DEVELOPMENTS IN AQUACULTURE AND FISHERIES SCIENCE

The following volumes are still available:

Developments in Aquaculture and Fisheries Science - 35

SCALLOPS: BIOLOGY, ECOLOGY AND AQUACULTURE

SECOND EDITION

Edited by

SANDRA E. SHUMWAY
Department of Marine Sciences
University of Connecticut
Groton, Connecticut
U.S.A.

G. JAY PARSONS
Fisheries and Oceans Canada
Aquaculture Science Branch
Ottawa, Ontario
Canada

ELSEVIER

Amsterdam – Boston – Heidelberg – London – New York – Oxford
Paris – San Diego – San Francisco – Singapore – Sydney – Tokyo

ELSEVIER B.V.
Radarweg 29
P.O. Box 211, 1000 AE
Amsterdam, The Netherlands

ELSEVIER Inc.
525 B Street
Suite 1900, San Diego
CA 92101-4495, USA

ELSEVIER Ltd.
The Boulevard
Langford Lane, Kidlington,
Oxford OX5 1GB, UK

ELSEVIER Ltd.
84 Theobalds Road
London WC1X 8RR
UK

First edition 2006

Library of Congress Cataloging in Publication Data
A catalog record is available from the Library of Congress.

British Library Cataloguing in Publication Data
A catalogue record is available from the British Library.

ISBN-13: 978-0-444-50482-1
ISBN-10: 0-444-50482-6
ISSN: 0167-9309 (Series)

∞ The paper used in this publication meets the requirements of ANSI/NISO Z39.48-1992 (Permanence of Paper).
Printed in The Netherlands.

Working together to grow
libraries in developing countries

www.elsevier.com | www.bookaid.org | www.sabre.org

ELSEVIER   BOOK AID
International   Sabre Foundation

Dedicated to K. S. (Sam) Naidu

Preface

From their continued appearance in art and culture to their savoured place at the dinner table, scallops have not lost their general appeal and continue to garner the attention of scientists, aquaculturists, chefs, shell collectors and others. Discussion during the 6th International Pectinid Workshop in 1987 provided the impetus for the first edition of *Scallops: Biology, Ecology and Aquaculture* which was published in 1991 and has been out of print for nearly a decade. Talk of a second edition began almost immediately, but was gently ignored for some years because of the time and effort required of all contributors and a sense on the part of the original editor that there was not enough new information to warrant a second volume.

Interest in all aspects of scallop biology and ecology has continued to increase, culture efforts have expanded globally and scallops continue to bring high prices from both the commercial fisheries and aquaculture ventures.

Potential authors began to inquire enthusiastically about the potential for a second edition, Elsevier Science Publishers were eager to add the volume to their series, and there was a renewed sense of need coupled with availability of new information to proceed with publishing a second edition.

The present volume represents an updated and revised version of the first edition. It is a comprehensive overview of the biology, ecology and aquaculture for scallop species worldwide. Some chapters have been completely rewritten by previous or new authors and two chapters have been reprinted as in the original volume (Chapter 5 - Neurobiology and Behaviour of the Scallop by Wilkens and Chapter 8 - Physiological Integrations and Energy Partitioning by Thompson and MacDonald) as they are as pertinent now as they were in 1991. Other chapters have been revised and new ones added (Nutrition and accounts by country of aquaculture and fisheries – Brazil, Venezuela and the Russian Federation now that the former Soviet Union has transformed since the first edition!).

This work would not have been possible without the assistance of several individuals. We thank Linda Kallansrude, Manon Chouinard and Dinah Helpert who provided secretarial and administrative support, Maille Lyons who took on the arduous task of developing the indices, Mara Vos-Sarmiento (Elsevier) for assistance, guidance and support and a special thanks to Sharon, Christopher, Michael and Max for their support and patience.

Finally, the authors are to be congratulated for their efforts and thanked profusely for their patience during the publication process.

We have dedicated this volume to Sam Naidu, friend and colleague, who devoted his career to working with fishermen, aquaculturists and government managers to ensure a sustainable fishery for future generations. Sam was a gentleman, generous with his knowledge and expertise, accepted no social barriers and was a pillar of the Pectinid Workshops. Moreover, he had perfected the art of living well and appreciating life.

Since publication of the first edition, there have been seven subsequent International Pectinid Workshops, the 15th recently held in Mooloolaba, Australia and the 16th planned for Halifax, Nova Scotia, Canada in 2007.

We hope this volume will continue to foster interest in scallops, especially aquaculture for a sustainable future, for years to come.

Sandra E. Shumway and G. Jay Parsons
Groton, Connecticut   Ottawa, Ontario

Preface from First Edition

Scallops are among the better known shellfishes and are widely distributed throughout the world. They are of worldwide economic importance and support both commercial fisheries and mariculture efforts. They have been the subject of numerous research efforts and their high economic stature encourages aquaculture efforts by both academicians and industrial researchers. The rapid growth, early maturity and high economic value of many scallop species offer special inducement to mariculturists. Scallops are already cultured successfully in a number of geographic locations and emphasis is increasing globally. Further, scallops play an important role in the structure and function of the communities of which they are a characteristic and important component.

Not since the publication of Marine Mussels (Bayne, 1975: IBP Handbook) has the current knowledge of a particular group of molluscs received a comprehensive coverage. His effort was a welcome addition to the scientific literature and provided an opportunity to synthesise a wealth of information. For successful culture and management, a good knowledge of the biology of the species is necessary. While scallops have been the subject of a considerable amount of research, to date no integrated account exists of the biology.

Discussions regarding the need for a comprehensive treatise on scallops first took place in 1987 during the 6th International Pectinid Workshop in Menai Bridge, Wales. The level of enthusiasm was high and convincing the individual authors to undertake writing the chapters was an easy task. The subsequent task of collecting and integrating those manuscripts proved to be considerably more arduous. There were further discussions regarding the final versions of manuscripts during the 7th International Pectinid Workshop held in Portland, Maine in 1989. These meetings provided a unique opportunity for specialists in the field of scallop biology representing over 20 countries to meet and discuss their individual views and I believe that the final product has benefited greatly from those open and often lively discussions.

The present volume presents a comprehensive overview of the biology, ecology and aquaculture prospects, both present and future, for scallops worldwide. Although the subject matter is specifically concerned with the biology, ecology and aquaculture of scallop species, the book should be of interest and relevance to a number of readers including advanced undergraduate/graduate students, mariculturists and research workers. Contributed chapters have been prepared by some of the foremost authorities in their respective fields and represents some 18 countries. Manuscripts submitted by authors whose first language is not English have been corrected for gross grammatical errors and clarity of presentation only. Their efforts to write a manuscript in a foreign language are to be applauded and should not be clouded by editorial license.

The book is divided into two main sections, the first dealing with general biology, ecology and physiology and the second dealing with fisheries and culture efforts in specific countries. Obviously, scallops are of more economic importance in some countries than in others and this is reflected in the specific coverage of specific species in Chapter 12.

The phylogenetic relationship of scallop species of commercial interest are examined by Waller in Chapter 1. The available information on larval requirements and ranges of tolerance of environmental variables, both prime importance to aquaculture efforts, are discussed by Cragg and Crisp in Chapter 2. Beninger and LePennec provide an elegant account of the internal anatomy of scallops in Chapter 3. The importance of scallop adductor muscles to the field of muscle physiology is clearly defined by Chantler in Chapter 4. Energy acquisition and utilisation are discussed by Bricelj and Shumway in Chapter 5 and the physiological integrations and partitioning of this energy are discussed in Chapter 6 by Thompson and MacDonald. Reproductive biology, another area of prime importance to culturists, is reviewed by Barber and Blake in Chapter 7. Two of the most striking features of scallops, eyes and their ability to swim, are discussed by Wilkens in Chapter 8: Neurobiology and Behavior. Diseases and parasites, major threats to aquaculture ventures, are discussed by Getchell in Chapter 9 and Gould and Fowler review the effects of pollutants on the commercial species of scallops in Chapter 10. General ecology, distribution and behavior are given in Chapter 11 by Brand. Beaumont and Zouros review the application of genetic tools to scallops and the importance of genetics to fisheries and aquaculture of scallops in Chapter 12. In Chapter 13, Orensanz, Parma and Iribane discuss population dynamics and the management of natural stocks. The commercial species of scallops are covered in Chapter 14 according to geographic location. Much of the data presented in Chapter 14 is being reported for the first time, including such elementary measures as growth, meat weight and landings.

This work would not have been possible without the efforts of Dr. Scott Siddall. His expertise in the field of desktop publishing coupled with his knowledge of shellfish biology made my task possible. His continued interest, support and interjections of humour when most needed made the task enjoyable. Jan Barter, Martha Hernandez-Davis, John Hedley and Jim Rollins helped in various ways and their willingness to undertake such arduous tasks as proofreading, crosschecking references, retyping of manuscripts and redrawing and repair of submitted diagrams are gratefully acknowledged. A special thanks must go to Bob Goodman, Elsevier Science Publishing Company, for his considerable patience in 'meeting' deadlines and his insight and support. Finally, I would like to thank the authors for their skill, patience and perseverance.

Sandra E. Shumway
Boothbay Harbor, Maine

List of Contributors

Alejandro Abarca	Pesquera San José, S.A., Cultivo de Ostiones, Casilla 342 Tongoy, Chile
Bruce J. Barber	Eckerd College, Galbraith Marine Science Laboratory, 4200 54th Avenue South, St. Petersburg, Florida, 33711 USA
Andy Beaumont	School of Ocean Sciences, University of Wales, Bangor, Menai Bridge, Gwynedd LL59 5EY Wales, UK
Peter G. Beninger	IsoMer, Faculté des Sciences, Université de Nantes, 44322 Nantes Cedex 3, France
Norman J. Blake	College of Marine Science, University of South Florida, 830 First Street South, St. Petersburg, Florida, 33701 USA
Eugenia Bogazzi	INIDEP (Instituto Nacional de Investigacion y Desarrollo Pesquero), Paseo Victoria Ocampo 1, CC 175, Mar del Plata (7600), Argentina
Neil F. Bourne	Fisheries and Oceans Canada, Pacific Biological Station, 3190 Hammond Plains Road, Nanaimo, British Columbia, Canada V9R 5K6
Susan M. Bower	Fisheries and Oceans Canada, Pacific Biological Station, 3190 Hammond Plains Road, Nanaimo, British Columbia, Canada V9R 5K6
Andrew R. Brand	Port Erin Marine Laboratory, University of Liverpool, Port Erin, Isle of Man, IM9 6JA
Claudia Bremec	INIDEP (Instituto Nacional de Investigacion y Desarrollo Pesquero), Paseo Victoria Ocampo 1, CC 175, Mar del Plata (7600), Argentina
V. Monica Bricelj	Institute for Marine Biosciences, National Research Council, 1411 Oxford Street, Halifax, Nova Scotia, Canada B3H 3Z1
Michael F. Bull	Deceased
Peter D. Chantler	Unit of Veterinary Molecular and Cellular Biology, Royal Veterinary College, University of London, Royal College Street, London NW1 0TU, United Kingdom
Néstor F. Ciocco	Centro Nacional Patagónico (CONICET), Bvd. Alte. Brown s/n, (9120) Puerto Madryn, Chubut, Argentina
Simon M. Cragg	Institute of Marine Sciences, School of Biological Sciences, University of Portsmouth, Ferry Road, Portsmouth PO4 9LY, United Kingdom
Peter J. Cranford	Fisheries and Oceans Canada, Marine Environmental Sciences Division, Bedford Institute of Oceanography, Dartmouth, Nova Scotia, Canada B2Y 4A2

Mike Dredge	Queensland Fisheries Service, 40 Hall Road, Narangba, Queensland 4504, Australia
Ana Farías	Instituto de Acuicultura, Universidad Austral de Chile. Av. Los Pinos, s/n, Balneario Pelluco, P. O. Box 1327, Puerto Montt, Chile
Esteban F. Félix-Pico	Centro Interdisciplinario de Ciencias Marinas, Departamento de Pesquerías y Biología Marina, Apartado Postal No. 592, La Paz, Baja California Sur, C.P. 23000, México
Luis Freites	Departamento de Biología Pesquera, Instituto Oceanográfico de Venezuela, Universidad de Oriente, Apdo. Postal 245, Cumaná 6101, Estado Sucre, Venezuela
Rodman G. Getchell	Aquatic Animal Health, Department of Microbiology and Immunology, College of Veterinary Medicine, Cornell University, Ithaca, New York, 14853–6401 USA
Ximing Guo	Haskin Shellfish Research Laboratory, Institute of Marine and Coastal Sciences, Rutgers, the State University of New Jersey, 6959 Miller Avenue, Port Norris, New Jersey, 08349 USA
John H. Himmelman	Département de Biologie, Université Laval, Quebec City, Quebec, Canada G1K 7P4
Hiroshi Ito	Hokkaido National Fisheries Research Institute, Fisheries Research Agency, Katsuarkoi, Kushiro, Hokkaido, 085–0802 Japan
Victor V. Ivin	Institute of Marine Biology, Far East State Branch of the Russian Academy of Sciences, Vladivostok 690041, Russia
Vasily Z. Kalashnikov	Institute of Marine Biology, Far East State Branch of the Russian Academy of Sciences, Vladivostok 690041, Russia
Yoshinobu Kosaka	Aomori Prefecture Fisheries Research Center, Aquaculture Institute, 10, Tsukidomari, Moura, Hiranai-cho, Aomori, 039–3381 Japan
Mario L. Lasta	INIDEP (Instituto Nacional de Investigacion y Desarrollo Pesquero), Paseo Victoria Ocampo 1, CC 175, Mar del Plata (7600), Argentina
Raymond B. Lauzier	Fisheries and Oceans Canada, Pacific Biological Station, 3190 Hammond Plains Road, Nanaimo, British Columbia, Canada, V9R 5K6
Marcel Le Pennec	Institut Universitaire Européen de la Mer, Université de Bretagne Occidentale, Place Nicolas Copernic, 29280 Plouzané, France

César J. Lodeiros — Departamento de Biología Pesquera, Instituto Oceanográfico de Venezuela, Universidad de Oriente, Apdo. Postal 245, Cumaná 6101, Estado Sucre, Venezuela

Yousheng Luo — Marine Fisheries Research Institute, No. 5–19 Xinghai Park, Dalian, Liaoning 116023, PRC

K. S. Naidu — Deceased

Maite Narvarte — Instituto de Biología Marina y Pesquera "Alte. Storni", Costanera s/n, CC 104, (8520) San Antonio Oeste, Río Negro, Argentina

Mark Norman — Taighde Mara Teo, Carna, Co. Galway, Ireland

Maximiano Nuñez — Departamento de Biología Pesquera, Instituto Oceanográfico de Venezuela, Universidad de Oriente, Apdo. Postal 245, Cumaná 6101, Estado Sucre, Venezuela

Bruce A. MacDonald — Biology Department and Centre for Coastal Studies and Aquaculture, University of New Brunswick, Saint John, New Brunswick, Canada E2L 4L5

Islay D. Marsden — School of Biological Sciences, University of Canterbury, Christchurch, New Zealand

Sergey I. Maslennikov — Institute of Marine Biology, Far East State Branch of the Russian Academy of Sciences, Vladivostok 690041, Russia

Sharon E. McGladdery — Fisheries and Oceans Canada, Aquaculture Science Branch, 200 Kent Street, Stn. 12W114, Ottawa, Ontario, Canada K1A 0E6

German E. Merino — Universidad Católica del Norte, Departamento de Acuicultura, Facultad de Ciencias del Mar, Casilla 117, Coquimbo, Chile

J.M. (Lobo) Orensanz — Centro Nacional Patagónico (CONICET), Bvd. Alte. Brown s/n, (9120) Puerto Madryn, Chubut, Argentina

Ana M. Parma — Centro Nacional Patagónico (CONICET), Bvd. Alte. Brown s/n, (9120) Puerto Madryn, Chubut, Argentina

G. Jay Parsons — Fisheries and Oceans Canada, Aquaculture Science Branch, 200 Kent Street, Stn. 12W114, Ottawa, Ontario, Canada K1A 0E6

G. Robert — Fisheries and Oceans Canada, Science Branch, P. 0. Box 1006, Dartmouth, Nova Scotia, Canada B2Y 4A2

Shawn M. C. Robinson — Fisheries and Oceans Canada, 531 Brandy Cove Road, Biological Station, St. Andrews, New Brunswick, Canada E5B 2L9

Guillermo Román — Instituto Espanol de Oceanografia, Centro Oceanografico de la Coruna, Muelle de Animas s/n, AP130, 15080 La Coruna, Spain

Guilherme S. Rupp   EPAGRI, Centro de Desenvolvimento em Aquicultura e Pesca, Rod. Admar Gonzaga 1188, Itacorubi, P. O. Box 502, Florianópolis, Santa Catarina, 88034–901, Brazil

Sandra E. Shumway   Department of Marine Sciences, University of Connecticut, 1080 Shennecossett Road, Groton, Connecticut, 06340 USA

Wolfgang Stotz   Universidad Católica del Norte, Departamento de Biologia Marina, Facultad de Ciencias del Mar, Casilla 117, Coquimbo, Chile

Øivind Strand   Department of Aquaculture, Institute of Marine Research, P. O. Box 1870 Nordnes, 5817 Bergen, Norway

Vitaly G. Tarasov   Institute of Marine Biology, Far East State Branch of the Russian Academy of Sciences, Vladivostok 690041, Russia

Raymond J. Thompson   Ocean Sciences Centre, Memorial University of Newfoundland, St. John's, Newfoundland and Labrador, Canada A1C 5S7

Teresa Turk   National Marine Fisheries Service, Northwest Fisheries Science Center, 2725 Montlake Blvd. E., Seattle, Washington, 98112 USA

Iker Uriarte   Instituto de Acuicultura, Universidad Austral de Chile. Av. Los Pinos, s/n, Balneario Pelluco, P. O. Box 1327, Puerto Montt, Chile

Juan Valero   School of Aquatic and Fishery Sciences, University of Washington, Box 355020, Seattle, Washington, 98195 USA

Anibal Vélez   Departamento de Biología Pesquera, Instituto Oceanográfico de Venezuela, Universidad de Oriente, Apdo. Postal 245, Cumaná 6101, Estado Sucre, Venezuela

Elisabeth von Brand   Universidad Católica del Norte, Departamento de Biologia Marina, Facultad de Ciencias del Mar, Casilla 117, Coquimbo, Chile

Thomas R. Waller   Department of Paleobiology, National Museum of Natural History, Smithsonian Institution, P. O. Box 37012, NHB MRC-121, Washington, D.C., 20013–7012 USA

Lon A. Wilkens   Department of Biology, University of Missouri-St. Louis, 8001 Natural Bridge Road, St. Louis, Missouri, 63121–4499 USA

Table of Contents

Chapter 1. New Phylogenies of the Pectinidae (Mollusca: Bivalvia): Reconciling Morphological and Molecular Approaches
Thomas R. Waller

Chapter 2. Development, Physiology, Behaviour and Ecology of Scallop Larvae
Simon M. Cragg

Chapter 3.   Structure and Function in Scallops
Peter G. Beninger and Marcel Le Pennec

Chapter 4.   Scallop Adductor Muscles: Structure and Function
Peter D. Chantler

Chapter 5.   Neurobiology and Behaviour of the Scallop
Lon A. Wilkens

Chapter 6.   Reproductive Physiology
Bruce J. Barber and Norman J. Blake

Chapter 7. Physiology: Energy Acquisition and Utilisation
Bruce A. MacDonald, V. Monica Bricelj and Sandra E. Shumway

Chapter 8. Physiological Integrations and Energy Partitioning
Raymond J. Thompson and Bruce A. MacDonald

Chapter 9.   Nutrition in Pectinids
Ana Farías and Iker Uriarte

Chapter 10. Genetics
Andy Beaumont

Chapter 11. Diseases and Parasites of Scallops
Sharon E. McGladdery, Susan M. Bower and Rodman G. Getchell

Chapter 12. Scallop Ecology: Distributions and Behaviour
    Andrew R. Brand

Chapter 13. Scallops and Marine Contaminants
Peter J. Cranford

Chapter 14. Dynamics, Assessment and Management of Exploited Natural Populations
J.M. (Lobo) Orensanz, Ana M. Parma, Teresa Turk and Juan Valero

Chapter 15. Fisheries Sea Scallop, *Placopecten magellanicus*
K.S. Naidu and G. Robert

Chapter 16. Sea Scallop Aquaculture in the Northwest Atlantic
G. Jay Parsons and Shawn M. C. Robinson

Chapter 17. Bay Scallop and Calico Scallop Fisheries, Culture and Enhancement in Eastern North America
Norman J. Blake and Sandra E. Shumway

Chapter 18. Scallops of the West Coast of North America
Raymond B. Lauzier and Neil F. Bourne

Chapter 19. The European Scallop Fisheries for *Pecten maximus, Aequipecten opercularis* and *Mimachlamys varia*
Andrew R. Brand

Chapter 20. European Aquaculture
Mark Norman, Guillermo Román and Øivind Strand

Chapter 21. Scandinavia
Øivind Strand and G. Jay Parsons

Chapter 22. Japan
Yoshinobu Kosaka and Hiroshi Ito

Chapter 23. Scallop Culture in China
Ximing Guo and Yousheng Luo

Chapter 24. Scallops Fisheries and Aquaculture of Northwestern Pacific, Russian Federation

Victor V. Ivin, Vasily Z. Kalashnikov, Sergey I. Maslennikov and Vitaly G. Tarasov

Chapter 25. Scallop Aquaculture and Fisheries in Brazil
Guilherme S. Rupp and G. Jay Parsons

Chapter 26. Argentina
Néstor F. Ciocco, Mario L. Lasta, Maite Narvarte, Claudia Bremec, Eugenia Bogazzi, Juan Valero and J.M. (Lobo) Orensanz

Chapter 27. Scallop Fishery and Aquaculture in Chile
Elisabeth von Brand, German E. Merino, Alejandro Abarca and Wolfgang Stotz

Chapter 28. Venezuela

César J. Lodeiros, Luis Freites, Maximiano Nuñez, Anibal Vélez and John H. Himmelman

Chapter 29. Mexico

Esteban Fernando Félix-Pico

Chapter 30. Scallop Fisheries, Mariculture and Enhancement in Australia
Mike Dredge

Chapter 31. New Zealand
Islay D. Marsden and Michael F. Bull

Scallops: Biology, Ecology and Aquaculture
S.E. Shumway and G.J. Parsons (Editors)
Published by Elsevier B.V.

Chapter 1

New Phylogenies of the Pectinidae (Mollusca: Bivalvia): Reconciling Morphological and Molecular Approaches

Thomas R. Waller

1.1 INTRODUCTION

Waller (1991) postulated evolutionary relationships among commercial scallops on the basis of shared derived morphological characters and produced a phylogeny scaled against geologic time using data from the fossil record. In the past decade, molecular genetic studies have tested most of the branching points in the original phylogeny. Although the morphological and molecular genetic trees have been largely concordant, there are some differences and some new findings that merit further testing and discussion. In the present study, I review the phylogenetic results of the molecular genetic studies and then present new data and new phylogenetic interpretations from an extensive study of the worldwide fossil record of the Pectinidae beginning at the dawn of the Cenozoic Era. The new phylogenies resolve at least some of the conflicts between morphological and molecular approaches. They also provide new hypotheses for further testing by both approaches.

1.1.1 Molecular Genetic Studies

Rice et al. (1993) analysed full sequences of the 18S rRNA gene from the pectinid species *Placopecten magellanicus* (Gmelin, 1791), *Chlamys islandica* (Müller, 1776), and *Argopecten irradians* (Lamarck, 1819) and seven species in the heterodont family Mactridae, with a branchiuran crustacean serving as an outgroup to the Bivalvia. Pectinid monophyly was supported by 100% bootstrap values for all methods of analysis. Parsimony and maximum-likelihood analyses placed either *Chlamys* or *Placopecten* first depending upon which gap value was chosen for the analysis. In contrast, Fitch-Margoliash distance analyses placed *Placopecten* first regardless of the gap values. *Argopecten irradians* consistently appeared as the latest branch among the three species, with either *Placopecten magellanicus* or *Chlamys islandica* as its sister species. With reference to these three species, Waller (1991, fig. 8) showed *Placopecten* (as a member of the "*Palliolum* group") branching off first and *Argopecten* (as a member of the "*Aequipecten* group") last. Waller (1984) had earlier concluded that the Pectinidae are monophyletic based upon the presence in all members of the group of a ctenolium, shown to be an autapomorphy for the family Pectinidae.

Kenchington and Roddick (1994) corroborated pectinid monophyly through analyses of the 18S rRNA gene in an expanded group of taxa that included additional bivalves (*Crassostrea, Mytilus*) as well as representatives of other molluscan classes (a gastropod and a chiton) and a spider used as a distant outgroup. Again, their pectinid phylogeny showed monophyly of the family Pectinidae and either *Chlamys* or *Placopecten* branching first among the same three pectinid species analysed by Rice et al. (1993).

Steiner and Müller (1996) analysed complete and partial 18S rDNA sequences from the same bivalve taxa plus two additional bivalves (*Atrina, Arca*), an additional gastropod and chiton, and a polychaete annelid as a distant outgroup. All methods of tree construction showed the same three species of Pectinidae referred to above to be a monophyletic group. In contrast to the preceding studies, however, their use of spectral analysis (Hendy and Penny 1993) found *Argopecten* branching first with no conflict but with little support. Among the most parsimonious trees in parsimony analysis of the sequence data, they found that either *Placopecten* or *Argopecten* branched first depending on whether *Crassostrea* is the sister group of the Pectinidae. All of their other tree reconstructions, however, indicate *Argopecten* as the first pectinid branch, i.e., as the sister group of *Placopecten magellanicus* plus *Chlamys islandica*, in contrast to Waller's (1991) phylogeny.

Frischer et al. (1998) used 18S rRNA gene sequence data, including alignments based on secondary structural constraints, from seven pectinid species: the three species mentioned above plus *Pecten maximus* (Linnaeus, 1758), *Chlamys hastata* (G.B. Sowerby, II, 1842), *Mimachlamys varia* (Linnaeus, 1758), and *Crassadoma gigantea* (Gray, 1825). Again, monophyly of the Pectinidae was firmly supported by both distance and parsimony methods. However, their analyses also provided very strong support for a closer relationship of *Pecten maximus* and *Argopecten irradians* to each other than to the other pectinid species. This is in contrast to Waller (1991), who placed *Pecten* on the same side of a deep split in the Pectinidae as *Placopecten* (in the *Palliolum* group) and placed *Argopecten* (in the *Aequipecten* group) on the other side of this split with the *Chlamys* and *Mimachlamys* groups (see Fig. 1.1 herein). Frischer et al. (1998) also found that the tribes Chlamydini and Mimachlamydini "generally group together, sharing a common ancestor with the Crassadomini and Palliolini" but with bootstrap values ranging widely depending upon the methods used.

Hayami and Matsumoto (1998) tested the phylogenies of Waller (1991, 1993) by analysing the mitochondrial cytochrome oxidase subunit I (COI) gene of seven Japanese scallop species: *Amusium japonicum* (Gmelin, 1791), *Cryptopecten vesiculosus* (Dunker, 1877), *Pecten albicans* (Schröter, 1802), *Mimachlamys nobilis* (Reeve, 1852), *Azumapecten farreri* (Jones and Preston, 1904), *Patinopecten yessoensis* (Jay, 1856), and *Swiftopecten swifti* (Bernardi, 1858). The oyster, *Crassostrea gigas* (Thunberg, 1793), was used as the outgroup. They found support for a clade containing *Amusium*, *Cryptopecten*, and *Pecten* forming a sister group of a clade containing *Mimachlamys*, *Azumapecten*, and *Patinopecten* + *Swiftopecten*, branching in that order. *Cryptopecten* is an aequipectinine genus closely related to *Argopecten*. Its sister-group relationship to *Pecten* shown by the COI gene analyses corroborates the close relationship of *Argopecten* and *Pecten* found by Frischer et al. (1998), in contrast to the well separated positions of these genera in the 1991 phylogeny of Waller (Fig. 1.1). Their phylogeny of the other genera, however, is similar to the relationships within the Chlamydinae indicated in Waller (1991, 1993).

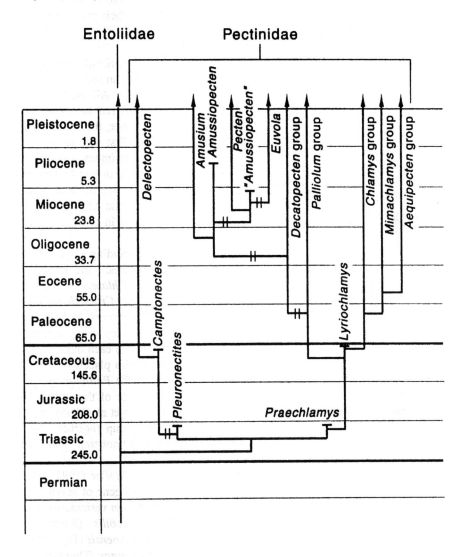

Figure 1.1. A phylogeny of groups of Pectinidae containing commercial species plotted against time reconstructed from Waller (1991, figs. 8 and 11 and text). Phyletic lines ending in arrows contain extant representatives, those ending in cross-bars are extinct. Double lines crossing a lineage indicate that its phylogenetic position has been changed in the present study. Dates of period or epoch boundaries, in millions of years, are from Berggren et al. (1995) for the Cenozoic Era and from Harland et al. (1990) for the Mesozoic Era.

Giribet and Carranza (1999) and Steiner (1999) found full support for monophyly of the Pectinidae in analyses of 18s rDNA sequences for the same seven pectinid species that were considered by Frischer et al. (1998). Giribet and Carranza's (1999: figs. 1a-c) most parsimonious trees using the complete alignment and unweighted parsimony gave identical arrangements of *Placopecten*, *Chlamys*, and *Argopecten* whether obtained by branch-and-bound search or by heuristic search. In both cases *Argopecten* is a sister group of *Chlamys* + *Placopecten*. Their strict concensus tree of 18 most parsimonious trees using heuristic search shows three sister groups: (a) *Argopecten* + *Pecten*, (b) species in the genera *Chlamys*, *Mimachlamys*, and *Crassadoma*, and (c) *Placopecten*. Steiner (1999) reanalysed the same data set and showed similar phylogenetic results for the Pectinidae.

Canapa et al. (1999, 2000a) determined a complete sequence of 18S ribosomal DNA for *Adamussium colbecki* (Smith, 1902), a large, smooth, thin-shelled scallop that is abundant around Antarctica. In analyses using sequences from the seven pectinid species mentioned above plus *Adamussium*, both neighbour-joining and maximum-parsimony methods yielded trees similar to those of Frischer et al. (1998). Here again, a close relationship between *Pecten maximus* and *Argopecten irradians* is supported by a high bootstrap value. *Adamussium* is shown as a sister group of this pair of taxa, but with a low bootstrap value. *Chlamys islandica*, *Chlamys hastata*, and *Mimachlamys varia* also form a clade with high bootstrap value. *Crassadoma gigantea* is close to these, but the position of *Placopecten* is uncertain. As the authors point out (p. 381), *Placopecten* appears close to *Chlamys* where the analysis includes other pteriomorphian families but "appears equidistant from *Chlamys* and the *Pecten-Aequipecten* groups when only pectinid sequences are considered."

Wilding et al. (1999) examined the genetic relationships of two European species, *Pecten maximus* and *P. jacobaeus* (Linnaeus, 1758), that have long been regarded as closely related but distinct species by malacologists and palaeontologists. Using allozyme electrophoresis and a type of mitochondrial DNA analysis (DNA PCR-RFLP) previously applied to populations, they found that genetic distances were of a magnitude "considerably lower than expected for congeneric species." However, rather than accepting the low genetic distances as evidence for conspecificity, the authors carefully reviewed the conflicting lines of evidence and suggested three possibilities for explaining the conflict: (a) The two are indeed valid species but this is not reflected in the particular genetic indices calculated. (b) These are not distinct species, meaning that both would carry the name *Pecten maximus*, the prior name in Linnaeus (1758). In this case the morphological differences would be ecophenotypic, i.e., reflecting local environmental conditions, and the allozyme differentiation "merely represents increased isolation by distance" between the two samples (*P. maximus* sampled from Britain, Ireland, and northern France, *P. jacobaeus* from southeastern Spain). This would require demonstration of range overlap and evidence of hybridisation in nature. Thus far, however, the authors point out that hybridisation to produce viable adults has been demonstrated only under laboratory conditions. The matter of range overlap is discussed further in the following section on new observations. (c) *P. jacobaeus* from southeastern Spain are interbreeding with *P. maximus* but interbreeding may not occur elsewhere, thus leading to an anomalously low genetic distance between the species. The authors discounted this possibility, however, in noting that there is an oceanographic front stretching across the Mediterranean from Almería, Spain, to Oran, Algeria, that has been associated with genetic breaks in the distributions of other marine species. Logically this front could separate *P. maximus*, a species of the northeastern Atlantic, from *P. jacobaeus*, a Mediterranean species. In this case, however, the population of *P. jacobaeus* sampled came from the Mediterranean side of the front, where interbreeding would not be expected. Wilding et al. (1999) concluded, after considering these possibilities, that the conflict between genetic-distance data that indicate conspecificity and morphological, biogeographical, and paleontological data that indicate species distinctness is still unresolved.

Canapa et al. (2000b, fig. 1) used partial sequences of the 16S rRNA gene to analyse the phylogenetic relationships of seven scallop species which they referred to as *Adamussium colbecki*, *Aequipecten opercularis* (Linnaeus, 1758), *Chlamys glabra* (Linnaeus, 1758), *Chlamys varia*, *Pecten jacobaeus*, and *Pecten maximus*. As indicated above, their two "*Chlamys*" species have been assigned to other genera and now bear the names *Mimachlamys varia* and *Flexopecten glaber* (see Waller 1986, 1991, 1993). The inclusion of a species of *Flexopecten* is of interest, because according to Waller (1991, but see below) it represents the "*Decatopecten* group", no representative of which had been previously analysed using molecular genetic information. The trees illustrated by Canapa et al. (2000b, figs. 1 and 2) are topologically identical to each other. With a *Mytilus* as the outgroup, these indicate the monophyly of the Pectinidae. Within the Pectinidae, they show *Aequipecten opercularis* and *Flexopecten glaber* as a sister group of the other five pectinid taxa. This is in contrast to Waller (1991), where *Flexopecten*, as a member of the *Decatopecten* group, is distant from *Aequipecten*. Within the five species that Canapa et al. (2000b) analysed, two European species of *Pecten* are joined by bootstrap values of 100%. Lower bootstrap values (93% and 95%) join *Adamussium colbecki* with *Chlamys islandica* and *Mimachlamys varia*. Their analysis, like that of Wilding et al. (1999), demonstrated a relationship between *Pecten maximus* and *P. jacobaeus* that is so close that it would appear to challenge the existing taxonomic concept that these taxa are distinct species. Although Canapa et al. (2000b) contend that this proximity is corroborated by evidence of morphological intergradation, the validity of this morphological evidence is questioned below.

Matsumoto and Hayami (2000) analysed sequences of the mitochondrial cytochrome c oxidase subunit I gene (COI) of 17 pectinid species, by far the most taxonomically extensive molecular genetic study of scallops performed thus far. The species are distributed among most of the clades discussed by Waller (1991), including the *Chlamys*, *Mimachlamys*, *Aequipecten*, *Decatopecten*, and *Pecten* groups. Groups not analysed are the *Palliolum* group and the pectinid genus *Delectopecten*, the latter hypothesised by Waller to be the sister group of all other extant Pectinidae. The authors also analysed the COI gene in two additional families of the superfamily Pectinoidea, Propeamussiidae (one species) and Spondylidae (four species) using a limid bivalve (*Limnaria*) and an arcoidean bivalve (*Scapharca*) as outgroups of the pectinoidean families.

The phylogenetic results of the COI analysis are remarkably concordant with the morphological phylogenies of Waller (1978, 1984, 1991, 1993). The Spondylidae and Propeamussiidae, as predicted, fall outside of the family Pectinidae. Within the Pectinidae, the subfamily Chlamydinae is a sister group of the family Pectininae. The former contains the tribes Chlamydini and Mimachlamydini, whereas the latter contains the Decatopectinini and Pectinini as sister groups. At a finer level within the Chlamydini, the bizarre, coral-inhabiting species *Pedum spondyloideum* Lamarck, 1819, is shown to be a sister group of *Laevichlamys squamosa* (Gmelin, 1791) + *Semipallium dianae* (Crandall, 1979). Waller (1993) had predicted a close relationship between *Pedum* and his new genus *Laevichlamys* based upon the early development of shell morphology and microsculpture in the two groups. Waller's (1993: 27) placement of *Patinopecten* close to *Chlamys*, in contrast to its previous position close to *Pecten*, is also corroborated, as is also a close relationship between the Decatopectinini and Pectinini. Waller's separation of the Propeamussiidae from the Pectinidae (Waller 1972, 1978, 1984) and placement of true *Amusium* close to *Pecten* is also corroborated, although in the COI phylogeny, *Amusium* is more basal in the Pectininae, branching off before the split of the *Decatopecten* and *Pecten* groups (but with low bootstrap value).

There are also, however, some conflicts between the COI phylogeny and the previous morphological phylogenies. As in the other molecular genetic studies mentioned above, an aequipectinine genus, *Cryptopecten*, is shown to have a closer relationship to *Pecten* than to any other genus that they analysed. Also, there is also no clear indication in the COI analysis that the subfamily Pectininae is closer to the tribe Mimachlamydini than to other members of the Chlamydinae. Matsumoto and Hayami (2000) did not analyse the COI gene in any member of the *Palliolum* group of Waller and thus did not address the persistent problem of the relationship of *Placopecten* to other pectinids.

Steiner and Hammer (2000) examined phylogenetic relationships throughout the Pteriomorphia by aligning new, nearly complete 18S rDNA sequences with those previously published. For the Pectinidae, they included the seven species analysed by Frischer et al. (1998) but added sequences from three more species: *Flexopecten glaber* (Linnaeus, 1758), *Excellichlamys spectabilis* (Reeve, 1853), and *Pedum spondyloideum* (Gmelin, 1791). Parsimony and maximum-likelihood trees have *Crassadoma*, *Placopecten*, *Chlamys*, and *Pedum* at the base and a series represented successively by *Mimachlamys*, *Excellichlamys*, *Pecten*, and *Argopecten* + *Flexopecten* at higher levels. These results corroborate those of the earlier molecular genetic studies. The position of the newly analysed *Excellichlamys*, however, is of considerable interest, because it is well separated from *Flexopecten* and is located below the *Pecten-Argopecten-Flexopecten* clade. This is in contrast to Waller (1986, 1991), who included *Excellichlamys* with *Flexopecten* in the same tribe Decatopectinini.

Campbell (2000) used 18S gene sequences to examine phylogeny through the entire class Bivalvia. The relationships of four pectinid taxa (*Placopecten magellanicus*, *Chlamys islandica*, *Argopecten irradians*, and *Pecten maximus*) were similar to those of previous studies, with *Placopecten* branching off first, then *Chlamys*, and then *Pecten* + *Argopecten*.

The issue of whether western *Pecten jacobaeus* and *P. maximus* are distinct at the species level was re-examined by Ríos et al. (2002). Using allozyme electrophoresis, they analysed variation in 15 allozyme loci of *Pecten maximus* and *P. jacobaeus*, respectively, on the west and east sides of the Almería-Oran Oceanographic Front in the western Mediterranean. They found evidence for a major genetic discontinuity between these taxa where they are separated by the oceanographic front.

The principal uncertainties or conflicts in comparisons of the morphological and molecular phylogenies outlined above raise a number of questions: (a) Is *Placopecten* representative of a deep split in the Pectinidae? If so, when did this split occur, and what are the paleontological roots of *Placopecten*? (b) If *Pecten* and the aequipectinine genera *Argopecten* and *Cryptopecten* are as closely related as indicated by molecular genetic studies, can this be reconciled with the fossil record of the origins of these groups? (c) If *Flexopecten* is closely related to *Aequipecten* and *Argopecten* and is separated from *Excellichlamys*, is it still a member of the *Decatopecten* group, or do morphology and the fossil record indicate that *Flexopecten* and the *Decatopecten* group have independent origins? (d) Molecular genetic studies indicate that two long-recognised species of the eastern Atlantic and Mediterranean, *Pecten maximus* and *P. jacobaeus*, respectively, are so genetically close that it is possible that they belong to the same species. A recent allozyme study, however, discovered a major genetic continuity between these species in the western Mediterranean. What more do morphology and the fossil record have to say about the distinction between these taxa?

1.1.2 Methods and Materials

Morphological terms used herein are explained in Waller (1991). Particularly important are the three terms describing the orientation of sculptural features: "antimarginal", referring to sculptural features that trace sweeping lines that are approximately perpendicular to the margin; "radial", referring to features that radiate from the beak and are therefore approximately perpendicular to the margin only in the central radial sector; and "commarginal" describing features that parallel the margin. Terms for hinge teeth with reference to the right valve were introduced in Waller (1986) and were further refined in Waller (1991: 8). Terms for adaptive shell shapes (chlamydoid, aequipectinoid, pectinoid, amusioid), discussed in Waller (1991: 10), refer to forms that have evolved over and over again in the history of the Pectinidae. These forms may be taxonomically important within clades defined on the basis of other characters but do not in themselves denote high-level taxa. Morphological clues to high-level pectinid relationships lie in the development of microsculpture, megasculpture, shell microstructure, and hinge dentition.

The phylogenies resulting from the present study (Figs. 1.2 and 1.3) are scaled against geologic time calibrated in millions of years and compartmentalised into standard European stages. For convenience, these European stage/age names have been used throughout the text rather than local names. The basis for calibration is the set of charts in Berggren et al. (1995) that relate global zones based on planktonic foraminifera, calcareous nannoplankton, magnetic polarity data, and radiometric data to European

8

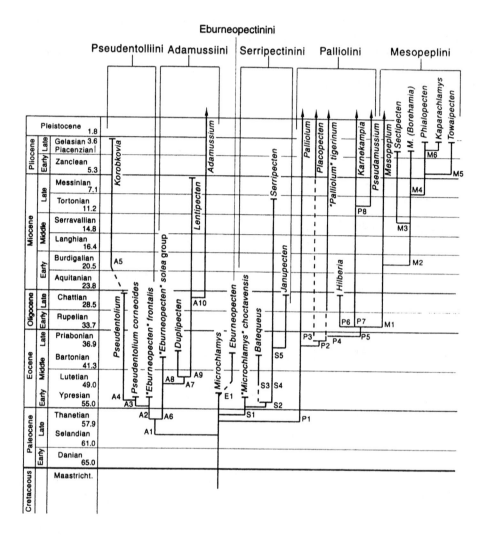

Figure 1.2. A phylogeny of the six tribes of the subfamily Palliolinae. Dashed lines indicate missing fossil records. Letter-number labels of nodes refer to characters discussed in text. Numbers on stage boundaries refer to time in millions of years ago. Positions of origins of lineages within stages are not significant. Other conventions are as in Figure 1.1.

stages. In general, the process of placing an age on a local fossil occurrence has involved (a) determining the geological formation in which the fossil occurs, either from original collection data or later stratigraphic updates in the literature, (b) determining from the literature whether there is microfossil zonal information for that formation or for formations that are in stratigraphic proximity, and (c) using the microfossil zonal determinations for entry into the Berggren et al. (1995) charts. Particularly useful sources of information on fossil occurrences and stratigraphic correlation via microfossil zones have been Palmer and Brann (1965), Ward and Blackwelder (1980), Blackwelder (1981), Siesser (1984), Ward (1985, 1992), Carter and Rossbach (1989), Harland et al. (1990), and Beu and Maxwell (1990).

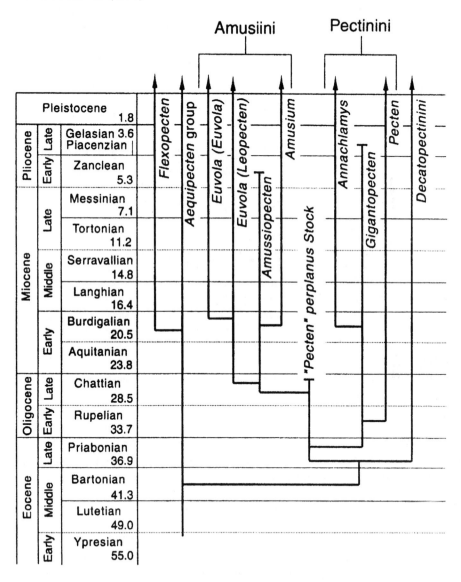

Figure 1.3. A phylogeny of three tribes of the subfamily Pectininae. See Figures 1.1 and 1.2 for explanation of numbers and symbols.

Almost all of the taxa referred to in this study have been examined first-hand by means of dissecting microscopes. High-intensity oblique lighting and, in some cases, coating with ammonium chloride, have been useful for bringing out details of sculptural features of low relief. European fossil taxa have been studied at major natural history museums, particularly in London, Brussels, Paris, and Lyon. Fossil taxa from the Americas have been studied mainly at the National Museum of Natural History, Smithsonian Institution, Washington. Information on the morphology of extant species is mainly from the extensive collections of the Department of Systematic Biology (Mollusk Division) at the National Museum of Natural History, Smithsonian Institution.

1.2 NEW OBSERVATIONS

1.2.1 Subfamily Camptonectinae Habe, 1977

Waller (1991, fig. 8) and Waller and Marincovich (1992: 219) indicated that the extant cosmopolitan genus *Delectopecten* Stewart, 1930, is a descendant of the extinct Jurassic-Cretaceous genus *Camptonectes* Agassiz in Meek, 1864, which in turn is derived from *Pleuronectites* von Schlotheim, 1820, of the Triassic. By this phylogenetic scheme, *Delectopecten* is a sister taxon of all other extant Pectinidae, with the last common ancestor in the early Middle Triassic (early Anisian), approximately 240 million years ago. New research by the author on Middle Triassic pectinids has led to a new concept of *Pleuronectites* as an evolutionary dead-end that underwent its main evolutionary change within the Muschelkalk Basin of southwestern Europe. In the course of its evolution within this basin in the Triassic, the genus reached gigantic size and became more inequivalved, with the right valve nearly flat and the left valve becoming greatly inflated (Hagdorn 1995). All *Pleuronectites* thus far examined lack antimarginal microsculpture throughout ontogeny, indicating that these are an unlikely phylogenetic source for the earliest *Camptonectes*, which have fine but persistent antimarginal microsculpture. The earliest *Camptonectes* thus far reported are Late Triassic (Carnian-Norian) in age and are from the Boreal-East Asian Province of Japan and eastern Siberia (Hayami 1989: 6). Their likely origin is in the Triassic *Chlamys* group, possibly in the genus *Praechlamys* Allasinaz, 1972. *Delectopecten*, as a phyletic descendant of *Camptonectes*, thus still remains as the sister group of all other extant Pectinidae, but the divergence point should now be dated from the early Late Triassic (Carnian), in the time range of approximately 225 to 230 million years ago (Harland et al. 1990).

1.2.2 Subfamily Palliolinae Korobkov in Eberzin, 1960

The subfamily Palliolinae is used here in a phylogenetic sense for the monophyletic assemblage of genera (Fig. 1.2) rooted in *Dhondtichlamys* Waller, 2001 (replacement name for *Microchlamys* Sobetski, 1977; see Appendix Note 1). This genus, based on the type species *Pecten pulchellus* Nilsson, 1827, originated in the mid-Cretaceous from a *Chlamys*-like ancestor probably in the genus *Lyriochlamys* Sobetski, 1977 (Waller 1991: 14). The time of origin is not certain but may have been Albian, based on the oldest record of *Dhondtichlamys* given by Dhondt (1972: 8, for *Pecten* (*Aequipecten*) *arlesiensis* Woods, 1902).

The major morphological innovations in the shell of *Dhondtichlamys* involve shell microstructure, shell sculpture, and shell shape. (See Dhondt 1972, for good figures of *Dhondtichlamys* species under the combination "*Lyropecten* (*Aequipecten*)"). Compared to *Lyriochlamys*, the leading edge of the inner crossed-lamellar aragonitic ostracum of *Dhondtichlamys* is withdrawn from the shell margins and hinge region and is restricted to an area that is mainly inside the pallial line. The microstructure of the foliated calcite of *Dhondtichlamys* between the pallial line and the margin is irregularly foliated. In this configuration, small patches of laths of uniform orientation produce tiny, reflective surfaces on the inner surface of well-preserved shells in contrast to the distinctly more uniform reflective surface of the corresponding region in most members of the *Chlamys* group. The contrast between these types of reflective surfaces may be observed by comparing palliolinine-type shells such as those of *Placopecten magellanicus* or "*Palliolum*" *tigerinum* (Gmelin, 1791) with shells of the *Chlamys* group such as those of *Chlamys islandica*, *Aequipecten opercularis*, or *Argopecten irradians*. Carter (1990: 297, and references therein) provided a detailed descriptive terminology for shell microstructures in the Bivalvia. For the purposes of the study, however, these two types of foliated calcite distinguished by the complexity of their reflective surfaces on the inner shell surface will be referred to as irregularly foliated and regularly foliated.

Both shell shape and radial macrosculpture are highly variable in Cretaceous *Dhondtichlamys*. In general, however, this genus, compared to coeval Cretaceous *Lyriochlamys*, is characterised by its smaller size, somewhat broader umbonal angle, smaller auricles relative to size of disk, and stronger radial plicae that are wider than interspaces on the right valve and narrower than interspaces on the left valve due to interlocking of the plicae at the ventral margin. After an ontogenetically early scaly or reticulate phase in which scales form atop ribs and commarginal lirae cross both interspaces and ribs, the rib crests become smooth. Antimarginal microsculpture that is relatively coarse for the small shell size forms in a uniform continuous antimarginal pattern in the rib interspaces but is absent from the rib crests. The antimarginal ridglets strengthen considerably where they cross the disk flanks and extend onto the bases of the auricles. The auricles themselves vary in relative lengths from nearly equal to unequal with the anterior auricle the longer.

Dhondtichlamys survived the end-Cretaceous mass extinction and is the earliest coarsely ribbed pectinid to be found in the Paleocene. The specimens that I have been able to examine directly belong to two possibly intergradational species, *Dhondtichlamys johnsoni* (Clark, 1895) and *D. greggi* (Harris, 1897), from the Upper Paleocene (Thanetian) and Lower Eocene (Ypresian) of the Gulf Coastal Plain of southeastern United States (see Palmer and Brann 1965, for stratigraphic and geographic data). These species resemble Cretaceous *Dhondtichlamys* in shell microstructure, shape, ribbing style, and microsculpture. Like some Cretaceous species, the lengths of the anterior and posterior auricles of the Paleocene forms are nearly equal, and the posterior margin of the posterior auricles is convex outward.

In Figure 1.2, *Microchlamys* (= *Dhondtichlamys*) is shown as the ancestral group for four major evolutionary radiations that comprise single tribes or pairs of tribes as follows: (a) Pseudentoliini, new tribe, plus Adamussiini Habe, 1977; (b) Eburneopectinini, new tribe; (c) Serripectinini, new tribe, and (d) Palliolini Korobkov in Eberzin, 1960, plus Mesopeplini, new tribe. The evidence for this phylogeny draws heavily on changing morphology within stratigraphic successions within geographic regions. Clear separation of the origins of similar morphologies in space and time provides strong evidence that these morphologies are convergent. This permits the disassembly of certain overly broad form genera, the outstanding example in this subfamily being *Eburneopecten* Conrad, 1865. Discussion of these clades begins at the left side of Figure 1.2 and proceeds to the right. In some cases, I continue to use a current genus-group name in quotation marks, indicating that this taxon can no longer be regarded as that genus but has not yet been renamed.

1.2.2.1 Tribe Pseudentoliini, new tribe

The new tribe Pseudentoliini is defined as a monophyletic group of genera rooted in "*Eburneopecten*" *frontalis* (Dall, 1898) (Appendix Note 2) of late Paleocene (Selandian, Thanetian) age in the North Atlantic. As presently understood, evolution from this root led successively through a series of morphologies represented by *Pseudentolium corneoides* (Harris, 1919), *Pseudentolium corneum* (J. Sowerby, 1818), and a succession of species in the genus *Korobkovia* Glibert and van de Poel, 1965. From the beginning of the Ypresian, the tribe was, so far as known, restricted to the seas of Europe and Asia and remained restricted to the Northern Hemisphere until its extinction at the end of the Pliocene.

At the base of this clade (A2 in Fig. 1.2), "*Eburneopecten*" *frontalis*, from the Upper Paleocene and Lower Eocene of Maryland, has the basic shape of Paleocene *Dhondtichlamys* and shares similar auricular costae and byssal fasciole. However, in contrast to the coarse plicae and strong antimarginal microsculpture of *Dhondtichlamys*, "*E.*" *frontalis* has nearly obsolete but ontogenetically continuous low, broad radial ribs and very fine antimarginal microsculpture. "*E.*" *frontalis* is closely followed stratigraphically by *Pseudentolium corneoides* in the early Ypresian of the North American Gulf Coast (A3, Fig. 1.2). This species, previously assigned to *Eburneopecten* (Palmer and Brann 1965: 134), is placed here in *Pseudentolium* because it displays distinctive features that characterise the taxa above Node A3. These features are associated with an evolutionary trend toward an amussioid adaptive form: a broad, low radial ridge on the inner surface of the disk inside each disk flank, disk gapes, auricular buttresses ending in low tubercles, a shallowing of the byssal notch, a reduction on the ontogenetic persistence of an active ctenolium, the formation of margins of the left auricles and of the right posterior auricle that form obtuse angles with the hinge line, and the complete loss of auricular costae. At Node A4, *Pseudentolium corneum* of the European Eocene and Oligocene evolved a distinctive flattened byssal fasciole and rounded shallow byssal notch with the functional ctenolium becoming obsolete before maturity. The nearly circular arc of the anterior margin of the right anterior auricle of this species is nearly identical to that of *Korobkovia*. Principal innovations at Node A5 that characterise *Korobkovia* are shortened auricles relative to disk length, a trend toward narrowing of the disk, increased size of disk gapes, obsolescence of the byssal notch, and low radial ribs expressed mainly on the interior surface the left valve. These features of *Korobkovia* developed progressively within the genus as exemplified by the European stratigraphic succession of *K. denudata* (Reuss, 1867) of the Burdigalian and Langhian, *K. woodi* (Nyst, 1861) of the Serravallian, and *K. oblonga* (Philippi, 1844) of the Upper Miocene and Pliocene.

1.2.2.2 Tribe Adamussiini Habe, 1977

Habe (1977) introduced the subfamily Adamussiinae to contain only the genus *Adamussium* Thiele, 1934, which he regarded as being different enough from all other Pectinidae to justify this elevated rank. In the present study the name is reduced in rank to the tribal level within the subfamily Palliolinae but is expanded to include phylogenetically related genera. The tribe is defined phylogenetically as rooted in the *"Eburneopecten" solea* group beginning in the Thanetian in Europe (A6, Fig. 1.2) and giving rise in the Southern Hemisphere to *Duplipecten* Marwick, 1928, and *Lentipecten* Marwick, 1928, in the Lutetian and to *Adamussium* in the Oligocene (possibly Chattian) (A7 to A10, Fig. 1.2). The last genus is extant in the antarctic region.

The oldest taxon included here in the *"E." solea* group is *"Eburneopecten" breviauritum* (Deshayes, 1824), from the Thanetian of the Paris Basin, France. In shape and size it resembles coeval *"Eburneopecten" frontalis* of the American Thanetian but with two important differences: (a) the ribbing of the European species is much finer and begins later in ontogeny after a non-costate post-prismatic stage; and (b) the antimarginal microsculpture of the European species is plesiomorphically coarser. Other species that are probably members of this group that have ontogenetically late-appearing radial costae are all Lutetian in age and are possibly synonymous. These are *"E." laekeniensis* (G. Vincent, 1881) from Belgium, *"E." paueri* (Frauscher, 1886) from Germany and France, and *"E." bourdoti* (Cossmann and Pissaro, 1903) from the Lutetian of France. All of these differ from Thanetian *"E." breviauritum* in having even later appearing radial costae, broader umbonal angles, steeper disk flanks, a deeper byssal notch with a persistent active ctenolium, antimarginal microsculpture that is finer in the centre of the disk with commarginal shingles or fine commarginal lirae of low relief. They also differ in having broader disk gapes and low auricular buttresses ending in tubercles. The tubercle of the right anterior buttress forms along the margin of the disk flank distal to the byssal notch, not proximal to the byssal notch as in the Pseudentoliini. Plesiomorphic features include nearly equal lengths of anterior and posterior auricles, and a posterior auricular margin that is outwardly convex with its overall trend either nearly at a right angle to the hinge line or forming a slightly obtuse angle. *"E." solea* (Deshayes, 1830) from the Lutetian of Belgium has this same suite of characters except for the late-appearing radial costae, which here are absent altogether, although its right anterior auricle still retains low radial costae.

The basis for the inference that the Southern Hemisphere genera *Duplipecten*, *Lentipecten*, and *Adamussium* share an ancestor rooted in the "*E.*" *solea* group is based on the presence of the characters mentioned above in the early ontogeny of both *Duplipecten* and *Lentipecten*. Both of these genera develop steeply inturned disk flanks in their right valves, with the point of inflection being a sharp edge separating the disk from the disk flank, and both show hypertrophy of a right anterior tubercle that forms distal to the byssal notch (not proximal as in *Pseudentolium*). It is likely, however, that these genera each originated from a common ancestor (A7, Fig. 1.2) rather than one from the other. Both appear at about the same time in the Bortonian stage (upper Lutetian) of New Zealand (Beu and Maxwell 1990). *Duplipecten* (A8, Fig. 1.2) evolved very deeply inturned disk flanks, particularly on its right valve, as well as autapomorphic low costae limited to the early ontogeny of its left valve. *Lentipecten* (Node A9), on the other hand, evolved amussioid features, with originally inturned, sharp-edged disk flanks reduced to mere sutures between the auricular and disk surfaces, more prominent internal auricular buttresses and tubercles, and restriction of the functional ctenolium to early ontogeny. *Adamussium* (A10, Fig. 1.2) is tied to the other members of the Adamussiini by its nearly equal auricles, the outwardly convex margins of its posterior auricles, and by its microsculpture, in which fine commarginal lirae interrupt the sweeping continuity of antimarginal striae. The regularity of this microstructural pattern is particularly close to that of *Lentipecten*, as shown by Beu and Dell (1989), who found a nearly morphologically intermediate fossil taxon preserved in the upper Oligocene (Chattian) level of a drill core from the Antarctic Ocean. Lastly, *Lentipecten* and *Adamussium* share a similar degree of extension of the prismatic stage of the right valve, to a height of 6 or 7 mm, that is higher than other members of this or of other clades in the subfamily.

Clearly many of the features of *Adamussium* are autapomorphies related to its extremely cold environment in Antarctica. Its shell is paper thin, and the low radial plicae are likely secondary features that serve to strengthen its shell. It lacks the amussioid auricular denticles of its hypothesised *Lentipecten* ancestor, but this may again be a secondary loss related to the difficulty of calcification in such a harsh environment.

1.2.2.3 Tribe Eburneopectinini, new tribe

Hertlein (1969: N352) used the term "*Eburneopecten* group" for an assemblage of 13 genera and subgenera having small, smooth shells, some without a ctenolium. He specifically stated that the relationships of these taxa "are not fully known". Indeed, the members of this heterogenous assemblage are now distributed among three families (Propeamussiidae, Entoliidae, and Pectinidae). Vaught (1989: 119), in a compendium of classifications of living Mollusca that provided no descriptions or analyses, used "Tribe Eburneopectenini" [sic] for Hertlein's polyphyletic group. However, because she was concerned only with living taxa, *Eburneopecten*, an extinct genus, was omitted from her list. Her use of a tribal name for this group is here regarded as a *nomen nudum* (ICZN 4th Edition: Article 13 and p. 111).

In the present study, the tribe Eburneopectinini is limited to the genus *Eburneopecten* because of morphological and stratigraphic evidence that this genus evolved from coarsely ribbed *Dhondtichlamys* in the early to mid-Eocene of eastern North America and is phylogenetically independent of so-called *"Eburneopecten"* in Europe (Fig. 1.2, Node E1 compared to A6). Well-preserved specimens of *E. scintillatus* consistently display fine radial costellae on the left beak that are ontogenetically decoupled from later costae that may appear near the ventral margin. The species retains costae throughout ontogeny on its right anterior auricle, whereas the costae on other auricles are absent or limited to early ontogeny. Similar decoupling of ribbing occurs within the range of variation of two older species that resemble *E. scintillatus*, and they are included here in *Eburneopecten*. One of these is *E. clarkeanus* (Aldrich, 1895) from the Cook Mountain Formation of Alabama of Bartonian age; the other is *E. burlesonensis* (Harris, 1919) from the Weches Formation of Texas of middle Lutetian age. The latter, older species resembles *Dhondtichlamys* in having strong steep-sided plicae but differs in having much finer antimarginal microsculpture in the rib interspaces. Both of these species exhibit great variation in the degree of rib interruption, with some specimens approaching the nearly smooth condition of *Eburneopecten scintillatus*. (See illustrations of these species in Harris 1919, pl. 14, figs. 11–13 and pl. 15, figs. 8–13.)

1.2.2.4 Tribe Serripectinini, new tribe

The new tribe Serripectinini is defined phylogenetically as a monophyletic group of genera beginning with a morphology represented by *"Chlamys" choctavensis* (Aldrich, 1895) of the early Ypresian of southeastern United States (S1, Fig. 1.2) and leading to the genera *Batequeus* Squires and Demetrion, 1990 (Lutetian, Baja California, Mexico; S3 in Fig. 1.2), *Serripecten* Marwick, 1928 (Ypresian to Tortonian of New Zealand; S4 in Fig. 1.2), and *Janupecten* Marwick, 1928 (Priabonian to Chattian, New Zealand; S5 in Fig. 1.2). The age ranges are taken from Beu and Maxwell (1990: 33–34) but translated into European stage names via their correlation chart (their fig. 1).

"Chlamys" choctavensis retains the shape and smoothness of primary plicae present in *Dhondtichlamys* and has the same pattern of coarse antimarginal striae restricted to rib interspaces. As in the *Dhondtichlamys greggi-johnsoni* lineage (see above), *"C." choctavensis* has anterior and posterior auricles of nearly equal length, a deep byssal notch, and an ontogenetically persistent active ctenolium. Morphological characters that tie *"C." choctavensis* with *Batequeus*, *Serripecten*, and *Janupecten* (Node S1, Fig. 1.2) are its broadened umbonal angle, lower convexity of both valves (still retaining a biconvex condition), incipient disk gapes, auricular tubercles (the left anterior tubercle present, other tubercles weak or absent), and, most importantly, the medial intercalation of secondary costae in rib interspaces in mid to late ontogeny on both valves. This species also shows the start of sculptural differences between the two valves in that the anterior and posterior ribs of the left valve develop erect cuspate scales beginning early in ontogeny, these being absent on the right valve. There is a very close resemblance between *"C." choctavensis* and *"Serripecten" tiorioriensis* Marwick, 1928, a species that occurs in rocks of approximately the same age (late Paleocene and/or earliest Eocene) in New Zealand. Beu and Maxwell (1990: 88) referred to this species as "the oldest known species of [*Serripecten*]", noting that it differs from later members of the genus by its

smaller size and its simpler radial sculpture, radial ribs that are smooth-crested, and the late appearance of secondary ribs. Other species that possibly represent early *Serripecten* have been described by Maxwell (1992) form the Bortonian (late Lutetian) stage of New Zealand.

Node S2 (Fig. 1.2) represents the start of extreme differential sculpture of the two valves, that of the left valve being densely costate, with all of the costae scaly, that of the right valve having only the coarser primary plicae and intercalated secondaries. *Batequeus* (Node S3) retains plesiomorphic biconvexity (Squires and Demetrion 1990, fig. 2(3)) while evolving medial bifurcation of its central right primary plicae. Thus far *Batequeus* is known only from its type species, *B. mezquitalensis* Squires and Demetrion, 1990, from the Bateque Formation of Baja California Sur, Mexico. Its age was originally given as middle Eocene based on planktonic foraminifera indicating zones P8 or P9 of Berggren et al. (1985). However, those zones now indicate a late Early Eocene age (late Ypresian) according to Berggren et al. (1995: 140). A second species of *Batequeus*, *B. ducenticostatus* Campbell, 1995, from the "Cross Formation" of probable Bartonian age of South Carolina, is not a *Batequeus* but rather is a member of the *Chlamys* group. Its radial sculpture is similar on the two valves, and its radial costae are more densely packed, more densely scaly, and with a more complex introduction pattern throughout ontogeny than in any member of the *Batequeus-Serripecten* clade.

Node S4 marks the start of an ontogenetic trajectory of increasing valve convexity so that maximum convexity is near the mid-point of the dorso-ventral dimension. This node also marks the steepening and actual inturning of the disk flanks, particularly those of the right valve. Both *Serripecten* and *Janupecten* have low, broadly trigonal plicae on the right valve. *Serripecten* develops distally pointed scales on these plicae, with the scales becoming distinctly skewed toward the outside near the anterior and posterior margins of the disk. Its differential sculpture, with costae more numerous and more scaly on the left valve than on the right, is shared with *Batequeus*.

Janupecten (S5, Fig. 1.2) was originally named as a subgenus of *Serripecten* by Marwick (1928: 455), but Hertlein (1969: N363) regarded the former as a junior synonym of the latter. The close relationship of *Janupecten* to *Serripecten* is indicated by remnant outwardly skewed scales atop the anterior and posterior ribs of the right valves observable on well-preserved specimens of *Janupecten polemicus* Marwick, 1943, and *J. uttleyi* (Marwick, 1928). On less well-preserved specimens, the exterior features on rib crests are obliterated by corrosion that is limited to the rib crests while the rib interspaces remain intact. Commonly, this corrosion is extensive enough so that the ribs may appear depressed below the level of the adjacent interspaces. This differential stability between rib crests and rib interspaces appears to be the result of a less stable shell microstructure present along the rib crests just interior to a very thin outer calcitic layer. This microstructure, which has not yet been investigated by scanning electron microscopy, turns chalky even before the thin exterior layer is breached, causing the ribs of some specimens to appear as white rays. Cross sections through the ribs show that the chalky material is restricted to a lenticular area beneath the rib crest. This unusual structure is regarded as an autapomorphy of *Janupecten*.

1.2.2.5 Tribe Palliolini Korobkov in Eberzin, 1960

The tribe Palliolini is rooted in *Palliolum* Monterosato, 1884, and contains the derivative genera *Placopecten* Verrill, 1897, *Karnekampia* Wagner, 1988, and *Pseudamussium* Mörch, 1853, all of which are endemic to the North Atlantic. Representative extant species are *Palliolum striatum* (Müller, 1776), *Placopecten magellanicus* (Gmelin, 1791), *Karnekampia sulcata* (Müller, 1776), and *Pseudamussium septemradiatum* (Müller, 1776). The tribe is paraphyletic because of the exclusion of the tribe Mesopeplini, new tribe, a closely knit group of five Southern Hemisphere genera with a morphological link to the Palliolini (see below).

A morphological definition of the Palliolini is difficult, because its constituent genera are among the most variable known in the Pectinidae. Like all members of the subfamily Palliolinae, members of the tribe Palliolini have irregularly foliated calcite in the ventral shell region outside the pallial line and differ in this respect from *Chlamys*. Perhaps the best morphological definition of the Palliolini that can be ventured is that the tribe consists of genera that have costate auricles but commonly either lack radial ribs on the disk, in which case the surface is dominated by coarse antimarginal microsculpture, or have primary ribbing in the form of numerous low costae. The higher magnitude plicae that occur in most taxa (above Node P2 in Fig. 1.2) are secondary, commonly arising by the merger of primary costae in early ontogeny. All but the more primitive members of the group have posterior auricles with an outwardly concave posterior margin at least in early ontogeny. Because of the secondary origin of the highest order ribs, ribbing style is enormously variable in spacing, ordering, and magnitude. Members of this group are generally chlamydoid in shape, with a deep byssal notch and persistent functional ctenolium, but some lineages evolved amussioid and pectinoid forms. The two best-known amusioid examples are *Palliolum gerardi* (Nyst 1835) in the Pliocene of the North Sea Basin in Europe (Janssen and Dijkstra 1996: 112), and the genus *Placopecten* in the Upper Miocene to Recent of eastern North America. Pectinoid form is exemplified by *Hilberia* von Teppner, 1922.

Palliolum is regarded here as the stem genus of the tribe. Among its oldest species are *Palliolum prestwichi* (Morris, 1852) from the Thanetian, *Palliolum vincenti* (Glibert 1933) from the lower Lutetian, *P. nysti* (G. Vincent, 1881) from the middle Lutetian, and *P. decussatum* (Münster in Goldfuss, 1833) from the Rupelian, all of which occur in northern Europe. Fossil species of the extant species *Palliolum incomparabile* (Risso, 1826) have been reported from rocks as old as Chattian (Báldi 1973: 183). Collectively these fossil species have posterior auricles that are shorter than the anterior ones in length and have convex to nearly straight posterior auricular margins that tend to form obtuse angles with the hinge. If radial costae are present on the disk, they are fine and numerous and tend to be continuous from early ontogeny. In these costate forms, coarse antimarginal striae commonly cross over the costae rather than being restricted to rib interspaces as in *Dhondtichlamys*. All auricles are multicostate regardless of whether costae are present on the disk. The extant species *Palliolum striatum* and *P. incomparabile* retain most of these ancestral features. The latter is the type species of the genus (subsequent designation by Crosse; see Hertlein 1969: N354).

All higher members of the Palliolini are united at Node P2 (Fig. 1.2) by (a) shared ribbing complexity, including the secondary major plicae mentioned above, and (b) increasing complexity of antimarginal striae in rib interspaces, commonly producing a "pseudoshagreen" microsculpture in very limited areas of the disk where not masked by other microsculpture. With the exception of *Hilberia*, all of the genera above this node also have at least some members that display third-order very fine costellae that are hardly greater in width than are the adjacent antimarginal striae. In the more primitive members of *Hilberia*, e.g., *H. hoeninghausi* (Defrance, 1825: 256), very coarse antimarginal ridgelets are present in a complex pattern in rib interspaces in early ontogeny.

At Node P4 (Fig. 1.2) the extant taxa "*Palliolum*" *tigerinum*, *Karnekampia*, and *Pseudamussium* are united by substantial auricular asymmetry, with the posterior auricles shortened relative to the length of the anterior auricles, and by the distinctly concave posterior margins of the posterior auricles, the presence of coarse radial costae on the left anterior auricle that produce a dentate anterior margin, and the presence of enlarged commarginal auricular lamellae on at least the basal part of the left anterior auricle. At least some populations in all of these groups also show the presence of large radial folds, although these may be highly variable in amplitude and number. The auricles of mature *Hilberia* are of nearly equal length, but they attain this symmetry secondarily during ontogeny. In juvenile shells of *Hilberia* the shorter posterior auricle has a distinctly concave posterior margin.

At Node P5 (Fig. 1.2) *Hilberia*, *Karnekampia*, and *Pseudamussium* are united by the presence of commarginal lirae in rib interspaces, these lirae being regularly spaced but of irregular commarginal trend. Their intersections with the fine radial costellae that are present in the earliest ontogeny of the left valve produce a finely reticulate pattern on the left umbo. The lirae tend to persist throughout ontogeny in *Karnekampia* and the more primitive species of *Hilberia*; in *Pseudamussium* the lirae disappear in later ontogeny either as part of a general smoothening of the shell or because they are masked by a complex microsculpture (see below). These genera also share a peculiar scale morphology. Each scale is enlarged but distally directed along the rib crest, commonly merging with the succeeded scale as well as extending laterally somewhat beyond the sides of each rib. These types of scales are present in *Hilberia hoeninghausi* (particularly on the right valve), in *Karnekampia*, and on the right valves of *Pseudamussium*.

At Node P6 (Fig. 1.2), *Hilberia*, comprising a cohesive set of species limited to the Nordic and Atlantic basins of Europe in the latest Eocene through Oligocene (Roger 1944; Anderson 1958), evolved a pectinoid form, with a convex right valve and flat left valve and a correlated set of changes leading to a non-functional ctenolium, very rounded byssal notch, equal length of auricles, basal auricular buttresses on the shell interiors, and inturned disk flanks. Unlike true *Pecten*, none of the species with strong radial plicae developed internal rib carinae and the hinge teeth are simple, with dorsal teeth dominant and resilial teeth poorly developed. As in all members of the subfamily Palliolinae thus far studied in detail, *Hilberia* has irregularly foliated calcite outside the pallial line in the central sector of the shell.

At Node P7 (Fig. 1.2), *Karnekampia* and *Pseudamussium* share the common presence of a pseudosurface formed by the side-to-side merger of fine third-order costellae that are

T-shaped in cross section. This microsculptural feature is common in the more southerly populations of species in both genera, whereas in more northerly populations its presence is variable within population samples. Although comparable costellae occur in both *Placopecten* and "*Palliolum*" *tigerinum*, their side-to-side merger has not been observed in *Placopecten* and occurs only very rarely in "*P.*" *tigerinum*. Lastly, at Node P8 *Pseudamussium*, represented in the present-day northeastern Atlantic and deep Mediterranean by *P. septemradiatum* and the possibly intergradational *P. clavatum* (Poli, 1795), is distinguished by its very shallow byssal notch (depth about one fourth or less the length of the anterior auricle compared to half the length of the auricle in *Karnekampia*), a ctenolium that is barely functional at maturity (compared to a strong and persistent functional ctenolium in *Karnekampia*), and five to seven large radial folds.

Except for the *Placopecten* clade (P3 in Fig. 1.2), the branching points of the phylogeny described above are corroborated by the relative times of known first occurrences in the fossil record. As stated above, the P1 lineage (*Palliolum*) is represented by fossils in the Thanetian and Lutetian. The main diversification of the tribe appears to have occurred in the Oligocene at a time of global cooling. A probable early member of the "*Palliolum*" *tigerinum* lineage in the Rupelian of Germany was originally described in the combination *Chlamys* (*Flexopecten*) *welschbergensis* Neuffer, 1973. But the published figures (Neuffer 1973, pl. 5, figs. 10–12) display the characters described above for Node P4. *Karnekampia* was also present by Rupelian time, an example being the species from the Rupelian of Germany identified by Neuffer (1973:37, pl. 5, figs. 18, 19) as *Chlamys* (*Chlamys*) *striatocostata* (Münster in Goldfuss, 1833). Specimens of this species reported by Glibert (1957:20, pl. 1, figs. 15a-d) from the Chattian of Belgium were examined by the author. They have the third-order T-shaped costellae and commarginal lirae that mark Node P6, and Glibert remarked on the probable relationship of this species to *Peplum* Bucquoy, Dautzenberg, and Dollfus, 1889, a genus now regarded by most malacologists as a junior synonym of *Pseudamussium*. *Karnekampia* proliferated during the Miocene in the North Sea Basin with the appearance of *Karnekampia lilli* (Pusch, 1837), a highly variable species with many subspecies, "varieties", and synonyms (Glibert 1945: 73; Marquet and Dijkstra 1999; and references therein). *Pseudamussium*, the youngest branch in the phylogeny, is not known in the fossil record before the Tortonian (Roger 1939: 214).

The position of *Placopecten* in the phylogeny of the Palliolinae is based on the demonstration of morphological transition between members of this genus and the fossil species "*Chlamys*" *decemnaria* (Conrad, 1834) by Gibson (1987: 68). At one end of its spectrum of variation "*C.*" *decemnaria* is chlamydoid in form, with a complex ribbing pattern beginning ontogenetically with fine costellae that merge and cluster into coarser secondary plicae. This ribbing pattern, coarse antimarginal microsculpture between the ribs and across the disk flanks, and limited areas of "pseudoshagreen" microsculpture on the disk resemble features that are present in "*Palliolum*" *tigerinum* and higher lineages in the phylogeny (Fig. 1.2). At the other end of the spectrum of variation, "*C.*" *decemnaria* approaches the shape and ribbing style of typical *Placopecten*, including the fossil species *P. clintonius* (Say, 1824), *P. princepoides* (Emmons, 1858), and *P.* sp. aff. *P. magellanicus* (Gibson 1987; see also Appendix Note 3). Given this transition, it is reasonable to place "*Chlamys*" *decemnaria* in the genus *Placopecten* (in the combination *Placopecten decemnarius*) and to expand the concept of *Placopecten* to include chlamydoid forms that show variation into *Placopecten*-type morphology. The reason that *Placopecten* in this expanded concept is placed as a sister group of the clade that originates at Node P4 (Fig. 1.2) rather than being a member of that clade is because of differences in auricular shape and auricular sculpture. The posterior margins of the posterior auricles of *Placopecten decemnarius* and other species of the genus are plesiomorphically straight or convex, not concave as at Node P4, and they lack the other auricular character states mentioned above for Node P4.

A difficulty with the phylogenetic position of *Placopecten* shown in Figure 1.2 involves the relative stratigraphic positions of *P. decemnarius* and the more typical members of the genus. The earliest species known is *P. princepoides*, which occurs in the Eastover Formation of Virginia in strata that are of late Tortonian age (Ward 1992: 4, 66). This species already has a broad nearly smooth shell with shallow byssal notch and obsolete ctenolium comparable to these features in extant *P. magellanicus*. Even though *Placopecten decemnarius* is more plesiomorphic, it has thus far been found only in the stratigraphically higher Yorktown Formation of Virginia of Pliocene age. A possible explanation of this discrepancy involves the completeness of the fossil record in the Atlantic Coastal Plain of eastern North America. Because *Placopecten* occurs near the northern limit of the Eastover and Yorktown Formations and because older formations to the north may represent shallower facies, it may be that its fossil history is simply not preserved. On the basis of the phylogeny as currently understood, *Placopecten* probably originated from an eastern Atlantic ancestral species. Curiously, one of the few living northeastern Atlantic species that occurs in the northwestern Atlantic is *Palliolum striatum*, based on USNM 764670, a specimen dredged from 192 m off Cape Cod, Massachusetts, in 1936.

1.2.2.6 Tribe Mesopeplini, new tribe

The new Tribe Mesopeplini is erected to contain a phylogenetically closely knit group of five palliolinine genera and one subgenus (Fig. 1.2, Nodes M1-M6), all of which are endemic to New Zealand and Australia (Beu 1978, 1995; Beu and Maxwell 1990): *Mesopeplum* (*Mesopeplum*) Iredale, 1929, Bartonian?, Chattian to Recent, New Zealand and Australia; *Mesopeplum* (*Borehamia*) Beu, 1978, Messinian to early Piacenzian, New Zealand only; *Sectipecten* Marwick, 1928, Serravallian to latest Pliocene, New Zealand and Chatham Islands; *Phialopecten* Marwick, 1928, Messinian to late Pliocene, New Zealand only; *Kaparachlamys* Boreham, 1965, late Pliocene, New Zealand and Chatham Islands; and *Towaipecten* Beu, 1995, latest Pliocene and earliest Pleistocene, New Zealand only. The only extant genus, *Mesopeplum* (*Mesopeplum*), is represented in present-day oceans by *M.* (*M.*) *caroli* Iredale, 1929, of southeastern and southern Australia and by *M.* (*M.*) *convexum* (Quoy and Gaimard, 1835) of New Zealand and the Chatham Islands. Reports of *Mesopeplum* in the Eocene and Oligocene of Argentina by Morra and Erdmann (1986) refer to giant members of the Chlamydinae that bear shagreen microsculpture and are therefore excluded from the Palliolinae.

Collectively, this group displays a higher level of variation in ribbing patterns and shell shapes than almost any other group of Pectinidae. What ties these genera together are demonstrations of morphological transitions in carefully collected stratigraphic sequences. These transitions and occurrences have been discussed in detail in a series of remarkable monographs by Beu (1978), Beu and Maxwell (1990), and Beu (1995). The relationships of these genera depicted in the phylogeny in Figure 1.2 (Nodes M1–M6) are based upon these works and my own examination of specimens representing all of the genera except *Kaparachlamys* and *Towaipecten*.

Mesopeplum (*Mesopeplum*) is the basal genus of the tribe (Beu 1995: 186), a conclusion that is supported by its morphology and its lowest stratigraphic occurrence. It shares with the *Karnekampia-Pseudamussium* clade (P5, Fig. 1.2) the presence of radial costae that are clustered in broad radial folds, posterior auricles with a distinctly concave posterior margin, auricular asymmetry (posterior auricle shorter than anterior auricle in primitive members), and commarginal lirae in costal interspaces. In all members of the Mesopeplini (Node M1), the commarginal lirae, which are present only in early ontogeny below Node M1, are present throughout ontogeny and in more derived genera may be extended into commarginal frills. A limited number of broad radial folds, generally five, is also present at least in the early ontogeny of primitive *Mesopeplum* (*Borehamia*), *Sectipecten*, and *Phialopecten* (Beu 1995: 32, 255) but more even ribbing patterns prevail in more derived forms. The evolution of *Sectipecten* from *Mesopeplum* (*Borehamia*) (Node M3) was discussed by Boreham (1961) and by Beu (1995: 55). All of the genera of Mesopeplini above Node M4 develop internal rib carinae, the only occurrence of this character within the subfamily Palliolinae. These carinae are readily apparent in *Phialopecten* but less so in *Towaipecten*, where plicae disappear in late ontogeny, and in *Kaparachlamys*, where the internal rib carinae are vestigial (limited mainly to the lateral areas of the inner disk surface). The secondary approach toward very fine external costation and smoothness, as well as shortening of auricles, and reduction in depth of the byssal notch, give *Kaparachlamys* a *Placopecten*-like appearance. Marwick (1928: 451), Boreham (1965: 25), and Beu (1995: 19) all attributed this similarity to evolutionary convergence.

If the *Karnekampia-Pseudamussium* clade of the Palliolini and the tribe Mesopeplini shared a common ancestor in the early Oligocene as indicated at Node P4 (Fig. 1.2), the diversification of these groups may be related to the cooling of the world's climate at that time. This may have allowed the spread of palliolinine taxa that had previously been confined to higher northern latitudes across the tropics into the Southern Hemisphere. The earliest occurrence of *Mesopeplum* in New Zealand, however, is still poorly known. Beu and Maxwell (1990: 338) reported that *M.* (*M.*) *convexum* (Quoy and Gaimard, 1835) may occur as early as Kaiatan Stage (= Bartonian, early Late Eocene) in New Zealand but also expressed uncertainty as to whether these early forms were a part of the same species lineage, the known stratigraphic range of which is from late Chattian to present.

1.2.3 Origins of the *Decatopecten* and *Pecten* Groups

Waller (1991: 15 and fig. 8; Fig. 1.1 herein) hypothesised a close relationship between *Pecten* Müller, 1776, and *Decatopecten* Rüppel in Sowerby, 1839, and suggested that the Decatopectinini may serve as a bridge between the *Pecten* group and the Palliolinae. *Amussiopecten* Sacco, 1897a, *sensu lato*, was shown as the basal member of the *Pecten* group in the Oligocene, leading to three separate clades represented by the extant genera *Amusium*, *Pecten*, and *Euvola* (Waller 1991, fig. 11). Subsequent studies of *Amussiopecten* and a search for its ancestors in the Eocene and Oligocene have led to new phylogenetic hypotheses (Fig. 1.3): Both the *Pecten* and *Decatopecten* groups are derived from an ancestral stock that in turn was derived from the *Aequipecten* group of Waller (1991). There is no close relationship between these groups and the Palliolinae. Furthermore, the *Decatopecten* group as originally defined is not monophyletic but rather consists of three extant components with separate origins: (a) *Decatopecten* and allied genera occurring mainly in the Indo-Pacific region (Waller 1986); (b) *Annachlamys* Iredale, 1939, occurring in the western Indo-Pacific region; and (c) *Flexopecten* Sacco, 1897a (including *Proteopecten* Monterosato, 1899, and *Lissopecten* Verrill, 1897), occurring in the Mediterranean and adjacent Atlantic. As indicated in Figure 1.3, the former *Decatopecten* and *Pecten* groups are now regarded as two separate monophyletic tribes, Decatopectinini and Pectinini, while the genera *Amussiopecten*, *Amusium*, and *Euvola* comprise a third monophyletic tribe Amusiini. Collective these three groups comprise the subfamily Pectininae.

A likely stem group of the Pectininae is the "*Pecten*" *perplanus* stock of Glawe (1969), an array of species and subspecies common in the Vicksburgian Stage of the Oligocene of southeastern U.S.A. Vicksburgian strata are now known to be equivalent in age to the Rupelian and early Chattian of Europe (Siesser 1984; Rossbach and Carter 1991). Both G.D. Harris (1951: 9) and Glawe (1969: 39) suggested that a probable ancestor of the *P. perplanus* stock in the Eocene is "*Pecten*" *elixatus* Conrad, 1844, said to be from the "Santee Limestone" of South Carolina (Glawe 1969: 39). The precise position of this fossil within limestone beds that have been called "Santee Limestone" is unknown and there is substantial confusion regarding the correlation of limestone units in this area (Campbell 1995). True Santee Limestone is generally regarded as Middle Eocene (middle Lutetian through Bartonian) in age (Palmer and Brann 1965; W.B. Harris et al. 1984), but it is possible that this species is from a later Eocene unit.

Glawe (1969), by plotting the frequency distribution of several morphotypes within stratigraphically successional population samples, showed that the "*Pecten*" *perplanus* stock during the early Oligocene split into two lineages, both derived from "*Pecten*" *perplanus perplanus* Morton, 1833. The first lineage evolved from this species to "*P.*" *perplanus poulsoni* Morton, 1834, to "*P.*" *perplanus byramensis* Gardner, 1945. The other evolved from "*P.*" *perplanus perplanus* to "*P.*" *howei mariannensis* Glawe, 1969, to "*P.*" *howei howei* (Mansfield 1940). The "*Pecten*" *perplanus* stock ended, so far as known, with the last occurrence of "*P.*" *howei howei* in the Chickasawhay Limestone and Paynes Hammock Formation of the American Gulf Coastal Plain in the late Rupelian or early Chattian (Glawe 1969: 57; Seisser 1984, fig. 3; Berggren et al. 1995: fig. 3).

Members of the *"Pecten" perplanus* stock have regularly foliated calcite in the midventral region outside of the pallial line, not the irregular foliation that is present throughout the Palliolinae. Their shell shape is generally biconvex and aequipectinoid but with the left valve distinctly less convex than the right, especially in the basal members and in the *byramensis* lineage, where some individuals may approach a pectinoid form. The byssal notch is only moderately deep in the geologically older members of the stock and becomes more shallow in both lineages. In all members of the group, the ctenolium terminates before maturity, and all have strong internal buttressing ridges aligned with the bases of the auricles and ending in tubercles. Auricular costae are strong in early members of the stock and tend to remain so in the *byramensis* lineage, but they become obsolete or disappear within the *howei* lineage. All members of the *perplanus* stock have strong commarginal lamellae in a characteristic pattern described below. Microsculpture of the left beak is seldom preserved, but it appears to be smooth in later members of the stock. A specimen identified as *"P." perplanus poulsoni* was found to have a transient pitted microsculpture in the late pre-radial stage of the left valve, suggesting that the smooth beaks of later members of the stock may be a derived condition. In the early radially plicate stage, very fine antimarginal striae may be present in rib interspaces between commarginal lamellae, but these disappear in later ontogeny. Finally, all members of the stock have well-developed internal rib carinae.

A link between the *"Pecten" perplanus* stock with *Aequipecten* is indicated not only by the presence of strong internal rib carinae, but also by the trend of commarginal lamellae on the flanks of radial plicae in early ontogeny. In the early ontogeny of *"P." perplanus perplanus* and some *"P." perplanus poulsoni*, before the unilirate phase (see below), these lamellae form tight distally concave arcs on steep rib flanks, a pattern that is also commonly present in the early ontogeny of aequipectinine genera such as *Aequipecten*, *Cryptopecten*, and some *Argopecten*.

Other characters found in the *"Pecten" perplanus* stock provide phylogenetic links to *Amussiopecten* and true *Pecten* (Fig. 1.3). Most members of the *"P." perplanus* stock have three pairs of hinge teeth on the right valve, an elongate dorsal tooth on each side of the resilifer, a single intermediate tooth that is shorter but equally strong at least on the anterior side, and a weakly developed resilial tooth. Ontogenetically, the intermediate tooth differentiates from an early out-turned resilial tooth, first on the anterior side and later on the posterior side. A similar ontogenetic pattern leading to a three-element hinge dentition occurs in some fossil *Mimachlamys* and *Aequipecten*, in some *Amussiopecten*, and in forerunners of true *Pecten*. In *Amussiopecten* this dentition pattern occurs mainly in the oldest members of this genus in the upper Oligocene, although it is retained in *Amussiopecten subpleuronectes* (d'Orbigny, 1852) and *A. burdigalensis* (Lamarck, 1806), which are, respectively, from the Aquitanian and Burdigalian of southern Europe. The latter species is the type species of the genus. Three-element dentition is also present in southern European species that have long been regarded as belonging in the ancestry of true *Pecten*: *"Pecten" arcuatus* Brocchi, 1814, from the late Eocene (Priabonian) and Oligocene (Rupelian and Chattian) (Depéret and Roman 1902: 10; Cossmann 1921: 167; Bongrain 1988: 189); *"P." carryensis* Gourret, 1890, of the Aquitanian and possibly Burdigalian (Depéret and Roman 1910: 135; Roger 1939: 250); and some members of the highly variable species *Oopecten rotundatus* (Lamarck, 1819) and *Pecten beudanti* Basterot, 1825, of the Aquitanian and Burdigalian (Roger 1939: 10, 240). Another character that ties the *perplanus* stock to these same taxa is unilirate rib sculpture (Glawe 1969), in which a single radial lira runs down the centre of each plica. The plicae are generally parabolic in cross section. Commarginal lamellae, which pass straight across the rib interspaces, pass up the curved flank of the rib in a broad distally concave arc, finally trending distally along the side of the narrow lira on the rib crest. This unilirate pattern may occur on either or both valves and may be limited to a short ontogenetic phase. A similar configuration is present in each of the extinct southern European species listed above as having three-element hinge dentition. Unilirate sculpture of the *perplanus* type seems to be absent from *Amussiopecten* and its derivatives, but presumably this absence is secondary, reflecting a general decrease in rib amplitude and a trend toward an amussioid adaptive form.

These characters provide the basis for the early part of the phylogeny of what will be called here the Tribe Pectinini (Fig. 1.3). "*Pecten*" *arcuatus* is regarded as the earliest known member of the tribe. This highly variable species of the Upper Eocene (Priabonian) and Oligocene (Rupelian and Chattian) of southern Europe has a thick, very inequivalved shell with strong simple plicae and a three-element hinge having prominent or even hypertrophied intermediate teeth. Its right valve is deeply convex; its left valve is highly variable in convexity, ranging from uniformly slightly concave, to overall convex but with a depressed ontogenetically early concavity, to uniformly concave (Bongrain 1988, pl. 17; Cossmann 1921). The forms that have ontogenetically early and temporary concavity of the left valve resemble later members of *Gigantopecten* Rovereto, 1899, that are well known for this feature, termed "coup de pouce" (thumb print) by Bongrain (1988: 56). "*Pecten*" *arcuatus* should probably be placed in this genus as its earliest member, a position implied in the detailed study of the genus by Bongrain (1988: 189). *Gigantopecten*, as its name implies (Appendix Note 4), reached gigantic size in its later phylogenetic history, becoming extinct in the late Pliocene. "*Pecten*" *arcuatus* had well developed commarginal lamellae, and on some specimens these combine with unilirate plicae to produce a sculptural pattern like that in the "*Pecten*" *perplanus* stock. As shown by Bongrain (1988), *Pecten beudanti* Basterot, a Burdigalian species that is regarded as the earliest true *Pecten*, resembles "*P.*" *arcuatus* in shell shape, and its left valve has coarse commarginal lamellae and incipient unilirate plicae. Unlike *arcuatus*, the hinge dentition of *P. beudanti* has weaker intermediate teeth that are commonly multiplied, a pattern that is found in extant *Pecten*. Beginning in the Burdigalian, true *Pecten* diversified rapidly and entered the Indian Ocean via the eastern gates of the Mediterranean before their initial closure at the end of the Burdigalian (Adams et al. 1990: 304; Rögl 1999: 343).

The diversification of true *Pecten* led repeatedly to forms with a right valve of only moderate convexity and a nearly flat left valve. These were assigned to a new subgenus *Pecten* (*Flabellipecten*) by Sacco (1897a, b) based on a Pliocene type species, *Pecten flabelliformis* (Brocchi, 1814). Depéret and Roman (1910: 106) mistakenly thought these later forms were the same as *Amussiopecten* Sacco, 1897, and chose to regard the latter subgenus as a junior synonym of the former even though *Amussiopecten* appeared first in Sacco (1897a).

Waller (1991: 38–39, fig. 11; Fig. 1.1 herein) emphasised the phylogenetic separation of true *Pecten* of the eastern Atlantic and western Indo-Pacific regions from *Euvola* of the Americas, both groups originating from the genus *Amussiopecten* in the Miocene. The new analysis outlined above increases the phylogenetic separation of these two groups by demonstrating an earlier origin for *Pecten* (Fig. 1.3). As in Waller (1991), both *Euvola* and *Amusium* are still regarded as derivatives from *Amussiopecten*, but new research on tropical American Tertiary pectinids now makes it possible to provide more detail on the *Amussiopecten-Euvola* relationship. The evidence lies in a stratigraphic succession of Central American species that lived during the Miocene when the Atlantic and Pacific were still connected by Central American seaways. Woodring (1982: 581) referred the taxa in this succession to his "*Flabellipecten gatunensis* group", claiming that in Panama, "The group of *Flabellipecten gatunensis* can be traced through nine stratigraphic units, ranging in age from late Oligocene to late Miocene or Pliocene." He regarded the first three taxa in this stratigraphic succession to be "an unbroken main lineage", these taxa being "*Flabellipecten*" *gatunensis protistus* Woodring, 1982, of the late Oligocene and early Miocene, followed by "*F.*" *gatunensis gatunensis* (Toula, 1909) of the early and middle Miocene, in turn followed by "*F*". *gatunensis macdonaldi* (Olsson, 1922) of the late Miocene or early Pliocene.

Examination of specimens of all of these taxa has led to a different interpretation. The early members of this succession lack auricular costae and are thus unlike the European subgenus *Pecten* (*Flabellipecten*), which has multiple auricular costae tending to be evenly distributed across the auricles as in *Pecten s.s.*. Later members of the Central American series exhibit a few weak, unevenly spaced auricular costellae, which may resemble in number and spacing the auricular costae in *Euvola*. Also unlike *Pecten* (*Flabellipecten*), all of the Central American taxa with the exception of the earliest one, "*Flabellipecten*" *gatunensis protistus*, lack intermediate hinge teeth. Instead, the dentition is dominated by dorsal teeth, thus resembling the hinge configuration of *Euvola* (Waller 1991: 38). It is only "*Flabellipecten*" *gatunensis protistus* that has intermediate hinge teeth, there being a single short but well differentiated one on each side of the right resilifer. The only significant difference between this taxon and typical Oligocene *Amussiopecten* is in ribbing persistence. In *Amussiopecten* the disk plicae during ontogeny fade out before reaching the margin of the disk (see Masuda 1971b). In "*F.*" *gatunensis protistus*, the distal fading of plicae is variable, as Woodring (1982: 583) indicated, and in some specimens the plicae persist throughout ontogeny.

Evolutionary trends within this Central American group include increasing persistence and amplitude of disk plicae and increase in adventitious auricular costae from a plesiomorphically non-costate state. These trends can be followed into the living eastern Pacific taxa "*Pecten*" *diegensis* Dall, 1898, "*P.*" *sericeus* Hinds, 1845, and "*P.*" *stillmani* Dijkstra, 1998, taxa that Woodring (1982: 584) also recognised as being closely related and in the same group. The same features are present in a group of fossil species in the Upper Miocene and Pliocene of southern California and Baja California, Mexico, commonly referred to *Flabellipecten* (e.g., Moore 1984 and Smith 1991). Masuda (1971a) had earlier erected a new genus, *Leopecten*, for this group, choosing as type species *Pecten* (*Patinopecten*) *bakeri* Hanna and Hertlein, 1927, from the Lower Pliocene of Baja California. This is an unusual endemic member of the genus because of its high degree of secondary radial costation, but clearly the name *Leopecten* should apply to the whole American "*Flabellipecten*" assemblage. Here I reduce *Leopecten* to the level of subgenus within *Euvola* because of the close phylogenetic linkage. Masuda (1971a: 171) anticipated Woodring's (1982) discovery of the succession of species linking *Amussiopecten* and *Leopecten* in stating, "... it is inferred that *Leopecten* may have descended from *Amussiopecten* before the Early Pliocene, but no intermediate form is known between the two."

Euvola (*Euvola*) differs from *Euvola* (*Leopecten*) in having a deeply convex right valve and a nearly flat to concave left valve, i.e., a pectinoid adaptive form. Early members of *Euvola s.s.* have either non-costate auricles or weak adventitious costae similar to those present on many *E.* (*Leopecten*); later members have two or three prominent costae on the dorsal regions of the left auricles. The earliest known *Euvola*, *sensu stricto*, is an unnamed species that occurs in the Baitoa Formation of the Dominican Republic in an age range that probably extends from late Burdigalian to Serravallian (Saunders et al. 1986: 37). This is possibly of about the same age as the oldest known occurrence of *Leopecten gatunensis gatunensis* in the lower member of the Alhajuela Formation of Panama (Woodring 1982: 584), thus indicating that the split between mildly right-convex *Euvola* (*Leopecten*) and deeply right convex *Euvola* (*Euvola*) had already occurred early in the history of this genus.

The new phylogenetic treatment of the *Decatopecten* group compared to that in Waller (1991) stems from new studies of the fossil history of this group. The tribe Decatopectinini Waller, 1986, as originally defined consisted of ten genera, all but one of which have distributions that are entirely or mainly in the tropical Indo-Pacific region. The exception is *Flexopecten*, which presently lives in the Mediterranean and adjacent eastern Atlantic with a single outlier species occurring in the western South Atlantic (Waller 1991: 36). In the original diagnosis of the Decatopectinini, four characters were used to define the group: (a) unusually fine and closely spaced regular commarginal lamellae, with a spacing density ranging from 30 to 70 lamellae per two-millimetre distance along a radius in center of disk at a distance of 10 mm from the beak; (b) very low inflation of the left beak, only slightly greater than that of the right beak; (c) antimarginal microsculpture very fine and restricted to early ontogeny of disk up to a point just distal to the origins of radial plicae; and (d) hinge dentition of the right valve dominated by dorsal and/or intermediate teeth, the latter sometimes multiple or absent, and resilial teeth low or absent.

Although *Flexopecten* is consistent with placement in the Decatopectinini in terms of characters (a) through (c), the extant species *Flexopecten glaber* (Linnaeus, 1758) and *F. flexuosus* (Poli, 1795) and their fossil counterparts in the European Pliocene and Pleistocene differ from other members of the tribe in hinge dentition. In *Flexopecten*, dorsal teeth are generally very weak, and resilial teeth are dominant, showing an out-turning and merger with the prominent ventral shelf of the hinge plate. Articulation is mainly the result of the resilial teeth of the right valve fitting into resilial sockets on the left and infradorsal teeth of the right valve resting in shallow infradorsal sockets on the left. Similar features are present in the closely related genus *Lissopecten*. This pattern is more like that in the *Aequipecten* group than that in the *Decatopecten* group. Furthermore, the left beak sculpture of *Flexopecten* is commonly pitted, unlike that of other Decatopectinini where the left beak is smooth. Collectively these features indicate that *Flexopecten* should be removed from the Decatopectinini and placed in the *Aequipecten* group. This would imply that the dense, closely spaced commarginal lamellae of *Flexopecten* developed independently of this feature in the Decatopectinini. If the genus *Antipecten* Cossmann and Peyrot, 1914, of the European Miocene (Burdigalian-Serravallian) is ancestral to *Flexopecten*, as indicated by Roger (1939: 69, 91), then convergent development of dense commarginal lamellae seems likely, because *Antipecten* has far-set rather than close-set lamellae. I found looped commarginal lamellae on steep rib flanks in the early ontogeny of the specimen of *Antipecten pseudopandorae* (Berkeley Cotter 1904) from the lower Burdigalian of Portugal figured by Roger (1939, pl. 5, fig. 7), suggesting descent from within the *Aequipecten* group.

Another of the ten genera originally included in the Decatopectinini may also have a different phylogenetic origin from other members of the tribe. All extant species of *Annachlamys* Iredale, 1939, begin ontogeny after the prismatic stage with close-set commarginal lamellae, although their spacing is somewhat greater than the lower density limit given for the Decatopectinini by Waller (1986: 30 lamellae per two-millimetres measured 10 mm from origin of growth). At a point in ontogeny that varies among species, the dense commarginal lamellae give way to far-set, projecting lamellae. The transition is similar in all species, with the far-set condition arising first on the rib crests because of the merger or termination of some lamellae as they pass from rib interspaces onto rib flanks. This stage is then followed by loss of close-set lamellae in the rib interspaces as well. These later, far-set projecting lamellae crossing both ribs and interspaces resemble closely the lamellae present in the *Pecten perplanus* stock. *Annachlamys reeve* (Adams and Reeve, 1850) develops a single lira on each rib crest at a stage in ontogeny just before the end of the close-set lamellae, and this lira then continues into the far-set stage before terminating. The resulting transient growth stage is remarkably similar to the unilirate sculpture of the *Pecten perplanus* stock. A similar sculptural ontogeny occurs in *Annachlamys iredalei* (Powell 1958). In contrast, the other Decatopectinini genera display persistent close-set lamellae throughout ontogeny, and the earliest lamellae are more closely spaced.

The fossil record of *Annachlamys* appears to be limited to the Neogene (Miocene to Pleistocene) of the Indo-Pacific region. Examples are *Annachlamys okinawaensis* Noda, 1991, from the Ryukyu Islands, southwest Japan, *Pecten (Chlamys) suvaënsis* Mansfield, 1926, and *Pecten (Chlamys) nausorensis* Ladd, 1934 from the late Miocene or early Pliocene of the Fiji Islands, and *Pecten murrayanus* Tate, 1886, from the Morgan Limestone of South Australia. The last species is possibly the oldest known *Annachlamys*. According to Ludbrook (1973: table 1) the Morgan Limestone contains planktic foraminiferal zones N7 and N8. According to Berggren et al. (1995) these are of late Burdigalian and Langhian age. Roger (1939: 16) noted that the extant species *Pecten leopardus*, which is now assigned to *Annachlamys*, is known from the Miocene of Java and that it is probably a living representative of his *Chlamys rotundata* group, a group that is closely allied to *Gigantopecten* (see Bongrain 1988). Indeed, some *Annachlamys* have an unusual ontogenetic shift in convexity of the left valve, with the proximal part flattened, recalling the more extreme expression of this condition in the *Gigantopecten* clade.

Removal of *Flexopecten* and *Annachlamys* from the tribe Decatopectinini leaves eight genera in the tribe: *Anguipecten* Dall, Bartsch, and Rehder, 1938; *Bractechlamys* Iredale, 1939; *Decatopecten* Rüppel in Sowerby, 1839; *Excellichlamys* Iredale, 1939; *Gloripallium* Iredale, 1939; *Juxtamusium* Iredale, 1939; *Mirapecten* Dall, Bartsch, and Rehder, 1938; and *Somalipecten* Waller, 1986. All of these have tropical Indo-Pacific distributions at present, and there are few unequivocal fossil records older than the Miocene. There is but one comparable genus in older rocks, *Anatipopecten* Hertlein, 1936, based on a type species, *Pecten anatipes* Morton, 1833, from the Oligocene of Alabama (Glawe 1974: 8). It resembles *Decatopecten* in having a low number of broad plicae and very dense, closely spaced commarginal lamellae, but it differs in lacking well-developed intermediate hinge teeth and in having only very weakly developed internal rib carinae. The oldest species assigned to *Anatipopecten* is *Chlamys* (*Lyropecten*) *incertae* Tucker-Rowland, 1938, which is common in the upper part of the Crystal River Formation of Florida (Nicol et al. 1989) of probable latest Eocene (latest Priabonian) age. To judge from figures of this species in Nicol et al. 1989: pl. 1), internal rib carinae are present at least in smaller individuals and intermediate hinge teeth are also present. This species, unlike *Annachlamys*, lacks sculptural features that would suggest a relationship to the *Pecten perplanus* stock, but the presence of internal rib carinae would suggest derivation from an aequipectinine ancestor.

1.2.4 Eastern Atlantic *Pecten*

As reviewed above, molecular genetic studies of Atlantic *Pecten maximus* and Mediterranean *Pecten jacobaeus* have produced conflicting results. Either the genetic difference is so small that it may not be significant at the species level, or, where the geographic ranges of these species are in contact along an oceanographic barrier in the western Mediterranean, "there is a major discontinuity in interpopulation allozyme variation" (Ríos et al. 2002: 237). Comments on morphological similarity made in connection with some of the genetic studies have also been conflicting. Canapa et al. (2000b) indicated that the very close genetic similarity between *Pecten maximus* and *Pecten jacobaeus* is corroborated by their strong morphological similarity and evidence of morphological intergradation. As evidence for morphological intergradation, they claimed that there is "a variety with intermediate features, classified as *P. intermedius* by Monterosato in 1878, which is found off the coast of Portugal (near Gibraltar) and everywhere the ranges overlap." Ríos et al. (2002: 223), in contrast, claimed that the major genetic discontinuity that separates populations on either side of Almería-Oran Oceanographic Front is "coincident in space with the morphological discontinuity between the two taxa."

As far as I am able to determine from the literature, solid evidence for morphological intergradation of these species has never been presented. *Pecten intermedius* Monterosato, 1899 (not 1878), mentioned by Canapa et al. (2000b), is not recognised today as a valid species. It was based on specimens of *Pecten jacobaeus* from the northern Adriatic, an area that today harbours populations of *P. jacobaeus* that are dense enough for commercial exploitation (Wilding et al. 1999). Purported zones of intergradation between *P. maximus* and *P. jacobaeus* along the southern Portugal coast mentioned by Canapa et al. (2000b) and along the Atlantic coast of northwest Africa have not been adequately documented and may not exist. In their extensive monographs of Neogene pectinids of Europe, Depéret and Roman (1902: 59, 97) did not report intergradation between these species, although they reported fossil *P. maximus* in Pleistocene terrace deposits in Atlantic Morocco to the south of Safi (about 32°N) and typical *P. jacobaeus* from Cádiz, on the Atlantic coast of Spain. Roger (1939: 246) reported that *P. maximus* or *maximus-jacobaeus* transitional forms occur rarely in Pleistocene deposits in Italy where *P. jacobaeus* is abundant. Although there were deep penetrations of *P. maximus* into the Mediterranean during cold periods of the Pleistocene (Raffi 1986; Malatesta and Zarlenga 1986), more detailed studies have failed to document any intergradation between these species. For example, Imbesi (1956), in a study of variations of *P. maximus* and *P. jacobaeus* from Pleistocene deposits in Sicily and Calabria, Italy, concluded that "although the two species show variability, no intermediate form exists."

There is, however, evidence for morphological separation. Along the Algerian coast, Pallary (1900: 377) found *Pecten maximus* both to the west and east of Oran, including specimens as well developed as those in the Atlantic, and was able to distinguish these from *P. jacobaeus* living in the same area. Apparently recognising morphological differences in the spat of the two species, Cano et al. (1989) reported settlement of *Pecten maximus* but not *P. jacobaeus* on spat collectors at Málaga, Spain, west of the oceanographic front. Rombouts (1991) and Wagner (1991) both described generalised morphological differences between the species, and Ríos et al. (2002: 229) were able to use these differences to corroborate their allozyme study of populations on either side of the oceanographic front in the western Mediterranean.

In the present study, examination of museum collections of *Pecten jacobaeus* and *P. maximus* failed to turn up evidence for intermediate forms in the critical areas of the western Algerian coast, southwestern Spain, and southern Portugal. Comparison of 75 valves of *P. jacobaeus* and about 100 valves of *P. maximus* from throughout their geographic ranges indicate that shells as small as 20 mm in height are clearly distinguishable. In right valves of *P. jacobaeus*, radial costellae begin to develop atop plicae within a range of shell heights of 11–23 mm, most commonly appearing in heights of 11–19 mm. In contrast, on right valves of *P. maximus*, radial costellae atop plicae begin to form in a range of heights from 26 to 47mm, mostly commonly in the range of 32–41 mm. Left valves differ more in the prominence of secondary costae atop plicae than in the timing of appearance. In late ontogeny, the left plicae of *P. jacobaeus* have rounded crests with secondary costae so weak that they are hardly discernible. In later ontogeny of the left valve of *P. maximus*, three to five well-developed radial costae appear. There is a greater difference in the presence and timing of appearance of secondary costae in the rib interspaces of right valves. In *P. jacobaeus*, these rib interspaces are non-costate (with a single exception, where costae appear very late in the ontogeny of a gerontic individual). In *P. maximus*, costellae begin to develop in the rib interspaces only a few millimetres later than do the costellae on the rib crests, and the interspace costellae attain the same amplitude and spacing as those on the rib crests.

Given the differences between these two species, the apparent lack of intermediate forms, their separate geographic ranges, and their long separation in the fossil record, it would seem premature to state that the very close genetic similarity indicated by partial sequences of the 16S rRNA gene indicates that they are parts of a single species. The same gene should be sequenced and compared in other species of *Pecten* that are morphologically more distant, such as *Pecten keppelianus* of the Canary and Cape Verd Islands and *Pecten sulcicostatus* G.B. Sowerby II, 1842, of southwester Africa, as well as Indo-Pacific *Pecten* species (see Waller 1991: 39).

1.3 CONCLUSIONS

New research on the morphology of fossil scallops of the family Pectinidae, particularly those from the early Cenozoic, has led to new phylogenetic hypotheses that are encapsulated in Figures 1.1–1.3. The new phylogenies indicate that extant pectinid genera comprise four subfamilies: Camptonectinae, Chlamydinae, Palliolinae, and Pectininae. Three of the four are monophyletic, the exception being the subfamily Chlamydinae, which is paraphyletic in the sense that it has provided the roots for the other three subfamilies. The Camponectinae, represented by the extant genus *Delectopecten*, is the oldest of the monophyletic subfamilies, having branched from a chlamydinine ancestor before the end of the Triassic, approximately 220 million years ago. Next oldest is the Palliolinae, with an origination in the mid-Cretaceous (Albian), about 100 million years ago. The Pectininae is the youngest monophyletic group, traceable to an aequipectinine subgroup of the Chlamydinae in the late Middle Eocene, about 40 million years ago.

One of the major results of the present study is a new phylogeny of the Palliolinae, demonstrating the importance of this group in Cenozoic pectinid evolution in both hemispheres. It is subdivided into five tribes, all but one of which are monophyletic: Adamussiini, Mesopeplini, Palliolini, Pseudentoliini, and Serripectinini. The exception is the tribe Palliolini, which is paraphyletic in the sense that it is the group from which the Mesopeplini, a Southern Hemisphere group, originated.

The stem genus of the Palliolinae, *Dhondtichlamys*, originated from a chlamydinine ancestor in the Cretaceous (Albian) and is traceable into the Paleocene and Lower Eocene. Diversification of the Palliolinae occurred in two major pulses (Fig. 1.2), one in the Paleocene and Early to Middle Eocene during a time of global warmth, the other in the latest Eocene and Early Oligocene, during a time of global cooling and readjustment of oceanic circulation (Flower and Kennett 1994; Zachos et al. 1994; Thomas et al. 2000). The warm-phase clades contain the longest ranging genera of the subfamily: (a) the European *Pseudentolium-Korobkovia* lineage, which lasted for about 51 million years, from the Early Eocene (Ypresian) to the Late Pliocene; (b) the New Zealand-Australian genus *Lentipecten* (shown here to be distinct from European *Pseudentolium*), which survived for about 40 million years, from the Middle Eocene (Lutetian) to the end of the Miocene; and (c) the New Zealand-Australian genus *Serripecten*, which persisted for about 46 million years, from the Early Eocene (Ypresian) into the Late Miocene (Tortonian). These long-ranging taxa may well have been eurythermal and may have achieved their longevity by being able to adapt to changing marine climate. Although the fossil record for the Eocene is not good enough to provide extensive distributional data, it is likely that these long-ranging genera were also geographically widespread. Two of the three clades that became restricted to the Southern Hemisphere, the Adamussiini and Serripectinini, originated during the warm phase of diversification.

Cooling-phase diversification during the latest Eocene through Early Oligocene (Rupelian, Chattian) includes the origin of the antarctic genus *Adamussium* from the long-ranging, probably eurythermal Southern Hemisphere genus *Lentipecten*. In the Northern Hemisphere, rapid diversification in the tribe Palliolini occurred in northern basins of Europe, leading to the extant northern Atlantic genera *Placopecten*, *Karnekampia*, and *Pseudamussium*, and to the Southern Hemisphere extant genus *Mesopeplum*. *Mesopeplum* is the stem as well as the only survivor of a monophyletic clade of genera that evolved in the Southern Hemisphere, mainly in New Zealand (Beu and Maxwell 1990; Beu 1995) during the Miocene and Pliocene (Fig. 1.3). The extinct derivative genera are *Mesopeplum* (*Borehamia*), *Phialopecten*, *Kaparachlamys*, and *Towaipecten*, all of which occur in an interval of time from the Late Miocene through the Pliocene.

The third clade to survive the end-Cretaceous mass extinction is the paraphyletic subfamily Chlamydinae. Its evolution was discussed in Waller (1991, 1993), where it was shown that the tribe Mimachlamydini developed during the late Paleocene or early Eocene two incipient morphological characters indicating that it is the root for the Aequipectini - a pitted microsculpture in the pre-radial stage of the left beak and internal rib carinae. New information on hinge dentition and trends of commarginal lamellae in Eocene and Oligocene pectinids indicates that the subfamily Pectininae, here redefined to include the tribes Pectini, Decatopectinini, and Amusiini, is rooted in the Aequipectinini, not in the Palliolinae contrary to Waller (1991).

Study of a stratigraphic succession of Pectininae in the Oligocene and Miocene of Central America reconfirms the phylogenetic independence of the American genus *Euvola* and the European-Indo-Pacific genus *Pecten*. *Euvola* is ultimately rooted in the widespread Oligocene-Miocene genus *Amussiopecten* and so also is *Amusium*. Collectively these three genera comprise the tribe Amusiini. In contrast, true *Pecten* as well as the extant genus *Annachlamys* are both derived, but at different times, from the Upper Eocene to Pliocene genus *Gigantopecten*, the basal member of which is the Upper Eocene-Lower Oligocene European species "*Pecten*" *arcuatus*. These three genera, *Pecten*, *Annachlamys*, and *Gigantopecten* comprise the tribe Pectinini. The tribe Decatopectinini of Waller (1986) is here reduced to eight genera with the removal of *Flexopecten*, assigned to the tribe Aequipectinini, and *Annachlamys*, assigned to the Pectinini. Thus restricted, the Decatopectinini is a sister group of the Pectinini + Amusiini.

These phylogenetic refinements based on morphology and the fossil record create substantial reconciliation with the results of molecular genetic analyses compared to the earlier phylogenies of Waller (1991). To return to the questions posed in the introduction, the important points are as follows:

a) Molecular studies have thus far been unable to resolve satisfactorily whether *Placopecten* or *Chlamys* branched off first in pectinid phylogeny. Although there is still lingering uncertainty because of the lack of suitable transitional fossils in the Miocene of northeastern North America, the present study reconfirms placement of *Placopecten* in the Palliolinae, meaning that it was a separate branch before the *Chlamys* group gave rise to *Aequipecten* and *Pecten*.

b) The molecular genetic results that have placed true *Pecten* close to the aequipectinine genera *Argopecten* and *Cryptopecten* are corroborated by the present study, which derives *Pecten* ultimately from an aequipectinine lineage.

c) The COI gene analysis of Matsumoto and Hayami (2000) that indicates a close relationship of *Flexopecten* and *Aequipecten* is also corroborated by the present study, in which *Flexopecten* is moved from the Decatopectinini to the Aequipectinini.

d) Molecular genetic studies that indicate that the extant species *Pecten jacobaeus* and *P. maximus* of the eastern North Atlantic and Mediterranean may actually be conspecific are not supported by the morphological study of specimens from present-day contact zones of these species or by studies of their fossil records. Reports of transitional forms may be spurious.

Thus far molecular genetic analyses have examined species distributed among 21 pectinid genera, but important gaps still remain for the testing of morphological phylogenies. These are *Delectopecten*, representing a very early branch of the Pectinidae, and the extant genera of the Palliolinae other than *Placopecten*, viz., the northeastern Atlantic genera *Palliolum*, *Karnekampia*, and *Pseudamussium*. Finally, the New Zealand-Australian genus *Mesopeplum* should be examined in order to verify its position within the Palliolinae.

ACKNOWLEDGMENTS

I thank Warren Blow, Barbara Bedette and Mary Parrish of the Smithsonian Institution for research assistance and artwork. Annie Dhondt, Koninklijk Belgisch Instituut voor Natuurwetenschappen, Brussels, Belgium, and David Jablonski, University of Chicago, provided useful information. This study has benefitted greatly from generous loans of European Cenozoic pectinids from the Département de Sciences de la Terre, Université Claude-Bernard, Villeurbanne, France, facilitated by Abel Prieur, and of New Zealand Cenozoic pectinids from the New Zealand Geological Survey, facilitated by Alan Beu.

REFERENCES

Adams, C.G., Lee, D.E. and Rosen, B.R., 1990. Conflicting isotopic and biotic evidence for tropical sea-surface temperatures during the Tertiary. Palaeogeography, Palaeoclimatology, Palaeoecology 77:289–313.

Adams, A. and Reeve, L.A., 1840–1850. Mollusca. In: A. Adams (Ed.). The Zoology of the Voyage of H.M.S. *Samarang*, under the Command of Captain Sir Edward Belcher, during the Years 1843–1846. Reeve et al., London. x + 87 p.

Aldrich, T.H., 1895. New or little known Tertiary Mollusca from Alabama and Texas. Bulletins of American Paleontology 1(2):1–30.

Allasinaz, A., 1972. Revisione dei Pettinidi triassici. Rivista Italiana di Paleontologia 78(2):189–428.

Anderson, H.-J., 1958. Die Pectiniden des niederrheinischen Chatt. Fortschritte in der Geologie von Rheinland und Westfalen 1:297–321.

Báldi, T., 1973. Mollusc fauna of the Hungarian Upper Oligocene (Egerian), studies in stratigraphy, palaeoecology, palaeogeography and systematics. Akadémiai Kiadó, Budapest. 511 p.

Basterot, M.B. de B., 1825. Mémoire geologique sur les environs de Bordeaux. Mémoires de la Société d'histoire naturelle de Paris 2:1–100.

Berggren, W.A., Kent, D.V., Flynn, J.J. and Van Couvering, J.A., 1985. Cenozoic geochronology. Geological Society of America Bulletin 96:1407–1418.

Berggren, W.A., Kent, D.V., Swisher III, C.C. and Aubry, M.-P., 1995. A revised Cenozoic geochronology and chronostratigraphy. In: W.A. Berggren, D.V. Kent, M.-P. Aubry and J. Hardenbol (Eds.). Geochronology, Time Scales and Global Stratigraphic Correlation. SEPM (Society for Sedimentary Geology) Special Publication No. 54. SEPM, Tulsa, Oklahoma. pp. 129–212.

Berkeley Cotter, J.C., 1904. Esquisse du Miocène marin portugais. In: G.F. Dollfus, J.C. Berkeley Cotter and J.P. Gomez. Mollusques tertiares du Portugal. Planches de Céphalopodes, Gastéropodes et Pélécypodes laissées par F.A. Pereira Da Costa. Imp. Acad. roy. Sci., Commission du service géologique du Portugal, Lisbon. 46 p.

Bernardi, M., 1858. Description d'espéces nouvelles. Journal de Conchyliogie 7:90–94.

Beu, A.G., 1978. Taxonomy and biostratigraphy of large New Zealand Pliocene Pectinidae (*Phialopecten* and *Mesopeplum*). New Zealand Journal of Geology and Geophysics 21(2):243–269.

Beu, A.G., 1995. Pliocene limestones and their scallops. Institute of Geological and Nuclear Sciences Monograph 10. Institute of Geological and Nuclear Sciences, Ltd., Lower Hutt, New Zealand. 243 p.

Beu, A.G. and Dell, R.K., 1989. Mollusca. In: P.J. Barrett (Ed.). Antarctic Cenozoic history from the CIROS-1 drillhole, McMurdo Sound. DSIR Bulletin (Department of Scientific and Industrial Research, Wellington, New Zealand) 245:135–141.

Beu, A.G., and Maxwell, P.A., 1990. Cenozoic Mollusca of New Zealand. New Zealand Geological Survey Paleontological Bulletin 58:1–518.

Blackwelder, B.W., 1981. Late Cenozoic stages and molluscan zones of the U.S. middle Atlantic Coastal Plain. Paleontological Society, Memoir 12 (Journal of Paleontology 55(5): supplement):1–34.

Bongrain, M., 1988. Les *Gigantopecten* (Pectinidae, Bivalvia) du Miocène Français. Cahiers de Paléontologie. Éditions du Centre National de la Recherche Scientifique, Paris. 230 p.

Boreham, A.U.E., 1961. The New Zealand Tertiary genus *Sectipecten* Marwick (Mollusca). Transaction of the Royal Society of New Zealand 88:665–688.

Boreham, A.U.E., 1965. A revision of F.W. Hutton's pelecypod species described in the Catalogue of Tertiary Mollusca and Echinodermata (1873). New Zealand Geological Survey Paleontological Bulletin 37:1–125.

Brocchi, G., 1814. Conchilogia fossile subappennina con osservazioni geologiche sugli Apennini e sul suolo adiacente, II. Stemperia Reale, Milan. pp. 241–712.

Bucquoy, E., Dautzenberg, Ph. and Dollfus, G., 1889. Les mollusques marins du Roussillon. Pelecypoda. Famille: Pectinidae - Genre *Pecten*. J.-B. Baillière & Fils et Chez l'Auteur Ph. Dautzenberg, Paris. pp. 61–112.

Campbell, D.C., 1995. New molluscan faunas from the Eocene of South Carolina. Tulane Studies in Geology and Paleontology 27:119–152.

Campbell, D.C., 2000. Molecular evidence on the evolution of the Bivalvia. In: E.M. Harper, J.D. Taylor and J.A. Crame (Eds.). The Evolutionary Biology of the Bivalvia. Geological Society of London Special Publication No. 177. The Geological Society, London. pp. 31–46.

Canapa, A., Barucca, M., Caputo, V., Marinelli, A., Nisi Cerioni, P. and Olmo, E., 2000a. A molecular analysis of the systematics of three Antarctic bivalves. Italian Journal of Zoology, Supplement 1: 127–132.

Canapa, A., Barucca, M., Marinelli, A. and Olmo, E., 1999. A molecular approach to the systematics of the Antarctic scallop *Adamussium colbecki*. Italian Journal of Zoology 66:379–382.

Canapa, A., Barucca, M., Marinelli, A. and Olmo, E., 2000b. Molecular data from the 16S rRNA gene for the phylogeny of Pectinidae (Mollusca: Bivalvia). Journal of Molecular Evolution 50:93–97.

Cano, J., Garcia, T. and Roman, G., 1989. Primeros resultados obtenidos con colectores de pectinidos en Málaga (S.E. de España). Boletín del Instituto de Estudios Almerienses 1988:259–267.

Carter, J.G., 1990. Shell microstructural data for the Bivalvia. Part 1. Introduction. In: J.G. Carter (Ed.). Skeletal Biomineralization: Patterns, Processes and Evolutionary Trends, Vol. 1. Van Nostrand Reinhold, New York. pp. 297–301.

Carter, J.G. and Rossbach, T.J., 1989. Summary of lithostratigraphy and biostratigraphy for the Coastal Plain of the southeastern United States. Biostratigraphy Newsletter, No. 3: fold-out correlation chart.

Clark, W.B., 1895. Contributions to the Eocene fauna of the middle Atlantic Slope. Johns Hopkins University Circular 15(121):3–6.

Clark, W.B., 1898. Collection of Eocene fossils. Johns Hopkins University Circular 18(137):18.

Clark, W.B. and Martin, G.C., 1901. Systematic paleontology, Eocene. Mollusca. Maryland Geological Survey, Eocene. The Johns Hopkins Press, Baltimore. 331 p.

Cockerell, T.D.A., 1911. The nomenclature of the Rhizopoda. Zoologischer Anzeiger 38:136–137.

Conrad, T.A., 1834. Description of new Tertiary fossils from the southern states. Journal of the Academy of Natural Sciences, Philadelphia, 1st Series 7:130–178.

Conrad, T.A., 1845. Descriptions of the eight new fossil shells of the United States. Proceedings of the Academy of Natural Sciences of Philadelphia, 1844, 2(6):174–175.

Conrad, T.A., 1865. Descriptions of new Eocene shells from Enterprise, Mississippi. American Journal of Conchology 1(2):137–141.

Cossmann, M., 1921. Synopsis illustré des Mollusques de l'Éocène et de l'Oligocène en Aquitaine. Mémoires de la Société Géologique de France, Paléontologie 24(1–2), Mémoire 55:1–220.

Cossmann, M. and Peyrot, A., 1914. Conchologie néogénique de l'Aquitaine. Actes de la Société Linneenne de Bordeaux 68:1–496.

Cossmann, M. and Pissaro, G., 1903. Faune Éocénique du Cotentin (Mollusques). Pélécypodes. Bulletin de la Société Géologique de Normandie 22:5–30.

Crandall, P.R., 1979. A new cone from off NE Taiwan and a new *Chlamys* from the Ryukyu Islands, Japan. Quarterly Journal of the Taiwan Museum 32(1–2):113–115.

Dall, W.H., 1898. Contributions to the Tertiary fauna of Florida. Part IV. I. Prionodesmacea, II. Teleodesmacea. Transactions of the Wagner Free Institute of Science of Philadelphia 3(4):571–947.

Dall, W.H., Bartsch, P. and Rehder, H.A., 1938. A manual of the Recent and fossil marine pelecypod mollusks of the Hawaiian Islands. Bernice P. Bishop Museum Bulletin 153:1–233.

Defrance, M.J.L., 1825. Dictionnaire des Sciences Naturelles par Plusieurs Professeurs du Jardin du Roi et des Principales Écoles de Paris, vol. 38. Paris and Strasbourg.

Depéret, Ch. and Roman, F., 1902. Monographie des Pectinidés néogénes de l'Europe et des régions voisines. Première partie: Genre *Pecten*. Mémoires de la Société Géologique de France, Paléontologie 10(1):1–74.

Depéret, Ch. and Roman, F., 1910. Monographie des Pectinidés néogénes de l'Europe et des régions voisines. II. Genre *Flabellipecten* Sacco, 1897. Mémoires de la Société Géologique de France, Paléontologie 18(2):105–139.

Deshayes, G.P., 1824. Description des coquilles fossiles des environs de Paris, vol. 1, part 1. Paris. pp. 1–80.

Deshayes, G.P., 1830. Description des coquilles fossiles des environs de Paris, vol. 1, part 4. Paris. pp. 239–332.

Dhondt, A.V., 1972. Systematic revision of the Chlamydinae (Pectinidae, Bivalvia, Mollusca) of the European Cretaceous. Part 2: *Lyropecten*. Bulletin de l'Institut Royal des Sciences Naturelles de Belgique, Sciences de la Terre 48(7):1–81.

Dijkstra, H.H., 1998. Notes on taxonomy and nomenclature of Pectinoidea (Mollusca: Bivalvia: Propeamussiidae, Pectinidae). 3. Nomina nova. Basteria 62(5–6):245–261.

Dunker, G., 1877. Mollusca nonnulla nova maaris japonici. Malakozoologische Blätter. 24:67–75.

Eberzin, A.G., 1960. Mollyuski – pantsirnye, dvustvorchatye, lopatonogie. In: Yu.A. Orlov (Ed.). Osnovy Paleontologii, USSR Academy of Science, Moscow. pp. 1–300.

Emmons, E., 1858. Report of the North Carolina Geological Survey. Agriculture of the eastern counties; together with descriptions of the fossils of the marl beds. Henry D. Turner, Raleigh. 314 p.

Flower, B.P. and Kennett, J.P., 1994. The middle Miocene climatic transition: East Antarctic ice sheet development, deep ocean circulation and global carbon cycling. Palaeogeography, Palaeoclimatology, Palaeoecology 108(3–4):537–555.

Frauscher, K.F., 1886. Das Unter-Eocän der Nordalpen und seine Fauna. I. Theil. Lamellibranchiata. Denkschriften der Kaiserlichen Akademie der Wissenchaften. Mathematisch-Naturwissenschaftliche Classe 51(2):37–270.

Frischer, M.E., Williams, J. and Kenchington, E., 1998. A molecular phylogeny of some major groups of Pectinidae inferred from 18S rRNA gene sequences. In: P.A. Johnston and J.W. Haggart (Eds.). Bivalves: an eon of evolution. University of Calgary Press, Calgary, Alberta, Canada. pp. 213–221.

Gardner, J., 1945. Mollusca of the Tertiary formations of northeastern Mexico. Geological Society of America Memoir 11:1–332.

Gibson, T.G., 1987. Miocene and Pliocene Pectinidae (Bivalvia) from the Lee Creek mine and adjacent areas. In: C.E. Ray (Ed.). Geology and Paleontology of the Lee Creek Mine, North Carolina, II. Smithsonian Contributions to Paleobiology 61:31–112.

Giribet, G. and Carranza, S., 1999. What can 18S rDNA do for bivalve phylogeny? Journal of Molecular Evolution 48(3):256–258.

Glawe, L.N., 1969. *Pecten perplanus* stock (Oligocene) of the southeastern United States. Geological Survey of Alabama Bulletin 91:1–179.

Glawe, L.N., 1974. Upper Eocene and Oligocene Pectinidae of Georgia and their stratigraphic significance. Geological Survey of Georgia, Information Circular 46:1–27.

Glibert, M., 1933. Monographie de la faune malacologique du Bruxellien des environs de Bruxelles. Mémoires du Musée Royal d'Histoire Naturelle de Belgique 53:1–214.

Glibert, M., 1945. Faune malacologique du Miocène de la Belgique. I. Pélécypodes. Mémoires du Musée Royal d'Histoire Naturelle de Belgique 103:1–263.

Glibert, M., 1957. Pélécypodes et gastropodes du Rupélien Supérieur et du Chattien de la Belgique. Mémoires du Musée Royal d'Histoire Naturelle de Belgique 137:1–98.

Glibert, M. and Van de Poel, L., 1965. Les Bivalvia fossiles du Cénozoïque étranger des collections de l'Institut Royal des Sciences Naturelles de Belgique. II. Pteroconchida, Colloconchida et Isofilibranchida. Mémoires du Musée Royal d'Histoire Naturelle de Belgique, 2nd Série. fasc. 78:1–105.

Gmelin, J.F., 1791. Caroli Linnaei systema naturae per tria naturae. Ed. 13, aucta, reformata, Vermes Testacea. Lipsiae 1(6):3021–3910.

Goldfuss, A., 1826–1844. Petrefacta Germaniae. Dusseldorf, Germany. 692 p.

Gourret, P., 1890. La faune tertiare marine de Carry, de Sausset et de Couronne (près Marseille). Facies des étages tertiaires dans la Basse-Provence. Mémoires de la Société Belge de Géologie, de Paléontologie et d'Hydrologie 4:73–143.

Gray, J.E., 1825. A list and description of some species of shells not taken notice of by Lamarck. Annals of Philosophy, n.s. 9(2):134–140.

Habe, T., 1977. Systematics of Mollusca in Japan: Bivalvia and Scaphopoda [in Japanese]. Hokuryukan, Tokyo. xiii + 372 p.

Hagdorn, H., 1995. Farbmuster und pseudoskulptur bei Muschelkalkfossilien. Neues Jahrbuch für Geologie und Paläontologie, Abhandlungen 195(1–3):85–108.

Hanna, G.D. and Hertlein, L.G., 1927. Expedition of the California Academy of Sciences to the Gulf of California in 1921. Proceedings of the California Academy of Sciences, Ser. 4, 16(6):137–157.

Harland, W.B., Armstrong, R.L., Cox, A.V., Craig, L.E., Smith, A.G. and Smith, D.G., 1990. A Geologic Time Scale 1989. Cambridge University Press, New York. xvi + 263 p.

Harris, G.D., 1897. The lignitic stage, Part I, stratigraphy and Pelecypoda. Bulletins of American Paleontology 2(9):1–102.

Harris, G.D., 1919. Pelecypoda of the St. Maurice and Claiborne stages. Bulletins of American Paleontology 6(31):1–260.

Harris, G.D., 1951. Preliminary notes on Ocala bivalves. Bulletins of American Paleontology 33(138):1–54.

Harris, W.B., Fullagar, P.D. and Winters, J.A., 1984. Rb-Sr glauconite ages, Sabinian, Claibornian and Jacksonian units, southeastern Atlantic Coastal Plain, U.S.A. Palaeogeography, Palaeoclimatology, Palaeoecology 47(1–2):53–76.

Hayami, I., 1989. Outlook on the post-Paleozoic historical biogeography of pectinids in the western Pacific region. In: H. Ohba, I. Hayami and K. Mochizuki (Eds.). Current Aspects of Biogeography in West Pacific and East Asian regions. Nature and Culture, No. 1. The University Museum, The University of Tokyo, Toyko. pp. 3–25.

Hayami, I. and Matsumoto, M., 1998. Phylogenetic classification of scallops and evaluation of their taxonomic characters. Fossils (Tokyo) 64:23–35.

Hendy, M.D. and Penny, D., 1993. Spectral analysis of phylogenetic data. Journal of Classification 10(1):5–24.

Hertlein, L.G., 1936. Three new sections and rectifications of some specific names in the Pectinidae. The Nautilus 50(1):24–27.

Hertlein, L.G., 1969. Family Pectinidae Rafinesque, 1815. In: R.C. Moore (Ed.). Treatise on Invertebrate Paleontology, Part N, Vol. 1, Mollusca 6, Bivalvia. Geological Society of America and University of Kansas, Lawrence, Kansas. pp. N348–N373.

Hinds, R.B., 1845. The zoology of the voyage of H.M.S. *Sulphur*, under the command of Capt. Sir E. Belcher. Mollusca, part 3. Smith and Elder, London. pp. 49–72.

Imbesi, M., 1956. Sulle forme intermedie fra il *"Pecten maximus* L." et il *"Pecten jacobaeus* L." nei giacimenti fossiliferi del quaternario in Sicilia e in Calabria. International Quaternary Association, 4th Congress, Rome-Pisa, Actes I:367–371.

International Commission on Zoological Nomenclature, 1999. International Code of Zoological Nomenclature, Fourth Edition. London, The International Trust for Zoological Nomenclature c/o The Natural History Museum, London. xxix + 306 p.

Iredale, T., 1929. Mollusca from the continental shelf of eastern Australia. Records of the Australian Museum 17(4):157–189.

Iredale, T., 1939. Mollusca, Part 1. Great Barrier Reef Expedition 1928–29, Scientific Reports, British Museum (Natural History), London 5(6):209–426.

Janssen, A.W. and Dijkstra, H.H., 1996. Morphological differences between two species of *Palliolum* (Bivalvia: Pectinidae). Basteria 59:107–113.

Jay, J.C., 1856. Report of the shells collected by the Japan Expedition, under the command of Commodore M.C. Perry, U.S.N., together with a list of Japan shells. In: M.C. Perry. Narrative of the expedition of an American squadron to the China seas and Japan, performed in the years 1853, 1853 and 1854, under the command of Commodore M.C. Perry, United States Navy, by order of the government of the United States, Vol. 2. Tucker, Washington, D.C. pp. 291–297.

Jones, K.H. and Preston, H.B., 1904. List of Mollusca collected during the expedition of H.M.S. *"Waterwitch"* in the China Seas, 1900–1903, with descriptions of new species. Proceedings of the Malacological Society of London 6:138–151.

Kenchington, E. and Roddick, D.L., 1994. Molecular evolution within the phylum Mollusca with emphasis on the class Bivalvia. In: N.F. Bourne, B.L. Bunting and L.D. Townsend (Eds.). Proceedings of the 9th International Pectinid Workshop, Nanaimo, B.C., Canada, April 22–27, 1993. Vol. 1. Canadian Technical Report of Fisheries and Aquatic Sciences 1994:206–213.

Ladd, H.S., 1934. Geology of Vitilevu, Fiji. Bernice P. Bishop Museum Bulletin 119:1–263.

Lamarck, J.B.P.A. de M. de, 1806. Suite des Mémoires sur les fossiles des environs de Paris. Annales du Muséum d-Histoire Naturelle 8(47):347–355.

Lamarck, J.B.P.A. de M. de, 1819. Histoire Naturelle des Animaux sans Vertèbres, vol. 6, pt. 1. Paris. 343 p.

Linnaeus, C., 1758. Systema Naturae per Regna Tria Naturae, Tenth Edition, vol. 1. Laurentii Salvii, Stockholm. iii + 824 p. [Facsimile reprints of whole work, Trustees of the British Museum (Natural History), London.]

Ludbrook, N.H., 1973. Distribution and stratigraphic utility of Cenozoic molluscan faunas in southern Australia. Science Reports of the Tohoku University, Sendai, Japan, Second Series (Geology), Special Volume No. 6 (Hatai Memorial Volume): 241–261.

Malatesta, A. and Zarlenga, F., 1986. Northern guests in the Pleistocene Mediterranean Sea. Geologica Romana 25:91–154.

Mansfield, W.C., 1926. Fossils from quarries near Suva, Viti Levu, Fiji Islands, and from Vavao, Tonga Islands, with annotated bibliography of the geology of the Fiji Islands. Papers from the Department of Marine Biology of the Carnegie Institution of Washington 23:87–103.

Mansfield, W.C., 1936. Stratigraphic significance of Miocene, Pliocene, and Pleistocene Pectinidae in the southeastern United States. Journal of Paleontology 10(3):168–192.

Mansfield, W.C., 1940. Mollusks of the Chickasawhay marl. Journal of Paleontology 14(3):171–225.

Marquet, R. and Dijkstra, H.H., 1999. Neogene species of *Pseudamussium* (Mollusca, Bivalvia, Pectinidae) from Belgium. Contributions to Tertiary and Quaternary Geology 36(1–4):45–57.

Marwick, J., 1928. The Tertiary Mollusca of the Chatham Islands including a generic revision of the New Zealand Pectinidae. Transactions of the New Zealand Institute 58:432–506.

Marwick, J., 1943. Some Tertiary Mollusca from North Otago. Transactions of the Royal Society of New Zealand 73(3):181–192.

Masuda, K., 1971a. On some *Patinopecten* from North America. Transactions and Proceedings of the Palaeontological Society of Japan, N.S. 83:166–178.

Masuda, K., 1971b. *Amussiopecten* from North America and northern South America. Transactions and Proceedings of the Palaeontological Society of Japan, N.S. 84:205–224.

Matsumoto, M. and Hayami, I., 2000. Phylogenetic analysis of the family Pectinidae (Bivalvia) based on Mitochondrial cytochrome C oxidase subunit 1. Journal of Molluscan Studies 66:477–488.

Maxwell, P.A., 1992. Eocene Mollusca from the vicinity of McCulloch's Bridge, Waihao River, South Canterbury, New Zealand: paleoecology and systematics. New Zealand Geological Survey Paleontological Bulletin 65:1–280.

Meek, F.B., 1864. Checklist of the invertebrate fossils of North America: Cretaceous and Jurassic. Smithsonian Miscellaneous Collections 177:1–40.

Monterosato, T.A. di, 1884. Nomenclatura Generica e Specifica di alcune Conchiglie Mediterranee. Palermo. 152 p.

Monterosato, T.A. di, 1899. Revision de quelques *Pecten* des mers d'Europe. Journal de Conchyliologie 47(3):182–193.

Moore, E.J., 1984. Tertiary marine pelecypods of California and Baja California: Propeamussiidae and Pectinidae. U.S. Geological Survey Professional Paper 118-B:1–112.

Mörch, O., 1852–1853. Catalogus Conchyliorum quae reliquit D. Alphonso, D. Aguirra et Gadea Comes de Yoldi, Pts. I and II. Havniae. 74 p.

Morra, G. and Erdmann, S., 1986. El genero *Mesopeplum* Iredale 1929 (Bivalvia: Pectinidae) en el Terciario marino Patagonico. Actas IV Congreso Argentino de Paleontología y Bioestratigrafia (Mendoza) 3:119–125.

Morris, J., 1852. Descriptions of some fossil shells from the Lower Thanet Sands. Quarterly Journal of the Geological Society of London 8:264–268.

Morton, S.G., 1833. Supplement to the "Synopsis of the Organic Remains of ferruginous sand formation of the United States" contained in vols. 17 and 18 of this Journal. American Journal of Science, 1st ser. 23(2):188–294.

Morton, S.G., 1834. Synopsis of the organic remains of the Cretaceous group of the United States. Appendix. Catalogue of the fossil shells of the Tertiary formation of the United States, by Timothy A. Conrad. Philadelphia. 88 p.

Müller, O.F., 1776. Zoologiae Danicae Prodromus, seu animalium Daniae et Norvegiae indigenarum characteres, nomina, et synonyma imprimis popularium. Copenhagen. xxxii + 281 p.

Neuffer, F.O., 1973. Die Bivalven des Unteren Meeressandes (Rupelium) im Mainzer Becken. Abhandlungen des Hessischen Landesamtes für Bodenforschung 68: 3–113.

Nicol, D., Jones, D.S. and Hoganson, J.W., 1989. *Anatipopecten* and the *Rotularia vernoni* Zone (Late Eocene) in peninsular Florida. Tulane Studies in Geology and Paleontology 22(2):55–59.

Nilsson, S., 1827. Petrificata Suecana Formationis Cretaceae, Descripta et Iconibus Illustrata. Pars Prior, Vertebrata et Mollusca Sistens. Londini Gothorum. pp. 1–39.

Noda, H., 1991. Molluscan fossils from the Ryukyu Islands, southwest Japan. Part 3. Gastropoda and Pelecypoda from the Yonabaru Formation in the southwestern part of Okinawa-jima. Science Reports of the Institute of Geoscience, University of Tsukuba, Section B, Geological Sciences 12:1–63.

Nyst, P.H., 1835. Recherches sur les coquilles fossiles de la province d'Anvers. Anvers, p. 19.

Nyst, P.H., 1861. Descriptions succinctes de dix espèces nouvelles de coquilles fossiles du Crag noir d'Edegem, près d'Anvers. Bulletins de l'Académie Royal des Sciences, des Lettres et des Beaux-Arts de Belgique, ser. 2, 12:188–197.

Olsson, A.A., 1922. The Miocene of northern Costa Rica. Bulletins of American Paleontology 9(39):1–309.

d'Orbigny, A., 1850–1852. Prodrome de Paléontologie Stratigraphique Universelle des Animaux Mollusques et Rayonnes Faisant Suite au cours Alementaire de Paléontologie et de Geologie Stratigraphiques. Victor Masson, Paris, 3 volumes.

Pallary, P., 1900. Coquilles marines du littoral du Département Doran. Journal de Conchyliologie 48(3):211–422.

Palmer, K.V.W. and Brann, D.C., 1965. Catalogue of the Paleocene and Eocene Mollusca of the southern and eastern United States. Part I. Pelecypoda, Amphineura, Pteropoda, Scaphopoda, and Cephalopoda. Bulletins of American Paleontology 48(218):1–466.

Philippi, R.A., 1844. Enumeratio Molluscorum Siciliae cum Viventium tum in tellure Tertiaria Fossilium quae in Itinere suo Observati, vol. 2. Halle.

Poli, J.X., 1795. Testaceae Utriusque Siciliae Eorumque Historia et Anatome, vol. 2. Parma. lxxvi + 264 p.

Powell, A.W.B., 1958. Mollusca of the Kermadec Islands, part 1. Records of the Auckland Institute and Museum 5(1–2):65–85.

Pusch, G.G., 1837. Plens Paläontologie oder Abbildung und Beschreibung der vorzüglichsten und der noch unbeschriebenen Petrefacten aus dem Gebirgsformationen in Polen, Volhynien und den Karpathen nebst einigen allgemeinen Beiträgen zur Petrefaktenkunde und einem Versuch zur Vervollstündigung der Geschichte des Europäischen Auer-Ochsen. E. Schweiizerbart's Verlagshandlung, Stuttgard. xiii + 218 p.

Quoy, J.C.R. and Gaimard, P., 1834–1835. Voyage de découvertes de l'Astrolabe, éxécuté par ordre du Roi pendant les années 1826–1827–1828–1829, soue le commandement de M.J. Dumont d'Urville. Mollusques. Zoologie Tome 3:1–954.

Raffi, S., 1986. The significance of marine boreal molluscs in the Early Pleistocene faunas of the Mediterranean area. Palaeogeography, Palaeoclimatology, Palaeoecology 52:267–289.

Reeve, L.A., 1852–1853. Monograph of the genus *Pecten*. Conchologia Iconica, vol. 8. L.A. Reeve, London, unnumbered pages are captions for 35 numbered plates.

Reuss, A.E., 1867. Die fossile Fauna der Steinsalzablagerung von Wieliczka in Galizien. Sitzungberichte der Mathematisch-Naturwissenschaftlichen Classe der Kaiserlichen Akaddemie der Wissenschaften, vol. 55, abt. 1, jahrgang 1867, heft 1–5:17–182.

Rice, E.L., Roddick, D. and Singh, R.K., 1993. A comparison of molluscan (Bivalvia) phylogenies based on palaeontological and molecular data. Molecular Marine Biology and Biotechnology 2(3):137–146.

Ríos, C., Sanz, S., Saavedra, C. and Peña, J.B., 2002. Allozyme variation in populations of scallops, *Pecten jacobaeus* (L.) and *P. maximus* (L.) (Bivalvia: Pectinidae), across the Almeria-Oran front. Journal of Experimental Marine Biology and Ecology 267(2):223–244.

Risso, A., 1826. Histoire naturelle des principales productions de l'Europe méridionale et particuliérement de celles des environs de Nice et des Alpes Maritimes, vol. 1. Chez F.-G. Levrault, libraire, Paris and Strasbourg. xii + 446 p.

Roger, J., 1939. Le genre *Chlamys* dans les formations Néogènes de l'Europe. Mémoires de la Société Géologique de France, N.S. 17(2–4), Mémoire 40:1–294.

Roger, J., 1944. Révision des Pectinidés de l'Oligocène du domaine nordique. Mémoires de la Société Géologique de France, N.S. 33(1), Mémoire 50:1–57.

Rögl, F., 1999. Mediterranean and Paratethys. Facts and hypotheses of an Oligocene to Miocene paleogeography (short overview). Geologica Carpathica [Bratislava] 50(4):339–349.

Rombouts, A., 1991. Guidebook to pecten shells: recent Pectinidae and Propeamussiidae of the world. Universal Book Services/Dr. W. Backhuys, Oegstgeest, The Netherlands. xiii + 157 pp.

Rossbach, T.J. and Carter, J.G., 1991. Molluscan biostratigraphy of the lower River Bend Formation at the Martin Marietta Quarry, New Bern, North Carolina. Journal of Paleontology 65(1):80–118.

Rovereto, G., 1899. Rectification de nomenclature. Revue critique paléozoologie 3(2):90.

Sacco, F., 1897a (June). I Molluschi dei terreni terziarii del Piemonte e della Liguria. Parte 24, Pectinidae. Bollettino dei Musei di Zoologia ed Anatomia comparata della R. Università di Torino 12(298):101–102.

Sacco, F., 1897b (December). I Molluschi dei terreni terziarii del Piemonte e della Liguria. Parte 24 (Pectinidae). Carlo Clausen, Torino, Italy. 116 p.

Saunders, J.B., Jung, P. and Biju-Duval, B., 1986. Neogene paleontology in the northern Dominican Republic. 1. Field surveys, lithology, environment, and age. Bulletins of American Paleontology 89(323):1–79.

Say, T., 1824. An account of some of the fossil shells of Maryland. Journal of the Academy of Natural Sciences of Philadelphia, First Series 4(1):124–155.

Schlotheim, E.F. von, 1820–1823. Die Petrefaktenkunde. Bekker, Gothenburg, Germany. 437 p.

Schröter, J.S., 1802. Neue cönchlienarten und wanderungen, ammerkungen und Berichtigungen nach dem Linneischen system der XII. Archiv für Zoologie und Zootomie 3(1):125–166.

Schuchert, C., 1905. Catalogue of the Type and Figured Specimens of Fossils, Minerals, Rocks and Ores in the Department of Geology, United States National Museum. Part I - Fossil Invertebrates. Government Printing Office, Washington, D.C. 704 p.

Siesser, W.G., 1984. Paleogene sea levels and climates: U.S.A. eastern Gulf Coastal Plain. Palaeogeography, Palaeoclimatology, Palaeoecology 47:261–275.

Smith, E.A., 1902. Report on the Collections of Natural History Made in the Antarctic Regions During the Voyage of the "Southern Cross". Part 7, Mollusca. British Museum (Natural History), London. pp. 201–213.

Smith, J.T., 1991. Cenozoic giant pectens from California and the Tertiary Caribbean Province: *Lyropecten*, "*Macrochlamis*", *Vertipecten*, and *Nodipecten* species. U.S. Geological Survey Professional Paper 1391:1–155.

Sobetski, V.A., 1977. "Bivalve mollusks of the Late Cretaceous platform seas" [in Russian]. Akademia Nauk SSSR, Trudy Paleontologicheskovo Instituta (Moscow) 159:3–155.

Sowerby II, G.B., 1839. A Conchological Manual. London. vi + 313 p.

Sowerby II, G.B., 1842. Thesaurus Conchyliorum, Vol. 1. Monography of the Genus *Pecten*. Sowerby, London. pp. 45–82.

Sowerby, J., 1818. The Mineral Conchology of Great Britain. vol. 3, part 36. W. Arding, London. pp. 1–16.

Squires, R.L. and Demetrion, R., 1990. New Eocene marine bivalves from Baja California Sur, Mexico. Journal of Paleontology 64(3):382–391.

Steiner, G., 1999. What can 18S rDNA do for bivalve phylogeny? Response. Journal of Molecular Evolution 48(3):258–261.

Steiner, G. and Hammer, S., 2000. Molecular phylogeny of the Bivalvia inferred from 18S rDNA sequences with particular reference to the Pteriomorphia. In: E.M. Harper, J.D. Taylor and J.A. Crame (Eds.). The Evolutionary Biology of the Bivalvia. Geological Society of London Special Publication No. 177. The Geological Society, London. pp. 11–29.

Steiner, G. and Müller, M., 1996. What can 18S rDNA do for bivalve phylogeny? Journal of Molecular Evolution 43(1):58–70.

Stewart, R.B., 1930. Gabb's California Cretaceous and Tertiary type lamellibranchs. Academy of Natural Sciences of Philadelphia Special Publication 3:1–314.

Tate, R., 1886. The lamellibranchs of the older Tertiary of Australia (Part 1). Transactions and Proceedings and Report of the Royal Society of South Australia 8:96–158.

Teppner, W. von, 1922. Pars 15. Lamellibranchiata tertiaria. "Anisomyaria". In: C. Diener (Ed.). Fossilium Catalogus. I: Animalia. W. Junk, Berlin. pp. 67–296.

Thiele, J., 1934–1835. Handbuch der Systematischen Weichtierkunde, Vol. 2. Gustav Fischer, Jena. pp. 779–1154.

Thomas, E., Zachos, J.C. and Bralower, T.J., 2000. Deep-sea environments on a warm earth: latest Paleocene-early Eocene. In: B.T. Huber, K.G. MacLeod and S.L. Wing (Eds.). Warm climates in earth history. Cambridge University Press, New York. pp. 132–160.

Thunberg, C.P., 1793. Tekning och beskrifning på en stor ostronsort ifrån Japan. Kongliga Svenska Vetenskaps-Akademiens, Handllingar 14(2):140–142.

Toula, F., 1909. Eine jungtertiäre Fauna von Gatun am Panama-Kanal. Geologischen Reichsanstalt, Jahrbuch 58:673–760.

Tucker-Rowland, H.I., 1938. The Atlantic and Gulf Coast Tertiary Pectinidae of the United States. Sec. III. Systematic descriptions. Musée Royal d'Histoire Naturelle de Belgique, Mémoire, 2nd ser., fasc. 13:1–76.

Vaught, K.C., 1989. A classification of the living Mollusca. American Malacologists, Inc., Melbourne, Florida. xii + 195 p.

Verrill, A.E., 1897. A study of the family Pectinidae, with a revision of the genera and subgenera. Transactions of the Connecticut Academy of Arts and Sciences 10:41–95.

Vincent, G., 1881. Description de deux peignes nouveaux du système Laekenien. Annales de la Société Royal Malacologique de Belgique 16:7–9.

Wagner, H.P., 1988. The status of four scallop species (Mollusca; Bivalvia; Pectinidae), with description of a new genus. Basteria 32:41–44.

Wagner, H.P., 1991. Review of the European Pectinidae (Mollusca: Bivalvia). Vita Marina 41(1):1–48.

Waller, T.R., 1972. The functional significance of some shell microstructures in the Pectinacea (Mollusca: Bivalvia). International Geological Congress, 24th Session, Section 7, Paleontology. Montreal, Canada. pp. 48–56.

Waller, T.R., 1978. Morphology, morphoclines and a new classification of the Pteriomorphia (Mollusca: Bivalvia). Philosophical Transactions of the Royal Society of London, B 284:345–365.

Waller, T.R., 1984. The ctenolium of scallop shells: functional morphology and evolution of a key family-level character in the Pectinacea (Mollusca: Bivalvia). Malacologia 25(1):203–219.

Waller, T.R., 1986. A new genus and species of scallop (Bivalvia: Pectinidae) from off Somalia, and the definition of a new tribe Decatopectinini. The Nautilus 100(2):39–46.

Waller, T.R., 1991. Evolutionary relationships among commercial scallops (Mollusca: Bivalvia: Pectinidae). In: S.E. Shumway (Ed.). Scallops: Biology, Ecology and Aquaculture. Elsevier, New York. pp. 1–73.

Waller, T.R., 1993. The evolution of "Chlamys" (Mollusca: Bivalvia: Pectinidae) in the tropical western Atlantic and eastern Pacific. American Malacological Bulletin 10(2):195–249.

Waller, T.R., 1996. Bridging the gap between the eastern Atlantic and eastern Pacific: a new species of Crassadoma (Bivalvia: Pectinidae) in the Pliocene of Florida. Journal of Paleontology 70(6):941–946.

Waller, T.R., 2001. Dhondtichlamys, a new name for Microchlamys Sobetski, 1977 (Mollusca: Bivalvia: Pectinidae), preoccupied by Microchlamys Cockerell, 1911 (Rhizopoda: Arcellinida). Proceedings of the Biological Society of Washington 114: 858–860.

Waller, T.R., and Marincovich, L., Jr., 1992. New species of Camptochlamys and Chlamys (Mollusca: Bivalvia: Pectinidae) from near the Cretaceous/Tertiary boundary at Ocean Point, North Slope, Alaska. Journal of Paleontology 66(2):215–227.

Ward, L.W., 1985. Stratigraphy and characteristic mollusks of the Pamunkey Group (Lower Tertiary) and the Old Church Formation of the Chesapeake Group - Virginia Coastal Plain. U.S. Geological Survey Professional Paper 1346:1–78.

Ward, L.W., 1992. Biostratigraphy of the Miocene, middle Atlantic Coastal Plain of North America. Virginia Museum of Natural History, Memoir 2:1–159.

Ward, L.W. and Blackwelder, B.W., 1980. Stratigraphic revision of Upper Miocene and Lower Pliocene beds of the Chesapeake Group, middle Atlantic Coastal Plain. U.S. Geological Survey Bulletin 1482-D:D1–D61.

Wilding, C.S., Beaumont, A.R. and Latchford, J.W., 1999. Are *Pecten maximus* and *Pecten jacobaeus* different species? Journal of the Marine Biological Association of the United Kingdom 79(5):949–952.

Woodring, W.P., 1982. Geology and paleontology of Canal Zone and adjoining parts of Panama. Description of Tertiary mollusks (Pelecypods: Propeamussiidae to Cuspidariidae; additions to families covered in P306-E; additions to gastropods; cephalopods). U.S. Geological Survey Professional Paper 306-F:541–759.

Woods, H., 1902–1903. A monograph of the Cretaceous Lamellibranchia of England. Palaeontographical Society, Monograph: Pectinidae: 145–232.

Zachos, J.C., Stott, L.D. and Lohmann, K.C., 1994. Evolution of early Cenozoic marine temperatures. Paleoceanography 9(2):353–387.

AUTHOR'S ADDRESS

Thomas R. Waller - Department of Paleobiology, National Museum of Natural History, Smithsonian Institution, P. O. Box 37012, NHB MRC-121, Washington, D.C. 20013–7012, USA (E-mail: waller.thomas@nmnh.si.edu)

APPENDIX

Note 1.

Dhondtichlamys Waller, 2001, is a replacement name for *Microchlamys* Sobetski, 1977. The latter name is a junior primary homonym of *Microchlamys* Cockerell, 1911, a rhizopodan protozoan (Waller 2001).

Note 2.

An explanation of the usage of the name "*Eburneopecten*" *frontalis* for a species occurring in the Paleocene and Lower Eocene of Virginia and Maryland is necessary because of errors in the literature. The name *Pecten rogersi* Clark, 1895, was given by Clark to a species from the Paleocene Aquia Formation of Virginia. Dall (1898, April), recognising that Clark's name is a junior homonym (non *Pecten rogersi* Conrad, 1834), renamed Clark's species *Pecten frontalis* and expanded the concept of the species to include younger material from the lower Jackson group of Mississippi (Moodys Branch Formation, early Bartonian in age; Palmer and Brann 1965: 135). Seven months later,

Clark (1898, November), apparently unaware of Dall's action, introduced another new name, *Pecten dalli*, for the same species. Clark and Martin (1901, caption for Pl. 44, fig. 7) designated a type specimen for *P. rogersi* Clark from the Aquia Formation at Potomac Creek, Virginia (now USNM 207171), and recognised that the species extends into the lower Eocene Nanjemoy Formation of Maryland and Virginia. Dall in Schuchert (1905: 487), apparently overlooking the fact that a type had already been designated, incorrectly referred to specimens of *Pecten* (*Pseudamussium*) *frontalis* from the later Eocene at Garlands Creek, Clark County, Mississippi, as "cotypes", believing that these specimens are conspecific with the older specimens from Maryland and Virginia. The Internal Code of Zoological Nomenclature, however, clearly mandates (ICZN 4th Edition, Article 72.7) that a new replacement name has the same type(s) as the original name. Furthermore, specimens in the Smithsonian collections identified by Dall as "*Pecten frontalis*" from Mississippi are within the range of variation of *Eburneopecten scintillatus* (Conrad, 1865) from the same locality and are not the same as "*Eburneopecten*" *frontalis* from Virginia and Maryland. Palmer and Brann (1965: 135) incorrectly accepted Dall's "cotypes" as valid types and therefore chose to limit application of Dall's name to Upper Eocene specimens from Mississippi. They further suggested that the name *Pecten dalli* Clark, 1898, in the combination *Eburneopecten dalli*, should be reinstated, even though it is an objective junior synonym of *P. frontalis*, and should be used for the Virginia and Maryland species. This is clearly not in accord with the ICZN code.

Note 3.

Ward (1992: 66) determined that *Placopecten princepoides* (Emmons, 1858) is a senior synonym of *P. clintonius rappahanockensis* (Mansfield, 1936).

Note 4.

The name *Gigantopecten* Rovereto, 1899, is used herein as a full genus rather than as a subgenus of *Pecten* as was done by Bongrain (1988). There is perhaps no other group of pectinids for which a genus-group name is more contentious. In recent years, there has been a nearly even split among taxonomists as to whether the name of this taxon should be *Gigantopecten* or *Macrochlamis* Sacco, 1897a. Some authors regard the name *Macrochlamis* to be a misspelling of *Macrochlamys*, the suffix of the latter reflecting the correct spelling of the name *Chlamys*. In fact Sacco (1897b) himself corrected the spelling of *Macrochlamis* to *Macrochlamys* three months later in the same year in a monograph, of which the earlier paper was only a summary. If an automatic emendation of *Macrochlamis* to *Macrochlamys* is allowed, then the name *Macrochlamys* Sacco, 1897, falls because it is a junior primary homonym of *Macrochlamys* Benson, 1832, a gastropod. It was for that reason that Rovereto (1899) introduced *Gigantopecten* as a new name. Even here, however, the matter does not end, depending on how many other genera are regarded as synonyms of this genus. If the list includes *Oopecten* Sacco, 1897, as in Hertlein (1969: N358), then this name has priority over *Gigantopecten*.

Scallops: Biology, Ecology and Aquaculture
S.E. Shumway and G.J. Parsons (Editors)
© 2006 Elsevier B.V. All rights reserved.

Chapter 2

Development, Physiology, Behaviour and Ecology of Scallop Larvae

Simon M. Cragg

2.1 INTRODUCTION

There is a range of larval types in the Bivalvia, with the veliger larva being characteristic for the subclasses Pteriomorphia and Heterodonta (Zardus and Martel 2002). Veliger larvae have a remarkably well-developed anatomy with shared features of shell, velum, musculature, gut, nervous system and, in the pediveliger stage, foot (Cragg 1996; Morse and Zardus 1997). This probably reflects not just homology, but also the fact that during their planktonic life, veliger larvae are exposed to the same selection pressures – those relating to planktonic life in the euphotic zone. Scallop larval anatomy fits the normal bivalve pattern, but certain features may be used to distinguish scallops from other bivalves, or from each other.

This chapter presents information on larvae of the Superfamily Pectinoidea, but virtually all information relates to the Pectinidae, particularly the shallow water commercially-exploited species. Nonetheless, there is value in investigating non-commercial, often small scallops in this era of molecular biology in which the genetic potential of the whole family or superfamily may be exploited.

The chapter is a substantial revision of the review of Cragg and Crisp (1991) with about two fifths of the references cited being new, often updating and sometimes replacing sources cited in that review. Some of the areas covered have had an explosion of new information due to advances in techniques. For example, field and mesocosm studies have offered fresh perspectives on larval distribution and behaviour. Also, fluorescence microscopy techniques have provided a new understanding of the larval nervous system. Experimental investigation of sensory function, until recently not feasible on such small organisms, is yielding insights into settlement behaviour. Advances in larval biochemistry relate particularly to lipid metabolism and to the role of neurotransmitters and neuromodulators, especially in metamorphosis control. Larval rearing techniques have been refined with much attention being paid to bacteria as sources of disease and as protection against disease.

2.2 SCALLOP LIFE HISTORY CHARACTERISTICS

Studies of scallop development have focused on species with commercial potential from shallow temperate to tropical waters, which, with one exception, show typical planktotrophic development. Species with smaller adults and from deeper waters have received less attention: evidence suggests that lecithotrophic development is more common in such cases. Direct development from egg to juvenile has not been reported in the Pectinidae, but does occur in the related Propeamussidae. In cases where direct study of development is difficult, evidence from egg and shell dimensions can be used to provide evidence of mode of development. Table 2.1 summarises the information available on these key dimensions.

The eggs of most species listed in Table 2.1 range from 60 to 85 µm in diameter, thus falling within the range characteristic for planktotrophic species in the Mytilidae (Ockelmann 1965). Mackie (1983) stated that bivalves with planktotrophic development generally have eggs with a diameter of less than 85 µµm, while the eggs of those with lecithotrophic development range in size from 90–140 µm. Zardus and Martel (2002) confirmed the planktotrophic range, but revised the range for lecithotrophic development for non-protobranch bivalves to 90–300 µm. Eggs in the predicted lecithotrophic range have been reported from the scallops *Equichlamys bifrons* – a continental shelf species, *Bathypecten vulcani* – a hydrothermal vent species and *Pseudohinnites levii* – an abyssal species. Larvae of *E. bifrons* provided with a microalgal diet were reared to metamorphosis by Dix (1976), but whether these larvae ingested the algae is not stated.

Mytilid species with larger eggs also have larger prodissoconch I shells (Ockelmann 1965) and parental stocks of *Mercenaria* producing larger eggs give rise to larger prodissoconch I shells (Goodsell and Eversole 1992). Egg size data for lecithotrophic scallops is limited, but *Equichlamys* and *Pseudohinnites*, follow the pattern of larger shells from larger eggs (Table 2.1). In planktotrophic scallop larvae, there is a weak negative correlation between egg size and prodissoconch II size at metamorphosis, but no clear trend between egg size and prodissoconch I size (Fig. 2.1). Stocks of *Pecten maximus* that produce larger eggs, in some cases produce larger prodissoconch I shells (Paulet et al. 1988) and in other cases do not (Cochard and Devauchelle 1993).

Planktotrophic larvae have a much larger prodissoconch II (PII) than PI, while there is little larval shell growth after formation of PI in lecithotrophic species. Berkmann et al. (1991) report a PI/PII ratio in planktotrophic larvae of 0.25 to 0.4 and of 0.6 or greater in lecithotrophic larvae. (To facilitate comparisons with the literature, the PI/PII ratio is used here, but it would perhaps provide a clearer insight into the difference between lecithotrophic and planktotrophic shells if data was presented in the form of percent growth from PI to PII.) Ratios in Table 2.1 range from 0.24 to 0.98 with two groupings: known or likely planktotrophic species ranging from 0.24 to 0.55; known or likely lecithotrophic species ranging from 0.87 to 0.98. A single specimen of *Crassodoma pusio* with a ratio of 0.63 falls between these. In addition to these species, there are a number of species reported by Schein (1989) and Kase and Hayami (1992) (though full data are not available) as having shells characteristic of non-feeding larvae (*Bathypecten eucymatus*, *Chlamydella* spp., *Cyclopecten* spp., *Hyalopecten frigidus*, *H. undatus*, *Parvamussium* spp., *Propeamussium centobi* and *Swiftopecten swifti*) or of planktotrophic larvae (*Chlamys bruei*).

Table 2.1

Key dimensions and shell features during ontogeny of scallops.

Species	Dimensions (µm)[1,2,3]				shell outline	Features[4]		Reference
	egg	PI	P II	PI/PII		ant. teeth	post. teeth	
Adamussium colbecki	55	124	346	0.36				Berkman et al. 1991
Aequipecten opercularis	68							Fullarton 1890
" "			260		+			Jørgensen 1946
" "			210	0.45	+			LePennec 1982
" "	63	95	180–229	0.44	+	3/4	3/4	Sasaki 1979
" "		80	191–265		+			Peña et al. 1998
Amusium balloti	63/76	110	200	0.55		3–4	3–4	Rose et al. 1988
Amusium pleuronectes	72	80	175–250	0.46				Belda & Del Norte 1988
Argopecten circularis	60	90	228	0.38	+			Aviles-Quevado & Muncino-Diaz 1988
Argopecten gibbus	60	85	235		+	3	3	Costello et al. 1973
Argopecten irradians		80	175	0.49		3	3	Chanley & Andrews 1971
" "		78	175	0.46				Loosanoff & Davies 1963
" "	62		190	0.41				Sastry 1965
Argopecten purpuratus		70	231	0.32				Bellolio et al. 1993
" "			222					Bellolio et al. 1994
" "			240					Illanes-Butcher 1987
" "			270					Padilla 1974 in Bellolio et al. 1994
" "								Farias et al. 1998
" "	67							Piquimil et al. 1991
Bathypecten eucymatus	59–65	100						Schein 1989
Bathypecten vulcani	100	148*	150–220	0.98				Le Pennec & Beninger 2000
Bractechlamys vexillum	82	90	170	0.53				Lefort 1992
Camptochlamys sp. A		73	231	0.32		>8	>8	Malchus 2000
Camptochlamys sp. B		65	246	0.26				Malchus 2000
Caribachlamys imbricata		157*	172*	0.91				Waller 1993
Caribachlamys ornata		156*	173*	0.90				Waller 1993
Caribachlamys sentis		152*	175*	0.87				Waller 1993
Chlamydella incubata		246–279	absent					Hayami & Kase 1993
Chlamydella tenuissima		261–298	absent					Kase & Hayami 1992; Hayami & Kase 1993
Chlamys asperrimus	60	80	220	0.36	+	3/4	4/5	Rose & Dix 1984

Chlamys bruei								Schein 1989
Chlamys distorta					+		3	Le Pennec 1980
Chlamys farreri			340		+	3/4	3	Kulikova et al. 1981
" "			220–230		+	5	5	Tanaka 1984
Chlamys hastata	70	105	240	0.44	+	3/6	3/6	Hodgson & Burke 1988
Chlamys islandica	70	120	305	0.39	+			Gruffydd 1976
Chlamys patriae						6	5	Barichivith & Stefoni 1980
" "								Solis et al. 1976
" "			175–200		+		+	Jorgensen 1946
Chlamys striata			310		+		+	Jorgensen 1946
Chlamys tigrinus		92*	197*	0.47				Waller 1993
Crassodoma gigantea		96*	208*	0.46				Waller 1993
Crassodoma multistriata		121*	191*	0.63				Waller 1993
Crassodoma pusio			130–190					Schein 1989
Cyclopecten ambiannulatus			absent					Kase & Hayami 1992; Hayami & Kase 1993
Cyclopecten ryukyuensis		165–204	270–400					Ockelmann 1965
Delectopecten vitreus		88*	363*	0.24				Waller 1991
Delectopecten vitreus		70	275–385	0.25				Schein 1989
Delectopecten vitreus var. abyssorum			170–200	0.88	+	3–4	3–4	Dix 1976
Equichlamys bifrons	120	150			+			Peña et al. 1998
Flexopecten flexuosus			172–234					Peña et al. 1998
Flexopecten glaber			172–226					Schein 1989
Hyalopecten frigidus			220					Schein 1989
Hyalopecten undatus			180–200					Lefort 1992
Mimachlamys gloriosa	85	95	180	0.53				Le Pennec 1980
Mimachlamys varia			186–245		+	3	3	Peña et al. 1998
" "			208–230					De la Roche 2002
" "								Schein 1989
Nodipecten nodosus			180					Schein 1989
Parvamussium fenestratum			170					Schein 1989
Parvamussium lucidum			170					Schein 1989
Parvamussium permirum			absent					Kase & Hayami 1992; Hayami & Kase 1993
Parvamussium crypticum		181–204	absent					Hayami & Kase 1993
Parvamussium decoratum		140	270					Kulikova & Tabunkov 1974
Patinopecten yessoensis			263	0.38				Maru 1985a
" "		107	285		+	2/6	2/6	Maru 1972
" "			270–280					Kulikova et al. 1981
" "	53	72				5	5	Yamamoto & Nishioka 1943

Species	PI	PII	Dimension	PI/PII	Shape	Teeth	Teeth	Reference
" " "	76	115	271–286	0.42				Yoo 1969
" " "			230–350					Ito et al. 1965
" " "					+			Park et al. 2001
Pecten albicans	77	118			+			Hotta 1977
" "	79	104	230–250	0.45	+		3/4	Tanaka 1984
	70	105	210–240	0.50		3/4		Knudsen 1979
Pecten alcocki			280					Peña et al. 1998
Pecten commutatus			221–296					Heasman et al. 1996
Pecten fumatus			185					Heasman et al. 1996
" "					+			Odhner 1914
Pecten glaber					+			Peña et al. 1998
Pecten incomparabile			225–284					Peña et al. 1998
Pecten jacobaeus	90		250	0.36	+			Gruffydd & Beaumont 1972
Pecten maximus	66/70	80	240	0.33	+	2/5	2/4	LePennec 1974 & 1980
" "			210/237					Paulet et al., 1988
" "			190–249		+	3/5	3/5	Sasaki 1979
" "			230–260					Cabello & Camacho 1976
Pecten meridonalis	68				+	3	3	Dix & Sjardin 1975; Dix 1976
Pecten novaezelandiae[5]	70				+	3/4	3/4	Booth 1983; Bull, pers. comm.
Pecten ponticus			270–364		+	3	3	Zakhvatkina 1959
Pecten septemradiatus			250–325		+	+	+	Jorgensen 1946
Placopecten magellanicus	64	105	292		+	+	+	Culliney 1975
" "	70							Desrosiers et al. 1996
Propeamussium centobi			170–180					Schein 1989
Pseudamussium clavatum			242–343					Peña et al. 1998
Pseudohinnites levii	170–180	260	265	0.98				Dijkstra and Knudsen 1997
Similipecten similis			100–170					Schein 1989
Spondylus tenebrosus	64							Parnell 2002
Swiftopecten swifti			240–250					Kulikova et al. 1981

Footnotes

1. Egg diameter indicated; prodissoconch I (PI) shell width parallel to hinge; prodissoconch II (PII) shell width at metamorphosis; PI/PII is ratio of the two shell widths.
2. Where two dimensions are divided with a slash, figures relate to different parental stocks.
3. Asterisk denotes dimensions estimated from published micrographs.
4. Shape: image of shell profile provided if marked +; presence of hinge teeth noted if marked +; numbers of anterior and posterior teeth with range.
5. Identification tentative.

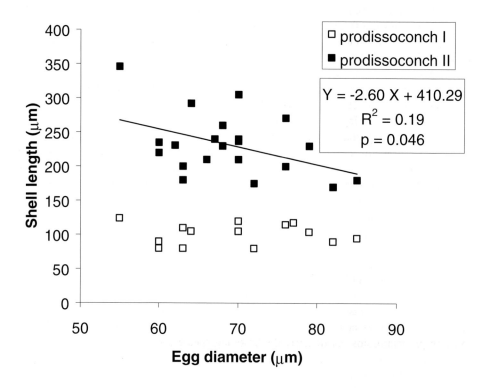

Figure 2.1. Relationship between egg diameter, and width of prodissoconch I and II shells in pectinid larvae with planktotrophic development.

By using evidence from egg size and larval shell ratios we can conclude that almost all shallow water scallop species have planktotrophic development, whether in the Antarctic (*Adamussium*), temperate waters (e.g., *Pecten*) or tropical waters (e.g., *Bractechlamys*) and at least two bathyal to abyssal species have this mode of development. Lecithotrophic development occurs in a single temperate water species (*Equichlamys*), several sublittoral cave-dwelling species (Kase and Hayami 1992), many bathyal and abyssal species (Schein 1989; Dijkstra and Knudsen 1997; Le Pennec and Beninger 2000), a hydrothermal vent species (*Bathypecten vulcani*) and species of *Caribachlamys* from coral reefs (Waller 1993). A further indication of the lecithotrophic, low dispersal mode of development for *Bathypecten* is the observation that larvae of a species of this genus only settle close to known parental stocks (Mullineaux et al. 1998). Gregarious settlement is reported in *Leptopecten* (Morton 1994), though whether this is due to having lecithotrophic larvae with limited dispersal capability, or to settlement cues from congeners remains to be established. Planktotrophy is the dominant development mode in euphotic waters, but where suspended particulate food is limited (e.g., in shallow caves and in the abyss) or where the parental environment is extremely patchy (coral reefs, mid-ocean ridges), lecithotrophy may improve the chances of survival to settlement in the parental environment. Lecithotrophic development may also correlate with small adult size (Kase and Hayami 1992, Schein 1989).

Ockelmann (1965) stated that there were no records of direct development or brood protection for any deep sea pectinids, the reviews of bivalve development by Sastry (1979) and Krasyanov et al. (1998) report none for pectinids in any environment and none have come to light during the preparation of this chapter. Brooded larvae lack a PII, and PI is often unusually shaped and textured (Berkmann et al. 1991). The propeamussiids examined by Hayami and Kase (1993) lacked a PII and in *Chlamydella incubata* and *C. tenuissima* the PI has radiating or diagonal ribs, and the shape is like a conical, broad-brimmed hat. These species were found to brood a few (commonly 3–7) embryos and to retain the juveniles in the suprabranchial chamber, attached by their byssus to the maternal gill. Mode of development in the genus is variable, however, as they point out that *C. favus* from lower neritic substrata off southern Australia has a smaller, normal-shaped PI and so is unlikely to be brooded. *Propeamussium imbriferum* and *P. groenlandicum* are also reported to be brooders (Wada 1953 in Kasyanov et al. 1998).

With their vast numbers of small eggs released per spawning and prolonged period of larval dispersal, the planktotrophic species appear to be *r*-strategists. Even a species which is mature at just over 10mm length and is thus less fecund and may not be planktotrophic shows characteristics of an *r*-strategist (Morton 1994). The small cave-dwelling propeamissids, particularly the brooding species, are *K*-strategists adapted for a stable low-productivity environment (Hayami and Kase 1993).

The timing of stages of larval development varies considerably from species to species, though information on early stages is sparse (Table 2.2). In addition to the putative constraints of food supply, depth and patchyness of adult environment, temperature also affects scallop life history characteristics.

Cell division during the cleavage stages is distinctly faster at 20°C than at 15 or 10°C (Zavarzina 1981). For a given species, time to metamorphosis varies with temperature across the physiologically tolerated temperature range: at 18°C, *Pecten maximus* larvae reached metamorphosis in 35 days, but at 15°C, 44 days were required and at 12°C, 51 days. At 9°C, only a small proportion of larvae metamorphosed after 65 days (Beaumont and Barnes 1992). Sublethal low temperatures arrest veliger growth and subsequent development when temperature rises to optimal temperatures is retarded (Beaumont and Budd 1982). Data from the planktotrophic pectinids on temperature and time to metamorphosis fits to an Arrhenius plot (Fig. 2.2) that also fits the data for individual species across the temperature range at which they can achieve metamorphosis. This suggests that the energy of activation of a critical step required for development to metamorphosis is remarkably constant within the family.

Size at metamorphosis of planktotrophic species shows a trend of decreasing size with increase in temperature evident particularly in data from laboratory rearing of species at close to their temperature optimum (Fig. 2.3). A negative correlation of this sort also to occurs in other bivalve families (Lutz and Jablonski 1978). Data from field collections of larvae generally lack sufficient information about temperatures during larval

Table 2.2

Timing of stages of development in scallops in laboratory culture.

Species	Temperature	Time (h or d)					Reference
		Blastula (h)	Gastrula (h)	Trochophore (h)	D-veliger (h)	Metamorphosis (d)	
Amusium balloti	18.5	7	16	28		22	Rose et al. 1988
Amusium pleuronectes	26					14	Belda & Del Norte 1988
Argopecten circularis	24					10	Aviles-Quevado & Muncino-Diaz 1988
" " "	20				36		Cruz & Ibarra 1997
Argopecten gibbus	23			24	48	16	Costello et al. 1973
Argopecten irradians	25			24	42		Gutsell 1930
" " "	21.5					14	Loosanoff & Davies 1963
" " "	24	5	9	24	48	13	Sastry 1965
" " "	24				18–28	10	Castagna & Duggan 1971
Argopecten purpuratus	20			12	24–36	16	Bellolio et al. 1993
" " "	23.5					12	Illanes in Bellolio et al. 1994
" " "	23					14	Illanes-Butcher 1987
" " "	17.5					31	Padilla 1974 in Bellolio et al. 1994
" " "	20					18	Farias et al. 1998
" " "	14					35	Piquimil et al. 1991
" " "	18				48		Piquimil et al. 1991
" " "	24				24		Piquimil et al. 1991
Bractechlamys vexillum	25					20	Lefort 1992
Chlamys asperrimus	17.5			24	48	20	Rose & Dix 1984
Chlamys hastata	12					42	Hodgson & Bourne 1988
" " "	16					34	Hodgson & Bourne 1988
" " "	16		18	30	50		Hodgson & Burke 1988
Chlamys islandica	7					70	Gruffydd 1976

Species							Reference
Chlamys nobilis	24.5				48	10	Ventilla 1982
Chlamys opercularis	17				72	17	LePennec 1982
Equichlamys bifrons	16					34	Dix 1976
Hinnites multirugosus	12.5				48	13	Olsen 1981
Mimachlamys gloriosa	26					10	Lefort 1992
Nodipecten nodosus	26.5		8		26		De la Roche 2002
Patinopecten yessoensis		18	25	18			Malakhov & Medvedeva 1986
"	12					36	Maru 1985a
"	13					32	Yoo & Imai 1968
"	15						Park et al. 2001
Pecten albicans	15	9	10	20	60	28	Hotta 1977
"	15				48		Tanaka 1984
"	19				40	21	Tanaka 1985
Pecten caurinus	12.5			48		34	Olsen 1981
Pecten fumatus	14				72	31	Dix & Sjardin 1974
"	14					15	Dix & Sjardin 1975
"	18.5					16	Frankish et al. 1990 in Heasman et al. 1998
"	18				48	11	Heasman et al. 1996
"	21				48		Heasman et al. 1996
Pecten maximus	15	12	12	24	48	30	Comely 1972
"	16			20	50	33	Gruffydd & Beaumont 1972
"	17				72		LePennec 1974
"	20				48		LePennec 1974
"	18					16	Paulet et al. 1988
"	20					21	Sasaki 1979
Placopecten magellanicus	15		30		96	35	Culliney 1975
"	13		24		96		Desrosiers et al. 1996
Spondylus tenebrosus	23	4	6	11	21	21	Parnell 2002

54

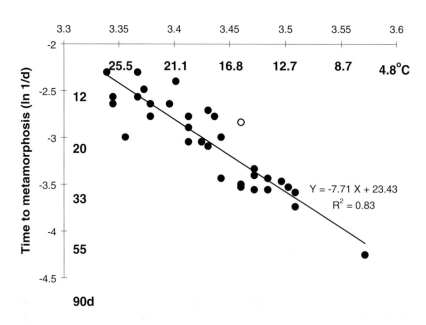

Figure 2.2. Arrhenius plot with units converted to days and degrees C of relationship between water temperature and time from fertilisation to metamorphosis for planktotrophic (•) and lecithotrophic (o) larvae. Sources are listed in Tables 2.1 and 2.3.

development to permit detailed analysis of this correlation. However, measurements from recently settled spat of eight pectinids at 20, 50 and 70 m depth by Peña et al. (1998) suggest that the correlation probably occurs under field conditions. The average PII width was 200 µm for species settling only at 20 m (*Flexopecten* spp.), was between 217 and 257 µm for species settling over the whole range (*Mimachlamys, Aequipecten,* and *Pecten* spp.) and was 291 µm for those that settled only at 70 m (*Pseudamussium*). The deeper settling species may well have experienced a colder temperature regime during larval development. The PII size/temperature trend of planktotrophic species may relate to the same development/temperature interaction that is manifest in the negative correlations between a) water temperature and length of the larval stage and b) egg size and size at metamorphosis.

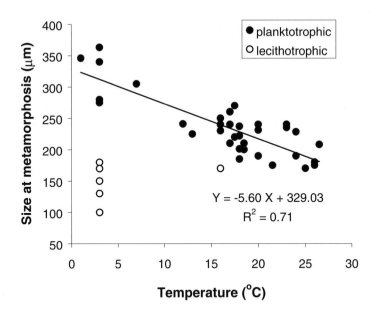

Figure 2.3. Relationship between water temperature and size at metamorphosis of planktotrophic and lecithotrophic larvae of pectinids. The regression line relates to the planktotrophic data. Abyssal and bathyal species have been assigned a very approximate water temperature of 3°C to permit display and comparison.

The data for lecithotrophic species is insufficient to investigate the relationships between temperature and PII size or time to metamorphosis. However, it seems likely that given the shorter developmental period for lecithotrophic species, size and time to metamorphosis is mainly regulated by egg reserves, which may in turn be directly affected by environmental constraints or by limitations imposed by adult size.

2.3 LARVAL DEVELOPMENT

2.3.1 Embryogenesis

In the Pectinidae, gametes are released into the sea and fertilisation occurs externally. Sperm penetration takes place while the egg is at the metaphase I stage of meiosis. Fertilisation and early cleavage stages have been investigated in detail in *Pecten* and *Placopecten* (Gruffydd and Beaumont 1970; Désilets et al. 1995). Temperature, salinity and pH all affect the timing of the stages of meiosis (Desrosiers et al. 1996). Subsequent cell divisions have been investigated in a number of species (Table 2.2). Cell lineages have been traced up to the 16-cell stage. Division is spiral, complete and heteroquadrantal; further cell divisions lead to the formation of a stereoblastula (Tanaka 1984; Malakhov and Medvedeva 1986): this blastula type is typical of bivalves with yolk-rich eggs (Sastry 1979). The blastula is not motile (Kulikova and Tabunkov 1974). A gastrula is then formed by epiboly and invagination (Gutsell 1930; Fullarton 1890; Drew 1906; Hodgson and Burke 1988).

The roughly spherical gastrula develops to form a trochophore larva. The end of the larva that leads as it swims - the apical end or pretrochal region - is rounded and surmounted with a tuft of long cilia. The other end of the larva is tapered with indentations on either side of the larva, corresponding to the shell gland and the archenteron (Gruffydd and Beaumont 1972; Malakhov and Medvedeva 1986; Hodgson and Burke 1988; Belolio et al. 1993).

2.3.2 Development of the Larval Shell

A surface infolding, the shell field invagination, develops on the dorsal surface of the trochophore. Within this is the shell gland (Fullerton 1890; Sastry 1965; Malakhov and Medvedeva 1986; Bellolio et al. 1993; Casse et al. 1998). This structure with its wide transverse opening consists of cells derived from blastomeres Xs and Xd (Malakhov and Medvedeva 1986). Cells with numerous yolk granules bear microvilli that line the gland and some of which partially close off the opening (Casse et al. 1998).

A thin pellicle elaborates across the opening, which allows the organic matrix of the shell to be deposited (Casse et al. 1998). Initially, a continuous sheet consisting of two discs joined by a bridge of the same material is formed (see Hodgson and Burke 1988 and Figs. 2.4 and 2.5), which together partially envelop the post-trochal region. Prodissoconch I shell secretion is initiated at the end of the trochophore stage after the gland everts (Malakhov and Medvedeva 1986, Casse et al. 1998). Casse et al. (1998) provide diagrams that show how identified cells are repositioned during eversion to form the edge of the mantle with its periostracal groove.

The appearance in SEM specimens of the pellicle and pre-prodissoconch I structure, suggests that it is thinner and less rigid than the larval shell in later stages (compare Figs. 2.4 and 2.7; see also Casse et al. 1998). Humphreys (1969) suggests that this material is the precursor of the periostracum and it is likely that there is initially little or no calcification, as is the case of *Tridacna squamosa* (LaBarbera 1974). In pectinids, the outer surface of the shell takes on a less flimsy appearance due to calcification within less than a day: the central region of the shell valves becomes covered with small pits and is surrounded by a rim which has no pits but is marked with radial striations (Lucas and Le Pennec 1976; Hodgson and Burke 1988; Bellolio et al. 1993; Casse et al. 1998; Fig. 2.6). This pattern of surface sculpture described by Carriker and Palmer as "punctate-stellate" also occurs in oyster larvae (Carriker and Palmer 1979; Waller 1981), but in oysters, the pitted (punctate) region is rather smaller.

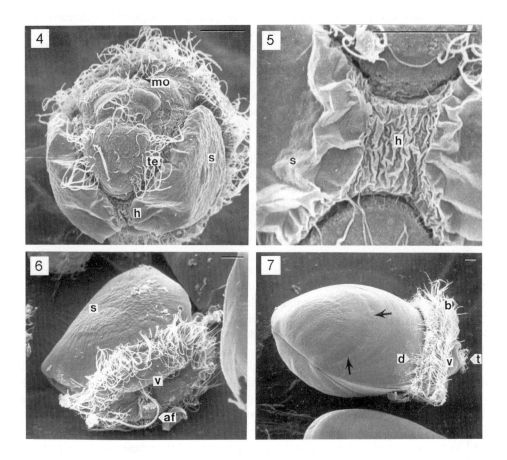

Figure 2.4. 42-hour larva of *Pecten maximus* viewed from opposite end to apical pole. Scale bar = 10 μm. **h** = developing hinge, **mo** = mouth, **te** = telotroch, **s** = non-calcified shell.

Figure 2.5. Detail of specimen in Fig. 2.4. Scale bar = 10 μm.

Figure 2.6. 50-hour D veliger of *Pecten maximus*. Shell (**s**) showing punctate stellate sculpturing of pits and radial striations. Non-retractable velum (**v**) with central apical flagellum (**af**). Scale bar = 10 μm.

Figure 2.7. Umboned veliger of *Pecten maximus*, showing concentric growth lines on the shell, the prodissoconch I/prodissoconch II boundary (**arrowheads**). Compound preoral cirri form a band (**b**) around the rim of the velum (**v**) and the apical tuft (**t**) protrudes from the centre of the velum. Scale bar = 10 μm.

The bridge between the two discs of the shell quickly develops into a straight-edged hinge. The discs become the shell valves and these take on the D-shaped outline characteristic of the prodissoconch I shell of early bivalve veligers (see Fig. 2.6 and Table 2.1). This shell consists of a single homogenous layer of vertical crystallites (Salaün 1994 in Casse et al. 1998)

A distinct change in shell growth occurs shortly after the formation of the D-shaped valves. Further shell-growth takes place at the margins of the prodissoconch I shell. This, the prodissoconch II, can be distinguished from the prodissoconch I by its concentric growth rings (Bellolio et al. 1993; Fig. 2.7). In *Placopecten* veligers there are major and minor growth rings correlating closely with the number of days growth after initiation of secretion of the prodissoconch II shell (Hurley et al. 1987). Ring deposition occurs at the same rate in constant conditions including constant illumination. Thus the possibility of an innate rhythm of secretion, perhaps on a tidal rather than daily cycle, such as that described in an adult bivalve, by Richardson et al. (1980) should be considered. However, Salaün and Le Pennec (1992) sought, but did not find any such correlation in their studies of *Pecten* veligers.

The punctate region of the prodissoconch I is secreted by the shell gland whereas the remainder of the shell is secreted by the rim of the mantle. The concentric growth rings appear once the larva becomes capable of closing the shell (Cragg 1985; Bellolio et al. 1993) and may be due to fluctuations in shell secretion caused by shell closure (Waller 1981). The larval hinge-line differentiates during the secretion of the prodissoconch II shell (Le Pennec 1980, Bellolio et al. 1994). At either end of the hinge there is a line of roughly rectangular teeth with matching depressions in the hinge of the opposite shell valve. These teeth resemble those on the hinges of adult taxodont bivalves. The middle part of the hinge-line is devoid of teeth: the ligament is located near the centre of this area (Zakhvatkina 1959; Waller 1976; Le Pennec 1980; Barichivith and Stefoni 1980; Kulikova et al. 1981; Tanaka 1984; Hodgson and Burke 1988). It is presumably due to the ligament that empty larval shell valves tend to gape and this provides the antagonistic mechanism opposing the adductor muscles (Cragg 1985). As the larva grows, the straight hinge line is gradually obscured by the developing umbos, though these never become as pronounced as in larvae of the other families of bivalves (Rees 1950; Chanley and Andrews 1971).

Scanning electron microscopy has proved particularly valuable in elucidating the disposition and number of the components of the hinge (Table 2.1). Changes in hinge morphology during development have been studied in various pectinids (Maru 1972; Le Pennec 1974 and 1980; Tremblay et al. 1987; Bellolio et al. 1994). Four-day-old *Placopecten magellanicus* veligers have no hinge teeth, but by seven days, six teeth are present; in later stages, up to ten and possibly more are present (Tremblay et al. 1987). D-veligers of *Patinopecten yessoensis* have only two teeth per shell valve at each end of the hinge line, rising to six by the end of the pediveliger stage (Maru 1972). In *Pecten ponticus, P. maximus, Chlamys hastata* and *Argopecten purpuratus* the number of hinge teeth also increase during larval development (Zakhvatkina 1959; Le Pennec 1974; Hodgson and Burke 1988; Bellolio et al. 1993), though no such increase is evident before metamorphosis in the micrographs of *Chlamys varia* (Le Pennec 1980). The amplitude of the teeth increases during larval development (Waller 1976; Bellolio et al. 1994).

The end of planktonic life is marked by a further distinct change in shell secretion. Shell continues to be added at the margin of the valves, but this juvenile (dissoconch) shell has a different appearance (Gruffydd and Beaumont 1972; Maru 1972; Le Pennec 1974; Dix 1976; Hodgson and Burke 1988; Waller 1991; Bellolio et al. 1993). This clearly marked boundary permits the estimation from spat of the size of pediveligers at metamorphosis (Jørgensen 1946; Merrill 1961). The difference in appearance reflects the chemical nature of the two shell types: the dissoconch has a higher calcium salt content and lower organic content than prodissoconch II (Merrill 1961), though the larval shell is sufficiently calcified to warrant decalcification before sectioning with glass knives for electron microscopy (Hodgson and Burke 1988; Cragg 1989). The boundary between prodissoconch II and the dissoconch is also a boundary between two crystalline forms of calcium carbonate: aragonite in prodissoconch II and calcite in the dissoconch (Waller 1976; Fatton and Bongrain 1980; Waller 1991). The dissoconch of pectinids is composed of two layers, the inner of which is formed of foliated calcite. The outer layer of the right valve is composed of prismatic calcite and that of the left is formed of a material that gives a pitted appearance to its surface (Fatton and Bongrain 1980 and authors reviewed therein; Bongrain and Fatton 1982).

Following metamorphosis, the hinge-line becomes straight again and the "wings" which are characteristic features of the hinge of adult scallops appear (Merrill 1961; Sastry 1965; Le Pennec 1974 and 1980). After reaching their maximum amplitude at metamorphosis, the larval hinge teeth slowly disappear (Le Pennec 1974; Hodgson and Burke 1988). The ligament grows along the dorsal edge of the ears and forms resilium in a deep triangular pit below the umbo (Le Pennec 1980: Bellolio et al. 1993).

2.3.3 Organogenesis

2.3.3.1 Development of ciliation, the prototroch and the velum

Apart from shell secretion, changes in the distribution of cilia on the larva are the most significant external evidence of the changes taking place during early larval development. Authors describing a number of pectinid species (see Table 2.3) are agreed that the first stage to possess cilia is the gastrula and that by the trochophore stage, the larva has a conspicuous apical flagellum composed of long cilia. They are less unanimous about the distribution of other cilia apart from the apical flagellum. The diagrams of Sastry (1965) and Costello et al. (1973) indicate that the remaining cilia of the trochophore are evenly distributed over the whole larva, whereas those of Drew (1906), Belding (1910) and Gutsell (1930) suggest that these cilia are restricted to the area extending from the apical tuft to the invagination of the shell gland. Malakhov and Medvedeva (1986) show these cilia only around the equatorial region of the larva. These observations used light microscopy with which, in such transparent organisms, it is difficult to determine the distribution of cilia. Using scanning electron microscopy, Hodgson and Burke (1988) found that in *Chlamys hastata* the ciliation of the gastrula is restricted to two groups of two to four primary trochoblasts on the perimeter of the blastopore and a further two groups of primary trochoblasts girdling the embryo. By the trochophore stage, the prototroch has developed from these primary trochoblasts and consists of a ring of cells bearing a band of sparsely distributed single cilia extending around the equator of the larva (the apical flagellum being situated at the anterior pole).

At the posterior pole, there is a small tuft or ring of cilia, the telotroch. SEM studies of *Pecten maximus* and *Argopecten purpuratus* larvae have revealed a similar prototroch (Cragg 1989, Bellolio et al. 1993) and, in the early shelled larva, a group of cilia posterior to the mouth (see Fig. 2.4), which may be homologous to the telotroch of Hodgson and Burke.

The most characteristic organ of bivalve veligers is the velum, which emerges as a distinct organ during the transition from trochophore to veliger. The pretrochal region, which is domed in the trochophore, becomes flattened, forming the roof of the velum: the prototroch becomes dilated, forming its rim. Initially the velum is an almost solid group of cells (Sastry 1965; Gruffydd and Beaumont 1972; Malakhov and Medvedeva 1986) which cannot be retracted, but the body cavity soon extends into the velum and the larva becomes capable of rapidly withdrawing the velum within the shell valves (Cragg 1985). For illustrations of the velum, see authors listed in Table 2.3. The roof of the velum is devoid of cilia apart from the centrally located apical flagellum (Fig. 2.6) or in later larvae, apical tuft (Hodgson and Burke 1988; Cragg 1989 and Figs. 2.7 and 2.8).

The rim of the velum, on the other hand, is profusely ciliated, with five bands or rings of cilia extending right around the rim (Hodgson and Burke 1988; Cragg 1989; Bellolio et al. 1993, Fig. 2.8). Closest to the roof of the velum is the inner preoral ring consisting of single cilia, below this are two rings of markedly longer cilia grouped into cirri, each composed of more than one row of cilia aligned at right angles to the rings. The arrangement of these cilia is best seen in the bases remaining in a deciliated velum (Bellolio et al. 1993). Below the preoral cirri is a tract of shorter single cilia - the adoral tract, and then a ring of postoral single cilia. The naming of the rings of cilia relates to their position relative to the mouth as the larva moves through the water. The mouth itself is located at the posterior end of the velum: it consists of an opening in the adoral tract with an associated tuft of cilia or compound cilia.

Viewed from above, the velum has a roughly oval outline, though with slight indentations on either side (Hodgson and Burke 1988) so that Culliney (1974) describes it as keyhole shaped and Cragg (1989) as pear-shaped. In *Chlamys hastata,* the preoral cirri appear to be borne on a ridge running around the rim of the velum, as are the postoral cirri, with the adoral tract running in a shallow gutter between the two ridges. The velum of *Pecten maximus* may have a similar topography (see Fig. 4 of Cragg 1989). When the velum retracts, it folds with a longitudinal furrow (Cragg 1985; Hodgson and Burke 1988). Apart from the region of the apical tuft, the roof of the velum consists of an extremely thin epithelium; so does the epithelium joining the velum to the mantle folds. The cells of the velum rim, on the other hand, are much larger, particularly those bearing the preoral cilia, which have numerous large mitochondria and an extensive ciliary rootlet system (Cragg 1989). Hodgson and Burke (1988) report the presence of secretory cells on the rim of the velum, among the cilia-bearing cells.

Table 2.3

Sources of information on pectinid larval anatomy.

Species	Features													Reference
	Embryonic cleavage	Trochophore	Velum	Gut	Musculature	Foot	Meta-morphosis	Hinge	Apical flagellum	Apical tuft	Eye	Statocysts	Gills	
Amusium balloti								+			+			Rose et al. 1988
Argopecten gibbus		+							+		+			Costello et al. 1972
Argopecten irradians		+							+					Belding 1910
" " " "	+	+							+					Gutsell 1930
" " " "	+	+							+			+		Sastry 1965
" " " "			+	+									+	Chanley & Andrews 1971
" " " "			+	+				+					+	Lutz et al. 1982
Argopecten purpuratus		+	+	+	+				+			+		Bellolio et al. 1993
Argopecten ventricosus											+			Sainz et al. 1998
Chlamys asperrimus											+			Rose & Dix 1984
Chlamys hastata	+	+	+				+	+	+		+			Hodgson & Burke 1988
Chlamys islandica			+								+			Gruffydd 1976
Chlamys opercularis	+		+											Fullarton 1898
" " "											+	+	+	Jorgensen 1946
Chlamys striata											+	+	+	Jorgensen 1946
Crassodoma gigantea											+			Whyte et al. 1992
Equipecten bifrons								+	+					Dix 1976
Patinopecten yessoensis	+	+							+					Malakhov & Medvedeva 1986
" " "											+			Maru 1972
" " "				+	+	+				+	+		+	Bower & Meyer 1990
Pecten albicans	+		+				+				+	+	+	Tanaka 1984
Pecten glaber			+	+							+	+	+	Jorgensen 1946

Species	1	2	3	4	5	6	7	Reference
Pecten maximus		+			+			Gruffydd & Beaumont 1972
" " "					+			Le Pennec 1974
" " "				+				Gruffydd et al. 1975
" " "	+							Cragg & Nott 1977
" " "						+	+	Cragg 1985
" " "							+	Cragg 1989
" " "			+			+		Cragg & Crisp 1991
" " "		+			+	+		Beninger et al. 1994
Pecten meridonalis				+				Dix 1976
Pecten novaezelandiae		+						Bull 1977
Pecten ponticus			+					Zakhvatkina 1959
Pecten septemradiatus	+		+					Jørgensen 1946
Placopecten magellanicus					+			Drew 1906
" " "				+				Tremblay 1987
" " "		+	+					Culliney 1974
unid. pectinid				+				Dominguez & Alcaraz 1983

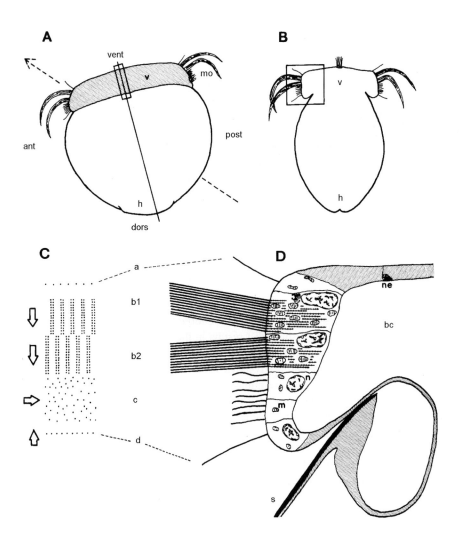

Figure 2.8. Disposition of cilia on the velum of veliger of *Pecten maximus* (from Cragg 1989 with minor modifications). A: lateral view of veliger in typical swimming attitude with swimming direction indicated by arrow. For the sake of clarity, only the cilia on anterior and posterior end of velum are shown. Solid line indicates plane of section shown in B, outlined area shown in C. B: Transverse section of veliger. Outlined area shown in D. C: Disposition of bases of cilia on the rim of the velum. **a**: inner preoral ring; **b1, b2**: upper and lower rings of preoral cirri; **c**: adoral tract; **d**: postoral ring. Arrows indicate direction of power stroke. D: Section of rim of velum. Note that upper and lower rows of preoral cirri are both shown, but as they are out of register, so they would not both appear in a true section.

2.3.3.2 Larval mantle

The inner surface of each shell valve is lined with an epithelium which turns back at the rim, forming two folds, then joins the epithelium of the velum or the foot or, at the anterior or posterior regions close to the hinge, forms a bridge between the valves. The epithelium is generally rather thin, being composed of cells that are much less tall from apical to basal surface than they are wide. Where it lines the shell, the apical surface lies directly against the shell: where it extends between the shell valves, it has a brush border of short microvilli at the apical (outward-facing) surface. The basal surface of the epithelium rests on a thin basal membrane. The cells of the folds are more columnar in character, with some much larger than the rest. A layer of periostracum curves back on itself to overlie the outer mantle fold and extends into the groove between the two folds. The cytology of the mantle folds of *Pecten maximus* is summarised in Fig. 2.9. The inner mantle fold bears cilia (Sastry 1965; Bellolio et al. 1993) which are grouped into a variety of formations described below in the section dealing with putative sense organs (section 2.3.3.5).

2.3.3.3 Musculature

No functional muscles are present in the trochophore, but rapid differentiation during the period of secretion of prodissoconch I results in the appearance in the early veliger of velar retractor muscles and the anterior adductor (Fullarton 1890; Sastry 1965; Maru 1972; Cragg 1985; Malakhov and Medvedeva 1986; Meyer and Bourne 1990; Bellolio et al. 1993). During development, the number of velar retractors increases and retractor muscles attach to the posterior body wall (Cragg 1985; Malakhov and Medvedeva 1986; Bellolio et al. 1993). The posterior adductor develops by the late veliger or pediveliger stage (Jørgensen 1946; Sastry 1965; Maru 1972; Tanaka 1984; Cragg 1985; Bellolio et al. 1993).

The foot of the pediveliger also has retractor muscles (Cragg 1985) perhaps developed from some of the retractors of the posterior body wall.

All retractor muscles are composed of striated muscle, but both anterior and posterior adductors have smooth and striated components. A strip of smooth muscle extends from the digestive mass and probably attaches at the rim of the shell. Smooth muscle fibres are also found near the mantle folds (Fig. 2.9) and associated with the gut (Cragg 1985). In *Pecten maximus,* the velar retractor muscles are attached to the shell near the hinge-line and branch profusely before attaching to the velum. They form a pattern that is symmetrical about the plane between the shell valves (Figs. 2.10 and 2.11). The pattern may vary in its details, but it has certain features that are constant for a given species. The posterior adductor consists of a single column, but the anterior adductor consists of two columns (Fig. 2.12), one of striated muscle and one of smooth muscle (Cragg 1985: Bellolio et al. 1993). The figures of Jørgensen (1946) suggest that *P. septemradiatus* and *Chlamys striata,* may also have a two-part anterior adductor, but that *Chlamys opercularis* does not.

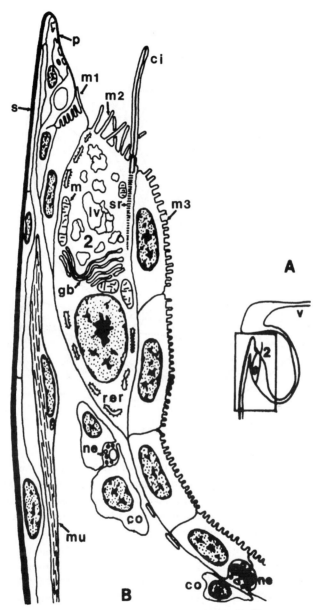

Figure 2.9. Section of mantle perpendicular to the ventral rim of the shell of a veliger of *Pecten maximus*. Location of detail shown in B is outlined in A. Plane of section similar to that in Fig. 2.5D. Labels: **ci** - cilium, **co** - companion cell for nerve, **gb** - Golgi body, **lv** - electron lucent vesicles, **m** - mitochondrion, **m1**, **m2**, **m3** - types of microvilli, **mu** - smooth muscle, **ne** - nerve profiles, **rer** - rough endoplasmic reticulum, **s** - shell, **sr** - striated rootlet, **v** - velum, **2** - type 2 ciliated mantle cell.

66

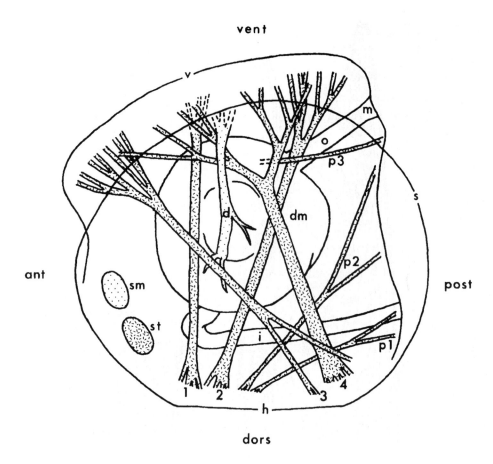

Figure 2.10. The major muscles of the veliger of *Pecten maximus* viewed from the left-hand side with smooth muscle indicated by spots and striated muscle by dashes (Cragg 1985). Only the left hand set of retractor muscles are shown. Labels: **d** – smooth muscle, **dm** - digestive mass (stomach plus digestive gland), **h** – hinge, **i** – intestine, **m** – mouth, **p1-p3** – posterior retractors, **sm** – smooth adductor, **st** – striated adductor, **1-4** – velar retractor muscles.

Where muscle is connected to the shell (Fig. 2.13), it attaches by means of modified mantle cells, whose structure of fibrils and hemidesmosomes resembles that of cells at muscle insertions on the larval shell of the oyster, *Crassostrea virginica* (Elston 1980) and of a nudibranch, *Phestilla sibogae* (Bonar 1978): where muscle attaches to the velum, there is no intervening cell layer and the structure of the point of attachment is less complex (unpublished obs.).

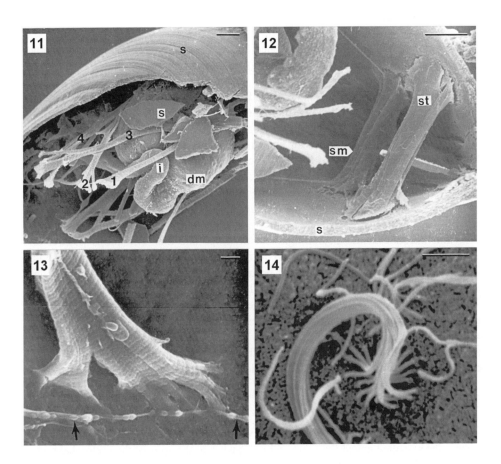

Figure 2.11. to 2.13. Veligers of *Pecten maximus*. Labels as for Fig. 2.10.

Figure 2.11. Umbone region broken away to reveal musculature. Scale bar 10 μm.

Figure 2.12. Shell broken open to show anterior adductor. Note fine strands (putative nerve cell processes) extending to either end of the smooth adductor. Scale bar 10 μm.

Figure 2.13. Velar retractor muscle at point of attachment to the shell. Note putative nerve process with varicosities (arrows). Scale bar 2 μm.

Figure 2.14. Apical flagellum of 42-h larva of *Pecten maximus*. Scale bar 2 μm.

2.3.3.4 Digestive tract

In addition to the shell gland mentioned above, the scallop trochophore has another invagination, the archenteron with its opening, the blastopore (Drew 1906; Belding 1910; Malakhov and Medvedeva 1986; Hodgson and Burke; 1988). The digestive tract develops from this, becoming open at both ends in the early veliger stage.

Development of the morphology of the tract from early veliger to post metamorphosis has been investigated in most detail in *Patinopecten yessoensis* (Bower and Meyer 1990). The mouth is situated at the posterior end of the velum, opening into a straight ciliated cylindrical oesophagus with actively beating cilia (Beaumont et al. 1987), which leads to the stomach. The stomach itself is ciliated and food can be seen rotating inside live larvae. A crystalline style with a style sac is present at the posterior end of the stomach (Sastry 1965; Tanaka 1984; Bower and Meyer 1990). A pair of digestive diverticulae opens off the stomach.

A thin intestine is initially straight but develops one then two loops (Bower and Meyer 1990, Figs. 2.10 and 2.11) leads to the anus which is located close to the hinge line in the posterior body wall, dorsal to the posterior adductor in pediveliger larvae. Light microscope studies indicate the presence of labial palps adjacent to the mouth in older veligers or pediveligers (Belding 1910; Jørgensen 1946; Sastry 1965; Maru 1972), but these have not been reported in SEM studies (Hodgson and Burke 1988; Cragg 1989; Bellolio et al. 1993). The diagrams showing the palps depict contracted larvae, whereas the SEM studies show expanded larvae. Further investigation is required to establish the location and nature of the palps, and their mode of development.

2.3.3.5 Sense organs

Scallop larvae develop a range of putative sense organs during larval life. First to appear is the so-called apical flagellum, which appears at the apex of the pretrochal region of the trochophore and is lost within a day or two of the development of the shell. In its place (now the centre of the velum) there remains a short apical tuft. During the veliger stage, a variety of ciliated tufts appear on the rim of the mantle. At about the time that the umbones become apparent, the veliger develops a pair of pigmented eye spots. By the pediveliger stage, a pair of statocysts has developed near the base of the foot.

2.3.3.6 Apical organ

The apical "flagellum" is in fact a group of long cilia adhering to one another to form a whip-like structure (Bellolio et al. 1993; Fig. 2.14). The apical tuft, which is found in the centre of the velum after the disappearance of the apical flagellum is composed of cilia which do not adhere together and which are situated in a shallow circular depression (Hodgson and Burke 1988; Fig. 2.15), which appears as a pit in contracted larvae (Bower and Meyer 1990).

The following description of the apical region in *Pecten maximus* is based on observations using preparation methods detailed in Cragg and Nott (1977) and Cragg

(1989). The apical flagellum generally protrudes stiffly from the larva with its tip bent back during swimming by the flow of water, but in some larvae it twitches intermittently and in one early veliger, it was seen to vibrate rapidly with small amplitude movements. It consists of up to 50 cilia with the normal 9 + 2 microtubule arrangement, which may be as much as 100 μm long (Figs. 2.6 and 2.14). It arises from a single cell (as was also reported in *Argopecten irradians* by Belding 1910). The rootlets of the cilia extend to the base of the cell. These rootlets are cross-striated by major bands and more numerous sub-bands with a striation pattern periodicity of approximately 50 nm (Fig. 2.16). The cirrus-bearing cell has a basally located nucleus, a Golgi body and, in the apical cytoplasm, numerous mitochondria that are larger than those of neighbouring cells. It also contains a number of spheroidal osmophilic granules resembling those found throughout the trochophore and early veliger, some as large as the nucleus. It is likely that these granules are the lipid droplets that act as the energy store for early larval stages (Whyte et al. 1987). The brush border of this cell is similar to that of neighbouring cells. The apical flagellum is lost at about the time the veliger becomes capable of closing the shell valves. Other ciliated cells appear adjacent to the cirrus cell, which they resemble but for the nature of their cilia (Fig. 2.17). These cilia lack striated rootlets and are shorter. They are obliquely rather than perpendicularly oriented to the free cell surface, or are associated with depressions in that surface, or are partly enveloped within the apical cytoplasm. Small cells appear at the base of the apical cirrus cell and its neighbours.

Further differentiation in the central region of the velum results in the formation of an apical organ directly overlying the connective between the two lobes of a cerebral ganglion. A similar arrangement is reported in *Patinopecten yessoensis* (Bower and Meyer 1990). The apical organ consists of a disc-shaped region of thickened epithelium formed from cell types not found elsewhere in the velum. It bears a number of short cilia and, in extended, fixed larvae, it lies in a depression at the apex of a volcano-like prominence in the centre of the velum (Figs. 2.7 and 2.15). A similarly shaped apical organ occurs in veligers of *Chlamys hastata* (Hodgson and Burke 1988). The shape of this region in live larvae has not been reported.

Much of the free surface of the apical organ is formed by cells bearing a particularly well-developed brush border (Fig. 2.18). The long microvilli of these cells rarely branch. In between the microvilli are layers of extracellular material resembling that observed in the brush border of the rim of the velum (Cragg 1989), but with a more extensive amorphous layer (Fig. 2.19). In the apical cytoplasm of these microvillus cells, there are membrane-bound osmophilic granules and electron-lucent vacuoles, and a compact Golgi body may be found. The nucleus is basally located. The cells bearing the cilia forming the tuft of the apical organ are rather elongated: nuclei belonging to them are found towards the base of the apical organ epithelium. They bear up to ten 9 + 2 type cilia attached to perpendicular striated rootlets (whose striation periodicity is about 50 nm) with associated horizontal rootlets which appear to stretch from cilium to cilium (Fig. 2.20). Horizontal microvilli arise from the free surface of these cells and interweave between the vertical microvilli of the brush border (Fig. 2.21). The cytoplasm of these cells contains elongated mitochondria and rough endoplasmic reticulum. Occasionally, dense cored vesicles 60–80 nm in diameter resembling those within the neuropil of the

70

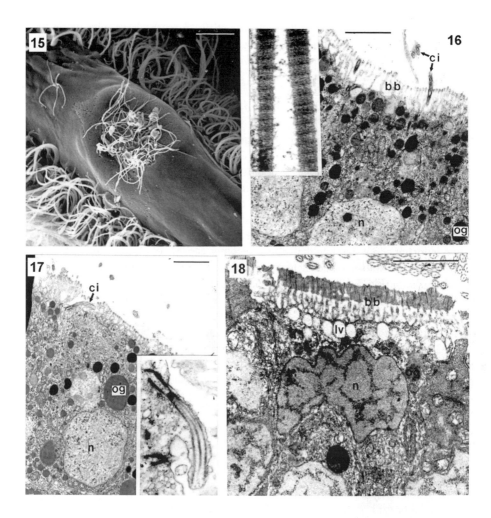

Figures 2.15–2.18. Electron micrographs of larvae of *Pecten maximus*. Labels: **bb** – brush border, **ci** – cilia, **lv** – electron lucent vacuole, **n** – nucleus, **og** – osmophilic granule.

Figure 2.15. Central region of velum of umboned veliger showing apical tuft arising from a circular depression in velum surface. Scale bar = 10 μm.

Figure 2.16. Cell bearing apical flagellum of early veliger. Scale bar = 2 μm. Inset shows striated rootlets of cilia forming apical flagellum.

Figure 2.17. Cell adjacent to the apical flagellum cell. Scale bar = 2 μm. Inset shows simple cilia bases and cilium.

Figure 2.18. Cell bearing tall brush border of the apical organ of later veliger larva. Scale bar = 2 μm.

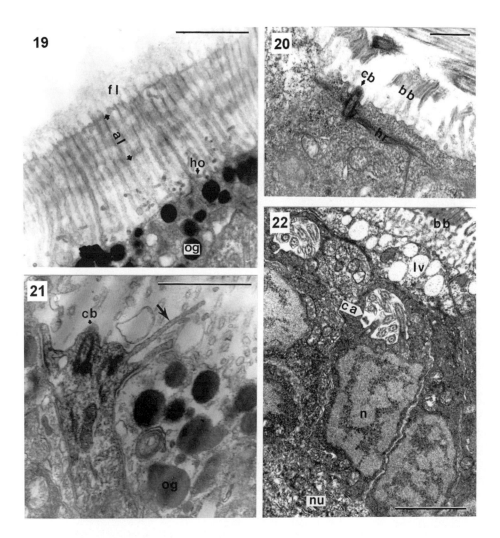

Figures 2.19–2.22. TEM images of apical organ of veligers of *Pecten maximus*. Labels as for Figs. 2.15–2.18 plus: **al** – amorphous extracellular layer, **ca** – ciliated cavity, **cb** – cilium base, **fl** – filamentous extracellular layer, **ho** – horizontal microvillous, **nu** – neuropil.

Figure 2.19. Complex brush border of apical organ. Scale bar = 1 μm.

Figure 2.20. Horizontal roots of apical tuft cilia. Scale bar = 1 μm.

Figure 2.21. Horizontal microvillus originating from cilium-bearing cell. Scale bar = 1 μm.

Figure 2.22. Ciliated cavity cell overlying neuropil of commissure between lobes of the cerebral ganglion. Scale bar = 2 μm.

72

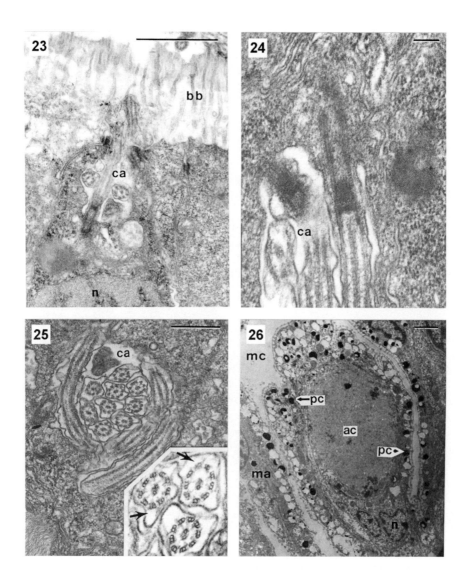

Figures 2.23–2.26. TEM images of veligers of *Pecten maximus*. Labels as for Figs. 2.15–2.22.

Figure 2.23. Ciliated cavity opening to the surface of the apical organ. Scale bar = 1 μm.

Figure 2.24. Base of cilium on ciliated cavity cell. Scale bar = 0.1 μm.

Figure 2.25. Cilia in cavity. Scale bar = 1 μm. Note in inset that side arms of cilia tubules point in clockwise direction in the case of two cilia (arrows) and anticlockwise in the other cilium.

Figure 2.26. Eye-spot with amorphous core (**ac**) which does not appear to be separated by a plasma membrane from the cytoplasm surrounding a basally located nucleus. A pigmented cup (**pc**) with osmophilic granules surrounds the core. Similar granules occur in the nearby mantle epithelium. Scale bar = 1 μm.

cerebral ganglion are found within these cells. The cytology of these cells resembles that of the apical cirrus cell from which they may be derived.

The third major component of the apical organ epithelium are irregularly-shaped cells with cilia projecting into an intracellular cavity (Fig. 2.22). These cells generally lie within the epithelium, but some have a small portion of the apical cytoplasm projecting between the other cell types to the free surface of the apical organ. Occasional examples with the cavity opening to this surface have been found (Fig. 2.23). No more than 15 cilia have been found in a single cell. The cilia lack striated roots (Fig. 2.24), but have the normal 9 + 2 configuration with side arms on the microtubules (Fig. 2.25). They extend into the cavity from more than one direction (Fig. 2.25). Intracellular axonemes have occasionally been found in these cells in pediveliger larvae. The cytoplasm of these cells contains multivesicular bodies, a Golgi body and dense-cored vesicles approximately 70 nm in diameter. The nucleus is generally located below the cavity. The basal portion of the cell lies close to the commissure between the two lobes of the cerebral ganglion with no intervening basement membrane. However, no extensions from these cells have been found in the neuropil and no synapses have been found in the epithelium.

2.3.3.7 Statocysts

Statocysts formed by invagination (Moor 1983) have been reported from several larval pectinids listed in Table 2.3. In *Pecten maximus,* the invagination of the epithelium of the foot remains, forming a spherical sac connected to the mantle cavity by a cylindrical ciliated canal (Cragg and Nott 1977). Openings to these canals have also been observed in *Argopecten purpuratus* (Bellolio et al. 1993). The cilia of the canal are directed inward. Specialised ciliated cells - hair cells - form the majority of the wall of the sac. The bases of the cilia belonging to one cell are arranged in a circle or a circular array, with basal feet pointing inwards and horizontal striated roots probably radiating outward. Several statoconia (dense non-cellular bodies) of varying shape, size and structure are found free in the lumen of the sac. These can be seen in jerky motion in live larvae. Similar bodies occur within a single non-ciliated cell in the wall of the sac. A small nerve stretches from the pleural ganglion along the wall of the statocyst canal towards the sac of the statocyst. Its fibres may connect with thin processes extending from the hair cells (Cragg and Nott 1977).

2.3.3.8 Eye spots

A pair of eye spots have been observed in older veligers and pediveligers of many pectinids (Table 2.3). Eye spots can be seen through the transparent shell, appearing roughly in the centre of the valve when the larva is viewed from the side. The actual location of the eye is (in *Chlamys hastata)* on the anterior aspect of the gill bar: it consists of one cell with pigmented granules forming an anteriorly directed cup and another cell within the cup (Hodgson and Burke 1988; Fig. 2.26). Eye spots are developed by larvae of the Pteriomorphia and Heterodonta (Cragg 1996) and appear to be retained in adults of some Arcoida and Pterioda (Morton 2001).

2.3.3.9 Mantle ciliation

A study of critical-point-dried specimens broken open to remove obscuring tissues (using the methods described by Cragg 1985) has shown that the inner mantle fold of *Pecten maximus* larvae, prior to metamorphosis, has five different sorts of groupings of cilia on it, referred to as types 1–5 in the following text (Figs. 2.9 and 2.27). At five days old, no groupings of cilia are evident on the inner mantle fold, but 13-day veliger larvae have type 2 groupings and type 1 groupings appear in umboned veligers. By the pediveliger stage, all five types of grouping are present.

As many as four type 1 groupings have been found on a valve, each with up to five short (less than 6 μm long) cilia protecting from a small circular depression (Fig. 2.28). Unlike cilia from some other grouping types, these cilia do not appear stiff in SEM specimens. These groupings are situated near the anterior margin of the shell. Six or less type 2 groupings (Fig. 2.29) were found in 13-day old larvae, but there were up to 13 of them in pediveliger specimens. They are each formed from a row of about 25 cilia (exceptionally as many as 70) arranged in a line. These cilia are comparatively long (up to 21 μm) and are generally straight in SEM specimens. In live retracted larvae, they can be seen beating with no metachronal pattern evident. Six or less type 3 groupings occur in an arc which runs around the posterior margin of the valve, but further from the rim of the valve than type 2 and 4 groupings. Type 3 groupings are stiff tufts of up to 14 cilia with a maximum length of about 9 μm (Fig. 2.30). The type 4 grouping is made up of a line of single cilia running around the posterior margin of the valve (Fig. 2.30). The type 5 grouping is found along the mid-line of the epithelial bridge which lies between the valves and dorsal of the anus. It consists of a row of about ten short (up to 8 μm) single cilia, each arising from a small hemispherical protruberance (Fig. 2.31).

Only type 2 and type 3 configurations have been identified in sections. Fig. 2.9 summarises what is known about the cytology of the cell-type bearing type 2 configurations. The cilia of such configurations arise from a single cell that is much longer than its neighbours. In SEM specimens, these cells can be distinguished not only by the line of cilia along their inner margin (i.e., the margin furthest from the shell rim), but also by their distinctive long unbranched microvilli. The basal part of these cells contains the nucleus and rough endoplasmic reticulum; the middle part is filled with a particularly large Golgi body; the apical cytoplasm is full of large electron-lucent vesicles. The cilia are attached to long striated rootlets. Simple nerves containing cell profiles, some with neurotransmitter-like vesicles (Fig. 2.32), have been observed in the mantle epithelium close to the type 2 cells, or in the space within the mantle folds (Fig. 2.9). These nerves generally have non-nerve companion cells closely associated with them. Croll et al. (1997) found cells in the mantle rim of *Placopecten* veligers, with fluorescence indicative of catecholamines, which can function as neurotransmitters.

The cells bearing the type 3 cilia have not been subjected to detailed study, but sections reveal that associated with these groupings there are one or two specialised cilia on the ventral side of the group, borne on a neighbouring cell which lacks microvilli. Such cilia have the normal 9 + 2 arrangement, but are surrounded by a corona of nine

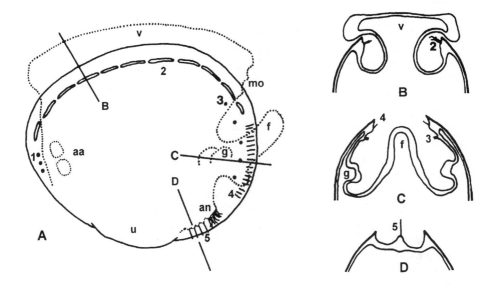

Figure 2.27.   Distribution of types of configuration of cilia on the inner mantle fold of the pediveliger. A: lateral view of left valve with soft tissues indicated by dotted line and shell by solid line. B, C, D: sections along the lines marked on A. Labels as for Figs 2.28–2.32 plus: **aa** – anterior adductor, **f** – foot, **g** – gill buds, **mo** – mouth.

microvilli that are wedge-shaped in cross section (Fig. 2.33). These microvilli project a small distance above the level of the microvilli of surrounding cells. Longitudinal fibrils extend from the microvilli into the apical cytoplasm. Extracellular filamentous material extends circumferentially from microvillus to microvillus, and radially from each microvillus to the cilium. Sections suggest that cilium and microvillar corona arise from a columnar extension from the surface of the cell which is otherwise devoid of cilia and microvilli. However, insufficient examples have been seen to determine whether this is always the case.

Patterns of cilia have also been found on the mantle fold of *Argopecten purpuratus* (Bellolio et al. 1993): groups of up to ten cilia emerging from protuberances and single cilia projecting from a circular depression. The groups and single cilia alternate around the rim. The groups appear similar to the type 3 cilia described above.

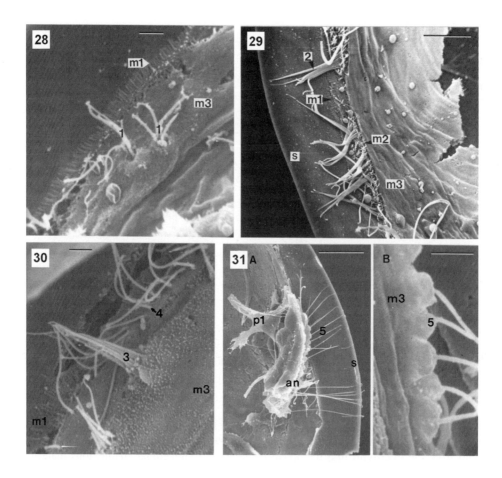

Figures 2.28.–2.31. SEM images of mantle of pediveligers of *Pecten maximus*. Labels: Labels: **an** – anus, **co** – companion cell for nerve, **gb** – Golgi body, **lv** – electron lucent vesicles, **m** – mitochondrion, **m1**, **m2**, **m3** – types of microvilli, **mu** - smooth muscle, **ne** – nerve profiles, **rer** – rough endoplasmic reticulum, **s** – shell, **1 - 5** – types 1–5 mantle cilia.

Figure 2.28. Type 1 cilia configurations on inner mantle fold. Scale bar = 2 μm.

Figure 2.29. Type 2 cilia configurations on inner mantle fold. Scale bar = 10 μm.

Figure 2.30. Type 3 and 4 configurations on inner mantle fold. Scale bar = 10 μm.

Figure 2.31. Type 5 cilia on mantle tissue around anus. Note edges of torn tissue. A: scale bar = 10 μm; B: scale bar = 2 μm.

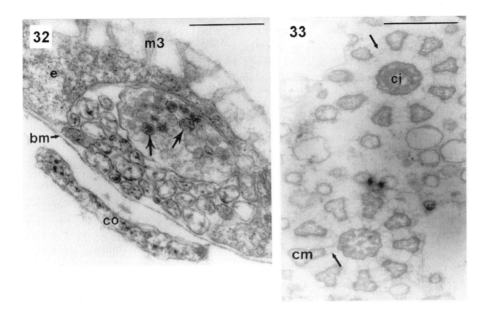

Figure 2.32. Bundle of nerve cell profiles in epithelium close to the inner mantle fold as shown in Fig. 2.9. Arrows indicate neurotransmitter granules. Scale bar = 0.5 μm.

Figure 2.33. Cilia from a type 3 configuration with coronas of microvilli. Note radial and circumferential arrangement of filamentous material (arrows). Scale bar = 0.5 μm.

2.3.3.10 Foot

The appearance of a foot in the later stages of larval development in pectinids has been noted by many of the authors listed in Table 2.3. Hodgson and Burke (1988) found that in *Chlamys hastata* the foot rudiment appears after 15 days development and becomes functional after 34 days, only a few days before metamorphosis. The functional foot in this species (and in *Argopecten* - Bellolio et al. 1993) is a somewhat sock- shaped organ, which is profusely ciliated on its "sole" (ventral surface), but sparsely ciliated on its lateral and dorsal surfaces. Culliney (1975) noted the presence of long mobile cilia on the toe of the foot of *Placopecten magellanicus.* These may correspond to the long cilia borne by processes from cell bodies with catecholamine-associated fluorescence found in this region by Croll et al. (1997). These processes extend from 15 to 20 cells, which seem to be connected to the pedal ganglion.

A byssal duct opens at the "heel" and a byssal grove extends forward along the "sole" from this opening (Gruffydd et al. 1975; Hodgson and Burke 1988). The duct has a bilobed blind ending *(Pecten maximus* - Gruffydd et al. 1975) or consists of two ciliated canals (*Chlamys hastata* - Hodgson and Burke 1988).

Five gland types can be distinguished in the foot of the pediveliger of *Pecten maximus*, characterised by their staining with Azan and toluidine blue. Equivalents of four of these have been identified in juveniles of *Patinopecten yessoensis* (Bower and Meyer 1990). The secretions of these glands are likely to resemble similarly staining secretions in homologous glands in *Ostrea* and *Mytilus* that have been examined with cytochemical techniques (Gruffydd et al. 1975). The primary byssus gland (also recognised by Hodgson and Burke) surrounds the bilobed blind ending to the byssal duct: it probably secretes highly sulphated acid mucopolysaccharides. The pedal mucus glands extend to the tip and anterior end of the sole of the foot. The secondary byssus glands are situated near the opening of the byssus duct and may produce a collagenous secretion. Two types of phenolic gland discharge a secretion likely to be phenolic protein into the anterior end of the byssal groove.

The gills are another organ that appears during the later stages of larval development and is retained through metamorphosis. Gill development in *Pecten* is of the papillary type (Raven 1958), in which a row of papillae is formed on either side of the foot in a postero-anterior sequence (Beninger et al. 1994). In studies of live pectinid pediveligers, gills have been detected in a number of species (Table 2.3). The observations of *Chlamys hastata* and *Pecten maximus* pediveligers using SEM and histological sections (Hodgson and Burke 1988; Beninger et al. 1993) reveal more detail: on either side of the foot there is a ridge of tissue extending from the mantle into the mantle cavity which bears three to five small lobes - the primary gill filaments - formed from cuboidal epithelium containing large vacuoles filled with granules. Belding (1910) reports that in *Argopecten irradians* there are initially two gill filaments on either side of the foot, but by metamorphosis there are four or five occasionally contractile filaments bearing active cilia. On each filament of *C. hastata* there is a single row of simple cilia that beat inwards in anteriorly-directed metachronal waves.

2.3.3.11 Nervous system

The nervous system of the pediveliger of *Ostrea* and *Mytilus* consist of a bilobed cerebral ganglion joined by connectives to paired pleural, visceral and pedal ganglia (Erdmann 1934; Ranieri 1995). Nerves run from the larval eyes to the same connective and the nerve from the statocyst connects to the visceral ganglion (Erdmann 1934). Most components of this system have been recognised in light microscope and TEM sections of larvae of *Patinopecten yessoensis* (Bower and Meier 1990), *Chlamys hastata* (Hodgson and Burke 1988) and *Pecten maximus*. The bilobed cerebral ganglion is described in the section of this chapter that deals with the apical organ. The visceral ganglion has been identified (pers. obs.) and the relationship between the pedal ganglion, the pleural ganglia and the statocysts is illustrated by Cragg and Nott (1975).

Hatschek (1880) described a plexus of nerves radiating to the velum rim from the cerebral ganglion in *Teredo*, while Raineri (1995) reported two rows of acetylcholinesterase-staining bipolar neurons at the base of the velum and connecting with the cerebral ganglion in *Mytilus*. Evidence for innervation of the velum has also been found in pectinids. Nerve profiles have been identified in TEM sections of the velum of

Pecten (Cragg 1989) and, in *Placopecten*, fluorescent staining for catecholamines has revealed nerve connections similar to those described by Raineri.   Four to eight catecholamine-staining cell bodies have been revealed on the outer rim of the velum. These are connected around the rim by two to three fibres (Croll et al. 1997).   No evidence of a radiating plexus was revealed by this approach, though a fibre connecting from a pair or more of flask-shaped, fluorescing cells on either side of the mouth was found to extend to the region of the cerebral ganglion. These cells may correspond to the nerve cell clusters with acetylcholinesterase activity near the oesophagus of *Mytilus* reported by Raineri (1995) and the catecholaminergic cells near the mouth of the prosobranch *Crepidula* (Dickinson et al. 1999).   The position of these cells and their connections is suggestive of control of feeding mediated by the cerebral ganglion, perhaps in response to sensory information.   Indeed, there is some evidence of nervous control of feeding: serotonin has been reported to inhibit the ciliary transport rate in the oesophagus of a nudibranch veliger (Pavlova et al. 1999).

Further studies combining aldehyde-induced fluorescence and immunofluorescence with confocal laser microscopy are needed to fully reveal the extent and nature of connections in the apical organ and velar nervous system.   A more detailed examination of distribution of neurotransmitters or neuromodulators needs to be considered as studies of a range of gastropod larvae have revealed staining for the peptide FMRFamide and for serotonin in cells of the apical organ and in fibres extending from the apical organ to the velum (Kempf et al. 1997; Dickinson et al. 1999).

There are nerves, which have not been mentioned in other descriptions of bivalve larvae - those associated with retractor and adductor muscles, and with the inner mantle folds, though their connections to ganglia remain to be clarified (Cragg 1985; this chapter). Those connecting to the retractor muscles have a constant diameter, while those connecting to the retractor muscles have varicosities, as has also observed in nerve processes in the velum (Croll et al. 1997).

2.3.3.12 Other organs

Anatomical and histological studies of veligers and pediveligers from other bivalve families (e.g., *Teredo* - Hatschek 1880; *Ostrea* - Erdmann 1934; *Crassostrea* - Galtsoff 1964; *Mytilus* - Bayne 1971) have revealed a few organs, which have either not been investigated in detail or not described at all from pectinid larvae. The larval heart or its precursor has been observed by Hatschek and Erdmann: it is functional in pediveligers of *Crassostrea virginica* (Galtsoff 1964). No equivalent of this has been reported from any pectinid larvae, nor is there any mention of the larval kidneys illustrated by Erdmann, Hatschek and Bayne.

2.3.4 Metamorphosis

Towards the end of larval life, pediveligers initiate settlement behaviour. This change is reversible – larvae can return to the previous behaviour patterns. However, at some stage during settlement, the irreversible morphogenetic changes involved in metamorphosis may be triggered. In scallop larvae, as in other bivalve larvae, the process of metamorphosis involves changes in the nature of shell secretion, loss of some organs, greater development and/or relocation of others. The changes in shell shape and structure, which are described in a section above, must result from changes in the mantle, but no information is available regarding the modifications of mantle fold morphology and cytology. The principal organs lost at metamorphosis are the velum, the velar retractor muscles and the anterior adductor.

Accounts of the loss of the velum differ markedly: whether this indicates that the process of velum loss varies between pectinid species or that different interpretations are put on the process is unclear. Most accounts state that the velum is lost at the end of the pediveliger stage: Culliney (1975) reports that the velum of *Placopecten magellanicus* is cast off in large pieces or even as a single piece, while in *Chlamys hastata* it is histolysed (Hodgson and Burke 1988) and in *Pecten maximus* it degenerates (Gruffydd and Beaumont 1972). The account of Sastry (1965) of metamorphosis of *Argopecten irradians* suggests a remarkable degeneration of the velum almost complete by the time the foot is functional. However, Gutsell (1930) and Chanley and Andrews (1971) provide illustrations of pediveligers of *A. irradians* that show no sign of degeneration of the velum. The loss of the anterior adductor at metamorphosis has been reported by Belding (1910) and Sastry (1965), but diagrams of Jørgensen show it to be retained in recently settled spat of *Pecten septemradiatus* and *Chlamys opercularis.*

There is a general migration, relative to the axes of the shell, of those organs that survive metamorphosis (Belding 1910; Jørgensen 1946; Sastry 1965; Hodgson and Burke 1988). This results in the mouth moving from its posterio-ventral larval location to the adult anterio-dorsal position, the foot becoming ventral rather than posterior and the posterior adductor migrating to the centre of the valve (Belding 1910). Sastry's diagrams indicate that the migration of the mouth commences during the veliger stage, but the illustrations of Chanley and Andrews (1971) and Gutsell (1930) show no evidence of such migration, even in pediveligers.

After metamorphosis, the gill filaments increase in length and number (Belding 1910; Gutsell 1930; Sastry 1965; Hodgson and Burke 1988). The outer filaments of the adult gill system appear after metamorphosis (Belding 1910), the cilia of the larval gills become the lateral ciliated band and the frontal cilia appear (Hodgson and Burke 1988). The gill becomes capable of suspension feeding within a day of metamorphosis, ensuring that there is little ingestion loss during the transition (Gruffydd and Beaumont 1972).

The shape of the foot gradually changes after metamorphosis (Sastry 1965). The glands present in the pediveliger foot become better developed (Gruffydd et al. 1975) and the byssus secreted changes abruptly to a more sticky form after metamorphosis (Culliney 1975). In *Argopecten irradians* the heart and pericardium are first apparent in the light microscope in spat (Belding 1910), with beating first evident two weeks after metamorphosis (Sastry 1965), but Fullarton (1896) reports a heart in a pediveliger, which may have been *Chlamys opercularis,* taken from the plankton. Of the putative sense organs of the pediveliger, the statocysts are retained (Belding 1910; Sastry 1965); so are the eye spots (Gruffydd 1976) though these probably have no association with the adult eyes which appear in much later stages on the rim of the mantle.

The fate of the groupings of cilia on the inner mantle fold is not known. However, Belding (1910) reports the presence of cilia on the mantle of spat and the region where the cilia occur on the mantle probably represents the precursor of the adult inner mantle fold, which bears sensory tentacles. These first appear a few days after metamorphosis (Sastry 1965). The fate of the apical organ is uncertain: in a number of species of bivalves, cells of the apical organ participate in the formation of the labial palps (Cragg and Crisp 1991). Sastry's comments on the formation of palps suggest that this may also be the case in *Argopecten irradians*. As there appear to be either genuine differences in palp development between different pectinid species or differences of interpretation, the question of palp development deserves closer scrutiny.

2.4 COMPARATIVE ANATOMY

The general anatomical layout of pectinid veligers - bivalved shell, velum, musculature, digestive tract, nervous system, certain sense organs and foot - is typical for planktotrophic bivalve larvae. The mantle structure is probably also typical. Two mantle folds have been reported from the larvae of a number of other bivalve species (Cragg and Crisp 1991): only the remarkably large, teleplanic pediveliger of *Planktomya* has been reported to have the adult complement of three folds (Allen and Scheltema 1972).

In the venerid *Ruditapes*, the non-motile flagellum borne by three central cells is retained through larval life and is surrounded by sub-epithelial cells with ciliated cavities (Tardy and Dongard 1993). Some form of apical ciliation, either in the form of a flagellum formed from numerous cilia or a ciliary tuft, has been reported for veligers of other bivalve taxa (Zardus and Martel 2002). Indeed, apical structures bearing tufts of cilia occur in planktonic larvae in many phyla (Nielsen 1995). The embryological origin of apical organ of larvae from a wide range of spiralian phyla is sufficiently similar to suggest homology (Hadfield et al. 2000). Furthermore, serotonin immunoreactivity has been detected in the apical organ of polychaetes, brachiopods, phoronids, nudibranchs and echinoids (Kempf et al. 1997). Additionally, the arrangement of a sensory ciliated epithelium overlaying the commissure between the lobes of the cerebral ganglion is common to gastropod and bivalve veligers. Indeed, the ciliated cavity cells of the apical organ of *Pecten maximus* resemble the cells termed ampullary neurons in the apical organ of the veligers of prosobranch and nudibranch gastropods, and annelid trochophores (see studies and reviews of Page 2002; Kempf et al. 1997). Thus, evidence for homology of elements of the apical sensory structures in molluscs is strong (Hadfield et al. 2000).

The opposed-band arrangement of cilia on the velum may be homologous with similar features in larvae from other molluscan classes and from other protostomian phyla (Strathmann et al. 1972; Strathmann and Leise 1979; Nielsen 1995), though independent evolution of the opposed-band mechanism has also been proposed (Rouse 1999).

The velar retractor muscles are lost at metamorphosis as are similar muscles in many gastropods. The question as to whether these larval muscles represent a synapomorphy for the entire phylum Mollusca has been investigated by Wanninger and Haszprunar (2002a). Their conclusion is that, as no such larval muscles occur in the Polyplacophora, they are not an ancestral feature for the phylum, and their observations of a lack of special larval muscles in scaphopods (Wanninger and Haszprunar 2002b) indicate that such retractor muscle systems rather than being an ancestral feature for the Conchifera (= Scaphopoda, Gastropoda, Bivalvia, Cephalopoda) could have evolved independently in the Gastropoda and Bivalvia.

The papillary mode of gill development occurs not only in pectinids, but also in *Arca, Anomia, Modiolus, Mytilus* and *Ostrea* of the Pteriomorphia and *Dreissena* of the Heterodonta (Raven 1958). Multiple statoconia have been found, not only in pectinid pediveligers, but also in other members of the Pteriomorphia: single statoconia on the other hand have been reported only from the Nuculoida, Veneroida and Myoida (Cragg and Nott 1977). A possible exception to this is the report of Stafford (1912) that veligers with single statoconia she had collected from the plankton could be identified *as Placopecten magellanicus.* However, Merrill (1961) doubted the precision of this identification and Stafford's illustrations do not resemble those of Culliney (1974), who reared larvae of this species. Indeed, the larvae illustrated differ sufficiently from the descriptions of the authors listed in Table 2.1 that it seems unlikely that they were pectinids. Gruffydd et al. (1975) found that there are homologues of certain of the pedal glands of *Pecten maximus* in two other members of the Pteriomorphia - *Mytilus edulis* and *Ostrea edulis.*

Tufts of cilia on the inner mantle fold, possibly of similar origin to those observed in *Pecten maximus,* have been reported in *Ostrea edulis* (Cranfield 1974; Waller 1981) and tufts of morphology and position similar to types 2 and 3 groupings of *P. maximus* were observed in another oyster, *Crassostrea gigas* (Cragg 1976). The cilia of the type 3 configurations with their corona of microvilli resemble structures on the mantle tentacles of *Placopecten magellanicus* adults (Moir 1977). They also resemble collar receptors found in organisms in several major taxa: gastropods (Crisp 1981), adult nuculid bivalves, turbellarians, (Haszprunar 1985) gastrotrichs (Teuchert 1976), kinorhinchs plus other invertebrates (Nebelsick 1992) and cnidarians (Golz and Thurm 1993). Several phylogenetic interpretations of these possible homologies are considered by Haszprunar.

Illustrations of the shell of pectinid veligers and pediveligers (citations in Table 2.1; Fig. 2.34) show that once the shell loses the "D" shape, it takes on a characteristic outline with rather indistinct umbones and an anterior margin which is more tightly curved than the posterior margin. None of the other bivalve families surveyed by Chanley and Andrews (1971) have such an outline, though mytilids are somewhat similar (see Fuller and Lutz 1989). The taxodont teeth at either end of the hinge-line in pectinid veligers are also similar to those in anomiids and mytilids, but unlike mytilids the region in between is devoid of teeth. The distribution and type of teeth distinguishes pectinid veligers from those of other bivalve families (Rees 1950; Le Pennec 1980). Lateral indentations to the outline of the velum (Culliney 1974; Hodgson and Burke 1988; Cragg 1989) may prove to be another characteristic feature of pectinids.

Features of the hinge, such as number of teeth, shape and position of ligament (Le Pennec 1980; Tanaka 1984), and of shell outline (Table 2.1, Fig. 2.34) may also permit species of pectinids to be distinguished from one another. Jørgensen (1946) noted species-specific shell pigmentation in larval pectinids. The pattern of retractor muscle distribution may also prove to be characteristic (Cragg 1985). Diagrams of Jørgensen

(1946) suggest that the degree of separation of the components of the anterior adductor and the pattern of coiling of the intestine varies from species to species. Further information relating to the features mentioned above would enable keys to pectinid larvae to be prepared.

Many descriptions of veligers list measurements of the dimensions of the shell and hinge line. Care should however be taken in using size data derived from laboratory-reared larvae as species-characteristic features, because Sasaki (1979) has shown that the size at metamorphosis can be affected by rearing conditions and the observations of Paulet et al. (1988) suggest that genetic factors may account for significant variation within a species of size at metamorphosis. Sasaki found that the mean width at metamorphosis of *Pecten maximus* was 247 µm in spat collected from the sea and 223 µm in hatchery-reared spat: the corresponding figures in *Chlamys opercularis* were 225 µm and 201 µm (the measurements for each species represent the means of 100 measurements of spat from the sea and of 60 spat from the hatchery). Paulet and his colleagues found that the mean size of metamorphosing larvae derived from one stock of *Pecten maximus* from a site in north western France differed significantly from those derived from another nearby stock with different spawning characteristics.

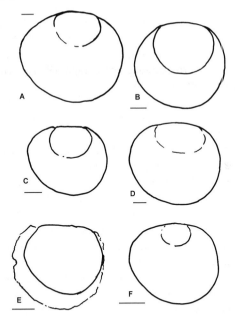

Figure 2.34. Outline of prodissoconch II shell in larval pectinids, with hinge above and anterior end to the left. A: *Pecten maximus*, scale bar 25 µm (Salaun and Le Pennec 1992); B: *Argopecten purpuratus*, scale bar 20 µm (Bellolio et al. 1994); C: *Bathypecten eucymatus*, scale bar 50 µm (Schein 1989); D: *Chlamys hastata*, scale bar 20 µm (Hodgson and Bourne 1988); E: *Equichlamys bifrons*, scale bar 50 µm (Dix 1976); F: *Adamussium holbecki*, scale bar 100 µm (Berkman et al. 1991).

2.5 FUNCTIONAL MORPHOLOGY

2.5.1 Locomotion

The trochophore moves by means of the simple, relatively short cilia of the prototroch. With the depletion of lipid reserves and the secretion of the shell, the larva becomes denser and requires a more effective means of propulsion. This is provided by an increase in the ciliated band circumference relative to body size due to the formation of the velum, and by the greater propulsive power of the long cirri of the main preoral ciliary bands. The cilia of the other bands on the velum are unlikely to contribute significantly to the propulsion of the larva. Though the main thrust of the preoral cirri tends to propel the larva velum first, the weight of the shell deflects the swimming path and a small tangential force generated either during the active beat or the recovery stroke probably accounts for the rotation of the larva as it swims. The density of the shell means that when the velum retracts, the direction of movement abruptly changes and the larva sinks more rapidly than it can swim (Cragg 1989).

The control of beating of the preoral cilia is probably due to a mechanism similar to that in gastropod veligers investigated by a number of authors reviewed by Arkett et al. (1987). Arkett and colleagues showed that in veligers of the gastropod *Calliostoma* the cells bearing the preoral cirri are connected by gap junctions and are thus electrically coupled. Propagated Ca^{++}-dependent action potentials are initiated from the cerebral ganglion, causing velum-wide arrests of ciliary beating, while reduction in beating rate is accompanied by a slow depolarisation of the cell membrane. Direct evidence of excitatory nervous control by sensory serotinergic cells in the probable homologue of the apical sense organ of marine gastropods has been obtained from laser ablation studies with pulmonate embryos (Kuang and Goldberg 2000). Experimentation has demonstrated that the velum of *Mytilus* larvae has dual innervation: excitatory serotinergic and inhibitory dopaminergic (Beiras and Widdows 1995). Acetylcholinesterase occurs in the ciliated cells of the velum in *Mytilus* (Raineri 1995). Immunofluorescence studies suggest nervous control of the ciliated cells with connected catecholaminergic cells on the rim of the velum, but no rim to apical organ connections yet located (Croll et al. 1997). The abrupt cessation of beating and changes in beating rate that occur in *Pecten maximus* veligers (Cragg 1989) suggest that nervous control similar to that revealed by Beiras and Widdows (1995) probably occurs in pectinids.

Precisely-ordered sequences of contraction of the various components of the musculature effects the body movements involved in velum contraction and protraction (Cragg 1985). In the pediveliger stage, crawling is accomplished by a combination of the creeping due to the cilia of the sole of the foot, followed by a sudden contraction of the foot presumably due to the pedal retractor muscles, which draw the shell towards the tip of the foot (Gruffydd et al. 1975 - *Pecten maximus*; Hodgson and Burke 1988 - *Chlamys hastata*). Mucus secreted by glands near the tip of the foot is likely to facilitate creeping, but no equivalents of the tip attachment glands of *Mytilus* and *Ostrea,* which probably temporarily anchor the tip of the foot during contractions, have been found in *Pecten maximus* pediveligers. The remaining pedal glands secrete various components of the byssus, which permit the larva to temporarily anchor itself (Gruffydd et al. 1975).

2.5.2 Feeding

The ciliation of the velum plays an important role in feeding. The arrangement of bands of cilia is comparable with those of organisms that feed by the "opposed band" filtering mechanism. In this mechanism, studied in veligers of *Crassostrea gigas* by Strathmann and Leise (1979) and of *Mercenaria mercenaria* by Gallager (1988), the long preoral cirri sweep particles such as algal cells towards the adoral tract. The postoral cilia may prevent particles from being swept further and the adoral tract carries the captured particles towards the mouth. Veligers can clear between 3.4 and 38.5 $\mu L\ h^{-1}$, when feeding in algal suspensions, with feeding rate increasing with increasing temperature (Lu and Blake 1997).

Bivalve larvae have been shown to be capable of taking up nutritionally-significant quantities of dissolved amino acids from seawater through the velum (Manahan and Crisp 1982) as can the gills of newly-metamorphosed spat of *Pecten maximus* (Manahan and Crisp 1983). Though the gills become the feeding organ in postlarvae, pediveliger gills are non-functional in this respect (Beninger et al. 1994). Gills become capable of filter-feeding a day after metamorphosis (Gruffydd and Beaumont 1972), so that the uptake of dissolved organic matter by the gills may help bridge the gap between the loss of velar filtering and the initiation of gill filtering (Manahan and Crisp 1983).

Particles captured by the velum enter the ciliated oesophagus. Le Pennec and Rangel-Davalos (1985) studied the fate of ingested unicellular algae in veligers of *Pecten maximus* using epifluorescence microscopy. The gut becomes functional shortly after shell secretion. At high densities of algal cells, ingestion and digestion are continuous and breakdown of ingested cells of readily digested algal species starts 5–6 hours after ingestion and is complete after 10–11 hours. Variations in ingestability of similar-sized algae (Lora-Vilchis and Maeda-Martinez 1997) suggest a selective capability, perhaps at a ciliated sphincter at the junction between the mouth and the oesophagus (Gallager 1988; Bower and Meyer 1990). The presence of catecholaminergic cells in this region, which have connections running to the apical organ (Croll et al. 1997), may be evidence of some form of nervous control of ingestion, with perhaps interaction through the apical organ to the swimming cirri. Field studies suggest that veligers of *Placopecten* are capable of selective feeding (Raby et al. 1997).

2.5.3 Respiration

The large surface-area-to-volume ratio of larvae, the currents generated by ciliary beating and the short internal diffusion distances mean that no separate respiratory organ is necessary of respiratory demands. No identified respiratory structures develop during the veliger stage and it is not certain whether the gills in the pediveliger function in gaseous exchange. The fluid of the body cavity is only separated from the surrounding seawater by the thin epithelia of the velum and mantle. The cells bearing the preoral cirri, which are directly in contact with a stream of water generated by ciliary activity, probably constitute the site of highest oxygen demand as evidenced by their large and numerous mitochondria. The mantle cavity is separated from this current by the velum, so perhaps the beating of the type 2 cilia of the mantle serve to prevent water from stagnating in the cavity.

2.5.4 Sensory Reception

The structure of the apical organ suggests that it is a sense organ, but a sensory role has not been demonstrated experimentally. Its role is particularly unclear in the trochophore and early veliger where the nervous system has yet to differentiate. In later veligers, broad conclusions about function can be drawn on the basis of structural similarities with more thoroughly investigated organisms. The complex brush border of the apical organ, with vertical and horizontal microvilli interwoven with extracellular material, is similar to that of molluscan chemosensory epithelia on the osphradium (Crisp 1973), olfactory organ (Yi and Emery 1991) and tentacles (Wright 1974). The cells bearing the cilia forming the apical tuft resemble sensory neurones in such epithelia. The ciliated cavity cells of the apical organ resemble cells in a range of putative or known chemosensory structures: the apical sensory organ of nudibranch veligers (Chia and Koss 1984), the rhinophore of *Nautilus* (Barber and Wright 1969), the lip of *Sepia* (Emery 1975), the olfactory epithelium of *Octopus* (Woodhams and Messenger 1974) and the ampullary system of chitons (Haszprunar et al. 2002). The sensory modality of this cell type remains to be established in many cases, though Lucero et al. (1992) have detected an electrical response to chemical stimuli from similar cells in squid. Detailed studies with the probably homologous (see section 2.4) apical sensory organ of gastropod larvae are also illuminating. Cells that correspond in location, size and number to cells with ciliated cavities in the apical sensory organ of a nudibranch veliger have been demonstrated to be responsible for detection of chemical settlement cues (Hadfield et al. 2000). Receptors for similar cues have been located on cilia harvested from metamorphically competent veligers of a prosobranch (Baxter and Morse 1992), though in this investigation, the location of the organ providing the cilia was not known.

The type 1, 3, 4 and 5 configurations of cilia on the mantle all consist of groups of cilia too small and too few to create significant currents by beating, so the possibility that they are sense receptors should be considered. As the mantle ciliation reaches greatest complexity in pediveligers, the sensory role may relate to cues for settlement. The cilia from the type 3 groupings with corona of microvilli resemble structures found in known mechanoreceptors (see for example Moir 1977; Haszprunar 1985; Golz and Thurm 1993). The orientation of the basal structures of the cilia from the cells forming the sac of the statocyst resembles that of cells demonstrated to be gravity receptors (Cragg and Nott 1977).

2.6 PHYSIOLOGY AND EFFECTS OF ENVIRONMENTAL VARIABLES

The processes of feeding, respiration and growth in bivalve veligers have been investigated in some detail, particularly in mussels and oysters (Bayne 1983; Crisp et al. 1985; Sprung 1984a, b, c), but until recently, little was known about such processes in pectinid larvae. The studies reviewed by Bayne (1983) show that energy is acquired initially from reserves contained in the egg, then from suspension feeding; the energy obtained is partitioned for maintenance, locomotion, growth and reserves for metamorphosis.

2.6.1 Energy Reserves and Feeding

The quantity and quality of egg-derived energy reserves is important for the early stages of larval development and can be modified by alterations to adult diet (Caers et al. 1999). Larvae originating from larger-than-average eggs have a higher survival rate over the first two days of development (Kraeuter et al. 1982) and take less time to reach metamorphosis (Paulet et al. 1988). Specific fatty acids are required in the maternal diet during maturation of eggs to ensure normal embryonic development, perhaps because of the role these molecules play in cell membrane synthesis (Le Pennec et al. 1998).

The energy reserves contained in scallop eggs consist of lipids (mainly neutral lipids), proteins and carbohydrate in the form of glycogen (Whyte et al. 1992; Lu et al. 1999b). In *Argopecten purpuratus* and *A. irradians* eggs, lipids represent, respectively 34% and 27% of ash-free dry weight (AFDW), proteins 38% and 64% and carbohydrates, 1.5% and 9.5% (Farias et al. 1998; Lu et al. 1999b). Vitamins B1 and B2 are also present in the eggs and are utilised during development (Seguineau et al. 2001). Early development is fuelled by energy reserves, especially glycogen and lipid, particularly neutral lipids. Soudant et al. (1998b) found that 20-carbon polyunsaturated fatty acid levels decrease during the lecithotrophic phase of larval development. Lipid consumption provides two thirds of the energy expenditure during embryogenesis (Lu et al. 1999b). Under optimal growth conditions, the proportions of different categories of lipid remain relatively constant during the planktotrophic phase (Delaunay et al. 1992; Soudant et al. 1998b), but total lipid accumulates during planktotrophy and then acts as an energy reserve for the process of metamorphosis (Whyte et al. 1987; Delaunay et al. 1992; Farias et al. 1998; Lu et al. 1999b). Lipid content is a significant predictor of metamorphosis success (Robert et al. 1999). The use of lipid reserves bridges the period between loss of velum, and gills becoming functional (Whyte et al. 1992). During metamorphosis, all three energy reserves decline by more than 50% of their premetamorphic mass, with protein and lipid breakdown accounting for over 90% of energy consumed (Lu et al. 1999b).

Planktotrophic feeding commences shortly after the development of the shell and velum. Veligers of *Patinopecten yessoensis* are capable of ingesting cells of the alga *Isochrysis* at a rate that ranges from 50 to 700 per hour depending on larval size. These veligers can filter 3–45 μL seawater per hour (MacDonald 1988). Like mussel veligers, they filter more efficiently than oyster veligers at low food concentrations, but are less efficient at high concentrations. When they are kept under crowded conditions, ingestion rates are reduced, perhaps due to frequent interruptions to feeding caused by collisions between larvae. Ingestion and filtration rates of *P. yessoensis* veligers are typical for bivalve veligers, as is assimilation efficiency. A food density of 6–10 cells mL^{-1} is required to support the metabolic demands of *P. yessoensis* veligers (MacDonald 1988). However, under natural conditions, some energy demands can be satisfied by dissolved organic matter (Manahan and Crisp 1982, 1983) or perhaps from bacteria associated with organic matter in particles large enough to be captured by the feeding apparatus.

Larvae are unable to convert or biosynthesise sterols essential for development. The sterol composition of larvae is strongly influenced by that of food algae, with sterols of some algae being more readily taken up than those of others (Soudant et al. 1998a). Studies of nutrition during the planktotrophic phase (see section 2.8.2) have concentrated on experiments with defined single or multiple species algal cultures. At sea, veligers will encounter a wide variety of algal species, often at suboptimal cell concentrations. Also, blooms of less nutritious species may prejudice larval survival (Nelson and Siddall 1988). In nearshore waters dominated by ultraplankton, stomach contents of *Placopecten magellanicus* veligers were mainly composed of small autotrophic flagellates and cyanobacteria (Raby et al. 1997), the nutritional value of which has yet to be investigated.

2.6.2 Respiration

Respiration rates for pectinid veligers are markedly higher than the 1.2–10 mL O_2 g (ash free dry weight)$^{-1}$ h^{-1} range for non-pectinid bivalve larvae (MacDonald 1988). Respiration rates in *Patinopecten yessoensis* veligers range from 13 to 29 mL O_2 g^{-1} h^{-1} and in *Argopecten irradians* veligers range from 4.6 to 15.2 mL O_2 g^{-1} h^{-1} (MacDonald 1988). In eyed veligers of *A. irradians* the range is 14.6–15.8 mL O_2 g^{-1} h^{-1} (Lu et al. 1999a). Higher respiration rates in pectinid larvae may be due to their being well fed while the other bivalves were not (Sprung, in MacDonald 1988). The high cost of respiration results in a growth efficiency, which is much lower than that of oysters or mussels (MacDonald 1988).

Respiration rates are depressed by crowding, presumably because during interruptions in swimming, the large preoral cirrus cells expend less energy. Swimming activity accounts for a considerable proportion of the energy demands of a veliger - 8–29% in the case of *Argopecten* larvae (Lu et al. 1999a). Further evidence of the energetic demands of veliger swimming is given by the much lower respiration rate of spat of *A. irradians* (1.8–5.3 mL O_2 g^{-1} h^{-1}) (Lu et al. 1999a).

Measurement of oxygen consumption of bivalve veligers in sealed chambers by polarography yields lower estimates than by coulometric respirometry (Hoegh-Guldberg and Manahan 1995), perhaps because the latter method operates at constant dissolved oxygen levels by replacing the oxygen consumed. Measurements with polarographic electrodes or Cartesian divers in which larvae were kept in a sealed chamber should therefore, perhaps be treated as underestimates.

Respiration rate increases with body weight. This can be expressed by an equation of the form: oxygen consumption = a (body weight) [b]. The value of b is close to 1 in bivalve veligers investigated to date, and is estimated at 0.92 for *Argopecten irradians* (Lu et al.

1999a) and 1.39 for *Patinopecten yessoensis* (where total rather than ash free dry weight was used: Macdonald 1988). The measured values of b are much higher in larvae than in adults (Lu et al. 1999a).

2.6.3 Growth

Growth of bivalve larvae can be described in terms of growth stages, cleavage rate, shell size increase or weight increase. Table 2.2 summarises the information available relating to the rate of embryological and larval development in the Pectinidae described in terms of growth stages. In most cases, this information relates to observations of larvae reared under laboratory conditions at close to optimum temperatures and feeding rates for the species in question. It is not known whether these species would develop at a similar rate in the sea, except in the case *of Patinopecten yessoensis* where development in the sea (Maru 1972) is somewhat slower than that in the laboratory (Yoo 1969). In the case of *Chlamys islandica,* the progress of a single cohort of larvae resulting from a brief spawning period in the sea near to Tromsø, northern Norway was followed (Gruffydd 1976). It was possible to estimate the length of larval life because 1) the brief spawning period was closely monitored; 2) the larvae were collected from the plankton from the earliest shelled stage until the numbers of pediveligers rapidly declined (presumably due to settlement and metamorphosis); 3) only three species of bivalve veliger were present in the plankton over the period in question and 4) the *Chlamys* larvae could be selected from these with some confidence, using features of shell morphology.

Larvae of *Patinopecten yessoensis* increase in body weight from 150 to 1,000 ng (ash free dry weight) during larval development (Whyte et al. 1987), which compares with a growth from 61 to 882 ng in *Mytilus* (Sprung 1984a). Though shell growth may not, under adverse conditions, be accompanied by growth of soft tissues (Bayne 1983), shell length measurements provide a convenient means of assessing growth. Information provided by the authors in Table 2.1 can be used to estimate shell growth rates. In general, under favourable conditions, the shells of scallop larvae grow at a rate of about 3–5 μm d^{-1}. Dissoconch growth rates are much faster (Gruffydd and Beaumont 1972; Rose et al. 1988). Larval shell growth rates give an indication of the effectiveness of feeding regimes and food species (Yoo 1969; Cary et al. 1981, 1982). The much slower rate of development of *C. islandica* larvae compared with the other species in Table 2.2 may partly be due to the lower temperatures at which these larvae were growing, but may also be due to the higher algal rations in the laboratory than in the sea where the *C. islandica* larvae were found.

Inherited factors also affect the rate of development. Different stocks of the same species may develop at different rates. In some cases, this correlates with differences in egg size (Paulet et al. 1988), but in other cases it does not (Cochard and Devauchelle 1993). The smaller larvae of a brood develop more slowly (Cabello and Camacho 1976). Experiments with reciprocal crosses between stocks with distinct larval growth rates implicate maternal effects (probably differences in quantity or quality of yolk reserves) in differences in early development rate (Cruz and Ibarra 1997). Larvae from cross-fertilised eggs develop significantly more rapidly than do those from self-fertilised eggs. This effect of heterosis only becomes evident after about two weeks or more of development (Beaumont and Budd 1983; Ibarra et al. 1995; Cruz and Ibarra 1997).

2.6.4 Biochemical Events during Metamorphosis

In *Pecten maximus*, the levels of serotonin and dopamine, which steadily increase during larval development, drop suddenly during metamorphosis (Robert et al. 1999). Whether this is due to these molecules being concentrated in tissues lost during metamorphosis or whether they are metabolised during the process remains to be determined. Noradrenaline levels do not fluctuate in the same way. These and other molecules that affect nerve cell function have been investigated for their role in settlement and/or metamorphosis of bivalve larvae. In experiments, the molecules are generally applied to the water containing the larvae, from which they can diffuse into the larval tissues. The site of action is therefore open to debate. The fact that ions such as K^+ can trigger metamorphosis (Martinez et al. 1999) suggests a nervous signalling mechanism of the sort investigated in gastropod veligers (Hadfield et al. 2000) is implicated. The neuroactive compounds could act at the apical (sensory) membrane of chemosensory cells, at the basal (signalling) membrane of these cells, at cells deeper within the nervous system or even on the target cells for metamorphosis signals from the nervous system.

Studies with oyster larvae suggest that a dopaminergic pathway promotes settlement behaviour patterns and an adrenergic pathway controls metamorphosis (Beiras and Widdows 1995). Scallop larvae, unlike oysters, do not undergo cementation at metamorphosis and the process of metamorphosis is more protracted in pectinids than oysters (Nicolas et al. 1998). Thus, the distinction between settlement and metamorphosis is less clear. The promotion of metamorphosis by L-DOPA in *Patinopecten* (Kingzett et al. 1990) and *Pecten* (Nicolas et al. 1998) suggests a dopaminergic pathway, though *Chlamys hastata* is not affected in the same way (Hodgson and Bourne 1988). Other tyrosine derivatives, especially a quinol extracted from a rhodophyte alga, *Delesseria sanguinea* have also been shown to promote pectinid metamorphosis (Yvin et al. 1985; Chevolot et al. 1991). The increased metamorphosis rates in larvae exposed to epinephrine or norepinephrine (Kingzett et al. 1990; Nicolas et al. 1996, 1998; Martinez et al. 1999) is perhaps evidence for an adrenergic pathway.

2.7 BEHAVIOUR AND LARVAL DISTRIBUTION

2.7.1 Characteristics of Locomotion during Larval Development

Scallop larvae first become capable of movement at the gastrula stage. Initially, movement consists of undirected rolling or spinning, but by the trochophore stage, the larva swims apical cirrus first, rotating about its long axis along an upward directed path (Drew 1906; Belding 1910; Gutsell 1930; Sastry 1965; Cragg 1980). By the veliger stage, larvae swim up paths, which describe vertically orientated cylindrical helices: they rotate about their dorso-ventral axis once per complete spiral (Belding 1910; Cragg 1980). Spiralling may be in either direction (Belding 1910), but in *Pecten maximus* the direction is clockwise in most larvae (Cragg 1980). Swimming is continuous until the veliger becomes capable of retracting the velum. Retraction may occur in response to an obnoxious stimulus (Drew 1906; Belding 1910; Comely 1972): this response (the "fright response" - La Barbera 1974) consists of an abrupt cessation of beating of the preoral cirri, followed by rapid retraction of the velum, then the closing of the shell valves (Cragg 1980). Sinking then occurs because the larval shell is much denser than water. This sequence, or the first and/or second part of it, also occurs at intervals, apparently spontaneously, so that an undisturbed larva alternates between rising and sinking. Such a pattern of swimming emerges after three days of development in *Pecten maximus*.

During the trochophore and very early veliger stages, the mean vertical velocity of *P. maximus* larvae is less than 0.5 mm s^{-1}, but later stages swim faster, with the highest mean velocity of about 1.2 mm s^{-1} recorded at the beginning of the pediveliger stage (Cragg 1980). The mean velocity of *Placopecten magellanicus* veligers ranges from 0.32 to 1.59 mm s^{-1} depending upon light, temperature and depth conditions, with the low velocities being recorded at low temperatures (Gallager et al. 1996). When the foot becomes functional, the larva sinks at increasingly frequent intervals and crawls for a while before returning to the water column (Gruffydd and Beaumont 1972; Cragg 1980; Chaitanawisuti and Menasveta 1992). The velum continues to be used for swimming during the pediveliger stage, though the foot may be extended during swimming (Hodgson and Burke 1988). Crawling is executed by a gliding motion of the foot due to its cilia, alternating with sharp jerks, which bring the shell close to the foot again (Hodgson and Burke 1988). Belding (1910) reports that after the degeneration of the velum, *Argopecten irradians* is capable of swimming by means of kicking motions assisted by the cilia of the foot. Young juveniles can enable themselves to be carried by water currents, by secreting a byssus thread that acts as a drogue (Beaumont and Barnes 1992). Swimming in the adult fashion by clapping the valves only appears some time after metamorphosis (Sastry 1965).

2.7.2 Responses to Stimuli

Locomotion alone is insufficient to keep the developing scallop in the appropriate environment. The series of types of locomotion described above are regulated by responses to environmental variables. Evidence of the sensitivity of larvae to stimuli is limited, though by the time of the pediveliger stage, the larva has a wide range of putative sense organs.

Though larval swimming is generally orientated upwards, this does not necessarily indicate the ability to detect the orientation of gravity. Several theories reviewed by Chia et al. (1984) suggest that small ciliated organisms with an anterior-posterior asymmetry and/or and asymmetrical distribution of density may move upwards, anterior end first, due to the forces generated by ciliary beating. It may be that the pectinid trochophore has an uneven distribution of lipid droplets less dense than the general cytoplasm, and the apical cirrus certainly provides a marked anterior-posterior asymmetry. When the shell is secreted, the difference in density between the shell and the velum ensures that the larva tends to swim velum uppermost (Cragg 1980). Larval shell density is between 1.20 and

1.34 depending on age (Gallager et al. 1996). The nature of the forces generated by the beating of the preoral cirri may be sufficient to account for the nature of the spiralling swimming path of the veliger without having to postulate any sensory response to environmental variables (Cragg 1989). Behavioural scientists rarely get the opportunity to vary gravity, but in an experiment in a jet travelling along a parabolic trajectory, Jackson et al. (1993) were able to do just that. They found that at the onset of microgravity conditions, the veligers of *Placopecten* change from swimming along tight spirals to swimming up steeper slopes with a wider diameter to the spiral. Given the lack of statocysts at this stage, the effect is likely to be a direct effect on the physics of the propulsion of the larva rather than a behavioural response to a detected change in the force of gravity. However, the structure of pediveliger statocysts does suggest that by this stage the larva is capable of detecting gravity, an ability which is likely to be used to orientate crawling movements, during which the shell will not impart a consistent orientation to the larva (Cragg and Nott 1977).

Sensitivity to pressure changes is widespread in marine animals without pressure-deformable structures (Knight-Jones and Morgan 1966). It has been observed in veligers and pediveligers of *Pecten maximus*, but other bivalve larvae of intertidal species appear to be more sensitive to this stimulus. The response to pressure increases was not detected in 3 day old *Pecten* larvae, but was found in larvae between 10 and 41 days old. The response consists of an increase in vertical velocity and an increased tendency to swim upwards rather than sink (Cragg 1980). Increased vertical velocity may be achieved by increasing velocity along the spiral path or by swimming up a steeper spiral (Cragg and Gruffydd 1975).

Kaartvedt et al. (1987) showed that diurnal variations in light intensity were correlated with changes in the vertical distribution of 4–20 day old *Pecten maximus* larvae in 4 m deep plastic enclosures suspended from the surface in the sea. Maru et al. (1973) examined the variation of vertical distribution of *Patinopecten yessoensis* veligers and changes in temperature, salinity and water currents throughout a 15 m water column in a shallow bay over several 24 hour cycles. They also measured light intensity at the surface. Their results indicate that veligers are distributed closer to the surface during the hours of darkness. No other correlation with environmental changes is evident in their data. They categorised larvae into size classes of less than 150 μm, 150–180 μm, 180–210 μm, and 210–240 μm (i.e., ranging from late D-veliger to umboned veliger - see Maru 1972). No difference between the light responses of the different size classes was detected. The observations of these and other (see next section) field and mesocosm studies indicate a response to light intensity changes not evident in trochophores of *Pecten maximus*, which accumulate at the surface whether they are illuminated or not (Cragg 1980). The orientation of illumination does not affect the orientation of the swimming path of veligers: the spiral path is vertically orientated whether the larva is illuminated from above, below or laterally (Cragg 1980).

Veligers swimming behaviour may be modified by changes in salinity. Gruffydd (1976) found that under laboratory conditions larvae of *Chlamys islandica* can detect salinoclines and respond by restricting their up and down movements to below the interface between a tolerated salinity and a lower salinity. In a vessel with a series of layers of salinity with 5% reductions of salinity in succeeding layers, larvae restricted their vertical movements to those layers with a salinity of 60% or more of that of the seawater in which they were reared. With 20% salinity steps, larvae restricted themselves to water of at least 80% of the rearing water salinity. Culliney (1974) found that veligers of *Placopecten magellanicus* reared at 32‰ and placed in waters of 16.9‰ or 21.5‰ ceased swimming, but recovered to swim close to the bottom of their containers. Salinities of 26.2‰ and 30.0‰ produced no detectable change in their normal up and down movements. Prior exposure for 6 hours to 20‰ salinity during the trochophore stage enables veligers of *Mizuhopecten yessoensis* reared at 32‰ to maintain activity in 14‰ seawater, whereas the activity of veligers without prior exposure is markedly lower (Yaroslavtseva et al. 1993). It is unclear whether this is a behavioural or a physiological response.

The fright reaction can be triggered by obnoxious stimuli. These may be physical shocks such as sharp taps on the culture container (Drew 1906; Belding 1910; Comely 1972; pers. obs.) or collisions with other larvae (Drew 1906; pers. obs.). Chemical stimuli may also act as a trigger, so that in order to prevent larvae from retracting when fixatives are applied, relaxants such as $MgCl_2$ need to be gently introduced into the water in which the larvae are swimming (Cragg and Nott 1977; Hodgson and Burke 1988). It is not known whether these relaxants interfere with sensory perception or block nervous control of the musculature.

The behaviour of pediveliger larvae suggests an exploration of potential sites for settlement. Golikov and Scarlatto (1970) report that on settlement on a plant or artificial substrate, larvae of *Patinopecten yessoensis* move downwards. No studies of pediveliger responses during crawling have been carried out so the only evidence of the stimuli to which they may respond is provided by environment in which settlement occurs. In the method of rearing for *Pecten maximus*, described by Beaumont et al. (1982), no settlement substrate is provided, as this would interfere with rearing procedures, and larvae metamorphose in glass containers. Small scale experiments reported by Culliney (1974) seem to indicate that metamorphosis of *Placopecten magellanicus* is promoted by the presence of fragments of shell, glass, or sand grains, with the metamorphosed larvae showing a preference for the underside of the pieces of these materials. The observations of Culliney (1974) and Golikov and Scarlatto (1970) suggest that on settlement the juvenile scallop is positively geotropic or geotactic, or negatively phototropic. Current velocity may also affect pectinid settlement behaviour, with higher velocities near the seabed promoting settlement (Eckmann 1987).

Recently-settled pectinid spat have also been collected from many natural substrates: *Pecten maximus* and *Chlamys opercularis* have been reported from stones, empty shells, bryozoans (particularly *Cellaria*), hydroids, and the algae *Laminaria saccharina* and *Desmarestia* (Brand et al. 1980; Mason 1958); *Patinopecten yessoensis* on the algae *Sargassum pallidum* and *S. kjellmanianum*, and the seagrass *Zostera asiatica* (Golikov and Scarlatto 1970); *Argopecten irradians* on seagrasses (Thayer and Stuart 1974; Eckmann 1987); *Argopecten ventricosus* on the seagrass *Zostera marina* (Santamaria et al. 1999); *Leptopecten latiauratus* on erect bryozoans and hydroids (Allen 1979); *Chlamys islandica* on hydroid perisarcs (Harvey et al. 1995b); *Adamussium colbecki* on shells of adult conspecifics (Cerrano et al. 2001). Pectinid larvae will settle on clean silt-free surfaces of many sorts (Brand et al. 1980), including a variety of artificial substrata made from polymers (section 2.8.4). The observation that artificial spat collectors are only effective for collecting pectinid spat for about one month after immersion (Wilson 1987), suggests that the development of the primary bacterial film may modify settlement behaviour. In some cases, this may involve masking the chemical cues emanating from the surface on which the film has developed (Harvey et al. 1995b). However, promotion rather than inhibition of settlement by the primary bacterial film has been reported in several pectinid species (Hodgson and Bourne 1988; Xu et al. 1991; Parsons et al. 1993; Pearce and Bourget 1996) and older bacterial films on natural surfaces are sometimes more attractive (Harvey et al. 1995a).

In their review of settlement in pectinids, Brand et al. (1980) found no evidence for species-specific chemical settlement cues. However, Cerrano et al. (2001) found that larvae of the Antarctic scallop *Adamussium* settle on adult shells in preference to rock, though settlement density is reduced by when epizoic hydroids are present on the shell. It is not clear whether this is due to avoidance behaviour prompted by a chemical signal from the hydroids or predation by the hydroids on settling larvae. *Chlamys islandica* spat settle preferentially on the perisarcs of dead rather than live hydroids or on filamentous red algae (Harvey et al. 1993). Choice experiments indicate that it is a high molecular weight protein or polysaccharide in the perisarcs of hydroids that promotes settlement (Harvey et al. 1995b). Also, purified chitin from perisarcs or from crustacean exoskeletons applied to artificial substrates increased the intensity of settlement on those substrates with no detectable difference between the attractiveness of different types of chitin (Harvey et al. 1997).

2.7.3 Distribution of Larvae

Laboratory experiments permit the role of individual factors in determining the distribution of larvae to be examined independently under controlled conditions, but the validity of extrapolation to field situations is constrained by the artificiality of experimental conditions (high larval densities, constraints on larval movements etc.). The use of mesocosms with much larger water volumes has improved the applicability of experimental findings, though artefacts may still complicate interpretation. For example, in mesocosms, as in smaller experimental vessels, circulation of dense aggregations of larvae may occur due to larva to larva interactions at the surface in still conditions (Gallager et al. 1996).

Field observations are constrained by the time required for counting mixed plankton samples and a reliable means of distinguishing them from larvae of other bivalves are required. The methods pioneered by Jørgensen (1946) and Rees (1950), which rely on characteristic shell outline and hinge features, have been refined by the use of SEM images (Tremblay et al. 1987). They are labour intensive to apply to large series of samples, so attempts have been made to develop alternatives. The most promising approach to date is the development of immunological probes that tag larvae of specific species or even genetically-distinct populations with fluorescent markers (Demers et al.

1993; Frischer et al. 2000; Paugam et al. 2000). Demers and colleagues used a monoclonal antibody with high specificity to the target species, but variable staining, while Paugam and colleagues used a polyclonal antibody that enabled laboratory cultures of mixed species to be separated, but did not prove effective with plankton samples. They suggest that the combination of immunological methods with minute magnetic beads may enable the sorting of plankton with a magnetic field.

Pectinid larvae have been collected in many plankton studies in neritic waters, which give information on horizontal and vertical distribution, and the temperature or season of appearance of the larvae (Table 2.4).

The timing of spawning is crucial to ensure temperatures and concentrations of phytoplankton adequate for larval development. Detailed analysis of correlations between mass spawning of *Placopecten magellanicus* and a wide range of environmental variables indicated that spawning coincided with either sharp temperature increases or with strong temperature fluctuations (Bonardelli et al. 1996). In either case, the gametes would be ejected into downwelled water masses likely to be suitable for larval development. A study of spawning of *Chlamys islandica* in the Gulf of St. Lawrence, Canada did not reveal a temperature cue, but phytoplankton blooms did appear to be associated with spawning events (Arsenault and Himmelman 1998).

Horizontal dispersal of larvae will be determined by water currents. Effects of currents on horizontal distribution have been investigated (Kulikova and Tabunkov 1974; Ito et al. 1975; Maru 1985a; Tremblay and Sinclair 1988, 1992) though conclusions are difficult to draw because of limited oceanographic information and/or because controlling factors are so complicated that there is no clear pattern. Nonetheless, there is evidence that oceanographic conditions on the continental shelf can keep different stocks of the same species of scallop separate by determining the horizontal dispersal of their larvae (Tremblay and Sinclair 1992). The extent to which larvae of shelf populations can be lost due to transport on currents from their parental area has been modelled (Tremblay et al. 1994). Tidal lagoons and estuaries offer a more constrained environment in which to investigate the effects of hydrographic conditions on dispersal. By examining the proportions and numbers of different larval stages, Maru (1985a) produced a budget for larval development within a lagoon, calculating egg production, death, loss due to outflow and settlement. Larvae of *Argopecten irradians* were found in greater numbers in the upstream part of an estuarine system, a distribution attributed by Moore and Marshall (1969) to asymmetry in the tidal circulation pattern, but also explicable in terms of vertical migration in response to salinity as inferred in the study of *Crassostrea* larvae by Wood and Hargis (1971).

Larvae may have some control over their vertical distribution. In the laboratory in the dark, trochophores of *Pecten maximus* accumulate in the upper few centimetres of a shallow vessel (as do trochophores of *Chlamys hastata* - Hodgson and Burke 1988), but once larvae develop a closable shell, they become distributed more evenly through the water column. By the late veliger stage, there is no concentration of larvae near the surface and, on reaching the pediveliger stage, most larvae accumulate just above the bottom (Cragg 1980). Early veligers of *Argopecten irradians* and *Placopecten*

Table 2.4

Plankton studies of pectinid larvae.

Species	Site	Reference
Argopecten circularis	Ensenada la Paz, Mexico	Quezada 1980
Argopecten irradians	Niantic Estuary, E. USA	Moore & Marshall 1967
Chlamys islandica	Balsfjord, Norway	Gruffydd 1976
Chlamys opercularis	western Scotland	Sasaki 1976
	North Sea	Rees 1951, 1954
	Öresund, Denmark	Jørgensen 1946
	Adriatic	Hrs-Brenko 1971
Chlamys patriae	Chiloe Is. Chile	Solis et al. 1976
Chlamys striata	North Sea	Rees 1951, 1954
	Öresund, Denmark	Jørgensen 1946
Pecten glaber	Adriatic Sea	Ohdner 1914
Pecten maximus	Irish Sea	Mason 1958
	North Sea	Rees 1951, 1954
	western France	Raby et al. 1994, 1997
Pecten ponticus	Black Sea	Zakhvatkina 1959
Pecten septemradiatus	Öresund, Denmark	Jørgensen 1946
Patinopecten yessoensis	Saroma Lagoon, Hokkaido, Japan	Maru et al. 1973
	Abashiri, Hokkaido, Japan	Hoshikawa & Kurata 1987
	Okunai, Japan	Kanno 1974
	Mutsu Bay, Japan	Ito et al. 1975
	Aniva Bay, E. Russia	Kulikova & Tabunkov 1974
	Peter the Great Bay, E. Russia	Kulikova et al. 1981
	Posyet Bay, Sea of Japan	Grigor'eva & Regulev 1992
	Yeong-il Bay, Korea	Yoo & Park 1979
Placopecten magellanicus	Bay of Fundy	Tremblay & Sinclair 1988
	Georges Bank, western N. America	Tremblay & Sinclair 1990a, b, 1992, 1994
pectinid	ria de Pontevedra, Spain	Dominguez & Alcaraz 1983

magellanicus also tend to be found near the surface while later stages occur near the bottom of their containers (Belding 1910; Sastry 1965; Culliney 1974).

In 9 m deep laboratory mesocosms with only limited water movement, *Placopecten* veligers from 6 to 49 days old tend to concentrate in the top 10 cm, especially at night, irrespective of the distribution of their food microalgae, except in the case of larvae larger than about 240 μm, which become evenly distributed through the water column (Gallager et al. 1996). A thermocline in the mesocosm of greater than 1 degree C may constrain the extent of vertical movement of larvae, leading to a concentration of larvae above the thermocline (Manuel et al. 2000).

Vertical distribution of larvae in the sea is more variable, partly because of the much greater water depths involved, but perhaps also because of vertical water movements (Maru 1985a). Developmental stages of *Chlamys islandica* from D-veligers to pediveligers were found in a fjord at the surface and at 40 m depth, but there were few larvae at 100 m (Gruffydd 1976). Veligers of *Pecten maximus* showed no clear vertical distribution in depths of up to 25 m (Sasaki 1979). However, most larvae of *Patinopecten yessoensis* were found within 8 m of the bottom in depths of 16–24 m in Yeong-il Bay, Korea (Yoo and Park 1979), while larvae of the same species from 170 to 275 μm in length were distributed from the surface to near the bottom at 14 m (Grigor'eva and Regulev 1992). Larvae of the same species in 15 m depth in a tidal lagoon in northern Japan with no marked temperature or salinity gradients, showed no clear difference between the distribution of different size classes from D-veliger to late veliger (Maru et al. 1973; Maru 1985a). No consistent pattern of *Placopecten magellanicus* veliger size distribution over the top 40 m in the Bay of Fundy was detected, but the mean size of veligers just above the seabed was slightly greater than that of larvae taken in vertical tows from 5 m above the seabed (Tremblay and Sinclair 1988). When the water column was well mixed, veligers were distributed almost evenly over the top 40 m, but when a thermocline was present, veligers were sometimes unevenly distributed with a peak associated with the thermocline (Tremblay and Sinclair 1990a). The vertical distribution of food organisms, whether at sea or in a mesocosm, rarely shows any evidence of affecting the distribution of veligers (Raby et al. 1994; Manuel et al. 1996 a, b).

Diurnal vertical migration by scallop veligers has been detected in mesocosms with artificial lighting providing a day-night cycle of illumination (Manuel et al. 1996 a, b), in mesocosms suspended in the sea under ambient lighting (Kaartvedt et al. 1987) and in the sea (Maru 1985a; Tremblay and Sinclair 1990b; Raby et al. 1994). Larvae from different parental stocks may show different intensities of vertical migration, even when kept in the same container (Manuel et al. 1996 a and b). Data from mesocosm and field studies of *Placopecten* veligers can be fitted to a tidal diel model, with tidal response varying from stock to stock in a way that ensures that pediveligers are adapted to the current regime of the larval and parental environment. There strong selection pressure imposed by the advective loss of larvae, which would favour the development of behavioural responses to diel and tidal cues appropriate to the particular environment (Manuel et al. 1997).

The vertical distribution of settlement of pectinids has also been examined. Spat may occur in surface plankton tows (Gruffydd 1976). The natural settlement of *Patinopecten yessoensis* on the western shores of the Sea of Japan occurs at depths of 1–3 m (Golikov and Scarlatto 1970), perhaps determined by the distribution of the settlement substrates - seagrass and algae. Patterns of vertical settlement distribution on artificial settlement substrates include: increasing with depth *(Chlamys islandica* - Wallace 1982; *Pecten maximus* - Sasaki 1979; *Patinopecten yessoensis* - Maru 1985a); increasing with depth but decreasing just above the bottom (Naidu and Scaplen 1976); fairly even, but less intense near the surface and bottom *(Pecten maximus* Brand et al. 1980; *Pecten fumatus* - Hortle and Cropp 1987); reducing with depth (*P. yessoensis* - Evans et al. in Brand et al. 1980). Different species of scallop at the same site have been shown to have different depth preferences at sampling depths of 20, 50 and 70 m (Peña et al. 1998). Factors controlling vertical distribution of settlement have been examined in experiments in 9.5 m deep mesocosms. In turbulent conditions, there was no effect of depth on settlement, while in low turbulence conditions settlement rate increased with depth and weak temperature stratification (approximately 1.5 degrees C) did not affect settlement distribution (Pearce et al. 1998). Strong stratification (approximately 4 degrees C) did limit settlement to above the thermocline and vertically stratified food distribution had an adverse effect on settlement (Pearce et al. 1996).

2.7.4 Development and the Larval Environment

Direct observation of scallop larvae in their natural environment presents considerable practical difficulties and plankton sampling provides only isolated snapshots of larval distribution. Nonetheless, the information reviewed above does permit predictions about the nature of larval life of pectinids in the sea to be made. The larval period can be considered to consist of three phases characterised by the nature of the energy source used and the nature of locomotion. The first phase is lecithotrophic, the second and third are planktotrophic with swimming the only type of locomotion in the second and crawling alternating with swimming in the third.

The embryonic phase is the most vulnerable to environmental conditions. The survival of larvae during this phase is dependent on spawning taking place when conditions are suitable for embryological development (see for example Broom and Mason 1978; Bonardelli et al. 1996). Passive orientation of swimming (see above) is probably sufficient to ensure that the early larvae move up into the water column from the adult environment to the larval environment. The apical flagellum may warn of the forces of surface tension that could rip the naked larva apart.

On becoming able to feed, the larva enters the dispersal phase. It moves up and down due to its own efforts, but lateral (dispersal) movements are due to water currents. The trigger that prevents the larva from sinking out of the water column is likely to be the resulting increase in pressure. The diurnal migrations observed by Maru et al. (1973), Kaartvedt et al. (1987) and Manuel et al. (1996a, b) may reduce the risk of predation by moving the larvae in response to rising light intensity into less well illuminated, deeper waters during the day. The fright reaction may also serve to protect larvae from predation and from surface tension effects in bursting bubbles. The response to lower salinity observed by Gruffydd (1976) would tend to prevent a larva passing up through a salinocline such as may occur in the enclosed waters in which the larvae of many pectinids are found (see for example, Moore and Marshall 1969). Temporary reductions in temperature may not be fatal at this stage as larvae can survive a temperature-induced arrest and continue development when the temperature rises again (Beaumont and Budd 1982). During this phase, larvae may be carried far from their parental stock (Tremblay and Sinclair 1988).

The third phase of larval life may be considered to serve to enable the larva to find a suitable settlement substrate. The fact that the larva does not abandon swimming altogether increases the area the larva can explore, because lateral dispersal may occur during each period of renewed swimming effort. Furthermore, the larva is able to feed during swimming so that energy reserves are maintained. During crawling movements, the sense organs developed in the late veliger larva may come into play: the statocysts to orientate crawling and perhaps some of the ciliated cells of the mantle to test the nature of the substrate.

During the embryonic phase, the chances of survival are affected by the prior spawning response of adults to environmental cues. During the dispersal phase, swimming behaviour has survival value; settlement behaviour permits the larva to be selective about the site for metamorphosis and juvenile life. Thus, the larva is not a passively dispersed "seed": the high risks of larval life are reduced by adult behaviour and the larva has some control over its own destiny.

Dispersal in species with small adult size and a patchy adult environment such as hydrothermal vents or submarine caves must require different mechanisms, particularly in brooding species. Hayami and Kase (1993) hypothesise that the colonisation of cave environments by non-planktotrophic species may occur as a result of pseudoplanktonic dispersal of juvenile attached to current-carried debris, though dispersal independent of attachment, by byssus drifting (Beaumont and Barnes 1992; Wang and Xu 1997) might also provide the necessary dispersal mechanism.

2.7.5 Tolerance of Environmental Variables

Light intensity is not normally considered as a factor limiting larval development or survival and light conditions are frequently not mentioned in experimental descriptions, but Krassoi et al. (1997) showed that high light intensities correlate with high levels of abnormal development in *Chlamys asperrima*.

Temperature is a key variable for larval development. Too low a temperature reduces or arrests growth (Beaumont and Budd 1982) while too high temperatures may permit rapid growth of harmful bacteria and ciliates in the culture, cause high mortality rates (Gruffydd and Beaumont 1972) or increased levels of abnormal embryos (Krassoi et al. 1997). Temperature optima may vary during embryonic and larval development (Heasman et al. 1996): this may represent adaptation to the differences in temperature between the embryonic, benthic environment and the larval environment – the euphotic zone. Optimum development of *Patinopecten yessoensis* embryos occurs at 20°C: development rate decreases with decreasing temperature, but still occurs even at 4°C; a decrease is also evident at 23°C and at 28°C no development takes place (Maru 1985b; Park et al. 2001). Temperature changes increase mortality of early cleavage stages, trochophores and D-veligers of *Argopecten irradians*, with mortality increasing with longer periods of exposure to suboptimal temperatures. Cleavage stages are most sensitive to thermal shock and veligers least (Wright et al. 1984). This species is more sensitive than non-pectinid bivalves (Wright et al. 1983). Temperature tolerance limits during the larval phase may determine adult distribution (Beu 1999).

Most rearing of pectinid larvae has been carried out using the seawater supply also used for the maintenance of the adults, with salinities within the range 30–35‰. Reduced salinities severely affect embryonic development and adversely affect growth of veligers (Gruffydd and Beaumont 1972; O'Connor and Heasman 1998). Embryonic development of *Patinopecten yessoensis* can take place over the range of about 14 to 21.5‰ salinity with a marked reduction in the rate of development at either end of this range (Maru

1985b). In *Chlamys asperrima*, percentage normal development declines sharply above and below the optimum salinity range of 31–33‰ (Krassoi et al. 1997). Gruffydd (1976) found that the survival of *Chlamys islandica* veligers over a 24-h exposure period was little affected by salinities as low as about 21‰, but markedly reduced at 14‰, with 100% mortalities at 7‰. Culliney (1974) noted that veligers of *Placopecten magellanicus* could survive for 48-h at salinities as low as 10‰, though there was evidence of tissue swelling and the larvae were incapable of normal swimming. The tolerance of reduced salinity varies according to whether larvae experience an abrupt reduction or a gradual transition (Davenport et al. 1975). If salinity fluctuates with a diurnal sinusoidal regime (as it might under an estuarine tidal regime) from a maximum of about 33.5 ‰, the salinity LC50 for *Pecten maximus* larvae is about 8‰, but abrupt changes give a LC50 of 16.6‰ and for constant low salinities the LC50 is 12‰ (Davenport et al. 1975). Also, the salinity tolerance range may be increased by prior exposure to suboptimal salinity (Yaroslavtseva et al. 1991).

The possibility of synergistic effects between sublethal salinities and temperatures has been investigated by mapping the tolerance of embryos and/or larvae of *Pecten maximus*, *Argopecten irradians* and *Mimachlamys asperrima* to a range of temperature/salinity combinations (Davenport et al. 1975; Tettelbach and Rhodes 1981; O'Connor and Heasman 1998). *P. maximus* veligers were exposed to the test temperature/salinity combination for a period of 24-h, returned to normal culturing conditions and after a recovery period were examined for activity of preoral cirri which was used as an index of survival. Their temperature tolerance was greatest at salinities close to 34‰. Prior to experimentation, *Argopecten irradians* embryos and larvae were kept at a salinity of 25‰ – close to that of the environment from which brood stock were obtained. Measurements were made of embryonic development, growth and survival of veligers at combinations of temperatures from 10 to 35°C and salinities ranging from 10 to 35‰. Normal embryonic development only occurred over a narrow range of salinities, while survival of veligers was markedly reduced at salinities of 15‰ or less, particularly at 10 and 20°C. Growth rates were greatest between 20 and 30‰ at 25–30°C, but were markedly reduced when both temperature and salinity fell outside these ranges. The tolerance of *Mimachlamys asperrima* to suboptimal temperatures and salinities varied over the period of development.

Pollutants affect the viability of scallop larvae in a number of ways. Parental exposure to heavy metals may result in gametes with elevated levels of phosphatases that give rise to embryos and early veliger larvae which also have elevated levels of these enzymes and which fail to develop normally (Karaseva and Medvedeva 1993). Larval exposure to organic pollutants and heavy metals also disrupts development (Liao and Li 1980; Beaumont et al. 1987; Krassoi et al. 1997). Settlement behaviour appears to be affected by organotin antifoulants (Minchin et al. 1987). Embryonic stages of bivalves are particularly sensitive to pollutants. This sensitivity has been exploited in the development of the D-larva bioassay in which the appearance of the characteristic D-shaped early veliger shell is used as a marker for normal development (Anonymous 1994; His et al. 1999). This ecotoxicological test is usually performed with oyster or mussel embryos, but *Chlamys asperrima* embryos show comparably high sensitivities to a range of pollutants (Krassoi et al. 1997; Pablo et al. 1997). Other bioassay tests have been developed with bivalve larvae. Changes in the parameters of larval swimming trajectories have been used to bioassay sublethal effects of pollutants (Praël et al. 2000) and of toxic algae (Yan et al. 2001, 2002). Improvements with cryopreservation techniques to permit storage of viable larvae or gametes would improve comparability of tests and would facilitate the use of pectinid larvae, which are more demanding of rearing conditions.

2.8 REARING METHODS

2.8.1 Spawning and Manipulation of Zygotes

Descriptions of larval rearing methods are available for most of the pectinid species listed in Table 2.1. Rearing of *Hinnites multirugosus* is described by Cary et al. (1981, 1982). Additional information on the rearing of *Patinopecten yessoensis* is provided by Yoo and Imai (1968) and on *Pecten maximus*, by Comely (1972) and Beaumont et al. (1982). The methods used are similar to those described by Loosanoff and Davis (1963), but pectinid larvae seem to be particularly sensitive to rearing conditions (Tettelbach and Rhodes 1981; Beaumont et al. 1987).

Some species are amenable to being brought into spawning condition in the laboratory so that larvae can be produced at most times of the year, while others seem restricted to the spawning period of the wild stocks (Beaumont et al. 1982; Gruffydd 1976). Subjecting ripe adults to a rise in temperature of about 5 degrees C is the spawning stimulus used for most pectinids, but Le Pennec (1982) found *Chlamys opercularis* to require more carefully regulated stimulation: he evaluated the efficacy of a variety of sequences of temperature change or UV stimulus, but concluded that the most important factor was the sexual cycle of this species. Induction of spawning in *Mimachlamys asperrima* by intragonadal injection of serotonin gives similar yields of larvae to spawnings induced by thermal shock (O'Connor and Heasman 1995). Stripped spermatozoa achieve lower fertilisation rates and produce more abnormalities than those obtained from induced spawning (Beaumont and Budd 1983). Most authors endeavoured to minimise self-fertilisation as larvae thus obtained grow more slowly and fewer survive to metamorphosis (Beaumont and Budd 1983; Ibarra et al. 1995). The proportion of sperm to eggs needs to be controlled so that the incidence of polyspermy is minimised (Gruffydd and Beaumont 1970; O'Connor and Heasman 1995; Desrosiers et al. 1996).

Treatment at meiosis I or II with heat or cold shock, pressure or cytochalasin B induces triploidy: higher heterozygosity is retained if meiosis I is targeted (Beaumont and Zouros 1991). Heat shock treatment does not affect larval growth rate (Toro et al. 1995). A brief treatment with the puromycin analogue 6-DMAP inhibits polar body extrusion and chromosome movement, but as it is water-soluble it can be washed out from cultures once triploidy has been induced, to avoid compromising embryo viability (Desrosier et al. 1993). This offers the possibility of rearing scallops in which growth is focused on somatic tissues.

Conditioning of adults for spawning is a protracted and not always reliable procedure. It should not only be designed to bring the adults into spawning condition, but also to ensure egg quality, which is a major determinant of larval survival and growth (Caers et al. 1999). When spawning is successful, it generally produces far more larvae than are required. Cryopreservation offers the possibility of storing excess larvae. Odintsova and Tsal (1995) investigated the efficacy of dimethyl sulphoxide (DMSO), glycerol, taurine and bovine serum albumin as cryoprotectants for cultures of bivalve cells. They found that the viability of cells from early veligers could be maintained if protected by 10% DMSO or 10% glycerol. Trochophore cells lost viability under the same conditions. Perhaps this approach could be tailored to protect whole larvae.

2.8.2 Feeding

Larvae have been cultured generally at densities of 10 per mL or lower, though Ibarra et al. (1997) found that densities up to 20 per mL gave satisfactory growth rates. Higher densities depress growth rates of *Pecten maximus* (Gruffydd and Beaumont 1972) and densities above two per mL are sufficient to markedly reduce long-term larval survival in *Patinopecten yessoensis* (MacDonald 1988).

Cultures are supplied with food in the form of single celled algae, which vary in their nutritional value. For example, *Isochrysis galbana,* particularly in the stationary phase of culture when the cells have a high lipid content, proved a superior food for *Hinnites balloti* larvae than *Pavlova lutheri, Rhodomonas* sp. and the chlorophyte, *Carteria pallida* (Cary et al. 1981). Though *Chaetoceros calcitrans* gives good growth rates, it does not support metamorphosis (Soudant et al. 1998a). Mixtures of two or three algal species may give higher growth rates (Gruffydd and Beaumont 1972; Narvarte and Pascual 2001). Mixed diets also ensure the uptake of a broader range of sterols (Soudant et al. 1998a). The algae used come from a range of taxa: diatoms (*Chaetoceros, Phaeodactylum, Thalassiosira*); chlorophytes *Dunaliella, Nanochloris, Tetraselmis*); chrysophytes (*Isochrysis, Pavlova*); cryptophytes (*Rhodomonas*) (see references in Table 2.1). Epifluorescence studies revealed that veligers of *Pecten maximus* and *Argopecten ventricosus* readily ingest and digest *Pavlova lutheri* and *Isochrysis galbana,* but that *Dunaliella promolecta* is less readily ingested and is poorly digested, and *Tetraselmis suecica* is poorly ingested if at all and is not digested (Le Pennec and Rangel Davalos 1985; Lora-Vilchis and Maeda-Martinez 1997). *A. ventricosus* ingests but does not digest *Nannochloris oculata* or *Phaeodactylum tricornutum.* The concentration of food organisms supplied is important: depending on size, *Patinopecten yessoensis* larvae require from 7 to 15 cells per µL to obtain sufficient energy to permit growth (MacDonald 1988). Growth rate of *Argopecten irradians* veligers increased with increasing concentration from one cell per µL up to an optimum of 20 cells μL^{-1} (Lu and Blake 1996a, b).

The culture water is changed every two or three days. Algal feed is added after the change, having been centrifuged (Gruffydd and Beaumont 1972) to prevent too much of the algal growth medium entering the culture.

Larvae are capable of ingesting and digesting bacteria. It is likely that though the estimated uptake of bacterial cells can be high (>10,000 cells $larva^{-1} d^{-1}$), the contribution to nutrition is qualitative rather than quantitative (Moal et al. 1996), with bacteria supplying vitamins absent in the food algae (Seguineau et al. 1993).

2.8.3 Control of Disease in Cultures

High mortalities in scallop cultures can be caused by viruses: a herpesvirus has been detected in cellular lesions in scallop and oyster larvae (Arzul et al. 2001). The causative agent of paralytic shellfish poisoning, the dinoflagellate *Alexandrium tamarense* depresses embryonic development and survival of *Chlamys farreri* larvae (Yan et al. 2001). However, bacteria are the main cause of mass mortalities in laboratory and hatchery culture (Nicolas et al. 1996; Robert et al. 1996). They may be brought into larval cultures via seawater, air, brood stock or algal cultures (in which they flourish due to the elevated nutrient levels) (Sainz et al. 1998). *Aeromonas hydrophila* and *Vibrio alginolyticus* have been associated with mass mortalities of *Argopecten purpuratus* larvae (Riquelme et al. 1996a). A *Pseudomonas* has been implicated in necrosis in *Euvola* larvae (Lodeiros et al. 1992). Velar necrosis has been regularly detected in cultures of *Pecten maximus* in which antibiotics were not used (Nicolas et al. 1996). The mechanism of infection in this case may be a combination of invasion of larval tissues and build up of toxins. Isolates from moribund larvae in these cultures include a vibrio that is specific to scallop larvae, *Vibrio pectinicida* (Lambert et al. 1998). A small, heat-stable and protease-resistant molecule that is toxic to *P. maximus* haematocytes has been extracted from the cytoplasm of *V. pectinicida* (Lambert et al. 2001). *Vibrio alginolyticus* has been found to infect *Argopecten purpuratus* larvae (Jorquera et al. 1999), causing diminished swimming, empty stomachs, formation of lipid granules in the gut, and velar necrosis or even detachment. These effects are due to an exotoxin protease that affects cilia and collagen (Sainz et al. 1998).

At each water change, culture containers can be sterilised with bleach, as can the larvae, provided care is taken to expose them briefly and only to low concentrations. The water used is filtered to 0.2 μm and UV sterilised (Gruffydd and Beaumont 1972). Antibiotics such as chloramphenicol, penicillin, sulphamethazine and streptomycin have been added to larval cultures to minimise bacterial build-up between water changes. In a closed system, the use of chloramphenicol has been demonstrated to improve growth and survival rates of veligers of *Argopecten purpuratus* (Uriarte et al. 2001). Torkildsen et al. (2000) established minimum inhibitory concentrations against bacteria isolated from scallop cultures for a range of antibiotics, with a view to providing alternatives to chloramphenicol, which is banned for use with animals intended for human consumption in many European countries.

The rearing method described by Beaumont et al. (1982) does not require antibiotics. Effective upwelling and downwelling flow-through systems have been developed which reduce the numbers of bacteria without compromising growth rate (Andersen et al. 2000), while enhancing metamorphosis rates (Robert and Nicolas 2000).

Some bacteria isolated from larval cultures produce substances that inhibit pathogenic bacteria (Riquelme et al. 1996b). These include members of the genera *Vibrio*, *Pseudomonas* and *Bacillus* (Riquelme et al. 2001). *Roseobacter* sp. produces a substance when in the presence of *Vibrio anguillarum* that enhances survival of scallop larvae: the substance is trypsin susceptible and stable at 100°C (Ruiz-Ponte et al. 1999). A substance produced by *Vibrio anguillarum* that has an inhibitory effect on the pathogen of *Argopecten purpuratus* has been characterised as an aliphatic hydroxyl ether (Jorquera et al. 1999). The potential of co-culturing such bacteria with food algae is under investigation (Riquelme et al. 1997; Avendaño and Riquelme 1999) and their addition to cultures has been shown to permit larvae to complete the planktotrophic phase without the need for antibiotics (Riquelme et al. 2001). Indeed, antibiotics might adversely affect the balance between probiotic and pathogenic bacteria. Veliger larvae have been shown to be capable of ingesting probiotic bacteria (Riquelme et al. 2000), which thus may provide a nutritional as well as bacteriostatic benefit.

2.8.4 Collection of Spat from Wild Stocks

The process of preparing brood stock, controlling spawning and rearing larvae through to metamorphosis requires considerable investment in equipment and skilled personnel. An alternative, in areas where sufficient numbers of larvae occur in the plankton, is to place collectors with a suitable substrate for larval settlement into the larval environment and to transplant the metamorphosed juveniles to aquaculture facilities. This approach has been used with considerable success in relatively enclosed waters in Japan (Ventilla 1977).

A variety of artificial settlement substrates have been successfully used: cotton nets, unravelled sisal ropes, birch brooms (Golikov and Scarlatto 1970), rice straw ropes, cedar twigs, scallop shells, monofilament net, Netlon (Ventilla 1977) or polyethylene film (Naidu and Scaplen 1976). *Aequipecten tehuelchus* settles more readily on monofilament Netlon inside a bag of 2 mm mesh than on monofilament onion bags or thin branches (Narvarte 2001). *Chlamys islandica* settles on nylon filaments of 0.15 mm diameter in preference to those of 0.8 mm or thicker (Wallace 1982). The rock scallop, *Hinnites multirugosus* settles in much greater numbers on gill net collectors than on stick collectors, but *Leptopecten latiauratus* settles readily on both types (Phleger and Cary 1983) and on virtually any artificial surface, and on live kelps (Morton 1994). Pearce and Bourget (1996) found that *Placopecten* settled in much greater numbers on polyester filter wool than on various diameters of monofilament nylon, onion bag material, polyethylene turf, acrylic sheets, or adult shells. Shape and dimensions rather than the chemical nature of the material appear to influence settlement, perhaps by modifying small-scale water flow patterns. Except under static hatchery conditions, larvae encountering a spat collector may be subject to turbulent flow in the immediate vicinity of the collector surfaces, which will affect the chances of the larva attaching. Miron et al. (1995, 1996) investigated the likely effect of collector design on larval velocity, trajectory within a collector by using an artificial flow regime in a flume and video endoscopy to follow passive movement of larva-sized particles.

Timing of the deployment of spat collectors to coincide with the availability of metamorphically competent larvae requires detailed information. Ito et al. (1975) give an example of the sort of information needed. They studied the relationship between appearance of *Patinopecten* larvae, settlement, and daily temperatures in the months proceeding spawning. Factors such as the time required for primary film formation on collector surfaces also have to be taken into account (see section 2.7.2 above).

REFERENCES

Allen, D.M., 1979. Biological aspects of the calico scallop, *Argopecten gibbus*, determined by spat monitoring. Nautilus 94:107–119.

Allen, J.A. and Scheltema, R.S., 1972. The functional morphology and geographical distribution of *Planktomya hensoni*, a supposed neotenous pelagic bivalve. J. Mar. Biol. Assoc. U.K. 52:19–32.

Andersen, S., Burnell, G. and Bergh, O., 2000. Flow-through systems for culturing great scallop larvae. Aquacult. Int. 8:249–257.

Anonymous, 1994. Standard guide for conducting acute toxicity tests starting with embryos of four species of saltwater bivalves. ASTM E 724–94.

Arkett, S.A., Mackie, G.O. and Singla, C.L., 1987. Neuronal control of ciliary locomotion in a gastropod veliger *(Calliostoma)*. Biol. Bull. 173:513–526.

Arsenault, D.J. and Himmelman, J.H., 1998. Spawning of the Iceland scallop (*Chlamys islandica* Müller, 1776) in the northern Gulf of St, Lawrence and its relationship to temperature and phytoplankton abundance. Veliger 41:180–185.

Arzul, I., Nicolas, J.L., Davidson, A.J. and Renault, T. 2001. French scallops: a new host for ostreid herpesvirus-1. Virology 290:342–349.

Avendaño, R.E. and Riquelme, C.E. 1999. Establishment of mixed culture probiotics and microalgae as food for bivalve larvae. Aquacult. Res. 30:893–900.

Barber, V.C. and Wright, D.E., 1969. The fine structure of the eye and optic tentacle of the mollusc *Cardium edule*. J. Ultrastruct. Res. 26:515–528.

Barichivith, J.U. and Stefoni D.L., 1980. Fijacion primaria y variaciones morphologicas durante le metamorphosis de algunos bivalvos Chilenos. Bol. Inst. Oceanogr. Sao Paulo. 29:367–369.

Bayne, B.L., 1971. Some morphological changes that occur at the metamorphosis of the larvae of *Mytilus edulis*. In: D.J. Crisp (Ed.). 4th European Marine Biology Symposium. Cambridge University Press. pp. 259–280.

Bayne, B.L., 1983. Physiological ecology of molluscan development. In: N.H. Verdonk, J.A.M. van den Biggelaar and A.S. Tompa (Eds.). The Mollusca, Vol. 3: Development. Academic Press, London. pp. 299–343.

Baxter, G. and Morse, D.E., 1992. Cilia from abalone larva contain a receptor-dependent G protein transduction system similar to that in mammals. Biol. Bull. 183:147–154.

Beaumont, A.R., 1986. Genetic aspects of the hatchery rearing of the scallop, *Pecten maximus* (L.). Aquaculture 57:99–110.

Beaumont, A.R. and Barnes, D.A., 1992. Aspects of veliger larval growth and byssus drifting of the spat of *Pecten maximus* and *Aequipecten opercularis*. ICES J. Mar. Sci. 49:417–423.

Beaumont, A.R. and Budd, M.D., 1982. Delayed growth of mussel *(Mytilus edulis)* and scallop (*Pecten maximus*) veligers at low temperatures. Mar. Biol. 71:97–100.

Beaumont, A.R. and Budd, M.D., 1983. Effects of self-fertilization and other factors on the early development of the scallop *Pecten maximus*. Mar. Biol. 76:285–289.

Beaumont, A.R. and Zouros, E., 1991. Genetics of scallops. In: S.E. Shumway (Ed.). Scallops: Biology, Ecology and Aquaculture of Scallops. Elsevier, Amsterdam. pp. 585–623.

Beaumont, A.R., Budd, M.D. and Gruffydd, Ll.D., 1982. The culture of scallop larvae. Fish Fmg. Int. 9:10–11.

Beaumont, A.R., Tserpes, G. and Budd, M.D., 1987. Some effects of copper on the veliger larvae of the mussel *Mytilus edulis* and the scallop *Pecten maximus* (Mollusca, Bivalvia). Mar. Env. Res. 21:299–309.

Beiras, R. and Widdows, J., 1995 Effects of the neurotransmitters dopamine, serotonin and norepinephrine on the ciliary activity of mussel (*Mytilus edulis*) larvae. Mar. Biol. 122:597–603.

Belda, C. and Del Norte, A., 1988. Notes on the induced spawning and larval rearing of the Asian moon scallop, *Amusium pleuronectes* (Linné), in the laboratory. Aquaculture 72:173–179.

Belding, D.L., 1910. A report on the scallop fishery of Massachusetts. Special Report Mass. Comm. Fish and Game. Wright and Potter Printing Co., State Printers, Boston (republished 1931).

Bellolio, G., Lohrmann, K. and Dupré, E., 1993. Larval morphology of the scallop *Argopecten purpuratus* as revealed by scanning electron microscopy. Veliger 36:332–342.

Bellolio, G., Toledo, P. and Campos, B., 1994. Morphología de la concha larval y postlarval del ostión *Argopecten purpuratus* (Lamarck, 1819) (Bivalvia, Pectinidae) en Chile. Revista Chilena de Historia Natural 67:229–237.

Beninger, P.G., Dwiono, S.A.P. and Le Pennec, M., 1994. Early development of the gill and implications for feeding in *Pecten maximus* (Bivalvia: Pectinidae). Mar. Biol. 119:405–412.

Berkman, P.A., Waller, T.R. and Alexander, S.P., 1991. Unprotected larval development in the Antarctic scallop *Adamussium colbecki*: Mollusca, Bivalvia, Pectinidae. Antarct. Sci. 3:151–158.

Beu, A.G., 1999. Fossil records of the cold-water scallop *Zygochlamys delicatula* (Mollusca : Bivalvia) off northernmost New Zealand: how cold was the Last Glacial maximum? N.Z. J. Geol. Geophys. 42:543–550.

Bonar, D.B., 1978. Fine structure of muscle insertions on the larval shell and operculum of the nudibranch *Phestilla sibogae* (Mollusca: Gastropoda) before and during metamorphosis. Tiss. Cell 10:143–152.

Bonardelli, J.C., Himmelman, J.H. and Drinkwater, K., 1996. Relation of spawning of the giant scallop, *Placopecten magellanicus*, to temperature fluctuations during downwelling events. Mar. Biol. 124:637–649.

Bongrain, M. and Fatton, E., 1982. Croissance et microstructure chez divers pectinides (bivalves) actuel et fossiles. Malacologia 22:333–336.

Booth, J.D., 1983. Studies on twelve common bivalve larvae, with notes on bivalve spawning seasons in New Zealand. N.Z. J. Mar. Freshwat. Res. 17:231–265.

Bower, S.M. and Meyer, G.R., 1990. Atlas of anatomy and histology of larvae and early juvenile stages of the Japanese scallop (*Patinopecten yessoensis*). Can. Spec. Publ. Fish. Aquatic Sci. 111:1–51.

Brand, A.R. Paul, J.D. and Hoogesteger, J.N., 1980. Spat settlement of the scallops *Chlamys opercularis* (L.) and *Pecten maximus* on artificial collectors. J. Mar. Biol. Assoc. U.K. 60:379–390.

Broom, M.J. and Mason, J., 1978. Growth and spawning in the pectinid *Chlamys opercularis* in relation to water temperature and phytoplankton concentration. Mar. Biol. 47:277–285.

Bull, M., 1977. Scallop research in Marlborough Sounds. Catch 77, 4:4–5.

Cabello, G.R. and Camacho, A.P., 1976. Cultivo do larvas de vieira, *Pecten maximus* (Linnaeus), en laboratoria. Bol. Inst. Esp. Oceanogr. 223:1–17.

Caers, M., Coutteau, P., Cure, K., Morales, V., Gajardo, G. and Sorgeloos, P., 1999. The Chilean scallop *Argopecten purpuratus* (Lamarck, 1819): II. Manipulation of the fatty acid composition and lipid content of the eggs via lipid supplementation of the broodstock diet. Comp. Biochem. Physiol. B-Biochem. Mol. Biol. 123:97–103.

Carriker, M.R. and Palmer, R.E., 1979. Ultrastructural morphogenesis of prodissoconch and early dissoconch valves of the oyster *Crassostrea virginica*. Proc. Natl. Shellfish. Assoc. 69:103–128.

Cary, S.C., Leighton, D.L. and Phleger, C.E., 1981. Food and feeding strategies in larval and early juvenile purple-hinge rock scallops *Hinnites multirugosus* (Gale). J. World Maricult. Soc. 12:156–169.

Cary, S.C., Phleger, C.E. and Leighton, D.L., 1982. Advances in the culture of the purple-hinged rock scallop. J. Shellfish. Res. 2:116–117.

Casse, N., Devauchelle, N. and Le Pennec, M., 1998. Embryonic shell formation in the scallop *Pecten maximus* (Linnaeus). Veliger 41:133–141.

Castagna, M.M. and Duggan, W., 1971. Rearing of the bay scallop, *Aequipecten irradians*. Proc. Natl. Shellfish. Assoc. 61:86–92.

Cerrano, C., Puce, S., Chiantore, M., Bavestrello, G. and Cattaneo-Vietti, R., 2001. The influence of the epizoic hydroid *Hydractinia angusta* on the recruitment of the Antarctic scallop *Adamussium colbecki*. Polar Biol. 24:577–581.

Chaitanawisuti, N. and Menasveta, P., 1992. Preliminary studies on breeding and larval rearing of the Asian moon scallop *Amusium pleuronectes*. J. Trop. Aquacult. 7:205–218.

Chanley, P. and Andrews, J.D., 1971. Aids for identification of bivalve larvae of Virginia. Malacologia 11:45–119.

Chevolot, L., Cochard, J. and Yvin, J., 1991. Chemical induction of larval metamorphosis of *Pecten maximus* with a note on the nature of naturally-occurring triggering substances. Mar. Ecol. Prog. Ser. 74:83–89.

Chia, E.S., Buckland, J. and Young, C.M., 1984. Locomotion of marine invertebrate larvae. Can. J. Zool. 62:1205–1222.

Cochard, J.C. and Devauchelle, N. 1993. Spawning, fecundity and larval survival and growth in relation to controlled conditioning in native and transplanted populations of *Pecten maximus* (L.): evidence for the existence of separate stocks. J. Exp. Mar. Biol. Ecol. 169:41–56.

Cole, J.A., 1938. The fate of larval organs in the metamorphosis of *Ostrea edulis*. J. Mar. Biol. Assoc. U.K. 22:469–484.

Comely, C.A., 1972. Larval culture of the scallop *Pecten maximus* (L). J. Cons. Int. Explor. Mer 34:365–378.

Costello, T.J., Hudson, J.H., Dupuy, J.L. and Rivkin, S., 1973. Larval culture of the calico scallop, *Argopecten gibbus*. Proc. Natl. Shellfish. Assoc. 63:72–76.

Cragg, S.M., 1976. Some aspects of the behaviour and functional morphology of bivalve larvae. PhD Thesis. University of Wales.

Cragg, S.M., 1980. Swimming behaviour of the larvae of *Pecten maximus* (L.) (Bivalvia). J. Mar. Biol. Assoc. U.K. 60:551–564.

Cragg, S.M., 1985. The adductor and retractor muscles of the veliger of *Pecten maximus (L.)* (Bivalvia). J. Moll. Stud. 51:276–283.

Cragg, S.M., 1989. The ciliated rim of the velum of *Pecten maximus* (Bivalvia, Pectinidae). J. Moll. Stud. 55:497–508.

Cragg, S.M., 1996. The phylogenetic significance of some anatomical features of bivalve veliger larvae. In: J.D. Taylor (Ed.). Origin and Evolutionary Radiation of the Mollusca. Oxford University Press. pp. 371–380.

Cragg, S.M. and Crisp, D.J., 1991. The biology of scallop larvae. In: S.E. Shumway (Ed.). Scallops: Biology, Ecology and Aquaculture. Elsevier, Amsterdam. pp. 72–132.

Cragg, S.M. and Nott, J.A, 1977. The ultrastructure of the statocysts in the pediveliger larvae of *Pecten maximus* (L.) (Bivalvia). J. Exp. Mar. Biol. Ecol. 27:23–36.

Cranfield, H.J., 1974. Observations on the morphology of the mantle folds of the pediveliger of *Ostrea edulis* L. and their function during settlement. J. Mar. Biol. Assoc. U.K. 54:1–12.

Crisp, D.J., Yule, A.B. and White, K.N., 1985. Feeding in oyster larvae. J. Mar. Biol. Assoc. U.K. 65:759–783.

Crisp, M., 1973. Fine structure of some prosobranch osphradia. Mar. Biol. 22(3):231–240.

Crisp, M., 1981. Epithelial sensory structures of trochids. J. Mar. Biol. Assoc. U.K. 61:95–106.

Croll, R.P., Jackson, D.L. and Voronezhskaya, E.E. 1997. Catecholamine-containing cells in larval and postlarval bivalve molluscs. Biol. Bull. 193:116–124.

Cruz, P. and Ibarra, A. 1997. Larval growth and survival of two catarina scallop (*Argopecten circularis*, Sowerby, 1835) populations and their reciprocal crosses. J. Exp. Mar. Biol. Ecol. 212:95–110.

Culliney, J.L., 1974. Larval development of the giant scallop, *Placopecten magellanicus* (Gmelin). Biol. Bull. 147:147–321.

Davenport, J., Gruffydd, L.D. and Beaumont, A.R., 1975. An apparatus to supply water of fluctuating salinity and its use in a study of larvae of the scallop *Pecten maximus* L. J. Mar. Biol. Assoc. U.K. 55:391–409.

De la Roche, J.P., Marin, B., Freites, L. and Velez, A., 2002. Embryonic development and larval and post-larval growth of the tropical scallop *Nodipecten* (=*Lyropecten*) *nodosus* (L. 1758) Mollusca: Pectinidae. Aquaculture Res. 33:819–827.

Delaunay, F., Marty, Y., Moal, J. and Samain, J.F., 1992. Growth and lipid class composition of *Pecten maximus* (L.) larvae grown under hatchery conditions. J. Exp. Mar. Biol. Ecol. 163:209–219.

Delaunay, F., Marty, Y., Moal, J. and Samain, J.F., 1993. The effect of monospecific algal diets on growth and fatty acid composition of *Pecten maximus* L. larvae. J. Exp. Mar. Biol. Ecol. 173:163–179.

Demers, A., Lagadeuc, Y., Dodson, J.J. and Lemieux, R., 1993. Immunofluorescence identification of early life history stages of scallops (Pectinidae). Mar. Ecol. Prog. Ser. 97:83–89.

Désilets, J., Gicquaud, C. and Dubé, F., 1995. An ultrastructural analysis of early fertilization events in the giant scallop *Placopecten magellanicus* (Mollusca, Pelecypoda). Invert. Reprod. Dev. 27:115–129.

Desrosiers, R.R., Désiltes, J. and Dubé, F., 1996. Early developmental events following fertilization in the giant scallop *Placopecten magellanicus*. Can. J. Fish. Aquat. Sci. 53:1382–1392.

Desrosiers, R., Gerard, A., Peignon, J., Naciri, Y., Dufresne, L., Morasse, J., Ledu, C., Phelipot, P., Guerrier, P. and Dubé, F., 1993. A novel method to produce triploids in bivalve molluscs by the use of 6-dimethylamineopurine. J. Exp. Mar. Biol. Ecol. 170:29–43.

Devauchelle, N., Dorange, G. and Faure, C., 1994. A technique for separating high-quality and low-quality embryos of the scallop *Pecten maximus* L. Aquaculture 120:341–346.

Dickinson, A.J.G., Nason, J. and Croll, R.P., 1999. Histochemical localization of FMRFamide, serotonin and catecholamines in embryonic *Crepidula fornicata* (Gastropoda, Prosobranchia). Zoomorphology 119:49–62.

Dijkstra, H.H. and Knudsen, J., 1997. The morphology and assignment of *Pseudohinnites levii* Dijkstra, 1989 (Bivalvia: Pectinoidea). Basteria 61:1–15.

Dix, T.G., 1976. Larval development of the queen scallop, *Equichlamys bifrons*. Aus. J. Mar. Freshwat. Res. 27:399–403.

Dix, T.G. and Sjardin, M.J., 1975. Larvae of the commercial scallop *Pecten meridionalis* from Tasmania, Aust. J. Mar. Freshwat. Res. 26:109–112.

Dominguez, M. and Alcaraz, M., 1983. Planktonic larvae of lamellibranch mollusks from the Ria-de-Pontevedra, northwestern Spain. Invest. Pesqueras 47:345–357.

Drew, G.A., 1906. The habits, anatomy and embryology of the giant scallop *(Pecten tenuicostatus,* Mighels). Univ. Maine Stud. 6:71 pp. + 17 plates.

Eckmann, J.E., 1987. The role of hydrodynamics in recruitment, growth and survival of *Argopecten irradians (L)* and *Anomia simplex* (D'Orbigny) within eelgrass meadows. J. Exp. Mar. Biol. Ecol. 106:165–191.

Elston, R., 1980. Functional anatomy, histology and ultrastructure of the soft tissues of the larval American oyster, *Crassostrea virginica*. Proc. Natn. Shellfish. Assoc. 70:65–93.

Emery, D.G., 1975. Ciliated sensory neurons in the lip of the squid *Loliguncula brevis* Blainville. Cell Tiss. Res. 157:323–329.

Erdmann, W., 1934. Untersuchungen über die Lebensgeschichte der Auster Nr. 5 Über die Entwicklung und die Anatomie der "ansatzreifen" Larve von *Ostrea edulis* mit Bemerkungen über die Lebensgeschichte der Auster. Wiss. Meersunters. N.F. Abt. Helgoland 19(6):1–24.

Farias, A., Uriarte, I. and Castilla, J.C., 1998. A biochemical study of the larval and postlarval stages of the Chilean scallop *Argopecten purpuratus*. Aquaculture 166:37–47.

Fatton, E. and Bongrain, M., 1980. Stades juvéniles de coquille de Pectinidés (Bivalves: observations au microscope électronique à balayage. Bull. Mus. natn. Hist. nat., Paris, 4e sèr., 2, sect C No. 4:291–319.

Frischer, M.E., Danforth, J.M., Tyner, L.C., Leverone, J.R., Marelli, D.C., Arnold, W.S. and Blake, N.J., 2000. Development of an *Argopecten*-specific 18S rRNA targeted genetic probe. Mar. Biotech. 2:11–20.

Fullarton, J.H., 1896. On the development of the common scallop *(Pecten opercularis* L.). Rep. Fish. Board Scot. No. 8, 6:290–299.

Fuller, S.C. and Lutz, R.A., 1989. Shell morphology of larval and post-larval mytilids from the north-western Atlantic. J. Mar. Biol. Assoc. U.K. 69:181–218.

Gallager, S.M., 1988. Visual observations of particle manipulation during feeding of a bivalve mollusc. Bull. Mar. Sci. 43:344–365.

Gallager, S.M., Manuel, J.L., Manning, D.A. and O'Dor, R., 1996. Ontogenetic changes in the vertical distribution of giant scallop larvae, *Placopecten magellanicus*, in 9-m deep mesocosms as a function of light, food, and temperature stratification. Mar. Biol. 124:679–692.

Golikov, A.N. and Scarlato, O.A., 1970. Abundance, dynamics and production properties of populations of edible bivalves *Mizuhopecten yessoensis* and *Spisula sachalinensis* related to the problem of organization of controllable submarine farms at the western shores of the Sea of Japan. Helgol. wiss. Meeresunters. 20:498–513.

Golz, R. and Thurm, U., 1993. Ultrastructural evidence for the occurrence of 3 types of mechanosensitive cells in the tentacles of the cubozoan polyp *Carybdea marsupialis*. Protoplasma 173:13–22.

Goodsell, J.G. and Eversol, A.G., 1992. Prodissoconch I and II length in *Mercenaria* taxa. Nautilus 106:119–122.

Grigor'eva, N.I. and Regulev, V.N., 1992. Vertical distribution of larvae of the scallop *Mizuhopecten yessoensis* and mussel *Mytilus trossulus* in Posyet Bay, Sea of Japan. Biologiya Morya – Marine Biology 2:64–70.

Gruffydd, L.D., 1976. The development of the larva of *Chlamys islandica* in the plankton and its salinity tolerance in the laboratory (Lamellibranchia, Pectinidae). Astarte 8:61–67.

Gruffydd, L.D. and Beaumont, A.R., 1970. Determination of the optimum concentration of eggs and spermatozoa for the production of normal larvae in *Pecten maximus* (Mollusca, Lamellibranchia). Helgol. wiss. Meersunters. 20:486–497.

Gruffydd, L.D. and Beaumont, A.R., 1972. A method of rearing *Pecten maximus* larvae in the laboratory. Marine Biology 15:350–355.

Gruffydd, L.D., Lane, D.J. and Nott, J.A., 1975. The glands of the larval foot in *Pecten maximus* L. and possible homologues in other bivalves. J. Mar. Biol. Assoc. U.K. 55:463–476.

Gutsell, J.S., 1930. Natural history of the bay scallop. Bull. U.S. Bur. Fish. 46:569–632.

Hadfield, M.G., Meleshkevitch, E.A. and Boudko, D.Y., 2000. The apical sensory organ of a gastropod veliger is a receptor for settlement cues. Biol. Bull. 198:67–76.

Harvey, M., Bourget, E. and Miron, G., 1993. Settlement of Iceland scallop *Chlamys islandica* spat in response to hydroids and filamentous red algae: Field observations and laboratory experiments. Mar. Ecol. Prog. Ser. 99:283–292.

Harvey, M., Bourget, E., Legault, C. and Ingram, R., 1995a. Short-term variations in settlement and early spat mortality of Iceland scallop, *Chlamys islandica* (O.F. Muller). J. Exp. Mar. Biol. Ecol. 194:167–187.

Harvey, M., Miron, G. and Bourget, E., 1995b. Resettlement of Iceland scallop (*Chlamys islandica*) spat on dead hydroids (*Tubularia larynx*): response to chemical cues from the protein-chitinous perisarc and associated microbial film. J. Shellfish Res. 14:383–388.

Harvey, M., Bourget, E. and Gagne, N., 1997. Spat settlement of the giant scallop, *Placopecten magellanicus* (Gmelin, 1791), and other bivalve species on artificial filamentous collectors coated with chitinous material. Aquaculture 148:277–298.

Haszprunar, G., 1985. On the anatomy and fine structure of a peculiar sense organ of *Nucula* (Bivalvia, Protobranchia). Veliger 28:52–62.

Haszprunar, G., Friedrich, S., Wanninger, A. and Ruthensteiner, B. 2002. Fine structure and immunocytochemistry of a new chemosensory system in the chiton larva (Mollusca: Polyplacophora). J. Morphol. 251:210–218.

Hatschek, B., 1880. Über Entwicklungsgeschichite *von Teredo.* Arb. Zool. Inst. Univ. Wien 3:1–44.

Hayami, I. and Kase, T., 1993. Submarine cave Bivalvia from the Ryukyu Islands: systematics and evolutionary significance. Tokyo University Museum Bulletin 35. 133 p.

Heasman, M.P., O'Connor, W.A. and Frazer, A.W.J., 1996. Ontogenetic changes in optimal rearing temperatures for the commercial scallop, *Pecten fumatus* Reeve. J. Shellfish Res. 15:627–634.

Hilbish, T.J., Sasada, K., Eyster, L.S. and Pechenik, J.A., 1999. Relationship between rates of swimming and growth in veliger larvae: genetic variance and covariance. J. Exp. Mar. Biol. Ecol. 239:183–193.

His, E., Beiras, R. and Seaman, M.N.L., 1999. The assessment of marine pollution – bioassays with bivalve embryos and larvae. Adv. Mar. Biol. 37:1–178.

Hodgson, C.A. and Burke, R.D., 1988. Development and larval morphology of the spiny scallop, *Chlamys hastata.* Biol. Bull. 174:303–318.

Hoegh-Guldberg, O. and Manahan, D.T., 1995. Coulometric measurement of oxygen consumption during development of marine invertebrate embryos and larvae. J. Exp. Biol. 198:19–30.

Hortle, M.E. and Cropp, D.A., 1987. Settlement of the commercial scallop *Pecten fumatus* (Reeve) on artificial collectors in eastern Tasmania. Aquaculture 66:79–95.

Hoshikawa, H. and Kurata, M., 1987. Spat settlement of the scallop *Patinopecten yessoensis* on artificial collectors off Abashiri. J. Hokkaido Fish. Exp. Stn. 44:1–12 (in Japanese).

Hotta, M., 1977. On rearing the larvae and young of Japanese scallop *Pecten* (*Notovola*) *albicans* (Shroter). Bull. Hiroshima Fisheries Exptl. Stn. 9:37–45.

Hrs-Brenko, M., 1971. Observations on the occurrence of planktonic larvae of several bivalves in the northern Adriatic Sea. In: D.J. Crisp (Ed.). Fourth European Marine Biology Symposium. Cambridge University Press. pp. 45–54.

Humphreys, W.J., 1969. Initial shell formation in the bivalve *Mytilus edulis.* Proc. Electr. Micr. Soc. Am. 27:272–273.

Hurley, G.V., Tremblay, M.J. and Couturier, C., 1987. Age estimation of sea scallop larvae *(Placopecten magellanicus)* from daily growth lines on shells. J. Northwest. Atl. Fish. Sci. 7:123–129.

Ibarra, A.M., Cruz, P. and Romero, B.A., 1995. Effects of inbreeding on growth and survival of self-fertilized catarina scallop larvae, *Argopecten circularis.* Aquaculture 134:37–47.

Ibarra, A.M., Ramirez, J.L. and Garcia, G.A., 1997. Stocking density effects on larval growth and survival of two Catarina scallop, *Argopecten ventricosus* (= *circularis*) (Sowerby II, 1842), populations. Aquacult. Res. 28:443–451.

Illanes-Bucher, J.E., 1987. Cultivation of the northern scallop of Chile *(Chlamys (Argopecten) purpurata)* in controlled and natural environments. In: A.R. Beaumont and J. Mason (Eds.). 6th International Pectinid Workshop. Menai Bridge, Wales. pp. xviii.

Imai, T., 1953. Mass production of molluscs by means of rearing in tanks. Venus Jap. J. Malacol. 25:159–167.

Imai, T. and Nishikawa, N., 1969. Artificial mass production of youngs of scallop and bloody clam. The Aquaculture 16:309–316.

Jablonski, D. and Lutz, R.A., 1980. Molluscan larval shell morphology: ecological and palaeontological applications. In: D.C. Rhoads and R.A. Lutz (Eds.). Skeletal Growth in Aquatic Organisms. Plenum Press, New York. pp. 323–378.

112

Jackson, D.L., O'Dor, R.K. and Gallager, S.M., 1993. Scallops in space: microgravity experiments with veliger locomotion. In: N.F. Bourne, B.L. Bunting and L.D. Townsend (Eds.). Proc. 9th Int. Pectinid Workshop, Nanaimo, BC, Canada, April 22–27, 1993. Can. Tech. Rep. Fish. Aquat. Sci. 1994(1):45–52.

Jørgensen, C.B., 1946. Lamellibranchia. In: G. Thorson (Ed.). Reproduction and larval development of Danish Marine Bottom Invertebrates. Ser. Plankton, Medd. Komm. Danmarks Fisk. Havunders. 4:277–311.

Jorquera, M.A., Riquelme, C.E., Loyola, L.A. and Munoz, L.F., 2000. Production of bactericidal substances by a marine vibrio isolated from cultures of the scallop *Argopecten purpuratus.* Aquacult. Int. 7:433–448.

Kaartvedt, S., Aksnes, D.L. and Egge, J.K., 1987. Effect of light on the vertical distribution of *Pecten maximus* larvae. Mar. Ecol. Prog. Ser. 40:195–197.

Kanno, H., 1970. On the relationship between the occurrence of pelagic larvae and attached spats in Okunai. Aquaculture 17:121–134.

Karaseva, E.M. and Medvedeva, L.A., 1993. Morphological and functional changes in the offspring of *Mytilus trossulus* and *Mizuhopecten yessoensis* after parental exposure to copper and zinc. Russian Journal of Marine Biology 19:276–280.

Kase, T. and Hayami, I., 1992. Unique cave mollusc fauna: composition, origin and adaptation. J. Moll. Studs. 58:446–448.

Kasyanov, V.L., Kryuchkova, G.A., Kulikova, V.A. and Medvedeva, L.A., 1998. Larvae of Marine Bivalves and Echinoderms (D.L. Pawson Translation Ed.). Science Publishers Inc., Enfield, New Hampshire, USA. 288 p.

Kempf, S.C., Page, L.R. and Pires, A., 1997. Development of serotonin-like immunoreactivity in the embryos and larvae of nudibranch mollusks with emphasis on the structure and possible function of the apical sensory organ. J. Comp. Neurol. 386:507–528.

Kingzett, B.C., Bourne, N. and Leask, K., 1990. Induction of metamorphosis of the Japanese scallop *Patinopecten yessoensis* Jay. J. Shellfish Res. 9:119–124.

Knight-Jones, E.W. and Morgan, E., 1966. Responses on marine animals to changes in hydrostatic pressure. In: H. Barnes (Ed.). Oceanogr. Mar. Biol. Ann. Rev., George Allen and Unwin Ltd., London 4:267–299.

Knudsen, J., 1979. Deep-sea bivalves. In: S. van der Spoel, A.C. van Bruggen and J. Lever (Eds.). Pathways in Malacology. Scheltema and Holkema, Utrecht. pp. 195–224.

Krassoi, R., Anderson, I. and Everett, D., 1997. Larval abnormalities in doughboy scallops *Chlamys (Mimachlamys) asperrima* L. in response to test conditions and six reference toxicants. Aust. J. Ecotox. 3:65–74.

Kraeuter, J.N., Castagna, M. and Van dessel, R., 1982. Egg size and larval survival of *Mercenaria mercenaria* (L.) and *Argopecten irradians* (Lamarck). J. Exp. Mar. Biol. Ecol. 56:3–8.

Kuang, S. and Goldberg, J.I. 2000. Laser ablation reveals regulation of ciliary activity by serotonergic neurons in molluscan embryos. J. Neurobiol. 47:1–15.

Kulikova, V.A. and Tabunkov, V.D., 1974. Ecology, reproduction, growth and productive properties of a population of the scallop *Mizuchopecten yessoensis* Dysodonta, Pectinidae in the Busset Lagoon, Aniva Bay. Zool. Zh. 53:1767–1774.

Kulikova, V.A., Medvedeva, L.A. and Guida, G.M., 1981. Morphology of pelagic larvae of three bivalve species of the family Pectinidae from Peter the Great Bay (Sea of Japan). Biologiya Morya 1981:75–77.

La Barbera, M., 1974. Calcification of the first larval shell of *Tridacna squamosa* (Tridacnidae: Bivalvia). Marine Biology 25:233–238.

Lambert, C., Nicolas, J.L., Cilia, V. and Corre, S., 1998. *Vibrio pectinida* sp. nov., a pathogen of scallop (*Pecten maximus*) larvae. Int. J. System. Bacteriol. 48:481–487.

Lambert, C., Nicolas, J.L. and Bultel, V., 2001. Toxicity to bivalve hemocytes of pathogenic *Vibrio* cytoplasmic extract. J. Invert. Pathol. 77:165–172.

Lefort, Y., 1992. Larval development of the scallop, *Mimachlamys gloriosa* (Reeve, 1853) from south-west lagoon of New Caledonia. C. R. Acad. Sci. Paris Sér III 314:601–607.

Le Pennec, M., 1974. Morphogenése de la coquille de *Pecten maximus* (L) élevé au laboratoire. Cah. Biol. Marine 15:475–482.

Le Pennec, M., 1980. The larval and post-larval hinge of some families of bivalve molluscs. J. Mar. Biol. Assoc. U.K. 60:601–617.

Le Pennec, M., 1982. Élevage experimental de *Chlamys opercularis* (L.) (Bivalvia, Pectinidae). Vie Mar. 4:29–36.

Le Pennec, M., 1998. Embryonic shell formation in the scallop *Pecten maximus* (Linnaeus). Veliger 41:133–141.

Le Pennec, M. and Rangel-Davalos, C., 1985. Observations by epifluorescence microscopy of ingestion and digestion of unicellular algae by young larvae of *Pecten maximus* Pectinidae Bivalvia. Aquaculture 47:39–52.

Le Pennec, M. and Beninger, P.G., 2000. Reproductive characteristics and strategies of reducing system bivalves. Comp. Biochem. Physiol. A 126:1–16.

Le Pennec, M., Robert, R. and Avendaño, M., 1998. The importance of gonadal development on larval production in pectinids. J. Shellfish Res. 17:97–101.

Liao, C. and Li, Y., 1980. Influence of Sheng-Li crude oil on the early development of the scallop *Chlamys farreri* (Jones et Preston). J. Shandong Coll. Ocean. 10:52–59.

Lodeiros, C., Freites, L. and Vélez, A., 1992. Necrosis bacilar en larvas del bivalvo *Euvola ziczac* (Linneo, 1758) causada por una *Pseudomonas* sp. Microbiología Acta Científica Vennezolana. 43:154–158.

Loosanoff, V.L. and Davis, H.C., 1963. Rearing of bivalve mollusks. Adv. Mar. Biol. 1:2–136.

Loosanoff, V.L., Davis, H.C. and Chanley, P.E., 1966. Dimensions and shapes of larvae of some marine bivalve mollusks. Malacologia 4(2):351–435.

Lora-Vilchis, M. and Maeda-Martinez, A., 1997. Ingestion and digestion index of catarina scallop *Argopecten ventricosus-circularis*, Sowerby II, 1842, veliger larvae with ten microalgae species. Aquacult. Res. 28:905–910.

Lu, Y. and Blake, N., 1996a. Optimum concentrations of *Isochrysis galbana* for growth of larval and juvenile bay scallops, *Argopecten irradians concentricus* (Say). J. Shellfish Res. 15:635–643.

Lu, Y. and Blake, N., 1996b. Clearance and ingestion rates of *Isochrysis galbana* by larval and juvenile bay scallops, *Argopecten irradians concentricus* (Say). J. Shellfish Res. 16:47–54.

114

Lu, Y.T., Blake, N.J. and Torres, J.J., 1999a. Oxygen consumption and ammonia excretion of larvae and juveniles of the bay scallop, *Argopecten irradians concentricus* (Say). J. Shellfish Res. 18:419–423.

Lu, Y.T., Blake, N.J. and Torres, J.J., 1999b. Biochemical utilization during embryogenesis and metamorphosis in the bay scallop, *Argopecten irradians concentricus* (Say). J. Shellfish Res. 18:425–429.

Lucas, A. and Le Pennec, M., 1976. Apports du microscope électronique a balayage dans l'étude de la morphogenese des coquilles larvaires de mollusques bivalves marins. 97 Congres nat. Soc. savantes, Nantes, 1972, Sciences, t. III:685–693.

Lucero, M.T., Horrigan, F.T. and Gilly, W.F., 1992. Electrical responses to chemical stimulation of squid olfactory receptor cells. J. Exp. Biol. 162:231–249.

Lutz, R.A. and Jablonski, D., 1978. Larval bivalve shell morphometry: a new paleoclimatic tool? Science 202:51–53.

Lutz, R., Goodsell, J., Castagna, M., Chapman, S., Newell, C., Hidu, H., Mann, R., Jablonski, D., Kennedy, V., Siddall, S., Goldberg, R., Beattie, H., Falmagne, C., Chestnut, A. and Partridge, A., 1982. Preliminary observations on the usefulness of hinge structures for identification of bivalve larvae. J. Shellfish Res. 2:65–70.

MacDonald, B.A., 1988. Physiological energetics of Japanese scallop *Patinopecten yessoensis* larvae. J. Exp. Mar. Biol. Ecol. 120:155–170.

Mackie, G.L., 1983. Bivalves. In: A.S. Tompa, N.H. Verdonk and J.A.M. Van den Biggelaar (Eds.). The Mollusca, Vol. 7: Reproduction. Academic Press.

Malakhov, V.V. and Medvedeva, L.A., 1986. Embryonic development in the bivalves *Patinopecten yessoensis* (Pectinida, Pectinidae) and *Spisula sachalinensis* (Cardiida, Mactridae). Zool. Zh. 65:732–740.

Malchus, N., 2000. Early shell stages of the Middle Jurassic bivalves *Camptochlamys* (Pectinidae) and *Atreta* (Dimyidae) from Poland. J. Moll. Stud. 66:577–581.

Manahan, D.T. and Crisp, D.J., 1982. The role of dissolved organic material in the nutrition of pelagic larvae: amino acid uptake by bivalve veligers. Amer. Zool. 22:635–646.

Manahan, D.T. and Crisp, D.J., 1983. Autoradiographic studies on the uptake of dissolved amino acids from sea water by bivalve larvae. J. Mar. Biol. Assoc. U.K. 63:673–682.

Manuel, J.M., Burbridge, S., Kenchington, E.L., Ball, M. and O'Dor, R.K., 1996a. Veligers from two populations of scallop *Placopecten magellanicus* exhibit different vertical distributions in the same mesocosm. J. Shellfish Res. 15:251–257.

Manuel, J.L., Gallager, S.M., Pearce, C.M., Manning, D.A. and O'Dor, R.K., 1996b. Veligers from different populations of sea scallop *Placopecten magellanicus* have different vertical migration patterns. Mar. Ecol. Prog. Ser. 142:147–63.

Manuel, J.L., Pearce, C.M. and O'Dor, R.K., 1997. Vertical migration for horizontal transport while avoiding predators: II. Evidence for the tidal/diel model from two populations of scallop (*Placopecten magellanicus*) veligers. J. Plankton Res. 19:1949–1973.

Manuel, J.L., Pearce, C.M., Manning, D.A. and O'Dor, R.K., 2000. The response of sea scallop (*Placopecten magellanicus*) veligers to a weak thermocline in 9-m deep mesocosms. Mar. Biol. 137:169–175.

Martinez, G., Aguilera, C. and Campos, E.O., 1999. Induction of settlement and metamorphosis of the scallop *Argopecten purpuratus* Lamarck by excess K$^+$ and epinephrine: energetic costs. J. Shellfish Res. 18:41–46.

Maru, K., 1972. Morphological observations on the veliger larvae of a scallop *Patinopecten yessoensis* (Jay). Scientific Rep. Hokkaido Fish. Expl. Stn. No. 1 14:55–62.

Maru, K., 1985a. Ecological studies on the seed production of scallop, *Patinopecten yessoensis* (Jay). Sci. Rep. Hokkaido Fish. Expl. Stn. 27:1–53.

Maru, K., 1985b. Tolerance of scallop, *Patinopecten yessoensis* (Jay) to temperature and specific gravity during early developmental stages. Sci. Rep. Hokkaido Fish. Expl. Stn. 27:55–64.

Maru, K., Obara, A., Kikuchi, K. and Okesaku, H., 1973. Studies on the ecology of the scallop *Patinopecten yessoensis* (Jay). 3. On the diurnal vertical distribution of scallop larvae. Sci. Rep. Hokkaido Fish. Exp. Stn. 15:33–52.

Mason, J., 1958. The breeding of the scallop, *Pecten maximus,* in Manx waters. J. Mar. Biol. Assoc. U.K. 37:653–671.

Merrill, A.S., 1961. Shell morphology in the larval and postlarval stages of the sea scallop, *Placopecten magellanicus* (Gmelin). Bull. Mus. Comp. Zool. Harvard 125:3–20.

Miron, G., Pelletier, P. and Bourget, E., 1995. Optimizing the design of giant scallop (*Placopecten magellanicus*) spat collectors – flume experiments. Mar. Biol. 123:285–291.

Miron, G., Ward, J., MacDonald, B. and Bourget, E., 1996. Direct observations of particle kinematics within a scallop (*Placopecten magellanicus*) spat collector by means of video endoscopy. Aquaculture 147:71–92.

Minchin, D., Duggan, C.B. and King, W., 1987. Possible effects of organotins on scallop recruitment. Mar. Pollut. Bull. 18:604–608.

Moal, J., Samain, J., Corre, S., Nicolas, J. and Glynn, A., 1996. Bacterial nutrition of great scallop larvae. Aquacult. Int. 4:215–223.

Moir, A.J.G., 1977. Ultrastructural studies on the ciliated receptors of the long tentacles of the giant scallop *Placopecten magellanicus* (Gmelin). Cell Tiss. Res. 184:359–366.

Moor, B., 1983. Organogenesis. In: N.H. Verdonk, J.A.M van den Biggelaar and A.S. Tompa (Eds.). The Mollusca, Vol. 3: Development. Academic Press, London. pp. 123–177.

Moore, J.K. and Marshall, N., 1967. The retention of lamellibranch larvae in the Niantic estuary. Veliger 10:10–12.

Morse, M.P. and Zardus, J.D., 1997. Bivalvia. In: F.W. Harrison and A.J. Kohn (Eds.). Microscopic Anatomy of Invertebrates. Volume 6A: Mollusca II. pp. 7–118.

Morton, B., 1994. The biology and functional morphology of *Leptopecten lauriaticus* (Conrad, 1837) – an opportunistic scallop. Veliger 37:5–22.

Morton, B., 2001. The evolution of eyes in the Bivalvia. Oceanogr. Mar. Biol. 39:165–205.

Mullineaux, L.S., Mills, S.W. and Goldman, E., 1998. Recruitment variation during a pilot colonization study of hydrothermal vents (9 degrees 50' N, East Pacific Rise). Deep-Sea Res. II Topical Stud. Oceanogr. 45:441–464.

Naidu, K.S. and Scaplen, R., 1976. Settlement and survival of the giant scallop, *Placopecten magellanicus,* larvae on enclosed polyethylene film collectors. In: T.V.R. Pillay and W.A. Dill (Eds.). Advances in Aquaculture. FAO, Rome. pp. 379–381.

Narvarte, M.A., 2001. Settlement of the tehuelche scallop, *Aequipecten tehuelchus* D'Orb, larvae on artificial substrata in San Matias Gulf (Patagonia, Argentina). Aquaculture 196:55–65.

116

Narvarte, M.A. and Pascual, M.S., 2001. Diet trials on tehuelche scallop *Aequipecten tehuelchus* (d'Orb) larvae. Aquacult. Int. 9:127–131.

Nebelsick, M., 1992. Sensory spots of *Echinoderes capitatus* (Zelinka, 1928) (Kinorhyncha, Cyclorhagida). Acta Zool. 73:185–195.

Nelson, C.L. and Siddall, S.E., 1988. Effects of an algal bloom isolate on growth and survival of bay scallop (*Argopecten irradians*) larvae. J. Shellfish Res. 7:683–694.

Nielsen, C., 1995. Animal Evolution: Interrelationships of the Living Phyla. OUP. 466 p.

Nicolas, L., Robert, R. and Chevolot, L., 1996. Effect of epinephrine and seawater turbulence on the metamorphosis of the great scallop. Aquacult. Int. 4:293–297.

Nicolas, L., Robert, R. and Chevolot, L., 1998. Comparative effects of inducers on metamorphosis of the Japanese oyster *Crassostrea gigas* and the great scallop *Pecten maximus*. Biofouling 12:189–203.

Nicolas, J., Corre, S., Gauthier, G., Robert, R. and Ansquer, D., 1996. Bacterial problems associated with scallop *Pecten maximus* larval culture. Dis. Aquat. Org. 27:67–76.

Ockelmann, K.W., 1965. Developmental types in marine bivalves and their distribution along the Atlantic coast of Europe. In: L.R. Cox and J.E. Peake (Eds.). Proc. 1st Europ. Malac. Congr., London. Conchological Society of Great Britain and Ireland and the Malacological Society of London. pp. 25–35.

O'Connor, W.A. and Heasman, M.P., 1995. Spawning induction and fertilization in the doughboy scallop *Chlamys* (*Mimachlamys*) *asperrima*. Aquaculture 136:117–129.

O'Connor, W.A. and Heasman, M.P., 1997. Diet and feeding regimens for larval doughboy scallops, *Mimachlamys asperrima*. Aquaculture 158:289–303.

O'Connor, W.A. and Heasman, M.P., 1998. Ontogenetic changes in salinity and temperature tolerance in the doughboy scallop, *Mimachlamys asperrima*. J. Shellfish Res. 17:89–95.

Odhner, N.H., 1914. Notizen über die Fauna bei Rovigno. Beiträge zur Kenntnis der marinen Molluskenfauna von Rovigno in Istrien. Zool. Anz. 44:156–170.

Odintsova, N. and Tsal, L., 1995. Cryopreservation of primary cell cultures of Bivalvia. Cryo-Letters 16:13–20.

Olsen, S., 1981. New candidates with aquaculture potential in Washington State: pinto abalone (*Haliotis kamtschatkana*), weathervane scallop (*Pecten caurinus*), and purple-hinge rock scallop (*Hinnites multirugosus*). J. Shellfish Res. 1:133.

Pablo, F., Buckney, R.T. and Lim, R.P., 1997. Toxicity of cyanide, iron-cyanide complexes and a blast furnace effluent to larvae of the doughboy scallop, *Chlamys asperrimus*. Bull. Env. Cont. Toxicol. 58:93–100.

Page, L.R., 2002. Comparative structure of the larval sensory organ in gastropods and hypotheses about function and developmental evolution. Invert. Reprod. Develop. 41:193–200.

Park, Y-J., Lee, J.-Y., Kim, W.-K. and Lee, C.-S., 2001. Egg development and larva growth of the scallop *Patinopecten yessoensis*. Korean J. Malacol.

Parnell, P.E., 2002. Larval development, precompetent period and a natural spawning event in the pectinacean bivalve *Spondylus tenebrosus* (Reeve, 1856). Veliger 45:58–64.

Parsons, G.J., Dadswell M.J. and Roff, J.C., 1993. Influence of biofilm on settlement of sea scallop, *Placopecten magellanicus* (Gmelin, 1791), in Passamaquoddy Bay, New Brunswick, Canada. J. Shellfish Res. 12:279–283.

Paugam, A., Le Pennec, M. and Genevieve, A.F., 2000. Immunological recognition of marine bivalve larvae from plankton samples. J. Shellfish Res. 19:325–331.

Paulet, Y.M., Lucas, A. and Gerard, G., 1988. Reproduction and larval development in two *Pecten maximus* (L.) populations from Brittany. J. Exp. Mar. Biol. Ecol. 119:145–156.

Pavlova, G.A., Willows, A.O.D. and Gaston, M.R. 1999. Serotonin inhibits ciliary transport in esophagus of the nudibranch mollusk *Tritonia diomedea*. Acta Biol. Hung. 50:175–184.

Pearce, C.M. and Bourget, E., 1996. Settlement of larvae of the giant scallop, *Placopecten magellanicus* (Gmelin), on various artificial and natural substrata under hatchery-type conditions. Aquaculture 141:201–221.

Pearce, C.M., Gallager, S.M., Manuel, J.L., Manning, D.A., O'Dor, R.K. and Bourget, E., 1996. Settlement of larvae of the giant scallop *Placopecten magellanicus* in 9-m deep mesocosms as a function of temperature stratification, depth, food, and substratum. Mar. Biol. 124:693–706.

Pearce, C.M., Gallager, S.M., Manuel, J.L., Manning, D.A., O'Dor, R.K. and Bourget E., 1998. Effect of thermoclines and turbulence on depth of larval settlement and spat recruitment of the giant scallop *Placopecten magellanicus* in 9.5 m deep laboratory mesocosms. Mar. Ecol. Prog. Ser. 165:195–215.

Peña, J.B., Ríos, C., Peña, S. and Canoles, J., 1998. Ultrastructural morphogenesis of pectinid spat from the western Mediterranean: a way to differentiate seven genera. J. Shellfish Res. 17:123–130.

Phleger, C.E. and Cary, S.C., 1983. Settlement of spat of the purple-hinge rock scallop *Hinnites multirugosus* (Gale) on artificial collectors. J. Shellfish Res. 3:71–73.

Piquimil, R.M., Figueroa, L.S., Contrera, O.C. and Avendaño, M.D., 1991. Chile. In: S.E. Shumway (Ed.). Scallops: Biology, Ecology and Aquaculture. Elsevier, Amsterdam. pp. 1001–1015.

Pouliot, F., Bourget, E. and Frechette, M., 1995. Optimizing the design of giant scallop (*Placopecten magellanicus*) spat collectors: Field experiments. Mar. Biol. 123:277–284.

Quezada, A-T., 1980. Densidad y fijacion de larves de lamellibranquios en la ensenada de La Paz, B.C.S., relacionados con factores fisico-quimicos. Memorias del 2 Simposio Latinamericano de Acuacultura. Dep. Pesca, Mexico City 1:787–821.

Raby, D., Lagadeuc, Y., Dodson, J.J. and Mingelbier, M. 1994. Relationship between feeding and vertical distribution of bivalve larvae in stratified and mixed waters. Mar. Ecol. Prog. Ser. 103:275–284.

Raby, D., Mingelbier, M., Dodson, J., Klein, B., Lagadeuc, Y. and Legendre, L.N., 1997. Food-particle size and selection by bivalve larvae in a temperate embayment. Mar. Biol. 127:665–672.

Raineri, M., 1995. Is a mollusc an evolved bent metatrochophore? A histochemical investigation of neurogenesis in *Mytilus* (Mollusca: Bivalvia). J. Mar. Biol. Assoc. U.K. 75:571–592.

Raven, C.P., 1958. Morphogenesis: the analysis of molluscan development. Pergamon Press, London. 311 p.

Rees, C,B., 1950. The identification and classification of Lamellibranch larvae. Hull Bull. Mar. Ecol. 19:73–104.

Rees, C.B., 1951. Continuous plankton records: first report on the distribution of Lamellibranch larvae in the North Sea. Hull Bull. Mar. Ecol. 3:105–134.

Rees, C.B., 1954. The distribution of lamellibranch larvae in the North Sea, 1950–51. Hull Bull. Mar. Ecol. 4:21–46.

Richardson, C.A., Crisp, D.J. and Runham, N.W., 1980. An endogenous rhythm in shell secretion in *Cerastoderma edule*. J. Mar. Biol. Assoc. U.K. 60:991–1004.

Riquelme, C., Hayashida, G., Toranzo, A., Vilches, J. and Chavez, P., 1995. Pathogenicity studies on a *Vibrio anguillarum*-related (var) strain causing an epizootic in *Argopecten purpuratus* larvae cultured in Chile. Diseases Aquat. Org. 22:135–141.

Riquelme, C., Toranzo, A., Barja, J., Vergara, N. and Araya, R., 1996a. Association of *Aeromonas hydrophila* and *Vibrio alginolyticus* with larval mortalities of scallop (*Argopecten purpuratus*). J. Invert. Path. 67:213–218.

Riquelme, C., Hayashida, G., Araya, R., Uchida, A., Satomi, M. and Ishida, Y., 1996b. Isolation of a native bacterial strain from the scallop *Argopecten purpuratus* with inhibitory effects against pathogenic vibrios. J. Shellfish Res. 15:369–74.

Riquelme, C., Araya, R., Vergara, N., Rojas, A., Guaita, M. and Candia, M., 1997. Potential probiotic strains in the culture of the Chilean scallop *Argopecten purpuratus* (Lamarck, 1819). Aquaculture 154:17–26.

Riquelme, C., Araya, R. and Escribano, R., 2000. Selective incorporation of bacteria by *Argopecten purpuratus* larvae: implications for the use of probiotics in culturing systems of the Chilean scallop. Aquaculture 181:25–36.

Riquelme, C.E., Jorquera, M.A., Rojas, A.I., Avendano, R.E. and Reyes, N., 2001. Addition of inhibitor-producing bacteria to mass cultures of *Argopecten purpuratus* larvae (Lamarck, 1819). Aquaculture 192:111–119.

Robert, R. and Nicholas, L., 2000. The effect of seawater flow and temperature on metamorphosis and postlarval development in great scallop. Aquacult. Int. 8:513–530.

Robert, R., Miner, P. and Nicolas, J., 1996. Mortality control of scallop larvae in the hatchery. Aquacult. Int. 4:305–313.

Robert, R., Nicolas, L., Moisan, C. and Barbier, G., 1999. Morphological and biochemical characterizations of the great scallop *Pecten maximus* metamorphosis. C. R. Acad. Sci. Paris, Sciences de la vie 322:847–853.

Rose, R.A. and Dix, T.G., 1984. Larval and juvenile development of the doughboy scallop, *Chlamys asperrimus*: Mollusca, Pectinidae. Aust. J. Mar. Freshwat. Res. 35:315–324.

Rose, R.A., Campbell, G.R. and Sanders, S.G., 1988. Larval development of the saucer scallop *Amusium balloti* (Bernadi) (Mollusca: Pectinidae). Aust. J. Mar. Freshwater Res. 39:153–160.

Rouse, G., 1999. Trochophore concepts: ciliary bands and the evolution of larvae in spiralian Metazoa. Biol. J. Linn. Soc. 66:411–464.

Ruiz Ponte, C., Cilia, V., Lambert, C. and Nicolas, J., 1998. *Roseobacter gallaeciensis* sp. nov., a new marine bacterium isolated from rearings and collectors of the scallop *Pecten maximus*. Int. J. Syst. Bacteriol. 48:537–542.

Ruiz-Ponte, C., Samain, J., Sanchez, J. and Nicolas, J., 1999. The benefit of a *Roseobacter* species on the survival of scallop larvae. Mar. Biotech. 1:52–59.

Sainz, J., Maeda-Martinez, A. and Ascencio, F., 1998. Experimental vibriosis induction with *Vibrio alginolyticus* of larvae of the catarina scallop (*Argopecten ventricosus* = *circularis*) (Sowerby II, 1842). Microb. Ecol. 35:188–192.

Salaün, M. and Le Pennec, M., 1992. La prodissoconque de *Pecten maximus* (mollusque, bivalve), outil potentiel pour l'estimation du recrutement. Ann. Inst. océanogr., Paris 68:37–44.

Santamaria, N., Felix-Pico, E., Sanchez-Lizaso, J., Palomares-Garcia, J. and Mazon-Suastegui, M., 1999. Temporal coincidence of the annual eelgrass *Zostera marina* and juvenile scallops *Argopecten ventricosus* (Sowerby II, 1842) in Bahia Concepcion, Mexico. J. Shellfish Res. 18:415–418.

Sasaki, R., 1979. A report of scallop and oyster in the course of Japan/Scotland exchange scholarship 1977/1978. Fisheries Div., Highlands and Islands Devp. Bd., Scotland, 24 p.

Sastry, A.N., 1965. The development and external morphology of pelagic larval and post-larval stages of the bay scallop, *Aequipecten irradians concentricus* Say, reared in the laboratory. Bull. Mar. Sci. Gulf Caribb. 15:417–435.

Sastry, A.N., 1979. Pelecypoda (excluding Ostreidae). In: A.C. Giese and J.S. Pearse (Eds.). Reproduction of Marine Invertebrates, Vol. 5. Molluscs: Pelecypods and Lesser Classes. Academic Press, New York. 1:13–292.

Seguineau, C., Laschi Loquerie, A., Moal, J. and Samain, J., 1993. Vitamines transfer from algal diets to *Pecten maximus* larvae. J. Mar. Biotechnology 1:67–71.

Seguineau, C., Laschi Loquerie, A., Moal, J. and Samain, J., 1996. Vitamin requirements in great scallop larvae. Aquacult. Int. 4:315–324.

Seguineau, C., Saout, C., Paulet, Y.M., Muzellec, M.L., Quere, C., Moal, J. and Samain, J.F., 2001. Changes in tissue concentrations of the vitamins B1 and B2 during reproductive cycle of bivalves. Part 1: the scallop *Pecten maximus*. Aquaculture 196:125–137.

Schein, E., 1989. Pectinidae (Mollusca, Bivalvia) bathyaux et abyssaux des campagnes Biogas (Golfe de Gascogne) systématique et biogéographique. Ann. Inst. océanogr., Paris 65:59–125.

Solis, U.I., Sanchez, A.P. and Navarette, V.S., 1976. Identification and description of larvae of bivalve mollusks in plankton from Estero de Castro, Chile. Bol. Soc. Biol. Concepcion 50:183–196.

Soudant, P., LeCoz, J., Marty, Y., Moal, J., Robert, R. and Samain, J., 1998a. Incorporation of microalgae sterols by scallop *Pecten maximus* (L.) larvae. Comp. Biochem. Physiol. A 119:451–457.

Soudant, P., Marty, Y., Moal, J., Masski, H. and Samain, J., 1998b. Fatty acid composition of polar lipid classes during larval development of scallop *Pecten maximus* (L.). Comp. Biochem. Physiol. A 121:279–288.

Sprung, M., 1984a. Physiological energetics of mussel larvae (*Mytilus edulis*). I. Shell growth and biomass. Mar. Ecol. Prog. Ser.17:283–293.

Sprung, M., 1984b. Physiological energetics of mussel larvae (*Mytilus edulis*). II Food uptake. Mar. Ecol. Prog. Ser. 17:295–305.

Sprung, M., 1984c. Physiological energetics of mussel larvae *(Mytilus edulis)*. IV. Efficiencies. Mar. Ecol. Prog. Ser. 18:179–186.

Stafford, J., 1912. On the recognition of bivalve larvae in plankton collections. Contr. Can. Biol. 1905–1910:221–242.

Strathmann, R.R., Jahn, T.L. and Fonseca, J.R.C., 1972. Suspension feeding by marine invertebrate larvae: clearance of particles by ciliated bands of a rotifer, pluteus and trochophore. Biol. Bull. 142:505–519.

Strathmann, R.R. and Leise, E., 1979. On feeding mechanisms and clearance rates of molluscan veligers. Biol. Bull. 157:524–535.

Tardy, J. and Dongard, S., 1993. The apical complex in the veliger of *Ruditapes philippinarum* (Adams and Reeve, 1850) Mollusca, Bivalvia, Veneridae. Comptes Rendu Acad. Sci. Ser. iii Sci. Vie 316:177–184.

Tanaka, Y., 1984. On the development of *Pecten albicans* Schroeter. Bull. Natl. Res. Inst. Aquacult. (Japan). 5:19–25 and Can. Transl. Fish. Aquat. Sci. 5321: 7p.

Tettelbach, S.T. and Rhodes, E.W., 1981. Combined effects of temperature and salinity on embryos and larvae of the northern bay scallop, *Argopecten irradians irradians*. Marine Biology 63:249–256.

Teuchert, G., 1976. Sinneseinrichtungen bei *Turbanella comuta* Remane (Gastrotricha). Zoomorphologie 83:193–207.

Thayer, G.W. and Stuart, H.H., 1974. The bay scallop makes its bed of seagrass. Mar. Fish. Rev. 36:27–30.

Torkildsen, L., Samuelsen, O.B., Lunestad, B.T. and Bergh, Ø., 2000. Minimum inhibitory concentrations of chloramphenical, florfenicol, trimethoprim/sulfadiazine and flumequine in seawater of bacteria associated with scallops (*Pecten maximus*) larvae. Aquaculture 185:1–12.

Toro, J., Sanheza, M., Paredes, L. and Canello, F., 1995. Induction of triploid embryos by heat-shock in the Chilean northern scallop *Argopecten purpuratus* Lamarck, 1819. NZ J. Mar. Freshwat. Res. 29:101–105.

Tremblay, M.J. and Sinclair, M.M., 1988. The vertical and horizontal distribution of sea scallop *(Placopecten magellanicus)* larvae in the Bay of Fundy in 1984 and 1985. J. Northw. Atl. Fish. Sci. 8:43–53.

Tremblay, M.J. and Sinclair, M.M., 1990a. Sea scallop larvae *Placopecten magellanicus* on Georges Bank: Vertical distribution in relation to water column stratification and food. Mar. Ecol. Prog. Ser. 61:1–15.

Tremblay, M.J. and Sinclair, M.M., 1990b. Diel vertical migration of sea scallop larvae *Placopecten magellanicus* in a shallow embayment. Mar. Ecol. Prog. Ser. 67:19–25.

Tremblay, M.J. and Sinclair, M.M., 1992. Planktonic sea scallop larvae (*Placopecten magellanicus*) in the Georges Bank region: Broadscale distribution in relation to physical oceanography. Can. J. Fish. Aquat. Sci. 49:1597–615.

Tremblay, M.J., Meade, L.D. and Hurley, G.V., 1987. Identification of planktonic sea scallop larvae *(Placopecten magellanicus)* (Gmelin). Can. J. Fish. Aquat. Sci. 44:1361–1365.

Tremblay, M.J., Loder, J., Werner, F., Naimie, C., Page, F. and Sinclair, M.M., 1994. Drift of sea scallop larvae *Placopecten magellanicus* on Georges Bank - a model study of the roles of mean advection, larval behavior and larval origin. Deep-Sea Res. II 41:7–49.

Uriarte, I., Farias, A. and Castilla, J.C., 2001. Effect of antibiotic treatment during larval development of the Chilean scallop *Argopecten purpuratus*. Aquacultural Engineering 25:139–147.

Ventilla, R.E., 1982. The scallop industry in Japan. Adv. Mar. Biol. 20:309–382.

Wallace, J.C., 1982. The culture of the Iceland scallop, *Chlamys islandica* (O. F. Müller). 1. Spat collection and growth during the first year. Aquaculture 26:311–320.

Waller, T.R., 1976. The development of the larval and early post-larval shell of the bay scallop, *Argopecten irradians*. Bull. Am. Malacol. Union Inc. 1976:46.

Waller, T.R., 1981. Functional morphology and development of veliger larvae of the European oyster, *Ostrea edulis* Linné. Smithson. Contrib. Zool. 328:1–70.

Waller, T.R., 1991 Evolutionary relationships among commercial scallops (Mollusca: Bivalvia: Pectinidae). In: S.E. Shumway (Ed.). Scallops: Biology, Ecology and Aquaculture. Elsevier, Amsterdam. pp. 1–55.

Waller, T.R., 1993. The evolution of *Chlamys* (Mollusca, Bivalvia, Pectinidae) in the tropical western Atlantic and eastern Pacific. Am. Malacol. Bull. 10:195–249.

Wang, W. and Xu, Z., 1997. Larval swimming and postlarval drifting behavior in the infaunal bivalve *Sinonovacula constricta*. Mar. Ecol. Prog. Ser. 148:71–81.

Wanninger, A. and Haszprunar, G., 2002a. Chiton morphogenesis: perspectives for the development and evolution of larval and adult muscle systems in molluscs. J. Morphol. 251:103–113.

Wanninger, A. and Haszprunar, G., 2002b. Muscle development in *Antalis entalis* (Mollusca, Scaphopoda) and its significance for scaphopod relationships. J. Morphol. 254:53–64.

Whyte, J., Bourne, N. and Ginther, N., 1991. Depletion of nutrient reserves during embryogenesis in the scallop *Patinopecten yessoensis* (Jay). J. Exp. Mar. Biol. Ecol. 149:67–79.

Whyte, J.N.C., Bourne, N. and Hodgson, C.A., 1987. Assessment of biochemical composition and energy reserves in larvae of the scallop *Patinopecten yessoensis*. J. Exp. Mar. Biol. Ecol. 113:113–124.

Whyte, J.N.C., Bourne, N. and Hodgson, C.A., 1990. Nutritional condition of rock scallop, *Crassodoma gigantea* (Gray), larvae fed mixed algal diets. Aquaculture 86:25–40.

Whyte, J.N.C., Bourne, N., Ginther, N.G. and Hodgson, C.A., 1992. Compositional changes in the larva to juvenile development of the scallop *Crassodoma giganteus* (Gray). J. Exp. Mar. Biol. Ecol. 163:13–29.

Wilson, J.H., 1987. Spawning of *Pecten maximus* Pectinidae and the artificial collection of juveniles in two bays in the West of Ireland. Aquaculture 61:99–112.

Wood, L. and Hargis, W.J., 1971. Transport of bivalve larvae in a tidal estuary. In: D.J. Crisp (Ed.). 4th European Marine Biology Symposium, C.U.P. pp. 29–44.

Woodhams, P.L. and Messenger, J.B., 1974. A note on the ultrastructure of the *Octopus* olfactory organ. Cell Tiss. Res. 152:253–258.

Wright, B.R., 1974. Sensory structure of the tentacles of the slug *Arion ater* (Pulmonata, Mollusca). 1. Ultrastructure of the distal epithelium, receptor cells and tentacular ganglion. Cell Tiss. Res. 151:229–244.

Wright, D.A., Roosenberg, W.H. and Castagna, M., 1984. Thermal tolerance in embryos and larvae of the bay scallop *Argopecten irradians* under simulated power plant entrainment conditions. Mar. Ecol. Prog. Ser. 14:269–273.

Wright, D.A, Kennedy, V.S., Roosenburg, W.H., Castagna, M. and Mihursky, J.A., 1983. Temperature tolerance of embryos and larvae of five bivalve species under simulated power entrainment conditions: a synthesis. Mar. Biol. 77:271–278.

Yamamoto, G. and Nishioka, C., 1943. On the development of the scallop by means of artificial fertilization. Bull. Jap. Soc. Sci. Fish. ll:219. (In Japanese)

Yan, T., Zhou, M.J., Fu, M., Wang, Y.F., Yu, R.C. and Li, J., 2001. Inhibition of egg hatching success and larvae survival of the scallop *Chlamys farreri*, associated with exposure to cells and cell fragments of the dinoflagellate *Alexandrium tamarense*. Toxicon 39:1239–1244.

Yaroslavtseva, L., Sergeeva, E. and Chan, G., 1991. The effect of short-term decrease in salinity on the adaptability of larvae of the Japanese scallop *Mizuhopecten yessoensis*. Biologiya Morya-Marine Biology 5:63–67.

122

Yi, H. and Emery, D.G., 1991. Histology and ultrastructure of the olfactory organ of the freshwater pulmonate *Helisoma trivolvis*. Cell Tiss. Res. 265:335–344.

Yoo, S.K., 1969. Food and growth of the larvae of certain important bivalves. Bull. Pusan Fish. Coll. 9:L65–87.

Yoo, S.K. and Imai, T., 1968. Food and growth of larvae of the scallop *Patinopecten yessoensis* (Jay). Bull. Pusan Fisheries Coll. 8:127–134.

Yoo, S.K. and Park, K.Y., 1979. Distribution of drifting larvae of scallop, *Patinopecten yessoensis*, in the Yeong-Il Bay. Journal Oceanol. Soc. Korea 14:54–60.

Yvin, J.C., Chevolot, L., Chevolot-Magueur, A.M. and Cochard, J.C., 1985. First isolation of jacaranone from an alga, *Delesseria sanguinea*, a metamorphosis inducer of *Pecten* larvae. J. Nat. Prod. 48:814–816.

Zakhvatkina, K.A., 1959. Larvae of bivalve mollusks of the Sevastopol region of the Black Sea. Akad. Nauk. SSSR Trudy Sevastopl'skei Biolog. Stantsii (Virginia Institute of Marine Science Translation Series No. 15: 41 p.) 11:108–152.

Zardus, J.D. and Martel, A.L., 2002. Phylum Mollusca: Bivalvia. In: C.M. Young (Ed.). Atlas of Marine Invertebrate Larvae. Academic Press, London. pp. 289–325.

Zavarzeva, E.G., 1981. A method of estimating the rate of development of mollusc embryos. Dokl. Biol. Sci. (Engl. Transl.) 266:527–528.

Xu, H., Xu, B. and Ji, W. 1991. Component of bacteria and their effects on settlement of larvae of the scallop. J. Fish. China 15:117–123.

AUTHOR'S ADDRESS

Simon M. Cragg - Institute of Marine Sciences, School of Biological Sciences, University of Portsmouth, Ferry Road, Portsmouth PO4 9LY, United Kingdom (E-mail: simon.cragg@port.ac.uk)

Scallops: Biology, Ecology and Aquaculture
S.E. Shumway and G.J. Parsons (Editors)
© 2006 Elsevier B.V. All rights reserved.

Chapter 3

Structure and Function in Scallops

Peter G. Beninger and Marcel Le Pennec

3.1 INTRODUCTION

At the beginning of the 1990's, the field of bivalve anatomy and function was characterised by hundreds of disparate papers, often based on extremely rudimentary techniques, in journals that often no longer exist. Since the publication of the first edition of this volume (Shumway 1991), two other works presenting modern compilations of bivalve anatomical and functional data have appeared, each focussing on a particular species: The Eastern Oyster *Crassostrea virginica* (Kennedy et al. 1996), and Biology of the Hard Clam (Kraeuter and Castagna 2001). A general presentation of bivalve micro-anatomy has appeared in the Microscopic Anatomy of Invertebrates series (Morse and Zardus 1997). In addition to the constitution of very important information resources, each of these works has considerably raised the profile of bivalve anatomy and function, while at the same time rendering the field accessible and comprehensible to non-specialists.

The common term 'scallop' may be used to designate members of the superfamily Pectinoidea, which includes the extant families Pectinidae, Entoliidae, Spondylidae, and Propeamussiidae (Waller 1991). Since the other families of this group either have alternate common names (the Spondylidae are also called 'thorny oysters') or are poorly known to non-specialists (the Propeamussiidae are small, very poorly-known deep-water species, while the Entoliidae are represented today by only one known species, described as 'very rare, cryptic, and surviving only in disjunct populations in the Caribbean and central and Western Pacific ocean' (Waller 1991)), we will herein restrict the term 'scallop' to members of the family Pectinidae, which are the subject of this chapter. The deep-sea *Bathypecten* genus will be excluded, because *Bathypecten vulcani* is, in fact, a propeamussiid (Dufour et al. in press).

3.1.1 An Overview of the Scallop Body

The family Pectinidae displays one of the greatest degrees of anatomical differentiation within the Bivalvia. Organs are quite distinct and easily located; although this facilitates the anatomical study of scallops, one is quickly impressed with the complexity of the body systems themselves.

Until the 1980's, much of the state of our knowledge concerning scallop anatomy had been derived from meticulous studies performed in the late nineteenth and early twentieth centuries, when observational techniques were still quite limited. Although a prodigious body of knowledge was built up by the work of Kellogg (1892, 1915), Drew (1906), and Dakin (1909, 1910a, b), it is appropriate to integrate more recent observations of scallop anatomy and function using the techniques now available. Accordingly, much of the emphasis in this chapter will be placed upon the results of more recent studies of scallop structure and function.

One of the earliest detailed descriptions of scallop general anatomy was that of Drew (1906). Other general descriptions of overall anatomy in various scallop species may be found in Kellogg (1892), Gutsell (1931), Pierce (1950), Yonge (1951), Bourne (1964), and Mottet (1979). As the structure of the shell, the adductor muscle and the larva have been treated elsewhere (Waller 1991; Chapters 2 and 4 in this work), only the postlarval soft body parts with the exception of the adductor muscle will be presented here (Fig. 3.1). Despite some inter-taxon anatomical variation, the following general overview of the scallop body may be considered representative.

The Pectinidae and other monomyarians are characterised by several important modifications of the more primitive isomyarian body form. The region corresponding to the isomyarian "anterior" is greatly reduced, such that the anterior adductor muscle is completely absent. Similarly, the isomyarian "ventral" region is shifted anteriorly, the foot being quite close to the mouth. The inadequacy of such conventional terms as anterior and ventral when describing scallop anatomy has been pointed out by Yonge (1951); Stasek (1963) has recommended that such terms not be used to refer to orientation, but rather to the relation of body parts with one another. While such an approach is technically justified, it would render the present text confusing to those less familiar with bivalve anatomy. For the sake of simplicity, the term "dorsal" will be considered to indicate those structures closest to the hinge line, while the term "anterior" will be used to designate those structures closest to the extreme point of the shell margin on the side where the mouth is located. The terms "ventral" and "posterior" will of course refer to the opposite directions, respectively, of the proceeding. The large posterior adductor muscle, situated near the centre of the shell, serves as a convenient reference point for the other body parts.

The internal shell faces are covered by a very thin, transparent mantle, which thickens and divides into distinct folds at the shell margin. The most evident of these folds is the mantle curtain, or velum, which modulates the exit of water from between the valves during valve adductions, such as the rejection of pseudofeces and the swimming response. The margin of the mantle carries a large and variable number of sensory tentacles and eyes.

The heart is contained within a thin, transparent pericardium immediately dorsal and slightly posterior to the adductor muscle. It is composed of two auricles and a single, large ventricle from which an anterior aorta extends antero-dorsally, while a posterior aorta arises from the ventricle and curves around the ventral margin of the adductor muscle. These two vessels then ramify to form the peripheral arterial system. The venous system is chiefly composed of a number of sinuses from which the blood flows through the kidneys to the gills. Blood from the gills enters the corresponding right or left auricle via an efferent branchial vessel.

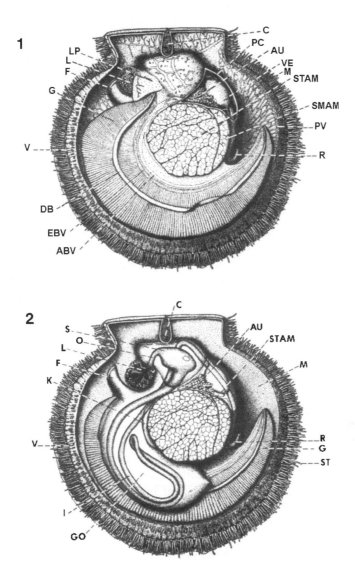

Figure 3.1. General anatomy of a large pectinid, *Placopecten magellanicus*. Figure 3.1.1. Left shell and mantle removed. Figure 3.1.2. Left shell, mantle and gill removed. Abbreviations: ABV, afferent branchial vessel; AU, auricle; C, chondrophore; DB, dorsal bend of gill filaments; EBV, efferent branchial vessel; F, foot; G, gill; GO, gonad; I, intestine; K, kidney; L, lips; LP, labial palp; M, mantle; O, oesophagus; PC, pericardium; PV, pallial vessels; R, rectum; S, stomach; ST, sensory tentacle; SMAM, smooth adductor muscle; STAM, striated adductor muscle; V, velum. After Drew (1906).

The paired kidneys are attached directly to the anterior margin of the adductor muscle. The kidneys open internally into the anterior region of the pericardial cavity, and externally into the common uro-genital pore at their ventral extremity.

The gonad of sexually mature scallops is also directly attached to the anterior margin of the adductor muscle, covering the kidneys. It curves around the ventral margin of the adductor muscle, where it is unattached. In some species, such as *Hinnites multirugosus* and *Placopecten magellanicus*, gonadal tissue may extend dorsally into the labial palps, digestive gland, or mantle from its anterior end (Yonge 1951; P.G. Beninger, unpublished personal observations). The foot and its associated byssal complex arise from the antero-dorsal extremity of the gonad, being attached to the left valve by a single pedal retractor muscle, which is variably developed in different species.

Ventral to the gonad are the two large gills, which are attached to the adductor muscle via suspensory membranes on the right and left insertions of the muscle to the shell. The relaxed gill follows the curvature of the shell margin, from the postero-dorsal level of the adductor muscle to a point just ventral to the anterior hinge margin. Each anterior gill extremity is enveloped on the right and left sides by the paired labial palps, which themselves become greatly folded to form the dorsal and ventral lip-apparatus, covering the mouth.

The oesophagus leads directly to the stomach, which is situated within the digestive gland, just ventral to the chondrophore. The intestine describes several loops in the digestive gland, which it exits posteriorly, descends ventrally and enters the gonad at its most antero-dorsal extremity. After a single elongated loop in the gonad, the intestine curves around the dorsal portion of the adductor muscle and up toward the hinge line, descends ventrally and traverses the pericardium and ventricle. The distal portion of the intestine is attached to the posterior side of the adductor muscle, becoming free in the region of the rectum and anus.

Although transparent and therefore somewhat difficult to see, the most conspicuous parts of the scallop nervous system are the two ganglionic concentrations. The parietovisceral ganglion is located on the antero-ventral surface of the adductor muscle. It innervates the adductor muscle, ventral mantle region, gills and some viscera. The fused pedal and cerebral ganglia are slightly ventral to the mouth, and are linked to the parietovisceral ganglion by a pair of cerebro-visceral connectives. These ganglia and connectives innervate the rest of the scallop body.

3.2 THE MANTLE AND ITS DERIVATIVES

The scallop mantle is a two-faced epithelial membrane, which intervenes in several key functions, such as secretion of the shell and ligament, sensory perception, and swimming response via the velum. Its role in pallial water circulation is probably minor, given its patchy cilia distribution (see below).

3.2.1 Gross Functional Anatomy

The mantle is composed of a sparsely-ciliated (Fig. 3.2) right and left lobe, which are fused dorsally along the cardinal plate and closely connected to the adductor muscle, digestive gland and pericardium. The two thin lobes separate from the visceral mass at the base of the gills, enclosing the large ventral pallial cavity. They are connected to the right and left shell valves by strands of connective tissue terminating in specialised tendon cells (Bubel 1984). At the mantle margin are found the various folds and grooves, including the numerous eyes and tentacles.

Although in the Pectinidae there are no specialised fused mantle regions to form siphons, the inhalant and exhalent currents follow definite trajectories due to the directed beating of the gill cilia. Thus, in *Pecten maximus* and *Chlamys opercularis* inhalant currents are created around the entire mantle, with the exception of a narrow postero-dorsal region where an exhalent current is evident (Dakin 1909; Ghiretti 1966). Similar observations have been made by Kellogg (1915) on *Placopecten magellanicus* and *Argopecten irradians*.

The free edges of the mantle lobes are well-developed and divided into three folds and two grooves. The terminology for these structures in the Pectinidae is derived from the studies of Drew (1906), Dakin (1909), Gutsell (1931) and summarised for bivalves in general by Petit (1978). The recent interest in bivalve functional anatomy has resulted in excellent presentations of the mantle lobes and periostracal formation in non-pectinid species (Morse and Zardus 1997; Eble 2001). Beginning with the most external formation, it is possible to identify the following (Fig. 3.3):

- an external or shell fold, subdivided into a primary shell fold, a periostracal groove, and a secondary shell fold;
- an external groove;
- a middle or sensory fold;
- an internal groove;
- an internal or velar fold;

The small shell fold contains the periostracal groove, which constitutes the glandular system responsible for secretion of the periostracum. Although Dakin (1909) states that this fold bears long, extensible tentacles in both *Pecten maximus* and *Chlamys opercularis*, none are reported for either *Argopecten irradians* (Gutsell 1931) or *Placopecten magellanicus* (Drew 1906; P.G. Beninger, personal observations). The shell fold of *P. magellanicus* does, however, bear numerous short tentacles (Moir 1977a) which respond to tactile stimulation (P.G. Beninger, unpublished observations), while long, extensible sensory tentacles are situated slightly medially to the eyes at the lateral margin of the more well-developed middle mantle fold (also called the ophthalmic or sensory fold). The eyes are considered in detail in Chapter 5 and the sensory tentacles in section 3.8.2.2 of the present chapter.

The internal fold (also called the velum, velar fold or mantle curtain) is the most conspicuous fold, consisting of a flap of tissue extending toward the pallial cavity at right angles to the shell in an undisturbed animal. In the dorsal-most anterior and posterior regions, this fold is progressively less developed toward its point of fusion at the hinge

128

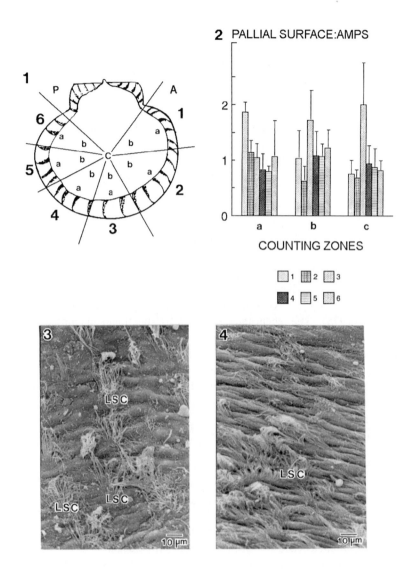

Figure 3.2. Mucocytes and cilia on the pallial surface of *Placopecten magellanicus*. Figure 3.2.1–3.2.2: Left mantle lobe divided into 6 regions within which semiquantitative mucocyte counts were effected centripetally (a → c). 1 = few mucocytes, 2 = mucocytes easily distinguished but distinct, 3 = mucocyte abundance makes individual mucocytes barely distinguishable, 4 = mucocyte abundance makes individual cells indistinguishable. Only acid mucopolysaccharide-secreting (AMPS) mucocytes were observed, and no distinct pattern (and therefore no distinct rejection pathway) was reported (Beninger and St-Jean 1997). 3.2.3–3.2.4: Sparse, randomly-located tufts of long simple cilia (LSC) on the posterior (3.2.3) and anterior (3.2.4) pallial surface of the right mantle lobe, indicating lack of distinct rejection pathway (Beninger et al. 1999).

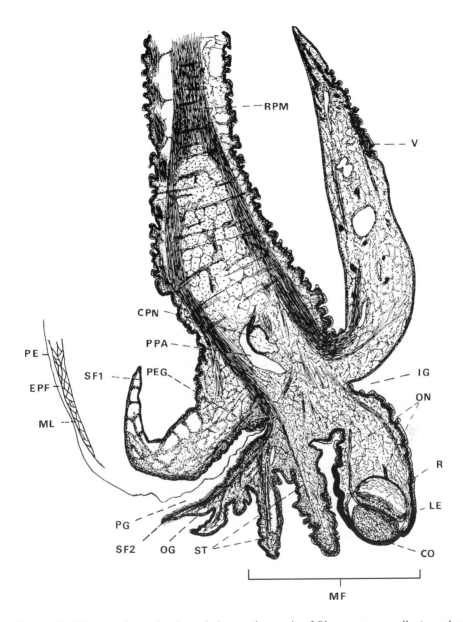

Figure 3.3. Diagram of a section through the mantle margin of *Placopecten magellanicus*, showing the secretion of the periostracum. Abbreviations: CO, cornea; CPN, circumpallial nerve; EPF, extrapallial fluid; IG, inner groove; LE, lens; MF, middle fold; ML, mineral layer; ON, optic nerves; OG, outer groove; PE, periostracum; PEG, periostracal groove; PPA, posterior pallial artery; R, retina; RPM, radial pallial muscles; SF1, primary shell fold; SF2, secondary shell fold; ST, sensory tentacles; V, velar fold. Modified after Drew (1906).

line. A row of guard tentacles is present on the inner-most margin of the velum in *Argopecten irradians* (Gutsell 1931) and in *Placopecten magellanicus* (Bourne 1964; Moir 1977a). The muscular velar fold plays an important role in water movement within the pallial cavity, notably during the swimming response, when it partially seals the pallial cavity, thereby directing jets of water which provide thrust. As the two valves close, the vela fold inward under the control of the parietovisceral ganglion and the radial pallial nerves, thus avoiding damage to the sensory structures of the mantle edge (Stephens 1978).

In the median dorsal region, the mantle is inserted into the ligament cartilage. There is no cardinal pallial crest (Ranson 1939) or pallial isthmus (Owen et al. 1953; Le Pennec 1978), and the hinge is almost totally reduced to the ligament. However, the cardinal region does show several long protuberances, which are in fact cardinal teeth or crura. The formation of the ligament and cardinal teeth is detailed in Chapters 1 and 2 of this volume.

The ligament is composed of two parts: the very prominent median internal cartilage or resilium resting upon a specialised shell formation called the resilifer, and a very fine external layer extending along the exterior dorsal shell margin. These two components of the ligament oppose an antagonistic force to that of the adductor muscle, so that in a relaxed state the two valves gape slightly, facilitating the circulation of water within the pallial cavity. Whereas in other bivalves the ligament is composed of calcium carbonate and a hydrated protein, in pectinids it is not calcified (Marsh et al. 1976). The rubbery consistency of the scallop ligament is more mechanically efficient than that of other bivalves, allowing the animal to swim and clear debris from the pallial cavity with a minimal energy cost (Trueman 1953). The ligament cartilage may be used to determine the age of individual scallops, as it often bears growth checks which are more distinct than those of the shell (Mottet 1979).

3.2.2 Microanatomy and Functions

Along the cardinal plate and in the regions where it adheres to the adductor muscle, digestive gland and pericardium, the mantle is simply composed of an internal epithelium and a thin subjacent conjunctive layer. In all other regions, the mantle is composed of three well-defined layers consisting of an internal and external epithelium separated by a thin layer of lacunar connective tissue. This middle layer is traversed by the pallial blood vessels, nerves, and muscles, which are particularly well-developed near the mantle margin.

Both the internal and the external epithelia rest upon a basal lamina composed of collagen fibres. The internal epithelium is composed of three cell types: columnar, microvilli-bearing cells, irregularly - distributed columnar cells bearing long ciliary tufts, and mucocytes. The secretions contained within the mucocytes are acid (high-viscosity) mucopolysaccharides. In contrast to those bivalves equipped with siphons and gill ventral particle grooves, neither the ciliary nor the mucocyte distributions present any pattern; this has been linked to the valve-clapping mode of pseudofeces evacuation from the pallial cavity (Beninger and St-Jean 1997; Beninger et al. 1999). It is likely that the radially ventral particle transport previously depicted for the mantle surface of *Pecten maximus* (Orton 1912) is actually the result of the ventralward beat of the frontal cilia of the gill ordinary filaments as viewed in a half-shell preparation, since the patchy distribution of mantle cilia would not be effective in particle transport. The high viscosity of the mucocyte secretions, in the absence of other mucociliary transport adaptations, suggests that they facilitate the passage of water over the mantle surface, as has been suggested for other pallial surfaces in bivalves (Beninger et al. 1997a; Beninger and St-Jean 1997; Beninger and Dufour 2000; Dufour and Beninger 2001). The mantle cilia of *Placopecten magellanicus* probably do not contribute much to the flow of water in the pallial cavity, due to their sparse distribution (Beninger et al. 1999). Additional studies on other pectinid species would be most welcome.

The large pallial surface area of the mantle is greatly multiplied by the apical microvilli; together with the finely-branched blood vessels of the connective layer, this indicates that the mantle may be an important site for the exchange of gases and other dissolved molecules. Wandering amoebocytes and mantle epidermal cells are capable of absorbing particulate matter from the pallial cavity of several non-pectinid bivalves, and the ultrastuctural characteristics of these cells indicate that they are capable of digesting such material (Bevelander and Nakahara 1966; Nakahara and Bevelander 1967; Machin 1977). As for the other surfaces of the pallial cavity, it is likely that direct exchange with the external medium may at least supplement the metabolic needs of the numerous epithelial cells.

We are unaware of data concerning the cell types of the external (shell side) epithelium in pectinids. In other bivalves, tall columnar cells presenting ultrastructural characteristics typical of translocation (e.g., apical microvilli and thick, convoluted basal lamina) have been observed (Neff 1972), and up to seven different cell types have been reported. Caecal cells extend mitochondria-rich processes into the shell itself, terminating on the outer shell surface (Reindel and Haszprunar 1996). This is clearly a very complex surface, which merits much further investigation.

Shell formation in bivalves generally follows two major steps: the cellular processes of ion transport and secretion of the organic matrix proteins; and the physico-chemical processes whereby calcium carbonate crystals are nucleated, oriented and assembled due to the presence of the organic matrix (for details, see Wilbur and Saleuddin 1983).

The detailed process of shell formation, and ontogenetic changes in shell microstructure and minerology, have been intensively studied by a number of workers in selected bivalve species (see reviews by Wilbur 1972, 1985; Crenshaw 1980; Wilbur and Saleuddin 1983; Watabe 1984), and for pectinids in particular (see Chapter 1 of this volume for review and references).

The shell is covered externally by a thin, fibrous organic layer called the periostracum, which may vary in thickness depending on the amount of mechanical abrasion and/or damage due to epizoonts. The periostracum serves as a matrix for the deposition of the calcium carbonate crystals of the shell (Taylor and Kennedy 1969; Saleuddin and Petit 1983). The formation of the periostracum has been studied in detail in the freshwater bivalve *Amblema plicata perplicata* (Petit 1978) as well as in several marine species (Saleuddin and Petit 1983; Petit et al. 1988). Glandular cells located at the base of the periostracal groove secrete the periostracal components, including quinone-tanned protein fibres (Petit 1978), polysaccharides and lipids (Brown 1952).

The outer edge of the shell fold contains numerous depressions, such that an extrapallial fluid is contained between the shell fold and the newly-formed periostracum (Fig. 3.3). Calcification is initiated at the small spicules of periostracum within this fluid-filled space, and progressively thickens (Wilbur and Saleuddin 1983). Extrapallial fluid may also be found in sub-periostracal depressions distributed within the shell. Mantle enzyme secretion has been studied in a number of bivalve species (Brown 1952; Waite and Wilbur 1976; Waite and Andersen 1978, 1980). Among the enzymes discovered, acid phosphatase and phenoloxidase are of particular interest. Acid phosphatase intervenes in the structural modification of the inner face of the periostracal layer (Chan and Saleuddin 1974), while phenoloxidase is involved in the tanning of the periostracal proteins (Bubel 1973a, b).

Between the outer (shell side) and inner (pallial side) epithelia, several cell types have been reported in non-pectinid bivalves: amoebocytes, porocytes/rhogocytes, nerve cells, haemocytes, gliointerstitial cells, as well as musculo-connective tissues (Reindl and Haszprunar 1996). Despite its deceptively simple gross appearance, the mantle is one of the most complex, multi-tasked organs in the Bivalvia.

3.3 PALLIAL ORGANS AND PARTICLE PROCESSING

The pectinid pallial organs are comprised of the mantle, gill, labial palps and lips. Studies to date have shown that, in pectinids, all of these structures save the mantle are involved in particle processing (Fig. 3.4), as outlined below. It is impossible to understand the mechanisms of particle processing without a good grasp of the underlying anatomy and micro-anatomy.

3.3.1 Gills

Much of the groundwork for bivalve gill anatomy was laid by Kellogg (1892, 1915), Janssens (1893), Ridewood (1903), Drew (1906), Setna (1930), and Atkins (1936; 1937a, b, c; 1938a, b, c; 1943). More recent studies on scallop gills in particular include the work of Owen and McCrae (1976), Morse et al. (1982), Reed-Miller and Greenberg (1982), Ciocco (1985), Beninger et al. (1988), Le Pennec et al. (1988), Beninger and Dufour (2000), and Dufour and Beninger (2001). Prior to the 1990's, function was largely inferred either from the possibilities afforded by the gill anatomy, by indirect means such as particle retention and clearance studies, or by observation of particle movement on opened animals. The resulting state of knowledge concerning particle processing mechanisms was eminently conjectural, and even an attempt to indicate the sizeable gaps and paradoxes concerning this subject in scallops was itself, in hindsight, somewhat naïve (although some of the conjectures were later borne out - Beninger 1991). From this point forward, an intense research effort was brought to bear on the problem, supplementing the existing techniques with tools specifically developed or adapted for the purpose: video endoscopy, mucocyte and cilia mapping, and confocal laser microscopy. The relevant

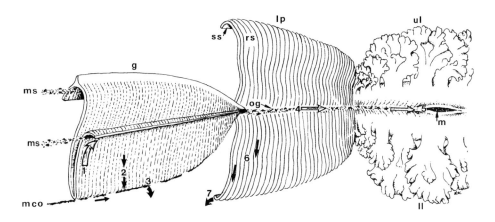

Figure 3.4. General organisation and function of the pallial organs involved in particle processing: gill (g), smooth (ss) or aboral surface, and ridged or oral sufrace of labial palps (lp), and upper (ul) and lower (ll) lips. Retained particles destined for further processing on the peribuccal organs are directed dorsally in the principal filament troughs (1), where they are incorporated into the anteriorward flow of mucus-particle slurry (ms) in the ciliated tracts of the gill arch and dorsal bend. Particles destined for rejection are directed ventrally in a viscous mucus on the ordinary filament plicae (2), where they are incorporated into mucus cords and masses of various length (mco), which are periodically expelled from the gill and pallial cavity (3) by valve adductions. Particles arriving from the dorsal ciliated tracts onto the labial palps (4) may continue along the oral groove (og), in a reduced-viscosity mucus, for ingestion (5) by the mouth (m). Particles rejected on the labial palps travel laterally in a viscous mucus within the troughs to the ventral margin (6), where they join pseudofeces moving posteriorly and are finally expelled from the postero-ventral tip (7). Modified from Beninger et al. (1990a).

studies concerning pectinids are, to date: Ward et al. 1991, 1993; Beninger et al. 1992, 1993, 1999; Beninger and St-Jean 1997a, b; Beninger and Dufour 2000; Dufour and Beninger 2001; Beninger et al. 2004). The following account of gill structure and function incorporates the information from all of these sources.

Adult scallops possess a heterorhabdic plicate gill, which means that the W-shaped left and right gills are composed of a series of two different types of filaments, suspended from the gill axis in a corrugated or plicate fashion. The principal filaments are situated in the troughs of the plicae, separated from each other by a variable number (11–20) of ordinary filaments (OF). The ascending branches of the filaments are approximately two-thirds the length of the descending branches (Figs. 3.5, 3.6.1).

The single reported exception to the heterorhabdic condition was the pectinid *Hemipecten forbesianus*, which was said to possess a homorhabdic gill (Yonge 1981). However, re-examination of the specimens studied by Yonge show unequivocally that the gill is heterorhabdic (Beninger, unpublished observations). There is thus no known exception to the rule of heterorhabdy in the pectinids.

134

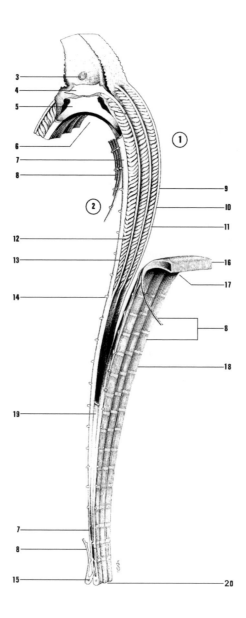

Figure 3.5. General organisation and macro-anatomy of the gill. 1 - suprabranchial chamber, 2 - infrabranchial chamber, 3 - branchial nerve, 4 - afferent branchial vessel, 5 - efferent branchial vessel, 6 - gill arch, 7 - lateral wall of principal filament, 8 - ordinary filaments, 9 - descending branch of principal filament, 10 - dorsal expansion, 11 - afferent vessel, 12 - interconnecting vessel, 13 - efferent vessel in principal filament, 14 - ciliated spur (cilifer), 15 - ciliated disc, 16 - dorsal bend, 17 - ciliated tract, 18 - ascending filaments, 19 - interlamellar junction, 20 - ventral bend.

3.3.1.1 Gill axis and arch

A thin suspensory membrane joins the gill axis to the adductor muscle. Within the gill axis is situated, in a dorso-ventrally aligned sequence, the branchial nerve, the afferent branchial vessel, and the efferent branchial vessel (Figs. 3.5, 3.35.1). On either side of the gill axis is situated the eversible osphradial ridge (Sec. 3.8.2.4). Relatively abundant, but distinct ciliary tufts are found from the dorsal extremity of the gill axis to the osphradial ridge; below this feature, the axis is abundantly covered with simple cilia.

The gill arch comprises the fused and staggered proximal extremities of the principal and ordinary filaments. Here the afferent and efferent blood vessels of the principal filaments join with their counterparts in the gill axis and arch (Fig. 3.5). The ventral surface of the arch forms a ciliated groove which collects and transports particles arriving in a mucus slurry from the principal filaments (Fig. 3.4); the arch itself contains very few mucocytes (Beninger et al. 1992, 1993).

3.3.1.2 Principal filaments and dorsal expansion

Principal filaments differentiate well after metamorphosis, at a size of about 4 mm in *Pecten maximus*, and 3.3–5.0 mm in *Placopecten magellanicus* (Veniot et al. 2003). This represents a change in both gill structure and function, and therefore in particle processing mechanisms. It may thus be a critical stage, and potential indirect cause of mortality, in the early life of pectinids (Beninger et al. 1994).

The proximal third of the abfrontal surface of the principal filament presents a rather complex structure historically termed the 'dorsal respiratory expansion' (Setna 1930). No studies have yet been performed to evaluate its contribution to gas exchange, or any other role. We will refer to it here simply as the *dorsal expansion*. The dorsal expansion consists of an abfrontal afferent vessel, an efferent vessel contained within the wall of the PF, and a number of variously-shaped interconnecting vessels (Figs. 3.5, 3.6.1–3.6.3). The afferent vessel is covered with patchy but abundant simple cilia, whereas the efferent and interconnecting vessels bear only tufts of simple long cilia (Beninger et al. 1988; Dufour and Beninger 2001). The convoluted basal lamina of the shallow, one-cell-thick layer of the interconnecting vessels, together with their extensive, ramified apical microvilli (Fig. 3.6.4) strongly indicate a specialised role in translocation of dissolved substances between the external medium and the gill (Le Pennec et al. 1988). In addition to evaluating the extent of gas exchange, future studies should also examine dissolved organic substances, which are actively taken up by the mussel gill (Manahan et al. 1982; Wright et al. 1984; Wright 1987).

The principal filament displays a dynamic anatomical organisation, due to the periodic contractions of portions of its lateral walls - the 'concertina' response (Kellogg 1915; Setna 1930; Owen and McCrae 1976; Beninger et al. 1988, 1992). In the resting state, the filament comprises a ciliated frontal surface within a trough formed by the lateral walls (Fig. 3.6.3). The margins of the lateral walls bear ciliated spurs or cilifers (Kellogg 1892; Reed-Miller and Greenberg 1982), which are directed out and away from the PF (Figs. 3.5, 3.6.1–3.6.3). The inner surface of the cilifer is ciliated, and

136

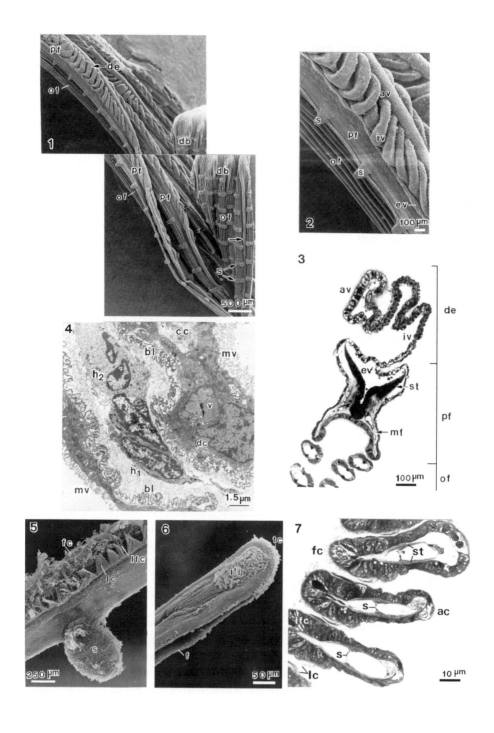

Figure 3.6. General organisation and microanatomy of the gills of *Placopecten magellanicus*. Figure 3.6.1. Low power scanning electron micrograph showing general organisation of principal (pf) and ordinary (of) filaments. Note rows of ciliated spurs (s) on abfrontal surface of ordinary filaments, responsible for structural integrity of plicae. de, dorsal expansion; db, dorsal bend; Arrow indicates abfrontal surface of ascending branch of principal filament. Ascending principal filaments are in contracted state. Figure 3.6.2. Scanning electron micrograph detail of dorsal expansion of principal filament (pf). av, afferent vessel; ev, efferent vessel; iv, interconnecting vessels; of, ordinary filaments; s, ciliated spurs. Figure 3.6.3. Thin-section light micrograph of a principal filament (pf) and dorsal expansion (de). Abbreviations: av, afferent vessel; ev, efferent vessel; iv, interconnecting vessel; of, ordinary filaments; st, supporting structures; mf, muscle fibres. Figure 3.6.4. Transmission electron micrograph of a longitudinal section of the dorsal expansion interconnecting vessels. Cells with dark nuclei and cytoplasm (dc), capable of containing large vacuoles (v). Cells with clear nuclei and cytoplasm (cc). Both haemocyte types I (h_1) and II (h_2) are present in the narrow lumen (see section 3.5.3.2). Note the apical microvilli (mv) and greatly-indented basal lamina (bl). Figure 3.6.5. Scanning electron micrograph detail of an ordinary filament. lc, lateral cilia; lfc, latero-frontal cilia; fc, frontal cilia; s, ciliated spur. Figure 3.6.6. Scanning electron micrograph of the interfilamentar junction (ifj) at ventral bend of an ordinary filament. Another ordinary filament (f) is attached behind. Note ciliated disc of interfilamentar junction, terminal cilia (tc) at ventral extremity, and absence of a particle groove. Figure 3.6.7. Thin section light micrograph of ordinary filaments. Rare abfrontal cilia (ac), lateral cilia (lc), latero-frontal (lfc) and frontal cilia (fc); supporting structures (st) and septum (s). Modified after Beninger et al. (1988) and Le Pennec et al. (1988).

interlocks with the cilia of similar spurs situated on the abfrontal surfaces of the ordinary filaments. The spurs of the ordinary filaments are ciliated on both sides, allowing adjacent filaments to adhere to each other at intervals along their lengths, thus preserving the structural integrity of the gill (Fig. 3.6.1). It appears that the orientation of the PF cilifers, which are ciliated on their inner side only, is responsible for the normal plicate arrangement of the gill filaments (Beninger et al. 1988). Contractions of the principal filament lateral walls modify the orientation of the cilifers, producing the concertina response, and is one of the mechanisms used to clear the gill and thus reduce ingestion volume when the stomach is full (Beninger et al. 1992, and section 3.3.1.5).

On the frontal surface of *Placopecten magellanicus*, the principal filament mucocytes contain a mixture of acid and neutral mucopolysaccharide secretions (Fig. 3.7), which would produce a mucus of relatively low viscosity, typical of enclosed or semi-enclosed transporting surfaces, which lead directly to other such surfaces, in bivalves (Beninger et al. 1993; Beninger and St-Jean 1997b).

3.3.1.3 Ordinary filament

The ordinary filaments have no dorsal expansions or lateral walls, and possess frontal, latero-frontal, and lateral ciliary tracts on their frontal surfaces (Figs. 3.6.5, 3.6.7). Contrary to many other bivalves, the scallop latero-frontal tract consists of only a single

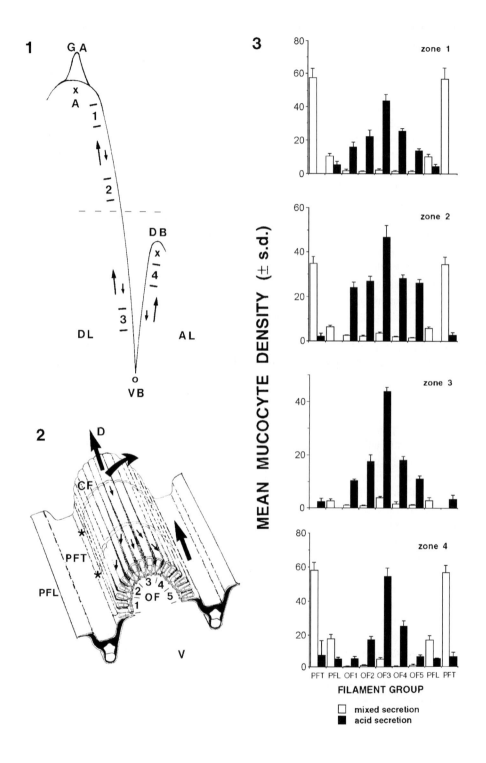

Figure 3.7. *Placopecten magellanicus* gill. Relationship between particle trajectories, filament types, dorsal (acceptance) and ventral (rejection) tracts, and mucocyte types and distributions. Figure 3.7.1. Schematic drawing of gill viewed along antero-posterior axis. Numbers designate mucocyte counting zones of Fig. 3.7.3, large arrows show particle movement on principal filaments (acceptance), small arrows show particle movement on ordinary filament plicae (rejection), and dashed line represents plane of section for Fig. 3.7.2. A, gill arch (ciliated tract); AL, ascending lamella; DB, dorsal bend (ciliated tract); DL, descending lamella; GA, gill axis; VB, ventral bend (ciliated tract); x, anteriorward direction of particles in dorsal ciliated tracts (A+DB = acceptance) perpendicular to plane of drawing; o, anteriorward direction of particles at ventral ciliated tract (VB; rejection) perpendicular to plane of drawing. The particle-mucus masses at the ventral bend are periodically dislodged by valve-clap movements. Figure 3.7.2. Schematic stereodiagram of descending lamella, showing filament groups of Figure 3.7.3 used for counting mucocytes. Large arrows indicate dominant currents and trajectories of particles destined for further processing on the labial palps (acceptance), small arrows indicate movement of particles destined for rejection beneath the dominant currents. Asterisks show dorso-ventral boundaries for a given count, corresponding to rows of cilifers (CF) visible through ordinary filaments. D, dorsal; OF 1–5, ordinary filament groups constituting a plica; PFL, principal filament lateral wall; PFT, principal filament trough; V, ventral. Figure 3.7.3. Mean mucocyte densities (± standard deviation) for each of the counting zones illustrated in Fig. 3.7.1 and each of the filament groups illustrated in Fig. 3.7.2. OF 1–5, ordinary filament groups 1–5; PFL, principal filament lateral wall; PFT, principal filament trough. Note association of acid mucopolysaccharide - secreting mucocytes with rejection function of ordinary filaments, mixed mucopolysaccharide -secreting mucocytes associated with transport of particles in semi-enclosed acceptance tracts of principal filaments. From Beninger et al. 1993.

row of cilia, which have been termed the pro-laterofrontal cilia (Owen and McCrae 1976). Rare cilia on the abfrontal surfaces are considered vestigial (Beninger et al. 1988; Beninger and Dufour 2000; Fig. 3.8), and it is interesting to contrast this with the dense abfrontal ciliation of the principal filaments. The presence of abundant cilia and mucocytes on the PF abfrontal surface (Fig. 3.8) has been related to the tardy evolutionary development of the PF's compared to the OF's; it has also been suggested that the mucocytes may play a role in preserving the structural integrity of the gill during valve adductions (Beninger and Dufour 2000).

The ordinary and principal filaments are organically joined only in the region of the dorsal bend (Morse et al. 1982; Le Pennec et al. 1988), which itself adheres to the mantle by means of interlocking cilia (Beninger et al. 1988). The ciliary connection to the mantle is easily broken when the gill retracts prior to valve adduction, or during a shunt response (Fig 3.9).

The ventral bend of each ordinary and principal filament bears a terminal ciliary tuft; the ventral bends are aligned via ciliary interfilamental junctions, forming a ciliated tract at the ventral bend, although there is no specialised trough (Fig. 3.6.6, 3.9.3). In addition, the ventral bends are imperfectly aligned along the dorso-ventral axis, such that this ciliated tract is in fact staggered (Beninger et al. 1988), probably aiding in the detachment of mucus-particle masses destined for rejection at this extremity.

140

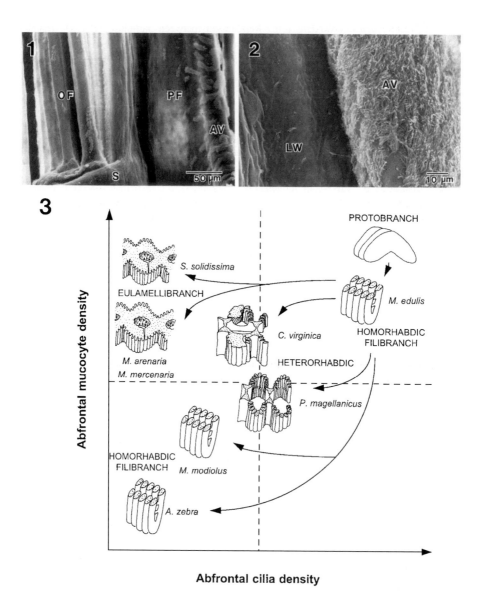

Figure 3.8. Cilia and mucocyte distributions on abfrontal surface of *Placopecten magellanicus* gill.
Figure 3.8.1. SEM of abfrontal surface, showing extremely sparse ciliation of ordinary filaments
(OF), much greater cilation of principal filament (PF) afferent vessel (AV). S, ciliated spur or
cilifer. Figure 3.8.2. SEM detail of abfrontal surface of principal filament. Note dense ciliation of
afferent vessel (AV), extremely sparse ciliation of lateral wall (LW). Figure 3.8.3. Relative
position of *P. magellanicus* gill compared to other major gill types with respect to abfrontal cilia and
mucocyte density. Note general tendency toward reduction of cilia on all ordinary filaments in more
recent gill types. From Beninger and Dufour (2000) and Dufour and Beninger (2001).

The gill filaments gradually decrease in length toward the anterior and posterior extremities. In the anteriormost region of the gill this shortening of the filaments results in the convergence of the dorsal feeding tracts with the oral groove at the base of each pair of labial palps (Fig. 3.4).

3.3.1.4 Haemolymph circulation in the gill

The gill filaments are essentially hollow tubes within which the haemolymph circulates. The internal walls of the filaments are strengthened by collagenous supporting structures (Fig. 3.6.7), which appear to be more numerous in the region of the ventral bend (Le Pennec et al. 1988). Blood enters the afferent vessel of the dorsal expansion, flows down the principal filament lumen and enters the organic junction between filaments in the dorsal bend. It then flows down the ordinary and principal filament lumina and rises dorsally to join the efferent vessel of the principal filament, which empties into the efferent branchial vessel (Morse et al. 1982). The lack of a distinct anatomical separation between afferent and efferent blood in the lumina of the filaments indicates that mixing of haemolymph probably occurs.

3.3.1.5 Particle processing on the gill

Owen (1978) originally proposed a non-'filter'-feeding mode of particle capture for several bivalve families, including the Pectinidae. In these species, the plical arrangement of the relaxed gill results in particle-laden water currents being directed toward the densely-ciliated troughs of the principal filaments and dorsally. Detailed *in vivo* endoscopic observations of the gills of *Placopecten magellanicus* and *Pecten maximus* support this proposed mechanism (Beninger et al. 1992; Beninger et al. 2004), but the microscopic events are not yet known. It is clear that the principal filaments send particles destined for secondary processing by the peribuccal organs dorsally to the gill arch and dorsal bend. A reduced-viscosity mucus accompanies these particles (Beninger et al. 1992; Fig. 3.7).

The ordinary filament plicae possess ventralward-beating frontal cilia, which direct mucus-particle masses to the ventral bend; this is probably accomplished by the phylogenetically ubiquitous mechanism of mucociliary transport, as has been recently documented *in vitro* for *Mytilus edulis* (Beninger et al. 1997b). These masses are characterised by a high-viscosity mucus, in keeping with the exposed surface and predominantly counter-current water flow (Beninger et al. 1992, 1993; Beninger and St-Jean 1997b). Most of these masses are rejected from the ventral bend, and exit the pallial cavity via valve adductions (Kellogg 1915; Owen and McCrae 1976; Beninger et al. 1992, 1999; Fig. 3.9).

3.3.1.6 Particle selection at the gill

As explained above, particles sent dorsally in the principal filament tracts are more likely to be ingested, whereas those sent ventrally are more likely to be rejected. It is

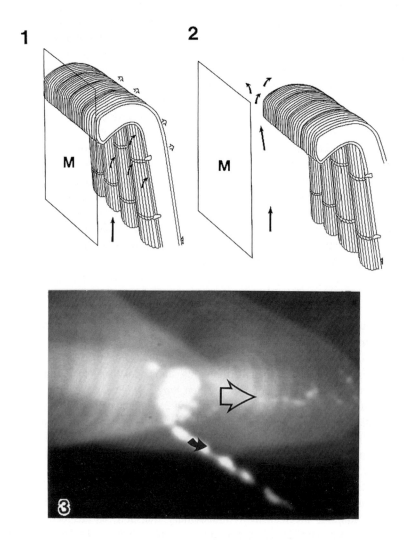

Figure 3.9. Ingestion volume regulation and pseudofeces formation on the *P. magellanicus* gill. Figure 3.9.1. Diagram showing dorsal bend in normal feeding position applied to the mantle surface (M). Solid arrows show direction of particle-laden water currents in infrabranchial cavity, open arrows show passage of particle-depleted water through the interfilamentar spaces. Figure 3.9.2. Dorsal bend detached from mantle surface, allowing particle-laden water to exit directly into the suprabranchial cavity. Figure 3.9.3. Video endoscopic micrograph showing mucus-particle mass detaching from ventral bend (solid arrow). Open arrow indicates anteriorward direction of pseudofeces movement along the ventral bend. From Beninger et al. (1992).

tempting to invoke a selective process for such a system, as has been suggested for the heterorhabdic gill in general (Atkins 1937a; Beninger and St-Jean 1997a), and the pseudolamellibranch oyster gill in particular (Ward et al. 1998); this would furnish a satisfying explanation of the heterorhabdic condition.   This question was settled by presenting known mixtures of intact and empty diatoms (*Coscinodiscus perforatus*) and *in vivo* sampling of the dorsal and ventral gill tracts, as well as palp pseudofeces, combined with microscopic determinations of particle proportions at each sampling site (Beninger et al. 2004).   Qualitative particle selection was indeed demonstrated on the gill, and a second level of particle selection was also found at the labial palps.   The organic casing and associated organic molecules of this diatom appear to constitute a key quality cue for *Pecten maximus* (Beninger and Decottignies, in press).

In contrast to the oyster gill, the width of the scallop principal filament trough (approx. 200 µm in *Placopecten magellanicus* and *Pecten maximus*), as well as its relative plasticity, allows the admission of most suspended particles (no doubt accounting for the presence of some very large particles in the stomachs of scallops (Mikulich and Tsikhon-Lukanina 1981; Shumway et al. 1987)).   The upper size constraint on the qualitative selection of particles at the scallop gill is thus much larger than that of the other major heterorhabdic gill type, the pseudolamellibranch gill of the oyster (Cognie et al. 2003).   At this point, however, the detectors, effectors, and microscopic sequence of events involved in qualitative selection are totally unknown.

3.3.1.7 Particle retention lower size limit

The lower size limit for the efficient retention of natural particles in scallops is approximately 5 µm.   Clearance studies indicate that pectinids, which possess only a single row of simple latero-frontal cilia, have a low retention efficiency for particles less than 5–7 µm in diameter, in contrast to the majority of bivalves, which possess well-developed, compound latero-frontal cilia called *cirri* (Møhlenberg and Riisgard 1978; Riisgard 1988).   It is difficult to make direct comparisons, however, since the vast majority of bivalves possessing cirri also have a homorhabdic gill.   The oyster (Fam. Ostreidae) is an interesting oddity in the bivalve world.   Kellogg (1892) called it a 'very degenerate form', which however possesses the most complex of gill types, with both latero-frontal cirri (albeit reduced in size) on its ordinary filaments, a heterorhabdic gill, and a ventral particle groove.   Much is still to be learned concerning gill function in this unique group.

3.3.1.8 Ingestion volume regulation on the gill

Five different mechanisms of ingestion volume regulation have been identified in the functioning scallop gill (Beninger et al. 1992): rejection of particles from the principal filament troughs onto the ordinary filament plicae, arrest of the anteriorward dorsal feeding currents, reduction of input from the principal filaments to the dorsal tracts, detachment of the dorsal bends from the mantle, allowing a flow-through shunt (Figs. 3.9.1, 3.9.2), and the concertina response described above.   The variety of such mechanisms indicate that this is an important process in scallops, particularly under high seston concentrations, and also that the scallop is probably able to fine-tune ingestion volume using a combination of these mechanisms.

3.3.2 Labial Palps and Lips

In addition to the paired labial palps found in all bivalves, pectinids possess a complex, ramified pair of lips covering the mouth. While considerable progress has been made in the understanding of their roles in particle processing, the extreme sensitivity of the scallop peribuccal structures to any local mechanical disturbances has rendered direct observation of particle processing impossible to date.

3.3.2.1 Labial palps

The labial palps of scallops are similar to those of the majority of bivalves, consisting of a right and left pair of tissue flaps, into which the anterior ends of the gills are inserted (Fig. 3.4). The detailed structure and ultrastructure of scallop labial palps has been reported for only two species, *Placopecten magellanicus* and *Chlamys varia* (Beninger et al. 1990 a). The two surfaces facing the gill consist of a ciliated epithelium organised into a series of troughs and ridges oriented at right angles to the gill axis (Figs. 3.4, 3.10.3). Each ridge consists of an anterior fold, a crest, and a posterior fold (Figs. 3.10.2, 3.11.3). At the base of the inner surfaces is a ciliated oral groove leading anteriorly to the mouth (Figs. 3.4, 3.11.1). The "smooth" outer or aboral surfaces of the palps possess tufts of cilia, which appear too sparse to be involved in mucociliary transport. The aboral surface is rich in mucocytes, however, probably facilitating the passage of pallial currents (Fig. 3.10.1). Between the inner and outer epithelia is a haemolymph sinus, within which numerous haemocytes may be observed, traversed by lacunar muscular-connective tissue (Figs. 3.10.1, 3.10.2).

The basal lamina of the labial palp oral surface is thick, complex, folded, and presents numerous indentations (Fig. 3.10.4). Together with the highly-branched microvilli and the observation of lysosomes in the apical region, these observations suggest that the palp oral surface is involved in absorption, although this requires much further study (Beninger et al. 1990a).

3.3.2.2 Particle processing on the labial palps

Pseudofeces are voided from both the labial palps and gill in the same manner, i.e., by valve adduction; it is thus difficult to determine the exact origin of the various pseudofeces masses. *In vivo* sampling (as in Beninger et al. 2004) allows us to at least sample pseudofeces observed to be rejected by the postero-ventral palp extremity (see below), so it is possible to affirm that the palps constitute a second site of qualitative particle selection. We do not yet know what the relative contributions of the gills and labial palps are to ingestion volume regulation and qualitative selection, and this is likely to be a very complex matter of investigation. Both of these processes involve rejection and acceptance, however, and mucocyte mapping data shed some light on the underlying mechanisms of each.

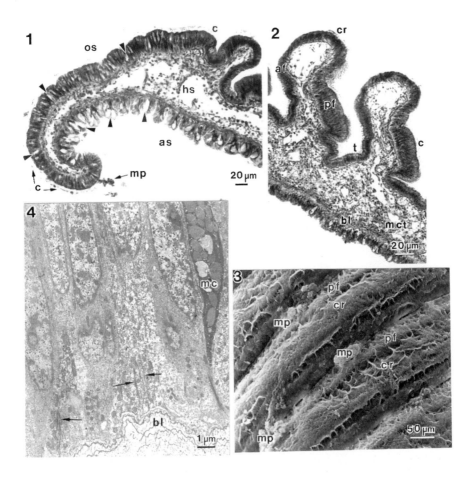

Figure 3.10. Labial palps of *Placopecten magellanicus*. Figure 3.10.1. Transverse histological section of palp margin. Note flattening of crests at the approach to the flat margin, and mucus-particle mass of pseudofeces (mp) at margin edge. Arrowheads indicate numerous narrow mucocytes on oral surface (os), goblet-cell type mucocytes on aboral surface (as). c, cilia; hs, haemolymphatic sinus. Modified Masson trichrome. Figure 3.10.2. Transverse histological section in mid-region of palp, showing structure of ridges and troughs. af, anterior fold; bl: basal lamina; c: cilia; cr, crest; mct, lacunar muscular-connective-tissue; pf, posterior fold; t, trough (rejection tract). Modified Masson trichrome. Figure 3.10.3. SEM of densely-ciliated palp oral surface, mid-region. cr, crest; mp, mucus-particle masses; pf, posterior fold. Figure 3.10.4. TEM detail of basal region of oral surface epithelium, showing part of elongated mucocyte (mc); ciliated and non-ciliated epithelial cells have indented basal membranes (arrowed); all cells rest upon multi-layered, irregular basal lamina (bl). From Beninger et al. (1990a).

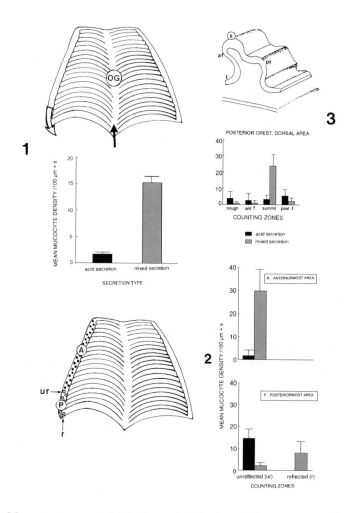

Figure 3.11. Mucocyte types and distribution on labial palps of *Placopecten magellanicus*. Figure 3.11.1. Labial palp mucocyte types and density throughout the oral groove (OG, means + standard deviations). Note dominance of lower-viscosity mixed secretions in this semi-enclosed surface. Bold arrow denotes arrival of mucus slurry from gill arch, hollow arrow shows rejection pathway from ventral margin. Figure 3.11.2. Labial palp mucocyte types and densities (means + standard deviations) on anterior (A) and posterior (P) ventral margin, on unreflected surface (UR) and reflected surface (R). Note absence of acid mucopolysaccharide-secreting mucocytes on reflected surface, facilitating discharge of pseudofeces. Figure 3.11.3. Mucocyte types and densities (means + standard deviations) on representative crest of palp oral surface (postero-dorsal region). Note dominance of acid mucopolysaccharides in trough (t), which is a semi-enclosed surface leading directly to the exposed surface of the palp ventral margin. Note also dominance of reduced-viscosity mucopolysaccharide-secreting mucocytes on crest summit (s), allowing extraction of particles from mucus for sorting. af, anterior fold; pf, posterior fold. From Beninger and St-Jean (1997a).

Particles destined for ingestion are accompanied in the oral groove by a medium-viscosity mucopolysaccharide (MPS), in keeping with the semi-enclosed nature of the transporting surfaces (Beninger and St-Jean 1997a, b; Fig. 3.11.1). Particles entering the palp troughs of several non-pectinid species have been observed via endoscopy to be destined for rejection, and this agrees with observations performed on dissected, live specimens of *Placopecten magellanicus* (Beninger and St-Jean 1997a; Beninger et al. 1997a). Particles destined for rejection are accompanied by acid (viscous) MPS in the palp troughs, to the lateral (ventral) margins which are also rich in acid MPS (Beninger and St-Jean 1997a; Figs. 3.11.2, 3.11.3), conforming to the rule of mucociliary transport on surfaces which are themselves exposed and subjected to a countercurrent water flow, or on surfaces which lead directly to same (Beninger and St-Jean 1997b). At the posterior-most tip of the palp ventral margin, the mucocyte viscosity again decreases; this probably assists in the release of the pseudofeces; Fig. 3.10.1), following which they are expelled from the pallial cavity by valve adduction (Beninger et al. 1999).

3.3.2.3 Lips

Despite their strategic position and visibly complex anatomy, the lips of pectinids have received relatively little attention. Drew (1906) described them briefly as ramifications of the labial palps, while Dakin (1909) also gave a summary account of these structures, suggesting that the ramified margins allow the lips to form a hood over the oral groove. Some observations of the lips of *Pecten maximus* were made by Gilmour (1964, 1974), although photomicrographs of these observations were largely lacking. Detailed structural and ultrastructural observations are only available for *Placopecten magellanicus* and *Chlamys varia* (Beninger et al. 1990b).

The lips arise as ramifications of the anterior margins of the paired labial palps (Fig. 3.4). Their dendritic configuration completely covers the mouth; upon separation they may be seen to comprise an upper and a lower lip (Fig. 3.4, 3.12.1). The cauliflower-like branches possess two distinct surfaces: a ciliated oral epithelium, and a non-ciliated aboral epithelium (Figs. 3.12.2–3.12.4). Both of these epithelia terminate in a covering of apical microvilli. Whereas the mucus secretion of the aboral surface probably facilitates the passage of pallial currents over this complex surface, the types and distribution of mucocytes on the oral surface presents a very definite gradient (Fig. 3.12.1), with a large number of acid-dominant MPS-secreting mucocytes close to the mouth, and a smaller number of acid and acid-dominant MPS - secreting mucocytes in the mid-region of the lips. It has been suggested that these characteristics function to prevent the removal of food particles destined for ingestion when the scallop initiates its frequent valve adductions for pseudofeces clearance (Beninger and St-Jean 1997b).

The internal structure of the lips consists of a haemolymphatic sinus containing numerous haemocytes, traversed by a lacunar muscular - connective tissue. Muscle fibres attached to the basal lamina are responsible for the rapid contraction observed in response to the slightest mechanical disturbance in the vicinity of the lips (Fig. 3.13.2). It may be presumed that the extension of the lips is accomplished via a forced flow of haemolymph into the lacunar connective tissue.

148

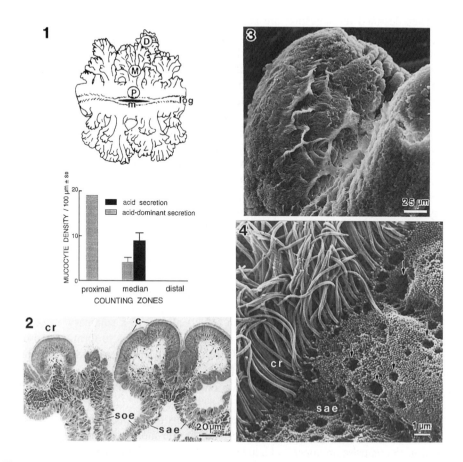

Figure 3.12. Lips of *Placopecten magellanicus*. Figure 3.12.1. Mucocyte types and distribution on the lips of *Placopecten magellanicus*. Mean counts ± standard deviation for epithelial mucocytes in three counting zones: D, distal; M, median; P, proximal to mouth (m). Note dominance of acid and acid-dominant mucopolysaccharide-secreting mucocytes in median region, facilitating trapping of particles dislodged from oral groove (see text). Standard deviation for proximal zone negligible. log, left oral groove. Figure 3.12.2. Semithin section of lips, showing ciliated (c) crests (cr) and intermediate sparsely-ciliated epithelium of oral surface (soe), and sparsely-ciliated aboral epithelium (sae). Figure 3.12.3. SEM of ciliated crest, showing dense ciliation and wave beat pattern. Figure 3.12.4. SEM of transition from ciliated epithelium of crest (cr, oral surface) to sparsely-ciliated epithelium of aboral surface (sae). Arrows indicate surface openings of mucocytes. From Beninger et al. (1990b).

As was observed for the scallop labial palps, the lip oral epithelium presents highly - branched apical microvilli, and a thick, greatly-indented basal lamina (Figs. 3.13.1, 3.13.2), suggestive of absorption and transport of dissolved or colloidal substances to the underlying haemolymphatic sinuses.

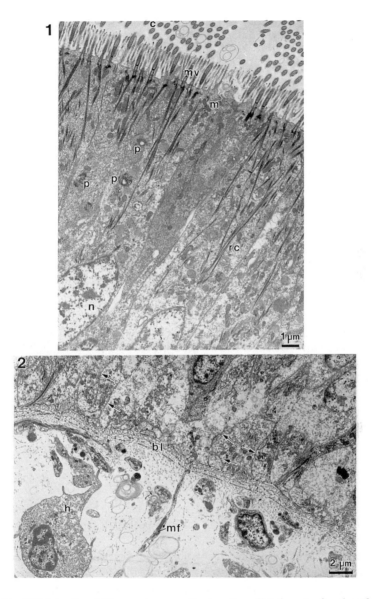

Figure 3.13. TEM details of *Placopecten magellanicus* lip, ciliated crest of oral surface. Figure 3.13.1. Apical region of epithelium. Note branched microvilli (mv) and phagosomes (p), indicating absorption of material from the epithelial surface. Ciliary roots (rc) extend beyond apical mitochondria (m) to the region of the nucleus (n). c, cilia. Figure 3.13.2. Basal region of epithelium. Note thick, multi-layered basal lamina (bl), with long indentations (arrows), indicative of transport to the underlying haemolymphatic sinus. h, haemocyte; mf, muscle fibre responsible for lip contraction.

3.4 DIGESTIVE SYSTEM AND DIGESTION

The scallop digestive system consists of the mouth, oesophagus, stomach and crystalline style, digestive gland and intestine. Its anatomy and function are similar to that of other bivalves (see Purchon 1977 and Morton 1983 for review), with several primitive features which are outlined below.

3.4.1 Mouth and Oesophagus

Particles and associated mucus enter the mouth from the oral groove at the base of the labial palps. The mouth is a simple ciliated opening to the narrow, ciliated oesophagus. The oesophageal epithelium is plicate, allowing accommodation, which may account for the presence of comparatively large food particles in the stomach (Mikulich and Tsikhon-Lukanina 1981; Shumway et al. 1987). Both the buccal and oesophageal epithelia are mucociliary, with a dominance of acid MPS-bearing mucocytes in the peribuccal region, and a mixture of acid and neutral MPS mucocytes in the oesophagus, as is the rule for exposed and enclosed mucociliary surfaces, respectively, in bivalves (Beninger et al. 1991; Beninger and St-Jean 1997b). The beating cilia of the mouth and oesophagus assist in transporting the relatively low-viscosity mucus - particle masses toward the stomach (Beninger and St-Jean 1997b). Apart from the epithelial mucocytes, no glands are present in the scallop oesophagus (in contrast to the 'salivary glands' found in other molluscs, including *Mytilus edulis* – Beninger and Le Pennec 1993).

The pectinid alimentary canal is characterised by a pseudostratified, microvillous-bearing epithelium with a dominance of tall, narrow, ciliated cells, interspersed with acid and neutral MPS - containing mucocytes (Figs. 3.14, 3.18), as well as non-ciliated absorptive cells. The epithelium rests upon a thick basal lamina, surrounded by a thick layer of connective tissue, occasionally traversed by strands of smooth muscle (Beninger et al. 1991; Le Pennec et al. 1991; and Fig. 3.18.4).

3.4.2 Stomach, Crystalline Style, and Gastric Shield

A detailed description of the pectinid stomach may be found in Purchon (1957). According to his system of classification, scallops possess a type IV stomach with the following salient features:

The oesophagus enters the stomach anteriorly and somewhat dorsally. The stomach is roughly oval in shape, with irregular folds and depressions delimiting different regions. A crystalline style projects across the stomach and into a well-developed dorsal hood, which is situated just above the gastric shield. The style originates in the style-sac, which is conjoined to the intestine in the Pectinidae. Depending on its degree of development, the style may occupy most of the lumen of the descending loop of the intestine. The historic term "crystalline style" is a misleading one, since the style is not crystalline, and it may only be presumed that the adjective refers to its translucent, pale coloration.

The exact composition of the crystalline style is not yet known, although it must be a stabilised mucopolysaccharide-type structure. Using very rudimentary methods, Dakin

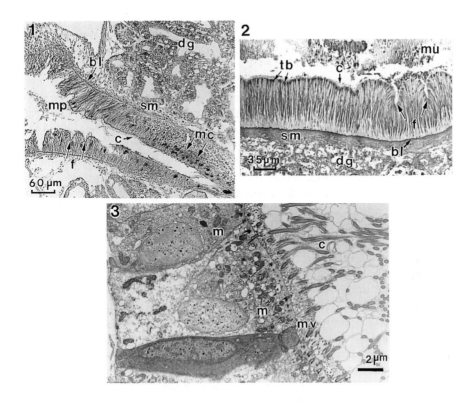

Figure 3.14. Structure and ultrastructure of the oesophagus in *Pecten maximus* and *Placopecten magellanicus*. Figure 3.14.1. Histological section of *Pecten maximus* oesophagus. Cilia (c) of ciliated pseudostratified epithelium, containing mucocytes (mc), beneath which are found a basal lamina (l), a smooth muscle layer (sm) and the digestive gland (dg). Note mucus-particle masses (mp) at the entry to the oesophagus, and folds (f) which allow ingestion of particles larger than the oesophageal diameter. PAS - alcian blue stain. Figure 3.14.2. Histological section of *Placopecten magellanicus* oesophagus. General structure as in Fig. 3.14.1, but note terminal bulbs (tb) of mucocytes, characteristic of this species. Abundant mucus (mu) is visible in the oesophagus, histochemically similar to that contained in the mucocytes. PAS-alcian blue stain. Figure 3.14.3. Transmission electron micrograph of the ciliated epithelium of the *Pecten maximus* oesophagus. Note the numerous mitochondria (m) at the bases of the cilia (c). The apical surface is covered with microvilli (mv).

(1909) detected unspecified proteinaceous substances in the style of *Pecten* sp. In a more modern study of the styles of twelve bivalve species, including *Pecten novaezelandiae*, Judd (1987) concluded that the crystalline style is probably composed of mucin type glycoproteins rather than glycosaminoglycans. The viscosity of these glycoproteins, or acid mucopolysaccharides, varies considerably among individuals and conditions, probably as a result of varying gut pH (Mathers and Colins 1979); in any event, the crystalline style can appear to be quite well-defined (but very easily broken or sectioned), or very poorly-defined (rather syrupy). It is presumably secreted by the walls of the style sac (Dakin 1909; Mathers 1976).

The gastric shield is situated at the head of the crystalline style. It covers the stomach epithelium, and is easily visualised in histological sections with either Fast Green or Alcian Blue. Shaw and Battle (1958) presented strong evidence that the oyster gastric shield may be composed of chitin (*N*-acetyl–D-glucosamine), and this has henceforth been supposed to be the case for bivalves in general (Morse and Zardus 1997). Chitin does not seem to be present anywhere else in the Bivalvia, so its presence in the gastric shield is intriguing. Obviously, the mechanical properties of chitin make it suitable for the grinding function it appears to perform (see below), and it may be that no other non-mineralised substance could perform such a task effectively. Although it contains other digestive enzymes (see below), the crystalline style does not appear to contain chitinases (Wojtowicz 1972); this is understandable, given the chitin composition of the gastric shield (Shaw and Battle 1958).

Although it was long assumed that one of the functions of the crystalline style was to drag mucus-particle 'strings' into the stomach (e.g., Purchon 1977), mucociliary transport via the ciliated epithelium of the oesophagus seems a much more plausible mechanism; moreover, as described above, particles appear to be ingested in a reduced-viscosity mucus, which would not lend itself to a 'capestan' model. However, the style rotation, effected by the co-ordinated ciliary beat of the stomach and intestinal epithelia, accomplishes two important functions: (1) it triturates the particles against the gastric shield, situated at the head of the style, and (2) it stirs the stomach contents, placing them in contact with plicate ciliary sorting areas located on the inner wall of the stomach (Purchon 1977). The crystalline style also undergoes periodic partial dissolution, liberating enzymes, which participate in extra-cellular digestion. In a study of *Placopecten magellanicus*, Wojtowicz (1972) identified relatively high α-amylase and laminarinase activity from the crystalline style. These enzymes originate in the style sac; they are incorporated into the style as it is secreted.

A cycle of dissolution - reconstitution of the crystalline style has been observed in both *Pecten maximus* and *Chlamys varia* subjected to diurnal changes in current flow (Fig. 3.15); this cycle has been attributed to variations in the pH of the style sac resulting from rhythmic activity in the digestive gland. This rhythmic activity in *P. maximus* and other nearshore bivalves appears to correspond to distinct feeding cycles as a function of the tides (Langton and Gabbott 1974; Mathers 1976; Mathers and Colins 1979).

The discovery of bacteriolytic activity in bacteria associated with the crystalline style of *Mytilus edulis* (Seiderer et al. 1987) raises the possibility that some extracellular digestion of bacteria occurs in the stomachs of bivalves. Such studies should be extended to the Pectinidae.

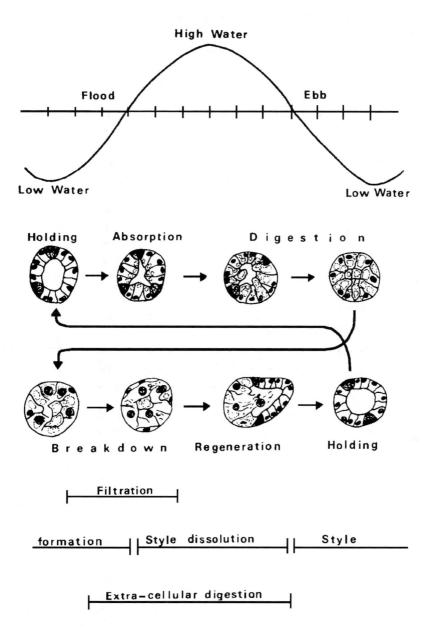

Figure 3.15. Diagrammatic illustration indicating the correlation of tidal cycle over a 12-h period with the cycles of feeding, extracellular digestion, and diphasic intracellular digestion in the tubules of a single individual of *Pecten maximus.* From Mathers (1976).

3.4.3 The Digestive Gland

To date, the terminology of the architecture of the digestive gland has been based on reconstructions of histological sections. Principal ducts from the gland open laterally into the plicated sorting regions of the stomach (two ducts in *Chlamys opercularis* and four in *Pecten maximus* - Purchon 1957). The principal ducts branch into secondary ducts, and finally into blind digestive 'tubules'. Recent studies using injected resin molds, however, show that the secondary ducts of *Pecten maximus* terminate in a short tubular region, followed by an acinus rather than a tubule (Le Pennec 2000 and Figs. 3.16.1, 3.16.2).

The probable mechanism of particle entry into the ducts and acini has been described by Owen (1955). Although the stomach and principal duct cilia create currents directed away from the principal ducts, a compensating current is created due to the distribution of cilia on only one side of the principal ducts (such a current was later demonstrated in *Ostrea edulis* by Mathers 1972). The movement of fluids into the non-ciliated secondary ducts and the acini is accomplished by aspiration, replacing the fluids absorbed by the digestive cells of the tubules. The weak current allows only the finest suspended particles to reach the acini (Owen 1955); larger particles are presumably digested in the stomach and intestine (see below).

The exact nature and number of different cell types in the bivalve digestive acini has been the object of considerable debate (see review by Morton 1983 and Weinstein 1995 for recent references). Although immature digestive cells are usually reported to be present in depressions or "cypts" of the acini (e.g., Henry et al. 1991), such crypts do not appear to be a constant feature in *Pecten maximus* (Henry et al. 1991; Le Pennec et al. 2001; and Fig. 3.16.3). Whether found in cypts or between and below the acinal cells, the immature cells are probably undifferentiated precursors for all of the other cell types observed in the tubule. The immature cells pass through a transient flagellated stage (Fig. 3.17.3), after which they differentiate either into absorptive (digestive) cells or secretory cells. The mature absorptive cells pinocytose particulate matter, which is digested within vacuoles called digestive spherules (Fig. 3.17.5). In *Pecten maximus*, the end product of this digestion appears as "residual bodies" (Mathers 1976 and Fig. 3.17.6). The absorptive cells eventually rupture, entraining the degeneration of the acinus (Mathers 1976; Mathers and Colins 1979), and the wastes are swept toward the stomach and ultimately the intestine. Acini are regenerated by immature cells.

The mature secretory cells possess an electron-dense cytoplasm rich in rough endoplasmic reticulum and Golgi bodies (Figs. 3.17.2, 3.17.4). These structures probably secrete digestive enzymes. Precise localisation of the enzymes is difficult, since cell type cannot be readily deduced from histochemical preparations. Whereas secretory cells would primarily secrete enzymes into the acinal lumen for extracellular digestion, intracellular digestion would be continued using the enzymatic equipment within the absorptive cells. A wide range of enzymes has been identified within the digestive gland (both the acini and the digestive ducts) of *Placopecten magellanicus* (Wojtowicz 1972), *Argopecten irradians* (Brock et al. 1986), and *Pecten maximus* (Henry et al. 1991), notably those involved in carbohydrate and peptide digestion. The absence of proteases (Henry et al. 1991) suggests that initial protein digestion occurs in the stomach.

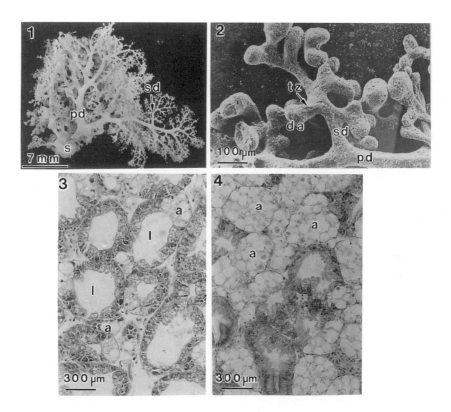

Figure 3.16. Topology and histology of the digestive gland in *Pecten maximus*. Figure 3.16.1. Internal mold of digestive gland, showing three-dimensional structure and anatomical relationships of point of attachment to stomach (s), principal duct (pd), and secondary duct (sd). Figure 3.16.2. Detail of internal mold, showing ramification of secondary duct (sd) from principal duct (pd), the short tubular zone (tz), terminating in the digestive acinus (da). Figure 3.16.3. Histological section of tubular and acinal regions in the Bay of Brest in April. Note large lumen (l), and few adipocyte-like digestive cells (a). Figure 3.16.4. Same type of section as in Fig. 3.16.3, but from a specimen sampled in the Bay of Brest in December. Note virtual absence of lumina and abundance of adipocyte-like digestive cells (a) in the digestive acini. Micrographs courtesy of G. Le Pennec, Université de Brctagne Occidentale.

Chitinase (N-acetylglucosaminase) is abundant in the digestive gland of *Pecten maximus*, however (Henry et al. 1991), and this probably enables scallops to utilise the energy-rich chitin contained in some diatom frustules (McLachlan et al. 1965; Jeuniaux 1982, as well as in the crustacean exoskeletons which may be abundant in bivalve stomachs, including those of pectinids (Shumway et al. 1987; Davenport and Lehane 2000; Lehane and Davenport 2002). Restriction of chitinases to the digestive gland (and intestine, see sections 3.4.4 and 3.4.5) may be sufficient to protect the gastric shield.

156

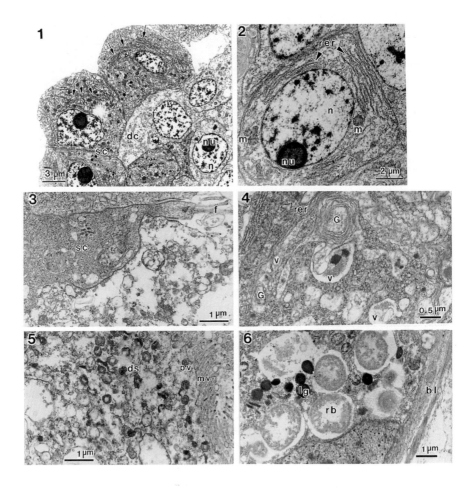

Figure 3.17. Transmission electron micrographs of digestive gland ultrastructure in *Pecten maximus*. Figure 3.17.1. Low-power micrograph showing general aspect of digestive (dc) and secretory (sc) cells. Arrows indicate rough endoplasmic reticulum in secretory cells. n, nucleus; nu, nucleolus. Figure 3.17.2. Detail of secretory cell, showing mitochondria (m) and rough endoplasmic reticulum (rer). Figure 3.17.3. Transmission electron micrograph of the apical region of a young secretory cell (sc) in the digestive gland. Note the presence of flagella (f). Figure 3.17.4. Transmission electron micrograph of the basal region of a mature secretory cell. Note the numerous vacuoles (v), extensive rough endoplasmic reticulum (rer), and the characteristic annular Golgi apparatus (G). Figure 3.17.5. Transmission electron micrograph of the apical region of a mature digestive cell, showing extensive microvilli (mv). Pinocytotic vesicles (pv) become digestive spherules (ds). Figure 3.17.6. Transmission electron micrograph of the basal region of an absorptive cell. Numerous residual bodies (rb) and lipid granules (lg) are present. Note the thick basal lamina (bl). Micrographs courtesy of A. Donval and G. Le Pennec, Université de Bretagne Occidentale, Brest, France.

Distinct digestive phases have been identified in several bivalve species (Morton 1977; Robinson and Langton 1980; Robinson et al. 1981), including *Pecten maximus* and *Aequipecten opercularis* (Mathers 1976, Mathers and Colins 1979). These phases are cyclic and linked to environmental factors such as the tidal cycle (Fig. 3.15), and may be somewhat complex, as in the case of *A. opercularis*, which displays co-dominant diphasic cycles as a function of tidal rhythm.

In addition to cyclicity on short time scales such as the tidal cycle, recent investigations have shown that ultrastructural characteristics of the digestive gland cells in *Pecten maximus* present long-term changes over the course of a year, notably with respect to the amount of lipid present in digestive cells within the acinus, which may resemble adipocytes at certain periods of the year (Le Pennec et al. 2001). The digestive cells of the tubular region store smaller lipid droplets, and these two types of lipid reserves are used differentially; it has been proposed that the lipid droplets of the tubular digestive cells are used for maintenance energy reserves, whereas those of the acinus are used for acute demand states such as gametogenesis (Le Pennec et al. 2001; Figs. 3.16.3, 3.16.4). Long-term variations in digestive gland lipid contents have also been observed in *Pecten maximus* using biochemical analyses (Strohmeier et al. 2000).

The pronounced endocytotic character of the digestive gland may account for the prevalence of prokaryotic infections of this organ by *Chlamidia*-like organisms (Morrison and Shum 1982).

3.4.4 Intestine, Rectum, and Anus

The remainder of the pectinid alimentary canal may be divided into three regions of similar length: the descending and ascending portions of the intestine, and the rectum. The descending portion of the intestine leaves the stomach mid-ventrally, passes through the digestive gland and into the gonad. In *Pecten maximus* and *Placopecten magellanicus*, the intestine continues almost to the distal extremity of the gonad before looping back, whereas in *Chlamys opercularis* the intestine curves back at about the level of the junction between the ovarian and seminal portions of the gonad. The ascending limb then returns through the digestive gland to follow the course previously described (Section 3.1).

The cell types, which form the intestine, are similar to those described for the alimentary canal in general. In the portion, which is exposed to the pallial cavity, a thin outer epithelium surrounds and protects the intestinal epithelium (Fig. 3.18.2). Although several studies of the anatomy and histology of the intestine have been performed on non-pectinid bivalves (see Morton 1983 for review), we are only aware of scattered information for a scallop species (*Pecten maximus* - Le Pennec et al. 1991a,b; Beninger et al. 2003). The intestinal epithelium consists of cells with apical microvilli and cilia, resting on a thick basal lamina (Figs. 3.18.1, 3.25). The underlying tissue is muscular-connective (Figs. 3.18.2, 3.18.4). Numerous mitochondria are present, and decomposing dead cells are frequently observed (Figs. 3.18.3–3.18.4). The intestinal cells themselves contain numerous enzymes: non-specific esterases, alkaline and acid phosphatases, chininase, and leucine aminopeptidase (Le Pennec et al. 1991a).

158

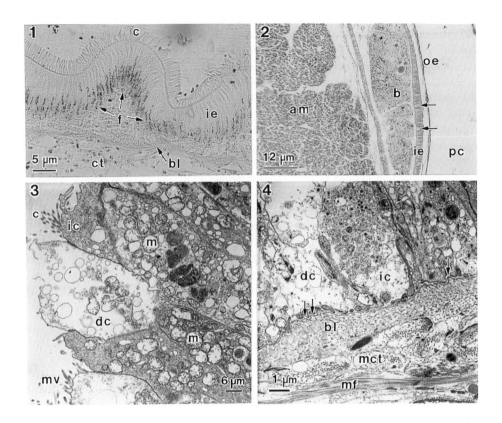

Figure 3.18. Histology, ultrastructure, and assimilation in the intestine of *Pecten maximus*. Figure 3.18.1. Portion of intestine passing through gonad. The intestinal epithelium (ie) is composed of tall, thin, pseudostratified, mainly ciliated (c) cells, resting on a thick basal lamina (bl) and surrounded by connective tissue (ct). Ferritin (f) injected into the intestinal lumen has been readily assimilated by the cells. The alimentary tract was allowed to purge for 48 h prior to injection. Prussian blue - nuclear red stain. Figure 3.18.2. Portion of intestine exposed to the pallial cavity (pc). Note thin outer epithelium (oe) surrounding intestinal epithelium (ie). Clear spaces (arrows) indicate acid mucopolysaccharide-secreting mucocytes. am, adductor muscle; b, bolus. Masson trichrome stain. Figure 3.18.3 Transmission electron micrograph of the apical region of intestinal epithelium cells. The intestinal cells (ic) bear cilia (c) and microvilli (mv). The numerous mitochondria (m) are dilated due to hypotonic fixation. Dead cells (dc) are frequently observed. Figure 3.18.4. Transmission electron micrograph of the basal region of the intestinal cells (ic). Note the thick basal lamina (bl), and the muscle fibres (mf) in the muscular-connective tissue (mct). Dead cells (dc) are again present. Micrographs courtesy of A. Donval, Université de Bretagne Occidentale, Brest, France.

The cellular structure of the rectum has been studied in *Patinopecten yessoensis*. This region consists mainly of ciliated cells, with few mucus cells. Some solitary cells of apparently neuroendocrine origin have also been observed, but their function has not been determined (Usheva 1983).

The scallop digestive system terminates in an anus, which is curved dorsally and away from the adductor muscle, such that the feces are voided in the excurrent water flow (Drew 1906). The anus is surmounted by a prominent lip (Dakin 1909), which in *Placopecten magellanicus* may function as a sphincter (P.G. Beninger, personal observations).

3.4.5 Digestive Sites and Postingestive Selection

As has been reported for some decades in various bivalves, the digestive gland is not the only site of extra-and intracellular digestion in bivalves. The intestine is also active in both types of digestion (Zacks 1955; Mathers 1973a,b; Stewart and Bamford 1976; Le Pennec et al. 1991a,b; Beninger et al. 2003; Fig. 3.18.1), and cannot simply be considered as a conduit for wastes and particles 'rejected' from the stomach and digestive gland (Brillant and MacDonald 2000). Indeed, as detailed in section 5, transfer of material from the intestine to developing oocytes suggests that in scallops the gonad intestinal loop may be an important site of both digestion and metabolite transfer to the gonad. As the majority of the intestine passes through the gonad and digestive gland, it is thus not possible to conclude that material therein is necessarily indigestible or that it has been 'rejected' following ingestion. While it is thus not evident what physiological significance may be attributed to the difference in location of digestion, it is clear that the presence of a given particle or fragment in the intestine does not signify that it will not be or has not been at least partially assimilated. Whereas the routing of particles to the digestive gland or to the intestine for eventual digestion indicates a difference in treatment, and hence a form of post-ingestive sorting, we do not know whether this reflects a difference in the percentage of each particle type assimilated by the digestive system. Rather, significantly different gut retention times may simply indicate that less nutritious particles spend less time in the digestive system (Brillant and MacDonald 2003).

3.5 CARDIO-VASCULAR SYSTEM

As in all bivalves, the pectinid circulatory system is said to be open, since venous hemolymph is collected chiefly in a number of well-developed sinuses. However, as Jones (1983) has pointed out, the term "open" should not be taken to mean that capillary vessels are absent. In fact, capillaries are visible in injected preparations (Drew 1906; Dakin 1909), and little is actually known concerning the microcirculation of bivalves.

3.5.1 General Circulation

The general circulation has been described in *Placopecten magellanicus* (= *Pecten tenuicostatus*) by Drew (1906), in *Pecten maximus* by Dakin (1909), and in *Argopecten irradians* by Gutsell (1931); this knowledge base has not since been significantly expanded. The following description is therefore based essentially on these sources. The general circulation may be divided into an arterial and a venous system (Figs. 3.19.1, 3.19.2).

3.5.1.1 The arterial system

Two main vessels constitute the central elements of the arterial system: the anterior and posterior aortas.

3.5.1.1.1 Anterior aorta

The anterior aorta supplies most of the visceral mass and is thus the more complex of the two vessel systems. It arises from the dorsal-most margin of the ventricle and curves dorsally around the digestive gland. Almost immediately after leaving the ventricle, it gives off a left and a right branch which supply blood to the digestive gland; these are visible at the surface of the gland as they begin to ramify and enter the gland itself.

The main branch of the aorta passes over the right side and into the anterior region of the digestive gland, producing ramifications, which pass through the gland to the left side and the left mantle lobe. Just before entering into the digestive gland, a branch separates from the aorta and continues dorsally and anteriorly. This is the anterior pallial artery; it bifurcates to form the left and right circumpallial arteries, which follow the margins of the left and right mantle lobes, branching off small connections to the mantle and becoming progressively smaller in diameter until they meet with the corresponding posterior circumpallial arteries. Drew (1906) states that in *Placopecten magellanicus* (= *Pecten tenuicostatus*) the posterior pallial vessel also arises from the anterior aorta, whereas Dakin (1909) affirms that in *Pecten maximus* it arises from the posterior aorta.

From within the digestive gland, the aorta gives off a branch, which passes anteriorly to supply the dorsal lip and the outer labial palps. As the aorta continues ventrally within the gland, it gives rise to another branch, which proceeds ventro-anteriorly and then curves dorsally into the foot. This is the pedal artery, and at its point of inflection it bifurcates, sending a vessel which supplies the lower lip and inner labial palps.

The aorta continues ventrally, giving rise to another branch running posteriorly to the digestive gland, then dividing into three vessels from within the gonad, supplying both the gonad and the convoluted intestine.

3.5.1.1.2 Posterior aorta

The posterior aorta chiefly supplies the rectum, the adductor muscle and the two mantle lobes. It leaves the ventral-most margin of the ventricle and proceeds ventrally along the rectum for a short distance before dividing into three branches. One of these continues along the rectum, supplying it with small branches to its extremity. A second curves dorsally to the fused mantle margin, where it bifurcates to become the right and left posterior circumpallial arteries. These vessels follow the mantle margin to meet with the anterior circumpallial arteries. The third branch of the posterior aorta constitutes the adductor artery. It runs antero-ventrally to the adductor muscle, ramifying most extensively from a point near the centre of the striated portion of the muscle.

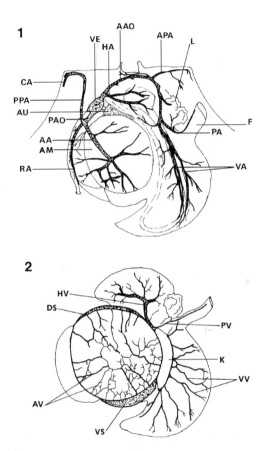

Figure 3.19. Schematic illustration of the arterial and venous systems in *Pecten maximus*. Figure 3.19.1. Arterial system. AA, adductor muscle artery; AAO, anterior aorta; AM, adductor muscle; APA, anterior pallial artery; AU, auricle; CA, circumpallial artery; F, foot; HA, hepatic artery; L, lips; PA, pedal artery; PAO, posterior aorta; PPA, posterior pallial artery; RA, rectal artery; VA, visceral arteries; VE, ventricle. Figure 3.19.2. Venous system. AV, adductor muscle veins; DS, dorsal sinus; HV, hepatic vein; K, kidney; PV, pedal vein; VS, ventral sinus; VV, visceral veins. After Gutsell (1931).

3.5.1.2 The venous system

In addition to blood vessels, the venous system comprises a number of sinuses, which collect blood from the various body parts. These sinuses are all located along the margin of the adductor muscle.

A large hepatic vein runs ventrally to join the pedal vein in a dorsal sinus, located between the pericardium and the adductor muscle. Venous blood from the muscle also flows into this sinus, which communicates with the dorsal extremities of the right and left kidneys. Two ventral sinuses situated on the antero-ventral margin of the adductor muscle extend antero-dorsally to enter the dorsal extremities of the respective right and left kidneys.

Most of the blood from the intestine and visceral mass flows through distinctly visible veins at the surface of the gonad to the external sides of the kidneys, where the veins communicate with the small kidney vessels. In summary, therefore, all returning blood thus far described passes into the kidneys. From there, the blood flows toward the right and left afferent branchial vessels in the right and left gill axes. After circulating in the gills (as described in Section 3.3), the blood joins the right and left efferent branchial vessels within the right and left gill axes. These vessels continue dorsally between the digestive gland and the adductor muscle, opening into the apices of the auricles. The right and left pallial veins join the corresponding efferent branchial vessels just prior to their entry into the right and left auricles; they drain blood from the very fine vessels and lacunae of the right and left mantle lobes.

3.5.2 The Heart

Scallops possess a typical bivalve heart composed of two auricles and one ventricle. In addition to its contractile role, the heart is also involved in excretion, as outlined below.

3.5.2.1 The ventricle

The ventricle is a smooth-walled chamber with a rather complex morphology (Dakin 1909). The rectum passes through the central portion of the ventricle, which is sac-shaped, while on either side joining the corresponding auricles are two pouches, thus creating a chamber with three very incompletely separated compartments. Both Drew (1906) and Dakin (1909) mentioned the existence of muscular sphincters or valves between the auricles and the ventricle; these structures may be seen in the histological section of Fig. 3.20.1. The valves presumably ensure a one-way flow of blood from the auricles to the ventricle.

The chamber of the ventricle is traversed by numerous muscle fibres, giving a distinct spongy appearance in section. This arrangement, also found in the auricles, is responsible for the great degree of contraction of the heart (Fig. 3.20.1).

3.5.2.2 The auricles and their excretory structures

The internal structure of the auricles resembles that of the ventricle, with criss-crossing muscle fibres and numerous lacunae (Figs. 3.20.1–3.20.3). The walls of the auricles are not smooth, as is the case for the ventricles, due to the presence of lateral, papillose outgrowths, which constitute the pericardial gland. This gland was observed by Dakin (1909) to be a site of nitrogenous waste uptake in *Pecten* spp. Although some differences of opinion exist, most recent studies on this system tend to confirm that haemolymph is ultrafiltered via numerous podocytes in this gland from the auricles to the pericardial cavity (see Jones 1983 for review). More recently, Morse and Zardus (1997) have confirmed this function in the pectinids *Chlamys hastata*, *Placopecten magellanicus*, and *Patinopecten caurinus*. In these pectinids, the ultrafiltrate passes to the kidneys via pedicels at the bases of the podocytes (Morse and Zardus 1997).

A second type of excretory cell, the *pore cell*, is found beneath the podocyte-containing outer epithelium of the pericardial gland, within the underlying muscular-connective tissue. These cells are capable of taking up large organic molecules (Morse and Zardus 1997), functioning to prefilter the haemolymph prior to ultrafiltration by the podocytes. They are naturally brown in colour, giving this region a brownish colour in living specimens. The pericardial cavity thus participates in at least two important physiological processes: refilling the heart (as mentioned in Section 3.5.2.4.1), and excretion (in which it acts as a reservoir for the auricular ultrafiltrate). More will be said of the excretory functions of this structure in Section 3.6.1.

A band of myocardial tissue situated on the ventral part of the pericardium connects the two auricles.

3.5.2.3 Structure and ultrastructure of heart cells

Scallop and other bivalve myocardial cells present some marked differences from vertebrate cardiac muscle cells. Sarcolemmic and T-tubules are absent in scallop myocardial cells, while the sarcoplasmic reticulum forms a network of tubules extending throughout the cell. This contrasts with adductor muscle cells, in which the sarcoplasmic reticulum is situated only under the sarcolemma. Stimulus for calcium release in scallop myocardial cells is presumably initiated at the sarcolemma and propagated along the sarcoplasmic reticulum to the interior of the cell (Sanger 1979).

Electron microscopic observations reveal the existence of both thin and thick filaments in the myocardial cells (Sanger 1979). The diameter of the thin filaments (6 nm) corresponds to that of actin, while the rather large diameter of the thick filaments (40 nm) led Sanger (1979) to suggest that they were composed of both myosin and paramyosin. Although Dakin (1909) believed that the myocardial muscle fibres of *Pecten maximus* were not striated, it is now known that scallop heart muscle fibres are striated, presenting distinct solid Z-bands (Sanger 1979).

Both intercalated discs and gap junctions are observed in scallop myocardial cells (Sanger 1979), as is true for oyster and mussel heart cells (Irisawa et al. 1973); hence, these cells are both mechanically and electrically coupled, in contrast to scallop adductor

164

muscle cells which present intercalated discs only and are thus mechanically but not electrically coupled (Sanger 1979). The scallop heart, like that of all other bivalves, therefore constitutes a functional syncytium.

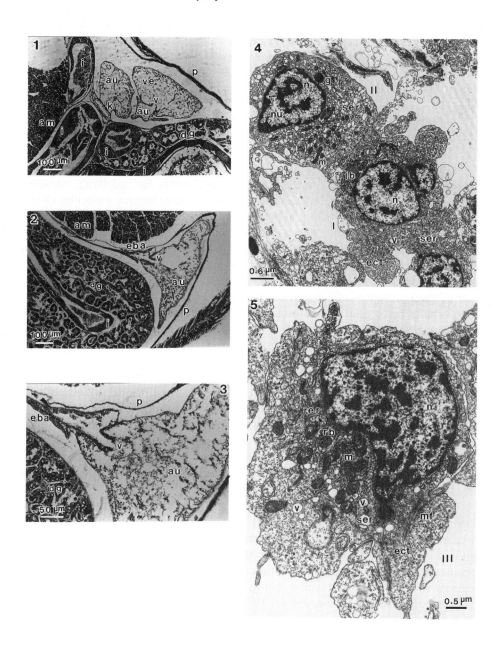

Figure 3.20. Structure of the pectinid heart and haemocytes. Figure 3.20.1. Histological section of a juvenile specimen of *Pecten maximus,* showing the anatomical relationships of the heart to surrounding structures. Note the sponge-like appearance of the heart due to the abundant crisscrossing muscle fibres. am, adductor muscle; au, auricle; dg, digestive gland; i, intestine; k, kidney; p, pericardium; v, valve-like structure between the auricle and ventricle (ve). Figure 3.20.2. Histological section through the auricle (au), efferent branchial artery (eba), and valve (v) of a juvenile specimen of *Pecten maximus.* Other abbreviations as in Fig. 3.20.1. Figure 3.20.3. Detail of Fig. 3.20.2 showing the auricular insertion of the efferent branchial artery and associated valve. Abbreviations as in Fig. 3.20.2. Figure 3.20.4. Transmission electron micrograph of haemocyte types I and II in *Chlamys varia.* Haemocyte type I: lobed nucleus (n) with associated juxtanuclear body (jb), around which are found numerous round mitochondria. The smooth endoplasmic reticulum (ser) is extremely well-developed and dilated, forming abundant saccules. The ectoplasm (ect) presents pseudopodial-type elongations, and vacuoles (v) of apparently endocytotic origin are visible. Haemocyte type II: the irregular nucleus (n) contains a small nucleolus (nu) close to the nuclear envelope. Numerous elongated mitochondria (m) are dispersed throughout the cytoplasm, which is chiefly occupied by dilated smooth endoplasmic reticulum (ser). A large, electron-dense granule (g) is present. Figure 3.20.5. Transmission electron micrograph of a type III haemocyte. Note the irregular nucleus (n), with its characteristic indentation near the cell centre. Numerous mitochondria (m) are grouped near this indentation. Rough endoplasmic reticulum (rer) and free ribosomes (rb) are abundant, while smooth endoplasmic reticulum (ser), showing dilated vesicles, is less common. Microfilaments (mf) are present at the boundary between the organite-rich endoplasm and the ectoplasm (ect). Vacuoles (v) are relatively abundant. Micrographs courtesy of P. Beninger and M. Auffret, Université de Bretagne Occidentale.

The fine innervation of the bivalve myocardium has not been extensively studied. Nerve endings with at least two different vesicle types are closely associated with the myocardium; these two vesicle types may correspond to different neurotransmitters (Sanger 1979).

3.5.2.4 Contraction

The contraction of the scallop heart may be divided into four related phenomena: refilling, coordination of the alternate auricular-ventricular beat, pacemaker mechanism, and regulation of the pacemaker mechanism.

3.5.2.4.1 Refilling

The passage of venous blood through large sinuses would intuitively result in low venous blood pressure, especially in light of the initially low systolic ventricular pressures of bivalve hearts, including those of scallops (see Jones 1983 for review). Ramsay (1952) proposed a mechanism of refilling in the hearts of invertebrates possessing a pericardial cavity. This mechanism was developed by Krijgsman and Divaris (1955) to explain the refilling of the bivalve heart; it has come to be known as the "constant volume" or "volume-compensating" mechanism. These authors concluded that contraction of the ventricle decreases the hydrostatic pressure in the pericardial cavity, thus forcing the auricle walls to expand; hence ventricular contraction automatically results in auricular expansion. This may be readily verified by cutting or piercing the pericardium at any point - the auricles and ventricle then fail to expand to any appreciable extent. The one-way valves at both extremities of the auricles ensure that auricular contraction will also automatically result in ventricular expansion.

Recent measurements indicate that the pericardium has a consistently lower pressure than either the auricles or the ventricles at all stages of the beat cycle of the bivalve heart; consequently it will promote the expansion of whichever chamber is in diastole (Jones 1983). Although there are some objections to this model, there is little doubt that it forms the basis for the refilling of the bivalve heart (see Jones 1983 for review).

3.5.2.4.2 Coordination of alternate AV beat

The volume-compensating mechanism provides an obligate mechanical link between the auricular and ventricular contractions, thereby ensuring a certain coordination of their respective beats. The actual coordinating stimulus is also purely mechanical, being mediated by alternate stretching of the auricular and ventricular myocardia. There appears to be no direct electrical interaction between the auricles and ventricle (Uesaka et al. 1987).

3.5.2.4.3 Pacemaker mechanism

The case for a diffuse myogenic pacemaker in the bivalve heart was put forward by Krijgsman and Divaris (1955); this has since become conventional wisdom (Brand and Roberts 1973; Jones 1983). Supporting evidence for the myogenic nature of the pacemaker comes from the electrocardiogram of the bivalve heart, which resembles a true myogram, with no trace of nervous impulses (Krijgsman and Divaris 1955). The pacemaker action potential is thus probably initiated by relatively unmodified myocardial cells.

Evidence for the diffuse nature of the myogenic pacemaker is much more equivocal. While denervated heart preparations amply demonstrate the existence of the pacemaker within the heart itself, contradictory results have accumulated which variously indicate a truly diffuse pacemaker, a relatively local pacemaker, and a "wandering" pacemaker in bivalve hearts (Jones 1983). We are unaware of any studies on the location of the pacemaker in pectinids.

Although no accounts of scallop heart action potential have been published to date, other bivalve species exhibit one of three different types of action potential: fast, slow, and spike-plateau. It has been suggested that these differences may be related to phylogeny (Jones 1983).

3.5.2.4.4 Regulation of pacemaker

The bivalve heart receives nerves from the cerebrovisceral connectives; hence, cardiac nerves enter both the right and left auricle (Jones 1983). These nerves originate in the parietovisceral ganglion (Krijgsman and Divaris 1955), and they contain both inhibitory and excitatory fibres (Jones 1983). As mentioned previously, the two different types of vesicles found in the nerve endings may correspond to different neurotransmitters; however, it should be noted that the neurochemistry of the bivalve heart is an evolving field and it is difficult to present a coherent synopsis at this point (see Jones 1983 for review).

On a functional level, it is clear that the pacemaker frequency may be modified by environmental or endogenous physiological factors. Brand and Roberts (1973) observed hypoxia-induced bradycardia followed by an overshoot when preparations of *Pecten maximus* were restored to normoxic conditions. Thompson et al. (1980) demonstrated a regulation of heartbeat during work in *Placopecten magellanicus*. The cardiac rhythm of this species was shown to increase two to three-fold during the rapid contractions of the adductor muscle valve-snap response.

Reported values for resting heartbeats in scallops maintained in well-oxygenated seawater, measured via electrodes passing through holes in the shell, are as follows: 15–20 beats min^{-1} in *Pecten maximus* at 12°C (Brand and Roberts 1973), 5–10 beats min^{-1} in *Placopecten magellanicus* at 5°C (Thompson et al. 1980). Earlier values of 25–30 beats min^{-1} reported for *Placopecten magellanicus* (Drew 1906) should not be retained, as the conditions of observation and measurement were not provided.

Cardiac output has been determined to be 64 mL h^{-1} in the previously-described preparations of *Placopecten magellanicus* (Thompson et al. 1980). More studies are required to determine such values in other species.

3.5.3 Haemolymph

The haemolymph of bivalves participates in a variety of physiological functions, including gas exchange, osmoregulation, nutrient distribution, elimination of wastes and internal defence. It also serves as a hydrostatic skeleton, such as during movements of the labial apparatus, tentacles, foot, and mantle edges.

Most bivalves lack circulating respiratory pigments (Booth and Mangum 1978; Barnes 1987), and there are no reports of such pigments in any scallop species. Certainly the low oxygen uptake efficiency of the haemolymph of *Placopecten magellanicus* (42% - Thompson et al. 1980) tends to confirm the lack of haemolymph respiratory pigments in this species. The essentially sedentary life habit, the enormous exposed body surface area, and the epibenthic habitat of scallops probably obviate the need for circulating respiratory pigments.

3.5.3.1 Plasma

Studies concerning the chemical composition of marine bivalve plasma have been reviewed by Burton (1983). The steady-state osmotic concentration of the plasma is equal to or marginally greater than that of the surrounding seawater. A slight hyperosmolarity may be necessary to the maintenance of urinal flow and mucus secretion. In any event, the metabolic cost of maintaining a significant long-term plasma-seawater osmotic gradient would be prohibitive in organisms such as bivalves which have vast surface areas exposed to the external medium (Burton 1983).

Most pectinid species are marine, some inhabit reduced-salinity waters, and none are adapted to freshwater (the current common name for *Placopecten magellanicus* - 'sea scallop'- is rather inappropriate, since there are no freshwater scallops). The ionic composition of the haemolymph in *Chlamys opercularis* has been studied by Shumway (1977); concentrations of Na^+, Mg^{2+} and Ca^{2+}, as well as overall osmotic concentration, followed that of the external medium in short-term fluctuating salinity regimes. The plasma K^+ concentration of bivalves is greater than that of seawater by a factor of 1–2 (Burton 1983); this probably reflects the normal intracellular-extracellular K^+ gradient (Natochin et al. 1979; Hochachka and Somero 1984).

Bivalve plasma contains numerous dissolved organic molecules. Little information is available on this subject for scallops, with the notable exception of *Placopecten magellanicus* (Thompson 1977). Plasma determinations performed immediately after field sampling yielded the following values (means and standard deviations): protein 153 ± 28, ammonia -N 0.24 ± 0.12, total carbohydrate 5.2 ± 1.2, lipid 13.7 ± 2.4 mg 100 mL^{-1}. It should be noted that plasma metabolite concentrations are influenced by physiological events such as short-term osmoregulation (see Pierce 1982) or the reproductive cycle (Thompson 1977). The subject of osmoregulation in bivalves is not within the scope of the present chapter; the reader is referred to Pierce (1982) and Beninger (1985) for a review and references.

3.5.3.2 Haemocytes

Although there are many descriptions of bivalve haemocytes in the literature, the lack of a universal classification scheme has greatly hindered attempts to draw together existing knowledge (Auffret 1988). Indeed, for a single bivalve species, a table of reported haemocyte types and names spans two full pages (Cheng 1996). In an extensive review paper, Cheng (1981) proposed a classification based on three categories of cell populations: hyalinocytes, granulocytes and serous cells. While this scheme may apply to most bivalves, Auffret (1985) did not observe any typical granulocytes in either *Pecten maximus* or *Chlamys varia*, and he did not consider the serous cells to be true haemocytes, since they have a different histological origin. As this appears to represent the only recent study of pectinid haemocytes, the cell types will be summarised here, with special reference to *Chlamys varia*.

3.5.3.2.1 Haemocyte types

Type I haemocytes contain central, lobed nuclei with abundant chromatin, partially condensed into *small* clots and a peripheral chromatin shell. Many *round* mitochondria are situated close to the nucleus; in *C. varia* these form a typical juxtanuclear body. The endoplasmic reticulum (ER) is smooth and often dilated to form numerous saccules. Electron-dense granules are rare (in contrast to the granulocytes of other bivalves), but there are numerous cytoplasmic inclusions and some glycogen particles, as well as vacuoles of apparently endocytotic origin (Fig. 3.20.4).

Type II haemocytes have non-lobed, eccentric nuclei containing abundant chromatin, either uncondensed or condensed in *large* masses of heterochromatin; a peripheral chromatin shell and nucleolus are also present. The numerous mitochondria are *elongated* and distributed throughout the cytoplasm. The smooth ER resembles that of the type I haemocytes except that it is even more vesiculated, filling most of the cytoplasm. Large electron-dense granulations are sometimes present (Fig. 3.20.4).

Type III haemocytes were observed in *C. varia* but not in *P. maximus*; inter-specific differences in haemocyte types may thus exist within the Pectinidae. Type III haemocytes have very irregular shapes, and a *polymorphic* nucleus indented near the cell centre, containing numerous clots of heterochromatin and a peripheral chromatin shell. The endoplasm is very dense, with many elongated mitochondria near the nuclear indentation at the cell centre; *rough* ER is well-developed while smooth ER is not; many free ribosomes are present. The ectoplasm is less dense than the endoplasm, containing large, possibly endocytotic vacuoles and microfilaments at the bases of the cell prolongations (Fig. 3.20.5).

3.5.3.2.2 Functions of haemocytes

Circulating haemocytes participate in five classes of physiological function in bivalves: wound repair, shell repair, nutrient digestion and transport, excretion, and internal defence. Details concerning the roles of haemocytes in these functions may be found in the review by Cheng (1981). Briefly, haemocytes intervene in wound repair by successive infiltration, clumping and plugging of the wound (without a fibrinogen-thrombin-fibrin system), followed by wound repair and phagocytosis of necrotic elements. Their known role in shell repair is that of calcium and organic matrix transfer from the digestive gland to the repair site. Haemocytes are present not only in the vascular system, but are also found wandering through tissues (an "open" circulatory system characteristic). They may thus absorb nutrients from the alimentary tract and pass them directly to other tissues, or to developing oocytes (see section 3.7.4.1.3). Thompson (1977) observed seasonal variations in the glycogen content of haemocytes of *Placopecten magellanicus*, with greater levels in summer than in winter.

Haemocytes may participate directly in excretion by absorbing wastes, passing directly across the epithelium of the nephridial tubules into the kidney lumen and thence to the exterior. Serous cells may also participate in excretion, since they originate as the pigmented cells in the pericardial gland. Although such cells were not observed in the haemolymph of the two pectinids examined by Auffret (1985), they have been observed in the pericardial gland of *Placopecten magellanicus* (Dakin 1909), as mentioned previously. These cells are capable of phagocytosing foreign particles such as carmine (Dakin 1909).

Most recent attention has been directed toward the roles of haemocytes in internal defence (Cheng 1981; Rodrick and Ulrich 1984). Although it is difficult to summarise the often contradictory observations, the following general points may be mentioned. Bivalve leucocytes are capable of phagocytosing foreign materials and degrading them via the lysosomes. The glycogen granules frequently found in bivalve haemocytes (including pectinids - Auffret 1985) may originate from digested bacteria (Rodrick and Ulrich 1984). In this respect, internal defence and nutrition are somewhat interrelated. Haemocytes also intervene in the extracellular lysis of bacteria through the production of circulating lysozymes (Cheng 1975).

3.6 EXCRETORY SYSTEM

The excretory system of bivalves comprises a number of diverse sites of excretory function (see review by Andrews 1988). The principal excretory organs are the auricular and pericardial glands, and the paired kidneys, also called nephridia or (historically) organs of Bojanus, who mentioned them first in 1819. The anatomical relationships of these structures to each other and to the gonad are illustrated in Fig. 3.21. Ubiquitous cells specialised in metal ion transport may also participate in excretory function.

3.6.1 Pericardial (Auricular) Glands

Gröbben (1888) was apparently the first to suggest an excretory function for the glands located in the auricular and pericardial walls of bivalves. However, it was not until 1942 that White demonstrated their excretory function and showed that their products were collected in the pericardial cavity, from whence they presumably exited via the reno-pericardial canal and kidney. This has since been confirmed for *Mytilus edulis* (Pirie and George 1979). Ultrastructural data show that these glands, also (historically) known as Keber's organs, appear to be the site of primary urine formation (i.e., ultrafiltration) in bivalves (Andrews and Jennings 1993; Morse and Zardus 1997).

There is some understandable confusion in the designation of auricular and pericardial glands. Glands, which originate in the outer auricle wall and project into the pericardial cavity, have been termed both 'auricular' and 'pericardial' glands. We will follow the terminology of Andrews and Jennings (1993): glands originating on the auricular wall are *auricular glands*, while those originating on the inner pericardial wall and projecting into the pericardial cavity are *pericardial glands*.

Auricular excretory glands appear to be the primitive condition in bivalves; their position in the auricle wall appears to impose a size constraint on the glands themselves, since an excessive development here would compromise auricular contraction. Most Pectinidae have auricular glands, but some species such as *Chlamys ruschenbergerii* have both auricular and pericardial glands (White 1942). The following description covers the general case (auricular glands); a good description of pericardial glands may be found in Eble (2001).

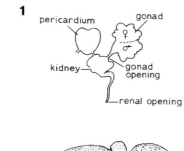

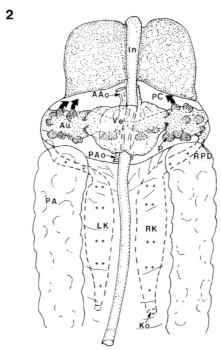

Figure 3.21. Schematic representation of the anatomical relationships between the pericardium, kidney and gonad in pectinids. Figure 3.21.1. The gonad opening to the kidney (modified after Mackie 1984). Figure 3.21.2. Dorsal view of the heart - pericardial cavity - kidney complex in *Chlamys hastata*. Primary urine formation occurs via ultrafiltration of haemolymph across podocytes (P) of auricular gland (curved arrows) into pericardial cavity (PC). Aao, anterior aorta; Au, auricle; In, intestine; Ko, kidney opening or nephrostome; LK, left kidney; PA, posterior adductor muscle; Pao, posterior aorta; RK, right kidney; RPD, right renopericardial duct. Modified from Morse and Zardus (1997).

The auricular glands of *Pecten maximus* and *Chlamys opercularis* were first described by Dakin (1909). More recent histological and ultrastructural studies of pectinid auricular glands include those of Andrews and Jennings (1993) and Morse and Zardus (1997). The general anatomical relationships of the scallop auricular glands are shown in Fig. 3.21.2. The ultrafiltration barrier is the greatly-folded basal lamina of the podocytes which cover the auricular glands. Within the underlying connective tissue are distinctive cells termed *pore cells* or brown cells (Morse and Zardus 1997), which do not appear to participate in ultrafiltration; more will be said of these below. Almost the entire cell volume is filled with secretory products; their numbers in the auricular wall give this structure a slightly brownish colour. Haszprunar (1996) proposed the name *rhogocyte* (= *Leydig's cell*), based on the diagnostic feature of cell membrane slits rather than pores; the semantic difference is perhaps somewhat tenuous. Although quite dense in the auricular wall, as mentioned above, these cells are ubiquitous in the bivalve body and can be found as solitary wandering cells in the haemolymph, and so may be also considered as a type of haemocyte. Although a partial homology to podocytes has been suggested (Haszprunar 1996), substantial ultrastructural differences exist (Morse and Zardus 1997). Pore cells appear to function in metal ion transport, and may have a role in metal detoxification, complementing that of the kidney (see below), and supporting the suggestion of a homology with nephrocytes (Haszprunar 1996).

Ultrafiltered haemolymph flows from the pericardial cavity to the kidneys via the renopericardial ducts or canals (Drew 1906; Pelseneer 1906; Dakin 1909; Potts 1967; and Fig. 3.21.2), where it is further processed as described below.

3.6.2 Kidney

The kidneys of scallops appear as brownish, flattened sacs attached to the adductor muscle, partially occluded by the gonad. As shown in Figs. 3.21.2 and 3.22.1, each kidney consists of a straight tube, surrounded by glandular tissue (as opposed to the U-shaped tube found in many other bivalves). The tube has classically been divided into a proximal or pericardial section, and a distal section or posterior sinus. The proximal sections of the right and left kidney tubules are connected by a transverse branch on their outer surfaces (Dakin 1909). The distal sections terminate in the renal or kidney openings, which open to the pallial cavity (Fig. 3.21). Given their dual function of evacuating kidney fluid and gametes, these openings are also called the reno-genital apertures.

Although historically considered to reflect separate secretory and excretory functions (Franc 1960), the division of the kidney tubule into proximal and distal sections is quite arbitrary, since it is lined by a single type of epithelial cell (Jennings 1984, cited by Andrews 1988). There exists a close anatomical relationship between the kidney and the parietovisceral ganglion (Fig. 3.22.1), and neurosecretory cells have been shown to penetrate the kidney tissue (Le Roux et al. 1987). The kidneys and the pericardial glands usually display a brownish coloration, due to the presence of granular concretions within their cells (Dakin 1909).

The kidney tubules open into numerous glandular ducts, giving the kidney tissue a spongy appearance (Turchini 1923; Andrews 1988), as in Fig. 3.22.2. The columnar external epithelium of the kidney rests upon a lacunar connective tissue, within which are elastic and muscular fibres (Fig. 3.22.2). The glandular cells (nephrocytes) lining the kidney tubules of *Pecten maximus* have a basal nucleus with dispersed chromatin and few organelles, with the exception of numerous mitochondria (Figs. 3.22.3–3.22.6). These cells contain many variously-sized vacuoles, which may fill most of the cell volume (Figs. 3.22.2, 3.22.3, 3.22.6, 3.22.7). Some vacuoles appear empty, while others contain variably electron-dense granules (Fig. 3.22.7) or myelin figures (Fig. 3.22.4). The cell membrane of the basal pole of the glandular cells presents numerous deep indentations (Fig. 3.22.5), while the apical pole is covered with microvilli (Fig. 3.22.3). These two characteristics indicate that these cells are sites of considerable exchanges with the renal fluid. Supporting cells are interspersed among the glandular cells of the kidney tubules (Fig. 3.22.4).

3.6.3 Functions of the Kidney and Pericardial Glands

Numerous authors have attempted to identify the composition of the urinary fluid of bivalves and elucidate the functions of their excretory system (see reviews by Franc 1960; Lucas and Hignette 1983; Martin 1983; Andrews 1988). Dakin (1909) states that uric acid is not found in bivalve kidneys, and that hippuric acid is eliminated by the pericardial glands of scallops. However, with the exception of Brunel's (1938) studies, which mention the existence of an allantoicase, a urease, an allantoinase and a uricase, little is known concerning the processes of organic nitrogen catabolism (Franc 1960).

The site of primary urine formation is now known to be the auricular or pericardial glands (Andrews and Jennings 1993). Further processing occurs within the kidney nephrocytes and pericardial gland pore cells, where various substances, notably metal ions, are accumulated, and which may combine with each other to form intracellular concretions or granules (Fig. 3.22.7). These concretions may either be excreted as solids, re-dissolved and eliminated as solutes, or grow until the intracellular fluid pressure ruptures the glandular cell, liberating the concretions (Franc 1960). Histochemical and ultrastructural studies in various bivalves suggest an apocrine-type secretion of renal concretions as residual bodies via the lysosomal-vacuolar system (Lowe and Moore 1979; Pirie and George 1979; George et al. 1980; Sullivan et al. 1988). In mussels, the concretions observed in the renal cells and in the pericardial gland pore cells are extruded to the renal tubules and excreted with the urine, which thus contains much particulate matter (Pirie and George 1979).

Mineral concretions were discovered very early in the kidneys of a variety of scallops, including *Argopecten irradians* (Kellogg 1892), *Placopecten magellanicus* (Drew 1906), *Pecten maximus* and *Chlamys opercularis* (Dakin 1909). Several of these authors also gave a qualitative chemical analysis of the concretions, showing the predominance of calcium phosphate, the nearly universal absence of urates, and traces of metals such as magnesium and iron (see review by Lucas and Hignette 1983). The accumulation of these metals was considered to represent either a long-term storage, or a

174

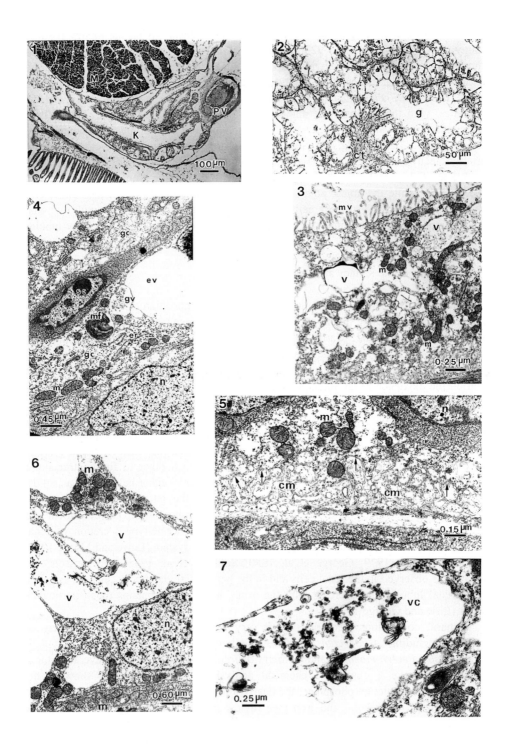

Figure 3.22 Tissue and cellular details of the kidney in *P. maximus*. Figure 3.22.1. Histological section of *Pecten maximus*, showing relationships between adductor muscle (M), kidney (K) and the parietovisceral ganglion (PV). Figure 3.22.2. Histological section of the kidney tubules, showing greatly-vacuolated gland (g) cells. ct, connective tissue. Figure 3.22.3. Transmission electron micrograph of apical pole of a glandular kidney cell. Note the numerous mitochondria (m), the apical microvilli (mv) and the variously-sized vacuoles (v). Figure 3.22.4. Transmission electron micrograph showing two glandular cells (gc) separated by a supporting cell (sc). An empty vacuole (ev), a granular vacuole (gv), and a myelin figure-containing vacuole (mf) are visible in one gland cell. er, endoplasmic reticulum; m, mitochondria; n, nucleus. Figure 3.22.5. Transmission electron micrograph of the basal poles of two glandular cells. Note the greatly-indented cell membrane (cm, arrows). m, mitochondria; n, nucleus. Figure 3.22.6. Transmission electron micrograph of a greatly-vacuolated glandular cell. The vacuoles (v) contain variously-sized particles. m, mitochondria; n, nucleus. Figure 3.22.7. Transmission electron micrograph showing detail of diverse vacuolar contents (vc). Photographs courtesy of A. Donval and P. Beninger, Université de Bretagne Occidentale, Brest, France.

means of detoxification-excretion of toxic elements (Simkiss 1976; Coombs and George 1977).

Using diverse techniques (differential centrifugation, X-ray diffraction, atomic absorption spectrophotometry, microanalysis), the quantitative composition of metals within the concretions has been determined, notably for *Pecten maximus* (George et al. 1980) and *Argopecten irradians* (Doyle et al. 1978; Carmichael et al. 1979). The concretions are chiefly composed of amorphous metal phosphates, which appear to be more closely related to the brushite-monetite series of phosphorites than to apatite (George et al. 1980). In *P. maximus,* for example, the elements Ca, Zn, Mn, and P comprise 50% of concretion dry weight, while organic material constitutes approximately 21–26% of concretion dry weight (George et al. 1980). The organic substances identified are proteins (6–9%) and oxalate (7% - Overnell 1981). In addition to these static observations, studies of the seasonal variations in composition of the concretions have been performed, as well as studies of the dynamics of their formation (see Bryan 1973 and reviews by Lucas and Hignette 1983; Yevich and Yevich 1985).

Through their accumulation of various trace metals in concretions (especially kidney granules), bivalves may act as biological concentrators (Martoja et al. 1975). The analysis of these metals or of their radioactive isotopes may thus be a means of monitoring environmental pollution (Chipman and Thommeret 1970; Masson and Ancellin 1976). The *in vitro* bioaccumulation of cadmium has been demonstrated in this way in the kidneys of *Argopecten irradians* (Carmichael and Fowler 1981). However, extreme care must be exercised in the interpretation of data from environmental monitoring studies, since the natural variation of granule elemental composition may be very large (George et al. 1980).

3.7 REPRODUCTIVE SYSTEM

A sound knowledge of the structure and function of the scallop reproductive system is obviously necessary for the understanding of their ecology, aquaculture and fisheries. For these reasons, the scallop reproductive system merits detailed consideration.

3.7.1 Sexuality: Gonochory, Hermaphroditism, and their Variants

Bivalves have long been considered gonochoric (sexes separate and invariable) in their overwhelming majority (de Lacaze-Duthiers 1854; Pelseneer 1894; Coe 1943b, 1945). The presence of a few hermaphroditic individuals in normally gonochoric species led Coe (1945) to conclude that primary germ cells are potentially ambisexual, cellular differentiation being based on an inhibition of one or the other potentiality via an unknown mechanism. An incomplete inhibition would thus produce animals with varying degrees of hermaphroditism.

Contrary to most bivalve families, the Pectinidae are predominantly hermaphroditic, with many specific and individual variations, and a general tendency towards protandry (producing first male and later female gametes - Coe 1945). Environmental factors may influence sex determination. In *Argopecten irradians*, low food and temperature levels result in male gametogenesis, whereas oogenesis is suppressed if food is absent (Sastry 1968).

Scallops cited as being predominantly gonochoric are *Chlamys tigerina, C. striata, C. furtiva* (Reddiah 1962), *Placopecten magellanicus* (Drew 1906; Posgay 1950; Naidu 1970), *Patinopecten yessoensis* (Yamamoto 1943) and *Chlamys islandica* (Sundet and Lee 1984). Even in strictly gonochoric species, a low frequency of hermaphrodites may be found (Naidu 1970; Merrill and Burch 1960; Wakui and Obara 1967; Ozanai 1975), while gonochoric features may be found in some normally hermaphroditic species (Mason 1958).

Most simultaneous hermaphroditic scallop species (male and female gametes developing at the same time) have gonads with distinct male and female portions, the male portion of the gonad being proximal and the female portion distal. However, a general tendency toward protandry precludes self-fertilisation in most cases (Sastry 1963; Fretter and Graham 1964). Although self-fertilised eggs are considered inferior to cross-fertilised ones in *Aequipecten irradians concentricus* (Sastry 1965), self-fertilisation may be a natural phenomenon in *Pecten maximus* (Wilkins 1978).

3.7.2 Origin and Formation of the Gonad

In the postlarval scallop, the gonad differentiates after most of the other organs; the organogenesis of this structure will therefore be described here. Primordial germ cells first appear in juvenile *Chlamys varia* in the region bounded dorsally by the pericardium, anteriorly by the developing kidney, and postero-ventrally by the margin of the adductor muscle. They arise from a group of mesodermal cells (Raven 1966; Wada 1968), and may be observed in *Chlamys varia* in animals as young as 6 months, measuring only 2–3 mm (Lucas 1965). In most pectinids, however, the gonad reaches its first maturity after one year (Reddiah 1962; Ozanai 1975).

Development of the gonad begins with the ventral extension of primordial germ cells in the form of tubules accompanied by loose connective tissue (Coe 1943a, 1945; Lucas 1965). The tubules then become folded, assuming the acinal-type structure of the gonad. The evacuating ducts differentiate as the germ cells multiply along the walls of the acini. This ontogenetic sequence suggests that a tubular gonad organisation is primitive. It is also very rare in the Bivalvia, being reported in the highly-modified, deep-sea endosymbiont-bearing protobranch *Acharax aline* (Beninger and Le Pennec 1997).

3.7.3 Anatomy, Histology, and Ultrastructure of the Adult Gonad

Since the first published anatomical description of a pectinid gonad (*Chlamys varia*) by de Lacaze-Duthiers (1854), numerous authors have used basic histological techniques to study the dynamic anatomy of this organ in various scallop species. Sastry (1979) provides a list of fifteen pectinid species whose spawning periods have been determined, usually by microscopic examination of the gonad. Surprisingly few ultrastructural studies of gonad structure and gametogenesis have been performed in bivalves, notably those of Pipe (1987a, b - *Mytilus edulis*), Eckelbarger and Davis (1996a, b - *Crassostrea virginica*), Beninger and Le Pennec (1997 - *Acharax alinae*), and Eckelbarger and Young (1999 - *Bathymodiolus childressi*). Fortunately, similar studies have also been performed in some detail for various pectinid species (Dorange et al. 1989a, b, c, d). The following account is based on this body of work, supported by micrographs.

In summary description, the gonad contains a large number of acini, whose walls are composed of connective tissue and primary germ cells. The lumen of the acinus is more or less filled with sex cells in varying stages of gametogenesis, depending on the reproductive state of the gonad. The acini empty into a network of ramified evacuating tubules, which consolidate as they approach the kidney (Figs. 3.21, 3.23.3, 3.27.1, 3.27.2). The gametes are shed into the pallial cavity via the renogenital openings of the right and left kidneys. The intestine and part of the crystalline style also occupy some of the gonad (see Section 3.4). Between these structures is a lacunar connective tissue, and the gonad is bounded by a thick outer epithelium (Figs. 3.23.1, 3.23.3).

Although in many species, the male gonad (or portion in simultaneous hermaphrodites) is yellowish-white and the female gonad (or portion) is variably reddish, sexing cannot be accomplished on the basis of gonad colour in all scallop species (Lucas 1965; Mottet 1979).

3.7.3.1 Outer epithelium

Apart from the brief descriptions of Mason (1958, 1963) for *Pecten maximus* and Lucas (1965) for *Chlamys varia,* most of our knowledge concerning the outer epithelium is derived from the electron microscopic studies of Dorange et al. (1989a) for *Pecten maximus.* The following description is based on the observations of the latter study.

178

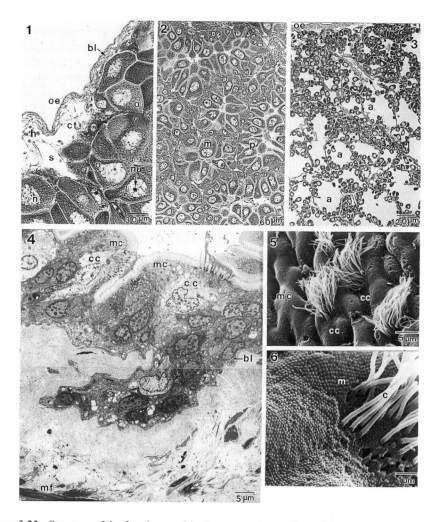

Figure 3.23. Structure of the female gonad in *Pecten maximus*. Figure 3.23.1. Histological section of the outer portion of a mature female gonad. bl, basal lamina; ct, perigonadal connective tissue; h, haemocytes; n, nucleus containing dispersed heterochromatin; nu, nucleolus; o, oocyte; oe, outer epithelium; s, haemolymph sinus; v, vitellus. Masson trichrome stain. Figure 3.23.2. Histological section of the interior of a mature female gonad, showing mature (m) and pedunculated (p) oocytes. Masson trichrome stain. Figure 3.23.3. Histological section of a partially-spawned female gonad. Most remaining oocytes are pedunculated. Abbreviations: a, acinus; e, evacuating duct opening into acinus; oe, outer epithelium. Masson trichrome stain. Figure 3.23.4. Transmission electron micrograph of the gonad outer epithelium. bl, basal lamina; cc, ciliated cells; mc, non-ciliated microvillous cells; mf, muscle fibres. Figure 3.23.5. Low-power scanning electron micrograph of the gonad outer epithelium. cc, ciliated cell; mc, non-cilated microvillous cell. Figure 3.23.6. Scanning electron micrograph detail of the gonad outer epithelium. m, microvilli; c, cilia. Micrographs courtesy of G. Dorange, Université de Bretagne Occidentale, France.

The outer epithelium is generally cuboidal, although some regions containing non-differentiated cells may show a pseudostratified appearance. Three cell types dominate in the epithelium: microvillous cells, ciliated cells, and mucocytes. Ultrastructural details for each cell type may be found in Dorange et al. (1989a). These cells rest upon a thick, often deeply-folded basal lamina, indicative of translocation (Figs. 3.23.1, 3.23.4).

3.7.3.2 Perigonadal connective tissue

The perigonadal connective tissue contains various cell types, as well as abundant collagen fibres. Smooth muscle cells are frequently observed (Figs. 3.23.1, 3.27.1); hence the designation "muscular-connective tissue" is in fact more appropriate. Mason (1958, 1963), suggests that this muscular-connective tissue may, through localised contractions, assist in the evacuation of gametes. The myocytes may also aid in the circulation of haemolymph within the gonad (Mason 1958, 1963).

Among the other cell types observed in the perigonadal connective tissue are macrophage-like cells, fibroblasts, and large vacuolar cells. The macrophage-like cells appear to correspond to "quiescent" type II haemocytes, which are not involved in internal defense, but rather in the assimilation and transport of the products of gamete lysis following atresia (Poder 1980; Dorange et al. 1989b). This topic will be dealt with in detail in Section 3.7.4.

3.7.3.3 Inter-acinal connective tissue

Since the gametes are produced in acini and not follicles, the name inter-acinal connective tissue is now widely used to describe the lacunar tissue found between the acini. It is structurally similar to the perigonadal connective tissue (Fig. 3.23.1). The amount of this tissue varies throughout the reproductive cycle, and it probably serves as an energy reserve for the developing gametes (Coe 1943a; Beninger 1987; Eckelbarger 1996a).

3.7.3.4 Haemolymph sinuses

These sinuses are present beneath the perigonadal connective tissue (Figs. 3.23.1, 3.27.1), and are particularly abundant in the inter-acinal connective tissue and at the periphery of the evacuating ducts; the intestinal loop is surrounded by a large sinus (Fig. 3.25). They are frequently associated with the infoldings of the outer epithelia, hence increasing the amount of surface area for exchange. Numerous haemocytes may be observed in these sinuses (Section 3.5), and macrophagous type II haemocytes are also present in the acini among the debris of oocyte lysis after spawning, during sexual resting periods, and in the later stages of atresia (Section 3.7.4).

180

3.7.3.5 Acini

The acini are bulb-shaped structures composed of an outer layer of connective tissue, from which presumably arise the inner layer of primordial germ cells; the gametes produced by these cells are eventually drained by the evacuating ducts. The developing gametes are readily visible along the inner margins of the acini, variably filling the lumina depending on the stage of gametogenesis (Figs. 3.23.2, 3.23.3, 3.27.1, 3.27.2, and Section 3.7.4).

3.7.3.6 Evacuating ducts

The ciliated columnar epithelium of the evacuating ducts rests upon a basal lamina, which is irregular at the junction of the gonoduct and acinus.

3.7.4 Gametogenesis

Gametogenesis and spawning are controlled by both endogenous and exogenous factors, among which the most important are temperature and food (Sastry 1966, 1968; Taylor and Capuzzo 1983). The process of gametogenesis in bivalves has been detailed in several works, such as those by Raven (1961), Sastry (1979), Dohmen (1983), Pipe (1987), Eckelbarger and Davis (1996a, b), and Morse and Zardus (1997). For the pectinid family, the principal histological studies are those of Coe (1945), Mason (1958, 1963), Lubet (1959), Reddiah (1962), Sastry (1963, 1966), Lucas (1965), Naidu (1970), Allarakh (1979). Ultrastructural studies have been performed on the gonads of *Pecten maximus* (Dorange et al. 1989a, b, c, d), allowing a better understanding of the events involved in the reproductive cycle. The following description is based largely on these ultrastructural studies.

3.7.4.1 Oogenesis

In pectinids, as in all other molluscs (and most other animals as well), complete meiosis of the female gametes is delayed until after spawning and fertilisation (see Longo 1983 for review). The mature scallop oocyte is blocked in the first metaphase of meiosis; the remaining meiotic stages are rapidly accomplished after fertilisation. This is readily observed *in vitro* by the appearance of polar bodies after fertilisation (Desilets et al. 1995).

Oogenesis may be divided into three distinct stages: premeiotic, previtellogenic, and vitellogenic.

3.7.4.1.1 Premeiotic stage

The stem cells, which constitute the boundaries of the acini, give rise to primary oogonia (8–10 μm), characterised by a high nuclear-cytoplasmic ratio. The nucleus contains clumped chromatin and occasionally a single, 3–4 μm nucleolus. Some mitochondria and endoplasmic reticulum cisternae are present in the cytoplasm. The germ cells divide mitotically to produce secondary oogonia. This stage is termed premeiotic because all divisions are purely mitotic.

3.7.4.1.2 Previtellogenic stage

The secondary oogonia enter into the first prophase of meiosis to give previtellogenic oocytes (Raven 1961). During the previtellogenic stage, the nucleus and cytoplasm of the oocyte increase in volume. The ultrastructural characteristics of the nucleus reveal young oocytes in leptotene, zygotene-pachytene and diplotene stages of prophase (Fig. 3.24.2). In leptotene, the chromosomes become visible whereas the nucleolus disappears. In zygotene-pachytene, the round or ovoid cells are easily identified due to the presence of the synaptonemal complex and the familiar lamp brush chromosomes.

As they enter diplotene, the oocytes elongate (Fig. 3.24.2). With the reappearance of the nucleus, dense aggregates (probably nuclear extrusions rich in ribonucleotides) accumulate in the cytoplasm. At this stage, auxiliary cells (Coe 1943a; Mason 1958, 1963; Raven 1966; Dorange et al. 1989c) migrate from the periphery of the acinus, establishing an intimate contact with the developing oocytes (Figs. 3.24.2, 3.24.4). The ultrastructural characteristics of the auxiliary cells in *Pecten maximus* are similar to those described for *Mytilus edulis* (Pipe 1987) and *Crassostrea gigas* (Eckelbarger and Davis 1996a). These cells participate in oocyte maturation as described below.

3.7.4.1.3 Vitellogenesis and metabolite transport to the oocyte

Knowledge of the exact pathways and mechanisms of metabolite transfer to the developing gametes is rather limited. A summary of known and proposed pathways may be found in Le Pennec et al. (1991), including a proposed pathway from the gonad-intestinal loop. Experimental support for this pathway has been provided using ferritin histochemistry (Beninger et al. 2003). Ferritin molecules are absorbed by the intestinal epithelial cells, transferred to the basal lamina, incorporated into vacuolated haemocytes, which migrate along connective-tissue fibres to the acini (Fig. 3.25). The haemocytes are able to penetrate the acinal basal lamina, and transfer the ferritin to developing oocytes. It is quite possible that similar haemocytes arrive in the acinus from the digestive gland. The relationship between these migrating haemocytes and auxiliary cells (see below) is not yet clear; the latter may serve as intermediaries.

Developing oocytes enter into vitellogenesis in the diplotene stage (Dorange et al. 1989b). Their size increases and they progressively become pedunculated (Figs. 3.24.1, 3.24.4). Various nuclear, cytoplasmic and membranal modifications occur before the oocyte detaches from the acinus wall.

Cortical glycoprotein granules become progressively denser in the cytoplasm while the oocyte is still largely attached to the underlying connective tissue. At this stage the ER is abundant. As the oocyte develops, the cortical granules and Golgi apparatus multiply. There is an intense production of vesicular and lamellar ER, as well as of various vitelline inclusions (Fig. 3.24.3). Three or four dictyosomes also appear at this point.

182

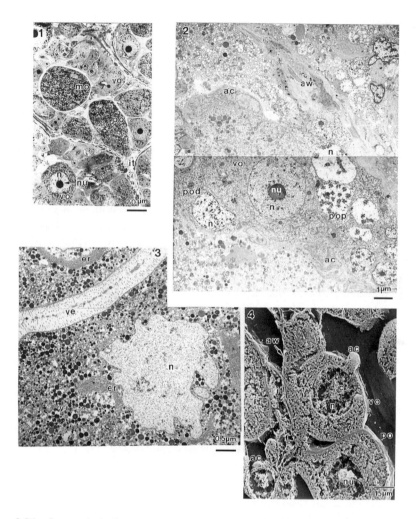

Figure 3.24. Oogenesis in *Pecten maximus*. Figure 3.24.1. Semi-thin resin section of oocytes in various stages of development. it, inter-acinal connective tissue; mo, mature oocyte; n, nucleus; nu, nucleolus; po, pedunculated oocyte; vo, oocyte at beginning of vitellogenesis. Figure 3.24.2. Transmission electron micrograph of oocytes in various stages of development. Note the auxiliary cells (ac) adhering to the vitelline envelope. aw, acinus wall; it, inter-acinal connective tissue. n, nucleus; nu, nucleolus; pod, previtellogenic oocyte in diplotene stage of first meiotic prophase; pop, previtellogenic oocyte in pachytene stage of first meiotic prophase; vo, vitellogenic oocyte. Figure 3.24.3. Transmission electron micrograph of part of vitellogenic oocyte. Note abundant endoplasmic reticulum (er), interspersed among the vitelline inclusions (vi) which they secrete. n, nucleus; ve, vitelline envelope. Figure 3.24.4. Scanning electron micrograph of oocytes in various stages of development. ac, auxiliary cell; aw, acinus wall; c, chromatin within nucleus (n); it, inter-acinal connective tissue; nu, nucleolus; po, pedunculated oocyte surrounded by vitelline envelope; vo, vitellogenic oocyte. Micrographs courtesy of G. Dorange, Université de Bretagne Occidentale.

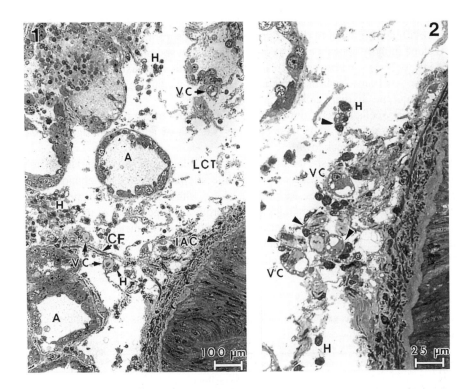

Figure 3.25. Anatomy of the proposed intestinal loop pathway of metabolite transfer to developing oocytes in *Pecten maximus*. Semi-thin sections of the epithelium of the gonad intestinal loop. Toluidine blue stain. Figure 3.25.1. Low-power micrograph showing proposed anatomical pathway for transfer of metabolites from intestine to haemocyte (H)-vesicular cell (VC) couples, and thence to acini (A) via intestine-acinus transfer complex (IAC), and connective fibres, (CF), within loose connective tissue (LCT). Figure 3.25.2. Detail of an intestinal-acinus transfer complex, showing greatly-vacuolated vesicular cells (VC) along pathway to acinus. Smaller haemocytes (H), with extensive filamentous cytoplasmic projections, can be seen in close contact with the vacuolated vesicular cells (arrows). Parts of another pathway may be observed in the upper right quadrant of the micrograph. From Le Pennec et al. (1991).

Throughout oogenesis, the attached auxiliary cells also present cytological modifications as their cytoplasmic volume increases. The granular ER multiplies, indicating intense protein synthesis. Mitochondria, smooth ER vesicles, α-glycogen vesicles and lipid-like globules appear in the cytoplasm. Similar ultrastructural changes have been observed in the auxiliary cells of *Mytilus edulis* (Pipe 1987) and *Crassostrea gigas* (Eckelbarger and Davis 1996a). Endocytotic figures may be seen between the developing oocyte and the auxiliary cell, indicating a transfer of nutrients.

The plasma membrane changes as the oocyte becomes peduncular. At the apical pole, microvilli appear and a fibrillar glycocalyx develops, forming the vitelline envelope. This envelope initially appears close to the auxiliary cell, and progressively surrounds the oocyte (Fig. 3.24.3). Close contact is maintained with the auxiliary cell via zonula adherens type junctions. Near the adherance zone, fibril-containing vacuoles are visible in the cytoplasm of the auxiliary cells.

The auxiliary cells detach from the oocytes at the end of the peduncular stage. The cytoplasm of the auxiliary cells then becomes vacuolated and myelin figures appear, indicative of membrane breakdown.

At the end of vitellogenesis, the thick vitelline envelope is slightly separated from the cytoplasmic membrane by a vitelline space, while vitelline reserves are abundant in the cytoplasm. ß-glycogen particles are present, along with typical ringed lamellae whose function is not known. Cortical granules are numerous at the periphery of the oocyte.

The mature oocytes of scallops measure 15–120 µm in diameter, depending on the species (Table 3.1). In a mature gonad, they often assume a polyhedral shape due to crowding in the acini, but resume a rounded shape when spawned.

Table 3.1

Dimensions (µm) of mature gametes in various pectinids. Dimensions are for histologically or electron microscopically-processed material.

Species	Oocytes	Spermatozoa		Reference
		Head	Flagellum	
Aequipecten opercularis	50–70			Amirthalingam (1928)
Chlamys varia	50–70	1.5–2		Reddiah (1962)
	45–50	2	45	Lucas (1965)
	65			LePennec (1978)
Chlamys distorta	58	2		Lucas (1965)
	60–70			LePennec (1978)
Chlamys tehuelcha	15–45			Christiansen and Olivier (1971)
Equichlamys bifrons	119.6			Dix (1976)
Aequipecten irradians	63			Sastry (1968)
Argopecten gibbus	60			Costello et al. (1973)
Placopecten magellanicus	80–90	1.5		Naidu (1970), Culliney (1974), MacDonald and Thompson (1986), Langton et al. (1987)
Pecten meridionalis	71.1			Dix and Jardin (1975)
Pecten maximus	70–80			Tang (1941)
	80–90			Mason (1958)
	70			LePennec (1978)
	75			Lubet et al. (1987)
	70			Dorange et al. (1987b)
	65–70			Paulet et al. (1988)
		1.7	45	Dorange (pers. comm.)

3.7.5 Oocyte Atresia

One of the first observations of oocyte atresia (necrosis) was that of Tang (1941) in *Pecten maximus.* This phenomenon was later observed in *Chlamys tehuelcha* (Christiansen and Olivier 1971) and *Patinopecten yessoensis* (Ozanai 1975). Although relatively common in bivalves, few ultrastructural studies of oocyte lysis have been performed (see Albertini 1985; Lubet et al. 1986; and Pipe 1987 for *Mytilus edulis;* and Dorange et al. l989b, c for *Pecten maximus).* Oocyte atresia is usually observed at the end of vitellogenesis, but even previtellogenic oocytes may be affected due to the presence of lytic enzymes, which spread throughout the acinus when other oocytes lyse.

The histological appearance of atresic oocytes is characterised by a modification of their staining affinities, beginning with the nucleus, which loses its basophilic properties. The peripheral cytoplasm then becomes clear, and the oocytes take on a much-deformed "jigsaw-puzzle" appearance (Fig. 3.26.2). Finally, the cell membranes rupture, discharging the nuclear and cytoplasmic contents. In this terminal stage, the oocytes appear as empty, deformed sacs piled against each other (Fig. 3.26.3).

In *Pecten maximus*, the first noticeable ultrastructural modification is the appearance of one or two nodules within the ER, consisting of concentric masses of cisternae. The rough and smooth ER then dilates, accompanied by a vacuolar degeneration of the cell. The mitochondria lose their cristae and become clear and deformed. The nuclear envelope expands into the cytoplasm, and the nucleus becomes multilobed prior to bursting (Fig. 3.26.5). A peripheral cytoplasmic necrosis then develops. Granules (probably from the rupture of the cortical granules) appear between the vitelline envelope and the plasma membrane. The perivitelline space increases, the microvilli detach and the plasma membrane ruptures. Glycogen granules accumulate in the lysed zones and in the vitelline envelope, which finally bursts, discharging the oocyte contents into the acinus (Fig. 3.26.4). Lytic debris spreads among the intact oocytes, while macrophage cells converge in the acinus (Dorange et al. 1989c).

Although the regulating mechanism of oocyte atresia is not well known, Lubet et al. (1986) demonstrated a neuroendocrine control *in vitro* for *Mytilus edulis;* it is probable that a similar mechanism exists in scallops. Environmental stimuli such as temperature may act as cues. In *Pecten maximus* from St.-Brieuc Bay (France), spawning only takes place if the water temperature reaches $15.5–16.0°C$; if not, pronounced atresia occurs in the mature gonads (Paulet et al. 1988). Since adductor muscle reserves are at their lowest at this time, it is possible that the lysed oocytes are recycled as maintenance substrates for the animal. Macrophagous type II haemocytes play an important role in the resorption of lytic debris and the redistribution of nutrients to the germ cells, both within the acinus and at the junction of the evacuating duct. Epithelial cells of the evacuating ducts also resorb lytic debris in *Mytilus edulis* (Pipe 1987).

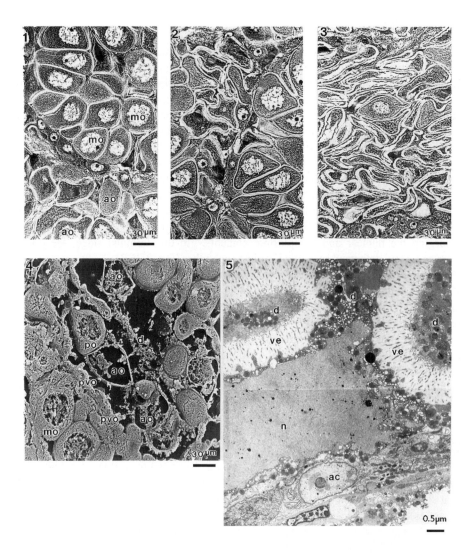

Figure 3.26. Oocyte atresia in *Pecten maximus*. Figure 3.26.1. Histological section of acini containing mature oocytes (mo) and atresic oocytes beginning lysis (ao). Figure 3.26.2. Histological section of gonad showing caracteristic "jigsaw puzzle" shape of oocytes in mid-atresia. Figure 3.26.3. Histological section showing advanced atresia. All oocytes are lysed. Masson Trichrome stain for Figs. 3.26.1–3.26.3. Figure 3.26.4. Scanning electron micrograph of atresic oocytes (ao), healthy mature oocytes (mo), pedunculated oocytes (po), and previtellogenic oocytes (pvo). Note lytic debris (d) releases fron atresic oocytes. Figure 3.26.5. Transmission electron micrograph of an atresic oocyte in which the nucleus (n), bereft of recognisable chromatin, has lysed, and is now in contact with lytic debris from the cytoplasm (d). Parts of two other atresic oocytes are visible, with vitelline envelopes (ve) surrounding lytic debria (d). ac, auxiliary cell. Micrographs courtesy of G. Dorange and Y. M. Paulet, Université de Bretagne Occidentale, France.

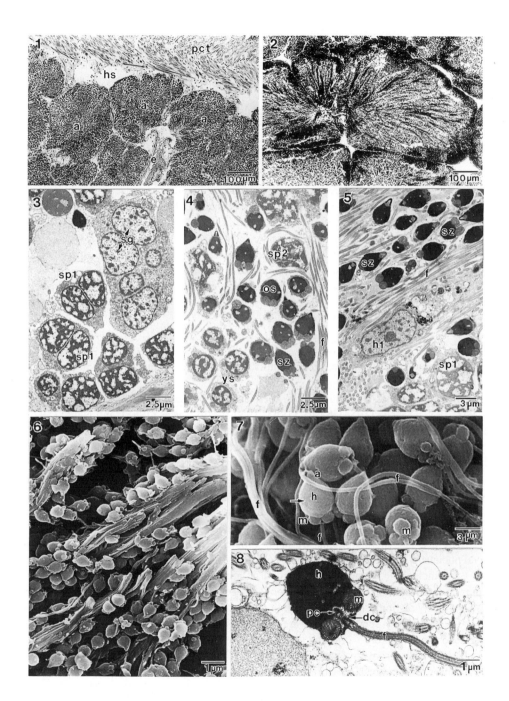

Figure 3.27. Spermatogenesis in scallops. Figure 3.27.1. Histological section of *Pecten maximus* gonad showing general organisation of acini (a) filled with developing gametes near perigonadal connective tissue (pct). An evacuating duct (e) drains two acini. A haemolymph sinus (hs) is visible. Figure 3.27.2. Histological section of a mature acinus in *Placopecten magellanicus*. Showing characteristic 'flow' pattern of mature spermatozoa toward the acinus lumen. Figure 3.27.3. Transmission electron micrograph of spermatogenesis in *P. maximus.* sg, spermatogonia dividing mitotically; sp1, primary spermatocytes near the end of the first meiotic division. Each sister cell will become a secondary spermatocyte. Figure 3.27.4. Transmission electron micrograph showing the development of the spermatids. f, flagellae; os, older spermatid; sp2, secondary spermatocyte, sz, spermatozoon; ys, young spermatid. Figure 3.27.5. Transmission electron micrograph of a relatively mature acinus in *P. maximus.* f, flagellae; h1, nucleus of macrophagous type I haemocyte; sp1, primary spermatocyte; sz, spermatozoa. Figure 3.27.6. Scanning electron micrograph showing orientation of mature spermatozoa within an acinus of *P. maximus.* Figure 3.27.7. Scanning electron micrograph of mature spermatozoa from *Chlamys varia*. The arrowed spermatozoon presents its acrosome (a), head (h), mitochondrial spheres (m), and flagellum (f). Figure 3.27.8. Transmission electron micrograph of a mature spermatozoan from *P. maximus.* f, flagellum; h, head; m, mitochondrial sphere; dc, distal centriole which has formed the flagellum; pc, proximal centriole. Micrographs courtesy of G. Dorange and P. Beninger, Université de Bretagne Occidentale, France.

3.7.6 Spermatogenesis, Spermatozoon Ultrastructure, and Taxonomy

Spermatogenesis in scallops has been studied by several authors, notably Lubet (1951, 1959), Mason (1958), Sastry (1963, 1966, 1979), Lucas (1965), and Naidu (1970), using light microscopy; and by Dorange et al. (1989d), using both light and electron microscopy. The following description is based on these studies. Dimensions are for histologically or electron microscopically-processed material.

The developing gametes are grouped in acini (Fig. 3.27.1, 3.27.2). Two cell types are visible in the acinus wall: stem cells, easily recognised by their finely granular nucleus and oval nucleolus, and spermatogonia, which are derived from these cells after several mitotic divisions (Fig. 3.27.3).

The primary spermatogonia are approximately 12 μm long and 7 μm wide. Their oval nucleus is about 5 μm in diameter, containing dispersed clumps of chromatin and one or two nucleoli. Numerous mitochondria are present in the cytoplasm. Primary spermatogonia divide mitotically to produce secondary spermatogonia, from which they are not easily distinguished under the light microscope.

The secondary spermatogonia are numerous at the start of sexual maturity. They measure approximately 7 μm long and 5 μm wide, with a rounded nucleus containing one nucleolus. The chromatin clumps are denser and the cytoplasm is reduced in comparison to primary spermatozoa. During sexual maturation, polynuclear spermatogonia are frequently observed in *Pecten maximus* (Dorange et al. 1989b). The secondary spermatogonia become spermatocytes, which detach from the acinus wall (Figs. 3.27.3–3.27.5). They are slightly smaller than the secondary spermatogonia. Primary spermatocytes present the various stages of the first meiotic division (Fig. 3.27.3). The first meiotic division products are the secondary spermatocytes, which are rarely observed because the second meiotic division occurs almost immediately. Their nucleus contains a dense network of chromatin (Lucas 1965).

The meiotic division of the secondary spermatocytes produces young spermatids, which differentiate to form older spermatids (Fig. 3.27.4) which are small (approximately 3 μm in diameter), containing a dense, round nucleus (approximately 1.8–2.4 μm). Their most characteristic feature is the 3–5 mitochondrial spheres, which form a collar at the base of the nucleus (Fig. 3.27.4). Several structural changes transform the spermatid into a mature spermatozoon. An annular Golgi structure, consisting of two concentric sacs, forms the acrosome, which elongates, invaginates, and becomes comma-shaped. The sacs then appear to fuse, giving rise to the definitive acrosome. The centriolar system migrates to the base of the nucleus and the distal-most centriole modifies to give rise to the flagellum. Throughout these latter stages of spermatogenesis there is a progressive reduction of the cytoplasm.

The ultrastructure of bivalve spermatozoa has been intensively studied, due to their usefulness in elucidating phylogenetic and taxonomic relationships (e.g., Popham 1979; Hodgson and Bernard 1986; Healy 1995, 1996; Eckelbarger and Davis 1996b; Garrido and Gallardo 1996; Le Pennec and Beninger 1996; Beninger and Le Pennec 1997; Morse and Zardus 1997; Kafanov and Drozdov 1998; Healy et al. 2000). As recently underscored by Healy et al. (2000), specific data for pectinids are scarce; however, the ultrastructural features of *Pecten maximus* have been detailed in Dorange and Le Pennec (1989d) and in the previous edition of this volume (Beninger and Le Pennec 1991). To date, the only comparative study of pectinid spermatozoon morphology and ultrastructure deals with ten species of confirmed pectinids (Le Pennec et al. 2002). The following description is based upon these works.

Two general categories of pectinid spermatozoon have been defined: the more primitive Category 1, characterised by a rounded shape, and the more advanced, elongated Category 2 spermatozoon (Le Pennec et al. 2002). Within these two categories are several groups, based on ultrastructural characteristics, which correspond well with the taxonomic affinities of Waller (1991): the *Pecten* group, the *Palliolium* group, the genus *Hinnites*, the *Aequipecten* group, the *Chlamys* group, and the *Mimachlamys* group (Fig. 3.28). The ultrastructural distinctions are based on the shape of the nucleus and the acrosomal depressions.

Notwithstanding the differences noted above, the mature scallop spermatozoan contains a dense, granular nucleus with two acrosomal depressions (Figs. 3.27.4–3.27.8, 3.28). The mid-piece presents four to five mitochondrial spheres (diameter 0.7–0.8 μm) (Fig. 3.27.7, 3.27.8), which surround the centriolar system and possess long crests. Together these sections measure approximately 3–6 μm, whereas the flagellum, of 9 + 2 structure, is approximately 45 μm long (Table 3.1 and Fig. 3.28), and 0.2–0.3 μm in diameter; these dimensions are typical of littoral bivalve species (Beninger and Le Pennec 1997). The spermatozoa develop in a centripetal manner, with their flagella toward the lumen of the acinus (Fig. 3.27.2).

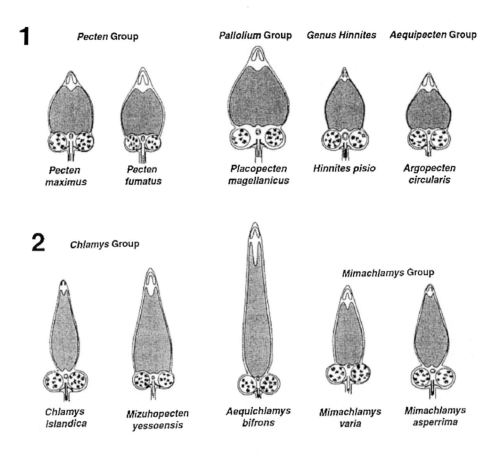

Figure 3.28. Schematic representations of pectinid spermatozoa, assembled into Categories 1 and 2, and into Groups (Waller 1991) corresponding to taxonomic affiliations (Le Pennec et al. 2002). For scale, *Pecten maximus* = 1.6 μm width.

3.7.7 Fertilisation

The events of fertilisation have been studied in detail in *Placopecten magellanicus*. Although similar to the accounts of other bivalve species, some differences were observed, notably the absence of a zygote nucleus (the male and female chromosomes align immediately on the mitotic. spindle of the first meiotic division). There does not seem to be a general rule for the stage of blockage of the first meiotic division in mature pectinid oocytes, as some species are blocked in prophase and others in metaphase (Desilets et al. 1995).

3.8 NERVOUS AND SENSORY SYSTEMS

The anatomical modifications of scallops (reduction of the anterior region, circumpallial mantle eyes and tentacles, central position of posterior organs such as the adductor muscle) are reflected in the general organisation of the nervous system (Fig. 3.29). Accordingly, the cerebral ganglia are reduced and closely related to the pedal ganglia, while the visceral ganglia are fused and complex, forming the distinctive pectinid parietovisceral ganglion. This structure innervates the majority of the scallop body (Fig. 3.29).

3.8.1 General Organisation of the Nervous System and Functional Anatomy of Principal Ganglia

The scallop nervous system consists of two main ganglionic concentrations and their nerves: the cerebral and pedal ganglia and the parietovisceral ganglion (Figs. 3.29, 3.30).

3.8.1.1 Cerebral and pedal ganglia

The paired cerebral and pedal ganglia are located quite close together beneath the integument, between the ventral lip and the foot. Wilkens (1981) shows them as contiguous structures in *Pecten ziczac*, and this agrees with our own histological observations in *Placopecten magellanicus* (Figs. 3.29, 3.32.1). However, Drew (1906) and Dakin (1909, 1928a) show these two paired structures as being somewhat more distinct in *Pecten maximus*, *Chlamys varia* and *Placopecten magellanicus*. They are linked to each other by a pair of cerebro-pedal connectives. The individual cerebral ganglia themselves are joined dorsally by a circumoesophageal cerebral commissure immediately posterior to the mouth (Fig. 3.30).

Also arising from the antero-dorsal portion of the cerebral ganglia are the right and left anterior pallial nerves, which run toward the oesophagus and through the digestive gland to enter the mantle above the dorsal lip. Here they split into several branches which join the circumpallial nerve (Figs. 3.29, 3.30).

Three smaller nerves also originate from the antero-dorsal region of the left and right cerebral ganglia (Fig. 3.30): the labial nerve, the palp nerve, and the statocystic nerve (the otocystic nerve of Drew 1906 and Dakin 1909).

The pedal ganglia are almost fused due to the massive pedal commissure. A large pedal nerve arises from each ganglion, ramifying and innervating the foot muscles (Figs. 3.30, 3.32.1).

3.8.1.2 The parietovisceral ganglion and its nerves

The paired visceral ganglia characteristic of other bivalve families have become greatly modified in the Pectinidae, being completely fused to form one large, complex parietovisceral ganglion composed of several distinct lobes (Fig. 3.31.1). This is the largest and most intricate ganglion found in the Bivalvia (Bullock and Horridge 1965) and

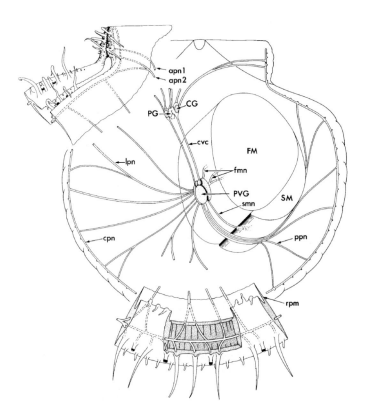

Figure 3.29. Diagram of the principal ganglia and their nerves in relation to body structure in *Pecten ziczac.* The upper left and lower insets illustrate the innervation of the mantle in the dorsal and ventral regions. Abbreviations: apn, anterior pallial nerve; CG, cerebral ganglion; cpn, circum-pallial nerve; cvc, cerebrovisceral connective; FM, fast striated muscle groups; fmn, fast motor nerve; lpn, lateral pallial nerve; PG, pedal ganglion; ppn, posterior pallial nerve; PVG, parietovisceral ganglion; rpm, radial pallial musculature; SM, smooth muscle groups; smn, slow motor nerve. From Wilkens (1981).

as such it merits special attention. The external anatomy of this structure has been described in *Pecten maximus* and *Chlamys* (= A*equipecten*) *opercularis* by Dakin (1910b, 1928a), in *Argopecten irradians* by Spagnolia and Wilkens (1983), and in *Patinopecten yessoensis* by Matsutani and Nomura (1984). It consists of a large ventrocentral (= posterior) lobe, flanked by two crescent-shaped lateral lobes. Its external anterior surface is surmounted by two bulbous dorsocentral (= anterior) lobes (Fig. 3.32.2). Near the point of insertion of the cerebro-visceral connectives are two spherical structures called accessory lobes (= accessory ganglia), which are considered to be part of the parietovisceral ganglion by Spagnolia and Wilkens (1983), but not by Matsutani and Nomura (1986).

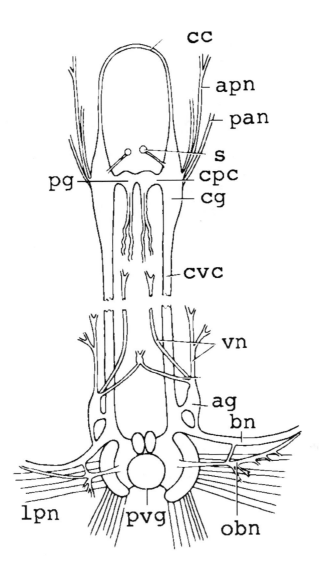

Figure 3.30. Diagram of the central nervous system of *Argopecten irradians* (rotated ventral view; see Fig. 3.29 for actual orientation). Abbreviations: ag, accessory ganglion; apn, anterior pallial nerve; bn, branchial nerve; cc, cerebral commissure; cg, cerebral ganglion; cpc, cerebro-pedal connective; cvc, cerebro-visceral connective; lpn, lateral pallial nerve; obn, osphradio-branchial nerve; pan, palp nerve; pg, pedal ganglion; pvg, parietovisceral ganglion; s, statocyst and nerve; vn, nerves to viscera. From Bullock and Horridge (1965), after Gutsell (1931).

1

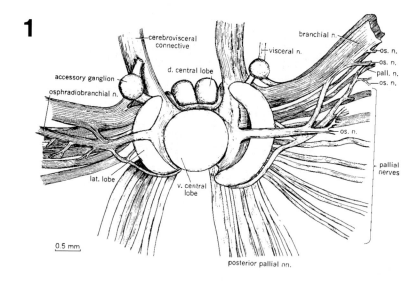

2

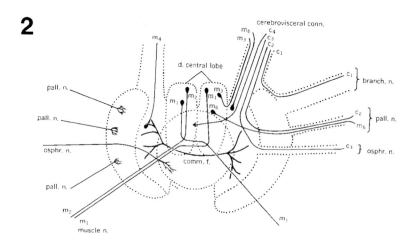

Figure 3.31. Anatomy and principal nerve tracts of the parietovisceral ganglion in *Pecten maximus* and *Chlamys varia*. Figure 3.31.1. Rotated ventral view of the ganglion. Abbreviations: d., dorsal; lat., lateral; n, nerve; nn, nerves; os. or osph., osphradial; pall., pallial; v, ventral. From Bullock and Horridge (1965) after Dakin (1910). Figure 3.31.2. Principal nerve tracts based on the observations of Dakin (1910). Afferent pathways are shown on the right, and efferent pathways on the left. Abbreviations: branch. n., branchial nerve; C 1–4, nerve fibres from cerebral ganglion; comm. f., commissural fibres; m 1–6, motor neurons; other abbreviations as in Fig. 3.32.1. From Bullock and Horridge (1965) after Dakin (1910).

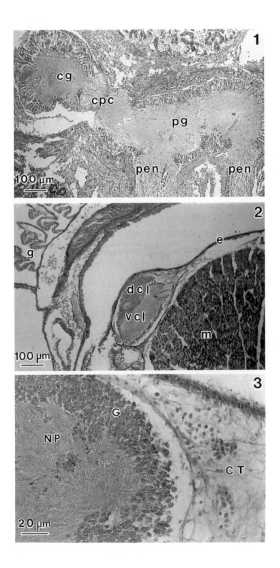

Figure 3.32. Structure and histology of the principal ganglia in *Placopecten magellanicus* and *Pecten maximus*. Masson Trichrome stain. Figure 3.32.1. Histological section of *Placopecten magellanicus* showing the anatomical relationship between one cerebral ganglion (cg) the fused pedal ganglia (pg), the cerebro-pedal connective (cpc), and the two pedal nerves (pen). Figure 3.32.2. Histological section of the parietovisceral ganglion in a juvenile specimen of *Pecten maximus*. Abbreviations: dcl, dorsocentral lobe; g, gill filaments arising from suspensory membrane; e, limiting epithelium; m, adductor muscle; vcl, vertrocentral lobe. Figure 3.32.3. Histological section of the ventrocentral lobe of the parietovisceral ganglion in a juvenile specimen of *Pecten maximus*. Abbreviations: CT, connective tissue; G, ganglion cells; NP, neuropile region. Micrographs courtesy of A. Donval and P. Beninger, Université de Bretagne Occidentale.

The parietovisceral ganglion is situated slightly postero-ventrally to the excretory pore of the right kidney, and is in close anatomical relation with this organ. It gives rise to most of the nerves of the viscera, adductor muscle, and mantle (along with its specialised sensory structures). Two cerebro-visceral connectives link the ventral parts of the cerebral ganglia to the parietovisceral ganglion (Figs. 3.29, 3.30). Close to the junction with the cerebro-visceral connectives are the large branchial nerves, which innervate the gills. Joining the ganglion on the right and left sides are the pallial nerves, adductor muscle nerves, osphradial nerves, and osphradio-branchial nerves. Ventrally on either side of the ganglion, large posterior pallial nerve tracts arise which curve around the adductor muscle, joining the circumpallial nerve by several branches, and ramifying to innervate most of the mantle. The circumpallial nerve innervates the mantle margin and its associated sensory structures (e.g., tentacles and eyes via the tentacle and optic nerves, respectively), as well as some of the pallial muscles (Figs. 3.29, 3.30, 3.33.1, 3.33.6).

An interesting aspect of the external anatomy of the parietovisceral ganglion are the relative sizes of the two lateral lobes. The left lateral lobe is larger than the right one in *Pecten maximus*; this has been related to the greater number of eyes on the left mantle margin in this species (Dakin 1910b). In *Argopecten irradians*, however, the two lateral lobes are symmetric, corresponding to an equal number of eyes on each side (Spagnolia and Wilkens 1983). The role of the lateral lobes in vision is dealt with in Chapter 5.

3.8.1.3 Histology and neurosecretions of the ganglia

The cerebral and pedal ganglia show similar histological organisation, with a cortex of ganglionic cells from which nerve fibres extend to the core or neuropile region (Fig. 3.32.1). The parietovisceral ganglion presents a more complex internal anatomy. While the ventro-central lobe is composed of a fibrillar neuropile and a peripheral ganglionic cortex, the two dorso-central lobes are almost entirely made up of the largest ganglion cells (Figs. 3.32.2, 3.32.3), whose nerve fibres ramify throughout other regions of the ganglion (Stephens 1978). Most of the pear-shaped ganglion cells are unipolar, but bipolar and multipolar cells also occur. Neuroglia cells and their fibres envelop the ganglion cells, and are also present in the neuropile and in the nerves. There are very few blind endings of nerve fibres in the neuropile (Dakin 1910b).

Much of our knowledge of the nerve tracts of the parietovisceral ganglion is based on the fragmentary observations of Dakin (1910b), presented by Bullock and Horridge (1965) and shown in Fig. 3.31.2. More recently, the routes of the radial pallial nerves were traced within the parietovisceral ganglion of *Argopecten* (= *Aequipecten) irradians* by Stephens (1978). The internal structure and nerve tracts of the lateral lobes have been studied in detail by Spagnolia and Wilkens (1983), and are presented in Chapter 5.

Details of the synaptic structure in bivalve ganglia are rather scant, with the notable exceptions of the study by Cobb and Mullins (1973) on the visceral ganglion of *Spisula solidissima*. In this species, most axons were shown to be varicose, containing several types of vesicles, while specialised synaptic contacts were rare or absent. These authors proposed that the varicose axon type may function in a similar manner as the

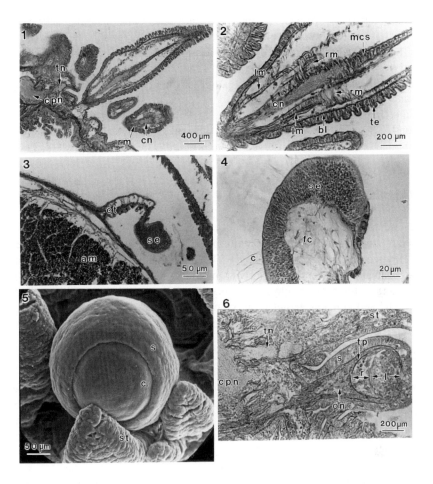

Figure 3.33. Structure and histology of various sensory organs in *Pecten maximus* and *Chlamys varia.* Masson trichrome except for Fig. 3.33.5. Figure 3.33.1. Longitudinal and cross-sections of a long tentacle (retracted) on the middle fold of the mantle of a juvenile *Pecten maximus.* cn, central tentacle nerve; cpn, circumpallial nerve; rm, radial muscles; tn, tentacle nerve. Figure 3.33.2. Detail of basal region of long sensory tentacle. bl, basal lamina; cn, central tentacle nerve; lm, longitudinal retractor muscle; mcs, musculo-connective sheath; rm, radial muscles; te, tentacle epithelium. Figure 3.33.3. Anatomical relationships of the abdominal sense organ in a juvenile *Pecten maximus.* am, adductor muscle; ct, connective tissue fold; se, sensory epithelium. Figure 3.33.4. Detail of the abdominal sense organ of a juvenile *Pecten maximus.* c, cilia; fc, fibrillar core; se, sensory epithelium. Figure 3.33.5. Scanning electron micrograph of an eye and sensory tentacles in *Chlamys varia.* c, cornea; s, eyestalk; st, sensory tentacle (retracted). Figure 3.33.6. Longitudinal section of the eye in a juvenile *Pecten maximus*, also showing the tentacle nerve (tn) in an adjacent tentacle, emanating from the circumpallial nerve (cpn). l, lens; on, optic nerve; r, retina; st, sensory tentacle; tp, tapetum. Micrographs courtesy of A. Donval and P. Beninger, Université de Bretagne Occidentale.

unspecialised varicose terminals of the vertebrate autonomic neurons. Such studies should be extended to the more complex parietovisceral ganglion of the Pectinidae.

The localisation of serotonin–like monoamines in the scallop ganglia has been performed in several studies (e.g., Chang et al.1984; Matsutani and Nomura 1986; Paulet et al. 1993). The relation between serotonin and spawning is well-known in many bivalve species, and it is not surprising that serotonin receptors are also situated in the walls of the gonad acini (Matsutani and Nomura 1986). This relation is explored further in Sections 3.8.2.4 and 3.8.3. Other secretions could include growth factor and gonial mitosis stimulating factor, as has been observed in *Mytilus edulis* (Mathieu et al. 1988; Toullec et al. 1988).

3.8.1.4 The circumpallial nerve

The circumpallial nerve is located just interior to the circumpallial artery. It is innervated by both the anterior and posterior pallial nerves, which originate in the cerebral and parietovisceral ganglia, respectively. The circumpallial nerves of the two mantle lobes are fused anteriorly and posteriorly at the hinge line. Both Drew (1906) and Dakin (1909) state that this nerve is physiologically a ganglion, since it is well supplied with ganglion cells and contains abundant nerve cells; however, integrative properties do not appear to exist in relation to visual stimuli (see Chapter 5).

3.8.2 Sensory Structures

Knowledge concerning the various sensory structures of scallops is quite unequal. While the visual system has been the object of continued study for over a century, the remaining structures have received much less attention and are still largely enigmatic.

3.8.2.1 Visual system

A conspicuous feature of scallops, the eyes have been the subject of some study from time to time in the literature (e.g., Dakin 1910a, 1928b; Land 1965, 1968; Morton 2000). The eyes are distributed around the margin of the middle (sensory) fold of the mantle, and originate at the base of the tentacles (Butcher 1930; Figs. 3.33.5, 3.33.6). The eye itself consists of a cornea, lens, double retina and tapetum (Fig. 3.33.6). Light is reflected off the tapetum to the inverted retina. The optic nerve from each eye joins the circumpallial nerve (Fig. 3.33.6), which is innervated by tracts from the lateral (optic) lobes of the parietovisceral ganglion. A detailed description of visual physiology is presented in Chapter 5.

3.8.2.2 Epithelial sensory cells and tentacles

Epithelial sensory cells are probably scattered over the scallop epidermis, but correlative ultrastructural and electrophysiological studies are lacking. Sensory cells are concentrated on papillae in the distal third of the long tentacles, situated at the margin of the middle mantle lobe (see Section 3.2). The haemocoel of the tentacle contains a central tentacular nerve, surrounded by a protective musculo-connective sheath, and flanked by two longitudinal retractor muscles (Figs. 3.33.1, 3.33.2). Extension of the tentacle is effected by hydrostatic pressure from the haemolymph, which fills the haemocoel; contraction of the longitudinal tentacle muscles causes the tentacles to retract to approximately one-tenth of the extended length (Fig. 3.33.1, 3.33.2). The central tentacular nerve is an extension of the tentacular nerve which radiates from the circumpallial nerve (Figs. 3.33.1, 3.33.2, 3.33.6).

Although histologically similar to ordinary epithelial cells, the tentacle sensory cells are often much narrower and bear cilia much longer than those of the other epithelial cells. The base of the sensory cell is joined to nerve fibres running to the central tentacular nerve (Dakin 1909). The structure and ultrastructure of the ciliated receptor cells on the papillae of the long tentacles of *Placopecten magellanicus* have been studied in detail by Moir (1977a). Each papilla bears three specialised cell types: a supporting cell associated with the putative sensory cell which carries up to five cilia, and a third cell type at the base of the papilla possessing many cilia as well as macrocilia (Fig. 3.34). The sensory cells are known to respond to mechanical stimulation.

3.8.2.3 Abdominal sense organ

The abdominal sense organ was first described in pectinids by Dakin (1909). It is visible as a small, yellowish fold of tissue situated on the adductor muscle near the anus (Fig. 3.33.3). Since Dakin's work this enigmatic organ was virtually ignored, until the detailed structural and ultrastructural study by Moir (1977b) and the electrophysiological studies of Zhadan and Semen'kov (1984) and Zhadan and Doroshenko (1985). Ciocco (1985) also described the histology of this structure. The following is a summary of these authors' work, as well as our own histological data presented here.

The sensory epithelium consists of two basic cell types: sensory cells and mucocytes. Zhadan and Semen'kov (1984) described two categories of sensory cells - those with single, long cilia, and those with multiple, short cilia. These two distinct cell categories are innervated by different nerve fibres, which descend toward the fibrillar inner core (Fig. 3.33.4). A third group of nerve fibres innervates the base of the organ. All three groups of nerve fibres respond to mechanical stimulation of the sensory cells, and are connected to the parietovisceral ganglion via one of the posterior pallial nerves. Each group of fibres exhibits different response characteristics.

Although the structural evidence would indicate a chemosensory role for the abdominal sense organ (Dakin 1909; Moir 1977b), it has also been suggested that it functions in mechanoreception (Charles 1966; Moir 1977b). Zhadan and Semen'kov (1984) and Zhadan and Doroshenko (1985) demonstrated the electrophysiological responses of this organ to mechanical stimulation, and suggested that cAMP and cAMP-dependent phosphorylation may be involved in mechanoreception. However, the hypothesis that it may function in the detection of vibrations such as the approach of a predator (Zhadan and Semen'kov 1984) is tenuous, given the position of the abdominal sense organ on one side of the body, well inside the pallial cavity which is almost totally

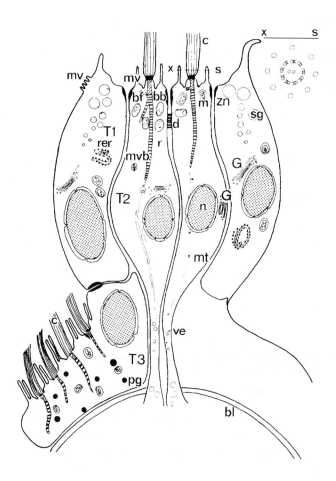

Figure 3.34. Composite diagram of a ciliated papilla on the distal third of a sensory tentacle in *Placopecten magellanicus.* bb, basal body; bf, basal foot of basal body; bl, basal lamina; c, cilia; d, septate desmosomes; G, Golgi apparatus; m, mitochondria; mt, microtubules; mv, microvilli; mvb, multivesicular bodies; n, nucleus; pg, pigment granules; r, ciliary root; rer, rough endoplasmic reticulum; sg, secretory granules; T1, Type I cell, (non-ciliated supporting cell); T2, Type II cell (ciliated sensory cell); T3, Type III cell (macrocilia-bearing cell); X-S, plane of cross-section shown in inset; zn, zonula adherentes. From Moir (1977a).

enclosed by the velum. It is more likely that such detection is effected by the extensile sensory tentacles, which are distributed around the mantle. To date, the suggestions of Charles (1966) and Moir (1977b) appear the most plausible: the abdominal sense organ, being situated directly in the path of exhalent water flow within the pallial cavity, probably functions in the regulation of water flow and, thus, in the regulation of feeding.

3.8.2.4 Osphradia

Sensory structures called osphradia are well known in gastropods (see Garton et al. 1984 for references), and zoologists have long supposed that similar structures exist in bivalves. Most authors agree that the osphradia of bivalves are small and difficult to detect (Drew 1906; Bullock and Horridge 1965; Charles 1966). Indeed, Gutsell (1931) was unable to observe osphradia in *Argopecten irradians*. To date, anatomical evidence indicates that in all probability the osphradia of gastropods and bivalves are not homologous (Kraemer 1979).

The early work of Dakin (1909) gives a description of putative osphradial ganglia in *Pecten maximus*, but does not pinpoint their location. Setna (1930) states that in several scallop species, the osphradia are paired sensory structures situated along the most lateral margins of the gill axis as raised ridges of tissue which are only visible in fresh specimens or in histological section. The osphradia are innervated by branches of the branchial nerve, and contain both ciliated sensory cells and bipolar neurons. Haszprunar (1985a,b; 1987a, b; 1992) did extensive electron microscopic investigations on the osphradia of various molluscan groups, but not the Pectinidae. Such a study was later performed in *Placopecten magellanicus* and *Pecten maximus* (Beninger et al. 1995a). The structure consists of an eversible ridge with subjacent muscle fibres, (Fig. 3.35.2), and a dorsal tuft cilia region. The eversible character of the ridge may account for the confusion concerning the location of the osphradium in species in which it is not pigmented (see below). The ridge epithelium contains many secretory cells, which appear to produce neurosecretions which are transported along microtubules to the bases of the cells, where axons join the osphradial nerve (Figs. 3.35.3–3.35.5). In *Pecten maximus*, pigment granules are also secreted by these cells, giving the osphradium a distinct orange pigmentation (and hence a sure way of locating the structure!). The tuft epithelium appears sensory in structure, with both free nerve fibres and ciliated cells (Fig. 3.35.6) similar to those described by Haszprunar (1987a). No specialised cilia form is observed, and it is probable that the 'paddle cilia' previously reported as diagnostic features of the osphradium (e.g., Haszprunar 1985) are artefacts of hypotonic fixation (Beninger et al. 1995b).

Numerous functions have been proposed for the bivalve osphradium, reviewed by Haszprunar (1987a). This author concluded that the most probable function is that of chemoreception of a gamete release signal. The histological and ultrastructural data for *Placopecten magellanicus* and *Pecten maximus*, together with data on the innervation of the osphradium and the location of monoamines in *Pecten maximus*, led Beninger et al. (1995a) to support this view, proposing a dual function of both stimulus reception (for the sensory cells), and of serotonin production (for the secretory cells of the osphradial ridge), which is then axonally transported (Fig. 3.35.5) and stored in the accessory lobe of the parietovisceral ganglion (Fig. 3.36). The rate of serotonin production and storage could depend on approximate timing cues, such as cyclic temperature changes or even more proximate changes due to downwelling events (Bonardelli et al. 1996). Release of the stored serotonin would induce gamete release in the nearby gonad, allowing vitally-precise synchronisation, enhancing the probability of fertilisation within a population.

202

The nature of the stimulus is not yet known, but is presumed to be a substance contained in male ejaculate.

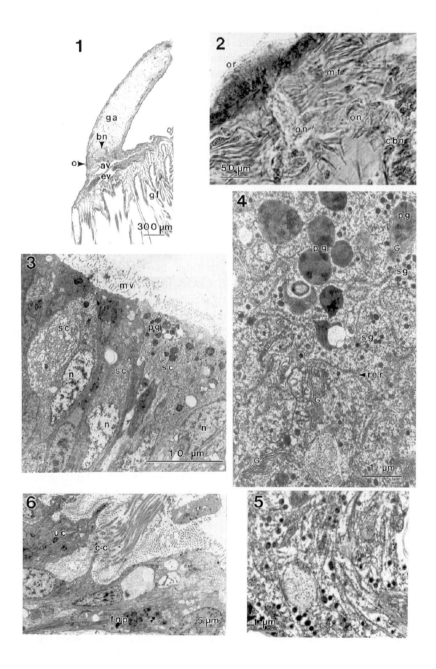

203

Figure 3.35. The osphradium in *Pecten maximus* and *Placopecten magellanicus*. Figure 3.35.1. Histological section showing general anatomical relationships: gill axis (ga), branchial nerve (bn), afferent vessel (av), efferent vessel (ev), osphradium (o), and gill filaments (gf). Figure 3.35.2. Histological section showing detail of osphradial ridge (or), an osphradial nerve (on), cortex of branchial nerve (cbn), and muscle fibres (mf) beneath osphradial epithelium. Figure 3.35.3. TEM showing secretory cells of osphradial ridge (sc), with large basal nuclei (n), and apical pigment granules (pg). mv, microvilli. Figure 3.35.4. TEM detail of apical region of a secretory cell in *Pecten maximus*. G, Golgi complexes; pg, pigment granules; rer, rough endoplasmic reticulum; sg, secretion granules. Figure 3.35.5. TEM detail of the basal region of a secretory cell, showing secretory granules aligned along microtubules, indicative of axonal transport. Figure 3.35.6 TEM detail of tuft cilia region, showing tuft cilia arising from ciliated cell (cc), adjacent free nerve process (fnp), and undifferentiated (supporting) cell (uc). From Beninger et al. (1995).

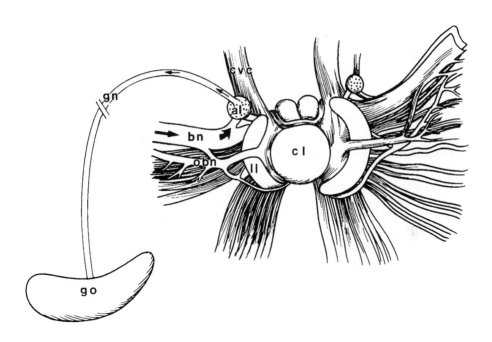

Figure 3.36. The proposed pathway for axonal transport of neurosecretions produced in the secretory cells of the pectinid osphradium (arrows). The dotted regions indicate sites of monoamine storage. Modified after Beninger et al. (1995).

3.8.2.5 Statoreceptors

Much of the remarkable early work concerning pectinid statoreceptors (Drew1906; Buddenbrock 1915) was confirmed and extended in Barber and Dilly (1969). The terminology used by the earlier authors is rather more complex; they distinguish *statocysts*, or closed statoreceptor sacs, and *statocrypts*, or statoreceptors, which communicate with the external medium. In addition, they distinguish *statoliths,* or single endogenously-formed crystals, and *statoconia*, or multiple small crystals. Statoconia of exogenous origin (in the case of statocrypts) are termed *pseudostatoconia.* The statoreceptors of scallops have been termed both statocrypts and statocysts by these authors. Regardless of semantics, the statoreceptors of scallops show a marked asymmetry, the left one being more developed than the right one. Each statoreceptor consists of a sac-shaped sensory epithelium, and a fluid-filled lumen containing one or several crystals. The sensory cells are inappropriately called *hair cells,* due to the sensory cilia at their apical extremities. The left statoreceptor hair cells present up to 30 9+2 type cilia each, with striated roots penetrating deeply into the cell body, whereas the right statocrypt hair cells only have about 10 such cilia. Each hair cell produces an axon at its basal extremity (Fig. 3.37). Both statoconia and statoliths have been reported in scallops (Drew 1906; Buddenbrock 1915; Barber and Dilly 1969). Buddenbrock (1915) believed the statoconia of *Pecten inflexus* to be exogenous, being predominantly formed of sponge spicules (which presumably enter via the statocystic canals) and cemented by mucus into a compact mass, whereas he reported other scallop species such as *Chlamys varia* to have true endogenous statoliths, the canal to the exterior being blind and vestigial (Buddenbrock 1915). The statoreceptor axons group to form the statocystic nerves leading to the cerebral ganglion, as mentioned previously.

The pronounced asymmetry of scallop statoreceptors may correspond to its sedentary life habit, with the animal always resting on the same valve (usually the right one). However, it should be noted that ablation of the statoreceptors in the scallop species studied by Buddenbrock (1915) does not affect the righting reflex. Furthermore, the vertical steering component of the swimming reflex is affected by removal of the left statoreceptor, though unaffected by removal of the right one (Buddenbrock 1915). The left statoreceptor may thus be an essential element in the control of the swimming reflex. This topic is covered in detail in Chapter 5.

3.8.3 Neurotransmitters and Neurohormones

Several different neurotransmitters have been partially identified in the scallop nervous system. The distribution of acetylcholinesterase and monoamine oxidase in the central nervous system of *Patinopecten yessoensis* indicates the presence of acetylcholine and a serotonin-like monoamine, respectively (Chang et al. 1984). A serotonin-like monoamine was also detected by Paulet et al. (1993) in the parietovisceral ganglion. Recent histochemical work has revealed the presence of FMRFamide (Phe-Met-Arg-Phe-NH$_2$) - like molecules in all of the ganglia of *Placopecten magellanicus* (Too and Croll 1995). Using HPLC, Pani and Croll (1995) detected the presence of the catecholamines

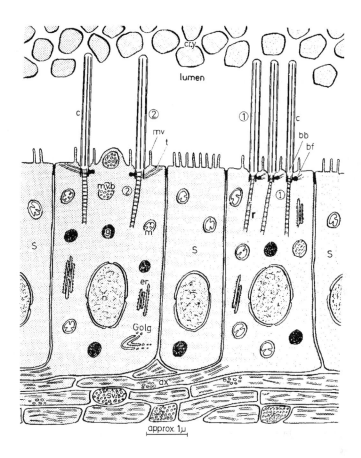

Figure 3.37. Diagram of the statoreceptor wall in *Pecten* sp. ax, sensory cell axons; bb, basal bodies (cinetosomes); bf, basal feet (in type 1 ciliary arrangement); c, cilia (putatively sensory); cry, statoreceptor crystals; er, endoplasmic reticulum; g, membrane-bound granules; Golg, Golgi bodies; m, mitochondria; mv, microvilli; mvb, multivesicular bodies; n, nuclei; r, ciliary roots; s, supporting cells; sv, synaptic vesicles; t, microtubular attachments (in type 2 ciliary arrangement). Reprinted from Barber and Dilly (1969).

3,4 dihydroxyphenylalanine, dopamine, norepinephrine, and epinephrine, as well as the indoleamine 5-hydroxytyptamine, in the ganglia, and in most organs of *Placopecten magellanicus*. The latter studies indicate that knowledge of pectinid and indeed bivalve neurophysiology, is quite embryonic at this stage. We are only now beginning to discover the types of neurotransmitters found in gross anatomical structures; precise mapping to the level of precision now available for the human nervous system is a seemingly distant goal.

The roles of the various bivalve neurotransmitters are even less well known. Serotonin has been implicated in the regulation of sodium transport (Dietz et al. 1984), gamete release (see Ram et al. 1997 for review), ciliary beat (e.g., Jørgensen 1976; Silverman et al. 1999), with cAMP mediation (Stommel and Stephens 1985), cardiac function (Welsh and Moorhead 1960) and probably also in the extremely ubiquitous and polyvalent function of mucus secretion (Lent 1974).

While all of the molecules mentioned above may function as neurotransmitters, it is equally likely that they also have roles as neurohormones. The existence of neurosecretory cells was demonstrated in all of the ganglionic concentrations of several bivalve species by Gabe (1955). Lubet (1955) showed the relationship between the secretory activity of these cells and the reproductive cycle of *Chlamys varia* and *Mytilus edulis*. In *Mytilus edulis*, the greatest concentration of such cells appears to be in the cerebral ganglia (Lubet and Mathieu 1982), from which a gonial mitosis stimulating factor has been isolated (Mathieu et al. 1988). Studies using organ cultures have shown that gametogenesis in *M. edulis* is influenced by nonspecies-specific neurosecretions of the cerebral ganglia (Lubet and Mathieu 1982). In addition, the gametogenesis of cultured *Chlamys opercularis* tissue may be triggered by *M. edulis* ganglia secretions (Allarakh 1979, cited by Toullec et al. 1988).

Since the experiments of Matsutani and Nomura (1982), it is known that serotonin stimulates the actual spawning in *Patinopecten yessoensis*. The same effect has also been obtained for a variety of other bivalves, including *Argopecten irradians* (Gibbons and Castanga 1984). Perikarya of serotonin-like monoamine-secreting neurons have been observed in the cerebral, pedal, and accessory ganglia of *Patinopecten yessoensis*, while serotonin-like, monoamine-secreting fibres derived mostly from the cerebro-visceral connective terminate in the acinal walls and gonoduct of the gonad (Matsutani and Nomura 1984, 1986). Beninger et al. (1995a) proposed that secretions formed in the scallop osphradium were monoamines axonally transported and stored in the accessory lobes of the parietovisceral ganglion; release to the gonad via the gonadal nerve would provoke gamete emission (Fig. 3.36).

In addition to neurosecretions involved in reproduction, the bivalve nervous system includes secretory neurons which regulate somatic growth. A nonspecies-specific growth factor isolated from the cerebral ganglia of *M. edulis* is active with *Pecten maximus* tissue, indicating that a similar secretion exists in pectinids (Toullec et al. 1988). A somatostatin-like substance has also been isolated from peripheral cells of the parietovisceral ganglion of *Pecten maximus* (Le Roux et al. 1987). Somatostatin-like receptor cells have been identified in the kidney (see Section 3.6.1) and in the digestive gland (Le Roux et al. 1987). Clearly, much further study needs to be done concerning the nature and roles of both neurotransmitters and neurohormones in bivalves, and notably in pectinids.

3.9 FOOT-BYSSAL COMPLEX

In contrast to mytilids (see Morse and Zardus 1997 for summary and references), relatively few studies have been performed on the foot and byssal complex of the Pectinidae. Nevertheless, a good account of the structure and function of this system may be derived from the work of Mahéo (1968, 1969), Gruffydd (1978) and Gruffydd et al. (1979).

3.9.1 External Morphology and Development of the Foot-byssal Complex

The pectinid foot arises out of the antero-dorsal surface of the gonad and is roughly cylindrical in shape, terminating in a sucker-like formation called the sole (Drew 1906; Gruffydd 1978), or, more appropriately, the cornet (Pelseneer 1906; Mahéo 1969). The cornet is wedge-shaped in longitudinal section, the dorsal surface being more prominent than the ventral surface. The pedal groove, which runs along the mid-ventral portion of the foot, is enclosed except for a small portion, which is visible as a slight depression just behind the cornet (Fig. 3.38).

The pectinid foot is small compared to that of the Mytilidae; however, it may be more or less developed depending on the adult life habit and the importance of the byssal gland in different species. Most scallops begin their postlarval life as byssally-attached juveniles; some species retain this capability while others grow into free-living adults. In byssally-fixed species such as *Chlamys varia*, the very active foot can be extended out through the byssal notch between the valves. In species in which the adult is free-living, such as *Pecten maximus*, the adult foot is a degenerate structure, which plays no role in locomotion or in fixation. Other species in which the foot-byssal complex regresses in the juvenile stage include *P. jacobeus, Chlamys islandica, C. squamosus,* and

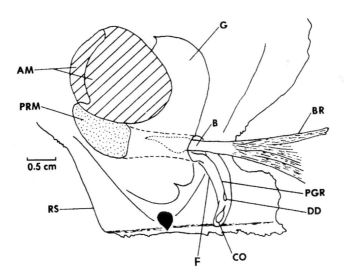

Figure 3.38. Schematic illustration of the anatomical relationships of the foot and byssal gland of *Chlamys varia*. AM, adductor muscle; B, byssus; BR, byssal ribbon; CO, cornet; DD, distal depression; F, foot; G, gonad; PGR, pedal groove; PRM, pedal retractor muscle; RS, right shell. After Mahéo (1969).

C. groenlandicus. In *Placopecten magellanicus* and other non-attached scallop species Drew (1906) suggests that the foot may be used to clean the labial palps and the anterior part of the gills, as observed in pearl oysters (*Pinctata* spp.). This interesting hypothesis merits verification.

3.9.2 Anatomy and Histology of the Foot-byssal Gland Complex

In most bivalves the foot is connected to the shell via an anterior and a posterior pair of pedal retractor muscles. In monomyarians, however, the anterior retractor muscles are generally absent, and in the Pectinidae only the left posterior retractor remains, being inserted on the left shell close to the dorsal margin of the adductor muscle. The degree of development of this muscle varies among species according to the degree of development of the foot, being rather well-developed in *Chlamys opercularis* and rudimentary in *Pecten maximus.*

The external covering of the foot consists of a typical pseudostratified ciliated epithelium (Mahéo 1969). Beneath this layer are found the glandular systems, nerve fibres from the pedal nerves, circular and longitudinal muscle fibres, and lacunar connective tissue filled with haemolymph (Fig. 3.40). The glandular systems, occupying nearly half the volume of the foot, consist of a protein gland, an enzyme gland, and a mucous gland.

3.9.2.1 The protein gland

The protein gland is the most voluminous gland of the foot-byssal complex. It is situated on both sides and above the pedal groove, and also extends into the proximal dorsal portion of the foot. In this region, the dorsal surface of the pedal groove becomes progressively folded, assuming an accordion-like appearance (Figs. 3.39, 3.40). The folds consist of connective tissue, and are simply called connective tissue folds or lamellae. The large (10–15 um) glandular cells secrete a proteinaceous substance either directly into the pedal groove (distal portion), or across the connective tissue lamellae in the proximal portion of the foot. In this region, the secretions traverse the lamellae at their ciliated crypts, and are then swept into the pedal groove by the ciliary action of the epithelial cells lining the groove. Each crypt or trough of the lamellae secretes its own individual ribbon (Gruffydd et al. 1979). Gruffydd (1978) divides the protein gland into two components: the primary (lamellar) byssal gland, and the secondary (distal) byssal gland running along the pedal groove.

3.9.2.2 The enzyme gland

The enzyme gland consists of a thin layer of tissue situated below the protein gland, on either side of the pedal groove (Figs. 3.39, 3.40). It does not extend into the proximal portion of the foot. The secretions of this gland include a polyphenoloxidase (Mahéo 1969; Gruffydd 1978).

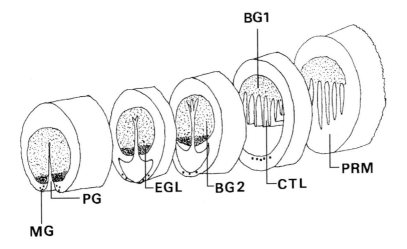

Figure 3.39. Schematic illustration of the glandular systems of the foot-byssal complex of *Chlamys varia*. BG1, primary byssus gland; BG2, secondary byssus gland; CTL, connective tissue lamellae; EGL, enzyme gland; MG, mucous gland; PG, pedal groove; PRM, pedal retractor muscle. Modified after Mahéo (1969).

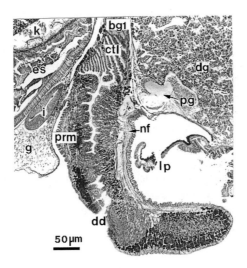

Figure 3.40. Oblique sagittal section of the foot and surrounding structures of a juvenile *Pecten maximus*. bg1, primary byssus gland; cs, crystalline style within the descending loop of the intestine; ctl, connective tissue lamellae; dd, distal depression; dg, digestive gland; es, oesophagus; g, developing gonad; i, ascending loop of the intestine; k, kidney; lp, sectioned labial palp; nf, nerve fibres; pg, pedal ganglion; prm, pedal retractor muscle. The Masson trichrome 1 stain does not allow the other glandular systems to be distinguished. Micrograph courtesy of P. Beninger and A. Donval, Université de Bretagne Occidentale.

3.9.2.3 The byssus

The byssus is formed of a root from which extend several relatively thick ribbons, in contrast to the many thin threads of the Mytilidae (Mahéo 1969; Gruffydd 1978; Gruffydd et al. 1979). It is composed of complex proteinaceous molecules (Bubel 1984). The root is formed of 30–50 fibrous ribbons formed from secretions of the primary byssus gland (Gruffydd 1978; Gruffydd et al. 1979). These fibres are covered by a thin matrix of amorphous material at their proximal extremities. The proximal portions are thick and unstructured, whereas the distal portions are finely folded and the matrix covering becomes progressively thinner (Gruffydd et al. 1979). The ribbons are bound together by a fibrous sheath, secreted by the secondary byssus gland (Gruffydd et al. 1979).

3.9.3 Functioning of the Foot-byssal Complex

Prior to fixation, the foot extends out from the byssal notch in the shell and actively explores the surrounding substrates. The cornet then presses against a substrate, and several minutes later (4–10 minutes in *Chlamys varia*) the foot retracts, leaving a filament which fixes the animal to the substrate (Mahéo 1969). Throughout the period of extension, protein granules are secreted by the primary and secondary byssus gland and are swept into the pedal groove. The secretions are presumably tanned by the products of the enzyme gland, although it is not yet known exactly how the proteinaceous secretions are formed into distinct ribbons bounded by a sheath. The ciliary beating of the pedal groove epithelial cells probably acts to mix the aromatic proteins and the polyphenoloxidase, promoting tanning. Gruffydd et al. (1979) propose that incompletely-tanned ribbons are stored in a special pouch of the primary byssus gland, being moved into the pedal groove and stretched outward by ciliary action.

Once the ribbon is tanned and anchored to the substratum, the sides of the pedal groove curl outward along the short non-enclosed length, leaving a ribbon, which extends into the primary byssus gland. A new thread may then be put in place, more or less fused with the previous one by incomplete tanning. In the proximal region of the foot, the accumulation of untanned and partially-tanned ribbons forms the byssal root. The root is held in place by the continuous contraction of numerous muscle fibres. Prior to the swimming response, these muscles relax, separating the untanned portions of the ribbons from the primary byssus gland. The byssus then detaches and the animal is freed of its connection to the substrate (Mahéo 1969).

ACKNOWLEDGMENTS

Photographic assistance was kindly provided by A. Le Mercier and L. Blanchard. Assistance with electron microscopy of previously unpublished work was provided by Dr. G. Sinquin. Mad. O. Aumaille assisted with some lettering and reference checks. Funding during the preparation of the second edition of this chapter was provided by the Syndicat Mixte pour le Développement de l'Aquaculture aux Pays de la Loire, although not for this project.

REFERENCES

Albertini, C., 1985. Recherches cytologiques et expérimentales sur l'ovogénèse chez la moule (*Mytilus edulis* L., Mollusque Bivalve). Thèse 3e cycle, Université de Caen, Caen, France. 117 p.

Allarakh, C., 1979. Recherches histologiques et expérimentales de la différenciation sexuelle et du cycle de reproduction de *Chlamys opercularis* L. (Mollusque Lamellibranche). Thèse 3e cycle, Université de Caen, Caen, France. 148 p.

Amirthalingam, C., 1928. On lunar periodicity in reproduction of *Pecten opercularis* near Plymouth in 1927–28. J. Mar. Biol. Assoc. U.K. 15:605–641.

Andrews, E.B., 1988. Excretory systems of molluscs. In: E.R. Trueman and M.R. Clarke (Eds.). The Mollusca, Vol. 11: Form and Function. Academic Press, San Diego. pp. 381–448.

Andrews, E.B., and Jennings, K.H., 1993. The anatomical and ultrastructural basis of primary urine formation in bivalve molluscs. J. Moll. Stud. 59:223–257.

Atkins, D., 1936. On the ciliary mechanisms and interrelationships of lamellibranchs. Part I: New observations on sorting mechanisms. Q. J. Microsc. Sci. 79:181–308.

Atkins, D., 1937a. On the ciliary mechanisms and interrelationships of lamellibranchs. Part II: Sorting devices on the gills. Q. J. Microsc. Sci. 79: 339–373.

Atkins, D., 1937b. On the ciliary mechanisms and interrelationships of lamellibranchs. Part III: Types of lamellibranch gills and their food currents. Q. J. Microsc. Sci. 79:375–421.

Atkins, D., 1937c. On the ciliary mechanisms and interrelationships of lamellibranchs. Part IV: Cuticular fusion. Q. J. Microsc. Sci. 79:423–445.

Atkins, D., 1938a. On the ciliary mechanisms and interrelationships of lamellibranchs. Part V: Note on gills of *Amussium pleuronectes* L. Q. J. Microsc. Sci. 80:321–329.

Atkins, D., 1938b. On the ciliary mechanisms and interrelationships of lamellibranchs. Part VI: Pattern of the lateral ciliated cells of the gill filaments. Q. J. Microsc. Sci. 80:331–344.

Atkins, D., 1938c. On the ciliary mechanisms and interrelationships of lamellibranchs. Part VII: Latero-frontal cilia of the gill filaments and their phylogenic value. Q. J. Microsc. Sci. 80:345–436.

Atkins, D., 1943. On the ciliary mechanisms and interrelationships of lamellibranchs. Part VIII: Notes on gill musculature in the microciliobranchia. Q. J. Microsc. Sci. 84:187–256.

Auffret, M., 1985. Morphologie comparative des types hémocytaires chez quelques Mollusques Bivalves d'intérêt commercial. Thèse de doctorat de spécialité, Université de Bretagne Occidentale, Brest, France. 155 p.

Auffret, M., 1988. Bivalve hemocyte morphology. In: S.G. Pal (Ed.). Disease Processes in Marine Bivalve Molluscs. American Fisheries Society, Special Publication Series 18, Bethesda, Maryland. pp. 169-177.

Barber, V.C. and Dilly, P.N., 1969. Some aspects of the fine structure of the statocysts of the molluscs *Pecten* and *Pterotrachea*. Z. Zellforsch. 94:462–478

Barnes, R.D., 1987. Invertebrate Zoology. Saunders College Publishing, Philadelphia. 893 p.

Beninger, P.G., 1985. Long-term variations in cation content of two populations of adult marine clam (*Tapes decussatus* L. and *T. philippinarum*) reared in a common habitat. Comp. Biochem. Physiol. 82A:945–949.

Beninger, P., 1987. A qualitative and quantitative study of the reproductive cycle of the giant scallop *Placopecten magellanicus,* in the Bay of Fundy (New Brunswick, Canada). Can. J. Zool. 65:495–498.

Beninger, P.G., 1991. Structures and mechanisms of feeding in scallops: paradigms and paradoxes. In: S.E. Shumway (Ed.). An International Compendium of Scallop Biology and Culture. World Aquaculture Workshops, Volume 1. World Aquaculture Society, Baton Rouge, LA. pp. 331–340.

Beninger, P.G., Auffret, M. and Le Pennec, M., 1990a. The peribuccal organs of *Placopecten magellanicus* and *Chlamys varia* (Mollusca : Bivalvia) : structure, ultrastructure and implications for feeding. I. The labial palps. Mar. Biol. 107:215–223.

Beninger, P.G. and Decottignies, P. What makes diatoms attractive for suspensivores? The organic casing and associated organic molecules of *Coscinodiscus perforatus* are quality cues for the bivalve, *Pecten maximus.* J. Plankton Res., in press.

Beninger, P.G., Decottignies, P. and Rincé, Y., 2004. Localization of qualitative particle selection sites in the heterorhabdic filibranch *Pecten maximus* (Bivalvia: Pectinidae). Mar. Ecol. Prog. Ser. 275:163–173.

Beninger, P.G., Donval, A. and Le Pennec, M., 1995a. The osphradium in *Placopecten magellanicus* and *Pecten maximus* (Bivalvia, Pectinidae): histology, ultrastructure and implications for spawning synchronisation. Mar. Biol. 123:121–129.

Beninger, P.G. and Dufour, S.C., 2000. Evolutionary trajectories of a redundant feature: lessons from bivalve gill abfrontal cilia and mucocyte distributions. In: E.M. Harper, J.D. Taylor and J.A. Crame (Eds.). The Evolutionary Biology of the Bivalvia. Geological Society, London, Special Publications 177:273–278.

Beninger, P.G., Dufour, S. and Bourque, J., 1997a. Particle processing mechanisms on the pallial organs of the eulamellibranchs *Spisula solidissima* and *Mya arenaria.* Mar Ecol. Prog. Ser. 150:157–169.

Beninger, P.G., Dwiono, S. and Le Pennec, M., 1994. Early development of the gill and implications for feeding in *Pecten maximus* (Bivalvia : Pectinidae). Mar. Biol. 119:405–412.

Beninger, P.G. and Le Pennec, M., 1993. Histochemistry of the bucco-oesophageal glands of *Mytilus edulis*: the importance of mucus in ingestion. J. Mar. Biol. Assoc. U.K. 73:237–240.

Beninger, P.G. and Le Pennec, M., 1997. Reproductive characteristics of a primitive bivalve from a deep-sea reducing environment: giant gametes and their significance in *Acharax alinae* (Cryptodonta: Solemyidae). Mar. Ecol. Prog. Ser. 157:195–206.

Beninger, P.G., Le Pennec, M. and Auffret, M., 1990b. The peribuccal organs of *Placopecten magellanicus* and *Chlamys varia* (Mollusca : Bivalvia): structure, ultrastructure, and implications for feeding. II. The lips. Mar. Biol. 107:225–233.

Beninger, P.G., Le Pennec, M. and Donval, A., 1991. Mode of particle ingestion in five species of bivalve molluscs. Mar. Biol. 108:255–261.

Beninger, P.G., Le Pennec, G. and Le Pennec, M., 2003. Demonstration of nutrient pathway from the digestive system to oocytes in the gonad intestinal loop of the scallop *Pecten maximus* L: the intestinal loop pathway. Biol. Bull. 205:83–92.

Beninger, P.G., Le Pennec, M. and Salaün, M., 1988. New observations of the gills of *Placopecten magellanicus* (Mollusca: Bivalvia), and implications for nutrition. Mar. Biol. 98:61–70.

Beninger, P.G., Lynn, J.W., Dietz, T.H. and Silverman, H., 1997b. Mucociliary transport in living tissue: the two-layer model confirmed in the mussel *Mytilus edulis* L. Biol. Bull. 193:4–7.

Beninger, P.G., Potter, T.M. and St-Jean, S.D., 1995b. Paddle cilia fixation artefacts in pallial organs of adult *Mytilus edulis* and *Placopecten magellanicus*. Can. J. Zool. 73:610–614.

Beninger P.G., and St-Jean, S.D., 1997a. Particle processing on the labial palps of *Mytilus edulis* and *Placopecten magellanicus* (Mollusca, Bivalvia). Mar. Ecol. Prog. Ser. 147:117–127.

Beninger, P.G. and St-Jean, S.D., 1997b. The role of mucus in particle processing by suspension-feeding marine bivalves: unifying principles. Mar. Biol. 129:389–397.

Beninger, P.G., St-Jean, S.D., Poussart, Y. and Ward, E., 1993. Gill function and mucocyte distribution in *Placopecten magellanicus* and *Mytilus edulis* (Mollusca : Bivalvia): the role of mucus in particle transport. Mar. Ecol. Prog. Ser. 98:275–282.

Beninger, P.G., Veniot, A. and Poussart, Y., 1999. Principles of pseudofeces rejection on the bivalve mantle: integration in particle processing. Mar. Ecol Prog. Ser. 178:259–269.

Beninger, P.G., Ward, J. E., MacDonald, B.A. and Thompson, R.J., 1992. Feeding processes of the gill in *Placopecten magellanicus* (Gmelin) (Mollusca : Bivalvia) as revealed using video endoscopy. Mar. Biol. 114:281–288.

Bevelander, G. and Nakahara, H., 1966. Correlation of lysosomal activity and ingestion by the mantle epithelium. Biol. Bull. 131:76–82.

Bojanus, L.H., 1819. Mémoire sur les organes respiratoires et circulatoires des coquillages bivalves en général, et spécialement sur ceux de l'Anodonte des Cygnes (*Anodon cygneum*). J. Physiol. 89:108–134.

Bonardelli, J.C., Himmelman, J.H. and Drinkwater, K., 1996. Relation of spawning of the giant scallop, *Placopecten magellanicus*, to temperature fluctuations during downwelling events. Mar. Biol. 124:637–649.

Booth, C.E. and Mangum, C.P., 1978. Oxygen uptake and transport in the lamellibranch mollusc *Modiolus demissus*. Physiol. Zool. 51:17–32.

Bourne, N., 1964. Scallops and the offshore fishery of the Maritimes. Bull. Fish. Res. Board Can. No. 145. 60 p.

Brand, A.R. and Roberts, D., 1973. The cardiac responses of the scallop *Pecten maximus* (L.) to respiratory stress. J. Exp. Mar. Biol. Ecol. 13:29–43.

Brillant, M.G.S. and MacDonald, B.A., 2000. Postingestive selection in the sea scallop, *Placopecten magellanicus* (Gmelin): the role of particle size and density. J. Exp. Mar. Biol. Ecol. 253:211–227.

Brillant, M.G.S. and MacDonald, B.A., 2003. Postingestive sorting of living and heat-killed *Chlorella* within the sea scallop, *Placopecten magellanicus* (Gmelin). J. Exp. Mar. Biol. Ecol. 290:81–91

Brock, V., Kennedy, V.S. and Brock, A., 1986. Temperature dependency of carbohydrate activity in the hepatopancreas of thirteen estuarine and coastal bivalve species from the North American east coast. J. Exp. Mar. Biol. Ecol. 103:87–101

Brown, C.H., 1952. Some structural proteins of *Mytilus edulis*. Q. J. Microsc. Sci. 4:487–502.

Brunel, A., 1938. Sur la dégradation des substances d'origine purique chez les Mollusques Lamellibranches. C. R. Acad. Sci. Paris Ser. III 206:858–860.

Butcher, E.O., 1930. The formation, regeneration, and transplantation of eyes in *Pecten* (*Gibbus borealis*). Biol. Bull. Mar biol. lab. Woods Hole 59:154–164.

Bryan, G.W., 1973. The occurrence and seasonal variation of trace metals in the Scallops *Pecten maximus* (L.) and *Chlamys opercularis* (L.). J. Mar. Biol. Assoc. U.K. 53:154–166.

Bubel, A., 1973a. An electron microscopic study of the periostracum formation in some marine bivalves. I. The origin of the periostracum. Mar. Biol. 20:213–221.

Bubel, A., 1973b. An electron microscopic study of the periostracum formation in some marine bivalves. II. The cell lining in the periostracal groove. Mar. Biol. 20:222–234.

Bubel, A., 1984. Mollusca. In: J. Bereiter-Hahn, A.G. Matoltsy and K.S. Sylvia Richards (Eds.). Biology of the Integument. I. Invertebrates. Springer-Verlag, Berlin. pp. 421–447.

Buddenbrock, W.V., 1915. Die Statocyste von *Pecten*, ihre Histologie und Physiologie. Zool. Jahrb. Abt. Allg. Zool. Physiol. Tiere 35:301–356.

Bullock, T.H. and Horridge, G.A., 1965. Mollusca: Pelecypoda. In: T.H. Bullock and G.A. Horridge (Eds.). Structure and Function in the Nervous Systems of Invertebrates, Vol. II. W.H. Freeman and Company, San Francisco. pp. 1390–1431.

Burton, R.F., 1983. Ionic regulation and water balance. In: A.S.M. Saleuddin and K. M. Wilbur (Eds.). The Mollusca. Physiology, Part 2. New York: Academic Press. pp. 291–347.

Carmichael, N.G. and Fowler, B.A., 1981. Cadmium accumulation and toxicity in the kidney of the bay scallop *Argopecten irradians*. Mar. Biol. 65:35–43.

Carmichael, N.G., Squibb, K.S. and Fowler, B.A., 1979. Metals in the molluscan kidney: a comparison of two closely related bivalve species (*Argopecten*), using X-ray microanalysis and atomic absorption spectroscopy. J. Fish. Res. Board. Can. 36:1149–1155.

Chan, J.F.Y. and Saleuddin, A.S.M., 1974. Acid phosphatase in the mantle edge of the shell-regenerating snail. Top. Curr. Chem. 64:1–112.

Chang, Y.G., Matsutani, T., Mori, K. and Nomura, T., 1984. Enzyme histochemical localization of monoamine oxydase and acetylcholinesterase in the central nervous system of the scallop, *Patinopecten yessoensis*. Mar. Biol. Lett. 5:335–345.

Charles, G.H., 1966. Sense organs (less cephalopods). In: K.M. Wilbur and C.M. Yonge (Eds.). Physiology of Mollusca, Vol. II. Academic Press, New York. pp. 465–521.

Cheng, T.C., 1975. Lysosomal and other enzymes in the hemolymph of *Crassostrea virginica* and *Mercenaria mercenaria*. Comp. Biochem. Physiol. 52B:443–447.

Cheng, T.C., 1981. Bivalves. In: N.A. Ratcliffe and A.F. Rowley (Eds.). Invertebrate Blood Cells, Vol. 1. Academic Press, London. pp. 223–300.

Cheng, T.C., 1996. Hemocytes: forms and functions. In: V.S. Kennedy, R.I.E. Newell and A.F. Eble (Eds). The Eastern Oyster, *Crassostrea virginica*. Maryland Sea Grant Books, College Park, Maryland USA. pp. 299-334.

Chipman, W. and Thommeret, J., 1970. Manganese content and the occurrence of the fallout of ^{54}Mn in some marine benthos of the Mediterranean. Bull. Inst. Oceanogr. Monaco 69, No. 1402: 15 p.

Christiansen, H.R. and Olivier, S.R., 1971. Sobre el hermaproditismo de *Chlamys tehuelcha* D'orb. 1846. (Pelecypoda, Filibranchia, Pectinidae). An. Soc. Cient. Argent. 3:115–127.

Ciocco, N.F., 1985. Biologia y ecologia de *Chlamys tehuelcha* d'Orbigny en el Golfo san José (Provincia del Chubut, Republica Argentina) Pelecypoda, Pectinidae. PhD Thesis. Universidad Nacional de La Plata, Argentina.

Cobb, J.L.S. and Mullins, P.A., 1973. Synaptic structure in the visceral ganglion of the lamellibranch mollusc, *Spisula solidissima*. Z. Zellforsch. 138:75–83.

Coe, W.R., 1943a. Development of the primary gonads and differentiation of sexuality in *Teredo navalis* and other Pelecypod Mollusks. Biol. Bull. 84: 178–187.

Coe, W.R., 1943b. Sexual differentiation in molluscs. I. Pelecypods. Q. Rev. Biol. 18:154–164.

Coe, W.R., 1945. Development of the reproductive system and variations in sexuality in *Pecten* and other Pelecypod Mollusks. Trans. Conn. Acad. Arts Sci. 36:673–700.

Cognie, B., Barillé, L., Massé, G. and Beninger, P.G., 2003. Selection and processing of large suspended algae in the oyster *Crassostrea gigas*. Mar. Ecol. Prog. Ser. 250:145–152.

Comely, C.A., 1972. Larval culture of the scallop, *Pecten maximus*. J. Cons. Int. Explor. Mer. 34:365–378.

Coombs, T.L. and George, S.G., 1977. Mechanisms of immobilization and detoxication of metals in marine organisms. In: D.S. McLusky and A.J. Berry (Eds.). Physiology and Behavior of Marine Organisms. Proceedings of the 12th European Marine Biology Symposium, Pergamon Press, Oxford. pp. 179–187.

Costello, T.J., Hudson, M.J., Dupuy, J.L. and Rivkin, S., 1973. Larval culture of the calico scallop *Argopecten gibbus*. Proc. Natl. Shellfish. Assoc. 63:72–76.

Crenshaw, M.A., 1980. Mechanisms of shell formation and dissolution. In: D.C. Rhoads and R.A. Lutz (Eds.). Skeletal Growth of Aquatic Organisms. Plenum, New York. pp. 67–86.

Culliney, J.L. 1974. Larval development of the giant scallop *Placopecten magellanicus*. Biol. Bull. 147:321–332.

Dakin, W.J., 1909. *Pecten*. Liverpool Marine Biology Committee Memoirs No. 17: 146 p.

Dakin, W.J., 1910a. The eye of *Pecten*. Q. J. Microsc. Sci. 55:49–112 + 2 pl.

Dakin, W.J., 1910b. The visceral ganglion of *Pecten*, with some notes on the physiology of the nervous system, and an inquiry into the innervation of the osphradium in Lamellibranchia. Mitt. Zool. Statz. Neapel 20:1–40 + 2 pl.

Dakin, W.J., 1928a. The anatomy and phylogeny of *Spondylus*, with a particular reference to the lamellibranch nervous system. Proc. R. Soc. Lond. B Biol. Sci. 103:337–354.

Dakin, W.J., 1928b. The eyes of *Pecten, Spondylus, Amussium*, and allied lamellibranchs, with a short discussion on their evolution. Proc. R. Soc. (B) 103:355–365

Davenport, J. and Lehane, C., 2000. Mussels *Mytilus edulis*: significant consumers and destroyers of mesozooplankton. Mar. Ecol. Prog. Ser. 198:131–137.

de Lacaze-Duthiers, H., 1854. Recherches sur les organes génitaux des acéphales lamellibranches. Annales de Sc. Nat. Paris, 4e série, Zoologie 3:155–248, Pl. 5–9.

Desilets, J., Gicquaud, C. and Dubé, F., 1995. An ultrastructural analysis of early fertilization events in the giant scallop, *Placopecten magellanicus* (Mollusca, Pelecypoda). Invert. Reprod. Dev. 27:115–129.

Dietz, T.H., Steffens, W.L., Kays, W.T. and Silverman, II., 1984. Serotonin localization in the gills of the freshwater mussel, *Ligumia subrostrata*. Can. J. Zool. 63:1237–1243.

Dix, T.G., 1976. Larval development of the Queen scallop *Equichlamys bifrons*. Aust. J. Mar. Freshwat. Res. 27:399–403.

Dix, T.G. and Jardin, S., 1975. Larvae of the commercial scallop, *Pecten meridionalis* from Tasmania, Australia. Aust. J. Mar. Freshwat. Res. 26:109–112.

Dohmen, M.R., 1983. Gametogenesis. In: N.H. Verdonk, J.A.M. van den Biggelaar and A.S. Tompa (Eds.). The Mollusca, Vol. 3: Development. Academic Press, London. pp. 1–48.

Dorange, G., Paulet, Y.M. and Le Pennec, M., 1989a. Etude cytologique de la partie femelle de la gonade de *Pecten maximus* récolté en baie de St. Brieuc. I. Caractéristiques ultrastructurales des tissus somatiques. Haliotis 19:287–297.

Dorange, G., Paulet, Y.M. and Le Pennec, M., 1989b. Etude cytologique de la partie femelle de la gonade de *Pecten maximus* récolté en baie de St. Brieuc. II. Ovogénèse et lyse ovocytaire. Haliotis 19:299–314.

Dorange, G. and Le Pennec, M., 1989c. Ultrastructural study of oogenesis and oocytic degeneration in *Pecten maximus* from the Bay of St. Brieuc. Mar. Biol. 103:339–348.

Dorange, G. and Le Pennec, M., 1989d. Ultrastructural characteristics of spermatogenesis in *Pecten maximus* (Mollusca, Bivalvia). Invert. Reprod. Devel. 15:109–117.

Doyle, L.J., Blake, N.J., Woo, C.C. and Yevich, P., 1978. Recent biogenic phosphorite: Concretions in Mollusk kidneys. Science 199:1431–1433.

Dufour, S.C. and Beninger, P.G., 2001. A functional interpretation of cilia and mucocyte distributions on the abfrontal surface of bivalve gills. Mar. Biol. 138: 295–309.

Dufour, S.C., Steiner, G. and Beninger, P.G. Phylogenetic analysis of the peri-hydrothermal vent bivalve *Bathypecten vulcani*, based on 18S rRNA and anatomical characters. Malacologia, in press.

Drew, G.A., 1906. The habits, anatomy, and embryology of the giant scallop (*Pecten tenuicostatus*, Mighels). Univ. Maine Studies Series, No. 6, Orono. 89 p.

Eble, A.F., 2001. Anatomy and histology of *Mercenaria mercenaria*. In: J.N. Kraeuter, and M. Castagna (Eds.). Biology of the Hard Clam. Developments in Aquaculture and Fisheries Science, Vol. 31. Elsevier, Amsterdam. pp. 117–220.

Eckelbarger, K.J., and Davis, C.V., 1996. Ultrastructure of the gonad and gametogenesis in the eastern oyster, *Crassostrea virginica*. I. Ovary and oogenesis. Mar. Biol. 127:79–87.

Eckelbarger, K.J. and Davis, C.V., 1996. Ultrastructure of the gonad and gametogenesis in the eastern oyster, *Crassostrea virginica*. II. Testis and spermatogenesis. Mar. Biol. 127:89–96

Eckelbarger, K.J. and Young, C.M., 1999. Ultrastructure of gametogenesis in a chemosynthetic mytilid bivalve (*Bathymodiolus childressi*) from a bathyal, methane seep environment (Northern Gulf of Mexico). Mar. Biol. 135:635–646.

Foster-Smith, R.L., 1975. A comparative study of the feeding mechanisms of *Mytilus edulis* L., *Cerastoderma edule* (L), and *Venerupis pullastra* (Montague) (Mollusca:Bivalvia). PhD Thesis, University of Newcastle-upon-Tyne, England. 153 p.

Franc, A., 1960. Classe des Bivalves. In: P.P. Grassé (Ed.). Traité de Zoologie, Tome 5, Fascicule 2. Librairie Masson, Paris. pp. 845–2133.

Fretter, V. and Graham, A., 1964. Reproduction. In: K.M. Wilbur and C.M. Yonge (Eds.). Physiology of Mollusca, Academic Press, New York 1:127–164.

Gabe, M., 1955. Particularités histologiques des cellules neurosécrétrices chez quelques Lamellibranches. C.R. Acad. Sci. Paris 240:1810–1812.

Garrido, O. and Gallardo, C.S., 1996. Ultrastructure of sperms in bivalve molluscs of the Mytilidae family. Invert. Reprod. Devel. 29:95–102.

Garton, D.W., Roller, R.A. and Caprio, J., 1984. Fine structure and vital staining of osphradium of the southern oyster drill, *Thais laemastoma canaliculata* (Gray) (Prosobranchia: Muricidae). Biol. Bull. 167:310–321.

George, S.G., Pirie, B.J.S. and Coombs, T.L., 1980. Isolation and elemental analysis of metal-rich granules from the kidney of the scallop *Pecten maximus* (L.). J. Exp. Mar. Biol. Ecol. 42:143–156.

Ghiretti, F., 1966. Respiration. In: K.W. Wilbur and C.M. Yonge (Eds.). Physiology of Mollusca Vol. II. Academic Press, New York. pp. 209–231.

Gibbons, M.C. and Castanga, M., 1984. Serotonin as an inducer of spawning in six bivalve species. Aquaculture 40:189–191.

Gilmour, T.H.J., 1964. The structure, ciliation and function of the lip-apparatus of *Lima* and *Pecten* (Lamellibranchia). J. Mar. Biol. Assoc. U.K. 44:485–498.

Gilmour, T.H.J., 1974. The structure, ciliation and function of the lips of some bivalve molluscs. Can. J. Zool. 52:335–343.

Gröbben, G., 1888. Die pericardialdruse der lamellibranchiaten. Arbeiten aus des Zoologischen Instituten der Universität Wien und der Zoologischen Station in Triest 7:355–444.

Gruffydd, Ll. D., 1978. The byssus and byssus glands in *Chlamys islandica* and other scallops (Lamellibranchia). Zool. Scr. 7:277–285.

Gruffydd, Ll. D., Budiman, A. and Nott, J.A., 1979. The ultrastructure of the byssus and the byssus glands in *Chlamys varia* L. (Lamellibranchia). J. Mar. Biol. Assoc. U.K. 59:597–603.

Gutsell, J.S., 1931. Natural history of the bay scallop. Bulletin of the Bureau of Fisheries (United States Bureau of Fisheries) 46:569–632.

Haszprunar, G., 1985a. The fine morphology of the osphradial sense organs of the Mollusca. I. Gastropoda, Prosobranchia. Phil. Trans. R. Soc. Lond. B 307:63–73.

Haszprunar, G., 1985b. The fine morphology of the osphradial sense organs of the Mollusca. II. Allogastropoda (Architectonicidae, Pyramidellidae). Phil. Trans. R. Soc. Lond. B 307:497–505.

Haszprunar, G., 1987a. The fine morphology of the osphradial sense organs of the Mollusca. IV. Caudofoveata and Solenogastres. Phil. Trans. R. Soc. Lond. B 315:63–73.

Haszprunar, G., 1987b. The fine morphology of the osphradial sense organs of the Mollusca. III. Placophora and Bivalvia. Phil. Trans. R. Soc. Lond. B 315:37–61.

Haszprunar, G., 1992. Ultrastructure of the osphradium of the Tertiary relict snail, *Campanile symbolium* Iredale (Mollusca, Streptoneura). Phil. Trans. R. Soc. Lond. B 337:457–469.

Haszprunar, G., 1996. The molluscan rhogocyte (pore cell, blasenzelle, cellule nucale), and its significance for ideas on nephridial evolution. J. Moll. Stud. 62:185–211.

Healy, J.M., 1995. Comparative spermatozoal ultrastructure and its taxonomic and phylogenetic significance in the bivalve order Veneroida. In: B.G.M. Jamieson, J. Ausio and J-L. Justine (Eds.). Advances in Spermatozoal Phylogeny and Taxonomy. Mem. Mus. Hist. Nat. 166:155–166.

Healy, J.M., 1996. Molluscan sperm ultrastructure: correlation with taxonomic units within the Gastropoda, Cephalopoda, and Bivalvia. In: J. Taylor (Ed.). Origin and Evolutionary Radiation of the Mollusca. Oxford University Press. pp. 99–113

Healy, J.M., Keys, J.L. and Daddow, L.Y.M., 2000. Comparative sperm ultrastructure in pteriomorphan bivalves with special reference to phylogenetic and taxonomic implications. In: E.M. Harper, J.D. Taylor and J.A. Crame (Eds.). The Evolutionary Biology of the Bivalvia. Geological Society, London, Special Publications 177:169–190.

218

Henry, M., Boucaud-Camou, E. and Lefort, Y., 1991. Functional micro-anatomy of the digestive gland of the scallop *Pecten maximus* (L.). Aquat. Living Res. 4:191–202.

Hochachka, P.W. and Somero, G.W., 1984. Biochemical Adaptation, Chapter Ten. Water Solute Adaptations: The Evolution and Regulation of Biological Solutions. Princeton University Press, Princeton, New Jersey. pp. 304–354.

Hodgson, A.N. and Bernard, R.T.F., 1986. Ultrastructure of the sperm and spermatogenesis of three species of Mytilidae (Mollusca, Bivalvia). Gamete Res. 15:123–135.

Irisawa, H., Irisawa, I. and Shigeto, N., 1973. Physiological and morphological correlation of the functional syncytium in the bivalve myocardium. Comp. Biochem. Physiol. 44A:207–219.

Janssens, F., 1893. Les branchies des acéphales. La cellule 9:6–90.

Jennings, K.H., 1984. The organization, fine structure and function of the excretory systems of the estuarine bivalve, *Scrobicularia plana* (da Costa) and the freshwater bivalve *Anodonta cygnea* (Linné) and other selected species. PhD Thesis, University of London, England. Cited by E.B. Andrews (1988).

Jeuniaux, C., 1982. La chitine dans le règne animal. Bull. Soc. Zool. Fr. 107:363–386.

Jones, H.D., 1983. The circulatory systems of gastropods and bivalves. In: A.S.M. Saleuddin and K.M. Wilbur (Eds.). The Mollusca, Vol. 5: Physiology, Part 2. Academic Press, New York. pp. 189–238.

Jørgensen, C.B., 1976. Comparative studies on the function of gills in suspension-feeding bivalves, with special reference to effects of serotonin. Biol. Bull. 151:331–343.

Judd, W., 1987. Crystalline style proteins from bivalve molluscs. Comp. Biochem. Physiol. 88B:333–339.

Kafanov, A.I. and Drozdov, A.L., 1998. Comparative sperm morphology and phylogenetic classification of recent Mytiloidea (Bivalvia). Malacologia 39:129–139.

Kellogg, J.L., 1892. A contribution to our knowledge of the morphology of the lamellibranchiate molluscs. Bull. U.S. Fish Commission 10:389–434 + Pl. 79–94.

Kennedy, V.S., Newell, R.I.E. and Eble, A.F. (Eds), 1996. The Eastern Oyster, *Crassostrea virginica*. Maryland Sea Grant Books, College Park, Maryland USA. 734 p.

Kellogg, J.L., 1915. Ciliary mechanisms of lamellibranchs with descriptions of anatomy. J. Morph. 26:625–701.

Kiørboe, T. and Møhlenberg, F., 1981. Particle selection in suspension-feeding bivalves. Mar. Ecol. Prog. Ser. 5:291–296.

Kraemer, L.R., 1979. Suprabranchial and branchial shelves of bivalved molluscs: structural/functional context of visceral ganglion, osphradium and branchial nerves. Am. Zool. 19:959.

Kraeuter, J.N. and Castagna, M. (Eds.), 2001. Biology of the Hard Clam. Developments in Aquaculture and Fisheries Science, Vol. 31. Elsevier, Amsterdam. 751 p.

Krijgsman, B.J. and Divaris, G.A., 1955. Contractile and pacemaker mechanisms of the heart of molluscs. Biol. Rev. Camb. Philos. Soc. 30:1–39.

Langton, R.W., Robinson, W.E. and Schick, D., 1987. Fecundity and reproductive effort of sea scallop *Placopecten magellanicus* from the Gulf of Maine. Mar. Ecol. Prog. Ser. 37:19–25.

Land, M.F., 1965. Image formation by a concave reflector in the eye of the scallop, *Pecten maximus*. J. Physiol. 179:138–153 + 1 pl.

Land, M.F., 1968. Functional aspects of the optical and retinal organization of the mollusc eye. Symp. Zool. Soc. Lond. 23:75–96.

Langton, R.W. and Gabbott, P.A., 1974. The tidal rhythm of extracellular digestion and the response to feeding in *Ostrea edulis*. J. Mar. Biol. Assoc. U.K. 24:181–187.

Lehane, C. and Davenport J., 2002. Ingestion of mesozooplankton by three species of bivalve: *Mytilus edulis*, *Cerastoderma edule*, and *Aequipecten opercularis*. J. Mar. Biol. Assoc. U.K. 82:615–619.

Le Pennec, G., 2000. La glande digestive de la coquille Saint Jacques *Pecten maximus*. Caractérisation fonctionnelle et mise en place d'un nouveau modèle *in vitro*. Doctoral thesis, Université de Bretagne Occidentale. 219 p.

Le Pennec, M., 1978. Genèse de la coquille larvaire et postlarvaire chez divers Bivalves marins. Thèse de Doctorat d'état, Université de Bretagne Occidentale, Brest, France. 216 p., 108 pl.

Le Pennec, M. and Beninger, P.G., 1996. Ultrastructural characteristics of spermatogenesis in three species of deep-sea hydrothermal vent mytilids. Can. J. Zool. 75:308–316.

Le Pennec, M., Beninger, P.G., Dorange, G. and Paulet, Y-M., 1991b. Trophic sources and pathways to the developing gametes of *Pecten maximus* (Bivalvia: Pectinidae). J. Mar. Biol. Assoc. U.K. 71: 451–463.

Le Pennec, M., Beninger, P.G. and Herry, A., 1988. New observations of the gill of *Placopecten magellanicus* (Mollusca: Bivalvia), and implications for nutrition. II. Internal anatomy and microanatomy. Mar. Biol. 98:229–238.

Le Pennec, M., Dorange, G., Beninger, P.G., Donval, A. and Widowati, I., 1991a. Les relations anse intestinale- gonade chez *Pecten maximus* (Mollusque, Bivalve). Haliotis 21: 57–69.

Le Pennec, G., Le Pennec, M. and Beninger, P.G., 2001. Seasonal digestive gland dynamics of the scallop *Pecten maximus* L in the Bay of Brest (France). J. Mar. Biol. Assoc. U.K. 81:663–671.

Le Pennec, G., Le Pennec, M., Beninger, P.G. and Dufour, S., 2002. Spermatogenesis in the archaic hydrothermal vent bivalve, *Bathypecten vulcani*, and comparison of spermatozoon ultrastructure with littoral pectinids. Invert. Reprod. Devel. 41:13–19.

Le Roux, S., Bellon-Humbert, C. and Lucas, A., 1987. Mise en évidence par immunocytochimie d'une substance apparentée à la somatostatine dans le système nerveux, le rein, et la glande digestive de juvéniles de *Pecten maximus* (Mollusque Bivalve). C. R. Acad. Sci. Paris Ser. III t 304:115–118.

Lent, C.M., 1974. Neuronal control of mucus secretion by leeches: toward a general theory for serotonin. Am. Zool. 14:931–942.

Longo, F.J., 1983. Meiotic Maturation and Fertilization. In: N.H. Verdonk, J.A.M. van den Bigelaar and A.S. Tompa (Eds.). The Mollusca, Vol. 3. Development. Academic Press, New York. pp. 49–89.

Lowe, D.M. and Moore, M.N., 1979. The cytochemical distribution of zinc (Zn II) and iron (Fe III) in the common mussel *Mytilus edulis* and their relationship with lysosomes. J. Mar. Biol. Assoc. U.K. 59:854–858.

Lubet, P., 1951. Sur l'émission des gamètes chez *Chlamys varia* L. (Moll. Lamellibr.). C. R. Hebd. Seances Acad. Sci. 233:1680–1681.

Lubet, P., 1955. Cycle neurosécrétoire de *Chlamys varia* (L.) et de *Mytilus edulis* L. C. R. Acad. Sci. Paris 241:119–121.

Lubet, P., 1959. Recherches sur le cycle sexuel et l'émission des gamètes chez les Mytilidés et les Pectinidés. Rev. Trav. Inst. Pêches Marit. 23:395–548.

Lubet, P., Albertini, L. and Robbins, I., 1986. Recherches expérimentales au cours des cycles annuels sur l'action gonadotrope exercée par les ganglions cérébroïdes sur la gamétogenèse femelle chez la moule *Mytilus edulis* L. (mollusque bivalve). C. R. Acad. Sci. Paris, Ser. III 303:575–580.

Lubet, P. and Mathieu, M., 1982. The action of internal factors on gametogenesis in pelecypod molluscs. Malacologia 22:131–136.

Lucas, A., 1965. Recherches sur la sexualité des Mollusques Bivalves. Bull. Biol. Fr. Belg. 99:115–247.

Lucas, A. and Hignette, M., 1983. Les concrétions rénales chez les bivalves marins: Etudes anciennes et récentes. Haliotis 13:99–113.

MacDonald, B.A. and Thompson, R.J., 1986. Production, dynamics and energy partitioning in two populations of the giant scallop *Placopecten magellanicus* (Gmelin). J. Exp. Mar. Biol. Ecol. 101:285–299.

Machin, J., 1977. Role of integument in molluscs. In: B.L. Gupta, R.B. Moreton, J.L. Oschman and B.J. Wall (Eds.). Transport of Ions and Water in Animals. Academic Press, London. pp. 735–762.

Mackie, G.L., 1984. Bivalves. In: A.S. Tompa, N.H. Verdonk and J.A.M. Van Den Biggelaar (Eds.). The Mollusca, Vol. 7: Reproduction. Academic Press, San Diego. pp. 351–403.

Mahéo, R., 1968. Observations sur l'anatomie et le fonctionnement de l'appareil byssogène de *Chlamys varia* L. Cah. Biol. Mar. 9:373–379.

Mahéo, R., 1969. Contribution à l'étude de l'anatomie et du fonctionnement du complexe byssogène de quelques Bivalves. Thèse de 3e cycle, Université de Rennes, France. 91 p. + 12 pl.

Manahan, D.T., Wright, S.H., Stevens, G.C. and Rice, M.A., 1982. Transport of dissolved amino acids by the mussel, *Mytilus edulis*: demonstration of net uptake from natural seawater. Science 215:1253–1255.

Marsh, H., Hopkins, G., Fisher, F. and Saas, R.L., 1976. Structure of the molluscan bivalve hinge ligament, a unique calcified elastic tissue. J. Ultrastruct. Res. 54:445–450.

Martin, A.W., 1983. Excretion. In: A.S.M. Saleuddin and K.M. Wilbur (Eds.). The Mollusca, Vol. 5: Physiology, Part 2. Academic Press, San Diego. pp. 353–405.

Martoja, R., Alibert, J., Ballan-Dufrancais, C., Jeantet, A.Y., Lhonore, D. and Truchet, M., 1975. Microanalyse et écologie. J. Microsc. Biol. Cell. 22:441–448.

Mason, J., 1958. The breeding of the scallop *Pecten maximus* L., in Manx waters. J. Mar. Biol. Assoc. U.K. 37:653–671.

Mason, J., 1963. Queen Scallop Fisheries in the British Isles. Fishing News Book, Farnham. 144 p.

Masson, M. and Ancellin, J., 1976. Aspects physiologiques et écologiques des contaminations des mollusques pélécypodes par les radionucléides et autres éléments à l'état de traces. Haliotis 7:123–130.

Mathers, N.F., 1972. The tracing of natural algal food labelled with a carbon 14 isotope through the digestive tract of *Ostrea edulis* L. Proc. Malac. Soc. Lond. 40:115–124.

Mathers, N.F., 1973a. A comparative histochemical survey of enzymes associated with the process of digestion in *Ostrea edulis* and *Crassostrea angulata* (Mollusca: Bivalvia). J. Zool. 169:169–179.

Mathers, N.F., 1973b. Carbohydrate digestion in *Ostrea edulis* L. Proc. Malacol. Soc. Lond. 40:359–367.

Mathers, N.F., 1976. The effects of tidal currents on the rhythm of feeding and digestion in *Pecten maximus* L. J. Exp. Mar. Biol. Ecol. 24:271–283.

Mathers, N.F. and Colins, N., 1979. Monophasic and diphasic digestive cycles in *Venerupis decussata* and *Chlamys varia*. J. Moll. Stud. 45:68–81.

Mathieu, M., Lenoir, F. and Robbins, I., 1988. A gonial mitosis - stimulating factor in cerebral ganglia and haemolymph of the marine mussel *Mytilus edulis* L. Gen. Comp. Endocrinol. 72:257–263.

Matsutani, T. and Nomura, T., 1982. Induction of spawning by serotonin in the scallop, *Patinopecten yessoensis* (Jay). Mar. Biol. Lett. 3:353–358.

Matsutani, T. and Nomura, T., 1984. Localization of monoamines in the central nervous system and gonad of the scallop *Patinopecten yessoensis*. Bull. Jap. Soc. Sci. Fish. 52:425–430.

Matsutani, T. and Nomura, T., 1986. Serotonin-like immuno-reactivity in the central nervous system and gonad of the scallop, *Patinopecten yessoensis*. Cell Tissue Res. 244:515–517.

McLachlan, J., McInnes, A.G. and Allen, J.A., 1965. Studies on the chitan (chitin: poly-n-acetylglucosamine) fibers of the diatom *Thalassiosira fluviatilis* Hustedt. Can. J. Bot. 43:707–713.

Merrill, A.S. and Burch, J.B., 1960. Hermaphroditism in the sea scallop, *Placopecten magellanicus* (Gmelin). Biol. Bull. 119:197–201.

Mikulich, L.V. and Tsikhon-Lukanina, Ye. A., 1981. Food of the scallop. Oceanol. 21:633–635.

Møhlenberg, F. and Riisgard, H.U., 1978. Efficiency of particle retention in 13 species of suspension feeding bivalves. Ophelia 17(2):239–246.

Moir, A.J.G., 1977a. Ultrastructural studies on the ciliated receptors of the long tentacles of the giant scallop, *Placopecten magellanicus* (Gmelin). Cell Tissue Res. 184:367–380.

Moir, A.J.G., 1977b. On the ultrastructure of the abdominal sense organ of the giant scallop, *Placopecten magellanicus* (Gmelin). Cell Tissue Res. 184:359–366.

Morrison, C. and Shum, G., 1982. Chlamydia-like organisms in the digestive diverticula of the bay scallop, *Argopecten irradians* (Link). J. Fish Dis. 5:173–184.

Morse, M.P., Robinson, W.E. and Wehling, W.E., 1982. Effects of sublethal concentrations of the drilling mud components attapulgite and Q-broxin on the structure and function of the gill of the scallop, *Placopecten magellanicus* (Gmelin). In: W.B. Vernberg, A. Calabrese, F.P. Thurberg and F.J. Vernberg (Eds.). Physiological Mechanisms of Marine Pollutant Toxicity. Academic Press, New York. pp. 235–259.

Morse, M.P. and Zardus, J.D., 1997. Bivalvia. In: F.W. Harrison and A.J. Kohn (Eds.). Microscopic Anatomy of Invertebrates, Vol. 6A. Mollusca II. Wiley-Liss, New York. pp. 7–118.

Morton, B.S., 1977. The tidal rhythm of feeding and digestion in the Pacific oyster, *Crassostrea gigas* (Thunberg). J. Exp. Mar. Biol. Ecol. 26:135–151.

Morton, B.S., 1983. Feeding and Digestion in Bivalvia. In: A.S.M. Saleuddin and K.M. Wilbur (Eds.). The Mollusca, Vol. 5. Physiology, Part 2. Academic Press, New York. pp. 65–147.

Morton, B.S., 2000. The function of pallial eyes within the Pectinidae, with a description of those present in *Patinopecten yessoensis*. In: E.M. Harper, J.D. Taylor and J.A. Crame (Eds.). Evolutionary Biology of the Bivalvia. Geological Society Special Publication 177, Geological Society Publishing House, Bath, U.K. pp. 247–255.

Mottet, M.G., 1979. A review of the fishery biology and culture of scallops. State of Washington Department of Fisheries Technical Report No. 39: 99 p.

Naidu, K.S., 1970. Reproduction and breeding cycle of the giant scallop *Placopecten magellanicus* (Gmelin) in Port au Port Bay, Newfoundland. Can. J. Zool. 48:1003–1012.

Nakahara, H. and Bevelander, G., 1967. Ingestion of particulate matter by the outer surface cells of the mollusc mantle. J. Morph. 122:139–146.

Natochin, Yu.V., Berger, V.Ya., Khlebovich, V.V., Lavrova, E.A. and Michailova, O. Yu., 1979. The participation of electrolytes in adaptation mechanisms of intertidal molluscs' cells to altered salinity. Comp. Biochem. Physiol. 63A:115–119.

Neff, J.M., 1972. Ultrastructure of the outer epithelium of the mantle in the clam *Mercenaria mercenaria* in relation to calcification of the shell. Tissue & Cell 4:591–600.

Orton, J.H., 1912. The mode of feeding of *Crepidula*, with an account of the current-producing mechanism in the mantle cavity, and some remarks on the mode of feeding in Gastropods and Lamellibranchs. J. Mar. Biol. Assoc. U.K. 9:444–478.

Overnell, J., 1981. Protein and oxalate in mineral granules from the kidney of *Pecten maximus* (L.). J. Exp. Mar. Biol. Ecol. 52:173–183.

Owen, G., 1955. Observations on the stomach and digestive diverticula of the Lamellibranchia. I. The Anisomyaria and Eulamellibranchia. Q. J. Microsc. Sci. 96:517–537.

Owen, G., 1966. Digestion. In: K.M. Wilbur and C.M. Yonge (Eds.). Physiology of Mollusca, Vol. II. Academic Press, New York. pp. 53–96.

Owen, G., 1978. Classification and the bivalve gill. Phil. Trans. Roy. Soc. Lond. B 284:377–385.

Owen, G. and McCrae, J.M., 1976. Further studies on the latero-frontal tracts of bivalves. Proc. R. Soc. Lond. B 194:527–544.

Owen, G., Trueman, E.R. and Yonge, C.M., 1953. The ligament in the Lamellibranchia. Nature 171:73–75.

Ozanai, K., 1975. Seasonal gonad development and sex alteration in the scallop, *Patinopecten yessoensis*. Bull. Mar. Biol. Stn. Asamushi 15:81–88.

Pani, A.M. and Croll, R.P., 1995. Distribution of catecholamines, indoleamines, and their precursors and metabolites in the scallop, *Placopecten magellanicus* (Bivalvia, Pectinidae). Cell. Molec. Neurobiol. 15:371–385.

Paulet, Y.M., Donval, A. and Bekhadra, F., 1993. Monoamines and reproduction in *Pecten maximus*, a preliminary approach. Invert. Reprod. Devel. 23:89–94.

Paulet, Y.M., Lucas, A. and Gerard, A., 1988. Reproduction and larval development in two *Pecten maximus* L. populations from Brittany. J. Exp. Mar. Biol. Ecol. 119:145–156.

Pelseneer, M.J., 1894. Hermaphroditism in Mollusca. Q. J. Microsc. Sci. 37:19–46.

Pelseneer, P., 1906. Mollusca. A Treatise on Zoology. Part V. Ray Lankester Ltd., London. 355 p.

Petit, H., 1978. Recherches sur des séquences d'évènements périostracaux lors de l'élaboration de la coquille d'*Amblema plicata perplicata* Conrad 1834. Thèse de doctorat d'Etat, Université de Bretagne Occidentale, Brest, France. 185 p.

Petit, H., Le Pennec, M. and Martin, J., 1988. Ultrastructure du sillon periostracal du Mytilidae des sources hydrothermales profondes du Pacifique oriental. Oceanol. Acta, Special Publication Number 8:191–194.

Pierce, M.E., 1950. *Pecten irradians*. In: F.A. Brown (Ed.). Selected Invertebrate Types. John Wiley and Sons Inc., New York. pp. 321–324.

Pierce, S.K., 1982. Invertebrate cell volume control mechanisms: a co-ordinated use of intracellular amino acids and inorganic ions as osmotic solute. Biol. Bull. 163:405–419.

Pipe, R.K., 1987. Oogenesis in the marine mussel *Mytilus edulis*: an ultrastructural study. Mar. Biol. 95:405–414.

Pirie, B.J.S. and George, S.G., 1979. Ultrastructure of the heart and excretory system of *Mytilus edulis* (L.). J. Mar. Biol. Assoc. U.K. 59:819–829.

Poder, M., 1980. Les réactions hémocytaires inflammatoires et tumorales chez *Ostrea edulis* (L.). Essai de classification des hémocytes des Mollusques Bivalves. Thèse de doctorat de 3e cycle, Université de Bretagne Occidentale, Brest, France. 95 p.

Popham, J.D., 1979. Comparative spermatozoan morphology and bivalve phylogeny. Malacol. Rev. 12:1–20.

Posgay, J.A., 1950. Investigations of the sea scallop, *Pecten grandis*. Massachusetts Dept. Nat. Resources, Div. Mar. Fish., Third report on investigations of methods of improving the shellfish resources of Massachusetts. pp. 24–30.

Potts, W.T.W., 1967. Excretion in the molluscs. Biol. Rev. Camb. Philos. Soc. 42:1–41.

Purchon, R.D., 1957. The stomach in the Filibranchia and Pseudolamellibranchia. Proc. Zool. Soc. Lond. 129:27–60.

Purchon, R.D., 1977. The Biology of the Mollusca (2nd ed.). Pergammon Press, Oxford. pp. 225–243.

Ram, G.L., Ram, M.L., Smith, S.S., Croll, R.P., 1997. Peptinergic and serotinergic mechanisms regulating spawning and egg-laying behavior in gastropods and bivalves. Malacol. Rev. 30:1–23.

Ramsay, J.A., 1952. A Physiological Approach to the Lower Animals. University Press, Cambridge. pp. 18–36.

Ranson, G., 1939. Le provinculum de la prodissoconque de quelques Ostréidés. Bull. Mus. Nat. Hist. Nat. (Paris) 2:318–332.

Raven, C., 1961. Oogenesis: the storage of development information. Pergamon Press, New York. 274 p.

Raven, C.P., 1966. Morphogenesis. The Analysis of Molluscan Development. 2nd ed. Pergamon Press, New York.

Reddiah, K., 1962. The sexuality and spawning of manx Pectinids. J. Mar. Biol. Assoc. U.K. 42:683–703.

Reed-Miller, C. and Greenberg, M.J., 1982. The ciliary junctions of scallop gills: the effects of cytochalasins and concanavalin A. Biol. Bull. (Woods Hole) 163:225–239.

Reindl, S. and Haszprunar, G., 1996. Fine structure of caeca and mantle of arcoid and limopsoid bivalves (Mollusca: Pteriomorpha). Veliger 39:101–116.

Ridewood, W.G., 1903. On the structure of the gills of the Lamellibranchia. Phil. Trans. Roy. Soc. Lond. B 195:147–284.

Riisgard, H.U., 1988. Efficiency of particle retention and filtration rate in 6 species of Northeast American bivalves. Mar. Ecol. Prog. Ser. 45:217–223.

Robinson, W.E. and Langton, R.W., 1980. Digestion in a subtidal population of *Mercenaria mercenaria* (Bivalvia). Mar. Biol. 58:173–179.

Robinson, W.E., Penington, M.R. and Langton, R.W., 1981. Variability of tubule types within the digestive glands of *Mercenaria mercenaria* (L.), *Ostrea edulis* L. and *Mytilus edulis* L. J. Exp. Mar. Biol. Ecol. 54:265–276.

Rodrick, G.E. and Ulrich, S.A., 1984. Microscopical studies on the hemocytes of bivalves and their phagocytic interaction with selected bacteria. Helgol. Wiss. Meeresunters. 37:167–176.

Saleuddin, A.S.M. and Petit, H., 1983. The mode of formation and the structure of the periostracum. In: A.S.M. Saleuddin and K.M. Wilbur (Eds.). The Mollusca, Vol. 4. Physiology, Part 1. Academic Press, New York. pp. 199–234.

Sanger, J.W., 1979. Cardiac structure in selected arthropods and molluscs. Am. Zool. 19:9–27.

Sastry, A.N., 1963. Reproduction of the bay scallop, *Aequipecten irradians* Lamark. Influence of temperature on maturation and spawning. Biol. Bull. 125:146–153.

Sastry, A.N., 1965. The development and external morphology of pelagic larval and post-larval stages of the bay scallop, *Aequipecten irradians concentricus* Say, reared in the laboratory. Bull. Marine Sci. 15:417–435.

Sastry, A.N., 1966. Temperature effects in reproduction of the bay scallops, *Aequipecten irradians* Lamarck. J. Mar. Biol. Assoc. U.K. 130:118–134.

Sastry, A.N., 1968. Relationships among food, temperature and gonad development of the bay scallop, *Aequipecten irradians concentricus* Say, reared in the laboratory. Bull. Mar. Sci. 15:417–435.

Sastry, A.N., 1979. Pelecypoda (excluding Ostreidae). In: A.C. Giese and J.S. Pearse (Eds.). Reproduction of Marine Invertebrates, Volume V, Molluscs: Pelecypods and Lesser Classes. Academic Press, New York. pp. 113–292.

Seiderer, L.J., Newell, R.C., Schultes, K., Robb, F.T. and Turley, C.M., 1987. Novel bacteriolytic activity associated with the style microflora of the mussel *Mytilus edulis* (L.). J. Exp. Mar. Biol. Ecol. 110:213–224.

Setna, S.B., 1930. Neuro-muscular mechanism of the gill of *Pecten*. Q. J. Microsc. Sci. 73:365–391.

Shaw, B.L. and Battle, H.I., 1958. The chemical composition of the gastric shield of the oyster *Crassostrea virginica* (Gmelin). Can. J. Zool. 37:214–215.

Shumway, S.E., 1977. Effect of salinity fluctuation on the osmotic pressure and Na^+, Ca^{2+} and Mg^{2+} ion concentrations in the hemolymph of bivalve molluscs. Mar. Biol. 41:153–177.

Shumway, S.E., Selvin, R. and Schick, D.F., 1987. Food resources related to habitat in the scallop *Placopecten magellanicus* (Gmelin, 1791): a qualitative study. J. Shellfish Res. 6:89–95.

Shumway, S.E. (Ed), 1991. Scallops: Biology, Ecology, and Aquaculture. Developments in Aquaculture and Fisheries Science, Vol. 21. Elsevier, Amsterdam. 1095 p.

Silverman, H.S., Lynn, J.W., Beninger, P.G. and Dietz, T.H., 1999. The role of latero-frontal cirri in particle capture by the gills of *Mytilus edulis*. Biol. Bull. 197:368–376.

Simkiss, K., 1976. Intracellular and extracellular routes in biomineralisation. In: J.C. Duncan (Ed.). Calcium in Biological Systems. Cambridge University Press, Cambridge. pp. 423–444.

Spagnolia, T. and Wilkens, L.A., 1983. Neurobiology of the scallop. II. Structure of the parietovisceral ganglion lateral lobes in relation to afferent projections from the mantle eyes. Mar. Behav. Physiol. 10:23–55.

Stasek, C.R., 1963. Orientation and form in the bivalved Mollusca. J. Morph. 112:195–214.

Stephens, P.J., 1978. The sensitivity and control of the scallop mantle edge. J. Exp. Biol. 75:203–221.

Stewart, M.G. and Bamford, D.R., 1976. Absorption of soluble nutrients by the mid gut of the bivalve *Mya arenaria*. J. Molluscan Stud. 42:63–73.

Strohmeier, T., Duinker, A. and Lie, Ø., 2000. Seasonal variations in chemical composition of the female gonad and storage organs in *Pecten maximus* (L.) suggesting that somatic and reproductive growth are separated in time. J. Shellfish Res. 19:741–747.

Stommel, E.W. and Stephens, R.E., 1985. Cyclic AMP and calcium in the differential control of *Mytilus* gill cilia. J. Comp. Physiol. A 157:451–459.

Sullivan, P.A., Robinson, W.E. and Morse, M.P., 1988. Isolation and characterization of granules from the kidney of the bivalve *Mercenaria mercenaria*. Mar. Biol. 99:359–368.

Sundet, J.H. and Lee, J.B., 1984. Seasonal variations in gamete development in the Iceland scallop, *Chlamys islandica*. J. Mar. Biol. Assoc. U.K. 64:411–416.

Tang, S.F., 1941. The breeding of the scallop (*Pecten maximus* L.) with a note on the growth rate. Proc. Liverpool Biol. Soc. 54:9–28.

Taylor, J.D. and Kennedy, W.J., 1969. The influence of the periostracum on the shell structure of bivalve molluscs. Calcif. Tissue Res. 3:274–283.

Taylor, R. and Capuzzo, J., 1983. The reproductive cycle of the bay scallop *Argopecten irradians irradians* (Lamarck), in a small coastal embayment on Cape Cod, Massachusetts. Estuaries 6:431–435.

Thompson, R.J., 1977. Blood chemistry, biochemical composition, and the annual reproductive cycle in the giant scallop, *Placopecten magellanicus*, from Southeast Newfoundland. J. Fish. Res. Board Can. 34:2104–2116.

Thompson, R.J., Livingstone, D.R. and de Zwaan, A., 1980. Physiological and biochemical aspects of the valve snap and valve closure responses in the giant scallop *Placopecten magellanicus*. I. Physiology. J. Comp. Physiol. 137:97–104.

Too, C.K.L. and Croll, R.P., 1995. Detection of FMRFamide-like immunoreactivities in the sea scallop *Placopecten magellanicus* by immunohistochemistry and Western Blot analysis. Cell Tissue Res. 281:295–304.

Toullec, J.Y., Lenoir, F., Van Wormhoudt, A. and Mathieu, M., 1988. Nonspecies-specific growth factor from cerebral ganglia of *Mytilus edulis*. J. Exp. Mar. Biol. Ecol. 119:111–117.

Trueman, E.R., 1953. The ligament of *Pecten*. Q. J. Microsc. Sci. 94:193–202.

Turchini, J., 1923. Contribution à l'étude de l'histologie comparée de la cellule rénale. L'excrétion rénale chez les mollusques. Arch. Morph. Gén. Exp. 18:7–253.

Uesaka, H., Yamagishi, H. and Ebara, A., 1987. Stretch-mediated interaction between the auricle and ventricle in an oyster *Crassostrea gigas*. Comp. Biochem. Physiol. 88A:221–227.

Usheva, L.N., 1983. Cell histomorphology and proliferation of the posterior intestine epithelium in the Yezo scallop *Patinopecten yessoensis*. Biol. Morya 3:17–24.

Veniot, A., Bricelj, V.M. and Beninger, P.G., 2003. Ontogenetic changes in gill morphology and potential significance for food acquisition in the scallop *Placopecten magellanicus*. Mar. Biol. 142:123–131.

Wada, S.K., 1968. Mollusca. 1. Amphineura, Gastropoda, Scaphopoda, Pelycypoda. In: M. Kume and K. Dan (Eds.). Invertebrate Embryology. Nolit Publishing House, Belgrade. pp. 485–525.

Waite, J.H. and Andersen, S.O., 1978. 3, 4-Dihydroxyphenylalanine in an insoluble shell protein of *Mytilus edulis*. Biochim. Biophys. Acta 541:107–114.

Waite, J.H. and Andersen, S.O., 1980. 3, 4-Dihydroxyphenylalaline and sclerotization of the periostracum in *Mytilus edulis.* Biol. Bull. 158:164–173.

Waite, J.H. and Wilbur, K.M., 1976. Phenoloxidase in the periostracum of the marine bivalve *Modiolus demissus* Dillwyn. J. Exp. Zool. 195:358–368.

Wakui, T. and Obara, A., 1967. On the seasonal change of the gonads of scallop, *Patinopecten yessoensis* (Jay), in lake Saroma, Hokkaido. Bull. Hokkaido Reg. Fish. Res. Lab. 32:15–32.

Waller, T.R., 1991. Evolutionary relationships among commercial scallops (Mollusca: Bivalvia: Pectinidae). In: S.E. Shumway (Ed.). Scallops: Biology, Ecology, and Aquaculture. Developments in Aquaculture and Fisheries Science, Vol. 21. Elsevier, Amsterdam. pp. 1–73.

Ward, J.E., Beninger, P.G., MacDonald, B.A. and Thompson, R.J., 1991. Direct observations of feeding structures and mechanisms in Bivalve molluscs using endoscopic examination and video image analysis. Mar. Biol. 111:287–291.

Ward, J.E., MacDonald, B.A., Thompson, R.J. and Beninger, P.G., 1993. Mechanisms of suspension-feeding in Bivalves: resolution of current controversies using endoscopy. Limnol. Oceanogr. 30:265–272.

Ward, J.E., Levinton, J.S., Shumway, S.E. and Cucci, T., 1998. Particle sorting in bivalves: *in vivo* determination of the pallial organs of selection. Mar. Biol. 131:283–292.

Watabe, N., 1984. Shell. In: J. Bereiter-Hahn, A.G. Matoltsy and K.S. Richards (Eds.). Biology of the Integument. I. The Invertebrates. Springer-Verlag, Berlin. pp. 448–485.

Weinstein, J.E., 1995. Fine structure of the digestive tubule of the Eastern oyster, *Crasssostrea virginica* (Gmelin, 1791). J. Shellfish Res. 14:97–103.

Welsh, J.H. and Moorhead, M., 1960. The quantitative distribution of 5-hydroxytryptamine in the invertebrates, especially in their nervous systems. J. Neurochem. 6:146–169.

White, K.M., 1942. The pericardial cavity and the pericardial gland of the Lamellibranchia. Proc. Malacol. Soc. Lond. 25:37–88.

Wilbur, K.M., 1972. Shell formation in molluscs. In: M. Florkin and B.T. Scheer (Eds.). Chemical Zoology, Vol. VII: Mollusca. Academic Press, New York. pp. 243–282.

Wilbur, K.M., 1985. Topics in molluscan mineralization: present status, future directions. Am. Malacol. Union Bull., Special Edition 1:51–58.

Wilbur, K.M. and Saleuddin, A.S.M., 1983. Shell formation. In: A.S.M. Saleuddin and K.M. Wilbur (Eds.). The Mollusca, Vol. 4. Physiology, Part 1. Academic Press, New York. pp. 235–287.

Wilkens, L.A., 1981. Neurobiology of the scallop. I. Starfish-mediated escape behaviors. Proc. R. Soc. Lond. B Biol. Sci. 211:241–372.

Wilkins, N.P., 1978. Length correlated changes in heterozygosity at an enzyme locus in the scallop (*Pecten maximus* L.). Anim. Blood Groups Biochem. Genet. 9:69–77.

Wojtowicz, M.B., 1972. Carbohydrases of the digestive gland and the crystalline style of the Atlantic deep-sea scallop (*Placopecten magellanicus*, Gmelin). Comp. Biochem. Physiol. 43A:131–141.

Wright, S.H., 1987. Alanine and taurine transport by the gill epithelium of a marine bivalve: effect of sodium on influx. J. Membr. Biol. 95:37–46.

Wright, S.H., Southwell, K.M. and Stevens, G.C., 1984. Autoradiographic analysis of amino-acid uptake by the gill of *Mytilus californianus.* J. Comp. Physiol. 154:249–256.

Yamamoto, G., 1943. Gametogenesis and the breeding season of the Japanese common scallop *Pecten (Patinopecten) yessoensis* Jay. Bull. Jap. Soc. Sci. Fish. 12:21–26.

Yevich, C.A. and Yevich, P.P., 1985. Histopathological effects of cadmium and copper on the sea scallop *Placopecten magellanicus*. In: F.J. Vernberg, F.P. Thurberg, A. Calabrese and W. Vernberg (Eds.). Marine Pollution and Physiology: Recent Advances. The Belle W. Baruch Library in Marine Science Number 13. pp. 187–198.

Yonge, C.M., 1951. Studies on Pacific coast mollusks. III. Observations on *Hinnites multirugosus* (Gale). Univ. Calif. Publications Zool. 56:409–420.

Yonge, C.M., 1981. On adaptive radiation in the Pectinacea, with a description of *Hemipecten forbesianus*. Malacologia 21:23–34.

Zacks, S.I., 1955. The cytochemistry of the amoebocytes and intestinal epithelium of *Venus mercenaria* (Lamellibranchiata), with remarks on a pigment resembling ceroid. Q. J. Microsc. Sci. 96:57–71

Zhadan, P.M. and Semen'kov, P.G., 1984. An electrophysiological study of the mechanoreceptory function of the abdominal sense organ of the scallop *Patinopecten yessoensis* (Jay). Comp. Biochem. Physiol. 78A:865–870.

Zhadan, R.M. and Doroshenko, P.A., 1985. Cyclic AMP-dependent processes and mechanosensitivity of the abdominal sense organ of the scallop *Patinopecten yessoensis* (Jay). Biol. Membr. 2:285–291.

AUTHORS ADDRESSES

Peter G. Beninger - IsoMer, Faculté des Sciences, Université de Nantes, 44322 Nantes Cedex 3, France (E-mail: Peter.Beninger@isomer.univ-nantes.fr)

Marcel Le Pennec - Institut Universitaire Européen de la Mer, Université de Bretagne Occidentale, Place Nicolas Copernic, 29280 Plouzané, France (E-mail: Marcel.Lepennec@univ-brest.fr)

Scallops: Biology, Ecology and Aquaculture
S.E. Shumway and G.J. Parsons (Editors)
© 2006 Elsevier B.V. All rights reserved.

Chapter 4

Scallop Adductor Muscles: Structure and Function

Peter D. Chantler

4.1 INTRODUCTION

Scallop aquaculture owes its commercial success to a global appreciation of the gastronomic delights of this animals' adductor muscles. For this reason alone, these muscles would have been considered worthy of study. Early physiologists were also impressed by the high muscle to body weight ratio, the apparent purity of the fibres and the clarity of function of the two main adductor muscles. More recently, biochemists have shown that these muscles possess a mode of regulation that is closely related to mechanisms found both in vertebrate smooth muscles and in non-muscle cells; furthermore, because scallop myosins have been shown to possess unique physical properties making them ideally suited for biochemical manipulation, they remain a key model for studies on the function and regulation of muscle contraction.

Typically, scallops possess two adductor muscles (Fig 4.1a). The phasic adductor, often the most dominant feature in the body of the animal, is the muscle sought by gourmets and gourmands alike. It is cross-striated and facilitates fast, repetitive opening and closing of the valves; contraction of this muscle causes a rapid ejection of water from the mantle cavity, enabling the animal to swim by jet propulsion (see Wilkens 1991) powered, in the main, by anaerobic metabolism (De Zwaan et al. 1980). The smaller tonic adductor is more commonly known as the catch muscle. It is a smooth muscle, lacking cross-striations; its slower contraction is capable of generating considerable force, keeping the animal's hinged shell closed for long periods of time with little expenditure of energy. In most scallops, the two types of adductor muscle lie closely apposed to one another and there is often a gradual transition of fibre type from one muscle to the next.

Molluscan muscles, like vertebrate skeletal muscles, contract by a sliding filament mechanism, the thin filaments sliding past the thick filaments through the cyclical action of crossbridges extending out from the myosin-containing thick filament to make contact with the actin-containing thin filaments (Fig. 4.1b; Hanson and Lowy 1959; Hanson and Lowy 1960). Our knowledge of the structure and biochemistry of scallop adductor muscles has largely been obtained from four species: *Pecten maximus*, from the shores of Britain, *Argopecten irradians* and *Placopecten magellanicus*, from the eastern seaboard of the United States, and *Patinopecten yessoensis* from the coastal waters off Japan.

The interested reader may supplement the following account with reviews which cover the regulation of contraction by the scallop striated adductor muscle (Szent-Gyorgyi and Chantler 1994), the comparative structure and biochemistry of molluscan muscles

230

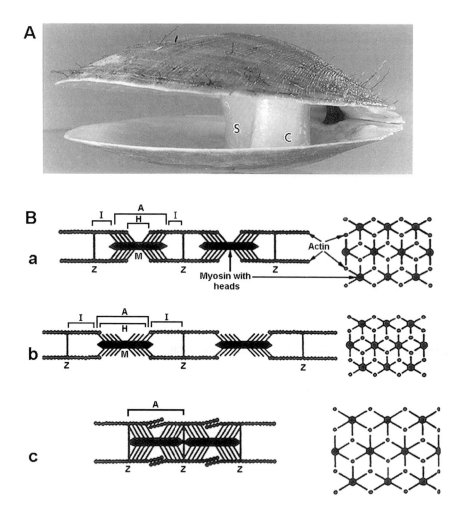

Figure 4.1 A. Photograph of *Placopecten magellanicus* after removal of all organs and tissues surrounding the adductor muscles. The striated (S) and catch (C) adductor muscles were left *in situ* and allowed to extend overnight under sustained pressure from the hinge (seen to the right of the muscles) in a solution containing 50% ethylene glycol. The catch muscle, which is the smaller of the two adductors, is seen to advantage in this perspective. B. Schematic of sarcomeric structure as seen in a typical striated muscle fibre. Longitudinal views through two sarcomeres (seen on the left) are compared with cross-sectional views through the A-band (seen on the right) at three different sarcomeric lengths. a). Sarcomeres at rest length (l_o). b). Sarcomeres stretched to 1.3 l_o. c). Sarcomeres contracted to 0.6 l_o. The locations of the A-band, I-band, H-zone, M-band and Z-line are indicated. Note the decrease in spacing between thin and thick filaments as a function of increasing sarcomere length. Adapted from "Muscles, molecules and movement", by J.R. Bendall (1969). Heinemann Educational Books Ltd.

(Chantler 1983), the structure of, and transmission across, molluscan neuromuscular junctions (Muneoka and Twarog 1983; Nicaise and Amsellem 1983), the general structure of muscle (Craig 1994) and the mechanism of contraction and motor function (Cooke 1997; Howard 1997; Geeves and Holmes 1999).

4.2 STRUCTURE OF THE STRIATED ADDUCTOR MUSCLE

4.2.1 Fibre Microanatomy

In scallops, the large adductor muscle is cross-striated. Cross-striations can be seen by light microscopy but are most clearly visualised by electron microscopy which reveals the sarcomeric structure, the alignments of thick and thin filaments and the Z-lines (Fig. 4.2). Each muscle cell of the scallop striated adductor muscle is relatively small and ribbon-like compared to vertebrate skeletal muscle cells. Individual muscle cells are shorter than the length of the muscle; for example, in *Argopecten irradians* the average fibre length for a muscle length of 2 cm was calculated to be close to 650 μm (Nunzi and Franzini-Armstrong 1981). The ribbon-like character of the muscle cells is best seen in cross-sections (Fig. 4.2a) where typically, the short and long diameters are of the order of 1 and 10 μm, respectively (Morrison and Odense 1974; Millman and Bennett 1976; Nunzi and Franzini-Armstrong 1981). Each cell contains a single, centrally-placed myofibril, sometimes partially or completely divided into submyofibrils (Nunzi and Franzini-Armstrong 1981) and is bounded by a surface membrane. There are no invaginations of this membrane that could be termed a transverse tubular system in this muscle (Sanger 1971; Nunzi and Franzini-Armstrong 1981), the small fibre diameter probably rendering such a system unnecessary. Nevertheless, there is a complex sarcoplasmic reticulum (Sanger 1971; Morrison and Odense 1974), the cisternae of which form peripheral couplings (junctional feet) with the surface membrane (Fig. 4.3), these structures being similar to those found in vertebrates (Nunzi and Franzini-Armstrong 1981). The composition of the sarcoplasmic reticulum membrane from *Placopecten magellanicus* appears to comprise a near-crystalline array of dimer ribbons of Ca^{2+}-ATPase molecules (Castellani and Hardwicke 1983; Castellani et al. 1985). Between cells, both in series and in parallel, are junctions similar to the *fasciae adherentes* of the intercalated discs of vertebrate myocardium (Fig. 4.3; Nunzi and Franzini-Armstrong 1981). Gap junctions are also fairly common.

In contrast to vertebrate fast twitch muscles, there is no evidence of M-bridges holding adjacent thick filaments together at the centre of a sarcomere. Each "A"-band within a sarcomcre typically has a somewhat ragged edge (Fig. 4.2e), possibly because of this absence of M-bridges necessary to hold the structure together and/or the presence of adjacent thick filaments of differing length. Twitchin-like mini-titins are present in the A-bands of both striated and smooth adductor muscles (Vibert et al. 1993) but it is not clear that these 0.2 μm long proteins connect the thick filaments to the Z-lines, as is found in vertebrate muscles. Indeed, the lateral alignment of Z-lines in adjacent myofibrils is very easily disturbed. Each myosin-containing thick filament (~20 nm backbone diameter and 1.7 μm in length in the Atlantic scallops) (Morrison and Odense 1974; Millman and

232

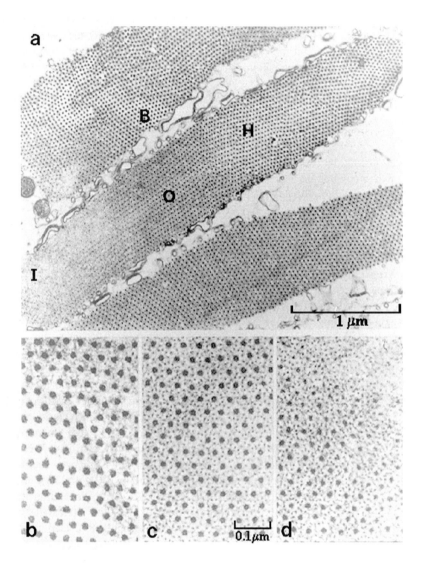

Figure 4.2. Structure of the striated adductor muscle (*Pecten maximus*). a). Low magnification transverse section showing the elongated profile of the muscle fibres and the appearance of the filaments at different regions in the sarcomere. B, bare zone; H, part of the H-zone where the thick filaments show projections; I, I-band; O, region where the thin and thick filaments overlap. b). High-magnification transverse section of the bare-zone region of the sarcomere. c). High-magnification transverse section through the overlap region. d). High-magnification transverse section through that part of the overlap region where the thick filaments taper. Magnification scale bars are seen in a) and c); magnifications of images seen in b), c) and d) are the same. Micrographs reproduced, with permission of the publisher, from Millman and Bennett (1976).

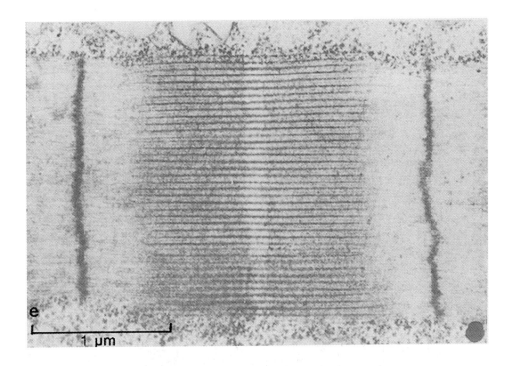

Figure 4.2e. Structure of the striated adductor muscle (*Pecten maximus*). e). Longitudinal section through ribbon-like muscle fibre. Magnification scale bars is included. Micrographs reproduced, with permission of the publisher, from Millman and Bennett (1976).

Bennett 1976) is packed in a hexagonal array, 60 nm from its neighbour, and is surrounded by approximately 12 actin-containing thin filaments, each 7–9 nm in diameter (Fig. 4.2c, d). The core of striated adductor thick filaments contains paramyosin (approximately 7% of the thick filament by weight) (Hardwicke and Hanson 1971; Levine et al. 1976; Millman and Bennett 1976) but this amount is small by contrast with thick filaments from the smooth adductor muscles.

4.2.2 Actin and Thin Filament Structure

The basic structure of the scallop striated adductor thin filament is very similar to that of its vertebrate counterpart. However, there is still controversy as to the amount and functionality of troponin in these filaments. Notwithstanding the variation in the amounts of troponin present within muscles of different molluscan species, the underlying structure of the thin filament is one with which we are familiar.

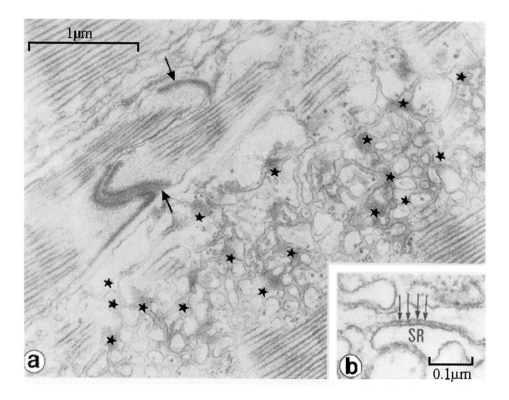

Figure 4.3. a). A fascia adherens joining two striated adductor muscle cells from *Argopecten irradians* in series (between arrows). Notice the dense material filling the junctional gap and the intermediate dense line between the two cell membranes. In the lower half of the image, a grazing view of the sarcoplasmic reticulum illustrates its anastomosing configuration. Cisternae formed at the crossover points (asterisks) form peripheral couplings with the surface membrane. b). Detail of a peripheral coupling. A strip of extracellular space crosses the image from left to right and the peripheries of two fibres are above and below. The junctional surface of a sarcoplasmic reticulum cisterna (SR) is coupled to the surface membrane by regularly spaced feet (arrows). Scale bars indicate the magnification. Micrographs reproduced, with permission of the publisher, from Nunzi and Franzini-Armstrong (1981).

Thin filament structure is based on the actin filament (Fig. 4.4). Tropomyosin, found in all molluscan thin filaments, is located within the major grooves on either side of the actin filament. Troponin, in those muscles where it is found, is located with a 38 nm periodicity along each thin filament, where it forms attachments to both tropomyosin and actin.

The structure of filamentous actin (F-actin) is well-known, considerable information coming from studies on scallop actin (Hanson and Lowy 1963; Millman and Bennett

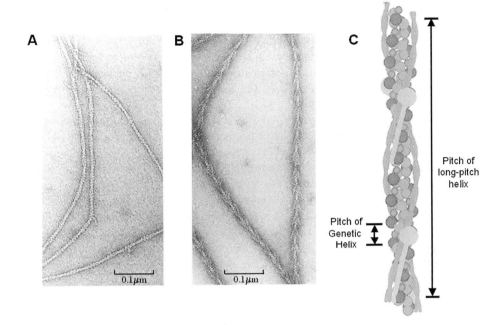

A   B   C

Figure 4.4. A. High-magnification micrograph of purified scallop F-actin, partially decorated with scallop myosin heads (S-1) (only a few S-1 molecules are visible).   Scale bar indicates magnification. B. High-magnification micrograph of purified scallop F-actin fully decorated with scallop S-1, illustrating arrowhead formation.   Scale bar indicates magnification.   C. Diagram illustrating the structure of an actin helix, composed of 28 subunits in 13 turns of the genetic helix, in association with tropomyosin and troponin.   Micrographs reproduced, with permission of the publisher, from Vibert and Craig (1982).   Diagram reproduced, with permission of the publisher, from Squire and Morris (1998).

1976; Craig et al. 1980; Egelman et al. 1982; Vibert and Craig 1982; Egelman et al. 1983).   F-actin is composed of a double string of globular subunits (G-actin protomers, each of 42 kD molecular weight), 9 ± 1 nm in diameter, wound as a right-handed helix about a central axis (Fig. 4.4c).   This structure may be described as a shallow helix of pitch 76.0 nm with crossovers every 38.0 nm along the filament – as determined for thin filaments from *Pecten maximus* and *Placopecten magellanicus*, using X-ray diffraction techniques (Millman and Bennett 1976).   The axial spacing of subunits in each strand of the helix is 5.5 nm, and there are 13–14 subunits per strand in one turn of each long-pitch helix.   The two chains are staggered, with respect to each other, by about half a subunit. Alternatively, F-actin structure may be described in terms of the "one-start", short-pitch (5.9 nm), left-handed "genetic" helix, formed as one moves about a unidirectional axis along a line passing through all protomers in turn (Fig. 4.4c). In the case of scallop F-actin, 28 actin protomers are present in 13 turns of the genetic helix (2.1538 protomers per turn) (Egelman et al. 1982). Despite this apparent order, individual subunits in the F-actin helix appear to exhibit considerable angular displacements with respect to adjacent subunits ($\pm$ 10°), imparting a random variable twist to the filament which manifests itself as a variation in pitch along the length of an individual filament (Egelman et al. 1982; Egelman et al. 1983); this rotation is thought to facilitate the interaction of actin with a great variety of actin-binding proteins.

F-actin can be depolymerised *in vitro* into its individual globular subunits by dialysis against ATP and trace amounts of calcium or magnesium, at low ionic strength. G-actin, so-formed, possesses one molecule of ATP and one tightly bound divalent cation per molecule, both of which are freely exchangeable. Although the crystal structure of G-actin has not been obtained directly, the structure of the actin monomer has been determined at atomic resolution through crystallisation as a complex with DNase I (Kabsch et al. 1990), gelsolin (McLaughlin et al. 1993) or profilin (Schutt et al. 1993) (Fig. 4.5a). The molecule displays two distinct domains separated by a cleft within which the nucleotide and divalent cation are bound. Each domain is further divided into a pair of subdomains, numbered from I through IV (Fig. 4.5a). Subdomain I contains both the amino and carboxyl termini of the molecule, as well as the primary site for interaction with myosin.

The particular isoform of actin found in scallop adductor muscles is β-actin (Margulis et al. 1982; Chantler 1983); this isoform appears to be distributed throughout all scallop tissues (Hue et al. 1989). Actin isoforms are distinguished by the identity of the four N-terminal amino acid residues (Vandekerckhove and Weber 1979); e.g., the N-terminal sequence from *Pecten maximus* is acetyl-Asp.Asp.Glu.Val... (Vandekerckhove and Weber 1984). Actin sequences are highly conserved: scallop actin and vertebrate actins differ in less than 6% of their residues; e.g., *Patinopecten yessoensis* actin has 21 (out of 374) amino acid substitutions with respect to vertebrate skeletal muscle actin, most changes occurring within subdomains I and III (Khaitlina et al. 1999). These structural differences confer subtle functional differences: scallop actin polymerises less readily than its vertebrate counterpart (Khaitlina 1986) and exhibits differing proteolytic susceptibility (Khaitlina et al. 1999) and antigenicity (Hue et al. 1988; Hue et al. 1989).

An increase in ionic strength causes G-actin to polymerise into F-actin, this process being accompanied by the hydrolysis of ATP to yield bound ADP. F-actin cannot be crystallised but its structure is amenable to analysis at low resolution by the combined techniques of optical diffraction and three dimensional reconstruction, which have been applied to the scallop thin filament (Craig et al. 1980; Vibert and Craig 1982; Egelman et al. 1983). More recently, the orientation of actin monomers within the filament structure has been established by modelling monomer structure against data obtained from fibre X-ray diffraction patterns from aligned gels of actin filaments (Holmes et al. 1990; Lorenz et al. 1993). In this structure (Fig. 4.5b), the long axis of the actin monomer passes through both domains and is arranged so that it is perpendicular to the axis of the F-actin filament, as originally envisaged from optical reconstructions (Egelman et al. 1983). Subdomains III and IV are shown to be positioned close to the helix axis and interact with subdomains

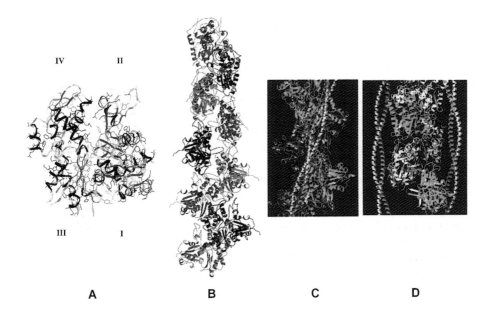

Figure 4.5. A. The structure of an actin monomer. This structure is shown as a wire-frame structure on which the α-carbon backbone has been highlighted. It is drawn in RasMol (version 2.5) using coordinates downloaded from the Protein Data Bank, Research Collaboratory for Structural Bioinformatics, Rutgers University (accession code 2BTF) for actin-profilin (Schutt et al. 1993); profilin has been rendered invisible so that the structure of the actin monomer can be seen clearly. α-helices are coloured green; component chains of β-sheets are coloured orange; the remainder of the α-carbon backbone is coloured grey. This globular structure is composed of four lobes (I, II, III and IV), II and IV being separated by a deep nucleotide binding cleft with ATP (red) positioned at the base. Note the position of Tyr-69 (blue), a residue which, when modified with a bulky group, affects the dynamic interaction between monomer and polymer (Chantler and Gratzer 1975). Both the N-terminus (yellow) and C-terminus (magenta) are located within lobe I. B. Ribbon structure of the actin monomer seen in A placed into a 13/6 helix (Holmes et al. 1990), the final structure being the best fit model for X-ray diffraction data from gels of oriented actin filaments. C. and D. Ribbon structures of the actin filament model seen in B together with tropomyosin (red) as seen in orthogonal views (Lorenz et al. 1995). Figures for B, C and D kindly provided by Prof. Ken Holmes and reproduced with permission of the publishers.

III or IV of adjacent actin monomers within the filament (Holmes et al. 1990). Small changes in the relative positions of subdomains II and IV occur upon polymerisation (Lorenz et al. 1993), accounting for the conformational changes known to occur when a G-actin monomer becomes a protomer in the F-actin structure, as deduced earlier by both functional (Offer et al. 1972; Chantler and Gratzer 1976) and spectroscopic (Higashi and

Oosawa 1965; Chantler and Gratzer 1975) techniques. Further conformational changes to actin protomers may be elicited upon myosin binding to F-actin when the latter is bound to tropomyosin (Tobacman and Butters 2000).

Tropomyosin has been found in all muscles without exception and is located within the two long-pitch grooves of the actin filament (Figs. 4.4c, 4.5c, d). It is a long, thin molecule, 41 nm in length and 2 nm in diameter (Cohen et al. 1971a), 66–70 kDa molecular weight (Woods 1969) and composed of two polypeptide chains arranged in parallel with each other, with no axial stagger, as α-helical coiled coils (Cohen and Holmes 1963; Lehrer 1975; Stewart 1975; Phillips et al. 1979). Tropomyosin can interact with seven actin protomers along each long-pitch helical groove of F-actin; it exhibits a 14-fold pseudo-repeat of charged residues which has led to suggestions that it can bind to each of the seven protomers at two different sites (Parry 1975; Stewart and McLachlan 1975; Hitchcock-DeGregori et al. 1990). Two different cDNAs, coding for the same tropomyosin sequence but possessing different polyadenylation signals, have been characterised from *Placopecten magellanicus* striated adductor muscle (Patwary et al. 1999). Scallop adductor muscles, similar to vertebrate skeletal muscles, express seven actin monomers for every tropomyosin (Chantler and Szent-Gyorgyi 1980). Using techniques of modelling fibre diffraction patterns similar to those described for actin above, the location of tropomyosin within the actin groove has been obtained with high precision (Lorenz et al. 1995) (Figs. 4.5c, d). Furthermore, direct evidence for the steric blocking model of tropomyosin action has been visualised by three-dimensional reconstruction of *Limulus* thin filaments, in the presence and absence of calcium (Lehman et al. 1994). The simplest interpretation is that tropomyosin can occupy either of two stable positions, one preventing access of myosin heads to subdomain I of actin (blocking state, observed in the absence of calcium) or, by moving deeper into the long-pitch actin groove, another stable position that facilitates strong, stereospecific myosin interaction (open state, observed in the presence of calcium) (Rayment et al. 1993b; Lehman et al. 1994). Other, more complex three-state interpretations are also possible and may be required to explain all aspects of the available data (Squire and Morris 1998; Tobacman and Butters 2000). Adjacent tropomyosin molecules within a groove of the long axis of the thin filament overlap each other in a head to tail fashion by nine residues (Phillips et al. 1979), a feature of possible significance for long range cooperativity in the thin filament (Brandt et al. 1984; Lehrer and Geeves 1998; Tobacman and Butters 2000).

Although scallop striated adductor tropomyosin may differ from its better known vertebrate counterpart in detail (Woods and Pont 1971; Woods 1976), the overall impression is that these striated muscle tropomyosins are very similar molecules. Many vertebrate muscles possess polymorphic forms of tropomyosin (Perry 1994). Despite evidence for more than one tropomyosin gene in *Placopecten magellanicus* (Patwary et al. 1999), a single isoform appears to be present within the striated adductor muscles of *Patinopecten yessoensis* and *Chlamys nobilis* (Ishimoda-Takagi et al. 1986; Ishimoda-Takagi and Kobayashi 1987). While it is easy to understand the role of tropomyosin in those scallop muscles possessing troponin, where it is essential for the steric blocking mechanism of thin filament-linked regulation, its role in scallop striated adductor muscles subject to thick filament-linked regulation alone is still not fully understood other than in its capacity of placing constraints on the rigidity and flexibility of the thin filament.

Troponin, the key regulatory complex isolated from muscles exhibiting thin filament-linked regulation, is composed of three subunits: troponin C (calcium binding subunit, 18–20 kDa), troponin I (inhibitory subunit, 19–40 kDa) and troponin T (tropomyosin binding subunit, 31–52 kDa). In such muscles, troponin is found associated with the thin filament in a stoichiometric ratio to tropomyosin. The individual troponin subunits are usually considered to be present in the ratio C:I:T = 1:1:1 (Potter 1974) although a ratio of C:I:T = 1:2:1 has been suggested for both vertebrate (Sperling et al. 1979) and invertebrate (Lehman et al. 1976; Ojima and Nishita 1986a; Ojima and Nishita 1992) muscles. In native thin filaments troponin attaches at regular intervals of 38.5 nm (Ohtsuki et al. 1967; Ebashi et al. 1969; Ohtsuki 1974; Ohtsuki 1975), the repeat being determined primarily by the elongated troponin T subunit (Phillips et al. 1979) (Fig. 4.4c). Several detailed reviews of troponin and its role are available (e.g., Perry 1994; Squire and Morris 1998).

As noted earlier (Chantler 1991), functional assays of scallop myofibrils, actomyosin (Lehman and Szent-Gyorgyi 1975) and fibres (Simmons and Szent-Gyorgyi 1978) have indicated a complete absence of thin filament regulation in striated adductor muscles from some scallops (e.g., *Placopecten magellanicus*). Consequently, the many reports of the existence of troponin in molluscan muscles have been regarded as somewhat problematic. The presence of troponin in molluscan muscles is variable and species-specific (Chantler 1983) but the case in favour of dual regulation in some adductor muscles has been made stronger in recent years. Whereas some striated adductor muscles appear to possess near normal levels of troponin (e.g., *Argopecten irradians*) (Goldberg and Lehman 1978; Lehman 1983a) others appear to possess only trace amounts (e.g., *Placopecten magellanicus*) (Szent-Gyorgyi 1976; Lehman 1981). Whole troponin has been isolated from the striated adductor of *Chlamys nipponensis akazara* and *Patinopecten yessoensis* (Ojima and Nishita 1986a; Ojima and Nishita 1992); striated adductor troponin I from *Argopecten irradians* hybridises with other vertebrate troponin subunits to form functional complexes (Goldberg and Lehman 1978) and antibodies to *Argopecten irradians* troponins I and C stain myofibrils from the same muscle with a 38 nm periodicity (Lehman 1981; Lehman 1983a). Amino acid sequences of troponin C from the striated adductor muscles of the Ezo-giant scallop, *Patinopecten yessoensis* (Nishita et al. 1994) and the Akazara scallop, *Chlamys nipponensis akazara* (Ojima et al. 1994) have been determined and shown to exhibit low (<30%) homology with their vertebrate counterparts; only one (site IV) of the four EF-hand motifs found in vertebrate skeletal muscle troponin retained calcium-specific binding functionality. Sequences of troponin I (Tanaka et al. 1998) and troponin T (Inoue et al. 1996) have also been obtained from the Akazara striated adductor and also shown to exhibit minimal (<30%) homology with vertebrate troponin subunits. Recently, the complete gene for troponin C from the striated adductor of *Patinopecten yessoensis* has been obtained (Yuasa and Takagi 2000) and shown to lack a fourth intron, found in vertebrates, a feature shared with *Drosophila* troponin C and calmodulin. However, even in the most favourable circumstances, scallop thin filament-linked calcium activation rarely achieves greater than a four-fold activation and elevated levels of magnesium are usually required for functional demonstration

(Goldberg and Lehman 1978; Lehman 1983b; Ojima and Nishita 1986b). In reconstitution experiments using *Patinopecten* striated adductor muscles, only readdition of light-chain was able to depress the actin-activated MgATPase rate in the absence of calcium (Shiraishi et al. 1999), the feature which most workers correlate with regulation; readdition of troponin C only restored the depressed activated rates to their intact values (Shiraishi et al. 1999). So, while some striated adductor muscles may exhibit dual-regulation it remains plausible that much of this troponin has little functional value and may represent vestigial components, molecular survivors of an earlier regulatory system superseded by the advent of regulatory myosins (Chantler 1983).

4.2.3 Thick Filament Structure

In order for muscle cells to achieve reliable, repetitive contractions, myosin molecules are organised into remarkable bipolar assemblages known as thick filaments. While the structure of striated muscle thick filaments is dictated to a large extent by associations between adjacent myosin tails, the heads being exposed on the thick filament surface, one must not ignore the contribution of paramyosin to molluscan thick filament structure, even though it is relatively low in the striated adductor (<7% by weight) (Szent-Gyorgyi et al. 1971; Levine et al. 1976). Thick filaments possess a central bare zone (Fig. 4.6a) where the myosin molecules are packed in an antiparallel manner. Farther away from the bare zone, on either side of it, myosin molecules are packed in a parallel arrangement. Short, beaded rods known as end-filaments have been observed to protrude from the tapering extremities of scallop thick filaments (Vibert and Castellani 1989).

Using optical diffraction and computer image reconstruction of relaxed thick filaments isolated from the striated adductor of *Placopecten magellanicus*, it has been possible to obtain direct information on thick filament structure in this muscle (Vibert and Craig 1983). Although the thick filaments taper towards their ends, there is a substantial constant region on which imaging techniques can be relied upon to give precise data. Order is replaced by disorder when calcium is added to the relaxed filaments, bringing about activation (Vibert and Craig 1985; Frado and Craig 1989). Image analysis of electron micrographs of relaxed scallop thick filaments revealed seven sets of surface projections at any one level (Vibert and Craig 1983) (Fig. 4.6c), a finding that has been substantiated further by rapid freezing and freeze-substitution techniques (Craig et al. 1991) and by helical reconstruction of frozen-hydrated scallop myosin filaments (Vibert 1992). Six or seven sets of surface projections had also been predicted earlier from X-ray diffraction studies of this muscle (Wray et al. 1975). Projections were found to lie close to the thick filament surface, extending along it axially, yet slewing slightly around the filament in an ordered array comprising seven long-pitch helices (Figs. 4.6b, c) (Vibert and Craig 1983; Levine 1993). Each surface projection is now thought to represent two heads arising from adjacent myosin molecules on adjacent crowns (Fig. 4.6b-III), which are in close association and related by an approximate two-fold rotation axis; the overlaps are arranged along the seven long-pitch helices (Vibert and Craig 1983; Levine et al. 1988; Levine 1993; Padron et al. 1997). The interaction between adjacent myosin heads

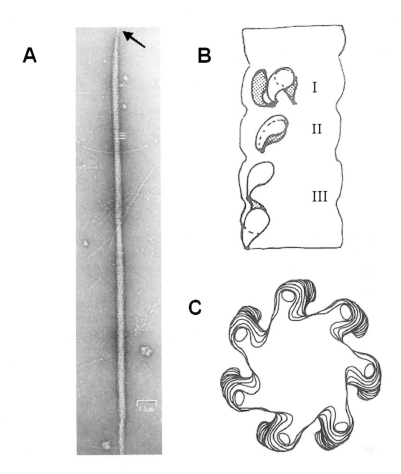

Figure 4.6. A. Electron micrograph of a negatively stained image of a thick filament isolated from the striated adductor muscle of *Placopecten magellanicus* in the relaxed state. Bar represents 0.1 μm. 14.5 nm periodicities are indicated (white bars). End-filaments are also indicated (black arrow). B. Interpretative diagram showing three plausible arrangements (I-III) of myosin heads on the surface of relaxed scallop striated adductor thick filaments taking account of the results of crosslinking experiments using bis_{22}ATP (Levine 1993). The bare zone is toward the bottom of the filament. Underlying heads are shaded except where overlapped. In I, heads are splayed circumferentially, overlapping adjacent heads within the same crown. In II, heads from each myosin molecule are not splayed. In III, heads are splayed axially along each of seven helices, leading to an overlap of adjacent heads from molecules within adjacent crowns; the latter is considered the most likely situation. C. Cross-section through the scallop thick filament seen in A as determined by three-dimensional reconstruction techniques, demonstrating the seven-fold symmetry of myosin heads at any crown. In this diagram, several sections covering a 14 nm axial extent (one crossbridge level), have been superposed. Material reproduced, with permission of the publisher, from Vibert and Craig (1983) (A and C) or from Levine (1993) (B).

along ordered helical tracks may be similar to that envisioned by Offer and Knight (1996) on isolated myosin molecules, and is responsible for stabilisation of the ordered, relaxed structure (Padron et al. 1997). As with most myosin thick filaments, sets of crossbridge projections indicate an axial periodicity along the filament of some 14.5 nm (Fig. 4.6a). The bare zone in these filaments was found to be 140 nm in length and the filaments had an overall length of 2.0 µm. This indicates 64 levels of crossbridges in each half of the thick filament and approximately 900 myosin molecules per filament. Fourier transforms calculated from images of the constant region confirmed the presence of this well-defined helical surface lattice and allowed precise measurement of the 48.0 nm right-handed helical pitch, consistent with that expected for a 7-start helix.

Realisation of the importance of head-head interactions has been key to establishing an atomic model of thick filament structure (Padron et al. 1997) but a detailed understanding of how myosin tails are arranged within the filament has yet to be achieved. Paramyosin-rich filaments are thought to possess a thin layer of myosin molecules wrapped around a dense paramyosin core (see below). However, in thick filaments from scallop striated muscle, where the molar ratio of paramyosin to myosin is less than 10%, it is unlikely that paramyosin is arranged as a central core. X-ray diffraction patterns from live scallop muscle do not exhibit reflections characteristic of the typical paramyosin 72.5 nm period (Wray et al. 1975; Millman and Bennett 1976). When native thick filaments, extracted from either *Argopecten* or *Placopecten* striated muscle, are exposed to low ionic strength, their structures fray into subfilaments, each approximately 10 nm in diameter (Vibert and Castellani 1989; Castellani and Vibert 1992); the most likely number of subfilaments per thick filament is seven. Treatment of native or frayed thick filaments with antibodies that recognise paramyosin provided no indication of paramyosin localisation, implying that paramyosin epitopes remain inaccessible even when subfilaments are exposed (Castellani and Vibert 1992). It is possible, therefore, that paramyosin is located at the core of each subfilament; indeed, each subfilament may in turn be composed of sub-subfilaments arranged in groups around a subfilament paramyosin core (Castellani and Vibert 1992). The exact arrangement remains unknown but it is possible that paramyosin could be restricted to a region on either side of the bare zone. In support of this, a 29 nm periodicity (corresponding to one of the molecular displacements found in paramyosin crystals) has been found to flank the bare zone by 150 nm in either direction (Vibert and Craig 1983). Alternatively, this could represent a periodic disturbance within myosin heads displayed on the surface of the filament (Wray et al. 1975). One consequence of the underlying subfilament structure within the thick filament backbone would be to cause individual crossbridges in relaxed muscle to take up precise yet varied orientations with respect to the filament axis, depending upon the particular rotational symmetry, thereby offering an explanation as to why filaments of muscles from different species that exhibit different rotational symmetries, helical symmetries and diameters possess crossbridges that appear to lie at similar radii from the filament centre (Wray et al. 1975; Wray 1979; Maw and Rowe 1980; Luther et al. 1981; Trinick 1981; Wray 1982).

Twitchins, or mini-titins, molecules in the size range 600–800 kDa, have been isolated from both *Placopecten* striated and smooth adductor muscles (Vibert et al. 1993).

Antibodies raised against scallop mini-titins localised the protein to the A-band, especially at the A-I junction. Mini-titins are giant kinases belonging to the immunoglobulin superfamily of which vertebrate titin and nematode twitchin are related members; antibodies to twitchin cross-react with scallop mini-titin (Vibert et al. 1993). Whereas titin is a much larger protein and extends throughout the I-band, connecting each thick filament to the Z-band (Trinick 1992), the restricted yet varied distribution of mini-titins currently limits our understanding of their function (but see below). Because insect muscle mini-titins connect the Z-band with the A-band (Nave and Weber 1990), it remains possible that the size distribution determined for scallop mini-titins is a gross underestimate (as a consequence of proteolysis or shearing during preparation) and that the intact structure could extend throughout the I-band to the Z-band, allowing mini-titins to act as shock absorbers by suspending thick filaments in the middle of the sarcomere. It remains unclear as to whether mini-titin is localised within the end-filaments, which protrude from the tips of thick filaments (Vibert et al. 1993) as originally suggested (Vibert and Castellani 1989) or is more extensive. Antibodies to twitchin kinase, a specific domain of the twitchin molecule, label scallop mini-titin close to the M-band. As noted above, scallop thick filament assemblies do not show any features which would correspond to the M-bridges seen in vertebrate skeletal muscle (Millman and Bennett 1976).

4.3 STRUCTURE OF THE SMOOTH ADDUCTOR MUSCLE

4.3.1 Fibre Microanatomy

The catch adductor muscle exhibits two major features which characterises it as a smooth muscle: firstly, it lacks striations and secondly, it possesses numerous dense bodies, both throughout the sarcoplasm and peripherally, which take the place of the missing Z-bands and provide anchoring sites for actin filaments. Additionally, this muscle exhibits a high proportion of paramyosin to myosin (2–10:1 by mass) and demonstrates the physiological property of catch; it is therefore categorised as a "paramyosin smooth" muscle (Hanson and Lowy 1960; Chantler 1983).

The scallop smooth adductor muscle has some similarities to the striated adductor. For example, the sarcoplasmic reticulum possesses the same components (Fig. 4.7), though both longitudinal tubules and cisternae are said to be less numerous (Nunzi and Franzini-Armstrong 1981). Intercalated discs and *fasciae adherentes* are also present. Like the striated adductor, the cells are also small, possessing a single fibril at the centre. In cross-section, however, the profile is more rounded (Fig. 4.7a), measuring about 3 x 5 μm in the case of *Argopecten irradians* (Nunzi and Franzini-Armstrong 1981). The average fibre length in this muscle was calculated to be close to 375 μm (Nunzi and Franzini-Armstrong 1981). Cross-sections through the smooth adductor reveal a variety of thick filament diameters: this is because individual thick filaments, arranged parallel to the fibre axis yet possessing no lateral alignment, are long and taper substantially towards their ends. The widest part is found throughout the centre of the filament, including the region comprising the bare zone. Mean filament diameters, measured in the smooth

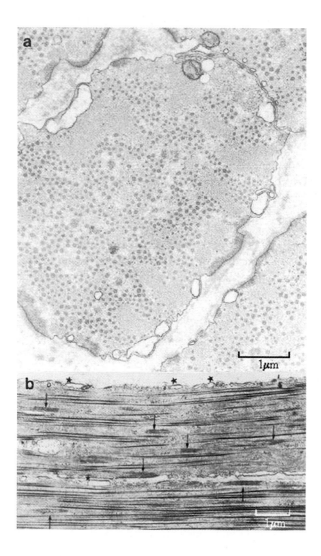

Figure 4.7. Scallop (*Placopecten magellanicus*) smooth adductor muscle. a). Low-magnification transverse section through catch adductor muscle fibres illustrating the variable diameter of thick filaments seen in a single cross-section, regions rich in thin filaments, a number of peripheral dense bodies as well as invaginations and vesicles of the sarcoplasmic reticulum. Micrograph courtesy of Dr. Clara Franzini-Armstrong. b). Longitudinal section through two adjacent smooth muscle fibres. Thin filaments are attached to elongated dense bodies both within the fibre and at the periphery (arrows). Elements of the sarcoplasmic reticulum are less numerous than in the striated portion, but have similar longitudinal profiles and peripheral cisternae (asterisks). Scale bars represent 1 μm. Micrograph reproduced, with permission of the publisher, from Nunzi and Franzini-Armstrong (1981).

adductors of the Atlantic scallops, peak in the range 20–50 nm (Philpott et al. 1960; Morrison and Odense 1974) though individual filaments may have diameters in excess of 100 nm (Elliott and Bennett 1982; Matsuno et al. 1996). Although it is generally accepted that paramyosin smooth filaments are very long (Fig. 4.7b), it would appear that no direct measurements have been made using filaments from scallop smooth adductor muscles. The anterior byssus retractor muscle (ABRM) of *Mytilus edulis* is often considered to be analogous to that of the scallop smooth adductor muscle (Nunzi and Franzini-Armstrong 1981) albeit with filaments having a larger average mean diameter (60–70 nm) (Sobieszek 1973). Here, the thick filaments have been estimated as being 20–30 μm in length (Lowy and Hanson 1962; Sobieszek 1973) and this is probably a reasonable guesstimate of filament length in the scallop smooth adductor. It may be noted that there is no necessary correlation between filament length and filament thickness (Nicaise and Amsellem 1983).

Figure 4.7a shows that in those regions of the cell possessing thick filaments, the filaments are approximately equally spaced from their neighbours, each surrounded by a halo of thin filaments. Nevertheless, it is clear that there are a number of regions in any cellular cross-section that are completely devoid of thick filaments and which possess huge numbers of thin filaments. It is possible that these areas represent regions adjacent to, but not coincident with, the dense bodies onto which the thin filaments are anchored. These dense bodies are fusiform structures, their shape readily envisioned by comparing their cross-sections (Fig. 4.7a) and longitudinal sections (Fig. 4.7b). Their sizes (1–2 μm long; 0.1–0.3 μm diameter) are comparable to those of dense bodies seen in the ABRM (Twarog 1967a; Sobieszek 1973). By analogy with the ABRM, 60–80 thin filaments will attach to opposite sides of each single dense body located in the body of the cell (Sobieszek 1973); peripheral dense bodies, attached to the sarcolemma in some way, only allow thin filament attachment on the cytoplasmic side.

4.3.2 Thin Filament Structure

Thin filaments from scallop smooth adductor muscles would appear to be very similar in structure to their striated muscle counterparts, described above and illustrated in Figures 4.4 and 4.5.

The structure of the underlying actin helix is virtually identical in both types of adductor muscle. An early X-ray diffraction study of molluscan muscles indicated that the structure of F-actin in the smooth adductor of *Pecten chlamys* was similar to that found in *Mytilus edulis* ABRM (Lowy and Vibert 1967), indicating a long-pitch crossover repeat of the order of 36–37 nm, as found in most muscles. Because the ABRM is often considered an experimental analog of the scallop smooth muscle, it is of interest to note that low-angle X-ray diffraction patterns of living ABRM muscle clearly show an intensity change on one of the actin layer lines upon activation of the muscle (Vibert et al. 1972). This is consistent with the movement of tropomyosin deeper into the actin groove upon activation of the muscle, and is suggestive of the presence of thin filament regulation in addition to myosin-linked control. Alternatively, it is also possible that tropomyosin movement occurs here in response to actin attachment by myosin crossbridges and is passive and inconsequential.

Scallop smooth adductor actin is also a β-isoform, similar to actin isolated from the striated adductor (Chantler 1983). Two isoforms of tropomyosin are found in the smooth adductors from *Patinopecten yessoensis* and *Chlamys nobilis*, both isoforms being different from the isoform of tropomyosin isolated from the striated adductor (Ishimoda-Takagi et al. 1986; Ishimoda-Takagi and Kobayashi 1987): one isoform is unique to the smooth adductor, the other appears to be identical to the tropomyosin isoform found in scallop cardiac muscle. Smooth muscle tropomyosin has also been shown to be an activating factor for actin-myosin interaction in the presence of ATP (Takahashi and Morita 1986). Troponin has been isolated from the smooth adductor muscles of *Chlamys nipponensis akazara* (Ojima and Nishita 1986b) and from the Ezo-giant scallop, *Patinopecten yessoensis* (Nishita et al. 1997). In the latter case, all three subunits were purified separately and shown to be functional, albeit restoring only 3-fold calcium sensitivity at best (Nishita et al. 1997). Nevertheless, these results are convincing and indicate that these proteins are not caldesmon degradation products, as had been suggested (Bennett and Marston 1990). Two isoforms of troponin C are expressed from a single gene in *Patinopecten yessoensis* through alternative splicing (Yuasa and Takagi 2000); both of these isoforms are found within the smooth adductor muscle. However, no troponin subunits were found in association with thin filaments prepared from smooth adductor muscles of the Atlantic scallops (Lehman and Szent-Gyorgyi 1975). Consequently, it remains a mystery as to why the scallop catch adductor muscles from some species possess functional troponins whereas others do not.

A second regulatory system involving the calmodulin binding protein, caldesmon (120–140 kDa molecular weight), is found in vertebrate smooth muscles where it dissociates from thin filaments in a calcium-dependent manner (Marston and Smith 1985). In the absence of calcium but the presence of tropomyosin, caldesmon binds to thin filaments and inhibits actomyosin interaction; this inhibition is relieved when calcium-calmodulin binds to caldesmon bringing about dissociation of the calcium-calmodulin-caldesmon complex from the thin filament. Furthermore, when caldesmon is phosphorylated by protein kinase C, it is unable to bind to the thin filament and consequently cannot inhibit the binding of myosin to actin (Yamakita et al. 1992). Caldesmon-like proteins of lower molecular mass (100–110 kDa) have been described in, and isolated from, the catch muscles of several molluscs (Bartegi et al. 1990; Bennett and Marston 1990), including scallops (Bartegi et al. 1990; Malnasi-Csizmadia et al. 1994) and, *in situ*, show a similar localisation to that found in vertebrates. Because a role for caldesmon in the maintenance of the latch state of vertebrate smooth muscle has been proposed (Marston 1989), the possibility that it may also be involved in maintaining the molluscan catch state is attractive (Bennett and Marston 1990). However, while some workers claim that caldesmon is only present in the smooth adductor muscle (Malnasi-Csizmadia et al. 1994), others claim that it shows the same abundance in striated adductor muscles (Bartegi et al. 1990). It has also been suggested that these invertebrate caldesmons are actually catchin molecules (Yamada et al. 2000), a newly reported protein found in smooth adductor muscles (see below). Until these issues are resolved, the role of scallop adductor muscle caldesmons will remain uncertain.

4.3.3 The Structure of Paramyosin-rich Thick Filaments

Paramyosin molecules are long rod-like coiled-coils (Cohen and Holmes 1963; Lowey et al. 1963; Kendrick-Jones et al. 1969) with a 124 nm contour length (Panté 1994); single paramyosin chains have a molecular mass of ~104 kDa (e.g., from *Chlamys nobilis*) (Matsuno et al. 1996); pairs of chains combine in parallel to form functional paramyosin molecules. Paramyosin is found within the core of molluscan thick filaments, exemplified by scallop smooth adductor muscles, hence their categorisation as "paramyosin smooth muscles" (Chantler 1983).

Paramyosin-rich thick filaments are large yet variable in size, >100 μm in length and can be more than 100 nm in diameter (Morrison and Odense 1974; Nunzi and Franzini-Armstrong 1981; Elliott and Bennett 1982); paramyosin typically makes up more than 80% of the thick filament mass (Millman 1967; Szent-Gyorgyi et al. 1971; Elliott 1974; Levine et al. 1976). *Pecten* smooth muscle thick filaments are bipolar in construction, similar to striated adductor thick filaments, but distinguished from them by their much greater size and by the arrangement of myosin with respect to paramyosin: in smooth muscle thick filaments a thin layer of myosin is weakly attached to the surface of the dense paramyosin core, surrounding it completely (Szent-Gyorgyi et al. 1971; Nonomura 1974; Cohen 1982). Comparatively little work has been done on structural aspects of smooth adductor muscles from scallop thick filaments. Although some differences between species have been reported (e.g., Matsuno et al. 1996), it is likely that similarities in the construction of all paramyosin-rich thick filaments transcend species boundaries (Cohen et al. 1971b; Elliott and Bennett 1982). Consequently, much of the information presented here was gained from studies of thick filaments from other molluscan muscles, especially the anterior byssus retractor muscle (ABRM) of *Mytilus edulis*.

Considerable effort has been expended in attempts to determine the structure of the paramyosin core (see Cohen and Szent-Gyorgyi 1971; Elliott and Bennett 1982; Chantler 1983; Bennett and Elliott 1989). Paramyosin cores, from which myosin has been removed, show a characteristic "checkerboard" appearance when seen by negative staining techniques in the electron microscope (Fig. 4.8e) (Hall et al. 1945; Cohen et al. 1971b; Elliott 1979; Matsuno et al. 1996). A unit cell of this array is seen in Figure 4.8g and is identical to the two-dimensional net deduced early-on from low angle X-ray diffraction patterns of molluscan smooth muscles (Bear 1944; Bear and Selby 1956) subsequently known as the "Bear-Selby net". Measurements of this net indicated a nodal repeat of 72.5 nm in one dimension (the vertical direction in Figure 4.8g) and a more variable distance between nodes at the same level in the second dimension (the horizontal direction in Figure 4.8g). This latter variability was shown to be mainly due to differences in the degree of hydration between samples used in different studies and, under carefully controlled conditions, this internodal distance was shown to be in the range 25–34 nm (Bennett and Elliott 1981).

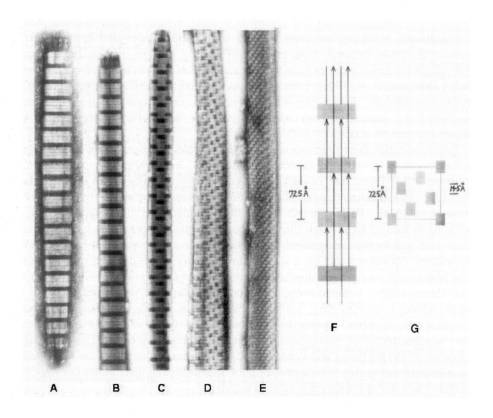

Figure 4.8. A). Bipolar paramyosin filament. Paramyosin from the white adductor muscle of the clam, *Mercenaria mercenaria*, precipitated from 50 mM Tris, pH 8.0, 1.0 M urea with 50 mM BaCl₂. The fine striations within the light bands show that the polarity of the arrays is opposite in the two halves of the filament. The band pattern is nonpolar (dihedral) at the centre. B). Paramyosin filament showing a simple polar array from the translucent adductor of *Mercenaria mercenaria* precipitated from 50 mM Tris, pH 8.0, with 50 mM MgCl₂. C). Paramyosin filament showing a transitional form from the smooth adductor muscle of the scallop, *Argopecten irradians*, precipitated with 50 mM BaCl₂ from 50 mM Tris, pH 8.0, 50 mM KSCN. The subfilament in the center is staggered by 2/5 of a period. This translation generates the net pattern. D). Paramyosin filament showing the synthetic net pattern from the smooth adductor of the scallop, *Placopecten magellanicus*, precipitated with 50 mM BaCl₂ from 50 mM Tris, pH 8.0, 0.1 M KSCN. E). Molluscan thick filament from which myosin has been removed. Paramyosin core from the white adductor of *Mercenaria mercenaria*. F). Two dimensional representation of the molecular packing of paramyosin molecules in the basic polar array. G). Unit cell of the paramyosin net pattern. Figures reproduced, with permission of the publisher, from Cohen and Szent-Gyorgyi (1971).

Paramyosin paracrystals can be obtained through precipitation by divalent cations in the presence of small quantities of chaotropic reagents (Fig. 4.8) (Kendrick-Jones et al. 1969; Cohen et al. 1971b; Cohen and Szent-Gyorgyi 1971). A variety of patterns were obtained when samples were negatively stained and viewed by electron microscopy, including the checkerboard array (Fig. 4.8d). Other arrangements included those with a banding pattern that lacked distinct nodes (Fig. 4.8a, b). These latter patterns could be arranged as non-polar (dihedral) structures (Fig. 4.8a) or as polar assemblies (Fig. 4.8b) (Kendrick-Jones et al. 1969), the polarity being seen by inspection of the minor striations within the main light band. The simplest interpretation of such a banding pattern is that paramyosin molecules do not bond end-to-end but engender gaps (where stain has penetrated) and overlaps (where stain is largely excluded) (Kendrick-Jones et al. 1969; Cohen and Szent-Gyorgyi 1971). Hydrodynamic and light-scattering data indicated that paramyosin molecules were about 135 nm long, as measured within the paracrystalline structure (Lowey et al. 1963). Thus, paramyosin was longer than a single banding repeat (72.5 nm) but shorter than the distance between alternate bands (2 x 72.5 nm). Hence, its dimensions were consistent with the gap-overlap model, adjacent molecules being staggered by 72.5 nm, the simplest arrangement being seen in Figure 4.8f. This interpretation of the regular paracrystalline banding pattern made it possible to determine the exact length of paramyosin with great precision, a value of 127.5 nm being ascertained (Kendrick-Jones et al. 1969).

Cohen et al. (1971b) suggested that the Bear-Selby net pattern (e.g., Fig. 4.8d, e) could be generated from the polar banding pattern (e.g., Fig. 4.8b) by unrestricted growth in the axial dimension yet restricted growth perpendicular to that axis, the nodal repeat shearing by 2/5 x 72.5 nm after only a few intermolecular spacings. An intermediate paracrystalline form that shows one such shear imposed on a polar banding pattern is seen in Figure 4.8c. Removal of surface myosin from thick filaments to yield paramyosin cores also yielded a variety of patterns including banding patterns as well as the Bear-Selby net (Nonomura 1974). That such patterns were manifestations of an underlying crystalline core was shown directly by tilting filament-bearing grids on a goniometer stage so as to view individual filaments after a rotation about their axis (Dover and Elliott 1979; Elliott 1979) or by performing a tilt series using embedded thick filaments cut in transverse section (Bennett and Elliott 1981).

The core of paramyosin-rich molluscan thick filaments is currently viewed as a layered, quasi-crystalline structure. In three dimensions the structure can be envisioned as being composed of two Bear-Selby planar arrays placed in nearly perpendicular apposition to each other (Elliott and Bennett 1984). In any one plane, adjacent paramyosin molecules are staggered by 72.5 nm. The array is not propagated throughout the filament but is sheared periodically in the direction of the filament axis by 2/5 x 72.5 = 29.0 nm. There appears to be between 4 and 10 parallel molecular arrays between shear lines, the exact number being species dependent (Elliott and Bennett 1984; Matsuno et al. 1996) and is variable in any one thick filament giving rise to the paracrystalline (as opposed to crystalline) nature of the core. The underlying reasons for the orderly shearing pattern are unknown. The paramyosin core must also be bipolar so as to maintain appropriate bonding with the bipolar arrangement on the surface. This bipolar arrangement has now been observed directly (Panté 1994), paramyosin assembling such that its amino terminal end points towards the midpoint of the filament (Fig. 4.9) (Weisel 1975; Panté 1994).

The arrangement of the surface layer of myosin which surrounds the paramyosin core is still open to question but is likely to be a monomolecular layer with from one to three myosin molecules associated with each node of the Bear-Selby net (Elliott 1974; Cohen 1982; Panté 1994). Based on sequence analyses of myosin and paramyosin from the nematode, *Caenorhabditis elegans*, a number of optimal charge pairings have been conjectured (Kagawa et al. 1989). Taking into account the known orientations of paramyosin and myosin within the bipolar filament, Panté (1994) has proposed a model (Fig. 4.9), based on an earlier version (Cohen 1982), in which myosin is arranged in an antiparallel manner to paramyosin with a stagger between their respective amino acid sequences of 196 residues – equivalent to 29 nm, a spacing which corresponds to the axial spacing of adjacent Bear-Selby nodes. Vital to this is an assembly competence domain, comprising a 29 amino acid residue region of unique charge profile, that is present at the carboxyl-termini of both molecules (Cohen and Parry 1998). Several models are possible and the one depicted in Figure 4.9c should not be thought of as final; the number of myosin molecules per node and the number of paramyosin molecules within each shear zone are still not known. However, this particular model (Panté 1994) does provide a clear insight into how myosin and paramyosin could be aligned, even to the extent of proposing a function for the 'skip' residues – single amino acid interruptions within the coiled-coil heptad repeat – which are also aligned within the two polypeptide sequences, suggesting they provide a form of "ball and socket" stabilisation.

It remains unclear that the proposed structure (Fig. 4.9) would provide for nodal radial symmetry (hence a radial distribution of myosin) around the filament. Such an arrangement is necessary because thin filaments have been observed to be distributed radially around these thick filaments (Lowy and Hanson 1962; Elliott 1964; Elliott and Bennett 1982; Lowy and Poulsen 1982) and radially arranged crossbridges can be visualised under rigor conditions (Sobieszek 1973; Bennett and Elliott 1981). It is also possible that myosin could be organised in a helical manner around the filament, as has been observed in some paramyosin-containing muscles where the molar ratio of paramyosin to myosin is close to unity (Castellani et al. 1983). This would provide a way for myosin to get around the "corners" and attain a radial distribution, the myosin coiled-coil possibly being placed at some, as yet unspecified, angle to the paramyosin coiled-coil. Note that stoichiometries obtained for the relative abundance of myosin and paramyosin in molluscan smooth muscle thick filaments (Szent-Gyorgyi et al. 1971; Elliott 1974; Levine et al. 1976) are not so precise as to place the concept of a single myosin surface layer beyond question. The essence of this concept is that from one to three myosin molecules, present within a surface area of 43,500 $\mathrm{\AA}^2$, would exhibit properties such that myosin-paramyosin interactions predominate (Cohen 1982); surface myosins present on the thick filaments of non-catch muscles, where myosin-myosin interactions predominate, are squeezed into a surface area of some 15,000 $\mathrm{\AA}^2$ (Cohen 1982). Small changes in the measurement of molecular stoichiometries, filament dimensions or molecular packing,

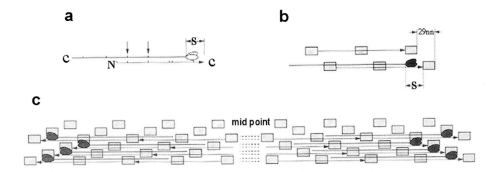

Figure 4.9. Possible molecular arrangement of myosin and paramyosin molecules in the native thick filament of molluscan smooth muscle. a). Antiparallel, lateral arrangement of myosin and paramyosin molecules for a stagger (s = -196 residues) in which the carboxyl terminal end of paramyosin is shifted away from the amino terminal end of the myosin rod by ~29 nm. Short vertical bars indicate skip residues of which the positions of two on either molecule (arrows) coincide. b). Schematic arrangement of paramyosin subfilaments illustrating relationship to 29 nm nodal gap stagger. In the lower subfilament, myosin (as in a)) is superposed. c). Schematic model to illustrate a possible arrangement of myosin within the thick filament relative to the Bear-Selby net. The latter is generated by shifting paramyosin subfilaments by 29 nm. Myosin heads are shown to coincide with nodal gaps, but other arrangements are possible. For simplification, only one paramyosin molecule is drawn per node and only four myosins are drawn on either side of the thick filament. Adapted from Pante (1994) and reproduced with permission of the publisher.

could have profound effects on the concept of whether myosin-paramyosin or myosin-myosin interactions predominate.

Myosin isolated from *Patinopecten yessoensis* smooth adductor possessing an isoform of the regulatory light chain, R-LCa, undergoes phosphorylation at Ser-11 of R-LCa (Sohma et al. 1985). This was shown using either gizzard myosin light chain kinase (MLCK) (Sohma et al. 1985) or an endogenous kinase purified from the smooth adductor (Sohma and Morita 1986; Sohma and Morita 1987; Sohma et al. 1988a). A calcium-dependent calcineurin-like phosphatase has been shown to dephosphorylate phosphorylated R-LCa (Inoue et al. 1990); whether this protein is related to the calcineurin-like phosphatase implicated in catch contraction (Castellani and Cohen 1992) is not yet known. These data are suggestive of a form of regulation more akin to vertebrate smooth muscle than to the regulation of scallop striated adductor myosin. Sequence analysis of the larger smooth muscle R-LC isoform (smoA) from *Placopecten* confirmed (Perreault-Micale et al. 1996a) the presence of a phosphorylation consensus sequence (Kemp and Pearson 1985); smoA could also be phosphorylated by MLCK (Perreault-Micale et al. 1996a). However, hybrid studies demonstrated that the ATPase obtained was characteristic of the parent myosin heavy chain (MHC) and the ATPase rates were not altered irrespective of R-LC phosphorylation (Perreault-Micale et al. 1996a). Thus, whether or not the specific phosphorylation of R-LCa or smoA portends a role in the regulation of catch contraction remains an open question.

Myosin from the ABRM of *Mytilus edulis* has been shown to be a substrate for heavy-chain phosphorylation by an endogenous heavy-chain kinase or by a heavy-chain kinase isolated from the slime mold, *Dictyostelium* (Castellani and Cohen 1987a; Castellani and Cohen 1987b). Heavy chain phosphorylation occurs at a serine residue within the tail-piece sequence, an extended region not found in striated adductor myosin (Castellani et al. 1988). A second phosphorylation site noted by these authors is actually found within the N-terminal sequence of catchin (Yamada et al. 2000) (see below). It was shown, *in vitro*, that this phosphorylation enhances the ability of the myosin molecule to fold at low ionic strength (Castellani and Cohen 1987b) suggesting that events in the neck region of the myosin molecule could be modulated by phosphorylation at the tip of the tail. Myosin heavy chain phosphorylation has also been demonstrated to occur in smooth adductor muscles of *Patinopecten yessoensis* using an endogenous kinase previously characterised as phosphorylating R-LCa (Sohma et al. 1988b).

The most important new thick filament protein to emerge recently, unrelated to myosin or paramyosin, is twitchin. As noted above, mini-titins, otherwise known as twitchins, are giant kinase molecules found in both scallop striated and smooth adductor muscles where they are located within the A-band and at the A-I junction (Vibert et al. 1993). Recently, Marion Siegman and collaborators (Siegman et al. 1997; Siegman et al. 1998) set out to find which proteins became phosphorylated when an ABRM (*Mytilus edulis*) muscle was released from the catch state, either by application of serotonin to the intact muscle or by the addition of cAMP to permeabilised muscles. Surprisingly, a 600 kDa protein exhibited the greatest increase in phosphorylation (Siegman et al. 1997) and was shown to be twitchin through a cDNA cloning strategy that used N-terminal amino acid sequence data to derive PCR primers (Siegman et al. 1998). The molar ratio of twitchin to myosin in these muscles was shown to be 1:15 and the level of cAMP-induced phosphorylation was somewhat greater than 0.5 mole/mole.

Catchin is the name given to a pair of new players on the scene – novel heat-stable, phosphorylated rod-shaped molecules, 112 kDa and 120 kDa in size, characterised from the catch adductor muscles of the scallop, *Argopecten irradians* and the mussel, *Mytilus galloprovincialis* (Yamada et al. 1997; Yamada et al. 2000). Often noted as an abundant contaminant of paramyosin preparations (Panté 1994), catchin had previously been identified as a proteolytic cleavage product of myosin (Castellani et al. 1988) through partial sequencing. Determination of the complete sequence (Yamada et al. 2000) revealed that catchin is an alternatively spliced product of the MHC gene with a unique non-helical N-terminal sequence comprising 156 amino acid residues, including a phosphorylatable serine; the remaining 830 C-terminal residues are identical to the MHC coiled-coil rod sequence distal to the S-2 junction and include the non-helical tail-piece. The unique N-terminus of catchin is encoded within the large intron following exon-18 of the smooth adductor MHC sequence (Nyitray et al. 1994). The contour length of catchin, 125 nm, is almost identical to that of paramyosin (Yamada et al. 1997; Yamada et al. 2000). While the presence of catchin (Yamada et al. 2000) and twitchin (Siegman et al.

1998) clearly have implications for the catch mechanism (see below under Catch Mechanism), nothing is known at present as to how these proteins are incorporated into, or interact with, the paramyosin smooth thick filament structure.

4.4 MYOSIN

Scallop adductor myosin is a regulatory myosin, possessing the necessary machinery within its own structure to control its interaction with actin. Similar to other conventional myosins it is composed of two heads (each ~16.5 nm in length, ~6.5 nm wide and ~4 nm thick) and a long tail (~150 nm in length) (Fig. 4.10). The N-terminal halves of each of the two myosin heavy chains (MHCs) (~200 kDa each) separately fold to form the heads of the molecule while the α-helical C-terminal halves wrap around each other to form the coiled-coil tail. Within each head the first ~780 residues of the MHC form the motor domain featuring the actin-binding and ATPase sites; each MHC then extends as a long α-helix stabilised by two light chains (LCs) (~20 kDa each) forming a distinct regulatory domain (RD) or lever arm. The essential light chain (E-LC) abuts the motor domain while the regulatory light chain (R-LC), furthest from the head, abuts the E-LC. Unlike their vertebrate counterparts, scallop striated and smooth muscle MHCs are expressed from a single gene as alternatively-spliced products (Nyitray et al. 1994). This gene, in *Argopecten irradians*, is 24 kilobases in size and is composed of 27 exons of which at least four are present in tandem pairs and are alternatively spliced in the striated and catch adductors (Nyitray et al. 1994). A minor catch isoform expressing a fifth alternatively-spliced exon (exon 13), that was noted in *Argopecten irradians* (Nyitray et al. 1994), was found to be the predominant isoform in *Placopecten magellanicus* catch myosin (Perreault-Micale et al. 1996b). Whereas the catch muscle MHC is also found in scallop foot, mantle, gonad and heart muscles, the striated muscle isoform appears to be expressed exclusively in the striated adductor muscle.

Complete primary amino acid sequences have been obtained for the MHCs (Nyitray et al. 1991; Perreault-Micale et al. 1996b; Janes et al. 2000), R-LCs (Kendrick-Jones and Jakes 1977; Collins et al. 1986; Goodwin et al. 1987; Perreault-Micale et al. 1996a; Janes et al. 2000) and E-LCs (Collins et al. 1986; Goodwin et al. 1987; Perreault-Micale et al. 1996a; Janes et al. 2000) of myosins from the striated adductor muscles of the Atlantic scallops *Argopecten irradians* (Collins et al. 1986; Goodwin et al. 1987; Nyitray et al. 1991), *Placopecten magellanicus* (Perreault-Micale et al. 1996a; Perreault-Micale et al. 1996b) and *Pecten maximus* (Kendrick-Jones and Jakes 1977; Janes et al. 2000). These are displayed and aligned in Figs. 4.11 and 4.12, where they are compared with the primary sequences of the catch adductor MHCs from *Argopecten irradians* (Nyitray et al. 1994) and *Placopecten magellanicus* (Perreault-Micale et al. 1996b).

Pecten maximus MHC cDNA was found to contain an open reading frame of 5,820 nucleotides encoding 1,940 amino acid residues as compared with 1,938, 1,941 and 1,935 amino acids which make up the MHCs of *Argopecten irradians*, *Placopecten magellanicus* and *Loligo pealei* (squid) striated muscle myosins, respectively (Fig. 4.11). Analysis of the *Argopecten* MHC sequence defined a number of residues required for regulation (Nyitray et al. 1991); these residues are conserved in all the regulatory MHC

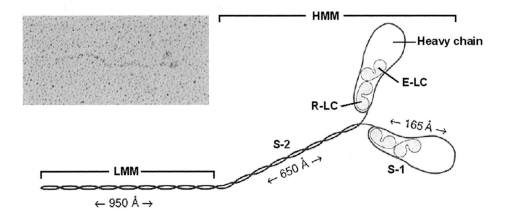

Figure 4.10. Schematic diagram of the myosin molecule (not drawn to scale) showing the organisation of the three pairs of polypeptide chains (myosin heavy chain; regulatory light chain (R-LC); essential light chain (E-LC)) and major proteolytic fragments (single headed subfragment-1 (S-1); double-headed heavy meromyosin (HMM); coiled-coil tail proximal to the heads (S-2); coiled-coil tail distal from the heads (LMM)). Modified from Rayment and Holden 1994 and reproduced with permission from the publisher. Inset: Rotary-shadowed electron microscopic image of scallop striated adductor myosin; Chantler and Kensler, unpublished.

sequences (Fig. 4.11). The three scallop MHC sequences are closely homologous to each other (Janes et al. 2000), the *Pecten* sequence exhibiting 97.5% and 95.6% identity to the *Argopecten* and *Placopecten* MHC sequences, respectively; by contrast, only 73.6% identity was found relative to a molluscan MHC sequence from squid (*Loligo pealei*) syphon retractor muscle (Matulef et al. 1998). *Placopecten* striated adductor myosin displays a higher actin-activated MgATPase rate than either *Argopecten* or *Pecten* myosins (Chantler and Szent-Gyorgyi 1980; Chalovich et al. 1984; Perreault-Micale et al. 1996b; Patel et al. 2000). Key sequence differences within a surface loop (Loop 1) (Spudich 1994) structure close to the nucleotide binding pocket, encoded by exon 5, are thought to account for these different turnover rates (Perreault-Micale et al. 1996b; Kurzawa-Goertz et al. 1998). Differences found within the catch muscle isoforms arise from the five alternatively-spliced exons. Three of these regions are found within the myosin head and encode the phosphate binding loop and Loop 1 (exon 5), part of the ATP binding cleft (exon 6) and a linking region between the ATP site and the actin-binding site (exon 13). The other two regions are within the tail and encode the S2 hinge region (exon 20) and lead to an additional ten amino acids within the catch isoform at the C-terminus of the non-helical tail-piece (exon 26) (Fig. 4.11) (Nyitray et al. 1994; Perreault-Micale et al. 1996b). The actin-activated MgATPase activity of myosin and myosin head (S-1) fragments isolated from the smooth adductor are two to three times lower than those measured for the striated adductor (Perreault-Micale et al. 1996a; Perreault-Micale et al.

1996b); these differing rates arise from differing values of V_{max} but not K_M (Perreault-Micale et al. 1996b) and are attributed to variations in the Loop 1 sequence because alternatively spliced sequences of exons 6 and 13 are identical within analogous muscles of the two species. No alternative splicing of Loop 2 has so far been detected in molluscan myosins. It is important to note that these reported sequence differences, while relevant to catch, cannot be the sole determinants of the catch mechanism because the smooth muscle isoform possessing all alternatively spliced regions was also detected in scallop mantle and heart muscle (Nyitray et al. 1994).

The deduced sequence of *Pecten* R-LC has 156 amino acids, exhibiting 97.4% identity to that found in *Argopecten* R-LC (Goodwin et al. 1987) and 90.4% identity to *Placopecten* R-LC (Perreault-Micale et al. 1996a) (Fig. 4.12a). Glycine-117, a residue previously shown to be critical for regulation (Jancso and Szent-Gyorgyi 1994), is conserved in all molluscan R-LCs so far sequenced: this residue is responsible for maintaining hydrogen bonding to the adjacent E-LC and is critical to the regulatory mechanism (see below). The deduced amino acid sequence of *Pecten* E-LC also has 156 amino acids, exhibiting 99.4% identity to the *Argopecten* E-LC (Goodwin et al. 1987) and 94.2% identity to *Placopecten* E-LC (Perreault-Micale et al. 1996a) (Fig. 4.12b). In addition, R-LC (Maita et al. 1984; Miyanishi et al. 1985) and E-LC (Maita et al. 1987) sequences have been determined for myosins from the striated adductor muscles of the Japanese scallops, *Patinopecten yessoensis* (Maita et al. 1984; Miyanishi et al. 1985; Maita et al. 1987) and *Chlamys nipponensis akazara* (Maita et al. 1984).

The sequence of the E-LC found in the smooth adductor muscle of *Placopecten magellanicus* is identical to its counterpart found in the striated adductor (Perreault-Micale et al. 1996a). However, two distinctive amino acid sequences have been determined for R-LCs isolated from the smooth adductor muscles of each of *Patinopecten yessoensis* (Miyanishi et al. 1985) and *Placopecten magellanicus* (Perreault-Micale et al. 1996a) which differ from their striated counterparts (Fig. 4.12a). Because the catch adductor muscle closely approximates the striated adductor muscle along its innermost surface (see Fig. 4.1), it is not too surprising that there is a gradual transition in fibre type from striated to smooth, moving from the inside surface to the outside of the catch adductor (Ruegg 1961). Consistent with this, there is a gradual transition in myosin type across the smooth adductor (Kondo and Morita 1981; Morita and Kondo 1982): the R-LC associated with the catch portion (R-LCa) of *Patinopecten* is found only around the outer perimeter of the muscle (Miyanishi et al. 1985); the larger of the two R-LCs (smoA) found in *Placopecten* is found only within the catch portion of the muscle whereas the other smooth muscle R-LC isoform (smoB) is present both in the intermediate and catch portion of the smooth muscle (Perreault-Micale et al. 1996a). Most amino acid differences occur within the N-terminal 52 residues; analysis of genomic DNA by Southern blotting and PCR suggests that all three isoforms (striated, smooth A and B) are generated by alternative splicing of a single gene (Perreault-Micale et al. 1996a).

```
                    *       *   *      SH3  *
1    MNIDYSDPDFQYLAVDRKKLMKEQTAAFDGKKNCWVPDEKEGFAPAEIQSSKGDEITVKITSDNSTRTVKKDDIQSMNPPKFEKLEDMAN   Pecten St
     ....F.................................S..............VA.S............................   Argopecten St
     ....FN........M.......P...........P...S..............VA.S.........Q..................   Placopecten St
     ....F.................................S.............E...V............................   Argopecten Ct
     ..TM.F....MEF.CLT.Q...EATSIP..........P...S..........E...V.....Q....................   Placopecten Ct
     ....F..........M...P...........PDF.VG........T...V....TDKTQE.V......GQR.....MNM.....   Loligo SR

                                                                          P-Loop
                **     *
91   MTYLNEASVLHNLRSRYTSGLLIYTYSGLFCIAVNPYRRLPIYTDSVISKYRGKRKTEIPPHLFSVADNAYQNMVTDRENQSCLITGESGA   Pecten St
     ..Y...............................A.......................................   Argopecten St
     ..N..G...A........................A.......................................   Placopecten St
     ..Y...............................A.......................................   Argopecten Ct
     ..N..G...A........................A.......................................   Placopecten Ct
     L.F....I.....E.F......I...........QGLVD......RA.M......I......Y.LQ......M....   Loligo SR

                                                                          Switch-I
         *    *                  (25/50kDa) Loop-1
181  GKTENTKKVIMYLAKVACAV--KK-KTDEEEASD-KKQGSLEDQIIQANPVLEAYGNAKTTRNNNSRFGKFIRIHFGPTGKIAGADIETY   Pecten St
     ...............-----------.E..............   Argopecten St
     ...............-----..S..EA.Q.............   Placopecten St
     .....F..ANL--Y.-QKQ..PTTHARASN............   Argopecten Ct
     .....S..ANL--Y.-QKE.PVPNLRA--SN....E.........V.......   Placopecten Ct
     ...Q.F.L.ASLAG..D.KE..KKK.-E.K.T....V.C......E.........TQ....   Loligo SR

268  LLEKSRVTYQQSAERNYHIFYQVCSNALPELNDIMLVTPDSGLYSFINQGCLITVDNIDDVEEFKLCDEAFDILGFTKEEKQSMFKCTASI   Pecten St
     ...I..I.....V............................................   Argopecten St
     ...I..I...EV..I.........................T................   Placopecten St
     ...I..H.................................................   Argopecten Ct
     ...I..H...EV.I.........................T................   Placopecten Ct
     ...LL.P.F..NIEKI.AV.P..G....T...G..E.MG.T.T..V...D..L.Y...GC.   Loligo SR
```

```
                   ACTIN BINDING
358  LHMGEMKFKQRPREEQAESDGTAEAEKVAFLCGINAGDLLKALLKPKVKVGTEMVTKGQNLNQVTNSVGALAKSLYDRMFNWLVKRVNKT  Pecten St
     ..........................................................M..V.......................R...  Argopecten St
     ...........................................Q..I.......S..............................R...  Placopecten St
     ..........................................................M..V.......................R...  Argopecten Ct
     ...........................................Q..I.....................................R...  Placopecten Ct
     ..L....W...--G....A.......L.V.......C.....I.......Y..Q.R.KD....IA...............R..Q...  Loligo SR

                      Switch-II                        Relay
448  LDTKAKRNYYIGVLDIAGFEIFDYNSFEQLCINYTNERLQQFFNHHMFILEQEEYKKEGIAWEFIDFGMDLQMCIDLIEKPMGILSILEE  Pecten St
     ..............F......................................................................  Argopecten St
     ..............F..................................V...................................  Placopecten St
     ..............F......................................................................  Argopecten Ct
     ..............F..................................V...................................  Placopecten Ct
     ........QFF......F........................V.............V..........L..A..E...........  Loligo SR

             ACTIN BINDING
538  ECMFPKADDKSFQDKLYQNHMGKNRMFTKPGKPTRPNQGPAHFELHHYAGNVPYSITGWLDKNKDPINENVVSLLSVSKEPLVAELFRAP  Pecten St
     ..................................................................E.......A..GA.......K...  Argopecten St
     .........YS....I...........................H...............................A.........A...  Placopecten St
     ..................................................................E.......A..GA.......K...  Argopecten Ct
     .........YS....M...........................H...............A.......E.QN...I.KM.TP....  Placopecten Ct
     ........S.T..KN..D.L..P..G...-PKAGCAE...C.......S..S..A..............................  Loligo SR

                                                            SH-
     (50/20kDa) Loop-2 ACTIN BINDING                   SH2 helix SH1
628  EEPVGGGGKKKKGKSSAFQTISAVHRESLNKLMKNLYSTHPSFVRCIIPNELKQPGLVDAELVLHQLCNGVLEGIRICRKGFPSRLIYS  Pecten St
     ..A.--...............................H.............................................  Argopecten St
     D..A..A.G...K........................H.............................................  Placopecten St
     ..A.--...............................H.............................................  Argopecten Ct
     D.A..A.G...K.........C..RR.N.H.......LE.D.....T..I.A.....R........N.I..............  Placopecten Ct
     RI-LTP....A........S..K...........................................................  Loligo SR
```

```
         converter                                                      ELC BINDING
                                                                                  *
718 EFKQRYSILAPNAIPQGFVDGKTVSEKILTGLQMDPAEYRLGTTKVFFKAGVLGNLEMRDERLSKIISMFQAHIRGYLIRKAYKKLQDQ  Pecten St
    ...................A.........................................................................  Argopecten St
    ...................A....S....................................................................  Placopecten St
    ...................A....S....................................................................  Argopecten Ct
    ...................A....S....................................................................  Placopecten Ct
    ....V.S..A...V.TD.V.SA.L.N...........M.D....................................M................  Loligo SR


            *
            RLC BINDING
                       **
808 RIGLSVIQRNIRKWLVLRNWQWKLYAKVKPLLSIARQEEEMKEQLKQMDKMKEDLAKTERIKKELEEQNVTLLEQKNDLFLQLQTIEDS  Pecten St
    ...S........................................................L...............................  Argopecten St
    ...S...................A....L............VE..................................................  Placopecten St
    ...............A.......L............VE.......................L...............................  Argopecten Ct
    ..TL...V..........E.R.FN....N....D.N.KAQEEFA....EF.SC.QMR.....TV.MQ..........................  Placopecten Ct
    ...TL...V.................E.R.FN....N....D.N.KAQEEFA....EF.SC.QMR.....TV.MQ.......VIAMSSG..A  Loligo SR


898 MGDQEERVEKLIMQKADFESQIKELEERLLDEDAAADLEGIKKMETDNSNLKKDIGDLENTLQKAEQDKAHKDNQISTLQGEMAQQDE  Pecten St
    ...D..................................A.A..................................IS.............  Argopecten St
    ...D...............................S.....G.A......E.HS..S.E..................IS.............  Placopecten St
    ...D...............................S.....A.A......E.HS..S.E...............................  Argopecten Ct
    ...D...............................S.....G.A......E.HS..S.E.......S.......................  Placopecten Ct
    I.A..KI.Q..K..S...T......DK.M......TE.SAQ...SDAEIGE...VE...AG.A...E.TT.....K..D............  Loligo SR


988 HIGKLNKEKKALEEANKKTSESLQAEEDKCNHLNKLKAKLEQALDELEDNLEREKKVRGDVEKAKRKVEQDLKSTQENVEDLERVKRELE  Pecten St
    ...D.........................................................................................  Argopecten St
    ...D.........................................................S...............................  Placopecten St
    ...D.........................................................................................  Argopecten Ct
    ...D.........................................................S...............................  Placopecten St
    .LS......N...VQ..L.D.........V..S..T..T......IP..D.........T..T......D.......................  Loligo SR
```

```
1078  ENVRRKEAEISTLNSKLEDEQNLVSQLQRKVKELQARIEELEEELEAERNARAKVEKQRAELNRELELGERLDEAGGATSAQIELNKKR   Pecten St
      ...........S.................I...............S.........................................   Argopecten St
      ...........T.................H...............S.........................................   Placopecten St
      ...........S..................................S.......................................   Argopecten Ct
      ...........T..................................S................A...M...................   Placopecten Ct
      DAG.K.DM.NG...........A..K.I........Q.T.....T.S...................                       Loligo SR

1168  EAELLKIRRDLEEASLQHEAQISALRKKHQDAANEMADQVDQLQKVKSLKEKDKKDIKREMDDLESQMTHNMKNKGCSEKVMKQFESQMS   Pecten St
      ......................................L...............................................   Argopecten St
      ....................................................D.................................V   Placopecten St
      .................I...................L........S.EQQLRS.VE..QA.IQ.IS....................   Argopecten Ct
      .................I...................S.ENNKMES.NE..QA.IQ.IS.........................V     Placopecten Ct
      .Q..RL.........TM...S.AT....N.E.T.LG..I.........R..E.TQLRA.....VQ..VE.AG..R......MS..M.A.L. Loligo SR

1258  DLNARLEDSQRSINELQSQKSRLQAENSDLSRQLEDAEHRVSVLSKEKSQLGSQLEDARRSLEDETRARSKLQNEVRNMHADMDAVREQL   Pecten St
      ........................T........................S..........E...............I.........   Argopecten St
      .................................................T..........E...T...I.......A.........   Placopecten St
      ...........................T.....................S..........E...T.......I.......I.....   Argopecten Ct
      ..............................................................E...T.....I...A.........   Placopecten Ct
      E...KID.QA..VS..T........T.AA..T....E..N.GQ.T.L..S..AS.....K....G.L.A...A...LNS.I.GI..S.   Loligo SR

1348  EEEQESKSDVQRQLSKANNEIQQWRSKFESEGANRTEELEDQKRKILGKLSEAEQTTEAANSKCSALEKAKSRLQQELEDMSIEVDRANA   Pecten St
      .........................................L...........A...............................     Argopecten St
      ...........................................NM....A.A..D................L.............     Placopecten St
      .............................................L....A.A..................L.............     Argopecten Ct
      ...........................................NM...A...D.....L..........G..LA.D.E.SS.......   Placopecten Ct
      .A.....L.A..R..A.V..........A.AD.....A..LQA.....ADTLH..AG...........G...LA.D.E.SS.       Loligo SR
```

260

```
1438 NVNQMEKKQRAFDKTTAEWQSKVNSLQSELENAQKESRGYSAELYRIKASVEEYQDSIGSLRRENKNLADEIHDLTDQLSEGRSTHELD  Pecten st
     S.........A..............S.............I......A.................................            Argopecten st
     S.........S..............S.............I.............S..........................            Placopecten St
     S.........A..............S.............I.....A.................................             Argopecten Ct
     S.........S..............S.............A.............S..........................            Placopecten Ct
     HA.NL.....N...VVS...H.C.D.A......F.VR.QC..VG.TVA..............G....N...E                     Loligo SR

1528 KARRRLEMEKEKLQAALEEAEGALEQEEAKVMRAQLEIATVRNEIDKRIQEKEEEFDNTRRNHQRALESMQASLEAEAKGKADAMRIKKK  Pecten st
     .................................................................                          Argopecten st
     .........................................L.......................                          Placopecten St
     .................................................................                          Argopecten Ct
     .........................................L.......................                          Placopecten Ct
     ...KH.AL.....................T...SQI.Q...R.L....I.........E.L.....                          Loligo SR

1618 LEQDINELEVALDASNRGKAEMEKTVKRYQQQIREMQTSIEEEQRDEARESYNMAERRCTLMSGEVEELRAALEQAERARKASDNELA    Pecten st
     ...............................................................                            Argopecten st
     ...........................................G.E.................                            Placopecten St
     ...............................................................                            Argopecten Ct
     ...........................................G.E.................                            Placopecten Ct
     .G.....T.....L.N.K.G....L.SQV....A.....K.H.Q.....AAIN.L....TI........AE....                  Loligo SR

1708 DANDRVNELTSQVSSVQGQKRKLEGDINAMQTDLDEMHGELKGADERCKKAMADAARLADELRAEQDHSSQVEKVRKNLESQVKEFQIRL  Pecten st
     .................N..............                                                           Argopecten st
     .................N..............                                                           Placopecten St
     .................N..............                                                           Argopecten Ct
     .................Q...GLS..M..S......L.V..                                                  Placopecten Ct
     ..S......QA...T.GS......VT...S......LNN..D....A.H....T......Q....                            Loligo SR
```

```
1798 DEAEAASSLKGGKKMIQKLESRVHELEAELDNEQRRHAETQKNMRKADRRLKELAFQADEDRKNQERLQELIDKLNAKIKTFKRQVEEAEE  Pecten St
     .........................................................................................  Argopecten St
     .........L...............................................................................  Placopecten St
     .........L...............................................................................  Argopecten Ct
     ..S..AA..........R.............S.........S...V...V...S..QE.....Y..M...V...QN.....Y.........  Placopecten Ct
                                                                                                  Loligo SR
```

```
                                       Tailpiece
1888 IAAINLAKYRKAQHELELAEEAERADTADSTLQKFRAKSRSSVSVSVQRSSVSVSA-N-----------  Pecten St
     ...................................................S.---------------  Argopecten St
     ...................................................A.---------------  Placopecten St
     .............S...........................S.AAHVAHHHVE  Argopecten Ct
     .............S...........................A.ASQAVHHHVE  Placopecten Ct
     .......F.V.Q...D.......QSEGA...L...N....AA.T.PM  Loligo SR
```

```
Pecten St        1940 residues
Argopecten St    1938 residues
Placopecten St   1941 residues
Argopecten Ct    1951 residues
Placopecten Ct   1950 residues
Loligo SR        1935 residue
```

Figure 4.11. Comparison of deduced sequences for myosin heavy chains from the striated (line 1: *Pecten maximus*; line 2: *Argopecten irradians*; line 3: *Placopecten magellanicus*) and smooth (line 4: *Argopecten irradians*; line 5: *Placopecten magellanicus*) scallop adductor muscles as well as from squid (*Loligo pealei*) syphon retractor muscle (line 6). Periods (.) indicate sequence as for *Pecten* striated muscle myosin heavy chain. Residues deduced to be involved in regulation (Nyitray et al. 1991; Matulef et al. 1998) are highlighted (blue) by asterisks. Attention is drawn to key features of the heavy chain sequence including SH3, the P-loop, the 25/50 kDa loop-1 junction, the 50/20 kDa loop-2 junction, Switch I and Switch II subdomains, the relay, converter and the fast-reacting thiols, SH1 and SH2, together with the non-helical tailpiece found in smooth myosin isoforms (all in green); as well as the malleable helix in-between SH1 and SH2 that is capable of melting upon nucleotide binding (bright green), the E-LC and R-LC binding regions (turquoise), residues involved in ATP binding (red), actin binding regions (violet), the "terminal" proline residue which marks the head/tail boundary (green and underlined), the four skip residues within the rod portion of the tail (pink and underlined). Attributions are: *Pecten maximus* striated myosin HC (Janes et al. 2000); *Argopecten irradians* striated myosin HC (Nyitray et al. 1991); *Argopecten irradians* smooth myosin HC (Nyitray et al. 1994); *Placopecten magellanicus* striated and smooth myosin HC isoforms (Perreault-Micale et al. 1996b); *Loligo pealei* syphon retractor muscle myosin HC (Matulef et al. 1998). Figure adapted and expanded from Janes et al. (2000). Single letter codes for the amino-acids are: D, aspartate; E, glutamate; N, asparagine; Q, glutamine; G, glycine; A, alanine; V, valine; L, leucine; I, isoleucine; S, serine; T, threonine; F, phenylalanine; W, tyrosine; Y, tryptophan; M, methionine; P, proline; K, lysine; R, arginine.

a) Regulatory Light Chain

```
                                  Ca²⁺-Mg²⁺ site
1    MADK*****AASGVLTKLPQKQIQEMKEAFSMIDVDRDGFVSKDD    Pecten striated
1    ....*****...............................E.       Argopecten striated
1    ....*****...................M.C.....IN...         Placopecten striated
1    ....ERAQR.T.N.FAR....LM.......T.I.QN....IDIN.     Placopecten smooth A
1    ....ERAQR...............T.I.QN....IDIN.           Placopecten smooth B

41   IKAISEQLGRTPDDKELTAMLKEAPGPLNFTMFLSIFSDKLSGTD    Pecten striated
41   .........A.................................       Argopecten striated
41   L.E.......................................        Placopecten striated
46   L.EMFSS...................................        Placopecten smooth A
46   L.EMFSS...................................        Placopecten smooth B

                              G117
86   SEETIRNAFAMFDEQENKKLNIEYIKDLLENMGDNFNKDEMRMTF    Pecten striated
86   .Q........................................       Argopecten striated
86   .........G....LD..........................       Placopecten striated
91   .........G....LD..........................       Placopecten smooth A
91   .........G....LD..........................       Placopecten smooth B

131  KEAPVEGGKFDYVKFTAMIKGSGEDEA                      Pecten striated
131  .....................E..                         Argopecten striated
131  ...........R.V.......D.D.                        Placopecten striated
136  ...........R.V.......D.D.                        Placopecten smooth A
136  ...........R.V.......D.D.                        Placopecten smooth B
```

b) Essential Light Chain

```
                    Ca²⁺-binding loop
1    MPKLSQDEIDDLKDVFELFDFWDGRDGAVDAFKLGDVCRCLGINP    Pecten
1    ..........................................       Argopecten
1    ............E...................I.........        Placopecten

46   RNEDVFAVGGTHKMGEKSLPFEEFLPAYEGLMDCEQGTFADYMEA    Pecten
46   ..........................................       Argopecten
46   ...................................Y......        Placopecten

91   FKTFDREGQGFISGAELRHVLTALGERLSDEDVDEIIRLTDLQED    Pecten
91   ...........................................K.......  Argopecten
91   ...................SG........E....N.......        Placopecten

136  LEGNVKYEDFVKKVMAGPYPDK                           Pecten
136  ....................                             Argopecten
136  ........E......T......                           Placopecten
```

Figure 4.12. Comparison of deduced sequences for the main isoforms of a) regulatory light chain (R-LC) and b) essential light chain (E-LC) from the striated muscles of each of three species of scallop: *Pecten maximus* (line 1); *Argopecten irradians* (line 2) and *Placopecten magellanicus* (line 3) and from the SmoA and Smo B isoforms of *Placopecten magellanicus* (R-LC, lines 4 & 5). E-LC isoforms are identical in striated and smooth muscles. The non-specific divalent cation binding site on the R-LC (red) and the Ca^{2+}-specific site on the E-LC (blue) are indicated; the R-LC glycine residue critical for regulation (Gly117) is also highlighted (blue). Single letter codes for the amino-acids are as defined in the legend to Figure 4.11. Periods (.) indicate sequence as for *Pecten* striated. Asterisks indicate extra residues present in the smooth R-LC isoforms but missing in their striated counterparts. Sequences determined by Janes et al. 2000 (*Pecten* striated R-LC and E-LC); Goodwin et al. 1987 (*Argopecten* striated R-LC and E-LC); Perreault-Micale et al. 1996a (*Placopecten* striated and smooth R-LC and E-LC). The N-terminus of each light-chain is most likely blocked by an acetyl group, as has been shown for *Pecten* R-LC (Kendrick-Jones and Jakes 1977). Figure adapted and expanded from Janes et al. 2000.

The determination of the structure of the isolated myosin head was a seminal *fin-de-siècle* achievement. First accomplished for chicken skeletal myosin (Rayment et al. 1993a), the effort has been repeated on myosin heads from *Dictyostelium* myosin II (Fisher et al. 1995), chicken smooth muscle myosin (Dominguez et al. 1998) and scallop (*Argopecten irradians*) striated adductor muscle myosin (Houdusse et al. 1999), each in the presence of various nucleotide analogs. The structure of the scallop myosin head, with ADP present at the active site, is seen in Figure 4.13 (Houdusse et al. 1999). Key landmarks (note locations shown within the sequence in Figure 4.11) on each globular head include the nucleotide binding pocket, the entrance to which is guarded by the 25/50 kDa loop (Loop 1); the actin binding site, which is closely associated with the 50/20 kDa loop (Loop 2); the SH3 domain, which has been suggested to limit the potential swing of the lever arm (Dominguez et al. 1998); the converter domain, important for communication between the motor domain and the lever arm (Dominguez et al. 1998); the regulatory domain (RD), where the long α-helix extending from either head is grasped by the two light chains, the E-LC being closest to the head. Both loop structures are variable in sequence and must be flexible because they cannot be visualised in the crystal structures; their function appears to be to control actin affinity (Loop 2), ATPase (Loops 1 and 2) and ADP release (Loop 1) rates (Spudich 1994; Kurzawa-Goertz et al. 1998; Murphy and Spudich 2000).

An understanding of the mechanism of myosin action has emerged from the crystallographic structures of myosin heads in the presence and absence of nucleotides. The conformation of the scallop myosin head, determined in the presence of MgADP, is unique among current structures: it has been referred to as a third state of myosin (state III), and has been compared critically to the other two key structural states (Houdusse et al. 1999; Houdusse et al. 2000). State I was first seen in chicken skeletal myosin in the absence of nucleotide (but with a sulphate ion within the phosphate binding pocket) (Rayment et al. 1993a) and more recently in scallop myosin in the absence of nucleotide (Houdusse et al. 2000); state II has been observed in chicken smooth muscle myosin in the presence of MgADP.AlF$_4$- (Dominguez et al. 1998) and in scallop myosin in the presence of MgADP.VO$_4$ (Houdusse et al. 2000). Similarities, as well as profound differences, between these three structures have been noted (Fig. 4.14). All three show conformationally invariant domains (the N-terminal domain, the upper 50 kDa domain; the lower 50 kDa domain and the converter domain) which are displaced relative to one another in each of the three states (Houdusse et al. 1999). To date, such a comparison within a single species is only possible for myosin from the scallop striated adductor muscle (Houdusse et al. 2000). In state III, the "jaws" of the 50 kDa domain are open (Houdusse et al. 1999), similar to the nucleotide-free structure (Rayment et al. 1993a; Houdusse et al. 2000); these differ from the partially closed form of the smooth myosin structure which is thought to represent the transition state complex (Fisher et al. 1995; Dominguez et al. 1998) (Fig. 4.14). The relative dispositions of the invariant domains with respect to each other appear to be determined by the presence of three flexible joints: switch II, the relay and the SH-1 helix (Fig. 4.14). Of these, small conformational changes in switch II, consequential to sensing the changing chemistry at the active site, lead to altered associations between the relay and the SH-1 helix (see Fig. 4.14).

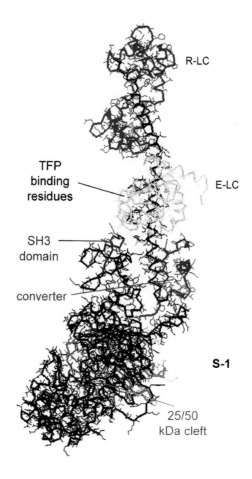

Figure 4.13. Structure of a scallop myosin head in the presence of MgADP. It is drawn using RasMol (version 2.4), coordinates being downloaded from the Protein Data Bank, Research Collaboratory for Structural Bioinformatics, Rutgers University (accession code 1b7t) (Houdusse et al. 1999). Illustrations are rendered as wire-frame structures on which the α-carbon backbone has been highlighted. Heavy-chain residues highlighted comprise: the converter region (residues Gly-695 - Phe-762; red); those amino acids which outline the 25/50 kDa cleft and the nucleotide binding cleft (residues Gln-170 - Lys-256; violet); surface loops (residues Lys-139 - Thr-146 and Asn-162 - Asn-169; orange), which have been shown to contact the E-LC in the chicken motor domain MgADP.AlF$_4$/E-LC structure (Dominguez et al. 1998); a region within the N-terminal SH3 domain (residues Ala-45 - Lys-60; light blue) which has been shown to be in close contact with the E-LC in the scallop MgADP S-1 structure (Houdusse et al. 1999). The R-LC is shaded red. The E-LC is shaded yellow. Residues shaded green within the E-LC structure indicate amino acids inferred to be involved in TFP binding (Patel et al. 2000). All other regions of the heavy-chain are shaded deep blue. Figure adapted from Patel et al. (2000) and reproduced with permission of the publisher.

Switch II also influences communication between the active site and the actin-binding surface and is critical to the putative release of inorganic phosphate through a separate exit at the base of the nucleotide cleft – known as the "back door" (Yount et al. 1995). The converter, which arises directly from the C-terminal end of the SH-1 helix, shows the greatest positional change in the three states (Houdusse et al. 2000); the lever arm is contiguous with the converter. Consequently, a small conformational change in switch II leads to altered inter-domain relationships, especially the relative orientations of the lower 50 kDa and converter domains, the lever arm effectively magnifying these changes by "wagging its tail".

The structure of the scallop striated adductor muscle myosin RD ($+Ca^{2+}$) has also been obtained, separately, and at high resolution (Xie et al. 1994; Houdusse et al. 1996) (Fig. 4.15). In regulatory myosins, activity of the motor domain is triggered directly or indirectly by the primary Ca^{2+} signal through a switch located within the RD. In the form of myosin regulation epitomised by vertebrate smooth muscle and non-muscle cells, Ca^{2+} triggers activation of myosin light chain kinase (MLCK) which catalyses the phosphorylation of a serine residue within each R-LC (Pfitzer 2001). By contrast, myosin-linked regulation is found in its simplest form in the striated adductor muscles of molluscs where the triggering ion binds directly and specifically to the first helix-loop-helix motif of the E-LC (Xie et al. 1994). Although the light chains belong to a family of proteins characterised by the possession of four potential EF hand motifs (Collins et al. 1986), X-ray structural analysis revealed that the calcium-specific binding site on scallop myosin E-LC does not conform to a classic EF-hand, as originally described for parvalbumin (Kretsinger and Nockolds 1973); instead, all seven liganding groups are provided by oxygen donors, including one water molecule and three backbone carbonyls, from within the tight horseshoe of a 9-residue loop (Houdusse and Cohen 1996). By contrast, the first helix-loop-helix motif of the scallop R-LC is a functional archetypal EF-hand, but can only bind divalent cations non-specifically (Bagshaw and Kendrick-Jones 1979; Bagshaw and Kendrick-Jones 1980). Neither light chain can bind calcium specifically or with high affinity when dissociated from myosin. The RD structure of scallop myosin reveals how the integrity of the E-LC calcium-specific horseshoe loop depends critically on contacts made between the R-LC and the E-LC over a very limited region. Two key hydrogen bonds originate from Gly-117 of the R-LC (Xie et al. 1994; Houdusse and Cohen 1996), a residue shown to be critical for stabilisation of the calcium-specific binding site (Jancso and Szent-Gyorgyi 1994), and make contact with the Phe-20 carbonyl oxygen and Arg-24 amido nitrogen of the E-LC. The side-chain of E-LC Arg-24 also makes two hydrogen bonds with R-LC Asp-118 (Houdusse and Cohen 1996).

The E-LC and R-LC structures are organised as a pair of lobes (Fig. 4.15); each lobe can adopt one of three conformations (Houdusse and Cohen 1995; Houdusse and Cohen 1996). The N-terminal lobe of the R-LC binds Mg^{2+} creating an 'open' state, facilitating non-polar interactions with the hydrophobic face of the MHC helix (Houdusse and Cohen 1996) whereas the N-terminal lobe of E-LC, when bound to the triggering Ca^{2+}, is in a 'closed' form and does not grip the MHC but is stabilised by interactions with the R-LC (Houdusse et al. 1996). These two conformations resemble those seen in calmodulin and troponin C which can form open, or closed, states in the divalent cation bound, or

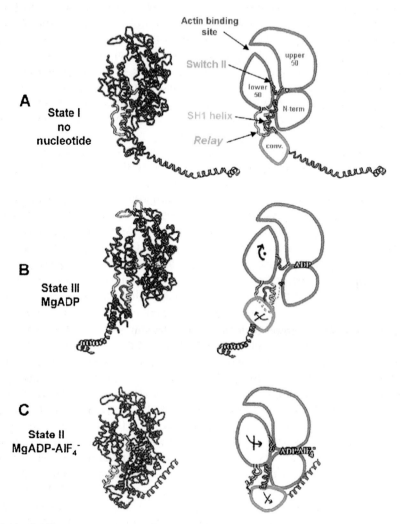

Figure 4.14. Subdomain arrangement in the three states of scallop myosin heads. All three states have been determined for scallop myosin (Houdusse et al. 2000) and have also been determined, individually, for chicken skeletal myosin heads in the absence of nucleotide (Rayment et al. 1993a) and *Dictyostelium* myosin heads in the presence of ADP.BeF₃ (Fisher et al. 1995) (state I); chicken smooth muscle myosin in the presence of ADP.AlF₄⁻ (Dominguez et al. 1998) and *Dictyostelium* myosin heads in the presence of ADP.VO₄ (Smith and Rayment 1996) (state II). Ribbon diagrams of the three states are shown on the left-hand side together with explanatory diagrams illustrating the relative dispositions of the invariant subdomains together with the flexible "joints" that link them together. Arrows indicate the direction of movement of the lower 50 kDa domain and the converter domain as compared with the state I. See text for details. Adapted from Houdusse et al. (1999) and reproduced here with permission of the publisher.

unbound forms, eliciting a significant change of conformation to create a hydrophobic cleft capable of binding a target α-helical peptide (Ikura et al. 1992; Vassylyev et al. 1998). Additionally, the C-terminal lobes of both the E-LC and R-LC (but not calmodulin or troponin C) display a third 'semi-open' state in which no metal is bound yet the MHC is gripped by the tips of these lobes (Houdusse and Cohen 1995; Houdusse et al. 1996). The same "IQ" motif (residues 788 to 798 of the *Pecten* MHC sequence) is gripped, from opposite sides, by the C-terminal and N-terminal lobes of the E-LC, whereas the C-terminal and N-terminal lobes of the R-LC grip tandem segments of the MHC (residues 814 to 836) (Xie et al. 1994; Houdusse et al. 1996). A ~40° bend in the MHC at the location of the E-LC-R-LC interface and an ~80° bend in the MHC located between the separate grips of the N and C-terminal lobes of the R-LC have been identified as sites of flexibility (Houdusse et al. 1996; Offer and Knight 1996).

When a regulatory myosin is in the off-state, its intrinsic MgATPase (Wells and Bagshaw 1985), actin-activated MgATPase (Kendrick-Jones et al. 1976; Chantler and Szent-Gyorgyi 1980) and actin-mediated motility (Vale et al. 1984) are inhibited. The relay of conformational changes, which enables ligand binding to the RD to switch on the motor, remains elusive. While the MHC converter region is critically placed, connecting a relatively rigid relay loop within the motor domain to the lever arm (Dominguez et al. 1998; Houdusse et al. 1999), it remains unclear how ligand binding to the LCs brings about conformational changes at the converter/E-LC interface necessary for activation. Trifluoperazine (TFP), a powerful inhibitor of calmodulin action, also binds to the C-terminal lobe of the scallop myosin E-LC and inhibits ATPase turnover (Patel et al. 2000). Regulation is not observed in S-1, the single head cleaved from myosin by proteolysis (Szent-Gyorgyi et al. 1973; Kalabokis and Szent-Gyorgyi 1997). The regulatory system thus transcends a single head and indeed, cooperative interactions between the heads underlie regulation. There is a requirement that both heads must bind Ca^{2+} (Chantler et al. 1981; Walmsley et al. 1990; Kalabokis and Szent-Gyorgyi 1997), in order to activate the molecule. The existence of head-head interactions is also shown by the negatively cooperative binding of R-LCs to R-LC-denuded myosin (Chantler and Szent-Gyorgyi 1980). However, a two-headed structure alone may not be sufficient for regulation since smooth myosin with a very short tail exhibits diminished regulation (Trybus et al. 1997); furthermore, single-headed scallop myosin retains a regulatory capability not seen in S-1 (Stafford et al. 1979), albeit lacking cooperativity (Kalabokis et al. 1996). Collectively, these observations suggest that head-head and head-tail interactions play a crucial role not only in producing cooperativity but also in the regulatory mechanism itself.

Despite the recent advances in myosin structure determination, atomic detail is currently confined to single-heads (Rayment et al. 1993a; Dominguez et al. 1998; Houdusse et al. 1999; Houdusse et al. 2000) or head fragments (Xie et al. 1994; Fisher et al. 1995; Houdusse and Cohen 1996). Two-headed fragments of myosin are too large for NMR structural determination and have not yet yielded diffractable crystals. As an alternative approach, an atomic model of the head-tail junction of scallop striated adductor muscle myosin was published (Offer and Knight 1996) derived by combining the known structure of key fragments through computer modelling. It predicted a number of specific interactions between the heads and between the heads and the tail. The construction of the

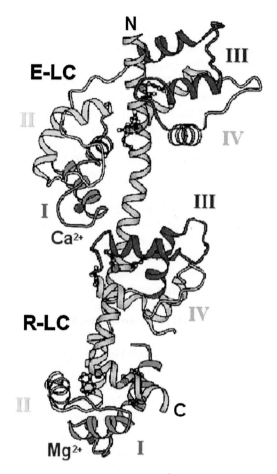

Figure 4.15. Structure of the regulatory domain ($+Ca^{2+}$) from scallop striated adductor myosin. This ribbon diagram demonstrates the distinctive interactions of the two lobes of each light chain with the underlying heavy chain (grey). All four subdomains (I through IV) are shown for both the regulatory (R-LC) and essential (E-LC) light chains. While each subdomain is capable of binding calcium in the related molecule, calmodulin (Houdusse and Cohen 1995), only subdomain I can bind divalent cations in the bound light chains; subdomain I of the R-LC possesses a non-specific divalent cation site (bound Mg^{2+} is shown as a red sphere); subdomain I of the E-LC possesses a calcium-specific site (bound Ca^{2+} also shown as red sphere) provided that both light chains remain bound. Note that the two lobes of the R-LC grasp the heavy chain in tandem whereas the two lobes of the E-LC interact with opposite sides of the same region of heavy chain - the E-LC N-terminal lobe making only surface contacts rather than tightly gripping the heavy chain. The long α-helix of the heavy chain shows a ~40° bend in the region between the two light chains and an ~80° bend in the region gripped by the N-terminal lobe of the RLC. Figure reproduced, with permission of the publisher, from Houdusse and Cohen (1996).

model was guided by an understanding of the side-chain packing in two-chain α-helical coiled-coils (Offer and Sessions 1995) leading to a model in which the two heads lay alongside one another, with their bases in contact but without steric clash (Fig. 4.16a). The principal interactions between the two heads were between the N-terminal lobe of a R-LC of one head and the C-terminal lobe of the R-LC on the other head; there were also head-tail interactions between each R-LC and its partnering MHC in the coiled-coil region (Offer and Knight 1996). Three pairs of residues within the R-LCs on either side of these interfaces (Ala42-Arg126, Glu45-Lys130, Gln46-Glu131) were in close proximity and demonstrated complementary charges suggesting the formation of salt links. The resulting model accounted for some of the conserved sequence features found in regulatory myosins and provided a structural basis for understanding head-head and head-tail interactions in regulated myosins. Functional consequences of this model were tested directly by site-directed mutagenesis: expressed R-LCs in which Arg126, Lys130 and Glu131 were converted to alanine residues, either singly, in pairs or as a triple mutant, then recombined with R-LC denuded scallop myofibrils (Colegrave et al. 2003) were analysed for regulation. If the symmetric model was correct in detail then each of these mutations, singly or in combination, should have disrupted normal regulation. However, it was found that regulation, as measured either through calcium binding, calcium dependent actin-activated MgATPase or actin sliding, was not impaired (Colegrave et al. 2003), suggesting that the symmetric model was inadequate to explain head-head cooperativity (Chantler et al. 1981) and calcium-dependent regulation (Chantler and Szent-Gyorgyi 1980).

An alternative model, derived from vertebrate smooth muscle (which is subject to regulation through R-LC phosphorylation), has also been proposed recently (Wendt et al. 1999; Wendt et al. 2001). It defines an asymmetric structure for the inhibited state, determined by electron microscope imaging of dephosphorylated myosin arrays within 2D lipid monolayers (Wendt et al. 1999; Wendt et al. 2001); the actin binding surface of one head abuts the converter domain of the adjacent head. This model is further supported by both crosslinking (Wu et al. 1999) and kinetic (Nyitai et al. 2002) data. Furthermore, the R-LC residues that were mutated in order to appraise the symmetric model (Colegrave et al. 2003) can be seen to be widely separated in the asymmetric model (Fig. 4.16b), explaining why these mutations had minimal effect on regulation. In this regard, it may be noted that while R-LC dissociation is a random event at temperatures ~20° (Chantler 1985) and is fairly fast, the rate being 0.014 s^{-1} (partially limited by the rate of dissociation of Mg^{2+} (0.058 s^{-1}) from the non-specific divalent cation site) (Bennett and Bagshaw 1986b), at temperatures <20°C dissociation is biphasic. This indicates that the binding constants for the R-LCs are not equivalent under these conditions (Chantler 1985), results that are consistent with an asymmetric structure. Further work needs to be done in order to verify the details of the asymmetric structure for the off-state of scallop myosin, but the ultimate goal of understanding regulation and cooperativity in the two-headed scallop myosin molecule appears to be attainable.

a)

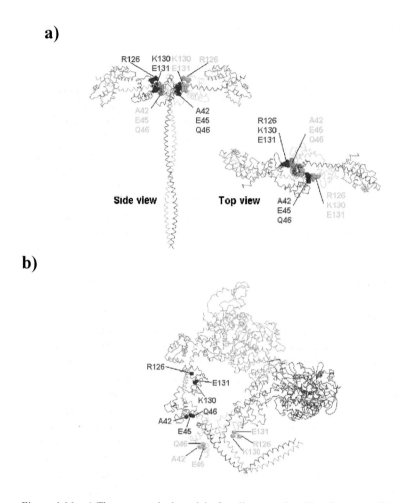

b)

Figure 4.16. a) The symmetrical model of scallop myosin. The views are of the N-terminal region of the coiled-coil as it joins the regulatory domain, seen from the side and from the top; the motor domain is not shown. The pair of backbone heavy chains are coloured in shades of blue; the pair of essential light chains are coloured in shades of green; red and yellow illustrate the pair of regulatory light chains. The critical residues that are proposed to participate in contacts between R-LCs are indicated. Image data derived from Offer and Knight 1996. Image shown copyright © of the Biochemical Society and reproduced here with their permission. b) The asymmetrical model of smooth-muscle myosin. Views of the asymmetrical model of smooth muscle myosin (Code 1I84, Protein Data Bank, Research Collaboratory for Structural Bioinformatics, Rutgers University) adapted to illustrate the location of the critical residues depicted for scallop myosin. Peptides colour-coded as in a). The residues in the R-LCs that are proposed to contact each other in the Offer & Knight model (a) are indicated; note that these sets of residues, on each of the regulatory light chains, are far apart from each other in this model. Image data derived from Wendt et al. (1999). Image shown copyright © of the Biochemical Society and reproduced here with their permission.

4.5 FUNCTION OF THE STRIATED ADDUCTOR

4.5.1 Mechanics

The cellular organisation of the scallop striated adductor muscle, described earlier, is designed to perform external work during a limited number of contractions rather than to maintain force during a single contraction. This is apparent form the work of Rall (1981), who compared the mechanics and energetics of contraction of the striated adductor muscle of *Placopecten magellanicus* to the more favoured experimental subject for physiology, the semitendinosus muscle of the frog, *Rana pipiens*. Such a conclusion also makes sense from what we know about the action of the striated adductor muscles *in vivo*: swimming by jet propulsion occurs because of the repeated expulsion of water by adduction of the shells, brought about through a series of contractions by the striated adductor muscle (Buddenbrock 1911; Marsh et al. 1992). After thirty or so such contractions the muscle becomes unexcitable and the shells remain closed through contraction of the smooth adductor muscle (Bayliss et al. 1930; Thompson et al. 1980).

Scallop and frog striated muscle exhibit similar isometric mechanical twitch kinetics, scallop fibres exhibiting mean half-times for tension rises of 43 msec (calcium release brought about by photolysis of nitr-5) (Lea et al. 1990) or 60 msec (electrical stimulation) (Rall 1981) as compared with 40 msec and 36 msec, respectively, for the frog. Twitch to tetanus ratios (Fig. 4.17) and heat coefficients are also similar (Rall 1981). This work suggested that the similarity of the isometric heat coefficients (force developed per unit energy liberated) indicates a basic similarity in mechanochemical conversion in scallop and frog striated muscle, assuming an unverified similarity between the *in vivo* enthalpy change for creatine phosphate and arginine phosphate hydrolysis (arginine phosphate, as opposed to creatine phosphate, is used as the back-up energy source in scallop striated muscle (De Zwaan et al. 1980)). Several distinct differences exist in the mechanics of these two muscles. The half-time taken for relaxation is significantly longer in scallop muscle (104 ± 2 msec) as opposed to the frog (66 ± 5 msec) under comparable conditions (Rall 1981). Figure 4.17 shows that scallop muscle develops about 50% less force than frog muscle, despite the similar twitch to tetanus ratio, an effect that may be more apparent than real: each small scallop muscle cell contains but a single myofibril surrounded by sarcoplasmic reticulum and sarcolemma, yet force is normalised per unit area; it may be that, in aggregate, a large area of scallop muscle is devoid of myofibrils, making the area comparison unrealistic. Scallop muscle also exhibits a dramatic fatigue during tetanus (Fig. 4.17; Rall 1981) and displays a greater post-tetanic twitch potentiation than frog muscle (Rall 1981), a feature of unknown significance.

For scallop striated muscle, the relationship between the development of isometric tension and calcium concentration, is very steep. In an activating solution containing 5 mM MgATP at pH 7.1 and 20°C, the rise in tension is completed between pCa values of 5.5 and 5.2 (Fig. 4.18a) giving rise to a Hill coefficient of 5.5 (Simmons and Szent-Gyorgyi 1985). When *in vitro* measurements of the relationship between actomyosin MgATPase and calcium concentration are compared under identical conditions with isometric tension measurements, the two sets of data superpose exactly (Fig. 4.18b;

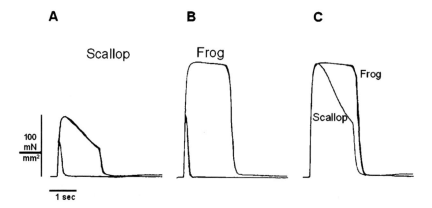

Figure 4.17. Twitch and tetanus records from scallop (*Placopecten magellanicus*) (A) and frog (*Rana pipiens*) (B) muscle. Tetanus was for 1.5 s at 50 Hz. In C, the tetanic contractions were scaled to the same peak value. Records reproduced, with permission of the publisher, from Rall 1981.

Chantler et al. 1981; Simmons and Szent-Gyorgyi 1985). The high value of the Hill coefficient obtained, as well as direct modelling of all possible ways calcium could activate a two-headed molecule (Chantler et al. 1981), indicated that an additional cooperative mechanism had to exist when myosin is associated within a thick filament, in order to account for the steepness of the activation curve with respect to calcium. It seemed likely that this was an intermolecular mechanism, propagated throughout the thick filament: full activation of one molecule allowing adjacent molecules to switch on both heads, even though only one head may have bound calcium (Chantler et al. 1981). This hypothesis has been substantiated further by structural studies on isolated scallop thick filaments which indicate that the ordered appearance, characteristic of the relaxed state, is lost over a very narrow range of free calcium concentrations upon activation (Vibert and Craig 1985). The structural basis for this intermolecular cooperativity may be related to a close interaction between pairs of heads arranged in helical strands along each half of the thick filament, one member of each pair coming from adjacent crossbridge levels (14.5 nm periodicity) (Levine et al. 1988; Levine 1993). Alternatively, it could be due to the interaction of heads at one crossbridge level with the underlying S-2 regions of myosin molecules located at the next crossbridge level along the filament.

4.5.2 The Interaction of Myosin with Actin

The interaction of myosin with actin in the presence of nucleotide is the process that lies at the core of muscle contraction: the crossbridge cycle. The *in vitro* equivalent of the crossbridge cycle is assumed, with appropriate caveats, to be the actomyosin kinetic cycle. A number of elementary steps in the scallop striated adductor myosin ATPase pathway

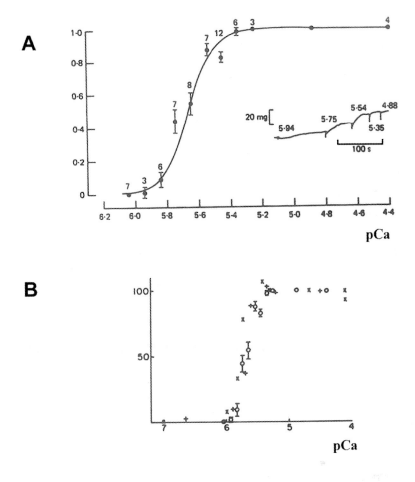

Figure 4.18. A). Relationship between isometric tension and calcium concentration (expressed as pCa = -log[Ca²⁺]). Tension was plotted relative to the value for pCa 4.88 . The number of fibre bundles tested is given for each calcium concentration. The error bars are ± 1 S.E. of the mean. Inset: tension record from one fibre bundle; numbers are the pCa values. Figure reproduced, with permission of the publisher, from Simmons and Szent-Gyorgyi (1985). B). The *in vitro* response of *Placopecten* myofibrillar ATPase to calcium was compared directly to tension development of fibre bundles under identical conditions. Conditions were: 3 mM $MgCl_2$, 2 mM ATP, 30 mM NaCl, pH 7.0. For ATPase studies, the pH was maintained at pH 7.0 in a pH-stat; free calcium levels were determined by incremental addition of calcium to the above mixture containing 0.1 mM EGTA (x). Alternatively, the pH was maintained by a 15 mM imidazole buffer and the use of 1 mM CaEGTA/EGTA mixtures to achieve a particular pCa level (+). For tension studies (o), the latter buffer was used and methods similar to those in A) were employed. Tension measurements were performed by Prof. R. M. Simmons. Figure reproduced, with permission of the publisher, from Chantler et al. (1981).

are nearly identical to analogous steps in the rabbit myosin ATPase cycle (Taylor 1979). The principle difference is that in the scallop ATPase pathway, there is a marked suppression of the product dissociation steps in the absence of calcium (Jackson and Bagshaw 1988; Nyitrai et al. 2002). Such calcium sensitivity, exhibited by myosin alone (i.e., in the absence of actin), was originally noted by Ashiba et al. (1980) and further characterised by Wells and Bagshaw (1985); the latter workers showed that it was possible to separate the long-lived intermediate, formed in the presence of ATP and the absence of calcium, from unbound nucleotide using column centrifugation techniques. The long-lived intermediate was identified as M.ADP.Pi and the rate limiting step in the absence of calcium has been assigned to the phosphate release step (Wells and Bagshaw 1985; Jackson and Bagshaw 1988; Nyitrai et al. 2002).

A simplified version of the scallop myosin ATPase scheme is seen in Figure 4.19a. There are two calcium-sensitive steps: the rate-limiting dissociation of phosphate and the slow release of ADP (Wells and Bagshaw 1985; Jackson and Bagshaw 1988; Nyitrai et al. 2002). The fact that there are two calcium-sensitive elementary steps may have some significance with regard the catch mechanism (see below). The dissociation of ADP is depicted as a single step in Figure 4.19a; in reality it will consist of a minimum of two

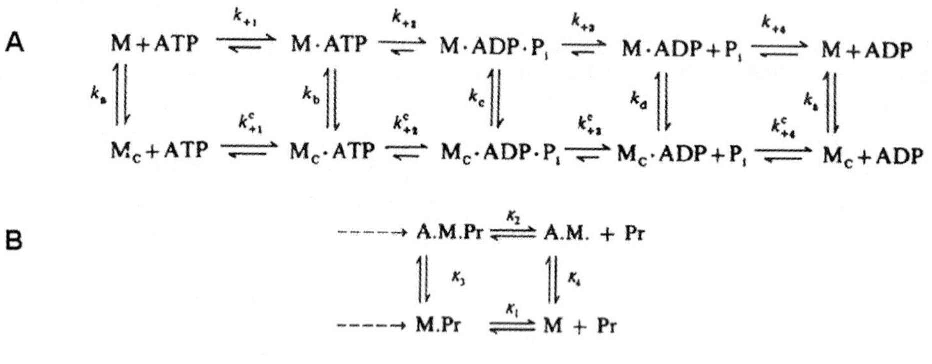

Figure 4.19. A. Elementary steps of the scallop myosin ATPase cycle both in the presence (Mc) and absence (M) of calcium. The majority of the data shown below were obtained for *Pecten maximus* HMM (Jackson and Bagshaw 1988) but where ranges are given they bracket data from *Pecten* as well as from *Argopecten irradians* and *Placopecten magellanicus* S-1 (Kurzawa-Goertz et al. 1998). Most rate constants were determined in 20 mM NaCl, 1 mM $MgCl_2$, 10 mM TES, 0.1 mM EGTA, pH 7.5, 20°C ± 0.2 mM $CaCl_2$ and were: $k_{+1} = 2$-5×10^6 $M^{-1}s^{-1}$; $k_{+2} = 200$ s^{-1}; $k_{+3} = 0.002$ s^{-1}; $k_{+4} = 0.01$ s^{-1}; $k_{c+1} = 2$-5×10^6 $M^{-1}s^{-1}$; $k_{c+2} = 200$ s^{-1}; $k_{c+3} = 0.2$ s^{-1}; $k_{c+4} = 6$ s^{-1}; $k_{-4} = k_{c-4} = 8 \times 10^5$ $M^{-1}s^{-1}$; $k_a = 25$ s^{-1}; $k_{-a} = 2.5 \times 10^8$ $M^{-1}s^{-1}$. B. Summary of product release steps for the acto-HMM ATPase. Conditions used were as for A. Of the equilibrium constants shown, K_3 and K_4 are calcium insensitive whereas K_1 and K_2 are calcium sensitive (Stanners and Bagshaw 1987). For S-1, $K_1 = 1$-2×10^5 M^{-1}; $K_2 = 5$-7×10^{-4} M; $K_3 \cong 2.5 \times 10^4$ M^{-1}; $K_4 = 4$-5×10^{-7} M (Kurzawa-Goertz et al. 1998), where the equilibrium constants are defined in a clockwise direction in scheme B (i.e., $K_1.K_2.K_3.K_4 = 1$).

steps, the dissociation ultimately occurring via a weakly-bound intermediate. Equilibrium binding studies have shown that ADP binds scallop HMM more strongly in the absence of calcium than in its presence (Wells et al. 1985; Kalabokis and Szent-Gyorgyi 1997) and the conformation of the myosin head upon binding ADP would also appear to be calcium-dependent (Jackson and Bagshaw 1988; Stafford et al. 2001).

The binding of ATP and its subsequent hydrolysis takes place in a series of fast, coupled steps that are similar in both vertebrate and invertebrate striated muscle. ATP binding is at least a two-step process involving the formation of a collision intermediate followed by a conformational change; all binding transitions are subsumed into a single step in Figure 4.19a. Experiments on *Dictyostelium* myosin constructs have shown that tryptophan fluorescence (Trp509 in the *Pecten* sequence) can be used to discriminate kinetically between the hydrolysis step and the closure of the 50 kDa jaws (see Fig. 4.14) (Malnasi-Csizmadia et al. 2000). The cleavage step is rapid and leads to the formation of enzyme-bound products (the so-called early phosphate burst). By analogy with the vertebrate myosin kinetic scheme (e.g., Adelstein and Eisenberg 1980), the hydrolysis step will have an equilibrium constant in the region of 1–10, ensuring a minimal decrease in free energy during nucleotide cleavage. The largest drop in free energy in the cycle accompanies phosphate release.

All the intermediates of the scheme depicted in Figure 4.19a, in the presence and absence of calcium, are expected to be in rapid equilibrium with actin at intermediate and high actin concentrations (Stein et al. 1979; Taylor 1979; Adelstein and Eisenberg 1980; Nyitrai et al. 2003). Calcium does not regulate the affinity of actin for scallop myosin heads in the absence of nucleotide (Nyitrai et al. 2003) or in the presence of excess ATP (Chalovich et al. 1984; Nyitrai et al. 2003). The latter results were shown by comparing actin-activated ATPase rates and the degree of binding of actin to HMM as a function of actin concentration, in the presence and absence of calcium, under identical conditions. At high actin concentrations, where most of the HMM was bound, the rate of ATP hydrolysis remained inhibited in the absence of calcium (Chalovich et al. 1984). These results were the first to imply that calcium must exert its control on attached intermediates of the kinetic cycle. Constants obtained (25°C, pH 7.0) were: for $HMM(+Ca^{2+})$, $V_{max} = 14$ s^{-1}, $K_{ATPase} = 3.1 \times 10^4$ M^{-1}, $K_{binding} = 2.1 \times 10^4$ M^{-1}; for $HMM(-Ca^{2+})$, $V_{max} = 3$ s^{-1}, $K_{ATPase} = 2.4 \times 10^4$ M^{-1}, $K_{binding} = 4.1 \times 10^3$ M^{-1}; for Ca.Mg S-1 (irrespective of calcium levels), $V_{max} = 15$ s^{-1}, $K_{ATPase} = 3.7 \times 10^4$ M^{-1}, $K_{binding} = 4.4 \times 10^4$ M^{-1}.

The 5-fold sensitivity to calcium exhibited by the $K_{binding}$ data compared with the virtual calcium insensitivity of the K_{ATPase} data led Chalovich et al. (1984) to discuss the possibility that the true value of K_{ATPase} was being obscured by a small population of calcium-insensitive molecules that possessed an elevated actin-activated ATPase at all times. This was proven to be the case by elegant studies involving measurements taken during single or limited turnovers of ATP hydrolysis whilst monitoring turbidity (Wells and Bagshaw 1984; Jackson et al. 1986; Ankrett et al. 1991b) or fluorescence (Wells et al. 1985; Jackson and Bagshaw 1988; Walmsley et al. 1990; Ankrett et al. 1991b) as a measure of complex formation between actin and HMM. These experiments established that the calcium sensitivity of fully regulated molecules could be as much as 650-fold at low ionic strength *in vitro* (Wells and Bagshaw 1984). As routinely prepared, between 10 and 30% of scallop HMM molecules are calcium insensitive, a fact which hampered earlier attempts to measure the rates of elementary steps of the scallop actomyosin ATPase scheme in the absence of calcium and has led to the exercise of considerable ingenuity in the design of experiments.

Direct interpretation of experiments using transient kinetics became possible with the advent of a method for preparing low yields of fully active, calcium-sensitive scallop HMM (Kalabokis and Szent-Gyorgyi 1997). These data, obtained using HMM prepared from *Argopecten irradians* (Nyitrai et al. 2002), yielded rate constants in agreement with the earlier studies on HMM from *Pecten maximus* (Jackson and Bagshaw 1988). Furthermore, apart from the calcium sensitive steps, individual rate constants were in agreement with those obtained from calcium insensitive *Argopecten irradians* S-1 (Kurzawa-Goertz et al. 1998) reinforcing the idea that individual activated heads act independently from each other. However, the data obtained from the improved HMM preparation also extended our knowledge in several important ways. As noted above, calcium does not regulate the affinity of actin for scallop myosin heads in the absence of nucleotide or in the presence of ATP; however, ADP binding (Nyitrai et al. 2002) and actin affinity in the presence of ADP (Nytrai et al. 2003) are calcium dependent. Detailed analysis indicates that k_{+4} is about 20-fold smaller than k^c_{+4} (Fig. 4.19a). Conversely, the reverse steps are calcium insensitive, k_{-4} taking the same value as k^c_{-4}. This, together with the calcium-sensitive phosphate dissociation step, accounts for the effect of calcium on scallop HMM kinetics. Access to the nucleotide pocket in the absence of calcium is not a problem because ATP was able to displace a fluorescent form of ADP (mant-ADP) with an identical dissociation constant to that in the presence of calcium; instead the off-state of the myosin head is stabilised by ADP in the absence of calcium. This off-state, at ADP concentrations below 50 μM for scallop HMM, requires only an occupancy of one ADP per two-headed molecule (Nyitrai et al. 2002), an observation consistent with an asymmetric structure during the maintenance of the off-state (see discussion above and Colegrave et al. 2003). Based on their detailed kinetic results, Geeves and coworkers proposed a classic Monod-Wyman-Changeux cooperative model (Monod et al. 1965) to account for the cooperative activation of scallop myosin by calcium in the presence of nucleotide (Nyitrai et al. 2002). In this model, the heads of scallop myosin exist in an equilibrium between "off" and "on" conformations, even in the absence of nucleotides and calcium where 30% of the heads are in the off-state. When one head is occupied by ADP, it spends 90% of its time in the off-state. Addition of one calcium ion to this ADP-bound state switches the equilibrium to 45% on; a full complement of bound calcium shifts the balance to 88% on (Nyitrai et al. 2002).

Using pyrene-labelled actin as a binding-sensitive fluorescence probe, the dissociation and association rate constants of actin and scallop HMM have been measured in the absence of nucleotide and shown to be calcium independent (equilibrium constant K_4 in Fig. 4.19b) (Wells and Bagshaw 1984; Stanners and Bagshaw 1987). ATP-induced dissociation of HMM from pyrene-actin was also calcium independent (Nyitrai et al. 2003). Titration of the fluorescence amplitude during binding indicated that the rigor complex of scallop HMM with actin requires both heads in a 1:1 head to actin stoichiometric ratio, each head tightly bound with an affinity <5 nM in the presence and absence of calcium (Nyitrai et al. 2003). In the absence of calcium and the presence of ADP, the heads do not bind actin to any significant degree (Nyitrai et al. 2003); this fact pre-empts determination of which head – the one with or without ADP – is able to interact with actin (see Fig. 4.16b). Because the binding of myosin heads to actin in the presence of ATP appears to be calcium insensitive (equilibrium constant K_3 in Fig. 4.19b) (Chalovich et al. 1984; Wells and Bagshaw 1984; Nyitrai et al. 2003), yet product dissociation from unbound myosin is calcium-sensitive (equilibrium constant K_1 in Figure 4.19b) (Wells and Bagshaw 1985; Nyitrai et al. 2002), it can be reasoned on thermodynamic grounds from the product release steps shown in Fig. 4.19b, that the dissociation of products from heads already attached to actin must be calcium-sensitive (equilibrium constant K_2 in Fig. 4.19b) (Stanners and Bagshaw 1987). The long-lived intermediate present in relaxed scallop striated adductor myofibrils is known to possess ADP (Marston and Lehman 1974; Wells and Bagshaw 1985) so, by analogy with the vertebrate actomyosin ATPase scheme (Adelstein and Eisenberg 1980; Taylor 1979) as well as the scallop myosin ATPase cycle (Jackson and Bagshaw 1988), it seems likely that the rate is limited by loss of products from the A.M.ADP.Pi state, phosphate dissociation being the favoured limiting factor. This rate, for HMM, is elevated >600-fold in the presence of calcium (Wells and Bagshaw 1985; Stanners and Bagshaw 1987); in the case of acto.S-1 (equivalent to the on-state of HMM), the rate of ADP release proved too fast to measure (>200 s^{-1}) (Kurzawa-Goertz et al. 1998). In the presence of calcium, ADP and actin bring about reciprocal lowering of the other's affinity for HMM by 30–50-fold, yet the heads still remain in the associated state (Nytrai et al. 2003). Effectively, ADP and actin mutually exclude each other from binding to HMM.

4.5.3 The Crossbridge Cycle

For several decades prior to atomic resolution of the structures involved, our working model of the crossbridge cycle was framed by a handful of influential contributions (Huxley 1957; Huxley 1969; Huxley and Simmons 1971). In the advent of structural knowledge at atomic resolution (Rayment et al. 1993a; Rayment et al. 1993b; Geeves and Holmes 1999; Houdusse et al. 2000) these ideas have been reformulated with increased precision. The crossbridge is defined as that working portion of the myosin molecule – the head (motor domain) plus neck (lever arm) and possibly a portion of the tail (HMM S-2) – capable of reaching out from the muscle thick filament so as to interact with actin in a stochastic and asynchronous manner, participating in a repetitive cycle comprising attached and detached states. Each crank of the cycle is thought to be powered by the hydrolysis of a single molecule of ATP (Toyoshima et al. 1990; Ishijima et al. 1998), though contrary views exist (Harada et al. 1990; Kitamura et al. 1999). Conventional muscle myosin is a non-processive motor with a low duty ratio (Howard 1997): i.e., the majority of the cycle time is spent during an obligatory period of detachment (Linari et al. 1998). Vertebrate myosin step sizes have been measured through a variety of techniques and the values obtained are likely to be similar for scallop myosins. Measurements vary from 4–15 nm (Uyeda et al. 1991; Molloy et al. 1995; Ishijima et al. 1998; Kitamura et al. 1999) to >100 nm (Yanagida et al. 1985; Harada et al. 1990; Sellers and Homsher 1991);

measurements made using compliant optical traps are likely to be most accurate and are in the range 4–15 nm (Mehta et al. 1999); within this spread, 4–6 nm is considered to be the most likely length of a single step. This narrow range encompasses results from mechanical experiments which show that tension in an active muscle can be reduced to zero upon rapid shortening by only 4–6 nm per half sarcomere (Huxley and Simmons 1971). Force is generated by the attached crossbridge pulling the thin filaments in the direction required for sarcomere/muscle shortening or for the maintenance of isometric force. Estimates of the average force generated by a single head under isometric conditions are in the range, 1–4 pN (Molloy et al. 1995; Block 1996). Both tilting of the myosin head and rotation of part of the lever arm have been observed during contraction through the application of spectroscopic techniques simultaneously with mechanochemical measurements (Irving et al. 1995; Baker et al. 1998; Corrie et al. 1999). In the case of smooth muscle myosin heads, structural analysis has provided evidence for a 3.5 nm movement upon ADP release (Whittaker et al. 1995). In general, the short range of the step size (<15 nm) together with the low duty ratio (<20%) suggests that individual myosin heads will only generate contractile force when appropriately oriented actin monomers become accessible, at integral values of the half pitch (36–37 nm) of the actin helix (Howard 1997). Thus, although the action of any single head may only be responsible for 4–6 nm of movement per half sarcomere, which would not in itself facilitate immediate rebinding, each head is placed within its appropriate target area with respect to the next actin binding site through the collective actions of hundreds of other heads within each bipolar filament.

The marriage of crossbridge mechanics and solution kinetics has illuminated our thinking on aspects of this cyclic mechanism (Eisenberg and Greene 1980; Eisenberg and Hill 1985; Hibberd and Trentham 1986; Cooke 1997; Goldman 1998; Geeves and Holmes 1999). Figure 4.20a depicts a clear, conceptual, one-to-one correlation between steps in the actomyosin ATPase cycle and possible orientations of the crossbridge. While oversimplistic, it has heuristic value for it illustrates a minimal structural scheme necessary to account for crossbridge cycling which, together with a minimal ATPase scheme (Fig. 4.20b), describe events that occur in solution at low actin concentration (Lymn and Taylor 1971). In this scheme, one ATP is hydrolysed per round of cycle; ATP binding causes detachment of the myosin head from actin followed by rapid hydrolysis of ATP on the detached head, bringing about a putative change in conformation which cocks the mechanism; the power-stroke (step 4) corresponds to the rate-limiting product release step(s) which are assumed to be accompanied by a large drop in free energy (Adelstein and Eisenberg 1980); in this scheme the cycle is completed by the generation of the 45° state rigor complex, poised to take up ATP once again.

In vitro experiments conducted at the high concentrations of actin approaching those present in muscle have suggested that all intermediates of the myosin ATPase pathway could be in rapid equilibrium with their actin-bound states and that hydrolysis of ATP could occur in the attached state (Stein et al. 1979; Adelstein and Eisenberg 1980; Eisenberg and Hill 1985). Although evidence in support of such rapid equilibria has been obtained in relaxed muscle (Brenner et al. 1982) myosin heads are known to be dissociated from actin for a considerable portion of the ATPase cycle (Linari et al. 1998).

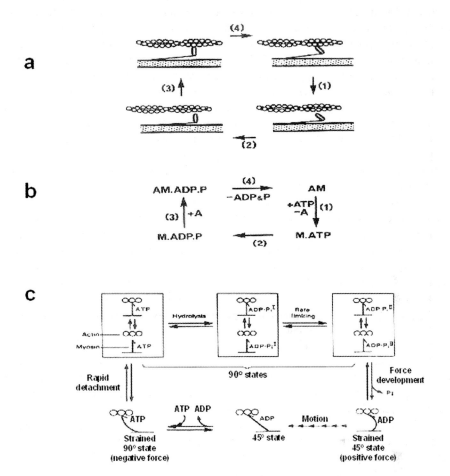

Figure 4.20. a). Schematic diagram of simplified structural features of the crossbridge cycle of muscle contraction (after Lymn and Taylor 1971). b). Basic kinetic scheme of the fundamental biochemical species that are thought to correlate with the structural events in a) (after Lymn and Taylor 1971). c). Schematic diagram combining both structural and biochemical features of the crossbridge cycle. Whilst more complicated than either a) or b), it is likely to be more realistic. The $90°$ states M.ATP, M.ADP.P_iI, and M.ADP.P_iII are in rapid equilibrium with A.M.ATP, A.M.ADP.P_iI, and A.M.ADP.P_iII, respectively. The motional transition does not have an activation energy or a rate constant. It is a continuous conformational change that occurs as the filaments slide past each other. It is the second step of a transition from the unstrained $90°$ conformation to the unstrained $45°$ conformation; the first step involves P_i release and has both an activation energy and a rate constant. a) and b) taken, with permission of the publisher, from Offer, G. 1974. In: A.T. Bull et al. (Eds.). Companion to Biochemistry: Selected Topics for Further Study. Longman, London. c) reproduced, with permission of the publisher, from Eisenberg and Hill (1985).

Consequently, hydrolysis is still envisaged as taking place on dissociated myosin heads and this step contributes to part of the cycle time (White et al. 1997). A more realistic scheme involving strongly bound states such as AM and AM.ADP, and weakly bound states such as AM.ATP and AM.ADP.Pi, is seen in Fig. 4.20c. In this scheme, different myosin intermediates will have distinct preferential angles of attachment to actin; two such angles (45° and 90°) are shown here and their structural counterparts have been identified in scallop myosin (Houdusse et al. 2000) (see Fig. 4.21). It is possible that each state of myosin can bind to actin through a variety of angles, the probability of attachment or detachment being determined by a free energy profile in each case. Indeed, a number of pre-power stroke attachment angles have been inferred from fluorescence energy transfer (Shih et al. 2000) and electron spin resonance (Roopnarine et al. 1999) spectroscopic approaches, the latter using adductor muscle fibres from *Placopecten magellanicus*. All rate constants in the cycle (Fig. 4.20c) are assumed to be sensitive to mechanical strain in the crossbridge (Eisenberg and Hill 1985).

As noted above (see Fig. 4.14) three conformational states of the scallop myosin head have been determined so far and these have provided mechanistic insights into the hydrolysis of ATP. Critical for the interaction with actin is the key observation that when all three structures are compared in a view where each actin-binding interface is constant, the structures differ dramatically with regard to the disposition of their lever arms – which could be taken as a glimpse of three still pictures at different stages within the crossbridge cycle (Houdusse et al. 2000). In state I, the lever arm appears to be an angle close to the rigor bond angle of 45° in the chicken skeletal structure (Rayment et al. 1993b) but is closer to 25° in scallop state I (Houdusse et al. 2000). In state II, this angle has changed to close to 90° (Dominguez et al. 1998; Houdusse et al. 2000). State III is different again, the lever arm making an angle of only around 15° to the actin filament (Houdusse et al. 1999) (Fig. 4.21a). State III is also remarkable in providing structural confirmation of a long-known biochemical observation (Reisler et al. 1974; Wells and Yount 1980; Huston et al. 1988) – the α-helix connecting the reactive thiols, SH1 (*Pecten* C706) and SH2 (*Pecten* C696), melts upon nucleotide hydrolysis, bringing these residues to within 0.7 nm of each other, facilitating chemical crosslinking (in states I and II these residues point in opposite directions at extreme ends of the SH α-helix, obviating crosslinking). There is also an overall rotation of the converter domain within the state III structure through angles of ~30° and 90°, as compared with the state I and state II structures, respectively (Houdusse et al. 1999; Houdusse et al. 2000). The step size, as deduced from these structural considerations, is ~12 nm (Houdusse et al. 2000).

While it is tempting to correlate the three states seen in Fig. 4.21a with the various stages of the crossbridge cycle, extreme caution must be taken. State I, while ostensibly free of nucleotide, has been attained with a sulphate ion at the active site (Rayment et al. 1993a); furthermore, the same structure has been observed in the presence of ATP analogs (Gulick et al. 1997) or ADP (Houdusse et al. 2000). Therefore, there is some ambiguity associated with this state – it has been referred to variously as a weak-binding, pre-hydrolysis state (Gulick and Rayment 1997) or, alternatively, as a near-rigor conformation (Houdusse et al. 2000). State II, crystallised in the presence of MgADP.AlF$_4$- (Fisher et al. 1995; Dominguez et al. 1998) or MgADP.VO$_4$ (Smith and Rayment 1996; Houdusse et

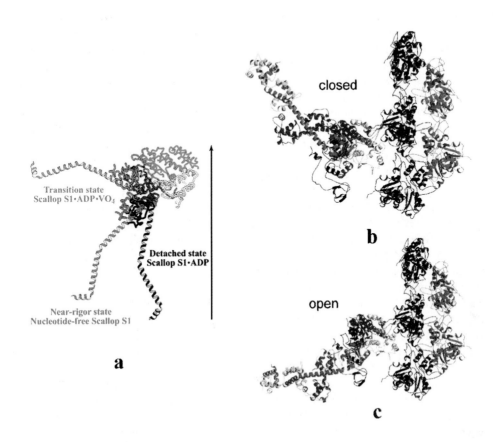

Figure 4.21. a) Illustration of the three distinct positions of the lever arm as defined from the three known conformational states of scallop myosin. These compare the nucleotide-free state (state I); the transition state (ADP.VO$_4$) (state II); the detached state (ADP) (state III). Ribbon diagrams oriented so as to superpose the conformation of the lower 50 kDa domain in all three states. Vertical arrow indicates the approximate direction of the actin axis. The end of the lever arm moves about 12 nm between states II and III and exhibits a small azimuthal displacement. Figure taken from Houdusse et al. (2000) and reproduced here with permission from the publisher. b) and c) Alternative analysis based on data from *Dictyostelium* myosin head structures showing the interaction of myosin heads, in either the closed (b) or open (c) conformations, with the actin filament structure. The CLOSED (ADP.VO$_4$) (Smith and Rayment 1996) and OPEN (ADP.BeF$_3$) (Fisher et al. 1995) conformations are thought to represent the pre-power-stroke and post-power-stroke states, respectively. The terms CLOSED and OPEN derive from whether or not the backdoor γ-phosphate release mechanism is closed or open and from the disposition of the switch II helix (Smith and Rayment 1996; Geeves and Holmes 1999). The end of the lever arm moves about 11 nm between the two states. Figures kindly provided by Prof. Ken Holmes and reproduced from Geeves and Holmes (1999) with permission of the publishers.

al. 2000) is a transition state intermediate and therefore represents a weak-binding, pre-hydrolysis state (Fig. 4.21a, b). Although state III was crystallised in the presence of ADP, the investigators suggest that it is actually another weak-binding state (Houdusse et al. 1999); in support of this, S-1 preparations in which SH1 and SH2 are crosslinked together with *para*-phenylene dimaleimide (*p*PDM) are well-characterised as having a weak affinity for actin (Chalovich et al. 1983). Consequently, strong binding state intermediates, responsible for actin-induced ADP release (see Fig. 4.20c) appear to be missing (Geeves and Holmes 1999; Houdusse et al. 2000). Because it will be difficult, if not impossible, to crystallise these fragments while bound to actin, the identification of strong binding states in isolation remains a problem. Crystals of *Dictyostelium* myosin heads possessing the bipyrimidal MgADP.VO$_4$ ion at the active site (Smith and Rayment 1996), together with the MgADP.AlF$_4$- containing fragments (Fisher et al. 1995; Dominguez et al. 1998), have been interpreted as representing pre-power-stroke structures (Geeves and Holmes 1999). The γ-phosphate binding pocket in these structures is in a "closed" state and the disposition of switch II is altered substantially relative to the nucleotide-free structure (Fig. 4.21b). Geeves and Holmes (1999) modelled this structure bound to an F-actin filament and showed that the displacement of the working stroke of the tail, from the "up" to the "down" position, occurred when the γ-phosphate binding pocket changed from the "closed" to an "open" configuration, consequential to an inward movement of switch II. The open configuration is deduced from myosin heads crystallised in the presence of MgADP.BeF$_3$ (Fisher et al. 1995) and is thought to define the post-power-stroke structure while bound to F-actin (Fig. 4.21c).

It is not entirely clear that all structures depicted in Fig. 4.21 are truly intermediates of the cycle. For example, state III could represent a weakened dragged-bridge that had failed to dissociate at the end of the cycle (Sellers and Homsher 1991; Cooke 1999). Also, the relaxed state may not necessarily be equivalent to a state within the crossbridge cycle: the time-course of activation of ABRM muscle suggests that during relaxation, the heads reside close to the thick filament and, upon activation, must first swing out towards the filament (Lowy and Poulsen 1987). Furthermore, under the conditions that pertain in muscle, it is likely that ATP can bind to myosin faster than the initial actomyosin state can revert to the rigor configuration. Consequently, the rigor state is not necessarily equivalent to a state within the crossbridge cycle. Clearly, although there has been much progress, there is more to learn.

4.5.4 Myosin-linked Regulation

In the absence of calcium, tension generation by the striated adductor muscle, as well as the Mg^{2+}-ATPase in the presence or absence of actin, and actin-dependent motility, are all repressed. An increase of calcium above micromolar levels elevates these activities by 100–1000 -fold. In scallop striated adductor muscle, this control is exerted through the action of regulatory myosin; upon activation calcium ions bind to specific sites, one on each head of myosin, and sensitise the molecule so that it can interact with actin (Kendrick-Jones et al. 1970; Kendrick-Jones et al. 1976; Chantler and Szent-Gyorgyi 1980; Chantler et al. 1981). In scallop smooth adductor muscle, myosin phosphorylation mechanisms may occur in addition to activation by direct calcium binding (Sohma et al. 1985; Castellani and Cohen 1987a; Takahashi et al. 1988). Myosin-linked regulation in the scallop contractile system has been demonstrated at the level of the myosin molecule alone (Ashiba et al. 1980; Wells and Bagshaw 1985; Nyitrai et al. 2002); actomyosin (or acto-HMM) in ATPase assays (Chantler et al. 1981; Wells and Bagshaw 1984; Ankrett et al. 1991b; Nyitrai et al. 2003) or motility assays (Vale et al. 1984; Colegrave et al. 2003); myofibrils (Kendrick-Jones et al. 1970; Kendrick-Jones et al. 1976; Chantler and Szent-Gyorgyi 1980) and muscle fibres (Simmons and Szent-Gyorgyi 1978; Kerrick et al. 1981; Lea et al. 1990). Myosin-linked control is considered to be the dominant form of regulation in scallop adductor muscles because specific removal of regulatory light chains (R-LCs) abolishes regulation in both myofibrils (Szent-Gyorgyi et al. 1973; Chantler and Szent-Gyorgyi 1980) and fibres (Simmons and Szent-Gyorgyi 1978; Simmons and Szent-Gyorgyi 1985); full regulation is restored upon readdition of a stoichiometric amount of pure R-LCs (Kendrick-Jones et al. 1976; Chantler and Szent-Gyorgyi 1980; Simmons and Szent-Gyorgyi 1985). Although it is plausible that the desensitisation procedure may also remove troponin components, it is known that troponin molecules are present in sub-stoichiometric amounts in muscles from many species of scallop and a functional "relaxing factor", isolated from purified adductor thin filaments, has been difficult to demonstrate in these species (see above). Furthermore, scallop myofibrils do not show a significant increase in ATPase activity in the presence of low concentrations of ATP and an absence of calcium (Knox et al. 1986), characteristics that would be expected of myofibrils extracted from muscles exhibiting thin-filament linked regulation. However, it is not possible to rule out a contribution from a small population of thin-filament regulated (or dually regulated) fibres, whose function is critically impaired by extraction procedures or EDTA treatment.

Calcium-specific binding to scallop striated adductor myosin, measured at 25° in the presence or absence of MgATP at low ionic strength, follows a simple binding isotherm and yields association constants close to 10^7 M^{-1} at pH 7.5 and 10^6 M^{-1} at pH 7.0 (Chantler et al. 1981). However, at physiological ionic strength or in the presence of MgADP or nucleotide analogs, calcium binds cooperatively, exhibiting Hill coefficients in the range 1.6–2.8 (Kalabokis and Szent-Gyorgyi 1997). In the presence of 1 μM calcium, 2 mM magnesium ± 1 mM ATP scallop myosin binds two calcium ions per molecule (Chantler and Szent-Gyorgyi 1980; Chantler et al. 1981; Kalabokis and Szent-Gyorgyi 1997). The association and dissociation steps for this process are extremely rapid, yielding rates of the order of $>10^8$ M^{-1} s^{-1} and 25 s^{-1}, respectively (Jackson and Bagshaw 1988). The measured dissociation rate constant is at least 30 x faster than the rate of dissociation of calcium ions from the non-specific divalent cation site located on each R-LC (Bagshaw and Reed 1977; Bennett and Bagshaw 1986a). Magnesium dissociates from the non-specific sites at an even slower rate (0.05 s^{-1}) (Bennett and Bagshaw 1986a). Consequently, it is only the calcium-specific site that displays appropriate kinetics to account for the rapid activation characteristic of scallop adductor muscles, which approach peak tension within 90 msec of stimulation (Rall 1981).

Isolation of the regulatory domain (RD) – the light chain binding fragment of the myosin heavy chain (MHC), together with associated light chains – has provided significant insights into myosin-linked regulation. Earlier small-scale tryptic preparations (Bennett et al. 1984; Szentkiralyi 1984) have been improved to generate high yield preparations, either through the direct use of trypsin (Kalabokis et al. 1994) or the sequential use of papain and clostripain (Kwon et al. 1990; Xie et al. 1994). More recently, both *Argopecten* (Malnasi-Csizmadia et al. 1999) and *Pecten* (Janes et al. 2000) RDs have been isolated following peptide expression from bacterial constructs. The RD complex exhibits calcium-specific binding, albeit of slightly lower affinity than the parent myosin molecule (Kwon et al. 1990) accompanied by a characteristic, reversible increase in intrinsic tryptophan fluorescence (Wells et al. 1985; Malnasi-Csizmadia et al. 1999; Janes et al. 2000), the increased emission arising mainly from Trp21 of the essential light chain (E-LC) (Malnasi-Csizmadia et al. 1999). Construction of hybrid RDs produced the first significant evidence in favour of a direct role for the E-LC in calcium-specific binding (Kwon et al. 1990) and showed that the primary calcium-binding capacity of this light chain could be modified by the source of heavy chain with which it was associated (Kalabokis et al. 1994). Crystallisation of the RD (Fig. 4.15) provided high resolution images of this critical regulatory region of the myosin molecule (Xie et al. 1994; Houdusse and Cohen 1996). The minimalist structures of the early proteolytic fragments (*Argopecten* MHC fragment 775–840) (Kalabokis et al. 1994) can now be extended at will in the amino-terminal (Janes et al. 2000) or carboxyl-terminal (Malnasi-Csizmadia et al. 1999) directions through recombinant techniques and bacterial expression. When the carboxyl terminal end is extended so as to include 48 heptad repeats of the myosin rod sequence (*Argopecten* MHC fragment 749–1175) dimerisation occurs leading to pairs of RD fragments joined together by a coiled-coil tail (Malnasi-Csizmadia et al. 1998). In the absence of RDs, the coiled-coil is unstable, a feature that may play a role during regulation (Malnasi-Csizmadia et al. 1998). Such lability may well underlie the generation of R-LC-R-LC dimers from *Mercenaria* R-LC hybrid scallop myosin molecules using the sulphydryl-specific reagent, 5,5'-dithiobis (2-nitrobenzoic acid) (DTNB) (Bower et al. 1992). These results imply a transient propinquity of 2 Å even though modelling studies (Offer and Knight 1996) (which assume no such lability), crosslinking (Chantler and Bower 1988) and FRET (Chantler and Tao 1986; Chantler et al. 1991) data place the distance between these residues (position 50 on the scallop R-LC) as >15 Å. A partial unwinding of the proximal coiled-coil, facilitating passage of one chain over the other in this neck region of myosin, would bring the two *Mercenaria* cysteine residues into the required proximity. Evidence for such dynamic flexibility within the proximal coiled-coil has been inferred from experiments on vertebrate muscle myosin (Knight 1996).

Head-head cooperativity during actin-activation can be demonstrated either in solution, using scallop striated adductor HMM, or in suspension, using myosin filaments (Chantler et al. 1981; Walmsley et al. 1990; Kalabokis and Szent-Gyorgyi 1997; Nyitrai et al. 2002). Early work (Chantler et al. 1981) showed that two calcium ions must bind to both heads of HMM in order to switch the entire molecule on. Under the conditions of these experiments (40 mM ionic strength; 0–2.0 mM ATP; pH 7.0–7.5) calcium binding was characterised by a single binding isotherm, making it possible to predict the activation curve in a model-dependent manner (Chantler et al. 1981). This demonstrated a cooperativity between the two heads of a single molecule during the activation process. As noted above, subsequent work has shown that if the ionic strength is raised to 170 mM or if the ATP is replaced by ADP, AMP-PNP or ADP-vanadate, calcium binding itself becomes cooperative (Kalabokis and Szent-Gyorgyi 1997), complicating interpretation of the activation curves under these conditions but now shown to be compatible with a classic Monod-Wyman-Changeux cooperative mechanism (Nyitrai et al. 2002). In turn, calcium decreases the affinity of ADP and AMP-PNP for myosin and HMM (Kalabokis and Szent-Gyorgyi 1997).

At low ionic strength, higher levels of cooperativity were detected for myosin filament activation than for soluble HMM (Hill coefficients ≥2.8) (Chantler et al. 1981; Kalabokis and Szent-Gyorgyi 1997). Furthermore, myosin ATPase data obtained *in vitro* as a function of free calcium (Chantler et al. 1981) superposed exactly on tension measurements made in the fibre under identical conditions (Fig. 4.18b) (Simmons and Szent-Gyorgyi 1985). Because the form of calcium dependence seen in the earlier work demonstrated a transition from a model where the binding of two calciums per myosin could switch the molecule on, to one where a single calcium would suffice (Chantler et al. 1981), the steep onset of activation was interpreted as indicating some form of intermolecular cooperativity operating between myosin molecules within each thick filament. This could be envisaged as occurring either through interactions between the heads of myosin molecules located at a single crown, or through interactions between heads in adjacent crowns (Fig. 4.6b) (Levine et al. 1988; Levine 1993). The recent development of HMM preparations demonstrating much greater levels of cooperativity (Kalabokis and Szent-Gyorgyi 1997), albeit still significantly less than myosin, raises the possibility that an intact HMM may be capable of duplicating the myosin data entirely. Unfortunately, attempts so far to prepare intact HMM through baculovirus expression have not yielded active preparations (Nyitray, unpublished – as quoted in Perreault-Micale et al. 1996b; Janes 2001). Structural data remain suggestive of an intermolecular role (Levine 1993). At minimum, regulation must require a cooperative interaction between the two heads of each myosin molecule, which is modulated by the level of occupancy of the calcium and nucleotide binding sites within a thick filament.

Global structural changes, consequential to calcium binding, have been demonstrated using scallop myosin but their relationship to the regulatory process remains obscure. Similar to vertebrate smooth and non-muscle myosin II molecules (Craig et al. 1983), myosins from the striated adductor muscles of *Pecten maximus* (Ankrett et al. 1991a) and from both the striated and smooth adductors of *Patinopecten yessoensis* (Takahashi et al. 1989) have been shown to fold as 10S monomers under relaxing conditions. This effect has also been demonstrated on *Argopecten irradians* HMM (Frado and Craig 1992; Stafford et al. 2001). At physiological ionic strength, calcium binding promoted unfolding of 10S myosin to yield the extended 6S conformation (Takahashi et al. 1989; Ankrett et al. 1991a) but this transition was much slower than activation of product release by calcium binding (Ankrett et al. 1991a). Also, the regular arrangement of myosin heads seen at each crown within the thick filament under relaxing conditions becomes disorganised upon addition of calcium (Vibert and Craig 1985; Frado and Craig 1989). It is plausible that the head-head interactions promoted by nucleotide binding under relaxing conditions become disrupted upon calcium binding, allowing the heads to act independently during the mechanochemical cycle (Kalabokis and Szent-Gyorgyi 1997).

What are the specific conformational changes that must supervene calcium-specific binding to a myosin head so as to trigger actin-activation? Although much detail is known regarding the structural changes undertaken at the active site during the catalytic cycle of an active myosin head (Rayment 1996; Houdusse et al. 2000), the structural changes involved in regulation remain subject to speculation for no myosin head has yet been crystallised in the "off-state". We do know that calcium does not control actin binding directly (Chalovich et al. 1984) but brings about an elevation of nucleotide hydrolysis at the active site (Ashiba et al. 1980; Wells and Bagshaw 1985). Hence, the structural consequences of calcium-specific binding must be first transmitted to the active site, influencing the rate of product release; structural transitions that accompany ATP cleavage will facilitate actin-activation. Unfortunately, a sufficient number of snapshots of this dynamic process have yet to be collected in order to construct a detailed picture of how calcium-specific binding couples to events at the active site. But some clues do exist.

The role of the E-LC may be key to the mechanism of regulation. Structural studies (Xie et al. 1994; Houdusse and Cohen 1996), together with functional dissection through light chain point mutation and chimera construction (Jancso and Szent-Gyorgyi 1994; Fromherz and Szent-Gyorgyi 1995), have pinpointed the exact residues involved in calcium-specific binding within the N-terminal domain of the E-LC (see Myosin, above, and Figs. 4.12 and 4.15). However, it is possible that calcium-triggering is not the only function of the E-LC – it may also be physically involved in controlling events at the active site. Early evidence in support of this hypothesis was from antibodies, produced against the scallop E-LC, that had no effect on calcium-specific binding yet specifically increased the actin-activated ($-Ca^{2+}$) ATPase activity leaving the actin-activated ($+Ca^{2+}$) ATPase rate unaffected (Wallimann and Szent-Gyorgyi 1981). More recently, limited physical interactions have been seen between the C-terminal domain of the E-LC and the motor domain (Houdusse et al. 1999). This interface has been suggested to be a target upon the application of micromolar amounts of TFP to *Pecten maximus* myosin, myosin S-1 or myofibrils (Patel et al. 2000), eliminating ATPase hydrolysis by locking myosin in a form which may mimic the off-state. However, a hydrophobic patch associated with the lower 50 kDa subdomain of the heavy chain has been shown to be responsible for abrogation of ATPase activity in both conventional and unconventional myosins (Sellers et al. 2003), and may well be the main site of TFP inhibition in scallop myosin. Higher (millimolar) concentrations of TFP lead to dissociation of the R-LC (Patel et al. 2000), reminiscent of the effect of EDTA (Chantler and Szent-Gyorgyi 1980). The rate limiting step of the scallop myosin MgATPase, in the absence of calcium, is one which involves a weak-binding $ADP.P_i$ intermediate (Marston and Lehman 1974; Wells and Bagshaw 1985; Jackson and Bagshaw 1988). Importantly, the 25/50 kDa loop, which guards the entrance to active site cleft, is known to be a major determinant in the rate of ADP release (Spudich 1994; Perreault-Micale 1996b; Kurzawa-Goertz et al. 1998).

It is plausible that an association between the two heads of myosin in the off-state could be maintained by salt-bridge interactions between R-LCs located on adjacent heads

(Offer and Knight 1996), or through similar interactions which arise through head-head overlap between myosins molecules situated within adjacent crowns along helical arrays spiralling around each thick filament (Levine 1993). Alternatively, and more likely, asymmetric structures within the filament may maintain the off-state, similar to those described for smooth muscle where intramolecular interactions between the actin binding surface of one head and the converter domain of the adjacent head have been observed (Wendt et al. 1999; Wendt et al. 2001). Although two heads are required, this may not be sufficient to ensure regulation. Short-tailed HMM molecules from smooth muscle exhibit diminished regulation (Trybus 1994; Trybus et al. 1997); two-headed fragments require at least twenty heptads of coiled-coil in order to resist chain dissociation in either smooth (Trybus et al. 1997) or scallop (Malnasi-Csizmadia et al. 1998) myosin fragments and thereby maintain regulation or calcium-specific binding, respectively. However, it is also possible that the conformational relay mechanisms differ between scallop adductor and vertebrate smooth muscles (Patel et al. 2000). For instance, smooth muscle myosin (Katoh and Morita 1996) and HMM (Trybus 1994) lacking E-LC has been shown to exhibit R-LC phosphorylation-dependent activation: an impossibility if the E-LC/motor domain interface is essential for regulation. Complete removal of E-LCs from either vertebrate (Wagner and Giniger 1981; Sivaramakrishnan and Burke 1982) or scallop (Ashiba and Szent-Gyorgyi 1985) myosins does not appear to perturb the intrinsic activity of the ATPase site – the presence of the E-LC is required in order to *curtail* activity in a regulatory myosin.

Clues to the next step in the conformational relay may be seen through examination of the structural locations of the regulation-sensitive residues identified within the primary scallop MHC sequence (Fig. 4.11) (Nyitray et al. 1991). From the scallop S-1.ADP structure (state III) (Houdusse et al. 1999) Asp54 (numbering as for *Pecten*) is located on the surface of the SH3 domain, close to the point of contact with the E-LC; Trp35 and Gly42, while closer in primary sequence to the N-terminus than SH3, are located deeper in the molecule along a shallow cleft that separates the SH3 domain from the rest of the motor domain (note that residues 16–26 are invisible in the crystal structure (Houdusse et al. 1999) therefore the location of this mobile region is not clear). On the other side of this cleft, facing Trp35, are Glu96 and Ala97; it seems likely that interactions across this cleft represent the next turn of the cogs in this calcium-dependent conformational relay. It is of interest to note that the myosin head is so compliant that this cluster of regulation-sensitive residues is located on the opposite side of the molecule relative to the E-LC in the transition state structure (state II) (Dominguez et al. 1998). It remains unclear from either the state II or the state III structures as to exactly how the above regulation-sensitive residues communicate with the remaining two in the head, Glu184 and Lys188, which are close to the active site but several peptide layers removed from the other regulation-sensitive residues.

Examination of myosin head structures (Rayment et al. 1993a; Houdusse et al. 1999) indicates that results from earlier photo-crosslinking studies, which purported to show that a region of the R-LC could move relative to the E-LC in a nucleotide-dependent manner in the absence of calcium yet in a calcium-dependent manner in the presence of nucleotide (Hardwicke et al. 1983; Hardwicke and Szent-Gyorgyi 1985), were likely to originate from intermolecular, rather than intramolecular events. Nevertheless, a calcium-dependent shift in the photolabelling of heavy chain residues by 2-[(4-azido-2-nitrophenyl)amino]ethyl diphosphate (NANDP), an ADP analog, has been observed (Kerwin and Yount. 1993). This altered labelling pattern is curious: Arg128, located close to the purine binding site, is labelled in the absence of calcium; yet in the presence of calcium Cys198 becomes labelled - a residue that is located some distance away from the nucleotide binding site, close to the surface of the molecule. If real, this suggests a dramatic and bizarre conformational change.

The functional link in the regulatory chain of command between the active site and the actin binding site appears to be the role of a single stretch of heavy-chain, switch II (Ile463-Asn472; numbering as for *Pecten*). Switch II lies at the base of the cleft that separates the 50 kDa region into upper and lower domains (Houdusse et al. 1999; Houdusse et al. 2000) and influences the shape of the cavity into which the nucleotide γ-phosphate fits. In both the nucleotide-free state I, and the detached state III, the jaws of the 50 kDa cleft are open (Fig. 4.14) yet the conformation of switch II remains unchanged. In the transition state II, switch II has moved towards the nucleotide binding site, partially closing the 50 kDa cleft and creating a distortion in the γ-phosphate bond, facilitating phosphate cleavage and possible subsequent departure by the "back door" (Fisher et al. 1995; Yount et al. 1995; Houdusse et al. 2000). All three flexible joints – switch II, the relay and the SH1 helix – respond in a coordinated manner to the cyclic hydrolysis of ATP during actin activation. Damage to the continuity of the α-carbon backbone in this critical region can abrogate actin activation, as demonstrated by proteolytic cleavage data. Digestion of scallop S-1(+LC) by trypsin leads to initial severance of the heavy-chain into 75 kDa N-terminal and 24 kDa C-terminal fragments (Szentkiralyi 1984; Szentkiralyi 1987), the cleavage probably occurring within the cluster of four lysine residues beginning at Lys636 (numbering as for *Pecten*). This uncouples actin-activation from the consequences of nucleotide cleavage at the active site, the former being abolished whilst the latter remains unaffected. If tryptic digestion is performed on scallop S-1(+LC) in the presence of excess actin, cleavage at this site is blocked. Instead, the primary cleavage site is found closer to the N-terminus, splitting the S-1 heavy-chain into 65 kD N-terminal and 31 kD C-terminal fragments (Szentkiralyi 1987), cleavage probably occurring at Lys547, Lys552, Lys560 or Arg562 (numbering as for *Pecten*). Such cleavage is unique to scallop myosin; severance of the S-1 heavy-chain at this site abolishes hydrolytic activity at the active site completely. Both sets of cleavage sites are within actin-associated regions of the lower 50 kDa domain; cleavage here would physically disconnect switch II and the relay from the SH1 helix.

In summary, regulation of scallop striated adductor contraction is accomplished by a complex allosteric mechanism. Myosin activation is triggered when calcium binds to calcium-specific sites located on the E-LC. The patency of this site is exquisitely maintained by the presence of the R-LC, hydrogen bonding from Gly117 being critical: when the R-LC is lost through divalent cation chelation, the E-LC can no longer bind calcium. Activation is cooperative; initially, calcium must bind to each site on *both* E-LCs of myosin in order to activate the molecule. The off-state is maintained through interactions between the two halves of the bipartite molecule, probably through an asymmetric conformation. When a sufficient number (~20%) of myosins become activated within a filament, cooperativity increases and the mechanism changes such that both heads are activated by a single bound calcium per molecule. At physiological ionic strength in the presence of ADP, calcium binding itself is also cooperative. Subtle changes in bonding between adjacent peptide chains, close to the calcium binding site on the E-LC, transmit this triggering signal to the nucleotide binding pocket. Although this transmission could be through the MHC, it is also possible that it occurs via the C-terminal domain of the E-LC. When the myosin head is in the detached state III conformation, refractive to actin binding, suggestive links in the chain of command are found between the E-LC and SH3 on the myosin head, and across the cleft between SH3 and the heart of the motor domain, regulation-sensitive residues lining the way. The ensuing changes lead to an increase in the rate of loss of nucleotide cleavage products, the loss of γ-phosphate possibly being rate-limiting. ATP hydrolysis, facilitated by this conformational relay, leads in turn to precise changes which now prime the myosin head into state II, the "closed" transition state capable of binding to actin. In this state, regulation-sensitive residues are remote from the E-LC, bound nucleotide is present as a transition-state complex and the lever arm is cocked for the power-stroke. Actin binding increases the flux of nucleotide hydrolysis through the active site by causing reciprocal movements in switch II and the converter; the γ-phosphate leaves by the back door and ADP dissociates from the nucleotide pocket, while the lever arm swings and remains attached to actin, generating the "open" post-power-stroke state, state I. This cycle repeats for as long as calcium remains bound. Calcium dissociation favours a return to the actin-dissociated state, a nucleotide-bound head conformation which becomes locked in an associated state with an adjacent myosin head – the relaxed state.

4.6 FUNCTION OF THE SMOOTH ADDUCTOR

4.6.1 Physiology

Molluscan paramyosin-containing smooth muscles are capable of both tonic and phasic contraction. When the active state of a tonic contraction decays, such muscles may still sustain large tensions for long periods of time with little expenditure of energy. This stretch-resistant state is known as "catch" and is characterised by economy of energy usage during a period of extremely slow force decay (Jewell 1959; Nauss and Davies 1966; Baguet and Gillis 1968). Although considerable progress has been made, we are still far from understanding the catch state. Much of our knowledge of molluscan catch muscle physiology comes from the study of the anterior byssus retractor muscle (ABRM) of *Mytilus edulis*, an experimental preparation introduced by Winton (1937); there are practical difficulties in performing many of the requisite tension measurements on the more fragile scallop fibre preparations. Consequently, in the absence of evidence to the contrary, I will assume that facts obtained from studies on the ABRM apply equally to catch contraction in the scallop. A number of prior reviews have emphasised various aspects of the subject (Ruegg 1971; Twarog 1976; Achazi 1982; Chantler 1983; Muneoka and Twarog 1983; Watabe and Hartshorne 1990; Ruegg 1992).

Catch muscle is innervated by at least two types of nerve: cholinergic excitatory and serotonergic relaxing nerves. The firing of the excitatory nerve releases acetylcholine, which brings about catch contraction by eliciting a brief active contraction which decays to the catch state. Phasic contraction can be brought about most readily through addition of acetylcholine to the bathing medium (Fig. 4.22a; Jewell 1959; Ruegg 1965) or upon stimulation by an alternating current (Winton 1937). The immediate action of acetylcholine is to mediate the release of calcium from internal stores (Sugi and Suzuki 1978).

If, while the muscle is undergoing isometric contraction during the active state, a "quick release" is performed (i.e., one end of the muscle is released and the muscle is allowed to shorten by 5% of its length), the tension will drop momentarily and completely, owing to relaxation of the series elastic elements. Nevertheless, tension recovery under these conditions is rapid (Fig. 4.22a), demonstrating that the muscle is in the active state. By contrast, the catch state is induced in muscle fibres by specifically washing acetylcholine away from the bathing medium (Fig. 4.22b; Ruegg 1965) or by direct current stimulation (Winton 1937). A quick release performed under these conditions results in a rapid and complete loss of tension followed by negligible tension recovery (Jewell 1959), tension only being restored if the muscle is stretched back to its original length. This shows that the stiffness of this muscle is high and similar to that attained in the rigor state, except that nucleotide is present here: such an experiment demonstrates the catch state. Relaxation from catch is brought about by application of serotonin to muscle fibres in the bathing medium (Fig. 4.22c; Twarog and Cole 1973); applied serotonin has no effect during phasic contraction. Simultaneous application of serotonin and acetylcholine results in the phasic contraction of muscle fibres (Muneoka and Twarog 1983). *In vivo*, relaxing nerves are thought to release serotonin in order to abolish catch.

It would appear that the catch state can only be entered provided the muscle is primed by first passing through the active state. Thus, a brief application of acetylcholine or neural stimulation in the presence of serotonin antagonists will induce catch (Muneoka and Twarog 1983). Once the excitatory signal is removed, the high calcium levels elicited by the signal drop to low levels comparable with the relaxed state (Twarog 1967b; Atsumi and Sugi 1976). In the intact muscle, therefore, catch is a low calcium ($<10^{-8}$ M) state (Ishii et al. 1989).

Skinned fibre preparations have been very useful in extending our knowledge of the catch state (Baguet 1973; Marchand-Dumont and Baguet 1975; Cornelius 1980; Cornelius 1982; Pfitzer and Ruegg 1982). Such preparations lack a functional plasma membrane and sarcoplasmic reticulum; they are freely permeable to proteins and ligands added at will. With skinned fibres catch can be induced by calcium chelation (e.g., by addition of 20 mM EGTA in the presence of MgATP) following initiation of the active state (Fig. 4.22d) (Baguet and Marchand-Dumont 1975; Cornelius 1982; Pfitzer and Ruegg 1982; Castellani and Cohen 1987a; Siegman et al. 1997), further suggesting that active state priming and an absence of calcium are required to induce the catch state. The slow decline in tension (Fig. 4.22d) has been shown to correlate with negligible actomyosin ATPase activity (Guth et al. 1984). Quick release experiments provide results analogous

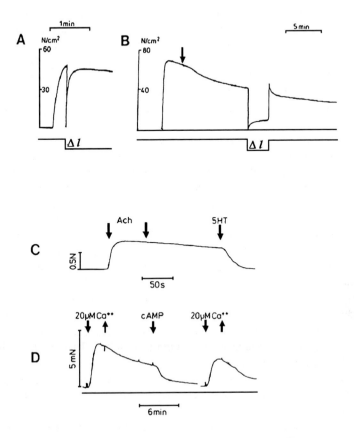

Figure 4.22. Physiological demonstration of the active and catch states in a molluscan muscle. A. Phasic isometric contraction of isolated anterior byssus retractor muscle (ABRM) of the bivalve, *Mytilus edulis*, after addition of 10 μM acetylcholine to the bathing medium (seawater). A quick release of the muscle by 5% of its length (Δl) causes an immediate drop in tension (due to discharge of the series elastic elements) followed by active redevelopment of tension by the contractile elements. B. Isometric contraction induced by 10 μM acetylcholine is terminated by washing out the drug (at arrow). During the very slow relaxation that follows, the ABRM is in catch. Tension is maintained passively. The active state is negligible, as indicated by the poor tension recovery following the quick release ($\Delta l = 5\% L_o$). Force is restored by restretching the fibres passively. C. Catch contraction of ABRM is induced by acetylcholine (first arrow) and then removed (second arrow). The slow relaxation which follows is accelerated by 10 μM 5-hydroxytryptamine (third arrow), which abolishes catch without inhibiting active contraction. D. A catch-like state can be induced in chemically skinned preparations of ABRM by the initial addition of 20 μM calcium (first arrow) followed by its removal (second arrow) upon addition of chelator. The slow relaxation that follows is accelerated by the addition of cAMP (10–100 μM). A subsequent calcium-induced active contraction is only slightly inhibited by cAMP. All solutions contained 5 mM MgATP, 20 mM imidazole, pH 6.7. Data reproduced, with permission of the publisher, from Ruegg (1986).

to those described above, obtained on intact fibres. Using permeabilised fibres, the relaxing action of serotonin has been shown to be mimicked by cAMP (Cornelius 1982; Pfitzer and Ruegg 1982), by the catalytic subunit of cAMP-dependent protein kinase (Pfitzer and Ruegg 1982; Castellani and Cohen 1987a), NaF (Castellani and Cohen 1987a) and by ATP-γ-S (Castellani and Cohen 1987a); furthermore, fibre perfusion with the calcium-activated phosphatase, calcineurin, maintains the catch state under conditions that would otherwise lead to its decay (Castellani and Cohen 1992). Such results cannot be attributed to perturbations of calcium or calcium sensitivity and we are left with the ineluctable conclusion that pathways involving phosphorylation/dephosphorylation reactions are involved in the control of catch. The prime suspects to be considered as the major substrates for phosphorylation are paramyosin (Achazi 1979; Watabe et al. 1989), twitchin (Siegman et al. 1997; Siegman et al. 1998), myosin (Castellani and Cohen 1987a; Castellani and Cohen 1987b) and catchin (Yamada et al. 2000).

4.6.2 Catch Mechanism

There are two main contenders as possible explanations of the catch mechanism. While one theory holds that the catch state involves formation of actomyosin links which are somehow preserved, in the presence of nucleotide (Lowy and Millman 1963; Baguet and Gillis 1968), an alternative explanation takes more graphic account of the unique structure of the paramyosin-rich thick filaments and suggests a mechanism involving direct interactions between adjacent thick filaments (Johnson et al. 1959; Ruegg 1961; Gilloteaux and Baguet 1977). Although the ability to produce catch is synonymous with the presence of paramyosin-rich thick filaments in invertebrates, catch-like states have been described in vertebrate smooth muscles (Murphy et al. 1983; Hai and Murphy 1988), where it is termed "latch" owing to its similarity to "catch"; latch and catch may be mechanistically similar. Indeed, this similarity has been consolidated, recently, with the observation that catch forces may exist at higher than resting calcium concentrations (Butler et al. 1998; Siegman et al. 1998), similar to latch. Vertebrate smooth muscles do not contain paramyosin. Consequently, if the key intermediate of catch or latch is confined to regulatory step(s) of the crossbridge cycle, then the role for paramyosin in invertebrate muscles may simply be structural, related to the large size of the thick filaments, rather than being a critical component necessary for the generation of a high passive tension.

Earlier work appeared to substantiate the "paramyosin model" because a merging, or fusion, of thick filaments was observed in cross-sections of *Mytilus edulis* ABRM undergoing catch contraction (Heumann and Zebe 1968; Gilloteaux and Baguet 1977; Hauck and Achazi 1987). However, these data have been criticised (Bennett and Elliott 1989) on two grounds – first, that events occurring within the muscle between the times of fixation and observation were largely uncontrolled, and second, that inter-filament crosslinking artefacts could occur during the fixation process. To overcome such drawbacks and arrest ABRM in a defined state of catch, Bennett and Elliott (1989) used freeze-substitution prior to electron microscopy, measuring mechanical properties of the muscle up until the instant of slam-freezing, thereby fixing the muscle instantaneously relative to the time-scale of the catch mechanism. Under these conditions, no evidence for thick filament clumping during catch was obtained; instead, well-separated thick filaments were found, surrounded by partial rings of thin filaments, the thick and thin filaments sometimes being closer to each other than during phasic contraction. These authors could only reproduce micrographs showing clumped thick filaments if cell membranes were badly damaged. The work of Bennett and Elliott (1989) represents the best ultrastructural evidence so far in favour of catch being a consequence of an altered actomyosin association rather than through thick filament interaction.

Although paramyosin may not play a direct role in the maintenance of catch, an indirect role in addition to its contribution to structure is more difficult to rule out. Paramyosin extracted from serotonin-treated muscles is able to incorporate two to four times as much covalently bound phosphate as paramyosin extracted from untreated or acetylcholine-treated muscles (Achazi 1979). This observation is consistent with a greater degree of paramyosin phosphorylation within the muscle during catch. These experiments, together with *in vitro* observations that paramyosin aggregates underwent a phase transition over a markedly small pH range (6.5–7.2) (Johnson et al. 1959; Ruegg 1961; Ruegg 1964), the extent determined by the degree of phosphorylation (Cooley et al. 1979), have implicated paramyosin phosphorylation in the regulation of the catch state. Because paramyosin smooth thick filaments are thought to have myosin arranged as a surface monolayer (Cohen 1982), it follows that only the outer layer of paramyosin may need to be phosphorylated in order to be functional. Hence, stoichiometries of less than 0.3 moles phosphate per mole paramyosin during catch would be entirely consistent with a putative role for paramyosin phosphorylation during the regulation of catch (Chantler 1983); values in the range 0.1–0.3 moles phosphate per mole paramyosin have been observed (Achazi 1979). Direct functional evidence for this form of regulation is still lacking however. Furthermore, in those instances where paramyosin phosphorylation was deduced from SDS gels, migrating at a position corresponding to paramyosin "aggregates" (Watabe et al. 1989), it is possible that the band observed originated from the recently identified twitchin-related protein (Siegman et al. 1997; Siegman et al. 1998) and may have been mistakenly identified in the earlier work.

As noted above (in Structure of the Smooth Adductor Muscle), a 600 kDa twitchin-like protein exhibited the greatest increase in phosphorylation when ABRM (*Mytilus edulis*) was released from the catch state (Siegman et al. 1997; Siegman et al. 1998). Most importantly, this twitchin-like protein undergoes phosphorylation exclusively during the cAMP dependent release from catch, no phosphate incorporation being detected in paramyosin, myosin light chains or heavy chains, or any other component (Siegman et al. 1998). cDNA sequencing of the N-terminal region of twitchin (from the mussel *Mytilus galloprovincialis*) reveals a putative kinase domain but so far the cAMP-dependent phosphorylation site remains unknown. If the kinase domain is operative, it does not appear to act downstream of cAMP activation because no other component of the muscle appears to be phosphorylated; consequently, it is postulated that phosphorylated twitchin releases the crossbridges from catch (Siegman et al. 1998). Phosphorylated twitchin has no effect at saturating calcium levels but requires some fraction of bridges that are not calcium-activated (Siegman et al. 1998; Butler et al. 1998).

The above results place severe constraints on the role of myosin phosphorylation mechanisms in catch. Nevertheless, one of two myosins found in the smooth adductor muscle of *Patinopecten yessoensis* (Kondo and Morita 1981; Morita and Kondo 1982), in that portion of the muscle furthest away from the striated adductor, is capable of undergoing phosphorylation on its R-LC (Sohma et al. 1985). This was shown using either vertebrate smooth muscle light-chain kinase or an endogenous kinase source (Sohma and Morita 1986; Sohma and Morita 1987; Sohma et al. 1985) and the phosphorylation site on the R-LC was demonstrated to be Ser-11 (Miyanishi et al. 1985). Such a property is more akin to vertebrate smooth muscle regulation than it is to the regulation of scallop striated muscle and raised the possibility that light-chain phosphorylation might play a role in the regulation of catch (Takahashi et al. 1988). Because the latch state of vertebrate smooth muscle correlates with light-chain dephosphorylation (Hai and Murphy 1988) a regulatory role for light-chain phosphorylation would, by analogy, be associated with release from the catch state, similar to the role of twitchin phosphorylation, described above. It may be noted, however, that the actin-activated MgATPase activities (Kendrick-Jones et al. 1970; Perreault-Micale et al. 1996a) of myosins isolated from the smooth adductor muscles of *Placopecten magellanicus* and *Argopecten irradians* are calcium-sensitive and exhibit rates that are independent of R-LC phosphorylation. Consequently, the functional significance of some isoforms of scallop smooth adductor R-LC to undergo reversible phosphorylation may have little relevance for the catch mechanism.

Heavy-chain phosphorylation of ABRM myosin was also observed to occur either in the presence of an endogenous heavy-chain kinase or in the presence of a heavy-chain kinase isolated from the slime mold, *Dictyostelium* (Castellani and Cohen 1987a; Castellani and Cohen 1987b). To date, no direct experiments have been published which would correlate this phosphorylation with specific steps in the catch mechanism. Using shadowed preparations visualised by electron microscopy, these authors were able to show *in vitro* that heavy-chain phosphorylation enhances the ability of the molecule to fold up at low ionic strength (Castellani and Cohen 1987b). These folded molecules are reminiscent of the asymmetric structures seen in the off-state of myosin from the striated adductor muscle (see above). Although they have not been observed *in vivo* as yet, these *in vitro* studies do suggest that events in the neck region of the myosin molecule could be modulated by phosphorylation at the tip of the tail (Castellani et al. 1988). It is plausible that tail phosphorylation could perturb events on the heads of myosin molecules several crossbridge crown repeats removed along the thick filament. Such heavy chain phosphorylation would be associated with release from the catch state (Castellani and Cohen 1987a; Castellani and Cohen 1992).

Catchin, expressed only in molluscan catch muscles, can also be phosphorylated on Ser-103 within its unique N-terminal sequence (Yamada et al. 2000). This was inferred for scallop (*Argopecten irradians*) catchin upon comparison of its sequence with that of a phosphorylated peptide sequence previously isolated from *Mytilus edulis* (Castellani et al. 1988). There is no current evidence for such phosphorylation upon initiation of, or exit from, the catch state. Nevertheless, it has been hypothesised that catchin may act in concert with twitchin during catch, or the unique N-terminus may be involved in tethering actin while the MHC-derived C-terminal coiled coil could interact with either myosin or paramyosin rod structures (Yamada et al. 2000).

Biochemical (Takahashi et al. 1988) and physiological (Butler et al. 1998; Galler et al. 1999) data have reinforced the emerging view that catch involves a myosin.ADP intermediate locked in an associated state with actin. Force decay in skinned ABRM fibres subject to flash photolysis in the presence of caged-ATP, is slowed approximately five-fold by the presence of MgADP (Galler et al. 1999). Force maintenance during the latch state of mammalian tonic smooth muscles also appears to require a delayed release of MgADP (Fuglsang et al. 1993; Khromov et al. 1995; Khromov et al. 1998). It is plausible, therefore, that cAMP-dependent phosphorylation of twitchin alters the configuration of catch crossbridges such that their affinity for ADP decreases (Butler et al. 1998; Galler et al. 1999). To explain why there was no effect of twitchin phosphorylation on either ATPase or force at saturating calcium levels, Butler and colleagues (1998) proposed that the structures affected by twitchin phosphorylation are those myosin heads that do not possess bound calcium. Their results have been summarised in a semi-quantitative model (Butler et al. 1998) in which the catch mechanism is simulated as a boxed set of reversible first order rate constants relating myosin and actomyosin to their calcium-liganded structures (Fig. 4.23a). Force generation is possible through actin binding but not by myosin alone; catch is represented by actin-bound, calcium-free, myosin. Twitchin phosphorylation is simulated by a 10-fold increase in dissociation rate constant (k_7) and the effects of increasing calcium concentration are simulated through the use of a range of rate constants (k_1) for calcium binding. The predicted tension curve derived from this model, including the relaxing effect of cAMP (Fig. 4.23b), is in good agreement with experimental data (Butler et al. 1998). Even better agreement (the calculated decay curve is exponential whereas the experimental data are biphasic) could be obtained if head-head cooperativity were taken into account and if precise binding data were known for β-actin and catch myosin, both in the presence and absence of calcium.

Many of the key pieces of the catch puzzle are now to hand. Twitchin, a molecule with a structure and periodicities related to both paramyosin and myosin, facilitates the ability of myosin heads to retain MgADP following prior activation (MgATP loading) and calcium dissociation, yet promotes MgADP dissociation upon phosphorylation. The critical differences evident between the primary sequences of smooth and striated adductor muscle MHCs (Fig. 4.11) must somehow facilitate an altered tertiary conformation within the nucleotide binding pocket such that myosin is locked in the actin-associated catch state, a consequence of the increased affinity for MgADP inherent to this conformation. It is likely that such changes must involve switch II (Fig. 4.14). Although the myosin head structure of a catch adductor myosin has not yet been determined, it is clear from other known structures (Dominguez et al. 1998; Houdusse et al. 1999) that entry to, or exit from, the nucleotide binding pocket may be guarded by the E-LC, itself the instigator of the conformational relay which ensues calcium triggering. Determination of the crystal structure of catch muscle myosin heads (or preferably two-headed fragments!) and computer modelling of these within the thick filament structure, paying particular attention to paramyosin, twitchin and catchin packing, should prove a suitable goal for the next decade.

296

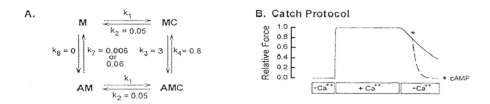

A.

$$M \underset{k_2 = 0.05}{\overset{k_1}{\rightleftarrows}} MC$$

$$k_8 = 0 \quad k_7 = 0.006 \atop \text{or} \atop 0.06 \quad k_3 = 3 \quad k_4 = 0.8$$

$$AM \underset{k_2 = 0.05}{\overset{k_1}{\rightleftarrows}} AMC$$

B. Catch Protocol

Figure 4.23. Catch simulation using a four-state model. A). MC and M represent detached, non-force producing states of myosin, with and without calcium, respectively. AMC and AM represent force producing states of myosin when attached to actin, with and without calcium, respectively. All rate constants are first order (units of s^{-1}). k_7 is taken to be 0.006 s^{-1} when twitchin is not phosphorylated and 0.06 s^{-1} when it is phosphorylated. Increases in calcium concentration may be simulated by increasing the value of k_1. B). Simulated force response curve during a catch protocol (continuous line) in which k_1 was increased from zero ($-Ca^{2+}$) to 1.5 s^{-1} ($+Ca^{2+}$) then decreased to zero once again. The effect of cAMP addition was simulated by increasing k_7 from 0.006 s^{-1} to 0.06 s^{-1} (dashed line beginning at asterisk). This quantitative model for catch, based on earlier consideration of the latch state in vertebrate smooth muscle (Hai and Murphy 1988), was devised by Butler et al. (1998) and is reproduced here with permission of the publishers.

4.7 ACHIEVEMENTS AND GOALS

Many of the earlier predictions (Chantler 1991) have been realised, not only enriching our understanding of the structure and function of scallop muscle but also providing insights into the wider field of contractility. The tools of molecular biology have been used, not only to determine a number of myosin primary sequences from several scallop species (e.g., Nyitray et al. 1991; Nyitray et al. 1994; Perreault-Micale et al. 1996b; Janes et al. 2000) but, with ever increasing sophistication, to facilitate the creation of mutant peptides which allow experimental dissection of the regulatory process (e.g., Reinach et al. 1986; Jancso and Szent-Gyorgyi 1994; Yang and Sweeney 1995; Trybus et al. 1997; Wu et al. 1999; Colegrave et al. 2003). Advances in data processing, structural analysis and the art of crystallography have yielded the first detailed structures of scallop myosin heads (Houdusse et al. 1999; Houdusse et al. 2000) and RDs (Xie et al. 1994; Houdusse and Cohen 1996) at atomic resolution. Improvements in sample preparation and image processing have led to advances in large-scale structure determinations of thick filaments and quasi-crystalline arrays by electron microscopic imaging (e.g., Levine 1993; Whittaker et al. 1995; Padron et al. 1997; Wendt et al. 1999). Our knowledge of the intricacies of scallop myosin ATPase kinetics (Kurzawa-Goertz et al. 1998; Nyitrai et al. 2002) and of cooperative aspects of regulation (Walmsley et al. 1990; Kalabokis and Szent-Gyorgyi 1997; Nyitrai et al. 2002; Nyitrai et al. 2003) has expanded. Our understanding of the catch mechanism has advanced dramatically through a series of new and significant findings (e.g., Siegman et al. 1997; Siegman et al. 1998; Butler et al. 1998; Galler et al. 1999; Yamada et al. 2000), though much remains to be learned.

Future goals must include a structural description at the atomic level of the catalytic events during each mechanochemical cycle of scallop myosin heads. With the use of appropriately characterised nucleotide analogs, there is no reason to doubt that this will not be accomplished within the next decade for this approach is already well advanced (Houdusse et al. 2000). Such a proposal might be aided through the defined expression of functional subfragments using appropriate cloning vectors. Myosin heads are too large to be expressed in bacterial expression systems and, so far, baculovirus expression systems have failed to yield functional products using scallop myosin (Nyitray, unpublished – as quoted in Perreault-Micale et al. 1996b; Janes 2001). Clearly, there is a need to try alternative eukaryotic expression systems. Because complete regulation can only be demonstrated in two-headed molecules, their functional expression or improved production quality (Kalabokis and Szent-Gyorgyi 1997) will be key to any crystallographic attempt to determine the structure of HMM. Herculean as such a task will be, its thunder could be stolen through refinement of other techniques such as computer modelling (e.g., Offer and Knight 1996) or electron microscope imaging techniques (e.g., Wendt et al. 1999) with respect to two-headed scallop myosin fragments. The latter approach may be the most direct way to probe the structure of the inhibited state, another "holy grail" that has so far proven elusive. Potential models can be tested for function using mutant light chains in scallop hybrid myofibrils or myosin (Sellers et al. 1980; Reinach et al. 1986; Goodwin et al. 1990; Colegrave et al. 2003). Alternative ways of achieving inhibition (Patel et al. 2000) may also offer a pragmatic solution to the quest for atomic resolution of the myosin conformation that prevents product release. An understanding of head-head cooperativity (Chantler et al. 1981; Walmsley et al. 1990; Kalabokis and Szent-Gyorgyi 1997; Nyitrai et al. 2002; Nyitrai et al. 2003) at the atomic level is likely to be another major benefit of such structural enterprises, complementing solution studies.

Physiological experiments, in conjunction with the use of specific molecular probes, will continue to be useful in the study of catch contraction. Such approaches could facilitate dissection of the roles of catchin (Yamada et al. 2000), twitchin (Butler et al. 1998) and myosin (Castellani and Cohen 1994) and decipher the meaning of their respective phosphorylation patterns during catch regulation. Physiological studies on the phasic adductor remains the "Achilles heel" of scallop muscle research. Although many pioneering experiments have been performed (e.g., Simmons and Szent-Gyorgyi 1980; Rall 1981; Simmons and Szent-Gyorgyi 1985; Lea et al. 1990), the delicate structure of scallop muscle fibres do not lend themselves to the sort of manipulations required during the demanding studies of crossbridge mechanics. However, we are now in the era of single molecule mechanics. Many elegant experiments have already been performed on various myosin motors through the use of optical tweezers or flexible microneedles (see e.g., Mehta et al. 1999). To date, these techniques have not been applied to components from scallop muscle. Application of laser trap technology to the scallop actomyosin interaction would allow a direct comparison of the step-size and force per head between motors isolated from both tonic and phasic adductor muscles, facilitating correlation between events at the nucleotide binding site and force production. Although the structural similarities between striated and catch adductor myosins are many, it may still be possible for catch muscle myosin to attain putative increased force simply by generating an increased angle of swing of the lever arm; this would lead to an increased step size and a higher duty ratio (Howard 1997). In principle, single molecule mechanics could determine, once and for all, whether or not there is any intrinsic difference between the force-producing states of striated and catch adductor muscle molecular motors. Another productive decade lies ahead!

ACKNOWLEDGMENTS

I thank the following for providing me with images of published and unpublished material: Drs. Pauline Bennett, Carolyn Cohen, Roger Craig, Arthur Elliott, Clara Franzini-Armstrong, Mike Geeves, Ken Holmes, Daniel Janes, Bob Kensler, Peter Knight, Rhea Levine, Barry Millman, Gerald Offer, Andrew Szent-Gyorgyi and Peter Vibert. I also thank Angie Poole and George Stodulski for help in preparing the figures, and a big thank-you to Sandy Shumway for supporting my request for colour plates. This review was written during the tenure of a grant from the Biotechnology and Biological Sciences Research Council in the U.K.

REFERENCES

Achazi, R.K., 1979. Phosphorylation of molluscan paramyosin. Pfluegers Arch. 379:197–201.

Achazi, R.K., 1982. Catch muscle. In: B.M. Twarog, R.J.C. Levine and M.M. Dewey (Eds.). Basic Biology of Muscles: A Comparative Approach. Raven Press, New York. pp. 291–308.

Adelstein, R.S. and Eisenberg, E., 1980. Regulation and kinetics of the actin-myosin-ATP interaction. Ann. Rev. Biochem. 49:921–956.

Ankrett, R.J., Rowe, A.J., Cross, R.A., Kendrick-Jones, J. and Bagshaw, C.R., 1991a. A folded (10S) conformer of myosin from a striated muscle and its implications for regulation of ATPase activity. J. Mol. Biol. 217:323–335.

Ankrett, R.J., Walmsley, A.R. and Bagshaw, C.R., 1991b. Kinetic analysis of regulated myosin ATPase activity using single and limited turnover assays. J. Cell Sci. Suppl. 14:1–5.

Ashiba, G., Asada, T. and Watanabe, S., 1980. Calcium regulation in clam foot muscle. J. Biochem. 80:837–844.

Ashiba, G. and Szent-Gyorgyi, A.G., 1985. Essential light-chain exchange in scallop myosin. Biochemistry 24:6618–6623.

Atsumi, S. and Sugi, H., 1976. Localization of calcium-accumulating structures in the anterior byssal retractor muscle of Mytilus edulis and their role in the regulation of active and catch contractions. J. Physiol. 257:549–560.

Bagshaw, C.R. and Kendrick-Jones, J., 1979. Characterization of homologous divalent metal ion binding sites of vertebrate and molluscan myosins using electron paramagnetic resonance spectroscopy. J. Mol. Biol. 130:317–336.

Bagshaw, C.R. and Kendrick-Jones, J., 1980. Identification of the divalent metal ion binding domain of myosin regulatory light-chains using spin-labelling techniques. J. Mol. Biol. 140:411–433.

Bagshaw, C.R. and Reed, G.H., 1977. The significance of the slow dissociation of divalent metal ions from myosin regulatory light-chains. FEBS Lett. 81:386–390.

Baguet, F., 1973. The catch-state in glycerol extracted fibers from a lamellibranch smooth muscle (ABRM). Pfluegers Arch. 340:19–34.

Baguet, F. and Gillis, J.M., 1968. Energy cost of tonic contraction in lamellibranch catch muscle. J. Physiol. 198:127–143.

Baguet, F. and Marchand-Dumont, G., 1975. The muscular membrane and calcium activation of the contractile system of a lamellibranch smooth muscle (ABRM). Pflügers Arch. 354:75–85.

Baker, J.E., Brust-Mascher, I., Ramachandran, S. and Thomas, D.D., 1998. A large and distinct rotation of the myosin light chain domain upon muscle contraction. Proc. Natl. Acad. Sci. USA 95:2944–2949.

Bartegi, A., Fattoum, A., Dagorn, C., Gabrion, J. and Kassab, R., 1990. Isolation, characterization and immunocytochemical localisation of caldesmon-like protein from molluscan striated muscle. Eur. J. Biochem. 185:585–595.

Bayliss, L.E., Boyland, E. and Ritchie, A.D., 1930. The adductor mechanism of *Pecten*. Proc. Roy. Soc. Ser. B 106:363–376.

Bear, R.S., 1944. X-ray diffraction on protein fibres. Feather rachis, porcupine quill tip and clam muscle. J. Amer. Chem. Soc. 66:2043–2050.

Bear, R.S. and Selby, C.C., 1956. The structure of paramyosin fibrils according to X-ray diffraction. J. Biophys. Biochem. Cytol. 2:55–69.

Bennett, A.J. and Bagshaw, C.R., 1986a. The kinetics of bivalent metal ion dissociation from myosin subfragments. Biochem. J. 233:173–177.

Bennett, A.J. and Bagshaw, C.R., 1986b. The mechanism of regulatory light-chain dissociation from scallop myosin. Biochem. J. 233:179–186.

Bennett, A.J., Patel, N., Wells, C. and Bagshaw, C.R., 1984. 8-Anilino-1-naphthalenesulphonate, a fluorescent probe for the regulatory light-chain binding site of scallop myosin. J. Mus. Res. Cell Mot. 5:165–182.

Bennett, P.M. and Elliott, A., 1981. The structure of the paramyosin core in molluscan thick filaments. J. Mus. Res. Cell Mot. 2:65–81.

Bennett, P.M. and Elliott, A., 1989. The 'catch' mechanism in molluscan muscle: an electron microscopy study of freeze-substituted anterior byssus retractor muscle of *Mytilus edulis*. J. Mus. Res. Cell Mot. 10:297–311.

Bennett, P.M. and Marston, S.B., 1990. Calcium regulated thin filaments from molluscan catch muscles contain a caldesmon-like regulatory protein. J. Mus. Res. Cell Mot. 11:302–312.

Block, S.M., 1996. Fifty ways to love your lever: Myosin motors. Cell 87:151–157.

Bower, S.M., Wang, Y. and Chantler, P.D., 1992. Regulatory light-chain Cys-55 sites on the two heads of myosin can come within 2Å of each other. FEBS Lett. 310:132–134.

Brandt, P.W., Diamond, M.S. and Schachat, F.H., 1984. The thin filament of vertebrate skeletal muscle cooperatively activates as a unit. J. Mol. Biol. 180:379–384.

Brenner, B., Schoenberg, M., Chalovich, J.M., Greene, L.E. and Eisenberg, E., 1982. Evidence for cross-bridge attachment in relaxed muscle at low ionic strength. Proc. Natl. Acad. Sci. 79:7288–7292.

Buddenbrock, W.V., 1911. Untersuchungen uber die schwimmbewegungen und die statocysten der gattung pecten. Sber. Heidelb. Acad. Wiss. 28:1–24.

Butler, T.M., Mooers, S.U., Li, C., Narayan, S. and Siegman, M.J., 1998. Regulation of catch muscle by twitchin phosphorylation: Effects on force, ATPase, and shortening. Biophys. J. 75:1904–1914.

Castellani, L. and Cohen, C., 1987a. Myosin rod phosphorylation and the catch state of molluscan muscles. Science 235:334–337.

Castellani, L. and Cohen, C., 1987b. Rod phosphorylation favors folding in a catch muscle myosin. Proc. Natl. Acad. Sci. USA 84:4058–4062.

Castellani, L. and Cohen, C., 1992. A calcineurin-like phosphatase is required for catch contraction. FEBS Lett. 309:321–326.

Castellani, L., Elliott, B.W. and Cohen, C., 1988. Phosphorylatable serine residues are located in a non-helical tailpiece of a catch muscle myosin. J. Mus. Res. Cell Mot. 9:533–540.

Castellani, L. and Hardwicke, P.M.D., 1983. Crystalline structure of sarcoplasmic reticulum from scallop. J. Cell Biol. 97:557–561.

Castellani, L., Hardwicke, P.M.D. and Vibert, P., 1985. Dimer ribbons in the three-dimensional structure of sarcoplasmic reticulum. J. Mol. Biol. 185:579–594.

Castellani, L. and Vibert, P., 1992. Location of paramyosin in relation to the subfilaments within the thick filaments of scallop striated muscle. J. Mus. Res. Cell Mot. 13:174–182.

Castellani, L., Vibert, P. and Cohen, C., 1983. Structure of myosin/paramyosin filaments from a molluscan smooth muscle. J. Mol. Biol. 167:853–872.

Chalovich, J.M., Chantler, P.D., Szent-Gyorgyi, A.G and Eisenberg, E., 1984. Regulation of molluscan actomyosin ATPase activity. J. Biol. Chem. 259:2617–2621.

Chalovich, J.M., Greene, L.E. and Eisenberg, E., 1983. Crosslinked myosin subfragment 1: a stable analogue of the subfragment-1.ATP complex. Proc. Natl. Acad. Sci. USA 80:4909–4913.

Chantler, P.D., 1983. Biochemical and structural aspects of molluscan muscle. In: A.S.M. Saleuddin and K.M. Wilbur (Eds.). The Mollusca. Vol. 4. Academic Press. pp. 77–154.

Chantler, P.D., 1985. Regulatory light-chains and scallop myosin. Form of light-chain removal or reuptake is dependent on the presence of divalent cations. J. Mol. Biol. 181:557–560.

Chantler, P.D., 1991. The structure and function of scallop adductor muscles. In: S.E. Shumway (Ed.). Scallops: Biology, Ecology and Aquaculture. Developments in Aquaculture and Fisheries Science, Vol. 21. Elsevier, Amsterdam. pp. 225–304.

Chantler, P.D. and Bower, S.M., 1988. Cross-linking between translationally equivalent sites on the two heads of myosin. Relationship to energy transfer between the same pair of sites. J. Biol. Chem. 263:938–944.

Chantler, P.D. and Gratzer, W.B., 1975. Effects of specific modification of actin. Eur. J. Biochem. 60:67–72.

Chantler, P.D. and Gratzer, W.B., 1976. The interaction of actin monomers with myosin heads and other muscle proteins. Biochemistry 15:2219–2225.

Chantler, P.D., Sellers, J.R. and Szent-Gyorgyi, A.G., 1981. Cooperativity in scallop myosin. Biochemistry 20:210–216.

Chantler, P.D. and Szent-Gyorgyi, A.G., 1980. Regulatory light-chains and scallop myosin: Full dissociation, reversibility and cooperative effects. J. Mol. Biol. 138:473–492.

Chantler, P.D. and Tao, T., 1986. Interhead fluorescence energy transfer between probes attached to translationally equivalent sites on the regulatory light-chains of scallop myosin. J. Mol. Biol. 192:87–99.

Chantler, P.D., Tao, T. and Stafford III, W.F., 1991. On the relationship between distance information derived from cross-linking and from resonance energy transfer, with specific reference to sites located on myosin heads. Biophys. J. 59:1242–1250.

Cohen, C., 1982. Matching molecules in the catch mechanism. Proc. Natl. Acad. Sci. USA 79:3176–3178.

Cohen, C., Caspar, D.L.D., Parry, D.A.D. and Lucas, R.M., 1971a. Tropomyosin crystal dynamics. Cold Spring Harbour Symp. Quant. Biol. 36:205–219.

Cohen, C. and Holmes, K.C., 1963. X-ray diffraction evidence for α-helical coiled-coils in native muscle. J. Mol. Biol. 6:423–432.

Cohen, C. and Parry, D.A.D., 1998. A conserved C-terminal assembly region in paramyosin and myosin rods. J. Struct. Biol. 122:180–187.

Cohen, C. and Szent-Gyorgyi, A.G., 1971. Assembly of myosin filaments and the structure of molluscan catch muscles. In: R.J. Podolsky (Ed.). Contractility of Muscle Cells and Related Processes. Prentice-Hall, Princeton, N.J. pp. 23–36.

Cohen, C., Szent-Gyorgyi, A.G. and Kendrick-Jones, J., 1971b. Paramyosin and the filaments of molluscan "Catch" muscle. J. Mol. Biol. 56:223–237.

Colegrave, M., Patel, H., Offer, G. and Chantler, P.D., 2003. Evaluation of the symmetric model for myosin-linked regulation: effect of site-directed mutations in the regulatory light chain on scallop myosin. Biochem. J. 374:89–96.

Collins, J.H., Jakes, R., Kendrick-Jones, J., Leszyk, J., Barouch, W., Theibert, J.L., Spiegel, J. and Szent-Gyorgyi, A.G., 1986. Amino acid sequence of myosin essential light-chain from the scallop *Aequipecten irradians*. Biochemistry 25:7651–7656.

Cooke, R., 1997. Actomyosin interaction in striated muscle. Physiol. Rev. 77:671–697.

Cooke, R., 1999. Myosin structure: Does the tail wag the dog? Curr. Biol. 9:R773–R775.

Cooley, L.B., Johnson, W.H. and Krause, S., 1979. Phosphorylation of paramyosin and its possible role in the catch mechanism. J. Biol. Chem. 254:2195–2198.

Cornelius, F., 1980. The regulation of tension in a chemically skinned molluscan smooth muscle. J. Gen. Physiol. 75:709–725.

Cornelius, F., 1982. Tonic contraction and the control of relaxation in a chemically skinned molluscan smooth muscle. J. Gen. Physiol. 79:821–834.

Corrie, J.E.T., Brandmeier, B.D., Ferguson, R.E., Trentham, D.R., Kendrick-Jones, J., Hopkins, S.C., van der Heide, U.A., Goldman, Y.E., Sabido-David, C., Dale, R.E., Criddle, S. and Irving, M., 1999. Dynamic measurement of myosin light-chain-domain tilt and twist in muscle contraction. Nature 400:425–430.

Craig, R., 1994. Structure of the contractile filaments. In: A.G Engel and C. Franzini-Armstrong (Eds.). Myology. Vol. 1, Second Edition. McGraw-Hill, Inc., New York. pp. 134–175.

Craig, R., Padron, R. and Alamo, L., 1991. Direct determination of myosin filament symmetry in scallop striated adductor muscle by rapid freezing and freeze substitution. J. Mol. Biol. 220:125–132.

Craig, R., Smith, R. and Kendrick-Jones, J., 1983. Light chain phosphorylation controls the conformation of vertebrate non-muscle and smooth muscle myosin molecules. Nature 302:436–439.

Craig, R., Szent-Gyorgyi, A.G., Beese, L., Flicker, P., Vibert, P. and Cohen, C., 1980. Electron microscopy of thin filaments decorated with a Ca^{2+}-regulated myosin. J. Mol. Biol. 140:35–55.

302

De Zwaan, A., Thompson, R.J. and Livingstone, D.R., 1980. Physiological and biochemical aspects of the valve snap and valve closure responses in the giant scallop *Placopecten magellanicus*. J. Comp. Physiol. 137:105–114.

Dominguez, R., Freyzon, Y., Trybus, K.M. and Cohen, C., 1998. Crystal structure of a vertebrate smooth muscle myosin motor domain and its complex with the essential light chain: Visualization of the pre-power stroke state. Cell 94:559–571.

Dover, S.D. and Elliott, A., 1979. Three-dimensional reconstruction of a paramyosin filament. J. Mol. Biol. 132:340–341.

Ebashi, S., Endo, M. and Ohtsuki, I., 1969. Control of muscle contraction. Q. Rev. Biophys. 2:351–384.

Egelman, E.H., Francis, N. and DeRosier, D.J., 1982. F-actin is a helix with a random variable twist. Nature (London) 298:131–135.

Egelman, E.H., Francis, N. and DeRosier, D.J., 1983. Helical disorder and the filament structure of F-actin are elucidated by the angle-layered aggregate. J. Mol. Biol. 166:605–629.

Eisenberg, E. and Greene, L.E., 1980. The relation of muscle biochemistry to muscle physiology. Ann. Rev. Physiol. 42:293–309.

Eisenberg, E. and Hill, T.L., 1985. Muscle contraction and free energy transduction in biological systems. Science 227:999–1006.

Elliott, A., 1974. The arrangement of myosin on the surface of paramyosin filaments in the white adductor of *Crassostrea angulata*. Proc. Roy. Soc. Ser. B 186:53–66.

Elliott, A., 1979. Structure of molluscan thick filaments: A common origin for diverse appearance. J. Mol. Biol. 132:323–340.

Elliott, A. and Bennett, P.M., 1982. Structure of the thick filaments in molluscan adductor muscle. In: B.M. Twarog, R.J.C. Levine and M.M. Dewey (Eds.). Basic Biology of Muscles: A Comparative Approach. Raven Press, New York. pp. 11–28.

Elliott, A. and Bennett, P.M., 1984. Molecular organization of paramyosin in the core of molluscan thick filaments. J. Mol. Biol. 176:477–493.

Elliott, G.F., 1964. Electron microscope studies of the structure of the filaments in the opaque adductor muscle of the oyster *Crassostrea angulata*. J. Mol. Biol. 10:89–104.

Fisher, A.J., Smith, C.A., Thoden, J.B., Smith, R., Sutoh, K., Holden, H.M. and Rayment, I., 1995. X-ray structures of the myosin motor domain of *Dictyostelium discoideum* complexed with MgADP.BeF$_x$ and MgADP.AlF$_4$-. Biochemistry 34:8960–8972.

Frado, L-L.Y. and Craig, R., 1989. Structural changes induced in Ca^{2+}-regulated myosin filaments by Ca^{2+} and ATP. J. Cell Biol. 109:529–538.

Frado, L-L.Y. and Craig, R., 1992. Structural changes induced in scallop heavy meromyosin molecules by Ca^{2+} and ATP. J. Mus. Res. Cell Mot. 13:436–446.

Fromherz, S. and Szent-Gyorgyi, A.G., 1995. Role of essential light chain EF hand domains in calcium binding and regulation of scallop myosin. Proc. Natl. Acad. Sci. USA. 92:7652–7656.

Fuglsang, A., Khromov, A., Torok, K., Somlyo, A.V. and Somlyo, A.P., 1993. Flash photolysis studies of relaxation and cross-bridge detachment: higher sensitivity of tonic than phasic smooth muscle to MgADP. J. Mus. Res. Cell Mot. 14:666–677.

Galler, S., Kogler, H., Ivemeyer, M. and Ruegg, J.C., 1999. Force responses of skinned molluscan catch muscle following photoliberation of ATP. Pflugers Arch - Eur. J. Physiol. 438:525–530.

Geeves, M.A., Goody, R.S. and Gutfreund, H., 1984. Kinetics of acto-S-1 interaction as a guide to a model for the crossbridge cycle. J. Mus. Res. Cell Mot. 5:351–361.

Geeves, M.A. and Holmes, K.C., 1999. Structural mechanism of muscle contraction. Ann. Rev. Biochem. 68:867–728.

Gilloteaux, J. and Baguet, F., 1977. Contractile filaments organization in function states of the anterior byssus retractor muscle (ABRM) of *Mytilus edulis* L. Cytobiologie 15:192–220.

Goldberg, A. and Lehman, W., 1978. Troponin-like proteins from muscles of the scallop, *Aequipecten irradians*. Biochem. J. 171:413–418.

Goldman, Y.E., 1998. Wag the tail: Structural dynamics of actomyosin. Cell 93:1–4.

Goodwin, E.B., Leinwand, L.A. and Szent-Gyorgyi, A.G., 1990. Regulation of scallop myosin by mutant regulatory light chains. J. Mol. Biol. 216:85–100.

Goodwin, E.B., Szent-Gyorgyi, A.G. and Leinwand, L.A., 1987. Cloning and characterization of the scallop essential and regulatory myosin light chain cDNAs. J. Biol. Chem. 262:11052–11056.

Gulick, A.M., Bauer, C.B., Thoden, J.B. and Rayment, I., 1997. X-ray structures of the MgADP, MgATP-γ-S and MgAMP.PNP complexes of *Dictyostelium discoideum* myosin motor domain. Biochemistry 36:11619–11628.

Gulick, A.M. and Rayment, I., 1997. Structural studies on myosin II: communication between distant protein domains. Bioessays 19:561–569.

Guth, K., Gagelmann, M. and Ruegg, J.C., 1984. Skinned smooth muscle: time course of force and ATPase activity during contraction cycle. Experientia 40:174–176.

Hai, C-M. and Murphy, R.A., 1988. Cross-bridge phosphorylation and regulation of latch state in smooth muscle. Am. J. Physiol. 254:C99–C106.

Hall, C.E., Jakus, M.A. and Schmitt, F.O., 1945. The structure of certain muscle fibrils as revealed by the use of electron stains. J. Appl. Phys. 16:459–465.

Hanson, J. and Lowy, J., 1959. Evidence for a sliding filament mechanism in tonic smooth muscles of lamellibranch molluscs. Nature 184:286–287.

Hanson, J. and Lowy, J., 1960. Structure and function of the contractile apparatus in the muscles of invertebrate animals. In: G.H. Bourne (Ed.). Structure and Function of Muscle. Academic Press. pp. 265–335.

Hanson, J. and Lowy, J., 1963. The structure of F-actin and actin filaments isolated from muscle. J. Mol. Biol. 6:46–60.

Harada, Y., Sakurada, K., Aoki, T., Thomas, D.D. and Yanagida, T., 1990. Mechanochemical coupling in actomyosin energy transduction studied by *in vitro* movement assay. J. Mol. Biol. 216:49–68.

Hardwicke, P.M.D. and Hanson, J., 1971. Separation of thick and thin myofilaments. J. Mol. Biol. 59:509–516.

Hardwicke, P.M.D. and Szent-Gyorgyi, A.G., 1985. Proximity of regulatory light-chains in scallop myosin. J. Mol. Biol. 183:203–211.

Hardwicke, P.M.D., Wallimann, T. and Szent-Gyorgyi, A.G., 1983. Light-chain movement and regulation in scallop myosin. Nature 301:478–482.

Hauck, R. and Achazi, R.K., 1987. The ultrastructure of a molluscan catch muscle during a contraction-catch-relaxation cycle. Eur. J. Cell Biol. 45:30–35.

Heumann, H-G. and Zebe, E., 1968. Uber die Funktionsweise glatter Muskelfasern. Zeitschrift f. Zellforsch. 85:534–551.

Hibberd, M.G. and Trentham, D.R., 1986. Relationships between chemical and mechanochemical events during muscle contraction. Ann. Rev. Biophys. Biophys. Chem. 15:119–161.

Higashi, S. and Oosawa, F., 1965. Conformational changes associated with polymerization and nucleotide binding in actin molecules. J. Mol. Biol. 12:843–865.

Hitchcock-DeGregori, S.E and Varnell, T.A., 1990. Tropomyosin has discrete actin-binding sites with sevenfold and fourteenfold periodicities. J. Mol. Biol. 214:885–896.

Holmes, K.C., Popp, D., Gebhard, W. and Kabsch, W., 1990. Atomic model of the actin filament. Nature 347:44–49.

Houdusse, A. and Cohen, C., 1995. Target sequence recognition by the calmodulin superfamily: Implications from light chain binding to the regulatory domain of scallop myosin. Proc. Natl. Acad. Sci. USA 92:10644–10647.

Houdusse, A. and Cohen, C., 1996. Structure of the regulatory domain of scallop myosin at 2A resolution: implications for regulation. Structure 4:21–32.

Houdusse, A., Kalabokis, V.N., Himmel, D., Szent-Gyorgyi, A.G. and Cohen, C., 1999. Atomic structure of scallop myosin subfragment S1 complexed with MgADP: A novel conformation of the myosin head. Cell 97:459–470.

Houdusse, A., Szent-Gyorgyi, A.G. and Cohen, C., 2000. Three conformational states of scallop myosin S1. Proc. Natl. Acad. Sci. USA 97:11238–11243.

Howard, J., 1997. Molecular motors: structural adaptations to cellular functions. Nature 389:561–567.

Hue, H.K., Benyamin, Y. and Roustan, C., 1989. Comparative study of invertebrate actins: antigenic cross-reactivity versus sequence variability. J. Mus. Res. Cell Mot. 10:135–142.

Hue, H.K., Labbe, J.P., Harricane, M.C., Cavadore, J.C., Benyamin, Y. and Roustan, C., 1988. Structural and functional variations in skeletal and *Pecten maximus* muscle actins. Biochem. J. 256:853–859.

Huston, E.E., Grammer, J.C. and Yount, R.G., 1988. Flexibility of the myosin heavy chain: direct evidence that the region containing SH1 and SH2 can move 10Å under the influence of nucleotide binding. Biochemistry 27:8945–8952.

Huxley, A.F., 1957. Muscle structure and theories of contraction. Prog. Biophys. Biophys. Chem. 7:255–318.

Huxley, A.F. and Simmons, R.M., 1971. Proposed mechanism of force generation in striated muscle. Nature 233:533–538.

Huxley, H.E., 1969. The mechanism of muscle contraction. Science 164:1356–1366.

Ikura, M., Clore, G.M., Gronenborn, A.M., Zhu, G., Klee, B.B. and Bax, A., 1992. Solution structure of a calmodulin-target peptide complex by multidimensional NMR. Science 256:632–638.

Inoue, A., Ojima, T. and Nishita, K., 1996. Cloning and sequencing of a cDNA for Akazara scallop troponin T. J. Biochem. 120:834–837.

Inoue, K., Sohma, H. and Morita, F., 1990. Ca2(+)-dependent protein phosphatase which dephosphorylates regulatory light chain-a in scallop smooth muscle myosin. J. Biochem. 107:872–878.

Irving, M., St. Claire Allen, T., Sabido-David, C., Craik, J.S., Brandmeir, B., Kendrick-Jones, J., Corrie, J.E.T., Trentham, D.R. and Goldman, Y.E., 1995. Tilting of the light-chain region of myosin during step length changes and active force generation in skeletal muscle. Nature 375:688–691.

Ishii, N., Simpson, A.W.N. and Ashley, C.C., 1989. Free calcium at rest and during "catch" in single smooth muscle cells. Science 243:1367–1368.

Ishijima, A., Kojima, H., Funatsu, T., Tokunaga, M., Higuchi, H., Tanaka, H. and Yanagida, T., 1998. Simultaneous observation of individual ATPase and mechanical events by a single myosin molecule during interaction with actin. Cell 92:161–171.

Ishimoda-Takagi, T. and Kobayashi, M., 1987. Molecular heterogeneity and tissue specificity of tropomyosin obtained from various bivalves. Comp. Biochem. Physiol. 88B:443–452.

Ishimoda-Takagi, T., Kobayashi, M. and Yaguchi, M., 1986. Polymorphism and tissue specificity of scallop tropomyosin. Comp. Biochem. Physiol. 83B:515–521.

Jackson, A.P. and Bagshaw, C.R., 1988. Transient-kinetic studies of the adenosine triphosphatase activity of scallop heavy meromyosin. Biochem. J. 251:515–526.

Jackson, A.P., Warriner, K.E., Wells, C. and Bagshaw, C.R., 1986. The actin-activated ATPase of regulated and unregulated scallop heavy meromyosin. FEBS Lett. 197:154–158.

Jancso, A. and Szent-Gyorgyi, A.G., 1994. Regulation of scallop myosin by the regulatory light chain depends on a single glycine residue. Proc. Natl. Acad. Sci. USA 91:8762–8766.

Janes, D.P., 2001. Structural and functional approaches to myosin-linked regulation using expressed protein fragments. Ph.D. Thesis. University of London.

Janes, D.P., Patel, H. and Chantler, P.D., 2000. Primary structure of myosin from the striated adductor muscle of the Atlantic scallop, *Pecten maximus*, and expression of the regulatory domain. J. Mus. Res. Cell Mot. 21:415–422.

Jewell, B.R., 1959. The nature of the phasic and the tonic responses of the anterior byssal retractor muscle of *Mytilus*. J. Physiol. 149:154–177.

Johnson, W.H., Kahn, J.S. and Szent-Gyorgyi, A.G., 1959. Paramyosin and contraction of "Catch Muscles". Science 130:160–161.

Kabsch, W., Mannherz, H.G., Suck, D., Pai, E.F. and Holmes, K.C., 1990. Atomic structure of the actin:DNase I complex. Nature 347:37–44.

Kagawa, H., Gengyo, K., McLachlan, A.D., Brenner, S. and Karn, J., 1989. Paramyosin gene (Unc-15) of *Caenorhabditis elegans*: Molecular cloning, nucleotide sequence and models for thick filament structure. J. Mol. Biol. 207:311–333.

Kalabokis, V.N., O'Neall-Hennessey, E. and Szent-Gyorgyi, A.G., 1994. Regulatory domains of myosins: influence of heavy chain on Ca^{2+}-binding. J. Mus. Res. Cell Mot. 15:547–553.

Kalabokis, V.N. and Szent-Gyorgyi, A.G., 1997. Cooperativity and regulation of scallop myosin and myosin fragments. Biochemistry 36:15834–15840.

Kalabokis, V.N., Vibert, P., York, M.L. and Szent-Gyorgyi, A.G., 1996. Single-headed scallop myosin and regulation. J. Biol. Chem. 271:26779–26782.

Katoh, T. and Morita, F., 1996. Roles of light chains in the activity and conformation of smooth muscle myosin. J. Biol. Chem. 271:9992–9996.

Kemp, B.E. and Pearson, R.B., 1985. Spatial requirements for location of basic residues in peptide substrates for smooth muscle myosin light chain kinase. J. Biol. Chem. 260:3355–3359.

Kendrick-Jones, J., Cohen, C., Szent-Gyorgyi, A.G. and Longley, W., 1969. Paramyosin: Molecular length and assembly. Science 163:1196–1198.

Kendrick-Jones, J. and Jakes, R., 1977. Myosin-linked regulation - A chemical approach. In: G. Rieker, A. Weber and J. Goodwin (Eds.). International Symposium on Myocardial Failure. Tergensee, Munich, W. Germany. Springer-Verlag. pp. 28–40.

Kendrick-Jones, J., Lehman, W. and Szent-Gyorgyi, A.G., 1970. Regulation in molluscan muscles. J. Mol. Biol. 54:313–326.

Kendrick-Jones, J., Szentkiralyi, E.M. and Szent-Gyorgyi, A.G., 1976. Regulatory light-chains in myosins. J. Mol. Biol. 104:747–775.

Kerrick, W.G.L., Hoar, P.E., Cassidy, P.S., Bolles, L. and Malencik, D.A., 1981. Calcium-regulatory mechanisms: Functional classification using skinned fibers. J. Gen. Physiol. 77:177–190.

Kerwin, B.A. and Yount, R.G., 1993. Photolabelling evidence for calcium-induced conformational changes at the ATP binding site of scallop myosin. Proc. Natl. Acad. Sci. USA 90:35–39.

Khaitlina, S.Y., 1986. Polymerisation of β-like actin from *Pecten maximus* adductor muscle. FEBS Lett. 198:221–224.

Khaitlina, S.Y., Antropova, O., Kuznetsova, I., Turoverov, K. and Collins, J., 1999. Correlation between polymerizability and conformation in scallop β-like actin and rabbit skeletal muscle α-actin. Arch. Biochem. Biophys. 368:105–111.

Khromov, A., Somlyo, A.V. and Somlyo, A.P., 1998. MgADP promotes a catch-like state developed through force-calcium hysteresis in tonic smooth muscle. Biophys. J. 75:1926–1934.

Khromov, A., Somlyo, A.V., Trentham, D., Zimmermann, B. and Somlyo, A.P., 1995. The role of MgADP in force maintenance by dephosphorylated crossbridges in smooth muscle: A flash photolysis study. Biophys. J. 69:2611–2622.

Kitamura, K., Tokunaga, M., Hikikoshi Iwane, A. and Yanagida, T., 1999. A single myosin head moves along an actin filament with regular steps of 5.3 nanometres. Nature 397:129–134.

Knight, P.J., 1996. Dynamic behaviour of the head-tail junction of myosin. J. Mol. Biol. 255:269–274.

Knox, M.K., Szent-Gyorgyi, A.G., Trueblood, C.E., Weber, A. and Zigmond, S., 1986. The effect of low ATP concentrations on relaxation in myosin regulated myofibrils from scallop. J. Mus. Res. Cell Mot. 7:110–118.

Kondo, S. and Morita, F., 1981. Smooth muscle of scallop adductor contains at least two kinds of myosin. J. Biochem. 90:673–681.

Kretsinger, R.H. and Nockolds, C.E., 1973. Carp muscle calcium binding protein. J. Biol. Chem. 248:3313–3326.

Kurzawa-Goertz, S.E., Perreault-Micale, C.L., Trybus, K.M., Szent-Gyorgyi, A.G. and Geeves, M.A., 1998. Loop 1 can modulate ADP affinity, ATPase activity and motility of different scallop myosins. Transient kinetic analysis of S1 isoforms. Biochemistry 37:7517–7525.

Kwon, H., Goodwin, E.B., Nyitray, L., Berliner, E., O'Neall-Hennessey, E., Melandri, F.D. and Szent-Gyorgyi, A.G., 1990. Isolation of the regulatory domain of scallop myosin: Role of the essential light chain in calcium binding. Proc. Natl. Acad. Sci. USA 87:4771–4775.

Lea, T.J., Fenton, M.J., Potter, J.D. and Ashley, C.C., 1990. Rapid activation by photolysis of nitr-5 in skinned fibres of the striated adductor muscle from the scallop. Biochim. Biophys. Acta 1034:186–194.

Lehman, W., 1981. Thin-filament-linked regulation in molluscan muscles. Biochim. Biophys. Acta. 668:349–356.

Lehman, W., 1983a. The distribution of troponin-like proteins on thin filaments of the bay scallop, *Aequipecten irradians*. J. Mus. Res. Cell Mot. 4:379–389.

Lehman, W., 1983b. The ionic requirements for regulation by molluscan thin filaments. Biochim. Biophys. Acta 745:1–5.

Lehman, W., Craig, R. and Vibert, P., 1994. Ca^{2+}-induced tropomyosin movement in *Limulus* thin filaments revealed by three-dimensional reconstruction. Nature 368:65–67.

Lehman, W., Regenstein, J.M. and Ransom, A.L., 1976. The stoichiometry of the components of arthropod thin filaments. Biochim. Biophys. Acta. 434:215–222.

Lehman, W. and Szent-Gyorgyi, A.G., 1975. Regulation of muscular contraction. Distribution of actin control and myosin control in the animal kingdom. J. Gen. Physiol. 66:1–30.

Lehrer, S.S., 1975. Intramolecular crosslinking of tropomyosin via disulfide bond formation: Evidence for chain register. Proc. Natl. Acad. Sci. USA 72:3377–3381.

Lehrer, S.S. and Geeves, M.A., 1998. The muscle thin filament as a classical cooperative/allosteric regulatory system. J. Mol. Biol. 277:1081–1089.

Levine, R.J.C., 1993. Evidence for overlapping myosin heads on relaxed thick filaments of fish, frog and scallop striated muscles. J. Struct. Biol. 110:99–110.

Levine, R.J.C., Chantler, P.D. and Kensler, R.W., 1988. Origin of myosin heads on *Limulus* thick filaments. J. Cell. Biol. 107:1739–1747.

Levine, R.J.C., Elfvin, M., Dewey, M.M. and Walcott, B., 1976. Paramyosin in invertebrate muscles. Content in relation to structure and function. J. Cell Biol. 71:273–279.

Linari, M., Dobbie, I., Reconditi, M., Koubassova, N., Irving, M., Piazzesi, G. and Lombardi, V., 1998. The stiffness of skeletal muscle in isometric contraction and rigor: the fraction of myosin heads bound to actin. Biophys. J. 74:2459–2473.

Lorenz, M., Poole, K.J.V., Popp, D., Rosenbaum, G. and Holmes, K.C., 1995. An atomic model of the unregulated thin filament obtained by X-ray fiber diffraction on oriented actin-tropomyosin gels. J. Mol. Biol. 246:108–119.

Lorenz, M., Popp, D. and Holmes, K.C., 1993. Refinement of the F-actin model against X-ray fiber diffraction data by the use of a directed mutation algorithm. J. Mol. Biol. 234:826–836.

Lowey, S., Kucera, J. and Holtzer, A., 1963. On the structure of the paramyosin molecule. J. Mol. Biol. 7:234–244.

Lowy, J. and Hanson, J., 1962. Ultrastructure of invertebrate smooth muscles. Physiol. Rev. Suppl. 42:34–47.

Lowy, J. and Millman, B.M., 1963. The contractile mechanism of the Anterior Byssus Retractor muscle of *Mytilus edulis*. Phil. Trans. Roy. Soc. Lond. Ser. B 246:105–148.

Lowy, J. and Poulsen, F.R., 1982. Time-resolved X-ray diffraction studies of the structural behaviour of myosin heads in a living contracting unstriated muscle. Nature 299:308–312.

Lowy, J. and Poulsen, F.R., 1987. Fast X-ray diffraction studies of muscle. In: J.M. Squire and P.J. Vibert (Eds.). Fibrous Protein Structure. Academic Press. pp. 451–494.

Lowy, J. and Vibert, P.J., 1967. Structure and organization of actin in a molluscan smooth muscle. Nature 215:1254–1255.

Luther, P.K., Munro, P.M.G. and Squire, J.M., 1981. Three-dimensional structure of the vertebrate muscle A-band. III. M-region and myosin filament symmetry. J. Mol. Biol. 151:703–730.

Lymn, R.W. and Taylor, E.W., 1971. Mechanism of adenosine triphosphate hydrolysis by actomyosin. Biochemistry 10:4617–4624.

Maita, T., Konno, K., Maruta, S., Norisue, H. and Matsuda, G., 1987. Amino acid sequence of the essential light-chain of adductor muscle myosin from Ezo giant scallop, *Patinopecten yessoensis*. J. Biochem. 102:1141–1149.

Maita, T., Konno, K., Ojuma, T. and Matsuda, G., 1984. Amino acid sequences of the regulatory light-chains of striated adductor muscle myosins from Ezo giant scallop and Akazara scallop. J. Biochem. 95:167–177.

Malnasi-Csizmadia, A.M., Bonet-Kerrache, A., Nyitray, L. and Mornet, D., 1994. Purification and properties of caldesmon-like protein from molluscan muscle. Comp. Biochem. Physiol. Biochem. Mol. Biol. 108:59–63.

Malnasi-Csizmadia, A., Hegyi, G., Tolgyesi, F., Szent-Gyorgyi, A.G. and Nyitray, L., 1999. Fluorescence measurements detect changes in scallop myosin regulatory domain. Eur. J. Biochem. 261:452–458.

Malnasi-Csizmadia, A., Shimony, E., Hegyi, G., Szent-Gyorgyi, A.G. and Nyitray, L., 1998. Dimerization of the head-rod junction of scallop myosin. Biochem. Biophys. Res. Comm. 252:595–601.

Malnasi-Csizmadia, A., Woolley, R.J. and Bagshaw, C.R., 2000. Resolution of conformational states of *Dictyostelium* myosin II motor domain using tryptophan (W501) mutants: Implications for the open-closed transition identified by crystallography. Biochemistry 39:16135–16146.

Marchand-Dumont, G. and Baguet, F., 1975. The control mechanism of relaxation in molluscan catch muscle (ABRM). Pflügers Arch. 354:87–101.

Margulis, B.A., Galaktionov, K.I., Podgornaya, O.I. and Pinaev, G.P., 1982. Major contractile proteins of mollusc: Tissue polymorphism of actin, tropomyosin and myosin light chains is absent. Comp. Biochem. Physiol. 72B:473–476.

Marsh, R.L., Olson, J.M. and Guzik, S.K., 1992. Mechanical performance of scallop adductor muscle during swimming. Nature 357:411–413.

Marston, S.B., 1989. What is latch? New ideas about tonic contraction in smooth muscle. J. Mus. Res. Cell Mot. 10:97–100.

Marston, S.B. and Lehman, W., 1974. ADP binding to relaxed scallop myofibrils. Nature 252:38–39.

Marston, S.B. and Smith, C.W.J., 1985. The thin filaments of smooth muscles. J. Mus. Res. Cell Mot. 6:669–708.

Matsuno, A., Kannda, M. and Okuda, M., 1996. Ultrastructural studies on paramyosin core filaments from native thick filaments in catch muscles. Tissue and Cell 28:501–505.

Matulef, K., Sirokman, K., Perreault-Micale, C.L. and Szent-Gyorgyi, A.G., 1998. Amino acid sequence of squid myosin heavy chain. J. Mus. Res. Cell Mot. 19:705–712.

Maw, M.C. and Rowe, A.J., 1980. Fraying of A-filaments into three subfilaments. Nature 286:412–414.

McLaughlin, P.J., Gooch, J.T., Mannherz, H.-G. and Weeds, A.G., 1993. Structure of gelsolin segment 1-actin complex and the mechanism of filament severing. Nature 364:685–692.

Mehta, A.D., Rief, M., Spudich, J.A., Smith, D.A. and Simmons, R.M., 1999. Single-molecule biomechanics with optical methods. Science 283:1689–1695.

Millman, B.M., 1967. Mechanism of contraction in molluscan muscle. Am. Zool. 7:583–591.

Millman, B.M. and Bennett, P.M., 1976. Structure of the cross-striated adductor muscle of the scallop. J. Mol. Biol. 103:439–467.

Miyanishi, T., Maita, T., Morita, F., Kondo, S. and Matsuda, G., 1985. Amino acid sequences of the two kinds of regulatory light chains of adductor smooth muscle myosin. J. Biochem. 97:541–551.

Molloy, J.E., Burns, J.E., Kendrick-Jones, J., Tregear, R.T. and White, D.C.S., 1995. Movement and force produced by a single head. Nature 378:209–212.

Monod, J., Wyman, J. and Changeux, J.-P., 1965. On the nature of allosteric transitions: a plausible model. J. Mol. Biol. 12:88–103.

Morita, F. and Kondo, S., 1982. Regulatory light-chain contents and molecular species of myosin in catch muscle of scallop. J. Biochem. 92:977–983.

Morrison, C.M. and Odense, P.H., 1974. Ultrastructure of some Pelecypod adductor muscles. J. Ultrastruct. Res. 49:228–251.

Muneoka, Y. and Twarog, B.M., 1983. Neuromuscular transmission and excitation-contraction coupling in molluscan muscle. In: A.S.M. Saleuddin and K.M. Wilbur (Eds.). The Mollusca. Vol. 4. Academic Press. pp 35–76.

Murphy, C.T. and Spudich, J.A., 2000. Variable surface loops and myosin activity: Accessories to a motor. J. Mus. Res. Cell Mot. 21:139–151.

Murphy, R.A., Aksoy, M.O., Dillon, P.F., Gerthoffer, W.T. and Kamm, K.E., 1983. The role of myosin light-chain phosphorylation in regulation of the cross-bridge cycle. Fed. Proc. 42:51–56.

Nauss, K.M. and Davies, R.E., 1966. Changes in inorganic phosphate and arginine during the development, maintenance and loss of tension in the anterior byssus retractor muscle of *Mytilus edulis*. Biochem. Z. 345:173–187.

Nave, R. and Weber, K., 1990. A myofibrillar protein of insect muscle related to vertebrate titin connects Z band and A band: Purification and molecular characterization of invertebrate mini-titin. J. Cell Sci. 95:535–544.

Nicaise, G. and Amsellem, J., 1983. Cytology of muscle and neuromuscular junction. In: A.S.M. Saleuddin and K.M. Wilbur (Eds.). The Mollusca. Vol. 4. Academic Press. pp. 1–33.

Nishita, K., Ojima, T., Takahashi, A. and Inoue, A., 1997. Troponin from smooth adductor muscle of Ezo-giant scallop. J. Biochem. 121:419–424.

Nishita, K., Tanaka, H. and Ojima, T., 1994. Amino acid sequence of troponin C from scallop striated adductor muscle. J. Biol. Chem. 269:3464–3468.

Nonomura, Y., 1974. Fine structure of the thick filaments in molluscan catch muscle. J. Mol. Biol. 88:445–455.

Nunzi, M.G. and Franzini-Armstrong, C., 1981. The structure of smooth and striated portions of the adductor muscle of the valves in a scallop. J. Ultrastruct. Res. 76:134–148.

Nyitrai, M., Szent-Gyorgyi, A.G. and Geeves, M.A., 2002. A kinetic model of the co-operative binding of calcium and ADP to scallop (*Argopecten irradians*) heavy meromyosin. Biochem. J. 365:19–30.

Nyitrai, M., Szent-Gyorgyi, A.G. and Geeves, M.A., 2003. Interactions of the two heads of scallop (*Argopecten irradians*) heavy meromyosin with actin: influence of calcium and nucleotides. Biochem. J. 370:839–848.

Nyitray, L., Goodwin, E.B. and Szent-Gyorgyi, A.G., 1991. Complete primary structure of a scallop striated muscle myosin heavy chain. J. Biol. Chem. 266:18469–18476.

Nyitray, L., Jancso, A., Ochiai, Y., Graf, L. and Szent-Gyorgyi, A.G., 1994. Scallop striated and smooth muscle myosin heavy-chain isoforms are produced by alternative RNA splicing from a single gene. Proc. Natl. Acad. Sci. USA 91:12686–12690.

Offer, G., Baker, H. and Baker, L., 1972. Interaction of monomeric and polymeric actin with myosin subfragment-1. J. Mol. Biol. 66:435–444.

Offer, G. and Knight, P., 1996. The structure of the head-tail junction of the myosin molecule. J. Mol. Biol. 256:407–416.

Offer, G. and Sessions, R., 1995. Computer modelling of the α-helical coiled-coil. Packing of side chains in the inner core. J. Mol. Biol. 249:967–987.

Ohtsuki, I., 1974. Localization of troponin in thin filament and tropomyosin paracrystal. J. Biochem. 75:753–765.

Ohtsuki, I., 1975. Distribution of troponin components in the thin filament studied by immunoelectron microscopy. J. Biochem. 77:633–639.

Ohtsuki, I., Masaki, T., Nonomura, Y. and Ebashi, S., 1967. Periodic distribution of troponin along the thin filament. J. Biochem. 61:817–819.

Ojima, T. and Nishita, K., 1986a. Troponin from Akazara scallop striated adductor muscles. J. Biol. Chem. 261:16749–16754.

Ojima, T. and Nishita, K., 1986b. Isolation of troponins from striated and smooth adductor muscles of Akazara scallop. J. Biochem. 100:821–824.

Ojima, T. and Nishita, K., 1992. Comparative studies on biochemical characteristics of troponins from Ezo-giant scallop (*Patinopecten yessoensis*) and Akazara scallop (*Chlamys nipponensis akazara*). Comp. Biochem. Physiol. 79B:525–529.

Ojima, T., Tanaka, H. and Nishita, K., 1994. Cloning and sequence of a cDNA encoding Akazara scallop troponin C. Arch. Biochem. Biophys. 311:272–276.

Padron, R., Alamo, L., Murgich, J. and Craig, R., 1997. Towards an atomic model of the thick filaments of muscle. J. Mol. Biol. 275:35–41.

Panté, N., 1994. Paramyosin polarity in the thick filament of molluscan smooth muscles. J. Struct. Biol. 113:148–163.

Parry, D.A.D., 1975. Analysis of the primary sequence of α-tropomyosin from rabbit skeletal muscle. J. Mol. Biol. 98:519–535.

Patel, H., Margossian, S.S. and Chantler, P.D., 2000. Locking regulatory myosin in the off-state with trifluoperazine. J. Biol. Chem. 275:4880–4888.

Patwary, M.U., Reith, M. and Kenchington, E.L., 1999. Cloning and characterization of tropomyosin cDNAs from the sea scallop, *Placopecten magellanicus* (Gmelin, 1791). J. Shellfish Res. 18:67–70.

Perreault-Micale, C.L., Jancso, A. and Szent-Gyorgyi, A.G., 1996a. Essential and regulatory light chains of *Placopecten* striated and catch muscle myosins. J. Mus. Res. Cell Mot. 17:533–542.

Perreault-Micale, C.L., Kalabokis, V.N., Nyitray, L. and Szent-Gyorgyi, A.G., 1996b. Sequence variations in the surface loop near the nucleotide binding site modulate the ATP turnover rates of molluscan myosins. J. Mus. Res. Cell Mot. 17:543–553.

Perry, S.V., 1994. Activation of the contractile mechanism by calcium. In: A.G. Engel and C. Franzini-Armstrong (Eds.). Myology, Vol. 1, Second Edition. McGraw-Hill, Inc., New York. pp. 529–552.

Pfitzer, G., 2001. Invited review: Regulation of myosin phosphorylation in smooth muscle. J. Appl. Physiol. 91:497–503.

Pfitzer, G. and Ruegg, J.C., 1982. Molluscan catch muscle: Regulation and mechanics in living and skinned anterior byssus retractor muscle of *Mytilus edulis*. J. Comp. Physiol. B 147:137–142.

Phillips, G.N., Lattman, E.E., Cummins, P., Lee, K.Y. and Cohen, C., 1979. Crystal structure and molecular interactions of tropomyosin. Nature 278:413–417.

Philpott, D.E., Kahlbrock, M. and Szent-Gyorgyi, A.G., 1960. Filamentous organization of molluscan muscles. J. Ultrastruct. Res. 3:254–269.

Potter, J.D., 1974. The content of troponin, tropomyosin, actin and myosin in rabbit skeletal muscle fibers. Arch. Biochem. Biophys. 162:436–441.

Rall, J.A., 1981. Mechanics and energetics of contraction in striated muscle of the sea scallop, *Placopecten magellanicus*. J. Physiol. 321:287–295.

Rayment, I., 1996. The structural basis of the myosin ATPase activity. J. Biol. Chem. 271:15850–15853.

Rayment, I. and Holden, H.M., 1994. The three-dimensional structure of a molecular motor. TIBS. 19:129–134.

Rayment, I., Holden, H.M., Whittaker, M., Yohn, C.B., Lorenz, M., Holmes, K.C. and Milligan, R.A., 1993b. Structure of the actin-myosin complex and its implications for muscle contraction. Science 261:58–65.

Rayment, I., Rypniewski, W.R., Schmidt-Base, K., Smith, R., Tomchick, D.R., Benning, M.M., Winkelmann, D.A., Wesenberg, G. and Holden, H.M., 1993a. Three-dimensional structure of myosin subfragment-1: a molecular motor. Science 261:50–58.

Reinach, F.C., Nagai, K. and Kendrick-Jones, J., 1986. Site-directed mutagenesis of the regulatory light-chain Ca^{2+}/Mg^{2+} binding site and its role in hybrid myosins. Nature 322:80–83.

Reisler, E., Burke, M., Himmelfarb, S. and Harrington, W.F., 1974. Spatial proximity of the two essential sulfhydryl groups of myosin. Biochemistry 13:3837–3840.

Roopnarine, O., Szent-Gyorgyi, A.G. and Thomas, D.D., 1998. Microsecond rotational dynamics of spin-labelled myosin regulatory light chain induced by relaxation and contraction of scallop muscle. Biochemistry 37:14428–14436.

Ruegg, J.C., 1961. On the tropomyosin-paramyosin system in relation to the viscous tone of lamellibranch catch muscle. Proc. Roy. Soc. Ser. B 154:224–249.

Ruegg, J.C., 1964. Tropomyosin-paramyosin system and "prolonged contraction" in a molluscan smooth muscle. Proc. Roy. Soc. Ser. B 160:536–542.

Ruegg, J.C., 1965. Physiologie und Biochemie des Sperrtonus. Helvetica Physiologica et Pharmacologica Acta (Suppl XVI):1–76.

Ruegg, J.C., 1971. Smooth muscle tone. Physiol. Rev. 51:201–248.

Ruegg, J.C., 1992. Calcium in muscle contraction. Second Edition. Springer-Verlag, Berlin.

Sanger, J.W., 1971. Sarcoplasmic reticulum of the cross-striated adductor muscle of the Bay scallop, *Aequipecten irradians*. Z. Zellforsch. 127:314–322.

Schutt, C.E., Myslik, J.C., Rozycki, M.D., Goonesekere, N.C. and Lindberg, U., 1993. Structure of crystalline profilin-β-actin. Nature 365:810–816.

Sellers, J.R., Wang, F. and Chantler, P.D., 2003. Trifluoperazine inhibits the MgATPase activity and *in vitro* motility of conventional and unconventional myosins. J. Mus. Res. Cell Mot. 24:579–585.

Sellers, J.R., Chantler, P.D. and Szent-Gyorgyi, A.G., 1980. Hybrid formation between scallop myofibrils and foreign regulatory light-chains. J. Mol. Biol. 144:223–245.

Sellers, J.R. and Homsher, E., 1991. A giant step for myosin. Curr. Biol. 1:347–349.

Shih, W.M., Gryczynski, Z., Lakowicz, J.R. and Spudich, J.A., 2000. A FRET-based sensor reveals large ATP hydrolysis-induced conformational changes and three distinct states of the molecular motor myosin. Cell 102:683–694.

Shiraishi, F., Morimoto, S., Nishita, K., Ojima, T. and Ohtsuki, I., 1999. Effects of removal and reconstitution of myosin regulatory light chain and troponin C on the Ca(2+)-sensitive ATPase activity of myofibrils from scallop striated muscle. J. Biochem. 126:1020–1024.

Siegman, M.J., Funabara, D., Kinoshita, S., Watabe, S., Hartshorne, D.J. and Butler, T.M., 1998. Phosphorylation of a twitchin-related protein controls catch and calcium sensitivity of force production in invertebrate smooth muscle. Proc. Natl. Acad. Sci. USA 95:5383–5388.

Siegman, M.J., Mooers, S.U., Li, C., Narayan, S., Trinkle-Mulcahy, L., Watabe, S., Hartshorne, D.J. and Butler, T.M., 1997. Phosphorylation of a high molecular weight (~600 kDa) protein regulates catch in invertebrate smooth muscle. J. Mus. Res. Cell Mot. 18:655–670.

Simmons, R.M. and Szent-Gyorgyi, A.G., 1978. Reversible loss of calcium control of tension in scallop striated muscle associated with the removal of regulatory light-chains. Nature 273:62–64.

Simmons, R.M. and Szent-Gyorgyi, A.G., 1980. Control of tension development in scallop muscle fibres with foreign regulatory light-chains. Nature 286:626–628.

Simmons, R.M. and Szent-Gyorgyi, A.G., 1985. A mechanical study of regulation in the striated adductor muscle of the scallop. J. Physiol. 358:47–64.

Sivaramakrishnan, M. and Burke, M., 1982. The free heavy-chain of vertebrate skeletal myosin subfragment-1 shows full enzymatic activity. J. Biol. Chem. 257:1102–1105.

Smith, C.A. and Rayment, I., 1996. X-ray structure of the magnesium (II).ADP.vanadate complex of the *Dictyostelium discoideum* myosin motor domain to 1.9Å resolution. Biochemistry 35:5404–5417.

Sobieszek, A., 1973. The fine structure of the contractile apparatus of the anterior byssus retractor muscle of *Mytilus edulis*. J. Ultrastruct. Res. 43:313–343.

Sohma, H., Inoue, K. and Morita, F., 1988a. A cAMP-dependent regulatory protein for RLC-a myosin kinase catalyzing the phosphorylation of scallop smooth muscle myosin light chain. J. Biochem. 103:431–435.

Sohma, H. and Morita, F., 1986. Purification of a protein kinase phosphorylating myosin regulatory light-chain-a (RLC-a) from a smooth muscle of scallop, *Patinopecten yessoensis*. J. Biochem. 100:1155–1163.

Sohma, H. and Morita, F., 1987. Characterization of regulatory light-chain-a myosin kinase. J. Biochem. 101:497–502.

Sohma, H., Sasada, H., Inoue, K. and Morita, F., 1988b. Regulatory light chain-a myosin kinase (aMK) catalyzes phosphorylation of smooth muscle myosin heavy chains of scallop, *Patinopecten yessoensis*. J. Biochem. 104:889–893.

313

Sohma, H., Yazawa, M. and Morita, F., 1985. Phosphorylation of regulatory light-chain-a (RLC-a) in smooth muscle myosin of scallop, *Patinopecten yessoensis*. J. Biochem. 98:569–572.

Sperling, J.E., Feldmann, K., Meyer, H., Jahnke, U. and Heilmeyer, L.M.G., 1979. Isolation, characterization and phosphorylation pattern of the troponin complexes TI2C and I2C. Eur. J. Biochem. 101:581–592.

Spudich, J.A., 1994. How molecular motors work. Nature 372:515–518.

Squire, J.M. and Morris, E.P., 1998. A new look at thin filament regulation in vertebrate skeletal muscle. FASEB J. 12:761–771.

Stafford, W.F., Jacobsen, M.P., Woodhead, J., Craig, R., O'Neall-Hennessey, E. and Szent-Gyorgyi, A.G., 2001. Calcium-dependent structural changes in scallop heavy meromyosin. J. Mol. Biol. 307:137–147.

Stafford, W.F., Szentkiralyi, E.M. and Szent-Gyorgyi, A.G., 1979. Regulatory properties of single-headed fragments of scallop myosin. Biochemistry 18:5273–5280.

Stanners, P.J. and Bagshaw, C.R., 1987. Interaction of scallop heavy meromyosin with pyrene-labelled actin. Biochem. Soc. Trans. 15:901–903.

Stein, L.A., Schwartz, R.P., Chock, P.B. and Eisenberg, E., 1979. Mechanism of actomyosin adenosine triphosphatase. Evidence that ATP hydrolysis can occur without dissociation of the actomyosin complex. Biochemistry 18:3895–3909.

Stewart, M., 1975. Tropomyosin: Evidence for no stagger between chains. FEBS Lett. 53:5–7.

Stewart, M. and McLachlan, A.D., 1975. Fourteen actin-binding sites on tropomyosin? Nature 257:331–333.

Sugi, H. and Suzuki, S., 1978. The nature of potassium- and acetylcholine-induced contractures in the anterior byssal retractor muscle of *Mytilus edulis*. Comp. Biochem. Physiol. C 61:275–279.

Szent-Gyorgyi, A.G., 1976. Comparative survey of the regulatory role of calcium in muscle. In: Calcium in Biological Systems, Symp. Soc. Exp. Biol. 30:335–347.

Szent-Gyorgyi, A.G. and Chantler, P.D., 1994. Control of contraction by calcium binding to myosin. In: A.G Engel and C. Franzini-Armstrong (Eds.). Myology. Vol. 1, Second Edition. McGraw-Hill, Inc., New York. pp. 506–528.

Szent-Gyorgyi, A.G., Cohen, C. and Kendrick-Jones, J., 1971. Paramyosin and the filaments of molluscan "Catch" muscles. J. Mol. Biol. 56:239–258.

Szent-Gyorgyi, A.G., Szentkiralyi, E.M. and Kendrick-Jones, J., 1973. The light-chains of scallop myosin as regulatory subunits. J. Mol. Biol. 74:179–203.

Szentkiralyi, E.M., 1984. Tryptic digestion of scallop S-1: Evidence for a complex between the two light-chains and a heavy-chain peptide. J. Mus. Res. Cell Mot. 5:147–164.

Szentkiralyi, E.M., 1987. An intact heavy-chain at the actin-subfragment-1 interface is required for ATPase activity of scallop myosin. J. Mus. Res. Cell Mot. 8:349–357.

Takahashi, M., Fukushima, Y., Inoue, K., Hasegawa, Y., Morita, F. and Takahashi, K., 1989. Ca^{2+}-sensitive transition in the molecular conformation of molluscan muscle myosins. J. Biochem. 105:149–151.

Takahashi, M. and Morita, F., 1986. An activating factor (tropomyosin) for the superprecipitation of actomyosin prepared from scallop adductor muscles. J. Biochem. 99:339–347.

Takahashi, M., Sohma, H. and Morita, F., 1988. The steady-state intermediate of scallop smooth muscle myosin ATPase and effect of light-chain phosphorylation. A molecular mechanism for catch contraction. J. Biochem. 104:102–107.

314

Tanaka, H., Ojima, T. and Nishita, K., 1998. Amino acid sequence of troponin-I from Akazara scallop striated adductor muscle. J. Biochem. 124:304–310.

Taylor, E.W., 1979. Mechanism of actomyosin ATPase and the problem of muscle contraction. Crit. Rev. Biochem. 6:103–165.

Thompson, R.J., Livingstone, D.R. and De Zwaan, A., 1980. Physiological and biochemical aspects of the valve snap and valve closure responses in the giant scallop, *Placopecten magellanicus*. J. Comp. Physiol. 137:97–104.

Tobacman, L.S. and Butters, C.A., 2000. A new model of cooperative myosin-thin filament binding. J. Biol. Chem. 275:27587–27593.

Toyoshima, Y.Y., Kron, S.J. and Spudich, J.A., 1990. The myosin step size: Measurement of the unit displacement per ATP hydrolyzed in an *in vitro* assay. Proc. Natl. Acad. Sci. USA 87:7130–7134.

Trinick, J.A., 1981. End-filaments: A new structural element of vertebrate skeletal muscle thick filaments. J. Mol. Biol. 151:309–314.

Trinick, J.A., 1992. Understanding the functions of titin and nebulin. FEBS Lett. 307:44–48.

Trybus, K.M., 1994. Regulation of expressed truncated smooth muscle myosins. J. Biol. Chem. 269:20819–20822.

Trybus, K.M., Freyzon, Y., Faust, L.Z. and Sweeney, H.L., 1997. Spare the rod, spoil the regulation: Necessity for a myosin rod. Proc. Natl. Acad. Sci. USA 94:48–52.

Twarog, B.M., 1967a. The regulation of catch in molluscan muscle. J. Gen. Physiol. 50:157–169.

Twarog, B.M., 1967b. Factors influencing contraction and catch in *Mytilus* smooth muscle. J. Physiol. 192:847–856.

Twarog, B.M., 1976. Aspects of smooth muscle function in molluscan catch muscle. Physiol. Rev. 56:829–838.

Twarog, B.M. and Cole, R.A., 1973. Relaxation of catch in a molluscan smooth muscle. II. Effects of serotonin, dopamine and related compounds. Comp. Biochem. Physiol. A. 46:831–835.

Uyeda, T.Q.P., Warrick, H.M., Kron, S.J. and Spudich, J.A., 1991. Quantized velocities at low myosin densities in an *in vitro* motility assay. Nature 352:307–311.

Vale, R.D., Szent-Gyorgyi, A.G. and Sheetz, M.P., 1984. Movement of scallop myosin on Nitella actin filaments: Regulation by calcium. Proc. Natl. Acad. Sci. USA 81:6775–6778.

Vandekerckhove, J. and Weber, K., 1979. The amino-acid sequence of actin from chicken skeletal muscle actin and chicken gizzard smooth muscle actin. FEBS Lett. 102:219–222.

Vandekerckhove, J. and Weber, K., 1984. Chordate muscle actins differ distinctly from invertebrate muscle actins. J. Mol. Biol. 179:391–413.

Vassylyev, D.G., Takeda, S., Wakatsuki, S., Maeda, K. and Maeda, Y., 1998. Crystal structure of troponin C in complex with troponin I fragment at 2.3 A resolution. Proc. Natl. Acad. Sci. USA 95:4847–4852.

Vibert, P., 1992. Helical reconstruction of frozen-hydrated scallop myosin filaments. J. Mol. Biol. 223:661–671.

Vibert, P. and Castellani, L., 1989. Substructure and accessory proteins in scallop myosin filaments. J. Cell Biol. 109:539–547.

Vibert, P. and Craig, R., 1982. Three-dimensional reconstruction of thin filaments decorated with a Ca^{2+}-regulated myosin. J. Mol. Biol. 157:299–319.

Vibert, P. and Craig, R., 1983. Electron microscopy and image analysis of myosin filaments from scallop striated muscle. J. Mol. Biol. 165:303–320.

Vibert, P. and Craig, R., 1985. Structural changes that occur in scallop myosin filaments upon activation. J. Cell Biol. 101:830–837.

Vibert, P., Edelstein, S.M., Castellani, L. and Elliott, B.W., 1993. Mini-titins in striated and smooth molluscan muscles: structure, location and immunological crossreactivity. J. Mus. Res. Cell Mot. 14:598–607.

Vibert, P., Haselgrove, J.C., Lowy, J. and Poulsen, F.R., 1972. Structural changes in actin-containing filaments of muscle. J. Mol. Biol. 71:757–767.

Wagner, P.D. and Giniger, E., 1981. Hydrolysis of ATP and reversible binding to F-actin by myosin heavy-chains free of all light-chains. Nature 292:560–562.

Wallimann, T. and Szent-Gyorgyi, A.G., 1981. An immunological approach to myosin light-chain function in thick filament linked regulation. 2. Effects of ant-scallop myosin light-chain antibodies. Possible regulatory role for the essential light-chain. Biochemistry 20:1188–1197.

Walmsley, A.R., Evans, G.E. and Bagshaw, C.R., 1990. The calcium ion dependence of scallop myosin ATPase activity. J. Mus. Res. Cell Mot. 11:512–521.

Watabe, S. and Hartshorne, D.J., 1990. Paramyosin and the catch mechanism. Comp. Biochem. Physiol. 96B:639–646.

Watabe, S., Tsuchiya, T. and Hartshorne, D.J., 1989. Phosphorylation of paramyosin. Comp. Biochem. Physiol. 94B:813–821.

Weisel, J.W., 1975. Paramyosin segments: Molecular orientation and interactions in invertebrate muscle filaments. J. Mol. Biol. 98:675–681.

Wells, C. and Bagshaw, C.R., 1984. The Ca^{2+} sensitivity of the actin-activated ATPase of scallop heavy meromyosin. FEBS Lett. 168:260–264.

Wells, C. and Bagshaw, C.R., 1985. Calcium regulation of molluscan myosin ATPase in the absence of actin. Nature 313:696–697.

Wells, C., Warriner, K.E. and Bagshaw, C.R., 1985. Fluorescence studies on the nucleotide and Ca^{2+}-binding domains of molluscan muscle. Biochem. J. 231:31–38.

Wells, J.A. and Yount, R.G., 1980. Reaction of 5-5'-dithiobis (2-nitrobenzoic acid) with myosin subfragment one: evidence for formation of a single protein disulfide with trapping of metal nucleotide at the active site. Biochemistry 19:1711–1717.

Wendt, T., Taylor, D., Messier, T., Trybus, K.M. and Taylor, K.A., 1999. Visualization of head-head interactions in the inhibited state of smooth muscle myosin. J. Cell Biol. 147:1385–1389.

Wendt, T., Taylor, D., Trybus, K.M. and Taylor, K.A., 2001. Three dimensional image reconstruction of dephosphorylated smooth muscle heavy meromyosin reveals asymmetry in the interaction between myosin heads and placement of subfragment 2. Proc. Natl. Acad. Sci. USA 98:4361–4366.

White, H.D., Belknap, B. and Webb, M.R., 1997. Kinetics of nucleoside triphosphate cleavage and phosphate release steps by associated rabbit skeletal actomyosin, measured using a novel fluorescent probe for phosphate. Biochemistry 36:11828–11836.

Whittaker, M., Wilson-Kubalek, E.M., Smith, J.E., Faust, L., Milligan, R.A. and Sweeney, H.L. 1995. A 35-Å movement of smooth muscle myosin on ADP release. Nature 378:748–753.

316

Wilkens, L.A., 1991. Neurobiology and behaviour of the scallop. In: S.E. Shumway (Ed.). Scallops: Biology, Ecology and Aquaculture. Developments in Aquaculture and Fisheries Science, Vol. 21. Elsevier, Amsterdam. pp. 429–469.

Winton, F.R., 1937. The changes in viscosity of an unstriated muscle (*Mytilus edulis*) during and after stimulation with alternating, interrupted and uninterrupted direct currents. J. Physiol. 88:492–511.

Woods, E.F., 1969. Comparative physiochemical studies on vertebrate tropomyosins. Biochemistry 8:4336–4344.

Woods, E.F., 1976. The conformational stabilities of tropomyosins. Aust. J. Biol. Sci. 29:405–418.

Woods, E.F. and Pont, M.J., 1971. Characterization of some invertebrate tropomyosins. Biochemistry 10:270–276.

Wray, J.S., 1979. Structure of the backbone in myosin filaments of muscle. Nature 277:37–40.

Wray, J.S., 1982. Organization of myosin in invertebrate thick filaments. In: B.M. Twarog, R.J.C. Levine and M.M. Dewey (Eds.). Basic Biology of Muscles: A Comparative Approach. Raven Press, New York. pp. 29–36.

Wray, J.S., Vibert, P. and Cohen, C., 1975. Diversity of crossbridge configurations in invertebrate muscles. Nature 257:561–564.

Wu, X., Clack, B.A., Zhi, G., Stull, J.T. and Cremo, C.R., 1999. Phosphorylation-dependent structural changes in the regulatory light chain domain of smooth muscle heavy meromyosin. J. Biol. Chem. 274:20328–20335.

Xie, X., Harrison, D.H., Schlichting, I., Sweet, R.M., Kalabokis, V.N., Szent-Gyorgyi, A.G. and Cohen, C., 1994. Structure of the regulatory domain of scallop myosin at 2.8Å resolution. Nature 368:306–312.

Yamada, A., Yoshio, M. and Nakayama, H., 1997. Bidirectional movement of actin filaments along long bipolar tracks of oriented rabbit skeletal muscle myosin molecules. FEBS Lett. 409:380–384.

Yamada, A., Yoshio, M., Oiwa, K. and Nyitray, L., 2000. Catchin, a novel protein in molluscan catch muscles, is produced by alternative splicing from the myosin heavy chain gene. J. Mol. Biol. 295:169–178.

Yamakita, Y., Yamashiro, S. and Matsumura, F., 1992. Characterization of mitotically phosphorylated caldesmon. J. Biol. Chem. 267:12022–12029.

Yanagida, T., Arata, T. and Oosawa, F., 1985. Sliding distance of actin filament induced by a myosin crossbridge during one ATP hydrolysis cycle. Nature 316:366–369.

Yang, Z. and Sweeney, H.L., 1995. Restoration of phosphorylation-dependent regulation to the skeletal muscle myosin regulatory light chain. J. Biol. Chem. 270:24646–24649.

Yount, R.G., Lawson, D. and Rayment, I., 1995. Is myosin a "back-door" enzyme? Biophys. J. 68:44S–49S.

Yuasa, H.J. and Takagi, T., 2000. The genomic structure of the scallop, *Patinopecten yessoensis*, troponin C gene: a hypothesis for the evolution of troponin C. Gene 245:275–281.

AUTHOR'S ADDRESS

Peter D. Chantler - Unit of Veterinary Molecular and Cellular Biology, Royal Veterinary College, University of London, Royal College Street, London NW1 0TU, United Kingdom (E-mail: pchant@rvc.ac.uk)

Scallops: Biology, Ecology and Aquaculture
S.E. Shumway and G.J. Parsons (Editors)
© 2006 Elsevier B.V. All rights reserved.

Chapter 5

Neurobiology and Behaviour of the Scallop

Lon A. Wilkens

5.1 INTRODUCTION

Perhaps the most striking feature of the scallop is the presence of eyes clearly visible around the margin of the shell. This is a somewhat unexpected feature for a bivalve representative of a molluscan class otherwise noted for the reduction of sense organs and cerebral nervous system. An additional 'surprise' is that scallops are able to swim by clapping the valves of the shell in rapid succession. Furthermore, swimming can be elicited by the touch of a starfish in what is now a classic illustration of an adaptive predator-prey interaction. Most of the scallop literature in the area of neurobiology and behaviour is related to vision and swimming, the topics around which this chapter will be cantered.

The behavioural repertoire of the scallop, which includes visual responses, and locomotion in conjunction with both escape and migration, is noteworthy among bivalves and corresponds with the free-living mode of existence of this animal. This contrasts sharply with the majority of bivalves either fixed more or less permanently to the substrate by cement or byssal formations, or buried in the substrate. The few additional bivalve species which exhibit overt behaviours, aside from simple siphon retraction and/or valve adduction, are also free living to varying extents. Notable examples include the "swash riding" coquina clam, *Donax* (Ellers 1987), active burrowers such as razor (Schneider 1982) and surf clams which, like cardiids, produce vigorous somersault manoeuvres to escape predation (Feder 1972; Prior et al. 1979), and the flame scallop, *Lima*, which also swims. The giant clam, *Tridacna*, is exceptional in exhibiting various visual and nonvisual behaviours even though sessile in habit (Stasek 1965; Wilkens 1987).

5.2 THE VISUAL SYSTEM

5.2.1 Functional Anatomy of the Eyes

Scallop eyes are located at the tip of a short stalk that extends outward from the middle fold of the mantle lining the circumference of the shell. The blue eyes, 1 to 1.5 mm in diameter, are found in association with both the upper (left) and lower (right) valves, although they occur in greater number and size in the upper mantle (Gutsell 1930). Differences in eye number are especially true of species in which the upper valve is flattened (slightly concave) (Dakin 1909), such as in *Pecten ziczac* where the eyes on the upper valve can number more than twice those on the lower valve (Wilkens 1981). In a detailed description of the eyes, Dakin (1910a) also points out that eyes vary in number from animal to animal, as well as between species. In *P. opercularis*, for example, he found no correlation between the size of the shell and the number of eyes. In fact, the largest eye count (111) was from one of the smaller scallops in his study. Also, eyes from the upper mantle tend to be slightly larger, and smaller eyes are spaced irregularly among those of full size in both regions (Dakin 1910a).

That there is no set pattern or number of eyes is supported by observations concerning the formation of eyes during development and regeneration. Butcher (1930) concludes that "new eyes . . . . make their appearance as space for them is provided". In his experiments on *Pecten* (= *Argopecten*) *gibbus*, eyes regenerated within forty days and their development, beginning as buds along the ophthalmic groove at the base of the tentacles, followed the same course as eyes appearing during growth. To determine the influence of nervous tissue on regeneration, Butcher performed a series of nerve-ganglion ablation and removal experiments and found that none of these procedures had much effect, that is, the eyes regenerated normally. In fact, eyes develop with little difficulty when transplanted at an early stage ("anlagen") onto the epithelial surface of the gonads. Growth of the optic nerve into the gonad also occurs when fully developed eyes are transplanted. It may be that nervous tissue, particularly from the eyes, inhibits the induction of anlagen during normal development, with new eyes forming only as the existing ones separate as the shell enlarges.

The eyes of scallops have attracted the attention of anatomists since the late 18th century primarily due to their resemblance (albeit superficial) to the camera eye of vertebrates (descriptive terms such as iris, rods, and ganglion cells appeared in the older literature reviewed by Dakin 1910a). The essential features of the eye were, in fact, clarified by Dakin and his diagram in sagittal section (Dakin 1910a; redrawn in 1928) has been widely reproduced. The following description is based on this model (Fig. 5.1) with ultrastructural details provided by Barber et al. (1967) (Fig. 5.2).

The functional components of the eye occupy an optic capsule of connective tissue and epithelial cells continuous with the stalk and mantle. Small longitudinal muscle fibres around the stalk allow the eyes to contract from a stimulus but diagonal fibres, present in the constantly moving tentacles, are absent. The epithelial cells are heavily pigmented on the sides of the eye, but distally they form a clear cornea with microvilli at the surface. A lens, comprised of irregular cells with few inclusions, lies beneath the cornea and is bounded laterally by a fluid-filled space.

Lining the back of the eye are the retina, the reflecting argentea and pigment layer. Unlike Dakin's illustration, the retina is not separated by the anterior fluid chamber but, rather, is in direct contact with the lens (*cf.* Figs. 5.2, 5.3A), separated only by a 2 μm-thick basement membrane. This is an important distinction when considering the optical properties of the eye (see below).

The retina itself is a highly specialised structure, the complexity of which would be puzzling with regard to the visual requirements of sessile bivalves (the thorny oyster *Spondylus* has eyes identical to those of the Pectinidae; Dakin 1928).

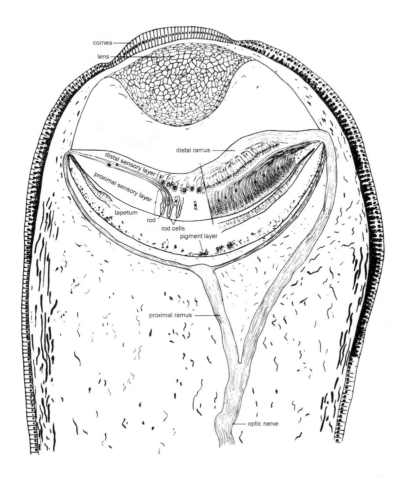

Figure 5.1. Diagram of the eye of *Pecten*. [From Bullock and Horridge 1965, after Dakin 1928]

Two distinct layers of retinal cells are present, the primary difference between which is the presence of ciliary versus microvillar sensory regions (supporting glial cells are also present). In the distal layer, each cell gives rise to approximately 100 modified cilia packed in a lamellar array on the distal surface facing the lens and cornea. Axons from these receptors arise lateral to the cilia and, with their fibres passing in front of the retina, collect at the retinal margin on the side of the shell to form the distal branch of the optic nerve. In the proximal layer, the outer segments of the receptors are covered with microvilli facing proximally, and contain one or two cilia with bundles of filaments projecting toward their base. The proximal receptors, with a rhabdom-like interdigitation of closely packed microvilli from adjacent cells, are similar to the photoreceptors of gastropod and cephalopod molluscs (see Barber et al. 1967, for references). The cilia of both proximal and distal receptors have a 9 + 0 fibril arrangement and basal bodies, but lack fibril arms and root systems. Axons from proximal receptors extend around the sides

320

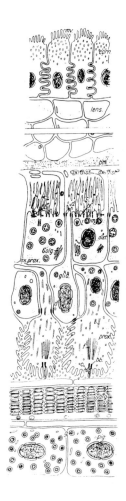

Figure 5.2. Diagrammatic sagittal section of the eye based on ultrastructural details. Beginning at the outer margin is the cornea (corn.), lens, basement membrane (bm), distal axon, (ax. dist.), distal cell (dist.), proximal axon (ax. prox.), glia, proximal cell (prox.), argentea (arg.), and pigment cell layer (pig.). [From Barber et al. 1967]

of the retina and join at the base of the capsule to form the proximal branch of the optic nerve. Thus, each retinal layer has a separate nerve branch, which extends 1 to 2 mm before joining to form the optic nerve (Hartline 1938b). Further, no synapses have been observed in the retina (Miller 1958; Barber et al. 1967) so that each receptor cell is represented by its own axon in the optic nerve.

The argentea (also tapetum), the reflecting layer responsible for the characteristic bright iridescence of the pupil, lies beneath the retina. This is a single layer of cells containing an array of flattened, membrane-bound guanine crystals and is the biological

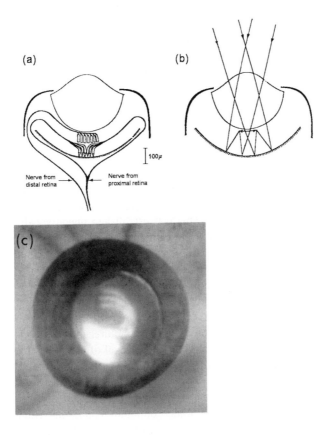

(a)

(b)

Nerve from distal retina

Nerve from proximal retina

I 100µ

(c)

Figure 5.3. Diagram of the eye and retinal layers (a) and in relation to the optical system (b). [From Land 1966] (c) Close-up of the eye in *P. maximus* illustrating focused image of a hand. [M.F. Land, unpublished photograph]

equivalent of a multilayered interference mirror (Land 1978). With its precise hemispheric shape, the argentea acts as a concave mirror with a focal length of approximately 200 µm (Land 1965). Thus, an inverted image is formed within the eye and at the level of the distal retina (Fig. 5.3b, c). The focal length of the lens is 1.5 mm and, since it lies in contact with the retina, its image would form behind the eye. The only optical function of the lens may be to correct slightly for aberration in the argentea. Thus, only the distal layer of the retina is capable of reacting to a focused image from a distant object, an important consideration for the physiological properties of the eye.

Finally, a layer of pigmented cells containing EM-dense granules lines the back of the eye.

5.2.2 Retinal Physiology

In view of the anatomic simplicity of the scallop retina, with its separate proximal and distal branches of the optic nerve, Hartline (1938b) believed that this preparation could provide insights into the mechanisms of 'off' discharges like those seen previously in the vertebrate retina (Hartline 1938a). An 'off' discharge in the scallop seemed likely since these animals, like many invertebrates, respond vigorously to shadows — in this case by shell closure.

The physiological mechanisms, which lead to spike discharges in the optic nerves of vertebrates, are actually quite different from those in the scallop. Nevertheless, Hartline (1938b) demonstrated clearly that the two layers of the scallop retina have distinct physiological properties; axons from the proximal layer produce a train of spikes when the eye is illuminated whereas axons from the distal layer respond vigorously when the light is turned off or reduced in intensity. This was established unequivocally by cutting the distal branch of the optic nerve to determine whether or not responses recorded from the common optic nerve had been eliminated (optic fibres are more accessible in the pallial nerves of the scallop; the relevant neural pathways are described briefly in a following section, and in more detail by Beninger and Le Pennec, this volume).

The 'on' response of the proximal retinal cells is typical of invertebrate photoreceptors in that spikes are elicited by light. The 'off' response, on the other hand, is representative of a new and unique type of primary photoreceptor which Hartline (1938b) described as "capable of being stimulated by the removal of an environmental agent." In fact, light itself is inhibitory, a physiological corollary to the activity that ensues at dimming or light offset. For example, re-illumination of the eye during an 'off' response rapidly terminates spiking with a latency of as little as 100 ms, approximately half the minimum latency for burst initiation under optimum conditions (Land 1966). The physiological complexity of 'off' receptor activity is suggested further by the fact that light itself is a prerequisite for the response, with burst strength and rapid onset both dependent on light adaptation (Land 1966). Thus, light has both excitatory and inhibitory effects, with inhibition being more direct in that it overrides the excitatory response. "Primary inhibition," as this phenomenon has been termed (Kennedy 1960), appears to be characteristic of all molluscan 'off' receptors, although the mechanisms may not be uniform. For example, the excitatory and inhibitory components of the 'off' response in *Spisula* are wavelength dependent with inhibition predominant at shorter wavelengths (540 nm) and excitation at longer wavelengths (600 nm) (Kennedy 1960). Whereas this indicates the presence of two photopigments within the same cell, a more recent study on *Mercenaria* (Wiederhold et al. 1973) finds evidence for only a single photopigment. The spectral properties of scallop photoreceptors are considered in the following section.

The functional importance of the 'off' receptors is suggested by the fact that only the distal retina lies in the plane of focus of the argentea. Land (1966) predicted that these receptors alone would respond to movement in the visual field. To test this hypothesis, he stimulated the eyes with a regular pattern of light/dark stripes so that movements produced no net change in overall stimulus intensity. Indeed, when the distal branch of the optic nerve was cut (Fig. 5.4), activity seen previously in response to movement was

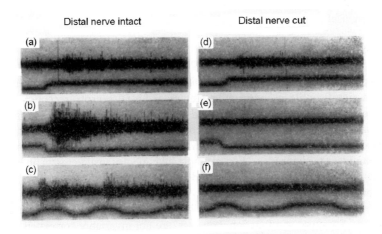

Distal nerve intact   Distal nerve cut

(a)   (d)

(b)   (e)

(c)   (f)

Figure 5.4. Abolition of 'off' and movement responses after cutting distal nerve. A photodiode (lower trace) monitors the stimulus: 'on' stimulus (a and d); 'off' stimulus (b and e); moving stripes (c and f), movement signalled by upward deflection. Screen brightness is 500 cd m^2. [From Land 1966]

abolished, even though an 'on' response persisted for control flashes. Tests using a single light or dark stripe produced similar results. A dark stripe, moved stepwise across the visual field triggered bursts of spikes that correspond to the shadowing of previously illuminated distal receptors; light stripes generated no response for similar patterns of movement unless left in one position for 10 s or more, in which case a sufficient level of light adaptation was achieved to trigger an 'off' response when the stripe was moved (Land 1966). The smallest stripe or degree of movement capable of producing a response subtended 2° at the eye, an angle represented by a distance of 9 μm in the retina. This approximates the spacing of the retinal photoreceptors.

Although the scallop eye is equipped with two layers of retinal cells, only the distal 'off' receptors are positioned for receiving a focused image. Thus, feature detection in the spatial environment is limited to the distal cell layer, with movement represented by the phasic activity of these photoreceptors. That 'off' cells perform this function is consistent with the fact that in bivalves, 'off' receptors featuring "primary inhibition" serve as shadow detectors (Kennedy 1960; Barber and Land 1967; Mpitsos 1973; Wiederhold et al. 1973). In the scallop, these receptors can signal movements in the environment at distances greater than required to cast a direct shadow on the animal, a distinct advantage for reacting to the approach of a fast moving predator. 'On' receptors, on the other hand, are subject to diffuse light and respond only to changes in overall intensity. The function of these receptors is less well understood although in species (*Argopecten irradians*, for example) where tonic background activity persists in the light (Hartline 1938b;

324

McReynolds and Gorman 1970a), these receptors could signal absolute levels of light intensity and be useful in migration and habitat selection.

5.2.3 Receptor Potential Biophysics

Scallop photoreceptors, especially the ciliary 'off' receptors of the distal retina, have been featured significantly in comparative studies of visual sensory receptors. Initially, interest in the biophysical properties stemmed from the fact that photoreceptors based on both ciliary and microvillar structures were contained in the same eye. The comparative significance of this unusual dichotomy relates to the evolution of photoreceptor cells and their physiological mechanisms. For example, the photosensitive structures of most invertebrate photoreceptors are composed of microvilli whereas all vertebrate photoreceptors are derived from a ciliary base (Eakin 1965). Furthermore, most invertebrate photoreceptors are depolarised by light whereas those of vertebrates are hyperpolarised. Structure and function, however, are not invariably correlated (Järvilehto 1979).

Intracellular recordings from scallop photoreceptors (Gorman and McReynolds 1969; McReynolds and Gorman 1970a) showed that cells in the proximal layer of the retina ('on' receptors) are depolarised by light and that cells in the distal retina ('off' receptors) are hyperpolarised (Fig. 5.5). Although the receptor potentials in both types are graded, with amplitudes proportional to the logarithm of intensity, depolarising 'on' receptors are at least 2 log units more sensitive to light. Maximum sensitivity in each receptor occurs at a wavelength of 500 nm, which suggests that the same photopigment is used by each cell despite their difference in response polarity. Axonal spikes invading the cell soma are seen occasionally and arise on the depolarising phase of the waveform, that is, at the onset in depolarising and at the offset in hyperpolarising receptor potentials (Fig. 5.6).

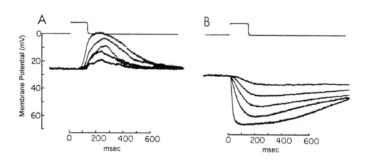

Figure 5.5. Responses of depolarising and hyperpolarising cells from the same eye to brief light flashes of varying intensity. Threshold intensity is log I = -5.7 for the depolarising cell (A) and log I = -3.0 for the hyperpolarising (B) cell. The upper trace is the output of a photocell monitoring the light flash. [From McReynolds and Gorman 1970a]

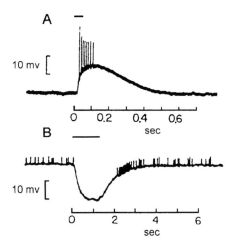

Figure 5.6. Axonal spikes recorded in depolarising (A) and hyperpolarising (B) cells. The duration of the light flash is indicated by horizontal lines. Intensity is log I = -4.2 (A) and log I = -2.4 (B). [From McReynolds and Gorman 1970a]

The receptor potentials in both cells are due to an increase in the ionic current, which crosses the cell membrane (McReynolds and Gorman 1970b). For depolarising receptor potentials, an increase in membrane permeability is consistent with the mechanisms of excitation in other invertebrate photoreceptors. An increase in membrane permeability for hyperpolarising potentials, however, differs from that of vertebrate photoreceptors where hyperpolarisation is due to a decrease in permeability. A detailed analysis of the ionic and spectral properties of the hyperpolarising distal receptors (Gorman and McReynolds 1978; Cornwall and Gorman 1979, 1983a, 1983b) has provided insights into the physiological mechanisms underlying "primary inhibition" in molluscan 'off' receptors. In addition, these results add substantially to our understanding of the mechanisms whereby photo-isomerisation of visual pigments is coupled to the electrical response.

The hyperpolarising response is primarily a light-activated increase in potassium current (Gorman and McReynolds 1978; Cornwall and Gorman 1983a). The reversal potentials for peak hyperpolarisation bear a Nernstian relationship with external potassium concentrations and attest to the specificity of the response as an outward potassium current. At light offset, repolarisation is due not only to a slow decay in potassium current but also includes a spike-like component (Fig. 5.7A). Various experiments (Cornwall and Gorman 1979) have established that this is a calcium spike and that it is partially obscured in normal circumstances by the residual voltage-dependent potassium current (cf. Fig. 5.7B, where potassium current has been blocked by hyperpolarising current injections). The voltage-dependent calcium current (spike) is also partially inactivated at low membrane potentials due to a sodium leakage current (cf. Fig. 5.7C, 7D). The existence of the calcium spike at light offset acts to increase the rate of repolarisation and may

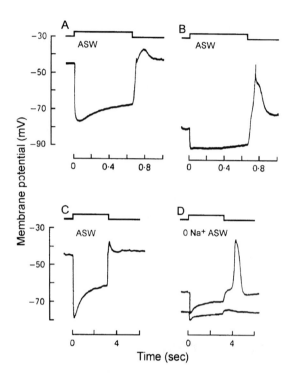

Figure 5.7. Calcium spike in hyperpolarising photoreceptors. The rapid repolarisation and small overshooting peak are indications of the spike in normal artificial seawater (ASW), (A) and (C). Hyperpolarisation of the membrane ($i = -2.0 \times 10^{-10}$ A) to near the reversal potential for potassium reduces the amplitude of the receptor potential and reveals a spike (B). Replacement of sodium ions with choline (D) also increases the membrane potential and, with the additive effect of depolarising current ($i = -3 \times 10^{-11}$ A), reveals a spike. [From Cornwall and Gorman 1979]

therefore be responsible for triggering the discharge of axonal spikes such as those recorded in earlier studies of optic nerve responses (Hartline 1938b; Land 1966). The calcium spike also provides a mechanism for understanding "primary inhibition" in that relative activation or inactivation of voltage-dependent calcium channels, as influenced by the graded amplitude of the receptor potential and the duration of light-mediated hyperpolarising potassium currents, determines the rate of rise and therefore the burst frequency.

Spectral properties similar to those first described in *Spisula* (Kennedy 1960), are also important considerations for understanding the physiology of 'off' receptors. Cornwall and Gorman (1983b) have demonstrated convincingly that shorter (blue) and longer (red) wavelengths, respectively, are responsible for the inhibitory and excitatory components of the receptor potential. These effects can be seen, independent of the ionic currents of the receptor potential, in the waveform of the early receptor potential (ERP), a small 2–4 mV potential associated directly with a charge shift in the pigment molecules resulting from an intense light flash. Blue light (500 nm) gives a negative ERP whereas red light (575 nm) gives a positive ERP (Fig. 5.8). The negative ERP, it can be concluded (Cornwall and Gorman 1983b), results from the photo-isomerisation of rhodopsin to metarhodopsin and the positive ERP in turn is the result of the regeneration of rhodopsin from metarhodopsin.

Spectral effects of the same polarity are seen in the receptor potential itself. A blue light flash initiates a prolonged hyperpolarising afterpotential (Fig. 5.9) which decays slowly, up to 5 min for an intense flash, and which can be terminated by a red light flash. Similarly, these hyperpolarising and depolarising responses can be attributed to the isomerisation of photopigment to metarhodopsin at short wavelengths, and the regeneration of rhodopsin at long wavelengths of light. Thus, light absorption by different photopigment states triggers opposite long-lasting, light-dependent shifts in membrane potential, both of which are mediated by currents carried through the potassium channels. This is the first demonstration in a photosensory system that light absorption by metarhodopsin directly affects ionic conductance.

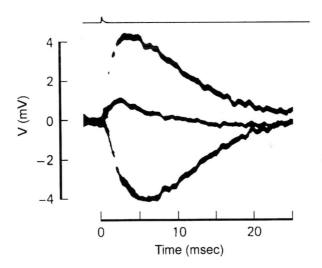

Figure 5.8. Effects of wavelength of light on the early receptor potential (ERP). White light, after preadaptation with white light, produces a small depolarisation (middle trace). Maximum responses are produced by red light (positive ERP), after preadaptation with blue light, and by blue light (negative ERP) after preadaptation with red light. Preadaptation with red and blue light, respectively, insures maximum concentrations of rhodopsin and metarhodopsin photopigments prior to the test flash. Response latency is negligible, as indicated by the light monitor (top trace). [From Cornwall and Gorman 1983b]

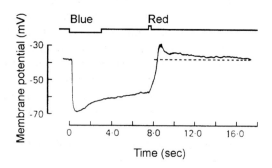

Figure 5.9. Receptor potential response to monochromatic light. A prolonged hyperpolarising afterpotential is produced by a 3 s blue flash and suppressed by a 200 ms red flash. A burst of small spikes riding the crest of the repolarising response and a prolonged depolarising afterpotential are also produced by the red flash. [From Cornwall and Gorman 1983b]

Cornwall and Gorman (1983b) also reported that shadow responses in the scallop, normally triggered by a decrease in illumination using white light, could be elicited instead by shining red light on two or three eyes when these had been preadapted by blue light. This is a behavioural confirmation of the spectral properties illustrated above (Fig. 5.9). Thus, we have a fairly complete description of the physiological properties of the 'off' receptors in the scallop distal retina, a system uniquely designed to signal behaviourally relevant information (shadows) without the interplay of synapses, as seen in other animals.

5.2.4 Anatomy and Physiology of Vision in the Central Nervous System

In contrast to the eyes, the remainder of the scallop visual system is little understood except for anatomical features of the 'optic lobes'. The relative lack of physiological information reflects the general unfavourability of bivalve ganglia for electrophysiological studies. There are no large cells, as in the gastropod CNS, and an insulating glial constituent is often present.

Visual activity travels to the CNS via the pallial nerves (see Beninger and Le Pennec, this volume), nerves which radiate peripherally from the parietovisceral ganglion to the circumpallial nerve at the margin of the mantle and which contain other sensory interneurons and motor fibres. The optic nerves initially contact the circumpallial nerve, raising the possibility that synaptic interactions occur at this level. However, available evidence suggests that the optic nerves are simply ensheathed in and 'follow' the pallial network into the ganglion. For example, while photoreceptor axons from adjacent eyes converge to a single pallial nerve, in my experience one never encounters activity from a single eye in more than one pallial nerve. Diverging pathways would be expected if synaptic interactions between optic nerve fibres and circumpallial neurons were present. In autoradiographic experiments (unpublished results) in which tritiated proline was injected into the eye, optic tracts have not been observed to enter the circumpallial nerve despite the intense labelling of optic nerve fibres as they skirt the circumpallial nerve (Fig. 5.10).

The optic fibres terminate primarily in the two lateral ('optic') lobes of the visceroparietal ganglion. Dakin (1910b) clearly established that these crescent-shaped lobes are specialised as visual centres noting that they are present only in scallops and that they vary in size corresponding to the number of eyes present. For example, the left lobe is distinctly smaller than the right lobe in species having fewer eyes in the upper mantle. In *Argopecten*, with its symmetrical valves, the lobes are nearly identical in size. Dakin also correctly interpreted the densely staining glomeruli characteristic of the lateral lobes as being specialised for visual function, and in a diagram has shown fibres from the pallial nerve branching to form one of these spherical structures. Autoradiographic experiments (Spagnolia and Wilkens 1983) have confirmed these observations (Fig. 5.11), and illustrate directly that pallial fibres originating in the eye are associated with the glomerular structures. However, this study suggests that the optic fibres actually terminate in a subcellular neuropile region above (ventral, anatomically) the glomerulus, in which case the glomerulus would consist primarily of fibrillae from neurons originating in the lateral lobe.

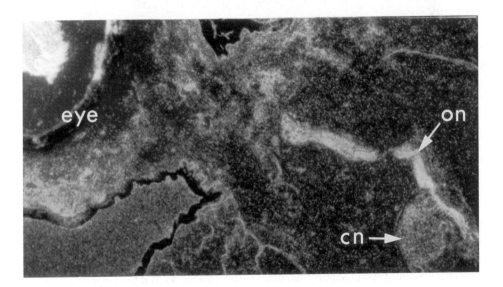

Figure 5.10. Darkfield micrograph of a radial section of the mantle showing the base of an eye (injected with tritiated proline), a segment of the optic nerve (o.n.), and a cross section of the circumpallial nerve (c.n.).

330

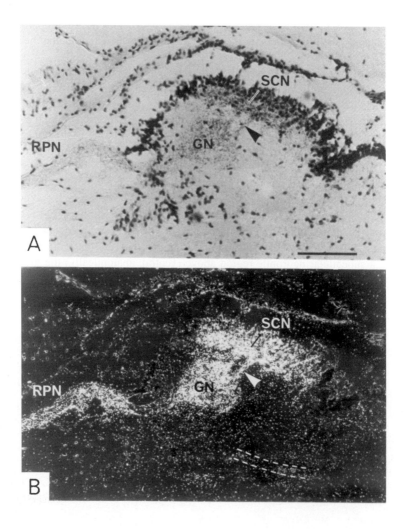

Figure 5.11. Autoradiograph of the lateral lobe two days after intraocular injection of labelled precursor. (A) Bright-field and (B) dark-field micrographs of a thionine-stained transverse section illustrate the incoming optic fibres of the radial pallial nerve (RPN), glomerulus (GN), and subcellular neuropile (SCN). Scale bar 100 μm. [From Spagnolia and Wilkens 1983]

Two or three eyes are located on the anterior- and posterior-most regions of the mantle in association with the wings of the shell (adjacent to the hinge). The pallial nerves innervating these segments of the mantle extend from the cerebral ganglion instead of the parietovisceral ganglion (see Beninger and Le Pennec, this volume). The central connections of the optic nerves from eyes in this region are unknown. Since glomerular structures have not been described in the cerebral ganglion, it is possible that optic fibres also bypass this ganglion enroute to the visceroparietal ganglion.

Only a brief electrophysiological study of visual activity in the central ganglia of the scallop has been reported (Wilkens and Ache 1977). Attempts to record intracellularly from cells of the lateral lobes have been unsuccessful to date. Thus, our information comes from recordings made at the surface of the lateral lobes by extracellular techniques.

Activity corresponding with both 'on' and 'off' retinal cell layers is present centrally (Fig. 5.12). Although it is not known whether this activity stems from afferent fibres terminating in the lateral lobes, or from neurons of the lateral lobe, there is a consistent difference between the light response recorded in the pallial nerve and the lateral lobe. The 'off' discharge predominant in pallial nerve recordings is relatively weak in the lateral lobe and can be elicited only after light adaptation at high intensities (cf. traces in Fig. 5.12A and B). The 'on' response, on the other hand, is present following either dark or light adaptation, bursts of spikes are distinguishable at stimulus frequencies up 5–6 Hz, and activity persists with little decrement after the initial burst. These results suggest that afferents from the two retinal layers terminate in different regions of the lateral lobe (tactile-evoked activity in the pallial nerve is also absent in recordings from the lateral lobe, Fig. 5.12C). It is tempting to speculate that the fibres traced autoradiographically to the subcellular neuropile (Spagnolia and Wilkens 1983) are responsible for the 'on' response recorded at the surface of the lateral lobe.

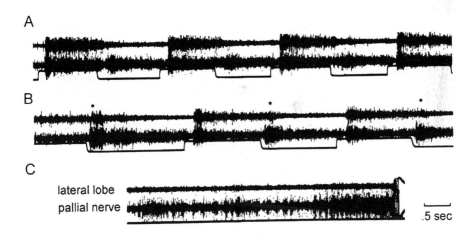

Figure 5.12. Light-evoked neural activity in the lateral lobes (upper traces) and pallial nerves (middle traces) of *Pecten ziczac*; stimulus monitor (lower traces) indicates onset of light with upward deflections. Records begin after 1 min of dark adaptation in (A) and 1 min of light adaptation in (B). The response to tactile stimulation is illustrated in (C). [From Wilkens and Ache 1977]

Both the anatomy and physiology of the lateral lobes indicate a consistent somatotopic organisation of the central visual structures, as might be expected from the arrangement of pallial nerves entering the ganglion. When two or more eyes are labelled autoradiographically, glomeruli in the lateral lobes vary in their position in direct relation to the location of the injected eyes around the shell margin. Similarly, electrical activity recorded from the anterior to posterior regions of the lateral lobes corresponds to stimulation of eyes at the respective anterior to posterior regions of the mantle. Based on the anatomy and physiology of the lateral lobes, it is clear that they are specialised exclusively for visual functions. Motor and nonvisual sensory elements of the pallial nerves form tracts at the base of the lateral lobes which communicate directly with central regions of the ganglion (Stephens 1978; Spagnolia and Wilkens 1983).

5.2.5 Visual Behaviours

Visual behaviours in the scallop include swimming and the orientation of shell position relative to the visual environment, reflex shell closure in response to shadows or movement, and extension of the tentacles toward regions of contrasting light intensity (Buddenbrock and Moller-Racke 1953). The familiar molluscan shadow response, in a strict sense, is probably of less importance for the scallop since image formation in the eye allows detection of movements well in advance of a direct shadow. Accordingly, scallops react more consistently to movement (referred to as the 'sight reaction', as in *Tridacna* by Stasek 1965) than to a general decrease in light intensity, a phenomenon apparent from cursory observations of animals in an aquarium.

The reflexive sight reaction is the most familiar visual behaviour and occurs only in response to decreases in light intensity (Gutsell 1930), or movements involving a dark object. A bright light stimulus will occasionally cause a delayed response (Wenrich 1916), but the circumstances where this occurs are probably not environmentally relevant. In scallops, the opposite behaviour of opening the shell has not been reported for visual stimuli.

The components of the sight reaction include a slight retraction of the mantle velum and valve adduction by activation of the smooth adductor muscle. In recordings of valve movements as well as by direct observation, the visual reflex lacks the twitch-like characteristics of striated adductor-mediated movements. This is confirmed by recording muscle potentials for various responses in the intact animal (Fig. 5.13). For example, the rate of valve adduction for the visual response (A) is lower and clearly distinguishable from the rapid closure triggered by mechanical and/or chemical stimuli (B); the latter type of response is invariably associated with large synchronous muscle potentials (small arrow) from the striated adductor. On occasion, rapid striated muscle contractions contribute to the visual response (see also Fig. 5.14), as reported previously by Land (1968) and Mellon (1969), but experimentally only when fully light adapted; twitch adductions are rarely observed in the field.

Thus, for a visual stimulus shell closure is a relatively 'smooth' response. The degree of closure, onset latency, and duration are variable, as illustrated in Fig. 5.14 by the responses of a scallop stimulated visually by a moving dark stripe. Response latency

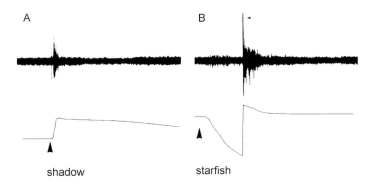

A   B

shadow   starfish

Figure 5.13. Adductor muscle EMGs and shell movements in a tethered scallop, *Pecten ziczac*, following visual (A) and mechanical (B) stimuli. EMG electrodes (7.5 µm Ag wire) were inserted into the smooth adductor muscle. The shadow was produced by an electromagnetic shutter. In (B) the starfish *Coscinasterias tenuaspina* was allowed to contact the mantle and tentacles.

varies from 0.25 s (Fig. 5.13A) to several seconds and valve adduction varies from near complete closure to only a slight decrease in shell gape. Reopening of the shell generally occurs in 5–10 s, although relaxation plateaus may last up to 30 s or more. Prolonged shell closure for several minutes, as occurs following handling, is not characteristic of visual responses.

An additional feature of the sight reaction that is readily apparent is the decrease in sensitivity to repetitive stimulation (Fig. 5.14). For example, a second stimulus within 10–15 s usually elicits no response. For stimulus intervals between 30 s and 5 min, a decrease in response amplitude occurs that is somewhat more rapid for the first few stimuli (6–7 at 2-min intervals) and which appears to be independent of other activities such as the spontaneous 'cough' after the sixth trial. Similar results were obtained by Ludel (1974) using a shadow as the stimulus. Summarising her results, the rate of response decrement was greatest for shorter stimulus intervals, as was the tendency for response failures, and neither effect varied as a function of light intensity. It can be concluded, after eliminating the possibilities of receptor adaptation and muscle fatigue, that response decrement is a central mechanism.

To further characterise this effect, that is, to determine whether the response decrement fulfills the criteria for habituation, starfish extract was used as a novel or dishabituating stimulus. As seen in Fig. 5.14, the sight reaction is potentiated after each application of the extract, an effect that was obtained in two of the four animals tested. However, as will be shown in the following section, starfish extract is far less effective as a releasor for escape behaviours than is contact with the intact animal. This suggests that dishabituation could be enhanced by the use of live starfish. Nonetheless, it appears that habituation of the sight reaction, and its dishabituation by the stimulus of a natural predator, is evidence of a simple form of nonassociative learning in the scallop.

334

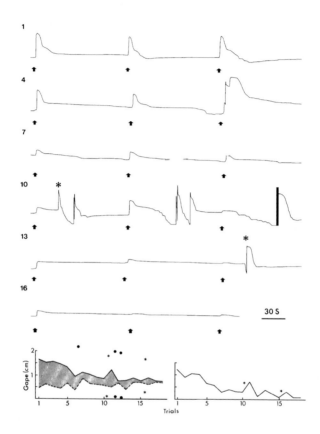

Figure 5.14. Visual responses of the bay scallop *Argopecten irradians* to movement of a dark stripe (2.5 cm wide) across the visual field at a distance of 0.5 m. The lower valve was clamped and shell movements were recorded by a force transducer attached to the upper valve. A photocell signalled the stripe as it entered the visual field (arrows). Responses to 18 consecutive stimuli at 2-min intervals are presented. Spontaneous 'coughs' occur after the 6th and 11th stimuli; a brief swim occurred after the 12th. Following the 10th and 15th stimuli a starfish extract was squirted onto the mantle from a syringe (*'s). Valve gape at rest (dashed line) and in response to the stimulus (solid line) is graphed in the insert (lower left) versus trial number. The asterisks indicate valve excursions for the starfish extract-mediated responses. Filled circles indicate spontaneous valve excursions. Net valve adduction is indicated in the graph insert at right.

The protective reflex behaviour is fully consistent with the physiological sensitivities of the distal photoreceptors which respond to dimming and which discriminate movement from a focused image. Thus, sight reactions need not rely on 'off' responses from numerous eyes. Indeed, Buddenbrock and Moller-Racke (1953) report that responses persist when only a single eye is present. In an extensive study of scallop visual behaviour, these authors also describe a variety of tentacle reflexes and orientation responses. They report, for example, that the long tentacles along a corresponding region of the mantle stretch out toward a slowly moving object. I have also observed tentacles in *P. ziczac* extending toward and 'following' an approaching starfish (within 1–2 cm). Although other sensory modalities could be involved, this behaviour and the related observations by Buddenbrock and Moller-Racke suggest that the eyes function in near-field predator detection. Since escape responses involving chemoreception are invariably dependent on contact with a starfish (avoidance reactions have not been reported for scallops), a visual role for the tentacle response is possible.

Buddenbrock and Moller-Racke (1953) also report that scallops swim in directions influenced by the relative brightness of their surroundings. For example, *P. jacobeus*, *P. opercularis*, and *P. flexuosus* swim toward light whereas *P. varius* swims toward dark regions, with orientation possible by shell rotations prior to or during swimming. The proximal receptors are believed to provide the sensory input for these behaviours since activity only from 'on' receptors is suited physiologically for discriminating static visual patterns (Land 1968). Although an image is not formed at the proximal retina, these photoreceptors are nonetheless capable of coarse spatial resolution, being directionally sensitive for the 50° visual field admitted to the eye (Land 1968). The overlap in visual field for the proximal receptors of adjacent eyes would be approximately 80%, an estimate based on 30 eyes arranged radially around the 180° shell circumference (hinge to hinge). Thus, at least 5 eyes would convey information from any point in the visual environment, with intensity weighted relative to the position of each eye and the extent of retinal illumination. Integration of such an overlapping mosaic centrally would provide the information necessary for orientation behaviours. The somatotopic organisation of glomeruli and the 'on' responses recorded in the lateral lobes of the nervous system (Wilkens and Ache 1977), are consistent with this interpretation of visual signal processing and orientation behaviour. However, 'off' receptors in the distal retina presumably would also be active in the case of swimming, especially so in light of the characteristic zig-zag motion of this behaviour.

5.3 THE LOCOMOTORY SYSTEM

5.3.1 Escape Responses and Swimming

In scallops, the methods of escape differ significantly from other bivalves, which either close the shell tightly for long periods of time, or depend on rapid burrowing and/or somersault behaviours utilising the foot. That these strategies are inappropriate is evident considering that the shell valves of the scallop, even when maximally adducted, do not seal tightly at or near the auricles and that the relatively small size of the smooth adductor muscle does not favour prolonged closure against the force of the hinge ligament. Also, the foot is greatly reduced, its use being to lay down byssal threads in some species, or when small in size.

Rather, escape mechanisms (other that the sight reaction described above) are manifestations of quick shell closure and the jet-like propulsion of water from the mantle

336

cavity. The corresponding structural modifications of the shell, mantle, and adductor muscle are considered to be derived from the adaptations of monomyarian bivalves (Drew 1906; Yonge 1936). Accordingly, changes in shell morphology accompanying the reduction or loss of the anterior adductor, attachment by the byssus — later by cementing, and an open mantle to increase feeding currents are modifications which gave lamellibranchs access to deeper and/or more turbid waters.

With increased turbidity and loss of the siphon, fouling was avoided by the enhancement of ciliary tracts on the palps and ctenidia for sloughing off sediments and a proportionate increase in the striated adductor for rapid valve adduction and currents which transport sediment out of the mantle cavity. In addition, the inner fold of the mantle — the velum or pallial curtain — was enlarged for use in the formation of inhalent and exhalent apertures.

Thus, in bivalves, basic structural modifications necessary for swimming and other escape behaviours are present as a result of changes which accompanied the colonisation of muddy habitats. In the scallop, sediment transported by ciliary tracts inside the mantle collects near the auricles and is ejected laterally and dorsally (Fig. 5.15). This pattern is not only efficient in eliminating sediments and waste, since there is no interference with incurrent streams elsewhere along the shell margin, but it matches the direction of currents used in swimming as well.

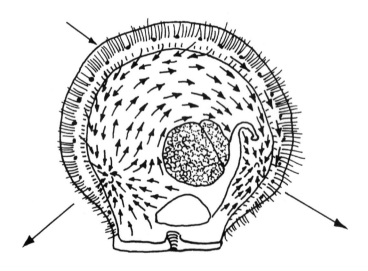

Figure 5.15. Currents in the scallop for elimination of sediment and waste materials. Small arrows, ciliary currents on the inner mantle surface, lower valve; bold arrows, currents due to valve adduction. [Redrawn from Yonge 1936; after Kellogg 1915]

Additional features acquired for locomotion include (a) lengthening of the adductor muscle to increase valve gape and the volume of the mantle cavity — and the propulsive force, (b) compression of the hinge ligament into a triangular pad, for maximum velocity of shell opening when the adductor is relaxed, and (c) development of the muscular and sensory components of the mantle which control current direction. The mechanics of escape behaviours, therefore, are cantered around currents jetted powerfully from the mantle cavity. The forces are considerable since they must overcome the drag and sinking of a shell that, although lightweight, is neither streamlined nor reduced to the extent found in other swimming molluscs (e.g., cephalopods).

Locomotion was first studied systematically by Buddenbrock (1911) and comprises three basic manoeuvres (Fig. 5.16), die Schwimmbewegungen (swimming), die Fluctbewegung (an escape or "jump" response; Thomas and Gruffydd 1971), and die Umkehrbewegung (the righting reflex); locomotory jet currents also are used by some species for excavation or burying activities. In swimming, water is taken into the mantle cavity ventrally and laterally during valve abduction and exits dorsally near the shell auricles during adduction to generate thrust for forward movement. Longitudinal muscles in the velum stiffen during adduction to prevent the forward escape of water, although some escape ventrally through the overhang of the upper mantle is thought to produce an upward force at the leading edge of the shell (Buddenbrock 1911). The dorsal jets produce an explosion of sand at the onset of swimming whereas little or no substrate disturbance is apparent elsewhere suggesting that the volume is proportionately small for the upward force. Hydrodynamic factors undoubtedly are involved in swimming as well, and these will vary somewhat depending on the shape of the shell (Stanley 1970). For example, the valves in *Placopecten* and *Amusium* are much less convex than in other scallops and these genera are among the most adept swimmers. The upper valve in *Placopecten* in fact is slightly more convex than the lower valve and may act as a hydrofoil in providing lift, a prediction correlated with the horizontal plane of the commissure during swimming in this genus (Stanley 1970). In scallops where the biconvex valves are equal, or nearly so, or where the upper valve is slightly concave (planoconvex), lift is predicted only when the plane of commissure is inclined during swimming, as it is in most species. However, from tests in flowing water (Gruffydd 1976), only *P. maximus* (planoconvex) produced lift when held with the shell commissure horizontal whereas *Placopecten*, *Chlamys islandica*, and *C. opercularis* required that the shell commissure be inclined. Lift, whether due to the inclination of the shell or to downward-directed currents, is not haphazard. Various reports indicate that swims of any distance begin with a steep climb after which movement is horizontal (Olsen 1955; Bourne 1964; Stanley 1970; Morton 1980). Swimming directed toward a visual pattern (Buddenbrock and Moller-Racke 1953) or away from a predator (Winter and Hamilton 1985) also suggests that currents jetted from the mantle are actively controlled by the velum.

For jump and righting behaviours, water is taken into the mantle cavity as in swimming. During the jump response, however, water is jetted back out of the ventral margin. Unlike the swim, in which the scallop appears to take successive bites out of the water (Dakin 1909), escape jumps characteristically involve 1–3 valve closures.

338

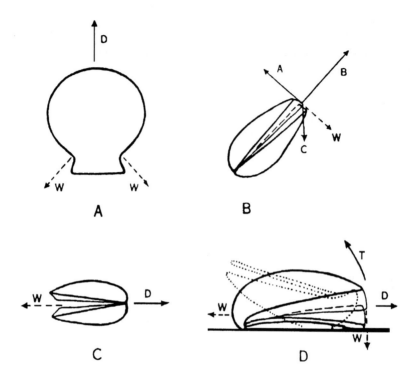

Figure 5.16. Jet-like currents involved in locomotion by the scallop. (A, B) Top and side view of swimming currents. (C) Currents used in jump responses. (D) Currents used in the righting reflex. Broken arrows indicate direction of current; solid arrows the direction of movement. (T) represents turning. [Yonge 1936; after Buddenbrock 1911]

Nonetheless, several behavioural responses are possible depending on the extent or strength of valve adduction and the direction of the jet current. For example, water jetted ventrally causes backward movement (hingeward), whereas water jetted laterally will result in movement in an anterior or posterior direction. In some instances water is jetted dorsally, as during swimming, but from only one side to effect a rotation of the shell. Alternatively, in planoconvex scallops recessed in the substrate, jump-like adductions do not result in movement of the scallop but instead dislodge, or possibly startle, potential predators.

Shell movements associated with locomotory and escape behaviours have been recorded by mechanoelectric devices attached to the upper valve, with the lower valve fixed to the chamber floor (Thomas and Gruffydd 1971; Stephens and Boyle 1978), or indirectly by monitoring the pressure in the mantle cavity (Moore and Trueman 1971).

Valve movements also have been recorded in tethered scallops (Wilkens 1981), a situation where the freely-moving animal is less likely to be influenced by the recording apparatus, particularly with regard to the dynamic forces associated with swimming (Vogel 1985). Various shell movements characteristic of the Bermuda sand scallop *P. ziczac* are illustrated in Figure 5.17. The resting position of the valves, as seen for these spontaneous behaviours, is characteristically about half the maximum shell gape, although some scallops may gape more widely. Coughs associated with irrigation of the mantle cavity are generally regular (see also Stephens and Boyle 1978), occurring singly (A) or in bouts (B); irregular activity may occur also, such as under crowded conditions in an aquarium (C), or after handling (Thomas and Gruffydd 1971). Invariably, each cough is preceded by an increase in shell gape, thereby increasing the volume of the mantle cavity. Likewise, increases in shell gape precede escape closures, a further indication of the relationship between cleansing mechanisms and swimming.

In coughs, jumps, and swims, valve adduction is rapid due to activation of the fast adductor (cf. Fig. 5.13). In contrast, the valves reopen gradually, except during swimming as seen more clearly at a faster time base (Fig. 5.17C). Thus, both fast and slow components of the adductor muscle are activated during these behaviours, slow relaxation of the smooth adductor being responsible for the gradual reopening of the shell. In swimming, activation of the smooth muscle is delayed until the end of the swim episode. Of equal importance to the twitch-like valve adductions, the shell must be able to reopen quickly during swimming, a requirement for the repetitive valve adductions (at 2–3 Hz) necessary for the scallop to remain waterborne. The primary opening force is supplied by the elasticity of the hinge ligament. However, Vogel (1985) has estimated that at a

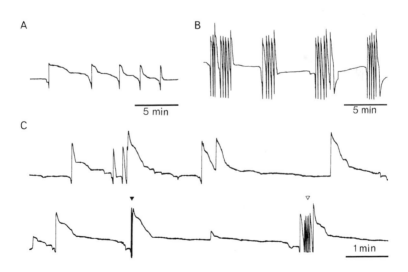

Figure 5.17. Spontaneous valve movements in *Pecten ziczac*. Records are from tethered animals partially recessed in the sand. Upward traces represent valve adduction. [From Wilkens 1981]

swimming velocity of 0.5 m s^{-1}, in *A. irradians* a force approximately one-quarter that of the hinge ligament is generated by movement of the shell through the water. In effect, an increase in internal pressure from the force of water at the ventral gape, and a decrease in outside pressure from water flowing over the convex surfaces of the shell, combines to assist the hinge in rapid valve abduction.

Scallops utilise their locomotory abilities primarily to escape predation and for habitat selection. Evidence of starfish-mediated swimming goes back at least to Dakin (1910b) who noted that the "close proximity" of *Asterias rubens* is the only stimulus effective in inducing swimming in *P. opercularis*. Indeed, a variety of predatory starfish and gastropods are reported to elicit swimming and/or jump escape responses when contact is made with the scallop mantle (Buddenbrock 1911; Lecomte 1952; Olsen 1955; Mellon 1969; Moore and Trueman 1971; Thomas and Gruffydd 1971; Bloom 1975; Stephens and Boyle 1978; Forester 1979; Ordzie and Garofalo 1980; Wilkens 1981; Peterson et al. 1982; McClintock 1983; Chernoff 1987; Pitcher and Butler 1987); translocations up to 3 m have been measured in the field for *A. irradians* (McClintock 1983), 4 m for *P. magellanicus* (Caddy 1968), and exceeding 10 m in *A. pleuronectes* under experimental conditions (Morton 1980). Were it not for these 'violent' reactions, which affect successful escapes by the scallop in many instances, predation levels would certainly be greater. As it is, laboratory estimates of successful attacks on *A. irradians* by the oyster drill *Urosalpinx cinera* are high (72.3%), with a mean density of attacks in the field of 1.7 m^{-2}. An estimated 75–80% destruction of Tasmanian scallop beds (Olsen 1955) and a mortality rate up to 25% for *P. magellanicus* in the Gulf of Saint Lawrence (Dickie and Medcof 1963) are effects attributed primarily to starfish predation. Escape locomotion is also thought to reduce predation in gastropod molluscs (Feder 1963; Margolin 1964).

In addition to locomotion, scallops (principally *Chlamys* sp.) derive protection from the encrustment of sponges on the upper valves (Bloom 1975; Chernoff 1987; Pitcher and Butler 1987). This mutualistic association is synergistic with the behavioural responses, increasing the probability of escape due to the interference of adhesion by starfish tube feet. Similarly, shell coloration, with the dorsal valve typically darker, and encrustment by benthic invertebrates other than sponges may offer protective camouflage.

Substrate and habitat characteristics are also major factors, which influence locomotory behaviour in scallops. Planoconvex scallops, *P. maximus* for example, which are normally recessed in sandy or sand-mud substrates swim frequently and disperse widely when placed on hard substrates (Baird 1958); when placed in sand little or no swimming occurs. An increase in swim frequency associated with hard substrates, and a greater tendency for small or unrecessed animals to swim has been noted for *P. maximus* by other investigators (Hartnoll 1967; Baird and Gibson 1956; Gruffydd 1976, for smaller *Chlamys*). Perhaps related to the life habits of 'recessing' scallops, visual stimuli or other disturbances generally do not induce swimming. For example, *P. maximus* reportedly does not swim in response to visual stimulation (Land 1968), the approach of divers (Baird 1958), or dragging operations (Baird and Gibson 1956); visual and tactile stimuli (except for starfish) are also ineffective for *P. opercularis* and *P. jacobeus* (Dakin 1910b). *A. pleuronectes* can be induced to swim only when picked up and drained of water (Morton 1980). Conversely, Caddy (1968) reports that *Placopecten* first faces away and then swims at the approach of a diver and that swimming accounts for a low drag capture efficiency (2.1%). However, large animals (>100 mm) again show a marked decrease in the tendency to swim.

Habitat preference is reversed for the bay scallop; *Argopecten* normally inhabits seagrass beds as opposed to sandy substrates. Experimentally, the probability of swimming is significantly greater when scallops are placed in a sandy area, and the distance covered by swimming increases for larger diameter sandy test sites (Winter and Hamilton 1985). Camouflage, food resources, and interference with the movements of predators are among the possible advantages afforded by seagrass beds.

Although it is not clear whether scallops routinely migrate long distances, circumstantial evidence suggests that swimming is used by some species for migration and that, in relation to swimming, community structure is further influenced by currents. For example, Morton (1980) argues that swimming in *Amusium* is sufficiently well developed as to make it "extremely doubtful that [it] is used merely for escape." The seasonal appearances of *Amusium* sp., based on the coastal scallop fisheries of Hong Kong, are also cited in support of mass migrations by this species. Yonge (1936) also states that *C. opercularis* and the bay scallop *A. irradians* make extensive migrations, a claim often made by fishermen and supported by my own collecting experience in the northern Gulf of Mexico (pers. obs.). Moore and Marshall (1967) also note that displacements of *Argopecten* are related to tidal currents. In *C. islandica*, the propensity to swim is augmented by current (Gruffydd 1976), a factor which may affect their distribution. For example, older scallops are found in the innermost regions of Norwegian fjords where inward-flowing currents are stronger than those flowing outward. Olsen (1955) also reports that adult scallops in the waters of Tasmania are consistently found in highest concentrations where tidal currents are strong and in beds parallel to the direction of flow, even though no differences in bottom characteristics were discernible. It is not surprising that swimming is effected by currents since the efficiency of directed locomotion would increase or decrease with current direction, and additional movement would continue after the swim as the scallop settles passively to the substrate.

5.3.2 Sensory and Motor Functions of the Mantle

The ciliated epithelial cells of the mantle, particularly those on the tentacles, serve as the primary sensory receptors for tactile and chemical stimuli. Two probable sensory structures of uncertain function, the abdominal sense organ and a structure reminiscent of the gastropod osphradium (Beninger and Le Pennec, this volume), are located inside the shell but the mantle edge is clearly preferable for most sensory features, as indicated further by the presence of the eyes.

The mantle tentacles have been associated primarily with the detection of predators but they undoubtedly contribute to a variety of activities. In fact, tentacles of different size and location are present; long tentacles which arise from the middle fold of the mantle in line with the eyes, and short tentacles on both the middle fold lateral to the eyes and lining the front edge of the velar curtain (inner mantle fold). There is no evidence, physiological or otherwise, that differences in sensitivity exist between the two types.

However, the short interlocking tentacles of the velum would seem appropriate for excluding unwanted particulate matter from the incurrent streams and may be sensitive primarily to tactile stimuli. The activity of the long tentacles suggests that they are involved in detecting either the suitability of the substrate or the posture of the shell in relation to it, or both. In *Argopecten*, for example, the long tentacles of the lower mantle typically extend downward, often in contact with the substrate. In the sand scallop *P. ziczac*, the lower tentacles curve tightly around the margin of the lower valve when the shell rests on the sand instead of extending outward as when recessed. The long tentacles therefore act as if they monitor substrate texture and shell position, but the chemical essence of the substrate also may be detected by this activity.

Chemosensitivity in the scallop is understood essentially from behavioural assays, that is, escape responses. Both swim and jump reflex behaviours are dependent on specific chemical releasing factors found in their predatory coinhabitants. In all instances, these responses appear to be contact dependent, an indication that the chemical stimulant is not released by the starfish or snail and/or that both tactile and chemical stimulus components are required to trigger an escape response. Peterson et al. (1982) also have shown that scallop emigration is significantly higher from test quadrants containing the predatory snail *Busycon* even though whether the scallops having fled actually came into contact with the snail was not determined. Scallops also discriminate between predatory and nonpredatory species. In *A. irradians*, the predatory *Asterias forbesi* and *Luidia clathrata* evoke swimming whereas the nonpredatory *Echinaster* sp. and *Henrica sanguinolenta* do not (Stephens 1978; McClintock 1983; Winter and Hamilton 1985). Predatory snails (*Urosalpinx cinerea, Eupleura caudata, Thais lapillus, B. canaliculatum*) are also distinguished from nonpredators (*Littorina littorea, Ilyanassa obsoleta*), as indicated by longer swims and shorter response latencies in the former group (Ordzie and Garofalo 1980). In *P. maximus* the predatory starfish *A. rubens, Marthasterias glacialis*, and *Astropecten irregularis* evoke swimming but the nonpredatory *Porania pulvillus* and *H. sanguinolenta* do not (Thomas and Gruffydd 1976). *P. jacobaeus* also distinguishes behaviourally between different asteroid species (Dakin 1910b; Lecomte 1952).

Active components isolated from the surface exudates of starfish extracts (*M. glacialis*) trigger escape responses like those evoked by the intact starfish (Mackie et al. 1968; Mackie 1970). These compounds have been identified as steriod glycosides (M1 and M2) and, like the synthetic surfactants Triton X-100 and Tergitol NPX, are effective escape stimulants for a variety of benthic invertebrates including the snail *Buccinum* and the brittle star *Ophiothrix*. In *P. maximus* and *C. opercularis*, threshold concentrations of starfish steroid glycosides cause tentacle withdrawal and contractions of the velum. Swimming is induced at higher concentrations (Mackie 1970; Stephens and Boyle 1978) but the frequency of swim responses was not addressed in these experiments. As observed by Thomas and Gruffydd (1971), and in my experience (Wilkens 1981), the pipetting of solutions of starfish extract onto the mantle is much less effective as a stimulus for swimming than direct contact with a live starfish.

The stimulus-response characteristics of the starfish-scallop interaction are highly stereotyped as well as species specific. Most stimuli, including those triggering the sight reaction, result in closure of the shell with valve adductions ranging from partial to complete; duration of shell closure generally reflects the intensity of stimulation. For example, any moving object large enough to cast a shadow or to create a near-field disturbance to which the animal is unaccustomed will cause the shell to close briefly. Direct probing of the mantle with any of a variety of glass, wood, or metallic objects generally causes rapid and sustained closure, sometimes followed by a delayed swim (Dakin 1910b; Thomas and Gruffydd 1971; Stephens and Boyle 1978; Wilkens 1981). However, contact with a predatory starfish invariably causes the shell to open widely, a response quite distinct from that of any other tactile stimulus. The behaviours described above are illustrated in Fig. 5.18. In each of these trials the scallop closed its shell when the starfish was introduced (by hand) into the aquarium. Even when completely adducted, as in 4 of the 6 trials shown here, contact by the starfish with only a few retracted tentacles produced an immediate (in a matter of seconds) opening of the valves, the biological equivalent of using a key to open a lock. A swim or jump response follows if the starfish remains in contact, generally for 5–10 s. In related experiments, responses were obtained by probing the mantle with a bare stainless steel wire and by a single tube foot into which the wire had been inserted (Wilkens 1981).

Again, the wire probe triggered shell closure whereas the tube foot caused shell opening and an escape response. Subsequent tests with *A. irradians*, *A. gibbus*, and *Notovola meridionalis* have produced similar results (unpublished observations). These results highlight the sensitivity as well as the combined chemo-tactile specificity of receptors in the scallop mantle to the stimulus of a natural predator. Scallops also differ somewhat in their responsiveness to starfish. Indeed, swimming is energetically expensive and Thompson et al. (1980) report that following a swim, *P. magellanicus* becomes refractory to further stimulation by starfish extract for 1–3 hr. *P. ziczac* will swim repetitively (Fig. 5.18) but the number of valve claps per swim decreases and the response latency may increase for successive swims.

The underlying sensory as well as motor organisation of the mantle, as it relates to various behaviours, has been described by Stephens (1978) and Stephens and Boyle (1978). With the mantle isolated from the central nervous system the tentacles become elongate and immobile and spontaneous activity in the afferents of the pallial nerves is absent. However, mechanical stimulation of a tentacle will cause it to retract and a broader stimulus will produce a localised retraction of the mantle edge, events associated respectively with single and multi-unit bursts of spikes in adjoining pallial nerves. Starfish extracts applied to the mantle also produce local retractions and are characterised by an initial multi-unit burst of activity followed by a low level afferent discharge for 30 s or more (Stephens 1978). If synaptic mechanisms are blocked by soaking in calcium-free, high-magnesium solutions, reflex movements and pallial nerve activity reversibly disappear. Thus, local reflex responses of the mantle, including those that resemble jet formation in the intact animal, are coordinated peripherally by synaptic interactions in the circumpallial nerve. These results also indicate that afferent activity in the pallial nerves, except for that in the optic nerves, is not from primary mechano- and chemoreceptors. Whether sensory blockade is peripheral to or at the level of the circumpallial nerve is unknown.

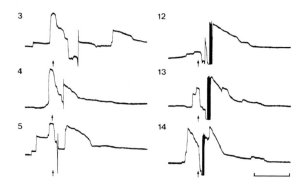

Figure 5.18. Starfish-mediated escape responses in *P. ziczac*. Valve movements are from an unrestrained scallop in response to contact by the starfish *Coscinasterias tenuaspina*. Consecutive stimuli were presented at 2-min intervals along the ventral edge (left-hand column), or at the posterior auricle (right-hand column) of the shell. These responses are characteristic of a sequence of 18 consecutive trials. Arrows indicate stimulus onset; adduction is represented by upward deflections. Calibration, 1 min. [Adapted from Wilkens 1981].

Jet apertures are formed by retraction of the radial muscles in corresponding regions of the upper and lower mantle velum, as illustrated in *P. ziczac* (Fig. 5.19). That jets are formed at the point of contact by the starfish is important for the escape jump in that currents are aimed directly at the predator. In an unrecessed scallop, *C. opercularis* for example, the resultant escape movement is in a direction opposite the point of attack (Stephens and Boyle 1978). For a recessed scallop the force of the current drives back the starfish. A similar directional, exhalent current is used by the giant clam *Tridacna* to startle and/or drive away potential predators (Stasek 1965).

The physiological mechanisms associated with jet formation and related behavioural responses can be investigated by direct activation of sensory tracts in the pallial nerves. These nerves are readily accessible after the careful removal of one of the valves. Weak en passant stimulation of a pallial nerve on one side of the animal (Fig. 5.20A) produces a threshold response in the corresponding contralateral nerve (Stephens 1978). This unilateral response is resistant to fatigue and calcium-free saline suggesting that reflex contractions of the type which form jets are coordinated by neural pathways involving electrical synapses which cross the visceroparietal ganglion to sites opposite in the mantle. Cobalt backfillings of pallial nerves also stain tracts through the ganglion and into the appropriate pallial nerves on the opposite side. Higher intensity stimulus pulses trigger a more sustained response in the corresponding nerves of both sides (Fig. 5.20B). This activity is correlated with rapid infolding of the velar curtain and is sensitive (reversibly)

Figure 5.19. Jet formation (arrow) in *Pecten ziczac* at the point of contact by a ray and tube feet of *Coscinasterias tenuaspina*. Note also the outstretching of tentacles toward the starfish.

to calcium-free saline and fatigue. With further increases in stimulus intensity, the smooth and then striated adductor muscles are activated (Fig. 5.20C, D); stimulus trains evoke swim-like rhythmic sequences of activity in the pallial nerves and striated muscle (Fig. 20E).

These results demonstrate at least a rough correlation between the stimulus intensity, as might range from lightly touching the mantle or tentacles to sustained contact by a predatory starfish, and neural responses, which correspond with jet formation, an escape jump, and a swim. The infolding of the velum, described as a protective response (Stephens 1978), represents the contractions of longitudinal muscles, which stiffen the velum and thus prevent water from escaping the mantle cavity other than at sites of jet formation. I have recorded equivalent ensembles of neural activity from the completely isolated ganglion of *Argopecten* (unpublished results) in response to a similar sequence of pallial nerve stimuli. An additional stage in the level of ganglionic excitation involves activity triggered in the ctenidial nerve. This response precedes activation of the striated muscle efferents and corresponds with an anterior retraction of the gill axis in anticipation of a jump or swim.

5.3.3 Innervation and Neuromuscular Physiology of the Adductor Muscle

As noted above, the adductor muscle of scallops and other monomyanarians is composed of two distinct types of muscle (Yonge 1936), a fast, twitch-like block of striated, somewhat translucent fibres and a slow, smooth muscle component of smaller, opaque fibres. In scallops the striated muscle constitutes 70–80% of the adductor mass.

346

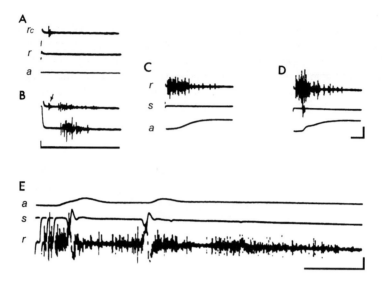

Figure 5.20. Efferent responses and mechanograms (a) from stimulation of the pallial nerves. (A) Threshold pulse evokes short-latency burst in the contralateral pallial nerve (rc); (B) Higher intensity pulse evokes a threshold response (arrow) followed by synchronous bursts on both sides. (C) Synchronous burst in pallial nerve and slow adductor contraction. (D) High stimulus intensity pulse triggers burst in pallial nerve and a fast muscle contraction (s, myograms in C and D). (E) Trains produce multiple cycles of activity. Calibration: 100 µV; 50 ms, A, B; 100 ms C-E. [Adapted from Stephens 1978]

The position of the adductor muscle in relation to the shell and nervous system is illustrated elsewhere in this volume (see Beninger and Le Pennec). The visceroparietal ganglion (VPG) lies on the anterior surface of the adductor and hence on the striated portion of the muscle.

Innervation of the smooth and striated muscle components by nerves from the VPG is equally distinct. A pair of bilaterally symmetrical nerves to the smooth muscle extend ventrally from beneath the VPG, course over the surface of the striated muscle and inward at the juncture of the two muscle blocks. The motor nerves innervating the striated muscle also emerge from the surface of the ganglion facing the muscle and branch, on each side, into three primary trunks – anterior, lateral, and posterior. Each trunk extends only a short distance before additional branches form and penetrate the striated muscle mass (unpublished results from *A. irradians*).

Despite the fact that the smooth muscle can be readily isolated with its efferent nerve supply intact, the efferent mechanisms which control this muscle are poorly understood. Stimulation of either the nerve or muscle produces contractions which decay slowly (5–10 s for return to baseline) and therefore summate readily into sustained contractures (Wilkens 1981). In contrast, the pattern of innervation and the junctional physiology of muscle fibres in the striated adductor is relatively well understood (Mellon 1968). By focal stimulation of small nerve branches and intracellular recording from single muscle fibres, a multiterminal pattern of muscle fibre innervation has been shown to exist. This is illustrated by a stepwise series of excitatory junction potentials (EJPs) whose amplitudes correspond reproducibly with stimulus thresholds for individual motor neurons (Fig. 5.21A). It was also shown that adjacent muscle fibres, and presumably muscle fibres within a motor field, are innervated in common by branches of the same motor neurons (Fig. 5.21B).

Muscle spikes arise from the EJPs, but typically only in response to more intense electrical stimulation and the temporal summation of junction potentials. Both overshooting (Fig. 5.21C) and undershooting spikes are seen although some fibres fail to spike. Thus it seems that spike threshold is high and/or that regenerative activity may be restricted to certain portions of the muscle fibre.

Graded contractions of the striated adductor may be possible as a result of high spike threshold and multiterminal innervation, although as yet no simultaneous recordings of mechanical tension have been obtained to test this hypothesis. For example, either spatial or temporal recruitment of EJPs could produce graded contractions, a mechanism similar to that of the multiply-innervated muscles in arthropods, as long as the spike threshold was not exceeded. In the intact animal the strength of contraction in the striated muscle varies, as indicated by rapid adductions ranging in extent from those of coughs to complete shell closure during swimming. Unlike arthropods, however, no inhibitory junction potentials have been observed, and the EJPs are non-facilitating and resistant to fatigue at stimulus frequencies up to 10–20 Hz, well above the rates necessary for swimming.

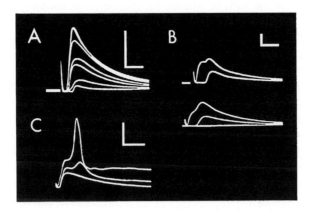

Figure 5.21. Junction potentials in striated muscle fibres. (A) Superimposed responses from a single fibre to stimuli of increasing intensity. (B) Similar response patterns from adjacent muscle fibres show common thresholds to nerve stimulation. (C) Summation of junction potentials with an over-shooting spike superimposed on the response at highest stimulus intensity. Calibration: (A, B) 20 mV, 20 ms; (C) 25 mV, 50 ms. [Adapted from Mellon 1968]

A striking feature of the motor system innervating the striated muscle is the close synchronisation of EJPs, a requirement for achieving maximum strength of contraction. Electromyograms (EMGs) recorded from the striated muscle of *A. irradians* during swimming characteristically show a single muscle spike for each adduction of the valves (Mellon 1969; Stephens and Boyle 1978), and the time course of the extracellularly recorded spike is comparable to that of the muscle action potential recorded intracellularly (Mellon 1968). There is also close synchronisation of the EMGs in different motor fields, including activity which originates from nerves on both sides of the ganglion. Thus, Mellon (1969) suggests that striated muscle motorneurons in the VPG may be electrically coupled; pools of electrically coupled motor neurons are well known from other animals. In other evidence, contractions in the striated muscle can be triggered by stimulating (antidromically) the cut proximal stumps of the lateral or posterior nerve trunks (unpublished results) if the anterior branch remains intact. These results are evidence for the electrical coupling of motor neurons although it is possible also that branches of the same motor neurons are found in each nerve trunk.

An interesting and as yet poorly understood feature of the scallop adductor concerns the regulation of tension in the smooth muscle. In the undisturbed animal, shell gape is maintained at or slightly greater than half maximum during normal feeding. Cutting the smooth adductor allows the valves to open widely whereas disruption of the striated muscle, in part due to trauma, results in partial smooth muscle contractures and smaller valve gapes (Mellon 1969; Stephens 1978; Wilkens 1981). Thus, the smooth muscle clearly exerts continual tension in opposing the opening force of the hinge resilium. The wide gaping of the valves triggered by a starfish or in advance of spontaneous cough and swimming behaviours is due therefore to relaxation of the smooth muscle.

It is unclear whether long-term tension in the smooth adductor depends on tonic excitatory input from the VPG or whether the muscle exists normally in a state of partial 'catch' tension. Lowy (1954) favours the tonic hypothesis, with relaxation resulting from a decrease in efferent activity. However, in electrical recordings from the nerve, and directly from the smooth muscle, tonic discharges consistent with activity that would be required for sustained tension have not been observed (Wilkens 1981). On the contrary, efferent discharges arise which correspond with muscle relaxation and increased valve gape (e.g., Fig. 5.13B). It is of interest, further, that contractions of the smooth muscle produced by ongoing electrical stimulation of the muscle itself can be effectively reversed by a starfish stimulus (Fig. 5.22). Thus, it seems certain that an active mechanism is responsible for relaxation of tension in the smooth adductor muscle and the corresponding valve abductions seen in various behaviours. Curiously, relaxation cannot be induced by electrical stimulation of the smooth muscle nerves in either the intact animal or in the isolated nerve-muscle preparation; electrical stimuli produce only muscle contractions.

5.3.4 Coordination of Locomotory Behaviour by the Central Nervous System

As previously described, swimming, escape, and visual behaviours are brought about by activity in the mantle and adductor muscle effectors. Local reflexes in the mantle are coordinated in part by the circumpallial nerve, in effect a circular ganglion, and by the

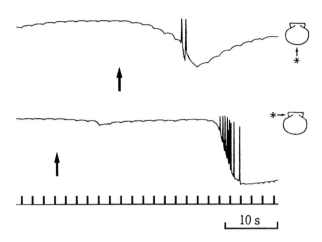

Figure 5.22. Tetanus-like contractions of the smooth adductor muscle evoked by direct electrical stimulation through electrodes implanted in the muscle. A starfish stimulus (arrows) triggers muscle relaxation as well as jumps (top record) and a swim (middle record); insets at right illustrate the site of contact by the starfish. Stimulus parameters: train rate, 0.5 Hz; train duration, 250 ms; pulses, 1.5 V and 1 ms. [Adapted from Wilkens 1981]

VPG by way of afferent and efferent tracts in the radiating pallial nerves. The adductor muscle is innervated entirely by efferent nerves from the VPG. Thus both sensory and motor functions of the scallop are dominated by the VPG, the largest of the central ganglia and most complex within bivalve molluscs (Dakin 1910b). However, not all pallial nerves connect centrally with the VPG; those which innervate dorsal segments of the mantle at the shell auricles originate from the cerebral ganglion instead.

Less is known about the role of the cerebral ganglion in the behaviour of scallops than is the case in other bivalves (reviewed by Bullock 1965) where cerebral functions include reflex control of the anterior adductor, coordination of visceral and pedal activities, and a dominant influence on behavioural rhythms. Nevertheless, several observations attest as well to the dominance of cerebral function in the scallop. Specifically, the cerebral ganglia are instrumental in behavioural decision making with respect to swimming or other escape responses and contain integral components of the neural circuits for these behaviours. For instance, a starfish stimulus along regions of the mantle innervated by pallial nerves from the cerebral ganglion triggers a significantly greater number of swim responses than equivalent stimulation at any other site; elsewhere an escape jump ensues (Wilkens 1981). This behavioural dichotomy is illustrated in Fig. 5.18 in data excerpted from a longer sequence of jump and swim responses. Equivalent results have been observed in *A. irradians* (unpub. obs.) and responses triggered by stimulation of the dorsal mantle, in this case by a predatory gastropod, also have a significantly shorter latency than equivalent stimuli along the ventral margin of the shell (Ordzie and Garofalo 1980). Thus, what otherwise may be equivalent sensory information, but from different regions of the mantle, triggers different behaviours depending on whether the information enters the cerebral or visceroparietal ganglion. Since jet formation is not possible along the hinge, or where the mantle is fused at the shell auricles, a swim is the appropriate response if a predator is detected at the dorsal shell margin. Whether or not susceptibility to predation is greater for attacks at the dorsal edge of the shell has not been determined.

The correlation between swimming and sensory input to the cerebral ganglion suggests that these ganglia are necessary for the initiation of swimming behaviour. Indeed, swimming does not appear to be possible if the cerebrovisceral connectives (CVCs) are cut (Stephens 1978; Wilkens 1981). Cutting the CVCs not only interrupts descending pathways otherwise activated by the dorsal mantle but also eliminates a component of the swim neural circuit in the cerebral ganglion since, under these circumstances, responses triggered from ventral regions of the mantle affect only weak jump-like valve adductions. It can be concluded that the occasional swims that result from stimulation of ventral mantle under experimental conditions, or at lower stimulus threshold in undisturbed animals, involve neural elements in the cerebral ganglia brought into play by activity in ascending pathways of the connectives. The circumpallial nerves do not substitute for or constitute pathways parallel to those in the connectives, as is evident from the behavioural disruption which follows transection of the CVCs. A functional pallial-circumpallial-pallial circuit between the cerebral and VPG is also unlikely in that the spread of electrical excitation decrements rapidly in the circumpallial nerve, as seen in experiments with the isolated mantle (unpublished results).

Isolation of the cerebral ganglia also profoundly alters the 'resting' posture of the animal by interfering with the ability of the smooth adductor muscle to relax, even in response to a starfish stimulus (Wilkens 1981). Similarly, in *Anodonta* the VPG can increase tonus in the posterior smooth adductor but cannot initiate relaxation in the absence of the cerebral ganglion (Barnes 1955). In the scallop, valve gape decreases following transection of the CVCs and any subsequent valve adductions decay with a long time course. Thus, even though the motor neurons innervating the smooth adductor muscle originate in the VPG, the cerebral ganglia are required for their normal function. The loss of the cerebral ganglia therefore inhibits swimming on two accounts. First, they appear to contain integral components of the swim motor circuit or to exert a required influence over its operation. Second, relaxation of the smooth muscle, which is responsible for the wide gape of the valves prior to a swim, and which must be sustained if the valves are to reopen rapidly, is blocked. That the smooth muscle does not relax following isolation of the cerebral ganglia by itself makes swimming impossible.

Earlier studies also emphasise the role of the cerebral ganglion. For example, Buddenbrock (1911) has shown that scallops respond asymmetrically to gravity due apparently to the function of the statocysts, sensory structures associated with the cerebral ganglia. Scallops suspended by a thread attached to either the dorsal or ventral margin of the shell generate movements that result in the left valve becoming uppermost, a similar result to that obtained by the righting reflex. This behaviour assumes differential control of the mantle curtain by the statocysts and cerebral ganglion, direct for dorsal regions and indirect via the VPG for the remainder of the mantle.

As indicated previously, the adductor muscle is innervated by motor neurons located in the VPG; efferent discharges coordinating mantle and gill movements also stem from this ganglion. The jump response is to an extent a reflex barrage of excitatory efferent activity in these motor pathways, although cerebral input is required as well for the release of tension in the smooth adductor muscle. The neural mechanisms are less certain for the rhythmic bursts of activity which underlie swimming. Mellon (1969) attributes swimming to the output of a relaxation oscillator in which stretch receptors located in the fast adductor, and activated when the muscle is stretched, in turn activate motor neurons to the striated muscle. Swimming begins when valve gape is adequate to discharge the stretch receptors and is maintained by subsequent afferent bursts triggered at each cycle of valve abduction; swimming is terminated either by stretch receptor adaptation or when the oscillator is damped by coactivation of the smooth adductor muscle.

The relaxation oscillator model of swimming is based on a simple myotatic reflex. However, evidence for a reflex mechanism is as yet circumstantial. For example, even though sudden stretch (controlled by a solenoid) elicits a transient electrical response in the striated muscle (Mellon 1969), the receptors which trigger this reflex have not been identified. The reflex activity, which appears as a synchronous muscle spike, is variable in number, amplitude, and latency but has been recorded during stretch sequences of up to 21 consecutive cycles.

More recently it has been shown that swimming sequences can be initiated without stretching the adductor muscle (Wilkens 1981). In the intact animal, for example, a swim sequence can be initiated by a starfish stimulus when the valves are near maximally adducted (Fig. 5.22). In related experiments in which the smooth muscle nerves had been cut, that is, where the smooth muscle relaxation phase of the response had been eliminated, a starfish stimulus applied to the dorsal mantle again triggered swimming. These results are interpreted as indicating that the rhythmic discharges which underlie swimming are evidence of a centralised neural circuit, specifically a central pattern generator whose neural components are derived from both the visceroparietal and cerebral ganglia and one which receives input from peripheral receptors with a sharply tuned specificity for natural predators of the scallop.

Although swim-like cyclical activity in the adductor muscle can be elicited under conditions that would fail to excite stretch receptors in the muscle, the frequency of swim adductions is less than the usual 3–5 Hz. It is likely therefore that sensory elements are incorporated at some level in the pattern generator for the rhythmical swim response. Attempts to record afferent activity in the motor nerves to the smooth and striated adductor muscles have been unsuccessful (unpublished results). In the absence of stretch receptors in the muscle the most likely source of afferent input is mechanoreceptors in the mantle. An additional candidate would be the abdominal sense organ, a mechanosensory structure located on the posterior face of the adductor muscle near the anal aperture. Although these receptors are known to be sensitive to vibrations (Zhadan and Semen'kov 1984), their low threshold sensitivity (0.006 μm) may also permit the monitoring of muscle movements. As with the source, the exact role of sensory elements in the pattern generator is unknown, although integration of sensory feedback as a timing mechanism seems most likely. Sensory, as well as premotor and motor neurons, are recognised components of oscillating neural circuits in other organisms (Delcomyn 1980; Grillner 1985; Reye and Pearson 1987).

The central role of the cerebral ganglion in escape responses and swimming is summarised by a diagram of the neural components relevant to these behaviours (Fig. 5.11, Wilkens 1981). Descending activity from the cerebral ganglia is essential not only for the rhythmical efferent discharges in VPG nerves to the striated muscle, but for the activation of afferent pathways innervating the smooth muscle, specifically those that are responsible for releasing tension in the smooth muscle prior to swims and jumps. In the latter instance, where the response is triggered by peripheral receptors that communicate directly with the VPG, neural pathways that ascend to the cerebral ganglia are required as well.

ACKNOWLEDGMENTS

Some of the unpublished results were obtained while the author was a Visiting Research Professor at the Whitney Laboratory for Experimental Marine Biology, and while supported by a grant from the Whitehall Foundation, Inc. A critique of the manuscript by Deforest Mellon, Jr., assistance with translation of reference material by Raissa Berg, and with photography by John Judd are greatly appreciated. A contribution (No. 280) from the Tallahassee, Sopchoppy and Gulf Coast Marine Biological Association.

REFERENCES

Baird, R.H., 1958. On the swimming behaviour of escallops (*Pecten maximus* L.). Proc. Malacol. Soc. Lond. 33:67–71.

Baird, T.H. and Gibson, F.A., 1956. Underwater observations on escallop (*Pecten maximus* L.) beds. J. Mar. Biol. Assoc. U.K. 35:555–562.

Barber, V.C., Evans, E. and Land, M.F., 1967. The fine structure of the eye of the mollusc *Pecten maximus*. Z. Zellforsch. 76:295–312.

Barber, V.C. and Land, M.F., 1967. Eye of the cockle, *Caridium edule*: anatomical and physiological investigations. Experientia 23:677–678.

Barnes, G.E., 1955. The behaviour of *Anodonta cygnea* L., and its neurophysiological basis. J. Exp. Biol. 32:158–174.

Bloom, S.A., 1975. The motile escape response of sessile prey: a sponge-scallop mutualism. J. Exp. Mar. Biol. Ecol. 17:311–321.

Bourne, N., 1964. Scallops and the offshore fishery of the Maritimes. Bull. Fish. Res. Board Can., No. 145. 60 p.

Buddenbrock, W. von, 1911. Untersuchungen über die Schwimmbewegungen und die Statocysten der Gattung Pecten. Sitzber. Heidelb. Akad. Wiss. 28:1–24.

Buddenbrock, W. von and Moller-Racke, I., 1953. Šber den Lichtsinn von *Pecten*. Pubbl. Zool. Stn. Napoli 24:217–245.

Bullock, T.H., 1965. Mollusca: pelecypoda and scaphopoda. In: T.H. Bullock and G.A. Horridge. Structure and Function in the Nervous System of Invertebrates. W.H. Freeman and Co., San Francisco/London. pp. 1387–1431.

Butcher, E.O., 1930. The formation, regeneration, and transplantation of eyes in *Pecten* (*gibbus, borealis*). Biol. Bull. 59:154–164.

Caddy, J.F., 1968. Underwater observations on scallop (*Placopecten magellanicus*) behaviour and drag efficiency. J. Fish. Res. Board Can. 25:2123–2141.

Chernoff, H., 1987. Factors affecting mortality of the scallop *Chlamys asperrima* (Lamarck) and its epizooic sponges in South Australian waters. J. Exp. Mar. Biol. Ecol. 109:155–171.

Cornwall, M.C. and Gorman, A.L.F., 1979. Contribution of calcium and potassium permeability changes to the off response of scallop hyperpolarizing photoreceptors. J. Physiol. 291:207–232.

Cornwall, M.C. and Gorman, A.L.F., 1983a. The cation selectivity and voltage dependence of the light-activated potassium conductance in scallop distal photoreceptor. J. Physiol. 340:287–305.

Cornwall, M.C. and Gorman, A.L.F., 1983b. Colour dependence of the early receptor potential and late receptor potential in scallop distal photoreceptor. J. Physiol. 340:307–334.

Dakin, W.J., 1909. Memoir on *Pecten*. In: L.M.B.C. Mem. Typ. Br. Mar. Pl. Anim., no. 17.

Dakin, W.J., 1910a. The eye of *Pecten*. Q. J. Microsc. Sci. 55:49–112.

Dakin, W.J., 1910b. The visceral ganglion of *Pecten*, with some notes on the physiology of the nervous system, and an inquiry into the innervation of the osphradium in the Lamellibranchiata. Mitt. Zool. Stn. Neapel 20:1–40.

Dakin, W.J., 1928. The eyes of *Pecten* and allied lamellibranchs. Proc. R. Soc. Lond. B 103:355–365.

Delcomyn, F., 1980. Neural basis of rhythmic behaviour in animals. Science 210:492–498.

Dickie, L.M. and Medcof, J.C., 1963. Causes of mass mortalities of scallops (*Placopecten magellanicus*) in the Southwestern Gulf of St. Lawrence. J. Fish. Res. Board Can. 20:451–482.

Drew, G.A., 1906. The habits, anatomy, and embryology of the giant scallop (*Pecten tenuicostatus* Mighels). Univ. Maine Studies, No. 6.

Eakin, R.M., 1965. Evolution of photoreceptors. Cold Spring Harbor Symp. Quant. Biol. 30:363–370.

Ellers, O., 1987. Passive orientation of benthic animals in flow. In: W.F. Herrnkind and A.B. Thistle (Eds.). Signposts in the Sea. Proceedings of a Multidisciplinary Workshop on Marine Animal Orientation and Migration. Florida State University, Tallahassee, Florida. pp. 45–68.

Feder, H.M., 1963. Gastropod defensive responses and their effectiveness in reducing predation by starfishes. Ecology 44:505–512.

Feder, H.M., 1972. Escape responses in marine invertebrates. Scient. Am. 227:93–100.

Forester, A.J., 1979. The association between the sponge *Halicondria panicea* (Pallas) and scallop *Chlamys varia* (L.): a commensal-protective mutualism. J. Exp. Mar. Biol. Ecol. 36:1–10.

Gorman, A.L.F. and McReynolds, J.S., 1969. Hyperpolarizing and depolarizing receptor potentials in the scallop eye. Science 165:309–310.

Gorman, A.L.F. and McReynolds, J.S., 1978. Ionic effects on the membrane potential of hyperpolarizing photoreceptors in scallop retina. J. Physiol. 275:345–355.

Grillner, S., 1985. Neurobiological bases of rhythmic motor acts in vertebrates. Science 228:143–149.

Gruffydd, L.D., 1976. Swimming in *Chlamys islandica* in relation to current speed and an investigation of hydrodynamic lift in this and other scallops. Norw. J. Zool. 24:365–378.

Gutsell, J.S., 1930. Natural history of the bay scallop (*Pecten irradians*). Bull. U.S. Bur. Fish. 46:569–632.

Hartline, H.K., 1938a. Response of single optic nerve fibers of the vertebrate eye to illumination of the retina. Am. J. Physiol. 121:400–415.

Hartline, H.K., 1938b. The discharge of impulses in the optic nerve of *Pecten* in response to illumination of the eye. J. Cell. Comp. Physiol. 11:465–478.

Hartnoll, R.G., 1967. An investigation of the movement of the scallop, *Pecten maximus*. Helgol. Wiss. Meeresunters. 15:523–533.

Järvilehto, M., 1979. Receptor potentials in invertebrate visual cells. In: H. Autrum (Ed.). Vision in Invertebrates. (Handbook of Sensory Physiology, VII/6A). Springer, Berlin/Heidelberg/New York. pp. 315–356.

Kellogg, J.L., 1915. Ciliary mechanisms of lamellibranchs with descriptions of anatomy. J. Morph. 26:625–701.

Kennedy, D., 1960. Neural photoreception in a lamellibranch mollusc. J. Gen. Physiol. 44:277–299.

Land, M.F., 1965. Image formation by a concave reflector in the eye of the scallop, *Pecten maximus*. J. Physiol. 179:138–153.

Land, M.F., 1966. Activity in the optic nerve of *Pecten maximus* in response to changes in light intensity and to pattern and movement in the optical environment. J. Exp. Biol. 45:83–99.

Land, M.F., 1968. Functional aspects of the optical and retinal organization of the mollusc eye. Symp. Zool. Soc. Lond. 23:75–96.

Land, M.F., 1978. Animal eyes with mirror optics. Sci. Am. 239:126–134.

Lecomte, J., 1952. Reactions de fuite des pectens en presence des asterides. Vie et Milieu 3:57–60.

Lowy, J., 1954. Contraction and relaxation in the adductor muscles of *Pecten maximus*. J. Physiol. 124:100–105.

Ludel, J., 1974. Behavioural responses to visual stimulation in the scallop (*Aequipecten irradians*). Florida Sci. 37:72–78.

Mackie, A.M., 1970. Avoidance reactions of marine invertebrates to either steroid glycosides of starfish or synthetic surface-active agents. J. Exp. Mar. Biol. Ecol. 5:63–69.

Mackie, A.M., Lasker, R. and Grant, P.T., 1968. Avoidance reactions of a mollusc *Buccinum undatum* to saponin-like surface-active substances in extracts of the starfish *Asterias rubens* and *Marthasterias glacialis*. Comp. Biochem. Physiol. 26:415–428.

Margolin, A.S., 1964. The running response of *Acmaea* to seastars. Ecology 45:191–193.

McClintock, J.B., 1983. Escape response of *Argopecten irradians* (Mollusca: Bivalvia) to *Luidia clathrata* and *Echinaster* sp. (Echinodermata: Asteroidea). Florida Sci. 46:95–100.

McReynolds, J.S. and Gorman, A.L.F., 1970a. Photoreceptor potentials of opposite polarity in the eye of the scallop, *Pecten irradians*. J. Gen. Physiol. 56:376–391.

McReynolds, J.S. and Gorman, A.L.F., 1970b. Membrane conductances and spectral sensitivities of *Pecten* photoreceptors. J. Gen. Physiol. 56:392–406.

Mellon, Jr., D., 1968. Junctional physiology and motor nerve distribution in the fast adductor muscle of the scallop. Science 160:1018–1020.

Mellon, Jr., D., 1969. The reflex control of rhythmic motor output during swimming in the scallop. Z. vergl. Physiol. 62:318–336.

Miller, W.H., 1958. Derivatives of cilia in the distal retina of *Pecten*. J. Biophys. Biochem. Cytol. 4:227–228.

Moore, J.D. and Trueman, E.R., 1971. Swimming of the scallop, *Chlamys opercularis* (L.). J. Exp. Mar. Biol. Ecol. 6:179–185.

Moore, J.K. and Marshall, N., 1967. An analysis of the movements of the bay scallop, *Aequipecten irradians*, in a shallow estuary. Proc. Natl. Shellfish Assoc. 57:77–82.

Morton, B., 1980. Swimming in *Amusium pleuronectes* (Bivalvia: Pectinidae). J. Zool. Lond. 190:375–404.

Mpitsos, G.J., 1973. Physiology of vision in the mollusk *Lima scabra*. J. Neurophysiol. 36:371–383.

Olsen, A.M., 1955. Underwater studies on the Tasmanian commercial scallop, *Notovola meridionalis* (Tate) (Lamellibranchiata: Pectinidae). Aust. J. Mar. Freshwater Res. 6:392–409.

Ordzie, C.J. and Garofalo, G.C., 1980. Behavioural recognition of molluscan and echinoderm predators by the bay scallop, *Argopecten irradians* (Lamarck) at two temperatures. J. Exp. Mar. Biol. Ecol. 43:29–37.

Peterson, C.H., Ambrose, W.G. and Hunt, J.H., 1982. A field test of the swimming response of the bay scallop (*Argopecten irradians*) to changing biological factors. Bull. Mar. Sci. 32:939–944.

Pitcher, C.R. and Butler, A.J., 1987. Predation by asteroids, escape response, and morphometrics of scallops with epizoic sponges. J. Exp. Mar. Biol. Ecol. 112:233–249.

Prior, D.J., Schneiderman, A.M. and Greene, S.I., 1979. Size-dependent variation in the evasive behaviour of the bivalve mollusc *Spisula solidissima*. J. Exp. Biol. 78:59–75.

Reye, D.N. and Pearson, K.G., 1987. Projections of the wing stretch receptors to central flight neurons in the locust. J. Neurosci. 7:2476–2487.

Schneider, D., 1982. Escape response of an infaunal clam *Ensis directus* Conrad 1843, to a predatory snail, *Polinices duplicatus* Say 1822. Veliger 24:371–372.

Spagnolia, T. and Wilkens, L.A., 1983. Neurobiology of the scallop. II. Structure of the parietovisceral ganglion lateral lobes in relation to afferent projections from the mantle eyes. Mar. Behav. Physiol. 10:23–55.

Stanley, S.M., 1970. Relation of shell form to life habits of the Bivalvia (Mollusca). Geol. Soc. Am. Mem. 125:1–296.

Stasek, C.R., 1965. Behavioural adaptation of the giant clam *Tridacna maxima* to the presence of grazing fishes. Veliger 8:29–35.

Stephens, P.J., 1978. The sensitivity and control of the scallop mantle edge. J. Exp. Biol. 75:203–221.

Stephens, P.J. and Boyle, P.R., 1978. Escape responses of the queen scallop *Chlamys opercularis* (L.) (Mollusca: Bivalvia). Mar. Behav. Physiol. 5:103–113.

Thomas, G.E. and Gruffydd, L.D., 1971. The types of escape reactions elicited in the scallop *Pecten maximus* by selected sea star species. Mar. Biol. 10:87–93.

Thompson, R.J., Livingstone, D.R. and DeZwann, A., 1980. Physiological and biochemical aspects of the valve snap and valve closure responses in the giant scallop *Placopecten magellanicus*. J. Comp. Physiol. B 137:97–104.

Vogel, S., 1985. Flow-assisted shell reopening in swimming scallops. Biol. Bull. 169:624–630.

Wenrich, D.H., 1916. Notes on the reactions of bivalve mollusks to changes in light intensity: image formation in *Pecten*. J. Anim. Behav. 6:297–318.

Wiederhold, M.L., MacNichol, Jr., E.F. and Bell, A.L., 1973. Photoreceptor spike responses in the hardshell clam, *Mercenaria mercenaria*. J. Gen. Physiol. 61:24–55.

Wilkens, L.A., 1981. Neurobiology of the scallop. I. Starfish-mediated escape behaviours. Proc. R. Soc. Lond. B 211:341–372.

Wilkens, L.A., 1987. The visual system of the giant clam *Tridacna*: behavioural adaptations. Biol. Bull. 170:393–408.

Wilkens, L.A. and Ache, B.W., 1977. Visual responses in the central nervous system of the scallop *Pecten ziczac*. Experientia 33:1338–1339.

Winter, M.A. and Hamilton, P.V., 1985. Factors influencing swimming in bay scallops, *Argopecten irradians* (Lamarck, 1819). J. Exp. Mar. Biol. Ecol. 88:227–242.

Yonge, C.M., 1936. The evolution of the swimming habit in the Lamellibranchia. Mém. Musée Royale D'Hist. Nat. Belg., Ser. II, 3:77–100.

Zhadan, P.M. and Semen'kov, P.G., 1984. An electrophysiological study of the mechanoreceptory function of abdominal sense organ of the scallop *Patinopecten yessoensis* (Jay). Comp. Biochem. Physiol. A 78:865–870.

AUTHOR'S ADDRESS

Lon A. Wilkens - Department of Biology, University of Missouri-St. Louis, 8001 Natural Bridge Road, St. Louis, Missouri 63121–4499 USA (E-mail: lon_wilkens@umsl.edu)

Scallops: Biology, Ecology and Aquaculture
S.E. Shumway and G.J. Parsons (Editors)
© 2006 Elsevier B.V. All rights reserved.

Chapter 6

Reproductive Physiology

Bruce J. Barber and Norman J. Blake

6.1 INTRODUCTION

Reproduction is an important aspect of the life history of any species and having an understanding of reproductive processes is central to the management of any commercial fishery. This is particularly true for the Pectinidae, which in many instances have meat (adductor muscle) weights (and thus yields) that vary seasonally in relation to gametogenesis and spawning. Thus, a trade-off is sometimes made between yield per unit effort and reproductive potential. For short-lived species in which annual recruitment is dependent on the reproductive success of the previous year class, it would seem prudent to regulate harvesting to maximise reproductive output. A good example of this is the bay scallop (*Argopecten irradians*) fishery along the east coast of the United States for which harvest each year is delayed until after spawning has been completed. The availability and quality of roe-on product obviously changes seasonally in conjunction with gonad maturity. Knowledge of spawning periods is important for understanding and predicting recruitment events and maximising the collection of scallop spat for enhancement or culture. Production of the Japanese scallop, *Patinopecten yessoensis*, in Mutsu Bay, Aomori Prefecture, is the ultimate example of this approach. An understanding of reproductive physiology is also critical for the successful hatchery production (especially conditioning and spawning of broodstock) of any species.

Several important reviews of molluscan reproduction currently exist. Fretter and Graham (1964) reviewed reproduction in the entire phylum, while Purchon (1977) concentrated on reproduction in freshwater bivalves. Seed (1976) reviewed reproduction in the Mytilidae. Sastry (1979) provided a comprehensive review of reproductive physiology in bivalves (excluding the Ostreidae). Andrews (1979) reviewed reproduction in the Ostreidae. Mackie (1984) summarised molluscan reproduction in both marine and freshwater environments. These reviews attempted the difficult task of describing a basic physiological process in a very diverse group of organisms. This often necessitated sacrificing either detail on a specific topic or completeness of coverage (either number of topics covered or number of species included).

This chapter is an update of Barber and Blake (1991), an attempt to summarise the current state of knowledge of reproductive physiology in the Pectinidae. The justifications for continuing this effort are just as valid today as they were ten years ago. Scallops are a diverse, but ecologically and commercially important group of marine bivalves that are sufficiently unique from a reproductive standpoint to deserve discrete attention. Second, significant contributions to the body of knowledge have been made during the last decade and need to be included. Third, much of what we have learned of the basic processes of reproductive physiology can be applied to the commercial culture of scallops, a rapidly expanding sector of the global aquaculture industry. Therefore, this chapter will concentrate on the description of gametogenic cycles, the factors (both endogenous and exogenous) that influence gametogenesis and spawning, the allocation of energy for gamete synthesis (energy metabolism), and the extrapolation of this information to critical aspects of commercial aquaculture, namely broodstock conditioning and spawning.

6.2 GAMETOGENIC CYCLES

6.2.1 Definition

Scallops, like other marine bivalves, have reproductive cycles that include periods of gamete formation, spawning, fertilisation, larval development, settlement and metamorphosis, and growth to reproductive maturity. Cycles of gametogenesis (the focus of this chapter) occur on a regular basis throughout the adult life of scallops and include a vegetative period followed by periods of differentiation, cytoplasmic growth, vitellogenesis (maturation), spawning (release of gametes), and resorption of unspawned gametes, all of which affect the relative size of the gonad (Fig. 6.1). Various exogenous and endogenous factors determine the timing and duration of these events for a particular species at a particular location and time. Thus, the reproductive cycle can be considered a genetically controlled response to the environment (Sastry 1979).

6.2.2 Means of Assessment

Several means of assessing gametogenesis in scallops have been utilised. Some of these are applicable to all bivalve species, and some are uniquely available to scallop biologists. The anatomy of the scallop is such that the gonad is easily visible and anatomically distinct from the rest of the visceral mass. As the gonad matures, macroscopic changes are readily visible. It increases in weight and size and becomes rounder in cross section as gametes become larger and more numerous. Also, as the sexes differentiate and gametogenesis proceeds, male gonads attain a cream colour and female gonads become reddish. After spawning and the release of gametes, gonads become smaller, flatter in cross section, colourless, and watery in appearance. Scallop gonads can thus be characterised as either empty, filling, full (mature), partially spawned, or spent, based on external appearance (Fig. 6.2). Similarly, microscopic changes are occurring within gonadal tissue as gametes develop, mature and are spawned. Both external and internal features have been used in varying degrees to assess reproduction in scallops (see Table 6.1).

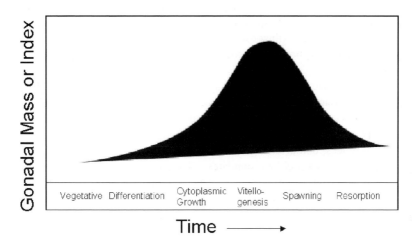

Figure 6.1. Generalised (annual) gametogenic cycle indicating various stages of development relative to gonad size.

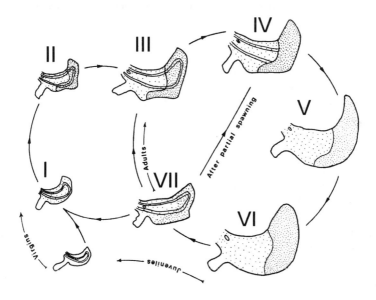

Figure 6.2. Macroscopic changes in the gonad of *Pecten maximus*: Stage I, developing; Stage II, differentiated; Stage III, recovering; Stage IV, filling; Stage V, half-full; Stage VI, full; Stage VII, spent (After Mason 1958a).

6.2.2.1 Visual observation

The most widely used (and simplest) method of assessing gametogenesis in scallops has been via gross visual examination of the gonads. Observations were made as to the relative size, shape ("condition"), and colour of the gonads of several species (Tables 6.2, 6.3 and 6.4). Mason (1958a) arbitrarily divided gonads of *Pecten maximus* into eight stages of maturity, on the basis of external appearance. Similarly, Naidu (1970) presented nine stages of gonad development for *Placopecten magellanicus*. Citing the difficulty in differentiating between undeveloped adults and virgin juveniles, Davidson and Worms (1989) recognised just four stages of gonad development for *P. magellanicus*.

In some instances, colour changes in the female gonad were assessed through more elaborate comparisons with colour standards. Tang (1941) used the "Colour Standards and Colour Nomenclatures" (Robert Ridgeway, Washington, 1912) to determine gonad development in *Pecten maximus*, and noted that as the gonad matured, the percentage of red hue gradually increased. The "Pantone Color Specifier" (Pantone Matching Systems of Pantone, Inc., New York) was used by Miller et al. (1979) to assess ovarian development in *Argopecten gibbus*. Gross visual examination of scallop gonads is the simplest method of assessing reproduction as it involves fresh material with no preparation. However, it only provides a rough estimate of overall gonadal development and no information on gamete development *per se*.

By examining fresh gonadal smears microscopically, more direct information can be obtained as to the development and viability of gametes. Spawning and fertilisation success is correlated with gamete maturity. The degree of motility of sperm and shape and size of ova was used to assess gamete maturity in *Chlamys* spp. (Reddiah 1962; Wiborg 1963) and *Placopecten magellanicus* (Naidu 1970). Bull (1976) also incorporated gonadal smears into his study of reproduction in *Pecten novaezelandiae*. Tang (1941) examined gamete viability in *P. maximus* by attempting artificial fertilisation in the laboratory. This approach provides a measure of gamete viability, but like macroscopic examination of the gonad, is only qualitative in nature.

6.2.2.2 Gonad mass and index

The anatomically distinct gonad of the scallop allows for easy removal from the rest of the soft tissue. Thus a fairly simple but effective means of assessing the timing, duration and extent of gametogenesis in scallops is to determine mean gonad mass (weight, g) on a regular basis throughout the year (e.g., Barber and Blake 1981; Barber et al. 1988). Dry weights are preferable to wet weights as water content varies seasonally. As gonad size is related to scallop (size) age (see Schick et al. 1992), it is necessary to sample individuals of a fairly discrete size range (e.g., Barber et al. 1988) or determine the gonad weight of a "standard" scallop by regressing gonad weight on shell height over a large size range (e.g., Dredge 1981; Robinson et al. 1981). Gametogenic cycles have been determined using gonad weights for several species of scallops, as indicated in Tables 6.2, 6.3 and 6.4. Bricelj et al. (1987) used ash-free dry weights to eliminate the contribution of inorganic intestinal components to gonad weight. Besides being useful for determining periods of gamete development, gonad weights provide an ecologically meaningful, quantitative estimate of fecundity or reproductive output.

Table 6.1

Macroscopic and microscopic characteristic of gonads of *Pecten maximus* in various stages of maturity (after Mason 1958a).

Gonad Stage	External appearance of gonad	Cytological details
Stage 0 Immature	Small, flat; transparent, colourless	Some connective tissue; narrow tubules with primary germ cells give rise to follicles
Stage I Developing	Growing; minute follicles visible; fawn-coloured; no sexual differentiation	Follicles growing; male follicles lined by spermatogonia; female follicles have oogonia; young oöcytes (up to 30 µm)
Stage II Differentiated	Growing; obviously differentiated into testis (white) and ovary (fawn)	Spermatazoa appear at centre of male follicles; many half-grown oöcytes (30–60 µm) in female follicles
Stage III Recovering	Larger, thicker; flabby, containing free water; testis white and ovary bittersweet orange	Male follicles contain more spermatozoa, but loosely packed; mostly growing oöcytes in female follicles
Stage IV Filling	Still larger and thicker; colouring brighter (testis white and ovary orange to pink)	Spermatogonia, spermatocytes and few spermatids; half-grown and few larger (60–80 µm) oöcytes; little connective tissue
Stage V Half-full	Thicker, firmer; brighter (testis white and ovary pink)	Lumina full of spermatozoa and almost fully grown oöcytes; very few oogonia and little connective tissue
Stage VI Full	Large, thick, firm; highly coloured (testis cream and ovary grenadine)	Follicles maximum size; male follicles packed with spermatozoa; female follicles packed with 80–90 µm polygonal oocytes
Stage VII Partially Spent or Spent	Dull, thin, collapsed; containing much free water; follicles empty; sexual differentiation may be lost	Follicles smaller, containing large spaces; few residual spermatocytes and oocytes

Table 6.2

Major spawning periods reported for various populations of pectinid species. Means of assessing reproductive status include **GO** (gross observation), **MO** (microscopic observation), **GW** (gonad weight), **GI** (gonad index), **H** (histology, with oocyte measurements (**Oo**) or sterological measurements (**s**)), **L** (larval abundances), and **S** (spat abundances).

Species	Location	Major spawning periods	Means of assessing	Source
Amusium japonicum balloti (Bernardi)	Queensland, Aust.	Aug. – Nov.	GO, GW, H	Dredge (1981)
	Shark Bay, Aust.	Dec. – Jan.	GO	Heald and Caputi (1981)
Amusium pleuronectes Linnaeus	Visayan Sea, Phil.	Jan., June, Oct.	GO, H	Llana and Aprieto (1980)
Argopecten circularis (Sowerby)	Panama	April	H	Villalaz and Gomez (1987)
	Mexico	Dec. – Mar.	GW, H (Oo)	Villalejo-Fuerte and Ochoa-Báez (1993)
Argopecten gibbus (Linnaeus)	Florida, USA	March – June	GO	Roe et al. (1971)
		Sept. – Oct.		
	Florida, USA	Jan. – May	GO	Miller et al. (1979, 1981)
		Oct. – Dec.		
	Florida, USA	Apr. – May		Moyer and Blake (1986)
		Aug. – Oct.		
Argopecten irradians irradians Lamarck	(See Table 6.3)			
Argopecten irradians concentricus (Say)	(See Table 6.3)			
Argopecten purpuratus (= *Chlamys purpuratus*) Lamarck	Peru	Feb. – May	GI, L, S	Wolff (1987, 1988)
Argopecten ventricosus (Sowerby)	Mexico	July – Dec.	GI, H	Luna-González et al. (2000)
Chlamys asperrimus Lamarck	Tasmania	Sept. – Oct.	GW, GI, S	Zacharin (1994)
		Dec. – Jan.		
	Jervis Bay, Aust.	Aug. – Sept.	GW, GI, MO	O'Connor and Heasman (1996)
Chlamys distorta (Da Costa)	Isle of Man, UK	May – Aug.	GO, H	Reddiah (1962)
Chlamys farreri (Kuroda)	Quingdao, China	May – June	GI, H	Liao et al. (1983)
		Sept. – Oct.		
	Possiet Bay, USSR	June – July	GI	Kalashnikova (1984)
Chlamys furtive (Loven)	Isle of Man, UK	June, Oct. – Nov.	MO, H	Reddiah (1962)

Species	Location	Season	Code	Reference
Chlamys islandica (O.F. Müller)	Norway	July – Aug.	GO, MO	Wiborg (1963)
	Norway	June – July	MO, GI	Skreslet and Brun (1969); Skreslet (1973)
	Norway	June – July	GW, H(s)	Sundet and Vahl (1981); Sundet and Lee (1984)
Chlamys opercularis (Linnaeus)	Iceland	July	GI, H	Thorarinsdóttir (1993)
	Firth of Forth, UK	Aug. – Sept.	GO, L, S	Fullarton (1889)
	Firth of Forth, UK	July – Aug.	GO	Dakin (1909)
	Plymouth, UK	June	GO	Amirthalingam (1928)
	Danish Waters	July – Aug.	GO	Ursin (1956)
	North Sea	Aug. – Sept.	GO	Ursin (1956)
	Faroe Waters	July – Sept.	GO	Ursin (1956)
	Hampshire, UK	June – July	GO	Broom and Mason (1978)
		Sept. – Oct.		
		Jan. – Feb.		
Chlamys septemradiata (O.F. Müller)	Clyde Sea, UK	June – July	GO, GW	Taylor and Venn (1979)
Chlamys striata (O.F. Müller)	Clyde Sea, UK	July – Aug.	GW	Ansell (1974)
Chlamys tehuelcha (d'Orbigny)	Isle of Man, UK	Aug. – Sept.	MO, H	Reddiah (1962)
	Chubut, Urug.	Oct. – Jan.	GO, GW, GI, H(Oo)	Lasta and Calvo (1978)
Chlamys tigerina (O.F. Müller)	Isle of Man, UK	June	MO, H	Reddiah (1962)
Chlamys varia (Linnaeus)	La Rochelle, Fra.	June – July, Sept. – Oct	H	Lubet (1959)
	Isle of Man, UK	June, Sept. – Nov.	MO, H	Reddiah (1962)
	Bay of Brest, Fra.	Sept. – Oct., May – June	GI, H	Shafee and Lucas (1980)
Chlamys zelandiae (= *Pecten zelandiae*) Gray	Brittany, Fra.	June, Aug. – Sept.	GI, S	Perodou and Latrouite (1981)
	Kilkieran Bay, Ire.	May – June, Aug.	GI, H, L, S	Burnell (1983)
	Bay of Islands, NZ	July – Aug.; Oct.	L	Booth (1983)
	Raumati Beach, NZ	Mar., June, July	L	Booth (1983)
	Wellington Hr., NZ	May – Dec.; Jan.	L	Booth (1983)
Hinnites giganteus (= *Crassadoma gigantea*) (Gray)	Washington, USA	June	Gi, H(Oo)	Lauren (1982)
	California, USA	May – Oct.	H	Malachowski (1988)

Species	Location		Code	Reference
Hinnites multirugosus (= *Crassodoma gigantean*) (Gray)				
Patinopecten caurinus (Gould)	California, USA	July – Aug., Oct. – Nov.	GI, H(Oo)	Jacobsen (1977)
	Alaska, USA	June – July	GO	Hennick (1970)
	B.C., Can. (inshore)	April – June	H(s)	MacDonald and Bourne (1987)
	(offshore)	July – Aug.		MacDonald and Bourne (1987)
	Oregon, USA	Jan. – June	H	Robinson and Breese (1984)
Patinopecten yessoensis	Mutsu Bay, Jpn.	March – May	H, L, S	Yamamoto (1943, 1950, 1953, 1968)
	Hokkaido, Jpn.	May – June	GI, H	Wakui and Obara (1967)
	Hokkaido, Jpn.	May – June	GI, H	Maru (1976, 1985)
	Hokkaido, Jpn.	April – June	GI	Kawamata (1983)
	Victoria, Aust.	Aug. – Oct.	GI, H	Sause et al. (1987)
Pecten alba (Tate)	Isle of Man, UK	April – Aug.	GO	Dakin (1909)
Pecten maximus (Linnaeus)	Isle of Man, UK	June	GO, MO, L	Tang (1941)
	Bantry Bay, Ire.	April – May, Sept.	GO	Gibson (1956)
	Isle of Man, UK	Aug. – Sept., April – May	H	Mason (1958a)
	La Rochelle, Fra.	April – May	H	Lubet (1959)
	Wales, UK	Spring, Autumn	GO	Baird (1966)
	Northern Ire.	May – June, Aug.	GW, GI	Stanley (1967)
	Clyde Sea, UK	June – July	GW	Comely (1974)
	Bay of Brest, Fra.	July – Aug.	GI, H(s)	Lubet et al. (1987)
	Connemara, Ire.	June	GI, H, S	Wilson (1987)
	Galway Bay, Ire.	April – May, July – Aug.	GI, H, S	Wilson (1987)
	Bay of Seine, Fra.	July – Aug.	GI, H(s), S	Lubet et al. (1991)
	Fosen, Nor.	June	GI, H	Strand and Nylund (1991)
Pecten nouvaezelandiae Reeve	Austevolbl, Nor.	July – Sept.	GI, H	Strand and Nylund (1991)
	Tasman Bay, NZ	March, Aug. – Sept.	GO	Choat (1960)
	Tasman Bay, NZ	March – April	GW	Tunbridge (1968)
	Marlborough Snd, NZ	Nov. – Dec.	GO, MO, GW, H, S	Bull (1976)
Placopecten magellanicus (Gmelin)	Aukland, NZ	Nov. – Dec.	GO, GI, H, S	Nicholson (1978)
	(See Table 6.4)			

Table 6.3

Major spawning periods reported for various populations of *Argopecten irradians*, in order of decreasing latitude. Means of assessing reproductive status include **GO** (gross observation), **MO** (microscope observation), **GW** (gonad weight), **H** (histology, with oöcyte measurements (**Oo**)), **L** (larval abundances), and **S** (spat abundances).

Sub-species	Location	Major spawning periods	Means of assessing	Source
Argopecten irradians irradians Lamarck	Mass., USA	May – July	H	Taylor and Capuzzo (1983)
	Mass., USA	June – Aug.	GO, MO	Belding (1910)
	Mass., USA	July – Sept.	GI, H	Sastry (1966b, 1970a)
	Rhode Is., USA	June – July	GO	Risser (1901)
	Conn., USA	June – July	GO, S	Marshall (1960)
	New York, USA	June	---	Hickey (1978)
	New York, USA	June – August	GW, GI	Bricelj et al. (1987)
	New York, USA	June – July, Aug. – Oct.	GW, GI, H	Tettelbach et al. (1999)
Argopecten irradians concentricus (Say)	N. Carolina, USA	July – Aug.	GO, L, S	Gutsell (1930)
	N. Carolina, USA	Sept. – Nov.	GI, H	Sastry (1966a, b, 1970a, b)
	Florida, USA	Aug. – Sept.	H	Sastry (1961)
	Florida, USA	Sept. – Nov.	GW, GI, H(Oo)	Barber and Blake (1981, 1983); Barber (1984)
	Florida, USA	Dec. – Feb.	GO, GW, GI	Bologna (1998)

Table 6.4

Major spawning periods reported for various populations of *Placopecten magellanicus* (Gmelin), in order of decreasing latitude. Means of assessing reproductive status include **GO** (gross observation), **MO** (microscope observation), **GW** (gonad weight), **H** (histology, with oocyte measurements (**Oo**) or sterological measurements (**s**)), **L** (larval abundances), and **S** (spat abundances).

Location	Major Spawning Periods	Means of Assessing	Source
Newfoundland, Canada	Sept. – Oct.; June	MO, H, L, S	Naidu (1970)
Newfoundland, Canada	Aug. – Sept.	GW, GI	Thompson (1977)
Newfoundland, Canada	Aug. – Sept.	H(s)	MacDonald and Thompson (1986, 1988)
Gulf of St. Lawrence, Canada	July – Sept.	GW, GI	Bonardelli et al. (1996)
Gulf of St. Lawrence, Canada	July – Aug.	GI, H	Davidson et al. (1993)
Gulf of St. Lawrence, Canada	Sept.	GI, H	Giguère et al. (1994)
Nova Scotia, Canada	June – Sept. / Dec. – Feb.	H	Borden (1928)
Nova Scotia, Canada	Aug. – Sept.	GO	Dickie (1955)
New Brunswick, Canada	Aug. – Sept.	GI, H(s)	Beninger (1987)
New Brunswick, Canada	July – Sept.	GI, H(Oo)	Parsons et al. (1992)
Maine, USA	Aug. – Sept.	GO, MO	Welch (1950)
Maine, USA	Aug. – Sept.	GW, GI, H(Oo)	Barber et al. (1988)
Maine, USA	Sept.	GW, GI, H	Robinson et al. (1981)
Georges Bank	Sept.	GO	Posgay and Norman (1958)
Georges Bank	Sept. – Oct.	GO	MacKenzie et al. (1978)
Georges Bank	Sept. – Oct.; May – June	GI, H(Oo, s)	Dibacco et al. (1995)
Mass., USA	Sept. – Oct.	GO	Posgay (1950)
New Jersey, USA	Oct. – Nov.	H(s)	MacDonald and Thompson (1988)
Mid-Atlantic, USA	April – May; Oct.	GW	DuPaul et al. (1989)
Mid-Atlantic, USA	April – June; Oct. – Nov.	GW	Kirkley and DuPaul (1991)
Mid-Atlantic, USA	May – June / Nov. – Dec.	GW, H(s)	Schmitzer et al. (1991)

The gonad index (GI) or gonosomatic index (GSI), which expresses gonadal mass as a proportion of total body mass is also used extensively to define gametogenic cycles in scallops:

$$\text{GI (GSI)} = [\text{Gonad Weight (g)} / \text{Total Tissue Weight (g)}] \cdot 100 \qquad (1)$$

Both wet and dry tissue weights can be used, but dry weights are preferable to eliminate the variability in water content encountered both seasonally and among the different tissues. Latrouite and Claude (1979) calculated GI as the percentage of gonad weight to remaining tissue weight rather than total tissue weight. For either ratio, GI is affected by changes in mass of either gonadal or somatic tissues. Gonad indices were used to assess gametogenesis in several species (Tables 6.2, 6.3 and 6.4).

The use of gonadal indices for assessment of relative gametogenic state assumes that the following conditions are met: 1) the allometric relationship between gonad and total tissue mass does not change over the size range of the population studied; 2) the slopes of the allometric growth equations for the gonad and total tissues are similar, and; 3) the mass of the non-gonadal tissue does not change over time. Bonardelli and Himmelman (1994) examined these assumptions for body component indices of *Placopecten magellanicus* ranging in shell height from 29–153 mm at two sites, and found that they were not met due to differences between maturing and fully mature individuals. Instead, it was recommended that gonadal mass be scaled to maximum shell height using a gonadal mass index (GMI), where b is the slope of the regression of gonadal mass to shell height for mature individuals and k is a constant to obtain a value greater than zero (Bonardelli and Himmelman 1994).

$$\text{GMI} = [\text{Gonadal Mass} / (\text{Shell Height})^b] \cdot k \qquad (2)$$

An index relating gonad weight to shell height (rapport gonado-hauteur, RGH) has been used to monitor reproduction in *Chlamys varia* (Perodou and Latrouite 1981; Burnell 1983) and *Pecten maximus* (Wilson 1987):

$$\text{RGH} = [\text{Gonad Weight (g)} / \text{Shell Height}^3 \text{ (mm)}] \cdot 10^3 \qquad (3)$$

Since scallops are not spherical, however, the use of the cube function in the denominator is questionable. It is important to note that both GI and RGH indices provide only relative estimates of gamete production. This subtle but important distinction separates gonad index from gonad mass as a means of assessing gametogenic stage and output.

6.2.2.3 Histology

The methods discussed so far, although simple, fast, and inexpensive, do not provide the ability to examine cytological changes occurring within the gonad. Histological preparation of gonadal tissue, although costly and time consuming, provides the means to examine and assess gamete development definitively. For example, a decrease in gonad mass or index, which may occur as the result of either spawning or resorption of gametes, can only be distinguished histologically. Although gametogenesis is a continual process, most investigators have ascribed developmental stages to certain cytological features, which are generally recognised in all bivalve molluscs. For example, Mason (1958a) arbitrarily divided gonads of *Pecten maximus* into eight stages of maturity, on the basis of external appearance and cytological details (Table 6.1; Fig. 6.2). Unfortunately, determination of these stages tends to be subjective and there is little agreement as to the number of stages that should be included. Complete cytological descriptions of gamete development (with micrographs) are also provided by Bull (1976) for *P. novaezelandiae*, Sastry (1961) for *Argopecten irradians*, Naidu (1970) and Davidson and Worms (1989) for *Placopecten magellanicus*, Dredge (1981) for *Amusium japonicum balloti*, and Burnell (1983) for *Chlamys varia*. Histology has been employed to varying degrees by other investigators, as indicated in Tables 6.2, 6.3 and 6.4.

In addition to qualitative information, histological preparations can also be used for generating quantitative data. For example, Burnell (1983), Sause (1987), and Wilson (1987) assigned numerical values to developmental stages so that a mean "histological index" could be calculated for each sample. Mean oöcyte diameter, obtained from histological sections, reflects the reproductive cycle, as oöcytes gradually increase in size as they develop, reaching a maximum size prior to spawning. Mean oöcyte diameter decreases sharply after spawning, as mostly larger, mature ova are released (Fig. 6.3). Thus mean oöcyte diameters were used to monitor gamete development in *Argopecten irradians* (Sastry 1970a; Barber and Blake 1981, 1983; Barber 1984), *Hinnites giganteus* (Lauren 1982), and *Placopecten magellanicus* (Barber et al. 1988). Lasta and Calvo (1978) measured follicle diameter as well as oöcyte diameter to follow the reproductive cycle of *Chlamys tehuelcha*. Mean oöcyte diameter, however, was not effective for defining gametogenic cycles in *C. varia*, as partial spawning and the development of new oöcytes occurred simultaneously (Burnell 1983). Since oöcytes are rarely circular in histological sections, measurement of mean area of oöcytes probably more accurately reflects oöcyte development than mean oöcyte diameter (e.g., Barber 1996). For either approach, however, there should be some standardisation of the number of microscopic fields observed from each section, how they are chosen, the number of oöcytes measured, and which oöcytes are measured (e.g., only those sectioned through the nucleus).

The application of stereological techniques to histological preparations can provide additional quantitative information in the form of volumetric ratios (Freere and Weibel 1966; Weibel et al. 1966). Thus gamete volume fractions were determined and applied to the study of gametogenesis in *Chlamys islandica* (Sundet and Lee 1984), *Patinopecten caurinus* (MacDonald and Bourne 1987), *Pecten maximus* (Lubet et al. 1987), and *Placopecten magellanicus* (MacDonald and Thompson 1986, 1989; Beninger 1987). This takes advantage of the fact that as gametogenesis proceeds, follicles become increasingly filled with gametes, and gonadal tissue as a whole is increasingly comprised of gametes. Estimation of volume fractions of multiple cytological features (e.g., connective tissue, follicle or lumen space, developing gametes, mature gametes, and resorbing gametes) over the course of an entire gametogenic cycle provides additional insights into gametogenic processes, including nutrient transfer and growth and development of gametes.

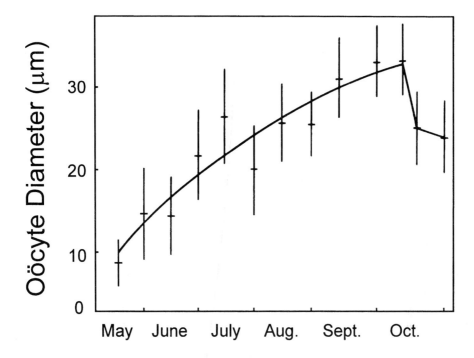

Figure 6.3. Reproductive cycle of *Argopecten irradians* as defined by mean (± 1 SD) oöcyte diameter. (From Barber and Blake 1981).

6.2.2.4 Abundance of larvae and spat

The presence and abundance of two other life stages of scallops have been used to indirectly estimate spawning periods. Seasonal peaks in larval abundance were used to determine spawning times for *Chlamys opercularis* (Fullarton 1889), *Argopecten irradians* (Gutsell 1930), *A. purpuratus* (Wolff 1988), *Pecten maximus* (Tang 1941), *Patinopecten yessoensis* (Yamamoto 1950, 1951b, 1953, 1968), *Placopecten magellanicus* (Naidu 1970), *Chlamys zelandiae* (Booth 1983), and *C. varia* (Burnell 1983). DiBacco et al. (1995) used the presence of larvae to confirm spring spawning of *P. magellanicus*. Similarly, the occurrence of spat (post-settlement juveniles) was used to infer spawning times for *C. opercularis* (Fullarton 1889), *C. varia* (Burnell 1983), *A. irradians* (Gutsell 1930; Marshall 1960), *A. gibbus* (Allen 1979), *A. purpuratus* (Wolff 1988), *P. yessoensis* (Yamamoto 1950), and *Pecten novaezelandiae* (Bull 1976). Although occurrence of larvae indicate spawning and fertilisation have occurred, absence of larvae does not necessarily indicate an absence of spawning activity. Also, larval settlement and metamorphosis, resulting in the occurrence of spat, is highly dependent on environmental conditions. Therefore, this approach, although not reflective of the events leading up to spawning, can provide some indication of spawning periods and the relative success of spawning and early life history stages.

All of the means of assessing reproduction in scallops have inherent advantages and disadvantages. The most complete approach would be to employ at least two methods, one quantitative (gonad weight or index) and one qualitative (histology). The histology is necessary for verification of reproductive events pertaining to gamete development not discernible by the other methods (see Beninger 1987). Total gamete weight can be calculated by multiplying gamete volume fraction by gonad weight (MacDonald and Thompson 1986).

6.2.3 Variations in Gametogenic Cycles

The events that occur in the reproductive cycle of a population include activation, growth and gametogenesis, ripening of gametes, spawning, and an inactive or "resting" period (Sastry 1979). Spawning periods, because they represent the culmination of the reproductive process and are definable by all the methods outlined above, provide a convenient focus for comparing gametogenic cycles. Within a population, scallops tend to develop and spawn synchronously. There are, however, inter- and intra-specific differences with respect to the frequency and timing of spawning as well as the duration of spawning activity (Tables 6.2, 6.3 and 6.4). There are also differences in the amount of gametogenic material produced (fecundity). Within a species, environmental differences between years and locations most likely result in the observed differences in reproductive success and recruitment. Genetic variation also contributes to the variability in gametogenesis between populations and species.

6.2.3.1 Intra-specific variations

Variations in the timing of gametogenic cycles between years have been observed for several species of scallops. The timing of gametogenic events was consistent between years for *Pecten novaezelandiae* (Tunbridge 1968; Bull 1976), *Amusium balloti* (Heald and Caputi 1981), *Argopecten irradians* (Barber and Blake 1983), *Pecten alba* (Sause 1987), *Pecten maximus* (Lubet et al. 1991), *Placopecten magellanicus* (Parsons et al. 1992), *Chlamys islandica* (Thorarinsdóttir 1993), and *Chlamys asperrima* (Zacharin 1994; O'Connor and Heasman 1996). Spawning occurred at similar times of the year, but differences in fecundity were observed between years for *C. islandica* (Skreslet 1973) and *Chlamys varia* (Shaffee and Lucas 1980). Differences between years in the timing or duration of spawning activity occurred for *P. maximus* (Mason (1958a), *P. magellanicus* (Naidu 1970; Davidson et al. 1993), and *Argopecten gibbus* (Miller et al. 1979). Differences between years in the timing of spawning and fecundity of *P. maximus* occurred at each of two sites (Wilson 1987; Strand and Nylund 1991). Spawning periods of *C. varia* (Lubet 1959) and *Pecten novaezelandiae* (Nicholson 1978) also varied between years. Egg diameter varied between years for *P. magellanicus* (MacDonald and Thompson 1986). On Georges Bank *P. magellanicus* was found to have a biannual spawning cycle, with a primary spawning event in the fall and a temporally erratic spring spawning event (DiBacco et al. 1995). In the mid-Atlantic, inter-annual variation in the timing, duration and magnitude of spawning in *P. magellanicus* also occurred, with a dominant spawning event in the spring and an unpredictable spawning event in the fall (Kirkley and DuPaul 1991; Schmitzer et al. 1991). Maximal gonad weights and indices of *A. irradians* were lower in a year when salinity was reduced as a result of a hurricane (Bologna 1998). Variations between years in spawning intensity (% gametes released) resulted in variable recruitment in populations of *Patinopecten yessoensis* (Yamamoto 1950), *P. maximus* (Lubet et al. 1991), and *C. asperrimus* (Zacharin 1994). There was an especially intense spawning event in 1983 in a population of *Argopecten purpuratus* that resulted in unusually high recruitment (Wolff 1987). Most year-to-year variation in gametogenesis can be related to environmental factors (see Regulation of Gametogenic Cycles, below).

Site-specific variations in scallop gametogenesis have been documented for several scallop species. Population differences in spawning intensity (% gametes released) or fecundity occurred for *Hinnites multirugosus* (Jacobsen 1977), *Amusium japonicum balloti* (Dredge 1981), and *Argopecten irradians* (Bricelj et al. 1987). The greatest difference in fecundity among four sites occurred over a distance of only 1.5 km (Bricelj et al. 1987). Differences between locations in the timing of spawning were seen for *Chlamys opercularis* (Fullarton 1889; Ursin 1956), *Chlamys varia* (Lubet 1959), *Pecten maximus* (Gibson 1956; Wilson 1987; Strand and Nylund 1991), *Argopecten circularis* (Villalejo-Fuerte and Ochoa-Báez (1993), and *Pecten novaezelandiae* (Nicholson 1978), but not for *Pecten alba* (Sause et al. 1987). Parsons et al. (1992) and Davidson et al. (1993) noted differences in the intensity and timing of spawning of *Placopecten magellanicus* between sites within Passamaquoddy Bay and the Gulf of St. Lawrence, respectively. The duration of spawning of *A. irradians* was found to extend into the autumn at one or more of four sites during three different years (Tettelbach et al. 1999).

Gametogenesis in scallops has been found to vary between sites of differing water depth. *Pecten novaezelandiae* spawned first in the deeper beds (28–40 m) among seven sites ranging in depth from 6–40 m (Tunbridge 1968). For *Chlamys islandica*, no differences in the onset of spawning occurred between sites having depths of 20, 40, and 50 m, but gonad index was significantly greater at the shallowest site (Skreslet and Brun 1969). In a separate study, Skreslet (1973) observed that spawning in *C. islandica* was delayed at the deepest (39 m) of three sites.

Gametogenesis in *Placopecten magellanicus* varies between sites having differing water depth. MacDonald and Thompson (1986) showed that scallops at 31 m had a reduced rate of gamete development compared to scallops at 10 m, but that the timing of spawning and size of spawned eggs was similar at both depths. Similarly, Schmitzer et al. (1991) found that *P. magellanicus* from a shallow (38–56 m) site exhibited larger gamete volume fractions than those from deeper (56 66 m). Barber et al. (1988) compared gametogenic cycles over much greater depth differences (13–20 m and 170–180 m), and found that gonad weight and gonad index for scallops of similar shell height was significantly greater at the shallower depth and that the deeper site was characterised by more resorption and less synchronous development of gametes (Fig. 6.4). Estimated fecundities ranged from 3.1–6.6 x 10^7 eggs per female at the shallow site and 1.4–2.4 x

10^7 eggs per female at the deep site, almost a threefold difference. It is not depth *per se*, but rather the factors that vary with depth, such as food and temperature that influence gametogenesis (Barber et al. 1988).

The wide geographic distribution of some scallop species makes comparisons of gametogenic cycles on a latitudinal basis possible. There are two species for which enough information exists to examine latitudinal trends in gametogenesis. *Argopecten irradians*, the bay scallop, is found in coastal bays and estuaries along the east coast of North America from Nova Scotia to Tampa, Florida (Gutsell 1930). Studies on the reproduction of this species from Massachusetts to Florida along the east coast of North America reveal latitudinal trends in the timing of gametogenic events (including spawning), the duration of spawning periods, and energetically related parameters such as fecundity and oöcyte size (Table 6.3). Spawning of *A. irradians* commences as early as May in Massachusetts populations (Taylor and Capuzzo 1983), June–July in New York (Tettelbach et al. 1999), July in North Carolina (Gutsell 1930), September in Florida (Barber and Blake 1981, 1983; Barber 1984), and December–February in the northern Gulf of Mexico (Bologna 1998). Exceptions to these general trends, however, do exist. In the Northern Gulf of Mexico, even though most spawning occurs over winter, spawning can occur as early as May (Bologna 1998). Tettelbach et al. (1999) demonstrated that scallops in New York are capable of spawning as late as September–November, even though peak spawning occurs in June–July. Sastry (1970a) compared populations from Woods Hole, Massachusetts and Beaufort, North Carolina and found that gametogenesis and spawning occurred earlier in the northern (Woods Hole) population and that the timing of maximum gonad response is significantly different between the two populations. The duration of the spawning period generally increases with decreasing latitude (Sastry 1979). The maximum mean gonad index and oöcyte diameter attained prior to spawning also decreased in a southerly direction (Sastry 1970a; Barber and Blake 1983). These trends were summarised by Barber and Blake (1983) as meaning that different populations of bay scallops have differing temperature requirements for the initiation of gametogenesis and spawning and that with decreasing latitude, a smaller portion of the overall energy budget is available for gametogenesis, resulting in reduced fecundity (Fig. 6.5). This was confirmed by Bricelj et al. (1987) who calculated that the size-specific fecundity of bay scallops is seven times greater in populations from New York than from Florida.

The sea scallop, *Placopecten magellanicus*, is distributed from the Gulf of St. Lawrence to Cape Hatteras in oceanic waters off the east coast of North America (MacKenzie et al. 1978). Both the timing and duration of spawning activity in this species vary latitudinally (Table 6.4). In locations north of Cape Cod (Newfoundland, Nova Scotia, New Brunswick, and Maine) spawning occurs as early as July, but primarily in August and September (Borden 1928; Welch 1950; Dickie 1955; Naidu 1970; Thompson 1977; Robinson et al. 1981; MacDonald and Thompson 1986, 1988; Beninger 1987; Barber et al. 1988; Parsons et al. 1992; Davidson et al. 1993). Spawning occurs primarily in September–October on Georges Bank, although a secondary spawning event in May–June may take place in some years (Posgay and Norman 1958; MacKenzie et al. 1978; DiBacco et al. 1995) and even later (October–November) off the coasts of

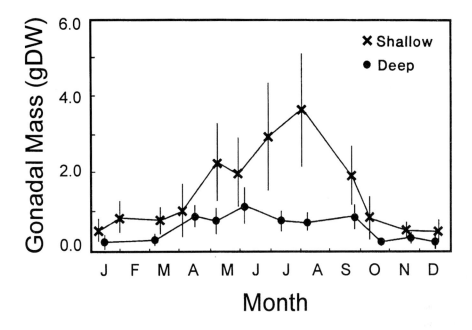

Figure 6.4. Reproductive cycles (mean ± 1 SD gonad weight) of scallops (*Placopecten magellanicus*) from shallow (13–20 m) and deep (170–180 m) sites. (From Barber et al. 1988).

Massachusetts and New Jersey (Posgay 1950; MacDonald and Thompson 1988). MacDonald and Thompson (1988) compared gametogenic cycles of *P. magellanicus* from sites off Newfoundland and the coast of New Jersey, USA. At the Newfoundland site, mature gametes appeared in April and spawning occurred two months earlier than at the New Jersey site. In the New Jersey population, mature gametes were present almost all year and a small decrease in gamete volume fraction in June and July suggested that a partial spawning occurred prior to final gamete development and complete spawning (MacDonald and Thompson 1988). Populations' further south (New Jersey to North Carolina) are characterised by biannual spawning periods, including dominant spring spawns and less predictable fall spawns (DuPaul et al. 1989; Kirkley and DuPaul 1991; Schmitzer et al. 1991). Thus, for both *Argopecten irradians* and *P. magellanicus*, as latitude decreases, primary spawning events occur later in the year and the duration of gametogenesis and spawning increases. This trend is supported further by the fact that tropical species, such as *Argopecten ventricosus*, have no seasonally defined gametogenic cycle (Luna-Gonzalez et al. 2000).

This raises the question as to whether observed differences in gametogenesis are due to environmental differences between years and locations, or genetic differences between populations. One way to distinguish between genetic and environmental effects is to mutually transplant individuals between two populations having differing gametogenic characteristics and observe the trends at the new locations. If the transplanted individuals

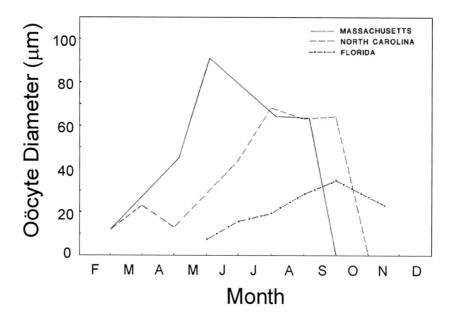

Figure 6.5. Reproductive cycles (mean oöcyte diameters) of scallops (*Argopecten irradians*) from latitudinally separated populations. (From Barber and Blake 1983).

assume the characteristics of the native population, the original differences were due to variation in environmental conditions. If, on the other hand, original differences in gametogenic cycles, especially the timing of events, are maintained at the new location, genetic differentiation would be supported.

Bay scallops, *Argopecten irradians irradians*, transplanted from Woods Hole, Massachusetts to Beaufort, North Carolina, where the local scallop population (*A. irradians concentricus*) spawns later in the year, failed to survive and develop gonads (Sastry 1966b). Subsequent laboratory studies in which scallops from both locations were collected prior to gonadal development and acclimated to identical temperatures and photoperiods, showed that maximum gonad growth occurred at 15°C in the Woods Hole scallops and at 23°C in the Beaufort scallops, indicating that these two populations of bay scallops did not acclimate for gametogenesis (Sastry 1966b). When broodstock from two populations of *Argopecten ventricosus* were spawned and offspring from each source population were transplanted to both source locations, differences between stocks in both age of sexual maturity and gonad index were found at one site, also suggestive of genetic differences (Cruz et al. 2000). Scallops (*Pecten maximus*) transplanted from St. Brieuc Bay to Quiberon Bay, France, underwent gonad development in the winter in conjunction with the local population (Latrouite and Claude 1979), indicating that the St. Brieuc scallops are able to acclimate for gametogenesis in Quiberon Bay. *Pecten maximus* transplanted as juveniles from the Bay of Brest, France and Scalpay, Scotland to the Bay of St. Brieuc, France, however, maintained cycles of gametogenesis and spawning characteristic of the site of origin and different from the native population (Mackie and Ansell 1993). Therefore, variation in scallop gametogenic cycles can be attributed to both environmental and genetic factors.

6.2.3.2 Inter-specific variations

The Pectinidae exhibit variation in gametogenic cycles characteristic of all marine bivalves. Gametogenesis occurs on an annual, semi-annual, or more continuous basis, and spawning takes place throughout the year (see Table 6.2). Species-specific variations in gametogenesis can only be determined by comparing populations of different species at a common site. Accordingly, Giguère et al. (1994) examined gametogenesis in *Placopecten magellanicus* and *Chlamys islandica* from the Îles-de-la-Madeleine, Canada, and found that *P. magellanicus* was characterised by a synchronous, annual gametogenic cycle, but that *C. islandica* had a more protracted spawning period, with considerable lysis of oöcytes.

Sause et al. (1987) suggested that pectinid species in the southern hemisphere spawn during the same range of calendar months as species in the northern hemisphere. In the southern hemisphere, however, these months (primarily June–October) correspond to winter-spring rather than summer-autumn periods. This would mean that scallops in the southern hemisphere initiate gamete development as water temperature is decreasing, and begin spawning when water temperature is at a minimum. This was the case for *Pecten alba*, in which peak spawning occurred 3–4 months before water temperature reached a maximum (Sause et al. 1987). *Pecten novaezelandiae* (Choat 1960), *Amusium japonicum* (Dredge 1981), and *Chlamys asperrima* (Zacharin 1994; O'Connor and Heasman 1996) also spawned primarily during the winter-spring (August–October) period. Spawning in *P. novaezelandiae* occurred in spring-early summer (Bull 1976), and both summer-autumn and spring-summer periods (Nicholson 1978). In addition, peak spawning of *Argopecten purpuratus* took place from February–May (summer-autumn) (Wolff 1988), spawning in *Chlamys tehuelcha* occurred over the summer (October–January) (Lasta and Calvo 1978), and spawning in *Chlamys zelandiae* was site specific and occurred throughout the year, especially in northern New Zealand (Booth 1983). The differences in the timing of spawning of species in the southern hemisphere just noted suggest that there is no consistent pectinid gametogenic response to temperature and related environmental factors. This will be discussed further below.

6.3 REGULATION OF GAMETOGENIC CYCLES

The gametogenic cycle of scallops (as for other bivalves) is a genetically controlled response to the environment (Sastry 1970a, 1979). Reproductive response is produced through an interaction of exogenous factors (temperature, salinity, light, food) and endogenous factors (neuronal, hormonal) within an organism. After attaining a certain physiological state, an organism exposed to the necessary environmental conditions, begins gonad growth and gametogenesis. Neuroendocrine activity plays a significant role in coordinating physiological processes within the organism to produce a reproductive response relative to the external environment. The successive events in the gametogenic cycle are affected differently by various factors (Sastry 1963, 1966a, 1968, 1970b; Sastry and Blake 1971; Blake and Sastry 1979). Although some of these factors have been investigated experimentally, most of the information has been obtained through observation, so is therefore correlative.

6.3.1 Gametogenesis and Fecundity

6.3.1.1 Exogenous regulation

Water temperature is the environmental factor most often cited as influencing bivalve reproduction (Sastry 1979). Gonad growth and gametogenesis have been correlated with seasonal variation in temperature for several pectinid species. For *Placopecten magellanicus*, gonadal differentiation began when water temperature was low (-1–0°C), but food level high (Thompson 1977). Schmitzer et al. (1991) noted that gametogenesis and spawning of *P. magellanicus* occurred twice, in the spring when water temperature was decreasing and in autumn when temperature was increasing. Low temperature was not inhibitory to gametogenesis in *Pecten novaezelandiae* (Bull 1976). *Patinopecten yessoensis* initiated gonad development when water temperature was minimal (-1.5°C), and the period of gonad growth coincided with increasing temperature (Wakui and Obara 1967; Maru 1976). Gonadal maturation in *Chlamys farreri* was closely related to temperature (Liao et al. 1983). For *Chlamys varia*, gametogenesis was initiated at a temperature of 4–5°C (a threshold of 70 degree-days) and the two periods of maximum gonad growth were correlated with rising temperatures and maximum chlorophyll-*a* levels; in addition, the rate of gamete development was correlated to the rate of increase in temperature (Burnell 1983). For *Argopecten irradians* from Woods Hole, Massachusetts, primary germ cells and gonial cells developed in winter and early spring, gamete differentiation began in April, and the population reached maturity in July in conjunction with an increase in temperature (Sastry 1970a). Bay scallops, *A. irradians*, from Beaufort, North Carolina, however, did not develop primary germ cells and gonial cells until spring, and differentiation of gametes occurred in conjunction with increasing temperature in May (Sastry 1966a, 1970a). A population of *A. irradians* from Tarpon Springs, Florida, underwent cytoplasmic growth in July when water temperature was near maximal (Barber and Blake 1983). Thus populations of bay scallops have differing minimum threshold temperatures for the initiation of gonad growth and gametogenesis (Sastry 1970a, 1979; Barber and Blake 1983). *Amusium japonicum balloti* responded differently in that gametogenesis started when temperature was near maximum (Dredge 1981). Similarly, the gonad index of *Argopecten purpuratus* was negatively correlated with water temperature (Wolff 1988). There was no relationship between temperature or salinity and gametogenesis for *Argopecten ventricosus* (Luna-Gonzalez et al. 2000). Thus the relationship between the timing and duration of gametogenic activity in scallops and water temperature is characteristic of a population under natural conditions. It follows that temperature will have less influence on gametogenesis in (tropical or polar) locations where seasonal variation in temperature is minimal.

It is possible to induce gametogenesis in scallops outside the normal reproductive period by exposing them to suitable temperatures. Maturation of eggs in excised gonads of *Patinopecten yessoensis* was accelerated with thermal stimulation (Yamamoto 1951c, 1968). Gametogenesis in *Argopecten irradians* was induced out of season by exposure to 23°C; failure to complete the process, however, was attributed to inadequate food supply (Turner and Hanks 1960). Mature reproductive state was also attained in *A. irradians* by exposure to elevated temperatures in winter (even though the natural population does not mature until August) (Sastry 1961, 1963). Development of gametes of *A. irradians* to maturity can be accelerated after gametogenesis has been initiated, and the rate of development to maturation is dependent on temperature Sastry (1963). Temperature alone did not induce gonad growth in *Argopecten ventricosus*, but rather interacted with food ration and time (Villalaz 1994).

The most complete examinations of the effects of temperature on successive events in the gametogenic cycle of any bivalve are those of Sastry (1966a, 1968, 1970a) and Sastry and Blake (1971) for *Argopecten irradians*. In Beaufort, North Carolina, gamete differentiation was initiated when the temperature exceeded 20°C. Prior to the initiation of gonad growth, scallops exposed to 10, 20, and 30°C without food resorbed the gonial cells. Individuals with a minimum amount of gonadal reserves and oöcytes in the beginning growth stages, developed to maturity at both 20 and 30°C. In scallops with accumulated gonad reserves and oöcytes in the growth phase, the time to completion of gametogenesis and release of gametes decreased with increasing temperature. At 10 and 15°C, scallops failed to complete gametogenesis; at 25°C, gametes were released in 8 days, and at 30°C, gametes were released in 5 days (Sastry 1966a).

Gonad growth and gametogenesis in *Argopecten irradians* occurred under temperature conditions at which nutrient mobilisation for the gonads took place (Sastry 1966a, 1968). In scallops in the resting stage, gametogenesis began upon exposure to a minimum threshold temperature of 20°C in the presence of ample food. At the subthreshold temperature of 15°C, scallops developed oögonia, but oöcyte growth did not occur even though food was present. Upon exposure to temperatures of 20 and 25°C, however oöcyte growth began immediately. Apparently scallops require a minimum threshold temperature for activation of the oöcyte growth phase, since oögonia can develop at subthreshold temperatures, but without further differentiation. If, however, scallops with oöcytes already in the cytoplasmic growth phase are maintained at a temperature below that necessary for initiating growth, oöcyte development continues until dissolution of the germinal vesicle (Sastry 1970b). Thus temperature acts as a triggering stimulus for initiation of the oöcyte growth phase (Sastry 1968, 1970b). The temperatures required for activating oöcyte growth at the beginning of oögenesis and for attaining maturity ultimately limit the annual period of gonad growth and gametogenesis in the natural environment.

Sastry and Blake (1971) demonstrated that temperature influences the initiation of oöcyte growth in *Argopecten irradians* by regulating the transfer of nutrient reserves to the gonads (Fig. 6.6). Scallops that were injected with ^{14}C-leucine (into the digestive gland) incorporated more labelled amino acid into the gonads at 15°C than at 5°C. Thus the rate of nutrient transfer to the gonads depends upon the stage of gametogenesis and the temperature of exposure (Sastry and Blake 1971).

Gametogenesis is an energy demanding process, as the mobilisation of nutrients to the gonad is essential for gamete development. It is still unclear, however, whether gonad development depends on recently ingested food, stored reserves, or some combination of the two (Sastry 1979; Barber 1984). Ultimately, all energy is acquired from ingested "food", the definition of which is varied and poorly defined. Food available to scallop populations has been measured as chlorophyll-*a* (Broom and Mason 1978; Illanes et al. 1985; MacDonald and Thompson 1985; Thorarinsdóttir 1993), suspended particulate matter (Nicholson 1978), and energy content of seston (MacDonald and Thompson 1985; Luna-Gonzalez et al. 2000). In most cases, however, food level was not directly obtained, but rather inferred from available records.

Gametogenesis and fecundity of several scallop populations have been related to seasonal food levels. The reproductive period of *Patinopecten yessoensis* coincided with increasing stomach content index (i.e., food) (Maru 1976). Gonadal differentiation in *Placopecten magellanicus* occurred when temperature was low but food was abundant (Thompson 1977). The high food levels that occurred at the end of spring were necessary for oöcyte maturation in *Pecten maximus* (Lubet et al. 1987). The three spawning periods observed for *Chlamys opercularis* were attributed to the abundant food supply at that site (Broom and Mason 1978). In contrast, gametogenesis in *C. opercularis* occurred in winter when food supply was low (Taylor and Venn 1979). For *Chlamys varia*, the onset of gametogenesis coincided with maximum available food (Burnell 1983). Similarly, gametogenesis in *Chlamys islandica* was initiated after spawning when food was abundant, was arrested in autumn as food supply decreased, and was completed the following spring as food supply reached a maximum (Thorarinsdóttir 1993). Cruz et al. (2000) attributed the lack of gametogenic development in *Argopecten ventricosus* at Bahia Concepción to low primary productivity and high average temperature.

The relationship between seasonal food abundance and gonadal development varies among populations of the bay scallop, *Argopecten irradians* (Sastry 1961, 1963, 1966a, 1970a). In Alligator Harbor, Florida, gametogenesis was correlated with seasonal changes in phytoplankton production in that both rapid gonadal growth in the spring and recovery after spawning in the fall coincided with phytoplankton blooms (Sastry 1961, 1963). In Beaufort, North Carolina, gonadal growth and gametogenesis in the spring coincided with peak food levels; although food concentrations remained high all summer, the gonad index showed no increase until early autumn (Sastry 1966a, 1968). In Woods Hole, Massachusetts, the period of gonadal development in *A. irradians* did not coincide with peak food levels (Sastry 1970a). Thus the amount of nutrients mobilised for the gonad appears to depend not only on food level, but also on temperature and the basic metabolic requirements of the animal; adaptational differences with respect to the relationship of food to reproduction exist between different populations of this species (Sastry 1970a, 1979).

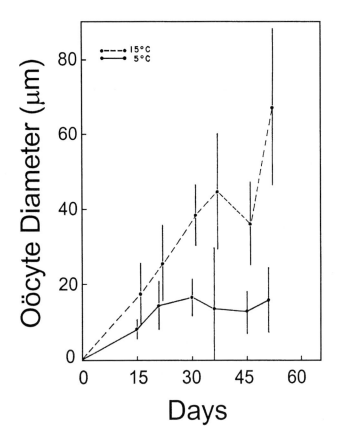

Figure 6.6. Oöcyte growth response in *Argopecten irradians* from Massachusetts as a function of temperature. (From Sastry and Blake 1971).

Stanley (1967) considered the relationships among food, temperature, and gametogenesis in several populations of *Pecten maximus*. Gonadal development in the spring was dependent on the rate of the winter recovery of weight that in turn was dependent on the magnitude of the food reserves accumulated during the previous summer and autumn. Over the winter, these reserves were presumably utilised for gametogenesis. Depending on the stage of recovery reached, spawning was triggered at some point during the spring rise in temperature, explaining both the differences in temperature recorded during this period and the earlier start and greater magnitude of the spring spawning period of the southern populations. The gonad then had sufficient time and food available to recover before the autumn spawning that occurred at all locations near the beginning of September, in conjunction with decreasing water temperature.

Elevated food (chlorophyll-*a*) levels and temperatures, the result of El Niño off the Chilean coast, greatly affect gametogenesis, fecundity, and recruitment in the local population of *Argopecten purpuratus*. In the years of El Niño, maximum mean gonad index was greater, spawning was synchronous and massive, and recruitment was over 3 times greater than in normal years (Illanes et al. 1985; Wolff 1987, 1988).

Sastry (1966a, 1968, 1970a) and Sastry and Blake (1971) experimentally examined the relationship between food supply and gametogenesis for *Argopecten irradians*. Scallops collected from the field at successive stages in their reproductive cycle were exposed to various temperatures and held without food in the laboratory (Sastry 1966a). During the initial stages of gonad growth and gametogenesis, starvation at 10, 20 and 30°C resulted in a decrease in digestive gland and gonad indices as well as resorption of oögonia and primary oöcytes. With minimal reserves in the gonad and oöcytes in the cytoplasmic growth phase, scallops held without food developed and released gametes at 25 and 30°C. At 15 and 20°C, however, digestive gland and gonad indices decreased and the oöcytes were resorbed. At 10°C, digestive gland and gonad indices remained the same, but the oöcytes failed to grow. Starved individuals having accumulated reserves and oöcytes in vitellogenesis released gametes at 20°C but not at 10°C. Scallops in the resting stage showed a decrease in both digestive gland and gonad indices and failed to initiate gametogenesis when starved at all experimental temperatures. Thus abundant food was necessary for gonad growth and gametogenesis in *A. irradians* since pre-stored nutrient reserves were not sufficient by themselves (Sastry 1966a). This is demonstrated by the decrease in tissue indices upon starvation, presumably as these nutrients are utilised for maintenance metabolism. Once the gonads have accumulated a minimal supply of energy reserves, however, gametes develop to maturity even though the animals receive no more food. For *A. irradians* it thus appears that gonad growth and gametogenesis occur under temperature conditions that allow nutrient mobilisation to the gonads in the presence of ample food, after basic metabolic requirements are satisfied (Sastry 1966a).

Sastry (1968) examined the relationships among food, temperature and gonad development for *Argopecten irradians* from Beaufort, North Carolina. The gonad index of scallops receiving food at 20°C increased, but the digestive gland index decreased. A slight increase in the gonad index and decrease in the digestive gland index occurred in scallops receiving food at 15°C. Both gonad and digestive gland indices decreased in starved animals at 15 and 20°C. Thus scallops held at temperatures below 20°C failed to accumulate gonad reserves, and pre-stored reserves were not sufficient to support gonad growth in the absence of food. It therefore appears that adequate food and a minimum threshold temperature (20°C) are necessary for the initiation of gonad growth and gametogenesis in *A. irradians* in North Carolina.

There have been no detailed laboratory studies examining the influence of salinity, light (day length), lunar phase, or tides on scallop gametogenesis. Sastry (1979), however, noted that gonad growth and gametogenesis in *Argopecten irradians* from Massachusetts were initiated during the spring in correlation with increasing day length, and maturity was attained when day length was maximal. In contrast, *A. irradians* from North Carolina initiated gonad growth and gametogenesis when the day length was about maximal, and matured and spawned in correlation with decreasing day length (Sastry

1979). Gonadal production in *A. irradians* was negatively impacted by a reduction in salinity related to a tropical storm (Bologna 1998). Amirthalingam (1928) noted that the percentage of ripe scallops (*Chlamys opercularis*) increased as the full lunar phase approached. A similar trend was noted for *Pecten maximus* (Tang 1941).

6.3.1.2 Endogenous regulation

Environmental factors are generally thought to stimulate an internal system regulating gametogenesis and sexual maturity. For invertebrates, an effective stimulation is received by receptors and transmitted to nerve ganglia. The neurosecretory cells in the stimulated ganglia secrete neurohormones, which regulate the physiological changes of the target organ, such as the gonad. Much of the knowledge of the role of the nervous system in bivalve reproduction has come from: 1) detection of monoaminergic and cholinergic neurons in the nervous system and gonads; 2) identification of nervous connections between the central nervous system and gonads; 3) analysis of the seasonal dynamics of monoamine levels in nervous ganglia, and; 4) experimental determination of the effects of monoamines and related compound on gametogenesis and spawning (Martínez and Rivera 1994).

The relationship between neurosecretion and gametogenesis in the Pectinidae is poorly understood, and few studies investigating these relationships exist. Based on a review of the histology of nerve cells in *Patinopecten yessoensis*, Osanai (1975) concluded that neuroproducts secreted by the large cells of the lower lobe of the cerebro-pleural ganglion accelerate gonad development and gametogenesis, and those from the small cells of the antero-lateral lobe of the visceral ganglion relate to gonad maturation. Accordingly, Osada and Nomura (1989) found that a seasonal variation in the level of catacholamines (noradrenaline and dopamine) in gonads of *P. yessoensis* was correlated to gametogenesis rather than water temperature, and was regulated by estrogen. Similarly, levels of prostaglandins in the hemolymph and ovary of *P. yessoensis* increased markedly during the spawning season; anti-estrogen inhibited this increase, suggesting that prostaglandins are involved in sexual maturation and that estrogen regulates production of prostaglandin in female scallops (Osada and Nomura 1990). Yamamoto (1951c, 1968), on the other hand, concluded that neither nervous nor circulatory systems were involved in the induction of maturation and ovulation in *P. yessoensis*, since only thermal stimulation was necessary for induction of these events outside the natural cycle. Substances secreted by the cerebropleural and visceral ganglia were found to be essential for controlling the annual gametogenic cycle of *Chlamys varia* (Lubet 1955). Levels of serotonin and noradrenalin increased in gonads, and levels of dopamine, noradrenalin and serotonin increased in adductor muscles of *Argopecten purpuratus* during active gametogenesis and decreased after spawning (Martínez and Rivera 1994; Martínez and Mettifogo 1998). Similarly, seasonal levels of serotonin in cerebral plus pedal ganglia and dopamine in the visceral ganglion were correlated with the gametogenic cycle of *Pecten maximus* (Paulet et al. 1993).

Blake (1972) and Blake and Sastry (1979) examined the neurosecretory cycle and its relationship to gametogenesis in *Argopecten irradians* (Fig. 6.7). Five neurosecretory stages were identified histologically (based on changes in size, granulation, and vacuolisation of neurons) that correlated with stages of oögenesis. In Stage I a neuroendocrine was secreted that allowed nutrient accumulation. Stages II and III acted as "on-off" mechanisms controlling the transfer of nutrients to the gonad. Stage IV corresponded with growth of oöcytes to maturity, and the secretion corresponding to Stage V initiated spawning.

Blake (1972) examined the effects of starvation and temperature on the gametogenic and neurosecretory cycles of *Argopecten irradians*. During the vegetative stage, starvation resulted in the resorption of germinal epithelium at both 5 and 15°C, but the neurosecretory cycle remained in Stage I. In scallops with oöcytes in the cytoplasmic growth phase and the neurosecretory cycle in Stage III, starvation effected no change at 5°C, but at 15°C, the oöcytes advanced to vitellogenesis with corresponding changes in neurosecretory stage. Scallops undergoing vitellogenesis in neurosecretory cycle stage IV were not affected by starvation at either 5 or 15°C. When partially spawned individuals were starved, the residual gametes were resorbed and the neurosecretory cycle returned to Stage II. Prolonged starvation at the time oöcytes were initiating cytoplasmic growth, or exposure of scallops to sub-threshold temperatures in the later stages of oögenesis, resulted in neuron degeneration and atrophy (Blake 1972).

The change in neurosecretory cycle from Stage II to Stage III in *Argopecten irradians* acted as a switching mechanism, and oöcyte growth was initiated or delayed depending on ambient temperature and food conditions (Sastry 1968, 1970b; Sastry and Blake 1971; Blake 1972; Blake and Sastry 1979). Both cycles advanced only under conditions of adequate food and threshold temperatures. Progression from Stage II to Stage III may have initiated the transfer of nutrients from the digestive gland to the gonad, which was necessary for the growth of oöcytes. After the neurosecretory cycle entered Stage III, exposure to sub-threshold temperature reversed the neurosecretory cycle to Stage II and delayed oöcyte growth. Scallops that experienced threshold temperatures for a prolonged period of time and were in Stage III of the neurosecretory cycle with oöcytes undergoing cytoplasmic growth, or scallops in Stage IV with oöcytes undergoing vitellogenesis, however, did not regress in neurosecretory stage upon exposure to sub-threshold temperatures. Prolonged exposure to sub-threshold temperature of scallops with oöcytes undergoing vitellogenesis resulted in vacuolisation of the cytoplasm and lysis of oöcytes (Sastry 1966a, 1968). Neuronal degeneration in *A. irradians* was in turn correlated with the extent of oöcyte disintegration and resorption (Blake 1972).

Thus, the gametogenic and neurosecretory cycles appear to be dynamically linked for *A. irradians*, with both cycles being regulated by changes in environmental temperature. Feedback controls through the mediating neuroendocrine substances may regulate metabolism, accumulation of reserves, mobilisation of nutrients for the gonads, and gametogenesis relative to changes in the environment (Gabbott 1975; Sastry 1979). The relationships between exogenous and endogenous factors and gametogenesis in *A. irradians* are summarised in Figure 6.7.

A minimum age (or size) is required before gametogenesis can be initiated. The age of sexual maturity in the Pectinidae varies between species from 71 days for *Argopecten gibbus* (Miller et al. 1979) to 3–5 years for *Chlamys islandica* (Wiborg 1963) (Table 6.5).

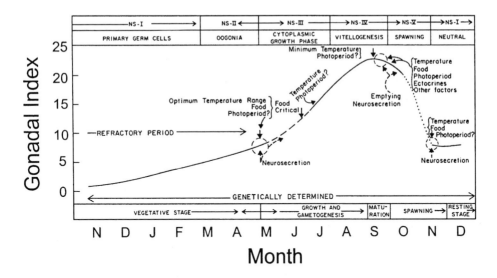

Figure 6.7. Summary of exogenous and endogenous regulation of the reproductive cycle of *Argopecten irradians.* (From Sastry 1979).

There may be a physiological explanation for a minimum age at which scallops initiate gametogenesis. Sundet and Vahl (1981) found that when food was abundant, immature *C. islandica* gave priority to fast growth while mature scallops stored nutrients for the production of gametes. This allocation of energy to somatic growth by young scallops agrees with the general life history theory of increasing reproductive effort with age, and as such is most likely species specific (genetically determined). Once the minimum "physiological age" has been reached, however, initiation of gametogenesis and the length of time required to attain sexual maturity may vary with environmental conditions (Cruz et al. 2000).

Blake (1972) examined the effect of age on gametogenic and neurosecretory cycles in *Argopecten irradians;* individuals collected from the field at different stages of the annual gametogenic cycle were held at 5 and 15°C and provided food. Three-month-old scallops having primary oöcytes (and in Stage I of the neurosecretory cycle) experienced no change in either gametogenic or neurosecretory stage at either temperature. Six-month-old scallops having oögonia and primary oöcytes showed no increase in oöcyte size or change in neurosecretory cycle stage at 5°C; but at 15°C, an increase in oöcyte diameter with a corresponding advance in the neurosecretory cycle to Stages III and IV was seen. At the age of 1 year, post-spawning scallops returned to Stage I of the neurosecretory cycle for a short time before advancing to Stage II with the development of oögonia. At this point, scallops held at 5°C did not progress in gametogenic or neurosecretory stage, while scallops held at 15°C progressed to later stages in both cycles. Although *A. irradians*, like other species, can be stimulated to mature faster by exposure to increasing

Table 6.5

Age and size of sexual maturity in various species of Pectinids.

Species	Age	Shell height	Source
Amusium japonicum		>90 mm	Dredge (1981)
Amusium pleuronectes		54 mm	Liano and Aprieto (1980)
Argopecten circularis		35 mm	Villalejo-Fuerte and Ochoa-Báez (1993)
Argopecten gibbus	71 days	20 mm	Miller et al. (1981)
Argopecten irradians	1 year	55–60 mm	Belding (1910); Gutsell (1930)
Argopecten ventricosus	4–5 months	20–33 mm	Cruz et al. (2000)
Chlamys islandica	3–5 year	35–40 mm	Wiborg (1963)
Patinopecten caurinus	3–4 year		Hennick (1970)
Pecten maximus	2–3 year		Tang (1941)
Pecten novaezelandiae	1 year	>50 mm	Choat (1960); Bull (1976); Tunbridge (1968
Placopecten	3 years	60 mm	Welch (1950)
magellanicus	1 year	23–75 mm	Naidu (1970)
		95 mm	Bonardelli and Himmelman (1994)
		>40 mm	Parsons et al. (1992)

temperatures (Sastry 1961; Turner and Hanks 1960), it appears that a minimum age has to be reached before cytoplasmic growth of oöcytes can be initiated (Blake 1972).

In summary, a combination of exogenous and endogenous factors is necessary before a gametogenic cycle can be initiated and completed. Fecundity, or the amount of gametogenic material produced, is dependent on the amount of energy available after maintenance requirements have been met. The portion of the energy budget allocated to gamete production ultimately depends on food availability. The relationship between fecundity and food supply has been examined for scallops in very few cases, as quantification of food level is beyond the scope of most studies. For *Placopecten magellanicus*, reproductive output (weight loss on spawning) was positively correlated with food concentration both between sites and between years at the same site (MacDonald and Thompson 1985). Similarly, Barber et al. (1988) found that fecundity (calculated number of ova) in *P. magellanicus* was reduced almost 3-fold in a deep-water (170–180 m) population compared to a shallow-water (13–20 m) population, presumably as the result of differing food levels at the two sites. For *Argopecten irradians*, Barber and Blake (1983) noted that maximum mean gonad indices decreased with decreasing latitude, from about 20% in Massachusetts, to 17% in North Carolina, to only 14% in Florida, in direct correlation to published values of food level (annual primary production) in the three locations. Bricelj et al. (1987) found differences in maximum gonad index and fecundity (weight loss on spawning) among four adjacent populations of *A. irradians* in New York and concluded that scallops are food limited in shallow, enclosed bays, where water circulation is restricted.

Metabolic stressors such as parasites and pollutants might reduce the amount of energy available for gamete production, negatively affecting fecundity. Parasitic castration of *Pecten alba* by a Bucephalid trematode was reported by Sanders (1966) and Sanders and Lester (1981). Total gamete weight in *Placopecten magellanicus* decreased by 60% in individuals exposed to both copper and cadmium, while doubling in control scallops over the same time period (Gould et al. 1988a).

6.3.2 Spawning and Spawning Synchrony

6.3.2.1 Exogenous regulation

Site-specific differences in the timing of spawning indicate that one or more environmental factors may act as a stimulus for the spawning process, once sexual maturation has been reached. Scallop life history strategies presumably have evolved such that for a given species in a particular location, spawning would occur at a time favourable for maximum rates of fertilisation and larval survival. Parameters potentially triggering gamete release include temperature, salinity, lunar phase, light, dissolved oxygen, pH, food level, mechanical shock, and ectocrines. These relationships have been examined by correlating spawning events to measured parameters in the field, as well as via controlled laboratory experiments.

Water temperature has been most frequently implicated in the initiation of spawning within the Pectinidae. Field studies have suggested that a minimum threshold temperature or critical temperature range is necessary for spawning to occur (Table 6.6). The temperature at which spawning is initiated, however, depends on acclimation history. Sastry (1961) acclimated mature *Argopecten irradians* to temperatures between 10.5 and 29.5°C at various times throughout the year and found that lower acclimation temperatures resulted in lower spawning temperatures. Thus the concept of a critical spawning temperature holds only for a population under a particular set of environmental conditions.

The initiation of spawning has been observed to occur with both increasing and decreasing temperatures. Spawning in *Chlamys opercularis* (Ursin 1956), *C. islandica* (Thorarinsdóttir 1993), *Patinopecten yessoensis* (Wakui and Obara 1967; Kawamata 1983), *Pecten novaezelandiae* (Bull 1976), *Amusium balloti* (Heald and Caputi 1981), and *Pecten alba* (Sause 1987) have all been associated with increasing water temperature. In contrast, *Hinnites multirugosus* (Jacobsen 1977) and *Argopecten gibbus* (Miller et al. 1981) spawned in conjunction with decreasing water temperature. In the case of *Pecten maximus* and *Placopecten magellanicus*, spawning in the spring occurred when water temperature was increasing and spawning in the fall occurred when water temperature was decreasing (Stanley 1967; Naidu 1970). *Argopecten irradians irradians* (the northern subspecies) spawned in response to increasing (Belding 1910; Sastry 1970a; Taylor and Capuzzo 1983; Bricelj et al. 1987) or decreasing (Tettelbach et al. 1999) water temperature, while *A. irradians concentricus* (the southern subspecies) spawned in conjunction with decreasing water temperature (Gutsell 1930; Sastry 1961, 1963; Barber and Blake 1981, 1983). Spawning of *P. yessoensis* was initiated by a sudden rise in water

Table 6.6

Temperatures at which spawning has occurred in wild populations of various Pectinids.

Species	Temperature (°C)	Source
Argopecten circularis	16–22	Villalejo-Fuerte and Ochoa-Báez (1993)
Argopecten gibbus	<22.5	Miller et al. (1981)
Argopecten irradians irradians	14–23	Bricelj et al. (1987)
	20–27	Tettelbach et al. (1999)
Argopecten irradians concentricus	25–30	Barber and Blake (1983)
Chlamys farreri	17–18	Kalashnikova (1984)
Chlamys islandica	8–10	Thorarinsdóttir (1993)
Chlamys opercularis	7–11	Amirthalingham (1928); Ursin (1956)
Chlamys varia	15	Burnell (1983)
Patinopecten yessoensis	8.0–8.5	Yamamoto (1951b, 1953, 1968)
Pecten maximus	15.5–16.0	Paulet et al. (1988)
	13–15	Strand and Nylund (1991)
	7–9	Strand and Nylund (1991)
Placopecten magellanicus	8–11	Posgay and Norman (1958)
	4–16	Bonardelli et al. (1996)

temperature caused by a strong warm-water current flowing into Mutsu Bay (Yamamoto 1950, 1951a, 1951b). In years that this occurred, larvae and spat were found in the water, but in years when these conditions did not occur, spawning did not take place and mortality resulted (Osanai 1975; Kawamata 1985). Maru (1985) suggested that *P. yessoensis* spawns when cumulative water temperature over 2.2°C reaches 285 degree-days. In Norway, the seasonal influx of meltwater establishes thermoclines over populations of *Chlamys islandica*; the movements of these differing water masses with the tides produce sudden temperature changes (either increases or decreases) that trigger spawning (Skreslet 1973; Skreslet and Brun 1969). In Iceland, *C. islandica* spawned when temperature was increasing from 8–10°C, in conjunction with maximum chlorophyll-*a* levels (Thorarinsdóttir 1993). The intrusion of colder water from offshore may initiate spawning of *Argopecten gibbus* in the vicinity of Cape Canaveral, Florida (Miller et al. 1981). Bonardelli et al. (1996) examined the relationships of spawning of *P. magellanicus* to environmental factors over eight years and found that all but one of 33 spawning events were associated with temperature change (25 coincided with an increase in water temperature associated with downwelling, seven coincided with fluctuations at high mean temperatures, and one coincided with decreasing water temperature). It thus appears that the influence of temperature on spawning in scallops depends more on the magnitude of change occurring over a short period of time period, rather than on the direction of change.

More precise information on the influence of temperature on spawning in scallops has been obtained in laboratory experiments. *Argopecten irradians* was induced to spawn by raising water temperature to 30–32°C and then lowering it to 22–28°C (Outten 1959; Sastry 1963). Spawning was also initiated in *Pecten albicans* by raising water temperature (Tanaka and Murakoshi 1985). *Patinopecten yessoensis* was induced to spawn by raising water temperature by 5°C, from 14–15°C to 19–20°C in 2–3 hr and raising the pH from 8.1 to 8.2–8.5 (Kinosita et al. 1943). Yamamoto (1951a, 1951b, 1953, 1968) found that spawning in *P. yessoensis* was induced by a sudden rise in temperature (as little as 0.5–1.0°C) and that raising the pH had no effect. After a series of trials involving several potential stimuli, Nicholson (1978) found that spawning in *Pecten novaezelandiae* was most effectively induced by slowly raising the temperature 5–7°C above ambient and then lowering the temperature to ambient.

In general, the strength of the stimulus needed to effect spawning decreases as maximum ripeness is attained. For *Argopecten irradians*, the minimum temperature required for spawning decreased as oöcytes become more developed (Sastry 1966a). Individuals with oöcytes in the beginning of the cytoplasmic growth phase failed to release gametes at 10, 15, and 20°C. At a later stage of development, individuals held at 20°C spawned, while those held at 15 and 10°C did not. Sastry (1966a) also noted that spermatozoa were released at a lower temperature than ova.

Limited information suggests that environmental parameters other than temperature can also influence spawning in scallops. Spawning in *Chlamys opercularis* (Amirthalingam 1928) and *Pecten maximus* (Tang 1941; Mason 1958a; Stanley 1967) has been associated with lunar phase. The lunar influence on spawning may be indirect, however, with the actual stimulus being provided by temperature changes associated with spring tides (Mason 1958a; Stanley 1967; Burnell 1983). Parsons et al. (1992) found a significant relationship between spawning of *Placopecten magellanicus* and lunar/tidal cycles which in turn would result in changes in temperature, food supply, and current speed, any of which could stimulate spawning. Naidu (1970) noted that spawning of *P. magellanicus* occurred in conjunction with periods of physical disturbance associated with high winds and rough seas. In support of this, Desrosiers and Dubé (1993) found that flowing seawater induced spawning in 93% and 100% of male and female *P. magellanicus,* respectively. Spawning of *Patinopecten yessoensis* was not related to lunar phase (Yamamoto 1952), but was correlated with the progression of the plankton community, following the abrupt seasonal change in water temperature (Yamamoto 1952, 1953). Yamamoto (1951a) found that decreasing the dissolved oxygen and pH of the water inhibited spawning of *P. yessoensis*. Lower salinity associated with the influx of meltwater may have provided a spawning stimulus for *Chlamys islandica*, but the effects of salinity changes could not be separated from the effects of simultaneous temperature changes associated with the meltwater (Skreslet 1973). Gametes of the opposite sex stimulated spawning in *Chlamys varia* (Lubet 1951). Thermal stimulation of spawning was more effective than mechanical stimulation of spawning in *C. varia* when combined with extracts of gametes (Lubet 1959). The presence of gametes in the water apparently provides a chemical stimulus (ectocrine) that triggers a spawning response in mature individuals.

Synchronous release of gametes of both sexes maximises the likelihood of successful fertilisation and the production of larvae. The degree of synchronisation of spawning, however, varies among species and within populations of the same species both between locations and years. The environmental factors responsible for bringing the population to a mature stage so that spawning can be as synchronous as possible have not received much attention. Exogenous factors may synchronise gonad development in members of a population at an early stage of gametogenesis so that spawning synchrony can be maximised, provided the proper spawning stimulus exists.

Experimental studies on environmental regulation of the neurosecretory cycle and gonad development in *Argopecten irradians* indicate that the controlling mechanisms within the organism may be responsive to environmental influences at the beginning of the cytoplasmic growth phase of oöcytes (Sastry 1966a, 1968, 1970b; Sastry and Blake 1971; Blake 1972; Blake and Sastry 1979). A population responding to environmental influences would initiate gonad development at the same time and progress synchronously toward maturity. However, oöcytes of different sizes occur within the gonads of individual scallops throughout the period of gonad development, both in the field and in the laboratory. Variation in the size of oöcytes size may be due to temporal differences in the initiation of cytoplasmic growth, which could be related to the amount of nutrients available for ovarian development. Although variation in development of oöcytes exists within individuals, the population as a whole is synchronous in development toward maturity. Even after spawning is initiated, there is a considerable range in oöcyte size, indicating that oöcyte development is continuing even at this point in the cycle. Thus over the approximately eight week spawning period, the release of gametes probably occurs in pulses as reflected by the stepwise decline in the gonad index and mean oöcyte diameter. If all members of the population are mature and react simultaneously to the factors inducing spawning, gametes will be released synchronously. Therefore, synchrony in spawning appears to depend on a critical state of physiological maturity of a population as a whole and on the responsiveness of the population to the exogenous factors inducing gamete release.

Variation in spawning synchrony in different years has been noted by Langton et al. (1987) for mature *Placopecten magellanicus*. They reasoned that synchronous spawning, which would allow the most effective fertilisation, might be the earliest indication of year class strength. Dickie (1955) reported that scallop abundance in any year class is correlated with water temperature six years previously. Higher temperatures might result in more favourable conditions for larvae or contribute to a more synchronous, intense spawning in *P. magellanicus*. If environmental conditions are not ideal for mass spawning, scallops may adopt a "dribble" spawning strategy thus ensuring that at least some larvae will survive (Langton et al. 1987). Similarly, early partial spawning may reflect a "bet-hedging" strategy in which at least some gametes would encounter environmental conditions favourable for development (Parsons et al. 1992).

Differences in synchrony of spawning occurred in two populations of *Pecten maximus* (Paulet et al. 1988). In the Bay of St. Brieuc, maturation was synchronous and spawning was synchronous and massive. In the Bay of Brest, mature individuals were present all year and partial spawnings occurred over a five month period. Massive, synchronous spawning might result in highly variable recruitment, depending on environmental conditions at the time of spawning. Spawning over a prolonged period decreases the likelihood of either very weak or very strong recruitment (Paulet et al. 1988), and could be an adaptation to an unpredictable environment.

Larval abundances have been related to spawning intensity and synchrony in populations of *Patinopecten yessoensis* (Yamamoto 1950, 1951b, 1968). In some years, mature gametes were retained, suggesting that a proper spawning stimulus (increased water temperature) was not received (Yamamoto 1950; Osanai 1975). Variation in local environmental conditions triggering the release of gametes in physiologically mature populations may thus play a major role in determining spawning synchrony and resultant year class success.

6.3.2.2 Endogenous regulation

Spawning is more difficult to induce in non-mature individuals than in mature individuals (e.g., Yamamoto 1968; Nicholson 1978; Malachowski 1988), suggesting that internal physiological conditions have to be met before external stimuli effectively induce spawning. Neuroendocrine control of spawning has been investigated for several species of scallops. A secretion associated with Stage V of the neurosecretory cycle was associated with spawning in *Argopecten irradians* (Blake 1972; Blake and Sastry 1979). Lubet (1955a) proposed a neurosecretory regulation of spawning in *Chlamys varia* based on the observed reduction in neurosecretory products of the cerebral ganglia in response to maximal environmental stimuli. Prior to spawning, neurosecretory products were released. Ablation of the cerebral ganglion accelerated spawning, while removal of the visceral ganglion retarded spawning (Lubet 1955b). It was thus suggested that the neurosecretion of cerebral ganglia has an inhibitory effect on spawning and that only after the dissipation of neurosecretory products does the individual become receptive to environmental factors stimulating spawning (Lubet 1955a, 1955b). Martínez et al. (1996) showed that levels of dopamine, noradrenaline and serotonin changed in nervous, muscle and gonadal tissue when spawning of *Argopecten purpuratus* was thermally induced.

Relationships between the presence of specific neurosecretory products and spawning have been investigated for *Patinopecten yessoensis*. Serotonin (5-hydroxytryptamine) effectively induced spawning of *P. yessoensis* (Matsutani and Nomura 1982; Matsutani and Nomura 1984), as well as of *Pecten albicans* (Tanaka and Murakoshi 1985) and *Hinnites giganteus* (Malachowski 1988). Methysergide (serotonin antagonist), lanthanum chloride and aspirin (inhibitor of prostaglandin synthesis) prevented induction of spawning with serotonin, providing further evidence that serotonergic mechanisms (as well as prostaglandin and dopaminergic mechanisms) are involved in scallop spawning (Matsutani and Nomura 1986). Osada et al. (1987) found significantly lower levels of dopamine in cerebral plus pedal ganglion, visceral ganglion, gills, and gonads of *P. yessoensis* after spawning was stimulated with UV irradiated seawater, suggesting that the release of dopamine is also important in the spawning of scallops. Prostaglandin levels decreased in ovaries but increased in testes of spawning *P. yessoensis*, suggesting inhibitory and stimulatory effects in female and male scallops, respectively (Osada et al. 1989).

Age may also influence the timing of spawning in scallops. Burnell (1983) found that for *Chlamys varia*, smaller (younger) individuals ripened and spawned earlier than larger (older) individuals. Also, in both the minor spring spawning and the major autumn spawning of *Pecten maximus*, individuals over the age of 8–9 years spawned later than the younger individuals (Stanley 1967). Thus, in spite of experiencing similar environmental stimuli, younger scallops in these populations spawned first, perhaps triggering spawning in the older individuals. Nicholson (1978) studied the effect of age-specific spawning responses for populations of *Pecten novaezelandiae* and found that the secondary spawn that occurred irregularly among the various sites and years involved primarily young (two year old) individuals; older individuals (>100 mm shell height) spawned less frequently and less completely. Thus the spawning characteristics exhibited by a particular population may depend to a large extent on the age structure of that population.

Research to date indicates that an external spawning stimulus is detected by specific receptors and communicated by a monoaminergic mechanism from the nervous system to the gonad, where the signal induces the action of some other compound or physiological event (e.g., adductor muscle contraction) that results in the release of gametes. There is also evidence that these neurohormonal mechanisms differ for male (primarily serotonin) and female (primarily catecholamines) scallops.

6.4 ENERGY METABOLISM

Gamete production is ultimately dependent on energy derived from ingested food. The energetic transformations that occur within a scallop after food energy is assimilated and before spawning occurs, however, are unclear. Like other marine bivalves, scallops exhibit cycles of energy storage and utilisation that are closely linked to annual cycles of gametogenesis (Gabbott 1975; Sastry 1979; Barber 1984). Although the specifics of these cycles are species and location dependent, energy is generally stored in one or more body components in the form of lipid, carbohydrate, or protein substrates when food intake exceeds basic maintenance requirements. These energy reserves are subsequently utilised to varying degrees (in conjunction with available food) to meet the energetic requirements of gametogenesis.

There are several types of information which when viewed in conjunction with gametogenesis, are relevant to the study of gametogenic energy metabolism in scallops. Changes in somatic tissue weights and indices indicate sites of energy storage. Changes in biochemical composition (lipid, carbohydrate, protein) of body components reveal which substrates contribute to energy metabolism at different times throughout the gametogenic cycle. Scallop physiological rates (oxygen consumption, carbon dioxide production, and ammonia excretion) are reflective of energy balance and substrate catabolism. The use of radiolabelled compounds helps define the energetic transformations and translocations occurring within the individual. Ultrastructural and histochemical approaches help visualise and verify specific routes and mechanisms of metabolite transfer. This section will attempt to assimilate available information that is relevant to gametogenic energy metabolism, or the manner in which assimilated energy is utilised with respect to gametogenesis in scallops.

6.4.1 Tissue Weights and Indices

Just as an increase in gonad mass (or index) is indicative of gametogenesis while a decrease signifies spawning, an increase in somatic body component weight indicates substrate accumulation or storage (growth), while a decrease is indicative of utilisation (degrowth). Changes in body component weights and indices thus have been useful for determining the relationships among growth and gametogenesis in several species of scallops. The adductor muscle of *Chlamys septemradiata* declined in weight over the winter, and reached a minimum prior to the period of gonad growth (Ansell 1974). For *Chlamys opercularis*, adductor muscle dry weight decreased in conjunction with gametogenesis (Taylor and Venn 1979). O'Connor and Heasman (1996) noted that the adductor muscle weight of *Chlamys asperrima* varied inversely with that of the gonad. The adductor muscle of *Pecten maximus* doubled in weight between March and November, but declined as the gonad increased in weight over the winter (Stanley 1967; Comely 1974). Faveris and Lubet (1991) found a similar inverse relationship between adductor muscle weight and gonad index for *P. maximus*. Weights of the digestive gland and adductor muscle of *Patinopecten yessoensis* decreased in conjunction with the increase in gonad weight (Fuji and Hashizume 1974). Lauren (1982) found that the adductor muscle index of *Hinnites giganteus* was inversely related to the gonad index. Digestive gland and adductor muscle indices in *Placopecten magellanicus* decreased as the gonad index increased (Robinson et al. 1981). Similarly, Schick et al. (1992) found an inverse relationship between weights of adductor muscle and gonad body components of *P. magellanicus*. For *Argopecten irradians*, Hickey (1978) noted that the adductor muscle index decreased during the period of gonad maturation and spawning and that adductor muscle weight gain occurred after spawning, while Sastry (1966a) found a reciprocal relationship between digestive gland and gonad indices. In the population of *A. irradians* studied by Barber and Blake (1981), adductor muscle dry weight decreased by two-thirds and adductor muscle index decreased by 18.7% over the periods of cytoplasmic growth and maturation of oöcytes (Fig. 6.8).

The importance of the scallop adductor muscle as a source of stored energy for gametogenesis and other metabolic demands becomes evident when gamete development is arrested. Scallop (*Argopecten irradians*) embryos treated with cytochalasin B to induce triploidy (effecting sterility), had adductor muscles that were 73% heavier and contained significantly more glycogen than diploid (control) scallops that underwent gametogenesis (Tabarini 1984). Similarly, Ruiz-Verdugo et al. (2000) found that triploid *Argopecten ventricosus* had significantly larger adductor muscle weights than diploids. In other words, when there is no energetic demand from the gametogenic process, energy reserves that are normally diverted to gamete synthesis remain in the adductor muscle. This approach may provide a more direct calculation of reproductive cost in scallops.

Simultaneous growth of gonadal tissue and a decrease in one or more somatic tissue weights or indices suggest that the utilisation of stored reserves is involved in the gametogenic process. For scallops, the adductor muscle appears to be the primary site of

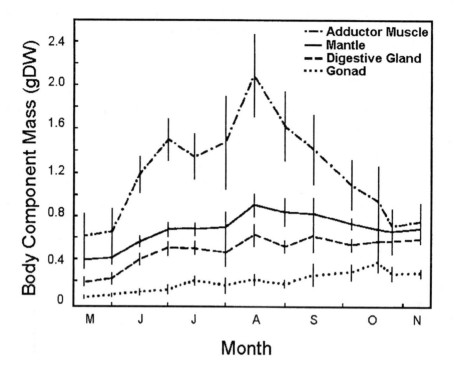

Figure 6.8. Energy storage cycle in *Argopecten irradians* as reflected by mean (±1 SD) body component dry weights. (From Barber and Blake 1981).

energy storage, with the digestive gland involved secondarily. The relative importance of the adductor muscle in scallop gametogenic energy metabolism, however, is site specific (Barber and Blake 1983; Bricelj et al. 1987) and may vary with environmental food levels.

6.4.2 Tissue Biochemical Composition

Biochemical analysis of scallop body components have revealed which substrates (protein, lipid, carbohydrate) are accumulated and utilised in conjunction with changes in weight of various body components with the increase in gonadal constituents associated with gametogenesis. For *Chlamys septemradiata* (Ansell 1974), *Chlamys opercularis* (Taylor and Venn 1979), and *Pecten maximus* (Stanley 1967; Comely 1974), declines in adductor muscle weight were accompanied by decreases in glycogen and protein contents and an increase in gonadal lipid content. Faveris and Lubet (1991) found an inverse relationship between adductor muscle lipid and protein (but not glycogen) contents and gonad index of *P. maximus*. Pollero et al. (1979) noted a reciprocal relationship between gonad and other tissue lipid contents in *Chlamys tehuelcha*. Glycogen stored in the adductor muscle of *Chlamys islandica* during the summer was subsequently depleted during the winter in conjunction with gametogenesis (Sundet and Vahl 1981). A loss of glycogen in somatic tissues was correlated to gonadal development in *Chlamys varia* (Shafee 1981). Mori (1975) found that digestive gland lipid and adductor muscle glycogen levels decreased prior to gonad development in *Patinopecten yessoensis*. For *Placopecten magellanicus*, digestive gland lipid and carbohydrate and adductor muscle carbohydrate were depleted as gametes matured and gonadal lipid content reached a maximum (Robinson et al. 1981). Couturier and Newkirk (1991), however, found that only adductor muscle protein content was inversely related to gonad weight of *P. magellanicus*. Lipid, protein and glycogen levels in plasma of *P. magellanicus* increased during the period of gonad growth (Thompson 1977), suggesting that any or all of these substrates were being mobilised for gamete synthesis. For *Argopecten irradians*, the utilisation of lipid from the digestive gland was associated with initiation of the oöcyte growth phase, and decreases in glycogen and protein from the adductor muscle occurred in conjunction with cytoplasmic growth and vitellogenesis and corresponding increases in protein and lipid in the gonad (Barber and Blake 1981). These relationships are shown in Figure 6.9. The authors also noted a significant correlation between decrease in adductor muscle glycogen content and increase in mean oöcyte diameter, indicating that oöcytes were growing as glycogen in the adductor muscle was being depleted (Fig. 6.10). A similar inverse relationship between adductor muscle glycogen content and gonad index occurred for *Argopecten purpuratus* (Martínez and Mettifogo 1998).

Additional support for the importance of adductor muscle glycogen as a source of energy for gametogenesis exists for *Placopecten magellanicus*. Gould et al. (1988b) reported that scallops from water over 110 m in depth had very low adductor muscle glycogen levels throughout the year compared to scallops from shallower depths. This corresponds with the finding of Barber et al. (unpublished) that scallops at depths of 170–180 m had smaller adductor muscles that showed little seasonal variation in weight compared to adductor muscles of scallops from depths of 13–20 m (Fig. 6.11). Similarly, there was a 66% reduction in fecundity in the deep-water population. Both of these studies supported the hypothesis that low food supply limits energy storage at greater depths. This would explain the reduced adductor muscle size and glycogen content and in turn, the reduced fecundity at greater depths. It follows that the greater the amount of energy available for gametogenesis (either from available food or reserves stored in the adductor muscle), the greater the fecundity will be.

Adductor muscle glycogen thus appears to be the energy substrate most readily stored and utilised for gametogenesis. Lipid stored in the digestive gland and lipid and protein stored in the adductor muscle may also be involved. Differences in the timing of utilisation of the various substrates suggest that they may be involved in different phases of the gametogenic cycle. Differences in the sources and timing of utilisation of various energy substrates are most likely due to either genetic differences between species or the quantity and quality of locally available food resources.

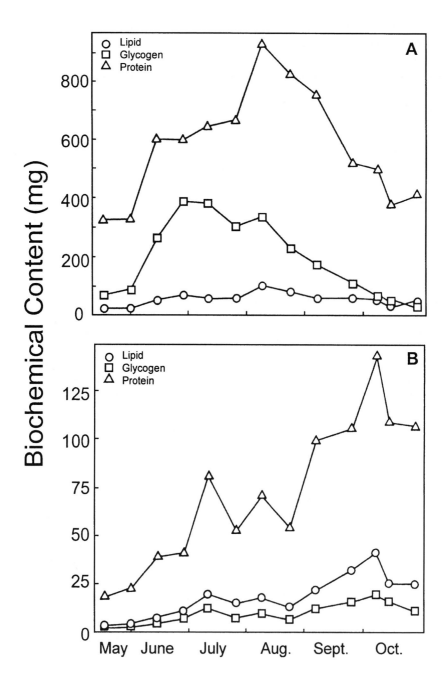

Figure 6.9. Energy storage cycle in *Argopecten irradians* as reflected by biochemical composition of (A) adductor muscle and (B) gonad. (From Barber and Blake 1981).

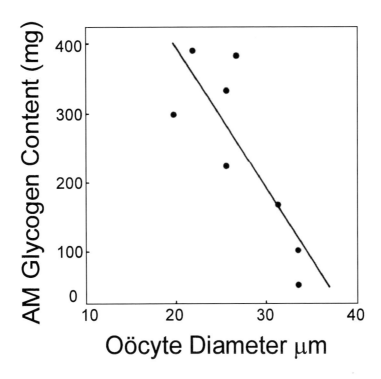

Figure 6.10. Relationship between glycogen content of adductor muscle and mean oöcyte diameter in *Argopecten irradians* (Y = 20.37X + 796.6; r = ⁻0.83). (From Barber 1984).

6.4.3 Physiological Indices

Rates of oxygen consumption, carbon dioxide production, and ammonia excretion, when converted to O/NH_3 and CO_2/O_2 molar ratios provide indices of the catabolic balance between protein, carbohydrate, and lipid substrates, and thus are pertinent to the study of gametogenic energy metabolism. The ratio of moles of oxygen consumed to moles nitrogen (ammonia) excreted provides an index of the relative amount of protein to non-protein catabolism occurring at a particular time (see Barber 1984). If the amino acids resulting from protein breakdown are deaminated and totally excreted as ammonia while the carbon skeletons are fully oxidised to carbon dioxide and water, the theoretically minimum O/NH_3 ratio (indicative of exclusive protein catabolism) is 9.3. Higher values indicate a greater contribution to energy metabolism from non-protein (lipid and carbohydrate) sources. Barber and Blake (1985b) examined trends in O/NH_3 ratios over the gametogenic cycle of *Argopecten irradians* and found an increase early in the gametogenic cycle, with values decreasing later in the cycle, as glycogen from the adductor muscle was being depleted (Fig. 6.12).

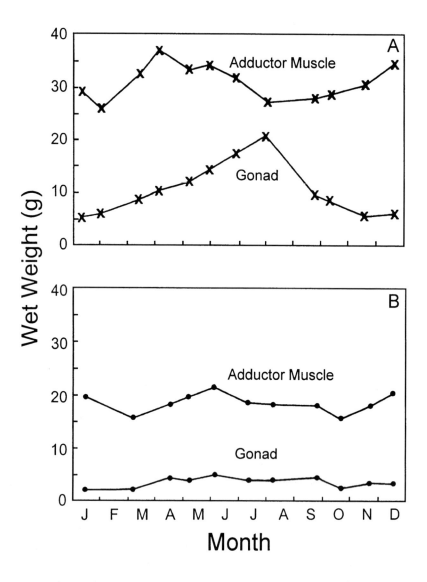

Figure 6.11. Comparison of adductor muscle and gonad wet weight of *Placopecten magellanicus* at (A) shallow (13–20 m) and (B) deep (170–180 m) sites. (From Barber et al. 1988, unpublished).

Additional information on the relative contribution of the various substrates to energy metabolism is provided by the respiratory quotient (RQ), defined as the ratio of the moles of carbon dioxide produced to the moles of oxygen consumed (see Barber 1984). When carbohydrate is oxidised, all of the oxygen utilised forms carbon dioxide, resulting in an RQ of 1.0. When protein and lipid are catabolised, some of the oxygen forms water,

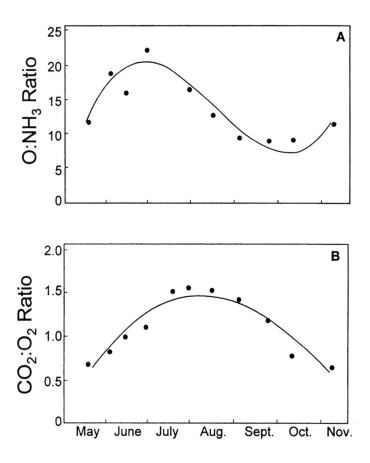

Figure 6.12. Seasonal variation in substrate catabolism of *Argopecten irradians* as reflected by (A) O/NH₃ and (B) RQ physiological indices. (From Barber and Blake 1985b).

resulting in RQ values of 0.79 for protein and 0.71 for lipid. An RQ greater than 1.0 indicates that carbohydrate is being converted to lipid. For *Patinopecten yessoensis*, the RQ of digestive gland tissue fell from 1.3 to 0.7 during the period of gonad development (Mori 1975). Barber and Blake (1985b) also measured RQ values of *Argopecten irradians* over the course of a gametogenic cycle and found that RQ increased from about 0.7 to 1.0 in the early stages of gametogenesis, exceeded 1.0 for most of the cytoplasmic growth period, then decreased to around 0.6 after spawning (Fig. 6.12).

The physiological indices thus reinforce the biochemical composition data by indicating relative contributions of the various substrates to gametogenic energy metabolism. In combination, these indices indicate that energy metabolism shifts from being lipid based during the initiation of gametogenesis to being carbohydrate based during the period of cytoplasmic growth of oöcytes. Additionally, RQ values during this period suggest that glycogen is being converted to lipid. Toward the time of spawning and especially after spawning, metabolism is fuelled primarily by protein.

6.4.4 Radiotracer Experiments

The use of radiolabelled compounds allows the movement and distribution of particular biomolecules to be followed over time. For studies pertaining to gametogenic energy metabolism, radiotracer experiments can provide more direct information with respect to the movement of particular substrates within the individual. Very few studies of this type have been undertaken for scallops. Sastry and Blake (1971) injected ^{14}C-leucine into the digestive gland of *Argopecten irradians* and found that in conjunction with oöcyte growth, incorporation of ^{14}C into gonads one week after injection was significantly higher at 15°C than at 5°C. Vassallo (1973) fed ^{14}C lipid (*Chlorella* extract) to *Chlamys hericia* via a cannula and found that lipid activity in the gonad increased as digestive gland lipid activity decreased.

Barber and Blake (1985a) fed ^{14}C-inoculated *Tetraselmis* sp. to *Argopecten irradians* at monthly intervals over the reproductive period and observed subsequent ^{14}C incorporation into lipid, carbohydrate and protein fractions of digestive gland, adductor muscle, and ovarian tissues. It was found that the pattern of ^{14}C incorporation varied seasonally in conjunction with oögenesis (Fig. 6.13). Resting stage scallops exhibited net radiocarbon losses in ovarian fractions and small radiocarbon losses in digestive gland fractions that were equalled by gains in adductor muscle carbohydrate and protein fractions. During the period of oöcyte growth, ^{14}C losses in digestive gland and adductor muscle fractions always exceeded progressive gains in the ovary. Increased carbon turnover (catabolism) of digestive gland lipid and adductor muscle carbohydrate fractions accompanied decreased turnover (anabolism) of ovarian lipid. After spawning, ^{14}C was rapidly lost from all body component fractions.

Radiotracer studies thus reinforce the patterns of energy storage and utilisation suggested by previous studies providing data on tissue weights and indices, biochemical composition, and substrate catabolism. Prior to gametogenesis, nutrients are stored in digestive gland and adductor muscle components. During cytoplasmic growth digestive gland lipid and adductor muscle carbohydrate is utilised in conjunction with an accumulation of lipid in the ovary.

6.4.5 Ultrastructure and Histochemistry

The approaches described so far are only gross indicators of nutrient storage and utilisation with respect to gamete formation, and provide little in the way of direct evidence as to how proposed metabolites are being moved within and between tissues of an animal undergoing gametogenesis. The use of histochemical techniques, in which thin tissue sections are stained to reveal the distribution of particular biomolecules (substrates and enzymes), in conjunction with histological and ultrastructural (electron microscopic) analyses, helps visualise and verify specific energetic pathways.

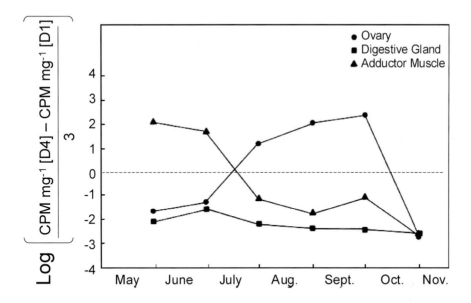

Figure 6.13. Relative seasonal incorporation of C^{14} by ovary, digestive gland, and adductor muscle body components of *Argopecten irradians*. (From Barber and Blake 1985a).

Gametogenesis of the scallop *Pecten maximus* is characterised by oöcyte atresia throughout the year, which reaches a maximum during vitellogenesis (Lubet et al. 1987; Dorange and Le Pennec 1989). Le Pennec et al. (1991) used ultrastructural, histological and histochemical approaches to identify sources, transfer mechanisms, and modes of incorporation of energy substrates to the developing oöcytes of *Pecten maximus*, paying particular attention to the fate of atretic oöcytes. They demonstrated that lysed material could be taken up directly by adjacent developing oöcytes via endocytotic vesicles or by gonoducts via ciliated epithelial cells with endocytic-like vacuoles. Lysed material that escapes the gonoducts through the gonopore enters the pallial cavity where it might also be taken up along the epithelia of the gonad itself by non-ciliated microvillous cells and transported to underlying tissue by amoebocyte transit cells.

It is also known that part of the scallop intestine loops through the gonad, and thus may play a role in digestion and assimilation of nutrients. The transfer of metabolites across the intestinal epithelium directly to developing gametes in the gonad was supported by ultrastructural and histochemical observations of *P. maximus* (Le Pennec et al. 1991) and *Argopecten ventricosus* (Luna-Gonzalez et al. 2000). Beninger et al. (unpublished) established that ferritin directly injected into the intestine of *P. maximus* was rapidly absorbed by the intestinal epithelium and transferred to hemocytes in the surrounding connective tissue. Ferritin-bearing hemocytes then migrated along connective tissue fibres to acini and appeared inside the acini of developing female gonads.

6.4.6 Mechanisms

Based on the available information on gametogenic energy metabolism in scallops, several mechanisms have been proposed for the cycling of available energy within an individual and its eventual transfer to and sequestration by developing gametes in the gonad. Barber (1984) suggested that three possible mechanisms are involved in the gametogenic energy metabolism of *Argopecten irradians concentricus*. The first mechanism involves the direct transfer of either stored or recently ingested fatty acids from the digestive gland to developing oöcytes in the gonad. Digestive gland lipid reaches high levels in some species prior to gametogenesis and decreases as oöcytes mature (Pollero et al. 1979; Mori 1975; Robinson et al. 1981; Barber and Blake 1981). Radioactive biomolecules move from digestive gland to gonad body components in conjunction with reproductive activity (Sastry and Blake 1971; Vassallo 1973; Barber and Blake 1985a). The existence of increased plasma lipid levels during oögenesis supports the feasibility of lipid transport between body components via the hemolymph (Thompson 1977). The second mechanism involves the conversion of carbohydrate to lipid in a manner similar to the glucose-fatty acid cycle in vertebrates. A general loss of glycogen reserves from adductor muscle tissue (which is not replaced until after spawning) often accompanies gonad lipid accumulation (Ansell 1974; Comely 1974; Mori 1975; Taylor and Venn 1979; Barber and Blake 1981; Robinson et al. 1981). High physiological indices during the early stages of gametogenesis indicate the predominance of carbohydrate catabolism and the conversion of glycogen to lipid (Mori 1975; Barber and Blake 1985b). The existence of a functional Kreb's cycle in bivalves is generally accepted, as evidenced by the presence of the carboxylic acid intermediates and enzymes involved in the cycle and the results of studies on the movement of ^{14}C from glucose through the cycle to $^{14}CO_2$ (Gabbott 1976). The third mechanism involves the breakdown of protein to indirectly support reproduction. Protein is accumulated as somatic tissue during periods of growth, and utilised over the gametogenic period, most commonly in species that undergo oögenesis over the winter when food levels are minimal and after lipid and carbohydrate reserves are depleted (Adachi 1979; Taylor and Venn 1979; Sundet and Lee 1984). In other species gametogenesis is primarily supported by available food and lipid and carbohydrate reserves so that protein is not catabolised to a great extent, or if it is, not until the other reserves are depleted (Barber and Blake 1981). This relationship is reflected by the low physiological index values obtained during later gametogenic and spawning periods (Mori 1975; Barber and Blake 1985b). Since gametogenesis is essentially complete at this point, energy derived from protein catabolism probably supports maintenance metabolism, rather that gametogenesis directly.

Le Pennec et al. (1991) proposed two additional mechanisms of energy transfer to developing oöcytes of *Pecten maximus*, including recycling of atretic oöcytes and intestinal loop transfer. Oöcytes themselves represent significant energy stores. In situations where oöcytes are lysed rather than spawned, ultrastructural and histochemical data have demonstrated that significant resorption of lytic debris occurs in the acini (via developing oöcytes), in gonoduct epithelium cells, and along gonadal epithelium (Le Pennec et al. 1991). Histochemical detection of injected ferritin in intestinal epithelium, hemocytes, connective tissue and acini of maturing ovaries demonstrates that digestion does occur along the portion of the intestine that runs through the gonad, and that energy assimilated is subsequently moved by hemocytes to developing oöcytes (Le Pennec et al. 1991; Beninger, unpublished). Figure 6.14 provides a schematic summary of all potential energy substrate sources and transfer mechanisms to developing gametes in scallops. The relative importance of these mechanisms varies between species and locations.

Additional neurohormonal control of the gametogenic cycle could occur indirectly, through control of the energy storage cycle. Martínez and Mettifogo (1998) found that when adductor muscle tissue of *Argopecten purpuratus* was incubated with dopamine, an increase in cyclic AMP occurred near the completion of oöcyte maturation. An increase in monoamine concentration in the adductor muscle would stimulate production of cyclic AMP, which in turn would induce the phosphorylation of glycolytic enzymes, thus increasing the catabolism of glycogen for gamete maturation.

6.5 APPLICATIONS TO AQUACULTURE

Successful culture of any species is dependent upon a comprehensive understanding of the reproductive process. As numerous wild stocks of scallops worldwide are becoming depleted, aquaculture is seen as a way of meeting a growing demand, either through stock enhancement efforts or commercial culture (see chapters relating to aquaculture, this volume). Much of the information presented in this chapter regarding the regulation of gametogenesis and spawning is relevant to the hatchery-based culture of scallops, including broodstock conditioning, induction of spawning, and the relationship between broodstock nutrition and larval growth and survival. Properly applied, a basic knowledge of gametogenesis will help commercial hatcheries produce a consistent, high quality, year round supply of seed organisms.

6.5.1 Broodstock Conditioning

It has long been known that scallops can be "conditioned" (induced to undergo gametogenesis) outside of normal seasonal cycles by exposure to adequate temperature and feeding regimes (Turner and Hanks 1960; Sastry 1961, 1963). Both the quantity and quality of gametes produced depend on adequate nutrition. For example, Martínez et al. (1992) found that *Argopecten purpuratus* fed a diet consisting of *Isochrysis galbana, Chaetoceros gracilis* and *C. calcitrans* for 48 days did not become fully ripe, but that individuals placed in the ocean reached maturity and spawned in the same period. The percentage of *A. purpuratus* reaching maturity was greater for individuals fed a microalgal diet supplemented with a lipid emulsion (Navarro et al. 2000). Monsalvo-Spencer et al. (1997) found that 95% of *Argopecten ventricosus* fed a diet consisting of a 6:3:1 ratio of *Isochrysis galbana: Chaetoceros* sp.: *Tetraselmis suecica* reached reproductive maturity in 27 days. Gonadal dry weight of *A. ventricosus* increased significantly after being fed a high ration for 40 days (Villalaz 1994). Not only are the contents of diets important, but the delivery of those diets may also have an impact on conditioning efficiency. Racotta et al. (1998) found that continuous feeding of *A. ventricosus* with equal cell numbers of

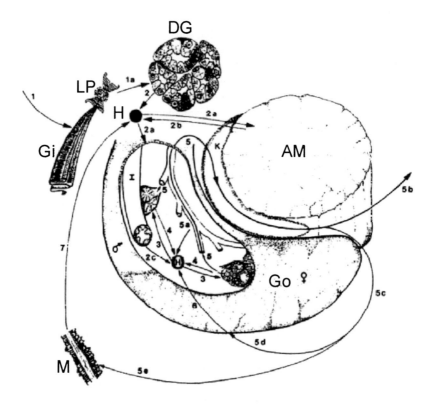

Figure 6.14.   Schematic summary of energy substrate sources and transfer mechanisms to developing gametes of scallops. 1 = ingestion of particulate matter captured and transported by the feeding epithelia (gills, Gi; labial palp complex, LP). 1a = entry of particles into diverticula of digestive gland (DG). 2 = diffusion of metabolites resulting from digestion into the hemolymph and hemocytes (H). 2a = distribution of these metabolites to various organs, including the adductor muscle (AM) and gonad (Go). 2b = hemolymphatic transfer of metabolites from the adductor muscle to the developing gametes. 2c = transfer of products of digestion within the gonad intestinal loop to the vesicular-cell hemocyte couples in intestine-acinus transfer complexes. 3 = transfer of metabolites from hemolymph (surrounding intestine, via vesicular cell-hemocyte couples and in sub-integumental hemolymphatic sinuses) to acini. 4, 5, 5a, 5b, 5d, 5e = lytic material resulting from gamete atresis (when this occurs) is incorporated into macrophagic hemocytes (4), or is expelled into the gonoducts and partially absorbed by the gonoduct epithelium (5). This material is also absorbed by macrophagous hemocytes within the gonoducts (5a = these may be expelled from the acini at the same time as the lytic debris). Any remaining lytic material is expelled through the gonopore and into the pallial cavity, where it may be lost to the external medium (5b), or directed toward the gonad integument (5d) or the mantle (M) epithelium (5e) by the circulation of the pallial fluid. 6, 7 = absorption of lytic material by the gonad epithelium and transfer to subjacent hemocytes (6), or by the mantle epithelium and transfer to subjacent hemocytes (7). (After Le Pennec et al. 1991).

Isochrysis galbana and *Chaetoceros muelleri* resulted in a higher gonadal index, a higher dry weight of spawned eggs, a greater concentration of acylglycerol in the gonad, and greater spawning synchrony compared to discontinuous (batch) feeding. Since the production of microalgal diets represents a major cost to commercial hatcheries, overfeeding of broodstock should be avoided. Paon and Kenchington (1995) found that unlike in wild populations, adductor muscles of *Placopecten magellanicus* conditioned in a hatchery did not decrease in weight while gonads were developing, suggesting that these animals were being given more food than they needed merely to complete gametogenesis. Similarly, Grant and Cranford (1989) found that *P. magellanicus* fed a diet of cultured phytoplankton had significantly greater somatic and gonad weights compared to scallops fed aged kelp powder or sediment diets, but that gonad index was not affected.

Cochard and Devauchelle (1993) found that scallops, *Pecten maximus*, originating from St. Brieuc were more difficult to artificially condition than scallops from Brest. Conditioning success (percentage of scallops spawning) varied between populations for differing conditions of photoperiod, food ration, and temperature (Devauchelle and Mingant 1991). These differences in response to conditioning between stocks of *P. maximus* suggest that there are genetic differences that could be commercially exploited.

6.5.2 Spawning

Once broodstock animals reach sexual maturity, synchronous and complete spawning of gametes is desired. For *Pecten ziczac*, thermal stimulation (rapid increase from 20 to 29°C) was found to be an effective method for inducing spawning in both sexes, but intragonadal injection of serotonin was effective only in inducing release of spermatozoa (Vélez et al. 1990). For maximum egg yield from *Chlamys asperrima*, a combination of temperature induction (raising water temperature 4°C in 30 min) and serotonin injection was suggested (O'Connor and Heasman 1995). Injections of serotonin, dopamine and noradrenaline into gonads of *Argopecten purpuratus* induced the release of sperm only, but injection of a combination of dopamine and prostaglandin induced spawning of both ova and spermatozoa (Martínez et al. 1996). For *Argopecten ventricosus*, serotonin injections effectively induced males to spawn, and thermal stimulation combined with exposure to sexual products resulted in a 50% spawning rate of both sexes (Monsalvo-Spencer et al. 1997). Desrosiers and Dubé (1993) found that exposure to flowing seawater was 10 times more effective at inducing spawning of *Placopecten magellanicus* than serotonin injection.

6.5.3 Larval Growth and Survival

Larval growth and survival depends to a large extent on the nutritional content of ova, which in turn depends on broodstock conditioning history. Inadequate levels of polyunsaturated fatty acids such as 20:4 (n-6), 22:6 (n-3), and 20:1 (n-9) in developing eggs are detrimental to larval development (Le Pennec et al. 1998). In support of this, Navarro et al. (2000) found that greatest survival rates of larval *Argopecten purpuratus* were achieved when broodstock were fed microalgal diets supplemented with a lipid emulsion containing essential fatty acids. Stocks of *Pecten maximus* conditioned and spawned in the laboratory produced larvae that performed similarly but exhibited differences in survival during and after metamorphosis (Cochard and Devauchelle 1993).

6.6 SUMMARY

6.6.1 Gametogenic Cycles

Gametogenic cycles in scallops include periods of inactivity (vegetative periods), cytoplasmic growth, vitellogenesis, spawning, and resorption of unspawned gametes. Several means of assessing gametogenesis in scallops exist, including visual observation of the gonad (both gross and microscopic examination of gonadal smears), gonad weights and indices, histology (including oöcyte diameter and stereology), and indirect means (larval and spat abundances). The most complete definition of a gametogenic cycle would include both quantitative estimates of fecundity and qualitative assessment of gametogenesis.

Considerable variation exists in pectinid gametogenic cycles. Intra-specific variations in the timing of spawning periods and the amount of gametogenic material produced occur between years and sites (location and depth), probably due to variations in environmental conditions that influence the gametogenic process. Aspects of gametogenesis vary along geographic ranges either as the result of environmental or genetic variability. Inter-specific differences in reproductive cycles also exist; the relative influence of genetic and environmental factors is unclear.

6.6.2 Regulation of Gametogenesis

Gametogenesis in scallops is regulated internally by neurosecretions and age and externally by environmental factors (primarily food and temperature), which synchronise the cycle. Regulation of gametogenesis is ultimately under genetic influence. Thus the factors regulating gametogenesis may vary between species or even within a species at different locations, resulting in variation in gametogenic cycles.

The regulation of gametogenesis is most completely understood for the bay scallop, *Argopecten irradians*. Once a minimum age is attained, a change in neurosecretory cycle stage (perhaps invoked by increased temperature) occurs. Scallops then become responsive to environmental stimuli for either initiation or delay of gonad growth. If minimum threshold day length, food, and temperature criteria are met, growth of oöcytes is initiated in conjunction with the transfer of nutrients from the digestive gland to the gonad. After a certain amount of gametogenic material is accumulated in the gonad, development of oöcytes to maturation is independent of food supply, providing minimal temperatures are maintained. This transition is marked by another change in neurosecretory cycle stage and effectively removes the gametogenic process from an uncertain environmental food supply. Fecundity, or the amount of gametogenic material produced, is determined by food availability above minimum threshold levels.

Spawning of gametes appears to rely on a combination of exogenous and endogenous factors. Once a state of gametogenic maturity is reached (through environmental and neurosecretory control), a variety of environmental stimuli can initiate spawning. The most universal of these is temperature. A rapid change in temperature appears to be more important than an absolute temperature or the direction of change. There is no critical, species-specific temperature for spawning. Physical disturbance (wind, tides) also stimulates spawning, as does the presence of gametes of the opposite sex (ectocrines). The influence of lunar periodicity on spawning is apparently indirect, resulting from temperature changes associated with spring tides. The effects of salinity and day length, are poorly understood. Changes in dissolved oxygen and pH are not relevant in seawater. Neurosecretory products such as serotonin and dopamine appear to be important in the spawning process. These chemicals or similar ones, released into the water with the gametes stimulate spawning in other members of the population, helping to synchronise the spawning process. Duration of spawning is limited by temperature and food supply and a return of the neurosecretory cycle to the stage coinciding with the resting stage.

Year class success might depend to a large extent on spawning intensity and synchrony, which in turn depends on gametogenic synchrony within the population and the presence of an effective spawning stimulus. The timing of spawning with respect to conditions suitable for larval growth and survival undoubtedly also contributes to recruitment success.

6.6.3 Energy Metabolism

The study of reproductive energy metabolism is concerned with the ways in which energy acquired from food is ultimately incorporated into gametogenic material. In scallops, nutrient reserves are stored primarily in digestive gland and adductor muscle body components during periods of somatic growth when food is abundant. Carbohydrate (mainly adductor muscle glycogen) appears to be of primary importance, but lipid stored in the digestive gland and protein stored in the adductor muscle also play important roles. These reserves may subsequently be catabolised to support maintenance metabolism when food is scarce and/or the production of gametes, depending on the timing of gametogenesis in relation to seasonal food supply.

Barber (1984) and Barber and Blake (1981, 1983, 1985a, 1985b) defined the gametogenic energy metabolism of *Argopecten irradians concentricus*. A period of growth and energy storage in which the adductor muscle increases in size and glycogen level occurs in the spring when food is abundant, prior to the initiation of gametogenesis. Metabolism is supported primarily by digestive gland lipid, which is declining in level at this time. Early in the gametogenic cycle maximum O/NH_3 and CO_2/O_2 (RQ) ratios indicate a shift to carbohydrate as the primary respiratory substrate, as the adductor muscle reaches maximal weight and its glycogen reserves begin to be utilised. As these glycogen stores are depleted in conjunction with gamete development, the adductor muscle decreases in weight and protein content. Gametes are mature by the end of the summer, as evidenced by maximum gonad weight and mean oöcyte diameter. Metabolism becomes primarily protein based as the adductor muscle reserves continue to be catabolised. After spawning commences in the fall (when food energy is less available), energy reserves are depleted and scallop physiological condition is poor, as indicated by rapid carbon turnover and increasing senescence. This cycle of energy storage and utilisation is an integral part of the reproductive process, acting as a buffer between the regulating factors and the actual production of gametes. The energy cycle itself is undoubtedly controlled by neuroendocrine and environmental factors but is critical to reproductive success in that it provides the energy and material required for gamete synthesis.

Several mechanisms for the transfer of energy to developing oöcytes have been proposed. These include: 1) direct transfer (via the hemolymph) of fatty acids from the digestive gland to developing ova; 2) the conversion of carbohydrate (stored in the adductor muscle) to fatty acids which are incorporated into developing ova; 3) the breakdown of protein (primarily adductor muscle tissue) to support energetic demands after other reserves are exhausted; 4) the recycling of energy from oöcyte atresis; and, 5) intestinal loop transfer.

6.6.4 Applications to Aquaculture

Compared to oysters, clams, and mussels, the commercial culture of scallops is relatively new. A greater understanding of the factors regulating gametogenesis and spawning of scallops will lead to a more consistent supply of seed organisms (Devauchelle and Mingant 1991). Like other bivalves, scallops can be conditioned out of season with a microalgal diet, although more research is needed in order to optimise specific dietary requirements and delivery systems. This will lead to improvements in conditioning time and subsequent health of larvae. Once scallops are fully ripe, spawning can be induced using thermal or mechanical stimulation, injection of serotonin, exposure to sexual products, or some combination.

REFERENCES

Allen, D.M., 1979. Biological aspects of the calico scallop *Argopecten gibbus*, determined by spat monitoring. Nautilus 94(4):107–119.

Amirthalingham, C., 1928. On lunar periodicity in reproduction of *Pecten opercularis* near Plymouth in 1927–1928. J. Mar. Biol. Assoc. U.K. 15(2):605–641.

Andrews, J.D., 1979. Pelecypoda: Ostreidae. In: A.C. Giese and J.S. Pearse (Eds.). Reproduction of Marine Invertebrates. Academic Press, New York. pp. 293–341.

Ansell, A.D., 1974. Seasonal changes in biochemical composition of the bivalve *Chlamys septemradiata* from the Clyde Sea area. Mar. Biol. 25:85–99.

Baird, R.H., 1966. Notes on an escallop (*Pecten maximus*) population in Holyhead Harbour. J. Mar. Biol. Assoc. U.K. 46:33–47.

Barber, B.J., 1984. Reproductive energy metabolism in the bay scallop, *Argopecten irradians concentricus* (Say). PhD Thesis, University of South Florida, Tampa. 122 p.

Barber, B.J., 1996. Gametogenesis of eastern oysters, *Crassostrea virginica* (Gmelin, 1791), and Pacific oysters, *Crassostrea gigas* (Thunberg, 1793) in disease-endemic lower Chesapeake Bay. J. Shellfish Res. 15(2):285–290.

Barber, B.J. and Blake, N.J., 1981. Energy storage and utilization in relation to gametogenesis in *Argopecten irradians concentricus* (Say). J. Exp. Mar. Biol. Ecol. 52:121–134.

Barber, B.J. and Blake, N.J., 1983. Growth and reproduction of the bay scallop, *Argopecten irradians* (Lamarck) at its southern distributional limit. J. Exp. Mar. Biol. Ecol. 66:247–256.

Barber, B.J. and Blake, N.J., 1985a. Intra-organ biochemical transformations associated with oögenesis in the bay scallop, *Argopecten irradians concentricus*, as indicated by ^{14}C incorporation. Biol. Bull. (Woods Hole) 168:39–49.

Barber, B.J. and Blake, N.J., 1985b. Substrate catabolism related to reproduction in the bay scallop, *Argopecten irradians concentricus*, as determined by O/N and RQ physiological indexes. Mar. Biol. 87:13–18.

Barber, B.J. and Blake, N.J., 1991. Reproductive Physiology. In: S.E. Shumway (Ed.). Scallops: Biology, Ecology and Aquaculture. Elsevier, New York. pp. 377–428.

Barber, B.J., Getchell, R., Shumway, S. and Shick, D., 1988. Reduced fecundity in a deep-water population of the giant scallop, *Placopecten magellanicus*, in the Gulf of Maine, U.S.A. Mar. Ecol. Prog. Ser. 42:207–212.

Belding, D.L., 1910. The scallop fishery of Massachusetts, including an account of the natural history of the common scallop. Mass. Div. Fish Game Mar. Fish. Ser. 5: 51 p.

Beninger, P.G., 1987. A qualitative and quantitative study of the reproductive cycle of the giant scallop, *Placopecten magellanicus*, in the Bay of Fundy (New Brunswick, Canada). Can J. Zool. 65(3):495–498.

Blake, N.J., 1972. Environmental regulation of neurosecretion and reproductive activity in the bay scallop, *Aequipecten irradians* (Lamarck). PhD Thesis, University of Rhode Island, Kingston. 161 pp.

Blake, N.J. and Sastry, A.N., 1979. Neurosecretory regulation of oögenesis in the bay scallop *Argopecten irradians irradians* (Lamarck). In: E. Naylor and R.G. Hartnoll (Eds.). Cyclic Phenomena in Marine Plants and Animals. Pergamon Press, New York. pp. 181–190.

Bologna, P.A.X., 1998. Growth, production, and reproduction in bay scallops *Argopecten irradians concentricus* (Say) from the northern Gulf of Mexico. J. Shellfish Res. 17(4):911–917.

Bonardelli, J.C. and Himmelman, J.H., 1995. Examination of assumptions critical to body component indices: application to the giant scallop *Placopecten magellanicus*. Can. J. Fish. Aquat. Sci. 52:2457–2469.

Bonardelli, J.C., Himmelman, J.H. and Drinkwater, K., 1996. Relation of spawning of the giant scallop, *Placopecten magellanicus*, to temperature fluctuations during downwelling events. Mar. Biol. 124:637–649.

Booth, J.D., 1983. Studies on twelve common bivalve larvae, with notes on bivalve spawning seasons in New Zealand. N.Z. J. Mar. Freshw. Res. 17:231–263.

Borden, M.A., 1928. A contribution to the study of the giant scallop, *Placopecten grandis* (S.). Fisheries Research Board of Canada, Manuscript Reports of the Biological Stations, No. 350. 24 p.

Bricelj, V.M., Epp, J. and Malouf, R.E., 1987. Intraspecific variation in reproductive and somatic growth cycles of bay scallops *Argopecten irradians*. Mar. Ecol. Prog. Ser. 36:123–137.

408

Broom, M.J. and Mason, J., 1978. Growth and spawning in the pectinid *Chlamys opercularis* in relation to temperature and phytoplankton concentration. Mar. Biol. 47:277–285.

Bull, M.F., 1976. Aspects of the biology of the New Zealand scallop, *Pecten novaezelandiae* Reeve 1853, in the Marlborough Sounds. PhD Thesis. Victoria University, Wellington. 175 p.

Burnell, G.M., 1983. Growth and reproduction of the scallop, *Chlamys varia* (L.), on the west coast of Ireland. PhD Thesis. National University of Ireland, Galway, Ireland. 295 p.

Choat, J.H., 1960. Scallop investigation Tasman Bay 1959–60. N.Z. Mar. Dept. Fish. Tech. Rep. 2: 51 p.

Cochard, J.C. and Devauchelle, N., 1993. Spawning, fecundity and larval survival and growth in relation to controlled conditioning in native and transplanted populations of *Pecten maximus* (L.): evidence for the existence of separate stocks. J. Exp. Mar. Biol. Ecol. 169:41–56.

Comely, C.A., 1974. Seasonal variations in the flesh weights and biochemical content of the scallop *Pecten maximus* (L.) in the Clyde Sea Area. J. Cons. Int. Explor. Mer 35:281–295.

Cruz, P., Rodriguez-Jaramillo, C. and Ibarra, A.M., 2000. Environment and population origin effects on first sexual maturity of catarina scallop, *Argopecten ventricosus* (Sowerby II, 1842). J. Shellfish Res. 19(1):89–93.

Dakin, W.J., 1909. *Pecten*, the edible scallop. Rep. Lancs. Sea Fish. Labs. No. 17. pp. 231–366.

Davidson, L.A., Lanteigne, M. and Niles, M., 1993. Timing of the gametogenic development and spawning period of the giant scallop *Placopecten magellanicus* (Gmelin) in the southern Gulf of St. Lawrence. Can. Tech. Rep. Fish. Aquat. Sci. 1935:14 p.

Davidson, L.A. and Worms, J., 1989. Stages of gonad development in the sea scallop *Placopecten magellanicus* (Gmelin) based on both macroscopic and microscopic observation of the gametogenic cycle. Can. Tech. Rep. Fish. Aquat. Sci. 1686:v + 20 p.

Desrosiers, R.R. and Dubé, F., 1993. Flowing seawater as an inducer of spawning in the sea scallop, *Placopecten magellanicus* (Gmelin, 1791). J. Shellfish Res. 12(2):263–265.

Devauchelle, N. and Mingant, C., 1991. Review of the reproductive physiology of the scallop, *Pecten maximus*, applicable to intensive aquaculture. Aquat. Living Resour. 4:41–51.

Dibacco, C., Robert, G. and Grant, J., 1995. Reproductive cycle of the sea scallop, *Placopecten magellanicus* (Gmelin, 1791), on northeastern Georges Bank. J. Shellfish Res. 14(1):59–69.

Dickie, L.M., 1955. Fluctuations in abundance of the giant scallop, *Placopecten magellanicus* (Gmelin), in the Digby area of the Bay of Fundy. J. Fish. Res. Bd. Can. 12:797–857.

Dorange, G. and Le Pennec, M., 1989. Ultrastructural study of oögenesis and oöcyte degeneration in *Pecten maximus* from the Bay of St. Brieuc. Mar. Biol. 103:339–348.

Dredge, M.C.L., 1981. Reproductive biology of the saucer scallop *Amusium japonicum balloti* (Bernardi) in central Queensland waters. Aust. J. Mar. Freshw. Res. 32:775–787.

DuPaul, W.D., Kirkley, J.E. and Schmitzer, A.C., 1989. Evidence of a semiannual reproductive cycle for the sea scallop, *Placopecten magellanicus* (Gmelin), in the Mid-Atlantic region. J. Shellfish Res. 8(1):173–178.

Faveris, R. and Lubet, P., 1991. Energetic requirements of the reproductive cycle in the scallop *Pecten maximus* (Linnaeus, 1758) in Baie de Seine (Channel). In: S.E. Shumway and P.A. Sandifer (Eds.). An International Compendium of Scallop Biology and Culture. The World Aquaculture Society, Baton Rouge. pp. 67–73.

Freere, R.H. and Weibel, E.R., 1967. Stereologic techniques in microscopy. Journal of the Royal Microscopical Society 87:25–34.

Fretter, V. and Graham, A., 1964. Reproduction. In: K.M. Wilbur and C.M. Yonge (Eds.). Physiology of Mollusca, Vol. 1. Academic Press, New York. pp. 127–164.

Fuji, A. and Hashizume, M., 1974. Energy budget for a Japanese common scallop, *Patinopecten yessoensis* (Jay), in Mutsu Bay. Bull. Fac. Fish. Hokkaido Univ. 25:7–19.

Fullarton, J.H., 1889. On the development of the common scallop (*Pecten opercularis*, L.). Fish. Board Scotland Annu. Rep. 8:290–299.

Gabbott, P.A., 1975. Storage cycles in marine bivalve mollusks: a hypothesis concerning the relationship between glycogen metabolism and gametogenesis. In: H. Barnes (Ed.). Ninth European Marine Biology Symposium. Aberdeen University Press, Aberdeen, Scotland. pp. 191–211.

Gabbott, P.A., 1976. Energy metabolism. In: B.L. Bayne (Ed.). Marine Mussels. Cambridge University Press, Cambridge, England. pp. 293–355.

Gibson, F.A., 1956. Escallops (*Pecten maximus* L.) in Irish waters. Sci. Proc. R. Dublin Soc. 27:253–271.

Giguère, M., Cliche, G. and Brulotte, S., 1994. Reproductive cycles of the sea scallop, *Placopecten magellanicus* (Gmelin), and the Iceland scallop, *Chlamys islandica* (O.F. Müller), in Îles-de-la-Madeleine, Canada. J. Shellfish Res. 13(1):31–36.

Gould, E., Rusanowsky, D. and Luedke, D.A., 1988b. Note on muscle glycogen as an indicator of spawning potential in the sea scallop, *Placopecten magellanicus*. Fish. Bull. 86:597–601.

Gould, E., Thompson, R.J., Buckley, L.J., Rusanowsky, D. and Sennefelder, G.R., 1988a. Uptake and effects of copper and cadmium in the gonad of the scallop *Placopecten magellanicus*: concurrent metal exposure. Mar. Biol. 97:217–223.

Grant, J. and Cranford, P.J., 1989. The effect of laboratory diet conditioning on tissue and gonad growth in the sea scallop *Placopecten magellanicus*. In: J.S. Ryland and P.A. Tyler (Eds.). Reproduction, Genetics and Distributions of Marine Organisms. Olsen and Olsen, Fredensborg, Denmark. pp. 95–105.

Gutsell, J.S., 1930. Natural history of the bay scallop. Bull. Bur. Fish. (U.S.) 46(193):569–632.

Heald, D.I. and Caputi, N., 1981. Some aspects of growth, recruitment and reproduction in the southern saucer scallop, *Amusium balloti* (Bernardi, 1861) in Shark Bay, Western Australia. Fish. Res. Bull. (West. Aust.) 25:1–33.

Hennick, D.P., 1970. Reproductive cycle, size at maturity, and sexual composition of commercially harvested weathervane scallops (*Patinopecten caurinus*) in Alaska. J. Fish. Res. Bd. Can. 27:2112–2119.

Hickey, M.T., 1978. Age, growth, reproduction and distribution of the bay scallop, *Aequipecten irradians irradians* (Lamarck), in three embayments of eastern Long Island, New York, as related to the fishery. Proc. Natl. Shellfish. Assoc. 68:80–81 (Abstract).

Illanes, B.J.E., Akaboshi, S. and Uribe, T.E., 1985. Effectos de la temperatura in la reprodución del ostion del norte *Chlamys (Argopecten) purpuratus* en la Bahía Tongoy Durant el fenómeno El Nino 1982–83. Invest. Pesq. (Chile) 32:167–173.

Jacobsen, F.R., 1977. The reproductive cycle of the purple-hinge rock scallop *Hinnites multirugosus* Gale 1928 (Mollusca: Bivalvia). PhD Thesis, San Diego State University, San Diego. 72 p.

Kalashnikova, C.A., 1984. Some biological features of the scallop *Chlamys farreri nipponesis* in the Possiet Bay of the Sea of Japan. Biol. Morya (Vladivostok) 1984(1):63–65.

Kawamata, K., 1983. Reproductive cycle of the scallop, *Patinopecten yessoensis* (Jay), planted in Funka Bay, Hokkaido. Sci. Rep. Hokkaido Fish. Exper. Stn. 25:15–20.

Kawamata, K., 1985. Abnormal gonad development observed in the scallops, *Patinopecten yessoensis* (Jay), cultured in Funka Bay, Hokkaido. Sci. Rep. Hokkaido Fish. Exper. Stn. 27:65–69.

Kinoshita, T., Shibuya, S. and Shimizu, Z., 1943. Induction of spawning of the scallop, *Pecten (Patinopecten) yessoensis* (Jay). Bull. Japan. Soc. Sci. Fish. 11:168–170.

Kirkley, J.E. and DuPaul, W.D., 1991. Temporal variations in spawning behavior of sea scallops, *Placopecten magellanicus* (Gmelin, 1791), in the mid-Atlantic resource area. J. Shellfish Res. 10(2):389–394.

Langton, R.W., Robinson, W.E. and Schick, D., 1987. Fecundity and reproductive effort of sea scallops *Placopecten magellanicus* from the Gulf of Maine. Mar. Ecol. Prog. Ser. 37:19–25.

Lasta, M.L. and Calvo, J., 1978. Ciclo reproductivo de la vieira (*Chlamys tehuelcha*) del golfo San José. Com. Soc. Malacologica Uruguay 5:1–42.

Latrouite, D. and Claude, S., 1979. Essai de culture suspendue de la coquille Saint Jacques *Pecten maximus*: influence de la transplantation sur le development gonadique. I.C.E.S., C.M. 1979/F:16: 10 p.

Lauren, D.J., 1982. Oögenesis and protandry in the purple-hinge rock scallop, *Hinnites giganteus*, in upper Puget Sound, Washington, U.S.A. Can. J. Zool. 60:2333–2336.

Le Pennec, M., Beninger, P.G., Dora, G. and Paulet, Y.M., 1991. Trophic sources and pathways to the developing gametes of *Pecten maximus* (Bivalvia: Pectinidae). J. Mar. Biol. Assoc. U.K. 71:451–463.

Le Pennec, M., Robert, R. and Avendaño, M., 1998. The importance of gonadal development on larval production in pectinids. J. Shellfish Res. 17(1):97–101.

Liao, C.-Y., Xu, Y.-F and Wang, Y.-L., 1983. Reproductive cycle of the scallop *Chlamys farreri*, (Jones et Preston) at Quindado. J. Fish. China 7(1):1–13.

Llana, M.E.G. and Aprieto, V.L., 1980. Reproductive biology of the Asian moon scallop *Amusium pleuronectes*. Fish. Res. J. Philipp. 5(2):1–10.

Lubet, P., 1951. Sur l'émission des gamètes chez *Chlamys varia* L. (Moll. Lamellibr.) C.R. Acad. Sci. 235:1680–1681.

Lubet, P., 1955a. Cycle neurosécrétoire chez *Chlamys varia* et *Mytilus edulis* L. (Mollusques Lamellibranches). C.R. Acad. Sci. 241:119–121.

Lubet, P., 1955b. Effets de l'ablation des centres nerveux sur l'émission des gamètes chez *Mytilus edulis* L. et *Chlamys varia* L. (Moll. Lamellibranches). Ann. Sci. Nat., Zool. Biol. Anim. 2:175–183.

Lubet, P., 1959. Recherches sur le cycle sexuel et l'émission des gamètes chez les mytilides et des pectinides (Mollusques bivalves). Rev. Trav. Inst. Pêches Marit. 23(4):387–548.

Lubet, P., Besnard, J., Faveris, R. and Robbins, I., 1987. Physiologie de la reproduction de la coquille Saint-Jacques (*Pecten maximus* L.). Oceanis 13:265–290.

Lubet, P., Faveris, R., Besnard, J.Y., Robbins, I. and Duval, P., 1991. Annual reproductive cycle and recruitment of the scallop *Pecten maximus* (Linnaeus, 1758) from the Bay of Seine. In: S.E. Shumway and P.A. Sandifer (Eds.). An International Compendium of Scallop Biology and Culture. The World Aquaculture Society, Baton Rouge. pp. 87–94.

411

Luna-González, A., Cáceres-Martínez, C., Zúñiga-Pacheco, C., López-López, S. and Ceballos-Vázquez, B.P., 2000. Reproductive cycle of *Argopecten ventricosus* (Sowerby 1842) (Bivalvia: Pectinidae) in the Rada del Puerto de Pichilingue, B.C.S., Mexico and its relation to temperature, salinity, and food. J. Shellfish Res. 19(1):107–112.

MacDonald, B.A. and Bourne, N.F., 1987. Growth, reproductive output, and energy partitioning in weathervane scallops, *Patinopecten caurinus*, from British Columbia. Can. J. Fish. Aquat. Sci. 44:152–160.

MacDonald, B.A. and Thompson, R.J., 1985. Influence of temperature and food availability on the ecological energetics of the giant scallop *Placopecten magellanicus*. II. Reproductive output and total production. Mar. Ecol. Prog. Ser. 25:295–303.

MacDonald, B.A. and Thompson, R.J., 1986. Influence of temperature and food availability on the ecological energetics of the giant scallop *Placopecten magellanicus*. III. Physiological ecology, the gametogenic cycle and scope for growth. Mar. Biol. 93:37–48.

MacDonald, B.A. and Thompson, R.J., 1988. Intraspecific variation in growth and reproduction in latitudinally differentiated populations of the giant scallop *Placopecten magellanicus* (Gmelin) Biol. Bull. (Woods Hole) 175:361–371.

MacKenzie, C.L., Jr., Merrill, A.S. and Serchuk, F.M., 1978. Sea scallop resources off the northeastern U.S. coast, 1975. Mar. Fish. Rev. 40(2):19–23.

Mackie, G.L., 1984. Bivalves. In: A.S. Tompa, N.H. Verdonk and J.A.M. van den Biggellaar (Eds.). The Mollusca, Vol. 7, Reproduction. Academic Press, New York. pp. 351–418.

Mackie, L.A. and Ansell, A.D., 1993. Differences in reproductive ecology in natural and transplanted populations of *Pecten maximus*: evidence for the existence of separate stocks. J. Exp. Mar. Biol. Ecol. 169:57–75.

Malachowski, M., 1988. The reproductive cycle of the rock scallop *Hinnites giganteus* (Grey) in Humboldt Bay, California. J. Shellfish Res. 7:341–348.

Marshall, N., 1960. Studies of the Niantic River, Connecticut with special reference to the bay scallop, *Aequipecten irradians*. Limnol. Oceanogr. 5:86–105.

Martínez, G., Garrote, C., Mettifogo, L., Pérez, H. and Uribe, E., 1996. Monoamines and prostaglandin E_2 as inducers of the spawning of the scallop, *Argopecten purpuratus* Lamarck. J. Shellfish Res. 15(2):245–249.

Martínez, G. and Mettifogo, L., 1998. Mobilization of energy from adductor muscle for gametogenesis of the scallop, *Argopecten purpuratus* Lamarck. J. Shellfish Res. 17:113–116.

Martínez, G. and Rivera, A., 1994. Role of monoamines in the reproductive process of *Argopecten purpuratus*. Invertebr. Reprod. Develop. 25(2):167–174.

Martínez, G., Saleh, Fl, Mettifogo, L, Campos, E. and Inestrosa, N., 1996. Monoamines and the release of gametes by the scallop *Argopecten purpuratus*. J. Exp. Zool. 274:365–372.

Martínez, G., Torres, M., Uribe, E., Díaz, M.A. and Pérez, H., 1992. Biochemical composition of broodstock and early juvenile Chilean scallops, *Argopecten purpuratus* Lamarck, held in two different environments. J. Shellfish Res. 11(2):307–313.

Maru, K., 1976. Studies on the reproduction of a scallop, *Patinopecten yessoensis* (Jay). 1. Reproductive cycle of the cultured scallop. Sci. Rep. Hokkaido Fish. Exp. Stn. 18:9–25.

Maru, K., 1985. Ecological studies on the seed production of scallop, *Patinopecten yessoensis* (Jay). Sci. Rep. Hokkaido Fish. Exp. Stn. 27:1–54.

Mason, J., 1958a. The breeding of the scallop (*Pecten maximus* L.) in Manx waters. J. Mar. Biol. Assoc. U.K. 37:653–671.

Mason, J., 1958b. A possible lunar periodicity in the breeding of the scallop, *Pecten maximus* (L.). Ann. Mag. Nat. Hist., Ser. 13 1:601–602.

Matsutani, T. and Nomura, T., 1982. Induction of spawning by serotonin in the scallop *Patinopecten yessoensis* (Jay). Mar. Biol. Lett. 3:353–358.

Matsutani, T. and Nomura, T., 1984. Localization of monoamines in the central nervous system and gonad of the scallop, *Patinopecten yessoensis*. Bull. Japan. Soc. Sci. Fish. 50:425–430.

Matsutani, T. and Nomura, T., 1986. Pharmacological observations on the mechanism of spawning in the scallop *Patinopecten yessoensis*. Bull. Japan. Soc. Sci. Fish. 52:1589–1594.

Miller, G.C., Allen, D.M. and Costello, T.J., 1981. Spawning of the calico scallop *Argopecten gibbus* in relation to season and temperature. J. Shellfish Res. 1:17–21.

Miller, G.C., Allen, D.M., Costello, T.J. and Hudson, J.H., 1979. Maturation of the calico scallop, *Argopecten gibbus*, determined by ovarian color changes. Northeast Gulf Sci. 3(2):96–103.

Monsalvo-Spencer, P., Maeda-Martínez, A.N. and Reynoso-Granados, T., 1997. Reproductive maturity and spawning induction in the catarina scallop *Argopecten ventricosus* (=circularis) (Sowerby II, 1842). J. Shellfish Res. 16 (1):67–70.

Mori, K., 1975. Seasonal variation in physiological activity of scallops under culture in the coastal waters of Sanriku District, Japan, and a physiological approach of a possible cause of their mass mortality. Bull. Mar. Biol. Sta. Asamushi 15:59–79.

Moyer, M.A. and Blake, N.J., 1986. Fluctuations in calico scallop production (*Argopecten gibbus*). In: Proceedings of the Eleventh Annual Tropical and Subtropical Fisheries Conference of the Americas. Texas Agricultural Extension Service, Texas A & M University. pp. 45–58.

Naidu, K.S., 1970. Reproduction and breeding cycle of the giant scallop *Placopecten magellanicus* (Gmelin) in Port au Port Bay, Newfoundland. Can. J. Zool. 4:1003–1012.

Navarro, J.M., Leiva, G.A., Martínez, G. and Aguilera, C., 2000. Interactive effects of diet and temperature on scope for growth for growth of the scallop *Argopecten purpuratus* during reproductive conditioning. J. Exp. Mar. Biol. Ecol. 247:67–83.

Nicholson, J., 1978. Feeding and reproduction in the New Zealand scallop *Pecten novaezelandiae*. MS Thesis, University of Auckland, Auckland, NZ. 75 p.

O'Connor, W.A. and Heasman, M.P., 1995. Spawning induction and fertilization in the doughboy scallop *Chlamys (Mimachlamys) asperrima*. Aquaculture 136:117–129.

O'Connor, W.A. and Heasman, M.P., 1996. Temporal patterns of reproductive condition in the doughboy scallop, *Chlamys (Mimachlamys) asperrima* Lamarck, in Jervis Bay, Australia. J. Shellfish Res. 15(2):237–244.

Osada, M., Matsutani, T. and Nomura, T., 1987. Implication of catecholamines during spawning in marine bivalve molluscs. Int. J. Invert. Reprod. Develop. 12:241–252.

Osada, M., Nishikawa, M. and Nomura, T., 1989. Involvement of prostaglandins in the spawning of the scallop, *Patinopecten yessoensis*. Comp. Biochem. Physiol. 94C:595–601.

Osada, M. and Nomura, T., 1989. Estrogen effect on the seasonal levels of catecholamines in the scallop *Patinopecten yessoensis*. Comp. Biochem. Physiol. 93C:349–353.

Osada, M. and Nomura, T., 1990. The levels of prostaglandins associated with the reproductive cycle of the scallop, *Patinopecten yessoensis*. Prostaglandins 40:229–239.

Osanai, K., 1975. Seasonal gonad development and sex alteration in the scallop, *Patinopecten yessoensis*. Bull. Mar. Biol. Stn. Asamushi 15(2):81–88.

Outten, L.M., 1959. Observations on the spawning and early development of the common scallop, *Pecten irradians*. J. Elisha Mitchell Scientific Soc. 75:73 (Abstract).

Paon, L.A. and Kenchington, E.L.R., 1995. Changes in somatic and reproductive tissues during artificial conditioning of the sea scallop, *Placopecten magellanicus* (Gmelin, 1791). J. Shellfish Res. 14(1):53–58.

Parsons, G.J., Robinson, S.M.C., Chandler, R.A., Davidson, L.A., Lanteigne, M. and Dadswell, M.J., 1992. Intra-annual and long-term patterns in the reproductive cycle of giant scallops *Placopecten magellanicus* (Bivalvia: Pectinidae) from Passamaquoddy Bay, New Brunswick, Canada. Mar. Ecol. Prog. Ser. 80:203–214.

Paulet, Y.M., Donval, A. and Bekhadra, F., 1993. Monoamines and reproduction in *Pecten maximus*, a preliminary approach. Invertebr. Reprod. Develop. 23:89–94.

Paulet, Y.M., Lucas, A. and Gerard, A., 1988. Reproduction and larval development in two *Pecten maximus* (L.) populations from Brittany. J. Exp. Mar. Biol. Ecol. 119:145–156.

Perodou, D. and Latrouite, D., 1981. Contribution à l'étude de la reproduction du pétoncle noir (*Chlamys varia*) de la baie de Quiberon. I.C.E.S. C.M. 1981/K:33. 10 p.

Pollero, R.J., Re, M.E., and Brenner, R.R., 1979. Seasonal changes of the lipids of the mollusc *Chlamys tehuelcha*. Comp. Biochem. Physiol. 64A:257–263.

Posgay, J.A., 1950. Investigations of the sea scallop, *Pecten grandis*. Third report on investigations of methods of improving the shellfish resources of Massachusetts. Div. of Mar. Fish., Dept. Conserv., Commonwealth of Massachusetts, Boston. 24–30 p.

Posgay, J.A. and Norman, K.D., 1958. An observation on the spawning of the sea scallop (*Placopecten magellanicus*) on Georges Bank. Limnol. Oceanogr. 3(4):478.

Purchon, R.D., 1977. The Biology of the Mollusca. Second Edition. Pergamon Press, Oxford.

Racotta, I.S., Ramirez, J.L., Avila, S. and Ibarra, A.M., 1998. Biochemical composition of gonad and muscle in the catarina scallop, *Argopecten ventricosus*, after reproductive conditioning under two feeding systems. Aquaculture 163:111–122.

Reddiah, K., 1962. The sexuality and spawning of Manx pectinids. J. Mar. Biol. Assoc. U.K. 42:683–703.

Risser, J., 1901. Habits and life-history of the scallop (*Pecten irradians*). 31st Annual Report to the Rhode Island Commissioners of Inland Fisheries: 47–55.

Robinson, A.M. and Breese, W.P., 1984. Spawning cycle of the weathervane scallop *Pecten (Patinopecten) caurinus* Gould along the Oregon coast. J. Shellfish Res. 4(2):165–166.

Robinson, W.E., Wehling, W.E., Morse, M.P. and McLeod, G.C., 1981. Seasonal changes in soft-body component indices and energy reserves in the Atlantic deep-sea scallop, *Placopecten magellanicus*. Fish. Bull. 79:449–458.

Roe, R.B., Cummins, R., and Bullis, H.R., 1971. Calico scallop distribution, abundance, and yield off eastern Florida, 1967–1968. Fish. Bull. U.S. 69:399–409.

Ruiz-Verdugo, C., Ramírez, J.L., Allen, S.K., Jr. and Ibarra, A.M., 2000. Triploid catarina scallop (*Argopecten ventricosus* Sowerby II, 1842): growth, gametogenesis, and suppression of functional hermaphroditism. Aquaculture 186:13–32.

Sanders, M.J., 1966. Parasitic castration of the scallop *Pecten alba* (Tate) by a Bucephalid trematode. Nature 212:307–308.

414

Sanders, M.J. and Lester, R.J.G., 1981. Further observations on a Bucephalid trematode infection in scallops (*Pecten alba*) in Port Phillip Bay, Victoria. Aust. J. Mar. Freshwater Res. 32:475–478.

Sastry, A.N., 1961. Studies on the bay scallop, *Aequipecten irradians concentricus* Say, in Alligator Harbor, Florida. PhD Thesis, Florida State University, Tallahassee. 118 p.

Sastry, A.N., 1963. Reproduction of the bay scallop, *Aequipecten irradians* Lamarck. Influence of temperature on maturation and spawning. Biol. Bull. (Woods Hole) 125:146–153.

Sastry, A.N., 1966a. Temperature effects in reproduction of the bay scallop, *Aequipecten irradians* Lamarck. Biol. Bull. (Woods Hole) 130:118–134.

Sastry, A.N., 1966b. Variation in reproduction of latitudinally separated populations of two marine invertebrates. Am. Zool. 6:374 (Abstract).

Sastry, A.N., 1968. The relationships among food, temperature, and gonad development of the bay scallop *Aequipecten irradians* Lamarck. Physiol. Zool. 41:44–53.

Sastry, A.N., 1970a. Reproductive physiological variation in latitudinally separated populations of the bay scallop, *Aequipecten irradians* Lamarck. Biol. Bull. (Woods Hole) 138:56–65.

Sastry, A.N., 1970b. Environmental regulation of oöcyte growth in the bay scallop *Aequipecten irradians* Lamarck. Experientia 26:1371–1372.

Sastry, A.N., 1979. Pelecypoda (excluding Ostreidae). In: A.C. Giese and J.S. Pearse (Eds.). Reproduction of Marine Invertebrates. Academic Press, New York. pp. 113–292.

Sastry, A.N. and Blake, N.J., 1971. Regulation of gonad development in the bay scallop, *Aequipecten irradians* Lamarck. Biol. Bull. (Woods Hole) 140:274–282.

Sause, B.L., Gwyther, D., Hanna, P.J. and O'Connor, N.A., 1987. Evidence for winter-spring spawning of the scallop *Pecten alba* (Tate) in Port Phillip Bay, Victoria. Aust. J. Mar. Freshw. Res. 38:329–337.

Schick, D.F., Shumway, S.E. and Hunter, M., 1992. Allometric relationships and growth in *Placopecten magellanicus*: The effects of season and depth. Proceedings of the Ninth International Malacological Congress. pp. 341–352.

Schmitzer, A.C., DuPaul, W.D. and Kirkley, J.E., 1991. Gametogenic cycle of sea scallops (*Placopecten magellanicus* (Gmelin, 1791)) in the mid-Atlantic Bight. J. Shellfish Res. 10(1):221–228.

Seed, R., 1976. Ecology. In: B.L. Bayne (Ed.). Marine Mussels: Their Ecology and Physiology. Cambridge University Press, Cambridge. pp. 13–65.

Shaffee, M.S., 1981. Seasonal changes in the biochemical composition and calorific content of the black scallop *Chlamys varia* (L.) from Lanveoc, Bay of Brest. Oceanol. Acta 4(3):331–341.

Shaffee, M.S. and Lucas, A., 1980. Quantitative studies on the reproduction of black scallop, *Chlamys varia* (L.) from Lanveoc area (Bay of Brest). J. Exp. Mar. Biol. Ecol. 42:171–186.

Skreslet, S., 1973. Spawning in *Chlamys islandica* (O.F. Müller) in relation to temperature variations caused by vernal meltwater discharge. Astarte 6(1):9–14.

Skreslet, S. and Brun, E., 1969. On the reproduction of *Chlamys islandica* (O.F. Müller) and its relation to depth and temperature. Astarte 2:1–6.

Stanley, C.A., 1967. The commercial scallop, *Pecten maximus* in Northern Irish waters. PhD Thesis, Queen's University of Ireland, Belfast. 111 p.

Strand, Ø. and Nylund, A., 1991. The reproductive cycle of the scallop *Pecten maximus* (Linnaeus, 1758) from two populations in Western Norway, 60° N and 64° N. In: S.E. Shumway and P.A.

Sandifer (Eds.). An International Compendium of Scallop Biology and Culture. The World Aquaculture Society, Baton Rouge. pp. 95–105.

Sundet, J.H. and Lee, J.B., 1984. Seasonal variations in gamete development in the Iceland scallop, *Chlamys islandica*. J. Mar. Biol. Assoc. U.K. 64:411–416.

Sundet, J.H. and Vahl, O., 1981. Seasonal changes in dry weight and biochemical composition of the tissues of sexually mature and immature Iceland scallops, *Chlamys islandica*. J. Mar. Biol. Assoc. U.K. 61:1001–1010.

Tabarini, C.L., 1984. Induced triploidy in the bay scallop, *Argopecten irradians*, and its effect on growth and gametogenesis. Aquaculture 42:151–160.

Tanaka, Y. and Murakoshi, M., 1985. Spawning induction of the hermaphrodite scallop, *Pecten albicans*, by injection with serotonin. Bull. Natl. Res. Inst. Aquacul. (Jpn.) 7:9–12.

Tang, S.F., 1941. The breeding of the escallop (*Pecten maximus* L.) with a note on the growth rate. Proc. Transl. Liverpool Biol. Soc. 54:9–28.

Taylor, A.C. and Venn, T.J., 1979. Seasonal variation in weight and biochemical composition of the tissues of the queen scallop, *Chlamys opercularis*, from the Clyde Sea area. J. Mar. Biol. Assoc. U.K. 59:605–621.

Taylor, R.E. and Capuzzo, J.M., 1983. The reproductive cycle of the bay scallop, *Argopecten irradians irradians* (Lamarck), in a small coastal embayment on Cape Cod, Massachusetts. Estuaries 6(4):431–435.

Tettelbach, S.T., Smith, C.F., Smolowitz, R., Tetrault, K. and Dumais, S., 1999. Evidence for fall spawning of northern bay scallops *Argopecten irradians irradians* (Lamarck 1819) in New York. J. Shellfish Res. 18(1):47–58.

Thompson, R.J., 1977. Blood chemistry, biochemical composition, and the annual reproductive cycle in the giant scallop, *Placopecten magellanicus*, from southeast Newfoundland. J. Fish. Res. Board Can. 34(11):2104–2116.

Thorarinsdóttir, G.G., 1993. The Iceland scallop, *Chlamys islandica* (O.F. Müller), in Breidafjördur, West Iceland II. Gamete development and spawning. Aquaculture 110:87–96.

Tunbridge, B.R., 1968. The Tasman Bay scallop fishery. N.Z. Mar. Dept. Fish. Tech. Rep. 18: 78 p.

Turner, H.J. and Hanks, J.E., 1960. Experimental stimulation of gametogenesis in *Hydroides dianthus* and *Pecten irradians* during the winter. Biol. Bull. Mar. Biol. Lab. Woods Hole 119:145–152.

Ursin, E., 1956. Distribution and growth of the queen *Chlamys opercularis* (Lamellibranchiata), in Danish and Faroese waters. Medd. Dan. Fisk. Havunders. 13: 32 p.

Vassallo, M.T., 1973. Lipid storage and transfer in the scallop *Chlamys hericia* Gould. Comp. Biochem. Physiol. 44A:1169–1175.

Vélez, A., Alifa, E. and Azuaje, O., 1990. Induction of spawning by temperature and serotonin in the hermaphroditic tropical scallop, *Pecten ziczac*. Aquaculture 84:307–313.

Villalaz, J.R., 1994. Laboratory study of food concentration and temperature effect on the reproductive cycle of *Argopecten ventricosus*. J. Shellfish Res. 13(2):513–519.

Villalaz, J.R. and Gomez, J.A., 1987. Esquema de cultivo de bivalves en Panama. In: J.A.J. Verreth, M. Carillo, S. Zanuy and E.A. Huisman (Eds.). Investigación Acuicola en América Latina, Pudoc Wageningen. pp. 310–319.

Villalejo-Fuerte, M. and Ochoa-Báez, R.I., 1993. The reproductive cycle of the scallop *Argopecten circularis* (Sowerby, 1835) in relation to temperature and photoperiod, in Bahia Concepcion, B.C.S., Mexico. Ciencias Marinas 19(2):181–202.

Wakui, T. and Obara, A., 1967. On the seasonal change of the gonads of scallop, *Patinopecten yessoensis* (Jay), in Lake Saroma, Hokkaido. Bull. Hokkaido Reg. Fish. Lab. 32:15–22.

Weibel, E.R., Kistler, G.S. and Scherle, W.F., 1966. Practical stereological methods for morphometric cytology. J. Cell Biol. 30:23–38.

Welch, W.R., 1950. Growth and spawning characteristics of the sea scallop, *Placopecten magellanicus* (Gmelin), in Maine waters. MA Thesis, University of Maine, Orono. 95 p.

Wiborg, K.F., 1963. Some observations on the Iceland scallop, *Chlamys islandica* (Müller), in Norwegian waters. Fiskeridir. Skr. (Havunders.) 13(6):38–53.

Wilson, J.H., 1987. Spawning of *Pecten maximus* (Pectinidae) and the artificial collection of juveniles in two bays in the west of Ireland. Aquaculture 61:99–111.

Wolff, M., 1987. Population dynamics of the Peruvian scallop *Argopecten purpuratus* during the El Niño phenomenon of 1983. Can. J. Fish. Aquat. Sci. 44(10):1684–1691.

Wolff, M., 1988. Spawning and recruitment in the Peruvian scallop *Argopecten purpuratus*. Mar. Ecol. Prog. Ser. 42:213–217.

Yamamoto, G., 1943. Gametogenesis and the breeding season of the Japanese common scallop *P. (Patinopecten) yessoensis* Jay. Bull. Japan. Soc. Sci. Fish. 12:21–26.

Yamamoto, G., 1950. Ecological note of the spawning cycle of the scallop (*Pecten yessoensis* (Jay)) in Mutsu Bay. Sci. Rep. Tôhoku Univ. (Ser. 4 Biol.) 18:477–481.

Yamamoto, G., 1951a. Induction of spawning in the scallop *Pecten yessoensis* Jay. Sci. Rep. Tôhoku Univ. (Ser. 4 Biol.), 19(1):7–10.

Yamamoto, G., 1951b. Ecological study on the spawning of the scallop *Pecten (Patinopecten) yessoensis* in Mutsu Bay. Bull. Japan. Soc. Sci. Fish. 17:53–56.

Yamamoto, G., 1951c. On acceleration of maturation and ovulation of the ovarian eggs in vitro in the scallop, *Pecten yessoensis* Jay. Sci. Rep. Tôhoku Univ. 4th Ser. (Biol.) 19(2):161–166.

Yamamoto, G., 1952. Further study on the ecology of spawning in the scallop in relation to lunar phases, temperature and plankton. Sci. Rep. Tôhoku Univ. (Ser. 4 Biol.) 19:247–254.

Yamamoto, G., 1953. Ecology of the scallop, *Pecten yessoensis* Jay. Sci. Rep. Tôhoku Univ. (Ser. 4 Biol.) 20(1):11–32.

Yamamoto, G., 1968. Studies on the propagation of the scallop, *Patinopecten yessoensis* (Jay), in Mutsu Bay. Fish. Res. Board Can. Transl. Ser. 1054: 68 p.

Zacharin, W., 1994. Reproduction and recruitment in the doughboy scallop, *Chlamys asperrimus*, in the D'Entrecasteaux Channel, Tasmania. Mem. Queensland Mus. 36(2):299–306.

AUTHORS ADDRESSES

Bruce J. Barber - Eckerd College, Galbraith Marine Science Laboratory, 4200 54th Avenue South, St. Petersburg, Florida, USA 33711 (E-mail: barberbj@eckerd.edu)

Norman J. Blake - College of Marine Science, University of South Florida, 830 First Street South, St. Petersburg, Florida, USA 33701 (E-mail: nblake@marine.usf.edu)

Scallops: Biology, Ecology and Aquaculture
S.E. Shumway and G.J. Parsons (Editors)
© 2006 Elsevier B.V. All rights reserved.

Chapter 7

Physiology: Energy Acquisition and Utilisation

Bruce A. MacDonald, V. Monica Bricelj and Sandra E. Shumway

7.1 INTRODUCTION

This chapter builds on a previous review of scallop physiology by Bricelj and Shumway (1991) included in an earlier edition of this text. The physiology of larval pectinids is not included, as the primary focus of this chapter is on post-settlement stages, juveniles and adults. Aspects related to nutrition (nutritional requirements of pectinids and substrate utilisation) and reproductive physiology are also excluded from this chapter. New, updated information is provided in areas that have received increasing attention since the publication of the first chapter, in particular the effects of flow, suspended sediments and of harmful algal blooms on scallop physiology. Readers are also referred to a comprehensive recent review of scallops in Iberoamerica, which includes several chapters on ecophysiology of scallops from that region (Maeda-Martínez 2002).

7.2 ENERGY ACQUISITION

7.2.1 Food Sources

Scallops are sublittoral, epifaunal, active suspension-feeding bivalves, which rely on suspended detrital material and phytoplankton as their food source. It has been suggested, however, that adult scallops are uniquely capable of exploiting food particles associated with surface sediments resuspended by their "shell clapping" activity (Davis and Marshall 1961). This ability has been demonstrated in the laboratory (Davis and Marshall 1961), but its significance in the natural environment is not known. Grant and Cranford (1989, 1991) and Cranford and Grant (1991) found that detrital diets alone, such as fresh and aged macroalgal detritus from the kelp *Laminaria longicruris*, or resuspended sediments containing benthic microalgae, were inadequate to support growth of adult *Placopecten magellanicus* under laboratory conditions. Only phytoplankton diets (*Isochrysis galbana* or *Chaetoceros gracilis*) fed at weight rations comparable to the detrital diets, sustained tissue and gonad growth in this species. Scope for growth (SFG) estimates derived from physiological data indicated that the phytoplankton diets yielded a positive SFG for both carbon and nitrogen, whereas aged kelp, a carbon-rich food source, yielded a positive (although relatively low) SFG for carbon but negative SFG for nitrogen (Grant and Cranford 1990). Aged kelp was readily absorbed (carbon absorption efficiency, at 87%, was comparable to that determined for the diatom diet) but resulted in markedly reduced clearance rates and nitrogen ingestion rates compared to algal diets. Fresh kelp resulted in lower and highly variable absorption efficiencies (0 to 60%) compared to aged kelp. Lack of growth of sea scallops on kelp detritus alone was thus attributed primarily to nitrogen deficiency and contrasts with findings in the mussel, *Aulacomya ater*, which can maintain a positive energy balance on a kelp diet (Stuart 1982). Studies using ^{15}N as a tracer (Alber et al. 1988) indicate that nitrogen released as dissolved organic matter by senescing macrophytes (*Enteromorpha, Gracilaria, Fucus* and *Spartina*) can be aggregated into amorphous particulate matter and ingested and assimilated by bay scallops, *Argopecten irradians*, under laboratory conditions. Bay scallops were capable of absorbing nitrogen and organic material from aggregates or flocs, although less efficiently than they were able to assimilate phytoplankton (Alber and Valiela 1996). These authors suggested that organic aggregates, or flocs, could potentially represent an important and nutritious food source for suspension-feeding bivalves. Furthermore, studies on the activity of digestive enzymes show that bivalves, including scallops, have the capacity to digest macroalgal detritus (Wojtowicz 1972; Brock et al. 1986). Overall, the above studies thus indicate that macrophyte detritus has the potential to supplement phytoplankton as a food source in the natural environment and may help to meet energy demands of scallops when the phytoplankton supply is low.

At least three studies based on gut content analysis (Davis and Marshall 1961; Vernet 1977 and Shumway et al. 1987) have independently shown that benthic and/or tychopelagic algae are an important component of the scallops' diet. Davis and Marshall (1961) found that benthic diatoms (e.g., *Melosira, Licomophora, Cocconeis*) were more abundant than planktonic diatoms in gut contents of *Argopecten irradians*. They also reported that the numerical abundance of live benthic algae in water samples increased with increasing proximity to the sediment surface (from 32% 30 cm above the bottom to 80% approximately 0.5 cm from the bottom). Shumway et al. (1987) compared gut contents of shallow (20 m) and deep water (180 m) populations of *Placopecten magellanicus* in the Gulf of Maine. Again, benthic algae such as *Melosira, Navicula* and *Pleurosigma* were found to outnumber pelagic forms in gut contents of the deep water population. Interestingly, both resting cysts of the toxic dinoflagellate *Protogonyaulax (=Alexandrium) tamarensis*, and *Dinophysis* cells, implicated in outbreaks of paralytic shellfish poisoning (PSP) and diarrhetic shellfish poisoning (DSP) respectively, were abundant in gut contents. Vernet (1977) determined the seasonal algal composition of gut contents of the Patagonian scallop, *Chlamys tehuelcha*, as well as that of bottom sediments and plankton samples collected immediately above scallop beds in the Gulf of San José, Argentina. Benthic algae such as *Synedra investens, Melosira sulcata, Grammathopora marina* and *Navicula* spp., were dominant in gut contents throughout most of the year (Fig. 7.1). Planktonic diatoms such as *Chaetoceros* and *Thalassiosira* species, which were the dominant component of the spring phytoplankton bloom, were rarely present in gut contents. Furthermore, benthic algae that attach to sand grains by gelatinous stalks, such as *Glyphodesmia distans, Glyphodesmia* spp. and *Plagiogramma interruptus*, and were abundant in sediments, were absent from gut contents, presumably because they are not readily resuspended. It is interesting to note that scallop eggs were relatively abundant in stomach contents during the spawning season.

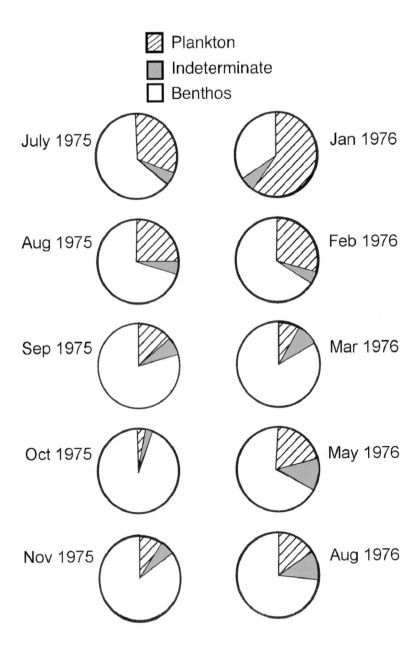

Figure 7.1. Seasonal changes in source of the diet (% numerical abundance of planktonic and benthic microalgae) as determined from stomach content analysis in the Patagonian scallop *Chlamys tehuelcha* (1975 cohort) (modified from Vernet 1977).

Adult scallops are capable of ingesting relatively large particles and this ability has been related to the absence of mechanisms for sorting particles on the basis of size by the gills of Pectinidae (Beninger et al. 1988). Particles up to 950 μm were described in the gut contents of *Patinopecten yessoensis* (Mikulich and Tsikhon-Lukamina 1981). Gut contents of *P. magellanicus* ranged from 10 to 350 μm (Shumway et al. 1987). By comparing the algal composition and size spectra of the plankton with stomach contents of *Chlamys tehuelcha*, Vernet (1977) found, however, that this scallop showed negative selection for particles exceeding 100 μm.

Changes in diet may be associated with major shifts in habitat of scallop species that move from an above-bottom, byssally attached juvenile stage to a free-living existence on the bottom as adults. Such changes have not yet been investigated.

7.2.2 Feeding Currents and Mechanisms of Particle Capture

Scallops are non-siphonate, ciliary suspension-feeders, which exhibit no fusion of the mantle edge. Water enters the mantle cavity along the ventral and anterior edge, and exits through the posterior exhalent opening (Hartnoll 1967). In *Argopecten irradians* this aperture is about five times larger than the anterior inhalent opening (Winter and Hamilton 1985). Scallops preferentially orient themselves by facing directly into the current (exhalent opening facing away from the direction of the flow) (Hartnoll 1967; Caddy 1968; Mathers 1976; Grant et al. 1992). Adults possess a plicate, heterorhabdic gill in which ordinary filaments and principal filaments form the crests and grooves respectively of the plicae.

Bivalves have been suggested to capture particles by direct interception with the ctenidial filament and transport the particles along the frontal surface of the filament by mucociliary processes (Ward et al. 1998a; See section below on "Particle retention efficiency" for alternative views). This mechanism for particle capture is consistent with theories of hydrosol filtration observed for many different groups of aquatic suspension-feeders (Shimeta and Jumars 1991). The low angle of particle approach to the filament increases the probability of encounter with frontal cilia and the vertical flow set up by the beating of the laterofrontal cilia or cirri reduces flow through the interfilamentary spaces and redirects particles to the frontal surface of the filament (Ward et al. 1998a). According to these authors, these two processes affecting flow patterns promote the retention of particles on the frontal surface by increasing the encounter efficiency with the frontal cilia.

While the mechanism of particle capture itself is likely to be similar for most species of bivalves, the velocity and sites of particle transport on the gill and the efficiency of retention vary among species including pectinids. Oysters (*Crassostrea virginica* and *Ostrea* sp.) and mussels (*Geukensia demissa*, *Mytilus edulis* and *Modiolus modiolus*) have well developed ventral ciliated grooves where particles are incorporated into a cohesive mucus string and transported at relatively low velocities anteriorly toward the labial palps (Ward 1996). In *C. virginica* the opposing surfaces of the ridged labial palps reduce the cohesiveness of the mucus strings and disperse entrapped particles for ingestion whereas in *M. edulis* the ridged surface of one palp and the action of the ciliated dorsal margin disperse particles (Ward 1996).

The scallop, *P. magellanicus,* and mussel, *Arca zebra,* however, lack ventral grooves and only small mucus aggregations or strings accumulate on the ventral margins which could be transported to the labial palps for rejection as pseudofeces (Ward 1996). Particles destined for ingestion by *P. magellanicus* are suspended in a slurry and transported at high velocities in the dorsal tracts rather than in a mucus-bound string along the ventral margin (Ward 1996).

7.2.3 Particle Retention Efficiency

The amount of food available to suspension-feeding bivalves is a function of the volume of water transported across the gills (pumping rate) as well as the efficiency with which particles are retained by the gill. Most juvenile and adult suspension-feeding bivalves are able to retain particles above 3–4 μm with 100% efficiency, and retention efficiency decreases with decreasing particle size (to between 35 to 90% for 2 μm particles) (Møhlenberg and Riisgård 1978; Riisgård 1988). This pattern is consistent with capture mechanisms other than sieving proposed by modern filtration theory (Rubenstein and Koehl 1977; Jørgensen 1981, 1983). In contrast, the limit for effective retention of particles in members of the Pectinidae studied to date, [including *Pecten opercularis, P. septemradiatus* (Møhlenberg and Riisgård 1978), *Chlamys opercularis* (Vahl 1972) and *C. islandica* (Vahl 1973a) and *Argopecten irradians concentricus* (Palmer and Williams 1980; Riisgård 1988)], is about 5–7 μm (Fig. 7.2). Free bacterioplankton, typically ranging in size from 0.3 to 1 μm, is therefore not available as a food source for pectinids unless bound in aggregates. The retention pattern for particles below 5–7 μm, however, varies somewhat between species. For example, *C. islandica* is more efficient in capturing small particles than *C. opercularis* (retention efficiencies for 2 μm = ca. 27 and 5%, respectively (Vahl 1973a).

It should be pointed out that retention efficiencies are generally determined using electronic particle counters. Jørgensen et al. (1984) found that these instruments tend to somewhat overestimate retention efficiencies of small particles (<2 μm in diameter) generally determined with a 50 μm aperture tube. This is partly attributed to interference by conductive colloidal particles in seawater, and electrical noise at the lower limit of resolution of the particle counter, although recent multisizers are capable of increased resolution at small sizes.

The influence of particle concentration on retention efficiency in bivalves is not fully resolved. Palmer and Williams (1980) found that *Argopecten irradians concentricus* were significantly more efficient in retaining small particles (<3.4 μm) at high algal concentrations that induce pseudofeces production (6.1 mg wet weight L^{-1}) than at low concentrations below the threshold for pseudofeces production (0.9 mg L^{-1}), whereas an inverse effect was observed in the oyster *Crassostrea virginica.* The addition of a higher concentration of bentonite clay to an algal suspension caused a reduction in the retention efficiency of small particles by *Placopecten magellanicus* (Cranford and Gordon 1992).

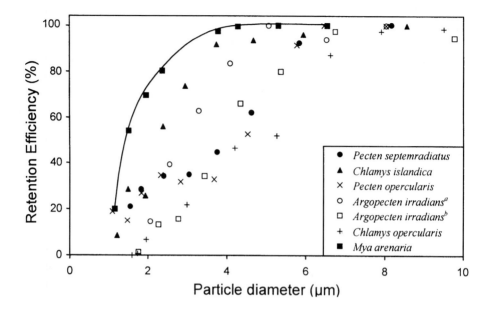

Figure 7.2. Retention efficiency (%) as a function of particle size in several pectinid species (redrawn from references cited in the text) compared to that of a bivalve with well developed eulaterofrontal cirri, e.g., the softshell clam, *Mya arenaria* (line fitted to the data from Møhlenberg and Riisgård 1978). [a]Riisgård 1988; [b]Palmer and Williams 1980, retention measured at low algal concentrations (0.9 mg wet weight L^{-1}).

Bivalves possess three types of ciliary tracts in their gill filaments: lateral (current producing), frontal (particle carrying), and laterofrontal tracts believed to play a role in particle retention (Owen and McCrae 1976). Although the mechanism of particle retention by the bivalve gill is not fully understood (see below) retention efficiency of small, bacteria-sized particles has been correlated with the development or structural complexity of laterofrontal tracts in both marine and freshwater bivalves (McHenery and Birbeck 1985; Silverman et al. 1997). Thus the poor retentiveness for small particles of the pectinid gill has been ascribed to the poor development of laterofrontal ciliary tracts in this group of bivalves (Owen and McCrae 1976; Møhlenberg and Riisgård 1978). The laterofrontal tracts consist only of small, simple pro-laterofrontal cilia, and lack the large, compound eulaterofrontal cirri characteristic of mussels (Owen and McCrae 1976). Thus, these authors suggest that the poorly developed laterofrontal cilia of pectinids play no role in particle retention. They suggest that, at low to moderate particle loads, the form of the plicae and U-shaped nature of principal filaments all combine to create a region of low pressure which tends to attract particles into the gutters formed by principal filaments.

Laterofrontal ciliary tracts, located across the entrance to interfilamentary spaces, are attributed a role in particle retention in numerous studies involving mussels, which are characterised by large, complex cirri. However, the exact mechanism of particle retention by laterofrontal cilia or cirri has remained controversial, ranging from their action as mechanical filters or sieves (direct mechanical interception) (Dral 1967) to a hydromechanical role in capture (Jørgensen 1990), or a combination of other multiple mechanisms (Ward et al. 1998a). A mechanical role of the cirri is supported by differential particle capture in freshwater bivalves with cirri of different sizes (Silverman et al. 1997). Capture via a hydrodynamic mechanism is presumably achieved by influencing the steepness and height of velocity gradients in the boundary zone between currents entering the gill interfilamentary space, and surface currents along the frontal surface of gill filaments (Jørgensen 1981). Jørgensen's velocity gradient hypothesis, developed for the mytilid filibranch gill, predicts that the critical size for efficient particle retention should be larger in bivalves (such as pectinids) with reduced laterofrontal ciliary tracts. Ward et al. (1998a) argued that low Reynolds numbers preclude a mechanical sieving role of laterofrontal cilia/cirri (also reviewed by Jørgensen 1996), and that hydrodynamic entrainment may occur on principal but not ordinary filaments. Based on endoscopic observations they concluded that particle capture occurs by direct interception at the frontal surface of ordinary gill filaments and indeed does not require the presence of laterofrontal tracts. They suggest that the main role of the laterofrontal tracts is to increase the probability of encounter of particles with the frontal surface of gill filaments, by inducing lateral flows that reduce particle loss through the interfilamentary spaces. However, the pectinid gill was not represented in their study and these findings remain controversial.

Despite the controversy over the mechanism of particle capture as determined by *in vitro* methods using isolated gill filament preparations or *in vivo* methods applied to intact animals (e.g., confocal microscopy and video endoscopy respectively), each of which has inherent limitations [see Comments by various authors in Limnol. Oceanogr. 45(5), 2000], several recent studies of mytilids may support the role of the laterofrontal cirri in capturing particles and directing them to the frontal surface of gill filaments for subsequent transport (Riisgård et al. 1996; Silverman et al. 1996, 1999, 2000). Using confocal laser scanning microscopy and high-speed video recordings of isolated gill sections, Silverman et al. (1999) concluded that the movement of cirri in *Mytilus edulis* was essential for successful capture of small, bacterial-sized (1 μm) particles although they could not discriminate between mechanical and hydromechanical interception. This was further supported by the fact that mussels treated with serotonin concentrations that stopped the beat of laterofrontal cirri, but not that of frontal or lateral cilia on the gill filaments, showed a 90% reduction in the rate of removal of bacteria from seawater relative to untreated controls.

Post-settlement scallops undergo major changes in the anatomy of feeding organs, especially the gill, during early development. Ontogeny recapitulates phylogeny as the postlarval gill experiences a gradual transition from a homorhabdic condition, with a single, non-plicate inner demibranch, to a heterorhabdic, plicate, reflected gill consisting of both inner and outer demibranchs (Kingzett 1993; Beninger et al. 1994; Veniot et al. 2003), characteristic of adult scallops (Beninger and Le Pennec, Chapter 3). Lateral and frontal cilia are well developed early on in the gills or postlarval stages, but the laterofrontal cilia are absent in these early stages (Veniot et al. 2003). It is expected that such profound anatomical changes are associated with differences in gill function: suspension-feeding appears to be relatively ineffective in postlarvae prior to reflection of the demibranchs, as evidenced in *Patinopecten yessoensis* by very low clearance rates at sizes <600 μm in shell height (SH) (Kingzett 1993). Yet little is known about the retention efficiency of the postlarval gill in any bivalve species.

From the gut contents of *P. yessoensis* offered suspensions of fluorescent polystyrene beads either 2, 6, 9, 23 or 55 μm in diameter, Kingzett (1993) determined that postlarvae ~ 300 and 1,000 μm SH were unable to ingest beads of the two largest bead sizes tested, suggesting that the upper limit for capture/ingestion of particles is approximately 20 μm. However, at 300 μm SH scallops were able to ingest approximately equal numbers of 2 and 6 μm beads when offered at the same concentration. Thus, retention efficiency differs between early postlarvae and adult scallops. Firstly, postlarvae appear to have a lower maximum size limit for particle retention, and secondly, despite the absence of laterofrontal cilia on gill filaments, postlarvae appear to be capable of effectively capturing small 2 μm particles. One of the limitations in Kingzett's study is that the different bead sizes were not fed to the animals simultaneously, thus precluding accurate measures of retention efficiency. It is important to note that measures of gut content do not allow differentiation of the particle retention/capture per se, but reflect the integration of feeding processes leading to ingestion: retention-transport-selection-ingestion. Confirmation of effective retention of small (1–2 μm) particles by the postlarval scallop gill, which lacks laterofrontal cilia, may challenge previous suggestions that these cilia are prerequisite for small-particle capture in suspension-feeding bivalves (Riisgård et al. 1996; Silverman et al. 1996, 1999, 2000). However, it is difficult to make direct comparisons between the mechanics of particle capture of early scallop postlarvae and adults, as the morphology of the gills differs markedly between the two stages.

7.2.4 Feeding Rates

The allometric relationship between clearance rate and body size (tissue dry weight) for several pectinid species is shown in Table 7.1. The weight exponent is variable, ranging from 0.58 to 0.94, with a mean of 0.7, which is within the range published for other bivalves (Bayne and Newell 1983). The allometric exponent was found to be considerably higher (mean = 0.92) in larvae and juveniles (up to ~10 mm) of the bay scallop, *Argopecten irradians concentricus* (Lu and Blake 1997) than that reported for adult pectinids. The relationship between clearance rate (CR, in μL hr^{-1}) and total body ash-free dry weight (AFDW, in mg) in these early development stages at 25°C and a near-optimum food concentration of 20 cells T-ISO μL^{-1} was described by the equation:

$$CR = 56.565 \cdot W^{0.931}, \text{ or when expressed in terms of shell height (H, in mm) as:}$$

$$CR = 1.051 \cdot H^{2.479} \tag{1}$$

Table 7.1

Parameters of the allometric relationship between clearance rate (CR; L h^{-1}) and tissue dry weight (W; g) according to the equation CR = aW^b, in several pectinid species.

Species	Size range (g flesh dry wt.)	Temp. (°C)	a	b	Source
Chlamys islandica	0.004–7.0	3.4	3.9	0.60	Yahl 1980
Chlamys hastata	1.8–2.2	12.8	0.145	0.943	Meyhofer 1985
Placopecten	1.8–42	5.5–8.5	0.616	0.76	MacDonald &
magellanicus		10–12	1.318	0.60	Thompson 1986
(10 m depth)		8–10	0.891	0.66	
"	0.016–0.6	10	3.435	0.855	Manning 1985[1]
Argopecten	0.05–4.2	10–26	5.827*	0.584	Kirby-Smith 1970
irradians	0.7–4.1	22–26	4.742	0.82	Chipman & Hopkins 1954[2]
concentricus					
Argopecten	0.1–5.7	12	2.45	0.80	Navarro & González 1998
purpuratus					

* Assuming 85% water content of tissues;
[1] Clearance rate determined with *Isochrysis galbana* (3.5 μm) and thus underestimates pumping rate;
[2] Calculated by Winter (1978).

Weight-normalised clearance rates (volume of water cleared of particles per unit time by an animal of standard tissue weight) of various pectinid species are shown in Table 7.2. It is generally difficult to make meaningful interspecies comparisons of feeding rates derived from studies differing in methodology and experimental conditions. This is especially true given that clearance rates are extremely sensitive to changes in food quality and quantity. Declining salinities (between 27 and 18 ppt) have also been shown to markedly reduce clearance rates in the Chilean scallop *Argopecten purpuratus* (Navarro and González 1998). Physiological rates of scallops are generally very susceptible to the negative effects of low salinities, as they are unable to maintain prolonged valve closure and thus cannot effectively isolate themselves from the environment. Interspecific comparisons in feeding rates are therefore best carried out from studies, which employ identical experimental protocols on a wide variety of bivalve species. For example, Meyhöfer (1985) found that weight-standardised pumping rates were highest for two filibranch species, the scallop *Chlamys hastata* (0.145 L h^{-1}) and mussel *Mytilus californianus* (0.133 L h^{-1}), followed by *Clinocardium nuttallii* (0.051 L h^{-1}) and finally by *Macoma nasuta* (0.0014 L h^{-1}), a deposit-feeder capable of facultative suspension-feeding. When *Mya arenaria* and *Placopecten magellanicus* were simultaneously exposed to various concentrations of particles scallops displayed higher weight specific

Table 7.2

Weight-standardised clearance rates (= pumping rates) (CR_s; L $h^{-1}g$ dry tissue weight^{-1}) of various pectinid species. $CR_s = CR_e \cdot (W_s/W_e)^b$, where CR_e and W_e are the clearance rate and tissue dry weight of the experimental animal, $W_s = 1$ g standardised weight and b is the exponent of the allometric relationship between CR and W.

Species	Temp. (°C)	Suspension	CR_s	Source
Chlamys hastata	12.8	Direct measurement w/ thermistor flowmeter	0.145	Meyhöfer 1985
Chlamys islandica	3.4	17 μm polysterene particles, 1,000–2,000 mL^{-1}	3.09	Vahl 1980
Chlamys opercularis	11–13	Natural seston + algae	13.589[a]	Vahl 1972
Chlamys opercularis	5 10 20	*Dunaliella euchlora*, 8,000– 10,000 cells mL^{-1}	1.64 3.23 5.90	McLusky 1973
Pecten (Chlamys) opercularis	10–13	Mixed algal suspension, 0.02–0.3 mg organic DW L^{-1}	14	Møhlenberg & Riisgård 1979
Pecten furtivus	10–13	”	31	”
Placopecten magellanicus 10–13 m	10–12	Natural seston, 5–10 mg DW L^{-1}	0.871–1.318	MacDonald & Thompson 1986
Argopecten irradians concentricus	22–26	*Nitzchia* (850–8,000 cells mL^{-1}) or *Chlamydomonas* (28,000 cells mL^{-1})	4.742	Chipman & Hopkins 1954
Argopecten i. concentricus	10–26 5	*Nitzchia* 1 x 10^5 – 5 x 10^5 cells mL^{-1}	5.82 1.75	Kirby-Smith 1970
Argopecten i. concentricus	21	*Thalassiosira pseudonana*, 50,000–340,000 cells mL^{-1} *Dunaliella tertiolecta*, 10,000–30,000 cells mL^{-1}	4.022 (0.31–8.78) 5.684 (0.65–11.90)	Palmer 1980
Argopecten i. irradians	22	*Thalassiosira weissflogii* 1,200 cells mL^{-1} 4,800 cells mL^{-1} 12,000 cells mL^{-1}	10.333 4.707 1.387	Kuenstner 1988
Argopecten purpuratus	12	*Isochrysis galbana* 30,000 cells mL^{-1}	2.6	Navarro & González 1998
Argopecten purpuratus	16–20	*Isochrysis galbana* + *Chaetoceros gracilis* 30,000 cells mL^{-1}	2.1[b]	Navarro et al. 2000

[a] Assuming 85% water content of tissues and weight exponent of 0.7
[b] Calculated based on a weight exponent of 0.7

clearance rates than clams, especially at concentrations less than 7.0 mg L^{-1} (Bacon et al. 1998). Møhlenberg and Riisgård (1979) compared feeding rates of thirteen bivalve species and reported clearance rates considerably higher than those from previous studies. Weight-standardised clearance rates of the two scallops, *Pecten furtivus* and *Pecten (Chlamys) opercularis*, based on only three measurements (Table 7.2), were comparable only to those of *Mytilus edulis* but were generally two to five times higher than those of other bivalves included in the study.

Feeding rates in bivalves, including pectinids, have also been shown to be influenced by chemical compounds such as metabolites extracted from cultured microalgae. There are several examples of bivalves exhibiting chemosensory abilities, with some dissolved phytoplankton metabolites typically producing inhibitory feeding responses as in the case of the blue mussel *Mytilus edulis* (Ward and Targett 1989). Stimulatory responses are much rarer but have been observed for clearance and ingestion rates in *Placopecten magellanicus* when scallops were exposed to metabolites from the diatom *Chaetoceros muelleri* Lemmermann (Ward et al. 1992). A microalgal diet supplemented with a lipid emulsion also had a stimulatory effect on clearance rates in *Argopecten purpuratus* (Navarro et al. 2000). A better understanding of the capability of bivalves to detect and respond to chemical cues has interesting implications for research on artificial diets possibly making them more palatable and more readily ingested by the bivalves.

It has recently been discovered that 6–15 μm ciliated and non-ciliated epithelial cells can be released from the pallial cavity of sea scallops *Placopecten magellanicus* (MacDonald et al. 1995; Potter et al. 1997). Release of a large number of these cells may lead to an estimate of negative clearance rates, i.e., higher concentrations of suspended particles after exposure to the bivalve than in the original suspension. Exfoliation of these cells has also been observed, although not quantified, for the blue mussel *Mytilus edulis* and the eastern oyster *Crassostrea virginica* (MacDonald, pers. obs.) and *Mercenaria mercenaria* (Bricelj, pers. obs.). Exfoliation of small numbers of cells may be a consequence of cellular turnover and normal physiological function. However, environmental stressors, such as unseasonally elevated water temperatures are known to increase the rate of cell release to the point that tissue damage results (Potter et al. 1997). It is not known what other stressors may cause increased exfoliation rates in bivalves or the short- and long-term consequences that any potential damage will have on the efficiency of particle retention and transport on the gill.

7.2.5 Clearance Rate in Relation to Food Concentration

When exposed to increasing suspended particulate loads, suspension-feeding bivalves are able to control the total amount of material ingested by: a) reducing the time spent pumping (discontinuous feeding behaviour), b) reducing their clearance rates, and/or c) increasing the amount of material rejected in pseudofeces (Foster-Smith 1975a, b; 1976). Palmer (1980) characterised the bay scallop, *Argopecten irradians*, as a continuous feeder, that showed no rhythmic cycles in clearance rates in response to tidal cycles or photoperiod. In contrast, Mathers (1976) suggested that the feeding activity of *Pecten maximus* was cyclical in response to changes in tidal flow. He indicated, however, that this was caused by strong reversible tidal currents and was probably not an endogenous feature of the scallop's feeding behaviour. Existing evidence, although limited to a few pectinid species (see below), suggests that scallops primarily regulate ingestion and compensate for short-term changes in food supply through fluctuations in clearance rate (mechanism b).

A strong inverse, linear relationship between clearance rate and algal concentration has been described for adult bay scallop *Argopecten irradians concentricus* (Palmer 1980; Fig. 7.3A). Clearance rates are reduced by 95% over the concentration range 0.94 to 9.4 mg dry wt. L^{-1} (1.23 to 12.3 x 10^6 μm^3). Therefore, above a threshold concentration of ca. 2 mg dry weight L^{-1}, algal ingestion rate becomes independent of concentration (Fig. 7.3B). Similarly, clearance rates of juvenile *A. i. irradians* were found to decline by 85% with a 10-fold increase in the concentration of *Thalassiosira weissflogii* (1.2 to 12 cells μL^{-1} = 0.83 to 8.3 x 10^6 μm^3 mL^{-1}) (Kuenstner 1988). Cahalan et al. (1989) reported a lower reduction (56%) for juvenile bay scallops exposed to concentrations between 7.5 and 68 cells μL^{-1} (0.25 to 2.28 x 10^6 μm^3 mL^{-1}) of *Isochrysis galbana*, an alga that is incompletely retained by the pectinid gill. Lu and Blake (1997) found that clearance rates of juvenile bay scallops, *A. i. concentricus* (~0.6 to 9.8 mm in shell height) declined exponentially with increasing algal concentration ranging between 5 and 50 *I. galbana* cells μL^{-1} (Fig. 7.3C). Ingestion rate, however, increased with increasing cell concentration, and approached an asymptote or maximum ingestive capacity at ~ 50 cells μL^{-1} (Fig. 7.3D). It is interesting to note that larvae reached this asymptote at a lower cell density (~ 20 cells μL^{-1}) than larger juveniles, suggesting that the latter are better able to handle dense algal assemblages than larval stages. An optimum cell density of 20 cells μL^{-1} was established for growth of juveniles of this species, as higher concentrations (up to 50 cells μL^{-1}) did not significantly increase growth rates (Lu and Blake 1996). Similarly, growth rates in shell height and tissue weight of juvenile *Nodipecten subnodosus* (lion's paw scallop) were maximised at the lowest algal concentration tested (33 T-ISO cells μL^{-1}). Growth was comparable at 33 and 66 T-ISO cells μL^{-1}, and declined at 100 cells μL^{-1} (García-Esquivel et al. 2000).

Clearance rates for *Placopecten magellanicus* and *Mya arenaria* both significantly declined as concentration and organic content of microalgae and silica mixtures increased between 1 and 14 mg L^{-1} (Bacon et al. 1998). However, other studies on *P. magellanicus* have reported clearance rates to be independent of particle concentration between 1–15 mg L^{-1} and this may be related to the type of particles used in the individual studies (Cranford and Grant 1990; Cranford and Gordon 1992; MacDonald and Ward 1994; Cranford et al. 1998). Several authors have emphasised the potential interactions between seston concentration and flow velocities on the feeding response of scallops (Wildish et al. 1992; Wildish and Saulnier 1993; Pilditch and Grant 1999a). How scallops respond to changes in particle concentration may not only depend on variations in flow speeds, but also on whether or not the flow is increasing or decreasing in velocity (Pilditch and Grant 1999a). These authors observed a decline in scallop clearance rates when phytoplankton concentration increased and flow velocities were increasing but found clearance rates to be independent of concentration, albeit reduced by 50%, when flow velocities were decreasing. They concluded that the scallop's short-term (hours) feeding history may be

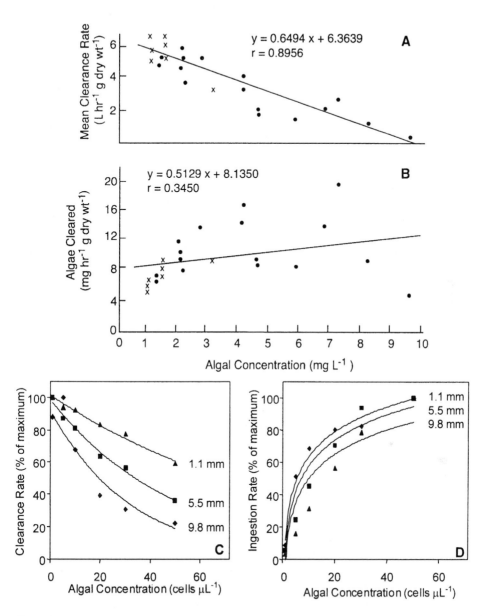

Figure 7.3. *Argopecten irradians concentricus* adults. Relationship between algal concentration (mg dry weight L^{-1}) and clearance rate (A) and amount of algae cleared (B). Each point represents the mean value for one experiment - that is for 5 to 32 hourly measurements; algae used were *Dunaliella tertiolecta* (x) or *Thalassiosira pseudonana* (●) (from Palmer 1980). *Argopecten irradians concentricus* juveniles. Relative clearance rate (C) and algal ingestion rate (D) as a % of the maximum of bay scallops ranging from 1.1 to 9.8 mm shell height at 25°C (calculated from data in Lu and Blake 1997). Data fitted to exponential (C) and logarithmic functions (D).

influential in determining its response to variations in the food supply and that concentration was the primary determinant of scallop clearance rate over a range of fluxes (Pilditch and Grant 1999a). See the section 7.2.10 on "Effects of flow on feeding and growth" for a more detailed discussion on the effects of flow on feeding activity.

Hawkins et al. (2001) have also emphasised the importance of the amount of suspended material in regulating clearance rates in the scallop *Chlamys farreri*, in particular the volume and composition of the seston. They found that clearance rate varied in a unimodal relation with short-term changes in seston abundance and was primarily dependent upon seston availability in terms of total volume rather than any gravimetric estimate of mass. They suggested morphometric limits to feeding behaviour, most likely constraints within the digestive system rather than on the gills. There is also an interaction between seston volume and seston quality with 58% of the variance in clearance explained by particle volume and chlorophyll concentration combined, in contrast to only 47% for seston volume alone (Hawkins et al. 2001). These authors observed over an order of magnitude variation in clearance rates and concluded that this potential for feeding adjustment would be of greatest physiological and ecological importance at low food availabilities because it would have proportionally more impact on net energy balance.

7.2.6 Influence of Temperature on Feeding Rates

The effect of temperature on feeding activity has been investigated in few pectinid species. Clearance rates of *Argopecten irradians concentricus* were independent of temperature between 10 and 26°C, but were markedly depressed at 5°C (Fig. 7.4B; Kirby-Smith 1970). Since the increase in metabolic expenditure between 10 and 26°C is not offset by a parallel increase in feeding activity, the irrigation efficiency (litres pumped per mL of O_2 consumed) decreases rapidly with increasing temperature above 10°C (Fig. 7.4A). Irrigation efficiency is generally assumed to be inversely related to the maintenance food requirement (that at which growth = 0) (Newell and Køfoed 1977). Therefore, it is expected that bay scallops will sustain a maximum energetic gain from the environment between 10 and 20°C, and that they will be severely stressed, and exhibit rapid weight loss, under suboptimal conditions of poor food supply and high temperature. Such conditions might arise due to an increase in water temperatures during winter (caused by thermal pollution) without a concomitant increase in food levels. The scallops' dependence on higher food levels at high temperatures is exemplified by the finding that growth rate of juveniles at lower temperatures (10 to 16°C) is independent of chlorophyll-*a* levels within the range of naturally occurring concentrations (0.5 to 5.5 µg L^{-1}), but becomes increasingly correlated with chlorophyll levels at higher temperatures (22 to 28°C) (Kirby-Smith 1970; Kirby-Smith and Barber 1974).

In contrast to the above findings for adults, clearance rate of juvenile *A. i. concentricus* was an increasing function of temperature between 10 and 30°C (Fig. 7.4D), which approximates the temperature range of this species in its natural habitat in Florida, USA (~12 to 32.5°C) (Lu and Blake 1997). Thus juveniles may be better able to exploit

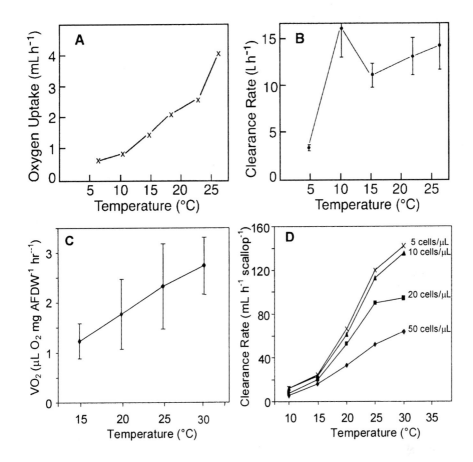

Figure 7.4. *Argopecten irradians concentricus* adults. Mean oxygen uptake (A) and clearance rate (B) of a standard animal 20 g in wet tissue weight (calculated from data by Kirby-Smith (1970; Tables VII and VIII for scallops fed *Nitzchia* sp.; vertical bars = standard errors). *Argopecten irradians concentricus* juveniles. C. Weight-specific mean oxygen uptake (± SD) vs. temperature of juveniles 3.5 to 5 mm in shell height (plotted from Lu et al. 1999). D. Clearance rate vs. temperature of 5 mm juveniles at different *Isochrysis galabana* cell concentrations (plotted from Table 3 in Lu and Blake 1997).

higher temperatures when these coincide with an abundant food supply than adults. As in adults, oxygen uptake in juveniles increased with increasing temperature (Fig. 7.4C).

In *Chlamys opercularis* the response of clearance rates to temperature is similar to that of adult bay scallops. Feeding rates are independent of acclimation temperature between 10 and 20°C, but drop markedly at 5°C (Q_{10} from 5 to 10°C = 3.87; Q_{10} from 10 to 20°C = 1.83; McLusky 1973). Unlike bay scallops, however, oxygen uptake remains constant between 10 and 20°C, allowing conservation of energy at higher temperatures. This strategy is similar to that displayed by *Mytilus edulis* (Bayne et al. 1976). In *Placopecten magellanicus* from Newfoundland, clearance rates were significantly correlated with ambient temperature in both shallow (10 m) and deep water (31 m) populations between 0 and 12°C. In the latter, clearance rates were also correlated with food availability (energy content of seston) (MacDonald and Thompson 1986). In juveniles of the subtropical scallop, *Argopecten ventricosus-circularis*, routine oxygen consumption rate increased steadily between 12 and 28°C, whereas clearance rate and algal ingestion rate were maximised between 19 and 20°C, and declined above and below this temperature range (Sicard et al. 1999). This physiological optimum temperature range was confirmed by determining that shell growth was also maximised at 19–22°C. Furthermore, scope for activity, defined as the difference between routine (active) VO_2 and standard (basal) VO_2 determined for fed and starved animals respectively, was also maximised at 19°C. This illustrates that physiological rates can be very useful in predicting optimum ranges of environmental factors for growth.

Marine suspension-feeders inhabiting coastal waters typically pump 15 L or more of water per equivalent mL of O_2 consumed (Jørgensen 1975). A mean irrigation efficiency or convection requirement of 17 (8 to 25) was reported for *Pecten latiauratus* (Jørgensen 1960), and values ranging from 15 (at 5°C) and 39 (at 20°C) were found for *Chlamys opercularis* acclimated to laboratory conditions (McLusky 1973). A considerably higher mean value of 79 (range = 63–97) was reported for the same species by Vahl (1972). A maximum irrigation efficiency of 8.8 L mL^{-1} O_2 was obtained for juvenile *A. ventricosus-circularis* at 19°C (Sicard et al. 1999).

7.2.7 Pseudofeces Production, Pre- and Post-ingestive Particle Selection

Pseudofeces production has been shown to be an important pre-ingestive mechanism because it facilitates the process of particle selection whereby less nutritious particles may be rejected and the quality of the ingested material improved proportionately. The labial palps are currently believed to be the principal site of pre-ingestive particle selection via pseudofeces production in bivalves, although the evidence remains inconclusive. Their role in particle selection is supported by Kiørboe and Møhlenberg's (1981) finding that selection efficiency is positively correlated with a relative index of palp size (ratio of palp area to clearance rate) and other indirect evidence. Direct endoscopic observations of the pectinid palps to confirm their role in particle selection have so far proved elusive due to their extreme sensitivity and obstruction by the complex, arborescent lips (Beninger et al. 1992). Morphological evidence of their role in selection, e.g., presence of sensory structures on the ridged surface, was found to be lacking in both *P. magellanicus* and *Chlamys varia* (Beninger et al. 1990). Selective rejection of particles may also occur on the bivalve heterorhabdic gill, e.g., via rejection of material from the ventral gill margin [most notably in oysters, *Crassostrea virginica* (Ward et al. 1997, 1998b)]. Endoscopic observations indicate that even in pectinids, which lack a ventral groove, particles bound in mucus may move ventrally along ordinary filaments, and break off as pseudofeces as they move anteriorly (Beninger et al. 1992). This type of transport is only observed at high particle concentrations or when the scallop's ingestive capacity is overloaded. Pseudofeces in adult scallops are ejected from the pallial cavity via valve clapping. Particle selection through the production of pseudofeces has been demonstrated in studies on feeding activity using laboratory diets and cultured algae mixed with sediments and assemblages of natural particles found in the environment (Kiørboe et al. 1980; Newell and Jordan 1983; Cranford and Gordon 1992; Iglesias et al. 1992; MacDonald and Ward 1994; Shumway et al. 1997; Bacon et al. 1998).

Scallops, however, typically do not produce copious amounts of pseudofeces, as compared to mussels and oysters. Bay scallops, *Argopecten irradians*, rejected only up to 25–35% of the algal cells filtered when exposed to bloom concentrations (0.55 to 1.46 x 10^6 cells mL^{-1} = 2.4 to 6.4 mg dry weight) of the pelagophyte *Aureococcus anorexefferens* (= *anophagefferens*) (Kuenstner 1988). MacDonald and Thompson (1986) reported that *Placopecten magellanicus* produced no pseudofeces when fed natural seston levels of 5 to 10 mg dry weight L^{-1}. In contrast, *Mytilus edulis* initiates pseudofeces production at seston concentrations between 2.6 and 5.0 mg L^{-1} depending on body size (Widdows et al. 1979). It has often been thought that bivalves only produce pseudofeces when exposed to concentrations of particles above a certain threshold value. However, studies by Ward et al. (1993) using endoscopic techniques have shown that pseudofeces are produced intermittently during the feeding process in many bivalves regardless of the concentration. It is likely that some pseudofeces are being produced even at low concentrations, although not in sufficient concentrations to be readily noticed or quantified unless specialised techniques are used.

The efficiency of particle selection in marine bivalves has been shown to be quite variable both within and among species, and is most likely related to palp size, seston conditions, or which indicator of seston quality is used to measure efficiency (Kiørboe and Møhlenberg 1981; Iglesias et al. 1992; Urrutia et al. 1996). Pectinids, like other marine bivalves have been shown to have the capability of rejecting poorer quality particles and improving the quality of the ingested ration. Recent studies on *Placopecten magellanicus* have confirmed the capability of this scallop species to reject particles of poorer quality preferentially based on both the organic composition of particles and the concentration of chlorophyll-*a* containing particles in the natural seston (Cranford and Gordon 1992; MacDonald and Ward 1994; Bacon et al. 1998). Results for *P. magellanicus* were consistent with those reported for other bivalve species such as *Mytilus edulis* and *Cardium edule* respectively, where the efficiency of selection diminished as the quality of the seston mixture declines (Bayne et al. 1993; Navarro et al. 1994) (Fig. 7.5). Both of the studies on *P. magellanicus* by MacDonald and Ward (1994) and Bacon et al. (1998) reported that the overall quantitative selection process, measured by the compensation index, was not effective at the lowest quality diets tested. Compensation index (CI) not only takes into consideration the quality of seston and pseudofeces, like most other indicators of sorting efficiency, but is unique because it takes into consideration quantities of material cleared and rejected. It is calculated as CI = (I_Q / SES$_Q$) -1, where I_Q is the quantity of chlorophyll-*a* ingested and SES$_Q$ is a measure of the quality of the seston, e.g., chlorophyll-*a* concentration. CI is the same as the index calculated by Navarro et al. (1992) and the benefit ratio (BR) given by Iglesias et al. (1992). In fact, for very poor-

434

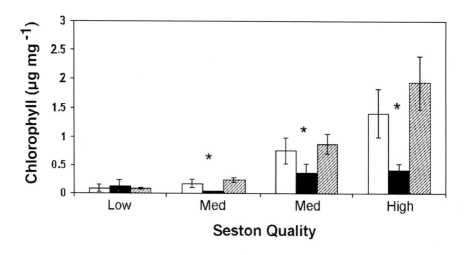

Figure 7.5. Comparison of the quality of material in the seston (SES_Q white box), with that in pseudofeces (PS_Q black box) and that ingested (I_Q striped box) by *Placopecten magellanicus* during each of the experimental periods. Data are from MacDonald and Ward (1994) and presented as mean $\pm$ SD, see their text for details.

quality diets, there may have been some evidence of organic enrichment of the pseudofeces rather than the ingested ration, or negative selection. Recognising that particle selection through the production of pseudofeces is a process of rejection MacDonald and Ward (1994) proposed the following explanation for negative selection. When poor-quality particles dominate the seston scallops may improve the quality of the material ingested by attempting to reject a high proportion of these particles. However, it may not be possible to retain the few relatively high-quality particles while rejecting the majority of low quality ones. This high rate of rejection may overwhelm the capacity of the system and result in non-selection or possibly even negative selection.

Both studies on particle selection in *Placopecten magellanicus* reported a negative relationship between compensation indices and clearance rates. This suggests that particle sorting on the labial palps is one of the limiting steps in the feeding process. In order for *Placopecten magellanicus* to increase the quality of the material ingested above 20% by particle selection through pseudofeces production they must reduce the volume of material entering the palps by reducing the clearance rate (MacDonald and Ward 1994). The strategy adopted by *P. magellanicus* when exposed to seston that varies in concentration and quality is to have relatively high clearance rates at low particle concentrations and as concentration increases significantly reduce clearance rate and increase the amount of pseudofeces produced. They rely on their pre-ingestive selection capabilities to reject poorer quality particles and enhance the quality of material ingested and maintain ingestion at higher concentrations.

It is well established that bivalves, including scallops, can selectively reject inorganic, sediment particles from mixed algal-sediment suspensions (see section on "Effects of suspended sediments on feeding and growth". Kiørboe and Møhlenberg (1981) determined the efficiency of particle selection of ten species of suspension-feeding bivalves fed algal-sediment mixtures from the ratio of chlorophyll-*a* to dry weight in the suspension and pseudofeces. The only pectinid tested, *Aequipecten opercularis*, showed an intermediate selection efficiency (= 5.4). A maximum efficiency of 15.8 was measured for *Spisula subtruncata*, and a minimum value of 2.9 for *Mytilus edulis* from a low turbidity environment. A low selection efficiency (SE = 2.7) was also determined for *Placopecten magellanicus* fed a mixture of algae and bentonite clay (Cranford and Gordon 1992). Values obtained in the two studies are not strictly comparable, however, as Cranford and Gordon (1992) based their SE estimates on data obtained from retained rather than suspended particles. Low capacity for particle selection between inorganic and organic particles was confirmed when sea scallops were fed natural resuspended sediment containing benthic diatoms (Cranford and Grant 1990).

Bivalves are also capable of discriminating among algal cells of similar size in their diet. Using flow cytometry, Shumway et al. (1985) showed that the diatom *Phaeodactylum tricornutum* was selectively rejected in pseudofeces of *Placopecten magellanicus* when this alga was fed in combination with the dinoflagellate *Prorocentrum* sp. and the cryptomonad *Chroomonas salina*. Similarly, *Thalassiosira pseudonana* was preferentially rejected in pseudofeces when present in a mixed suspension with *C. salina*, *Prorocentrum* and the toxic dinoflagellate *Protogonyaulax tamarensis* (clone GT429) (Shumway and Cucci 1987). Differential clearance of various algal clones in a mixed suspension was also demonstrated using flow cytometry in three species of 1–2 mm juvenile scallops (Shumway et al. 1997). Selective removal from suspension was greatest in *Patinopecten yessoensis*, which showed significantly higher clearance rates for *Chroomonas salina* (clone 3C, ~6.1 µm), *Pheodactylum tricornutum* (PHAEO, ~5.0 µm) and especially *Amphidinium carterae* (AMPHI, ~10 µm), than for *Isochrysis galbana* (T-ISO, 3–6 µm) or *Chaetoceros muelleri* (CHGRA, 4–9 µm). Lowest clearance rates were observed in all three scallop species with the smallest alga tested, a 2–3 µm unidentified prasinophyte (Omega 48). However, in *P. yessoensis*, algal size alone was not sufficient to explain the differences in clearance rate observed.

Selection in the gut or digestive gland-stomach complex is less well known, but it is believed that some bivalves can differentiate between particles within the gut and preferentially digest those particles that give the most nutritional benefit (Bricelj et al. 1984; Shumway et al. 1985; Lopez and Levinton 1987; Bayne et al. 1993; Wang and Fisher 1996). Postingestive selection may occur either by the retention of some particles longer than others in the stomach so that extracellular digestion has more time to act, or by directing some particles to the digestive gland for intracellular digestion. Cranford et al. (1998) found that two sizes of microspheres were passed at different rates through the gut of the sea scallop *Placopecten magellanicus*. While postingestive sorting in bivalves has been confirmed for several species, few studies have attempted to isolate the factors influencing selection among different particles presented simultaneously. *P. magellanicus* has been shown to be capable of distinguishing between particles of different size and density by retaining larger particles longer than smaller ones and lighter particles longer than denser ones (Brillant and MacDonald 2000). These authors further showed that

P. magellanicus has well-developed postingestive sorting capabilities and could sort organic from inorganic particles and some particles based solely on chemical properties (Brillant and MacDonald 2002). These scallops retained protein-coated beads in the gut longer than uncoated beads of identical diameter and density.

7.2.8 Absorption Efficiency

Few studies have attempted to determine the utilisation efficiency with which the ingested ration is absorbed of algal diets by pectinids. Available data suggest, however, that scallops do not differ markedly from other bivalves in their absorption capabilities. Using ^{14}C labelling techniques Peirson (1983) found that adult *Argopecten irradians concentricus* absorbed most algal species tested with absorption efficiencies (AE) ranging between 78.1 and 89.9%. These values represent maximum efficiencies since they were determined at relatively low mean algal rations (= 2 mm^3 L^{-1} = 468 µg C L^{-1}). Only the chlorophyte, *Chlorella autotrophica*, was inefficiently absorbed (AE = 17.4%), as previously observed both in adult and larval oysters (Floyd 1953; Babinchak and Ukeles 1979). Low absorption efficiencies are generally attributed to the indigestible cell wall of this alga. The diatom, *Thalassiosira pseudonana* used in Peirson's study yielded the highest absorption efficiencies (89.9%). This is almost identical to a mean value of 89.7% obtained for bay scallops fed *T. weissflogii* by Kuenstner (1988) with the twin ^{14}C-^{51}Cr radiotracer method. Veliger larvae of *Pecten maximus* inefficiently ingested and digested *Dunaliella primolecta*, and were unable to digest *Platymonas* (*Tetraselmis*) *suecica* (Le Pennec and Rangel-Dávalos 1985).

High absorption efficiencies are not sufficient to support growth of bivalves, and thus cannot be used as single predictors of food value. For example, Peirson (1983) reported a high absorption efficiency (83.3%) for *Dunaliella tertiolecta*, an alga known to support poor growth of oysters due to its deficiency in essential polyunsaturated fatty acids (Langdon and Waldock 1981). A high efficiency (90.6%) was also measured for *Aureococcus anophagefferens* (Kuenstner 1988); a species which caused starvation of bay scallops at bloom concentrations under field conditions (Bricelj et al. 1987a). Grant and Cranford (1989) reported that aged kelp (*Laminaria*) detritus was absorbed with high efficiency (70–80%) but did not support growth of adult *Placopecten magellanicus*.

A commonly used technique to estimate absorption efficiency is the Conover (1966) method which is based on the assumption that the organic and inorganic components must be ingested in the same proportions that occurs in the food supply (no ingestion selectivity). Adjustments must be made to the estimate of absorption efficiency if the bivalve is rejecting poorer quality particles through the production of pseudofeces otherwise the assumptions of the Conover ratio are violated. While some studies have shown that absorption efficiency increased with particle concentration, several recent studies on *Placopecten magellanicus* have shown absorption efficiency to be much better correlated to the quality than the concentration of the seston (Cranford 1995; Grant et al. 1997; Cranford et al. 1998; MacDonald et al. 1998). In many of these studies scallops were exposed to natural concentration of seston in the field or mixtures of microalgae and inorganic particles in the laboratory to mimic seston rather than various concentrations of pure cultured microalgae alone. Absorption efficiency in field and laboratory studies consistently increased with the quality of the seston whether percent organics or the concentrations of nitrogen or carbon were used as indicators of quality (Fig. 7.6). Similar relationships between absorption efficiency and diet quality have been reported for *Chlamys islandica, Placopecten magellanicus, Mytilus edulis, M. galloprovincialis, Cerastoderma edule, Crassostrea gigas* and *Mya arenaria* (Vahl 1980; Bayne et al. 1987; Cranford and Grant 1990; Iglesias et al. 1992; Navarro et al. 1992; Navarro and Iglesias 1993; Hawkins et al. 1998; MacDonald et al. 1998).

Absorption efficiencies in *Argopecten purpuratus* did not significantly increase when the quality of the microalgae was improved by the addition of a lipid emulsion and were observed to decrease from 85 to 27% when carbohydrates were added to the microalgal diet (Navarro et al. 2000). However, clearance rates were stimulated by the diets supplemented with lipids resulting in a significant increase in absorption rates and a several-fold increase in scope for growth. In postlarvae of *A. purpuratus* absorption efficiency (AE) was found to correlate positively with the protein content of a mixed algal diet (*Isochrysis galbana*, T-ISO, and *Chaetoceros neogracile*) which ranged between 48 and 27% of total organics depending on culture conditions (Uriarte and Farías 1995; 1999). Absorption efficiency (AE) was highest (74.6%) when scallop postlarvae were fed the high-protein diet, and dropped to only 30.8% on the low-protein diet of the same algal species. High-protein diets were found to yield higher growth rates and survival of 1.8 mm postlarvae (also associated with high metabolic rates) but not in 6 mm juveniles suggesting that protein requirements may be greater during early scallop ontogeny (Uriarte and Farías 1999). It was also suggested that this transition in substrate utilisation might be associated with the change from a sedentary, byssally attached habit in *A. purpuratus* <5 mm, to a free-swimming mode above this size threshold. With the possible exception of *A. purpuratus,* absorption efficiency in pectinids is very similar to that observed in many other bivalves. There was no significant difference in absorption efficiencies between *P. magellanicus* and *Mya arenaria* when they were exposed simultaneously to a range of particle concentrations and qualities (MacDonald et al. 1998).

A strong and consistent relationship between a physiological response, such as absorption efficiency, and some characteristic of the food supply, such as organic composition can be used to improve the predictive power of numerical models of feeding behaviour and estimate carrying capacity of different environments. Cranford (1995) found that between 74 and 84% of the variance in absorption efficiency could be explained by the variance in food quality, expressed as the percentage of organic matter, C or N. Complete digestive acclimation to dietary conditions in *P. magellanicus* takes several days but its food supply fluctuates considerably over much shorter time scales (Cranford 1995). Rather than adjusting digestive processes in response to frequent dietary fluctuations this species acclimated to the lower quality diet and maintained a high state of digestive acclimation for at least 12 h to enhance absorptive capabilities and energy gain (Cranford 1995). In a seasonal field study of *P. magellanicus* and *Mytilus edulis* Cranford and Hill (1999) did not find good relationships between absorption efficiency and seston organic composition and a suite of environmental conditions were only able to explain

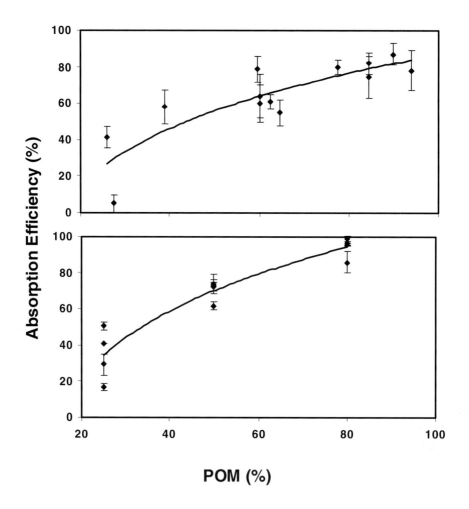

Figure 7.6. *Placopecten magellanicus.* Absorption efficiencies (mean ± SD) for scallops exposed to (A) natural seston in the field (from Cranford 1995) and (B) mixtures of microalgae and inorganic particles in the laboratory (from MacDonald et al. 1998).

28% of the variation in sea scallop and mussel ingestion rates. This work emphasises the complications associated with predicting feeding and digestion processes in bivalves and the need to have a better understanding of their previous history and other physiological processes such as metabolic demands associated with reproduction, for example (Kreeger 1993; Cranford and Hill 1999).

Absorption efficiency was not significantly influenced by salinity changes ranging from 30 to 18 ppt in *Argopecten purpuratus* (Navarro and González 1998). However, due to the strong inhibitory effect of low salinities on clearance and ingestion rates, absorption rate was highly reduced (by 96%) between 30 and 18 ppt.

7.2.9 Effects of Suspended Sediments on Feeding and Growth

Bivalve species differ considerably in their ability to cope with suspended sediment loads (see discussion by Bricelj and Malouf 1984), which may reduce growth rates by "diluting" the available food. Growth enhancement by low additions (<5–10 mg L^{-1}) of bottom sediments to algal diets were reported for *Mytilus edulis* (Kiørboe et al. 1981) and surf clams, *Spisula subtruncata* (Møhlenberg and Kiørboe 1981), but not for hard clams, *Mercenaria mercenaria* (Bricelj et al. 1984). Bricelj and Malouf (1984) hypothesised that bivalves which regulate ingestion primarily by a reduction in clearance rates (such as hard clams, softshell clams and scallops), are more likely to be vulnerable to high suspended sediment concentrations than bivalves such as mussels, oysters and surf clams which control ingestion mainly by increasing pseudofeces production (provided they are also capable of high ingestion selectivity). Although this simplified classification remains valid, recent work shows that intermediate strategies may also occur, and that the response to suspended sediment concentrations may also vary within a species depending on prevailing seston characteristics (Navarro and Iglesias 1993). A combination of both strategies occurs in *P. magellanicus* in response to mixtures of algae and inorganic, silicate particles (Bacon et al. 1998). Clearance rates declined sharply, and pseudofeces production increased with increasing seston concentration regardless of diet quality (% particulate organic matter, POM). However, sea scallops only rejected in pseudofeces 7 to 14% of the filtered ration, whereas rejection attained 40 to 93% in mussels, *Mytilus edulis* (Bayne 1993; Hawkins et al. 1996) and up to 58% in cockles, *Cerastoderma edule* (Navarro et al. 1994).

Shell growth rates of *Chlamys opercularis* were significantly depressed by moderate concentrations (11 to 37 mg dry weight L^{-1}) of iron ore particles (Richardson et al. 1984). Iron ore suspensions exceeding ca. 25 mg L^{-1} caused abnormal thickening on the interior surface of shell valves presumably due to failure of the mantle edge to extend fully in the presence of high densities of inorganic particles. Tissue and shell growth rates of juvenile bay scallops, *Argopecten irradians*, were unaffected, however, by natural sediment concentrations between 5 and 44 mg L^{-1} fed in combination with an algal diet (50 x 10^6 *Pseudoisochrysis paradoxa* cells L^{-1}) (Korol 1985). Duggan (1973) and Monical (1980) attributed reduced survival and growth of scallops (*Argopecten irradians* and *Hinnites multirugosus*) suspended near the bottom, relative to those at mid-depth, to increased suspended sediment concentrations. Depth-related differences in growth rates of *Argopecten irradians* were investigated by Korol (1985) in central Long Island Sound, N.Y., U.S.A. with contrasting results. Growth rates of bay scallops suspended within the turbidity zone, (1 m above a muddy bottom) were greater than at mid-depth and near the surface during the autumn. Growth enhancement near the bottom was attributed to 3–4 fold higher seston levels at depth relative to surface waters. In this study, chlorophyll and organic content of seston were relatively uniform throughout the water column in the fall, and near-bottom seston levels remained low, below 17 mg L^{-1}. Increased growth near the bottom presumably resulted from the combined effects of resuspension of bottom detrital material and increased availability of surface food particles due to breakdown of stratification at this time of the year. High-frequency characterisation of site-specific seston quality and quantity and other environmental variables is thus critical in interpreting the effects of resuspended bottom material. A fall-winter study in a nearshore environment in Nova Scotia, Canada which included extensive environmental measurements showed an inconsistent relationship between soft-tissue growth of *P. magellanicus* deployed in cages and distance from the bottom (0 to 200 cm) over time (Emerson et al. 1994). However, by the end of the study period, soft tissue weight was significantly less on the bottom than at ≥50 cm above bottom suggesting that high seston loads near-bottom inhibited growth. Temperature and the ratio of particulate to organic to inorganic matter, POM/PIM, were found to be the best predictors of growth.

A diet consisting only of intertidal resuspended sediments was unable to support growth of *Placopecten magellanicus* (Cranford and Grant 1990). Scallops showed lower clearance rates for the suspended sediment diet than for an algal diet offered at a comparable ration, and low absorption efficiency (29% for organic carbon). Although a higher absorption efficiency was measured for sedimentary organic nitrogen (50%), the nitrogen content of sediment was extremely low. Thus negative values of scope for growth were obtained when calculated in terms of both carbon and nitrogen (Grant and Cranford 1991).

Studies on the effects of montmorillonite clays, e.g., bentonite (~2 μm median particle diameter) and attapulgite, the main components of water-based muds used in oil and gas drilling activities, on adult *Placopecten magellanicus*, have contributed to our understanding of the effects of fine-grained sediments in this important commercial species (see also chapter by Cranford in this volume). These clays are biologically inert and thus can be used to mimic fine-grained suspended sediment. Significant reduction in somatic and reproductive tissue growth of sea scallops, and extensive mortalities (~15% and 45% after 15 and 30 days of exposure, respectively) were documented at only 10 mg bentonite clay L^{-1} in the presence of ambient food levels relative to controls with no clay addition (Cranford and Gordon 1992). Therefore *P. magellanicus*, which typically inhabits low turbidity environments with seston concentrations <2–5 mg DW L^{-1} (e.g., Emerson et al. 1994; Grant et al. 1997), is highly intolerant of sediment loading. Higher, transient levels occur due to storm events and tidal and wind-driven bottom resuspension and anthropogenic activities (oil drilling, dredging).

In the presence of an algal diet, clearance rate (CR) of *P. magellanicus* declined exponentially with increasing clay concentration, resulting in 50% reduction at ~6 mg L^{-1}, whereas CR remained constant over the same range of dry weight concentrations of a pure algal diet (Fig. 7.7A, B). Low clay concentrations, <1 mg L^{-1}, enhanced clearance rates relative to controls as observed in mussels and oysters. Increasing bentonite concentrations resulted in an exponential decline in ingestion rate of organic matter thus demonstrating a food dilution effect (Fig. 7.7C), and caused a decline in absorption efficiency when added to natural seston (Fig. 7.8A). Physiological responses returned immediately to control levels upon removal of bentonite from the suspension. Integration of physiological data allowed calculation of a negative scope for growth at bentonite concentrations ≥8 mg L^{-1} (Fig. 7.8B). Negative effects on growth of bentonite can thus be attributed to the combined effects of reduced ingestion rate of organic matter and reduced absorption efficiency, which lead to reduced absorption rate. Reduced metabolic rate,

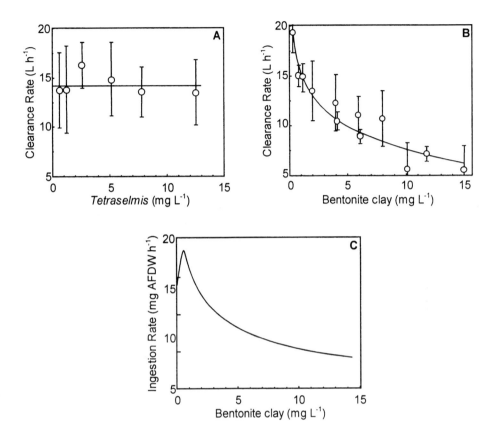

Figure 7.7. *Placopecten magellanicus* adults. Effects on clearance rates (mean ± SD) of increasing concentration of algae (*Tetraselmis suecica*) (A), compared to those of PIM (bentonite clay) added to a 1 mg L⁻¹ algal suspension (B); effects of clay on calculated algal ingestion rate (algal ash-free dry weight, AFDW, per hour) (C). Source: modified from Cranford and Gordon (1992).

documented at concentrations ≥4 mg L⁻¹ (Cranford et al. 1999) is insufficient to compensate for food dilution effects in this species.

Compensation of food dilution by inorganic particles may also occur via selective rejection of PIM in pseudofeces and can be quantified by the compensation index (CI), as selection efficiency alone does not take into account the amount of pseudofeces produced (see section 7.2.7 on particle selection). In *P. magellanicus* exposed to mixtures of algae and silicate, a positive CI, indicative of the capacity to enhance the quality of ingested vs. available particles, was obtained at seston concentrations 3–7 mg L⁻¹ and when diet quality (% POM) was ≥50% (Bacon et al. 1998). This compensation strategy was ineffective (negative CI values) when seston quality was very low (25% POM).

442

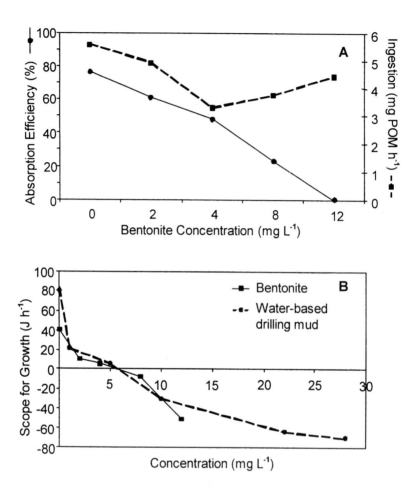

Figure 7.8. *Placopecten magellanicus* adults. Effects of increasing concentrations of PIM (bentonite clay) added to natural seston on A) ingestion rate and percent absorption efficiency (AE) of particulate organic matter (POM). B) Effects of bentonite and water-drilling muds (composed primarily of bentonite with some barite) on scope for growth (plotted from data in Cranford et al. 1999, for unconditioned scallops, with no prior exposure to clay).

It is now well recognised that functional responses of bivalves may vary considerably with the type of suspended particulate matter, i.e., artificial diets (e.g., algae and inert particles) vs. natural suspensions of low organic content (Bayne et al. 1993; Hawkins et al. 1996). In *P. magellanicus* the reduction in CR and in ingestion rate of organic matter caused by bentonite additions was greater when scallops were fed natural seston than a unialgal diet (Cranford and Gordon 1992). Increasing concentrations of natural, resuspended bottom sediment from Georges Bank (that passed through a 102 µm mesh sieve) resulted in an exponential decline in clearance rates, but in contrast to bentonite, this decline was only detected at concentrations ≥ 12 mg L^{-1} (Grant et al. 1997). The threshold concentration for visible pseudofeces production was also higher for resuspended sediment, (>10 mg L^{-1}) than that found for bentonite (~ 2 mg clay L^{-1}) by Cranford and Gordon (1992). Thus resuspended bottom sediment (≥ 5 μm median size) was less deleterious to adult *P. magellanicus* than bentonite and can even contribute a potential food source under conditions of food limitation. Grant et al. (1997) estimated that sea scallops could absorb organic matter ($\sim 30\%$ of dry weight) associated with resuspended bottom sediment with $\sim 40\%$ efficiency. Inorganic clay particles may also provide a potential food source via adsorption of dissolved organic matter. In this study sea scallop gill retention efficiency of resuspended sediments attained 100% at ~ 5 μm and did not decline with increasing sediment load (Grant et al. 1997), as was observed with clay (Cranford and Gordon 1992).

Effects on scallop feeding physiology can also vary with the degree of aggregation of individual sediment particles. Fine, clay-sized particles readily form larger flocs or aggregates in seawater, and it remains unclear whether they are again broken down to individual particles during bivalve feeding. Scallops are known to ingest relatively large individual particles up to 350 μm for *P. magellanicus* (Shumway et al. 1987) and 950 μm for *Patinopecten yessoensis* (Mikulich and Tsikhon-Lukamina 1981) and could thus potentially ingest large flocs even if these were not disaggregated on the gills. White (1997) found that flocculation increased the ability of adult *P. magellanicus* to retain small (<7 μm) clay particles and thereby increased the availability of clay for ingestion. For the preparation of feeding suspensions in this study rapid formation of large flocs was induced by the addition of gelatin, which may lead to formation of more robust flocs than those occurring in natural seston. Lyons and Ward (2003) also found that flocs (marine aggregates) increased the uptake of small (1 μm) fluorescent beads in *P. magellanicus*.

High concentrations of PIM not only result in food dilution but may also cause direct and irreversible physical damage to the bivalve gill thus impairing feeding capacity. The filibranch gill of scallops is more vulnerable to this type of sublethal structural damage than the eulamellibranch gill. In adult *Placopecten magellanicus* heavy damage resulting from the breakdown of the ciliary junctions (cilifers) between gill filaments occurred after 4 to 9 days of exposure to attapulgite clay at concentrations of 0.5 to 1 mg L^{-1} and moderate damage was observed at 0.1 mg L^{-1} (Morse 1982).

The contribution of PIM to total seston has been used as an index of the nutritional quality of available particles. Vahl (1980) found that absorption efficiencies of *Chlamys islandica* (ranging between 10 and 55%) were inversely related to the fraction of inorganic matter in the seston, and that the ratio of PIM/POM (particulate inorganic matter to organic matter) provided a good correlate of growth in this population. Enhanced growth rates of *C. islandica* held in suspended culture near the sea surface, compared to those held at 40 m, were also attributed to depth-related differences in PIM/POM (Wallace and Reinsnes 1985). These authors recommended use of the PIM/POM ratio as an index of the nutritional value of seston. Adverse effects on growth of scallops were predicted when the PIM/POM ratio exceeded a critical value of 3.5 (i.e., when inorganic material comprises more than 78% of the total seston). This supports Vahl's (1980) estimate, obtained by extrapolation, that no POM would be absorbed when PIM comprises more than 80% of the seston (PIM/POM = 4). Similarly, using controlled mixtures consisting of a fixed concentration (1 mg L^{-1}) of algae or natural seston and varying concentrations of PIM (bentonite clay) to manipulate diet quality, Cranford (1995) found an inverse logarithmic relationship between absorption efficiency (AE) of organic matter of the sea scallop, *P. magellanicus*, and increasing bentonite levels. The relationship between AE and diet quality (Q) whether expressed as the % of POM, particulate organic carbon, POC, or nitrogen, PON, per unit dry weight seston) was described (after Bayne et al. 1987) by the following exponential equation:

$$AE = a\,[1 - e^{-b(Q-c)}]$$
(2)

Where, a = the maximum value of AE, >92%, b = the rate of increase of AE with increasing diet quality, and c = diet quality at which AE = 0. This coefficient was estimated at 14% POM/DW (PIM/POM = 6) for *P. magellanicus* in this study, compared to 7% in *Mytilus edulis* (reviewed by Cranford 1995) and has been suggested as an index of a bivalve's capacity to survive in a nutritionally poor environment (Navarro and Iglesias 1993). Adverse effects on sea scallop growth were found at a relatively low PIM/POM ratio of 1.5 (addition of 10 mg L^{-1} bentonite) (Cranford and Gordon 1992, Cranford 1995), below the critical threshold proposed above for *C. islandica*. Interspecific comparisons from different studies must be interpreted with caution, however, as the coefficients in the equation can vary considerably within a species as a function of environmental conditions, and were shown to vary between laboratory and field studies (Cranford et al. 1998).

Scallops appear to be particularly susceptible to siltation both in the laboratory (Castagna 1975) and in the field (Duggan 1973; Tettelbach et al. 1985). In bay scallops, *A. irradians*, burial by shifting sediments has been suggested as a significant cause of mortalities during the winter when low temperatures result in reduced activity (Tettelbach et al. 1990). Laboratory studies confirmed that juveniles (19–25 mm in shell height) partially (2/3) covered by a 1 cm-deep sediment layer at temperatures <10°C showed significant mortalities relative to unburied controls by the end of 10 weeks. Therefore, byssal attachment to elevated substrates in juveniles and some adult scallops may provide a mechanism of avoiding burial by fine-grained sediments and exposure to high near-bottom turbidities, in addition to a predator refuge. Using ciliary activity of gill sections as a physiological index Yamamoto (1960) found that the tolerance of juvenile *Patinopecten yessoensis* to suspended silt and low oxygen tension increased with scallop size (over the range 17 to 100 mm). Since high turbidity and low oxygen levels are both conditions associated with uncompacted, fine-grained sediments, Yamamoto's findings explain the often observed low survival of juvenile scallops in muddy bottoms. Growth rates of adult scallops can also be adversely affected in soft, muddy substrates; Gruffydd (1974) found that the maximum shell size of *Pecten maximus* from the North Irish Sea decreased significantly with increasing mud content of sediments.

7.2.10 Effects of Flow on Feeding and Growth

Studies of natural populations of shallow water pectinids such as *Argopecten irradians*, have generally found that larger size and faster growth rates are associated with areas of relatively strong currents (Belding 1910; Gutsell 1930; Marshall 1960). Emigration rates of the bay scallop *Argopecten irradians concentricus* increased with scallop density but only in the presence of a high tidally oscillating flow regime of 28 cm s^{-1} but not at flows of 10 cm s^{-1} (Powers and Peterson 2000). This study illustrates how flow can facilitate biological interactions and influence density dependence. Reduced growth rates of suspension-feeding bivalves in areas of low current speeds and/or high population densities are attributed to a reduction in seston supply in the benthic boundary layer (Fréchette and Bourget 1985).

Excessively high current speeds could, however, potentially reduce growth by inhibiting the scallops' feeding activity. Kirby-Smith (1972) investigated the effect of current speed (within the range 0.2 to 12.8 cm s^{-1}) on growth of adult *Argopecten irradians concentricus* in an apparatus consisting of pipes with different outflow diameters. He found that shell growth rate and condition (muscle weight/shell height) decreased at current speeds exceeding about 6 cm s^{-1}. It has been noted however, that the current speeds reported in his study, estimated by dividing the observed volume discharge rate by the cross-sectional area of the pipe, underestimate the speeds actually experienced by the scallops (Eckman et al. 1989). Clearance rates in *Placopecten magellanicus* are a unimodal function of flow: they are positively related with flow speed below 5–10 cm s^{-1}, independent of flow at intermediate flows of 10 to 20 cm s^{-1} and decline when velocities exceed 15 to 20 cm s^{-1} (Wildish and Kristmanson 1985; Wildish et al. 1992; Wildish and Saulnier; 1993). Wildish et al. (1992) found that higher seston concentrations could offset the velocities at which feeding inhibition occurs.

Many studies on scallop feeding activity have been undertaken via short-term experiments in static chambers or in poorly defined flow fields. Feeding activity in suspension-feeders measured under laboratory conditions may differ from feeding rates observed in the dynamic boundary layer flows found in coastal waters (Rubenstein and Koehl 1977, LaBarbera 1981). In a unique flume study Pilditch and Grant (1999a) exposed the sea scallop *Placopecten magellanicus* to increasing and decreasing concentrations of food and incremental changes in flow speeds between 5 and 25 cm s^{-1}. These authors did not find a positive correlation between clearance and flow speeds but observed clearance to be independent of flow speed between 5–15 cm s^{-1} and inhibited at 25 cm s^{-1}. However, clearance rates were different in the ascending and descending phases of the experiments with a lower flow speed of 10–15 cm s^{-1} inhibiting feeding on the descending phase. They suggested that the response to flow speed may be a function of feeding history whereby clearance may be inhibited at lower flow velocities after feeding on high concentrations of phytoplankton. This research emphasises the direct effects of flow, its interaction with seston concentration and the significance of short-term feeding history in regulating scallop feeding rates.

Evidence of a growth limiting upper velocity obtained in experimental flume systems appears to contradict field observations. For example, Eckman et al. (1989) report that juvenile *Argopecten irradians* within dense eelgrass beds (1,100 shoots m^{-2}) in Back Sound, North Carolina, U.S.A. regularly experience current speeds as high as 5–28 cm s^{-1}. Adult giant scallops (*Placopecten magellanicus*) in the Bay of Fundy are found where depth-integrated maximum tidal velocities exceed 100 cm s^{-1} (Wildish and Peer 1983) and are thus well above the critical thresholds determined experimentally. Similarly, Bricelj et al. (1987a) found highest reproductive output and muscle condition of adult *A. irradians* in a site where surface currents can exceed 70 cm s^{-1}. Flow velocities within the benthic boundary layer will be lower, however, than depth integrated or free stream velocities. Furthermore, scallops often occupy depressions in bottom sediments (recessing) (Caddy 1968) where they may avoid high flow velocities. Wildish and Kristmanson (1988) attempted to reconcile this apparent contradiction in a study that examined the influence of periodic changes in flow regime on growth of *Placopecten magellanicus*. Scallops were found to maintain maximal rates of shell growth as long as they were exposed to growth limiting current velocities (>10 cm s^{-1}) less than a third of the time. In contrast to studies evaluating the effect of scallop orientation on growth in flumes using steady current velocities, Claereboudt et al. (1994) did not observe any differences in growth of *P. magellanicus* related to orientation relative to current direction in the field. However, these authors did find a reduction in tissue growth at their high velocity site (16 cm s^{-1}) consistent with other studies of sea scallops in flumes where growth was reduced at velocities exceeding 10 cm s^{-1} (Wildish et al. 1987; 1992). Pearl nets decreased water flow by 46–61% thereby reducing velocities that inhibit feeding and result in greater growth inside the nets than outside for scallops held at the high velocity site. This is in contrast to observations at the low velocity site (9 cm s^{-1}) where the pearl nets reduced flow inside to the point where growth inside the nets was significantly lower possibly due to seston depletion (Claereboudt et al. 1994).

In contrast with the results reported earlier, Cahalan et al. (1989) found that growth rates of juvenile (3–7 mm) bay scallops, *Argopecten irradians*, determined in a flume, were independent of current speed between 1 and 15 cm s^{-1}. Eckman et al. (1989) found that shell growth of juvenile *A. irradians* (8–14 mm) declined gradually with increasing flow velocity over the range 1.4 to 17.2 cm s^{-1}, irrespective of the scallops' orientation to the flow. At any given flow velocity, scallops grew more rapidly when oriented with anterior margins facing upstream, their preferential orientation. There was no evidence however, of a sharp decline in growth above some threshold flow velocity. Furthermore, a statistically significant effect of flow velocity on growth was only detected for scallops with their anterior opening oriented downstream. In this study responses of scallops to current speed and turbulence were determined in a series of pipes with outflows of different diameters, and low velocities were measured directly with a thermistor-bead flowmeter.

Juvenile bay scallops in their natural habitat live byssally attached to elevated substrates, and therefore routinely experience flows greater than 3 cm s^{-1}, and as high as ca. 17 cm s^{-1} (Eckman et al. 1989). Given their preferential orientation, they are unlikely to remain in an unfavourable orientation where growth is more strongly inhibited by increasing flow velocity. Eckman et al. (1989) additionally found that growth rates were independent of the presence or intensity of turbulence. Animals were tested in flows ranging from laminar to fully turbulent (Reynolds number = 840 to 5,600), with flow velocity held within a narrow range of 1.7 to 3.9 cm s^{-1}.

In a laboratory study that examined the effects of flow speed and orientation on growth of juvenile *Nodipecten subnodosus*, highest shell and tissue growth rates were obtained at flow speeds between 5 and 10 cm s^{-1} and when scallops were in their natural orientation (ventral, inhalant margin oriented towards the flow) (García-Esquivel et al. 2000). This study was conducted in tubes cut in half, in which flow was controlled via a recirculating pump, and current velocities were determined using rhodamine dye. Free scallops showed reduced tissue weight at the highest speed tested (15 cm s^{-1}) even though this scallop species is known to experience current velocities over 100 cm s^{-1} in its natural habitat in Baja California, Mexico. This high current speed caused detachment of juveniles as well as growth inhibition. Flow had no significant effect on growth when scallops were oriented horizontally against the flow, an orientation, which resulted in the slowest growth.

7.2.11 Effects of Harmful and Toxic Algae

Like other suspension-feeding bivalve molluscs, scallops accumulate toxins associated with harmful algal species. In most instances (but not all!), adductor muscles do not accumulate toxins and are thus safe for human consumption even in the presence of harmful algae. Whole scallops are a serious threat to public health, especially since many species bind the toxins in various tissues for extended periods of time and are rendered unsuitable for human consumption even when blooms are not evident (see Shumway and Cembella 1993; Bricelj and Shumway 1998 and references therein). In addition to becoming vectors for these algal toxins, many species of scallops are themselves adversely affected by the toxins (Table 7.3).

Perhaps the best understood responses of scallops to harmful algae are those associated with brown tides caused by the picoplanktonic alga *Aureococcus anophagefferens* in mid-Atlantic, U.S.A. estuaries. In eastern Long Island, NY, where blooms have recurred since 1985, brown tides have led to recruitment failure, growth inhibition and decimation of local bay scallop (*Argopecten irradians*) populations (reviewed by Bricelj and Lonsdale 1997) despite intensive reseeding efforts (Tettelbach and Wenczel 1993). The mid-summer, main spawning period of scallops has been shown to overlap with the period of occurrence of *A. anophagefferens* (Bricelj et al. 1987a). Due to their small size (~2 µm) *A. anophagefferens* cells are retained with only 36% efficiency by adult bay scallops (Cosper et al. 1987). However, their negative effects are attributed to an unknown, dopamine-mimetic, bioactive/toxic metabolite associated with brown tide cells, which suppresses the activity of gill lateral cilia (Gainey and Shumway 1991) and thus clearance rates of bivalves such as *Mytilus edulis* and *Mercenaria mercenaria* (Tracey 1988; Bricelj et al. 2001). These effects are observed even when non-toxic algae of high food value are present in a mixed phytoplankton assemblage with *A. anophagefferens*. Toxic effects require direct contact with brown tide cells and do not appear to be associated with the production of dissolved toxic exudates (Ward and Targett 1989). It is of interest to note that *in vitro* trials by Gainey and Shumway (1991) did not document gill ciliary inhibition by brown tide in adult *A. irradians*, although natural populations of this species are known to be adversely affected by brown tides.

Although *A. anophagefferens* bloom densities in the field can attain 1–2 x 10^6 cells mL^{-1}, the inhibitory effect on bivalve feeding is concentration-dependent and in short-term studies of juvenile hard clams occurred above a threshold cell density of $\geq$35 x10^3 cells mL^{-1} (Bricelj et al. 2001). Densities from 1.9 x 10^5 to 7.5 x 10^5 *A. anophagefferens* cells mL^{-1} were found to significantly reduce survival and growth of bay scallop larvae in laboratory studies, even in the presence of nutritious algae (Gallager et al. 1989). Scallop larvae were able to maintain ciliary-driven swimming activity, as well as capture of nutritious algae in the presence of *Aureococcus*. However, the latter caused increased rejection of non-toxic algae from the oesophagus and thus reduced ingestion rate. Thus, Gallager et al. (1989) suggested that the mechanism of action of brown tide cells in larvae made differ from that of adults, in that *Aureococcus* may interfere with the chemosensory function of larvae rather than ciliary activity.

The ichthyotoxic flagellate *Prymnesium* spp. (isolate 97–20–1) at 10^5 cells mL^{-1} was found to elicit toxic effects in juvenile bay scallops, *A. irradians*: copious mucus and pseudofeces production and valve gape not responsive to stimulation, necrosis of the digestive gland and gill as revealed by histopathology and presence of moribund or dead animals within 24 h of exposure (Wikfors et al. 2002).

Several species of dinoflagellates, the most common components of toxic algal blooms, have been shown to have negative impacts on scallops. Yan et al. (2001) reported inhibition of egg hatching success and larval survival of *C. farreri* when exposed to intact cells and cell fragments of *Alexandrium tamarense* (isolate from Daya Bay, PRC). Hatching rate of *C. farreri* decreased to 30% of controls after exposure to *A. tamarense* cells or cellular fragments at 100 cells mL^{-1}, and was only 5% after exposure at 500 cells mL^{-1}. Larval (D-stage) survival rates decreased significantly after exposure for 6 d at 3000 cells mL^{-1} and above; no larvae survived after 14 d exposure at 10,000 cells mL^{-1} or 20 days at 5,000 cells mL^{-1}. However, these experimental concentrations exceed environmental levels by several orders of magnitude. Exposure to saxitoxin (STX) standard did not inhibit egg hatching, suggesting that an unknown metabolite, rather than PSP toxins, was the causative agent of hatching failure. These results may have far reaching implications for devastation of field or hatchery populations exposed to this common dinoflagellate.

While heart rate, feeding rate, and irrigation rates of *P. magellanicus* were unaffected by the presence of PSP-producing *A. tamarense* (at 10^4 cells L^{-1}) (Shumway and Cucci 1987; Gainey and Shumway 1988a, b), scallops did produce copious amounts of a white, mucus-like substance, and the animals exhibited a striking escape response with either violent swimming activity, partial, sustained shell-valve closure, or a combination of the two. It was suggested that this increased metabolic output might prove detrimental to the scallops if it continued for any extended length of time. This type of avoidance response was not described in other scallop species fed PSP-producing dinoflagellates, e.g., *Pecten maximus* fed *A. minutum* (Bougrier et al. 2001). Bivalve species that are resistant to the effects of PSP toxins can assimilate and grow on toxic *Alexandrium* spp.. Absorption efficiency of organic matter (OM) [(OM absorbed/OM filtered) x 100] of *P. maximus* fed

Table 7.3

A summary of toxic and noxious algal species associated with scallops (after Shumway 1990; Shumway and Cembella 1993 and Landsberg 2002). Where known, strain is given below the algal species name followed by cell toxicity and concentration. Algal species are given as in the original publications with current taxonomic status added. See text for details.

Algal Species	Scallop Species	Effects	Location	Reference
Dinoflagellate Species				
Alexandrium tamarense (ATHK; 1.175 x 10⁻⁹g STX eq cell⁻¹; see text for cell concentrations)	*Chlamys farreri* (larvae and juveniles)	Inhibition of egg hatching success after 36 h exposure at 100 cells mL⁻¹; decreased larval survival after 6 d at 3000 cells mL⁻¹	Laboratory	Yan et al. 2001
Protogonyaulax tamarensis (= *A. tamarense*) (GT429; 5x10⁵ cells L⁻¹)	*Placopecten magellanicus* (adults)	Violent swimming and mucus production	Laboratory	Shumway and Cucci 1987; Gainey and Shumway 1988b
Ptychodiscus brevis (= *Karenia brevis*) (5–500 x 10³ cells L⁻¹)	*Argopecten irradians*	Adult scallop mortality (21%); almost total recruitment failure	North Carolina (Field)	Summerson and Peterson 1990
Karenia brevis (<900 cells mL⁻¹)	*Argopecten irradians* (larvae)	Delayed metamorphosis and larval development, decreased filtration and mortality	Laboratory	Leverone and Blake 2001
Gymnodinium veneficum	*Pecten maximus* (adults)	100% mortality within 1 h of exposure	Laboratory	Abbott and Ballantine 1957
Gyrodinium aureolum (PLY 497; 10⁵ cells L⁻¹)	*Argopecten irradians* *Placopecten magellanicus* (juveniles <15mm)	Some evidence of poor food quality, no obvious toxic effects in either species at cellular level; 100% mortality in *A. irradians*; no mortality but copious mucus production in *P. magellanicus*	Laboratory	Smolowitz and Shumway 1997; Lesser and Shumway 1993
G. cf. aureolum (up to 5 x 10⁵ cells L⁻¹)	*Pecten maximus*	High mortality in postlarvae and juveniles; cessation of feeding in larvae; inhibition of reproduction and growth	Bay of Brest, France and laboratory	Lassus and Berthome 1988; Erard-LeDenn et al. 1990
Heterocapsa circularisquama (10⁵–10⁶ cells L⁻¹)	*Chlamys nobilis* (adults)	Mass mortalities	Ago Bay, Japan	Matsuyama et al. 1995; 1999; personal communication

Species	Test organism	Effects	Location	Reference
Lingulodinium polyedrum (= *Gonyaulax polyhedra*) (as *Gonyaulax* sp.) (10^7 cells L^{-1})	*Leptopecten* [as *Pecten*] *L. latiauratus*	Mortalities (probably a result of anoxia and high temperature, not toxins)	Ensenada, Baja California, Mexico	Stohler 1959
Prorocentrum lima	*Argopecten irradians* (juveniles; pre-reproductive adults)	Reduced absorption efficiency relative to *Thalassiosira weissflogii*	Laboratory	Bauder et al. 2001
Prorocentrum minimum	*Argopecten irradians* (post-set; 0.61g mean live weight)	Mortality, atrophy and necrosis of the digestive gland absorptive cells, and systemic effects	Laboratory	Wikfors and Smolowitz 1993, 1995; personal communication
Prymnesium sp. (10^5 cells mL^{-1})	*Argopecten irradians* (juveniles, 10 and 50 mm)	Violent twitching, abundant mucus production; valve gape; all moribund or dead within 24h; severe, acute total or near-total necrosis of digestive gland and ducts, gill and other tissues; dead hemocytes	Laboratory	Wikfors et al. 2002
Pfiesteria shumwayae (100–2,500 cells mL^{-1})	*Argopecten irradians*	Rapid mortality caused by aggressive feeding on larvae by *Pfiesteria*; no mortality when cells were isolated from scallops in dialysis tubing	Laboratory	Shumway and Springer 1996; unpublished
Pfiesteria piscicida (150–3,500 cells mL^{-1})	*Argopecten irradians*	Decrease or cessation of clearance rate in adults; mortality in larvae, spat and adults	Laboratory	Springer et al. 2000, 2002
Pelagophyte Species				
Aureococcus anophagefferens	*Argopecten irradians*	Larval shell growth reduced and increased mortalities	Laboratory	Gallager et al. 1989
A. anophagefferens	*Argopecten irradians*	Mass mortalities	Long Island, NY	Cosper et al. 1987
A. anophagefferens	*Argopecten irradians*	76% reduction in adductor weights; recruitment failure	Long Island, NY	Bricelj et al. 1987b
Diatom Species				
Rhizosolenia chunii ($56–187 \times 10^3$ cells mL^{-1})	*Pecten alba* (adult)	Digestive gland lesions and subsequent mortalities possibly caused by *R. chunii*	Port Phillip Bay, Australia	Parry et al. 1989
Rhizosolenia delicatula Ceratulina pelagica (7.3×10^5 cells L^{-1})	*Pecten maximus* (adult)	Decline in growth rate; clogged gills affecting feeding and respiration	Bay of Brest, France	Lorrain et al. 2000; Chauvaud et al. 1998

A. minutum was 42% (Bougrier et al. 2001) and absorption efficiency [(OM absorbed /OM ingested) x 100] of a highly toxic strain of *A. fundyense* by *Mytilus edulis* was 60–63% (Bricelj et al. 1990).

Very limited information is available on the effects of the brevetoxin-producer *Ptychodiscus brevis* (= *Karenia brevis*) on bivalve molluscs, although this species is known to cause mass fish mortalities. Bay scallops, *Argopecten irradians*, appear to be particularly sensitive to the toxic effects of this dinoflagellate. A mortality of 21% and almost total (98%) recruitment failure of *A. irradians* was reported by Summerson and Peterson (1990) for scallops exposed to a renegade bloom of this dinoflagellate in North Carolina. Clearance rates of juvenile bay scallops were significantly reduced at concentrations ≥ 50 intact cells *K. brevis* mL^{-1} as well as in the presence of lysed cells (Leverone and Blake 2002). These authors also found that *K. brevis* inhibited metamorphosis and caused mortalities of bay scallop larvae above 500 cells mL^{-1}. While this dinoflagellate species is usually restricted to the Gulf of Mexico, this example is a clear indication of the devastating impacts that harmful algal blooms can have on scallop populations and a warning with regard to potential impacts at aquaculture sites.

In complementary studies (Lesser and Shumway 1993; Smolowitz and Shumway 1997), juvenile *Argopecten irradians* and *Placopecten magellanicus* were exposed to bloom conditions (10^5 cells L^{-1}) of *Gyrodinium aureolum* for a week. These studies clearly demonstrated a species-specific impact of this alga on the scallops. Mortality was 100% in *A. irradians* and zero in *P. magellanicus*. Reduced clearance rates were noted in *A. irradians* and production of copious amounts of mucus were noted in *P. magellanicus*. No mortalities were noted in either species when exposed to *Alexandrium tamarense*. At the cellular level, *G. aureolum* had a marked effect on the digestive gland of *A. irradians*. There was a significant decrease in the height of absorptive cells and increased lumen diameter which, at least, suggest poor food quality of *Gyrodinium*. There was no direct evidence of toxic effects such as necrosis and sloughing of digestive gland epithelial cells in the digestive gland. Some animals showed inflammation in the kidney but the cause was not delineated. No such impacts were noted in *P. magellanicus*. Erard-LeDenn et al. (1990) noted minor mortalities in 2-year old *P. maximus* exposed to *G. aureolum* and also noted a 'stress ring' on shells of exposed animals in the field, clearly indicating growth inhibition. Postlarvae (1 mm) experienced major die-offs (100% at one site, 85% at another) at cell concentrations of 2 x 10^5 cells L^{-1}. These blooms were extensive and lasted several weeks during the summer period and caused mass mortalities of the individuals in the nurseries and trays. Growth of juveniles (5–30 mm shell height) stopped completely for one month and resumed when the red tide vanished. In addition, the adult scallops exhibited abnormally low gonadal indices. Laboratory studies revealed a decreased filtration rate of *P. maximus* exposed to *G. aureolum* and also demonstrated differences between toxicity of different strains of *G. aureolum*.

Heterocapsa circularisquama, a toxic dinoflagellate from Japanese waters has been implicated in mass mortalities of several species of bivalve molluscs, including scallops (*Chlamys nobilis;* Matsuyama 1999; Matsuyama et al. 1996, 2001). While empirical data on scallop mortality are not available, this alga is known to be a molluscicide and it has had devastating impacts on aquaculture for various species including mussels, oysters and pearl oysters in Japanese waters. Establishment of scallop culture in areas prone to this alga would be very risky business.

Impacts of the dinoflagellate, *Prorocentrum minimum,* on postlarval scallops, *A. irradians,* were studied by Wikfors and Smolowitz (1993) as part of an investigation of observed poor growth rates in caged hard clams, *Mercenaria mercenaria.* They provided scallops with mixed diets of T-ISO and *P. minimum* in varying ratios. Daily ration was equalised to 0.012 mL of packed cells per individual scallop. No algal diet supported good, consistent growth of scallops and survival varied between treatments, with 100% mortality noted in one week in one trial and four weeks in a second trial. Histological observations after one week of exposure to the mixed diet revealed poorly developed digestive diverticula and attenuation of the epithelium with abnormal vacuolation and necrosis. They also noted the presence of large thrombi in the open vascular system of the mantle, digestive diverticula, heart, gill, and kidney tissues, suggesting systemic effects of a toxin. Since *P. minimum* is a common component of the phytoplankton during the summer in scallop growing areas, there is considerable potential for detrimental impacts on both natural and cultured populations of *A. irradians.* Exposure of juvenile and adult *Argopecten irradians,* to *Prorocentrum lima,* a producer of diarrhetic shellfish poisoning (DSP) toxins which can act as potent phosphatase inhibitors, had no adverse effects on survival of scallops over a 2-wk period (Bauder et al. 2001). In this study there was no evidence of abnormal behaviour (shell clapping or avoidance by swimming) and byssal attachment and climbing of juveniles remained unaffected. Feeding (clearance rates) were comparable to those of controls offered an equal biovolume of non-toxic diatoms.

Scallops exposed to the dinoflagellate, *Pfiesteria* spp. are impacted by both toxins and physical attack by *Pfiesteria* (Springer et al. 2002; Shumway, unpublished). Bay scallop larvae (*A. irradians*) are aggressively fed upon by *Pfiesteria piscicida* zoospores (all functional types) within minutes of zoospore introduction. Left in direct contact with the zoospores, the larvae are dead within 30–45 minutes. High rates of larval mortality have also been observed in experiments where toxic zoospores were constrained within dialysis tubing. These results suggest the presence of dissolved toxin(s). *P. shumwayae* zoospores are even more aggressive in their feeding response towards *A. irradians* larvae (death within 10–15 minutes) although the same cultures cannot induce larval mortality when constrained within dialysis tubing.

Juvenile and adult bay scallops rapidly cleared *P. piscicida* from suspension at concentrations ranging from 150–3,500 cells mL^{-1} (Springer et al. 2002; Shumway unpublished). Toxic zoospores were cleared from suspension at progressively slower rates then nontoxic (algal fed; TOX-B, NON-IND) zoospores. A significant difference in grazing rates also exists between juvenile scallops presented with freshly isolated *P. piscicida* (TOX-B) versus those fed non-inducible cultured isolates (3+ months in culture). This is presumably due to residual toxicity in the TOX-B cultures. *P. shumwayae* is cleared from solution at a relatively lower rate than observed for *P. piscicida* (concentrations ranging from 100–2,500 cells mL^{-1}) but the cause for these observed differences has yet to be determined. The presence of *Pfiesteria* spp. has already had devastating impacts on at least one bivalve hatchery in North Carolina.

Unlike dinoflagellates, diatoms have rarely been attributed harmful effects on shellfish in general or scallops specifically (see Shumway 1990; Shumway and Cucci 1987; Gainey and Shumway 1988b). Unusually large blooms of the diatom, *Rhizosolenia chunii*, in Port Phillip Bay, Australia in 1987–1988, initially resulted in a bitter taste in mussels, oysters and scallops (*Pecten alba*) (Parry et al. 1989). Abnormally high mortalities were noted in mussels and oysters, which were followed histologically during the post-exposure period. While scallops exhibited increased levels of mortality post-bloom, it is not clear whether these mortalities were caused by parasites, diseases or exposure to *R. chunii*.

In a more recent study, Chauvaud et al. (1998) suggested that large populations of a diatom, *Rhizosolenia delicatula*, could have led to reduction in feeding due to gill clogging, and cessation of growth in *Pecten maximus* in the Bay of Brest, France. In a later study of the same area, Lorrain et al. (2000) showed a decreased growth rate from 180 to 80 μm d^{-1} in *P. maximus* associated with large blooms of *R. delicatula* and another diatom, *Ceratulina pelagica*. Again, they postulated that the reduction in growth rate was directly caused by the presence of these diatoms but no cause and effect was demonstrated and no laboratory studies accompanied the field work.

7.3 ENERGY UTILISATION: METABOLIC EXPENDITURE

Metabolic rate of bivalves, as measured by the rate of oxygen consumption (VO_2), is known to be influenced by a number of variables, including temperature, body size, oxygen tension, food concentration, salinity, reproductive state, activity level and physiological condition. The effect of salinity can be illustrated in *Argopecten purpuratus*, a euryhaline species, in which VO_2 increased with a decrease in salinity from 30 to 24 ppt, but declined with a further reduction in salinity from (24 to 18 ppt) (Navarro and González 1998). This pattern was similar to that observed for excretion rate (see section 7.4.1 on "Excretion and byssus secretion").

The allometric relationship between body mass (tissue dry weight) and VO_2 has been determined for several pectinid species (Table 7.4). The coefficient (b) of this relationship varies between 0.486 and 0.986 [average = 0.75 excluding Vahl and Sundet (1985) values for reproductively mature individuals]. This mean value closely approximates that of 0.727 estimated for bivalves (Bayne and Newell 1983) and the value of 0.75 estimated for poikilotherms in general (Hemmingsen 1960). The biological significance of this parameter has been discussed by numerous authors (e.g., Zeuthen 1953; Hemmingsen 1960). A higher allometric coefficient (b) has been obtained in both larval and juvenile *Argopecten irradians concentricus* (Lu and Blake 1999), as has also been found for clearance rate. In juveniles (0.5 to 7 mm in shell height) the allometric equation is:

$$VO_2 \ (\mu L \ O_2 \ h^{-1}) = 2.142 \times AFDW^{0.905} \text{ where AFDW in mg} \tag{3}$$

Table 7.4

Parameters of the allometric relationship between oxygen consumption (VO_2; mL O_2 h^{-1}) and tissue dry weight (W; g), following the equation $VO_2 = a\ W^b$, in several pectinids. (Unless specified, regressions were selected from seasons during which the animals are not at the peak of their reproductive development).

Species, location	Size range (g tissue DW)	Temp. (°C)	a	b	Source
Argopecten	0.47–2.99	17.4	0.931	0.725	Bricelj et al. 1987a
irradians irradians	0.84–2.86	10.5	0.368	0.733	
New York, U.S.A.	0.87–4.37	1.5	0.065	0.986	
Argopecten circularis,	0.04–1.78	20	0.479	0.715	Silva Loera 1986*
Mexico					
Placopecten	0.01–18	19	0.399	0.837	Shumway et al.
magellanicus		10	0.363	0.838	1988
Maine U.S.A.					
P. magellanicus - 10 m	1.8–42	5.5–8.5	0.447	0.79	MacDonald &
Newfoundland,		10–12	0.339	0.78	Thompson 1986
Canada - 31 m	0.5–25	1.8–3.5	0.214	0.76	
		5.5–7.2	0.234	0.79	
Argopecten purpuratus	0.1–5.7	12	0.30	0.55	Navarro & González 1998
Chlamys islandica,	0.05–2.6	3.8	0.098	0.87	Vahl 1978
Norway					
Chlamys, Immature	0.02–0.9	5.7	0.145	0.486	Vahl & Sundet
islandica Mature	0.5–5		0.251	0.567	1985
Norway Mature	0.4–6		0.242	0.759	
Chlamys delicatula,	0.01–0.82	10	0.147	0.527	Mackay &
New Zealand					Shumway 1980
Chlamys opercularis,	0.3–3.0	10	0.385	0.63	Mackay &
Denmark					Shumway 1980[+]
Chlamys varia,	0.1–2.3	10	0.334	0.70	Shafee 1982
France		15	0.344	0.77	
Patinopecten		22.4	0.579	0.817	Fuji & Hashizume
yessoensis,	0.5–15	14.8	0.398	0.777	1974
Japan		5.8	0.181	0.862	

* Animals starved for 36 h prior to measurements.
[+] Calculated from data in McLusky (1973).

This agrees with results by Riisgård (1998), who found that larval and early juvenile stages of bivalves are characterised by a higher exponent (b = 0.9–1.0) than larger animals (b = 0.6–0.7) and thus challenged the validity of a constant allometric exponent for VO_2.

Metabolic rate (VO_2) is also highly responsive to food levels: VO_2 is markedly reduced to basal levels during starvation as in other bivalves (Grant and Cranford 1991), and in *P. magellanicus* at 10°C, VO_2 was found to be significantly higher (by 31%) at high food levels (ambient seston supplemented with cultured algae) than at low food levels (ambient seston) (Pilditch and Grant 1999b). Respiration rate was also significantly greater (by 36%) in *P. magellanicus* fed cultured phytoplankton compared to detrital diets containing fresh or aged kelp, or resuspended sediment (Grant and Cranford 1991). The increase in metabolic rate of feeding scallops is expected to be primarily associated with the post-ingestive costs of digestion and growth rather than the mechanical cost of filtration.

7.3.1 Metabolic Rate and Oxygen Availability

Aquatic organisms have been characterised as oxyregulators or oxyconformers depending on their ability to maintain a VO_2 independent of declining oxygen tension (PO_2) over some range of PO_2 values, or one that conforms (declines) with PO_2. The critical oxygen tension (P_c) is given by the inflection point of the function relating VO_2 and PO_2, beyond which VO_2 becomes dependent of ambient oxygen tension. A critical value of about 48–56% oxygen saturation was determined for *Pecten maximus* at 10°C (Brand and Roberts 1973), and a low value of ca. 20% for the oxyregulators *Argopecten irradians* and the deep sea scallop *Pecten grandis* (van Dam 1954). Other scallop species are poor oxygen regulators: in *Chlamys islandica* (Vahl 1972, 1978) and *Placopecten magellanicus* (Shumway, unpublished). VO_2 is independent of oxygen tension only to approximately 60–70% oxygen saturation, and in *C. delicatula* VO_2 declines gradually with decreasing oxygen tension, with no clear inflection point (Mackay and Shumway 1980) (Fig. 7.9). Similarly, in *Argopecten ventricosus* VO_2 remains independent of PO_2 only between 100 and 76% of oxygen saturation (Sicard et al. 1999 in Maeda-Martínez et al. 2000).

The qualitative distinction between oxyregulators and conformers has been criticised by Mangum and Van Winkle (1973), who point out that few species exhibit perfect regulation or strict oxyconformity over a wide range of external oxygen conditions, and that these conditions represent only extremes in a continuum. Bayne (1971) suggested the adoption of the ratio K_1/K_2 (where K_1 is the intercept, and K_2 is the slope of a plot of PO_2 against PO_2/VO_2) as a more relevant quantitative index of an organism's oxygen dependence. A relatively higher K_1/K_2 value indicates a reduced ability to regulate VO_2. The response to declining oxygen tension of bivalve mollusks can vary with environmental conditions such as temperature and salinity (Shumway and Koehn 1982), nutritional condition, and body size (Shumway 1983). For example, Silva Loera (1986) found that small bay scallops, *Argopecten circularis*, (0.1 g in dry tissue weight) show a more marked dependency of VO_2 on PO_2 and therefore a higher K_1/K_2 ratio (565), than large individuals (1.0 g; K_1/K_2 = 137.6). In bivalve molluscs the relationship between the

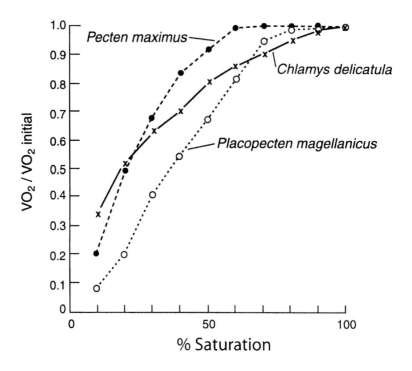

Figure 7.9. The response of three oxyconforming pectinid species to declining oxygen tension (VO_2 = weight-specific oxygen uptake). Data for *Chlamys delicatula* from Mackey and Shumway 1980; *Pecten maximus* from Brand and Roberts 1974; *Placopecten magellanicus* from Shumway, unpublished data.

K_1/K_2 index and weight-specific VO_2 (mL h^{-1} g^{-1}) is described by an exponential equation. For the deep water scallop *Chlamys delicatula* (Mackay and Shumway 1980) this relationship is given by:

$$K_1/K_2 = 115.78 \times VO_2^{0.769} \tag{4}$$

The relatively high value of the constant in this equation (= 115.78) compared to other bivalves species again indicates that *C. delicatula* is a poor oxyregulator. *Artica islandica* is an extreme oxyconformer, as indicated by an intercept value of 1,000, whereas the mussels *Geukensia demissa* show intermediate values (76 and 63, respectively).

In scallops, as observed in other bivalves, variations in metabolic rate are generally reflected in changes in heart rate. In littoral molluscs, respiratory stress caused by sudden aerial exposure typically results in rapid bradycardia (reduction in heart rate) and valve closure. On the other hand, *Pecten maximus* initially responds to air exposure by violent adductions of the shell and tachycardia, followed by gradual bradycardia, accompanied by wide gaping of valves. Inability to control air gaping and consequent vulnerability to desiccation is a characteristic feature of many sublittoral species including scallops (Brand and Roberts 1973). The inability to remain closed for extended periods of time also makes scallops highly vulnerable to low salinity stress (Stockton 1984; Tettelbach et al. 1985), especially at higher temperatures (Mercaldo and Rhodes 1982). Little is known, however, about the air-breathing capacity of scallops during emersion, which may occur naturally during stranding by storms, during seeding and shipping activities, or harvesting and return to the water of sublegal scallops (Duncan et al. 1994; Dredge 1997). *Pecten maximus* suffered respiratory acidosis and died after 72 h of emersion (Duncan et al. 1994). Acidosis was largely attributed to aerobic metabolism (CO_2 production) rather than accumulation of anaerobic end products. The decline in pH was not buffered by mobilisation of Ca from shell dissolution and resulting increase in HCO_3^-, as observed in intertidal bivalves that are more tolerant of emersion. Negligible survival of *Argopecten ventricosus* spat (3.5 mm shell height) was obtained after 26 h of air-exposure at temperatures ranging from 17 to 28°C (Maeda-Martínez et al. 2000). At a shorter exposure time (19 h) survival rate, measured after 2 h of re-immersion, was inversely related to temperature. Survival could be further enhanced during shipping by packaging animals tightly and thus preventing shell gaping, and by using pure oxygen. Predator avoidance response of juvenile *P. magellanicus* was not affected by 4 h air exposure at 18°C, but was adversely affected by the combined stress of air exposure and cold shock (18 to 8°C) (Lafrance et al. 2002). Thus both duration of air exposure and temperature (absolute value and change in temperature) are critical determinants of scallop survival during emersion.

A reduced capacity for oxyregulation, or increased VO_2 dependence on oxygen tension, may be a characteristic feature of species which are unlikely to experience low oxygen levels in their natural habitat (Bayne 1973). Scallops, as sublittoral, epifaunal bivalves, which are incapable of sustaining prolonged valve closure and are relatively intolerant of aerial exposure, are unlikely to experience low oxygen levels in the environment. In their analysis of 31 species of marine invertebrates, however, Mangum and Van Winkle (1973) found no correlation between the response of VO_2 to declining oxygen tension, and environmental oxygen level.

Hochachka and Somero (1984, Table 5-1) correlated the tolerance to anoxia in both terrestrial and aquatic organisms with the amount of glycogen stores, since glycogen is the main respiratory substrate during anaerobiosis. In this respect it is noteworthy that scallops, which are relatively intolerant of anoxia, contain relatively low levels of glycogen in the adductor muscle, the main storage organ (attaining maximum values of 23–25% of muscle dry weight in first-year *Argopecten irradians* (Epp et al. 1988) and 18% in *Chlamys islandica* (Sundet and Vahl 1981)). In contrast, *Mytilus edulis*, an intertidal bivalve that commonly utilises anaerobic pathways during prolonged valve closure induced by aerial exposure, attains high maximum seasonal glycogen levels of 42 to 53% in the mantle, the principle long term storage organ in mytilids (de Zwaan and Zandee 1972; Gabbott 1983).

7.3.2 Metabolic Cost of Reproduction

An increase in VO_2 associated with the metabolic cost of reproduction has been documented in the bay scallop, *Argopecten irradians irradians* (Bricelj et al. 1987b), giant scallop, *Placopecten magellanicus* (Shumway et al. 1988), Iceland scallop, *Chlamys islandica* (Vahl 1978) and black scallop *Chlamys varia* (Shafee 1982). Figure 7.10 illustrates the relationship between oxygen uptake, environmental temperature and gametogenic stage in three of these pectinid species. In contrast, MacDonald and Thompson (1986) found no significant correlation between oxygen uptake and gametogenic activity in *P. magellanicus* from Newfoundland populations at depths of 10 and 31 m. In bay scallops, gametogenesis (gonadal growth) was associated with a 50% increase in the routine rate of oxygen uptake relative to that predicted on the basis of seasonal temperature (Bricelj et al. 1987b). Vahl and Sundet (1985) found that sexually mature *C. islandica* have a higher size-specific metabolic rate during the period when they experience intense gamete differentiation, than immature scallops. For a scallop of comparable size (e.g., 0.8 g in tissue dry weight), the VO_2 of mature males and females is 57% and 70% higher respectively than that of immature scallops. They also found that the metabolic rate of sexually mature males increases at a faster rate with body size than that of mature females, and that for size classes greater than 1.2 g in tissue weight (ca. 50 mm in shell height) mature males have a higher metabolic rate than females. These authors suggest that the high cost of sperm production, involving protein synthesis, cannot be met by stored glycogen reserves, thereby restricting spermatogenesis to periods of high food availability (March to October), while oögenesis can continue throughout the fall and early winter at Balsfjord, Norway.

Bivalves typically show a pattern of increasing reproductive output with increasing age/size (Peterson 1983). Two pectinid species, however, the iteroparous *Chlamys islandica* (Vahl 1984), and the semelparous scallop *Argopecten irradians irradians* (Bricelj et al. 1987b, and Bricelj and Krause 1992), exhibit reproductive senility (*sensu* Peterson 1983) in that size-specific reproductive output declines in older individuals. Reproductive output declined in senescent *Placopecten magellanicus* from low food environments but was maintained in larger senescent scallops from adjacent higher food environments (MacDonald and Bayne 1993). The combination of seasonal availability of the food supply and the timing of the reproductive cycle permitted reproductive output to be maintained during a period of the life cycle when maintenance requires all the available resources. Fecundity was sacrificed for maintenance when food resources were limited but when food was more readily available fecundity in the large senescent individuals continued to increase (MacDonald and Bayne 1993). In the bay scallop, post-reproductive individuals approaching the end of their lifespan, exhibit a significantly lower weight-normalised VO_2 than young individuals which have not yet undergone reproduction (Bricelj et al. 1987b). Thus senescence is associated with lowered metabolic expenditure, as found for senescent gastropods *Ancylus fluviatilis* and *Planorbis contortus* (Calow 1975) and the limpet *Ferrissia rivularis* (Burky 1971).

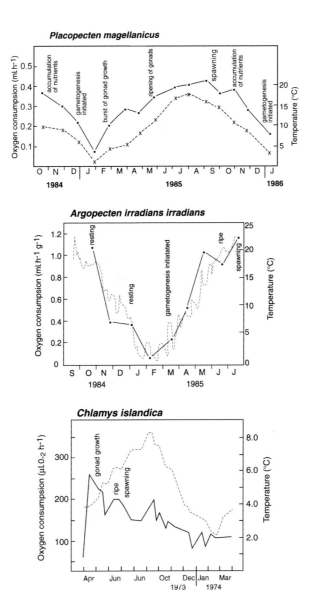

Figure 7.10. Seasonal changes in oxygen consumption of three pectinid species in relation to environmental temperature and reproductive stage. Weight-normalised oxygen uptake of *Chlamys islandica* and *Placopecten magellanicus* from Vahl (1978) and Shumway et al. (1988), respectively. Reproductive stages of *C. islandica* from Sundet and Vahl (1981). Weight-specific oxygen uptake of *Argopecten irradians irrandians* from Bricelj et al. (1987a).

7.3.3 Metabolic Rate in Relation to Temperature and Latitude

Spärck (1936) observed that scallop species with a more northerly distribution, e.g., the arctic scallop, *Pecten groenlandicus*, had a relatively higher VO_2 than those of southern latitudes, such as the boreal *P. varius* and Mediterranean *P. flexuosus*, when compared at the same temperature. Bricelj et al. (1987b) also found that the northern bay scallop, *Argopecten irradians irradians*, had a higher metabolic rate than the southern Florida subspecies, *A. i. concentricus* (Barber and Blake 1985), at a comparable temperature range (20–25°C). Since these two studies examined VO_2 at environmental temperatures, however, the possibility that the latitudinal differences observed simply reflect a lower degree of acclimatisation to higher temperatures in the northerly scallop population cannot be ruled out. In *A. i. irradians* seasonal changes in VO_2 closely track changes in temperature, which can explain 93% of the seasonal variation in metabolic rate (Fig. 7.10). In the southerly bay scallop, although oxygen uptake increases significantly with environmental temperature, the latter explains only 23% of the variation in metabolism (Barber and Blake 1985). The temperature range over which VO_2 was determined, however, was considerably narrower for the Florida population (21.5–31.7°C) than for the New York population (1.5–22.9°C). Barber and Blake (1985) suggested that the combined effects of lower food supply and higher temperature-dependent metabolic rates at lower latitudes may limit this species' southern distribution along the Atlantic coast.

A positive, significant correlation between VO_2 and both seasonal and laboratory acclimation temperature has been described in *Placopecten magellanicus* from Maine at a depth of 20 m (Shumway et al. 1988; Fig. 7.10). A strong correlation between VO_2 and ambient temperature was found in *P. magellanicus* from Newfoundland at 10 m, but not in those from 31 m (MacDonald and Thompson 1986). In *Patinopecten yessoensis* oxygen uptake closely tracks seasonal temperature changes, except during the winter months when weight-standardised VO_2 increases by 13% while water temperatures are still declining (5.8 to 4.6°C) (Fuji and Hashizume 1974). This period, extending between December and February, coincides with that of maximal gonadal growth, suggesting that the increase in metabolic rate is related to reproductive cost, as found in other scallop species. The above examples serve to illustrate that the degree to which seasonal changes in metabolic rate are determined by temperature can vary considerably both among pectinid species and within the same species.

In a unique study, Pilditch and Grant (1999b) compared the effects on VO_2, NH_4^+ and growth rate of *P. magellanicus* exposed to a fluctuating temperature regime (6 to 15°C over an 8-day cycle, averaging 10°C) to those of scallops held at a constant temperature of 10°C. The fluctuating regime was intended to simulate the relatively high-frequency temperature changes occurring in coastal waters due to upwelling. They found a small (15%) but significant increase in VO_2 of scallops exposed to the fluctuating temperature treatment compared to those held at constant temperature, although this was not detected as a significant difference in growth of shell or total soft tissue dry weight. More importantly, they found no evidence of compensatory acclimation of sea scallops to fluctuating temperatures, as VO_2 remained dependent on ambient temperature even after 48 days of exposure.

Shafee (1982) developed a predictive multiple regression equation for *Chlamys varia*, which allows calculation of the metabolic expenditure for this species at any time of the year. Oxygen consumption (VO_2; $\mu L\ h^{-1}$) was related to ambient temperature (°C), tissue dry weight (W; g), food supply (F), and reproductive condition (a) as follows: VO_2 = (-89.92 + 23.37 T + 1, 183.76 a + 134.47 F) $W^{0.72}$ (r^2 = 0.96). Food supply was defined as F = 0 or F = 1, for starved and fed animals respectively. The gonad index was defined as the constant "a" in the allometric equation relating gonad dry weight (G) to total tissue dry weight (G = aW^b).

Seasonal Q_{10} values of oxygen consumption for several pectinid species are shown in Table 7.5. Values are typically >1, indicating that most of the scallop species listed, unlike the blue mussel, *Mytilus edulis*, have a limited ability to acclimatise their metabolic rate to seasonal temperature changes, and are thus unable to conserve energy at higher temperatures. In *Chlamys varia* standard and routine oxygen uptake rates were temperature dependent (Q_{10} >2), while active VO_2 (that of a starved animal suddenly exposed to food) showed a weak response to temperature (Q_{10} = 1.0) (Table 7.5).

7.3.4 Metabolic Rate in Relation to Activity Levels

Early work by Spärck (1936) indicated that bivalves capable of swimming such as members of the Pectinidae and Limidae families, were characterised by higher metabolic rates, as well as a greater increase in VO_2 with increasing temperature than other more sedentary bivalves of the Astartidae, Veneridae, and Tellinidae families. Van Dam (1954) however, reported VO_2 rates for *Pecten grandis* and *P. irradians*, which were within the range of values published for non-swimming boreal species.

Table 7.6 compares routine rates of oxygen uptake of pectinids with other bivalves at a common temperature (ca. 10°C). Values were carefully selected from studies in which animals were fed prior to measurements, a relatively wide size range was used, weight standardisation was carried out, and experimental animals were not experiencing rapid gonadal growth. Analysis of this table yields no evidence of increased metabolic rates in scallops when compared to a wide variety of more sedentary and sessile bivalves. Values shown, however, reflect metabolic rates of quiescent, undisturbed animals confined in experimental respirometers. They do not reflect active metabolic rates elicited during swimming or "clapping" activity of scallops. The cost of such activity cannot be incorporated in the energy budget of scallops without further understanding of its size and temperature dependence, and its contribution to overall energy expenditure. Decomposition of total metabolic rate into several functional components (Clarke 1987), partially achieved by measuring standard, routine and active metabolic rates (Bayne et al. 1976) has been carried out in considerable detail for *Mytilus edulis*, but has rarely been attempted for pectinids.

Limited information is available on the metabolic rate of scallops during locomotory activity, e.g., crawling of juveniles or swimming activity. Mackay and Shumway (1980) induced vigorous swimming of *C. deliculata* using starfish foot or extract, before placing

Table 7.5

Temperature coefficient (Q_{10}) for oxygen consumption in several pectinid species. $Q_{10} = (K_1/K_2)^{[10/(T1-T2)]}$ where K_1 and K_2 = oxygen uptake at temperatures T_1 and T_2.

Species	Temperature (°C)		Q_{10}	Source
Chlamys opercularis	5–10		4.42	McLusky 1973[a]
	10–20		0.78	
Chlamys varia	10–15	standard:	2.19–6.05	Shafee 1982
		routine:	1.75–4.67	
		active:	1.08–1.62	
Placopecten magellanicus	5–15		2.38	Shumway et al. 1988
	10–20		1.58	
Patinopecten yessoensis	5.8–14.8		1.09	Fuji and Hashizume
	8.9–16.9		1.78	1974
	14.8–22.4		1.64	
Argopecten irradians	5–15		3.36	Bricelj et al. 1987a
irradians	10–20		2.15	
A. i. concentricus	6.5–10.2		2.27	Kirby-Smith 1970
	10.2–18.0		3.16	
	18.0–25.5		2.39	
Argopecten circularis	20–30		1.94[b]	Silva Loera 1986
			2.77[c]	

[a] Laboratory acclimated scallops;

[b, c] for scallops 1 g and 0.1 g dry tissue weight, respectively.

animals into respirometers. They found that the VO_2 of active animals was 2.4 times higher than that of resting (starved) animals, while feeding did not cause an increase in oxygen consumption above the standard rate of starved scallops. This observation was related to the fact that scallops are sublittoral, continuous feeders. An elevated VO_2 in response to feeding was suggested to be more typical of intertidal, discontinuous feeders such as *Mytilus edulis*. Swimming to exhaustion of *Chlamys hastata* confined within a respirometer, induced by contact with starfish, caused a significant, 3- to 4-fold increase in VO_2 (Donovan et al. 2003). On the other hand, seasonal temperature changes did not greatly influence the metabolic rate of *C. islandica*, and Vahl (1978) suggested that a significant portion (34%) of the seasonal variability in VO_2 in this population could be explained by changes in food availability. Oxygen uptake was also significantly correlated with food levels in both shallow and deep water populations of *Placopecten magellanicus* (MacDonald and Thompson 1986). In *Chlamys varia* routine VO_2 was on the average 1.6 to 1.9 times greater than standard oxygen uptake rates determined for starved individuals (Shafee 1982).

Table 7.6

Weight-standardised routine rates of oxygen consumption (VO_2; mL O_2 h^{-1} g dry tissue weight^{-1}) of pectinid species compared to other bivalves.

Species	Temp. (°C)	VO_2	Source	Notes
Chlamys varia	10	0.339	Shafee 1982	A
Chlamys islandica	8	0.195	Vahl 1978	A
Chlamys delicatula	10	0.147	Mackay & Shumway 1980	B
Chlamys opercularis	10	0.229	McLusky 1973	B
	10–13	0.182	Vahl 1972	B, C
Argopecten irradians irradians	10	0.425	Bricelj et al. 1987a	A
A. i. concentricus	10	0.249–0.357	Kirby-Smith 1970	A, C
A. purpuratus	12	0.54	Navarro & González 1998	B
Placopecten magellanicus	10	0.244	Shumway et al. 1988	B
	10–12	0.339	MacDonald & Thompson 1986	A
Patinopecten yessoensis	9	0.293	Fuji & Hashizume 1974	A
Donax vittatus	10	0.238–0.275	Ansell 1973	A
Arctica islandica	10	0.317	Taylor & Brand 1975	B
Cerastoderma edule	10	0.200	Newell & Bayne 1980	A
Mytilus edulis	10	0.370	Vahl 1973b	B
	15	0.381	Bayne 1973	B, D
Geukensia demissa	10	0.185	Hilbish 1987	B, E
Choromytilus meridionalis	12	0.430	Griffiths 1980	A
Ostrea edulis	10	0.059	Calculated from Newell et al. 1977	A

Notes:
- A. VO_2 measured at ambient, seasonal temperature;
- B. Measured following laboratory acclimation;
- C. Assuming a tissue water content of 85% and slope b = 0.75;
- D. *M. edulis* can acclimate its routine VO_2 between 10 and 20°C;
- E. Calculated by extrapolation from data at 5 and 15°C.

7.3.5 Anaerobic Metabolism

Bivalves may experience functional anaerobiosis (e.g., following vigorous activity such as burst swimming in scallops) or environmental anaerobiosis, as occurs under hypoxic or anoxic environmental conditions. Recent work has established that in many bivalves, anaerobic metabolism can contribute significantly to the total metabolic rate, particularly in intertidal species, which suffer periodic aerial exposure (Pamatmat 1980; Shick et al. 1983). Therefore oxygen consumption does not always provide a reliable measure of total metabolism or heat loss as measured by direct calorimetry. Physiological and biochemical studies conducted on *Placopecten magellanicus* indicate that in scallops, anaerobic pathways are predominantly utilised for energy production during sudden bursts of activity (swimming or valve snapping escape response induced experimentally by predator stimulation) (Thompson et al. 1980; de Zwaan et al. 1980). Swimming activity is exhibited by members of at least four bivalve families: Pectinidae, Amussidae, Limidae and Cardidae. Scallops differ, however, from actively swimming bivalves such as the fileshell *Limaria fragilis* (family Limidae), which display slower sustained swimming activity fueled predominantly by aerobic mechanisms of ATP production (Baldwin and Lee 1979).

Scallops swim by means of jet propulsion in which water is expelled from the mantle cavity by repeated muscle contractions (valve snapping). This activity is powered by the phasic (striated or fast) portion of the adductor muscle, and to a lesser extent by the smaller catch or smooth portion, while more prolonged valve closure is exclusively powered by the catch muscle. Thompson et al. (1980) showed that in *Placopecten magellanicus* rapid shell valve adductions result in accelerated heart rate (2 to 3 fold increase), enhanced stroke volume and 5-fold increase in cardiac output, and a decrease in the PO_2 of post-branchial blood to a low value of 15 mm Hg. Although these physiological responses increase the supply of oxygen to the adductor muscle, they are insufficient to meet this organ's high energy demand during vigorous activity. The effectiveness of O_2 uptake by the scallops' blood (42% in *P. magellanicus*) and the supply of oxygen to the muscle are limited by the lack of respiratory pigments, and generally poor development of the bivalve open circulatory system. De Zwaan et al. (1980) suggested that different scallop species may vary in the relative contribution of aerobic and anaerobic metabolism to the total energy demand during swimming. In *P. magellanicus* these authors estimated that the former represents only 3% of the anaerobic contribution. However, due to the higher ATP yield of aerobic versus anaerobic glycogen utilisation, this translates into as much as 30% of the total ATP provided to the muscle by glycogen catabolism. The energy demand during valve snapping and the subsequent recovery phase is thus largely met by anaerobic glycolysis, and by the breakdown of high-energy phosphagen compounds (arginine phosphate, Arg P) in muscle tissue (arginine phosphate + ADP = arginine + ATP). The relative contribution of these two processes to the total energy demand during valve snapping varies between muscle parts, with phosphoarginine hydrolysis contributing most (72%) of the ATP requirement in the phasic muscle and only 34% in the catch muscle (de Zwaan et al. 1980).

The amino acid octopine is the main end product of anaerobic glycolysis accumulated in muscle tissue (both phasic and catch portions) as a result of exhaustive swimming and valve clapping in scallops (e.g., Grieshaber and Gade 1977; Baldwin and Opie 1978; Chih and Ellington 1983), as well as cephalopods, *Nautilus pompilus*, *Loligo vulgaris* and *Sepia officinalis*, and in *Cardium tuberculatum*, following vigorous jumping (reviewed by

Zandee et al. 1980). Octopine synthesis is catalysed by the enzyme octopine dehydrogenase (Odh) according to the reaction:

Pyruvate + arginine + NADH = octopine + NAD^+

Other metabolic end products such as propionate and succinate are more typical of sedentary or sessile bivalves exposed to low oxygen tensions for prolonged periods. Breakdown of glycogen to octopine is associated with relatively low ATP yield per fuel equivalent (i.e., low efficiency of energy production) and rapid fatigue, but yields a relatively high rate of energy production or ATP output per unit time compared to propionate and succinate (Zandee et al. 1980; Livingstone 1991). Thus glycogen conversion to octopine is adaptive during temporary muscle anoxia associated with short term burst activity, such as swimming in scallops. Octopine production provides active marine molluscs (scallops, cephalopods) with an anaerobic pathway functionally analogous, although evolutionarily more primitive than that of lactate production, characteristic of vertebrates as well as many marine, freshwater and terrestrial invertebrates including gastropods (Livingstone 1991). Octopine production, however, differs from lactate production in that it results in a less acidic end product, has a lower energy yield, and requires an aminoacid (arginine) as well as carbohydrate as substrates.

Thus, swimming to exhaustion, induced experimentally by contact with starfish, caused a significant (~2-fold) increase in octopine levels, as well as a significant decrease in Arg P levels in the scallop *Chlamys hastata* (Donovan et al. 2003). In this study, encrustation of scallops with barnacles negatively affected their swimming capacity and resulted in increased anaerobic energy expenditure. Octopine levels of scallops exhausted by swimming were higher in encrusted than unencrusted scallops. The latter also consumed approximately twice as much Arg P as unencrusted scallops. In *Argopecten irradians concentricus* octopine production in the adductor muscle was restricted to periods of contractile activity (burst swimming) but did not occur during environmental hypoxia (Chih and Ellington 1983). Following 4 h of hypoxia there was a decline in Arg P and accumulation of succinate but no detectable octopine production.

A 5-fold increase in ODH activity in juvenile *Argopecten irradians* was related to a decrease in their percent attachment to the eelgrass canopy and thus increase in swimming activity between 6.7 and 29 mm SH (García-Esquivel and Bricelj 1993). On this basis, these authors proposed Odh activity as an index of the scallops' capacity for burst swimming activity. A significant positive correlation was described between the degree of individual multilocus heterozygosity and octopine accumulation after burst activity in *P. magellanicus* (Volckaert and Zouros 1989), and between heterozygosity and Odh and pyruvate kinase activities in the adductor muscle of the scallop *Euvola ziczac* (Alfonsi et al. 1995). In this context, Pérez et al. (2000) found that in *E. ziczac* the affinity of the Odh enzyme for pyruvate was greater in heterozygous than homozygote individuals at the Odh locus. These findings support the suggestion made by Volckaert and Zouros (1989) that the fitness advantage of heterozygosity in motile bivalves such as scallops may be related to their increased metabolic capacity for burst swimming involved in predator escape.

Repeated valve snapping can only be maintained for a few minutes and is followed by a period of apparent exhaustion, evidenced by the scallops' lack of response to further stimulation and valve closure. During this period blood PO_2 remains low, and scallops consume no oxygen, yet an increased heart rate is maintained. Thus recovery takes place under hypoxic/anoxic conditions. Utilisation of phosphoarginine and accumulation of octopine appear to occur more or less sequentially in scallops, the former providing the main energy source during initial valve snapping, and the latter accumulating mainly during later stages of swimming (Chih and Ellington 1983) or during recovery following exhaustion (Gäde 1980). Thus octopine formation serves primarily to replenish cytoplasmic NAD^+ required to maintain a high glycolytic flux (Baldwin and Opie 1978; Zandee et al. 1980). Upon reopening of valves, scallops exhibit a transient increase in VO_2 above normal, resting levels (Thompson et al. 1980). This increase probably reflects the repayment of an oxygen debt, i.e., the oxidation of anaerobic end products (octopine) accumulated in the tissues, although it could also reflect reoxygeneration of the hemolymph. Restoration of physiological functions to resting values takes several hours in exhausted scallops. In adult *Argopecten irradians* a considerably shorter recovery period of 90 s was required before 100% of bay scallops tested experimentally could attempt a second swim following a prior swim, but the swimming distance was significantly shorter than on the first attempt, suggesting that full physiological recovery had not been achieved (Winter and Hamilton 1985).

Scallops do not normally experience prolonged periods of valve closure. However, Thompson et al. (1980) were able to induce valve closure in *Placopecten magellanicus* for up to 90 min by occasional tapping of the shell margin. Under these conditions, physiological responses differed from those elicited during valve snapping activity. The PO_2 of blood from the adductor muscle sinus remained relatively high (40 mm Hg), there was no increase in heart rate, and scallops continued to take up oxygen although at a reduced rate. Scallops are able to ventilate through the gape which occurs where the valves do not closely oppose each other. Thus although anaerobic pathways are invoked in the catch adductor, metabolism during valve closure remains largely aerobic. Furthermore, during valve closure phosphagen contributes less energy than anaerobic glycolysis to the total energy demand of the muscle, and octopine is no longer the sole end product of anaerobic metabolism, as observed during valve snapping (de Zwaan et al. 1980). Thus, both alanine and succinate were produced during valve closure.

A study using *in vivo* nuclear magnetic resonance to measure changes in phosphorus-containing metabolites showed that laboratory-induced hypoxia in juvenile *Placopecten magellanicus* collected in winter, when glycogen stores are low, caused a marked reduction in Arg P concentrations and intracellular pH, with an associated increase in inorganic phosphate (P_i) levels (Jackson et al. 1994), thus indicating a reliance on this phosphagen to meet the energy demands during environmental anaerobiosis. A similar response to hypoxia was observed in excised adductor muscle from adults, but was not apparent in live juvenile scallops collected in the summer. On this basis the authors suggested that changes in Arg P and P_i, or in the ratio Arg P/P_i might provide a useful and more sensitive indicator of stress and energy status of scallops than the adenylate energy charge (AEC).

7.4 ENERGY UTILISATION

7.4.1 Excretion and Byssus Secretion

The excretion of nitrogenous excreta represents a potentially significant loss of energy in bivalves, particularly during conditions of severe nutritive stress when reliance on protein catabolism to support metabolic demand increases (Gabbott and Bayne 1973). Although ammonia is the major nitrogenous excretory product in bivalves, under certain circumstances dissolved organic nitrogen (primary amines) may comprise a significant fraction of total nitrogen excretion. In *Mytilus edulis* for example, the excretory loss represents up to 31% of the respiratory energy demand during the winter (calculated from Hilbish and Koehn 1985). At this time, amine and ammonia excretion contribute about 76 and 24%, respectively, to the total energy loss as nitrogenous excreta. Very limited data are available on excretion rates of nitrogenous products in pectinids. It is of particular interest to obtain information on the relative loss of primary amines and ammonia in this bivalve group, since several scallop species are known to rely heavily on protein catabolism during gametogenesis as well as during periods of negative energy balance (e.g., overwintering conditions) (Epp et al. 1988).

Ammonium excretion rates (VNH_4) have been determined for adult *Argopecten irradians concentricus* (Barber and Blake 1985), and are estimated to represent about 14% (range = 8.5 to 18.5%) of the respiratory energy loss. A similar contribution of VNH_4 to the respiratory loss (= 13.5%) was obtained for juveniles of this subspecies (Lu and Blake 1999). Ammonia excretion of adults increased significantly with decreasing salinity, a finding consistent with the role of excretory products (free amino acids) in cell volume regulation of bivalves (Deaton et al. 1984). Similarly, in *A. purpuratus* VNH_4 increased between 30 and 24 ppt, but declined with a further decrease in salinity (24 to 18 ppt), a range over which these scallops are physiologically impaired and show negative scope for growth (Navarro and González 1998).

The ratio of oxygen consumed to NH_4 - N excreted (O:N, calculated in atomic equivalents) was used in conjunction with respiratory quotients ($CO_2:O_2$) to identify seasonal changes in the dominant catabolic substrates in *A. irradians concentricus* (Barber and Blake 1985). Oxygen:N values, which ranged from about 6 to 22, were generally lower than those reported for mussels, *Mytilus edulis*. In this species values of 30 or below are generally indicative of a stressed animal with relatively high protein catabolism, and typically exceed 50 during periods of tissue growth (range = 17 to 120; Fig. 14 in Bayne and Newell 1983). In *Argopecten purpuratus* O:N values ranged between 9.5 and 32.6 over a salinity range of 30 to 18 ppt, showing a consistent decline with decreasing salinity in larger individuals (5 to 10 g dry soft tissue weight) (Navarro and González 1998). Values <15 were associated with unsuitable conditions for growth in larger scallops. There was also a negative relationship between scallop size and the O:N ratio, indicative of higher protein catabolism in younger individuals. In juvenile (57 mm SH) *P. magellanicus* exposed to varying experimental conditions (constant or fluctuating temperature regime; high or low food supply), O:N ratios ranged from 12 to 20 (Pilditch and Grant 1999b). In this study sea scallops showed positive shell and soft tissue growth rates under all experimental conditions tested. A lower value of 8.9 was obtained for this species under thermal stress (Grant and Cranford 1991). Starvation of juvenile *P. magellanicus* yielded a C:N ratio of 12.7 compared to a value of 31 for scallops fed algal diets (Volckaert 1988 in Grant and Cranford 1991). Lower O:N ratios in scallops may be related to the more significant contribution of protein catabolism to the total energy metabolism in this group of bivalves (Epp et al. 1988) compared to *M. edulis*, which preferentially utilises glycogen stores to fuel gametogenesis (Gabbott 1975, 1983). Utilisation of the O:N ratio as a generalised index of physiological condition in scallops thus requires additional information on suitable reference values under normal as well as stressed conditions in each species.

Byssus secretion persists into the adult phase only in some pectinid species, such as *Chlamys islandica* and *C. opercularis*. In *Placopecten magellanicus* byssus formation (number of threads secreted) and rate of byssus attachment decline with increasing body size (Caddy 1972). The rate of byssus formation also increases with temperature, while percent attachment is temperature independent. The cost of byssus production as a function of scallop size/age has been determined for adult *Chlamys islandica* by measuring the energy content of byssal threads, which are composed mainly of quinone tanned protein (Vahl 1981). Byssus secretion represents only a minor component of the energy budget, ranging between 4 and 14% of somatic production.

7.4.2 Growth

Growth represents the integrated response of physiological processes of energy acquisition and expenditure detailed in the previous sections. Scope for growth and resource allocation between somatic and reproductive tissue production in pectinids are examined in a separate chapter (Thompson and MacDonald 1991; Chapter 8, this volume). Seasonal growth in bivalves, including pectinids, is influenced by the interaction of several environmental variables, particularly water temperature and food supply (e.g., Broom and Mason 1978; Bayne and Newell 1983). Food availability has often been found to exert a greater influence on growth rate than temperature in temperate scallop species (Orensanz 1984; MacDonald and Thompson 1985), as well as in species inhabiting regions of continually low temperatures, such as the subarctic scallop *Chlamys islandica* (Vahl 1978) and Antarctic scallop, *Adamussium colbecki* (Stockton 1984).

Scallops are generally very vulnerable to the effects of low salinities, as they are unable to maintain prolonged valve closure. In *Argopecten purpuratus* positive scope for growth (determined from physiological rate measures) was only obtained at 27 to 30 ppt. At lower salinities (24 to 18 ppt) scope for growth was negative as a result of the combined effects of reduced clearance and ingestion rates and high excretion and aerobic respiration rates (Navarro and González 1998).

In pectinids, intraspecific variability in growth rates and tissue weight for a given shell height has most frequently been correlated with differences in water depth. Scallops from inshore, shallower waters typically display higher growth rates and maximum sizes than those from deeper waters (*Placopecten magellanicus*: MacDonald and Thompson 1985; Schick et al. 1988; *Patinopecten caurinus*: Haynes and Hitz 1971; MacDonald and

Bourne 1987; *Pecten maximus*: Mason 1957; *Hinnites multirugosus*: Leighton 1979; see also Chapter 28 Lodeiros et al. this volume and Lodeiros et al. 1998). Depth *per se* is not the limiting factor, but, as demonstrated for *P. magellanicus* by MacDonald and Thompson (1985), growth is promoted by relatively higher temperatures and more importantly, higher food levels in shallow waters lying within the productive euphotic zone. For *Placopecten* from Canadian waters, differences in somatic weight between depths were more pronounced than differences in shell height. The negative correlation between maximum size and depth reported for many scallop populations thus supports Sebens' (1982) theoretical growth model, which predicts that maximum size attained by animals with indeterminate growth increases with habitat suitability (lower physiological stress).

Field experiments on the effect of depth on scallop growth convey mixed results. Comparisons between studies are complicated due to differences in species, starting sizes, stocking densities, gear, site, time of year and environmental conditions. Some reports document a decrease in growth with an increase in culture depth while others show no significant differences or even an increase in growth with depth. Côté et al. (1993) examined the influence of depth on the growth of juvenile sea scallops *Placopecten magellanicus* using pearl nets with 6-mm mesh and 0.16 m^2 bottom surface area. Nets were filled with 50, 100, 200, or 300 scallops (mean height 14.4 mm) and suspended from the surface at 9, 15, and 21 m. Growth, measured as dry mass of the shell, muscle and other soft tissues, tended to decrease with depth although the differences were only significant for soft tissue. At all depths, after 1 year, growth (measured as mean shell height) of scallops stocked at 50 individuals per net was significantly larger than of scallops stocked at 200 and 300 individuals per net. Focusing on *P. magellanicus* spat, Grecian et al. (2000) also evaluated the effects of culture depth. Four size classes (1.4–1.69 mm, 1.70–1.99 mm, 2.00–2.99 mm, ≥3.0 mm) of scallop spat were sorted and deployed in different types of nursery-based equipment including collector bags (1.2 and 2.0 mm) and pearl nets (1.5 and 3.0 mm) at two depths (5 and 10 m) for 10 months. Highest growth rates were found for the largest size class, in pearl nets and suspended in shallower waters. Rupp et al. (2004) examined depth and density effects on the lion's paw scallop, *Nodipecten nodosus*, in Southern Brazil. Postlarvae (0.4 mm) were deployed in collector bags at two densities (340 and 150 scallops/spat bag) at two depths (4 and 12 m) for 26–27 days. Growth rates were higher in shallower waters (4 m) during the summer (March-April) when South Atlantic Central Water (SACW) creates temperature and food quality differences with depth. In the winter (June-July), without SACW intrusions there were no depth related growth differences. Results from a computer model of the effects of depth and density on the growth of the Icelandic scallop, *Chlamys islandica* predict no effect of depth as long as no food limitation occurs (Fréchette and Daigle 2002). The model suggests that if food limitation occurs, growth rates would decrease with depth.

Emerson et al. (1994) also evaluated the effects of depth on the growth of *P. magellanicus*. Mesh cages (10 x 50 x 100 cm; 6-mm mesh) were filled with 100 juvenile scallops (mean shell height 39.7 ± 5.0 mm) and secured to a frame at 0, 20, 50, 100, or 200 cm above the bottom in approximately 7-m deep water. By the second period of the growth experiment (October-November), scallops on the bottom had significantly larger shell heights than scallops at any other depth, but by the end of the experiment (March) it was determined that water depth had no consistent effect on shell height. Depth also had no significant effect on the ash content of soft tissues or the weight of the adductor muscle. The weight of soft tissues (excluding muscle) of scallops on the bottom was 40% less than that of scallops growing more than 50 cm off the bottom. Emerson et al. (1994) concluded that high seston concentrations near the bottom inhibited growth rather than providing an energetic benefit. Parrish et al. (1995) also reported that growth rates for *P. magellanicus* were similar over a range of depths. In this study, 200 scallops approximately 40 mm in shell height were ear hung on 1.2 x 1.2 m frames at 20, 50, and 300 cm above the bottom in 10-m water. Mean shell height (mm) and mean dry body mass (mg) were similar at all three depths over the course of the experiment (May-December) but appeared to be highest at 300 cm off the bottom during the latter part of the sampling period although the seston at that depth had the lowest concentration of microplankton, chlorophyll-*a*, AFDW and total lipids. Parrish et al. (1995) concluded that other factors than simply the availability of food, or high energy compounds are important to the growth of scallops. Kleinman et al. (1996), using *P. magellanicus* (mean shell height 22.2 ± 0.1 mm) compared growth in uncontained bottom cultures to growth in suspended pearl nets (100 scallops per net). During most months of the survey, significant differences were found between their three study sites for both culture methods. Although the average soft tissue condition index was significantly higher in the suspended cultures, shell growth rates, average adductor muscle condition index and average whole dry weights were all significantly higher for scallops in bottom cultures. Water temperature, total particulate matter, and chlorophyll concentration explained up to 66% of the variation in shell growth rates in suspended culture and up to 78% in bottom culture, and between 55% and 80% variation in soft tissue growth in suspended and bottom culture, respectively (Kleinman et al. 1996).

The depth at which pearl nets are suspended in the water column can also result in substantial growth differences for juvenile sea scallops. In an experiment, in which pearl nets were variously suspended from 5 m off-bottom to 0.25 m off-bottom, Dadswell and Parsons (1991) found that the mean size of juveniles ranged from 54 to 42 mm, respectively. Growth differences of this magnitude were similar to the effects of stocking density as determined by Parsons and Dadswell (1992). The effect of suspension depth on growth in sea scallops and other species of scallops has been attributed to differences in food level (Wallace and Reinsnes 1984, 1985; Côté et al. 1993). In a study contrasting an open and sheltered location in the Iles-de-la-Madeleines, scallop spat were found to grow and survive better in pearl nets that were grown near the surface (Gaudet 1994). Côté et al. (1993) found that spat grew better at 9 m than at 21 m and related this to the levels of food in the 0.7 to 5.0 μm size range, although density effects masked some of the trends. From the above observations we can see there is likely no one depth that will be suitable in all locations. The optimum culture depth for a particular location will depend on the distribution of the food in the water column and the depth to which wave action will impact the suspended culture gear. There may also be interactions with other fouling species that may be depth dependent.

Intraspecific variation in growth between localities at similar depths has also been related to food limitation associated with high scallop densities (Gruffydd 1974; Orensanz 1984) and/or reduced flow (Cooper and Marshall 1963; Eckman 1987). Only a few studies have evaluated the effects of density on scallop growth. Côté et al. (1993) investigated the effects of stocking density on the growth of juvenile sea scallops *Placopecten magellanicus*. Using 6-mm mesh pearl nets with a bottom surface area of 0.16 m^2 and scallops with a mean shell height of 14.4 mm, nets were filled with 50, 100, 200, or 300 scallops and suspended from the surface at 9, 15, and 21 m. Growth, measured as dry mass of the shell, muscle and other soft tissues, decreased with increasing density but the differences were only significant between 50 and 300 scallops per pearl net. After 1 year, at all depths, the mean shell heights of scallops stocked at 50 individuals per net were significantly larger than of scallops stocked at 200 and 300 individuals per net. Using the same style of pearl nets, Côté et al. (1994) stocked 50, 75, 100, 150, 200, or 250 scallops (shell height 29.6 ± 0.3 mm) per net and anchored the nets 9 m above the bottom (approximately 15 m from the surface). In one set of nets all scallops were alive while in the other set of nets only 25 scallops were alive and the others were scallop shells glued together as "dummies". After three months (July-October), at densities of 100 scallops/net and greater, mean growth rates of both shell and dry mass were always significantly lower in nets without dummies suggesting that the effects of density on growth were due to food limitations and not the lack of space. Growth models for Iceland scallops, *Chlamys islandica* also support the importance of food depletion on scallop growth (Fréchette and Daigle 2002).

The results for scallop spat are not as clear. Grecian et al. (2000) found no significant difference in growth rates between *P. magellanicus* spat (2.0–3.3 mm shell height) stocked at 2,600 and 5,200 scallops per collector bag deployed at 5-m for 8 months. They used "onion" collector bags, measuring 40 x 80 cm with a 3-mm nominal mesh size, filled with 1 meter of Netron®. The authors suggest that their relatively low stocking densities may not have created limitations of food or space for the spat. Also working with scallop spat, Rupp et al. (2004) examined the effect of density on the lion's paw scallop, *Nodipecten nodosus*, in Southern Brazil. Postlarvae (0.4 mm) were deployed in collector bags at two densities (340 and 150 scallops/spat bag) at two depths (4 and 12 m) for 26–27 days. A slight, but statistically significant decrease in shell heights was observed at both depths for postlarvae stocked at the higher density.

Due to the high energetic cost of reproduction, shell and somatic tissue growth in pectinids may also be greatly influenced by reproductive events. Cessation or retardation of shell growth during the reproductive period (gonadal growth and spawning), and resumption of growth following spawning have been reported in *Argopecten irradians* (Bricelj et al. 1987a, Bricelj and Krause 1992), *Notovola meridionalis* (Fairbridge 1953) and *Patinopecten yessoensis* (Maru and Obara 1967), whereas shell and gonadal growth coincide in other pectinids such as *Chlamys opercularis* (Broom and Mason 1978) and *C. islandica* (Vahl 1978; 1981). In pectinids in which gametogenesis involves the utilisation of energy stores (reviewed by Barber, this volume), growth of somatic tissues may not follow the same pattern as shell growth, and is expected to be more strongly influenced by the reproductive cycle than in species which meet their energy demand primarily from the external food supply.

Growth curves shown in Figure 7.11a, b, c illustrate the remarkable variability in growth rates, longevity and maximum size displayed by the Pectinidae. Shell growth has commonly been described by the von Bertalanffy model, and age information obtained from external growth rings in the shell. The largest scallop species, which attain asymptotic heights (H_∞) of up to 160–170 mm, are generally long-lived, with lifespans ranging between 18 and 23 years. These species are often characteristic of deeper waters (up to 100–200 m), such as *Chlamys islandica* (Vahl 1981), *Patinopecten caurinus* (MacDonald and Bourne 1987) and *Placopecten magellanicus* (MacDonald 1986; Schick et al. 1988), or of moderate depths (up to 50–60 m) such as *Pecten maximus* (Mason 1957) and *Crassadoma gigantea* (MacDonald and Bourne 1989). In contrast, species restricted to shallow coastal waters (<10 m) are generally characterised by shorter lifespans (2–8 years) and smaller asymptotic size. For example, *Argopecten irradians irradians* reaches a maximum size of 80 mm at 2 years (Bricelj and Krause 1992), *Chlamys varia* attains ca. 54 mm at 7–8 years (Conan and Shafee 1978) and *C. tehuelcha* (H_∞ = 83–91 mm) lives only about 6 years (Orensanz 1984).

Large size, protracted lifespan and deep water habitat are however, not always correlated with slow growth. The Bertalanffy growth coefficient (k) provides a measure of the rate of which animals reach their asymptotic shell height. The purple hinge rock scallop *Crassadoma gigantea* is slow growing, as reflected by its low k value (= 0.17) (MacDonald and Bourne 1989), while the giant Pacific sea scallop *Patinopecten caurinus*

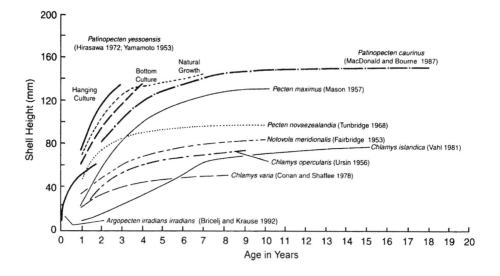

Figure 7.11a. Comparative growth curve of various pectinid species.

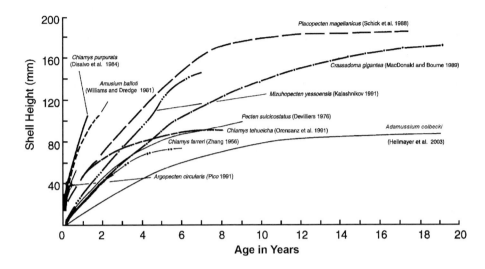

Figure 7.11b. Comparative growth curves of various pectinid species.

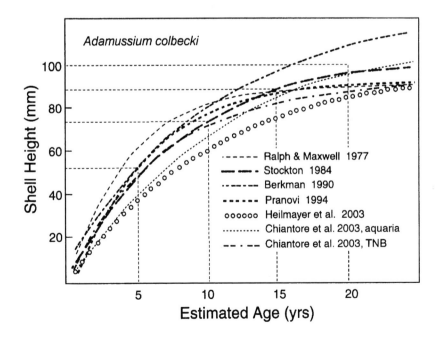

Figure 7.11c. Von Bertalanffy growth functions of Antarctic scallop *Adamussiun colbecki* populations from various studies. TNB = Terra Nova Bay. After Heilmayer et al. (2003).

grows fairly rapidly (k = 0.39) (MacDonald and Bourne 1987). A literature review of growth parameters for the giant scallop *Placopecten magellanicus* (Schick et al. 1988) provides an average k value of 0.24 (n = 74; SD = 0.06), which ranges widely between 0.17 and 0.38 depending on habitat suitability.

Finally, scallops grown in suspended culture generally achieve faster shell and somatic tissue growth rates, and a greater tissue weight and thinner shell for a given height than wild scallops of comparable age growing on the bottom (Ventilla 1982; MacDonald 1986; MacDonald and Bourne 1989). This agrees with the observation made earlier for natural scallop populations, that within a species, growth rate in shallow water is often greater than in deeper waters. In *Placopecten magellanicus* MacDonald (1986) showed that accelerated growth in above-bottom culture was also accompanied by a reduction in longevity and maximum size (from 174 to 128 mm in shell height). Thus, provided that fouling is controlled, enhanced growth of scallops suspended in surface waters results from the exploitation of more favourable seston and temperature conditions in the overlying water column relative to those present on the bottom (Leighton 1979; Wallace and Reinsnes 1985).

REFERENCES

Abbott, B.C. and Ballantine, D., 1957. The toxin from *Gymnodinium veneficum* Ballantine. J. Mar. Biol. Assoc. U.K. 36:169–189.

Alber, M. and Valiela, I., 1996. Utilization of microbial aggregates by bay scallops, *Argopecten irradians* (Lamarck). J. Exp. Mar. Biol. Ecol. 195:71–89.

Alber, M., D'Avanzo, C. and Valiela, I., 1988. Assimilation of organic aggregates by bay scallops, *Argopecten irradians*. Eos. Trans. Am. Geophysical Union 69(44):1093 (abstract).

Alfonsi, C., Nussetti, O. and Pérez, J.E., 1995. Heterozygosity and metabolic efficiency in the scallop *Euvola ziczac* (L. 1748). J. Shellfish Res. 14:389–393.

Ansell, A.D., 1973. Oxygen consumption by the bivalve *Donax vittatus* (da Costa). J. Exp. Mar. Biol. Ecol. 11:311–328.

Babinchak, J. and Ukeles, R., 1979. Epifluorescence microscopy, a technique for the study of feeding *Crassostrea virginica* veliger larvae. Mar. Biol. 87:69–76.

Bacon, G.S., MacDonald, B.A. and Ward, J.E., 1998. Physiological responses of infaunal (*Mya arenaria*) and epifaunal (*Placopecten magellanicus*) bivalves to variations in the concentration and quality of suspended particles I. Feeding activity and selection. J. Exp. Mar. Biol. Ecol. 219:105–125.

Baldwin, J. and Lee, A.K., 1979. Contributions of aerobic and anaerobic energy production during swimming in the bivalve mollusk *Limaria fragilis* (family Limidae). J. Comp. Physiol. 129:361–364.

Baldwin, J. and Opie, A.M., 1978. On the role of octopine dehydrogenase in the adductor muscle of bivalve molluscs. Comp. Biochem. Physiol. 61B:85–92.

Barber, B.J. and Blake, N.J., 1985. Substrate catabolism related to reproduction in the bay scallop *Argopecten irradians concentricus*, as determined by O/N and RQ physiological indexes. Mar. Biol. 87:13–18.

Bauder, A.G., Cembella, A.D., Bricelj, V.M. and Quilliam, M.A., 2001. Uptake and fate of diarrhetic shellfish poisoning toxins from the dinoflagellate *Prorocentrum lima* in the bay scallop *Argopecten irradians*. Mar. Ecol. Prog. Ser. 213:39–52.

Bayne, B.L., 1971. Oxygen consumption by three species of lamellibranch mollusc in declining ambient oxygen tension. Comp. Biochem. Physiol. A 40:955–970.

Bayne, B.L., 1973. The responses of three species of bivalve mollusc to declining oxygen tension at reduced salinity. Comp. Biochem. Physiol. A 45:793–806.

Bayne, B.L., 1993. Feeding physiology of bivalves: time dependence and compensation for changes in food availability. In: R.F. Dame (Ed.). Bivalve Filter Feeders in Estuarine and Marine Ecosystem Processes. NATO ASI series, Vol. G33, Springer-Verlag, Berlin. pp. 1–24.

Bayne, B.L. and Newell, R.C., 1983. Physiological energetics of marine molluscs. In: A.S.M. Saleuddin and K.M. Wilbur (Eds.). The Mollusca, 4(1). Academic Press, New York. pp. 407–515.

Bayne, B.L., Thompson, R.J. and Widdows, J., 1976. Physiology: I. In: B.L. Bayne (Ed.). Marine Mussels, Their Ecology and Physiology. Cambridge University Press, London. pp. 121–206.

Bayne, B.L., Hawkins, A.J.S. and Navarro, E., 1987. Feeding and digestion by the mussel *Mytilus edulis* L. (Bivalvia: Mollusca) in mixtures of silt and algal cells at low concentrations. J. Exp. Mar. Biol. Ecol. 111:1–22.

Bayne, B.L., Iglesias, J.I.P., Hawkins, A.J.S., Navarro, E., Heral, M. and Deslous-Paoli, J.M., 1993. Feeding behaviour of the mussel, *Mytilus edulis*: responses to variations in quantity and organic content of the seston. J. Mar. Biol. Assoc. U.K. 73:813–829.

Belding, D.L., 1910. A report upon the scallop fishery of Massachusetts, including the habits, life history of *Pecten irradians*, its rate of growth, and other facts of economic value. Boston, Commonwealth of Massachusetts. 150 p.

Beninger, P.G., Le Pennec, M. and Salaun, M., 1988. New observations of the gills of *Placopecten magellanicus* (Mollusca: Bivalvia), and implications for nutrition. Mar. Biol. 98:61–70.

Beninger, P.G., Auffret, M. and Le Pennec, M., 1990. Peribuccal organs of *Placopecten magellanicus* and *Chlamys varia* (Mollusca: Bivalvia): structure, ultrastructure and implications for feeding. I. The labial palps. Mar. Biol. 107(2):215–223.

Beninger, P.G., Ward, J.E., MacDonald, B.A. and Thompson, R.J., 1992. Gill function and particle transport in *Placopecten magellanicus* (Mollusca: Bivalvia) as revealed using video endoscopy. Mar. Biol. 114:281–288.

Beninger, P.G., Dwiono, S.A. P. and Le Pennec, M., 1994. Early development of the gill and implications for feeding in *Pecten maximus* (Bivalvia: Pectinidae). Mar. Biol. 119:405–412.

Berkman, P.A., 1990. The population biology of the Antarctic scallop, *Adamussium colbecki* (Smith, 1902) at New Harbor, Ross Sea. In: K.R. Kerry and G. Hempel (Eds.). Antarctic Ecosystems. Ecological Change and Conservation. Springer, Berlin, NY. pp. 281–288.

Bougrier, S., Lassus, P., Beliaeff, B., Bardouil, M., Masselin, P., Truquet, P., Matignon, F., Mornet, F. and Le Baut, C., 2001. Feeding behavior of individuals and groups of king scallops (*Pecten maximus*) contaminated experimentally with PSP toxins and detoxified. In: G.M. Hallegraeff, S.I. Blackburn, C.J. Bolch and R.J. Lewis (Eds.). Harmful Algal Blooms 2000. IOC of UNESCO, Paris. pp. 407–410.

Brand, A.R. and Roberts, D., 1973. The cardiac responses of the scallop *Pecten maximus* (L.) to respiratory stress. J. Exp. Mar. Biol. Ecol. 13:29–43.

Bricelj, V.M. and Krause, M.K., 1992. Resource allocation and population genetics of the bay scallop, *Argopecten irradians irradians*: effects of age and allozyme heterozygosity on reproductive output. Mar. Biol. 113:253–261.

Bricelj, V.M. and Malouf, R.F., 1984. Influence of algal and suspended sediment concentrations on the feeding physiology of the hard clam *Mercenaria mercenaria*. Mar. Biol. 85:155–165.

Bricelj, V.M. and Lonsdale, D.J., 1997. *Aureococcus anophagefferens*: causes and ecological consequences of brown tide in U.S. mid-Atlantic coastal waters. Limnol. Oceanogr. 42 (5, part 2):1023–1038.

Bricelj, V.M. and Shumway, S.E., 1991. Physiology: Energy acquisition and utilization. In: S.E. Shumway (Ed.). Scallops: Biology, Ecology and Aquaculture. Elsevier, Amsterdam. pp. 305–346.

Bricelj, V.M. and Shumway, S.E., 1998. Paralytic shellfish toxins in bivalve molluscs: occurrence, transfer kinetics and biotransformation. Rev. Fisheries Science 6(4):315–383.

Bricelj, V.M., Malouf, R.E. and de Quillfeldt, C. 1984. Growth of juvenile *Mercenaria mercenaria* and the effect of resuspended bottom sediments. Mar. Biol. 84:167–173.

Bricelj, V.M., Epp, J. and Malouf, R.E., 1987a. Intraspecific variation in reproductive and growth cycles of bay scallops *Argopecten irradians*. Mar. Ecol. Prog. Ser. 36:123–137.

Bricelj, V.M., Epp, J. and Malouf, R.E., 1987b. Comparative physiology of young and old cohorts of bay scallop *Argopecten irradians irradians* (Lamarck): mortality, growth, and oxygen consumption. J. Exp. Mar. Biol. Ecol. 112:73–91.

Bricelj, V.M., Lee, J.H., Cembella, A.D. and Anderson, D.M., 1990. Uptake kinetics of paralytic shellfish toxins from the dinoflagellate *Alexandrium fundyense* in the mussel *Mytilus edulis*. Mar. Ecol. Prog. Ser. 63:177–188.

Bricelj, V.M., MacQuarrie, S.P. and Schaffner, R.A., 2001. Differential effects of *Aureococcus anophagefferens* isolates "brown tide" in unialgal and mixed suspensions on bivalve feeding. Mar. Biol. 139:605–615.

Brillant, M.G.S. and MacDonald, B.A., 2000. Postingestive selection in the sea scallop, *Placopecten magellanicus* (Gmelin): the role of particle size and density. J. Exp. Mar. Biol. Ecol. 253:211–227.

Brillant, M.G.S. and MacDonald, B.A., 2002. Postingestive selection in the sea scallop (*Placopecten magellanicus*) on the basis of chemical properties of particles. Mar. Biol. 141:457–465.

Brock, V., Kennedy, V.S. and Brock, A., 1986. Temperature dependency of carbohydrase activity in the hepatopancreas of thirteen estuarine and coastal bivalve species from the North American east coast. J. Exp. Mar. Biol. Ecol. 103:87–101.

Broom, M.J. and Mason, J., 1978. Growth and spawning in the pectinid *Chlamys opercularis* in relation to temperature and phytoplankton concentration. Mar. Biol. 47:277–285.

Burky, A.J., 1971. Biomass turnover, respiration and interpopulation variation in the stream limpet *Ferrissia rivularis* (Say). Ecol. Monogr. 41:235–251.

Caddy, J.F., 1968. Underwater observations on scallop (*Placopecten magellanicus*) behaviour and drag efficiency. J. Fish. Res. Bd. Canada 25(10):2123–2141.

Caddy, J.F., 1972. Progressive loss of byssus attachment with size in the sea scallop, *Placopecten magellanicus* (Gmelin). J. Exp. Mar. Biol. Ecol. 9:179–190.

Cahalan, J.A., Siddall, S.E. and Luckenbach, M.W., 1989. Effects of flow velocity, food concentration, and particle flux on the growth rates of juvenile bay scallops, *Argopecten irradians*. J. Exp. Mar. Biol. Ecol. 129:45–60.

Calow, P., 1975. The respiratory strategies of freshwater gastropods (*Ancylus fluviatilis* Mull. and *Planorbis contortus* Linn.) in relation to temperature, oxygen concentration, body size and season. Physiol. Zool. 48:114–129.

Castagna, M., 1975. Culture of the bay scallop, *Argopecten irradians*, in Virginia. Mar. Fish. Rev. 37:19–24.

Chauvaud, L., Thouzeau, G. and Paulet, Y.M., 1998. Effects of environmental factors on the daily growth rate of *Pecten maximus* juveniles in the Bay of Brest (France). J. Exp. Mar. Biol. Ecol. 227:83–111.

Chiantore, M., Cattaneo-Vietti, R. and Helimayer, O., 2003. Antarctic scallop (*Adamussium colbecki*) annual growth rate at Terra Nova Bay. Polar Biol. 26:416–419.

Chih, P.C. and Ellington, W.R., 1983. Energy metabolism during contractile activity and environmental hypoxia in the phasic adductor muscle of the bay scallop *Argopecten irradians concentricus*. Physiol. Zool. 56(4):623–631.

Chipman, W.A. and Hopkins, J.G., 1954. Water filtration by the bay scallop, *Pecten irradians* as observed with the use of radioactive plankton. Biol. Bull. Mar. Biol. Lab., Woods Hole 107:80–91.

Claereboudt, M.R., Himmelman, J.H. and Côté, J.,1994. Field evaluation of the effect of current velocity and direction on the growth of the giant scallop, *Placopecten magellanicus*, in suspended culture. J. Exp. Mar. Biol. Ecol. 183:27–39.

Clarke, A., 1987. Temperature, latitude and reproductive effort. Mar. Ecol. Prog. Ser. 38:89–99.

Conan, G. and Shafee, M.S., 1978. Growth and biannual recruitment of the black scallop *Chlamys varia* (L.) in Lanvéoc area, Bay of Brest. J. Exp. Mar. Biol. Ecol. 35:59–71.

Conover, R.J., 1966. Assimilation of organic matter by zooplankton. Limnol. Oceanogr. 11:338–354.

Cooper, R.A. and Marshall, N., 1963. Conditions of the bay scallop, *Aequipecten irradians*, in relation to age and the environment. Chesapeake Sci. 4:126–134.

Cosper, E.M., Dennison, W.C., Carpenter, E.J., Bricelj, V.M., Mitchell, J.G., Kuenstner, S.H., Coleflesh, D. and Dewey, M., 1987. Recurrent and persistent brown tide blooms perturb coastal marine ecosystem. Estuaries 10:284–290.

Côté, J., Himmelman, J., Claereboudt, M. and Bonardelli, J., 1993. Influence of density and depth on the growth of juvenile sea scallops (*Placopecten magellanicus*) in suspended culture. Can. J. Fish. Aquat. Sci. 50(3):1857–1869.

Côté, J., Himmelman, J. and Claereboudt, M., 1994. Separating effects of limited food and space on growth of the giant scallop *Placopecten magellanicus* in suspended culture. Mar. Ecol. Prog. Ser. 106:85–91.

Cranford, P.J., 1995. Relationships between food quantity and quality and absorption efficiency in sea scallops *Placopecten magellanicus* (Gmelin). J. Exp. Mar. Biol. Ecol. 189:123–142.

Cranford, P.J. and Grant, J., 1990. Particle clearance and absorption of phytoplankton and detritus by the sea scallop *Placopecten magellanicus* (Gmelin). J. Exp. Mar. Biol. Ecol. 137:105–121.

Cranford, P.J. and Gordon, D.C. Jr., 1992. The influence of dilute clay suspensions on sea scallop (*Placopecten magellanicus*) feeding activity and tissue growth. Neth. J. Sea Research 30:107–120.

Cranford, P.J. and Hill, P.S., 1999. Seasonal variation in food utilization by the suspension-feeding bivalve molluscs *Mytilus edulis* and *Placopecten magellanicus*. Mar. Ecol. Prog. Ser. 190:223–239.

Cranford, P.J., Emerson, C.W., Hargrave, B.T. and Milligan, T.G., 1998. In situ feeding and absorption responses of sea scallops *Placopecten magellanicus* (Gmelin) to storm-induced changes in the quantity and composition of the seston. J. Exp. Mar. Biol. Ecol. 219:45–70.

Cranford, P.J., Gordon Jr, D.C., Lee, K., Armsworthy, S.L. and Tremblay, G.H., 1999. Chronic toxicity and physical disturbance effects of water- and oil-based drilling fluids and some major constituents on adult sea scallops (*Placopecten magellanicus*). Mar. Environm. Res. 48:225–256.

Dadswell, M.J. and Parsons, G.J., 1991. Potential for aquaculture of sea scallop, *Placopecten magellanicus* (Gmelin, 1791) in the Canadian Maritimes using naturally produced spat. In: S.E. Shumway and P.A. Sandifer, (Eds.). An International Compendium of Scallop Biology and Culture. World Aquaculture Workshops, No. 1. The World Aquaculture Society, Baton Rouge, LA. pp. 300–305.

Davis, R.L. and Marshall, N., 1961. The feeding of the bay scallop, *Aequipecten irradians*. Proc. Nat. Shellfish. Assoc. 52:25–29.

Deaton, L.E., Hilbish, T.J. and Koehn, R.K., 1984. Protein as a source of amino nitrogen during hyperosmotic volume regulation in the mussel *Mytilus edulis*. Physiol. Zool. 57:609–619.

DeVilliers, G., 1976. Exploratory fishing for and growth of scallop *Pecten sulcicostatus* off the Cape South Coast. Rep. South Africa Dept. Industries, Sea Fisheries Branch Investigational Report, No. 112:23 p.

Disalvo, L.H., Alarcon, E., Martinez, E. and Uribe, E., 1984. Progress in mass culture of *Chlamys Argopecten purpurata* Lamarck (1819) with notes on its natural history. Revista Chilena de Historia Natural 57:35–45.

Donovan, D.A., Bingham, B.L., From, M., Fleisch, A.F. and Loomis, E.S., 2003. Effects of barnacle encrustation on the swimming behaviour, energetics, morphomtery, and drag coefficient of the scallop *Chlamys hastata*. J. Mar. Biol. Assoc. U.K. 83:1–7.

Dral, A.D.G., 1967. The movements of the laterofrontal cilia and the mechanism of particle retention in the mussel (*Mytilus edulis* L.). Neth. J. Sea Res. 3:391–422.

Dredge, M.C.L., 1997. Survival of saucer scallops, *Amusium japonicum balloti*, as a function of exposure time. J. Shellfish Res. 16:63–66.

Duggan, W.P., 1973. Growth and survival of the bay scallop, *Argopecten irradians*, at various locations in the water column and at various densities. Proc. Nat. Shellfish. Assoc. 63:68–71.

Duncan, P., Spicer, J.I., Taylor, A.C. and Davies, P.S., 1994. Acid-base disturbances accompanying emersion in the scallop *Pecten maximus* (L.). J. Exp. Mar. Biol. Ecol. 182:15–25.

Eckman, J.E., 1987. The role of hydrodynamics in recruitment, growth, and survival of *Argopecten irradians* (L.) and *Anomia simplex* (D'Orbigny) within eelgrass meadows. J. Exp. Mar. Biol. Ecol. 106:165–191.

Eckman, J.E., Peterson, C.H. and Cahalan, J.A., 1989. Effects of flow speed, turbulence, and orientation on growth of juvenile bay scallops, *Argopecten irradians concentricus* (Say). J. Exp. Mar. Biol. Ecol. 132:123–140.

Emerson, C.W., Grant, J., Mallet, A. and Carver, C., 1994. Growth and survival of sea scallops *Placopecten magellanicus*: effects of culture depth. Mar. Ecol. Prog. Ser. 119:119–132.

Epp, J., Bricelj, V.M. and Malouf, R.E., 1988. Seasonal partitioning and utilization of energy reserves in two age classes of the bay scallop *Argopecten irradians irradians* (Lamarck). J. Exp. Mar. Biol. Ecol. 121:113–136.

Erard-LeDenn, E., Morlaix, M. and Dao, J.C., 1990. Effects of *Gyrodinium* cf. *aureolum* on *Pecten maximus* (post larvae, juveniles and adults). In: E. Granéli, E.L. Sundström and D.M. Anderson (Eds.). Toxic Marine Phytoplankton. Elsevier/North Holland. pp. 132–136.

Fairbridge, W.S., 1953. A population study of the Tasmanian "commercial" scallop, *Notovola meridionalis* (Tate) (Lamellibranchiata, Pectinidae). Aust. J. Mar. Freshw. Res. 4:1–40.

Floyd, D.J., 1953. Foods and feeding of oysters as observed with the use of radioactive plankton. Proc. Natl. Shellfish. Assoc., Convention Addresses. pp. 171–180.

Foster-Smith, R.L., 1975a. The effect of concentration of suspension on the filtration rates and pseudofaecal production for *Mytilus edulis* L., *Cerastoderma edule* (L.) and *Venerupis pullastra* (Montagu). J. Exp. Mar. Biol. Ecol. 17:1–22.

Foster-Smith, R.L., 1975b. The effect of concentration of suspension and inert material on the assimilation of algae by three bivalves. J. Mar. Biol. Assoc. U.K. 55:411–418.

Foster-Smith, R.L., 1976. Some mechanisms for the control of pumping activity in bivalves. Mar. Behav. Physiol. 4:4160.

Fréchette, M. and Bourget, E., 1985. Food limited growth of *Mytilus edulis* L. in relation to the benthic boundary layer. Can. J. Fish. Aquat. Sci. 42:1166–1170.

Fréchette, M. and Daigle, G., 2002. Reduced growth of Iceland scallops *Chlamys islandica* (O.F. Müller) cultured near the bottom: a modeling study of alternative hypotheses J. Shellfish Res. 21:87–91.

Fuji, A. and Hashizume, M., 1974. Energy budget for a Japanese common scallop, *Patinopecten yessoensis* (Jay), in Mutsu Bay. Bull. Fac. Fish. Hokkaido Univ. 25:7–19.

Gabbott, P.A., 1975. Storage cycles in marine bivalve molluscs: a hypothesis concerning the relationship between glycogen metabolism and gametogenesis. Proc. 9th Europ. Mar. Biol. Symp. pp. 191–211.

Gabbott, P.A., 1983. Developmental and seasonal metabolic activities in marine molluscs. In: P.W. Hochachka (Ed.). The Mollusca, 2, Environmental Biochemistry and Physiology. Academic Press, New York. pp. 165–219.

Gabbott, P.A. and Bayne, B.L., 1973. Biochemical effects of temperature and nutritive stress on *Mytilus edulis* L. J. Mar. Biol. Assoc. U.K. 53:269–286.

Gäde, G., 1980. Biological role of octopine formation in marine mollusks. Mar. Biol. Lett. 1:121–135.

Gainey Jr, L.G. and Shumway, S.E., 1988a. Physiological effects of *Protogonyaulax tamarensis* on cardiac activity in bivalve molluscs. Comp. Biochem. Physiol. C 91:159–164.

Gainey Jr, L.F. and Shumway, S.E. 1988b. A compendium of the responses of bivalve molluscs to toxic dinoflagellates. J. Shellfish Res. 7:623–628.

Gainey Jr, L.F. and Shumway, S.E. 1991. The physiological effect of *Aureococcus anophagefferens* ("brown tide") on the lateral cilia of bivalve mollusks. Biol. Bull. 181:298–306.

Gallager, S.M., Stoecker, D.K. and Bricelj, V.M., 1989. Effects of the brown tide alga on growth, feeding physiology and locomotory behavior of scallop larvae. In: E.M. Cosper, E. Carpenter and V.M. Bricelj (Eds.). A Novel Phytoplankton Bloom. Causes and Impacts of Recurrent Brown Tides and Other Unusual Blooms. Lecture Notes on Coastal and Estuarine Studies. Springer-Verlag, New York. pp. 511–542.

García-Esquivel, Z. and Bricelj, V.M., 1993. Ontogenic changes in microhabitat distribution of juvenile bay scallops, *Argopecten irradians irradians* (L.), in eelgrass beds, and their potential significance to early recruitment. Biol. Bull. 185:42–55.

García-Esquivel, Z., Parés-Sierra, G. and García-Pámanes, L., 2000. Effect of flow speed and food concentration on the growth of juvenile scallops *Nodipecten subnodosus*. Ciencias Marinas 26(4):621–641.

Gaudet, M., (1994) Intermediate culture strategies for sea scallop (*Placopecten magellanicus*) spat in Magdalen Islands, Quebec. Bull. Aquacul. Assoc. Canada 94–2:22–28.

Grant, J. and Cranford, P., 1989. The effect of laboratory diet conditioning on tissue and gonad growth in the sea scallop *Placopecten magellanicus*. In: J.F. Ryland and P.A. Tyler (Eds.). Reproduction, Genetics and Distributions of Marine Organisms. 23rd Eur. Mar. Biol. Symp., Olsen and Olsen, Denmark. pp. 95–105.

Grant, J. and Cranford, P.J., 1991. Carbon and nitrogen scope for growth as a function of diet in the sea scallop *Placopecten magellanicus*. J. Mar. Biol. Assoc. U.K. 71:437–450.

Grant, J., Cranford, P. and Emerson, C., 1997. Sediment resuspension rates, organic matter quality and food utilization by sea scallops (*Placopecten magellanicus*) on Georges Bank. J. Mar. Res. 55:965–994.

Grant, J., Emerson, C.W. and Shumway, S.E., 1992. Orientation, passive transport, and sediment erosion features of the sea scallop *Placopecten magellanicus* in the benthic boundary layer. Can. J. Zool. 71:953–959.

Grecian, L.A., G.J. Parsons, P. Dabinett and C. Couturier. 2000. Influence of season, initial size, depth, gear type and stocking density on the growth rates and recovery of sea scallop, *Placopecten magellanicus*, on a farm-based nursery. Aquaculture International 8:183–206.

Grieshaber, M. and Gade, G., 1977. Energy supply and the formation of octopine in the adductor muscle of the scallop *Pecten jacobaeus*. Comp. Biochem. Physiol., B 58:249–252.

Griffiths, R.J., 1980. Filtration, respiration and assimilation in the black mussel *Choromytilus meridionalis*. Mar. Ecol. Prog. Ser. 3:63–70.

Gruffydd, L.I.D., 1974. The influence of certain environmental factors on the maximum length of the scallop, *Pecten maximus* L. J. Cons. Int. Explor. Mer 35(3):300–302.

Gutsell, J.S., 1930. Natural history of the bay scallop. U.S. Bur. Fish. Bull. 46:569–632.

Hartnoll, R.G., 1967. An investigation of the movement of the scallop, *Pecten maximus*. Helgol. Wiss. Meeresunters 15:523–533.

Hawkins, A.J.S., Smith, R.F.M., Bayne, B.L. and Héral, M., 1996. Novel observations underlying the fast growth of suspension-feeding shellfish in turbid environments: *Mytilus edulis*. Mar. Ecol. Prog. Ser. 131:179–190.

Hawkins, A.J.S., Bayne, B.L., Bougrier, S., Heral, M., Iglesias, J.I.P., Navarro, E., Smith R.F.M. and Urrutia, M.B., 1998. Some general relationships in comparing the feeding physiology of suspension-feeding bivalve molluscs. J. Exp. Mar. Biol. Ecol. 219:87–103.

Hawkins, A.J.S., J.G. Fang, P.L. Pascoe, Z.H. Zhang, X.L. Zhang, and M.Y. Zhu, 2001. Modelling short-term responsive adjustments in particle clearance rate among bivalve suspension-feeders: separate unimodal effects of seston volume and composition in the scallop *Chlamys farreri*. J. Exp. Mar. Biol. Ecol. 262:61–73.

Haynes, E.B. and Hitz, C.R., 1971. Age and growth of the giant Pacific sea scallop, *Patinopecten caurinus*, from the Strait of Georgia and outer Washington coast. J. Fish. Res. Bd. Canada 28:1335–1341.

Heilmayer, O., Brey, T., Chiantore, M., Cattaneo-Vietti, R. and Arntz, W.E., 2003. Age and productivity of the Antarctic scallop, *Adamussium colbecki*, in Terra Nova Bay (Ross Sea, Antarctica). J. Exp. Mar. Biol. Ecol. 288:239–256.

Hemmingsen, A.M., 1960. Energy metabolism as related to body size and respiratory surfaces and its evolution. Rep. Steno. Mem. Hosp. Copenh. 9:7–110.

Hilbish, T.J., 1987. Response of aquatic and aerial metabolic rates in the ribbed mussel *Geukensia demissa* (Dillwyn) to acute and prolonged changes in temperature. J. Exp. Mar. Biol. Ecol. 105:207–218.

Hilbish, T.J. and Koehn, R.K., 1985. The physiological basis of natural selection at the LAP locus. Evolution 39(6):1302–1317.

Hirasawa, Y., 1972. The rapid increase in the scallop fishery of Mutsu Bay and the future management of the fishing ground. Gyogo Keizai Kenkyu 19:1–36.

Hochachka, P.W. and Somero, G.N., 1984. Limiting oxygen availability. In: Biochemical Adaptation. Princeton University Press, Princeton, New Jersey. pp. 145–181.

Iglesias, J.I.P., Navarro, E., Alvarev, P. and Armentia, I., 1992. Feeding, particle selection and absorption in cockles *Cerastoderma edule* (L.) Exposed to variable conditions of food concentration and quality. J. Exp. Mar. Biol. Ecol. 162:177–198.

Jackson, A.E., deFreitas, A.S.W., Hooper, L., Mallet, A. and Walter, J.A., 1994. Phosphorus metabolism monitored by ^{31}P NMR in juvenile sea scallop (*Placopecten magellanicus*) overwintering in pearl nets at a Nova Scotian aquaculture site. Can. J. Fish. Aquat. Sci. 51:2105–2114.

Jørgensen, C.B., 1960. Efficiency of particle retention and rate of water transport in undisturbed lamellibranches. J. Cons. Perm. Int. Explor. Mer. 26:94–116.

Jørgensen, C.B., 1975. Comparative physiology of suspension feeding. Ann. Rev. Physiol. 37:57–79.

Jørgensen, C.B., 1981. A hydromechanical principle for particle retention in *Mytilus edulis* and other ciliary suspension feeders. Mar. Biol. 61:277–282.

Jørgensen, C.B., 1983. Fluid mechanical aspects of suspension feeding. Mar. Ecol. Prog. Ser. 11:89–103.

Jørgensen, C.B., 1990. Bivalve Filter-feeding: Hydrodynamics, Bioenergetics, Physiology and Ecology. Olsen and Olsen.

Jørgensen, C.B., 1996. Bivalve filter feeding revisited. Mar. Ecol. Prog. Ser. 142:287–302.

Jørgensen, C.B., Kiørboe, T., Møhlenberg, F. and Riisgård, H.U., 1984. Ciliary and mucus-net filter feeding, with special reference to fluid mechanical characteristics. Mar. Ecol. Prog. Ser. 15:283–292.

Kalashnikov, V.Z., 1991. Soviet Union. In: S.E. Shumway (Ed.). Scallops: Biology, Ecology and Aquaculture. Elsevier, Amsterdam. pp. 1057–1082.

Kingzett, B.C., 1993. Ontogeny of suspension feeding in post-metamorphic Japanese scallops, *Patinopecten yessoensis* (Jay). M.S. Simon Fraser University, British Columbia, Canada. 117 p.

Kiørboe, T. and Møhlenberg, F., 1981. Particle selection in suspension-feeding bivalves. Mar. Ecol. Prog. Ser. 5:291–296.

Kiørboe, T., Møhlenberg, F. and Nøhr, O., 1980. Feeding, particle selection and carbon absorption in *Mytilus edulis* in different mixtures of algae and resuspended bottom material. Ophelia 19:193–205.

Kiørboe, T., Møhlenberg, F. and Nøhr, O., 1981. Effect of suspended bottom material on growth and energetics in *Mytilus edulis*. Mar. Biol. 61:283–288.

Kirby-Smith, W.W., 1970. Growth of the scallops, *Argopecten irradians concentricus* (Say) and *Argopecten gibbus* (Linne), as influenced by food and temperature. Ph.D. Thesis, Duke University, Durham, North Carolina. 126 p.

Kirby-Smith, W.W., 1972. Growth of the bay scallop: the influence of experimental water currents. J. Exp. Mar. Biol. Ecol. 8:7–18.

Kirby-Smith, W.W. and Barber, T.R., 1974. Suspension-feeding aquaculture systems: effects of phytoplankton concentration and temperature on growth of the bay scallop. Aquaculture 3:135–145.

Kleinman, S., Hatcher, B., Scheibling, R., Taylor, L. and Hennigar, A., 1996. Shell and tissue growth of juvenile sea scallops (*Placopecten magellanicus*) in suspended and bottom culture in Lunenburg Bay, Nova Scotia. Aquaculture 142:75–97.

Korol, B.L., 1985. The effect of suspended sediment on the feeding and growth of the bay scallop, *Argopecten irradians*. M.S. thesis. State University of New York at Stony Brook, New York. 66 p.

Kreeger, D.A., 1993. Seasonal patterns in the utilization of dietary protein by the mussel *Mytilus trossulus*. Mar. Ecol. Prog. Ser. 95:215–232.

Kuenstner, S.H., 1988. The effects of the "Brown Tide" alga on the feeding physiology of *Argopecten irradians* and *Mytilus edulis*. M.S. thesis, State University of New York at Stony Brook. 84 p.

LaBarbera, M., 1981. Water flow patterns in and around three species of articulate brachiopods. J. Exp. Mar. Biol. Ecol. 55:185–206.

Lafrance, M.L., Guderly, H. and Cliche, G., 2002. Low temperature, but not air exposure slows the recuperation of juvenile scallops, *Placopecten magellanicus*, from exhausting escape responses. J. Shellfish Res. 21:605–618.

Landsberg, J.H. 2002. The effects of harmful algal blooms on aquatic organisms. Rev. Fish. Sci. 10(2):113–390.

Langdon, C.J. and Waldock, M.J., 1981. The effect of algal and artificial diets on the growth and fatty acid composition of *Crassostrea gigas* spat. J. Mar. Biol. Assoc. U.K. 61:431–448.

Lassus, P. and Berthome, J.P. 1988. Status of the 1987 algal blooms in IFREMER. ICES/annex III C.M. 1988/F; 33A:5–13.

Leighton, D.L., 1979. A growth profile for the rock scallop *Hinnites multirugosus* held at several depths off La Jolla, California. Mar. Biol. 51:229–232.

Le Pennec, M. and Rangel-Dávalos, C.D., 1985. Observations en microscopie à épifluorescence de l'ingestion et de la digestion d'algues unicellulaires chez des jeunes larves de *Pecten maximus* (Pectinidae, Bivalvia). Aquaculture 47:39–51.

Lesser, M.J. and Shumway, S.E., 1993. Effects of toxic dinoflagellates on clearance rates and survival in juvenile bivalve molluscs. J. Shellfish Res. 12:377–381.

Leverone, J.R. and Blake, N.J., 2002. Effects of the toxic dinoflagellate, *Karenia brevis*, on larval mortality and juvenile feeding behavior in the bay scallop, *Argopecten irradians*. J. Shellfish. Res. 21(1):396–397 (abstract).

Livingstone, D.R., 1991. Origins and evolution of pathways of anaerobic metabolism in the animal kingdom. Amer. Zool. 31:522–534.

Lodeiros, C., Rengel, J., Freites, L., Morales, F. and Himmelman, J., 1998. Growth and survival of the tropical scallop *Lyropecten* (*Nodipecten*) *nodosus* maintained in suspended culture at three depths. Aquaculture 165:41–50.

Lopez, G.R., and Levinton, J.S., 1987. Ecology of deposit-feeding animals in marine sediments. Q. Rev. Biol. 62:235–260.

Lorrain, A., Paulet, Y-M., Chauvaud, L., Savoye, N., Nezan, E. and Guerin, L., 2000. Growth anomalies in *Pecten maximus* from coastal waters (Bay of Brest, France): relationship with diatom blooms. J. Mar. Biol. Assoc. U.K. 80(4):667–673.

Lu, Y.T. and Blake, N.J., 1996. Optimum concentrations of *Isochrysis galbana* for growth of larval and juvenile bay scallops, *Argopecten irradians concentricus* (Say). J. Shellfish Res. 15(3):635–643.

Lu, Y.T. and Blake, N.J., 1997. Clearance and ingestion rates of *Isochrysis galbana* by larval and juvenile bay scallops, *Argopecten irradians concentricus* (Say). J. Shellfish Res. 16(1):47–54.

Lu, Y.T. and Blake, N.J. 1999. Oxygen consumption and ammonia excretion of larvae and juveniles of the bay scallop, *Argopecten irradians concentricus* (Say). J. Shellfish Res. 18(2):419–423.

Lyons, M.M. and Ward, J.E., 2003. Suspension-feeding bivalves, marine aggregates and the accessibility of small particles. J. Shellfish Res. 22(1):342. (abstract).

MacDonald, B.A., 1986. Production and resource partitioning in the giant scallop *Placopecten magellanicus* grown on the bottom and in suspended culture. Mar. Ecol. Prog. Ser. 34:79–86.

MacDonald, B.A. and Thompson, R.J., 1985. Influence of temperature and food availability on the ecological energetics of the giant scallop *Placopecten magellanicus*. I. Growth rates of shell and somatic tissue. Mar. Ecol. Prog. Ser. 25:279–294.

MacDonald, B.A. and Thompson, R.J., 1986. Influence of temperature and food availability on the ecological energetics of the giant scallop *Placopecten magellanicus*. III. Physiological ecology, the gametogenic cycle and scope for growth. Mar. Biol. 93:37–48.

MacDonald, B.A. and Bourne, N.F., 1987. Growth, reproductive output, and energy partitioning in weathervane scallops, *Patinopecten caurinus*, from British Columbia. Can. J. Fish. Aquat. Sci. 44:152–160.

484

MacDonald, B.A. and Bourne, N.F., 1989. Growth of the purple-hinge rock scallop, *Crassadoma gigantea* Gray, 1825 under natural conditions and those associated with suspended culture. J. Shellfish Res. 8:179–186.

MacDonald, B.A. and Bayne, B.L., 1993. Food availability and resource allocation in senescent *Placopecten magellanicus*: evidence from field populations. Func. Ecol. 7:40–46.

MacDonald, B.A. and Ward, J.E., 1994. Variation in food quality and particle selectivity in the sea scallop *Placopecten magellanicus* (Mollusca: Bivalvia). Mar. Ecol. Prog. Ser. 108:251–264.

MacDonald, B.A., Ward, J.E. and McKenzie, C.H., 1995. Exfoliation of epithelial cells from the pallial organs of the sea scallop, *Placopecten magellanicus*. J. Exp. Mar. Biol. Ecol. 191:151–165.

MacDonald, B.A., Bacon, G.S. and Ward, J.E., 1998. Physiological responses of infaunal (*Mya arenaria*) and epifaunal (*Placopecten magellanicus*) bivalves to variations in the concentration and quality of suspended particles II. Absorption efficiency and scope for growth. J. Exp. Mar. Biol. Ecol. 219:127–141.

Mackay, J. and Shumway, S.E., 1980. Factors affecting oxygen consumption in the scallop *Chlamys delicatula* (Hutton). Ophelia 19(1):19–26.

Maeda-Martínez, A.N., Sicard, M.T. and Reynoso-Granados, T., 2000. A shipment method for scallop seed. J. Shellfish Res. 19(2):765–770.

Malouf, R.E. and Bricelj, V.M., 1989. Comparative biology of clams: environmental tolerance, feeding and growth. In: J. Manzi and M. Castagna (Eds.). Clam Mariculture in North America. Elsevier, New York. pp. 23–73.

Mangum, C. and Van Winkle, W., 1973. Responses of aquatic invertebrates to declining oxygen conditions. Amer. Zool. 13:529–541.

Marshall, N., 1960. Studies of the Niantic River, Connecticut with special reference to the bay scallop, *Aequipecten irradians*. Limnol. Oceanogr. 5:86–105.

Maru, H. and Obara, A., 1967. Studies on the ecology of the scallop *Patinopecten yessoensis* (Jay). II. On the seasonal variation on the fatness of the soft body. Hokkaidoritsu Suisan Shikenjo Hokoku 15:23–32.

Mason, J., 1957. The age and growth of the scallop, *Pecten maximus* (L.) in Manx waters. J. Mar. Biol. Assoc. U.K. 36:473–492.

Mathers, N.F., 1976. The effects of tidal currents on the rhythm of feeding and digestion in *Pecten maximus* L. J. Exp. Mar. Biol. Ecol. 24:271–283.

Matsuyama, Y., 1999. Harmful effect of dinoflagellate *Heterocapsa circularisquama* on shellfish aquaculture in Japan. Jap. Aquacult. Res. Quart. 22:283–293.

Matsuyama, Y., Uchida, T., Honjo, T. and Shumway, S.E., 2001. Impacts of the harmful dinoflagellate, *Heterocapsa circularisquama*, on shellfish aquaculture in Japan. J. Shellfish Res. 20:1269–1272.

Matsuyama, Y., Nagai, K., Mizuguchi, T., Fujiwara, M., Ishimura, M., Yamaguchi, M., Uchida, T. and Honjo, T., 1995 Ecological features and mass mortality of pearl oysters during the red tide of *Heterocapsa* sp. in Ago Bay in 1992. Nippon Suisan Gakkaishi 61:35–41 (In Japanese).

Matsuyama, Y., Uchida, T., Nagai, K., Ishimura, M., Nishimura, A., Yamaguchi, M. and Honjo, T., 1996. Biological and environmental aspects of noxious dinoflagellate red tides by *Heterocapsa circularisquama* in the west Japan. In: T. Yasumoto, Y. Oshima and Y. Fukuyo, (Eds.). Harmful and Toxic Algal Blooms. IOC of UNESCO, Paris. pp 247–250.

McHenery, J.G. and Birbeck, T.H., 1985. Uptake and processing of cultured microorganisms by bivalves. J. Exp. Mar. Biol. Ecol. 90:145–163.

McLusky, D.S., 1973. The effect of temperature on the oxygen consumption and filtration rate of *Chlamys (Aequipecten) opercularis* (L.) (Bivalvia). Ophelia 10:114–154.

Mercaldo, R.S. and Rhodes, E.W., 1982. Influence of reduced salinity on the Atlantic bay scallop *Argopecten irradians* (Lamarck) at various temperatures. J. Shellfish Res. 2(2):177–182.

Meyhöfer, E., 1985. Comparative pumping rates in suspension feeding bivalves. Mar. Biol. 85:137–142.

Mikulich, L.V. and Tsikhon-Lukanina, A., 1981. Food of the scallop. Oceanology 21(5):633–635.

Møhlenberg, F. and Riisgård, H.U., 1978. Efficiency of particle retention in 13 species of suspension feeding bivalves. Ophelia 17(2):239–246.

Møhlenberg, F. and Riisgård, H.U., 1979. Filtration rate, using a new indirect technique, in thirteen species of suspension-feeding bivalves. Mar. Biol. 54:143–147.

Møhlenberg, F. and Kiørboe, T., 1981. Growth and energetics in *Spisula subtruncata* (Da Costa) and the effect of resuspended bottom material. Ophelia 20(1):79–90.

Monical, J.B. Jr., 1980. Comparative studies on growth of the purple-hinge rock scallop *Hinnites multirugosus* (Gale) in the marine waters of southern California. Proc. Nat. Shellfish. Ass. 70(1):14–21.

Morse, P. 1982. Effects of sublethal concentrations of the drilling mud components attapulgite and Q-broxin on the structure of the gill of the scallop, *Placopecten magellanicus*. In: W.B. Vernberg, A. Calabrese, F.P. Thurberg and F.J Vernberg (Eds.). Physiological Mechanisms of Marine Pollutant Toxicity. Academic Press, New York. pp. 235–259.

Navarro, E. and Iglesias, J.I.P., 1993. Infaunal filter-feeding bivalves and the physiological response to short-term fluctuations in food availability and composition. In: R.F. Dame (Ed.). Bivalve Filter Feeders in Estuarine and Coastal Ecosystem Processes. NATO ASI Series, Vol. G33, Springer-Verlag, Berlin. pp. 25–56.

Navarro, E., Iglesias, J.I.P. and Ortega, M.M., 1992. Natural sediment as a food source for the cockle *Cerastoderma edule* (L.): Effect of variable particle concentration on feeding, digestion and scope for growth. J. Exp. Mar. Biol. Ecol. 156:69–87.

Navarro, E., Iglesias, J.I.P., Ortega, M.M. and Larretxea, X., 1994. The basis for a functional response to variable food quantity and quality in cockles *Cerastoderma edule* (Bivalvia Cardiidae). Physiol. Zool. 67:468–496.

Navarro, J.M., and González, C.M., 1998. Physiological responses of the Chilean scallop *Argopecten purpuratus* to decreasing salinities. Aquaculture 167:315–327.

Navarro, J.M., Leiva, G.E., Martinez, G. and Aguilera, A., 2000. Interactive effects of diet and temperature on the scope for growth of the scallop *Argopecten purpuratus* during reproductive conditioning. J. Exp. Mar. Biol. Ecol. 247:67–83.

Newell, R.C. and Kofoed, L.H., 1977. Adjustment of the components of energy balance in the gastropod *Crepidula fornicata* in response to thermal acclimation. Mar. Biol. 44:275–286.

Newell, R.I.E. and Bayne, B.L., 1980. Seasonal changes in the physiology, reproductive condition and carbohydrate content of the cockle *Cardium* (= *Cerastoderma*) *edule* (Bivalvia: Cardiidae). Mar. Biol. 56:11–19.

Newell, R.I.E. and Jordan, S.J., 1983. Preferential ingestion of organic material by the American oyster *Crassostrea virginica*. Mar. Ecol. Prog. Ser. 13:47–53.

Newell, R.I.E., Johnson, L.G. and Kofoed, L.H., 1977. Adjustment of energy balance in response to temperature change in *Ostrea edulis*. Oecologia 30:97–110.

Orensanz, J.M., 1984. Size, environment, and density: the regulation of a scallop stock and its management implications. In: G.S. Jamieson and N. Bourne (Eds.). North Pacific Workshop on Stock Assessment and Management of Invertebrates. Can. Sp. Publ. Fish. Aquat. Sci. 92:195–227.

Orensanz, J.M., Parma, A.M. and Iribarne, O.O., 1991. Population dynamics and management of natural stocks. In: S.E. Shumway (Ed.). Scallops: Biology, Ecology and Aquaculture. Elsevier, Amsterdam. pp. 625–714.

Owen, G. and McCrae, J.M., 1976. Further studies on the laterofrontal tracts of bivalves. R. Soc. Lond. B. 194:527–544.

Palmer, R.E., 1980. Behavioral and rhythmic aspects of filtration in the bay scallop, *Argopecten irradians concentricus* (Say), and the oyster, *Crassostrea virginica* (Gmelin). J. Exp. Mar. Biol. Ecol. 45:273–295.

Palmer, R.E. and Williams L.G., 1980. Effect of particle concentration on filtration efficiency of the bay scallop *Argopecten irradians* and the oyster *Crassostrea virginica*. Ophelia 19(2):163–174.

Pamatmat, M.M., 1980. Facultative anerobiosis of benthos. In: K.R. Tenore and B.C. Coul (Eds.). Marine Benthic Dynamics. Belle W. Baruch Library in Marine Science, 11, Univ. of South Carolina Press, Columbia, South Carolina. pp. 69–90.

Parrish, C.C., McKenzie, C.H., MacDonald, B.A. and Hatfield, E.A., 1995. Seasonal studies of seston lipids in relation to microplankton species composition and scallop growth in South Broad Cove, Newfoundland. Mar. Ecol. Prog. Ser. 129:151–164.

Parry, G.D., Langdon, J.S. and Huisman, J.M., 1989. Toxic effects of a bloom of the diatom *Rhizosolenia chunii* on shellfish in Port Phillip Bay, southeastern Australia. Mar. Biol. 102:25–41.

Parsons, G. J. and Dadswell M. J., 1992. Effect of stocking density on growth, production, and survival of the giant scallop, *Placopecten magellanicus*, held in intermediate suspension culture in Passamaquoddy Bay, New Brunswick. Aquaculture 103:291–309.

Peirson, W.M., 1983. Utilization of eight algal species by the bay scallop, *Argopecten irradians concentricus* (Say). J. Exp. Mar. Biol. Ecol. 68:1–11.

Pérez, J.E., Nusetti, O., Ramirez, N. and Alfonsi, C., 2000. Allozyme and biochemical variation at the octopine dehydrogenade locus in the scallop *Euvola ziczac*. J. Shellfish. Res. 19(1):85–88.

Peterson, C.H., 1983. A concept of quantitative reproductive senility: application to the hard clam, *Mercenaria mercenaria* (L.). Oecologia (Berlin) 58:164–168.

Pico, E.F.F., 1991. Mexico. In: S.E. Shumway (Ed.). Scallops: Biology, Ecology and Aquaculture. Elsevier, Amsterdam. pp. 943–980.

Pilditch, C.A. and Grant, J., 1999a. Effects of variations in flow velocity and phytoplankton concentration on sea scallop (*Placopecten magellanicus*) grazing rates. J. Exp. Mar. Biol. Ecol. 240:111–136.

Pilditch, C.A. and Grant, J., 1999b. Effects of temperature fluctuations and food supply on the growth and metabolism of juvenile sea scallops (*Placopecten magellanicus*). Mar. Biol. 134:235–248.

Potter, T.M., MacDonald, B.A. and Ward, J.E., 1997. Exfoliation of epithelial cells by the scallop *Placopecten magellanicus:* seasonal variation and the effects of elevated water temperatures. Mar. Biol. 127:463–472.

Powers, S.P. and Peterson, C.H., 2000. Conditional density dependence: The flow trigger to expression of density-dependent emigration in bay scallops. Limnol. Oceanogr. 45:727–732.

Pranovi, R., Marcato, S. and Zanellato, R., 1994. Analisi biometriche e biologia di popolazione del mollusco anatartico *Adamussium colbecki* a Baia terra Nova, Mare di Ross, Atti Dell 'Istituto Veneto Di Scienze Lettere Ed Arti Tomo CLII (1993–1994) 152:123–136.

Ralph, R., Maxwell, J.G.H., 1977. Growth of two Antarctic lamellibrancsh: *Adamussium colbecki* and *Laternula elliptica*. Mar. Biol. 42:171–175.

Richardson, C.A., Gale, G.F. and Venn, T.J., 1984. The effect of iron ore suspensions and food on the shell growth rates and tissue weights of queen scallops *Chlamys opercularis* held in the laboratory. Mar. Environ. Res. 13:1–31.

Riisgård, H.U., 1988. Efficiency of particle retention and filtration rate in 6 species of northeast American bivalves. Mar. Ecol. Prog. Ser. 45:217–223.

Riisgård, H.U., 1998. No foundation of "a ¾ power scaling law" for respiration biology. Ecology Letters 1:71–73.

Riisgård, H.U., Larsen P.S. and Nielsen, N.F., 1996. Particle capture in the mussel, *Mytilus edulis*: the role of latero-frontal cirri. Mar. Biol. 127:259–266.

Rubenstein, D.L. and Koehl, M.A.R., 1977. The mechanisms of filter feeding: some theoretical considerations. Amer. Nat. 111:981–994.

Rupp, G.S., Parsons, G.J., Thompson, R.T., and de Bem, M.M., 2004. Effect of depth and stocking density on growth and retrieval of the postlarval lion's paw scallop, *Nodipecten nodosus* (Linnaeus, 1758). J. Shellfish Res. 23(2):473–482.

Schick, D.F., Shumway, S.E. and Hunter, M.A., 1988. A comparison of growth rate between shallow water and deep water populations of scallops, *Placopecten magellanicus* (Gmelin, 1971), in the Gulf of Maine. Am. Malac. Bull. 6:1–8.

Schick, J.M., Zwaan de, A. and de Bont, A.M.T., 1983. Anoxic metabolic rate in the mussel *Mytilus edulis* L. estimated by simultaneous direct calorimetry and biochemical analysis Physiol. Zool. 56(1):56–63.

Sebens, K.P., 1982. The limits to indeterminate growth: an optimal size model applied to passive suspension feeders. Ecology 63(1):209–222.

Shafee, M.S., 1982. Variations saisonnières de la consommation d'oxygène chez le pétoncle noir *Chlamys varia* (L.). de Lanvéoc (rade de Brest). Oceanol. Acta 5(2):189–197.

Shimeta, J. and Jumars, P.A., 1991. Physical mechanisms and rates of particle capture by suspension-feeders. Oceanogr. Mar. Biol. Annu. Rev. 29:191–257.

Shumway, S.E., 1983. Factors affecting oxygen consumption in the coot clam, *Mulinia lateralis* (Say). Ophelia 22:143–171.

Shumway, S.E., 1990. A review on the effects of algal blooms on shellfish and aquaculture. J. World. Aquat. Soc. 21:65–105.

Shumway, S.E. and Koehn, R.K., 1982. Oxygen consumption in the American oyster *Crassostrea virginica*. Mar. Ecol. Prog. Ser. 9:59–68.

Shumway, S.E. and Cucci, T.L., 1987. The effects of the toxic dinoflagellate *Protogonyaulax tamarensis* on the feeding and behaviour of bivalve mollusks. Aquat. Toxicol. 10:9–27.

Shumway, S.E., and Cembella, A.D., 1993. The impact of toxic algal on scallop culture and fisheries. Rev. in Fisheries Sci. 1(2):121–150.

Shumway, S.E., Cucci, T.L., Newell, R.C. and Yentsch, C.M., 1985. Particle selection, ingestion and absorption in filter feeding bivalves. J. Exp. Mar. Biol. Ecol. 91:77–92.

Shumway, S.E., Selvin, R., Schick, D.F., 1987. Food resources related to habitat in the scallop *Placopecten magellanicus* (Gmelin, 1791): a qualitative study. J. Shellfish Res. 6:89–95.

Shumway, S.E., Barter, J. and Stahlnecker, J., 1988. Seasonal changes in oxygen consumption of the giant scallop, *Placopecten magellanicus* (Gmelin). J. Shellfish Res. 7:77–82.

Shumway, S.E., Cucci, T.L., Lesser, M.P., Bourne, N. and Bunting, B., 1997. Particle clearance and selection in three species of juvenile scallops. Aquaculture International. 5:89–99.

Sicard, M.T., Maeda-Martinez, A.N., Ormart, P., Reynoso-Granados, T. and Carvalho L., 1999. Optimum temperature for growth of the catarina scallop (*Argopecten ventricosus-circularis*, Sowerby II, 1842). J. Shellfish. Res. 18(2):385–392.

Silva Loera, H.A., 1986. Efecto del tamaño corporal, tensión de oxígeno y temperatura sobre la tasa de consumo de oxígeno en la escalopa *Argopecten circularis* Sowerby) (Molusca: Lamellibranchia). M.S. Thesis, Instituto Tecnológico y de Estudios Superiores de Monterrey, Mexico. 93 p.

Silverman, H., Lynn, J.W. and Dietz, T.H., 1996. Particle capture by the gills of *Dreissena polymorpha, Corbicula fluminea*, and *Carunculina texasensis*. Biol. Bull. 189:308–319.

Silverman, H., Nichols, S.J., Cherry, J.S., Achberger, E., Lynn, J.W. and Dietz, T.H., 1997. Clearance of laboratory-cultured bacteria by freshwater bivalves: differences between lentic and lotic unionids. Can. J. Zool. 75(11):1857–1866.

Silverman, H., Lynn, J.W., Beninger, P.G. and Dietz, T.H., 1999. The role of latero-frontal cirri in particle capture by the gills of *Mytilus edulis*. Biol. Bull. 197:368–376.

Silverman, H., Lynn, J.W. and Dietz, T.H., 2000. *In vitro* studies of particle capture and transport in suspension-feeding bivalves. Limnol. Oceanogr. 45(5):1199–1203.

Smolowitz, R. and Shumway, S.E., 1997. Possible cytotoxic effects of the dinoflagellate, *Gyrodinium aureolum*, on juvenile bivalve molluscs. Aquaculture International 5:291–300.

Spärck, R., 1936. On the relation between metabolism and temperature in some marine lamellibranchs, and its zoogeographical significance. K. Danske Vidensk, Selfk. Skr., Biol. Med. 13:1–27.

Springer, J., Shumway, S.E. and Burkholder, J., 2000. Behavioral variability of the toxic dinoflagellate *Pfiesteria piscicida*, when introduced to larval and adult shellfish. J. Shellfish Res. 19:637–638.

Springer, J., Shumway, S.E., and Burkholder, J., 2002. Interactions between the toxic estuarine dinoflagellate, *Pfiesteria piscicida,* and two species of commercially important bivalve molluscs. Mar. Ecol. Prog. Ser. 245:1–10

Stockton, W.L., 1984. The biology and ecology of the epifaunal scallop *Adamussium colbecki* on the west side of McMurdo Sound, Antarctica. Mar. Biol. 78:171–178.

Stohler, R. 1959. The red tide of 1958 at Ensenada, Baja California, Mexico. The Veliger 2:32–35.

Stuart, V., 1982. Absorbed ration, respiratory costs and resultant scope for growth in the mussel *Aulacomya ater* (Molina) fed on a diet of kelp detritus of different ages. Mar. Biol. Lett. 3:289–306.

Summerson, H.C. and Peterson, C.H. 1990. Recruitment failure of the bay scallop *Argopecten irradians concentricus*, during the first red tide, *Ptychodiscus brevis*, outbreak recorded in North Carolina. Estuaries 13:322–331.

Sundet, J.H. and Vahl, O., 1981. Seasonal changes in dry weight and biochemical composition of the tissues of sexually mature and immature Iceland scallops, *Chlamys islandica*. J. Mar. Biol. Assoc. U.K. 61:1001–1010.

Taylor, A.C. and Brand, A.R., 1975. Effects of hypoxia and body size on the oxygen consumption of the bivalve *Arctica islandica* (L.). J. Exp. Mar. Biol. Ecol. 19:187–196.

Tettelbach, S.T. and Wenczel, P., 1993. Reseeding efforts and the status of bay scallop *Argopecten irradians* (Lamarck, 1819) populations in New York following the occurrence of "brown tide" algal blooms. J. Shellfish Res. 12:423–431.

Tettelbach, S.T., Auster, P.J., Rhodes, E.W. and Widman, J.C., 1985. A mass mortality of northern bay scallops, *Argopecten irradians irradians*, following a severe spring rainstorm. The Veliger 27(4):381–385.

Tettelbach, S.T., Smith, C.F., Kaldy III, J.E., Arroll, T.W. and Denson, M.R., 1990. Burial of transplanted bay scallops *Argopecten irradians irradians* (Lamarck, 1819) in winter. J. Shellfish Res. 9(1):127–134.

Thompson, R.J. and MacDonald, B.A., 1991. Physiological integrations and energy partitioning. In: S.E. Shumway, (Ed.). Scallops: Biology, Ecology and Aquaculture. Elsevier, Amsterdam. pp. 347–376.

Thompson, R.J., Livingstone, D.R. and Zwaan de, A., 1980. Physiological and biochemical aspects of the valve snap and valve closure responses in the giant scallop, *Placopecten magellanicus*. I. Physiology. J. Comp. Physiol. 137:97–104.

Tracey, G.A., 1988. Feeding reduction, reproductive failure, and mortality in *Mytilus edulis* during the 1985 "brown tide" in Narragansett Bay, Rhode Island. Mar. Ecol. Prog. Ser. 50:73–81.

Tunbridge, B.R., 1968. The Tasmanian bay scallop fishery. N.Z. Mar. Dep. Fish. Tech. Rep. 18:1–92.

Uriarte, I. and Farías, A., 1995. Effect of broodstock origin and postlarval diet on postlarval performance of the Chilean scallop *Argopecten purpuratus*. In: P. Lavens and I. Roelants (Eds.). Larvi'95-Fish and Shellfish Larviculture Symposium. European Aquaculture Society, Sp. Publ. 24, Gent, Belgium. pp. 69–72.

Uriarte, I. and Farías, A., 1999. The effect of dietary protein content on growth and biochemical composition of Chilean scallop *Argopecten purpuratus* (L.) postlarvae and spat. Aquaculture 180:119–127.

Urrutia, M.B., Iglesias, J.I.P., Navarro, E. and Prou, J., 1996. Feeding and absorption in *Cerastoderma edule* under environmental conditions in the Bay of Marennes-Oleron (Western France). J. Mar. Biol. Assoc. UK. 76:431–450.

Ursin, E., 1956. Distribution and growth of the queen, *Chlamys opercularis* (Lamellibranchiata) in Danish and Faroese waters. Medd. Dan. Fish.-Havunders. 1:1–31.

Vahl, O., 1972. Particle retention and relation between water transport and oxygen uptake in *Chlamys opercularis* (L.) (Bivalvia). Ophelia 10:67–74.

Vahl, O., 1973a. Efficiency of particle retention in *Chlamys islandica* (O. F. Müller). Astarte 6:21–25.

490

Vahl, O., 1973b. Pumping and oxygen consumption rates of *Mytilus edulis* L. of different sizes. Ophelia 12:45–52.

Vahl, O., 1978. Seasonal changes in oxygen consumption of the Iceland Scallop (*Chlamys islandica* (O.F. Müller)) from 70°N. Ophelia 17(1):143–154.

Vahl, O., 1980. Seasonal variations in seston and the growth rate of the Iceland scallop, *Chlamys islandica* (O.F. Müller) from Balsfjord, 70°N. J. Exp. Mar. Bio. Ecol. 48:195–204.

Vahl, O., 1981. Energy transformations by the Iceland scallop, *Chlamys islandica* (O.F. Müller), from 70°N. I. The age-specific energy budget and net growth efficiency. J. Exp. Mar. Biol. Ecol. 53:182–196.

Vahl, O., 1984. Size-specific reproductive effort in (*Chlamys islandica*): reproductive senility or stabilizing selection? In: P.E. Gibbs (Ed.). Proceedings of the Nineteenth European Marine Biology Symposium. Cambridge University Press, Plymouth/Devon. pp. 521–527.

Vahl, O. and Sundet, J.H., 1985. Is sperm really so cheap? In: J.S. Gray and M.E. Christiansen (Eds.). Marine Biology of Polar Regions and Effects of Stress on Marine Organisms. John Wiley and Sons, New York. pp.281–285.

van Dam, L., 1954. On the respiration in scallops (Lamellibranchiata). Biol. Bull. 107:192–202.

Veniot, A., Bricelj, V.M. and Beninger, P.G., 2003. Ontogenetic changes in gill morphology and potential significance for food acquisition in the scallop *Placopecten magellanicus*. Mar. Biol. 142: 123–131. Erratum in Mar. Biol. 142:827–832 (2003).

Ventilla, R.F., 1982. The scallop industry in Japan. Adv. Mar. Biol. 20:309–382.

Vernet, M., 1977. Alimentación de la vieira tehuelche (*Chlamys tehuelchus*). Comisión Nacional de Estudios Geo-Heliofísicos. Centro Nacional Patagónico, Puerto Madryn, Argentina, unpublished report. 26 p.

Volckaert, F., 1988. The implications of heterozygosity in the scallop *Placopecten magellanicus*. PhD dissertation, Dalhousie University, Halifax, Canada.

Volckaert, F. and Zouros, E., 1989. Allozyme and physiological variation in the scallop *Placopecten magellanicus*, and a general model for the effects of heterozygosity on fitness in marine mollusks. Mar. Biol. 101:1–11.

Wallace, J. C. and T. G. Reinsnes. 1984. Growth variation with age and water depth in the Iceland scallop (*Chlamys islandica*, Pectinidae). Aquaculture 41:141–146.

Wallace, J.C. and Reinsnes, T.G., 1985. The significance of various environmental parameters for growth of the Iceland scallop, *Chlamys islandica* (Pectinidae), in hanging culture. Aquaculture 44:229–242.

Wang, W.-X., and Fisher, N.S., 1996. Assimilation of trace elements and carbon by the mussel *Mytilus edulis*: effects of food composition. Limnol. Oceanogr. 41:197–207.

Ward, J.E., 1996. Biodynamics of suspension-feeding in adult bivalve molluscs: particle capture, processing and fate. Invert. Biol. 115:218–231.

Ward, J.E., and Targett, N.M., 1989. Influence of microalgal metabolites on feeding behavior of the blue mussel *Mytilus edulis*. Mar. Biol. 101:313–321.

Ward, J.E. and Targett, N.M., 1989. Are metabolites from the brown tide alga, *Aureococcus anophagefferens*, deleterious to mussel feeding behavior? In: E. Cosper, V.M. Bricelj and E.J. Carpenter (Eds.). Novel Phytoplankton Blooms: Causes and Impacts of Recurrent Brown Tides and Other Unusual Blooms. Springer Verlag, New York. pp. 543–556.

Ward, J.E., Cassell, H.K. and MacDonald, B.A., 1992. Chemoreception in the sea scallop *Placopecten magellanicus* (Gmelin). I Stimulatory effects of phytoplankton metabolites on clearance and ingestion rates. J. Exp. Mar. Biol. Ecol. 163:235–250.

Ward, J.E., Levinton, J.S., Shumway, S.E. and Cucci, T.L., 1997. Linking individual function to ecosystem control: Identification of the gills as the locus of particle selection in a bivalve mollusk. Nature 390:131–132.

Ward, J.E., Sanford, L.P., Newell, R.I.E. and MacDonald, B.A., 1998a. A new explanation of particle capture in suspension-feeding bivalve molluscs. Limnol. Oceanogr. 43:741–752.

Ward, J.E., Levinton, J.S., Shumway, S.E. and Cucci, T., 1998b. Particle sorting in bivalves: in vivo determination of the pallial organs of selection. Mar. Biol. 131:283–292.

Ward, J.E., MacDonald, B.A., Thompson, R.J. and Beninger, P.G., 1993. Mechanisms of suspension-feeding in bivalves: resolution of current controversies using endoscopy. Limnol. Oceanogr. 38:265–272.

White, M.J., 1997. The effect of flocculation on the size-selective feeding capabilities of the sea scallop *Placopecten magellanicus*. MSc Thesis, Dalhousie University, Halifax, Nova Scotia, Canada.

Widdows, J., Fieth, P. and Worrall, C.M., 1979. Relationships between seston, available food and feeding activity in the common mussel *Mytilus edulis*. Mar. Biol. 50:195–207.

Wikfors, G.H. and Smolowitz, R.M., 1993. Detrimental effects of a *Prorocentrum* isolate upon hard clams and bay scallops in laboratory feeding studies. In: T. Smayda and Y. Shimizu (Eds.). Toxic Phytoplankton Blooms in the Sea. Elsevier/North Holland, New York. pp. 447–452

Wikfors, G.H. and Smolowitz, R.M., 1995. Do differences in dinoflagellate damage depend upon digestion? J. Shellfish. Res. 14(1):282.

Wikfors, G.H., Alix, J.H., Smolowitz, R.M., Wallace, L. and Hégaret, H., 2002. Detrimental effects of a recent *Prymnesium* isolate from Boothbay Harbor, Maine (USA) upon juvenile bay scallops, *Argopecten irradians*. J. Shellfish Res. 21(1):397 (abstract).

Wildish, D.J. and Peer, D., 1983. Tidal current speed and production of benthic macrofauna in the lower Bay of Fundy. Can. J. Fish. Aquat. Sci. 40 (Suppl. 1):309–321.

Wildish, D.J. and Kristmanson, D.D., 1985. Control of suspension feeding bivalve production by current speed. Helgolander Meeresunters. 39:237–243.

Wildish, D.J. and Kristmanson, D.D., 1988. Growth response of giant scallops to periodicity of flow. Mar. Ecol. Prog. Ser. 42:163–169.

Wildish, D.J. and Saulnier, A.M., 1993. Hydrodynamic control of filtration in *Placopecten magellanicus*. J. Exp. Mar. Biol. Ecol. 174:65–82.

Wildish, D.J., Kristmanson, D.D. and Saulnier, A.M., 1992. Interactive effect of velocity and seston concentration on giant scallop feeding inhibition. J. Exp. Mar. Biol. Ecol. 155:161–168.

Wildish, D.J., Kristmanson, D.D., Hoar, R.L., DeCoste, A.M., McCormick, S.D. and White, A.W., 1987. Giant scallop feeding and growth response to flow. J. Exp. Mar. Biol. Ecol. 113:207–220.

Williams, M.J. and Dredge, M.C.L., 1981. Growth of the saucer scallop, *Amusium japonicum balloti* Habe in central eastern Queensland. Aust. Mar. Freshw. Res. 32:657.

Winter, J.E., 1978. A review of the knowledge of suspension feeding in lamellibranchiate bivalves, with special reference to artificial aquaculture systems. Aquaculture 13:1–33.

Winter, M.A. and Hamilton, P.V., 1985. Factors influencing swimming in bay scallops, *Argopecten irradians* (Lamarck, 1819). J. Exp. Mar. Biol. Ecol. 88:227–242.

Wojtowicz, M.B., 1972. Carbohydrases of the digestive gland and the crystalline style of Atlantic deep-sea scallop (*Placopecten magellanicus*, Gmelin). Comp. Biochem. Physiol. 43A:131–141.

Yamamoto, G., 1960. Ecology of the scallop, *Pecten yessoensis* Jay. Sci. Rep. Tohoku Univ. Fourth Ser. (Biol.) 20:11–32.

Yan, T., M. Zhou, M. Fu, Y. Wang and J. Li., 2001. Inhibition of egg hatching success and larvae survival of the scallop, *Chlamys farreri*, associated with exposure to cells and cell fragments of the dinoflagellate *Alexandrium tamarense*. Toxicon. 39(8):1239–1244.

Zandee, D.I., Holwerda, D.A. and Zwaan de, A., 1980. Energy metabolism in bivalves and cephalopods. In: R. Gilles (Ed.). Animals and Environmental Fitness, 1. Pergamon, Oxford. pp. 185–206.

Zeuthen, E., 1953. Oxygen uptake as related to body size in organisms. Quat. Rev. Biol. 28:1–12.

Zhang, X., 1956. Observation on the reproduction and growth of scallop, *Chlamys farreri*, J. Zool. China 8:235–249.

Zwaan de, A. and Zandee, D.I., 1972. Body distribution and seasonal changes in the glycogen content of the common sea mussel *Mytilus edulis*. Comp. Biochem. Physiol. A 43:53–58.

Zwaan de, A., Thompson, R.J. and Livingstone, D.R., 1980. Physiological and biochemical aspects of the valve snap and valve closure responses in the giant scallop *Placopecten magellanicus*. II Biochemistry. J. Comp. Physiol. 137:105–114.

AUTHORS ADDRESSES

Bruce A. MacDonald - Biology Department and Centre for Coastal Studies and Aquaculture, University of New Brunswick, Saint John, New Brunswick, Canada E2L 4L5 (E-mail: bmacdon@unbsj.ca)

V. Monica Bricelj - Institute for Marine Biosciences, National Research Council, Halifax, 1411 Oxford Street, Nova Scotia, Canada B3H 3Z1 (E-mail: Monica.Bricelj@nrc-cnrc.gc.ca)

Sandra E. Shumway - Department of Marine Sciences, University of Connecticut, Groton, Connecticut, USA 06340 (E-mail: Sandra.Shumway@uconn.edu)

Scallops: Biology, Ecology and Aquaculture
S.E. Shumway and G.J. Parsons (Editors)

Chapter 8

Physiological Integrations and Energy Partitioning

Raymond J. Thompson and Bruce A. MacDonald

8.1 INTRODUCTION

In the Pectinidae, the measurement of shell growth rate has received a great deal of attention owing to its importance of fisheries management and aquaculture. There have also been numerous studies on reproductive cycles in several species of scallops to evaluate gametogenic activity and the time of spawning. However, much less emphasis has been placed on physiological energetics, which is concerned with the study of energy balance within individuals, not only in terms of the acquisition and expenditure of energy but also the efficiency with which it is converted from one form to another (Bayne and Newell 1983). This topic has recently been reviewed for marine molluscs in general (Bayne and Newell 1983) and for bivalves specifically (Griffiths and Griffiths 1987). In this chapter, we describe the influence of environmental factors on energy flow and the partitioning of energy within individuals and populations of pectinids. With the exception of some studies on two or three species, most investigators have not determined the amount of energy invested in somatic tissue growth or spent on gametes. The degree to which available resources are allocated to reproduction will strongly influence future form and function, including survivorship, fecundity and energy available for subsequent growth, yet until recently intraspecific variation in resource partitioning resulting from environmental influences has received little attention in scallops and has not been well understood.

8.2 ENERGY BALANCE, PHYSIOLOGICAL INTEGRATIONS AND THE PARTITIONING OF ENERGY BETWEEN GROWTH AND REPRODUCTION

8.2.1 Energy Budgets

The most common method of assessing energy balance in an individual organism is to measure the components of the energy budget equation where:

$$C = P + R + F + U \tag{1}$$

The primary source of energy is food consumption (C), whereas energy expenditures include faecal losses (F), excretory products (U), respiratory heat loss (R) and energy invested in production (P) (Vahl 1981a; Bayne and Newell 1983). By rearranging this equation it is possible to calculate several parameters relevant to energy balance, particularly absorbed ration AB = C - F or AB = C x AE, where AE is absorption efficiency. The proportion of energy potentially available for physiological activity is referred to as the assimilated ration A and is given by A = AB - U or A = R + P. Assimilated energy is used to support the requirements of routine and active metabolism (i.e., respiration R), the remainder being available for production (P).

Fuji and Hashizume (1974), working on the Japanese scallop *Patinopecten yessoensis*, provided some of the earliest information on energy balance in marine bivalves, and combined seasonal measurements of growth, respiration and faecal production to illustrate energy flow through young scallops. In their study and many others on bivalves, no distinction was made between absorbed and assimilated ration, because bivalves are primarily ammonotelic and nitrogenous losses usually represent a small component of the total budget, e.g., Less than 5% in the giant scallop *Placopecten magellanicus* (MacDonald and Thompson 1986a). Excreted nitrogen, however, may constitute a significant metabolic loss for some species, such as the blue mussel *Mytilus edulis,* under certain conditions (Bayne and Scullard 1977). More than half of the assimilated energy in young *Patinopecten yessoensis* and *Choromytilus meridiionalis* (black mussel) is used to meet metabolic requirements, the remainder being divided almost equally between soft tissue growth and gamete synthesis (Fuji and Hashizume 1974; Griffiths 1981a).

8.2.2 Scope for Growth

An organism can only allocate energy to growth or reproduction while it remains in positive energy balance i.e., if absorbed ration exceeds total metabolic losses. If the latter exceed the former, then energy reserves are utilised for maintenance purposes (Newell 1979). The energy available for production after respiration and excretion have been subtracted from absorption [AB - (R + U)] is referred to as scope for growth (Warren and Davis 1967; Bayne and Newell 1983; Griffiths and Griffiths 1987), which in a bivalve is the physiological equivalent of the sum of reproductive output, somatic tissue growth and shell production. Scope for growth is often measured in short term laboratory studies to predict the response of the whole organism to a variety of conditions in the natural environment, and there has been reasonable agreement between growth predicted from physiological measurement (scope for growth) and growth determined directly (Dame 1972; Bayne and Worrall 1980). A better understanding of the physiological mechanisms which ultimately influence energy balance in pectinids will require measurements of the integrated response (scope for growth) under natural temperature and food conditions.

8.2.3 Growth Efficiency and Turnover Ratio

The efficiency with which food ration is converted to production (P/C) is referred to as total gross growth efficiency, whereas the proportion of absorbed ration converted to production (P/AB) is termed total net growth efficiency. Production of somatic tissue alone as a fraction of ingested ration (Pg/C) and absorbed ration (Pg/AB) are referred to as gross and net somatic growth efficiency respectively. Not only do these efficiencies vary with age but they may also fluctuate in response to changes in the environment.

The ratio of production to biomass (P/B), also known as the turnover ratio, has been widely used to estimate the production of natural populations from biomass estimates when mortality, growth or age composition are unknown (Banse and Mosher 1980), and has also been employed to compare energy flow through different populations in relation to the standing stock. It is important to distinguish biomass, which is an accumulated quantity of mass or energy, from production, which represents the rate of incorporation of organic matter or energy into the tissues (Davis 1963; Crisp 1984). In this review we shall discuss turnover ratios for individual scallops as well as for populations, and where appropriate we shall also take into account the reproductive component of production. Many of the P/B values recorded in the literature are based on somatic production rather than total production (Warwick 1980).

8.2.4 Growth of Shell and Somatic Tissue

In many studies, growth in bivalves has been described in terms of an increase in some dimension of the shell, usually length, the maximum distance between the anterior and posterior margins (Seed 1980). For the Pectinidae, shell height, which is the maximum distance between the dorsal (hinge) and ventral margins, has frequently been substituted for length. Growth rate may be expressed as an increase in size per unit time (usually years), i.e., the absolute growth rate, or as a proportional increase per unit time, i.e., the relative growth rate (Seed 1976). The rate of change in one variable relative to another, e.g., shell volume to shell length, has also been used to describe growth rates using allometric equations. The most popular method of describing growth, however, is to establish the relationship between shell size and age in order to construct a growth curve, which is often fitted to a non-linear mathematical function such as the von Bertalanffy model for predicative comparative purposes. In order for such models to be used effectively it is essential that certain criteria be met, including representation from all age classes, especially the youngest ones, and that the limitations and significance of the parameters should be recognised. For example, the parameter K, or Brody growth coefficient, is not a measure of absolute growth rate *per se* but rather an indicator of relative growth, representing the rate at which the asymptotic size is attained.

The majority of studies describing growth in pectinids have dealt exclusively with growth rates of the shell, but in order to measure production and energy flow it is necessary to record the weight of the soft tissue. Body weights are then converted to energy equivalents after biochemical analysis. Short-term variation is tissue weight is not always reflected in a change in shell dimensions, e.g., in the mussel *Mytilus edulis,* in which shell and somatic growth are not necessarily seasonally coupled (Hilbish 1986). Shell growth may continue throughout the entire lifespan in long-lived pectinids, whereas growth may continue throughout the entire lifespan in long-lived pectinids, whereas growth of the soft tissues of the soma often ceases completely and may become negative (Vahl 1981a; Macdonald and Thompson 1985a). Short-term variation in somatic tissue weight or energy content is a more sensitive indicator of growth fluctuations within the individual than is a change in shell dimensions, because unlike the soft tissue the shell cannot shrink rapidly under adverse environmental conditions e.g., in *Placopecten magellanicus* (Macdonald and Thompson 1985a). For describing the growth of shell or soft tissue, polynomial equations provide an alternative to the non-linear models such as the von Bertalanffy, and possess the advantage of not imposing asymptotic behaviour (Roff 1980), as well as being easier to deal with statistically when making comparisons between groups. The disadvantage of the polynomial is that the coefficients have no biological significance, unlike the parameters of the von Bertalanffy function. MacDonald and Thompson (1985a) found that shell growth in *P. magellanicus* was described well by a von Bertalanffy treatment, whereas a polynomial was required for the analysis of somatic growth, owing to the lack of an asymptotic weight as a result of the resorption of somatic tissue in senescent individuals. The selection of an appropriate model to fit a growth curve will largely depend on the properties of the data.

8.2.5 Reproductive Effort

Physiological indices such as the scope for growth afford one means by which the response of the whole organism to environmental change may be described, but a useful alternative is to establish the partitioning of assimilated energy between growth and reproduction. Of particular significance is the reproductive effort, defined as that proportion of the energy available to the individual which is channelled into reproduction. Approximately twenty years ago considerable interest arose among life history theorists in the concept of reproductive effort as a fitness correlate, and in the last decade several workers have responded to the challenge of producing empirical data to test some of the demographic theory. Estimates of reproductive effort have now been made for a variety or organisms from diverse taxa, and included experimental observations as well as determinations for natural populations. In many instances the objective in measuring reproductive effort has been to provide an insight into the ecology of the species or population rather than to test the predications of life-history theory. There is now a considerable body of data dealing with the reproductive energetics of bivalve molluscs (reviewed by Bayne and Newell 1983; Griffiths and Griffiths 1987), and much of this work has been undertaken on four pectinid species, the Iceland scallop *Chlamys islandica* (Vahl 1981b), the black scallop *Chlamys varia* (Shafee and Lucas 1982), the giant or deep-sea scallop *Placopecten magellanicus* (MacDonald 1986; MacDonald et al. 1987; MacDonald and Thompson 1988) and the weathervane scallop *Patinopecten caurinus* (MacDonald and Bourne 1987). For other pectinids, notably the bay scallop *Argopecten irradians*, information is available on reproductive output and the production of adductor muscle tissue (Bricelj et al. 1987a) but the absence of data sets containing both gamete production and total production of somatic tissue precludes any analysis of reproductive effort.

The measurement of reproductive effort presents some difficulties. There are two approaches which can be taken, both of which require an estimate or reproductive output, usually obtained by the weight loss on spawning. The first procedure is to express reproductive output as a proportion or percentage of total production by the individual, for which the somatic component is generally determined from the annual increment in tissue weight for each age class. Thus we have RE = $Pr/(Pg + Pr)$, where Pr is gamete production and Pg is somatic production over the same time period (usually one year). If significant, the production or organic material such as conchiolin deposited in the shell matrix (Ps) should be included in the denominator. When appropriate, byssus production (Pby) may also be considered (Vahl 1981a; Shafee and Lucas 1982). Data for weight are usually converted to common energy units. This is a useful means of calculating reproductive effort when the ages of individual animals can be determined, as in many scallop species, and is particularly relevant ecologically, since it considers the partitioning of energy by the organism in terms of net production.

In the second approach, suggested by Hirshfield and Tinkle (1975) and developed largely by Calow (1978; 1979), reproductive effort is derived by relating gonad output to some measure of energy input, so that RE = Pr/I, where I is absorbed ration, or RE = $Pr/(I - R)$, where R is metabolic rate. Other authors (e.g., Vahl 1981b) have proposed various alternatives based on these fundamental equations. Such physiological indices have proved particularly useful in experimental work in which food ration is controlled (Calow and Woollhead 1977; Thompson 1983), because they do not depend on the determination of somatic growth over long periods. In field studies, however, it is usually too difficult or time consuming to measure food availability, ingestion rate, absorption efficiency and metabolic rate under natural conditions, and the physiological approach to the measurement of reproductive effort is rarely adopted. There are, nevertheless, three studies on natural populations of bivalves (including two pectinid species) in which both the physiological and the production methods have been used (Vahl 1981b; Bayne et al. 1983; Macdonald et al. 1987). In each study the two indices of reproductive effort showed similar trends.

Other alternatives for the estimation of reproductive effort have been employed, particularly the expression of gamete production as a proportion of body weight. This is an attractive option because it circumvents the necessity of measuring somatic production, for which a more convenient quantity (biomass) is substituted, but it can lead to serious underestimates of reproductive effort, especially in long-lived species, because reproductive output in a single year is divided by somatic production accumulated over several years. The procedure is of limited use in comparative studies. Some authors have also used reproductive output as an index of reproductive effort, which has led to unnecessary semantic difficulties and has contributed to the lack of a general understanding or agreed definition of the latter term. Although there is usually a positive correlation between reproductive output and effort, there are circumstances in which a reduction in reproductive output is associated with enhanced reproductive effort, especially during poor conditions (Hirshfield 1980; Thompson 1983). In this review we shall use the term reproductive effort to refer only to those indices in which gonad production is expressed in terms of total production (Pr + Pg) or of some estimate of energy input into the organism (e.g., I or I - R).

8.3 CHANGES RELATED TO AGE OF INDIVIDUALS

8.3.1 Scope for Growth

Very little information exists on variation in scope for growth as bivalves get larger or older. In the mussel *Aulacomya ater*, smaller individuals gain more energy than larger ones at the same food ration because the latter must expend a greater proportion of their absorbed energy to offset metabolic costs (Griffiths and Griffiths 1987). According to MacDonald and Thompson (1985b, 1986a), annual gamete production, assimilated energy and respiratory losses are increasing functions of age in the giant scallop *Placopecten magellanicus* (Fig. 8.1a). These curves were derived by combining physiological measurements (metabolic expenditure, clearance rate, absorption efficiency and energy content of the natural seston) for each month of the year to obtain an estimate of cumulative annual scope for growth (A - R). Unlike instantaneous estimates of scope for growth, which are often greater for smaller individuals than for larger ones, cumulative annual values increase with age for the first five years of life, after which scope remains constant. This is consistent with direct measurements, which demonstrate that annual production in a 15-year old scallop is almost twice that of a five-year old. When expressed in terms of energy per unit body weight, however, annual cumulative scope for growth decreases with age (Fig. 8.1b). In the population of *P. magellanicus* studied by MacDonald and Thompson (1986a,b), scope for growth underestimated measured production by about 25–60% in scallops from shallow water, whereas in those from deeper water the predicted values (not shown) exceeded measured production by approximately the same amount.

8.3.2 Growth Efficiency and Turnover Ratio

Gross and net somatic growth efficiencies for *Placopecten magellanicus* (MacDonald and Thompson, unpublished), together with net somatic growth efficiencies for *Chlamys islandica* (Vahl 1981a), are presented in Fig. 8.2. Both species invest similar percentages of the absorbed ration in somatic growth, although this may vary between populations. As scallops grow older an increasing amount of assimilated energy is required to support respiratory demands, and more of the non-respired assimilation is spent on reproduction, resulting in decreased growth efficiencies.

Data for *Chlamys varia* (Conan and Shafee 1978; Shafee and Lucas 1980, 1982) and *Placopecten magellanicus* (MacDonald and Thompson 1985b; 1988) confirm the generally held view that the turnover ratio decreases with age, indicating that older individuals are less productive per gram of body tissue than younger individuals (Fig. 8.3). When total production (Pt) rather than somatic production (Pg) is used as the numerator, the P/B ratio may become independent of age in the last few age classes, despite increasing body mass (Fig. 8.3). This is because reproductive output often continues to increase throughout the lifetime of the scallop, whereas somatic production remains constant or declines in older individuals (MacDonald and Thompson 1985b; 1988).

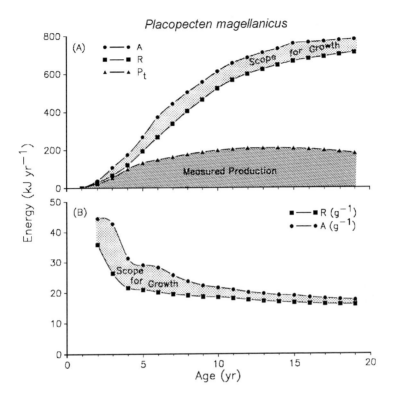

Figure 8.1. (A) Scope for growth per scallop compared with measured production (Pt) in the giant scallop *Placopecten magellanicus.* Scope = A - R, where A is assimilated ration and R is respiratory loss. (B) Scope for growth per gram of dry tissue (less shell). Data from MacDonald and Thompson (1985b; 1986a).

8.3.3 Somatic Growth and Maximum Size

The growth rate of an individual is the product of resource availability and the efficiency with which it is converted into body tissue (Calow and Townsend 1981). Relationships between the rates of feeding or oxygen uptake and body size are often described using the allometric equation $y = aW^b$ where a and b are fitted parameters, W is body weight and y is the predicted rate. The exponent (b) for respiratory energy loss is usually higher than the exponent describing energy gain from feeding (Bayne and Newell 1983). This results in a reduction in the energy available for growth as the organism increases in size, and provides a physiological basis for the decrease in growth by older individuals (Sebens 1979). Another factor which contributes to a decrease in somatic growth with increasing age is the gradual redistribution of this available energy to favour gamete production rather than growth as the animal becomes larger (see "Somatic

500

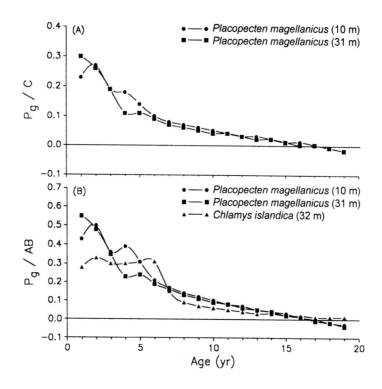

Figure 8.2. (A) Gross growth efficiency Pg/C (where Pg is annual somatic production and C is annual food intake) in giant scallops *Placopecten magellanicus* from 10 m and 31 m depths at Sunnyside, Trinity Bay, Newfoundland. (B) Net growth efficiency Pg/AB (where AB is the absorbed ration over one year) in *P. magellanicus* from 10 m and 31 m at Sunnyside, and in *Chlamys islandica* from Balsfjord, Norway. Data from MacDonald and Thompson (unpublished) and Vahl (1981a).

production and reproductive output"). According to Bayne and Newell (1983) and Bayne (1984) it is this process which may ultimately set the limit on size. Although smaller individuals are more efficient in acquiring energy relative to their metabolic expenses than are larger ones, this advantage is partially offset by associated disadvantages such as a higher maintenance ration per unit body weight, a more rapid weight loss during starvation and a higher predation pressure (Bayne 1985; Paine 1976).

8.3.4 Somatic Production and Reproductive Output

Most of the studies on reproductive energetics in bivalves demonstrate that gonad output Pr increases as the animal becomes older, whereas somatic production Pg reaches a maximum early in life and remains constant or declines thereafter (Bayne and Newell

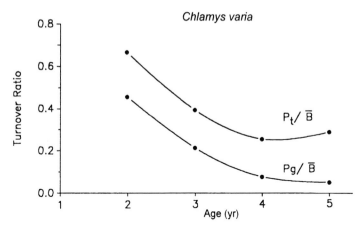

Figure 8.3. Turnover ratio (production: biomass ratio) for individual black scallops *Chlamys varia*. Pg is somatic production, Pt total production, and B mean biomass. Data from Conan and Shafee (1978), Shafee and Lucas (1980; 1982).

1983; Griffiths and Griffiths 1987). Certain long-lived pectinid species, such as *Placopecten magellanicus* and *Chlamys islandica*, are ideal for such studies, especially since the age of an individual can often be determined by counting annual growth rings on the shell or the ligament. Thus in one population of *Placopecten magellanicus* studied by MacDonald and Thompson (1985b), Pg increased with age until it peaked at approximately five years, after which Pg fell steadily (Fig. 8.4a). In senescent individuals, Pg became negative i.e., somatic tissue was resorbed, a phenomenon which has been recorded for *Chlamys islandica* (Vahl 1981c) and other aquatic invertebrates (Calow and Woollhead 1977) but dues not seem to be general. In *Placopecten magellanicus* Pr was very low in young individuals but gradually increased throughout the life of the scallop (Fig. 8.4a), so that eventually almost all the surplus energy was spent on gametes. The same trends have been observed in other populations of the giant scallop (MacDonald and Thompson 1985b; 1988; MacDonald 1986; Langton et al. 1987), and similar relationships among Pg, Pr and age have been described for various pectinids (see Figs. 8.4b, 8.5a, 8.6a), including *C. islandica* (Vahl 1981a), *Patinopecten caurinus* (MacDonald and Bourne 1987), and *C. varia* (Shafee and Lucas 1982). Maximum Pr exceeded maximum Pg in *Placopecten magellanicus* by at least a factor of two, but they were approximately equal in *C. islandica* (Fig. 8.4). Iceland scallops appeared to mature later than giant scallops, and in the former Pr did not become a detectable part of production until the scallops were at least five years old. Nevertheless, the transition from an emphasis on growth to enhanced reproduction occurred at approximately the same age (7–8 yrs) in both species. Shell production Ps was in low *Placopecten magellanicus* (Fig. 8.4a) and *Patinopecten caurinus* (Fig. 8.5a), but relatively high in young individuals of the short-lived *C. varia* (Fig. 8.6a). In *Patinopecten caurinus* and *C. varia* Ps decreased with age, but no clear trend was observed in *Placopecten magellanicus* or *C. islandica*.

502

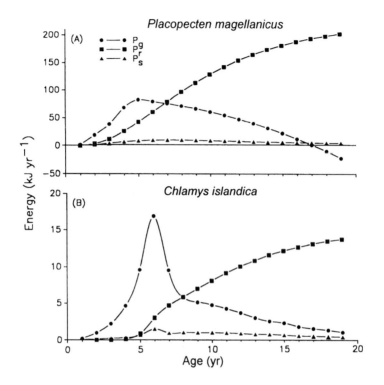

Figure 8.4. Somatic production (Pg) reproductive output (Pr) and shell production (Ps) in (A) the giant scallop *Placopecten magellanicus* (B) the Iceland scallop *Chlamys islandica*. Data from MacDonald and Thompson (1985b) and from Vahl (1981a).

8.3.5 Reproductive Effort

These age-related changes in Pg and Pr in several species of scallops are reflected in an increase in reproductive effort as the individual grows older (Figs. 8.5b, 8.6b, 8.7). This has been demonstrated in *Placopecten magellanicus* (MacDonald 1986, MacDonald et al. 1987, Langton et al. 1987, MacDonald and Thompson 1988), *Patinopecten caurinus* (MacDonald and Bourne 1987), *Chlamys islandica* (Vahl 1981b) and *C. varia* (Shafee and Lucas 1982), as well as in other bivalves (Bayne and Newell 1983; Griffiths and Griffiths 1987). Reproductive effort (expressed as 100 x Pr/(Pg + Pr)) is usually high in old individuals and may approach 100% in senescent animals of some species, e.g., *Patinopecten caurinus* and *Placopecten magellanicus* (Figs. 8.5, 8.7). Thus, there is a gradual transition from an early phase in which growth is favoured to a later phase in which reproductive output greatly exceeds somatic production. This may be a result of selective pressure for early increase in size as a refuge from predation (Paine 1976), coupled with selection for high fecundity (i.e., greater fitness) in larger individuals.

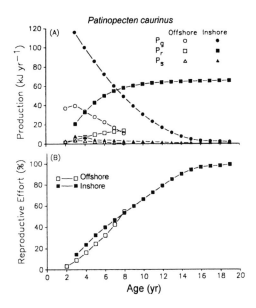

Figure 8.5. Somatic production (Pg), reproductive output (Pr), shell production (Ps) and reproductive effort in offshore and inshore populations of the weathervane scallop *Patinopecten caurinus* from British Columbia. Data from MacDonald and Bourne (1987).

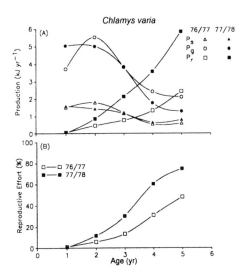

Figure 8.6. Somatic production (Pg), reproductive output (Pr), shell production (Ps) and reproductive effort in the black scallop *Chlamys varia* during two consecutive years. Data from Shafee and Lucas (1982).

The data for scallops and other bivalves are consistent with most life-history theory, which predicts that reproductive effort should increase with age in species which experience low mortality in the reproductive phase and which continue to grow after maturation (Charlesworth and Leon 1976; MacDonald et al. 1987).

8.4 ENVIRONMENTAL INFLUENCE ON PRODUCTION

8.4.1 Scope for Growth and Growth Efficiency

Any environmental factor that alters the rates of ingestion, absorption, excretion or oxygen uptake will influence scope for growth in either a positive or negative fashion. The blue mussel *Mytilus edulis,* for example, is capable of remaining in positive energy balance over a wide temperature range (5–20°C), whereas the European oyster *Ostrea edulis* is not (Widdows 1978; Newell et al. 1977). Scope for growth in *M. edulis* and *Aulacomya ater* fed with algal cultures increases with cell concentration until absorption efficiency declines, eventually resulting in negative value for scope (Bayne and Newell 1983). The relationship between scope for growth and the availability of seston to natural populations of bivalves is much more complex, owing to the production of pseudofaeces, fluctuations in food quality, and the difficulty in estimating absorption efficiency.

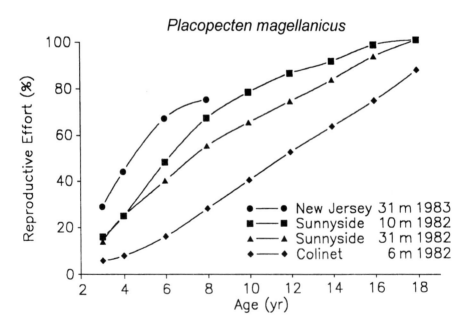

Figure 8.7. The relationship between reproductive effort and age in several populations of the giant scallop *Placopecten magellanicus.* Modified from MacDonald and Thompson (1988) and MacDonald et al. (1987).

Seasonal values of scope for growth in scallops feeding on natural seston at ambient temperature have been reported for *Placopecten magellanicus* from a population in Newfoundland (MacDonald and Thompson 1986a). Scope was lowest during the winter, often approaching zero or becoming negative, and maximum values were observed in May (≈1°C), August and September (12°C) when food was abundant (Fig. 8.8). The excess energy available in May was spent on gamete production rather than somatic growth, because somatic weight actually decreased during this period. Small scallops were almost always in positive energy balance and had the potential to grow or synthesise gametes throughout the year, whereas the largest scallops may only have been capable of significant growth or gamete production in a few months of the year. Similar seasonal trends in scope for growth have been reported for other bivalve species, such as *Chlamys islandica* (Vahl 1980), *Mytilus edulis* (Bayne and Widdows 1978; Thompson 1984a) and *Mya arenaria* (Gilfillan et al. 1976), although *Mya arenaria* apparently does not take advantage of the abundant food supply during the spring bloom in March. The combined effects of ration and temperature on energy balance are considerable, ration often being the most influential single factor (Bayne and Newell 1983).

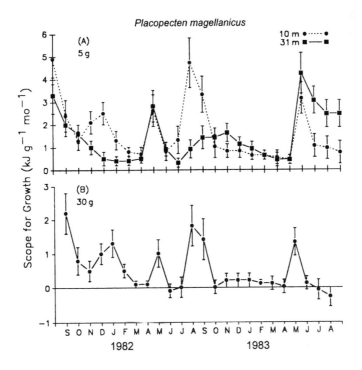

Figure 8.8. The seasonal cycle in scope for growth per unit dry weight of soft tissues in the giant scallop *Placopecten magellanicus* from 10 m and 31 m depth at Sunnyside, Trinity Bay, Newfoundland. (A) scallop of dry tissue weight 5 g. (B) scallop of dry weight 30 g. Modified from MacDonald and Thompson (1986a).

Estimates of growth efficiency based on physiological measurements have been made for *Chlamys islandica* (net efficiency; Vahl 1981a) and for two groups of *Placopecten magellanicus* from different water depths at one location (gross and net efficiencies; MacDonald and Thompson 1986a; b). Although ingested ration for larger *P. magellanicus* was greater in scallops from the more favourable shallow water environment than in those from deeper water, there were no differences between groups in the proportion of ingested or of absorbed ration which was invested in somatic production at any given age (Fig. 8.2).

8.4.2 Growth of Shell and Somatic Tissue

The influence of environmental conditions such as food availability and ambient temperature on growth and production has been the central theme of many studies on marine bivalves. The aim of such studies has often been to understand how a species contributes to energy flow in the ecosystem, or to evaluate the suitability of habitats for species under consideration for mariculture. The approach often adopted has been to compare individuals collected from various points along an environmental gradient over which food and temperature conditions may be expected to vary, such as the intertidal zone, different water depths, polluted habitats or sites separated geographically. Although food ration has been recognised as the major factor regulating growth and production, very few studies have actually quantified the food resources available to suspension feeders at various times of the year (Bayne and Newell 1983). In the few studies in which seston has been measured, growth rates have been well correlated with food availability for *Mytilus edulis* (Rodhouse et al. 1984, Page and Hubbard 1987), *Placopecten magellanicus* (MacDonald and Thompson 1985a), *Macoma balthica* (Hummel 1985; Thompson and Nichols 1988) and *Crassostrea gigas* (Brown 1988).

In populations of *Placopecten magellanicus* from shallow water habitats in southeastern Newfoundland, better food and temperature conditions are reflected in faster shall and somatic growth rates (Fig. 8.9). Somatic growth (less shell) is more sensitive to environmental conditions that is shell growth, the environmental effects being more evident in the somatic growth curves than in those for the shell. In the Bay of Fundy, however, scallops from deep water do not grow more slowly than those from shallow water, probably because the water column is vertically mixed and the food supply more uniformly distributed (MacDonald and Thompson 1985a). Local oceanographic conditions determine whether growth declines with increasing water depth. Not only do scallops from shallow water environments in Newfoundland often grow faster buy they also reach a much greater asymptotic height and maximum somatic weight. Somatic weight actually decreases in the oldest individuals, a phenomenon which is revealed by fitting a polynomial but is obscured by the forced asymptote if the more traditional von Bertalanffy equation is used. Maximum size is reached when energy input from feeding equals metabolic losses, so that there is no surplus energy available for further growth. These findings are consistent with Sebens' (1979; 1982; 1987) theoretical growth model that predicts an increase in asymptotic size if ration is increased or metabolic demand

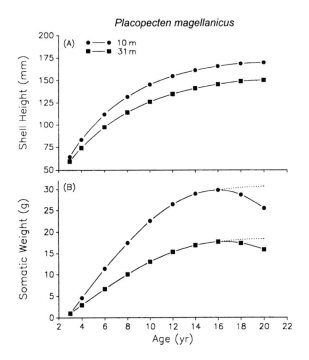

Figure 8.9. Growth curves for the giant scallop *Placopecten magellanicus* from 10 m and 31 m depth at Sunnyside, Trinity Bay, Newfoundland. (A) shell growth fitted by the von Bertalanffy model. (B) somatic tissue growth fitted by a polynomial, the dashed lines showing the asymptotes that would be produced by a von Bertalanffy equation. Modified from MacDonald and Thompson 1985a).

decreased (Fig. 8.10), and they provide evidence for the role of ration in establishing the maximum size which the scallop may attain in any particular environment.

Whereas the majority of studies have concentrated on annual growth, others have been concerned with the growth rates or physiological condition of individual animals at different times of the year. The instantaneous growth rate is dependent upon seasonal cycles of food availability, water temperature and possibly gametogenesis. A great deal of energy is required for the production of gametes in marine bivalves, which often results in close coupling between the reproductive cycle and the energy available for growth (Bayne 1985). In *Macoma balthica*, somatic tissue growth and gamete development often occur simultaneously, with food availability limiting growth in some months, depending upon the location of the year (Thompson and Nichols 1988). In *Argopecten irradians irradians,* most adductor muscle growth takes place in the autumn after gamete development and spawning in spring/summer (Bricelj et al. 1987a). Gamete development in the bay scallop sometimes requires the utilisation of stored muscle reserves, although in some locations ambient food resources are adequate (Bricelj et al. 1987b). In *Placopecten*

508

magellanicus from southeast Newfoundland, gametogenic activity proceeds very rapidly in the spring as food availability and temperature increase simultaneously (MacDonald and Thompson 1986a). Somatic weight decreases as the gonad grows during gametogenesis, although outside the reproductive period the giant scallop is capable of growing in any month despite very low temperatures (Fig. 8.11).

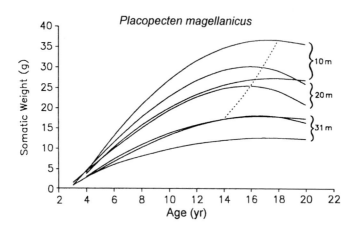

Figure 8.10. Somatic growth curves (polynomials) for giant scallops *Placopecten magellanicus* from various depths at Sunnyside, Trinity Bay, Newfoundland in different years. The dashed line joins points on each curve at which somatic weight is either a maximum or becomes asymptotic. Modified from MacDonald and Thompson (1985a).

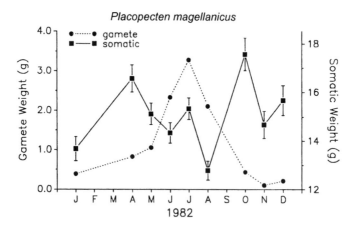

Figure 8.11. Seasonal changes in dry weight of the somatic tissue and the dry weight of gametes in the gonad for the giant scallop *Placopecten magellanicus* from 10 m depth at Sunnyside, Trinity Bay, Newfoundland. Modified from MacDonald and Thompson (1986a).

8.4.3 Reproductive Effort

The growth and fecundity of any organism are clearly influenced by the quality of the environment, and pectinids are no exception. One question which has preoccupied life-history theorists has been the effect of environmental variation on fitness characteristics such as reproductive effort within single species. Much of the experimental and ecological work on the reproductive energetics of bivalves and other aquatic invertebrates has been stimulated by the need to test empirically some of the predictions of demographic theory, although these studies have also provided useful insights into the ecology of several species, particularly in determining the degree of intraspecific variation in growth and reproductive output. Such research on pectinids has often focussed on water depth as an environmental variable (MacDonald et al. 1987; Barber et al. 1988), although there are studies in which inshore and offshore populations have been compared (MacDonald and Bourne 1987) or scallops in suspended culture compared with those growing on the bottom (MacDonald 1986). In all cases variation between populations is attributable not to factors such as depth *per se*, but to differences in food, temperature, salinity or other characteristics which define a particular water mass. In *Patinopecten caurinus*, for example, MacDonald and Bourne (1987) found that somatic production, gonad output and reproductive effort were all greater in animals from the productive inshore waters of the Georgia Strait, British Columbia, than in those from the cool, deeper water below the euphotic zone off the west coast of Vancouver Island (Fig. 8.5).

The most comprehensive information on intraspecific variation in energy partitioning by a pectinid has been obtained for the giant scallop *Placopecten magellanicus*, which occurs in the northwest Atlantic. MacDonald and Thompson (1985b) studied several populations of this species distributed along depth gradients in shallow inlets on the southeast coast of Newfoundland and in Passamaquoddy Bay, near St. Andrews, New Brunswick. For two of the Newfoundland locations, Sunnyside and Colinet, temperature and food availability (energy content of the seston) were measured throughout the year at selected depths from which scallops were sampled, whereas for other sites only temperature data were obtained (MacDonald and Thompson 1985a). At all locations except St. Andrews, the number of annual day degrees was a decreasing function of depth. Annual food availability was negatively correlated with water depth at Sunnyside in 1982, but not in the subsequent year, whereas at Colinet no variation in food ration was observed between depths in 1982 or 1983. These differences in environmental conditions at various depths were reflected in the values obtained for somatic production (Pg) and reproductive output (Pr) by individual scallops, because at all locations except Colinet and St. Andrews there was a negative correlation between age-specific Pg and depth, and in general Pr at any given age was also greater in scallops from shallower water. Thus for at least one site (Sunnyside) a direct relationship was demonstrated between favourable environmental conditions and enhanced Pg and Pr (especially the latter), and in other (Colinet) no differences were observed in age-specific Pg and Pr between scallops from various depths, owing to the absence of a depth gradient in food availability at this location. Age-related reproductive effort, whether expressed as Pr/(Pg + Pr) or as Pr per unit ingested ration, was also greater in scallops from shallow water at Sunnyside, but there was no such variation at Colinet. The lack of a correlation between Pr or Pg and depth at St. Andrews was attributed to unusual hydrographic conditions in the Bay of Fundy, where there is extreme vertical mixing as a result of tidal forces. Reduced fecundity has also been observed in a population of *P. magellanicus* from deep water in the Gulf of Maine (Barber et al. 1988), and Skreslet and Brun (1969) recorded a higher gonadosomatic index in Iceland scallops *Chlamys islandica* from shallow water than those from greater depths.

In addition to variation in Pg, Pr and reproductive effort between conspecifics from different depths or localities, comparisons of observations made in different years on one or more populations provide further evidence of intraspecific variation. Shafee and Lucas (1982) found much greater differences in Pr than in Pg or Ps between consecutive years in *Chlamys varia,* resulting in large interannual variation in reproductive effort (Fig. 8.6). In several populations of *Placopecten magellanicus*, MacDonald and Thompson (1985b) and MacDonald et al. (1987) also recorded small differences in Pg between years, but greater variation in Pr and reproductive effort (Fig. 8.7), especially in shallow water. There was a good correlation between reproductive output and food availability for scallops from shallow water in each of the two years for which data were obtained at Sunnyside, although in deeper water this trend was not observed. Annual variation in total production of individual scallops was largely a function of variability in gamete output. These results are consistent with life-history theory, which predicts that reproductive effort should increase when conditions are favourable (Goodman 1979). MacDonald and Thompson (1985b) suggested that the investment of surplus energy in gamete production at favourable locations or in good years may be an appropriate strategy for a species like *P. magellanicus* which experiences a variable, unpredictable environment. The highly variable recruitment which is characteristic of *P. magellanicus* in Newfoundland, and perhaps elsewhere, may be a consequence of a conservative strategy of controlled growth and opportunistic reproduction.

In a study of growth and reproduction in populations of the giant scallop *Placopecten magellanicus* from southeast Newfoundland to northern New Jersey, MacDonald and Thompson (1988) found considerable intraspecific variation in the production of somatic and germinal tissue, but only in Pr was there any evidence of latitudinal differentiation, both Pr and reproductive effort (Fig. 8.7) being greater in New Jersey scallops than in those from more northerly locations. This enhanced reproductive output was associated with reduced longevity, which is consistent with theory discussed by Calow (1979). Variation on a microgeographic scale, such as that observed between depths at Sunnyside and other locations in Newfoundland, may exceed variation on a latitudinal scale, owing to the sensitivity of the animal to local environmental conditions. A considerable body of evidence has accumulated which demonstrates great phenotypic plasticity in the reproductive output of scallops and other bivalves over small geographic ranges (Bayne et al. 1983; Bricelj et al. 1987a; MacDonald and Bourne 1987; MacDonald and Thompson 1988). The resulting variability in reproductive effort is illustrated in Fig. 8.7, which presents only some of the available data for *P. magellanicus* but includes the maximum and minimum values recorded, i.e., covers the entire range for the species. Although there is little information on the degree to which variation in growth and reproduction is genetically controlled, there is clearly a strong environmental influence. Thus in the Newfoundland populations of *P. magellanicus* studied by MacDonald and Thompson (1985a; b) not only were large variations in reproductive output observed along depth gradients over horizontal distances of approximately 100 m, which makes genetic differentiation extremely unlikely, but the values for Pr were also correlated with food availability. There is a large environmental component in the determination of somatic growth, fecundity and reproductive effort in bivalves which operates in the short-term and allows opportunistic exploitation of a variable food resource, but there must also be a genetic component which is less well documented and more difficult to quantify, and which presumable provides a mechanism for adaptation to long-term changes in the environment.

8.5 REPRODUCTIVE VALUE AND COST

8.5.1 Residual Reproductive Value

The concept of reproductive value was introduced by Fisher (1930) as a measure of the average production of young expected from a female over her lifetime. This was modified by Williams (1966), who considered reproductive value in terms of two components, current fecundity and future reproductive potential or residual reproductive value (RRV):

$$RRV = \sum_{t = x + 1}^{\omega} \frac{l_t}{l_x} m_t \tag{2}$$

Where l_t/l_x is the probability of survival from age x to age t, ω is the age of last reproduction and m_t is gamete production at age t. This index appeals to life-history theorists because it incorporates two characteristics, fecundity and mortality, which are likely to be under selective pressure. Thus RRV is probably a better fitness correlate than reproductive effort, but from the point of view of the empiricist it is less attractive because it requires an estimate of mortality for each age class. In many bivalve species this is difficult and tedious, with the result that there are few treatments of RRV in the literature. A uniform mortality rate has often been assumed for all age classes, but this is a questionable practice, since in some populations of at least two species, *Mytilus edulis* and *Placopecten magellanicus,* mortality is greater in very young and very old individuals than in those of intermediate age (Thompson 1984b; MacDonald and Thompson 1986b).

Vahl (1981b) demonstrated that in *Chlamys islandica* RRV is low in young individuals, but increases as they grow older, reaching a poorly-defined maximum at about 13 years before falling to zero at 23 years. In the mussel *Mytilus edulis* maximum RRV occurs at 4–6 years (Bayne et al. 1983; Thompson 1984b), and there are differences between populations in age related RRV which correlate well with environmental quality (Bayne et al. 1983). According to MacDonald et al. (1987), RRV in *Placopecten magellanicus* from various depths at Sunnyside and Colinet, Newfoundland, attains maximum values at 4–5 years (Fig. 8.12). These authors also found differences in the magnitude of RRV which, like reproductive effort, was greater in scallop from shallow

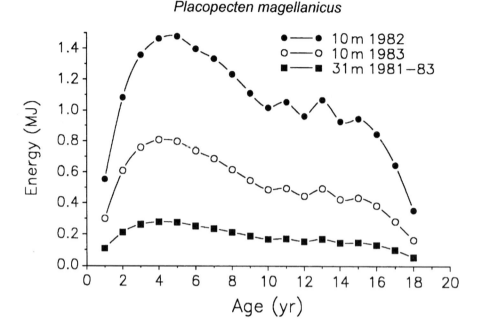

Figure 8.12. Age-related residual reproductive value in the giant scallop *Placopecten magellanicus* from 10 m and 31 m depth at Sunnyside, Trinity Bay, Newfoundland during consecutive years. Modified from MacDonald et al. (1987).

water at Sunnyside that in those collected from deeper water, and which also varied between years in the former group (Fig. 8.12). No consistent differences in RRV were observed between years or between depths at Colinet, where depth gradients in temperature or food availability were not apparent.

8.5.2 Reproductive Cost

Calow (1979) argued that an index of reproductive effort should take into account the degree to which reproduction represents a drain on the non-reproductive energy requirements of the animal. He proposed a physiological index of reproductive cost (C) expressed in terms of energy input I, reproductive output Pr and the metabolic demand R^* of the somatic tissue:

$$C = 1 - [(I - Pr)/R^*] \qquad (3)$$

The measurement of absorbed ration I over an entire reproductive cycle is difficult, and the separation of the measured metabolic rate into reproductive and non-reproductive components in natural populations presents a further challenge. Consequently this index has not been used as extensively as others, and the only estimates of reproductive cost in bivalves are those which have been made by Bayne et al. (1983) and Thompson (1984b) for the blue mussel *Mytilus edulis,* and by MacDonald et al. (1987) for the giant scallop *Placopecten magellanicus.* In both species the cost of reproduction was low in small individuals, but increased as they grew, reaching maximum values in the largest, most fecund animals (Fig. 8.13). In the population of *P. magellanicus* studied by MacDonald et al. (1987) at Sunnyside, Newfoundland, reproductive cost was lower in scallops from deep water than in those from shallow water (Fig. 8.13), owing largely to reduced fecundity in the former group. In general, values for reproductive cost were negative, suggesting a "restrained" reproductive strategy (Calow 1979), whereas positive values, indicating a diversion of energy from somatic maintenance to gamete production (i.e., a "reckless" strategy), were observed only in large individuals from shallow water, where conditions were more favourable.

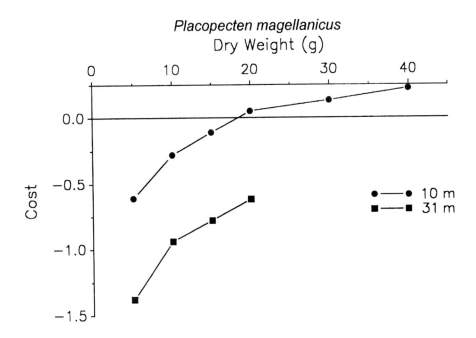

Figure 8.13. Age-related reproductive cost in the giant scallop *Placopecten magellanicus* from 10 m and 31 m depth at Sunnyside, Trinity Bay, Newfoundland. Modified from MacDonald et al. (1987).

8.6 POPULATION PRODUCTION

Energy flow in populations can be evaluated by expanding the concept of an energy budget for individuals to include parameters relevant to populations. Production (estimated as $P = \Delta B + E$) includes the change in biomass (ΔB) which results not only from the growth of individuals but also from the addition of new individuals through recruitment and immigration, and must consider energy losses or elimination (E) through mortality and emigration. In a stable population, biomass does not vary annually because energy gains through production are counterbalanced by mortality losses. Although expenditure on gametes is a major component of individual production, there is often a major loss to the population when the animals spawn. Because the partitioning of this production into somatic growth and gamete output is age dependent, estimates of population production not only require information on the number of individuals per unit area, but also their age structure, which is often described by an age-frequency distribution. Indices of population production or turnover ratios (P/B) depend upon the balance between young, fast-growing animals and older, slow-growing, fecund individuals.

One means of estimating population energy flow requires the integration of physiological measurements such as ingestion and metabolic rates made under natural conditions with population characteristics such as density, growth rate and reproductive output. The limitations of extrapolating short-term laboratory studies of individual animals to whole populations under field conditions have been discussed by Bayne and Newell (1983) and some investigators have gone to considerable lengths to measure physiological activity under as realistic conditions as possible, especially in the field (Bayne et al. 1977). The classical procedure for the assessment of population energy flow, however, is to estimate production either from the increase in biomass per unit time, or by the application of "standard" P/B ratios to measured biomass. This is a direct approach, since production represents the net result of physiological activity in the organism, and it eliminates the need for physiological measurements and the extrapolation of data. The use of a P/B ratio is particularly attractive, because it provides an estimate of production from a single biomass measurement without the need for repeated sampling, but the difficulty here is that most bivalves undergo pronounced seasonal changes in body weight and energy content, so that the production estimate is sensitive to the time of sampling. One major problem with direct estimates of population production is that they often consider only the somatic growth component and omit reproductive output, which in many bivalves can result in an underestimate of total production by as much as tenfold, depending on the species and the age composition of the population (Bayne and Worrall 1980; Griffiths 1981b; Thompson 1984b). The magnitude of the underestimate will also depend on the stage of gonad development at the time of sampling. Wherever possible, sampling should be carried out immediately before and immediately after spawning, so that gamete production may be determined as well as somatic growth.

Despite the frequent use of energy units to express production per unit area, meaningful comparisons between populations remain difficult because methodology is often not uniform and there are spatial and temporal differences in population structure which are seldom quantified (Griffiths 1981b; Broom 1983). Pectinid populations appear to be less productive than those of many other bivalves (Table 9 in Broom 1983). This is presumably the result of differences in population structure, especially the relatively low densities characteristic of many scallop species, or of behavioural patterns, particularly migration after settlement, rather than of any physiological properties of individual animals. For example, allocations of assimilated energy to respiration in the pectinids *Chlamys islandica, C. varia* and *Placopecten magellanicus* are similar to those of several other bivalves (Vahl 1981c; Shafee and Conan 1984; MacDonald and Thompson 1986a). For these three species, information is available on the relevant population characteristics, including physiological measurements and estimates of the various components of production (Vahl 1981a; c; Shafee and Conan 1984; MacDonald and Thompson 1986a; b). Despite some major differences in growth rate, maximum size, longevity and population density between these species, there are many similarities in their production and patterns of energy flow, e.g., as seen in the P/B ratios (Table 8.1). Although the metabolic demand represents a similar proportion of production in all three scallops, they differ greatly in the proportion which is lost through mortality (25–30% in *P. magellanicus,* 41% in *C. islandica,* and more than 100% in a population of *C. varia* which may be heavily fished). The lower values for *P. magellanicus* may be attributable to the predominance of older (larger) individuals, providing a size related refuge from predation similar to that described for the horse mussel *Modiolus modiolus* (Seed and Brown 1978).

Neither *C. islandica* nor *C. varia* grow as large as *P. magellanicus,* and may therefore be susceptible to predation throughout their life spans, as in the cockle *Cerastoderma edule* (Seed and Brown 1978). Between 15 and 25% of total production in *C. varia* and *C. islandica* is spent on gametes, compared with 46–72% in *P. magellanicus,* depending upon the population studied. The reproductive output of a 19-year-old giant scallop (approximately 170 mm in height) is equivalent to the combined output of 14 Iceland scallops of this age.

The significance of age structure and environmental conditions becomes apparent when intraspecific comparisons are made between populations from different locations. MacDonald and Thompson (1986b) found that the proportion of large individuals of *Placopecten magellanicus* was greater at Sunnyside, Newfoundland, than at Colinet, where the younger age classes predominated. As a result the Sunnyside population, with its high reproductive potential, spent almost three times more energy on gametes than on somatic tissue. *P. magellanicus* from Colinet is similar to populations of the hard clam *Mercenaria mercenaria* (Hibbert 1977) and the oyster *Ostrea edulis* (Rodhouse 1978) which also partition their production almost equally between growth and reproduction, although there is very little information available on intraspecific variation in population production for other bivalve species. The rapid growth and high fecundity of individual scallops from shallow water at Sunnyside has been attributed to favourable food and temperature conditions (MacDonald and Thompson 1985a,b), but the Colinet population was almost twice as productive as the Sunnyside one, owing to the greater density of the former. Some of the physical and biological factors that determine production operate at the level of the individual animal (e.g., the effects of food and temperature on scope for growth), others at the population level (e.g., the influence of substrate and ocean currents

516

Table 8.1

Population characteristics for three pectinid species: P_g = production due to somatic growth; P_r = reproductive output; P_s = organic production by the shell; P_bY = byssal production: A = assimilated energy; a, Sunnyside; b, Colinet; c, age range; d, net production after subtracting degrowth; e, calculated from instantaneous mortality rate. From MacDonald and Thompson (1986b).

	Chlamys islandica (Vahl 1981a;c)	*Chlamys varia* (Shafee and Conan 1984)	*Placopecten magellanicus* (MacDonald and Thompson 1986a;b)
Density (scallops m^{-2})	64	2.3	0.19[a], 0.84[b]
Age range (yr)	1–20	0–4	3–19
Biomass (kJ m^{-2})	1662	40.1–52.5	97.6[a], 196[b]
Annual mortality	0.17	0.35–0.93[e]	(0.03–0.38)[b, c]
Production (kJ m^{-2} yr^{-1})	771[d]	20.8–22.6	31.8[a], 59.5[b]
Assimilation (kJ^{-2} yr^{-1})	2772	191–242	120[a]
Production/biomass	0.46	0.43–0.52	(0.92–0.25)[a,c], (0.66–0.19)[b,c]
Respiration/assimilation	0.72	0.89–0.91	0.74[a]
Reproductive output/ assimilation	0.07	0.02–0.03	0.19[a]
$(P_g + P_s + P_{by})$/ assimilation	0.21	0.07–0.08	0.07[a]
Net production efficiency $[(P_g+P_r+P_{by})/A \times 100]$	(18–42)[c]	(4–22)[c]	(20–46)[a,c]

on recruitment, and of predation on mortality rate). Despite evidence that there may be considerable intraspecific variation in population production, very little is known about the range of responses under different environmental conditions.

ACKNOWLEDGMENTS

The work was supported by NSERC operating grants to both authors. The figures were prepared by Ms. Margaret Miller.

REFERENCES

Banse, K. and Mosher, S., 1980. Adult body mass and annual production/biomass relationships of field populations. Ecol. Monogr. 50:355–379.

Barber, B.J., Getchell, R., Shumway, S. and Schick, D., 1988. Reduced fecundity in a deep-water population of the giant scallop *Placopecten magellanicus* in the Gulf of Maine, USA. Mar. Ecol. Prog. Ser. 42:207–212.

Bayne, B.L., 1984. Aspects of reproductive behaviour within species of bivalve molluscs. Adv. Invert. Reprod. 3:357–366.

Bayne, B.L., 1985. Responses to environmental stress: tolerance, resistance and adaptation. In: J.S. Gray and M.E. Christiansen (Eds.). Marine Biology of Polar Regions and Effects of Stress on Marine Organisms. Wiley and Son, Chichester. pp. 331–349.

Bayne, B.L. and Newell, R.C., 1983. Physiological energetics of marine molluscs. In: A.S.M. Saleuddin and K.M. Wilbur (Ed.). The Mollusca, Vol. 4. Academic Press Inc., New York. pp. 407–515.

Bayne, B.L., Salkeld, P.N. and Worrall, C.M., 1983. Reproductive effort and value in different populations of the marine mussel, *Mytilus edulils* L. Oecologia 59:18–26.

Bayne, B.L. and Scullard, C., 1977. Rates of nitrogen excretion by species of *Mytilus* (Bivalvia: Mollusca). J. Mar. Biol. Assoc. U.K. 57:355–369.

Bayne, B.L. and Widdows, J., 1978. The physiological ecology of two populations of *Mytilus edulis* L. Oecologia 37:137–162.

Bayne, B.L., Widdows, J. and Newell, R.I.E., 1976. Physiological measurements on estuarine bivalve molluscs in the field. In: B.F. Keegan, P. O'Ceidigh and P.J.S. Boaden (Eds.). Biology of Benthic Organisms. Pergamon Press, New York. pp. 57–68.

Bayne, B.L. and Worrall, C.M., 1980. Growth and production of mussels *Mytilus edulis* from two populations. Mar. Ecol. Prog. Ser. 3:317–328.

Bricelj, V.M., Epp, J. and Malouf, R.E., 1987a. Intraspecific variation in reproductive and somatic growth cycles of bay scallops *Argopecten irradians*. Mar. Ecol. Prog. Ser. 36:123–137.

Bricelj, V.M., Epp, J. and Malouf, R.E., 1987b. Comparative physiology of young and old cohorts of bay scallops *Argopecten irradians irradians* (Lamarck): mortality, growth and oxygen consumption. J. Exp. Mar. Biol. Ecol. 112:73–91.

Broom, J.J., 1983. Mortality and production in natural, artificially-seeded an experimental populations of *Anadara granosa* (Bivalvia: Arcidae). Oecologia 58:389–397.

Brown, J.R., 1988. Multivariate analyses of the role of environmental factors in seasonal and site-related growth variation in the Pacific oyster *Crassostrea gigas*. Mar. Ecol. Prog. Ser. 45:225–236.

Calow, P., 1978. The evolution of life-cycle strategies in freshwater gastropods. Malacologia 17:351–364.

Calow, P., 1979. The cost of reproduction - a physiological approach. Biol. Rev. 54:23–40.

Calow, P. and Townsend, C.R., 1981. Resource utilization in growth. In: C.R. Townsend and P. Calow (Eds.). Physiological Ecology: An Evolutionary Approach to Resource Use. Blackwell Scientific Publications, Oxford. pp. 220–244.

Calow, P. and Woollhead, A.S., 1977. The relationship between ration, reproductive effort and age-specific mortality in the evolution of life-history strategies - some observations on freshwater triclads. J. Anim. Ecol. 46:765–781.

Charlesworth, B. and Leon, J.A., 1976. The relation of reproductive effort to age. Am. Nat. 110:449–459.

Conan, G. and Shafee, M.S., 1978. Growth and biannual recruitment of the black scallop *Chlamys varia* (L.) in Lanvaeoc area, Bay of Brest. J. Exp. Mar. Biol. Ecol. 35:59–71.

Crisp, D.J., 1984. Energy flow measurements. In: N.A. Holme and A.D. McIntyre (Eds.). Methods for the Study of Marine Benthos. Blackwell Scientific Publications, Oxford. pp. 284–372.

Dame, R.F., 1972. The ecological energies of growth, respiration and assimilation in the intertidal American oyster *Crassostrea virginica*. Mar. Biol. 17:243–250.

Davis, C.C., 1963. On questions of production and productivity in ecology. Arch. Hydrobiol. 59(2):145–161.

Fisher, R.A., 1930. The Genetical Theory of Natural Selection. Oxford University Press. 272 p.

Fuji, A. and Hashizume, M., 1974. Energy budget for a Japanese common scallop, *Patinopecten yessoensis* (Jay), in Mutsu Bay. Bull. Fac. Fish. Hokkaido Univ. 25:7–19.

Gilfillan, E.S., Mayo, D., Hanson, S., Donovan, D. and Jiang, L.C., 1976. Reduction in carbon flux in *Mya arenaria* caused by a spill of no. 6 fuel oil. Mar. Biol. 37:115–123.

Goodman, D., 1979. Regulating reproductive effort in a changing environment. Am. Nat. 113:735–748.

Griffiths, C.L. and Griffiths, R.J., 1987. Bivalvia. In: T.J. Pandian and F.J. Vernberg (Eds.). Animal Energetics. Vol. 2. Bivalvia through Reptilia. Academic Press, New York. pp. 1–87.

Griffiths, R.J., 1981a. Aerial exposure and energy balance in littoral and sublittoral *Choromytilus meridionalis* (Kr.) (Bivalvia). J. Exp. Mar. Biol. Ecol. 52:231–241.

Griffiths, R.J., 1981b. Production and energy flow in relation to age and shore level in the bivalve *Choromytilus meridionalis* (Kr.). Estuar. Coast. Shelf Sci. 13:477–493.

Hibbert, C.J., 1977. Energy relations of the bivalve *Mercenaria mercenaria* on an intertidal mudflat. Mar. Biol. 44:77–84.

Hilbish, T.J., 1986. Growth trajectories of shell and soft tissue in bivalves: seasonal variation in *Mytilus edulis*, L. J. Exp. Mar. Biol. Ecol. 96:103–113.

Hirshfield, M.F., 1980. An experimental analysis of reproductive effort and cost in the Japanese medaka, *Oryzias latipes*. Ecology 61:282–292.

Hirshfield, M.F. and Tinkle, D.W., 1975. Natural selection and the evolution of reproductive effort. Proc. Nat. Acad. Sci. 72:2227–2231.

Hummel, H., 1985. Food intake of *Macoma balthica* (Mollusca) in relation to seasonal changes in its potential food on a tidal flat in the Dutch Wadden Sea. Neth. J. Sea Res. 19:52–76.

Langton, R.W., Robinson, W.E. and Schick, D., 1987. Fecundity and reproductive effort of sea scallops *Placopecten magellanicus* from the Gulf of Maine. Mar. Ecol. Prog. Ser. 37:19–25.

MacDonald, B.A., 1986. Production and resource partitioning in the giant scallop *Placopecten magellanicus* grown on the bottom and in suspended culture. Mar. Ecol. Prog. Ser. 34:79–86.

MacDonald, B.A. and Bourne, N.F., 1987. Growth, reproductive output, and energy partitioning in weathervane scallops, *Patinopecten caurinus* from British Columbia. Can. J. Fish. Aquat. Sci. 44:152–160.

MacDonald, B.A. and Thompson, R.J., 1985a. Influence of temperature and food availability on the ecological energetics of the giant scallop *Placopecten magellanicus*. I. Growth rates of shell and somatic tissue. Mar. Ecol. Prog. Ser. 25:279–294.

MacDonald, B.A. and Thompson, R.J., 1985b. Influence of temperature and food availability on the ecological energetics of the giant scallop *Placopecten magellanicus*. II. Reproductive output and total production. Mar. Ecol. Prog. Ser. 25:295–303.

MacDonald, B.A. and Thompson, R.J., 1986a. Influence of temperature and food availability on the ecological energetics of the giant scallop *Placopecten magellanicus*. III. Physiological ecology, the gametogenic cycle and scope for growth. Mar. Biol. 93: 37–48.

MacDonald, B.A. and Thompson, R.J., 1986b. Production, dynamics and energy partitioning in two populations of the giant scallop *Placopecten magellanicus* (Gmelin). J. Exp. Mar. Biol. Ecol. 101:285–299.

MacDonald, B.A. and Thompson, R.J., 1988. Intraspecific variation in growth and reproduction in latitudinally differentiated populations of the giant scallop *Placopecten magellanicus* (Gmelin). Biol. Bull. 175:361–371.

MacDonald, B.A., Thompson, R.J. and Bayne, B.L., 1987. Influence of temperature and food availability on the ecological energetics of the giant scallop *Placopecten magellanicus*. IV. Reproductive effort, value and cost. Oecologia 72:550–556.

Newell, R.C., 1979. Biology of Intertidal Animals. Marine Ecological Surveys Ltd., Faversham. 781 p.

Newell, R.C., Johnson, L.G. and Kofoed, L.H., 1977. Adjustment of the components of energy balance in response to temperature in *Ostrea edulis*. Oecologia 30:97–110.

Page, H.M. and Hubbard, D.M., 1987. Temporal and spatial patterns of growth in mussels *Mytilus edulis* on an offshore platform: relationship to water temperature and food availability. J. Exp. Mar. Biol. Ecol. 111:159–179.

Paine, R.T., 1976. Size-limited predation: An observational and experimental approach with the *Mytilus-Pisaster* interaction. Ecology 57:858–873.

Rodhouse, P.G., 1978. Energy transformations by the oyster *Ostrea edulis* L. in a temperature estuary. J. Exp. Mar. Biol. Ecol. 34:1–22.

Rodhouse, P.G., Roden, C.M., Burnell, G.M., Hensey, M.P., McMahon, T., Ottway, B. and Ryan, T.H., 1984. Food resource, gametogenesis and growth of *Mytilus edulis* on the shore and in suspended culture: Killary Harbour, Ireland. J. Mar. Biol. Assoc. U.K. 64:513–529.

Roff, D.A., 19890. A motion for the retirement of the von Bertalanffy function. Can. J. Fish. Aquat. Sci. 37:127–129.

Sebens, K., 1979. The energetics of asexual reproduction and colony formation in benthic marine invertebrates. Am. Zool. 19:683–697.

Sebens, K.P., 1982. The limits to indeterminate growth: an optimal size model applied to passive suspension feeders. Ecology 63:209–222.

Sebens, K.P., 1987. The ecology of indeterminate growth in animals. Ann. Rev. Ecol. Syst. 18:371–407.

Seed, R., 1976. Ecology. In: B.L. Bayne (Ed.). Marine Mussels: Their Ecology and Physiology. Cambridge University Press, Cambridge. pp. 13–65.

Seed, R., 1980. Shell growth and form in the Bivalvia. In: D.C. Rhoads and R.A. Lutz (Eds.). Skeletal Growth in Aquatic Organisms: Biological Records of Environmental Change. Plenum Press, New York. pp. 23–67.

Seed, R. and Brown, R.A., 1978. Growth as a strategy for survival in two marine bivalves, *Cerastoderma edule* and *Modiolus modiolus*. Journal of Animal Ecology 47:283–292.

Shafee, M.S. and Conan, G., 1984. Energetic parameters of a population of *Chlamys varia* (Bivalvia: Pectinidae). Mar. Ecol. Prog. Ser. 18:253–262.

Shafee, M.S. and Lucas, A., 1980. Quantitative studies on the reproduction of black scallop, *Chlamys varia* (L.) from Lanveoc area (Bay of Brest). J. Exp. Mar. Biol. 42:171–186.

Shafee, M.S. and Lucas, A., 1982. Variations saisonnie res du bilan energetique chez les individus d'une population de *Chlamys varia* (L.): Bivalvia, Pectinidae. Oceanol. Acta 5:331–338.

520

Skreslet, S. and Brun, E., 1969. On the reproduction of *Chlamys islandica* (O.F. Muller) and its relation to depth and temperature. Astarte 2:1–6.

Thompson, J.K. and Nichols, F.H., 1988. Food availability controls seasonal cycle of growth in *Macoma balthica* (L) in San Francisco Bay, California. J. Exp. Mar. Biol. Ecol. 116:43–61.

Thompson, R.J., 1983. The relationship between food ration and reproductive effort in the green sea urchin, *Strongylocentrotus droebachiensis*. Oecologia 56:50–57.

Thompson, R.J., 1984a. The reproductive cycle and physiological ecology of the mussel *Mytilus edulis* in a subarctic, non-estuarine environment. Mar. Biol.79:277–288.

Thompson, R.J., 1984b. Production, reproductive effort, reproductive value and reproductive cost in a population of the blue mussel *Mytilus edulis* from a subarctic environment. Mar. Ecol. Prog. Ser. 16:249–257.

Vahl, O., 1980. Seasonal variations in seston and in the growth rate of the Iceland scallop, *Chlamys islandica* (O.F. Müller) from Balsfjord, 70°N. J. Exp. Mar. Biol. Ecol. 48:195–204.

Vahl, O., 1981a. Energy transformations by the Iceland scallop, *Chlamys islandica* (O.F. Müller), from 70°N. I. The age-specific energy budget and net growth efficiency. J. Exp. Mar. Biol. Ecol. 53:281–296.

Vahl. O., 1981b. Age-specific residual reproductive value and reproductive effort in the Iceland scallop, *Chlamys islandica* (O.F. Müller). Oecologia 51:53–56.

Vahl, O., 1981c. Energy transformations by the Iceland scallop, *Chlamys islandica* (O.F. Müller), from 70°N. II. The population energy budget. J. Exp. Mar. Biol. Ecol. 53:297–303.

Warren, C.E. and Davis, G.E., 1967. Laboratory studies on the feeding, bioenergetics, and growth of fish. In: S.D. Gerking (Ed.). The Biological Basis of Freshwater Fish Production. Blackwell Scientific Publications, Oxford. pp. 175–214.

Warwick, R.M., 1980. Population dynamics and a secondary production of benthos. In: K.R. Tenore and B.C. Coull (Eds.). Marine Benthic Dynamics. University of South Carolina Press, Columbia. pp. 1–24.

Widdows, J., 1978. Physiological indices of stress in *Mytilus edulis*. J. Mar. Biol. Assoc. U.K. 58:125–142.

Williams, G.C., 1966. Natural selection, the costs of reproduction, and a refinement of Lack's principle. Am. Nat. 100:687–690.

AUTHORS ADDRESSES

Raymond J. Thompson - Ocean Sciences Centre, Memorial University of Newfoundland, St. John's, Newfoundland, Canada A1C 5S7 (E-mail: Thompson@mun.ca)

Bruce A. MacDonald - Biology Department and Centre for Coastal Studies and Aquaculture, University of New Brunswick, Saint John, New Brunswick, Canada E2L 4L5 (E-mail: bmacdon@unbsj.ca)

Scallops: Biology, Ecology and Aquaculture
S.E. Shumway and G.J. Parsons (Editors)
© 2006 Elsevier B.V. All rights reserved.

Chapter 9

Nutrition in Pectinids

Ana Farías and Iker Uriarte

9.1 INTRODUCTION

Over the last decade, the area of nutrition in larviculture of bivalve molluscs has undergone a period of great development, primarily due to increased implementation of controlled cultures under hatchery conditions. These operations extend from broodstock conditioning through production of small juveniles, with a larval culture phase of as long as four weeks between spawning and metamorphosis. One of the greatest costs involved in these types of culture is that of the food required, principally based on microalgae (Coutteau and Sorgeloos 1992; Uriarte et al. 2001). Much research has thus been centred on more efficient and cost-effective means for obtaining foods for each stage of the culture process, both in microalgal culture technology and in formulation of microalgal substitutes. This area has promoted the development of nutritional studies in bivalve molluscs for broodstock conditioning and larval culture, as well as for postlarval culture after metamorphosis until the progeny reach 2 to 10 mm in size.

9.2 PECTINID FEEDING

The pectinids are sestotrophic bivalve filter-feeders, which are capable of ingesting living and inert particles suspended in the water column (Lucas 1982). It has been observed that in bottom dwelling pectinids, inert organic material in the seston may be responsible for most of the energetic input (Hunauld et al. 2002), while it has been shown in other species that detritus may contribute to the diet during periods in which the environmental offering of phytoplankton is unable to satisfy their energetic requirements (Cranford and Grant 1990). Organic materials from discarded feed and faeces from salmon culture contribute to the diet of pectinids maintained in culture in southern Chile (42°S; Seguel et al. 1998). Phytoplankton is, however, the main nutrient source for pectinids as the nutritional value of organic detritus becomes degraded by bacterial decomposition, and leakage of its contained micronutrients.

Use of microalgae cultured in support of intensive bivalve culture, including pectinids, has promoted active monitoring of the nutritive characteristics of these algae in order to satisfy the nutritional requirements of broodstock, larvae and postlarvae. Of the traditionally used microalgal species in controlled bivalve culture, two species stand out, *Isochrysis* aff. *galbana* (clone *T-Iso*) and *Chaetoceros neogracile* (Coutteau and Sorgeloos 1992).

Based on the premise that utilisation of protein in food was most efficient when the protein contained essential amino acids in a ratio compatible with tissue formation, Webb and Chu (1983) proposed that the low nutritional value of some microalgae such as *Phaedactylum tricornutum* was due to an inadequate amino acid profile, as this alga lacked tryptophan. Microalgae currently considered to be of high nutritional value have similar essential amino acid profiles as the larvae in culture (Brown 1991), and contain higher concentrations of these amino acids than do the larvae. From this it is concluded that the protein in favourable microalgal diets presently used in molluscan hatcheries is of adequate quality, and that continued research on this protein should be directed to the area of its proportional quantity in culture diets.

Relevant to lipids, the pioneering work of Chu and Webb (1984) demonstrated the importance of essential fatty acids and the relation between n-3 and n-6 fatty acids, which promote growth in bivalves, and thus act as a quality indicator for microalgae. The survey mentioned above (Coutteau and Sorgeloos 1992) indicated that the preference for *T-Iso* and *C. neogracile* was associated with their high levels of total lipids and content of docosahexanoic acid (DHA) in *T-Iso*, and elevated levels of carbohydrates, eicosapentanoic acid (EPA) and riboflavin in *C. neogracile*. In general, varied diets favour controlled molluscan culture, with use of a DHA-enriched alga, such as *T-Iso* or other prasinophyceaen, and an alga high in EPA such as *Chaetoceros* or other diatoms (Table 9.1).

Cultures of microalgae may be manipulated in order to increase their content of proteins, lipids, and carbohydrates, as well as increasing their levels of n-3 HUFA. Increases in algal biomass need to be achieved in order to make such microalgal production methods cost-effective. For example, methodology for increasing the protein content of microalgae involves increasing the nitrogen content of the algal culture medium, achieving a 50% rise above the normal cellular protein content (Fig. 9.1) and produces a significant increase in the growth rate of the microalgal culture.

In general, the majority of algal species used in molluscan cultures contain high levels of vitamins. The vitamin C found in the majority of microalgae fulfils the requirements of pectinids and other bivalves as the average values for this compound in these microalgae is between 0.11 and 1.62% of their dry weight and it is known that in fishes and crustaceans these requirements may be as high as 0.02% of the diet. Brown and Miller (1992) suggested the vitamin C found in the microalgae is efficiently incorporated into filter feeders. As with vitamin C, little is known concerning the riboflavin requirements of molluscs, although fishes and crustaceans may have requirements as high as 6 μg g^{-1} of the diet, and the microalgae used in molluscan cultures contain amounts of this vitamin generally greater than 20 μg g^{-1}. *Chaetoceros gracilis* is one of the species with the highest content of this vitamin (Brown and Farmer 1994).

Adult pectinids often contain large numbers of their own eggs in their digestive tracts during the reproductive season, as well as particles of up to 950 μm (Shumway et al. 1987; Bricelj and Shumway 1991). They normally carry out negative selection of particles over 100 μm and have a high efficiency for retention of particles of less than 40 μm (Vernet 1977). Pectinid larvae efficiently retain particles from 0.5 to 8 μm depending on the size of the larva. The rate of consumption of food particles depends on a number

Table 9.1

Microalgal diets used in the feeding of Latin American pectinids. Data are listed in the same units as used in the references. (After Farías 2001).

Species	Stage of Development	Microalgal diet	Ration	Author
Argopecten purpuratus	Larva	*Pavlova lutheri, Isochrysis galbana, Isochrysis aff galbana (T-Iso), Chaetoceros gracilis*	>20 cells μL^{-1}	Valdivieso et al. 1988
	Larva	*Nanochloropsis oculata + T-Iso + C. gracilis* (ratio 1:1:1)	45 cells μL^{-1}	Martinez 1991
	Larva to postlarva of 14 mm	*T-Iso + P. lutheri + Chaetoceros calcitrans + Thalassiossira minima* (ratio 1:1:1:1)	30–120 cells μL^{-1}	Farías et al. 1998
	Postlarva 400 μm	*I. galbana + C. calcitrans + C. gracilis* (ratio?)	?	Martinez et al. 1992
	Postlarva 18 mm	*T-Iso + C. gracilis* (ratio 1:1)	$8 \bullet 10^{8}$ cells $tank^{-1}$ d^{-1}	Uriarte and Farías 1995
	Brood stock 3 years	*T-Iso + P. lutheri + C. gracilis + Tetraselmis suecica* (ratio 1:1:1:1)	0.5–1.5% dry biomass of microalga per wet wt. minus shell	Caers et al. 1999
	Brood stock 24 months	*T-Iso + C. gracilis* (ratio 1:1)	$2.5 \bullet 10^{9}$ cells $pectinid^{-1}$ d^{-1}	Farías et al. 1997; Farías and Uriarte 2001
	Brood stock 8–10 cm	*I. galbana + C. calcitrans + C. gracilis* (ratio?)	?	Martinez et al. 1992
Argopecten ventricosus = circularis	Larva	*I. galbana + Tetraselmis chuii +* Sp. X (ratio 1:1.5:3)	$5.5 \bullet 10^{4}$ cells d^{-1}	Aviles 1990
	Larva	*T-Iso*		Fundación CENAIM-ESPOL 1997a
	Pre-seed 2 mm	*T-Iso*	8% dry biomass of microalgae per live weight	Fundación CENAIM-ESPOL 1997b

Juveniles	I. galbana + T. chuii + Sp. X (ratio 2:1:2)	$2.5 \cdot 10^5$ cells d^{-1}	Aviles 1990
Brood stock	I. galbana	$9.9 \cdot 10^5$ cells mL^{-1} d^{-1}	Aviles 1990
Brood stock of 8–16 months	T-Iso + C. gracilis	$1.25 \cdot 10^4$–$5 \cdot 10^4$ cells mL^{-1} d^{-1}	Villalaz 1994a, b
Brood stock 50 mm	I. galbana + Chaetoceros muelleri (ratio 1:1)	$4 \cdot 10^9$ cells pectinid^{-1} d^{-1} = 0.48% dry biomass microalga per live weight 1.55% dry biomass microalga per wet weight minus shell	Racotta et al. 1998
Brood stock 50 mm	I. galbana + C. muelleri (ratio 1:1)	$3 \cdot 10^9$ cells pectinid^{-1} d^{-1}	Ramirez et al. 1999
Nodipecten subnodosus			
Larva	?	3,750–10,000 cells larva^{-1} d^{-1}	Ortiz 1994
Larva	T-Iso	?	Villavicencio 1997
Larva	T-Iso + P. lutheri (ratio 1:1)	10,000 cells mL^{-1}	Lora-Vilchis and Maeda-Martinez 1997
Brood stock	T-Iso + P. lutheri + Thalassiossira pseudonana + C. gracilis (ratio 3:3:2:2)	150,000 cells mL^{-1}	Villavicencio 1997
Pecten maximus			
Larva	I. galbana + P. lutheri + Rhodomonas baltica (ratio 1.25:1.25:1)	50 cells µL^{-1}	Román, pers. comm.
Brood stock	Skeletonema costatum + T-Iso	$4.6 \cdot 10^8$ cells pectinid^{-1} h^{-1}	Román and Campos 1993

? data not given in reference >< Data not exact in the reference, d = day

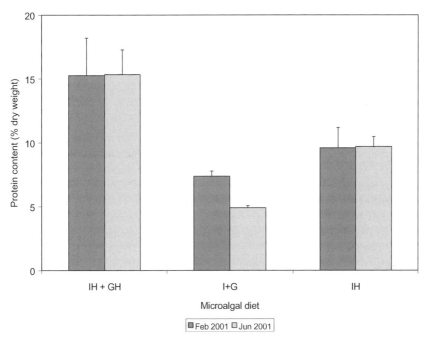

Figure 9.1. Protein content of different mass cultures of a microalgal mixture consisting of *I. aff. galbana (T-Iso)* and *Chaetoceros neogracile* conditioned in media with high (IH + GH) and normal (I + G) nitrogen concentrations, or with *T-Iso* alone, conditioned in a high nitrogen concentration (IH). Each value is the mean of seven replicates. Bar = 1 SE.

of variables, including temperature, food concentration, nutritive value of the food, and water flow (Wildish et al. 1987; Bricelj and Shumway 1991). Absorption both in larvae and in juveniles and adults is controlled by the levels of digestive enzymes in their digestive tracts (Samain et al. 1992). Digestion in pectinids occurs partially in the stomach and crystalline style sac, and partially within the digestive diverticula in the digestive gland. Since there are no salivary glands in filter feeding bivalves, the carbohydrases important in digestion occur in the digestive gland and crystalline style. Amylase has been detected in the stomach in a selection area composed of B type cells containing large granules (Benlimame et al. 1993). Amylase and laminarinase activity have been detected in the crystalline style, and act extracellularly at an acid pH in the crystalline style sac and stomach. The digestive gland is rich in intracellular carbohydrases such as glucosidases, galactosidases, and chitobiases, and intracellular peptidases have been detected, which may characterise them as "lysosomal" (Wojtowickz 1972; Boucaud-Camou et al. 1993). High extracellular amylase activity has also been detected in channels within the digestive gland without proteolytic activity. Aside from the absorption of macronutrients, larvae, juveniles and adults have mechanisms for the absorption of micronutrients, which is an energy-requiring process. These processes may be saturable as a function of dose, such as in the case in the absorption of vitamin C in larvae of *Pecten maximus* (Seguineau et al. 1993).

Pectinids may be fed using different microalgal cultures under experimental conditions or in intensive culture. Most feeding and nutritional studies have been carried out with these types of diet, and although positive empirical results have been obtained, there has been little explanation of how natural diets fulfil the nutritional needs of these bivalves. Also, little is known concerning how the pectinids respond to the complex mixtures of bacteria, phytoplankton, and chemical signals, which affect them in their natural environment. Variability in nutritional quality of seston may range from 5 to 80% in organic content, with a C:N ratio from <4 to >26 (Bayne and Hawkins 1990). This spectrum may vary with the length of day, season of the year, or geographic locality, which make it difficult to define a "typical natural diet", or reasonably define the "normal" feeding behaviour of a bivalve. In *Placopecten magellanicus* a chemoreceptive mechanism has been described which operates at low concentration levels which allows the species to discriminate between types of particles prior to their capture and ingestion (Ward et al. 1992).

Given the high variability inherent in living cultures of microalgae, much research has been carried out on identifying potential microalgal substitutes. Pure-cultured microalgae is now available as a frozen paste, or in dried form, as are heterotrophic cultures of microalgae. These products may be used as supplements or even substitutes for traditional microalgal cultures. Yeasts have been tested with some success in clam cultures, but these have not been found useful for the pectinids due to their low nutritional value and their high rates of sedimentation. Tests made by substituting 50% of the traditional microalgal diet of juvenile *A. circularis* (Fundación CENAIM-ESPOL 1998) with yeast or dry microalgal powder produced a significant decrease in their growth in both cases when compared with normally fed controls. In tests where diets containing 100% yeast or dry microalgae were given to *A. circularis* juveniles, they failed to grow at all and almost all the individuals died by the fourth week of the experiment (Table 9.2).

Some thought was given in the past to the possibility of producing a well balanced diet for filter feeders by means of microencapsulation technology. These methods have still not emerged from the experimental stage; however, they are potentially useful for the study of nutritional requirements in bivalves in general (Kreeger and Langdon 1993; Seguineau et al. 1993).

Development of lipid emulsions for the feeding of bivalves has given good results at the experimental level. Coutteau et al. (1996a) and Caers et al. (1999) have shown that both pectinid juveniles and broodstock were capable of incorporating portions of the fatty acids present in lipid emulsions into their triglycerides. Lipid emulsions given as a 50% (w/w) supplement in the microalgal diet of *A. purpuratus* broodstock significantly increased lipid levels in their eggs (Caers et al. 1999). In mixed diets of 70% microalgae and 30% lipid emulsions, the broodstock of *A. purpuratus* reached better conditioning than those with a 100% microalgal diet (Martínez et al. 2000a). In diets for larval *A. purpuratus*, lipid emulsions may be substituted at up to 40% of the microalgal diet without affecting their growth and survival when compared with a diet of *I. galbana* (*T-Iso*) and *C. neogracile* (Uriarte et al. 2003). The microalgal diet for *A. purpuratus*

Table 9.2

Use of artificial diets in feeding *A. circularis*. The microalgal mixture was composed of *T-Iso* and *C. neogracile*. Results after four weeks of experimentation. Each value represents the mean and standard deviation of three replicates. (After CENAIM-ESPOL Foundation 1998).

Treatment	Increase in wet weight (mg)	Increase in dry weight (mg)	Increase in length (mm)	Survival (%)
100% algae	0.79 ± 0.06 [a]	0.4 ± 0.03 [a]	6.3 ± 0.4 [a]	98.3 ± 1.4 [a]
50% algae + 50% yeast	0.4 ± 0.05 [b]	0.2 ± 2.02 [b]	3.9 ± 0.45 [b]	95.0 ± 4.3 [a]
100% yeast	0.05 ± 0.07 [c]	-0.01 ± 0.05 [c]	0.2 ± 1.7 [c]	8.3 ± 10.1 [b]
50% algae + 50% dry algae	0.38 ± 0.05 [b]	0.19 ± 0.02 [b]	3.8 ± 0.2 [b]	95.8 ± 5.2 [a]
50% dry algae	0.01 ± 0 [c]	0.02 ± 0 [c]	0.1 ± 0.7 [c]	2.5 ± 4.3 [b]
Control (not fed)	-----	-----	-----	0 ± 0 [b]
50% algae	0.31 ± 0.01 [b]	0.16 ± 0.01 [b]	3.6 ± 0.3 [b]	95.8 ± 3.8 [a]

Values with the same exponent within each column not significantly different ($P > 0.05$).

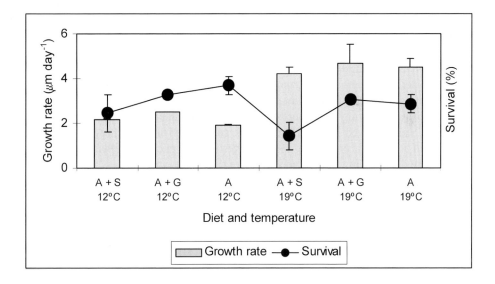

Figure 9.2. Experiment in 20-L tanks with *Argopecten purpuratus* larvae grown at 12°C and 19°C with three different diets: (A) = microalgae alone; (A + G) = microalgae plus 20% glucose; (A + S) = microalgae plus 20 starch. Each value is the mean of three replicates. Bar = 1 SE. (After Uriarte 2000).

juveniles may be substituted up to 20% with lipid emulsions enriched with EPA and DHA, but at a substitution level of 40% growth decreases (Fig. 9.2). At present there is at least one commercial mollusc hatchery which has incorporated routine use of lipid emulsions into its culture procedure (Sorgeloos 1998), and appears to be a useful alternative for replacing a portion of the microalgal diet in pectinids (Uriarte et al. 2003).

Use or not of substitutes for microalgae is highly dependent on the quality of the microalgae utilised in the culture procedure. Prior to the consideration of such substitutes, the high variability in nutritional value of cultured microalgae needs to be eliminated by standardisation of mineral nutrients (fertilisers), lighting conditions, photoperiod, temperature, and stage of harvest of the cultures. The concept of adding microalgal substitutes is only valid when there is a widely accepted "standard microalgal diet", a known portion of which can then be experimentally replaced by food supplements.

9.3 NUTRITIONAL REQUIREMENTS DURING REPRODUCTIVE CONDITIONING

The composition of diets administered to molluscan broodstock has been discussed as having a major influence on their reproductive capacity and on the quality of their gametes, (Robinson 1992a; Samain et al. 1992). Protein increase in the microalgal diet decreases the time to maturity, increases the fecundity of the females, and improves the energy balance in pectinid broodstock (Farías and Uriarte 2001). Results of physiological analyses have shown that diets low in protein produce malnutrition in *A. purpuratus* broodstock characterised by a high degree of excretion of endogenous proteins to make up for lack of dietary proteins during reproductive conditioning. High respiratory rates are also observed under these conditions, probably due to an elevated level of carbohydrates in protein-deficient diets (Farías and Uriarte 2001). Villalaz (1994a) suggest that the adductor muscle stores protein in support of reproductive activity on the gonad of *A. ventricosus*, reiterating the importance of proteins in the broodstock diet. A significant difference was observed between broodstock conditioned in a hatchery and broodstock conditioned at sea (Martínez et al. 1992) suggesting that microalgal diets produced in the hatchery are deficient in proteins, and therefore need to be enriched.

One of the nutritional factors that most affects the quality of spawnings in molluscs is the content of EPA in eggs, which is also the most abundant in larvae during metamorphosis (Robinson 1992b). The EPA/DHA ratio varies between species, which have been conditioned in the field and those conditioned in the hatchery (Utting and Millican 1997). The pattern of fatty acids in the lipids of the eggs reflects the lipids in the diet, where in *Chlamys tehuelcha* a strong correlation was observed between the fatty acid composition of the dominant phytoplankton and the lipid composition of the gonads (Pollero et al. 1979). Lipids in the eggs of marine molluscs are rich in n-3 HUFA, the origin of which is probably from diatoms and dinoflagellates (Pollero et al. 1979; Trider and Castell 1980; Napolitano et al. 1992; Robinson 1992b). This observation implies that developing embryos and emerging larvae have high n-3 HUFA requirements, which need to be derived from the diet of the broodstock. Martínez et al. (2000b) demonstrated that a 30% substitution of emulsions enriched with n-3 HUFA into the microalgal diet during reproductive conditioning of *A. purpuratus* broodstock produced higher fecundity and better responses to spawning induction. Supplementing the broodstock diet with n-3 HUFA induced effects in the fatty acid profiles of the triglycerides in the eggs, but did not produce variation in the composition of their phospholipids (Caers et al. 1999). In comparing high-protein microalgal diets with diets normal in protein but 30% substituted with a lipid emulsion enriched with n-3 HUFA, *A. purpuratus* broodstock produced significantly better quality progeny on the high protein diet (Hernández 2001; Uriarte et al. 2004).

No requirements for carbohydrates have been demonstrated during the reproductive conditioning of pectinids, although accumulation of glycogen reserves has been observed in the muscle for subsequent use in gonadal maturation. Increase in the feeding ration has been shown to produce increases in carbohydrates in the gonad (Martínez et al. 2000b). In cultures of *A. purpuratus* grown in polyculture systems with salmon in Chile's X^{th} Region (42°S) accumulations of carbohydrate occurred in both muscle and gonad during observations made both in tanks and in the sea. This accumulation of carbohydrates arose from the availability of protein-lipid particles coming from artificial feeding of cultured salmonids, and produced a reproductive event in the pectinids in winter, which was not observed in control cultures of *A. purpuratus* distant from the salmon cultures (Seguel et al. 1998).

In addition to provitamin A activities, carotenoids are required for gonadal development and maturation, fertilisation, eclosion, and larval viability, as well as for other functions. Since these pigments originate in the phytoplankton, it is expected that filter feeders such as the pectinids might not be limited with regard to these pigments when fed on microalgal diets. Carotenoids observed in the gonad of *A. circularis* have included lutein and astaxanthin at a level of 2×10^5 mg g gonad^{-1} (Escarria et al. 1989).

9.4 NUTRITIONAL REQUIREMENTS OF THE LARVAE

Larvae of *A. ventricosus* demonstrated better growth in a mixture of *I. galbana* and *C. gracilis* than when cultured separately with each of these algae; using the mixture, values for the daily deposition of protein were increased, as well as the apparent use of protein (Millán 1997). The factorial effects of the protein levels using microalgal and lipid-substituted diets on growth and survival, and energetic reserves in larvae have been compared for *A. purpuratus* (Machulas 2000; Uriarte 2000; Nuñez 2002). In all cases the best growth of the larvae was obtained using high-protein diets, while there were no detectable effects attributable to the lipid-substituted diet. Larval survival appeared to be independent of both protein and lipid dietary levels. Metabolism of energetic reserves observed in pectinid larvae fed with different levels of protein and n-3 HUFA (Uriarte 2000) showed that the protein content of the larvae did not vary due to type of diet, although the opposite was true for the lipids and the carbohydrates. Lipid levels in larvae increased in proportion to increase in lipid levels in the microalgal diet ($F_{lipids} = 6.09$, df = 1,6, $P = 0.05$) and the carbohydrates in the larvae were affected by the protein level in the diet, but not when protein was replaced by lipids in the diet ($F_{protein} = 9.93$, df = 1,4, $P = 0.03$). These results indicate a greater requirement for proteins during the larval phase, which could be satisfied with a high protein microalgal diet. Microalgal growth in a high-nitrogen medium probably not only increases in protein content, but also may improve in availability of essential amino acids which could be limiting at lower protein concentrations.

There is considerable information available on the lipid requirements of larvae of marine organisms, primarily the fishes. These studies have centred primarily on the optimum requirements for highly unsaturated fatty acids (n-3 and n-6 HUFA) in early fish developmental stages. HUFA identified as essential for marine organisms include docosahexanoic acid (DHA), 22: 6n-3, eicosapentanoic acid (EPA), 20:5n-3, and arachidonic acid (ARA), 20:4n-6. These lipids are normally produced by phytoplankton and reach the larvae via the food chain. Only a portion of the fatty acids required in the diet serve in functions requiring esterified essential fatty acids in membrane phospholipids. The majority serves an energetic functions, such as a portion of the triglycerides. The remaining portions of the non-essential fatty acids are the saturated, monounsaturated, and remaining polyunsaturates (PUFA). Less than a decade ago, it was thought that only crustaceans had requirements for phospholipids in their diets, and that these could be biosynthesised by fishes. However, the use of artificial diets has shown that larvae of marine fish have a limited capacity for the biosynthesis of phospholipids. Phospholipids are required in aquatic feeds to form the chylomicrons in order to transport triglycerides across the intestinal mucosa to the lymph and low-density lipoproteins (Sargent et al. 1999b). Phospholipid requirements have not been determined for molluscs; most studies have indicated a requirement for fatty acids. Sargent et al. (1999a) proposed that the ideal lipid composition in the diet of marine larvae would be one which duplicated the lipid composition of the vitellum in the eggs based on it's content of phospholipids and triglycerides, where the phospholipids are naturally rich in HUFA. This suggests that larvae would be saved the task of biosynthesis of phospholipids and PUFAs, and the triglyceride content would provide the energy sufficient for larval development. Experimental diets might be useful for definition of optimal levels of lipids consistent with those hypothesised over the last three years (Geurden et al. 1997, Fontagne et al. 1998), but technological developments so far do not allow preparation of a diet of this type (Sargent and Tacon 1999).

During molluscan larval development a greater percentage of n-3 HUFA are observed in the phospholipid fraction than in that of the triglycerides due to a greater selective insertion of n-3 than n-6 in the membrane phospholipids (Chu and Webb 1984). Larvae of *P. maximus* show a restricted capacity for desaturation of fatty acids, although this capacity is sufficient for regulation and stabilisation of the PUFA content of the membrane phospholipids (Soundant et al. 1998). The n-3 HUFA composition of phosphatidylcholine in larvae of *A. purpuratus* varies in relation to the diet administered, while the composition of the phosphatidylethanolamine is highly conservative and is maintained independent of the diet (Farías et al. 2001).

Early studies of larval bivalve molluscs demonstrated that their triglycerides were independent of the diet, while phospholipids were highly conservative and tended to be independent of the composition from that of the microalgae used as food (Waldock and Nascimento 1979, Napolitano et al. 1988). Studies using lipid emulsions enriched in n-3

HUFA as supplements to the microalgae, however, have shown that the composition of the polyunsaturated fatty acids in the diet affected both the larval triglycerides and their phospholipids, and that their EPA/DHA became the same as that of the emulsions offered (Coutteau et al. 1996a). It has also been found that the PUFA in *P. maximus* in the phospholipids are highly deficient in EPA, DHA, or ARA (Soudant et al. 1998), and results of studies of premetamorphic larvae have shown that the microalgal diets considered to be of high nutritive value can be further improved by adding n-3 HUFA emulsions. These provide better growth and survival in the early postlarval stages (Coutteau et al. 1994). Marty et al. (1992) found that larvae of *P. maximus* had high dietary requirements for DHA, while Delaunay et al. (1993) detected a requirement for EPA and ARA in larvae of this species. Coutteau et al. (1996b) showed that larvae of *Placopecten magellanicus*, although able to efficiently incorporate n-3 HUFA from lipid emulsions, did not demonstrate improved growth or survival during the larval period, which may be because the microalgal diet satisfied their requirements for EPA and DHA. Uriarte et al. (2003) have shown that at 19°C, using diets of *T-Iso* and *C. neogracile*, larvae of *A. purpuratus* did not require n-3 HUFA lipid supplementation. They also showed that the microalgal diets of *A. purpuratus* larvae can be 40% substituted with lipid emulsions without affecting larval growth and survival. This may be because the PUFA stored in the neutral lipids of the larvae are utilised during metamorphosis for supplying energy, and perhaps as well for synthesising new membrane phospholipids (Delaunay et al. 1993). The studies of Thompson et al. (1993, 1994) have shown that in at least *C. gigas* and *Patinopecten yessoensis* the increase of dietary EPA and DHA was not beneficial. These authors suggested that the best diets for bivalve larvae were those high in saturated lipids, which only contained moderate quantities of polyunsaturated lipids, because the energy from the food was more efficiently utilised when it is liberated via β oxidation of saturated fatty acids. The studies of Sargent et al. (1999b) point rather to threshold concentrations of PUFA dependent on the interrelations with the essential n-3 HUFA. Larvae of *A. purpuratus* show better growth and survival with raised amounts of DHA/EPA in the diet, together with low values for EPA/HUFA, suggesting that the dietary EPA is less important, probably because the larvae may obtain it by desaturation of 18:3n-3 (Farías et al. 2001). Also in *A. purpuratus* larvae, the levels of ARA remain relatively constant even when they are fed diets not containing this fatty acid (Farías et al. 2001), thus suggesting it is not a dietary requirement for development of these larvae. An elevated content of EPA was observed in D-larvae of laboratory reared *A. purpuratus* fed with a mixture of *Nannochloropsis oculata* + *T-Iso* + *C. gracilis* at a 1:1:1 cell dry wt. ratio (Martínez 1991), which became reduced by up to 27% in pediveliger larvae at 16 days. Concurrently, the levels of DHA and ARA showed stable values throughout the entire larval development. The EPA/DHA ratio declined from 1.5 to 0.5 and the DHA/ARA ratio remained at about 12 during the entire period (Table 9.3). These results may indicate a use of the EPA in energy metabolism while the DHA tended to be conserved and maintained at high levels during the entire larval period. Farías et al. (2003) suggested that *A. purpuratus* larvae may be capable of selectively retaining DHA or converting sufficient 18:3n-3 and 20:5n-3 to 22:6n-3 for fulfilling their requirements.

Table 9.3

Compositions of fatty acids in larvae of *Argopecten purpuratus* fed with a 1:1 mixture of *T-Iso* and *C. calcitrans*. Each value is the mean of 2 to 6 replicates. (After Martínez 1991).

Fatty acid	Days of larval life					
	0	2	7	12	16	24 (post- larvae)
C16:0	15.65	14.50	12.00	12.05	12.85	13.00
C17:0	0.87			1.25	1.35	1.60
C18:0	4.63	5.85	5.65	4.45	4.25	3.90
C16:1	8.30	6.75	7.60	10.70	11.85	12.40
C18:1	5.25	6.20	10.45	10.35	11.60	9.50
C22:1	0.73	0.50		0.30	0.75	
C24:1	0.70	0.85		1.75	1.60	
C18:2	1.17	1.20	1.75	2.60	2.50	2.70
C18:3n-6		1.20	2.20	0.60		
C18:3n-3	1.50			2.95	2.85	3.30
C18:4 + C20:1	6.13	6.60	7.60	6.95	7.95	9.10
C20:2	0.57	0.70	0.60	0.70	0.70	
C20:3	0.25	0.30	0.40	0.40	0.30	
C20:4	1.33	1.35	1.45	1.60	1.15	1.30
C20:5	20.48	12.15	5.35	7.20	5.70	5.60
C24:0 + c22:4	1.33	2.35	1.50	0.70	0.60	
C22:5	1.30	1.30	1.40	0.60	0.40	
C22:6	13.10	16.50	17.45	13.45	11.80	15.70
C20:5 / C22:6	1.56	0.74	0.31	0.54	0.48	0.36
C22:6 / C20:4	9.84	12.22	12.03	8.41	10.26	12.08

Nevejan et al. (2002) observed a high level of survival and development to the premetamorphic phase in larvae fed with *Dunaliella tertiolecta* supplemented with coconut oil. This microalga is deficient in HUFA, which provides additional evidence that pectinid larvae have a high capacity for the desaturation of linolenic and linoleic fatty acids. These studies did not, however, control for the possibility that bacterial flora of the culture tanks or in the digestive tracts of the larvae might be active in the synthesis of n-3 HUFA, or detect if true *de novo* synthesis of these fatty acids was accomplished by the larvae.

Larval requirements for fatty acids may vary significantly with variations in culture temperature. There are currently debated results concerning the effects of diet, temperature and their interactions on larval fatty acids, phospholipids and triglycerides. Also controversial have been comparisons made between similar species with different geographical origins which have been cultivated in a common habitat (Beninger and Stephan 1985). Studies on eggs of *P. magellanicus* from two geographically separated populations showed that eggs from broodstock originating at 31 m depth had a high content of DHA in their phospholipids while those from 10 m depth had significantly lower concentrations of this lipid. The lower temperature of the deeper waters compared with that of shallower waters during the last phases of gametogenesis may have caused the variation in the fatty acid profile of this species, discarding diet as the determining factor in this variation, given that the fatty acids in the triglycerides was similar in the scallops from both habitats (Napolitano et al. 1992).

On theoretical grounds, it can be supposed that lipids (rather than carbohydrates) would have evolved as the primary energy source in tiny larvae of molluscs such as those of bivalves, as they are not only a more concentrated energy source, but also aid in the floatability of these planktonic organisms. The carbohydrates assume importance during the benthic phase of bivalve existence, with feeding habits adapted to ingestion of carbohydrate-rich particulate food of plant origin. Within a month after metamorphosis, tissue carbohydrates become more abundant than lipids in *A. purpuratus* tissues (Farías et al. 1998). For this reason it is probable that in larvae the hemolymph glucose levels are maintained by gluconeogenesis, to be replaced by glucogenesis and glucogenolysis only after metamorphosis. The addition of carbohydrates as glucose and soluble starch to *A. purpuratus* larvae cultured with *T-Iso* and *C. gracilis* produced no increase in their growth, while their survival was negatively affected by the use of starch (Uriarte 2000; Fig. 9.2). So, carbohydrates are not important determining factors of quality in the diet of larval bivalves. Whyte et al. (1989), however, found that quantity of dietary carbohydrates significantly increased the nutritional condition of larvae of *Patinopecten yessoensis*, suggesting an important role for carbohydrates in conserving nutrients such as proteins and fatty acids for tissue synthesis rather than as energy sources.

Vitamin requirements in diets of marine species vary with the species in culture, growth rate, capacity for vitamin synthesis by symbiotic microorganisms in the digestive tract, and the composition of ingested food. For species such as the pectinids which feed on phytoplankton, it is unlikely that vitamin deficiencies occur since the microalgae, particularly in the exponential phase of growth, contain sufficiently high levels of vitamins to satisfy the known requirements of various marine species (Seguineau et al. 1993). Accumulations of vitamins C and E have been observed in *P. maximus* during larval development, accompanied by a large reduction in riboflavin (B_2); the latter vitamin was the least abundant among those studied by Seguineau et al. (1993).

The mineral requirements of larval pectinids are probably at trace levels except for calcium required for production of the shell. Since seawater contains appreciable levels of calcium and trace minerals, uptake of these substances from this medium undoubtedly relies on ingestion and active transport of needed minerals across membranes of marine animals (Tacon 1990).

9.5 POST-METAMORPHIC NUTRITIONAL REQUIREMENTS

Increase in growth rate from metamorphosis to the 2-mm size as a result of administration of high-protein diets has been observed in pectinids, where the growth period between settlement and transfer of "pre-seed" to sea was reduced from 50 to 35 days (Uriarte and Farías 1995). This was due, at least in part, to the fact that in postlarvae of *A. purpuratus* food absorption efficiencies may vary between 74.6% (± 2.7) for a high protein *T-Iso-C. gracilis* mixture to 30.8% for a mixture of the same microalgae cultured to have a low protein content. Intermediate values for absorption were observed using this mixture when the algae were cultivated in traditional medium (Uriarte and Farías 1995).

The above cited stimulative effect of the high protein diet could not be duplicated with "seed" scallops 6 mm in length, suggesting that there was a greater requirement for proteins in the diet at the postlarval stage than in more advanced stages of growth in these scallops (Uriarte and Farías 1999). Low-protein diets produce reduced growth which was only slightly superior to that observed when the postlarvae were held without food, and began to experience high mortality. Postlarvae maintained on low-protein microalgae show high respiratory rates, probably associated with the higher relative carbohydrate content of this diet. The significant decrease of protein in hatchery-produced juveniles of *A. purpuratus* compared to those maintained at sea as observed by Martínez et al. (1992) reiterates the importance of enrichment of proteins in the microalgae in controlled cultures.

Regarding lipids, the existence of mechanisms in juvenile bivalves have been proposed for the desaturation and elongation of fatty acids which allow them to synthesise EPA and DHA when these are unavailable from microalgae. Experiments with seed-sized *C. gigas*, *Ostrea edulis*, and *Venerupis pullastra* showed that their growth was less when they were fed microalgae deficient in n-3 HUFA than when they were fed with mixtures of microalgae which contained EPA and DHA, or when these lipids were added as dietary supplements (Langdon and Waldock 1981; Waldock and Holland 1984; Uriarte et al. 1990; Albentosa et al. 1994). This information has served as empirical evidence supporting the concept that fatty acids are necessary in the diets of juvenile molluscs. The importance of ARA as an essential fatty acid was demonstrated by Caers et al. (1998) working with juvenile *Tapes* sp. clams. She also indicated that the EPA/DHA and DHA/ARA ratios were those definitively affecting the degree of impact that supplementing these fatty acids could have on the growth of juvenile molluscs. Martínez et al. (1992) found that early juveniles of *A. purpuratus* had a higher content of DHA when they were cultured at sea than when cultured in the hatchery, suggesting that the hatchery diet used (*T-Iso*, *C. gracilis* and *C. calcitrans*) was deficient in this fatty acid.

In the culture of *Nodipecten subnodosus*, Velasco (1997) found that larvae fed with a mixture of *I. galbana* (*T-Iso*) and *C. calcitrans*, or with *T-Iso* alone, were able to develop to metamorphosis, while the same larval development could not be obtained with *C. calcitrans* alone. Although this author did not discuss the nutritional quality of the diets, it is probable that a deficiency of DHA in *C. calcitrans* was one of the principal factors responsible for this result. Postlarvae of *A. purpuratus* obtained in the laboratory by Martínez (1991) had a 25% lower content of EPA after metamorphosis of the larvae, while the levels of DHA and ARA remained high.

EPA and DHA are essential lipids in postlarvae and juvenile oysters, and deficiencies in these compounds provoke retardation in their growth (Chu and Webb 1984). Lipid emulsions enriched with EPA and DHA may efficiently replace up to 20% of a microalgal diet for *A. purpuratus* although not producing a significant increase in their growth rate (Uriarte et al. 2003). Nuñez (2002) studied the joint effect of protein in the microalgae and substitution of a portion of the microalgae by lipid emulsions enriched with EPA and DHA using *A. purpuratus* seed of 4 mm in length. Growth of the seed was not affected by the protein levels in the microalgal diet, but it was affected by the degree of replacement of the lipids; i.e., when 40% of the microalgae were replaced by lipids, there was a negative effect on the growth of the juveniles, while no differences were noted at replacement levels from zero to 20% (Fig. 9.3). Survival of these postlarvae was neither affected by the protein levels or by replacement of the lipids.

Carbohydrates may improve the growth of juvenile bivalves when their levels are raised in a microalgal diet, and once adequate levels of proteins and essential fatty acids have been provided (Enright et al. 1986). Increase in dietary carbohydrates may be achieved by cultivation of microalgae deficient in nitrogen, and adding adequate fractions to mixtures containing microalgae, which contain high levels of protein. Carbohydrase-type enzymes have been found in juvenile and adult bivalves (Samain et al. 1992), and for juvenile *P. maximus* the carbohydrate digestive capacity can limit the growth, mainly in the groups of 4-mm scallops (Økland et al. 2001). The greater accumulation of carbohydrates in juvenile scallops cultured under hatchery conditions compared with congeners cultured at sea demonstrated that microalgal diets produced in the hatchery contained higher energetic levels than those available in the environment (Martínez et al. 1992).

Although there are no studies on the vitamin requirements of postlarval pectinids, in accord with Seguineau et al. (1993) microalgal diets are capable of supplying sufficient levels of vitamin C and other vitamins required by most of the marine species.

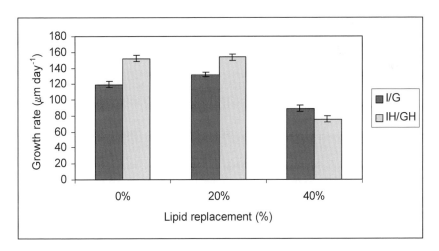

Figure 9.3. Growth rates of 4 mm *Argopecten purpuratus* seed fed with high protein (IH/GH) *T-Iso* + *C. neogracile* and normal protein (I/G) microalgae, and levels of substitution of microalgae by lipid emulsions EmDHA and EmEPA of 0, 20 and 40% of the dry weight of the microalgae. Each value is the mean of four replicates. Bar = 1 SE. (After Nuñez 2002).

9.6 SUMMARY

Fulfilment of the nutritional requirements of pectinids in the reproductive conditioning, larval culture, and postlarval culture phases is determined by the levels and quality of the proteins and lipids in the diets utilised in their culture. The protein content of the microalgal diet affects the time to reach maturity, the fecundity of the females, and the energy balance in the broodstock, showing that diets high in protein favour all these parameters and also improve the quality of the larvae. An increase in the protein content of microalgal food by manipulation of their culture medium significantly increases larval growth and also favours improved growth of the postlarval juveniles to the 5 mm size.

Pectinid larvae show greater growth and survival when the diets contain high levels of DHA/PA together with low values for EPA/HUFA, with EPA being less important because the larvae are efficient in desaturation beginning with the linolenic acid (18:3n-3), and also seem to have the capacity for obtaining ARA from linoleic acid (18:3n-6). The microalgal diet for pectinids in the larval phase may be substituted up to 40% by lipid emulsions enriched with EPA and DHA, while in the postlarval phase to 2 mm this substitution may be up to 20% of the microalgal diet without damaging growth and survival in this phase.

The carbohydrates begin to be important in late juvenile stages, which may be related to availability of enzymes in the digestive system where amylases only appear after the scallops reach the 5 mm size. Pectinids at all stages are assured of adequate levels of vitamins and pigments when fed with microalgae. Micronutrients in the marine environment of the scallops readily provide their mineral requirements by direct absorption from seawater.

REFERENCES

Albentosa, M., Labarta, U., Perez-Camacho, A., Fernández-Reiriz, M.J. and Beiras, R., 1994. Fatty acid composition of *Venerupis pullastra* spat fed on different microalgae diets. Comp. Biochem. Physiol. 108A:639–648.

Aviles, M., 1990. Crecimiento de la almeja catarina (*Argopecten circularis*) en función del alimento, con anotaciones sobre su biología y desarrollo. Tesis de Maestro en Ciencias, Instituto Politécnico Nacional, La Paz, México.

Bayne, B.L. and Hawkins, A.J.S., 1990. Filter-feeding in bivalve molluscs: controls on energy balance. In: Animal nutrition and transport processes. 1. Nutrition in wild and domestic animals. Comparative Physiology. Basel, Karger 5:70–83.

Beninger, P.G. and Stephan, G., 1985. Seasonal variations in the fatty acids of the triacylglycerols and phospholipids of two populations of adult clam (*Tapes decussatus* L. and *T. philippinarum*) reared in a common habitat. Comp. Biochem. Physiol. 81B:591–601.

Benlimame, N., Henry, M., Boucaud-Camou, E., Donval, A., Van Wormhoudt, A. and Mathieu, M., 1991. Scanning and transmission electron microscopic study of the stomach of *Pecten maximus* (L.): ultrastructural localization of amylase secreting cells. 9th International Pectinid Workshop. Cherbourg 1991.

Boucaud-Camou, E., Henry, M., Lefort, Y. and Mathieu, M., 1991. Fine structure and enzymes of the digestive gland of *Pecten maximus* (L.). 9th International Pectinid Workshop. Cherbourg 1991.

Bricelj, V.M. and Shumway, S., 1991. Physiology: Energy acquisition and utilization. In: S.E. Shumway (Ed.). Scallops: Biology, Ecology and Aquaculture. Developments in Aquaculture and Fisheries Science, Vol. 21. Elsevier, Netherlands. pp. 305–346.

Brown, M.R., 1991. The amino-acid and sugar composition of 16 species of microalgae used in mariculture. J. Exp. Mar. Biol. Ecol. 145:79–99.

Brown, M.R. and Farmer, C.L., 1994. Riboflavin content of six species of microalgae used in mariculture. J. Appl. Phycol. 6:61–65.

Brown, M.R. and Miller, K.A., 1992. The ascorbic acid content of eleven species of microalgae used in mariculture. J. Appl. Phycol. 4:205–215.

Caers, M., Coutteau, P., Cure, K., Morales, V., Gajardo, G. and Sorgeloos, P., 1999. The Chilean scallop *Argopecten purpuratus* (Lamarck, 1819); II. Manipulation of the fatty acid composition and lipid content of the eggs via lipid supplementation of the broodstock diet. Comp. Biochem. Physiol. 123B:97–103.

Caers, M., Coutteau, P., Lombeida, P. and Sorgeloos, P., 1998. The effect of lipid supplementation on growth and fatty acid composition of *Tapes philippinarum* spat. Aquaculture 162:287–299.

Chu, F.E. and Webb, K.L., 1984. Polyunsaturated fatty acids and neutral lipids in developing larvae of the oyster, *Crassostrea virginica*. Lipids 19:815–820.

Committee on Animal Nutrition, Board on Agriculture, National Research Council, 1993. Nutrient requirements of fish. National Academy Sciences. 114 p.

Coutteau, P., Caers, M., Curé, K. and Gajardo, G., 1996b. Supplementation of lipid emulsions to algal diets in the hatchery rearing of bivalves. In: G. Gajardo and P. Coutteau (Eds.). Improvement of the Commercial Production of Marine Aquaculture Species. Proceedings of a Workshop on Fish and Mollusc Larviculture. Impresora Creces, Santiago, Chile. pp. 145–154.

Coutteau, P., Caers, M., Malle, A., Moore, W., Manzi, J.J. and Sorgeloos, P., 1994. Effect of lipid supplementation on growth, survival and fatty acid composition of bivalve larvae. In: P. Kestemont, J. Muir, F. Sevila and P. Williot (Eds.). Measures for Success. CEMAGREF. pp. 213–218.

Coutteau, P., Castell, J.D., Ackman, R.G. and Sorgeloos, P., 1996a. The use of lipid emulsions as carriers for essential fatty acids in bivalves: a test case with juvenile *Placopecten magellanicus*. J. Shellfish Res. 15:259–264.

Coutteau, P. and Sorgeloos, P., 1992. The use of algal substitutes and the requirement for live algae in the hatchery and nursery rearing of bivalve molluscs: an international survey. J. Shellfish Res. 11:467–476.

Cranford, P.J. and Grant, J., 1990. Particle clearance and absorption of phytoplankton and detritus by the sea scallop *Placopecten magellanicus* (Gmelin). J. Exp. Mar. Biol. Ecol. 137:105–121.

Delaunay, F., Marty, Y., Moal, J. and Samain, J.F., 1993. The effect of monospecific algal diets on growth and fatty acid composition of *Pecten maximus* (L.) larvae. J. Exp. Mar. Biol. Ecol. 173:163–179.

Enright, C.T., Newkirk, G.F., Craigie, J.S. and Castell, J.D., 1986. Growth of juvenile *Ostrea edulis* L. fed *Chaetoceros gracilis* Schütt of varied chemical composition. J. Exp. Mar. Biol. Ecol. 96:15–26.

Escarria, S., Reyes, C. and Rohan, D., 1989. Estudio bioquímico de la escalopa *Argopecten circularis.* Ciencias Marinas 15:63–72.

Farías, A., 2001. Capítulo 5: Nutrición en Moluscos Pectínidos. In: A.N. Maeda-Martínez (Ed.). Los Moluscos Pectínidos de Iberoamérica: Ciencia y Acuicultura. Editorial Limusa, S.A. de C.V., Mexico City, Mexico. pp. 89–104.

Farías, A., Bell, J.G., Uriarte, I. and Sargent, J.R., 2003. Polyunsaturated fatty acids in total lipid and phospholipids of Chilean scallop *Argopecten purpuratus* (L.) larvae: Effects of diet and temperature. Aquaculture 228:289–305.

Farías, A. and Uriarte, I., 2001. Effect of microalgae protein on the gonad development and physiological parameters for the scallop *Argopecten purpuratus* (Lamarck, 1819). J. Shellfish Res. 20:97–105.

Farías, A., Uriarte, I. and Castilla, J.C., 1998. A biochemical study of the larval and postlarval stages of the Chilean scallop *Argopecten purpuratus.* Aquaculture 166:37–47.

Farías, A., Uriarte, I. and Varas, P., 1997. Estudio de los requerimientos nutricionales de ostión del norte *Argopecten purpuratus* (Lamarck, 1819) durante el acondicionamiento reproductivo. Rev. Biol. Mar. Oceanogr. Valparaíso 32:127–136.

Fontagne, S., Geurden, I., Escaffre, A. and Bergot, P., 1998. Histological changes induced by dietary phospholipids in intestine and liver in common carp (*Cyprinus carpio*, L.) larvae. Aquaculture 161:213–223.

Fundación CENAIM-ESPOL, 1997a. Cultivo de especies no tradicionales (moluscos). Informe técnico no publicado Proyecto n°P-BID-198. 1er semestre.

Fundación CENAIM-ESPOL, 1997b. Cultivo de especies no tradicionales (moluscos). Informe técnico no publicado Proyecto n°P-BID-198. 2° semestre.

Fundación CENAIM-ESPOL, 1998. Cultivo de especies no tradicionales (moluscos). Proyecto n°P-BID-198, Informe técnico tercer semestre.

Gardner, D. and Riley, J.P., 1972. The component fatty acids of the lipids of some species of marine and freshwater molluscs. J. Mar. Biol. Assoc. U.K. 52:827–838.

Geurden, I., Coutteau, P. and Sorgeloos, P., 1997. Effect of dietary phospholipid supplementation on growth and fatty acid composition of European sea bass (*Dicentrarchus labrax*) and turbot (*Scophthalmus maximus* L.) juveniles from weaning onwards. Fish Physiol. Biochem. 16:259–272.

Hernández, J., 2001. Efecto de la alimentación en reproductores de ostión del norte (*Argopecten purpuratus*) y ostra del Pacífico (*Crassostrea gigas*) sobre la maduración y la calidad de la progenie. Tesis de Ingeniería Pesquera, Universidad Austral de Chile. 73 p.

Hernández, J., Uriarte, I. and Farías, A., 2000. Estudio comparativo de las progenies del ostión del norte (*Argopecten purpuratus*) y la ostra del Pacífico (*Crassostrea gigas*), en reproductores acondicionados con distintas calidades de dieta. In: Abstracts of XX Congreso Ciencias del Mar, Concepción, Chile. 136 p.

Hunauld, P., Vélez, A., Jordan, N., Himmelman, J.H., Morales, F., Freites, L. and Lodeiros, C., 2002. Contribution of food availability to the more rapid growth of the scallop, *Euvola ziczac*, in bottom than in suspended culture. Rev. Biol. Trop. 50(4) (in press).

Langdon, C.J. and Waldock, M.J., 1981. The effect of algal and artificial diets on the growth and fatty acid composition of *Crassostrea gigas* spat. J. Mar. Biol. Assoc. U.K. 61:431–448.

Lora -Vilchis, M.C. and Maeda-Martínez, A.N., 1997. Ingestion and digestion index of catarina scallop *Argopecten ventricosus* = *circularis* (Sowerby II, 1842) veliger larvae with ten microalgae species. Aquaculture Research 28:905–910.

Lucas, A., 1982. La nutrition des larves de bivalves. Oceanis 8(5):363–388.

Machulas, J.P., 2000. Cálculo para la construcción de un sistema de suministro de agua de mar en el Hatchery UACh y Desarrollo del cultivo de ostión del norte, *Argopecten purpuratus* (Lamarck, 1819) utilizando emulsiones enriquecidas en lípidos. Tesis de Ingeniería Pesquera, Universidad Austral de Chile. 131 p.

Martínez, G., 1991. Estudios nutricionales en larvas de ostión *Argopecten purpuratus*. Informe Técnico, Fondo Desarrollo Productivo, CORFO, Chile. 40 p.

Martínez, G., Aguilera, C. and Mettifogo, L., 2000a. Interactive effects of diet and temperature on reproductive conditioning of *Argopecten purpuratus* broodstock. Aquaculture 183:149–159.

Martínez, G., Brokordt, K., Aguilera, C., Soto, V. and Guderley, H., 2000b. Effect of diet and temperature upon muscle metabolic capacities and biochemical composition of gonad and muscle in *Argopecten purpuratus* Lamarck 1819. J. Exp. Mar. Biol. Ecol. 247:29–49.

Martínez, G., Torres, M., Uribe, E., Díaz, M.A. and Pérez, H., 1992. Biochemical composition of broodstock and early juvenile Chilean scallops, *Argopecten purpuratus* Lamarck, held in two different environments. J. Shellfish Res. 11(2):307–313.

Marty, Y., Delaunay, F., Moal, J. and Samain, J.F., 1992. Changes in the fatty acid composition of the scallop *Pecten maximus* (L.) during larval development. J. Exp. Mar. Biol. Ecol. 163:221–234.

Millán, M.M., 1997. Experimentos de inducción a la maduración gonádica de *Argopecten ventricosus* (Sowerby, 1842) y estudio del valor nutricional de *Isochrysis galbana* y *Chaetoceros gracilis*, durante su crianza larvaria. Tesis de Maestría en Ciencias en Acuacultura, Universidad Autónoma de Baja California Sur. La Paz, México.

Napolitano, G.E., MacDonald, B.A., Thompson, R.J. and Ackman, R.G., 1992. Lipid composition of eggs and adductor muscle in giant scallops (*Placopecten magellanicus*) from different habitats. Marine Biology 113:71–76.

Napolitano, G.E., Ratnayake, W.M.N. and Ackman, R.G., 1988. Fatty acid components of larval *Ostrea edulis* (L.): importance of triacylglycerols as a fatty acid reserve. Comp. Biochem. Physiol. 90B:875–883.

Nevejan, N., Saez, I., Gajardo, G. and Sorgeloos, P., 2002. Relative importance of DHA and EPA for *Argopecten purpuratus* larvae (Lamarck, 1819). In: Taller Internacional de Larvicultura de Invertebrados Marinos: Estado del Arte y sus proyecciones. Universidad Austral de Chile, Puerto Montt. pp. 21–26.

Nuñez, M.J., 2002. Sustitución y suplementación de dieta viva con dieta inerte (emulsiones lipídicas) y su interacción con la temperatura en fundión del rendimiento larvario y postlarvario del ostión del norte, *Argopecten purpuratus*, en condiciones controladas. Tesis de Ingeniería Pesquera, Universidad Austral de Chile. 80 p.

Økland, S.N., Hoehne Reitan, K. and Reitan, K.I., 2001. Lipid and carbohydrate digestion in scallop (*Pecten maximus*) juveniles. In: C.I. Hendry, G. Van Stappen, M. Willie and P. Sorgeloos (Eds.). Larvi'01-Fish and Shellfish Larviculture Symposium. European Aquaculture Society, Special Publication No. 30, Oostende, Belgium. pp. 445–448.

Ortiz, G., 1994. Efecto de la ración alimenticia y la densidad de cultivo sobre el desarrollo larval de *Nodipecten subnodosus*. Tesis de Oceanologo, Universidad Autónoma de Baja California, México.

Pollero, R.J., Ré, M.E. and Brenner, R.R., 1979. Seasonal changes of the lipids of the mollusc *Chlamys tehuelcha*. Comp. Biochem. Physiol. 64A:257–263.

Racotta, I.S., Ramírez, J.L., Avila, S. and Ibarra, A., 1998. Biochemical composition of gonad and muscle in the catarina scallop, *Argopecten ventricosus*, after reproductive conditioning under two feeding systems. Aquaculture 163:111–122

Robinson, A., 1992a. Dietary supplements for reproductive conditioning of *Crassostrea gigas kumamoto* (Thunberg). I. Effects on gonadal development, quality of ova and larvae through metamorphosis. J. Shellfish Res. 11:437–441.

Robinson, A., 1992b. Dietary supplements for reproductive conditioning of *Crassostrea gigas kumamoto* (Thunberg). II. Effects on glycogen, lipid and fatty acid content of broodstock oyster and eggs. J. Shellfish Res. 11:443–447.

Román, G. and Campos, M.J., 1993. Acondicionamiento de *Pecten maximus*. In: A. Cervino, A. Landin, A. de Coo, A. Guerra and M. Torre (Eds.). Actas IV Congreso Nac. Acuicult. Centro de Investigaciones Marinas, Pontevedra, Spain. pp. 305–310.

Samain, J.F., Séguineau, C., Cochard, J.C., Delaunay, F., Nicolas, J.L., Marty, Y., Galois, R., Mathieu, M. and Moal, J., 1992. What about growth variability for *Pecten maximus* production? Oceanis 18:49–66.

Sargent, J., Bell, G., McEvoy, L., Tocher, D. and Estevez, A., 1999a. Recent developments in the essential fatty acid nutrition of fish. Aquaculture 177:191–199.

Sargent, J., McEvoy, L., Estevez, A., Bell, G., Bell, M., Henderson, J. and Tocher, D., 1999b. Lipid nutrition of marine fish during early development: current status and future directions. Aquaculture 179:217–229.

Sargent, J.R. and Tacon, A.G.J., 1999. Development of farmed fish: a nutritionally necessary alternative to meat. Proceedings of the Nutrition Society 58:377–383.

Seguel, M., Osses, J., Farías, A., Avila, M., Plaza, H. and Pavez, H., 1998. Evaluación del efecto de mitigación de aporte de nutrientes al medio al desarrollar policultivos marinos de salmónidos. Informe final FIP 94–01. Instituto de Fomento Pesquero. 192 p.

Seguineau, C., Laschi-Loquiere, A., Leclerq, M., Samain, J.F., Moal, J. and Fayol, V., 1993. Vitamin transfer from algal diet to *Pecten maximus* larvae. J. Mar. Biotechnol. 1:67–71.

Shumway, S.E., Selvin, R. and Schick, D.F., 1987. Food resources related to habitat in the scallop Placopecten magellanicus (Gmelin, 1791): A quantitative study. J. Shellfish Res. 6:89–95.

Sorgeloos, P., 1998. Progress in larviculture nutrition of fish and shellfish. XXXIII International Symposium of New Species for Mediterranean Aquaculture. Alghero, 22–24 April 1998. pp. 41–43.

Soudant, P., Marty, Y., Moal, J., Masski, H. and Samain, J.F., 1998. Fatty acid composition of polar lipid classes during larval development of scallop *Pecten maximus* (L.). Comp. Biochem. Physiol. 121A:279–288.

Tacon, A., 1990. Standard methods for the nutrition and feeding of farmed fish and shrimp. Volume I. The essential nutrients. Argen Laboratory Press, Redmond, Washington, USA. 117 p.

Thompson, P.A., Guo, M.X. and Harrison, P.J., 1993. The influence of irradiance on the biochemical composition of three phytoplankton species and their nutritional value for larvae of the Pacific oyster (*Crassostrea gigas*). Mar. Biol. 117:259–268.

Thompson, P.A., Guo, M.X. and Harrison, P.J., 1994. Influence of irradiance on the nutritional value of two phytoplankton species fed to larval Japanese scallops (*Patinopecten yessoensis*). Mar. Biol. 119:89–97.

Trider, D.J. and Castell, J.D., 1980. Influence of neutral lipid on seasonal variation of total lipid in oysters, *Crassostrea virginica*. Proceedings of the National Shellfisheries Association 70:112–118.

Uriarte, I., 2000. Informe de Avance n°3, FONDECYT 1970807, Chile. 11 p.

Uriarte, I. and Farías, A., 1995. Effect of broodstock origin and postlarval diet on postlarval performance of the Chilean scallop *Argopecten purpuratus*. In: P. Lavens and I. Roelants (Eds.). Larvi'95- Fish and Shellfish Larviculture Symposium. European Aquaculture Society, Special Publication 24, Gent, Belgium. pp. 69–72.

Uriarte, I. and Farías, A., 1999. The effect of dietary protein on growth and biochemical composition of Chilean scallop *Argopecten purpuratus* (L.) postlarvae and spat. Aquaculture 180:119–127.

Uriarte, I., Farías, A., Hernández, J., Schäfer, C. and Sorgeloos, P., 2004. Reproductive conditioning of Chilean scallop (*Argopecten purpuratus*) and the Pacific oyster (*Crassostrea gigas*): effects of enriched diets. Aquaculture 230:349–357.

Uriarte, I., Farías, A., Navarro, J.M., Cancino, J.M., Gajardo, G. and Nevejan, N., 2003. The effects of lipid emulsions and temperature on the hatchery performance of Chilean scallop *Argopecten purpuratus* (Lamarck, 1819) larvae. Aquaculture Research 34:899–902.

Uriarte, I., Farías, A., Peña, J.B. and Mestre, S., 1990. Efecto de diferentes dietas microalgales en la composición de ácidos grasos y aminoácidos de *Ostrea edulis* L. Iberus 9:261–267.

Uriarte, I., Rupp, G. and Abarca, A., 2001. Capítulo 8: Producción de juveniles de pectínidos iberoamericanos bajo condiciones controladas. In: A.N. Maeda-Martínez (Ed.). Los Moluscos Pectínidos de Iberoamérica: Ciencia y Acuicultura. Editorial Limusa, S.A. de C.V., Mexico City, Mexico. pp. 147–171.

Utting, S. and Millican, P.F., 1997. Techniques for the hatchery conditioning of bivalve broodstocks and the subsequent effect on egg quality and larval viability. Aquaculture 155:45–54.

Valdivieso, V., Loyola, C. and Gallegos, F., 1988. Producción experimental en semillas de concha de abanico en ambiente controlado, Callao-Perú. In: Rev. Pacífico Sur (número especial), Comisión Permanentes del Pacífico Sur (CPPS). Memorias del Simposio Internacional de los Recursos Vivos y las Pesquerías en el Pacífico sudeste. Viña del Mar. pp. 601–606.

Velasco, G., 1997. Cultivo larvario a nivel piloto del callo de hacha *Atrina maura* Sowerby y de la almeja mano de león *Nodipecten subnodosus* Sowerby con dos especies de microalgas. Tesis de Maestro en Ciencias, Centro de Investigación Científica y Educación Superior de Ensenada, Baja California, México.

Vernet, M., 1977. Alimentación de la vieira tehuelche (*Chlamys tehuelchus*). Comisión Nacional de estudios Geo-Heliofísicos, Centro Nacional Patagónico. Chubut, Argentina.

Villalaz, J.R., 1994a. Morphometric and biochemical changes in two age classes of the tropical scallop, *Argopecten ventricosus*, under laboratory conditions. Amer. Malac. Bull. 11:67–72.

542

Villalaz, J.R., 1994b. Laboratory study of food concentration and temperature effect on the reproductive cycle of *Argopecten ventricosus*. J. Shellfish Res. 13:513–519.

Villavicencio, G., 1997. Acondicionamiento gonadal, desarrollo embrionario y cultivo de larvas de la almeja mano de león, *Nodipecten subnodosus* (Sowerby, 1835), alimentadas con cuatro especies de microalgas. Tesis de Ingeniero en Acuicultura, Secretaría de Educación Pública, Instituto tecnológico del Mar, La Paz, México.

Waldock, M.J. and Holland, D.L., 1984. Fatty acid metabolism in young oysters, *Crassostrea gigas*: polyunsaturated fatty acids. Lipids 19:332–336.

Waldock, M.J. and Nascimento, I.A., 1979. The triacylglycerol composition of *Crassostrea gigas* larvae fed on different algal diets. Marine Biology Letters 1:77–86.

Ward, J.E., Cassell, H.K. and MacDonald, B.A., 1992. Chemoreception in the sea scallop *Placopecten magellanicus* (Gmelin). I. Stimulatory effects of phytoplankton metabolites on clearance and ingestion rates. J. Exp. Mar. Biol. Ecol. 163:235–250.

Webb, K.L. and Chu, F.E., 1983. Phytoplankton as a food source for bivalve larvae. In: G.D. Pruder, C.J. Langdon and D.E. Conklin (Eds.). Proceedings of the Second International Conference on Aquaculture Nutrition and Physiological Approaches to Shellfish Nutrition, Oct 27–29, 1981. Louisiana State Univ. pp. 272–336.

Whyte, J.N.C., Bourne, N. and Hodgson, C.A., 1989. Influence of algal diets on biochemical composition and energy reserves in *Patinopecten yessoensis* (Jay) larvae. Aquaculture 78:333–347.

Wildish, D.J., Kristmanson, D.D., Hoar, R.L., DeCoste, A.M., McCormick, S.D. and White, A.W., 1987. Giant scallop feeding and growth responses to flow. J. Exp. Mar. Biol. Ecol. 113:207–220.

Wojtowickz, M.B., 1972. Carbohydrases of the digestive gland and the crystalline style of the Atlantic deep-sea scallop (*Placopecten magellanicus*, Gmelin). Comp. Biochem. Physiol. 43A:131–141.

AUTHORS ADDRESSES

Ana Farías - Instituto de Acuicultura, Universidad Austral de Chile. Av. Los Pinos, s/n, Balneario Pelluco, P. O. Box 1327, Puerto Montt, Chile (E-mail: afarias@uach.cl)

Iker Uriarte - Instituto de Acuicultura, Universidad Austral de Chile. Av. Los Pinos, s/n, Balneario Pelluco, P. O. Box 1327, Puerto Montt, Chile (E-mail:iuriarte@uach.cl)

Scallops: Biology, Ecology and Aquaculture
S.E. Shumway and G.J. Parsons (Editors)
© 2006 Elsevier B.V. All rights reserved.

543

Chapter 10

Genetics

Andy Beaumont

10.1 INTRODUCTION

Before protein electrophoresis was developed and applied to the study of population genetics (Harris 1966; Lewontin and Hubby 1966) little was known of the genetics of any scallop species, or indeed any other bivalve mollusc. Naturally, early use of protein electrophoresis to detect allozyme loci concentrated on those species such as humans and fruit flies about which much was already known, but as the methodology became more easily accessible and cheaper, allozyme polymorphisms were investigated in a wider range of species. A glance at the references for this chapter will illustrate that the earliest publications on scallop allozymes appeared some ten years after the method had been developed (Beaumont and Gruffydd 1975; Mathers 1975). Although chromosome genetics had been studied in many organisms much earlier than this, descriptions of chromosome numbers and karyotypes of scallops were also only appearing in the literature at around this time (Beaumont and Gruffydd 1974; Ieyama 1975).

In general, genetic studies in scallops have not been aimed at the furtherance of the science of genetics, but rather at the application of genetic methods to problems relating to the biology of the species. These problems include the phylogenetic relationships among species, the degree of gene flow among conspecific populations, the degree to which life history parameters or physiological and adaptational responses may be determined by genetic differences among individuals within populations. Of particular importance, also, is the increasing awareness of the relevance of genetics to the fishery and aquaculture of scallops.

Because this chapter appears in a volume intended as a general introduction to the biology and ecology of scallops and their potential for aquaculture, the assumption is made that not all readers will be familiar with the genetic methodology or terminology described. Therefore short explanations and discussions of relevant techniques and concepts are provided at appropriate points together with further references to the wider literature. For a general introduction to biotechnology and genetics in fisheries and aquaculture see Beaumont and Hoare (2003).

For the purpose of review the various studies of the genetics of scallop species could be classified in at least three different ways: according to the genetic approaches employed, according to species studies, or according to the question being addressed. The first approach has been adopted here and the chapter is divided into three main sections. The first covers chromosomes and ploidy manipulation, the second introduces genetic markers (both protein and DNA-based) and their use in population genetic and systematics, and the final section deals with hybridisation, inbreeding and quantitative genetics. Because of the extent of the information available within the second section, a part of it is subdivided on a species by species basis.

10.2 CHROMOSOMAL GENETICS AND PLOIDY MANIPULATION

10.2.1 Chromosome Numbers and Karyotypes

The chromosomes are normally invisible in the cell nucleus but become visible during the process of cell division when they coil and contract and become associated with a microfibrillar structure, the spindle. The diploid (2N) number of chromosomes present in each somatic cell consists of a haploid (N) number of homologous pairs of chromosomes of identical size and shape except where sex chromosomes are present. During mitotic cell division each chromosome is replicated to form 2 chromatids which divide and separate into the daughter cells. During the production of gametes from the germ cells, however, two maturation divisions take place, meiosis I and meiosis II. The first is a reduction division in which each resultant nucleus receives one chromosome from each of the homologous pairs and is therefore haploid. The second maturation division, meiosis II, is effectively a mitotic division of the chromosomes into chromatids. The consequences of meiosis I and II are therefore the production, from one germ cell, of four haploid nuclei. In males, four spermatozoa are produced from each germ cell but in females unequal cytoplasmic division during meiosis I results in one haploid set of chromosomes remaining in the egg while the other set is extruded within the first polar body. Similarly, the second polar body contains one of the sets of chromosomes derived from meiosis II. The first polar body may occasionally undergo its own meiosis II giving three polar bodies.

In scallops, as in most other bivalves, which release their eggs directly into the sea, the chromosomes in the egg are initially uncontracted and contained within the nuclear membrane, the germinal vesicle. In laboratory spawnings, both *P. magellanicus* and *P. maximus* eggs are initially non-spherical but soon round off (Gruffydd and Beaumont 1970; Desilets et al. 1995). According to Desilets et al. (1995) the germinal vesicle has broken down by the time *P. magellanicus* eggs are released, but in *P. maximus* it is possible to identify the germinal vesicle in the non-spherical eggs on their release and the breakdown of the vesicle is coincident with the eggs becoming spherical. At some point, which may therefore vary between species, the germinal vesicle breaks down and the homologous pairs of chromosomes come to lie, as bivalents, on the metaphase plate of the meiosis I spindle (Metaphase I). Scallop eggs then remain at this stage until activated by sperm attachment. Desilets et al. (1995) report that *P. maximus* oocytes are fertilised at the germinal vesicle stage, but all our observations on *P. maximus* indicate that the germinal vesicle has almost always broken down before egg activation. Nevertheless, one of the difficulties in establishing the earliest possible timing of egg activation in laboratory trials is that eggs are usually kept separate from spermatozoa before mixing, or samples are not taken immediately on release. In the case of *P. maximus*, which is a hermaphrodite, a small proportion of eggs are usually found to be self-fertilised. In this case, it is possible that spermatozoa attach to the egg during passage along the gonoduct, or immediately on release, and eggs at this point would still have an intact germinal vesicle.

Once activated by the attachment of spermatozoa, meiosis I and II are completed and the first and second polar bodies are extruded sequentially at the same position on the egg surface. In the laboratory, in 32‰ filtered seawater at 15°C, the first polar body is usually extruded from *P. maximus* eggs within 25 minutes and the second within 50 minutes from activation (Gruffydd and Beaumont 1970). The process seems to be a little slower in *Placopecten magellanicus*, with anaphase I at 30 minutes and anaphase II at 60 minutes after activation of the egg by spermatozoa (Desilets et al. 1995). However, the trials with *P. magellanicus*, although at the same temperature (15°C), were carried out in pH-adjusted seawater at 28‰ salinity and this may have affected the speed of development. Lower temperatures clearly slow the process because, in *Patinopecten yessoensis*, anaphase I and II occur at around 60 and 100 minutes respectively at a temperature of 9°C (Li et al. 2000). While the meiotic divisions are taking place, the sperm head enters the egg and the male pronucleus migrates to join the female pronucleus on the metaphase plate of the first cleavage spindle and, as such, there is no true zygotic nucleus formed before syngamy (Desilets et al. 1995; Li et al. 2000).

Chromosome counts and karyotypes in early studies on bivalves were usually obtained by observing the chromosomes at meiosis I and II either in released eggs or in male gonad, or during the early mitotic divisions of the embryo (e.g., Beaumont and Gruffydd 1974). However, Thiriot-Quievreux and Ayraud (1982) developed a reliable method of obtaining metaphase chromosomes from bivalve gill tissue, which is now widely used. In order to obtain mitotic chromosomes at the metaphase stage when they are most contracted and most easily identifiable, embryos or gill tissues are treated with colchicine, which arrests mitosis at metaphase. In a brief review, Menzel (1968) gave chromosome numbers of 23 species in 9 families of bivalves and in a more extensive paper Thiriot-Quievreux (1994) reported on the chromosome numbers of all aquatic groups. Generally, chromosome numbers tend to be relatively constant within families and Thiriot-Quievreux (1994) observed the following modal haploid numbers for oysters (N = 10), mussels (N = 14) and scallops (N = 19).

The chromosome numbers of seventeen pectinid species and sub-species are given in Table 10.1 and although many species have the modal number n = 19 (Thiriot-Quievreux 1994) there is considerable variation in chromosome number within the group. Some species in the genus *Chlamys* have n = 19, (*C. farreri* and sub-spp., *C. distorta* and *C. islandica*) while *C. nobilis* has n = 16 and *C. glabra* has n = 14. Both *Argopecten* species studied have n = 16. Beaumont and Gruffydd (1974) first reported n = 13 for *Aequipecten opercularis* but Rasotto et al. (1981) observed n = 14 for this species. Insua et al. (1998) confirm without doubt that n = 13 for Atlantic coast *A. opercularis*. There are reports of differing chromosome numbers within a single mollusc species (*Nucella lapillus*, Hoxmark 1970) although this is a very unusual occurrence and it is possible that Mediterranean populations of *A. opercularis* could have a different chromosome number from other populations. However, it is more likely that the n = 14 reported for

Table 10.1

Chromosome numbers and karyotypes in Pectinidae. Centromere positions t/a = telocentric or acrocentric, s = sub-telocentric, sm = sub-metacentric, m = metacentric. * = described in Brand et al. (1990) as " ... 7 metacentric (M) or sub-metacentric (SM) pairs ... "

Species	N	2N	t/a	t/st	st	st/sm	sm	sm/m	m	Reference
Placopecten magellanicus	19	38								Beaumont and Gruffydd 1974
Patinopecten (Mizuhopecten) yessoensis	19	38	3		3	6	4	1	2	Komaru and Wada 1985; Yabu and Maru 1987
Nodipecten nodosus	19	38			7		6		6	Pauls and Affonso 2000
Pecten maximus	19	38	14		1		2		2	Beaumont and Gruffydd 1974
Pecten jacobeus	19	-								Rasotto et al. 1981
Pecten albicans	19	38	12		4			3		Ieyama 1975
Pecten albicans	-	38	14			3		1	1	Komaru and Wada 1985 Beaumont and Gruffydd 1974;
Mimachlamys varia	19	38			4		10		5	Rasotto et al. 1981; Baron et al. 1989
Chlamys farreri	19	38			11		5		3	Wang et al. 1990; Yang et al. 2000
Chlamys farreri farreri	19	38		2	7	6	1		3	Komaru and Wada 1985
Chlamys farreri nipponensis	19	38								Yabu and Maru 1987
Chlamys distorta	19	38								Beaumont and Gruffydd 1974
Chlamys islandica	19	-								Beaumont and Gruffydd 1974
Chlamys nobilis	16	32	13						3	Komaru and Wada 1985; Komaru et al. 1988
Chlamys glabra	14	-								Rasotto et al. 1981
Argopecten irradians	16	32								Wada 1978
Argopecten purpuratus	16	32	4		3			7*	2	Brand et al. 1990
Argopecten purpuratus	16	32	11		5					Gajardo et al. 2002
Aequipecten opercularis	14	-								Rasotto et al. 1981
Aequipecten opercularis	13	26								Beaumont and Gruffydd 1974
Aequipecten opercularis	13	26	7					4	2	Insua et al. 1998

Mediterranean *C. opercularis* (Rasotto et al. 1981) was incorrect and was due to chromosome breakage, a feature noted by the authors.

How might the variant chromosome numbers in the group have evolved from the modal number? Evolution of chromosome numbers generally depends upon breakage and/or fusion of the chromosomes but these events must involve the centromere, the non-staining, constricted region of the chromosome, which forms the attachment point of the chromosome onto the spindle (White 1973). A chromosome portion with no centromere will generally be lost during cell division since it cannot attach to the spindle.

As the evolutionary relationships between various scallop groups have become better understood (Waller 1991), genus names have been changed to reflect these relationships. For example, *A. opercularis* was previously included in the genus *Chlamys*, but evidence from various sources supported its reassignment to the genus *Aequipecten* (Beaumont 1991a; Waller 1991). More recently, using molecular data from the 16S rRNA molecule, Canapa et al. (2000b) have shown that *Chlamys glabra* clusters with *A. opercularis* (with which it shares a similar chromosome number) and is completely separate from the other *Chlamys* species included in their study (*C. islandica*). It is probable that the inclusion of *C. glabra* in the genus *Chlamys* will be reassessed in the light of these observations.

Species karyotypes are determined by examination of the morphology of the chromosomes, principally by their length and by the position of the centromere which can be terminal or almost terminal (telocentric or acrocentric), median (metacentric) or somewhere in between (sub-telocentric, sub-metacentric). In addition, different staining methods have been employed to identify certain chromosomal regions, for example, silver staining for the nucleolar organising region (NOR), C-banding for localisation of heterochromatin, or G-banding (Thiriot-Quievreux 1994). Modern DNA technology has made possible the localisation of satellites, telomeric sequences and individual genes (e.g., ribosomal RNA genes) using the technique of fluorescent in-situ hybridisation (FISH) (Insua et al. 1998).

Karyotypes have been studied in several scallop species (Table 10.1). The first point to note is that two different authors present two different karyotypes for *P. albicans*. Ieyama (1975) reported 12 telocentric and 7 non-telocentric pairs while Komaru and Wada (1985) identified 14 telocentric and 5 non-telocentric pairs. These differences are most likely due to differences in chromosome preparation methods although different karyotypic races can occur within species. The second observation that can be made is that there is no consistency of centromeric position in the karyotype of those species containing the modal number of 19 chromosome pairs. For example, *P. maximus* has 14 telocentric pairs while *P. yessoensis* has only 3 and *C. farreri farreri* apparently has none. Therefore, evolution of scallop chromosomes appears to have involved considerable changes in chromosome arm lengths without loss of whole chromosomes or centromeres. Where loss of chromosomes has occurred, for example, in *C. nobilis* where N = 16, it can be postulated that 'Robertsonian' centric fusion (White 1973) has taken place. This is the process whereby the centromeres of two telocentric or acrocentric chromosomes fuse to produce a single non-telocentric chromosome. Three such fusions within an ancestral karyotype of 19 telocentric pairs would produce 13 telocentric and 3 metacentric pairs of chromosomes which is the precise karyotype reported for *C. nobilis* by Komaru and Wada (1985). Of the two other species having n = 16, no karyotype data are published for the bay scallop *A. irradians* and the karyotype for Chilean scallop, *A. purpuratus* (Brand et al. 1990) is very different with only 4 telocentric chromosome pairs. However, bearing in mind that there is evidence of considerable evolutionary changes in chromosome arm lengths in scallop species with the modal number, it is not surprising to observe no clear evidence of centric fusion in other species.

The most detailed karyotypic study in any of the pectinids is that by Insua et al. (1998) on *A. opercularis*. They demonstrated that the C-banding pattern showed heterochromatin blocks of differing staining intensity in various regions of all the chromosomes and this assisted in the identification of some specific chromosomes. In contrast, C-banding in mussels and oysters is usually observed in telomeric or metameric positions, and not on every chromosome. A single strong consistent NOR was located to a particular telocentric chromosome (No. 7) and this was confirmed by FISH localisation of an 18S-28S rRNA probe to the same position. Finally, in the first such study in molluscs, the 5S rRNA gene was located in one arm of the largest metacentric chromosome (No. 1) (Insua et al. 1998). Telomeric sequences have now been isolated and sequenced from the bay scallop (*A. i. irradians*) by Estabrooks (1999) and these will make useful tools for the study of the karyotype of this species. In addition, because the number of repeats of telomeric sequences are related to senescence, and the bay scallop is essentially an annual species, these sequences will be valuable in the study of longevity in scallops.

10.2.2 Ploidy Manipulation

Artificial changes in ploidy can be induced during the egg maturation divisions in fish and shellfish. The principle technique used in ploidy manipulation involves allowing chromosome division (karyokinesis) but preventing cytoplasmic division (cytokinesis) at meiosis I, meiosis II or first cleavage division. This can be achieved by subjecting eggs to a temperature shock, a pressure shock or to chemical action (cytochalasin B (CB) or 6-dimethyl amino-purine, (6-DMAP)) at the appropriate time during the maturation or first cleavage divisions. Triploids are produced by suppression of either meiosis I or meiosis II and tetraploids by suppression of first cleavage. A further technique is the use of irradiated spermatozoa, together with suppression of meiosis I, meiosis II or first cleavage, to produce gynogenetic diploid offspring. Four main methods have been employed to identify polypoid individuals; direct chromosome counting in early embryos, flow cytometry of larvae or adults, DAPI staining or electrophoresis of juveniles or adults. For a review of ploidy manipulation in molluscan shellfish, see Beaumont and Fairbrother (1991) where methods and consequences are more fully discussed.

Triploid shellfish exhibit features of value to the aquaculturist. During gametogenesis in diploid individuals, homologous pairs of chromosomes synapse early during meiosis I, but in triploids, pairing of the chromosomes is likely to be interfered with by the presence of a third homologous chromosome. For this reason, triploid individuals are expected to be sterile and evidence from triploid finfish and bivalves indicate that the sterility generally manifests itself early during gonad development, producing animals with small or very reduced gonads (Beaumont and Fairbrother 1991). Because less energy is utilised in gonad production, adult triploids are expected to have a higher somatic production than mature diploids and this extra meat weight is of economic importance to aquaculture. In the case of oysters, the lack of extensive reproductive tissue also enables year-round marketing (Allen 1988). A further advantage of sterility is that triploids of exotic species can be cultivated without danger of spawning under cultivation and competing with indigenous species. Recently, evidence of reversion to diploidy in older oysters, which were initially triploids has caste doubt on the value of triploids for this purpose (Shatkin et al. 1997; S. K. Allen, pers. comm.).

Methods for inducing triploidy by physical or chemical shock are seldom 100% effective. Therefore the potential value of producing tetraploid individuals is that they can be used as broodstock to provide diploid gametes which, when used with haploid gametes from diploids, will give 100% triploid offspring. In principle, tetraploids can be produced by suppression of the first cleavage division and this has proven to be a practical method for producing tetraploid fish. However, attempts to use this method in bivalves have always failed: although small numbers of tetraploid embryos can be produced, these do not survive to maturity. Guo and Allen (1994) have nevertheless developed a method for production of tetraploid oysters (C. gigas), which relies upon suppression of meiosis I in eggs produced by triploid females. Although only a few eggs are produced, and only a small proportion of the treated eggs produce tetraploid embryos, many of these tetraploid embryos do survive to maturity and are effective as broodstock (Guo et al. 1996).

Triploids have been produced in a number of pectinid species: A. irradians (Tabarini 1984), P. maximus (Beaumont 1986; Clarke 1991), C. nobilis (Komaru et al. 1988; Komaru and Wada 1989, 1991; Zeng et al. 1995), M. varia (Baron et al. 1989), A. purpuratus (Toro et al. 1995) and Argopecten ventricosus (= circularis) (Ruiz-Verdugo et al. 2000).

The cytochalasin B (CB) chemical method was used in most triploid induction studies on scallops. Toro et al. (1995) used heat shock successfully on A. purpuratus and Komaru and Wada (1989, 1991) tried pressure treatment on C. nobilis, but with limited success. As with other bivalves (Beaumont and Fairbrother 1991), a major side effect of the use of CB was a significant reduction in the number of larvae produced (Tabarini 1984; Beaumont 1986; Baron et al. 1989; Ruiz-Verdugo et al. 2000).

Higher overall heterozygosity is predicted in triploids compared with diploids and, because multi-locus heterozygosity (MLH — see later) is often positively correlated with size at age in many bivalves, then we might expect to detect faster growth in triploids compared with diploids. In studies where reliable data on larval growth of triploids have been collected, results are equivocal. For P. maximus, Beaumont (1986) and Clarke (1991, see Beaumont 2000a) found that 3N larvae grew significantly faster than 2N larvae, but no such trend was evident when comparing growth of 3N and 2N larvae of M. varia (Baron et al. 1989) or A. purpuratus (Toro et al. 1995). There has been some uncertainty in the bivalve literature about whether meiosis I triploids might have higher heterozygosity than meiosis II triploids (Stanley et al. 1984; Beaumont and Kelly 1989; Beaumont and Fairbrother 1991; Hawkins et al. 1994), but this actually depends on recombination rates of the loci which are being scored (Beaumont 2000a). Across the genome as a whole there is probably little difference between the heterozygosity of meiosis I and meiosis II triploids. Also, the detection of slight, and barely significant differences in growth rates between cohorts of 3N and 2N larvae will depend on the proportion of the 3N cohort, which are actually triploid (often <80%). Finally, speculation about any marginal relationship between heterozygosity and growth in triploids, relative to diploids, may be outweighed by the potential effect of polyploid gigantism as proposed by Guo and Allen (1994) and Guo et al. (1996).

Twelve month old triploid *A. irradians* had a significantly higher mean adductor muscle weight and total body tissue weight than their diploid counterparts (Tabarini 1984). Although shell height and length were not significantly different, shell inflation was significantly greater in triploids compared to diploids and, as observed in other bivalves, the majority of triploid scallops failed to ripen or spawn. The effect of triploidy on the adductor muscle index was not significant at 6 months but only became significant after gonad growth had occurred in the diploids. A similar pattern was seen in triploid *C. nobilis* (Komaru and Wada 1989) with significant differences in meat weight and shell width between 3N and 2N scallops only becoming evident at 14 months. In the case of *A. ventricosus* (= *circularis*) (Ruiz-Verdugo et al. 2000), performance of 4 to 14 month old triploid scallops was significantly better than diploids in all traits measured (shell dimensions, tissue weights, muscle index) except for gonad index, which was significantly depressed in triploids as the scallops reached maturity. Interestingly, the normal hermaphrodite status of this species was radically distorted with all triploids eventually becoming female (Ruiz-Verdugo et al. 2000).

These studies have demonstrated that significant increases in weight or size of triploid over diploid scallops are only detectable when triploid sterility becomes evident. This suggests that neither increased heterozygosity, nor polyploid gigantism play a major role in increased size of triploid scallops. Considering the impressive effect of triploidisation on muscle weight and size, the aquacultural potential of triploids scallops is high if it is the "roe-off" market (mainly the USA), which is targeted. However, there is no advantage to culturing triploid scallops for the "roe-on" or whole-body market, in spite of their greater muscle weight, because of the underdeveloped gonad and its potential discolouration in triploids (Ruiz-Verdugo et al. 2000).

10.3 GENETIC MARKERS AND POPULATION GENETICS

10.3.1 Genetic Markers

In order to study genetic variation at the individual or population level it is necessary to use markers of known genes or of portions of the genome. The development of markers of protein variation — allozymes — in the 1960's paved the way for the first extensive studies of the genetics of non-human organisms, which like the scallop, were neither important agricultural crops (e.g., maize) nor amenable to genetic studies by virtue of ease of handling and short generation time (e.g., fruit flies).

Part of the motivation for population genetic studies in scallops came from the relative success that allozyme studies were having in addressing population and species level problems in other marine bivalves, particularly mussels and oysters. Moreover, in several bivalve species, certain interesting genetic phenomena such as heterozygote deficiency and the correlation between degree of heterozygosity and growth or viability, had become apparent. These studies not only increased our understanding of the biology of the species concerned, but also made substantial contributions to several fundamental problems in population genetics and evolution, such as the nature of heterosis, the genetic control of variation of flux along enzymatic pathways, the physiological basis of variable growth. For a wider discussion of these phenomena see reviews by Zouros and Pogson (1994), Gaffney (1994) and Britten (1996) as well as papers by Koehn et al. (1988), Hawkins et al. (1994) and Hedgecock et al. (1996).

In the last 25 years of the 20th Century there have been several significant developments in DNA technology which have led to the routine use of several novel genetic marker systems based directly on variation at the level of the DNA molecule (Beaumont and Hoare 2003). The ways in which allozymes, and these new markers, have been used to address questions on the genetics of scallops will be presented and discussed following an outline of the principle techniques, which have been employed.

10.3.1.1 Allozymes

The demonstration of electrophoretic separation and identification of proteins — usually enzymes — on gels introduced a novel and simple methodology for investigating genetic variability in individuals and populations. Soluble proteins from tissue samples are placed at an origin point on a flat bed gel of starch or polyacrylamide and an electric current passed through the gel. The proteins, which are charged molecules, migrate along the gel and are separated according to their different mobilities. Identification of the final position of the proteins or enzymes is achieved by the use of general protein or specific enzyme stains.

Mutations of a gene at a locus may alter the charge, size or configuration of the protein product of that locus and these alterations will generally affect the mobility of the protein on a gel. Thus, the protein products of different alleles at a locus can be detected. Homozygotes will exhibit a single band on the gel and heterozygotes of monomeric enzymes, two bands. Banding patterns for heterozygotes of dimeric and other oligomeric enzymes reflect the random association of sub-units and consist of three or more bands. Thus, electrophoresis of a sample of individuals from a population can provide an estimate of genotype frequencies, and, by simple calculation, allele frequencies at protein encoding loci.

10.3.1.2 DNA methods

Routine procedures exist for the extraction of DNA from animals and its separation from other cellular constituents (Sambrook et al. 1989). Isolated DNA can be "cut and pasted" using restriction enzymes (REs), which cut at specific nucleotide sequence sites, and DNA ligases, which enable fragments of DNA to be joined. DNA can be stained by ethidium bromide or by radiolabelling, which is much more sensitive. However, in order to visualise a fragment of DNA, or to sequence it, the fragment needs to be replicated and millions of copies produced. This can be achieved by two main methods. Firstly, DNA can be cloned in bacteria by inserting the DNA fragment into the DNA molecule of a vector. The vector can be either a plasmid or a virus (phage), which can enter bacterial cells and be replicated along with the bacteria during its normal growth. Thus, bacterial colonies containing millions of copies of particular fragments can be produced. The second method for copying, or amplifying DNA is the polymerase chain reaction (PCR) technique, which uses a pair of short (15–30 bp) oligonucleotide primers each complementary to one end of the target DNA. These are extended towards each other by a thermostable DNA polymerase in a reaction cycle of three parts: denaturation, primer annealing and polymerization. The DNA is first denatured at 95°C, an annealing step follows at 55°C, which allows the primers to bind, and a final polymerization step takes place at 72°C. The reaction mixture, held in a microtube and placed in the PCR machine, contains the template DNA, primers, DNA polymerase, Mg^{2+}, a mixture of oligonucleotides (dNTPs) all suspended in a buffer. In each cycle, which takes only a few minutes, the number of copies is doubled and so millions of copies are produced after 30 to 40 cycles. For an overview of these methods see Turner et al. (1997).

Mitochondrial (mt)DNA is a circular molecule present in all mitochondria. The basic advantages of mtDNA for population and evolutionary studies are the ease with which it can be isolated from the nuclear DNA, its uniformity in the individual organism, its relatively small size in animals (16–17 kb — but see later), and its uniparental inheritance (Meyer 1994). The finding that animal mtDNA evolves at a much higher rate than single copy nuclear DNA adds to this list of desired properties. Simple tissue extractions will already contain millions of copies of mtDNA and because mtDNA can be separated from nuclear DNA by centrifugation methods, subsequent whole molecule mtDNA amplification is not usually required.

10.3.1.3 DNA based markers

10.3.1.3.1 Restriction fragment length polymorphism (RFLP)

Any amplified length of DNA can be incubated with a suite of REs each of which has a different target sequence. Depending on the RE cut sites present, a number of fragments will result, which can be separated on the basis of length using gel electrophoresis. Mutations, which affect the sequence at any RE cut site, will change the length of fragments produced and these polymorphisms are RFLPs.

10.3.1.3.2 Random amplified polymorhic DNA (RAPDs)

PCR primers for amplifying known sequences of DNA are usually 15–30 bp long. If the primers are much shorter than this then there is the chance that other random regions of the genome will be amplified in addition to the targeted sequence. The RAPD method exploits this by subjecting template DNA to PCR with a single primer of 10 bp in length. Wherever this primer is complementary to opposite strands of DNA separated by less than 2–3 kb, an amplified fragment will be produced so the final PCR product will contain from none to several fragments of different sizes. Fragments are separated by agarose gel electrophoresis and bands can be sized and scored. Screening DNA with a range of 10 bp primers enables large numbers of randomly amplified fragments to be scored. Where a mutation occurs at a primer site, preventing the primer from annealing, the normal fragment product will be missing and a polymorphism is recognised by the absence of a band on the gel.

10.3.1.3.3 Microsatellites

Microsatellites are short (<1 kb) regions of the genome containing a repeated motif consisting of 1 to 5 base pairs. Genetic variability results from the variation in the number of these repeats and the detection of this variation depends on sizing the fragment of DNA (containing the microsatellite region), which is PCR-amplified from the raw DNA of an individual. Identification and isolation of microsatellite loci is a complex procedure (see Estoup and Angers (1998) for details) but once primers have been developed for the loci then the routine screening of genotypes is rapid, and can be automated to some extent by using an automatic sequencer to size the alleles.

10.3.1.3.4 Amplified fragment length polymorphism (AFLP)

The AFLP method (Vos et al. 1995) can provide a large number of reliably reproducible size fragment-markers from DNA of unknown sequence. Extracted DNA is cut using a pair of restriction enzymes, which leave cohesive (sticky) ends, one a frequent (4 base) cutter, the other a rare (6 base) cutter. The majority of fragments produced will have 4 base-cut ends, the minority will have 6 base-cut ends and the remainder will have 6 base - 4 base ends. Two oligonucleotide "adapters" designed to attach to the cohesive cut ends of both the 6-base and 4 base fragment terminals are then ligated to the resulting fragments. These adapters are used as the basis for polymerase chain reaction (PCR) primers, which are also designed to overlap with the DNA fragment by one, two or three specified bases. The primer for the less frequent (6 base) cut adapter is radiolabelled before PCR amplification and the resulting fragments are separated on polyacrylamide gel and visualised by autoradiography or fluorescence. By varying the number of specified bases on the primers the number of fragments amplified can be controlled. Typically 50-100 restriction fragments can be identified and separated in a single run of the AFLP method. Although there are currently no publications on the use of AFLP in scallops, this method is included here because it has been used in China to investigate the reduction of genetic variation in 2 selected populations of *C. farreri* compared with a wild population (Z. Yu, pers. comm.) and there is little doubt that AFLPs markers will be used extensively in the near future in the search for quantitative trait loci (see section on quantitative genetics).

10.3.1.4 Mitochondrial DNA in scallops

An initial study on *P. magellanicus* showed that the mtDNA molecule was much larger than the expected 16–17 kb, reaching 31–42 kb in length, and it was also noted that there was considerable variation in size between the mtDNA from different individuals (Snyder et al. 1987). It was discovered that the main cause of this variation was the variable number of copies of an element that occurred in direct tandem repeats in all mtDNA molecules of this species. This 1.47 kb repeat element was sequenced by La Roche et al. (1990) and it was shown to occur as 2 to 8 copies per molecule. More recent studies have revealed further complexity with 3 separate regions in *P. magellanicus* mtDNA which exhibited tandem length repeat polymorphisms: locus I (1450 bp), locus II (250–1000 bp) and locus III (100 bp) (Fuller and Zouros 1993). Locus I and II overlap slightly, which complicates analysis, but locus III occurs on the opposite side of the molecule (Cook and Zouros 1994). Although mtDNA size variation has been identified in other invertebrates (Moritz et al. 1987; Gjetvaj et al. 1992), the presence of at least 2 (possibly 3) independent repeat sequence polymorphisms located in different regions of the mtDNA is a very rare phenomenon (Cook and Zouros 1994) and has been demonstrated only in *P. magellanicus* and one other scallop species, *C. islandica* (Gjetvaj et al. 1992). Using mtDNA extracted from 2 parents and their 1 year old offspring, Cook and Zouros (1994) demonstrated that there was a very high mutation rate associated with locus 1 and locus II because 15% of the offspring had mtDNA haplotypes which differed from what could be detected in the mother. Although mtDNA is expected to be under strong selection for economy of size, there is no evidence that the extensive size polymorphisms in *P. magellanicus* is anything other than selectively neutral (Zouros et al. 1992).

MtDNA repeat sequence variation has been identified in several other scallop species, namely *P. maximus, Crassodoma gigantea, A. opercularis, Chlamys hastata, C. islandica* and *P. yessoensis* (Gjetvaj et al. 1992; Rigaa et al. 1993, 1995; Boulding et al. 1993). Gjetvaj et al. (1992) used the *P. magellanicus* repeat motif to test for its presence in other species, but found no hybridisation product. Rigaa et al. (1993, 1995, 1997) studied length variation in the mtDNA of *P. maximus* in detail and identified a sequence of 1,586 bp, which occurs as 2 to 5 repeated units. They compared this repeat region with the locus I (1.6 kb) repeat in *P. magellanicus* (Fuller and Zouros 1993) and, like Gjetvaj et al. (1992), found no clear similarity. Nevertheless, some homology was evident when comparing the nucleotide content of different domains: both species exhibit an Adenine+Thymine-rich region, and an upstream Guanine-rich region preceded by a potential hairpin structure (Rigaa et al. 1995). Furthermore, this bears some similarity to the 1.2 kb non-coding region of the *Mytilus edulis* mitochondrial genome (Hoffman et al. 1992) and suggests that there may be systematic value in further studies of these repeat regions in scallop mtDNA (Rigaa et al. 1995).

From the point of view of using mtDNA as a genetic marker the extensive length variation of this molecule in scallops presents difficulties. The numbers of repeats do not appear to be population specific in *P. magellanicus* (Fuller and Zouros 1993) or in *P. yessoensis* (Boulding et al. 1993), although when the structure of the molecule and the position and size of the repeat motif is better known some direct RFLP population analysis is possible as demonstrated by Rigaa et al. (1997) in *P. maximus*. The main method employed to get usable genetic markers out of scallop mtDNA has been to clone and sequence a part of the molecule, which is not in the repeat region and use PCR to amplify this fragment directly (Boulding et al. 1993; Wilding et al. 1997; Heipel et al. 1999).

Although most scallop species so far investigated have mtDNA repeat sequence variation, *A. irradians* and *Argopecten gibbus* have a fixed length mtDNA molecule of around 16.5 kb without repeat sequences (Gjetvaj et al. 1992; Blake and Graves 1995) and therefore are amenable to standard mtDNA RFLP analysis. Seyoum et al. (2003) have developed a very effective mtDNA marker to be used as a genetic tag for *A. irradians* in Florida.

10.3.1.5 Types of data produced by different markers

Allozymes and microsatellites produce co-dominant genotypic data such that homozygotes and heterozygotes can be unambiguously identified (except in the case of null alleles, see later). On the other hand, RAPDs and AFLPs produce dominant data where one homozygote (band-band) and the heterozygote (band-no band) are indistinguishable. RFLP data provide information on sequence differences between variants based on RE cut sites and can be quantified as nucleotide differences. Because mtDNA is essentially a haploid clone (as opposed to the diploid, sexually reproduced nuclear DNA) analysis is based on nucleotide, or haplotype divergence.

10.3.1.6 Analysis of data

Once data on the numbers and proportions of different genotypes at a co-dominant marker locus (e.g., allozyme or microsatellite locus) are available for a sample of organisms, the frequencies of the alleles at that locus can be estimated. The Hardy–Weinberg model predicts the expected genotype frequencies based on these estimated allele frequencies. For the simplest case, consider a locus with two co-dominant alleles A and B present at frequencies p and q where $p + q = 1$. The Hardy–Weinberg model states that the frequency of the genotypes AA, AB and BB in the next generation will be p^2, $2pq$ and q^2, respectively. Significant deviations from genotype frequencies predicted by the Hardy–Weinberg model may be produced by a number of factors such as natural selection, migration into, out of or between populations, mating structure, or the presence of non-staining protein products of an allele (null allele). Early studies generally used χ^2 Goodness of Fit tests to detect significant deviations from the model, but nowadays exact tests are used (Raymond and Rousset 1995).

In early literature relating to bivalves, the degree of deviation of genotype frequencies from Hardy–Weinberg expectations at a locus was commonly denoted by Selander's (1970) index "D" where:

$$D = (H_o - H_e)/H_e \qquad (1)$$

Where, H_o = observed numbers of heterozygotes and H_e = expected numbers of heterozygotes according to the Hardy–Weinberg model and

$$H_e = 1 - \Sigma x^2_i \qquad (2)$$

Where, x_i is the frequency of the ith allele. A negative value of Selander's (1970) D indicates a deficiency of heterozygotes. Although Selander's (1970) D has been used in some bivalve literature, the much better and more widely used statistic is Wright's (1951) F, sometimes called the inbreeding coefficient.

Wright's F statistics (Wright 1951) provide a measure of deviation from the Hardy–Weinberg model for each allele at a locus, in individual sub-populations (Fis) and for the total population (Fit) and are calculated as:

$$\text{Fis or Fit} = (H_e - H_o) / H_e \qquad (3)$$

Positive values of Fis and Fit indicate a deficiency of heterozygotes against the model; negative values an excess of heterozygotes. For each locus, if all the sub-populations are in agreement with the Hardy–Weinberg model, and share the same alleles at the same frequencies, then the mean of Fis will equal Fit. However, if one or other sub-population differs in allele frequency, or is in disagreement with the Hardy–Weinberg model, then mean Fis will differ from Fit. This feature is use to calculate the statistic Fst, which is a measure of the degree of differentiation between sub-populations where:

$$\text{Fst} = (\text{Fit} - \text{mean Fis}) / (1 - \text{mean Fis}) \qquad (4)$$

When there is no sub-population differentiation, Fst is zero. Values of Fst above zero indicate differentiation and the significance of this differentiation can be calculated. Fst can further be related to the size of the sub-populations and the migration rate between them such that:

$$\text{Fst} = 1 / (1 + 4Nm) \qquad (5)$$

Where N is the effective sub-population size and m is the proportion of that effective population size that migrates between sub-populations each generation. The effective population size is the number of individuals, which actually make a genetic contribution to the next generation and therefore excludes immature or otherwise infertile organisms.

It is important to note that the assumptions made in the development of the F statistics are, amongst others that all sub-populations are of equal size, and that they each exchange equal numbers of migrants with all other sub-populations at each generation (Wright 1951). Clearly, this is seldom the case in the wild and particular caution is required in extrapolation of Fst to statements about Nm, the number of migrants between sub-populations.

Although the significance of allele frequency differences between populations can be simply tested by Contingency Table χ^2 tests (more recently, more accurately, by exact tests), and Fst can provide an estimate of potential sub-population structure, Nei's (1978) genetic identity (I) and distance (D) provide further insight into the relative degree of differentiation among populations. Genetic identity is computed for any pairwise combination of populations according to:

$$I = \Sigma(P_{ix} \cdot P_{iy}) / \sqrt{\Sigma(P_{ix}^2 \cdot P_{iy}^2)} \tag{6}$$

where P_{ix} is the frequency of allele i in population x and P_{iy} is the corresponding frequency in population y. Genetic distance, D is related to genetic identity (I) according to the formula:

$$D = - \ln I \tag{7}$$

Values of Nei's (1978) D for allozyme data are interpreted as the net number of codon substitutions per locus that have accumulated since separation of any two populations. The rate at which this "molecular clock" accumulates mutations is still a matter of debate (Avise 1994) but Nei (1978) suggests that a value of D = 1.0 implies a separation of 5 million years. In the *Drosophila willistoni* group of fruit flies, for example, the mean genetic distance is 0.031 for conspecific populations, 0.229 for semi-species, 0.660 for sibling species and 1.044 for full species (Ayala et al. 1974).

10.3.2 Overall Genetic Variation

Two indices of genetic variation are generally employed for allozyme data; P, the proportion of polymorphic loci and H_o, the mean observed heterozygosity per locus. These indices are also relevant for microsatellite loci. Table 10.2 lists values of P and H_o for a number of pectinid species. In species such as *P. yessoensis*, where some estimates show low values of P and H_o (Nikiforov and Dolganov 1982; Balakirev et al. 1995) this is possibly due to the inclusion in these studies of a high proportion of non enzymatic loci, which are generally less variable than enzymatic loci (Smith and Fujio 1982). In the case of *Lyropecten nodosa* and *Euvola ziczac* (Coronado et al. 1991), which have the lowest values, the authors suggest that gene duplication may have increased the number of loci, but that these loci are mainly monomorphic. An alternative suggestion is that the populations from which these hermaphroditic species were sampled were inbred due to extensive self-fertilisation. The values of P and H_o for the microsatellite loci in *P. magellanicus* illustrate the very high level of variation characteristic of these genetic markers compared with allozymes.

Considering all the scallop species for which we have allozyme data (Table 10.2), the mean of P is 0.49 and the mean heterozygosity is 0.164. These values are higher than for other bivalves (Ahmad et al. 1977; Buroker 1980; Buroker et al. 1983) and for invertebrates in general (Hedrick 1985). It is important to bear in mind that these estimates of genetic variability are taken from studies based on relatively few gene loci (10–25) and may therefore be rather unreliable estimates of overall genomic variability.

558

Table 10.2

Estimates of proportion of allozyme loci (except [b]), which are polymorphic (P) and average levels of observed heterozygosity (Ho) in scallop species and other bivalves. [a] = values calculated from data in paper; [b] = microsatellite loci; [c] = probably *Argopecten irradians concentricus* - collected from North Carolina.

Species	P	Ho	Reference
Placopecten magellanicus	0.38	0.139	Foltz and Zouros 1984
Placopecten magellanicus	1.00	0.802	Gjetvaj et al. 1997 [b]
Patinopecten yessoensis	0.41	0.158	Kijima et al. 1984
Patinopecten yessoensis	0.33	0.094	Nikoforov and Dolganov 1992
Patinopecten yessoensis	0.35	0.087	Balakirev et al. 1995
Lyropecten nodosa	0.02	0.016	Coronado et al. 1991
Euvola ziczac	0.18	0.074	Coronado et al. 1991
Swiftopectn swifti	0.31	0.051	Balakirev et al. 1995
Pecten maximus	0.72	0.215	Beaumont and Beveridge 1984
Pecten jacobaeus	0.77	0.235 [a]	Pena et al. 1995
Mimachlamys varia	0.72	0.284	Beaumont and Beveridge 1984
Chlamys farreri	0.78	0.284	Zhang and Zhang 1997
Chlamys farreri nipponensis	0.49	0.225	Balakirev et al. 1995
Chlamys distorta	0.73	0.321	Beaumont and Beveridge 1984
Chlamys islandica	0.21	0.116	Fevolden 1989
Chlamys islandica	0.35 [a]	0.182 [a]	Fevolden 1992
Argopecten irradians [c]	0.33	0.116	Wall et al. 1976
Argopecten irradians irradians	0.35	0.129 [a]	Bricelj and Krause 1992
Argopecten gibbus	0.64	0.103	Krause et al. 1994
Aequipecten opercularis	0.59	0.155	Beaumont and Beveridge 1984
Mytilus edulis	0.38	0.095	Ahmad et al. 1977
Oyster spp.	0.38	0.125	Buroker et al. 1983
Invertebrates	0.40	0.112	Hedrick 1985

Theoretically from an aquaculture point of view, high genetic variability should be an advantage since overall genomic heterozygosity is associated with increased vigour (Lerner 1954). Nevertheless, in practice, certain bivalves, such as the European mussel, which appear to have a low overall level of genetic variability, are able to survive and grow vigorously both in culture and in a wide range of environmental conditions. By comparison, European scallop species tend to grow less vigorously and to be more problematical to rear in culture than mussels.

10.3.3 Genetic Differentiation of Populations

One of the more valuable contributions of electrophoretic assays of genetic variation is in the area of taxonomy and population subdivision. Before the advent of allozyme electrophoresis, taxonomic studies relied almost exclusively on morphological characters. These characters are useful for discriminating among species, but are of little use for the study of differences between geographic populations or for the study of the related problems of hybridisation and gene flow. Examples of the use of allozyme data for such applications in marine bivalves can be found in Skibinski et al. (1983), Koehn et al. (1984), Varvio et al. (1988), McDonald et al. (1991) and Gosling (1992).

10.3.3.1 Placopecten magellanicus

Zouros and Gartner–Kepkay (data cited in Beaumont and Zouros 1991) used variation at 5 polymorphic allozyme loci to assess the level of sub-population structure of *P. magellanicus* with samples taken throughout its range extending from the Bay of Fundy and the coast off Nova Scotia to the Southern end of Georges Bank. Contingency Table χ^2 tests across all populations revealed significant allele frequency differences at the octopine dehydrogenase (*Odh*) and the 6–phosphogluconic acid dehydrogenase (*Pgd*) loci but not at the other three loci. However, as in other scallop allozyme studies published before the mid 1990s, no Bonferroni correction for multiple tests of the same hypothesis (type I error) was applied (Lessios 1993). When such a correction is applied to these data, only the *Odh* value remains significant.

Nei's (1978) genetic distances (*D*) for all populations are given in Table 10.3 and the last column of the table estimates the 'average' distance of each population. This represents the distance that one would obtain for this population if all other populations were combined into one. The sample from Georges Bank South (number 4), which is the southernmost sample, showed the largest genetic distances in most pairwise comparisons (average distance $D = 0.0078$). Even so, these genetic distances are typical of conspecific populations of other species and they provide no evidence for severe or prolonged isolation of these populations from each other. Nevertheless, the significant differentiation in allele frequencies at one of the five loci suggests that the various geographic populations of this species should not be considered as parts of a single randomly mating population. However, the only locus responsible for the significant differentiation among geographic populations is *Odh* for which, as discussed later, there are good reasons to suspect strong selection pressures.

P. magellanicus allozyme data from Volckaert et al. (1991) also show low values of Nei's (1978) genetic distance, ranging from 0.002 to 0.004, when comparing a shallow water (Damariscotta River) population, a neighbouring deep water (Gulf of Maine) population, and two Canadian populations from Nova Scotia (Foltz and Zouros 1984) and New Brunswick (Volckaert and Zouros 1989). Beaumont and Zouros (1991) later noted earlier studies confirmed that there is very little identifiable allozyme-genetic variation between populations of *P. magellanicus*, and they also provide no evidence for an allozyme-genetic basis for the physiological and biochemical differences exhibited by

Table 10.3

Genetic distances (D, Nei 1978) among eight populations of *Placopecten magellanicus* based on allele frequencies at five enzyme loci (modified from Beaumont and Zouros 1991). The last column gives the average genetic distance of each population from all others. Populations: 1 = Middle Grounds, 2 = Browns Bank, 3 = Bay of Fundy, 4 = Georges Bank (South), 5 = Second Peninsula, 6 = Georges Bank (North), 7 = Grand Manan, 8 = Georges Bank (North East).

	2	3	4	5	6	7	8	mean D
1	0.005	0.007	0.007	0.002	0.002	0.003	0.008	0.004
2		0.007	0.003	0.006	0.003	0.005	0.005	0.004
3			0.013	0.006	0.004	0.004	0.004	0.006
4				0.013	0.008	0.008	0.011	0.008
5					0.004	0.002	0.005	0.005
6						0.004	0.004	0.004
7							0.006	0.004
8								0.006

shallow water and deep water populations in Maine. These authors report a significant *Odh* allele frequency difference between the deep and shallow water populations, but this is not significant following Bonferroni correction.

The population structure of *P. magellanicus*, deduced from allozyme studies, is therefore one of a geographically fragmented species whose constituent populations exchange individuals at a rate that prevents the accumulation of marked genetic differentiation. What genetic heterogeneity there is probably results from a combination of reduced migration rates over geographic distance, and varying selection pressures.

Patwary et al. (1994) developed a suite of RAPD markers for *P. magellanicus* and tested them for Mendelian inheritance, but did not carry out an extensive study of population structure. Differences in frequencies of bands in samples from different locations were noted but not quantified or further analysed. They did establish that great care was required initially to ensure that scored bands could be reliably repeated and pointed out that though the RAPD technique was, in principle, a speedy method for estimating genetic structure, the time taken to ensure reliability, and the fact that the data are not co-dominant, were major drawbacks. Gjetvaj et al. (1997) published primers for seven microsatellite DNA markers in *P. magellanicus* and confirmed their Mendelian inheritance using a full-sib family. They did not use the markers to investigate population structure, but pointed out the power of these markers as tools for such studies by virtue of their high variation and the fact that individual larvae could be genotyped. One of the loci has already been used in a deep mesocosm study of larval behaviour (Manuel et al. 1996).

10.3.3.2 Patinopecten (= Mizuhopecten) yessoensis

Both Russian and Japanese groups have studied the population structure of *P. yessoensis* across much of its range using allozymes (Kijima et al. 1984; Balakirev et al. 1995; Dolganov 1995; Pudovkin and Dolganov 1995; Dolganov and Pudovkin 1997, 1998). Data from some of these studies, brought together and re-analysed by Dolganov and Pudovkin (1998), have indicated that there are several relatively distinct populations of *P. yessoensis* around the Sea of Japan and the Sea of Okhotsk. A single sample taken from Furunhelm Island, just south of Posjet Bay, Primorye, appeared different from all other groups, while an essentially panmictic group (the Primorye group) extends from Posjet Bay along the Russian coast to Chikhachev Bay in the North of Primorye. Two distinct groups inhabit Terpenija Bay and Aniva Bay in Sakhalin Island, to the North of Japan. Earlier analysis by Kijima et al. (1984), which apparently showed genetic differentiation of samples from the North coast of Hokkaido, was criticised by Dolganov and Pudovkin (1998). In this later analysis, Kijima et al.'s (1984) samples from the North coast of Hokkaido grouped together with Dolganov and Pudovkin's (1998) Kuril Island samples with characteristic allele frequencies at most loci to form the fifth population group of *P. yessoensis*. Although there are significant allele frequency differences between these 5 groups, Nei's (1978) mean pairwise genetic distances are not particularly large and range from 0.003 to 0.037. It is therefore probable that there is a certain amount of larval interchange between some of these groups. Following the expectation that significant linkage disequilibrium might be detectable in populations, which were sharing recruitment from two different sources, Pudovkin et al. (1998) tested the Primorye group and the Furenhelm Island sample for linkage disequilibrium at six polymorphic loci. They found a significant disequilibrium in the Furenhelm sample when only the Promorye data were used, but this disequilibrium was not significant when analysed in the context of data from all five groups of scallops (Dolganov and Pudovkin 1998).

Boulding et al. (1993) used analysis of mtDNA to investigate population differences between a stock of *P. yessoensis* imported into, and bred in Canada for 3 generations and two populations in Japan, one of which was the source of the Canadian importation. As noted earlier, length variation caused by tandem repeat regions is common in the mtDNA of most scallop species so these authors used PCR amplification of length-invariant regions of the mtDNA to provide fragments for RFLP analysis. Pure mtDNA was extracted, cut into fragments, and then the fragments were cloned. Clones were sequenced and tested against known (at that time) conserved regions of the mtDNA of other animal groups. Four sequences were selected as having no potential length-variant component and pairs of primers were designed for these sequences. Following RFLP analysis of the PCR amplified products of these sequences, 11 mtDNA haplotypes were identified. Comparison of the frequencies of these haplotypes revealed no significant difference between either of the Japanese populations or the imported Canadian sample. Although the main purpose of Boulding et al.'s (1993) approach was to investigate potential bottleneck effects from the importation of the Japanese scallop into Canada (Beaumont 2000b), the results support the expectation that *P. yessoensis* from Mutsu Bay

(Aomori) and Uchiura Bay (Hokkaido), because of their geographical closeness, are likely to be part of a single panmictic population.

Using 3 of Boulding et al.'s (1993) mtDNA markers, Wilbur et al. (1997) analysed genetic variation in a *P. yessoensis* sample from Amursky Bay in Peter the Great Bay (Siberia, Primorye) and compared this with Boulding et al.'s (1993) data on Japanese populations and the Canadian imported stock. The results suggested two rather contrasting situations. Basically, there was very little divergence between the 26 mtDNA haplotypes identified, that is, all haplotypes differed from each other by no more than one or two site changes. Thus, there is no evidence for a deep split at some time in the past between the groups. On the other hand, a significant haplotype frequency difference was evident between the Primorye sample and all others suggesting that there has been little recent mixing between the Japanese and the Primorye population. This result supports the findings of allozyme studies (Dolganov and Pudovkin 1998) and adds strength to the idea that the Primorye population should be considered as a separate stock to populations in Japan.

10.3.3.3 Pecten maximus

There have been two extensive allozyme population genetic studies of *P. maximus* but neither reveals any evidence of sub-population structure. Using 8 polymorphic loci across 13 populations from the UK and France, Beaumont et al. (1993) reported a range of pairwise Nei's (1978) genetic distance values of 0.003 to 0.031. Similar values ranging from 0.007 to 0.045 were found by Wilding et al. (1998) using 7 polymorphic loci across 9 populations.

Wilding et al. (1997) developed genetic markers from the mitochondrial DNA of *P. maximus* by extracting and cloning fragments of the mtDNA, which did not contain length-variable regions (Rigaa et al. 1993, 1995). Following primer development, the marker, *Pma1* (a 2kb *Eco*RI - *Hind*III fragment), was amplified by PCR from samples of *P. maximus* DNA from 9 sites. Following cutting with 6 restriction enzymes 59 different composite haplotypes were identified and sequence divergence values derived from these were used to estimate genetic difference between the populations. Incomplete data from a second marker (*Pma2*, a 3.8kb *Hind*III - *Hind*III fragment) were also analysed. All the UK and French coast samples were essentially indistinguishable from one another and only one site, Mulroy Bay, Ireland, showed a significant difference from all other sites. Earlier studies on the different physiology and reproductive patterns of *P. maximus* from St. Brieuc and the Rade de Brest had suggested that these populations might be genetically distinct (Cochard and Devauchelle 1993; Mackie and Ansell 1993). These sites were included in Beaumont et al.'s (1993) and Wilding et al.'s (1997, 1998) studies and showed no differentiation. Rigaa et al. (1997) also compared populations of *P. maximus* from St. Brieuc, the Rade de Brest, and the Glenan Islands using whole mtDNA RFLP analysis. No significant differences were detected between the populations.

Heipel et al. (1999) used Wilding et al.'s (1997) marker *Pma1*, and a second marker developed from another non-length-variable region of the mtDNA, to investigate population differentiation amongst more localised populations around the Isle of Man

(UK) with outlying samples from Plymouth and Mulroy Bay. Their nucleotide divergence results showed Mulroy Bay and one of the Isle of Man samples clustering separately from the rest, but these two samples nevertheless each exhibited several unique and rare haplotypes implying that they were actually quite genetically distinct. A parallel study on these same population using RAPDs (Heipel et al. 1998) showed a clear distinction between the Mulroy Bay population and all others. In addition, the data show some structuring amongst geographically close *P. maximus* samples taken from around the Isle of Man, something which had not been detected in mtDNA or allozyme studies for this species. As previously mentioned (Patwary et al. 1994), concerns are often expressed about the confidence with which RAPD data can be accepted due to the difficulties in reproducibility of banding patterns in such low stringency PCR with relatively short primers. However, in this case, only bands, which were reproducible (in initial duplicate samples from a few individuals) were scored and the data were recorded as phenotypes (presence or absence of bands) rather than as allele frequencies (which requires the assumption of Hardy–Weinberg equilibrium at a locus).

Two populations of *P. maximus* along the Norwegian coast showed negligible genetic differentiation at Wilding et al.'s (1997) *Pmal* mtDNA marker (Ridgeway and Dahle 2000). However, there was a significant difference between the pooled haplotype frequencies of the two Norway populations and the combined haplotype frequencies of all the populations studied by Wilding et al. (1997) which included samples from the UK, France and Ireland. Direct comparison with those from Scotland to the closest Norway populations was not made.

In conclusion, there is little evidence of substantial genetic differentiation of *P. maximus* populations throughout its range. Apart from the possible differences of the Norwegian populations, the only population, which appears to be genetically isolated from others is Mulroy Bay. Because of the enclosed nature of Mulroy Bay, self-recruitment of the scallop population is probably the rule. Isolated in this way for around 10,000 years since the last Ice Age, the forces of random genetic drift and natural selection for local adaptation can readily explain the genetic differences observed. This genetic isolation may have important consequences for aquaculture of *P. maximus*. Mulroy Bay has been a regular source of scallop spat for aquaculture enterprises elsewhere although the genetic integrity of the Mulroy Bay scallops has probably not yet been compromised by the transfer of scallops into the Bay from other sites (Beaumont 2000b).

10.3.3.4 Aequipecten opercularis

Mathers (1975) investigated variation at the glucose phosphate isomerase (*Gpi*) locus in populations of *A. opercularis* and the black scallop *M. (Chlamys) varia*. He showed that *Gpi* allele frequencies differed significantly between populations of *A. opercularis* on the West and on the East coasts of Ireland. In a more extensive study, Beaumont (1982a) proposed at least four relatively genetically isolated regions of *A. opercularis* around the British Isles, namely (a) the North and West Scottish coast, (b) the Irish Sea, (c) the West Brittany and West Irish coast and (d) the English Channel. These conclusions were based on data from 3 enzyme loci (leucine amino peptidase (*Lap*), phosphoglucomutase (*Pgm*) and *Odh*) and one non-enzymatic protein locus (*Pt-A*, Beaumont and Gruffydd 1975) but the enzyme loci contributed little towards population differentiation. The most striking geographical differences were in the frequencies of the three *Pt-A* alleles: $Pt-A^3$ was most frequent in the Irish Sea and some English Channel populations, $Pt-A^2$ was commonest in the North and high frequency of $Pt-A^1$ characterised populations on the West coast of France and Ireland (Beaumont 1982a). Mean values of Nei's (1978) *D* for pairwise comparisons between these regions range from 0.178 to 0.233, while values within regions are generally less than 0.02. Allowing for the fact that the *A. opercularis* genetic distance values are derived from only three loci, and that it is only the *Pt-A* locus which shows big differences in allele frequency, the data nevertheless do suggest that the samples come from distinct populations within a single species.

Almost all the *A. opercularis* samples from all populations were in good agreement with Hardy–Weinberg expectations which reduces the likelihood that strong genotype dependent selection is changing allele frequencies each generation and the conclusion was therefore drawn that restricted gene flow caused the observed population differentiation. Data on currents, to some extent, support this conclusion but no similar population differentiation was evident in *P. maximus* (Beaumont et al. 1993; Wilding et al. 1998), which is surprising given the similarity of their distributions, their larval longevity and dispersal capacity (Beaumont and Barnes 1992).

Neither Beaumont (1982a) nor Macleod et al. (1985) were able to detect significant genetic differentiation amongst *A. opercularis* stocks within the Irish Sea although the latter authors did show that the source of recruitment of some Manx stocks may differ from year to year. Lewis and Thorpe (1994a, 1994b) explored this observation further by analysis of 4 allozyme loci in 12 UK populations each of which were sub-divided into year classes based on shell ring counts. They were unable to detect any significant allele frequency differences between year classes within any population, but did demonstrate significant geographic differentiation between some of the populations. However, because Lewis and Thorpe (1994a) used a discriminant function method to analyse their allele frequency data, neither F statistics, nor genetic distance values are provided which makes it difficult to relate their results to other studies.

10.3.3.5 Chlamys islandica

Fevolden (1989, 1992) analysed genetic differentiation in populations of *C. islandica* from Anddamen (Norway), Moffen (Spitzbergen), Jan Mayen Island and Bear Island using 6 polymorphic loci. Values of Nei's *D* ranged from 0.005 to 0.02 with the greatest genetic separation evident between the Norwegian mainland sites (Anddamen and Tromso) and all other sites (Fevolden 1992). Fevolden (1992) discusses the various mechanisms, which might be active in maintaining these minor genetic differences between populations and concludes that both adaptive selection and genetic drift may contribute to the observed population differentiation.

There is a small North American fishery for *C. islandica* as well as the major fishery for *P. magellanicus*. One management tool of the fishery is to set the minimum number of meats per 500 g and this has worked well for the *P. magellanicus* fishery. However, *C. islandica* has a smaller meat at commercial size than *P. magellanicus* and, the scallops are shucked at sea, and the meats are impossible to separate visually. This leaves open the opportunity for vessels supposedly working the *C. islandica* fishery to land large numbers of meats from undersize *P. magellanicus*. Kenchington et al. (1993) identified diagnostic allozyme differences and specific18S rRNA gene RFLP differences, which enable unequivocal genetic identification of muscle tissue from the two species and provides a very valuable management tool.

10.3.3.6 Chlamys farreri

Zhang and Zhang (1997) examined differentiation at a number of variable allozyme loci between 5 *C. farreri* populations from China. The sample from Haiyangdao (Liaoning Province) differed most from all others with a mean value of Nei's (1978) *D* of 0.068, and this is consistent with the separation of this population from the others by a low salinity coastal current. More recent studies based at the University of Qingdao, China, have used AFLPs to demonstrate the reduction of genetic variation in 2 selected populations of *C. farreri* compared with a wild population (Z. Yu, pers. comm.).

10.3.3.7 Mimachlamys varia

In an experiment where samples of *M. varia* were reciprocally transplanted between sites on the West and South coasts of Ireland, Gosling and Burnell (1988) demonstrated significant genetic differences between sites. However, some of these differences were reduced by selective mortalities in transplanted animals and they proposed that both genetic drift and selection were important in maintaining allele frequency differences between sites.

10.3.3.8 Euvola (Pecten) ziczac

Samples from the southern and northern extremes of the fishery for *E. ziczac* along the Brazilian coast were analysed at 2 polymorphic loci and revealed alleles unique to each extreme (Wanguemert et al. 2000). Although a wider study is clearly needed, this initial investigation does suggest that there is population differentiation of significance to the fishery for this species.

10.3.3.9 Argopecten gibbus

The calico scallop *A. gibbus* occurs on the East coast of the USA, from North Carolina down to Florida, and in the Gulf of Mexico and samples taken from these three regions were investigated by Krause et al. (1994) to test for population differentiation. They used morphological traits (shell shape characters) and allozymes (7 polymorphic loci) but revealed relatively low levels of morphological and genetical differences between the samples. Nei's (1978) *D* was smallest between the two Atlantic coast populations (D = 0.009) and greatest across the Florida peninsula (D = 0.070) indicating at least some genetic differentiation. In contrast, Blake and Graves (1995) found insignificant mtDNA restriction site differences between Atlantic and Gulf coast populations of the calico scallop.

10.3.3.10 The Argopecten irradians species complex

Strong population sub-structure of *A. irradians*, based on morphological differences between populations on the east coast of the USA, was noted and quantified in the 1960s. The species was classified by Clarke (1965) and Waller (1969), based on morphological data, into three sub-species: *A. irradians irradians* (northern bay scallop - from Massachusetts to New Jersey), *A. irradians concentricus* (Atlantic bay scallop - from New Jersey to South Carolina, and along the Gulf coast of Florida) and *A. irradians amplicostatus* (to the west of Galveston, Texas along the Gulf coast). Furthermore, strong shell size differences in scallops from the Florida bay and Florida Keys area, encouraged Petuch (1987) to propose an additional, very localised, sub-species, *A. irradians taylorae*. Kraeuter et al. (1984) demonstrated that hybridisation was possible between *A. i. amplicostatus* and *A. i. concentricus* and that hybrids had intermediate rib numbers to those diagnostic of the parental sub-species.

Using morphological and allozyme data (Marelli et al. 1997a, 1997b) and mtDNA (RFLP) data (Blake and Graves 1995), Petuch's (1987) claim for sub-specific status of *A. irradians taylorae* has now been effectively dismissed. In addition, these studies agree that samples of North Carolina bay scallops differ more strongly from Florida scallops (supposedly the same sub-species) than they do from samples from Massachusetts (classified as a separate sub-species). Wilbur and Gaffney (1997) collected bay scallops from four regions — the North Atlantic, the East Atlantic, the Gulf coast of Florida and of Texas — and spawned each group and reared the offspring in a common environment. They measured and compared shell parameters between the parental and the offspring groups and concluded that the morphological differences between all four source groups of bay scallops had a genetic basis. Therefore, their results would suggest that the northern *A. a. irradians* and the southern *A. a. amplicostatus* are probably safe distinct groups, but the sub-species *A. a. concentricus* is split into two geographically distinct, and probably genetically distinct groups, one on the East Atlantic and the other from the tip of the Florida Cape round into the Gulf of Mexico. Marelli et al. (1997b) reported a Nei's (1978) D value of around 0.14 between the Florida bay scallops and those from the east coast of the USA. A much smaller D value, around 0.06, separated putative *A. a. irradians* and *A. a. concentricus* populations. While the sub-structuring of the *A. irradians* species complex is becoming clearer, the precise classification of the groups still remains to be resolved. Adamkewicz and Castagna (1998) published evidence for genetic control of shell colour and pattern in *A. irradians*, but whether this will be of value in determining systematic relations between the sub-species is not known.

There is an added complication. According to Rhodes (1991), because bay scallops are an important fishery and aquaculture resource, there has almost certainly been extensive unrecorded mixing of populations of the two Atlantic coast sub-species,

A. a. irradians and *A. a. concentricus.* There is no apparent barrier to hybridisation between the sub-species (Kraeuter et al. 1984) and any hybridisations within mixed populations will make it increasingly difficult to establish the true original taxonomic status of the *A. irradians* complex. The recent development of two very variable mtDNA sequence markers (471 bp and 450 bp) by Seyoum et al. (2003), designed to identify the contribution of hatchery-produced *A. irradians* in the natural populations on the Florida coast, provides an extremely sensitive method for investigation of the wider systematics of this sub-species group.

Samples of a mixed stock of *A. irradians* sub-spp. were introduced into China from the USA during the 1980s and 1990s and these introductions have since resulted in very extensive cultured production of bay scallops (Beaumont 2000b). Blake et al. (1997) investigated mtDNA variation between the original stocks and the Chinese scallops and discovered very significant differences. The potential genetic consequences of introductions such as these are discussed by Beaumont (2000b).

10.3.4 Genetic Differences at Species Level and above

There have been several studies on relationships at the species or genus level based on allozyme data in oysters (e.g., Buroker 1979a, 1979b; Buroker et al. 1983) and mussels (McDonald et al. 1991) but few such studies have been carried out in pectinids. Based on allozyme data, L. Woodburn (in Beaumont and Mason 1987) proposed that the Australian *Pecten* 'species' *P. fumatus* (New South Wales), *P. albus* (South Australia and Victoria), *P. meridionalis* (Tasmania) and possible also *P. modestus* (Western Australia) should be considered as populations of a single species rather than as separate species. Beaumont (1991a) used allozyme data to investigate the systematic relationships between *P. maximus* and three species, which had previously been placed in the genus *Chlamys*. Pairwise genetic distances (*D*) between all these species ranged from 1.53 to 2.75 (mean 1.99) which is outside the range for species differences within a genus (0.118–1.273) suggested by Ferguson (1980). These data added strong support to the decision of Waller (1991) to place *Chlamys opercularis* into the genus *Aequipecten* and *Chlamys varia* into the genus *Mimachlamys*.

In more recent years, sequence data from mitochondrial and nuclear genes have been used as systematic tools. Pectinid phylogeny has been investigated principally using sequence data from the 18S rRNA gene (Rice et al. 1993; Kenchington et al. 1994; Kenchington and Roddick 1994; Frischer et al. 1999, 2000; Canapa et al. 1999). The distinction proposed by Waller (1991) between the sub-families Pectininae (containing the *Pecten, Decatopecten* and *Pallollium* groups) and Chlamydinae (containing the *Chlamys, Mimachlamys* and *Aequipecten* groups) is generally supported by these analyses. However, using partial sequence data from another mitochondrial gene, 16S rRNA, Canapa et al. (2000b) concluded that the *Chlamys* group is not monophyletic, implying that not all *Chlamys* spp. can be assigned to a single sub-family. The two *Chlamys* species, which seemed so different were *C. islandica* and *C. glabra*, while *Mimachlamys varia* (which they named *Chlamys varia*) clustered with *C. islandica*. It is worth noting (Table 10.1) that chromosome numbers vary between these species such that *C. islandica* and *M. varia* both have the modal chromosomal number of N = 19, while *C. glabra* has N = 14, which is very close to N = 13 for *A. opercularis* with which it clusters in Canapa et al.'s (2000b) study. Perhaps some early chromosomal rearrangement was a defining event in the evolution of the pectininds producing two lineages, which would then begin to diverge at both the mitochondrial and nuclear genome. Robertsonian-type chromosomal rearrangements (White 1973) involve changes in centromere position, but are not expected to substantially increase or decrease the amount of chromosomal material. However, the chromosome number differences between *P. maximus* (N = 19, the modal number for scallops) and *A. opercularis* (N = 13) are reflected in the amount of DNA present in the cell nucleus because the *P. maximus* genome is 27% heavier than that of *A. opercularis* (Rodriguez-Juiz et al. 1996). This suggests that chromosomal changes in the pectinid lineage were not simple Robertsonian events.

Several studies have now questioned the separate specific status of *P. maximus* and *P. jacobaeus*. Allozyme data from Huelvan (1985), incorporated into Beaumont's (1991a) analysis, showed that these two taxa were effectively equivalent to sub-populations of a single species. A further allozyme analysis, supplemented by mtDNA PCR-RFLP, confirmed the very close similarity between them (Wilding et al. 1999). Finally, Canapa et al. (2000a) found only a single nucleotide difference (in 548 bp) in the 16S rRNA gene between the taxa. The ranges of the two taxa overlap, according to Canapa et al. (2000b) an intermediate variety (*P. intermedius*) was once reported, and there is ample evidence that hybridisation between the taxa is possible in the laboratory (Wilding et al. 1999). Although there are morphological distinctions, the consensus is that these two taxa should be regarded as sub-populations of the same species, or possibly as sub-species.

10.3.5 Heterozygote Deficiency

Many allozyme studies on marine bivalves have shown that levels of heterozygosity in populations are often considerably lower than expected under Hardy–Weinberg equilibrium (Zouros and Foltz 1984; Gaffney 1994). Scallops are no exception to this trend and heterozygote deficiencies have been reported at a number of loci in *P. magellanicus* (allozyme loci: Folz and Zouros 1984; Volckaert and Zouros 1989; microsatellite loci: Gjetvaj et al. 1997), *P. yessoensis* (Fujio and Brand 1991), *P. maximus* (Beaumont and Beveridge 1984; Beaumont et al. 1993; Wilding et al. 1998), *P. jacobaeus* (Pena et al. 1995), *A. opercularis* (Beaumont 1982a, 1982b), *C. islandica* (Fevolden 1992), *C. farreri* (Zhang and Zhang 1997), *M. varia* (Beaumont and Beveridge 1984; Gosling and Burnell 1988), *C. distorta* (Beaumont and Beveridge 1984), *A. i. irradians* (Bricelj and Krause 1992), *A. purpuratus* (Galleguillos and Troncoso 1991) and *E. ziczac* (Coronado et al. 1991).

A common pattern exhibited in studies involving several loci and several populations is one where a small proportion of the loci studied show a significant deviation from the Hardy–Weinberg model, and in most such cases, the deviation is towards too few heterozygotes (positive value of Selander's (1970) D, or negative value of Fis). The second frequent observation is that significantly more than half of the tests of Hardy–Weinberg equilibrium will have a negative Fis (or positive D).

Data from Beaumont et al. (1993) are presented in Table 10.4 to illustrate these characteristics and it can be seen that overall, 61 of the 99 individual test Fis values are positive, and 7 of the 8 mean Fis values are positive. Both these are significantly different (p <0.05 by Sign Test: Sokal and Rohlf 1991) from the equal numbers of positive and negative predicted by the Hardy–Weinberg model. Furthermore, only one of the 19 tests showing significant values of Fis (without Bonferroni correction) is for a negative Fis. The pattern differs between loci: in some cases such as *Me-2*, the heterozygote deficiency is very pronounced in all populations, in others, for example *Pgm*, the effect is negligible.

Is such a level of heterozygote deficiency important, and what could be its cause? According to population genetics theory, heterozygote deficiency may appear in the population as a consequence of various processes. One obvious explanation is that heterozygotes may suffer higher mortality at an early stage of the life cycle, so that when the population is sampled after selective mortality has occurred there are fewer heterozygotes than expected. However, several other processes not involving selection may act on the population and cause heterozygote deficiency. One such process is inbreeding, i.e., when matings among relatives occur at frequencies higher than expected by chance alone. In marine bivalves, this may occur either as a consequence of self-fertilisation or when sperm from a given male more often fertilise eggs from related females.

Several scallop species such as *P. maximus*, *A. irradians* and *A. ventricosus* (= *circularis*) are hermaphroditic and, in hatchery situations, some self-fertilisation is the rule rather than the exception (Beaumont et al. 1982; Wilbur and Gaffney 1991; Ibarra et al. 1995). Whether this is the case in the wild is not known, but occasional instances of significant self-fertilisation in natural populations, particularly involving isolated individuals, must at least be a possibility. Perhaps of more relevance is the observation that self-fertilised *P. maximus* or *A. ventricosus* (= *circularis*) larvae do not grow as rapidly as cross-fertilised ones and are therefore less likely to survive to adulthood (Beaumont and Budd 1983; Ibarra et al. 1995).

Although rare hermaphroditic specimens of *P. magellanicus* have been reported (Naidu 1970), self-fertilisation is also an unlikely explanation of heterozygote deficiency in this species because of the rarity of hermaphrodites and the lack of a consistent trend toward heterozygote deficiency at all loci. Excessive fertilisations among relatives could occur only if the spatial distribution of adults is such that close relatives settle next to each other. It is difficult to see how this can happen given that scallops have a planktonic larval stage that experiences wide dispersal. Nevertheless, very interesting evidence is emerging that cohorts of *P. magellanicus* larvae may be able to position themselves in the water column in a non-random way (Manuel et al. 1996) and this opens up the possibility that related individuals could settle close to each other more frequently than would be expected under truly random dispersal.

If heterozygote deficiencies were being produced by self-fertilisation, or by mating of close relatives, then levels would tend to be similar across loci but this does not appear to be the case in most studies (as illustrated in Table 10.4) thus ruling out self-fertilisation as a major cause of heterozygote deficiency in these species.

Table 10.4

Deficiencies of heterozygotes against Hardy-Weinberg equilibrium at 8 loci in 13 populations of *P. maximus*. N+ve = number of positive Fis values, N-ve = number of negative Fis values, *+ve = number of significant +ve Fis values (without Bonferroni correction), *-ve = number of significant -ve Fis values (without Bonferroni correction). Numbers of tests for some loci do not sum to 13 because these loci were monomorphic in some populations. *Mdh* = malate dehydrogenase, *Me* = Malic enzyme, *Idh* = isocitrate dehydrogenase, *Gdp* = glycerophosphate dehydrogenase. Data modified from Beaumont et al. (1993).

locus	*Odh*	*Mdh-2*	*Me-1*	*Me-2*	*Idh-1*	*Pgd*	*Gdp*	*Pgm*	mean totals
Mean Fis	0.078	0.068	0.237	0.288	0.075	0.039	-0.034	0.015	0.096
N+ve	9	7	5	13	6	8	6	7	61
N-ve	4	5	5	0	6	5	7	6	38
* +ve	1	1	5	8	2	1	0	0	18
* -ve	0	0	0	0	0	0	1	0	1

The next hypothesis to be considered is population subdivision and mixture. This hypothesis (also known as the Wahlund effect) assumes that the species exists as a collection of sub-populations and that matings occur randomly within sub-populations, but matings between sub-populations do not occur or are severely restricted. The resulting zygotes may, however, disperse and be mixed so that, upon settlement, the sampled population will not consist of individuals resulting from one but from several random mating pools of individuals. In such cases genotypic frequencies will not be in Hardy–Weinberg proportions, but will exhibit heterozygote deficiencies. The level of heretozygote deficiency produced will depend on the degree of difference in allele frequencies between breeding sub-populations and considerable allele frequency heterogeneity is needed to generate significant deficiencies. Although population subdivision and mixture is an attractive hypothesis it has proven elusive to rigorous test. One consequence of the hypothesis is that it should produce non-random association between non-allelic genes creating two-locus disequilibria (Nei and Li 1973; Li and Nei 1974) but Gaffney (1994) failed to find any evidence of this in data sets from *Crassostrea virginica, Mulinia lateralis* and *A. irradians*. Further evidence against this hypothesis is the consistent lack of significant allele frequency heterogeneity in scallop species: there is, in effect, no sub-population structuring.

One source of heterozygote deficiency at allozyme loci could be the existence of one or more null alleles (alleles which code for a non-active protein product). In the case of a locus coding for a monomeric protein (heterozygotes produce 2 bands on the gel) individuals which are heterozygous for a null allele and an active allele will be mis-scored on electrophoretic gels as homozygotes for the active allele producing an apparent deficiency of heterozygotes. At dimeric protein loci (where heterozygotes exhibit 3 bands on the gel) it is theoretically possible to identify null / active allele heterozygotes, but often individuals exhibiting unexpected banding patterns are excluded from analysis. Because null allele homozygotes are expected to be lethal, null allele frequencies are not likely to be high in natural populations but there may be several such alleles at a locus and their detection is difficult. In pair matings, however, null heterozygote parents can be detected when offspring genotypes are scored. There have been several studies on bivalves which have detected a surprisingly high frequency of null alleles at a number of loci (see review by Gaffney 1994; Hoare and Beaumont 1995; Del Rio-Portilla and Beaumont 2000) and it is therefore probable that some of the heterozygote deficiencies detected in scallops could be the result of mis-scoring of genotypes due to the presence of null alleles.

One interesting observation concerning the heterozygote deficiency in populations of oysters and mussels is that the deficiency is more pronounced in young cohorts and declines as the cohort becomes older (Zouros et al. 1983; Diehl and Koehn 1985). Foltz and Zouros (1984) examined this phenomenon in a large sample of *P. magellanicus*, which had been aged by counting the annual rings on the shell and resilium (Merrill et al. 1966). The mean number of heterozygous loci per individual within each age group was determined by electrophoresis at six loci and this number increased with age. This suggests that heterozygous scallops survived better than homozygous ones and would indicate a reduction in any heterozygote deficiency with age. Fujio and Brand (1991) also found higher heterozygosity in larger *P. yessoensis*. However, contradictory results have been reported for other scallop species. Larger individuals of *P. maximus* were, on average, less heterozygous at the *Gpi* locus than smaller ones (Wilkins 1978). Also, heterozygosity at the protein *Pt–A* locus was lower in larger than in smaller size classes of *A. opercularis*, while significant variation in *Pgm* heterozygosity occurred between year classes rather than simply between size classes (Beaumont 1982b). Fevolden (1992) tested cohorts of *C. islandica*, aged by ring counts, but found no effect of single locus, or multiple locus heterozygosity on longevity.

10.3.6 Heterozygosity and Growth

Growth is a critical trait in marine molluscs because faster growth in the wild translates into higher viability. Larger animals are more likely to escape predators, to survive periods of starvation due to their larger energy reserve, to enter reproductive age earlier and, once mature, to produce larger gonads. From the aquaculture point of view, faster growth is a self evident objective. For these reasons, some of the genetic studies on bivalves have addressed, directly or indirectly, the relationship between allozyme polymorphisms and growth. Bivalves are good models for such studies because they are relatively easy to collect in large numbers, they can be marked and studied in situ, many species can be aged, and certain species are exposed to environmental conditions that may vary markedly on a micro- or macro-geographic scale. In addition, they have indeterminate growth (unlike, for example, insects that undergo metamorphosis and cease to grow as adults) and shell length is a permanent record of growth (it cannot regress with time like, for example, the weight of mammals). It is well documented that shell dimensions exhibit high variance, even in cohorts of the same age collected from the same site, and much of this variance must have a genetic basis.

The first studies that looked for a relationship between individual growth and allozyme variation in bivalves were on oysters (Zouros et al. 1980; Koehn and Shumway 1982). They were later widened in scope and extended to mussels (Koehn and Gaffney 1984; Gentili and Beaumont 1988), clams, (Koehn et al. 1988; Gaffney 1990) and scallops (Beaumont et al. 1985; Volckaert and Zouros 1989; Fevolden 1992; Pogson and Zouros 1994; Dolganov and Pudovkin 1995; Rios et al. 1996). To test for this relationship, samples of scallops of known age (either a single cohort, or aged by ring counts) are obtained and scored at a number of allozyme loci. Individuals will be heterozygous at none, one, two, etc. of the scored loci and can therefore be classed into groups according to their multi-locus heterozygosity (MLH). Shell size, or data from some other growth parameter, within each of these groups can then be tested against the null hypothesis that there is no relationship between MLH class and the growth parameter measured. Several (although not all) studies on bivalves have proved the null hypothesis incorrect. This is the phenomenon of the MLH - growth correlation, which can be stated as follows: the expected shell length of an individual is larger, the higher is its degree of its allozyme heterozygosity.

The MLH - growth correlation is an interesting observation and has been the subject of several reviews (Mitton and Grant 1984; Zouros and Foltz 1987; Zouros and Pogson 1994; Britten 1996; Mitton 1998). However, the phenomenon is neither general nor predictable and the number of studies that have failed to detect significant correlations are nearly as numerous as those that have reported positive correlations. Nevertheless, there appear to be certain conditions that favour the correlation and others that do not. Firstly, the correlation does not appear to hold among individuals that have been grown in the laboratory, or are the products of single matings or even mass matings in which the vast contribution of gametes come from few individuals (Beaumont 1991b). Secondly, the MLH - growth correlation is more frequently observed in juveniles, but seldom detected in adults. This may be due to the allocation of energy to gonad production, rather than growth, in mature individuals (Rodhouse et al. 1986).

In contrast to oysters and mussels, studies on scallops have generally failed to find any significant MLH - growth correlation. Foltz and Zouros (1984) examined about 750 *P. magellanicus*, aged by ring counts, but no correlation between MLH and shell length was found, within each age class. This negative study does not have the statistical power of the corresponding studies in other bivalves where sample sizes were much larger, the animals were at the juvenile stage, and there were no uncertainties concerning the age of scored specimens. Volckaert and Zouros (1989) later attempted a more exhaustive study of the correlation in *P. magellanicus*. They examined several age cohorts for the same set of loci as Foltz and Zouros (1984) but again, there was no evidence of a significant correlation (either positive or negative) between MLH and shell size. Beaumont et al. (1985) used both hatchery reared single families and wild caught spat of *P. maximus*, on-grown in lantern nets to the juvenile stage. No MLH - growth correlation was detected. It is perhaps not surprising that this study failed to detect an MLH - growth correlation in offspring of single matings but lack of correlation in the wild caught juveniles was unexpected. Fevolden (1992) tested cohorts of *C. islandica*, aged by ring counts, but found no MLH - growth correlation in any year class. Studies on *A. i. irradians* and *E. ziczac* have also failed to find MLH - growth correlations (Bricelj and Krause 1992; Alfonsi et al. 1995).

Assuming that heterozygosity at the few loci used in these studies cannot realistically reflect the overall genomic heterozygosity of individuals (Mitton 1998), two main hypotheses have been advanced to explain the MLH - growth correlation. One school of thought holds the view that the enzymes recorded by electrophoresis are the casual agents of the correlation, i.e., heterozygosity for these enzymes directly affect growth rate (the "direct involvement hypothesis", Koehn et al. 1988). Another school (the "associative overdominance hypothesis", Zouros et al. 1988) advances the view that the enzyme variants act as mere markers of genetic abnormalities that are responsible for the genetic variation in growth, but which cannot be detected by the electrophoretic assay involved. Such abnormalities may include null alleles (i.e., variants that do not produce active enzymes), chromosomal losses or homozygosity for linked deleterious genes. An individual that is multiply heterozygous for the marker genes will have a much lower chance of carrying one or more of these abnormalities compared to an individual that is multiply homozygous for the marker genes. This can be equated with the effect of inbreeding (Mitton 1998). Thus, the correlation between electrophoretic heterozygosity and growth results more from the fact that there is a correlation between homozygosity for electrophoretic markers and the presence of a hidden genetic defect than from the direct involvement of the scored allozymes in the animal's physiology.

Testing between these two hypotheses has proved difficult: evidence for the direct involvement hypothesis was provided by Koehn et al. (1988) working on *M. lateralis*, while Beaumont et al. (1995) reported evidence for the associative overdominance theory using cohorts of triploid mussels. Scallops have also featured in the search for confirmation of one or other theory. A critical evaluation requires that the test of the competing hypotheses make opposing predictions. One such test is to compare MLH - growth correlations in a sample of organisms using both allozyme loci and neutral loci. If the direct involvement hypothesis is correct, the correlation should be detected for the allozyme loci, but not for the neutral loci. If associative overdominance is the main causal factor, both sets of loci should show correlations. Pogson and Zouros (1994) developed a specific set of cDNA probes to detect single locus RFLP polymorphisms at sites close to transcribed regions of the genome in *P. magellanicus*. Eight of these RFLP loci were scored, together with 9 allozyme (7 enzyme, 2 non-enzymatic protein) loci, in a large sample of 21 month-old scallops, which had been collected as a single cohort of spat. The shell height of each scallop was accurately measured and tests were made between MLH and shell size for all loci, for the enzymatic allozyme loci alone, and for the non-enzymatic and RFLP loci alone. The results supported the direct involvement hypothesis because a significant MLH - growth correlation was found for the enzymatic loci, but not for the neutral loci. Therefore there is conflicting evidence in the literature supporting both hypotheses. Of course, neither theory need be exclusive. Indeed, there is no reason why both causes cannot be operating in the same genome and the prominence of one or other cause may depend to a great extent on which particular allozyme loci are scored.

Whatever the genetic cause of the MLH — growth correlation, the physiological interpretation seems clear. Koehn and Shumway (1982) and Hawkins et al. (1986) have shown that the degree of heterozygosity relates to basal metabolism. The physiological interpretation of the correlation is that the increased level of heterozygosity (either for the scored enzyme loci or for hidden detrimental genes) enables the individual to sustain its basal metabolism with lower expenditures of ATP. Thus, after meeting their catabolic needs (and provided that everything else is equal), heterozygous individuals are left with higher amounts of energy to fuel other functions. It is suggested that the organism should use this energy in whatever function will provide the maximum returns in terms of Darwinian fitness. For sedentary animals such as mussels and oysters, and for as long as the organism is in the juvenile stage, the best investment is growth. In the mature stages of these same organisms, the best investment will be production of larger gonads. This could explain why the MLH — growth correlation is not found as frequently among cohorts of adults as among juveniles (Diehl and Koehn 1985) and why MLH might correlate with gonadal size (Rodhouse et al. 1986). Garton (1984) proposes that in the predatory snail *Thais heamastoma* the higher energetic efficiency of heterozygotes results in higher predatory activity.

These considerations do not directly help us to explain why MLH — growth correlations are not detected in scallops. In those species of scallop which swim, one may be tempted to suggest that active swimming for the purpose of escaping predators will be the expenditure of surplus ATP favoured by natural selection. If this is the case, then one would not expect to find a correlation between shell size and heterozygosity. One would expect, however, to find a correlation between heterozygosity and survival, since the essence of the hypothesis is that heterozygotes, through their enhanced ability for active swimming, escape predation more effectively than homozygotes. The results of Volckaert and Zouros (1989) are consistent with this explanation, but as noted earlier, evidence from other studies does not show differential survival of heterozygotes (Wilkins 1978; Beaumont 1982b).

Volckaert and Zouros (1989) attempted to obtain more direct evidence for the hypothesis that heterozygous scallops may have higher scope for movement. Animals were placed, one at a time, in specially designed chambers and excited by injecting starfish extract in the water with which the chamber was filled. The time to exhaustion was measured and examined in relation to the genotype at the octopine dehydrogenase locus (*Odh*) which is a key enzyme involved in energy metabolism under anaerobic conditions (Beaumont et al. 1980; Gäde 1980). The study revealed that different *Odh* genotypes have different mean times to exhaustion, a relationship, which was not observed at any of the other loci scored. It is possible therefore that the *Odh* locus is under strong selective pressure in the natural environment, most likely because of its importance in the animal's energy availability under stress. If the demands on active and sustained swimming are higher in one population than in another, this may result in one *Odh* allele being favoured in one population and another allele in another population. This could explain why allele frequencies at the *Odh* locus vary among populations more extensively than allele frequencies at other loci (Beaumont and Zouros 1991). If a certain amount of mixture occurs among these populations the Wahlund effect will predict that the *Odh* locus will also show a higher deficiency of heterozygotes, which was indeed observed. Note, however, that in some cases (e.g., Volckaert et al. 1991) Bonferroni corrections of probabilities question the significance of *Odh* heterozygote deficiencies. In addition, there is mounting evidence that high frequencies of null alleles can be present at the *Odh* locus (as well as other loci) and these could account for observed heterozygote deficiencies (Gaffney 1994; Hoare and Beaumont 1995).

Others have studied the *Odh* locus in scallops in relation to its biochemistry. Alfonsi et al. (1995) reported that MLH at 6 loci in *E. ziczac* was correlated with the maximal activity of the enzymes pyruvate kinase and octopine dehydrogenase from its adductor muscle. These enzymes are involved in the anaerobic metabolism of adductor muscle swimming and therefore impact on the fitness of the animal. Following from this work Perez et al. (2000) demonstrated that *Odh* heterozygotes have a higher K_m for pyruvate and that this would be a fitness advantage during swimming when pyruvate levels are low. Such apparent overdominance is rather different to the situation in *P. magellanicus* reported by Volckaert and Zouros (1989) where differences between populations were proposed to be due to allele specific selection. Indeed, overdominance for *Odh* would be expected to produce the opposite of heterozygote deficiency, a heterozygote excess.

The search for links between single locus allozyme polymorphisms and specific factors upon which natural selection can act has been undertaken with several species (Mitton 1998) and the demonstration of selection for salinity in *M. edulis* involving the *Lap* [94] allele by Hilbish and Koehn (1985) is one of the few well documented examples. Krause and Bricelj (1995) investigated the relationship between variation at the glucose phosphate isomerase (*Gpi*) locus and reproduction and growth in *A. i. irradians* from Connecticut, USA. They showed that *Gpi* genotype could be linked with up to 15% of the size of a full grown scallop, but that this effect was not heterotic: it was not necessarily heterozygous genotypes which were exclusively scored in larger individuals. Krause (1995) attempted to explore the rates of flux through the biochemical pathway containing GPI but with limited success. Nevertheless, here is another example of a scored locus, or some other locus in linkage disequilibrium with it, which may affect the fitness of scallops and genotypes at this locus therefore have potential for use in marker assisted selection (MAS) in breeding programmes.

10.4 QUANTITATIVE GENETICS

10.4.1 Heritability and Artificial Selection

Quantitative genetics is the study of characteristics or traits of populations, which are influenced by many genes and also by environmental factors. Examples of such traits are size, weight and growth rate and the phenotypic values of such characteristics can generally be measured on a quantitative scale. A quantitative trait for a population can be described by the mean and the variance of the measurements of the trait from a sample of the population. Phenotypic variance can be partitioned most simply into genetic and environmental components such that:

$$Vp = Vg + Ve \qquad (8)$$

where, the total phenotypic variance is Vp, the variance due to genetic differences among individuals is Vg and the component of variance due to environmental differences is Ve. The genetic component may be further partitioned into that due to additive (Va), dominance (Vd) or epistatic effects (Vi) such that:

$$Vg = Va + Vd + Vi \qquad (9)$$

The term for additive genetic variance describes how much of the total phenotypic variation is passed from generation to generation in a predictable way and the ratio of additive variance to the total phenotypic variance is called the "narrow-sense" heritability (h^2 = Va / Vp) while "broad-sense" heritability (h^2 = Vg / Vp) estimates the proportion of phenotypic variation due to all genetic effects.

Values of h^2 may theoretically vary from zero, where phenotypic variance is entirely due to environmental effects, to 1.00, where all the variance is due to genetic effects. The higher the heritability, the greater will be the resemblance between parents and offspring and other related individuals and it is this resemblance between related individuals which is used experimentally to estimate heritabilities of important traits. A number of methods can be used such as regression of full-sib means on the mid-parent mean or comparison of full-sibs, half-sibs or parents (dams and sires) and offspring using analysis of covariance. In addition to the experimental estimation of heritability for a trait, a "realised" heritablity can be calculated following artificial selection for the trait in the parents, and its assessment in offspring.

Once values of h^2 have been estimated, programmes of selection involving breeding from the best individuals ("mass" or "individual" selection) or from the best families (family selection) can be undertaken in order to improve the mean phenotypic value of the offspring compared to the parents. For full details of methods of analysis, selection programmes and other aspects of quantitative genetics the reader is referred to Falconer (1981) and Kearsey and Pooni (1996).

Until recently there were no published data on heritabilities of any traits in scallops and no artificial selection programmes had been instigated although some quantitative genetic investigations had been carried out on oysters and mussels (Beaumont 1994). The last decade has seen an increase in interest in quantitative genetics and selection programmes for bivalve shellfish (Beaumont 2000a; Ward et al. 2000) and there are now some published data on quantitative genetics in scallops.

Jones et al. (1996) collected growth data from 20 full-sibling families of *P. magellanicus* and estimated broad-sense heritabilities of larval growth ranging from 1.10 (SE ± 0.17) to 1.24 (SE ± 0.40). Strictly speaking, h^2 cannot exceed 1.0, but estimates of h^2 are often imprecise and have high standard errors. Jones et al.'s (1996) results suggest that much, if nor all of larval growth rate is controlled by genetic factors but being broad-sense h^2 estimates these include genetic variation due to genotypic combinations within loci and interaction between loci, neither of which are amenable to direct artificial selection. In addition, heritability estimates from full-sib families are subject to maternal effects and, for example, the quantity and quality of lipids present in the eggs will influence the performance of larvae and may inflate estimates of h^2. In their study Jones et al. (1996) recorded lipid weight egg^{-1} for all the dams used and, as might be expected, detected a relationship between larval performance and lipid content of the eggs at day 4, but not at later larval stages.

Crenshaw et al. (1991) published a realised heritability value of 0.206 for growth rate in *A. irradians concentricus* based on a single generation of selection. Broodstock scallops were chosen on the basis of their growth rate and only those which grew faster than 1 standard deviation from the mean were used as parents in the selected group. Mating between a random selection of scallops from the whole broodstock provided a control population. Following the observation of reduced larval growth in similarly selected groups of the clam *Mercenaria mercenaria* (Heffernan et al. 1991), Heffernan et al. (1992) analysed embryonic viability and larval growth rate of the offspring from the Crenshaw et al. (1991) selection trials. Offspring from selected lines had significantly lower embryonic survival and larval growth rate than controls. This demonstrated that, as for clams, there was a negative larval response following selection for faster juvenile and adult growth. These results question the merits of the normal hatchery practice of artificial selection of the fastest growing larvae. Further support for this view comes from the demonstration that faster growing *Mytilus edulis* larvae are not guaranteed to become rapidly growing juveniles, and also that faster growing larvae are no more heterozygous at allozyme loci than slower growing larvae (Del Rio-Portilla and Beaumont 2000).

Although the genetic basis of the yellow colour morph of the Chilean (= Peruvian) scallop, *A. purpuratus*, has not been determined, this qualitative trait is associated with reduced growth rate (Wolff and Garrido 1991) and therefore will be a useful marker for selection trials; notably, yellow shelled *A. purpuratus* should not be used as broodstock in trials designed to improve growth rate.

10.4.2 Inbreeding and Hybridisation

One technique, which has proved to be of value in animal and plant breeding, is the production of inbred lines, which are homozygous at the majority of their gene loci. Different inbred lines can be produced to ensure homozygosity for different alleles at the various gene loci and crosses between these pure lines produce hybrid offspring, which are heterozygous at many loci and exhibit 'hybrid vigour' or heterosis.

Many scallop species are hermaphroditic and the potential for the rapid production of inbred lines by the use of self-fertilisation is a feature of importance in such species. Repeated selfings over several generations will produce highly inbred lines more quickly than matings between close relatives because the inbreeding coefficient in selfings (up to a maximum of 0.5) is greater than that for matings between sibs (0.25). As indicated earlier, induced gynogenomes may be almost 100% inbred and could provide an even more rapid method for the production of highly inbred lines. Nevertheless, the higher the degree of inbreeding, the greater is the likelihood of producing deleterious homozygous combinations of alleles at more loci, and thus causing inbreeding depression.

An early study investigated possible effects of inbreeding due to self-fertilisation in *P. maximus* (Beaumont and Budd 1983). Eggs from two adult *P. maximus* were crossed or selfed with sperm from the same parent or from other individuals in a 2 x 5 design. Assessments were made of yield, normality and size of first 'D' larvae on the third day after fertilisation, and of larval size after 2 weeks. Genetic variation was evident at all stages with males and females generally having a significant interactive effect. It was demonstrated that selfing did not significantly effect normality, yield or larval size at day 3 but did cause a significant reduction in growth rate of larvae over 2 weeks. The conclusion from this study was that there were no physiological barriers to self-fertilisation in *P. maximus*, but that inbreeding depression was recognisable in larval growth rate after only one generation of selfing.

Ibarra et al. (1995) have demonstrated similar results on larval growth in selfed *A. ventricosus* (= *circularis*). Larvae were produced from three sources: exclusive self-fertilisations, pair matings or mass spawnings. The authors note that because of the occasional simultaneous release of the eggs and sperm in these hermaphrodites, some selfing may have occurred in the pair matings and the mass mating. In spite of this, significant depression of both larval growth and survival was observed in the exclusively selfed larvae compared with the other groups.

Thus, the production of highly homozygous lines as broodstock by the use of self-fertilisation does not look particularly promising. Nevertheless, as hatchery rearing and on-growing techniques improve, selfing may yet become a feasible strategy in scallop broodstock management.

In principle, offspring with hybrid vigour (heterosis) can be produced by crossing individuals from closely related species, sub-species or different populations. Kraeuter et al. (1984) crossed *A. i. amplicostatus* and *A. i. concentricus* from the east coast of the USA and demonstrated that although rib number and shell colour were principally under genetic control, there was no evidence for heterosis in growth rate. They do point out, however, that all the offspring were reared at the temperature appropriate to only one of the sub-species and that growth differences between the two pure lines and the hybrids may have been due to genes controlling temperature adaptation rather than growth rate.

Cruz and Ibarra (1997) reciprocally crossed two populations of *A. ventricosus* (= *circularis*) from either side of the Baja California peninsula, Mexico, and compared hatchery performance of larvae from the crosses and the pure lines. They discovered a clear maternal effect on early larval growth. Larvae from the Bahia Magdalena population (M) and the reciprocal cross produced with those eggs (MC) had a significantly better survival and faster early larval growth rate than any larvae produced from Bahia Concepcion females (C or CM). This early maternal effect, probably due to higher general productivity in Bahia Magdalena, became non-significant after 11 days. By day 17, there was evidence of heterosis because the mean size of the reciprocal cross larvae was greater than the mean size of pure line larvae. However, it was really the poor performance of the Bahia Concepcion larvae which depressed the mean size of the pure line larvae and neither of the reciprocal cross larval groups were significantly better than the Bahia Magdalena pure line larvae.

The catarina scallop larvae produced by Cruz and Ibarra's (1997) experimental cross (M, MC, CM and C) were settled and grown to 1.5 mm in shell length in the hatchery before being placed out in bags at sites in Bahia Magdalena and Bahia Concepcion (Cruz et al. 1998). As the scallops grew they were transferred from suspended bags to suspended trays and finally to bottom anchored trays. Data on growth and mortality were tested for the effect of age, environment, genetic group and their interactions using analysis of variance. Growth and survival of all genetic groups was significantly lower in Bahia Concepcion, a high temperature, low food environment, than in Bahia Magdalena where temperatures are lower and productivity is higher. Only for one growth character (shell width) was the native population the best population in that environment. No overall significant heterosis of any trait was evident in the reciprocal crosses (MC, CM) over the pure line crosses (M, C). In fact, heterosis was negative for survival and the conclusion of the authors was that the Bahia Concepcion catarina scallops were adapted to their (more stressful) environment and that these adaptations were not present in the Bahia Magdalena scallops.

The poor growth potential of the Bahia Concepcion environment was reinforced by the discovery that none of the scallops in this environment had reached sexual maturity by 7 months (Cruz et al. 2000). In contrast, in Bahia Magdalena, the M and pooled MC - CM populations had a significantly higher gonad index and matured at 4 months compared with the C group, which matured at 5 months. This suggests that there is some genetic element to the control of time of maturation and that this element is dominant because its effect is seen in both reciprocal crosses. This study on catarina scallops by Ibarra's group (Cruz and Ibarra 1997; Cruz et al. 1998; Cruz et al. 2000) is an important demonstration of the need to be wary of population transfers, often warned of by geneticists (e.g., Beaumont 2000b), but seldom tested so clearly in scallops.

Studies on inbreeding effects and heterosis in oysters have produced conflicting results (Lannan 1980; Longwell and Stiles 1973; Mallet and Haley 1983; Hedgecock et al. 1995, 1996; Bierne et al. 1998). Some authors report reduced larval viability due to inbreeding but others, while demonstrating increased growth rate in outbred offspring have shown an increased larval survival in inbred offspring. In trials with *C. gigas*, Hedgecock et al. (1995) demonstrated both positive and negative heterosis in larval and juvenile traits, which were later confirmed by physiological measurements (Hedgecock et al. 1996). Further research is required before the advantages and disadvantages of inbreeding and hybridisation as strategies in bivalve hatcheries can be fully assessed.

10.5 GENOME MAPPING AND GENE SEQUENCES

There are very few published data on genome mapping or linkage groups in bivalves (Foltz 1986; Beaumont 1994; McGoldrick and Hedgecock 1997; Ward et al. 2000) but this aspect of genetics in aquaculture is likely to increase in importance. The use of highly variable markers like microsatellite loci (Gjetvaj et al. 1997), and the ability to rapidly screen large numbers of amplified fragment polymorphism (AFLP) loci for use as genetic markers enables a number of approaches. Linkage of such markers to growth rate or disease resistance can be explored and their relative positions on the chromosomes can be used to develop a genome map of the species. Such information, together with estimated or realised heritabilities (h^2) for important traits, can be used to design and initiate selective breeding programmes based on marker assisted selection (MAS) and the use of family, individual or mixed family/individual selection (Beaumont and Hoare 2003).

There are an increasing number of publications reporting sequences of specific genes or chromosomal regions in scallops. Examples include an actin gene in *P. magellanicus* (Patwary et al. 1996), telomeric DNA in *A. irradians* (Estabrooks 1999), *Adamussium colbecki* satellite DNA with homology to the mammalian CENP-B box (Canapa et al. 2000a), ribosomal internal transcribed spacers (ITS) in *C. farreri* (Y. Yu, pers. comm.), the troponin C and calcineurin genes in *P. yessoensis* (Yuasa et al. 2000; Uryu et al. 2000), histone gene in *P. magellanicus* and *A. irradians* (Brown et al. 2000), expressed sequence tags (ESTs) and the metallothionein gene in *A. irradians* (Picozza et al. 2000), the scallop ATPase molecule (Ryan et al. 2000) and the fructose-1,6-biphosphate aldolase gene in *P. magellanicus* (Patwary et al 2002). Such information, although mainly of a descriptive nature at present, is of great potential importance in the understanding of the genome of scallops and in the emerging technology of gene transfer and the production of genetically modified organisms (GMOs) (Cadoret et al. 2000; Hackett and Alvarez 2000; Moav 2000; Powers and Kirby 2000).

10.6 CONCLUSION

Genetic studies on scallops have helped to advance knowledge in a number of areas of scallop biology notably, stock structure, taxonomy, evolution, fisheries and aquaculture. The main focus of studies using allozymes and more recently developed DNA markers has been on the search for population differentiation, species level differences and higher level systematics. The identification of genetically distinct populations such as the Mulroy Bay *P. maximus* population, and the differentiation of Russian and Japanese populations of *P. yessoensis* are of fundamental value to the fishery for these species, and their management for aquaculture. Studies on chromosome numbers and karyotypes have provided data for many scallop species, which are relevant to their taxonomy and these more traditional approaches have been supplemented by the use of sequence data from rRNA genes. It is clear that the classification of scallops, and ideas about their evolution based on paleontological studies have been strongly influenced by these new data. Continuing discrepancies between various approaches illustrated in this chapter show that there is more of the story still to be told.

The contribution of studies on scallops to the phenomena of the MLH - growth correlation and heterozygote deficiency has been important and scallops have been used as experimental organisms to elucidate the possible causes of these phenomena. In addition, studies on the relationship between heterozygosity and metabolic parameters such as growth and viability are of increasing relevance to the aquaculture of scallops as well as other bivalves. Other areas of scallop genetics, such as the use of selfing to produce inbred lines, the production of triploids, and the estimation of heritabilities of quantitative traits are all fields of research, which could make important contributions towards scallop aquaculture.

As hatchery production of scallops takes on increasing importance, developments currently emerging in the domestication of oysters will be, or are already being employed to enhance scallop aquaculture. Examples include the development of genome maps with linkages to quantitative trait loci (Ward et al. 2000) and, assuming that ethical issues surrounding GMOs can be satisfactorily resolved, the use of gene transfer technology (Cadoret et al. 2000; Lin et al. 2000) to enhance performance and develop commercial strains of scallops.

ACKNOWLEDGMENTS

My thanks are due to the many scientists who have attended Pectinid Workshops over the years, who contributed to my understanding of genetics and scallops, and who often allowed me access to unpublished data. My particular thanks are due to Dr. Ziniu Yu and Dr. Zhaoping Wang for their help in accessing work being carried out in China.

REFERENCES

Adamkewicz, L. and Castagna, M., 1988. Genetics of shell colour and pattern in the bay scallop, *Argopecten irradians*. J. Hered. 79:14–17.

Ahmad, M., Skibinski, D.O.F. and Beardmore, J.A., 1977. An estimate of the amount of genetic variation in the common *Mytilus edulis*. Biochem. Genet. 15:833–846.

Alfonsi, C., Nussetti, O. and Perez, J.E., 1995. Heterozygosity and metabolic efficiency in the scallop *Euvola ziczac* (L. 1748). J. Shellfish Res. 14:389–393.

Allen, S.K., 1988. Triploid oysters ensure year-round supply. Oceanus 31:58–63.

Avise, J.C., 1994. Molecular Markers, Natural History and Evolution. Chapman and Hall, New York.

Ayala, F.J., Tracey, M.L., Hedgecock, D. and Richmond, R., 1974. Genetic differentiation during the speciation process in *Drosophila*. Evolution 28:576–592.

Balakirev, E.S., Dolganov, S.M. and Priima, T.F., 1995. The level of genetic variation in *Swiftopecten swifti, Mizuhopecten (Patinopecten) yessoensis*, and *Chlamys farreri nipponensis* from the Sea of Japan. In: P. Lubet, J. Barret and J-C Dao (Eds.). Fisheries, Biology and Aquaculture of Pectinids: 8th International Pectinid Workshop. IFREMER, Actes de Colloques 17:235–238.

Baron, J., Diter, A. and Bodoy, A., 1989. Triploidy induction in the black scallop (*Chlamys varia* L.) and its effect on the larval growth and survival. Aquaculture 77:103–111.

Beaumont, A.R., 1982a. Geographic variation in allele frequencies at three loci in *Chlamys opercularis* from Norway to the Brittany coast. J. Mar. Biol. Assoc. U.K. 62:243–261.

Beaumont, A.R., 1982b. Variations in heterozygosity at two loci between year classes of a population of *Chlamys opercularis* (L) from a Scottish sea-loch. Mar. Biol. Letts. 3:25–34.

Beaumont, A.R., 1986. Genetic aspects of hatchery rearing of the scallop, *Pecten maximus* (L). Aquaculture 57:99–110.

582

Beaumont, A.R., 1991a. Genetic distances between some scallop species. In: S.E. Shumway and P.A. Sandifer (Eds.). An International Compendium of Scallop Biology and Culture. World Aquaculture Workshops, No. 1. World Aquaculture Society, Baton Rouge. pp. 151–155.

Beaumont, A.R., 1991b. Genetic studies of laboratory reared mussels, *Mytilus edulis*: heterozygote deficiencies, heterozygosity and growth. Biol. J. Linn. Soc. 44:273–285.

Beaumont, A.R., 1994. The application and relevance of genetics in aquaculture. In: A.R. Beaumont (Ed.). Genetics and Evolution of Aquatic Organisms. Chapman and Hall, London. pp. 467–486.

Beaumont, A.R., 1994. Linkage studies in *Mytilus edulis*, the mussel. Heredity 72:557–562.

Beaumont, A.R., 2000a. Genetic considerations in hatchery culture of bivalve shellfish. In: M. Fingerman and R. Nagabhushanam (Eds.). Recent Advances in Marine Biotechnology. Volume IV, Aquaculture, Part A, Seaweeds and Invertebrates. Oxford and IBH, New Delhi. pp. 87–109.

Beaumont, A.R., 2000b. Genetic considerations in transfers and introductions of scallops. Aquaculture International 8(6):493–512.

Beaumont, A.R. and Barnes, D.A., 1992. Aspects of veliger larval growth and byssus drifting of the spat of *Pecten maximus* and *Aequipecten (Chlamys) opercularis*. ICES J. Mar. Sci. 49:417–423.

Beaumont, A.R. and Beveridge, C.M., 1984. Electrophoretic survey of genetic variation in *Pecten maximus, Chlamys opercularis, C. varia* and *C. distorta* from the Irish Sea. Mar. Biol. 81:299–306.

Beaumont, A.R. and Budd, M.D., 1983. Effects of self fertilisation and other factors on the early development of the scallop *Pecten maximus*. Mar. Biol. 76:285–289.

Beaumont, A.R., Budd, M.D. and Gruffydd, Ll.D., 1982. The culture of scallop larvae. Fish Farm. Int. 9:10–11.

Beaumont, A.R., Day, T.R. and Gäde, G., 1980. Genetic variation at the octopine dehydrogenase locus in the adductor muscle of *Cerastoderma edule* (L) and six other bivalve species. Mar. Biol. Letts. 1:137–148.

Beaumont, A.R. and Fairbrother, J.E., 1991. Ploidy manipulation in molluscan shellfish: a review. J. Shellfish Res. 10:1–18.

Beaumont, A.R., Fairbrother, J.E. and Hoare, K., 1995. Multilocus heterozygosity and size: a test of hypotheses using triploid *Mytilus edulis*. Heredity 75:256–266.

Beaumont, A.R., Gosling, E.M., Beveridge, C.M., Budd, M.D. and Burnell, G., 1985. Studies on heterozygosity and growth rate in the scallop, *Pecten maximus* (L). In: P.E. Gibbs (Ed.). Proc. 19th Eur. Mar. Biol. Symp., Plymouth. Cambridge University Press. pp. 443–455.

Beaumont, A.R. and Gruffydd, Ll.D., 1974. Studies on the chromosomes of the scallop *Pecten maximus* (L) and related species. J. Mar. Biol. Assoc. U.K. 54:713–718.

Beaumont, A.R. and Gruffydd, Ll.D., 1975. A polymorphic system in the sarcoplasm of *Chlamys opercularis*. J. Cons. Int. Explor. Mer. 36:190–192.

Beaumont, A.R. and Hoare, K., 2003. Biotechnology and Genetics in Fisheries and Aquaculture. Blackwell Science, Oxford.

Beaumont, A.R. and Kelly, K.S., 1989. Production and growth of triploid *Mytilus edulis* larvae. J. Exp. Mar. Biol. Ecol. 132:69–84.

Beaumont, A.R. and Mason, J., 1987. The Sixth International Pectinid Workshop, Menai Bridge, Wales. 9–14 April, 1987. ICES Mimeo. CM 1987/K:3. 39 p.

Beaumont, A.R., Morvan, C., Huelvan, S., Lucas, A. and Ansell, A.D., 1993. Genetics of indigenous and transplanted populations of *Pecten maximus*: no evidence for the existence of separate stocks. J. Exp. Mar. Biol. Ecol. 169:77–88.

Beaumont, A.R. and Zouros, E., 1991. Genetics of scallops. In: S.E. Shumway (Ed.). Scallops: Biology, Ecology and Aquaculture. Developments in Aquaculture and Fisheries Science, Vol. 21. Elsevier, Amsterdam. pp. 585–623.

Bierne, N., Launey, S., Naciri-Graven, Y. and Bonhomme F., 1998. Early effect of inbreeding as revealed by microsatellite analyses on *Ostrea edulis* larvae. Genetics 148:1893–1906.

Blake, S.G., Blake, N.J., Oesterling, M.J. and Graves J.E., 1997. Genetic divergence and loss of diversity in two cultured populations of the bay scallop, *Argopecten irradians* (Lamark, 1819). J. Shellfish Res. 16:55–58.

Blake, S.G. and Graves, J.E., 1995. Mitochondrial DNA variation in the bay scallop, *Argopecten irradians* (Lamark, 1819), and the Atlantic calico scallop, *Argopecten gibbus* (Linnaeus, 1758). J. Shellfish Res. 14:79–85.

Boulding, E.G., Boom, J.D.G. and Beckenbach, A.T., 1993. Genetic variation in one bottlenecked and two wild populations of the Japanese scallop (*Patinopecten yessoensis*): empirical parameter estimates from coding regions of mitochondrial DNA. Can. J. Fish. Aquat. Sci. 50:1147–1157.

Bricelj, V.M. and Krause, M.K., 1992. Resource allocation and population genetics of the bay scallop, *Argopecten irradians irradians* - effects of age and allozyme heterozygosity on reproductive output. Mar. Biol. 113:253–261.

Britten, H.B., 1996. Meta-analysis of the association of multilocus heterozygosity and fitness. Evolution 50:2158–2164.

Brown, M.V., Strausbaugh, L. and Stiles, S., 2000. Methodology for the generation of molecular tags in *Placopecten magellanicus* (sea scallop) and *Argopecten irradians* (bay scallop). J. Shellfish Res. 19:569 (abstract only).

Buroker, N.E., 1980. An examination of the trophic resource stability theory using oyster species of the family Ostreidae. Evolution 34:204–207.

Buroker, N.E., Chanley, P., Cranfield, H.J. and Dinamani, P., 1983. Systematic status of two oyster populations of the genus *Tiostrea* from New Zealand and Chile. Mar. Biol. 77:191–200.

Buroker, N.E., Hershberger, W.K. and Chew, K.K., 1979a. Population genetics of the family Ostreidae I. Intraspecific studies of *Crassostrea gigas* and *Saccostrea commercialis*. Mar. Biol. 54:157–170.

Buroker, N.E., Hershberger, W.K. and Chew, K.K., 1979b. Population genetics of the family Ostreidae II. Interspecific studies of the genera *Crassostrea* and *Saccostrea*. Mar. Biol. 54:171–184.

Cadoret, J-P., Bachere, E., Roch, P., Mialhe, E. and Boulo, V., 2000. Genetic transformation of farmed marine bivalve molluscs. In: M. Fingerman and R. Nagabhushanam (Eds.). Recent Advances in Marine Biotechnology. Volume IV, Aquaculture, Part A, Seaweeds and Invertebrates. Oxford and IBH, New Delhi. pp. 111–126.

Canapa, A., Barucca, M., Cerioni, P.N. and Olmo, E., 2000a. A satellite DNA containing CENP-B box-like motifs is present in the Antarctic scallop *Adamussium colbecki*. Gene 247:175–180.

584

Canapa, A., Barucca, M., Marinelli, A. and Olmo, E., 2000b. Molecular data from the 16S rRNA gene for the phylogeny of the Pectinidae (Mollusca: Bivalvia). Journal of Molecular Evolution 50:93–97.

Canapa, A., Barucca, M., Marinelli, A. and Olmo, E., 1999. A molecular approach to the systematics of the Antarctic scallop *Adamussium colbecki*. Ital. J. Zool. 66:379–382.

Clarke, A.H. Jr., 1965. The scallop superspecies *Aequipecten irradians* (Lamark). Malacologia 2:161–188.

Clarke, D.F., 1991. Ploidy manipulation in the great scallop, *Pecten maximus* L. M.Sc. thesis, University of Wales.

Cochard, J.C. and Devauchelle, N., 1993. Spawning, fecundity and larval growth in relation to controlled conditioning in five populations of *Pecten maximus*: evidence for the existence of separate stocks. J. Exp. Mar. Biol. Ecol. 169:41–56.

Cook, D.I. and Zouros, E., 1994. The highly variable and highly mutable mitochondrial DNA molecule of the deep-sea scallop *Placopecten magellanicus*. Nautilus 108:85–90.

Coronado, C., Gonzalez, P. and Perez, J.E., 1991. Genetic variation in Venezuelan molluscs *Pecten ziczac* and *Lyropecten nodosa* (Pectinidae). Carib. J. Sci. 27:71–74.

Crenshaw, J.W. Jr., Heffernan, P.B. and Walker, R.L., 1991. Heritability of growth rate in the southern bay scallop, *Argopecten irradians concentricus* (Say, 1822). J. Shellfish Res. 10:55–63.

Cruz, P. and Ibarra, A.M., 1997. Larval growth and survival of two catarina scallop (*Argopecten circularis*, Sowerby, 1835) populations and their reciprocal crosses. J. Exp. Mar. Biol. Ecol. 212:95–110.

Cruz, P., Ramirez, J.L., Garcia, G.A. and Ibarra, A.M., 1998. Genetic differences between two populations of catarina scallop (*Argopecten ventricosus*) for adaptations for growth and survival in a stressful environment. Aquaculture 166:321–335.

Cruz, P., Rodriguez-Jaramillo, C. and Ibarra, A.M., 2000. Environment and population origin effects on first sexual maturity of catarina scallop, *Argopecten ventricosus* (Sowerby II, 1812). J. Shellfish Res. 19:89–93.

Del Rio-Portilla, M. and Beaumont, A.R., 2000. Larval growth, juvenile size and heterozygosity in laboratory reared mussels, *Mytilus edulis*. J. Exp. Mar. Biol. Ecol. 254:1–17.

Desilets, J., Gicquaud, C. and Dubé, F., 1995. An ultrastructural analysis of early fertilization events in the giant scallop, *Placopecten magellanicus* (Mollusca, Pelecypoda). Invert. Reprod. Dev. 27(2):115–129.

Diehl, W.J. and Koehn, R.K., 1985. Multiple locus heterozygosity, mortality, and growth in a cohort of *Mytilus edulis*. Mar. Biol. 88:265–271.

Dolganov, S.M., 1995. Allozyme markers in scallop *Mizuhopecten yessoensis* Jay. Russian Journal of Genetics (Genetika) 31:705–711.

Dolganov, S.M. and Pudovkin, A.I., 1995. Lack of correlation between allozyme heterozygosity and size variation in juveniles of the scallop *Mizuhopecten yessoensis*. In: P. Lubet, J. Barret and J-C. Dao (Eds.). Fisheries, Biology and Aquaculture of Pectinids: 8th International Pectinid Workshop. IFREMER, Actes de Colloques 17:239–242.

Dolganov, S.M. and Pudovkin, A.I., 1997. Genetic diversity of the scallop *Mizuhopecten (Patinopecten) yessoensis* Jay, 1856 from Primorye. Russian Journal of Genetics (Genetika) 33:1187–1194.

Dolganov, S.M. and Pudovkin, A.I., 1998. Population genetic structure of the Japanese scallop *Mizuhopecten (Patinopecten) yessoensis* from Sakhalin Island and the southern Kuril Islands. Russian Journal of Genetics (Genetika) 34:1196–1204.

Estabrooks, S.L., 1999. The telomeres of the bay scallop, *Argopecten irradians* (Lamark). J. Shellfish Res. 18:401–404.

Estoup, A. and Angers, B., 1998. Microsatellites and minisatellites for molecular ecology: theoretical and empirical considerations. In: G.R. Carvalho (Ed.). Advances in Molecular Ecology. IOS Press. pp. 55–86.

Falconer, D.S., 1981. Introduction to Quantitative Genetics. Second Edition. Longman Inc., New York. 340 p.

Ferguson, A., 1980. Biochemical Systematics and Evolution. Blackie, London.

Fevolden, S.E., 1989. Genetic differentiation of the Iceland scallop, *Chlamys islandica*, in the northern Atlantic Ocean. Mar. Ecol. Prog. Ser. 51:77–85.

Fevolden, S.E., 1992. Allozymic variability in the Iceland scallop *Chlamys islandica* - geographic variation and lack of growth-heterozygosity correlations. Mar. Ecol. Prog. Ser. 85:259–268.

Foltz, D.W., 1986. Segregation and linkage studies of allozyme loci in pair crosses of the oyster *Crassostrea virginica*. Biochem. Genet. 24:941–956.

Foltz, D.W. and Zouros, E., 1984. Enzyme heterozygosity in the scallop *Placopecten magellanicus* (Gmelin) in relation to age and size. Mar. Biol. Letts. 5:255–263.

Frischer, M.E., Danforth, J.M., Tyner, L.C., Leverone, J.R., Marelli, D.C., Arnold, W.S. and Blake, N.J., 2000. Development of an *Argopecten* - specific 18SrRNA targeted genetic probe. Mar. Biotech. 2:11–20.

Frischer, M.E., Williams, J. and Kenchington, E., 1998. A molecular phylogeny of the major groups of pectinidae inferred from 18SrRNA gene sequences. In: P.A. Johnston and J.W. Haggart (Eds.). Bivalves: an Eon of Evolution - Paleobiological Studies Honouring Norman D Newell. University of Calgary Press, Calgary, Alberta. pp. 213–221.

Fujio, Y. and von Brand, E., 1991. Differences in degree of homozygosity between seed and sown populations of the Japanese scallop, *Patinopecten yessoensis*. Nippon Suisan Gakkaishi 57(1):45–50.

Fuller, K.M. and Zouros, E., 1993. Dispersed discrete length polymorphism of mitochondrial DNA in the scallop *Placopecten magellanicus* (Gmelin). Current Genetics 23:365–369.

Gäde, G., 1980. Biological role of octopine formation in marine molluscs. Mar. Biol. Letts. 1:121–135.

Gaffney, P.M., 1990. Enzyme heterozygosity, growth rate and variability in *Mytilus edulis*: Another look. Evolution 44:204–210.

Gaffney, P.M., 1994. Heterosis and heterozygote deficiencies in marine bivalves: more light? In: A.R. Beaumont (Ed.). Genetics and Evolution of Aquatic Organisms. Chapman and Hall, London. pp. 146–153.

Gajardo, G., Parraguez, M. and Colihueque, N., 2002. Karyotype analysis and chromosome banding of the Chilean-Peruvian scallop *Argopecten purpuratus* (Lamark, 1819). J. Shellfish Res. 21: 585-590.

Galleguillos, R.A. and Troncoso, L.S., 1991. Protein variation in the Chilean-Peruvian scallop *Argopecten purpuratus* (Lamark, 1819). In: S.E. Shumway and P.A. Sandifer (Eds.). An

International Compendium of Scallop Biology and Culture. World Aquaculture Workshops, No. 1. World Aquaculture Society, Baton Rouge. pp. 146–150.

Garton, D.W., 1984. Relationship between multiple locus heterozygosity and physiological energetics of growth in the estuarine gastropod *Thais haemastoma*. Physiol. Zool. 57:530–543.

Gentili, M.R. and Beaumont, A.R., 1988. Environmental stress, heterozygosity and growth rate in *Mytilus edulis*. J. Exp. Mar. Biol. Ecol. 120:145–153.

Gjetvaj, B., Ball, R.M., Burbridge, S., Bird, C.J., Kenchington, E. and Zouros, E., 1997. Amounts of polymorphism at microsatellite loci in the sea scallop *Placopecten magellanicus*. J. Shellfish Res. 16:547–553.

Gjetvaj, B., Cook, D.I. and Zouros, E., 1992. Repeated sequences and large-scale size variation of mitochondrial DNA: a common feature among scallops (Bivalvia: Pectinidae). Mol. Biol. Evol. 9:106–124.

Gosling, E.M., 1992. Genetics of *Mytilus*. In: E.M. Gosling (Ed.). The Mussel *Mytilus*: Ecology, Physiology, Genetics and Culture. Elsevier, Amsterdam. pp. 309–382.

Gosling, E.M. and Burnell, G.M., 1988. Evidence for selective mortality in *Chlamys varia* (L) transplant experiments. J. Mar. Biol. Assoc. U.K. 68:251–258.

Gruffydd, Ll.D. and Beaumont, A.R., 1970. Determination of the optimum concentration of eggs and spermatozoa for the production of normal larvae in *Pecten maximus* (Mollusca, Lamellibranchia). Helgolander wiss. Meeresunters 20:486–497.

Guo, X. and Allen, S.K. Jr., 1994. Sex determination and polyploid gigantism in the dwarf-surf clam, *Mulinia lateralis* Say. Genetics 138:1199–1206.

Guo, X., DeBrosse, G.A. and Allen, S.K. Jr., 1996. All-triploid Pacific oysters (*Crassostrea gigas* Thunberg) produced by mating tetraploids and diploids. Aquaculture 142:149–161.

Hackett, P.B. and Alkvarez, M.C., 2000. The molecular genetics of transgenic fish. In: M. Fingerman and R. Nagabhushanam (Eds.). Recent Advances in Marine Biotechnology. Volume IV, Aquaculture, Part B, Fishes. Oxford and IBH, New Delhi. pp. 77–145.

Harris, H., 1966. Enzyme polymorphisms in man. Proc. Roy. Soc. Lond. B 164:298–310.

Hawkins, A.J.S., Bayne, B.L. and Day, A.J., 1986. Protein turnover, physiological energetics and heterozygosity in the blue mussel, *Mytilus edulis*: the basis of variable age – specific growth. Proc. R. Soc. Lond. B 229:161–176.

Hawkins, A.J.S., Day, A.J., Gerard, A., Naciri, Y., Ledu, C., Bayne, B.L. and Heral, M., 1994. A genetic and metabolic basis for faster growth among triploids induced by blocking meiosis I but not meiosis II in the larviparous European flat oyster, *Ostrea edulis* L. J. Exp. Mar. Biol. Ecol. 184:21–40.

Hedgecock, D., McGoldrick, D.J. and Bayne, B.L., 1995. Hybrid vigour in Pacific oysters: an experimental approach using crosses among inbred lines. Aquaculture 137:285–298.

Hedgecock, D., McGoldrick, D.J., Manahan, D.T., Vavra, J., Appelmans, N. and Bayne B.L., 1996. Quantitative and molecular analysis of heterosis in bivalve molluscs. J. Exp. Mar. Biol. Ecol. 203:49–59.

Hedrick, P.W., 1985. Genetics of Populations. Jones and Bartlett Inc., Boston, USA. 629 p.

Heffernan, P.B., Walker, R.L. and Crenshaw, J.W. Jr., 1991. Negative larval response to selection for increased growth rate in *Mercenaria mercenaria*. J. Shellfish Res. 10:199–202.

Heffernan, P.B., Walker, R.L. and Crenshaw, J.W. Jr., 1992. Embryonic and larval responses to selection for increased rate of growth in adult bay scallops, *Argopecten irradians concentricus* (Say, 1822). J. Shellfish Res. 11:21–25.

Heipel, D.A., Bishop, J.D.D., Brand, A.R. and Thorpe, J.P., 1998. Population differentiation of the great scallop *Pecten maximus* in western Britain investigated by randomly amplified polymorphic DNA. Mar. Ecol. Prog. Ser. 162:163–171.

Heipel, D.A., Bishop, J.D.D. and Brand, A.R., 1999. Mitochondrial DNA variation among open-sea and enclosed populations of the scallop *Pecten maximus* in western Britain. J. Mar. Biol. Assoc. U.K. 79:687–695.

Hilbish, T.J. and Koehn, R.K., 1985. Dominance in physiological phenotypes and fitness at an enzyme locus. Science 229:52–54.

Hoffman, R., Boore, J.L. and Brown, W.M., 1992. A novel mitochondrial genome organisation in the blue mussel, *Mytilus edulis*. Genetics 131:397–412.

Hoare, K. and Beaumont, A.R., 1995. Effects of an *Odh* null allele and GPI low activity allozyme on shell length in laboratory-reared *Mytilus edulis* L. Mar. Biol. 123:775–780.

Hoxmark, R.C., 1970. The chromosome dimorphism of *Nucella lapillus* (Prosobranchia) in relation to wave exposure. Nytt. Mag. Zool. 18:229–238.

Huelvan, S., 1985. Variabilité Génétique de Populations de *Pecten maximus* (L) en Bretagne. Thése du Doctorat de 3eme Cycle, Université de Bretagne Occidentale, Brest, France. 196 p.

Ibarra, A.M., Cruz, P. and Romero, B.A., 1995. Effects of inbreeding on growth and survival of self-fertilized catarina scallop larvae, *Argopecten circularis*. Aquaculture 134:37–47.

Ieyama, H., 1975. Chromosome numbers of three species in three families of *Pteriomorpha* (Bivalvia). Jap. J. Malac. (Venus) 34:26–32.

Insua, A., Lopez-Pinon, M.J. and Mendez, J., 1998. Characterization of *Aequipecten opercularis* (Bivalvia: Pectinidae) chromosomes by different staining techniques and fluorescent in situ hybridization. Genes Genet. Syst. 74(4):193–200.

Jones, R., Bates, J.A., Innes, D.J. and Thompson, R.J., 1996. Quantitative genetic analysis of growth in larval scallops (*Placopecten magellanicus*). Mar. Biol. 124:671–678.

Kearsey, M.J. and Pooney, H.S., 1996. The Genetical Analysis of Quantitative Traits. Chapman and Hall, London.

Kenchington, E., Naidu, K.S., Roddick, D.L., Cook, D.I. and Zouros, E., 1993. Use of biochemical genetic markers to discriminate between adductor muscles of the sea scallop (*Placopecten magellanicus*) and the Iceland scallop (*Chlamys islandica*). Can. J. Fish. Aquat. Sci. 50:1222–1228.

Kenchington, E. and Roddick, D.L., 1994. Molecular evolution within the phylum mollusca with emphasis on the class Bivalvia. In: N.F. Bourne, B.L. Bunting and L.D. Townsend (Eds.). Proceedings of the 9th International Pectinid Workshop, Nanaimo, British Columbia, 1993. Can. Tech. Rep. Fish. Aquat. Sci. 1994(1):206–213.

Kenchington, E.L., Roddick, D.L., Singh, R.K. and Bird, C.J., 1994. Analysis of small-subunit rRNA gene sequences from six families of molluscs. J. Mar. Biotech. 1:215–217.

Kijima, A., Mori, K. and Fujio, Y., 1984. Population differences of the Japanese scallop *Patinopecten yessoensis* on the Okhotsk Sea coast of Hokkaido. Bull. Jap. Soc. Sci. Fish. 50(2):241–248.

588

Koehn, R.K., Diehl, W.J. and Scott, T.M., 1988. The differential contribution by individual enzymes of glycolysis and protein catabolism to the relationship between heterozygosity and growth rate in the coot clam, *Mulinia lateralis*. Genetics 118:121–130.

Koehn, R.K. and Gaffney, P.M., 1984. Genetic heterozygosity and growth rate in *Mytilus edulis*. Mar. Biol. 82:1–7.

Koehn, R.K., Hall, J.G., Innes, D.J. and Zera, A.J., 1984. Genetic differentiation of *Mytilus edulis* in eastern North America. Mar. Biol. 79:117–126.

Koehn, R.K. and Shumway, S.E., 1982. A genetic/physiological explanation for differential growth rate among individuals of the American oyster, *Crassostrea virginica* (Gmelin). Mar. Biol. Letts. 3:35–42.

Komaru, A., Uchimura, Y., Ieyama, H. and Wada, K.T., 1988. Detection of induced triploid scallops *Chlamys nobilis*, by DNA microfluorometry with DAPI staining. Aquaculture 69:210–209.

Komaru, A. and Wada, K.T., 1985. Karyotypes of four species in the Pectinidae (Bivalvia: Pteriomorphia). Jap. J. Malac. (Venus) 44:249–259.

Komaru, A. and Wada, K.T., 1989. Gametogenesis and growth of induced triploid scallops, *Chlamys nobilis*. Nippon Suisan, Gakkaishi 55:447–452.

Komaru, A. and Wada, K.T., 1991. Different processes of pronuclear events in pressure-treated and CB-treated zygotes at the 2nd meiosis in scallop. Nippon Suisan Gakkaishi 57:1219–1223.

Kraeuter, J., Adamkewicz, L., Castagna, M., Wall, R. and Karney, R., 1984. Ridge number and shell colour in hybridised subspecies of the Atlantic bay scallop, *Argopecten irradians*. Nautilus 98:17–20.

Krause, M.K., 1995. The role of *Gpi* polymorphism in glycolytic flux variations and its effect on genotype dependent variability in the bay scallop. In: P. Lubet, J. Barret and J-C Dao (Eds.). Fisheries, Biology and Aquaculture of Pectinids: 8th International Pectinid Workshop. IFREMER, Actes de Colloques 17:243–247.

Krause, M.K. and Bricelj, V.M., 1995. Genotypic effect on quantitative traits in the northern bay scallop, *Argopecten irradians irradians*. Mar. Biol. 123:511–522.

Krause, M.K., Arnold, W.S. and Ambrose, W.G., 1994. Morphological and genetic variation among three populations of calico scallops *Argopecten gibbus*. J. Shellfish Res. 13:529–537.

Lannan, J.E., 1980. Broodstock management of *Crassostrea gigas*. IV. Inbreeding and larval survival. Aquaculture 21:353–356.

La Roche, J., Snyder, M., Cook, D.I., Fuller, K. and Zouros, E., 1990. Molecular characterization of a repeat element causing large-scale size variation in the mitochondrial DNA of the sea scallop *Placopecten magellanicus*. Mol. Biol. Evol. 7:45–64.

Lerner, I.M., 1954. Genetic Homeostasis. Oliver and Boyd, Edinburgh/London. 134 p.

Lessios, H.A., 1992. Testing electrophoretic data for agreement with Hardy–Weinberg expectations. Mar. Biol. 112:517–523.

Lewis, R.I. and Thorpe, J.P., 1994a. Temporal stability of gene frequencies within genetically heterogeneous populations of the queen scallop *Aequipecten (Chlamys) opercularis*. Mar. Biol. 121:117–126.

Lewis, R.I. and Thorpe, J.P., 1994b. Are queen scallops, *Aequipecten (Chlamys) opercularis* (L.), self recruiting? In: N.F. Bourne, B.L. Bunting and L.D. Townsend (Eds.). Proceedings of the 9th International Pectinid Workshop, Nanaimo, British Columbia, 1993. Can. Tech. Rep. Fish. Aquat. Sci. 1994(1):214–221.

Lewontin, R.C. and Hubby, J.L., 1966. A molecular approach to the study of genetic heterozygosity in natural populations II. Amount of variation and degree of heterozygosity in natural populations of *Drosophila pseudoobscura*. Genetics 54:595–609.

Li, W–H. and Nei, M., 1974. Stable linkage disequilibrium with epistasis in sub-divided populations. Theor. Pop. Biol. 6:173–183.

Li, Q., Osada, M., Kashihara, M., Hirohashi, K. and Kijima, A., 2000. Meiotic maturation, fertilization and effect of ultraviolet irradiation on the fertilizing sperm in the Japanese scallop. Fisheries Sci. 66:403–405.

Lin, C-M., Siri, S., Stiles, S. and Chen, T., 2000. Production of transgenic mollusks and crustaceans. J. Shellfish Res. 19:576 (abstract only).

Longwell, A.C. and Stiles, S., 1973. Gamete cross incompatibility and inbreeding in the commercial oyster *Crassostrea virginica* Gmelin. Cytologia 38:521–533.

Mackie, L.A. and Ansell, A.D., 1993. Differences in reproductive ecology in natural and transplanted populations of *Pecten maximus*: evidence for the existence of separate stocks. J. Exp. Mar. Biol. Ecol. 169:57–75.

Macleod, J.A.A., Thorpe, J.P. and Duggan, N.A., 1985. A biochemical genetic study of population structure in queen scallop (*Chlamys opercularis*) stocks in the Northern Irish Sea. Mar. Biol. 87:77–82.

Mallet, A.L. and Haley, L.E., 1983. Effects of inbreeding on larval and spat performance in the American oyster. Aquaculture 33:229–235.

Manuel, J.L., Burbridge, S., Kenchington, E.L., Ball, M. and O'Dor, R.K., 1996. Veligers from two populations of scallop *Placopecten magellanicus* exhibit different vertical distributions in the same mesocosm. J. Shellfish Res. 15:251–257.

Marelli, D.C., Krause, M.K., Arnold, W.S. and Lyons, W.G., 1997a. Systematic relationships among Florida populations of *Argopecten irradians* (Lamark 1819) (Bivalvia: Pectinidae). The Nautilus 110:31–41.

Marelli, D.C., Krause, M.K., Lyons, W.G. and Arnold, W.S., 1997b. Subspecific status of *Argopecten irradians concentricus* (Say, 1822) and the bay scallops of Florida. The Nautilus 110:42–44.

Mathers, N.F., 1975. Environmental variability at the phosphoglucose isomerase locus in the genus *Chlamys*. Biochem. Syst. Ecol. 3:123–127.

McDonald, J.H., Seed, R. and Koehn, R.K., 1991. Allozymes and morphometric characters of three species of *Mytilus* in the Northern and Southern Hemispheres. Mar. Biol. 111:323–333.

McGoldrick, D.J. and Hedgecock, D., 1997. Fixation, segregation and linkage of allozyme loci in inbred families of the Pacific oyster *Crassostrea gigas* (Thunberg): implications for the causes of inbreeding depression. Genetics 146:321–334.

Menzel, R.W., 1968. Chromosome number in nine families of marine pelecypod molluscs. Nautilus 82:45–58.

Merrill, A.S., Posgay, J.A. and Nichy, F.E., 1966. Annual marks on shell and ligament of sea scallop (*Placopecten magellanicus*). Fisheries Bull. 65:299–311.

Meyer, A., 1994. Molecular phylogenetic studies of fish. In: A.R. Beaumont (Ed.). Genetics and Evolution of Aquatic Organisms. Chapman and Hall, London. pp. 219–249.

Mitton, J.B., 1998. Molecular markers and natural selection. In: G.R. Carvalho (Ed.). Advances in Molecular Ecology. IOS Press, Amsterdam. pp. 225–241.

590

Mitton, J.B. and Grant, M.C., 1984. Association among protein heterozygosity growth rate and developmental homeostasis. Ann. Rev. Ecol. Syst. 15:479–499.

Moav, B., 2000. The application of gene manipulation to aquaculture. In: M. Fingerman and R. Nagabhushanam (Eds.). Recent Advances in Marine Biotechnology. Volume IV, Aquaculture, Part B, Fishes. Oxford and IBH, New Delhi. pp. 147–187.

Moritz, C., Dowling, P.E. and Brown, W.M., 1987. Evolution of animal mitochondrial DNA: relevance for population biology and systematics. Ann. Rev. Ecol. Syst. 18:269–292.

Naidu, K.S., 1970. Reproduction and breeding cycle of the giant scallop, *Placopecten magellanicus* (Gmelin) in Port-au-Port Bay, Newfoundland. Can. J. Zool. 48:1003–1012.

Nei, M., 1978. Estimation of average heterozygosity and genetic distance from a small number of individuals. Genetics 89:583–590.

Nei, M. and Li, W.S., 1973. Linkage disequilibrium in subdivided populations. Genetics 75:213–219.

Nikiforov, S.M. and Dolganov, S.M., 1982. Genetic variation of the Japanese scallop *Patinopecten yessoensis* from the Vostok Bay, Sea of Japan. Marine Biology, Vladivostok 2:46–50.

Patwary, M.U., Kenchington, E.L., Bird, C.J. and Zouros, E., 1994. The use of random amplified polymorphic DNA markers in genetic studies of the sea scallop *Placopecten magellanicus* (Gmelin, 1791). J. Shellfish Res. 13:547–553.

Patwary, M.U., Reith, M. and Kenchington, E.L., 1996. Isolation and characterization of a cDNA encoding actin gene from sea scallop. J. Shellfish Res. 15:265–270.

Patwary, M.U., Wauchope, A., Short, T.W. and Catapane, E.J., 2002. Molecular cloning and characterisation of a fructose-1,6-biphosphate aldolase cDNA from the deep-sea scallop *Placopecten magellanicus*. J. Shellfish Res. 21:591–596.

Pauls, E. and Affonso, P.R.A.M., 2000. The karyotype of *Nodipecten nodosus* (Bivalvia: Pectinidae). Hydrobiologia 420:99–102.

Pena, J.B., Moraga, D., Mestre, S. and Le Pennec, M., 1995. Biochemical genetics of the Mediterranean scallop *Pecten jacobaeus*. In: P. Lubet, J. Barret and J-C. Dao (Eds.). Fisheries, Biology and Aquaculture of Pectinids. 8th International Pectinid Workshop. IFREMER, Actes de Colloques 17:249–252.

Perez, J.E., Nusetti, O., Ramirez, N. and Alfonsi, C., 2000. Allozyme biochemical variation at the octopine dehydrogenase locus in the scallop *Euvola ziczac*. J. Shellfish Res. 19:85–88.

Petuch, E.J., 1987. New Caribbean Molluscan Faunas. The Coastal Education and Research Foundation, Charlottesville, Virginia.

Picozza, E., Crivello, J., Brown, M.V., Strausbaugh, L. and Stiles, S., 2000. Status report for the characterization of the bay scallop, *Argopecten irradians*, genome. J. Shellfish Res. 19:578–579 (abstract only).

Pogson, G.H. and Zouros, E., 1994. Allozyme and RFLP heterozygosities as correlates of growth rate in the scallop *Placopecten magellanicus*: a test of the associative overdominance hypothesis. Genetics 137:221–231.

Powers, D.A. and Kirby, V.L., 2000. Gene transfer in marine finfish and shellfish. In: M. Fingerman and R. Nagabhushanam (Eds.). Recent Advances in Marine Biotechnology. Vol. IV, Aquaculture, Part B, Fishes. Oxford and IBH, New Delhi. pp. 229–248.

Pudovkin, A.I. and Dolganov, S.M., 1995. Population genetic structure of the scallop *Mizuhopecten yessoensis* in the northern part of its geographic area. In: P. Lubet, J. Barret and J-C. Dao

(Eds.). Fisheries, Biology and Aquaculture of Pectinids. 8th International Pectinid Workshop. IFREMER, Actes de Colloques 17:253–256.

Pudovkin, A.I., Zaykin, D.V. and Dolganov, S.M., 1998. Interlocus association of allozyme genotypes in settlements of scallop *Mizuhopecten (Patinopecten)* in coastal waters of Primorye. Russian Journal of Genetics (Genetika) 34:299–305.

Rasotto, M., Altieri, D. and Colombera, D., 1981. I Chromosomi spermatocitari di 16 species appartenenti alla classe Pelycypoda. Atti. Congr. Soc. Malac. Ital, Salice Terme 1198:113–127.

Raymond, M. and Rousset, F., 1995. An exact test for population differentiation. Evolution 49:1280–1283.

Rhodes, E.W., 1991. Fisheries and aquaculture of the bay scallop, *Argopecten irradians*, in the eastern United States. In: S.E. Shumway (Ed.). Scallops: Biology, Ecology and Aquaculture. Developments in Aquaculture and Fisheries Sciences, Vol. 21. Elsevier, Amsterdam. pp. 913–924.

Ridgeway, G.M. and Dahle, G., 2000. Population genetics of *Pecten maximus* of the northeast Altantic coast. Sarsia 85:167–172.

Rice, E.L., Roddick, D. and Singh, R., 1993. A comparison of molluscan (Bivalvia) phylogenies based on paleontological and molecular data. Mol. Mar. Biol. Biotech. 2:137–146.

Rigaa, A., Le Gal, Y., Sellos, D. and Monnerot, M., 1993. Mapping and repeated sequence organisation of mitochondrial DNA in scallop, *Pecten maximus*. Mol. Mar. Biol. Biotech. 2:218–224.

Rigaa, A., Monnerot, M. and Sellos, D., 1995. Molecular cloning and complete nucleotide sequence of the repeated unit and flanking gene of the scallop *Pecten maximus* mitochondrial DNA: putative replication origin features. J. Mol. Evol. 41:189–195.

Rigaa, A., Sellos, D. and Monnerot, M., 1997. Mitochondrial DNA from the scallop *Pecten maximus*: an unusual polymorphism detected by restriction fragment length polymorphism analysis. Heredity 79:380–387.

Ríos, C., Canales, J. and Peña, J.B., 1996. Genotype-dependent spawning: evidence from a wild population of *Pecten jacobaeus* (L.) (Bivalvia: Pectinidae). J Shellfish Res. 15:645–651.

Rodhouse, P.G., McDonald, J.H., Newell, R.I.E. and Koehn, R.K., 1986. Gamete production, somatic growth and multiple-locus enzyme heterozygosity in *Mytilus edulis*. Mar. Biol. 90:209–214.

Rodriguez-Juiz, A.M., Torrado, M. and Mendez, J., 1996. Genome-size variation in bivalve molluscs determined by flow cytometry. Mar. Biol. 126(3):489–497.

Ruiz-Verdugo, C.A., Ramirez, J.L., Allen, S.K. and Ibarra, A.M., 2000. Triploid catarina scallop (*Argopecten ventricosus* Sowerby II, 1842): growth, gametogenesis, and suppression of functional hermaphroditism. Aquaculture 186:13–32.

Ryan, C., Chen, M. and Hardwicke, P.M.D., 2000. Two Ca2+ dependent tryptic cleavage sites in the cytoplasmic domain of scallop Ca ATPase. Biophys. J. 78:435.

Sambrook, J., Fritsch, E.F. and Maniatis, T., 1989. Molecular Cloning: a Laboratory Manual, Second Edition, 3 volumes. Cold Spring Harbour Laboratory, New York.

Selander, R.K., 1970. Behaviour and genetic variation in natural populations. Amer. Zool. 10:53–66.

592

Seyoum, S., Bert, T.M., Wilbur, A., Arnold, W.S. and Crawford, C., 2003. Development, evaluation and application of a mitochondrial genetic tag for the bay scallop, *Argopecten irradians*. J. Shellfish Res. 22:111–117.

Shatkin, G., Shumway, S.E. and Hawes, R., 1997. Considerations regarding the possible introduction of the Pacific oyster (*Crassostrea gigas*) to the Gulf of Maine: a review of global experience. J. Shellfish Res. 16:463–477.

Skibinski, D.O.F., Beardmore, J.A. and Cross, T.F., 1983. Aspects of the population genetics of *Mytilus* (Mytilidae; Mollusca) in the British Isles. Biol. J. Linn. Soc. 19:137–183.

Smith, P.J. and Fujio, Y., 1982. Genetic variation in marine teleosts: high variability in habitat specialists and low variability in habitat generalists. Mar. Biol. 69:7–20.

Snyder, M., Fraser, A.R., La Roche, J., Gartner–Kepkay, K.E. and Zouros, E., 1987. A typical mitochondrial DNA from the deep-sea scallop *Placopecten magellanicus*. Proc. Natl. Acad. Sci., USA 84:7595–7599.

Sokal, R.R. and Rohlf, F.J., 1995. Biometry. Third edition. W. Freeman and Company, New York. 887 p.

Stanley, J.G., Hidu, H. and Allen, S.K., 1984. Growth of American oysters increased by polyploidy induced by blocking meiosis I but not meiosis II. Aquaculture 37:147–155.

Tabarini, C.L., 1984. Induced triploidy in the bay scallop, *Argopecten irradians*, and its effect on growth and gametogenesis. Aquaculture 42:151–160.

Thiriot-Quievreux, C., 1994. Advances in cytogenetics of aquatic organisms. In: A.R. Beaumont (Ed.). Genetics and Evolution of Aquatic Organisms. Chapman and Hall. pp. 369–388.

Thiriot-Quievreux, C. and Ayraud, N., 1982. Les caryotypes de quelques especes de Bivalves et de Gasteropodes marins. Mar. Biol. 70:165–172.

Toro, J.E., Sanhueza, M.A., Paredes, L. and Canello, F., 1995. Induction of triploid embryos by heat shock in the Chilean northern scallop *Argopecten purpuratus* Lamark, 1819. N. Z. J. Mar. Freshwat. Res. 29:101–105.

Turner, P.C., McLennan, A.G., Bates, A.D. and White, M.R.H., 1997. Instant Notes in Molecular Biology. BIOS Scientific Publishers, Oxford.

Uryu, M., Nakatomi, A., Watanabe, M., Hatsuse, R. and Yazawa, M., 2000. Molecular cloning of cDNA encoding two subunits of calcineurin from scallop testis: demonstration of stage-specific expression during maturation of the testis. J. Biochem. 127:739–746.

Varvio, S.L., Koehn, R.K. and Väinölä, R., 1988. Evolutionary genetics of the *Mytilus edulis* complex in the North Atlantic region. Mar. Biol. 98:51–60.

Volckaert, F., Shumway, S.E. and Schick, D.F., 1991. Biometry and population genetics of deep- and shallow-water populations of the sea scallop *Placopecten magellanicus* (Gmelin, 1791) from the Gulf of Maine. In: S.E. Shumway and P.A. Sandifer (Eds.). An International Compendium of Scallop Biology and Culture, World Aquaculture Workshops, No. 1. World Aquaculture Society, Baton Rouge. pp. 156–163.

Volckaert, F. and Zouros, E., 1989. Allozyme and physiological variation in the scallop *Placopecten magellanicus* and a general model for the effects of heterozygosity on fitness in marine molluscs. Mar. Biol. 103:51–61.

von Brand, E., Belloglio, G. and Lohrmann, K., 1990. Chromosome number of the Chilean scallop *Argopecten purpuratus*. Tohoku Journal of Agriculture Research 40:91–95.

Vos, P., Hogers, R., Bleeker, M., Reijans, M., van de Lee, T., Hornes, M., Fritjers, A., Pot, J., Peleman, J., Kuiper, M. and Zabeau, M., 1995. AFLP: a new technique for DNA fingerprinting. Nucleic Acids Res. 23:4407–4414.

Wada, K., 1978. Chromosome karyotypes of three bivalves: the oysters *Isognomon alatus* and *Pinctada imbricata* and the bay scallop, *Argopecten irradians irradians*. Biol. Bull. 155:235–245.

Wall, J.R., Wall, S.R. and Castagna, M., 1976. Enzyme polymorphisms and genetic variation in the bay scallop, *Argopecten irradians*. Genetics 83:1, s81.

Waller, T.R., 1969. The evolution of the *Argopecten gibbus* stock (Mollusca: Bivalvia), with emphasis on the Tertiary and Quaternary species of eastern North America. J. Paleont. 43:1–125.

Waller, T.R., 1991. Evolutionary relationships among commercial scallops (Mollusca: Bivalvia: Pectinidae). In: S.E. Shumway (Ed.). Scallops: Biology, Ecology and Aquaculture. Developments in Aquaculture and Fisheries Sciences, Vol. 21. Elsevier, Amsterdam. pp. 1–73.

Wang, M., Zheng, J. and Yu, H., 1990. The karyotype of Zhikong scallop *Chlamys farreri*. Journal of the Oceanology University of Qingdao 20:81–85 (in Chinese).

Wanguemert, M.G., Petuzzo, P.R. and Borzone, C.A., 2000. Preliminary analysis of the genetic variability of two natural beds of the scallop *Euvola ziczac* (Linnaeus, 1758) in Brazil. Brazil. Arch. Biol. Technol. 43:235–240.

Ward, R.D., English, L.J., McGoldrick, D.J., Maguire, G.B., Nell, J.A. and Thompson, P.A., 2000. Genetic improvement of the Pacific oyster *Crassostrea gigas* (Thunberg) in Australia. Aquaculture Res. 31:35–44.

White, M.J.D., 1973. Animal Cytology and Evolution. Third Edition. Cambridge University Press. 961 p.

Wilbur, A.E. and Gaffney, P.M., 1991. Self-fertilization in the bay scallop *Argopecten irradians*. J. Shellfish Res. 10:274 (abstract).

Wilbur, A.E. and Gaffney, P.M., 1997. A genetic basis for geographic variation in shell morphology in the bay scallop, *Argopecten irradians*. Mar. Biol. 128:97–105.

Wilbur, A.E., Orbacz, E.A., Wakefield, J.R. and Gaffney, P.M., 1997. Mitochondrial genotype variation in a Siberian population of the Japanese scallop, *Patinopecten yessoensis* (Jay). J. Shellfish Res. 16:541–545.

Wilding, C.M., Beaumont, A.R and Latchford, J.W., 1997. Mitochondrial DNA variation in the scallop *Pecten maximus* (L.) assessed by a PCR-RFLP method. Heredity 79:178–189.

Wilding, C.M., Beaumont, A.R. and Latchford, J.W., 1999. Are *Pecten maximus* and *Pecten jacobaeus* different species? J. Mar. Biol. Assoc. U.K. 79:949–952.

Wilding, C.M., Latchford, J.W. and Beaumont, A.R., 1998. An investigation of possible stock structure in *Pecten maximus* (L.) using multivariate morphometrics, allozyme electrophoresis and mitochondrial DNA polymerase chain reaction-restriction fragment length polymorphism. J. Shellfish Res. 17:131–139.

Wilkins, N.P., 1978. Length-correlated changes in heterozygosity at an enzyme locus in the scallop (*Pecten maximus* L). Anim. Blood Grps. biochem. Genet. 9:69–77.

Wolff, M. and Garrido, J., 1991. Comparative study of growth and survival of two colour morphs of the Chilean scallop *Argopecten purpuratus* (Lamark 1819) in suspended culture. J. Shellfish Res. 10:47–53.

594

Wright, S., 1951. The genetical structure of populations. Ann. Eugen. 15:323–354.

Yabu, H., and Maru, K., 1987. Chromosome number of *Patinopecten yessoensis* and *Chlamys farreri nipponensis*. Nippon suisan Gakkaisha 53:319.

Yang, H.P., Que, H.Y., He, Y.C. and Zhang, F.S., 2000. Chromosome segregation in fertilized eggs from zhikong scallop *Chlamys farreri* (Jones and Preston) following polar body 1 inhibition with cytochalasin B. J. Shellfish Res. 19:101–105.

Yuasa, H.J. and Takagi, T., 2000. The genomic structure of the scallop, *Patinopecten yessoensis*, troponin C gene: a hypothesis for the evolution of troponin C. Gene 245:275–281.

Zeng, Z., Chen, M., Lin, Q., Chen, P., Liu, W. and Chen, Y., 1995. Induced triploidy in scallop, *Chlamys nobilis*. J. Oceanogr. (Taiwan Strait, Taiwan Haixia) 14:155–162.

Zhang, G. and Zhang, F., 1997. The genetic structure and variation of five populations in the Chinese scallop, *Chlamys farreri*. Proc. 4th Asian Fisheries Forum, China Ocean Press, Beijing. pp. 422–425.

Zouros, E., 1987. On the relation between heterozygosity and heterosis: an evaluation of the evidence from marine molluscs. In: Isozymes: Current Topics in Biological and Medical Research, Vol. 15: Genetics Development and Evolution. Alan R. Liss, New York. pp. 255–270.

Zouros, E. and Foltz, D.W., 1984. Possible explanations of heterozygote deficiency in bivalve molluscs. Malacologia 25:583–591.

Zouros, E.W. and Foltz, D., 1987. The use of allelic isozyme variation for the study of heterosis. In: M.C. Rattazzi, J.G. Scandalios and G.S. Witt (Eds.). Isozymes: Current Topics in Biological and Medical Research, Vol. 13. Liss, NY.

Zouros, E. and Pogson, G.H., 1994. The present status of the relationship between heterozygosity and heterosis In: A.R. Beaumont (Ed.). Genetics and Evolution of Aquatic Organisms. Chapman and Hall, London. pp. 135–146.

Zouros, E., Pogson, G.H., Cook, D.I. and Dadswell, M.J., 1992. Apparent selective neutrality of mitochondrial DNA size variation: a test in the deep-sea scallop *Placopecten magellanicus*. Evolution 46:1466–1476.

Zouros, E., Romero-Dorey, M. and Mallet, A.L., 1988. Heterozygosity and growth in marine bivalves: further data and possible explanations. Evolution 42:1332–1341.

Zouros, E., Singh, S.M., Foltz, D.W. and Mallet, A.L., 1983. Post-settlement viability in the American oyster (*Crassostrea virginica*): an overdominant phenotype. Genet. Res., Camb. 41:259–270.

Zouros, E., Singh, S.M. and Miles, H.E., 1980. Growth rate in oysters – an overdominant phenotype and its possible explanation. Evolution 34:856–867.

AUTHOR'S ADDRESS

Andy Beaumont - School of Ocean Sciences, University of Wales, Bangor, Menai Bridge, Gwynedd LL59 5EY Wales, UK (E-mail: a.r.beaumont@bangor.ac.uk)

Scallops: Biology, Ecology and Aquaculture
S.E. Shumway and G.J. Parsons (Editors)

595

Chapter 11

Diseases and Parasites of Scallops

Sharon E. McGladdery, Susan M. Bower and Rodman G. Getchell

11.1 INTRODUCTION

Much has changed with respect to the knowledge of diseases and parasites of scallops since the first edition of this book (Getchell 1991). Many observations made in that review, however, set the stage for the advancements in knowledge described in the present review. Significant progress has been made in diagnostic capability, experience and the ability to distinguish between infectious pathogens and cryptic histopathological lesions in scallops. This has paralleled significant advances in understanding of scallop physiology and behaviour (as described in detail elsewhere in this volume), both under wild and culture conditions. Evidence of this is the progress made in combating opportunistic infections of larvae and juveniles of several scallop species reared under hatchery conditions. Last, but not least, there has been a massive expansion in the diversification of scallop species being brought into stock enhancement and aquaculture programs on a global scale. Parallel development of remote underwater photography technology, for surveillance and stock assessment purposes, as well as, suspension culture techniques, have also greatly improved our capability to detect health challenges in many scallop species - challenges which previously evaded direct observation in the scallop's benthic domain.

The present review will follow that of Getchell (1991), dividing sections by taxonomic group of pathogen. However, a listing by scallop species has also been compiled in Table 11.1, to facilitate scallop specific cross-referencing. The infectious agents described in detail by Getchell (1991) remain, to ensure as comprehensive a reference as possible.

11.2 MICROBIAL DISEASES

11.2.1 Viruses

Only two viral observations from scallops have been reported to date. The first was a salmonid pathogen, Infectious Pancreatic Necrosis Virus (IPNV). This unenveloped, double-strand RNA Birnavirus was found in the King scallop (Coquille St.-Jacques) *Pecten maximus* (Mortensen 1993a; Mortensen et al. 1992, 1998) from Norway. The IPNV particles were isolated from hepatopancreas, gonad, kidney, mantle, gill, rectum and haemolymph preparations of scallops within 24-h of exposure to bath challenges (Mortensen et al. 1992). The virus could still be detected in the hepatopancreas 50 days

Table 11.1

Diseases and parasites of scallops.

Scallop Species	Parasite or Disease	Effect on Scallop	References
Adamussium colbecki	*Cibicides refulgens* (Foraminifera)	Normally superficial fouling, but penetration of shell can lead to feeding on mantle fluids and scallop weakening	Alexander & DeLaca 1987
Aequipecten (Chlamys) opercularis	Rickettsial-like organism (Bacteria)	None reported	LeGall et al. 1992
"	Unidentified microsporidian (Protista)	No adverse effects observed, under natural conditions	Lohrmann et al. 1999, 2000b
"	*Licnophora auerbachii* (Protista)	Histological damage to the eyes	Harry 1980
"	*Paranthessius pectinis* (Copepoda, Crustacea)	None reported	Reddiah & Williamson 1958
"	*Modiolicola inermis* (Copepoda, Crustacea)	None reported - common in mantle cavity	Reddiah & Williamson 1958
"	*Cibicides lobatulus* (Formanifera)	None reported	Howard & Haynes 1976
"	*Suberites ficus* ssp. *rubrus* (Porifera)	None reported - associated with reduced fouling by other colonies	Armstrong et al. 1999
Amusium balloti	*Sulcascaris sulcata* (Nematoda)	Associated with brown lesions in adductor muscle	Cannon 1978; Lester et al. 1980
Amusium pleuronectes	*Pinnotheres* sp. (Crustacea)	None	Llana 1979
Argopecten gibbus	*Marteilia* sp. (Ascetospora, Protista)	Associated with extreme mortalities in SE USA	Moyer et al. 1993, 1995
"	*Echeneibothrium* sp. (Cestoda)	Associated with gonad atrophy	Singhas et al. 1993
"	*Sulcascaris sulcata* (Nematoda)	None reported - associated primarily with the gonad	Lichtenfels et al. 1978, 1980; Blake et al. 1984
"	*Porrocaecum pectenis* (Nematoda)	Brown discolouration of the adductor muscle (associated with haplosporidian hyperparasite of the encysted nematode)	Cheng 1967, 1973, 1978; McLean 1983
"	*Ceratonereis tridentata* (Polychaeta)	Pest	Wells and Wells 1962

"	*Tumidotheres* (*Pinnotheres*) *maculatus* (Crustacea)	Mechanical compression of soft-tissues (gills, mantle, gonad)	Getchell 1991
"	*Odostomium seminuda* (Pyramidellidae: Gastropoda)	Mantle retraction	Wells and Wells 1961
Argopecten irradians	*Vibrio natriegens* (Vibrionaceae, Bacteria)	Associated with mass mortalities in adult scallops at a hatchery in China	Zhang et al. 1998
"	Chlamydia-like organisms (Rickettsiales, Bacteria)	Mass mortalities of larvae associated with infections in US hatchery. Mass mortalities of adult scallop in China also associated with digestive gland infections. Infections elsewhere show no clinical effects	Morrison and Shum 1982; Leibovitz 1989; Wang et al. 1998
"	Rickettsial-like organisms (Rickettsiales, Bacteria)	None reported except hypertrophy of the infected cell	Morrison and Shum 1983; Leibovitz et al. 1984; McGladdery et al. 1993a
"	Unidentified haplosporidian (Haplosporidia, Protista)	Associated with high post spawning mortalities in China, but infection levels were relatively low	Chu et al. 1996
"	*Nematopsis duorari* & *N. ostrearum* (Coccidia, Protista)	None reported	Léger and Dubosqc 1925 (cited in Sprague 1970); Kruse 1966; Sprague 1970
"	*Pseudoklossia*-like sp. (Coccidia)	Associated with adult mortalities under unnatural holding conditions	Leibovitz et al. 1984; McGladdery 1990, 1993a,b; Karlsson 1991, Getchell 1991; Cawthorn et al. 1992; Whyte et al. 1994
"	*Perkinsus karlssoni* (taxonomic affinity under reinvestigation)	Associated with adult mortalities under hatchery holding conditions	McGladdery et al. 1991; Whyte et al. 1993; Goggin et al. 1996
"	*Proctoeces maculatus* (Digenea, Platyhelminthes)	None reported	Karlsson 1991
"	Unidentified metacercaria (Digenea)	None reported	Karlsson 1991
"	Unidentified Echinostome (Digenea)	Focal haemocyte infiltration only	McGladdery et al. 1993a
"	*Himasthla quissetensis* (Digenea)	None reported	Stunkard 1983
"	Sanguilicolid sporocyst (Digenea)	Castration	Linton 1915

598

Host	Organism	Effect	Reference
"	*Parachristianella* sp. (Cestoda, Platyhelminthes)	None reported	Cake 1976, 1977
"	*Rhinebothrium* sp. (Cestoda)	None reported	Cake 1976, 1977
"	*Acanthobothrium* sp. (Cestoda)	None reported	Cake 1976, 1977
"	*Anthobothrium* sp. (Cestoda)	None reported	Cake 1976, 1977
"	*Eutetrarhynchus* sp. (Cestoda)	None reported	Cake 1976, 1977
"	*Tylocephalum* sp. (Cestoda)	None reported	Cake 1976, 1977
"	*Polypocephalus* sp. (Cestoda)	None reported	Cake 1976, 1977, 1979
"	Unidentified turbellarian (Turbellaria)	None reported	Leibovitz et al. 1984
"	*Sulcascaris sulcatus* (Nematoda)	None reported	Lichtenfels et al. 1978
"	*Porrocaecum pectenis* (Nematoda)	Brown discolouration of the adductor muscle (associated with haplosporidian hyperparasite of the encysted nematode)	Cheng 1967, 1973, 1978; McLean 1983
"	*Polydora ciliata, Polydora websteri,* *Polydora* spp (Polychaeta)	Extensive mud blisters associated with shell weakening	Turner & Hanks 1959; Russel 1973; Leibovitz et al. 1984; Karlsson 1991
"	*Tumidotheres (Pinnotheres) maculatus, Pinnotheres* spp. (Brachyura, Crustacea)	Mechanical compression of soft-tissues (gills, mantle, gonad), associated with weakening and weight loss	Cheng 1967; Kryczynski 1972; Karlsson 1991
"	*Odostomium seminuda* (Pyramidellidae: Gastropda)	Mantle retraction	Wells & Wells 1961; Leibovitz et al. 1984
"	*Prorocentrum* sp.	Dinoflagellate toxicity and mechanical damage under hatchery conditions	Leibovitz et al. 1984
"	*Zoochlorella* sp. (Algae)	Green proliferative granulomas in the mantle and tentacles	Leibovitz et al. 1984
"	*Scypha* sp. (Porifera)	Induced shell deformities	Leibovitz et al. (1984)
"	*Sirolpidium zoophthorum* (Oomycete, Eumycota)	Pathogenic to larval scallops	Martin et al. 1997
Argopecten (Chlamys) purpuratus	*Vibrio* spp. (Vibrionaceae, Bacteria)	Hatchery-related mortalities	DiSalvo 1994; Chavez & Riquelme 1994; Riquelme et al. 1995, 1996, 1997, 2000 Lohrmann et al. 2000a, 2002
"	Rickettsial-like organisms (Rickettsiales, Bacteria)	None reported	Lohrmann et al. 2000a, 2002
"	*Trichodina* sp. (Oligohymenophora, Protista)	None reported	Lohrmann et al. 2000a, 2002

Host	Organism	Effect	Reference
"	Kidney coccidia (Coccidia, Protista)	Associated with poor-quality gamete production	DiSalvo 1994
"	Hemiurid (Platyhelminthes)	Associated with castration	Mateo et al. 1975
"	Unidentified Digenean metacercariae	Encysted in the palps - no pathology observed	Lohrmann et al. 1991
"	Unidentified Phyllobothriidae (Cestoda, Platyhelminthes)	None reported	Oliva et al. 1986
"	Unidentified Oncobothriidae (Cestoda, Platyhelminthes)	None reported	Oliva et al. 1986
"	Unidentified larval cestode	None reported	Lohrmann & Smith 1993
"	Ciona intestinalis (Chordata, Ascidiacea)	Smothering related mortalities by fouling coverage	Uribe & Etchepare 1999
Argopecten ventricosus (circularis) ssp. aequisulcatus)	Stephanostomum sp. (Digenea, Platyhelminthes)	Black spot lesions - no adverse clinical affects but impacts marketability	Pérez-Urbiola & Martínez-Díaz (2001)
"	Echinocephalus pseudouncinatus (Nematoda)	None reported	McLean 1983
Chlamys (Mimachlamys) asperrina	Bucephalus sp. (Digenea, Platyhelminthes)	Reduced reproductive potential in heavily infected scallops	Heasman et al. 1996
Chlamys farreri (nipponensis)	Trichodina jadranica (Oligohymenophora, Protista)	None reported	Xu et al. 1995
"	Pectenophilus ornatus (Copepoda, Crustacea)	Gill hypertrophy at the attachment site of the copepod along with blood-feeding is associated with chronic weakening	Nagasawa et al. 1988, 1991; Nagasawa & Nagata 1992
"	Pinnotheres sp. (Brachyura, Crustacea)	None reported	Cheng 1967
Chlamys islandica	Gymnophallus sp. (Digenea, Platyhelminthes)	None reported	Chubrick 1966 (cited in Lauckner 1983)
Chlamys nobilis	Tylocephalum sp. (Cestoda, Platyhelminthes)	None reported	Sakaguchi 1973 (cited in Lauckner 1983)
Chlamys (Pecten) varia	Nematospis pectinis (Porosporidae, Protista)	None reported	Léger & Duboscq 1925 (cited in Sprague 1970)
Euvola (Pecten) ziczac	Flavobacterium sp. (Gracilicutes, Bacteria)	Associated with larval mortalities	Lodeiros et al. 1989
"	Pseudomonas sp. (Bacteria)	Associated with larval mortalities	Lodeiros et al. 1992

Host	Agent	Effects	References
Mizuhopecten (Patinopecten) yessoensis	Intracellular bacterium (possible Mycoplasma) (Tenericutes, Bacteria)	Associated with weakening and pinkish-orange pustules in adult scallops.	Bower & Meyer 1991; Bower et al. 1992; Bower & Meyer 1994
"	Rickettsial-like organisms (Rickettsiales, Bacteria)	None reported	Elston 1986; Friedman 1994
"	*Trichodina pectinis, T.* sp. (Oligohymenophora, Protista)	None reported	Lauckner 1983; Kurochkin et al. 1986
"	*Nematopsis* sp. (Porosporidae, Protista)	None observed	Bower & Meyer (pers. comm.)
"	*Perkinsus* sp. (Apicomplexa, Protista)	None reported	Kurochkin et al. 1986
"	*Perkinsus qugwadi* (Scallop Protistan Unknown (SPX)) (Perkinsea, Protista)	Associated with mass mortalities and creamy-white pustules (especially in gonad, digestive gland & mantle)	Bower et al. 1990, 1992, 1995, 1997, 1998, 1999; Bower & Meyer 1994; Blackbourn et al. 1998
"	Scallop Protistan 'Ghost' (SPG) (Protista)	Focal lesions with epithelial ulceration in gut, gonoduct, gill and mantle	Bower et al. 1992 1994
"	*Pseuodostylochus ostreophagus* (Turbellaria)	Predation-related mortalities	Bower & Meyer 1994
"	*Polydora ciliata* (Polychaeta)	None reported	Mori et al. 1985
"	*Polydora concharum* (Polychaete)	Effects to shell microstructure only reported	Mori et al. 1985; Sato-Okoshi & Okoshi 1993
"	*Polydora convexa* (Polychaete)	Effects to shell microstructure only reported	Sato-Okoshi & Okoshi 1993
"	*Polydora variegata* (Polychaete)	Effects to shell microstructure only reported	Mori et al. 1985; Sato-Okoshi & Okoshi 1993
"	*Polydora websteri* (Polychaete)	Heavy shell infestation can produce clinical soft-tissue effects, but infestation usually light.	Kurochkin et al. 1986; Sato-Okoshi & Okoshi 1993; Bower & Meyer 1994
"	*Pectenophilus ornatus* (Copepoda, Crustacea)	Gill hypertrophy at copepod attachment, plus blood-feeding associated with chronic weakening	Nagasawa et al. 1988, 1991; Nagasawa & Nagata 1992
"	*Cliona* sp. (Porifera)	Heavy shell infestation can produce clinical soft-tissue effects, but infestation usually light.	Kurochkin et al. 1986

"	Pollutant effects	Weakening and muscle atrophy, necrosis. Gametogenesis abnormalities	Syasina et al. 1996; Usheva 1999
"	Copper and Zinc toxicity	Embryo-toxicity reported	Karaseva & Medvedeva 1994
"	Mussel fouling	Smothering suspended scallops	Kurata et al. 1996
"	Solidobalanus hesperius (Cirrepeda, Crustacea) fouling	Smothering	Ovsyannikova & Levin 1982
Pecten alba	Bucephalus-like trematode (Digenea, Platyhelminthes)	Castration by gonad displacement	Sanders & Lester 1981
Pecten (Patinopecten) caurinus	Nematopsis sp. (Porosporidae, Protista)	None observed	Bower & Meyer (pers. comm.)
Pecten fumatus	Bucephalus-like trematode (Digenea, Platyhelminthes)	Castration by gonad displacement	Heasman et al. 1996
"	Sabella spallanzanii (Polychaeta)	Fouling-related mortalities	Clapin & Evans 1995; O'Connor et al. 1999
Pecten maximus	Infectious Pancreatic Necrosis Virus (Finfish virus)	None reported	Mortensen 1993a; Mortensen et al. 1992, 1998
"	Vibrio pectenicida (Vibrionaceae, Bacteria)	Larval mortalities	Lambert & Nicolas 1998; Nicolas et al. 1996; Lambert et al. 1999a
"	Vibrio splendidus (Bacteria)	Brown inner shell deposits in broodstock	Lambert et al. 1999b
"	Rickettsial-like organisms (Rickettsiales, Bacteria)	Associated with mass mortalities in France	Comps 1983; LeGall et al. 1988, 1991, 1992; LeGall & Mialhe 1992; Kellner-Cousin et al. 1993
"	Pseudoklossia pectinis (Coccidia, Protista)	None reported	Léger and Duboscq 1917
"	Polydora sp. (Polychaete)	Shell damage and associated spat mortalities	Mortensen et al. 1999
"	Pomatoceros triqueto (Polychaete)	None reported	Burnell et al. 1995
"	Modiolicola maxima (Copepoda, Crustacea)	None reported	Reddiah & Williamson 1958
"	Modiolicola sp. (Copepoda, Crustacea)	None reported	Mortensen 1993b
"	Mussel fouling	Smothering related mortalities	Minchin & Duggan 1989

Host	Organism/Disease	Effect	Reference
"	Unknown environmental disturbances	Abnormal melanisation and microstructural distortions of the shell	Larvor et al. 1996
Pecten novaezelandiae	Virus Like Particles (VPL) (Virus)	Necrotic digestive gland epithelial cells with associated mortalities	Hine & Wesney 1997
"	*Paravortex* sp. (Turbellaria)	Associated with weak and dying scallop but direct clinical effect unclear	Woods & Hayden 1998
Placopecten magellanicus	Bacterial Abscess Disease or brown spot disease (various Bacteria)	Abscess lesions in adductor muscle	Sherburne & Bean 1986; McGladdery 1990; Getchell 1991
"	Rickettsial like organism (RLO) (Rickettsiales, Bacteria)	Associated with adult mortalities in the eastern US but none reported in other cases	Ballou 1984; Gulka et al. 1983; Gulka & Chang 1984; Leibovitz et al. 1984
"	Chlamydial like organism (Rickettsiales)	None reported apart from slight cell hypertrophy	Morrison & Shum 1982
"	*Trichodina* sp. (Oligohymenophora, Protista)	None reported	Beninger et al. 1988; McGladdery et al. 1993a; McGladdery & Stephenson (unpubl. data)
"	Unidentified digenean (Platyhelminthes)	None reported	McGladdery et al. 1993a
"	*Urastoma*-like turbellarians (Turbellaria)	None reported	McGladdery et al. 1993a
"	*Polydora concharum* (Polychaete)	Shell damage with severe infestations	Evans 1969; Blake 1969
"	*Polydora socialis* (Polychaete)	None reported	Blake 1969
"	*Polydora websteri* (Polychaete)	Shell erosion and weakening in extreme infestations may render scallop more susceptible to predation	Evans 1969; Bergman et al. 1982
"	*Dodecaceria concharum* (Polychaete)	Enlarges *Polydora* tunnels, increasing shell damage	Evans 1969; Leibovitz et al. 1984
"	*Pseudopotamilla reniformis*	Secretion of large ridges of shell material to contain worms up to 100 mm in length	Blake 1969
"	Pea Crabs (Crustacea)	Compression of gonad possibly affecting gonadal development	Cheng 1967; Karlsson 1991; Getchell 1991

"	*Coccomyxa parasitica*	Green discolouration of the mantle and associated granulomas in scallops over 3 yrs old.	Naidu 1971; Stephenson & South 1975
"	*Cliona celata; Cliona vastifica* (Porifera)	None reported	Medcof 1949; Evans 1969; Sindermann 1971; McGladdery et al. 1993a
	Hydractinia echinata (Cnidaria)	If attached to internal edge of shell may induce shell deformities	Merrill 1967
"	Mussels	Fouling-related mortalities	Claereboudt et al. 1994; Gryska et al. 1996
"	Cadmium & Copper	Larval toxicity	Yevich & Yevich 1985
Placopecten meridionalis	*Polydora websteri*	Formed mud blisters between the shell and mantle	Skeel 1979

later, suggesting possible sequestering of the viral particles in this organ. Titres dropped below detectable levels after day 8 in the haemolymph and day 30 from the rectum. No evidence of viral replication or pathological effects was found in the scallop tissues, however, it is believed that these scallops may play a significant role as reservoirs of viable IPNV in the natural environment (Mortensen et al. 1992). Other challenges using an Orthomyxovirus-like (enveloped single-strand RNA) virus, responsible for Infectious Salmon Anaemia (ISA) showed no evidence of uptake or carriage by the same scallop species (Bjoershol et al. 1999).

The other observation of "virus-like particles" was reported from the New Zealand scallop *Pecten novaezelandiae*, and was associated with mortalities of up to 39% (Hine and Wesney 1997). The toheroa clam *Paphies ventricosum*, from independent mortalities, was found to carry similar viral-like particles. Light microscopy of diseased scallops revealed necrotic digestive gland epithelial cells, associated with sloughed or pyknotic cells in the tubule lumens. Some sloughed cells contained abnormal granular cytoplasmic inclusions which were DNA-negative. Transmission electron microscopy revealed electron dense, unenveloped virus-like particles (22–30 nm in diameter) in the cytoplasm adjacent to the outer nuclear membrane, or in orderly arrays along the cisternae of highly modified endoplasmic reticulum (ER), of digestive cells. Some cisternae were dilated to form vacuolar inclusions. Secretory cells showed similar viral-like particle arrays with dilated ER but rarely vacuolar inclusions. The ribonucleic acid component, along with the close association with the rough endoplasmic reticulum of infected cells, suggests an enterovirus-like (Picornaviridae) or calicivirus affiliation. However, these characteristics are indicative of a new group of viruses from molluscs, and require more examination in order to determine viral family affinities and their precise role in the mortalities observed.

No other viruses or viral-like particles have been described from scallops to date, although they have been documented from a number of other bivalve species, with increasing frequency, over the last decade (Hine et al. 1992; Comps and Cochennec 1993; Norton et al. 1993; Elston 1997; Hine et al. 1998; Miyazaki et al. 1999; Comps et al. 1999). Many of these more recent investigations have even been successful in determining biochemical characteristics which aid in classification of the viruses under investigation, e.g., the *Herpes*-like viruses of the Pacific oyster *Crassostrea gigas* (Le Deuff and Renault 1999). Viral detection is likely to increase with better access to transmission electron microscopy, however, the persistent lack of self-replicating cell-lines for isolation of marine bivalve (and other invertebrate) viruses continues to hamper investigation of diseases with possible viral aetiology. As noted by Getchell (1991) "*With the imminent development of marine invertebrate cell cultures, the coming decade should be one of rapid expansion of knowledge about viruses of marine animals (Sindermann 1984), including scallops.*" Research into such techniques has been limited, compared to that for finfish and terrestrial cell-lines, thus the need still exists. Until such techniques are developed, scallop (and other aquatic invertebrate) viruses will remain difficult to detect, and many infectious diseases of 'unknown aetiology' will continue to defy accurate diagnosis. Use of finfish cell-lines to isolate viruses from molluscs must also be treated with due caution, since these have been shown to detect pathogens of finfish (and other contaminants) sequestered by the molluscs, rather than obligate viruses requiring molluscan cells for replication (Hill et al. 1986; Mortensen et al. 1992).

11.2.2 Prokaryota

The most commonly recorded bacterial infections and diseases of scallops are attributed to members of the Gram-negative Vibrionaceae and Rickettsiales (intracellular Rickettsias and Chlamydias). As noted by Getchell (1991), these organisms are not usually harmful to adult scallops, except where bacterial concentrations exceed levels found under natural conditions. This is consistent with the effect of the same and related bacteria on other bivalve species, especially under culture conditions (Elston 1984).

11.2.2.1 Vibrionaceae

Leibovitz et al. (1984) list three bacterial diseases of cultured *Argopecten irradians*: 1) bacterial swarming, 2) bacillary necrosis, and 3) vibriosis. The overwhelming majority of bacteria associated with marine bivalves are Gram-negative. Due to their efficient filtering mechanism, bivalves are able to accumulate large numbers of microorganisms from the surrounding seawater and thus harbour a rich bacterial flora. The role of bacteria as organisms causing disease in bivalves is not always clear. Species of *Pseudomonas* and *Vibrio* (*Listonella* and *Photobacterium*) are natural constituents of the molluscan digestive tract. Yet, species of these genera have been shown to cause most bacterial diseases of affected pelecypods. The pathogenicity of bacteria for bivalves appears to be negatively correlated to age of the mollusc, decreasing as they grow older (Lauckner 1983). Recent studies by Lambert and Nicolas (1998) have clearly demonstrated a direct inhibition of haemocyte chemiluminescent (CL) activity (associated with intracellular degradation of phagocytosed particles) in both *Pecten maximus* and the Pacific oyster, *Crassostrea gigas*, when challenged with several strains of Vibrionaceae and *Altermonas* spp. Bacterial strains associated with mortalities of oyster larvae produced less CL inhibition in scallop haemocytes than in oysters and visa versa. This is consistent with other observations where bacterial strain and species can be clearly identified and host specificity examined (Nicolas et al. 1996). Although still recognised as principally a larval problem, exceptions are now beginning to emerge for adult bivalves, including scallops (Zhang et al. 1998).

Under hatchery conditions, scallops are most commonly attacked by vibrionacaea at the velar stage of their development. The role of similar infections in wild population dynamics is poorly understood; however, hatchery holding of high concentrations of larvae in flow-through systems, often fed by waters at higher than ambient temperature to accelerate growth and feeding, appear conducive to bacterial "blooms" and subsequent mortality events (Elston 1984). Any physiological stress due to crowding, feed detritus accumulation, rapid water quality changes (pH, oxygen, turbidity, etc.) should, therefore, be regarded as a potential trigger for bacterial proliferation and bacillary necrosis.

Tubiash et al. (1965) determined the pathogenicity of various bacteria by inoculating 24-h suspensions of the test organism into 400-mL cultures of 2 to 7 day old larvae. Host susceptibility was determined in a similar fashion, with *Argopecten irradians* larvae as well as other bivalves being challenged with an isolated bacterial strain of high virulence. The Gram-negative bacilli were obtained from dead and moribund bivalve larvae. The disease noted was termed "bacillary necrosis".

The course of the disease was swift and dramatic. Early signs, 4 to 5 h after inoculation with the bacterial pathogen, included a reduction of motility and the presence of many larvae lying on the bottom with either their rudimentary foot or velum extended. With this quiescent behaviour came the spreading of 'swarms' of bacteria from separate foci within the culture. By 8 h, death began to appear, with widespread granular necrosis of the tissues. Examination by light microscopy revealed massive invasion by bacteria throughout the larval tissues. In a heavily infected larval culture, mortality often reached 100% within 18 h. Ciliated protistans appeared as secondary invaders at the height of the epizootic, possibly feeding on the bacteria. *A. irradians* larvae experienced 100% mortality after 12 h when challenged with Tubiash's pathogen M 17, presumably a species of *Vibrio*.

Guillard (1959) also described the course of bacterial infection of bivalve larval cultures. The mechanism by which the bivalve larvae were destroyed was suggested to be by invasion or at least contact, rather than by a bacterial exotoxin. Tubiash et al. (1965, 1970) concluded that the entry of the pathogen was via the alimentary tract because malformed, non-feeding larvae were the last to show signs of the infection. And finally Brown and Losee (1978) have suggested that bacterial pathogens attach to and penetrate the velum and presumably spread from this site.

More recent examples of hatchery-based vibriosis problems have been described by Di Salvo (1994), Chavez and Riquelme (1994) and Riquelme et al. (1995, 1996, 1997, 2000) in the Chilean scallop, *Argopecten* (*Chlamys*) *purpuratus*, from northern Chile. These authors found the predominant bacteria to be *Vibrio anguillarum*-like strains. Challenges with cell-free supernatant prepared from bacterial cultures reduced larval scallop survival indicating a possible endotoxin effect, rather than bacterial penetration of the tissues. Mortalities were most rapid and significant at water temperatures of 25°C. Similar results were found with *Pseudomonas* sp. and *V. anguillarum* infections of larval *Euvola* (*Pecten*) *ziczac* scallops from Venezuela (Lodeiros et al. 1989, 1992).

Pecten maximus larvae in scallop hatcheries in France have also suffered mass mortalities due to vibrios. Although vibriosis has been a common feature of *P. maximus* production for over 15 years in Brittany hatcheries, only recently has differentiation and identification of a new species of *Vibrio* (related to the *Vibrio splendidus* group) been possible via immunoassays and nucleic acid sequencing (Nicolas et al. 1996; Lambert and Nicolas 1998; Lambert et al. 1999a). Interestingly, another *V. splendidus* related strain was isolated from brown inner shell deposits in 3-year-old broodstock *P. maximus*, showing signs of debilitating disease. Although challenges with isolates from these shell deposits reproduced the disease, identical isolates from control scallops as well as inlet water precluded determination of the exact cause-effect relationship (Lambert et al. 1999b). Infection by another vibrio, *Vibrio tapetis* is also associated with brown shell deposits and high levels of mortality in the clam species *Tapes decussatus* and *T. philippinarum* in Atlantic Europe (Nicolas et al. 1992; Borrego et al. 1996).

The only other documented case of mortalities of adult scallops attributed to *Vibrio* infection, was a recent report from China, where adult *Argopecten irradians* suffered mass mortalities in a hatchery in Shandong Province in 1996. Bacteria isolated from the kidney of one of the affected scallops was characterised serologically as *Vibrio natriegens* and challenge experiments using this isolate reproduced the clinical signs observed during the initial outbreak, namely digestive tract and kidney haemocyte infiltration, along with gonad and mantle atrophy (Zhang et al. 1998). No other vibrios have been reported as causing disease problems in adult scallops, in culture facilities or under open-water conditions.

11.2.2.2 Intracellular Prokaryotes (Rickettsiales; Chlamydiales and Mycoplasma)

Obligate intracellular parasites, notably the Rickettsiales and Chlamydiales, are commonly found in the epithelial cells of the gills and digestive diverticula of a wide range of bivalves, including scallops (Chang et al. 1980). Most infections appear benign, despite relatively dense colonisation. However, Leibovitz (1989) reported chlamydiosis as a serious disease of larval and postmetamorphic bay scallops, *Argopecten irradians*, in a hatchery and Morrison and Shum (1982) described a chlamydia-like organism from the digestive diverticula of older juveniles and adults of the same species. The presence of reticulate and elementary bodies within the inclusions in infected epithelial cells indicated that the pathogen was more related to the chlamydiales than rickettsiales, since the latter possess a single developmental stage compared with the at least two forms that comprise the chlamydial life-cycle. About 40% of the bay scallops examined were infected, as well as many giant sea scallops, *Placopecten magellanicus* (Morrison and Shum 1982). Basophilic staining (haematoxylin and eosin) spherical, or sub-spherical, inclusion bodies were seen within intracellular vacuoles in the epithelial cells of the blind-ending tubules of the digestive diverticula. The inclusions varied in size and in number. Heavy infections were associated with similar inclusion bodies in the kidney epithelia. Morrison and Shum (1982) suggested that the initial invasion may have occurred by endocytosis of the elementary bodies by the tubule epithelial cells (which are responsible for nutrient absorption and waste disposal via endocytosis and pinocytosis, respectively). Since the infections were sometimes heavy and some degeneration of host cells was present, these authors believed the pathogen should have some deleterious effect on the scallop host, but none was evident. Other bivalves present had a much lower prevalence of infection than either of the two scallop species. A recent investigation of mass mortalities of bay scallops, introduced to Japan for culture purposes, revealed high levels of a similar chlamydial-like infection of the digestive diverticula. This was directly attributed to the mass mortalities by the investigating authors (Wang et al. 1998); however, the precise linkage between infection and pathogenicity remains unclear.

Later work by Morrison and Shum (1983) showed a similar infection in the kidney epithelia of bay scallops (*Argopecten irradians*), which they identified as a rickettsial-like organism, due to the lack of different development stages characteristic of chlamydiales. The kidney rickettsial-like colonies were characterised by separate membrane-bound vacuoles within the infected cell cytoplasm. The prokaryotes multiplied by binary fission and could expand colony size to a scale that induced hypertrophy of the infected cell.

In 1983, Gulka et al. reported a mass mortality of giant sea scallops, (*Placopecten magellanicus*), in Narragansett Bay, Rhode Island. Gross clinical signs of infection included retracted mantles and greyish, flaccid, adductor muscles, similar to the pathology described by Medcof (1949) which he had associated with ageing scallops and heavy shell infestation by clionid sponges and *Polydora* polychaete worms. In the Gulka et al. (1983) study, however, affected scallops were of varying ages and demonstrated little shell infestation. Histologically there were myodegenerative changes in the adductor muscle and intracellular basophilic inclusion bodies within the epithelial cells of the gill, plicate membrane, and other body surfaces. Heavily infected tissues, especially the gills, often lost normal tissue architecture. Ultrastructural investigation showed that the infectious prokaryote cells possessed a thin cell wall, and measured 1.9–2.9 µm x 0.5 µm. Uniformity of structure supported tentative assignation to the rickettsial-like intracellular organisms. There was strong correlation between heavy infections and adductor muscle degeneration, described as "gray muscle disease" (Ballou 1984), which the authors postulated was due to gill dysfunction precipitating metabolic stress and pathological changes in the muscle tissue.

In their second examination of this prokaryotic infection, Gulka and Chang (1984) transplanted uninfected *Placopecten magellanicus* to cages in Narragansett Bay. Within 2.5–3.5 months, up to 53% of the transplants became infected. Scallops held in aquaria using the same water were 100% infected after 5 months. Mortality in the caged population was substantially higher than in scallops maintained in the laboratory. Inoculation challenges using homogenates of infected gill tissue resulted in heavy infections of gill tissue within 20–25 days. Attempts to infect the blue mussel, *Mytilus edulis*, and soft-shell clam, *Mya arenaria*, were unsuccessful. Attempts to culture the pathogen also failed, except in gill tissue preparations or via direct inoculation of homogenates into *P. magellanicus*. However, both field and laboratory challenges during this study failed to reveal the adductor muscle degenerative changes associated with the earlier investigation. This weakens the suggested linkage between the rickettsial infections and mortalities of *P. magellanicus* in Narragansett Bay. McGladdery et al. (1993a) also reported intracellular basophilic inclusion bodies in other giant sea scallops, with no apparent clinical effect.

Light to moderate rickettsial-like infections of the gill (Fig. 11.1) have also been found in wild, captive and cultured adult bay (*Argopecten irradians*) and sea scallops (*Placopecten magellanicus*) by Leibovitz et al. (1984). In these cases, no significant mortality was detected and larval and juvenile scallops remained free of infection. The organisms were morphologically similar to the prokaryotes in the gills of *P. magellanicus* in Rhode Island (Gulka et al. 1983) and those found in other molluscs by Comps (1983), but differed from the rickettsial-like organisms described from the kidney of bay scallops by Morrison and Shum (1983).

The apparent lack of disease in some cases of rickettsial-like infections in giant sea scallops, compared with extreme pathogenicity in other cases is a conundrum, which occurs in other pectinids; such as *Aequipecten* (*Chlamys*) *opercularis* (LeGall et al. 1992),

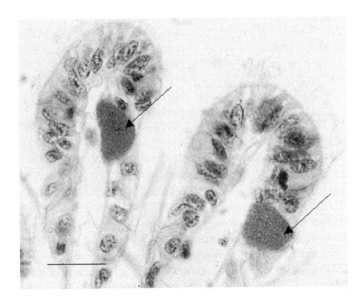

Figure 11.1.   Rickettsial-like microcolonies (arrows) in the gill epithelia of *Placopecten magellanicus* (giant sea scallop).   Hematoxylin and eosin stain; scale bar = 30 μm.

Argopecten purpuratus (Lohrmann et al. 2000a, 2002), *Mizuhopecten* (*Patinopecten*) *yessoensis* (Elston 1986; Friedman 1994) and *Pecten maximus* (Comps 1983; LeGall et al. 1988, 1991, 1992; LeGall and Mialhe 1992; Kellner-Cousin et al. 1993).   However, McGladdery (1998) noted that similar organisms have been *clearly* linked to pathogenic infections in other molluscan species.   An exacerbatory effect under sub-optimal growing conditions, as suggested by some larval mortalities under hatchery conditions, is possible. In addition, distinguishing between species, with no marine invertebrate cell-line to assist in isolation and purification, complicates differentiation between pathogenic and non-pathogenic strains.   Recent advances with immuno- and nucleic acid-labeling, however, show great promise for overcoming this diagnostic challenge (LeGall et al. 1992).

   The occurrence of an interesting third group of intracellular prokaryotes - the Mycobacteria (formerly known as the Mollicutes) has been speculated in association with a pustule disease affecting Japanese scallops (*Mizuhopecten* (*Patinopecten*) *yessoensis*) on the Pacific coast of Canada (Bower and Meyer 1991, 1994; Bower et al. 1992; Bower, S.M., 1998. Synopsis of infectious diseases and parasites of commercially exploited shellfish: Intracellular bacterial disease of scallops. URL: http://www.pac.dfo-mpo.gc.ca/sci/shelldis/pages/ibdsc_e.htm. Date last revised December 18, 2002). Complaints of pinkish-orange pustules up to 10 mm in diameter in the soft-tissues (including the adductor muscle) and conchiolin lined shell erosions along the edge of the shell were investigated and revealed the presence of intracellular prokaryotes in a few haemocytes within abscess lesions or other foci of haemocyte infiltration (Fig. 11.2).

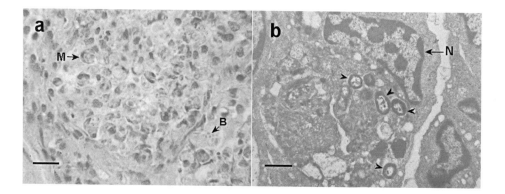

Figure 11.2. (a) Histological section through a pustule in *Mizuhopecten* (*Patinopecten*) *yessoensis* consisting of haemocytes at various stages of necrosis including a haemocyte (M) that appears to be engulfing necrotic tissue and another associated with a colony of intracellular prokaryotes (B). Haematoxylin and eosin stain; scale bar = 15 μm. (b) Electron micrograph of a pustule showing a healthy haemocyte (N indicates its nucleus) which appears to have engulfed a necrotic haemocyte that is infected with prokaryotes (arrows). Uranyl acetate and led citrate stain; scale bar = 1 μm.

Pustule lesions may become encapsulated and necrotic. Under ultrastructural examination, a few infected haemocytes and autophagy of infected haemocytes was detected (Bower, S.M., 1998. Synopsis of infectious diseases and parasites of commercially exploited shellfish: Intracellular bacterial disease of scallops. URL: http://www.pac.dfo-mpo.gc.ca/sci/shelldis/pages/ibdsc_e.htm. Date last revised December 18, 2002). Diagnosis is speculative because intracellular prokaryotes could result from the phagocytosis of bacteria from secondary infections and such bacteria could resemble Mycobacteria especially because few infected haemocytes were available for examination. However, the lesions appeared to be aseptic apart from the few haemocytes containing prokaryotic organisms. The lesions occurred in Japanese scallops from six grow out localities experiencing poor growth and mortalities in 1989. The condition appeared to be related to sub-optimal culture practices and has not re-appeared in subsequent years.

11.2.2.3 Other bacterial pathogens of scallops

Another adductor muscle abscess condition, called bacterial abscess disease or brown spot disease, has been reported from giant sea scallop (*Placopecten magellanicus*) from several beds along the Maine and Atlantic Canadian coasts. First observed from Harpswell Sound in 1977, this disease was noted in Muscongus Bay in 1979, Damariscotta River in 1980, and St. Croix River in 1981 (Sherburne and Bean 1986). Scallops examined from the Damariscotta River from 1985–1987 had similar mucoid abscess lesions 1–2 mm in diameter scattered throughout the muscle (Fig. 11.3).

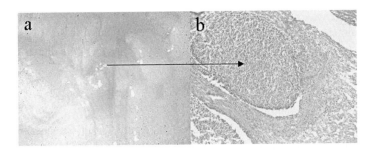

Figure 11.3. (a) *Placopecten magellanicus* (giant sea scallop) adductor muscle with "brown spot" abscess, x 10; (b) Corresponding histopathology consisting of normal muscle fibres, haemocyte infiltration, and myotome necrosis. Hematoxylin and eosin stain, x 250.

Five to ten per cent of the larger specimens (mean shell height = 124 mm) collected at two sampling stations in the Damariscotta were found to be infected. Gram positive pleomorphic bacteria were consistently observed from the necrotic foci and morphologically similar bacteria were seen in histological sections prepared from infected adductor muscle, gonad, and kidney tissue (Getchell, unpubl. data). Attempts to culture this probable pathogen have failed. Similar observations in scallops collected off southwest Nova Scotia and Cape Breton yielded bacterial cultures containing both Gram-positive, as well as Gram-negative bacterial species (McGladdery et al. 1993a; McGladdery 1998). Secondary infections probably complicated isolation and identification of the primary cause. Similar to the Japanese scallop pustule disease case, there is an apparent correlation to sub-optimal growing conditions that may trigger development of the "brown spot disease" condition in giant sea scallop. Many giant sea scallops showed extensive shell damage by *Polydora* spp. (mainly *P. concharum*) polychaetes and clionid sponges, including perforation of the adductor muscle attachment site. In Maine, many affected scallops came from estuarine beds subject to wide salinity fluctuations that could compromise scallop condition and their ability to remove entrained sediment resulting in irritation of the soft-tissues. Obviously, this condition requires more investigation.

11.2.2.4 Bacterial management under hatchery conditions

Antibiotic preparations may offer a practical short term means of control for bacillary necrosis in cultured scallop larvae, however, such practices are not recommended for flow-through systems or chronic/prophylactic bacterial management. In addition to the fact that scallop larvae may use many bacteria as a nutritional supplement (Samain et al. 1989; Moal et al. 1996), rapid development of resistance to many antibiotics is now well-documented for a number of aquatic bacteria - especially *Vibrio* spp. (Alderman and Barker 1997). Revisions to government approved antibiotic use in seawater facilities has further restricted their application (Robert et al. 1996). In addition, seawater appears to have properties, which are antagonistic to the efficacy of some antibiotics (Torkildsen et al. 2000). Another way to avoid difficulties with bacillary necrosis is to optimise water quality and maintain vigilant monitoring of influent pipes, algal food sources and tank surfaces. Every hatchery and laboratory has its own unique set of challenges and must establish its own controls, standards, and preventative measures. However, any evidence of reduced feeding should be treated as an alarm for potential bacterial build up in the system. Feeding of non-feeding larvae rapidly provides an ideal nutrient medium for bacterial proliferation, especially under warm water conditions. Brown and Russo (1979) found that a combination of filtration and ultraviolet irradiation of seawater reduced the occurrence of bacterial diseases. Brown (1981) noted, however, that many shellfish hatcheries do not use any disinfection methods and continue to be plagued by intermittent occurrences of bacteria related diseases that often destroy larval cultures around the sixth day of development. Sanitation and quality of food cultures may be more important than sanitation of the larval cultures themselves (Elston 1984); however, as discussed for vibriosis in *Pecten maximus* larvae, recent work on probiotic bacteria (Ruiz-Ponte et al. 1999; Riquelme et al. 2000) also shows promise for controlling bacterial proliferation.

11.3 MYCOTA

Few cases of fungal infection of scallops or other bivalves have been reported despite the ubiquitous distribution of marine fungi. Lauckner (1983) mentions a shell disease where fungi biodegrade bivalve shells and describes mycotic infections in the soft parts of larval and adult pelecypods, but de-emphasises the dangers fungi pose to bivalve health. The Phycomycete, *Sirolpidium zoophthorum* has been reported to attack cultured larvae of several species of bivalves and was implicated in severe epizootic mortalities in cultivated oyster larvae (Davis et al. 1954). Bay scallops (*Argopecten irradians*) at the same hatchery, however, rarely developed epizootic infections. Larvae of all ages were affected, with branched mycelia evident inside the shell. Infection occurs by the release of zoospores from sporangia that protrude from the shell. In a heavy infection, most of the larvae die within 2 to 4 days, but a small percentage may survive. The *S. zoophthorum* zoospores can be cultured on nutrient agar incubated between 20 to 30°C. Septate thalli show branched development (Loosanoff and Davis 1963). In a recent abstract, Martin et al. (1997) outline the continued occurrence of *S. zoophthorum* infections in the soft-tissues of cultured bay scallop larvae, but no further details on the significance of this infection to health or subsequent survival of cultured bay scallops is provided.

Getchell (1991) reported a single case of a fungal infection in the adductor muscle of *Placopecten magellanicus* from the Sheepscot River, Maine. Examination by light microscopy revealed typical fungi with both mycelial growth and clusters of spores within the lesion (Fig. 11.4) (Sherburne 1982). Hyphae with septate walls and apparent budding were purplish-red in colour by PAS staining. Initial work placed the fungus in the genus *Hormodendrum*. While studying another fungus (*Hormoconis resinae* Deuteromycetes) from lesions in American plaice (*Hippoglossoides platessoides*) collected from the Sable-Western Bank complex off the Scotian Shelf of Nova Scotia, Strongman et al. (1997)

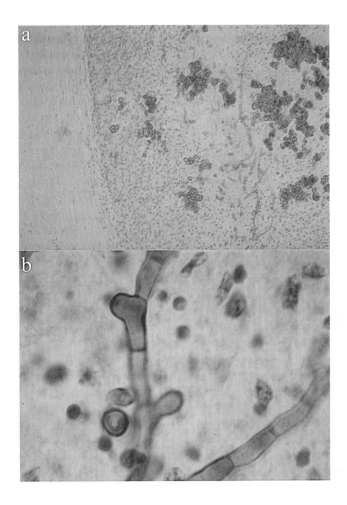

Figure 11.4. (a) *Placopecten magellanicus* (giant sea scallop) histological section of adductor muscle. Mycelial growth and clusters of spores are present within the lesion. Normal tissue is on the left side. (b) Same specimen with fungus showing septate walls and apparent budding from several hyphae. Two rounded spores are evident to the left of the hyphae. Hematoxylin and eosin stain.

investigated complaints of greenish black adductor muscle nodules in giant sea scallop meats from the same area. In addition to culturing a *Cladosporium* sp., the authors also isolated a *Penicillium* sp. from the lesions. No further reports of this condition have been recorded. As with "brown spot" and other infections of this nature - collection of samples from fisheries is often complicated by offshore discard of grossly affected adductor muscle "meats". No similar conditions have been reported from cultured scallops.

614

11.4 PROTISTA

11.4.1 Sarcomastigophorea (Amoebae and Flagellates)

Several flagellate and amoeboid infections have been reported from a variety of marine bivalve and crustacean species, however, none have been reported to date from scallops. An interesting possible exception is an unidentified protistan, described as "scallop protistan with a ghosty appearance" (SPG, Fig 11.5), which infects Japanese scallop, *Mizuhopecten* (*Patinopecten*) *yessoensis* along the Pacific coast of Canada (Bower et al. 1992, 1994). Although not conclusively identified to Phylum, the parasite does demonstrate characteristics typical of the Amoebida, namely translucent cytoplasmic inclusions, pleomorphic body shape, no surface ornamentation or evidence of flagellae/cilia. The mononuclear cells measure 10–15 µm in diameter, and are found within focal haemocyte aggregations located close to the epithelia of the gut, gonad, gill and mantle. Histological lesions are frequently associated with grossly visible ulcerative lesions in the mantle surface, but infections are normally light. The significance of these infections is not clearly understood and further study is complicated by the sporadic occurrence of the infection. A similar looking organism was also observed in *M. yessoensis* broodstock imported into Ireland from Japan (Bower, unpublished data).

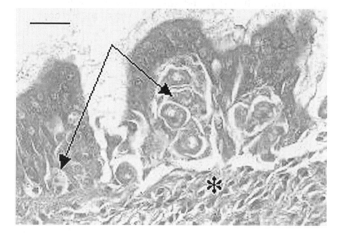

Figure 11.5. "SPG" lesions in *Mizuhopecten* (*Patinopecten*) *yessoensis* (Japanese or Yesso scallop) adjacent to the epithelium of the intestinal tract containing abnormally vacuolated cells (arrows) surrounded by intense focal haemocyte infiltration (*). Hematoxylin and eosin stain; x 630; scale bar = 10 µm.

11.4.2 Labyrinthomorpha (Thraustochytrids and Labythinuloids)

No labyrinthuloids have been reported from scallops, to date. However, the infectious agent of bay scallops, *Argopecten irradians*, initially described by McGladdery et al. (1991) as *Perkinsus karlssoni* (Fig. 11.6), but subsequently recognised as not being a perkinsiid (Goggin et al. 1996), shows ultrastructural features that may place it closer to the thraustochytrid/ labyrinthuloid complex, than the apicomplexa.

11.4.3 Apicomplexa

Included within the phylum Apicomplexa are three groups of molluscan pathogens that have been identified in scallops, the perkinsiids (Perkinsorida), coccidians (Eucoccidiia) and gregarines (Eugregarinida).

11.4.3.1 Perkinsorida

The first perkinsiid reported from scallops was a *Perkinsus* sp. from *Mizuhopecten* (*Patinopecten*) *yessoensis* imported to the Popov Islands on east coast of Russia from Japan (Kurochkin et al. 1986). The pathological effects of the lesions described were not elucidated and no further reports on this infection have been published.

Another perkinsiid was encountered in the same scallop species cultured in British Columbia, Canada. Initially, the infectious agent was reported as Scallop Protistan Unknown (SPX) (Bower et al. 1990, 1992, 1995, 1997, 1998; Bower and Meyer 1994), however, detailed taxonomic investigation has confirmed SPX as a new species of perkinsiid, *Perkinsus qugwadi* Blackbourn, Bower and Meyer, 1998 (Blackbourn et al. 1998) (Fig. 11.7). This infection is believed to be native to BC, and is characterised by

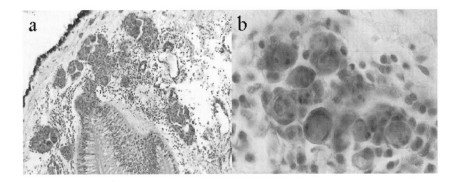

Figure 11.6. (a) '*Perkinsus*' *karlssoni* lesions in the connective tissue and stomach epithelium of a pre-spawning *Argopecten irradians* (bay scallop). Hematoxylin and eosin, x 160; (b) High power (x 630) of protistan-like inclusions within lesions and surrounding hyalinocyte response. Hematoxylin and eosin.

616

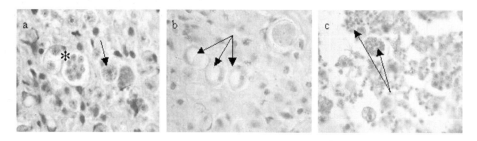

Figure 11.7. (a) *Perkinsus qugwadi* trophozoites (arrow) and tomites (*) in connective tissue of *Mizuhopecten* (*Patinopecten*) *yessoensis* (Japanese or Yesso scallop); (b) vacuolated trophozites ('signet-ring' form) of *Perkinsus qugwadi*; and (c) zoospore stage of *Perkinsus qugwadi* (arrows). Hematoxylin and eosin stain; x 1000 (oil).

massive proliferation throughout the connective tissues of all organs, leading to weakening and death. Infected scallops may demonstrate creamy-white pustules up to 5 mm in diameter in many tissues, especially the gonad, but these appear to bear no direct correlation to level of infection. The disease is not easily transmitted from scallop to scallop, with most transmission being confined to a few heavily infected juvenile scallops (<40 mm shell height) within which zoospores develop (a characteristic, which differs from all other described *Perkinsus* species where sporocyst development is external (Perkins 1996)). Progeny of *M. yessoensis* that survived an epizootic outbreak of *P. qugwadi* had a significant increase in resistance to infection and resulting mortalities. Hybrid scallops, resulting from a cross between *M. yessoensis* females (from the same group of scallops that survived an epizootic outbreak of *P. qugwadi*) and *Patinopecten caurinus* males (native to British Columbia), had similar resistance to *P. qugwadi*. The identification of scallop stocks that are resistant to *P. qugwadi* has facilitated the development of a scallop culture industry in British Columbia. No other native sympatric scallop species (*Chlamys rubida* or *C. hastata*) appear to be susceptible (Bower et al. 1999).

As discussed under Section 11.4.2, the taxonomic status of a bay scallop parasite identified as *Perkinsus karlssoni* by McGladdery et al. (1991), is now recognised to be invalid, based on the morphology of the biflagellate stage isolated from infected scallops by culture in fluid thioglycollate media (Whyte et al. 1993a) and by DNA analysis of cultured isolates (Goggin et al. 1996). Nevertheless, the possibility that the biflagellate isolates analysed to date were not free-living contaminants from scallop tissues but are representatives of the pathogen observed histologically remains to be confirmed. The infection was originally described by Karlsson (1991) from bay scallops in Rhode Island. At about the same time, McGladdery and co-workers were investigating bay scallops introduced for culture purposes into Atlantic Canada (McGladdery et al. 1993b, Whyte et al. 1993b). Due to the bay scallops being an exotic species to Atlantic Canadian waters, there was concern over the potential to introduce new infections into

Canadian bivalve population, thus, these investigations concentrated on host-specificity and transmission potential. Results indicated that the infection is specific to bay scallops (Bower, S.M. and McGladdery, S.E., 1996. Synopsis of infectious diseases and parasites of commercially exploited shellfish: *Perkinsus karlssoni* of scallops. URL: http://www.pac.dfo-mpo.gc.ca/sci/shelldis/pages/perkarsc_e.htm. Date last revised September 18, 1996). The infection is characterised by intense swirl-like haemocyte aggregations, within which the parasite is frequently engulfed and masked (McGladdery et al. 1991). Infections were boosted under warm water (20°C) recirculating holding conditions where the parasites became more evident in proliferative masses which distended the basal membranes of digestive tubule and duct epithelia (Whyte et al. 1993b). During these investigations, another parasite infection was also stimulated - a kidney coccidian, described further under Section 11.4.3.2. Interestingly, the coccidian induced a similar haemocyte encapsulation response to that caused by the *Perkinsus*-like parasite, making it difficult to distinguish the two species in advanced cases (Whyte et al. 1994).

11.4.3.2 Eucoccidiia

The first report of a coccidian infection in scallops was that of Léger and Duboscq (1917) who described a new species, *Pseudoklossia pectinis*, from the kidney of European great scallop (King scallop, Coquille St.-Jacques), *Pecten maximus*, from Roscoff, France. Oocysts of this species measured 32 to 35 um in diameter and were found solely in the kidney epithelial cells, where gamonts were also detected. A recent review by Desser and Bower (1997), however, questions the assignation of this species to the genus *Pseudoklossia* due to the description by Léger and Duboscq (1917) of syzygous division in the gamont stage. Desser and Bower (1997) note that this is a taxonomic feature of the family Adeloridae, rather than the Aggregatidae within which *Pseudoklossia* currently occurs. No meronts (schizonts) were found which is consistent with the belief that most eimeriid coccidians have a heteroxenous life-cycle. Schizogony (merogony) takes place in one host and gametogony and sporogeny take place in another host. No further descriptions, reports or investigation of the significance of this parasite to *P. maximus* have been reported.

Leibovitz et al. (1984) were the first to describe heavy renal coccidial infections in captive and cultured adult bay scallops, *Argopecten irradians* (Fig. 11.8). The disease consisted of coccidia at different life stages infecting the epithelium of the kidney, causing tissue destruction within the kidney, and renal tubule impaction by coccidia and epithelial debris. In cultured scallops, mortalities often exceeded 80% and coccidia also occurred in the digestive diverticula, gills, and gonad. Similar infections were also found in wild bay scallops, but less frequently and with limited pathology. Similar observations were subsequently reported from bay scallops held in captivity in Atlantic Canada as part of a host specificity study for an unrelated infection (Whyte et al. 1994). Although more than one type of oocyst was described by Leibovitz et al. (1984), suggesting the possible presence of more than one species, only one coccidian species is suspected in the

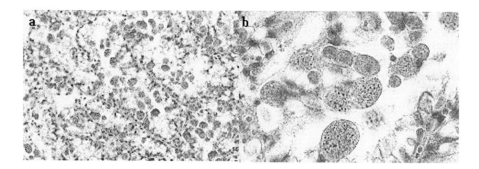

Figure 11.8. (a) Extensive proliferation of an unidentified coccidian throughout the renal tubules of *Argopecten irradians* (bay scallop). Hematoxylin and eosin stain, x 160; (b) Meront development and release from epithelial cells of the renal tubule. Hematoxylin and eosin, x 630.

Canadian studies (McGladdery 1990, 1993b; Cawthorn et al. 1992; Whyte et al. 1994), as well as independent eastern US observations (Getchell 1991; Karlsson 1991).

Another coccidian was reported by DiSalvo (1994) from the kidneys of broodstock Chilean scallops, *Argopecten purpuratus*. This was related to production of poor quality gametes, but no further descriptions of the infection have been reported. Other studies of this scallop species have not found this parasite (Lohrmann et al. 1991, 2000a; Lohrmann and Smith 1993) and it is speculated that the infection detected by DiSalvo (1994) may have been opportunistic.

11.4.3.3 Eugregarinida

Most gregarines complete their development within an arthropod host. Members of the family Porosporidae, however, undergo intermediate developmental stages in marine bivalves. Lauckner (1983) summarises the life cycle of these gregarines. At the mid-point in the cycle gymnospores are drawn into the mantle cavity of a compatible bivalve host where they are engulfed by phagocytes on the surface of the gill or mantle epithelium. It is within these cells that the sporozoites form. The genus *Nematopsis*, which includes most species infecting bivalves, is characterised by the sporozoites being encapsulated by a resistant spore (actually the oocyst stage, according to Lauckner (1983)).

Nematopsis pectinis was identified in *Chlamys* (*Pecten*) *varia* from Europe (Léger and Duboscq 1925). Although originally assigned to the genus *Porospora*, Sprague (1970) reviewed the taxonomic description and based on the encapsulation of the sporozoites, transferred them to the genus *Nematopsis*. Their final decapod host was unknown. Similar gregarines have been observed in the Japanese scallop *Mizuhopecten* (*Patinopecten*) *yessoensis* and *Pecten* (*Patinopecten*) *carinus* (Bower and Meyer pers. comm.). The bay scallop, *Argopecten irradians* has been found to be susceptible to

N. ostrearum, which has an oocyst size of approximately 14 x 10 μm (Sprague 1970). *Nematopsis duorari*, with spores 19 x 10 μm, has also been found in *A. irradians*. The final host is the shrimp, *Penaeus duorarum*. Kruse (1966) found that *A.* (*Aequipecten*) *irradians* shed *Nematopsis* spores in their mucous strings. These, with other bottom debris, are then ingested by the shrimp. Based on results of infection studies with oysters, Sprague and Orr (1955) decided that the parasites were detrimental to their intermediate hosts, but did not believe that *Nematopsis* was a significant mortality factor in the wild. No pathogenicity has been described for *Nematopsis* infections of scallops.

11.4.4 Microspora

Although microsporidians are obligate intracellular parasites of a wide variety of vertebrate and invertebrate hosts, few bivalves have been found to be infected and only one case in scallops has been reported (Lohrmann et al. 1999, 2000b). The infection was found in *Aequipecten opercularis* (the queen scallop) collected from several sites around the English and Welsh coasts of the UK between 1997 and 1998 (Fig. 11.9). Prevalences ranged from 4.5–20%. Two stages of spore development were observed within the blood

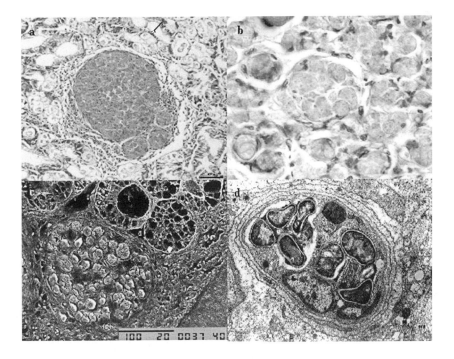

Figure 11.9. (a) Microsporidian colony in the digestive diverticula of *Aquipecten opercularis* (queen scallop) Hematoxylin and eosin, scale bar = 50 μm. (b) Individual sporocysts within colony, Feulgen stain. (c) Microsporidian colony within digestive gland diverticula, SEM. (d) Individual sporocyst ultrastructure, TEM. (Images kindly provided by Dr. K. Lohrmann, Universidad Católica del Nord, Coquimbo, Chile)

sinuses and tissues of the digestive gland, but no other developmental stages were detected. The authors speculate that other developmental stages may be present in other tissues of the same host, or an alternate host may be involved in the life cycle. Mature spores measured 2.3 x 1.2 μm, contained polar bodies with 7–9 coils and were found within the vascular sinuses. Immature spores measuring 2.3 x 1.2 μm were found within hypertrophied host cells (up to 300 μm in diameter) in the digestive gland. The polar tube of some spores had 10–12 coils, while others showed only 7–8 coils. There was no overt pathology associated with these infections. Cells infected by immature spores showed only minor focal infiltration and fibroblast like encapsulation layers.

11.4.5 Ascetospora

11.4.5.1 Marteiliida

Several species of *Marteilia* have been well-described from European populations of Flat oysters (*Ostrea edulis*), mussels (*Mytilus edulis*, *M. galloprovincialis*) and a pelecypod, *Scrobicularia piperata*. In addition, the Sydney rock oyster (*Saccostrea commercialis*) and possibly the black-lip oyster (*Crassostrea echinata*) are also host of the marteiliad, *Marteilia sydneyi*. Only one marteiliad has been reported from scallops and its impact was so severe and rapid that no material was available post-epizootic to permit detailed taxonomic investigation. The *Marteilia* sp. was found in mass mortalities of Calico scallop, *Argopecten gibbus*, off the Atlantic coast of Florida in late 1989–90 (Moyer et al. 1993, 1995). The marteiliad overwhelmed the digestive tubules of the scallops, which showed rapid weakening and death. No further infections of calico scallop have been detected since the original reports.

11.4.5.2 Balanosporida

The Balanosporida contain some of the most devastating parasites affecting bivalve molluscs: *Haplosporidium nelsoni* and *Haplosporidium costale*, the causative agents of MSX and SSO disease, respectively, of American oysters (*Crassostrea virginica*); and *Bonamia ostreae* and *Bonamia exitiosus* of European and New Zealand flat oysters (*Ostrea edulis* and *Ostrea lutaria*), respectively (OIE 2003). To date, however, only one unidentified species has been reported from scallops (Chu et al. 1996). More than 83% of bay scallops, *Argopecten irradians* from two stocks being cultured in the Yellow Sea area of China were found infected. Most infections found were light (Fig. 11.10), but heavy post-spawning mortalities in the same area led the authors to suggest that this haplosporidian may be a potential causative factor.

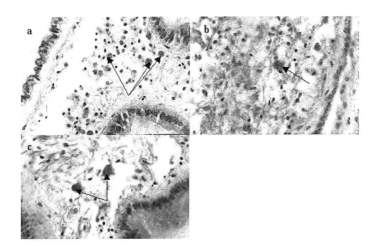

Figure 11.10 (a) Multinucleated schizont stage of an unidentified Haplosporidian parasite infecting the connective tissues of *Argopecten irradians* (bay scallop) collected from China. Hematoxylin and eosin, x 100; (b) and (c) denser plasmodial stages of the same parasite in the mantle and digestive diverticula, respectively. Hematoxylin and eosin, x 250 (images kindly provided by Dr. E.M. Burreson, Virginia Institute of Marine Science, VA, USA).

11.4.6 Ciliates

The majority of ciliates associated with bivalves are, like the flagellates described under Section 11.4.1, probably harmless commensals. More than 150 species of ciliates have been found in the mantle cavity, on the gills, or in the digestive diverticula of marine bivalves. Most are primarily commensal, but can become pathogenic if their numbers become unusually high, the physiological state of the host is compromised, or an environmental stress factor shifts the equilibrium (Lauckner 1983).

Trichodinids are peritrichous ciliates that are common symbionts of amphibians, fishes and bivalves (Fig. 11.11). Most are commensals feeding on bacteria and occurring at low prevalences and intensity of infestation within the mantle cavity. Some infections of trichodinids in bivalves, however, have been linked to tissue damage and mortalities (Lauckner 1983). Trichodinids are easily recognised by their dome shape, rows of cilia, conspicuous circle of hooklets and horse-shoe shaped macronucleus.

In scallops, *Trichodina pectenis* has been reported in the mantle cavity of *Mizuhopecten* (*Patinopecten*, *Pecten*) *yessoensis* and *Trichodina polandiae* from *Chlamys* sp. collected from the Gulf of Peter the Great (Sea of Japan) (Stein 1974 cited in Lauckner 1983). The former species is described as having a ring of 22–31 denticles, each denticle having 7–9 radial rods. The species described from *Chlamys* sp. has 20–26 denticles with 7–10 radial rods on each (see Table 13–9 in Lauckner 1983). Another, possibly identical, trichodinid was reported on *M. yessoensis* by Kurochkin et al. 1986. Xu et al. (1995)

622

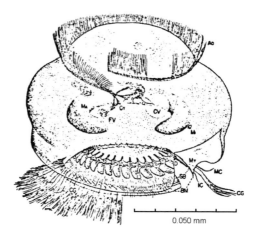

0.050 mm

Figure 11.11. *Trichodina myicola* from the clam *Mya arenaria* showing rows of cilia and typical circle of hooklets. (After Uzmann and Stickney 1954 J. Protozool. 1:152).

reported another species, *Trichodina jadranica* from the gills of *Chlamys farreri*. This ciliate averaged 37.8 μm in diameter with 21–24 denticles and an adoral ciliated membrane spiral of approximately 400°. Lohrmann et al. (2000a, 2002) reported a *Trichodina* sp. from the gills of the Chilean scallop, *Argopecten purpuratus* and trichodinids are also commonly found in the mantle cavity of giant sea scallops (*Placopecten magellanicus*) from Atlantic Canada (Beninger et al. 1988; McGladdery et al. 1993a) (Fig. 11.12). In some *P. magellanicus*, numbers may exceed over 100 per 5 μm thick tissue section of gill, but no tissue damage has been found to date.

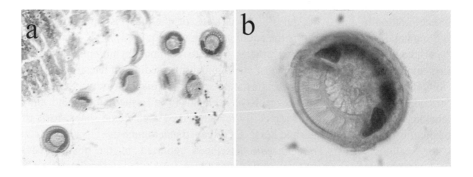

Figure 11.12. (a) Unidentified trichodinid ciliates on the surface of giant sea scallop (*Placopecten magellanicus*) from Atlantic Canada. Hematoxylin and eosin; x 160. (b) High power magnification (x 630) of denticle arrangement of one of the giant sea scallop trichodinids. Hematoxylin and eosin stain.

Whilst most trichodinids usually occupy many areas within the mantle cavity of a bivalve, the heterotrich, *Licnophora auerbachi* resides in a more unusual niche, the eyes of the scallop. These ectoparasites are highly motile and very difficult to detach from the eye surface. Harry (1977) found that 85 out of 88 queen scallops, *Aequipecten* (*Chlamys*) *opercularis*, collected from County Down, Ireland, harboured *L. auerbachi*. Only a few *Pecten maximus* and *Chlamys varia* were found infested, indicating a degree of host specificity that Harry (1977) was able to confirm with *in vitro* investigations. One specimen of *A. opercularis* was parasitised by ciliates on 94 out of 103 eyes. Loss of pigment from the iris and signs of disintegration were two of the prominent pathological features of this heavily infected individual. Harry (1980) suggested that the action of the basal disk as it attaches caused damage to the epidermis of the eye. Like *Trichodina*, *Licnophora auerbachi* is normally a filter-feeder probably thriving on bacteria (Fig. 11.13).

The eyes of scallops are not able to form focused images, so it is unlikely that the presence of these ciliates affect their visual response to stimuli. However, *L. auerbachi* might have an effect on young developing scallops by interfering with their light-dark shadow detection escape response (Harry 1977, 1980).

11.5 PLATYHELMINTHS

11.5.1 Trematodes

Bucephalid and Fellodistomid digenean trematodes parasitise many species of marine bivalves, and some are known to cause significant damage to the host's reproductive tissue. Sanders and Lester (1981) documented a bucephalid trematode infection of *Pecten alba* from Bass Strait, Australia, which turned the scallops gonad bright red or orange in colour. Histological observations showed the gonads contained sporocysts with cercariae throughout most of the year. The sporocysts occupied the bulk of the gonad as the scallop reached a state of sexual maturation. Over the 6-month period during which sporocysts produced cercariae, the average weight of the adductor muscle fell below that of unparasitised scallops. The loss of reproductive potential and energy reserves stored in the adductor muscle reflect the strain this trematode has on its scallop host. Other *Bucephalus* sp. sporocyst infections of Australian scallops were reported by Heasman et al. (1996). These authors followed infections in two species of scallops (*Chlamys asperrima* and *Pecten fumatus*) from New South Wales, Australia, and documented significant increases in prevalence of infection. Over a two year period prevalences increased to 66% in *P. fumatus* over 80 mm in length and 40% in *C. asperrima* over 75 mm in length, from 5.1% and 4.3%, respectively in 1991/92.

Mateo et al. (1975) reported castration of *Argopecten purpuratus* from Peru in association with sporocysts showing hemiuroid characteristics.

624

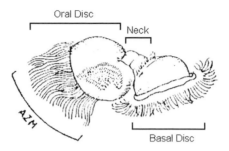

Figure 11.13. *Licnophora auerbachi* from the eye of *Chlamys opercularis* showing: 1) oral disc with cilia making up the adoral zone of membranelle (AZM); 2) neck region; and 3) basal disc. (After Harry 1980).

Fellodistomid digeneans belonging to the genus *Proctoeces*, are also well-documented as causing varying degrees of tissue damage in bivalves (Bray 1983). The most commonly reported species is *Proctoeces maculatus*, however, the broad geographic and host ranges reported for this species, have lead some authors to speculate that this name contains more than one species and requires re-investigation (Lauckner 1983; Bower 1995). The metacercarial stage of *P. maculatus* is reported from various Mollusca, Polychaeta and Echinoidea. Adults occur in mollusc-eating fishes such as labrids and sparids, however, progentic development is also documented in mussels, which normally host the sporocyst developmental stage (Bray 1983). Karlsson (1991) described adult worms, tentatively identified as *Proctoeces* sp., from the kidney and opening of the gonoduct into the kidney of bay scallops from Rhode Island. Some trematodes showed attachment to the kidney epithelia via the oral sucker. Although most tissue sections revealed only single worm infections, one scallop kidney section harboured over 50 *Proctoeces*. Bray (1983) also reported *P. maculatus* in bay scallops collected from Rhode Island.

Unidentified metacercariae belonging to the Echinostomatidae, Fellodistomidae or Gymnophallidae have also been reported from bay scallops from Rhode Island. Karlsson (1991) noted varying numbers of metacercariae encysted in the labial palps, digestive gland and gill tissues. Despite numbers exceeding 50 metacercariae per tissue section in some specimens, there was little evidence of a strong host response. Similar metacercariae have been reported from the labial palps of *Argopecten purpuratus* in Chile (Lohrmann et al. 1991), *Chlamys islandica* from the Barents and White Seas (Chubrick 1966, cited in Lauckner 1983), and in *Placopecten magellanicus* from Atlantic Canada (McGladdery et al. 1993a). The echinostomatid *Himasthla quissetensis* uses *Argopecten irradians* as a second intermediate host, before final maturation in the intestine of the herring gull, *Larus argentatus*. The metacercariae, 140–190 um in diameter, occur in the mantle, gills, and foot, but do not undergo any further development in the cyst (Stunkard

1938). A similar lack of growth occurs when metacercariae of *Renicola thoidus* parasitise *Argopecten irradians*. The adult trematode infects the kidney of shore birds (Getchell 1991). Adult trematodes are rare in scallops and possibly only occur as accidental infections (Fig. 11.14).

In a recent description by Pérez-Urbiola and Martìnez-Diaz (2001), a condition known as "pimientilla" disease in catarina scallop *Argopecten ventricosus (circularis)* from Baja California Sur, México, has been found to be caused by *Stephanostomum* sp., a species of Acanthocolpid digenean. This is the first description of this group of digeneans using molluscs as second intermediate hosts. Most other species of *Stephanostomum* encyst in the body cavity of benthic fish and mature in the guts of piscivorous fish. The metacercariae encyst throughout the soft tissues of the scallops, eliciting a strong melanisation response, which gives the scallop a peppery ("pimientilla") appearance and renders it unmarketable. Except for signs typical for metacercarial infections (i.e., a focal haemocyte infiltration and encapsulation), the effect on the health of the scallop was not reported, despite extremely high levels of infection (Pérez-Urbiola and Martìnez-Diaz 2001).

Unidentified larval sanguinicolids have been recorded from *Argopecten irradians* from Woods Hole, Massachusetts (Linton 1915). Interestingly, this group of digeneans develop into adults directly from cercariae, omitting the usual metacercarial encystment stage. Within the bay scallop, the only report of pathology was castration caused by sporocyst proliferation throughout the gonadal tissues. There have been no further reports of sanguilicolid sporocyst infections of bay scallops (or any other scallop) since Linton's 1915 note, however, Lauckner (1983) provides a thorough review of infections of other molluscan groups.

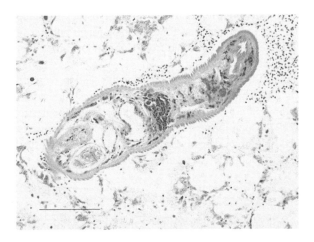

Figure 11.14. Unidentified digenean embedded in the connective tissue of the digestive gland of *Placopecten magellanicus*. Hematoxylin and eosin stain; x 250. Scale bar = 0.1mm.

11.5.2 Cestodes

Among the helminth parasites of shellfish, cestodes, particularly larval forms (metacestodes), are common inhabitants within the digestive tract of Atlantic bay scallops (*Argopecten irradians*). Seven species of larval cestode were identified in the eastern Gulf of Mexico by Cake (1977, 1979). These include the four most prevalent, *Parachristianella* sp. which encysts singly along the walls of the intestine, *Tylocephalum* sp. in the digestive tract walls and digestive gland, *Rhinebothrium* sp. in the stomach and digestive diverticula, and *Polycephalus* sp. in the visceral masses and digestive gland (Fig. 11.15). The other single isolations were *Acanthobothrium* sp., *Anthobothrium* sp., and *Eutetrarhynchus* sp. *Tylocephalum* sp. metacestodes were also reported from five of 25 *Chlamys nobilis* from Japan (Sakaguchi 1973 cited in Lauckner 1983). The scallops are initially infected by ingesting eggs, gravid proglottids or free swimming coracidia. Because the plerocercoids lack taxonomic characteristics, most larval forms can be identified only to genus.

None of these genera have been associated with overt pathology, most infections being light and eliciting a focal haemocyte response similar to that described for metacercarial digeneans. High levels of infection, however, may cause physiological stress that can affect growth, reproduction as well as edibility. In calico scallops, *Argopecten gibbus*, from North Carolina, high levels of infection by metacestodes belonging to the genus *Echeneibothrium* have been associated with significant gonad atrophy (Singhas et al. 1993). Infections by cestodes of the same genus have also been associated with aberrant behaviour in littleneck clams (*Protothaca staminea* and *P. laciniata*) from California (Sparks and Chew 1966; Warner and Katkansky 1969).

The Chilean scallop, *Argopecten purpuratus* is also known to be infected by larval cestodes. Oliva et al. (1986) found metacestodes, which they tentatively assigned to the Phyllobothriidae and Oncobothriidae, in the gonadal tissues of samples collected from Antofagasta, Chile; and Lohrmann and Smith (1993) found another unidentified larva in the intestine of the same species collected from Coquimbo, Chile. None of the infections was associated with overt pathology.

Because the larval forms infect the viscera of the scallop, which is discarded during processing, there is little chance of humans consuming cestodes with the edible adductor muscle. Any cestodes that are not discarded, however, would be destroyed by the cooking process and Cake (1977) mentions experimental evidence that suggests that *Tylocephalum* larvae are destroyed by human digestive acids and enzymes. The final hosts for all these cestode genera are elasmobranch fish (sharks, skates and rays)

11.5.3 Turbellaria

Turbellarian flatworms are commonly found on the gills or within the guts of a wide range of bivalves (Lauckner 1983), however, few have been reported from scallops. An unidentified turbellarian was reported from the digestive diverticula of bay scallop *Argopecten irradians* (Leibovitz et al. 1984), with no apparent pathological effect.

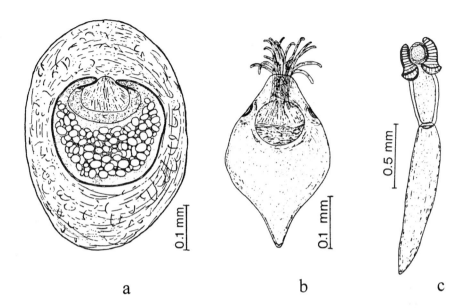

Figure 11.15. Larval cestodes from the scallop, *Argopecten irradians concentricus*. (a) Encapsulated, acaudate glando-procercoid of *Tylocephalum* sp. (b) Tentaculo-plerocercoid of *Polypocephalus* sp. (c) Bothridio-plerocercoid of *Rhinebothrium* sp. (After Cake 1976).

A turbellarian resembling *Urastoma cyprinae*, a common inhabitant of the gills of American oysters (*Crassostrea virginica*) in Atlantic Canada, has been reported from the gills and body surfaces of giant sea scallop from the same area (McGladdery et al. 1993a) (Fig. 11.16). Compared with oyster infestation levels, those found on the scallops occur in much lower numbers (1–2 per tissue section and maximum prevalences of 10%) than occur on oysters (up to 1000's of *U. cyprinae* and 100% prevalence). In both scallops and oysters from Atlantic Canada, there was no obvious histopathological effect on the gills, however, hypertrophy of gill epithelial cells in mussels from Spain has been attributed to the same turbellarian (Robledo et al. 1994) and there is some evidence for biochemical alteration in the mucoid chemistry of heavily infected oysters (Brun et al. 2000).

A recent report by Woods and Hayden (1998) associated the presence of a *Paravortex* sp. of turbellarian in the digestive diverticula with weakening and mortalities in the New Zealand scallop, *Pecten novaezelandiae*. Clinically, the scallops showed mantle retraction, gaping and impaired swimming behaviour prior to death. Prevalences of the *Paravortex* were greater in moribund scallops, but intensity of infections were low (<3 turbellarians per scallop) and not all affected scallops were infected. Other diseases, such as Malpeque disease of American oysters (*C. virginica*) have also been found to show a positive correlation with turbellarian proliferation (Drinnan, R.E., unpubl. data); however,

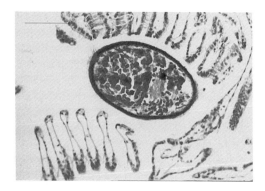

Figure 11.16. *Urastoma cyprinae* on the gills of giant sea scallop (*Placopecten magellanicus*) from Newfoundland. Hematoxylin and eosin stain. Scale bar = 1 mm.

this is believed to be a secondary, opportunistic proliferation, rather than a primary cause, due to inconsistencies between manifestation of disease signs and presence of the turbellaria. The ongoing work of Brun et al. (2000) may be able to shed some light on bivalve-turbellarian interactions at a physiological level. Documentation of effects at the predation level are more clear, for example *Pseudostylochus ostreophagus* predation on small juveniles of Japanese scallops (*Mizuhopecten* (*Patinopecten*) *yessoensis*) (Bower and Meyer 1994).

11.5.4 Nematodes

There is a lack of published information on parasitic larval nematodes of marine molluscs because they are difficult to identify and require special procedures, such as enzyme digestion, to isolate them from their embedded locations within bivalve tissues (Cheng 1978).

An abundant literature exists on the ascarid nematode, *Sulcascaris sulcata*, which infests larval scallops and, as an adult, inhabits the stomach of sea turtles, *Chelonia mydas* and *Caretta caretta* (Sprent 1977). Two scallops that serve as intermediate hosts are *Amusium balloti* and *Chlamys* sp. in Australian waters (Cannon 1978; Lester et al. 1980). Lichtenfels et al. (1978, 1980) and Blake et al. (1984) identified larval *S. sulcata* from the Atlantic scallops *Argopecten gibbus* and *A. irradians*. Parasitised green and loggerhead turtles occur in the same general area as the infected scallops.

The life cycle of *S. sulcata* starts with the release of negatively buoyant eggs from adult worms living in the intestine of the sea turtles. The eggs are shed with the turtle faeces, sink and adhere to the sea floor. After about 1 week, third stage larvae hatch and infect their molluscan host. Fourth stage larvae within the molluscs are ingested by the sea turtles and attach to the esophageal-gastric junction where they moult into adults within 7 to 21 days. Subsequent time to maturity takes at least 5 more months (Berry and Cannon 1981).

In Australia, scallop adductor muscle was infected, while all *S. sulcata* occurred in the gonad of *Argopecten gibbus* off the Atlantic coast of the United States. The worms are white and inconspicuous when small, and yellow to pale orange or brown when larger. There is cellular infiltration around the larvae, which usually lie loosely coiled near the surface of the adductor muscle. The presence of many larvae can damage the muscle enough to cause a loss of tonicity (Cannon 1978, cited in Berry and Cannon 1981).

The presence of nematodes, and the brownish capsule in which they are found, make some of the processed adductor muscles unsuitable for export. Up to 64% of commercially harvested saucer scallops, *Amusium balloti*, have been found to be infected by *S. sulcata* (Lester et al. 1980). Depending on whether an infected scallop is consumed cooked or raw, *S. sulcata* may be regarded as an aesthetic problem, or as a threat to public health. However, infectivity trials have failed to infect teleost or elasmobranch fish, or chickens or cats, thus, there is no evidence that they are a threat to human health (Berry and Cannon 1981) and no human cases of ascaridosis have been reported from consumption of infected scallops.

A small percentage of *Amusium balloti* were also infested with a larval gnathostome, *Echinocephalus* sp., and two other scallop species, *Anachlamys leopardus* and *Chlamys asperrimus*, were found to contain *S. sulcata* (Lester et al. 1980).

McLean (1983) identified second stage juveniles of *Echinocephalus pseudouncinatus* from the adductor muscles of *Argopecten aequisulcatus* (*A. ventricosus*) from Baja California, Mexico. The worms measured 14 to 17 mm in length and were located within yellow or brown spots on the surface of the adductor muscle. The same species of nematode infects pink abalone (*Halitois corrugata*) from California, and causes blistering of the foot. Heavy infections are associated with weakening of the foot, holdfast attachment and grazing capability (Sindermann 1990). The final hosts of *E. pseudouncinatus* are elasmobranch fish.

Another nematode *Porrocaecum pectenis* reported in *Argopecten irradians* and *A. gibbus* is usually invisible macroscopically. However, brown discolouration of the worm occurs when it becomes infected by an unidentified haplosporidian hyperparasite (Cheng 1967, 1973, 1978; McLean 1983). This renders the worm visible and reduces the quality of the scallop meat.

11.6 POLYCHAETES

Even though polychaete worms do not generally penetrate the soft tissues of the bivalves that host their refugia, the excavation of their tunnels in the matrix of the bivalve shell can have a significant indirect impact on physiology and overall survival of the bivalve. In addition, the mud substrate used to line the tunnels created by the most common family of shell-dwellers, the polydoriids, can exacerbate the erosion effect. Subsequent enlargement of the tunnels creates opportunities for other polychaetes, which, unlike the relatively innocuous *Polydora* spp., are easily visible by consumers, e.g., *Ceratonereis tridentata* (Wells and Wells 1962) and *Nereis diversicolor* (McGladdery et al. 1993a). Shell damage and lively polychaete occupants can significantly impact live mollusc markets. Although this is not usually a factor for scallop meat markets, many species are marketed in shell. Another effect of shell excavation is tunnel perforation of the inner shell surface - especially at the adductor muscle attachment site - which can create access for soft-tissue irritants, abrasion and secondary opportunistic microbial infection (see Section 11.2.2.3). Weakened adductor muscle attachment also impairs shell closure, swimming, feeding and escape behaviour in scallops. Bergman et al. (1982) reported that scallop shells weakened by *P. websteri* boring are more vulnerable to crushing by predatory decapods. The patterns of breakage among scallop (*Placopecten magellanicus*) shells damaged by rock crabs and lobsters suggest that successful predation is often accomplished by fracture of the upper valve near the umbo where the level of polydorid infestation was generally high. It was unclear what effect shell infestation had on scallop natural mortality (Elner and Jamieson 1979). Turner and Hanks (1959) considered *P. websteri* infestation a contributing factor to unusually high mortalities of bay scallops in Fairhaven, Massachusetts. Polychaete burrows close to the scallop hinge may have interfered with the function of the adductor muscle, which in many cases was in poor condition. Similar damage to the same scallop species have been linked to *Polydora ciliata* and *Polydora* spp. by Russel (1973), Leibovitz et al. (1984) and Karlsson (1991).

The harmful effects of polychaete shell damage vary with the intensity of infestation and the type of burrow formed. The most obvious chronic effect is development of mud blisters. These form as sand or mud breach the nacre layer of the shell. Continuous erosion of non-peripheral surfaces is progressively walled off by conchiolin and nacre secretion from the epidermal tissues in contact with the irritant. Such effects from *Polydora concharum, P. socialis* and *P. websteri* are well documented for the giant sea scallop *Placopecten magellanicus* (Blake 1969; Evans 1969; Bergman et al. 1982; McGladdery et al. 1993b). Wells and Wells (1962) and Blake and Evans (1973) describe effects in bay scallops (*A. irradians*) from the eastern US, and Skeel (1979) described *Polydora* infections in four species of cultivated bivalves in Australia. *Placopecten meridionalis*, was infected with *P. websteri* which formed mud blisters between the shell and mantle. An alternative route for inducing such damage is polychaete larvae swimming into the mantle cavity and burrow between the shell and mantle, however, most evidence supports externally driven penetration to the inner surface. Shell damage at the peripheral edges can also induce mantle retraction and growth deformities (indentations, overgrown or double lipped edges, etc.). Regardless of route and site of irritation, this challenge can assume extreme proportions if it persists and the resultant mud blisters take over a significant percentage of the inner shell cavity (Fig. 11.17).

Mori et al. (1985) investigated the polydorid infestation of scallops, in this case *Mizuhopecten* (*Patinopecten*) *yessoensis* in Japan. The boring polychaetes were identified as *Polydora variegata, P. ciliata* and *P. concharum* (Fig. 11.18). The worms settled almost exclusively on the left valve of these cultured scallops. Two years after seeding, the shell region nearest the attachment of the adductor muscle was heavily penetrated by polydoriid tunnels, which frequently contacted the muscle attachment surface. These authors suggested that these polydoriids could have a significant influence on the growth of scallops in that area. Similar observations have also been reported for this scallop species in Japan (Sato-Okoshi and Okoshi 1993), Russia (Kurochkin et al. 1986) and Pacific Canada (Bower and Meyer 1994). On the west coast of Canada shell damage

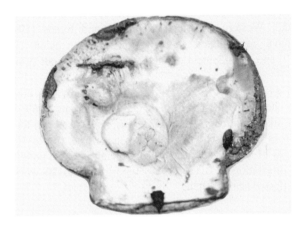

Figure 11.17. *Placopecten magellanicus* with extensive mud blisters occupying one third of the inner surface of the valve. Note poor condition of adductor muscle attachment site.

(thickening, growth stunting, impairment of adductor muscle function and associated mortalities) from a related species *P. websteri* is so severe that it renders culture of *M. yessoensis* impractical in some locations (Bower et al. 1994). Similar observations have recently been reported for polydoriid infestation of great scallop (Coquille St.-Jacques, *Pecten maximus*), spat from scallop nurseries in Norway (Mortensen et al. 1999). The spat measuring only 5–7 mm were found to have shells heavily infested by a *Polydora* sp. This infestation was associated with losses representing a third of Norway's cultured scallop production in 1997. Another polychaete, the tubeworm *Pomatoceros triqueto*, has been associated with fouling-related mortalities of up to 65% in the same scallop species in Ireland (Burnell et al. 1995).

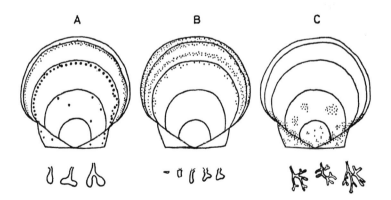

Figure 11.18. Comparison of burrowing regions and morphology of burrows in three species of *Polydora*. (a) *P. variegata*; (b) *P. ciliata*; and (c) *P. concharum*. (After Mori et al. 1985).

A fouling spionid, *Sabella spallanzanii* has been reported on *Pecten fumatus* shells from Cockburn Sound, Western Australia by Clapin and Evans (1995). Resultant smothering losses have necessitated re-evaluation of suspension culture methods (O'Connor et al. 1999) and development of shell attachment methods (gluing) rather than cage-culture.

Cirratulids, such as *Dodecaceria concharum*, may occupy the empty burrows of *Polydora spp.* in the shells of *Placopecten magellanicus* (Evans 1969; Leibovitz et al. 1984). And sabellids, like *Pseudopotamilla reniformis*, have also been found in the shells of *P. magellanicus* from Maine. The large size of the worms (up to 100 mm) and the secretion of large ridges of shell material to contain them may be a potential problem for scallops with heavy infestations. Blake (1969) reported up to 30 worm tubes from a single *P. reniformis* infested scallop shell.

11.7 CRUSTACEA

11.7.1 Pinnotheriidae

Many species of pinnotherid crabs (also known as "pea crabs") live symbiotically in the mantle cavity of marine bivalves. The scallops, *Argopecten irradians concentricus*, *A. gibbus*, and *Placopecten magellanicus*, are hosts for the species *Tumidotheres* (*Pinnotheres*) *maculatus* along the eastern US (Getchell 1991). Disease conditions associated with pinnotherid crabs include emaciation, reduced filtering capacity, damage to gills, palps, and mantle and compression of the gonad which may affect gonadal development. There are no observations of crabs feeding directly on host tissues, but their location within the mantle cavity inflicts a variety of host conditions ranging from slight irritation to severe structural alterations and pathology (Lauckner 1983).

Infestation with *T. maculatus* was found to cause stunting in *Argopecten irradians concentricus* from the eastern US. Bay scallops containing adult female pea crabs were slightly smaller than uninfested scallops. Growth of 3 size groups of scallops was measured over a 3 month period. Smaller scallops hosting pea crabs grew significantly less than uninfested scallops, although the larger scallops showed no significant difference between infested and uninfested groups (Kruczynski 1972).

Other scallops documented as hosting pinnotheriid crabs include *Amusium pleuronectis* (Llana 1979) and *Chlamys farreri* (Cheng 1967).

11.7.2 Copepodidae

Several copepod parasites have been reported from scallops. Reddiah and Williamson (1958) recorded *Paranthessius pectinis* and *Modiolicola inermis* from *Aequipecten opercularis* collected from around the Isle of Man, off the UK, as well as *Modiolocola maxima* from *Pecten maximus* in the same area. The effects of these copepod infestations were not reported as being significant. A similar *Modiolicola* sp. infestation of *Pecten maximus* from Norway was, likewise, not associated with pathological effect (Mortensen 1993b).

In contrast, *Pectenophilus ornatus*, a copepod parasite on the gills of the scallops, *Mizuhopecten (Patinopecten) yessoensis, Chlamys farreri nipponensis* and *C. f. farreri* from Japan (Nagasawa et al. 1988, 1991; Nagasawa and Nagata 1992) is associated with evident hypertrophy of the host tissue at the attachment site of the highly modified copepod (Fig. 11.19). The female assumes a modified sac-like morphology reaching 8 mm in diameter, resembling a rhizocephalan barnacle (Elston et al. 1985). The adult copepods are brilliant orange except for the area immediately surrounding the male vesicle which is reddish (Nagasawa et al. 1988). The parasite is devoid of appendages of any sort and is completely unsegmented. Its body is attached to the host by means of a tapering stalk (Fig. 11.19 and 11.20). The eggs are incubated within a spacious cavity communicating with the exterior by an unpaired birth pore through which the larvae are emitted. The nauplii closely resemble other copepod nauplii (Fig. 11.21). *Pectinophilus ornatus* feeds on the haemolymph of the scallop and may infest up to 100% of the cultured populations. It causes reduced marketability and decreased physiological condition. Combatting the parasite or controlling its distribution will depend on improved knowledge of its life cycle, reproductive dynamics and longevity.

11.8 GASTROPODS

The Family Pyramidellidae comprises a few small (3–5 mm in length) Odostomid gastropods which parasitise marine bivalves by feeding on haemolymph drawn from near the edge of the mantle or, in the case of clams, the siphon (Lauckner 1983; McGladdery et al. 1993a). Leibovitz et al. (1984) observed large numbers of the gastropod, *Odostomia* sp., on the sides of holding tanks, and on the valves, mantle, and pallial cavity of captive

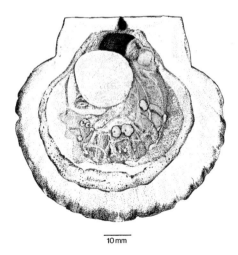

10 mm

Figure 11.19. *Patinopecten yessoensis* left valve with 15 small and large specimens of *Pectenophilus ornatus* attached to the left demibranchs. (After Nagasawa et al. 1988).

634

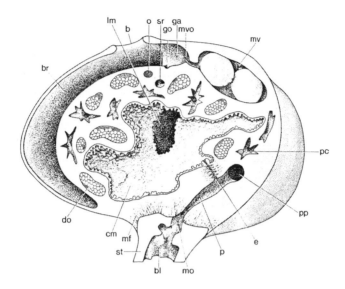

Figure 11.20. *Pectenophilus ornatus* sectioned in median plane. Microanatomy of female labelled as follows: b, birth pore; bl, blood lacuna; br, brood pouch; cm, midgut, central part; do, diverticulum of ovary; e, esophagus; ga, genital atrium; go, genital orifice; lm, lateral diverticulum; og midgut; mf, muscle fibres; mo, mouth; mv, male vesicle with two males; mvo, opening for male vesicle; o, unpaired part of ovary; p, pharynx; pc, pharyngeal channel; pp, pharyngeal pouch; sr, seminal receptacle; st, stalk. (After Nagasawa et al. 1988).

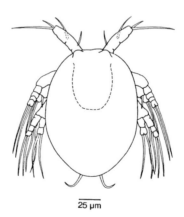

25 μm

Figure 11.21. *Pectenophilus ornatus* nauplius in dorsal view. (After Nagasawa et al. 1988).

bay scallops (*Argopecten irradians*). Developing embryos of the gastropod were also found within the soft tissues. These authors associated these gastropods with high mortalities among their broodstock and experimental bay scallops, however, *Odostomium* spp. are considered to be relatively benign haemolymph predators in other species.

Odostomium seminuda was common on the upper valve of *Argopecten irradians* and on the 'ears' of *A. gibbus* from North Carolina (Wells and Wells 1961). Constant irritation of the calico scallop's mantle margin by penetration of the pyramidellids' proboscis during feeding reportedly resulted in retraction of the mantle in affected areas.

Another pyramidellid, *Turbonilla interrupta*, is known to attack scallops along the eastern North American coast (Morton 1967), however, no further details of pathology or distribution have been reported.

The common oyster drill, *Urosalpinx cinerea*, has demonstrated predatory behaviour in a laboratory study of attacks on bay scallop, *Argopecten irradians*. Predation resulted in 72.3% mortality. The snails also demonstrated a positive chemotactic response to scallop effluent in a choice chamber (Ordzie and Garofalo 1980).

11.9 ALGAE

Many algae have been labelled as symbionts of marine bivalves, e.g., *Zoochlorella* spp., which colonise the mantle, eyes, and tentacles of wild, captive and cultured adult bay (*Argopecten irradians*) and sea scallops (*Placopecten magellanicus*) south of Cape Cod. The infections produced green proliferative granulomatous lesions of the mantle and tentacles and haemocyte infiltration into the eyes. The disease was not seen in young scallops (Leibovitz et al. 1984).

Early work by Naidu (1971) showed another alga, *Coccomyxa parasitica*, to be parasitic on *Placopecten magellanicus* from Newfoundland. The peripheral mantle tissue, as well as portions of the gonad and adductor muscle when exposed to ambient light were affected. When infestation levels were high, there was a decline in the condition and weight of the viscera, including the adductor muscle. The mantle had a dark green colour, was slippery due to a layer of mucus, and had a musty odour. Histological work demonstrated the encapsulation of the algal colonies by a network of connective tissue fibres. Because the normal mantle morphology is disrupted, shell deformities consisting of extra shell margins and grossly warped valves may occur. Of almost 3,000 scallops sampled by Naidu (1971), 17.6% were found to be infested. Scallops 8–10 years of age had the highest prevalences, while no algal cells were found in scallops younger than 3 years of age. Levels of infestation were positively correlated with light intensity.

Stevenson and South (1975) have described the causative agent in detail. They also showed that endocytosis occurs throughout the infestation and contributes to the spread of the algal cells. The algae also enters through normal feeding by the scallop, but seems highly resistant to digestion. The algal cells do not appear to be highly pathogenic or cause a rapid death of the host. *Coccomyxa parasitica* infestations are restricted to shallow water populations of *Placopecten magellanicus* inhabiting the more northern latitudes of its range (Stevenson and South 1975).

Leibovitz et al. (1984) described a die-off of post-metamorphic juvenile cultured bay scallops (*Argopecten irradians*) in association with a bloom of the dinoflagellate, *Prorocentrum* sp. in a Long Island Sound hatchery's water supply. The juvenile scallops showed signs of distress including gaping shells, injuries to the soft tissues by the dinoflagellates' minute spines, and impaction of the pallial cavity. Melanisation of the resulting lesions in the mantle followed, with secondary invasion of bacteria and ciliates.

Around the same time many coastal embayments of Long Island, New York, experienced algal blooms during the summers of 1985 and 1986 that were unprecedented in their persistence and cell density ($>10^9$/litre). Small microalga decreased light penetration to less than 1 meter, severely affecting eel grass (*Zostrea*) beds. The 1985 bloom coincided with the spawning period of the bay scallop (*Argopecten irradians*) and was blamed for the massive recruitment failure of that year-class (Cosper et al. 1987). Transmission electron microscopy indicated that the responsible microalga was a previously undescribed species, similar to the chrysophyte species tentatively designated *Aureococcus anorexefferens*, which also bloomed in Narragansett Bay, Rhode Island in 1985 (Sieburth et al. 1986). Grazing experiments revealed that bay scallops were inefficient at retaining the 'brown tide' cells, but this alone does not account for the impact that was observed. Possible mechanisms investigated were toxicity effects, reduced ingestion rates, reduced absorption efficiencies at high densities and poor nutritional qualities of the brown tide alga (Cosper et al. 1987).

Following a picoplankton bloom in Long Island Sound, NY, 1985, Bricelj et al. (1987) measured a 76% reduction in adductor muscle weight of adult bay scallops compared with 1984 data. However, survivors rebounded in September to triple their mean muscle weight. Mortality rates were not documented during either 1985 or 1986, thus, impact on survival is unknown.

11.10 FORAMINIFERANS

The calcareous foraminiferan *Cibicides refulgens* is a conspicuous and abundant component of the epifaunal community, living on the valves of the Antarctic scallop, *Adamussium colbecki*. One of its modes of nutrient acquisition is parasitism by eroding through the scallop's shell, and using dissolved free amino acids from the highly concentrated pool in the extrapallial cavity (Alexander and DeLaca 1987). These authors showed that 50% of attached *C. refulgens* significantly erode the surface of the scallop's shell to excavate channels into the extrapallial cavity. Whether the cumulative effects of *C. refulgens* induced shell erosions have a detrimental influence on the bivalve in the marginal Antarctic environment remains unknown, but seems likely. Hayward and Haynes (1976) reported the association of other foraminiferans including *Cibicides lobatulus* on the queen scallop *Aequipecten* (*Chlamys*) *opercularis*.

11.11 PORIFERA

As with polydoriid polychaetes, clionid sponges are also known as ubiquitous shell colonising organisms in both live and dead marine bivalves (Lauckner 1983). Shell destruction follows excavation through etching and discarding of shell fragments to permit sponge proliferation and development of inhalent and exhalent passages throughout the shell.

Evans (1969) detailed the extent of shell destruction by *Cliona vastifica* in *Placopecten magellanicus* from Newfoundland, Canada. The radiographs of infested shells showed how little shell material was left between the inner and outer surfaces of heavily colonised individuals. Externally, clionid invasion of the shell can be identified by small holes which open into the inhalent and exhalent tunnels, usually starting at the hinge end of the shell with predominant damage to the ventral valve (Warburton 1958).

Clionid invasion of the shell may also effect the soft parts of scallops and cause a condition Medcof (1949) termed 'dark-meat'. The adductor muscles of affected individuals have a greyish brown tinge, are flaccid and stringy, and reduced in size. The meat yield of lightly infested scallops was about 59%, while that of heavily infested scallops was only 46% of the yield of healthy individuals. Off Digby, Nova Scotia, Canada, only those scallops 8 to 9 years or older were affected by clionid sponges. Invasion by the pest starts near the hinge and spreads until the entire shell is honeycombed with burrows. The shell can reach over 3 times its normal thickness because of repeated nacre deposition on the inner shell surface to attempt shell repair. Several other reports of *Cliona celata* and *C. vastifica* shell damage have also been reported from *P. magellanicus*, but most are not associated with adductor muscle discolouration (Evans 1969; Sindermann 1971; McGladdery et al. 1993a). Kurochkin et al. (1986) report clionid damage to the shells of *Patinopecten (Mizuhopecten) yessoensis* in the Sea of Japan/Peter the Great Sea. These infestations were also associated with soft-tissue retraction and reduction of physiological condition.

Leibovitz et al. (1984) described clionid induced shell deformities in juvenile and adult bay scallops (*Argopecten irradians*), probably caused by a *Scypha* sp. These are tubular sponges that attach to the internal surface of the valve, as well as the ventral margins and external surface. No significant mortalities were seen, so the fate of these deformed scallops is unknown. More recently, a reciprocal shell colonisation relationship was reported on *Aequipecten opercularis*, where the sponge *Siberites ficus* ssp. *rubrus* was associated with reduced shell colonisation by other fouling agents (Armstrong et al. 1999). Within the geographic range of this species, this provides a natural alternative to chemical/salinity/shock treatments for problem colonisation and may provide potential for aquaculture use to control shell matrix colonisation.

Several species of scallop have evolved a mutualism with epizooic sponges on their shells. The sponge protects the scallop from predatory starfish, while being protected from nudibranchs and/or deriving additional nutrients generated by the host's inhalant current (Bloom 1975; Forester 1979; Chernoff 1987; Pitcher and Butler 1987).

11.12 CNIDARIA

The colonial hydroid, *Hydractinia echinata*, a common inhabitant on the external surface of the shell of the sea scallop, *Placopecten magellanicus*, sometimes attaches to the internal shell surface. This intrusion interferes with the normal activities of the scallop's mantle, often causing shell deformities. The scallop reacts by secreting a new shell edge within the existing perimeter and bypassing the hydroid colony. Batteries of nematocysts are abundant at the leading edge of the colony and the mantle withdraws presumably because of the discharge of the nematocysts into it. Successful control of this intrusion is accomplished when the scallop overgrows the impinging hydroid colony. No evidence has been found that *H. echinata* may ultimately cause the death of the scallop (Merrill 1967).

11.13 NON-INFECTIOUS DISEASES

In addition to the organisms that colonise scallops or cause infectious diseases, scallops also suffer from diseases that are not infectious. The most familiar cases are associated with aquaculture when animals are kept in suboptimal conditions, especially in land based systems, where diseases caused by bacteria (often *Vibrio* spp. as described in Section 11.2.2.1) can be encountered. Scallops cultured in open-water systems, however, have also encountered problems such as smothering by various organisms, including mussels (Minchin & Duggan 1989; Kurata et al. 1996) and barnacles (Ovsyannikova and Levin 1982). In wild stocks, abiotic factors, such as unknown environmental disturbances, were associated with abnormal melanisation and microstructural distortions of the shell of *Pecten maximus* in Brittany, France (Larvor et al. 1996). Anthropogenic pollutants have also been attributed to diseases affecting both cultured and wild stocks. For example, Syasina et al. (1996) and Usheva (1999) have described weakening muscle atrophy associated with necrosis and gametogenesis abnormalities in *Mizuhopecten yessoensis* from polluted areas of Peter the Great Bay, Sea of Japan. Also, Karaseva and Medvedeva (1994) reported embryo toxicity in the same species of scallop after parental exposure to copper and zinc.

11.14 SUMMARY

With growing interest in scallop enhancement, stock regeneration and molluscan aquaculture species diversification, comes a growing opportunity for improved health and disease understanding for these pectinids. Suspension culture, hatchery manipulated reproduction and increased proprietary observation is bound to reinforce the many gaps evident in this parasite, pest and disease summary. Many parasites, which have been reported in wild populations and subsequently forgotten, may become more readily available for study, or more problematic under culture conditions. In addition, the new tools being developed for other mollusc groups (primarily oysters) will be more accessible in the near future for distinguishing between significant scallop pathogens and innocent commensals - especially at the microbial level. With this in mind it is clearly anticipated that this chapter will also be superceded in the next 10 years - these authors look forward to reading it.

ACKNOWLEDGMENTS

The authors gratefully acknowledge the kind assistance of everyone who submitted images to illustrate the diseases described. Special thanks are due to Dr. Karin Lohrmann, Universidad Catolica del Norte, Facultad de Ciencias del Mar, Cas. 117, Coquimbo, Chile, for her valuable manuscript input as well as many images supplied during her doctoral thesis work at CEFAS, UK.

REFERENCES

Alderman, D. and Barker, G., 1997. Antibiotic resistance. Trout News 24:23–27.

Alexander, S.P. and DeLaca, T.E., 1987. Feeding adaptations of the foraminifera *Cibicides refulgens* living epizoically and parasitically on the antarctic scallop *Adamussium colbecki*. Biol. Bull. 173:136–159.

Armstrong, E., McKenzie, J.D. and Goldsworthy, G.T., 1999. Aquaculture of sponges on scallops for natural products research and antifouling. J. Biotechnol. 70:163–174.

Ballou, R., 1984. Gray muscle disease of scallops identified. Maritimes 28(2):14.

Beninger, P.G., LePennec, M. and Salaün, M., 1988. New observations of the gills of *Placopecten magellanicus* (Mollusca: Bivalvia), and implications for nutrition. Mar. Biol. 98:61–70.

Bergman, K.M., Elner, R.W. and Risk, M.J., 1982. The influence of *Polydora websteri* borings on the strength of the shell of the sea scallop, *Placopecten magellanicus*. Can. J. Zool. 60:2551–2556.

Berry, G.N. and Cannon, L.R., 1981. The life history of *Sulcascaris sulcata* (Nematoda: Ascaridoidea), a parasite of marine molluscs and turtles. Int. J. Parasit. 11:43–54.

Blackbourn, J., Bower, S.M. and Meyer, G.R., 1998. *Perkinsus qugwadi* sp. nov. (incertae sedis), a pathogenic protozoan parasite of Japanese scallops, *Patinopecten yessoensis*, cultured in British Columbia, Canada. Can. J. Zool. 76:942–953.

Blake, J.A., 1969. Systematics and ecology of shell-boring polychaetes from New England. Am. Zool. 9:813–820.

Blake, J.A. and Evans, J.W., 1973. *Polydora* and related genera as borers in molluscs shells and other calcareous substrates (Polychaeta: Spionidae). Veliger 15:235–249.

Blake, N.J., Barber, B.J. and Rodrick, G.E., 1984. Occurrence levels of the adductor muscle parasite *Sulcascaris sulcata* Rudolphi in the calico scallop *Argopecten gibbus* Linne. J. Shellfish Res. 4(1):82 (abstract)

Bloom, S.A., 1975. The motile escape response of a sessile prey: a sponge-scallop mutualism. J. Exp. Mar. Biol. Ecol. 17:311–321.

Bjoershol, B., Nordmo, R., Falk, K. and Mortensen, S., 1999. Cohabitation of Atlantic salmon (*Salmo salar* L.) and scallop *Pecten maximus*) - challenge with Infectious Salmon Anaemia (ISA) Virus and *Aeromonas salmonicida* subsp. *salmonicida*. 12th International Pectinid Workshop, Bergen, Norway, May 5–11, 1999 (abstract)

Borrego, J.J., Castro, D., Luque, A., Paillard, C., Maes, P., Garcia, M.T. and Ventosa, A., 1996. *Vibrio tapetis* sp. nov., the causative agent of the brown ring disease affecting cultured clams. Int. J. Syst. Bacteriol. 46(2):480–484

640

Bower, S.M., 1995. Parasitic Diseases of Shellfish. In: P.T.K. Woo (Ed.). Fish Diseases and Disorders. Vol. I, Protozoan and Metazoan Infections. CAB International, Oxon UK. pp. 673–728.

Bower, S.M., Blackbourn, J. and Meyer, G.R., 1997. A new and unusual species of *Perkinsus* pathogenic to cultured Japanese scallops, *Patinopecten yessoensis*, in British Columbia, Canada. J. Shellfish Res. 16:333. (abstract)

Bower, S.M., Blackbourn, J. and Meyer, G.R., 1998. Distribution, prevalence and pathogenicity of the protozoan *Perkinsus qugwadi* in Japanese scallops, *Patinopecten yessoensis*, cultured in British Columbia, Canada. Can. J. Zool. 76:954–959.

Bower, S.M., Blackbourn, J., Meyer, G.R. and Nishimura, D.J.H., 1992. Diseases of cultured Japanese scallops (*Patinopecten yessoensis*) in British Columbia, Canada. Aquaculture 107:201–210.

Bower, S.M., Blackbourn, J., Meyer, G.R. and Welch, D.W., 1999. Effect of *Perkinsus qugwadi* on various species and strains of scallops. Dis. Aquat. Org. 36:143–151.

Bower, S.M., Blackbourn, J., Nishimura, D.J.H. and Meyer, G.R., 1990. Diseases of cultured Japanese scallops (*Patinopecten yessoensis*) in British Columbia, Canada. In: A. Figueras (Ed.). 4th International Colloquium on Pathology in Marine Aquaculture. Vigo, Spain. pp. 67–68.

Bower, S.M., McGladdery, S.E. and Price, I.M., 1994. Synopsis of infectious diseases and parasites of commercially exploited shellfish. Ann. Rev. Fish Dis. 4:1–199.

Bower, S.M. and Meyer, G.R., 1991. Disease of Japanese scallops (*Patinopecten yessoensis*) caused by an intracellular bacterium. J. Shellfish Res. 10(2):513. (abstract)

Bower, S.M. and Meyer, G.R., 1994. Causes of mortalities among cultured Japanese scallops, *Patinopecten yessoensis*, in British Columbia, Canada. In: N.F. Bourne, B. Bunting and L.D. Townsend (Eds.). Proceedings of the 9th International Pectinid Workshop, Nanaimo, B.C., Canada. Can. Tech. Rep. Fish. Aquat. Sci. 1994:85–94.

Bower, S.M., Meyer, G.R. and Blackbourn, J., 1995. The enigmatic scallop protozoan pathogen (SPX) of cultured Japanese scallops, *Patinopecten yessoensis*, in British Columbia, Canada. J. Shellfish Res. 14:261. (abstract)

Bray, R.A., 1983. On the fellodistomid genus *Proctoeces* Odhner, 1911 (Digenea), with brief comments on two other fellodistomid genera. J. Nat. Hist. 17:321–339.

Bricelj, V.M., Epp, J. and Malouf, R.E., 1987. Comparative physiology of young and old cohorts of bay scallop *Argopecten irradians irradians* (Lamarck): mortality, growth, and oxygen consumption. J. Exp. Mar. Biol. Ecol. 112:73–91.

Brown, C., 1981. A study of two shellfish-pathogenic *Vibrio* strains isolated from a Long Island hatchery during a recent outbreak of disease. J. Shellfish Res. 1:83–87.

Brown, C. and Losee, E., 1978. Observation on natural and induced epizootics of vibriosis in *Crassostrea virginica* larvae. J. Invertebr. Pathol. 31:41–47.

Brown, C. and Russo, D.J., 1979. Ultraviolet light disinfection of shellfish hatchery sea water. I. Elimination of five pathogenic bacteria. Aquaculture 17:17–23.

Brun, N.T., Ross, N.W. and Boghen, A.D., 2000. Change in the electrophoretic profiles of gill mucus proteases of the eastern oyster *Crassostrea virginica* in response to infection by the turbellarian *Urastoma cyprinae*. J. Invert. Pathol. 75(2):163–170.

Burnell, G.M., Barnett, M., O'Carroll, T. and Roantree, V., 1995. Scallop spat collection and ongrowing trials in south-west Ireland. Actes. Colloq. IFREMER 17:139–144.

Cake, E.W., 1976. A key to larval cestodes of shallow-water, benthic mollusks of the northern Gulf of Mexico. Proc. Helminthol. Soc. Wash. 43(2):160–171.

Cake, E.W., 1977. Larval cestode parasites of edible molluscs of the northeastern Gulf of Mexico. Gulf Res. Rep. 6:1–8.

Cake, E.W., 1979. *Polypocephalus* sp. (Cestoda: Lecanicephalidae): a description of tentaculoplerocercoids from bay scallops of the northeastern Gulf of Mexico. Proc. Helminthol. Soc. Wash. 46(2):165–170.

Cannon, L.R.G., 1978. A larval ascaridoid nematode from Queensland scallops. Int. J. Parasitol. 8:75–80.

Cawthorn, R.J., MacMillan, R.J. and McGladdery, S.E., 1992. Epidemic of *Pseudoklossia* sp. (Apicomplexa) in bay scallops *Argopecten irradians* maintained in a warm water recirculating facility. Bull. Can. Soc. Zoologists 23(2):37–3 (abstract)

Chang, S.C., Harshbarger, J.C. and Otto, S.V., 1980. Status of cytoplasmic prokaryote infections and neoplasms in bivalve molluscs. Sixth Food and Drug Science Symposium on Aquaculture: Public Health, Regulatory and Management Aspects, February 12–14, 1980.

Chavez, C.P. and Riquelme, S.C., 1994. Analysis of the bacteriological quality of *Argopecten purpuratus* broodstock, for their use in aquaculture. Rev. Latinoam. Acuicult. 43:96–99 (in Spanish)

Cheng T.C., 1967. Marine molluscs as host for symbioses. Adv. Marine Biol. 5:1–424.

Cheng T.C., 1973. Human parasites transmissible by seafood- and related problems. In: C.O. Chichester and H.D. Graham (Eds.). Microbial Safety of Fishery Products. Academic Press, New York. p. 163–189.

Cheng, T.C., 1978. Larval nematodes parasitic in shellfish. Mar. Fish. Rev. 40:39–42.

Chernoff, H., 1987. Factors affecting mortality of the scallop *Chlamys asperrima* (Lamarck) and its epizooic sponges in South Australian waters. J. Exp. Mar. Biol. Ecol. 109:155–171.

Chu, F.-L., Burreson, E.M., Zhang, F. and Chew, K.K., 1996. An unidentified haplosporidian parasite of bay scallop *Argopecten irradians* cultured in the Shandong and Liaoning provinces of China. Dis. Aquatic Org. 25:155–158.

Chubrick, G.K., 1966. Fauna and ecology of larval trematodes in molluscs of the Barents and White Seas (in Russian). Trudy murmansk. biol. Inst. 10:78–166.

Claereboudt, M.R., Bureau, D., Côté, J. and Himmelman, J.H., 1994. Fouling development and its effect on the growth of juvenile giant scallops (*Placopecten magellanicus*) in suspended culture. Aquaculture 121:327–342.

Clapin, G. and Evans, D.R., 1995. The status of the introduced marine fanworm *Sabella spallanzanii* in Western Australia: a preliminary investigation. Crimp Tech. Tasmania, No. 2, 34 p.

Comps, M., 1983. Infections rickettsiennes chez les mollusques bivalves des cotes francaises. Rapp. P.-v. Reun. Cons. Int. Explor. Mer. 182:134–136.

Comps, M. and Cochennec, N., 1993. A *herpes*-like virus from the European oyster *Ostrea edulis* L. J. Invertebr. Pathol. 62(2):201–203.

Comps, M., Herbaut, Ch. and Fougerouse, A., 1999. Virus-like particles in pearl oyster *Pinctada margaritifera*. Bull. Eur. Assoc. Fish. Pathol. 19(2):85–88.

Cosper, E.M., Dennison, W.C., Carpenter, E.J., Bricelj, V.M., Mitchell, J.G., Kuenstner, S.H., Colflesh, D. and Dewey, M., 1987. Recurrent and persistent brown tide blooms perturb coastal marine ecosystem. Estuaries 10:284–290.

Davis, H.C., Loosanoff, V.L., Weston, W.H. and Matin, C., 1954. A fungus disease in clam and oyster larvae. Science 120:36–38.

Desser, S.S. and Bower, S.M., 1997. *Margolisiella kabatai* gen. et sp. n. (Apicomplexa: Eimeriidae), a parasite of native littleneck clams, *Protothaca staminea*, from British Columbia, Canada, with a taxonomic revision of the coccidian parasites of bivalves (Mollusca: Bivalvia). Folia Parasit. 44:241–247.

DiSalvo, L., 1994. Chronic infection of broodstock as a potential source of substandard gametes and larval infection in Chilean scallop hatcheries. In: N.F. Bourne, B.L. Bunting and L.D. Townsend (Eds.). Proceedings of the 9th International Pectinid Workshop, Nanaimo, BC Canada. Can. Tech. Rep. Fish. Aquat. Sci. 1994:107–111.

Elner, R.W. and Jamieson, G.S., 1979. Predation of sea scallops, *Placopecten magellanicus*, by the rock crab, *Cancer irroratus*, and the American lobster, *Homarus americanus*. J. Fish. Res. Board Can. 36:537–543.

Elston R.A., l984. Prevention and management of infectious diseases in intensive mollusc husbandry. J. World Maricult. Soc. 15:284–300.

Elston, R.A., 1986. Occurrence of branchial rickettsialies-like infections in two bivalve molluscs, *Tapes japonica* and *Patinopecten yessoensis*, with comments on their significance. J. Fish. Dis. 9:69–71.

Elston, R.A., 1997. Special topic review: Bivalve mollusc viruses. World J. Microbiol. Biotechnol. 13:393–403.

Elston, R.A., Wilkinson, M.T. and Burge, R., 1985. A rhizocephelan-like parasite of a bivalve mollusc, *Patinopecten yessoensis*. Aquaculture 49:359–361.

Evans, J.W., 1969. Borers in the shell of the sea scallop, *Placopecten magellanicus*. Am. Zool. 9:775–782.

Forester, A.J., 1979. The association between the sponge *Halichondris panicea* (Pallas) and scallop *Chlamys varia* (L.): A commensal-protective mutualism. J. Exp. Mar. Biol. Ecol. 36:1–10.

Friedman, C.S., 1994. Rickettsiales-like infection of the Japanese scallop, *Pationpecten yessoensis*. In: N.F. Bourne, B. Bunting and L.D. Townsend (Eds.). Proceedings of the 9th International Pectinid Workshop, Nanaimo, B.C., Canada. Can. Tech. Rep. Fish. Aquatic Sci. 1994:115.

Getchell, R.G., 1991. Diseases and parasites of scallops. In: S.E. Shumway (Ed.). Scallops: Biology, Ecology and Aquaculture. Development in Aquaculture and Fisheries Science, Vol. 21. Elsevier, Netherlands. pp. 471–494.

Goggin, C.L., McGladdery, S.E., Whyte, S.K. and Cawthorn, R.J., 1996. An assessment of lesions in bay scallops *Argopecten irradians* attributed to *Perkinsus karlssoni* (Protozoa, Apicomplexa). Dis. Aquat. Org. 24:77–80.

Gryska, A., Parsons, G.J., Shumway, S.E., Geib, K.P., Emery, I. and Kuenstner, S., 1996. Polyculture of sea scallops suspended from salmon cages. J. Shellfish Res. 15:454.

Guillard, R.R.H., 1959. Further evidence of the destruction of bivalve larvae by bacteria. Biol. Bull. Mar. Biol. Lab., Woods Hole 117:258–266.

Gulka, G., Chang, P.W. and Marti, K.A., 1983 Prokaryotic infection associated with a mortality of the sea scallop, *Placopecten magellanicus*. J. Fish Dis. 6:355–364.

Gulka, G. and Chang, P.W., 1984. Pathogenicity and infectivity of a rickettsia-like organism in the sea scallop, *Placopecten magellanicus*. J. Fish Dis. 8:309–318.

Harry, O.G., 1977. Observations on *Licnophora auerbachi*, a ciliate from the eyes of the Queen scallop *Chlamys opercularis* (Mollusca: Bivalvia). J. Protozool. 24:48A.

Harry, O.G., 1980. Damage to the eyes of the bivalve *Chlamys opercularis* caused by the ciliate *Licnophora auerbachi*. J. Invertebr. Pathol. 36:283–291.

Hayward, J.J. and Haynes, J.R., 1976. *Chlamys opercularis* (Linneaus) as a mobile substrate for a foraminifera. J. Foraminiferal Res. 6:30–38.

Heasman, M.P., O'Connor, W.A. and Frazer, A.W.J., 1996. Digenean (Bucephalidae) infections in commercial scallops, *Pecten fumatus* Reeve, and doughboy scallops *Chlamys (Mimachlamys) asperrima* (Lamarck) in Jervis Bay, New South Wales. J. Fish Dis. 19(5):333–339.

Hill, B.J., Way, K. and Alderman, D.J., 1986. IPN-like birnaviruses in oysters: infection or contamination? In: C.P. Vivares, J.-R. Bonami and E. Japers (Eds.). Pathology in Marine Aquaculture (*Pathologie en Aquaculture Marin*). European Aquaculture Society, Special Publication No. 9, Bredene, Belgium. pp. 297. (abstract)

Hine, P.M. and Wesney, B., 1997. Virus-like particles associated with cytopathology in the digestive gland epithelium of scallops *Pecten novaezelandiae* and toheroa *Paphies ventricosum*. Dis. Aquat. Org. 29:197–204.

Hine, P.M., Wesney, B. and Hay, B.E., 1992. Herpesviruses associated with mortalities among hatchery-reared larval Pacific oysters *Crassostrea gigas*. Dis. Aquat. Org. 12(3):135–142.

Hine, P.M., Wesney, B and Besant, P., 1998. Replication of a herpes-like virus in larvae of the flat oyster *Tiostrea chilensis* at ambient temperatures. Dis. Aquatic. Org. 32(3):161–171.

Karaseva, E.M. and Medvedeva, L.A., 1994. Morphological and functional changes in the offspring of *Mytilus trossulus* and *Mizuhopecten yessoensis* after parental exposure to copper and zinc. Russ. J. Mar. Biol. 19(4):83–89.

Karlsson, J.D., 1991. Parasites of the bay scallop, *Argopecten irradians* (Lamarck, 1819). In: S.E. Shumway and P.A. Sandifer (Eds.). International Compendium of Scallop Biology and Culture. World Aquaculture Society, Baton Rouge, LA. p. 180–190.

Kellner-Cousin, K., LeGall, G., Déspres, B., Kaghad, M., Legoux, P., Shire, D. and Mialhe, E., 1993. Genomic DNA cloning of rickettsia-like organisms (RLO) of Saint-Jacques scallop *Pecten maximus*: Evaluation of prokaryote diagnosis by hybridisation with non-isotopically labeled probe and by polymerase chain reaction. Dis. Aquat. Org. 15(2):145–152.

Kruczynski, W.L., 1972. The effect of the pea crab, *Pinnotheres maculatus* Say, on the growth of the bay scallop, *Argopecten irradians concentricus* Say. Chesapeake Sci. 13:218–220.

Kruse, D.N., 1966. Life cycle studies on *Nematopsis duorari* n. sp. (Gregarina: Poropsoridae), a parasite of the pink shrimp (*Penaeus duorarum*) and pelecypod molluscs. Diss. Abstr. 27B:2919-B.

Kurochkin, Y.V., Tsimbalyuk, E.M. and Rybakkov, A.V., 1986. Parazitî i bolyezni. In: P.A. Motavkin (Ed.). Primorskii grebeshok (The Yezo scallop or Japanese common scallop *Mizuhopecten yessoensis* (Jay). Institute of Marine Biology, Far East Science Centre, Academy of the USSR, Vladivostok. pp. 174–182. (in Russian)

Kurata, M., Nishida, Y. and Mizushima, T., 1996. Fouling organisms attached to cultured scallops, *Patinopecten yessoensis*, in Funka Bay. Sci. Rep. Hokkaido Fish. Exp. Stn. 49:15–22.

644

Lambert, C. and Nicolas, J.L., 1998. Specific inhibition of chemiluminescent activity by pathogenic vibrios in haemocytes of two marine bivalves: *Pecten maximus* and *Crassostrea gigas*. J. Invert. Pathol. 71(1):53–63.

Lambert, C., Nicolas, J.L. and Cilla, V., 1999b. *Vibrio splendidus*-related strain isolated from brown deposit in scallop (*Pecten maximus*) cultured in Brittany (France). Bull. Eur. Assoc. Fish Pathol. 19(3):102–106.

Lambert, C., Nicolas, J.-L., Cilla, V. and Corre, S., 1999a. *Vibrio pectenicida* sp. nov., a pathogen of *Pecten maximus* larvae. Book of Abstracts, 12th International Pectinid Workshop, Bergen, Norway.

Larvor, H., Cuif, J.P. and Devauchelle, N., 1996. Abnormal melanisation and microstructural distortions of the shell of the great scallop living in shallow water. Aquacult. Int. 4(3):237–252.

Lauckner, G., 1983. Diseases of mollusca: bivalvia. In: O. Kinne (Editor), Diseases of Marine Animals. Biologische Anstalt Helgoland, Hamburg 2:477–961.

Le Deuff, R.M. and Renault, T., 1999. Purification and partial genome characterization of a herpes-like virus infecting the Japanese oyster, *Crassostrea gigas*. J. Gen. Virol. 80(5):1317–1322.

LeGall, G., Chagot, D., Mialhe, E. and Grizel, H., 1988. Branchial Rickettsiales-like infection associated with mass mortality of sea scallop *Pecten maximus*. Dis. Aquat. Org. 4(3):229–232.

LeGall, G. and Mialhe, E., 1992. Purification of Rickettsiales-like organisms associated with *Pecten maximus* (Mollusca: Bivalvia): Serological and biochemical characterisation. Dis. Aquat. Org. 12(3):215–230.

LeGall, G., Mialhe, E., Chagot, D. and Grizel, H., 1991. Epizootiological study of rickettsiosis of the Saint-Jacques scallop *Pecten maximus*. Dis. Aquat. Org. 10(2):139–145.

LeGall, G., Mourton, C., Boulo, V., Paolucci, F., Pau, B. and Mialhe, E., 1992. Monoclonal antibody against a gill Rickettsiales-like organism of *Pecten maximus* (Bivalvia): Application to indirect immunofluorescence diagnosis. Dis. Aquat. Org. 14:213–217.

Léger, L. and Duboscq, O., 1917. *Pseudoklossia pectinis* n. sp. et l'origine des adéleidées. Arch. Zool. Exptl. Gen. 56:88–95.

Léger, L. and Duboscq, O., 1925. Les Porosporides et leur évolution. Trav. Sta. Zool. Wimereux 9:126–139.

Leibovitz, L., 1989. Chlamuydiosis: a newly reported serious disease of larval and postmetamorphic bay scallops, *Argopecten irradians* (Lamarck). J. Fish Dis. 12:125–136.

Leibovitz, L., Schott, E.F. and Karney, R.C., 1984. Diseases of wild, captive and cultured scallops. J. World Maricul. Soc. 15:269–283.

Lester, R.J., Blair, D. and Heald, D., 1980. Nematodes from scallops and turtles from Shark Bay, western Australia. Aust. J. Mar. Freshwat. Res. 31:713–717.

Lichtenfels, J.R., Bier, J.W. and Madden, P.A., 1978. Larval anisakid (*Sulcascaris*) nematodes from Atlantic molluscs with marine turtles as definitive hosts. Trans. Am. Microsc. Soc. 97:109–207.

Lichtenfels, J.R., Sawyer, T.K. and Miller, G.C., 1980. New hosts for larval *Sulcascaris* sp. (Nematoda, Anisakidae) and prevalence in the calico scallop (*Argopecten gibbus*). Trans. Am. Microsc. Soc. 99(4):448–451.

Linton, E., 1915. Note on trematode sporocysts and cercariae in marine molluscs of the Woods Hole region. Biol. Bull. Mar. Biol. Lab., Woods Hole 28:198–209.

Llana, E.G., 1979. Notes on the occurrence of the pea crab (*Pinnotheres* sp.) in the Asian moon scallop (*Amusium pleuronectes* Linne). Fish. Res. J. Philipp. 4(2):41–43.

Lodeiros, C., Freites, L., Fernandez, E., Velez, A. and Bastardo, J., 1989. Antibiotic effect of three marine bacteria in the larval survival of the infected scallop *Pecten ziczac* with *Vibrio anguillarum*. Bol. Inst. Oceanogr. Venez. 28(1–2):165–129. (in Spanish)

Lodeiros, C., Freites, L. and Velez, A., 1992. Bacillary necrosis in larvae of the bivalve *Euvola ziczac* (Linneo, 1758) caused by a *Pseudomonas* sp. Acta Cient. venez. 43(3):154–158.

Lohrmann, K.B., A.R. Brand and S.W. Feist. 2002. Comparison of the parasites and pathogens present in a cultivated and in a wild population of scallops (Argopecten purpuratus Lamarck, 1819) in Tongoy Bay, Chile. J. Shellfish Res. 21:557–561.

Lohrmann, K.B., Feist, S.W. and Brand, A.R., 1999. Presence of a microsporidian parasite in the queen scallop *Aequipecten opercularis* (L.). 12th Int. Pectinid Workshop, Book of Abstracts, Bergen, Norway, 5–11 May 1999.

Lohrmann, K.B., Feist, S.W. and Brand, A.R., 2000a. The parasite fauna of a cultivated and a wild population of *Argopecten purpuratus* from Tongoy Bay, Chile: a comparison. Proceedings XVIIIth Int. Congress Zool., Athens, Greece. pp. 85.

Lohrmann, K.B., Feist, S.W. and Brand, A.R., 2000b. Microsporidiosis in queen scallops (*Aequipecten opercularis* L.) from U.K. waters. J. Shellfish Res. 19(1):71–75.

Lohrmann, K.B. and Smith, Y., 1993. Platelmintos parásitos en el ostión del Norte, *Argopecten purpuratus* Lamarck, 1918. XIII Jornadas de Ciencias del Mar. pp. 119.

Lohrmann, K.B., Smith, Y., Díaz, S., Bustos, M. and Cortés, C., 1991. Presencia de un tremátodo digeneo en *Argopecten purpuratus* provenientes de poblaciones naturales. IV Congreso Latinoamericano de Ciencias del Mar. pp. 104 (in Spanish)

Loosanoff, V.L. and Davis, H.C., 1963. Rearing of bivalve molluscs. Adv. Mar. Biol. 1:1–136.

Martin, C., Stiles, S., Choromanski, J., Widman, J.C. Jr., Schweitzer, D. and Cooper, C., 1997. *Sirolpidium zoophthorum*, lethal fungus parasites of bivalve larvae: Recent observations in bay scallop cultures. J. Shellfish Res. 16:291. (abstract)

Mateo, E.S., Peña, C.D., Guzmán, E.L. and López, R.C., 1975. Parásito causante de castración de la concha de abanico *Argopecten purpuratus*. Biol. Lima 18(40):81–86. (in Spanish)

McGladdery, S.E., 1990. Shellfish parasites and diseases of the east coast of Canada. Bull. Aquaculture Assoc. Can. 90–3:14–18.

McGladdery, S.E., 1998. Microbial diseases of shellfish. In: D. Bruno and P.T.K. Woo (Eds.). Fish Diseases and Disorders, Volume III. CAB International, Oxon, UK. pp. 723–811.

McGladdery, S.E., Bradford, B.C. and Scarratt, D.J., 1993b. Investigations into transmission of parasites of bay scallop, *Argopecten irradians* (Lamarck, 1819) during quarantine introduction into Canadian waters. J. Shellfish Res. 12:49–58.

McGladdery, S.E., Cawthorn, R.J. and Bradford, B.C., 1991. *Perkinsus karlssoni* n. sp. (Apicomplexa) in bay scallops *Argopecten irradians*. Dis. Aquat. Org. 10:127–137.

McGladdery, S.E., Drinnan, R.E. and Stephenson, M.F., 1993a. A manual of the parasites, pests and diseases of Atlantic Canadian bivalves. Can. Tech. Rep. Fish. Aquat. Sci. No. 1931: 121 p.

McLean, N., 1983. An echinocephalid nematode in the scallop *Argopecten aequisulcatus* (Mollusca: Bivalvia). J. Invertebr. Pathol. 42:273–276.

Medcof, J.C., 1949. Dark-meat and the shell disease of scallops. Prog. Rep. Atlant. Cst. Stns. 45:3–6.

Merrill, A.S., 1967. Shell deformity of molluscs attributable to the hydroid, *Hydractinia echinata*. Fish. Bull. Fish Wildl. Serv. U.S. 66:273–279.

Minchin, D. and Duggan, C.B., 1989. Biological control of the mussel in shellfish culture. Aquaculture 81:97–100.

Miyazaki, T., Goto, K., Kobayashi, T., Kageyama, T. and Miyata, M., 1999. Mass mortalities associated with a virus disease in Japanese pearl oysters *Pinctada fucata martensii*. Dis. Aquatic. Org. 37(1):1–12.

Moal, J., Samain, J.F., Corre, S., *Nicolas*, J.L. and Glynn, A., 1996. Bacterial nutrition of great scallop larvae. Aquacult. Int. 4(3):215–223.

Mori, K., Sato, W., Nomura, T. and Imajima, M., 1985. Infestation of the Japanese scallop *Patinopecten yessoensis* by boring polychaetes, Polydora, on the Okhotsk Sea coast of Hokkaido, especially in Abashiri waters. Bull. Jap. Soc. Sci. Fish. 51:371–380.

Morrison, C. and Shum, G., 1982. Chlamydia-like organisms in the digestive diverticula of the bay scallop, *Argopecten irradians* (Lmk). J. Fish Dis. 5:173–184.

Morrison, C. and Shum, G., 1983. Rickettsias in the kidney of the bay scallop, *Argopecten irradians* (Lamarck). J. Fish Dis. 6:537–541.

Mortensen, S.H., 1993a. Passage of infectious pancreatic necrosis (IPNV) through invertebrates in the aquatic food chain. Dis. Aquat. Org. 16:41–45.

Mortensen, S.H., 1993b. A health survey of selected stocks of commercially exploited Norwegian bivalve molluscs. Dis. Aquat. Org. 16(2):149–156.

Mortensen, S.H., Bachère, E., LeGall, G. and Mialhe, E., 1992. Persistence of infectious pancreatic necrosis virus (IPNV) in scallops *Pecten maximus*. Dis. Aquat. Org. 12(3):221–227.

Mortensen, S.H., Nilsen, R.K. and Hjeltnes, B., 1998. Stability of an infectious pancreatic necrosis virus (IPNV) isolate stored under different laboratory conditions. Dis. Aquat. Org. 33(1):67–71.

Mortensen, S.H., Torkildsen, L., Hernar, I., Harkestad, L., Fosshagen, A and Bergh, Oe., 1999. One million scallop, *Pecten maximus*, spat lost due to brittle worm, *Polydora* sp., infestation. Book of Abstracts. 12th International Pectinid Workshop, Bergen, Norway.

Morton, J.E., 1967. Molluscs. Hutchinson, London. 244 p.

Moyer, M.A., Blake, N.J. and Arnold, W.S., 1993. An ascetosporan disease causing mass mortalities in the Atlantic calico scallop, *Argopecten gibbus* (Linnaeus, 1758). J. Shellfish Res. 12(2):305–310.

Moyer, M.A., Blake, N.J., Darden, R.L. and Arnold, W.S., 1995. Mass mortality in the calico scallop *Argopecten gibbus* caused by a *Marteilia*. In: J. Barret, J-C. Dao and P. Lubet (Eds.). Pêches, Biologie et Aquaculture des Pectinides: 8e Atelier International Sur Les Pectinides, Cherbourg (France), 22–29 Mai 1991. Institut Francais de Recherche Pour l'Exploitation de la Mer, Dcom/Se, Plouzane (France), Actes Colloq. IFREMER, no. 17. pp. 41–44.

Nagasawa, K., Bresciani, J. and Lutzen, J., 1988. Morphology of *Pectinophilus ornatus*, new genus, new species, a copepod parasite of the Japanese scallop *Patinopecten yessoensis*. J. Crust. Biol. 8:31–42.

Nagasawa, K. and Nagata, M., 1992. Effects of *Pectenophilus ornatus* (Copepoda) on the biomass of cultured Japanese scallop *Patinopecten yessoensis*. J. Parasitol. 78:552–554.

Nagasawa, K., Takahashi, S., Tanaka, S. and Nagata, M., 1991. Ecology of *Pectenophilus ornatus*, a copepod parasite of the Japanese scallop *Patinopecten yessoensis*. Bull. Plankton Soc. Japan 1991:495–502.

Naidu, K.S., 1971. Infection of the giant scallop *Placopecten magellanicus* from Newfoundland with an endozoic alga. J. Invertebr. Pathol. 17:145–157.

Nicolas, J.L., Ansquer, D., Cochard, J.C., 1992. Isolation and characterisation of a pathogenic bacterium specific to Manila clam *Tapes philippinarum* larvae. Dis. Aquat. Org. 14(2):153–159.

Nicolas, J.L., Corre, S., Gauthier, G., Robert, R. and Ansquer, D., 1996. Bacterial problems associated with scallop *Pecten maximus* larval culture. Dis. Aquat. Org. 27(1):67–76.

Norton, J.H., Shepherd, M.A. and Prior, H.C., 1993. Papovavirus-like infection of the golden-lipped pearl oyster *Pinctada maxima*, from the Torres Strait, Australia. J. Invertebr. Pathol. 62(2):198–200.

O'Connor, S.J., Heasman, M.P. and O'Connor, W.A., 1999. Evaluation of alternative suspended culture methods for the commercial scallop, *Pecten fumatus* Reeve. Aquaculture 171:237–250.

OIE, 2003. Manual of Diagnostic Tests for Aquatic Animal Diseases. Office International des Epizooties, Paris, France.

Oliva, M., Herrera, H., Matulic, J. and Severino, B., 1986. Parasitismo en el ostión del norte *Chlamys (Argopecten) purpuratus* (Lamarck, 1819). Parasitología al Día 10:83–86. (in Spanish)

Ordzie, C.J. and Garofalo, G.C., 1980. Predation, attack success, and attraction to the bay scallop, *Argopecten irradians* (Lamarck) by the oyster drill, *Urosalpinx cinerea* (Say). J. Exp. Mar. Biol. Ecol. 47:95–100.

Ovsyannikova, I.I. and Levin, V.S., 1982. Growth dynamics of the barnacle *Solidobalanus hesperius* on valves of Yeso scallop in bottom culture conditions. Sov. J. Mar. Biol. 8(4):224–230.

Pérez-Urbiola, J.C. and S.F. Martínez-Díaz. 2001. *Stephanostomum* sp. (Trematoda: Acanthocolpidae), the cause of "pimientilla" disease in catarina scallop *Argopecten ventricosus* (*circularis*) (Sowerby II, 1842) in Baja California Sur, México. Journal of Shellfish Research 20: 107–109.

Perkins, F.O., 1996. The structure of *Perkinsus marinus* (Mackin, Owen ad Collier, 1950) Levine, 1978 with comments on taxonomy and phylogeny of *Perkinsus* spp. J. Shellfish Res. 15(1):67–87.

Pitcher, C.R. and Butler, A.J., 1987. Predation by asteroids, escape response, and morphometrics of scallops with epizoic sponges. J. Exp. Mar. Biol. Ecol. 112:233–249.

Reddiah, K. and Williamson, D.I., 1958. On *Modiolicola inermis* Canu and *Modiolicola maxima* (Thompson), lichomolgid copepods associated with pectinid lamellibranchs. Ann. Magazine Nat. Hist. Ser.13, Vol. I:689–701.

Riquelme, S.C., Araya, R. and Escribano, R., 2000. Selective incorporation of bacteria by *Argopecten purpuratus* larvae: implications for the use of probiotics in culturing systems of the Chilean scallop. Aquaculture 181:25–36.

Riquelme, S.C., Araya, R., Vergara, N., Rojas, A., Guaita, M. and Candia, M., 1997. Potential probiotic strains in the culture of the Chilean scallop *Argopecten purpuratus* (Lamarck, 1819). Aquaculture 154:17–26.

Riquelme, C., Hayashida, G., Vergara, N., Vasquez, A., Morales, Y. and Chavez, P., 1995. Bacteriology of the scallop *Argopecten purpuratus* (Lamarck, 1819) cultured in Chile. Aquaculture 138:49–60.

Riquelme, S.C., Toranzo, A.E., Barja, J.L., Vergara, N. and Araya, R., 1996. Association of *Aeromonas hydrophila* and *Vibrio alginolyticus* with larval mortalities of scallop (*Argopecten purpuratus*). J. Invert. Pathol. 67:213–218.

Robert, R., Miner, P and Nicolas, J.L., 1996. Mortality control of scallop larvae in the hatchery. Aquacult. Int. 4(4):305–313.

Robledo, J.A., Càceres-Martinez, J., Sluys, R. and Figueras, A., 1994. The parasitic turbellarian *Urastoma cyprinae* (Platyhelminthes: Urastomidae) from blue mussel *Mytilus galloprovincialis* in Spain: Occurrence and pathology. Dis. Aquatic Org. 18:203–210.

Ruiz-Ponte, C., Samain, J.F., Sanchez, J.L. and Nicolas, J.L., (1999) The benefit of a *Roseobacter* species on the survival of scallop larvae. Mar. Biotech. 1(1):52–59.

Russell, H.J., Jr., 1973. An experimental seed bay scallop stocking of selected Rhode Island waters. Rhode Island Division of Fish and Wildlife. 64 p.

Samain, J.F., Ansquer, D., Cochard, J.C., Daniel, J.Y., Delaunay, F., Jasq, E., Le Coz, J.R., Nicolas, J.L., Marty, Y and Moal, J., 1989. Growth, digestive enzymes and fatty acid patterns of *Pecten maximus* larvae, some food quality and bacteria interactions. Program of the First International Marine Biotechnology Conference, Japanese Society for Marine Biotechnology, Tokyo, Japan. pp. 84. (abstract only)

Sanders. M.J. and Lester, R.J., 1981. Further observations on a bucephalid trematode infection in scallops (*Pecten alba*) in Port Phillip Bay, Victoria. Aust. J. Mar. Freshwat. Res. 32:475–478.

Sato-Okoshi, W. and Okoshi, K., 1993. Microstructure of scallop and oyster shells infested with boring *Polydora*. Bull. Jap. Soc. Sci. Fish. 59(7):1243–1247.

Sherburne, S.W., 1982. First record of a fungus disease in any scallop - found in a deep-sea scallop, *Placopecten magellanicus*, from the Sheepscot River, Maine. Maine Dept. Mar. Resources, Res. Ref. Doc. 82/16. 4 pp.

Sherburne, S.W. and Bean, L.L., 1986. A synopsis of the most serious diseases occurring in Maine shellfish. American Fisheries Society, AFS Fish Health Section Newsletter 14:5.

Sieburth, J.M., Johnson, P.W. and Hargraves, P.E., 1986. Characterization of *Aureococcus anorexefferens gen. et sp. nov.* (Chrysophyceae): The dominant picoplankter during the summer 1985 bloom in Narragansett Bay, Rhode Island. Proceedings of Emergency Conference on "Brown Tide" Long Island, New York, 1986. State Dep. New York State, Albany, New York. pp. 5.

Sindermann, C.J., 1971. Predators and diseases of commercial marine Mollusca of the United States. Rep. Am. Malacological Union 1970:35–36.

Sindermann, C.J., 1984. Disease in marine aquaculture. Helgolander Meeresunters. 37:505–532.

Sindermann, C.J., 1990. Principal Diseases of Marine Fish and Shellfish. Vol. 2, Diseases of Marine Shellfish, 2nd edition. Academic Press, San Diego.

Singhas, L.S., West, T.L. and Ambrose, W.G. Jr., 1993. Occurrence of *Echeneibothrium* (Platyhelminthes, Cestoda) in the calico scallop *Argopecten gibbus* from North Carolina. Fish. Bull. 91(1):179–181.

Skeel, M.E., 1979. Shell-boring worms (Spionidae: Polychaeta) infecting cultivated bivalve molluscs in Australia. Proc. World Maricul. Soc. 10:529–533.

Sparks, A.K. and Chew, K.K., 1966. Gross infestation of the littleneck clam, *Venerupis staminea*, with a larval cestode (*Echeneibothrium* sp.). J. Invert. Pathol. 8:413–416.

Sprague, V., 1970. Some protozoan parasites and hyperparasites in marine bivalve molluscs. In: S.F. Sniesko (Ed.). A Symposium on Diseases of Fishes and Shellfishes. Am. Fish. Soc., Wash., Spec. Publ. No. 5. pp. 511–526.

Sprague, V. and Orr, P.E., 1955. *Nematopsis ostrearum* and *N. prytherchi* (Eugregarinina: Porosporidae) with special reference to the host-parasite relations. J. Parasit. 41:89–104.

Sprent, J.F., 1977. Ascaridoid nematodes of amphibians and reptiles: *Sulcascaris*. J. Helminth. 51:379–387.

Stevenson, R.N. and South, G.R., 1975. Observations on phagocytosis of *Coccomyxa parasitica* (Coccomyxaceae: Chloroccales) in *Placopecten magellanicus*. J. Invertebr. Pathol. 25:307–311.

Strongman, D.B., Morrison, C.M. and McClelland, G., 1997. Lesions in the musculature of captive American plaice *Hippoglossoides platessoides* caused by the fungus *Hormoconis resinae* (Deuteromycetes). Dis. Aquat. Org. 28:107–184.

Stunkard, H.W., 1938. The morphology and life cycle of the trematode *Himasthla quissentensis* (Miller and Northrup, 1926). Biol. Bull. Mar. Biol. Lab., Woods Hole 75:145–164.

Syasina, I.G., Vashchenko, M.A., Zhadan, P.M. and Karaseva, E.M., 1996. The state of gonads and development of progeny in the scallop *Mizuhopecten yessoensis* from polluted areas of Peter the Great Bay, Sea of Japan. Biol. Morya/Mar. Biol. 22(4):255–262.

Torkildsen, L., Samuelsen, O.B., Lunestad, B.T. and Berghe, O.E., 2000. Minimum inhibitory concentrations of chloramphenicol, florfenicol, trimethoprim/sulfadiazine and flumequin in seawater of bacteria associated with scallop (*Pecten maximus*) larvae. Aquacult. 185(1–2):1–12.

Tubiash, H.S., Chanley, P.E. and Leifson, E., 1965. Bacillary necrosis, a disease of larval and juvenile bivalve molluscs. J. Bact. 90:1036–1044.

Tubiash, H.S., Colwell, R.R. and Sakazaki, R., 1970. Marine vibrios associated with bacillary necrosis, a disease of larval and juvenile bivalve molluscs. J. Bact. 103:271–272.

Turner, H.J. Jr. and Hanks, J.E., 1959. Infestation of *Pecten irradians* by *Polydora*. Nautilus 72(4):109–111.

Uribe, E. and Etchepare, I., 1999. Effects of biofouling by *Ciona intestinalis* on suspended culture of *Argopecten purpuratus* in Baia Inglesa, Chile. Book of Abstracts. 12th International Pectinid Workshop.

Usheva, L.N., 1999. Histopathology of the adductor muscle in the scallop *Mizuhopecten yessoensis* from polluted areas of Peter the Great Bay, Sea of Japan. Russ. J. Mar. Biol. 25(5):412–416.

Wang, W., Luo, W., Xue, Q., Song, O., Zhu, J., Tan, J., Hou, Y and Zou, G., 1998. Pathological research on chlamydia-like organisms in the hepatopancreatic gland of the bay scallop, *Argopecten irradians* (Lamarck). Mar. Sci./Haiyang Kexue 3:23–25.

Warburton, F.E., 1958. Boring sponges, *Cliona* species, of eastern Canada, with a note on the validity of *C. lobata*. Can. J. Zool. 36:123–125.

Warner, R.W. and Katkansky, S.C., 1969. Infestation of the clam *Protothaca staminea* by two species of tetraphyllidean cestodes (*Echeneibothrium* spp.). J. Invert. Pathol. 13:129–133.

Wells, H.W. and Wells, M.J., 1961. Three species of *Odostomia* from North Carolina, with description of new species. Nautilus 74:149–157.

Wells, H.W. and Wells, M.J., 1962. The polychaete *Ceratonereis tridentata* as a pest of the scallop *Aequipecten gibbus*. Biol. Bull. Mar. Biol. Lab., Woods Hole 122:149–159.

Whyte, S.K., Cawthorn, R.J., MacMillan, R.J. and Despres, B., 1993a. Isolation and purification of developmental stages of *Perkinsus karlssoni* (Apicomplexa: Perkinsea), a parasite affecting bay scallops *Argopecten irradians*. Dis. Aquat. Org. 15:199–205.

Whyte, S.K., Cawthorn, R.J., McGladdery, S.E., MacMillan, R.J. and Montgomery, D.M., 1993b. Cross-transmission studies of *Perkinsus karlssoni* (Apicomplexa) from bay scallops *Argopecten irradians* to native Atlantic Canadian shellfish species. Dis. Aquat. Org. 17:33–39

Whyte, S.K., Cawthorn, R.J. and McGladdery, S.E., 1994. Co-infection of bay scallops *Argopecten irradians* with *Perkinsus karlssoni* (Apicomplexa, Perkinsea) and an unidentified coccidian parasite. Dis. Aquat. Org. 18:53–62.

Woods, C.M.C. and Hayden, B.J., 1998. An observation of the turbellarian *Paravortex* sp. i the New Zealand scallop *Pecten novaezelandiae* (Bivalvia: Pectinidae). New Zealand J. Mar. Freshwater Res. 32:551–553.

Xu, K., Lei, Y. and Song, W., 1995. Morphological studies on *Trichodina jadranica* Raabe, 1958, a scallop ciliate parasite (Protozoa: Ciliophora: Pertitricha). J. Ocean Univ. Qingdao 25(3):321–326.

Yevich, C.A. and Yevich, P.P., 1985. Histopathological effects of cadmium and copper on the sea scallop *Placopecten magellanicus*. Marine Pollution and Physiology: Recent Advances. Belle W. Baruch Libr. Mar. Sci. No. 13:187–198.

Zhang, X., Liao, S., Li, Y., Ji, W. and Xu, H., 1998. Studies on pathogenic bacteria (*Vibrio natriegens*) of *Argopecten irradians* Lamark. J. Ocean Univ. Qingdao/Qingdao Haiyang Daxue Xuebao 28(3):426–432.

AUTHORS ADDRESSES

Sharon E. McGladdery - Department of Fisheries and Oceans Canada, Aquaculture Science Branch, 200 Kent Street, Stn. 12W114, Ottawa, Ontario, Canada K1A 0E6 (E-mail: McgladderyS@dfo-mpo.gc.ca)

Susan M. Bower - Pacific Biological Station, Department of Fisheries and Oceans Canada, 3190 Hammond Plains Road, Nanaimo, British Columbia, Canada V9R 5K6 (E-mail: BowerS@dfo-mpo.gc.ca)

Rodman G. Getchell - Aquatic Animal Health, Department of Microbiology and Immunology, College of Veterinary Medicine, Cornell University, Ithaca, New York, USA 14853–6401 (E-mail: rgg4@cornell.edu)

Scallops: Biology, Ecology and Aquaculture
S.E. Shumway and G.J. Parsons (Editors)
© 2006 Elsevier B.V. All rights reserved.

Chapter 12

Scallop Ecology: Distributions and Behaviour

Andrew R. Brand

12.1 INTRODUCTION

There are some 400 known living species in the bivalve family Pectinidae, commonly known as scallops. They occur in all the seas of the world from polar regions to the tropics. Most of the commercially important species occur in the inshore waters of the continental shelves but scallops are found in waters of all depths from the intertidal zone down to some 7,000 m.

Scallops are perhaps best known for their ability to swim – an unusual adaptation for a bivalve mollusc - and the swimming escape responses of species like *Argopecten irradians* (Lamarck), *Aequipecten opercularis* (L.), *Pecten maximus* (L.) and *Placopecten magellanicus* (Gmelin) have long been the subject of investigation (e.g., Verrill 1897; Buddenbrock 1911; Bayliss et al. 1930; Belding 1931; Gutsell 1931; Baird 1958; Caddy 1968; Waller 1969; Moore and Trueman 1971; Thomas and Gruffydd 1971; Thorburn and Gruffydd 1979). The swimming habit of scallops is thought to have evolved through modification of pre-existing mantle cavity cleansing mechanisms in byssally attached monomyarian ancestors (Drew 1906; Yonge 1936). Probably all scallops form a byssus for attachment when young (Verrill 1897) but many of the larger and more active species (e.g., *Mizuhopecten yessoensis* (Jay), *Pecten fumatus* (Reeve), *Pecten maximus, Placopecten magellanicus*) later become free-living on the seabed and lose the ability to secrete a byssus when they reach a larger size. Many small and delicate-shelled forms (e.g., *Chlamys vitrea* (Gmelin), *Chlamys furtiva* (Lovén)) and species living in areas of strong current flow (e.g., *Chlamys islandica* (O.F. Müller), *Mimachlamys varia* (L.)) are capable of releasing the byssus and making active swimming movements but generally remain attached throughout life. Some species, like the small European *Chlamys distorta* (da Costa) or the massive *Hinnites multirugosa* (Gale) of the Pacific coast of North America, swim freely in early life but later becomes cemented by the right valve to the rock and subsequent growth is irregular, the shell taking on the configuration of the surface to which it is attached (Yonge 1951). The scallops then, by virtue of the number of species, their geographical distribution and the range of habitats they occupy, are a supremely successful group of bivalve molluscs, and important members of a number of benthic communities. They are extremely attractive and prominent animals with many species of considerable commercial importance, the flesh being considered a luxury food in many countries. For these various reasons, scallops have been the subject of much scientific research. However, despite the vast literature, detailed knowledge of the ecology is restricted to a few species and there are some surprising gaps in our understanding of other common species.

This chapter describes the geographical and local distribution of scallop populations, concentrating on the species of commercial importance. It also discusses various factors that affect distribution, including environmental conditions, ecological interactions with other species and relevant aspects of the behaviour of post-larval scallops, such as byssus attachment, orientations and movements. In reviewing current knowledge of these topics the aim is to present a comparative account of all species. The usual common names and geographical region of occurrence of the species referred to most frequently are given in Table 12.1. Of these, *Argopecten gibbus* (L.), *A. irradians*, *Aequipecten opercularis*, *Chlamys islandica*, *Pecten maximus* and *Placopecten magellanicus* are considered in the greatest detail because these commercially important species have received the most study.

Table 12.1

Common names and geographical region of occurrence of some commercially important scallop species.

Species	Common name	Geographical region
Aequipecten opercularis	Queen scallop	E. North Atlantic
Aequipecten tehuelchus	Vieira; Vieira tehuelche	W. South Atlantic
Amusium balloti	Southern saucer scallop	Indo-West Pacific
Amusium pleuronectes	Moon scallop	Indo-West Pacific
Argopecten gibbus	Calico scallop	W. North Atlantic, Gulf of Mexico
Argopecten irradians	Bay scallop	W North Atlantic, E. coast U.S.A
Argopecten purpuratus	Ostion	E. South Pacific
Chlamys islandica	Iceland scallop	Sub-arctic
Equichlamys bifrons	Queen scallop	S.W. South Pacific
Mimachlamys asperrima	Doughboy scallop	S. Australia
Mimachlamys varia	Black; variegated scallop	E. North Atlantic
Mizuhopecten yessoensis	Yezo scallop; hotate gai	W. North Pacific
Patinopecten caurinus	Weathervane scallop	E. North Pacific
Pecten fumatus	Commercial scallop	S. Australia
Pecten jacobaeus	Pilgrims scallop	Mediterranean
Pecten maximus	Great scallop; escallop; Coquille St. Jacques	E. North Atlantic
Pecten novaezelandiae	New Zealand scallop	New Zealand
Placopecten magellanicus	Sea scallop; giant scallop	W. North Atlantic

12.2 GEOGRAPHICAL DISTRIBUTION

12.2.1 Geographical Distribution of Commercially Important Species

   While pectinids occur commonly in all seas of the world, the species that are of large enough body size and occur in sufficiently dense aggregations to be commercially exploited are found mostly in high latitudes, between about 30° and 55° in both northern and southern hemispheres. For the main commercially exploited species, the distribution of fishable concentrations has generally been well established, but even for these species, reliable information on the full geographical distribution is often remarkably scant. For each species there is a geographical and bathymetric range where environmental conditions are generally suitable for survival. Towards the extremities of their range, environmental factors such as water temperatures become limiting and scallops are scarce. Where conditions are favourable, scallops frequently occur in dense local populations, which may be of sufficient extent and density to support commercial fisheries. In the following account the geographical distributions of the most important commercial species are considered in an order based on geographical regions.

12.2.1.1 North Atlantic species

12.2.1.1.1 Pecten maximus and P. jacobaeus

   The geographical distribution of the great scallop, *Pecten maximus*, is shown in Figure 12.1. This species occurs along the eastern coast of the North Atlantic from northern Norway south to the Iberian peninsula (Tebble 1966), and has been reported also off West Africa, the Azores, Canary Islands and Madiera (Mason 1983). It also extends a short distance into the Mediterranean for in some years there has been a small fishery for *P. maximus* off the Costa del Sol, extending from the mouth of the river Guadiaro as far east as La Caleta de Valez in the province of Malaga (Cano and Garcia 1985). Further east it is replaced by the closely related *P. jacobaeus,* which is heavily exploited in many places throughout the Mediterranean and Adriatic (Piccinetti et al. 1986; Mattei 1995). The bathymetric range of *P. maximus* is from just below the low water mark (Tebble 1966) to about 183 m (Forbes and Hanley 1853) but it is most common in water of 20–45 m. It is essentially a coastal species and occurs on bottoms of clean firm sand, fine gravel or sandy gravel, sometimes with an admixture of mud (Mason 1983). It occurs in sufficient densities for commercial exploitation in many relatively small locations off the coasts of France and the British Isles (Figure 12.1), but the principal fisheries are in the Baie de St. Brieuc, Baie de la Seine, eastern and western English Channel, the north Irish Sea all around the Isle of Man, the Clyde, West of Kintyre, Orkney, Shetland and the Moray Firth.

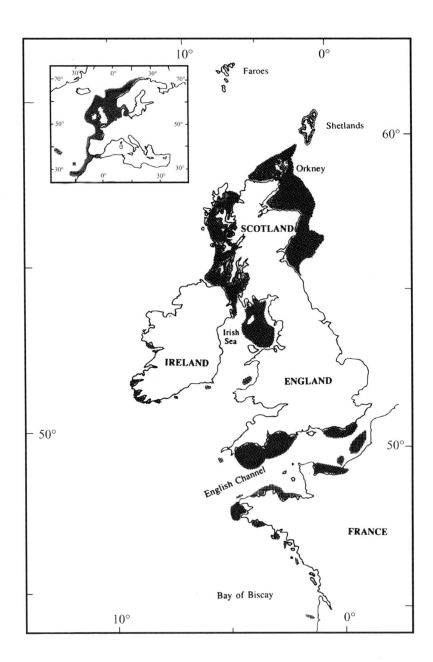

Figure 12.1. *Pecten maximus*. Principal fishing grounds around Britain and France (redrawn and modified from Mason 1983), together with the approximate geographical distribution of the species (inset).

12.2.1.1.2 Aequipecten opercularis

The other commonly exploited species in the eastern North Atlantic is the queen scallop, *Aequipecten opercularis*. This has a similar geographical range to *Pecten maximus* (Figure 12.2), having an approximate latitudinal range of 30°N to 70°N, but it extends further into the Mediterranean and Adriatic (Tebble 1966; Broom and Mason 1978; Piccinetti et al. 1986). It has a similar bathymetric range and is frequently found coincident with *Pecten maximus* on the same grounds but can live also on harder gravel and shelly bottoms because, unlike *Pecten*, it does not usually recess in the seabed (Mason 1983). The major aggregations that support commercial fisheries around the British Isles are relatively few in number and are widely separated (Rolfe 1973; Sinclair et al. 1985), notably in the western English Channel, Kish Bank, the north Irish Sea, Clyde, Orkney and Shetland, but it is abundant also in places in the North Sea, the Kattegat and around the Faroes (Ursin 1956).

12.2.1.1.3 Mimachlamys varia

A third European species, of some minor commercial value in France, is the black or variegated scallop, *Mimachlamys varia* L. (Letaconnoux and Audouin 1956; Shafee and Conan 1984). This small scallop is very common around the British Isles and has been reported from Denmark to the Iberian peninsula, in the Mediterranean and off the coast of West Africa to Senegal (Tebble 1966). It is widely distributed along the west coast of Ireland and is locally abundant within sheltered bays where it is found byssally attached to shells, stones, boulders or rock faces (Rodhouse and Burnell 1979). In the north Irish Sea it is a prominent member of the epifauna on banks of the horse mussel, *Modiolus modiolus*, together with *Chlamys distorta* (Jones 1951).

12.2.1.1.4 Placopecten magellanicus

On the western side of the North Atlantic the sea scallop, *Placopecten magellanicus* (Gmelin), has a geographical range that extends from Pistolet Bay, Newfoundland and the north shore of the Gulf of St Lawrence to Cape Hatteras, North Carolina (Figure 12.3) (Posgay 1957; Squires 1962; Bourne 1964). The usual depth range for this species is from 18–110 m but it occurs in shallower water in the northern part of the range and has been reported from depths as shallow as 2 m (Naidu and Anderson 1984). At the southern end of the range it is found in much deeper water, usually in excess of 55 m (Bourne 1964), while the known bathymetric range extends down to 384 m (Mcrrill 1959). Sea scallop populations of sufficient extent and density to support commercial fisheries occur from Virginia Capes (36°50'N) to Port au Port Bay, Newfoundland (48°40'N), with fishing grounds off Virginia Capes; off New York City; around Block Island, Rhode Island; on Georges Bank; off Cape Cod; along the coast of Maine; the Bay of Fundy, particularly off Digby, Nova Scotia; the southern Gulf of St Lawrence; Scotia Shelf (Browns Bank, Emerald Bank, Sable Island Bank, Banquereau); Port au Port Bay and St Pierre Bank (Bourne 1964; Naidu and Anderson 1984). Of these, the Georges Bank fishing grounds are by far the most productive and have been exploited consistently for more than 50 years (Serchuk et al. 1979; Sinclair et al. 1985; Murawski et al. 2000).

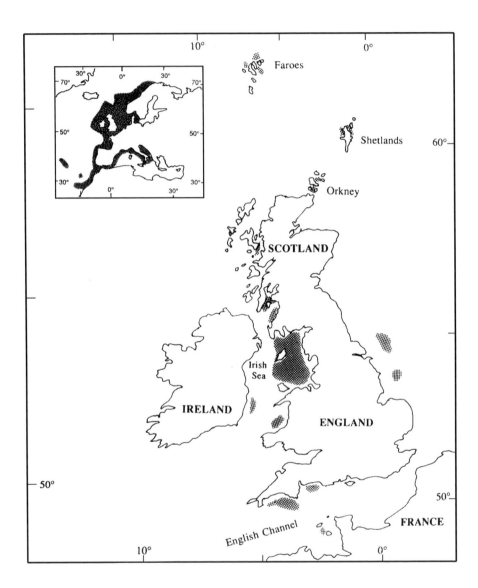

Figure 12.2. *Aequipecten opercularis.* Principal fishing grounds around Britain (redrawn and modified from Mason 1983), together with the approximate geographical distribution of the species (inset).

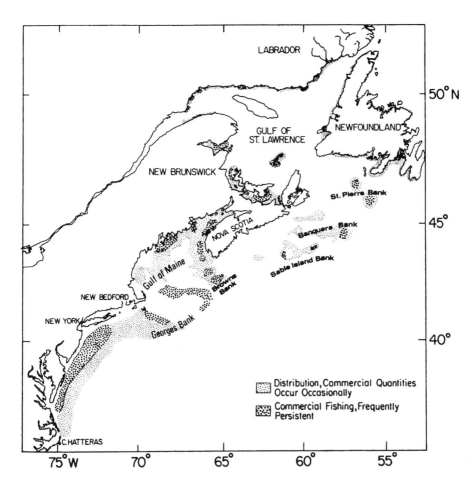

Figure 12.3. *Placopecten magellanicus*. Principal fishing grounds (map kindly provided by S. E. Shumway).

12.2.1.1.5 Argopecten irradians

The bay scallop occurs along the eastern coast of the U.S.A. and is recorded from Cape Cod, Massachusetts to the Laguna Madre, Southern Texas (Clarke 1965) (Figure 12.4), but present day distributions are not as extensive as previously reported. Three subspecies have long been recognised, based on shell morphometrics (Clarke 1965; Waller 1969). The geographical variation in morphology would appear to have a strong genetic basis (Wilbur and Gaffney 1997) but genetic variation may not coincide with variations in the morphological criteria used in subspecies designation (Bricelj et al. 1987b; Blake 1994; Blake and Graves 1995) and further studies are required to clarify the taxonomic status of some populations. The northern subspecies, *A. i. irradians*, extends

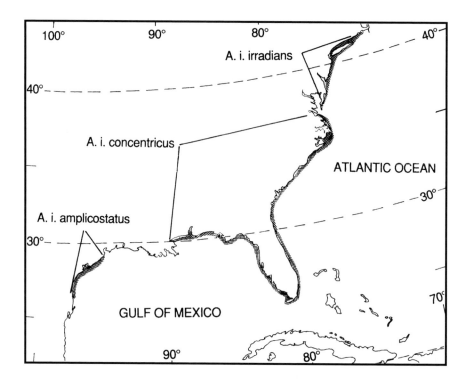

Figure 12.4. *Argopecten irradians*. Approximate geographical distribution of the three subspecies along the eastern coast of the U.S.A. (Redrawn and modified from Broom 1976).

from the north shore of Cape Cod, Massachusetts (42°N) to the area between New Jersey and Maryland (39–40° N) where it intergrades with *A. i. concentricus* (Clarke 1965). This subspecies then extends south, but is now uncommon on the Atlantic coast of Florida, and into the Gulf of Mexico where the most reliable westerly record of its presence is around the Chandeleur Islands of Louisiana (29°N 88°W) (Waller 1969; Broom 1976). Surprisingly for such a prominent and easily cultured bivalve, the western limit of *A. i. concentricus*, and the degree to which it intergrades with the most southerly subspecies, *A. i. amplicostatus*, is not well known. *A. i. amplicostatus* itself has been reported to occur from Galverston, Texas (26°N 95°W) to the Laguna Madre, southern Texas (26°N 97°W) (Waller 1969), but the southern limit of this subspecies is not known. A fourth subspecies, *Argopecten irradians taylorae*, with a restricted distribution in Florida Bay and west of the Florida Keys, has been proposed (Petuch 1987), but subsequent genetic and morphometric analyses have failed to separate this from Florida populations of *A. i. concentricus* (Marelli et al. 1997). As the common name implies, the bay scallop is found in protected coastal bays, sounds, estuaries and the inshore sides of barrier islands. It generally occurs very close inshore, in water of less than 12 m, typically on sandy substrates amongst the eelgrass, *Zostera*, though it does extend down to about

20 m (Belding 1931; Gutsell 1931; Marshall 1960; Sastry 1962; Cooper and Marshall 1963; Bologna and Heck 1999; Peterson et al. 2001).

12.2.1.1.6 Argopecten gibbus

The calico scallop, *Argopecten gibbus*, is closely related to the bay scallop; it is found also in the western North Atlantic and the Gulf of Mexico but in deeper water further offshore. It has a more southerly distribution, with a geographical range extending from the northern side of Greater Antilles (about 18°N) and throughout the Gulf of Mexico to Bermuda, and slightly north of Cape Hatteras, North Carolina (36°N) (Allen and Costello 1972) (Figure 12.5). Its precise southerly limit and distribution in the West Indies is unknown due to taxonomic confusion with other species, such as *A. nucleus*, which has an overlapping distribution extending from southeastern Florida to the coasts of Colombia and Venezuela (Waller 1969). *Amusium papyraceum, A. laurenti* and *Euvola zic-zac* are also commercially exploited at different places within the Caribbean (Penchaszadeh and Salaya 1985) but these species are of limited local importance and are not the subject of major fisheries, like *Argopecten gibbus*. Calico scallops are most abundant near coastal prominences; the largest, commercially fished concentrations occur over a large area off the east coast of Florida, near Cape Kennedy, though the location of individual scallop beds varies from year to year.

Lesser concentrations are found near Cape Lookout, North Carolina and in the northwest Gulf of Mexico near Cape San Blas, Florida. More sporadically fished grounds have also been reported off Savannah, Georgia and along the west coast of Florida from off Egmont Key to Key West (Cummins 1971; Allen and Costello 1972; Allen 1979). The calico scallop is found in warm, open marine waters and has been classified as a subtropical tolerant species (Miller and Richards 1980). It has been reported to occur in depths of less than 2 m down to 277 m but the commercial concentrations generally occur from about 20–70 m (Waller 1969; Allen and Costello 1972), with the largest concentrations off Cape Kennedy at depths of 33–42 m (Miller and Richards 1980).

12.2.1.2 Sub-arctic species

12.2.1.2.1 Chlamys islandica

The distribution of the Iceland scallop overlaps with that of *Placopecten magellanicus* in the western North Atlantic, though the two species apparently occur together on the same beds only on St. Pierre Bank, the westward extension of the Grand Banks of Newfoundland (Naidu and Anderson 1984). *C. islandica*, however, is a sub-arctic species, with its main distribution in the sub-arctic transition zone (Ekman 1953). On the western side of the Atlantic *C. islandica* occurs in greatest abundance in the Straits of Belle Isle, along the Labrador shelf, St Pierre Bank and over most of the Grand Banks (Naidu et al. 1982), but it has also been reported in fishable concentrations from as far south as Nantucket, Massachusetts (41° 32'N) (Naidu and Anderson 1984) and to occur as far north as the Parry Islands (75°N) (Wiborg 1963). It is found along East and West

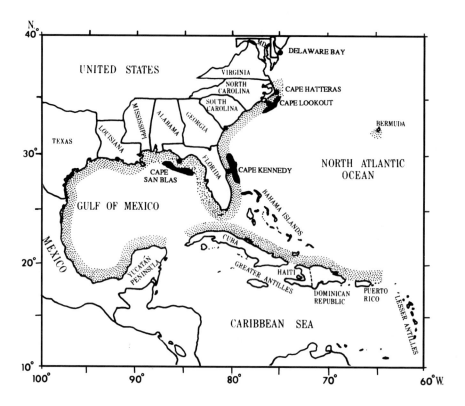

Figure 12.5. *Argopecten gibbus.* Approximate geographical distribution (stippled areas) and principal fishing grounds (black areas). (From Allen 1979).

Greenland, around Jan Mayen and Iceland, and in the eastern North Atlantic and Arctic oceans Iceland scallops occur along the Norwegian coast, mainly north of the Arctic Circle, western and northern Spitzbergen, Bear Island, Hopen Island, Murman coast, White Sea, west coast of Novaja Zemlja and the Kara Sea (Ockelmann 1958; Wiborg 1963; Lubinsky 1980; Hansen and Nedreaas 1986; Sundet and Fevolden 1994). *C. islandica* has also been recorded from the Bering Strait and extends into the Pacific Ocean, the Okhotsk Sea and down the eastern Pacific coast as far south as Puget Sound (Mottet 1979). However, Waller (1991) considers that *C. islandica* is limited to the sub-arctic regions of the North Atlantic and the records from the North Pacific region are due to taxonomic confusion with closely related species. *C. islandica* is found mainly in depths of 10–110 m, though it has been taken in trawl catches down to 525 m (Hansen and Nedreaas 1986). In Newfoundland, fishable concentrations occur mainly in waters deeper than 55 m, on hard bottoms with variable substrate composition consisting of sand, gravel, shell fragments and stones (Naidu and Anderson 1984). Wiborg (1963) comments that the densest populations along the Norwegian coast are found in fjords with one or more sills at the entrance, in places with comparatively strong currents. This latter observation was noted also by Naidu (1982) for Iceland scallop populations in the Gulf of St. Lawrence.

12.2.1.3 North Pacific species

12.2.1.3.1 Patinopecten caurinus

The weathervane scallop, *Patinopecten caurinus* Gould, is found in the eastern North Pacific. It occurs along the west coast of North America from San Francisco (37°N) to the Gulf of Alaska (60°N) and extends also into the Bering Sea (Grau 1959; Mottet 1979; Bernard 1983). It has been fished commercially off the coast of Alaska since 1967 (Haynes and Powell 1968; Haynes and McMullin 1970), initially off Kodiak and Yakutat, but extending later to grounds off Dutch Harbor and other areas such as south-east Alaska, Cook Inlet, Alaska Peninsula, the Bering Sea and Prince William Sound (Kruse and Shirley 1994; Shirley and Kruse 1995; Kruse et al. 2000). Large commercial concentrations were found off Coos Bay, Oregon in 1981 but were depleted in one year (Starr and McCrea 1983). Small beds have occasionally been exploited in the region of Puget Sound and it is frequently taken incidentally in other fisheries off the coasts of Canada, Washington and Oregon. *P. caurinus* has been reported from the intertidal zone down to depths of 300 m, but is most abundant between 45–130 m on beds of mud, clay, sand and gravel (Hennick 1973; Kruse and Shirley 1994).

12.2.1.3.2 Mizuhopecten yessoensis

On the western side of the North Pacific several species of scallops are the subject of important commercial fisheries off the coasts of Japan, Korea and Russia. Of these, *Mizuhopecten yessoensis* is by far the most important and, with the decline of natural fisheries in the 1960's, enormous quantities of this species have since been produced in both hanging and seabed culture (Ventilla 1982; Ito 1991). *M. yessoensis* is a cold water species that occurs around Hokkaido and the northern part of Honshu Island, with a natural distribution that extends as far south as Tokyo Bay on the Pacific coast and Toyama Bay in the Sea of Japan. It is found also in the Kurile Islands, the Sea of Japan along the coasts of Korea and the USSR, off Sakhalin Island and extending north into the Sea of Okhotsk (see distribution maps in Ito 1991). The Japanese commercial fisheries were mainly confined to the Okhotsk Sea coast of Hokkaido, Volcano Bay and Mutsu Bay in Aomori but, with the development of culture techniques, scallops are now produced in other areas further south (Motoda 1977). Fishing and cultivation of *M. yessoensis* in Russia is mainly in the Primorye region, south Sakhalin and the southern Kurile Islands (Kalashnikov 1991). *M. yessoensis* is a coastal species, commonly found in water of up to 40 m in depth but a maximum depth of 82 m has been reported in Peter the Great Bay (Skarlato 1981). It occurs where the bottom is fairly hard, with little mud (Imai 1980).

12.2.1.3.3 Pecten albicans, Mimachlamys nobilis and Chlamys farreri

These three species are of commercial value for fishing or cultivation in Japan and elsewhere in the western North Pacific, but they are not as abundant as *Mizuhopecten yessoensis* (Ventilla 1982). *P. albicans* Schröter and *Mimachlamys nobilis* (Reeve) are predominantly southern species found in shallow inshore reef areas, from the southern part of Hokkaido in Japan to Korea, Taiwan and China, with *M. nobilis* extending as far south as Indonesia (Ito 1991; Lou 1991). *Chlamys farreri* (Jones & Preston) has been reported from the north of the Japanese island of Honshu (Ventilla 1982) to the Fujian province in China at latitude 25° N (Lou 1991). Its distribution is patchy along the coast of China but it is fished and cultivated in many locations.

12.2.1.4 Southern hemisphere species

12.2.1.4.1 Pecten fumatus

Compared with the northern hemisphere, the geographical distribution of scallop species in the southern hemisphere is not so well known. There are established fisheries for several species of scallops around the coasts of Australia, in both the temperate and tropical regions (Figure 12.6) (Gwyther et al. 1991). Large commercial scallops of the genus *Pecten* occur from Newcastle, New South Wales, southward to Bass Strait, around Tasmania, all along the south coast and up the west coast as far north as Shark Bay (Fairbridge 1953; Olsen 1955; Young and Martin 1989). In the past these were considered to be four different species on the basis of shell morphology: *Pecten fumatus* (Reeve) occurring in New South Wales, *P. alba* (Tate) in Victoria and South Australia, *P. (= Notovola) meridionalis* (Tate) in Tasmania and *P. modestus* (Reeve) in Western Australia. However, electrophoretic studies of protein variation suggest that at least the first three of these are synonymous and should be included in the single species complex, *P. fumatus* (Woodburn 1987, 1989; Gwyther et al. 1991), known locally as the commercial scallop. The western limit of the distribution of *P. fumatus* and the southern and eastern limits of *P. modestus* are unclear. Throughout its range *P. fumatus* occurs in enclosed bays like Jervis Bay and Port Phillip Bay, as well as in exposed oceanic situations like the Banks Strait. It occurs on bottom substrates varying from muddy sand to coarse sand, in depths of 7–60 m. *P. modestus* occupies similar habitats in Western Australia, where it is found in 14–22 m depth in Cockburn Sound and 16–35 m in Geographe Bay. Important fishing grounds for *P. fumatus* have been located in Port Phillip Bay, off Lakes Entrance, in the Bass Strait, off the Furneaux Group and in several areas around the coast of Tasmania, including D'Entrecasteaux Channel. None of these grounds have supported a fishery for a very long period before the stocks have been depleted (Anonymous 1984b; Pontin and Millington 1985; Gwyther 1989; Young and Martin 1989) and, despite management efforts to stabilise these fisheries, annual production is very variable due to the irregularity of recruitment (Gwyther et al. 1991; Coleman 1998). There have also been small fisheries for *P. fumatus* in southern New

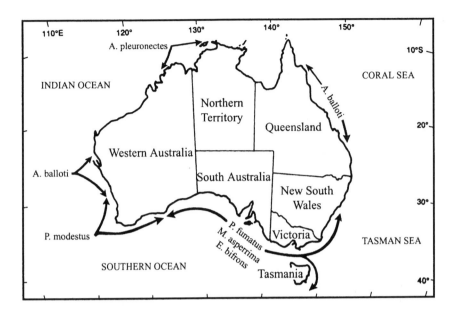

Figure 12.6. Scallop species exploited around the coasts of Australia.

South Wales (Jervis Bay and Bermagui) and central South Australia (Coffin Bay and Boston Bay), and for *P. modestus* in Cockburn Sound in Western Australia.

12.2.1.4.2 Mimachlamys asperrima and Equichlamys bifrons

In some areas of Australia commercial scallops are fished together with the doughboy scallop, *Mimachlamys asperrima* (Lamarck), and the queen scallop, *Equichlamys bifrons* (Lamarck), but these scallops are of minor commercial importance (Pontin and Millington 1985; Young and Martin 1989; Wolf and White 1995). *Mimachlamys asperrima* is by far the most abundant of these two species and in many areas it is more common than the commercial scallop, *Pecten fumatus*. It is distributed from Geographe Bay in Western Australia, around Southern Australia and Tasmania to the New South Wales coast (Figure 12.6) (Young and Martin 1989). It occurs in similar habitats to *Pecten fumatus*, in depths ranging from 7–69 m, but is usually byssally attached throughout life to solid objects within the bottom substrate. *Equichlamys bifrons* is found in waters of 2–40 m deep off New South Wales, Victoria, South Australia and Tasmania. It occupies a similar habitat to the other two species but is most common on coarse bottom substrates. In shallow bays of the D'Entrecasteaux Channel and Huon river estuary, *E. bifrons* is almost completely restricted to seagrass beds and rarely occurs on adjacent sandy areas (Wolf and White 1997). Unlike *M. asperrima*, the adults do not remain byssally attached but recess into shallow depressions in the seabed.

12.2.1.4.3 Amusium balloti and A. pleuronectes

In the tropical and subtropical waters of the Indo-Pacific several species of the genus *Amusium* are fished commercially. *A. balloti* and *A. pleuronectes* are common species with wide distributions throughout the region, the latter being recorded from the Indian Ocean, South China Sea, Indo-China, Japan, the Philippines, New Guinea, Indonesia, Java and Australia (Morton 1980). The Australian stocks of each have been assigned to two subspecies: the southern saucer scallop *Amusium japonicus balloti* (Bernardi) and the moon scallop *Amusium pleuronectes australiae* Habe.

The geographical range of *A. j. balloti* in Australia extends from Hervey Bay in southern Queensland, around the north and west of tropical Australia to Esperance on the south coast of Western Australia (Young and Martin 1989). It is found in water of 10–75 m depth but is most abundant at 25–55 m. It is fished commercially by otter trawlers in Queensland in the region from Mackay to the southern end of Fraser Island (21°S to 25°S) (Dredge 1985a; Dichmont et al. 2000). The same species is also fished in Western Australia, where it occurs in commercial densities in Shark Bay, around the Abrothos Islands and as far south as Rottnest Island (32°S) (Anonymous 1984a; Anonymous 1984b; Joll and Caputi 1995).

Amusium pleuronectes has a wide distribution around northern and western Australia but is found in shallower water than *A. balloti*. In Northern Queensland it is most abundant on firm, sandy, carbonate bottoms in depths of 12–24 m. It is fished commercially over a wide area, mainly as a by-catch of prawn trawlers, but particularly off Darwin in the Northern Territory and Admiralty Gulf in Western Australia (Sanders 1970; Young and Martin 1989). *A. pleuronectes* is the most common and abundant species of scallop in the Philippines where it occurs on sandy-mud to mud bottoms at 18–40 m depth (Llana 1983, 1985; Del Norte 1991). It is also common along the continental shelf of southern China, and Morton (1980) suggests that it migrates into the shallow coastal waters of Hong Kong to breed since it occurs in the markets only in the winter months when it is sexually mature.

12.2.1.4.4 Pecten novaezelandiae

In the temperate waters of New Zealand, *Pecten novaezelandiae* occurs sporadically around the entire coastline including Stewart Island and the Chatham Islands (34°S – 47°S). It is found on a wide variety of substrates in semi-estuarine and coastal waters from low tide to at least 90 m (Bull 1991). It is fished commercially in various locations but the main beds are at the north end of South Island (Golden Bay, Tasman Bay and the Marlborough Sound) and the north of North Island (Bream Bay, Spirits Bay, the Bay of Plenty, Hauraki Gulf) (Tunbridge 1968; Bull 1989, 1991). Since 1989, a rotational fishing regime, coupled with extensive stock enhancement, has been operated with great success in Tasman Bay and Golden Bay to stabilise the fishery and achieve environmental and commercial sustainability (Arbuckle and Metzger 2000; Dredge et al. 2002).

12.2.1.4.5 Argopecten purpuratus and A. ventricosus

The ostion, *Argopecten purpuratus* Lamarck, is a scallop of considerable commercial importance in the eastern Pacific, its range extending on both sides of the equator. It is fished and cultivated in Chile and Peru (Wolff and Wolff 1983; Wolff 1988; Piquimil et al. 1991; Stotz 2000; Wolff and Mendo 2000). The natural distribution of this species is from Corinto, Nicaragua (12°30'N) to Valparaiso, Chile (33°S), occurring in semi-protected bays with sedimentary substrates (Grau 1959; Waller 1969; Piquimil et al. 1991), however it has been introduced for cultivation into sheltered bays much further south (Gonzalez et al. 1999). According to Olsson (1961), it is a typical cold-water scallop of the Humboldt current, occurring most abundantly along the coast of Peru, southward from the Bay of Sechura. The distribution of *A. purpuratus* overlaps with that of the more northerly *Argopecten ventricosus* Sowerby (previously but erroneously called *Argopecten circularis* (Waller 1995)), which extends from Paita, Peru (5°S) to Monterey Bay, California (37°N) (Grau 1959; Waller 1969). *A. ventricosus* is particularly abundant in some sheltered locations on both the Pacific and Gulf coasts of Baja California, where it is fished by divers who also exploit *Pecten vogdesi* and *Nodipecten subnodosus* in some areas (Félix-Pico 1991; Félix Pico et al. 1999). This latter species is now considered to have great potential for cultivation in Mexico, due to its large size and fast growth rate (Félix-Pico et al. 2003; Racotta et al. 2003)

12.2.1.4.6 Aequipecten tehuelchus

Aequipecten tehuelchus (d'Orbigny) was for many years the only scallop in the southwestern Atlantic that was commercially exploited. It is a warm temperate species that ranges from Rio de Janeiro, Brazil (23°S) to Camarones, Argentina (45°S), with commercial densities occurring mostly at depths greater than 15 m (Ciocco 1991; Orensanz et al. 1991b). The fishery is mostly confined to the San Matías and San José Gulf's and has developed strongly since 1968. However, annual landings are highly variable due to irregular recruitment, intense harvesting and sporadic catastrophic stranding mortalities caused by storms (Iribarne et al. 1991; Orensanz et al. 1991b).

12.2.1.4.7 Zygochlamys patagonica

Z. patagonica is a cold temperate species of the continental shelf of Patagonia (Orensanz et al. 1991b), with an extensive distribution off the coasts of Chile (Piquimil et al. 1991), Argentina (Waloszek 1991; Lasta and Bremec 1997) and Uruquay (Gutierrez and Defeo 2003). It extends from Puerto Montt (41°S) in the Pacific Ocean south to Cape Horn and northwards in the Atlantic Ocean from the Burwood Bank and the Falkland Islands to the estuary of the Rio de la Plata (36°S) (Waloszek 1991). Off the coast of Chile it occurs from 2–40 m in protected bays, inlets and channels (Piquimil et al. 1991), but on the Argentinian and Uruquayan shelf it has a very extensive offshore distribution with the main concentrations from 60–80 m depth (Orensanz et al. 1991b; Lasta and Bremec 1997). There is a long-established dive fishery for *Z. patagonica* in Chile

(Piquimil et al. 1991), but the extensive offshore populations on the Argentinean shelf have only been exploited by dredging since 1996 (Lasta and Bremec 1999).

There are other species of scallop that occur in abundance but their geographical distribution is not well known as they have not been commercially exploited or are remote from areas of regular scientific activity. For example, in the southeastern Atlantic, stocks of *Pecten sulcicostatus* have been surveyed off South Africa that may prove to be worth exploiting (de Villiers 1976). Further south, the Antarctic scallop, *Adamussium colbecki*, has a circumpolar distribution with a depth range that extends from just below the sea ice to 1500 m (Dell 1972; Berkman 1991). The highest densities are usually found in shallow water near to shore (Stockton 1984), but other factors such as the persistency of sea ice also play a part (Chiantore et al. 2001). Antarctic scallops have never been exploited and, as pointed out by Berkman (1991), this makes them useful baseline measures for assessing the potential impacts of human activities.

12.2.2 Factors Affecting Geographical Distribution

The geographical distribution of a species depends on many interrelated factors but for most marine animals temperature is the primary factor that limits the overall geographical range. Temperature can directly or indirectly affect the survival of larvae and adults, and is known to influence reproduction though its effects on gamete maturation, spawning, embryonic development, the length of planktonic life and larval settlement (Kinne 1970). There is a very large literature concerning the effects of temperature on the ecology, physiology and distribution of marine organisms (e.g., see Kinne 1970; Newell 1970; Hawkins et al. 2003), including many studies on scallops.

Placopecten magellanicus is a cold-water species with a temperature optimum of about 10°C (Posgay 1953). Dickie (1958) found the upper lethal temperature for this species to range from 20–24°C, depending on the acclimation temperature. He also found that the upper lethal temperature is raised quite rapidly (1.7°C per day) by acclimation to higher temperature, but the loss of acclimation to high temperature is very slow and appears to take as long as 3 months. Sudden increases or decreases in temperature, too small to be lethal, both caused debility, the scallops gaping, secreting large quantities of mucus and becoming quiescent. Dickie and Medcof (1963) and Bourne (1964) discuss the ecological relevance of these findings. At the northern end of the range *Placopecten* occurs in shallow depths, where the water is warmest. The distribution in this part of the range is apparently determined largely by low summer temperatures, which fail to reach the spawning threshold or prolong larval development, resulting in recruitment failure. In the more southerly part of the range, sea scallops live at greater depths where the water remains cold, separated from the warm surface layers by a thermocline. It has been suggested that many of the sudden mass mortalities of sea scallops that have been reported on beds at intermediate or shallow depths in the south-western Gulf of St Lawrence result from perturbations to the position of the thermocline which subject scallops to sudden large changes in temperature (Dickie and Medcof 1963). These may be of sufficient extent and duration to be the direct cause of death but, more often, thermal shock increases mortality indirectly by inhibiting the scallop's normal escape reactions and making them more vulnerable to predation. Such reactions will cause complex interactions between predator and prey species, depending on their respective thermal tolerances.

A similar experimental study of upper temperature tolerance was carried out by Paul (1980b) on the queen scallop, *Aequipecten opercularis*. He also found that the upper lethal temperatures (LD50-48h) increased markedly in the range 19–25°C depending on the acclimation temperature. By investigating the responses of three size groups he was also able to show that very young queens were more resistant to high temperatures than animals of intermediate or large size (Figure 12.7) and showed less effect of acclimation. The greater thermal resistance of young animals agrees with previous findings for some other groups (Huntsman and Sparks 1924; Brett 1956).

Paul (1980b) also carried out experiments to study the effects of temperature on the rate of byssus formation in *Aequipecten opercularis*, which he found to be at a maximum at 18°C. This corresponds fairly closely with the results of McLusky (1973), who found that the rates of oxygen consumption and filtration were at a maximum at 15°C but fell sharply at 20°C, which he suggested was near the upper lethal temperature for this species. Ursin (1956) made a detailed study of the distribution of *A. opercularis* populations in relation to environmental temperatures. He suggests that temperatures below 2 or 3°C are lethal, since North Sea populations were severely affected when winter temperatures fell to this level in 1947. In the area he investigated, *Aequipecten* was not found at localities where summer temperatures normally exceed 17°C. The temperature range of 2–17°C is characteristic of a boreal species but *Aequipecten opercularis* extends also into the Mediterranean and Adriatic, where it is widely distributed in shallow water and must experience temperatures of 13–26°C. The Mediterranean form appears to be morphologically different to the boreal and the environmental differences suggest that they may also be physiologically distinct.

Environmental temperature appears also to play an important part in determining the geographical distribution of *Chlamys islandica*. This sub-arctic species generally occurs in temperatures ranging from -1.3°C to 8°C and Wiborg (1963) considers that its disappearance from areas where it was once abundant is due to the steady rise in temperature in northern waters, which began in about 1930. Similarly, the geographical distribution of *Argopecten gibbus* is probably limited primarily by temperature (Waller 1969). This species is one of a faunal assemblage in the northern Gulf of Mexico that occurs in waters deep enough to avoid winter cooling (Parker 1956). The temperature range for *A. gibbus* is given as 9.9°C to 33°C (Waller 1969; Allen and Costello 1972), but Vernberg and Vernberg (1970) found that calico scallops collected off North Carolina, near the northern end of their range, did not survive 48 hours at experimental temperatures of 10°C and suggest that the species has tropical affinities.

The geographical distribution of many scallop species is clearly associated with the thermal characteristics of the major ocean current systems. Thus, the cold Labrador current allows the southward extension of the sub-arctic *Chlamys islandica* down the western Atlantic coast, while its junction with the Gulf Stream in the region of Cape Hatteras marks the southern limit of *Placopecten magellanicus* and the northern limit of *Argopecten gibbus*. Similarly, in the western Pacific, the current systems determine the

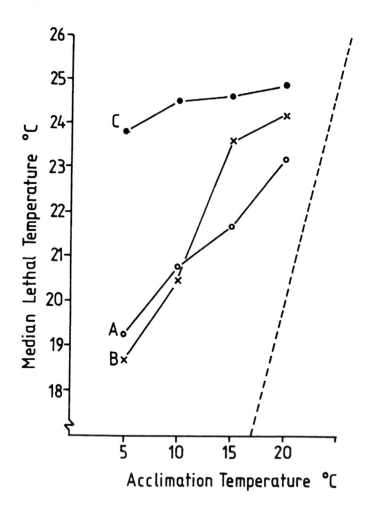

Figure 12.7. Temperature tolerance of *Aequipecten opercularis*. Median lethal temperatures (LD$_{50}$ 48-h) for three size groups (A, >60 mm; B, 30–40 mm; C, 5–10 mm shell height), acclimated to different temperatures. The dashed line shows where acclimation temperature = median lethal temperature. (From Paul 1980b).

geographical distribution of scallop species around the coast of Japan. *Mizuhopecten yessoensis* occurs naturally in the regions influenced by cold currents in the winter and warm currents in the summer (for maps showing these currents see Ito 1991). This species cannot tolerate high temperatures and the temperature range for survival appears to be 5–23°C (Ventilla 1982). The ecological effect of temperature as a limiting factor has been dramatically demonstrated by the very high mortalities obtained during attempts to cultivate *M. yessoensis* in the south of Honshu, where water temperatures in May and June exceed 22°C. In contrast, high mortalities of the warm water *Chlamys nobilis* can occur in winter if the temperature falls below the low temperature survival threshold of 8°C.

While temperature is obviously a major factor determining the overall geographical range of a scallop species, it interacts with numerous other factors such as water depth, substrate type, food availability, salinity, turbidity and the occurrence of competitors and predators in determining the local distribution. For species, like scallops, in which the mobility of the adult is very limited, the geographical distribution of a population is likely to be governed mainly by barriers to the dispersal of the larvae (Macleod et al. 1985). Patterns of water movement and other hydrographic features are therefore likely to be of major importance. Furthermore, since scallops have relatively long-lived pelagic larvae, the spatial and temporal stability of scallop populations will be dependent on the nature and stability of the hydrographic regime. The interplay of these various factors results in the local aggregation of scallops into particular areas of high density within their geographical range, which is a characteristic feature of scallop populations.

12.3 LOCAL DISTRIBUTION

12.3.1 Spatial Distribution

All scallop species have an aggregated distribution within their geographical range but before discussing the nature of spatial distribution patterns it is necessary to define the general terms used here to describe the scale or magnitude of aggregations. As shown in Figures 12.1, 12.2, 12.3 and 12.5, for four commercially fished species, there are a limited number of major areas within the geographical range of each species where the population is sufficiently abundant to support a commercial fishery. Such areas will be referred to as 'grounds' and are usually widely separated by areas that are environmentally unsuitable for the species. For example, the Western Channel, the north Irish Sea, the Clyde and West of Kintyre are *Pecten maximus* grounds, which are typically large areas separated by tens or even hundreds of kilometres. The absolute size of a ground may vary enormously, from an isolated bay of a few km^2, to the Georges Bank ground for *Placopecten magellanicus* which has an area of more than 31,000 km^2. Within each ground there is usually a number of regions, typically of an area of several km^2, where scallop abundance is higher than elsewhere, which will be referred to as 'beds'. Beds may be permanent aggregations, precise in their location and separated by clearly demarked areas that are unsuitable for scallops, or they may be temporary aggregations that vary in their location from year to year, resulting from uneven settlement or early survival. Finally, within each bed, the distribution of scallops may be aggregated into 'patches', the scale of which is generally measured in terms of tens or hundreds of m^2. It must be stressed that the terms 'grounds', 'beds' and 'patches' are used here simply to indicate the general scale of aggregation and are not intended to be precise definitions of the scale, absolute density or nature of spatial pattern, all of which may differ considerably between species or between locations. More detailed consideration of the spatial structure of scallop populations is given elsewhere (Orensanz 1986; Caddy 1989; Orensanz et al. 1991a), with some variation in terminology. The terms used here are obviously convenient for describing aggregations of exploited species but are not so appropriate for the less abundant non-commercial species.

For each species the major grounds are generally widely separated geographically, and environmental conditions are usually sufficiently different to produce marked and frequently consistent differences in population parameters. For example, differences in growth rate between grounds have been reported for *Pecten maximus* (Mason 1983), *Placopecten magellanicus* (Bourne 1964), *Aequipecten opercularis* (Taylor and Venn 1978), *Aequipecten tehuelchus* (Ciocco 1991) and *Mizuhopecten yessoensis* (Yamamoto 1975), and differences in breeding cycles between grounds have been described for many species (Sastry 1970, 1979; Taylor and Venn 1979; Mason 1983; Paulet et al. 1988). There are also frequently differences in population size- and age-structure that arise partly from differences in the regularity of recruitment. This raises the important question of whether the populations of scallop grounds are self-sustaining. Scallops have fairly long-lived pelagic larvae and it has been often considered that dispersion is an important adaptive function of the larval phase of relatively sedentary benthic invertebrates (e.g., Thorson 1946; Thorson 1950). For scallops, however, the extent of larval dispersal and the origin of spat settling on a ground are not known for any species and are difficult to investigate. The general assumption that recruitment to scallop populations is derived mainly from larvae transported, often large distances, from surrounding areas (Posgay 1979b) has important implications to the strategy of fishery management, but is based on little evidence. These questions have been discussed in detail by Sinclair et al. (1985), who review the available evidence for *Aequipecten opercularis*, *Pecten maximus* and *Placopecten magellanicus*. They note that, for each of these species, the major fishing grounds are relatively few in number, of characteristic absolute abundance, precise in their geographical location and have been persistent in these locations for very long periods, all of which, they believe, strongly implies that the populations on these grounds are self-sustaining. They go on to point out that many of these scallop aggregations are found in areas with tidally-induced oceanographic features, relatively persistent gyres or characterised by two-layer circulation, all of which could provide mechanisms for larval retention. Naidu and Anderson (1984) consider that estuarine entrainment-type circulation systems (Tully and Barber 1960) are responsible for retaining scallop larvae over the parent grounds in many coastal populations around Newfoundland. For the calico scallop, gyres created by coastal prominences are considered to be responsible for retaining larvae (Allen 1979; Miller et al. 1981). The effects of water circulation and oceanographic fronts on larval transport and recruitment of *Pecten maximus* and *Aequipecten opercularis* in the north Irish Sea has been frequently discussed (e.g., Beaumont 1982a; Macleod et al. 1985; Duggan 1987), and areas with particularly good recruitment and growth of *Pecten* appear to correspond with the position of major fronts (Murphy 1986). Elsewhere, potential larval dispersal pathways and recruitment links have been predicted by hydrographic modelling (Darby and Durance 1989; Dare et al. 1994), and these studies conclude that localised scallop fisheries vary in the extent to which they are likely to be self-sustaining. In recent years the techniques of biochemical genetics have been brought to bear on these problems, initially with studies of allozyme variation and, more recently, using DNA-based molecular methods, and there is now increasing evidence to suggest that many scallop populations may be largely self-recruiting.

For *Aequipecten opercularis*, the extent to which populations are reproductively isolated has been studied in a number of investigations of intraspecific biochemical genetic variation (Beaumont and Gruffydd 1975; Mathers 1975; Beaumont 1982a, b; Beaumont and Beveridge 1984; Macleod et al. 1985; Duggan 1987; Lewis and Thorpe 1994a, b). On the basis of mean genetic identities at nine sample sites, together with a consideration of current systems, Beaumont (1982a) suggested that there are at least four relatively genetically isolated populations of *A. opercularis* throughout its geographical range: a North Atlantic population extending from Northern Ireland through the north and west coasts of Scotland to Norway; a Celtic Sea population extending from the west coast of Ireland to the west coast of Brittany; an English Channel population and an Irish Sea population. Subsequent studies by Macleod et al. (1985) and Duggan (1987) generally agreed with Beaumont's suggestion of Irish Sea homogeneity in showing an apparent uniformity of allele frequencies in samples from different *A. opercularis* beds around the Isle of Man. However, the appearance of a rare allele in the 1979 year-class (*Pgi* allele 4), which later appeared also in spat of the 1984 year-class and very rarely in some other year-classes, suggests that there may be occasional recruitment of larvae from elsewhere. Interestingly, Mathers (1975) found significant differences in allele frequencies between queen scallop populations from the east and west coasts of Ireland, and his west coast sample may have contained the same rare allele that appeared in the 1979 and 1984 year-classes in the Isle of Man, indicating potential gene flow between these sites. Beaumont and Beveridge (1984) studied a sample of *A. opercularis* taken from an area near Anglesey, N. Wales. While the allele frequencies they obtained did not differ significantly from those obtained for populations around the Isle of Man (Duggan 1987), they found an additional allele at one locus that was not found in the Isle of Man samples. This suggests that the Anglesey ground may be to some extent genetically isolated from the grounds further north. In a more recent study, Lewis and Thorpe (1994a, b) investigated allele frequencies for each year-class in 12 populations of *A. opercularis* from around the UK. They found highly significant variations between sites, but allele frequencies between year-classes were predominantly homogeneous within sites. This temporal genetic stability within geographically highly heterogeneous populations suggests that queen scallops may be predominantly self-recruiting.

Studies of genetic variation have also been carried out on various other scallop species including *Argopecten irradians* and *Argopecten gibbus* (Blake 1994; Krause et al. 1994; Blake and Graves 1995), *Chlamys islandica* (Fevolden 1989, 1992), *Mizuhopecten yessoensis* (Kijima et al. 1984; Wilbur et al. 1997) and *Pecten maximus* (Beaumont et al. 1993; Wilding et al. 1997, 1998; Heipel et al. 1998, 1999; Ridgway and Dahle 2000).

In the work on *Pecten maximus*, initial allozyme studies failed to show any differentiation between 13 widely scattered populations in the British Isles and France, suggesting that these stocks might be regarded as a panmictic unit (Beaumont et al. 1993).

However, more recent studies using high-resolution DNA-based methods have shown that the British and Norwegian scallop populations belong to different gene pools (Ridgway and Dahle 2000), and that the Mulroy Bay population is significantly different from other UK populations (Wilding et al. 1997, 1998; Heipel et al. 1998). Mulroy Bay is a very enclosed body of water with restricted tidal transfer of water in and out, which could potentially cause the retention of larvae and self-recruitment. In the mid 1980's there was a dramatic decrease in scallop spat settlement in Mulroy Bay, linked to the use of tributyltin (TBT) on salmon nets. While the scallop populations have subsequently recovered, following a ban on the use of TBT (Minchin et al. 1987), this event could have been responsible for the low genetic diversity of the Mulroy Bay scallops. In addition to these more pronounced differences, a low but significant degree of genetic differentiation has also been found between commercially fished beds of *Pecten maximus* around the Isle of Man (Heipel et al. 1998, 1999), indicating some restriction to genetic exchange in relatively closely-located, open-sea populations.

Comparative studies of reproduction and larval development of *Pecten maximus* in the Bay of Brest and the Bay of St. Brieuc, France, including transplant experiments, have shown differences in reproductive pattern between the two bays that are genetically determined (Paulet et al. 1988). Similarly, scallops transplanted from Scotland also retained critical features of their reproductive cycle after growing in the Bay of St Brieuc for two years (Ansell et al. 1988; Cochard and Devauchelle 1993; Mackie and Ansell 1993). These transplant studies provide strong evidence for the existence of different stocks, despite the failure of allozyme studies to confirm genetic differentiation (Beaumont et al. 1993). In this case allozyme methods, which examine only a small proportion of the entire genome, were less able to detect stock differentiation then studies of the reproductive physiology. Despite these limitations, the techniques of biochemical genetics, particularly the high-resolution DNA-based methods, clearly have much to offer and more detailed surveys covering larger geographical areas, including populations with different patterns of recruitment, will be especially valuable for assessing the extent to which populations on different grounds are self-sustaining.

Different beds within a scallop ground would not normally be expected to be genetically isolated, though in inshore areas there may well be hydrographic mechanisms that cause larval retention within a very small area. Environmental conditions such as water depth, substrate type and food availability can vary considerably between beds, even within a small area, leading to differences in growth rates and breeding cycles. *Pecten maximus* and *Aequipecten opercularis* populations on different beds in the north Irish Sea have been frequently shown to differ in various population parameters (e.g., Mason 1957, 1958; Gruffydd 1974a, b; Murphy 1986; Allison 1993), and similar results have been obtained for many other species.

The shape and size of a bed may be constant, defined by physical or topographic factors such as the distribution of a suitable substrate, or they may be variable from year to year, depending on both the spatial distribution of settlement within an otherwise suitable area and the subsequent effects of fishing. On heavily exploited grounds, scallop aggregations are of temporary duration as fishing both reduces scallop abundance and causes scallops to disperse (Robert and Jamieson 1986). Caddy (1970), surveying scallop grounds in the Northumberland Strait from a submersible, noted high-density aggregations of *Placopecten magellanicus* of at least 0.25–1.0 km in diameter, but on some grounds scallop beds can be very much larger. On Georges Bank, for example, *P. magellanicus* beds can be more than 30 km long yet often only 4 or 5 km wide (Jamieson 1978). This long and thin shape appears to be characteristic of scallop aggregations in several species, for it has been described also for *Amusium balloti* (Heald and Caputi 1981; Dredge 1985b), *Argopecten gibbus* (Cummins 1971; Roe et al. 1971; Allen and Costello 1972), *Mizuhopecten yessoensis* (Goshima and Fujiwara 1994), *Pecten fumatus* (Olsen 1955) and *Pecten novaezelandiae* (Tunbridge 1968). These authors all note that the long axis of these aggregations is parallel to the flow of tidal currents or to the depth contours, and the suggestion has frequently been made that beds of this shape arise by the concentration of larvae settling out in regions of strong current flow. There are some similar observations for other benthic invertebrates. Fager (1964), for example, reported an elliptical-shaped bed of the polychaete *Owenia fusiformis* and attributed this to the passive accumulation of larvae by hydrodynamic processes associated with a local rip current. Scheltema (1986) also illustrates how an elliptical distribution of settling larvae could result from the process of eddy-diffusion, acting together with horizontal advection in a current. There is evidence for the importance of hydrodynamic processes that result in passive accumulation and deposition of benthic invertebrate larvae at large spatial scales (Butman 1987). Such processes may well operate to produce large scallop aggregations. However, the species of scallop for which long, thin beds have been recorded are all very active swimmers. Various studies have shown a predominant downstream dispersal during swimming movements (e.g., Belding 1931; Bourne 1964; Moore and Marshall 1967; Gruffydd 1976b; Imai 1980; Posgay 1981; Minchin and Mathers 1982; Melvin et al. 1985; Goshima and Fujiwara 1994) and this would also lead to elongated patterns of distribution in areas of strong current flow.

The location of beds within large fishing grounds, and changes from year to year, can be determined by mapping the distribution of catch per unit effort as an index of scallop abundance. This requires detailed information from commercial fishing boats on the location of catches, or very extensive research vessel surveys. There are very few fisheries for which sufficiently detailed catch data of either type are available. The large fisheries for *Pecten maximus* and *Aequipecten opercularis* in the north Irish Sea have been studied in this way since 1981, based on catch and effort data reported by commercial fishermen for a 5 x 5 nautical mile (9.3 x 9.3 km) grid (Murphy 1986; Allison 1993; Beukers-Stewart et al. 2003). More detailed studies of the spatial distribution of *Placopecten magellanicus* over Georges Bank have been carried out using Canadian fishermen's logbook data, which allows catches to be attributed to one-minute squares of latitude and longitude (1.38 x 1.85 km at 42°N). Isopleth mapping techniques have been used on these, and on research vessel survey data, to produce contour maps of scallop abundance (Jamieson and Chandler 1981; Robert and Jamieson 1986). More recently, video survey techniques have also proved to be a fast and accurate way of determining scallops abundance and mapping the location of beds (Thouzeau et al. 1991; Stokesbury 2002).

The distribution of scallops within a bed has been frequently described as patchy for most species of scallop including *Adamussium colbecki* (Berkman 1991), *Aequipecten opercularis* (Soemodihardjo 1974; Mason et al. 1979b; Mason 1981), *Amusium balloti* (Heald and Caputi 1981), *Argopecten gibbus* (Cummins 1971; Allen and Costello 1972), *A. irradians* (Belding 1931; Tettelbach 1991), *Chlamys islandica* (Vahl 1981), *Mimachlamys varia* (Shafee 1979), *Mizuhopecten yessoensis* (Goshima and Fujiwara 1994), *Pecten fumatus* (Olsen 1955), *P. maximus* (Baird and Gibson 1956; Mason 1983), *P. novaezelandiae* (Tunbridge 1968) and *Placopecten magellanicus* (Bourne 1964; Caddy 1970, 1975; Stokesbury and Himmelman 1993; Stokesbury 2000). The densities of individuals in a patch can be exceptionally high: Mason (1983), for example, once found *Aequipecten opercularis* in layers three or four deep at a density of 500–1,000 scallops m^{-2}. Such dense patches are exceptional, however, even for this species. Some examples of the densities reported for different species of scallop are given in Table 12.2. Because of the aggregated distribution patterns, at different spatial scales, density estimates depend very much on the area over which they are measured. Methods for estimating densities on the scale of fishing grounds therefore generally give much lower estimates than the techniques appropriate for measuring small-scale patches (Orensanz et al. 1991a). However, average densities for many exploited species are usually in the range 0.1–10 scallops m^{-2} on a good fishing ground.

The spatial distribution of scallop species has rarely been investigated using methods appropriate for the study of biological pattern (see Andrew and Mapstone 1987), mainly because of inherent problems in the methods for capturing or observing scallops in their natural habitat. Dredge surveys do not provide sufficient resolution for studies of small-scale spatial distribution patterns (Caddy 1970; Mason et al. 1982). To do this it is necessary to use more quantitative techniques, such as diver quadrant counts (Olsen 1955; Vahl 1981; Caddy 1968) or underwater photography or television, taken from a towed sledge (Caddy 1975; Franklin et al. 1980b; Thouzeau and Hily 1986; Thouzeau et al. 1991; Goshima and Fujiwara 1994) or from a submersible (Caddy 1970; Langton and Robinson 1990). These methods all have some advantages and disadvantages, depending on the aim of the study (Mason et al. 1982). Vahl (1981) used diver quadrant counts to study the spatial distribution pattern of *Chlamys islandica* on a bed in Balsfjord, North Norway. He used the variance to mean ratio test (Elliot 1971) to show that the distribution was not random, but highly contagious, and went on to demonstrate a good agreement with the negative binomial distribution. Similar statistical techniques have been applied to photographic surveys of *Placopecten magellanicus* and have also shown contagious distributions, even at low densities (Caddy 1970; Langton and Robinson 1990). Stokesbury and Himmelman (1993, 1995a) and Stokesbury (2000) have carried out the most detailed investigations of scallop distribution pattern, at different spatial scales, in studies of two *Placopecten magellanicus* beds in Port Daniel Bay, eastern Canada. At a large scale (km), scallop distribution was contagious and strongly associated with gravel or gravel-sand substrates, although not all gravel substrates contained high densities of scallops. At a small scale (cm), distribution was also contagious, with the scallops aggregated into small clumps. The average clump size was 1.13 m^2 in one bed and 4.5 m^2 in another, and the majority of scallops (79% in one bed

Table 12.2

Some estimates of density recorded for different scallop species.

Species	Location	Density (numbers m^{-2})	Reference
Aequipecten opercularis	English Channel	mean 0.46; max. 4.46	Franklin et al. 1980b
	Guernsey	5.7	Askew et al. 1974 (unpubl.).
	W. Scotland	500–1000	Mason 1983
Argopecten gibbus	Florida east coast	mean 43; max. 108	Cummins 1971; Roe et al. 1971
A. irradians	Long Island N.Y.	0.2–4.4	Bricelj et al. 1987b
	Niantic River, Conn.	up to 75	Cooper and Marshall 1963
	Newport river N.C.	19.5	Thayer and Stuart 1974
	- after fishing	3.0	
A. purpuratus	Chile	0.2–5.0	Illanes-Bucher 1987
Chlamys islandica	Balsfjord, Norway	means 56.0–69.9	Vahl 1981
Mimachlamys varia	Ireland – various areas	means 1.8–28.0	Rodhouse and Burnell 1979
	Bay of Brest, France	mean 2.3	Shafee and Conan 1984
Mizuhopecten yessoensis	Mutsu Bay – adults	1.5–6.0	Imai 1980
	- spat	50–250	
Pecten fumatus	Tasmania	mean 5.0*; max. 25–30*	Olsen 1955
P. maximus	Strangford Lough	0.67*	Hartnoll 1967
	Claonaig Bay, Scotland -	0.12*	Mason et al. 1979a
	after fishing	0.03*	
	English Channel	mean 0.16; max. 2.33	Franklin et al. 1980b
	Isle of Man	0.04–0.13*	Murphy 1986
		0.01–0.04	Wilson and Brand 1995
Placopecten magellanicus	Digby	max. 6.3*	Dickie 1955
	Georges Bank – recruits	4–5	Caddy 1975
	- older scallops	0.5	
	Georges Bank – spat	1.7–123	Larsen and Lee 1978
	Gulf of Lawrence	1.4–6.6	Caddy 1968
	Fippennies Ledge	0.98	Langdon and Robinson 1987 (unpubl.)
	Newfoundland	0.19–0.86	MacDonald and Thompson 1986b
	Georges Bank	0.25–0.59	Stokesbury 2002

* These values were originally recorded in other units but have been converted here to number per m^2

and 87% in the other) were in clumps. In the two beds, the average number of scallops per clump was about three, both sexes were present in 79% of clumps of adult scallops, and 91% and 75% of nearest neighbour pairs were 100 cm or closer. The authors suggest that this small-scale aggregation is an adaptation to increase fertilisation success, and warn that disturbance to these aggregations, as occurs during fishing (Langton and Robinson 1990), may decrease reproductive success. A small-scale random distribution, in agreement with the Poisson distribution, has been reported for some scallop populations (e.g., Mason et al. 1982; Thouzeau et al. 1991; Goshima and Fujiwara 1994) but, as pointed out by Andrew and Mapstone (1987), the spatial pattern observed using these methods depends on the scale at which the population is sampled, relative to the scale at which it is aggregated. In contrast to the findings of Stokesbury and Himmelman (1993), MacDonald and Bajdik (1992) reported a small-scale contagious distribution in a low-density bed of *Placopecten magellanicus* and a random distribution in a high-density bed. They suggest that sea scallops form local aggregations at low densities if food supplies are not limiting, but distributions become random at high densities as scallops attempt to reduce competition for resources by spacing themselves out. The mechanisms responsible for the establishment and maintenance of such small-scale aggregations are not known, though Stokesbury (2000) considers that swimming in *Placopecten magellanicus* may have evolved so that individuals could form clumps in order to enhance reproductive success, as well as to escape from predators. Clearly, more research is required on this subject.

12.3.2 Year-class Separation

Scallop beds frequently show a regional separation of year-classes that is probably largely a consequence of spatial differences in settlement and early survival. In heavily exploited species, this can lead to the concentration of the fishery on aggregations of the recruiting year-class (Caddy 1975; Murphy 1986) and the subsequent rapid depletion of stocks in that area. Year-class separation has been reported for *Amusium balloti* (Heald and Caputi 1981), *Aequipecten opercularis* (Soemodihardjo 1974), *Pecten maximus* (Gruffydd 1974a; Minchin and Mathers 1982; Mason 1983; Murphy 1986; Brand and Murphy 1992; Allison 1993), *Pecten fumatus* (Olsen 1955) and *Placopecten magellanicus* (Bourne 1964; Caddy 1975; Fritz and Haven 1981; Robert and Jamieson 1986; Thouzeau et al. 1991). Gruffydd (1974a) found regional variations in population age structure of *Pecten maximus* in unexploited areas of the north Irish Sea which indicated uneven recruitment, though taken over the whole area investigated recruitment was remarkably consistent from year to year. Subsequently, Murphy (1986) and Allison (1993) noted differences in the strength of particular year-classes between grounds to the east and to the west of the Isle of Man which could result from annual variations in the northerly flowing residual currents affecting larval dispersal to these grounds. For *Placopecten magellanicus*, isopleth mapping of survey catch rates for scallops of different ages has allowed the spatial distribution pattern of different year-classes on Georges Bank to be determined (Robert and Jamieson 1986; Thouzeau et al. 1991).

It has been suggested that scallop larvae may be gregarious (Mason 1983), that settlement patterns may depend on the distribution of suitable primary settlement sites like hydroids and bryozoans (Larsen and Lee 1978; Brand et al. 1980; Dare and Bannister 1987; Minchin 1992), and that density dependent mechanisms can affect recruitment success (Vahl 1982). However, the recruitment processes leading to spatial differences in year-class distribution are poorly understood, for any scallop species, due to the difficulties of studying the larval and pre-recruit stages in the field. Pectinid larvae are difficult to identify to species (Rees 1950; Le Pennec 1974, 1980; Gruffydd 1976a; Slater 2003), and are not easy to distinguish from the larvae of some other bivalve families, like the anomiids and the mytilids, which are often abundant in the plankton. Settled spat and pre-recruit juvenile scallops are difficult to collect in representative abundance and, in accounting for their distribution, it is generally impossible to distinguish between differential larval settlement and differential post-settlement mortality. Following primary settlement from the plankton, juvenile pectinids can detach the byssus and swim or crawl short distances. This may be to escape from predators, or in some species it could be a re-dispersion strategy used by spat to find more favourable microhabitats after settling initially on unsuitable substrata (Wallace 1982). Experiments carried out on *Chlamys islandica* suggest that resettling spat show an aggregation response to spat that has already settled (Harvey et al. 1993). The possibility of scallops settling in nursery areas, where conditions are particularly favourable for settlement and early survival, and migrating or dispersing later to adult beds has often been suggested but until recently there was little evidence of this occurring in any species. Of the earlier work, Vahl (1981) showed delayed recruitment to a population of *Chlamys islandica,* and Allen (1953) reported inshore settlement followed by a possible offshore movement in the deepwater *Chlamys septumradiata.* Thouzeau et al. (1991) found that juvenile (age 1) *Placopecten magellanicus* were concentrated on gravel and pebble sediments in the northern half of Georges Bank and there was a clear dispersion to the south (age 2) and the south-west (age 3+) as they grow older. More recently, a tagging study of juvenile *Chlamys islandica* demonstrated movement into deeper water (Arsenault et al. 2000). This provides the first experimental evidence of juvenile scallops swimming from a nursery area to recruit into the adult populations, for the scallop movements in this study were predominantly downslope and could not be attributed to downstream advection. Recent studies by Kamenos and Moore (2003) suggest that maerl beds may act as nursery areas for juvenile *Aequipecten opercularis.*

12.3.3 Factors Affecting Local Distribution

Many of the factors affecting local distribution have been mentioned above in accounting for the observed patterns of spatial distribution. Environmental factors such as temperature, depth, food availability, substrate type, water currents, turbidity, oxygen concentration and salinity are all likely to be important, as are ecological interactions with competitors and predators. There have been very few experimental studies carried out under controlled conditions to determine the effects of any of these factors on scallop growth, reproduction or survival, so it is generally impossible to assess the ecological significance of particular factors. Indeed, since all are interrelated in a complex manner it is of limited value to study single factors in isolation.

12.3.3.1 Depth

Depth is an important factor influencing many aspects of the biology of scallop species (Shumway and Schick 1987) but its ecological effects are usually associated with variations in temperature, food availability or substrate type (Dickie and Medcof 1963; Gruffydd 1974b; Leighton 1979; Posgay 1979a; Wallace and Reinsnes 1984, 1985; MacDonald and Thompson 1985a, b, 1986a). *Argopecten gibbus*, for example, has been reported in waters varying from 9.9–33.0°C (Waller 1969; Allen and Costello 1972) but temperatures outside the range 15–27°C are probably lethal to the major portion of the population (Miller et al. 1981). This restricts the species to a well-defined depth range off Cape Kennedy, with the largest concentrations occurring in 33–42 m (Miller and Richards 1980). However, irregular intrusions of deep, cold, water on to the Florida-Hatteras shelf affect both the local distribution and the abundance of calico scallop stocks (Miller et al. 1981). Similarly, temperature variation with depth accounts for the different depth distribution of *Placopecten magellanicus* along its geographical range (Bourne 1964), and populations in areas with unstable hydrographic conditions may suffer sudden mass mortalities when exposed to rapid temperature changes caused by oscillations in the depth of the thermocline (Dickie and Medcof 1963).

Reduced food availability with increasing depth from 19 to 31 m was considered to be a primary factor regulating growth, reproductive output and reproductive effort of *P. magellanicus* in Newfoundland (MacDonald and Thompson 1985a, b, 1986a, b, 1987). Populations of sea scallops living at much greater depths (170–180 m) in the Gulf of Maine have greatly reduced shell and tissue growth, compared with populations in shallow water (Schick et al. 1988). Barber et al. (1988) also found lower fecundities in these populations. This suggests that the deepwater populations are not self-sustaining but depend instead on sporadic recruitment from populations in shallower water (Shumway and Schick 1987). Nutrient stress was also considered to be a main factor accounting for the absence of *Aequipecten opercularis* from areas below 200 m (Ursin 1956).

12.3.3.2 Substrate type

Most scallop species are found on a range of bottom substrates but the important commercial species typically occur on harder substrates of gravel and coarse to fine sand (e.g., Mason 1983; Young and Martin 1989; Thouzeau et al. 1991; Stokesbury and Himmelman 1993; Dare et al. 1994). Indeed, the association of scallops with such deposits is so strong that high-resolution acoustic bottom discrimination techniques are increasingly being used to map scallop grounds and provide more precise estimates of scallop abundance (Burns et al. 1995; Magorrian et al. 1995; Kaiser et al. 1998; Kostylev et al. 2003). Many species, such as *Amusium pleuronectes* (Young and Martin 1989), *Argopecten gibbus* (Allen and Costello 1972), *A. irradians* (Belding 1931; Gutsell 1931), *Pecten maximus* (Mason 1983), *P. fumatus* (Young and Martin 1989) and *Placopecten magellanicus* (Bourne 1964) are able to tolerate some silt or mud in the substrate but, even for these species, the areas of highest abundance and the fastest growth rates are normally areas with little mud. Such bottom substrates typically occur in areas of strong current flow.

12.3.3.3 Currents

Mason (1983) comments that *Pecten maximus* and *Aequipecten opercularis* are most abundant just inside or just away from areas of strong currents. There are very strong tidal currents in the English Channel and Dare et al. (1994) found a clear inverse relationship between *Pecten maximus* distribution and the magnitude of frictional bottom stress, with virtually all commercial scallop grounds restricted to areas where bottom stress is low (less than 10 dynes cm^{-2}). Ursin (1956) classifies *A. opercularis* as a current dependent species, and associations with strong currents have been noted for many other species including *Argopecten irradians* (Belding 1931), *A. gibbus* (Allen and Costello 1972), *Chlamys islandica* (Wiborg 1963; Naidu et al. 1982), *Mizuhopecten yessoensis* (Imai 1980), *Pecten fumatus* (Fairbridge 1953; Olsen 1955) and *Placopecten magellanicus* (Bourne 1964).

Areas with strong currents generally provide favourable conditions for benthic filter feeders but there is now much experimental evidence to show that feeding and growth of *Placopecten magellanicus*, *Argopecten irradians* and *Chlamys islandica* are inhibited at very high rates of flow (Kirby-Smith 1972; Wildish et al. 1987, 1992; Wildish and Kristmanson 1988; Eckman et al. 1989; Wildish and Saulnier 1992, 1993; Claereboudt et al. 1994; Arsenault et al. 1997). Pilditch and Grant (1999) carried out a detailed investigation of the effects of variable food supply on *P. magellanicus*, measuring clearance rates while manipulating both flow velocity and phytoplankton concentration. They found that, with constant phytoplankton concentration, clearance rates during declining flow rates (15–5 cm s^{-1}) were 30% higher than during increasing flow rates (5–15 cm s^{-1}) and suggest that these elevated clearance rates during declining flow may compensate for the period of feeding inhibition they found at peak flow velocity (25 cm s^{-1}). The ecological implications of feeding inhibition in high current flows are not clear, however, for the behaviour of recessing into depressions in the seabed may enable *Placopecten magellanicus* and other scallop species to take advantage of boundary layer effects and avoid the high current velocities in apparently unsuitable areas (Yager et al. 1993; Stokesbury and Himmelman 1995b). In *Chlamys islandica*, Arsenault et al. (1997) found that the use of crevices to avoid predation resulted in higher growth rates for small individuals (15–30 mm shell height), but had no detectable effect on larger individuals (30–60 mm). They concluded that the markedly lower current velocities within crevices increased the feeding efficiency of small animals, since the velocity threshold at which feeding is inhibited is generally lower in small individuals. For *Argopecten irradians*, swimming responses operate to maintain position within eelgrass beds (Winter and Hamilton 1985), where current speeds are significantly reduced (Eckman 1987).

Like feeding, larval settlement and byssal attachment of spat are also likely to be physically limited, at some level, by high water current velocity but, while there are some field observations, there are few experimental data for this for any scallop species. The settlement, growth and early survival of *A. irradians* in eelgrass beds have been studied by Eckman (1987) in relation to hydrodynamics. He found that newly settled larvae were significantly more abundant in areas with a lower density of eelgrass shoots, and consequently with higher rates of seawater flux. The subsequent growth and survival of *Argopecten* was more difficult to interpret but indicated the importance of hydrodynamic processes. Survival was generally low in regions with strong currents, suggesting that young *Argopecten* are dislodged from their byssal attachment to the eelgrass blades at higher current velocities. However, in this and numerous other field studies of bay scallop survival in seagrass beds (e.g., Pohle et al. 1991; Ambrose and Irlandi 1992; Ambrose et al. 1992; Bologna and Heck 1999), it is difficult to distinguish any effects of water currents from the effects of predation. Recently, Irlandi et al. (1999) showed that neither seagrass patch size nor plant density – both factors that affect the rate of current flow over the seabed – had any relationship with the survival of *Argopecten* over periods of 3–7 weeks, though there were significant short-term (i.e., 24 h) effects. They suggested that this lack of a longer-term relationship was due to differential responses to patch size by the many potential predators of juvenile scallop. Current speed and turbulent mixing have been shown to affect the efficiency with which predators prey on scallops. Thus, Powers and Kittinger (2002) found that whelk predation on bay scallops increased with increases in water flow, while blue crab predation decreased. They conclude that current speed and turbulent mixing affect the chemosensory abilities of both the predator to locate the scallop, and the scallop to detect and avoid approaching predators, with predator success depending on whether the predator or the prey is most affected.

12.3.3.4 Turbidity

Living in open coastal environments with coarse bottom substrates and relatively high current flow, most commercially important scallop species are not normally confronted with water of high turbidity. Other species, like the inshore, estuarine, *Argopecten irradians* and the many deepwater species, like *Chlamys septumradiata* (Allen 1953), commonly live on soft muddy bottoms and must be adapted for dealing with more turbid conditions. The shallow, subtidal, *Minnivola pyxidatus* is able to survive in dredged and trawled waters with high levels of settling and resuspended silt off the coast of Hong Kong, and elsewhere in the Western Pacific, but its numbers are declining locally because of seabed perturbations (Morton 1996). Increases in the amount of particulate inorganic matter in the water have been shown to have a detrimental affect on absorption efficiency and growth of *Chlamys islandica* (Vahl 1980; Wallace and Reinsnes 1985), and ciliary activity of *Mizuhopecten yessoensis* (Yamamoto 1956b), but there have been no experimental studies of the effects of turbidity on survival for these or any other scallop species. It seems likely, however, that turbidity is an important ecological factor that limits the occurrence and distribution of both adults and newly settled larvae. Scallop spat is very susceptible to silt and to low oxygen tension (Belding 1931; Yamamoto 1956b, 1960; Motoda 1977). The reduced water current velocity in the boundary layer just above the surface of the sediment allows the deposition of fine particles and organic detritus; the oxidation of this organic material can then lead to the local depletion of dissolved oxygen in this layer. Yamamoto (1960) considered that such conditions are a major cause of mortality of the spat of *Mizuhopecten yessoensis* when they first release the byssus and turn to bottom life at a size of 6–10 mm shell length. He reported high mortalities of spat, amounting to 82–100% in some habitats, occurring in the summer and early autumn when the water temperature is high but mortality decreasing in the winter when the water temperature falls and the spat have reached a shell length of 30–40 mm. He concluded that mortality during this stage of the life history is most important in determining the size of the adult population, and that the recruitment of young scallops in any year depends on the presence or absence of an anaerobic layer. The habit of settling on algae, erect bryozoans or other surfaces would appear to provide scallop spat with a clean, silt-free surface raised above the boundary layer (Brand et al. 1980; Paul 1981; Dare and Bannister 1987; Harvey et al. 1993, 1994). The distribution of suitable sites for larval settlement and early survival may therefore be a major factor determining the local distribution of adult scallops.

12.3.3.5 Salinity

Salinity is another ecological factor that clearly acts to limit distribution, but there have been few studies of distribution in relation to salinity or of the salinity tolerance of scallop species. Most scallops live in fully saline water and species such as *Argopecten gibbus*, *Argopecten purpuratus* and *Aequipecten opercularis* do not survive well in reduced salinities. *Argopecten gibbus* occurs where the salinity is fairly stable and ranges from 31–37‰ (Allen and Costello 1972). *Argopecten purpuratus* is stenohaline and natural beds of this species occur in bays along the Chilean coast where salinity fluctuations are generally low (34.5–35.2‰). Uribe et al. (2003) believe that the scarcity of *A. purpuratus* at the extremes of its range results from lower salinities to the north of Paita, Peru (5°S), due to the presence of Equatorial water, and south of Valparaiso, Chile (33°S), due to the strong presence of lower salinity sub-Antarctic water together with freshwater run-off from rivers entering the bays. The shoreward limit of distribution of *Aequipecten opercularis* in the transition area between the Skagerrak and the Baltic follows closely the 30‰ isohaline and Ursin (1956) considered that queen scallops do not tolerate salinities below 25‰, based on their distribution. This was essentially confirmed by the detailed studies of Paul (1980a) on the salinity-temperature relationships of *A. opercularis*. He found that salinities of 16–28‰ were lethal after a 24 h experimental exposure, depending on the temperature and size of the scallop, with mortalities increasing at the extremes of temperature, and spat having a slightly greater tolerance than larger individuals (Figure 12.8). For the Australian doughboy scallop, *Mimachlamys asperrima*, spat are also more tolerant of salinity reductions than the adults (O'Connor and Heasman 1998) while, conversely, spat of *Mizuhopecten yessoensis* appear to be less tolerant than larger individuals. *M. yessoensis* spat is seriously affected by slightly reduced salinities and ciliary activity stops in 75% normal seawater salinity (Yamamoto 1956a), but adults can survive in 60% seawater, although the LT50 in 40% seawater is only 3 hours (Ventilla 1982). In contrast to these open-sea species, *Argopecten irradians*

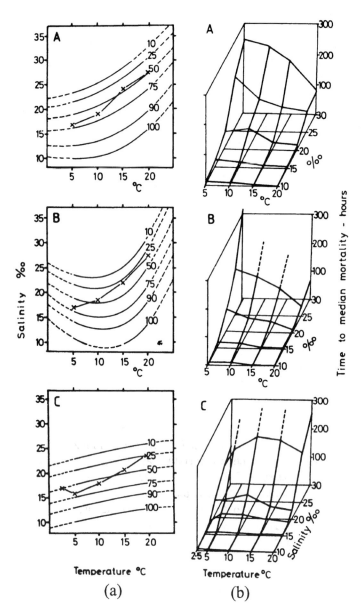

Figure 12.8. Salinity temperature tolerance of *Aequipecten opercularis*. (a). Regression surfaces for % mortality of three size groups (A. >60 mm; 30–40 mm; 5–10 mm shell height) at different salinity-temperature combinations. Isopleths are produced at intervals of mortality between 10 and 100%. The broken lines represent extrapolations outside the test area; crosses show estimated median lethal salinities. (b). Time to median mortality for the three size groups (A, B, C), plotted as heights above a grid for the various salinity-temperature combinations. Broken lines are extrapolations where median mortality was not reached. (From Paul 1980a).

lives in very shallow bays and estuaries and has been reported in salinities ranging from 10‰ to 38‰ (Belding 1931; Gutsell 1931; Castagna and Chanley 1973; Duggan 1975). This is one of the few scallop species likely to be regularly exposed to very low salinities in its natural habitat during periods of heavy rain and freshwater run-off from the land. Mercaldo and Rhodes (1982) found that adult bay scallops are able to tolerate exposure to low salinities for varying periods of time depending on the temperature. They recorded excellent survival for up to 48 hours in freshwater at 10°C, but not at 24°C, suggesting that this species is better able to withstand exposure to low salinities in the winter months. Nevertheless, Gutsell (1931) reported that periods of heavy freshwater run-off and severe cold weather can cause heavy mortalities in bay scallop populations, and Tettelbach et al. (1985) recorded nearly 100% mortality of northern bay scallops at the highest locations in the Poquonock River, Connecticut, following a very severe rainstorm in June 1982. Similarly, Stockton (1984) states that the formation of a hyposaline lens of seawater under the sea ice during the summer melt is the main cause of mortality of *Adamussium colbecki* in McMurdo Sound, Antarctica.

Various experimental studies of the combined effects of temperature and salinity have also been made on the larval stages of *Argopecten irradians,* both in the U.S.A. (e.g., Tettelbach and Rhodes 1981), and in China (Lin et al. 1989 1991; He and Zhang 1990), where hatchery-reared bay scallops are now the basis of a massive cultivation industry. Tettelbach and Rhodes (1981) showed that normal embryonic development of the northern bay scallop *A. i. irradians* takes place only over a very narrow range of temperature and salinity, from 20–27°C at 25±1‰ salinity. Survival and growth of larvae occur over wider ranges, though temperatures greater than 35°C and salinities less than 10‰ are lethal to all life-history stages. The optimum conditions for larval development of the southern bay scallop, *A. i. concentricus,* occur at rather higher levels of both temperature and salinity. Castagna (1975) gives optimal values of 26–28°C at 28–30‰ and the minimum salinity necessary for survival to the straight-hinged larval stage is 25.5‰. The higher temperature optimum for larval development of *A. i. concentricus* corresponds with a slightly higher temperature at which spawning is initiated in this southern subspecies (Sastry 1963, 1970). These experimental determinations of temperature and salinity tolerance limits, and the optimal requirements of the larvae of the two subspecies, were carried out independently using adult scallops acclimated, conditioned and spawned under different environmental conditions. It is not known if the differences observed are due to genetic or environmental factors. The narrow salinity range for normal larval development of the bay scallop would appear to be an important factor limiting larval dispersal and thereby restricting gene flow between populations in bays and estuaries separated by waters of full salinity. Lin et al. (1989) note that *A. irradians* larvae will not attach when the salinity is below 11.7‰ or above 35‰, and this will also limit their distribution.

Salinity tolerance experiments have also been performed on the larvae of *Chlamys islandica,* which may experience reduced salinities in the fjords of northern Norway (Gruffydd 1976a). These larvae are unable to tolerate salinities much lower than 80% seawater for any length of time and possess behavioural responses that prevent them from swimming through vertical salinity gradients into unfavourable conditions. Larvae of

Pecten maximus are unlikely to experience salinity levels much below those of the open sea but are able to tolerate reduced salinities for a short time. Laboratory experiments showed that tidal sinusoidal fluctuations in salinity were much less harmful than steady exposure to reduced salinities, or to abrupt changes in salinity, indicating the special danger of sudden osmotic shocks (Davenport et al. 1975). Recent studies of the effects of reduced salinity on *P. maximus* spat by Christophersen and Strand (2003) found reduced growth, increased mortality and no byssal attachment when salinity was reduced to 20‰. They concluded that locations exposed to salinities below 25‰ for extended periods should be avoided for nursery cultivation of *P. maximus*.

12.3.3.6 Competitors and predators

The ecological role of competitors and predators in determining the local distribution of scallops is poorly understood. Bivalves and other benthic filter feeders may compete for suspended food particles and are thought to filter out and destroy large numbers of planktonic larvae (Thorson 1946; Kristensen 1957); dense assemblages may therefore prevent settlement of larvae in the area (Woodin 1976). This interaction can occur between adults and larvae of the same species, providing a density dependent population regulatory mechanism which could account for the inverse correlations frequently found between settlement density and adult population density for various species of bivalve (Fitch 1965; Hancock 1973; Williams 1980). Such inverse correlations have not been found in any scallop populations, however, and few scallop species would appear to occur in sufficiently high densities for this mechanism to play an important part in regulating the population. Vahl (1982) found evidence for the density dependent control of recruitment to a population of *Chlamys islandica* in Balsfjord, Northern Norway but he considered that this was due to the primary settlement of larvae away from the adult population, followed by a later movement of juveniles on to the adult beds.

Spat and juvenile scallops are very vulnerable to predation by certain crabs, starfish, gastropods and bottom-feeding fish (e.g., Medcof and Bourne 1964; Kim 1969; Buestel and Dao 1979; Elner and Jamieson 1979; Minchin 1981; Volkov et al. 1982; Lake et al. 1987; Lake and McFarlane 1994; Strieb et al. 1995; Arsenault and Himmelman 1996b; Barbeau et al. 1996; Peterson et al. 2001), and in areas of high predator abundance this may well limit scallop distribution. Few attempts have been made to determine predation rates on scallop spat in the natural environment but studies of the mortalities of spat within the confines of collector bags (e.g., Brand et al. 1980; Paul et al. 1981; Thouzeau 1991; Zhang 1996), and of juvenile scallops relayed at higher than normal densities on the seabed (e.g., Volkov et al. 1982; Minchin 1991; Barbeau and Scheibling 1994b; Lake and McFarlane 1994; Barbeau et al. 1996, 1998; Hatcher et al. 1996; Fleury et al. 1997; Barbeau and Caswell 1999; Strand et al. 1999), suggest that predation rates are very variable but can be very high when certain predators are abundant. Eckman (1987) considered that predation was less important than water movements in determining the abundance of juvenile *Argopecten irradians* but numerous subsequent studies have stressed the importance of predation on this species (Prescott 1990; Bricelj et al. 1991; Pohle et al. 1991; Garcia-Esquivel and Bricelj 1993; Strieb et al. 1995; Irlandi et al. 1999;

Peterson et al. 2001). The vulnerability of young scallops to predation obviously varies between species, due to different behavioural and morphological adaptations. This is well illustrated where two scallop species live together in the same habitat. Spat and juveniles of *Placopecten magellanicus* and *Chlamys islandica*, for example, are both preyed upon by the American plaice (*Hippoglossoides platessoides*) and to a lesser extent by the yellowtail flounder (*Limanda ferruginea*) on St. Pierre Bank, but the far greater vulnerability of *Chlamys islandica* is probably due to its greater tendency to remain byssally attached and its weaker swimming escape response (Naidu and Meron 1986). No quantitative data are available for the two common European species but spat of *Pecten maximus* is extremely thin-shelled and fragile compared with *Aequipecten opercularis* of the same size and is likely to be susceptible to a wider range and size of predators. However, in older animals, tethering experiments have shown that the major scallop predators (*Asterias rubens* and *Cancer pagurus*) prey preferentially on *A. opercularis*, rather than *P. maximus*, despite its better swimming escape response (Whittington 1993).

Predation on *Pecten maximus* is reduced when the spat leave sites of byssus attachment and start to recess in the seabed (Buestel and Dao 1978). As scallops get bigger and the shell thickens they become less vulnerable to predation, at least by crabs, which usually feed by breaking the shell open. In experiments on predation of *Placopecten magellanicus* by the rock crab, *Cancer irroratus*, and the American lobster, *Homarus americanus*, Elner and Jamieson (1979) found that preferred prey size increased with predator size, but smaller scallops were particularly vulnerable as predators of all sizes could eat them. Scallops larger than a critical size of about 70 mm shell height were immune to predation by even comparatively large rock crabs and lobsters. Lake et al. (1987) obtained similar results in experiments on predation of *Pecten maximus* by four species of crab. Predation varied with the size of both the crab and the scallop, and they too found that predation was negligible for scallops >70 mm shell height (Figure 12.9). Pennington (1999) confirmed these findings in field trials using tethered scallops and time-lapse photography, backed up by laboratory experiments. However, he found that the common starfish (*Asterias rubens*) was by far the main predator of scallops in the area of the Irish Sea where he carried out his studies. Although starfish appeared to be more flexible than crabs in their ability to deal with larger scallops, the scallops still achieved a size refuge from starfish predation at around 75 mm shell length for starfish of up to 100 mm arm length. Starfish predation was also strongly seasonal, following a bimodal pattern over two years that was associated with the breeding cycle. Interestingly, temperature was not a major factor in determining levels of predation on tethered scallops in this study. Starfish predation showed no correlation with temperature, and while there was a consistent positive association between crab predation and temperature, only one of the statistical tests proved significant. In contrast, Barbeau et al. (1994c) found that both starfish and crab predation on the sea scallop *Placopecten magellanicus* increased significantly with temperature, albeit over a rather greater temperature range. They also found that predation fate by sea stars increased with temperature because of the decreased effectiveness of the scallops' swimming escape response. Barbeau and co-workers have carried out a series of studies on the interactions between juvenile *P. magellanicus* and its predators (Barbeau and Scheibling 1994a, b, c; Barbeau et al. 1994, 1996, 1998), and have

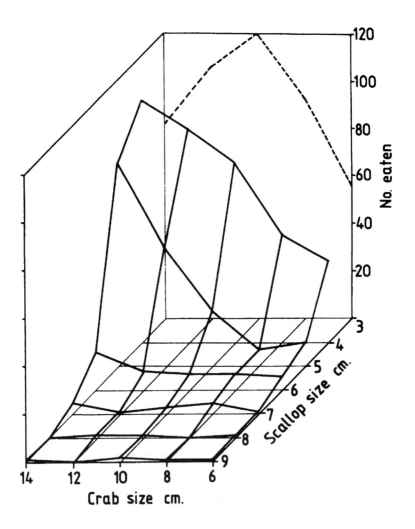

Figure 12.9. Scallop predator-prey size relationships. The total numbers of *Pecten maximus* of different size groups eaten during a 96 h period are plotted in relation to the size of the crab *Cancer pagurus*. The broken line indicates experiments with 3 cm scallops when the time was limited to 48 h. (From Lake et al. 1987).

recently summarised this information into a matrix model to predict scallop survival to commercial size in seabed reseeding operations (Barbeau and Caswell 1999). There is also a very substantial literature on the interactions of the bay scallop, *Argopecten irradians*, and its predators (e.g., Ordzie and Garofalo 1980a, b; Peterson et al. 1989; Prescott 1990; Bricelj et al. 1991; Pohle et al. 1991; Strieb et al. 1995; Bologna and Heck 1999; Irlandi et al. 1999; Powers and Kittinger 2002), and growing information for other species such as *Chlamys islandica* (Brun 1968; Arsenault and Himmelman 1996b), *Mizuhopecten yessoensis* (Volkov et al. 1982; Kalashnikov 1991; Gabaev and Kolotukhina 1999) and *Argopecten purpuratus* (Wolff and Alarcon 1993; Marahrens 1995; Ortiz and Wolff 2002a, b; Ortiz et al. 2003). Many of the experimental studies of predation on scallops have been based on laboratory studies or tethering experiments, but the relevance of such findings to the situation in the natural environment, where the scallop is free to swim away from the predator, is not always clear (however, see Barbeau and Scheibling 1994b).

Numerous authors have listed the natural predators of each species of scallop but it has been rarely possible to make any quantitative assessment of the mortality rates they generate. Of the few estimates available, most involve predation on adult scallop stocks by large predatory starfish. Thus, Olsen (1955) describes the destruction of a stock of *Pecten fumatus* on one bed, which he considered was primarily due to predation by the large eleven-armed spiny starfish *Coscinasterias calamaria* (Gray). From the decline in scallop densities he estimated that the mortality of scallops amounted to 75–80% over a four-year period. Dickie and Medcof (1963) reviewed the causes of mass mortalities of *Placopecten magellanicus* in the southwestern Gulf of St. Lawrence, estimating mortality as the % of 'cluckers' (empty shells still attached by the hinge) in the total catch. They considered that the largest and most devastating mass mortalities, with up to 80% cluckers, were caused by sudden temperature fluctuations resulting from hydrographic phenomena, but that several of the slower and less spectacular mortalities, with up to 25% cluckers, were caused by predators, either acting alone or, together with temperature debility of the scallops, making them more vulnerable to predation. These higher than usual mortalities were generally associated with an unusual abundance of predators, especially the starfish, *Asterias vulgaris*.

Dickie (1955) described a more precise method of calculating the rate of natural mortality, based on the proportion of cluckers in the catch and the average time for the shell valves to disarticulate in the sea. His results showed that annual natural mortality rates for the same bed were relatively constant from year to year, with average values of 4.5%, 13.8% and 15.7% for three groups of beds in progressively deeper water. Taken over all beds in the Digby area of the Bay of Fundy, he calculated the average annual rate of natural mortality to be about 10% (instantaneous coefficient, $M = 0.1054$). This method has been frequently used in subsequent studies of sea scallop populations on Georges Bank and elsewhere (e.g., Merrill and Posgay 1964; Naidu et al. 1983; Naidu and Cahill 1985; Robinson et al. 1994), and although there is some variability, the estimated values are generally low, in the region of 10–15% ($M = 0.10$–0.16). Estimates of natural mortality for other scallops species, made using this or other techniques, have generally produced values of a similar magnitude for unfished populations of long-lived species such as *Pecten maximus* (Baird 1966; Gruffydd 1974a; Brand et al. 1991) and *Chlamys islandica* (Naidu et al. 1982, 1983; Naidu 1988), though values for short-lived species such as *Aequipecten tehuelchus* (Ciocco 1996), *Argopecten gibbus* (Roe et al. 1971; Allen and Costello 1972), *A. irradians* (Belding 1931; Bricelj et al. 1987a), *Comptopallium radula* (Linne) and *Mimachlamys gloriosa* (Reeve) (Lefort 1994), and *Amusium balloti* (Dredge 1985a) are naturally considerably higher. Higher values are also frequently recorded for exploited populations (e.g., Naidu and Cahill 1984; Naidu 1988; Allison 1993; Allison and Brand 1995) because estimates of natural mortality include scallops damaged but not caught in dredges (Caddy 1968, 1973; Gruffydd 1972). Naidu (1988) comments that, depending on the gear used, 4–8 times as many *Chlamys islandica* die from encounters with fishing gear than through natural causes. While natural mortality is notoriously one of the most difficult population parameters to estimate accurately, and published values show considerable variation from area to area and year to year, there is little evidence to suggest that predation rates are normally high for any scallop species. There are, however, two observations worth noting. Over a 7-year study period, Petersen et al. (2001) recorded an annual virtual extinction of *Argopecten irradians* in one densely occupied seagrass bed, which they showed was due to predation by migrating cownose rays. In an earlier study Brun (1968) found an extremely dense belt of the starfish *Asterias rubens*, about 10 m wide and at least 100 m long, with densities of up to 97 starfish m^{-2}, moving on a broad front across a bed of *Chlamys islandica*. In front of the *Asterias*, scallop densities were about 60 scallops m^{-2}, while behind the belt of starfish practically all the scallops were eaten. Such dense aggregations of predators are not normal and this example of an extremely high level of predation would appear to be exceptional. From all available evidence it would seem that predation on adult scallops reduces population density but, for the longer-lived species at least, is not usually sufficiently high to control distribution, except on a very local scale.

Most scallop species, however, respond to contact with their common predators by a swimming escape response (e.g., Buddenbrock 1911; Mellon 1969; Moore and Trueman 1971; Thomas and Gruffydd 1971; Bloom 1975; Ordzie and Garofalo 1980a; Wilkens 1981; McClintock 1983; Dautov and Karpenko 1984; Winter and Hamilton 1985; Pitcher and Butler 1987; Joll 1989; Morton 1996). Swimming by one individual frequently stimulates other scallops nearby and results in mass swimming activity (Chapman et al. 1979; Vahl and Clausen 1980; Minchin and Mathers 1982; Howell and Fraser 1984). Scallops may also be stimulated to swim by other environmental factors, or as an avoidance reaction to fishing gear (Moore and Marshall 1967; Caddy 1968; Gruffydd 1976b; Chapman et al. 1979; Morton 1980; Ansell et al. 1998; Jenkins et al. 2003). In areas of strong current flow, swimming activity can lead to downstream dispersal (Bourne 1964; Moore and Marshall 1967; Gruffydd 1976b; Imai 1980; Posgay 1981; Minchin and Mathers 1982; Melvin et al. 1985; Goshima and Fujiwara 1994; Sakurai and Seto 2000). It is also possible that some species of scallop undergo directed migrations (Allen 1953; Morton 1980; Vahl 1981, 1982; Arsenault et al. 2000). Such movements could account for some of the many reports by fishermen of large commercial scallop stocks disappearing overnight from an area (e.g., Belding 1931; Rees 1957; Rolfe 1973). Whatever the method of stimulation, swimming activity, coupled with downstream dispersal or directed migrations, may therefore play an important part in determining the local distribution of scallop populations. These aspects of scallop behaviour, and their ecological significance, will therefore be considered in more detail.

12.4 SCALLOP BEHAVIOUR - ASPECTS AFFECTING DISTRIBUTION

As might be expected for a group of animals possessing a heavy calcified shell, active swimming behaviour is relatively unusual among the bivalve molluscs, though it has been reported to occur in some members of the families Pectinidae, Amussiidae, Limidae, Solenidae, Solemyidae and Cardiidae (Stanley 1970). Of these, it is best developed in the Pectinidae and the Amussiidae, the only families that swim with the commissure plane in a non-vertical position. Almost all species in these two families have the capacity for free swimming movements for at least part of their life history. Scallops swim by a type of jet propulsion, the powerful water currents for which are generated by rapid adductions that clap the shell valves shut. This is thought to have evolved from the efficient mantle cavity cleansing mechanisms of their byssally attached monomyarian ancestors (Drew 1906; Yonge 1936). Scallops display a range of adaptations associated with swimming and this is reflected in the varying swimming abilities of different species. Post-larval scallops pass through a period of byssal attachment and some species retain the ability to secrete a byssus throughout life. Other species lose the byssus and become free living, using swimming reactions to control their orientation, to recess into the seabed and to escape from predators. The behavioural mechanisms associated with byssus attachment, orientations and movements in post-larval scallops are therefore very important aspects of scallop ecology, affecting both distribution and population dynamics. Behavioural mechanisms associated with larval settlement are also most important but these are considered elsewhere in this book.

12.4.1 Byssus Attachment

The early post-settlement stages of most bivalves secrete a byssus for attachment and a functional byssus persists into the adult in several taxonomic groups, particularly those with a sessile mode of life (Yonge 1962). The neotenous retention of this post-larval structure is thought to have had a profound affect on the evolution of the superfamily Anisomyaria, leading to the suppression of the anterior adductor muscle and the evolution of the monomyarian condition of the Pectinacea (Yonge 1953, 1962; Yonge and Thompson 1976).

Probably all scallops secrete a byssus when young (Verrill 1897) and many species such as *Chlamys islandica* (Gruffydd 1976b, 1978; Vahl and Clausen 1980; Arsenault and Himmelman 1996a), *C. vitrea* (Verrill 1897), *C. furtiva, C. tigerina, Mimachlamys varia* (Reddiah 1959; Briggs 1983) and *M. asperrima* (Young and Martin 1989) retain the ability to form a byssus throughout life. These are generally species of small body size, or those living in areas of strong current flow, and attachment to rocks or other firmly fixed substrates is obviously advantageous for maintaining position in these species. Other scallops cease to form a byssus and live free of attachment on the seabed at some stage in their life history. The large *Pecten maximus* generally loses the byssus soon after metamorphosis and few larger than 15 mm shell length are found attached (Dakin 1909; Eggleston 1962; Mason 1969, 1983; Minchin 1978, 1992; Franklin et al. 1980a). *Mizuhopecten yessoensis* appears to detach at a similar size (Yamamoto 1956a, 1964;

Motoda 1977; Ventilla 1982), while *Pecten novaezelandiae* detaches at a much smaller size of 4–5 mm (Bull 1976). Compared with other scallops the byssus of *Amusium balloti* is a very weak, transient, structure (Wang et al. 2001); it does not appear until after the prodissoconch shell is produced, well after metamorphosis, and is lost when they reach a shell height of 4–5 mm. In *Placopecten magellanicus* (Caddy 1968, 1972; Manuel and Dadswell 1994) and *Aequipecten opercularis* (Soemodihardjo 1974; Paul 1980b), the proportion of the population that is byssally attached at any one time declines with age so that most individuals large enough to be commercially exploited are unattached. *Argopecten gibbus* apparently behaves similarly (Allen 1979), with byssus attachment declining from a shell height of 6–10 mm, but some individuals occasionally attach up to 38 mm. *Argopecten irradians* has a life-span of only 20–26 months and is seldom byssally attached after the first year of life, but individuals of 15–16 months have been observed occasionally fastened to eelgrass (Belding 1931). Byssus attachment inhibits swimming in *Placopecten magellanicus* (Caddy 1972) and *Mimachlamys varia* (Soemodihardjo 1974) but in these, and virtually all other species, contact with predators stimulates attached scallops to release the byssus and make swimming escape responses. The ability to swim efficiently declines in older individuals of some species (Olsen 1955; Caddy 1968, 1972; Gould 1971; Mathers 1976b; Dautov and Karpenko 1984; Parsons and Dadswell 1992; Cheng and DeMont 1996b) but is only lost altogether in the few species, like *Chlamys distorta* or *Hinnites multirugosus*, which become cemented to rock or other hard surfaces later in life (Yonge 1951; Tebble 1966; Soemodihardjo 1974).

Considering the evolutionary significance and great ecological importance of the byssus it is surprising that there have been so few field or laboratory studies of byssus attachment in any scallop species. A number of experiments have been carried out to study the affects of body size and temperature on byssus formation in *Placopecten magellanicus* (Caddy 1972) and *Aequipecten opercularis* (Soemodihardjo 1974; Paul 1980b). In both species, individuals alternate between attached and unattached phases, and the proportion attached at any time decreases progressively with increasing shell size (Figure 12.10). Small scallops are more active in byssus formation than larger scallops: small individuals secrete more byssus threads and re-attach much more rapidly than larger individuals. Thus, Caddy (1972) noted that the time to 50% attachment for *Placopecten* increased from 7 minutes for juveniles of 2–5 mm shell height to 140 minutes for animals larger than 120 mm. The number of byssus threads secreted per scallop per week also increased markedly with temperature but there was no corresponding increase in the proportion attached at the end of the experiment at temperatures above 5°C (Figure 12.11). This suggests that temperature affects the rate of attachment and detachment equally, so that the proportion attached at equilibrium remains independent of temperature over the normal temperature range of the species. For *Aequipecten opercularis* the rate of byssus attachment was highest and least variable at 18°C, declining and becoming more variable at both higher and lower temperatures (Paul 1980b) (Figure 12.12). Byssus formation in this species appears to be extremely sensitive to environmental conditions. Roberts (1973) found that byssogenesis in small *A. opercularis* is twice as sensitive to the insecticide Endosulphan as it is in the mussel *Mytilus edulis* and suggested that byssus attachment could make a very useful and rapid bioassay for potentially toxic compounds.

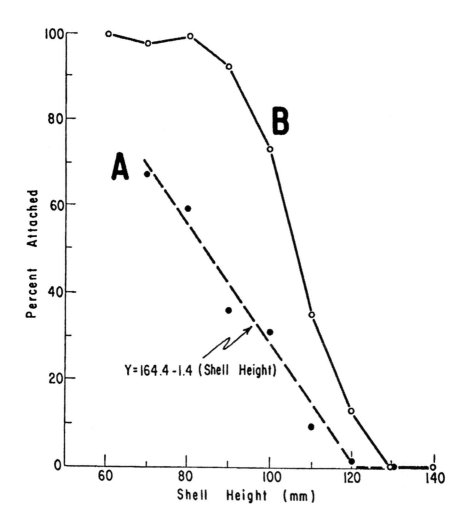

Figure 12.10. The % of *Placopecten magellanicus* byssally attached in relation to shell height. A, under natural conditions: regression calculated for 493 scallops in the size range 70–120 mm shell height, collected from L'Etang Harbour in August, 1967. B, under laboratory conditions: cumulative attachment of 424 scallops to a fibreglass surface during 40 days in the laboratory. (From Caddy 1972).

This has subsequently been confirmed in several studies using scallop species (e.g., Pesch and Stewart 1980; Martoja et al. 1989).

In the early stages after larval settlement the byssus functions to anchor spat in a clean, silt-free environment. In some species, like *Chlamys islandica*, which live in areas where the maximum current velocity above the seabed may exceed the swimming speed of the scallop (Gruffydd 1976b), it continues to serve this function throughout life. Since

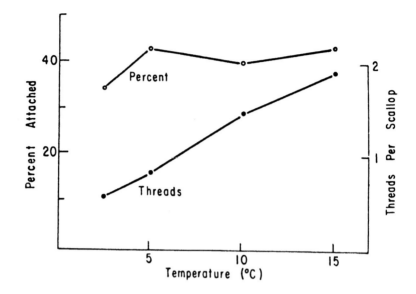

Figure 12.11. The effect of temperature on byssal attachment of *Placopecten magellanicus* (size range 60–100 mm shell height). Percent - % scallops byssally attached to gravel after 7 days. Threads - the number of byssus threads secreted per scallop in 7 days. (From Caddy 1972).

byssus attachment inhibits swimming (Caddy 1972; Soemodihardjo 1974), the retention of a functional byssus in some of the more mobile species may serve to reduce dispersal of juvenile scallops away from favourable settlement areas (Minchin 1992).

12.4.2 Recessing

After releasing the byssus, many scallops recess into saucer-shaped depressions in the seabed, typically so that the upper (left) valve is level with or just below the surface of the sediment. The process of recessing in *Pecten maximus* was first described in detail and photographed by Baird (1958) and his observations have subsequently been confirmed in numerous other studies (e.g., Minchin 1978, 1992; Fleury et al. 1996; Maguire et al. 1999a, 2002b; Minchin et al. 2000). In *P. maximus* a series of powerful adductions eject water from the mantle cavity in the region of the posterior auricle, the water jets being directed downwards by muscular control of the velum. This may cause the scallop to swing round but also lifts the shell at an angle to the seabed so that subsequent water jets blow a hollow in the sediment. When this is of sufficient depth, a more powerful adduction lifts the scallop and it lands precisely in the recess. Sediment disturbed in the process settles on top of the shell and more accumulates from water jets directed downwards around the edge of the shell. After a few days a layer 5 mm thick may be present, making the position of the scallop difficult to detect when the valves are closed.

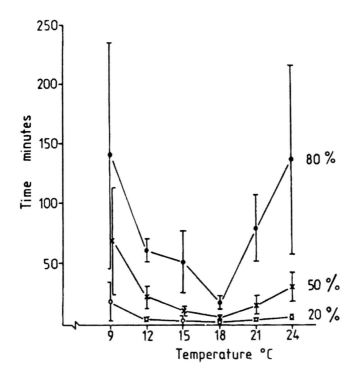

Figure 12.12. The effect of temperature on byssus attachment of *Aequipecten opercularis*. Time taken to reach 20, 50 and 80% byssal attachment at different temperatures. Mean values with 95% confidence limits. (From Paul 1980b).

The precision with which the recessing process is carried out indicates that recessing is a complex, closely-coordinated, behaviour pattern in *Pecten maximus*. Recessing usually starts soon after the scallop is exposed on the seabed. Once recessing commences, it takes from 2–20 minutes for *P. maximus* to complete the process and spat as small as 6–10 mm shell length have been seen to recess by this method (Minchin 1978, 1992; Minchin et al. 2000).

There are widespread reports of many other unattached scallops living in similar depressions on suitable soft substrates, including *Adamussium colbecki* (Berkman 1994), *Aequipecten opercularis* (Chapman et al. 1979), *Amusium pleuronectes* (Morton 1980), *Argopecten gibbus* (Allen and Costello 1972), *A. purpuratus* (Illanes-Bucher 1987), *Mizuhopecten yessoensis* (Tyurin 1990; Sakurai and Seto 2000), *Pecten fumatus* and *Equichlamys bifrons* (Olsen 1955; Young and Martin 1989), *Pecten novaezelandiae* (Tunbridge 1968) and *Placopecten magellanicus* (Caddy 1968; Stokesbury and Himmelman 1995b), but none of these authors provide a good description of the recessing process. Indeed, although *Pecten fumatus* and *Equichlamys bifrons* may occur in typical saucer-shaped depressions, Olsen (1955) is emphatic in stating that these scallops do not bury themselves. An alternative mechanism to account for the formation of recesses in *Placopecten magellanicus* is given by Grant et al. (1993). In flume experiments, these authors found that for scallops with the ventral margin facing upstream, sediment was eroded from a horseshoe-shaped trough around the ventral margin of the shell, and deposited in two elongated crests at each end of the trough and on the auricles and dorsal margin of the scallop shell. Based on these observations they suggest that recesses are formed by the interaction between passive sediment transport and active excavation of sediment by the scallop. In view of the enormous number of observations that have been made on different scallop species worldwide, the absence in the literature of any good description of a rapid, closely-coordinated, recessing process, similar to that repeated observed in *Pecten maximus,* suggests that passive sediment transport may be an important component of recess formation in many species, particularly those that do not normally recess. For example, *Aequipecten opercularis* does not normally recess but some individuals, heavily encrusted with epifauna, have been observed in typical depressions (Chapman et al. 1979). Similarly, *Argopecten irradians* usually remains on the surface of the substrate but has been reported to recess in coarse sand, especially in winter (Belding 1931; Waller 1969) when scallop activity is low and water currents may be elevated. Recessed scallops are generally unattached but about 40% of recessed *Placopecten magellanicus* are anchored by a byssus to shell fragments in the sediment (Caddy 1968), and recessed spat of *Pecten maximus* is often similarly attached (Minchin 1992). Tank experiments have shown that *Pecten maximus* move frequently and disperse widely when placed on a concrete bottom, but recess rapidly, with little movement, when on sand (Baird 1958). In Strangford Lough, unrecessed *Pecten* moved about more regularly but recessed individuals stayed in the same position for an average of 27 days, probably moving only when disturbed (Hartnoll 1967). It is of interest to note that this average duration of stay in one position is remarkably similar to that of 31 days calculated for byssally attached *Chlamys islandica* in a Norwegian fjord (Vahl and Clausen 1980).

Following disturbance by predators or fishing gear, and after reseeding in culture operations, the speed of recessing of *Pecten maximus* is an important factor in reducing vulnerability to predation by visual predators such as crabs, fish and octopus. Recessing may reduce predation by camouflaging the scallop against visual predators but this is of limited value on grounds where the major predators of adult scallops are starfish, which detect their prey by chemosensory mechanisms. However, the layer of sediment on the upper valve may confer some protection from starfish by restricting contact with the tube feet, and it also reduces the settlement of epifauna on the shell, which may both reduce competition with epifaunal filter feeders and improve swimming performance. In recent years the rate of recessing of *Pecten maximus* has been frequently used as a quantitative test of scallop vitality in studies of the ecological impact of dredging (Maguire et al. 1997, 1999a, b, 2002a, b), and to assess methods for reseeding scallops on the seabed in aquaculture (Fleury et al. 1996, 1997).

Whatever the mechanism of recess formation, for many scallop species there may be important benefits of living in a recess associated with feeding. Studies of the gut contents of various scallops, including *Argopecten irradians* (Gutsell 1931; Davis and

Marshall 1961), *Aequipecten opercularis* (Hunt 1925; Aravindakshan 1955; Christensen and Kanneworff 1985; Lawrence 1993), *Mizuhopecten yessoensis* (Imai 1980; Mikulich and Tsikhon-Lukanina 1981), *Pecten maximus* (Lawrence 1993) and *Placopecten magellanicus* (Shumway et al. 1987), have generally found both pelagic and benthic food species, together with variable quantities of detritus. Scallops live at the sediment-water interface and appear to be opportunistic filter feeders, taking advantage of benthic organisms and material deposited on the surface of the sediment, as well as particles suspended in the near-bottom water (Shumway et al. 1987). When scallops recess, the inhalant water current is lowered to at least the level of the sediment surface and benthic material can be more easily drawn into the mantle cavity. Detritus and material falling from the water column will tend to collect in the depressions excavated by the scallops, particularly in areas with strong current flow, and this will increase the supply of available food. In a comparative study of the feeding mechanism of *Pecten maximus* and *Aequipecten opercularis* in a strongly tidally-mixed area of the north Irish Sea, Lawrence (1993) found that resuspended benthic diatoms were present in the benthic boundary layer throughout the year and comprised the major dietary item of both species. Planktonic diatom species were only available, and present in the gut contents, in spring and early summer. She concluded that *P. maximus*, which usually recesses, probably receives a lower quality diet than the unrecessed *A. opercularis* since *P. maximus* draws in water from nearer the sediment surface and take in more detritus. Comparative laboratory experiments showed that *P. maximus* has higher clearance rates and exhibits greater particle selection than *A. opercularis*, but is more sensitive to resuspended sediment in the diet. *Argopecten irradians* has been reported to re-suspend the surface sediments by vigorous shell clapping activity (Davis and Marshall 1961). This mechanism would operate more efficiently in recessed scallops and may be widespread in other species. Filter feeding in scallops, as in all bivalves, depends on relatively weak currents through the mantle cavity generated by the ctenidia, and is inhibited in strong water currents (e.g., Kirby-Smith 1972; Wildish et al. 1987, 1992; Wildish and Kristmanson 1988; Wildish and Saulnier 1993; Pilditch and Grant 1999). This suggests that, in areas with very strong current flow, efficient feeding may only be possible for recessed scallops. In such areas, recessing will also provide a limited form of anchorage, preventing the scallop from being washed away by the current.

12.4.3 Orientation

Scallops on the seabed live with the commissure plane horizontal and the morphological right valve underneath. If they get turned over they are able to regain the normal position by a special movement known as the righting reflex (Dakin 1909; Buddenbrock 1911). In this, water is ejected downwards by a powerful adduction in such a way that the shell flips over into the correct orientation. Young scallops generally right themselves with one adduction but older scallops may require several successive attempts (Minchin 1992). The speed of righting of *Pecten maximus* has been used as a measure of stress in recent studies of reseeding and the ecological impact of fishing (Minchin et al. 1999, 2000; Maguire et al. 2002a, b).

Some scallops appear to control their orientation with regard to the direction the shell is facing. Hartnoll (1967) found that the orientation of *Pecten maximus* in Strangford Lough was non-random, with recessed scallops tending to orientate so that the ventral shell gape faced directly into the tidal flow, which in this area is virtually unidirectional (Figure 12.13a). A similar orientation with respect to water currents was noted by Caddy (1968) for a population of *Placopecten magellanicus* in the Gulf of St Lawrence. In two more typical locations, where the direction of tidal flow reverses, Mathers (1976a) found a tendency for *Pecten maximus* to orientate to either of two directions (Figure 12.13b and c). Approximately half the population faced the ebb current while the other half faced the flood current, and individual scallops did not change position when the tidal flow changed direction. This apparently imposed a rhythm of feeding and digestion in these scallops, phased to the tidal cycle, with scallops facing the ebb current being 6 h out-of-phase with those facing the flood current. Such orientation to a water current probably increases the efficiency of filter feeding. With the ventral shell gape facing into the flow, tidal currents will assist mantle cavity ventilation because the water is moving past the scallop in a direction that complements both the inhalant and the exhalant ventilatory stream of the scallop (Figure 12.14). However, various more recent studies have often failed to find clear orientation in the field or have produced conflicting results. Thus, MacDonald and Bajdik (1992) studied two populations of *Placopecten magellanicus* and found that one population oriented so that at least a portion of the inhalant margin was directed into the prevailing current, while the other population was randomly orientated. Stokesbury and Himmelman (1995b) found that sea scallop orientation was random in two natural beds and at three sites used for tag-release experiments, two of which were on gravel substrates and one on sand. However, tagged scallops released on bedrock were significantly orientated so they concluded that, although sea scallops have the ability to orient, they do not do so on gravel and sand substrates if current velocities are less than 18.3 cm s^{-2}. More complex relationships between scallop orientation and current speed have also been found for the Japanese scallop, *Mizuhopecten yessoensis,* by Sakurai and Seto (2000), who concluded that orientation to flow maximised filtration and avoided dislodgement.

Scallops that do not recess may also orientate to currents but there have been few observations. Chapman et al. (1979) and Chapman (1981) report underwater observations that unattached *Aequipecten opercularis* also orientate so that the ventral gape faces into the current. Thorburn and Gruffydd (1979) found no evidence for this in experimental studies in a flume, but found instead that *A. opercularis,* made lethargic by high temperatures, tended to be passively orientated with the hinge facing upstream at higher water velocities. In similar flume experiments on *Chlamys islandica* (Gruffydd 1976b) and *Placopecten magellanicus* (Grant et al. 1993), the hydrodynamically stable passive orientation at higher current flows was also with the hinge facing into the current. When the tentacles of *C. islandica* were fully extended, the dorso-ventral axis of the scallops in these experiments was directly aligned in the direction of the current, but with the tentacles withdrawn the shell rotated in an anticlockwise direction by about 40°. When byssally attached in this orientation the elongated anterior auricle would make it impossible for the shell to be overturned by strong currents (Stanley 1970), but it would appear to be difficult for the scallop to feed efficiently as the water currents would directly

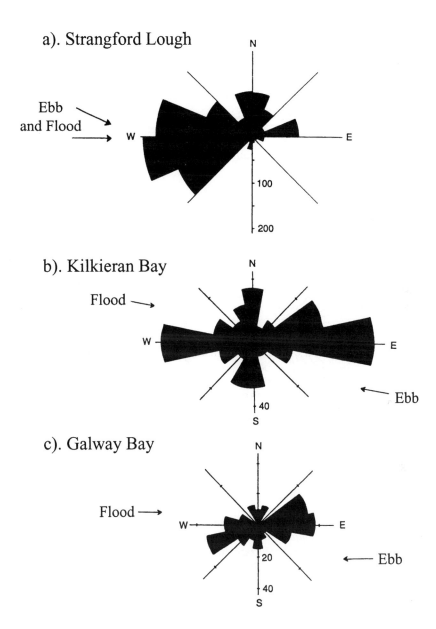

Figure 12.13. The orientation of *Pecten maximus* in relation to water currents. Diagrams showing the direction of orientation of the ventral shell margin of scallop populations in a) Strangford Lough, b) Kilkieran Bay, c) Galway Bay. The directions of ebb and flow tidal currents at each location are indicated by arrows. (Redrawn and modified from a) Hartnoll 1967 and b), c) Mathers 1976).

698

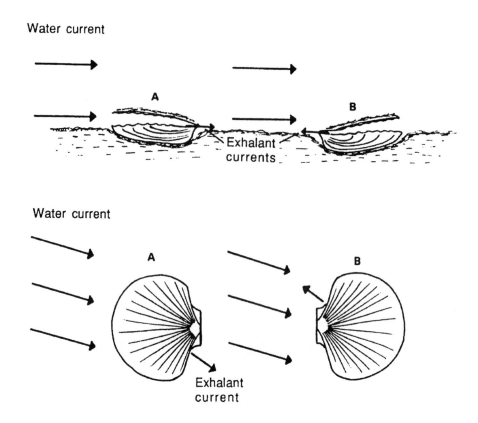

Figure 12.14. The life position and orientation of *Pecten maximus* in relation to water currents. In A), the direction of the water current complements the inhalant and exhalant streams of the scallop, in B), the directions are opposed. (From Mathers 1976).

oppose the exhalant ventilatory stream. The underwater observations of Olsen (1955), however, suggest that *Pecten fumatus* orientates with the hinge facing into the current in a strong tide but that orientation is haphazard in weaker tides. The occurrence of both active and passive components of scallop orientation (Grant et al. 1993), and the ecological significance of orientation in species with different life habits, is therefore not well understood and more detailed studies are required.

12.4.4 Swimming

The swimming behaviour of scallops is considered in some detail in Chapter 5 of this book, approached from a physiological viewpoint; this chapter concentrates instead on the more ecological aspects. The mechanism of swimming has been studied frequently in *Pecten maximus* (Antony 1906; Dakin 1909; Buddenbrock 1911; Bayliss et al. 1930;

Buddenbrock and Möller-Racke 1953; Baird 1958; Thomas and Gruffydd 1971; DeMont 1990; Fleury et al. 1996; Jenkins and Brand 2001) and *Placopecten magellanicus* (Caddy 1968; Dadswell and Weihs 1990; Manuel and Dadswell 1991, 1993, 1994; Parsons and Dadswell 1992; Parsons et al. 1992; Bowie et al. 1993; Carsen et al. 1994, 1995, 1996; Cheng et al. 1996; Cheng and DeMont 1996a, b; Kleinman et al. 1996; Lafrance et al. 2002). Other species have been studied in varying detail including *Adamussium colbecki* (Ansell et al. 1998), *Aequipecten opercularis* (Rees 1957; Moore and Trueman 1971; Stephens and Boyle 1978; Chapman et al. 1979; Thorburn and Gruffydd 1979; Chapman 1981; Ansell and Ackerly 1994; Jenkins et al. 2003), *Amusium balloti* (Joll 1989), *Amusium pleuronectes* (Morton 1980), *Argopecten gibbus* (Allen and Costello 1972), *A. irradians* (Belding 1931; Gutsell 1931; Moore and Marshall 1967; Ordzie and Garofalo 1980a; Peterson et al. 1982; Winter and Hamilton 1985; Guzik and Marsh 1989), *Chlamys hastata* (Guzik and Marsh 1990; Donovan et al. 2002), *C. islandica* (Gruffydd 1976b; Vahl and Clausen 1980; Brokordt et al. 2000a), *Euvola ziczac* (Brokordt et al. 2000b), *Mimachlamys asperrima* and *Equichlamys bifrons* (Pitcher and Butler 1987), *Mizuhopecten yessoensis* (Dautov and Karpenko 1984; Sakurai and Seto 2000) and *Pecten fumatus* (Fairbridge 1953; Olsen 1955).

There have been a number of important comparative studies on the functional morphology and evolution of the swimming habit in scallops (Yonge 1936, 1953; Waller 1969; Stanley 1970), and special emphasis has been placed on allometric changes in muscular mechanics during growth (Gould 1971; Soemodihardjo 1974) and adductor muscle obliquity (Thayer 1972). These latter, largely theoretical, studies made inferences about the swimming abilities of different species that have been tested and discussed in subsequent experimental investigations of scallop hydrodynamics and behaviour (e.g., Moore and Trueman 1971; Gruffydd 1976b; Thorburn and Gruffydd 1979; Morton 1980; Vogel 1985; Dadswell and Weihs 1990; Hayami 1991; Manuel and Dadswell 1991 1993, 1994; Millward and Whyte 1992; Ansell and Ackerly 1994; Cheng et al. 1996; Sakurai and Seto 2000). Most investigations of swimming have been carried out in laboratory tanks but particularly valuable observational and experimental studies have been carried out in the natural environment to determine reactions to environmental conditions, fishing gear and contact with predators in *Aequipecten opercularis* (Chapman et al. 1979; Jenkins et al. 2003), *Amusium balloti* (Joll 1989), *Argopecten irradians* (Moore and Marshall 1967; Peterson et al. 1982; Winter and Hamilton 1985), *Pecten maximus* (Baird 1958; Hartnoll 1967; Minchin 1992; Fleury et al. 1996) and *Placopecten magellanicus* (Caddy 1968; Parson and Dadswell 1992; Carsen et al. 1994, 1996).

Swimming responses appear to be very similar in all species. In addition to the righting reflex mentioned above, scallops utilise two locomotory escape responses, the jumping and swimming reactions (Figure 12.15) (Buddenbrock 1911; Thomas and Gruffydd 1971). During jumping, water is ejected from the ventral mantle margin and usually propels the animal hinge-line (dorsal) foremost. The shell adductions that produce these jets are few in number (usually 1–3), generate relatively low pressures in the mantle cavity (Moore and Trueman 1971), and recur at low frequencies so that the scallop, if it takes off at all, returns to the seabed between each adduction. During swimming, the scallop moves ventral edge foremost, propelled from the seabed by water ejected dorsally

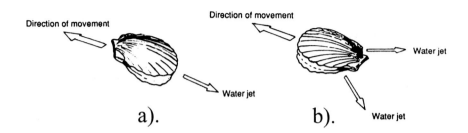

Figure 12.15. (a) Scallop jumping and (b) swimming movements, indicating the direction of movement, and the position and direction of water jets.

on either side of the hinge in a rapid series of adductions that recur at a high frequency (2–5 adductions s⁻¹) and generate relatively high pressures in the mantle cavity. A wide gaping of the valves for 5–15 seconds characteristically precedes the swimming response, whereas the valves gape only momentarily before jumping. Swimming is therefore a more vigorous response and moves the scallop a greater distance than jumping.

In a typical burst of swimming (Figure 12.16a) the adult scallop takes off from the seabed and rises up into the water at an angle of about 30–50°, the trajectory then levels off and the scallop may swim more-or-less horizontally for a short distance before active swimming stops and the animal sinks passively to the seabed in a series of side-slips (Caddy 1968; Chapman et al. 1979; Morton 1980).

This type of jet propelled swimming is very effective in adult scallops, but has limitations in juveniles due to problems of scaling. In a detailed analysis of swimming in juvenile sea scallop of 5–30 mm shell height, Manuel and Dadswell (1993) describe three types of swimming. Very small sea scallops predominantly use stepwise swimming, which involves the alternation of two types of valve clap cycle, the up-clap and the over-clap, that alternate more vertical and more horizontal body angles in a step-like pattern. Small scallops develop so much lift that they ascend from the seabed at a very steep angle and often follow a spiral flight path with little horizontal displacement. However, since the maximum velocity that small scallops can attain by swimming is lower that the water currents they often experience on the seabed, staying up in the water column may be advantageous for maximising downstream movement. As sea scallops increase in size the stable swimming type progressively replaces stepwise swimming, with variable swimming occurring in a small proportion of animals in the intermediate size range. In stable swimming the shell is maintained at a relatively stable body angle, the angle of ascent declines, swimming velocity increases and the flight path levels out so that the scallops become increasingly capable of horizontal displacement under their own power. During stable swimming over-claps predominate but some scallops may execute an up-clap every fourth or fifth adduction (Figure 12.16b) to produce an undulating flight similar to that observed in some adult sea scallops by Caddy (1968). From their experimental and modelling studies Manuel and Dadswell (1993) conclude that sea scallops pass

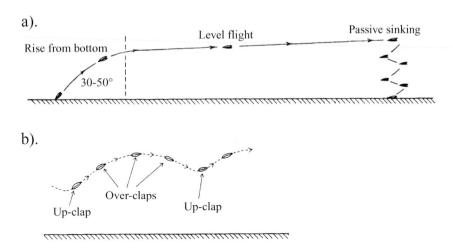

Figure 12.16. Swimming in *Placopecten magellanicus*. a). The path taken during a typical single burst of swimming (From Caddy 1968). b). Undulating level swimming, showing up-claps and over-claps (From Manuel and Dadswell 1994).

a hydrodynamic threshold at a shell height of 12–16 mm, which they believe corresponds to the change from turbulent to laminar flow around the shell. Below this threshold swimming is energetically very costly, which may explain why very small sea scallops tend to remain byssally attached.

Some comparative data on the swimming performance of different scallops is given in Table 12.3. These values may not all be strictly comparable, as they were obtained by different workers using a variety of methods, but they do allow some assessment of swimming ability and of factors affecting swimming. *Amusium* was long presumed to be the most proficient bivalve swimmer, based on its morphological adaptations (Yonge 1936; Stanley 1970; Gould 1971; Thayer 1972), but until the investigations by Morton (1980) on *Amusium pleuronectes*, swimming had not actually been observed in any species. His results showed that, compared with the scallops previously studied, *A. pleuronectes* swam further, faster and appeared also to have the greatest endurance, as it made the largest number of adductions in one swimming burst. However, subsequent studies of *A. balloti* by Joll (1989) showed that this is an even better swimmer, for it swims substantially faster and more than double the maximum single-swim distance reported for *A. pleuronectes*. *Placopecten magellanicus* has an excellent shell shape for producing hydrodynamic lift (Gruffydd 1976b; Thorburn and Gruffydd 1979) and is an excellent swimmer: it swims at a similar speed to *A. pleuronectes* but makes fewer adductions per swim and travels a much shorter distance. *Aequipecten opercularis* is also relatively fast and has been frequently shown to be a very good swimmer. The information available for other species is more limited but *Pecten fumatus* and *Argopecten irradians*, in particular, also appear to be powerful swimmers.

Table 12.3

Swimming performance of some scallop species during one (or more) swimming burst.

Species	Distance m	Height m	Speed cm·s^{-1}	Number of Adductions	Reference
Aequipecten opercularis					
Claonaig Bay	1.5	-	35–40		Chapman et al. 1979
Loch Ewe	1.1	0.6–1.2	29–36	-	
maximum	-	1.5	56	-	
maximum – multiple swims	6.6	-	-	-	
65–75 mm	-	-	25	12	Moore and Trueman 1971
60–70 mm	-	-	18–34	5–19	
Amusium balloti					
<60 mm		0.2–0.5	80–100		Joll 1989
>60 mm		0.5–1.0	20–70		
maximum – 1 swim	23.1		160		
maximum – 2 swims	30.8				
Amusium pleuronectes					
60 mm	-	-	45	19	Morton 1980
100 mm	-	-	37	25	
maximum	>10	-	73	>50	
Argopecten irradians					
Average 64 mm	0.97	0.67	-	-	Winter and Hamilton 1985
maximum	7.6	-	-	-	Belding 1931
Argopecten gibbus					
40 mm	0.3	0.46	-	-	Allen and Costello 1972
Chlamys islandica					
20 mm	-	-	15	-	Gruffydd 1976
40 mm	-	-	25	-	
70 mm	-	-	ca. 35		
Pecten maximus					
12 mm spat	0.45	0.2	-	-	Minchin 1978
adults	-	0.8–0.9	-	-	
maximum	ca. 3	0.8–0.9	-	-	Minchin and Mathers 1982
Pecten fumatus	4.6	1.8	-	-	Olsen 1955
Placopecten magellanicus					
mean	2.3	0.4	-	12	Caddy 1968
>100 mm	0.9	0.07–0.1	-	6–8	
maximum	4.3	0.9	>67.7	21	
5–30 mm – range		0.05–0.60	1.2–46.6		Manuel and Dadswell
– Mean		0.195	17.3		1993

Various factors affect swimming performance. Chapman et al. (1979) attributed the relatively poorer performance of the queen scallop *Aequipecten opercularis* in Loch Ewe, compared with queens in Claonaig Bay, to the presence of shell-fouling epifauna in Loch Ewe, which increased both weight and drag. This conclusion is supported by the experimental studies of Winter and Hamilton (1985), who found that weights added to the shell of *Argopecten irradians* caused a marked reduction in the distance travelled in a burst of swimming. They calculated that a weight equivalent to the maximum weight of shell epifauna (6 g) caused a 28% decrease in the distance travelled. In contrast, Jenkins and Brand (2003) found that the epifaunal load on the shell had no effect on the ability of *Aequipecten opercularis* to swim above the dredge mouth. Swimming performance also varies with body size but the relationships are not straightforward and vary among species. Several species that swim well as juveniles appear to lose the ability to swim during ontogeny (Yonge 1936; Olsen 1955; Baird and Gibson 1956; Caddy 1968; Waller 1969; Minchin and Mathers 1982). Gould (1971) attributes this to the inherent problems of scaling, which he and Thayer (1972) believe is partly counteracted in older scallops by allometric growth of shell and muscle. Morton (1980) refutes much of this analysis for *Amusium pleuronectes,* pointing out that although small individuals swim faster, the large individuals swim for longer periods of time and hence go farther. Distance travelled per swim also increases with body size in *Amusium balloti* (Joll 1989), *Aequipecten opercularis* (Chapman et al. 1979) and *Argopecten irradians* (Winter and Hamilton 1985), but decreases markedly in large *Placopecten magellanicus* (Caddy 1968) and *Pecten maximus* (Minchin and Mathers 1982). In contrast to *Amusium pleuronectes,* the swimming speed of *Amusium balloti* (Joll 1989), *Aequipecten opercularis* (Chapman et al. 1979) and *Chlamys islandica* (Gruffydd 1976b) increases with body size and, up to a certain size, it probably does also in *Placopecten magellanicus* for Caddy (1968) comments that sea scallops of 70 mm shell length and over can keep pace with fishing gear towed at 1–2 knots, while smaller individuals fall behind. However, sea scallops over 100 mm shell length are recessed and do not respond to approaching gear. Larger animals require an increased swimming speed if hydrodynamic lift is to compensate for increasing body weight (Gould 1971). Irrespective of differences in swimming performance, it would seem that, for most species, large individuals swim less often and require a higher threshold stimulus to elicit the swimming reaction (Baird and Gibson 1956; Caddy 1968; Allen and Costello 1972; Mathers 1976a; Morton 1980; Minchin and Mathers 1982). At the other end of the size range, swimming is also problematic, and energetically expensive, for very small scallops. This has been clearly demonstrated for *Placopecten magellanicus* (Manuel and Dadswell 1993, 1994) but the same factors may also apply in other species.

Taken overall, Table 12.3 shows that few scallops swim more than 1 m off the seabed or greater than 5 m distance in one swimming burst. On repeated stimulation the distance travelled in successive bursts of swimming falls sharply and an increasing proportion of scallops refuse to swim (Chapman et al. 1979; Winter and Hamilton 1985; Joll 1989). Chapman et al. (1979) found that very few *Aequipecten opercularis* made more than 8 consecutive swims, and none more than 11, and the mean cumulative distance travelled for all swims was only 6.6 m in Claonaig Bay and 3.5 m in Loch Ewe. For *Amusium balloti* only one individual out of 55 tested could be stimulated to make 4 consecutive swims (Joll 1989), but this individual (93 mm shell height) achieved a cumulative distance of 27.4 m, while another individual (109 mm shell height) covered 30.8 m in two swims. Biochemical evidence suggests that the inability to undertake multiple swims is due to fatigue, rather than habituation to the stimulus; scallops take several hours to recover normal physiological functions after swimming to exhaustion (Gäde et al. 1978; Grieshaber 1978; de Zwaan et al. 1980; Thompson et al. 1980; Livingstone et al. 1981; Brokordt et al. 1999, 2000a, b; Jenkins and Brand 2001). It must therefore be concluded that, with the possible exception of *Amusium*, scallop swimming reactions involve rapid movements over very short distances. Such responses are well suited for escape reactions but are inappropriate for efficient long-distance swimming.

In addition to predators, scallop swimming reactions are known to be evoked or modified by certain environmental factors. *Argopecten irradians* swims more often, and for greater distances, when released on sand than in eelgrass patches (Winter and Hamilton 1985). In addition, over short distances (25 cm), bay scallops appear to be able to swim towards seagrass beds using visual information (Hamilton and Koch 1996). The mechanisms controlling these reactions are not known but such behaviour serves to aggregate bay scallops in their preferred habitat of seagrass beds, where conditions are possible better for feeding and avoiding predators. Water currents also affect swimming activity. Experimental studies of *Chlamys islandica* in a flume showed that all sizes swam more often at higher current speeds, or when current speed was raised, but swimming activity was suppressed when current speed decreased (Gruffydd 1976b). At all current speeds a greater proportion of the smallest size group became byssally attached and swam less frequently than the larger size groups. Gruffydd considered that these reactions to currents may be of ecological significance, leading to the inward movement of the larger animals in Norwegian fjords. However, observations in one fjord by Vahl and Clausen (1980) showed that a very high proportion of *C. islandica* is normally byssally attached (97%) and the frequency of movement is very low. They also comment that swimming reactions to the presence of a diver are less evident in periods when the currents are strong. *Aequipecten opercularis* may be induced to swim by stimulating the dorsal velum with a water jet from a syringe (Chapman et al. 1979), but no relationship was found between current speed and swimming frequency in laboratory experiments with this species (Thorburn and Gruffydd 1979). In contrast, Moore and Marshall (1967) recorded increased swimming activity in *Argopecten irradians* 1–2 hours after tidal reversals, which suggests that this was a current response.

Many scallop species react by swimming when approached by divers (Gibson 1956; Baird 1958; Moore and Marshall 1967; Vahl and Clausen 1980; Minchin and Mathers 1982; Parsons and Dadswell 1992) and by fishing gear (Olsen 1955; Caddy 1968; Chapman et al. 1979; Jenkins et al. 2003). When one scallop swims it often induces swimming in others nearby, setting off a chain-reaction (Chapman et al. 1979; Vahl and Clausen 1980; Minchin and Mathers 1982; Howell and Fraser 1984). These may be responses to stimulation by light, water currents or vibrations but there have been few experimental studies. Scallops have complex and sensitive eyes that can detect movements and shadows (Buddenbrock and Möller-Racke 1953; Land 1966).

Underwater observations of *Aequipecten opercularis* in Claonaig Bay, Scotland, showed that queen scallops responded to fishing gear by swimming at distances of 1–1.5 m and were only caught in trawls when off the seabed (Chapman et al. 1979). The importance of vision in stimulating swimming is evident from the fact that daylight catches in this fishery generally exceed those taken at night. Conversely, the active-swimming *Pecten fumatus* is more easily caught at night (Olsen 1962), but this difference is probably more a reflection of the different type of fishing gear used, than of differences in the reactions of the scallops, for the dredges used in this Tasmanian fishery capture scallops only when they are inactive. To fish successfully with dredges the boats generally work in formation in this fishery, and good catches are taken from a bed only after several hours dredging, when the scallops have become too fatigued to swim.

Surprisingly little is known about the seasonal variability of scallop swimming, though various studies in both the field and the laboratory have noted correlations between temperature and different components of swimming performance. Pitcher and Butler (1987), for example, recorded fewer escape responses by *Mimachlamys asperrima* in August at 11.4°C, than in April at 19°C. Scheibling et al. (1995) noted that active swimming of *Placopecten magellanicus* to approaching divers was observed in November at 6°C, but not in January and February when water temperature was ~0°C, and swimming started again as temperature increased to 5°C in March. Parsons and Dadswell (1992) also describe a very clear seasonal pattern of swimming activity for *Placopecten magellanicus*. They found that very few scallops swam in the winter months at temperatures of <4°C, but swimming increased in spring and summer to a maximum in late summer and autumn that corresponded with the annual temperature maxima. They also found that the clap rate during swimming correlated with temperature over the range 3–12°C. This confirmed previous laboratory studies on juvenile sea scallops by Manuel and Dadswell (1991), who also reported that mean and maximum swimming velocities correlated with temperature. Recently, Lafrance et al. (2002) investigated the effect of a thermal shock, such as might occur during a scallop reseeding operation. They found that clap rate and maximum number of claps per swim were significantly reduced following the sudden transfer of juvenile sea scallops from 18°C to 8°C, and although there was some subsequent acclimation, clap rate did not fully acclimate within 6 days. Clearly, many of the physiological properties of scallop adductor muscle are temperature dependent, as shown for *Argopecten irradians* by Olson and Marsh (1993) who reported that maximum velocity at zero force, and therefore maximal power output, increased significantly with temperature over the range 10–20°C. As well as swimming activity, the time taken to respond to stimuli before swimming can also vary with season. Ordzie and Garofalo (1980a) found that the response time of *Argopecten irradians* before swimming was significantly greater at 12°C than at 20°C, but that both predator recognition and swimming performance were temperature independent. In a more detailed study of seasonal variability in the swimming behaviour of *Aequipecten opercularis*, Jenkins et al. (2003) recorded the proportion of queen scallops that evaded dredge capture by swimming up into the water column to be captured in a net deployed above the dredge. They found a strong positive correlation with seawater temperature, and laboratory experiments indicated that the observed seasonal variability in dredge avoidance was probably the result of changes in the time taken to respond to stimuli, rather than changes in the ability to perform a large number of valve adductions. In contrast, Jenkins and Brand (2001) found no seasonal variability in the response time of *Pecten maximus*.

Although the results of these various studies have been inconsistent, temperature clearly has a direct affect on scallop swimming behaviour through its effects on metabolic activity and nerve transmission. However, in all studies involving seasonal comparisons, the direct effects of temperature are confounded with other factors that vary seasonally, such as reproductive state, food supply and stored energy reserves. Seasonal cycles of reproduction, and energy storage and utilisation, are well known in scallops (see Thompson and MacDonald 1991), and the experimental manipulation of temperature in the laboratory has shown a positive relationship between temperature and the metabolic condition of the adductor muscle in *Placopecten magellanicus* (Kleinman et al. 1996). Brokordt et al. (2000a) showed that reproductive stage had no effect on the number and rate of valve claps performed by *Chlamys islandica*, although there was a reduction in the ability to recover from exhausting swimming activity during gonadal maturation and spawning. Similar results were obtained for *Euvola ziczac* (Brokordt et al. 2000b). Jenkins and Brand (2003) also found no effect of reproductive state on avoidance of dredge capture in *Aequipecten opercularis*. In *Pecten maximus* the reduction in adenylic energetic charge of the adductor muscle in response to the stress of dredge capture varied seasonally and was lowest in February, when glycogen levels in the muscle were at a minimum (Maguire et al. 2002c). Despite these various findings, confirmation of the direct role of temperature in causing seasonal variability in swimming performance requires further laboratory experiments to unconfound variability in seawater temperature with seasonal variability in the metabolic status of scallops, and to clarify which components of swimming behaviour contribute to the observed differences between seasons.

A proper understanding of seasonal variability in scallop swimming has important implications for both fisheries and aquaculture. In fisheries, the precise description of seasonal variations in swimming activity may allow more efficient use of different gear types and more effective gear to be developed. Furthermore, quantitative estimates of seasonal changes in catchability, such as those made by Jenkins and Brand (2003), will allow more reliable stock assessments to be made in studies using commercial or scientific catch rates as indices of population abundance. In aquaculture, knowledge of seasonal variations in swimming is important for successful bottom culture or sea ranching of scallops. Numerous studies have shown that juvenile sea scallops, reseeded on the seabed at high densities, disperse at a greater rate in the warmer months (e.g., Carsen et al. 1995; Hatcher et al. 1996), and similar results have been obtained for *Pecten maximus* (Fleury et al. 1996). Seasonal variability in scallop escape responses also affect the vulnerability of reseeded scallops to predation, and since predators are also affected by seasonal factors, the interactions are complex (Barbeau and Scheibling 1994c; Barbeau et al. 1994, 1996).

12.4.5 Reactions to Predators

It is very well known that contact with predators can induce an escape reaction in many scallops, and this behaviour has received increasing attention as the reseeding of juvenile scallops has become more frequent. Most reports of escape reactions have involved responses to asteroid starfish (e.g., Dakin 1909; Buddenbrock 1911; Lecomte 1952; Mellon 1969; Moore and Trueman 1971; Thomas and Gruffydd 1971; Soemodihardjo 1974; Bloom 1975; Stephens and Boyle 1978; Wilkens 1981; Dautov and Karpenko 1984; Chernoff 1987; Pitcher and Butler 1987; Jenkins and Brand 2001; Ortiz et al. 2003), or gastropod molluscs (Ordzie and Garofalo 1980a, b; Peterson et al. 1982; Powers and Kittinger 2002), and there is little evidence that strong reactions are induced by other important predators such as crustaceans and fish.

Scallops react to contact by a predator by closing the shell, jumping or swimming, three distinct responses of increasing intensity (Thomas and Gruffydd 1971). The response of *Pecten maximus, P. jacobaeus* and *Aequipecten opercularis* varies greatly with different species of starfish: a high proportion make vigorous escape movements (jumping or swimming) after contact by predatory starfish species such as *Asterias rubens, Astropecten irregularis* and *Marthasterias glacialis,* whereas non-predatory starfish like *Solaster papposus, Porania pulvillus* and *Henricia sanguinolenta* evoke more minor reactions, or none at all (Lecomte 1952; Thomas and Gruffydd 1971; Soemodihardjo 1974). A similar ability to discriminate between predatory and non-predatory gastropod molluscs has been demonstrated for *Argopecten irradians* (Ordzie and Garofalo 1980a). Such behavioural recognition of potential predators has been described for other molluscs groups (Bullock 1953) and may be a common feature of molluscan behaviour. To evoke the swimming reaction it is generally necessary for the starfish or gastropod to touch the mantle margin or tentacles of the scallop, which suggests that the response is chemosensory. The reactions to starfish are possibly mediated by surface-active agents, secreted by the tube-feet of the starfish (Feder and Lasker 1964), which have been identified as 'saponin-like' or, more specifically, as steroid glycosides (Mackie 1970; Mackie and Turner 1970). Nothing is known of the mechanism by which the gastropod predators of *Argopecten irradians* are selectively identified. It is likely that some scallop escape reactions to predators are also induced by visual stimuli (Uexküll 1912; Buddenbrock and Möller-Racke 1953; Morton 2001) or by tactile stimuli such as vibrations, but this has been mostly inferred from scallop escape reactions to divers and fishing gear (Olsen 1955; Caddy 1968; Chapman et al. 1979; Joll 1989; Jenkins et al. 2003). For most species, reactions to tactile stimuli are not as vigorous or as predictable as the response to physical contact by starfish. Since many scallop species appear to swim less readily as they get older (or bigger) (Baird and Gibson 1956; Caddy 1968; Allen and Costello 1972; Mathers 1976b; Morton 1980; Minchin and Mathers 1982; Parsons and Dadswell 1992; Jenkins et al. 2003) it might be expected that escape reactions to predators would also change during ontogeny. Morton (1980) postulates that such behavioural changes occur in *Amusium pleuronectes*, and that larger scallops rely more on sustained shell closure, rather than escape movements, to avoid predators. However, this conclusion is based mostly on functional interpretations of the morphology, together with knowledge of changes in swimming performance with increasing size, and there have been few direct investigations. Thomas and Gruffydd (1971) concluded that the escape response to stimulation by *Asterias rubens* was independent of age for *Pecten maximus* from 3–8 years old, although their data suggests a decline in the response of the 7 and 8 year olds. More recently, Jenkins and Brand (2001) found that the time taken for *P. maximus* to respond to a starfish stimulus was significantly longer in larger (105–110 mm shell length), compared with smaller (80–95 mm) scallops. For *Aequipecten opercularis,* Soemodihardjo (1974) found no difference in the proportion of animals of different age groups that responded to *Asterias rubens* by swimming, but Jenkins et al. (2003) found a lower propensity to swim to avoid gear capture in larger individuals. Variations in swimming behaviour and ability with age and body size therefore differ among species and the relationships are not straightforward. Escape reactions are of considerable advantage to scallops in reducing predation. The ability to differentiate between potential predators and harmless species is clearly of great adaptive value for it minimises feeding interruptions and prevents energy being wasted in making unnecessary escape responses. This is particularly advantageous for species that recess (Thomas and Gruffydd 1971), or byssally attach (Vahl and Clausen 1980), for it saves these species the considerable additional energy cost of making a new depression or re-forming the byssus.

In recent years a number of studies have described mutualistic associations between species of *Chlamys* and various epizoic sponges, which act, together with the scallop escape reactions, to reduce predation by starfish. Such associations are amenable to manipulative experiments in laboratory tanks, or enclosures in the sea, in order to test the nature and ecological significance of the interaction between the species (Bloom 1975; Forester 1979; Chernoff 1987; Pitcher and Butler 1987; Pond 1992; Burns and Bingham 2002; Donovan et al. 2002). Epizoic sponges on the upper valve of the scallop greatly inhibit predation by starfish but some sponges are more effective than others (Pitcher and Butler 1987). The scallop probably derives benefit mainly because the presence of the sponges greatly reduces adhesion of the starfish tube-feet to the shell, but tactile camouflage may also be involved (Bloom 1975; Forester 1979; Chernoff 1987; Pond 1992). The interactions between species are clearly complex for the presence of sponges on the shell reduces the responsiveness of *Mimachlamys asperrima* to contact by predatory starfish (Pitcher and Butler 1987), but does not appear to inhibit swimming in *Chlamys hastata* (Donovan et al. 2002). In addition to providing protection from predators, the presence of sponges on the shell may confer additional selective advantages for *Mimachlamys asperrima* with crellid sponges on the shell have higher growth rates of shell and tissue than those without sponges (Pitcher and Butler 1987). Similar associations between scallops and other taxa, such as the epibiotic relationship between the hydroid *Hydractinia angusta* and the Antarctic scallop *Adamussium colbecki,* may also defend the scallop from predators (Cerrano et al. 2001).

12.4.6 Dispersal and Migrations

The final aspect of scallop behaviour to consider is the extent and nature of any population movements. Many fishermen all over the world believe that scallops migrate from one area to another and this view has occasionally received scientific support. Yonge (1936), for example, stated that *Aequipecten opercularis* and *Argopecten irradians* both 'make extensive migrations from time to time'. Numerous studies have recorded individual scallops, and occasionally whole populations, moving for distances that vary from a few metres to many kilometres, but there is, in fact, very little evidence to show that these movements are active, directed, migrations.

Tagging experiments have been carried out on several populations of *Pecten maximus* (Gibson 1953, 1956; Baird and Gibson 1956; Mason 1957; Murphy 1986; Brand and Murphy 1992; Allison 1993). They all show very little movement, with nearly all returns from the same bed or within 1–2 km of the point of release. However, the analysis of tag returns from a commercial fishery is not a very accurate technique for assessing such small movements as it is generally impossible to determine the precise position of recapture. Howell and Fraser (1984) made a more precise study by transplanting tagged scallops into a small area at very high density and recording subsequent dispersal from a fixed point by a series of diving surveys. They found that some 60% of the live tagged scallops were still within 30 m of the release point after 18 months, confirming earlier views that this species moves very little (Hartnoll 1967). More active swimming species such as *Placopecten magellanicus* may move rather greater distances but there is no evidence of extensive migration (Baird 1954; Dickie 1955; Posgay 1963, 1981; Melvin et al. 1985). Tag returns from Georges Bank, analysed by Posgay (1981), showed that 85% of recaptures were within 10 miles (16.1 km) of the release point. He suggested that adult *Placopecten* move very little, that the direction of swimming is probably random and any net movement with time is brought about by tidal currents. The larger tagging study by Melvin et al. (1985) confirmed and amplified these findings. They reported 52% of recaptures within 5 km, 77% within 10 km and 94% within 25 km of the release point; found that mid-sized scallops (60–89 mm) moved greater distances than smaller or larger size groups and used vector analysis to show that scallop movements followed closely the complex patterns in direction and velocity of water movements over Georges Bank, Browns Bank and the Great South Channel. Carsen et al. (1995) found that tagged juvenile *P. magellanicus* were much more mobile in the autumn but the net daily displacement was <3 m for more than 60% of the scallops recorded. *Mizuhopecten yessoensis* tagged in Japan moved an average of 5–8 nautical miles (9.3–14.8 km) in 59 days and the direction of this movement was attributed to strong currents (Imai 1980). Similarly, movements of *Argopecten irradians* have also been considered to be due to current transport (Belding 1931; Moore and Marshall 1967), though one recent study has shown that over very short distances (<25 cm) bay scallops are able to use visual information to swim towards seagrass beds (Hamilton and Koch 1996). *Chlamys islandica* swims more frequently in strong currents (Gruffydd 1976b) and current transport could account for the unusual distribution of this species in Norwegian fjords, with progressively older age distributions towards the upper reaches (Wiborg 1963).

Although byssally attached throughout life, this species does move about (Vahl and Clausen 1980) and recruitment to some populations is delayed for several years (Vahl 1981, 1982), but until recently there was little evidence that this movement of older scallops is a directed migration.   However, Arsenault et al. (2000) tagged juvenile *C. islandica* on an inshore bed in the northern Gulf of St Lawrence and found that movement was consistently towards increasing depth, and not in the direction of the tidal currents. Although the mechanism controlling this downslope movement is not clear, this would appear to be the first experimental evidence indicating the directed movement of a scallop population away from a nursery area.

Most scallop species are not well-adapted for making extensive migrations: they swim for very short distances (see Table 12.3), become rapidly fatigued and take a long time to recover after swimming.   When they swim in response to predators, divers or fishing gear the direction of swimming is usually directed initially away from the source of stimulation but the course is erratic (Caddy 1968; Chapman et al. 1979; Winter and Hamilton 1985; Joll 1989; Carsen et al. 1996).   This leads to a random dispersal, but in areas of strong current flow there will be a tendency for swimming scallops to be carried downstream.   Such random or downstream dispersal probably account for the majority of the scallop movements reported in the literature.   When one scallop swims it often stimulates others nearby and this can lead to mass swimming activity in a population (Chapman et al. 1979; Vahl and Clausen 1980; Minchin and Mathers 1982; Howell and Fraser 1984).   In strong currents these scallops may be carried for some distance downstream, stimulating each other to swim several times before they become exhausted. This could explain some of the examples of mobile species like *Argopecten irradians* and *Aequipecten opercularis* suddenly disappearing from a fishing ground (Belding 1931; Rees 1957; Rolfe 1973).   Commercial fishermen often pull their gear along very precise 'tows', where the bottom topography is known to be suitable, so scallop populations may not need to be carried very far to become suddenly unavailable to fishing.

In scallops transplanted at high densities the aggregation of predators increases swimming escape responses and, together with the tendency to stimulate one another to swim, leads to a gradual dispersal.   For adult *Pecten maximus* this dispersal is random, or downstream, and continues until the distance between individuals is sufficient to reduce these interactions to more natural levels (Howell and Fraser 1984).   Several studies have investigated movements of juvenile *P. magellanicus* in bottom culture trials (e.g., Cliche et al. 1994; Giguère and Cliche 1994; Scheibling and Hatcher 1994; Carsen et al. 1995; Scheibling et al. 1995), and while dispersal is usually rapid, the general conclusion is that local hydrodynamics is not a reliable predictor of dispersion in natural habitats.   Large numbers of juvenile *Mizuhopecten yessoensis* are regularly transplanted to prepared beds for cultivation in Japan (Ito 1991).   These scallops generally remain in the area where they are released, but movements both inshore and offshore have been recorded, though the mechanisms responsible for these movements are not clear (Yamamoto 1951, 1953), or have been attributed to downstream advection (Goshima and Fujiwara 1994).   Volkov et al. (1982) had similar difficulty accounting for the direction of movement of juvenile *M. yessoensis* transplanted in Vityaz Bay, Sea of Japan, but concluded that starfish played an indirect role by stimulating swimming and the scallops then tended to move into deeper water 'for purely mechanical reasons'. They do not mention the direction of water currents in the area.

Dispersal into deeper water may also occur in populations of *Chlamys septumradiata* in the Clyde Sea area for young individuals have been found byssally attached in shallow areas and older animals live on the deep-water muds where there are few surfaces for attachment (Allen 1953). Golikov and Scarlato (1970) state that pre-spawning aggregations of *Mizuhopecten yessoensis* in the southern Primorje (Sea of Japan) move into very shallow water when temperatures increase in the spring and disperse again into deeper water in the winter. Insufficient information is available to explain these, or many other examples of apparent movements in scallop populations, but there is currently little direct evidence that any of these are active, directed, migrations. Minchin (1987) noted that *Pecten maximus*, unable to recess, consistently moved up the sloping sides of a large pond. Apart from the observations of Arsenault et al. (2000) on *Chlamys islandica* and Hamilton and Koch (1996) on *Argopecten irradians*, this would appear to be one of the only records of a directed movement in any scallop species but the ecological relevance of Minchin's observation is doubtful for the scallops were in very unnatural conditions, in much shallower water than they normally live. There is experimental evidence to support the movement of *C. islandica* into deeper water as they increase in size (Arsenault et al. 2000), but this is local-scale movement leading to spatial size partitioning within the same inshore bed, and the mechanism to account for this downslope movement is not known.

The strongest case for any large-scale migration in scallops is advocated by Morton (1980) for some members of the genus *Amusium*. Pointing out the many morphological adaptations of *Amusium* for efficient swimming, he puts the persuasive argument that 'it would seem ridiculous to suggest that such modifications arose simply to remove the scallop from the immediate proximity of a predator'. Instead, he argues, there is circumstantial evidence to suggest that two species of *Amusium* undergo active, seasonal migrations, coming in to the shallow coastal waters of Hong Kong to breed. This is based on the observation that both species appear in Hong Kong markets only when they are sexually mature, *A. pleuronectes* in the winter months and *A. japonicum* in the summer months, though the beam trawlers that catch these scallops fish in the same areas throughout the year. Morton himself admits that this is weak evidence of migration and fails to consider the alternative possibility that the seasonal appearance of mature scallops in the catch could result from a reduced ability to avoid capture when the gonads are full (however, recent studies have shown that reproductive state does not appear to affect swimming performance in *Aequipecten opercularis* (Jenkins et al. 2003)). There is also no support for seasonal migration in *Amusium* from tagging studies carried out in Australia on the related *A. balloti;* of 677 tagged scallops returned, only 14 had moved more than 10 km in up to 121 weeks at liberty (Williams and Dredge 1981). Despite the lack of sound evidence for migration, members of the genus *Amusium* have wide distributions throughout the Indo-Pacific region and are superbly adapted for a highly mobile mode of life, so extensive movements remain a possibility and field studies are urgently required to determine if these are directed movements that could be termed a migration.

12.5 FURTHER STUDY

Since the first edition of this book was published in 1991, knowledge of factors affecting the occurrence, spatial distribution and abundance of scallop populations has expanded very rapidly. As expected, the worldwide development of scallop culture, based on either the collection of naturally-produced larvae settling on artificial collectors placed in the sea after spawning, or on hatchery culture, has provided the impetus, the opportunity, and some useful methods for studying the biology of the larval and early post-larval stages of many species of scallop, and knowledge of these life-history stages has therefore increased enormously. Similarly, the development of bottom culture as a low-cost cultivation technique in many countries has stimulated extensive research on the ecology, physiology and behaviour of juvenile scallops, and on predator/prey interactions. Despite the wide scope and volume of this research there are inevitably many topics that require further study. Many of these have already been pointed out in the preceding sections but special mention should be made of some areas for research that are both important and timely.

Scallops have distinct habitat preferences, live in aggregated distributions and are relatively immobile so they are easily detected and caught in commercial or recreational fisheries. Ease of capture, combined with variable recruitment patterns, make scallops very vulnerable to overfishing and lead to 'boom and bust' fisheries. The high market value of scallops, and the decreasing availability of alternative fisheries as other stocks are overfished, result in high fishing effort and make it possible for boats to continue to fish for scallops at very low densities. In doing so, the heavy dredge gear used in most fisheries kills and damages many other animals on the seabed, changes the physical nature of sediments and modifies benthic habitats and communities (e.g., Dayton et al. 1995; Bradshaw et al. 2000; Brand 2000; Collie et al. 2000; Hall-Spencer and Moore 2000; Veale et al. 2000). In order to limit the widespread ecological impact of scallop dredging on benthic communities, and to provide more stable, economically efficient fisheries, there is growing evidence to suggest that the future of scallop fisheries may lie with closed area management (Horwood 2000). Rotational fishing of small closed areas, combined with predator clearance and stock enhancement, has long been part of the highly successful Japanese seabed cultivation of scallops (Ventilla 1982), and seasonal, rotational or permanent closures to bottom fishing, with or without stock enhancement, are increasingly being proposed or introduced to restore habitat, rebuild populations and manage scallop fisheries elsewhere (e.g., Dao and Carval 1999; Arbuckle and Metzger 2000; Murawski et al. 2000; Myers et al. 2000; Arnold 2001; Bradshaw et al. 2001). There have been some notable successes. Short rotational closures (<4 years), together with re-seeding, have been used with great successful to increase yield-per-recruit, maintain a stable fishery and sustain full-time employment for scallop fishers in New Zealand (Arbuckle and Metzger 2000). On Georges Bank, scallop biomass increased 14-fold following a 4-year closure to mobile fishing gear, without stock enhancement (Murawski, et al. 2000). The success of these and other closures, together with pressure from conservation bodies to protect endangered species and habitats, will undoubtedly lead to the introduction of other closed areas in the coming years. These closures will be fruitful fields for research. There has been a massive research effort on the ecological impacts of scallop dredging in recent years (reviewed in Jennings and Kaiser 1998; Bradshaw et al 2000), but there is relatively little information available on the recovery of benthic organisms, communities or habitats after closure to fishing (Collie et al. 2000; Bradshaw et al. 2001). Properly replicated large-scale experiments examining the time scale of recovery within closed areas are now urgently required so that closures can be designed to effectively manage fisheries and conserve benthic habitats. Research to determine the provenance of larvae and recruitment mechanisms is also essential so that closed areas are sited to protect sources of scallop recruitment, rather than sinks. Much of this knowledge is likely to come from the rapidly developing fields of biochemical and molecular genetics (Sweijd et al. 2000). Molecular genetic and immunological techniques are currently available that allow individual bivalve veliger larvae to be identified to species (e.g., Hu et al. 1992; Frischer et al. 2000; Paugam, et al. 2000), and with rapid developments in methods such as microsatellites, minisatellites and DNA sequencing it is becoming increasing possible to carry out high resolution studies on sub-populations or stocks (Thorpe et al. 2000). Such techniques could be used to evaluate the extent to which closed areas export larvae, juvenile or adult scallops, and to assess the effectiveness of stock enhancement, particularly when the stock enhancement involves the transfer of scallops from other locations or from hatcheries. The molecular genetic techniques will also be available to be applied more widely, together with ecological investigations and oceanographic modeling, in multidisciplinary studies of recruitment mechanisms, which has long been one of the most challenging and intractable problems in scallop fisheries biology.

ACKNOWLEDGMENTS

I am grateful for the assistance and support of many colleagues but particularly my research assistants Bryce Beukers-Stewart, Stuart Jenkins, Belinda Vause and Mog Mosley. Chris Bridge assisted with the illustrations. I would also like to commend Sandy Shumway and Jay Parsons for their patience in waiting for the manuscript. It is a pleasure also to acknowledge the generous financial support provided by the Isle of Man Department of Agriculture, Fisheries and Forestry for much of my research on scallops.

REFERENCES

Allen, D.M., 1979. Biological aspects of the calico scallop, *Argopecten gibbus*, determined by spat monitoring. Nautilus 93:107–119.

Allen, D.M. and Costello, T.J., 1972. The calico scallop, *Argopecten gibbus*. Special Scientific Report-Fisheries Series 656, National Marine Fisheries Service, National Oceanic and Atmospheric Administration. 19 p.

Allen, J.A., 1953. Observations on the epifauna of the deep-water muds of the Clyde Sea area, with special reference to *Chlamys septemradiata* (Müller). Journal of Animal Ecology 22:240–260.

714

Allison, E.H., 1993. The dynamics of exploited populations of scallops (*Pecten maximus* (L.)) and queens (*Chlamys opercularis* (L.)) in the North Irish Sea. PhD Thesis. University of Liverpool. 254 p.

Allison, E.H. and Brand, A.R., 1995. A mark-recapture experiment on queen scallops, *Aequipecten opercularis*, on a North Irish Sea fishing ground. Journal of the Marine Biological Association of the United Kingdom 75:323–335.

Ambrose, W.G. and Irlandi, E.A., 1992. Height of attachment on seagrass leads to trade-off between growth and survival in the bay scallop *Argopecten irradians*. Marine Ecology Progress Series 90:45–51.

Ambrose, W.G.J., Peterson, C.H., Summerson, H.C. and Lin, J., 1992. Experimental tests of factors affecting recruitment of bay scallops (*Argopecten irradians*) to spat collectors. Aquaculture 108:67–86.

Andrew, N.L. and Mapstone, B.D., 1987. Sampling and the description of spatial pattern in marine ecology. Oceanography and Marine Biology Annual Review 25:39–90.

Anonymous, 1984a. Abrolhos Islands scallop bonanza - but fears for the future. Australian Fisheries 43:4–5.

Anonymous, 1984b. Scallops and abalone support valuable fisheries. Australian Fisheries 43:44–50.

Ansell, A.D. and Ackerly, S.C., 1994. Swimming in *Aequipecten opercularis*: preliminary scaling considerations. In: N.F Bourne, B.L. Bunting and L.D. Townsend (Eds.). Proceedings of the 9th International Pectinid Workshop, Nanaimo, B. C. Canada, April 22–27, 1993. Canadian Technical Report of Fisheries and Aquatic Sciences 1994 (1): 3–11.

Ansell, A.D., Cattaneo-Vietti, R. and Chiantore, M., 1998. Swimming in the Antarctic scallop *Adamussium colbecki*: analysis of *in situ* video recordings. Antarctic Science 10:369–375.

Ansell, A.D., Dao, J.D., Lucas, A., Mackie, A.L. and Morvan, C., 1988. Reproductive and genetic adaptations in natural and transplant populations of the scallop, *Pecten maximus*, in European waters. Report to the European Commission, EEC Scientific Cooperation Contract, No. ST2J - 0058-1-UK (CD). 50 p.

Antony, R., 1906. Contribution á l'étude de la mode de vie et de la locomotion de *Pecten*. Bulletin du Musée Océanographique de Monaco 85:1–11.

Aravindakshan, I., 1955. Studies on the biology of the queen scallop, *Chlamys opercularis* (L.). PhD Thesis. University of Liverpool. 79 p.

Arbuckle, M. and Metzger, M., 2000. Food for thought, a brief history of the future of fisheries management. Challenger Scallop Enhancement Company, Nelson, New Zealand. 25 p.

Arnold, W.S., 2001. Bivalve enhancement and restoration strategies in Florida, USA. Hydrobiologia 465:7–19.

Arsenault, D.J., Giasson, M.C. and Himmelman, J.H., 2000. Field examination of dispersion patterns of juvenile Iceland scallops (*Chlamys islandica*) in the northern Gulf of St Lawrence. Journal of the Marine Biological Association of the United Kingdom 80:501–508.

Arsenault, D.J., Girard, P. and Himmelman, J.H., 1997. Field evaluation of the effects of refuge use and current velocity on the growth of juvenile Iceland scallops, *Chlamys islandica* (O.F. Muller, 1776). Journal of Experimental Marine Biology and Ecology 217:31–45.

Arsenault, D.J. and Himmelman, J.H., 1996a. Ontogenic habitat shifts of the Iceland scallop, *Chlamys islandica* (Muller, 1776), in the northern Gulf of St Lawrence. Canadian Journal of Fisheries and Aquatic Sciences, 53:884–895.

Arsenault, D.J. and Himmelman, J.H., 1996b. Size related changes in vulnerability to predators and spatial refuge use by juvenile Iceland scallops *Chlamys islandica*. Marine Ecology Progress Series 140:115–122.

Baird, F.T., 1954. Migration of the deep sea scallop (*Placopecten magellanicus*). Maine Department of Sea and Shore, Fisheries Circular 14: 8 p.

Baird, R.H., 1958. On the swimming behaviour of escallops (*Pecten maximus* L.). Proceedings of the Malacological Society of London 33:67–71.

Baird, R.H., 1966. Notes on an escallop (*Pecten maximus*) population in Holyhead harbour. Journal of the Marine Biological Association of the United Kingdom 46:33–47.

Baird, R.H. and Gibson, F.A., 1956. Underwater observations on escallop (*Pecten maximus* L.) beds. Journal of the Marine Biological Association of the United Kingdom 35:555–562.

Barbeau, M.A. and Caswell, H., 1999. A matrix model for short-term dynamics of seeded populations of sea scallops. Ecological Applications 9:266–287.

Barbeau, M.A., Hatcher, B.G., Scheibling, R.E., Hennigar, A.W. and Risk, A.C., 1996. Dynamics of juvenile sea scallops (*Placopecten magellanicus*) and their predators in bottom seeding trials in Lunenburg Bay, Nova Scotia. Canadian Journal of Fisheries and Aquatic Sciences 53:2494–2512.

Barbeau, M.A. and Scheibling, R.E., 1994a. Behavioural mechanisms of prey size selection by sea stars (*Asterias vulgaris* Verrill) and crabs (*Cancer irroratus* Say) preying on juvenile sea scallops (*Placopecten magellanicus* (Gmelin)). Journal of Experimental Marine Biology and Ecology 180:103–136.

Barbeau, M.A. and Scheibling, R.E., 1994b. Procedural effects of prey tethering experiments: predation of juvenile scallops by crabs and sea stars. Marine Ecology Progress Series 111:305–310.

Barbeau, M.A. and Scheibling, R.E., 1994c. Temperature effects on predation of juvenile sea scallops (*Placopecten magellanicus* (Gmelin)) by sea stars (*Asterias vulgaris* Verrill) and crabs (*Cancer irroratus* Say). Journal of Experimental Marine Biology and Ecology 182:27–47.

Barbeau, M.A., Scheibling, R.E. and Hatcher, B.G., 1998. Behavioural responses of predatory crabs and sea stars to varying density of juvenile sea scallops. Aquaculture 169:87–98.

Barbeau, M.A., Scheibling, R.E., Hatcher, B.G., Taylor, L.H. and Hennigar, A.W., 1994. Survival analysis of tethered juvenile sea scallops *Placopecten magellanicus* in field experiments - Effects of predators, scallop size and density, site and season. Marine Ecology Progress Series 115:243–256.

Barber, B.J., Getchell, R., Shumway, S. and Schick, D., 1988. Reduced fecundity in a deep-water population of the giant scallop *Placopecten magellanicus* in the Gulf of Maine, USA. Marine Ecology Progress Series 42:207–212.

Bayliss, L.E., Boyland, E. and Ritchie, A.D., 1930. The adductor mechanism of *Pecten*. Proceedings of the Royal Society of London 106B:363–376.

Beaumont, A.R., 1982a. Geographic variation in allele frequencies at three loci in *Chlamys opercularis* from Norway to the Brittany coast. Journal of the Marine Biological Association of the United Kingdom 62:243–261.

Beaumont, A.R., 1982b. Variations in heterozygosity at two loci between year classes of a population of *Chlamys opercularis* (L.) from a Scottish sea-loch. Marine Biology Letters 3:25–33.

Beaumont, A.R. and Beveridge, C.M., 1984. Electrophoretic survey of genetic variation in *Pecten maximus*, *Chlamys opercularis*, *C. varia* and *C. distorta* from the Irish Sea. Marine Biology 81:299–306.

Beaumont, A.R. and Gruffydd, L.D., 1975. A polymorphic system in the sarcoplasm of *Chlamys opercularis*. Journal du Conseil International pour l'Exploration de la Mer 36:190–192.

Beaumont, A.R., Morvan, C., Huelvan, S., Lucas, A. and Ansell, A.D., 1993. Genetics of indigenous and transplanted populations of *Pecten maximus*: no evidence for the existence of separate stocks. Journal of Experimental Marine Biology and Ecology 169:77–88.

Belding, D.L., 1931. The scallop fishery of Massachusetts - including an account of the natural history of the common scallop. Commonwealth of Massachusetts, Division of Fisheries and Game, Marine Fisheries Series 3: 51 p.

Berkman, P.A., 1991. Spatial distribution of an unexploited nearshore Antarctic scallop population. In: S.E. Shumway and P.A. Sandifer (Eds.). An International Compendium of Scallop Biology and Culture. World Aquaculture Society, Baton Rouge, LA. pp. 176–179.

Berkman, P.A., 1994. Epizoic zonation on growing scallop shells in McMurdo Sound, Antarctica. Journal of Experimental Marine Biology and Ecology 179:49–67.

Bernard, F.R., 1983. Catalogue of the living Bivalvia of the eastern Pacific Ocean: Bering Strait to Cape Horn. Canadian Special Publications in Fisheries and Aquatic Sciences 61: 102 p.

Beukers-Stewart, B.D., Mosley, M.W.J. and Brand, A.R., 2003. Population dynamics and predictions in the Isle of Man fishery for the great scallop, *Pecten maximus* (L.). ICES Journal of Marine Science 60:224–242.

Blake, S.G., 1994. Mitochondrial DNA variation in three populations of the bay scallop, *Argopecten irradians*. In: N.F Bourne, B.L. Bunting and L.D. Townsend (Eds.). Proceedings of the 9th International Pectinid Workshop, Nanaimo, B.C. Canada, April 22–27, 1993. Canadian Technical Report of Fisheries and Aquatic Sciences 1994 (1):200–204.

Blake, S.G. and Graves, J.E., 1995. Mitochondrial DNA variation in the bay scallop, *Argopecten irradians* (Lamarck, 1819), and the Atlantic calico scallop, *Argopecten gibbus* (Linnaeus, 1758). Journal of Shellfish Research 14:79–85.

Bloom, S.A., 1975. The motile escape response of sessile prey: a sponge-scallop mutualism. Journal of Experimental Marine Biology and Ecology 17:311–321.

Bologna, P.A.X. and Heck, K.L. Jr, 1999. Differential predation and growth rates of bay scallops within a seagrass habitat. Journal of Experimental Marine Biology and Ecology 239:299–314.

Bourne, N., 1964. Scallops and the offshore fishery of the Maritimes. Bulletin of the Fisheries Research Board of Canada 145: 60 p.

Bowie, M.A., Layes, J.D. and DeMont, M.E., 1993. Damping in the hinge of the scallop *Placopecten magellanicus*. Journal of Experimental Biology 175:311–315.

Bradshaw, C., Veale, L.O., Hill, A.S. and Brand, A.R., 2000. The effects of scallop dredging on gravelly seabed communities. In: M.J. Kaiser and S.J. de Groot (Eds.). Effects of Fishing on Non-Target Species and Habitats. Blackwell Science, Oxford. pp. 83–104.

Bradshaw, C., Veale, L.O., Hill, A.S. and Brand, A.R., 2001. The effect of scallop dredging on Irish Sea benthos: experiments using a closed area. Hydrobiologia 465:129–138.

Brand, A.R., 2000. North Irish Sea scallop (*Pecten maximus* and *Aequipecten opercularis*) fisheries: effects of 60 years of dredging on scallop populations and the environment. Alaska Department of Fish and Game, Special Publication 14:37–43.

Brand, A.R. and Murphy, E.J., 1992. A tagging study of North Irish Sea scallop (*Pecten maximus*) populations: comparisons of an inshore and an offshore fishing ground. Journal of Medical and Applied Malacology 4:153–164.

Brand, A.R., Allison, E.H. and Murphy, E.J., 1991. North Irish Sea scallop fisheries: a review of changes. In: S.E. Shumway and P.A. Sandifer (Eds.). An International Compendium of Scallop Biology and Culture. World Aquaculture Society, Baton Rouge, LA. pp. 204–218.

Brand, A.R., Paul, J.D. and Hoogesteger, J.N., 1980. Spat settlement of the scallops *Chlamys opercularis* (L.) and *Pecten maximus* (L.) on artificial collectors. Journal of the Marine Biological Association of the United Kingdom 60:379–390.

Brett, J.R., 1956. Some principles in the thermal requirements of fishes. Quarterly Review of Biology 31:75–87.

Bricelj, V.M., Epp, J. and Malouf, R.E., 1987a. Comparative physiology of young and old cohorts of bay scallop *Argopecten irradians irradians* (Lamarck): mortality, growth, and oxygen consumption. Journal of Experimental Marine Biology and Ecology 112:73–91.

Bricelj, V.M., Epp, J. and Malouf, R.E., 1987b. Intraspecific variation in reproductive and somatic growth cycles of bay scallops *Argopecten irradians*. Marine Ecology Progress Series 36:123–137.

Bricelj, M., Garcia-Esquivel, Z. and Strieb, M., 1991. Predatory risk of juvenile bay scallops, *Argopecten irradians*, in eelgrass habitat. Journal of Shellfish Research 10:271.

Briggs, C.F., 1983. A study of some sublittoral populations of *Asterias rubens* and their prey. PhD Thesis. University of Liverpool. 193 p.

Brokordt, K., Lafrance, M., Himmelman, J., Nusetti, O. and Guderley, H., 1999. The use of escape responses to evaluate the physiological status of scallops; experiments with temperate and tropical species. Book of Abstracts, 12th International Pectinid Workshop, Bergen, Norway, 5–11 May, 1999. pp. 90–91.

Brokordt, K.B., Himmelman, J.H. and Guderley, H.E., 2000a. Effect of reproduction on escape responses and muscle metabolic capacities in the scallop *Chlamys islandica* Muller 1776. Journal of Experimental Marine Biology and Ecology 251:205–225.

Brokordt, K.B., Himmelman, J.H., Nusetti, O.A. and Guderley, H.E., 2000b. Reproductive investment reduces recuperation from exhaustive escape activity in the tropical scallop *Euvola ziczac*. Marine Biology 137:857–865.

Broom, M.J., 1976. Synopsis of biological data on scallops *Chlamys* (*Aequipecten*) *opercularis* (Linnaeus), *Argopecten irradians* (Lamarck), *Argopecten gibbus* (Linnaeus). Food and Agriculture Organisation, Fisheries Synopsis No 114, FIRS/S114. 44 p.

Broom, M.J. and Mason, J., 1978. Growth and spawning in the pectinid *Chlamys opercularis* in relation to temperature and phytoplankton concentration. Marine Biology 47:277–285.

Brun, E., 1968. Extreme population density of the starfish *Asterias rubens* L. on a bed of Iceland scallop *Chlamys islandica* (O.F. Müller). Astarte 1 (32):1–4.

Buddenbrock, W. von, 1911. Untersuchungen über die Schwimmbewegungen und die Statocysten der Gattung *Pecten*. Sitzungsberichte der Heidelberger Akademie der Wissenschaften 28:1–24.

Buddenbrock, W.V. and Möller-Racke, I., 1953. Über den Lichtsinn von *Pecten*. Publicazione della Stazione Zoologica di Napoli 24:217–245.

Buestel, D. and Dao, J.C., 1978. Aquaculture extensive de la coquilles St. Jacques: resultats d'un semis experimental. Départment Scientifique du Centre Oceanologique du Brest, Contribution No. 624: 10 p (mimeo).

Buestel, D. and Dao, J.C., 1979. Aquaculture extensive de la coquille Saint Jacques: resultats d'un semis experimental. Pêche Maritime 1215:361–365.

Bull, M.F., 1976. Aspects of the biology of the New Zealand scallop, *Pecten novaezealandiae* Reeve 1853, in the Marlborough Sounds. PhD Thesis. Victoria University, Wellington, New Zealand. 175 p.

Bull, M.F., 1989. The New Zealand scallop fishery: a brief review of the fishery and its management. In: M.C.L. Dredge, W.F. Zacharin and L.M. Joll (Eds.). Proceedings of the Australasian Scallop Workshop, Hobart, Australia. Tasmanian Government Printer, Hobart. pp. 42–50.

Bull, M.F., 1991. New Zealand. In: S.E. Shumway (Ed.). Scallops: Biology, Ecology and Aquaculture. Developments in Aquaculture and Fishery Science, Volume 21. Elsevier, Amsterdam. pp. 853–859.

Bullock, T.H., 1953. Predator recognition and escape responses of some intertidal gastropods in presence of starfish. Behaviour 5:130–140.

Burns, D., Egan, T., Seward, E. and Naidu, K.S., 1995. Identification of scallop beds using an acoustic ground discrimination technique. Book of Abstracts, 10th International Pectinid Workshop, Cork, Ireland, 26 April – 2 May, 1995. pp. 11–12.

Burns, D.O. and Bingham, B.L., 2002. Epibiotic sponges on the scallops *Chlamys hastata* and *Chlamys rubida*: increased survival in a high-sediment environment. Journal of the Marine Biological Association of the United Kingdom 82:961–966.

Butman, C.A., 1987. Larval settlement of soft sediment invertebrates: the spatial scales of pattern explained by active habitat selection and the emerging role of hydrodynamical processes. Oceanography and Marine Biology Annual Review 25:113–165.

Caddy, J.F., 1968. Underwater observations on scallop (*Placopecten magellanicus*) behaviour and drag efficiency. Journal of the Fisheries Research Board of Canada 25:2123–2141.

Caddy, J.F., 1970. A method of surveying scallop populations from a submersible. Journal of the Fisheries Research Board of Canada 27:533–549.

Caddy, J.F., 1972. Progressive loss of byssus attachment with size in the sea scallop, *Placopecten magellanicus* (Gmelin). Journal of Experimental Marine Biology and Ecology 19:179–190.

Caddy, J.F., 1973. Underwater observations on the tracks of dredges and trawls and some effects of dredging on a scallop ground. Journal of the Fisheries Research Board of Canada 30:173–180.

Caddy, J.F., 1975. Spatial model for an exploited shellfish population, and its application to the Georges Bank scallop fishery. Journal of the Fisheries Research Board of Canada 32:1305–1328.

Caddy, J.F., 1989. A perspective on the population dynamics and assessment of scallop fisheries, with special reference to the sea scallop, *Placopecten magellanicus* Gmelin. In: J.F. Caddy (Ed.). Marine Invertebrate Fisheries: Their Assessment and Management. John Wiley & Sons, New York. pp. 559–589.

Carsen, A., Hatcher, B.G. and Scheibling, R.E., 1994. Displacement patterns of sea scallops, *Placopecten magellanicus*: a comparison of flume and short and long-term field experiments. In: N.F Bourne, B.L. Bunting and L.D. Townsend (Eds.). Proceedings of the 9th International

Pectinid Workshop, Nanaimo, B.C. Canada, April 22–27, 1993. Canadian Technical Report of Fisheries and Aquatic Sciences 1994 (1):19.

Carsen, A.E., Hatcher, B.G. and Scheibling, R.E., 1996. Effect of flow velocity and body size on swimming trajectories of sea scallops, *Placopecten magellanicus* (Gmelin): a comparison of laboratory and field measurements. Journal of Experimental Marine Biology and Ecology 203:223–243.

Carsen, A.E., Hatcher, B.G., Scheibling, R.E., Hennigar, A.W. and Taylor, L.H., 1995. Effects of site and season on movement frequencies and displacement patterns of juvenile sea scallops *Placopecten magellanicus* under natural hydrodynamic conditions in Nova Scotia, Canada. Marine Ecology Progress Series 128:225–238.

Castagna, M., 1975. Culture of the bay scallop, *Argopecten irradians*, in Virginia. Marine Fisheries Review 37:19–24.

Castagna, M. and Chanley, P., 1973. Salinity tolerance of some marine bivalves from inshore and estuarine environments in Virginia waters on the western mid-Atlantic. Malacologia 12:57–96.

Cerrano, C., Puce, S., Chiantore, M., Bavestrello, G. and Cattaneo-Vietti, R., 2001. The influence of the epizoic hydroid *Hydractinia angusta* on the recruitment of the Antarctic scallop *Adamussium colbecki*. Polar Biology 24:577–581.

Chapman, C.J., 1981. The swimming behaviour of queens in relation to trawling. Scottish Fisheries Bulletin 46:7–10.

Chapman, C.J., Main, J., Howell, T. and Sangster, G.I., 1979. The swimming speed and endurance of the queen scallop *Chlamys opercularis* in relation to trawling. In: J.C. Gamble and J.D. George (Eds.). Progress in Underwater Science, Volume 4. Pentech Press, London. pp. 57–72.

Cheng, J.Y., Davison, I.G. and DeMont, M.E., 1996. Dynamics and energetics of scallop locomotion. Journal of Experimental Biology 199:1931–1946.

Cheng, J.Y. and DeMont, M.E., 1996a. Hydrodynamics of scallop locomotion: unsteady fluid forces on clapping shells. Journal of Fluid Mechanics 317:73–90.

Cheng, J.Y. and DeMont, M.E., 1996b. Jet-propelled swimming in scallops: swimming mechanics and ontogenic scaling. Canadian Journal of Zoology 74:1734–1748.

Chernoff, H., 1987. Factors affecting mortality of the scallop *Chlamys asperrima* (Lamarck) and its epizooic sponges in South Australian waters. Journal of Experimental Marine Biology and Ecology 109:155–171.

Chiantore, M., Cattaneo-Vietti, R., Berkman, P.A., Nigro, M., Vacchi, M. and Sciaparelli, S., 2001. Antarctic scallop (*Adamussium colbecki*) spatial population variability along the Victoria Land Coast, Antarctica. Polar Biology 24:139–143.

Christensen, H. and Kanneworff, E., 1985. Sedimenting phytoplankton as a major food source for suspension and deposit feeders in the Øresund. Ophelia 24:223–244.

Christophersen, G. and Strand, Ø., 2003. Effect of reduced salinity on the great scallop (*Pecten maximus*) spat at two rearing temperatures. Aquaculture 215:79–92.

Ciocco, N.F., 1991. Differences in individual growth rate among scallop (*Chlamys tehuelcha* (d'Orbigny)) populations from San José Gulf (Argentina). Fisheries Research 12:31–42.

Ciocco, N.F., 1996. "In situ" natural mortality of the tehuelche scallop, *Aequipecten tehuelchus* (d'Orb, 1846), from San José Gulf (Argentina). Scientia Marina 60:461–468.

Claereboudt, M.R., Himmelman, J.H. and Côté, J., 1994. Field-evaluation of the effect of current velocity and direction on the growth of the giant scallop, *Placopecten magellanicus*, in suspended culture. Journal of Experimental Marine Biology and Ecology 183:27–39.

Clarke, A.H., 1965. The scallop superspecies *Aequipecten irradians* (Lamarck). Malacolgia 2:161–188.

Cliche, G., Giguère, M. and Vigneau, S., 1994. Dispersal and mortality of sea scallops, *Placopecten magellanicus* (Gmelin 1791), seeded on the sea bottom off Îles-de-la-Madeleine. Journal of Shellfish Research 13:565–570.

Cochard, J.C. and Devauchelle, N., 1993. Spawning, fecundity and larval survival and growth in relation to controlled conditioning in native and transplanted populations of *Pecten maximus* (L.): evidence for the existence of separate stocks. Journal of Experimental Marine Biology And Ecology 169:41–56.

Coleman, N., 1998. Counting scallops and managing the fishery in Port Phillip Bay, south-east Australia. Fisheries Research 38:145–157.

Collie, J.S., Hall, S.J., Kaiser, M.J. and Poiner, I.R., 2000. A quantitative analysis of fishing impacts on shelf-sea benthos. Journal of Animal Ecology 69:785–798.

Cooper, R.A. and Marshall, N., 1963. Condition of the bay scallop *Aequipecten irradians* in relation to age and environment. Chesapeake Science 4:126–134.

Cummins, R.J., 1971. Calico scallops of the southeastern United States, 1959–69. U.S. Department of Commerce, Special Scientific Report – Fisheries No. 627: 22 p.

Dadswell, M.J. and Weihs, D., 1990. Size-related hydrodynamic characteristics of giant scallop, *Placopecten magellanicus* (Bivalvia: Pectinidae). Canadian Journal of Zoology 68:778–785.

Dakin, J., 1909. *Pecten*. Liverpool Marine Biology Committee Memoirs on Typical British Marine Plants and Animals, No. 17. William & Norgate, London. 136 p.

Darby, C.D. and Durance, J.A., 1989. Use of the North Sea water parcel following model (NORSWAP) to investigate the relationship of larval source to recruitment for scallop (*Pecten maximus*) stocks of England and Wales. ICES CM 1989, Doc No. K:28. 19 p.

Dare, P.J. and Bannister, R.C.A., 1987. Settlement of scallop, *Pecten maximus*, spat on natural substrates off south west England: the hydroid connection. 6th International Pectinid Workshop, Menai Bridge, Wales, 9–14 April, 1987. 12 p. (mimeo).

Dare, P.J., Darby, C.D., Durance, J.A. and Palmer, D.W., 1994. The distribution of scallops, *Pecten maximus*, in the English Channel and Celtic Sea in relation to hydrographic and substrate features affecting larval dispersal and settlement. In: N.F. Bourne, B.L. Bunting and L.D. Townsend (Eds.). Proceedings of the 9th International Pectinid Workshop, Nanaimo, B.C., Canada, April 22–27, 1993. Canadian Technical Report of Fisheries and Aquatic Sciences 1994 (1):20–27.

Dautov, S.S.H. and Karpenko, A.A., 1984. Behaviour and mechanisms of locomotion of two species of scallops from the Sea of Japan. Soviet Journal of Marine Biology 9:271–275.

Davenport, J., Gruffydd, L.D. and Beaumont, A.R., 1975. An apparatus to supply water of fluctuating salinity and its use in a study of the salinity tolerences of larvae of the scallop *Pecten maximus* L. Journal of the Marine Biological Association of the United Kingdom 55:391–409.

Davis, R.L. and Marshall, N., 1961. The feeding of the bay scallop, *Aequipecten irradians*. Proceedings of the National Shellfish Association 52:25–29.

Dayton, P.K., Thrush, S.F., Tundi Agardy, M. and Hofman, R.J., 1995. Environmental effects of marine fishing. Aquatic Conservation 5:205–232.

Del Norte, A.G.C., 1991. Philippines. In: S.E. Shumway (Ed.). Scallops: Biology, Ecology and Aquaculture. Developments in Aquaculture and Fishery Science, Volume 21. Elsevier, Amsterdam. pp. 825–834.

Dell, R.K., 1972. Antarctic benthos. Advances in Marine Biology 10:1–216.

DeMont, M.E., 1990. Tuned oscillations in the swimming scallop *Pecten maximus*. Canadian Journal of Zoology 68:786–791.

Dichmont, C.M., Dredge, M.C.L. and Yeomans, K., 2000. The first large-scale fishery-independent survey of the saucer scallop, *Amusium japonicum balloti* in Queensland, Australia. Journal of Shellfish Research 19:731–739.

Dickie, L.M., 1955. Fluctuations in abundance of the giant scallop *Placopecten magellanicus* (Gmelin), in the Digby area of the Bay of Fundy. Journal of the Fisheries Research Board of Canada 12:797–857.

Dickie, L.M., 1958. Effects of high temperature on survival of the giant scallop. Journal of the Fisheries Research Board of Canada 15:1189–1211.

Dickie, L.M. and Medcof, J.C., 1963. Causes of mass mortalities of scallops (*Placopecten magellanicus*) in the southwestern Gulf of St. Lawrence. Journal of the Fisheries Research Board of Canada 20:451–482.

Donovan, D.A., Bingham, B.L., Farren, H.M., Gallardo, R. and Vigilant, V.L., 2002. Effects of sponge encrustation on the swimming behaviour, energetics and morphometry of the scallop *Chlamys hastata*. Journal of the Marine Biological Association of the United Kingdom 82:469–476.

Dredge, M.C.L., 1985a. Estimates of natural mortality and yield-per-recruit for *Amusium japonicum balloti* Bernardi (Pectinidae) based on tag recoveries. Journal of Shellfish Research 5:103–109.

Dredge, M.C.L., 1985b. Growth and mortality in an isolated bed of saucer scallops, *Amusium japonicum balloti* (Bernard). Queensland Journal of Agricultural Animal Science 42:11–21.

Dredge, M., Duncan, P., Heasman, M., Johnston, B., Joll, L., Mercer, J., Souter, D. and Whittingham, T., 2002. Feasibility of scallop enhancement and culture in Australian waters. Queensland Government, Department of Primary Industries. 124 p.

Drew, G.A., 1906. The habits, anatomy and embryology of the giant scallop (*Pecten tenuicostatus*, Mighels). University of Maine Studies Series 6: 89 p.

Duggan, N.A., 1987. Recruitment in North Irish Sea scallop stocks. PhD Thesis. University of Liverpool. 142 p.

Duggan, W.P., 1975. Reactions of the bay scallop, *Argopecten irradians*, to gradual reductions in salinity. Chesapeake Science 16:284–286.

Eckman, J.E., 1987. The role of hydrodynamics in recruitment, growth, and survival of *Argopecten irradians* (L.) and *Anomia simplex* (D'Orbigny) within eelgrass meadows. Journal of Experimental Marine Biology and Ecology 106:165–191.

Eckman, J.E., Peterson, C.H. and Cahalan, J.A., 1989. Effects of flow speed, turbulence, and orientation on growth of juvenile bay scallops *Argopecten irradians concentricus* (Say). Journal of Experimental Marine Biology and Ecology 132:123–140.

Eggleston, D., 1962. Spat of the scallop (*Pecten maximus* L.) from off Port Erin, the Isle of Man. Annual Report of the Marine Biological Station, Port Erin 74:29–32.

Ekman, S., 1953. Zoogeography of the Sea. Sidgwick and Jackson, London. 417 p.

Elliot, J.M., 1971. Some methods for the statistical analysis of samples of benthic invertebrates. Freshwater Biology Association, Publication No. 25. 144 p.

Elner, R.W. and Jamieson, G.S., 1979. Predation of sea scallops, *Placopecten magellanicus*, by the rock crab, *Cancer irroratus* and the American lobster, *Homarus americanus*. Journal of the Fisheries Research Board of Canada 36:537–543.

Fager, E.W., 1964. Marine sediments: effects of a tube-building polychaete. Science 143:356–359.

Fairbridge, W.S., 1953. A population study of the Tasmanian 'commercial' scallop, *Notovola meridionalis* (Tate) (Lamellibranchia, Pectinidae). Australian Journal of Marine and Freshwater Research 4:1–40.

Feder, H.M. and Lasker, R., 1964. Partial purification of a substance from starfish tube feet which elicits escape responses in gastropod molluscs. Life Sciences 3:1047–1051.

Félix-Pico, E.F., Villalejo-Fuerte, M., Tripp-Quezada, A. and Holguin-Quinones, O., 1999. Growth and survival of *Lyropecten subnodosus* (Sowerby, 1835) in suspended culture at the National Marine Park of Bahia de Loreto, B.C.S., Mexico. Book of Abstracts, 12th International Pectinid Workshop, Bergen, Norway, 5–11 May 1999. pp. 39–40.

Félix-Pico, E.F., 1991. Mexico. In: S.E. Shumway (Ed.). Scallops: Biology, Ecology and Aquaculture. Developments in Aquaculture and Fishery Science, Volume 21. Elsevier, Amsterdam. pp. 943–979.

Félix-Pico, E.F., Arellano-Martínez, M., Ceballos-Vázquez, B.P., Armenta, M., Domíngues-Valdéz, P.M. and García-Aguilar, A.M., 2003. Growth of *Nodipecten subnodosus* (Sowerby, 1835) in suspended culture in the Guerrero Negro lagoon, Baja California Sur México. Book of Abstracts, 14th International Pectinid Workshop, St. Petersburg, Florida, USA, 23–29 April, 2003. pp. 11–12.

Fevolden, S.E., 1989. Genetic differentiation of the Iceland scallop *Chlamys islandica* (Pectinidae) in the northern Atlantic Ocean. Marine Ecology Progress Series 51:77–85.

Fevolden, S.E., 1992. Allozymic variability in the Iceland scallop *Chlamys islandica* - geographic variation and lack of growth-heterozygosity correlations. Marine Ecology Progress Series 85:259–268.

Fitch, J.E., 1965. A relatively unexploited population of Pismo clams, *Tivela stultorum* (Mawe, 1923) (Veneridae). Proceedings of the Malacological Society of London 36:309–312.

Fleury, P.G., Dao, J.-C., Mikolajunas, J.P., Minchin, D., Norman, M. and Strand, Ø, 1997. Concerted action on scallop seabed cultivation in Europe (1993–1996). Specific Community Programme for Research, Technological Development and Demonstration in the Field of Agriculture and Agro-Industry, Inclusive Fisheries, AIR 2-CT 93–1647. 118 p.

Fleury, P.G., Mingant, C. and Castillo, A., 1996. A preliminary study of the behaviour and vitality of reseeded juvenile great scallops, of three sizes in three seasons. Aquaculture International 4:325–337.

Forbes, E. and Hanley, S., 1853. A History of the British Mollusca and their Shells, vols. 1 and 2. Van Voorst, London. 557 p.

Forester, A.J., 1979. The association between the sponge *Halichondria panicea* (Pallas) and scallop *Chlamys varia* (L.): a commensal-protective mutualism. Journal of Experimental Marine Biology and Ecology 36:1–10.

Franklin, A., Pickett, G.D. and Conner, P.M., 1980a. The escallop (*Pecten maximus*) and its fishery in England and Wales. Ministry of Agriculture Fisheries and Food, Laboratory Leaflet 51: 19 p.

Franklin, A., Pickett, G.D., Holme, N.A. and Barrett, R.L., 1980b. Surveying stocks of scallops *Pecten maximus* and queens *Chlamys opercularis* with underwater television. Journal of the Marine Biological Association of the United Kingdom 60:181–192.

Frischer, M.E., Danforth, J.M., Tyner, L.C., Leverone, J.R., Marelli, D.C., Arnold, W.S. and Blake, N.J., 2000. Development of an *Argopecten*-specific 18S rRNA targeted genetic probe. Marine Biotechnology 2:11–20.

Fritz, L.W. and Haven, D.S., 1981. An investigation of sea scallops (*Placopecten magellanicus*) of the mid-Atlantic from commercial samples in 1979. Annual Meeting of the National Shellfisheries Association, Hyannis, USA, 9–12 June 1980. 1:14.

Gabaev, D.D. and Kolotukhina, N.K., 1999. The effect of predation by *Nucella (Thais) heyseana* on population of Japanese scallop *Mizuhopecten yessoensis* (Jay). Russian Journal of Ecology 30:133–135.

Gäde, G., Weeda, E. and Gabbott, P.A., 1978. Changes in the level of octopine during the escape responses of the scallop, *Pecten maximus* (L.). Journal of Comparative Physiology 124:121–127.

Garcia-Esquivel, Z. and Bricelj, V.M., 1993. Ontogenic changes in microhabitat distribution of juvenile bay scallops, *Argopecten irradians irradians* (L), in eelgrass beds, and their potential significance to early recruitment. Biological Bulletin 185:42–55.

Gibson, F.A., 1953. Tagging of escallops (*Pecten maximus* L.) in Irish waters. Journal du Conseil International pour l'Exploration de la Mer 19:204–208.

Gibson, F.A., 1956. Escallops (*Pecten maximus* L.) in Irish waters. Scientific Proceedings of the Royal Dublin Society 27:253–271.

Giguère, M. and Cliche, G., 1994. Dispersal of sea scallop, *Placopecten magellanicus*, juveniles seeded on the bottom off the Îles-de-la-Madelaine, Québec, Canada. In: N.F. Bourne, B.L. Bunting and L.D. Townsend (Eds.). Proceedings of the 9th International Pectinid Workshop, Nanaimo, B.C., Canada, April 22–27, 1993. Canadian Technical Report of Fisheries and Aquatic Sciences 1994 (2): p 34.

Golikov, A.N. and Scarlato, O.A., 1970. Abundance, dynamics and production properties of populations of edible bivalves, *Mizuhopecten yessoensis* and *Spisula sachalinensis* related to problems of organisation of controllable submarine farms at the western shores of the Sea of Japan. Helgoländer wissenschaftliche Meeresuntersuchungen 20:498–513.

Gonzalez, M.L., Lopez, D.A., Perez, M.C., Requelme, V.A., Uribe, J.M. and Le Pennec, M., 1999. Growth of the scallop, *Argopecten purpuratus* (Lamarck, 1819), in southern Chile. Aquaculture, 175:307–316.

Goshima, S. and Fujiwara, H., 1994. Distribution and abundance of cultured scallop *Patinopecten yessoensis* in extensive sea beds as assessed by underwater camera. Marine Ecology Progress Series 110:151–158.

Gould, S.J., 1971. Muscular mechanics and the ontogeny of swimming in scallops. Palaeontology 14:61–94.

Grant, J., Emerson, C.W. and Shumway, S.E., 1993. Orientation, passive transport, and sediment erosion features of the sea scallop *Placopecten magellanicus* in the benthic boundary layer. Canadian Journal of Zoology 71:953–959.

Grau, G., 1959. Pectinidae of the eastern Pacific. Allen Hancock Pacific Expeditions 23, 308 p.

Grieshaber, M., 1978. Breakdown and formation of high-energy phosphates and octopine in the adductor muscle of the scallop, *Chlamys opercularis* (L.), during escape swimming and recovery. Journal of Comparative Physiology 126 B:269–276.

Gruffydd, L.D., 1972. Mortality of scallops on a Manx bed due to fishing. Journal of the Marine Biological Association of the United Kingdom 52:449–455.

Gruffydd, L.D., 1974a. An estimate of natural mortality in an unfished population of the scallop *Pecten maximus* (L.). Journal du Conseil International pour l'Exploration de la Mer 35:209–210.

Gruffydd, L.D., 1974b. The influence of certain environmental factors on the maximum length of the scallop, *Pecten maximus* L. Journal du Conseil International pour l'Exploration de la Mer 35:300–302.

Gruffydd, L.D., 1976a. The development of the larva of *Chlamys islandica* in the plankton and its salinity tolerance in the laboratory (Lamellibranchia, Pectinidae). Astarte 8:60–67.

Gruffydd, L.D., 1976b. Swimming in *Chlamys islandica* in relation to current speed and an investigation of hydrodynamic lift in this and other scallops. Norwegian Journal of Zoology 24:365–378.

Gruffydd, L.D., 1978. The byssus and byssus glands in *Chlamys islandica* and other scallops (Lamellibranchia). Zoologica Scripta 7:277–285.

Gutierrez, N. and Defeo, O., 2003. Development of a new scallop *Zygochlamys patagonica* fishery in Uruguay: latitudinal and bathymetric patterns in biomass and population structure. Fisheries Research 62:21–36.

Gutsell, J.S., 1931. Natural history of the bay scallop (*Pecten irradians*) - reproduction and development. Bulletin of the U.S. Bureau of Fisheries 46:599–632.

Guzik, S.K. and Marsh, R.L., 1989. Thermal effects on the kinematics of swimming in the scallop *Argopecten irradians*. American Zoologist 29:30A.

Guzik, S.K. and Marsh, R.L., 1990. Thermal effects on the swimming performance of the scallop *Chlamys hastata*. American Zoologist 30:132 A.

Gwyther, D., 1989. History of management in the Victorian scallop industry. In: M.C.L. Dredge, W.F. Zacharin and L.M. Joll (Eds.). Proceedings of the Australasian Scallop Workshop, Hobart, Australia. Tasmanian Government Printer, Hobart, pp. 12–20.

Gwyther, D., Cropp, D.A., Joll, L.M. and Dredge, M.C.L., 1991. Australia. In: S.E. Shumway (Ed.). Scallops: Biology, Ecology and Aquaculture. Developments in Aquaculture and Fishery Science, Volume 21. Elsevier, Amsterdam. pp. 835–851.

Hall-Spencer, J.M. and Moore, P.G., 2000. Scallop dredging has profound, long-term impacts on maerl habitats. ICES Journal of Marine Science 57:1407–1415.

Hamilton, P.V. and Koch, K.M., 1996. Orientation toward natural and artificial grassbeds by swimming bay scallops, *Argopecten irradians* (Lamarck, 1819). Journal of Experimental Marine Biology and Ecology 199:79–88.

Hancock, D.A., 1973. The relationship between stock and recruitment in exploited invertebrates. Rapport et Procès-Verbaux des Réunions, Conseil International pour l'Exploration de la Mer 164:113–131.

Hansen, K. and Nedreaas, K., 1986. Measurements of Iceland scallop (*Chlamys islandica* Müller) in the Spitzbergen and Bear Island regions. International Council for the Exploration of the Sea, CM 1986/K:26. 19 p.

Hartnoll, R.G., 1967. An investigation of the movement of the scallop *Pecten maximus*. Helgoländer wissenschaftliche Meeresuntersuchungen 15:523–533.

Harvey, M., Bourget, E. and Miron, G., 1993. Settlement of Iceland scallop *Chlamys islandica* spat in response to hydroids and filamentous red algae: field observations and laboratory experiments. Marine Ecology Progress Series 99:283–292.

Harvey, M., Bourget, E., Miron, G. and Legault, C., 1994. Settlement of Iceland scallop, *Chlamys islandica*, spat on natural substrata: relationship between hydroids and scallops. In: N.F. Bourne, B.L. Bunting and L.D. Townsend (Eds.). Proceedings of the 9th International Pectinid Workshop, Nanaimo, B.C. Canada, April 22–27, 1993. Canadian Technical Report of Fisheries and Aquatic Sciences 1994 (1): p. 160.

Hatcher, B.G., Scheibling, R.E., Barbeau, M.A., Hennigar, A.W., Taylor, L.H. and Windust, A.J., 1996. Dispersion and mortality of a population of sea scallop (*Placopecten magellanicus*) seeded in a tidal channel. Canadian Journal of Fisheries and Aquatic Sciences 53:38–54.

Hawkins, S.J., Southward, A.J. and Genner, M.J., 2003. Detection of environmental change in a marine ecosystem - evidence from the western English Channel. The Science of the Total Environment 310:245–256.

Hayami, I., 1991. Living and fossil scallop shells as airfoils - an experimental study. Paleobiology 17:1–18.

Haynes, E.B. and McMullin, J.C., 1970. Relationship between meat weight and shell height of the giant Pacific sea scallop, *Patinopecten caurinus*, from the Gulf of Alaska. Proceedings of the National Shellfisheries Association 60:50–53.

Haynes, E.B. and Powell, G.C., 1968. A preliminary report on the Alaska sea scallop fishery - exploration, biology, and commercial processing. Department of Fish and Game (Alaska), Information Leaflet 125: 20 p.

He, Y. and Zhang, F, 1990. The influence of environmental salinity on various development stages of the bay scallop *Argopecten irradians* Lamarck. Oceanologia et Limnologia Sinica 21:197–204.

Heald, D.I. and Caputi, N., 1981. Some aspects of growth, recruitment and reproduction in the southern saucer scallop *Amusium balloti* (Bernard 1861) in Shark Bay, Western Australia. Fisheries Research Bulletin, Department of Fisheries and Wildlife (West Australia) 25:1–33.

Heipel, D.A., Bishop, J.D.D. and Brand, A.R., 1999. Mitochondrial DNA variation among open-sea and enclosed populations of the scallop *Pecten maximus* in western Britain. Journal of the Marine Biological Association of the United Kingdom 79:687–695.

Heipel, D.A., Bishop, J.D.D., Brand, A.R. and Thorpe, J.P., 1998. Population genetic differentiation of the great scallop *Pecten maximus* in western Britain investigated by randomly amplified polymorphic DNA. Marine Ecology Progress Series 162:163–171.

Hennick, D.P., 1973. Sea scallop, *Patinopecten caurinus*, investigations in Alaska. Alaska Department of Fish and Game, Division of Commercial Fisheries, Completion Report, 5–23-R, Juneau.

Howell, T.R.W. and Fraser, D.I., 1984. Observations on the dispersal and mortality of the scallop, *Pecten maximus* (L.). International Council for the Exploration of the Sea, Shellfish Committee, CM1984, Doc No K:35. 13 p.

Hu, Y.-P., Lutz, R.A. and Vrijenhoek, R.C., 1992. Electrophoretic identification and genetic analysis of bivalve larvae. Marine Biology 113:227–230.

Hunt, O.D., 1925. The food of the bottom fauna of the Plymouth fishing grounds. Journal of the Marine Biological Association of the U.K. 13:560–599.

Huntsman, A.G. and Sparks, M.I., 1924. Limiting factors for marine animals. 3. Relative resistance to high temperatures. Contributions to Canadian Biology and Fisheries 2:95–114.

Illanes-Bucher, J.E., 1987. Cultivation of the northern scallop of Chile (*Chlamys* (*Argopecten*) *purpurata*) in controlled and natural environments. 6th International Pectinid Workshop, Menai Bridge, Wales, UK, April 1987. pp. 13 (mimeo).

Imai, T., 1980. Aquaculture in Shallow Seas. Progress in Shallow Sea Culture, Part III. A. A. Balkena, Rotterdam. 615 p.

Iribarne, O.O., Lasta, M.I., Vacas, H.C., Parma, A.M. and Pascual, M.S., 1991. Assessment of abundance, gear efficiency and disturbance in a scallop dredge fishery: results of a depletion experiment. In: S.E. Shumway and P.A. Sandifer (Eds.). An International Compendium of Scallop Biology and Culture. World Aquaculture Society, Baton Rouge, LA. pp. 242–248.

Irlandi, E.A., Orlando, B.A. and Ambrose, W.G., 1999. Influence of seagrass habitat patch size on growth and survival of juvenile bay scallops, *Argopecten irradians concentricus* (Say). Journal of Experimental Marine Biology and Ecology 235:21–43.

Ito, H., 1991. Japan. In: S.E. Shumway (Ed.). Scallops: Biology, Ecology and Aquaculture. Developments in Aquaculture and Fishery Science, Volume 21. Elsevier, Amsterdam. pp. 1017–1055.

Jamieson, G.S., 1978. Identification of offshore scallop (*Placopecten magellanicus*) concentrations and the importance of such procedures in stock assessment and population dynamics. 2nd International Pectinid Workshop. 8–13 May 1978, Brest, France. pp. 16 (mimeo).

Jamieson, G.S. and Chandler, R., 1981. The potential for research and fishery performance data isopleths in population assessment of offshore, sedentary, contagiously distributed species. Canadian Atlantic Fisheries Scientific Advisory Committee, Document 80/77. 32 p.

Jenkins, S.R. and Brand, A.R., 2001. The effect of dredge capture on the escape response of the great scallop, *Pecten maximus* (L.): implications for the survival of undersized discards. Journal of Experimental Marine Biology and Ecology 266:33–50.

Jenkins, S.R., Lart, W., Vause, B.J. and Brand, A.R., 2003. Seasonal swimming behaviour in the queen scallop (*Aequipecten opercularis*) and its effect on dredge fisheries. Journal of Experimental Marine Biology and Ecology 289:163–179.

Jennings, S. and Kaiser, M.J., 1998. The effects of fishing on marine ecosystems. Advances in Marine Biology 34:201–351.

Joll, L.M., 1989. Swimming behaviour of the saucer scallop *Amusium balloti* (Mollusca: Pectinidae). Marine Biology 102:299–305.

Joll, L.M. and Caputi, N., 1995. Geographic variation in the reproductive cycle of the saucer scallop, *Amusium balloti* (Bernardi, 1861) (Mollusca: Pectinidae) along the Western Australian coast. Australian Journal of Marine and Freshwater Research 46:779–792.

Jones, N.S., 1951. The bottom fauna off the south of the Isle of Man. Journal of Animal Ecology 20:132–144.

Kaiser, M.J., Armstrong, P.J., Dare, P.J. and Flatt, R.P., 1998. Benthic communities associated with a heavily fished scallop ground in the English Channel. Journal of the Marine Biological Association of the United Kingdom 78:1045–1059.

Kalashnikov, V.Z., 1991. Soviet Union. In: S.E. Shumway (Ed.). Scallops: Biology, Ecology and Aquaculture. Developments in Aquaculture and Fishery Science, Volume 21. Elsevier, Amsterdam. pp. 1057–1082.

Kamenos, N.A. and Moore, P.G., 2003. Attachment of juvenile queen scallop (*Aequipecten opercularis*) to live maerl and other substrata in laboratory conditions. Book of Abstracts, 14th International Pectinid Workshop, St. Petersburg, Florida, 23–29 April, 2003. pp. 201.

Kijima, A., Mori, K. and Fujio, Y., 1984. Population differences in gene frequency of the Japanese scallop *Patinopecten yessoensis* on the Okhotsk Sea coast of Hokkaido. Bulletin of the Japanese Society of Science and Fisheries/Nissuishi 50:241–248.

Kim, Y.S., 1969. Selective feeding on several bivalve molluscs by starfish, *Asterias amurensis* Lütken. Bulletin of the Faculty of Fisheries, Hokkaido University 19:244–249.

Kinne, O., 1970. Temperature - animals, invertebrates. In: O. Kinne (Ed.). Marine Ecology, a Comparative Treatise on Life in Oceans and Coastal Waters. Volume 1, Environmental Factors. Wiley-Interscience, New York. pp. 821–995.

Kirby-Smith, W.W., 1972. Growth of the bay scallop: the influence of experimental water currents. Journal of Experimental Marine Biology and Ecology 8:7–18.

Kleinman, S., Hatcher, B.G. and Scheibling, R.E., 1996. Growth and content of energy reserves in juvenile sea scallops, *Placopecten magellanicus*, as a function of swimming frequency and water temperature in the laboratory. Marine Biology 124:629–635.

Kostylev, V.E., Courtney, R.C., Robert, G. and Todd, B.J., 2003. Stock evaluation of giant scallop (*Placopecten magellanicus*) using high-resolution acoustics for seabed mapping. Fisheries Research 60:479–492.

Krause, M.K., Arnold, W.S. and Ambrose, W.G.J., 1994. Morphological and genetic variation among three populations of calico scallops, *Argopecten gibbus*. Journal of Shellfish Research 13:529–537.

Kristensen, I., 1957. Differences in density and growth in a cockle population in the Dutch Wadden Sea. Archives Néerlandaises de Zoologie 12:351–453.

Kruse, G.H., Barnhart, J.P., Rosenkranz, G.E., Funk, F.C. and Pengilly, D., 2000. History and development of the scallop fishery in Alaska. Alaska Department of Fish and Game, Special Publication 14:6–12.

Kruse, G.H. and Shirley, S.M., 1994. The Alaskan scallop fishery and its management. In: N.F Bourne, B.L. Bunting and L.D. Townsend (Eds.). Proceedings of the 9th International Pectinid Workshop, Nanaimo, B.C. Canada, April 22–27, 1993. Canadian Technical Report of Fisheries and Aquatic Sciences 1994 (2):170–177.

Lafrance, M., Guderley, H. and Cliche, G., 2002. Low temperature, but not air exposure slows the recuperation of juvenile scallops, *Placopecten magellanicus*, from exhausting escape responses. Journal of Shellfish Research 21:605–618.

Lake, N.C.H., Jones, M.B. and Paul, J.D., 1987. Crab predation on scallop (*Pecten maximus*) and its implication for scallop cultivation. Journal of the Marine Biological Association of the United Kingdom 67:55–64.

Lake, N.C.H. and McFarlane, C.L., 1994. Crab and starfish predation of the scallop *Pecten maximus* (L.) during seabed cultivation. In: N.F. Bourne, B.L. Bunting and L.D. Townsend (Eds.). Proceedings of the 9th International Pectinid Workshop, Nanaimo, B.C., Canada, April 22–27, 1993. Canadian Technical Report of Fisheries and Aquatic Sciences 1994 (2):56.

Land, M.F., 1966. Activity in the optic nerve of *Pecten maximus* in response to changes in light intensity, and to pattern and movements in the optical environment. Journal of Experimental Biology 45:83–99.

Langton, R.W. and Robinson, W.E., 1990. Faunal associations on scallop grounds in the western Gulf of Maine. Journal of Experimental Marine Biology and Ecology 144:157–171.

Larsen, P.F. and Lee, R.M., 1978. Observations on the abundance, distribution and growth of postlarval sea scallops, *Placopecten magellanicus,* on Georges Bank. Nautilus 92:112–116.

Lasta, M. and Bremec, C., 1997. *Zygochlamys patagonica* (King & Broderip,1832): development of a new scallop fishery in the southwestern Atlantic ocean. Book of Abstracts, 11th International Pectinid Workshop, La Paz, Mexico, 10–15 April, 1997. pp. 138–139.

Lasta, M.L. and Bremec, C.S., 1999. Development of the scallop fishery (*Zygochlamys patagonica*) in the Argentine Sea. Book of Abstracts, 12th International Pectinid Workshop, Bergen, Norway, 5–11 May, 1999. pp. 154–155.

Lawrence, S.J., 1993. The feeding ecology and physiology of the scallops *Pecten maximus* (L.) and *Aequipecten opercularis* (L.) in the North Irish Sea. PhD Thesis. University of Liverpool. 225 p.

Le Pennec, M., 1974. Morphogenése de la coquille de *Pecten maximus* (L.) élevé au laboratoire. Cahiers de Biologie Marine 15:475–482.

Le Pennec, M., 1980. The larval and post-larval hinge of some families of bivalve molluscs. Journal of the Marine Biological Association of the United Kingdom 60:601–617.

Lecomte, J., 1952. Réactions de fuite des pectens en présence des astérides. Vie Milieu 3:57–60.

Lefort, Y., 1994. Growth and mortality of the tropical scallops: *Anachlamys flabellata* (Bernardi), *Comptopallium radula* (Linne) and *Mimachlamys gloriosa* (Reeve) in southwest lagoon of New Caledonia. Journal of Shellfish Research 13:539–546.

Leighton, D.L., 1979. A growth profile for the rock scallop *Hinnites multirugosus* held at several depths off La Jolla, California. Marine Biology 51:229–232.

Letaconnoux, R. and Audouin, J., 1956. Contribution à l'étude du pétoncle (*Chlamys varia* L.). Revue de Travaux de l'Institute des Pêches Maritimes, Nantes 20:133–155.

Lewis, R.I. and Thorpe, J.P., 1994a. Are queen scallops, *Aequipecten* (*Chlamys*) *opercularis* (L.), self recruiting? In: N.F Bourne, B.L. Bunting and L.D. Townsend (Eds.). Proceedings of the 9th Pectinid Workshop, Nanaimo, B. C. Canada, April 22–27, 1993. Canadian Technical Report of Fisheries and Aquatic Sciences 1994 (1):214–221.

Lewis, R.I. and Thorpe, J.P., 1994b. Temporal stability of gene-frequencies within genetically heterogeneous populations of the queen scallop, *Aequipecten (Chlamys) opercularis*. Marine Biology 121:117–126.

Lin, R., Chen, M. and Lin, B., 1989. Effects of temperature and salinity on attachment and metamorphosis of bay scallop larvae *Argopecten irradians* (Lamarck). Journal of Oceanography, Taiwan Straight/Taiwan Haixia 8:60–67.

Lin, R., Lin, B. and Chen, M., 1991. Study on effects of temperature and salinity on migrating behaviour, growth and survival of juvenile bay scallop *Argopecten irradians* (Lamarck). Journal of Oceanography, Taiwan Straight/Taiwan Haixia 10:133–138.

Livingstone, D.R., Zwaan, A. de and Thompson, R.J., 1981. Aerobic metabolism, octopine production and phosphoarginine as sources of energy in the phasic and catch adductor muscles of the giant scallop *Placopecten magellanicus* during swimming and the subsequent recovery period. Comparative Biochemistry and Physiology 70B:35–44.

Llana, M.E.G., 1983. Size composition, occurrence, distribution and abundance of scallops in Visayan Sea. Phillipines Journal of Fisheries 16:75–94.

Llana, M.E.G., 1985. Some aspects of the biology of the scallop *Amusium pleuronectes* in the Phillipines. Book of Abstracts, 5th International Pectinid Workshop, La Coruña, Spain, 6–10 May, 1985. pp. 21.

Lou, Y., 1991. China. In: S.E. Shumway (Ed.). Scallops: Biology, Ecology and Aquaculture. Developments in Aquaculture and Fishery Science, Volume 21. Elsevier, Amsterdam. pp. 809–824.

Lubinsky, I., 1980. Marine bivalve molluscs of the Canadian central and eastern Arctic: faunal composition and zoogeography. Canadian Bulletin of Fisheries and Aquatic Sciences 207:1–111.

MacDonald, B.A. and Bajdik, C.D., 1992. Orientation and distribution of individual *Placopecten magellanicus* (Gmelin) in two natural populations with differing production. Canadian Journal of Fisheries and Aquatic Sciences 49:2086–2092.

MacDonald, B.A. and Thompson, R.J., 1985a. Influence of temperature and food availability on the ecological energetics of the giant scallop *Placopecten magellanicus* 1. Growth rates of shell and somatic tissue. Marine Ecology Progress Series 25:279–294.

MacDonald, B.A. and Thompson, R.J., 1985b. Influence of temperature and food availability on the ecological energetics of the giant scallop *Placopecten magellanicus* 2. Reproductive output and total production. Marine Ecology Progress Series 25:295–303.

MacDonald, B.A. and Thompson, R.J., 1986a. Influence of temperature and food availability on the ecological energetics of the giant scallop *Placopecten magellanicus* 3. Physiological ecology, the gametogenic cycle and scope for growth. Marine Biology 93:37–48.

MacDonald, B.A. and Thompson, R.J., 1986b. Production, dynamics and energy partitioning in two populations of the giant scallop *Placopecten magellanicus* (Gmelin). Journal of Experimental Marine Biology and Ecology 101:285–299.

MacDonald, B.A., Thompson, R.J. and Bayne, B.L., 1987. Influence of temperature and food availability on the ecological energetics of the giant scallop *Placopecten magellanicus* 4. Reproductive effort, value and cost. Oecologia (Berlin) 72:550–556.

Mackie, A.M., 1970. Avoidance reactions of marine invertebrates to either steroid glycosides of starfish or synthetic surface active agents. Journal of Experimental Marine Biology and Ecology 5:63–69.

Mackie, A.M. and Turner, A.B., 1970. Partial characterization of a biologically active steroid glycoside isolated from the starfish *Marthasterias glacialis*. Journal of Biochemistry 117:543–550.

Mackie, L.A. and Ansell, A.D., 1993. Differences in reproductive ecology in natural and transplanted populations of *Pecten maximus* - evidence for the existence of separate stocks. Journal of Experimental Marine Biology and Ecology 169:57–75.

Macleod, J.A.A., Thorpe, J.P. and Duggan, N.A., 1985. A biochemical genetic study of population structure in queen scallop (*Chlamys opercularis*) stocks in the Northern Irish Sea. Marine Biology 87:77–82.

Magorrian, B.H., Service, M. and Clarke, W., 1995. An acoustic bottom classification survey of Stranford Lough, Northern Ireland. Journal of the Marine Biological Association of the United Kingdom 75:987–992.

Maguire, J.A., Fleury, P.G. and Burnell, G., 1997. A method for quantifying scallop quality. Book of Abstracts, 11th International Pectinid Workshop, La Paz, Mexico, 10–15 April, 1997. pp. 50–52.

Maguire, J.A., Fleury, P.G. and Burnell, G.M., 1999a. Some methods for quantifying quality in the scallop *Pecten maximus* (L.). Journal of Shellfish Research 18:59–66.

Maguire, J.A., O'Connor, D.A. and Burnell, G.M., 1999b. An investigation into behavioural indicators of stress in juvenile scallops. Aquaculture International 7:169–177.

Maguire, J.A., Coleman, A., Jenkins, S. and Burnell, G.M., 2002a. Effects of dredging on undersized scallops. Fisheries Research 56:155–165.

Maguire, J.A., Jenkins, S. and Burnell, G.M., 2002b. The effects of repeated dredging and speed of tow on undersized scallops. Fisheries Research 58:367–377.

Maguire, J.A., O'Donoghue, M., Jenkins, S., Brand, A.R. and Burnell, G.M., 2002c. Temporal and spatial variability in dredging induced stress in the great scallop, *Pecten maximus* (L.). Journal of Shellfish Research 21:81–86.

Manuel, J.L. and Dadswell, M.J., 1991. Swimming behaviour of juvenile giant scallop, *Placopecten magellanicus*, in relation to size and temperature. Canadian Journal of Zoology 69:2250–2254.

Manuel, J.L. and Dadswell, M.J., 1993. Swimming of juvenile sea scallops, *Placopecten magellanicus* (Gmelin): a minimum size for effective swimming? Journal of Experimental Marine Biology and Ecology 174:137–175.

Manuel, J.L. and Dadswell, M.J., 1994. Scale and hydrodynamics change the swimming behaviour of juvenile sea scallops, *Placopecten magellanicus*. In: N.F Bourne, B.L. Bunting and L.D. Townsend (Eds.). Proceedings of the 9th International Pectinid Workshop, Nanaimo, B. C. Canada, April 22–27, 1993. Canadian Technical Report of Fisheries and Aquatic Sciences 1994 (1):59–65.

Marahrens, M., 1995. Estimating the habitat specific natural mortality rates of the Chilean scallop (*Argopecten purpuratus*). Book of Abstracts, 10th International Pectinid Workshop, Cork, Ireland, 27 April – 2 May, 1995. pp. 2–3.

Marelli, D.C., Krause, M.K., Arnold, W.S. and Lyons, W.G., 1997. Systematic relationships among Florida populations of *Argopecten irradians* (Lamarck, 1819) (Bivalvia: Pectinidae). Nautilus 110:31–41.

Marshall, N., 1960. Studies of the Niantic River, Connecticut with special reference to the bay scallop *Aequipecten irradians*. Limnology and Oceanography 5:86–105.

Martoja, M., Truchet, M. and Berthet, B., 1989. Effets de la contamination experimentale par l'argent chez *Chlamys varia* L. (Bivalve, Pectinidae). Donnees quantitatives, histologiques et microanalytiques. Annales Institut Oceanography Paris (Nouv. Ser.) 65:1–13.

Mason, J., 1957. The age and growth of the scallop, *Pecten maximus* (L.), in Manx waters. Journal of the Marine Biological Association of the United Kingdom 36:473–492.

Mason, J., 1958. The breeding of the scallop, *Pecten maximus*, in Manx waters. Journal of the Marine Biological Association of the United Kingdom 37:653–671.

Mason, J., 1969. The growth of spat of *Pecten maximus* (L.). International Council for the Exploration of the Sea, Shellfish and Benthos Committee Meeting Paper, CM 1969/ K: 32. 3 p.

Mason, J., 1981. The Scottish queen fishery. Scottish Fisheries Bulletin 46:3–7.

Mason, J., 1983. Scallop and Queen Fisheries in the British Isles. Fishing News Books Ltd (Buckland Foundation), Farnham, UK. 144 p.

Mason, J., Chapman, C.J. and Kinnear, J.A.M., 1979a. Population abundance and dredge efficiency studies on the scallop, *Pecten maximus* (L.). Rapport et Procès-Verbaux des Réunions. Conseil International pour l'Exploration de la Mer 175:91–96.

Mason, J., Drinkwater, J., Howell, T.R.W. and Fraser, D.I., 1982. A comparison of methods of determining the distribution and density of the scallop, *Pecten maximus* (L.). International Council for the Exploration of the Sea, Shellfish Committee, CM 1982/K:24. 5 p.

Mason, J., Shanks, A.M., Fraser, D.I. and Shelton, R.G.J., 1979b. The Scottish fishery for the queen, *Chlamys opercularis* (L.). International Council for the Exploration of the Sea, CM 1979/ K:37. 6 p.

Mathers, N.F., 1975. Environmental variability at the phosphoglucose isomerase locus in the genus *Chlamys*. Biochemical Systematics and Ecology 3:123–127.

Mathers, N.F., 1976a. *In situ* studies on the scallop, *Pecten maximus*, with reference to the association between feeding rhythms and tidal currents. Journal of Molluscan Studies 42:455–456.

Mathers, N.F., 1976b. The effects of tidal currents on the rhythm of feeding and digestion in *Pecten maximus*. Journal of Experimental Marine Biology and Ecology 24:271–283.

Mattei, N., 1995. Situation of pectinid aquaculture (*Pecten jacobaeus*) in Italy and some information on Italian pectinid fishery. Book of Abstracts, 10th International Pectinid Workshop, Cork, Ireland, 26 April–2 May, 1995. pp. 59–60.

McClintock, J.B., 1983. Escape response of *Argopecten irradians* (Mollusca: Bivalvia) to *Luidia clathrata* and *Echinaster* sp. (Echinodermata: Asteroidea). Florida Scientist 46:95–101.

McLusky, D.S., 1973. The effect of temperature on the oxygen consumption and filtration rate of *Chlamys (Aequipecten) opercularis* (L.) (Bivalvia). Ophelia 10:114–154.

Medcof, J.C. and Bourne, N., 1964. Causes of mortality of the sea scallop, *Placopecten magellanicus*. Proceedings of the National Shellfisheries Association 53:33–50.

Mellon, D.J., 1969. The reflex control of rhythmic motor output during swimming in the scallop. Zeitschrift für Vergleichende Physiologie 62:318–336.

Melvin, G.D., Dadswell, M.J. and Chandler, R.A., 1985. Movement of scallops *Placopecten magellanicus* (Gmelin, 1971) (Mollusca: Pectinidae) on Georges Bank. Canadian Atlantic Fisheries Scientific Advisory Committee, Research Document 85/30. 29 p.

Mercaldo, R.S. and Rhodes, E.W., 1982. Influence of reduced salinity on the Atlantic bay scallop *Argopecten irradians* (Lamarck) at various temperatures. Journal of Shellfish Research 2:177–181.

Merrill, A.S., 1959. A comparison of *Cyclopecten nanus* (Verrill & Bush) and *Placopecten magellanicus* (Gmelin). Bulletin of the Museum of Comparative Zoology at Harvard College 2:209–228.

Merrill, A.S. and Posgay, J.A., 1964. Estimating the natural mortality rate of the sea scallop, *Placopecten magellanicus*. Research Bulletin of the International Commission for the Northwest Atlantic Fisheries 1:88–107.

Mikulich, L.V. and Tsikhon-Lukanina, E.A., 1981. Food composition of the yeso scallop. Oceanology 21:894–897.

Miller, G.C., Allen, D.M. and Costello, T.J., 1981. Spawning of the calico scallop *Argopecten gibbus* in relation to season and temperature. Journal of Shellfish Research 1:17–21.

Miller, G.C. and Richards, W.J., 1980. Reef fish habitat, faunal assemblages and factors determining distributions in the south Atlantic Bight. Proceedings of the Gulf and Caribbean Fisheries Institute 32:114–130.

Millward, A. and Whyte, M.A., 1992. The hydrodynamic characteristics of six scallops of the Superfamily Pectinacea, Class Bivalvia. Journal of Zoology, London 227:547–566.

Minchin, D., 1978. The behaviour of young escallops (*Pecten maximus* (L.) (Pectinidae). 2nd International Pectinid Workshop, Brest, France, May, 1998. pp. 11 (mimeo).

Minchin, D., 1981. The escallop *Pecten maximus* in Mulroy Bay. Fisheries Bulletin (Dublin) 1:13 p.

Minchin, D., 1991. Decapod predation and the sowing of the scallop *Pecten maximus* (Linnaeus 1758). In: S.E. Shumway and P.A. Sandifer (Eds.). An International Compendium of Scallop Biology and Culture. World Aquaculture Society, Baton Rouge, LA. pp. 191–197.

Minchin, D., 1992. Biological observations on young scallops, *Pecten maximus*. Journal of the Marine Biological Association of the United Kingdom 72:807–819.

Minchin, D., Duggan, C.B. and King, W., 1987. Possible effects of organotins on scallop recruitment. Marine Pollution Bulletin 18:604–608.

Minchin, D., Haugum, G.A., Skjaeggestad, H. and Strand, O., 1999. Effect of air exposure on scallop behaviour and subsequent survival. Book of Abstracts, 12th International Pectinid Workshop, Bergen, Norway, 5–11 May, 1999. pp. 101–102.

Minchin, D. and Mathers, N.F., 1982. The escallop, *Pecten maximus* (L.), in Killary Harbour. Irish Fisheries Investigations, Series B Marine 25:13 p.

Minchin, D., Skjaeggestad, H., Haugum, G.A. and Strand, O., 2000. Righting and recessing ability of wild and naive cultivated scallops. Aquaculture Research 31:473–474.

Moore, J.D. and Trueman, E.R., 1971. Swimming of the scallop, *Chlamys opercularis* (L.). Journal of Experimental Marine Biology and Ecology 6:179–185.

Moore, J.K. and Marshall, N., 1967. An analysis of the movements of the bay scallop *Aequipecten irradians* in a shallow estuary. Proceedings of the National Shellfisheries Association 57:77–82.

Morton, B., 1980. Swimming in *Amusium pleuronectes* (Bivalvia: Pectinidae). Journal of Zoology 190:375–404.

Morton, B., 1996. The biology and functional morphology of *Minnivola pyxidatus* (Bivalvia: Pectinoidea). Journal of Zoology 240:735–760.

Morton, B., 2001. The evolution of eyes in the Bivalvia. Oceanography and Marine Biology Annual Review 39:165–205.

Motoda, S., 1977. Biology and artificial propagation of Japanese scallop (General review). Proceedings of the 2nd Soviet-Japan Joint Symposium on Aquaculture, Moscow, USSR, November 1973, Tokai University. pp. 75–120.

Mottet, M.G., 1979. A review of the fishery biology and culture of scallops., State of Washington Department of Fisheries, Technical Report 39: 292 p

Murawski, S.A., Brown, R., Lai, H.-L., Rago, P.J. and Hendrickson, L., 2000. Large-scale closed areas as a fishery-management tool in temperate marine systems: the Georges Bank experience. Bulletin of Marine Science 66:775–798.

Murphy, E.J., 1986. An investigation of the population dynamics of the exploited scallop, *Pecten maximus* (L.), in the North Irish Sea. PhD Thesis. University of Liverpool. 155 pp.

Myers, R.A., Fuller, S.D. and Kehler, D.G., 2000. A fisheries management strategy robust to ignorance: rotational harvest in the presence of indirect fishing mortality. Canadian Journal of Fisheries and Aquatic Sciences 57:2357–2362.

Naidu, K.S., 1988. Estimating mortality rates in the Iceland scallop, *Chlamys islandica* (O.F. Müller). Journal of Shellfish Research 7:61–71.

Naidu, K.S. and Anderson, J.T., 1984. Aspects of scallop recruitment on St. Pierre Bank in relation to oceanography and implications for resource management. Canadian Atlantic Fisheries Scientific Advisory Committee, Research Document 84/29. 15 p.

Naidu, K.S. and Cahill, F.M., 1984. Status and assessment of St. Pierre bank scallop stocks, 1982–1983. Canadian Atlantic Fisheries Scientific Advisory Committee, Research Document 84/69. 54 p.

Naidu, K.S. and Cahill, F.M., 1985. Mortality associated with tagging in the sea scallop, *Placopecten magellanicus* (Gmelin). Canadian Atlantic Fisheries Scientific Advisory Committee, Research Document 85/21. 10 p.

Naidu, K.S., Cahill, F.M. and Lewis, D.B., 1982. Status and assessment of the Iceland scallop *Chlamys islandica* in the northeastern Gulf of St. Lawrence. Canadian Atlantic Fisheries Scientific Advisory Committee, Research Document 82/02. 66 p.

Naidu, K.S., Lewis, D.B. and Cahill, F.M., 1983. St. Pierre Bank: an offshore scallop buffer zone. Canadian Atlantic Fisheries Scientific Advisory Committee, Research Document 83/16. 48 p.

Naidu, K.S. and Meron, S., 1986. Predation of scallops by American plaice and yellowtail flounder. Canadian Atlantic Fisheries Scientific Advisory Committee, Research Document 86/62. 25 p.

Newell, R.C., 1970. Biology of intertidal animals. Paul Elek (Scientific Books) Ltd., London. 555 p.

O'Connor, W.A. and Heasman, M.P., 1998. Ontogenetic changes in salinity and temperature tolerance in the doughboy scallop, *Mimachlamys asperrima*. Journal of Shellfish Research 17:89–95.

Ockelmann, W.K., 1958. The zoology of East Greenland. Meddelelser om Grønland 122 (4):1–256.

Olsen, A.M., 1955. Underwater studies on the Tasmanian commercial scallop, *Notovola meridionalis* (Tate), (Lamellibranchiata: Pectinidae). Australian Journal of Marine and Freshwater Research 6:392–409.

Olsen, A.M., 1962. The commercial scallop fishery of the D'Entrecasteaux Channel, Tasmania. Fisheries Management Seminar, Sydney, New South Wales, 21 September, 1962. pp. 147–148.

Olson, J.M. and Marsh, R.L., 1993. Contractile properties of the striated adductor muscle in the bay scallop *Argopecten irradians* at several temperatures. Journal of Experimental Biology 176:175–193.

Olsson, A.A., 1961. Molluscs of the Tropical Eastern Pacific, Particularly from the Southern Half of the Panamic-Pacific Faunal Province (Panama to Peru). Ithaca, New York. 574 p.

Ordzie, C.J. and Garofalo, G.C., 1980a. Behavioural recognition of molluscan and echinoderm predators by the bay scallop, *Argopecten irradians* (Lamark), at two temperatures. Journal of Experimental Marine Biology and Ecology 43:29–37.

Ordzie, C.J. and Garofalo, G.C., 1980b. Predation, attack success, and attraction to the bay scallop, *Argopecten irradians* (Lamarck) by the oyster drill, *Urosalpinx cinerea* (Say). Journal of Experimental Marine Biology and Ecology 47:95–100.

Orensanz, J.M., 1986. Size, environment and density: the regulation of a scallop stock and its management implications. Canadian Special Publication in Fisheries and Aquatic Sciences 92:195–227.

Orensanz, J.M., Parma, A.M. and Iribarne, O.O., 1991a. Population dynamics and management of natural stocks. In: S.E. Shumway (Ed.). Scallops: Biology, Ecology and Aquaculture. Developments in Aquaculture and Fishery Science, Volume 21. Elsevier, Amsterdam. pp. 625–713.

Orensanz, J.M., Pascual, M. and Fernandez, M., 1991b. Argentina. In: S.E. Shumway (Ed.). Scallops: Biology, Ecology and Aquaculture. Developments in Aquaculture and Fishery Science, Volume 21. Elsevier, Amsterdam, pp. 981–999.

Ortiz, M., Jesse, S., Stotz, W. and Wolff, M., 2003. Feeding behaviour of the asteroid *Meyenaster gelatinosus* in response to changes in abundance of the scallop *Argopecten purpuratus* in northern Chile. Archiv für Hydrobiologie 157:213–225.

Ortiz, M. and Wolff, M., 2002a. Dynamical simulation of mass-balance trophic models for benthic communities of north-central Chile: assessment of resilience time under alternative management scenarios. Ecological Modelling 148:277–291.

Ortiz, M. and Wolff, M., 2002b. Trophic models of four benthic communities in Tongoy Bay (Chile): comparative analysis and preliminary assessment of management strategies. Journal of Experimental Marine Biology and Ecology 268:205–235.

Parker, R.H., 1956. Macro-invertebrate assemblages as indicators of sedimentary environments in east Mississippi Delta region. Bulletin of the American Association of Petroleum Geologists 40:295–376.

Parsons, G.J. and Dadswell, M.J., 1992. Seasonal and size-related swimming behaviour in the giant scallop, *Placopecten magellanicus*. Journal of Shellfish Research 11:559.

Parsons, G.J., Warren-Perry, C.R. and Dadswell, M.J., 1992. Movements of juvenile scallops *Placopecten magellanicus* (Gmelin, 1791) in Passamaquoddy Bay, New Brunswick. Journal of Shellfish Research 11:295–297.

Paugam, A., Le Pennec, M. and Genevieve, A.F., 2000. Immunological recognition of marine bivalve larvae from plankton samples. Journal of Shellfish Research 19:325–331.

Paul, J.D., 1980a. Salinity-temperature relationships in the queen scallop *Chlamys opercularis*. Marine Biology 56:295–300.

Paul, J.D., 1980b. Upper temperature tolerance and the effects of temperature on byssus attachment in the queen scallop *Chlamys opercularis* (L.). Journal of Experimental Marine Biology and Ecology 46:41–50.

Paul, J.D., 1981. Natural settlement and early growth of spat of the queen scallop *Chlamys opercularis* (L.) with reference to the formation of the first growth ring. Journal of Molluscan Studies 47:53–58.

Paul, J.D., Brand, A.R. and Hoogesteger, J.N., 1981. Experimental cultivation of the scallops *Chlamys opercularis* (L.) and *Pecten maximus* (L.) using naturally produced spat. Aquaculture 24:31–44.

Paulet, Y.M., Lucas, A. and Gerard, A., 1988. Reproduction and larval development in two *Pecten maximus* (L.) populations from Brittany. Journal of Experimental Marine Biology and Ecology 119:145–156.

Penchaszadeh, P.E. and Salaya, J.J., 1985. Ecology, fisheries and reproductive cycles of Caribbean Venezuelan scallops (*Amusium papyraceum*, *Amusium laurenti* and *Aequipecten lineolaris*). Book of Abstracts, 5th International Pectinid Workshop, La Coruña, Spain, 6–10 May, 1985. pp. 28.

Pennington, D., 1999. Studies of aspects of predation on the Manx scallop, *Pecten maximus* (L.), populations. PhD Thesis. University of Liverpool. 188 p.

Pesch, G.G. and Stewart, N.E., 1980. Cadmium toxicity to three species of estuarine invertebrates. Marine Environmental Research 3:145–156.

Peterson, C.H., Ambrose, W.G. Jr. and Hunt, J.H., 1982. A field test of the swimming response of the bay scallop (*Argopecten irradians*) to changing biological factors. Bulletin of Marine Science 32:939–944.

Peterson, C.H., Fodrie, F.J., Summerson, H.C. and Powers, S.P., 2001. Site-specific and density-dependent extinction of prey by schooling rays: generation of a population sink in top-quality habitat for bay scallops. Oecologia 129:349–356.

Peterson, C.H., Summerson, H.C., Fegley, S.R. and Prescott, R.C., 1989. Timing, intensity and sources of autumn mortality of adult bay scallops *Argopecten irradians concentricus* Say. Journal of Experimental Marine Biology and Ecology 127:121–140.

Petuch, E.J., 1987. New Caribbean Molluscan Fauna. The Coastal Education and Research Foundation, Charlottesville, Virginia.

Piccinetti, C., Simunovic, A. and Jukic, S., 1986. Distribution and abundance of *Chlamys opercularis* (L.) and *Pecten jacobaeus* L. in the Adriatic Sea. FAO Fisheries Report No. 345, FIPL/R345. pp. 99–105.

Pilditch, C.A. and Grant, J., 1999. Effect of variations in flow velocity and phytoplankton concentration on sea scallop (*Placopecten magellanicus*) grazing rates. Journal of Experimental Marine Biology and Ecology 240:111–136.

Piquimil, R.N., Figueroa, L.S., Contreras, O.C. and Avendaño, M.D., 1991. Chile. In: S.E. Shumway (Ed.). Scallops: Biology, Ecology and Aquaculture. Developments in Aquaculture and Fishery Science, Volume 21. Elsevier, Amsterdam. pp. 1001–1015.

Pitcher, C.R. and Butler, A.J., 1987. Predation by asteroids, escape response and morphometrics of scallops with epizoic sponges. Journal of Experimental Marine Biology and Ecology 112:233–249.

Pohle, D.G., Bricelj, V.M. and García-Esquivel, Z., 1991. The eelgrass canopy: an above–bottom refuge from benthic predators for juvenile bay scallops *Argopecten irradians*. Marine Ecology Progress Series 74:47–59.

Pond, D., 1992. Protective commensal mutualism between the queen scallop *Chlamys opercularis* (Linnaeus) and the encrusting sponge *Suberites*. Journal of Molluscan Studies 58:127–134.

Pontin, C. and Millington, P., 1985. Management of the Bass Strait scallop fishery. Australian Fisheries, September 1985. pp. 2–5.

Posgay, J.A., 1953. Sea scallop investigations. Sixth Report on Investigations of the Shellfisheries of Massachusetts. Massachusetts Department of Natural Resources, Division of Marine Fisheries. pp. 9–24.

Posgay, J.A., 1957. The range of the sea scallop. Nautilus 71:55–57.

Posgay, J.A., 1963. Tagging as a technique in population studies of the sea scallop. International Commission for Northwest Atlantic Fisheries, Special Publication 4:268–271.

Posgay, J.A., 1979a. Depth as a factor affecting the growth rate of the sea scallop. International Council for the Exploration of the Sea. ICES CM/K:27. 4 p.

Posgay, J.A., 1979b. Population assessment of the Georges Bank sea scallop stocks. Rapport et Procès-Verbaux des Réunions, Conseil International pour l'Exploration de la Mer 175:109–113.

Posgay, J.A., 1981. Movements of tagged sea scallops of Georges Bank. Marine Fisheries Review 43:19–25.

Powers, S.P. and Kittinger, J.N., 2002. Hydrodynamic mediation of predator-prey interactions: differential patterns of prey susceptibility and predator success explained by variation in water flow. Journal of Experimental Marine Biology and Ecology 273:171–187.

Prescott, R.C., 1990. Sources of predatory mortality in the bay scallop *Argopecten irradians* (Lamarck): interactions with seagrass and epibiotic coverage. Journal of Experimental Marine Biology and Ecology 144:63–83.

Racotta, I.S., Ramirez, J.L., Ibarra, A.M., Rodriguez-Jaramillo, M.C., Carreno, D. and Palacios, E., 2003. Growth and gametogenesis in the lion-paw scallop *Nodipecten (Lyropecten) subnodosus*. Aquaculture 217:335–349.

Reddiah, K., 1959. Studies on the biology of Manx pectinids (Lamellibranchia) and on the copepods associated with some invertebrates. PhD Thesis. University of Liverpool. 118 p.

Rees, C.B., 1950. The identification and classification of lamellibranch larvae. Hull Bulletins of Marine Ecology 3:73–104.

Rees, W.J., 1957. The living scallop. In: I. Cox (Ed.). The Scallop. Shell Transport and Trading Co. Ltd., London. pp. 15–32.

Ridgway, G.M.I. and Dahle, G., 2000. Population genetics of *Pecten maximus* of the northeast Atlantic coast. Sarsia 85:167–172.

Robert, G. and Jamieson, G.S., 1986. Commercial fishery data isopleths and their use in offshore sea scallop (*Placopecten magellanicus*) stock evaluations. Canadian Special Publications in Fisheries and Aquatic Sciences 92:76–82.

Roberts, D., 1973. Some sub-lethal effects of pesticides on the behaviour and physiology of bivalved molluscs. PhD Thesis. University of Liverpool. 148 p.

Robinson, S.M.C., Martin, J.D., Chandler, R.A. and Parsons, G.J., 1994. The use of video in the assessment of high mortality in a population of the giant scallop in the Bay of Fundy, Canada. In: N.F Bourne, B.L. Bunting and L.D. Townsend (Eds.). Proceedings of the 9th International Pectinid Workshop, Nanaimo, B.C. Canada, April 22–27, 1993. Canadian Technical Report of Fisheries and Aquatic Sciences 1994 (1):116.

Rodhouse, P.G. and Burnell, G.M., 1979. *In situ* studies on the scallop *Chlamys varia*. In: J.C. Gamble and J.D. George (Eds.). Progress in Underwater Science. Pentech Press, Plymouth. pp. 87–98.

Roe, R.B., Cummins, R. and Bullis, H.R., 1971. Calico scallop distribution, abundance and yield off eastern Florida, 1967–1968. National Marine Fisheries Service, Fisheries Bulletin 69:399–409.

Rolfe, M.S., 1973. Notes on queen scallops and how to catch them. Ministry of Agriculture Fisheries and Food, Shellfish Information Leaflet 27:13 p.

Sakurai, I. and Seto, M., 2000. Movement and orientation of the Japanese scallop *Patinopecten yessoensis* (Jay) in response to water flow. Aquaculture 181:269–279.

Sanders, M.F., 1970. The Australian scallop industry. Australian Fisheries Newsletter 29:2–11.

Sastry, A.N., 1962. Some morphological and ecological differences in two closely related species of scallop *Aequipecten irradians* Lamarck and *Aequipecten gibbus* Dall from the Gulf of Mexico. Quarterly Journal Florida Academy of Science 25:89–95.

Sastry, A.N., 1963. Reproduction of the bay scallop *Aequipecten irradians* Lamarck. Influence of temperature on maturation and spawning. Biological Bulletin of the Marine Biological Laboratory, Woods Hole 125:146–153.

Sastry, A.N., 1970. Reproductive physiology variation in latitudinally separated populations of the bay scallop *Aequipecten irradians* Lamarck. Biological Bulletin of the Marine Biological Laboratory, Woods Hole 138:56–65.

Sastry, A.N., 1979. Pelecypoda (excluding Ostreidae). In: A.C. Giese and J.S. Pearse (Eds.). Reproduction of Marine Invertebrates. Academic Press, New York. pp. 113–292.

Scheibling, R. and Hatcher, B.G., 1994. Movement and mortality of juvenile scallops released in bottom culture trials. In: N.F Bourne, B.L. Bunting and L.D. Townsend (Eds.). Proceedings of the 9th International Pectinid Workshop, Nanaimo, B. C. Canada, April 22–27, 1993. Canadian Technical Report of Fisheries and Aquatic Sciences 1994 (2):97.

Scheibling, R.E., Hatcher, B.G., Taylor, L. and Barbeau, M.A., 1995. Seeding trial of the giant scallop (*Placopecten magellanicus*) in Nova Scotia. In: P. Lubet, J. Barret and J-C. Dao (Eds.). Fisheries, Biology and Aquaculture of Pectinids. 8th International Pectinid Workshop, Cherbourg, France, 22nd-29th May, 1991, IFREMER, Actes de Colloques 17:123–129.

Scheltema, R.S., 1986. On dispersal and planktonic larvae of benthic invertebrates: an eclectic overview and summary of problems. Bulletin of Marine Science 39:290–322.

Schick, D.F., Shumway, S.E. and Hunter, M.A., 1988. A comparison of growth rate between shallow water and deep water populations of scallops, *Placopecten magellanicus* (Gmelin, 1791), in the Gulf of Maine. American Malacological Bulletin 6:1–8.

Serchuk, F.M., Wood, P.W., Posgay, J.A. and Brown, B.E., 1979. Assessment and status of sea scallop (*Placopecten magellanicus*) populations off the northeast coast of the United States. Proceedings of the National Shellfisheries Association 69:161–191.

Shafee, M., 1979. Underwater observation to estimate the density and spatial distribution of black scallop, *Chlamys varia* (L.) in Lanveoc (Bay of Brest). Bulletin Office National Pêches Tunisie 3:143–156.

Shafee, M.S. and Conan, G., 1984. Energetic parameters of a population of *Chlamys varia* (Bivalvia: Pectinidae). Marine Ecology Progress Series 18:253–262.

Shirley, S.M. and Kruse, G.H., 1995. Development of the fishery for weathervane scallops, *Patinopecten caurinus* (Gould, 1850), in Alaska. Journal of Shellfish Research 14:71–78.

Shumway, S.E. and Schick, D.F., 1987. Variability of growth, meat count and reproductive capacity in *Placopecten magellanicus*: are current management policies sufficiently flexible? International Council for the Exploration of the Sea, CM 1987/K:2. 15 p.

Shumway, S.E., Selvin, R. and Schick, D.F., 1987. Food resources related to habitat in the scallop *Placopecten magellanicus* (Gmelin, 1791): a qualitative study. Journal of Shellfish Research 6:89–95.

Sinclair, M., Mohn, R.K., Robert, G. and Roddick, D.L., 1985. Considerations for the effective management of Atlantic scallops. Canadian Technical Report in Fisheries and Aquatic Sciences 1382:113 p.

Skarlato, O.A., 1981. Bivalve molluscs of the temperate latitudes of the western part of the Pacific Ocean. Akademiia Nauka SSSR, Zoological Institute, Leningrad. 461 p.

Slater, J., 2003. Forecasting the scallop spatfall using morphological identification of scallop larvae. Book of Abstracts, 14th International Pectinid Workshop, St. Petersburg, Florida, 23–29 April, 2003. pp. 24–26.

Soemodihardjo, S., 1974. Aspects of the biology of *Chlamys opercularis* (L.) (Bivalvia) with comparative notes on four allied species. PhD Thesis. University of Liverpool. 110 p.

Squires, H.T., 1962. Giant scallops in Newfoundland coastal waters. Bulletin of the Fisheries Research Board of Canada 135: 29 p.

Stanley, S.M., 1970. Relation of shell form to life habitats in the Bivalvia (Mollusca). Memoranda of the Geological Society of America 125:1–296.

Starr, R.M. and McCrea, J.E., 1983. Weathervane scallop (*Patinopecten caurinus*) investigations in Oregon 1981–1983. Oregon Department of Fisheries and Wildlife, Information Report, 83–10: 55 p.

Stephens, P.J. and Boyle, P.R., 1978. Escape responses of the queen scallop *Chlamys opercularis* (L.) (Mollusca: Bivalvia). Marine Behaviour and Physiology 5:103–113.

Stockton, W.L., 1984. The biology and ecology of the epifaunal scallop *Adamussium colbecki* on the west side of McMurdo Sound, Antarctica. Marine Biology 78:171–178.

Stokesbury, K.D.E., 2000. Physical and biological variables influencing the spatial distribution of the giant scallop *Placopecten magellanicus*. Alaska Department of Fish and Game, Special Publication 14:13–19.

Stokesbury, K.D.E., 2002. Estimation of sea scallop abundance in closed areas of Georges Bank, USA. Transactions of the American Fisheries Society 131:1081–1092.

Stokesbury, K.D.E. and Himmelman, J.H., 1993. Spatial distribution of the giant scallop *Placopecten magellanicus* in unharvested beds in the Baie des Chaleurs, Québec. Marine Ecology Progress Series 96:159–168.

Stokesbury, K.D.E. and Himmelman, J.H., 1995a. Biological and physical variables associated with aggregations of the giant scallop *Placopecten magellanicus*. Canadian Journal of Fisheries and Aquatic Sciences 52:743–753.

Stokesbury, K.D.E. and Himmelman, J.H., 1995b. Examination of orientation of the giant scallop, *Placopecten magellanicus*, in natural habitats. Canadian Journal of Zoology 73:1945–1950.

Stotz, W., 2000. When aquaculture restores and replaces an overfished stock: is the conservation of the species assured? The case of the scallop *Argopecten purpuratus* in Northern Chile. Aquaculture International 8:237–247.

Strand, Ø., Haugum, G.A., Hansen, E. and Monkan, A., 1999. Fencing scallops on the seabed to prevent intrusion of the brown crab *Cancer pagurus*. Book of Abstracts, 12th International Pectinid Workshop, Bergen, Norway, 5–11 May, 1999. pp. 58–59.

Strieb, M.D., Bricelj, V.M. and Bauer, S.I., 1995. Population biology of the mud crab, *Dyspanopeus sayi*, an important predator of juvenile bay scallops in Long Island (USA) eelgrass beds. Journal of Shellfish Research 14:347–357.

Sundet, J.H. and Fevolden, S.E., 1994. Distribution of the Icelandic scallop, *Chlamys islandica*, in the northeast Atlantic. In: N.F Bourne, B.L. Bunting and L.D. Townsend (Eds.). Proceedings of the 9th International Pectinid Workshop, Nanaimo, B.C., Canada, April 22–27, 1993. Canadian Technical Report of Fisheries and Aquatic Sciences 1994 (2):179.

Sweijd, N.A., Bowie, R.C.K., Evans, B.S. and Lopata, A.L., 2000. Molecular genetics and the management and conservation of marine organisms. Hydrobiologia 420:153–164.

Taylor, A.C. and Venn, T.J., 1978. Growth of the queen scallop, *Chlamys opercularis*, from the Clyde Sea area. Journal of the Marine Biological Association of the United Kingdom 58:687–700.

Taylor, A.C. and Venn, T.J., 1979. Seasonal variation in weight and biochemical composition of the tissues of the queen scallop, *Chlamys opercularis*, from the Clyde Sea area. Journal of the Marine Biological Association of the United Kingdom 59:605–621.

Tebble, N., 1966. British Bivalve Seashells. British Museum, London. 212 p.

Tettelbach, S.T., 1991. Seasonal changes in a population of northern bay scallops, *Argopecten irradians irradians* (Lamarck, 1819). In: S.E. Shumway and Sandifer (Eds.). An International Compendium of Scallop Biology and Culture. The World Aquaculture Society, Baton Rouge, LA. pp. 164–175.

Tettelbach, S.T. and Rhodes, E.W., 1981. Combined effects of temperature and salinity on embryos and larvae of the Northern Bay scallop *Argopecten irradians irradians*. Marine Biology 63:249–256.

Tettelbach, S.T., Auster, P.J., Rhodes, E.W. and Widman, J.C., 1985. A mass mortality of northern bay scallops, *Argopecten irradians irradians*, following a severe spring rainstorm. Veliger 27:381–385.

Thayer, C.W., 1972. Adaptive features of swimming monomyarian bivalves (Mollusca). Forma et Functio 5:1–32.

Thayer, C.W. and Stuart, H.H., 1974. The bay scallop makes its bed of seagrass. Marine Fishery Review 36:27–30.

Thomas, G.E. and Gruffydd, L.D., 1971. The types of escape reactions elicited in the scallop *Pecten maximus* by selected sea star species. Marine Biology 10:87–93.

Thompson, R.J., Livingstone, D.R. and de Zwaan A., 1980. Physiological and biochemical aspects of the valve snap and valve closure responses in the giant scallop *Placopecten magellanicus*. 1. Physiology. Journal of Comparative Physiology 137:97–104.

Thompson, R.J. and MacDonald, B.A., 1991. Physiological integrations and energy partitioning. In: S.E. Shumway (Ed.). Scallops: Biology, Ecology and Aquaculture. Developments in Aquaculture and Fisheries Science, Volume 21. Elsevier, Amsterdam. pp. 347–372.

Thorburn, I.W. and Gruffydd, L.D., 1979. Studies of the behaviour of the scallop *Chlamys opercularis* (L.) and its shell in flowing sea water. Journal of the Marine Biological Association of the United Kingdom 59:1003–1023.

Thorpe, J.P., Solé-Cava, A.M. and Watts, P.C., 2000. Exploited marine invertebrates: genetics and fisheries. Hydrobiologia 420:165–184.

Thorson, G., 1946. Reproduction and larval development of Danish marine bottom invertebrates. Meddelelser fra Danmarks Fiskeri – og Havundersøgelser 4:1–523.

Thorson, G., 1950. Reproductive and larval ecology of marine bottom invertebrates. Biological Revue of the Cambridge Philosophical Society 25:1–45.

Thouzeau, G., 1991. Experimental collection of postlarvae of *Pecten maximus* (L) and other benthic macrofaunal species in the Bay of Saint-Brieuc, France 1. Settlement patterns and biotic interactions among the species collected. Journal of Experimental Marine Biology and Ecology 148:159–179.

Thouzeau, G. and Hily, C., 1986. A.QUA.R.E.V.E.: Une technique nouvelle d'echantillonnage quantitatif de la macrofaune epibenthique des fonds meubles. Oceanologica Acta 9:509–513.

Thouzeau, G., Robert, G. and Smith, S.J., 1991. Spatial variability in distribution and growth of juvenile and adult sea scallops *Placopecten magellanicus* (Gmelin) on eastern Georges Bank (Northwest Atlantic). Marine Ecology Progress Series 74:205–218.

Tunbridge, B.R., 1968. The Tasman Bay scallop fishery. New Zealand Marine Department of Fisheries, Report No. 18. 78 p.

Tully, J.P. and Barber, F.G., 1960. An estuarine analogy in the sub-arctic Pacific Ocean. Journal of the Fisheries Research Board of Canada 17:91–112.

Tyurin, A.N., 1990. Behavioural reactions of scallop *Mizuhopecten yessoensis* and mussel *Crenomytilus grayanus* to the reduction of salinity, oxygen content, and to the effect of synthetic detergents. Zoologicheskii Zhurnal 69:31–37.

Uexküll, J. von., 1912. Studien über den Tonus. VI. Die Pilgermuschel. Zeitschrift für Biologie 58:305–332.

Uribe, E., Blanco, J.L. and Yamashiro, C., 2003. Effect of salinity on the distribution of *Argopecten purpuratus* on the SW Pacific coast. Book of Abstracts, 14th International Pectinid Workshop, St. Petersburg, Florida, USA, 23–29 April, 2003. pp. 125–126.

Ursin, E., 1956. Distribution and growth of the queen, *Chlamys opercularis* (Lamellibranchiata) in Danish and Faroese waters. Meddelelser fra Danmarks Fiskeri-og Havundersøgelser, Ny Serie 1:1–31.

Vahl, O., 1980. Seasonal variations in seston and in the growth rate of the Iceland scallop, *Chlamys islandica* (O.F. Müller) from Balsfjord, 70°N. Journal of Experimental Marine Biology and Ecology 48:195–204.

Vahl, O., 1981. Energy transformations by the Iceland scallop, *Chlamys islandica* (O.F. Müller), from 70°N. II. The population energy budget. Journal of Experimental Marine Biology and Ecology 53:297–303.

Vahl, O., 1982. Long-term variation in recruitment of the Iceland scallop, *Chlamys islandica*, from northern Norway. Netherlands Journal of Sea Research 16:80–87.

Vahl, O. and Clausen, B., 1980. Frequency of swimming and energy cost of byssus production in *Chlamys islandica* (O.F. Müller). Journal du Conseil International pour l'Exploration de la Mer 39:101–103.

Veale, L.O., Hill, A.S., Hawkins, S.J. and Brand, A.R., 2000. Effects of long-term physical disturbance by commercial scallop fishing on subtidal sedimentary habitats and epifaunal assemblages. Marine Biology 137:325–337.

Ventilla, R.F., 1982. The scallop industry in Japan. Advances in Marine Biology 20:310–390.

Vernberg, F.J. and Vernberg, W.B., 1970. Lethal limits and the zoogeography of the faunal assemblages of coastal Carolina waters. Marine Biology 6:26–32.

Verrill, A.E., 1897. A study of the family Pectinidae, with a revision of the genera and subgenera. Transactions of the Connecticut Academy of Arts and Sciences 10:41–96.

Villiers, G. de, 1976. Exploratory fishing for and growth of scallop *Pecten sulcicostatus* off the Cape south coast. Republic of South Africa Department of Industries, Sea Fisheries Branch, Investigational Report 112:1–23.

Vogel, S., 1985. Flow-assisted shell reopening in swimming scallops. Biological Bulletin Marine Biological Laboratory, Woods Hole 169:624–630.

Volkov, Y.P., Dadaev, A.A., Levin, V.S. and Murakhveri, A.M., 1982. Changes in the distribution of yezo scallop and starfishes after mass planting of scallops at the bottom of Vityaz Bay (Sea of Japan). Soviet Journal of Marine Biology 8:216–223.

Wallace, J.C., 1982. The culture of the Iceland scallop, *Chlamys islandica* (O.F. Müller). 1. Spat collection and growth during the first year. Aquaculture 26:311–320.

Wallace, J.C. and Reinsnes, T.G., 1984. Growth variation with age and water depth in the Iceland scallop (*Chlamys islandica*, Pectinidae). Aquaculture 41:141–146.

Wallace, J.C. and Reinsnes, T.G., 1985. The significance of various environmental parameters for growth of the Iceland scallop, *Chlamys islandica* (Pectinidae), in hanging culture. Aquaculture 44:229–242.

Waller, T.R., 1969. The evolution of the *Argopecten gibbus* stock (Mollusca: Bivalvia), with emphasis on the Tertiary and Quaternary species of eastern North America. Paleontological Society Memoir 3: 125 p.

Waller, T.R., 1991. Evolutionary relationships among commercial scallops (Mollusca: Bivalvia: Pectinidae). In: S.E. Shumway (Ed.). Scallops: Biology, Ecology and Aquaculture. Developments in Aquaculture and Fisheries Science, Volume 21. Elsevier, Amsterdam, pp. 1–73.

Waller, T.R., 1995. The misidentified holotype of *Argopecten circularis* (Bivalvia, Pectinidae). Veliger 38:298–303.

Waloszek, D., 1991. *Chlamys patagonica* (King and Broderip 1832), a long "neglected" species from the Patagonian coast. In: S.E. Shumway and P.A. Sandifer (Eds.). An international compendium of scallop biology and culture. World Aquaculture Society, Baton Rouge, LA. pp. 256–263.

742

Wang, S., Duncan, P.F., Knibb, W. and Degnan, B.M., 2001. Successful hatchery production and the first report of byssal attachment in the 'unusual' scallop, *Amusium balloti*. Book of Abstracts, 13th International Pectinid Workshop, Coquimbo, Chile, 18–24 April, 2001. pp. 71–72.

Whittington, M.W., 1993. Scallop aquaculture in Manx waters: spat collection and the role of predation in seabed cultivation. PhD Thesis. University of Liverpool. 199 p.

Wiborg, K.F., 1963. Some observations on the Iceland scallop *Chlamys islandica* (Müller) in Norwegian waters. Fiskeridirecktoratets Skrifter. Serie Havundersøkelser 13:38–53.

Wilbur, A.E. and Gaffney, P.M., 1997. A genetic basis for geographic variation in shell morphology in the bay scallop, *Argopecten irradians*. Marine Biology 128:97–105.

Wilbur, A.E., Orbacz, E.A., Wakefield, J.R. and Gaffney, P.M., 1997. Mitochondrial genotype variation in a Siberian population of the Japanese scallop, *Patinopecten yessoensis*. Journal of Shellfish Research 16:541–545.

Wilding, C.S., Beaumont, A.R. and Latchford, J.W., 1997. Mitochondrial DNA variation in the scallop *Pecten maximus* (L.) assessed by a PCR-RFLP method. Heredity 79:178–189.

Wilding, C.S., Latchford, J.W. and Beaumont, A.R., 1998. An investigation of possible stock structure in *Pecten maximus* (L.) using multivariate morphometrics, allozyme electrophoresis and mitochondrial DNA polymerase chain reaction restriction fragment length polymorphism. Journal of Shellfish Research 17:131–139.

Wildish, D.J. and Kristmanson, D.D., 1988. Growth responses of giant scallops to periodicity of flow. Marine Ecology Progress Series 42:163–169.

Wildish, D.J., Kristmanson, D.D., Hoar, R.L, DeCoste, A.M., McCormick, S.D. and White, A.W., 1987. Giant scallop feeding and growth responses to flow. Journal of Experimental Marine Biology and Ecology 113:207–220.

Wildish, D.J., Kristmanson, D.D. and Saulnier, A.M., 1992. Interactive effect of velocity and seston concentration on giant scallop feeding inhibition. Journal of Experimental Marine Biology and Ecology 155:161–168.

Wildish, D.J. and Saulnier, A.M., 1992. The effect of velocity and flow direction on the growth of juvenile and adult giant scallops. Journal of Experimental Marine Biology and Ecology 155:133–143.

Wildish, D.J. and Saulnier, A.M., 1993. Hydrodynamic control of filtration in *Placopecten magellanicus*. Journal of Experimental Marine Biology and Ecology 174:65–82.

Wilkens, L.A., 1981. Neurobiology of the scallop. I. Starfish-mediated escape behaviours. Proceedings of the Royal Society of London, Series B 211:341–372.

Williams, J.G., 1980. The influence of adults on the settlement of spat of the clam, *Tapes japonica*. Journal of Marine Research 38:729–741.

Williams, M.J. and Dredge, M.C.L., 1981. Growth of the saucer scallop, *Amusium japonicum balloti* Bernardi in central eastern Queensland. Australian Journal of Marine and Freshwater Research 32:657–666.

Wilson, U.A.W. and Brand, A.R., 1995. Variations in commercial scallop (*Pecten maximus*) density in a seasonal fishery. Book of Abstracts, 10th International Pectinid Workshop, Cork, Ireland, 26 April–2 May, 1995. pp. 13–14.

Winter, M.A. and Hamilton, P.V., 1985. Factors influencing swimming in bay scallops, *Argopecten irradians* (Lamark, 1819). Journal of Experimental Marine Biology and Ecology 88:227–242.

Wolf, B.M. and White, R.W.G., 1995. Age and growth of the queen scallop, *Equichlamys bifrons*, in the D'Entrecasteaux Channel and Huon River estuary, Tasmania. Marine and Freshwater Research 46:1127–1135.

Wolf, B.M. and White, R.W.G., 1997. Movements and habitat use of the queen scallop, *Equichlamys bifrons*, in the D'Entrecasteaux channel and Huon river estuary, Tasmania. Journal of Shellfish Research 16:533–539.

Wolff, M., 1988. Spawning and recruitment of the Peruvian scallop *Argopecten purpuratus*. Marine Ecology Progress Series 42:213–217.

Wolff, M. and Alarcon, E., 1993. Structure of a scallop (*Argopecten purpuratus*) dominated subtidal macro-invertebrate assemblage in Northern Chile. Journal of Shellfish Research 12:295–304.

Wolff, M. and Mendo, J., 2000. Management of the Peruvian bay scallop (*Argopecten purpuratus*) metapopulation with regard to environmental change. Aquatic Conservation-Marine and Freshwater Ecosystems 10:117–126.

Wolff, M. and Wolff, R., 1983. Observations on the utilization and growth of the pectinid *Argopecten purpuratus* (L.) in the fishing area of Pisco, Peru. Boletin Instituto del Mar del Peru 7:6 p.

Woodburn, L., 1987. Variation in Australian *Pecten*. Book of Abstracts, 6th International Pectinid Workshop, Menai Bridge, Wales, 9th-14th April, 1987. p. 34.

Woodburn, L., 1989. Genetic variation in southern Australasian *Pecten*. In: M.C.L. Dredge, W.F. Zacharin and L.M. Joll (Eds.). Proceedings of the Australasian Scallop Workshop, Hobart, Australia. Tasmanian Government Printer, Hobart. pp. 226–240.

Woodin, S.A., 1976. Adult-larval interactions in dense infaunal assemblages: patterns of abundance. Journal of Marine Research 34:25–41.

Yager, P.L., Nowell, A.R.M. and Jumars, P.A., 1993. Enhanced deposition to pits: a local food source for benthos. Journal of Marine Research 51:209–236.

Yamamoto, G., 1951. Ecological note on the transplantation of the scallop *Patinopecten yessoensis* Jay in Mutsu Bay, with specific reference to the succession of the benthic communities. Scientific Reports of the Tôhoku University, 4th Series (Biology) 19:11–16.

Yamamoto, G., 1953. Ecology of the scallop, *Pecten yessoensis* Jay. Science Reports of the Tôhoku University, 4th Series (Biology) 20:11–32.

Yamamoto, G., 1956a. Habitats of spats of the scallop, *Pecten yessoensis* Jay, which turned to bottom life. Bulletin of the Marine Biological Station of Asamushi, Tokyo University 22:149–156.

Yamamoto, G., 1956b. On the behaviour of the scallop under some environmental conditions with special reference to effects of suspended silt, lack of soluble oxygen and others on ciliary movement of gill pieces. Japanese Journal of Ecology 5:172–175.

Yamamoto, G., 1960. Mortalities of the scallop during its life cycle. Bulletin of the Marine Biological Station of Asamushi, Tôhoku University 10:149–152.

Yamamoto, G., 1964. Studies on the propagation of the scallop, *Patinopecten yessoensis* Jay, in Mutsu Bay. Fisheries Research Board of Canada Translation Series, No. 1054: 64 p.

Yamamoto, G., 1975. Recent advances in the ecological studies on the Japanese scallop. Bulletin of the Marine Biological Station of Asamushi 15:53–58.

Yonge, C.M., 1936. The evolution of the swimming habit in the Lamellibranchia. Mémoires du Musée Royal d'Histoire Naturelle de Belgique (2me series) 3:77–100.

Yonge, C.M., 1951. Studies on the Pacific coast mollusks III: observations on *Hinnites multirugosus* (Gale). University of California Publications in Zoology 55:409–420.

Yonge, C.M., 1953. The monomyarian condition in the Lamellibranchiata. Transactions of the Royal Society of Edinburgh 62:443–478.

Yonge, C.M., 1962. On the primitive significance of the byssus in the Bivalvia and its effects in evolution. Journal of the Marine Biological Association of the United Kingdom 42:113–125.

Yonge, C.M. and Thompson, T.E., 1976. Living Marine Molluscs. Collins, London. 288 p.

Young, P.C. and Martin, R.B., 1989. The scallop fisheries of Australia and their management. CRC Critical Reviews in Aquatic Sciences 1:615–638.

Zhang, J.Z., 1996. Further studies of scallop spat settlement and early survival in Isle of Man coastal waters. PhD Thesis. University of Liverpool. 254 p.

Zwaan, A de., Thompson, R.J. and Livingstone, D.R., 1980. Physiological and biochemical aspects of the valve snap and valve closure response in the giant scallop *Placopecten magellanicus* II: Biochemistry. Journal of Comparative Physiology 137:105–114.

AUTHOR'S ADDRESS

Andrew R. Brand - Port Erin Marine Laboratory, University of Liverpool, Port Erin, Isle of Man, IM9 6JA (E-mail: arbrand@liverpool.ac.uk)

Scallops: Biology, Ecology and Aquaculture
S.E. Shumway and G.J. Parsons (Editors)
© 2006 Elsevier B.V. All rights reserved.

Chapter 13

Scallops and Marine Contaminants

Peter J. Cranford

13.1 INTRODUCTION

The marine environment receives a wide range of industrial, agricultural and domestic wastes. Marine contaminants may have an immediate impact on local coastal communities, and may further interact with living resources in shelf waters as the contaminants are transported seaward. Coastal and shelf waters also receive direct inputs of contaminants from the atmosphere and from offshore industries. Contaminant exposure can induce a wide range of biological effects in organisms including biochemical, immunological, physiological, and bioenergetic responses to stress, altered growth and reproductive output, and histopathological effects. Depending on the exact nature and temporal and spatial scale of the biological response, impacts on individuals may have implications at population, community, and ecosystem levels. The Pectinidae have received considerable attention in studies of contaminant distributions and biological effects. Reasons for this attention include the fact that they are largely sedentary and unable to escape contaminant exposure, have a wide geographic distribution, appear to be relatively sensitive to a wide range of pollutants, and are commercially important.

The risk of any contaminant impacting epibenthic organisms (e.g., post-settlement scallops) is related to numerous physical, chemical and biological processes that are responsible for the contaminant reaching, and accumulating in the near seabed region. An important mechanism by which contaminants are transported in the marine environment is by scavenging (uptake) onto particulate matter and particle settling. As a result, marine sediments are a major sink for many trace contaminants. The slow release of contaminants from sediments can result in scallops being subjected to chronic exposure long after specific anthropogenic discharges cease. Once subjected to an acute or chronic exposure to a specific contaminant, the impact on scallops is related to contaminant bioavailability and the potential for bioaccumulation (amount accumulated in tissues) of toxicants. The former requires a process by which the contaminant is taken up and the latter is related to the rates of uptake and loss of contaminants that can vary greatly for different contaminants and organisms. Scallops, as suspension feeders, may be exposed directly to dissolved and particulate contaminants in the water column, or indirectly through their food sources. As a result of scavenging onto particulate matter, concentrations of contaminants in solid form are often several orders of magnitude greater than the solute concentration. Particle ingestion is therefore an important means for the uptake of particle-reactive contaminants (including most metals and organic contaminants) by scallops. With large volumes of water entering the mantle cavity and passing the gills during feeding, scallops are also highly susceptible to the passive assimilation of bioavailable contaminants. This direct uptake of soluble contaminants is generally orders of magnitude greater than the amount assimilated indirectly with solid food.

This paper reviews research on different classes of anthropogenic contaminants as they relate to the Pectinidae, but is not meant to be an exhaustive literature review. Owing to the large number of potential pollutants, each section of this review focuses on specific contaminants that have received a relatively high degree of research interest over approximately the past decade. Much of the earlier research on scallops was reviewed by Gould and Fowler (1991). In order to avoid simply identifying numerous *perceived* risks to scallops from a very wide range of human activities, an attempt has been made, where possible, to discuss the *real* risk to populations by (1) integrating laboratory biological effects data with knowledge of the capacity of scallops to adapt to contaminant exposure, (2) identifying major oceanic processes known to enhance or reduce chronic contaminant exposure, (3) providing information on environmentally relevant exposure conditions, and (4) comparing the perceived zone of impact with the distribution of local scallop stocks. This approach, which relies on specific geographical case studies, is intended to emphasise contaminant issues of foremost importance to scallop stocks.

13.2 TRACE METALS

Even though many trace metals are essential to organism metabolism, all are capable of causing a wide range of deleterious effects in marine organisms and populations at concentrations that are only slightly higher than typically found dissolved in seawater (Rainbow 1993). Metals can be taken up by scallops passively through diffusion from the surrounding water and from food ingestion and absorption. Protein binding of metals prevents the outward diffusion from tissue, leading to accumulation in the body if metal uptake exceeds excretion. There is considerable variability in biological responses (harmfulness degree, type of response and degree of bioaccumulation) to metallic contaminants, even among similar zoological groups and life-stages (Berthet et al. 1992). All bivalve molluscs have the ability to accumulate large amounts of trace metals in their tissues. As a result, exposure of scallop stocks to trace metals could make them toxic to their predators, including humans.

Cadmium, a nonessential trace metal, is recognised as one of the most deleterious metallic contaminants and has somewhat dominated the inorganic trace metals literature pertaining to scallops in recent years. Much of the oceanic input of cadmium from rivers and atmosphere is believed to be from anthropogenic origin (Neff 2002). Cadmium tends to bind weakly to particles and much of the oceanic pool is bioavailable. Members of the Pectinidae appear to accumulate cadmium to a higher degree than other organisms studied (Gould et al. 1985; Evtushenko et al. 1990; Lukyanove et al. 1993; Viarengo et al. 1993), indicating the absence of an effective mechanism for regulating tissue concentrations. Cadmium levels in the hepatopancreas and kidney were observed to increase with age in the Japanese scallop *Mizuhopecten yessoensis* (Evtushenko et al. 1990; and Lukyanove et al. 1993), and in tissues associated with the digestive gland of the Antarctic scallop

Adamussium colbecki (Mauri et al. 1990; Viarengo et al. 1993), and the sea scallop *Placopecten magellanicus* (Uthe and Chou 1987). The digestive gland was found to contain 91% of the total cadmium accumulated by *P. magellanicus* (Uthe and Chou 1987). The very high cadmium content stored by *A. colbecki* from isolated regions around Antarctica appears to be related to the relatively high cadmium concentrations in seawater upwelled from the deep Southern Ocean (Viarengo et al. 1993). It is interesting to note that the amount of cadmium in *P. magellanicus* tissues was higher for animals from offshore sites (Georges Bank and Browns Bank, Canada) than at coastal sites receiving substantial anthropogenic cadmium input (Uthe and Chou 1987). This illustrates the importance of natural factors such as the local food supply or primary productivity in contributing to tissue cadmium levels in scallops (Bryan 1973).

There are marked differences in the ability of scallops to regulate the concentration of different metals stored in tissues. While cadmium is not effectively regulated by scallops and continues to accumulate with age (above), many other trace elements remain at constant levels as the animal grows, an indication that these metals are highly regulated by the scallop. Lukyanove et al. (1993) provided evidence to support homeostasis of the essential trace elements zinc and copper in *M. yessoensis,* and Gould et al. (1988) provided support for the tight regulation of magnesium in *P. magellanicus* gonad. Cadmium exposure and accumulation has been shown to cause the redistribution of essential trace metals (e.g., copper, zinc and iron) present in *M. yessoensis* tissues, and this disruption to metals homeostasis may interfere with metabolism and enzymatic activity (Chelomin et al. 1995). Similarly, *P. magellanicus* exposed to cadmium altered zinc, manganese, and copper homeostasis in the kidney, potentially permitting toxic responses to these previously sequestered metals (Fowler and Gould 1988).

Fishery products can be screened for potential hazard to humans by comparing metals content in edible tissues with calculated risk-based concentrations (RBC). The RBC for cadmium in tissues, calculated based on consumption of 54 g day^{-1} for 175 days year^{-1} by an adult, is 13.5 μg g^{-1} dry weight (Neff 2002). This RBC is generally less than the majority of values detected in scallop tissue, which ranges from 0.58 to 43.6 μg Cd g^{-1} dry weight (Berkman and Nigro 1992; Neff 2002, and calculated from data in Uthe and Chou 1987). For these cadmium stores to be a hazard to consumers, including humans, they must be in a bioavailable form. Cadmium stored in tissue as solid concretions are not bioavailable to consumers and are nontoxic to the scallop (Neff 2002). Such concretions have been described in the kidney of *Argopecten irradians* and in *P. magellanicus* following cadmium exposure (Carmichael and Fowler 1981; Fowler and Gould 1988), but were not observed in *A. colbecki* that are known to accumulate large amounts of cadmium (Nigro et al. 1992). The other major tissue store for metals is binding to soluble ligands such as metallothioneins. The fraction of total accumulated cadmium in the digestive gland of *Pecten jacobaeus* and *A. colbecki* that was bound to metallothionein was reported to be 60% and 30%, respectively (Viarengo et al. 1993). Lukyanove et al. (1993) and Stone et al. (1986) reported that metallothionein does not play a significant role in cadmium accumulation in *M. yessoensis* and *Pecten maximus.* Most of the cadmium accumulated (72–99%) in *M. yessoensis* was found to bind to high-molecular-weight proteins (Evtushenko et al. 1986; Evtushenko et al. 1990; Syasina et al. 1994). Maximal cadmium accumulations in male *M. yessoensis* gonad were determined to be associated with high-molecular-weight proteins, while accumulations in female gonad were primarily associated with medium-molecular-weight proteins (Syasina et al. 1994). Evtushenko et al. (1986) noted that after 30 days of exposure of *M. yessoensis*, cadmium was redistributed from high-molecular-weight proteins to metallothionein-like proteins. The binding of cadmium to these macromolecules renders them nontoxic to the scallop and to predators (Viarengo et al. 1993; Neff 2002).

Although scallops are strong accumulators of potentially hazardous metals, total body concentration is not a good indicator of either metals toxicity to the scallop, or of the danger to consumers. The detoxification of metals by binding to proteins or to insoluble granules greatly reduces levels of harmful metabolically available metals in scallop tissues, and limits the potential for biomagnification up the food chain. There is little evidence to support biomagnification of cadmium, or other inorganic trace metals, in the marine environment (Gray 2002; Neff 2002).

13.3 TRACE ORGANICS

Trace organic contaminants entering the sea from anthropogenic sources include a diverse range of compounds including organometals, organophosphates, organohalogens, and other compounds with endocrine-disrupting properties (Moore et al. 2002). Many of the compounds of concern are relatively lipophilic, resulting in a high propensity for the contaminant to leave the aqueous environment and accumulate in biota or the organic fraction of sediments. The primary means of uptake of lipophilic compounds is directly from the surrounding water, rather than from food, and a major route of intake by bivalves is through the gill. An important concern associated with trace organics is the effect on biota of long-term, low level exposure. The potential for chronic effects is particularly high for those toxic compounds that are mobile and persist in the environment for many years. Such compounds include many organochlorine insecticides (DDT, dieldrin, toxaphene and chlordane), PCBs, dioxins, brominated flame retardants, polychlorinated napthalenes, and chlorinated paraffins (Moore et al. 2002). Unlike trace metals, the degree of accumulation of organic contaminants in bivalve tissues is generally closely linked to biological responses. Stress from organic contaminant accumulations in tissues may cause immunotoxicity, impaired reproduction, carcinogenicity, increased susceptibility to disease and parasites, and eventually mortality (Moore et al. 2002). In addition to biological damage, there are also potential hazards associated with contaminant biomagnification and health concerns for humans ingesting toxic organic bioaccumulations in seafood.

The treatment of vessels and other marine structures with anti-fouling biocides has resulted in the accumulation of toxic contaminants in scallops, leading to impacts at the population level. Organotin paint, mainly bis-(tri-*n*-butyltin)-oxide (TBT), has been shown to be detrimental to the growth and survival of juvenile *Pecten maximus*, but not to adults (Paul and Davies 1986). Juvenile *P. maximus* have been observed to progressively increase TBT accumulations during chronic exposure, and depuration is slow with 60–80% remaining after eight weeks (Davies et al. 1986). In contrast, oysters tend to accumulate less TBT than scallops and rapidly eliminate the body burden soon after the exposure ceases (Davies et al. 1986). The adductor mussel appears to be the primary storage site for TBT in scallops (Davies et al. 1986). Minchin et al. (1987) monitored the settlement of bivalves and the organotin content of adult scallops in different bays in Ireland during an eight-year period. The body burden of TBT in adult scallop tissue was higher than in other species tested and was greatest near known sources of TBT. The settlement of the scallops *P. maximus*, *Lima hians* and *Chlamys varia* onto collectors in Mulroy Bay either failed or was greatly reduced during years when TBT was used to treat salmonid cage netting (Minchin et al. 1987). Although other factors may also be implicated in the observed temporal variations in scallop recruitment off Ireland (Minchin et al. 1987), these observations appear to indicate that the known effect of TBT on juvenile scallops resulted in changes in scallop populations prior to the ban on the use of TBT paint for many applications. The TBT issue did not disappear with the more limited allowable application of this product, but remains a concern for organisms residing near major shipping lanes, offshore installations and shipyards.

Bivalve filter-feeders are known to be highly sensitive to hydrocarbon contaminants over a concentration range as low as 0.03–3.0 mg L^{-1} (Widdows et al. 1981, 1987, 1995; Axiak and George 1987; Stekoll et al. 1980). Two- and three-ring aromatic hydrocarbons, with a molecular weight cut-off at fluoranthene, and aliphatic hydrocarbons of less than n-C_{11} are the primary sources of hydrocarbon toxicity to *Mytilus edulis* (Widdows and Donkin 1989, 1992). Cranford et al. (1999) showed *P. magellanicus* to be highly sensitive to a low aromatic content mineral oil composed of aliphatics less than n-C_{18}. High mortality and negative growth (resorption of gonad) were observed during chronic exposure to just 0.03 mg L^{-1} mineral oil. The total hydrocarbon body burden at the end of this 37-day exposure was relatively high at 1,143 µg g^{-1} dry tissue weight (i.e., strong accumulator) and the digestive gland was the primary hydrocarbon storage site (Cranford et al. 1999).

Hydrocarbon-growth relationships indicate growth reductions in *M. edulis*, the most extensively studied bivalve species in this regard, occur when tissue dose exceeds 0.1–1 µg g^{-1} dry tissue weight (Widdows and Donkin 1989; Widdows et al. 1995). Reductions in scope-for-growth during exposure to hydrocarbons appear to be attributed primarily to reductions in particle clearance rate, but increased energy utilisation is also a significant contributing factor (Axiak and George 1987; Widdows et al. 1987; Cranford et al. 1999). Total polyaromatic hydrocarbons (PAHs) measured in the pink scallop, *Chlamys rubida,* from bays "oiled" by the *Exxon Valdez* crude oil spill in Alaska in March 1989 averaged 0.29 µg g^{-1} (maximum of ~1.2 µg g^{-1}) in 1989, but decreased by a factor of 15 to background levels by 1990 (Armstrong et al. 1995). Based partially on these scallop data, these authors concluded that large marine oil spills do not negatively affect fauna residing in shelf areas deeper than 20 m.

Orimulsion®, a newly developed bitumen-based fuel is used by thermal power plants as a lower cost alternative to traditional heavy fuels. This product has a relatively high density, very high PAH content, and forms fine bitumen droplets when added to seawater. Based on the belief that an accidental spill of this product posed a relatively high threat to benthic suspension-feeding organisms, laboratory experiments were conducted to determine the effects of this fuel on adult *P. magellanicus* (Armsworthy et al. 1999). Mortalities during 43-day chronic exposures to Orimulsion® increased with increasing fuel concentration, but were generally low with 82% survivorship at the highest concentration tested (10 mg L^{-1}). Sublethal effects were more pronounced with scope-for-growth reduced by 80% at just 0.01 mg L^{-1}. Growth reductions resulted mainly from effects on clearance rate (EC_{50} value was between 0.01 and 0.1 mg L^{-1}), but respiration rate also increased significantly relative to controls. Further research was conducted on newer formulations of Orimulsion®, to assess manufacturer claims of reduced toxicity (Armsworthy et al. 2000). Laboratory tests with *P. magellanicus* indicated that chronic exposure to Orimulsion-100 and Orimulsion-400 had no significant sublethal effects on adult sea scallops. One of the few effects observed was a significant 30% increase in clearance rate in scallops exposed to the 0.1 mg L^{-1} treatment. Orimulsion dose levels in sea scallops were as high as 56 µg PAH g^{-1}, with no apparent growth inhibition. The large multi-ring aromatics in the bitumen are less toxic than the PAHs frequently found in crude and heavy fuel oil. The dramatic reduction in toxicity of the newer Orimulsion® formulations appears to be related to changes in the surfactant used to maintain the bitumen/water emulsion. The general lack of observed detrimental biological effects suggests that the impact on scallop stocks of a limited Orimulsion® spill will probably be minimal (Armsworthy et al. 2000).

Biomagnification of organic chemicals in marine systems has been convincingly shown for only a few compounds, including methylmercury, some PCBs and PCDs, and DDT (Neff 2002). Organic mercury is presently the only trace metal (and physicochemical form) for which evidence exists to support biomagnification in the marine environment (Gray 2002). Although scallops accumulate methylmercury, they are of the second trophic level, and are expected to accumulate less methylmercury and be of less of a risk to consumers than predatory fish.

13.4 OFFSHORE OIL AND GAS OPERATIONS

The interaction between scallops and offshore oil and gas operations has been the topic of considerable research over the past decade. Much of this work is related to the rapid expansion of offshore oil and gas operation off the east coast of Canada, including areas supporting significant commercial sea scallop (*P. magellanicus*) and Icelandic scallop (*Chlamys islandica*) stocks. A major concern related to offshore oil and gas operations has been the potential environmental impacts associated with the exposure of marine organisms to operational waste discharges. Drilling wastes (spent drilling mud and well cuttings) are the primary concern during exploration and development drilling operations, while produced water recovered from the hydrocarbon bearing strata is the highest volume waste generated during production. Produced water may contain elevated concentrations of metals, nutrients, radionuclides, hydrocarbons, and trace amounts of chemical agents. Discharged drilling fluids (muds) have been the primary focus of environmental impact studies because their drift, dispersion, and dilution are generally lower than those of dissolved or buoyant discharges such as produced water.

Drilling mud formulations are highly variable containing any combination of over a thousand minor components (GESAMP 1993). Barite ($BaSO_4$) and bentonite (montmorillonite clay) together comprise the majority of the insoluble components of most drilling fluids (GESAMP 1993). Early studies of the fate and effects of drilling and production wastes assumed that the fine drilling mud particles and dissolved contaminants would readily dissipate resulting in low concentrations on energetic offshore banks (Neff 1987). However, it is now known that flocculation (the adhesion of smaller particles to form large particles) and surface adsorption (the adhesion of small particles to larger particles and/or droplets) are important processes in the transport of wastes from oil and gas operations (Fig. 13.1). Drilling wastes readily flocculate in seawater to form aggregates on the order of 0.5–1.5 mm in diameter with high settling velocities (Milligan and Hill 1998) such that the bulk of drilling mud discharges settle rapidly and can accumulate on the seabed (Muschenheim et al. 1995; Muschenheim and Milligan 1996). Potentially toxic metals in produced water have also been observed to transform from dissolved to particulate forms that settled rapidly. Resuspension/deposition processes in the benthic boundary layer tend to concentrate particulate wastes in suspension near the seabed before eventually being dispersed by currents and waves (Muschenheim and Milligan 1996). These observations are consistent with the conclusion that impacts from drilling operations are most severe on benthic communities (National Research Council 1983).

Early studies on drilling waste toxicity to scallops, and other species, focused on the biological effects of relatively concentrated waste suspensions (100–1,000 mg L^{-1}). Multi-organism studies of this type identified scallops as one of only a few organisms that appear to be sensitive to impacts from drilling wastes (Neff 1987). As a result of high waste dispersion processes in the marine environment, waste concentrations greater than about 10 mg L^{-1} are only relevant to a small region in the water column near the discharge point (Neff 1987). For a significant impact to occur at the population level, a long-term biological response would have to be caused by relatively low waste concentrations, a large portion of the population would have to be exposed, and the observed biological effects would have to be relevant to scallop survival, growth and/or reproduction. A preliminary study was conducted in 1988 to assess the potential hazard to commercial sea scallop stocks on Georges Bank of proposed oil and gas drilling activities. Chronic exposure to well cuttings containing two types of mineral oil-based drilling mud resulted in increased mortalities, cessation of reproductive and somatic tissue growth, and a decrease in condition index and metabolic reserves (Cranford and Gordon 1991). The observed impact was greater than had been observed for other bivalve filter-feeders previously tested, raising concerns about the impact on scallop stocks residing in the immediate vicinity of a drilling platform.

Cranford and Gordon (1992) developed a more ecologically relevant laboratory methodology for studying the chronic effects of different types of drilling wastes on scallops. This included the ability to conduct chronic exposures under realistic flow conditions, and the ability to expose scallops to dilute concentrations (<10 mg L^{-1}) of suspended drilling wastes. Exposure of *P. magellanicus* to suspended bentonite clay revealed a low tolerance to this material, with a 50% reduction in clearance rate

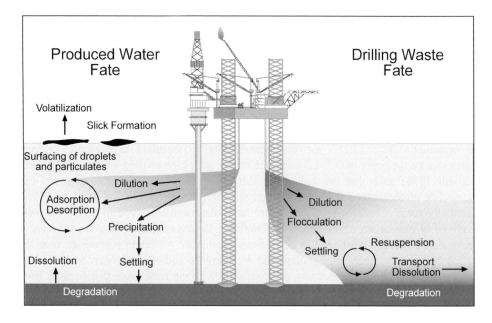

Figure 13.1. Major processes controlling the environmental fate (i.e., exposure of scallop stocks) of wastes from offshore oil and gas drilling and production activities.

(EC_{50}) observed at 2 mg clay L^{-1} (Cranford and Gordon 1992). Additional significant effects detected during chronic exposure to 10 mg L^{-1} bentonite included high mortality, tissue weight loss, and cessation of gonad development. The scallops immediately employed several compensatory mechanisms in an attempt to adapt to the presence of the suspended clay particles, including altering retention and selection efficiencies (Cranford and Gordon 1992) and absorption efficiency (Cranford 1995). These mechanisms allowed the scallops to maintain normal growth rates (relative to the control group) in the 2 mg L^{-1} bentonite treatment, but physiological compensations were largely ineffective at adapting to higher clay concentrations.

The high sensitivity of sea scallops to the suspended clay was not expected as only about 40% of this material is of a size that can be retained by the feeding apparatus (Cranford and Gordon 1992). Fine inorganic particles (<10 µm) have been observed to have a greater impact on gill ciliary activity in *Pecten novaezelandiae* than larger (10–30 µm) particles (Stevens 1987). Although the mechanism for this effect is unknown, the importance of gill ciliary activity in scallop feeding may explain changes in the feeding response observed by Cranford and Gordon (1992). High concentrations of attapulgite clay, also found in some drilling muds, has been shown to have an additional impact on gill ciliary functions in *P. magellanicus*. Morse et al. (1982) observed that 100–1,000 mg L^{-1} suspension of attapulgite caused an "unzipping" of the ordinary filaments of the gill at the ciliary interfilamentary junctions. Clay has been used to displace harmful algal blooms and spilled oil by spraying it over affected areas (Bragg and Yang 1995; Wood et al. 1997; Frank et al. 2000). The clay promotes the flocculation process, leading to settlement of the algae and oil on the seabed. Bay scallops (*Argopecten irradians*) exposed to clay/algal cell suspensions decreased clearance rate (Frank et al. 2000), but the effect was largely limited to concentrations greater than 100 mg L^{-1}. The effect of kaolin clay on bay scallops was not as great as was observed for *P. magellanicus* exposed to bentonite (Cranford and Gordon 1992) and it may be speculated that the presence of clay in a flocculated form reduces its impact on scallops. White (1997) showed that when bentonite was incorporated into large robust flocs in the laboratory, it no longer affected sea scallop clearance rates. However, Cranford et al. (1998) studied sea scallops exposed in the field to natural flocs and found these to be readily disrupted into component particles during feeding.

Additional studies with *P. magellanicus* were conducted to determine chronic biological effects for a wide range of operational drilling muds and their major constituents, and to identify the cause of any effect. Types of drilling muds tested are characterised as water-, oil-, or synthetic-based (WBM, OBM, and SBM, respectively) depending on the fluid component of the drilling mud. An OBM containing a "low-toxicity" (<0.5% aromatic hydrocarbon content) mineral oil had the greatest impact and was fatal to all sea scallops at 2 mg L^{-1} within 11 days (Cranford et al. 1999). The lowest concentration tested (0.5 mg L^{-1}) caused cessation of feeding and somatic tissue growth, and resorption of gonad tissue. Three types of SBM were tested, including a low viscosity ester-based fluid and two synthetic oil-based formulations from oil and gas production sites on the Grand Banks (Hibernia) and Scotian Shelf (Sable Offshore Energy Project). All three SBMs were not lethal to *P. magellanicus* during the chronic exposures, but caused a significant reduction in tissue growth at the 1.0 mg L^{-1} concentration (P.J. Cranford and S.L. Armsworthy, unpublished data). One test detected a significant growth effect at 0.1 mg L^{-1}. Growth reductions observed during exposure to all waste types resulted primarily from reduced energy intake through feeding (clearance rate) and digestion (absorption efficiency), but increased energy losses to respiration and excretion also contributed to reduced growth (Cranford et al. 1999).

Barite particles found in most drilling fluids have been characterised as being toxicologically inert (Neff 1987), but chronic exposures of sea scallops to barite resulted in the complete cessation of gonad growth at the lowest concentration tested (0.5 mg L^{-1}) (Cranford et al. 1999). More recent exposures show this impact to occur at concentrations as low as 0.05 mg L^{-1} (P.J. Cranford and S.L. Armsworthy, unpublished data). SEM observations of the ctenidia of suspension-feeding bivalves have shown gill damage after exposure to barite (Barlow and Kingston 2001). It has also been suggested that barite particles cause nutritional stress by adsorption of mucus secretions, which are critical to normal feeding and digestion processes (Cranford et al. 1999).

In the past, the biological effects of drilling wastes have been grouped as being caused primarily by a combination of (1) chemical toxicity from hazardous pollutants and biodegradation products, (2) organic enrichment of sediment that may produce anoxia, and (3) physical smothering. The above studies with sea scallops show that an additional, and potentially more important cause for concern exists that could have consequences at the population level if drilling wastes are discharged in marine areas supporting significant scallop stocks. Physical effects caused by chronic exposure to very low concentrations (>0.05 mg L^{-1}) of chemically inert components found in all drilling muds (e.g., bentonite and barite) are capable of causing a significant impact on growth and reproduction, at least in *P. magellanicus*. These results may help to explain why observed impacts on benthic communities around drilling platforms tend to be greater than predicted based on the measured distribution of toxic contaminants (hydrocarbons and metals).

Cranford et al. (2003) employed a set of numerical models to evaluate the potential risk to *P. magellanicus* stocks on northeastern Georges Bank during exploratory oil and gas drilling. Georges Bank, which straddles the United States-Canadian boundary, is one of the most productive fishing banks in the North Atlantic Ocean, and supports the largest offshore scallop fishery in the world. Hydrocarbon resources on Georges Bank are believed to be extensive. Numerical models are valuable tools for evaluating the potential environmental impact of marine contaminants. These models provide a quantitative framework for integrating knowledge on the intrinsic physico-chemical properties of the different contaminants, the extrinsic processes that control their transport and fate in the environment, and information on the biological effects. The models used to predict the drilling waste zone of influence and the impact of chronic exposure on scallop growth indicated a potential for 0–48 days of growth inhibition over a 92-day drilling scenario. The wide variation in growth inhibition predictions was related primarily to specific oceanographic conditions at the site and the settling velocity of the mud (Cranford et al. 2003). Impacts were predicted to be greatest in the vertically stratified region around the side of the Bank that supports relatively small scallop stocks, but dense aggregations are found in some areas (Fig. 13.2). Impacts are expected to be more localised in the tidal front region, which has the densest scallop stocks, and minimal in the central, vertically well mixed region of Georges Bank (Fig. 13.2). The implications of such growth inhibition at the population level are not believed to be significant for a single exploration well (Cranford et al. 2003). However, the risk to scallop stocks would be much greater during oil and gas development operations where multiple wells are drilled almost continuously, sometimes at multiple sites, over many years. Drilling for oil and gas on Georges Bank is currently prohibited by legislation in both Canada and the US.

13.5 SCALLOPS AS SENTINEL ORGANISMS

The strategy for monitoring the health of the oceans is evolving from a focus on chemical contaminant distributions to pollution assessments that emphasise the measurement of meaningful biological responses to available contaminants (Bayne et al. 1988; Stagg 1998). Information obtained on contaminant concentrations during field monitoring programs is greatly enhanced by the availability of biological effects data and it has been suggested that the analysis of chemical contaminants is only necessary if a biological effect is detected (e.g., Gray 1999). An advantage of biological effects monitoring programs is that they can indicate the presence of a contaminant that was not previously identified as being of concern. Data on biological responses to contaminants have often come from laboratory studies conducted to reflect acute or relatively short-term

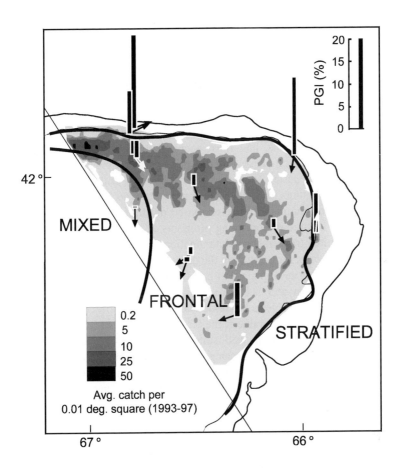

Figure 13.2. Potential growth inhibition (PGI) predicted at model application sites on Georges Bank during drilling of a typical exploratory oil and gas well, in relation to the distribution of *P. magellanicus* stocks and oceanographic zones. PGI estimates (vertical bars) are shown as the percent of time that tissue growth is predicted to be inhibited by the effects of drilling waste discharges. PGI estimates are averaged over a radius of 10 km from the discharge point during the first 62 days of drilling activities. Stock distributions were calculated based on average catch statistics from 1993–1997. The net direction of the drilling waste plume at each site during summer is indicated by the arrows. Adapted from Cranford et al. (2003).

chronic responses to highly contaminated conditions. It is more important to know the effects of much longer-term exposures to very small, and often fluctuating quantities of chemicals. Extrapolation of laboratory results to natural conditions through numerical modelling or environmental risk assessment approaches contain, and must account for, uncertainties regarding prediction of a wide range of environmental processes affecting contaminant availability to organisms (e.g., adsorption, desorption, vertical flux, resuspension, dissolution, and degradation). Although it is important to continue to quantify exposure-response relationships for different contaminants under controlled laboratory conditions for predictive purposes, the long-term monitoring of biological responses to ambient contaminant concentrations would help to reduce uncertainty by directly looking for impacts of xenobiotic substances on populations.

A number of prerequisites have been developed for the selection of biological indicators of contaminant exposure and bioavailability. The selected indicator species should be sedentary, abundant, widely distributed, long-lived, easily obtained, of sufficient size for tissue analysis, and must accumulate the contaminant with a high concentration factor. Based on these criteria, mussels and oysters have been widely employed for monitoring coastal contaminant distributions, and there is now a good appreciation of what constitutes low to high contaminant levels in their tissues. Scallops are also well suited as contaminant indicators based on these criteria, but they have not been widely employed in coastal monitoring programs. Some exceptions include the use of *A. colbecki* to monitor trace metal contamination around Antarctica (Berkman and Nigro 1992), and *P. maximus* for estimating the degree of coastal PAH pollution in France (Friocourt et al. 1985). For an organism to be of use for assessing and monitoring *both* environmental contamination and subsequent biological effects, the indicator species should be susceptible and sensitive to the toxic and/or physical effects of a wide range of contaminants. It is also highly desirable if the target organism is economically important to the study area. Scallops appear well suited to the role as a sentinel species for contaminant and biological effects monitoring owing to the combination of being a strong accumulator of many contaminants and exhibiting a relatively high sensitivity to the toxic and physical effects of many contaminants (this review). They are also resident in commercial quantities in deep shelf waters where more traditional bioindicator species would require transplantation.

An integrated approach to environmental effects monitoring, known as the exposure-dose-response triad has been developed that is based on weight-of-evidence from the results of caged bivalve studies (Salazar and Salazar 1998). This method combines measurements of environmental contaminant concentrations in water and/or sediment (exposure) with measurements of bioaccumulation (dose), and biological responses. It is important to measure both exposure and dose as the response may be related primarily to the dose (trace organics), or to the exposure (inorganic metals). No direct relationships exist between total bioaccumulation of metals and toxicity due to the physico-chemical forms of sequestration (Berthet et al. 1992). A wide range of potentially relevant biological variables (bioindicators) are available for use in monitoring programs, with some indicating that the organism was exposed to a pollutant, and others showing the biological consequence of contamination. The most useful types of measures provide an early warning of the onset of toxicity and/or provide information that is directly relevant to effects at the population to ecosystem levels. The latter include measures of effects on growth, reproductive potential, and survival. Such effects are clearly relevant to population sustainability as they determine the individual's ability to contribute to the population gene pool.

Scallop populations have been employed on several occasions in experimental biological effects monitoring programs. Commercial stocks of *P. maximus* and *Aequipecten opercularis* were employed for detecting insecticide exposure in the Bay of Brest area of France (Bocquene 1997), and *Euvola (Pecten) ziczac* populations were sampled for the same purpose at two sites in Bermuda (Owen et al. 2002). Both studies used acetylcholinesterase activity (AChE) as a biomarker of organophosphorus and carbamate pesticide exposure. AChE activity was shown to be substantially inhibited at environmentally relevant concentrations (Owen et al. 2002). Bocquene (1997) found no evidence of AChE inhibition in adductor muscle extracts that would indicate the presence of insecticide contaminants, but Owen et al. (2002) showed seasonal AChE inhibition in scallop hemolymph that appears to reflect contamination. The hemolymph assay was shown to be of greater value for biomarker applications than the muscle extract method (Owen et al. 2002). Regoli et al. (1997, 1998) proposed using lysosomal and antioxidant responses in *A. colbecki* as an early warning tool for detecting environmental disturbance in the Antarctic environment. Although laboratory studies by these authors demonstrated significant lysosomal and antioxidant responses to copper and mercury exposure, only a limited effect was observed in scallops residing in cages in a bay receiving discharges from the Italian Antarctic Base (Regoli et al. 1998). It was subsequently concluded that the Base should not have a major impact on benthic communities in the bay. It should be noted that the sensitivity of the Antarctic scallop biomarkers to detect an impact in the field has not been established as only a single, and extremely high concentration was used to document the response to the different metals.

In response to laboratory observations of the high sensitivity of sea scallop growth and physiological responses to very dilute concentrations of operational drilling wastes (Cranford and Gordon 1991, 1992; Cranford et al. 1999), two biological effects monitoring tools were developed using scallops as sentinel organisms. These tools were tested at 80 m depth at the Hibernia oil production site on the Grand Banks off Newfoundland, Canada. The first approach is based on the *in situ* sediment trap method of Cranford and Hargrave (1994) for continuously and autonomously monitoring feeding and digestion responses of bivalve filter feeders. In this method, sediment traps (HABITRAPs) are used to quantitatively collect biodeposits at regular programmed intervals from a cohort of scallops feeding under natural conditions in the field (Fig. 13.3). This method has been used extensively for measuring scallop functional responses to natural environmental variations over different time scales (Cranford and Hargrave 1994; Cranford et al. 1998; Cranford and Hill 1999). The rationale for applying this method as a monitoring tool is that food utilisation by bivalves is highly sensitive to the presence of many contaminants, including dilute concentrations of drilling wastes (Axiak and George 1987; Widdows et al. 1987; Cranford et al. 1999). The provision of biological response measurements over short time intervals (days) permits the identification of waste release

Figure 13.3. Biological effects monitoring tools using scallops as sentinel organisms for measuring effects of offshore oil and gas drilling operations. Cages (top) and sediment traps (bottom) were deployed in 80 m of water near the Hibernia platform off Newfoundland. The cage is placed on the seabed using an acoustic release (top left photo) and is recovered after the subsurface floats are released by a timed mechanism. *Placopecten magellanicus, Chlamys islandica* and *Mytilus edulis* were held separately in 'pocket' nets (top right) behind a predator mesh on each wing of the cage (top left). Time-series sediment traps called HABITRAPs (bottom left) were deployed near the seabed with and without *C. islandica* (bottom right) to monitor scallop biodeposition and sedimentation rates over 3-day intervals for a total of 72 days (photos by the author).

practices (e.g., rate and type of waste discharge) having the greatest impact on scallops. The HABITRAP approach is unique in that sequential sampling is automated over user defined time scales, while providing information on all three elements of the exposure-dose-response triad. Measurement of contaminants or tracers collected in the sediment traps define the exposure, analysis of contaminants in scallop tissues at the end of the exposure defines the dose, and the time series of feeding, absorption, and/or biodeposition rate defines the response.

A second tool was developed and tested for biological effects monitoring at offshore oil and gas operations that consisted of a series of *in situ* moorings containing caged *C. islandica*, *P. magellanicus* and *M. edulis* (Fig. 13.3). Six identical moorings were placed on the seabed between 0.5 to 6 km from the Hibernia platform. Data on lethal and sublethal effects and hydrocarbon body burdens were collected to help define the spatial extent of benthic impacts from oil drilling activities. Changes in survivorship, somatic and reproductive growth, and condition index were selected as biomonitoring parameters because they are relatively easy and inexpensive to measure (i.e., practical) and are relevant to population dynamics. Mortality (<10%) of scallops was low during the study, and no significant difference in mean tissue dry weight (gonad, adductor, digestive, and total) was detected between the six sites at the end of the three-month exposure (P.J. Cranford and S.L. Armsworthy, unpublished data). HABITRAP data showed that drilling wastes (barium) were only in suspension for a brief period during the study and that concentrations were below the threshold known to impact feeding and growth rates.

With many countries presently attempting to establish ecosystem-based approaches to the management of marine resources, the objectives and methodologies of current environmental and fisheries surveillance programs need to be re-evaluated. Any assessment of the cumulative effect of all activities potentially impacting an environment that is prone to large natural variations is a formidable task. A pragmatic approach to cumulative effects monitoring may be to integrate environmental quality and fisheries management programs based, in part, on a similar suite of core biomonitoring parameters. Routine stock assessments measure changes in scallop population dynamics (e.g., abundance, distribution, age structure, growth rate and reproductive effort) that can be related directly to the fishery and to other human-induced stresses and natural variations. For regions having significant scallop resources and which already conduct stock assessments, a cumulative effects monitoring approach could be implemented by conducting additional diagnostic bioassay and biomarker tests (including contaminant bioaccumulations in scallops) to identify probable causes for observed changes in wild scallop populations. Interpretation of data from such a biological monitoring program would have to be based on obtaining a thorough understanding of contaminant accumulation dynamics and contaminant-response relationships for the target species.

ACKNOWLEDGMENTS

I am grateful to B.T. Hargrave and S.L. Armsworthy for providing valuable comments on the manuscript. Unpublished results are presented from studies funded by Fisheries and Oceans Canada, and the Program for Energy Research and Development.

REFERENCES

Armstrong, D.A., Dinnel, P.A., Orensanz, J.M., Armstrong, J.L., McDonald, T.L., Cusimano, R.F., Nemeth, R.S., Landolt, M.L., Skalski, J.R., Lee, R.F. and Huggett, R.J., 1995. Status of selected bottomfish and crustacean species in Prince William Sound following the Exxon

Valdez oil spill. In: P.G. Wells, J.N. Butler and J.S. Hughes (Eds.). Exxon Valdez Oil Spill: Fate and Effects in Alaskan Waters. ASTM, Philadelphia, PA, USA. pp. 485–547.

Armsworthy, S.L., Cranford, P.J. and Lee, K., 1999. Effects of a new bitumen fuel source on the growth and energetics of sea scallops. J. Shellfish Res. 18:312.

Armsworthy, S.L., Cranford, P.J., Lee, K. and Tremblay, G.H., 2000. Effects of Orimulsion on food acquisition and growth of sea scallops. Proceedings of the 23rd Arctic and Marine Oilspill Program (AMOP) Technical Seminar, Environment Canada, June 14–16, Vancouver, British Columbia. pp. 1003–1022.

Axiak, V. and George, J.J., 1987. Effects of exposure to petroleum hydrocarbons on the gill functions and ciliary activities of a marine bivalve. Mar. Biol. 94:241–249.

Barlow, M.J. and Kingston, P.F., 2001. Observations on the effect of barite on gill tissues on the suspension feeder Cerastoderma edule (Linné) and the deposit feeder Macoma balthica (Linné). Mar. Poll. Bull. 42:71–76.

Bayne, B.L., Clarke, K.R. and Gray, J.S., 1988. Background and rationale to a practical workshop on biological effects of pollutants. Mar. Ecol. Prog. Ser. 46:1–5.

Berkman, P.A. and Nigro, M., 1992. Trace metal concentrations in scallops around Antarctica: Extending the Mussel Watch Programme to the Southern Ocean. Mar. Poll. Bull. 4:322–323.

Berthet, B., Amiard, J.C., Amiard-Triquet, C., Martoja, M. and Jeantet, A.Y., 1992. Bioaccumulation, toxicity and physico-chemical speciation of silver in bivalve molluscs: Ecotoxicological and health consequences. Sci. Total Environ. 125:97–122.

Bocquene, G., 1997. Cholinesterase activity in the common scallop and the queen scallop from the Bay of Brest (France): a tool for the detection of effects of organophosphorous and carbamate insecticides. Ann. Inst. Oceanogr. Paris 73:59–68.

Bragg, J.R. and Yang, S.H., 1995. Clay-oil flocculation and its role in natural cleansing in Prince William Sound following the Exxon Valdez oil spill. In: P.G. Wells, J.N. Butler and J.S. Hughes (Eds.). Symposium on Environmental Toxicology and Risk Assessment—Exxon Valdez Oil Spill, Atlanta, GA (USA), 26–28 April 1993. Exxon Valdez Oil Spill: Fate and Effects in Alaskan Waters. ASTM, Philadelphia, PA, USA. pp. 178–214.

Bryan, G.W., 1973. The occurrence and seasonal variation of trace metals in the scallops Pecten maximus (L.) and Chlamys opercularis (L.). J. Mar. Biol. Assoc. U.K. 53:145–166.

Carmichael, N.G. and Fowler, B.A., 1981. Cadmium accumulation and toxicity in the kidney of the bay scallop Argopecten irradians. Mar. Biol. 65:35–43.

Chelomin, V.P., Bobkova, E.A., Lukyanova, O.N. and Chekmasova, N.M., 1995. Cadmium-induced alterations in essential trace element homoeostasis in the tissues of scallop Mizuhopecten yessoensis. Comp. Biochem. Physiol., C. 110C:329–335.

Cranford, P.J., 1995. Relationships between food quantity and quality and absorption efficiency in sea scallops Placopecten magellanicus (Gmelin). J. Exp. Mar. Biol. Ecol. 189:123–142.

Cranford, P.J., Emerson, C.W., Hargrave, B.T. and Milligan, T.G., 1998. In situ feeding and absorption responses of sea scallops Placopecten magellanicus (Gmelin) to storm-induced changes in the quantity and composition of the seston. J. Exp. Mar. Biol. Ecol. 219:45–70.

Cranford, P.J. and Gordon, D.C., 1991. Chronic sublethal impact of mineral oil-based drilling mud cuttings on adult sea scallops. Mar. Poll. Bull. 22:339–344.

Cranford, P.J. and Gordon, D.C. Jr., 1992. The influence of dilute clay suspensions on the sea scallop (*Placopecten magellanicus*) feeding activity and tissue growth. Neth. J. Sea Res. 30:107–120.

Cranford, P.J., Gordon, D.C. Jr., Hannah, C.G., Loder, J.W., Milligan, T.G., Muschenheim, D.K. and Shen, Y., 2003. Modelling potential effects of petroleum exploration drilling on northeastern Georges Bank scallop populations. Ecol. Modelling 166:19–39.

Cranford, P.J., Gordon, D.C. Jr., Lee, K., Armsworthy, S. and Tremblay, G.-H., 1999. Chemical toxicity and physical disturbance effects of water- and oil-based drilling fluids and major additives on adult sea scallops (*Placopecten magellanicus*). Mar. Env. Res. 48:225–256.

Cranford, P.J. and Hargrave, B.T., 1994. *In situ* time-series measurements of ingestion and absorption rates of suspension-feeding bivalves: *Placopecten magellanicus* (Gmelin). Limnology and Oceanography 39:730–738.

Cranford, P.J. and Hill, P.S., 1999. Seasonal variation in food utilization by the suspension-feeding bivalve molluscs *Mytilus edulis* and *Placopecten magellanicus*. Mar. Ecol. Prog. Ser. 190:223–239.

Davies, I.M., McKie, J.C. and Paul, J.D., 1986. Accumulation of tin and tributyltin from anti-fouling paint by cultivated scallops (*Pecten maximus*) and Pacific oysters (*Crassostrea gigas*). Aquaculture 55:103–114.

Evtushenko, Z.S., Belcheva, N.N. and Lukyanova, O.N., 1986. Cadmium accumulation in organs of the scallop *Mizuhopecten yessoensis*. 2. Subcellular distribution of metals and metal-binding proteins. Comp. Biochem. Physiol., C. 83C:377–383.

Evtushenko, Z.S., Lukyanova, O.N. and Belcheva, N.N., 1990. Cadmium bioaccumulation in organs of the scallop *Mizuhopecten yessoensis*. Mar. Biol. 104:247–250.

Fowler, B.A. and Gould, E., 1988. Ultrastructural and biochemical studies of intracellular metal-binding patterns in kidney tubule cells of the scallop *Placopecten magellanicus* following prolonged exposure to cadmium or copper. Mar. Biol. 97:207–216.

Frank, D., Ewert, L., Shumway, S. and Ward, J.E., 2000. Effect of clay suspensions on clearance rate in three species of benthic invertebrates. J. Shellfish Res. 19:663.

Friocourt, M.P., Bodennec, G. and Berthou, F., 1985. Determination of polyaromatic hydrocarbons in scallops (*Pecten maximus*) by UV fluorescence and HPLC combined with UV and fluorescence detectors. Bull. Environ. Contam. and Toxicol. 34:228–238.

GESAMP, 1993. Impact of oil and related chemicals and wastes on the marine environment. GESAMP Report and Studies No. 50: 180 p.

Gould, E. and Fowler, B.A., 1991. Scallops and pollution. In: S.E. Shumway (Ed.). Scallops: Biology, Ecology and Aquaculture. Developments in Aquaculture and Fisheries Science, No. 21. Elsevier, Amsterdam. pp. 495–515.

Gould, E., Greig, R.A., Rusanowsky, D. and Marks, B.C., 1985. Metal-exposed sea scallops, *Placopecten magellanicus* (Gmelin): A comparison of the effects and uptake of cadmium and copper. In: F.J. Vernberg, F.P. Thurberg, A. Calabrese and W.B. Vernberg (Eds.). Marine Pollution and Physiology: Recent Advances. South Carolina Press, Columbia. pp. 157–186.

Gould, E., Thompson, R.J., Buckley, L.J., Rusanowsky, D. and Sennefelder, G.R., 1988. Uptake and effects of copper and cadmium in the gonad of the scallop *Placopecten magellanicus*: Concurrent metal exposure. Mar. Biol. 97:217–233.

Gray, J.S., 1999. Using science for better protection of the marine environment. Mar. Poll. Bull. 39:3–10.

Gray, J.S., 2002. Biomagnification in marine systems: the perspective of an ecologist. Mar. Poll. Bull. 45:46–52.

Lukyanova, O.N., Belcheva, N.N. and Chelomin, V.P., 1993. Cadmium bioaccumulation in the scallop *Mizuhopecten yessoensis* from an unpolluted environment. In: R. Dallinger and P.S. Rainbow (Eds.). Ecotoxicology of Metals in Invertebrates. Lewis Publishers, Boca Raton. pp. 25–35.

Mauri, M., Orlando, E., Nigro, M. and Regoli, F., 1990. Heavy metals in the Antarctic scallop *Adamussium colbecki* . Mar. Ecol. Prog. Ser. 67:27–33.

Milligan, T.G. and Hill, P.S., 1998. A laboratory assessment of the relative importance of turbulence, particle composition, and concentration in limiting maximal floc size and settling behaviour. J. Sea Res. 39:227–241.

Minchin, D., Duggan, C.B. and King, W., 1987. Possible effects of organotins on scallop recruitment. Mar. Poll. Bull. 18:604–608.

Moore, M.R., Vetter, W., Gaus, C., Shaw, G.R. and Müller, J.F., 2002. Trace organic compounds in the marine environment. Mar. Poll. Bull. 45:62–68.

Morse, M.P., Robinson, W.E. and Wehling, W.E., 1982. Effects of sublethal concentrations of the drilling mud components attapulgite and Q-broxin on the structure and function of the gill of the scallop *Placopecten magellanicus* (Gmelin). In: W.B. Vernberg, A. Calabrese, F.P. Thurberg and F.J. Vernberg (Eds.). Physiological Mechanisms of Marine Pollutant Toxicity. Academic Press, New York. pp. 235–259.

Muschenheim, D.K. and Milligan, T.G., 1996. Flocculation and accumulation of fine drilling waste particulates on the Scotian Shelf (Canada). Mar. Pollut. Bull. 32:740–745.

Muschenheim, D.K., Milligan, T.G. and Gordon, D.C. Jr., 1995. New technology and suggested methodologies for monitoring particulate wastes discharged from offshore oil and gas drilling platforms and their effects on the benthic boundary layer environment. Can. Tech. Rep. Fish. Aquat. Sci. 2049:x + 55.

National Research Council, 1983. Drilling discharges in the marine environment. Panel on Assessment of Fates and Effects of Drilling Fluids and Cuttings in the Marine Environment. Marine Board, National Research Council. National Academy Press, Washington, D.C. 180 p.

Neff, J.M., 1987. The potential effects of drilling effluents on marine organisms on George Bank. In: Georges Bank. MIT Press. pp. 551–559.

Neff, J.M., 2002. Bioaccumulation in Marine Organisms: Effects of Contaminants from Oil Well Produced Water. Elsevier, Amsterdam. 452 p.

Nigro, M., Orlando, E. and Regoli, F., 1992. Ultrastructural localization of metal binding sites in the kidney of the Antarctic scallop *Adamussium colbecki*. Mar. Biol. 113:637–643.

Owen, R., Buxton, L., Sarkis, S., Toaspern, M., Knap, A. and Depledge, M., 2002. An evaluation of hemolymph cholinesterase activities in the tropical scallop, *Euvola* (*Pecten*) *ziczac*, for the rapid assessment of pesticide exposure. Mar. Poll. Bull. 44:1010–1017.

Paul, J.D. and Davies, I.M., 1986. Effects of copper- and tin-based anti-fouling compounds on the growth of scallops (*Pecten maximus*) and oysters (*Crassostrea gigas*). Aquaculture 54:191–203.

Rainbow, P.S., 1993. The significance of trace metal concentrations in marine invertebrates. In: R. Dallinger and P.S. Rainbow (Eds.). Ecotoxicology of Metals in Invertebrates. Lewis Publishers, Boca Raton. pp. 3–23.

Regoli, F., Nigro M. and Orlando, E., 1998. Lysosomal and antioxidant responses to metals in the Antarctic scallop *Adamussium colbecki*. Aquat. Toxicol. 40:375–392.

Regoli, F., Principato, G.B., Bertoli, E., Nigro, M. and Orlando, E., 1997. Biochemical characterization of the antioxidant system in the scallop *Adamussium colbecki*, a sentinel organism for monitoring the Antarctic environment. Polar Biol. 17:251–258.

Salazar, M.H. and Salazar, S.M., 1998. Using caged bivalves as part of an exposure-dose-response triad to support an integrated risk assessment strategy. In: A. de Peyster and K. Day (Eds.). Proceedings-Ecological Risk Assessment: A Meeting of Policy and Science. SETAC Press. pp. 167–192.

Stagg, R.M., 1998. The development of an international programme for monitoring the biological effects of contaminants in the OSPAR convention area. Mar. Env. Res. 46:307–313.

Stekoll, M.S., Clement, L.E. and Shaw, D.G., 1980. Sublethal effects of chronic oil exposure on the intertidal clam *Macoma balthica*. Mar. Biol. 57:51–60.

Stevens, P.M., 1987. Response of excised gill tissue from the New Zealand scallop *Pecten novaezelandiae* to suspended silt. N.Z. J. Mar. Fresh. Res. 21:605–614.

Stone, H.C., Wilson, S.B. and Overnell, J., 1986. Cadmium-binding components of scallop (*Pecten maximus*) digestive gland. Partial purification and characterization. Comp. Biochem. Physiol. 85C:259–268.

Syasina, I.G. and Lukyanova, O.N., 1994. Ultrastructural and biochemical changes in gonads of the scallop *Mizuhopecten yessoensis*: Results of cadmium accumulation. Russ. J. Mar. Biol. 19:265–271.

Uthe, J.F. and Chou, C.L., 1987. Cadmium in sea scallop (*Placopecten magellanicus*) tissues from clean and contaminated areas. Can. J. Fish. Aquat. Sci. 44:91–98.

Viarengo, A., Canesi, L., Mazzucotelli, A. and Ponzano, E., 1993. Cu, Zn and Cd content in different tissues of the Antarctic scallop *Adamussium colbecki*: Role of metallothionein in heavy metal homeostasis and detoxication. Mar. Ecol. Prog. Ser. 95:163–168.

Widdows, J. and Donkin, P., 1989. The application of combined tissue residue and physiological measurements of mussels (*Mytilus edulis*) for the assessment of environmental pollution. Hydrobiologia 188/189:455–461.

Widdows, J. and Donkin, P., 1992. Mussels and environmental contaminants: bioaccumulation and physiological aspects. In: E. Gosling (Ed.). The Mussel *Mytilus*: Ecology, Physiology, Genetics and Culture. Developments in Aquaculture and Fisheries Sciences, Vol. 25. Elsevier, Amsterdam. pp. 383–424.

Widdows, J., Donkin, P., Brinsley, M.D., Evans, S.V., Salkeld, P.N., Franklin, A., Law, R.J. and Waldock, M.J., 1995. Scope for growth and contaminant levels in north sea mussels *Mytilus edulis*, Mar. Ecol. Prog. Ser. 127:131–148.

Widdows, J., Donkin, P. and Evans, S.V., 1987. Physiological responses of *Mytilus edulis* during chronic oil exposure and recovery. Mar. Env. Res. 23:15–32.

Widdows, J., Phelps, D.K. and Galloway, W., 1981. Measurement of physiological condition of mussels transplanted along a pollution gradient in Narragansett Bay. Mar. Env. Res. 4:181–194.

White, M.J., 1997. The effect of flocculation on the size-selective feeding capabilities of the sea scallop *Placopecten magellanicus*. MSc Thesis. Dalhousie University, Halifax, Nova Scotia.

Wood, P., Lunel, T., Lee, K. and Bailey, N., 1997. Investigation into role of clay-oil-flocculation— final report. AEA Technology PLC, Abingdon, UK. 110 p.

AUTHOR'S ADDRESS

Peter J. Cranford - Fisheries and Oceans Canada, Marine Environmental Sciences Division, Bedford Institute of Oceanography, Dartmouth, Nova Scotia, Canada B2Y 4A2 (E-mail: CranfordP@mar.dfo-mpo.gc.ca)

Scallops: Biology, Ecology and Aquaculture
S.E. Shumway and G.J. Parsons (Editors)
© 2006 Elsevier B.V. All rights reserved.

765

Chapter 14

Dynamics, Assessment and Management of Exploited Natural Populations

J.M. (Lobo) Orensanz, Ana M. Parma, Teresa Turk and Juan Valero

14.1 INTRODUCTION

Since the first edition of this book, published a decade ago, there has been substantial progress in our understanding of scallop population dynamics. Most significant are new results on the structure of metapopulations, the mechanisms involved in larval retention/dispersal, and the relationship between aggregate stock abundance and subsequent settlement/recruitment. There has been a renewed interest in the fishing process, particularly regarding the behaviour of fishers and fleets. Methods used for estimating abundance have paid increasing attention to spatial structure, including both design-based approaches (e.g., adaptive sampling) and model-based methods (e.g., geostatistics). Yet, among the subjects covered in this chapter, it is perhaps in the management arena where changes have been most radical. This is in part a result of the debates over sustainability and governance originating in the so-called "World Fishing Crisis". Management of benthic resources has begun to emerge with an identity of its own, differentiated from conventional (finfish) management by a new and growing emphasis on spatially explicit approaches, and on the role of stakeholders in co-management.

In revising this chapter we have tried to do justice to these fundamental developments, which required a complete rewriting of all sections. We have virtually dropped the entire section on the estimation of growth and mortality parameters, which is covered at length in the first edition. The readers are referred to other chapters of this book for new developments on these topics.

14.2 SPATIAL SCALES AND SCALE-SPECIFIC PROCESSES

All scallop stocks are structured as "metapopulations"[1] in which subpopulations of sedentary[2] post-larval individuals are connected with each other through the dispersal of pelagic larvae. The metapopulation is a useful conceptual framework for the analysis and management of benthic fisheries in general, as indicated by trends in the literature over the last decade (Orensanz and Jamieson 1998). This emerging paradigm emphasises the spatial structure of benthic stocks, in contrast with the fin-fishery-inspired approaches that once used to dominate the scene. Emphasis on spatial structure brings with it the need to identify appropriate spatial scales for the observation, analysis and management of exploited scallop stocks. Here we adopt the hierarchy of spatial scales presented by Orensanz and Jamieson (1998).

Megascale, at which there is detectable geographic genetic differentiation between populations.

Macroscale (large scale): metapopulations composed of spatially disjoint subpopulations connected through the dispersal of pelagic larvae.

Mesoscale (intermediate scale): subpopulations within a metapopulation. This is typically the scale of fishing *beds*, usual units in traditional stock assessment.

Microscale (small scale): neighbourhoods of individuals, within which a certain interaction between the individual and its environment is effective.

Nanoscale, characterised by processes taking place at low Reynolds numbers.

Relevant processes taking place at each of these scales are summarised in Table 14.1. In this review, we will emphasise macro, meso and microscale phenomena:

Macroscale: source-sink (SS) dynamics in metapopulations. The pattern of connectivity[3] between components of a metapopulation is the scaffolding underlying SS dynamics, ranging from the ideal model of a well mixed larval pool (maximal connectivity) at one extreme, to a collection of closed self-sustaining populations (minimal connectivity) at the other. Connectivity is never symmetrical: larval flows between two subpopulations will always be stronger in one direction. Maximum asymmetry is found in non-reproductive *pseudo-populations* (absolute sinks), which depend exclusively on imported recruits and do not contribute larvae to other populations.

Mesoscale: contraction-expansion (CE) dynamics in subpopulations. The mesoscale is the domain of traditional fisheries models, in which emphasis is on change in population size over time. In recent years, there has been increasing interest in the CEs of geographic range that accompanies change in aggregate[4] abundance. In fishes and other mobile organisms CEs of subpopulations are mediated by the active displacements of individual organisms along environmental gradients; changes in the range occupied by a subpopulation happen within the time-frame of a generation and can be rapid. In scallops and other sedentary invertebrates, CEs are mediated mostly by the advection of larvae; changes in the range occupied by a subpopulation involve more than one generation and can be slow and noisy.

Microscale: Density-dependent (DD) processes within neighbourhoods. The DD mechanisms that control recruitment[5] to benthic stocks, whether compensatory or depensatory, always have small operational spatial scales. The effects of those processes on recruitment, however, may occur at varying spatial scales depending on whether DD occurs before or after dispersal (Botsford and Hobbs 1995). *Pre-dispersal* DD processes includes reduction of growth and fecundity at high density due to competition for a limited food supply, and the DD of fertilisation rate in broadcast spawners. Typical *post-dispersal* DD processes involve various forms of inhibition of settlement or recruitment by high densities of local residents, as well as gregarious settlement of larvae in the neighbourhood of adult conspecifics.

The scales introduced above are defined in terms of processes of interest, not of dimensional bounds. Roughly, in the case of scallops, typical metric scales are in the order of mm for the nanoscale, cm to m for the microscale, km for the mesoscale, tens to hundreds of km for the macroscale, and thousands of km for the megascale. We use the words "site", "bed" and "ground" for, loosely, regions commensurate with micro, meso and macroscale processes, respectively. A "fishing ground" is typically occupied by a metapopulation. "Beds" (equivalent to the Spanish "bancos") within a ground are more or less discrete patches with high (fishable) density, generally occupied by a subpopulation. Scallop fishing grounds and beds are identified by name in most scallop fisheries. Sites are small partitions of the bottom (in the order of tens of m^2), commensurate with the typical experimental sites of ecologists.

In traditional finfish fishery models, there is no explicit consideration of spatial structure or spatial scales. Basic models of the dynamics of exploited stocks are constructed around two tenets (Caddy 1975):

The unit stock concept: A "unit stock" is defined as a self-sustaining unit of population, closed to immigration or to recruits originated by other stocks. In its more general form, this concept makes no explicit allowance for spatial distribution of abundance and fishing effort; fishing mortality is assumed to be proportional to fishing effort.

The dynamic pool assumption: Real populations are contagiously distributed, and fishing effort tends to concentrate on dense areas. Active dispersal by the fish is invoked to spread the local effects of fishing events ("redistribution"), so that locally depleted areas are replenished by immigration. Without this assumption "catchability" would be unevenly distributed, *CPUE* ("catch per unit of effort") would not be proportional to abundance, and fishing mortality would not be proportional to fishing effort.

It is immediately obvious that the very nature of scallop stocks is at odds with these basic notions:

The unit stock assumption rarely holds. Subpopulations in a metapopulation are generally "open" to recruitment, and connectivity between subpopulations is never symmetrical.

The dynamic pool assumption never holds. Fishing intensity tends to match the distribution of density. Adult scallops, being sedentary, do not "redistribute" themselves following a spatially localised fishing event. Although scallops are able to move, the scale of their displacements is very small relative to the scale of the fishing process. Fishing mortality varies spatially; the chances of being fished vary amongst individuals, and are related to density in their neighbourhoods. The relations between *CPUE* and abundance, and between fishing mortality and effort, depart from simple proportionality (Section 14.5).

Table 14.1

A hierarchy of relevant spatial scales, typical methods of pattern description, and processes commensurate with each scale. In this overview we emphasise macro, meso and microscale phenomena.

Spatial scale	Definition	Typical method of pattern observation or description	Examples of processes commensurate with scale
MEGA	Zoogeographic range of an Artenkreis or Rassenkreis, respectively collections of closely related species or genetically differentiable stocks	Large-scale maps	Speciation
MACRO (= large)	Metapopulations, or "populations of populations"	GIS	Dispersal of pelagic larvae among populations
MESO (= intermediate)	Populations (or subpopulations) within a metapopulation. Typically the scale of individual fishing beds	Geostatistical methods, GAMS	Tactical decisions by fishers
MICRO (= small)	Neighbourhoods of individuals, the pertinent scale to describe interactions taking place in dense "patches". The "neighbourhood" is a region within which a certain interaction between the individual and its environment is effective	Spatial point process methods; concentration profiles	Competition Fertilisation rate Sensorial perception
NANO	Processes taking place at low Reynolds numbers (predominance of viscous over inertial forces)	Microphotography; flow chambers	Particle collection in the feeding structures of suspensivores. Interaction between eggs and sperm in broadcast spawners. Settlement of pelagic larvae in the boundary layer

Throughout this chapter we discuss each subject (structure, dynamics, the fishing process, assessment and management) with reference to the pertinent spatial scale. Whenever we refer to aggregate meta- or subpopulations, this is made explicit.

14.3 POPULATION STRUCTURE AND DYNAMICS

14.3.1 Aggregate Stocks

The primary factors tuning year-class-strength are those that affect advection and survival of pelagic larvae. These factors pertain largely to oceanographic conditions. Thus, it is natural that research on modulation of year-class-strength and the search for signs of climatic forcing should focus on abundance indices aggregated at relatively large spatial scales. Usually, the longest time series of data available correspond to the catches. These are limited because the contribution of individual year classes is blurred by the dynamics of the stock and the fishing process (recruitment fisheries provide some exceptions; Section 14.6.2.3). Yet, catch statistics are informative about the broadest patterns of fluctuation in population abundance. We address this subject in Section 14.3.1.1, below. In Section 14.3.1.2, we consider empirical evidence (mostly of correlational nature) about the influence of climate on larval availability and recruitment. The transition between these two stages – pelagic larvae and recruits – is, however, modulated by factors other than larval availability. A rich literature has accumulated over the last decade on the relative importance of pre- vs. post-settlement processes in regulating recruitment; the "recruitment limitation hypothesis", claiming that recruitment is limited by larval availability, has been scrutinised. We present empirical evidence about correlation between abundance of different life-history stages in Section 14.3.1.3. Finally, in Section 14.3.1.4 we discuss the elusive relationship between aggregate spawning stock and recruitment. In the case of scallop-like organisms the recruitment process involves a sequence of processes with different operating spatial scales, which are considered in Sections 14.3.2–14.3.4.

14.3.1.1 Patterns of fluctuation

Caddy and Gulland (1983) classified harvested stocks in four groups, according to their pattern of fluctuation:

(1) Steady stocks remain at about the same level year after year, with variations that generally do not exceed 20–30% of the long-term average. Some boreal or arctic populations of long-lived scallop species may fall into this group. Golikov and Scarlato (1970) characterised the *Patinopecten yessoensis* population from Posjet Bay as "stationary", although they did not give quantitative information on long-term population size or year-class strength. Density in a bed of *Chlamys islandica* studied by Vahl (1982) in North Norway showed insignificant change over 9 years.

(2) Cyclical stocks show a repeated alternation of periods of high and low abundance. Although strict periodic behaviour is possible, fluctuations with uneven period (quasi-periodic) are more common. Cycles can result from predator-prey interactions, nonlinear renewal, or fluctuations in an external (climatic) variable modulating year-class strength. The most dramatic scallop cycle was that documented by Caddy (1979a) for a *Placopecten* stock of the Bay of Fundy (Nova Scotia). Annual landings from the Digby inshore fishery fluctuated cyclically between 1922 and 1971. Correlograms for catch, fleet size and inshore catch per boat showed pronounced maxima at lags of 8–10 years, suggesting an underlying periodicity of 9 years. Annual catch and fleet size were strongly cross-correlated and in phase. Caddy (1979a) considered several alternative hypotheses to explain the cycle, and concluded that climatic forcing (perhaps fine-tuned by density dependence and harvesting) was the most consistent with the pattern. Inshore catches of *Placopecten* from Maine between 1897–1980 also show quasi-cyclic fluctuations (Dow 1969). Average biomass and year class strength of *Aequipecten tehuelchus* in the San Roman bed (Patagonia) showed a 5-year cycle over the period 1969–1983 (Orensanz 1986). It was produced by three exceptionally strong year-classes (1970, 1975, 1980), and may have resulted, to some extent, from density-dependent settlement (Section 14.5).

(3) Irregular stocks fluctuate greatly from year to year without any clear pattern, as is frequently the case when recruitment is strongly dependent on hydrographic conditions. Irregular stocks are exemplified by the Georges Bank and St. Pierre Bank stocks of *Placopecten magellanicus* (Mohn et al. 1988; Naidu and Anderson 1984).

(4) Spasmodic stocks show irregular pulses of high abundance followed by periods of scarcity or collapse. Many scallop stocks fit this type. Exceptional environmental conditions underlie spasmodic pulses in three well-documented and dramatic cases:

i) The *Patinopecten yessoensis* stock of Mutsu Bay, Japan (Yamamoto 1964; Yamamoto in Imai 1971; Aoyama 1989; Anonymous 1990), before 1970. Kishigami (cited by Yamamoto in Imai 1971) informed of pulses in 1861–62, the late 1860's and the late 1880's. In the present century, there were pulses in 1909–1910, 1926–1929; 1950, 1959, and 1970–1973. Yamamoto (1960, 1964) attributed them to "abnormal fecundity" due to exceptionally high spat survival. Under normal conditions, plankton (including scallop larvae) is concentrated by eddies (Section 14.3.1.1) in areas 30–70 m deep, that are not the normal habitat of the scallop. Flow in eddies stop during the summer, producing stratification and anoxic conditions near the bottom, and mass mortality of scallop spat. In some years, however, stratification does not occur, anoxic conditions do not develop, and sinking plankton does not accumulate. Survival of spat is then high, resulting in the "abnormal fecundity" phenomenon (Yamamoto 1964; Imai 1971).

ii) In the *Amusium pleuronectes* stock of Shark Bay (Western Australia), a large recruitment in 1990 resulted in spawning stocks in 1991–1992 about ten times larger than previous levels (Caputi et al. 1998). Yet, subsequent recruitments failed due to the Leeuwin Current strength (Joll and Caputi 1995; Caputi et al. 1998; Section 14.3.1.2).

iii) The population explosions of *Argopecten purpuratus* in Perú and northern Chile following El Niño events (Wolff 1987; Navarro-Piquimil et al. 1991; Wolff and Mendo 2000). Following the 1983 event, stock size exceeded more than 60 times that of "normal" years. Factors involved in these pulses include increased individual growth rate, increased larval survival and/or intensified spawning activity associated with temperature 6–8°C higher than normal, and higher survival due to mortality or emigration of predators and competitors.

14.3.1.2 Climatic forcing

The most dramatic examples of climatic forcing are provided by some spasmodic populations, discussed above. Climatic forcing is most often revealed by cross-correlation between year-class strength and environmental variables. Even in the absence of long time series of data, geographic coherence in recruitment (large-scale spatial correlation in year-class-strength) is often a signature of climatic forcing, as this is in most cases the only large-scale factor that can synchronise the pattern. Orensanz (1986) and Ciocco and Aloia (1991) reported that year classes of the tehuelche scallop (*Aequipecten tehuelchus*) are weak or strong across the entire metapopulation of San José Gulf, pointing to large-scale climatic forcing as the likely cause.

Climate drives scallop recruitment, and consequently the dynamics of scallop populations, in two main ways:

(1) Variation in reproductive schedules: Like other sedentary organisms (and unlike pelagic or demersal fishes) scallops are not able to migrate tracking their environmental preferences. While highly mobile organisms are able to adjust the location of their spawning grounds in response to year-to-year variation in oceanographic conditions, scallops respond only by "tuning" their schedule of reproductive events (timing and magnitude), eventually resulting in the failure of spawning in a given season. Failed spawning has been well documented for *Patinopecten yessoensis* (Yamamoto 1964), and for *Pecten maximus* in the Bay of St. Brieuc (France), the latter following an unusually cold year. The impact of spawning failures is most significant in the case of short-lived semelparous species, as illustrated by the collapse of the *Argopecten gibbus* fishery off Cape Canaveral (east Florida) in 1985 (Moyer and Blake 1986). This population apparently has two spawning periods: a main one in spring (average size of spawning scallops small), and a smaller one in the fall (average size of spawning scallops large). In the autumn of 1984, however, no reproduction was observed. This resulted in a small spawning population of small scallops in the next spring. The main spawning stock was composed of the few large scallops that over-wintered; the fall spawning, although small, may be very important for the persistence of this stock.

A sudden change in temperature is the major trigger for spawning in scallops and other broadcast spawners with external fertilisation. This points towards strong selective pressure favouring synchrony of reproductive events. There are three well-documented cases in which high abundance or larval settlement apparently results from temperature-mediated synchrony in spawning:

i) Successful spawning of *Patinopecten yessoensis* in Mutsu Bay (Japan) requires a sudden rise in temperature (to a minimum of 8–8.5°C) within the spawning season, between March and May (Yamamoto 1964). Yamamoto (1964) investigated different hypotheses to explain a series of exceptional sets (1948, 1952, 1954, 1957 and 1960). High abundance of pelagic larvae was observed in four years (1948, 1952, 1956 and 1957), in all of which temperature rose abruptly in April.

ii) In *Pecten maximus* from Brittany, reproductive success increases with spawning synchrony. When temperature fluctuates near the spawning trigger (16°C), partial asynchronous spawning prevails, and relative larval density is low (Boucher 1985).

iii) In *Placopecten magellanicus* of the Gulf of St. Lawrence (Canada), an 8-yr study by Bonardelli et al. (1996) indicated that spawning was not controlled by critical temperature or cumulative degree-days thresholds, but was strongly associated with sudden changes in temperature caused by downwelling of warm surface water. Delay between locations coincided with the propagation rate of downwelling events. Strong cohorts (1988, 1990) resulted from abrupt, synchronous spawning (Claereboudt and Himmelman 1996, discussed by Bonardelli et al. 1996). Weakest cohorts (1986, 1989) resulted from seasons in which multiple (asynchronous) spawning prevailed. A relationship between year-class strength and spawning synchrony had been suggested before for other populations of the same species by Dickie (1955) and Langton et al. (1987).

Although temperature is the variable most often invoked, other factors may be significant as well in controlling the timing of spawning events, e.g., phytoplankton (Arsenault and Himmelman 1998) and lunar or tidal phases (Parsons et al. 1992). Bonardelli et al. (1996), however, found no indication of these factors being triggers for spawning in *Placopecten magellanicus* from the Gulf of St. Lawrence (Canada).

(2) Variation in larval advection: Dickie (1955) found that abundance and recruitment in the Digby area (Bay of Fundy) stock of *Placopecten magellanicus* was positively correlated with water temperature 6 years before. Spawning in the Digby area begins by mid July, peaks in August-September, and terminates by late September. Lagged cross-correlation between year-class strength and water temperature was highest for the month of October, pointing to mechanisms operating once spawning was over and larvae were in the plankton. Dickie proposed a composite hypothesis involving temperature-mediated rate of larval development and larval advection. Low temperatures are indicative of great exchange between the Bay of Fundy and outside water masses. Slower larval development and increased duration of larval stages in cold years would result in a high proportion of larvae being advected off the bay. In contrast, high temperatures would lead to higher retention of larvae in the vicinity of the parental beds, good settlement, and high recruitment six years later. Experimental studies (MacDonald 1988; Tettelbach and Rhodes 1981) have confirmed that survival of scallop larvae does not decrease (and often increases) at low temperatures (within normal range), but that development rate is depressed. Caddy (1979a) later expanded Dickie's time series to 50 years, and found significant cross-correlation between temperature and *CPUE* with lags of up to 5 years.

The cyclic pattern that he found (Section 14.3.1.1) indicates that long-period tidal oscillations might be responsible for fluctuations in both temperature and (directly or indirectly) recruitment. In the saucer scallop (*Amusium balloti*) of Shark Bay (Western Australia) the strength of recruitment is significantly and negatively correlated with the strength of the offshore Leeuwin Current (which is under the influence of the ENSO events) during the spawning season (Joll and Caputi 1995). Proximate mechanisms possibly involved are reduced spawning (discussed above), and/or flushing of larvae away from suitable grounds.

14.3.1.3 Correlation between consecutive life-history stages

Stages involved in the recruitment process include spawning, fertilisation, larval survival and dispersal, settlement, and post-settlement growth and survival (Hancock 1976). Correlation between consecutive stages is informative for the identification of recruitment bottlenecks. Strong correlation between larval availability in the water column and recruitment, for example may give support to the "recruitment limitation hypothesis" (Ólafsson et al. 1994), stressing the importance of larval supply to explain recruitment success. Alternatively, lack of correlation may point to the significance of processes affecting the settlement/post-settlement stages, or it may simply reflect substantial process and/or observation errors, or a mismatch between the operating spatial scales of process and observation. Indeed, a cautionary note is needed here. Larval availability may appear to be "limiting" or not depending on the scale of observation. Suppose, as a simplification, that density of conspecifics was a major factor controlling post-dispersal survival (pure post-dispersal compensation; Section 14.3.4.3). A bed is usually a collection of sites, some "saturated" by residents and/or settlers, others not (Sections 14.3.1.4 and 14.3.4). Observations from saturated sites would stress the significance of post-settlement survival. Yet, at the larger scale of the bed or ground recruitment could still increase with increased larval availability, provided that many unsaturated sites were left. This would highlight the relative importance of larval availability (the "recruitment limitation hypothesis"). The two perspectives are relevant, but must be properly specified. The astounding increase in recruitment in Mutsu Bay that accompanied the development of aquaculture and consequent stabilisation of an abundant spawning stock (Section 14.3.1.4) may point to the existence of unsaturated sites (even at high aggregate levels abundance). In the San José Gulf metapopulation of tehuelche scallops, large-scale forcing of settlement/recruitment was detected in lightly-populated beds as well as in beds that showed unequivocal signs of overcompensation (Orensanz 1986).

Long-term studies conducted in Brittany (France) by the *Centre Oceanologique de Bretagne* on *Pecten maximus* (Buestel et al. 1979; Boucher 1985), under the framework of the "*Programme National sur le Determinisme du Recrutement*" (Thouzeau 1987, 1991a), and in Mutsu Bay (Japan) by the *Aquaculture Center of the Aomori Prefecture* on *Patinopecten yessoensis* (Ventilla 1982; Aoyama 1989), are unique in having addressed simultaneously the multiple stages involved in scallop recruitment. These programs combined observations of gonadal condition, abundance of pelagic larvae, and settlement on artificial collectors. In Brittany (Boucher 1985) correspondence was (i) weak between spawning intensity and larval density, (ii) good between larval density and settlement (as has been also observed in *Patinopecten yessoensis*; Kanno (1970) cited by Ventilla (1982), his fig. 12), and (iii) very good between settlement and recruitment to age class 2+. Correspondence between spawning activity and density of larvae of *Pecten maximus* in the plankton improved when precocious and late cohorts were discriminated. The spawning schedule (rather than larval transport, starvation or predators) seemed to be the primary factor determining differences in larval density between cohorts and year-classes in Brittany. For a given amount of sexual products, according to Boucher (1985), late spawning bursts produce more larvae than early ones. Interestingly, Bonardelli et al. (1996) observed that synchronous spawning occurring in late August, coinciding with high mean gonadal mass, gave rise to strong cohorts in *Placopecten magellanicus* in the Gulf of St. Lawrence (Canada).

Good correlation between settlement indices (e.g., from spat collectors) and pre-recruits or recruits has been reported in other cases as well, e.g., in *Pecten fumatus* from Port Phillip Bay (southeastern Australia) (Sause et al. 1987), and *Aequipecten opercularis* and *Pecten maximus* from Scotland (Fraser 1991).

Not surprisingly, significant lagged cross-correlation between pre-recruits and recruits has been observed on many occasions (e.g., Coleman et al. 1988), allowing short-term forecasts that are valuable to managers and fishers. This is particularly so in the case of "recruitment fisheries", where the abundance of pre-recruits is strongly correlated with the catch in the following year. Examples have been presented by Joll and Caputi (1995), for *Amusium balloti* of Western Australia, and by Thouzeau (1991b) for the *Pecten maximus* stock of the Bay of St. Brieuc (France). The most impressive example is the correlation between seed input and scallop production in Japan (Ito 1991, f.15). Beyond their practical significance, these "mega-experiments" strengthen the notion that many sites within scallop beds are well below carrying capacity under natural conditions. These observations, however, are not conclusive with respect to the relative significance of pre- vs. post-settlement processes under natural conditions, as scallop production involves both sowing and control of predators and competitors.

14.3.1.4 The relation between aggregate spawning stock and settlement/recruitment

The relation between aggregate stock size and recruitment is the result of diverse processes, decoupled in their temporal and spatial scales. Spawning and larval survival depend on large-scale conditions (climate, spawning stock size), while receptiveness of sites to settlement depends on local (small-scale) conditions. Recruitment to one bed is affected by conditions prevailing in other (eventually distant) beds. Young et al. (1989) captured this notion clearly: *"the number of larvae competent to settle is related to the parent stock, but the number actually settling is dependent on unknown physical or biological factors at the settlement site"*. Decoupling of spatial scales relevant to the different life history stages involved in the recruitment process implies that, in principle, aggregate stock-recruitment relationships (e.g., as described by the Beverton-Holt and Ricker stock-recruitment models) should not be expected in any subpopulation open to settlers originated in other subpopulations. At that spatial scale, lack of pattern is something to be expected *a priori*, not to be explained *a posteriori*.

Good settlement may occur occasionally at times when the spawning stock is at a very low abundance (Joll 1989; Román 1991; Naidu 1991, p. 869; Kenchington and Lundy 1996, p. 3; Ciocco and Monsalve 1999a, 1999b). Observation of these compelling cases, in isolation, may give a false impression of either large-scale over-compensation (strong negative relation) or no relation between spawning stock and recruitment. On average, however, evidence suggests that pulses of recruitment are more frequent (in time and/or space) when spawning stock is higher. There is a handful of informative cases in which observations on recruitment were made in assumably self-sustaining stocks. These fall into three groups:

1) Recruitment after spawning biomass was dramatically enhanced by massive aquaculture:

i) In Mutsu Bay (Japan), before the 1970s, average settlement of *Patinopecten yessoensis* per standard collector was in the order of a few hundred spat (Ventilla 1982, his fig. 12; Aoyama 1989). After scallop aquaculture multiplied and stabilised the size of the spawning stock, settlement rose to more than 10,000 spat per collector (improved forecasting of settlement peaks also contributed). After years of bottom seeding in Sarutsu, the practice was stopped during 1980–1985 because of the high availability of wild scallops from natural recruitment (Ito 1991, p. 1044).

ii) Chilean scallop (*Argopecten purpuratus*) harvests from natural populations sharply peaked at *ca.* 5,000 t in 1984, following the 1982/83 ENSO event (Navarro-Piquimil et al. 1991); stocks were overfished afterwards, and the fishery was subsequently closed in 1988 (Stotz and González 1997a). During the 1990s a parcel of 1,900 ha of seabed in Tongoy Bay (North-Central Chile) was destined to scallop aquaculture. Production rose fast, soon reaching an all-time maximum of *ca.* 11,000 t in 1994. During the same period, average density in a TURF granted to an adjacent fishing village rose gradually six-fold over a three-year period. While this may reflect in part the positive effects of co-management (Section 14.6.3.4), increased recruitment from the new, stable adjacent source is likely to have had a major effect.

iii) Recovery of Jicon scallop (*Chlamys farreri*) stocks in Liaoning and Shandong Provinces (China) has been attributed to the development of raft aquaculture for that species (Luo 1991).

2) Recruitment after spawning biomass declined to very low levels due to either natural causes or overfishing:

i) The age structure of the tehuelche scallop (*Aequipecten tehuelchus*) population of San Matías Gulf (Argentina) indicates that recruitment was stable before the stocks were overfished during 1969–1971, but turned to spasmodic (two pulses of recruitment over a 30-year period) afterwards (Ciocco et al., Chapter 26).

ii) Dredge (1988) observed weak recruitment of *Amusium japonicum balloti* in depleted beds of Queensland.

iii) Young and Martin (1989) noticed that recruitment failure of most beds of *Pecten fumatus* in southern Australia occurred after a period of commercial fishing (ending by the mid-1980s) and suggested that a certain minimum adult density was required to ensure successful spawning and high levels of recruitment. Young et al. (1990) conducted a two-year study of six sites in Bass Strait. Settlement rate (spat collectors) was correlated with an index of parental stock abundance (*CPUE*). Evidence synthesised by Young (1994) indicates that, before the fisheries started, at least some populations were composed of several consecutive cohorts. Recruitment occurred all years to the eastern Banks Strait region during 1979–1983, although not always in the same beds. By contrast, no significant recruitment was observed between 1985 and 1989.

iv) Marelli et al. (1999) monitored density and recruitment of a collapsed bay scallop stock from Florida; lack of recovery was attributed to a combination of natural and anthropogenic causes, and depensatory dynamics. They suggested that recovery could take at least a decade.

v) Summerson and Peterson (1990) presented compelling evidence supporting the hypothesis that recruitment of bay scallops failed in 1987–1988 in Bogue and Back Sounds (North Carolina) due to the first recorded red tide of *Ptychodiscus* (= *Karenia*) *brevis*. Recruitment failed again during the next four years. Peterson and Summerson (1992) observed a positive relation between settlement rate (measured with spat collectors), recruitment and adult abundance across 9 beds, and showed that reduction in recruitment (relative to pre-disturbance levels) conformed to the documented pattern of immediate direct effects after the red tide. In subsequent years (1992–1994), Peterson et al. (1996) transplanted spawners from a donor site where scallops were abundant to four receiver sites where populations had not recovered. Adult density in western Bogue Sound (receiver sites) increased by 258% following spawner transplantation as compared to a non-significant change of 8% in control sounds (Back and Core Sounds), although settlement indices (spat collectors) could not confirm that the transplants succeeded through the enhancement of larval abundance.

(3) Stock-recruitment relationship in the absence of exceptional changes of spawning biomass:

i) McGarvey et al. (1993) analysed the stock-recruitment relationship in a self-sustaining stock, *Placopecten magellanicus* from Georges Bank (Sections 14.3.2.1 and 14.3.2.3). Their study was based on 11 years of survey data, including information on age structure. Time series of recruitment and annual egg production (based on age-specific fecundities) were correlated in two areas, which may imply that they constitute a self sustaining stock (see also Section 14.3.2.1).

ii) Fifas et al. (1990) analysed recruitment of the *Pecten maximus* stock from the Bay of St. Brieuc (also believed to be self-sustaining) using a multiplicative model incorporating spawning biomass, thermal anomalies during critical periods of the reproductive cycle, and incidental mortality of pre-recruits due to fishing. The assumption of a linear relation between spawning biomass and recruitment could not be rejected in favour of models incorporating compensatory density-dependence.

This collection of cases appears to reveal some characteristics of scallop population dynamics:

i) Recruitment is positively correlated with stock size, even at moderate to high levels of abundance. "Under-saturation" of beds may point to the existence of many unsaturated sites, even at high levels of aggregated abundance.

ii) Steady or irregularly fluctuating iteroparous stocks tend to shift to a spasmodic pattern after their abundance has been reduced (the *Pecten fumatus* and tehuelche scallop cases); alternatively, spasmodic stocks have become steady following enhancement of the spawning biomass (the Mutsu Bay case). This indicates that at low density levels strong recruitment is exceptional, probably resulting from fortuitous combinations of circumstances (e.g., right thermal triggering of spawning in a small source or reproductive hotspot).

iii) In the absence of such fortuitous combinations of circumstances, recovery of depleted beds can be lengthy and erratic (e.g., *Pecten fumatus* and bay scallop stocks); depletion events (natural or due to overfishing) tend to have long-lasting effects.

Overall, these observations are consistent with the "recruitment-limitation hypothesis" (Peterson and Summerson 1992), which may hold even at moderate to high levels of abundance. Existence of under-saturated sites, even at high levels of aggregated abundance, is expected to occur due to the contagious spatial distribution of scallop populations. Caputi et al. (1998), for example, showed that, following a spasmodic recruitment pulse in the *Amusium balloti* stock from Shark Bay (Western Australia), the area of the bottom covered by scallops was less than 10%, even at peak observed density (7 scallops m^{-2}). Compensatory density-dependence ("inhibition") has been shown to occur in very dense patches of scallops and other benthic bivalves (Section 14.3.4.3), e.g., in the tehuelche scallop stock of San José Gulf (Argentina). But even in the latter case Orensanz (1986) reported that the 1975 year-class was relatively strong in the entire region, both in lightly populated beds and in others in which compensation was concurrently demonstrated by small-scale observation of crowded patches. Ciocco and Aloia (1991) made comparable observations on other year classes.

14.3.2 Macroscale

14.3.2.1 Patterns of connectivity: metapopulation structure

Although the term "metapopulation" became fashionable only in recent years, the conceptual model associated with it has long been implicit in scallop population dynamics. Fairbridge (1953) clearly conceptualised the scallop metapopulation from the D'Entrecasteaux Channel, Tasmania (itemisation is ours):

i) "Larvae originating from any one area in the Channel [are] distributed throughout most, if not all, of the scallop-carrying parts of the Channel.

ii) Although the adult swims, it does not move in numbers from one area to another.

iii) There are, therefore, more or less clearly delimited subpopulations in the various bays and straits of the Channel which must be treated, for certain purposes, as entities.

iv) Scallops in the Channel are [most probably] a genetically homogeneous [meta]population."

Metapopulations vary in the pattern and degree of connectivity among their component subpopulations (Fig. 14.1); some examples follow:

1) Subpopulations that share a common larval pool. Fairbridge's model is one of high connectivity. The tehuelche scallop metapopulation from San Jose Gulf (Patagonia) is similarly structured (Orensanz 1986); high density beds are spread over a sparsely populated background, interconnected by larval dispersal. Thouzeau and Lehay (1988) investigated connectivity in the metapopulation of *Pecten maximus* from the Bay of Saint-Brieuc (Brittany) using a numerical simulation. Their results are consistent with the observed distribution of the reproductive stock and of areas with strong settlement, supporting the idea of intense larval flow between subpopulations. These three examples of high connectivity correspond to metapopulations confined to semi-enclosed bays with significant tidal circulation.

(2) Self-sustaining subpopulations. Sinclair et al. (1985; see also Sinclair 1987), probably looking at a larger geographical scale, suggested that many discrete aggregations of scallops are self-sustaining. This viewpoint stresses the importance of mechanisms of larval retention in the vicinity of the parental stock. They found support for this view in the correlation between observed patterns of distribution and some oceanographic features ("inductive approach" *sensu* Sinclair et al. 1985, p. 5), including shelf fronts and gyres. Tremblay et al. (1994) investigated larval dispersion on Georges Bank using a circulation model. In the simulations, the dispersing particles mimicked the behaviour of sea scallop larvae. Results indicate significant larval exchange among the three major aggregations, with self-seeding possible for two of them, but unlikely for the third. Retention of particles on Georges Bank as a whole was 10–73% before mortality; the authors concluded that Georges Bank should be considered self-sustaining. Young (1994) suggested that subpopulations in the *Pecten fumatus* metapopulation from Bass Strait (Australia) are self-sustained. Arnold et al. (1998) compared abundance and settlement indices (spat collectors) among four geographically separated bay scallop populations along the west coast of Florida. Observed lack of coherence was thought to indicate that the populations are self-sustaining, even if occasional larval exchanges between populations are sufficient to maintain genetic homogeneity (Section 14.3.2.2).

(3) Asymmetric source-sink links. Connectivity between two subpopulations is always asymmetric. Exchange of larvae between beds in the Georges Bank model, discussed above, provides an example. Extreme asymmetry is found in the case of "absolute sinks"

Closed Populations

Single Source

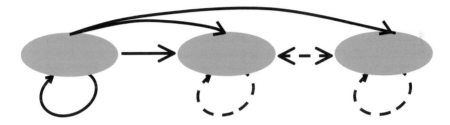

Multiple Sources

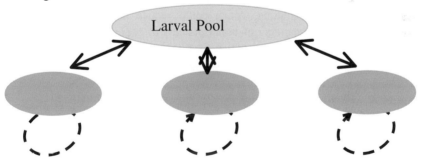

Larval Pool

Figure 14.1. Schematic examples of metapopulation structure (connectivity between subpopulations). *Top*: closed, self-sustaining populations, is a good model for bay scallops from bays and estuaries along the east coast of the US and the west coast of Florida. *Middle*: single source and multiple sinks, is best exemplified by the Peruvian Bay scallop. *Bottom*: multiple sources contributing to a common larval pool, is a reasonable model for metapopulations confined to semi-closed bodies of water (e.g., the tehuelche scallop stock from San José Gulf, Argentina). See Section 14.3.2 for further information on these and other scallop cases.

(pseudopopulations). Dredge (1988) hypothesised that a spawning ground of *Amusium balloti* in Queensland is the source for a sink population located to the southwest (Hervey Bay). The most dramatic example is the metapopulation of the Peruvian bay scallop from north Chile and Peru, which extends over thousands of kilometres of coastline (Wolff and Mendo 2000), a range over which no genetic variation has been detected (Galleguillos and Troncoso 1991). The Independence Bay (Peru) subpopulation, near the center of the latitudinal range of distribution, appears to be the main source of larvae to the whole metapopulation. Some formerly extinct beds were recolonised following the 1983 El Niño event.

Most metapopulations (as exemplified by the Georges Bank's) are a mix of these basic patterns. Although life-history traits relevant to connectivity may be species- or population-specific, patterns of connectivity are always contingent upon the local oceanography. In the case of *Aequipecten opercularis* from Europe, for example, the Gulf Stream drift facilitates high connectivity along the west coast of Scotland, while the stock of the Bay of St. Brieuc (Brittany) appears to be relatively isolated, "locked" by an oceanographic eddy.

Patterns of connectivity have important implications for management. In general, the lower the connectivity, the more conservative the management should be in order to avoid the serial overfishing of the population units. Small, isolated beds are particularly vulnerable; this is the case of the small, discrete populations of *Pecten maximus* confined to bays along the west coast of Ireland (Ansell et al. 1991). Fevolden (1989) recommended that a conservative approach to the management of the Iceland scallop should treat each of the main areas (Jan Mayen, Spitsbergen, Bear Island and northern Norway) as a separate stock unit (Parsons et al. 1991). A different management issue addressed by Lewis and Thorpe (1994) is the possible effect of stock enhancement through extensive transplantation, which could reduce the fitness of receptor local populations. Some understanding of broad patterns of connectivity is also important in the design of reproductive refuges, which ideally should be located "upstream" and be self-sustaining to some extent (Section 14.6.3.2).

14.3.2.2 Metapopulation structure and geographic genetic differentiation

Low connectivity does not necessarily result in detectable geographic genetic differentiation. Since small levels of gene flow may be sufficient to erase differences, subpopulations that are isolated from a population dynamics viewpoint may not be isolated in the sense of population genetics. In the case of *Placopecten magellanicus*, for example, there appears to be enough gene flow to erase differentiation in the geographically fragmented metapopulation (Herbinger et al. 1998; Zouros and Gartner-Kepkay in Beaumont and Zouros 1989; Naidu 1991), yet there is compelling evidence indicating that subpopulations are self-sustaining (Sinclair 1987; Tremblay et al. 1994; Manuel et al. 1996, discussed below). Krause et al. (1994) estimated gene flow between three populations of the calico scallop (Marquesas Keys and Cape Canaveral, Florida, and Cape Lookout, North Carolina) based on electrophoretic study of seven polymorphic loci. They concluded that gene flow between populations was significant enough to erase any detectable differences, yet the populations are distant enough from each other (*ca.* six degrees of latitude) to suspect that their dynamics are uncoupled (see also Arnold et al. 1998). Furthermore, results on restriction site variation of mitochondrial DNA in populations from the Atlantic and Gulf coasts of Florida were consistent with the null hypothesis of a common genetic pool (Blake and Graves 1995). The same study could not detect differences between four bay scallops populations spread over a wide geographic range (Gulf of Mexico to New England). In the latter two cases there is little doubt that the populations sampled are dynamically independent from each other.

While low connectivity does not imply limited gene flow, the opposite is true: geographic genetic differentiation is strong evidence of low connectivity. Studies of genetic variation of *Pecten maximus* around the British Islands using mDNA and RAPD banding patterns indicate that, while there is little geographic variation overall, a stock from a semi-enclosed sea lough from Ireland (Mulroy Bay) differed significantly from populations in open water (Heipel et al. 1998, 1999); this result is consistent with other pieces of evidence indicating that populations in semi-enclosed coastal systems tend to be self-sustaining (Sections 14.3.2.1 and 14.3.2.3). Geographic genetic differentiation was early documented among scallops for *Aequipecten opercularis* (Beaumont 1991); Lewis and Thorpe (1994) found highly significant inter-site genetic heterogeneity in a study of twelve populations of *A. opercularis* around the British Islands. They suggested that, in spite of having pelagic larvae, local populations appear to be self-sustaining. On the other hand, Macleod et al. (1985) identified one allele that was consistently unique to a single year class in all populations of *A. opercularis* around the Isle of Man. They interpreted this result as indicating that larvae settling in the area (at least that year) were not the offspring of the local adult populations, and speculated that the populations were maintained largely or entirely by immigration of larvae from elsewhere. Brand et al. (1991) interpreted this as indication of a source-sink system in the North Irish Sea, and hypothesised that this was the reason for recruitment being strongest in the most heavily fished beds around the Isle of Man.

Evidence of differentiation may be found also in the comparative study of life history, including transplant experiments. Ansell et al. (1991) concluded that differences in the reproductive ecology of different populations of *Pecten maximus* indicate relative genetic isolation among stocks (a hypothesis substantiated by Mackie and Ansell (1993), and Cochard and Devauchelle (1993)), although this is not supported by allozyme polymorphisms (Beaumont et al. 1993; Wilding et al. 1998).

Manuel et al. (1996) kept larvae of *Placopecten magellanicus* spawn by parents from different areas (Georges Bank, Passamaquoddy Bay and Mahone Bay) in a large experimental mesocosm. The larvae from the different stocks had different vertical distributions, identifiable by means of a microsatellite probe (see also Section 14.3.2.3). The source populations experienced contrasting hydrographic regimes, suggesting that genetically fixed behavioural differences might have resulted from natural selection for increased retention near the parental location (Section 14.3.2.3). These results are consistent with the very low density of larvae observed by Tremblay and Sinclair (1992) in the waters between Georges Bank and the Scotian Shelf.

Other transplant experiments have shown that differences in life-history parameters have no genetic basis. For example, cross-transplants of *Aequipecten tehuelchus* in San José Gulf indicate that differences in growth rate among subpopulations are ecophenotypic (Ciocco 1992). *Placopecten magellanicus* transferred from locations where spawning is annual to others where spawning is biennial, shift to the pattern of the receiving location (Dadswell and Parsons 1994). Physiologically different shallow and deep-water populations of *Placopecten magellanicus* from Newfoundland did not differ genetically based on cDNA and microsatellite markers (Herbinger et al. 1998).

The subject of genetic diversification is treated in detail in Chapter 10 of this book.

14.3.2.3 Larval retention/dispersal: mechanisms and oceanographic scenarios

Larval dispersal has been often inferred from information on the duration of the larval stage (6–70 days in different scallop species; Orensanz et al. 1991) and vectors of residual circulation estimated from drift card data or circulation models (e.g., Roberts 1997; Thouzeau 1991b; Craig and McLoughlin 1994). Larvae, however, are not drift-cards or passive particles; larval dispersal is influenced by factors other than water circulation: diffusion, survival, and active behaviour in the water column (Cowen et al. 2000). As pointed by Craig and McLoughlin (1994), Tremblay et al. (1994) and others, the appropriate use of simulation models is to assess the sensitivity of scallop larval drift to various factors. Even subtle responses to environmental cues can have a strong influence on larval retention/dispersal. Manuel and O'Dor (1997) and Manuel et al. (1997), for example, synthesised information on vertical migration of *Placopecten magellanicus* veligers from mesocosm experiments and field studies. Following evidence suggesting that they migrate in response to both tidal and diurnal stimuli, they proposed a tidal/diel conceptual model, and applied it to the Passamaquoddy Bay and Georges Bank populations (eastern Canada). Both respond to the daily light cycle (a phenomenon observed also in other species, e.g., in *Patinopecten yessoensis*; Ito (1964); Maru et al. (1973)), but differ in their response to tidal cues. Georges Bank veligers appear to utilise baroclinic differences in tidal transport to maintain themselves on the bank. Passamaquoddy Bay veligers respond by swimming up at slack water (high and low tides) and down when currents are strongest. Such behaviour would minimise dispersal on the strong tidal currents in the Bay of Fundy. Horizontal transport resulting from vertical migration was considered by the authors as the most likely selective pressure to create and maintain these different behaviours against the homogenising effects of larval dispersal.

Coincidentally, several modelling and empirical studies (Craig and McLoughlin 1994; Swearer et al. 1999; Jones et al. 1999; Cowen et al. 2000) have emphasised the importance of larval retention near the parental ground and "self-recruitment" in populations with pelagic larvae. Ultimately, as pointed by Cowen et al. (2000), the relative significance of dispersal and retention is largely one of scale. Good settlement in the vicinity of Maria Island, Great Oyster Bay (Tasmania), in December 1987, appears, based on modelling results, to have been spawned about 15 km away from the collectors (Craig and McLoughlin 1994). This could be claimed to support larval dispersal or retention, depending on the scale of specific interest.

Association between scallop concentrations and oceanographic features is generally believed to reflect, at least to some extent, retention mechanisms. Such features include:

(1) Shelf fronts. Sinclair et al. (1985) pointed to the fact that many scallop aggregations occur in zones of transition between tidally mixed (inshore) and stratified (offshore) waters. In some documented cases recruitment appears to be associated with the seasonal development of an upper layer of warm water, as indicated by Tremblay and Sinclair (1991):

i) All areas of high concentration of *Zygochlamys patagonica* from the Argentine Shelf are associated with different types of frontal systems (Ciocco et al., Chapter 26). One of those aggregations matches a surface warm water pool (delimited by the 18°C surface isotherm), 20–30 m deep that starts to develop in spring and matures during the summer (Orensanz et al. 1991).

ii) Robinson et al. (1999) conducted a 10-year study in Passamaquoddy Bay (*Placopecten magellanicus*), in which spat collectors were deployed annually on a grid and environmental data were recorded concurrently. High-density of settlers corresponded to warmer, stratified waters that generally had higher concentration of chlorophyll-*a*.

Gallager et al. (1996) and Pearce et al. (1996) studied the ontogeny of behaviour and the settlement of *Placopecten magellanicus* larvae kept in experimental mesocosms. Consistent with the field observations mentioned above, they found that in stratified water larvae are confined to the upper layer, above the thermocline. According to these results, settlement should increase in areas where, and/or at times when, water stratification is disrupted.

(2) Frontal systems along the edge of gyres. Premetz and Snow (1953) attributed concentration of *Placopecten magellanicus* on Georges Bank to a large circular gyre that forms during the summer months, when larvae are in the plankton. The clockwise circulation gyre over the bank narrows into an intense current jet along the bank's northern edge. The shallow central part of the bank remains vertically well mixed all year, primarily due to strong tidal currents. A frontal zone separates the mixed area from the surrounding stratified waters (Tremblay and Sinclair 1991, 1992).

(3) Retention/trapping by gyres and eddies. Dickie (1955) emphasised retention in the Digby area (Bay of Fundy) in warm years, and Caddy (1979a) suggested that a counterclockwise gyre in the center of the Bay of Fundy might retain scallop larvae and influence recruitment to grounds in its vicinity (Tremblay and Sinclair 1988). Yamamoto (1964) observed enhanced settlement of *Patinopecten yessoensis* in Mutsu Bay in areas underneath eddies and slow-running water. Both are characterised by "swirling currents" that concentrate large quantities of planktonic organisms and suspended silt. "Marine snow" forms underneath during the spring, and heavy settlement of scallops and other organisms was observed when oxygen was not depleted during the summer. Heald and

Caputi (1981) found that juveniles of *Amusium balloti* settle regularly in some areas of Shark Bay (Western Australia), presumably reflecting the existence of persistent eddy systems. Thouzeau and Lehay (1989) found indication that larvae of *Pecten maximus* are retained in areas of the Bay of St. Brieuc in which gyre-clockwise residual currents were measured. Aggregation underneath gyres does not necessarily imply retention of larvae generated *in situ*. Larvae trapped into gyres could come from other areas, being concentrated (rather than retained) by the gyre. Dredge (1988) suggested that a gyre in Hervey Bay (Queensland) might act as a trap for larvae of *Amusium balloti* spawned in grounds located to the northwest.

(4) Semi-enclosed bays. Larval retention in semi-enclosed bays was stressed by Boucher (1985) for St. Brieuc Bay (*Pecten maximus*) and by Fairbridge (1953) for the D'Entrecasteaux Channel (*Pecten fumata*). Semi-enclosed, non-estuarine bays are often areas of good settlement. This is the case of Mutsu Bay in Japan (Yamamoto 1964), San José Gulf in Patagonia (Orensanz 1986), Bogue Sound in North Carolina (Gutsell 1931), and the Bays of Brest and Saint-Brieuc in Brittany (Boucher 1985). Stocks within these bays are, in general, metapopulations with strong connectivity that may play an important role supplying larvae to adjacent open grounds.

(5) Coastal topography. Good settlement has often been observed in association with headlands and prominent features of the seabed, possibly as a result of interaction between coastal topography and wind- or tidally-induced circulation. Good recruitment of *Aequipecten tehuelchus* off the south coast of San Jose Gulf (Patagonia) almost always occurs at leeward of prominent coastal features, perhaps due to eddies generated by strong SW winds (Orensanz 1986). Particular areas adjacent to coastal headlands or prominent bottom features consistently attract settlement of *Pecten fumatus* in Bass Strait (Young 1994, p. 337) and *Amusium balloti* in Queensland (Gwyther et al. 1991). *Patinopecten yessoensis* in the Okhots Sea settles preferentially at the foot of newly-built pier walls or beneath kelp beds (Kalashnikov 1991, p. 1059).

(6) Estuarine-like circulation, which provides a mechanism of larval retention well documented for many invertebrates, has been invoked for scallops by Moore and Marshall (1967) and Naidu and Anderson (1984).

Although relationships between stock location and oceanographic features are frequently invoked, they have been rarely substantiated by process-oriented studies (the "deductive approach" *sensu* Sinclair et al. 1985: p. 5).

14.3.2.4 Metapopulation models

Metapopulation models have been discussed by Hanski (1998). Botsford (1995) and Botsford and Hobbs (1995) described the population dynamic theory appropriate to harvested metapopulations of marine invertebrates, in which subpopulations are linked by meroplanktonic larvae. Their modelling approach, which explicitly considers the spatial structure of these stocks, was discussed using crustacean fisheries and the tehuelche scallop stock of San José Gulf (Argentina) as examples. This is a promising direction for the modelling of scallop metapopulation dynamics; models of larval dispersion (Section 14.3.2.3) are a first step.

14.3.3 Mesoscale

14.3.3.1 Persistence, extinction and resurgence of subpopulations

In the preceding section we addressed the general arrangement and connectivity of subpopulations in a metapopulation. Here we focus on the individual subpopulations. The latter tend to be more or less discrete units ("beds") in the case of bivalves (e.g., Lasta and Bremec 1999). Sinclair et al. (1985) emphasised the unappreciated significance of the well-known fact that many commercial scallop fishing beds are recurrent in location and spatially discontinuous. In the northwestern Atlantic some have been observed and exploited in precisely the same locations for more than a century. They speculated that this supports a connection between scallop fishing beds and oceanographic features (shelf fronts, gyres, et al.) facilitating larval retention.

Bed discreteness is well known to fishers, who identify them by name in virtually every scallop fishery. They are recognised *de facto* when sampling surveys are designed, or when the origin of the catch is recorded. Yet, the dynamics of beds has received little attention from scientists. Information of the type *"Bed A has not been fished since xxx"*, *"We found a new bed off xxx"* or *"Bed B used to have three times its present extension"* is rarely treated formally in the technical or scientific literature. Over recent years, however, the origination, extinction, persistence, expansion, contraction and shape of beds have started to attract due attention, facilitated by the expanding capabilities to analyse geo-referenced data such as GIS and spatial statistics (see Section 14.4.3). The significance of this subject for the analysis of the fishing process and for the implementation of spatially-explicit harvesting strategies should be obvious.

14.3.3.2 Relation between abundance and area occupied by a subpopulation

As the size of a subpopulation (*N*) changes over time, the area that it occupies (i.e., the extension of the bed, *A*) and the overall density (*N/A*) are also likely to change. Analysing this process, the most basic in bed dynamics, requires the development of formal methods to draw the boundary and estimate the extension of the bed (Section 14.5.2). A power function relation between area occupied (*A*) and abundance (*N*), i.e., $A \propto N^\beta$, where β is a constant, is more or less implicit in several models, for example in the "proportional density" and "constant density" models (β respectively equals 0.5 and 1) discussed by Hilborn and Walters (1992).

Caddy (1989b) introduced a graphical model of the spatial expansions and contractions of a sedentary population with pelagic larvae (Fig. 14.2). Habitats within the maximal range of distribution rank from "favourable" to "unfavourable". When total population abundance increases, the range of distribution expands into unfavourable habitats, and when abundance declines the range contracts to the most favourable habitats. Suitability of a site, defined in terms of persistence of occupation through time, has two components:

i) Larval availability, determined, for example, by the existence of oceanographic systems that facilitate the retention of larvae.

ii) Availability of suitable benthic habitat, determined by factors affecting post-larval growth and survival, e.g., adequate substrate, good food supply, low density of predators, etc. (Thouzeau 1991b).

In Caddy's (1989b) and other idealised models of geographic range dynamics (e.g., McCall 1990) suitability gradually vanishes away from a central core; range contraction is visualised as a gradual retraction of the periphery. In more realistic scenarios, patterns associated with stock contractions more likely involve the "vacuolation" of dense cores, fragmentation of beds, and extinction of some beds.

Another popular model of CE dynamics was McCall's (1990) extension of the Ideal Free Distribution (Fretwell and Lucas 1970). He developed the model with pelagic fishes in mind, but presented an extension for the case of benthic shellfish. The assumptions of the latter are not reasonable for the case of scallop stocks.

The relationships between the area occupied by a subpopulation, its size, and catchability have been investigated for several finfish stocks during recent years (Crecco and Overholtz 1990; Rose and Leggett 1991; Swain and Wade 1993; Marshall and Frank 1994; Swain and Morin 1996; Rose and Kulka 1997). The spatial dynamics of finfish populations differs from that of sedentary organisms in that changes in the area occupied are often a short-term result of individual movements in the first, but are mediated by intergenerational processes (renewal) in the second (at least in the absence of fishing or catastrophic mortality). Yet, conceptual and formal models introduced recently for finfish stocks are applicable to scallop bed dynamics.

14.3.4 Microscale

In this section, we consider the small-scale spatial structure of scallop populations, and its implications for density-dependent processes. We do not discuss the factors and processes that underlay those patterns (microhabitat selection, movements, etc.), which were reviewed in the first edition of the book by Brand (1991, pp. 550–564) and Orensanz et al. (1991, pp. 662–667). See also Chapter 12 (Brand).

14.3.4.1 Density, neighbourhoods and concentration

Among benthic ecologists there is perhaps no notion better established than "*density*", the number of individuals found per unit of area. Its analysis is central to the assessment of stocks of benthic sedentary organisms. In a typical study, "quadrats" are sampled according to some pre-established design (typically at random locations, eventually stratified), density is estimated, and departure from spatial randomness is investigated

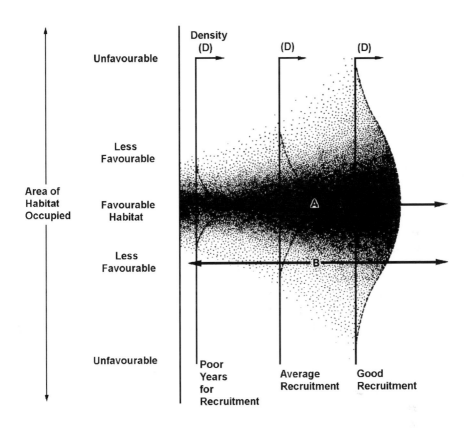

Figure 14.2.   Caddy's graphical model of the geographical contraction/expansion of a subpopulation.   Habitats within the maximal range of distribution rank from "favourable" to "unfavourable".   When total population abundance increases, the range of distribution expands into unfavourable habitats, and when abundance declines the range contracts to the most favourable habitats. (Modified from Caddy 1989b).

using the frequency distribution of quadrat counts (e.g., Elliott 1977; Section 14.5.3). Yet, Orensanz et al. (1998) argued that emphasis on the observation of density at arbitrary locations, together with reliance on quadrats for its measurement, have obscured the significance of a related but distinct concept: *concentration* or *crowding*.   The latter is density experienced by an individual within a certain *neighbourhood* (as opposed to density measured by an observer at arbitrary locations).   If the location of the individuals in a given region is conceived as a spatial process, then density and concentration capture the first and second order properties of that process (analogous to familiar notions of mean and variance).

The analysis of concentration requires meaningful conceptual constructs, and formal models to render these notions tractable:

(1) The notion of "neighbourhood" provides a conceptual framework. Defined with respect to an individual, it is a spatial region within which a certain interaction between the individual and its environment (including other individuals of its own kind) is effective (or, from the viewpoint of the observer, within which its effectiveness is evaluated) (Orensanz 1986). Various neighbourhoods can be defined, each relative to a specific process:

i) *Trophic*, relative to resource utilisation, eventually leading to competitive interactions.

ii) *Reproductive*, related, in the case of broadcast spawners, to gamete dilution away from the individual.

iii) *Sensory*, relative to perception of environmental conditions, including the presence of conspecific, predators, etc.

(2) The theory of stochastic spatial point processes provides an appropriate formal framework (Ripley 1977, 1981; Diggle 1983; Cressie 1991). Observed patterns of distribution of individual animals in space are modelled as realisations of spatial point processes; in the simplest case neighbourhoods are modelled as circular isotropic regions centred on the individuals. Most individuals in populations with an aggregated spatial distribution experience a high concentration, even if high-density patches are scarce and small (Orensanz 1986). This has important and self-evident implications in the case of density-dependent processes (resource depletion, fertilisation rate in broadcast spawners, predator-prey interactions, spatial allocation of fishing effort, etc.), that may affect a large fraction of the population even if they operate only over comparatively small areas (Orensanz et al. 1998).

"Neighbourhoods" are defined by interactions, not by an absolute metric ruler. For example, trophic neighbourhoods of suspension feeders may be extensive, particularly in the presence of turbulence or flow in the benthic boundary layer. The neighbours of a spawning individual, on the other hand, are the conspecific located in close proximity, at a distance such that gametes can cross-fertilise.

14.3.4.2 Concentration profiles

While ecologists have focused mostly on mean concentration or crowding for assessing spatial pattern (e.g., Elliott 1977), interest in the *distribution* of concentration in a population arose in the 1980's among natural resource scientists. First, Clark (1982, 1985) introduced such distributions (as *concentration profiles*) as a way of describing the amount of resource found at different densities, and to investigate the response of harvesters to such patterns. The observational scale of interest for the analysis of these profiles is necessarily large, commensurate with the spatial scale of the fishing process (i.e., the observational scale upon which harvesters make tactic decisions). Profiling of concentration at smaller scales was first proposed by Orensanz (1986, as distributions of "*neighbourhood density*", his fig. 3:e). Concentration profiles are commonly classified into "types" (Clark 1982, 1985; Hilborn and Walters 1992, pp.184–188; Orensanz et al. 1998, their fig. 8) generally associated with density frequency distributions at the same scale of observation (Fig. 14.3); the definition of types is not consistent in the literature.

In traditional fishery science there is an emphasis in abundance, but in the case of sedentary resources the concentration profile for a given abundance level is an important consideration as well. For example, the rates of fishing mortality and of fertilisation are both concentration-dependent (Sections 14.3.4.3 and 14.4.3): divers only harvest scallops in dense patches, and the chances of fertilisation are higher for crowded scallops. In other words, it is not the same thing to have 100 scallops spread over one km^2, or to have them crowded in 10 m^2: In the first case they will be unattractive to divers, and unlikely to get their eggs fertilised; the opposite holds in the second case. This creates a delicate situation in which divers target those individuals that are most likely to reproduce successfully. This phenomenon (first addressed by Gross and Smyth 1946) is one of the most likely causes of recruitment overfishing in benthic fisheries. The problem is well illustrated by the dramatic contrast between the concentration profiles of the tehuelche scallop stock of the San José Gulf before and after the fishery collapsed (Fig. 14.4). A survey of the entire metapopulation was conducted in 1975 (Orensanz 1986; Prince and Hilborn 1998, their fig. 2) and again 20 years later (Ciocco et al. 1997) after the fishery had collapsed, presumably due to overfishing. While a large fraction of the scallops lived at high concentrations in 1975, the scallops left after the collapse were not only few but also highly diluted (Orensanz et al. 1997). The vertical lines in the figure show the density threshold for commercial diving harvesting in 1975 (*ca.* 25 scallops m^{-2}). Between then and 1996, mostly due to a series of economic crises and a devalued peso (scallops are an export product) the density threshold dropped to very low levels.

14.3.4.3 Density- vs. concentration-dependence; compensation vs. depensation

Density-dependence in scallop metapopulations is likely to be significant only during benthic stages (larval settlement through fertilisation). Most density-dependent interactions have small operational scales, as individuals are influenced only by their neighbours (Orensanz et al. 1998). Density-dependence can be compensatory or depensatory. In the first the *per capita* reproductive contribution decreases at higher densities, a result of increased competition for resources, slower growth, less energy available for reproduction and increased mortality. In depensatory density-dependence (also known as inverse density-dependence or "Allee effects") the reproductive contribution per capita decreases as density decreases. This is the case of fertilisation rate in broadcast spawners: the less neighbours around, the lower the chances of fertilisation.

The effects of localised (small-scale) density-dependence on the (large-scale) dynamics of the metapopulation are mediated by larval dispersal. Thus, it is convenient to distinguish between pre- and post-dispersal density-dependent mechanisms (Botsford and Hobbs 1995), whether these are compensatory or depensatory. Pre-dispersal density-

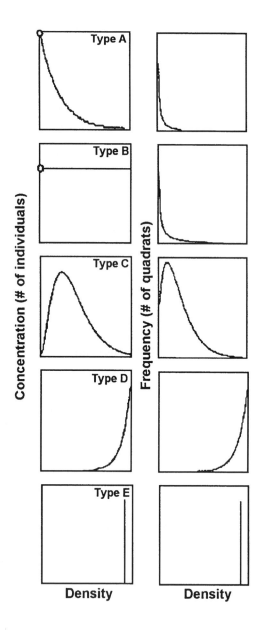

Figure 14.3. Five general types of concentration profile. *First column*: concentration profiles. *Second column*: corresponding distributions of density (concentration - the y-axis in figures of the first column - is the product of the two axes of the figures of the second column). (Modified from Orensanz et al. 1998).

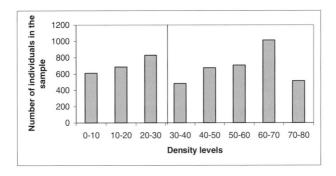

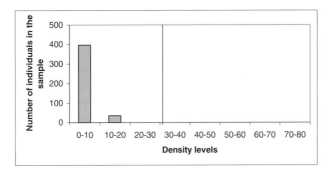

Figure 14.4. Concentration profiles for two surveys of the tehuelche scallop stock from San José Gulf conducted in 1975 and 20 years later, after the fishery had collapsed. Notice that the scallops left are highly diluted. The vertical lines show the density threshold for commercial diving harvesting in 1975. Between then and 1996 the density threshold dropped to very low levels, mostly due to a series of economic crises and a devalued peso (scallops are an export product). (Modified from Orensanz et al. 1997).

dependent processes are those that affect the reproductive output either directly (e.g., fertilisation success), or indirectly (e.g., density-dependent growth). They depend on the concentration profile of the populations, which summarise the density conditions experienced by individuals (concentration). Post-dispersal processes affect the rate of recruitment through the effect of resident densities on larval settlement and survival of pre-recruits. Figure 14.5 summarises the types of density-dependence most frequently considered for benthic sedentary invertebrates with broadcast spawning, external fertilisation, and pelagic larvae, such as scallops. Notice that in the lower-left panel of Figure 14.5 the solid line represents an "envelope" or "boundary" indicating the *maximum* density of new settlers expected for a given density. In all the other panels the solid line corresponds to the expected level (*mean*) of the response variable (gametes produced per local resident, fertilisation rate, density of new settlers) conditioned on the density of resident adults.

792

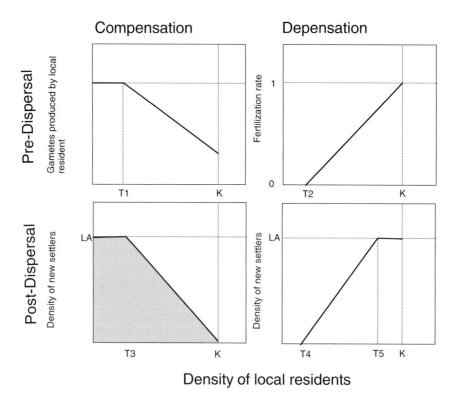

Figure 14.5. Types of density-dependence most frequently presumed and/or investigated for benthic sedentary invertebrates, which, like scallops, have broadcast spawning, external fertilisation, and pelagic larvae. Notice than in the lower-left panel the solid line represents an "envelope", indicating the *maximum* density of new settlers expected for a given density of residents. In all the other panels the solid line corresponds to the expected value (*mean*) of the response variable, conditioned on the density of resident adults. K: local carrying capacity; T1-T5: density threshold levels; LA is an index of larval availability.

(1) Pre-dispersal depensation: fertilisation rate. Pennington (1985) demonstrated that per-capita fertilisation rate (the fraction of the eggs produced by each individual that get fertilised) increases with density in sea urchins (as scallops, external fertilisers). This effect was already foreseen by Belding (1910: p. 26): "there are a great many chances in nature against fertilisation of the egg. Scallops may be some distance apart and the spermatozoa must travel far before they can meet the egg [, and] the male cell can only be attracted to the egg from a short distance". Gross and Smyth (1946) first suggested that the elimination of the densest segments of a bivalve metapopulation may have a disproportionate effect on reproductive success and renewal (Fairbridge 1953: p. 38; Orensanz 1986), and that this should be a major concern in bivalve fisheries (Levitan and Sewell 1998).

(2) Post dispersal depensation: attraction of settling larvae to conspecifics. Evidence for scallops (unlike other epibenthic bivalves, like oysters) is very limited. Vahl (1982) suggested that recruitment success of the Icelandic scallop is dependent on the density of resident adults.

(3) Pre-dispersal compensation: competition. Competition for limited resources has a hierarchical sequel of compensatory effects:

  i) *Depletion of food resources.* Density-dependent depression of growth and survival rates are commonly assumed to be linked to *depletion of food resources* and intraspecific competition for food (Frechette and Bourger 1985; Orensanz 1986). This implies that the resource in question is in short supply, or that feeding individuals can deplete the resource in their neighbourhood. Seston depletion by suspension feeding bivalves has been invoked or demonstrated at different scales, ranging from the neighbourhood of individuals kept in still water to large water bodies. Dense scallop populations have the capacity to deplete phytoplanktonic resources in confined coastal water bodies (Mori 1975; Vahl 1979; Orensanz 1986; Bricelj et al. 1987).

  ii) *Individual growth.* Depressed individual growth at high experimental density was shown by Belding (1910) for *Argopecten irradians*. Similar results have been obtained in other studies (Orensanz et al. 1991). Inferential evidence of density-dependent growth in natural field populations usually consists of negative correlation between density (or biomass) and size-at-age (Cooper and Marshall 1963; Gruffydd 1974; Orensanz 1986). Orensanz (1986) found significant differences in growth among four year-classes of *Aequipecten tehuelchus* in the San Roman bed (Patagonia); growth rate was negatively correlated with density. Density-dependent depression of growth rate became evident in several indices of condition when average local biomass exceeded 1 kg m^{-2}. At larger scales, long-term trends in growth rate have also been attributed to density-dependent effects. *Patinopecten yessoensis* sown on the seabed of Mutsu Bay grew twice as fast in the sixties than in the seventies (Ventilla 1982: his fig. 29, p. 367). Most areas were overstocked in the interim period, when the bay apparently reached its carrying capacity.

  iii) *Recruitment to the Fishery.* Density-dependent effects on growth can be strong enough to affect the age of entry to the fishery, to the extent that they are worth considering by managers. The best known case is that of *Patinopecten yessoensis* from Mutsu Bay (Ventilla 1982: his fig. 28). For sown scallop to reach commercial size within 2.5–3.5 years, they must be distributed at a density of 5–6 m^{-2}. Overstocking, resulting in slower growth, delays significantly the time at which scallops can be harvested. The Tehuelche scallop population from San Roman provides an example involving a natural stock (Orensanz 1986). The 1974 year-class that grew under low density conditions, reached commercial size (on average) at the age of 1.4 years, while the 1975 year-class, that grew under high density conditions, reached the same size at the age of 2.2 years. *Pecten fumatus* from Port Phillip Bay (Australia) attained a shell length of 70–80 mm at the age of 2.5–3.5 years during the sixties, at the onset of the fishery. Twenty years later, the same size was reached at age 1+ (i.e., more than one year sooner). This was interpreted by Gwyther and McShane (1988) as a density-dependent effect.

iv) *Mortality.* Density-dependent mortality of suspension-feeding bivalves attributable to starvation has been rarely demonstrated for natural populations. Crowding has been found to be the cause of mass mortality events under exceptionally high density conditions, like in *Patinopecten yessoensis* cultivated in Japan (Mori 1975; Ventilla 1982; Aoyama 1989). Intermediate rearing in Mutsu Bay (Japan) has been limited to 700 million spat in order to avoid the deleterious effects of density-dependence. Caddy (1989a) speculated that density-dependent predation may be effective in maintaining low-density levels in *Placopecten magellanicus.* Noel (1997) reported extremely high mortality during years of exceptional abundance in *Pecten fumatus* from Port Philip Bay (Australia), which could perhaps be attributed to compensatory density-dependence. Besides direct effects through depletion of food or oxygen, and attraction of predators, the effect of density on mortality can be indirect or lagged: (i) retardation of growth due to starvation could protract exposure to some predators, and (ii) periods of environmental stress could be more harmful to individuals previously exposed to food shortages (Orensanz et al. 1991).

v) *Production* should reflect the aggregated effects of density-dependence on growth, survival and recruitment. The San Roman (Patagonia) population of *Aequipecten tehuelchus* (Orensanz 1986) provides a well-documented example. Mean biomass at the core of the bed showed cyclical variation over a 15-year period, with maxima of 1.3 kg m^{-2} reached every 5 years. This ceiling was set by compensatory mechanisms that became detectable above a mean biomass of 1.0 kg m^{-2}. While average biomass almost doubled from 1975 to 1976 (521 to 980 g m^{-2}), annual production remained constant (762 to 767 g m^{-2}), even when there was not a remarkable change in age structure. This was indicative of an effective local carrying capacity. Meat production was 382 g m^{-2} yr^{-1}. Coincidentally, Charpy-Roubaud et al. (1982) estimated (independently) that phytoplankton production in the same area could support an annual production of 379 g m^{-2} yr^{-1} of herbivore bivalve meats. The bed was at its carrying capacity during 1974–1976, with scallops monopolising suspended food resources. Estimated filtration rate for the dense core bed was 14,400 L d^{-1} of water, roughly equivalent to the overlying water column (8,000–17,000 L). In an empirical study conducted on the same bed a few years later, Ciocco (1991) estimated values remarkably similar: 359 g m^{-2} yr^{-1} and 368 g m^{-2} yr^{-1}, respectively. The relationship between stocking density and production has been well established for the bottom culture of *Patinopecten yessoensis* in Mutsu Bay, Japan (Aoyama 1989). Maximum yield is 1.2 kg m^{-2} (less than 6 scallops of age 3+ per m^{2}). At a larger scale, it is thought that the bay as a whole reached its carrying capacity during the late seventies (Ventilla 1982). Mass mortality events in 1975–1977 were attributed to overstocking. It is estimated now that the bay can support a maximum of 700 million spat yr^{-1} under intermediate culture, keeping the annual production (crop) at 50,000–60,000 tons.

vi) *Reproductive Potential.* Gonad weight is - roughly - a power function of size, and potential fecundity is proportional to gonad weight. Thus, density-dependent depression of growth rate and production can result in reduced fecundity. Orensanz (1986) found that the relative gonad weight of *Aequipecten tehuelchus* showed the effects of density-dependent growth before any other index of condition. Reproductive output during a high-density year (1975) was 80% smaller for age class 1+, and 25% smaller for age class 2+ than during a low-density year (1974). *Patinopecten yessoensis* from Mutsu Bay showed a 20–25% reduction in ovary weight between the sixties and the seventies, when growth rate declined due to overstocking (Ventilla 1982).

(4) Post-dispersal compensation: inhibition. A negative relationship between the density of adult residents and settlement/early survival rate has been observed, among suspension-feeding bivalves, in field populations of cockles (Hancock 1976; André and Rosenberg 1991; Bachelet et al. 1992) and scallops (Vahl 1982; Orensanz 1986). The functional form of the relationship takes the form of an "envelope": given a certain density of residents there is a *maximum* possible number of settlers, but that number can be as low as 0 if settlers are not available. It is unclear whether the main inhibitory effect involves pediveligers being inhaled by resident adults, or juveniles being out-competed by residents.

Two factors contribute to the complexity of any postulated mechanism of negative adult-larval (or juvenile) interaction. First, refuge provided by primary settlement substrates elevated from the bottom or located away from adult habitat (Minchin 1992) may alleviate negative interactions. Second, post-settlement dispersal of bivalves (Sigurdsson et al. 1976) may result in different areas for settlement and recruitment. Juvenile scallops can disperse attached to algae, seagrass, etc. (Ingersoll 1887; Orensanz 1986), drifting suspended from a byssal filament (Sigurdsson et al. 1976), or by means of their own movements. Vahl (1982) studied recruitment to a bed of *Chlamys islandica* over a 9-year period. Juveniles settled in unidentified areas outside the bed, and immigrated at the age of 2 years or older.

(5) Tradeoffs between compensation and depensation occur, for example, when the reproductive contribution of a spawner increases with density due to increased fertilisation rate (pre-dispersal depensatory density-dependence), but decreases due to reduced growth rate (pre-dispersal compensatory density-dependence). Such a balance was hypothesised for scallops by Orensanz (1986) and Kalashnikov (1991, p. 1061). Levitan (1991) investigated the same phenomenon in sea urchins, and estimated that the reproductive contribution per capita was very stable over a wide density range due to the balance between the two antagonistic processes.

14.4 THE FISHING PROCESS

The *fishing process* is the sequence of actions through which a resource is located, harvested, and depleted by a *fishing force* ("*the fleet*") composed of discrete *fishing units*. The latter range from individual rakers and divers, to offshore industrial dredgers and trawlers (Section 14.4.1). The *fishing mortality* imposed on the population is related to the amount of *fishing effort* exerted by the fleet (Section 14.4.2). If fishing operations were spread at random and all scallops were identical to each other, all the scallops in the harvested population would have the same chance of being caught by a unit of effort. Under this assumption (*homogeneity*) fishing mortality would be simply proportional to fishing effort[6]. In reality, however, beds with higher density or closer to port tend to be harvested more intensely. The analysis of fleet behaviour and the spatial allocation of fishing effort (Section 14.4.3), depletion (Section 14.4.4) and variable vulnerability (Section 14.4.5) are important to understand where fishers fish, and why they fish where they fish. This is a central issue in the assessment and management of benthic stocks. The assessment of the fishing process attends also to several aspects of gear performance: *area fished* with a unit of effort, *gear efficiency* and *gear selectivity* (Section 14.4.6).

14.4.1 Types of Fishing Gear used in Scallop Fisheries

Scallop harvesting techniques fall into four main groups (see regional chapters in this book for further information), each of them presenting particular problems for the measurement and standardisation of fishing effort (Caddy 1979b):

(1) Dredging. Dredges and dredge behaviour have been described for many fisheries, including those in Britain (Baird 1959; Mason 1970; Chapman et al. 1977; Noel 1980; Franklin et al. 1980; Howell 1983), France (Dupouy 1978; Noel 1982; Fifas 1991), Ireland (Gibson 1956), eastern Canada and New England (MacPhail 1954; Bourne 1964, 1966; Caddy 1968; Cook 1983; Robert and Lundy 1988), southeastern US (Cummins 1971), Japan (Ito 1964; Yamamoto in Imai 1971; Ventilla 1982), Australia (Fairbridge 1953; Young and Martin 1989; Cover and Sterling 1994), and Argentina (Olivier and Capitoli 1980; Orensanz et al. 1991). Scallop dredges range from rigid-framed, with or without teeth, to essentially modified beam trawls with upper and lateral rigid supports and a lower sweep chain. Cover and Sterling (1994) investigated scallop dredges from an engineering perspective in a comparative experimental study.

(2) Trawling, employed in the US calico scallop fishery (Rivers 1962), in tropical *Amusium* fisheries (Heald and Caputi 1981; del Norte et al. 1988; Young and Martin 1989; Dredge 1994), in the *Pecten ziczac* fishery of South Brazil (Pezzuto and Borzone 1997), in the Patagonian scallop fishery of the Argentine Shelf (Ciocco et al., Chapter 26), and in the *Pecten jacobaeus* fishery of the Adriatic (Hall-Spencer et al. 1999).

(3) Diving is utilised, at least to some extent, to catch almost every commercial scallop species, including those that are harvested primarily with dredges, like *Placopecten magellanicus* in Maine (Dow 1969; Tamanini 1985), or *Pecten maximus* and *Aequipecten opercularis* in Britain (Hardy 1981). Many coastal scallop fisheries are caught exclusively by divers (often because of regulations banning dredges and trawls), e.g., *Pecten vogdesi* in the Gulf of California, *Aequipecten tehuelchus* in San José Gulf (Argentina; Ciocco et al., Chapter 26), *Zygochlamys patagonica* in southern Chile, and *Argopecten purpuratus* in northern Chile and Perú (Stotz and González 1997b).

(4) Hand-operated devices, including rakes and the brideog, a long-poled hand net employed in Ireland (Gibson 1956).

Gear evolves as fisheries develop, often through technological improvements that result in punctuated increments in efficiency, as is well illustrated by commercial scallop diving. Women divers of the Russian far east ("baba seals", Kalashnikov 1991; p. 1073), who picked *Patinopecten yessoensis* with no diving gear at all, were among the first to do it commercially. The first compressors introduced were hand-operated (Luo 1991). During the second half of the century, the "hookah" became the basic system utilised world-wide. A hookah team involves fishers with skin-diving gear getting air from on-board compressors through flexible hoses.

14.4.2 Effort and Fishing Mortality

14.4.2.1 Effort data

Effort data are most often provided by the fishers themselves (catch slips, logbooks, interviews) or the administrative track of the catch (catch/sales slips). On-board observer programs are a possibility in the case of industrial fisheries, like the Argentine offshore fishery for Patagonic scallops (Ciocco et al., Chapter 26) and the Alaska fishery for weathervane scallops (Turk 2000). Aerial surveys are an option in the case of small-scale or artisanal fisheries where the effort, spread over extensive and intricate coastal regions, is difficult to monitor with the methods listed above.

Fishing intensity (defined as the amount of effort *applied per unit of area* over a given time interval) can be assessed indirectly in the case of trawls and dredges, since the gear leaves readable marks on the substrate and on shelled organisms encountered by the gear but not caught. Depending on the nature of the substrate, dredging or trawling intensity can be mapped with the help of side-scan sonar. Encounters with the gear are often recorded in the shell of surviving scallops as *shock marks*, which provides a way to map fishing intensity with fine spatial resolution (Caddy 1972, 1989a; Naidu 1988).

14.4.2.2 Effort units

The operational time of a fishing unit can be partitioned into a number of components (Hilborn and Walters 1992): *travel time* to fishing grounds (which can take up a considerable fraction of the time budget), *search time, setting time* (spent in shooting and hauling the gear), *fishing/catching time* (the period during which the gear effectively operates), and *processing time* (handling of the catch on the deck). Detailed analysis of these components may prove informative. The ratio between handling and search times, for example, has been used as an index to describe stages in the development of a fishery (Caddy 1979a). Handling time should be relatively more important at early stages, whereas search time is expected to increase as the resource is depleted. In the case of dredgers and trawlers a unit of effort (e.g., a fishing day) often includes several dredge or trawl hauls. Catch and handling time per haul (which includes time spent culling and processing the catch on the deck) are positively correlated with abundance (Caddy 1975, 1979a, 1989b). Search time has decreased consistently in many scallop fisheries during recent years as a result of improved navigational and hydroacoustic equipment. Search efficiency is also related to the power of the engines; Fifas (1991) incorporated this relationship in his analysis of catchability in the St. Brieuc Bay (France) scallop fishery.

The choice of units to express effort is more often than not determined by the nature of the data available. Sources of variation increase for aggregated effort units like fishing days or trips, vs. - for example - hours dredged. However, even coarse effort information (for example fleet size) can be useful in the absence of more refined data. Caddy (1979a) found good correlation between estimated annual fishing mortality and fleet size for the Digby fishery for the period 1941–1951 (effort data and estimates of fishing mortality rate from Dickie 1955). This permitted the estimation of fishing mortality rate using fleet size data for other years, assuming that the fishing power of the boats was constant. Estimates can be improved with additional information, as Caddy did with the duration of the fishing season, and the proportion of "fishable days" (when weather allows fishing) during the open season.

14.4.2.3 Components of fishing mortality under homogeneity assumptions

Under homogeneity assumptions, fishing mortality is decomposed into the number of effort units applied per unit of time and the *catchability*, approximately the fraction of the population removed by a unit of effort. The catchability can be calculated from the area fished by a unit of effort (e.g., the *area swept* by a dredge haul), the total area occupied by the stock, and the efficiency of the gear (the fraction of the scallops in the path of the gear that are caught). The homogeneity assumptions, however, do not hold in real scallop fisheries (Caddy 1975, 1979b, 1989a). The spatial allocation of fishing effort is random or haphazard only in controlled fishing experiments (e.g., Iribarne et al. 1991), and organisms are never randomly distributed at the operating spatial scale of the fishing process (random patterns can occur at smaller scales, see Section 14.5.3.3). Besides problems derived from nonrandom effort allocation, catchability varies in relation to an assortment of variables that can be roughly classified into:

i) *Technological factors*. Fishing power of effort units depend on many factors such as boat design, quality of the equipment, type of gear, fisher's experience, etc. Trawling speed, for example, affects catchability in the fisheries for highly mobile *Amusium* species (Joll and Caputi 1995). Gear efficiency is discussed in Section 14.4.6.

ii) *Environmental factors* affecting the performance of fishing units or fishing gear: bottom quality, weather conditions, water visibility, depth, etc. Catchability increases after the first passes with the Australian dredge, a result of flattening of the substrate (Young and Martin 1989; Currie and Parry 1999).

iii) *Biological factors*, mostly related with behavioural aspects related to gear avoidance.

The effect of all these factors may vary over time, at various scales:

i) *Within-year fluctuations* can be associated with environmental conditions affecting fishing power (e.g., seasonality of weather conditions), or with biological factors.

Fluctuations in vulnerability due to behavioural cycles range in frequency from diel to annual.

ii) *Long-term trends* often reflect the evolution of gear design (Section 14.6.1). Contractions and expansions of the geographic range of the stock (or of the area harvested) can also have a substantial effect on catchability.

14.4.2.4 Relative fishing power and standardisation of effort

Stock assessment requires aggregate estimates of fishing effort. Since the fishing force is always heterogeneous, the aggregation of effort requires some form of standardisation based on the relative fishing power (*RFP*) of the fishing units. The *RFP* of the fishing units can be assessed experimentally, or through the analysis of *CPUE* data. Experimental studies are often impractical, and have been used mostly to assess one particular component of *RFP*: gear performance (Section 14.4.6).

CPUE analysis requires stratification of the data by meaningful categories or factors, such as statistical area, time (year, season, month), time of the day (day, night), and type of fishing unit, defined according to various attributes. The latter may include fishing gear (when several fishing styles coexist in the fishery), size of the vessels (measured in tonnage or length) or power of their engines (measured in HP, often used in dredge and trawl fisheries), skill and experience (important in the case of commercial divers). The intuitively appealing model elaborated by Robson (1966) and Gavaris (1980) assumes that the effects of various sources of variation are *multiplicative*. If, for example, abundance increases from one year to the next (year factor), the *CPUE* for all vessel or gear types may reflect the change similarly and increase by a fixed proportion (as opposed to a fixed amount). This model can be fitted to stratified *CPUE* data with the help of any statistical package that handles multiple regression. Alternatively, the problem can be approached through GLIM (Generalised Linear Models). Sources of variation in *CPUE* may interact with each other. For example, the *RFP* of different boats may vary over space: some crews may have a better knowledge of certain regions, the *RFP* of small and large vessels may rank reversibly in inshore and offshore grounds, etc. Interaction terms can be added to the multiplicative model to account for these effects. One problem, however, is often insurmountable: the unbalanced distribution of the information, particularly when there are many empty cells in the data matrix (combinations of factor levels for which there is no information).

14.4.3 Spatial Patterns of Effort Allocation

The dynamic implications of the spatial structure of the population and effort allocation were first explored formally by Caddy (1975) by means of a simulation model of the Georges Bank scallop fishery. Overall, fishing mortality estimated on the basis of area swept (homogeneity assumptions) were shown to underestimate grossly the actual mortality, due to the contagious nature of fishing effort. When homogeneity assumptions do not hold, the spatial allocation of fishing effort is controlled by a hierarchy of factors:

i) *The spatial distribution of the resource.*

ii) *Location-specific effects*, generally related to accessibility and operational costs: distance from port, exposure to weather, depth, type of bottom, etc.

iii) *Performance of the fishing unit at a given location*, which may or may not vary significantly and consistently among locations.

Simply stated, the first condition for fishing to occur at a location is that there is something substantial to be fished, the second that the location is accessible, and the third that the fishing gear has the appropriate power at that location.

Fishers travel from port to fishing beds, search for fishable concentrations, probe along the way, deploy the gear, fish, process the catch, re-deploy the gear, etc. Elaborate search behaviour, density thresholds at which divers move to a new location, and sequential depletion of beds are easiest to observe in diving fisheries (Orensanz 1986), but also in trawl and dredge fisheries when detailed data are available. Turk (2000) investigated the behaviour of individual vessels in the Gulf of Alaska using a GIS platform and logbook information, providing the most detailed study case available on the behaviour of a scallop fleet. As an illustration, Figures 14.6 and 14.7 show serial effort allocation to six beds near Kodiak Island, plus two periods of search. Effort shifted to a new bed after a *CPUE* threshold of *ca.* 2,000 kg km^{-2} was reached.

Various sequential patterns of effort allocation have been mentioned for other scallop fisheries:

i) Expansion of the area fished, typical of developing fisheries in which effort extends progressively to beds more distant from port, a case nicely illustrated by Brand et al. (1991) with the scallop fisheries of the North Irish Sea.

ii) Contraction of the area fished. Boats may be scattered at the opening of the fishery, and aggregated once dense beds are located (Fairbridge 1953).

iii) Serial depletion and shifts, described above for the Alaska fishery, have been reported in many other cases (Fairbridge 1953; Dickie 1955; Gibson 1956; Dredge 1986; Ansell et al. 1991; Briggs 1991). Renzoni (1991) described this pattern for the Italian scallop fishery: "an exploitable bank is discovered and the word spreads among the fishermen. The bank is exploited completely or until it is no longer 'economic' to exploit the area".

iv) Spontaneous rotation was described by Blake and Moyer (1991, p. 899) for the calico scallop fishery. Beds off Cape San Blas (Northern Gulf of Mexico) are harvested only every 2–3 years, after other stocks have been depleted.

Problems of interest in the analysis of effort allocation involve the identification of the components of suitability and other determinants of fisher's behaviour. Observed patterns can be compared with predictions of simple models. Caddy (1975), for example, proposed a model of *proportional effort allocation* (PEA), under which the intensity of fishing effort over a season is proportional to abundance at the beginning of the fishery (usually estimated from initial *CPUE*). Observed departures from predictions made by the PEA model may result from various factors:

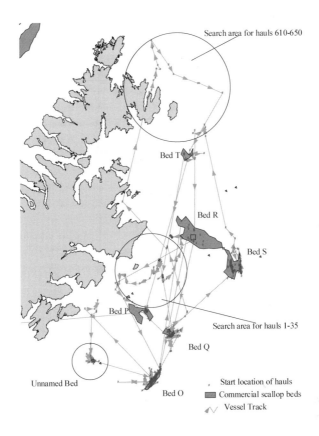

Figure 14.6. Behaviour of an individual fishing vessel in the Alaska weathervane scallop fishery in 1993 (see also Figure 14.8). The area shown corresponds to the north of Kodiak Island, in the Gulf of Alaska. (From Turk 2000).

   i)   If there is a *density threshold* below which fishing stops.

   ii)   The relation between *CPUE* and abundance is non-linear (Section 14.5.2.3).

   iii)   Depletion proceeds serially, bed after bed.

   iv)   Suitability has components other than *CPUE*.   Location-specific factors contributing to suitability include:

802

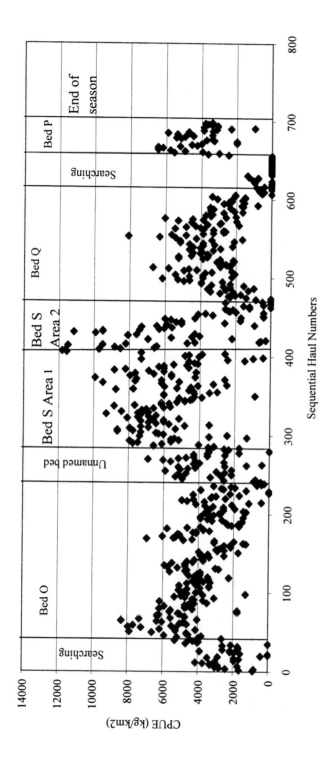

Figure 14.7. *CPUE* for the sequence of hauls made along the trajectory shown in Figure 14.7. The vessel moved often during a two-month period. As the *CPUE* declined on a given bed, the vessel usually began searching for a new bed or moved to another known bed. Usually mean *CPUE* dropped to less than 2,000 kg km^{-2} before the vessel abandoned each area. (From Turk 2000).

a) Hazard factors, for example related to depth in the case of commercial divers, to exposure in the case of small vessels, and to bottom texture in the case of dredgers and trawlers.

b) Differential costs of effort units, most often associated with distance from port.

c) Uneven value of the catch from different beds (related to the bed-specific composition of the catch by sizes, quality grading, etc).

v) *Omniscience*: Simple models assume that fishers have *a priori* knowledge about the distribution of suitability. Such information is, however, incomplete. Uncertainty about density and location-specific factors leads to time expenditures in exploration and probing. Fishermen give some weight to experience from previous fishing seasons ("*memory*"), often "survey" the fishing grounds at the beginning of the season, and frequently alternate fishing with probing and exploration. Memory can be effectual in the short or long term. Caddy (1975) noticed that in the Georges Bank dredge scallop fishery the areas most heavily fished in previous years received more effort than expected under proportional allocation, reflecting the "inertial" effect of historical effort distribution.

vi) *Interactions* between fishing or gear units, including two forms with opposite signs: *interference* and *cooperation*.

Walters et al. (1993) extended Caddy's PEA model to account for location-specific factors using an "attractiveness index", and for past experience using a "memory weighting factor."

14.4.4 The Depletion Process

Depletion, the reduction of stock size through the action of fishing, is the most immediate consequence of the fishing process. In the case of serial depletion, fishers target the densest patches available first, shifting to the next in the rank as the first are depleted. As a result, *CPUE* is not a good index of stock abundance (Section 14.5.2.3). In a simplified scenario (Orensanz et al. 1998), the relation between *CPUE* and abundance depends on the concentration profile of the stock (Section 14.3.4.2; Fig. 14.3) and the gear efficiency (Section 14.4.6). Three patterns are possible (Hilborn and Walters 1992; Orensanz et al. 1998):

i) *Proportionality*, shown only in the case of concentration Profile Type B.

ii) *Hyperdepletion*, in which *CPUE* declines faster than abundance. This is the case for Profile Type A: *CPUE* declines rapidly at the beginning of the season as the few high-density quadrats are depleted.

iii) *Hyperstability*, in which *CPUE* is stable or declines more slowly than abundance. This is the case for Profile Type E (constant concentration, strictly regular spatial pattern), in which *CPUE* does not change until it suddenly drops to zero when the entire stock has been depleted. Profile Type D leads to a less extreme case of hyperstability.

804

The interpretation of depletion data requires caution. In the weathervane scallop fishery of Alaska, management areas are closed once a catch quota is reached. Figure 14.8 shows the trend of daily *CPUE* during a five-month fishing season. Based on their previous knowledge, fishers tend to target the densest management areas first, shifting to lower-density beds after each bed is closed. Notice that *CPUE* did not change much while each area was harvested, presumably due to hyperstability (serial depletion within each management area). If the complete series was examined without discriminating the areas, the result would be an *apparent* decline of *CPUE* due to the sequence in which the areas were harvested. Application of fishing success methods (Section 14.5.2.3) to the aggregate series would result in statistically significant parameters (initial biomass, fishing mortality), but these would be meaningless.

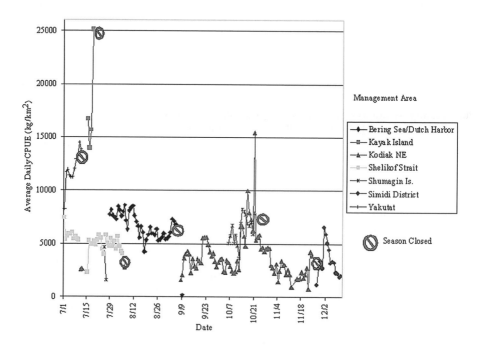

Figure 14.8. Behaviour of the fleet in the weathervane scallop fishery of Alaska during the summer and fall of 1993. When management areas were closed, the scallop fleet generally moved as a group from the highest producing beds to the next highest producing beds. The Yakutat, Kayak Island, and Kodiak NE area beds were closed when the upper bound of the guideline harvest range was reached, whereas the Shelikof Strait area was closed due to a significant drop in *CPUE*. All other areas were closed when the crab bycatch cap was reached. (From Turk 2000).

14.4.5 Vulnerability and Selectivity

Not all the individual scallops in a stock are equally vulnerable to fishing. We just addressed one case: scallops that are close to other scallops (i.e., live at high densities) are most likely to be caught. *Vulnerability* is determined by two families of factors, plus their interactions:

i) *Intrinsic attributes* of the individuals, such as size, age, behaviour, phenotype or condition (e.g., physiological or reproductive condition, load of parasites and epibionts, etc).

ii) *Contingencies* or circumstantial factors such as location, depth, local density, availability of refuges, water visibility, weather, etc.

Vulnerability of scallops varies through time. The level of activity is influenced by temperature, visibility, and fatigue. Temperature-dependent activity probably underlies the seasonal trends in catch-per-haul (highest in winter) observed in some fisheries (Dow 1962, 1969). Vulnerability can increase at night, due to reduced visibility of the gear (Olsen 1955), and after repeated dredging, due to fatigue of the scallops (Olsen 1953, 1955; Fairbridge 1953) and/or levelling off of the substrate by the gear (Hardy 1981: pp. 29–30). Swimming capability increases with age/size in *Amusium*, reducing the vulnerability of older/larger individuals (Joll and Caputi 1995).

Vulnerability of individual scallops is determined by:

i) Fishers' decisions about where to fish (pertaining mostly to contingent factors).

ii) Avoidance of the gear by individuals with certain attributes (e.g., largest scallops swimming away from the path of an approaching trawl).

iii) Retention of individuals that have entered the gear, often related to the mesh size of trawls and dredges, or that are in the path of a diver, mostly related to size.

The term *selectivity* refers to the relative vulnerability as a function of some intrinsic attribute, generally size. We explicitly distinguish *global selectivity*, (resulting from the composite effects of (i) through (iii)), from *gear selectivity* (resulting from (ii) and (iii)), typically assessed experimentally (Section 14.4.6). The hierarchical components of selectivity that are actually assessed depend on the nature of the data. Most of the literature on selectivity is narrowly focused on modelling *gear* selectivity (Section 14.4.6), although factors not related to the gear can be more important in the case of scallop stocks (e.g., consider the case of "blending", Section 14.6.2.1). Fifas (1991) described an interesting case of global selectivity for the Bay of St. Brieuc (France). The fishery targets two age groups (2+ and 3+). Before a reinforced gear design was introduced, scallops had a refuge in rocky areas. Effort in the flats targeted concentrations of age-class 2+, but was random relative to the distribution of age-class 3+.

806

14.4.6 Gear Performance

When fishing power is assessed experimentally, results apply only to (i) the gear utilised, (ii) the conditions prevalent at the experimental site, and (iii) the time when the experiment was conducted. "Gear" is defined here in a broad sense, eventually including a vessel and its crew (e.g., Dare et al. 1994). Components relative to effort allocation (for example ability to locate dense patches) are not considered, or included only at a very restricted spatial scale. Estimated parameters may not represent gear performance of commercial operations, as efficiency and selection are affected by the duration of the tows, towing speed, etc., which are usually standardised in experiments. Problems relative to the specificity of conditions under which experiments are conducted are well illustrated by the systematic differences between identically designed dredges or trawls towed simultaneously on port and starboard of the vessel (DuPaul et al. 1988a, b), well known to commercial fishermen. Experimental results on gear performance are most sensibly applied to the assessment of the gear and sampling units used in sampling surveys of abundance (for example the efficiency of a survey trawl or dredge). The experimental assessment of area fished, gear efficiency and selectivity are discussed below.

1. Area fished. Area swept by a dredge or trawl haul is calculated directly from the length of the haul and the width of the gear's mouth. By contrast, it is very difficult to define the area effectively fished by a commercial diver, which in some cases allocate effort selectively even at the smallest spatial scales.

2. Efficiency. Efficiency can be affected by the nature of the bottom, operation conditions, gear leaping, gear saturation, skill of the crew, size and behaviour of the scallops, etc. (Baird and Gibson 1956; Chapman et al. 1977; Caddy 1979b; Shafee 1979; Currie and Parry 1999; etc.). *CPUE* increases during April-June in the Lingayen Gulf (Philippines) fishery for *Amusium pleuronectes* (del Norte et al. 1988), perhaps reflecting the influence of calm waters on gear performance. Dredges can saturate if they are towed to fullness with scallops or debris (Bourne 1965; Sanders 1966; Millar and Naidu 1991), or if the effect of "bulldozing" of sediments in front of the frame is substantial (Serchuk and Smolowitz 1980; Caddy 1989a; Iribarne et al. 1991). Efficiency is also affected by the intrinsic factors affecting vulnerability, and their seasonal variation (Section 14.4.5).

Methods that have been utilised to estimate the efficiency of scallop trawls and dredges are summarised in Table 14.2. Values estimated for dredges are often below 20%: 5-12% in the Digby fishery for *Placopecten magellanicus* (Dickie 1955), 13-25% in the British fishery for *Pecten maximus* (Chapman et al. 1977; Mason et al. 1979, 1982), 7-28% for *Chlamys varia* in the Bay of Brest (Shafee 1979), 15-21% for *Aequipecten tehuelchus* in San Matías Gulf (Iribarne et al. 1991), 44% for the "rapido" scallop dredge used in the Adriatic (Hall-Spencer et al. 1999), etc. Efficiency of all dredges vary greatly between different bottom types. A highly efficient French dredge ("Erquy" type) reached 30% on hard bottoms and 67% on soft bottoms (Buestel et al. 1985). Estimated efficiency of the Australian dredge was only 12% in the hard bottoms of Bass Strait (McLoughlin et

Table 14.2

Methods that have been utilised to estimate the efficiency of scallop trawls and dredges.

Method	Comments	Scallop examples
Concurrent estimation of density (or density known), CPUE, and area fished by a unit of effort		
Overlapping measurements.	Density is assessed before or after fishing in the site swept by the gear.	Buestel et al. (1985); Hall-Spencer et al. (1999)
Non-overlapping measurements, density estimated.	Non-overlapping hauls allocated randomly within each experimental site. Density in background, or background and dredged areas, can be estimated through a sample of quadrats, underwater TV, or a concurrent mark-recapture experiment.	Dickie (1955), Gruffydd (1972), Caddy (1968, Shafee (1979), Mason et al. (1982); Guguère and Brulotte (1994)
Non-overlapping measurements, density known.	Density manipulated either by placing animals in an unpopulated experimental site or by releasing marked specimens in an area where there is already some background density. In the latter case the experiment can be designed to yield a simultaneous estimate of the background density.	Dickie (1955)
Scallops "planted" randomly or regularly over the entire site.	Can be logistically complicated because large numbers of individuals may have to be released to obtain significant recaptures.	McLoughlin et al. (1991)
Scallops "planted" along parallel stripes, stretching the entire width of a rectangular experimental plot.	Fishing is conducted along parallel non-overlapping trajectories (e.g., hauls) stretching the entire length of the plot. In this way every haul encounters the same average density of released individuals, but it is not necessary to "plant" the entire plot.	Schneider (1992)
Depletion experiments		
Random or haphazard.	Compared to other methods, this setup has some advantages: it is logistically easier to implement in relatively large experimental plots, and (partly for that reason) captures more realistically the *modus operandi* of commercial fishing gear.	Iribarne et al. (1991); Currie and Parry (1999)
Systematic, covering the whole experimental plot at each sampling date.	The method is constrained to a relatively small plot size (so that the plot can be swept quickly), accurate positioning (so that there is no overlap between adjacent sweeps), and the assumption of no small-scale movements. The slope of the Leslie or DeLury equations is an estimate of efficiency.	Joll and Penn (1990)

al. 1991), but reached 38–44% on firm sand bottoms and 51–56% on soft, flat muddy bottoms of Port Phillip Bay (Currie and Parry 1999).

3. Gear selectivity. Although gear efficiency can vary in relation to many intrinsic attributes (Section 14.4.5), size is the one commonly considered. Interest in assessing the size-selectivity of fishing gear stems from the fact that "size-limits" are the most widespread tactic in the management of scallop fisheries (Section 14.6.4.1). The formal assessment of gear size-selection has received most attention in the case of trawls and dredges, largely because it is assumed that selectivity can be controlled by regulating mesh size.

The literature on scallop gear selectivity is a bit intricate, because different things are estimated with different experimental set ups. We introduce a minimal formalism to facilitate discussion. The selectivity of dredges and trawls can be partitioned into components related to capture (S^C_L) and retention (S^R_L). If the gear is size-selective, then there is some size or size interval for which efficiency (f_L) is maximal (f_{max}). Then,

$$f_L = f_{max}\ S^{C^*}_L\ S^{R^*}_L$$

$$S^C_L = \frac{N(L)_{entering\ gear}}{N(L)_{path\ of\ gear}}$$

$$S^R_L = \frac{N(L)_{retained\ by\ gear}}{N(L)_{entering\ gear}}$$

where $N(L)$ denotes number of scallops in the L size category, C is capture, and R retention; "*" indicates that figures are normalised so that they are within the range [0,1]. The retention component is generally related to the mesh size of the bag. Ideally, for any given mesh size there is a lower size threshold below which a negligible fraction is retained, and an upper size threshold above which virtually all individuals are retained. Methods used to estimate the selectivity of scallop trawls and dredges are summarised in Table 14.3. Selection curves for dredges and trawls take the form of sigmoid ogives (Dickie 1955; Caddy 1972; Naidu et al. 1982; Millar and Naidu 1991; Fifas and Berthou 1992, 1999). There is an extensive literature dealing with models of gear selectivity.

14.4.7 Incidental Fishing Mortality and Sub-lethal Damage

Incidental fishing mortality is very significant in many scallop fisheries, particularly dredging, because of substantial disturbance of the substrate and the associated community (Section 14.6.3.3), as well as damage to individuals left behind by the gear or culled on board and returned to the sea. This results in immediate incidental mortality, as well as various degrees of sub-lethal damage, which can eventually increase "natural" mortality.

Table 14.3

Methods used to estimate the different components of the selectivity of scallop trawls and dredges. See text (Section 3.6) for notation.

Type of observation	Component assessed	Assumptions and/or problems	Examples
Catch at size and numbers of scallops at size left in the area swept	f_L	Assumes no gear avoidance through lateral escape; poor assumption in the case of highly mobile species	McLoughlin et al. (1991)
Catch at size; SFD and density of the unfished population known (seeding)	f_L	Estimation of f_{max} assumes that catchability is the same for sown and undisturbed individuals	Dickie (1955)
Mark-recapture experiments (essentially similar to the preceding)	f_L	Estimation of f_{max} assumes that catchability is the same for tagged and untagged individuals	Fifas and Berthou (1992, 1999); Iribarne et al. (1991); McLoughlin et al. (1993)
SFDs of the catch and of a *sample* of scallops left in the area swept	$S^C_L S^R_L$		
SFDs of the catch and of a sample from the unfished population	$S^C_L S^R_L$		
Double-bag end	S^R_L		Caddy (1972b)
Simultaneous tows with different mesh	S^{R*}_L	Same caveats as above regarding capture selectivity.	Dickie (1955); Serchuk and Smolowitz (1980); Millar and Naidu (1991); Naidu et al., (1992); Giguère and Brulotte (1994)
Samples obtained with gear rigged with different meshes; selectivity unknown for both	Only the relative efficiency of one gear relative to the other can be assessed	Results are contingent upon the composition of the population (which is unknown)	DuPaul et al. (1988a, 1988b)

(1) Factors involved. Damaging impacts include:

i) Mechanical shock, including squashing of individuals run over by dredges or passing through the mesh, damage of undersized animals discarded on board, etc. Incidental damage depends on the type of substrate. In the case of the Australian dredge it is highest in the hard bottoms of Bass Strait (60% of the catch smashed; McLoughlin et al. 1993), but very low in the sandy and muddy bottoms of Port Phillip (less than 1% of the catch damaged; Currie and Parry 1999). Shepard and Auster (1991) observed a comparable pattern in an experimental study of the commercial "rock drag" used in Maine: 25.5% of incidental damage on rocky bottoms vs. 7.7% on sand substrate.

ii) Prolonged air exposure.

iii) Turbidity and siltation. Direct observations of dredge operations showed a dense layer of liquid silt drifting close to the bottom and the scallops containing large quantities of sediment (Caddy 1989a). Using "crawl" velocity of pieces of gill tissue as an indicator of response to suspended silt, Stevens (1987) concluded that the juveniles of *Pecten novaezelandiae* may be less tolerant to suspended silt than adults, as indicated by Imai (1971) for *Patinopecten yessoensis*. High concentration and low size of suspended silt particles resulted in reduced piece crawl velocity and increased scallop mortality. Turbidity may also deplete dissolved oxygen in the water, reducing ciliar activity. The effect is more pronounced in juveniles because they cannot respire anaerobically by means of the metabolism of the crystalline style (Stevens 1987). Dredging has been prohibited in some cases when silt-sensitive spat are present on the bottom (Imai 1971).

iv) Increased exposure to predators. Active muscle adduction resulting from the escape response elicited by the gear (Dickie 1955; Caddy 1968, 1973) leads to fatigue, reducing the capability to escape from predators. Scavengers and predators have been seen on the path left by the gear feeding on damaged scallops (Caddy 1968; Chapman et al. 1977; Ramsay et al. 1998).

v) Mortality due to contamination by *Vibrio* has been was associated with the very large numbers of decomposing scallops left on the sea bed after intensive dredging (McLoughlin et al. 1991).

(2) Assessment of incidental mortality requires identification of the factors and stages of the fishing process involved, and, importantly, whether short-term and/or delayed effects are under consideration. Several approaches are outlined below:

i) *Direct observation of the area swept.* Caddy (1973) used a submersible to observe lethal damage caused by dredges to sea scallops left along the tract over rough bottoms; estimated incidental mortality, 13% to 17%, was comparable to the efficiency of the dredge. By contrast, in a similar study conducted on smooth sand/mud bottoms Murawski and Serchuk (1989) observed negligible dredge-induced incidental mortality (less than 5%) in the same species. This illustrates the importance of local variation due to the nature of the substrate.

ii) *Modelling the dynamics of live bivalves and empty shells during a depletion episode.* McLoughlin et al. (1991) investigated the incidental mortality caused by the

Australian scallop dredge during a fishing season in the Bass Strait area (Tasmania). They classified the harvested population into "live", "damaged" and "dead" (empty shell) components, and recorded the catch and *CPUE* by component for a specific fishing bed. Assuming equal catchability of all fractions, they estimated that only 12% to 22% of the initial stock in the Banks Strait ground was landed as catch, the rest being wasted as a result of dredging-induced mortality, either direct (mechanical damage) or indirect (increase of environmental stress, disease and/or predation due to the decay of dead individuals). Modelling exercises like this require some *ad-hoc* ingenuity to make the best possible use of the information available.

iii) *Correlation between apparent natural mortality and fishing effort.* Naidu (1988, 1992) suggested using estimates of apparent natural mortality based on ratios of cluckers-to-live scallops to assess incidental fishing mortality of Icelandic scallops around Newfoundland (Canada). Apparent natural mortality was higher in harvested compared to unfished beds.

iv) *Controlled survival experiments,* consisting of comparing survival of undamaged scallops (controls) with that of, depending upon the aspect of incidental mortality;

a) Individuals recovered from the area swept by the gear.

b) Individuals from the catch that would be normally discarded (culling mortality).

Experimental specimens can be kept in tanks on board, laboratory aquaria (Gruffydd 1972), or cages on the sea bottom (Murawski and Serchuk 1989). The controls can involve animals collected with non-damaging methods from unfished areas, or even individuals from the catch judged "undamaged". Gruffydd (1972), for example, kept scallops caught with a Manx dredge in aquaria for one month, and found that the mortality of the ones that were severely damaged (broken hinge or mantle disattached) was 5 to 13 times higher than that of the undamaged group. Even animals that are apparently "undamaged", however, can experience delayed effects and suffer a higher mortality than undisturbed individuals. Murawski and Serchuk (1989) observed a high survival rate (91% over several days) after placing sea scallops caught with a dredge into cages on bottom. This figure may be an underestimate because of possible effects of caging, which were not controlled with undisturbed scallops. This result strongly suggests that culling of undersized scallops from commercial dredge catches is a prudent conservation measure.

14.5 ASSESSMENT OF ABUNDANCE AND ITS SPATIAL DISTRIBUTION

14.5.1 Macroscale

The most important objective of scallop stock assessment at the macroscale level is to map the grounds and beds (usually subpopulations of a metapopulation), as well as other variables that are significant for habitat classification, survey design, modelling of the fishing process, etc. Caddy and Garcia (1986) presented an overview of "thematic mapping" applications to fisheries, anticipating issues and directions that became tractable after Geographic Information Systems (GIS) and spatial statistics software became widely available (e.g., Stolyarenko 1995). The volume of geo-referenced information has exploded over the last decade. New techniques for the continuous recording of bottom properties and resources have been developed, and techniques involving underwater TV and small submersibles have been improved (Goshima and Fujiwara 1994). Recent applications to the survey of benthic resources and their habitats include acoustic bottom-type classification systems (Greenstreet et al. 1997; Kaiser et al. 1998; Hamilton et al. 1999), high-resolution sidescan sonar (McRea et al. 1999), and underwater laser line scan technology (Traccy et al. 1998). While some (like the latter) are still too costly, hydroacoustic bottom-type classification technology has been incorporated by industrial fishing vessels on both coasts of the US. In a typical example of the macro-scale level of assessment, Smith and Greenhawk (1998) used sub-bottom profiling equipment, side-scan sonar, underwater video and seafloor classification systems to assess and map oyster habitat in Chesapeake Bay. Information was geo-referenced with GPS and integrated in a GIS.

Caddy and Caroci (1999) applied a GIS system to the analysis of spatial effort allocation, which illustrates the importance of modelling processes at the appropriate scale, and the value of geo-referenced information. The study by Turk (2000) described earlier on the mapping of fishing beds of *Patinopecten caurinus* in Alaska (Section 14.4.3) is another example.

14.5.2 Mesoscale

Methods for the estimation of the abundance of a scallop bed or ground can be divided in two broad categories: (a) those based on direct sampling (survey-sampling, line-transects, mark-recapture), and (b) those based on modelling the depletion process. Information on the spatial distribution of abundance of sessile organisms is often as important as (or even more important than) knowledge of total abundance. A main difference between various methods is that some (e.g., survey-samples) can be used to assess total abundance and spatial pattern, while others address only the first aspect (e.g., simple-random-sample survey designs, mark-recapture and depletion-based methods). The spatial distribution of abundance in scallop populations has important implications for modelling and management.

14.5.2.1 Estimation of aggregated abundance: methods based on sampling

(1) Types of data used in scallop surveys. The best data are perhaps quadrat samples obtained by divers (Orensanz 1986; Ansell et al. 1991; Noel 1997; Young and Martin 1989), although this alternative can be impracticable due to logistical constraints (e.g., depth) or cost. In those cases, the option is dredge or trawl surveys, which may require a concurrent analysis of gear efficiency and selectivity (Section 14.4.6). Photo-quadrats may be valuable in many cases (e.g., Goshima and Fujiwara 1994), but are inefficient for "recessing" species or when the bottom is covered by algae, eelgrass, etc. Underwater TV and other devices used to make continuous observations along transects or stripes are discussed later in this section.

(2) Sampling-surveys: design- vs model-based approaches. Smith and Mohn (1987) distinguished two basic types of approaches that have been followed in the assessment of scallop stocks by direct sampling surveys:

i) *"Design-based inference"*, in which estimation of stock abundance is based on some randomised sampling design. The population is assumed to be composed of a finite number of sample units, and only the process of selection among those units is assumed random. Optimally, a sampling scheme is designed *a priori* on the basis of available information, so as to minimise the variance of the abundance estimator given logistic constraints. The most common scheme is stratified-random (Smith and Robert 1998), where strata are delimited on the basis of some auxiliary variable (depth, *CPUE*, bottom type, etc.) correlated with scallop density. Research surveys based on stratified-random schemes have been conducted annually by Canada and USA to estimate standing stock of *Placopecten magellanicus* of the Georges Bank (Mohn et al. 1987; Serchuk and Wigley 1987; Smith and Robert 1998). In the USA survey, strata are defined on the basis of depth contours, and number of tows are allotted to strata in proportion to stratum area. By contrast, in Canada contours of commercial *CPUE* are used, and sampling effort is concentrated in areas of high commercial *CPUE*.

ii) *"Model-based inference"*, in which a probability model of the distribution of abundance is fitted to the survey data. Abundance is estimated as a function of the estimated model parameters. This approach results in more precise estimators than those based on a randomised design. The caveat is that the estimates relay on the specific model used, whose selection is usually *ad hoc*. The simplest design-based approach is to fit a model to the frequency distribution of catch per tow data, which is commonly highly skewed. Several skewed probability distributions exist that can be used to fit *CPUE* data (for a review, see Elliott 1977; Smith 1988). A more complex approach is to use models that explicitly account for the spatial distribution of the variable under study. Catch per tow data are often more similar between close samples than between tows farther apart from each other. Geostatistical methods explicitly use the structure of spatial autocorrelation of the data to construct the estimators. Other models, e.g., contouring and Delaunay triangles, are implicit (Heald 1979, cited by Young and Martin 1989, p. 629; Robert et al. 1990, 1994).

Design and model based (kriging) approaches were discussed by Warren (1998); he used a scallop data set for illustration. While some proponents of model-based approaches have pointed to the fact that they produce estimates with smaller standard errors than design-based methods, Warren (1998) emphasised that variances obtained by the two methods, strictly speaking, cannot be compared, and that application of model-based methods requires subjective judgement. From the perspective of the assessment and management of sedentary resources; however, geostatistics and other model-based approaches are often a choice not because of the reduction in the variances, but because they are most suitable for mapping abundance, often more significant that the estimation of abundance itself.

The spatial allocation of survey samples defines the sampling design. Systematic grids are preferred in the case of model-based approaches (Conan 1985; Nicolajsen and Conan 1987), and simple or stratified random schemes in the case of the design-based approach. Annual surveys of *Pecten fumatus* in Port Phillip Bay (Australia) changed from a systematic to a stratified-random design in 1982 (Young and Martin 1989). Systematic grids (long disregarded because of the traditional dominance of design-based approaches) have some intuitive merits: they avoid the cost of redundant information collected when stations end up being close to each other, provide a more even coverage of the fishing bed, and are logistically easier to conduct.

Adaptive sampling (Thompson 1992; Thompson and Seber 1996), a promising development among design-based approaches, is particularly well suited for the assessment of contagiously distributed stocks, especially when information about the location of aggregations is lacking. Stolyarenko (1995), Ciocco et al. (1997) and Bechtol and Bue (1998) applied adaptive sampling designs to the assessment of scallop stocks.

(3) Line transects and stripe counts. Continuous recording of scallop density along transects has been done using submersibles (Caddy 1970, 1976; Cummins 1971), and television and photographic cameras towed on a sledge or dredge (McIntyre 1956; Cummins 1971; Holme and Barrett 1977; Franklin et al. 1980; Merrien 1980; Mason et al. 1982). Problems related to the application of these methodologies are discussed in detail in Holme (1984). The AQUAREVE ("Application QUAntitative d'un Robot Epibenthique avec controle Video d'Echantillonnage"; Thouzeau and Hily 1986) has been successfully utilised in the assessment of *Pecten maximus* in the Bay of St. Brieuc, France (Thouzeau and Lehay 1988). Efficiency has been reported to be higher than 80% under appropriate meteorological conditions (Thouzeau and Lehay 1988). Giguère and Brulotte (1994) compared the sampling performance of video and dredges with lined and unlined buckets; they concluded that video surveys are the best method for estimating size frequency distribution and density, but are time costly.

(4) Mark-recapture methods. Abundance estimation through mark-recapture methods rests on a number of assumptions, which are particularly restrictive in the case of sessile animals because (i) marked and unmarked individuals do not mix randomly after release, and (ii) effort allocation is unlikely to be random relative to the proportion of marked animals. Dickie (1955) conducted some mark-recapture experiments to estimate abundance of *Placopecten magellanicus* in the Bay of Fundy. He noticed that the non-random pattern of effort allocation, in conjunction with high mortality rate of marked animals and low reporting rates of marks in the catch (especially for empty valves), resulted in overestimation of scallop abundance. Small-scale marking experiments can be valuable for assessing movements (e.g., Stokesbury and Himmelman 1996), growth and mortality parameters (e.g., Dredge 1988) or vulnerability to fishing gear (Section 14.4.6).

14.5.2.2 Mapping

Different methods have been used for mapping scallop abundance based on research survey and/or geo-referenced catch and effort data:

(1) Designed-based methods: stratification. Standard stratified estimates yield the simplest type of "map": all unsampled locations in the strata are assigned the strata mean. Smith and Robert (1998) discussed stratified designs using the *Placopecten magellanicus* stock from Georges Bank as an example. The design of the annual Georges Bank scallop survey is based on commercial *CPUE* during the previous months (Jamieson et al. 1981; Robert and Jamieson 1986). Several studies, however, have shown that strata and (particularly) allocation based on *CPUE* do not provide in this case gains in precision as compared to a simple random design (Mohn et al. 1987; Smith and Mohn 1987; Smith and Robert 1998). Smith and Robert (1998) proposed, alternatively, a two-tier approach in which *CPUE*-based strata are partitioned into sediment-based strata; the latter were (as a first approach) based on coarse sedimentary-ecological units defined by Thouzeau et al. (1991). As an improvement, sediment-based strata could be defined in the course of the survey using hydroacoustic bottom-type classification systems (Simon et al. 1997; Hamilton et al. 1999).

(2) Model-based methods incorporating survey design. Smith and Robert (1998) illustrated two approaches that incorporate the survey design as well as auxiliary information (e.g., sediment, depth, etc.) using also the sea scallop. Sediment type information was utilised to define strata within existing strata (post-stratification), and as factors in a simple linear model. Both methods resulted in smaller variances compared to the standard method, but in addition yielded maps of the distribution of abundance with a considerable level of spatial resolution (24 bins).

(3) Geostatistical methods have been employed in the analysis of surveys of *Placopecten magellanicus* (Conan 1985; Ecker and Heltshe 1994), *Chlamys islandica* (Nicolajsen and Conan 1987), and *Pecten maximus* (Buestel et al. 1985), among others. It must be emphasised that "kriging", often applied in fisheries stock assessment, is just one method among others in the vast realm of geostatistics. The reader is referred to Rossi et al. (1992) for a series of interesting examples applicable to the modelling of ecological spatial dependence in scallop beds.

(4) General Additive Models (GAMs). The GAM (Hastie and Tibshirani 1990; Kaluzny 1987) is a nonparametric generalisation of multiple linear regression; both relate the dependent variable (density in this case) to possibly important covariates. In GAM, however, the covariates are assumed to affect the dependent variable through additive, unspecified (non linear) functions. Scatterplot smooths in GAM replace least square fits in regression. This method was applied by Swartzman et al. (1992) to the analysis of the spatial distribution of five flatfish species of the Bering Sea, based on survey data. The model is applicable to a variety of scallop stocks.

(5) Mapping with input from the commercial fleet. Over recent years there has been a growing appreciation of data generated by fishers themselves, which at least should complement survey data. Beyond contour mapping (Robert and Jamieson 1986), geostatistics and other methods have been proposed for mapping abundance with geo-referenced *CPUE* data, e.g., Vignaux et al. (1997). Figure 14.9 (from Turk 2000) illustrates the importance of considering information from the commercial fleet. The US National Marine Fisheries Service (NMFS) has been conducting a trawl survey of benthic and demersal sources of the eastern Bering Sea since 1976. The design is a systematic grid, with stations 20 nautical miles apart from each other. Turk (2000) mapped commercial beds of *Patinopecten yessoensis* using catch and effort information from logbook records. When the two maps (surveys and commercial beds) were superimposed, it became clear that the surveys had never hit the beds; scallop beds went undetected by the survey for *ca.* 25 years.

14.5.2.3 Methods based on modelling the depletion process

These methods are the core of stock-assessment in traditional fishery science. In their simplest form, the depletion of a closed population is modelled to estimate initial biomass and catchability. In catch-at-age methods, the most complex members of the family, the depletion of a series of year classes is tracked through time. The related catch-at-size models translate recruitment (or age composition) and growth rates into size structure to fit the available data. In most cases *CPUE* is used as an index of abundance, which presents problems in the case of sedentary organisms. These methods were developed to estimate total stock abundance; they are not informative about the spatial structure of the stock, which is at least as important as total abundance for the management of sedentary resources. These models are comprehensively covered in fishery stock assessment textbooks (e.g., Hilborn and Walters 1992). We only discuss some of their limitations, and mention scallop examples.

(1) CPUE: limitations as an index of abundance. *CPUE* has well-understood limitations as an index of abundance (Hilborn and Walters 1992), but those are particularly severe in the case of sedentary organisms, as discussed in Section 14.4.4. At the same time *CPUE* is extremely valuable because it is a primary determinant of the spatial allocation of fishing effort (Section 14.4.3). Departure of proportionality between *CPUE* and abundance, and consequently problems in the use of *CPUE* as an index of abundance, has several origins:

i) *Correlation between the geographic range of the stock and stock size* implies a non-linear relation between stock size and catchability, even in homogeneous populations
ii) *Sequential effort allocation along density gradients and serial depletion*, discussed in Sections 14.4.3 and 14.4.4, imply that abundance is not reflected by *CPUE*. This is one of the most important notions to be understood in the assessment of stocks of sedentary invertebrates.

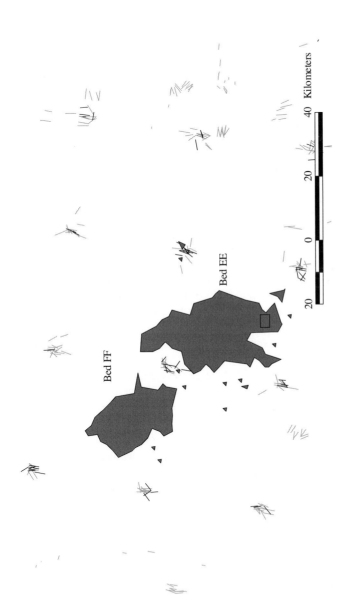

Figure 14.9. Trawl tracks from Bering Sea surveys (1978–1997). *Dark lines*: scallops present; *light lines*: no scallops. Scallop beds were delineated with logbook data from the commercial fleet (1993–1997). (From Turk 2000).

iii) *Saturation*: handling time, gear capacity, holding capacity and some management controls (e.g., trip limits) put ceilings on *CPUE* (e.g., in the scallop fishery of the Bay of St. Brieuc, France; Fifas 1991) and result in density-dependent catchability.

iv) *Observation errors* can generate spurious density-dependence trends in catchability.

Problems [i]-[iii] reflect real and serious non-linearities in the relation between *CPUE* and density; problem [iv] can make non-linearities apparent even if they do not exist, due to difficulties with the identity or measurement of variables. All these problems are likely to be inextricably confounded in most data sets.

(2) Fishing Success Methods. Catch and effort data can be utilised to estimate initial abundance (i.e., at the beginning of the fishing season) in removal experiments, provided that the quantity of animals removed over the season (or experiment) is large enough to produce a detectable decline in abundance. *CPUE* is commonly used as an index of abundance. "Fishing success methods" are treated in detail by Ricker (1975: chapter 6) and Seber (1982: chapters 7 and 8). Two main families are of common use in fishery research: regression of *CPUE* on cumulative catch ("Leslie method") and of log(*CPUE*) on cumulative effort ("DeLury method"). Dickie (1955), in the best-known scallop application, obtained yearly estimates of the size of the Digby stock of *Placopecten magellanicus* over 10 years (1941–1951), using a modified version of the Leslie method. Standard fishing success methods assume closed populations (no migration, recruitment or natural mortality), no competition between effort units and constant catchability. Models, however, can be modified in a number of ways for specific purposes, as is well illustrated by Dickie's (1955) pioneering study. Fishing success methods have been utilised in a variety of scallop stock assessments (Ito 1964; Joll 1987, cited by Young and Martin 1989; Wolff 1987; Turk 2000).

There is a strong warning on the use of fishing success methods in shellfish stock assessment, as *CPUE* is rarely a good index of total stock abundance (Turk 2000). They should be avoided whenever the fishermen are able to locate, relocate and selectively deplete dense patches within a fishing bed (Section 14.4.3), as best exemplified by commercial diving fisheries. Parameters estimated under those circumstances are likely to be meaningless, unless individual areas of relatively homogeneous density can be analysed independently and total stock size is estimated as the sum of the individual estimates. In dredge fisheries these methods tend to be adequate when (1) dense patches are small compared to the length of fishing tows (i.e., when they are "invisible" to fishers), and (2) the boundaries of the fishing bed are definable and data are collected at the appropriate spatial scale.

(3) Catch-at-size and catch-at-age methods. A variety of methods collectively known as "catch-at-age analyses" are central to fish stock-assessment. They provide estimates of abundance and fishing mortality by age and year from information on total catch-at-age taken over a number of years. In standard virtual population analysis (VPA; Ricker 1975), natural mortality is assumed known, as well as the terminal fishing rates producing

the latest catches from each of the cohorts involved in the analysis. This method has been applied to *Pecten maximus* (Mason et al. 1980, 1981, 1991; Fifas 1991), *Chlamys opercularis* (Mason et al. 1979) and *Placopecten magellanicus* (Kenchington et al. 1995). VPA tuned with various other indexes of abundance and fishing effort have been the standard procedure in the Georges Bank scallop stocks (Mohn et al. 1988). In the past (Mohn et al. 1984), the separable VPA (SVPA) of Pope and Shepherd (1982) was employed. A key assumption of SVPA is that fishing mortalities by age and year are a product of an age-specific factor (selectivity or partial recruitment) and a time-dependent factor, proportional to fishing effort. This reduces substantially the number of parameters needed to construct the matrix of fishing mortalities, at the expense of some loss of model flexibility. This loss of flexibility may seriously limit the applicability of the method to scallop fisheries: the separability assumption is not likely to hold when age classes are spatially segregated, specially if recruitment is highly variable and fishing effort is directed to the most abundant cohorts. When this is the case, the pattern of age-specific allocation of effort may shift over time as a result of changes in the relative abundance of the age components. Thus, time- and age-dependent effects determining fishing mortalities interact. As noticed by Mohn et al. (1984) the analysis of residuals from the fitting may point to these temporal shifts in the age-specific fishing pattern.

14.5.3 Microscale

14.5.3.1 Small-scale spatial data

The data collected to investigate the small-scale spatial distribution and abundance of scallops belong to three basic types:

i) *Counts of individuals within quadrats* are the most popular technique in the study of the spatial distribution of scallops at small scales. Spatial information about the individuals within quadrats is lost, however, and so nothing can be learned for scales below the quadrat size. Quadrat samples can be obtained directly by divers (e.g., Orensanz 1986) or remotely with underwater photography (e.g., Goshima and Fujiwara 1994). The popular book of Elliott (1977) is perhaps the best available introduction to the treatment of this type of data (see Shafee 1979, Mason et al. 1982, Thouzeau and Lehay 1988, and Goshima and Fujiwara 1994, for detailed scallop examples).

ii) *Distances between individuals and their nearest neighbours, or between points in space and the nearest individuals*, which have been rarely used in scallop studies. These data are not further considered below.

iii) *Complete maps with the coordinates of all the individuals in the study area*. A vector of data (size, age, sex, phenotype, etc.) can be associated with each individual. Maps are relatively easy to obtain (at least for small study areas) in the case of epibenthic invertebrates like scallops. Underwater photography and diving were used in two studies of giant scallops (*Placopecten magellanicus*) from Eastern Canada, in study areas ranging from 9 to 256 m^{-2} (MacDonald and Bajdik 1992; Stokesbury and Himmelman 1993). The use of quadrats is so deeply rooted that mapped data are often reduced to quadrat counts before analysis (e.g., MacDonald and Bajdik 1992; Stokesbury and Himmelman 1993), with a consequent loss of costly information.

14.5.3.2 Spatial pattern

Three basic types of spatial pattern - random, regular and clustered - are generally recognised in ecological research. Spatial point patterns are often compared to a realisation of a completely spatially random point process. Strictly defined, *complete spatial randomness* (the "white noise" of spatial point processes) is equivalent to a *homogeneous* Poisson process; given an area A where N individuals are found, the N individuals are distributed independently (and with uniform probability) over A. Intuitively, this means that individuals are equally likely to occur anywhere within A and that they do not interact with each other, either repulsively or attractively. Complete spatial randomness characterises the absence of structure (or signal) in the data. As such, it is often the null hypothesis in statistical tests to determine whether there is spatial structure in a given point pattern. Clustered patterns (most common in nature) can reflect environmental heterogeneity (e.g., a mosaic of bottom types) or interactions among individuals (e.g., reproductive aggregations). Detection and description of spatial patterns (including their change with scale) is of significance in the estimation of abundance, the assessment of fishing strategies, and inference about ecological processes. Small-scale pattern has been investigated in scallops by means quadrat counts and mapped data:

(1) Quadrat data: pattern at a single scale. Under complete spatial randomness the number of organisms per quadrat has a Poisson distribution, which has been observed within homogeneous habitats, e.g., as defined by sediment type (Thouzeau et al. 1991). Complete spatial randomness can be tested with a goodness-of-fit test (Elliott 1977); if rejected, and since scallops always tend to show a clustered pattern, the density frequency distribution is commonly modelled with the negative binomial distribution (e.g., Shafee 1979; Vahl 1981a, 1981b; Orensanz 1986; Thouzeau and Leahy 1988; Thouzeau 1995). The intuitive notion of aggregation entails a relation between concentration and density, used to capture pattern in indices like Lloyd's "patchiness" and Morisita's "index of dispersion" (Goshima and Fujiwara 1994). These indices cannot capture the complexity and multiplicity of scales of most spatial patterns. Location of quadrats within the study area and of individuals within quadrats are not considered, and consequently most of the spatial information is effectively lost.

(2) Mapped data: scale dependency of pattern and the K-function. While mean concentration at a single scale has been often used to calculate indices of aggregation, more significant aspects, like the spatial dependence between different regions of the spatial process (Cressie 1991, p. 615) or the distribution of concentration in a population at a scale of interest ("concentration profiles", Section 14.3.4.2), have received little attention (Orensanz et al. 1998). Concentration at different scales is best captured by means of the K-function. This, the point process analog of geostatistics' variogram, allows assessment of departures from complete spatial randomness at different spatial

scales. A scallop example is in point. MacDonald and Bajdik (1992) collected data on the individual location of sea scallops in two 15 m × 15 m plots at two unfished sheltered sites (Colinet and Sunnyside) off Newfoundland (eastern Canada) (Colinet data shown in Fig. 14.10, left). The authors partitioned the study region into quadrats using various quadrat sizes, and calculated the mean/variance ratio and a transformation of mean crowding for each partition size. They concluded that the pattern was "haphazard" at Colinet (Fig. 14.10) and aggregated at Sunnyside, and speculated about the meaning of this result. The K-function analysis shows, however, that both patterns fall within the range of 100 realisations of a completely spatially random process, for all the scales about which the data are informative.

14.5.3.3 Scaling problems

Concentration analysis allows and forces focus on scale. Below we discuss some pertinent examples.

i) *Poor small-scale resolution.* Abundance data routinely collected during sampling-surveys of scallop socks (Section 14.5.2.1) are generally not informative about concentration at small scales, yet it is at these small scales that density-dependent processes governing stock dynamics are operationally effective (Section 14.3.4.3). The assessment of concentration at low scales for a whole stock, however, is a difficult proposition. One possibility is "disjoint windowing". Stokesbury and Himmelman (1993), for example, mapped the location of individual scallops within disjoint windows (square plots) in two unharvested beds from the Baie des Chaleurs, Gulf of St. Lawrence (eastern Canada). A total of 37 9-m^2 disjoint windows were marked at haphazardly selected sites. Such data could be utilised in the analysis of concentration profiles at scales smaller than those defined by the quadrat size, across an entire bed. Pattern could be assessed *within each* of the windows by means of the K-function or related methods. Analyses of pattern within the disjoint windows, however, could not be simply combined to provide a consolidated description of pattern across the entire region without making the additional assumption that the point process is stationary across the whole region under study. Such assumption would be difficult to meet (notice, however, that concentration profiles are free of stationary assumptions). Mapped data collected from relatively small disjoint windows should not be combined, *prima facie*, to make statements about small-scale pattern across the entire region surveyed.

ii) *Poor large-scale resolution.* Problems of large-scale resolution, even though pervasive, are often less apparent. Consider the example introduced earlier (Section 14.4) about sequential depletion for different types of concentration profiles. Frequency distributions of density reported for benthic invertebrates (usually estimated with a sample of quadrats) are most often described by the negative binomial distribution (Type C in Fig. 14.3; Section 14.3.4.2). Under sequential depletion, such distributions lead to *hyperdepletion* (at least at the beginning of the harvest), as initial effort targets the small areas that concentrate much of the resource. This entails an apparent paradox: while the overwhelming majority of investigated spatial patterns would suggest hyperdepletion as the most frequent type of *CPUE* trajectory, *hyperstability* seems to prevail in shellfish fisheries. Intriguingly, patterns of spatial distribution associated with hyperdepletion under strict sequential allocation (e.g., regular or quasi-regular patterns) are rarely (if ever) reported by shellfish biologists. The reason may be a mismatch between the spatial scale used in the assessment of pattern and the operational scales of the fishing process. Patterns of spatial distribution of benthic shellfish are generally analysed using small units (e.g., one-m^2 quadrats). The fishing process (including perception and harvesting) operates at larger scales, even in the case of hand-gatherers. A hypothetical diver fishing an epibenthic species will never harvest small (say one-m^2) patches in a perfect sequence of descending density, even in the unlikely case that he had a complete map of the fine-scale distribution of the resource. Pairs of small patches that rank next to each other in density will be variably (sometimes widely) separated in space. Since moving from patch to patch is costly, harvesting will depart from a simple response to small-scale pattern. If entire beds (rather than small patches) are the subject of a sequential harvest, and each bed is harvested systematically (as is often the case in diving shellfisheries), the result is hyperstability, no matter what the internal spatial pattern within each bed looks like. Surveys of spatial pattern at a small scale may be of little significance to the understanding of processes with a larger operational scale.

The distribution and pattern of concentration has generally been investigated at small scales by ecologists, and at large scales by fishery scientists. Yet, both are meaningful in the assessment of stocks of sedentary invertebrates.

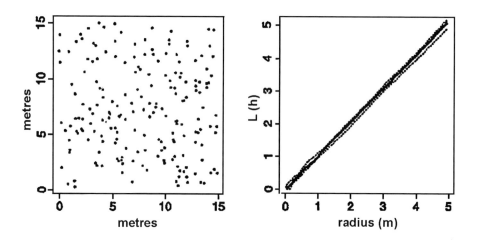

Figure 14.10. Small-scale spatial pattern. Left: individual location of sea scallops in a plot at Colinet, an unfished sheltered site off Newfoundland (eastern Canada). The observed L-function (solid line, a variant of the K-function used for graphical presentation) falls within the range of 100 realisations of a completely spatially random process (dashed envelopes), for all the scales about which the data are informative. (Data from MacDonald and Bajdik 1992; figure modified from Orensanz et al. 1998).

14.6 MANAGEMENT

Fishery science and management have changed dramatically over the last ten years. The term "collapse" became a frequent key-word in the literature. A global debate took momentum on how to manage fisheries sustainably. The Precautionary Approach to fisheries management, ecosystem management, Marine Protected Areas (MPAs) and governance, rarely heard of a decade ago, took center stage. Prominent fishery collapses motivated scientific scrutiny of various hypotheses (e.g., climate *vs.* overfishing), examination of the eventual failure of the management systems involved, cross-incrimination among the various sectors in those systems (fishermen, scientists, managers, politicians), *exposes* by the press, outrage by the public, and finger-pointing by the conservation movement. It has been a most eventful decade. Thus, this Section required re-writing. Below we address aspects of the debate that pertain most immediately to the management of scallop fisheries.

14.6.1 Types of Scallop Fisheries

Scallops from shallow water environments have been harvested since pre-historic times, as shown by the presence of bay scallops (*Argopecten irradians*) in shell middens along the coasts of the eastern USA (Rhodes 1991). Yet, most scallop beds are deep enough that they were reached by fishers only after the introduction of dredges and diving. Other groups of bivalves (e.g., mussels and clams) were (world-wide) more significant to early fishing communities.

Modern scallop fisheries belong to two major groups:

i) Small-scale and artisanal operations, characterised by small vessels (usually less than 15 m long) operating in inshore waters, making daily fishing trips, and often landing at small harbours along the coast. Common methods include dredging, trawling (e.g., *Amusium* fisheries) and commercial diving. Examples are most scallop fisheries from Australia, Latin America and Europe, as well as the bay scallop (*Argopecten irradians*) fishery of eastern USA states.

ii) Large-scale industrial operations, involving larger dredgers or trawlers that make trips of several days of duration, and land catches at one or a few harbours with appropriate logistic facilities. Examples are the offshore fisheries for sea scallop (*Placopecten magellanicus*) in eastern North America, calico scallops (*Argopecten gibbus*) in the southeast USA, the Patagonic scallop (*Zygochlamys patagonica*) in Argentina and the weathervane scallop (*Patinopecten caurinus*) of Alaska.

Distinction between these two groups is significant because the most appropriate or viable management options (tactic and strategic) are different. Scallop resources, taken as a whole, are not a significant resource for traditional or subsistence fishing communities. Generally, scallop fisheries have at least some level of technology and are oriented to the commercial exploitation of a highly prised product.

14.6.2 Overfishing

An exploited population can be viewed as composed of a number of cohorts, whose individuals grow and suffer natural mortality. Recruitment of new cohorts to the exploitable stock occurs when animals become vulnerable to the fishing gear, and available and subject to fishing. The total yield obtained from a given cohort depends on its initial abundance (R), and on the mean yield (Y) obtained per recruit (Y/R). A harvest strategy based on purely biological considerations should therefore contemplate two different problems: (a) the *recruitment overfishing problem*, in which the depletion of the reproductive population results in poor recruitment, and (b) the *growth overfishing problem*, in which animals are caught when they are too small. If recruitment is assumed independent of parent stock size (at least over some range of stock sizes), fishery regulations can be aimed only at optimising the harvest from a Y/R perspective, this is, taking the best advantage of any given level of available recruitment.

14.6.2.1 Growth overfishing: yield-per-recruit (Y/R) analysis

Orensanz et al. (1991) made a general conceptual presentation of the growth overfishing problem in scallop stocks, to which we address the interested reader. Mohn (1986) reviewed some of the approaches used to conduct Y/R analysis in shellfish stocks. The Beverton and Holt (1957) formulation has been utilised to evaluate the effects of different size limits and fishing intensities on the sea scallop fishery of the Georges Bank since the 1960s (Caddy 1989a; Serchuk et al. 1979; Sinclair et al. 1985), and other stocks of *Placopecten magellanicus* (e.g., Kenchington et al. 1995), and to *Chlamys islandica* (Lanteigne et al. 1986). With different modifications, Y/R analysis has been applied to several scallop stocks: the giant scallops of the Georges Bank (Mohn et al. 1988), *Chlamys opercularis* in Scotland (Mason et al. 1979), *C. islandica* in the Gulf of St. Lawrence (Naidu et al. 1982), *Pecten maximus* in Europe (Ansell et al. 1991*), Amusium balloti* and *Pecten fumatus* in Australia (Dredge 1985; Gwyther and McShane 1988), etc. Scallop Y/R is frequently expressed in terms of mean muscle weight per recruit, although any other variable can be chosen depending on which portion of the animal is commercialised. Dao et al. (1975) used, for *Pecten maximus*, the weight of the muscle plus the gonad, which is highly valued in the French market. A Thompson-Bell analysis was conducted on a monthly base in order to account for seasonal changes in the quality of the meats.

Y/R analysis, which in its simplest form deals with single cohorts or equilibrium conditions, can be extended to address the problem of maximising the yield (or some other output) from any standing stock, whatever its age structure. This is most easily done with book-keeping algorithms, where total yield from the standing stock under a given fishing strategy results from summing up the separate contributions of the cohorts, each weighed by its relative abundance. Mohn (1986), for example, applied one such procedure to the Georges Bank scallop stock. Detection of strong recruitment pulses has often prompted *ad hoc* management recommendations, intended to delay harvesting until recruits reach larger sizes (see for example Robert et al. 1987; Giguère and Legare 1988a, b).

Murawski and Fogarty (1984) extended the Thompson-Bell model to incorporate the spatial dimension, assuming that year-classes are segregated in space and can be managed independently. A simulation was used to evaluate different rules of spatial allocation of fishing effort. Although cohorts may be to some extent spatially segregated (and, therefore, they can be managed independently), the behaviour of the fishers may result in suboptimal harvesting of more than one cohort simultaneously. In the Georges Bank, for example, when meat-count regulations were enforced, the fishing fleet commonly targeted the densest beds, which usually correspond to young animals, and only secondarily the older, more scarce animals so as to comply with meat count regulations. This behaviour, commonly known as "blending", results in growth overfishing, since harvesting of scallops below the stipulated meat standards is legalised (Naidu 1984; Smith and Robert 1991). The global selectivity of the fishery reflects these practices: fishing mortalities at age estimated from cohort analysis for the Georges Bank stock decline subsequent to the age groups targeted (3–4 ring scallops) (Caddy and Jamieson 1977). This contrasts markedly with the age-specific selectivity of the fishing gear, which increases monotonically with size. Mohn et al. (1984) extended the Thompson-Bell Y/R analysis to account for this (or any other) type of behaviour of the fishing fleet (see also Mohn 1986). An optimal level of (nominal) fishing effort can be computed conditioned on the assumptions about effort allocation by age. For example, fishing effort can be allocated to the different age classes in proportion to their abundance, unless meat counts fall below the enforced level.

Wolff and Mendo (2000) presented a very interesting case involving a non-stationary system: the Independencia Bay (Perú) fishery of *Argopecten purpuratus* (Sections 14.3.2.1). Population parameters differ sharply between normal and El Niño years. Besides, fishers converge to Independencia Bay from other regions during scallop booms; the increase in effort lags behind the increase in population size, with the result that the fishing force is still increasing after the stock has started to decline. The estimated optimal ages at entry for "normal" and "El Niño" years, based on a Thompson and Bell's Y/R analysis, were 50 and 70 mm, respectively. A different legal size was recommended for the two states, given that effort cannot be realistically enforced. The size limits suggested should allow for at least one reproductive event in this short-lived species.

A modest first step to bringing the recruitment overfishing problem into Y/R analysis has been to tie the Y/R figures to some index of mean reproductive potential per recruit, and then evaluate the trade-offs between gains in yield and reproductive losses for the different harvest regimes considered. A standard index of reproductive output for scallops (Mohn et al. 1984) is the sum of age-specific mean gonad weights across all age components, each weighted by the fraction of animals surviving up to that age. Mason et al. (1981), for example, found that for *Pecten maximus* an increase in fishing effort would only slightly increase Y/R, at the expense of a considerable decrease in spawning biomass per recruit. In the *Amusium japonicum* fishery of Queensland (Australia) Dredge (1994) concluded that increasing the size limit would increase spawners per recruit marginally, while decreasing yield per recruit.

14.6.2.2 Recruitment overfishing

Many scallop fisheries have collapsed, some shortly after their onset. A collection of examples extracted from the literature is illustrative:

i) *Aequipecten opercularis* of Spain, after the 1960s (Román 1991).

ii) *Aequipecten tehuelchus* of San Matias Gulf in 1972 (Olivier and Capitoli 1980; Orensanz 1986), and San José Gulf in 1997 (Ciocco and Orensanz 1997) (Argentine Patagonia).

iii) *Argopecten gibbus* off Cape Canaveral (east Florida) in 1985 (Moyer and Blake 1986).

iv) *Patinopecten caurinus* beds off Oregon (USA) were depleted during a year of intense fishing (1981); the beds have not recovered after 20 years (Turk 2000).

v) *Pecten fumatus* near Hobart (Tasmania) in 1908, the D'Entrecasteaux Channel (Tasmania) in the late 1940's (Fairbridge 1953) and later in 1963 (Gwyther et al. 1991, p. 838), and many fishing beds from Tasmania, Bass Strait, South Australia and New South Wales (Australia) during the mid 1980s (Gwyther et al. 1991; Young and Martin 1989; McLoughlin 1994; Young 1994; Zacharin 1994).

vi) *Pecten jacobaeus* stocks from the Adriatic (Italy) were on the brink of commercial extinction during the early 1990s (Mattei and Pellizzato 1996).

vii) *Pecten maximus* of Cardigan Bay (UK), in 1980 (Ansell et al. 1991, p. 732), and Málaga (Spain), in 1978 (Ansell et al., p. 735; Román 1991).

viii) *Pecten modestus* of Cockburn Sound, West Australia (Hancock 1979; Young and Martin 1989; Young 1994).

ix) *Pecten vogdesi* of the Gulf of California in 1973 (Félix-Pico 1991).

x) *Pecten ziczac* of southern Brazil in 1982 (Pezzuto and Borzone 1997; Perez and Pezzuto 1998).

xi) *Placopecten magellanicus* of Mahone Bay (Sinclair et al. 1985, p. 9) and Nantucket (Sinclair, in Kenchington et al. 1995, p. 18).

Many collapse and recovery events are not systematically documented in the technical literature, in spite of the information that they contain about the dynamics of harvested scallop stocks (Section 14.3.1.4). A comparative examination of documented cases points to several roads to recruitment overfishing:

i) No *apparent* relationship between spawning stock and recruitment (Hancock 1973), an illusion magnified by occasional pulses of recruitment when the spawning stock is at a low level (Section 14.3.1.4). In these cases it is difficult to know if the population would have vanished even in the absence of harvest (albeit after a longer period).

ii) Serial depletion of beds in metapopulations with low connectivity (Section 14.3.2.1), a case of hyperstability. Hyperstability is a frequent road to overfishing in sedentary resources. Although rarely investigated in scallop fisheries, it is implicit in verbal depictions of the depletion process, e.g., "[there is] little evidence of over-exploitation or depletion of grounds, perhaps masked by the ability of the fleet to locate and move to new areas" (Ansell et al. 1991, p. 735). The best-documented case in the literature is the serial depletion of *Pecten fumatus* stocks in Australia (Young and Martin 1989, particularly their fig. 3).

iii) Depletion of source subpopulations, a hypothesis frequently advanced but difficult to substantiate.

iv) Absence of a low-density threshold at which fishing stops ("break-even point"). This occurs when scallops are fished together with species of higher value, as shrimp (Perez and Pezzuto 1998).

v) Reduction of abundance to a level at which fertilisation rate is reduced (pre-dispersal depensation), a problem that can be anticipated from changes in the concentration profile (Section 14.3.4.2). This has been suggested as a factor preventing recovery of depleted stocks of bay (Marelli et al. 1999) and sea scallops (Kenchington and Lundy 1996).

vi) Illusory high productivity due to the harvest of the biomass accumulated in a virgin stock targeted by a new fishery, documented for high-latitude long-lived species (Orensanz et al. 1998). Naidu et al. (1995, 1998) discussed the development of the Icelandic scallop (a long-lived species) in the Grand Banks of Newfoundland (eastern Canada). They predicted that at the effort levels of 1998 all known commercial aggregations could be severely depleted before significant recovery occurs.

vii) Elimination of natural refuges. Experience tells that natural refuges (Section 14.6.3.2) vanish as a result of technological improvements and changing economical incentives. The first is exemplified by the introduction of a reinforced dredge design in St. Brieuc Bay, which allowed fishing in previously inaccessible rocky areas (Fifas 1991). A case of the second is the willingness of commercial divers to risk fishing at greater depths in the Argentine tehuelche scallop fishery prompted by deteriorating economic conditions during the early 1990's (Ciocco et al., Chapter 26).

viii) Habitat destruction, principally the removal of primary settlement substrates.

ix) Economical incentives to harvest scallops smaller than the size at first spawning, a case observed in some fisheries for short-live, fast-growing tropical species (Dredge 1988; Young and Martin 1989, p. 634).

The tehuelche scallop fishery of San José Gulf provides an informative study case (Orensanz et al. 1997). During the 1970s, when divers harvested dense, shallow beds, the fishery had some features that favoured self-regulation (Orensanz 1986). Divers abandoned a fishing bed at relatively high-density thresholds (*ca.* 20 individuals m^{-2}), and selected scallops larger than the legal size (allowing for one or two spawning seasons) because smaller scallops did not have a market. Patches below the density threshold, deep beds beyond the reach of divers, and individuals below legal/commercial size provided for a significant spawning stock. In fact, the fishery proved to be sustainable for *ca.* 20 years in the virtual absence of effective regulations or enforcement. During the early 1990s, however, a new market developed for small (pre-reproductive) scallops, and a hard economic crisis led the fishers to lower their density threshold and to risk fishing in deeper waters. The fishery collapsed in 1996, remaining closed for three years (it has reopened but for small catches in recent years).

Among the roads to overfishing listed above, (1) has occasionally supported a dangerous rationale. If scallops fisheries depend upon sporadic and localised recruitment bursts, it has been reasoned, management strategies aimed at maintaining some optimal stock size for maximum long-term productivity may be futile (Fairbridge 1953, p. 37; Dredge 1986: p. 174; Perrin and Hay 1987). Gulland (pers. comm. in Hancock 1979), for example, considered that the high fecundity of scallops would result in recruitment being independent of stock size. Hancock (1979) went on to suggest that "some stocks of scallops can almost be treated as a nonrenewable resource"; harvesting strategies would be designed by comparison to "extermination". Young (1994) summarised and criticised the (often implicit) rationales behind such propositions with relation to *Pecten fumatus*, but his synthesis is applicable to all scallop stocks:

"The economics of fishing will not reduce the stock below a minimum level that will provide sufficient breeding population to ensure recruitment and maintain the viability of the fishery. [...] Recruitment [is] unrelated to resident stock in any given bed, but derived from contiguous areas as a function of larval drift in relation to residual circulation. [Observed fluctuations in the catches] merely reflect recruitment variability which is unaffected by fishing pressure, and the stocks would be assumed to replenish from unfished areas of low density at such time as oceanic conditions are suitable to produce a 'random' pulse of recruits."

In most cases, unfortunately, the relationship between stock and recruitment is difficult to specify. Under such circumstances, as Sinclair et al. (1985, p. 4) clearly put it, "since quantitative criteria to monitor [recruitment overfishing] are generally lacking, simple qualitative guidelines may have to suffice". Preventive conservation measures that have been suggested or implemented include:

i) A legal size that is above the size at sexual maturity (Section 14.6.4.1).

ii) Setting apart geographic segments of a metapopulation as reproductive refuges. Knowledge about biology and oceanography are always incomplete and the results of closures are difficult to assess (Kenchington and Lundy 1996). Yet, some general criteria have been advanced for their design (Section 14.6.3.2).

iii) Assuming that subpopulations are self-sustaining when recommending regulatory measures. Sinclair et al. (1985) tentatively concluded that "many of the scallop aggregations that support commercial fisheries are self-sustaining populations. Thus, some minimum spawning biomass must be maintained *in each fishery area* to ensure long-term harvesting potential".

iv) Stopping the harvest when a certain threshold density is reached, the latter perhaps tuned to the concentration-dependency of fertilisation rate (Section 14.3.4.3).

v) Consolidation of a broodstock (Heath 1999; Marelli et al. 1999). This is an option in the case of small confined stocks in which creation of high-density "reproductive hotspots", even if small, could produce recruitment pulses in the event of favourable conditions.

A crisis in the Bay of Fundy fishery prompted recommendations that incorporated some of these guidelines (Kenchington et al. 1995, p. 18): 20% of the Digby beds had to be closed for 10 years, closed areas should match historical nursery beds as much as possible, and a rotational program was to be considered.

14.6.2.3 A special case: recruitment fisheries

In "recruitment fisheries" all or most of the individuals that grow above legal size during the closed season (new recruits) are caught during the next fishing season. The fraction left behind by the fishery at the end of the season (usually known as "residual stock") is often determined by the *CPUE* level at which the vessels break even in economic terms. Legal size and timing of the season are generally tuned to allow scallops to reproduce at least once in their life-times. This strategy is well suited to short-lived or quasi-semelparous species, in which most of the potential reproductive output is produced before the scallops enter the fishery. In these scenarios, in principle, there is no risk of recruitment overfishing; stock fluctuations are attributable to natural causes only. Examples include the fisheries for *Amusium balloti* of Abrolhos Island, (Joll 1989), Shark Bay (Western Australia) and Queensland (Gwyther et al. 1991). In the latter, a density of one scallop per 150 m^2 has been determined as the breakeven threshold. In some cases harvest has to be delayed to allow for spawning at the cost of a lower yield. This is exemplified by fast-growing tropical species like *Amusium balloti* from Queensland, in which maximum yield corresponds to an age of 6–8 months (Dredge 1988), when the scallops have not yet reproduced. Based on these calculations and on trends in *CPUE*, Dredge (1988) concluded that this stock had been recruitment overfished.

Another example is the bay scallop (*Argopecten irradians*) fishery of New England. Spawning of a cohort takes place in summer, fall or early winter (one year of age), and few individuals survive to spawn twice. Thus, if the fishery is confined to the winter there is no risk of recruitment overfishing (Belding 1910). In this fishery, scallops can be retained when they have marked the one annual growth ring that most individuals would lay down in their lifetime, an interesting alternative to a legal size (Rhodes 1991). In the calico scallop (*Argopecten gibbus*) industrial fishery off southeastern USA, also a recruitment fishery, not even a legal size or a season are enforced. The industry's self-regulating association has agreed to limit fishing in a bed until at least 75% of the individuals exceed 38 mm; the latter has a commercial motivation but is also a conservation measure, as it allows scallops to undergo their first reproductive event (Blake and Moyer 1991).

While harvesting the post-reproductive recruits is a sensible option for short-lived, quasi-semelparous populations, the practice is troubling when longer-lived iteroparous species are turned into recruitment fisheries by excessive fishing effort. This becomes apparent when landings start to track abundance (catch being an estimate of recruitment), which generally points to a loss of potential yield (growth overfishing). This situation is exemplified by the competitive inshore fishery for sea scallops of the Bay of Fundy (Canada) and the Georges Bank US fishery, both managed through effort regulations (Section 14.6.4.4). In the latter standardised dredge surveys conducted by the National

Marine Fisheries Service generally track landings closely, indicating the high overall harvest rate and the level of dependence on incoming recruitment (Murawski et al. 2000). Beyond growth overfishing, this type of recruitment fisheries may be prone to recruitment overfishing. The decline of the *Pecten maximus* fisheries of St. Brieuc and Seine Bays (France) during the 1980s has been attributed to this kind of situation (Ansell et al. 1991; Thouzeau 1995). By contrast, in the Port Phillip Bay (southeastern Australia) fishery for *Pecten fumatus* the residual stock left after one season would not suffice to sustain the fishery during the following year, but is assumed to be the minimum needed to give rise to good recruitment (Gwyther et al. 1991).

14.6.3 Sustainability

In the search for sustainable fisheries, overfishing has repeatedly been identified as the primary problem (Botsford et al. 1997; NRC 1999). Solutions often advanced include precautionary reductions in catch levels (Section 14.6.3.1), establishment of marine reserves (Section 14.6.3.2) and ecosystem management (Section 14.6.3.3). These prescriptions, however, do not address the key failings in most fisheries systems; overfishing is a symptom of poor governance systems rather than the disease to be treated. Use and property rights, the template that shapes governance systems, are discussed in Section 14.6.3.4.

14.6.3.1 The Precautionary Approach to Fisheries Management (PAFM) and Risk Assessment (RA)

While mostly everybody in the fisheries community has an intuitive idea of what the PAFM is about, several subtly distinct interpretations coexist (Hilborn et al. 2000):
    i) Application of the Precautionary Principle, implying a reversal of the burden of proof: no fishing is allowed if the harvest policy has not been demonstrated to be sustainable (Garcia 1994).
    ii) Adoption of cautious harvest levels, guided by "biological reference points".
    iii) An emphasis on institutional arrangements that provide for monitoring of the fishery, feedback on regulations and effective implementation (Hilborn et al. 2000).

The Precautionary Principle is significant in environmental policy, but not in fisheries management. Institutions and feedback have a prominent role in the PAFM as interpreted in a well-known technical consultation conducted by FAO (FAO 1995). Yet, the PAFM is often (and narrowly) identified with biological reference points, a result of fishery regulations articulated after the Sustainable Fisheries Act (USA), several international treaties, and other regulatory frameworks. Following interpretation (ii) of the PAFM, generic harvest control rules based on target and limit reference points have been proposed as default harvest guidelines that could perform adequately for most fisheries and circumstances (Restrepo et al. 1998). Generic harvest rules specify the desired harvest rate or catch as a function of the actual stock size; their implementation requires the use of specific harvest regulations such as a total allowable catch (TAC) or an effort control. Management of the US fishery for sea scallops (*Placopecten magellanicus*) is a case in point. The level of fishing effort that maximises yield for a given selectivity function (F_{max}), and the equilibrium biomass (constant recruitment) reached with that effort (B_{max}), are calculated through yield-per-recruit analysis (see NEFSC 1999, for a description of the models and the estimation protocol). F_{max} and B_{max} are then used as proxies for F_{msy} and B_{msy} (biological reference points for surplus-production models), as B_{msy} is the target biomass according to the US Sustainable Fisheries Act (SFA). The stock is considered "overfished" whenever biomass drops below $B_{msy}/4$; overfishing occurs when $F > F_{msy}$, independently of biomass level. This can lead to a mind-boggling situation in which a stock that is not overfished is being overfished. The overfishing F-threshold decreases linearly to zero as biomass decreases to $B_{msy}/4$. When biomass falls in the range $[B_{msy}/4 < B < B_{msy}]$, F should be reduced to allow the stock to be rebuilt within 10 years.

Endorsement of the PAFM (however interpreted) during the 1990s gave prominence to "risk assessment" (RA), usually in the form of the calculation of the probabilistic consequences of alternative management actions (Hilborn et al. 2000). Applications of risk analyses in fisheries followed two fundamentally different approaches:

i) The first is associated with interpretation (ii) of the PAFM and the calculation of "reference points". Given that stock size is always estimated with error, and that the relationship between management regulations and fishing mortality is always uncertain, RA is used to quantify the probability that various TACs or levels of fishing effort meet the adopted harvest rule. Examples of this approach are found in Gavaris and Sinclair (1998), Patterson (1999) and Rosenberg and Restrepo (1994).

ii) In the second approach, RA attempts to evaluate the longer-term consequences of management actions without resorting to reference points *per se*. Because management consequences generally depend on a series of actions, not just a single annual limit placed on catches or effort, future management actions need to be evaluated as a consistent set of feedback rules or management procedures used to derive the future controls. The possible outcomes of alternative management procedures are evaluated using models that are tailored to specific situations (e.g., Ianelli and Heifetz 1995; McAllister et al. 1994; Geromont et al. 1999; Parma 2001).

The first approach to RA presents several shortcomings. First, the performance of alternative management choices, or "immediate management actions" (Gavaris and Sinclair 1998), are only evaluated with respect to the adopted reference points or harvest rules and not in terms that are more meaningful to the managers and stakeholders involved with the fishery in question. Secondly, the risks of exceeding the target or limit harvest rates are evaluated for a single annual decision (the annual TAC or limit on effort) and this fails to account for the compounding effect of, say, always choosing the most liberal control within the range considered permissible (e.g., upper bound of a TAC range). Finally, discussions about the appropriateness of specific reference points or their proxies may attract more attention than the actual evaluation of the consequences of alternative actions. In other words, it is often more difficult to decide what is the appropriate value for F_{MSY} or for B_{MSY} than it is to evaluate the possible consequences of alternative, nameless, harvest strategies.

While this is valid in general for any fishery, it is particularly relevant in the case of stocks of sedentary organisms, such as scallops, for which stock-aggregate reference points based on MSY may not be appropriate for management. Focus on MSY-related reference points may distract attention away from other strategies, e.g., spatially-explicit ones, which would be a more natural approach to controlling harvest rates.

The second approach to RA requires the use of an operating model to simulate the stock and fishery dynamics (Butterworth and Punt 1999). This is a preferred option for the *design* of stock-specific harvesting strategies, which may include different sorts of spatially-explicit regulations: territorial use rights (Section 14.6.3.4), MPAs (Section 14.6.3.2), rotation (Section 14.6.4.5), feedback from experimental management with spatial treatments and controls (Section 14.6.4.7), and localised interventions to enhance productivity (Section 14.6.4.6).

14.6.3.2 Marine Protected Areas (MPAs)

Spatial reproductive refuges, a particular form of marine protected areas (MPAs), have attracted attention as a ultimate safeguard against recruitment overfishing (NRC 2000), and may be particularly well suited to benthic resources (Orensanz and Jamieson 1998). Natural refuges, for example rocky areas where dredging is not possible (Sinclair et al. 1985, p. 9; Naidu 1991, p. 869) or depths beyond the reach of commercial divers, play a significant role in many fisheries. According to Carl Walters:

"If we look at fisheries that have been successful over the long term, the reason for their success is not to be found in assessment, learning and management models, but in the existence of a *spatial accident*, something about the spatial structure of population dynamics interacting with regulatory systems, or about the behaviour of the species and fishers, that creates a *large scale refuge* for a substantial segment of the spawning population." (*North Pacific Symposium on Invertebrate Stock Assessment and Management*, Nanaimo, Canada, 1995; quoted by Orensanz and Jamieson 1998)

When such "natural accidents" do not exist, a priority of any sensible management plan should be to create their equivalence: reproductive refuges that are off limits to fishery operations. Reproductive refuges were implemented early in the management of bivalve stocks; the "reproductive reserves" introduced by Washington State (USA) for oysters, early in the century, are just an example. They have received renewed attention in recent years in situations that combine sedentary resources with complex spatial structure, little available information, and difficulties of enforcement. Criteria prescribed for the design of systems of reproductive refuges (Fairbridge 1953; Caddy 1989b) should attend to:

i) The location relative to current systems and potential mechanisms for larval advection.

ii) The spatial pattern of the spawners within refuges.

iii) Areas of high scallop density that have consistently experienced good settlement over the years. Recurrence of recruitment successes (as evidenced by a persistent high density of scallops of multiple ages) must be indicative of hydrographic conditions that favour larval retention, a desirable attribute of a refuge.

In response to an unprecedented crisis in the Bay of Fundy scallop fishery in 1995, Kenchington and Lundy (1996) suggested that density had dropped to levels low enough that depensation (concentration-dependent fertilisation rate) was a concern. Following recommendations from a working group that 20% of each scallop bed should be closed to all forms of dredging during ten years, Kenchington and Lundy (1996) made specific suggestions about possible sittings of refuges based on location of persistent beds and circulation patterns (inferred mostly from drift cards).

Even small but dense refuges (reproductive "hotspots") may prove effective in some cases, particularly in semi-closed bodies of water. Empirical evidence indicates that in broadcast spawners with external fertilisation, conveniently located reproductive "hotspots" can contribute to the replenishment or expansion of populations. Peterson et al. (1996) provided empirical support to this contention, pointing to the possibility of manipulating hotspots as part of restocking programs (Section 14.6.4.6).

In general, MPAs are part of strategic plans implemented to satisfy the complex societal agenda for the use of the maritime space, of which fisheries management objectives are just a component. Thus, the establishment of an MPA often has multiple objectives, some related to management of one or more stocks, others to conservation, others to the provision of different ecosystem services to society. In Alaska, for example, vast expanses of seabed are closed to dredging for protection of crab recruitment habitat (Turk 2000). The multiplicity of goals has led to considerable confusion regarding the implementation of MPAs. The Chilean fishing act of 1992 (LGPA, the Spanish acronym) is remarkable in that it makes a clear distinction between "marine reserves" and "marine parks", the latter conceived for the preservation of marine biodiversity. Chile's first reproductive reserve, La Rinconada, was implemented in the north-central coast for the conservation of the *Argopecten purpuratus* stocks.

Recent interest in scallops reproductive reserves arose as a by-product of groundfish management in the US portion of Georges Bank, where three groundfish closures were implemented in 1994. Scallop dredging was banned within the closures because of groundfish bycatch, particularly flounders (NEFSC 1999; Murawski et al. 2000). As a result, fishing effort was re-directed to the Mid-Atlantic Region; in 1998 two additional areas with substantial abundance of young scallops were closed there for three years to increase yield per recruit and spawning biomass. The biomass of sea scallops (*Placopecten magellanicus*) within the Georges Bank closed areas increased 14-fold between 1994–1998. In July, 1998, total and harvestable scallop densities (expressed as biomass) were 9 and 14 times higher in closed than in adjacent open areas.

14.6.3.3 Ecosystem management concerns

Benthic shellfish harvesting devices, unlike pelagic gear or longlines, may affect the substrate and/or the biological assemblages of harvested patches (Auster 1998; Kaiser 1998; Raloff 1996). Local recovery may be related to changes in the benthic ecosystem rather than to the removal of the target species (Caddy 1979a; Fonseca et al. 1984; Orensanz 1986). Decline in *CPUE* of *Pecten fumatus* in Tasmania during the late forties (Fairbridge 1953) was blamed on the "constant stirring or plowing of the bottom by the dredges." Caddy (1979a) noticed that common overfishing criteria could not explain the declining trend observed in the Digby fishery since the mid-thirties. He suggested that habitat deterioration caused by fisheries operation could be the cause. The collapse of the fishery for *Aequipecten tehuelchus* in the NW of the San Matías Gulf in 1971 caused much concern about the ecological impact of dredging. It was estimated that each site of a fishing bed was dredged 7 times - on average - during a regular fishing season (Iribarne et al. 1991), implying a high level of disturbance. Auster et al. (1996) estimated that between 1984 and 1990 the entire US side of the Gulf of Maine was impacted annually (on average), while Georges Bank was impacted 3–4 times during the same period.

The ecological effects of dredging are, however, difficult to identify with certainty. Fairbridge (1953) for example, observed that settlement was poor in both dredged and undredged beds. Hill et al. (1999) re-sampled the benthos in 200 sites in a region of the Irish Sea that had been first sampled in 1946–1951; some of them were subject to heavy scallop dredging afterwards. Clear changes were apparent, but regardless of dredging. Laurenson et al. (1993) collected a wide array of evidence on the impact of trawling for saucer scallops off southwestern Australia. Comparison of commercially trawled and untrawled beds, combined with visual observation, indicated no significant impact on the benthic community, and short-lived physical effects on the bottom. Early experimental studies were qualitative (Caddy 1973; Butcher et al. 1981), or involved the disturbance of small experimental plots (Eleftheriou and Robertson 1992). Currie and Parry (1996) investigated changes induced by scallop dredging in the benthos of Port Phillip Bay (Australia) following a BACI (before, after, control, impact) design; disturbance mimicked commercial fishing operations in pattern and scale. Abundance of most species decreased by 20–30% after dredging, but the impact became undetectable for most of them following their next recruitment pulse; few had not yet recovered after 14 months. Changes in community structure caused by dredging were smaller than those that occur naturally between seasons or years.

Biogenic substrate (shell hash, sponge and coral bottoms, beds of epibenthic bivalves, seagrass meadows, etc) are particularly vulnerable, and their recovery from disturbance is often slow (if they recover at all). These substrates are important systemic components, often providing structural refuges or an adequate supply of primary settlement substrate, generally considered an important factor for successful scallop recruitment. Dredging or raking often removes or destroys such substrates (Auster et al. 1996), including shell hash (Caddy 1979a; Olivier and Capitoli 1980), hydroids, *Atrina* mussels (Hall-Spencer et al. 1999), and eelgrass (Belding 1910; Marshall 1947; Thayer and Stuart 1974). Catch is landed unsorted in the San Matías Gulf dredge fishery (Argentina); Olivier and Capitoli

(1980) suggested that removal of thousands of tons of shell hash (the primary settlement substrate in that ground) during the years 1968–1971 was the cause of several years of poor settlement (1971–1979). A similar concern has been recently raised with regards to the Icelandic scallop (*Chlamys islandica*) fishery (Naidu 1997). Caddy (1979a) observed that empty shells from at-sea shucking accumulated upstream and downstream the main beds in the Digby scallop beds, probably transported by tidal currents. Some components of the epifauna (mostly hydrozoans and bryozoans), known to be important substrates for primary settlement of *Placopecten*, were scarce in the fishing beds, but abundant further up the bay (Caddy 1970). He noted also (Caddy 1989a) that the removal of epifauna leaves the sediment unprotected from laminar water flow on the bottom; resulting changes in sediment composition may affect long-term productivity. Experimental studies conducted in North Carolina indicate that commercial scallop dredging and trawling for fish in shallow waters disrupt the vegetation (mostly eelgrass) and the bottom. This may impede the regrowth of the seagrass, to which scallops spat attaches. Thayer and Stuart (1974) suggested a 2-year harvest rotation period to allow for the recovery of seagrass beds.

Increased turbidity and siltation constitutes a second family of ecological side effects of scallop dredging. Black and Parry (1994) found that dredging may cause re-distribution of fine sediments in Port Phillip Bay (Australia). For typical currents, suspended sediment at above natural concentrations was confined to a region within about 54 m of the dredge.

Bans on dredging were early imposed in many fisheries because of concerns about their ecological side effects. In Chile, for example, only diving has been allowed in the shellfisheries for decades. In spite of the country being one of the world's main producers of molluscs, dredges are unheard of along its 38 latitudinal degrees of coastline. Scallop dredging was banned and replaced by commercial divers in the San José Gulf (Argentina) fishery for tehuelche scallops during the early 1970s. Dredges were replaced by trawls (generally regarded as less offensive) in the *Amusium* fisheries of Shark Bay (late 1960s) and Abrolhos Bank (early 1980s) (Young and Martin 1989, p. 630).

A long-ignored problem, impacts of dredging and trawling on benthic ecosystems are a subject of much current concern (Malakoff 1998). Reflecting this change of attitude, the USA Magnuson-Stevens Act now mandates that any fisheries management plan must include a provision to minimise, to the extent practicable, adverse effects on such habitat caused by fishing, and identify other actions to encourage the conservation and enhancement of such habitat (Schmidt 1998; Watling and Norse 1998).

14.6.3.4 Strategic structures: use and property rights

Fishers, like everybody else, behave rationally: they act in their best interest given certain conditions, among them the framework created by the management system. For a strategy to work, it is necessary that the interest of the fishers coincide as much as possible with the interest of society. This is particularly true in small-scale fisheries spread over long stretches of coast, in which traditional controls (size, quotas, effort) are often unenforceable. For a system to work, fishers (and other participants in the management system as well) should be offered the appropriate incentives. Use and/or property rights are a step in that direction. Scallop stocks are generally state property in the sense of Feeny et al. (1998). Three general forms of use rights have are found in the management systems of scallop fisheries. The natural form of fishing rights for stocks of sedentary organisms in coastal regions are co-management agreements through which *communities* are granted territorial use rights ("Territorial Use Rights for Fishers", or TURFs). Territories are difficult to demarcate and enforce in the case of offshore fishing grounds, where the obvious counterpart to communal territorial rights are *individual* rights, either to participate in the fishery or to have a certain share of effort or quota. Examples are presented below.

(1) *TURFs*. Territorial use rights over benthic resources involve a variety of formats: leases, quasi-property, property; also, they can be derived from traditional marine tenure, or introduced as a *de novo* management system. In an example of the latter, the Chilean Fisheries Act of 1992 formally introduced the right of fishing villages ("caletas") to claim tenure over portions of the seabed (AMERBs, Spanish acronym for "Areas for the Management and Exploitation of Benthic Resources"). These TURFs are granted after presentation of a base-line study and a management plan, and are effectively co-managed between fishers organisations in the caletas and the central fisheries authority. Stotz and González (1997a) described the effect of the implementation of a TURF for scallop (*Argopecten purpuratus*) exploitation by the Puerto Aldea Caleta (Tongoy Bay, Central-North Chile). The beds had been severely overexploited during the 1980s (largely through illegal fishing), and closed to legal fishing in 1988 (Stotz and González 1997a, 1997b). Following the claim of *de facto* use rights in 1991 (even before TURFS were enacted), fishers began taking care of the beds. Subsequently, average density increased six-fold over a three-year period, from 0.45 m^{-2} in 1992 to more than 3 m^{-2} in 1995. Although recovery of unfished beds is rapid, probably due to spawning from massive aquaculture in the bay (Section 14.3.1.4), illegal fishing continues to be intense in open-access areas. Production from fishers' organisations within the granted TURFs is likely to evolve from the exploitation of natural stocks to aquaculture (Chavez 1999).

Communal use rights have been well documented in Japanese fisheries. Following a ten-year post-war period (1943–1953) of disorderly fishing and declining yield, the drive to return to orderly catching practices resulted in the adoption of communal operation systems in about 1952 (Anonymous 1990). Ito (1964) described a case in the Okhotsk Sea Japanese scallop (*Patinopecten yessoensis*) fishery in which a strict management program (quotas, seasonal closures, minimum size, restocking, control of predators, fishers cooperative organisation) was voluntarily implemented in the Abashiri District starting in 1957, while no such a program was adopted in the neighbouring Soya District. The level of the catches was stabilised in Abashiri, in contrast to Soya, where catches plummeted to about 5% of the pre-war mean annual catch.

(2) *Limited entry*. Licensing programs, a tool for direct effort regulation (Section 14.6.4.4), have often evolved into limited entry programs, and then to making permits transferable (e.g., the scallop fisheries of southeastern Australia and Tasmania). In the initial stages of implementation of limited entry programs (or the moratoria on fleet size that often precede them) tough decisions have always to be made about who has the right to stay, which can create much turmoil in the industry. Limited entry was introduced in the Georges Bank Canadian fishery for *Placopecten magellanicus* in 1973 (Naidu 1991), while on the US side only a permit is required. In Australia, limited entry systems were introduced in the *Pecten fumatus* fisheries of Victoria (1971) and Tasmania (1985), and in the *Amusium balloti* fisheries of Queensland (1979), Shark Bay (1985) and Abrolhos Bank (1986) (Gwyther et al. 1991). Limited entry programs are often implemented in the wake of (or as reaction to) more entrants than the resource is presumed to sustain in the long term. Attempts to downsize the fisheries from East Coast of the US in 1994 left many displaced scallop boats looking for new opportunities, Alaska being a possibility (Shirley and Kruse 1995; Turk 2000). Alaska scallop fishers reacted persuading managers (the North Pacific Fishery Management Council) to introduce a system of limited entry, which froze at 18 the number of participants (Gay 1994).

(3) *Individual Transferable Quotas (ITQs)*. Brander and Burke (1995) compared two fisheries for sea scallops (*Placopecten magellanicus*) from Nova Scotia (eastern Canada): the inshore fishery of the Bay of Fundy and the offshore fishery of Georges Bank. The first is managed almost exclusively by traditional measures of input control, while the second has evolved to management by individual quotas in the form of "Enterprise Allocations", or EAs. These are given to license holders who may utilise one or more vessels to catch their quota allocation, thus differing from individual quota allocation systems (IVQs, ITQs) in which quotas are allocated to individual boat licenses. The behaviour of the two fleets has been radically different. Offshore harvesters, with secure catch shares, set conservative TACs to build up biomass. This resulted in an increase of *CPUE*, and consequently in lower effort to catch the TAC. In the inshore fishery, on the other hand, competitive license-holders seek improvement through campaigns to increase access to other fishing areas. With neither TACs nor individual entitlements, the harvest of a good year-class over time could not be managed without complex effort controls, which the fisheries authority (DFO) cannot afford to implement. In the EA program, by contrast, each enterprise manages its quota within the annual fishing plan and has a major interest in the longer-term management of bumper year-classes, and of the fishery in general. This case shows the advantages of quasi-property management regimes. ITQs or ITEs (Individual Transferable Effort) have been advocated as an option for scallop fisheries in other industrial offshore fisheries, e.g., in the northeastern USA (Kirkley et al. 1997) and Alaska (Gay 1996).

14.6.3.5 Diversification of the resource-base

Whichever the nature of the management system, diversification of the resource-base available to a fishing fleet is a way to spread the risks associated with individual resources due to uncertainty about recruitment and markets (Hilborn et al. 2001). A multiplicity of "methiers" ("polyvalence") is thought to be behind the relative success of the small-scale fleet of Saint-Brieuc, for which the coquille Saint-Jacques (*Pecten maximus*) is a key element (Anonymous 1995). Institutional frameworks, however, have often pushed fisheries towards specialisation. In Canada, specialisation is a byproduct of attempts by managers to stabilise catches within individual fisheries, fleets ending up "locked" by limited entry. As is the case of most aspects of management, problems associated with specialisation can be aggravated when allocation criteria are inconsistent or ambiguous. A mixture of scallop specialists and a dual-license fleet, the latter switching between scallops and groundfish, participate in the Bay of Fundy fishery (Brander and Burke 1995; Kenchington et al. 1995). During a bonanza in 1988–1990 boats in the dual-license fleet shifted to scallop fishing, doubling the number of scallop vessels in two years. This eroded the limited entry program in the scallop fishery, removing the incentives for spreading the bonanza over a longer time horizon. Limited entry in each of the fisheries might have resulted in a more stable fishery, which should not necessarily detract from the possibility of a boat stacking licenses for different fisheries, each of them with its own cap on entry.

14.6.4 Tactics

Whichever the strategic arrangement or management system implemented, any management plan will include tactics (and "controls" associated with them), in various combinations (e.g., see Young and Martin 1989, for Australian scallop fisheries). Size limits (Section 14.6.4.1, below) form part of almost every scallop management plan. Catch (quotas), escapement (thresholds) and effort, the traditional core of fisheries regulation, have been discussed by Hilborn (1986) and Perry et al. (1999) for invertebrate fisheries, and are exemplified in Sections 14.6.4.2–14.6.4.4 with scallop cases. Stocks of sedentary organisms present the opportunity to implement other tactics, many of which are spatially explicit (Orensanz and Jamieson 1998; Perry et al. 1999). A typical example is various combinations of rotation (Section 14.6.4.5), enhancement (Section 14.6.4.6), experimental management with spatially defined treatments and controls (Section 14.6.4.7), and reproductive refuges (Section 14.6.3.2).

14.6.4.1 Size limits

Age-at-harvest has been rarely used as a control. As an example, bay scallops harvested in Massachusetts are required to have one annual growth ring (MacFarlane 1991). Usually, it is the size of harvested scallops that is commonly regulated by means of (i) areal/seasonal closures, (ii) absolute size limits, and/or (iii) meat counts (maximum number of meats per pound).

Legal size limits are established to allow scallops to reproduce at least once before they are caught. Minimum legal sizes are popular because they are relatively easy to enforce, their rationale is easily understood by fishers and managers (Fairbridge 1953; Gibson 1956; Imai 1971; Ito 1964; Olivier and Capitoli 1980, etc.) and, unlike TACs or effort controls, their effectiveness does not depend on having reliable assessments. In Victoria (Australia), however, minimum size regulations were abandoned in favour of allowing the industry to determine an acceptable commercial size in order to avoid excessive handling-related damage to discarded scallops.

Because of geographic gradients in growth rate, size limits may need to be regionally adjusted. Two bays (Saint-Brieuc and Seine) produce 90–95% of the scallop (*Pecten maximus*) harvests from France. Abundance has been depressed in Seine in recent years, while the Saint-Brieuc fishery appears comparatively healthy. Growth rate and the reproductive cycle are very different between the two stocks; transplant experiments support the hypothesis that differences have a genetic basis (Section 14.3.2.2.). In Seine, scallops reach commercial size (110 mm) after their second winter, which is the age of first spawning. Thus, many of the scallops fished are pre-reproductive. In St. Brieuc, by contrast, scallops reach a smaller commercial size (102 mm) at an older age (most of the recruits have lived through their third winter). Considering that these two stocks appear to be self-sustaining and that both are now "recruitment fisheries" (Section 14.6.2.2), differences in the average number of reproductive opportunities provided by the respective size-specific harvesting regimes may explain their different success (Kiesel 1996).

Size limits are regulated through the selectivity of the fishing gear. This can be done in a number of ways:

i) Capture characteristics of the gear, for example through spacing of teeth in front of the dredge (Imai 1971; Dupouy 1978), or development of a "search image" for size in commercial divers (Orensanz 1986);

ii) Retention characteristics of the gear, for example through ring size in the New Bedford scallop dredge (Bourne 1965; Caddy 1971, 1972; Smolowitz 1979, in Sinclair et al. 1985) or mesh size in the "Keta-ami" scallop dredge in Japan (Imai 1971); or

iii) Selection of the catch on board of the vessel (practised in most scallop fisheries).

Regulation of selectivity based on mesh size may be reasonable in the case of dredges with a rigid mesh (e.g., dredges built of linked rigid rings). Unfortunately, selectivity is difficult to control through mesh size with other gear designs because the mesh (no matter how large) soon gets clogged with debris and non-target organisms (Medcof 1952; Fairbridge 1953; Baird and Gibson 1956; Bourne 1965; Orensanz 1986). Culling the catch on board or landing the gross catch unsorted (Olivier and Capitoli 1980) may worsen the problem. Under such circumstances, incidental mortality (Section 14.6.6.5) and disturbance of the benthic ecosystem are more important considerations than size-selectivity.

Meat counts may be preferable to size limits when the size of the shell is not a good predictor of muscle weight (Caddy 1989a), such as when there is great variability among localities, or because of seasonal changes in condition. Meat-count regulations, however, may lead to growth overfishing, as discussed below due to the "blending" practice (Section 14.6.2.1; Naidu 1991). In the US fishery for sea scallops meat counts were abandoned in 1994 in favour of effort controls, with majority support from the industry (Griffin 1993).

14.6.4.2 Quota regulations

Quota systems have been used mostly in industrial fisheries (e.g., the eastern Canada fisheries for sea scallops), but rarely in the case of coastal, small-scale scallop fisheries. One case is the St. Brieuc fishery for *Pecten maximus*, in which a quota system was introduced in 1974 (Fifas 1991). In the *Pecten fumatus* fishery from Port Phillip Bay (Australia) tentative quotas, based on a pre-season index of abundance, are adjusted during the season through consultation between fishers and managers (Noel 1997).

14.6.4.3 Escapement regulations

Thresholds (expressed in terms of some index of abundance, e.g., *CPUE*) are a member of the constant escapement family (Hilborn 1986; Quinnn et al. 1990). In the Port Phillip Bay (Australia) the *Pecten fumatus* season is closed when catch rate falls to a prescribed level which corresponds to an estimated residual population of 60 million scallops (Gwyther 1990). The Pertuis Charentais fishery for *Pecten maximus* is also managed with *CPUE* threshold regulations (Ansell et al. 1991). Sometimes fishing becomes unprofitable below some threshold density ("break-even" point) in the absence of regulations, but this may be too low for successful renewal (e.g., the *Pecten fumatus* fisheries of Bass Strait; Young 1994). In the *Amusium balloti* fishery of Queensland (Australia), for example, the cut off point in 1986–1987 corresponded to a density of one scallop per 120–150 m^{-2} (Dredge 1988). Maintaining a certain threshold density, qualitatively related to density dependence of fertilisation rate, has been occasionally proposed as a rationale (Fairbridge 1953, p. 38; Orensanz 1986; Caddy 1989a). Thresholds and harvest rates have been occasionally combined in the form of mixed strategies. In the recovery plan for the *Argopecten purpuratus* fishery of Tongoy Bay (Chile) an exploitation rate of 10% was proposed once the stock reaches a threshold average density of 10 scallops m^{-2}, while the exploitation rate during the recovery period would be only 5% (Stotz and González 1997a).

14.6.4.4 Direct effort regulations

(1) Restrictions on the efficiency of effort. Effort regulations are at least one among other components in the regulation of most managed scallop fisheries. Specific effort control measures include:

i) Restrictions on the size and/or power of the boats (Imai 1971; Dupouy 1978; Ansell et al. 1991).

ii) Size of gear units, and number of units (Fairbridge 1953; Dickie 1955; Dupouy 1978; Perrin and Hay 1987; Gwyther and McShane 1988; Ansell et al. 1991; Gwyther et al. 1991).

iii) Closed seasons and temporary spatial closures (Belding 1910; Sinclair et al. 1985; Gibson 1956; Fairbridge 1953; Ito 1964; Imai 1971; Dupouy 1978; Perrin and Hay 1987, etc.).

iv) Number of fishing days, or duration of the fishing day (Dupouy 1978; Gwyther and McShane 1988; Gwyther et al. 1991; Dredge 1994). In an extreme case, the Saint-Brieuc (France) fishery for *Pecten maximus* is opened during the fishing season in two slots of ½ hr each per week (Kiesel 1996).

The most prominent example is the US fishery for sea scallops (*Placopecten magellanicus*), the largest scallop fishery in the world, where the management strategy changed in 1994 from "meat count regulation" (Section 14.6.2.1) to effort regulation (NEFSC 1999). Fishing mortality (F) is calculated in two ways: (1) from year-to-year changes in an index of abundance of pre- and post-recruits (which provides an estimate of total mortality, Z), and a 'guesstimate' of natural mortality (M), and (2) combining a swept-area estimate of abundance, an experimental estimate of dredge efficiency, and catch. Effort is then regulated to satisfy management goals (see Section 14.6.4.4). Assessment and management with this protocol has become questionable after the introduction of large closed areas (Section 14.65.3.2) (NEFSC 1999, pp. 111–112).

Individual limits on catch per day or per trip have been used for *Pecten maximus* in France (Ansell et al. 1991), for *Pecten fumatus* in Australia (Gwyther et al. 1991), for almeja catarina (*Argopecten circularis*) in Bahia Concepción (México; León-Carballo et al. 1991), for *Aequipecten tehuelchus* in San José Gulf (Argentina), etc. These limits are, in general, just one more way to keep a handle on effort, not a real quota system designed to achieve a desired harvest rate.

(2) Seasons. Closed seasons in scallop fisheries have followed a number of rationales: as one mean among others to control effort (the most common case), minimising disturbance during critical stages (for example the settlement season), allowing short-lived species to spawn at least once, because of product quality (e.g., the *Pecten fumatus* fishery of Victoria, SE Australia; Gwyther et al. 1991, p. 839) or because of red tides or other environmental conflicts. An interesting case of a faulty rationale that has been invoked from time to time is that of protection of spawning individuals in the case of iteroparous species (see Shepherd 1993, for a concise commentary).

(3) Licensing. Many regulated scallop fisheries have incorporated licensing as just one more option in the menu available to control effort (Dupouy 1978; Gwyther and McShane 1988). In France, for example, a cap was placed on the number of licenses issued each year in some of the locally-managed fisheries for *Pecten maximus* (Ansell et al. 1991). Sometimes, however, licensing programs have been a first step in the direction of limited entry (Section 14.6.3.4).

14.6.4.5 Rotation

Rotating fishing areas is a natural option for the management of sessile species. Indeed, rotational strategies have been developed spontaneously for many invertebrate fisheries (e.g., in the calico scallop fishery; Blake and Moyer 1991), as an option that is intuitively sustainable even in the absence of reliable stock assessments. Local populations can be partitioned into areas or plots, each subjected to periodic pulse fishing (Munch-Petersen 1973; Sluczanowski 1984; Botsford et al. 1993). The optimal rotation problem (Clark 1976) is an analogue of Y/R analysis (Section 14.6.2.1), but here the control is the time interval between successive fishing pulses instead of a combination of age/size at recruitment and effort. An important conservation benefit of rotational strategies is that they may result in increased spawning biomass per recruit compared to traditional F-based strategies, for comparable yield levels. This advantage is exacerbated when there is substantial incidental mortality, as is the case of scallop dredge fisheries (Section 14.4.7) (Myers et al. 2001). Enforcement of spatial closures may be easier than other controls in nearshore fisheries, and also in offshore fisheries if vessels are required to have electronic location beacons (as in the US side of Georges Bank and in the offshore Argentine scallop fishery). Rotational strategies have been considered in New Zealand (Bull 1991), in the *Chlamys varia* fishery of the Rade de Brest (Ansell et al. 1991), in the sea scallop fishery of North America (Brander and Burke 1995; Kenchington et al. 1995; Myers et al. 2001), etc. The main obstacle for their implementation is opposition from other users of the seabed. Recovery of scallop abundance in areas closed to sea scallop (*Placopecten magellanicus*) dredging in Georges Bank, initially for groundfish protection, have encouraged managers to contemplate a formal rotation program in the scallop fishery (Murawski et al. 2000).

14.6.4.6 Direct interventions to enhance productivity

The management of marine [fin]fisheries relays largely upon regulations of effort, catch and minimum size. Benthic resources offer, instead, the possibility of more active interventions, destined to enhance productivity at the scale of patches of the seabed. Interventions fall into three broad groups: habitat manipulation, restocking of depleted areas, and control of predators. There is, obviously, a continuum between enhancing the productivity of natural stocks and aquaculture. One has often led to the other, as has been the case in Japan (Yamamoto 1964; Aoyama 1989). Spatially defined use rights (Section 14.6.3.4) provide incentives for investing in the enhancement of productivity, as seen in the cases of Chile and Japan.

(1) Habitat manipulation. This refers largely to the provision of primary settlement substrate. Seagrass meadows, important settlement substrate for bay scallops (Stauffer 1937; Dreyer and Castle 1941; Marshall 1947), have been decimated along the east coast of the US by different agents (anthropogenic and natural), "brown tides" being the newest among them. Replanting programs have often been implemented to palliate the effect of these catastrophes on scallop recruitment (Kelley 1988). "Shellstocking", the practice of returning empty shells to the fishing beds, was advocated for the tehuelche scallop fishery of Patagonia (Olivier and Capitoli 1980) and for the Icelandic scallop fishery of eastern Canada (Naidu 1997).

(2) Restocking. In 1906 Belding (1910) transplanted bay scallops to an area where they had virtually disappeared. Strong settlement in the following year was attributed to the

spawn of the transplanted lot. This and other incidental observations prompted him to suggest transplants of "seed" scallops from sites exposed to winter mortality to others where survival and growth were known to be good (Belding 1910, p. 123), but this was ruled out due to high costs and political problems. Peterson et al. (1996) and Blake (1998) showed that, as anticipated by Belding (1910), restocking may help the recovery of bay scallops from many isolated estuaries along the east coast of the US and the west coast of Florida, from which they disappeared due to environmental deterioration (pollution, loss of sea grass, brown and tides, etc.).

Stocking effectively started in Japan in the thirties, and developed through successive stages (Mottet 1979; Ventilla 1982; Aoyama 1989; Kitada and Fujishima 1997): from transplants between beds (Yamamoto 1950, 1953, 1964), to the use of artificial spat collectors, to - finally - massive sowing of the seabed. The latter involves several stages: preparation of the seabed by removing predators and competitors, sowing, thinning, and fishing (Anonymous 1990; Ito 1991). The spectacular increase in settlement rate in Mutsu Bay after 1975 (Ventilla 1982; Aoyama 1989) is the most dramatic example of the effects of massive restocking (Section 14.3.1.4).

Experimental seeding programs, often combined with rotation, have been conducted for several shallow water species, including *Pecten maximus* and *Chlamys varia* in the British Islands (Buestel et al. 1985; Ansell et al. 1991; Brand et al. 1991; Heath 1999) and France (Dao and Carval 1999), *Placopecten magellanicus* in eastern Canada (Cliche et al. 1994; Barbeau et al. 1996; Hatcher et al. 1996), *Pecten fumatus* in Australia (Cropp 1986; Perrin and Hay 1987: p. 7), *Pecten meridionalis* in New Zealand (Bull 1991), *Argopecten irradians* along the east coast of the US (Middleton 1983; Goldberg et al. 1999), and *A. purpuratus* in Chile (Munita 1988). Restocking, however, is still an unrealistic option in the management of many fisheries (see for example Dredge 1988: p. 239). A major problem with restocking/rotation programs are legal limitations to closing of beds (Bull 1991; Ansell et al. 1991), which is of course not the case in systems structured around territorial use rights (Section 14.6.3.4). Managers and fishers from Iles-de-la-Madeleine (Quebec) are collaborating since 1993 in a unique large-scale restocking program ("REPERE"); 1.7 million juveniles were seeded, 77,000 of them tagged. Fishing was closed initially, and reopened in 1997 (Nadeau and Cliche 1997, 1999; Cliche and Giguère 1999; Nadeau et al. 1999). Annual landings from the Bay of Brest fishery (*Pecten maximus*) have increased from 50 tons during the early 1980s to 200 tons in recent years as a result of the combined rotational harvest of natural and seeded beds (Dao and Carval 1999).

(3) Control of Predators. Elimination of predators has been frequently recommended, sometimes even enforced, as part of scallop management programs (Belding 1910; Ito 1964; Golikov and Scarlato 1970). Starfish have been the most frequent target. Active control techniques, like applications of calcium hydroxide to kill starfish (Arima et al. 1975), have been tried in some cases. Most frequently, however, only predators caught in the fishing gear are destroyed. Fishermen in the Okostsk Sea Japanese fishery are sometimes required to retain and land starfish (Ito 1964). Abundance of starfish is assessed during the scallop surveys, allowing an estimate of the effect of removal. Every

year fishermen killed 1.5–5.0% of the starfish present on the beds (Ito 1964). Control measures can be more indirect, as was the case with the prohibition of cleaning the catch at sea in Tasmania (Fairbridge 1953), in the belief that throwing scallop waste on the beds would enhance starfish populations. The effect of predator removal on managed scallop stocks remains – to our best knowledge – unassessed.

14.6.4.7 Experimental management

Benthic resources offer the opportunity to implement experimental management programs with spatially distributed treatments and controls. One limitation in the case of forms with pelagic larvae is that only measures associated with pre-dispersal biological may be amenable to this type of experimentation, as larvae generated in an experimental plot settle elsewhere. Yet, Peterson et al. (1996) showed that experiments can be informative even in the case of measures relative to pre-dispersal processes (enhancement of the spawning stock in their case). In most cases, however, experimentation has dealt with control of predators, the secondary effects of harvesting techniques, density-dependence of post-settlement growth and mortality, etc.

Ito (1964) described a situation in the Okhotsk Sea fishery that came close to experimentation with the entire management system, following a ten-year post-war period (1943–1953) of disorderly fishing and declining yield. A strict management program (quotas, seasonal closures, minimum size, restocking, control of predators, cooperative organisation of the fishers) was voluntarily implemented in the Abashiri District starting in 1957, while no such a program was adopted in the neighbouring Soya District. The level of the catches was stabilised in Abashiri, in contrast to Soya, where catches plummeted to about 5% of the pre-war mean annual catch.

ACKNOWLEDGMENTS

We dedicate this contribution to our colleague and friend John F. Caddy, always the first to think in the right direction.

REFERENCES

André, C. and Rosenberg, R., 1991. Adult-larval interactions in the suspension-feeding bivalves *Cerastoderma edule* and *Mya arenaria*. Mar. Ecol. Progr. Ser. 71:227–234.

Anonymous, 1990. Scallop aquaculture - Solving the mystery of good and poor harvests. Yamaha Fishery Journal 34:8 p.

Anonymous, 1995. Baie de Saint-Brieuc - La coquille sauve les meubles. France-Eco-Peche December 1995:2831.

Ansell, A.D., Dao, J.-C. and Mason, J., 1991. Three European scallops: *Pecten maximus*, *Chlamys (Aequipecten) opercularis* and *C. (Chlamys) varia*. In: S.E. Shumway (Ed.). Scallops: Biology, Ecology and Aquaculture. Vol. 21. Elsevier, Amsterdam. pp. 715–751.

Aoyama, S., 1989. The Mutsu Bay scallop fisheries: scallop culture, stock enhancement, and resource management. In: J.F. Caddy (Ed.). Marine Invertebrates Fisheries: Their Assessment and Management. John Wiley & Sons. pp. 525–539.

Arima, K., Shimazaki, H., Matsui, T., Sugiya, T. and Kamada, R., 1975. [The results of mass-planting scallop seed to Mori and Yakumo in Funka Bay]. Hokusuishi Geppo 32(10):1–21. (In Japanese)

Arnold, W.S., Marelli, D.C., Bray, C.P. and Harrison, M.M., 1998. Recruitment of bay scallops *Argopecten irradians* in Floridan Gulf of Mexico: scales of coherence. Mar. Ecol. Progr. Ser. 170:143–157.

Arsenault, D.J. and Himmelman, J.H., 1998. Spawning of the Iceland scallop (*Chlamys islandica* Muller, 1776) in the northern Gulf of St. Lawrence and its relationship to temperature and phytoplankton abundance. The Veliger 41:180–185.

Auster, P.J., 1998. A conceptual model of the impacts of fishing gear on the integrity of fish habitats. Cons. Biol. 12:1198–1203.

Auster, P.J., Malatesta, R.J., Langton, R.W., Watling, L., Valentine, P.C., Donaldson, C.L.S., Langton, E.W., Shepard, A.N. and Babb, I.G., 1996. The impacts of mobile fishing gear on seafloor habitats in the Gulf of Maine (northwest Atlantic): implications for conservation of fish populations. Rev. Fish. Sci. 4:185–202.

Bachelet, G., Desprez, M., Ducrotoy, J.-P., Guillou, J., Labourg, P.-J., Rybarczyk, H., Sauriau, P.-G., Elkaim, B. and Glémarec, M., 1992. Rôle de la compétition intraspécifique dans la régulation du recrutement chez la coque, *Cerastoderma edule* (L.). Ann. Inst. Océanogr. (Paris) 68:75–87.

Baird, R.H., 1959. Factors affecting the efficiency of dredges. In: H. Kristjonsson (Ed.). Modern Fishing Gear of the World. Fishing News (Books) Ltd., London. pp. 222–224.

Baird, R.H. and Gibson, F.A., 1956. Underwater observations on escallop (*Pecten maximus* L.) beds. J. Mar. Biol. Assoc. U.K. 35:555–562.

Barbeau, M.A., Hatcher, B.G., Scheibling, R.E., Hennigar, A.W., Taylor, L.H. and Risk, A.C., 1996. Dynamics of juvenile sea scallops (*Placopecten magellanicus*) and their predators in bottom seeding trials in Lunenburg Bay, Nova Scotia. Can. J. Fish. Aquat. Sci. 53:2494–2512.

Beaumont, A.R., 1991. Allozyme data and scallop stock identification. J. Cons. 47:333–338.

Beaumont, A.R., Morvan, C., Huelvan, S., Lucas, A. and Ansell, A.D., 1993. Genetics of indigenous and transplanted populations of *Pecten maximus*: no evidence of the existence of separate stocks. J. Exp. Mar. Biol. Ecol. 169:77–88.

Beaumont, A.R. and Zouros, E., 1991. Genetics of scallops. In: S.E. Shumway (Ed.). Scallops: Biology, Ecology and Aquaculture. Vol. 21. Elsevier, Amsterdam. pp. 585–624.

Bechtol, W.R. and Bue, B., 1998. Weathervane scallops, *Patinopecten caurinus* near Kayak Island, Alaska, 1996. Alaska Dept. of Fish and Game, Regional Information Report No. 2A98–20.

Belding, D.L., 1910. A report upon the scallop fishery of Massachusetts; including the habits, life history of *Pecten irradians*, its rate of growth and other facts of economic value. The Commonwealth of Massachusetts. 150 p.

Beverton, R.J.H. and Holt, S.J., 1957. On the dynamics of exploited populations. U.K. Min. Agr. Fish., Fish. Invest. (Ser. 2) 19:533 p.

Black, K.P. and Parry, G.D., 1994. Sediment transport rates and sediment disturbance due to scallop dredging in Port Phillip Bay. 2nd Australasian Scallop Workshop, Triabunna, Tasmania, 23 March 1993. Mem. Queensl. Mus. 36:315–326.

Blake, N.J., 1998. The potential for reestablishing bay scallops to the estuaries of the west coast of Florida. Trans. 63rd No. Am. Wild. and Natur. Resour. Conf. pp. 184–189.

Blake, N.J. and Moyer, M.A., 1991. The calico scallop, *Argopecten gibbus*, fishery of Cape Canaveral, Florida. In: S.E. Shumway (Ed.). Scallops: Biology, Ecology and Aquaculture. Vol. 21. Elsevier, Amsterdam. pp. 899–911.

Blake, S.G. and Graves, J.E., 1995. Mitochondrial DNA variation in the bay scallop, *Argopecten irradians* (Lamarck, 1819), and the Atlantic calico scallop, *Argopecten gibbus* (Linnaeus, 1758). J. Shellfish Res. 14:79–85.

Bonardelli, J.C., Himmelman, J.H. and Drinkwater, K., 1996. Relation of spawning of the giant scallop, *Placopecten magellanicus*, to temperature fluctuations during downwelling events. Mar. Biol. 124:637–649.

Botsford, L.W., 1995. Population dynamics of spatially distributed, meroplanktonic, exploited marine invertebrates. ICES Mar. Sci. Symp. 199:118–128.

Botsford, L.W., Castilla, J.C. and Peterson, C.H., 1997. The management of fisheries and marine ecosystems. Science 277:509–515.

Botsford, L.W. and Hobbs, R.C., 1995. Recent advances in the understanding of cyclic behavior of Dungeness crab (*Cancer magister*) populations. ICES Mar. Sci. Symp. 199:157–166.

Botsford, L.W., Quinn, J.F., Wing, S.R. and Brittnacher, J.G., 1993. Rotating spatial harvest of a benthic invertebrate, the red sea urchin, *Strongylocentrtus franciscanus*. Proceedings of the International Symposium Management Strategies for Exploited Fish Populations. Alaska Sea Grant College Program. AK–SG–93–02:409–428.

Boucher, J., 1985. Caracteristiques dynamiques du cycle vital de la coquille Saint-Jacques (*Pecten maximus*). Hypotheses sur les stades critiques pour le recrutement. CIEM CM 1985/K23:11 p. (mimeo)

Bourne, N., 1964. Scallops and the offshore fishery of the Maritimes. Bull. Fish. Res. Bd. Canada 145:60 p.

Bourne, N., 1965. A comparison of catches by 3- and 4-inch rings on offshore scallop drags. J. Fish. Res. Bd. Canada 22:313–333.

Bourne, N., 1966. Relative fishing efficiency and selection of three types of scallop drags. ICNAF Res. Bull. 3:15–25.

Brand, A.R., 1991. Scallop ecology: distributions and behaviour. In: S.E. Shumway (Ed.). Scallops: Biology, Ecology and Aquaculture. Vol. 21. Elsevier, Amsterdam. pp. 517–584.

Brand, A.R., Allison, E.H. and Murphy, E.J., 1991. North Irish Sea scallop fisheries: a review of changes. In: S.E. Shumway and P.A. Sandifer (Eds.). An International Compendium of Scallop Biology and Culture. World Aquaculture Workshops No. 1. World Aquaculture Society, Baton Rouge, LA. pp. 204–218.

Brand, A.R., Wilson, U.A.W., Hawkins, S.J., Allison, E.H. and Duggan, N.A., 1991. Pectinid fisheries, spat collection, and the potential for stock enhancement in the Isle of Man. ICES Mar. Sci. Symp. 192:79–86.

Brander, L., and Burke, D.L., 1995. Rights-based vs. competitive fishing of sea scallops *Placopecten magellanicus* in Nova Scotia. Aquat. Living Resour. 8:279–288.

Bricelj, V.M., Epp, J. and Malouf, R.E., 1987. Intraspecific variation in reproductive and somatic growth cycles of bay scallops *Argopecten irradians*. Mar. Ecol. Progr. Ser. 36:123–137.

Briggs, R.P., 1991. A study of the Northern Ireland fishery for *Pecten maximus* (L.). In: S.E. Shumway and P.A. Sandifer (Eds.). An International Compendium of Scallop Biology and Culture. World Aquaculture Workshops No. 1. World Aquaculture Society, Baton Rouge, LA. pp. 249–255.

Buestel, D., Dao, J.C. and Gohin, F., 1985. Estimation d'un stock naturel de coquilles Saint Jacques par une methode combinant les dragages et la plongee. Traitment des resultants par une methode geostatistique. ICES CM 1985/K:18:19 p. (mimeo)

Buestel, D., Dao, J.C. and Lemarie, G., 1979. Collecte de naissain de Pectinides en Bretagne. Rapp. Pv. Réun. CIEM 175:80–84.

Bull, M.F., 1991. New Zealand. In: S.E. Shumway (Ed.). Scallops: Biology, Ecology and Aquaculture. Vol. 21. Elsevier, Amsterdam. pp. 853–859.

Burns, D., Egan, T., Seward, E. and Naidu, K.S., 1995. Identification of scallop beds using an acoustic ground discrimination technique (abstract only). Book of Abstracts, 10th International Pectinid Workshop, Cork, Ireland.

Butcher, T., Matthews, J., Glaister, J. and Hamer, G., 1981. Study suggests scallop dredges causing few problems in Jervis Bay. Aust. Fish. 40:9–12.

Butterworth, D.S. and Punt A.E., 1999. Experiences in the evaluation and implementation of management procedures. ICES J. Mar. Sci. 56:985–998.

Caddy, J.F., 1968. Underwater observations on scallop (*Placopecten magellanicus*) behaviour and drag efficiency. J. Fish. Res. Bd. Canada 25:2123–2141.

Caddy, J.F., 1970. A method of surveying scallop populations from a submersible. J. Fish. Res. Bd. Canada 27:535–549.

Caddy, J.F., 1971. Efficiency and selectivity of the Canadian offshore scallop dredge. ICES CM 1971/K:25:8 p. (mimeo)

Caddy, J.F., 1972. Size selectivity of the Georges Bank offshore dredge and mortality estimate for scallops from the northern edge of Georges in the period June 1970 to 1971. ICNAF Redbook 1972:79–85.

Caddy, J.F., 1973. Underwater observations on tracks of dredges and trawls and some effects of dredging on a scallop ground. J. Fish. Res. Bd. Canada 30:173–180.

Caddy, J.F., 1975. Spatial model for an exploited shellfish population, and its application to the Georges Bank scallop fishery. J. Fish. Res. Bd. Canada 32:1305–1328.

Caddy, J.F., 1976. Practical considerations for quantitative estimations of benthos from a submersible. In: E.A. Drew, J.N. Lythgoe and J.D. Woods (Eds.). Underwater Research. Academic Press, New York, NY. pp. 285–298.

Caddy, J.F., 1979a. Long-term trends and evidence for production cycles in the Bay of Fundy scallop fishery. Rapp. P.-v. Réun. CIEM 175:97–108.

Caddy, J.F., 1979b. Some considerations underlying definitions of catchability and fishing effort in shellfish fisheries, and their relevance for stock assessment purposes. Fish. Mar. Serv. (Canada) MS Rep. No. 1489:19 p.

Caddy, J.F., 1989a. A perspective on the population dynamics and assessment of scallop fisheries, with special reference to the sea scallop *Placopecten magellanicus* Gmelin. In: J.F. Caddy

(Ed.). Marine Invertebrate Fisheries: Their Assessment and Management. John Wiley & Sons, New York, NY. pp. 559–574.

Caddy, J.F., 1989b. Recent developments in research and management for wild stocks of bivalves and gastropods. In: J.F. Caddy (Ed.). Marine Invertebrate Fisheries: Their Assessment and Management. John Wiley & Sons, New York, NY. pp. 665–700.

Caddy, J.F. and Caroci, F., 1999. The spatial allocation of fishing intensity by port-based inshore fleets: a GIS application. ICES J. Mar. Sci. 56:388–403.

Caddy, J.F. and Garcia, S., 1986. Fisheries thematic mapping - A prerequisite for intelligent management and development of fisheries. Océanogr. Trop. 21:31–52.

Caddy, J.F. and Gulland, J.A., 1983. Historical patterns of fish stocks. Mar. Pol. Oct. 1983:267–278.

Caddy, J.F. and Jamieson, G.S., 1977. Assessment of Georges Bank (ICNAF subdivision 5Ze) scallop stock 1972–76 incorporated. CAFSAC Res. Doc. 77/32:22 p.

Caputi, N., Penn, J.W., Joll, L.M. and Chubb, C.F., 1998. Stock-recruitment-environment relationships for invertebrate species of Western Australia. In: G.S. Jamieson and A. Campbell (Eds.). Proceedings of the North Pacific Symposium on Invertebrate Stock Assessment and Management. Can. Spec. Publ. Fish. Aquat. Sci. 125:247–256.

Chapman, C.J., Mason, J. and Kinnear, J.A.M., 1977. Diving observations on the efficiency of dredges used in the Scottish fishery for the scallop, *Pecten maximus* (L.). Scott. Fish. Res. Rep. 10:16 p.

Charpy-Roubaud, C.J., Charpy, L.J. and Maestrini, S.Y., 1982. Fertilité des eaux côtieres nord - patagoniques: facteurs limitant la production du phytoplancton et potentialités d'exploitation mytilicole. Oceanol. Acta 5:179–188.

Chavez, L.P., 1999. Conversion of artesanal fishermen to artesanal scallop culturists. 12th International Pectinid Workshop, Bergen (Norway), 5–11 May 1999, book of abstracts. (abstract only)

Ciocco, N.F., 1991. Differences in individual growth rate among scallop (*Chlamys tehuelcha* (d'Orb.)) populations from San José Gulf (Argentina). Fisheries Research 12:31–42.

Ciocco, N.F., 1992. Differences in individual growth rate among scallop (*Chlamys tehuelcha* (d'Orb.)) populations from the San José Gulf (Argentina): experiments with transplanted individuals. J. Shellfish Res. 11:27–30.

Ciocco, N.F. and Aloia, D.A., 1991. La pesquería de vieira tehuelche, *Chlamys tehuelcha* (d'Orb., 1846), del golfo San José (Argentina): abundancia de clases anuales. Scientia Marina 55:569–575.

Ciocco, N.F. and Monsalve, M.A., 1999a. The tehuelche scallop, *Aequipecten tehuelchus*, from San José Gulf (Argentina): spats settlement during the collapse of the fishery. 12th International Pectinid Workshop, Bergen (Norway), 5–11 May 1999, book of abstracts. (abstract only)

Ciocco, N.F. and Monsalve, M.A., 1999b. La vieira tehuelche, *Aequipecten tehuelchus* (d'Orb., 1846), del golfo San José (Argentina): captación de postlarvas durante el colapso de la pesquería. Biología Pesquera (Chile) 27:23–36.

Ciocco, N.F., Monsalve, M.A., Diaz, M.A., Vera, R., Signorelli, J.C. and Diaz, O., 1997. La vieira tehuelche del golfo San José: primeros resultados de la campaña de relevamiento SANJO/96. LAPEMAR, Centro Nacional Patagónico, Puerto Madryn, Argentina, Technical Report 1:30 p.

Ciocco, N.F. and Orensanz, J.M., 1997. Collapse of the Tehuelche scallop (*Aequipecten tehuelchus*) fishery from San José Gulf (Argentina). In: E.F. Félix Pico (Ed.). 11th International Pectinid

Workshop, La Paz, Baja California Sur, México, April 10–15, 1997), Book of Abstracts. CICIMAR, La Paz, México.

Claereboudt, M.R. and Himmelman, J.H., 1996. Recruitment, growth and production of giant scallops (*Placopecten magellanicus*) along an environmental gradient in Baie des Chaleurs, eastern Canada. Mar. Biol. 124:661–670.

Clark, C., 1976. Mathematical Bioeconomics: The Optimal Management of Renewable Resource. John Wiley & Sons, New York, NY. 352 p.

Clark, C., 1982. Concentration profiles and the production and management of marine fisheries. In: W. Eichlorn, R. Henn, K. Neumann and R.W. Shephard (Eds.). Economic Theory of Natural Resources. Physica-Verlag, Wurzburg-Wien, Heidelberg, Germany. pp. 97–112.

Clark, C., 1985. Bioeconomics Modelling and Fisheries Management. John Wiley & Sons, New York, NY. 291 p.

Cliche, G. and Giguère, M., 1999. Research program (REPERE) on scallop (*Placopecten magellanicus*) culture and restocking in Iles-de-la-Madeleine (Quebec, Canada). 12th International Pectinid Workshop, Bergen (Norway), 5–11 May 1999, book of abstracts. (abstract only)

Cliche, G., Giguère, M. and Vigneau, S. 1994. Dispersal and mortality of sea scallops, *Placopecten magellanicus* (Gmelin, 1791), seeded on the sea bottom of Iles-de-la-Madeleine. J. Shellfish Res. 13:565–570.

Cochard, J.C. and Devauchelle, N., 1993. Spawning, fecundity and larval survival and growth in relation to controlled conditioning in native and transplanted populations of *Pecten maximus* (L.): evidence for the existence of separate stocks. J. Exp. Mar. Biol. Ecol. 169:41–56.

Coleman, N., Gwyther, D. and Burguess, D., 1988. Prospects for the Victorian scallop fishery. Australian Fisheries 47:16–17.

Conan, G.Y., 1985. Assessment of shellfish stocks by geostatistical techniques. ICES CM 1985/K:30:24 p.

Cook, D., 1983. Designs of the drag. Virginia Sea Grant Mar. Res. Bull. 15:6–8.

Cooper, R.A. and Marshall, N., 1963. Condition of the bay scallop, *Argopecten irradians*, in relation to age and the environment. Chesapeake Sci. 4:126–134.

Cover, P. and Sterling, D., 1994. Scallop dredging - an engineering approach. 2nd Australasian Scallop Workshop, Triabunna, Tasmania, Australia. Mem. Queensland Mus. 36:343–349.

Cowen, R.K., Lwiza, K.M.M., Sponaugle, S., Paris, C.B. and Olson, D.B., 2000. Connectivity of marine populations: open or closed? Science 287:857–859.

Craig, P.D. and McLoughlin, R.J., 1994. Chapter 16: modelling scallop larvae movement in Great Oyster Bay. In: P.W. Sammarco and M.L. Heron (Eds.). The Bio-Physics of Marine Larval Dispersal. American Geophysical Union, Washington, DC. pp. 307–326.

Crecco, V. and Overholtz, W.J., 1990. Cause of density-dependent catchability for Georges Bank haddock *Melanogrammus aeglefinus*. Can. J. Fish. Aquat. Sci. 47:385–394.

Cressie, N., 1991. Statistics for Spatial Data. John Wiley & Sons, New York, NY. 900 p.

Cropp, D.A., 1986. Re-seeding scallops on the sea bed. Australian Fisheries November 1986:16.

Cummins, R., Jr., 1971. Callico scallops of the southeastern United States, 1959–1969. NOAA Spec. Scient. Rep. (Fisheries) 627:22 p.

Currie, D.R. and Parry, G.D., 1996. Effects of scallop dredging on a soft bottom sediment community: a large-scale experimental study. Mar. Ecol. Progr. Ser. 134:131–150.

850

Currie, D.R. and Parry, G.D., 1999. Impacts and efficiency of scallop dredging on different soft substrates. Can. J. Fish. Aquat. Sci. 56:539–550.

Dadswell, M.J. and Parsons, G.J., 1994. Reproduction and recruitment dynamics of the sea scallop, *Placopecten magellanicus*, in the Canadian Maritimes. In N.F. Bourne, B.L. Bunting and L.D. Townsend (Eds.). Proceedings of the 9th International Pectinid Workshop, Nanaimo, B.C., Canada, April 22–27, 1993. Can. Tech. Rep. Fish. Aquat. Sci. 1994 (1):147. (abstract only)

Dao, J.C. and Carval, J.P., 1999. Present status of the scallop production (*Pecten maximus*) in Bay of Brest (France) combining aquaculture and fishery. 12th International Pectinid Workshop, Bergen (Norway), 5–11 May 1999, book of abstracts. (abstract only)

Dao, J.C., Laurec, A. and Buestel, D., 1975. Application de la dynamique des populations au gisement de coquilles St. Jacques de la Baie de St. Brieuc. Recherche d'un modele bio-economique. In: Economie et Problemes des Peches Locales, OCDE. pp. 165–171.

Dare, P.J., Palmer, D.W., Howell, M.L. and Darby, C.D., 1994. Experiments to assess the relative dredging performances of research and commercial vessels for estimating the abundance of scallops (*Pecten maximus*) in the western English Channel fishery. MAFF (Lowestoft), Fish. Res. Tech. Rep. 96:9 p.

Del Norte, A.G.C., Capuli, E.C. and Mendoza, R.A., 1988. The scallop fishery of Lingayen Gulf, Philippines. Asian Fish. Sci. 1:207–213.

Dickie, L.M., 1955. Fluctuations in abundance of the giant scallop, *Placopecten magellanicus* (Gmelin), in the Digby area of the Bay of Fundy. J. Fish. Res. Bd. Canada 12:797–856.

Diggle, P.J., 1983. Statistical Analyses of Spatial Point Patterns. Academic Press, London. 148 p.

Dow, R.L., 1962. A method of predicting fluctuations in the sea scallop populations of Maine. Comm. Fish. Rev. 24 (10):1–4.

Dow, R.L. 1969. Sea scallop fishery. In: F.E. Firth (Ed.). The Encyclopedia of Marine Resources. Van Nostrand Reinhold Co. pp. 616–623.

Dredge, M.C.L., 1985. Estimates of natural mortality and yield-per-recruit for *Amusium japonicum balloti* Bernardi (Pectinidae) based on tag recoveries. J. Shellfish Res. 5:103–109.

Dredge, M.C.L., 1986. The effect of variation in prawn and scallop stocks on the behaviour of a fishing fleet. In: T.J.A. Hundloe (Ed.). Fisheries Management, Theory and Practice in Queensland. Griffith Univ. Press. pp. 167–177.

Dredge, M.C.L., 1988. Recruitment overfishing in a tropical scallop fishery? J. Shellfish Res. 7:233–239.

Dredge, M.C.L., 1994. Modelling management measures in the Queensland scallop fishery. Mem. Queensl. Mus. 36:307–314.

Dreyer, W.A. and Castle, W.A., 1941. Occurrence of the bay scallop, *Pecten irradians*. Ecology 28:425–427.

DuPaul, W.D., Heist, E.J. and Kirkley, J.E., 1988a. Comparative analysis of sea scallop escapement/retention and resulting economic impacts. Virginia Sea Grant College Program. 150 p.

DuPaul, W.D., Heist, E.J., Kirkley, J.E. and Testaverde, S., 1988b. A comparative analysis of the effects on technical efficiency and harvests of sea scallops (*Placopecten magellanicus*) by otter trawls of various mesh sizes. East Coast Fisheries Association. 70 p.

Dupouy, H., 1978. L'exploitation de la coquille Saint-Jacques *Pecten maximus* L. en France. 1er Partie: Presentation des pecheries. Science et Peche 276:1–11.

Ecker, M.D. and Heltshe, J.F., 1994. Geostatistical estimates of scallop abundance. In N. Lange, L. Ryan, L. Billard, D. Brillinger, L. Conquest and J. Greenhouse (Eds.). Case Studies in Biometry. John Wiley & Sons Inc., New York. pp. 125–144.

Eleftheriou, A. and Robertson, M.R., 1992. The effects of experimental scallop dredging on the fauna and physical environment of a shallow sandy community. Neth. J. Sea Res. 30:289–299.

Elliott, J.M., 1977. Some Methods for the Statistical Analysis of Samples of Benthic Invertebrates, 2nd edition. Freshwater Biological Association, Scientific Publication 25:160 p.

FAO, 1995. Precautionary approach to fisheries. Part 1: guidelines on the precautionary approach to capture fisheries and species introductions. FAO Fish Tech Pap. 350/1:52 p.

Fairbridge, W.S., 1953. A population study of the Tasmanian "commercial" scallop Notovola meridionalis (Tate) (Lamellibranchia, Pectinidae). Austr. J. Mar. Freshw. Res. 4:1–40.

Feeny, D., Berkes, F., McCay, B.J. and Acheson, J.M., 1998. The tragedy of the commons: twenty-two years later. In: J.A. Baden and D.S. Noonan (Eds.). Managing the Commons. Indiana University Press.

Félix-Pico, E.F., 1991. México. In: S. Shumway (Ed.). Scallops: Biology, Ecology and Aquaculture. Vol. 21. Elsevier, Amsterdam. pp. 943–980.

Fevolden, S.E., 1989. Genetic differentiation of the Iceland scallop Chlamys islandica (Pectinidae) in the northern Atlantic Ocean. Mar. Ecol. Progr. Ser. 51:77–85.

Fifas, S., 1991. Un modéle empirique de capturabilité pour le stock de coquilles Saint-Jacques (Pecten maximus L.) en Baie de Saint-Brieuc (Manche, France). CIEM, Shellfish Committee, CM 1991/K:37:20 p.

Fifas, S. and Berthou, P., 1992. Un modele d'efficacité d'une drague experimentale a coquilles Saint Jacques (Pecten maximus L.). Exemple de la pecherie de la Baie de Saint-Brieuc (Manche, France). Etude des sensibilites. CIEM, Statistics Committee, CM 1992/D:27:20 p.

Fifas, S. and Berthou, P., 1999. An efficiency model of a scallop dredge: sensitivity study. ICES J. Mar. Sci. 56:489–499.

Fifas, S., Dao, J.-C. and Boucher, J., 1990. Un modèle empirique du recrutement pour le stock de coquilles Saint-Jacques, Pecten maximus (L.) en baie de Saint-Brieuc (Manche, France). Aquat. Living Resour. 3:13–28.

Fonseca, M.S., Thayer, G.W. and Chester, A.J., 1984. Impact of scallop harvesting on eelgrass (Zostera marina) meadows: implications for management. North Am. J. Fish. Manag. 4:286–293.

Franklin, A., Pickett, G.D, Holme, N.A. and Barret, R.L., 1980. Surveying stocks of scallops (Pecten maximus) and queens (Chlamys opercularis) with underwater television. J. Mar. Biol. Assoc. U.K. 60:181–191.

Fraser, D.I., 1991. Settlement and recruitment in Pecten maximus (Linnaeus, 1758) and Chlamys (Aequipecten) opercularis (Linnaeus, 1758). In: Shumway, S.E. and Sandifer, P.A. (Eds.). An International Compendium of Scallop Biology and Culture. World Aquaculture Workshops No. 1. World Aquaculture Society, Baton Rouge, LA. pp. 28–35.

Frechette, M. and Bourger, E., 1985. Food-limited growth of Mytilus edulis L. in relation to the benthic boundary layer. Can. J. Aquat. Fish. Sci. 42:1166–1170.

Fretwell, S.D. and Lucas, H.L. 1970. On territorial behavior and other factors influencing habitat distribution in birds. I. Theoretical development. Acta Biotheoretica 19:16–36.

Gallager, S.M., Manuel, J.L., Manning, D.A. and O'Dor, R., 1996. Ontogenetic changes in the vertical distribution of giant scallop larvae, *Placopecten magellanicus*, in 9-m deep mesocosms as a function of light, food, and temperature stratification. Mar. Biol. 124:679–692.

Galleguillos, R.A. and Troncoso, L.S., 1991. Protein variation in the Chilean-Peruvian scallop *Argopecten purpuratus* (L.). In: S.E. Shumway and P.A. Sandifer (Eds.). An International Compendium of Scallop Biology and Culture. World Aquaculture Workshops No. 1. World Aquaculture Society, Baton Rouge, LA. pp. 146–150.

Garcia, S., 1994 The Precautionary Principle: its implications in capture fisheries management. Ocean & Coastal Man. 22:99–125.

Gavaris, S., 1980. Use of a multiplicative model to estimate catch rate and effort from commercial data. Can. J. Fish. Aquat. Sci. 37:2272–2275.

Gavaris, S. and Sinclair, A., 1998. From fisheries assessment uncertainty to risk analysis for immediate management actions. In: F. Funk, T.J. Quinn II, J. Heifetz, J.N. Ianelli, J.E. Powers, J.F. Schweigert, P.J. Sullivan and C.I. Zhang (Eds.). Fishery Stock Assessment Models. Lowell Wakefield Fisheries Symposium. University of Alaska Sea Grant College Program Rep. 98–01.

Gay, J. 1994. Limited entry for Alaska scallops. National Fisherman Nov. 1994:22–23.

Geromont, H.F., De Oliveira, J.A., Johnston, S.J. and Cunningham, C.L., 1999. Development and application of management procedures for fisheries in southern Africa. ICES J. Mar. Sci. 56:952–966.

Gibson, F.A., 1956. Escallops (*Pecten maximus* L.) in Irish waters. Scient. Proc. R. Dublin Soc. 27:253–270.

Gibson, F.A., 1959. Notes on the escallop (*Pecten maximus* L.) in three closely associated bays in the west of Ireland. J. Cons. int. Explor. Mer 24:366–371.

Giguère, M. and Brulotte, S., 1994. Comparison of sampling techniques, video and dredge, in estimating sea scallop (*Placopecten magellanicus* (Gmelin)) populations. J. Shellfish Res. 13:25–30.

Giguère, M. and Legare, B., 1988a. Exploitation du petoncle aux Iles-de-la-Madeleine en 1987. CAFSAC Res. Doc. 88/16:38 p. (mimeo)

Giguère, M. and Legare, B., 1988b. Exploitation du petoncle Cote-Nord du Golfe St-Laurent et Gaspesie-1987. CAFSAC Res. Doc. 88/17. (mimeo)

Goldberg, R., Pereira, J., Clark, P. and Faber, B., 1999. A critical evaluation of strategies for stock enhancement of the bay scallop, *Argopecten irradians*. 12th International Pectinid Workshop, Bergen (Norway), 5–11 May 1999, book of abstracts. (abstract only)

Golikov, A.N. and Scarlato, O.A., 1970. Abundance, dynamics and production properties of populations of edible bivalves *Mizuhopecten yessoensis* and *Spisula sachalinensis* related to the problem of organization of controllable submarine farms at the western shores of the Sea of Japan. Helg. wiss. Meeresunters. 20:498–513.

Goshima, S. and Fujiwara, H., 1994. Distribution and abundance of cultured scallop *Patinopecten yessoensis* in extensive sea beds as assessed by underwater camera. Mar. Ecol. Progr. Ser. 110:151–158.

Greenstreet, S.P.R., Tuck, I.D., Grewar, G.N., Armstrong, E., Reid, D.G. and Wright, P.J., 1997. An assessment of the acoustic survey technique, RoxAnn, as a means of mapping seabed habitat. ICES J. Mar. Sci. 54:939–959.

Griffin, N., 1993. Scallopers react to new plan. National Fisherman May 1993:19.

Gross, F. and Smyth, J.C., 1946. The decline of oyster populations. Nature 157:540–542.

Gruffydd, Ll.D., 1972. Mortality of scallops on a Manx scallop bed due to fishing. J. Mar. Biol. Assoc. U.K. 52:449–455.

Gruffydd, Ll.D., 1974. The influence of certain environmental factors on the maximum length of the scallop, *Pecten maximus* L. J. Cons. Int. Explor. Mer 35:300–302.

Gutsell, J.S., 1931. Natural history of the bay scallop (*Pecten irradians*). U.S. Bur. Fish. Bull. 46:569–632.

Gwyther, D., 1990. Yield assessments in the Port Phillip Bay scallop fishery. In: M.C.L. Dredge, W.F. Zacharin and L.M. Joll (Eds.). Proceedings of the Australasian Scallop Workshop, Taroona, Tasmania, 1988. Tasmanian Government Printer. Hobart, Australia, pp. 111–121.

Gwyther, D., Cropp, D.A., Joll, L.M. and Dredge, M.C.L., 1991. Australia. In: S.E. Shumway (Ed.). Scallops: Biology, Ecology and Aquaculture. Vol. 21. Elsevier, Amsterdam. pp. 835–851.

Gwyther, G. and McShane, P.E., 1988. Growth rate and natural mortality of the scallop *Pecten alba* Tate in Port Phillip Bay, Australia, and evidence for changes in growth rate after a 20-year period. Fish. Res. 6:347–361.

Hall-Spencer, J.M., Froglia, C., Atkinson, R.J.A. and Moore, P.G., 1999. The impact of rapido trawling for scallops, *Pecten jacobaeus* (L.), on the benthos of the Gulf of Venice. ICES J. Mar. Sci. 56:111–124.

Hamilton, L.J., Mulhearn, P.J. and Poeckert, R., 1999. Comparison of RoxAnn and QTC-View acoustic bottom classification system performance for the Cairns area, Great Barrier Reef, Australia. Cont. Shelf Res. 19:1577–1597.

Hancock, D.A., 1973. The relationship between stock and recruitment in exploited invertebrates. Rapp. Pv. Reun. CIEM 164:113–131.

Hancock, D.A., 1979. Population dynamics and management of shellfish stocks. Rapp. Pv. Reun. CIEM 175:8–19.

Hanski, I., 1998. Metapopulation dynamics. Nature 396:41–49.

Hardy, D., 1981. Scallops and the Diver-Fisherman. Fishing News Books, London. 132 p.

Hardy, D., 1991. Scallop Farming. Fishing News Books. Oxford, England. 237 p.

Hastie, T. and Tibshirani, R., 1990 Generalized Additive Models. Chapman & Hall, London.

Hatcher, B.G., Scheibling, R.E., Barbeau, M.A., Hennigar, A.W., Taylor, L.H. and Windust, A.J., 1996. Dispersion and mortality of a population of sea scallop (*Placopecten magellanicus*) seeded in a tidal channel. Can. J. Fish. Aquat. Sci. 53:38–54.

Heald, D.I. and Caputi, N., 1981. Some aspects of growth, recruitment and reproduction in the southern saucer scallop, *Amusium balloti* (Bernardi, 1861) in Shark Bay, Western Australia. Dept. Fish. Wildl. (West Australia), Fish. Res. Bull. 25:33 p.

Heath, P., 1999. Scallop stock enhancement initiative for the Irish Sea and Mulroy Bay, EU. 12th International Pectinid Workshop, Bergen (Norway), 5–11 May 1999, book of abstracts. (abstract only)

Heipel, D.A., Bishop, J.D.D. and Brand, A.R., 1999. Mitochondrial DNA variation among open-sea and enclosed populations of scallop *Pecten maximus* in western Britain. J. Mar. Biol. Assoc. U.K. 79:687–695.

Heipel, D.A., Bishop, J.D.D., Brand, A.R. and Thorpe, J.P., 1998. Population genetic differentiation of the great scallop *Pecten maximus* in western Britain investigated by randomly amplified polymorphic DNA. Mar. Ecol. Progr. Ser. 162:163–171.

Herbinger, C.M., Vercaemer, B.M., Gjetvaj, B. and O'Dor, R.K., 1998. Absence of genetic differentiation among geographically close sea scallop (*Placopecten magellanicus* G.) beds with cDNA and microsatellite markers. J. Shellfish Res. 17:117–122.

Hilborn, R., 1986. A comparison of alternative harvest tactics for invertebrate fisheries. In G.S. Jamieson and N. Bourne (Eds.). North Pacific Workshop on Stock Assessment and Management of Invertebrates. Can. Spec. Publ. Fish. Aquat. Sci. 92:313–317.

Hilborn, R., Maguire, J.-J., Parma, A.M. and Rosenberg, A.A., 2001. The precautionary approach and risk management. Can they increase the probability of success in fishery management? Can. J. Fish. Aquat. Sci. 58:99–107.

Hilborn, R. and Walters, C.J., 1992. Quantitative Fisheries Stock Assessment: Choice, Dynamics, and Uncertainty. Chapman and Hall, London. 570 p.

Hill, A.S., Veale, L.O., Pennington, D., Whyte, S.G., Brand, A.R. and Hartnoll, R.G., 1999. Changes in Irish Sea benthos: possible effects of 40 years of dredging. Est. Coast. Shelf Sci. 48:739–750.

Holme, N.A., 1984. Photography and television, pp. 66–98. In: N.A. Holme and A.D. McIntyre (Eds.). Methods for the Study of Marine Benthos, 2nd edition, IBP Hand Book 16. Blackwell Scientific Publications, Oxford. 387 p.

Holme, N.A. and Barrett, R.L., 1977. A sledge with television and photographic cameras for quantitative investigation of the epifauna on the continental shelf. J. Mar. Biol. Assoc. U.K. 57:391–403.

Howell, T.R.W., 1983. A comparison of the efficiencies of two types of Scottish commercial scallop dredge. ICES CM 83/K:38:5 p. (mimeo)

Ianelli, J.N. and Heifetz, J., 1995. Decision analysis of alternative harvest policies for the Gulf of Alaska Pacific Ocean Perch fishery. Fish. Res. 24:35–63.

Imai, T., 1971. (The evolution of scallop culture). In: Aquaculture in Shallow Seas: Progress in Shallow Sea Culture. pp. 261–364. In Japanese, translated by the US Dept. of Commerce (NTIS). 615 p.

Ingersoll, E., 1887. The scallop and its fishery. Am. Nat. 20:1001–1006.

Iribarne, O.O., Lasta, M.I, Vacas, H.C., Parma, A.M. and Pascual, M.S., 1991. Assessment of abundance, gear efficiency and disturbance in a scallop dredge fishery: results of a depletion experiment. In: S.E. Shumway and P.A. Sandifer (Eds.). An International Compendium of Scallop Biology and Culture. World Aquaculture Workshops No. 1. World Aquaculture Society, Baton Rouge, LA. pp. 242–248.

Ito, H., 1991. Japan. In: S.E. Shumway (Ed.). Scallops: Biology, Ecology and Aquaculture. Vol. 21. Elsevier, Amsterdam. pp. 1017–1055.

Ito, S., 1964. (On the scalloping in Okhotsk Sea). Ser. Propag. Mar. Prod. 7:2–40. In Japanese, Translation of the Fish. Res. Bd. of Canada No. 1115:65 p.

Jamieson, G.S., 1993. Marine invertebrate conservation: evaluation of fisheries over-exploitation concerns. Amer. Zool. 33:551–567.

Jamieson, G.S., Lundy, M.J., Kerr, G.L. and Witherspoon, N.B., 1981. Fishery characteristics and stock status of Georges Bank scallops. CAFSAC Res. Doc. No. 81/70:35 p.

Joll, L.M., 1989. Recruitment variation in stocks of the saucer scallop *Amusium balloti* in the Abrolhos Bank area. In: M.C.L. Dredge, W.F. Zachrin and L.M. Joll (Eds.). Proceedings of the Australasian Scallop Workshop. Tasmanian Government Printer, Hobart, Australia. pp. 61–67.

Joll, L.M. and Caputi, N., 1995. Environmental influences on recruitment in the saucer scallop (*Amusium balloti*) fishery of Shark Bay, Western Australia. ICES Mar. Sci. Symp. 199:47–53.

Jones, G.P., Millicich, M.J., Emslie, M.J. and Lunow, C., 1999. Self-recruitment in a coral reef fish population. Nature 402:802–804.

Kaiser, M.J., 1998. Significance of bottom-fishing disturbance. Cons. Biol. 12:1230–1235.

Kaiser, M.J., Armstrong, P.J., Dare, P.J. and Flatt, R.P., 1998. Benthic communities associated with a heavily fished scallop ground in the English Channel. J. Mar. Biol. Assoc. U.K. 78:1045–1059.

Kalashnikov, V.Z., 1991. Soviet Union. In: S.E. Shumway (Ed.). Scallops: Biology, Ecology and Aquaculture. Vol. 21. Elsevier, Amsterdam. pp. 1057–1082.

Kaluzny, S., 1987. Estimation of trends in spatial data. PhD Thesis Dissertation. University of Washington, Seattle. 165 p.

Kelley, K., 1988. Seagrass replanting efforts may improve fisheries. National Fisherman March 1988:14–16.

Kenchington, E. and Lundy, M.J., 1996. An assessment of areas for scallop broodstock protection in the approaches to the Bay of Fundy. DFO Atlantic Fish. Res. Doc. 96/13:21 p.

Kenchington, E., Lundy, M.J. and Roddick, D.L., 1995. An overview of the scallop fishery in the Bay of Fundy 1986 to 1994 with a report on fishing activity trends amongst the dual license holders in the full bay fleet. DFO Atlantic Fish. Res. Doc. 95/126:40 p.

Kenchington, E., Roddick, D.L. and Lundy, M.J., 1995. Bay of Fundy scallop analytical stock assessment and data review 1981–1994: Digby Grounds. DFO Atlantic Fish. Res. Doc. 95/10:70 p.

Kiesel, A., 1996. Coquille Saint-Jacques - Une question de baies. France-Eco-Peche Dec. 1996:43–46.

Kirkley, J.E., DuPaul, W.D. and Moore Niels, E., 1997. Rights-based strategies for managing the US Northwest Atlantic sea scallop. In: E.F. Félix Pico (Ed.). 11th International Pectinid Workshop, La Paz, Baja California Sur, México, April 10–15, 1997. Book of Abstracts: 49. CICIMAR, La Paz, México.

Kitada, S. and Fujishima, H., 1997. The stocking effectiveness of scallops in Hokkaido. Nippon Suisan Gakkaishi 63:686–693.

Krause, M.K., Arnold, W.S. and Ambrose, W.G., Jr., 1994. Morphological and genetic variation among three populations of calico scallops, *Argopecten gibbus*. J. Shellfish. Res. 13:529–537.

Langton, R.W., Robinson, W.R. and Schick, D., 1987. Fecundity and reproductive effort of sea scallops *Placopecten magellanicus* from the Gulf of Maine. Mar. Ecol. Progr. Ser. 37:19–25.

Lanteigne, M., Davidson, L. and Worms, J., 1986. Status of the Icelandic scallop, *Chlamys islandica*, in the northeastern Gulf of St. Lawrence, 1985. CAFSAC Res. Doc. 86/76:20 p.

Lasta, M.L. and Bremec, C.S., 1999. Development of the scallop fishery (*Zygochlamys patagonica*) in the Argentine Sea. 12th International Pectinid Workshop, Bergen (Norway), 5–11 May 1999, book of abstracts. (abstract only)

Laurenson, L.J.B., Unsworth, P., Penn, J.W. and Lenanton, R.C.J., 1993. The impact of trawling for saucer scallops and western king prawns on the benthic communities in coastal waters off south-western Australia. Fish. Res. Rep. Fish. Dept. West. Aust. 100:93 p.

856

León-Carballo, G., Reinecke-Reyes, M. and Ceseña-Espinoza, N., 1991. Abundancia y estructura poblacional de los bancos de almeja catarina *Argopecten circularis* (Sowerby, 1835) durante abril de 1988, en Bahía Concepción, BCS. Ciencia Pesquera (México) 8:125 p.

Levitan, D., 1991. Influence of body size and population density on fertilization success and reproductive output in a free-spawning invertebrate. Biol. Bull. 181:261–268.

Levitan, D.R., and Sewell, M.A., 1998. Fertilization success in free-spawning marine invertebrates: review of the evidence and fisheries implications. In: G.S. Jamieson and A. Campbell (Eds.). Proceedings of the North Pacific Symposium on Invertebrate Stock Assessment and Management. Can. Spec. Publ. Fish. Aquat. Sci. 125:159–164.

Lewis, R.I. and Thorpe, J.P., 1994. Are queen scallops, *Aequipecten (Chlamys) opercularis* (L.), self recruiting? In: N.F. Bourne, B.L. Bunting and L.D. Townsend (Eds.). Proceedings of the 9th International Pectinid Workshop, Nanaimo, B.C., Canada, April 22–27, 1993 Can. Tech. Rep. Fish. Aquat. Sci. 1994:(1):214–221.

Luo, Y., 1991. China. In: S.E. Shumway (Ed.). Scallops: Biology, Ecology and Aquaculture. Vol. 21. Elsevier, Amsterdam. pp. 809–824.

MacDonald, B.A., 1988. Physiological energetics of Japanese scallop *Patinopecten yessoensis* larvae. J. Exp. Mar. Biol. Ecol. 120:155–170.

MacDonald, B.A. and Bajdik, C.D., 1992. Orientation and distribution of individual *Placopecten magellanicus* (Gmelin) in two natural populations with differing production. Can. J. Fish. Aquat. Sci. 49:2086–2092.

MacFarlane, S.L., 1991. Managing scallops *Argopecten irradians irradians* (Lamarck, 1819) in Pleasant Bay, Massachusetts; large is not always legal. In: S.E. Shumway and P.A. Sandifer (Eds.). An International Compendium of Scallop Biology and Culture. World Aquaculture Workshops No. 1. World Aquaculture Society, Baton Rouge, LA. pp. 264–272.

Macleod, J.A.A., Thorpe, J.P. and Duggan, N.A., 1985. A biochemical genetic study of population structure in queen scallop (*Chlamys opercularis*) stocks in the Northern Irish Sea. Mar. Biol. 87:77–82.

MacPhail, J.S., 1954. The inshore scallop fishery of the Maritime Provinces. Fish. Res. Bd. Canada, Atlantic Biol. Sta. Circ. (General Ser.) 22:4 p.

Mackie, L.A. and Ansell, A.D., 1993. Differences in reproductive ecology in natural and transplanted populations of *Pecten maximus*: evidence for the existence of separate stocks. J. Exp. Mar. Biol. Ecol. 169:57–75.

Malakoff, D., 1998. Papers posit grave impact of trawling. Science 282:2168–2169.

Manuel, J.L., Burbridge, S., Kenchington, E.L., Ball, M. and O'Dor, R.K., 1996. Veligers from two populations of scallop *Placopecten magellanicus* exhibit different vertical distributions in the same mesocosm. J. Shellfish Res. 15:251–257.

Manuel, J.L., Gallager, S.M., Pearce, C.M., Manning, D.A. and O'Dor, R.K., 1996. Veligers from different populations of sea scallop *Placopecten magellanicus* have different vertical migration patterns. Mar. Ecol. Progr. Ser. 142:147–163.

Manuel, J.L. and O'Dor, R.K., 1997. Vertical migration for horizontal transport while avoiding predators: I. A tidal/diel model. J. Plankton Res. 19:1929–1948.

Manuel, J.L., Pearce, C.M. and O'Dor, R.K., 1997. Vertical migration for horizontal transport while avoiding predators: II. Evidence for the tidal/diel model from two populations of scallop (*Placopecten magellanicus*) veligers. J. Plankton Res. 19:1494–1973.

Marelli, D.C., Arnold, W.S. and Bray, C., 1999. Levels of recruitment and adult abundance in a collapsed population of bay scallops (*Argopecten irradians*) in Florida. J. Shellfish Res. 18:393–399.

Marshall, C.T. and Frank, K.T., 1995. Density-dependent habitat selection by juvenile haddock (*Melanogrammus aeglefinus*) on the southwestern Scotian Shelf. Can. J. Fish. Aquat. Sci. 52:1007–1017.

Marshall, N., 1947. Abundance of bay scallops in the absence of eelgrass. Ecology 28:321–322.

Maru, K., Obara, A., Kikuchi, K. and Okesaku, H., 1973. Studies on the ecology of the scallop, *Patinopecten yessoensis* (Jay). 3. On the diurnal vertical distribution of scallop larvae. Scient. Rep. Hokkaido Fish. Exper. Sta. 15:33–47. (in Japanese, with English abstract)

Mason, J., 1970. A comparison of various gears used in catching queens and scallops in Scottish waters. ICES CM 1970/K:19:3 p. (mimeo)

Mason, J., Chapman, G.J. and Kinnear, J.A.M., 1979. Population abundance and dredge efficiency studies on the scallop, *Pecten maximus* (L.). Rapp. Pv. Reun. CIEM 175:91–96.

Mason, J., Cook, R.M., Bailey, N. and Fraser, D.I., 1991. An assessment of scallops, *Pecten maximus* (Linnaeus, 1758), in Scotland west of Kintyre. In: S.E. Shumway and P.A. Sandifer (Eds.). An International Compendium of Scallop Biology and Culture. World Aquaculture Workshops No. 1. World Aquaculture Society, Baton Rouge, LA. pp. 231–241.

Mason, J., Drinkwater, J., Howell, T.R.W. and Fraser, D.I., 1982. A comparison of methods of determining the distribution and density of the scallop, *Pecten maximus* (L.). ICES CM 1982/K:24:5 p., figs. (mimeo)

Mason, J., Nicholson, M.D. and Shanks, A.M., 1979. A comparison of exploited populations of the scallop *Pecten maximus* (L.). Rapp. Pv. Reun. CIEM 175:114–120.

Mason, J., Shanks, A.M. and Fraser, D.I., 1980. An assessment of scallop, *Pecten maximus* (L.), stocks off South-west Scotland. ICES CM 1980/K:27:4 p. and tables. (mimeo)

Mason, J., Shanks, A.M. and Fraser, D.I., 1981. An assessment of scallop, *Pecten maximus* (L.), stocks at Shetland. ICES CM 1981/K:19 p. (mimeo)

Mason, J., Shanks, A.M., Fraser, D.I. and Shelton, R.G.J., 1979. The Scottish fishery for the queen, *Chlamys opercularis* (L.). ICES CM 1979/K:37:6 p., figs., tables. (mimeo)

Mattei, N. and Pellizzato, M., 1996. A population study of three stocks of a commercial Adriatic pectinid (*Pecten jacobaeus*). Fish. Res. 26:49–65.

McAllister, M.K., Pikitch, E.K., Ount, A.E., and Hilborn, R., 1994. A Bayesian approach to stock assessment and harvest decisions using the sampling/importance resampling algorithm. Can. J. Fish. Aquat. Sci. 51:2673–2687.

McCall, A.D., 1990. Dynamic Geography of Marine Fish Populations. Washington Sea Grant Program. Univ. of Washington Press , Seattle. 153 p.

McGarvey, R., Serchuk, F.M. and McLaren, I.A., 1993. Spatial and parent-age analysis of stock-recruitment in the Georges Bank sea scallop (*Placopecten magellanicus*) population. Can. J. Fish. Aquat. Sci. 50:564–574.

McIntyre, A.D., 1956. The use of trawl, grab and camera in estimating marine benthos. J. Mar. Biol. Assoc. U.K. 35:419–429.

McLoughlin, R.J., 1994. Sustainable management of Bass Strait scallops. Mem. Queensl. Mus. 36:307–314.

McLoughlin, R., Young, P.C., Martin, R.B. and Parslow, J., 1991. The Australian scallop dredge: estimates of catching efficiency and associated indirect fishing mortality. Fish. Res. 11:1–24.

McLoughlin, R., Zacharin, W., Cartwright, I., Gwyther, D., Probestl, M. and Sterling, D., 1993. New harvesters for the scallop industry in south east Australia. Australian Fisheries August 1993:12–14.

McRea, J.E., Jr., Greene, H.G., O'Connell, V.M. and Wakefield, W.W., 1999. Mapping marine habitats with high resolution sidescan sonar. Oceanol. Acta 22:679–686.

Medcof, J.C., 1952. Modification of drags to protect small scallops. Fish. Res. Bd. Canada, Atlantic Progr. Rep. 52:9–14.

Merrien, A., 1980. Utilisation de la television sous-marine pour l'inventaire et l'estimation directe des ressources en coquilles Saint-Jacques Ex: la Baie de Saint-Brieuc. ICES CM 1980/B:22 p. (mimeo)

Middleton, K., 1983. Bay scallop seed-release program could increase native stock levels. National Fisherman March 1983:26–28.

Minchin, D., 1992. Biological observations on young scallops, *Pecten maximus*. J. Mar. Biol. Assoc. U.K. 72:807–819.

Millar, R.B. and Naidu, K.S., 1991. The Iceland scallop (*Chlamys islandica*) size selectivity of an offshore scallop survey dredge. CAFSAC Res. Doc. 91/81:17 p.

Mohn, R.K., 1986. Generalizations and recent usages of yield per recruit analysis. Can. Spec. Publ. Fish. Aquat. Sci. 92:318–325.

Mohn, R.K., Robert, G. and Roddick, D.L., 1984. Status and harvesting strategies for Georges Bank scallop stock (NAFO SA 5Ze). ICES CM 1984/K:15:31 p. (mimeo)

Mohn, R.K., Robert, G. and Roddick, D.L., 1987. Research sampling and survey design of Georges Bank scallops (*Placopecten magellanicus*). J. Northw. Atl. Fish. Sci. 7:117–121.

Mohn, R.K., Robert, G. and Roddick, D.L., 1988. Georges Bank scallop stock assessment-1987. CAFSAC Res. Doc. 88/3:29 p. (mimeo)

Moore, J.K. and Marshall, N., 1967. The retention of lamellibranch larvae in the Niantic estuary. The Veliger 10:10–12.

Mori, K., 1975. Seasonal variation in physiological activity of scallops under culture in the coastal waters of Sanriku District, Japan, and a physiological approach of a possible cause of their mass mortality. Bull. Mar. Biol. Stat. Asamushi 15:59–79.

Mottet, M., 1979. A review of the fishery biology and culture of scallops. State of Washington, Dep. of Fisheries, Tech. Rep. 39:100 p.

Moyer, M.A. and Blake, N.J., 1986. Fluctuations in calico scallop production (*Argopecten gibbus*). 11th Tropical and Subtropical Fisheries Conference of the Americas, Proc. Texas A&M Univ. Dept. Anim. Science. pp. 45–57.

Munch-Petersen, S., 1973. An investigation of a population of the soft clam (*Mya arenaria* L.) in a Danish estuary. Medd. Danmarsk Fisk. Havunders. (NS) 7:47–73.

Munita, C., 1988. Repoblacion de recursos marinos. Chile Pesquero Marzo–Abril. pp. 23–25.

Murawski, S.A., Brown, R., Lai, H.L., Rago, P.J. and Hendrickson, L., 2000. Large-scale closed areas as a fishery-management tool in temperate marine systems: the Georges Bank experience. Bull. Mar. Sci. 66:775–798.

Murawski, S.A. and Fogarty, M.J., 1984. A spatial yield model for bivalve populations accounting for density dependent growth and mortality. ICES, Shellfish Committee, CM 1984/K:26:13 p. (mimeo)

Murawski, S.A. and Serchuk, F.M., 1989. Environmental effects of offshore dredge fisheries for bivalves. ICES, Shellfish Committee, CM 1989/K:27,:11 p, figures.

Myers, R.A., Fuller, S.D. and Kehler, D.G., 2000. A fisheries management strategy robust to ignorance: rotational harvest in the presence of indirect fishing mortality. Can. J. Fish. Aquat. Sci. 57:2357–2362.

Nadeau, M. and Cliche, G., 1997. Scallop (*Placopecten magellanicus*) seeding trials in Iles-de-la-Madeleine, Québec, 1993 to 1996. Bull. Aquacul. Assoc. Canada 97–2:69–71.

Nadeau, M. and Cliche, G., 1999. Recapture rate of seeded scallops (*Placopecten magellanicus*) during commercial fishing activity in Iles-de-la-Madeleine (Quebec, Canada). 12th International Pectinid Workshop, Bergen (Norway), 5–11 May 1999, book of abstracts. (abstract only)

Nadeau, M., Cliche, G. and Hebert, D., 1999. Experimental dredging of starfishes and crabs before commercial seeding of sea scallops in Magdalen Islands (Quebec, Canada). J. Shellfish Res. 18:332 (abstract).

Naidu, K.S., 1984. An analysis of the scallop meat count regulation. CAFSAC Res. Doc. 84/73:17 p. (mimeo)

Naidu, K.S., 1988. Estimating mortality rate in the Iceland scallop, *Chlamys islandica* (O.F. Muller). J. Shellfish Res. 7:61–71.

Naidu, K.S., 1991. Sea scallop: *Placopecten magellanicus*. In: S.E. Shumway (Ed.). Scallops: Biology, Ecology and Aquaculture. Vol. 21. Elsevier, Amsterdam. pp. 861–897.

Naidu, K.S., 1997. Scalping scallops: ecological considerations relative to the large-scale removal without replacement of shells from scallop beds. In: E.F. Félix Pico (Ed.). 11th International Pectinid Workshop, La Paz, Baja California Sur, México, April 10–15, 1997. Book of Abstracts. CICIMAR, La Paz, México. p. 53.

Naidu, K.S. and Anderson, J.T., 1984. Aspects of scallop recruitment on St. Pierre Bank in relation to oceanography and implications for resource management. CAFSAC Res. Doc. 84/29:15 p. (mimeo)

Naidu, K.S., Cahill, F.M. and Lewis, D.B., 1982. Status and assessment of the Iceland scallop *Chlamys islandica* in the northeastern Gulf of St. Lawrence. CAFSAC Res. Doc. 82/02:66 p. (mimeo)

Naidu, K.S., Cahill, F.M. and Seward, E.M. 1998. The Iceland scallop: a fishery under siege in Newfoundland. Canadian Stock Assessment Secretariat, Res. Doc. 98/149.

Naidu, K.S., Cahill, F.M., Veitch, P.J. and Stansbury, E.E., 1995. Status of the Iceland scallop *Chlamys islandica* on the Grand Banks of Newfoundland (NAFO Division 3LNO), 1994. DFO Atl. Fish. Res. Doc. 95/136.

Navarro-Piquimil, R., Sturla-Figueroa, L., Cordero-Contreras, O. and Avendaño-D., M., 1991. Chile. In: S.E. Shumway (Ed.). Scallops: Biology, Ecology and Aquaculture. Vol. 21. Elsevier, Amsterdam. pp. 1001–1015.

NEFSC (Northeast Fisheries Science Center). 1999. B-Sea Scallops (*Placopecten magellanicus*). In: 29th Northeast Regional Stock Assessment Workshop (29th SAW), SARC (Stock Assessment Review Committee) Consensus Summary of Assessment. NEFSC Reference Document 99–14:91–172.

Nicolajsen, A. and Conan, G.Y., 1987. Assessment by geostatistical techniques of populations of Iceland Scallop (*Chlamys islandica*) in the Barent Sea. ICES CM 1987/K:14:17 p. (mimeo)

Noel, C., 1997. Estimating scallop abundance and management of the fishery in Port Philip Bay, South-East Australia. In: E.F. Félix Pico (Ed.). 11th International Pectinid Workshop, La Paz, Baja California Sur, México, April 10–15, 1997. Book of Abstracts. CICIMAR, La Paz, México. pp. 44-46.

Noel, H.S., 1980. Britains scallop industry marked by impressive gear development. National Fisherman November 1980:56–57.

Noel, H.S., 1982. French scallop dredge outfishes Scottish and New Bedford types. National Fisherman October 1982:44 p.

NRC (National Research Council), 1999. Sustaining Marine Fisheries. NRC Press. Washington, DC.

Ólafsson, E.B., Peterson, C.H. and Ambrose, W.G. Jr., 1994. Does recruitment limitation structure populations and communities of macro-invertebrates in marine soft sediments: the relative significance of pre- and post-settlement processes. Oceanogr. Mar. Biol. Ann. Rev. 32:65–109.

Olivier, S.R. and Capitoli, R., 1980. Edad y crecimiento en *Chlamys tehuelcha* (d'Orbigny) (Mollusca, Pelecypoda, Pectinidae) del golfo San Matias (Pcia. de Rio Negro, Argentina). An. Centro. Cienc. Mar Limnol. UNAM (Mexico) 7:129–140.

Olsen, A.M., 1953. Diving investigations on scallops. Fish. News Lett. Aust. 12(7):5–7.

Olsen, A.M., 1955. Underwater studies on the Tasmanian commercial scallop, *Notovola meridionalis* (Tate) (Lamellibranchiata: Pectinidae). Aust. J. Mar. Freshwat. Res. 6:392–409.

Orensanz, J.M., 1986. Size, environment and density: the regulation of a scallop stock and its management implications. In: G.S. Jamieson and N. Bourne (Eds.), North Pacific Workshop on Stock Assessment and Management of Invertebrates. Canada Spec. Publ. Fish. Aquat. Sci. 92:195–227.

Orensanz, J.M. and Jamieson, G.S., 1998. The assessment and management of spatially structured stocks: an overview of the North Pacific Symposium on Invertebrate Stock Assessment and Management. In: G.S. Jamieson and A. Campbell (Eds.). Proceedings of the North Pacific Symposium on Invertebrate Stock Assessment and Management. Can. Spec. Publ. Fish. Aquat. Sci. 125:441–460.

Orensanz, J.M., Parma, A.M. and Ciocco, N.F., 1997. Reproductive reserves and zoning of uses as the only framework to prevent overfishing and protect wildlife in San José Gulf Marine Park (Argentine Patagonia). Fisheries Centre Research Report Series, University of British Columbia, Canada 5(1):21–22. (abstract)

Orensanz, J.M., Parma, A.M., and Hall, M.A., 1998. The analysis of concentration and crowding in shellfish research. In: G.S. Jamieson and A. Campbell (Eds.). Proceedings of the North Pacific Symposium on Invertebrate Stock Assessment and Management. Can. Spec. Publ. Fish. Aquat. Sci. 125:143–158.

Orensanz, J.M., Parma, A.M. and Iribarne, O., 1991. Population dynamics and management of natural stocks. In: S.E. Shumway (Ed.). Scallops: Biology, Ecology and Aquaculture. Vol. 21. Elsevier, Amsterdam. pp. 625–713.

Orensanz, J.M., Pascual, M. and Fernández, M., 1991. Argentina. In: S.E. Shumway (Ed.). Scallops: Biology, Ecology and Aquaculture. Vol. 21. Elsevier, Amsterdam. pp. 981–999.

Parma, A.M., 2002. In search of robust harvest rules for Pacific halibut in the face of uncertain assessments and decadal changes in productivity. Bull. Mar. Sci. 70(2):423-453.

Parsons, G.J., Dadswell, M.J. and Rodstrom, E.M., 1991. Scandinavia. In: S.E. Shumway (Ed.). Scallops: Biology, Ecology and Aquaculture. Vol. 21. Elsevier, Amsterdam. pp. 763–775.

Parsons, G.J., Robinson, S.M.C., Chandler, R.A., Davidson, L.A., Lanteigne, M. and Dadswell, M.J., 1992. Intra-annual and long-term patterns in the reproductive cycle of giant scallops *Placopecten magellanicus* (Bivalvia: Pectinidae) from Passamaquoddy Bay, New Brunswick, Canada. Mar. Ecol. Progr. Ser. 80:203–214.

Patterson, K.R., 1999. Evaluating uncertainty in harvest control law catches using Bayesian Markov Chain Monte Carlo virtual population analysis with adaptive rejection sampling and including structural uncertainty. Can. J. Fish. Aquat. Sci. 56:208–221.

Pearce, C.M., Gallager, S.M., Manuel, J.L., Manning, D.A., O'Dor, R.K. and Bourget, E., 1996. Settlement of larvae of the giant scallop, *Placopecten magellanicus*, in 9-m deep mesocosms as a function of temperature stratification, depth, food, and substratum. Mar. Biol. 124:693–706.

Pennington, J.T., 1985. The ecology of fertilization of echinoid eggs: the consequences of sperm dilution, adult aggregation and synchronous spawning. Biol. Bull. 169:417–430.

Perez, J.A.A. and Pezzuto, P.R., 1998. Valuable shellfish species in the by-catch of shrimp fishery in southern Brazil: spatial and temporal patterns. J. Shellfish Res. 17:303–309.

Perrin, R.A. and Hay, P.R., 1987. A history of the Tasmanian scallop industry: diagnosing problems of management. Pap. Proc. R. Soc. Tasm. 121:1–14.

Perry, R.I., Walters, C.J. and Boutillier, J.A., 1999. A framework for providing advice for the management of new and developing invertebrate fisheries. Rev. Fish Biol. Fish. 9:125–150.

Peterson, C.H. and Summerson, H.C., 1992. Basin-scale coherence of population dynamics of an exploited marine invertebrate, the bay scallop: implications of recruitment limitation. Mar. Ecol. Progr. Ser. 90:257–272.

Peterson, C.H., Summerson, H.C. and Luettich, R.A., Jr., 1996. Response of bay scallops to spawner transplants: a test of recruitment limitation. Mar. Ecol. Progr. Ser. 132:93–107.

Pezzuto, P.R. and Borzone, C.A., 1997. The scallop *Pecten ziczac* (Linnaeus, 1758) fishery in Brazil. J. Shellfish Res. 16:527–532.

Pope, J.G. and Shepherd, J.G., 1982. A simple method for the consistent interpretation of catch-at-age data. J. Cons. int. Explor. Mer 40:176–184.

Premetz, E.D. and Snow, G.W., 1953. Status of the New England sea-scallop fishery. Comm. Fish. Rev. 15:1–17.

Prince, J. and Hilborn, R., 1998. Concentration profiles and invertebrate fisheries management. In: G.S. Jamieson and A. Campbell (Eds.). Proceedings of the North Pacific Symposium on Invertebrate Stock Assessment and Management. Can. Spec. Publ. Fish. Aquat. Sci. 125:187–196.

Quinn, T.J., II, Fagen, R. and Zheng, J., 1990. Threshold management policies for exploited populations. Can. J. Fish. Aquat. Sci. 47:2016–2029.

Raloff, J., 1996. Fishing for answers. Deep trawls leave destruction in their wake - but for how long? Science News 150:268–271.

Ramsay, K., Kaiser, M.J. and Hughes, R.N., 1998. Responses of benthic scavengers to fishing disturbance by towed gears in different habitats. J. Exp. Mar. Biol. Ecol. 224:73–89.

862

Restrepo, V.R., Thompson, G.G., Mace, P.M., Gabriel, W.L., Low, L.L., MacCall, A.D., Methot, R.D., Powers, J.E., Taylor, B.L., Wade, P.R. and Witzig, J.F., 1998. Technical Guidance on the Use of Precautionary Approaches to Implementing National Standard 1 of the Magnuson-Stevens Fishery Conservation and Management Act. NOAA Technical Memorandum NMFS-F/SPO-31.

Renzoni, A., 1991. Italy. In: S.E. Shumway (Ed.). Scallops: Biology, Ecology and Aquaculture. Vol. 21. Elsevier, Amsterdam. pp. 777–788.

Rhodes, E.W., 1991. Fisheries and aquaculture of the bay scallop, *Argopecten irradians*, in the eastern United States. In: S.E. Shumway (Ed.). Scallops: Biology, Ecology and Aquaculture. Vol. 21. Elsevier, Amsterdam. pp. 913–924.

Ricker, W.E., 1975. Computation and interpretation of biological statistics of fish populations. Can. Dep. Env., Fish. Mar. Serv. Bull. 191:382 p.

Ripley, B.D., 1977. Modelling spatial patterns. J. Roy. Stat. Soc. (B) 39:172–212. (includes a multi-authored Discussion section)

Ripley, B.D., 1981. Spatial Statistics. John Wiley & Sons. 252 p.

Rivers, J.B., 1962. A new scallop trawl for North Carolina. Comm. Fish. Rev. 24:11–14.

Robert, G., Black, G.A.P. and Butler, M.A.E., 1994. Georges Bank scallop stock assessment-1994. DFO Atlantic Fish. Res. Doc. No. 94/97:42 p.

Robert, G., Buttler-Connolly, M.A.E. and Lundy, M.J., 1987. Perspectives on the Bay of Fundy scallop stock and its fishery. CAFSAC Res. Doc., 87/27:30 p. (mimeo)

Robert, G., Buttler-Connolly, M.A.E. and Lundy, M.J., 1990. Bay of Fundy scallop stock assessment-1989. CAFSAC Res. Doc. 90/31:35 p.

Robert, G. and Jamieson, G.S., 1986. Commercial fishery data isopleths and their use in offshore sea scallop (*Placopecten magellanicus*) stock evaluations. In: G.S. Jamieson and N. Bourne (Eds.). North Pacific Workshop on Stock Assessment and Management of Invertebrates. Can. Spec. Publ. Fish. Aquat. Sci. 92:76–82.

Robert, G. and Lundy, M.J., 1988. Gear performance in the Bay of Fundy scallop fishery. I. Preliminaries. CAFSAC Res. Doc. 88/19. (mimeo)

Roberts, C.M., 1997. Connectivity and management of Caribbean coral reefs. Science 278:1454–1457.

Robinson, S.M.C., Thomas, A., Martin, J.D., Page, F.H., Cliche, G. and Giguère, M., 1999. Using remote sensing satellite technology to study the early life history ecology of scallop larvae. 12th International Pectinid Workshop, Bergen (Norway), 5–11 May 1999. Book of abstracts. (abstract only)

Roe, R.B., Cummins, R., Jr. and Bullis, H.R., 1971. Calico scallop redistribution, abundance, and yield off eastern Florida, 1967–1968. Fish. Bull. 69:399–409.

Román, G., 1991. Spain. In: S.E. Shumway (Ed.). Scallops: Biology, Ecology and Aquaculture. Vol. 21. Elsevier, Amsterdam. pp. 753–762.

Rose, G.A. and Kulka, D.W., 1999. Hyperaggregation of fish and fisheries: how catch-per-unit-of-effort increased as the northern cod (*Gadus morhua*) declined. Can. J. Fish. Aquat. Sci. 56 (Suppl. 1):118–127.

Rose, G.A. and Leggett, W.C., 1991. Effects of biomass-range interactions on catchability and migratory demersal fish by mobile fisheries: an example of Atlantic cod (*Gadus morhua*). Can. J. Fish. Aquat. Sci. 48:843–848.

Rosenberg, A.A. and Restrepo, V.R., 1994. Uncertainty and risk evaluation in stock assessment advice for U.S. marine fisheries. Can. J. Fish. Aquat. Sci. 51:2715–2720.

Rossi, R.E., Mulla, D.J., Journel, A.G. and Franz, E.H., 1992. Geostatistical tools for modeling and interpreting ecological spatial dependence. Ecol. Monogr. 62:277–314.

Roughgarden, J. and Iwasa, Y., 1986. Dynamics of a metapopulation with space-limited subpopulations. Theor. Pop. Biol. 29:235–261.

Sanders, M.J., 1966. History of Victorian scallop fishery. Aust. Fish. 25:11.

Sauce, B.L., Gwyther, D. and Burgess, D., 1987. Larval settlement, juvenile growth and the potential use of spatfall indices to predict recruitment of the scallop *Pecten alba* Tate in Port Phillip Bay, Victoria, Australia. Fish. Res. 6:81–92.

Schmidt, K., 1998. Ecology's catch of the day. Science 281:192–193.

Schneider, D.C., 1992. Gear efficiency trials with Icelandic scallops on St. Pierre Bank. Technical Report prepared for Clearwater Fine Foods, Inc., Bedford, Nova Scotia, Canada. Unpubl. MS. 15 p.

Seber, G.A.F., 1982. The Estimation of Animal Abundance and Related Parameters, 2nd ed. Charles Griffin and Company Ltd. 654 p.

Serchuk, F.M. and Smolowitz, R.J., 1980. Size selection of sea scallops by an offshore survey dredge. ICES, Shellfish Committee CM/K:24:38 p.

Serchuk, F.M. and Wigley, S.E., 1987. Evaluation of USA and Canadian research vessel surveys for sea scallops (*Placopecten magellanicus*) for Georges Bank. J. Northw. Atl. Fish. Sci. 7:1–13.

Serchuk, F.M., Wood, P.W., Posgay, J.A. and Brown, B.E., 1979. Assessment and status of sea scallop (*Placopecten magellanicus*) populations off the Northeast coast of the United States. Proc. Nat. Shellfish Assoc. 69:161–191.

Shafee, M.S., 1979. Underwater observations to estimate the density and spatial distribution of black scallop, *Chlamys varia* (L.) in Lanveoc (Bay of Brest). Bull. Off. Natl. Peches Tunisie 3:143–156.

Shepard, A.N. and Auster, P.J., 1991. Incidental (non-capture) damage to scallops caused by dragging on rock and sand substrates. In: S.E. Shumway and P.A. Sandifer (Eds.). An International Compendium of Scallop Biology and Culture. World Aquaculture Workshops No. 1. World Aquaculture Society, Baton Rouge, LA. pp. 219–230.

Shepherd, J.G. 1993. Key issues in conservation of fisheries. Ministry of Agriculture, Fisheries and Food. Directorate of Fisheries Research, Lowestoft. Laboratory leaflet Number 72:19 p.

Shirley, S.M. and Kruse, G.H., 1995. Development of the fishery for weathervane scallop. *Patinopecten caurinus* (Gould, 1850) in Alaska. J. Shellfish Res. 14:71–78.

Sigurdsson, J.D., Titman, C.W. and Davies, B.A., 1976. The dispersal of young post-larval bivalve molluscs by byssus threads. Nature 262:386–387.

Sinclair, M., 1987. Marine Populations - An Essay on Population Regulation and Speciation. University of Washington Press, Seattle. 252 p.

Sinclair, M., Mohn, R.K., Robert, G. and Roddick, D.L., 1985. Considerations for the effective management of Atlantic scallops. Can. Tech. Rep. Fish. Aquat. Sci. 1382:97 p.

Sluczanowski, P.R., 1984. A management oriented model of an abalone fishery whose substocks are subject to pulse fishing. Can. J. Fish. Aquat. Sci. 41:1008–1014.

Smith, G.F. and Greenhawk, K.N., 1998. Shellfish benthic habitat assessment in the Chesapeake Bay: progress toward integrated technologies for mapping and analysis. Workshop on Spatial

Data and Remote Sensing in Invertebrate Fisheries, Fort Walton (Florida). J. Shellfish Res. 17:1433–1437.

Smith, S.J., 1988. Evaluating the efficiency of the distribution mean estimator. Biometrics 44:485–493.

Smith, S.J. and Mohn, R.K., 1987. Considerations on the representation and analysis of a spatially aggregated resource: Georges Bank scallops. ICES CM 1987/K:26:12 p., 7 figs. (mimeo)

Smith, S.J. and Robert, G., 1991. Scallops, sampling and the law. 1991 Proceedings of the Section on Statistics and the Environment, American Statistical Association. pp. 102–109.

Smith, S.J. and Robert, G., 1998. Getting more out of your survey designs: an application to Georges bank scallops (*Placopecten magellanicus*). In: G.S. Jamieson and A. Campbell (Eds.). Proceedings of the North Pacific Symposium on Invertebrate Stock Assessment and Management. Can. Spec. Publ. Fish. Aquat. Sci. 125:3–14.

Stauffer, R.C., 1937. Changes in the invertebrate community of a lagoon after disappearance of the grass. Ecology 18:427–431.

Stevens, P.M., 1987. Response to excised gill tissue from the New Zealand scallop *Pecten novaezelandiae* to suspended silt. N.Z. J. Mar. Freshwat. Res. 21:605–614.

Stokesbury, K.D.E. and Himmelman, J.H., 1993. Spatial distribution of the giant scallop *Placopecten magellanicus* in unharvested beds in the Baie des Chaleurs, Quebec. Mar. Ecol. Progr. Ser. 96:159–168.

Stokesbury, K.D.E. and Himmelman, J.H., 1994. Biological and physical variables associated with aggregations of the giant scallop *Placopecten magellanicus*. Can. J. Fish. Aquat. Sci. 52:743–753.

Stokesbury, K.D.E. and Himmelman, J.H., 1996. Experimental examination of movement of the giant scallop, *Placopecten magellanicus*. Mar. Biol. 124:651–660.

Stolyarenko, D.A., 1995. Methodology of shellfish surveys based on a microcomputer geographic information system. In: D.E. Aiken, S.L. Waddy and G.I. Conan (Eds.). Shellfish Life Histories and Shellfishery Models- Selected papers from a symposium held in Moncton, New Brunswick, 25–29 June 1990. Rapp. Proc. Verb. CIEM 199:259–266.

Stotz, W. and González, S.A., 1997a. Recovery of an overfished scallop bed (*Argopecten purpuratus* (Lamarck, 1819)) in Tongoy Bay (Chile) by the use of a new management tool. In: E.F. Félix Pico (Ed.). 11th International Pectinid Workshop, La Paz, Baja California Sur, México, April 10–15, 1997). Book of Abstracts. CICIMAR. La Paz, México. pp. 54–55, 2 figs.

Stotz, W. and González, S.A., 1997b. Abundance, growth, and production of the sea scallop *Argopecten purpuratus* (Lamarck 1819): bases for sustainable exploitation of natural scallop beds in north-central Chile. Fish. Res. 32:173–183.

Summerson, H.C. and Peterson, C.H., 1990. Recruitment failure of the bay scallop, *Argopecten irradians concentricus*, during the first red tide, *Ptychodiscus brevis*, outbreak recorded in North Carolina. Estuaries 13:322–331.

Swain, D.P. and Morin, R., 1996. Relationships between geographic distribution and abundance of American plaice (*Hippoglossoides platessoides*) in the southern Gulf of St. Lawrence. Can. J. Fish. Aquat. Sci. 53:106–119.

Swain, D.P. and Wade, E.J., 1993. Density-dependent geographic distribution of Atlantic cod (*Gadus morhua*) in the southern Gulf of St. Lawrence. Can. J. Fish. Aquat. Sci. 50:725–733.

Swartzman, G., Huang, C. and Kaluzny, S., 1992. Spatial analysis of Bering Sea groundfish survey data using generalized additive models. Can. J. Fish. Aquat. Sci. 49:1366–1378.

Swearer, S.E., Caselle, J.E., Lea, D.W. and Warner, R.R., 1999. Larval retention and recruitment in an island population of a coral-reef fish. Nature 402:799–802.

Tamanini, B., 1985. Diving for dollars in Maines ice-cold waters. National Fisherman April 1985:24–26.

Tettelbach, S.T. and Rhodes, E.W., 1981. Combined effects of temperature and salinity on embryos and larvae of the northern bay scallop *Argopecten irradians irradians*. Mar. Biol. 63:249–256.

Thayer, G.W. and Stuart, H.H., 1974. The bay scallop makes its bed of seagrass. Mar. Fish. Rev. 36:27–30.

Thompson, S.K., 1992. Sampling. John Wiley & Sons, Inc., New York. 343 p.

Thompson, S.K. and Seber, G.A.F., 1996. Adaptive Sampling. John Wiley & Sons Inc., New York. 265 p.

Thouzeau, G., 1987. Facteurs responsables de la repartition spatiale des post-larves et juveniles de *Pecten maximus*. Devenir des pre-recrues issues des pontes des etes 1985 et 1986. Programme National sur le Determinisme du Recrutement (PNDR), Informations 7 (Nantes, France):46. (abstract only)

Thouzeau, G., 1991a. Experimental collection of postlarvae of *Pecten maximus* (L.) and other benthic macrofaunal species in the Bay of Saint-Brieuc, France. I. Settlement patterns and biotic interactions among the species collected. J. Exp. Mar. Biol. Ecol. 148:159–179.

Thouzeau, G., 1991b. Déterminisme du pré-recrutement de *Pecten maximus* (L.) en baie de Saint-Brieuc: processus régulateurs de l'abondance, de la survie et de la croissance des post-larves et juvéniles. Aquat. Living Resour. 4:77–99.

Thouzeau, G., 1995. Aspects de la dynamique spatio-temporelle du pré-recrutement de *Pecten maximus* L., en baie de Saint-Brieuc. ICES Mar. Sci. Symp. 199:31–39.

Thouzeau, G. and Hily, C., 1986. AQUAREVE: une technique nouvelle d'echantillonage quantitatif de la macrofaune epibenthique des fonds meubles. Oceanol. Acta 9:509–513.

Thouzeau, G. and Lehay, D., 1988. Variabilite spatio-temporelle de la distribution, de la croissance et de la survie des juveniles de *Pecten maximus* (L.) issus des pontes 1985, en baie de Saint-Brieuc. Oceanol. Acta 11:267–283.

Thouzeau, G., Robert, G. and Smith, S.J., 1991. Spatial variability in distribution and growth of juvenile and adult sea scallops *Placopecten magellanicus* (Gmelin) on eastern Georges Bank (Northwest Atlantic). Mar. Ecol. Progr. Ser. 74:205–218.

Thouzeau, G., Robert, G. and Ugarte, R., 1991. Faunal assemblages of benthic megainvertebrates inhabiting sea scallop grounds from eastern Georges Bank, in relation to environmental factors. Mar. Ecol. Progr. Ser. 74:61–82.

Tracey, G.A., Saade, E., Stevens, B., Selvitelli, P. and Scott, J., 1998. Laser line scan survey of crab habitats in Alaskan waters. J. Shellfish Res. 17:1483–1486.

Tremblay, M.J., Loder, J.W., Werner, F.E., Naimie, C.E., Page, F.H. and Sinclair, M.M., 1994. Drift of sea scallop larvae *Placopecten magellanicus* on Georges Bank: a model study of the roles of mean advection, larval behavior and larval origin. Deep Sea Res. II. Top. Stud. Oceanogr. 41:7–49.

Tremblay, M.J. and Sinclair, M.M., 1988. The vertical and horizontal distribution of sea scallop (*Placopecten magellanicus*) larvae in the Bay of Fundy in 1984 and 1985. J. Northw. Atl. Fish. Sci. 8:45–53.

Tremblay, M.J. and Sinclair, M.M., 1991. Inshore-offshore differences in the distribution of sea scallop larvae: implications for recruitment. ICES Mar. Sci. Symp. 192:39. (abstract only)

Tremblay, M.J. and Sinclair, M.M., 1992. Planktonic sea scallop larvae (*Placopecten magellanicus*) in the Georges Bank region: broadscale distribution in relation to physical oceanography. Can. J. Fish. Aquat. Sci. 49:1597–1615.

Turk, T., 2000. Distribution, abundance and spatial management of the weathervane scallop (*Patinopecten caurinus*) fishery in Alaska. MSc Thesis. University of Washington, Seattle.

Vahl, O., 1979. Volume of water pumped and particulate matter deposited by the Iceland scallop (*Chlamys islandica*) in Balsfjord, northern Norway. In: H.J. Freeland, D.M. Farmer and C.D. Levings (Eds.). Fjord Oceanography. Plenum Press. pp. 639–644.

Vahl, O., 1981a. Energy transformations by the Iceland scallop, *Chlamys islandica* (O.F. Muller), from 70° N. II. The population energy budget. J. Exp. Mar. Biol. Ecol. 53:297–303.

Vahl, O., 1981b. Age-specific residual reproductive value and reproductive effort in the Iceland scallop, *Chlamys islandica* (O.F. Muller). Oecologia 51:53–56.

Vahl, O., 1982. Long term variation in recruitment of the Iceland scallop, *Chlamys islandica* (O.F. Muller), from Northern Norway. Neth. J. Sea Res. 16:80–87.

Ventilla, R.F., 1982. The scallop industry in Japan. Adv. Mar. Biol. 20:309–382.

Walters, C.J., Hall, N., Brown, R. and Chubb, C., 1993. Spatial model for the population dynamics and exploitation of the Western Australian rock lobster, *Panulirus cygnus*. Can. J. Fish. Aquat. Sci. 50:1650–1662.

Warren, W.G., 1998. Spatial analysis for marine populations: factors to be considered. In: G.S. Jamieson and A. Campbell (Eds.). Proceedings of the North Pacific Symposium on Invertebrate Stock Assessment and Management. Can. Spec. Publ. Fish. Aquat. Sci. 125:21–28.

Watling, L. and Norse, E.A., 1998. Disturbance of the seabed by mobile fishing gear: a comparison to forest clearcutting. Cons. Biol. 12:1180–1197.

Wilding, C.S., Latchford, J.W. and Beaumont, A.R., 1998. An investigation of possible stock structure in *Pecten maximus* (L.) using multivariate morphometrics, allozyme electrophoresis and mitochondrial DNA polymerase chain reaction-restriction fragment length polymorphism. J. Shellfish Res. 17:131–139.

Wolff, M., 1987. Population dynamics of the Peruvian scallop *Argopecten purpuratus* during the El Nino phenomenon of 1983. Can. J. Fish. Aquat. Sci. 44:1684–1691.

Wolff, M. and Mendo, J., 2000. Management of the Peruvian bay scallop (*Argopecten purpuratus*) metapopulation with regard to environmental change. Aquatic Conser. Mar. Freshw. Ecosyst. 10:117–126.

Yamamoto, G., 1950. Ecological note on transplantation of the scallop, *Pecten yessoensis* Jay, in Mutsu Bay, with special reference to the succession of the benthic communities. Sci. Rep. Tohoku Univ. 19:11–16.

Yamamoto, G., 1953. Ecology of the scallop, *Pecten yessoensis* Jay. Sci. Rep. Tohoku Univ. (4) 20:11–32.

Yamamoto, G., 1960. Mortalities of the scallop during its life cycle. Bull. Mar. Biol. Station, Asamushi, Tohoku Univ. 10:149–152.

Yamamoto, G., 1964. (Studies on the propagation of the scallop, *Patinopecten yessoensis* (Jay), in Mutsu Bay). Japan Mar. Res. Prot. Assoc., Booklet 6:77 p. In Japanese. Translation of the Fish. Res. Bd. Canada No. 1054.

Young, P.C., 1994. Chapter 17. Recruitment variability in scallops: potential causes for the loss of Bass Strait populations. In: P.W. Sammarco and M.L. Heron (Eds.). The Bio-Physics of Marine Larval Dispersal. American Geophysical Union, Washington, DC. pp. 327–342.

Young, P.C. and Martin, R.B., 1989. The scallops fisheries of Australia and their management. Rev. Aquat. Sci. 1:615–638.

Young, P.C., Martin, R.B., McLoughlin, R.J. and West, G., 1990. Variability in spatfall and recruitment of commercial scallops (*Pecten fumatus*) in Bass Strait. In: M.C.L. Dredge, W.F. Zachrin and L.M. Joll (Eds.). Proceedings of the Australasian Scallop Workshop. Tasmanian Government Printer, Hobart, Australia. pp. 80–91.

Zacharin, W., 1994. Scallop fisheries in southern Australia: managing for stock recovery. Mem. Queensl. Mus. 36:241–246.

AUTHORS ADDRESSES

J.M. (Lobo) Orensanz - Centro Nacional Patagónico (CONICET), Bvd. Alte. Brown s/n, (9120) Puerto Madryn, Chubut, Argentina (E-mail: lobo@cenpat.edu.ar)

Ana M. Parma - Centro Nacional Patagónico (CONICET), Bvd. Alte. Brown s/n, (9120) Puerto Madryn, Chubut, Argentina (E-mail: parma@cenpat.edu.ar)

Teresa Turk - National Marine Fisheries Service, Northwest Fisheries Science Center, 2725 Montlake Blvd. E., Seattle, Washington, USA 98112 (E-mail: teresa.turk@noaa.gov)

Juan Valero - School of Aquatic and Fishery Sciences, University of Washington, Box 355020, Seattle, Washington, USA 98195 (E-mail: juan@u.washington.edu)

ENDNOTES

[1] The concept was introduced in marine population dynamics simultaneously in 1986 by Roughgarden and Iwasa (a model inspired by barnacles) and by Orensanz (a scallop study), who used the slightly different term "megapopulation". In subsequent years the term "metapopulation" became the standard in population dynamics. The two are equivalent, at least in this context.

[2] By "sedentary" we do not mean sessile or immobile. Rather, we indicate that the movements of post-settlers (benthic juvenile and adult scallops) have a small spatial scale compared to processes relevant to large-scale dynamics (e.g., larval dispersal and fleet behaviour).

[3] In the first edition of this book we introduced the word "connectedness" which (as our spell-check now indicates) is not part of the English language. "Connectivity", an appropriate alternative with the same meaning, has become the standard term (e.g., Roberts 1997).

[4] Throughout this chapter we use the term "aggregate" (as in "aggregate abundance") in reference to an entire metapopulation or subpopulation, with no consideration of its internal spatial structure.

[5] "Recruitment" is used here in the sense of recruitment to the fishable stock. In benthic ecology the word is often used to denote a combination of settlement and early post-larval stages.

[6] The homogeneity assumption is also met if the individuals of the harvested population are mobile and redistribute themselves after a fishing operation. This is never the case for scallops.

Scallops: Biology, Ecology and Aquaculture
S.E. Shumway and G.J. Parsons (Editors)

869

Chapter 15

Fisheries Sea Scallop, *Placopecten magellanicus*

K. S. Naidu[†] and G. Robert

15.1 INTRODUCTION

From an economic viewpoint, the sea scallop, *Placopecten magellanicus* (also called giant scallop, smooth scallop, ocean scallop or Atlantic deep sea scallop) is one of the most important pectinid species in the world. Between 1990 and 1999, it alone accounted for some 28% of the mean annual global production of all scallop species combined (Table 15.1). Sporadic booms in natural production associated with temporal fluctuations in abundance in some species (e.g., calico scallop) and manipulated production through enhancement in some others, particularly the Japanese scallop, *Patinopecten yessoensis*, have in recent years relegated sea scallop landings to a seemingly secondary role, spuriously depressing the sea scallop contribution to world tonnage. Increasingly it has become difficult to distinguish as distinct and separate scallop landings based on wild populations (capture fisheries) from those involving production involving some form of manipulation (e.g., sea ranching).

The Atlantic sea scallop is a relatively large mollusc commonly reaching sizes between 10–15 cm and frequently beyond. While large as contrasted with several other scallop species, the implied gigantism is not always characterised by unusual or disproportionate shell size. The largest sea scallop ever recorded measured 211 mm (shell height, tangential dorso-ventral measurement), a size a little larger than the previous recorded of 208 mm (Norton 1931) and had an adductor muscle (meat) weight of 231 g (0.51 lb.) (Naidu, unpubl.). Rock scallops, for example, are better endowed with shell heights approaching 250 mm (Hennick, cited in Kaiser 1986). Maximum age recorded for sea scallops is 29 years (Naidu, unpubl.). The shell of the sea scallop is almost circular in outline with symmetrical wings at the hinge (Fig. 15.1). Whereas the lower right valve is white, flat and smooth, the left valve is usually light to pale brown, convex and delicately ribbed. Occasionally, both shell valves are white. Concentric rings on the delicately ribbed surface of the left valve have been verified to be annual (Stevenson and Dickie 1954; Posgay 1962; Naidu 1969) and are commonly used for age determinations. Oxygen isotope records have also confirmed that growth lines are in fact annual events, consistent with biological interpretation (Tan et al. 1988). Hurley et al. (1987) have shown that the number of growth lines in laboratory reared post-larval shells is related to the actual age in days. Growth rings are especially pronounced in northern shallow-water populations (Naidu 1975). Repeated encounters with fishing gear in heavily fished aggregations and the haphazard deposition of shock rings makes interpretation of annual growth rings

Table 15.1

Nominal landings (MT, round) of scallop species. Percentages represent species contribution to global production in any given year. Adapted from Yearbook of Fishery Statistics, Capture Production, FAO, Rome, Vol. 88, 1999.

Species	1990	1991	1992	1993	1994	1995	1996	1997	1998	1999	Average
Argopecten gibbus (Calico scallop)	11,220 / 2.0%	0 / 0.0%	0 / 0.0%	0 / 0.0%	74,325 / 12.2%	10,003 / 2.0%	0 / 0.0%	0 / 0.0%	0 / 0.0%	0 / 0.0%	9,555 / 1.8%
Argopecten irradians (Atlantic bay scallop)	2,596 / 0.5%	2,062 / 0.4%	1,564 / 0.3%	2,670 / 0.5%	524 / 0.1%	1,593 / 0.3%	230 / 0.0%	452 / 0.1%	690 / 0.1%	216 / 0.0%	1,260 / 0.2%
Argopecten purpuratus (Peruvian calico scallop)	900 / 0.2%	1,632 / 0.3%	5,446 / 1.1%	2,732 / 0.6%	842 / 0.1%	3,113 / 0.6%	2,095 / 0.4%	4,013 / 0.8%	23,546 / 4.5%	30,141 / 5.3%	7,446 / 1.4%
Argopecten circularis (Pacific calico scallop)	29,471 / 5.2%	19,428 / 4.0%	5,290 / 1.0%	5,883 / 1.2%	8,570 / 1.4%	1,256 / 0.2%	17,290 / 3.5%	2,320 / 0.5%	2,726 / 0.5%	1,864 / 0.3%	9,410 / 1.8%
Chlamys islandica[1] (Iceland scallop)	12,712 / 2.3%	11,509 / 2.4%	19,714 / 3.8%	15,072 / 3.1%	15,076 / 2.5%	17,440 / 3.4%	20,242 / 4.1%	21,614 / 4.5%	16,668 / 3.2%	11,874 / 2.1%	16,192 / 3.1%
Chlamys opercularis (Queen scallop)	16,778 / 3.0%	14,263 / 2.9%	16,027 / 3.1%	15,263 / 3.1%	10,538 / 1.7%	8,040 / 1.6%	7,183 / 1.5%	11,443 / 2.4%	14,486 / 2.8%	15,260 / 2.7%	12,928 / 2.5%
Patinopecten yessoensis (Japanese scallop)	231,770 / 41.2%	180,998 / 37.1%	196,570 / 38.3%	226,780 / 45.9%	274,290 / 45.2%	279,382 / 54.9%	276,406 / 56.0%	266,957 / 56.0%	294,211 / 56.3%	305,510 / 53.8%	253,287 / 48.4%
Patinopecten caurinus (weathervane scallop)	2,415 / 0.4%	3,649 / 0.7%	6,884 / 1.3%	6,224 / 1.3%	4,944 / 0.8%	1,950 / 0.4%	2,372 / 0.5%	2 / 0.0%	3,228 / 0.6%	2,642 / 0.5%	3,431 / 0.7%
Pecten maximus (Great Atlantic scallop)	12,384 / 2.2%	15,001 / 3.1%	21,212 / 4.1%	20,429 / 4.1%	24,994 / 4.1%	23,471 / 4.6%	31,521 / 6.4%	35,657 / 7.5%	35,527 / 6.8%	35,411 / 6.2%	25,561 / 4.9%
Pecten jacobeus (Great Mediterranean scallop)	1 / 0.0%	0 / 0.0%	0 / 0.0%	202 / 0.0%	308 / 0.1%	23 / 0.0%	52 / 0.0%	95 / 0.0%	50 / 0.0%	68 / 0.0%	80 / 0.0%
Pecten novaezelandiae (New Zealand scallop)	563 / 0.1%	1,032 / 0.2%	1,255 / 0.2%	1,116 / 0.2%	1,277 / 0.2%	1,905 / 0.4%	759 / 0.2%	2,557 / 0.5%	574 / 0.1%	769 / 0.1%	1,181 / 0.2%
Placopecten magellanicus (Sea scallop)	217,153 / 38.6%	211,132 / 43.3%	192,822 / 37.6%	143,921 / 29.1%	146,032 / 24.0%	121,230 / 23.8%	109,382 / 22.2%	102,028 / 21.4%	99,432 / 19.0%	131,962 / 23.3%	147,509 / 28.2%
other Pectinidae	24,374 / 4.3%	27,435 / 5.6%	46,165 / 9.0%	53,582 / 10.8%	45,558 / 7.5%	39,779 / 7.8%	26,231 / 5.3%	29,523 / 6.2%	31,155 / 6.0%	31,667 / 5.6%	35,547 / 6.8%
Totals	562,337	488,141	512,949	493,874	607,278	509,185	493,763	476,661	522,293	567,384	523,387

[1] modified from Naidu et al. (2001).

Figure 15.1. Left and right valves from a 5-year old sea scallop showing prominent concentric growth rings. The right valve is flatter and usually lighter in colour than the left valve.

sometimes difficult and frequently impossible. Under these circumstances, it may be necessary to utilise growth bands on the resilium (Merrill et al. 1966).

15.2 DISTRIBUTION

Confined to the Northwest Atlantic, the sea scallop has a geographical range from the north shore of the Gulf of St. Lawrence (Squires 1962) to Cape Hatteras, North Carolina (Posgay 1957a, Fig. 15.2). Although primarily a continental shelf species, and found usually in depths ranging from about 10–100 m (5–55 fm), it may be found in shallow water just below low tide (Verrill and Smith 1873; Read 1967). There is even a record of it occurring intertidally in the Gulf of Maine (MacKenzie 1979) but this must be considered anomalous and needs verification. Farther south, because it is restricted to areas with maximum water temperatures less than about 20°C, they are usually found at depths exceeding 55 m (30 fm). Merrill (1959) extends its bathymetric range down to 384 m (210 fm). As an environmental variable, depth has a pervasive influence on nearly all aspects of the growth and biology of the species (Posgay and Merrill 1979; Shumway and Schick 1987; Schick et al. 1988).

Sea scallop beds of sufficient extent and density to support commercial fisheries occur from Virginia Capes (latitude 36°50'N) to Port au Port Bay, Newfoundland, Canada (latitude 48°40'N). Offshore, sea scallops have been exploited commercially on Georges Bank, the Mid-Atlantic Shelf, Browns Bank, German Bank, Lurcher Shoals, Grand

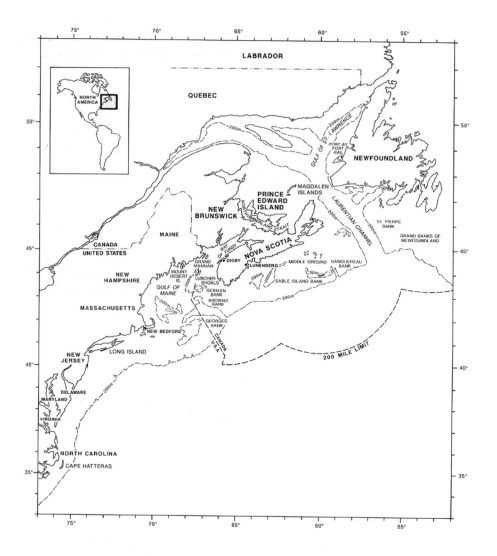

Figure 15.2. Principal sea scallop fishing grounds.

Manan, around Sable Island, Middle Ground, Banquereau Bank, and on St. Pierre Bank (Fig. 15.2). Georges Bank consistently rates as the world's largest producer of scallop. The Bay of Fundy (especially off Digby) and Gulf of Maine also have had a long history of production. For Canada, the Bay of Fundy landings have been second to Georges Bank except for very recently (Table 15.2). In the center of its range (Georges Bank and Middle Atlantic Shelf), scallops have been quite successful and have withstood moderate to heavy exploitation. The Mid-Atlantic area off Long Island and New Jersey (New York Bight) and Delmarva and Virginia-North Carolina regions has become more important in recent years sometimes contributing to more than half of the USA total scallop production. The recent closure of areas on Georges Bank has also contributed to the rise for the Mid-Atlantic area. In the Gulf of Maine, the majority of catches come from inshore U.S. territorial waters. Georges Bank where most of the offshore effort is directed constitutes the world's largest single natural scallop resource (Caddy 1989). Scallop production on Georges Bank has been attributed to the presence of a large gyre which forms during the summer and later helps to retain planktonic scallop larvae within the area until they are ready to metamorphose and settle to the sea bottom (Larsen and Lee 1978). Towards the extremes of their range sea scallops generally have been less successful and have not withstood continued heavy exploitation (Dickie and Medcof 1963). Fisheries in fringe areas such as the northeast coast of the United States (Serchuk et al. 1979) and St. Pierre Bank (Naidu et al. 1983b) typically are characterised by a disproportionate dependence on sporadic recruitment of a single or a few intermittent and, sometimes, well-spaced year-classes. Consequently, in fringe areas, fisheries must cope with wide and, sometimes, catastrophic temporal fluctuations.

As in most scallop fisheries, sea scallop recruitment, even in the center of its range frequently is irregular and pose undue problems to an industry that is typically overcapitalised. Even in industries that are no longer overcapitalised (Canadian deep sea fleet), the irregular recruitment plays havoc in the supply and demand sides of the market. In some other areas such as the Bay of Fundy (Caddy 1979; Dadswell et al. 1984; Robert et al. 1984), there is evidence of cyclicity in production which appears to be associated with periodic tidal phenomena. These departures from "steady-state" have wide-ranging implications for the orderly development and judicious management of scallop fisheries.

Table 15.2

Relative (percent) contribution of recent sea scallop landings from principal scallop producing areas of Northwest Atlantic (Canada).

Year	Georges Bank 5Z	St. Pierre Bank 3P	Sable Island and Western Banks 4W	Browns Bank 4Xo,p	German/ Lurcher 4Xq	Bay of Fundy and approaches 4Xr,s, 5Y
1990	57.3	1.7	4.5	2.0	0	34.5
1991	65.9	1.5	4.0	2.3	0	26.3
1992	64.0	0.7	5.0	4.6	1.5	24.2
1993	63.7	0.9	2.1	5.9	6.9	20.5
1994	52.9	0.5	1.1	14.8	9.3	21.4
1995	31.3	0.6	2.1	31.6	9.7	24.6
1996	60.1	0.4	3.3	14.9	3.5	17.7
1997	73.7	0	3.0	8.7	5.8	8.8
1998	66.6	0	4.4	10.0	7.6	11.4
1999	62.6	0	4.7	8.3	13.8	10.6

15.3 HISTORY OF FISHERY

The sea scallop is the most important commercial bivalve in North America. In the United States, the fishery is over 100 years old (Table 15.3), and second in value only to the American lobster. The fishery developed inshore off New England, but was of little commercial significance until 1883, when several scallop beds near Mt. Desert Island, Maine were discovered (Smith 1891; Premetz and Snow 1953) (Fig. 15.2). Ports in Maine were the leading producers. Landings between 1889 and 1924 were characterised by considerable fluctuations with maximum landings recorded in 1905 (195 t meats). With the discovery of scallop beds off Long Island in the early 1920's, Middle Atlantic ports temporarily became the center of U.S. scallop operations. New England regained leadership in sea scallop production in the mid 1930's with the discovery of offshore scallop beds on Georges Bank (O'Brien 1961). New Bedford, Mass., soon emerged and continues to be the leading U.S. port for the sea scallop industry. By the early 1930's the U.S. fishery had already spread offshore to Georges Bank. Between 1975 and 1979, more than 50% of the total U.S. scallop production has been from the middle Atlantic Shelf (Anonymous 1977).

In Canada, first landings recorded in the fisheries statistics of Canada were made in 1886 when 300 dozen were sold from Lunenburg County (Bourne 1964). A fishery developed along the southwest coast of New Brunswick in 1889. The Digby scallop beds were discovered in 1920 and the fishery there expanded rapidly into the Bay of Fundy. Landings here have fluctuated widely and reached a record of 4,600 t meats in 1989. The Digby fishery also initiated the practice of shucking scallops at sea. While minor fisheries for the mollusc were reported inshore elsewhere in the Maritimes, including the Gulf of St. Lawrence, the Digby fishery continued to be the mainstay of the Canadian fishery until 1945 when a long-standing interest in the offshore banks began to materialise. Encouraged by the relative success of the initial forays offshore, Canadian authorities, both public and private, conducted systematic explorations off the coast of Nova Scotia and the vast apron of shelf to the south and southeast of Newfoundland commonly referred to as the Grand Banks of Newfoundland (Dickie and Chiasson 1955; Somerville and Dickie 1957; MacPhail and Muggah 1965; Rowell et al. 1966a, 1966b). Beds of sea scallops on St. Pierre Bank, off the south coast of Newfoundland (Fig. 15.2), for example, were discovered during one such exploratory mission in 1953. Many of the earlier surveys were not fully documented (Caddy and Chandler 1968). Industry frequently adopted a proprietorial attitude towards newly discovered scallop beds and were loathe to publicise new finds. By 1954, landings by the Canadian offshore fleet surpassed those of the inshore fleet for the first time (Bourne 1964). Canadian scallop landings nearly doubled to 1044 t meats within six years after a concerted attempt had begun in 1956 to prosecute stocks offshore. By 1965, 75 percent of the annual removals from Georges Bank was taken by the Canadian fleet. The Canadian effort on Georges Bank was primarily focused on the Northern Edge and Peak, which encompass some of the most productive areas of the Bank. From the outset the Canadian sea scallop industry was and continues to be centered in the Province of Nova Scotia.

Table 15.3

Sea scallop landings (metric tons, meats) from the Northwest Atlantic, 1887–1998.

Year	USA[1] (t, meats)	Year	USA[1] (t, meats)	Can[2] (t, meats)	Total (t, meats)
1887	112	1951	8.503	91	8.594
1888	91	1952	8,451	91	8,542
1889	141	1953	10,713	242	10,955
1892	53	1954	7,997	234	8,231
1897	435	1955	10,036	289	10,325
1898	156	1956	9,102	424	9,526
1899	24	1957	9,523	841	10,364
1900	79	1958	8,608	1,181	9,789
1901	286	1959	11,178	2,646	13,824
1902	61	1960	12,065	3,680	15,745
1903	62	1961	12,456	4,906	17,362
1904	216	1962	11,174	6,089	17,263
1905	200	1963	9,038	7,314	16,352
1906	255	1964	7,704	7,538	15,242
1907	236	1965	9,105	8,257	17,362
1908	834	1966	7,237	8,284	15,521
1909	843	1967	4,646	6,033	10,679
1910	919	1968	5,473	7,049	12,522
1911	663	1969	3,362	6,227	9,589
1912	842	1970	2,613	5,845	8,458
1913	353	1971	2,593	4,995	7,588
1914	386	1972	2,655	5,118	7,773
1916	266	1973	2,401	4,788	7,189
1919	89	1974	2,721	6,478	9,199
1921	38	1975	4,421	7,962	12,383
1924	154	1976	8,721	11,392	20,113
1926	506	1977	11,104	14,090	25,194
1928	216	1978	14,483	13,111	27,594
1929	1,130	1979	14,256	10,716	24,972
1930	1,111	1980	12,566	8,357	20,923
1931	1,058	1981	11,742	10,638	22,380
1932	1,517	1982	9,044	7,798	16,842
1933	2,009	1983	8,707	6,133	14,840
1935	1,955	1984	7,739	4,160	11,899
1937	3,989	1985	6,742	5,333	12,075
1938	4,041	1986	8,661	6,577	15,238
1939	4,440	1987	13,534	8,749	22,282
1940	3,467	1988	13,414	9,285	22,699
1941	3,622	1989	14,917	11,079	25,996
1942	3,258	1990	17,449	9,955	27,404
1943	2,508	1991	17,280	9,443	26,723
1944	2,209	1992	14,296	10,205	24,501
1945	2,590	1993	7,389	10,473	17,863
1946	5,236	1994	7,641	9,316	16,957
1947	6,647	1995	8,030	7,138	15,168
1948	7,546	1996	7,921	5,800	13,721
1949	8,299	1997	6,027	6,464	12,490
1950	9,063	1998	5,586	6,467	12,052

[1] U.S.A. landings from Anon.(1989), F. M. Serchuk (pers. comm.) and NAFO Statistical Bulletins.
[2] Canadian landings modified from Anon.(1989), and NAFO Statistical Bulletins and Naidu et al. (2001).

With the fuller development of the Canadian and U.S. fleet after World War II, Georges Bank saw a dramatic increase in annual landings and quickly became established as the world's largest producer of scallops. Concomitant with the success of the then competitive Georges Bank fishery and the discovery of new beds elsewhere along the Atlantic seaboard came inevitable overcapitalisation. Frequently, industry is unable to respond in sympathy to the downturns associated with the extreme dynamism exhibited. As in other fisheries, correction to overcapitalisation is never as rapid as the initial response to a perceived new opportunity and is nearly always fraught with difficulty (Caddy and Gulland 1983). Industry generally has been slow to respond to these fluctuations in abundance.

The shared jurisdiction of this resource by Canada and the U.S. prior to ratification of the Convention for the Law of the Sea (October 1984) necessitated a joint management regime under the auspices of the International Commission for the Northwest Atlantic Fisheries (ICNAF). Attempts to increase yield per recruit by controlling the size of recruitment by imposing a size limit, in the case of scallops a meat count (number of meats/unit weight), were met with only partial success. The meat count was too high to be effective. A competitive fishery, although Canada had limited entry since 1973, led to very high levels of fishing effort and fleet overcapacity on both sides. As a consequence of the World Court decision on the fisheries jurisdiction of the United States and Canada on Georges Bank, the offshore fleets became restricted to their respective national zones (Fig. 15.2). Canada and the United States continue to be the sole participants of this lucrative fishery.

With the advent of the 200-mile Fisheries Conservation Zone in U.S. waters, the New England and mid-Atlantic Fisheries Management Councils developed and implemented the Sea Scallop Fisheries Management Plan (FMP) to regulate the fishery. Under the FMP, sea scallop fisheries for the Gulf of Maine, Georges Bank, and the mid-Atlantic are managed to address concerns of growth and recruitment overfishing via biological reference points such as yield per recruit. Over the last 25 years, there have been 2 periods of landings over 100,000 metric tons (round weight), the late 1970's and from 1987 to 1992.

Landings in the latter period peaked near 150,000 metric tons in 1990 (Table 15.4). Between the 2 peaks, most of the 1980's produced considerably less with a minimum slightly over 50,000 metric tons in 1985. After the last peak, annual landings have declined to lower values (1997–1998) than the minimum encountered during the 1980's. On Georges Bank landings peaked in 1990 and decreased thereafter due to poor stock conditions and area closures. In the mid-Atlantic region, landings peaked in 1989 and then fluctuated around a lower average estimate in later years. In the U.S. Gulf of Maine the 1998 landings reached a record high; there might have been problems in assigning catches to fishing areas.

After the jurisdiction for fisheries on Georges Bank had been settled by the World Court in October 1984, the Canadian offshore scallop industry focused on stock rehabilitation through better harvesting of the resource. Important reductions in fishing effort and the adoption of a new management regime were to rebuild the stock while optimising the yield, stabilise catches from year to year, and rationalises the fleet.

Table 15.4

Quantity and landed value of sea scallops in the United States and Canada, 1978–1998. (Quantity in metric tons round weight, value in millions of U.S. and Canadian dollars).

Year	USA[1] Quantity	Value	Canada[2,3] Quantity	Value
1978	120,209	78.6	109,402	63.5
1979	118,325	105.4	89,491	74.5
1980	104,298	107.2	70,475	68.5
1981	97,459	105.3	89,897	99.6
1982	75,065	74.6	65,097	60.5
1983	72,326	107.9	51,286	70.8
1984	64,234	94.3	36,470	56.4
1985	56,125	71.8	47,210	62.2
1986	71,886	91.2	57,063	74.5
1987	109,784	124.7	73,903	95.0
1988	109,543	123.0	77,854	85.9
1989	124,284	130.0	92,188	93.1
1990	145,341	149.1	83,278	86.9
1991	142,818	153.8	79,538	81.4
1992	117,918	153.4	92,163	100.7
1993	60,540	97.1	90,749	120.1
1994	63,819	84.8	91,626	139.0
1995	65,620	90.0	68,670	103.2
1996	65,537	98.3	59,987	90.7
1997	51,502	89.8	66,023	100.9
1998	45,658	75.1	63,030	95.9

[1] 1978–1988 U.S.A. landings from ICNAF/NAFO Statistical Bulletins and values from the U.S. Dept of Commerce and National Marine Fisheries Service. 1989–1998 U.S.A. landings and values from the NMFS web site.
http://www.st.nmfs.gov/webplcomm/plsq1/webst1.MF_ANNUAL_LANDINGS.RESULTS
[2] Canadian landings and values aggregate scallop species. Sea scallops and Icelandic scallops are included, with sea scallops the more prominent.
[3] 1978–1988 Canadian landings and values from Canadian Fisheries Statistical Review. 1989–1998 Canadian landings and values from the Department of Fisheries and Oceans web site.
http://www.dfo-mpo.gc.ca/communic/statistics/landings/land_e.htm

Landings from Georges Bank, the area contributing the most, peaked in 1987 and in 1992–93 at over 50,000 metric tons. Landings have been on a declining trend since the last peak. The range of landing figures since 1986 (16,500–56,500 metric tons) when the management plan was significantly altered is more narrow than the pre-1986 period when the fishery was competitive. Other offshore areas on the Scotian Shelf (German Bank, Browns Bank, Sable Island and Western Banks) may be exploited on a somewhat sustainable basis but do not produce landings comparable to Georges Bank (range of 1,700 to 17,000 metric tons). However, the same strict management measures generate comparable catch rates (Table 15.5). Less productive areas like St. Pierre Bank are exploited occasionally with a maximum under 1,500 metric tons.

Table 15.5

Sea scallop catch-per-unit-effort (kg meats/crew-hour-metre) from NAFO[1] areas fished by the Canadian deep-sea fleet.

Year	St. Pierre Bank[2] 3P	Banquereau Bank[3] 4V	Sable Island and[3] Western Banks 4W	Browns Bank[4] 4Xo,p	German Bank[3] 4Xq	Georges Bank[5] 5Z
1990	0.417	-	0.240	0.538	-	0.494
1991	0.358	0.240	0.284	0.557	-	0.599
1992	0.298	0.303	0.242	0.757	0.998	0.561
1993	0.278	0.364	0.192	0.829	0.756	0.627
1994	0.211	-	0.266	1.688	0.833	0.523
1995	0.184	-	0.302	1.292	0.428	0.349
1996	0.155	-	0.288	0.757	0.252	0.617
1997	0.133	-	0.305	1.013	1.509	0.744
1998	-	0.626	0.239	0.904	0.983	0.518
1999	-	0.594	0.207	0.886	1.524	0.700

[1] Northwest Atlantic Fisheries Organization
Source:
[2] G. Robert, Department of Fisheries and Oceans, Dartmouth, Nova Scotia, Canada.
[3] Robert, G. and M.A.E. Butler. 1997. Scallop stock status for 1996 – Eastern Scotian Shelf and German Bank. DFO Atl. Fish. Res. Doc. 97/49. After 1996, G. Robert.
[4] Robert, G. and M.A.E. Butler. 1998. Browns Bank north scallop stock assessment – 1997. DFO Atl. Fish. Res. Doc. 98/70. After 1997, G. Robert.
[5] Robert, G., G.A.P. Black, M.A.E. Butler, and S.J. Smith. 2000. Georges Bank scallop stock assessment – 1999. DFO Can. Stock Assess. Sec. Res. Doc. 2000/016.

15.4 POPULATION BIOLOGY

Unambiguous identification and separation of stocks within this species has been problematical as has been the assumption, sometimes justifiable, that fisheries are based on self-sustaining megapopulations (Caddy 1975; Orensanz 1986). Although phenotypic differences abound throughout its geographic range, genetic distances appear to be relatively small. Overall, the evidence for the existence of discrete stock units is not persuasive (Zouros and Gartner-Kepkay 1985; Volckaert et al. 1990) and suggests sufficient larval exchange between adjacent areas even if sometimes only unidirectional (Squires 1962; Naidu and Anderson 1984). Conceptually, it is tempting to reconcile the larval phase in this relatively sedentary species with its adaptive function of dispersal. A recent review (Sinclair et al. 1985) of important scallop fisheries for which long-term data are available, on the other hand, suggest that the main fishable concentrations are associated with strong tidal circulations (Tremblay and Sinclair 1988) or located in areas of persistent gyres (Caddy 1979; Posgay 1979a). Sinclair et al. (1985) argue that self-sustaining characteristics are borne out by the persistence in widely separated areas of scallop aggregations. But this view does not easily explain the persistence of "terminal" scallop populations with reduced reproductive output (estimated gamete production) (MacDonald and Thompson 1985b, 1986a) and reduced reproductive effort (proportion of non-respired assimilated energy allocated to reproduction (MacDonald et al. 1987; Robert and Lundy 1985). Knowledge of the source(s) of recruitment for a given aggregation and the relative larval contribution from within and or without is important to a clearer understanding of the overall population dynamics of that aggregation. Knowledge is quite limited in this area. Modelling exercises of scallop larval transport on Georges Bank support the theory of larval retention where gyral circulation exists (Lewis et al. 1999). The advent of remotely sensed satellite data to delineate ocean features such as thermal fronts could illustrate their influence on scallop settlement.

In exploited populations, it is also necessary to be concerned with important concepts of overfishing (recruitment overfishing versus growth overfishing). Allegations of recruitment overfishing notwithstanding, empirical data supporting its possible occurrence in sea scallops is weak, if not non-existent. Such data that are available seem to indicate that reproductive success in this highly fecund species is not wholly constrained by a low biomass. A healthy ten-year-old female may release up to 90–95 million eggs during a single spawning season (MacDonald 1984). A plot of age 4+ biomass against recruits (lagged by three years) for Georges Bank, for example, show that above-average recruitment occurred even when biomass was at historically low levels during the 1970's (Mohn et al. 1987). The same type of data with more recent information, including the 1997 biomass, confirmed that low or high levels of spawning stock biomass do not necessarily relate to high recruitment. It would appear that moderate biomass levels produce higher recruitment (Robert et al. 2000). The apparent lack of a linear stock recruit relationship suggests that conventional finfish assessment models may not be easily extrapolated to sea scallops. Refugia of unexploitable scallops, either at very low densities and or inaccessible to traditional fishing gear must necessarily safeguard the self-reproducing capacity of the megapopulation. This would be the case particularly in an area where water circulation is conducive to retaining the otherwise fugitive larvae within a given area for at least as long for them to settle on suitable substrates and complete metamorphosis (approx. 1–2 months) (Culliney 1974; Larsen and Lee 1978). As such, recruitment overfishing is seldom, if ever, specifically addressed in the protection and husbanding of the resource. Even if not completely understood, the apparent resilience to recovery in this species appears to be fairly widespread however. Also, the spawning potential is implicitly safeguarded by regulations which seek to maximise yield per recruit (older individuals contribute little to somatic production but a great deal to reproductive output (MacDonald and Thompson 1986b)). The authors are not aware of an exploited population of sea scallops with a significant production history whose imminent biological collapse has been somehow prevented through regulation of fishing activity or one that has not self-rehabilitated from near commercial extinction. Ongoing investigations into the culture and resource enhancement of this species in Newfoundland have shown that concentrations of spat on artificial substrates are sometimes disproportionately high (Naidu et al. 1983a) and not commensurate with spawning scallop biomass in the area from where seed were collected. Similarly, in Port au Port Bay, Newfoundland settlement of up to 6,500 seed/collector (x = 4,000) has been recorded even when mean annual production during the five years preceding was only 7.2 t meats (M. Lanteigne, pers. comm. [Science Branch, Department of Fisheries and Oceans, Moncton, New Brunswick, Canada ElC 9B6]). Until recently, research on the sea scallop has been confined primarily to the adult phase. The dearth of information concerning larval biology of the species in the open sea remains a conspicuous gap in our knowledge of the life history of this species (see Tremblay et al. 1987; Tremblay and Sinclair 1988). Field studies on spawning, larval distribution, and spat settlement now have become major concerns in Canada (Naidu and Scaplen 1979; Sinclair et al. 1985; Naidu et al. 1989). The question of whether scallop aggregations are self-sustaining has not been unequivocally resolved. Neither has a relationship been shown to exist between source of recruitment for a given scallop aggregation and the degree of temporal stability of that aggregation.

On the basis of historical patterns of production, Caddy and Gulland (1983) attempted to broadly classify fisheries as being a) steady, b) cyclical, c) irregular, and d) spasmodic. They proposed that approaches to management of fish stocks must recognise this departure from the often-assumed equilibrium or steady state conditions. For sea scallops it is apparent that while the biomass in each persistent self-reproducing aggregation has shown considerable fluctuations, not all of which are ascribable to fishing, the order of magnitude of average population size has remained more or less constant (Sinclair et al. 1985). On the basis of this observation, Caddy (1989) concludes that sea scallop fisheries are better described as cyclical (Caddy 1979; Dadswell et al. 1984; Robert et al. 1984), irregular or spasmodic, as suggested for populations in the northern (Naidu and Cahill 1984) and southern (Serchuk et al. 1979) limits of their distribution range. Four strong year-classes, those of 1957, 1972, 1977 and 1982 for example, have produced major peaks in landings of the Georges Bank fishery prior to 1990. The 1972 year-class resulted in peak landings of nearly 18,000 t in each of 1977 and 1978. Consequently, optimising the yield per recruit has been a major management objective of this recruitment-driven fishery, at least over the short term. In fact, more recent strong year-classes are not directly reflected in peak annual landings of the Canadian deep sea fleet because quota management spreads the benefits of an abundant year-class over 2–3 years.

15.5 GROWTH AND YIELD PER RECRUIT

Sea scallop growth rates are highly variable, depending on location. While much data on sea scallop growth have been assembled over the years, much remain in a form not amenable to a formal publication (see for example, Posgay and Merrill 1979). Growth rates calculated for various aggregations are reported usually in the form of the von Bertalanffy growth equation. Considerable differences in the growth pattern of scallops is evident for different populations of scallops (Table 15.6). Naidu (1969) reported that scallops from more northern latitudes generally have larger H_∞ values; conversely, K values, generally, are smaller. Posgay (1953) reported that scallops living in an area with a narrow temperature range near 10°C exhibited the best growth. Growth rates have been shown to be negatively correlated to a combination of temperature and food availability with low temperature and low food levels producing the slowest growing scallops

Table 15.6

Summary of estimated von Bertalanffy parameters for selected populations of the sea scallop, *Placopecten magellanicus.*

H∞	Parameter k	t₀	Location	Source
Port au Port, Bay, Newfoundland, Canada				
152.4	0.21	-0.48	Boswarlos	Naidu 1975
160.5	0.19	-0.88	West Bay	Naidu 1975
139.9	0.27	0.11	Fox Is. River	Naidu 1975
St. Pierre Bank, Canada				
146.9	0.22	0.35	St. Pierre Bank, Newfoundland	Naidu and Cahill 1984
Gulf of St. Lawrence, Canada				
103.7–108.8	0.33–0.37	0.46–0.67	Tormentine Bed, Northumberland St. P.E.I.	Chouinard 1984
114.8	0.28	-0.28	Central Strait, Northumberland St. P.E.I.	Jamieson 1979
128.0	0.19	-1.10	Eastern Region, Northumberland St., P.E.I.	Jamieson 1979
127.8	0.22	-0.50	Western Region, Northumberland St., P.E.I.	Jamieson 1979
126.2	0.21	-0.40	Northumberland St. (all regions)	Jamieson 1979
149.7	0.14	0.71	Basse-Cote Nord	D'Amours and Pilote 1982
147.2	0.25	0.44	Iles de la Madeleine	D'Amours and Pilote 1982
Scotian Shelf, Canada				
108.8	0.36	1.60	Browns Bank	Jamieson et al. 1981
113.5	0.27	1.31	Browns Bank	Robert et al. 1985
124.6	0.28	1.60	German Bank	Jamieson et al. 1981
130.6	0.23	1.39	German Bank	Robert et al. 1985
160.3	0.18	1.28	Middle Ground	Robert et al. 1985
161.5	0.18	1.34	Middle Ground	Robert et al. 1986
156.0	0.20	1.35	Sable, Western Bank	Robert et al. 1985
139.1	0.21	1.38	Sable, Western Bank	Robert et al. 1986
156.0	0.18	1.28	Lurcher Shoals	Robert et al. 1985
155.8	0.18	1.22	Lurcher Shoals	Robert et al. 1986
Gulf of Maine, U.S.A				
174.3	0.22	-1.24	Gulf of Maine	Serchuk et al. 1982
207.0	0.14	-0.23	Damariscotta River	Langton et al. 1987
218.0	0.13	0.17	Jericho Bay	Schick et al. 1987
148.0	0.27	0.10	Ringtown Island	Schick et al. 1987
116.0	0.28	-0.01	20 mi. S. of Booth Bay Harbor	Schick et al. 1987
223.0	0.09	-0.37	W. Jeffreys Lodge	Schick et al. 1987
Georges Bank (Canada/U.S.A.)				
148.9	0.26	1.00	Georges Bank	Posgay 1962
145.5	0.38	1.50	Georges Bank	Brown et al. 1972
146.4	0.35	1.40	Georges Bank	Posgay 1976
143.6	0.37	1.00	Georges Bank	Posgay 1979a
152.5	0.34	-1.45	Georges Bank	Serchuk et al. 1982
161.4	0.18	1.20	Georges Bank	Roddick and Mohn 1985
146.5	0.30	1.32	Northeast Peak	Posgay 1962
141.8	0.28	1.00	Northern Edge	Posgay 1962
Mid-Atlantic (U.S.A.)				
151.8	0.30	-1.13	Mid-Atlantic Bight, U.S.A.	Serchuk et al 1982

(MacDonald and Thompson 1985a). Furthermore, under the same general conditions of temperature and food availability in the Bay of Fundy, Smith et al. (2000) have found growth-rates to be spatially specific. Length-weight relationship of sea scallop populations have been described by numerous authors (e.g., Haynes 1966; Worms and Davidson 1986b; Naidu et al. 1983b; Naidu and Cahill 1984; Robert and Lundy 1987). While sexual maturity may be attained as early as age 1, initial spawning does not occur until the second year after the deposition of the first growth ring when scallops typically are between 23–75 mm (Naidu 1970; Posgay 1979b). Fecundity of the younger age-groups contributes little to total egg production however (MacDonald and Thompson 1986b).

Knowledge of the age structure of populations, growth rate and longevity are essential for effective management of the fishery. Aspects of the general population dynamics of scallop species are considered in Chapter 14. Recently, Caddy (1989) has succinctly summarised some of the approaches utilised in the assessment of scallop fisheries with particular reference to the sea scallop, *Placopecten magellanicus*. The assumption that the main scallop fisheries are self-sustaining megapopulations (Orensanz 1986) is reasonable and could be the basis, as it frequently is, for management measures instituted for their rational exploitation (Caddy 1989). Within each superpopulation gradients of scallop density exist, often culminating in a patch containing scallops of above-average density. A fishing ground consists of areas where scallops occur in high densities (beds), interspersed with areas where densities are as low as 1/40 of that in the center of the bed (Caddy 1970). The high density contagions are actively sought by fishermen. Once located, they are systematically harvested until yield/tow becomes uneconomic. Sea scallop catch-per-unit-effort (CPUE) is reported usually in kg/crew-hour-meter (kg crhm^{-1}). For the offshore beds catch rates typically range from about 0.2 to 0.6 kg crhm^{-1} (Table 15.5). The 1990 catch rate series for Browns Bank is well above the norm although it is regularly exploited (200–2,000 metric tons of meats annually). They correspond to a fishery developed on the northern part of Browns Bank in areas not previously fished. Right from the start in 1989, the fishery has been managed with a quota and a meat count (Robert and Butler 1998). Good densities and good yields equate above average catch rates. As in *Patinopecten yessoensis* (Querellou 1975), postsettlement dispersal from within area(s) of primary settlement sometimes occurs through random swimming activity of juveniles (Caddy 1968; Morton 1980) resulting in their being further dispersed usually along tidal axes (Posgay 1981; Melvin et al. 1985) to form typically elongated or elliptical scallops beds (Posgay 1981; Robert et al. 1982; Melvin et al. 1985; Caddy 1989). Caddy (1989) has postulated the existence of intraspecific density regulation and is of the view that this may account for the low densities with size (age) observed in nature. Densities within a 117 km^2 scallop bed on Georges Bank in 1970, for example, were estimated to be between 2 and 4 scallops m^{-2}. Elsewhere, in the Northumberland Strait scallop densities have been variously reported to be between 4 and 6 scallops m^{-2} (Caddy 1968) and up to 4–7 scallops m^{-2} (Caddy 1970). Similarly, densities of between 2 and 4 scallops m^{-2} have been reported from a northern shallow-water population in Port au Port Bay (Naidu 1969). The high density contagions frequently are too small to be adequately addressed in random stratified sampling schemes particularly when employed over large areas and pose undue problems for estimation of resource abundance. Notions of scallop distribution and densities in a scallop bed will likely be challenged by the recent availability of multibeam sonar backscatter bottom maps. Preliminary investigations on Browns Bank strongly correlate habitat types defined by sediment with scallop densities (Kostylev et al. 2003).

Although there is some overlapping of areas of settlement, sea scallop beds characteristically are dominated by one or few year-classes (Caddy 1968, 1970, 1971b, 1972). Not surprisingly, it is not uncommon to see dredge-sampled catch to comprise of similar-sized individuals (Fig. 15.3). This homogeneity in size composition is particularly striking in fringe areas where the fishery may sometimes rely almost entirely on a single age cohort.

15.6 GEAR AND BOATS

The sea bottom where scallops are found is highly variable, usually consisting of mud, sand, pebbles, rocks, and even boulders, but highest densities tend to be on bottoms where byssal attachment is likely, e.g., glacial till (Caddy 1989). Extremely rugged gear may be required depending on the type of sea bottom. Improvised rakes of unspecified configuration have been used from the early days with varying success. With the advent of motorised vessels, greater complexity in gear design and fishing techniques became possible. Majority of landings comes from vessels using scallop rakes (dredges or drags). Specially modified groundfish trawls are sometimes used (Dr. F.M. Serchuk, pers. comm.). For a summary of scallop gear used in various fisheries, including those commonly used for harvesting sea scallops, the reader should consult Stone and Hurley (1987). These include the Digby-type rake (Fig. 15.4), the Green sweep drag, Cayenne drag, Alberton drag, the Japanese (Keta-ami) dredge, and the Scottish spring-toothed drag. Because of their relatively small size the foregoing gear types are usually used in shallow waters nearshore. Inshore vessels sometimes use up to 9 Digby-type rakes individually shackled to a single tow bar. Each component of the gang of drags operates independently of one another. This allows the fleet of drags to conform closely to the bottom contours so typical of the eulittoral zone. Overall efficiency for the various inshore rakes is typically low. Efficiency of Digby drags for capturing sea scallops on rocky and smooth bottoms, for example, has been estimated to vary from about 5 to 12%, respectively (Dickie 1955). Higher efficiency on smooth grounds is attributed to improved gear performance through more consistent contact with the substrate. The offshore scallop rake (New Bedford scallop drag) is one of the sturdiest fishing gears used in the capture of sea scallops (Fig. 15.5). Widely used in the offshore scallop fishery in Canada and the United States, it has undergone little change since it was first introduced. Essentially, the drag consists of a heavy metal frame and a bag knit with steel rings 3 or 4 inches in diameter and variously inter-connected with links. Rubber chafing gear is sometimes used to extend the life of the rake, particularly when fishing hard bottom. Chafing gear can be of two types: shingles which are attached in rows to the belly of the rake to decrease wear on links and rings and rubber rollers which are fitted to the full length of the club stick. When fishing smooth bottom, a bag may last for as long as five

884

Figure 15.3. Typical haul of sea scallops from St. Pierre Bank showing single-cohort recruitment.

Figure 15.4. Gang of 4 Digby scallop rakes commonly used in the inshore sea scallop fishery.

Figure 15.5. New Bedford scallop rake used in deeper, offshore waters.

trips; but on rocky bottom, it may not last for even one trip. For a detailed description of the New Bedford scallop drag the reader is referred to Bourne (1964). The efficiency of the offshore rake is somewhat higher than for inshore drags and estimated to be 15.4% (Caddy 1971a). Progressively higher capture efficiencies were observed with increasing shell size with highest efficiency recorded for the larger, non-motile scallops (Caddy 1968). Swimming activity rather than selection by the drag appears to be responsible for the low drag efficiency for the capture of scallops smaller than 100 mm (Caddy 1968). Also, overall efficiency of the offshore dredge was higher over gravel on Georges Bank than on sand in the Gulf of St. Lawrence (Caddy 1968).

Factors known to affect the size selectivity of various types of gear commonly used in capturing sea scallops include inter-ring spacing (Medcof 1952; Caddy 1971a); ring size (Medcof 1952; Dickie 1955; Bourne 1960, 1962, 1965; Posgay 1958; Caddy 1971a; Serchuk and Smolowitz 1980; Smolowitz and Serchuk 1987); mesh size used in the rope back (Caddy 1968); rock chains (Bourne 1965; Smolowitz et al. 1985); tooth length, width and spacing (MacPhail 1954; Caddy 1973); chafing gear (Smolowitz and Serchuk 1987); substrate composition, and escape responses. Repeated towing over the same ground may also improve overall efficiency due to smoothing effects of the gear on the bottom (Caddy 1977). Most of the selection, particularly for larger sea scallops, is passive.

The majority of selection studies for the sea scallop have attempted to increase escapement of small scallops by increasing the internal diameter of the rings. While selective size retention favouring larger scallops has generally been a major concern, its resolution has been somewhat problematical because of the accumulation of trash, including disarticulated shells and progressive masking with tow time of ring size and ring numbers, resulting in the impairment of selectivity particularly during saturation tows. It is apparent, however, that much of the selection takes place through the inter-ring spaces through the belly than through the back.

Overall efficiency of capture is determined by gear configuration, behaviour (example, propensity to being bysally attached, swimming ability, and endurance), substrate composition and consistency, and speed of tow. Sea scallops are among the most efficient swimmers in the Family (Pectinidae) (Morton 1980). The low capture rate for small sea scallops is no doubt related to their ability to elude capture gear by swimming (Caddy 1968).

Two size-classes of vessels are usually recognised. Typically, inshore components are under 19.8 m (65 ft, LOA) while the offshore or deep sea fleet consists of larger vessels, sometimes approaching 46 m (150 ft, LOA) (Fig. 15.6). Vessels not only employ rugged gear but also must be quite powerful and seaworthy. Once built of wood (Royce 1946; Posgay 1957b), the tendency now is towards steel hulls. The smaller, wooden boats average about 40–60 gross tons and are powered by 200–300 hp diesel engines. They have extensive sheathing and protection against chafing by the warps and rakes. While some of them continue to make excursions offshore, particularly during the summer, they are normally restricted to operations nearshore. Their normal crew complement is between 11 and 13 men, including skipper, mate, cook, and engineer. The larger steel vessels averaging between 280 and 360 gross tons are powered by 650–1,200 hp diesel engines. Offshore vessels operate 24 hours a day and are capable of fishing year-round. Typically they carry up to 24 men for a 12-day excursion. Modern vessels are equipped with navigational gear, including DGPS (Differential Global Positioning System), satellite receiver-computers, with plotters, radar, depth sounders, automatic pilots, single and side-band, and VHF radios. Since 1998 the deep sea fleet is equipped with 'black boxes' (satellite-based monitoring systems) for tracking vessel activity and secure messaging.

15.7 EXPLOITATION AND RESOURCE MANAGEMENT

The Atlantic sea scallop fishery is of major importance to both the United States and Canada. In the United States, responsibility for the control and administration of all aspects of fishery resources is shared between public Fishery Management Councils (Regional), the Federal Department of Commerce and State Governments; in Canada, it is with the Minister, Federal Department of Fisheries and Oceans. The status of scallop resources is assessed periodically, particularly for the more important stocks. An extensive literature exists on the state of exploited scallop aggregations throughout its range (see other chapters in this volume). Exploitation of most important stocks like Georges Bank is evaluated on a regular basis in the Unites States or annually in Canada. On the American side of Georges Bank research vessel surveys play an important role to provide estimates of fishing mortality and stock biomass. There are concerns that the modified Delury model used so far underestimated stock size; a proposed two-stage dynamic model in a Bayesian framework might help (NEFSC 1999). There are difficulties with the collection of commercial fishery data. The stock evaluation is further complicated by the existence of closed areas; for example, fishing mortality rates are estimated for the stock as a whole but they actually apply to the portion of the stock from the open areas only. In Canada, stock assessment and short-term projections are based on biomass estimates from research vessel surveys, relative age frequencies and temporal changes in the commercial catch and catch per unit of effort. Sequential population analysis (SPA) has been carried out on scallop populations on Georges Bank (e.g., Mohn et al. 1989, Robert et al. 2000). General production models are not easily reconciled with the severe dynamism exhibited by the species. A Collie-Sissenwine production model (Collie-Sissenwine 1983) was applied to Georges Bank data in the last assessment with mixed results. But the two types of models revealed the same trends for fishing mortality rates and biomass estimates.

Figure 15.6. Typical offshore scallop vessel.

Contagious distribution of young scallops poses special problems in designing research surveys, as do sometimes the smallness of the aggregations themselves. The U.S. survey on Georges Bank, for example, uses a stratified-random design with scallop-sampling strata based on water depth and latitude (Serchuk and Wigley 1986; Lai and Hendrickson 1997). To reflect the impacts of closed areas implemented in the Georges Bank region in 1995 and in the mid-Atlantic region in 1998 sampling stations were reallocated among open and closed strata and the total number of sampling stations increased starting in 1997. The Canadian survey design uses a stratified-random scheme with sampling strata based on past commercial catch-per-unit-effort contours derived prior to each survey (Caddy and Chandler 1979; Jamieson and Chandler 1980; Robert and Jamieson 1986). The survey design was refined in 1997 to increase the precision of the estimated number of scallops either adults (Smith and Robert 1998) or juveniles. Juvenile scallops were poorly represented by the commercial data stratification due to small size; an adaptive strategy is reducing the variance by 30% (Robert et al. 2000). Whereas the former attempts to estimate total biomass, the latter methodology recognises the fishable area to be limited and, consequently, is more concerned with the fishable biomass (Mohn et al. 1985). Resource managers use the scientific advice to manage the fishery in their respective jurisdictions frequently preferring a holistic approach by taking into account prevailing socio-economic conditions.

Historically, regulations in the American offshore scallop fishery had been imposed by industry; they included crew size, trip restrictions such as time spent at sea and time between trips, etc. With the advent of the 200-mile Fisheries Conservation Zone, the New England and mid-Atlantic Fisheries Management Councils developed sea scallop Fisheries Management Plan (FMP) to regulate the fishery in 1982. Initially the FMP had a 40 meats per pound size limit with a minimum shell height of 3 inches. A year later, the Regional Director invoked the temporary adjustment provision in the FMP by setting a 35 meats per pound and a 3 3/8 inch minimum shell height. The minimum shell height was subsequently raised to 3 ½ inches in 1988. These measures met with resistance from industry and did not entirely fulfil the intended purpose of improving yield per recruit. The single set of measures applied to the whole area where sea scallops occur and did not allow for local differences in biological growth-rates (for example). Also, the size limit stating the number of meats in a pound allows mixing very small scallops with a few large ones as long as the limit is met. The FMP Amendment #4 (NEFMC 1993) introduced in 1994 shifted scallop fishery management from meat count regulation to effort control for the entire American conservation zone. New measures included a day-at-sea reduction schedule, an increase in minimum ring size, a decrease in crew members, and several restrictions on dredge gears. The days-at-sea for full time participants declined from 204 in 1994 to 120 in 2000. The minimum ring size went from 3 inches to 3 ½ in 1996. Maximum crew size went from 9 to 7 shortly thereafter. Dredge gears could not have chafing gear; double links, triple links in belly allowed only; no less than 5 inches mesh twine top. In December 1994, 3 groundfish closed areas were implemented on the U.S. side of Georges Bank. Scallop fishing was also prohibited in the closed areas to reduce bycatch and protect groundfish habitat. Subsequently, scallop fishing effort moved from Georges Bank to the mid-Atlantic region increasing exploitation and decreasing stock abundance in the latter region. In March 1998, 2 areas of the mid-Atlantic region were closed to scallop fishing for 3 years to increase yield per recruit and spawning biomass. Substantial abundance of young scallops had been observed in the 2 areas (NEFSC 1999). Amendment #7 to the FMP implemented a definition for overfishing in 1998 via biological reference points (F_{max}, B_{max} as a proxy to B_{MSY}); this was required by the Sustainable Fishery Act. Over the last 2 years the FMP has been modified to allow for limited exploitation in portions of 2 of the Georges Bank closed areas under strict rules (possession limits, gear restrictions and stowage while transiting, days-at-sea accrual, in-season catch adjustments, etc.). For commentaries on this very complex fishing plan, see Kirkley and DuPaul (1998).

Scallop dredges used in the sea scallop fishery are characterised by relatively poor species-specific and size selectivity. Bycatch issues revolve around the capture of undersized scallops and groundfish that are either retained or discarded and the ancillary damage to bycatch animals in the gear and on deck. Gear modifications like increasing dredge ring size have reduced the collection of small scallops but damage still occur through gear handling on deck. Mortality rates on groundfish bycatch are perceived to be high and wasteful if retention is not allowed or desirable (DuPaul et al. 1996). Modest success in groundfish escapement has been reported by changing the mesh in the dredge twine top. Changes in harvesting practices like taking steps to discard unwanted animals overboard immediately at haul-back could make a big difference with little effort.

Regulations affecting the Canadian offshore scallop fishery were developed primarily for Georges Bank (Caddy and Sreedharan 1971). Then, as now, Georges Bank accounted for most of the landings. It is also the area with the best growth-rate and good recruitment patterns. As other offshore aggregations became exploited, some regulations such as the size limit were modified to reflect different biological environments. Mandatory logs for vessels 19.8 m or more in overall length were introduced in Canada in 1973 (Sinclair et al. 1985). In 1976, the Canadian offshore fishery became designated as one of limited entry to control total effort expended. Regulations also specified trip limits (13,700 kg) and trip duration (not exceeding 12 consecutive 24-hr periods) in 1977. In many areas, sea scallop fishing regulations are primarily aimed at optimising the age of first capture (Posgay 1958). This is achieved through the imposition of a size limit that seeks to maximise yield per recruit from the population. Most important in terms of addressing total fishery yield and maximum yield/recruit is the meat count, which specifies the number of scallop meats allowed per unit weight. As the average scallop size increases, the meat count declines. It is an average count with attendant difficulties. The Canadian offshore fishery has been under meat count regulation since 1973. When first introduced the count was 60 meats per pound. From a yield point of view, this meat count was set too high to be effective but it introduced the regulation measure. The meat count was lowered to 50 meats per pound in 1974, then to 45 in 1975 and 40 in 1976. The next decrease, to 35 meats per pound or 39 meats per 500 grams, waited until 1983.

When both Canada and the United States declared a 200-mile fishing zone extended jurisdiction in 1977, most of Georges Bank became a disputed zone claimed by both countries. Competitive fishing by both the Canadian and American scallop fleets continued in the disputed zone and grew more intensive until 1984, when the International

Court of Justice established an international boundary in the Gulf of Maine. The northeast section of Georges Bank was awarded to Canada. As a result of the no-rules fishery leading up to the Court decision, the 1984 Canadian scallop landings from Georges Bank were less than 2,000 tonnes of meats or 16,000 tonnes round weight, the lowest catch on record.

After the dispute affecting fisheries jurisdiction on Georges Bank had been settled, the Canadian scallop industry focused on stock rehabilitation through better harvesting of the resource. There were 3 goals: optimise the yield, reduce the fishing effort, and inject a degree of stability in landings. Catch quotas (TAC's) were imposed on the offshore scallop fleet as enterprise allocations (EA's). Each of the offshore companies holding fishing licences receives a percentage share of the annual TAC; a company may utilise one or more vessels to catch its share. The percentage shares were negotiated between licence holders in 1986; they were based on historical fishing performance on Georges Bank and the number of licences held by each company. At first, only Georges Bank was managed under EA's; gradually all offshore fishing areas were integrated under TAC's, the last one in 1994. Since 1986, annual removals have generally stabilised (Fig. 15.7). There were 77 active licence holders in 1984. Only one third the initial number of licence holders are actively involved in the fishery today. But the fishing power has not necessarily been reduced by the same ratio. A quasi-property management regime appears to succeed where other regimes such as input control in the inshore fishery have been far from successful (Brander and Burke 1995). The meat count for Georges Bank was lowered to 33 meats per 500 grams in 1986 to direct exploitation toward slightly larger scallops. The meat count for the western areas of the Scotian Shelf is set at 40 per 500 grams and 45 for the eastern Scotian Shelf.

Compliance with the meat count is readily achieved through blending large meats with smaller ones, the latter frequently being predominant in the landed catch (Naidu 1984). This strategy legitimises the harvesting of considerable numbers of small scallops. As the regulation was intended primarily for the protection of young scallops and delaying capture of those only recently recruited to maximise yield per recruit, the meat count in its present form has not had the full intended regulatory effect to total management of the resource. This has been demonstrated for St. Pierre Bank where the proportion of young scallops in the catch has been shown to sometimes exceed 65% (Naidu 1984). An attempt to partially redress the effects of blending in the meat count measure was the voluntary introduction of 50-count tolerance in the catch. In 1995, a voluntary monitoring program to discourage the presence of small meats in the catch (50+ meats per 500 grams) was implemented. A low tolerance level (10% by number of meats 10 grams or less) added more restriction to the regulatory 33 meats per 500 grams. The tolerance measure on small meats had its greatest effects on the Georges Bank fishery because of its recruitment driven nature. Trends in exploitation rates on age 3 scallops, the youngest age recruiting to the gear, show exploitation rates over 40% in 1981 when the meat count regulation had been relaxed and a strong year class was recruiting to the fishery (Robert et al. 2000) (Fig. 15.8). The meat count reduction to 33 meats per 500 grams in 1986, shifted age at first capture to a slightly older scallop and exploitation on age 3 became minimal at 5%. It has been reduced to almost nil with the monitoring of small meats in the catch starting in

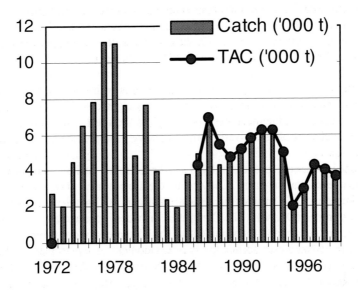

Figure 15.7. Profile of catches and TAC (since 1986) in tonnes of meats from the offshore scallop fleet for the Canadian side of Georges Bank, 1972–1999.

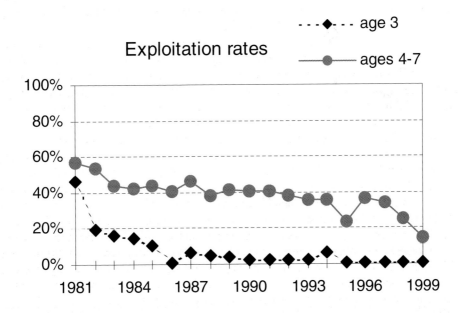

Figure 15.8. Exploitation rates on age 3 and the directed age group (ages 4–7) for Georges Bank offshore scallops since 1981.

1995. The aggregation of scallops into beds allows the fishery to direct its activities toward a sedentary target of specific size. The exploitation rate on the age group most important to the fishery (ages 4–7) has experienced less variability since the implementation of EA's. It varied little from 40% between 1988 and 1994. Recently, exploitation rates have generally been lower, in the 20% range. An annual time series of average meat count and percentage of meats 10 grams or less in the catch presents the same key elements (Fig. 15.9). While the meat weight distribution in the catch is a reflection of 3 variables: the meat size distribution in the stock, the quota, and meat count restrictions, the count and its derivatives may improve yield provided that it is low enough to be effective.

The decision-making process related to the offshore scallop management regime operates on a consultative basis. Advice to the Department of Fisheries and Oceans (DFO), Government of Canada (the fishery manager) is through the Offshore Scallop Advisory Committee (OSAC). The Committee provides input and advice to DFO on the conservation, protection, and management of the offshore scallop resource. It serves as the pre-eminent consultative forum for the development of the annual offshore scallop fishing plan (Anonymous 1999). There is no hard and fast rule to TAC setting. Overfishing concerns are considered in the light of rational harvesting and biological advice. The Committee's other terms of reference deal with regulatory measures, the administration of the EA programmes, and the introduction of new fishing technologies.

Catch size distribution

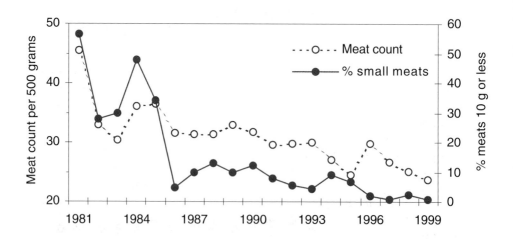

Figure 15.9. Distribution of scallop meat sizes in the commercial catch from Georges Bank, 1981 to 1999. The meat count is the number of scallop meats per 500 grams. The percentage of meats weighting 10 grams or less from commercial catch samples is also shown.

Bycatch issues have been the focus of discussions at OSAC. Since 1996, all incidental catches of all groundfish species, except monkfish, are returned to the water. Monkfish have accounted for approximately 85% of the groundfish bycatch landed by offshore scallop vessels (Anonymous 1999). The scallop industry has been proactive in testing gear modifications to reduce bycatch of groundfish and undersized scallops (Anonymous 1995). The use of large square mesh in the twine top of the dredge led to a decrease in the capture of groundfish like cod and haddock but flatfish like flounders are more difficult to deal with. Additional research is needed not only in gear modifications but also in harvesting practices to curtail bycatch mortalities.

Regional or District closures are dictated by the perceived incompatibility of recreational pursuits with commercial fishing activity and or the alleged interaction between scallop dragging and its effect on lobster grounds and lobster trapability (Jamieson and Campbell 1985). Lobsters are known to avoid moving drag rakes (Scarratt 1972). In one study (Pringle and Jones 1971), only one lobster was captured during approximately 30 hrs of fishing with a rock drag. With some clear exceptions (e.g., the Green sweep drag), lobster bycatch appears to be negligible. Neither is there evidence to suggest that scallop drags modify permanently the preferred habitat of lobsters. Physical gear conflicts with passive lobster gear, however, is likely and, presumably, is the rationale behind the institution of selective closures where the two activities must coexist. Management measures sometimes are developed and applied on a regional basis. In the Bay of Fundy, for example, closures are instituted that maintain some of the more productive grounds close to shore for exploitation during the winter months. The offshore drag and the Green Sweep scallop drag are banned. Regulations controlling minimum ring size (82 mm, inside diameter) and effective total width (5.5 m or 16.4 ft) of fishing gear are also in effect.

Provisions for temporary closure of areas, even if seldom enforced, is nearly always included in the Fishing Regulations. This facilitates the issuance of variation orders as and when the need arises. Management measures in the U.S., for example, allow for temporary adjustment of the meat count and shell height standards. The Canadian roe-on fishery for adductor muscles with gonads attached recognises the sporadic occurrence of paralytic shellfish poisoning (PSP) and domoic acid in scallops. While the toxins have little or no visible effect on the shellfish, they have proven to be poisonous to humans. Regulatory agencies, both in Canada and the United States, have developed necessary safeguards and protocol for the harvesting and marketing of roe-on scallops. All product is subject to detention, sampling, analyses, and certification prior to marketing. Toxicity levels are rigorously monitored throughout the fishing season. Should the results of the analyses exceed established standards (in Canada, PSP under 80 μg 100 g^{-1}; domoic acid under 20 μg g^{-1}), the product is subject to forfeiture and the source area and all adjacent grounds are immediately closed. Three consecutive acceptable levels over a 14-day period are required before the suspect area may be reopened. As only scallop roes become toxic, the problem is restricted to areas where meats are kept with the roe attached. The adductor muscle does not accumulate the toxin. The roe-on fishery is limited to Georges Bank, Western and Sable Banks, and the bulk of landings occurs between May and June when the roe is well developed. The rather limited Canadian roe-on fishery commenced in 1987 and has continued off and on to the present. Between 10 and 100 metric tons of roe are harvested annually, primarily for the European market. As only a small percentage of sea scallops are visibly hermaphroditic (Naidu 1970; Worms and Davidson 1986a), it has been necessary to duplicate the dioecious condition (as in *Pecten maximus*) by mixing approximately the same numbers of testes and ovaries in order to commercialise the product.

Inshore stocks in Canada are generally not regulated by a minimum meat count or minimum shell size. When in force it varies with location and, for a given area, it may even change seasonally. In the Northumberland Strait, for example, it remains at 52/500 g throughout the year. In the Bay of Fundy, the count used to vary from 55/500 g (October to April) to 72/500 g (May-September); currently it is 45 meats per 500 grams throughout the Bay. Elsewhere, including St. Pierre Bank the meat count remains at 33/500 g (30/lb) year round. Maine is the only State in the United States that imposes a season (November 1 to April 14) in its territorial waters (Mullen and Moring 1986). In the United States, removals are subject to both State and Federal Regulations. All removals are subject to the meat count, currently set at 30/lb (or 3 1/2 in or 89 mm shellstock). Although the minimum shell size stipulated is smaller than the average size corresponding to the blended count of 30/lb, it represents the absolute minimum size required to produce a nonviolative legal count. Consequently, enterprises in the United States sometimes find it advantageous to land scallops in the shell. A recently introduced regulation now requires scallop fishermen to unload their catches during specified daylight hours.

In some areas scallops may be in close proximity to industrial centers and consequently exposed to contaminants. Heavy metal (Pb, Zn, and Cu) levels in scallop tissues may become elevated in scallops found close to areas where storage and or shipboard ore loading had taken place over extended periods (Barrie 1984). Ray et al. (1984) have shown cadmium burdens in scallops from some areas offshore (example, Browns and Georges Bank) to be generally higher than from other areas such as Chaleur Bay, New Brunswick (Canada). Later, Uthe and Chou (1987) found that the higher levels were not due to contamination from anthropogenic or natural sources, but rather indicate that scallops were feeding on a nutritionally marginal diet containing relatively large amounts of cadmium.

Recreational fishing activity include the use of SCUBA (self-contained underwater breathing apparatus) and the use of rakes, tongs and dip-netting. In Canada, an individual fishing for scallops under a recreational fishing license is allowed no more than 100 scallops in any one day. In the United States, a non-commercial license permits a daily bag limit of two bushels of shellstock, or 4 quarts of meat. Also, it is unlawful to fish for or take scallops during night-time within Maine's territorial waters. Recreational activity is restricted to shallow water throughout the Gulf of Maine, Nova Scotia, north shore along Quebec, and around Newfoundland. While relatively insignificant in terms of volume removed, its per-unit-weight contribution to the economy is sometimes quite considerable.

Because of the high unit value, the species continues to be fully exploited. Frequently, they are harvested before reaching the size corresponding to maximum yield per recruit. While increasing age at first capture and decreasing fishing mortality are widely recognised as desirable management objectives in increasing the long-term yield from the resource as demonstrated by the Canadian Georges Bank fishery, yield-per-recruit and stock projections show that the species is in some cases overexploited, sometimes at a level higher than F_{max}.

15.8 OUTLOOK

The extreme dynamism of this species precludes long-term catch projections. Production will depend on recruitment patterns and the nature of management regime selected particularly for the offshore fishing areas. Fishery performance on Georges Bank will continue to influence patterns of exploitation elsewhere on the Atlantic seaboard. Over the short term, exploitation scenarios of a very different nature could take place. For example, accumulated biomass in the closed areas of Georges Bank (U.S. side) could provide levels of catches in relation with the opening of areas. Exploitation strategies could help along to optimise the catch. It would also relieve some of the fishing pressure exerted on the mid-Atlantic sea scallop beds. In Canada, sound management practices in the offshore and good incoming recruitment on Georges Bank should maintain landings at average historical levels or better. Production in the Bay of Fundy reached 38,000 metric tons in 1989, the highest in its 100-year history. Since then, lack of important recruitment and very high levels of effort have kept the fishery from performing. A pulse of good recruitment has been observed recently but the success of the fishery hinges in improving the management of the resource. The Northumberland Strait and other areas of the Gulf of St. Lawrence will continue to have only a marginal input to the overall Canadian landings. Inshore, because of wide fluctuations in recruitment, sustainable yields will continue to be highly variable; pulse fishing related to stock abundance and economic factors will dictate the pattern of exploitation.

In Canadian waters, two recent introductions will change the course of the offshore scallop fishery, satellite-based vessel monitoring and 3-D bottom imaging. Monitoring of vessel positions by satellite tracking has been introduced in early 1998. The monitoring provides improved knowledge of effort distribution both in time and space. While conventional logbooks give an estimated fishing location on a daily basis, satellite polling allows for multiple, at random if so wished, queries on vessel's location during the course of the day on a real time basis. Objectivity no doubt is, the second advantage of the monitoring system. Contrary to positions recorded in logbooks by fishers, there is no interpretation involved with the satellite-based monitoring system. Accuracy in location of fishing activities, especially with a sedentary target like sea scallop, is much improved with Global Positioning System. Applied to the offshore scallop fleet, the technology has allowed micro-management of fishing areas to become a reality. Most of the scallop grounds are prime habitats for the sea scallop and have traditionally been fished to this day. At times, scallop beds form in marginal habitats and although shell growth is similar to the prime areas, meats are small and slow growing. Scallop abundance in marginal habitats could be high but the size of the meat would make it difficult to fish at legal count and these grounds would be more or less ignored. If and when warranted, a management area is divided into the traditional area where the mainstay of the fishery continues to be managed as usual and a marginal area with a separate management regime where the meat count may be set higher. The fishery management by EA's also requires a quota for such a marginal area. The lack of historical fishery data or survey information precludes the use of most stock assessment tools. Rolling TAC's have been used in a few offshore areas over the last 3 years; the success of the measure rests mostly with the full co-operation of the scallop fleet. A rolling TAC has 2 key elements: a small quota (100 or 200 tonnes meats) to be fished over a short period, say 6 weeks at a set meat count. At the end of the first 6 weeks period, catch rates and meat counts are evaluated. Quota increments are repeated provided catch rates do not drop significantly and meat counts are met without difficulties (Robert and Butler 1997). Fishing areas can now undergo effective closures without the expense of enforcement patrols. Closures may be invoked to protect broodstock, nursery areas, sensitive habitats, etc. A seed closure was implemented on Browns Bank in late 1997 after the annual stock survey found low density levels of juvenile scallops throughout the Bank except for a patch of very high densities on the northern edge. The closure was lifted only after the year class of juveniles had reached commercial size.

The application of 3-D bottom imaging (Fig. 15.10) is changing fishing strategies into harvesting practices (Manson and Todd 2000). Using the latest technology in multibeam sonar, maps detailing seabed habitat, topography, and geology have been produced for the main offshore scallop fishing areas. The mapping information offers powerful new tools to improve the management of both the fishery and the habitat. Mapping scallop grounds has improved catch rates, 2.5X in one area; it reduces bycatch and bottom disturbance; it improves vessel operating efficiency with reduced gear costs and lowered fuel consumption. Less fishing effort means reduced incidental fishing mortality on scallops and other fauna. Maps of scallop habitat laid over survey and commercial catch data allow the harvesting of commercial size scallops while avoiding concentrations of juveniles. The ecosystem may become better protected. Once sensitive habitats are identified and mapped, managers could introduce protective measures more easily.

Interest has been manifested in conducting exploration activities for oil and gas on Georges Bank. In the early 1980's, exploration wells were drilled on the U.S. side of the Bank. In response to concerns about potential risks to the Georges Bank ecosystem from petroleum exploration a first series of moratoria on oil and gas activities ensued on both the American and Canadian sides of the Bank. A condition of the Canadian moratorium was a review, by 1999, of the environmental and socio-economic impacts of exploration and drilling activities. Through the Georges Bank Review Panel the ecological status of Georges Bank was reviewed for potential impacts of petroleum exploration activities (Boudreau 1998). In 1998, the U.S. President extended the moratorium over U.S. waters to the year 2012. Although actual impacts of exploratory drilling would be time and area-specific, there were uncertainties associated with various other aspects of the review. In the end, the Panel recommended the moratorium to be continued. It also, has been extended until the year 2012.

Figure 15.10. Multibeam bathymetric map showing rough terrain, dome-shaped sand dunes called barchans, and a gravel bed on the left near the top of the picture.

REFERENCES

Anonymous, 1977. Provisional nominal catches in Northeast Atlantic, 1976. Int. Comm. North Atl. Fish. Summary Doc. 77/VI/29. 52 p.

Anonymous, 1995. Fisheries Management Plan. 1995 Scotia-Fundy Region Offshore Scallops. Communications Branch, Maritimes Region, Department of Fisheries and Oceans, Halifax, Nova Scotia, Canada. 11 p.

Anonymous, 1999. Scotia-Fundy Offshore Integrated Fisheries Management Plan. Communications Branch, Maritimes Region, Department of Fisheries and Oceans, Dartmouth, Nova Scotia, Canada. 35 p.

898

Barrie, J.D., 1984. Heavy metals in marine sediments and biota at three ore loading sites in Newfoundland. Report prepared for Env. Prot. Ser. by LGL Ltd., Environmental Research Associates, St. John's, Newfoundland. 87 p.

Bourne, N., 1960. Selection of Georges Bank scallops by Canadian draggers. ICNAF Standing Comm. Res. and Stat. Ann. Mtg. pp. 62–65.

Bourne, N., 1962. Offshore scallop gear research. ICNAF Ann. Mtg. Doc. No. 56, Ser. No. 999. 11 p.

Bourne, N., 1964. Scallops and the offshore fishery of the Maritimes. Fish. Res. Bd. Canada, Bull. No. 145. 60 p.

Bourne, N., 1965. A comparison of catches by 3- and 4-inch rings on offshore scallop drags. J. Fish. Res. Bd. Canada 22(2):313–333.

Boudreau, P.R. (Ed.), 1998. The possible environmental impacts of petroleum exploration activities on Georges Bank ecosystem. DFO Can. Stock Assess. Res. Doc. 98/170.

Brander, L. and Burke, D.L., 1995. Rights-based vs. competitive fishing of seal scallops *Placopecten magellanicus* in Nova Scotia. Aquat. Living Resource 8:279–288.

Brown, B.E., Parrack, M. and Flescher, D.D., 1972. Review of the current status of the scallop fishery in International Commission for the Northwest Atlantic Fisheries (ICNAF) Division 5Z. ICNAF Research Document 72/113, Serial No. 2829. 13 p.

Caddy, J.F., 1968. Underwater observations on scallop (*Placopecten magellanicus*) behaviour and drag efficiency. J. Fish. Res. Board Can. 25(10):2123–2141.

Caddy, J.F., 1970. A method of surveying scallop populations from a submersible. J. Fish. Res. Board Can. 27(3):535–549.

Caddy, J.F., 1971a. Efficiency and selectivity of the Canadian offshore scallop dredge. ICES C.M. 1971/K:25. 5 p.

Caddy, J.F., 1971b. Recent scallop recruitment and apparent reduction in cull size by the Canadian fleet on Georges Bank. Int. Comm. Northwest Atl. Fish. Redbook Part 111:147–155.

Caddy, J.F., 1972. Size selectivity of the Georges Bank offshore dredge and mortality estimate for scallops from the Northern Edge of Georges in the period June 1970 to 1971. Int. Comm. Northwest Atl. Fish. Redbook Part III: 79–85.

Caddy, J.F., 1973. Underwater observations on the track of dredges and trawls and some effects of dredging on a scallop ground. J. Fish. Res. Bd. Canada 30(2):173–180.

Caddy, J.F., 1975. Spatial model for an exploited shellfish population and its application to the Georges Bank scallop fishery. J. Fish. Res. Board Can. 32(8):1305–1328.

Caddy, J.F., 1977. Some considerations underlying definitions of catchability and fishing effort in shellfish fisheries, and their relevance for stock assessment purposes. ICES C.M.1977/K:18. 22 p.

Caddy, J.F., 1979. Long-term trends and evidence for production cycles in the Bay of Fundy. Rapp. P.-v. Reun., Const. int. Explor. Mer 175:97–108.

Caddy, J.F., 1989. A perspective on the population dynamics and assessment of scallop fisheries, with special reference to the sea scallop, *Placopecten magellanicus* (Gmelin). In: J.F. Caddy (Ed.). Marine Invertebrate Fisheries: Their Assessment and Management. John Wiley and Sons, Inc. pp. 559–589.

Caddy, J.F. and Chandler, R.A., 1968. Lurcher scallop survey, March 1967. Fish. Res. Bd. Canada, MS Rept., Ser. No. 965. 36 p.

Caddy, J.F. and Chandler R.A., 1979. Georges Bank scallop survey, August 1966: A preliminary study of the relationship between research vessel catch, depth and commercial effort. Fish. Res. Bd. Can. MS Rept. 1054. 13 p.

Caddy, J.F. and Gulland, J.A., 1983. Historical patterns of fish stocks. Marine Policy 7(4):267–278.

Caddy, J.F. and Sreedharan, A., 1971. The effect of recruitment to the Georges Bank scallop fishery on meat sizes landed by the offshore fleet in the summer of 1970. Fish. Res. Bd. Canada. Tech. Rept. No. 256. 35 p.

Chouinard, G.A., 1984. Growth of the sea scallop (*Placopecten magellanicus*) on the Tormentine Bed, Northumberland Strait. CAFSAC Res. Doc. 84/61. 16 p.

Collie, J.S. and Sissenwine, M.P., 1983. Estimating population size from relative abundance data measured with error. Canadian Journal of Fisheries and Aquatic Sciences 40:1871–1879.

Culliney, J.L., 1974. Larval development of the giant scallop, *Placopecten magellanicus* (Gmelin). Biol. Bull. 147(2):321–332.

D'Amours, D. and Pilote, S., 1982. Données biologiques sur le Pétoncle D'islande (*Chlamys islandica*) et le Pétoncle Géant (*Placopecten magellanicus*) de la Basse-Côte-Nord du Québec (Secteur de la Tabatière). Cahier d'information No. 99. 47 p.

Dadswell, M.J., Chandler, R.A., Robert, G. and Mohn, R., 1984. Southwest New Brunswick and Grand Manan scallop stock assessment, 1983. CAFSAC Res. Doc. 84/28. 21 p.

Dickie, L.M., 1955. Fluctuations in abundance of the giant scallop, *Placopecten magellanicus* (Gmelin), in the Digby area of the Bay of Fundy. J. Fish. Res. Bd. Canada 12(6):797–857.

Dickie, L.M. and Chiasson, L.P., 1955. Offshore and Newfoundland scallop explorations. Fish. Res. Bd. Canada, Gen. Ser. Circular No. 25. 4 p.

Dickie, L.M. and Medcof, J.C., 1963. Causes of mass mortalities of scallops (*Placopecten magellanicus*) in the southwestern Gulf of St. Lawrence. J. Fish. Res. Bd. Canada 20(2):451–482.

DuPaul, W.D., Brust, J.C. and Kirkley J.E., 1996. Bycatch in the United States and Canadian sea scallop fisheries. Solving bycatch: Considerations for today and tomorrow. Solving bycatch workshop, Seattle, Wash. U.S.A. Alaska Sea Grant College Program Report No. 96–03:175–181.

Haynes, E.S., 1966. Length-weight relation of the sea scallop, *Placopecten magellanicus* (Gmelin). ICNAF Res. Bull. No. 3, p. 32–48.

Hurley, G.V., Tremblay, M.J. and Couturier, C., 1987. Age estimation of sea scallop larvae (*Placopecten magellanicus*) from daily growth lines on shells. J. Northw. Atl. Fish. Sci. 7(2):123–130.

Jamieson, G.S., 1979. Status and assessment of Northumberland Strait scallop stocks. Fish. Mar. Serv. Tech. Rept. No. 904. 34 p.

Jamieson, G.S. and Campbell, A., 1985. Sea scallop fishing impact on American lobsters in the Gulf of St. Lawrence. Fish. Bull. 83(4):575–586.

Jamieson, G.S. and Chandler, R.A., 1980. The potential for research and fishery performance data isopleths in population assessment of offshore, sedentary, contagiously distributed species. CAFSAC Res. Doc. 80/77. 32 p.

Jamieson, G.S., Kerr, G. and Lundy, M.J., 1981. Assessment of scallop stocks on Browns and German Banks - 1979. Can. Tech. Rept. Fish. Aquat. Sci. No. 1014. 17 p.

Kaiser, R.J., 1986. Characteristics of the Pacific weathervane scallop (*Pecten* (*Patinopecten*) *caurinus*, Gould 1850) fishery in Alaska. Alaska Dept. Fish. Game, Div. Comm. Fish., Kodiak, Alaska. 100 p.

Kirkley, J. and DuPaul, W., 1998. The U.S. Northwest Atlantic sea scallop fishery: an overview of problems and potential solutions. Virginia Sea Grant Marine Resource Advisory No. 71, VSG–98–10.

Kostylev, V.E., Courtney, R.C., Robert, G. and Todd B.J., 2003. Stock evaluation of giant scallop (*Placopecten magellanicus*) using high-resolution acoustics for seabed mapping. Fish. Res. 60:479–492.

Lai, H.L. and Hendrickson, L., 1997. Current resource conditions in USA Georges Bank and mid-Atlantic sea scallop populations: Results of the 1997 NMFS sea scallop research vessel survey. NMFS, NEFC, Woods Hole Lab. Ref. Doc 97–09. 73 p.

Langton, R.W., Robinson, W.E. and Schick, D., 1987. Fecundity and reproductive effort of sea scallops, *Placopecten magellanicus*, from the Gulf of Maine. Mar. Ecol. Prog. Ser. 37:19–25.

Larsen, P.F. and Lee, R.M., 1978. Observations on the abundance, distribution and growth of postlarval sea scallops, *Placopecten magellanicus*, on Georges Bank. Nautilus 92:112–116.

Lewis, C.V.W., Lynch, D.R., Fogarty, M.J. and Mountain, D., 1999. Effect of area closures on Georges Bank bivalves: Larval transport and population dynamics. http://asio.org/santafe99/abstracts/SS09WE0245S.html.

MacDonald, B.A., 1984. The partitioning of energy between growth and reproduction in the giant scallop, *Placopecten magellanicus* (Gmelin). PhD Thesis. Memorial University of Newfoundland. 202 p.

MacDonald, B.A. and Thompson, R.J., 1985a. Influence of temperature and food availability on the ecological energetics of giant scallop, *Placopecten magellanicus* (Gmelin). I. Growth rates of shell and somatic tissue. Mar. Ecol. Prog. Ser. 25:279–294.

MacDonald, B.A. and Thompson, R.J., 1985b. Influence of temperature and food availability on the ecological energetics of the giant scallop, *Placopecten magellanicus*. II. Reproductive output and total production. Mar. Ecol. Prog. Ser. 25:295–303.

MacDonald, B.A. and Thompson, R.J., 1986a. Influence of temperature and food availability on the ecological energetics of the giant scallop, *Placopecten magellanicus*. III. Physiological ecology, the gametogenic cycle and scope for growth. Mar. Biol. 93:37–48.

MacDonald, B.A. and Thompson, R.J., 1986b. Production, dynamics and energy partitioning in two populations of the giant scallop, *Placopecten magellanicus* (Gmelin). J. Exp. Mar. Biol. Ecol. 101(3):285–299.

MacDonald, B.A., Thompson, R.J. and Bayne, B.L., 1987. Influence of temperature and food availability on the ecological energetics of the giant scallop, *Placopecten magellanicus*. IV. Reproductive effort, value and cost. Oecologia (Berlin) 72(4):550–556.

MacKenzie, C.L. Jr., 1979. Biological and fisheries data on sea scallop *Placopecten magellanicus* (Gmelin). NOAA Fish. Cent., Sandy Hook Lab., New Jersey. Tech. Ser. Rep. 19. 34 p.

MacPhail, J.S., 1954. The inshore scallop fishery in the Maritime Provinces. St. Andrews Biol. Stn. Circular Gen. Ser. No. 22. 4 p.

MacPhail, J.S. and Muggah, H., 1965. Offshore scallop explorations. Nova Scotia Dept. Fisheries Project 204.

Manson, G. and Todd, B.J., 2000. Mapping the grounds: Seabed maps turn hunting into harvesting. Fishing News International February 2000:20–22.

Medcof, J.C., 1952. Modification of drags to protect small scallops. Fish. Res. Bd. Canada, Atl. Prog. Rept. 52:9–14.

Melvin, G.D., Dadswell, M.J. and Chandler, R.A., 1985. Movement of scallops, *Placopecten magellanicus* (Gmelin 1791) (Mollusca: Pectinidae) on Georges Bank. CAFSAC Res. Doc. 85/30. 29 p.

Merrill, A.S., 1959. A comparison of *Cyclopecten nannus* Verrill and Bush and *Placopecten magellanicus* (Gmelin). Occ. Papers on Mollusks, Harvard Univ. 2(25):209–228.

Merrill, A.S., Posgay, J.A. and Nichy, F.E., 1966. Annual marks on shell and ligament of sea scallop (*Placopecten magellanicus*). U.S. Fish. and Wildlife Ser., Fish. Bull. 65(2):299–311.

Mohn, R.K., Robert, G. and Black, G.A.P., 1989. Georges Bank scallop assessment - 1988. CAFSAC Res. Doc. 89/21. 26 p.

Mohn, R.K., Robert, G. and Roddick, D.L., 1985. Research sampling and survey design of Georges Bank scallops. NAFO SCR Doc. 85/97. 17 p.

Mohn, R.K., Robert, G. and Roddick, D.L., 1987. Georges Bank scallop assessment - 1986. CAFSAC Res. Doc. 87/9. 23 p.

Morton, B., 1980. Swimming in *Amusium pleuronectes* (Bivalvia: Pectinidae). J. Zool., Lond. 190(3):375–404.

Mullen, D.M. and Moring, J.R., 1986. Species profiles: Life histories and environmental requirements of coastal fishes and invertebrates (North Atlantic). Sea scallop. Biol. Rep. U. S. Fish Wildl. Serv. 1986. 21 p.

Naidu, K.S., 1969. Growth, reproduction, and unicellular endosymbiotic alga in the giant scallop, *Placopecten magellanicus* (Gmelin) in Port au Port Bay, Newfoundland. MSc Thesis. Memorial University of Newfoundland. 181 p.

Naidu, K.S., 1970. Reproduction and breeding cycle of the giant scallop, *Placopecten magellanicus* (Gmelin) in Port au Port Bay, Newfoundland. Can. J. Zool. 48(1):1003–1012.

Naidu, K.S., 1975. Growth and population structure of a northern shallow-water population of the giant scallop, *Placopecten magellanicus* (Gmelin). ICES C.M. 1975/K:37. 17 p.

Naidu, K.S., 1984. An analysis of the scallop meat count regulation. CAFSAC Res. Doc. 84/73. 18 p.

Naidu, K.S. and Anderson, J.T., 1984. Aspects of scallop recruitment on St. Pierre Bank in relation to oceanography and implications for resource management. CAFSAC Res. Doc. 84/29. 15 p.

Naidu, K.S. and Cahill, F.M., 1984. Status and assessment of St. Pierre Bank scallop stocks 1982–83. CAFSAC Res. Doc. 84/69. 56 p.

Naidu, K.S., Cahill, F.M. and Lewis, D.B., 1983a. Relative efficacy of two artificial substrates in the collection of sea scallop (*Placopecten magellanicus*) spat. J. World Mar. Soc. 12(2):165–171.

Naidu, K.S., Cahill, F.M. and Seward, E.M., 2001. The Scallop Fishery in Newfoundland and Labrador Becomes Beleaguered. CSAS Res. Doc. 2001/064. 35 p.

Naidu, K.S., Fournier, R., Marsot, P. and Worms, J., 1989. Culture of the sea scallop, *Placopecten magellanicus*: Opportunities and constraints. In: A.D. Boghen (Ed.). Cold-water Aquaculture in Atlantic Canada. The Canadian Institute for Research and Development. Tribune Press Ltd. Sackville, NB, Canada. pp. 211–239.

902

Naidu, K.S., Lewis, D.B. and Cahill, F.M., 1983b. St. Pierre Bank: An offshore scallop buffer zone. CAFSAC Res. Doc. 83/16. 48 p.

Naidu, K.S. and Scaplen, R., 1979. Settlement and survival of giant scallop, *Placopecten magellanicus*, larvae on enclosed polyethylene film collectors. In: T.V.R. Pillay and W.A. Dill (Eds.). Advances in Aquaculture. Fishing News Book Ltd. England. pp. 379–381.

NEFMC, 1993. Amendment #4 and supplemental environmental impact statement to the sea scallop fishery management plan. NEFMC, Saugus, MA.

NEFSC, 1999. 29th Northeast Regional Assessment (29th SAW), Stock Assessment Review Committee (SARC) Consensus summary of assessments. Northeast Fisheries Science Center Reference Document 9–14.

Norton, A.H., 1931. Size of the giant scallop (*Pecten grandis* Sol., *P. magellanicus* Gmelin). Nautilus 44(3):99–100.

O'Brien, J.J., 1961. New England sea scallop fishery, and marketing of sea scallop meats, 1939–60. U.S. Bur. of Comm. Fish., Market News Service, Boston. 48 p.

Orensanz, J.M., 1986. Size, environment, and density: the regulation of a scallop stock and its management implications. In: G.S. Jamieson and N. Bourne (Eds.). North Pacific Workshop on Stock Assessment and Management of Invertebrates. Can. Spec. Publ. Fish. Aquat. Sci. 92:195–227.

Posgay, J.A., 1953. The sea scallop fishery. In: Sixth Report on Investigations of Methods of Improving the Shellfish Resources of Massachusetts. Commonwealth of Massachusetts, Dept. Nat. Res., Div. Mar. Fish. pp. 9–24.

Posgay, J.A., 1957a. The range of the sea scallop. The Nautilus 71(2):55–57.

Posgay, J.A., 1957b. Sea scallop boats and gear. U.S. Fish and Wildlife Ser., Fishery Leaflet 442. 11 p.

Posgay, J.A., 1958. Maximum yield in the sea scallop fishery. ICNAF Doc. No. 28, Ser. No. 554. 17 p.

Posgay, J.A., 1962. Maximum yield per recruit of sea scallops. ICNAF Doc. No. 73, Ser. No. 1016. 20 p.

Posgay, J.A., 1976. Population assessment of the Georges Bank sea scallop stocks. ICES C.M. 1976/K:34. 6 p.

Posgay, J.A., 1979a. Population assessment of the Georges Bank sea scallop stocks. Rapp. P.-v. Reun. Cons. int. Explor. Mer 175:109–113.

Posgay, J.A., 1979b. Sea scallop, *Placopecten magellanicus* (Gmelin). In: M.D. Grosslein and T. Azarovitz (Eds.). Fish Distribution. MESA New York Bight Monograph No. 15, New York Sea Grant Institute. New York.

Posgay, J.A., 1981. Movement of tagged sea scallops on Georges Bank. Mar. Fish. Rev. 43(4):19–25.

Posgay, J.A. and Merrill, A.S., 1979. Age and growth data for the Atlantic coast sea scallop, *Placopecten magellanicus*. Nat. Mar. Fish. Ser. Northeast Fishery Center, Woods Hole Laboratory. Lab. Ref. Doc. 79–58. 97 p.

Premetz, E.D. and Snow, G.H., 1953. Status of New England sea-scallop fishery. Comm. Fish. Rev. 15(5):1–17.

Pringle, J.D. and Jones, D.J., 1971. The interaction of lobster, scallop, and Irish moss fisheries off Borden, Prince Edward Island. CAFSAC Res. Doc. 80/71. 17 p.

Querellou, J., 1975. Exploitation des coquilles Saint-Jacques, *Patinopecten yessoensis* Jay, au Japon. Milieux, methodes, resultats, organisation de la production. Publ. Assoc. Dev. Aquicult. 75:82.

Ray, S., Woodside, M., Jerone, V.E. and Akagi, H., 1984. Copper, zinc, cadmium and lead in scallops (*Placopecten magellanicus*) from Georges and Browns Banks. Chemosphere 13(11):1247–1254.

Read, K.R.H., 1967. Thermal tolerance of the bivalve mollusc, *Lima scabra* Born, in relation to environmental temperature. Proc. Malacol. Soc., Lond. 37(4):233–241.

Robert, G., Black, G.A.P., Butler, M.A.E. and Smith, S.J., 2000. Georges Bank scallop stock assessment - 1999. DFO Can. Stock Assess. Sec. Res. Doc. 2000/016.

Robert, G. and Butler, M.A.E., 1997. Scallop stock status for 1996 - Eastern Scotian Shelf and German Bank. DFO Can. Stock Assess. Sec. Res. - Doc. 97/49.

Robert, G. and Butler, M.A.E., 1998. Browns Bank north scallop stock assessment - 1997. DFO Can. Stock Assess. Sec. Res. Doc. 98/70.

Robert, G. and Jamieson, G.S., 1986. Commercial fishery data isopleths and their use in offshore sea scallop (*Placopecten magellanicus*) stock evaluations. Can. Spec. Publ. Fish. Aquat. Sci. 92:76–82.

Robert, G., Jamieson, G.S. and Lundy, M.J., 1982. Profile of the Canadian offshore scallop fishery on Georges Bank, 1978 to 1981. CAFSAC Res. Doc. 82/15. 33 p.

Robert, G. and Lundy, M.J., 1985. Reproductive aspects of the deep sea scallop (*Placopecten magellanicus*) in the Bay of Fundy near Digby, Nova Scotia. 5th. Pectinid Workshop, La Coruna, Spain, 6–10 May, 1985.

Robert, G. and Lundy, M.J., 1987. Shell height-meat weight allometry of Georges Bank scallop (*Placopecten magellanicus*) stocks. CAFSAC Res. Doc. 87/40. 39 p.

Robert, G., Lundy, M.J. and Butler-Connolly, M.A.E., 1985. Scallop fishing grounds on the Scotian Shelf. CAFSAC Res. Doc. 85/28. 45 p.

Robert, G., Lundy, M.J. and Butler-Connolly, M.A.E., 1986. Scallop fishing grounds on the Scotian Shelf - 1985. CAFSAC Res. Doc. 86/41. 43 p.

Robert, G., Lundy, M.J. and Connolly, M.A.E., 1984. Recent events in the scallop fishery of the Bay of Fundy and approaches. CAFSAC Res. Doc. 84/71. 41 p.

Roddick, D.L. and Mohn, R.K., 1985. Use of age-length information in scallop assessments. CAFSAC Res. Doc. 85/37, 16 p.

Rowell, T.W., Lord, E.I. and Somerville, G.M., 1966a. Scallop explorations - 1954. Offshore Newfoundland and Nova Scotia, Inshore Newfoundland and Magdalen Islands. Fish. Res. Bd. Canada, MS Rept. Ser. (Biological) No. 880.

Rowell, T.W., Somerville, G.M. and Lord, E.I., 1966b. Offshore scallop explorations - 1957. Fish. Res. Bd. Canada, MS Rept. Ser. (Biological) No. 881.

Royce, W.F., 1946. Gear used in the sea scallop fishery. Comm. Fish. Rev. 8(12):7–11.

Scarratt, D.J., 1972. Investigations into the effects on lobsters of raking Irish moss, 1970–71. Fish. Res. Bd. Can. Tech. Rep. 329. 20 p.

Schick, D.F., Shumway, S.E. and Hunter, M., 1988. A comparison of growth rate between shallow water and deep water populations of scallops, *Placopecten magellanicus* (Gmelin, 1791) in the Gulf of Maine. American Malacol. Bull. 6(1):1–8.

904

Serchuk, F.M. and Smolowitz, R.J., 1980. Size selection of sea scallops by an offshore scallop survey dredge. ICES C.M. 1980/K:24. 38 p.

Serchuk, F.M. and Wigley, S.E., 1986. Evaluation of USA and Canadian research vessel surveys for sea scallops (*Placopecten magellanicus*) on Georges Bank. J. Northw. Atl. Fish. Sci. 7:1–13.

Serchuk, F.M., Wood, P.W. Jr., Posgay, J.A. and Brown, B.E., 1979. Assessment and status of sea scallop (*Placopecten magellanicus*) populations off the Northeast coast of the United States. Proc. Natl. Shellfish Assoc. 69:161–191.

Serchuk, F.M., Wood, P.W. Jr. and Rak, R.S., 1982. Review and assessment of the Georges Bank, mid-Atlantic and Gulf of Maine Atlantic sea scallop (*Placopecten magellanicus*) resources. Nat. Mar. Fish. Serv., Woods Hole Oceanographic Inst. Ref. Doc. No. 82–06. 132 p.

Shumway, S.E. and Schick, D.F., 1987. Variability of growth, meat count and reproductive capacity in *Placopecten magellanicus*: Are current management policies sufficiently flexible? ICES C.M. 1987/K:2. 26 p.

Sinclair, M., Mohn, R.K., Robert, G. and Roddick, D.L., 1985. Considerations for the effective management of Atlantic scallops. Can. Tech. Rept. Fish. Aquat. Sci. 1382:vii + 113 p.

Smith, H.M., 1891. The giant scallop fishery of Maine. Bull. U.S. Fish. Comm. 4:313–335.

Smith, S.J. and Robert, G., 1998. Getting more out of your survey design: an application to Georges Bank scallops (*Placopecten magellanicus*). In: G.S. Jamieson and A. Campbell (Eds.). Proceedings of the North Pacific Symposium on Invertebrate Stock Assessment and Management. Spec. Publ. Can. Fish. And Aquat. Sci. 125:3–13.

Smith, S.J., Kenchington, E.L., Lundy, M.J., Robert, G. and Roddick, D., 2000. Spatially specific growth-rates for sea scallops (*Placopecten magellanicus*). In: G.H. Kruse, A.B.N. Benz, M. Dorn, S. Hills, R. Lipcius, D. Pelletier, C. Roy, S.J. Smith and D. Whiterell (Eds.). Spatial Processes and Management of Marine Populations. University of Alaska Sea Grant, AK–SG–00–04, Fairbanks. pp. 211–231.

Smolowitz, R.J. and Serchuk, F.M., 1987. Current technical concerns with sea scallop management. In: Proceedings of the Ocean - An international workplace. IEEE Service Center. Piscataway, N.J. 08854.

Smolowitz, R.J., Serchuk, F.M., Nicholas, J. and Wigley, S.E., 1985. Performance of an offshore survey dredge equipped with rock chains. NAFO SCR Doc. 85/89.

Somerville, G.M. and Dickie, L.M., 1957. Offshore scallop explorations, 1957. Fish. Res. Bd. Canada, Gen. Ser. Circular. No. 30. 4 p.

Squires, H.J., 1962. Giant scallops in Newfoundland coastal waters. Bull. Fish. Res. Bd. Canada No. 135. 29 p.

Stevenson, J.A. and Dickie, L.M., 1954. Annual growth rings and rate of growth of the giant scallop, *Placopecten magellanicus* (Gmelin) in the Digby area of the Bay of Fundy. J. Fish. Res. Board Can. 11(5):660–671.

Stone, H.S. and Hurley, G.V., 1987. Scallop behaviour/fishing gear interactions. Project Report No. 123. Hurley Fisheries Consulting Ltd., Dartmouth, Nova Scotia. 84 p.

Tan, F.C., Cai, D. and Roddick, D.L., 1988. Oxygen isotope studies on sea scallops, *Placopecten magellanicus*, from Browns Bank, Nova Scotia. Can. J. Fish. Aquat. Sci. 45(8):1378–1386.

Tremblay, M.J., Meade, L.D. and Hurley, G.V., 1987. Identification of planktonic sea scallop larvae (*Placopecten magellanicus*) (Gmelin). Can. J. Fish. Aquat. Sci. 44(7):1361–1366.

Tremblay, M.J. and Sinclair, M.M., 1988. The vertical and horizontal distribution of sea scallop (*Placopecten magellanicus*) larvae in the Bay of Fundy in 1984 and 1985. J. Northw. Atl. Fish. Sci. 8:43–53.

Uthe, J.F. and Chou, C.L., 1987. Cadmium in sea scallop (*Placopecten magellanicus*) tissues from clean and contaminated areas. Can. J. Fish. Aquat. Sci. 44(1):91–98.

Verrill, A.E. and Smith, S.I., 1873. Report upon the invertebrate animals of Vineyard Sound and adjacent waters, with an account of the physical features of the region. Report of the U.S. Fish. Comm. for 1871–72.

Volckaert, F., Shumway, S.E. and Schick, D.F., 1990. Biometry and population genetics of deep- and shallow-water populations of the sea scallop, *Placopecten magellanicus* (Gmelin, 1791) from the Gulf of Maine. In: S.E. Shumway (Ed.). An International Compendium of Scallop Biology and Culture. World Aquaculture Workshops No. 1. World Aquaculture Society, Baton Rouge, LA. pp. 156–163.

Worms, J.M. and Davidson L.A., 1986a. Some cases of hermaphroditism in the sea scallop, *Placopecten magellanicus* (Gmelin), from the southern Gulf of St. Lawrence, Canada. Jap. Jour. Malacol. 45(2):116–126.

Worms, J.M. and Davidson, L.A., 1986b. The variability of southern Gulf of St. Lawrence sea scallop meat weight-shell height relationships and its implications for resource management. ICES C.M.1986/K:24. 33 p.

Zouros, E. and Gartner-Kepkay, K., 1985. Influence of environment and human selection on the genetic structure of some economically important marine animal species. Rept. Canada Dept. Supply and Services (DSS) Contract 08 SC FP–101–3–0301. 73 p.

AUTHORS ADDRESSES

K. S. Naidu[†] - Science Branch, Department of Fisheries and Oceans, P. 0. Box 5667, St. John's, Newfoundland A1C 5X1 Canada

G. Robert - Science Branch, Department of Fisheries and Oceans, P. 0. Box 1006, Dartmouth, Nova Scotia B2Y 4A2 Canada (E-mail: RobertG@mar.dfo-mpo.gc.ca)

[†] Deceased.

Chapter 16

Sea Scallop Aquaculture in the Northwest Atlantic

G. Jay Parsons and Shawn M. C. Robinson

16.1 INTRODUCTION

Scallop culture in the Northwest Atlantic primarily involves the sea or giant scallop *Placopecten magellanicus* (Gmelin, 1791) whose culture is presently the basis of a slowly expanding and evolving industry. Within the last two decades, the aquaculture potential of the sea scallop has been widely reported (Naidu 1978; Naidu and Cahill 1986; Young-Lai and Aiken 1986; Gaudet 1989, Couturier 1988; Tremblay 1988; Wildish et al. 1988; Dadswell 1989; Dadswell and Parsons 1989; Couturier 1990; Robinson 1993; Dabinett and Couturier 1994; Couturier et al. 1995; Cliche et al. 1997; Neima 1997; Dadswell 2001; Grecian et al. 2003). The attractive characteristics for the sea scallop can include good spat supply, fast growth, large meat size, high fecundity, reliance on natural sources for food supply, high market value, available markets and consumer acceptance. The biological feasibility of producing a market sized scallop within two to three years has been demonstrated (Dadswell and Parsons 1991) but early reports of its economical feasibility were not as favourable (Frishman et al. 1980; Gilbert and Cantin 1987; Wildish et al. 1988), mainly due to projected slow growth and high costs of equipment and labour. New technology and research on optimising the methodology and handling of scallops has helped to change the economic outlook to that of a viable venture (Dadswell and Parsons 1992a; Ford 1997; Penney and Mills 2000). Innovative culture-based methods (to Atlantic Canada) have also been suggested to increase the production of existing scallop beds (Robinson 1993), which are now being implemented.

Scallop culturing technology and methodology are derived from the Japanese experience, which first started on a large scale in the 1950's (Imai 1977; Ventilla 1982; Aoyama 1989). The basic procedure involves four steps: 1) spat or seed procurement, 2) intermediate culture, 3) final grow-out, and 4) harvesting and marketing (for general scallop culturing manuals see Mottet 1979; Bourne et al. 1989; Neima 1997). Spat procurement relies on either hatchery-produced spat or obtaining wild spat using artificial collectors, which usually consists of "onion" bags filled with monofilament gillnetting or other materials. Intermediate culture of juveniles consists of growing scallops in pyramidal or conical shaped "pearl" nets until they are large enough for the next stage. For final grow-out to market size there are several different types of technology including pearl nets, lantern nets (a multi-tiered accordion style net), trays, large cages, ear hanging (where a hole is drilled in the "ear" or auricle of the scallop and they are hung on suspended ropes) or bottom culture (enhancement) (where scallops are seeded directly on the bottom). The traditional product is the meat, but higher value roe-on meat and whole scallop ("princess") markets do exist.

The bay scallop *Argopecten irradians* is an introduced species to Atlantic Canada and interest in culturing this species in this region has declined of late. For a description of earlier efforts, see Couturier (1990) and Couturier et al. (1995). The Iceland scallop *Chlamys islandica* also exists in parts of Atlantic Canada and there has been minor interest in culturing this species in Québec (lower north shore), but these efforts are not reported here (see Frechette et al. 2000; Frechette and Daigle 2002; Thomas et al. 2002).

16.2 HISTORY OF SEA SCALLOP CULTURE

The history of sea scallop culture can be traced to the early efforts of Newfoundland researchers in the 1970's. With the decline in the once lucrative fishery, especially in the Port au Port area, research was initiated to examine potential for restocking and enhancing the local populations (Naidu et al. 1989; Naidu 1991). These early efforts focussed primarily on spat collection techniques as a means of securing a large supply of juveniles (Naidu and Scaplen 1979). It was a similar decline in the fishery of the Magdellen Islands and lower north shore of Québec that stimulated interest among Québec researchers to investigate techniques of scallop enhancement (Fournier and Marsot 1987). Their initial work concentrated primarily on hatchery rearing techniques but has expanded to wild spat collection and grow-out (Cliche et al. 1994; Giguère et al. 1995; Cliche et al. 1997; Nadeau and Cliche 1997, 1998). In the 1980's, scallop aquaculture research in Passamaquoddy Bay, New Brunswick and Mahone Bay, Nova Scotia evolved from a research program originally established to examine sea scallop recruitment processes (Dadswell et al. 1987; Dadswell and Parsons 1991; Parsons 1994). The research in New Brunswick and Nova Scotia has encompassed spat collection, intermediate grow-out and several methods of final grow-out. A large-scale project in Newfoundland in the 1990's under the Canada/Newfoundland Agreement on Economic Renewal Aquaculture Component (ACERA) resulted in the establishment of commercial-scale hatchery production and intermediate and final grow-out and equipment trials (Dabinett et al. 1998; Parsons et al. 1998).

There have been also two additional, large-scale research programs established to examine many aspects of sea scallop culture. The first was a multi-university program called OPEN (Ocean Productivity and Enhancement Network) and the second was a provincial/federal government program in Québec called REPERE (REcherche sur le Pétoncle à des fins d'Élevage et de REpeuplement). From these initial modest beginnings twenty-five years ago, operations have been established in Newfoundland, Prince Edward Island, New Brunswick, Nova Scotia and Québec with mixed degrees of success. Currently, there are small to medium scale commercial operations in Nova Scotia and Québec and a large-scale enhancement enterprise in Québec. As well, sea scallop culture efforts using both suspended and bottom culture techniques have been examined in Maine and off Martha's Vineyard in Massachusetts (Goudey and Smolowitz 1996; Karney 1996; Karney et al. 1996; Langan et al. 1997; Parsons et al. 2002). For a comprehensive description of the history of scallop culture in eastern Canada, see Couturier et al. (1995).

16.3 HATCHERY SPAT PRODUCTION

Large-scale hatchery production offers several advantages to potential aquaculturists including genetic selection of a regular and plentiful supply of scallop seed at a relatively low price. Viable larval rearing, using hatchery technology, has been reported for many species of bivalves including scallops, clams and oysters (Loosanoff and Davis 1963; Gruffydd and Beaumont 1972) and several comprehensive hatchery manuals have been written (Dupuy 1977; Castagna and Kraeuter 1981; Bourne et al. 1989; Neima 1997).

Early work on larval rearing of the sea scallop was conducted mainly to describe the development of the early life-history stages (Baird 1953, Bourne 1964) and Culliney (1974) was the first to ascribe the potential of artificial larval rearing for aquaculture purposes. The successful rearing of the sea scallop larvae through to settlement and metamorphosis had eluded researchers until relatively recently (Fournier and Marsot 1986; Cliche and Beaulieu 1989; Dabinett 1989; Mallet 1989; Karney 1996; Karney et al. 1996; Dabinett et al. 1998). These breakthroughs have stimulated much research on hatchery-related topics with the aim of producing an efficient and economical source of sea scallop spat (Manning 1985; Couturier 1986; Gillis and Dabinett 1989; Kean-Howie et al. 1991; Jackson 1992; Dabinett et al. 1998; Ryan 2000).

16.3.1 Broodstock Conditioning

The spawning season of most sea scallop populations occurs annually between July and October (Naidu 1970; Robinson et al. 1981; Barber et al. 1988; MacDonald and Thompson 1988; Parsons et al. 1992a; Grecian et al. 2001a). A semi-annual reproductive cycle has been described for *Placopecten magellanicus* in some areas (Dupaul et al. 1989; Dibacco 1991; Schmitzer et al. 1991, Dadswell and Parsons 1992a). The presence of two annual spatfalls has important implications for scallop aquaculture in enhanced spat collection and accelerated growth (Dadswell and Parsons 1992a). To assure a continuous supply of spat throughout the year, however, spawning has to be induced outside the normal spawning period. This can be accomplished by "conditioning" the adult scallops in the hatchery by manipulating their temperature and feeding regime in order to stimulate gonad production and ripeness (Neima 1997; Dabinett et al. 1998).

Successful attempts to condition sea scallops have been realised in the laboratory, in commercial hatcheries and by holding adults in suspended culture (Manning 1985; Couturier 1986; Neima 1997; Dabinett et al. 1998). Couturier and Aiken (1989) demonstrated a photoperiod effect on the reproductive cycle of the sea scallop, which appears to be key to conditioning scallops for successful spawnings throughout the year. Couturier and Newkirk (1991) found, however, that seasonal differences in spawning times can have an effect on larval vigour. Broodstock nutrition is also an important consideration; Napolitano et al. (1992) found that the allocation of lipids to eggs of wild scallops from well and poorly-fed populations were the same suggesting that the adults maintain the quality of eggs and vary the amount produced. Pernet et al. (2003a, b) and Pernet and Tremblay (2004) examined the link between lipid class variation and sea scallop gonad and reproductive state and larval development and performance.

16.3.2 Spawning Methods

The three methods most often used to induce spawning in bivalves include varying water temperature, increased water circulation and injection of serotonin. All three methods have been used singly or in combination to induce spawning successfully in sea scallops (Culliney 1974; Manning 1985; Couturier 1986; Fournier and Marsot 1986; Dabinett 1989; Marshal and Lee 1991; Dabinett et al. 1998).

Temperature cycling has been used to induce spawning in a number of bivalves (Loosanoff and Davis 1963). Generally, the technique involves either gradually or suddenly increasing or decreasing the temperature. Culliney (1974) was able to stimulate sea scallops to spawn by increasing temperatures by 3–5°C and Fournier and Marsot (1986) induced spawning by decreasing temperatures from 10°C to 4°C. A variation of this technique involves removing scallops from the water for 1 to 2 hours prior to re-immersion followed by thermal stimulation or serotonin injection (Gruffydd and Beaumont 1970, Young-Lai and Aiken 1986; Marshall and Lee 1991).

Dabinett et al. (1998) stated that scallops placed in shallow trays and subjected to increased water circulation (20 L min^{-1}) are usually induced to spawn within an hour.

Serotonin (a neurotransmitter) injection is a reliable method for inducing spawning in scallops. The procedure is to inject about 0.5 mL of 2 mM serotonin into the adductor muscle and spawning usually follows within a half hour (Matsutani and Nomura 1982; Gibbons and Castagna 1984; Couturier 1986). Other biochemicals that have been used for inducing spawning in other scallop species are potassium chloride (Beaumont and Hall 1999) and monoamines and prostaglandin (Martinez et al. 2000).

16.3.3 Egg and Larval Development

Eggs incubated at temperatures of 12 to 16°C reach the "D" larval (shelled) stage in 2 to 4 days. Growth rates of D larvae to day 28 larvae, which have been fed on a mixed algal diet, range from 2.0 to 6.4 μm d^{-1} (Culliney 1974; Fournier and Marsot 1986; Couturier and Newkirk 1987; Fournier and Marsot 1987; Cliche and Beaulieu 1989; Mallet 1989). Larvae fed on mixed algal diets have better growth rates than those fed on unialgal diets (Mallet 1989; Dabinett et al. 1998; Ryan 2000). Marshall and Lee (1991) reported that larval sea scallops can use dissolved organic carbon (DOC), in the form of glycine, as a source of food. They further suggest that larvae are capable of competing with bacteria in the uptake of DOC. Larvae were also found to grow better when they were fed on algae in the log phase rather than the 'stationary' phase (Ryan et al. 1998; Ryan 2000). There was a positive correlation between growth of larvae and spat and sterol and methyl ketone content of the algae. In a similar study, larval scallop growth was found to be independent of the lipid content of the eggs, but was highly correlated with the genetic lines of heritability (Jones et al. 1996).

Larval survival rates are quite variable, yields from the egg to D larvae range from 30 to 76% and survival from egg to pediveliger (about day 28) range from 4.6 to 85.4% (Fournier and Marsot 1986; Couturier and Newkirk 1987; Fournier and Marsot 1987; Cliche and Beaulieu 1989; Dabinett 1989); Mallet (1989) reported an average daily mortality of 3.2%. Couturier and Newkirk (1987) found a positive relationship between initial larval survival and adult gonad lipid levels, but egg size was found to be a poor indicator of larval survival (Couturier 1991).

16.3.4 Settlement and Metamorphosis

Larval settlement and metamorphosis represent a period of high mortality since it is a period of their life-history in which larvae undergo behavioural changes associated with their search for an appropriate substrate upon which to settle. Early attempts to induce settlement in sea scallops were either unsuccessful or larvae were reluctant to settle (Culliney 1974). If the proper cue or stimulus is not received, settlement can be delayed and high mortalities may result. With improved handling and rearing techniques some success in settling sea scallop spat have been realised at several hatcheries in Québec, Newfoundland and Nova Scotia over the last fifteen years (Fournier and Marsot 1986, 1987; Cliche and Beaulieu 1989; Dabinett 1989; Mallet 1989; Dabinett et al. 1998).

Sea scallop settlement typically occurs about 35 to 45 days after fertilisation at a size of about 250 μm (Culliney 1974; Fournier and Marsot 1986; Cliche and Beaulieu 1989). Fournier and Marsot (1987) and Cliche and Beaulieu (1989) reported survival rates of 0.5 to 10.9% from egg to postlarvae. Mallet (1989) found a percent metamorphosis of 2 to 5% and Dabinett (1989) reported 0.75 to 22.2% survival from the egg stage to 3 mm spat.

A variety of cultch (artificial substrate) materials have been used as settlement surfaces for the sea scallop (Tremblay 1988). Culliney (1974) found scallops to settle predominantly on the underside of small pebbles and fragments of glass and shell. Fournier and Marsot (1986) and Dabinett (1989) reported scallops settling onto Nitex™, scallop shells, and fibreglass tanks and panels. The use of chemicals, such as potassium chloride, L-Dopamine, epinephrine, GABA, etc., to induce settlement in sea scallops have been ineffective (Tremblay 1988). However, chitin from several sources (hydroids, crabs, shrimp, lobster) was found to be very effective at increasing larval settlement in both laboratory and field experiments (Harvey et al. 1997). Bacterial films have also been found to increase the larval scallop settlement rates (Parsons et al. 1993a; Pearce and Bourget 1996).

16.3.5 Spat Husbandry

Recent work has focused on the dietary requirements and feeding regimes of raising post-metamorphic spat in the nursery phase. Metamorphosis results in structural changes including the loss of the velum, which is used in larval feeding, and the development of a filter-feeding mechanism involving the new gills and cilia. The ingestion rate of the alga *Isochrysis galbana* by juvenile scallops was described by the relationship IR = 1.5×10^4 W x 1.22 where IR = ingestion rate (cells min^{-1}) and W = body weight (g) (Strickland and Dabinett 1993). However, Lesser et al. (1991) demonstrated that juvenile scallops feed selectively. Juveniles fed an unialgal diet selected particles based on size, whereas for those fed multialgal diets, selection did not appear to be based on size but rather on other characteristics of the algae or pre-ingestive sorting. A mixed algal diet, of

3 species, resulted in better growth than an unialgal or paired algal diet for 2 mm spat (Gillis and Dabinett 1989). Similar findings were reported by Parrish et al. (1999) and Milke et al. (2004) who found that mixed species diets worked better for growing juveniles and that omega 3 and omega 6 fatty acids were essential to providing an optimal diet. Mallet (1989) found analogous results for larvae (see 16.3.3). Hollett and Dabinett (1989) have found that a ration of 45 cells·μL^{-1} of a multialgal diet provided the best growth and highest spat growth efficiencies. Their experimental range of algal densities was 12 to 87 cells·μL^{-1}.

Several recent studies have examined the physical conditions under which juveniles should be grown. Frenette and Parsons (2001) and Frenette (2004) showed that lethal temperatures were above 18°C and lethal salinities below 25‰. The highest filtering rates for juveniles were found at 13°C and 32‰ (Frenette et al. 2002; Frenette 2004). Un-ionised ammonia levels in the water were found to inhibit feeding of juvenile scallops at levels at 0.54 mg·L^{-1} and above (Dabinett et al. 1998; Grecian et al. 2001b). There was some evidence of size-related tolerances of ammonia as well as temperature-dependent sensitivity (Abraham et al. 1996).

Kean-Howie et al. (1989, 1991) studied the nutritional physiology of juvenile sea scallops and successfully developed a microparticulate food capsule upon which juveniles will feed. These studies have provided the basis for a more complete understanding of the specific dietary requirements of scallops, which have enabled researchers to manipulate the different dietary components of the microparticles. Coutteau et al. (1996) have used lipid emulsions as feed additives to increase the levels of n-3 highly unsaturated fatty acids in *Placopecten* and resulting in 20% more lipid being incorporated into juvenile sea scallops compared to the algal-fed controls.

16.4 WILD SPAT COLLECTION

16.4.1 Spat Collection Techniques

An alternative to producing spat in a hatchery is collecting wild spat using artificial substrates consisting of "onion" bags containing monofilament gillnetting. This technology or variations thereof, which are based on Japanese designs (Ventilla 1982; Ito 1990), has been successfully employed for many scallop species, including the sea scallop, *Placopecten magellanicus* (Buestel et al. 1979; Brand et al. 1980; Naidu et al. 1981; Grecian et al. 2001a). Spat collectors are suspended in the water column in an area that has a local population of sea scallops. The bags are generally deployed just after the scallops have spawned, which generally occurs from mid-summer to autumn (Naidu 1970; Robinson et al. 1981; MacDonald and Thompson 1988; Parsons et al. 1992a). Sea scallop larvae remain planktonic for about a month (Culliney 1974), after which they develop an affinity to settle on filamentous substrates (Larsen and Lee 1978) such as the monofilament gillnetting within the collectors.

Several factors can contribute to successful spat collection. Often, surveys for scallop larvae are conducted during the initial stages of spat collection programs to help choose potential spat collection areas (Parsons 1994; Giguère et al. 1995; Dadswell 2001).

Several studies have been conducted to determine the efficacy of different artificial settlement substrates (Naidu 1979; Parsons et al. 1996; Niles and Davidson 2002). Naidu (1979) found no differences in settlement of sea scallops on variously coloured substrates, but Naidu et al. (1981) did find that monofilament gillnetting was better than polyethylene sheets in obtaining sea scallop spat. They also determined that 540 g of monofilament gillnetting per "onion" bag (42 x 75 cm) was the optimum amount of material for this type of collector. It would appear that medium densities of medium diameter monofilament work better than very dense concentrations likely because of the water-flow characteristics created within the bag (Miron et al. 1995; Pouliot et al. 1995). Niles and Davidson (2002) found that black fuzzy rope was as successful as Netron for collecting spat. They also noticed that there were differences in collection efficiencies of ropes with different colours and suggested more work should be done on this aspect. Settlement intensity also varies with depth within the water column. Studies have shown settlement to be minimal near the surface and near the bottom and to be highest near the mid-depths (Naidu and Scaplen 1979; Dadswell et al. 1987; Bonardelli 1988).

16.4.2 Settlement Intensity

In Atlantic Canada, spat settlement studies have shown settlement to vary temporally and spatially, both on a local and regional level. From 1977 to 1987, spat studies at Port au Port, Newfoundland (NL) reported mean settlement to range from 5 to 600 spat per bag (Naidu and Cahill 1986; Naidu et al. 1989; Naidu 1991). Since 1988, with improved techniques and a better understanding of the vagaries of the bay, spat collection has been consistently greater than 3,000 spat per bag (Lanteigne et al. 1991).

In the Baie de Chaleurs (Québec/New Brunswick) and the lower north shore of Québec (QC), where both the Iceland scallop, *Chlamys islandica*, and the sea scallop, *Placopecten magellanicus*, occur, settlement has been found to be about 2,500 and 165 total spat per bag, for each area respectively (Bonardelli 1987; Chislett 1989). Settlement season occurred during September and October.

Spat collection in the Northumberland Strait occurs from late August to late October and average peak settlement varied annually from 400 to 1,100 spat per collector (Grecian et al. 2001a).

In Passamaquoddy Bay, New Brunswick (NB), spat collection has been monitored at a scallop aquaculture site since 1983 and average settlement has varied annually from 100 to 400 spat per bag (Dadswell et al. 1988; Parsons 1989; Dadswell and Parsons 1991). Settlement season was protracted, occurring from late August through to October, with a peak settlement period of 2 to 3 weeks (Dadswell et al. 1987). This site was also shown to be increasing the recruitment of scallops to the bottom directly underneath, likely from drop-off from the collection bags (Parsons et al. 1994). In a study examining the spatial distribution of spat within Passamaquoddy Bay, NB, Robinson et al. (1991, 1992) found the number of spat varied greatly throughout the bay with maximum settlement >3,000 spat per bag. Over the duration of this study, annual site specific settlement rates varied only by a factor of 2 to 3 times and spatial patterns were consistent interannually (Robinson et al. 1999).

In Mahone Bay, Nova Scotia (NS), there are two spat settlement periods each year, one in July and the other in September and October. Settlement rates initially varied from about 10 to 80 spat per collector (Dadswell and Parsons 1991, 1992a), but with the development of a commercial scallop farm in the area, spat collection has increased to over 500 spat per bag (Dadswell 2001).

16.4.3 Maximising Spat Collection

In order to minimise the uncertainty associated with the magnitude of the annual spatfall, there are several important factors, which should be considered if scallop settlement is to be maximised. First, it is axiomatic that spat collectors should be placed in an area near to or downstream from where scallops occur. Second, knowing the spawning period and duration is essential for the timing of spat collector deployment. Third, knowledge of local oceanographic currents and the settlement period of predators, such as sea stars, is an aid. Selecting areas with gyres or regions around shoals seems to assist with spat settlement (Robinson et al. 1992, 1999). Spat settlement can be further enhanced by avoiding the deployment of collectors at the surface or near the bottom (Naidu and Scaplen 1979; Bonardelli 1987; Dadswell et al. 1987). Studies have shown that biological films on the settlement substrate can enhance spat settlement (Hodgson and Bourne 1988, Parsons et al. 1990, 1993a) and that the spat bags should be pre-soaked for a couple of weeks prior to peak spatfall (Bonardelli 1987, 1988). There is, however, a trade-off in that the spat bags should not be pre-soaked for too long a period as heavy fouling will reduce settlement activity and fouling organisms will compete with scallops for food and space (e.g., blue mussels, *Mytilus edulis*). Harvey et al. (1997) demonstrated that larval scallops are attracted to chitin coated surfaces for settlement in a field experiment where they coated collectors with chitin. There were no observed differences in the sources of chitin (i.e., prawn, hydroids, crab or lobster). An additional factor to consider is the timing of the starfish (*Asterias* spp.) and gastropod (*Astyris mitrella*) settlement as they are voracious predators of the spat (Naidu and Scaplen 1979; Tetu and Davidson 2002). In Passamaquoddy Bay, NB peak starfish settlement occurs about 2 to 3 weeks prior to peak scallop settlement, hence spat collector deployment should be delayed until the peak starfish period has passed (Parsons et al. 1990; Parsons 1994) (Fig. 16.1). In Port au Port Bay, NL Naidu and Scaplen (1979) have shown that starfish settlement overlaps temporally with scallop settlement but are spatially different in their depth distribution. Thus, starfish settlement can be avoided by deploying spat bags at a greater water depth. In the Baie de Chaleurs (QC/NB) and Québec lower north shore starfish do not present a problem to aquaculturists (Tremblay 1988).
Prediction of the timing and rate of annual spatfall can be achieved with various types of monitoring protocols. The Japanese routinely monitor the occurrence of the larval pediveliger stage in order to predict settlement time (Ventilla 1982; Ito 1990). They have also found a relationship between larval abundance and spat settlement and can, based on larval abundance, predict potential number of spat per bag and determine how many spat collectors will be required to be deployed, in order to catch the quantity of scallops required each year. A larval abundance-spatfall index has yet to be developed for sea

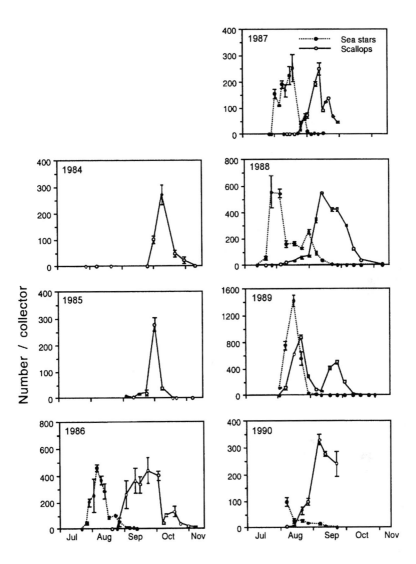

Figure 16.1. Mean number of sea scallops per collector per two-week sampling period for 1984 to 1990 and mean number of sea stars per collector per two-week sampling period for 1986 to 1990 in Passamaquoddy Bay, New Brunswick. Vertical bars ± 1 SE. (After Parsons 1994).

scallops, but the monitoring program employed in Passamaquoddy Bay, NB entails routine sampling of the spawning stage of adults, using the gono-somatic index (ratio of gonad weight to soft body weight) as an index of spawning. Once adults have started to spawn, starfish and scallop settlement is monitored by the weekly deployment of two spat collectors. The collectors are left in the water for two weeks after which they are returned to the lab, the contents of the bags are washed and the early settlement stages of the starfish and scallops are identified and counted. When the starfish settlement is decreasing, the annual deployment of spat collectors occurs.

16.4.4 Spat Growth

Parsons et al. (1993b) have validated an ageing technique of newly settled sea scallop spat based on a daily growth ridge formation. They reported growth rates of 32 to 57 μm d^{-1} for postlarval scallops up to 30 days old.

Naidu and Scaplen (1979) reported a mean spat size of 12.4 mm after 10 months of growth in spat bags deployed in Port au Port, Newfoundland. Dadswell et al. (1987) have monitored the monthly growth of spat from Passamaquoddy Bay, NB. Mean shell height after 10 months of growth in suspension was between 10 to 15 mm and spat growth was found to be correlated with temperature during this period (Parsons et al. 1990). Both Naidu and Scaplen (1979) and Dadswell et al. (1987) found spat size to increase with depth of water.

A semi-annual reproductive cycle of sea scallops in Mahone Bay results in the presence of two annual spatfalls with overall enhanced spat collection and accelerated growth of the early cohort compared to the latter (Dadswell and Parsons 1992a) (Fig. 16.2).

16.4.5 Spat Sorting

After spat collectors have been left in suspension for 9 to 11 months, the scallops are large enough that they will not fall through the mesh of the bags or through the mesh of the pearl nets (intermediate culture) and their shells are hard enough that they can be handled and sorted (Dadswell and Parsons 1991). Sorting the collectors is a labour-intensive technique, which requires scallops to be hand-picked or sorted away from the rest of the settled invertebrates (other molluscs, echinoderms, worms, tunicates, etc.) (Naidu and Cahill 1986). To assist further in reducing sorting time, several companies have developed prototype spat graders (Anonymous 1991a, 1991b, 1991c), which should help to semi-automate the sorting procedure and reduce a portion of the labour component.

16.5 INTERMEDIATE CULTURE

Intermediate culture, which typically involves the use of pyramidal shaped pearl nets, trays or small mesh lantern nets, is a necessary step for raising all scallop species (Ventilla 1982; Paul et al. 1981; Wallace and Reinsnes 1984; Naidu and Cahill 1986; Parsons and Dadswell 1994, Levy 1999). Spat produced from a hatchery, due to their small size, must first go through a nursery and intermediate culture stage before final grow-out. Grecian et al. (2000, 2003) has examined several strategies for the deployment of hatchery-reared sea scallops to sea-based grow-out sites. When scallops are harvested from spat collectors, generally at an average size of about 10 mm, they are also too small for

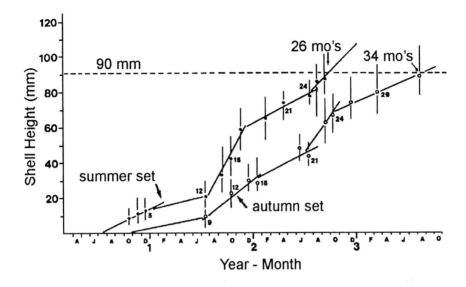

Figure 16.2. Shell height growth of sea scallops from 1989 in Mahone Bay, Nova Scotia summer (closed circles) and autumn (open circle) cohorts grown in suspension culture. Vertical bars are ranges. Numbers beside means (circles) represent months of growth since spat settlement. (After Dadswell and Parsons 1992a).

hanging culture (ear hanging or cages), while releasing juvenile scallops directly on the bottom will result in large losses (Scheibling et al. 1991; Hatcher et al. 1996). Thus, intermediate culture is important for increasing both survival and initial growth (Imai 1977).

16.5.1 Growth

Juvenile scallops grown in pearl nets are usually held for about a year during which they attain sizes of 40 to 60 mm. Field experiments have shown that a number of factors can influence growth, namely stocking density, size, gear type, deployment time, depth, season, and location (Grecian et al. 2000, 2003).

16.5.2 Stocking Density

Growth of juveniles in pearl nets has been found to be inversely related to stocking density for sea scallops, *Placopecten magellanicus* (Dadswell and Parsons 1991; Côté et al. 1993; Parsons and Dadswell 1992; Penney 1995), bay scallops, *Argopecten irradians* (Duggan 1973; Widman and Rhodes 1991), and Japanese scallops, *Patinopecten yessoensis* (Ventilla 1982). Parsons and Dadswell (1992) found that juvenile sea scallops grown over a period of a year at densities ranging from 136 to 818 per m^2 had mean growth rates ranging from 0.08 to 0.11 mm d^{-1}. Dadswell and Parsons (1991) reported

918

similar findings for sea scallops held over shorter growing periods. Grecian et al. (2000) found no stocking density effects for very small hatchery-reared scallops; however stocking density per unit area remained low throughout their study.

A general caveat for growing scallops in pearl nets or cage culture has been to never grow scallops at densities occupying more than a third of the floor space (Taguchi 1978; Mottet 1979; Paul et al. 1981), the implication being that growth will be compromised under crowded conditions. Parsons and Dadswell (1992) found, however, that juvenile sea scallops grew at densities equivalent to 115% of the floor area and exhibited substantial growth at densities of up to 50% or 60% of the floor area (Fig. 16.3). Densities of 80–90 11-mm spat per pearl net were recommended by Gaudet (1994) for large scale intermediate culture operations in the Iles-de-la-Madeleine. A study by Côté et al. (1994) demonstrated that reduced growth in suspended culture is due to the effects of food depletion by the scallops in the crowded conditions and not the physical effect of the crowding. In a related study, Claereboudt et al. (1994) also showed that the decrease in flow within pearl nets resulted in an increase of growth and somatic tissue of the caged scallops due to the more optimal flow regime. This effect probably persists until density dependent factors begin to predominate.

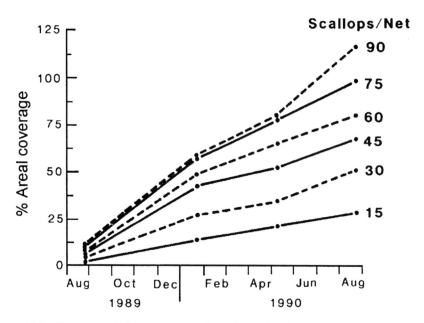

Figure 16.3. Percent areal floor coverage of pearl nets by the sea scallop at various stocking densities over a year's growth in intermediate suspension culture (After Parsons and Dadswell 1992).

16.5.3 Depth

The depth at which pearl nets are suspended in the water column can also result in substantial growth differences for juvenile sea scallops. In an experiment in which pearl nets were variously suspended from 5 m off-bottom to 0.25 m off-bottom Dadswell and Parsons (1991) found that the mean size of juveniles ranged from 54 to 42 mm, respectively. Growth differences of this magnitude were similar to the effects of stocking density as determined by Parsons and Dadswell (1992). The effect of suspension depth on growth in sea scallops and other species of scallops has been attributed to differences in food level (Wallace and Reinsnes 1984; Côté et al. 1993; Frechette et al. 2000). In a study done in both an open and sheltered location in the Iles-de-la-Madeleines, scallop spat were found to grow and survive better in pearl nets that were grown near the surface (Gaudet 1994). Côté et al. 1993) found that spat grew better at 9 m than at 21 m and was related to the levels of food in the 0.7 to 5.0 μm size range, although density effects masked some of the trends. From the above observations we can see there is likely no one depth that will be suitable in all locations. The optimum culture depth for a particular location will depend on the distribution of the food in the water column and the depth to which wave action will impact the suspended culture gear. There may also be interactions with other fouling species that may be depth dependent (Claereboudt et al. 1994; Grecian et al. 2000).

16.5.4 Season

Parsons and Dadswell (1992) and Grecian et al. (2000) reported a decreased growth rate in juvenile sea scallops during the winter. Seasonal effects on growth were, as would be expected, related to the annual variation in temperature and food levels. In many areas within the distribution range of *Placopecten*, winter water temperatures can drop to 1°C.

16.5.5 Location

A number of studies throughout the east coast of Canada have examined growth rates of juvenile scallops held in suspension. Naidu and Cahill (1981) reported growth rates of 0.096 mm d^{-1} for Spencer's Cove, NL. Dadswell and Crawford-Kellock (1989) and Dadswell and Parsons (1991, 1992b) found growth rates of 0.12 to 0.18 mm d^{-1} during the autumn for scallops of age 1+ grown in Mahone Bay, NS and Dadswell and Parsons (1992a) reported a remarkable 0.33 mm d^{-1} for juvenile scallops from the summer settlement period (i.e., the June-July spawning) during the first six months of the second year. Growth rates of 0.098 to 0.117 mm d^{-1}, 0.12 mm d^{-1} and 0.084 to 0.111 mm d^{-1} have been reported for Passamaquoddy Bay, NB by Wildish et al. (1988), Dadswell and Parsons (1991) and Parsons and Dadswell (1992), respectively. These growth rates are remarkably similar given the broad geographic area. One of the reasons for this might be that certain density thresholds of scallops were not exceeded that could have resulted in food depletion of the local seston fields. There is an optimal relationship between flow rate and seston concentration in the feeding dynamics of sea scallops (Wildish and

Saulnier 1992). As with stocking density and depth, spatial differences in growth of scallops, held in suspension and grown on the bottom, have been attributed to variations in food and temperature (MacDonald and Thompson 1985a, 1985b, 1986).

16.5.6 Survival

Survival of scallops was independent of stocking density for sea scallops (Côté et al. 1993; Parsons and Dadswell 1992). Parsons and Dadswell (1992) found a mean survival of 91% and attributed the 9% mortality to initial handling methods. In another study, Dadswell and Parsons (1991) reported a survival of 57%; the higher mortalities may have been due to the repeated handling of scallops. Survival rates have been linked to initial spat size at seeding (Penney and Mills 1996) for scallops grown to 60 mm in lantern nets; mesh size was not found to be a significant factor.

16.5.7 Gear Selection

Pearl nets have almost exclusively been used for the intermediate culture of all scallop species (Taguchi 1978; Ventilla 1982; Naidu and Cahill 1986; Bourne et al. 1989). Dadswell and Parsons (1992b) have evaluated the handling performance and cost associated with several different net designs (Table 16.1). The total cost associated with raising 100,000 scallops ranged from $1,684 to $3,666. Two of the gear types (Shibetsu and Lantern nets) that were the easiest to handle were also the most expensive to purchase. This preliminary study did not, however, examine the long-term growth and survival of scallops for the different nets.

Table 16.1

Handling times and costs associated with growing 100,000 sea scallops, *Placopecten magellanicus*, in intermediate culture for several different net designs (from Parsons and Dadswell 1994).

Net Type	Loading Time (hr)	Unloading Time (hr)	Total Time (hr)	Labour Cost ($)[1]	Cost of Gear ($)[2]	Total Cost ($)[3]
Pearl-square (9 mm)	104.2	33.8	138.0	966	1,860	2,826
Pearl-square (6 mm)	104.2	42.6	146.8	1,028	1,810	2,838
Pearl-square (3 mm)	132.3	64.8	197.1	1,380	2,050	3,430
Pearl-round (6 mm)	125.3	37.0	162.3	1,136	2,530	3,666
Superlantern	133.3	25.0	158.3	1,108	2,530	3,638
Shibetsu	47.9	17.5	65.4	458	3,090	3,548
Lantern	92.6	7.7	100.3	702	2,734	3,436
Oyster Tray	83.3	14.4	97.7	684	1,000	1,684

1. Based on an hourly rate of $7.00.
2. Based on a stocking density of approximately 180 scallop m^{-2} and amortised over 5 years.
3. Total cost = labour cost + amortised cost of gear.

16.5.8 Strategies for Intermediate Culture

There are a number of competing factors to consider when developing an overall strategy for intermediate culture. The type of gear chosen appears to be a trade-off between ease of handling (labour) and the cost and durability of the gear. Handling of nets (single or repeated) must be weighed against cleaning the nets to reduce fouling (increased growth and labour) (Cole et al. 1996; Devaraj and Parsons 1997) and increased mortality associated with repeated handling. Some of the handling due to fouling can be reduced by avoiding areas that are conducive to higher settlement rates by undesirable species. Culturing techniques in Japan recommend limiting handling nets to two times a year compared to the traditional four times a year (Taguchi 1976; Ventilla 1982). As well, the deployment time of gear is critical to maximising growth and avoiding predators such as starfish (Levy 1999, Grecian et al. 2000; Grecian et al. 2003) (Fig. 16.4).

The 'best' stocking density for growing scallops in intermediate culture is also a trade-off between reduced growth rates at high densities and a large number of nets (high cost) at lower stocking densities (Parsons and Dadswell 1992). Growth rates can be intentionally manipulated to ensure a continuous supply of the desired size of scallops in order to maximise efficiency of gear use and or to support continuous market supply. It may also be desirable to manipulate growth rates based on the choice of stocking density as influenced by the depth / food levels (Frechette et al. 2000), timing of transfer to final grow-out (ear hanging or cages), desired final market product (meat or whole animal) and or market size (e.g., princess scallop versus meats only).

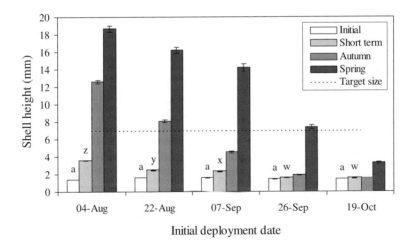

Figure 16.4. Mean shell height of scallops deployed over five consecutive two-week intervals in 1997 (short-term) and on November 8, 1997 (Autumn) and June 24, 1998 (spring) at Poole's Cove, Newfoundland. The initial date of an interval was the final date of the previous short-term interval. Common letters denote no significant difference among shell heights for each sampling period (Tukey's b test). Vertical bars are ± SE. (After Grecian et al. 2003).

16.6 FINAL GROW-OUT

Scallops are removed from intermediate culture at a size of about 40 to 60 mm (from 16 to 26 months of age) for final grow-out to market size. There are two basic methods for the final grow-out of scallops, suspension culture and bottom seeding (also known as enhancement, sowing, planting or relaying) (Mottet 1979; Ventilla 1982). Suspension culture takes about 0.5 to 2 additional years to reach market size, whereas bottom grown scallops require a further 2 to 3 years.

16.6.1 Suspension Culture

Suspension culture can entail the use of cages (e.g., multi-tiered Lantern nets) or ear hanging. Traditionally, these types of nets were relatively small so that they could be handled easily by a single person, although they could have up to 20 different levels within. More recently in eastern Canada, larger scale cages have been designed to grow larger numbers of animals in a single unit in an attempt to design a system that was more applicable to automation (Parsons et al. 1998). These larger cages are either larger versions of lantern nets or they are stackable plastic trays mass produced through injection moulding systems. There is a significant relationship between the mesh-size of the tray and the flow rates through the tray (Brake and Parsons 1998) so growers should carefully match the trays to their existing conditions. To date, however, the performance of these new trays or large cages has not been fully evaluated over several growth cycles.

Ear hanging consists of hanging scallops (>50 mm shell height) by their "ear" or auricle. A small hole is drilled into one of their ears or in the byssal notch and the scallops are then either threaded onto loop cord or stuck on a barbed plastic "aggie" pin, which has been inserted through a piece of braided rope (Bourne et al. 1989). A commercial ear hanging machine has become available for semi-automating the ear hanging procedure (Anonymous 1991a, 1991c). Loop cord consists of a piece of braided of rope, which contains loops of 1 to 2 cm which are evenly spaced about 10 to 15 cm apart (Dadswell and Parsons 1991). Scallops are then threaded onto the loop with a piece of nylon monofilament. Separation of the "aggie" pins is usually about 15 cm.

16.6.2 Bottom Culture

Bottom culture involves seeding scallops directly on the bottom and is also known as "enhancement". The type of bottom required for this type of operation should be consistent with the physical and biological qualities that comprise traditional scallop beds (see Stokesbury and Himmelman 1995). As a result, this implies that the traditional fishermen typically will be involved with most of these broad-scale initiatives. Ultimately, some form of property rights or limited access to the resource will have to be established if private investment is being sought. A large leased area is necessary if a substantial number of scallops are to be cultured. For example, if 1,000,000 scallops are to be seeded annually on the bottom at a density of $5 \cdot m^{-2}$, 20 ha yr^{-1} would be required. In Mutsu Bay, Japan 23,000 ha are used for bottom culture by co-operatives (Ventilla 1982).

The one main advantage of bottom culture over suspension culture is its low capital cost (Frishman et al. 1980; Wildish et al. 1988). There are, however, several disadvantages; growth rates are slower, sea scallops are capable of swimming and could potentially swim off a lease site (Caddy 1968; 1989; Parsons et al. 1992b), and scallops are highly susceptible to predation (Elner and Jamieson 1979; Jamieson et al. 1982; Barbeau and Scheibling 1994a, b; Barbeau et al. 1998). Losses of 20 to 50% are routinely experienced in Japanese scallop culture after 2 to 3 years of growth on bottom (Taguchi 1976; Ventilla 1982). These losses are due primarily to predation, although dispersion of a seeded population was relatively rapid in a shallow water site in Nova Scotia (Scheibling et al. 1991; Hatcher et al. 1996). Several studies have shown that predation rates are highest shortly after the seeding operation and that as prey density increases so does the predation rate, although not predator numbers (Barbeau et al. 1996; Barbeau et al. 1998; Wong et al. 2001). Crabs and starfish have been the major predators identified to date, each having different behaviour patterns that are also affected by temperature (Barbeau and Scheibling 1994a, 1994b) and substrate (Wong and Barbeau 2003). Enhancement efforts in the Iles-de-la-Madeleine, Québec have shown that 94–99% of the scallops seeded on the bottoms in some areas have been lost (Cliche et al. 1997; Nadeau and Cliche 1998). The seeding strategy, time of year, scallop condition and predators all played a role in the loss of the seeded scallops. At this site, there appeared to be relatively little dispersal of scallops from the general area of the seeded sites based on a combination of diving and video techniques on tagged scallops (Cliche et al. 1994). To help understand the population dynamics of the seeded juveniles, some mathematical models have been developed for the process (Barbeau and McDowell 1998; Barbeau and Caswell 1999).

16.6.3 Growth

Both the shell and meat growth rates of scallops grown in suspension are faster compared to bottom culture. Both Naidu and Cahill (1978) and MacDonald (1986) have reported final annual growth rates of 0.06 mm·d^{-1} for sea scallops from Newfoundland whereas MacDonald (1986) reported a growth rate of 0.044 mm·d^{-1} for bottom grown wild scallops. Bottom cultured and wild scallops from Passamaquoddy Bay, NB and the Bay of Fundy, NB each had annual growth rates of 0.06 mm·d^{-1} (Wildish et al. 1988; Chandler et al. 1989). In Passamaquoddy Bay, scallops grown in cages and by ear hanging had annual growth rates of 0.072 mm·d^{-1} and 0.08 mm·d^{-1}, respectively (Wildish et al. 1988; Dadswell and Parsons 1991). Grant et al. (2003) also found scallops grown by ear hanging out performed those grown in cage culture.

Shell growth rate height was inversely related to stocking density. Naidu and Cahill (1978) reported shell heights of 93.3 to 71.7 mm for scallops grown in cage culture at densities of 10 to 50 scallops. In an experiment in which scallops were cultured by ear hanging on loop cord, Dadswell and Parsons (1991) found no significant difference in growth between scallops, which were spaced 5 and 10 cm apart.

Growth of scallops, as measured by meat yield, was also greater in scallops cultured in suspension compared to bottom grown ones; 90 mm scallops from Passamaquoddy

Bay, NB and the Bay of Fundy, NB had meat weights of 10.4 g and 5.0 to 8.1 g, respectively (Jamieson and Lundy 1979; Chandler et al. 1989). Scallops cultured in cages and by ear hanging in Passamaquoddy Bay, NB had meat yields of 11.2 g and 16.3 g, respectively (Wildish et al. 1988; Dadswell and Parsons 1991) and in Mahone Bay, NS, suspended scallops averaged 15.0 g for meat weight (Dadswell and Parsons 1992a). This represents a 7.7 to 47.1% increase in meat weight for a similar size scallop grown in suspension versus a bottom grown animal. MacDonald (1986) reported similar differences in the somatic weight of suspended and bottom grown scallops from Newfoundland. Seasonal differences are also found in the relative weight of the different organs (i.e., gonad, adductor muscle) in cultured scallops (Penney and McKenzie 1996).

From a comparison of growth rates and meat yields of sea scallops by ear hanging (Dadswell and Parsons 1991) and cage culture (Wildish et al. 1988), it would appear that the ear hanging technique produces faster growing scallops with a greater meat yield.

16.7 SCHEDULING AND STRATEGIES

The timing and co-ordination of the different culturing stages is contingent upon cost factors and equipment availability but also type of products to be marketed (e.g., whole animal versus meat only) and continuity of supply. This requires that different sized scallops (e.g., 50 mm and 90 mm) be available throughout the year. Since the availability of spat is presently restricted to the summer (Dadswell and Parsons 1992a) or autumn (Dadswell et al. 1987), ensuring a continuous supply of certain size scallops entails manipulating their stocking density at intermediate and or final grow-out stages to enhance or slow down their growth, which will result in a staggered supply of appropriate sizes.

The typical scheduling of events includes in "Year 1" monitoring of spawning and starfish and spat settlement in the summer and spat collector deployment in the autumn (~30 days after spawning). "Year 2" involves sorting spat in the early summer when they are large enough to handle and immediately placing them in pearl nets. In the autumn the nets will be required to be thinned and cleaned, especially if the pearl nets were stocked with high densities of scallops (100 to 200/net). In the summer of "year 3", when the scallops are 24 months old and have reached about 50 mm, they will be ready to either market as whole animals (princess scallops) or be placed in final grow-out (ear hanging or cage culture). In the summer of "year 4" at approximately 36 months of age, the scallops can be harvested for roe-on meats and in the autumn for meats only.

16.8 SITE SELECTION

Site selection criteria are an integral component to the success of any aquaculture venture. Criteria can be viewed on two different scales, a large and small scale. In general, features on a large scale share characteristics in the order of 10 to 100 km, whereas small scale features vary on the order of 0.1 to 1 km. In other words, some bodies of water share common traits, such as a bay being completely covered with ice in the winter, whereas within a bay, a local promontory may have better current characteristics for spat collection and scallop growth than an open water site. Site selection criteria can also be generally applied to all aspects of scallop aquaculture or be specific to certain phases, such as spat collection.

16.8.1 Environmental Criteria

16.8.1.1 Temperature and salinity regime

Fluctuations in temperature and salinity for a particular site should be within acceptable ranges as obviously there will be seasonal cycles present. Sea scallops tolerate temperatures of 0 to 18°C and salinities of 25 to 32‰ (Dickie 1958; Frenette et al. 2002; Frenette 2004), but some acclimation is possible by the animals (Dickie 1958). Bergman et al. (1996) reported that salinities at 16‰ and below were lethal and that salinities between 16 and 21‰ put scallops into catatonic shock at 1°C. Preferred temperatures are likely close to 12°C and salinities over 31‰. Particular attention should be paid to the extreme values, i.e., the temperature should not exceed 18°C during the summertime nor should salinity drop below 25‰, especially during spring run-off. Temperature can affect the severity of the physiological response to environmental stressors such as heavy metals (Nelson et al. 1977) and also behavioural escape responses (Dickie and Medcof 1963).

16.8.1.2 Ice coverage

Ice coverage may restrict access to sites during the winter and all lines and floats will have to be submerged. Similarly, if ice flows are present, lines and floats will have to be submerged. Generally, it is a good practice to submerge sites during the winter to avoid storm damage and wear on gear.

16.8.1.3 Wind

The seasonal prevailing wind pattern and fetch should be considered, as certain exposures are more susceptible to wind generated wave and turbulence. Scallops do not grow well under strong wave action or in areas with high concentrations of suspended sediment (particulate inorganic matter) (Frenette 2004). Submerging the head ropes to 2 to 4 m below the surface will generally decrease turbulence. On very exposed sites, even benthic cultured scallops can suffer abrasion due to the movement of gravel and sand etc. (Miyake at al. 1995).

16.8.1.4 Bottom type

Bottom type is especially important if bottom culture is being considered. Scallops live mainly on gravely or sandy substrates and not on muddy bottoms (Stokesbury and Himmelman 1995). Selection of a poor bottom will result in emigration of scallops, particularly sand (Stokesbury and Himmelman 1996). Movement of 10 to 15 km are not uncommon (Melvin et al. 1985).

16.8.1.5 Turbidity

Scallops are sensitive to the amount of suspended particulate matter in the water, but are capable of maintaining high growth rates by increasing their feeding rates for short periods of time (Cranford 1995). Obviously, benthic animals have evolved to deal with a certain amount of re-suspension events due to bottom turbulence. Several different species of both terrestrial and marine sources of seston have been identified from scallop gut contents (Shumway et al. 1987) and several studies have showed that scallops are able to actively select high-quality chlorophyll-containing particles (Shumway et al. 1985; MacDonald and Ward 1994). However, if the inorganic component of the seston reaches levels in excess of 2 mg dm^3, then labial sorting begins and the production of pseudofaeces was initiated followed by reductions in growth (Cranford and Gordon 1992). Phytoplankton has been demonstrated to be a better food source than "aged" kelp particles in laboratory studies (Cranford and Grant 1990; Grant and Cranford 1991) and therefore, scallops often have better growth rates higher up in the water column (Emerson et al. 1994). However, mixed results were found in Lunenberg Bay in Nova Scotia where bottom cultured scallops sometimes showed greater growth rates than those in suspension culture (Kleinman et al. 1996). They also found spatial differences between site in growth rates. Frenette et al. (2001) and Frenette (2004) found poor scallop growth in shallow-water embayments of southeastern New Brunswick, which had high turbidity.

16.8.2 Biological Constraints

16.8.2.1 Predators

The presence of predators is again of primary interest if bottom culture is being considered. The main species to consider are starfish (*Asterias* spp.), wolffish (*Anarhichas lupus*), lobsters (*Homarus americanus*) and rock crabs (*Cancer irroratus*) (e.g., Dickie and Medcof 1963; Elner and Jamieson 1979) (see also section 16.6.2). Predators may be reduced or avoided altogether in suspension culture by good husbandry practises (i.e., keeping nets and ear hanging lines off the bottom).

16.8.2.2 Fouling organisms

Fouling organisms are generally ubiquitous and have to be tolerated. The best control against fouling is to determine settlement times of the main fouling organisms and then plan the scheduling of the different culturing phases accordingly (pearl net and cages). Some degree of control can also be gained by deploying nets deeper in the water. In some regions, settlement of fouling organisms is negligible at depths greater than 10 m.

16.8.2.3 Phycotoxins

When considering the production of princess scallops, the presence and seasonality of phycotoxins need to be taken into account. Toxins in scallops are a risk to consumers and testing is required to ensure product safety (Shumway et al. 1988). The occurrence of toxic algae are generally known for certain bodies of water (e.g., Bay of Fundy - Gulf of Maine, Gulf of St. Lawrence) although the species may vary as well as their mode of action (Prakash et al. 1971; Medcof 1985; Gainey and Shumway 1988; Martin and White 1988). The consumption of high levels of toxic algae has been shown to negatively affect the growth rate of scallops (Chauvaud et al. 1998). The implications of toxic algae on culture and fisheries have attracted several reviews (i.e., Shumway et al. 1988; Desbiens et al. 1990; Shumway 1990; Shumway and Cembella 1992).

Paralytic shellfish poison (PSP) toxin has been reported in *Placopecten magellanicus* for several many years (i.e., Bourne 1965; Hsu et al. 1979; Jamieson and Chandler 1983; Cembella et al. 1994) and is a feature that must be accommodated in the culture protocols. However, small-scale differences in the occurrence and concentration of toxin in bivalves do occur (White 1988; Robinson et al. 1999; Parsons et al. 2002). *P. magellanicus* has been categorised as a slow detoxifier of the saxitoxin (PSP) from the dinoflagellate *Alexandrium fundyensis* taking from months to years to detoxify.

Amnesic shellfish poison (ASP) or domoic acid is a neuro-toxic amino acid that can be produced by the marine diatoms *Nitzchia* (= *Psuedo-nitzschia*) *pungens* (Subba Rao et al. 1988; Maranda et al. 1990) and *Nitzschia pseudodelicatissima* (Martin et al. 1990). Gilgan et al. (1990) reported on the distribution and magnitude of domoic acid in Atlantic Canada in 1988. Haya et al. (1991) reported on the domoic acid levels in shellfish and phytoplankton from the Bay of Fundy. It has been reported from scallops (Bird and Wright 1989).

Another potential toxin, diarrehetic shellfish toxins (DSP) cause gastrointestinal distress in humans and scallops can acquire DSP toxins by ingesting toxic dinoflagellates (such as *Prorocentrum lima*) from the water column or from benthic seston (Yasumoto et al. 1984). These species are recognised to be a hazard to consumers and therefore culture and fisheries operations (i.e., Dahl and Yndestad 1985; Freudenthal and Jijina 1988).

16.8.2.4 Diseases and parasites

The scarcity of parasite and disease related problems in the sea scallop bodes well for aquaculturists (see Chapter 11). However, because scallop culturing can be intensive, the culture conditions may amplify any of the natural diseases present. At the present time, there is not a good understanding of the endemic parasites and diseases that currently affect the local scallop populations (*Placopecten*, *Chlamys* and *Argopecten*) (Ball and McGladdery 2001), but databases are being developed. For example, rickettsia-like organisms were found to infect the gills and plicate membranes of sea scallops near Rhode Island and were found to be quite contagious and species-specific with uninfected individuals (Gulka and Chang 1985). However, the health impacts to the scallop could not be determined. Maintaining an awareness of any potential causative agents and eliciting the co-operation with various agencies (government, universities and industry) may ensure a relatively disease free operation (McGladdery 1990). In addition, certain challenge tests are being developed to evaluate the responses of scallops to potential pathogens (Belvin et al. 2001; Roussy et al. 2001).

16.9 PRODUCTS AND MARKETING

Of the approximately 350 species of extant scallops, only about twelve are commercially fished or cultured on a large scale and the bulk of the world output is consumed by only a few countries, mainly, USA, Japan and Western Europe (De Franssu 1990). In North America, the principal product of the commercial sea scallop fishery has been the meat only. The landed processed product is then either sold fresh or frozen, with the USA being the largest consumers of this product (Department of Fisheries and Oceans 1989). Individual Quick Frozen (IQF) meats are rapidly dominating the market in North America. In Canada the bulk of the scallops from the commercial fishery are exported to the USA as meats, but throughout the last forty years, scallops with the roe-on have been periodically marketed (Bourne and Read 1965; Jamieson 1979; Department of Fisheries and Oceans 1989). The primary destination of this product has been to France (Dupouy 1983; De Franssu 1990).

Sea scallop culturists have shown a desire to produce three products: meats, roe-on meats and whole animals (~50 to 70 mm). Taste panel assessments could not differentiate between cultured and commercial caught scallops, thus a cultured sea scallop product is expected to receive consumer acceptance (Naidu and Botta 1978). Thus far, whole animals have been marketed in Québec, Ontario and New England (USA) and fresh meats have been sold locally or regionally. Regulatory agencies have established a protocol for testing whole animals and roe-on meats to ensure that these products are free of phycotoxins and bacterial pollution. If growers can successfully market whole animals and roe-on meats, cash flow can be increased through the sale of smaller animals; however, shelf life of live whole scallops is limited.

The market for meats is of sea scallops demands a quality product with prices primarily determined by the size of the commercial catch (i.e., supply) (Lines 1988). Aquaculturists can deliver a high quality, fresh product as access to farms and market-size scallops are generally close to the main markets and the product is available year-round.

16.10 ECONOMICS

The main constraint to larger scale development is that it is a capital and labour intensive venture with relatively long grow-out period (>3 years from egg to minimum meat market size). The main capital costs are primarily associated with the gear, i.e., pearl nets and lantern cages. Certain husbandry practices, however, can reduce costs. For spat collection, used monofilament gillnetting is usually readily available from the commercial fishery and for final grow-out using ear hanging, the main cost is associated with purchasing the "aggie" pins and rope (exclusive of the drill or ear hanging machine).

Labour, another large component of a culture operation, is centred on each of the culturing steps and occurs at different times of the year. A large amount of labour is require to sort spat, load and thin pearl nets and lantern nets, clean nets, ear hang the scallops, and harvest the product. Savings can be gained depending upon the type of gear used (e.g., easy loading nets, (Dadswell and Parsons 1992b)) and strategy employed (e.g., using a low initial density for intermediate culture thereby reducing the need for thinning

(Parsons and Dadswell 1992)). Any form of mechanisation that a grower can utilise will assist the operation (e.g., pressure cleaning, spat sorter, spat grader).

Several economic analyses have been conducted on the different grow-out strategies (e.g., bottom culture, cage culture and ear hanging). Based on the high cost of labour and gear and the price for meat only, the early economic scenarios were marginal at best (Frishman et al. 1980; Gilbert and Cantin 1987; Wildish et al. 1988).

An economic assessment by Ford (1997), however, highlighted that scallops grown in suspension had a larger meat yield to shell size than allowed for in earlier meat market assessments. This adjustment to the economic model, combined with growing scallops to a larger meat size (weight), indicated that an economically viable industry could be realised based on a meats only product; a less risky market in North America than a roe-on or whole animal market. If a consistent whole animal or roe-on meat product could be established in the market, then yield per scallop, cash flow and revenues would increase (De Franssu 1990; Dadswell and Parsons 1992a; Ford 1997). Studies in Passamaquoddy Bay, NB and Mahone Bay, NS have also shown that with experience both growth rates and efficiency have improved over time (Dadswell and Parsons 1991, 1992a).

The modelling exercise conducted by Penney and Mills (2000) reported on a 2-year pilot-scale study and concluded that vertically-integrated farms seeding 1 million spat annually were economically viable. Sensitivity analyses indicated that sale price for the harvested product was key. Other variables such as capital costs, labour and mortality were not as sensitive.

16.11 SOCIAL ISSUES

There has been some reluctance on the part of the regulating agencies and the general coastal communities in eastern Canada to embrace the concept of scallop aquaculture. This occurs at a number of administrative levels and for a number of reasons; not the least being an active wild fishery that is concerned about competition for markets and for space. In a report on the environmental requirements for sea scallops in eastern Canada, the limiting factor for scallop culture in the coastal areas was reported to be coastal sewage/faecal pollution and toxic phytoplankton (Stewart and Arnold 1994). While these reasons have hindered scallop culture and can be limiting, depending upon the product being produced (i.e., meats vs. roe-on), it is only part of the story. In general, there are only two marine species that are being grown in large amounts in Eastern Canada, the blue mussel (*Mytilus edulis* and *M. trossulus*) and the Atlantic salmon (*Salmo salar*). The production levels of both of these species have grown quickly and are now nearing peak capacity in many areas. This has resulted in resource use conflicts with other users such as the fishery, recreation and tourism industries. There are also concerns about the potential environmental effects that may be occurring at some of the culture sites. All of these issues can influence the perception of local coastal residents as to how other culture industries may evolve and has sometimes increased the level of local concern. As the existing industries look for ways of continually expanding and with requests for marine leases from a new culture candidate, one solution to address public concern is to do a better job at integrating the culture into the local economy and society (e.g., Parsons et al.

2002). Starting small and scaling up is probably a good approach with the appropriate checks and balances built in to the development plan. Bivalve culture often takes a significant amount of space (10–20 ha per operation) so the challenge will be to devise methods to intensify the culture while maintaining the long-term sustainability of the site.

16.12 FUTURE PROSPECTS

There is a strong demand for scallops (De Franssu 1990). However, the economic feasibility of sea scallop culture is marginal at best. As the commercial culturing of sea scallops develops and increases in size, it will move further along the "learning scale" and become more efficient. As operations become larger they will benefit from the "economy of scale", i.e., increased volume/decreased costs, and with the development of new technologies or approaches, such as the ear hanging machine and spat sorters or new approaches (e.g., triploidy (Jackson et al. 2003) or low cost, reliable hatchery production), significant savings on labour may accrue. The medium term outlook is a modest commercial industry in Atlantic Canada that could increase if it can realise a premium product for the marketplace.

ACKNOWLEDGMENTS

Our thanks to Sharon Ford for reviewing the chapter.

REFERENCES

Abraham, A., Couturier, C. and Parsons, J., 1996. Toxicity of un-ionized ammonia to juvenile giant scallops, *Placopecten magellanicus*. Bull. Aquacul. Assoc. Canada 96–3:68–70.

Anonymous, 1991a. Dutch ear hanger goes to Japan. Fish Farming International 18(11):42.

Anonymous, 1991b. Baby giant scallops: grading before going back to sea. Technomar News Canada, Rimouski, Québec No. 1: 8 p.

Anonymous, 1991c. New developments for the shellfish industry from Franken. Aquanotes, Aquaculture Association of Nova Scotia 14:45.

Aoyama, S., 1989. The Mutsu Bay scallop fisheries: scallop culture, stock enhancement, and resource management. In: J.F. Caddy (Ed.). Marine Invertebrate Fisheries: Their Assessment and Management. John Wiley & Sons, New York. pp. 525–539.

Baird, F.T., 1953. Observations on the early life history of the giant scallop (*Pecten magellanicus*). Maine Dept. Sea Seashore Fisheries, Res. Bull. No. 14, 7 p.

Ball, M.C. and McGladdery, S.E., 2001. Scallop parasites, pests and diseases: implications for aquaculture development in Canada. Bull. Aquacult. Assoc. Can. 101–3:13–18.

Barbeau, M.A. and Caswell, H., 1999. A matrix model for short-term dynamics of seeded populations of sea scallops. Ecol. Appl. 9:266–287.

Barbeau, M.A., Hatcher, B.G., Scheibling, R.E., Hennigar, A.W., Taylor, L.H. and Risk, A.C., 1996. Dynamics of juvenile sea scallops (*Placopecten magellanicus*) and their predators in bottom seeding trials in Lunenburg Bay, Nova Scotia. Can. J. Fish. Aquat. Sci. 53:2494–2512.

Barbeau, M.A. and McDowell, J.E., 1998. Improving the production of scallops from bottom culture. World Aquacult. 29:12–16.

Barbeau, M. and Scheibling, R., 1994a. Temperature effects on predation of juvenile sea scallops *Placopecten magellanicus* (Gmelin) by sea stars (*Asterias vulgaris* Verrill) and crabs (*Cancer irroratus* Say). *J. Exp. Mar. Biol. Ecol.* 182:27–47.

Barbeau, M.A. and Scheibling, R.E., 1994b. Behavioral mechanisms of prey size selection by sea stars (*Asterias vulgaris* Verrill) and crabs (*Cancer irroratus* Say) preying on juvenile sea scallops (*Placopecten magellanicus* (Gmelin)). J. Exp. Mar. Biol. Ecol. 180:103–136.

Barbeau, M.A., Scheibling, R.E. and Hatcher, B.G., 1998. Behavioural responses of predatory crabs and sea stars to varying density of juvenile sea scallops. Aquaculture 169:187–98

Barber, B.J., Getchell, R., Shumway, S. and Schick, D., 1988. Reduced fecundity in a deep water population of the giant scallop *Placopecten magellanicus* in the Gulf of Maine, USA. Mar. Ecol. Prog. Ser. 42:207–212.

Beaumont, A. and Hall, V., 1999. Getting the queen to spawn: Experiments with *Aequipecten opercularis*. 12th Int. Pectinid Workshop, Bergen (Norway), Book of Abstracts: p. 114.

Belvin, S., Tremblay, R., Roussy, M. and McGladdery, S.E., 2001. Investigating the cause of episodic mortalities in the giant scallop, *Placopecten magellanicus*, in the Gulf of St. Lawrence. Bull. Aquacult. Assoc. Can. 101–3:32–35.

Bergman, C., Parsons, J. and Couturier, C., 1996. Tolerance of the giant sea scallop, *Placopecten magellanicus*, to low salinity. Bull. Aquacul. Assoc. Canada 96–3:62–64.

Bird, C.J. and Wright, J.L.C., 1989. The shellfish toxin, domoic acid. World Aquaculture 20:40–41.

Bonardelli, J.C., 1987. Collection of scallop spat in the Baie des Chaleurs. Proc. Ann. Meet. Aquacul. Assoc. Canada 1:18–19.

Bonardelli, J.C., 1988. Optimizing collection of pectinid spat on collectors. J. Shellfish Res. 7:150.

Bourne, N., 1964. Scallops and the offshore fishery of the Maritimes. Bull. Fish. Res. Board Can. 145: ix + 60 p.

Bourne, N., 1965. Paralytic shellfish poison in sea scallops (*Placopecten magellanicus*, Gmelin). J. Fish. Res. Board Can. 22:1137–1149.

Bourne, N., Hodgson, C.A. and Whyte, J.N.C., 1989. A manual for scallop culture in British Columbia. Can. Tech. Rep. Fish. Aquat. Sci. 1694: 215 p.

Bourne, N. and Read, F.C., 1965. Fuller utilization of sea scallops (*Placopecten magellanicus*, Gmelin). Fish. Res. Bd. Canada MS. Rept. 806: 24 p.

Brand, A.R., Paul, J.D. and Hoogester, J.N., 1980. Spat settlement of the scallops *Chlamys opercularis* (L.) and *Pecten maximus* (L.) on artificial collectors. J. Mar. Biol. Assoc. U.K. 60:379–390.

Brand, A.R., Wilson, U.A.W., Hawkins, S.J., Allison, E.H. and Duggan, N.A., 1991. Pectinid fisheries, spat collection, and the potential for stock enhancement in the Isle of Man. ICES mar. Sci. Symp. 192:79–86.

Brake, J. and Parsons, G.J., 1998. Flow rate reduction in scallop grow-out trays. Bull. Aquacult. Assoc. Canada 98–2:62–64.

Buestel, D., Dao, J.C. and Lemarie, G., 1979. Collecte de naissain de pectinidés en Bretagne. Rapp. P.-v. Réun. Cons. int. Explor. Mer. 175:80–84.

Caddy, J.F., 1968. Underwater observations on scallop (*Placopecten magellanicus*) behaviour and drag efficiency. J. Fish. Res. Board Can. 25:2123–2141.

Caddy, J.F., 1989. A perspective on the population dynamics and assessment of scallop fisheries, with special reference to the sea scallop, *Placopecten magellanicus* Gmelin. In: J.F. Caddy (Ed.). Marine Invertebrate Fisheries: Their Assessment and Management. John Wiley & Sons, New York. pp. 559–589.

Castagna, M. and Kraeuter, J.N., 1981. Manual for growing the hard shell clam, *Mercenaria*. Spec. Rep. Appl. Mar. Sci. Ocean Eng. No. 249. Virginia Institute of Marine Science, Gloucester Point, VA 110 p.

Cembella, A.D., Shumway, S. and Larocque, R., 1994. Sequestering and putative biotransformation of paralytic shellfish toxins by the sea scallop, *Placopecten magellanicus*: Seasonal and spatial scales in natural populations. J. Exp. Mar. Biol. Ecol. 180:1–22.

Chandler, R.A., Parsons, G.J. and Dadswell, M.J., 1989. Upper and northern Bay of Fundy scallop surveys, 1986–1987. Can. Tech. Rept. Fish. Aquat. Sci. 1665: iii + 37 p.

Chauvaud, L., Thouzeau, G. and Paulet, Y.M., 1998. Effects of environmental factors on the daily growth rate of *Pecten maximus* juveniles in the Bay of Brest (France). J. Exp. Mar. Biol. Ecol. 227:83–111.

Chislett, R.R., 1989. Scallop culturing on the Quebec lower north shore (Baie de la Bussiere) 1986 to 1988. Atelier sur l'Elevage du Pétoncle Géant, Gapsé (Québec). Ministère de l'Agriculture, des Pêcheries et de l'Alimentation, Québec. Cahier Spécial d'Information No. 12:67–70.

Claereboudt, M., Bureau, D., Côté, J. and Himmelman, J., 1994. Fouling development and its effects on the growth of juvenile giant scallops (*Placopecten magellanicus*) in suspended culture. Aquaculture 121:327–342.

Cliche, G. and Beaulieu, J.L., 1989. Compte rendu des activités de recherche en écloserie-nurserie de pétoncle géant aux Îles-de-la-Madeline. Atelier sur l'Elevage du Pétoncle Géant, Gaspé (Québec). Ministère de l'Agriculture, des Pêcheries et de l'Alimentation, Québec. Cahier Spécial d'Information No. 12:39–41.

Cliche, G., Giguère, M. and Vigneau, S., 1994. Dispersal and mortality of sea scallops, *Placopecten magellanicus* (Gmelin 1791), seeded on the sea bottom off Iles-de-la-Madeleine. Journal of Shellfish Research 13:565–570.

Cliche, G., Vigneau, S. and Giguère, M., 1997. Status of a commercial sea scallop enhancement project in Iles-de-la-Madeleine (Quebec, Canada). Aquaculture International 5:259–266.

Cole, F., Parsons, J. and Couturier, C., 1996. Flow dynamics in and around pearl nets of various mesh sizes. Bull. Aquacul. Assoc. Canada 96–3:77–79.

Côté, J., Himmelman, J.H. and Claereboudt, J., 1994. Separating effects of limited food and space on growth of the giant scallop *Placopecten magellanicus* in suspended culture. Mar. Ecol. Prog. Ser. 106:85–91.

Côté, J., Himmelman, J., Claereboudt, M. and Bonardelli, J., 1993. Influence of density and depth on the growth of juvenile sea scallops (*Placopecten magellanicus*) in suspended culture. Can. J. Fish. Aquat. Sci. 50(3):1857–1869.

Coutteau, P., Castell, J.D., Ackman, R.G. and Sorgeloos, P., 1996. The use of lipid emulsions as carriers for essential fatty acids in bivalves: A test case with juvenile *Placopecten magellanicus*. Journal of Shellfish Research 15:259–264.

Couturier, C.Y., 1986. Aspects of reproduction and larval production in *Placopecten magellanicus* held in a semi-natural environment. MSc Thesis. Dalhousie University, Halifax, Nova Scotia. 108 p.

Couturier, C.Y., 1988. Shellfish toxins aplenty. Bull. Aquacul. Assoc. Canada 88–2:11–25.

Couturier, C.Y., 1990. Scallop aquaculture in Canada: fact or fantasy. World Aquaculture 21(2):54–62.

Couturier, C.Y., 1991. Egg size and larval viability in sea scallops. Bull. Aquacul. Assoc. Canada 91–3:91–93.

Couturier, C.Y. and Aiken, D.E., 1989. Possible role of photoperiod in sea scallop reproduction. Bull. Aquacul. Assoc. Canada 89–3:65–67.

Couturier, C., Dabinett, P. and Lanteigne, M., 1995. Scallop culture in Atlantic Canada. In: A. Boghen (Ed.) Cold-water Aquaculture. The Tribune Press, Sackville, NB. pp. 297–340.

Couturier, C.Y. and Newkirk, G.F., 1987. Aspects of reproduction and larval production in *Placopecten magellanicus* held in a semi-natural environment. Proc. Ann. Meet. Aquacul. Assoc. Canada 1:26–27.

Couturier, C.Y. and Newkirk, G.F., 1991. Biochemical and gametogenic cycles in scallops, *Placopecten magellanicus* (Gmelin, 1791), held in suspension culture. In: S. E. Shumway and P.A. Sandifer (Eds.). An International Compendium of Scallop Biology and Culture. World Aquaculture Workshops, No. 1. The World Aquaculture Society, Baton Rouge, LA. pp. 107–117.

Cranford, P.J., 1995. Relationships between food quantity and quality and absorption efficiency in sea scallops *Placopecten magellanicus* (Gmelin). J. Exp. Mar. Biol. Ecol. 189:123–142.

Cranford, P.J. and Gordon, D.C., Jr., 1992. The influence of dilute clay suspensions on sea scallop (*Placopecten magellanicus*) feeding activity and tissue growth. Neth. J. Sea Res. 30:107–120.

Cranford, P.J. and Grant, J., 1990. Particle clearance and absorption of phytoplankton and detritus by the sea scallop *Placopecten magellanicus* (Gmelin). J. Exp. Mar. Biol. Ecol. 137:105–121.

Culliney, J.L., 1974. Larval development of the giant scallop *Placopecten magellanicus* (Gmelin). Biol. Bull. 147:321–332.

Dabinett, P.E., 1989. Hatchery production and grow-out of the giant scallop *Placopecten magellanicus*. Bull. Aquacul. Assoc. Canada 89–3:68–70.

Dabinett, P., Caines, J., Crocker, K., Levy, L. and Ryan, C., 1998. A comprehensive research and development plan for scallop aquaculture in Newfoundland: Technical Components. Final Report. Part A: Belleoram scallop hatchery and nursery. Canada/Newfoundland Agreement on Economic Renewal Aquaculture Component. St. John's, Newfoundland. 49 p.

Dabinett, P. and Couturier, C., 1994. Scallop culture in Newfoundland. Bull. Aquacult. Assoc. Can. 94–3:8–11.

Dadswell, M.J., 1989. Potential for giant scallop (*Placopecten magellanicus*) aquaculture in Atlantic Canada. Bull. Aquacul. Assoc. Canada 89:19–22.

Dadswell, M.J., 2001. A review of the status of sea scallop (*Placopecten magellanicus*) aquaculture in Atlantic Canada in the year. Aquacult. Assoc. Can. Spec. Pub. 4:72–75.

Dadswell, M.J., Chandler, R.A. and Parsons, G.J., 1987. Spat settlement and early growth of *Placopecten magellanicus* in Passamaquoddy Bay, Canada. Sixth International Pectinid Workshop, March 1987, Menai Bridge, Wales. 14 p.

Dadswell, M.J., Chandler, R.A. and Parsons, G.J., 1988. Spat settlement and early growth of *Placopecten magellanicus* in Passamaquoddy Bay, Canada. J. Shellfish Res. 7:153–154.

Dadswell, M.J. and Crawford-Kellock, P., 1989. Growth of giant scallop juveniles (*Placopecten magellanicus*) at four sites on the Atlantic coast of Nova Scotia. Nova Scotia Department of Fisheries ERDA Report 18: 56 p.

Dadswell, M.J. and Parsons, G.J., 1989. Potential for giant scallop (*Placopecten magellanicus*) aquaculture in Nova Scotia. Aquanotes, Aquaculture Assoc. Nova Scotia 11:17–21.

Dadswell, M.J. and Parsons, G.J., 1991. Potential for aquaculture of sea scallop, *Placopecten magellanicus* (Gmelin, 1791) in the Canadian Maritimes using naturally produced spat. In: S.E. Shumway and P.A. Sandifer (Eds.). An International Compendium of Scallop Biology and Culture. World Aquaculture Workshops, No. 1. The World Aquaculture Society, Baton Rouge, LA. pp. 300–307.

Dadswell, M.J. and Parsons, G.J., 1992a. Exploiting life-history characteristics of the sea scallop, *Placopecten magellanicus*, (Gmelin, 1791), from different geographic locations in the Canadian maritimes to enhance suspended culture grow-out. J. Shellfish Res. 299–305.

Dadswell, M.J. and Parsons, G.J., 1992b. Evaluation of intermediate culture techniques for the giant scallop, *Placopecten magellanicus*, in Passamaquoddy Bay, New Brunswick. Atlantic Fisheries Adjustment Program final report, Canada Department of Fisheries and Oceans, 28 p.

Dahl, E. and Yndestad, M., 1985. Diarrhetic shellfish poisoning (DSP) in Norway in the Autumn 1984 related to the occurrence of *Dinophysis* spp. In: D.M. Anderson, A.W. White and D.G. Baden (Eds.). Toxic Dinoflagellates. Elsevier, New York. pp. 495–500.

De Franssu, L., 1990. The world market for bivalves - oyster-mussel - clam - scallop. FAO Globefish Research Programme 4: 117 p.

Department of Fisheries and Oceans, 1989. Scallop market outlook, March 1989. Can. Economic and Commercial Analysis Report 12: 17 p.

Desbiens, M., Coulombe, F., Gaudreault, J., Cembella, A.D. and Laroucque, R., 1990. PSP toxicity of wild and cultured blue mussels induced by *Alexandrium exacavatum* in Gaspé Bay (Canada): implications for aquaculture. In: E. Graneli, D.M. Anderson, L. Edler and B. Sunström (Eds.). Toxin Marine Phytoplankton. Elsevier, New York. pp. 459–462.

Devaraj, M. and Parsons, G.J., 1997. Effect of fouling on current velocities in pearl nets of various mesh sizes. Bull. Aquacult. Assoc. Canada 97–2:72–74.

Dibacco, C., 1991. Considering a semi-annual reproductive cycle for the sea scallop (*Placopecten magellanicus*) on Georges Bank. J. Shellfish Res. 10: 271–272. (Abstract)

Dickie, L.M., 1958. Effects of high temperature on survival of the giant scallop. J. Fish. Res. Bd. Canada 15(6):1189–1211.

Dickie, L.M. and Medcof, J.C., 1963. Causes of mass mortalities of scallops (*Placopecten magellanicus*) in the South-western Gulf of St. Lawrence. J. Fish. Res. Bd. Canada 20(2):451–482.

Duggan, W.P., 1973. Growth and survival of the bay scallop, *Argopecten irradians*. Proc. Natl. Shellfisheries Assoc. 63:68–71.

Dupaul, W.D., Kirkley, J.E. and Schmitzer, A.C., 1989. Evidence of a semiannual reproductive cycle for the sea scallop *Placopecten magellanicus* (Gmelin, 1791), in the mid-Atlantic region. J. Shellfish Res. 8:173–178.

Dupouy, H., 1983. Bilan et perspectives de la pêche et de la culture des Pectinides (coquille Saint-Jacques et pétoncle) dans le Monde. La Pêche Maritime 1983:704–712. (Can. Transl. Fish. Aquat. Sci. 5417: 38 p., 1989).

Dupuy, J.L., Windsor, N.T. and Sutton, C.F., 1977. Manual for design and operation of an oyster hatchery. Spec. Rep. Appl. Mar. Sci. Ocean Eng. No. 142. Virginia Institute of Marine Science, Gloucester Point, VA. 104 p.

Elner, R. and Jamieson, G.S., 1979. Predation of sea scallops, *Placopecten magellanicus*, by the rock crab, *Cancer irroratus* and the American lobster, *Homarus americanus*. Can. J. Fish. Aquat. Sci. 36:537–543.

Emerson, C.W., Grant, J., Mallet, A. and Carver, C., 1994. Growth and survival of sea scallops *Placopecten magellanicus* - effect of culture depth. Mar. Ecol. Prog. Ser. 108:119–132.

Ford, E.S., 1997. An assessment of the financial feasibility of framing the sea scallop, *Placopecten magellanicus*, in Atlantic Canada. MBA project. University of New Brunswick (Saint John Campus library), Saint John, New Brunswick, Canada. 72 p.

Fournier, R. and Marsot, P., 1986. Écloserie expérimentale de larves du pétoncle géant (*Placopecten magellanicus*). Colloque sur l'aquiculture. Conseil des productions animales du Québec. Ministère de l'Agriculture, des Pêcheries et de l'Alimentation, Québec. pp. 163–166.

Fournier, R. and Marsot, P., 1987. La production artificielle du naissain de pétoncle géant. Proc. Ann. Meet. Aquacul. Assoc. Canada 1:22–23.

Frechette, M. and Daigle, G., 2002. Reduced growth of Iceland scallops *Chlamys islandica* (O.F. Muller) cultured near the bottom: A modelling study of alternative hypotheses. J. Shellfish Res. 21(1):87–91.

Frechette, M., Gaudet, M. and Vigneau, S., 2000. Estimating optimal population density for intermediate culture of scallops in spat collector bags. Aquaculture 183:105–124.

Frechette, M., Giguere, M. and Daigle, G. 2000. Etude de l'effet du site d'elevage et de la provvenance des specimens sur le potentiel aquicole du petoncle d'islande (*Chlamys islandica*) en Cote-Nord. Rapp. Can. Ind. Sci. Haul. Aquat. No. 258. 32 p.

Frenette, B., 2004. Environmental factors influencing the growth and survival of juvenile sea scallops, Placopecten magellanicus (Gmelin, 1791). MSc Thesis. Memorial University of Newfoundland, St. John's, Newfoundland, Canada. 142 p.

Frenette, B. and Parsons, G.J., 2001. Salinity-temperature tolerance of juvenile giant scallops, *Placopecten magellanicus*. Aquacult. Assoc. Can. Spec. Pub. 4:76–78.

Frenette, B., Parsons, G.J. and Davidson, L.A., 2001. Growth and survival of giant sea scallops (*Placopecten magellanicus*) in shallow water embayments of southeastern New Brunswick. Aquacult. Assoc. Can. Spec. Pub. 4:79–81.

Frenette, B., Parsons, G.J. and Davidson, L.A., 2002. Influence of salinity and temperature on clearance rate and oxygen consumption of juvenile sea scallops *Placopecten magellanicus* (Gmelin). Aquacult. Assoc. Can. Spec. Pub. 2002:17–19.

Freudenthal, A.R. and Jijina, J.L., 1988. Potential hazards of *Dinophysis* to consumers and shellfisheries. J. Shellfish Res. 7:695–701.

Frishman, Z., Nooman, A., Naidu, K.S. and Cahill, F.M., 1980. Farming scallops in Newfoundland Canada: A cost-benefit analysis. Third Scallop Workshop, Port Erin, Isle of Man. 28 p.

Gainey, L.F., Jr., and Shumway, S.E., 1988. A compendium of the responses of bivalve molluscs to toxic dinoflagellates. J. Shellfish Res. 7:623–628.

Gaudet, M., 1989. La pectiniculture au Québec: réalités et potentialités. Atelier sur l'Elevage du Pétoncle Géant, Gapsé (Québec). Ministère de l'Agriculture, des Pêcheries et de l'Alimentation, Québec. Cahier Spécial d'Information No. 12:7–8.

Gaudet, M., 1994. Intermediate culture strategies for sea scallop (*Placopecten magellanicus*) spat in Magdalen Islands, Quebec. Bull. Aquacul. Assoc. Canada 94–2, 22–28.

Gibbons, M.C. and Castagna, M., 1984. Serotonin as an inducer of spawning in six bivalve species. Aquaculture 40:189–191.

Giguère, M., Cliche, G. and Brulotte, S., 1995. Review of research done between 1986 and 1994 of wild scallop spat collection in the Iles-de-la-Madeleine. Can. Tech. Rep. Fish. Aquat. Sci. 2061:83 p.

Gilbert, E. and Cantin, C., 1987. Scallop culture in Quebec, description of the production cycle and financial analysis of a culture method. Fisheries and Oceans Canada. Economic Services Division. Cat. No. FS23-118/1978E. 50 p.

Gilgan, M.W., Burns, B.G. and Landry, G.J., 1990. Distribution and magnitude of domoic acid contamination of shellfish in Atlantic Canada during 1988. In: E. Graneli, D.M. Anderson, L. Edler and B. Sunström (Eds.). Toxin Marine Phytoplankton. Elsevier, New York. pp. 469–474.

Gillis, C.A. and Dabinett, P.E., 1989. A comparison of unialgal and mixed algal diets on growth of spat of *Placopecten magellanicus*. Bull. Aquacul. Assoc. Canada 89–3:77–79.

Goudey, C.A. and Smolowitz, R.J., 1996. Open-ocean culture of sea scallops off New England. Int. Conf. on Open Ocean Aquaculture, Portland, ME (USA), 8–10 May 1996 (Unhmp-cp-sg-96-9). pp. 179–192.

Grant, J. and Cranford, P.J., 1991. Carbon and nitrogen scope for growth as a function of diet in the sea scallop *Placopecten magellanicus*. J. Mar. Biol. Assoc. U.K. 71:437–450.

Grant, J., Emerson, C.W., Mallet, A. and Carver, C., 2003. Growth advantages of ear hanging compared to cage culture for sea scallops, *Placopecten magellanicus*. Aquaculture 217:301–323.

Grecian, L.A., Davidson, L.A. and Parsons, G.J., 2001a. Temporal monitoring of sea scallop, *Placopecten magellanicus*, spawning and spat settlement in the Northumberland Strait, New Brunswick. Aquacult. Assoc. Can. Spec. Pub. 4:82–84.

Grecian, L.A., Parsons, G.J., Dabinett, P. and Couturier, C., 2000. Influence of season, initial size, depth, gear type and stocking density on the growth rates and recovery of the sea scallop on a farm-based nursery. Aquaculture International 8:183–206.

Grecian, L.A., Parsons, G.J., Dabinett, P. and Couturier, C., 2001b. Toxicity of un-ionized ammonia to nursery-sized sea scallop, *Placopecten magellanicus*, spat. Bull. Aquacult. Assoc. Can. 4:85–88.

Grecian, L.A., Parsons, G.J., Dabinett, P. and Couturier, C., 2003. Effect of deployment date and environmental conditions on growth rate and retrieval of hatchery-reared sea scallops, *Placopecten magellanicus* (Gmelin, 1791), at a sea-based nursery. J. Shellfish Res. 22(1):101–109.

Gruffydd, Ll.D. and Beaumont, A.R., 1970. Determination of the optimum concentration of eggs and spermatozoa for the production of normal larvae in *Pecten maximus* (Mollusca, Lamellibranchia). Helgol. wiss. Meersunters. 20:486–497.

Gruffydd, Ll.D. and Beaumont, A.R., 1972. A method for rearing *Pecten maximus* larvae in the laboratory. Mar. Biol. 15:350–355.

Gulka, G. and Chang, P.W., 1985. Pathogenicity and infectivity of a rickettsia-like organism in the sea scallop, *Placopecten magellanicus*. J. Fish Diseases 8:309–318.

Harvey, M., Bourget, E. and Gagne, N., 1997. Spat settlement of the giant scallop, *Placopecten magellanicus* (Gmelin, 1791), and other bivalve species on artificial filamentous collectors coated with chitinous material. Aquaculture 148:277–298.

Hatcher, B.G., Scheibling, R.E., Barbeau, M.A., Hennigar, A.W., Taylor, L.H. and Windust, A.J., 1996. Dispersion and mortality of a population of sea scallop (*Placopecten magellanicus*) seeded in a tidal channel. Can. J. Fish. Aquat. Sci. 53:38–54.

Haya, K., Martin, J.L., Burridge, L.E., Waiwood, B.A. and Wildish, D.J., 1991. Domoic acid in shellfish and plankton from the Bay of Fundy, New Brunswick, Canada. J. Shellfish Res. 10:113–118.

Hodgson, C.A. and Bourne, N., 1988. Effect of temperature on larval development of the spiny scallop, *Chlamys hastata* Sowerby, with a note on metamorphosis. J. Shellfish Res. 7:349–357.

Hollett, J. and Dabinett, P.E., 1989. Effect of ration on growth and growth efficiency of spat of the giant scallop, *Placopecten magellanicus* (Gmelin). Bull. Aquacul. Assoc. Canada 89–3:71–73.

Hsu, C.P., Marchand, A. and Shimizu, Y., 1979. Paralytic shellfish toxins in the sea scallop, *Placopecten magellanicus*, in the Bay of Fundy. J. Fish. Res. Bd. Canada 36:32–36.

Imai, T., 1977. Aquaculture in shallow seas: Progress in shallow sea culture. Part II. The evolution of scallop culture. Translation from original Japanese. National Technical Information Service, Springfield VI. pp. 261–364.

Ito, H., 1990. Some aspects of offshore spat collection of Japanese scallop. In: A.K. Sparks (Ed.). Marine Farming and Enhancement. NOAA Tech. Rept. NMFS 85:35–48.

Jackson, D.L., 1992. Physiological ecology of *Placopecten magellanicus*: Food availability and larval growth response. MSc Thesis. Dalhousie University, Halifax, Nova Scotia. 73 p.

Jackson, D.L., MacDonald, B.W., Vercaemer, B., Guo, X., Mallet, A. and Kenchington, E.L., 2003. Investigations with triploid Atlantic sea scallops, *Placopecten magellanicus*, at the Bedford Institute of Oceanography, 2000–2003. Can. Tech. Rep. Fish. Aquat. Sci. no. 2460. 53 p.

Jamieson, G.S., 1979. Status and assessment of Northumberland Strait scallop stocks. Can. Fish. Mar. Serv. Tech. Rept. 904: 37 p.

Jamieson, G.S. and Chandler, R.A., 1983. Paralytic shellfish poison in sea scallops (*Placopecten magellanicus*) in the west Atlantic. Can. J. Fish. Aquat. Sci. 40:313–318.

Jamieson, G.S. and Lundy, M.J., 1979. Bay of Fundy scallop stock assessment - 1978. Fisheries and Marine Service Technical Report 915:vii + 14 p.

Jamieson, G.S., Stone, H. and Etter, M., 1982. Predation of sea scallops (*Placopecten magellanicus*) by lobsters (*Homarus americanus*) and rock crabs (*Cancer irroratus*) in underwater cage enclosures. Can. J. Fish. Aquat. Sci. 39:499–505.

Jones, R., Bates, J.A., Innes, D.J. and Thompson, R.J., 1996. Quantitative genetic analysis of growth in larval scallops (*Placopecten magellanicus*). Mar. Biol. 124:671–677.

Karney, R.C., 1996. Hatchery culture techniques for the giant sea scallop *Placopecten magellanicus*. J. Shellfish Res. 15(2):455–456. (Abstract)

Karney, R.C., Dutra, F.A., Dutra, D. and Dutra, J., 1996. Hatchery and field culture techniques for the giant sea scallop *Placopecten magellanicus*. J. Shellfish Res. 15(2):482. (Abstract)

Kean-Howie, J.C., O'dor, R.K. and Grant, J., 1989. The development of a microparticulate diet for nutrition physiology research with the sea scallop, *Placopecten magellanicus*. Bull. Aquacul. Assoc. Canada 89–3:74–76.

938

Kean-Howie, J.C., O'dor, R.K., Grant, J. and Castell, J.D., 1991. The effects of current velocity and food concentration on the ingestion of a microparticulate diet by juvenile sea scallops (*Placopecten magellanicus*). In: S.E. Shumway and P.A. Sandifer (Eds.). An International Compendium of Scallop Biology and Culture. World Aquaculture Workshops, No. 1. The World Aquaculture Society, Baton Rouge, LA. pp. 347–352.

Kleinman, S., Hatcher, B., Scheibling, R., Taylor, L. and Hennigar, A., 1996. Shell and tissue growth of juvenile sea scallops (*Placopecten magellanicus*) in suspended and bottom culture in Lunenburg Bay, Nova Scotia. Aquaculture 142:75–97.

Langan, R., Kuenstner, S., Parsons, G.J., Shumway, S.E. and Simonitsch, M., 1997. Sea scallop enhancement and culture in New England. J. Shellfish Res. 16(1):355. (Abstract)

Lanteigne, M., Davidson, L.-A. and Andrews, J., 1991. Collecting juvenile sea scallops (*Placopecten magellanicus*) with artificial collectors, in Port au Port Bay, Newfoundland (Canada). J. Shellfish Res. 10:297. (Abstract)

Larsen, P.F. and Lee, R.M., 1978. Observations on the abundance, distribution and growth of postlarval sea scallops, *Placopecten magellanicus*, on Georges Bank. Nautilus 92:112–116.

Lesser, M.P., Shumway, S.E., Cucci, T., Barter, J. and Edwards, J., 1991. Size specific selection of phytoplankton by juvenile filter-feeding bivalves: Comparison of the sea scallop *Placopecten magellanicus* (Gmelin, 1791) with *Mya arenaria* Linnaeus, 1758 and *Mytilus edulis* Linnaeus, 1758. In: S.E. Shumway and P.A Sandifer (Eds.). An International Compendium of Scallop Biology and Culture. World Aquaculture Workshops, No. 1. The World Aquaculture Society, Baton Rouge, LA. pp. 341–346.

Levy, L.A., 1999. Growth rates and recovery of hatchery-reared sea scallop, *Placopecten magellanicus* (Gmelin, 1791) spat under a variety of nursery conditions. MSc Thesis. Memorial University of Newfoundland, St. John's, Newfoundland. 209 p.

Lines, R., 1988. Markets for Norwegian scallops. Fiskeriteknologisk Forskninst. Rapp. A48:107 p. (Can. Transl. Fish. Aquat. Sci. 5440: 135 p., 1989).

Loosanoff, V.L. and Davis, H.C., 1963. Rearing of bivalve mollusks. Adv. Mar. Biol. 1:1–136.

MacDonald, B.A., 1986. Production and resource partitioning in the giant scallop *Placopecten magellanicus* grown on the bottom and in suspended culture. Mar. Ecol. Prog. Ser. 34:79–86.

MacDonald, B.A. and Thompson, R., 1985. Influence of temperature and food availability on the ecological energetics of the giant scallop *Placopecten magellanicus*. I. Growth rates of shell and somatic tissue. Mar. Ecol. Prog. Ser. 25:279–294.

MacDonald, B. and Thompson, R., 1985b. Influence of temperature and food availability on the ecological energetics of the giant scallop *Placopecten magellanicus*. II. Reproductive output and total production. Mar. Ecol. Prog. Ser. 25:295–303.

MacDonald, B.A. and Thompson, R.J., 1986. Influence of temperature and food availability in the ecological energetics of the giant scallop *Placopecten magellanicus*. III. Physiological ecology, gametogenic cycle, and scope for growth. Mar. Biol. 93:36–48.

MacDonald, B.A. and Thompson, R., 1988. Intraspecific variation in growth and reproduction in latitudinally differentiated populations of the giant scallop *Placopecten magellanicus* (Gmelin). Biol. Bull. 175: 361–371.

MacDonald, B. and Ward, J., 1994. Variation in food quality and particle selectivity in the sea scallop *Placopecten magellanicus* (Mollusca:Bivalvia). Mar. Ecol. Prog. Ser. 108:251–264.

Mallet, A.L., 1989. Larval growth, larval mortality and metamorphosis success of the giant scallop, *Placopecten magellanicus*. Atelier sur l'Elevage du Pétoncle Géant, Gapsé (Québec). Ministère de l'Agriculture, des Pêcheries et de l'Alimentation, Québec. Cahier Spécial d'Information No. 12:49–51.

Manning, D.A.-M., 1985. Aquaculture studies of the giant scallop *Placopecten magellanicus* (Gmelin): conditioning of broodstock and energy requirements of the larvae and juveniles. MSc Thesis. Memorial University of Newfoundland, St. John's, Newfoundland. 201 p.

Maranda, L., Wang, R., Masuda, K. and Shimizu, Y., 1990. Investigation of the source of domoic acid in mussels. In: E. Graneli, D.M. Anderson, L. Edler and B. Sunström (Eds.). Toxin Marine Phytoplankton. Elsevier, New York. pp. 300–304.

Marshall, C.T. and Lee, K., 1991. Uptake of dissolved glycine by sea scallop (*Placopecten magellanicus* (Gmelin, 1791)) larvae. In: S.E. Shumway and P.A. Sandifer (Eds.). An International Compendium of Scallop Biology and Culture. World Aquaculture Workshops, No. 1. The World Aquaculture Society, Baton Rouge, LA. pp. 60–66.

Martin, J.L., Haya, K., Burridge, L.E. and Wildish, D.J., 1990. *Nitzschia pseudodelicatissima* - a source of domoic acid in the Bay of Fundy, eastern Canada. Mar. Ecol. Prog. Ser. 67:177–182.

Martin, J.L. and White, A.W., 1988. Distribution and abundance of the toxic dinoflagellate *Gonyaulax excavata* in the Bay of Fundy. Can. J. Fish. Aquat. Sci. 45:1968–1975.

Martinez, G., Olivares, A.Z. and Mettifogo, L., 2000. In vitro effects of monoamines and prostaglandins on meiosis reinitiation and oocyte release in *Argopecten purpuratus* Lamarck. Invertebr. Reprod. Dev. 38:61–69.

Matsutani, T. and Nomura, T., 1982. Induction of spawning by serotonin in the bay scallop *Patinopecten yessoensis* (Jay). Mar. Biol. Lett. 3:353–358.

McGladdery, S., 1990. Shellfish parasites and diseases on the east coast of Canada. Bull. Aquacul. Assoc. Canada 90–3:14–18.

Medcof, J.C., 1985. Life and death with *Gonyaulax*: an historical perspective. In: D.M Anderson, A.W. White and D.G. Baden (Eds.). Toxic Dinoflagellates. Elsevier, New York. pp. 1–8.

Melvin, G.D., Dadswell, M.J. and Chandler, R.A., 1985. Movement of scallops *Placopecten magellanicus* (Gmelin, 1791) (Mollusca: Pectinidae) on Georges Bank. Can. Dept. Fish. Oceans CAFSAC Res. Doc. 85/30.

Milke, L.M., Bricelj, V.M. and Parrish, C.C., 2004. Growth of postlarval sea scallops, *Placopecten magellanicus*, on microalgal diets, with emphasis on the nutritional role of lipids and fatty acids. 234:293–317.

Miron, G., Pelletier, P. and Bourget, E., 1995. Optimizing the design of giant scallop (*Placopecten magellanicus*) spat collectors: Flume experiments. Marine Biology 123:285–291.

Miyake, H., Matsuoka, M. and Furuya, K., 1995. Loss of and damage to scallops due to storms in the Sea of Okhotsk. Fish. Oceanogr. 4:293–302.

Mottet, M.G., 1979. A review of the fishery biology and culture of scallops. Washington Dept. Fish. Tech. Rep. 39:100 p.

Nadeau, M. and Cliche, G., 1997. Scallop (*Placopecten magellanicus*) seeding trials in Iles-de-la-Madeleine, Quebec, 1993 to 1996. Bull. Aquacult. Assoc. Can. 97–2:69–71.

Nadeau, M. and Cliche, G., 1998. Evaluation of the recapture rate of seeded scallops (*Placopecten magellanicus*) during commercial fishing activity in Iles-de-la-Madeleine, Québec. Bull. Aquacult. Assoc. Can. 98–2:79–81.

Naidu, K.S., 1970. Reproduction and breeding cycle of the giant scallop *Placopecten magellanicus* (Gmelin) in Port au Port Bay, Newfoundland. Can. J. Zool. 48:1003–1012.

Naidu, K.S., 1978. Culture of the sea scallop, *Placopecten magellanicus* (Gmelin). Proc. Natl. Shellfish. Assoc. 68:83–84. (Abstract)

Naidu, K.S., 1979. Preliminary observations on the effect of substrate colour on larval settlement in the sea scallop, *Placopecten magellanicus* (Gmelin). Proc. Natl. Shellfish. Assoc. 70: 128.

Naidu, K.S., 1991. Sea scallop, *Placopecten magellanicus*. In: S.E. Shumway (Ed.). Scallops: Biology, Ecology and Aquaculture. Developments in Aquaculture and Fisheries Science, Vol. 21. Elsevier Science Publishing Co. Inc., New York, N.Y. pp. 861–897.

Naidu, K.S. and Botta, J.R., 1978. Taste panel assessment and proximate composition of cultured and wild sea scallops, *Placopecten magellanicus* (Gmelin). Aquaculture 15:243–247.

Naidu, K.S. and Cahill, F.M., 1978. Scallop culture in Newfoundland: A progress report for 1977–1978. Report to the Department of Fisheries, Government of Newfoundland and Labrador. 32 p. (Unpublished)

Naidu, K.S. and Cahill, F.M., 1981. Scallop culture report, 1980–81. Report to the Department of Fisheries, Government of Newfoundland and Labrador. 46 p. (Unpublished)

Naidu, K.S. and Cahill, F.M., 1986. Culturing giant scallops in Newfoundland waters. Can. MS Rept. Fish. Aquat. Sci. 1876:iv + 23 p.

Naidu, K.S., Cahill, F.M. and Lewis, D.B., 1981. Relative efficacy of two artificial substrates in collection of sea scallop (*Placopecten magellanicus*) spat. J. World Mariculture Soc. 12:165–171.

Naidu, K.S., Fournier, R., Marsot, P. and Worms, J., 1989. Culture of the sea scallop, *Placopecten magellanicus*: Opportunities and constraints. In: A.D. Boghen (Ed.). Cold-water Aquaculture in Atlantic Canada. The Canadian Institute for Research on Regional Development, Université de Moncton, Moncton, New Brunswick, Canada. pp. 210–239.

Naidu, K.S. and Scaplen, R., 1979. Settlement and survival of giant scallop, *Placopecten magellanicus*, larvae on enclosed polyethylene film collectors. In: T.V.R. Pillay and W.A. Dill (Eds.). Advances in Aquaculture. Fishing News (Books) Ltd., England. pp. 379–381.

Napolitano, G.E., MacDonald, B.A., Thompson, R.J. and Ackman, R.G., 1992. Lipid composition of eggs and adductor muscle in giant scallops (*Placopecten magellanicus*) from different habitats. Mar. Biol. 113:71–76.

Neima, P.G., 1997. Report on commercial scallop hatchery design. Can. Tech. Rep. Fish. Aquat. Sci. 1997:62 p.

Nelson, D.A., Calabrese, A. and McInnes, J.R., 1977. Mercury stress on juvenile bay scallops, *Argopecten irradians*, under various salinity-temperature regimes. Mar. Biol. 43:293–297.

Niles, M.M. and Davidson, L.A., 2002. The use of fuzzy rope as sea scallop spat collectors. Aquacult. Assoc. Can. Spec. Pub. 5:20–22.

Parrish, C.C., Wells, J.S., Yang, Z. and Dabinett, P., 1999. Growth and lipid composition of scallop juveniles, *Placopecten magellanicus*, fed the flagellate *Isochrysis galbana* with varying lipid composition and the diatom *Chaetoceros muelleri*. Mar. Biol. 133:461–471.

Parsons, G.J., 1989. Spat settlement and growth of the giant scallop, *Placopecten magellanicus,* in Passamaquoddy Bay, New Brunswick. Atelier sur l'Elevage du Pétoncle Géant, Gapsé (Québec). Ministère de l'Agriculture, des Pêcheries et de l'Alimentation, Québec. Cahier Spécial d'Information No. 12:71.

Parsons, G.J., 1994. Reproduction and recruitment of the giant scallop *Placopecten magellanicus* and its relationship to environmental variables. PhD Thesis. University of Guelph, Guelph, Ontario. 196 p.

Parsons, G.J. and Dadswell, M.J., 1991. Stocking density effects on growth and survival of the giant scallop, *Placopecten magellanicus*, held in intermediate suspension culture. Bull. Aquacul. Assoc. Canada 91-3:94–96.

Parsons, G.J. and Dadswell, M.J., 1992. Effect of stocking density on growth, production, and survival of the giant scallop, *Placopecten magellanicus*, held in intermediate suspension culture in Passamaquoddy Bay, New Brunswick. Aquaculture 103:291–309.

Parsons, G.J. and Dadswell, M.J., 1994. Evaluation of intermediate culture techniques, growth, and survival of the giant scallop, *Placopecten magellanicus*, in Passamaquoddy Bay, New Brunswick. Can. Tech. Rep. Fish. Aquat. Sci. 2012:vii + 29 p.

Parsons, G.J., Dadswell, M.J. and Roff, J.C., 1990. Influence of environmental factors on the maximization of spat settlement in the giant scallop, *Placopecten magellanicus*. J. Shellfish Res. 8: 458. (Abstract)

Parsons, G.J., Dadswell, M.J. and Roff, J.C., 1993a. Effect of biofilm on settlement of sea scallop, *Placopecten magellanicus* in Passamaquoddy Bay, New Brunswick, Canada. J. Shellfish Res. 12(2):279–283.

Parsons, G.J., Robinson, S.M.C., Chandler, R.A., Davidson, L.A., Lanteigne, M. and Dadswell, M.J., 1992a. Intra-annual and long-term patterns in the reproductive cycle of the giant scallop, *Placopecten magellanicus*, (Bivalvia: Pectinidae) from Passamaquoddy Bay, New Brunswick, Canada. Mar. Ecol. Prog. Ser. 80:203–214.

Parsons, G.J., Robinson, S.M.C. and Martin, J.D., 1994. Enhancement of a giant scallop bed by spat from a scallop aquaculture site. Bull. Aquacul. Assoc. Can. 94-3:21–23.

Parsons, G.J., Robinson, S.M.C., Martin, J.D. and Chandler, R.A., 1996. Comparative collection of scallop spat using different types of artificial collectors. Bull. Aquacul. Assoc. Canada 96-1:70–72.

Parsons, G.J., Robinson, S.M.C., Roff, J.C. and Dadswell, M.J., 1993b. Daily growth rates as indicated by valve ridges in postlarval giant scallop *Placopecten magellanicus* (Bivalvia: Pectinidae). Can. J. Fish. Aquat. Sci 50(3):456–464.

Parsons, G.J., Shumway, S.E., Kuenstner, S. and Gryska, A., 2002. Polyculture of sea scallops (*Placopecten magellanicus*) suspended from salmon cages. Aquaculture Int. 10:65–77.

Parsons, G.J., Struthers, A. and Yetman, G., 1998. A comprehensive research and development plan for scallop aquaculture in Newfoundland: Technical Components. Final Report. Part B: Intermediate and final growout research and development and equipment development research and development. Canada/Newfoundland Agreement on Economic Renewal Aquaculture Component. St. John's, Newfoundland. 18 p.

Parsons, G.J., Warren-Perry, C.R. and Dadswell, M.J., 1992b. Movements of juvenile sea scallops *Placopecten magellanicus* (Bivalvia: Pectinidae) in Passamaquoddy Bay, New Brunswick. J. Shellfish Res. 11(2):295–297.

Paul, J.D., Brand, A.R. and Hoogesteger, J.N., 1981. Experimental cultivation of the scallops *Chlamys opercularis* (L.) and *Pecten maximus* (L.) using naturally produced spat. Aquaculture 24:31–44.

942

Pearce, C.M. and Bourget, E., 1996. Settlement of larvae of the giant scallop, *Placopecten magellanicus* (Gmelin), on various artificial and natural substrata under hatchery-type conditions. Aquaculture 141:201–221.

Penney, R., 1995. Effect of gear type and initial stocking density on production of meats and large whole scallops (*Placopecten magellanicus*) using suspension culture in Newfoundland. Can. Tech. Rep. Fish. Aquat. Sci. 2079.

Penney, R. and McKenzie, C., 1996. Seasonal changes in the body organs of cultured sea scallop, *Placopecten magellanicus*, and coincidence of spawning with water temperature, seston, and phytoplankton community dynamics. Can. Tech. Rep. Fish. Aquat. Sci. 2104: 26 p.

Penney, R.W. and Mills, T.J., 1996. Effect of spat grading and net mesh size on the growth and survival of juvenile cultured sea scallop, *Placopecten magellanicus*, in Newfoundland. Bull. Aquacult. Assoc. Canada 96–3:80–83.

Penney, R.W. and Mills, T.J., 2000. Bioeconomic analysis of a sea scallop, *Placopecten magellanicus*, aquaculture production system in Newfoundland, Canada. J. Shellfish Res. 19:113–124.

Pernet, F. and Tremblay, R., 2004. Effect of varying levels of dietary essential fatty acid during early ontogeny of the sea scallop *Placopecten magellanicus*. J. Exp. Mar. Biol. Ecol. 310:73–86.

Pernet, F., Tremblay, R. and Bourget, E., 2003. Biochemical indicator of sea scallop (*Placopecten magellanicus*) quality based on lipid class composition. Part II: Larval growth, competency and settlement. J. Shellfish Res. 22(2):377–388.

Pernet, F., Tremblay, R. and Bourget, E., 2003. Biochemical indicator of sea scallop (*Placopecten magellanicus*) quality based on lipid class composition. Part I: Broodstock conditioning and young larval performance. J. Shellfish Res. 22(2):365–375.

Pouliot, F., Bourget, E. and Frechette, M., 1995. Optimizing the design of giant scallop (*Placopecten magellanicus*) spat collectors: Field experiments. Mar. Biol. 123:277–284.

Prakash, A., Medcof, J.C. and Tenant, A.D., 1971. Paralytic shellfish poisoning in eastern Canada. Bull. Fish. Res. Board Can. No. 177:vii + 87 p.

Robinson, S.M.C., 1993. The potential for aquaculture technology to enhance wild scallop production in Atlantic Canada. World Aquacult. 24:61–67.

Robinson, S.M.C., Haya, K., Martin, J.L., Martin, J.D. and LeGresley, M., 1999. Spatial distributions of PSP within the Quoddy Region of the Bay of Fundy, measured in the giant scallop, *Placopecten magellanicus*. Can. Tech. Rep. Fish. Aquat. Sci. 2261:87. (Abstract)

Robinson, S.M.C., Martin, J.D., Chandler, R.A. and Parsons, G.J., 1991. Spatial patterns of spat settlement in the sea scallop, *Placopecten magellanicus*, compared to hydrographic conditions in Passamaquoddy Bay, New Brunswick, Canada. J. Shellfish Res. 10:272–273. (Abstract)

Robinson, S.M.C., Martin, J.D., Chandler, R.A. and Parsons, G.J., 1999. An examination of the linkage between the early life history processes of the sea scallop and local hydrographic characteristics. J. Shellfish Res. 18(1):314. (Abstract)

Robinson, S.M.C., Martin, J.D., Chandler, R.A., Parsons, G.J. and Couturier C., 1992. Spatial patterns of spat settlement in the sea scallop, *Placopecten magellanicus*, compared to hydrographic conditions in Passamaquoddy Bay, New Brunswick, Canada. C.A.F.S.A.C. Res. Doc. 92/115: 26 p.

Robinson, W.E., Wehling, W.E., Morse, M.P. and McLeod, G.C., 1981. Seasonal changes in soft-body component indices and energy reserves in the Atlantic deep-sea scallop, *Placopecten magellanicus*. Fish. Bull. 79:449–458.

Roussy, M., Tremblay, R. and McGladdery, S., 2001. Scallop (*Placopecten magellanicus*) mortality: Challenge test 1. Aquacult. Assoc. Can. Spec. Pub. 4:122–124.

Ryan, C.M., 2000. Effect of algal cell density, dietary composition, growth phase and macronutrient concentration on growth and survival of giant scallop *Placopecten magellanicus* (Gmelin, 1791) larvae and spat in a commercial hatchery. MSc Thesis. Memorial University

Ryan, C.M., Parsons, J. and Dabinett, P., 1998. Effect of algal harvest phase on larval and post-larval growth of giant scallops (*Placopecten magellanicus*) in a commercial hatchery. Bull. Aquacul. Assoc. Canada 98–2:65–67.

Scheibling, R., Hatcher, B., Taylor, L. and Barbeau, M., 1991. Seeding trials of the giant scallop (*Placopecten magellanicus*) in Nova Scotia. In: J. Barret, J. Dao and P. Lubet (Eds.). Fisheries, Biology, and Aquaculture of Pectinids. 8th International Pectinid Workshop. Cherbourg, France, 22–29 Mai 1991. Institut Francais de Recherche Pour le Mer, DCOM/SE. Plouzane, France. pp. 123–130.

Schmitzer, A.C., Dupaul, W.D. and Kirkley, J.E., 1991. Gametogenic cycle of sea scallops (*Placopecten magellanicus* (Gmelin, 1791)) in the mid-Atlantic Bight. J. Shellfish Res. 10:221–228.

Shumway, S.E. (Ed.), 1988. Toxic algal blooms: hazards to shellfish industry. J. Shellfish Res. 7:587–705.

Shumway, S.E., 1990. A review of the effects of algal blooms on shellfish and aquaculture. J. World Aquacult. Soc. 21:65–104.

Shumway, S.E. and Cembella A.D., 1992. Toxic algal blooms: Potential hazards to scallop culture and fisheries. Bull. Aquacult. Assoc. Can. 92–4:59–68

Shumway, S.E., Cucci, T.L., Newell, R.C. and Yentsch, C.M., 1985. Particle selection, ingestion, and absorption in filter-feeding bivalves. J. Exp. Mar. Biol. Ecol. 91:77–92.

Shumway, S.E., Selvin, R. and Schick, D.F., 1987. Food resources related to habitat in the scallop *Placopecten magellanicus* (Gmelin, 1791): A quantitative study. J. Shellfish Res. 6:89–95.

Shumway, S.E., Sherman-Caswell, S. and Hurst, J.W., 1988. Paralytic shellfish poisoning in Maine: monitoring a monster. J. Shellfish Res. 7:643–652.

Stokesbury, K.D.E. and Himmelman, J.H., 1995. Biological and physical variables associated with aggregations of the giant scallop *Placopecten magellanicus*. Can. J. Fish. Aquat. Sci. 52:743–753.

Stokesbury, K.D.E. and Himmelman, J.H., 1996. Experimental examination of movement of the giant scallop, *Placopecten magellanicus*. Mar. Biol. 124:651–660.

Strickland, J. and Dabinett, P., 1993. Rates of feeding, oxygen consumption and ammonia excretion as a function of size for hatchery reared spat of the sea scallop *Placopecten magellanicus*. Bull. Aquacult. Assoc. Can. 93–4:125–127.

Subba Rao, D.V., Quilliam, M.A. and Pocklington, R., 1988. Domoic acid - a neurotoxic amino acid produced by the marine diatom *Nitzschia pungens* in culture. Can. J. Fish. Aquat. Sci. 45:2076–2079.

Taguchi, K., 1976. Japanese scallop culture techniques boost yield. Austral. Fish. 35:20–23.

Taguchi, K., 1978. A manual of scallop culture methodology and management. Fish. Mar. Serv. Can. Transl. Ser. 4198: 416 p.

Tetu, C. and Davidson, L.A., 2002. Recruitment of predators in sea scallop, *Placopecten magellanicus*, collectors in southeastern Northumberland Strait 1999–2000. Aquaculture Association of Canada Special Publication 5:23–26.

Thomas, B., Giguere, M. and Brulotte, S., 2002. Success of natural capture of giant scallop (*Placopecten magellanicus*) on Gaspe peninsula (Quebec). Aquacul. Assoc. Canada Spec. Publ. 5:13–16.

Tremblay, M.J. (Ed.), 1988. A summary of the proceedings of the Halifax sea scallop workshop, August 13–14, 1987. Can. Tech. Rep. Fish. Aquat. Sci. 1605: 12 p.

Ventilla, R.F., 1982. The scallop industry in Japan. Adv. Mar. Biol. 20:309–382.

Wallace, J.C. and Reinsnes, T.G., 1984. Growth variation with age and water depth in the Iceland scallop (*Chlamys islandica*, Pectinidae). Aquaculture 41:141–146.

White, A.W., 1988. PSP: poison for Fundy shellfish culture. World Aquacult. 19:23–26.

Widman, J.C. and Rhodes, E.W., 1991. Nursery culture of the bay scallop, *Argopecten irradians irradians*, in suspended mesh nets. Aquaculture 99:257–267.

Wildish, D.J. and Saulnier, A.M., 1992. The effects of velocity and flow direction on growth of juvenile and adult giant scallops. J. Exp. Mar. Biol. Ecol. 133:133–143.

Wildish, D.J., Wilson, A.J., Young-Lai, W., Decoste, A.M., Aiken, D.E. and Martin, J.D., 1988. Biological and economic feasibility of four grow-out methods for the culture of giant scallops in the Bay of Fundy. Can. Tech. Rept. Fish. Aquat. Sci. 1658:iii + 21 p.

Wong, M.C. and Barbeau, M.A., 2003. Effects of substrate on interactions between juvenile sea scallops (*Placopecten magellanicus* Gmelin) and predatory sea stars (*Asterias vulgaris* Verrill) and rock crabs (*Cancer irroratus* Say). J. Exp. Mar. Biol. Ecol. 287:155–178.

Wong, M.C., Davidson, L.A., Grecian, L.A. and Barbeau, M.A., 2001. Predation on seeded juvenile sea scallops (*Placopecten magellanicus*) in the Northumberland Strait. Bull. Aquacult. Assoc. Can. 2001:93–96.

Yasumoto, T., Murata, M., Oshima, Y., Matsumoto, G.K. and Clardy, J., 1984. Diarrhetic shellfish poisoning. In: E.P. Ragelis (Ed.). Seafood Toxins. American Chemical Society, Washington, D.C. pp. 207–214.

Young-Lai, W.W. and Aiken D.E., 1986. Biology and culture of the giant scallop, *Placopecten magellanicus*: a review. Can. Tech. Rep. Fish. Aquat. Sci. 1478:iv + 21 p.

AUTHORS ADDRESSES

G. Jay Parsons - Marine Institute, Memorial University, P. O. Box 4920, St. John's, Newfoundland, Canada A1C 5R3 [Current Address: Fisheries and Oceans Canada, Aquaculture Science Branch, 200 Kent St., Ottawa, Ontario, Canada K1A 0E6] (E-mail: ParsonsJa@dfo-mpo.gc.ca)

Shawn M. C. Robinson - Fisheries and Oceans Canada, 531 Brandy Cove Road, Biological Station, St. Andrews, New Brunswick, Canada E5B 2L9 (E-mail: RobinsonSM@mar.dfo-mpo.gc.ca)

Chapter 17

Bay Scallop and Calico Scallop Fisheries, Culture and Enhancement in Eastern North America

Norman J. Blake and Sandra E. Shumway

17.1 INTRODUCTION

Scallops are commercially important shellfish worldwide (Table 17.1) with the average total catch between 1988 and 1997 at 522,894 pounds of meats (FAO 1997). Five species (sea scallop, *Placopecten magellanicus*; Icelandic scallop, *Chlamys islandica*, calico scallop, *Argopecten gibbus*; bay scallop, *Argopecten irradians*; weathervane scallop, *Patinopecten caurinus*) contribute to the major wild fisheries in North America with minor fisheries for two other species (pink scallop, *Chlamys rubida* and spiny scallop, *Chlamys hastata*).

Aquaculture and enhancement efforts are still limited activities in North America but world aquaculture of all scallops exceeds 1.7 million metric tonnes (O'Bannon 1999). Where scallop aquaculture activities do occur in North America, they contribute substantially to the local economies. Further, production from domestic activities (fisheries, aquaculture and enhancement) does not totally meet supply requirements and scallops are regularly imported from other countries and comprise about 40–80% of the total United States supply (Table 17.2).

This chapter is intended to present an overview of calico and bay scallop fisheries, aquaculture and enhancement efforts in eastern North America. Sea scallops are covered in Chapters 15 and 16 and west coast species are covered in chapter 18.

17.2 FISHERIES

17.2.1 Bay Scallop, *Argopecten irradians*

17.2.1.1 Distribution

The bay scallop, *Argopecten irradians* (Lamarck), occurs in the shallow (<10 m), protected estuarine habitats along the east coast of the United States (Fig. 17.1) from Cape Cod, Massachusetts to Texas (Clarke 1965). Three subspecies have been described based upon shell morphometrics (Waller 1969). The northern subspecies, *Argopecten irradians irradians*, extends from Cape Cod to approximately New Jersey and Maryland where it may hybridise with the southern bay scallop, *Argopecten irradians concentricus* (Clarke 1965). Populations of the southern bay scallop of sufficient size to harvest become discontinuous about North Carolina. It is absent along the Georgia coast (Walker et al.

Table 17.1

Nominal landings (MT) of major scallop species and % contribution to global production by year (Source FAO).

Species	1988	1989	1990	1991	1992	1993	1994	1995
x *Argopecten gibbus* (Calico scallop)	121,720 14%	67,330 8%	11,220 1%	0 0%	0 0%	0 0%	74,325 4%	10,003 1%
x *Argopecten irradians* (Atlantic bay scallop)	2,329 0%	1,360 0%	2,596 0%	2,062 0%	1,564 0%	2,670 0%	524 0%	1,593 0%
Argopecten purpuratus (Peruvian calico scallop)	7,828 1%	4,012 0%	2,212 0%	2,920 0%	7,758 1%	7,715 1%	12,073 1%	11,808 1%
Argopecten circularis (Pacific calico scallop)	9,220 1%	12,520 1%	29,501 3%	19,438 2%	5,350 0%	5,943 0%	8,611 1%	1,270 0%
Chlamys islandica (Iceland scallop)	10,059 1%	10,772 1%	12,117 1%	11,107 1%	21,269 2%	16,798 1%	17,230 1%	23,570 1%
Chlamys opercularis (Queen scallop)	15,613 2%	14,533 2%	16,832 2%	14,324 2%	16,089 2%	15,325 1%	10,576 1%	8,086 0%
Patinopecten yessoensis (Japanese scallop)	466,530 53%	502,163 59%	570,939 63%	558,683 64%	742,792 69%	1,196,713 81%	1,300,314 79%	1,423,869 85%
x *Patinopecten caurinus* (weathervane scallop)	961 0%	1,398 0%	2,415 0%	3,649 0%	6,884 1%	6,224 0%	4,944 0%	1,950 0%
Pecten maximus (Great Atlantic scallop)	15,962 2%	12,656 1%	12,517 1%	15,159 2%	21,391 2%	20,598 1%	25,158 2%	23,657 1%
Pecten jacobeus (Great Mediterranean scallop)	4 0%	1 0%	1 0%	0.5 0%	0.5 0%	202 0%	308 0%	23 0%
Pecten novaezelandiae (New Zealand scallop)	5,784 1%	4,264 0%	4,504 0%	8,256 1%	10,040 1%	8,928 1%	9,088 1%	14,160 1%
x *Placopecten magellanicus* (Sea scallop)	193,487 22%	206,898 24%	217,153 24%	211,132 24%	192,822 18%	143,921 10%	146,032 9%	121,230 7%
other Pectinidae	32,863 4%	18,117 2%	24,041 3%	28,949 3%	46,014 4%	53,577 4%	46,230 3%	38,843 2%
World Totals	882,360	856,024	906,048	875,679	1,071,973	1,478,614	1,655,413	1,680,062

x Species fished commercially in the United States.

Table 17.1 continued

Species	1996	1997	1998	1999	2000	2001	Average
x *Argopecten gibbus* (Calico scallop)	0 / 0%	0 / 0%	0 / 0%	0 / 0%	0 / 0%	0 / 0%	20,328 / 2%
x *Argopecten irradians* (Atlantic bay scallop)	230 / 0%	452 / 0%	690 / 0%	216 / 0%	154 / 0%	30 / 0%	1,176 / 0%
Argopecten purpuratus (Peruvian calico scallop)	12,275 / 1%	15,806 / 1%	41,041 / 3%	52,394 / 3%	33,125 / 2%	28,991 / 2%	17,140 / 1%
Argopecten circularis (Pacific calico scallop)	17,300 / 1%	2,330 / 0%	2,729 / 0%	1,864 / 0%	6,287 / 0%	3,227 / 0%	8,971 / 1%
Chlamys islandica (Iceland scallop)	22,752 / 1%	24,673 / 1%	18,946 / 1%	12,018 / 1%	13,471 / 1%	9,469 / 1%	16,018 / 1%
Chlamys opercularis (Queen scallop)	7,234 / 0%	11,489 / 1%	14,633 / 1%	15,860 / 1%	14,949 / 1%	20,511 / 1%	14,004 / 1%
Patinopecten yessoensis (Japanese scallop)	1,541,660 / 87%	1,523,784 / 86%	1,150,430 / 82%	1,234,235 / 81%	1,442,970 / 81%	1,489,403 / 79%	1,081,749 / 75%
x *Patinopecten caurinus* (weathervane scallop)	2,372 / 0%	2 / 0%	3,228 / 0%	2,642 / 0%	2,012 / 0%	1,052 / 0%	2,838 / 0%
Pecten maximus (Great Atlantic scallop)	31,955 / 2%	36,072 / 2%	35,908 / 3%	36,626 / 2%	37,851 / 2%	30,809 / 2%	25,451 / 2%
Pecten jacobeus (Great Mediterranean scallop)	52 / 0%	95 / 0%	50 / 0%	68 / 0%	570 / 0%	150 / 0%	109 / 0%
Pecten novaezelandiae (New Zealand scallop)	5,080 / 0%	18,848 / 1%	4,592 / 0%	6,152 / 0%	2,912 / 0%	6,792 / 0%	7,814 / 1%
x *Placopecten magellanicus* (Sea scallop)	109,382 / 6%	102,028 / 6%	99,432 / 7%	131,962 / 9%	196,993 / 11%	254,196 / 14%	166,191 / 183%
other Pectinidae	23,896 / 1%	26,438 / 2%	28,872 / 2%	27,250 / 2%	26,940 / 2%	34,202 / 2%	32,588 / 3%
World Totals	1,774,188	1,762,017	1,400,551	1,521,287	1,778,234	1,878,832	1,394,377

x Species fished commercially in the United States.

947

Table 17.2

US supply of scallops, 1976–2000 (Source FAO).

Year	Import	Export	Total
1976	11,455	0	11,550
1977	13,511	0	13,561
1978	12,867	0	14,279
1979	11,410	0	11,326
1980	9,473	0	9,743
1981	11,896	0	16,572
1982	9,462	0	11,504
1983	15,549	0	9,591
1984	12,372	0	24,039
1985	19,067	275	10,508
1986	21,734	370	6,281
1987	15,392	385	8,897
1988	14,533	303	10,348
1989	18,601	1,174	7,049
1990	18,156	3,272	5,350
1991	13,438	3,197	3,456
1992	17,622	1,634	3,166
1993	23,613	1,884	4,961
1994	25,766	2,755	5,217
1995	21,981	2,707	5,378
1996	26,684	2,808	8,073
1997	27,366	4,501	7,340
1998	24,132	3,484	9,374
1999	20,231	3,367	6,640
2000	24,530	4,164	8,774

1991) and from the east coast of Florida (unpub. data). It reappears on the southwest coast of Florida in Florida Bay and continues north along the west coast to Pensacola. Although not common, the southern subspecies reportedly is found as far west as Louisiana (Waller 1969; Broom 1976). The third subspecies, *Argopecten irradians amplicostatus*, occurs in the western Gulf of Mexico (Waller 1969). The degree to which these subspecies are physiologically different is not well understood (Bricelj et al. 1987) but Blake et al. (1997) have shown that there are significant genetic differences.

17.2.1.2 Biology

Bay scallops are functional hermaphrodites, which reproduce essentially once during their 12–24 month life span. Spawning is largely catastrophic, although the northern populations are more synchronous than the southern subspecies (Barber and Blake 1991).

The northern subspecies from New York spawn from June through August (Bricelj et al. 1987) while the Massachusetts populations spawn May through July (Taylor and Capuzzo 1983). In North Carolina, bay scallops spawn from July through August (Sastry 1968). On the west coast of Florida, the southern subspecies spawns in early August in the north (Sastry 1961) and October in the south (Barber and Blake 1983). Little has been reported about the spawning of the subspecies *A. irradians amplicostatus* in the western Gulf of Mexico. See Chapter 6 for a more detailed discussion of scallop reproduction.

The planktonic larval stage lasts 12–14 days (Sastry 1965), and it is during this planktonic stage that the larval distribution and eventual recruitment may be controlled by the hydrodynamics of the estuary (Eckman 1987). As the larvae metamorphose, the prodissoconchs (190 μm) typically attach to blades of seagrass. Growth during the winter is slow but by early spring scallops of 20–25 mm can be seen on the seagrass (Barber and Blake 1983). By late spring scallops become unattached and settle to the bottom at which time growth becomes rapid. Scallops of 40–50 mm occur by summer and may reach 60–90 mm by early winter (approximately 16 months of age). Growth is continuous in the Florida subspecies even at 33°C, only 2°C below the upper lethal temperature of the

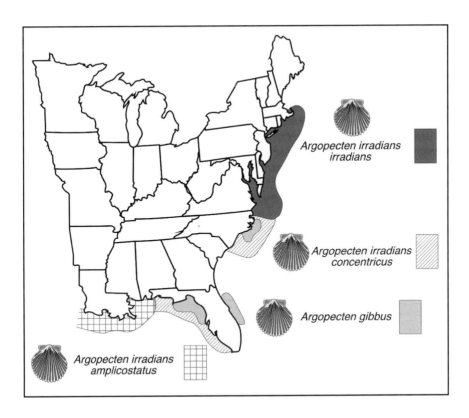

Figure 17.1. Distribution of scallops (*Argopecten* sp.) in eastern North America.

adults (Sastry 1961). Natural mortality after 12 months of age is high due to senescence, and only a few survive into the winter months of their second year.

17.2.1.3 Fishery

Landings are highly variable among years and among the various sections of the eastern United States (Table 17.3). The populations are dependent upon natural recruitment for continuation, although some enhancement efforts have been attempted. In 1985, bay scallop populations in the northeast were decimated by blooms of a previously unknown microalga, *Aureococcus anophagefferens* ("brown tide") (Tettelbach and Wenczel 1993). Three successive years of algal blooms resulted in virtually all-native stock in the Peconic Bays and the New York fishery being essentially eliminated. In addition, eelgrass beds were also depleted, thus reducing the total area of suitable habitat for scallop settlement. Landings for 1998 were 105,000 pounds of meats valued at $US 368,000 (Fig. 17.2). This represents an increase of 33,000 pounds (46%) and $58,000 (19%) compared with 1997 (O'Bannon 1999). The average ex-vessel price per pound of bay scallop meats was $3.50 in 1998 compared with $4.31 in 1997. Meat counts range between 50 and 100 per pound. Total catch and landed values from 1950–1999 are presented in Table 17.3 and Fig. 17.2.

Commercial fishing records for *A. irradians* date back to 1858 (Ingersoll 1886) and the introduction of the dredge in 1874. Commercial fishing for *A. irradians* is strictly limited and there is a large recreational fishery. In most areas, harvest is usually limited to the months of August-December. The bay scallop fishery is a protected resource throughout the range of the species. Commercial harvest of the species was prohibited in Florida waters in 1995.

Scallops are usually collected by diver, handpicking or rake. Some fishermen use small boats equipped with outboard engines and one or two small dredges. Scallops are culled on board and only the meats are presently harvested. Catch limits are determined on a season-by-season basis by fisheries officials in accordance with population fluctuations (Rhodes 1990; 1991).

Table 17.3

Historical catch statistics (total catch by regions) for bay scallops, (*Argopecten* sp.) for the period 1950–1999 (numbers in thousands). (NMFS, pers. comm.).

Year	New England		Middle Atlantic		South Atlantic		GULF		Grand Total	
	Lbs	$	Lbs	$	Lbs	$	Lbs	$	Lbs	$
1950	1,376	1,130	27	32	72	39	125	63	1,600	1,264
1951	1,253	959	101	121	183	96	252	161	1,789	1,337
1952	1,188	913	182	255	254	126	210	48	1,834	1,342
1953	2,397	1,222	162	102	65	33	229	53	2,853	1,410
1954	987	688	127	110	52	26	43	10	1,209	834
1955	1,070	837	226	210	78	39	223	53	1,597	1,139

1956	433	433	464	426	125	63	278	70	1,300	992
1957	1,230	880	674	447	109	37	315	91	2,328	1,455
1958	1,013	680	688	413	169	58	401	75	2,271	1,226
1959	591	700	385	386	128	51	82	19	1,186	1,156
1960	1,063	759	843	674	69	27	56	14	2,031	1,474
1961	704	671	862	621	106	42	36	14	1,708	1,348
1962	1,425	1,081	1,353	851	168	67	213	68	3,159	2,067
1963	391	492	577	404	321	122	228	59	1,517	1,077
1964	466	595	1,063	886	340	173	18	14	1,887	1,668
1965	459	562	982	766	379	196	39	24	1,859	1,548
1966	880	1,076	492	408	399	184	9	4	1,780	1,672
1967	455	579	248	258	387	211	7	5	1,097	1,053
1968	491	776	218	374	639	422	143	122	1,491	1,694
1969	1,172	1,592	249	377	613	383	80	61	2,114	2,413
1970	1,101	1,704	365	470	130	91	104	56	1,700	2,321
1971	2,063	3,531	144	234	60	42	48	39	2,315	3,846
1972	1,776	3,407	93	215	128	110	35	40	2,032	3,772
1973	694	1,462	230	467	37	33	53	63	1,014	2,025
1974	567	1,014	694	872	220	199	16	18	1,497	2,103
1975	1,054	2,568	444	713	135	105	14	16	1,647	3,402
1976	890	1,973	438	816	248	194	14	24	1,590	3,007
1977	1,044	3,085	199	489	257	509	46	58	1,546	4,141
1978	1,521	4,982	280	837	221	393	49	91	2,071	6,303
1979	1,382	5,967	346	1,243	193	514	62	137	1,983	7,861
1980	1,356	6,671	431	1,840	328	1,107	11	29	2,126	9,647
1981	964	4,630	244	891	189	656	22	62	1,419	6,239
1982	2,022	8,949	500	1,809	137	352	13	35	2,672	11,145
1983	1,083	6,491	167	992	205	509	22	75	1,477	8,067
1984	808	4,573	279	1,264	384	876	10	26	1,481	6,739
1985	958	5,812	174	828	456	1,072	4	10	1,592	7,722
1986	509	3,797	13	65	306	838	27	86	855	4,786
1987	341	2,813	(2)	3	155	501	19	80	515	3,397
1988	530	3,339	(2)	2	39	73	39	73	608	3,487
1989	215	1,494	2	22	84	214	57	162	358	1,892
1990	254	1,683	11	132	62	128	56	204	383	2,147
1991	191	1,363	16	117	45	100	0.2	0.6	252	1,580
1992	569	4,079	25	182	22	54	2	11	618	4,326
1993	145	1,502	15	143	153	365	9	37	322	2,047
1994	1	1	272	1,765	73	133	0	0	346	1,899
1995	8	66	26	203	201	401	0	0	235	670
1996	1	3	1	0.4	29	113	0	0	31	116
1997	0	0	7	66	64	214	0	0	71	280
1998	0	0	2	22	103	289	0	0	105	311
1999	0	0	6	96	30	103	0	0	36	199

952

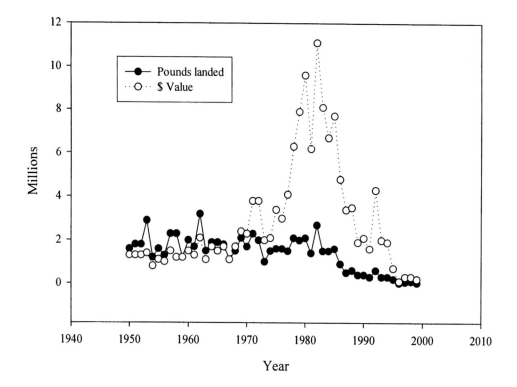

Figure 17.2. Landings and value of bay scallops, *Argopecten irradians*. Data from NMFS (pers. comm.).

17.2.2 Calico Scallop, *Argopecten gibbus*

17.2.2.1 Distribution

The calico scallop, *Argopecten gibbus,* is largely restricted to the sub-temperate and tropical waters of western North Atlantic (Fig. 17.3) with the major stocks distributed from Cape Hatteras, North Carolina to the Cape San Blas area of the northeastern Gulf of Mexico (Waller 1969). Calico scallops have also been collected from the Greater Antilles, Bermuda, and the western portions of the Gulf of Mexico (Waller 1969) but stocks of commercial importance have not been reported from these areas. At least one other species, *Argopecten circularis*, is commonly called a calico scallop and is harvested in the Republic of Panama.

Argopecten gibbus, inhabits the relatively shallow waters of the continental shelf and upper slope out to reported depths of 370 m (Waller 1969). Major commercial stocks along the Atlantic and Gulf coasts are usually located in depths of 20–50 m on hard sand bottom.

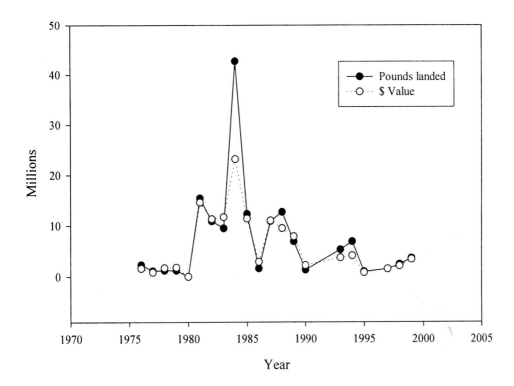

Figure 17.3. Landings and value of calico scallops, *Argopecten gibbus*. Data from NMFS (pers. comm.) and Florida Marine Research Institute, Fishery Dependent Monitoring Program.

17.2.2.2 Biology

The calico scallop, like the bay scallop, is a functional hermaphrodite. The species exhibits two spawning periods during each calendar year when optimal environmental conditions occur (Moyer and Blake 1986). The first and predominant spawn of the calendar year occurs in late spring from April to June. This follows the accumulation of gonadal reserves during the winter months. When spawning does occur it is rapid and often complete. A large percentage of any subpopulation normally spawns over a 1–3 week period.

The second reproductive cycle of the calendar year begins with an accumulation of gonadal reserves in July and August with spawning occurring in late summer to early autumn. The "fall spawn" apparently occurs only when environmental conditions are optimal and does not occur every year. Only a portion of the total population is involved in this minor spawn; however, the "fall spawn" appears to be important to the maintenance of the commercial stocks into the spring and summer months.

The individual scallop normally reproduces for the first time at an approximate age of either 6–12 months depending upon season of birth and the presence or absence of a fall spawn. Age, however, appears to be less important than environmental conditions since calico scallops as young as 3 months occasionally spawn. Whether or not there are any differences in the viability of larvae produced by scallops of different ages has not been examined. A scallop will normally spawn either 2 or 3 times during its 18–24 month life span, although the number of scallops reaching a third spawn in usually very low.

Like the bay scallop (Sastry 1968; Blake and Sastry 1979; Barber and Blake 1985; Barber and Blake 1991), the environmental controls for reproduction in the calico scallop include food, temperature, and age (Moyer 1997). The current dynamics in the region of Cape Canaveral may also be important to the maintenance of the population and these variables have received considerable study (Lee and Brooks 1979; Leming 1979; Blanton et al. 1981; Lee and Atkinson 1983). During the summer months, coastal upwelling brings deep Gulf Stream water onto the shelf resulting in lowered bottom temperatures. This nutrient rich, colder water may be important in allowing the scallops to accumulate nutrients rapidly.

Miller et al. (1981) have suggested that the cold water intrusions may have both favourable and unfavourable results. On the favourable side, the intrusion may initiate spawning, provide food through phytoplankton blooms resulting from nutrient-rich water, and provide an optimal temperature range of 15–27°C for survival. The intrusions may be unfavourable if bottom temperatures are lowered below 15°C where mortality may occur.

Growth rates are not constant but vary as environmental conditions change (Blake and Moyer 1991; Moyer 1997). Growth rates of greater than 1.5 mm per week can occur during upwelling events and can drop to 0.3–0.8 mm per week as bottom water temperatures increase.

Costello et al. (1973) estimated that calico scallops are planktonic for 16 days under laboratory conditions and set at about 0.25 mm shell height. Therefore, Allen (1979) estimated that 10 mm scallops would be about 51 days old and 28 mm in about 3 months. Considering a two-month interval from spawning to 23 mm, a total of 8 months are required for the scallops to reach 38–40 mm, the commercial size desired by the industry (Blake and Moyer 1991). This period could of course be increased or decreased depending upon environmental conditions, particularly bottom water temperatures.

The calico scallop lives a maximum of two years (Roe et al. 1971). However, since post-spawning mortality may be high, only a small percentage of a year class may survive after 18 to 20 months. Roe et al. (1971) estimated monthly mortality rates ranging from 1% to 31% depending upon the age composition of the population. High mortalities may also occur, even in the young of the population, if bottom water temperatures decrease below 15°C (Miller et al. 1981).

In late 1988 and again in 1991, an ascetosporan parasite, *Marteilia* sp., virtually eliminated the calico scallop population in the vicinity of Cape Canaveral (Moyer et al. 1993). The parasite had not been previously reported in the western Atlantic and the calico scallop population has not rebounded to pre-parasite levels.

17.2.2.3 Fishery

The calico scallop, *Argopecten gibbus*, supports a highly variable fishery (Fig. 17.3) off the coast of Florida and North Carolina. Although the exact locations of commercial stocks vary from year to year, the Cape Lookout, Cape Canaveral, and Cape San Blas areas (Fig. 17.4) are particularly important to the fishery. The North Carolina fishery is small (Table 17.4) and extremely transient (Allen and Costello 1972) but the scallop meats produced are often large for the species and are good quality. The Cape San Blas fishery of Florida is also highly variable and commercial fishing efforts may occur only every 2–3 years and only after other stocks have been depleted.

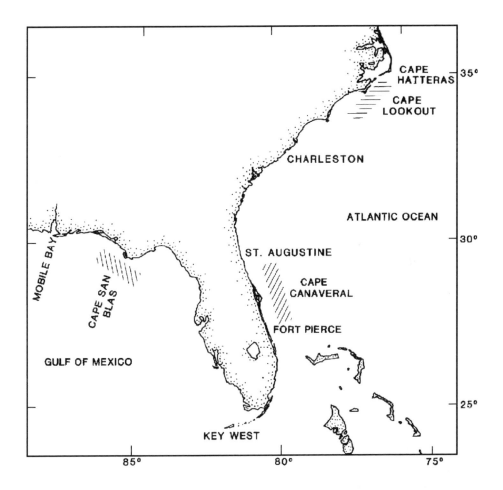

Figure 17.4. Commercial fishing grounds for the calico scallop, *Argopecten gibbus* (after Blake and Moyer 1991).

Table 17.4

Historical catch statistics (total catch by state) for calico scallops, (*Argopecten gibbus*) for the period 1976–1999 (numbers in thousands). *(NMFS, pers. comm.)

Year	Florida		Georgia		South Carolina		North Carolina		Grand Total	
	Lbs	$	Lbs	$	Lbs	$	Lbs	$	Lbs	$
1976	2,269	1,622	0	0	0	0	0	0	2,269	1,622
1977	1,113	837	0	0	0	0	0	0	1,113	837
1978	478	752	87	118	611	803	0	0	1,176	1,673
1979	1,157	1,711	33	45	0	0	43	81	1,233	1,837
1980	0.4	1.0	0	0	0	0	0	0	0.4	1.0
1981	15,171	14,277	0	0	0	0	244	307	15,415	14,585
1982	10,842	11,274	41	41	0	0	0	0	10,883	11,315
1983	9,352	11,556	0	0	0	0	102	179	9,453	11,735
1984	42,742	23,246	0	0	0	0	0	0	42,742	23,246
1985	12,283	11,450	0	0	0	0	0	0	12,283	11,450
1986	1,566	2,861	0	0	0	0	0	0	1,566	2,861
1987	10,933	11,001	0	0	0	0	0	0	10,933	11,001
1988	12,039	8,752	0	0	0	0	668	702	12,707	9,454
1989	6,611	7,439	0	0	0	0	336	469	6,946	7,907
1990	874	1,711	0	0	0	0	385	531	1,259	2,241
1991	39		0	0	0	0	0	0	39	
1992	200		0	0	0	0	0	0	200	
1993	5,307	3,660	0	0	0	0	0	0	5,307	3,660
1994	6,879	4,072	0	0	0	0	0	0	6,879	4,072
1995	945	805	0	0	0	0	0	0	945	805
1996	0	0	0	0	0	0	0	0	0	0
1997	1,545	1,489	0	0	0	0	0	0	1,545	1,489
1998	2,397	2,065	0	0	0	0	0	0	2,397	2,065
1999	3,572	3,411	0	0	0	0	0	0	3,572	3,411

* 1991, 1992 and 1993 values for Florida are from the Florida Marine Research Institute, Fishery Dependent Monitoring Program.

The most significant and persistent commercial stock occurs in the vicinity of Cape Canaveral, Florida. Commercial stocks of calico scallops were first observed off Cape Canaveral in 1960 (Bullis and Cummins 1961). Investigations by the Bureau of Commercial Fisheries (now the National Marine Fisheries Service) from 1960 to 1968 showed that commercial concentrations varied greatly along the coast from the St. Johns River near the Florida/Georgia boarder to Fort Pierce, Florida in depths of 19 to 74 m (Bullis and Cummins 1961; Drummond 1969; Cummins 1971; Roe et al. 1971). More recently the most productive beds have been found to be located in 30 to 50 m of water extending from about St. Augustine, Florida to Fort Pierce, Florida, making a harvestable area of about 2,000 square miles (Fig. 17.4).

The calico scallop grows to less than 3 inches and the adductor muscle (meat) is small and white to pink in colour (meat count range 100–300; normally 150–200). Hand-shucking is not economically feasible; thus, even though large stocks of the calicos scallops were known as early as 1949, the species was not harvested commercially prior to automation in the late 1960's.

Although the current commercial fishery began in 1967, significant landings did not begin until 1980 (Fig. 17.3). This was due to an increase in stocks and to the refinement of automated steam processing equipment. The equipment allowed shellstock representing up to 800 gallons of shucked meats to be processed and packaged within an hour.

Calico scallops are fished using a modified Gulf shrimp trawler. Typically a scallop net, similar to a shrimp net, is towed from each side of the boat. Towing time is regulated and the nets must be equipped with "turtle exclusion devices". No culling occurs and unless processing occurs at sea, the entire catch is returned to shore.

Processing begins as the shellstock is removed from the boat deck by a "knuckle boom". The scallops are then placed in a "batching hopper" which relays the unsorted load to an automated shaker. The shaker removes the broken and dead shell, sand, and unwanted species. Steam is then used to "shock" the scallop meats and to separate the meat and other tissue from the shell. A series of rollers then eviscerates the scallops separating the visceral mass from the adductor muscle. The visceral mass, commonly eaten in other countries, is discarded while the adductor muscle is rapidly cooled in a series of water baths and chill tanks. Quality control includes "hand culling" of any shell fragments or other imperfections before entry into the final series of chill tanks prior to packaging. With the exception of the "hand culling" the entire process can occur without the meats touching human hands and since the product is shipped immediately to a wholesaler or retailer, it can reach the consumer within 24 to 48 hours without being frozen.

The volatile nature of production of this species is clearly demonstrated in Table 17.4 and Fig. 17.3. During the peak in 1984, landings were almost 43 million pounds and reduced to 39,000 pounds in 1991. Annual variations in production of calico scallops impact not only the total catch in the U.S., they also determine the position of the U.S. among other world scallop producers. Landings of calico scallops were 3,572,000 pounds of meats valued at $US 3,411,000 in 1999. This represented an increase of 1,175,000 pounds (33%) and $US 1,346,000 (65%) compared with 1998. All calico scallops were landed in Florida in 1998 and 1999. The average ex-vessel price was $0.86 per pound of meats in 1998 compared to $0.95 in 1999 (Florida Marine Research Institute, Fishery Dependent Monitoring Program, pers. comm.).

Since stocks of *A. gibbus* are annual, over-fishing has not been considered a problem; thus, there are no state or federal fishery management programs. The fishery is totally dependent upon the natural population and regulation of landings has been attempted by the industry. Fishing efforts are limited until at least 75% of the stock at a particular location reaches a shell height of at least 38 mm, the point at which much of the population will have undergone their first spawning event. A second spawning is not guaranteed and only takes place when environmental conditions are optimal. Concerns by regulators about the quantity and disposition of by-catch may eventually lead to more formal regulations.

17.3 AQUACULTURE AND ENHANCEMENT

During the years between 1920 and 1926, William Firth Wells carried out some bivalve culture investigations, which he reported in his annual reports to the New York State Conservation Commission. His reports show that besides propagating the eastern oyster, *Crassostrea virginica*, he also cultured quahogs, *Mercenaria mercenaria*, soft clams, *Mya arenaria*, mussels, *Mytilus edulis* and bay scallops, *Argopecten irradians* (State of New York Conservation Department 1969). Wells used a milk separator to clarify his culture water and to collect larvae from cultures for transfer. One of the earliest species he was able to culture was the bay scallop. This was perhaps the first bivalve cultured in the manner similar to what we think of today as aquaculture in the United States (personal communication, Joseph Glancy, now deceased). Most scallop culture in the United States currently utilises the bay scallop, *A. irradians irradians* or *A. irradians concentricus*. No attempts have been made to culture *Argopecten gibbus* in the United States because of its small size and the relatively large natural populations.

The high market value of bay scallops combined with rapid growth and short life span of the species has led to the development and refinement of hatchery, nursery, and grow-out techniques (Castagna and Duggan 1971; Castagna 1975; Rhodes and Widman 1980; Mann and Taylor 1981; Karney 1991; Oesterling and DuPaul 1993; Lu and Blake 1997). Using the information about the natural reproductive cycle investigators have used either artificial conditioning and various stimuli to spawn adults for larval production in a hatchery or relied upon the harvest of reproductively mature adults from the natural environment for spawning stock.

Embryonic and larval stages of marine invertebrates are often the most sensitive to the environmental conditions and extreme mortality of the embryos and larvae can occur if a strict environment is not maintained in a hatchery. Temperature, salinity, and food supply appear to be the most important environmental variables regulating the growth and reproduction of the adults as well as the survival and growth of the embryos and larvae. The combined effects of temperature and salinity on the development of embryos and survival and growth of the larvae of the northern bay scallop, *A. irradians irradians*, was extensively studied by Tettelbach and Rhodes (1981). Only embryos cultured at 20°C–25‰ and 25°C–25‰ showed a greater than 70% normal development. At salinities greater than or lesser than 25‰, normal development declined markedly. Larval survival of this northern subspecies for 2–5 days after fertilisation occurred over a much wider temperature-salinity range with the optimal combination being 18.7°C and 28.1‰. Maximal larval survival at time of settlement occurred at 20°C–25‰. Lu (1996) has shown that this same temperature and salinity combination is optimum for the Florida bay scallop larvae and juveniles and that maximum growth of Florida bay scallop larvae and juveniles occurs when a food supply of *Isochrysis galbana* is maintained at a density of 10–40 thousand cells per mL of culture.

A number of different techniques have been used in hatcheries to maintain larvae and early juveniles. These techniques have included static, aerated water tanks (Castagna and Duggan 1971; Castagna 1975; Karney 1991; Oesterling and DuPaul 1993; Lu and Blake 1996) for early larval development. Sieves suspended in down-flowing aerated conicals (Karney 1991), raceways or troughs fitted with inserts (Oesterling and DuPaul 1993), and plastic strips suspended in aerated tanks (Lu and Blake 1996) have been used for pediveliger and post-set animals. Flow-through upwelling systems (Oesterling and DuPaul 1993) and nylon-meshed bags suspended in the natural environment (Lu and Blake 1997) are two of the more commonly employed techniques for the nursery of early juveniles from 1–8 mm.

For grow-out, Castagna and Duggan (1971) used large pens in order to demonstrate that a market-sized scallop could be attained in 5–7 months. Early releases of juvenile scallops into the natural environment for commercial harvest proved unsuccessful due to heavy predation (Morgan et al. 1980). Rhodes and Widman (1980) utilised Japanese pearl nets and lantern nets to determine the optimum scallop density for maximum growth in Long Island Sound. Oesterling and DuPaul (1993) used polyethylene trays or plastic mesh cages placed directly on the bottom, placed on cinder blocks, or suspended in the water column on a rack system to demonstrate an optimum density of about 500–800 scallops per square meter for attaining a market-size scallop of 40 mm in 7 months. More recently Oesterling (pers. comm.) employed a floating pen system to suspend the plastic cages in the water column.

Heavy mortalities can occur at all three points in the aquaculture process. Pathogens associated with scallop diseases have been reviewed by Leibovitz et al. (1984). Pathogenic *Pseudomonas* can bloom on the solid surfaces of culture vessels and kill larvae or early juveniles but can be minimised with regular cleaning and water changes (Karney 1991). In the nursery and grow-out stages fouling of the nets or cages can reduce flow rates which leads to reduced growth and increased mortality but this can also be minimised with frequent cleaning (Karney 1991). In the final stages of grow-out, especially during the summer, mortality can occur for unknown reasons (Oesterling, pers. comm.; Lu and Blake 1997) although fouling of the shells by tunicates, barnacles and oysters may be one of the primary causes but frequent cleaning of the shells to decrease the degree of fouling may also lead to mortality.

A number of companies have attempted to culture scallops but have been unsuccessful economically and there is currently no profitable, private aquaculture industry for bay scallops in the United States. Some recent attempts to market the whole bay scallop product to a speciality market have had mixed success due largely to production problems.

The best success with the aquaculture of the species has been in China (Ximing et al. 1999) where in 1982, 26 individuals of *Argopecten irradians irradians* transplanted from the United States were spawned successfully (Fusui Zhang, pers. comm.). Using techniques largely developed in the U.S., the annual bay scallop production in China steadily increased to over 50,000 tons live weight in 1988. The shucked muscles are imported into the U.S. at $2.00 per pound (Zhang 1995). This success can be partially attributed to the low cost of abundant labour as well as to the legal utilisation of the total marine habitat.

A few organisations have been involved in enhancement programs, also utilising bay scallops. Perhaps the most successful program of shellfish enhancement is carried out by the Martha's Vineyard Shellfish Group. This is a consortium of 5 towns (Chilmark, Gay Head, Oak Bluffs, Tisbury and West Tisbury) on the island of Martha's Vineyard off the coast of Massachusetts. This group, using a number of federal and state grants, built a solar-assisted hatchery to produce bay scallops, *A. irradians irradians* and clams, *Mercenaria mercenaria* (Karney 1978). The hatchery methods used at Martha's Vineyard are fairly standard except that the seawater is partially warmed in a passive solar system within the solarium-type building. The post-set scallops are held in an indoor, semi-closed nursery system supplemented with cultured algae until the juveniles are 3–5 mm height, then moved out to a small embayment in burlap bags with a brick anchor and a plastic cola bottle inside the bag for a float. Several hundred to a few thousand seed are placed in each bag, which is then anchored over the submerged vegetation found in the bay. This allows the seed to grow to a size that offers sanctuary from some predators before the bag rots away allowing the juveniles to escape a few at a time and spread into the vegetation (Karney, pers. comm.). Each township has legal jurisdiction over its own shellfish waters, sale of harvesting licenses and control of the harvest. Each township supporting the hatchery buys seed at about cost for replenishment or enhancement of an area. The effect of scallop enhancement has been to add a degree of stability to the harvest in the area that is seeded (Karney 1978).

Another enhancement program was carried out in the Long Island Sound area after heavy mortalities of native scallops were observed (see above). The mortality and subsequent recruitment failure was caused by a form of picoplankton, *Aureococcus annophagefferens*. Extensive reseeding of hatchery-reared scallops was initiated in the Peconic Bays by the Long Island Green Seal Committee in 1986 (see Tettelbach and Wenczel 1993). In the following two years, seed scallops (*A. irradians)* were purchased from a number of hatcheries and released in selected areas to enhance or replace the natural populations, which were lost. The effects of this enhancement effort were not quantified in all areas, but a number of scientists involved in this experiment initially believed the effects of the seed planting were minimal (Bricelj et al. 1987; Tettelbach and Wenczel 1993). Recent work by Krause, however, showed about 25% of the scallops in the area were survivors of those released (Krause 1992). Subsequent reseeding efforts were further hampered by the presence of a shell-boring parasite, *Polydora* sp. and yet another "brown tide". While the enhancement efforts are encouraging, the status of the New York bay scallop fishery is still precarious.

Over 100,000 adult bay scallops were transplanted to western Bogue Sound, North Carolina in each of three successive years (1992, 1993, and 1994). Recruitment density at experimental sites increased by as much as 568% while adult densities after recruitment increased by as much as 258%. Recruitment appeared to limit population size in Bogue Sound which implied that larval subsidy through increased spawning stock may enhance bay scallops in the system (Peterson et al. 1996).

In 1996, the state of Florida began an enhancement program along the west coast of Florida from Tampa Bay to Crystal River. Scallops reared in a hatchery and nursery to about 20 mm in size were placed at selected locations in wire cages at a density of 300 per square meter. The growth and mortality of these were followed until spawning. Fouling and mortality of the scallops were severe in the late summer but spawning of the survivors is thought to have contributed to an increased population in the Crystal River area.

17.4 FUTURE

Bay scallops in the eastern United States have historically been considered a delicacy, especially in the local markets where the fresh product is consumed. Its high demand in these markets guarantees a continued high value. However, problems associated with the fishery will need to be addressed.

In addition to showing sharp natural seasonal and annual fluctuations in stock densities, bay scallops along the east coast of North America will continue to be influenced by other pressures that will limit harvests. Loss of habitat through the decline of seagrass beds and the degradation of habitat through algal blooms such as "brown tide" may contribute to the further decline of bay scallop stocks. Also the increase in the public awareness about the occurrence of "harmful algal blooms" may also lead to a decline in the consumption of bay scallops. These problems may eventually lead to the profitability of bay scallop aquaculture, especially if the whole-scallop market develops further. However, the success of bay scallop aquaculture will also be influenced by stringent, governmental leasing regulations, grow-out problems such as fouling, and market development.

The annual landings of the calico scallop fishery will probably continue at relatively low levels (7–10 million pounds) compared to 1984. Although the calico scallop has less value than the bay scallop, the continued landings will ensure its place in the scallop market. It is unlikely that demand for the product will lead to aquaculture ventures in North America.

REFERENCES

Allen, D.M., 1979. Biological aspects of the calico scallop, *Argopecten gibbus*, determined by spat monitoring. The Nautilus 94(4):107–119.

Allen, D.M. and Costello, T.J., 1972. The calico scallop, *Argopecten gibbus*. NOAA Technical Report NMFS SSRF-656. 19 p.

Barber, B.J. and Blake, N.J., 1983. Energy storage and utilization in relation to gametogenesis in *Argopecten irradians concentricus* (Say). J. Exp. Mar. Biol. Ecol. 52:121–134.

Barber, B.J. and Blake, N.J., 1985. Substrate catabolism related to reproduction in the bay scallop, *Argopecten irradians concentricus*, as determined by O/N and RQ physiological indexes. Mar. Biol. 87:13–18.

Barber, B.J. and Blake, N.J., 1991. Reproductive physiology. In: S.E. Shumway (Ed.). Scallops: Biology, Ecology and Aquaculture. Developments in Aquaculture and Fisheries Science, Vol. 21. Elsevier, New York. pp. 377–420.

Blake, N.J. and Moyer, M.A., 1991. The calico scallop, *Argopecten gibbus*, fishery of Cape Canaveral, Florida. In: S.E. Shumway (Ed.). Scallops: Biology, Ecology and Aquaculture. Developments in Aquaculture and Fisheries Science, Vol. 21. Elsevier, New York. pp. 899–924.

Blake, N.J. and Sastry, A.N., 1979. Neurosecretory regulation of oogenesis in the bay scallop, *Argopecten irradians irradians* (Lamarck). In: E. Naylor and R.G. Hartnol (Eds.). Cyclic Phenomena in Marine Plants and Animals. Proceedings of 13th European Marine Biology Symposium, Sept. 27–Oct. 4, 1978, Isle of Man. Oxford, Pergamon Press. pp. 181–190.

Blake, S.G., Graves, J.E., Blake, N.J. and Oesterling, M.J., 1997. Genetic divergence and loss of diversity in two cultured populations of the bay scallop, *Argopecten irradians* (Lamarck, 1819). J. Shellfish Res. 16(1):55–58.

Blanton, J.L., Atkinson, L.P., Pietrafesa, L.J. and Lee, T.N., 1981. The intrusion of Gulf Stream Water across the continental shelf due to topographically-induced upwelling. Deep Sea Research 28(4):393–405.

Bricelj, V.M., Epp, J. and Malouf, R.E., 1987. Intraspecific variation in reproductive and somatic growth cycles of bay scallops *Argopecten irradians*. Mar. Ecol. Prog. Ser. 36:123–137.

Broom, M.J., 1976. Synopsis of biological data on scallops (*Chlamys* (*Aequipecten*) *opercularis* (Linnaeus), *Argopecten irradians* (Lamarck), *Argopecten gibbus* (Linnaeus)). FAO Fisheries Synopsis No. 114. 44 p.

Bullis, H.R. and Cummins, R., 1961. An interim report of the Cape Canaveral scallop bed. Commercial Fisheries Review 23(10):1–8.

Castagna, M., 1975. Culture of the bay scallop, *Argopecten irradians* in Virginia. Marine Fish. Rev. 37(1):19–24.

Castagna, M. and Duggan, W.P., 1971. Rearing of the bay scallop, *Aequipecten irradians*. Proc. Natl. Shellfish. Assoc. 61:80–85.

Clarke, A.H. Jr., 1965. The scallop superspecies *Aequipecten irradians* (Lamarck). Malacologia 2:161–188.

Costello, T.J., Hudson, J.H., Dupay, J.L. and Rivkin, S., 1973. Larval culture of the calico scallop, *Argopecten gibbus*. Proc. Natl. Shellfish. Assoc. 63:72–76.

Cummins, R., 1971. Calico scallops of the southeastern United States, 1959–1969. NOAA Technical Report NMFS SSRF-657. 22 p.

Drummond, S.B., 1969. Explorations for calico scallop, *Pecten gibbus*, in the area off Cape Kennedy, Florida, 1960–66. U.S. Fish Wildl. Serv., Fish. Ind. Res. 5:85–101.

Eckman, J.E., 1987. The role of hydrodynamics in recruitment, growth, and survival of *Argopecten irradians* (L.) and *Anomia simplex* (D'Orbigny) within eelgrass meadows. J. Exp. Mar. Biol. Ecol. 106:165–191.

FAO, 1997. Yearbook of Fishery Statistics, Capture Production. Vol. 84. FAO, Rome.

Ingersoll, E., 1886. The scallop and its fishery. Am. Naturalist 20(12):1001–1006.

Karney, R.C., 1978. A program for the development of the shellfisheries of five towns on Martha's Vineyard. First Annual Report to the Economic Development Administration, U.S. Dept. Commerce Tech. Asst. Grant No. 01–6–01519.

Karney, R.C., 1991. Ten years of scallops culture on Martha's Vineyard. In: S.E. Shumway and P.A. Sandifer (Eds.). Scallop Biology and Culture. World Aquaculture Workshops, No. 1. World Aquaculture Society, Baton Rouge, LA. pp. 308–312.

Krause, M.K., 1992. Use of genetic markers to evaluate the success of transplanted bay scallops. J. Shellfish Res. 11:199.

Lee, T.N. and Atkinson, L.P., 1983. Low-frequency current and temperature variability for Gulf Stream frontal eddies and atmospheric forcing along the southwest U.S. outer continental slope. J. Geophys. Res. 88:4541–4567.

Lee, T.N. and Brooks, D., 1979. Initial observations of current, temperature and coastal sea level response to atmospheric and Gulf Stream forcing. J. Geophys. Res. Lett. 6:321–324.

Leibovitz, L.E., Schott, F. and Karney, R.C., 1984. Diseases of wild, captive, and cultured scallops. Journal of the World Mariculture Society 15:269–283.

Leming, T.D., 1979. Observations of temperature, current and wind variations off the central eastern coast of Florida. NOAA Technical Memorandum NMFS-SEFC-6 (U.S. Dept. of Commerce). 172 p.

Lu, Y., 1996. The physiological energetics of larvae and juveniles of the bay scallop, *Argopecten irradians concentricus* (Say). Ph.D. dissertation, University of South Florida, St. Petersburg, Florida. 160 p.

Lu, Y. and Blake, N.J., 1997. The culture of the southern bay scallop, *Argopecten irradians concentricus*, in Tampa Bay, an urban Florida estuary. Aquaculture International 5:439–450.

Mann, R. and Taylor, R.E. Jr., 1981. Growth of the bay scallop, *Argopecten irradians*, in a waste recycling aquaculture system. Aquaculture 24:45–52.

Miller, G.C., Allen, D.M. and Costello, T.J., 1981. Spawning of the calico scallop *Argopecten gibbus* in relation to season and temperature. J. Shellfish Res. 1(1):17–21.

Morgan, D.E., Goodsell, J. and Matthiessen, G., 1980. Release of hatchery-reared bay scallops (*Argopecten irradians*) onto shallow coastal bottom in Waterford, Connecticut. Proc. World Mariculture Society 11:247–261.

Moyer, M.A., 1997. The reproductive ecology of the calico scallop, *Argopecten gibbus* (Linnaeus), and mass mortality linked to a protistan. Ph.D. dissertation, University of South Florida, St. Petersburg, Florida. 175 p.

Moyer, M.A. and Blake, N.J., 1986 Fluctuations in calico scallop production (*Argopecten gibbus*). Proc. 11th Annual Tropical and Subtropical Fisheries Conference of the Americas. pp. 45–58.

Moyer, M.A., Blake, N.J. and Arnold, W.S., 1993. An ascetosporan disease causing mass mortality in the Atlantic calico scallop, *Argopecten gibbus* (Linnaeus, 1758). J. Shellfish Res. 12(2):305–310.

O'Bannon, B.K. (Ed.). 1999. Fisheries of the United States, 1998. Current Fishery Statistics No. 9800. NOAA/NMFS. 130 p.

Oesterling, M.J. and DuPaul, W.D., 1994. Shallow water bay scallop (*Argopecten irradians*) culture in Virginia. In: N.F. Bourne, B.L. Bunting and L.D. Townsend (Eds.). Proceedings of the 9th International Pectinid workshop, Nanaimo, B.C., Canada, April 22–27, 1993. Volume 2. Can. Tech. Rep. Fish. Aquat. Sci. 1994:217.

Peterson, C.H., Summerson, H.C. and Luettich, R.A. Jr., 1996. Response of bay scallops to spawner transplants: a test of recruitment limitation. Mar. Ecol. Prog. Ser. 132:93–107.

Rhodes, E.W., 1991. Fisheries and aquaculture of the bay scallop, *Argopecten irradians*, in the eastern United States. In: S.E. Shumway (Ed.). Scallops: Biology, Ecology and Aquaculture. Developments in Aquaculture and Fisheries Science, Vol. 21. Elsevier, New York. pp. 913–921.

964

Rhodes, E.W. and Widman, J.C., 1980. Some aspects of the controlled production of the bay scallop (*Argopecten irradians*). Proc. World Mariculture Soc. 11:235–246.

Roe, R.B., Cummins, R. Jr. and Bullis, H.R. Jr., 1971. Calico scallop distribution, abundance, and yield off eastern Florida, 1967–1968. Fish. Bull., U.S. 69:399–409.

Sastry, A.N., 1961. Studies on the bay scallop, *Aequipecten irradians concentricus* Say, in Alligator Harbor, Florida. Ph.D. thesis, Florida State University, Tallahassee. 118 p.

Sastry, A.N., 1968. The relationships among food, temperature, and gonad development of the bay scallop, *Aequipecten irradians* Lamarck. Physiol. Zool. 41:44–53.

Shumway, S.E. and Castagna, M., 1994. Scallop fisheries, culture and enhancement in the United States. Mem. Queensland Museum 36(2):283–298.

State of New York Conservation Department. 1969. Early oyster culture investigations by the New York State Conservation Commission (1920–1926).

Taylor, R.E. and Capuzzo, J.M., 1983. The reproductive cycle of the bay scallop, *Argopecten irradians*, (Lamarck), in a small coastal embayment on Cape Cod, Massachusetts. Estuaries 6(4):431–435.

Tettelbach, S.T. and Rhodes, E.W., 1981. Combined effects of temperature and salinity on embryos and larvae of the northern bay scallop, *Argopecten irradians irradians*. Mar. Biol. 63(3):249–256.

Tettelbach, S.T. and Wenczel, P., 1993. Reseeding efforts and the status of bay scallop *Argopecten irradians* (Lamarck, 1819) populations in New York following the occurrence of "brown tide" algal blooms. J. Shellfish Res. 12:423–431.

Walker, R.L., Heffernan, P.B., Crenshaw, J.W. Jr. and Hoats, J., 1991. Effects of mesh size, stocking density and depth on the growth and survival of pearl net cultured bay scallops, *Argopecten irradians concentricus*, in shrimp ponds in South Carolina, U.S.A. J. Shellfish Res. 10:465–469.

Waller, T.R., 1969. The evolution of the *Argopecten gibbus* stock (Mollusca: Bivalvia), with emphasis on the tertiary and quartenary species of eastern North America. J. Paleontology 43: Suppl. 5. 125 p.

Ximing, G., Ford, S.E. and Zhang F., 1999. Molluscan. aquaculture in China. J. Shellfish Res. 18(1):19–31.

Zhang, F., 1995. The rise of the bay scallop industry in China. In: P. Lubet and J.-C. Dao (Eds.). Fisheries, Biology and Aquaculture of Pectinids. Actes de Colloques No. 17. IFREMER, Brest, France. pp. 131–138.

AUTHORS ADDRESSES

Norman J. Blake - College of Marine Science, University of South Florida, St. Petersburg, Florida 33701 USA (E-mail: nblake@marine.usf.edu)

Sandra E. Shumway - Dept of Marine Sciences, UCONN, 1080 Shennecossett Road, Groton, CT 06340 USA (E-mail: Sandra.Shumway@uconn.edu)

Scallops: Biology, Ecology and Aquaculture
S.E. Shumway and G.J. Parsons (Editors)
© 2006 Elsevier B.V. All rights reserved.

Chapter 18

Scallops of the West Coast of North America

Raymond B. Lauzier and Neil F. Bourne

18.1 INTRODUCTION

The west coast of North America, defined here as extending from the Mexico-California border to coastal Alaska and into the Bering Sea (Fig. 18.1), is an extensive area that appears to have considerable habitat suitable to support large scallop populations. Although much of the coastal area is mountainous and considerable parts of the coast fall to deep depths within a few km of shore, there are extensive areas with water depths less than 100 m that appear to have good scallop habitat. However, scallop landings from this area have never been large and in 1997 were only 3,181 t, less than 0.2% of world landings for that year (Anonymous 1999).

Although scallop landings from commercial fisheries have been minor, the west coast of North America has a rich scallop fauna. A recent publication (Coan et al. 2000) divided the Suborder Pectinoidea into two Superfamilies, Pectinoidea and Anomoidea (Fig. 18.2). The superfamily Pectinoidea has four living families world wide, two of which, Family Pectinidae and Family Propeasmussidae are well represented along the west coast of North America. Thirteen species of the Family Propeasmussidae have been found in this area, but all are small and most are rare and have a local distribution in deep water. Fifteen species of the Family Pectinidae occur in this area and seven have been or are currently utilised in commercial fisheries: weathervane, *Patinopecten caurinus*, Japanese weathervane, *Mizuhopecten yessoensis*, rock, *Crassadoma gigantea*, reddish or pink, *Chlamys rubida*, and spiny, *Chlamys hastata* (Starr and McCrae 1983; Bourne 1991; Sizemore and Palensky 1993; Shirley and Kruse 1995; Shumway and Castagna 1995; Lauzier and Parker 1999). The Pacific calico scallop, *Argopecten ventricosus*, was harvested in southern California but has been protected since 1954 (Bourne 1991). Minor landings of the Bering scallop, *Chlamys behringiana* (*=pseudoislandica*) were reported from Alaska in 1991 and 1993 (Shirley and Kruse 1995). The other species may be taken in recreational fisheries or for souvenirs but landings are minor.

The absence of sizeable scallop populations is real since numerous surveys have been undertaken to investigate scallop resources in these waters (Quayle 1961, 1963; Rathjen and Rivers 1964; Haynes and Powell 1968; Ronholt and Hitz 1968; Bourne 1969). Other surveys investigated invertebrate resources off the west coast of North America (Grau 1959; Bernard et al. 1967, 1968, 1970). In addition thousands of tows have been made with otter trawls and other dragging gear throughout the area by both research and commercial vessels. No large unknown or unexploited beds of scallops are believed to exist in this area.

966

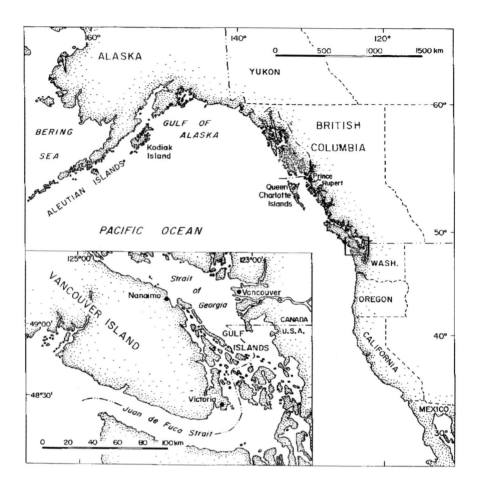

Figure 18.1. Map of the northeast Pacific showing the location of some places mentioned in the text.

Although the fisheries are small they are important to local economies and biological studies have been undertaken on several species to develop management policies for fisheries. Further, some species can be readily cultured and initial studies show potential for establishment of a significant aquaculture industry in the area.

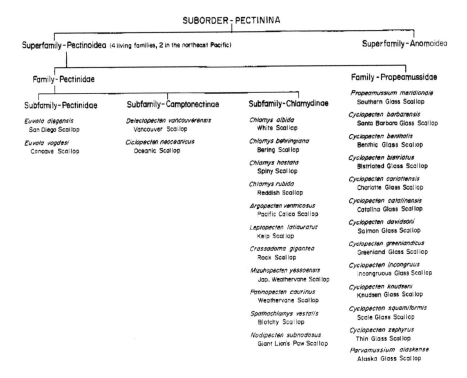

SUBORDER - PECTININA

Superfamily - Pectinoidea (4 living families, 2 in the northeast Pacific)

Superfamily - Anomoidea

Family - Pectinidae

Family - Propeamussidae

Subfamily - Pectinidae	Subfamily - Camptonectinae	Subfamily - Chlamydinae
Euvola diegensis San Diego Scallop	*Delectopecten vancouverensis* Vancouver Scallop	*Chlamys albida* White Scallop
Euvola vogdesi Concave Scallop	*Ciclopecten neoceanicus* Oceanic Scallop	*Chlamys behringiana* Bering Scallop
		Chlamys hastata Spiny Scallop
		Chlamys rubida Reddish Scallop
		Argopecten ventricosus Pacific Calico Scallop
		Leptopecten latiauratus Kelp Scallop
		Crassadoma gigantea Rock Scallop
		Mizuhopecten yessoensis Jap. Weathervane Scallop
		Patinopecten caurinus Weathervane Scallop
		Spathochlamys vestalis Blotchy Scallop
		Nodipecten subnodosus Giant Lion's Paw Scallop

Propeamussium meridionale
Southern Glass Scallop

Cyclopecten barbarensis
Santa Barbara Glass Scallop

Cyclopecten benthalis
Benthic Glass Scallop

Cyclopecten bistriatus
Bistriated Glass Scallop

Cyclopecten carlottensis
Charlotte Glass Scallop

Cyclopecten catalinensis
Catalina Glass Scallop

Cyclopecten davidsoni
Salmon Glass Scallop

Cyclopecten greenlandicus
Greenland Glass Scallop

Cyclopecten incongruus
Incongruous Glass Scallop

Cyclopecten knudseni
Knudsen Glass Scallop

Cyclopecten squamiformis
Scale Glass Scallop

Cyclopecten zephyrus
Thin Glass Scallop

Parvamussium alaskense
Alaska Glass Scallop

Figure 18.2. Scallop species found in coastal waters off the west coast of North America. After Coan et al. (2000).

18.2 FISHERIES

18.2.1 Weathervane Scallop

The weathervane scallop (*Patinopecten caurinus*) provides the largest scallop landings on the west coast and is the species generally referred to when people enquire about the feasibility of establishing a large scallop fishery in the area (Table 18.1).

18.2.1.1 Biology

Weathervane scallops are large and can attain a shell height of 250 mm (Hennick 1973). The lower right valve is larger and more convex than the left and the external colour is yellowish-white to light brown. There are 18–22 squarish flat-topped ribs that radiate from the umbo to the ventral margin. The upper left valve is pale brown to brick red in colour and has 18–22 low and gently rounded ribs. Interior surfaces of both valves

Table 18.1

Landings of scallops in tonnes (whole weight) from the west coast of North America 1967–1999. Unless noted otherwise, landings from Oregon, Washington and Alaska are weathervane scallops, *Patinopecten caurinus*; landings from British Columbia are pink and spiny scallops *Chlamys rubida* and *C. hastata*.

Year	Oregon	Washington Pink & Spiny	Total	Weathervane	British Columbia	Alaska
1967	0		-		0	3
1968	0		-		0	6,466
1969	0		-		0	7,132
1970	0		-		0	5,552
1971	0		-		0	3,589
1972	0		-		0	4,499
1973	0		-		0	4,277
1974	0		-		0	1,945
1975	0		-		0	1,680
1976	0		-		0	Confidential
1977	0		-		0	Confidential
1978	0		-		0	0
1979	0		-		0	96
1980	0		1		0	2,377
1981	7,645		47		0	3,564
1982	675		18		8	3,523
1983	715		Tr		11	750
1984	1,430		142		17*	1,503
1985	371		23		53*	2,418
1986	48		20		62	2,631
1987	6		16		66	2,319
1988	13		0		57	1,315
1989	Tr		0		66	2,061
1990	Tr	23		0	69	5,680
1991	Tr	24		Tr	82	4,382
1992	130	12		49	91	6,714
1993	70	16		98	90	5,901
1994	Tr	27		1	104	4,862
1995	Tr	38		35	96	1,583
1996	0	16		3	102	2,823
1997	4	5		0	73	3,099
1998	22	4		0	54	3,220
1999	0	Tr		0	?	3,230

* Includes 1.4 t and 0.3 t of weathervane scallops in 1984 and 1985, respectively.
Tr = Trace, under 1 t

are smooth and glossy white (Coan et al. 2000).

Weathervane scallops are found from central California (37°N) to the northern part of the Gulf of Alaska (60° N), west in the Aleutian Islands to Amlia Island (53°N 174°W) and into the Bering Sea on Petral Bank in depths of 10–300 m, generally on sand or mud bottom (Grau 1959; Bernard 1983; Kaiser 1986; Shirley and Kruse 1995; Foster 1997; Coan et al. 2000). Distribution is patchy throughout the range and no extensive beds occur; e.g., in British Columbia (B.C.) the two largest centres of population (both relatively small) are in the Gulf Islands region in the southern part of the Strait of Georgia and in Dixon Entrance off the north coast of the Queen Charlotte Islands with small local populations reported from a few other locations. In areas where commercial fisheries occur, distribution is sporadic and local (Starr and McCrae 1983; Kaiser 1986; Shirley and Kruse 1995). In Alaska the fishery occurs on many small beds scattered throughout the vast coastal waters of the southern part of the State but mainly in the Yakutat, Kodiak Island and Dutch Harbour areas (Shirley and Kruse 1995; Kruse et al. 2000). Maximum abundance off Oregon was in 80–100 m and in 70–100 m in Alaska (Starr and McCrae 1983; Kaiser 1986; Shirley and Kruse 1995).

It is interesting that although B.C. is in the centre of distribution of weathervane scallops, only minor populations occur there. The reason for this lack of extensive populations is unknown. It may be due to environmental conditions, excessive predation or perhaps diseases (Bower et al. 1999).

Sexes are separate although a single occurrence of hermaphrodism was reported (Hennick 1971). Sexual maturity is attained at a shell height of 70 mm and spawning occurs at various times. Off Oregon, spawning extended from mid January through June (Robinson and Breese 1984). In the Gulf Islands population of southern B.C., gonads began to fill out in late autumn and by January, it was possible to sex individuals macroscopically by colour. By early April, the gonads were full and spawning occurred from late April to mid June (MacDonald and Bourne 1987). In Alaska, gonads developed from October to May and a single spawning occurred from early June to mid July (Hennick 1970).

Larval development was studied in the laboratory. At 14°C, larvae reached maturity, 250–270 μm, in 28–32 days. Large mortalities were experienced at metamorphosis and few juveniles were raised in the laboratory (Bourne, unpublished data).

Age and growth have been determined for several weathervane scallop populations along the west coast. In B.C., significant differences were observed in growth of populations in the Strait of Georgia and off the west coast of Vancouver Island (MacDonald and Bourne 1987). Mean shell height at annuli and asymptotic shell heights were greater for inshore scallops where they attained a shell height of 100 mm in three years compared to five years in the offshore population. This difference in growth rate was probably due to differences in environmental conditions between the two locations. Another factor causing the difference in growth may be that offshore scallops were generally riddled with sponges and annelids that may cause a diversion of energy for growth to shell repair (Haynes and Hitz 1971; MacDonald and Bourne 1987).

Off Oregon, differences in growth were reported for two populations, one required almost five years to attain a shell height of 100 mm and the other six years (Starr and McCrae 1983). Scallops off Oregon also tended to be riddled with invasive organisms.

In Alaska, growth was found to vary among areas but the asymptotic shell height was similar to that for inshore populations in B.C. (Kaiser 1986; Shirley and Kruse 1995). Scallops attained a shell height of 100 mm in less than two years around Kodiak and in about five years in the Cape Fairweather to Cape Yakutat area.

Weathervane scallops are reasonably long lived. In the Strait of Georgia, animals that measured 230 mm shell height were over 20 years old. In Alaska, a scallop that measured 250 mm shell height was 28 years of age (Hennick 1973).

Few studies have been undertaken to assess causes of mortality in weathervane scallops. Kruse (1994) estimated the instantaneous natural mortality (M) of weathervane scallops was between 0.04 and 0.25, corresponding to annual mortality rates of 4–22%.

Recruitment has been assessed only briefly along the coast. In Oregon, one or two strong years classes supported the brief intense fishery that began in 1981, but since then recruitment has been low. In B.C. and Alaska, recruitment appears to be consistent but low. One indication of low recruitment in B.C. is that size distribution in the Gulf Islands population has varied little over years of sampling with no indication of strong year classes (Bourne, unpublished data).

Limited assessments have been made of populations along the coast (Bourne 1991). A study in Alaska calculated scallop biomass ranged from 12,335 t to 17,445 t in the Cape Spencer-Cape St. Elias area (Ronholt et al. 1977) but this estimate should be regarded as minimal since inefficient gear was used in the survey (Kaiser 1986). Studies are currently underway in Alaska to attempt estimation of scallop populations there (Kruse et al. 2000). Estimates of the small population in the Gulf Islands area of southern B.C. from both dragging operations and observations from the underwater submersible PISCES showed the density was about 1 scallop per 63 square meters (Bourne, unpublished MS).

Size frequency distribution in most populations shows a skew towards larger animals. When liners were used in drags, few small scallops (under 100 mm shell height) were caught. Off Oregon, most scallops caught in surveys were under 130 mm shell height. At two sites, most were between 100–130 mm shell height but at another location most were under 100 mm. Larger animals tended to occur in shallower waters (Starr and McCrae 1983). In Alaska, scallops were considerably larger than reported off Oregon and in commercial samples, most were over 120 mm shell height (Kaiser 1986). The age composition of scallops from commercial catches from 1968 through 1972 showed most were 7–11 years in age in 1968–1970 catches but scallops 2–6 and 7–11 years in age were about equal in 1971 and 1972 samples, indicating the effect of the fishery (Hennick 1973; Shirley and Kruse 1995). In B.C., in spite of using a variety of sampling gear, most animals harvested in the Gulf Islands area have been over 125 mm shell height.

An interesting phenomenon of weathervane scallops occurs in B.C. on the Masset beaches located on the northeast coast of Graham Island in the Queen Charlotte Islands. Generally each winter, live adult weathervane scallops are washed ashore and stranded on these beaches after severe storms. Such strandings may occur three to four times during some winters. Numbers washed ashore may amount to several thousand and they are collected by residents for local consumption (R. Schatz, pers. comm.). The reason for these strandings is unknown. Several scallop surveys in these waters have failed to locate any significant populations (Bourne 1969).

18.2.1.2 Fishery

Minor landings of weathervane scallops occurred sporadically along the coast until the late 1950's. The only recorded landings for this period were from Washington and between 1936–1952, they averaged about 360 t (Cheney and Mumford 1986).

Explorations in the 1960's led to development of a commercial fishery in Alaska that began in 1967 and has continued to the present (Kaiser 1986; Kruse and Shirley 1994; Shirley and Kruse 1995;. Foster 1997; Kruse et al. 2000). Landings have fluctuated for several reasons from a peak of 7,132 t (whole weight) in 1969 to nil in 1978 (Table 18.1). After 1980 they began to increase again and reached another peak of 6,714 t in 1992 but have stabilised at about 3,000 t after 1996 because of management policies (Kruse et al. 2000) (Table 18.1). The number of boats in the fishery has fluctuated from a high of 19 to nil, from 1979 to 1993 they ranged from 1–18 (Kaiser 1986; Shirley and Kruse 1995). Size of boats in the fishery has varied from a keel length of 18 m to over 31, most are between 24–38 m. In recent years, scalloping has become the sole source of income for many of the vessels.

Hennick (1970) described three main fishing areas in the early years of the Alaskan fishery: 1) eastern Gulf of Alaska between Cape Spencer and Cape St. Elias; 2) west side of Kodiak and Afognak Islands and the Shelikof Strait including mainland areas of the Alaska Peninsula; 3) east side of Kodiak and Afognak Islands including Albatross and Portlock Banks. These areas have continued to be important harvesting areas for the fishery, but as the fishery expanded in the 1980's, scallopers explored and exploited new scallop grounds. Landings began and have continued in the Dutch Harbour area, and harvesting began in the Bering Sea in 1986 and extended as far west as Amlia Island along the Aleutian Islands chain. In 1992, scallops were harvested from Prince William Sound for the first time (Shirley and Kruse 1995; Kruse et al. 2000).

In B.C., only minor harvesting has occurred on two small populations in Dixon Entrance and in the Gulf Islands area of the Strait of Georgia. Peak landings were 1.4 t in 1984 (Table 18.1) (Farlinger and Bates 1985).

In Washington, a sporadic fishery has continued on beds located in the coastal area where landings have ranged from a peak of 142 t in 1984 to 0 (Table 18.2) (R.E. Sizemore, pers. comm.). This small sporadic fishery will probably continue and depend on the size of the stocks, economics of harvesting and success in other fisheries.

In the local intense fishery off Oregon, landings peaked at 7,645 t in 1981 and 1,430 t in 1984, but then declined (Table 18.1) (Starr and McCrae 1983). Small landings have continued in a sporadic fishery that depends partly on size of populations and interest in the fishery (Robinson 1997; R. Hannah, pers. comm.).

A small one or two boat fishery harvested weathervane scallops in northern California in 1981 that was an extension of the intense Oregon fishery (Bourne 1991).

Table 18.2

Annual landings (tonnes) and effort in the pink and spiny scallop fishery in British Columbia, 1982–1998, as reported on harvest logs.

Year	Type and # of Licences Issued Dive	Trawl	# of Vessels with Landings From Logs Dive	Trawl	Total Vessel Fishing Hours From Logs Dive	Trawl	Dive Gear Landings From Logs (t)	Trawl Gear Landings From Logs (t)
1982			1	1	5	157	#	#
1983	Z 11		4	2	196	N/A	8.0	#
1984	Z 17		5	4	198	115	15.6	4.3
1985	Z 22		4	3	43	272	10.8	10.5
1986	Z 24		7		542	0	35.7	0
1987	ZI 29		8		1,317	0	69.0	0
1988	ZI 17		9	4	645	318	48.9	9.1
1989	ZI 43		7	5	382	917	32.4	29.1
1990	ZI 57		9	6	639	356	64.2	10.0
1991	ZI 61		7	12	966	550	47.6	23.1
1992	ZI 83		8	10	640	625	38.4	21.1
1993	ZI 35	ZR 44	8	12	741	453	77.3	9.9
1994	ZI 37	ZR 32	15	4	767	109	73.4	4.2
1995	ZI 29	ZR 43	14	9	652	843	76.1	9.8
1996	ZI 39	ZR 40	14	6	1,073	404	94.5	7.4
1997	ZI 41	ZR 38	9	4	1,244	248	73.6	4.0
1998	ZI 20	ZR 34	9	5	1,104	266	54.6	4.7
1999	ZI 22	ZR 16	8	3	824	181	36.7	2.2
2000*	XZI 12	XZR 2	10	-	1,096	-	53.2	-
2001	XZI 10	XZR 2	7	1	710	#	43.5	#
2002	ZXI 10	XZR 6	9	5	794	144	49.4	2.7

* In 2000 both Dive and Trawl commercial fisheries were discontinued and converted to experimental status; 2001 and 2002 are Preliminary Data. # Landings not reported due to confidentiality requirements.

Limited populations of weathervane scallops have been found off Pelican Bay but they are too small to support commercial harvesting.

In the Alaskan fishery, the catch is shucked at sea, adductor muscles bagged and sold when the vessel lands. Attempts were made to use mechanical shuckers but they are now prohibited in the fishery (Shirley and Kruse 1995; Kruse et al. 2000). Most of the catch in the Oregon fishery was shucked at sea but some catches of whole scallops were landed, particularly by smaller boats and shucking was done at shore facilities.

18.2.1.3 Gear

Several types of gear have been used to harvest weathervane scallops on the west coast. Prior to development of fisheries in the 1960's, most of the gear was made locally and frequently old shrimp trawls were used. With development of the Alaskan fishery, most of the gear used was the typical east coast offshore drag (Bourne 1964). Initially many of vessels in the fishery were east coast scallopers that went to Alaska because of the depressed fishing conditions on the east coast (Kaiser 1986). Some Alaskan boats converted to scallopers. The fishery off Oregon was started by east coast scallopers on their way to Alaska (Starr and McCrae 1983; Robinson 1997). Some local boats converted to offshore scallopers and used east coast type drags but other boats, particularly smaller ones, used a variety of gear including shrimp trawls (Starr and McCrae 1983).

18.2.1.4 Management

Several policies have been established to manage weathervane scallop fisheries, the most extensive are in Alaska where the major fishery occurs.

Shirley and Kruse (1995) and Kruse et al. (2000) have described management of the weathervane scallop fishery in Alaska in detail. In the initial stages of the fishery, management was by a set of passive regulations that included gear restriction, type of gear, minimum inside diameter of rings on drags, closed areas to protect crab resources and fishing seasons. With resurgence of the fishery in the 1990's, there was a shift to a more active management of the fishery and the policy now includes number and size of drags, minimum inside diameter of rings (10.2 cm), establishment of nine registration areas with established harvest limits, an overall limit on landings of 4,700 t (whole weight) (1.24 million lbs. shucked meats), area closures to protect crab habitat, limited entry, limit on crew size to 12, ban on automatic shucking machines and establishment of an onboard observer plan to collect data and monitor incidental catches.

When the commercial fishery was open in B.C., weathervane scallops could only be caught with dragging gear with a maximum width of 2 m and a size limit of 120 mm shell height is also in place. Under new management policies, if large quantities of weathervane scallops were found off the B.C. coast it would be classed as a new and developing fishery and biological information would have to be gathered before a commercial fishery was permitted (Perry et al. 1999).

In Washington, the commercial fisheries are regulated by gear size and mesh or ring size. Harvest is also regulated by the tribal rights treaty agreement.

In Oregon, the scallop fishery is regulated by limited entry, a permit system, type of gear and mesh or ring size.

18.2.2 Rock Scallop

Rock scallops (*Crassadoma gigantea*) are unique to the Pacific coast of North America and have an interesting life history.

18.2.2.1 Biology

Rock scallops are large and massive and can attain a shell height of 250 mm (Bourne 1969; Coan et al. 2000). The valves are irregular in shape. In younger animals, the external surface of the upper left valve is sculptured with strong unequal radiating ribs studded with short spines. The valves become massive and are brown or green in colour and are frequently pitted with holes of invasive organisms or encrusted by plant and animal growth. Internally the shell is smooth and white except for the deep purple colour at the hinge area.

An interesting feature of rock scallops is that until they are 2–3 cm in shell height they are free swimming. Auricles on the upper left valve are unequal, the scallop has a roundish appearance and the valves are usually orange in colour. When 2–3 in cm shell height they attach to a rock or other hard substrate by the lower right valve and remain there for life, hence the name purple-hinge rock scallop or rock oyster. In laboratory and hatchery studies, it was found that a small percentage of juveniles ($\approx$ 10%) did not attach to a substrate but remained unattached on the bottom of tanks.

Rock scallops occur from Mexico to the Aleutian Islands in Alaska, Lat. 25–60° N, from the lower intertidal zone (1 m intertidal) to subtidal depths of 80 m (Grau 1959; Bernard 1983; Coan et al. 2000). They occur mostly on subtidal cliffs in areas with strong currents or oceanic surges. Distribution is patchy and they are not particularly abundant in any one locality. No attempts have been made to estimate population size or structure in any area. Generally, there is a preponderance of older animals in most areas.

Rock scallops cannot be aged by measuring annual rings, since the valves become so massive that rings are obliterated. Estimates of age and growth have been made by counting rings on the resilium and from observations in culture work (Leighton 1979; MacDonald and Bourne 1989; MacDonald et al. 1991). In California, *C. gigantea* attained a shell height of 120 mm in two years (Leighton 1979). In B.C., rock scallops in natural populations attained a shell height of 100 mm in about 5 years but a shell height of 90–100 mm, mean 80 mm in three years when grown in suspended culture (MacDonald and Bourne 1989; MacDonald et al. 1991). Rock scallops of 27 years of age were recorded in B.C.

Sexes of rock scallops are separate although one individual hermaphrodite was reported in California (Jacobsen 1977).

Breeding was studied in California and in Puget Sound, Washington. In southern California spawning occurred over a protracted period in two populations (Jacobsen 1977). Both populations spawned from October to January, one population spawned again from March to June and the other from June to August. In Puget Sound, rock scallops spawned from June to August (Lauren 1982). Observations in B.C. indicated that rock scallops have a protracted spawning period from June to October although there may be one or two pulses during this period. The gonad never completely emptied and animals could be sexed macroscopically by colour throughout the year. In experimental work in B.C., rock scallops from wild populations were injected with seratonin biweekly to stimulate spawning, and viable eggs and sperm were obtained from June to January (Bourne, unpublished data).

Larval development has been studied in the laboratory during culture work and also in commercial hatcheries (Cary 1982; Bourne et al. 1989; Agoni, pers. comm.). Larvae reached a mature size of 220 μm in 18–20 days at 21°C.

Observations of populations along the coast indicate that recruitment is probably low but consistent; however, spatfall may be heavier than indicated by recruitment. Quantities of juvenile rock scallops have been collected on oyster strings grown in suspended culture in B.C. and elsewhere. In attempts to collect scallop spat on the west coast of Vancouver Island, about 25% of the collected juveniles were rock scallops, the remainder were pink and spiny scallops, *Chlamys rubida* and *C. hastata* (Bourne 1988; Bourne et al. 1989). Whether sufficient quantities of juvenile rock scallops could be collected consistently along the coast to support large-scale culture operations is not known.

18.2.2.2 Fishery

Attempts were made by scuba divers to harvest rock scallops commercially but they failed because it was too difficult and costly to chisel them off rocks. At present, there is no commercial fishery for rock scallops along the west coast.

Minor landings (less than 1 t annually) reported from California are from aquaculture operations (Shaw 1997; R. Hulbrock, pers. comm.).

Rock scallops are harvested in recreational fisheries along the coast and the catch is regulated by daily harvest limits.

18.2.3 Pink and Spiny Scallops

Two *Chlamys* species, pink scallops and spiny scallops occur widely with overlapping distributions, on the west coast of North America. Fisheries for these two species occur mainly in the inside waters of Washington and B.C.

18.2.3.1 Biology

The pink scallop, *Chlamys rubida* (Hinds, 1845), also known as the smooth pink scallop, reddish scallop, and the swimming scallop (Harbo 1997), was once considered to be a subspecies of *C. islandica* (Waller 1991). A number of subspecies have been named for Asiatic populations; however, they do not appear to have any morphological features that are outside the range of variation for eastern Pacific populations (Waller 1991).

The spiny scallop, *C. hastata* is also known as the pink scallop, Pacific pink scallop, and the swimming scallop (Harbo 1997). There are number of subspecies; *C. hastata hastata*, which ranges furthest south; and *C. hastata hericia* and *C. hastata pugetensis*, which are the two more northerly subspecies (Waller 1991). Here, all are considered as *C. hastata*.

Grau (1959) and Coan et al. (2000) give detailed descriptions of both species. Left valves of pink scallops have 20–42 finely imbricated primary ribs, interspaced with one or more strongly imbricated riblets and reticulated microsculpture. The large anterior auricle has 5 to 12 imbricated radial riblets. The right valve has 18–30 broad flat radial ribs, bifurcating ventrally, interspaced with a few intercalary riblets, and strongly reticulated microsculpture. The medium-size anterior auricle has 5–6 low smooth radial ribs. The right valve is generally white, and the left valve may be red, pink, orange, or rarely white. Maximum shell height is 71 mm.

Left valves of the spiny scallop have 8–20 primary ribs with strong imbricated spines, interspaced with 1–10 coarsely scaled intercalary riblets and fine striose microsculpture. The anterior auricle has many strongly imbricated radial ribs. The right valve has 12–24 rounded primary ribs, similar to the left valve, but with only 3–4 intercalary riblets. The large anterior auricle has 7–10 imbricated radial riblets. The colour is usually pink to yellow, and often streaked. Maximum shell height is 80 mm.

Shells of pink and spiny scallops are usually encrusted with one of two species of symbiotic sponge, *Myxilla incrustans* and *Mycale adherens* (Harbo 1997). Barnacles, 25 to 30 mm in height, are often attached (Grau 1959).

Pink scallops are found from 33°N – 58°N, San Diego, California to Kodiak Island, Alaska (Bernard 1983). It is not common south of Puget Sound, Washington (Harbo 1997). The spiny scallop is found from 33°N – 60°N, San Diego, California to the Gulf of Alaska (Bernard 1983). Both species have a discontinuous distribution throughout their range along the coast, but can occur in small dense beds (Bourne 1991).

Pink and spiny scallops are epibenthic swimming scallops that typically lie with the right (more convex) valve on the bottom. Like many other *Chlamys* species, they can form a byssus throughout their life, and retain the ability to alternate between attached and detached phases, a characteristic of species with a small body size or species that inhabit relatively high current areas (Brand 1991).

Pink scallops are found at depths ranging from 1 to 200 m in water temperatures ranging from 1°C to 17°C (Bernard 1983). Pink scallops are usually found on softer substrates than spiny scallops (Bourne 1991). Ellis (1967) found only pink scallops on soft harbour sediments in the southern Gulf Islands of British Columbia. Grau (1959) reported pink scallops as usually found off rocky shores. Quayle (1963) found pink scallops throughout coastal B.C. on various substrates, predominantly in sandy and mud substrates, but occasionally on gravel and rocky bottoms.

Spiny scallops are found at depths ranging from 2 to 150 m in water temperatures ranging from 0°C to 23°C (Bernard 1983). They usually occur on firm gravel or rocky substrates (Grau 1959; Bourne 1991) and rocky reefs (Harbo 1997) in areas of strong current (Bourne 1991) but have been found on muddy and sand substrates (Ellis 1967; Quayle 1963). Current was found to be a stimulus to larval settlement (Cooke 1986).

Both pink and spiny scallops are dioecious, and display asynchronous seasonal reproductive activity. Their gonads may not completely empty during spawning, so these species can be sexed throughout the year by the colour of the gonads (Bourne and Harbo 1987). Both species mature at 2 years (Bourne and Harbo 1987) and 25–35 mm shell height. Pink scallops spawn twice a year, first in March and again in September or October, spiny scallops spawn once a year in July or August (Macdonald et al. 1991). In pink scallops, reproductive effort slowly increases with age to an asymptotic maximum that does not exceed its somatic production, but in spiny scallops, reproductive effort increases sharply after three years and exceeds somatic production in its fifth year

(Macdonald et al. 1991). This pattern of increasing reproductive effort is characteristic of most iteroparous pectinids.

Settlement of spiny scallop larvae occurred in 34 days in 16°C water and 42 days in 12°C water (Cooke 1986; Hodgson and Burke 1988). Settlement and metamorphosis was only seen on fouled surfaces. A flow of water also enhanced settlement and metamorphosis.

Growth of both pink and spiny scallops was determined for animals collected from commercial catches (Bourne and Harbo 1987) and during experimental fisheries (Lauzier et al. In Prep). Spiny scallops attain a shell height of 60 mm in 3½ years, pink scallops in 5 years. MacDonald et al. (1991) showed differences in growth and reproductive strategies between pink and spiny scallops. Until recently, the maximum age for both species was thought to be 6 years (Macdonald et al. 1991). However, a small number of individuals (~1%) in biological samples collected from harvested populations in southern B.C. waters were aged at 7, 8 and 9 years (Lauzier, unpubl. data). Preliminary estimates of population size or structure, recruitment, natural and fishing mortality have been undertaken in the first two years of experimental fisheries on pink and spiny scallops (Lauzier et al. In Prep).

18.2.3.2 Fisheries

A small experimental fishery has existed in Washington since 1990 (Sizemore and Palensky 1993) (Table 18.1). There was a sharp decline in landings after 1996 due to large decrease in effort. The fishery is confined to a few vessels certified by the Washington State Department of Health operating in restricted areas and subject to weekly PSP (paralytic shellfish poisoning) monitoring. This fishery is managed by Department of Health actions on PSP monitoring and closures. The fishery is closed annually between July 1 and October 1 due to consistently elevated PSP levels during this period. This fishery is limited by PSP closures rather than stock abundance (R.E. Sizemore, pers. comm.).

Pink and spiny scallops have been commercially harvested in B.C. from 1982 to December 1999 and from 2000 to date under experimental licence. The fishery was confined to inshore waters using two different harvest techniques: a dive fishery and a trawl fishery. Total scallop landings in B.C. from both the dive fishery and the trawl fishery from 1982 to 2002 are shown in Fig. 18.3 and Table 18.2; landings for 2001 and 2002 are preliminary. Landings in the dive fishery comprise 86.8% of the cumulative commercial and experimental scallop fisheries to date. There was a decline in number of vessels as well as vessel days in the trawl fishery in the early to mid 1990's, and a concurrent increase in effort in the dive fishery (Fig. 18.3, Table 18.2).

During the commercial fishery prior to 1999, there was very little available information on the status of pink and spiny scallop stocks in B.C., the only available data were from reported effort and landings. However, fluctuations in effort and landings may have been due to the lack of markets and PSP closures. Pink and spiny scallops are marketed whole, either fresh or frozen. Harvesters claim that limited markets have impeded development of this fishery in the past. Some areas were closed to the fishery

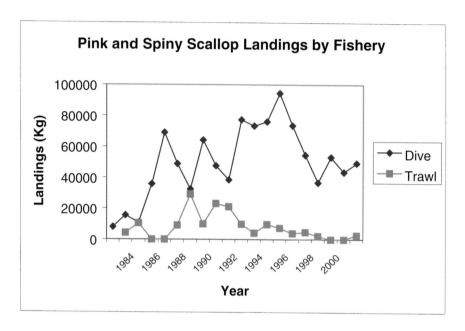

Figure 18.3. Total pink and spiny scallop landings by fishery (trawl and dive) from British Columbia waters, 1982–1999.

because of concerns based on anecdotal information. Experimental fisheries and fishery-independent surveys were implemented in 2000 to provide sufficient information for scientifically-based assessments.

The trawls used in the fishery have evolved since 1990 to produce a marketable product, which is competitive with that in the dive fishery, and to reduce bycatch. The trawls, consisting of a steel or aluminium frame on runners fitted with a rope mesh net bag 2 m wide by 1.4 m deep, were designed to capture scallops as they are swimming, since the crossbar and the bottom of the trawl net is usually 20 cm off the bottom. Only one trawl with a maximum 2 m width is permitted per boat. One particular model, the butterfly scallop trawl, is most commonly used in the trawl fishery. Tests conducted during the experimental fisheries program show considerable variation in butterfly scallop trawl efficiency, but there is virtually no bycatch, and negligible habitat damage.

The trawl fishery occurs generally in depths that are greater than diveable depths, where pink scallops occur more commonly. There is also no size selectivity in the trawl fishery, especially as the trawl bag begins to fill, resulting in a higher proportion of smaller pink scallops in the trawl fishery, as compared to catches in the dive fishery.

Most of the catch in the dive fishery is spiny scallops. Analysis of samples from the dive fishery showed that 81–99% of the catch was spiny scallops, showing divers target on this species. Pink scallops typically occur deeper than the spiny scallops, generally beyond the diveable depths. Divers in the dive fishery select for larger scallops because of higher economic return.

18.2.3.3 Management

Pink and spiny scallops were passively managed as one species, one stock in B.C. with unlimited entry, minimum size limits (55 mm shell height) as well as seasonal and area closures. Evidence indicates that this management policy was insufficient to sustain stocks in heavily harvested areas. Effort in the B.C. scallop fishery was largely market driven and limited by PSP closures. As markets developed it was felt that a more active management policy was needed to allow for orderly and sustainable development of the resource. Data limited fisheries, e.g., the pink and scallop fisheries, are now considered to be new and developing fisheries, and a process has been initiated, which includes an extensive review of existing information, followed by a framework for experimental fisheries in order to collect additional information identified as missing but necessary for ongoing assessments and a biologically based management policy (Perry et al. 1999).

After an initial review, it was concluded the overall status of pink and spiny scallops in B.C. was unknown, as there was insufficient data for assessment (Lauzier and Parker 1999) and the commercial scallop fisheries were closed on December 31, 1999. Information gaps identified through this review included the distribution of pink and spiny scallop stocks, estimates of biomass, availability of suitable stocks to the fishery, recruitment, natural and total fishing mortality.

B.C. pink and spiny scallop fisheries are now considered as experimental fisheries (Perry et al. 1999). A framework for assessment and management of B.C. pink and spiny scallop fisheries was prepared and approved that included identification and delineation of scallop assessment and management units, along with a systematic plan of biomass surveys (Lauzier et al. 2000). Experimental fisheries and biomass surveys that collect biological samples to assess age structure, growth and natural mortality have been undertaken since 2000 in collaboration with the dive and trawl fleets. There is also an ongoing re-evaluation of minimum shell height and its applicability to each species. The present plans include progressing towards limited dive and trawl commercial fisheries, with continued regular biomass surveys and biological data collection for scientifically based assessment programs and management plans.

18.3 AQUACULTURE

Scallop populations along the west coast of North America are limited. No large unknown harvestable beds are believed to exist in the area and hence future landings from fisheries will remain similar to those at present. Current management policies are designed to maintain these levels of landings. Occasional dense sets of weathervane scallops may occur in some areas that could support periodic intense local fisheries as occurred in Oregon (Starr and McCrae 1983) but this would be infrequent. If scallop landings are to increase significantly on the Pacific coast they will have to result from development of scallop aquaculture operations.

There is considerable interest in scallop culture along the west coast for several reasons, among others; the emerging acceptance of aquaculture as a reliable method to increase aquatic production, the long success of the oyster industry and recent development of Manila clam, *Venerupis philippinarum*, culture in the area, strong markets for scallops, but above all the outstanding success of scallop culture in Japan (Ventilla 1989; Aoyama 1989; Ito 1991) and more recently in China (Guo et al. 1999). Scallop production from Japan and China is entirely from culture operations and accounts for about 90% of present world landings. The west coast has considerable areas of protected waters and the belief is they could support extensive scallop culture operations.

Initial attempts have been made to investigate the feasibility of scallop culture in several areas along the west coast and a small industry exists at present (Bourne and Bunting 1995). In any culture operation a primary requisite is an abundant, reliable and inexpensive supply of juveniles (seed). Much of the scallop culture research along the west coast has focused on developing methods to provide this consistent supply of seed. Juveniles for culture operations can be obtained in two ways; from natural sets, or from production in hatcheries. There are advantages and disadvantages to both methods. The large culture operation in Japan obtains its entire supply of juveniles from natural sets (Ventilla 1982). Where large consistent natural sets do not occur, juveniles must be produced in hatcheries but this can add significantly to costs of culture operations.

18.3.1 Pink and Spiny Scallops

Attempts were made by research workers and industry in Washington and B.C. to culture pink and spiny scallops (Bourne 1991). An advantage of pink and spiny scallop culture is that seed can be collected from natural sets, up to 2,000 juveniles per collector bag. Species composition of sets was not determined although it was believed most of the sets were spiny scallops. Some sets had rock scallops and at one site approximately 25% of the set was rock scallops. A problem in using juveniles from natural sets could be separating species.

The conclusion from culture studies with pink and spiny scallops was they are too small and growth is too slow (MacDonald et al. 1991) to support an economically viable culture operation. No further work was done with either species other than perhaps as a hobby farm operation.

Another problem with pink and spiny scallop culture is that as mentioned previously, the whole scallop is marketed. Outbreaks of shellfish toxins could cause periodic closures of pink and spiny scallop culture operations (Shumway and Cembella 1993).

18.3.2 Weathervane Scallop

The weathervane scallop appears to be an ideal candidate for culture on the west coast. It is native, large, has a rapid growth rate and it should be possible to produce commercial sized animals (100 mm shell height) within a period of 2–3 years. Experimental work was undertaken in Washington, B.C. and Alaska to investigate the potential for weathervane scallop culture.

Attempts to collect juveniles from natural sets in all three areas failed. In Washington, no juveniles were caught. In B.C. only 10 were collected in one year and none in two other years, probably because of heavy sets of the crab, *Cancer oregonensis* in the collector bags. In Alaska, a joint project between the State of Alaska and the Japanese Overseas Fishery Cooperation Foundation (OFCF) attempted to collect juvenile weathervane scallops but few were caught. The conclusion from this work was that no areas are known along the coast where large quantities of juvenile weathervane scallops can be caught consistently to support a large culture operation as occurs in Japan.

Since juvenile weathervane scallops cannot be obtained from natural sets the alternative is to produce them in hatcheries. Work was undertaken in Washington (Olsen 1983; 1984) and in B.C. by both research workers and industry to develop methods to breed weathervane scallops (Bourne et al. 1989; R. Saunders, pers. comm.). The species has proven extremely difficult to breed successfully in labs or hatcheries and few juveniles were produced, too few to use to develop nursery methods or to use in extensive growout trials. Research studies are needed to solve the bottleneck of producing large quantities of weathervane scallop seed consistently in hatcheries before the species can be considered for culture operations along the coast. At present, there is no current work on weathervane scallop culture on the west coast.

Weathervane scallops were crossed with Japanese weathervane scallops to determine if such a hybrid had advantages for culture in B.C. The cross was successful and DNA testing was undertaken and continues to show this is a successful cross. Breeding of this cross has continued and offspring are used in culture operations in B.C. (R. Saunders, pers. comm.).

18.3.3 Rock Scallops

There was considerable interest in rock scallop culture in California, Washington and B.C. in the 1970's and early 1980's resulting from work in California where methods were developed to produce juveniles in hatcheries and grow them to commercial size in the open environment (Leighton and Phleger 1977; Phleger and Leighton 1980; Phleger et al. 1981; Cary 1982; Phleger and Cary 1983). Results showed it was possible to culture rock scallops to a shell height of 100 mm in two years in California.

Experimental rock scallop breeding was undertaken in Washington (Olsen 1983, 1984) and in B.C. and methods were developed to breed rock scallops and grow them to commercial size (Bourne et al. 1989). In B.C., growth was reasonable and a shell height of 80 mm was attained in three years (MacDonald et al. 1991). Rock scallops have an advantage over other species in that as mentioned above, juveniles attach themselves to a substrate and remain there the rest of their lives. In culture operations, juveniles were attached to a variety of substrates, growth was good and mortalities were low; however, a problem was experienced in growout. Rock scallops tend to assume the shape of the substrate they are attached to and frequently were thin, hence the adductor muscle was also thin. Several substrates were investigated as a surface for attachment but none was widely accepted. Interest in rock scallop culture generally declined.

Recently there has been renewed interest in rock scallop culture along the Pacific coast. Small production has occurred in California incidental to an oyster culture operation (Shaw 1997) and limited harvest was made by divers from offshore oil rigs (C. Culver, R. Hulbrock, pers. comm.). The possibility of using offshore platforms for rock scallop culture is being investigated. Experimental breeding of rock scallops has also occurred in Oregon (A. Robinson, pers. comm.).

In Washington, limited rock scallop culture was undertaken by one company incidental to oyster culture operations; methods were developed to breed rock scallops in a hatchery and grow them to commercial size (J. Agosti, pers. comm.). Annual production was small, under 1 t, but rock scallops from this operation were on the menu of a restaurant in Seattle in March 2000 (Bourne, pers. obs.). Recently another company has initiated a project to produce seed in their hatchery and attempt growout in exposed areas of the coast (J. Davis, pers. comm.).

Significant developments have occurred in rock scallop culture in Alaska (J. Agosti, pers. comm.). Seed is now produced routinely in a hatchery and there are at least three farms licensed to grow and sell rock scallops in the southeast part of the state and more are expected to begin culture operations. It is expected that by the end of 2000 there will be several farms culturing rock scallops in the state. Rock scallop culture could become a significant industry in the state in the next ten to twenty years.

Rock scallop culture has considerable potential along the west coast of North America. The bottleneck of a consistent seed supply has been solved with production from hatcheries. The problem will be to develop nursery and growout systems to permit economically viable production of deeply cupped marketable size rock scallops. However, with continued research and development it is expected this can be achieved in the near future permitting development of a significant rock scallop industry.

18.3.4 Japanese Weathervane Scallop

In 1981, a research project was initiated to determine the economic feasibility of scallop culture in B.C. Initial work focussed on four native species, weathervane, rock, pink and spiny but none proved to be a suitable species. A decision was made to import the Japanese weathervane scallops (*Mizuhopecten yessoensis*) and determine if it could be cultured in B.C. Considerable biological information existed on this species, broodstock was available and a large culture industry existed in Japan (Ventilla 1982). Broodstock was imported under quarantine conditions and methods were developed to breed them in a hatchery, culture juveniles in a nursery and grow juveniles to commercial size. The species proved to be a suitable candidate for culture in B.C.; seed could be readily produced in a hatchery and they could be grown to commercial size (100 mm shell height) at various locations in B.C. within two years (Bourne et al. 1989; Bourne and Hodgson 1991).

Commercial scallop culture began in B.C. in 1989 with construction of a hatchery by a private company using technology developed in the research program although the company has developed considerable new technology since then. Scallops can be bred routinely over an extended period and large quantities of juveniles produced in a land built nursery. Production could be easily expanded to meet a much larger demand. Methods have been developed to avoid predation of juveniles by a flat worm, *Pseudostylochus orientalis*, and to avoid high mortalities in growout that occurred at some locations because of a protozoan disease, *Perkinsus gugwadi* (SPX) (Blackbourn et al. 1998; Bower et al. 1998, 1999). Using offspring of the weathervane-Japanese weathervane cross, a strain of Japanese weathervane scallops has been produced that appears to be immune to this disease and mortality rates are greatly reduced. Scallops measuring 100 mm shell height can be produced within a 2 y period at several sites in B.C. Maximum annual production has been about 200 t from a maximum of ten farms (R. Saunders, pers. comm.).

The company has developed other techniques for scallop culture, among them the ability to ship eyed (mature) larvae and juveniles to all parts of the world. They have successfully shipped them to Morocco, where a Japanese weathervane scallop industry is beginning. Eyed larvae and juveniles have also been successfully shipped to Japan.

The future of scallop culture in B.C. continues to be bright but continued improvements are needed in all phases of culture to make scallop farming more economically attractive. It is estimated that annual production of juveniles will need to be about 40 million to make the hatchery-nursery complex economically viable. To some extent it is a problem of critical mass that requires a sufficient number of people to become involved in scallop culture to make operations of a scallop hatchery economically viable. Scallop culture has been proven in B.C. and the industry must now develop to a sufficient level to attract investment to achieve its potential.

18.3.5 Pacific Calico Scallop

Mention should be made of a Pacific calico scallop, *Argopecten ventricosus*, culture operation. A Washington State company has initiated a project to rejuvenate a scallop culture industry in Baja California, Mexico, using the Pacific calico scallop (Felix-Pico 1991). Some seed may be produced in their Washington hatchery, but it is expected that most if not all operations will be based eventually in Mexico (K. Johnson, pers. comm.).

18.4 FUTURE

Scallop landings from fisheries for natural stocks on the west coast of North America will probably continue at present levels in the foreseeable future. No large unknown populations of any harvestable species are believed to exist in the area that could add significantly to landings and management policies for pink, spiny and weathervane scallops are in place to assure continued existing sustainable production.

Landings of weathervane scallops will continue to be virtually all from Alaska unless dense year classes settle in other areas, but this will be infrequent. Weathervane scallop resources are now actively managed in Alaska with an annual overall cap of 4,600 t that will maintain landings at or below this level. If occasional dense populations are found in other areas, management policies are in place to assure optimum utilisation of the resource.

Landings of pink and spiny scallops are expected to remain low, about 200–300 t annually and will be mostly from B.C. and Washington. A problem in this fishery could be outbreaks of shellfish toxins that appear to be increasing in frequency in recent years. Since the entire soft parts of these species are consumed, such outbreaks could lead to more frequent closures in these fisheries. Another limiting factor in this fishery in B.C. has been market development. Harvesters claim that limited markets have impeded development of this fishery in the past. However, as markets develop and strengthen, there is a need for biologically-based assessment and management systems that allow for orderly and sustainable development of the resource. Resource inventory surveys are being conducted for B.C. pink and spiny scallop stocks, as a first step towards structured stock assessments and an active management system.

Scallop production on the west coast of North America could increase in the future through aquaculture.

There is little, if any, potential for culture of pink and spiny scallops and considerable research and development is required to solve the problem of a large, consistent supply of weathervane scallop seed for culture operations. If juvenile weathervane scallops can be produced at reasonable cost then the species could have potential for culture on the west coast.

A small industry culturing Japanese weathervane scallops exists at present in B.C. that could be greatly expanded. Large quantities of seed can now be produced consistently in a hatchery. The efficiency of all phases of culture need to be improved and strong support is needed from governments to involve more people in scallop culture and make it an economically attractive business. It is expected that production in this culture industry will continue to increase in the future.

There has been a resurgence of interest in rock scallop culture along the entire west coast and the seed supply problem has been solved, at least to some degree, with production from hatcheries. Continuing research and development is required to improve the efficiency of culture operations to make rock scallop culture an economically viable industry. Results of such studies could result in establishment of a rock scallop culture industry on the west coast in the near future.

ACKNOWLEDGMENTS

Many people have aided us in preparation of this paper. In particular we would like to thank the following people for sharing information and data on scallops with us; J. Agosti, J. Barnhart, B. Bunting, C. Carver, J. Davis, R. Hannah, R. Hulbroch, K. Johnson, G. H. Kruse, A. Robinson, R. Saunders, R. Schatz, and R. E. Sizemore.

REFERENCES

Anonymous, 1999. FAO statistics.

Aoyama, S., 1989. The Mutsu Bay scallop fisheries: scallop culture, stock enhancement and resource management. In: J.F. Caddy (Ed.). Marine Invertebrate Fisheries; Their Assessment and Management. J. Wiley and Sons. pp. 525–539.

Bernard, F.R., 1983. Catalogue of the living Bivalvia of the eastern Pacific Ocean: Bering Strait of Cape Horn. Can. Spec. Publ. Fish. Aquat. Sci. 61. 102 p.

Bernard, F.R., Bourne, N. and Quayle, D.B., 1967. British Columbia faunistic survey. A summary of dredging activities in western Canada 1878–1966. Fish. Res. Bd. Can. MS Rep. 920. 61 p.

Bernard, F.R., Bourne, N. and Quayle, D.B., 1968. British Columbia faunistic survey. A summary of dredging activities 1966–1967. Fish. Res. Bd. Can. MS Rep. 975. 6 p.

Bernard, F.R., Bourne, N. and Quayle, D.B., 1970. British Columbia faunistic survey. A summary of dredging activities 1967–1969. Fish. Res. Bd. Can. MS Rep. 1082. 7 p.

Blackbourn, J., Bower, S.M. and Meyer, G.R., 1998. *Perkinsus gugwadi* sp. nov. (incertae sedis), a pathogenic protozoan parasite of Japanese scallops, *Patinopecten yessoensis*, cultured in British Columbia, Canada. Can. J. Zool. 76:954–959.

Bourne, N., 1964. Scallops and the offshore fishery of the Maritimes. Fish. Res. Bd. Can. Bull. 145. 61 p.

Bourne, N., 1969. Scallop resources of British Columbia. Fish. Res. Bd. Can. Tech. Rep. 60 p.

Bourne, N., 1988. Scallop culture in British Columbia. In: S. Keller (Ed.). Proc. Fourth Alaska Aquaculture Conference. Alaska Sea Grant Rep. 88–4. Univ. Alaska. pp. 35–41.

Bourne, N., 1991. Scallops of the west coast of North America. In: S.E. Shumway (Ed.). Scallops: Biology, Ecology and Aquaculture. Developments in Aquaculture and Fisheries Science, Vol. 21. Elsevier, Amsterdam. pp. 925–942.

Bourne, N.F. and Bunting, B.L., 1995. The potential for scallop culture in British Columbia, Canada. In: P. Lubet, J. Barret and J.C. Dao (Eds.). Fisheries, Biology and Aquaculture of Pectinids; 8th International Pectinid Workshop. IFREMER. pp. 75–77.

Bourne, N.F. and Harbo, R., 1987. Size limits for pink and spiny scallops. In: Status of Invertebrate Fisheries off the Pacific Coast of Canada (1985/86). Can. Tech. Rep. Fish. Aquat. Sci. No. 1576. pp. 113–122.

Bourne, N., Hodgson, C.A. and Whyte, J.N.C., 1989. A manual for scallop culture in British Columbia. Can. Tech. Rep. Fish. Aquat. Sci. 1694. 215 p.

Bourne, N. and Hodgson, C.A., 1991. Development of a viable nursery system for scallop culture. In: S.E. Shumway and P.A. Sandifer (Eds.). Scallop Biology and Culture. World Aquaculture Workshops, No. 1. World Aquaculture Society, Baton Rouge, LA. pp. 273–286.

Bower, S.M., Blackbourn, J. and Meyer, G.R., 1998. Distribution, prevalence and pathogenicity of the protozoan *Perkinsus gugwadi* in Japanese scallops, *Patinopecten yessoensis*, cultured in British Columbia, Canada. Can. J. Zool. 76:954–959.

Bower, S.M., Blackbourn, J., Meyer, G.R. and Welch, D.W., 1999. Effect of *Perkinsus gugwadi* on various species and strains of scallops. Dis. Aquat. Org. 36:143–151.

Brand, A.R., 1991. Scallop ecology: distributions and behaviour. In: S.E. Shumway (Ed.). Scallops: Biology, Ecology and Aquaculture. Developments in Aquaculture and Fisheries Science, Vol. 21. Elsevier, Amsterdam. pp. 517–584.

Cary, S.C., 1982. Food and feeding strategies in early purple-hinge rock scallops, *Hinnites multirugosus* (Gale). MS Thesis. San Diego State Univ., San Diego, California. 100 p.

Cary, S.C., Leighton, D.L. and Phleger, C.F., 1981. Food and feeding strategies in culture of larval and early purple-hinge rock scallops, *Hinnites multirugosus* (Gale). J. World Mar. Soc. 12(1):156–169.

Cheney, D.P. and Mumford, T.F., 1986. Shellfish and seaweed harvests of Puget Sound. Washington Sea Grant Program. Univ. Washington Press, Seattle, Wash. 164 p.

Coan, E.V., Scott, P.V.V. and Bernard F.R., 2000. Bivalve seashells of western North America. Marine bivalve mollusks from Arctic Alaska to Baja California. Santa Barbara Mus. of Nat. History. Nat. His. Monograph 2, Studies in Biodiversity 2. 764 p.

Cooke, C.A., 1986. Embryogenesis and morphology of larval structures in *Chlamys hastata*, with an examination of the effect of temperature on larval development and factors affecting settlement and metamorphosis. MSc Thesis. University of Victoria, Victoria, B.C. 143 p.

Ellis, D.V., 1967. Quantitative benthic investigations. II. Satellite Channel species data, February 1965 – May 1965. Fish. Res. Bd. Can. Tech Rep No. 35.

Farlinger, S. and Bates, K.T., 1985. Review of shellfish fisheries in northern British Columbia to 1984. Can. MS Rep. Fish. Aquat. Sci. 1841. 33 p.

Felix-Pico, E.F., 1991. Scallops in Mexico. In: S.E. Shumway (Ed.). Scallops: Biology, Ecology and Aquaculture. Developments in Aquaculture and Fisheries Science, Vol. 21. Elsevier, Amsterdam. pp. 943–980.

Foster, N.R., 1997. The molluscan fisheries of Alaska. In: C.L. Mackenzie, V.G. Burrell, A. Rosenfield and W.L. Hobart (Eds.). The History, Present Condition and Future of the Molluscan Fisheries of North and Central America and Europe. (2). Pacific Coast and Supplemental Topics. NOAA Tech. Rep. NMFS 128. pp. 131–144.

Grau, G., 1959. Pectinidae of the eastern Pacific. Univ. Calif. Publ. Allan Hancock Found. Pac. Exp. 23. Viii. 208 p.

Guo, X., Ford, S.E. and Zhang, F., 1999. Molluscan aquaculture in China. J. Shellfish Res. 18(1):19–31.

Harbo, R.M., 1997. Shells and Shellfish of the Pacific Northwest. Harbour Publishing Madeira Park, B.C. 270 p.

Haynes, E.G. and Hitz, C.R., 1971. Age and growth of the giant Pacific sea scallop, *Patinopecten caurinus*, from the Strait of Georgia and outer Washington coast. J. Fish. Res. Bd. Can. 28:2112–2119.

Haynes, E.B. and Powell, G.C., 1968. A preliminary report on the Alaska sea scallop; Fishery exploration, biology and commercial processing. Alaska Dept. Fish. Game Info. Leaflet 125. 20 p.

Hennick, D.P., 1970. Reproductive cycle, size at maturity, and sexual composition of commercially harvested weathervane scallops, *Patinopecten caurinus*, in Alaska. J. Fish. Res. Bd. Can. 27:2112–2119.

Hennick, D.P., 1971. A hermaphroditic specimen of weathervane scallop, *Patinopecten caurinus*, in Alaska. J. Fish. Res. Bd. Can. 28:608–609.

Hennick, D.P., 1973. Sea scallop, *Patinopecten caurinus*, investigations in Alaska. Final Rep. July 1, 1969-June 30, 1972. Comm. Fish. Res. Develop. Act. Project No. 5–23–R.

Hodgson, C.A. and Burke, R.D., 1988. Development and larval morphology of the spiny scallop, *Chlamys hastata*. Biol. Bull. 174:303–318.

Jacobsen, F.R., 1977. The reproductive cycle of the purple-hinge rock scallop *Hinnites multirugosus* Gale 1928 (Mollusca: Bivalvia). PhD Thesis. San Diego State Univ., San Diego, Calif. 72 p.

Ito, H., 1991. Scallop culture in Japan. In: S.E. Shumway (Ed.). Scallops, Biology, Ecology and Aquaculture. Developments in Aquaculture and Fisheries Science, Vol. 21. Elsevier, Amsterdam. pp. 1017–1055.

Kaiser, R.J., 1986. Characteristics of the Pacific weathervane scallop [*Pecten (Patinopecten) caurinus*, Gould 1850] fishery in Alaska. Alaska Dep. Fish Game, Div. Comm. Fish., Kodiak, Alaska. 100 p.

Kruse, G.H., 1994. Fishery management plan for commercial scallop fisheries in Alaska. Draft Special Pub. 5, Alaska Dep. Fish and Game, Comm. Fish. Manag. and Dev. Div., Juneau, Alaska.

Kruse, G.H., Barnhart, J.P., Rosenkranz, G.E., Funk, F.C. and Pengilly, D., 2000. History and development of the scallop fishery in Alaska. In: Alaska Department Fish and Game and University of Alaska Fairbanks A Workshop Examining Potential Fishing Effects on Population Dynamics and Benthic Community Structure of Scallops with Emphasis on the Weathervane Scallop *Patinopecten caurinus* in Alaskan Waters. Alaska Dep. Fish and Game, Div. Commer. Fish. Special Pub. 14. Juneau. pp. 6–12.

Kruse, G.H. and Shirley, S.M., 1994. The Alaskan scallop fishery and its management. In: N.F. Bourne, B.L. Bunting and L.D. Townsend (Eds.). Proc. 9th International Pectinid Workshop. Can. Tech. Rep. Fish. Aquatic Sci. 1994(2):170–177.

Lauren, D.J., 1982. Oogensis and protandry in the purple-hinge rock scallop, *Hinnites multirugosus*, in upper Puget Sound, Washington, U.S.A. Can. J. Zool. 60:2333–2336.

Lauzier, R.B., Campagna, S. and Hinder, R., 2000. Framework for pink (*Chlamys rubida*) and spiny (*C. hastata*) scallop fisheries in waters off the west coast of Canada. Can. Stock Ass. Secretariat Res. Doc. 2000/123. 36 p.

Lauzier, R.B. and Parker, G., 1999. A review of the biology and fisheries of the pink scallop and spiny scallop. Can. Stock Ass. Secretariat, Res. Doc. 99/153. 46 p.

Lauzier, R.B., Walther, L.C, Hajas, W., Lessard, J., and Wylie, E., In Prep. Progressing to a scientifically-based assessment and management system for renewed commercial pink and spiny scallop fisheries off the British Columbia coast. Can. Sci. Advice Secretariat Res. Doc.

Leighton, D.L., 1979. A growth profile for the rock scallop, *Hinnites multirugosus* held at several depths off La Jolla, California. Mar. Biol. 51:229–232.

Leighton, D.L. and Phleger, C.P., 1977. A growth profile for the rock scallop, *Hinnites multirugosus*, held at several depths off La Jolla, California. Mar. Biol. 51:229–232.

MacDonald, B.A. and Bourne, N.F., 1987. Growth, reproductive output, and energy partioning in weathervane scallops, *Patinopecten caurinus*, from British Columbia. Can. J. Fish. Aquat. Sci. 44:152–160.

MacDonald, B.A. and Bourne, N.F., 1989. Growth of the purple-hinge rock scallop, *Crassadoma gigantea*, Gray, 1825, under natural conditions and those associated with suspended culture. J. Shellfish Res. 8(1):179–186.

MacDonald, B.A., Thompson, R.J. and Bourne, N.F., 1991. Growth and reproductive energetics of three scallop species from British Columbia (*Chlamys hastata, Chlamys rubida*, and *Crassadoma gigantea*). Can. J. Fish. Aquat. Sci. 48:215–221.

Olsen, S.L., 1983. Abalone and scallop culture in Puget Sound. J. Shellfish Res. 3(1):113.

988

Olsen, S.L., 1984. Completion report on invertebrate aquaculture, shellfish enhancement Project 1978–1983. Shellfish enhancement project, final report, Oct. 1, 1978-Mar. 30, 1983. Washington Dept. Fish. Olympia, Wash. 85 p.

Perry, R.A., Walters, C.J. and Boutillier, J.A., 1999. A framework for providing scientific advice for the management of new and developing fisheries. Rev. Fish. Biol. and Fisheries 9:125–150.

Phleger, C.F. and Cary, S.C., 1983. Settlement of spat of the purple-hinge rock scallop, *Hinnites multirugosus* (Gale) on artificial collectors. J. Shellfish Res. 3(1):71–73.

Phleger, C.F., Cary, S.C. and Leighton, D.L., 1981. Algae and detrital matter as food for juvenile purple-hinge rock scallops, *Hinnites multirugosus*, Gale. J. World Mar. Soc. 12(2):180–185

Phleger, C.F. and Leighton, D.L., 1980. Refinements in culture technology for the Purple-hinge rock scallop. Univ. Cal. Sea Grant Project R/A–31. Tech. Rep. 1978–1980. San Diego State Univ., San Diego, CA. 22 p.

Quayle, D.B., 1961. Deep water clam and scallop survey in British Columbia, 1960. Fish. Res. Bd. Can. MS Rep. Ser. (Biol) 717. 87 p.

Quayle, D.B., 1963. Deep water clam and scallop survey in British Columbia, 1961. Fish. Res. Bd. Can. MS Rep. Ser. (Biol) 746. 39 p.

Rathjen, W.F. and Rivers, J.B., 1964. Gulf of Alaska scallop explorations, 1963. Comm. Fish. Rev. 26(3). 7 p.

Robinson, A.M., 1997. Molluscan fisheries in Oregon: past, present and future. In: C.L. MacKenzie, V.G. Burrell, A. Rosenfield and W.L. Hobart (Eds.). The History, Present Condition, and Future of the Molluscan Fisheries of North and Central America and Europe. (2) Pacific Coast and Supplemental Topics. NOAA Tech. Rep. NMFS 128. pp. 75–87.

Robinson, A.M. and Breese, W.P., 1984. Spawning cycle of the weathervane scallop *Pecten* (*Patinopecten*) *caurinus* Gould along the Oregon coast. J. Shellfish Res. 4(2):165–166.

Ronholt, L.L. and Hitz, C.R., 1968. Scallop explorations off Oregon. Comm. Fish. Rev. 30(7):42–49.

Ronholt, L.L., Shippen, H.H. and Brown, E.S., 1977. Demersal fish and shellfish resources of the Gulf of Alaska from Cape Spencer to Unimak Pass, 1948–1976: A historical review. NOAA/OCSEAP, Final Rep. (2). 955 p.

Sizemore, R.E. and Palensky, L.Y., 1993. Fisheries management implications of new growth and longevity data for pink (*Chlamys rubida*) and spiny scallops (*C. hastata*) from Puget Sound, Washington. J. Shellfish Res. 12(1):145–146.

Shaw, W.N., 1997. The shellfish industry of California-past, present and future. In: C.L. MacKenzie, V.G. Burrell, A. Rosenfield and W.L. Hobart (Eds.). The History, Present Condition, and Future of the Molluscan Fisheries of North and Central America and Europe. (2) Pacific Coast and Supplemental Topics. NOAA Tech. Rep. NMFS 1128. pp. 57–74.

Shirley, S.M. and Kruse, G.H., 1995. Development of the fishery for weathervane scallops, *Patinopecten caurinus* (Gould, 1850), in Alaska. J. Shellfish Res. 14(1):71–78.

Shumway, S.E. and Castagna, M., 1994. Scallop fisheries, culture and enhancement in the United States. Memoirs Queensland Museum (Brisbane, Australia) 36(2):283–298.

Shumway, S.E. and Cembella, A.D., 1993. The impact of toxic algae on scallop fisheries and culture. Rev. Fish. Sci. 1(2):121–150.

Starr, R.M. and McCrae, J.E., 1983. Weathervane scallop (*Patinopecten caurinus*) investigations in Oregon, 1981–1983. Oregon Dept. Fish. Wildlife. Info. Rep. 83–10. 55 p.

Ventilla, R.F., 1982. The scallop industry in Japan. Adv. Mar. Biol. 20:309–338.

Waller, T.R, 1991. Evolutionary relationships among commercial scallops (Mollusca: Bivalvia: Pectinidae). In: S.E. Shumway (Ed.). Scallops: Biology, Ecology and Aquaculture. Developments in Aquaculture and Fisheries Science, Vol. 21. Elsevier, Amsterdam. pp. 1–55.

AUTHORS ADDRESS

Raymond B. Lauzier - Fisheries and Oceans Canada, Science Branch, Pacific Region, Pacific Biological Station, Nanaimo, British Columbia, Canada, V9R 5K6 (E-mail: LauzierR@dfo-mpo.gc.ca)

Neil F. Bourne - Fisheries and Oceans Canada, Science Branch, Pacific Region, Pacific Biological Station, Nanaimo, British Columbia, Canada, V9R 5K6 (E-mail: BourneN@dfo-mpo.gc.ca)

Scallops: Biology, Ecology and Aquaculture
S.E. Shumway and G.J. Parsons (Editors)
© 2006 Elsevier B.V. All rights reserved.

991

Chapter 19

The European Scallop Fisheries for *Pecten maximus*, *Aequipecten opercularis* and *Mimachlamys varia*

Andrew R. Brand

19.1 INTRODUCTION

Although 28 species of scallop have been recorded in European waters (Nordsieck 1969), relatively few are of large enough body size and occur in sufficiently dense concentrations to be commercially exploited. Of these, the 'great scallop', 'coquille Saint-Jacques' or simply 'scallop', *Pecten maximus* (L.) is the most important on account of its wide distribution and high market value. In recent years, the common name 'king scallop' has also been applied to this species, a name introduced by the U.K. industry to distinguish it from the second most important species, the 'queen scallop', 'queen' or 'queenie', *Aequipecten* (formerly *Chlamys*) *opercularis* (L.). The Iceland scallop, *Chlamys islandica* (Müller) is also of considerable economic importance, but is a sub-arctic species and has a more restricted European distribution. Finally, several other species are of more limited local importance including the 'St. James' or 'pilgrim's scallop', *Pecten jacobaeus* (L.), which replaces *P. maximus* in most of the Mediterranean, the 'black' or 'variegated scallop', *Mimachlamys varia* (L.), and the 'smooth scallop', *Flexopecten glaber* (L.).

Pecten maximus and *A. opercularis* both have a long history of exploitation and in recent years have been the subjects of important commercial fisheries by boats from England, Wales, Scotland, Northern Ireland, the Republic of Ireland, the Isle of Man, the Channel Islands, France and Spain. These fisheries, and the stocks that support them, have been occasionally investigated by ICES Working Groups (Anonymous 1979, 1989) and other workers. Mason (1983) reviewed the history of the fisheries around the British Isles and gave a general account of the catching, processing and marketing of these species, and there have been more specific studies of the fisheries of England and Wales (Franklin et al. 1980a), Northern Ireland (Briggs 1980a; Briggs 1987), the Isle of Man (Brand et al. 1991a; Brand and Prudden 1997; Brand 2000) and Guernsey (Jory 2000). Scallop fisheries in France have been described by Dupouy (1978), Dupouy et al. (1983) and Dao and Carval (1999), and in Spain by Cano and Garcia (1985) and Román (1991). More limited information is also available for various pectinid fisheries in Italy (Renzoni 1991; Mattei and Pellizzato 1996), Jugoslavia (Margus 1990; Margus 1991) and Greece (Lykakis and Kalathakis 1991). There is a very extensive literature on the biology and ecology of *P. maximus and A. opercularis*, but that for *M. varia* is more limited. Previous work on these three species has been reviewed by Mason (1983) and more recently by

Ansell et al. (1991). This section aims to bring the latter work up-to-date by incorporating studies on the biology of these three species that have come out over the last ten years and presenting, where possible, the latest information on their fisheries. Although *C. islandica* is a commercially important European scallop species, it is not included in this section, nor is any information given on the Scandinavian fisheries for *P. maximus* or *A. opercularis,* as these topics are covered elsewhere in this volume.

Pecten maximus and *Aequipecten opercularis* have similar geographical ranges along the east coast of the North Atlantic from northern Norway to West Africa, the Azores, Canary Islands and Madiera (Tebble 1966; Mason 1983; Brand 1991) but *A. opercularis* extends farther into the Mediterranean and the Adriatic (Broom 1976; Cano and Garcia 1985; Piccinetti et al. 1986). They are found from just below the low water mark to a depth of more than 150 m (Forbes and Hanley 1853; Tebble 1966) but are essentially coastal species, most common at 20–45 m. *P. maximus* occurs on bottoms of clean firm sand, fine gravel or sandy gravel, sometimes with an admixture of mud; *A. opercularis* is frequently found on the same grounds but it can also live on harder gravel and shelly bottoms (Mason 1983). *P. maximus* is predominantly an Atlantic species but it does extend along the south-eastern coast of Spain into the Mediterranean and a bed off Malaga has been commercially fished since 1982 (Cano and Garcia 1985). It has long been considered that *P. maximus* is replaced throughout most of the Mediterranean by the morphologically distinct *P. jacobaeus.* A number of genetic studies using allozyme electrophoresis (Huelvan 1985), mitochondrial DNA (Wilding et al. 1999) and partial sequences of the 16S rRNA gene (Canapa et al. 2000) have suggested that the genetic distance between *P. maximus* and *P. jacobaeus* is too small to consider them as separate species. However, the recent studies by Ríos et al. (1999, 2002), based on allozyme frequencies, found a major genetic discontinuity in the area of the Almeria-Oran oceanographic front that was coincident with the morphological discontinuity between the two taxa and appears to confirm the specific status. The main areas of commercial exploitation of *P. maximus* and *A. opercularis* are along the western coasts of the British Isles and France, but include the Moray Firth off the north-east coast of Scotland, the Shetland Islands and both sides of the English Channel (Brand 1991).

The smaller *Mimachlamys varia* does not extend quite as far north as the other two species but it has a wide distribution from Denmark, along the western coasts of the British Isles, France and the Iberian peninsula, throughout most of the Mediterranean and as far south as Senegal on the coast of west Africa (Tebble 1966). Light (1988) considers that *Mimachlamys nivea* is a geographical sub-species of *M. varia,* confined to the west coast of Scotland. *M. varia* occurs at depths ranging from the low intertidal to some 80 m, usually on shelly or sandy-gravel sediments where it byssally-attaches to shells, stones and boulders (Le Pennec and Diss-Mengus 1985) or rock faces (Rodhouse and Burnell 1979). It is also often a prominent member of the epifauna on banks of the horse mussel, *Modiolus modiolus* (Jones 1951). The main commercial exploitation of *M. varia* is along the French Atlantic coast, in recent years particularly the eastern part of the Bay of Brest.

19.2 BIOLOGY AND ECOLOGY

19.2.1 The Great Scallop, *Pecten maximus* (L.)

Pecten maximus is a large scallop. In many areas it can grow to a shell length (the largest dimension) of 150 mm but some even larger individuals have been recorded, including one with a shell length of 210 mm (Minchin 1978). It is inequivalve, with a flat upper (morphological left) valve and a deeply cupped lower (right) valve that generally overlaps the left at its margin. Both valves bear 15–17 broad, radiating ribs and numerous concentric corrugations and fine striae. The left valve is generally reddish brown in colour but can vary from light pink to almost black. The right valve is usually off-white, yellowish or light brown. Both valves can be marked with bands, spots or zigzags of red, pink or bright yellow. Growth rings are often very clear on both valves, particularly the left.

P. maximus is usually found recessed into shallow depressions in the seabed so that the upper valve is level with, or just below, the surface of the sediment (Baird 1958). Recessing is a complex, closely coordinated, behaviour pattern involving powerful jetting of water from the mantle cavity. The entire process usually takes from 2–20 minutes to complete, but depends on the nature of the sediment (Maguire et al. 1999), and scallops as small as 6–10 mm recess (Minchin 1992; Minchin et al. 2000). Particles of sand disturbed by the water jets resettle on the shell so that recessed scallops are difficult for visual predators or divers to detect. On settling from the plankton, and for some time after, juvenile scallops are byssally-attached but they can release the byssus to move to a new location where they secrete a new byssus. Scallops cease to form a byssus at a small size and few larger than 15 mm are attached (Eggleston 1962; Franklin et al. 1980a; Minchin 1992).

Whether recessed or byssally-attached, young scallops swim readily when disturbed. Like most pectinids, *P. maximus* appears to have three responses to disturbance (Thomas and Gruffydd 1971; Brand 1991): closing the valves, a jumping reaction that propels the scallop hinge foremost and a more vigorous swimming reaction that propels the scallop ventral edge foremost. Movement is achieved by powerful water jets from the mantle cavity, generated by shell adductions. Scallops are good swimmers when young but swimming performance and willingness to swim decline with age (Mathers 1976a; Minchin and Mathers 1982; Jenkins and Brand 2001). Jumping and swimming provide an escape reaction from predators, divers and fishing gear. Contact with predatory starfish elicits vigorous escape reactions, whereas non-predatory starfish evoke more minor reactions or none at all (Thomas and Gruffydd 1971; Soemodihardjo 1974). Many of these responses are chemosensory. The ability to discriminate between potential predators and harmless species minimises interruptions to feeding and prevents energy being wasted making unnecessary escape reactions, resecreting the byssus or reconstructing a recess (Thomas and Gruffydd 1971). Reactions to divers or fishing gear involve responses to light, water currents or vibrations (Baird 1958; Minchin and Mathers 1982) and have important implications for commercial fishermen (Chapman et al. 1979). The fast phasic adductor muscles of *P. maximus*, like other scallops, store glycogen as an energy reserve and contain high levels of adenosine nucleotides and arginine phosphate (Ansell 1977; Grieshaber and Gäde 1977). Swimming adductions are fuelled, at least initially, by the anaerobic metabolism of the arginine phosphate; this is subsequently regenerated aerobically from the resulting octopine when swimming ceases (Gäde et al. 1978; Grieshaber 1978). As a consequence of these physiological mechanisms, scallops are well adapted for short distance escape responses but they become rapidly fatigued and take a long time to recover after swimming (Grieshaber and Gäde 1977; Gäde et al. 1978; de Zwaan et al. 1980). Such characteristics are inappropriate for long-distance swimming and there is no sound evidence, from direct observations or tagging experiments, of *P. maximus* or any other scallop species undergoing extensive, directed migrations (Brand 1991). However, when one scallop swims it often induces swimming in others nearby, leading to mass swimming activity (Minchin and Mathers 1982; Howell and Fraser 1984), and repeated swimming can result in downstream dispersal in areas with strong water currents.

After disturbance, scallops attempt to recess by repeated ejection of water jets directed at the sediment (Baird 1958). Swimming movements may be used to move short distances if the sediment is unsuitable for recessing. The speed of recessing (Maguire et al. 2002) and swimming ability (Jenkins and Brand 2001) are greatly reduced after the stress of dredge capture, so undersized discards may be less able to escape from predators. Once recessed, scallops are less vulnerable to predation, and probably feed more efficiently (Brand 1991). Recessed scallops orientate to water currents so that mantle cavity ventilation is assisted (Hartnoll 1967; Mathers 1976a). This probably increases the efficiency of filter feeding and imposes rhythms of feeding and digestion, phased with the tidal cycle (Mathers 1976b). Living at the interface between sediment and water column, scallops encounter sedimenting phytoplankton, suspended benthic diatoms and resuspended detritus (Christensen and Kanneworff 1985). In a comparative study of the feeding ecology of *P. maximus* and *A. opercularis* in a strongly tidally mixed area of the north Irish Sea, Lawrence (1993) found that resuspended benthic diatoms were present in the benthic boundary layer throughout the year and comprised the major dietary item of both species. Planktonic diatom species were only available, and present in the gut contents, in spring and early summer. She concluded that *P. maximus* probably receive a lower quality diet than *A. opercularis* since they draw in water from nearer the sediment surface and take in more detritus. Laboratory experiments showed that *P. maximus* has higher clearance rates and exhibits greater particle selection than *A. opercularis*, but is more sensitive to resuspended sediment in the diet. Pseudofaeces production occurs at algal concentrations exceeding 10 cells μL^{-1} so enhanced food levels in the sea do not necessarily increase scallop growth (Skjaeggestad et al. 1999).

Within their geographical range scallops have aggregated distributions at different spatial scales. For various scallop population's scales of aggregation have been defined, with some variation in terminology, by a number of workers (Orensanz 1986; Caddy 1989a; Brand 1991; Orensanz et al. 1991). These range from the largest, which is the 'stock', 'reproductively isolated population' or 'megapopulation', down through the 'fishing ground', the 'bed' and the 'patch'. Since the degree to which the scale of aggregation can be perceived depends on the size of the sampling unit (Andrew and

Mapstone 1987), reliable estimates of scallop density and stock size are difficult to obtain, and the comparison of estimates based on different survey methods is problematic. Stock assessments of *P. maximus,* based on dredging, have been carried out regularly in several areas, particularly around the coasts of Scotland (Mason and Drinkwater 1976; Mason et al. 1980, 1981, 1984), and the north Irish Sea (Murphy 1986; Brand et al. 1991a; Allison 1993; Brand and Allison 1994; Beukers-Stewart et al. 2001a). Scallop density and catch per unit effort (CPUE) show a long-term decline on all Isle of Man fishing grounds, with densities of <0.03 scallops m^{-2} now typical of the inshore fishing grounds and many of the offshore grounds (Brand and Prudden 1997). There is a closed season for scallop fishing throughout the north Irish Sea (June–October inclusive), and for the most heavily exploited fishing grounds, where the populations are dominated by young scallops, there are large seasonal variations in density. Estimates of density of legal-sized scallops (>109 mm shell length) on all the heavily fished grounds are now usually around 0.03 scallop m^{-2}, at the start of the fishing season, a level clearly perceived by fishermen to be economically viable and dependent largely on the strength of summer recruitment (Wilson and Brand 1995). By the end of the season, however, densities have fallen to around 0.01 scallops m^{-2}, which probably approximates to the density that the fishermen currently consider to be uneconomic. When densities fall to this level the fishermen move off and fish elsewhere. These values may be compared with earlier density estimates for the same grounds of 0.09–0.12 scallops m^{-2} for 1954/5 and 0.2 for 1965/6 (based on tagging), 0.04–0.11 for 1981/4 and 0.03–0.04 for 1986/90 (based on dredge CPUE) and 0.01–0.03 for 1989/92 (based on diving) (Mason and Colman 1955; Gruffydd 1972; Murphy 1986; Allison 1993; Wilson 1994). The density estimates based on dredge CPUE assumed a dredge efficiency of 20%, a value derived from earlier studies (Chapman et al. 1977; Mason et al. 1979a). However, in a recent study, Beukers-Stewart et al. (2001a) reported rather higher dredge efficiencies for one Irish Sea fishing ground. Using two independent methods of assessment, they calculated 'Newhaven' spring-tooth dredge efficiencies for legal sized scallops (>109 mm shell length) of 29.5% from a depletion experiment and 40.7% from a concurrent diver survey of dredge tracks. Other work has shown that the efficiency of spring-tooth dredges varies on different substrates from 6% on stony ground to nearly 40% on smooth grounds (Dare et al. 1993), so some of the densities quoted above are likely to be underestimated, while others are overestimated. An even greater variation in dredge efficiency with substratum type was reported in the study by Buestel et al. (1985), carried out by divers following 'French' dredges (with a diving plane) in the Bay of St. Brieuc, France. They found an average density of 0.6 scallops m^{-2} over an area of some 33 km^2 of the bay, with dredge efficiency ranging from 25–74% on different sediments.

The use of diving, dredging and towed television camera surveys to assess the distribution and density of *P. maximus* was compared by Mason et al. (1982). They found that, compared with divers, dredges recovered only 18% of the scallops in their path and 36.3% were observed using underwater video. Diver surveys record the highest numbers of scallops but can only cover a relatively small area and are restricted in the water depth in which the divers can operate effectively. Divers also often select areas with abnormally high local densities, and diver estimates of up to 8 scallops m^{-2} have been recorded

(Minchin and Mathers 1982). Underwater video can survey larger areas, operate in deeper water and give information on large-scale patchiness in scallop distributions (Giguère and Brulotte 1994; Hall-Spencer et al. 1999). Franklin et al. (1980b) used underwater television to survey stocks of P. maximus and A. opercularis in the English Channel off Devon and Cornwall. While the mean densities over the whole area surveyed were 0.16 scallops m^{-2} and 0.46 queens m^{-2}, they found maximum densities for two-minute videotape segments (approximately 20 m^2) of 2.33 scallops m^{-2} and 4.46 queens m^{-2}. Thouzeau and Hily (1986) developed the A.QUA.R.E.V.E. system that combines a video camera with a dredge and this has been very successful for surveying juvenile scallop abundance (Thouzeau and Lehay 1988). Recently, Hall-Spencer et al. (1999) concluded that video surveys underestimate the density of recessed scallops because they are difficult to discern on the screen. In a survey of P. jacobaeus in the Gulf of Venice they recorded an average density of only 0.023 scallops m^{-2} from the videotapes but subsequently caught 0.025–0.034 scallops m^{-2} in the same area using a 'Rapido' trawl for which they had calculated a capture efficiency of 44%.

Growth in P. maximus is strongly seasonal and most populations can be easily aged by counting the rings laid down on the shell when growth resumes in spring (Mason 1957). Problems can be encountered with the identification of the first ring, and with the later rings in scallops aged more than 10 years (Mason 1983; Allison 1993). However, validation by repeated seasonal observations of the position of the most recently formed ring in relation to the edge of the shell, tagging and recovery, growth experiments with spat of known ages (Paul 1981; Allison et al. 1994) and oxygen isotope analysis (Dare and Deith 1991) have all shown that this traditional method of ageing P. maximus can be cheap, rapid and reliable. In addition to the annual growth rings, the shells of P. maximus also bear fine concentric ridges or striae that have been considered to represent daily growth increments (Antoine 1978). Gruffydd (1981) cast doubts on the reliability of this method when he reported that the frequency of formation of striae of juvenile P. maximus in good culture conditions was very variable, but always less than one per day. Deviation from a daily deposition of growth rings have been confirmed in all the more recent studies of P. maximus, and other scallop species, but various mechanisms have been proposed to account for these deviations and the technique would appear to be very valuable for investigating seasonal and environmental factors affecting growth (Chauvaud et al. 1998; Chauvaud and Strand 1999).

The growth pattern of P. maximus, in terms of both shell length and meat weight, generally conforms to the von Bertalanffy growth formula (Broom 1976; Murphy 1986; Allison 1993), but the Gompertz equation may provide a better fit for the younger age classes (Pope and Mason 1080; Mason 1983). Other model-free methods of analysis have also been used (Aldrich and Lawler 1996). Seasonally oscillating growth models that account for the winter cessation of growth have also been fitted (Allison 1994). The latter are of particular value for calculating potential meat yields from the fishery, and investigating exploitation strategies, because the edible portions of pectinids (the adductor muscles and gonads) show such large seasonal cycles. There have been numerous studies of the growth rates of P. maximus at different locations around the British Isles and France, and the older studies are reviewed in Mason (1983), Buestel and Laurec (1976) and Antoine et al. (1979a). Growth rates on different fishing grounds have been compared from growth curves of mean length or weight-for-age (e.g., Mason 1957; Gruffydd 1974a), from the parameters of the von Bertalanffy growth formula (Antoine et al. 1979a; Murphy 1986; Acosta and Román 1994) or from growth performance indices derived from these parameters (Allison 1993). Large differences in growth rate have been found between different populations, both on a wide geographical scale and between relatively closely situated locations, and these differences have been attributed to various environmental and other factors. Acosta and Roman (1994) conclude that, with some exceptions, growth rate is higher, the asymptotic length is smaller, and the life span in shorter in southern populations. In the north Irish Sea, growth rates of scallops are generally highest on the inshore fishing grounds off the west and north-eastern coasts of the Isle of Man and lowest on grounds farther offshore to the south and east of the Island (Allison 1993). Thouzeau and Lehay (1988) also found that the growth of juvenile *P. maximus* was highest at onshore locations in the Bay of St. Brieuc.

Differences in temperature, organic carbon content of the seston and current speeds at the different sites were considered to explain the differences in growth rate of *P. maximus* in suspended culture (Wilson 1987b), and similar factors were suggested to explain growth differences in juvenile scallops in the Bay of St. Brieuc (Thouzeau and Lehay 1988). Chauvaud et al. (1998), however, concluded that normal growth of *P. maximus* is regulated mainly by bottom-water temperature, salinity and, to a lesser extent, river flow, rather than food, but they found that toxic algal blooms led to major reductions in shell growth rate. Laing (2002) found that food cell clearance rates and mean growth rate of *P. maximus* spat were usually significantly lower at a salinity of 28 psu than at 28–30 psu, while Laing (2000) also suggests that temperature rather than food is the factor most often limiting growth in the sea. Food can, however, be a limiting factor in both natural (Gruffydd 1974a) and cultivated (Norman and Minchin 1999; Maguire and Burnell 2001) populations. It would seem that genetic factors play only a minor part in determining shell growth characteristics. Thus, the native growth characteristics were not retained when *P. maximus* spat were transplanted from the Bay of St. Brieuc to the Bay of Brest (Antoine et al. 1979b), or when Irish and Scottish scallops were transplanted to sites in Brittany (Dao et al. 1985a; Huelvan 1985). However, scallops transplanted from Mulroy Bay had significantly lower growth rates and carbohydrate content than native Bantry Bay scallops grown under the same conditions (Maguire and Burnell 2001), so further studies are necessary. Unlike many other bivalves there appears to be no association between heterozygosity and size in either natural populations or hatchery-reared families of *P. maximus* (Beaumont et al. 1985), and this has been confirmed for other pectinids (Fevolden 1992). In *P. jacobaeus*, the level of heterozygosity has been related to the timing of spawning (Ríos et al. 2002).

Size and age distributions of natural populations have been extensively studied for populations around the coasts of Scotland (reviewed in Mason 1983), the Isle of Man (Brand et al. 1991a; Allison 1993; Brand and Prudden 1997; Brand 2000), Northern Ireland (Briggs 1991; Briggs 1995) and France (Dupouy 1978; Dupouy et al. 1983; Lubet et al. 1991). In all fished areas there has been a shift in the population age structure towards the younger age groups as the older and larger scallops are progressively removed by size-selective fishing. This is particularly evident on the most heavily fished grounds, like the Bradda Head ground off the south west of the Isle of Man, where for the last 15–20 years up to 70% of the catch has been 4 years old or less, and mainly below the minimum legal landing size (110 mm shell length) (Brand et al. 1991b; Brand and Prudden 1997). The success of the fishery on such grounds is very dependent on the strength of annual recruitment. Annual recruitment has been remarkably regular for many years on inshore grounds in the north Irish Sea (Brand et al. 1991a; Beukers-Stewart et al. 2001b), while elsewhere it can be very variable. For example, Dupouy et al. (1983) cited a ratio of 20:1 between good and bad year-classes in the Bay of St. Brieuc. For many larger fishing grounds, recruitment may be temporally regular but with considerable spatial variability (Gruffydd 1974b; Bannister 1986). The strength of annual recruitment can be predicted by pre-recruitment surveys for 1 and 2 year old scallops using small mesh dredges (Thouzeau and Lehay 1988; Thouzeau 1991a; Allison 1993; Palmer 1995; Beukers-Stewart et al. 2001b). Future recruitment of each year-class may also be predicted at an early age from the numbers of spat settling in artificial collectors (Buestel et al. 1979; Fraser 1991; Beukers-Stewart et al. 2001b), particularly for very strong year-classes.

Pecten maximus is a long-lived scallop, with a life span exceeding 20 years in extreme cases (Tang 1941). However, the average life span is much less, even in unexploited populations (Dupouy et al. 1983). In the north Irish Sea the populations on all grounds were dominated by old scallops before fishing began, with up to 50% of the population 9 years old or older (Brand et al. 1991a). By 1995, after 15–50 years of exploitation, the populations on all the inshore grounds and many of the offshore grounds were dominated by young scallops, with up to 70% of the catch less than 4 years old and scallops older than 7 years were rare (Brand and Prudden 1997). There have been relatively few studies of the mortality rates of *P. maximus* in either unexploited or exploited populations. The most detailed estimates have come from two large tagging experiments carried out on the stocks around the Isle of Man (Brand and Murphy 1985; Murphy 1986; Allison et al. 1989; Brand and Murphy 1992; Allison 1993), and from analyses of CPUE and population age structure data (Murphy 1986; Wilson 1994) for the same stocks. Annual exploitation rates from the tagging experiments carried out in 1982/3 and 1987/8 were very high for the inshore grounds, with values of up to 55%, while for some of the offshore grounds exploitation rates increased from 1–5% in 1982/3 to 5–20% in 1987/8 (Brand et al. 1991a). Fishing mortality (F) increased considerably between the 1982/3 and 1983/4 fishing seasons on all grounds and has been maintained in subsequent years. While natural mortality (M) increases with age, F appears to be similar for all recruited age-classes, implying that all fully recruited age-classes are equally vulnerable to fishing. The estimates of natural mortality ($M = 0.31$–0.61) derived from tagging experiments (Brand et al. 1991a) are considerably higher than the assumed value of $M = 0.15$ commonly used in *P. maximus* stock assessments (Mason et al. 1979b; Mason et al. 1981), but these values include incidental or non-capture mortality (F_i) caused by dredge damage. Recent studies have shown there is the potential for high levels of mortality in undersized discards and scallops that encounter the dredges but are not captured (Jenkins and Brand 2001). It is most important that F_i can be separated from M, for M is used as a constant in stock assessment models and must not vary with the amount of fishing. Incidental mortality is an important, but often neglected, factor and can considerably exceed M in some other scallop fisheries (e.g., Naidu 1988; Shepard and Auster 1991). An estimate of M = 0.12 for one north Irish Sea ground in the absence of fishing (Allison 1993) is similar to the value of 0.164 previously calculated for 7 year old scallops from an un-fished Manx population (Gruffydd 1974b), suggesting that the conventionally assumed value of 0.15 is reasonable for these grounds.

The main predators of *P. maximus* are decapod crustaceans (*Cancer pagurus, Carcinus maenas, Liocarcinus puber*) and starfish (*Asterias rubens, Marthasterias glacialis*), but certain gastopods, octopus and fish can be important in some areas (Minchin 1991; Halary et al. 1994; Lake and McFarlane 1994; Grall et al. 1996; Veale et al. 2000a). The range of potential predators is much greater for spat and juvenile scallops (Brand et al. 1980; Minchin 1984; Fleury et al. 1996), including unexpected predators like the sea anemone *Anthopleura ballii* (Minchin 1983). Predation on *P. maximus* is reduced when the spat leave sites of byssal attachment and start to recess in the seabed (Buestel and Dao 1979). Older scallops appear to have a refuge in size from many predators (Paul et al. 1986; Lake et al. 1987; Lake and McFarlane 1994) but this can sometimes be overcome (Wilson and Brand 1994). The vulnerability of small scallops to predation has important implications for stock enhancement or bottom culture and has been the subject of much research in recent years (e.g., Dao et al. 1985a; Paul et al. 1986; Halary et al. 1994; Lake and McFarlane 1994; Maguire et al. 2002). A survival rate of 75–90% per year was reported for juvenile scallops of greater than 25mm shell height experimentally seeded in the Bay of Brest (Dao et al. 1985a), but scallops up to 60mm or more have been highly vulnerable to predation elsewhere (Paul et al. 1986; Lake et al. 1987; Lake and McFarlane 1994), making site selection, predator removal and control, or enclosure of the scallops in nets, cages or corrals, important for economic success. In addition to predators, periods of prolonged cold weather, like the very cold winter of 1962/3, may also cause mortality of scallops in shallow water populations (Boucher and Fifas 1997). Although blooms of toxic or noxious algae appear to be coming more frequent occurrences in European waters, as elsewhere, and have led to fisheries closures (Andersen 1996; Campbell et al. 2001), they do not generally cause the mortality of adult scallops, though growth and reproduction may be inhibited (Shumway and Cembella 1993). Algal blooms can, however, cause serious mortalities of *P. maximus* larvae and juveniles (Minchin 1984; Lassus and Berthome 1988; Erard-Le Denn et al. 1990; Glemarec 1997).

Diseases do not appear to be a major cause of scallop mortality in the sea. Apart from mass mortalities in the Bay of St. Brieuc, France, in 1985–6, caused by a rickettsiales-like microorganism (RLO) in the gills (Le Gall et al. 1988; Le Gall et al. 1991), there are few records of disease in natural populations of *Pecten maximus*. Mortensen et al. (1992) isolated infectious pancreatic necrotic virus from scallops in Norway but in a later health survey (Mortensen 1993) found no serious pathological agents. Similarly, a recent detailed survey found very few pathogens or parasites in two U.K. scallop populations (Lohrmann 2000). The RLO described by Le Gall (1988) was common in *P. maximus* from the English Channel, with a prevalence of 33.3%, but was completely absent from

Irish Sea samples, although RLOs were present in *Aequipecten opercularis* taken from the same grounds (prevalence 15%), but this was probably a different species. The only parasite found in *P. maximus* from the Irish Sea was a coccidean present in the kidney (prevalence 5.3%) and this was absent from the Channel population. An unusual mass-mortality of *P. maximus* spat in nursery cultivation near Bergen, Norway, was caused by massive settlements of the polychaete *Polydora* sp., which bores into the shell (Mortensen et al. 1999), but it is not known if *Polydora* causes mortalities in natural populations. Under hatchery conditions, bacterial and viral infections of larvae and brood stock remain a serious problem for *P. maximus* culture, but this is considered elsewhere in this volume.

Environmental pollution has only rarely been reported to be a problem for natural populations of *Pecten maximus*, or any other European scallop. Routine monitoring for metals, organochlorine pesticide and PCB residues in the U.K. (e.g., Portmann 1979; Franklin 1987; Franklin and Jones 1995) have shown no cause for concern. Scallops accumulate various metals, often to many times their environmental concentrations (Bryan 1973). Many of these accumulate in the kidney as metal-phosphate granules, prior to excretion into the urinary tract (George et al. 1980). Cadmium is also bound to proteins in the digestive gland, accumulating to a level of approximately 100 ppm wet weight (Stone et al. 1986). The affects of various anti-fouling compounds have been tested on juvenile and adult scallops in cultivation. High levels of copper accumulated in scallop tissue from copper oxide-based anti-fouling paint and especially from copper-nickel alloy trays (Davies and Paul 1986) but there was no significant accumulation of nickel. There was some increase in the growth of spat with the copper oxide treatment, but growth of the adult scallops was not affected; the copper-nickel trays, however, caused high mortalities and inhibited growth in adult scallops (Paul and Davies 1986). Bis-(tri-n-butyltin) oxide (TBT)-based anti-fouling paint was detrimental to the growth and survival of juvenile scallops but had no effect on adult scallops. There was, however, evidence for the operation of a storage/detoxification mechanism in adult scallops that involves the progressive transfer of TBT to the adductor muscle (Davies et al. 1986). There is also evidence that very poor spat settlement and decline of the adult population of scallops in Mulroy Bay, Ireland, in the period 1982–85, was caused by the use of TBT-based anti-fouling net-dips using in salmonid farming (Minchin et al. 1987a; Minchin and Ni Donnachada 1995; Slater 1995). There is very little evidence that environmental pollution from the large European oil and gas exploration and production industry, or tanker spillage, has had much impact on *P. maximus* populations (Grainger et al. 1984). The recent development of a method for maintaining scallop digestive acini alive *in vitro* opens up new possibilities for using *P. maximus* in ecotoxicology research (Le Pennec and Le Pennec 2001).

Pecten maximus is a simultaneous hermaphrodite, with the prominent mature gonad containing a proximal, creamy-coloured testis and a distal, orange-coloured ovary, of approximately equal size. The first detailed work on scallop reproductive cycles was that of Mason (1958a b), who combined descriptions of the histology of the gonad structure with subjective assessments of the stage of development, based on the external appearance of the gonad. This work subsequently became the model on which virtually all studies of seasonal scallop reproduction world-wide were based, although the range of study techniques subsequently broadened to include various weight-based gonad indices and stereological analyses of cell volume fractions. Mason's study was on scallops from the inshore fishing grounds around the Isle of Man, where adult scallops usually have two peaks of spawning each year: a partial spawning in spring (April or May) and a more complete and closely synchronised spawning in autumn (late August or September). There may also be a minor partial summer spawning in July or early August. Virgin and juvenile scallops differ in having only one major spawning, in autumn. Mason (1958b) also provided evidence of a possible lunar periodicity in spawning, with peaks of spawning at full moons. A similar annual pattern, with two main peaks of spawning each year for adult scallops, has also been reported in other areas, including south-west Ireland, Galway Bay on the west coast of Ireland (Wilson 1987a), Strangford Lough in Northern Ireland (Stanley 1967), Loch Torridon on the west coast of Scotland (Mason 1969) and, with some variation in the extent of the summer recovery, in the Clyde (Comely 1974). Other studies, however, have shown both latitudinal and local variations in the timing and number of annual spawning peaks, and the degree of synchronisation between individuals.

At the northern end of the geographical range, in western Norway, Strand and Nylund (1991) recorded single spawning peaks for two populations, the most northerly having a closely synchronised spawning in June and the more southerly a less well-synchronised spawning in late July-August or August-September. In Loch Creran, on the west coast of Scotland, there is also only one main spawning period each year, but the timing can vary from year to year and spawning is not closely-synchronised between all individuals in the population (Fegan et al. 1985; Mackie 1986). While two main spawning peaks are usual for most Irish Sea populations, Wanninayake (1994) found different seasonal cycles in two closely located populations. The Laxey Bay population in 20–30 m water depth conforms to the usual seasonal pattern for the inshore grounds (Mason 1958a; Duggan 1987; Allison 1993), with distinct spawning peaks in spring or early summer (April–June) and again in autumn (August–September). In contrast, the offshore Port St. Mary population, 30 km farther south and in 50–70 m water depth, has only a single summer peak (June–July). In addition, although summer spawning starts later in the Port St. Mary population, the recovery of the gonads after spawning occurs 1–2 months earlier. Thus full gonads are present for a longer period in the Port St. Mary population, and gonad index and gonad weight are higher for much of the year. In the absence of any late summer or autumn spawning peak, this deeper water Irish Sea population therefore resembles populations farther north in the geographical range.

Populations of *P. maximus* farther south in the geographical range show further variation in seasonal patterns of gametogenesis, spawning and recovery (Lubet et al. 1991). This is particularly apparent for populations in different parts of the English Channel. In the Bay of St. Brieuc, on the northern coast of Brittany, native scallops show highly synchronised maturation of all individuals in spring, leading to a massive spawning in July; this is followed by secondary spawnings until the end of August but, uniquely, recovery of the gonad is then delayed until the following spring, when they mature rapidly (Paulet et al. 1988; Mackie and Ansell 1993). This cycle differs markedly from that in the nearby Bay of Brest, where the scallops show little synchrony between individuals and there are repeated cycles of maturation of cohorts of gametes throughout the year (Cochard 1985; Paulet et al. 1988). Many of the oocytes produced during the winter months, however, become atretic, undergo cytolysis and are probable resorbed, rather than being spawned (Dorange and Le Pennec 1989; Le Pennec et al. 1991). This results in considerable seasonal variation in the quality of the gametes produced (Cochard 1985; Paulet et al. 1992). There are also differences in the egg size, larval life span and duration of metamorphosis, highlighting differences in reproductive strategy between the two bays (Paulet et al. 1988). Much research on the reproductive cycle has also been carried out on the scallop population in the Bay of Seine (Lubet et al. 1987a, b; Lubet et al. 1991) where the sexual cycle is continuous throughout the year but the spawning season is restricted to the period July–October, and oocyte lysis occurs from November to the following July. The most southerly investigations of *P. maximus* have been for a natural shallow water population (Acosta and Román 1994) and stocks in suspended culture in the Ria de Arousa, NW Spain (Pazos et al. 1995; Pazos et al. 1996; Pazos et al. 1997a). Here, there are two main spawning peaks, one in winter (February–March) and one in late spring and early summer (May–July), but minor spawnings are possible in late summer and early autumn, giving a very long spawning season. Some authors have suggested that the presence of large numbers of atretic oocytes at certain times of the year appears to be characteristic of shallow, warmer water populations towards the southerly end of the geographical range (Ansell et al. 1991; Besnard 1991; Lubet et al. 1991). However, this conclusion was drawn before oocyte lysis had been reported in some of the most northerly (Strand and Nylund 1991), and the most southerly (Pazos et al. 1995), *P. maximus* populations. The adaptive significance of this seemingly energetically wasteful reproductive strategy in some populations – and its apparent absence from populations around Britain – has not been explained, although it has been suggested that oocyte lysis and resorbtion may provide energy for metabolism when the energy supplies from the adductor muscle and ingested food are insufficient (Lubet et al. 1987c; Strand and Nylund 1991), or to develop new oocyte cohorts (Paulet and Boucher 1991). Lubet et al. (1991) conclude that it is 'an adaptive response to environmental conditions'. Of these, temperature and food supply are likely to be the most important.

Differences in reproductive strategy are usually closely linked with seasonal cycles of energy acquisition and storage. In scallops, metabolic reserves of protein and glycogen are stored mainly in the adductor muscle, and lipids are stored mainly in the digestive gland. Seasonal cycles of energy storage and utilisation have been studied in various tissues in relation to the spawning cycle for populations of *P. maximus* in Norway (Strohmeier et al. 2000), Scotland (Comely 1974; Mackie 1986), the Irish Sea (Wanninayake 1994), France (Faveris 1987; Lubet et al. 1987b; Faveris and Lubet 1991) and Spain (Acosta and Román 1995; Pazos et al. 1997a, b; Roman et al. 1999). During the seasonal cycle there are periods of net accumulation of reserves, when the protein and glycogen content of the adductor muscle and the lipid content of the digestive gland increase, and periods of net depletion when these reserves are utilised to contribute to metabolic requirements for maintenance, or for the development of the gonad. In Norway, spawning coincides with the spring phytoplankton bloom and switches in energy allocation appear to divide the year clearly into two parts: priority is given to somatic growth and energy storage in the period from June to October, while reproductive growth occurs only from October to June (Strohmeier et al. 2000). On the west coast of Scotland, metabolic reserves increase through the spring and summer, with only a slight reduction during the main spawning season, to reach a peak in autumn (Comely 1974; Mackie 1986). The reserves in the adductor muscle and digestive gland then decline during the winter, contributing to maintenance and the requirements of the rapidly developing gonad. A very similar cycle of energy storage and utilisation has been reported for scallops in the Ria de Arousa, Spain, the most southern population so far investigated (Pazos et al. 1997b). In Irish Sea populations the seasonal pattern of change is also generally similar, but there is some variation between fishing grounds. Energy reserves build up to higher levels in the deeper water Port St. Mary offshore population and, with no autumn spawning on this ground, the recovery of the gonads and the depletion of stored reserves proceeds 1–3 months earlier than in the shallow water Laxey Bay population (Wanninayake 1994). For both grounds the fall in lipid contents in the digestive gland during the period of gametogenesis, plus a small contribution from the adductor muscle, was more than adequate to account for the lipid increase in the ovary, but specific fatty acids may need to be synthesised from other substrates. In the French populations the seasonal cycles of biochemical reserves are more complex, with significant falls in glycogen and lipid in summer, coinciding with spawning (Faveris and Lubet 1991; Mackie and Ansell 1993). Hence, in Bay of Seine scallops, there are two major peaks in energy reserves, in December and in July (Lubet et al. 1987a, b).

Many of the differences in the reproductive and energy storage cycles that have been reported for different populations of *P. maximus* undoubtedly represent responses to differing environmental conditions, particularly variations in temperature, food supply and day length. However, for other populations, particularly those in the closely located Bays of Brest and St. Brieuc, there is evidence that the differences are determined by genetic factors, acting independently of the environment (Paulet et al. 1988; Mackie and Ansell 1993). This implies the existence of genetically distinct stocks in different localities and hence some degree of genetic isolation between stocks. Strong support for this hypothesis has come from a series of transplant experiments carried out in the Bay of St. Brieuc, by scientists from IFREMER, Centre de Brest, in association with local fishermen's organisations. This work was undertaken as part of a restocking programme that aimed to reinstate commercial stocks in the Bay by introducing juvenile scallops from other localities (Dao et al. 1985a, b). Batches of young scallops were taken from spat collectors near the Isle of Scalpay (Scotland) and hatchery reared spat from the Bay of Brest, and grown alongside native St. Brieuc scallops. The reproductive and storage cycles of the native stocks in these two areas differ in well-defined ways from the native stocks of the Bay of St. Brieuc. Unfortunately, a full reciprocal transplant experiment was not possible because national legislation prohibits transfers of shellfish between some areas. The results, never-the-less, provide very strong evidence that genetic factors control features of the reproductive cycle (Ansell et al. 1988; Mackie and Ansell 1993). Both Scottish and the Bay of Brest transplants retained critical features of their reproductive cycle after growing in the Bay of St. Brieuc for two years. In both these stocks, recovery of the gonad began in late summer, soon after spawning, and there was no post-spawning refractory period that is so characteristic of St. Brieuc native scallops. As a result, the mean state of maturity of the transplants diverged progressively from that of the St. Brieuc natives, and the reproductive cycle was less well synchronised between individuals. Since the three stocks were growing under identical environmental conditions, the study strongly suggests that some features of the reproductive cycle are under genetic control, and additional evidence to support this has come from hatchery conditioning trials carried out on the same native and transplant stocks (Cochard and Devauchelle 1993). At the same time, other aspects of the reproductive cycle, such as the time and duration of spawning, are clearly subject to environmental influence (Mackie and Ansell 1993). Thus spawning and recovery of the gonads started later, and with a longer spawning period, in Scottish native scallops growing on the west coast of Scotland, than in Scottish transplants in the Bay of St. Brieuc. The seasonal cycles of energy storage and utilisation also appear to be under environmental, rather than genetic, control. There was no detectable difference in the annual pattern of accumulation and utilisation of reserves between the Scottish and Bay of Brest transplants and the native Bay of St. Brieuc stock, growing under the same conditions, despite the different reproductive cycles. Further evidence comes from the experimental studies of Saout et al. (1999a) who found that energy partitioning between somatic growth and gonad growth was strongly influenced by the experimental manipulation of temperature and day length. The relative influence of genetic and environmental factors therefore differs for the reproductive cycle and the storage cycle. This finding is potentially important, for the two cycles might become de-coupled in transplanted populations (Mackie and Ansell 1993) and this could have significant, but unpredictable, biological and commercial consequences.

While the studies of reproductive adaptations from transplant experiments and hatchery conditioning strongly support the existence of genetically distinct stocks, at least for the native scallops from the Bay of St. Brieuc, direct studies of genetic variability, using the techniques of gel electrophoresis, failed to find any genetic differentiation between any of the scallop populations studied (Beaumont et al. 1993). This agreed with earlier studies comparing *P. maximus* populations from Scotland, Ireland and Brittany (Huelvan 1985), and from different grounds around the coast of Scotland (Morvan in (Ansell et al. 1988)). The high values for the mean genetic identity between populations shown in these three studies suggested that the European stocks of *P. maximus* might be a single panmictic unit, in keeping with the relatively long pelagic life of the species. However, genetic differentiation has been detected in similar studies of the queen scallop, *Aequipecten opercularis* (Mathers 1975; Beaumont 1982b; Beaumont and Beveridge 1984; Lewis 1992; Lewis and Thorpe 1994a; Lewis and Thorpe 1994b), which has a very similar geographical distribution and a comparable larval life span (Le Pennec 1982). Since the failure to detect small-scale differentiation between *P. maximus* populations could be a consequence of the limitations of the allozyme methodology, in which only a few loci were available for analysis, some more recent studies have used high-resolution DNA-based methods. Wilding et al. (1997) used PCR-RFLP of mitochondrial DNA to investigate population structure of

P. maximus from the U.K., Ireland and Brittany. They found no evidence of differentiation within and between the U.K. and Brittany populations (including Bay of St. Brieuc scallops), but the population from Mulroy Bay, Ireland, was distinct from the others. This latter finding was confirmed by Heipel et al. (1998), using RAPD (predominantly nuclear DNA) markers. They found significant variation between populations from Mulroy Bay, Plymouth (southwest England) and the Irish Sea. In addition, they reported a low but significant degree of differentiation between five commercially fished grounds around the Isle of Man. A subsequent study of restriction-site variation in two PCR-amplified mitochondrial DNA fragments, carried out on the same samples (Heipel et al. 1999), also showed evidence of population differentiation. Mulroy Bay was again distinct from the other populations, and had the lowest genetic variability of all the sites, reflecting the relative isolation of this very enclosed water body. There was also slight but significant differentiation among the Isle of Man populations, with the East Douglas fishing ground appearing distinct from the other Manx locations. These two papers provide the first evidence of population genetic structuring in exploited open-water stocks of *P. maximus*. Recently, Ridgway and Dahle (2000) have also used restriction analysis of PCR-amplified mtDNA fragments to show that the Norwegian and British scallop populations belong to different gene pools. The identification of genetically relatively isolated stocks in different areas, from studies of reproductive ecology or genetic differentiation, is an essential element in the management of fisheries for it has important implications for stock assessments, stock enhancement and cultivation projects involving the transfer of stocks from one area to another.

When *Pecten maximus* spawns the eggs and sperm are released into the water column where fertilisation takes place. The fertilised egg then develops through the usual trochophore and shelled veliger stages, during which it filter-feeds on phytoplankton, and grows to a length of about 250 μm before settlement and metamorphosis occurs (Le Pennec 1974). There is a large and rapidly growing literature on the development, growth, survival and nutritional requirements of *P. maximus* larvae under hatchery conditions (reviewed elsewhere in this volume) but studies under natural conditions are few. In the sea, the length of larval life depends on environmental factors like temperature and food availability, but genetic factors may also be important (Paulet et al. 1988). The larvae are probably in the plankton for 3–8 weeks before settlement, during which time they are carried with the currents (Le Pennec 1974; Paulet et al. 1988; Beaumont and Barnes 1991). Survival in the planktonic phase is strongly influenced by hydrographic conditions, the availability of food, the abundance of predators and competitors, and the presence of unusual conditions such as toxic algal blooms or local depletions of oxygen (Minchin 1985; Erard-Le Denn et al. 1990).

Natural settlement of scallop larvae is mainly on emergent filamentous substrates such as bryozoans and hydroids that provide an erect, silt-free, surface (Eggleston 1962; Soemodihardjo 1974; Mason 1983; Dare and Bannister 1987; Minchin 1992). After initial settlement, small *P. maximus* spat (250–530 μm shell length) appear to be capable of secondary dispersal by bysso-pelagic drifting (Beaumont and Barnes 1991; Beaumont and Barnes 1992). There is also experimental evidence that quinone or quinone-like compounds, such as occur in aqueous extracts of certain red algae like *Delesseria sanquinea*, can act as chemical inducers of metamorphosis (Chevolot et al. 1991). Around the Isle of Man most pectinid spat have been found attached to bryozoans of the genus *Cellaria* (Brand et al. 1980) but they will settle on a wide range of natural and artificial surfaces including ropes and plastic netting. This behaviour provides the means of capturing large numbers of naturally produced spat for cultivation (see elsewhere in this volume). Since spat of *P. maximus* is notoriously difficult to find on the seabed, artificial spat collectors provide scientists with a convenient technique to study seasonal and spatial patterns of settlement (Brand et al. 1980; Paul et al. 1981; Wilson 1987c; Fraser 1991; Thouzeau 1991b, c; Zhang 1996; Nance 2001) and a possible means of making early predictions of year-class strength in the commercial fishery (Fraser and Mason 1987; Burnell 1989; Brand and Prudden 1997; Beukers-Stewart et al. 2001b). There is recent evidence that settlement peaks off the west coast of Scotland occur on, or close to, the dates of full moons (Nance 2001). The results of spat collector studies can be difficult to interpret, for in such studies it is often impossible to distinguish between differences in initial settlement and differences in early survival (Zhang 1996). This could account for some of the apparent discrepancies noted between the seasonal pattern of settlement in spat collectors and seasonal patterns of spawning in the local populations shown by studies of gonad maturation stages (Brand et al. 1980; Wilson 1987a), rather than attributing these discrepancies to large-scale larval transport mechanisms. The relationships between gonad maturation, spawning, larval abundance, spat collection and recruitment in the Bay of St. Brieuc were reviewed by Boucher (1985). He concluded that, for this area, the influence of environmental factors such as temperature leads to synchrony of gonad maturation and mass spawning, which is a major determinant of recruitment. Despite numerous other studies in France (Fifas et al. 1990; Thouzeau 1991a; Paulet et al. 1992; Boucher and Fifas 1997; Saout et al. 1999b) and elsewhere (Fraser 1991; Allison 1993; Dare et al. 1994), factors affecting recruitment to commercial stocks remains one of the least well understood aspects of *P. maximus* ecology. While discussing the general problem of scallop recruitment in relation to management, Sinclair et al. (1985) suggested that the main commercially fished grounds *of Pecten maximus, Aequipecten (Chlamys) opercularis* and *Placopecten magellanicus* are self-sustaining. Subsequent hydrographic (Darby and Durance 1989; Dare et al. 1994) and genetic (Lewis and Thorpe 1994a; Lewis and Thorpe 1994b; Heipel et al. 1998; Heipel et al. 1999) studies are providing increasing support for this conclusion.

19.2.2 The Queen Scallop, *Aequipecten opercularis* (L.)

The queen scallop *Aequipecten opercularis* is smaller than *Pecten maximus,* growing to a maximum shell height of about 90 mm. The shell is inequivalve but, unlike *P. maximus*, both valves are cupped. The more deeply cupped upper (left) valve can be of a range of colours, including yellow, orange, red, brown and purple; the lower (right) valve is generally lighter in colour, frequently off-white, buff or brown. Both valves can have a conspicuous pattern of lines, spots or other markings in various colours. The shell colour is often much brighter and more variable in the first few months, after which it frequently becomes orange-brown but the first growth zone retains its original colour, albeit fading throughout life. There are some 19–22 broad radiating ribs on both valves, together with numerous concentric growth striae that aggregate to form growth rings. The visibility of the growth rings varies considerably from ground to ground in this species but they are seldom clear.

As with *Pecten maximus*, the first settled stages of *Aequipecten opercularis* are byssally attached to stones, shells or epibenthic organisms. The morphological characteristics for identifying small spat of *A. opercularis* have been described, together with comparative descriptions of six other pectinids from the western Mediterranean, including *Pecten jacobaeus, Mimachlamys varia, Flexopecten glaber and Flexopecten flexuosus* (Peña et al. 1998). *A. opercularis* spat settles very readily on artificial collectors, so the distribution, abundance and various aspects of the early biology of queens have been frequently studied (e.g., Brand et al. 1980; Paul 1981; Brand et al. 1991b; Thouzeau 1991b, c; Chauvand et al. 1996; Román et al. 1996; Zhang 1996; Nance 2001). Queen scallops retain the ability to attach until quite a large size, although the proportion that is byssally-attached falls with age and most queens larger than 50 mm are free-living (Soemodihardjo 1974; Paul 1980a). Small queens secrete more byssus threads and re-attach much more rapidly than larger individuals. The rate of attachment is highest at 18°C and declines, and becomes more variable, at both higher and lower temperatures (Paul 1980a). Byssus production in *A. opercularis* is extremely sensitive to environmental conditions, such as the presence of pollutants in the water, so it makes a very sensitive bioassay for marine contaminants (Roberts 1973). Although the spat of *A. opercularis* retain the ability to byssally attach to a larger size than *P. maximus*, the detached spat are much more active and show a preference to climb up and attach to vertical surfaces. This behaviour provides the basis for a simple but effective method of separating spat of the two species taken from collectors (Paul 1985). Once they cease to form a byssus, adult *A. opercularis* remain much more active than *P. maximus*. They do not recess but rest on the seabed, with the more deeply cupped left shell uppermost, and are excellent swimmers throughout life (Ansell and Ackerly 1994), though swimming ability varies seasonally. In the warmer summer and autumn months, queens swim readily in response to approaching divers and fishing gear (Chapman et al. 1979), and to escape from predatory starfish (Stephens and Boyle 1978). Underwater observations have shown that queens swim when fishing gear is 1–1.5 m away, and are only caught in bottom trawls when they are off the seabed (Chapman et al. 1979; Chapman 1981). The importance of vision in stimulating swimming in *A. opercularis* is supported by the fact that daylight catches in trawl fisheries greatly exceed those taken at night. Although swimming can be an effective short distance escape response, queens are not capable of swimming large distances. In Loch Ewe, Scotland, queens could be stimulated to swim for up to 11 times in succession, during which they travelled a maximum distance of only 6.6 m before becoming exhausted (Chapman et al. 1979). In areas with strong water currents, repeated swimming may result in downstream dispersal, and this could account for some of the many fishermen's reports of queen scallops suddenly disappearing from a fishing ground.

A. opercularis has not been studied as extensively as *P. maximus*, probably because of its lower market value. However, some aspects of the physiology and genetics have received considerable attention, and there have been several comparative studies of the two species. Feeding mechanisms of queens have been investigated, both in the field and in the laboratory (Vahl 1972; Mathers et al. 1979a; Lawrence 1993). While queens are unable to efficiently filter particles smaller than about 7 μm in diameter, they are able to pump large volumes of water at relatively low metabolic cost and show a flexible feeding response to changing dietary quantity and quality. The effects of temperature acclimation on the rates of oxygen consumption and filtration were studied by McLusky (1973), who suggested that 20°C was near the upper lethal temperature for this species. However, Paul (1980a) found the upper temperature tolerance (LT_{50}-48h) for large and intermediate sized queens was between 19–24°C, depending on the acclimation temperature, while for spat it was 24–25°C, and byssus production was at its maximum at 19°C. These experiments were all carried out on Scottish and Irish Sea populations so it would be interesting to investigate temperature tolerances in Mediterranean or Adriatic populations that are acclimated to higher temperatures. Reduced salinities ranging from 16–28‰ were lethal to *A. opercularis* after a 24 h exposure period, depending on the temperature and size of the animal (Paul 1980b). Mortality increased at extremes of temperature and spat had a slightly greater tolerance of low salinity than larger individuals.

Studies of the distribution and abundance of *A. opercularis* populations have been carried out in many areas throughout the geographical range (e.g., Brienne 1954; Ursin 1956; Pickett and Franklin 1975; Mason et al. 1979c; Franklin et al. 1980b; Piccinetti et al. 1986) but the Irish Sea stocks have been particularly well studied (Aravindakshan 1955; Watson 1971; Soemodihardjo 1974; Allison 1993; Veale et al. 2001). *A. opercularis* has a patchy, aggregated distribution, like *P. maximus,* but it occurs on a wider range of substrates and, when present, can occur at very much higher densities (Mason 1983). The life span of *A. opercularis* is shorter than that of *P. maximus,* with a maximum probably not exceeding 8–10 years and generally much less. In the Irish Sea queens reach a commercially acceptable size (50mm shell height) at an age of 14–18 months; beyond 4–5 years old the natural mortality rate is high and individuals more than 6 years old are rare on all fishing grounds (Brand and Prudden 1997). With so few year-classes present in the exploited populations, the success of the fishery in any year is highly dependent on the strength of recruitment. The spatial and temporal variability in year-class strength is considerably greater for *A. opercularis* populations than for *P. maximus* populations (Allison 1993). However, the main north Irish Sea queen fishing grounds have remained spatially persistent since the fishery began in 1969, although the areas of good recruitment within each ground vary from year to year.

Age and growth has been studied in many areas and the earlier data have been reviewed by Taylor and Venn (1978) and Mason (1983). Since then, Richardson et al. (1982) studied seasonal variations in growth of queens suspended at different depths, and Richardson et al. (1984) carried out an experimental study of the effects of iron ore suspensions and food on shell and tissue growth rates. More recently, Allison (1993) compared the growth rates of queens on different fishing grounds around the Isle of Man. He found them to be very variable, both spatially and temporally, and went on to develop a growth model that accounted for seasonal oscillations in the edible yield (Allison 1994). There have also been a number of studies of the growth of *A. opercularis* in suspended culture in Galicia, NW Spain (Román and Acosta 1991; Lowe 1996; Román et al. 1999). The high growth rates achieved at high densities indicate that this species has good potential for aquaculture in the Galician rias.

Other aspects of the population dynamics of queen scallop stocks are generally poorly understood compared with the state of knowledge of *P. maximus*. Mortality rates have rarely been determined. Allison and Brand (1995) calculated instantaneous rates of total (Z), fishing (F) and natural (M) mortality from tagging for an Irish Sea queen population and found them to be 0.41, 0.21 and 0.20 per month, respectively. The calculated value of M is very high but includes incidental fishing mortality (F_i) resulting from gear damage. An estimate of M for an adjacent, largely un-fished area, based on age-frequency analysis, gave a value of only 0.036 per month (0.43 per year). If natural mortality on the fished ground was at a similar level, then F_i would account for most of the mortality calculated as M in the tagging study. Incidental mortality may therefore approach F in this fishery. Very high levels of F_i, often exceeding F and M, have been reported for some other scallop fisheries (e.g., Naidu 1988; McLoughlin et al. 1991) and highlighted the need to assess the full impact of gear damage on both target stocks and the environment. In recent years, this has been frequently investigated for various European scallop and queen scallop fisheries (Hall-Spencer et al. 1999; Bradshaw et al. 2000, 2001, 2002; Hall-Spencer and Moore 2000a, b; Veale et al. 2000b, 2001; Pranovi et al. 2001).

The main predators of *A. opercularis* are mostly the same species that prey on *P. maximus*, notably decapod crabs and starfish prey on adult queens, and a wider range of animals prey on the spat and juveniles. There is evidence from field and laboratory experiments that the two most important predators in the Irish Sea, *Cancer pagurus* and *Asterias rubens,* both feed preferentially on *A. opercularis*, rather than *P. maximus*, despite the better escape response of the queen scallop (Whittington 1993; Veale et al. 2000a). Pond (1992) has shown that the encrusting *Suberites* sponges that frequently colonise the shell of *A. opercularis* (Ward and Thorpe 1991) forms a protective-commensal relationship with the queen scallop, in which the sponge protects the scallop from predation by *Asterias rubens*, probably by reducing tube-foot adhesion. Diseases do not appear to be an important cause of mortality in *A. opercularis*. In a recent survey of healthy stocks, Lohrmann (2000) found few pathogens present in *A. opercularis* populations from five locations around the U.K., though there were more different parasites, and they generally occurred at high prevalences, than in *P. maximus* taken concurrently from some of the same sites. Rickettsiales-like micro-organisms were present in gill epithelial cells but the prevalence was much higher in queens from the English Channel (85%) than from those taken in the Irish Sea and the North Sea (<20%). Microsporidian spores were present in queens from all locations, but with a much higher prevalence (40%) at one Irish Sea site (Lohrmann et al. 2000). This is the first time a microsporidian infection has been reported in any scallop species. There have been some studies of heavy metal accumulation in *A. opercularis* (Bryan 1973; Topping 1973; Howard and Nickless 1978) and of the effects of iron ore suspensions on feeding and growth (Richardson et al. 1984) but there is no evidence of mortalities caused by environmental pollution.

Like *P. maximus*, *A. opercularis* is a simultaneous hermaphrodite with the prominent gonad containing a proximal creamy-coloured testis and a distal bright red ovary. The onset of sexual maturity occurs very early in this species, when they are about one year old (Soemodihardjo 1974). The annual reproductive cycle has been studied at various locations throughout the geographical range (Aravindakshan 1955; Ursin 1956; Soemodihardjo 1974; Broom and Mason 1978; Taylor and Venn 1979; Castagnolo 1991; Canales et al. 1995; Román et al. 1995; Peña et al. 1996). The general trend appears to be for a single annual spawning peak at the northern end of the latitudinal range and multiple peaks at the southern end, but local conditions can result in different spawning patterns in some closely located populations. For example, in the north Irish Sea, where the most detailed studies have been carried out, three more-or-less distinct peaks of spawning each year are characteristic of the inshore populations (Aravindakshan 1955; Duggan 1987; Wanninayake 1994); they generally occur in February–March, June–July and September–October, with the autumn spawning appearing to be the most important. However, for a deeper-water population farther offshore, Wanninayake (1994) found only two peaks, with no evidence of the autumn spawning. Instead, the spawning peaks were less well-defined and the redevelopment of the gonad through the autumn and winter proceeded some two months earlier than for the inshore population studied. In the smaller number of annual spawning peaks and the absence of an autumn spawning in the deeper water population, *A. opercularis* therefore resembles *P. maximus* populations studied concurrently on the same two fishing grounds. Despite the apparent importance of the autumn spawning on most north Irish Sea grounds, as indicated by studies of the reproductive cycle, most spat settlement on benthic epifauna and artificial spat collectors in this area over many years has come from spring or early summer spawnings (Brand et al. 1980; Paul 1981; Brand and Prudden 1997). Beaumont and Hall (1999), discussing the difficulty of artificially inducing *A. opercularis* to spawn in the hatchery, suggest that it may be a 'dribble' spawner, rather than a complete spawner, under natural conditions. This may well be the case with some populations where spawning and partially spawned gonads are present in most months of the year (Canales et al. 1995). Seasonal cycles of energy storage and utilisation have been described for queen populations from the west coast of Scotland (Taylor and Venn 1979) and the Irish Sea (Wanninayake 1994), and estimates of reproductive effort made for Bay of Brest populations (Lucas et al. 1978; Lucas 1982).

The population genetics of *A. opercularis* has received considerable attention. Unlike similar studies on *P. maximus*, allozyme electrophoresis has readily demonstrated genetic differentiation between queen populations on a relatively small geographical scale, and few studies have used DNA-based molecular techniques. After an early study by Mathers (1975) reported significantly different *Gpi* allele frequencies in queen populations from the west and east coasts of Ireland, Beaumont (1982b) carried out a more extensive study of queens from some 20 locations ranging from Norway to Brittany. Of the four loci investigated, one locus was monomorphic, and two showed little variation, but there were large differences in the frequencies of three alleles at a protein locus, *Pt-A*. From these data Beaumont proposed that there were at least four relatively genetically isolated populations of *A. opercularis* around the British Isles, namely (1) the north and west

Scottish coast, (2) the Irish Sea, (3) the west Brittany and west Irish coast, and (4) the English Channel. More recent workers have shown highly significant differences in allele frequencies between sites, both within and around the Irish Sea (Lewis 1992; Lewis and Thorpe 1994b). These, and several other studies, have also taken advantage of the fact that queens can be aged from shell rings to study the stability of gene frequencies between year-classes. The predominant situation appears to be one of consistency of allele frequencies within sites from one year to the next. This temporal genetic stability, coupled with highly significant geographical heterogeneity, suggests the possibility that populations could be predominantly self-recruiting (Lewis and Thorpe 1994a; Lewis and Thorpe 1994b), despite having a planktonic larval stage lasting several weeks. For many *A. opercularis* and *P. maximus* populations it seems likely that hydrographic mechanisms resulting in larval retention may restrict local genetic exchange (Heipel et al. 1998). Unusual hydrographic events may, however, bring an occasional influx of larvae from further afield, as occurred in the Irish Sea in 1979 when an uncommon allele appeared in only one year-class of queens (Macleod et al. 1985) following a short period of exceptional storm-force winds. Some other aspects of genetic variability have also been studied, notably variations in the level of heterozygosity. Beaumont (1982a) concluded that differences in heterozygote deficiency between year-classes of queens in a Scottish sea-loch were due mainly to selection acting at the larval and/or the juvenile stage. Beaumont and Beveridge (1984) and Beaumont and Zouros (1991) effectively rule out self-fertilisation as a major cause of heterozygote deficiency, while Lewis and Thorpe (1995) consider that allozyme heterozygosity may act as a buffer against developmental abnormality in queen scallop spat.

19.2.3 The Black or Variegated Scallop, *Mimachlamys varia* (L.)

Mimachlamys varia (L.) is a small scallop, rarely growing to a shell height exceeding 60 mm. The shell is inequivalve, with both valves convex. It has protruding ears, the anterior of which is much larger than the posterior, and there is a prominent byssal notch in the right anterior ear. The shell colour is very variable, ranging through off-white, yellow, orange, pink, red, brown, purple, dark green and grey to almost black, often with prominent irregular patterns in many colours. Both valves have 25–35 prominent radiating ribs, each bearing spatulate spines that are particularly well-developed near the strongly crenulated margin. There are also numerous concentric growth striae, with annual growth lines sometimes clear.

Mimachlamys varia usually lives attached by the byssus to shells, stones or boulders on coarse sand or gravel substrates, but it also attaches to rock faces (Rodhouse and Burnell 1979) and is frequently associated with dense banks of the horse mussel, *Modiolus modiolus* (Jones 1951; Roberts 1975) and the oyster, *Ostrea edulis* (Forester 1979; Burnell 1991). Unlike *Pecten maximus* and *Aequipecten opercularis*, adult *Mimachlamys varia* are usually byssally attached (Mahéo 1968; Soemodihardjo 1974), but they can detach the byssus and utilise similar escape responses to the other scallops, although they are not such powerful swimmers (Millward and Whyte 1992). The shells are often heavily colonised by epizoites, including algae, sponges and serpulids

(Rodhouse and Burnell 1979; Burnell 1983). Forester (1979) noted that dense growths of the sponge *Halichondria panicea* were common epizoites on *M. varia* on the Atlantic coast of Ireland. He showed that this partnership was a commensal-protective mutualism, the sponge benefiting from the scallop's feeding currents while protecting it from predatory starfish. Although *M. varia* can extend into deeper water (Tebble 1966), abundant populations appear to occur only within sheltered bays, in shallow water down to 10 m (Rodhouse and Burnell 1979). *M. varia* contributes little to European fisheries, except in some parts of France. Despite this, there has been much research on some aspects of its biology.

The most detailed series of studies have been carried out on a population of *M. varia* from Lanvéoc, Bay of Brest, France (Shafee 1979 1980a, b, 1981, 1982; Shafee and Lucas 1980, 1982), covering metabolism, growth and reproduction. Studies of growth are complicated by the presence of clear biannual recruitment, and the growth curves of spring and autumn spawned cohorts remain different throughout life (Conan and Shafee 1978). When various growth models were fitted to data for annual and seasonal growth, linear day degree models were found to be capable of predicting growth in shell height (Shafee 1980a). Empirical models using temperature, food availability and gonad index as independent variables showed that temperature and food together were decisive factors determining growth rates. There were also clear seasonal variations in the growth of body tissues, but this was not closely correlated with changes in shell height. Seasonal changes in gonad index and other quantitative aspects of the reproductive cycle were studied (Shafee and Lucas 1980), together with related studies on seasonal changes in biochemical composition and calorific content (Shafee 1981), and oxygen consumption (Shafee 1982). Data from these various studies were integrated to assess seasonal variation in the overall energy budget (Shafee and Lucas 1982), and to calculate reproductive effort (Shafee and Lucas 1980; Lucas 1982).

Growth and reproduction has also been studied for populations on the west coast of Ireland (Rodhouse and Burnell 1979; Burnell 1983; Burnell 1991; Burnell 1995). The life span of *M. varia* in this area is estimated to be 9–10 years. Growth characteristics, as shown by the parameters of the von Bertalanffy growth equation, were markedly different for *M. varia* in Lough Ine and Inner Roskeeda Bay. Seasonal growth at these sites was best described by a model incorporating temperature (as day degrees >9°C) and standing crop of phytoplankton (as chlorophyll-*a*). A population model was fitted and reproductive effort calculated for each age group in the population.

The reproductive biology of *M. varia* differs considerably from *P. maximus* and *A. opercularis* because *M. varia* is a successive protrandric hermaphrodite (Lubet 1959; Reddiah 1962; Lucas 1965; Rodriguez et al. 1991). There is usually an imbalance in the sex ratio, with males predominating among small scallops and females among the larger animals. At any one time an individual may function as either a male or a female, but can change sex after spawning. For the Lanvéoc population, change of sex may occur within a single season or less commonly between successive seasons, and individuals may undergo several successive changes (Lucas 1965). In most areas where the reproductive cycle has been studied there appear to be two main periods of spawning each year, but the timing can differ between areas, and between years. For Bay of Brest and other French populations there is a partial spawning in May-June and a complete spawning in September–October, but other intermittent spawnings can occur between May and September (Dalmon 1935; Lubet 1959; Buestel et al. 1979; Shafee and Lucas 1980; Dao et al. 1985c). In the northern Irish Sea, around the Isle of Man, the first spawning takes place about a month later than in French waters, while the second appears to be at about the same time or slightly later, extending into November (Reddiah 1962). On the west coast of Ireland the first peak of spawning activity occurs in late May or early June, at a similar time to the French populations, but the second is earlier, in early August, although an additional minor spawning occurs in September in some years (Burnell 1983, 1991). Burnell noted that the major spawnings coincide with peaks in chlorophyll *a* and appear to be triggered by temperature fluctuations of 1–2°C around 15°C that occur at spring tides.

The biannual pattern of spawning in *M. varia* is generally reflected in the pattern of spat settlement in collectors (Conan and Shafee 1978; Burnell 1991; Peña et al. 1996), although there was only one peak in settlement off the Isle of Man in 1993 and 1994 (Zhang 1996). However, good settlement in collectors has not been found to be a reliable predictor of good recruitment in the natural population (Buestel 1978). Furthermore, spat settlement does not appear to be closely correlated with the intensity of spawning as measured by the gonad index (Lucas 1965; Buestel et al. 1978), or with the population reproductive effort (Shafee and Lucas 1980). Larvae of *M. varia* have been reared under hatchery conditions (Le Pennec and Diss-Mengus 1985, 1987) and spat settlement occurs in about 25 days at 18°C (Burnell 1983). On the west coast of Ireland, where the second spawning usually results in the highest settlements, the higher mean temperatures in July and August appear to be more conducive to larval survival and settlement than the conditions in May and June (Burnell 1991). Burnell found that spat collectors need a period of 2–3 weeks immersion to develop a biofilm before they become attractive for larval settlement of *M. varia*, after which they continue to attract settlement for several weeks. He also noted that spat often settled on the calcareous tubes of *Sabella pavonina* that had settled on the collectors and suggested that this association may be important to *M. varia* spat.

There have been a few studies on aspects of the genetics of *M. varia* (Mathers 1975; Beaumont and Beveridge 1984), including a reciprocal transplant experiment (Gosling and Burnell 1988). In this latter work, there was genetic evidence of selective mortality when scallops from Loch Hyne on the south coast of Ireland were transplanted into Roskeeda Bay on the west coast. Baron et al. (1989) investigated the induction of triploidy in hatchery cultured *M. varia* with the aim of producing a strain that requires less energy and food for growth in areas of poor nutritional capacity. Although they obtained a high percentage of triploids (78.5%), their triploid larvae had a significantly higher mortality before metamorphosis than untreated diploid controls, and there was no difference in growth.

Other aspects of the biology of *M. varia* that have been investigated include studies of digestive cycles (Mathers et al. 1979b), feeding, and the structure and function of the labial palps (Beninger et al. 1990, 1991). The toxicity of silver and its bioaccumulation has also been investigated (Metayer et al. 1990; Berthet et al. 1992). Martoja et al. (1989) found that silver accumulated in the glandular cells of the primary byssus gland and inhibited byssus secretion. *M. varia* was also one of several species of bivalve mollusc affected by organotin pollution in Mulroy Bay, Ireland (Minchin et al. 1987b).

Mimachlamys varia is a valuable commercial species in France and attempts have been made to manage natural stocks and replenish those depleted by overfishing (Letaconnoux and Audouin 1956; Dao et al. 1985c). Since very large numbers of *M. varia* spat often settle on artificial collectors, much effort has gone into cultivation from naturally collected spat (Buestel et al. 1978; Latrouite 1978; Latrouite et al. 1980; Dao et al. 1985c; Burnell 1991), but spat have also been reared in a commercial hatchery (Le Pennec and Diss-Mengus 1985, 1987). More recently, interest has been growing into the possibilities of cultivating this species in Spain (Cancelo et al. 1992; Peña et al. 1995; 1996; Cano et al. 1999).

19.3 FISHERIES

Scallops have been landed in Europe for more than 100 years but although directed fishing in most countries started in the 1930s, modern dredge fisheries only really started to develop in the 1950s and 1960s, based mainly on stocks of *Pecten maximus* and *Aequipecten opercularis* around the coasts of the British Isles and France (Mason 1983; Dupouy and Latrouite 1976; Dupouy 1978). Over the last 30 years the fisheries have expanded enormously, and other areas and other species have been exploited.

In recent years three species of scallop, *Pecten maximus, Aequipecten opercularis* and *Chlamys islandica*, have accounted for most of the catches in European waters. The fishery for *Mimachlamys varia* is small and concentrated in two areas in France, the eastern part of the Bay of Brest and the Perthuis Charentes, near La Rochelle. Catches of the St. James' scallop, *Pecten jacobaeus*, make only a small contribution to European scallop landings, but this species is landed in many Mediterranean, Adriatic and Aegean ports and is undoubtedly under-reported or not identified to species in the official statistics. In some years more than 1,000 t has been landed in Chioggia market, Italy (Renzoni 1991), but the stocks declined in the 1990s due to over-fishing and the fishery was on the brink of commercial extinction (Mattei 1995; Mattei and Pellizzato 1996). In recent years landings of *P. jacobaeus* in Turkey have increased and 570 t was reported for 2000.

Total European landings of *Pecten maximus, Aequipecten opercularis* and *Chlamys islandica,* from 1985 to 2000 are shown in Fig. 19.1. This, and other figures in this chapter, is based on statistical data taken from the FAO Fishstat Plus database and from ICES Bulletin Statistique, supplemented with published and unpublished data from national sources. Where major discrepancies occur between the statistical sources, these have been resolved by discussions with colleagues involved in fisheries research. One problem in determining changes in the fisheries is the increasing proportion of scallops that are not attributed to species in the official statistics. This category, recorded in the statistics as 'scallops nei' (not elsewhere included), has increased steadily over the last 10 years and now accounts for nearly 20% of the total landings.

Catches of *Pecten maximus* declined slowly until 1990 but have since risen steadily and the 37,000 t catch for 2000 represents about 50% of total European scallop landings.

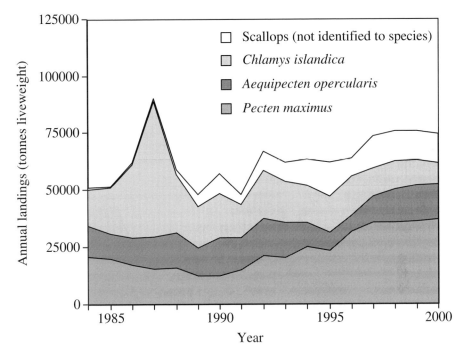

Figure 19.1. Total annual European landings (tonnes live weight) of *Pecten maximus, Aequipecten opercularis, Chlamys islandica* and other scallops (not identified to species in the statistics), 1984–2000. Based on data from FAO Fishstat Plus database, with additional data from Ø. Strand (personal communication).

Annual *Aequipecten opercularis* landings have fluctuated between 7–16,000 t, but after falling to a low level in 1995 and 1996, following poor recruitment in the Irish Sea, catches have increased again to some 15,000 in 1999 and 2000, representing 20% of European scallop landings. After the development of the large Norwegian fishery for *Chlamys islandica* around Jan Mayen, Bear Island and Spitzbergen in the mid-1980s, catches of this species peaked in 1987 but subsequently declined rapidly and now make up <15% of European scallop landings. Overall, total European scallop landings dropped sharply from 1987 to 1989, with the demise of the Norwegian *C. islandica* fishery, but since then there has been a fairly steady increase, due mainly to the increasing landings of *P. maximus*.

Catches of *Pecten maximus* from each of the main fishing areas between 1974 and 1998 are shown in Fig. 19.2. From 1974 to 1989 there was a marked decline in catches from both the east and west Channel grounds that was only partly offset by steadily increasing catches from the Irish Sea, Scottish west coast and northern North Sea grounds. Since 1990, catches have increased again on both the east and west Channel grounds, and with steady increases elsewhere, particularly on grounds to both the east and west of Scotland, the total *P. maximus* landings have increased greatly.

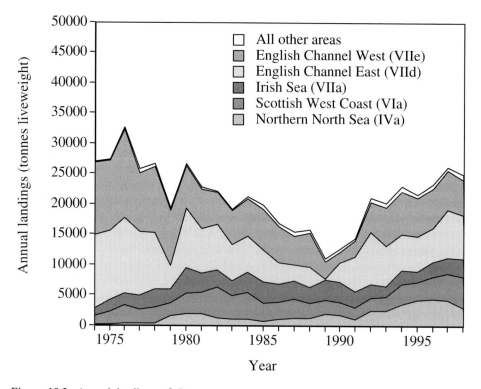

Figure 19.2. Annual landings of *Pecten maximus* (tonnes live weight) from European waters, 1974–1998, showing the landings taken from each major sea area fished. Roman numerals refer to ICES fishing areas.

Landings of *Aequipecten opercularis,* from the main fishing areas between 1974 and 1998 are shown in Fig. 19.3. The Irish Sea has always been the most important queen fishing area, contributing up to 80% of total annual landings. However, landings from the Faroes have increased steadily since 1975, with one massive fishery in 1990, and have contributed more than 30% to total queen landings in recent years. Elsewhere, queens are fished in many areas but these fisheries are generally sporadic. Although the main market for processed queen scallops is now in France, very few queens are caught in French waters. Overall, queen landings have been very variable from year to year, depending mainly on the success of recruitment. In the Irish Sea, poor recruitment, particularly of the 1992 year-class, resulted in a sharp drop in catches in the period 1994–1996, but the fishery has since recovered.

Gear, boats, fishing practices, areas fished and approaches to management differ in many respects from country to country. The development of the fisheries and their present status is described below for the main scallop and queen fishing countries. Although considered country by country, there is in fact considerable interaction between the fisheries due to the mobile nature of many modern scallop dredgers, and the traditional

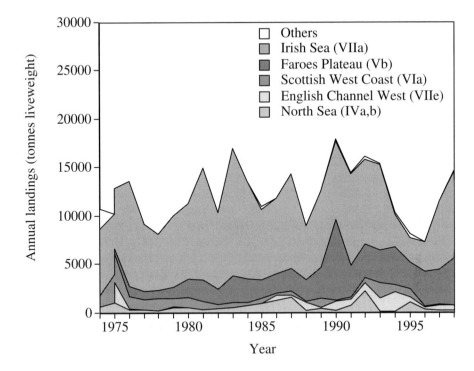

Figure 19.3. Annual landings of *Aequipecten opercularis* (tonnes live weight) from European waters, 1974–1998, showing the landings taken from each major area fished. Roman numerals refer to ICES fishing areas.

lack of effort regulation in U.K. scallop fisheries. The Isle of Man fishery is considered first, and in the most detail. Though not the largest or most economically important fishery, there is a long, well-documented, history of both scallop and queen fishing in the waters round the Isle of Man, and many of the features described for the Manx fishery apply also to other fisheries.

19.3.1 Isle of Man

Fishing for *P. maximus* started close inshore off the west coast of the Isle of Man in 1937. After the war the fishery developed rapidly, particularly in the 1960s, as more and larger boats joined the fishery and gear developments enabled the boats to exploit rougher areas of seabed. The start of the *A. opercularis* fishery in 1969 was also a major influence on the exploitation of new scallop grounds. The distributions of the two species overlap in many areas, so scallops have subsequently been fished in many areas where they occur in densities that would not be viable, were it not for the additional queen by-catch. Since the mid-1980s scallop fishing has taken place on a number of more-or-less discrete fishing grounds all around the Isle of Man, while the more restricted queen fishing grounds are mostly to the north, east and south of the Island (for maps see Brand et al. 1991a; Brand 2000).

In the early years, scallops were fished with relatively large (3 ft 6 ins–6 ft wide; 107–183 cm) dredges with a fixed tooth-bar, grouped together in 'gangs' on a heavy steel towing bar. However, 'Newhaven' spring-toothed dredges were introduced in 1972 and rapidly replaced fixed tooth-bar dredges as the normal gear. Then, in the early 1980s, individual dredge size was reduced to 2 ft 6 ins (76 cm), or sometimes 2 ft (61 cm), and these small spring-toothed dredges have subsequently become more-or-less universal throughout the British Isles, rigged to fish for either scallops or queens (Fig. 19.4a). Modern 'Newhaven' dredges fishing for scallops have 9 teeth on the standard 76 cm wide dredge, set 75 mm apart. Behind the tooth-bar the collecting bag is made up of a belly of steel rings (>75 mm internal diameter) and a netting back (100 mm stretched mesh) that extends forward ahead of the tooth-bar. As queen scallops do not recess into the seabed, and swim actively to avoid capture, they can be caught with a wider range of fishing gear (Mason 1983). In the early days of the Isle of Man queen fishery various types of dredge without teeth were tried but 'Newhaven' spring-toothed dredges rigged with shorter, more closely set teeth, and smaller ring diameter bellies were found to be efficient on the rough fishing grounds around the Isle of Man and were subsequently generally adopted. Newhaven spring-tooth dredges fishing for queens usually have 10 teeth on the standard 76 cm wide dredge, and belly rings of about 57 mm internal diameter (Fig. 19.4a). However, in the last few years a new type of toothless dredge, known as the 'skid' dredge, has been introduced and is gaining popularity. Skid dredges are a modification of toothed dredges, with each dredge frame mounted on ski-like runners and the tooth-bar replaced with a tickler chain (Fig. 19.4b). When fishing for scallops or queens, the number of dredges varies with the power of the boat, and while 4–8 per side is usual for the Manx fleet (Fig. 19.5), some very big vessels elsewhere pull much larger spreads of gear, up to 20 per side. Because of the large amount of bottom debris and invertebrate by-catch retained by the small-mesh bellies, boats dredging for queens use mechanical riddles on deck to sort the catch. On some fishing grounds bottom trawls are very efficient at catching queens in the summer months, when the queens swim actively to avoid capture, and some Manx boats trawl for queens for a few months each year. The trawls used are similar to groundfish nets but with heavier bobbin wheels or rock-hoppers on the groundline (Chapman et al. 1979; Brand et al. 1991a).

Management of fisheries in the north Irish Sea is complicated by the number of national jurisdictions involved. Management out to 12 miles around the Isle of Man is the responsibility of the Isle of Man Government, while the Republic of Ireland Government, and the U.K. Departments for England and Wales, Northern Ireland and Scotland all have jurisdiction in their respective areas. The Isle of Man Government has always taken a leading role in scallop fisheries management and maintained the strictest regime. Some of the measures introduced by the Manx Government in 1943, such as the closed season and minimum legal landing size (MLLS), were subsequently adopted by the U.K. Government and operate throughout the north Irish Sea (ICES area VIIa). The current closed season

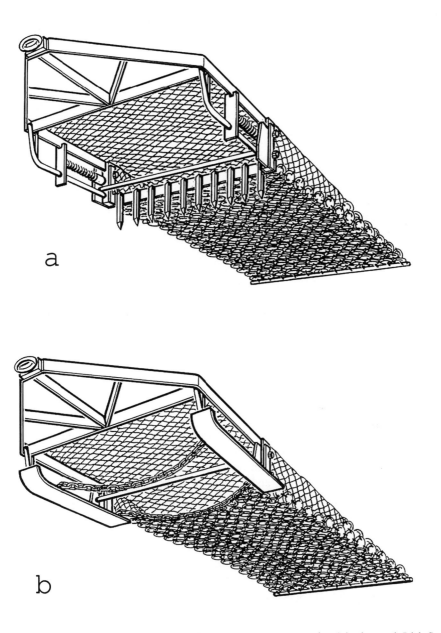

Figure 19.4. Dredges used to catch queen scallops (*Aequipecten opercularis*) in the north Irish Sea. a). Newhaven spring tooth-bar dredge, 76 cm wide, with 10 teeth and 57 mm internal diameter belly rings. b). 'Skid' dredge with ski-like runners and tickler chains instead of a tooth-bar. Diagrams kindly supplied by Sea Fish Industry Authority.

Figure 19.5. A gang of four 'Newhaven', 76 cm wide, spring tooth-bar dredges attached to a towing bar with large rubber bobbins at each end. The dredges are rigged to fish for queen scallops with 10 teeth and 57 mm internal diameter belly rings.

runs from 1 June to 31 October inclusive (apart from Cardigan Bay off the coast of Wales where fishing is allowed to continue throughout June) and the MLLS is 110 mm shell length. Additional Manx legislation restricts fishing within the 3-mile limit to vessels under 50 ft (15.24 m) in length, and fishing for scallops is prohibited after 18.00 hours GMT or before 06.00 hours GMT on any day. The possession or landing of shucked scallop meat by any fishing boat is also prohibited and vessels dredging for species other than scallops are restricted to only a 1% by-catch of scallops. There are also a series of technical measures governing the gear. The aggregate width of dredges that a boat may use is limited to 25 ft (762 cm) in the 3-mile area and 40 ft (1,219 cm) in the 3–12 mile area. Within the 12-mile territorial sea it is prohibited to use scallop dredges that have more than 9 teeth, a tooth spacing between the internal edges of less than 75 mm, belly rings with a clear opening of less than 75 mm internal diameter, and a mesh size of less than 100 mm in the netting cover. French dredges with a diving plane are also banned. In view of the number of measures governing the scallop fishery it is perhaps surprising that there are no specific regulations restricting queen fishing, other than the overall gear size restrictions and the prohibition on boats >50 ft (15.24 m) within the 3-mile limit. However, processors do not usually accept catches that contain a high proportion of queens less than about 55 mm shell height because it is not economic to process them. Since 1989, Isle of Man legislation has also prohibited dredging and trawling (but not static gear) within a small closed area of about 2 km² within the Bradda Inshore fishing ground off Port Erin. This area was closed to allow scientists from the Port Erin Marine Laboratory to assess scallop cultivation and stock enhancement, but has also proved an invaluable facility for studying the impact of dredging on benthic communities, and their subsequent recovery (Bradshaw et al. 2000, 2001).

The numbers of Isle of Man registered boats fishing for scallops and queens has fallen substantially from a maximum of about 70 in 1984 to about 30 in recent years. However, the average boat length has increased and more than 50% of active vessels are now >50 ft (15.24 m) which disqualifies them for fishing inside the 3-mile limit. Average engine power has increased more dramatically than length so the remaining vessels are able to tow more gear and exploit the offshore grounds more safely. However, this is not a modern, well-kept fleet, for many of the vessels are over 30 years old.

In addition to the Manx boats, there is a variable, but unknown, number of U.K. and Irish boats fishing scallops and queens in the north Irish Sea each year. The most important of these is a fleet of Scottish vessels operating from Kirkudbright; this includes several purpose-built dredgers with modern gear and catch-handling facilities that are probably the most efficient scallop and queen dredgers in the U.K. The Kirkudbright boats, and many other U.K. boats from other areas, are nomadic and fish wherever the best opportunities arise. The number of boats fishing in the north Irish Sea can therefore vary enormously, both within and between fishing seasons. No reliable data on fishing effort are available but there is no doubt that, overall, U.K. fishing effort on scallops has risen greatly in recent years as many more boats have turned to scalloping for at least part of the year when whitefish quotas are exhausted, or as a more economic alternative to *Nephrops* trawling in the winter months. Catches by the Manx fleet are landed daily, in the shell, processed on the Isle of Man and mostly exported fresh to continental Europe. Other boats operate from various ports in England, Wales, Scotland, Northern Ireland and the Republic of Ireland, landing daily or from 2–3 day trips, with the catches transported overland to shore-based processors.

The annual landings of scallops (*P. maximus*) from the north Irish Sea (Area VIIa), by country, over the period 1969–2001 are shown in Fig. 19.6. Overall, catches have been quite variable from year to year but rose to a high level of some 4000 t in 1980, fell to less than 2000 t by 1993, and rose again to nearly 4000 t in 2001. In the early part of this period, boats from the Isle of Man took the majority of the catch, but since 1980 boats from elsewhere, particularly Scotland and England and Wales, have taken an increasing proportion of the catch and the Manx share has declined to <30%.

Annual landings of queens (*A. opercularis*) by country, over the same period, are shown in Fig. 19.7. Overall, landings have been very variable from year to year with peaks in 1972, 1976, 1981, 1983, 1987, 1991, 1998 and 2001 and very poor fisheries in the period 1994–1996. These fluctuations in catch mainly reflect variable recruitment, but annual variations in fishing effort by the mobile U.K. fleet, and market demand, also have an effect. When the queen fishery first started in 1969 only Manx boats took part, but since the early 1970s increasing numbers of other vessels, particularly Scottish, have

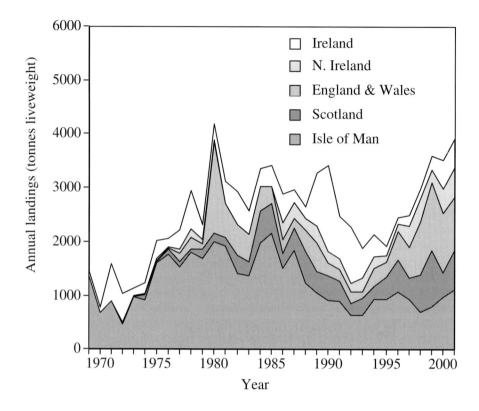

Figure 19.6. Annual landings of scallops (*Pecten maximus*) (tonnes live weight) in the Irish Sea (ICES Area VIIa), by country, 1969–2001.

progressively taken the bulk of the catch and the Manx share has declined to less than 20%.

The proximity of these fisheries to the Port Erin Marine Laboratory has ensured that they are amongst the most thoroughly studied and best-documented scallop fisheries in the world (see, for example, Tang 1941; Mason 1957, 1958a; Gruffydd 1972, 1974a, 1974b; Murphy 1986; Brand et al. 1991a; Allison 1993, 1994; Brand and Allison 1994; Allison et al. 1994; Brand 2000; Beukers-Stewart et al. 2001a; Jenkins et al. 2001; Jenkins and Brand 2001). Biometric data is available, dating back to the start of the scallop fishery in 1937, and commercial catch and effort data from fishermen's logbooks, recorded on an unusually small spatial scale (5 x 5 nautical miles) has been collected for both scallops and queens since 1981. Two, large-scale, tagging experiments have been carried out on scallops (Brand and Murphy 1985, 1992; Allison 1993) and a smaller study on queens (Allison and Brand 1995), and these have provided estimates of spatial variation in exploitation and mortality rates. Stock assessments have been made using yield per

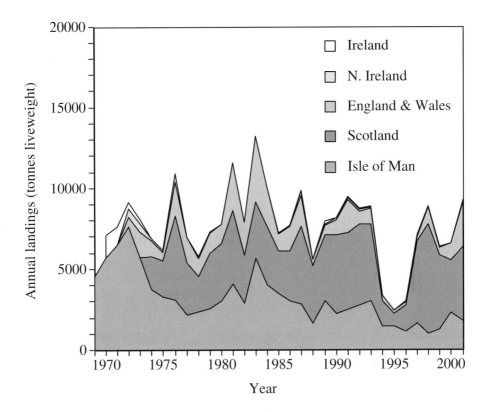

Figure 19.7. Annual landings of queen scallops (*Aequipecten opercularis*) (tonnes live weight) in the Irish Sea (ICES Area VIIa), by country, 1969–2001.

recruit models for a number of fishing grounds (Murphy 1986; Murphy and Brand 1987; Allison 1993) and annual reports on the fisheries and the state of the stocks are made to the Isle of Man Government (e.g., Brand and Beukers-Stewart 2002). In recent years there has been a large programme of research to investigate the environmental impact of scallop dredging (Brand et al. 1997; Brand 2000; Bradshaw et al. 2000, 2001, 2002; Veale et al. 2000a, b, 2001; Jenkins and Brand 2001; Jenkins et al. 2001).

19.3.2 Scotland

Commercial fishing for *Pecten maximus* in Scotland started in the Clyde sea area in the 1930s but it was not until the 1960s that it started to spread to other areas (Mason et al. 1991). As new grounds were discovered, more boats joined in and fishing spread first to the west of Kintyre in 1963, then to the north-west coast and Orkney and Shetland in 1967, followed by the Moray Firth in 1977. Since the early 1990s larger vessels have also exploited extensive scallop grounds well offshore in the central North Sea, up to 200 km east of the Scottish coastline. Scallop dredging is therefore a significant fishery around almost the entire coast of Scotland, including the Western and Northern Isles. Total Scottish landings have averaged about 8,000 t in recent years, with a first sale value of some £13 million. Fishing for queens began in the Clyde in 1967, and in the Shetlands and the north Irish Sea in 1970, and these three areas, particularly the latter, continue to dominate queen landings in Scotland. The boats dredge for scallops using standard 2 ft 6 in (76 cm) 'Newhaven' spring-tooth dredges. Smaller vessels up to 16 m in length dominate the west coast fleet and generally tow 2–10 dredges per side. The Moray Firth and North Sea grounds are exploited mainly by larger vessels towing 10–14 dredges per side. In the early years the Scottish scallop fishery was dominated by dedicated scallop dredgers, but over the last decade increasing numbers of *Nephrops* trawlers have converted to scallop dredging in the winter months, particularly on the west coast grounds. Queens are also generally fished with spring-tooth dredges, with smaller belly rings, netting mesh and tooth spacing. Otter trawls have also been used in the Clyde and beam trawls have proved effective in Orkney and Shetland (Mason 1983). In some areas of Scotland, particularly the west coast, Orkney and Shetland, there are significant diver caught landings of scallops from shallow, inshore beds.

For management purposes Scottish fisheries have been divided into 8 management areas, namely the Irish Sea, Clyde, West Kintyre, North West, Orkney, Shetland, North East and East Coast. These areas differ considerably in size, exploitation history, stock structure and annual recruitment patterns. In recent years the North East management area has provided the largest landings, followed by the North West, East Coast, Irish Sea and West Kintyre. All Scottish waters are subject to the EU MLLS of 100 mm shell length and shucking at sea is banned. Since January 2000, EU rules have also required a minimum mesh size of >100 mm in the dredge netting back, and restricted the whitefish by-catch to 5% when dredging for scallops. In the Clyde area there is a statutory weekend ban on mobile gear, while a voluntary weekend ban operates to the west of Kintyre. A seasonal winter ban also operates from October to March in a number of west coast sea lochs and adjacent waters to protect nursery grounds for herring and plaice.

Since 1990, Scottish scallop fisheries have been increasingly affected by algal toxins that have resulted in area closures for food safety reasons, particularly in the most productive summer months. Paralytic shellfish poisoning (PSP) closures occurred in the 1980s but became more frequent in the 1990s on both the east and west coasts. Closures have generally been of 1–3 months and most commonly affected Orkney and the Foray Firth. However, in 1995, much of the eastern North Sea, including the Moray Firth was closed for over 2 months in mid-summer. Even more serious, however, was the accumulation of amnesic shellfish poisoning (ASP) toxins in 1999 that resulted in large-scale closures of Orkney, the Moray Firth and almost the entire west coast grounds (Campbell et al. 2001), including Northern Irish waters. Large areas of the west coast were closed in July 1999 and some remained closed to scallop and queen fishing for more than 9 months. As well as affecting the boats, the regularity and duration of these closures has become a serious problem to the Scottish processing industry, leading to the inability to service debt, the loss of skilled processing staff and disruption of established markets

(Denton 2000). Once lost, these markets may be difficult to recover in the face of cheap, but reliable, imports of farmed scallops into continental markets. It has been estimated that the direct cost to the industry of the ASP closure was about £10 million (Denton 2000). However, to date, no recorded case of any human illness in the U.K. has been attributed to ASP (Campbell et al. 2001).

The collection of statistical and biometric data for the Scottish fisheries is the responsibility of the Scottish Office Agriculture Fisheries and the Environment (SOAEFD), who hold long datasets for the main areas fished. These have been used periodically for stock assessments for the main fished areas (e.g., Mason et al. 1979b, c, 1980, 1981, 1991).

19.3.3 England and Wales

Scallops have been landed in England and Wales for over a century but it was not until the mid-1970s that a significant dredge fishery developed (Mason 1983), firstly in the western English Channel, then in the eastern Channel and later in the Irish Sea. The scallop fishery in the English Channel and western approaches remains the most important for England and Wales. This fishery has developed strongly since the late 1980s (Fig. 19.8), due mainly to increased exploitation of offshore grounds by large (28–35 m) ex-Dutch beam trawlers towing 14–20 standard 76 cm 'Newhaven' spring-toothed dredges per side. These vessels are owned by financial interests throughout the U.K. and form a mobile fleet, prepared to dredge for scallops and queens wherever the best opportunities arise. In response to this, local authority Sea Fisheries Committees, with the power to manage fisheries out to six miles, have introduced by-laws limiting the number and type of dredges that may be used in their districts. This has provided some protection to the inshore fishery, so vessels in the 8–15 m size range have continued to exploit the traditional inshore grounds successfully, while the numbers of 15–25 m vessels have fallen. Landings of scallops into England and Wales from the Irish Sea were high in the early 1980s, then declined, but have risen sharply since 1996, although a substantial proportion of these landings are from Scottish registered vessels. Queens are fished mostly in the Irish Sea, though there is a small fishery off the Yorkshire coast, landing mainly at Whitby.

The England and Wales scallop fisheries are subject to the EU MLLS of 100 mm shell length, apart from the Irish Sea and the eastern English Channel, where the MLLS is 110 mm shell length. From January 2000, the EU permitted fish by-catch when dredging for scallops was reduced from 10% to 5%. National legislation has also banned the use of French dredges with a diving plane, which were used extensively on certain grounds in the western Channel; these were banned because they retain high by-catches of protected fish species, particularly Dover sole. Proposed national legislation will also restrict the number and size of dredges to be used in the 6–12 mile zone. In response to growing concern about the displacement of effort from whitefish stocks into scallop fisheries, restrictive licensing for scallops was introduced throughout the U.K. in April 1999 for vessels over 10 metres. Under this scheme licenses were granted to any vessel that had caught more than 1 t of scallops in any year between 1 January 1994 and 31 May 1998.

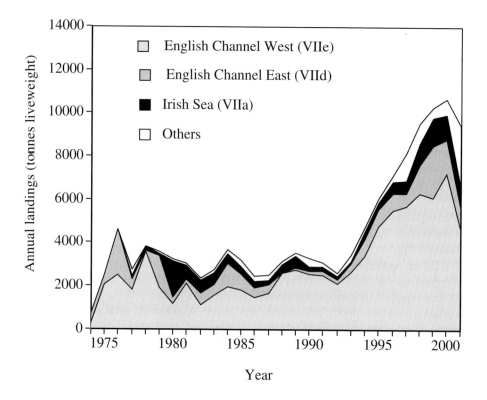

Figure 19.8. Annual landings of *Pecten maximus* (tonnes liveweight) in England and Wales, 1974–2001, from each major fishing area. Roman numerals refer to ICES fishing areas. Data from DEFRA.

As a result of this low entry qualification, 454 U.K. vessels >10 m are currently entitled to fish for scallops. A process of consultation with the industry is now taking place to determine how restrictive licensing and further technical measures can be used to control the exploitation of scallops.

The Department for Environment, Food and Rural Affairs (DEFRA) is responsible for the collection of commercial catch statistics for the England and Wales fisheries. Scientists at the Centre for Environment, Fisheries and Aquaculture Science, an agency of DEFRA, are carrying out research aimed at assessing the state of scallop stocks in the English Channel (Dare and Deith 1991; Dare et al. 1993, 1994; Dare and Palmer 1994; Palmer 1995), but are hampered by the sparse nature of the biological data for this area.

19.3.4 Northern Ireland

Scallops have been landed in Northern Ireland since the 1920s but a commercial fishery was first established in 1933 and provided winter employment for up to 90 boats (Mason 1983). After the war the fishery remained at a low level until the 1980s but has since supported a small but locally important fishery with some 10–20 boats fishing mainly from Kilkeel, Portavogie, Donaghadee and Bangor. The grounds fished are mainly in the Irish Sea (ICES area VIIa), off the east coast of County Down between Ardglass and Larne (Briggs 1995), including Belfast Lough and parts of Strangford Lough. However, since the early 1990s there has also been a small summer fishery off the Londonderry and Antrim coasts, in ICES area Via, which is not subject to a summer closed. For the Irish Sea grounds, the dates of the closed season (currently 1 June to 31 October) have generally followed those in operation elsewhere in area VIIa in order to prevent the diversion of effort onto Northern Ireland grounds. Similarly, the MLLS of 110 mm applies throughout Northern Irish waters. Since March 1997, the 'Conservation of Scallops Regulations (Northern Ireland) 1997' has also controlled the number, size, tooth number, tooth spacing, belly ring diameter and net mesh of dredges, banned French dredges, prohibited diving for scallops during the closed season and restricted the permitted by-catch of scallops to <1% when fishing for sea fish other than scallops. As Strangford Lough is an important marine conservation area, scallop dredging in the Lough is limited to the southern end, and only during the period 1 November–30 April.

Pecten maximus landings in Northern Ireland have generally been only a few hundred tonnes per year. However, they increased steadily in the late 1990s and in 2001 some 650 t were landed with a first sale value of £791,000 (Fig. 19.9). Most of this catch was taken from the Irish Sea (VIIa) grounds (shown in Fig. 19.6) but in recent years some 100 t per annum have been taken from the northern grounds in area VIa. In the first few years of the Irish Sea queen scallop fishery (1972–1976) some 500 t of *A. opercularis* per year were landed in Northern Ireland, but landings have since have been low, generally less than 100 t per year. Harmful algal blooms have also affected the Northern Ireland scallop fisheries in recent years, with short public health closures due to DSP and ASP toxins above the permitted levels.

The collection of data relevant to management of the Northern Ireland fisheries is the responsibility of the Department of Agriculture and Rural Development (DARD). Surveys of stock structure are carried out annually and scientific assessments of the fishery have been reported in a series of studies (Briggs 1980a, b, 1987, 1991, 1995).

19.3.5 Republic of Ireland

There is a very long history of scallop fishing in Ireland but the fisheries have been mostly on a very small scale, based mainly on small boats fishing numerous inshore beds in the bays around the south east, south and west coasts (Mason 1983). In recent years the main beds fished have been off Kilmore Quay, Dunmore East and the Saltee Islands on the south east coast, and in Roaringwater Bay, between Cape Clear and Sherkin Island,

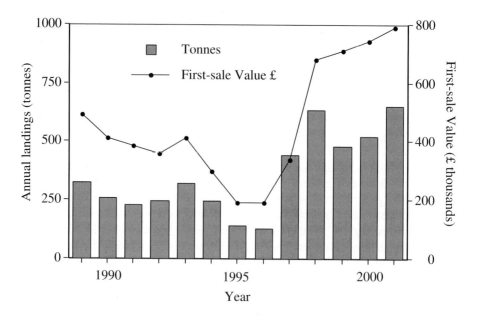

Figure 19.9. Annual landings (tonnes liveweight) and first-sale value (£1000s) of *Pecten maximus* landed in Northern Ireland, 1989–2001 (from ICES areas VIa +VIIa). Data from DARD.

Dunmanus Bay, Bantry Bay, Berehaven Sound, Whiddy Island and Glengariff on the south west coast. Further up the west coast there are also large beds in Kilkieran Bay and Betraghboy Bay, near Galway. The fisheries can be fairly clearly divided into two distinct types: inshore and offshore (Fleury et al. 1997). The inshore grounds are exploited by boats of 7–10 m length and 40–120 hp, typically using two dredges 70–130 cm wide. The design of the dredges varies considerably and many are used without steel ring bellies, and are often towed by rope and hauled with pot haulers. These local fisheries are often exploited to the point of extinction, as the small multi-purpose boats will fish for very small daily returns during the winter months when no viable alternatives exist. The offshore fisheries off the south-east and east coasts are fished seasonally by part of the Irish beam-trawl fleet. These vessels are 18–25 m and 300–500 hp, and tow 10–16 standard 76 cm wide 'Newhaven' spring-toothed dredges. Many of the local fishermen exploiting the inshore grounds may not report their landings, so the official statistics are probably greatly underestimated. Reported annual landings of *P. maximus* in Ireland have rarely been more than a few hundred tonnes. Landings for the north Irish Sea (ICES area VIIa) by the offshore fishery are probably more reliable; since 1969 they have varied between 86–1611 t, with high landings in the period 1989–1992 (Fig. 19.6). A dense bed of queen scallops on the Kish Bank off the east coast produced a substantial fishery in 1970 (1,367 t) but was quickly fished out and queen landings since 1974 have been negligible (Fig. 19.7). Irish scallop fisheries are subject to the EU MLLS of 100 mm shell length, and seasonal closures, a weekend ban on fishing and restricted daily fishing hours are also applied in some areas. Scallop dredging is also banned in parts of Mulroy Bay known as the North Water and Moross Channel in order to protect spawning stocks in this important spat collecting area.

19.3.6 France

With the main European market for scallops in France, fishing for scallops along the French coast has been an important economic activity for many years. In recent years some French 800 boats have fished seasonally for scallops (Dao et al. 1997). The French fishery for *Pecten maximus* can be divided into a number of different regions: the eastern Channel inshore grounds, particularly the Bay of Seine; the eastern Channel offshore grounds; the Bay of St. Brieuc and other western Channel grounds; and the Atlantic coast from the Bay of Brest to Pertuis Charentes. The fishery developed first on the Atlantic coast grounds in the 1950s but there was also fishing at this time along the Channel coast between Dieppe, Boulogne and the English inshore grounds (Ansell et al. 1991). Fishing then extended into the Bay of St. Brieuc in 1962 and the Bay of Seine a few years later. These two bays subsequently became the mainstay of the French scallop industry. In the Bay of St. Brieuc, catches increased steadily to 12,000 t in 1972 and there was a similar expansion in the Bay of Seine, while the traditional Atlantic fisheries declined to a very low level. In the Bay of Brest, for example, annual production dropped from 1,500–2,600 t in the 1950s to only 100–300 t after 1963. However, with strong landings in the western and eastern Channel, total landings of *P. maximus* in France rose to a peak of 25,300 t in 1973, but then declined steadily to a low of 5,077 t in 1990. Since then they have increased again, and for the period 1992–2000 have been fairly stable at 12–14,000 t. The detailed history of each fishery and the state of the stocks has varied from region to region, but all French fisheries now depend more-or-less exclusively on the exploitation of recruitment that occurs when the scallops are 2–3 years old (Dao et al. 1993).

In the Bay of St. Brieuc, the increase in scallop abundance that brought the start of the fishery in 1963 has been linked with the collapse of *Octopus* populations. *Octopus* were considered to be the main predators of scallops during the 1950s (Piboubés 1973), but they died off during the very cold winter of 1962/3 and never returned. Strong scallop recruitment between 1968 and 1973 resulted in a large increase in stock size, which led to the peak in productivity in the mid-1970s. Since then stock size and yield have varied with recruitment, though recent declines are considered to be due to over-fishing. Studies of water currents, and features of the reproductive cycle (Paulet et al. 1988), suggest that the scallop stock in the Bay of St. Brieuc is self-sustaining (Dao et al. 1993). Stocks in the eastern English Channel are spread over a large area but have followed broadly similar trends to the Bay of St. Brieuc. The only fishery in this area for which detailed knowledge is available is the Bay of Seine. Here the declines through the 1980s were followed by variable but increasing catches in the 1990s. Water current circulation modelling suggests that the Bay of Seine may be important in supplying larvae to the offshore grounds in the eastern Channel (Darby and Durance 1989; Dare et al. 1994). In the Bay of Brest, massive mortalities of scallops during the severe winter of 1962/3

(Boucher and Fifas 1997), followed by poor recruitment for nearly 30 years, resulted in low stocks until the 1990s, when a period of stronger recruitment, augmented by an extensive programme of cultivation (Dao et al. 1997; Fleury et al. 1997; Dao and Carval 1999), has brought about some increase in the stock. However, catches are still very low compared with historic levels. There have been very similar trends in the stocks on other French Atlantic scallops beds off Iroise, Ar Men, Concarneau, Quiberon and Belle Ile.

In the main fishing areas, annual stock assessments are made by IRFEMER to evaluate recruit and pre-recruit abundance, and spatial distribution. This is used as the basis for management strategies, which are decided between fishing representatives, administrators and scientists (Morin and Vigneau 1995). For the Bay of St. Brieuc, which has been most intensively studied, stock assessments are made in two ways (Dao et al. 1993): an annual evaluation of biomass in June using an experimental dredge with a diving plate and small rings in the belly (50 mm instead of the 85 mm used in the commercial fishery) and indirect methods including cohort analysis, catchability models (Fifas 1991), and equilibrium modelling of yield and fecundity per recruit. Since steady state models cannot take inter-annual recruitment variability into account, an empirical model of recruitment has been developed that includes the level of spawning stock biomass, the effects of fishing on the 0-age group, and the effects on recruitment of climatic conditions during the maturation and pre-spawning periods (Fifas et al. 1990).

French scallop fisheries are subject to EU and national regulations that restrict the MLLS to 100 mm, except for the eastern Channel where it is 110 mm, the maximum allowed fishing season to the period 1 October–15 May, and the minimum belly ring size to 72 mm. These regulations are strengthened by a complex series of local regulations that apply to so-called 'listed' beds. Fishing effort is controlled in various ways. Thus, the fishing season is shortened to run from November–April in the Bay of St. Brieuc and November–March in the Bay of Brest; the maximum permitted number of fishing days per week varies between two in the Bay of St. Brieuc and four in the Bay of Seine; and the number of fishing hours per day also varies from three-quarter to one hour in the Bay of St. Brieuc, two hours in the Bay of Brest and four to six hours on the inshore eastern Channel grounds. Total catch or individual daily quotas are also applied in some areas. Licences to fish are required on all grounds and the numbers of licences issued annually are restricted to 90 in the Bay of Brest, 259 in the Bay of St. Brieuc and 268 in the Bay of Seine. Boat size is also restricted to 12 m in the Bay of Brest, 13 m in Bay of St. Brieuc and 16 m in the Bay of Seine, and maximum engine power in these three fisheries is limited to 150 kW, 185 kW and 330 kW, respectively. Various types of dredge are used in the French fisheries including the typical 'French' dredge with a diving plane and fixed teeth, the 'heavy Breton dredge' and spring-toothed dredges. The permitted number, weight, width, tooth spacing and belly ring diameter of the dredges all vary between the different fishing grounds. Dredges with a diving plane are officially banned from all areas except the Bay of St. Brieuc, though they are apparently still used, while spring-toothed dredges are banned in the Sea of Iroise. The overall aim of these complex management regimes is to ensure regular landings from year to year, rather than to maximise yield per recruit.

Fisheries for *Aequipecten opercularis* and *Mimachlamys varia* are not so important or so well developed as those for *Pecten maximus*. The queen scallop fishery is sporadic and depends on good recruitment. Fishing is normally based on a single year-class and starts when a high-density aggregation is detected. The main grounds are in the western part of the Bay of Brest, the Bay of Camaret and the western Channel. The *Mimachlamys varia* fishery is more stable and the same boats fish for the two species, often at the same time. However, the main fishing areas for black scallops are in the eastern part of the Bay of Brest and in the Perthuis Charentes. In the Bay of Brest the dredges are restricted to a maximum width of 1.8 m, a maximum weight of 120 kg and belly rings with an internal diameter of no less than 42 mm. Catches of *M. varia* in the Bay of Brest fluctuate between 200–400 t per year; in the Perthuis Charentes they were historically some 800 t per year but the fishery collapsed between 1969 and 1991, though there has since been some recovery and annual catches were up to 200 t by 1997.

The management of the French fisheries has been supported in recent years by a substantial programme of research on recruitment determinism in scallops (e.g., Thouzeau and Lehay 1988; Thouzeau 1991a, b, c; Paulet et al. 1992). This includes work done within a broadly based series of studies in the Bay of Brest to explain inter-annual fluctuations of abundance of *Pecten maximus*, and with the ultimate aim of restoring water quality and marine resources in this heavily impacted ecosystem (Thouzeau and Chauvaud 1995; Paulet et al. 1997; Thouzeau et al. 2000; Chauvaud et al. 1996, 1997).

19.3.7 Spain

There has been a scallop fishery in the Galician Rias since at least 1870 but no landings data were available until 1962 (Román 1991). For *P. maximus*, annual catches of up to 300 t per year were recorded in the 1970s, but these declined to less than 100 t throughout the 1980s, before increasing again to a maximum of 675 t in 1996. In the 1960s there were also substantial catches of *Aquipecten opercularis* and some *Mimachlamys varia*, with combined totals of up to 700 t per year, but these subsequently dropped to such low levels that they are no longer recorded. There was also a short 'boom and bust' fishery for *P. maximus* in the Mediterranean, following the discovery of beds off Malaga in 1978 (Cano and Garcia 1985); catches in this fishery reached nearly 1,000 t in 1983, but by 1985 had declined to less than 100 tonnes. With such limited and unreliable supplies of pectinids from Spanish capture fisheries, and very strong market demand, the attention of scientists has increasingly turned to the possibilities of cultivation (Román and Acosta 1991; Román and Fernandez 1991; Román et al. 1996, 1999; Campos et al. 1999; Cano et al. 1999).

19.4 THE FUTURE

The history of scallop fisheries round the world indicates that natural scallop fisheries are very difficult to manage successfully. The reasons for this are numerous, but relate to three features of scallop fisheries: the biological characteristics of scallops themselves, the lack of appropriate management models and the common property nature of wild scallop resources (Perrin and Hay 1987). Scallops have distinct habitat preferences, live in aggregated distributions and are relatively immobile. This means that they are predictably located by fishermen and easily caught. In addition, their method of reproduction, long planktonic larval life and environmental sensitivity results in large annual and spatial variations in spat settlement and survival to recruitment. In consequence, large fluctuations in stock abundance are characteristic of scallops and this is compounded in some species by the short natural life-span. These features of scallop populations make traditional fishery assessment methods that were mainly developed for finfish difficult to apply or inappropriate (Caddy 1989b; Orensanz et al. 1991). The common property nature of wild scallop stocks applies to most marine fish stocks but is of particular relevance to immobile shellfish species. Unless there is the political will to enforce strict limited entry restrictions, scallop stocks in the open sea are common property resources and, as stocks decline, it is in the rational self-interest of individual fishermen to increase, rather than decrease, their fishing effort, since the consequences of over-fishing are shared by all participants (Berkes 1985). This conflict between personal and communal benefit has been termed 'the tragedy of the commons' (Hardin 1968) and is an important factor leading to over-capitalisation and over-fishing in scallop fisheries (Perrin and Hay 1987).

At present, natural fisheries for *Pecten maximus* in Europe are nearly all fully exploited or over-exploited. With increasing exploitation, the older year-classes have progressively disappeared and all fisheries in France, and most in the U.K. and Ireland, are now dependent on fishing the recruiting year-classes. With the natural variability in recruitment, this has led to 'boom and bust' fisheries and instability of supplies for fishermen and processors. Scallop densities have also fallen markedly so that, in the north Irish Sea for example, the density of scallops on most grounds is now less than 3 per $100m^2$ at the start of the fishing season and less than half that at the end of the fishing season (Wilson and Brand 1995). As stocks declined, fishermen have purchased larger, more powerful, boats capable of operating farther offshore and pulling larger gangs of dredges in order to maximise catches. In consequence, European fishermen now use large quantities of diesel fuel, pulling large spreads of gear over enormous areas of seabed, in order to gather up the very sparsely-distributed stocks of scallops. The high prices paid for scallops have made it possible to continue fishing on low densities but, in doing so, the toothed dredges used throughout Europe kill and damage many other animals on the seabed, change the physical nature of the sediments and modify benthic habitats and communities (Bradshaw et al. 2000; Brand 2000; Hall-Spencer and Moore 2000b; Veale et al. 2000b).

In order to preserve scallop stocks in the face of increasing exploitation, fishery management bodies throughout Europe have introduced progressively more strict regulatory regimes. European capture fisheries are also facing increasing competition from cultivated scallops. To date, the generally low-value species of cultivated scallop that have been imported, frozen, into European markets from countries such as China and Chile have competed directly with queen scallops (*Aequipecten opercularis*), but have not made a great impact on the market for fresh, high-value 'coquille St. Jacques' (*Pecten* spp.). The main supplies of *P. maximus* continue to come from capture fisheries, but this situation is likely to change in the next few years. Despite early demonstrations of the feasibility of collecting the spat and culturing scallops and queens in France, Ireland, Scotland and the Isle of Man (e.g., Minchin 1976; Ventilla 1977a; Ventilla 1977b; Buestel et al. 1979; Buestel and Dao 1979; Brand et al. 1980; Paul et al. 1981; Wieland and Paul 1983; Norman and Ludgate 1995), commercial cultivation has been slow to develop and total European production of farmed pectinids is still under 500 t per year. Production is steadily increasing but large-scale production has been inhibited mainly by the shortage of spat. However, this is likely to change in the next few years for there are now several well-funded projects to develop hatchery culture of *P. maximus* in France, the U.K., Ireland and Norway. In addition, a brood stock of *P. maximus* has been imported into Chile (Illanes et al. 1999) with the aim of eventually supplying high value 'coquille St. Jacques' to the French market.

With the problems of managing natural stocks in the face of increasing exploitation, together with increasing competition from cultivation, European capture fisheries face an uncertain future. In all fisheries, the boats now fish under strict management controls that greatly restrict their fishing effort and prevent them operating with high economic efficiency. As cultivated scallops become available in quantity it will become increasing difficult for traditional capture fisheries to compete economically. In order to promote more productive, economically efficient fisheries – and to prevent the widespread ecological impact of scallop dredging on seabed communities – there is growing evidence to suggest that the future of scallop fisheries may lie with closed area management (Horwood 2000). Seasonal closure of scallop fishing grounds have been applied for many years in Europe, and elsewhere, but the conservation value of seasonal closures for scallops and benthic communities is limited unless they are combined with strict control of fishing effort. Rotational closures, however, with a cycle of several years, can be used to prevent both growth and recruitment over-fishing (Myers et al. 2000, National Research Council 2001), and appear to be particular suitable for scallop fisheries management, with or without stock enhancement. The Japanese have long used rotational fishing of small closed areas, in combination with predator clearance and stock enhancement, as part of their seabed cultivation technique (Ventilla 1982). These techniques have subsequently been used elsewhere, most notably with Japanese assistance in Tasmania (e.g., Thompson 1994; Thompson et al. 1995) and New Zealand, where short rotations (<4 years) have been successfully employed, together with re-seeding, to increase yield per recruit, maintain a stable fishery and sustain full-time fishermen (Arbuckle and Metzger 2000). Developments have been slower in Europe, but a collaborative scheme between government and fishermen's organisations in France, involving stock enhancement and rotational fishing, is now contributing to the recovery of scallop production in the Bay of Brest (Dao and Carval 1999). If this proves to be economically viable, similar ventures are likely to follow elsewhere, particularly on grounds close inshore.

Of more interest to capture fisheries, however, are the possibilities of using large-scale closed areas as a fishery-management tool, without stock enhancement. This was recently given a major boost by the report of a 14-fold increase in scallop (*Placopecten magellanicus*) biomass on Georges Bank that followed a 4-year closure to any gears capable of retaining groundfish (Murawski et al. 2000). This increase in scallop biomass was, in fact, a fortuitous outcome of large-scale area closures designed to protect depleted groundfish stocks, but it was remarkable because it took place during a period of only moderate to below-average scallop recruitment. Even greater increases in *Pecten maximus* biomass have been recorded in the small Isle of Man closed area, compared with the fished area outside, albeit after a substantially longer period of closure (Brand and Beukers-Stewart 2002). In addition to increases in scallop density and biomass, rotational closures for <5 years can increase habitat complexity, and the production and diversity of benthic communities (Collie et al. 2000; Bradshaw et al. 2001), though the communities are likely to be dominated by mobile, robust, fecund and fast-growing opportunists. Longer rotational closures (5–10 years) would be needed to allow slower growing, less fecund species to proliferate (Collie et al. 1997; Bradshaw et al. 2002). However, to protect the most vulnerable and long-lived species, and the complex communities that develop on biogenic reefs and beds of horse mussels (*Modiolus modiolus*), serpulid worms (*Serpula vermicularis*) and maerl (e.g., *Lithothamnion corallioides* and *Phymatolithon calcareum*), permanent closures are generally necessary (Magorrian 1995; Moore et al. 1998; Collie et al. 2000; Hall-Spencer and Moore 2000a, b). Although permanently closed areas deprive fishermen of grounds, there can be some benefits to adjacent fishing grounds through the export of pelagic larvae and the 'spill-over' of juveniles and adults dispersing around the edges (Horwood et al. 1998; Hall-Spencer et al. 1999; National Research Council 2001). It therefore seems likely that the future management of near-shore waters will involve series or networks of closed areas of different sizes, closed for different periods and with differing objectives concerned with the conservation of fishery resources and marine communities.

Although there has been a substantial amount of research on short and long-term ecological impacts of scallop dredging (for example, European studies include Eleftheriou and Robertson 1992; Hall-Spencer et al. 1999; Brand 2000; Hall-Spencer and Moore 2000a; Bradshaw et al. 2000, 2002; Veale et al. 2000b, 2001), comparatively little information is available on the recovery of benthic organisms or habitats after closure to fishing (Collie et al. 2000; Bradshaw et al. 2001). As closed areas are introduced, supporting research will be required in many topics. For example, properly replicated large scale experiments will be required to evaluate the effectiveness of closed areas and determine the time scale of recovery for important species and habitats (Jennings 2000; National Research Council 2001). Studies will also be needed to evaluate factors such as the size, shape, connectivity and overlap of closed areas (Jennings 2000; Roberts and Hawkins 2000). Diversion of fishing effort could have major effects on scallop stocks and the benthos outside closed areas (Bradshaw et al. 2001), and this will also need to be evaluated in order to prepare schemes to minimise harmful impacts. To inform decisions on the siting of closed areas, a better understanding of scallop recruitment mechanisms is essential so that sources of scallop recruitment, rather than sinks, are conserved. Research is also needed to devise methods for assessing the extent to which closed areas export larvae, juveniles and adults to surrounding beds. Finally, it will be important to seek ways of limiting fishing effort and of modifying gear to reduce the damage it causes when closed areas are re-opened.

ACKNOWLEDGMENTS

I am very grateful to many European colleagues who have provided data and permission to quote from unpublished reports, particularly Richard Briggs, Dave Palmer and Guillermo Roman, together with Bill Lart, Partick Berthou, Ross Campbell, Julie Maquire and other members of the ECODREDGE research project. My thanks are due to Christine Drewery of the Sea Fish Industry Authority for her superb drawings of the dredges in Fig. 19.4. It is a pleasure to acknowledge also the continued financial support of the Isle of Man Department of Agriculture, Fisheries and Forestry for much of my research on scallops.

REFERENCES

Acosta, C.P. and Román, G., 1994. Growth and reproduction in a southern population of scallop *Pecten maximus*. In: N.F. Bourne, B.L. Bunting and L.D. Townsend (Eds.). Proceedings of the 9th International Pectinid Workshop, Nanaimo, B.C., Canada, April 22–27, 1993. Volume 1. Canadian Technical Report of Fisheries and Aquatic Sciences 1994:119–126.

Acosta, C.P. and Román, G., 1995. Reproductive and reserve storage cycles in *Pecten maximus*, reared in suspension. 2. Energy storage cycle. In: P. Lubet, J. Barret and J. Dao (Eds.). Fisheries, biology and aquaculture of pectinids. 8th International Pectinid Workshop, Cherbourg, France, 22–29 May, 1991. IFREMER, Actes de Colloques No 17, 1995. pp. 189–192.

Aldrich, J.C. and Lawler, I.F., 1996. The model-free analysis of growth curves. Marine Ecology-Pubblicazioni della Stazione Zoologica di Napoli I. 17:491–500.

Allison, E.H., 1993. The dynamics of exploited populations of scallops (*Pecten maximus* L.) and queens (*Chlamys opercularis* L.) in the North Irish Sea. PhD Thesis. University of Liverpool.

Allison, E.H., 1994. Seasonal growth models for great scallops (*Pecten maximus* (L.)) and queen scallops (*Aequipecten opercularis* (L.)). Journal of Shellfish Research 13:555–564.

Allison, E.H. and Brand, A.R., 1995. A mark-recapture experiment on queen scallops, *Aequipecten opercularis*, on a North Irish Sea fishing ground. Journal of the Marine Biological Association of the U.K. 75:323–335.

Allison, E.H., Brand, A.R. and Murphy, E.J., 1989. Mortality rates in North Irish Sea *Pecten maximus*, from tagging experiments. 7th International Pectinid Workshop, Portland, Maine, U.S.A., 20–25 April, 1989. 7 p. (mimeo)

Allison, E.H., Wilson, U.A.W. and Brand, A.R., 1994. Age determination and the first growth ring in North Irish Sea populations of the scallop, *Pecten maximus* (L.). Journal of Molluscan Studies 60:91–95.

Andersen, P., 1996. Design and implementation of some harmful algal monitoring systems. Intergovernmental Oceanographic Commission Technical Series 44. 102 p.

Andrew, N.L. and Mapstone, B.D., 1987. Sampling and the description of spatial pattern in marine ecology. Oceanography and Marine Biology Annual Review 25:39–90.

Anonymous, 1979. Report on the Working Group on assessment of scallop stocks, Dublin 4–6 April 1979. ICES CM 1979, Doc No K:6. 25 p.

Anonymous, 1989. Report of the Working Group on pectinid stocks, Aberdeen, 22–25 November, 1988. ICES CM 1989, Doc No. K:12. 60 p.

Ansell, A.D., 1977. The adenosine triphosphate content of some marine bivalve molluscs. Journal of Experimental Marine Biology and Ecology 28:269–283.

Ansell, A.D. and Ackerly, S.C., 1994. Swimming in *Aequipecten opercularis*: Preliminary scaling considerations. In: N.F. Bourne, B.L. Bunting and L.D. Townsend (Eds.). Proceedings of the 9th International Pectinid Workshop, Nanaimo, B.C., Canada, April 22–27, 1993. Volume 1. Canadian Technical Report of Fisheries and Aquatic Sciences 1994:3–11.

Ansell, A.D., Dao, J.-C. and Mason, J., 1991. Three European scallops: *Pecten maximus*, *Chlamys (Aequipecten) opercularis* and *C. (Chlamys) varia*. In: S.E. Shumway (Ed.). Scallops: Biology, Ecology and Aquaculture. Elsevier, Amsterdam. pp. 715–751.

Ansell, A.D., Dao, J.D., Lucas, A., Mackie, A.L. and Morvan, C., 1988. Reproductive and genetic adaptations in natural and transplant populations of the scallop, *Pecten maximus*, in European waters. Report to the European Commission, EEC Scientific Cooperation Contract, No. ST2J-0058-1-UK (CD). 50 p.

Antoine, L., 1978. La croissance journaliere chez *Pecten maximus* (L.) (Pectinidae, Bivalvia). Haliotis 9:627–636.

Antoine, L., Arzel, P., Laurec, A. and Morize, E., 1979a. La croissance de la coquille Saint-Jaques (*Pecten maximus* (L.)) dans les divers gisements francais. Rapport et Procès-Verbaux des Réunions. Conseil International pour l'Exploration de la Mer 175:85–90.

Antoine, L., Garen, P. and Lubet, P., 1979b. Conséquences sur la maturation et la croissance d'une transplantation de naissain de *Pecten maximus* (L.). Cahiers de Biologie Marine 20:139–150.

Aravindakshan, I., 1955. Studies on the biology of the queen scallop, *Chlamys opercularis* (L.). PhD Thesis. University of Liverpool.

Arbuckle, M. and Metzger, M., 2000. Food for thought, a brief history of the future of fisheries management. Challenger Scallop Enhancement Company, Nelson, New Zealand. 25 p.

Baird, R.H., 1958. On the swimming behaviour of escallops (*Pecten maximus* L.). Proceedings of the Malacological Society of London 33:67–71.

Bannister, R.C.A., 1986. Assessment and population dynamics of commercially exploited shellfish in England and Wales. In: G.S. Jamieson and N. Bourne (Eds.). North Pacific Workshop on Stock Assessment and Management of Invertebrates, Canadian Technical Special Publication of Fisheries and Aquatic Sciences. pp. 182–194.

Baron, J., Diter, A. and Bodoy, A., 1989. Triploidy induction in the black scallop (*Chlamys varia* L.) and its effect on larval growth and survival. Aquaculture 77:103–111.

Beaumont, A.R., 1982a. Variations in heterozygosity at two loci between year classes of a population of *Chlamys opercularis* (L.) from a Scottish sea–loch. Marine Biology Letters 3:25–33.

Beaumont, A.R., 1982b. Geographic variation in allele frequencies at three loci in *Chlamys opercularis* from Norway to the Brittany coast. Journal of the Marine Biological Association of the U.K. 62:243–261.

Beaumont, A.R. and Barnes, D.A., 1991. Larval growth and byssal drifting of scallop spat. In: P. Lubet, J. Barret and J.C. Dao (Eds.). Fisheries, Biology and Aquaculture of Pectinids. 8th International Pectinid Workshop, Cherbourg, France, 22–29 May, 1991. IFREMER, Actes de Colloques No 17. p. 207.

Beaumont, A.R. and Barnes, D.A., 1992. Aspects of veliger larval growth and byssus drifting of the spat of *Pecten maximus* and *Aequipecten* (*Chlamys*) *opercularis*. ICES Journal of Marine Science 49:417–423.

Beaumont, A.R. and Beveridge, C.M., 1984. Electrophoretic survey of genetic variation in *Pecten maximus*, *Chlamys opercularis*, *C. varia* and *C. distorta* from the Irish Sea. Marine Biology 81:299–306.

Beaumont, A.R., Gosling, E.M., Beveridge, C.M., Budd, M.D. and Burnell, G.M., 1985. Studies on heterozygosity and size in the scallop, *Pecten maximus*. In: P.E. Gibbs (Ed.). Proceedings of the 19th European Marine Biology Symposium, Plymouth, Devon (UK), Cambridge University Press. pp. 443–455.

Beaumont, A. and Hall, V., 1999. Getting the queen to spawn: experiments with *Aequipecten opercularis*. Book of Abstracts, 12th International Pectinid Workshop, Bergen, Norway, 5–11 May, 1999. p. 114.

Beaumont, A.R., Morvan, C., Huelvan, S., Lucas, A. and Ansell, A.D., 1993. Genetics of indigenous and transplanted populations of *Pecten maximus*: no evidence for the existence of separate stocks. Journal of Experimental Marine Biology and Ecology 169:77–88.

Beaumont, A.R. and Zouros, E., 1991. Genetics of scallops. In: S.E. Shumway (Ed.). Scallops: Biology, Ecology and Aquaculture. Vol. 21. Elsevier, Amsterdam. pp. 585–617.

Beninger, P.G., Auffret, M. and Le Pennec, M., 1990. Peribuccal organs of *Placopecten magellanicus* and *Chlamys varia* (Mollusca: Bivalvia): Structure, ultrastructure and implications for feeding. 1. The labial palps. Marine Biology 107:215–223.

Beninger, P.G., Le Pennec, M. and Donval, A., 1991. Mode of particle ingestion in 5 species of suspension-feeding bivalve mollusks. Marine Biology 108:255–261.

Berkes, F., 1985. Fishermen and the 'tragedy of the commons'. Environmental Conservation 12:199–205.

Berthet, B., Amiard, J.C., Amiardtriquet, C., Martoja, M. and Jeantet, A.Y., 1992. Bioaccumulation, toxicity and physicochemical speciation of silver in bivalve mollusks - ecotoxicological and health consequences. Science of the Total Environment 125:97–122.

Besnard, J.-Y., 1991. Seasonal variations in the lipids and fatty acids of the female gonad of the scallop, *Pecten maximus* (Linnaeus 1758) in the Bay of Seine (French Channel). In: S.E. Shumway and P.A. Sandifer (Eds.). An International Compendium of Scallop Biology and Culture. World Aquaculture Workshops No. 1. World Aquaculture Society, Baton Rouge, LA. pp. 74–86.

Beukers-Stewart, B.D., Jenkins, S.R. and Brand, A.R., 2001a. The efficiency and selectivity of spring-toothed scallop dredges: a comparison of direct and indirect methods of assessment. Journal of Shellfish Research 20:121–126.

Beukers-Stewart, B.D., Mosley, M.W.J. and Brand, A.R., 2001b. Predicting fluctuations in scallop catches: how far can you go? Book of Abstracts, 13th International Pectinid Workshop, Coquimbo, Chile, 18–24 April, 2001. pp. 20–22.

Boucher, J., 1985. Caracteristiques dynamiques du cycle vital de la coquille Saint-Jacques (*Pecten maximus*): Hypotheses sur les stades critiques pour le recruitement. ICES CM 1985, Doc No. K:23. 6 p.

Boucher, J. and Fifas, S., 1997. Scallop (*Pecten maximus*) population dynamics in the Bay of Brest: was yesterday different from today? Annales de l'Institut Océanographique 73:89–100.

Bradshaw, C., Veale, L.O. and Brand, A.R., 2002. The role of scallop-dredge disturbance in long-term changes in Irish Sea benthic communities: a re-analysis of an historical dataset. Journal of Sea Research 47:161–184.

Bradshaw, C., Veale, L.O., Hill, A.S. and Brand, A.R., 2000. The effects of scallop dredging on gravelly seabed communities. In: M.J. Kaiser and S.J. de Groot (Eds.). Effects of Fishing on Non-Target Species and Habitats. Blackwell Science, Oxford. pp. 83–104.

Bradshaw, C., Veale, L.O., Hill, A.S. and Brand, A.R., 2001. The effect of scallop dredging on Irish Sea benthos: experiments using a closed area. Hydrobiologia 465:129–138.

Brand, A.R., 1991. Scallop ecology: distributions and behaviour. In: S.E. Shumway (Ed.). Scallops: Biology, Ecology and Aquaculture. Elsevier, Amsterdam. pp. 517–584.

Brand, A.R., 2000. North Irish Sea scallop (Pecten maximus and Aequipecten opercularis) fisheries: effects of 60 years of dredging on scallop populations and the environment. Alaska Department of Fish and Game, Special Publication 14:37–43.

Brand, A.R. and Allison, E.H., 1994. Estimating abundance on north Irish Sea scallop, Pecten maximus (L.), fishing grounds. In: N.F. Bourne, B.L. Bunting and L.D. Townsend (Eds.). Proceedings of the 9th International Pectinid Workshop, Nanaimo, B.C., Canada, April 22–27, 1993. Volume 2. Canadian Technical Report of Fisheries and Aquatic Sciences 1994:121–130.

Brand, A.R., Allison, E.H. and Murphy, E.J., 1991a. North Irish Sea scallop fisheries: a review of changes. In: S.E. Shumway and P.A. Sandifer (Eds.). An International Compendium of Scallop Biology and Culture. World Aquaculture Workshops No. 1. World Aquaculture Society, Baton Rouge, LA. pp. 204–218.

Brand, A.R. and Beukers-Stewart, B.D., 2002. Shellfish Research Report to the Isle of Man Government, Department of Agriculture, Fisheries and Forestry, June 2002. Port Erin Marine Laboratory, University of Liverpool. 38 p.

Brand, A.R., Hill, A.S., Veale, L.O. and Hawkins, S.J., 1997. The environmental impact of scallop dredging. Book of Abstracts, 11th International Pectinid Workshop, La Paz, Baja California Sur, Mexico, 10–15 April, 1997. pp. 40–43.

Brand, A.R. and Murphy, E.J., 1985. Estimation of mortality rates by a mark, release, recapture experiment for the Isle of Man scallop stocks. 5th Pectinid Workshop, La Coruña, Spain, 6–10 May, 1985. 4 p. (mimeo)

Brand, A.R. and Murphy, E.J., 1992. A tagging study of North Irish Sea scallop (Pecten maximus) populations: comparisons of an inshore and an offshore fishing ground. Journal of Medical and Applied Malacology 4:153–164.

Brand, A.R., Paul, J.D. and Hoogesteger, J.N., 1980. Spat settlement of the scallops Chlamys opercularis (L.) and Pecten maximus (L.) on artificial collectors. Journal of the Marine Biological Association of the U.K. 60:379–390.

Brand, A.R. and Prudden, K.L., 1997. The Isle of Man scallop and queen fisheries: past, present and future. Report to Isle of Man Department of Agriculture, Fisheries and Forestry by Port Erin Marine Laboratory, University of Liverpool. 101 p.

Brand, A.R., Wilson, U.A.W., Hawkins, S.J., Allison, E.H. and Duggan, N.A., 1991b. Pectinid fisheries, spat collection, and the potential for stock enhancement in the Isle of Man. ICES Marine Science Symposia 192:79–86.

Brienne, H., 1954. Le vanneau (Chlamys opercularis L.) en manche orientale. Science et Pêche 1:9–11.

Briggs, R.P., 1980a. The Northern Ireland fishery for the escallop *Pecten maximus* (L.). 3rd Pectinid Workshop, Port Erin, Isle of Man, 13–16 May, 1980. 7 p. (mimeo)

Briggs, R.P., 1980b. Shell morphology and growth in relation to minimum legal size of the escallop, *Pecten maximus* L. from Northern Ireland waters. Record of Agricultural Research 28:41–48.

Briggs, R.P., 1987. An appraisal of the Northern Ireland scallop stocks. 6th International Pectinid Workshop, Menai Bridge, Wales, 9–14 April, 1987. (mimeo)

Briggs, R.P., 1991. A study of the Northern Ireland fishery for *Pecten maximus* (L.). In: S.E. Shumway and P.A. Sandifer (Eds.). An International Compendium of Scallop Biology and Culture. World Aquaculture Workshops No. 1. World Aquaculture Society, Baton Rouge, LA. pp. 249–255.

Briggs, R.P., 1995. A review of research on the Northern Ireland scallop stocks Book of Abstracts, 10th International Pectinid Workshop, Cork, Ireland, 26 April–2 May, 1995. pp. 16–18.

Broom, M.J., 1976. Synopsis of biological data on scallops *Chlamys (Aequipecten) opercularis* (Linnaeus), *Argopecten irradians* (Lamarck), *Argopecten gibbus* (Linnaeus). Food and Agriculture Organisation, Rome. 44 p.

Broom, M.J. and Mason, J., 1978. Growth and spawning in the pectinid *Chlamys opercularis* in relation to temperature and phytoplankton concentration. Marine Biology 47:277–285.

Bryan, G.W., 1973. The occurrence and seasonal variation of trace metals in the scallop *Pecten maximus* (L.) and *Chlamys opercularis* (L.). Journal of the Marine Biological Association of the U.K. 53:145–166.

Buestel, D., 1978. Comparison de l'importance des fixations du naissain de *Pecten maximus* sur les collecteurs et de l'importance du recrutement dans le millieu naturel en Baie de St. Brieuc. 2nd Pectinid Workshop, Brest, France, 8–13 May 1978. 6 p. (mimeo)

Buestel, D. and Dao, J.C., 1979. Aquaculture extensive de la coquille Saint Jacques: resultats d'un semis experimental. Pêche Maritime 1215: 361–365.

Buestel, D., Dao, J.C. and Golin, F., 1985. Estimation d'un stock naturel de coquilles Saint Jacques par une methode coulinant les dragages et la plongee. Traitement des resultats par une methode geostatistique. ICES CM 1985, Doc. No. K:18. 9 p.

Buestel, D., Dao, J.-C. and Lemarie, G., 1979. Collection of pectinid spat in Brittany. Rapport et Procès-Verbaux des Réunions. Conseil International pour l'Exploration de la Mer 175:80–84.

Buestel, D. and Laurec, A., 1976. La croissance de la coquille Saint Jacques (*Pecten maximus* L.) en rade de Brest et en Baie de Saint-Brieuc. Haliotis 5:173–177.

Buestel, D., Muzellec, M.L. and Bergere, A., 1978. Captages de naissain de petoncles noirs *Chlamys varia* en rade de Brest. 2nd Pectinid Workshop, 8–13 May, 1978, Brest, France. 5 p. (mimeo)

Burnell, G.M., 1983. Growth and reproduction of the scallop *Chlamys varia* (L.) on the west coast of Ireland. PhD Thesis. National University of Ireland, Galway.

Burnell, G.M., 1989. Scallop culture in Ireland - fact or fantasy. Scottish Shellfish Grower 3:5.

Burnell, G.M., 1991. Annual variations in the spawning and settlement of the scallop *Chlamys varia* (L.) on the west coast of Ireland. In: S.E. Shumway and P.A. Sandifer (Eds.). An International Compendium of Scallop Biology and Culture. World Aquaculture Workshops No. 1. World Aquaculture Society, Baton Rouge, LA. pp. 47–59.

Burnell, G.M., 1995. Age-related protandry in the scallop, *Chlamys varia* (L.) on the west coast of Ireland. ICES Marine Science Symposia 199:26–30.

1040

Caddy, J.F., 1989a. A perspective on the population dynamics and assessment of scallop fisheries, with special reference to the sea scallop, *Placopecten magellanicus* Gmelin. In: J.F. Caddy (Ed.). Marine Invertebrate Fisheries: Their Assessment and Management. John Wiley & Sons, New York. pp. 559–589.

Caddy, J.F., 1989b. Recent developments in research and management for wild stocks of bivalves and gastropods In: J.F. Caddy (Ed.). Marine Invertebrate Fisheries: Their Assessment and Management. John Wiley & Sons, New York. pp. 665–700.

Campbell, D.A., Kelly, M.S., Busman, M., Bolch, C.J., Wiggins, E., Moeller, P.D.R., Morton, S.L., Hess, P. and Shumway, S.E., 2001. Amnesic shellfish poisoning in the king scallop, *Pecten maximus*, from the west coast of Scotland. Journal of Shellfish Research 20:75–84.

Campos, M.J., Roman, G., Cano, J., Vazquez, M.C., Garcia, T., Fernandez, L. and Presasa, M.C., 1999. Scallop *Pecten maximus* culture in Fuengirola, Malaga, south Spain. A first attempt. Book of Abstracts, 12th International Pectinid Workshop, Bergen, Norway, 5–11 May, 1999. pp. 41–42.

Canales, J., Rios, C. and Peña, J.B., 1995. The reproductive cycle of the queen scallop *Aequipecten opercularis* (L). in the coast of Oropesa (Castellón), East of Spain. Book of Abstracts, 10th International Pectinid Workshop, Cork, Ireland, 27 April-2 May, 1995. pp. 87–88.

Canapa, A., Barucca, M., Marinelli, A. and Olmo, E., 2000. Molecular data from the 16S rRNA gene for the phylogeny of Pectinidae (Mollusca : Bivalvia). Journal of Molecular Evolution 50:93–97.

Cancelo, M.J., Guerra, A., Fernandez, A., Gabin, C. and Fernandez, J., 1992. La culture suspendue de *Chlamys varia*, de la nourricerie a la taille commerciale, en Galice (Espagne). International Symposium on the Marine Molluscs: Biology and Aquaculture, Brest (France), 9 Nov 1990. pp. 119–126.

Cano, J., Campos, M.J., Roman, G., Vazquez, M.C., Garcia, T., Fernandez, L. and Presasa, M.C., 1999. Pectinid settlement on collectors in Fuengirola, Malaga, south Spain. Book of Abstracts, 12th International Pectinid Workshop, Bergen, Norway, 5–11 May, 1999. pp. 70–71.

Cano, J. and Garcia, T., 1985. Scallop fishery in the coast of Malaga, S.E. Spain 5th Pectinid Workshop, La Coruña, Spain, 6–10 May, 1985. 8 p. (mimeo)

Castagnolo, L., 1991. La pesca e la riproduzione di *Pecten jacobaeus* L. e di *Aequipecten opercularis* (L.) nell'alto Adriatico. Bollettino Malacolgico Italiano 27:39–48.

Chapman, C.J., 1981. The swimming behaviour of queens in relation to trawling. Scottish Fisheries Bulletin 46:7–10.

Chapman, C.J., Main, J., Howell, T. and Sangster, G.I., 1979. The swimming speed and endurance of the queen scallop *Chlamys opercularis* in relation to trawling. In: J.C. Gamble and J.D. George (Eds.). Progress in Underwater Science. Pentech Press, London. pp. 57–72.

Chapman, C.J., Mason, J. and Kinnear, J.A.M., 1977. Diving observations on the efficiency of dredges used in the Scottish fishery for the scallop, *Pecten maximus* (L.). Scottish Fisheries Research Report 10. 16 p.

Chauvaud, L. and Strand, O., 1999. Growth traits in three populations of *Pecten maximus*. Book of Abstracts, 12th International Pectinid Workshop, Bergen, Norway, 5–11 May, 1999. pp. 166–167.

Chauvaud, L., Thouzeau, G. and Grall, J., 1996. Experimental collection of great scallop postlarvae and other benthic species in the Bay of Brest: settlement patterns in relation to spatio-temporal variability of environmental factors. Aquaculture International 4:263–288.

Chauvaud, L., Thouzeau, G. and Paulet, Y.M., 1997. Effects of algal blooms of *Gymnodinium* cf. Nagasakiense on the daily growth of *Pecten maximus* juveniles in the Bay of Brest (France). Book of Abstracts, 11th International Pectinid Workshop, La Paz, Baja California, Mexico, 10–15 April, 1997. pp. 70–72.

Chauvaud, L., Thouzeau, G. and Paulet, Y.M., 1998. Effects of environmental factors on the daily growth rate of *Pecten maximus* juveniles in the Bay of Brest (France). Journal of Experimental Marine Biology and Ecology 227:83–111.

Chevolot, L., Cochard, J.C. and Yvin, J.C., 1991. Chemical induction of larval metamorphosis of *Pecten maximus* with a note on the nature of naturally-occurring triggering substances. Marine Ecology Progress Series 74:83–89.

Christensen, H. and Kanneworff, E., 1985. Sedimenting phytoplankton as a major food source for suspension and deposit feeders in the Øresund. Ophelia 24:223–244.

Cochard, J.C., 1985. Observation sur la viabilité des oeufs de coquille Saint-Jacques en Rade de Brest. 5th Pectinid Workshop, La Coruña, Spain, 6–10 May, 1985. 8 p. (mimeo)

Cochard, J.C. and Devauchelle, N., 1993. Spawning, fecundity and larval survival and growth in relation to controlled conditioning in native and transplanted populations of *Pecten maximus* (L.) - evidence for the existence of separate stocks. Journal of Experimental Marine Biology and Ecology 169:41–56.

Collie, J.S., Escanero, G.A. and Valentine, P.C., 1997. Effects of bottom fishing on the benthic megafauna of Georges Bank. Marine Ecology Progress Series 155:159–172.

Collie, J.S., Hall, S.J., Kaiser, M.J. and Poiner, I.R., 2000. A quantitative analysis of fishing impacts on shelf-sea benthos. Journal of Animal Ecology 69:785–798.

Comely, C.A., 1974. Seasonal variations in the flesh weights and biochemical content of the scallop *Pecten maximus* (L.) in the Clyde Sea Area. Journal du Conseil international pour l'Exploration de la Mer 35:281–295.

Conan, G. and Shafee, M.S., 1978. Growth and biannual recruitment of the black scallop *Chlamys varia* (L.) in Lanvéoc area, Bay of Brest. Journal of Experimental Marine Biology and Ecology 35:59–71.

Dalmon, J., 1935. Mollusques. Note sur la biologie du pétoncle (*Chlamys varia* L.). Revue des Travaux de L'Institut des Pêches Maritimes, Nantes, France 8:268–281.

Dao, J.C., Barret, J. and Fleury, P.G., 1997. Scallop (*Pecten maximus*) sea-bed cultivation in the Bay of Brest in France. Book of Abstracts, 11th International Pectinid Workshop, La Paz, Baja California Sur, Mexico, 10–15 April, 1997. pp. 5–6.

Dao, J.-C., Buestel, D., Gerard, A., Halary, C. and Cochard, J.-C., 1985b. Le programme de repeuplement de coquilles St. Jacques (*Pecten maximus* L.) en France: finalité, resultats et perspectives. Colloques Franco-Japonais d'Océanographie, Marseille, France 16–21 September, 1985. 7:67–82.

Dao, J.C., Buestel, D. and Halary, C., 1985a. Note sur l'evolution comparee des coquilles Saint-Jacques d'origine differente en Bretagne Nord. 5th Pectinid Workshop, La Coruña, Spain, 6–10 May, 1985. pp. 12.

Dao, J.C. and Carval, J.P., 1999. Present status of the scallop production (*Pecten maximus*) in Bay of Brest (France) combining aquaculture and fishery. Book of Abstracts, 12th International Pectinid Workshop, Bergen, Norway, 5 –11 May, 1999. pp. 145.

Dao, J.C., Gerard, A. and Buestel, D., 1985c. Acquis biologique sur le petoncle noir (*Chlamys varia*) en rade de Brest 1973–1983. Consequences sur l'amenagement de la resource. 5th Pectinid Workshop, La Coruña, Spain, 6–10 May 1985. p. 11. (mimeo)

Dao, J.-C. et al., 1993. Concerted action on scallop seabed cultivation in Europe. Specific Community Programme for Research, Technological Development and Demonstration in the Field of Agriculture and Agro-Industry, Inclusive Fisheries, AIR 2-CT 93–1647. 24 p.

Darby, C.D. and Durance, J.A., 1989. Use of the North Sea water parcel following model (NORSWAP) to investigate the relationship of larval source to recruitment for scallop (*Pecten maximus*) stocks of England and Wales. ICES CM 1989, Doc No. K:28. 19 p.

Dare, P.J. and Bannister, R.C.A., 1987. Settlement of scallop, *Pecten maximus*, spat on natural substrates off south west England: the hydroid connection 6th International Pectinid Workshop, Menai Bridge, Wales, 9–14 April, 1987. 12 p. (mimeo)

Dare, P.J., Darby, C.D., Durance, J.A. and Palmer, D.W., 1994. The distribution of scallops, *Pecten maximus*, in the English Channel and Celtic Sea in relation to hydrographic and substrate features affecting larval dispersal and settlement. In: N.F. Bourne, B.L. Bunting and L.D. Townsend (Eds.). Proceedings of the 9th International Pectinid Workshop, Nanaimo, B.C., Canada, April 22–27, 1993. Volume 1. Canadian Technical Report of Fisheries and Aquatic Sciences 1994:20–27.

Dare, P.J. and Deith, M.R., 1991. Age determination of scallops, *Pecten maximus* (Linnaeus, 1758), using stable oxygen isotope analysis, with some implications for fisheries management in British waters. In: S.E. Shumway and P.A. Sandifer (Eds.). An International Compendium of Scallop Biology and Culture. World Aquaculture Workshops No. 1. World Aquaculture Society, Baton Rouge, LA. pp. 118–133.

Dare, P.J., Key, D. and Connor, P.M., 1993. The efficiency of spring-loaded dredges used in the western English Channel fishery for scallops, *Pecten maximus* (L.). ICES CM 1993, Doc No. B:15, Ref. K. 17 p .

Dare, P.J. and Palmer, D.W., 1994. The use of dredge efficiency factors for estimating indirectly population composition and abundance of scallops, *Pecten maximus*. In: N.F. Bourne, B.L. Bunting and L.D. Townsend (Eds.). Proceedings of the 9th International Pectinid Workshop, Nanaimo, B.C., Canada, April 22–27, 1993. Volume 2. Canadian Technical Report of Fisheries and Aquatic Sciences 1994:137–142.

Davies, I.M., McKie, J.C. and Paul, J.D., 1986. Accumulation of tin and tributyltin from anti-fouling paint by cultivated scallops (*Pecten maximus*) and Pacific oysters (*Crassostrea gigas*). Aquaculture 55:103–114.

Davies, I.M. and Paul, J.D., 1986. Accumulation of copper and nickel from anti-fouling compounds during cultivation of scallops (*Pecten maximus* L.) and Pacific oysters (*Crassostrea gigas* Thun.). Aquaculture 55:93–102.

de Zwaan, A., Thompson, R.J. and Livingstone, D.R., 1980. Physiological and biochemical aspects of the valve snap and valve closure response in the giant scallop *Placopecten magellanicus* II: Biochemistry. Journal of Comparative Physiology 137:105–114.

Denton, W., 2000. Proposals for dealing with the algal toxin problem that has hit the scallop industry. Sea Fish Executive Report, Sea Fish Industry Authority, Hull. 13 p. (mimeo)

Dorange, G. and Le Pennec, M., 1989. Ultrastructural study of oogenesis and oocytic degeneration in *Pecten maximus* from the Bay of St. Brieuc. Marine Biology 103:339–348.

Duggan, N.A., 1987. Recruitment in North Irish Sea scallop stocks. PhD Thesis. University of Liverpool.

Dupouy, H., 1978. L'exploitation de la coquille Saint-Jacques, *Pecten maximus* L., en France. 1ere partie: presentation des pecheries. Science et Pêche, Bulletin d'Institute des Pêches maritimes 276:1–11.

Dupouy, H., de Kergariou, G. and Latrouite, D., 1983. L'exploitation de la coquille Saint-Jacques *Pecten maximus* (L.) en France. 2. Evaluation et gestion du stock de la baie de Saint-Brieuc. Science et Pêche, Bulletin d'Institut Scientifiques des Pêches Maritimes 331: 5–11.

Dupouy, H. and Latrouite, D., 1976. Scallop fisheries in France. 1st Scallop Workshop, Baltimore, Ireland, 11–16 May, 1976. 14 p. (mimeo)

Eggleston, D., 1962. Spat of the scallop (*Pecten maximus* L.) from off Port Erin, Isle of Man. Annual Report of the Marine Biological Station, Port Erin 74:29–32.

Eleftheriou, A. and Robertson, M.R., 1992. The effects of experimental scallop dredging on the fauna and physical environment of a shallow sandy community. Netherlands Journal of Sea Research 30:289–299.

Erard-Le Denn, E., Morlaix, M. and Dao, J.C., 1990. Effects of *Gyrodinium cf. aureolum* on *Pecten maximus* (post larvae, juveniles and adults). Proceeding of the 4th International Conference on Toxic Marine Phytoplankton, Lund (Sweden), 26–30 June 1989. pp. 132–136.

Faveris, R., 1987. Studies on the evolution of glycogen content of somatic and germinal tissues during the annual reproductive cycle in *Pecten maximus* L. (Bay of Seine) 6th International Pectinid Workshop, Menai Bridge, Wales, UK, 9th–14th April, 1987. 13 p. (mimeo)

Faveris, R. and Lubet, P., 1991. Energetic requirements of the reproductive cycle in the scallop *Pecten maximus* (Linnaus, 1758) in Baie de Seine (Channel). In: S.E. Shumway and P.A. Sandifer (Eds.). An International Compendium of Scallop Biology and Culture. World Aquaculture Workshops No. 1. World Aquaculture Society, Baton Rouge, LA. pp. 67–73.

Fegan, D.F., Mackie, L.A. and Ansell, A.D., 1985. Reproduction, larval development and settlement of the scallop *Pecten maximus* in the Firth of Lorn and Loch Creran, West Scotland. 5th Pectinid Workshop, La Coruña, Spain, 6–10 May 1985. 12 p. (mimeo)

Fevolden, S.E., 1992. Allozymic variability in the Iceland scallop *Chlamys islandica* - Geographic variation and lack of growth-heterozygosity correlations. Marine Ecology Progress Series 85:259–268.

Fifas, S., 1991. An empirical catchability model for the scallop stock, *Pecten maximus* (L.) in the Saint-Brieuc Bay (English Channel, France). ICES CM 1991, Doc. No. K: 37, Ref. D. 20 p.

Fifas, S., Dao, J.C. and Boucher, J., 1990. Un modele empirique du recrutement pour le stock de coquilles Saint-Jacques, *Pecten maximus* (L.) en baie de Saint-Brieuc (Manche, France). Aquatic Living Resources 3:13–18.

Fleury, P.G., Mingant, C. and Castillo, A., 1996. A preliminary study of the behaviour and vitality of reseeded juvenile great scallops, of three sizes in three seasons. Aquaculture International 4:325–337.

1044

Fleury, P.G. et al., 1997. Concerted action on scallop seabed cultivation in Europe (1993–1996). Specific Community Programme for Research, Technological Development and Demonstration in the Field of Agriculture and Agro-Industry, Inclusive Fisheries, AIR 2 - CT 93 –1647. 118 p.

Forbes, E. and Hanley, S., 1853. A History of the British Mollusca and Their Shells. Van Voorst, London. 557 p.

Forester, A.J., 1979. The association between the sponge *Halichondria panicea* (Pallas) and scallop *Chlamys varia* (L.): a commensal-protective mutualism. Journal of Experimental Marine Biology and Ecology 36:1–10.

Franklin, A., 1987. The concentration of metals, organochlorine pesticide and PCB residues in marine fish and shellfish: results from MAFF fish and shellfish monitoring programmes. MAFF Aquatic Environment Monitoring Report 16:38.

Franklin, A. and Jones, J., 1995. Monitoring and Surveillance of Non-radioactive Contaminants in the Aquatic Environment and Activities Regulating the Disposal of Wastes at Sea, 1993. MAFF Aquatic Environment Monitoring Report 44. 68 p.

Franklin, A., Pickett, G.D. and Conner, P.M., 1980a. The escallop (*Pecten maximus*) and its fishery in England and Wales. Ministry of Agriculture Fisheries and Food, Laboratory Leaflet 51. 19 p.

Franklin, A., Pickett, G.D., Holme, N.A. and Barrett, R.L., 1980b. Surveying stocks of scallops *Pecten maximus* and queens *Chlamys opercularis* with underwater television. Journal of the Marine Biological Association of the U.K. 60:181–192.

Fraser, D.I., 1991. Settlement and recruitment in *Pecten maximus* (Linnaeus 1758) and *Chlamys (Aequipecten) opercularis* (Linnaeus 1758). In: S.E. Shumway and P.A. Sandifer (Eds.). An International Compendium of Scallop Biology and Culture. World Aquaculture Workshops No. 1. World Aquaculture Society, Baton Rouge, LA. pp. 28–36.

Fraser, D.I. and Mason, J., 1987. Pectinid spat studies in selected areas of the west coast of Scotland, 1982–1986. ICES CM 1987, Doc. No K:4. 9 p.

Gäde, G., Weeda, E. and Gabbott, P.A., 1978. Changes in the level of octopine during the escape responses of the scallop, *Pecten maximus* (L.). Journal of Comparative Physiology 124:121–127.

George, S.G., Pirie, B.J.S. and Coombs, T.L., 1980. Isolation and elemental analysis of metal-rich granules from the kidney of the scallop, *Pecten maximus* (L.). Journal of Experimental Marine Biology and Ecology 42:143–156.

Gibson, F.A., 1953. Tagging of escallops (*Pecten maximus* L.) in Irish waters. Journal du Conseil International pour l'Exploration de la Mer 19:204–208.

Giguère, M. and Brulotte, S., 1994. Comparison of sampling techniques, video and dredge, in estimating sea scallop (*Placopecten magellanicus*, Gmelin) populations. Journal of Shellfish Research 13:25–30.

Glemarec, M., 1997. The Bay of Brest: environmental disturbance and its impact on the biota. Annales de l'Institut Oceanographique 73:113–122.

Gosling, E.M. and Burnell, G.M., 1988. Evidence for selective mortality in *Chlamys varia* (L.) transplant experiments. Journal of the Marine Biological Association of the U.K. 68:251–258.

Grainger, R.J.R., Duggan, C., Minchin, D. and O'Sullivan, D., 1984. Investigations in Bantry Bay following the Betelgeuse oil tanker disaster. Irish Fisheries Investigation (B Marine) 27. 24 p.

Grall, J., Chauvaud, L., Thouzeau, G., Fifas, S., Glemarec, M. and Paulet, Y.-M., 1996. Distribution of *Pecten maximus* (L.) and its main potential competitors and predators in the Bay of Brest. Comptes Rendus de l'Academie des Sciences Serie III-Sciences de la Vie 319:931–937.

Grieshaber, M., 1978. Breakdown and formation of high-energy phosphates and octopine in the adductor muscle of the scallop, *Chlamys opercularis* (L.), during escape swimming and recovery. Journal of Comparative Physiology B 126:269–276.

Grieshaber, M. and Gäde, G., 1977. Energy supply and the formation of octopine in the adductor muscle of the scallop, *Pecten jacobaeus* (Lamarck). Comparative Biochemistry and Physiology B 58:249–252.

Gruffydd, L.D., 1972. Mortality of scallops on a Manx bed due to fishing. Journal of the Marine Biological Association of the U.K. 52:449–455.

Gruffydd, L.D., 1974a. The influence of certain environmental factors on the maximum length of the scallop, *Pecten maximus* L. Journal du Conseil International pour l'Exploration de la Mer 35:300–302.

Gruffydd, L.D., 1974b. An estimate of natural mortality in an unfished population of the scallop *Pecten maximus* (L.). Journal du Conseil International pour l'Exploration de la Mer 35:209–210.

Gruffydd, L.D., 1981. Observations on the rate of production of external ridges on the shell of *Pecten maximus* in the laboratory. Journal of the Marine Biological Association of the U.K. 61:401–411.

Halary, C., Royer, Y., Corlouer, J.P. and Dao, J.-C., 1994. Effects of predation and competition on scallop, *Pecten maximus*, seabed cultivation in Saint Brieuc Bay: preliminary results. In: N.F. Bourne, B.L. Bunting and L.D. Townsend (Eds.). Proceedings of the 9th International Pectinid Workshop, Nanaimo, B.C., Canada, April 22–27, 1993. Volume 2. Canadian Technical Report of Fisheries and Aquatic Sciences 1994:39–49.

Hall-Spencer, J.M., Froglia, C., Atkinson, R.J.A. and Moore, P.G., 1999. The impact of Rapido trawling for scallops, *Pecten jacobaeus* (L.), on the benthos of the Gulf of Venice. ICES Journal of Marine Science 56:111–124.

Hall-Spencer, J.M. and Moore, P.G., 2000a. Scallop dredging has profound, long-term impacts on maerl habitats. ICES Journal of Marine Science 57:1407–1415.

Hall-Spencer, J.M. and Moore, P.G., 2000b. Impact of scallop dredging on maerl grounds. In: M.J. Kaiser and S.J. de Groot (Eds.). Effects of Fishing on Non-target Species and Habitats. Blackwell Science Ltd, Oxford. pp. 105–117.

Hardin, G., 1968. The tragedy of the commons. Science 162:1243–1248.

Hartnoll, R.G., 1967. An investigation of the movement of the scallop *Pecten maximus*. Helgoländer wissenschaftliche Meeresuntersuchungen 15:523–533.

Heipel, D.A., Bishop, J.D.D. and Brand, A.R., 1999. Mitochondrial DNA variation among open-sea and enclosed populations of the scallop *Pecten maximus* in western Britain. Journal of the Marine Biological Association of the U.K. 79:687–695.

Heipel, D.A., Bishop, J.D.D., Brand, A.R. and Thorpe, J.P., 1998. Population genetic differentiation of the great scallop *Pecten maximus* in western Britain investigated by randomly amplified polymorphic DNA. Marine Ecology Progress Series 162:163–171.

Horwood, J.W., 2000. No-take zones: a management context. In: M.J. Kaiser and S.J. de Groot (Eds.). Effects of Fishing on Non-Target Species and Habitats. Blackwell Science, Oxford. pp. 302–311.

Horwood, J.W., Nichols, J.H. and Milligan, S., 1998. Evaluation of closed areas for fish stock conservation. Journal of Applied Ecology 35:893–903.

Howard, A.G. and Nickless, G., 1978. Heavy metal complexation in polluted molluscs. 3. Periwinkles (*Littorina littorea*), cockles (*Cardium edule*) and scallops (*Chlamys opercularis*). Chemical-Biological Interactions 23:227–231.

Howell, T.R.W. and Fraser, D.I., 1984. Observations on the dispersal and mortality of the scallop, *Pecten maximus* (L.). ICES CM 1984, Doc No K:35. 13 p.

Huelvan, S., 1985. Variabilité génétique de populations de *Pecten maximus* L. en Bretagne (Genetic variation in natural populations of *Pecten maximus*). Thèse 3ème cycle. Université Bretagne Occidentale, Brest, France.

Illanes, J., Uribe, E. and Solar, C., 1999. First attempt of introduction of *Pecten maximus* (L.) to Chile. Book of Abstracts, 12th International Pectinid Workshop, Bergen (Norway), 5 –11 May 1999. pp. 25–26.

Jenkins, S.R., Beukers-Stewart, B.D. and Brand, A.R., 2001. Impact of scallop dredging on benthic megafauna: a comparison of damage levels in captured and non-captured organisms. Marine Ecology Progress Series 215:297–301.

Jenkins, S.R. and Brand, A.R., 2001. The effect of dredge capture on the escape response of the great scallop, *Pecten maximus* (L.): implications for the survival of undersized discards. Journal of Experimental Marine Biology and Ecology 266:33–50.

Jennings, S., 2000. Patterns and prediction of population recovery in marine reserves. Reviews in Fish Biology and Fisheries 10:209–231.

Jones, N.S., 1951. The bottom fauna off the south of the Isle of Man. Journal of Animal Ecology 20:132–144.

Jory, A.M., 2000. The fishery and ecology of the scallop *Pecten maximus* (L.) in Guernsey. PhD Thesis. University of Southampton.

Laing, I., 2000. Effect of temperature and ration on growth and condition of king scallop (*Pecten maximus*) spat. Aquaculture 183:325–334.

Laing, I., 2002. Effect of salinity on growth and survival of king scallop spat (*Pecten maximus*). Aquaculture 205:171–181.

Lake, N.C.H., Jones, M.B. and Paul, J.D., 1987. Crab predation on scallop (*Pecten maximus*) and its implication for scallop cultivation. Journal of the Marine Biological Association of the U.K. 67:55–64.

Lake, N.C.H. and McFarlane, 1994. Crab and starfish predation of the scallop *Pecten maximus* (L.) during seabed cultivation. In: N.F. Bourne, B.L. Bunting and L.D. Townsend (Eds.). Proceedings of the 9th International Pectinid Workshop, Nanaimo, B.C., Canada, April 22–27, 1993. Volume 2. Canadian Technical Report of Fisheries and Aquatic Sciences 1994:56.

Lassus, P. and Berthome, J.P., 1988. Status of 1987 algal blooms in IFREMER. ICES/annex III C.M.1988/F. pp. 5–13.

Latrouite, D., 1978. Captage de *Chlamys varia* en baie de Quiberon (Bretagne-Sud). Resultats de 1976 et 1977. 2nd Pectinid Workshop, Brest, France, 8–13 May, 1978. 16 p. (mimeo)

Latrouite, D., de Kergariou, G., Claude, S. and Perodou, D., 1980. Le captage de pectinides en Baie de Quiberon; filieres de production. 3rd Pectinid Workshop, Port Erin, Isle of Man, 13–16 May, 1980. 8 p. (mimeo)

Lawrence, S.J., 1993. The feeding ecology and physiology of the scallops *Pecten maximus* (L.) and *Aequipecten opercularis* (L.) in the North Irish Sea. PhD Thesis. University of Liverpool.

Le Gall, G., Chagot, D., Mialhe, E. and Grizel, H., 1988. Branchial rickettsiales-like infection associated with a mass mortality of sea scallop *Pecten maximus*. Diseases of Aquatic Organisms 4:229–232.

Le Gall, G., Mialhe, E., Chagot, D. and Grizel, H., 1991. Epizootiological study of rickettsiosis of the Saint-Jacques scallop *Pecten maximus*. Diseases of Aquatic Organisms 10:139–145.

Le Pennec, G. and Le Pennec, M., 2001. Evaluation of the toxicity of chemical compounds using digestive acini of the bivalve mollusc *Pecten maximus* (L.) maintained alive in vitro. Aquatic Toxicology 53:1–7.

Le Pennec, M., 1974. Morphogenése de la coquille de *Pecten maximus* (L.) élevé au laboratoire. Cahiers De Biologie Marine, 15:475–482.

Le Pennec, M., 1982. L'elevage experimental de *Chlamys opercularis* (L.), (Bivalvia, Pectinidae). Vie Marine 4: 29–36.

Le Pennec, M., Beninger, P.G., Dorange, C. and Paulet, Y.-M., 1991. Trophic sources and pathways to the developing gametes of *Pecten maximus* (Bivalvia: Pectinidae). Journal of the Marine Biological Association of the U.K. 71:451–463.

Le Pennec, M. and Diss-Mengus, B., 1985. Rearing of *Chlamys varia* in commercial hatchery. 5th Pectinid Workshop, La Coruña, Spain, 6–10 May 1985. 9 p. (mimeo)

Le Pennec, M. and Diss-Mengus, B., 1987. Aquaculture of *Chlamys varia* (L.): data on the biology of larvae and postlarvae. Vie Marine 8: 37–42.

Letaconnoux, R. and Audouin, J., 1956. Contribution à l'étude du pétoncle (*Chlamys varia* L.). Revue de Travaux de l'Institute des Pêches Maritimes, Nantes 20:133–155.

Lewis, R.I., 1992. Population genetics of the queen scallop, *Chlamys opercularis* (L.). PhD Thesis. University of Liverpool.

Lewis, R.I. and Thorpe, J.P., 1994a. Are queen scallops, *Aequipecten* (*Chlamys*) *opercularis* (L.), self recruiting? In: N.F. Bourne, B.L. Bunting and L.D. Townsend (Eds.). Proceedings of the 9th International Pectinid Workshop, Nanaimo, B.C., Canada, April 22–27, 1993. Volume 1. Canadian Technical Report of Fisheries and Aquatic Sciences 1994:214–221.

Lewis, R.I. and Thorpe, J.P., 1994b. Temporal stability of gene-frequencies within genetically heterogeneous populations of the queen scallop, *Aequipecten (Chlamys) opercularis*. Marine Biology 121:117–126.

Lewis, R.I. and Thorpe, J.P., 1995. Docs allozyme heterozygosity provide a buffer against developmental instability in wild queen scallop spat? Book of Abstracts, 10th International Pectinid Workshop, Cork, Ireland, April 26–May 2, 1995. pp. 47–48.

Light, J.M., 1988. The status of *Chlamys varia* (L., 1758) and *Chlamys nivea* (MacGillivary, 1825); an appraisal using biometrics and geographical distribution. Journal of Conchology 33:31–41.

Lohrmann, K., 2000. Scallops disease survey. PhD Thesis. University of Liverpool.

Lohrmann, K.B., Feist, S.W. and Brand, A.R., 2000. Microsporidiosis in queen scallops (*Aequipecten opercularis* (L.)) from UK waters. Journal of Shellfish Research 19:71–75.

Lowe, E., 1996. Growth, gametogenesis, settlement and culture of the scallop, *Aequipecten opercularis* in Galicia, North West Spain. MSc Thesis. University College Cork.

Lubet, P., 1959. Recherches sur le cycle sexuel et l'émission des gametes chez les mytilides et des pectinides (Mollusques bivalves). Revue de Travaux de l'Institute des Pêches Maritimes, Nantes 23:387–548.

Lubet, P., Besnard, J.Y. and Faveris, R., 1987a. Competition énergétique entre tissus musculaire et gonadique chez la coquille St. Jacques (*Pecten maximus* L.). Haliotis 16:173–180.

Lubet, P., Besnard, J.Y., Faveris, R. and Robbins, I., 1987b. Physiologie de la reproduction de la coquille Saint-Jacques (*Pecten maximus* (L.)). Oceanis 13:265–290.

Lubet, P.E., Dorange, G. and Robbins, I.,1987c. Cytological investigations on the annual reproductive cycle of the scallop (*Pecten maximus* L.). 6th International Pectinid Workshop, Menai Bridge, Wales, UK, 9–14 April, 1987. 11 p. (mimeo)

Lubet, P., Faveris, R., Besnard, J.Y., Robbins, I. and Duval, P., 1991. Annual reproductive cycle and recruitment of the scallop, *Pecten maximus* (Linnaeus 1758) from the Bay of Seine (France). In: S.E. Shumway and P.A. Sandifer (Eds.). An International Compendium of Scallop Biology and Culture. World Aquaculture Workshops No. 1. World Aquaculture Society, Baton Rouge, LA. pp. 87–94.

Lucas, A., 1965. Recherches sur la sexualité des Mollusques Bivalves. Bulletin Biologique Francaise et Belgique 99:115–247.

Lucas, A., 1982. Evaluation of reproductive effort in bivalve molluscs. Malacologia 22:183–187.

Lucas, A., Calvo, J. and Trancart, M., 1978. L'effort de reproduction dans la stratégie demographique de six bivalves de l'Atlantique. Haliotis 9:107–116.

Lykakis, J.J. and Kalathakis, M., 1991. Greece. In: S.E. Shumway (Ed.). Scallops: Biology, Ecology and Aquaculture. Elsevier, Amsterdam. pp. 795–808.

Mackie, L.A., 1986. Aspects of the reproductive biology of the scallop *Pecten maximus*. PhD Thesis. Heriot-Watt University, Edinburgh.

Mackie, L.A. and Ansell, A.D., 1993. Differences in reproductive ecology in natural and transplanted populations of *Pecten maximus* - evidence for the existence of separate stocks. Journal of Experimental Marine Biology and Ecology 169:57–75.

Macleod, J.A.A., Thorpe, J.P. and Duggan, N.A., 1985. A biochemical genetic study of population structure in queen scallop (*Chlamys opercularis*) stocks in the Northern Irish Sea. Marine Biology 87:77–82.

Magorrian, B., 1995. The impact of commercial trawling on the benthos of Strangford Lough. PhD Thesis. The Queen's University of Belfast.

Maguire, J.A. and Burnell, G.M., 2001. The effect of stocking density in suspended culture on growth and carbohydrate content of the adductor muscle in two populations of the scallop (*Pecten maximus* (L.)) in Bantry Bay, Ireland. Aquaculture 198:95–108.

Maguire, J.A., Coleman, A., Jenkins, S. and Burnell, G.M., 2002. Effects of dredging on undersized scallops. Fisheries Research 56:155–165.

Maguire, J.A., O'Connor, D.A. and Burnell, G.M., 1999. An investigation into behavioural indicators of stress in juvenile scallops. Aquaculture International 7:169–177.

Mahéo, R., 1968. Observations sur l'anatomie et le functionnement de l'appareil byssogéne de *Chlamys varia* (L.). Cahiers de Biologie Marine 9:373–379.

Margus, D., 1990. The scallop (*Pecten jacobaeus* L.) in the Krka river estuary. Acta Biologica Iugoslavica, Serija E. Ichthyologica 22:69–77.

Margus, D., 1991. Yugoslavia. In: S.E. Shumway (Ed.). Scallops: Biology, Ecology and Aquaculture. Vol. 21. Elsevier, Amsterdam. pp. 789–793.

Martoja, M., Truchet, M. and Berthet, B., 1989. Effets de la contamination experimentale par l'argent chez *Chlamys varia* L. (Bivalve, Pectinidae). Donnees quantitatives, histologiques et microanalytiques. Annales Institut Oceanography Paris (Nouv. Ser.) 65:1–13.

Mason, J., 1957. The age and growth of the scallop, *Pecten maximus* (L.), in Manx waters. Journal of the Marine Biological Association of the U.K. 36:473–492.

Mason, J., 1958a. The breeding of the scallop, *Pecten maximus*, in Manx waters. Journal of the Marine Biological Association of the U.K. 37:653–671.

Mason, J., 1958b. A possible lunar periodicity in the breeding of the scallop, *Pecten maximus* (L.). Annals and Magazine of Natural History, Series B 1:601–602.

Mason, J., 1969. The growth of spat of *Pecten maximus* (L.). ICES, CM 1969, Doc No K: 32. 3 p.

Mason, J., 1983. Scallop and queen fisheries in the British Isles. Fishing News Books, Farnham (UK). 144 p.

Mason, J., Chapman, C.J. and Kinnear, J.A.M., 1979a. Population abundance and dredge efficiency studies on the scallop, *Pecten maximus* (L.). Rapport et Procès-Verbaux des Réunions. Conseil International pour l'Exploration de la Mer 175:91–96.

Mason, J. and Colman, J.S., 1955. Note on a short-term marking experiment on the scallop *Pecten maximus* (L.) in the Isle of Man. Annual Report of the Marine Biological Station, Port Erin 67:34–35.

Mason, J., Cook, R.M., Bailey, N. and Fraser, D.I., 1991. An assessment of scallops, *Pecten maximus* (Linnaeus 1758), in Scotland west of Kintyre. In: S.E. Shumway and P.A. Sandifer (Eds.). An International Compendium of Scallop Biology and Culture. World Aquaculture Workshops No. 1. World Aquaculture Society, Baton Rouge, LA. pp. 231–241.

Mason, J. and Drinkwater, J., 1976. The stocks of scallops, *Pecten maximus* in the Clyde Sea area and west of Kintyre in 1973–1974. Annales Biologiques, Copenhagen 31:183–184.

Mason, J., Drinkwater, J., Howell, T.R.W. and Fraser, D.I., 1982. A comparison of methods of determining the distribution and density of the scallop, *Pecten maximus* (L.). ICES CM 1982/K:24. 5 p.

Mason, J., Nicholson, M.D. and Shanks, A.M., 1979b. A comparison of exploited populations of the scallop, *Pecten maximus* (L.). Rapport et Procès-Verbaux des Réunions. Conseil International pour l'Exploration de la Mer 175:114–120.

Mason, J., Shanks, A.M. and Fraser, D.I., 1980. An assessment of scallop, *Pecten maximus* (L.), stocks off southwest Scotland. ICES CM 1980, Doc No. K:27. 11 p.

Mason, J., Shanks, A.M. and Fraser, D.I., 1981. An assessment of scallop, *Pecten maximus* (L.), stocks at Shetland. ICES CM 1981/K:19. 4 p.

Mason, J., Shanks, A.M., Fraser, D.I. and Shelton, R.G.J., 1979c. The Scottish fishery for the queen, *Chlamys opercularis* (L.). ICES CM 1979, Doc No. K: 37. 6 p.

Mason, J., Shelton, R.G.J., Chapman, C.J. and Howard, F.G., 1984. The state of Scottish shellfish stocks. Fishing Prospects 1984:38–44.

Mathers, N.F., 1975. Environmental variability at the phosphoglucose isomerase locus in the genus *Chlamys*. Biochemical Systematics and Ecology 3:123–127.

Mathers, N., 1976a. *In situ* studies on the scallop, *Pecten maximus*, with reference to the association between feeding rhythms and tidal currents. Journal of Molluscan Studies 42:455–456.

Mathers, N.F., 1976b. The effects of tidal currents on the rhythm of feeding and digestion in *Pecten maximus*. Journal of Experimental Marine Biology and Ecology 24:271–283.

Mathers, N.F., Lee, J. and Rankin, A., 1979a. *In situ* SCUBA studies of the queen scallop, *Chlamys opercularis* , in an enclosed sea loch. 1. Population structure and feeding cycles. 12th Symposium of the Underwater Association, 10 March, 1978, British Museum (Natural History), London. pp. 73–86 p.

Mathers, N.F., Smith, T. and Colins, N., 1979b. Monophasic and diphasic digestive cycles in *Venerupis delussuta* and *Chlamys varia*. Journal of Molluscan Studies 45:68–81.

Mattei, N., 1995. Situation of pectinid aquaculture (*Pecten jacobaeus*) in Italy and some information on Italian pectinid fishery. Book of Abstracts, 10th International Pectinid Workshop, Cork, Ireland, April 26-May 2, 1995. pp. 59–60.

Mattei, N. and Pellizzato, M., 1996. A population study on three stocks of a commercial Adriatic pectinid (*Pecten jacobaeus*). Fisheries Research 26:49–65.

McLoughlin, R.J., Young, P.C., Martin, R.B. and Parslow, J., 1991. The Australian scallop dredge: estimates of catching efficiency and associated indirect mortality. Fisheries Research 11:1–24.

McLusky, D.S., 1973. The effect of temperature on the oxygen consumption and filtration rate of *Chlamys (Aequipecten) opercularis* (L.) (Bivalvia). Ophelia 10:114–154.

Metayer, C., Amiard-Triquet, C. and Baud, J.P., 1990. Variations inter-specifiques de la bioaccumulation et de la toxicite de l'argent a l'egard de trois mollusques bivalves marins. Water Research 24:995–1001.

Millward, A. and Whyte, M.A., 1992. The hydrodynamic characteristics of six scallops of the Super Family Pectinacea, Class Bivalvia. Journal of Zoology, London 227:547–566.

Minchin, D., 1976. Spawning and rearing of *Pecten maximus* at Lough Hyne (Ine), 1st Scallop Workshop, Baltimore, Ireland, 11–16 May, 1976. 13 p. (mimeo)

Minchin, D., 1978. An exceptionally large escallop (*Pecten maximus* (L.)) from west Cork. Irish Naturalists' Journal 19:202.

Minchin, D., 1983. Predation on young *Pecten maximus* (L.) (Bivalvia), by the anemone *Anthopleura ballii* (Cocks). Journal of Molluscan Studies 49:228–231.

Minchin, D., 1984. Aspects of the biology of young escallops *Pecten maximus* (Linnaeus) (Pectinidae: Bivalvia) about the Irish coast. PhD Thesis. Trinity College, University of Dublin.

Minchin, D., 1985. Possible effects of an intense algal bloom of *Gyrodinium aureolum* on a year class of escallops (*Pecten maximus*). 5th Pectinid Workshop, La Coruña, Spain, 6th–10th May, 1985. 14 p. (mimeo)

Minchin, D., 1991. Decapod predation and the sowing of the scallop *Pecten maximus* (Linnaeus 1758). In: S.E. Shumway and P.A. Sandifer (Eds.). An International Compendium of Scallop Biology and Culture. World Aquaculture Workshops No. 1. World Aquaculture Society, Baton Rouge, LA. pp. 191–197.

Minchin, D., 1992. Biological observations on young scallops, *Pecten maximus*. Journal of the Marine Biological Association of the U.K. 72:807–819.

Minchin, D., Duggan, C.B. and King, W., 1987a. Possible effects of organotins on scallop recruitment. Marine Pollution Bulletin 18:604–608.

Minchin, D., Duggan, C.B. and King, W., 1987b. Possible influence of organotins on the Pectinacea. 6th International Pectinid Workshop, Menai Bridge, Wales, 9–14 April, 1987. 8 p. (mimeo)

Minchin, D. and Mathers, N.F., 1982. The escallop, *Pecten maximus* (L.) in Killary Harbour. Irish Fisheries Investigations B (Marine) 25. 13 p.

Minchin, D. and Ni Donnachada, C., 1995. Optimising scallop sowings for the restocking of an adult population in Mulroy Bay, Ireland. In: P. Lubet, J. Barret and J.C. Dao (Eds.). Fisheries, Biology and Aquaculture of Pectinids. 8th International Pectinid Workshop, Cherbourg, France, 22–29 May, 1991. IFREMER, Actes de Colloques No 17. pp. 175–182.

Minchin, D., Skjaeggestad, H., Haugum, G.A. and Strand, O., 2000. Righting and recessing ability of wild and naive cultivated scallops. Aquaculture Research 31:473–474.

Moore, C.G., Saunders, G.R. and Harries, D.B., 1998. The status and ecology of reefs of *Serpula vermicularis* L. (Polychaeta : Serpulidae) in Scotland. Aquatic Conservation-Marine and Freshwater Ecosystems 8:645–656.

Morin, J. and Vigneau, J., 1995. Eastern Channel scallop stock management. In: P. Lubet, J. Barret and J.C. Dao (Eds.). Fisheries, Biology and Aquaculture of Pectinids. 8th International Pectinid Workshop, Cherbourg, France, 22–29 May, 1991. IFREMER, Actes de Colloques No 17. pp. 35–40.

Mortensen, S.H., 1993. A health survey of selected stocks of commercially exploited Norwegian bivalve mollusks. Diseases of Aquatic Organisms 16:149–156.

Mortensen, S.H., Bachere, E., Le Gall, G. and Mialhe, E., 1992. Persistence of infectious pancreatic necrosis virus (IPNV) in scallops *Pecten maximus*. Diseases of Aquatic Organisms 12:221–227.

Mortensen, S., Torkildsen, L., Hernar, I., Harkestad, L., Fosshagen, A. and Bergh, Oe., 1999. One million scallop, *Pecten maximus*, spat lost due to a bristle worm, *Polydora* sp. infestation. Book of Abstracts, 12th International Pectinid Workshop, Bergen, Norway, 5 –11 May 1999. p. 96.

Murawski, S.A., Brown, R., Lai, H.-L., Rago, P.J. and Hendrickson, L., 2000. Large-scale closed areas as a fishery-management tool in temperate marine systems: the Georges Bank experience. Bulletin of Marine Science 66:775–798.

Murphy, E.J., 1986. An investigation of the population dynamics of the exploited scallop, *Pecten maximus* (L.), in the North Irish Sea. PhD Thesis. University of Liverpool.

Murphy, E.J. and Brand, A.R., 1987. Yield per recruit analysis for North Irish Sea scallop (*Pecten maximus* (L.)) stocks. 6th International Pectinid Workshop, Menai Bridge, Wales, 9–14 April 1987. 16 p. (mimeo)

Myers, R.A., Fuller, S.D. and Kehler, D.G., 2000. A fisheries management strategy robust to ignorance: rotational harvest in the presence of indirect fishing mortality. Canadian Journal of Fisheries and Aquatic Sciences 57:2357–2362.

Naidu, K.S., 1988. Estimating mortality rates in the Iceland scallop, *Chlamys islandica* (O.F. Müller). Journal of Shellfish Research 7:61–71.

Nance, D.A., 2001. Settlement of the scallops *Pecten maximus* (L.) and *Aequipecten opercularis* (L.) and their predators: the starfish *Asterias rubens* L. and the crabs *Necora puber* (L.) and *Cancer pagurus* L. on the west coast of Scotland. PhD Thesis. University of Aberdeen.

National Research Council, 2001. Marine protected areas: tools for sustaining ocean ecosystems. National Academy Press, Washington. 279 p.

1052

Nordsieck, F., 1969. Die europaischen Meeresmuscheln (Bivalvia) vom Eismeer bis Kapverden, Mittelmeer and Schwarzes Meer. Gustav Fisher Verlag, Stuttgart. 256 p.

Norman, M. and Ludgate, R., 1995. Initial survival of juvenile king scallops (*Pecten maximus*) after seeding out in Connemara Bays. Book of Abstracts, 10th International Pectinid Workshop, Cork, Ireland, April 26-May 2, 1995. pp. 80–81.

Norman, M. and Minchin, D., 1999. Comparison of the growth of wild and sown king scallops (*Pecten maximus*) in two Connemara Bays (Ireland). Book of Abstracts, 12th International Pectinid Workshop, Bergen, Norway, 5–11 May 1999. pp. 54–55.

Orensanz, J.M., 1986. Size, environment and density: the regulation of a scallops stock and its management implications. Canadian Special Publication Fisheries and Aquatic Sciences 92:195–227.

Orensanz, J.M., Parma, A.M. and Iribarne, O.O., 1991. Population dynamics and management of natural stocks. In: S.E. Shumway (Ed.). Scallops: Biology, Ecology and Aquaculture. Vol. 21. Elsevier, Amsterdam. pp. 625–713.

Palmer, D.W., 1995. Interpreting catch and effort data from the western Channel fishery for the scallop *Pecten maximus*. (L.) Book of Abstracts, 10th International Pectinid Workshop, Cork, Ireland, 26 April-May, 1995. pp. 9–10.

Paul, J.D., 1980a. Upper temperature tolerance and the effects of temperature on byssus attachment in the queen scallop *Chlamys opercularis* (L.). Journal of Experimental Marine Biology and Ecology. 46:41–50.

Paul, J.D., 1980b. Salinity-temperature relationships in the queen scallop *Chlamys opercularis*. Marine Biology 56:295–300.

Paul, J.D., 1981. Natural settlement and early growth of spat of the queen scallop *Chlamys opercularis* (L.) with reference to the formation of the first growth ring. Journal of Molluscan Studies 47:53–58.

Paul, J.D., 1985. Scallop cultivation: a simple method of separating spat of *Pecten maximus* (L.) and *Chlamys opercularis* (L.) by means of behavioural differences. Aquaculture 50:161–167.

Paul, J.D., Brand, A.R. and Hoogesteger, J.N., 1981. Experimental cultivation of the scallops *Chlamys opercularis* (L.) and *Pecten maximus* (L.) using naturally produced spat. Aquaculture 24:31–44.

Paul, J.D. and Davies, I.M., 1986. Effects of copper- and tin-based anti-fouling compounds on the growth of scallops (*Pecten maximus*) and oysters (*Crassostrea gigas*). Aquaculture 54:191–203.

Paul, J.D., Lake, N.C.H. and Jones, M.B., 1986. Scallop predation by crabs, in relation to the development of on-bottom cultivation. Sea Fish Industry Authority, Technical Report, 289. 17 p.

Paulet, Y.M., Bekhadra, F., Devauchelle, N., Donval, A. and Dorange, G., 1997. Seasonal cycles, reproduction and oocyte quality in *Pecten maximus* from the Bay of Brest. Annales de l'Institut Oceanographique 73:101–112.

Paulet, Y.M. and Boucher, J., 1991. Is reproduction mainly regulated by temperature or photoperiod in *Pecten maximus*? Invertebrate Reproduction & Development 19:61–70.

Paulet, Y.M., Dorange, G., Cochard, J.C. and Le Pennec, M., 1992. Reproduction and recruitment in *Pecten maximus* (L.). Annales de l'Institut Oceanographique 68:45–64.

Paulet, Y.M., Lucas, A. and Gerard, A., 1988. Reproduction and larval development in two *Pecten maximus* (L.) populations from Brittany. Journal of Experimental Marine Biology and Ecology 119:145–156.

Pazos, A.J., Román, G., Acosta, C.P., Abad, M. and Sánchez, J.L., 1995. Influence of gametogenic cycle on female gonad biochemical composition in the wild scallop *Pecten maximus* from suspended culture in the Ria de Arousa (Galicia, N. W. Spain). Book of Abstracts, 10th International Pectinid Workshop, Cork, Ireland, April 26–May 2, 1995. pp. 109–110.

Pazos, A.J., Román, G., Acosta, C.P., Abad, M. and Sanchez, J.L., 1996. Influence of the gametogenic cycle on the biochemical composition of the ovary of the great scallop. Aquaculture International 4:201–213.

Pazos, A.J., Román, G., Acosta, C.P., Abad, M. and Sanchez, J.L., 1997b. Seasonal changes in condition and biochemical composition of the scallop *Pecten maximus* (L.) from suspended culture in the Ria de Arousa (Galicia, NW Spain) in relation to environmental conditions. Journal of Experimental Marine Biology and Ecology 211:169–193.

Pazos, A.J., Román, G., Acosta, C.P., Sanchez, J.L. and Abad, M., 1997a. Lipid classes and fatty acid composition in the female gonad of *Pecten maximus* in relation to reproductive cycle and environmental variables. Comparative Biochemistry and Physiology B 117:393–402.

Peña, J.B., Canales, J., Adsuara, J.M. and Sos, M.A., 1995. Settlement of pectinid spat on filamentous collectors in Torre de la Sal beach, Castellon, E. Spain. Book of Abstracts, 10th International Pectinid Workshop, Cork, Ireland, April 26–May 2, 1995. pp. 115–116.

Peña, J.B., Canales, J., Adsuara, J.M. and Sos, M.A., 1996. Study of seasonal settlements of five scallop species in the western Mediterranean. Aquaculture International 4:253–261.

Peña, J.B., Ríos, C., Peña, S. and Canales, J., 1998. Ultrastructural morphogenesis of pectinid spat from the western Mediterranean: a way to differentiate seven genera. Journal of Shellfish Research 17:123–130.

Perrin, R.A. and Hay, P.R., 1987. A history of the Tasmanian scallop industry: diagnosing problems of management. Proceedings of the Royal Society of Tasmania 121:1–14.

Piboubés, R., 1973. Pêches et conchyliculture en Bretagne-Nord. 1. Bulletin du Centre des Etudes et Recherches Scientifiques, Biarritz 9:30–150.

Piccinetti, C., Simunovic, A. and Jukic, S., 1986. Distribution and abundance of *Chlamys opercularis* (L.) and *Pecten jacobaeus* L. in the Adriatic Sea. FAO Fisheries Report No. 345, FIPL/R345. pp. 99–105.

Pickett, G.D. and Franklin, A., 1975. Techniques for surveying queen scallop populations off southwest England in May-June 1974. Ministry of Agriculture, Fisheries and Food, Fisheries Laboratory, Technical Report 14. 20 p.

Pond, D., 1992. Protective commensal mutualism between the queen scallop *Chlamys opercularis* (Linnaeus) and the encrusting sponge *Suberites*. Journal of Molluscan Studies 58:127–134.

Pope, J.A. and Mason, J., 1980. The fitting of growth curves for *Pecten maximus* (L.). ICES CM 1980, Doc No. K: 28. 5 p.

Portmann, J.E., 1979. Chemical monitoring of residue levels in fish and shellfish landed in England and Wales during 1970–73. MAFF Aquatic Environment Monitoring Report 1. 70 p.

Pranovi, F., Raicevich, S., Franceschini, G., Torricelli, P. and Giovanardi, O., 2001. Discard analysis and damage to non-target species in the "rapido" trawl fishery. Marine Biology 139:863–875.

Reddiah, K., 1962. The sexuality and spawning of Manx pectinids. Journal of the Marine Biological Association of the U.K. 42:683–703.

Renzoni, A., 1991. Italy. In: S.E. Shumway (Ed.), Scallops: Biology, Ecology and Aquaculture. Elsevier, Amsterdam. pp. 777–788.

Richardson, C.A., Gale, G.F. and Venn, T.J., 1984. The effect of iron ore suspensions and food on the shell growth rates and tissue weights of queen scallops Chlamys opercularis held in the laboratory. Marine Environmental Research 13:1–31.

Richardson, C.A., Taylor, A.C. and Venn, T.J., 1982. Growth of the queen scallop Chlamys opercularis in suspended cages in the Firth of Clyde. Journal of the Marine Biological Association of the U.K. 62:157–169.

Ridgway, G.M.I. and Dahle, G., 2000. Population genetics of Pecten maximus of the Northeast Atlantic coast. Sarsia 85:167–172.

Ríos, C., Sanz, S. and Peña, J.B., 1999. Genetic relationships between populations of Pecten jacobaeus and Pecten maximus. Book of Abstracts, 12th International Pectinid Workshop, Bergen, Norway, 5–11 May, 1999. pp 109–110.

Ríos, C., Sanz, S., Saavedra, C. and Peña, J.B., 2002. Allozyme variation in populations of scallops, Pecten jacobaeus (L.) and P. maximus (L.) (Bivalvia : Pectinidae), across the Almeria-Oran front. Journal of Experimental Marine Biology and Ecology 267:223–244.

Roberts, C.D., 1975. Investigations into a Modiolus modiolus (L.) (Mollusca; Bivalvia) community in Strangford Lough, N. Ireland. Report of the Underwater Association 1:27–45.

Roberts, C.M. and Hawkins, J.P., 2000. Fully-protected marine reserves: a guide. World Wildlife Fund, York.

Roberts, D., 1973. Some sub-lethal effects of pesticides on the behaviour and physiology of bivalved molluscs. PhD Thesis. University of Liverpool.

Rodhouse, P.G. and Burnell, G.M., 1979. In situ studies on the scallop Chlamys varia. In: J.C. Gamble and J.D. George (Eds.). Progress in Underwater Science. Pentech Press, Plymouth. pp. 87–98.

Rodriguez, X., Espinos, F.J., Otero, F. and Silveiro, N., 1991. Sexual inversion in Chlamys varia: preliminary results. In: S.E. Shumway and P.A. Sandifer (Eds.). An International Compendium of Scallop Biology and Culture. World Aquaculture Workshops No. 1. World Aquaculture Society, Baton Rouge, LA. p. 106.

Román, G., 1991. Spain. In: S.E. Shumway (Ed.). Scallops: Biology, Ecology and Aquaculture. Vol. 21. Elsevier, Amsterdam. pp. 753–762.

Román, G. and Acosta, C.P., 1991. Growth of Chlamys (Aequipecten) opercularis (Linnaeus 1758) reared in experimental rafts. In: S.E. Shumway and P.A. Sandifer (Eds.). An International Compendium of Scallop Biology and Culture. World Aquaculture Workshops No. 1. World Aquaculture Society, Baton Rouge, LA. pp. 140–141.

Román, G., Campos, M.J. and Acosta, C.P., 1995. Reproduction and settlement of C. opercularis in Galicia, NW Spain. Book of Abstracts, 10th International Pectinid Workshop, Cork, Ireland, April 26-May 2, 1995. pp. 119–120.

Román, G., Campos, M.J. and Acosta, C.P., 1996. Relationships among environment, spawning and settlement of queen scallop in the Ria de Arosa (Galicia, NW Spain). Aquaculture International 4:225–236.

Román, G., Campos, M.J., Acosta, C.P. and Cano, J., 1999. Growth of the queen scallop (*Aequipecten opercularis*) in suspended culture: influence of density and depth. Aquaculture 178:43–62.

Román, G., Cano, J., Campos, M.J., Vázquez, C., García, T., Fernández, L. and Presas, C., 1999. The reproductive and reserve storage cycles of *Pecten maximus* in Malaga, south Spain. Book of Abstracts: 12th International Pectinid Workshop, Bergen, Norway, 5–11 May 1999. pp. 125–126.

Román, G. and Fernández, I., 1991. Ear hanging culture of scallop (*Pecten maximus* (Linnaeus 1758) in Galicia. In: S.E. Shumway and P.A. Sandifer (Eds.). An International Compendium of Scallop Biology and Culture. World Aquaculture Workshops No. 1. World Aquaculture Society, Baton Rouge, LA. pp. 322–330.

Saout, C., Paulet, Y.-M. and Muzellec, M.L., 1999b. Relationship between gonad growth and hatching success in the scallop *Pecten maximus* L. from the Bay of Brest (France): an experimental three-year study. Book of Abstracts, 12th International Pectinid Workshop, Bergen, Norway, 5–11 May, 1999. pp. 120–121.

Saout, C., Quere, C., Donval, A., Paulet, Y.M. and Samain, J.F., 1999a. An experimental study of the combined effects of temperature and photoperiod on reproductive physiology of *Pecten maximus* from the Bay of Brest (France). Aquaculture 172:301–314.

Shafee, M.S., 1979. Underwater observation to estimate the density and spatial distribution of black scallop, *Chlamys varia* (L.) in Lanvéoc (Bay of Brest). Bulletin Office National Pêches (Tunisie) 3:143–156.

Shafee, M.S., 1980a. Application of some growth models to the black scallop, *Chlamys varia* (L.) from Lanvéoc, Bay of Brest. Journal of Experimental Marine Biology and Ecology 43:237–250.

Shafee, M.S., 1980b. Ecophysiological studies on a temperate bivalve *Chlamys varia* (L.) from Lanvéoc, Bay of Brest. PhD Thesis. Université de Bretagne Occidentale.

Shafee, M.S., 1981. Seasonal changes in the biochemical composition and calorific content of the black scallop *Chlamys varia* (L.) from Lanvéoc, Bay of Brest. Oceanologica Acta 4:331–341.

Shafee, M.S., 1982. Variations saisonnières de la consommation d'oxygène chez le petoncle noir, *Chlamys varia* (L.) de Lanvéoc (rade de Brest). Oceanologica Acta 5:189–197.

Shafee, M.S. and Lucas, A., 1980. Quantitative studies on the reproduction of black scallop, *Chlamys varia* (L.) from Lanvéoc area (Bay of Brest). Journal of Experimental Marine Biology and Ecology 42:171–186.

Shafee, M.S. and Lucas, A., 1982. Variations saisonnières du bilan énergétique chez les individus d'une population de *Chlamys varia* (L.): Bivalvia, Pectinidae. Oceanologica Acta 5:331–338.

Shepard, A.N. and Auster, P.J., 1991. Incidental (non-capture) damage to scallops caused by dragging on rock and sand substrates. In: S.E. Shumway and P.A. Sandifer (Eds.). An International Compendium of Scallop Biology and Culture. World Aquaculture Workshops No. 1. World Aquaculture Society, Baton Rouge, LA. pp. 219–230.

Shumway, S.E. and Cembella, A.D., 1993. The impact of toxic algae on scallop culture and fisheries. Reviews in Fisheries Science 1:121–150.

Sinclair, M., Mohn, R.K., Robert, G. and Roddick, D.L., 1985. Considerations for the effective management of Atlantic scallops. Canadian Technical Report in Fisheries and Aquatic Sciences 1382. 113 p.

1056

Skjaeggestad, H., Andersen, S. and Strand, O., 1999. The capacity of the great scallop, *Pecten maximus*, to exploit enhanced food levels. Book of Abstracts, 12th International Pectinid Workshop, Bergen, Norway, 5–11 May, 1999. pp. 49–50.

Slater, J., 1995. Scallop spat collection in Mulroy Bay, Ireland, 1979–1994. Book of Abstracts, 10th International Pectinid Workshop, Cork, Ireland, April 26–May 2, 1995. pp. 68–69.

Soemodihardjo, S., 1974. Aspects of the biology of *Chlamys opercularis* (L.) (Bivalvia) with comparative notes on four allied species. PhD Thesis. University of Liverpool.

Stanley, C.A., 1967. The commercial scallop, *Pecten maximus* (L.) in Northern Irish waters. PhD Thesis. The Queens University, Belfast.

Stephens, P.J. and Boyle, P.R., 1978. Escape responses of the queen scallop *Chlamys opercularis* (L.) (Mollusca: Bivalvia). Marine Behaviour and Physiology 5:103–113.

Stone, H.C., Wilson, S.B. and Overnell, J., 1986. Cadmium-binding proteins in the scallop *Pecten maximus*. Environmental Health Perspectives 65:189–191.

Strand, O. and Nylund, A., 1991. The reproductive cycle of the scallop *Pecten maximus* (Linnaeus 1758) from two populations in Western Norway, 60°N and 64°N. In: S.E. Shumway and P.A. Sandifer (Eds.). An International Compendium of Scallop Biology and Culture. World Aquaculture Workshops No. 1. World Aquaculture Society, Baton Rouge, LA. pp. 95–105.

Strohmeier, T., Duinker, A. and Lie, O., 2000. Seasonal variations in chemical composition of the female gonad and storage organs in *Pecten maximus* (L.) suggesting that somatic and reproductive growth are separated in time. Journal of Shellfish Research 19:741–747.

Tang, S.F., 1941. The breeding of the scallop (*Pecten maximus* (L.) with a note on growth rate. Proceeding and Transactions of the Liverpool Biological Society 54:9–28.

Taylor, A.C. and Venn, T.J., 1978. Growth of the queen scallop, *Chlamys opercularis*, from the Clyde Sea area. Journal of the Marine Biological Association of the U.K. 58:687–700.

Taylor, A.C. and Venn, T.J., 1979. Seasonal variation in weight and biochemical composition of the tissues of the queen scallop, *Chlamys opercularis*, from the Clyde Sea area. Journal of the Marine Biological Association of the U.K. 59:605–621.

Tebble, N., 1966. British bivalve seashells. British Museum, London. 212 p.

Thomas, G.E. and Gruffydd, L.D., 1971. The types of escape reactions elicited in the scallop *Pecten maximus* by selected sea star species. Marine Biology 10:87–93.

Thompson, J.D., 1994. The development of scallop husbandry fishing in Tasmania through technology transfer from Hokkaido. In: N.F. Bourne, B.L. Bunting and L.D. Townsend (Eds.). Proceedings of the 9th International Pectinid Workshop, Nanaimo, B.C., Canada, April 22–27, 1993. Volume 2. Canadian Technical Report of Fisheries and Aquatic Sciences 1994:109.

Thompson, J.D., Fujimoto, T., Moriya, H. and Ikeda, T., 1995. Enhancement of the scallop *Pecten fumatus* in Tasmania - Japanese technology transfer down under. Journal of Shellfish Research 14:280.

Thouzeau, G., 1991a. Determination of *Pecten maximus* (L.) pre-recruitment in the Bay of Saint-Brieuc (France, Brittany): Processes regulating the abundance, survival and growth of post-larvae and juveniles. Aquatic Living Resources 4:77–99.

Thouzeau, G., 1991b. Experimental collection of postlarvae of *Pecten maximus* (L.) and other benthic macrofaunal species in the Bay of Saint-Brieuc, France. 1. Settlement patterns and biotic interactions among the species collected. Journal of Experimental Marine Biology and Ecology 148:159–179.

Thouzeau, G., 1991c. Experimental collection of postlarvae of *Pecten maximus* (L.) and other benthic macrofaunal species in the Bay of Saint-Brieuc, France 2. Reproduction patterns and postlarval growth of 5 mollusk species. Journal of Experimental Marine Biology and Ecology 148:181–200.

Thouzeau, G. and Chauvaud, L., 1995. Experimental collection of benthic post-larvae in the Bay of Brest: 3-dimensional vs. 2-dimensional artificial substratum collectors. Book of Abstracts, 10th International Pectinid Workshop, Cork, Ireland, April 26–May 2, 1995. pp. 128–129.

Thouzeau, G., Chauvaud, L., Grall, L. and Guerin, L., 2000. Do biotic interactions control pre-recruitment and growth of *Pecten maximus* (L.) in the Bay of Brest? Comptes Rendus de l'Academie Des Sciences Serie III-Sciences de la Vie 323:815–825.

Thouzeau, G. and Hily, C., 1986. A.QUA.R.E.V.E: Une technique nouvelle d'echantillonnage quantitatif de la macrofaune epibenthique des fonds meubles. Oceanologica Acta 9:509–513.

Thouzeau, G. and Lehay, D., 1988. Variabilite spatio-temporelle de la distribution, de la croissance et de la survie des juveniles de *Pecten maximus* issus des pontes de 1985, en baie de Saint-Brieuc. Oceanologica Acta 11:267–283.

Topping, G., 1973. Heavy metals in shellfish from Scottish waters. Aquaculture 1:379–384.

Ursin, E., 1956. Distribution and growth of the queen, *Chlamys opercularis* (Lamellibranchiata) in Danish and Faroese waters. Meddelelser fra Danmarks Fiskeri-og Havundersøgelser, Ny Serie 1:1–31.

Vahl, O., 1972. Particle retention and relation between water transport and oxygen uptake in *Chlamys opercularis* (L.) (Bivalvia). Ophelia 10:67–74.

Veale, L.O., Hill, A.S. and Brand, A.R., 2000a. An *in situ* study of predator aggregations on scallop (*Pecten maximus* (L.)) dredge discards using a static time-lapse camera system. Journal of Experimental Marine Biology and Ecology 255:111–129.

Veale, L.O., Hill, A.S., Hawkins, S.J. and Brand, A.R., 2000b. Effects of long-term physical disturbance by commercial scallop fishing on subtidal sedimentary habitats and epifaunal assemblages. Marine Biology 137:325–337.

Veale, L.O., Hill, A.S., Hawkins, S.J. and Brand, A.R., 2001. Distribution and damage to the by-catch assemblages of the northern Irish Sea scallop dredge fisheries. Journal of the Marine Biological Association of the U.K. 81:85–96.

Ventilla, R.F., 1977a. A scallop spat collector trial off the Northern Ardnamurchan coast. White Fish Authority, Field Report 485. 16 p.

Ventilla, R.F., 1977b. A preliminary investigation of the growth potential of collected scallop spat. White Fish Authority, Field Report 502. 16 p.

Ventilla, R.F., 1982. The scallop industry in Japan. Advances in Marine Biology 20:310–390.

Wanninayake, T., 1994. Seasonal cycles of two species of scallop (Bivalvia: Pectinidae) on an inshore and an offshore fishing ground. PhD Thesis. University of Liverpool.

Ward, M.A. and Thorpe, J.P., 1991. Distribution of encrusting bryozoans and other epifauna on the subtidal bivalve *Chlamys opercularis*. Marine Biology 110:253–259.

Watson, P.S., 1971. Survey off north and north-east coast of Ireland for queen scallop (*Chlamys opercularis* (L.)). Ministry of Agriculture (Government of Northern Ireland), Fisheries Research Leaflet No. 1. 13 p.

Whittington, M.W., 1993. Scallop aquaculture in Manx waters: spat collection and the role of predation in seabed cultivation. PhD Thesis. University of Liverpool.

Wieland, T. and Paul, J.D., 1983. Potential for scallop culture in the U.K. Sea Fish Industry Authority, Technical Report, 224. 40 p.

Wilding, C.S., Beaumont, A.R. and Latchford, J.W., 1997. Mitochondrial DNA variation in the scallop *Pecten maximus* (L.) assessed by a PCR-RFLP method. Heredity 79:178–189.

Wilding, C.S., Beaumont, A.R. and Latchford, J.W., 1999. Are *Pecten maximus* and *Pecten jacobaeus* different species? Journal of the Marine Biological Association of the U.K. 79:949–952.

Wilson, J.H., 1987a. Spawning of *Pecten maximus* (Pectinidae) and the artificial collection of juveniles in two bays in the West of Ireland. Aquaculture 61:99–111.

Wilson, J.H., 1987b. Environmental parameters controlling growth of *Ostrea edulis* L. and *Pecten maximus* L. in suspended culture. Aquaculture 64:119–131.

Wilson, J.H., 1987c. Spawning of *Pecten maximus* (Pectinidae) and the artificial collection of juveniles in 2 bays in the West of Ireland. Aquaculture 61:99–111.

Wilson, U.A.W., 1994. The potential for cultivation and restocking of *Pecten maximus* (L.) and *Aequipecten* (*Chlamys*) *opercularis* (L.) on Manx inshore fishing grounds. PhD Thesis. University of Liverpool.

Wilson, U.A.W. and Brand, A.R., 1994. Nowhere to hide? Predator impact on seeded scallops. In: N.F. Bourne, B.L. Bunting and L.D. Townsend (Eds.). Proceedings of the 9th International Pectinid Workshop, Nanaimo, B.C., Canada, April 22–27, 1993. Volume 2. Canadian Technical Report of Fisheries and Aquatic Sciences 1994:118.

Wilson, U.A.W. and Brand, A.R., 1995. Variations in commercial scallop (*Pecten maximus*) density in a seasonal fishery. Book of Abstracts, 10th International Pectinid Workshop, Cork, Ireland, 26 April-2 May, 1995. pp. 13–14.

Zhang, J.Z., 1996. Further studies of scallop spat settlement and early survival in Isle of Man coastal waters. PhD Thesis. University of Liverpool.

AUTHOR'S ADDRESS

Andrew R. Brand - Port Erin Marine Laboratory, University of Liverpool, Port Erin, Isle of Man, IM9 6JA (E-mail: arbrand@liverpool.ac.uk)

Scallops: Biology, Ecology and Aquaculture
S.E. Shumway and G.J. Parsons (Editors)

Chapter 20

European Aquaculture

Mark Norman, Guillermo Román and Øivind Strand

20.1 INTRODUCTION

The success of scallop aquaculture development in Japan created interest in scallop culture in Europe in the 1970–80s (Buestel 1981; Dao et al. 1999). The main species of interest were *Pecten maximus* and *Aequipecten* (*Chlamys*) *opercularis,* while *Chlamys varia, Chlamys islandica* and *Flexopecten flexuosus* have later also been considered for farming. These species are distributed along most Atlantic, Mediterranean and North sea coastlines (Figure 20.1). Today scallop culture is on an industry developmental level in Britain, France, Norway, Ireland, Italy and Spain. The commercial production of cultured scallops is still low, less than 500 tonnes a year. The culture of *P. maximus* in Norway and *C. islandica* in Iceland and Norway are covered in the Chapter 20 (Scandinavia).

Most of the work on scallop culture reviewed in the early 1980's focused on natural spat collection and farming by suspended culture operations (Berry and Burnell 1981; Paul et al. 1981; Wieland and Paul 1983). With the Japanese scallop aquaculture model and technology as a basis, several adaptations and experimental developments were undertaken with the aim of reducing the high cost of production associated with a long grow out, European labour costs and the high cost of materials.

Most of the experimental and development work was centred on areas where wild spat could be collected. In Scotland this was in the Rassay Sound off the Isle of Skye, while in Ireland it was in Mulroy Bay, County Donegal where significant catches of *P. maximus* were reported in the late 1970's (Minchin 1981, 1983). Other focus areas for aquaculture development were those areas with declining traditional fisheries such as the Rade de Brest, France and Connemara, Ireland.

In Spain, research is in progress to determine the feasibility of cultivating scallops found in two regions, the galician Rías Baixas and the coast of Malaga. From an historic point of view, the most important areas for mollusc production are the Rías Baixas in Galicia, NW of Spain. The Rías Baixas are wind protected, deep and narrow bays, where high primary productivity (250–537 g C m^{-2} y^{-1}) (Bode and Varela 1998) and mild temperatures (12–20°C) are recorded. A small-scale culture in this area supplies the market with about 10–20 tonnes of *P. maximus* annually. Experiments are being performed with *C. varia* and *A. opercularis.*

The highest *P. maximus* landings in Spain are recorded in Malaga. The region has an open shore with frequent easterly and westerly winds, which hinder the use of rafts. Recently, mooring of long-lines for mussel culture allowed the start of experiments for hanging culture of *P. maximus* and *C. varia.* Other species, such as *A. opercularis* and

1060

Spat Collection Sites ▯

Hatcheries ●

On-growing Areas ✪

Map Showing Main Scallop Aquaculture Locations in Europe

Figure 20.1 Map showing main scallop aquaculture locations in Europe.

Flexopecten flexuosus are present in the Malaga area, but presently are not considered as research targets for commercial culture.

20.2 PECTEN MAXIMUS

The first aquaculture operations in Europe in the 1970's started with attempts to collect natural spat, following the Japanese model (Ventilla 1982). In Ireland and Britain, locations with a regular spat-fall were identified (Brand et al. 1980; Minchin 1981; Paul et al. 1981; Fraser 1983) and commercial collection programmes established. These methods are currently used to supply culture operations. Based on an extended study of settlement patterns of the great scallop, *P. maximus*, and other benthic species in relation to spatio-temporal variability of environmental factors in the Bay of Brest, Chauvaud et al. (1996) recommended improvements on the efficiency of spat collectors (physical and structural criteria), to collector placement, and on the intensity of biotic interactions at the postlarval stage.

Induction of synchronised spawning by rising broodstock held in suspended nets to shallower depths of warmer water was proposed by Minchin (1993) in order to improve conditions for collection of scallop spat within sea-loughs. However, this is not a technique currently used by industry.

The amount of spat from natural collection programmes in Scotland and Ireland has varied each year for the past twenty years, from several thousand to several million (Scottish Executive 2001; J. Slater, pers. comm.). This irregular supply of spat has been caused by a range of factors including; organo-tin pollution; insufficient brood stock; lack of commercial markets for spat (Minchin et al. 1987).

In France, the failure of natural spat collection forced the restocking programme (Buestel et al. 1979) to develop hatcheries, nurseries and intermediate culture. The development of commercial scaled spat production in hatchery started in 1983, resulted in a production of annual production between 1988 and 1995 of 10 million spat, and has increased in more recent years to 25 million spat (Dao et al. 1999). Production of larvae and spat were previously done in laboratory scale (Comely 1972; Beaumont et al. 1982), but the French experience has been a reference for most other hatcheries in Europe (Fleury et al. 1997; Dao et al. 1999; Strand and Parsons, Chapter 21). The genetic aspects of hatchery rearing of the scallop *P. maximus* were reviewed by Beaumont (1986).

Devauchelle and Mingant (1993) reviewed information of reproductive biology of the scallop, *P. maximus*, and summarised the experience during ten years of an intensive aquaculture program in France for the development of spat production in hatchery. They presented data on how environmental factors sustain gametogenic activity at different times of the year, biochemical changes associated with gametogenesis, the consistency of egg quality and development of a simple test to predict egg quality.

After the production of postlarvae (2 mm) in hatchery-nursery, intermediate culture of spat in cages in the sea (2–30 mm) and seabed culture at densities of 10 m^2 until dredging 30–36 months later when they reach the marketable size of 100 mm shell length. Production cost calculations of the French production model showed that a survival rate of 30% after sowing was required to gain economic feasibility (Paquotte and Fleury 1994).

Studies on culture of juvenile great scallops held in suspended nets in Bertraboy Bay, Ireland (Wilson 1987) and of spat (5–14 mm shell height) held indoor and fed cultured algae (Laing 2000) suggest that growth in culture conditions is mainly regulated by water temperature, rather than the amount of food available. Laing (2000) showed that as temperature increased, the scallops consumed a bigger ration, and the highest ration that they consumed efficiently determined their maximum growth rate. This is supported by Chauvaud et al. (1998) studying effects of environmental factors on the daily shell growth of juvenile *P. maximus* from the Bay of Brest. Growth was mainly regulated by bottom-water temperature, salinity and, to a lesser extent, river flows, rather than food. However, scallop food intake and growth were high when the algal concentration did not exceed a critical threshold, dependent on the dominant species and sedimentation rate of diatoms (Chauvaud et al. 2001). They observed events of slow growth during and after diatom blooms (*Rhizosolenia delicatula* and *Chaetoceros sociale*) and dinoflagellate (*Gymnodinium* cf. *nagasakiense*) blooms.

Intermediate culture is a necessary stage in *P. maximus* culture, which is usually performed in suspension. However, suspension cage culture of the pectinoid form of pectinids (Waller 1992) shows delayed growth after some time or when reaching a certain size (Paul et al. 1981; Slater 1995). This is perhaps due to changes in the natural habitat, lack of sediment for recessing, or to the negative wave effect as Lodeiros et al. (1999) observed in *Euvola ziczac*. This is one reason why the trend is towards bottom culture of the pectinoid form of pectinids. However in some regions the environmental conditions or techniques allow suspension culture to commercial size in suspended cages as described in Malaga (Cano et al. 2000) or by ear-hanging (Paul 1988; Román and Fernández 1991; Gallagher 1999). In Ireland, stocking density and type of nets was shown to have a clear effect on growth and carbohydrate content in scallops (50 mm shell height) grown in suspended culture (Maguire and Burnell 2001).

Generally, scallops are not ear-hung until they have reached a shell height of 55 mm or more (Ventilla 1982; Dadswell and Parsons 1991; O'Connor et al. 1999). According to Gallagher (1999), the minimum size should be higher than 50 mm, excessive fouling should be avoided and culture duration should not be more that two years. Several studies have reported heavy fouling on the scallops in ear hanging (Minchin 1975; Cano et al. 2000) and particularly the settlement of *Balanus* sp. have caused high mortality.

In Galicia, Román et al. (2001) ear-hung juveniles of 66.6 mm mean shell height in September. The scallops were detached from collectors as 26.6 mm spat by the middle of November the previous year. In spite of the fact that no cleaning operations were performed at all during the growth period, the scallops grew well. By the end of May the next year, they were 81.2±9.3 mm in shell height (14.5% commercial size), and by mid October they had reached 91.0±9.7 mm height and the meat content (muscle plus gonad) was 24.4±9.1 g. Roughly two years after detachment from collectors, 60.9% of the scallops had reached commercial size. Survival was 75.8% by May and 50.5% by October.

Stocking density and type of nets was shown to have a clear effect on growth and carbohydrate content in scallops (50 mm shell height) grown in suspended culture (Maguire and Burnell 2001).

The great scallop, *Pecten maximus*, is considered to be one of the most precious members of the scallops, and there is a large European marked for live large size animals. Given the problems related to mortality, shell deformation and low growth, and high costs of farming this scallop in suspended culture, seabed culture has been regarded as the most feasible culture method in central and northern Europe (Paul et al. 1981; Lake et al. 1987; Minchin 1991; Dao et al. 1999). Cano et al. (2000) considered Fuengirola, Malaga, in southern Spain as the only place in Europe where suspended great scallop culture would be profitable by producing commercial scallop of 10 cm in shell length in approximately 18 months.

20.3 *AEQUIPECTEN OPERCULARIS*

In Galicia the most abundant scallop species obtained on collectors is *Aequipecten opercularis* at a rate of 25–200 per bag (Ramonell and Malvar 1993; Acosta et al. 1999; Román et al., unpublished data). In Malaga collection has been unsuccessful.

Spat collection in Scotland typically yields a mixture of *A. opercularis* and *P. maximus* spat (Paul et al. 1980). *A. opercularis* ongrowing trials using standard

Japanese equipment and various adaptations demonstrated the suitability of these species for suspended culture. Farming this species in Scotland has been technically successful but production costs at times exceed the market price of wild caught scallops. There have been several marketing innovations including the marketing of "Princess scallops", partially grown (15–18 months from settlement), that are sold whole to be prepared in a similar manner to mussels. There are currently a number of farms using hanging culture techniques to produce *A. opercularis* in the sea lochs of western Scotland (Scottish Executive 2001).

20.4 *CHLAMYS VARIA*

In Spain, small numbers of *C. varia* has been reported to settle on collectors in Galicia (Román et al. 1987; Ramonell et al. 1990; Ramonell and Malvar 1993; Román et al. 1996).

In the Rade de Brest, Brittany there is a traditional dredge fishery for *C. varia*. Stock enhancement (bottom culture) using hatchery-produced seed (Louro et al. 2003) has been attempted. Small seed were released but with very poor survival. However, subsequent trials using similar size seed that had attached to substrate (broken roof tiles) in tanks prior to seeding significantly improved survival (J-C. Dao, pers. comm.).

20.5 SUMMARY

All aspects of scallop culture have been the subject of a considerable amount of research effort in several European countries. This research and development effort has been in a range of economic and industry climate; from Norway with little or no traditional or modern shellfish farming to France and Spain with large, well established industries for other species of bivalves.

Despite many technological advances, a large-scale scallop culture industry has not been successfully developed in Europe.

A number of factors have been identified as constraints to the emergence of such an industry. These include; the relatively long grow out period for king scallops; the high production cost of hatchery seed; an inconsistent supply of wild seed; low market prices for queen scallops (*Aequipecten* spp.); and high labour costs associated with hanging culture.

These constraints are the subject of research and development programmes in Norway, Ireland, France, Spain and the U.K.

REFERENCES

Acosta, C.P., Cano, J., Campos, M.J., Román, G. and Lowe, E., 1997. Effect of depth on reproductive cycles and reserves in suspension cultures of *Aequipecten opercularis*. 11th International Pectinid Workshop 10–15 April 1997. La Paz, BCS, México. Book of Abstracts.

Acosta, C.P., Román, G. and Alvarez, M.J., 1990. Cultivo de zamburiña (*Chlamys varia*) II. Preengorde en batea. Actas del III Congreso Nacional de Acuicultura Acosta, C.P. pp. 533–538.

Bode, A. and Varela, M., 1998. Primary production and phytoplankton in three Galician Rías Altas (NW Spain): seasonal and spatial variability. Sci. Mar. 62(4):319–330.

Brand, A.R., Paul, J.D. and Hoogeteger, J.N., 1980. Spat settlement of the scallops *Chlamys opercularis* (L.) and *Pecten maximus* (L.) on artificial collectors. J. Mar. Biol. Assoc. U.K. 60:379–390.

Buestel, D., 1981. L'exploitation de la coquille St. Jaques *Patinopecten yessoensis* L. au Japon. Possibilités d'application du modéle de développement Japonais a l'espése francaise *Pecten maximus* L. Aquaculture Extensive et Repeuplement. Publ. Cnexo, France. Actes Colloques No. 12. pp. 15–32.

Buestel, D., Dao, J.C. and Lemarie, G., 1979. Collecte de naissain de pectinides en Bretagne. Rapp. et P.-v. Reun. Cons. Int. Explor. Mer. 175:80–84.

Cancelo, M., Fernández, A., Gabin, C. and Guerra, A., 1990. Analisis de crecimientode semilla de zamburiña (*Chlamys varia* L) cultivada en batea. Actas del III Congreso Nacional de Acuicultura. pp. 527–532.

Cano, J., Campos, M.J. and Román, G., 2000. Growth and mortality of the king scallop grown in suspended culture in Málaga, Southern Spain. Aquaculture International 8:207–225.

Cano, J., Campos, M.J., Román, G., Vázquez, M.C., García, T., Fernández, L. and Presas, M.C., 1999. Pectinid settlement on collectors in Fuengirola, Málaga, South Spain. 12th International Pectinid Workshop, 5–11 May 1999. Bergen, Norway. Book of Abstracts.

Cano, J. and García T., 1985. 5th International Pectinid Workshop. La Coruña, Spain.

Chauvaud, L., Donval, A., Thouzeau, G., Paulet, Y.-M. and Nézan, E., 2001. Variation in food intake of *Pecten maximus* (L.) from the Bay of Brest (France): Influence of environmental factors and phytoplankton species composition. C.R. Acad. Sci. Paris, Life Science, Ecology 324:743–755.

Chauvaud, L., Thouzeau, G. and Grall, J., 1996. Experimental collection of the great scallop postlarvae and other benthic species in the Bay of Breast: settlement patterns in relation to spatio-temporal variability of environmental factors. Aquaculture International 4:263–288.

Chauvaud, L., Thouzeau, G. and Paulet, Y.-M., 1998. Effects of environmental factors on the daily growth rate of *Pecten maximus* juveniles in the Bay of Brest. J. Exp. Mar. Biol. Ecol. 227(1):83–111.

Comely, C.A., 1972. Larval culture of the scallop *Pecten maximus* (L.). J. Cons. Int. Explor. Mer. 34:365–378.

Devauchelle, N. and Mingant, C., 1991. Review of the reproductive physiology of the scallop, *Pecten maximus*, applicable to intensive aquaculture. Aquat. Living Resour. 4:41–51.

Fraser, D.I., 1983. Observations on the settlement of pectinid spat off the west coast of Scotland in 1982. Int. Counc. Explor. Sea, C.M. K:40. 5 p.

Gallagher, J., 1999. Survival and growth of ear-hung scallop (*Pecten maximus*) in Mulroy Bay. 12th International Pectinid Workshop, 5–11 May 1999, Bergen, Norway. Book of Abstracts.

Grall, J., Chauvaud, L., Thouzeau, G., Fifas, S., Glémarec, M. and Paulet, Y-M., 1996. Distribution of *Pecten maximus* (L.) and its main potential competitors and predators in the Bay of Brest. C.R. Acad. Sci. Paris, Life Science, Ecology 319:931–937.

Laing, I., 2000. Effect of temperature and ration on growth and condition of king scallop (*Pecten maximus*) spat. Aquaculture 183:325–334.

Louro, A., de la Roche, J.P., Campos, M.J. and Román, G., 2003. Hatchery rearing the black scallop, *Chlamys varia* (L.). J. Shellfish Res. 22:95–99.

Maguire, J. and Burnell, G., 2001. The effect of stocking density in suspended culture on growth and carbohydrate content of the adductor muscle in two populations of the scallop (*Pecten maximus* L.) in Bantry Bay, Ireland. Aquaculture 198:95–108.

Maguire, J., Cashmore, D. and Burnell, G., 1999. The effect of transportation on the juvenile scallop *Pecten maximus* (L.). Aqua. Res. 30:325–333.

Maguire, J., Fleury, P.G. and Burnell, G., 1999. Some methods for quantifying quality in the scallop *Pecten maximus* (L.). J. Shellfish Res. 18:59–66.

Malvar, M. and Ramonell, R., 1993. Cultivo de vieira (*Pecten maximus*) en el medio natural - fase piloto - Ría de Pontevedra, 1989–1991. Actas del IV Congreso Nacional de Acuicultura. pp. 395–400.

Minchin, D., 1975. Experimental hanging culture of *Pecten maximus* in the west of Ireland, with a note on tagging. Int. Counc. Explor. Sea, C.M. K:3. 5 p.

Minchin, D., 1981. The escallop, *Pecten maximus* in Mulroy Bay. Fisheries Bulletin No. 1. The Stationary Office, Dublin.

Minchin, D., 1983. The escallop, *Pecten maximus* in Mulroy Bay-Part 2. 4th International Pectinid Workshop Aberdeen 10–13 May 1983.

Minchin, D., 1993. Scallop spat production within sea-loughs by means of induced synchronised spawnings – a possible solution. Int. Counc. Explor. Sea, C.M. F:32. 6 p.

Minchin, D., Duggan, C.B. and King, W., 1987. Possible effects of organotins on scallop recruitment. Mar. Poll. Bull. 18:604–608.

Parsons, G.J. and Dadswell, M.J., 1992. Effect of stocking density on growth, production and survival of the giant scallop, *Placopecten magellanicus*, held in intermediate suspension culture in Passamaquoddy Bay, New Brunswick. Aquaculture 103:291–309.

Paul, J.D., Brand, A.R and Hoogeteger, J.N., 1981. Experimental cultivation of the scallops *Chlamys opercularis* (L.) and *Pecten maximus* (L.) using naturally produced spat. Aquaculture 24:31–44.

Paquotte, P. and Fleury, P.G., 1994. Production costs in French scallop culture. In: N.F. Bourne, B.L. Bunting and L.D. Townsend (Eds.). Proceedings of the 9th International Pectinid Workshop, Nanaimo, B.C., Canada, April 22–27, 1993. Vol. 2. Can. Tech. Rep. Fish. Aquat. Sci. 1994:66–72.

Ramonell, R. and Malvar, M., 1993. Captación natural de semilla de volandeira *Chlamys opercualris* (Linné 1767) los años 1989, 1990 y 1991 y su crecimiento en batea. Actas del IV Congreso Nacional de Acuicultura. pp. 371–376.

Ramonell, R., Román, G., Acosta, C.P. and Malvar, M., 1990. Captación de semilla de pectínidos en colectores. Resultado de la campaña de prospección en Bueu (Ría de Pontevedra, Galicia) en 1988. Actas del III Congreso Nacional de Acuicultura. pp. 439–444.

Rhodes, E.W. and Widman, J.C., 1984. Density-dependent growth of the bay scallop *Argopecten irradians irradians*, in suspension culture. Int. Counc. Explor. Sea 18: 8 p.

Román, G. and Acosta, C.P., 1990. Cultivo de vieira en batea. I. Crecimiento. Actas del III Congreso Nacional de Acuicultura. pp. 539–544.

Román, G., Campos, M.J. and Acosta, C.P., 1996. Relationships among environment, spawning and settlement of Queen scallop in the Ría de arosa (Galicia, Spain). Aquaculture International 4:225–236.

Román, G., Campos, M.J. and Cano, J., 1999. On some factors affecting *Aequipecten* settlement. 12th International Pectinid Workshop. 5–11th May 1999. Bergen, Norway. Book of abstracts.

Román, G., Campos, M.J. and Cano, J., 1999. Cultivo de vieiras en Málaga en sistemas flotantes en mar abierto. Informe de progreso.

Román, G. and Cano, J., 1987. Pectinid settlement in Málaga, S. Spain in 1985. Sixth International Pectinid Workshop. Menai Bridge, Wales.

Román, G., Cano, J., Campos, M.J. and López-Linares, J.I., 2001. Biología y Cultivo de la Vieira en Málaga. Junta de Andalucía (Ed.). Consejería de Agricultura y Pesca. Colección: Pesca y Acuicultura. Serie: Recursos pesqueros. J. de Haro Artes gráficas, Sevilla. 75 p.

Román, G., Cano, J. and García-Jiménez, T., 1985. A first trial with pectinid collectors in Málaga, S. Spain. 5th International Pectinid Workshop, La Coruña, Spain.

Román, G., Fernández-Cortés, F., Acosta, C.P. and Rodriguez-Moscoso, E., 1987. Primeras experiencias con colectores de pectínidos en las rías de Arosa y Aldán. Actas del II Congreso Nacional de Acuicultura. Cuad. Marisq. Publ. Téc. No. 12. pp. 375–380.

Scottish Executive, 2001. Scottish Shellfish Farm Production Survey 2001. Scottish Executive Environmental and Rural Affairs Department.

Slater, J., 1995. An initial ear hanging trial with *Pecten maximus* in Mulroy Bay. 10th International Pectinid Workshop. April 27–May 2. Cork, Ireland.

Ventilla, R.F., 1982. The scallop industry in Japan. Adv. Mar. Biol. 20:309–382.

Waller, T.R., 1991. Evolutionary relationships among commercial scallops (Mollusca: Bivalvia: Pectinidae). In: S.E. Shumway (Ed.). Scallops: Biology, Ecology and Aquaculture. Developments in Aquaculture and Fisheries Science, Vol. 21. Elsevier, Amsterdam. pp. 1–74.

Wieland, T. and Paul, J., 1983. Potential for scallop culture in the U.K. Sea Fish Industry Authority Technical Report 224. 23 pp.

Wilson, J.H., 1987. Environmental parameters controlling growth of *Ostrea edulis* L. and *Pecten maximus* L. in suspended culture. Aquaculture 64:119–131.

AUTHORS ADDRESSES

Mark Norman - Taighde Mara Teo, Carna, Co. Galway, Ireland (E-mail: mnorman@taighde-mara.iol.ie)

Guillermo Román - Instituto Espanol de Oceanografia, Centro Oceanografico de la Coruna, Muelle de Animas s/n, AP130, 15080 La Coruna, Spain (E-mail: guillermo.roman@co.ieo.es)

Øivind Strand - Department of Aquaculture, Institute of Marine Research, P. O. Box 1870 Nordnes, 5817 Bergen, Norway (E-mail: oivind.strand@imr.no)

Scallops: Biology, Ecology and Aquaculture
S.E. Shumway and G.J. Parsons (Editors)

Chapter 21

Scandinavia

Øivind Strand and G. Jay Parsons

21.1 INTRODUCTION

The commercial exploitation of scallops in Scandinavia, which includes Norway, Denmark, Sweden, Faroes, Iceland, and Greenland (Fig. 21.1) is moderately recent. Commercial fishing activity began in Iceland in 1969, the Faroes in 1971, and in the mid 1980's it began in Greenland and Norway (Table 21.1). In Denmark, only by-catches of scallops have been reported. From a modest start of 402 metric tonnes (t) in 1969 (Table 21.1), representing 0.3% of the global scallop landings (FAO 2000), the fishery for this region increased to 60,580 t, which corresponded to 8.1% of the world's total scallop harvest in 1987. This position was rather short-lived and in 1996, the catch of 20,596 t corresponded to 1.2% of the global scallop landings.

Three species of commercial interest *Chlamys islandica* (Muller), *Aequipecten opercularis* (L.) and *Pecten maximus* (L.) occur in the waters of northern Europe. The Iceland scallop (*C. islandica*) has a northern boreal distribution. The other two species have a boreal-temperate distribution. The queen scallop (*A. opercularis*) is fished in the Faroes and a fishery for the great scallop (*P. maximus*) exists in Norway. *C. islandica* is found in concentrations great enough to be exploited on a large scale in Iceland, Greenland and Norway. All three species have a commercial potential for aquaculture in Scandinavian waters, and considerable effort is being conducted on aquaculture development of *P. maximus* in Norway. *C. islandica* and *P. maximus* will be the main species dealt with in this chapter.

21.2 CHLAMYS ISLANDICA

21.2.1 Biology

Chlamys islandica is the northernmost species of the family Pectinidae and is distributed in the northern boreal or subarctic transition zone (Ekman 1953). It has been described as circumpolar in its distribution and its occurrence in the north Pacific from Alaska to Puget Sound, Washington (Wiborg 1963; Abbott 1974; Bernard 1983) has been questioned by Waller (1991). In the western Atlantic, it is distributed from the Arctic seas to southern Massachusetts (Abbott 1974; Serchuk and Wigley 1984). In the eastern Atlantic it is found from Spitsbergen south to Lofoten Islands, Norway (Wiborg 1973; Wiborg et al. 1974) and from Greenland, Iceland and Jan Mayen (Ockelmann 1958; Wiborg 1963; Kjolholt and Hansen 1986) east to the Barents and White seas (Fig. 21.1)

1068

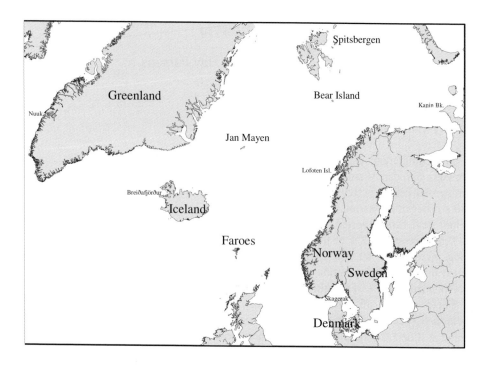

Figure 21.1. Scandinavia and Greenland waters.

(Wiborg 1963; Zenkevitch 1963). In Norway relict populations are found in fjords along the coast south from Lofoten Islands to the southern part of the west coast (Greve and Samuelsen 1970).

The Iceland scallop is found off the western coast of Greenland (Pedersen 1988a), ranging from Qaanaaq (Thule) (76°N) in the north (Lubinsky 1980; Pedersen 1988b) to Nanortalik on the southernmost coast (Fig. 21.1) (Pedersen 1988b). Its distribution off eastern Greenland is restricted to the southeastern coast and a relict population in the Franz Josephs Fjord (Ockelmann 1958). *C. islandica* is distributed all around Iceland except the south coast (Eiríksson 1997). However, the main distribution is rather discontinuous and almost entirely limited to fjord areas. The major Norwegian distribution is located north of the Lofoten Islands where it occurs in coastal and fjord areas, in the deeper waters of Bear Island, Spitsbergen and Jan Mayen (Wiborg 1963; Wiborg 1973; Wiborg et al. 1974; Rubach and Sundet 1988; Strand and Vølstad 1997).

Near the coast it is found on the inner sides of fjords with one or two shallow sills and low bottom temperatures of about -1.5°C to 9°C (Wiborg 1962; Wiborg 1963; Wiborg and Bøhle 1974; Eiríksson 1997). It is found in depths of 10–250 m; however, the largest concentrations occur in waters <100 m (Wiborg 1963; Venvik and Vahl 1985; Eiríksson

Table 21.1

Scallop landings (metric tonnes) for Scandinavia and Greenland from 1969–1999. J.M. = Jan Mayen; B.I. = Bear Island; Sp. = Spitsbergen, C. = coastal waters of north Norway.

Year	Iceland[1]	Norway[2] J.M, B.I. and Sp.	Norway[2] C.	Norway[2]	Greenland[3]	Faroes[4]
	C. islandica	C. islandica	C. islandica	P. maximus	C. islandica	A. opercularis
1969	402	-	-	-	-	-
1970	2,432	-	-	-	-	594
1971	3,658	-	-	-	-	414
1972	7,349	-	-	-	-	530
1973	4,848	-	-	-	-	817
1974	2,851	-	-	-	-	287
1975	2,784	-	-	-	-	542
1976	3,669	-	-	-	-	811
1977	4,427	-	-	-	-	403
1978	8,719	-	-	-	-	849
1979	7,800	-	-	-	-	1,426
1980	9,079	-	-	-	-	2,017
1981	10,186	-	-	-	-	1,762
1982	12,076	-	-	-	-	1,873
1983	15,181	-	-	-	-	2,256
1984	15,583	-	-	-	410	2,167
1985	17,068	1,192	-	-	1,693	1,793
1986	16,429	14,499	124	-	864	1,891
1987	13,272	44,098	849	-	1,086	1,275
1988	10,058	14,189	688	-	720	1,978
1989	10,772	6,098	760	-	NA	2,700
1990	12,380	6,750	584	-	NA	3,767
1991	10,297	4,022	214	-	NA	4,199
1992	12,443	6,200	400	-	1,995	3,846
1993	11,870	5,290	21	-	1,403	4,037
1994	8,401	6,200	34	4	1,876	3,282
1995	8,382	5,010	-	59	2,342	3,109
1996	8,873	7,000	-	27	1,176	3,520
1997	10,432	-	-	45	1,944	3,797
1998	10,038	-	-	112	2,200	5,367
1999	8,800	-	-	421	2,134	5,507

Sources; [1] Anonymous (2000) [2] Norwegian Directorate of Fisheries [3] Grønland Licenskontrol [4] A. Nicolajsen, Fisheries Laboratory of the Faroes

1997). *C. islandica* occurs on hard coarse sediments, consisting of sand, gravel, shells, and occasionally clay and fine sand (Wiborg 1962; Wiborg 1963; Vahl and Clausen 1980; Eiríksson 1997). These areas are often associated with strong currents and the scallops are invariably attached to stones or shells (Wiborg 1963; Vahl and Clausen 1980).

The Iceland scallop is a dioecious species (Wiborg 1963). Ripe female gonads are a pinkish hue and the male gonads are white. Male gonads are heavier than females, yet this has not been shown to occur throughout the year (Wiborg et al. 1974; Thorarinsdóttir 1993). Scallops are sexually mature at two and a half to six years of age (Sundet and Vahl 1981; Vahl 1981a; Vahl 1982; Thorarinsdóttir 1993), faster growing individuals reaching maturity first (Vahl 1981b). Wiborg (1963) and Pedersen (1988a) found the sex ratio to be 1:10 for Iceland scallops from Norway and Greenland, respectively. Scallops from Jan Mayen had a ratio of males to females of 0.8:1 (Rubach and Sundet 1988). The onset of spawning in northern Norway and in Breiðafjörður in Iceland varies annually from the end of June to the beginning of July (Skreslet and Brun 1969; Skreslet 1973; Gruffydd 1976a; Sundet and Vahl 1981; Sundet and Lee 1984; Thorarinsdóttir 1993). Furthermore, spawning occurs over a very short duration (about two weeks) (Skreslet 1973; Gruffydd 1976a; Sundet and Lee 1984; Thorarinsdóttir 1993). Gruffydd (1976a) observed shallow water scallops to spawn first, quickly followed by animals in deeper regions. Skreslet and Brun (1969) reported no difference with depth yet suggest that larger scallops may spawn earlier than smaller animals. Spawning appears to be triggered by short-term variations in temperature brought about by vernal meltwater discharge (Skreslet 1973).

Gonads start to recover immediately after spawning. In northern Norway spermatogenesis occurs only from March to October and oogenesis occurs throughout the winter months as well (Sundet and Lee 1984). The gonad weight of both sexes increases most rapidly in the spring from March to May (Skreslet and Brun 1969; Sundet and Vahl 1981; Sundet and Lee 1984). Thorarinsdóttir (1993) found that spermatogenesis in scallops from Breiðafjörður occurs during two periods of the year, from July to October and from February to June. However, in contrast to the Iceland scallops from northern Norway oogenesis did not continue throughout the winter.

This species is highly fecund and eggs are externally fertilised. The planktotrophic larvae spend up to 10 weeks, from late June to the end August, in the plankton before settling out (Gruffydd 1976a; Wallace 1982). Scallop larvae can detect salinity gradients and avoid lower salinity waters (Gruffydd 1976a). They can tolerate levels <80% ambient seawater. Settlement, onto artificial spat collectors, commences in September in both Norway (Wallace 1982) and Iceland (Thorarinsdóttir 1991). The metamorphosed larvae, which are about 305 µm, are capable of byssal attachment (Gruffydd 1976a).

Growth rates in *C. islandica* vary seasonally. The highest annual growth rate and net growth efficiency occurs in the spring (Sundet and Vahl 1981; Vahl 1981b) when the scallop utilises a large portion of the annual spring phytoplankton bloom (Vahl 1980a; Vahl 1980b; Thorarinsdóttir 1994). This is the period when Iceland scallops are also metabolically most active (Vahl 1978). Seasonal growth rates can be explained by variations in the proportion of particulate organic matter (POM) to particulate inorganic matter (PIM) (Vahl 1980a). Iceland scallops can effectively retain particles as small as 7 µm (Vahl 1973) but they can not preferentially select POM over PIM. As the ratio of

PIM increases, the scallops are not able to obtain sufficient food to sustain high growth rates. *C. islandica* has its greatest growth efficiency during the first six years of life (Vahl 1981b) and age-specific growth rates were highest for three and four year olds (10–11 mm yr^{-1} in scallops from Balsfjord, Norway) (Brun 1971).

Growth rates are available for several populations of scallops off northern Norway, Bear Island banks (Bjørnøya), Spitsbergen, Iceland and Greenland. Icelandic populations of *C. islandica* have the fastest growing scallops. A five-year-old scallop from northwestern Iceland was 50 mm in shell height and a ten-year-old 84-mm (Eiríksson 1988). The fastest growth reported in Norway occurred off Andamsfjord, northern Norway, where a five-year-old was 39-mm shell height and a ten-year-old 82-mm (Wiborg 1962; Wiborg 1963). Bear Island banks, Balsfjord and Korsfjord (northern Norway) have approximately the same growth rates, a five year old being 35–40 mm and a ten year old 70 mm (Wiborg 1963; Wiborg 1973; Vahl 1981b). Growth around Bear Island is quite variable, the northern beds having greater growth than the southern or eastern areas, respectively (Wiborg 1973). Growth in scallops from northern Spitsbergen was 30 mm and 60–65 mm for five and ten year old scallops, respectively (Wiborg et al. 1974). The growth rates of *C. islandica* on the Mitra bank (northwest Spitsbergen) are the lowest, 25 mm for a five year old and 50 mm for a ten year old (Wiborg et al. 1974). Growth of scallops from western Greenland was slower than Iceland, yet comparable to Spitsbergen (Pedersen 1988a).

The adductor muscle (meat) weight for a given size scallop is larger in areas with faster growth rates (Wiborg 1963; Wiborg et al. 1974). Growth rates are influenced by depth; Wiborg (1973) showed that scallops found at 75–105 m grew better than at depths >105 m. The Iceland scallop is capable of limited swimming. Wiborg (1963) and Gruffydd (1976b) found that medium sized scallops (30–40 mm) tend to swim most often, but Vahl and Clausen (1980) report that all sizes (20–80 mm) move equally often in the field. Swimming speed is proportional to size and all sizes swam more often at higher current velocities (Gruffydd 1976b). Scallops moved an average of once every thirty-one days (Vahl and Clausen 1980) and are invariably attached by their byssal threads (Wiborg 1963; Gruffydd 1978). Although swimming is infrequent, there is some evidence that this species migrates (Wiborg 1963; Vahl 1982) especially into inner fjords (Gruffydd 1976b). A survey along a transect of Ulfsfjord showed an increase in the size of scallops from the outer to the inner fjord (Wiborg 1963).

The natural predators of *Chlamys islandica* include starfish, *Asterias rubens* (Brun 1968), cod (Wiborg 1973) and eider ducks, *Somateria* spp. (Brun 1971).

In the early 1960's, Wiborg (1962) and Wiborg (1963) were the first to survey scallop beds to determine their population structure. Several northern Norwegian fjords from Balsfjord to Neidenfjord were surveyed using a six foot Baird scallop dredge and a four foot Baird mussel dredge. A wide size range of scallops (10–100 mm) were present; however the majority of animals were greater than 60 mm. While there probably was some gear selectivity against the smaller scallops, the population was still dominated by larger individuals. Wiborg and Bøhle (1974) reported that several other populations of Iceland scallops in the northern Norway region were also composed of larger scallops, e.g., at Kvænangen, 80% were between 55–75 mm. At Kvænangen and Kraksund, they found that the size distribution of scallops remained unchanged when surveyed in 1963 and in 1967. The distribution of Iceland scallops off Bear Island in 1968 and 1969 were also mainly composed of larger scallops, 91% and 85%, respectively, in the 55–85 mm range (Wiborg 1973). The average size distribution of scallops decreased with depth on beds in this area. Wiborg et al. (1974) re-surveyed this area again in 1973 using a larger commercial scallop dredge and found that all the scallops were greater than 60 mm. The size frequencies of C. islandica off Spitsbergen were predominately composed of scallops in the 60–80 mm range (Wiborg et al. 1974). The size range was smaller for scallops from Mitra bank (40–70 mm). This reflects the slow growth for this area as mentioned above. In Jan Mayen, 64% of the scallops caught were >65 mm (Rubach and Sundet 1988). The size distribution of scallops from eight locations in Breiðafjörður, western Iceland and five different depths at Sletta, northwestern Iceland all had a similar pattern with respect to location and depth (Eiríksson 1988). The majority of scallops were >65 mm. In Greenland the size distribution of scallops was strongly skewed to the larger scallops in the range of 60–110 mm with the majority being from 80–100 mm (Pedersen 1988b).

Wiborg (1963) determined the age of C. islandica by counting the alternating light and dark bands on the shell surface. However, Johannessen (1973) found that the lines produced on the ligament (resilium) were a more reliable and accurate method of ageing. Maximum longevity in the Iceland scallop is reported to be over twenty years (Vahl 1981a; Vahl 1981b; Vahl 1982).

The age composition of C. islandica populations generally corresponds to their size distribution (Wiborg 1973). Iceland scallops from northern Norway ranged from two to twelve years, with the majority being eight or older (Wiborg 1963). At Bear Island banks and off Spitsbergen, the majority of C. islandica were between the ages of ten and sixteen (Wiborg et al. 1974). In western Greenland, 40% of the scallops are over 21 years, which led Pedersen (1988a) to conclude that it is an accumulated virgin population.

Few data are available concerning Iceland scallop population densities. Wiborg (1973) estimated densities, from scallop dredge tows, to be as high as 80 m^{-2} for Bear Island, with an average of about 10 m^{-2}. Some areas around Spitsbergen, which can be as high as Bear Island, also average 10 scallops m^{-2} (Wiborg et al. 1974). Densities in inner Porsangerfjord and outer Balsford can reach as high as 60 m^{-2} and 50 m^{-2}, respectively (Wiborg and Bøhle 1974). Densities at Balsford decline towards the inner fjord (Wiborg 1963). A detailed study of scallop density in Balsford using SCUBA and randomly placed quadrats by Vahl (1981c) demonstrated that the spatial distribution is highly contagious and fits the negative binomial distribution. Estimates of density of scallops from all age classes ranged from 56 m^{-2} to 70 m^{-2}. In a later study, Venvik and Vahl (1985) found a mean density of 75 m^{-2}. Earlier studies on the distribution of Iceland scallops around Bear Island and Spitsbergen noted a low and irregular recruitment (Wiborg 1973; Wiborg et al. 1974). They based their observations on the low abundance of small scallops. Vahl (1981c) reports that C. islandica are not fully recruited in to the adult population until four years of age. Furthermore, recruitment can be highly variable, success being dependent on the density of the adult population already established (Vahl 1982). For existing beds, this suggests that recruitment will be low. However, if a favourable habitat is made available, e.g., by an attack of predators (Brun 1968), then a massive colonisation of juvenile scallops could take place (Vahl 1982; Vahl 1985). Pedersen (1988a) has estimated a recruitment of 3–4% for scallops from western Greenland.

The patterns of variation that have been found at selected allozyme loci support the suggestion that populations of *C. islandica* in the Northeast Atlantic are genetically structured (Fevolden 1989; Fevolden 1992; Galand and Fevolden 2000). These studies include scallops from Jan Mayen, northern Norway, Spitsbergen, Bear Island (Fevolden 1989; Fevolden 1992), Kap Kanin (Russia), Breiðafjörður (Iceland) and Fauskangerpollen (southwestern Norway) (Fig. 21.1) (Galand and Fevolden 2000). The relict population from Fauskangerpollen was genetically most distinct from any of the other populations. In contrast with findings for several other bivalves, but in accordance with observations for other pectinids, Fevolden (1992) were not able to show any increase in heterozygosity with age, nor were positive growth-heterozygosity correlations demonstrated.

Little information is available on the mortality rates of Iceland scallops. Vahl (1981c) estimated the natural instantaneous mortality rate (M) of fully recruited scallops from Norway to be 0.17 yr^{-1}. This was calculated from the slope of the regression of the natural logarithm of mean density of year classes over time. From an estimate of egg production (no. m^{-2}) and back calculating the density of age-group four to density at settlement, Vahl (1982) obtained an estimate of larval survival to be 0.00005%. This assumes the same mortality rate in all age classes and an instantaneous population growth rate of zero.

21.2.2 Fishery

The commercial exploitation of *Chlamys islandica* in Iceland, started in 1969 and quickly reached a peak total catch of 7,400 t in 1972 (Table 21.1). The harvest in subsequent years declined to between 2,784 t and 4,427 t due to a decline in the price of scallops in the U. S. market. In 1978 after the introduction of mechanical shuckers, the catch rebounded and increased to the historical maximum of 17,068 t in 1985 (Table 21.1), of which 12,128 t were caught in Breiðafjörður (Fig. 21.1) (Eiríksson 1997). After 1985–1986 the annual landings decreased to a stable level of 8,000–12,000 t. Fishing effort in the 1980s and early 1990s was high, resulting in a 30–40% decline in stock abundance indices, and a 25% decrease in CPUE in Breiðafjörður between 1983–1993 (Eiríksson 1997; Anonymous 2000). Since 1993, the older component of the Breiðafjörður stock has seemingly remained stable, accompanied by an increase in CPUE due to improved fishing gear.

Scallops are caught, for the most part, inshore in fjords and bays. When the fishery commenced it was centred off the northwest coast but quickly shifted to the western coast after rich new beds were discovered. The main fishing grounds are concentrated on scallop beds in Breiðafjörður at depths of 18 to 46 m (Eiríksson 1997). Catches from this area now account for 85–90% of the total landings. The other main fishing areas include Arnafjörður, Ísafjörður and Húnaflói in northwestern Iceland (Eiríksson 1997; Anonymous 2000).

The vessels formerly used in the Iceland fishery are traditional boats (10–20 m), which were converted over from shrimp trawlers and gill-netters (Venvik and Vahl 1985; Eiríksson 1997). The Breiðafjörður fleet is composed of multipurpose trawlers or gill-netters with a crew of 4–7 and one 1.5–2.7 m dredge, depending of size of boat (Eiríksson 1997). Most dredges are British types introduced during the 1970's and modified, especially in connection with the rapid expansion of the Breiðafjörður fishery (Eiríksson 1997). The dredges are similar in size as the original British prototypes, but it is up to three times heavier.

The Icelandic dredge was developed in the 1980's, primarily for use in the Ísafjörður and Húnaflói area (Eiríksson 1997). It is equipped to fish on both sides, and with rolling bars this dredge slides more easily over rough bottom, large stones and boulders.

Tows are made from the stern at a speed of about 2 knots and are maintained for a five to ten minute duration in the Breiðafjörður area where catch rates have generally been high. Tows of up to 30 minutes are quite regular, especially in the smaller fjord fisheries (Eiríksson 1997). The dredge is hauled in on the side. They use a warp ratio of 5–10:1 in order to increase the stability of the tow (Eiríksson 1988). The catch is machine washed and sorted before it is landed in 300–500 kg lots in aluminium tanks or bags for machine shucking and processing the next day. The scallop processing plants are located at the ports of landing.

In the plant, mechanical shuckers de-meat the shells and then clean the rim and gonad from the adductor muscle. Most scallops were destined for the United States market until 1988, after which they were increasingly exported to France, including the roe-on French market (Eiríksson 1997). The meats are size and quality graded prior to packing. The most common method of preservation of scallops to be stored is by quick-freezing. Processing for the roe-on market involves increased manual handling following mechanical shucking, though this production increases the yield from 10–12% up to 15–18%.

The scallop fishery is strictly managed by both seasonal and area quotas, and a series of resource and exploratory surveys (see Eiríksson 1997). These measures have prevented the stock from being over-exploited and resulted in a well-managed and stabilised scallop fishery.

A fishery for the Iceland scallop commenced in western Greenland in the autumn of 1983 (Pedersen 1988a; Pedersen 1988c; Pedersen 1988d). The fishery increased from 410 t in 1984 to 1,693 t in 1985, and until 1992 the annual catch was lower than 1,086 t (Table 21.1). After the implementation of quota regulation for the main fishing areas, the catches have been from 1,176 t to 2,342 t. There are seven main fishing areas in western Greenland, including: Nuuk, Sisimiut, Attu/Tugtulik, Diskofjord, Mellemfjord, Mudderbugten and Sønder Upernavik. Based on exploratory surveys, the fishery is managed by quotas in Nuuk, Attu/Tugtulik and Mudderbugten. The scallop fishery is regulated by licensing, minimum shell size and a yearly quota (Pedersen 1988a). Resource (stock assessment) surveys are carried out in order to assess the stock biomass and to make recommendations on catch levels (Pedersen, pers. comm.). In 1995 stock assessment in the Nuuk area (Figure 21.1) estimated the biomass to be 4,400 t, a reduction from an estimated 8,500 t in 1988.

Scallops are mainly harvested for the roe-on market (Pedersen 1988c). The smaller boats land the scallops whole and deliver the scallops to fish plants in Nuuk and Paamiut where they were manually shelled. This represented 45% and 6% of the catch in 1985 and 1986, respectively. The rest of the landings were caught by larger vessels where the scallops were shucked onboard (Pedersen, pers. comm.). In western Greenland scallops are harvested throughout the year but most of the fishery takes place during the months of November to May (Pedersen 1988c).

Iceland scallops were found in substantial quantities in some fjords in northern Norway already about a century ago (in Strand and Vølstad 1997). The scallop was used as bait for local fisheries (Wiborg 1963; Wiborg and Bøhle 1974), but no data are available on the size of this fishery. A small fishery for human consumption existed near Kvænangen in the early 1960's and Richardsen (in Venvik and Vahl 1985) estimated the total catch to be approximately 500 t. The fishery did not persist due to a poor market and underdeveloped fishing and processing technologies, but the scallop beds at depths of 15 to 60 m in Balsfjord, Kvænangen and Porsanger were considered to have fishing grounds of commercial potential. Wiborg (1973) and Wiborg et al. (1974) found that even larger beds occurred in the Bear Island, Spitsbergen and Jan Mayen areas in depths of 40 to 100 m. However, cost-efficient gear for harvesting scallops in offshore areas did not exist at the time.

In 1985, the large scale commercial harvest of scallops began with 1,192 t being harvested that year (Table 21.1), and it increased rapidly as ocean-going ships discovered large quantities of scallops at Jan Mayen (Strand and Vølstad 1997). The gear and techniques were borrowed from Canada, the United States, Iceland, and the Faeroes. Fishing effort was subsequently switched to scallop grounds off Bear Island and Spitsbergen. However, scallop shells from this area are encrusted with barnacles, which caused problems in the mechanised processing of the scallops. During the first year only 3–4 vessels participated in the fishery, but effort quickly expanded, peaking at 27 vessels during 1986 and 1987. Subsequent participation dropped to 13 vessels in 1988, 3–4 in 1989–1990, 2 vessels in 1991 and 1 vessel in 1996. Since 1997, Iceland scallops have not been landed in Norway. The numbers of fishermen varied from about 10–12 on the smaller vessels to 36 on the larger ones. The daily catch often exceeded 600 t for the most efficient vessels.

A total of 11 vessels, 10–14 m in length, were licensed for fishing in the coastal areas of northern Norway, while 34 vessels participated in the offshore fishery between 1985 and 1992. Ocean-going vessels, ranging in length from 29 to 69 m, were mostly modified factory trawlers, fresh fish and shrimp trawlers, purse seiners, longline vessels and supply ships from the oil industry. The larger of the specially designed ships typically operated three dredges simultaneously, and sophisticated instrumentation to optimise fishing effort (see Strand and Vølstad 1997). The boats had up to three mechanised production lines for onboard processing and freezing of scallops. Crews onboard these boats could fish for 24 hours a day.

When the fishery began, the boats used a single-side action dredge (2.5 m wide) from Iceland, that was towed at 2 knots. Afterward, catch efficiency was increased through use of a double-action dredge and higher towing speed. The dredge was a modification of US and Canadian dredge types and was towed at 4–5 knots with equal efficiency on both sides (Strand and Vølstad 1997).

During the fishery for Iceland scallops few regulations existed to protect the stock from over-exploitation. Inside the straight protection line along the coast of northern Norway, the fishing for Iceland scallops was restricted to 1 August–1 March. For 1985–1994, the total catch quota was 500–700 t round weight. Since 1995 the catch quota has been 250 t round weight, but the catch has been minimal (Table 21.1). In 1986, bounded catch reporting by logbooks was implemented. In 1987, the first coastal scallop bed, south of Tromsø, was closed. Fishing was prohibited inside a protection zone extending 4 nmi from the coastline of Spitsbergen. Outside this zone, no regulations were in effect for registered vessels participating in the Barents Sea scallop fishery in 1985. In 1986 restrictions preventing new boats from entering the fishery were implemented. In 1987 a lower size limit of 65 mm was introduced for all areas, and, because stocks were depleted, scallop beds were closed at Jan Mayen. In 1989, a limited area of these beds was reopened for fishing, while the scallop beds at Bear Island and Moffen (Spitsbergen) were closed. Subsequent catches have been from reopened areas regulated by a limited fishing period and a total catch quota. In 1993 exploratory fishing conducted at Kanin Bank (Russia) (Fig. 21.1) in collaboration with the Polar Research Institute of Marine and Oceanography in Murmansk, revealed scallop beds of commercial interest. Of the catch landed in Norway in 1996, 5,500-t whole weight were taken on the Kanin Bank.

Fevolden (1989) cautioned that a conservative approach to management of the Iceland scallop should be taken by considering each of the main areas (Jan Mayen, Spitsbergen, Bear Island and northern Norway) as if they were separate genetic units.

Norway exported most of their scallops, as a frozen product, to French and U.S. markets, but since 1987–1988 exports to France dominated.

21.2.3 Aquaculture

The cultivation of *Chlamys islandica* has been considered in Iceland, Greenland and Norway as a possible method of increasing the production of scallops in the area. Techniques for collecting Iceland scallop spat using artificial techniques have been tested successfully in Balsfjord, northern Norway (Wallace 1982) and in Breiðafjörður, Iceland (Thorarinsdóttir 1991). Using bags filled with nylon monofilament line, Wallace (1982) collected moderate numbers of *C. islandica* spat (average about 500 spat/net) in Balsfjord, Norway. Improved collection of spat occurred by placing the bags within 10 m of the bottom. Thorarinsdóttir (1991) collected 100–150 spat/net at a depth of 10 m off the bottom in Breiðafjörður, and observed heavier settlements on 0.2-mm than on 0.4-mm monofilament. Settlement, onto artificial spat collectors, commences in September in both Norway (Wallace 1982) and Iceland (Thorarinsdóttir 1991). After the first year, spat grew to an average size of 7 mm and 9 mm in Norway and Iceland, respectively, with the greatest rate of growth occurring from May to September. Larvae of *C. islandica* have been raised under laboratory conditions (Gruffydd 1976a) using the standard methods of Gruffydd and Beaumont (1972). However, spawning of adults has been difficult until the period in which natural spawning occurs. In 1986–1988 the University of Tromsø conducted research on artificial spat production of Iceland scallops in a laboratory-scale hatchery, but it was not successful.

Subsequent to spat collection is the culturing of this species by either suspension or bottom culture. Growth of *C. islandica* in "Netlon" mesh bags held in suspension in northern Norwegian waters has been shown to be depth dependant (Reinsnes 1984; Wallace and Reinsnes 1984); growth being fastest at about mid-depth. Wallace and Reinsnes (1985) attributed these depth differences to variations in seston content. The greater amount of food in the upper water column leads to a higher growth rate for scallops held in suspension compared to bottom populations (Wallace and Reinsnes 1984; Thorarinsdóttir 1994). The market size of 60–70 mm was reached in 3–4 years in northern Norway (Wallace and Reinsnes 1985) and estimated to require 4 years in Breiðafjörður (Thorarinsdóttir 1994), compared with 6–8 years for scallops growing in the wild. Thorarinsdóttir (1994) considered the 4 years required to reach market size to be too long to warrant the additional costs associated with the suspended culture. There are currently no efforts to cultivate *C. islandica* in Iceland.

The first commercial-size farm culturing scallops was established in northern Norway in 1985 (Strand and Vølstad 1997). In 1987–1989 a total of 14,000 spat collectors were set out by this farm, and as many as 20,000 spat/collector bag were obtained. This farm was intended as a model for the development of an industry based on cultivating Iceland scallops in northern Norway. Farms conducting growth trials were subsequently established along the coast. Afterward, spat production declined as methods of handling the collectors were insufficient, but some higher productivity was obtained by removing spat from the collectors after 2 years instead of 1 year. In early 1990′s the model farm was shut down due to low productivity, and farming activities along the coast declined. Spurred by the recent effort to develop shellfish farming in Norway, Iceland scallops are again being considered for cultivation.

Bottom culture is not considered to be a practical option due to the low natural growth rates, taking at least six years to reach commercial size (60 mm) (Wallace and Reinsnes 1985). Furthermore, while *C. islandica* is not a highly mobile species (Gruffydd 1976b; Gruffydd 1978), natural predators (Brun 1968; Brun 1971) also present an added problem to a concentrated bottom grow-out program. In Greenland, experimental transfers of *C. islandica* from slow growth areas to fast growth areas were performed to evaluate the feasibility to regenerate overfished scallop grounds (Engelstoft 2000). It was concluded that the possible benefit of transferring scallops did not warrant the cost associated with the operation.

21.3 PECTEN MAXIMUS

21.3.1 Biology

In Scandinavian waters *Pecten maximus* is found from Lofoten Islands in northern Norway (69°N) down to Kattegat (58°N) (Fig. 21.1) (Jagerskiold 1971; Wiborg and Bøhle 1974; Høysæter 1986). It also occurs around the Faroes (Nicolajsen 1997). On the west coast of Norway, *P. maximus* is mainly found between depths of 5 to more than 30 m.

Along the southern coast the scallops are more scattered and appears to have a deeper distribution. In the Skagerak-Kattegat area it is mainly found between depths of 25–50 m, and is not considered to be a common species (Christensen 1979). The distribution in Skagerak and Kattegat is believed to be limited by hydrographical conditions, particularly large variations in temperature and salinity resulting from an influx of cold brackish water from the Baltic sea during late winter and early spring. They are more common along the western coast of Norway, northwards to about 67°N. The largest concentrations are found at 15–30 m. In Froan (64°N) maximum densities of 5–6 scallops m^{-2} have been observed.

The reproductive cycle of *P. maximus* varies along the coast of Norway (Strand and Nylund 1991). In a population from the western coast (60°N) they observed spawning activity from July to September at a water temperature of 13–15°C, while a population from mid Norway (64°N) showed a synchronised spawn in June when water temperature was 7–9°C. Strohmeier et al. (2000) observed spawning from late March until September in the western population (60°N). In the northern population the gonad recovery was initiated in August when the temperature was at maximum, and was followed by a rapid increase in the gonadosomatic index through the autumn (Strand and Nylund 1991). In the southern population rebuilding of the gonad was found from October–December (Strand and Nylund 1991; Strohmeier et al. 2000). Strohmeier et al. (2000) studied the seasonal variations in chemical composition of the gonad and storage organs in *P. maximus* and suggests that switches in energy allocation divide the season in two parts in scallops from western Norway: (1) a period of reproductive growth from October to June and (2) a period of somatic growth and storage from June to October.

Scallops transferred from the northern population to be used as broodstock at the hatchery outside Bergen (60°N) seem to maintain the reproductive cycle (T. Magnesen, pers. comm.). The difference in the reproductive cycle along the coast and their maintenance after transfer suggests that there is a genetic difference among populations. However, comparisons among these populations, using allozyme frequencies (Igland and Nævdal 1995) and restriction fragment analysis of PCR amplified mtDNA fragments (Ridgeway and Dahle 2000), show no evidence of differentiation. Ridgeway and Dahle (2000) argue that this may be explained by the high dispersal capability of the pelagic stage combined with high rate of transport in the coastal currents. They also compared the Norwegian data with data from British populations and concluded that they apparently belong to separate gene pools.

21.3.2 Fishery

Exploitation of the great scallop in Norway has primarily been impeded by the unfavourable bottom conditions for dredging. Early dredging attempts yielded low returns, due mainly to rough bottoms and an abundance of seaweed that filled the dredges in only short tows (Wiborg and Bøhle 1974). Since the 1960's, SCUBA diving has been the common harvesting method. The harvest by leisure divers has probably been extensive in some areas along the coast (Strand and Vølstad 1997). The harvest of the great scallop is not regulated, while selling scallops is regulated through licensed dealers. In the beginning of the 1990's this statutory marketing was implemented, and figures on catch appeared (Table 21.1). Prompted by efforts on developing a shellfish industry in Norway the catches in recent years have increased up to 421 t with a value of 8.3 million NOK in 1999 (Table 21.1), when 40–50 divers were registered by the Regional fishermen's sales organisations in mid Norway. A diver may catch 150–250 kg scallops per day (3–4 scallops per kg), 3–5 days a week (T. Strohmeier, pers. comm.). The possibility of over-exploitation of the harvestable stock has been an issue also among many harvesters, particularly in relation to the possibilities of stock enhancement through cultivation. The increase of divers participating in this fishery in 2000, has led the Norwegian Labour Inspection Authority to observe the requirements on diver certification for scallop harvesters in the main harvesting areas. This has reduced the number of diver-fishermen and consequently the supply of scallops to the markets.

21.3.3 Aquaculture

Spurred by the increasing European scallop cultivation interest in the 1980's, the feasibility of commercial culture of *P. maximus* in Norway was considered. Extensive areas along the coast might provide suitable habitat, and growth studies from western Norway suggested that scallops attain commercial size (100 mm) in 4–5 years after spawning (Strand 1986). During the late 1980's, focus was on the development of hatchery technology and cultivation methods appropriate for Norwegian waters (see Strand and Vølstad 1997). Attempts have been made to collect spat in Norway, but results have so far been poor (Strand and Vølstad 1997; H. Skjæggestad, pers. comm.). The large-scale pilot commercial hatchery, which was built in Øygarden, north of Bergen, in 1987, was closed in 1991, mainly due to problems in scaling up the production. However, based on promising biological results the hatchery was reopened in 1993, and became a core activity in the Norwegian Scallop Programme, which was established in 1994 by farmers, regional authorities and research institutions. The initial challenge in this programme was to link the expertise in the research sector with the private interests and experience in shellfish farming. The participants consisted of many small enterprises, which alone were not able to make the required financial contribution to the development. In 1996 the hatchery was taken over by a private company where 40 participants in the programme became shareholders. The programme was extended and embraced all research and industry development activities on scallop farming. The main activities were to develop and distribute knowledge and expertise, develop networks, to co-ordinate the activities and in collaboration with the governmental authorities to adjust regulations allowing this development activity in the coastal zone. In 1995 the Directorate of Fisheries issued a regulation to obtain a license for intermediate culture trials at up to 10 sites granted for 15 months. This regulation was extended in 1997 to include license for seabed culture trials at three sites granted for 5 years. Attraction of larger investment to the shellfish industry development was another goal in the programme, and this was realised in 1997–1998 when several companies together with governmental funding made assets in scallop farming development. Still, a long term and extensive effort to develop this industry is required. In 2000, the parliament forwarded a law proposal regulating sea-ranching of sedentary species, whereby a statutory right will be implemented allowing the recapture of released individuals.

The developments of scallop cultivation in Norway focus on spat production in intensive hatchery and nursery, intermediate hanging culture in sea and finally seabed culture for growth to market size. Faced with the high costs of farming scallops in suspended culture, seabed culture is considered to have a greater potential for commercial development (Strand and Mortensen 1995). At present there is one hatchery in Norway. The production of postlarvae (2 mm) in the hatchery has steadily increased from 1 million in 1993 to 15 million in 1999, and since 1997 the production of 15 mm spat have stabilised at about 2 million spat (Fig. 21.2) (Magnesen 2000). Production capacity is now largely forced by demand from farmers. Development of methods and technology have significantly improved the efficiency of scallop spat production in the hatchery. In particular, during 1998 and 1999 a better efficiency was obtained in all stages up to postlarvae, and in 1999 a better strategy for broodstock conditioning provided that only half the number of broodstock was needed, compared to previous years, to produce the 2.4 billion eggs (Fig. 21.2) (Magnesen 2000). The spat originates from broodstock collected from populations in Hordaland and Trøndelag, in order to supply local stock for the main seeding areas. Scallops from these areas have a different reproductive cycle (Strand and Nylund 1991) and stock-specific conditioning regimes have been established (Magnesen et al. 1999). Broodstock conditioned at 12–13°C and fed flagellates and diatoms at a 1:1 mixture for up to sixteen weeks are spawned from December to June. The problem of low larvae survival during February–April in previous years has been resolved by improved conditioning in 2000 (T. Magnesen, pers. comm.). This has significantly increased the production efficiency of spat for optimal transfer to nurseries and further growth before the first winter. Andersen and Ringvold (2000) indicated that broodstock diets richer in the Tahitian *Isochrysis galbana* improves larval yield during spring-summer, while diets richer in diatoms may result in higher larval yield in winter. Duinker et al. (1999) exposed adult scallops to simulated natural (decreasing), constant and increasing photoperiods and results showing significantly higher gonad indexes and larger oocyte diameters in scallops given the latter two photoperiod regimes suggests that photoperiod affects early rebuilding of gonads in scallops from western Norway.

In the hatchery, larvae grow at 18°C to settlement in 3–4 weeks, and postlarvae grow to 2 mm size in 6–8 weeks (Magnesen 2000). Since 1996 there have been occasional high mortalities of larvae, suspected to have a bacterial aetiology. Since the use of the antibacterial agent chloramphenicol is not accepted in a commercial industry, considerable effort was made to find alternative agents or prophylactic methods to solve the bacterial problem. The minimum inhibitory concentrations for several antibacterial agents have been studied to develop optimal therapeutic procedures (Torkildsen et al. 2000), but large scale trials of alternative antibacterial agents have so far not been successful (L. Torkildsen and Ø. Bergh, pers. comm.). Studies on the potential use of probiotic bacterial strains show promising results (L. Torkildsen and Ø. Bergh, pers. comm.), but this concept is still prospective. The use of flow-through systems for the larvae culture is promising since bacterial numbers are low and survival of larvae is

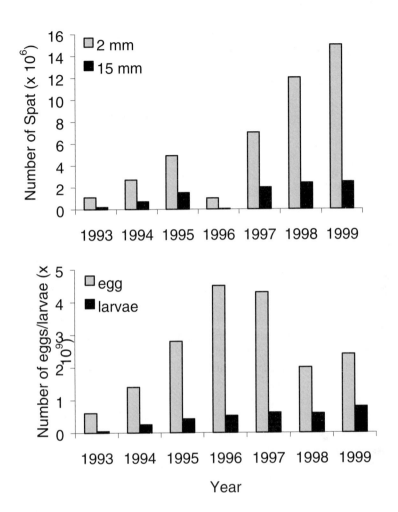

Figure 21.2. Hatchery production of *Pecten maximus*; egg, larvae, postlarvae (2 mm) and spat (15 mm) during 1993–1999 (from Magnesen 2000).

acceptable (Andersen et al. 2000; Magnesen 2000), but further studies have shown significant seasonal variation (T. Magnesen, pers. comm.). Andersen et al. (2000) also suggest that large flow-through silos (developed for rearing fish larvae) should be further developed with the aim of avoiding prophylactic use of antibacterial agents.

The first transfers of postlarvae (2 mm) from the hatchery to nursery in the sea during early spring resulted in high mortality, and hindered the extension of seasonal deployments, which is needed for scaling up the production. High survival rates (20–50%) of postlarvae (2 mm) to a saleable spat of 15 mm were obtained only by transferring spat after the seawater temperature increased to 8°C in May–June (Christophersen and Magnesen 1996). For larger spat (>4 mm) high survival rates were obtained at temperatures higher than 6–7°C. Between 1995 and 1997, a traditional Norwegian oyster poll (landlocked marine basin) (Strand 1996) was used as a water source for a landbased nursery. The temperature of the basin water can reach 10°C by April, facilitating an early transfer of postlarvae from the hatchery, and providing efficient predator control and high survival compared to the culture in sea conditions (Christophersen and Magnesen 1996). However, during the summer of 1997 scallop spat suffered high mortality. The scallops were heavily infested with tube dwelling polychaetes (*Polydora* sp.) which were considered the main cause for this mortality (Mortensen et al. 2000). In total, approximately one million spat — representing one third of Norway's intensive scallop spat production in 1997, was lost and scallop cultivation at this site was terminated. Transfer of the postlarvae to the sea in June–July has so far proved to be the most reliable method for grow-out to a saleable size. At present postlarvae are also transferred to a landbased nursery using outdoor tanks and recycling of heated water, from February and onwards (Magnesen 2000).

Scallop spat produced in hatcheries for continued on-growing normally require transportation between sites, and the associated stresses may cause mortalities. This situation has been particularly apparent in Norway since spat have originated from only one hatchery, and only a few nurseries and on-growing locations are found along much of the southwestern coastline (Brynjelsen and Strand 1996). While transport methods have been established, there is a need to assess the impact of transport on subsequent survival and growth, and to develop methods to assess the vitality of spat in advance of transfer. Christophersen (2000) investigated the effects of simulated transport (i.e., spat held in coffee filters at a temperature of 10°C) between hatchery and nursery on survival and growth of scallop spat of 2 mm size. To ensure high survival and growth rates, transportation is recommended to last no more than 12 h, which is within the travelling time commonly practised in Norway. Christophersen (2000) states that transportation up to 24 h may be recommended when spat in good condition are transferred to favourable environments, if these variables are known. Based on the need to develop simple methods to assess the vitality in advance of transfer, Minchin et al. (2000) studied the effects of air exposure on the righting and movement behaviour of scallops and related this to subsequent survival. As for scallop spat of 2 mm size (Christophersen 2000), Minchin et al. (2000) suggested that emersed transfer of juvenile scallops lasting more than 12 h should be avoided. Scallops emersed for 18–24 h showed subsequent (10 days) mortality that was 10–30%. They also showed that scallops did not show significantly different behaviour whether they were emersed upright or inverted. However, chilling (<10°C) during transport may prolong scallop vitality. It was demonstrated that at temperatures >9°C, behavioural responses like measures of movement and righting, may be a simple and effective method to assess vitality which can assist in the management of scallop culture.

Intermediate culture trials have been carried out along the western coast (59–65°N) by farmers participating in the Norwegian Scallop Programme aiming at farmer training and increased understanding of criteria for site selection along the coast. Scallops cultured at selected locations grow from 15–20 mm to 45–55 mm in one year (Strand and Mortensen 1995; Brynjelsen and Strand 1996). During summer growth is favourable in shallow depths (5–10 m) to depths below 15 m (Lund 1991; Brynjelsen and Strand 1996), while survival during winter is generally higher below 15 m (Brynjelsen and Strand 1996). During the winter of 1996, scallops suffered high mortalities at many sites. At all sites where minimum temperature was below 2°C, scallops experienced total mortality (Brynjelsen and Strand 1996). For acceptable levels of survival, temperatures above 4°C are required. The sites where scallops had total mortality were all located in the southernmost (Rogaland) and northernmost (Trøndelag) regions, and historical temperature data from the critical northern coastal areas have been used in identifying areas of high risk for scallop cultivation. Subsequent experiments showed that survival and behaviour were significantly affected by temperature differences between 3°C and 2°C (Strand, unpublished data).

Large variations in seawater parameters occur in Norwegian coastal areas. Temperature and salinity variations are caused by variations in fresh water supply and meteorological conditions, but may be exceeded by seasonal variations. The tolerance to reduced salinity in scallops is highly temperature dependent (Strand et al. 1993), and at 5°C a high mortality of scallops occurs at salinities lower than 26‰. Strand et al. (1993) recommended that scallop farms in western Norway should be located in areas where salinity seldom drops below 29 ppt. Also filtration rate and behaviour of the scallops were severely affected by lowering the temperature from 9°C to 5°C (Strand et al. 1993). Along the southwestern coast, upwelling may rapidly develop during summer after the onset of northwesterly winds (Sætre et al. 1988). Chauvaud and Strand (1999) demonstrated that such upwelling events succeeded by a small drop in temperature, from 8°C to 6°C, negatively affected the shell growth of scallops; up to a 50% decrease in daily growth rate.

Motivated by the use of landlocked marine basins for extensive marine fish larvae culture in Norway, Andersen and Naas (1992) compared growth of juvenile scallops in a basin loaded with fertilisers with growth at two locations at sea: one being potentially loaded with organic materials from a fish farm, and the other an unloaded site. The results indicated that fertilised basins are not useful locations for farming scallops unless the temperature, and phytoplankton density and composition are carefully controlled. Reitan (2000) showed that scallop spat held in a landlocked marine basin enhanced by artificial fertiliser attained a bigger size than scallops held outside the basin at an unfertilised site, but growth of scallops in this experiment was inferior to what could be expected in selected locations for scallop farming along the coast (Lund 1991; Strand and Mortensen 1995; Brynjelsen and Strand 1996).

Releases of scallops on to the seabed for cultivation during the mid 1990′s revealed heavy predation from the brown crab *Cancer pagurus*, and this has been a major impediment to the development of scallop seabed cultivation in Norway. Seeding trials during autumn showed extensive attraction of crabs to the sites, and high mortality was shown for scallops up to about 50–60 mm shell height (Strand et al. 1998). This immense and unacceptable level of crab predation on some of the first seeded scallops prompted an interest from the industry to develop fences in order to prevent the crab from accessing the scallops (Strand et al. 1999). Plates of steel (0.6 mm) coated by PVC, with a smooth surface, erected 30 cm height above the sediment and dug 20 cm into the sediment are now used by farmers to prevent the brown crab access to fenced scallops. Further technical improvements and biological assessments are being made to consider the economical viability of fenced scallop cultivation on the seabed. Other strategies now being considered are scaling up the number of scallops released to the seabed, seeding at low densities and active predator control.

Predation experiments have shown that the crab *C. pagurus* prefer cultivated scallops to wild scallops (Haugum et al. 1999). This is probably due to weaker shells in cultivated scallops. Shell strength increase with scallop size, but wild scallops were twice as strong as cultivated scallops. A better knowledge on the variation in shell strength of scallops and the determinant factors affecting the mechanical structure of scallops shells are assumed to be of major importance for the development of scallop seabed cultivation in Norway.

Co-location or duo-culture of Atlantic salmon and scallops is being considered in Norway in order to benefit from higher efficiency on labour, production facilities, management and even biological production (i.e., sustainable integrated aquaculture). Bivalve harvest through cultivation is also argued to balance the nutrient input to the marine environment from fish farming (Reitan 2000), and concepts are launched of using energy from the petroleum industry, as heated water and bioprotein, in an integrated system for marine aquaculture including bivalves (Åsgård et al. 1999; Jacobsen 1999). However, at present the distance between aquaculture sites is regulated to be at least one kilometre, which is a preventive measure to avoid disease transmission between bivalves and fish (e.g., Atlantic salmon). Laboratory tank cohabitation experiments of Atlantic salmon (*Salmo salar*) with scallops *P. maximus* challenged with the bacterium *Aeromonas salmonicida*, causing the disease furunculosis in salmonid fish species, showed that furunculosis was transmitted to susceptible salmon via the scallops (Bjørshol et al. 1999). Studies of the fate of the fish pathogenic infectious pancreatic necrosis virus (IPNV) in scallops (Mortensen et al. 1992; Mortensen 1993a) showed that bivalves probably should be considered potential vectors of fish pathogenic viruses. These studies have been guidelines in the management of production and movements of cultured bivalves and fish in coastal areas of Norway (Mortensen 2000). Nordtug et al., (1999) showed in a field study of scallop cohabitation with salmon that scallop growth and survival was not affected, and laboratory tests with adequate medical and chemical substances used in fish farming indicated only an acute negative effect in the very close vicinity of the salmon farm.

A restricted health survey on selected stocks of *P. maximus* has been performed in Norway since 1989 (Mortensen 1993b). To date, no serious infectious diseases or parasites have been detected (Mortensen, pers. comm.). There is a high priority on bivalve disease research and management in order to protect the industry from disease introductions (Mortensen 2000).

21.4 *AEQUIPECTEN OPERCULARIS*

The queen scallop *A. opercularis* has almost the same distribution as *P. maximus*. It is found from the Lofoten Islands (69°N) in northern Norway, the Faroes and south to the Danish Sound (56°N) (Figure 21.1) (Ursin 1956; Broom 1976; Høysæter 1986). In the Faroes it is found on sandy, rocky or soft bottom from 50 to more than100 m depth, being most abundant around 100 m (Ursin 1956, Nicolajsen 1997). In Skagerak-Kattegat queen scallops are generally found on fine sand and gravel from 20 m down to 100 m. *A. opercularis* prefer flowing water, salinities higher than 25‰ and temperatures in the range of 3–16°C. These requirements are thought to determine their distribution pattern in this area (Ursin 1956).

On the west coast of Denmark mudflats dominate the coastline. Queen scallops are rather scarce in this area since it appears that *A. opercularis* avoids soft bottoms. High numbers of *A. opercularis* are found between Doggers Bank and the coast of northern England and Scotland (Broom 1976).

In Scandinavian waters there is probably one spawning period per year, and it lasts from June (Faroes) to October (Skagerak-Kattegat) (Ursin 1956). Seasonal growth rates are not well documented from this area. Ursin (1956) found that *A. opercularis* from Danish and Faroese waters grows rapidly early in life, 40 mm in 2 years, after which growth slows reaching about 60 mm in 4 years, and ceases finally after about 6 years at a length of 70 mm. The density and distribution varies from time to time and unfortunately there exists no actual estimates of densities. The most dense populations are found in the southern Kattegat, in the Sound and around the Faroes.

In the past, *A. opercularis* have been locally fished in small amounts (27 t, 1954) in the Danish sound and in Faroese waters (Ursin 1956; Broom 1976). Since 1970 a small but stable commercial fishery for queen scallop in the Faroes has developed with an average catch of about 500 t, 2,000 t and 3,000–4,000 t per year during the 1970's, 1980's and 1990's, respectively (Table 21.1). In recent years the catches have increased to more than 5,000 t per year. Until 1988 the fishing fleet comprised from 2 to 9 ships (Nicolajsen 1997). After 1988, only one boat has been dredging on the main traditional beds 1–15 nmi. from the eastern coast. The catch season is from August to February, and the fishery uses two 12-foot modified Scottish type dredges. The fishery is managed by licence and bounded catch-reporting by logbooks. A large factory trawler, originally built for a Faroes company for the Iceland scallop fishery in the Barents Sea in 1987–1989, was allowed to fish queen scallop beds north from the Faroes. This fishery was closed in 1991 (Nicolajsen 1997). Before 1990, almost all the production went to US markets. At present the main market is France. Exploratory fishing for queen scallops in southwestern Norway in 1998–1999, using dredges from the fishery in Shetland islands, has been unsuccessful (E. Aarseth, pers. comm.).

At present, there are no attempts at trying to cultivate *A. opercularis*. Some trials have been made to collect spat and grow them in suspended culture in Norway with some success, but no real effort has been made. Growth studies in pearl nets suspended at 5-m depth-intervals from 5 to 25 m at locations along a gradient from protected fjords to exposed coastal areas were performed during May to September (Lund 1990).

Temperature rather than food quantity, measured as chlorophyll, POM and POM/PIM ratio, was the most important factor explaining variation in growth of soft parts and shell. Production of *A. opercularis* in suspended nets was higher in shallow depths than deeper, while there were no differences between the locations. In Denmark and Sweden the hydrographical conditions, influenced by brackish water, restrict the distribution of *A. opercularis*. *A. opercularis* would be more suitable for aquaculture development in the Norway and the Faroes islands.

ACKNOWLEDGMENTS

We would like to thank J. Bates, Ø. Bergh, R. A. Chandler, J. Cleghorn, M. J. Dadswell, J. J. Engelstoft, S. E. Fevolden, T. Magnesen, S. Mortensen, K. S. Naidu, A. Nicolajsen, S. A. Pedersen, E. M. Rödström, G. G. Thorarinsdóttir, T. R. Waller, D. Warren, and E. Aarseth for their assistance in the preparation of this chapter.

REFERENCES

Abbott, R.T., 1974. American seashells. New York. Van Nostrand, 541 p. (second rev. ed.).

Andersen, S., Burnell, G. and Bergh, Ø., 2000. Flow-through systems for culturing great scallop larvae. Aquaculture International 8(2/3):249–257.

Andersen, S. and Naas, K.E., 1992. Shell growth and survival of scallop (*Pecten maximus* L.) in a fertilized, shallow seawater pond. Aquaculture 110:71–86.

Andersen, S. and Ringvold, H., 2000. Seasonal differences in effect of broodstock diet on spawning success in the great scallop. Aquaculture International 8(2/3):259–265.

Anonymous, 2000. Nytjastofnar sjávar 1999/2000. Aflahorfur fiskveiðiárið 2000/2001. [State of Marine Stocks in Icelandic Waters 1999/2000. Prospects for the Quota year 2000/2001.] Marine Research Institute. Reykjavík 2000. 176 p.

Åsgård, T., Austreng, E., Holmefjord, I. and Hillestad, M., 1999. Resource efficiency in the production of various species. In: N. Svenning, H. Reinertsen and M. New (Eds.). Sustainable Aquaculture. Food for the future? Proceedings of the Second International Symposium on Sustainable Aquaculture, Oslo, 2–5 November 1997. Balkema, Rotterdam. pp. 171–183.

Bernard, F.R., 1983. Catalogue of the living Bivalvia of the eastern Pacific Ocean: Bering Strait to Cape Horn. Can. Spec. Publ. Fish. Aquat. Sci. 61: 102 p.

Bjørshol, B., Nordmo, R., Falk, K. and Mortensen, S., 1999. Cohabitation of Atlantic salmon (*Salmo salar*) and scallop (*Pecten maximus*) - challenge with Infectious Salmon Anemia (ISA) Virus and *Aeromonas salmonicida* subsp. *salmonicida*. Twelfth International Pectinid Workshop, May 5–11, 1999, Bergen, Norway (abstract).

Broom, M.J., 1976. Synopsis of biological data on scallops (*Chlamys (Aequipecten) opercularis* (Linnaeus), *Argopecten irradians* (Lamark), *Argopecten gibbus* (Linnaeus)). FAO Fish. Synop. 114: 44 p.

Brun, E., 1968. Extreme population density of the starfish *Asterias rubens* L. on a bed of Iceland scallop, *Chlamys islandica* (O.F. Muller). Astarte No.32:1–3.

Brun, E., 1971. Predation of *Chlamys islandica* (O. F. Muller) by eiders *Somateria* spp. Astarte 4 (1):23–29.

Brynjelsen, E. and Strand Ø., 1996. Prøvedyrking av stort kamskjell i mellomkultur - 1995–1996. [Growth trials with great scallops in intermediate culture – 1995–96] Fisken og Havet, nr 18. 28 p.

Chauvaud, L. and Strand Ø., 1999. *Pecten maximus*: Memory of upwelling events in Norway. Twelfth International Pectinid Workshop, May 5–11, 1999, Bergen, Norway (abstract).

Christensen, J.M., 1979. Musslor i havet [Bivalves in the Sea]. Wahlstrom and Widstrand, Stockholm. 119 p.

Christophersen, G., 2000. Effects of air emersion on survival and growth of hatchery reared great scallop spat. Aquaculture International 8(2/3):159–168.

Christophersen, G. and Magnesen, T., 1996. Yngelproduksjon av stort kamskjell 1995. [Scallop spat production 1995] SMR-rapport, 22/96. University of Bergen. 21 p.

Duinker, A., Saut, C. and Paulet, Y.M., 1999. Effect of photoperiod on conditioning of the great scallop. Aquaculture International 7:449–457.

Eiríksson, H., 1988. Horpudiskurinn, *Chlamys islandica*, Muller. Hafrannsoknir, 35. hefti, bls. 5–40.

Eiríksson, H., 1997. The Molluscan fisheries of Iceland. In: C.L. MacKenzie, Jr., V. Burrell, A. Rosenfield and W.L. Hobart (Eds.). The History, Present Condition, and Future of the Molluscan fisheries of North America and Europe. U.S. Dep. Commerce, NOAA Tech. Rep. NMFS 129:39–47.

Ekman, S., 1953. Zoogeography of the Sea. Sigwick and Jackson, London. 417 p.

Engelstoft, J.J., 2000. Omplantning av kammuslinger, *Chlamys islandica*, ved Nuuk. Teknisk rapport nr. 30. Pinngortitaleriffik, Grønlands Naturinstitut. 18 p.

FAO. 2000. FAO web site http://apps.fao.org/fishery/fprod1-e.htm.

Fevolden, S.E., 1989. Genetic differentiation of the Iceland scallop *Chlamys islandica* (Pectinidae) in the northern Atlantic Ocean. Mar. Ecol. Prog. Ser. 51:77–85.

Fevolden, S.E., 1992. Allozymic variability in the Iceland scallop *Chlamys islandica* – geographic variation and lack of growth-heterozygosity correlations. Mar. Ecol. Prog. Ser. 85:259–268.

Galand, P.E. and Fevolden, S-E., 2000. Population structure of *Chlamys islandica* in the Northeast Atlantic - northern stocks compared with a southern relict population. Sarsia 85:183–188.

Greve, L. and Samuelsen, T.J., 1970. A population of *Chlamys islandica* (O.F. Muller) found in Western Norway. Sarsia 45:17–24.

Gruffydd, Ll.D., 1976a. The development of the larvae of *Chlamys islandica* in the plankton and its salinity tolerance in the laboratory (Lamellibranchia, Pectinidae). Astarte 8:61–67.

Gruffydd, Ll.D., 1976b. Swimming in *Chlamys islandica* in relation to current speed and an investigation of hydrodynamic lift in this and other scallops. Norw. J. Zool. 24:365–378.

Gruffydd, Ll.D., 1978. The byssus and byssus glands in *Chlamys islandica* and other scallops (Lamellibranchia). Zool. Scr. 7 (4):277–285.

Gruffydd, Ll.D. and Beaumont, A.R., 1972. A method for rearing *Pecten maximus* larvae in the laboratory. Mar. Biol. 15:350–355.

Haugum, G.A., Strand, Ø. and Minchin, D., 1999. Are cultivated scallops wimps? Twelfth International Pectinid Workshop, May 5–11, 1999, Bergen, Norway (abstract).

Høysæter, T., 1986. An annotated check-list of marine molluscs of the Norwegian coast and adjacent waters. Sarsia 71:73–145.

Igland, O.T. and Nævdal, G., 1995. Analysis of genetic differentiation between samples of the scallop *Pecten maximus* (Linnaeus, 1758) from two areas in Norway; Hordaland and Trøndelag. University of Bergen, Senter for Miljø og Ressursstudier, SMR-rapport 18/95. 11 p.

Jagerskiold, L.A., 1971. A survey of the marine benthonic macro-fauna along the Swedish west coast 1921–1938. Act. Reg. Soc. Sci. Litt. Goth. Zoologica 6: 146 p.

Johannessen, O.H., 1973. Age determinations in *Chlamys islandica* (O.F. Muller). Astarte 6:15–20.

Kjolholt, J. and Hansen, M.M., 1986. PCBs in scallops and sediments from North Greenland. Mar. Pollut. Bull. 17 (9):432–434.

Lubinsky, I., 1980. Marine bivalve molluscs of the Canadian central and eastern Arctic: Faunal composition and zoogeography. Can. Bull. Fish. Aquat. Sci. 207:1–111.

Lund, B., 1990. Vekst av *Chlamys opercularis* og *Pecten maximus* i hengende kultur.[Growth of *Chlamys opercularis* and *Pecten maximus* in hanging culture.] Cand. scient. thesis, University of Bergen. 86 p.

Magnesen, T., 2000. Yngelproduksjon av stort kamskjell. [Great scallop spat production] Norsk Fiskeoppdrett 2:24–26.

Magnesen, T., Brynjelsen, E. and Knutsen, I., 1999. Yngelproduksjon av stort kamskjell 1998. [Scallop spat production 1998] SMR-rapport, 29/99. University of Bergen. 33 p.

Minchin, D., Haugum, G.A., Skjæggestad, H. and Strand, Ø., 2000. Effect of air exposure on scallop behaviour, and the implications for subsequent survival in culture. Aquaculture International 8(2/3):169–182.

Mortensen, S.H., 1993a. Passage of infectious pancreatic necrosis virus (IPNV) through invertebrates in an aquatic food chain. Dis. Aquat. Org. 16:41–45.

Mortensen, S.H., 1993b. A health survey of selected stocks of commercially exploited Norwegian bivalve molluscs. Dis. Aquat. Org. 16:149–156.

Mortensen, S., 2000. Scallop introductions and transfers, from an animal health point of view. Aquaculture International 8(2/3):267–271.

Mortensen, S.H., Bachere, E., LeGall, G. and Mialhe, E., 1992. Persistence of infectious pancreatic necrosis virus (IPNV) in scallops (*Pecten maximus*). Dis. Aquat. Org. 12:221–227.

Mortensen, S., Van Der Meeren, T., Fosshagen, A., Hernar, I., Harkestad, L., Torkildsen, L. and Bergh, Ø., 2000. Mortality of scallop spat in cultivation, infested with tube dwelling bristle worms, *Polydora* sp. Aquaculture International 8(2/3):267–271.

Nicolajsen, Á., 1997. The history of the Queen scallop fishery of the Faroe Islands. In: C.L. MacKenzie, Jr., V. Burrell, A. Rosenfield and W.L. Hobart (Eds.). The History, Present Condition, and Future of the Molluscan fisheries of North America and Europe. U.S. Dep. Commerce, NOAA Tech. Rep. NMFS 129:49–56.

Nordtug, T., Lundheim, R. and Myhren, H., 1999. Cohabitation of Atlantic salmon (*Salmo salar*) and scallop (*Pecten maximus*) - field studies and uptake of medicines and copper. Twelfth International Pectinid Workshop, May 5–11, 1999, Bergen, Norway (abstract).

Ockelmann, W.K., 1958. The zoology of east Greenland. 4. Marine Lamellibranchiata. Medd. Grønl. Kom. Ledel. Geol. Geogr. Unders. Grønl. 122(4):1–256.

Pedersen, S.A., 1988a. Inshore scallop resources, *Chlamys islandica*, in the Nuuk area West Greenland. ICES C. M. 1988/k:17.

Pedersen, S.A., 1988b. Kammuslinger, *Chlamys islandica*, ved vestgrønland. [Icelandic scallops, *Chlamys islandica*, at West Greenland] Rapport til Grønlands Hjemmestyr fra Grønlands Fiskeriundersøgelser-December 1988.

Pedersen, S.A., 1988c. The Icelandic scallop off western Greenland. Can. Transl. Fish. Aquat. Sci. No. 5379: 34 p.

Pedersen, S.A., 1988d. The Icelandic scallop off western Greenland: trial fishery for the Icelandic scallop at the western Greenland banks-1986. Can. Transl. Fish. Aquat. Sci. No. 5382: 9 p.

Reinsnes, T.G., 1984. Miljøets betydning for vekst hos haneskjell. [Environmental impact on growth of Iceland scallop] Norsk Fiskeoppdrett 9(4):33–36.

Reitan, K.I., 2000. Resultater fra en gjødslet fjord. Kan økt primærproduksjon i fjord gi økt vekst av skjell? [Results from a fertilised fjord.] Norsk Fiskeoppdrett 12:30–31.

Ridgeway, G.M.I. and Dahle, G., 2000. Population genetics of *Pecten maximus* of the Northeast Atlantic coast. Sarsia 85:167–172.

Rubach, S. and Sundet, J.H., 1988. Resource inventory of Iceland scallops (*Chlamys islandica* (O.F. Muller)) in the Jan Mayen and Spitsbergen areas during 1986. Can. Transl. Fish. Aquat. Sci. 5352: 65 p.

Serchuk, F.M. and Wigley, S.E., 1984. Results of the 1984 sea scallop research vessel survey: status of sea scallop resources in the Georges Bank, Mid Atlantic, and Gulf of Maine Regions and abundance and distributions of Iceland scallops off the southeastern coast of Cape Cod. Nat. Mar. Fish. Ser., Northeast Ref. Doc. No. 83–84. 74 p.

Skreslet, S., 1973. Spawning in *Chlamys islandica* (O.F. Muller) in relation to temperature variations caused by vernal meltwater discharge. Astarte 6(1):9–14.

Skreslet, S. and Brun E., 1969. On the reproduction of *Chlamys islandica* (O.F. Muller) and its relation to depth and temperature. Astarte 2(1–2):1–6.

Strand, Ø., 1986. Vekst, aldersfordeling og gytetidspunkt hos *Pecten maximus* L. i Sørværet, Ytre Sunnfjord (Growth, age and spawning time of *Pecten maximus* L. in the Ytre Sunnfjord, western Norway). Cand. scient. thesis, University of Bergen. 81 p.

Strand, Ø., 1996. Enhancement of the bivalve production capacity in a landlocked heliothermic marine basin. Aquaculture Research 27:355–373.

Strand, Ø., Haugum, G.A., Hansen, E. and Monkan, A., 1999. Fencing scallops on the seabed to prevent intrusion of the brown crab *Cancer pagurus*. Twelfth International Pectinid Workshop, May 5–11, 1999, Bergen, Norway (abstract).

Strand, Ø., Haugum, G.A. and Skjæggestad, H., 1998. Strategi for utsetting av Stort kamskjell bunnkultur - eller må de kastes på sjøen? [A strategic seeding of the Great scallops to the seabed – or should they just be thrown over board]. Norsk Fiskeoppdrett nr. 17, s. 40–41.

Strand, Ø. and Mortensen, S., 1995. Stort kamskjell. Biologi og dyrking. (Great scallop. Biology and culture.) Kystnæringen Forlag, ISBN 82–7595–013–9. 84 p.

Strand, Ø. and Nylund, A., 1991. The reproductive cycle of the scallop *Pecten maximus* (L.) from two populations in Western-Norway, 60N and 64N. In: S.E. Shumway and P. A. Sandifer (Eds.). An International Compendium of Scallop Biology and Culture. World Aquaculture Workshop No. 1. World Aquaculture Society. pp. 95–105.

Strand, Ø., Solberg, P.T., Andersen, K.K. and Magnesen, T., 1993. Salinity tolerance of juvenile scallops (*Pecten maximus*) at low temperature. Aquaculture 115:169–179.

1090

Strand, Ø. and Vøllstad, J.H., 1997. The Molluscan fisheries and culture of Norway. In: C.L. MacKenzie, Jr., V. Burrell, A. Rosenfield and W.L. Hobart (Eds.). The History, Present Condition, and Future of the Molluscan fisheries of North America and Europe. U.S. Dep. Commerce, NOAA Tech. Rep. NMFS 129:7–24.

Strohmeier, T., Duinker, A. and Lie, Ø.,2000. Seasonal variations in chemical composition of the female gonad and storage organs in *Pecten maximus* (L.) – suggesting that somatic and reproductive growth are separated in time. J. Shellfish Res. 19(2):741–747.

Sundet, J.H. and Lee, J.B., 1984. Seasonal variations in gamete development in the Iceland scallop, *Chlamys islandica*. J. Mar. Biol. Assoc. U.K. 64:411–416.

Sundet, J.H. and Vahl, O., 1981. Seasonal changes in the dry weight and biochemical composition of the tissues of sexually mature and immature Iceland scallops, *Chlamys islandica*. J. Mar. Biol. Assoc. U.K. 61:1001–1010.

Sætre, R., Aure, J. and Ljøen, R., 1988. Wind effects on the lateral extension of the Norwegian coastal water. Cont. Shelf Res. 8(3):239–253.

Thorarinsdóttir, G.G., 1991. The Iceland scallop, *Chlamys islandica* (O.F. Müller) in Breiðafjörður, west Iceland. I. Spat collection and growth during the first year. Aquaculture 97:13–23.

Thorarinsdóttir, G.G., 1993. The Iceland scallop, *Chlamys islandica* (O.F. Müller), in Breiðafjörður, West Iceland II. Gamete development and spawning. Aquaculture 110:87–96.

Thorarinsdóttir, G.G., 1994. The Iceland scallop, *Chlamys islandica* (O.F Müller), in Breiðafjörður, west Iceland. III. Growth in suspended culture. Aquaculture 120:295–303.

Torkildsen, L., Samuelsen, O.B., Lunestad, B.T. and Bergh, Ø., 2000. Minimum inhibitory concentrations of chloramphenicol, florfenicol, trimethoprim/sulfadiazine and flumequine in seawater of bacteria associated with scallops (*Pecten maximus*) larvae. Aquaculture 185:1–12.

Ursin, E., 1956. Distribution and growth of the Queen, *Chlamys opercularis* (Lamellibranchiata), in Danish and Faroese waters. Medd. Danm. Fisk. Havsunders 13: 32 p.

Vahl, O., 1973. Efficiency of particle retention in *Chlamys islandica* (O.F. Muller). Astarte 6:21–25.

Vahl, O., 1978. Seasonal changes in oxygen consumption of the Iceland scallop (*Chlamys islandica* (O.F. Muller)) from 70°N. Ophelia 17:143–154.

Vahl, O., 1980a. Seasonal variations in seston and in the growth rate of the Iceland scallop, *Chlamys islandica* (O.F. Muller) from Balsfjord, 70°N. J. Exp. Mar. Biol. Ecol. 48:195–204.

Vahl, O., 1980b. Volume of water pumped and particulate matter deposited by the Iceland scallop (*Chlamys islandica*) in Balsfjord, northern Norway. In: H.J. Freeland, D.M. Farmer and C.D. Levings (Eds.). Fjord Oceanography. NATO Conf. Ser. 4 Mar. Sci. pp. 639–644.

Vahl, O., 1981a. Age specific residual reproductive value and reproductive effort in the Iceland scallop, *Chlamys islandica* (O.F. Muller). Oecologia 51:53–56.

Vahl, O., 1981b. Energy transformations by the Iceland scallop, *Chlamys islandica* (O.F. Muller), from 70°N. I. The age-specific energy budget and net growth efficiency. J. Exp. Mar. Biol. Ecol. 53:281–296.

Vahl, O., 1981c. Energy transformations by the Iceland scallop, *Chlamys islandica* (O.F. Muller), from 70°N. II. The population energy budget. J. Exp. Mar. Biol. Ecol. 53:297–303.

Vahl, O., 1982. Long term variation in recruitment of the Iceland scallop, *Chlamys islandica* from northern Norway. Neth. J. Sea Res. 16:80–87.

Vahl, O., 1985. Size specific reproductive effort in (*Chlamys islandica*): Reproductive senility or stabilizing selection? In: P.E. Gibbs (Ed.). Proc. 19th European Mar. Biol. Symp., Plymouth, Devon, U.K., 16–21 September 1984. Cambridge University Press, N.Y. pp. 521–527.

Vahl, O. and Clausen, B., 1980. Frequency of swimming and energy cost of byssus production in *Chlamys islandica* (O.F. Muller). J. Cons. int. Explor. Mer. 39(1):101–103.

Venvik, T. and Vahl, O., 1985. Opportunities and limitations for harvesting and processing Iceland scallops. Can. Transl. Fish. Aquat. Sci. 5191: 76 p.

Wallace, J.C., 1982. The culture of the Iceland scallop, *Chlamys islandica* (O.F. Muller). I. Spat collection and growth during the first year. Aquaculture 26:311–320.

Wallace, J.C. and Reinsnes, T.G., 1984. Growth variation with age and water depth in the Iceland scallop (*Chlamys islandica*, Pectinidae). Aquaculture 41:141–146.

Wallace, J.C. and Reinsnes, T.G., 1985. The significance of various environmental parameters for growth of the Iceland scallop, *Chlamys islandica* (Pectinidae), in hanging culture. Aquaculture 44:229–242.

Waller, T.R., 1991. Evolutionary relationships among commercial scallops (Mollusca; Bivalvia; Pectinidae). In: S.E. Shumway (Ed.). Scallops: Biology, Ecology, and Aquaculture. Developments in Aquaculture and Fisheries Sciences Vol. 21. Elsevier, Amsterdam. pp. 1–73.

Wiborg, K.F., 1962. Haneskjellet, *Chlamys islandica* (O.F. Muller) og dets utbredelse i noen nordnorske fjorder. [Iceland scallop, *Chlamys islandica* (O.F. Muller), in northern Norwegian fjords.] Fisket Gang 47 (22):640–646.

Wiborg, K.F., 1963. Some observations on the Iceland scallop *Chlamys islandica* (O.F. Muller) in Norwegian waters. Fiskeridir. Skr. Ser. Havunders. 13(6):38–53.

Wiborg, K.F., 1973. Distribution of *Chlamys islandica* on Bear Island Banks. Transl. Fish. Res. Board Can., No. 2314: 7 p.

Wiborg, K.F. and Bøhle, B., 1974. Occurrences of edible shellfish (bivalves) in Norwegian coastal waters (with a selection of marine gastropods). Transl. Ser. Fish. Mar. Serv. Can. No. 2978: 34 p.

Wiborg, K.F., Hansen, K. and Olsen, H.E., 1974. Iceland scallop (*Chlamys islandica*) (O.F. Muller) at Spitsbergen and Bear Island - Investigation in 1973. Transl. Ser. Environ. Can. Fish. Mar. Serv. No. 3131: 18 p.

Zenkevitch, L., 1963. Biology of the Seas of the U.S.S.R. George Allen and Unwin Ltd., London. 955 p.

AUTHORS ADDRESSES

Øivind Strand - Department of Aquaculture, Institute of Marine Research, P. O. Box 1870 Nordnes, 5817 Bergen, Norway (E-mail: oivind.strand@imr.no)

G. Jay Parsons - Marine Institute, Memorial University, P. O. Box 4920, St. John's, Newfoundland, Canada A1C 5R3 [Current Address: Fisheries and Oceans Canada, Aquaculture Science Branch, 200 Kent St., Ottawa, Ontario Canada K1A 0E6] (E-mail: ParsonsJa@dfo-mpo.gc.ca)

Chapter 22

Japan

Yoshinobu Kosaka and Hiroshi Ito

22.1 INTRODUCTION

The scallop industry supports Japan's coastal fisheries and cultures with an annual increase in economic activity in shellfish producing communities (Fig. 22.1, Table 22.1). In 2000, the gross fisheries production of Japan totalled 6,384,100 metric tons (mt) and 1,875,290 million yen; these statistics include coastal fisheries of 1,576,140 mt (576,449 million yen) and marine cultures of 1,230,783 mt (527,230 million yen). Molluscan shellfishies were 838,450 mt (127,214 million yen), which was 13.1% (6.8%) of gross fisheries production, and various shellfish, such as scallop, oyster, short neck clam, surf clam, abalone and ark shell were utilised. Japanese scallop landings reached 514,989 mt (84,146 million yen), which was 61.4% (66.1%) of the molluscan shellfisheries production. The scallop was found to be the most valuable, with the oyster coming second; the oyster landing being 37,950 million yen in value and 221,252 mt in weight. The pearl oyster produces a gem, and its culture has become a big industry; however, it is separated from the shellfish fisheries statistics.

'Yoshoku', possessive culture in the demarcated sea, is held by the demarcated fishery, its produce contributing to the stabilisation of the supply of good quality seafood. 'Yoshoku' areas are separated by caging or fencing from the common sea or culture facilities that may be set up in these areas. Scallop hanging culture, belonging to this culture category, is a high activity in northern Japan. Concerning the other culture system, 'zoshoku' or 'saibai', sea farming or marine ranching is a project in the common sea by the right of the common fishery. 'Zoshoku' areas are designed by a consensus in the common sea. 'Zoshoku' has been conducted recently at every stage from feasibility studies to economic enterprises on the marine species of fish and shellfish. Of the seeds released for sea farming in Japan, the largest numbers are scallops (Table 22.2). The scallop industry is said to be the honour student of sea farming in Japan. The high industrial activity of the Japanese scallop currently leads all other developing mariculture efforts.

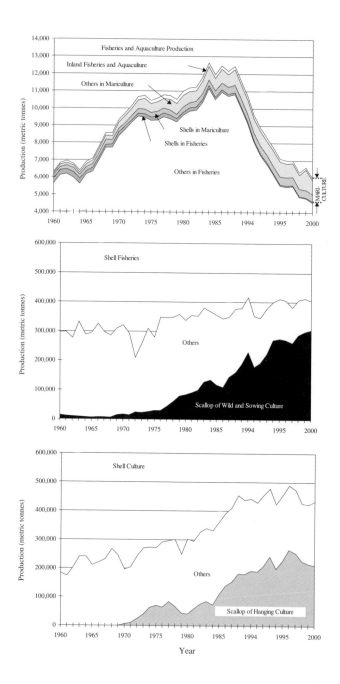

Figure 22.1. Changes in annual production of Japan's fisheries and aquaculture 1960–2000. The middle and lower graph show shell production. Source: Ministry of Agriculture, Forestry and Fisheries, Government of Japan, 'Annual report on statistics of fishery and aquaculture production'.

Table 22.1

Annual production of Japan's fisheries and Japanese scallop, *Patinopecten (Mizuhopecten) yessoensis* Jay, 1910–2000.

Year	Marine, ×10³ metric tons			Inland, ×10³ metric tons				Marine, ×10³ metric tons			Scallop, metric tons			
	Catch	Harvest	Sub-total	Catch	Harvest	Sub-total	Total	Distant	Offshore	Coastal	Hokkaido	Aomori	Other	Sub-total
1910											12,163	11,907	0	24,070
1911											9,844	2,343	0	12,187
1912	1,622	23	1,645		3		1,649				13,773	965	0	14,738
1913	1,943	26	1,969		3		1,974				12,545	2,146	0	14,691
1914	1,947	25	1,972		3		1,977				22,103	1,746	0	23,849
1915	1,999	28	2,027		4		2,032				45,554	261	0	45,815
1916	2,171	29	2,200		5		2,206				8,343	412	0	8,755
1917	1,932	31	1,963		5		1,969				3,683	347	0	4,030
1918	1,806	31	1,837		4		1,842				5,246	1,038	0	6,284
1919	2,205	38	2,243		4		2,249				13,520	361	0	13,881
1920	2,433	42	2,475		5		2,482				14,124	68	0	14,192
1921	2,128	41	2,169		7		2,177				12,000	36	0	12,036
1922	2,403	39	2,442		6		2,540				15,609	88	0	15,697
1923	2,435	30	2,465		8		2,475				11,694	72	0	11,766
1924	2,579	37	2,616		8		2,626				26,444	23	0	26,467
1925	2,794	39	2,833		8		2,843				36,525	57	0	36,582
1926	3,020	40	3,060		9		3,071				28,122	8,575	0	36,697
1927	3,191	44	3,235		10		3,248				28,071	13,128	0	41,199
1928	3,038	45	3,083		11		3,096				26,564	28,308	0	54,872
1929	3,067	49	3,116		12		3,129				25,544	19,504	0	45,048
1930	3,135	37	3,172		13		3,186				36,219	1,444	0	37,663
1931	3,308	62	3,360		14		3,376				31,270	1,386	0	32,656
1932	3,491	49	3,540		15		3,556				20,573	1,013	0	21,586
1933	3,996	49	4,045		17		4,064				40,392	155	0	40,547
1934	4,177	74	4,251		19		4,272				78,674	134	0	78,808
1935	3,863	93	3,956		19		3,977				68,012	134	0	68,146
1936	4,217	90	4,307		20		4,330				45,195	227	0	45,422

1937	3,927	91	4,018		21		4,041				42,840	223	0	43,063
1938	3,580	76	3,656		19		3,677				24,738	392	0	25,130
1939	3,583	76	3,659		20		3,681				23,055	382	0	23,437
1940	3,427	78	3,505		19		3,526				46,573	148	0	46,721
1941	3,702	109	3,811		20		3,833				17,020	410	0	17,430
1942	3,480	94	3,574		26		3,601				62,775	556	0	63,331
1943	3,236	87	3,323		31		3,356				43,512	191	0	43,703
1944	2,375	61	2,436		21		2,458				16,189	127	0	16,316
1945	1,750	61	1,811		12		1,824				1,008	40	0	1,048
1946	2,075	24	2,099		6		2,107				9,888	133	0	10,021
1947	2,257	23	2,280		4		2,285				9,187	166	0	9,353
1948	2,477	36	2,513		4		2,518				7,376	121	0	7,497
1949	2,666	53	2,719	37	3	40	2,761				14,071	131	0	14,202
1950	3,255	48	3,303	63	5	68	3,373				9,978	9,296	0	19,274
1951	3,774	88	3,862	60	6	66	3,390				5,502	4,000	0	9,502
1952	4,646	113	4,759	53	9	62	4,823				9,486	1,927	0	11,413
1953	4,387	144	4,531	57	8	65	4,598				11,320	1,220	0	12,540
1954	4,304	145	4,449	82	9	91	4,541				16,782	1,196	0	17,978
1955	4,658	154	4,812	82	11	93	4,907				14,555	821	0	15,376
1956	4,488	180	4,668	90	13	103	4,772				11,621	1,484	0	13,105
1957	5,067	244	5,311	81	14	95	5,407				15,257	1,029	0	16,286
1958	5,198	214	5,412	78	15	93	5,506				14,917	5,060	0	19,977
1959	5,568	225	5,793	75	15	90	5,884				11,541	9,991	0	21,532
1960	5,817	284	6,102	74	15	90	6,192				10,098	5,196	0	15,294
1961	6,287	322	6,609	81	18	100	6,710				8,962	1,613	0	10,743
1962	6,397	362	6,760	84	20	104	6,864				9,326	741	0	10,066
1963	6,200	389	6,590	84	23	108	6,698				8,523	437	0	8,956
1964	5,868	362	6,231	89	29	118	6,350				6,600	215	0	6,800
1965	6,382	380	6,761	113	33	146	6,908				5,524	283	0	5,742
1966	6,558	405	6,963	103	36	140	7,103	1,912	2,773	1,872	6,691	715	0	7,370
1967	7,241	470	7,712	97	42	139	7,851	2,403	2,828	2,011	5,095	1,658	66	6,819
1968	7,993	522	8,515	103	52	155	8,670	2,830	3,158	2,004	3,841	1,125	23	4,989
1969	7,976	473	8,449	112	52	164	8,613	3,165	2,949	1,862	8,620	5,936	88	14,644
1970	8,598	549	9,147	119	48	168	9,315	3,429	3,279	1,889	10,478	10,412	1,439	22,329
1971	9,149	609	9,575	101	50	151	9,909	3,674	3,541	1,934	13,292	8,619	2,888	24,799

Year														
1972	9,400	648	10,048	109	56	165	10,213	3,905	3,594	1,902	15,283	23,768	7,454	46,505
1973	9,793	791	10,584	114	64	179	10,763	3,989	3,984	1,820	20,114	33,557	7,832	61,503
1974	9,749	880	10,629	112	67	179	10,808	3,698	4,178	1,874	32,041	46,773	9,045	87,859
1975	9,573	773	10,346	127	72	199	10,545	3,168	4,470	1,935	44,902	46,538	9,147	100,587
1976	9,605	850	10,455	124	77	201	10,656	2,969	4,637	2,000	61,887	26,874	6,453	95,214
1977	9,688	861	10,659	126	82	208	10,757	2,683	4,898	2,107	101,134	21,885	3,696	126,715
1978	9,683	917	10,600	138	90	228	10,828	2,175	5,518	1,990	98,938	25,568	2,908	127,414
1979	9,477	883	10,359	136	95	231	10,590	2,066	5,458	1,953	93,786	26,868	2,701	123,355
1980	9,909	992	10,900	128	94	221	11,122	2,167	5,705	2,037	88,429	31,609	3,500	123,538
1981	10,143	960	11,103	124	92	216	11,319	2,165	5,939	2,038	113,728	33,205	3,312	150,245
1982	10,231	938	11,170	122	96	219	11,388	2,089	6,070	2,072	131,801	41,192	3,386	176,379
1983	10,697	1,060	11,756	117	94	211	11,967	2,132	6,428	2,137	165,668	43,624	3,977	213,269
1984	11,501	1,111	12,612	107	97	204	12,816	2,280	6,956	2,266	167,395	39,595	2,230	209,220
1985	10,877	1,088	11,965	110	96	206	12,171	2,111	6,498	2,268	177,547	44,245	5,002	226,794
1986	11,341	1,198	12,540	106	94	200	12,739	2,336	6,792	2,213	183,755	57,388	8,457	249,600
1987	11,129	1,137	12,267	101	97	198	12,465	2,344	6,634	2,151	219,392	71,616	6,773	297,781
1988	11,259	1,327	12,586	99	99	198	12,785	2,247	6,896	2,114	248,698	82,913	10,021	341,632
1989	10,440	1,272	11,712	103	99	202	11,913	1,976	6,340	2,123	290,306	70,107	8,978	369,391
1994	9,570	1,273	10,843	112	97	209	11,052	1,496	6,081	1,992	336,361	76,051	9,316	421,728
1993	8,511	1,262	9,773	107	97	204	9,978	1,179	5,438	1,894	274,231	80,222	13,478	367,931
1992	7,771	1,306	9,077	97	91	188	9,266	1,270	4,534	1,968	300,495	84,993	16,061	401,549
1991	7,256	1,274	8,530	91	86	177	8,707	1,139	4,256	1,861	356,058	89,810	19,402	465,270
1990	6,590	1,344	7,934	93	77	170	8,103	1,063	3,720	1,807	366,568	83,730	19,956	470,254
1995	6,007	1,315	7,322	92	75	167	7,489	917	3,260	1,831	403,466	82,017	17,220	502,703
1996	5,974	1,276	7,250	94	73	167	7,417	817	3,256	1,901	426,571	88,567	21,540	536,678
1997	5,985	1,273	7,258	86	67	153	7,411	863	3,343	1,779	388,004	101,064	26,182	515,250
1998	5,315	1,227	6,542	79	64	143	6,684	809	2,924	1,582	398,727	91,090	24,119	513,936
1999	5,239	1,605	6,844	71	63	134	6,626	834	2,800	1,605	407,363	86,840	23,696	517,899
2000	5,022	1,231	6,253	71	61	132	6,384	855	2,591	1,576	405,068	90,684	19,236	514,988

Table 22.2

Seed output (1,000 individuals) of main species for sea farming in Japan, 1977–1984. (Source Japan Sea-Farming Association Survey).

Year	Japanese scallop	Kuruma shrimp	Blue crab	Abalone	Sea bream
1977	2,139,363	255,515	6,917	7,015	4,667
1978	1,566,655	280,075	7,870	7,143	5,109
1979	1,699,127	337,229	12,171	8,462	8,600
1980	1,525,333	297,843	11,519	10,560	10,358
1981	2,127,447	302,138	11,212	12,074	12,044
1982	1,647,327	275,402	14,997	12,279	12,866
1983	1,607,213	300,584	19,523	18,334	15,519
1984	1,776,130	293,620	19,972	19,014	16,176
1985	2,009,893	291,909	27,101	19,431	12,655
1986	2,192,170	306,882	30,067	21,278	17,248
1987	2,926,131	336,933	20,639	22,398	23,712
1988	3,027,969	323,964	26,296	20,579	17,376
1989	3,231,386	288,747	29,722	22,659	16,080
1990	2,888,058	315,427	28,880	23,111	16,777
1991	2,824,520	317,878	35,594	25,839	16,803
1992	3,348,166	322,565	28,115	25,758	20,558
1993	3,123,771	304,235	27,562	23,911	20,610
1994	2,960,325	277,866	30,985	23,124	21,309
1995	2,989,328	275,192	34,919	24,975	22,395
1996	2,961,393	257,709	33,649	27,020	22,414
1997	2,849,530	265,002	41,247	26,406	23,643
1998	2,755,286	225,129	35,924	28,045	22,853
1999	2,805,597	206,835	34,225	28,523	24,296
2000	2,990,244	183,140	30,726	30,255	20,912

Two scallop species, Japanese and Bay, *Patinopecten* (*Mizuhopecten*) *yessoensis* (Jay, 1857) and *Pecten* (*Notovola*) *albicans* (Schröeter, 1802), have mariculture potential. Even so, there are several other species of scallops, which are also utilised for food in Japan, e.g., Noble or 'Hiougi-gai', *Chlamys* (*Mimachlamys*) *nobilis* (Reeve, 1852); Swift's or 'Ezokinchaku-gai', *C.* (*Swiftopecten*) *swiftii* (Bernardi, 1858); Farrer's or 'Azumanishiki-gai', *C.* (*Azumapecten*) *farreri* (Jones et Preston, 1904); and Japanese Moon or 'Tsukihi-gai', *Amusium japonicum* (Gmelin, 1791).

Cultures of Noble and Farrer's scallops have been attempted, though for these production levels remain low. In this paper, the mariculture and fisheries of Japanese, Bay and Noble scallops are described.

22.2 *PATINOPECTEN (MIZUHOPECTEN) YESSOENSIS*

The Japanese scallop, *Patinopecten (Mizuhopecten) yessoensis* Jay, is called 'Hotate-gai' in Japanese and has other names such as Japanese Common, Yesso or Giant Ezo scallop (Fig. 22.2). Jay (1856) obtained the Japanese scallop specimen from Hakodate of south Hokkaido, northern Japan, and reported it as *Pecten yessoensis* n. sp. Dall (1898) called *Patinopecten*, as one section of *Pecten*, and Arnold (1906) separated it as a subgenus. Masuda (1963) called the subgenus *Mizuhopecten* separating it from *Patinopecten* and Habe (1977) adopted this subgenus name.

22.2.1 Biology

The Japanese scallop is a cold water species distributed in the coastal areas of the sub arctic zones such as the North Pacific Ocean, the south Okhotsk Sea, the Japan Sea, and along the coasts of the Kurile Islands, Sakhalin, Hokkaido, and northern Honshu and North Korea. The southern limit of the natural distribution in Japan is Toyama Bay on the Japan Sea coast and Tokyo Bay on the Pacific Ocean coast (Fig. 22.3). The main mariculture areas are Hokkaido, north Honshu, north Japan, and some of the surrounding areas, which change seasonally. These coasts are influenced alternately by both warm currents in summer and cold ones in winter alternately from year to year (Fig. 22.4). Aberrant water masses also (Komaki 1975; Fujii and Sato 1977) appear in each season (Fig. 22.5).

Shell valves are loosely shut together, the right valve being larger than the left one. The shell shapes are slightly convex in the right valve with a yellow-white surface. In the left valve the shapes are either quite flat or slightly convex with a purplish brown surface or an occasionally with an orange yellow surface. Both interiors of the valves are white. The anterior ear and the posterior one have symmetrical beaks at both sides of umbones and make hinge-margins straight; however, the right valve has a shallow byssal notch below the front ear. Hinges are united by a narrow ligament. Valves are ornamented with radiating ribs of between15 and 32 ribs on the right and 13 to 31 on the left. It has been found that the number of ribs varies geographically and is maintained hereditarily (Kinoshita 1935b). The number of ribs of a scallop in Mutsu Bay and from the northern part of Iwate Prefecture is between 19 and 22 on the left valve, 18 and 21 on the right. The number in Japan Sea, Hokkaido and the Okhotsk Sea is between 23 and 24 on the left valve, 22 and 23 on the right. The number in Funka Bay is between that of the numbers in Mutsu bay and in the Okhotsk Sea (Kawamata 2002).

The life cycle begins when the demersal eggs are fertilised in the sea after spawning (Fig. 22.6). Fertilised eggs start cleavage and reach the subsequent development of the trochophore at four days. The early veliger larva stage with a fully formed prodissoconch shell, (D-larva) is reached 5–7 days after fertilisation. The umbones of the late veliger larvae are fully-grown and overhang the straight hinge by 30–35 days. This pediveliger attaches to the substratum with byssal threads by 40 days. Immediately after settlement, rapid changes in the shell morphology of the dissoconch (spat shell) and growth of internal organs take place, leading to the adult scallop form.

1100

Figure 22.2. A: *Patinopecten* (*Mizuhopecten*) *yessoensis* (Jay, 1857). B: *Chlamys* (*Azumapecten*) *farreri* (Jones et Preston, 1904). C: *Pecten* (*Notovola*) *albicans* (Schröeter, 1802). D: *Chlamys* (*Swiftopecten*) *swiftii* (Bernardi, 1858).

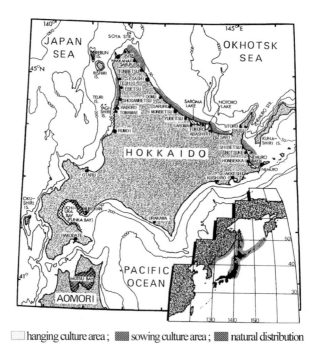

☐ hanging culture area ; ▨ sowing culture area ; ▨ natural distribution

Figure 22.3. Main mariculture areas in Japan and the natural distribution in East Asia of the Japanese scallop, *Patinopecten* (*Mizuhopecten*) *yessoensis* Jay.

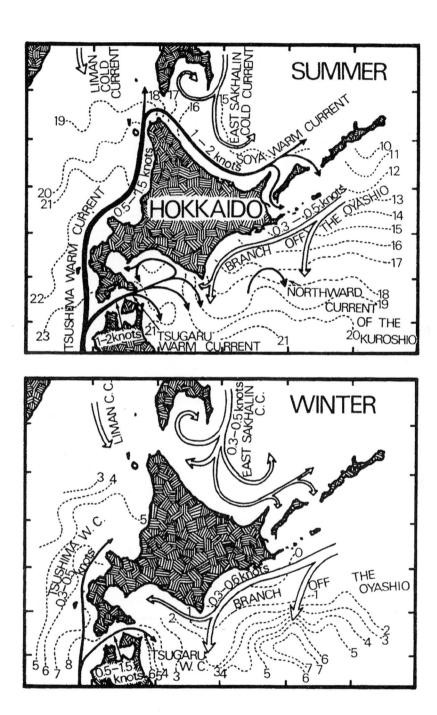

Figure 22.4. Main currents around Hokkaido and north Honshu in summer and winter.

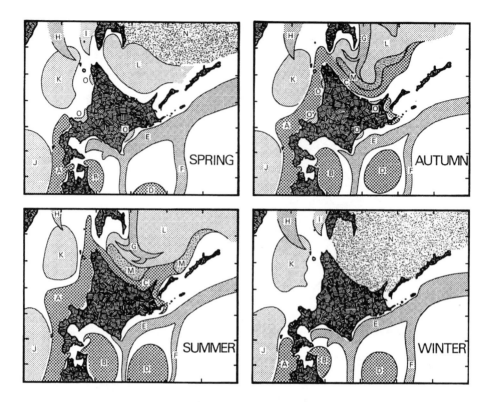

Figure 22.5. Water masses around Hokkaido and northern Honshu at each season. A: Tsushima warm current, B: Tsugaru warm current, C: Soya warm current, D: Northward current of Kuroshio, E: Coastal branch off the Oyashio (Kuril cold current), F: Offshore branch off the Oyashio (Kuril cold current), G: East Sakhalin cold current, H: Liman cold current, I: West Sakhalin coastal water, J: Cold water west off the Tsugaru Straits, K: Cold water west off Thugaru Straits, K: Cold water west off Musashi-tai, L: Inter-cool water, M: Mixed water areas of circulating current, N: Floating ice area, 0: Coastal water area (after Komaki 1975; Fujii and Sato 1977).

After being attached for 4–5 months, the spat grows to approximately one centimetre (cm) shell length, starts to fall off the substratum and the free spat inhabits the sea bottom. The bottom dwelling spat grows into a juvenile of 2–5 cm shell length by the next spring. The scallop grows to 5–9 cm shell length and 16–80 g body weight at two years old, 8–12 cm and 60–170 g at three years old, 10–15 cm and 110–300 g at four years old, and then larger with a reduction in growth rate as it grows older. Some scallops reach 20 cm and 1 kg at more than ten years old.

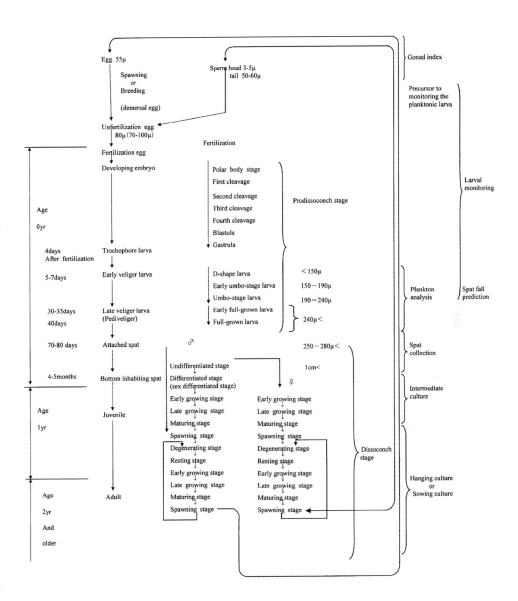

Figure 22.6. Life cycle of Japanese scallop, *Patinopecten* (*Mizuhopecten*) *yessoensis* Jay, with a note on the culture process. (modification from Yamamoto 1964; Maru 1972, 1976, 1978; Kawamata et al. 1981; Kosaka et al. 1997).

The scallops dig shallow depressions on the sandy bottom, feeding on phytoplankton and detritus collected. The feeding and growth of scallop are obstructed when the flow surpasses 20 cm s^{-1} (Miyazono 2000).

The optimum temperature for feeding is within the range of 5–20°C. The lower and upper limits are 4 and 22–23°C, respectively. The optimum temperature for their maximal growth is between 10 and 15°C (Yamamoto 1964). A spat is capable of growing at even 25°C though the upper lethal temperature of a 1-year-old scallop is 22–23°C (Maru 1985; Kosaka 1996).

An energy flow diagram for scallop, *Patinopecten* (*Mizuhopecten*) *yessoensis* (Jay), in Mutsu Bay, has been outlined as follows. About 20–30% of energy is voided as egesta, while most of the rest is assimilated. About 50% (1-year-old scallops) to 70% (3-year-old scallops) of assimilated energy is lost to the surroundings as heat (respiration). The net growth efficiency ranged from 50% (1-year-old scallops) to 29% (3-year-old scallops). The ratio of energy for the development of gonad tissue to the energy assimilated is about 2% in one-year-old scallops to12% in 3-year-old ones (Fuji et al. 1974). Assimilation efficiency was high from May to July and low from November to January (Kurata et al. 1991). Oxygen consumption increased with temperature, excluding April-May when a water temperature was 0–18°C. On the other hand, oxygen consumption decreases when the water temperature is higher than 18°C (Kurata 1996a).

Juveniles differentiate sexually after little over one year (about 15 months). This scallop is gonochoristic. When mature, the gonad of the female scallop turns reddish orange in colour and that of the male turns pale yellow. Fecundity varies with age, but is usually in the range of 8 to 18 X 10^{7} eggs per female (Yamamoto 1943). Previously, the scallop matured at more than two years old (Wakui and Obara 1967; Maru 1978a; Osanai et al. 1980; Kawamata, Tamaoki and Fuji 1981). Recently, the sexual maturation or differentiation begins at one-year-old and spermatozoa and so ovaries of one-year-old scallops are spawned from mid February to late April (Kosaka et al. 1996a). Hermaphroditic scallops are rarely found (Yamamoto 1964; Maru 1978b). In Saroma Lake the gonad develops from autumn to winter, becoming mature by May, with the spawning occurring from May to June (Maru 1976, 1985). In Funka Bay the gonads mature from March to April and spawning occurs from May to June (Kawamata 1981, 1983). In the Japan Sea coast of Hokkaido scallop become mature from March to April and spawn from April to May (Kawamata 1988). The gonads of scallop in Mutsu Bay develop from October to February, mature from January to February and the gametes were spawned from February to June (Kosaka et al. 1996b, 1997a).

The temperature in spawning season is 7–12°C in Saroma Lake, 4–12°C in Funka Bay, 6–10°C in Mutsu Bay and 8–12°C in Toni Bay, Iwate (Maru 1976; Kawamata 1983; Kosaka et al. 1997a; Mori 1975).

22.2.2 Fishery

The scallop had been the main important export item to China in the mid-Tokugawa period (Tokugawa jidai 1600–1868) of Japan (Thubata 1982). At that time the scallop was fished by a hand trawl or 'Teguri-ami' in Mutsu Bay, and was landed mainly in the regions of Hiranai and Kominato, Aomori. The fishery around the east Hokkaido coasts, facing the Okhotsk Sea was started in the late 1800's (Ito 1964). Use of the powered trawl or 'Gangara-biki' the scraper, scallop dredge, otherwise known as 'Hotategai-keta-ami', was extended in the 1940's because of a food production increase and a labour force decrease during the World War II. Catch statistics are available from the late 1800's for Mutsu Bay and the early 1900's for Hokkaido.

Until 1968, the scallop fishery had been dependent on natural resources. At that time fishing efforts concentrated on the high density populations which appeared by excellent natural recruitment of a year class. Scallop fishing for these resources would cease within 2–3 years because populations were completely exhausted. This was a normal fishing style for the natural resources. There was almost no regulation except the simple control of the shell size and the fishing period. Thus, catch statistics fluctuated in relation to the natural biomass changes (Fig. 22.7).

The history of the Japanese scallop fishery from 1910–1987 is classified into six periods as follows (modified from Tanaka 1963; Kanno and Wakui 1978; Thubata 1982): (1) 1910–1923: temporally high catch period, annual catch fluctuated with one peak of high landing in 1915, annual average catch was 16,000 mt; (2) 1924–1932: stable high level catch period, annual average catch was 37,000 mt; (3) 1933–1944: extreme high catch period, average was 43,000 mt; (4) 1945–1968: stable low level catch period, catch levelled down to average of 11,000 mt without peak; (5) 1969–1976: hanging culture development period, hanging culture production rapidly increased, average of 57,000 mt (Fig. 22.8); (6) 1977–1996: sowing and hanging culture mass production period, the production passed 100,000 mt mark in 1977 and 200,000 mt mark in 1983, 400,000 mt mark in 1992, and then it got to over 500,000 mt in 1995 at last. (7) 1997–2000: More than 500,000 mt is being produced steadily after 1997. The scallop industry in Japan develops into a mass production epoch.

Hokkaido has yielded a major catch in from times of old to date. The scallops caught around Hokkaido represented 77% of cumulative scallop landings in Japan during the period 1910–2000 (Table 22.1). The sowing culture and wild production from the sea bottom around Hokkaido has contributed 50% to the total amounts in Japan, 1981–2000. This especially so in 2000 when the production of sowing culture and wild production in the Okhotsk Sea came to occupy 53% of the total amounts in Japan (Fig. 22.9, Table 22.3). The mass mortality that happened in 1975–1977 in Mutsu Bay, Aomori and in 1977–1980 in Funka Bay, Hokkaido, resulted in the hanging culture production immediately decreasing in the late 1970's (Kanno et al. 1980). After that the hanging culture production has been extended smoothly and reached its highest level, 265,553 mt in 1995. However, production level has decreased since 1996 due to mortality and the growth defect in the Funka Bay. The hanging culture production in Funka Bay and Mutsu Bay in 2000 occupied 44% and 42% of the total harvest in Japan, respectively (Table 22.4). Thus, the Hokkaido region is the home of the Japanese scallop in Japan.

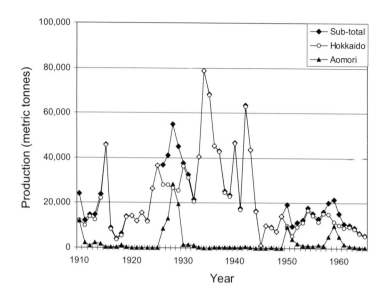

Figure 22.7. Changes in annual production of Japanese scallop, *Patinopecten* (*Mizuhopecten*) *yessoensis* Jay, in Japan, 1910–1965.

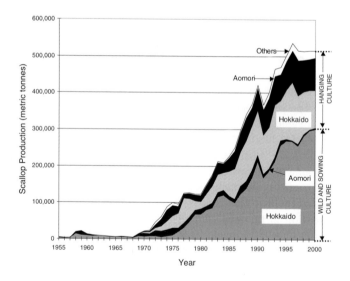

Figure 22.8. Changes in annual productions of harvest and catch of Japanese scallop, *Patinopecten* (*Mizuhopecten*) *yessoensis* Jay, in Japan, 1955–2000. (Source: Minister of Agriculture, Forestry and Fisheries, Government of Japan, "Annual report on statistics of fishery and aquaculture production").

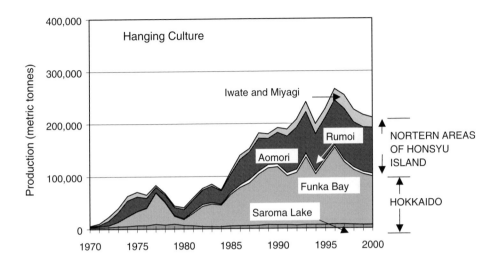

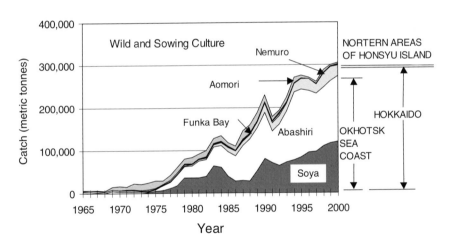

Figure 22.9. Changes in annual production of harvest and catch of Japanese scallop, *Patinopecten* (*Mizuhopecten*) *yessoensis* Jay, in every district of Japan, 1965–2000. Source: Ministry of Agriculture, Forestry and Fisheries, Government of Japan, 'Annual report of statistics of fishery and aquaculture production'.

Table 22.3

Annual catch (metric tons) of Japanese scallop, *Patinopecten* (*Mizuhopecten*) *yessoensis* Jay, in Japan, 1952–2000. Appending harvest statistics. (Source: Japan Ministry of Agriculture, Forestry and Fisheries).

Column groups: Hokkaido — Okhotsk Sea (Soya, Abashiri), Nemuro, Pacific Ocean (Kushiro, Tokachi, Hidaka), Funka Bay (Iburi, Osima, Rumoi), Japan Sea (Isikari, Shiri-beshi, Hiyama), Sub-total. Honsyu — Tohoku (Aomori, Iwate, Miyagi, Sub-total), Others (Others, Sub-total). Catch Total, Harvest Total, Grand Total.

Year	Soya	Abashiri	Nemuro	Kushiro	Tokachi	Hidaka	Iburi	Osima	Rumoi	Isikari	Shiri-beshi	Hiyama	Hokkaido Sub-total	Aomori	Iwate	Miyagi	Tohoku Sub-total	Others	Honsyu Sub-total	Catch Total	Harvest Total	Grand Total
1952											1	0								3,232		3,232
1953	457	1,917	528			16	20	17					2,960	325			325		325	3,287		3,287
1954	561	2,565	1,115			23	58	12					4,336	317			317		317	4,654		4,654
1955	892	2,101	1,068			1	55	13					4,133	219			219		219	4,352		4,352
1956	599	1,282	936		0	9	109	47					2,986	369			369		369	3,355		3,355
1957	80	2,757	783			9	113	16					3,760	275			275		275	4,036		4,036
1958	108	12,447	1,559	11			334	71					14,533	5,060			5,061	12	5,073	19,607		19,607
1959	586	8,550	1,528				794	85					11,547	9,991	0		9,992		9,992	21,539		21,539
1960	718	6,014	1,891			90	707	39					9,462	4,407	0		4,407		4,407	13,870		13,870
1961	540	7,156	886			80	285	13					8,962	1,780			1,780		1,780	10,743		10,743
1962	397	7,566	1,108	39			294	9					9,326	737			737	2	739	10,066		10,066
1963	370	6,872	1,037	11			194	37					8,523	432			432		432	8,956		8,956
1964	0	4,500	1,900	0			100	0					6,600	200			200		200	6,800		6,800
1965	68	2,722	2,575	24			126	10					5,524	214	4		218		218	5,742		5,742
1966	77	3,966	2,538	14			90	6					6,691	654	25		680		680	7,370		7,370
1967	61	1,732	3,104		42			140	15				5,095	1,719	5		1,724		1,724	6,819		6,819
1968		525	3,118				143	55					3,841	1,102	46		1,148		1,148	4,989		4,989
1969	90	5,272	3,091	16			92	59					8,620	5,527	496		6,022	1	6,023	14,644		14,644
1970	51	3,064	3,174	9			400	489					7,187	9,048	242		9,290		9,290	16,477	5,853	22,330
1971		3,457	2,867			0	618	1,290					8,232	6,199	9		6,208		6,208	14,440	10,361	24,801
1972	78	3,899	2,658				1,058	798					8,491	14,874	9	35	14,918	45	14,963	23,454	23,049	46,503
1973	122	3,191	1,714				766	675	1				6,469	15,340	22	3	15,365	374	15,739	22,208	39,297	61,505

Year																				
1974	1,674	2,294	1,204	83				2,392	443	26		8,116	16,925	6		160	17,091	25,207	62,651	87,858
1975	4,680	3,577	868	109				1,476	299	16	11	11,036	19,063	6		169	19,238	30,274	70,313	100,587
1976	7,169	10,789	800	214				1,705	536	83	4	21,300	8,662	8	9	291	8,970	30,270	64,946	95,216
1977	11,429	15,294	1,425	98	14			1,970	787	599		31,602	11,699	4	7	190	11,900	43,502	83,213	126,715
1978	18,746	23,626	1,457	125	34	40		2,309	1,010	699		47,426	12,131	15	9	83	12,238	59,664	67,750	127,414
1979	35,387	26,873	1,563	71	45	36		2,593	1,351	134	206	68,248	11,448	21	1	15	11,486	79,734	43,622	123,356
1980	35,955	28,421	2,284		0	0	3	784	515	764		68,771	14,290	11	1	62	14,364	83,134	40,403	123,537
1981	36,444	39,295	3,320		0	0		246	1,043	106	106	80,560	10,568	6	2	3	10,579	91,139	59,106	150,245
1982	38,679	39,916	3,116		0	0		884	618	2,393	45	85,651	13,845	6	1	1	13,854	99,505	76,876	176,381
1983	63,286	44,736	5,406	92	0		2	286	740	1,251	117	115,914	12,059	20	5	138	12,222	128,136	85,134	213,270
1984	60,424	51,686	6,014					957	698	589	327	120,697	14,194	13	1	334	14,542	135,239	73,981	209,220
1985	38,945	57,590	9,822	66			3	935	138	1,502	204	109,205	9,047	18	3	4	9,072	118,277	108,517	226,794
1986	28,677	57,590	11,615	26				1,065	658	1,173		100,504	9,082	15	7	123	9,227	109,731	139,869	249,600
1987	29,336	81,325	11,261	29				995	554	1,929	1	125,430	19,728	15	9	176	19,928	145,358	152,423	297,781
1988	27,483	97,952	10,813					1,131	517	1,243	301	139,139	20,532	7	4	7	20,550	159,689	181,935	341,624
1989	52,420	100,021	13,234	85				796	140	1,840	257	168,837	20,272	7		2	20,281	189,118	180,265	369,383
1990	80,953	108,122	19,997	19				1,251	701	2,141	118	213,441	16,213	9		4	16,226	229,667	192,044	421,711
1991	70,879	73,746	19,783	58				2,864	212	725		168,385	10,685	6		1	10,692	179,077	188,858	367,935
1992	63,458	97,068	20,558	49				4,304	63	1,928	449	187,877	5,593	4		1	5,598	193,475	208,063	401,538
1993	71,647	119,399	18,093					1,742	458	1,657	112	213,108	10,733	3			10,736	223,844	241,439	465,283
1994	76,378	159,182	19,584	34				1,380	282	902		257,742	13,117	8		23	13,148	270,890	199,372	470,262
1995	83,765	159,017	23,984	26				1,913	291	334	150	269,480	5,354	18		26	5,399	274,879	227,823	502,702
1996	95,418	143,775	27,626				6	888	200	17	29	267,959	3,136	28		1	3,165	271,124	265,553	536,677
1997	97,222	136,289	21,351	48				1,160	195	232		256,497	4,656	11			4,667	261,164	254,086	515,250
1998	109,121	138,900	32,761	69				1,303	351	35		282,540	5,261				5,261	287,801	226,133	513,934
1999	116,489	146,624	32,263	49				1,748	200			297,373	2,255				2,255	299,628	216,018	515,646
2000	119,188	153,979	26,000	30				1,940	194			301,331	2,955				2,955	304,286	210,703	514,989

Table 22.4

Annual harvest (mt) of *Patinopecten* (*Mizuhopecten*) *yessoensis* Jay, in Japan, 1970–2000. (Source: Japan Min. Agriculture, Forestry and Fisheries).

| | Hokkaido | | | | | | | | | | | | | Honsyu | | | | | Harvest |
| | Okhotsk Sea | | Nemuro | Pacific Ocean | | | Funka Bay | | Japan Sea | | | | | Tohoku | | | Others | | |
Year	Soya	Abashiri	Nemuro	Kushiro	Tok-achi	Hidaka	Iburi	Osima	Rumoi	Isikari	Shiri-beshi	Hiyama	Sub-total	Aomori	Iwate	Miyagi	Others	Sub-total	Total
1970		2,488					681	121			1		3,291	1,364	996	190	11	2,562	5,853
1971		2,802					1,656	598			4		5,060	2,419	2,383	486	12	5,301	10,361
1972		3,486					2,244	1,021			41		6,792	8,894	6,459	870	35	16,258	23,049
1973		4,665					5,200	3,752	12		16		13,645	18,217	6,928	458	49	25,652	39,297
1974		5,100				6	7,729	10,936	40		114		23,925	29,848	7,501	1,270	107	38,726	62,651
1975		6,618				1	8,898	17,969	156		225		33,866	27,475	7,902	967	104	36,448	70,313
1976		6,579				6	7,780	25,449	248		525		40,587	18,212	5,474	640	32	24,359	64,946
1977		9,027				9	11,821	48,053	356		266		69,532	10,186	3,016	432	47	13,681	83,213
1978		7,616				2	2,653	40,245	509		487		51,512	13,437	2,577	210	14	16,238	67,750
1979		9,309				1	1,213	13,956	829		230		25,538	15,420	2,429	202	33	18,084	43,622
1980	49	6,771				0	2,683	8,822	1,139		194		19,658	17,259	3,303	154	29	20,745	40,403
1981	28	5,916					5,178	18,830	2,984		232		33,168	22,637	3,143	142	16	25,938	59,106
1982	17	4,278				0	5,726	32,623	3,180	170	156		46,150	27,347	3,176	180	23	30,726	76,876
1983	6	3,832					7,212	35,986	1,997	294	427		49,754	31,429	3,665	263	23	35,380	85,134
1984	36	3,678		40		3	7,587	33,715	911	366	367		46,698	25,068	1,922	293		27,283	73,981
1985	32	5,090		41		15	10,314	49,506	2,037	440	867		68,342	35,198	4,113	864		40,175	108,517
1986	86	5,246		25		15	14,439	58,550	3,746	231	913		83,251	48,306	7,227	1,085		56,618	139,869
1987	52	5,426				40	15,599	65,862	5,172	644	1,163	4	93,962	51,888	5,287	1,283	3	58,461	152,423
1988	32	6,014				220	15,926	81,926	3,647	612	1,164	4	109,545	62,380	6,958	3,040	12	72,390	181,935
1989	14	7,445					18,240	89,616	4,308	800	1,023	5	121,451	49,835	6,215	2,744	20	58,814	180,265
1990	9	6,823					17,716	92,888	3,864	812	767	22	122,901	59,838	5,817	3,484	4	69,143	192,044
1991		6,791					16,230	77,142	3,775	676	1,174	58	105,846	69,536	7,722	5,746	8	83,012	188,858
1992		5,902				60	17,970	82,466	4,462	949	701	83	112,593	79,399	9,724	6,319	28	95,470	208,063
1993	4	7,011				60	20,634	108,055	5,171	1,028	871	116	142,950	79,076	11,814	7,573	26	98,489	241,439
1994	8	6,642				58	17,942	78,852	3,867	631	699	126	108,825	70,613	10,466	9,450	18	90,547	199,372
1995		6,687		0			21,654	100,672	3,788	576	508	100	133,985	76,663	8,493	8,671	11	93,838	227,823
1996		7,664		1			21,519	124,492	3,704	564	593	74	158,611	85,431	9,633	11,868	10	106,942	265,553
1997		7,512		2		0	21,067	97,319	4,073	723	689	122	131,507	96,408	11,867	14,298	6	122,579	254,086
1998		6,605		6		36	19,836	84,680	3,358	919	616	130	116,186	85,828	9,262	14,852	5	109,947	226,133
1999		6,340		4		21	17,825	78,929	3,258	611	612	137	107,737	84,585	8,781	14,910	5	108,281	216,018
2000		6,360		5		7	14,665	77,768	3,617	611	601	104	103,738	87,729	5,755	13,476	5	106,965	210,703

22.2.3 Culture

Scallop mariculture developed rapidly due to the technical invention of successful methods for natural spat collection and the intermediate culture in embayment seas in the mid-1960's after 30 years of investigative trials. Scallop spat collection in Japan was first attempted with successful results at an experimental stage in Saroma Lake in 1934 (Kinoshita 1935a). Industrial results of the spat collection fluctuated greatly because the technical methods were still undeveloped. After many experiments, a successful spat collector was invented in 1964, by Mr. Toyosaku Kudo, a fisherman in Mutsu Bay, at Aomori Prefecture (Yamamoto et al. 1971; Tsubata 1982). The scallop collector is a double structure and is composed of a mesh bag filled with substratum. The result of the use of this type of collector has been that in embayment seas relatively stable spat collection has been achieved. After the mid-1970's, offshore spat collection in open sea coastal areas was developed because of a lack of seed for sowing (Ito 1984).

The spat are caged for seed production by a long line system for several months. The caged culture for seed production is called "chukan-ikusei" or intermediate culture. The intermediate cultured seed are used for both hanging culture in exclusive sea areas and sowing on prepared sea bottoms.

Scallops in hanging culture grow more rapidly than in sowing culture (Fig. 22.10). The weight of a 2-year-old scallop grown by hanging culture has the same weight as that of a 3-year-old scallop grown by sowing culture in Mutsu Bay, and the weight of a 2.5 years old scallop grown by hanging culture in Saroma Lake is the same as a 4.5-year-old scallop grown by sowing culture in the Okhotsk Sea. In addition, the growth of scallop in hanging culture is greatly different between Saroma Lake and Mutsu Bay. The scallops in hanging culture in Saroma Lake grow to only about 90 g in 2 years, though ones in Mutsu Bay grow to over 120 g in 2 years. This difference is caused by differences in the environment, the culture technique, the culture density and the food supply.

22.2.4 Seed Production

Recently all scallop seed has been produced from a natural spat collection. Until the natural seed production became a mass production stage, the artificial seed production in land tanks had been examined by the mid-1970's (Kinoshita et al. 1943; Yamamoto and Nishioka 1943; Tanaka et al. 1977; Tsubata 1982).

The mass and cost of natural seeds gain an advantage of those of artificial seeds. Seed production takes the sequential process from spat collection to intermediate culture.

22.2.5 Spat Collection

A process of spat collection and its research is as follows (Ito 1986; Ito et al. 1988a; Aoyama 1989): (1) check the spawning: to estimate the timing of spawning use changes in the gonad index, weight ratio of gonad to soft body; (2) search for the larvae: determine the set timing and location of spat collectors by larval monitoring; (3) set the collectors: collectors are set where the larvae are common; (4) check the attached spat: count spat

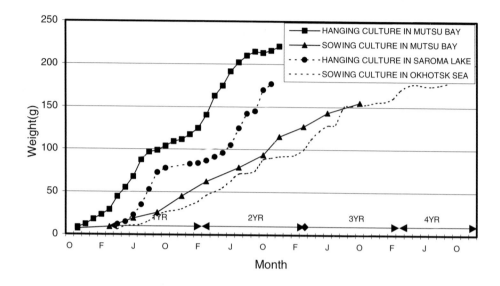

Figure 22.10. Growth curves of Japanese scallop, *Patinopecten* (*Mizuhopecten*) *yessoensis* Jay, in hanging culture and sowing culture (modification from Aoyama 1989; Sato 1993; Nishihama 1994).

and measure shell size to determine the intermediate culture management planning and timing. Although this process has been directly derived from the spat collection data, other information supports it as well (Ito and Hamaoka 1988b).

Since the 1970's, surveys and research have been carried out with the objective of providing forecasts of available spat collection in Mutsu Bay. Every year from December to July, studies are conducted on the gonad index of broodstock, on larval density, spat, and environmental conditions, using automatic monitoring buoy robots (Aoyama 1989). In the Okhotsk and Japan Seas, Hokkaido, studies on gonad index of broodstock, the larval density, the spat, and temperature are done from April to September (Kurata 2001) and similar studies are done in Funka Bay, Iwate and Miyagi annually.

More than 10,000 spat per collector can now be collected consistently in Mutsu Bay. In strong years, more than 100,000 spat can be collected on average, the record catch being more than 1,000,000 spat in one collector. Because, the production of spat fluctuates in Funka Bay greatly (Table 22.5); spat are imported from the Okhotsk Sea or the Japan Sea, Hokkaido (Fig. 22.11: Kurata et al. 2001). Juvenile seeds of six months to one-year-old are imported from the Japan Sea or other regions and then released in the Okhotsk Sea and the Nemuro Strait. The annual volume of sales of seed reaches more than 2 billion individuals in Hokkaido.

Collectors for natural spat collection vary with conditions and purposes as follows (Fig. 22.12).

The collectors being used in Saroma Lake and Notoro Lake are made of Netron, with a large mesh size, which is loosely stuffed into the outer net of a netron bag. The mesh size of the outer net is 5 mm, larger than the mesh size of an onion bag. Worn-out salmon and trout gill nets are sometimes used as substratum.

Worn-out salmon and trout gill net, or netron is being used as substratum in Mutsu Bay. Though the onion bag of 1×5-mm mesh size is commonly used for outer nets, bags of 1×1-mm, 1×2-mm and 2×2-mm mesh sizes are also used. It varies due to the conditions, or the preference of the fisherman.

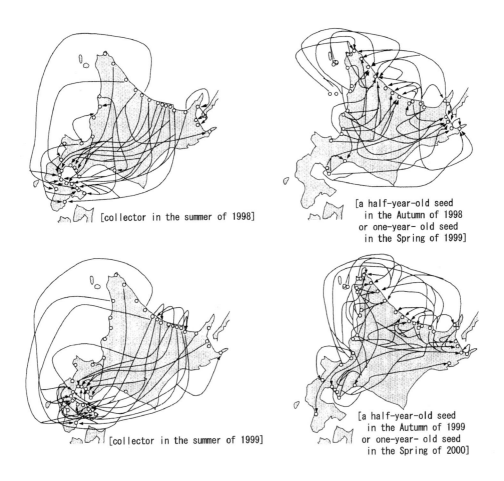

[collector in the summer of 1998]

[a half-year-old seed in the Autumn of 1998 or one-year- old seed in the Spring of 1999]

[collector in the summer of 1999]

[a half-year-old seed in the Autumn of 1999 or one-year- old seed in the Spring of 2000]

Figure 22.11. Transport routes of imported scallop seeds in Hokkaido.

Table 22.5

Number of spat attached to collector (Source: Hokkaido Abashiri Fisheries Experimental Station, Hokkaido Hakodate Fisheries Experimental Station and Aomori Aquaculture Research Center survey). The numbers for Funka Bay are individuals per 100 g net weight and others are individuals per collector. The numbers for Mutsu Bay are average numbers; others are the minimum and maximum numbers.

Year	Hokkaido Okhotsk Sea Soya min	max	Abashiri min	max	Nemuro Nemuro min	max	Pacific Ocean Hidaka min	max	Funka Bay Iburi min	max	Oshima min	max	Japan Sea Rumoi min	max	Ishikari min	max	Shiribeshi min	max	Honshu Tohoku Aomori mean
1976																			4,021
1977																			15,917
1978																			39,637
1979																			34,600
1980																			30,610
1981																			59,200
1982																			1,416
1983																			35,111
1984																			10,053
1985												2,563	7,563							35,377
1986												935	9,100							7,117
1987	0	935	78	10,019	131	800	128	250	1,848	14,192	44	54,186	0	920	0	481	0	415	62,058
1988	0	1,799	22	3,400	40	955	40	185	9,720	33,787	1,536	26,765	0	1,768	442	2,849	33	1,506	32,605
1989	113	1,551	317	10,644			179	558	1,250	224,520	2,284	847,872	0	1,866	470	2,369	3	725	18,282
1990	197	4,251	43	3,600	1,445	3,634	255	2,283	3,188	168,716	1,084	26,985	0	1,631	323	3,212	28	2,290	16,347
1991	0	2,613	1,303	7,339			435	2,523	2,839	663,552	1,934	136,413	0	5,877	807	2,704	260	1,700	133,771
1992	802	5,470	2,500	14,022	2,123	5,086	467	3,776	20	1,324	15	1,429	0	5,034	1,208	4,610	72	4,663	40,647
1993	4,174	11,000	1,879	14,544	117	1,430	22	2,341	167	1,680	185	3,167	340	4,556	1,047	4,660	22	3,323	67,444
1994	868	4,297	680	22,209	1,848	5,460			9,500	220,000	48,000	570,400	0	5,456	2,374	5,195	104	3,088	155,283
1995	28	4,101	829	23,500	418	960			93,148	1,650,447	9,750	582,229	0	3,995	1,219	1,484	311	4,821	38,585
1996	992	8,360	461	16,919			808	9,248	967	281,098	5,902	89,771	891	6,521	1,135	2,668	183	2,500	115,277
1997	520	4,353	244	22,347	808	2,169	4	76	8,302	70,473	465	5,061	1,426	9,847	1,816	4,970	21	759	95,813
1998	1,520	5,129	34	64,930	1,527	5,329	7	169	18	275	3	300	0	579	2,413	5,848	3	6,270	59,304
1999	42	4,184	31	19,721			205	784	10	1,028	54	10,162			507	5,168	0	6,411	67,033
2000	395	11,282	5	8,000	1,435	1,938	2,633	3,258	126	4,937	497	3,309	0	1,210	1,085	2,486	0	5,650	91,368

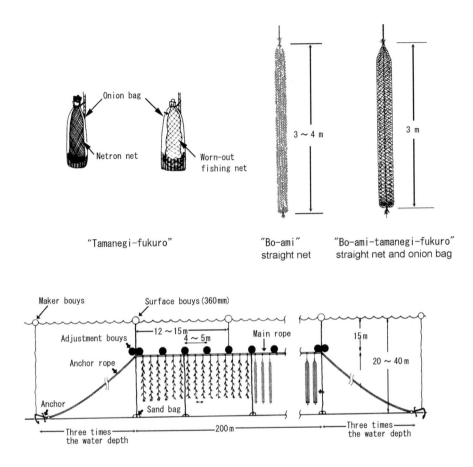

Figure 22.12. Facilities for spat collection for Japanese scallop, *Patinopecten* (*Mizuhopecten*) *yessoensis* Jay.

Juvenile starfishes stick to collectors every year in Funka Bay and they grow inside the bag, and prey upon the scallop spat. Generally, the 'Straight net' style collector is being used at present as a countermeasure to this. The 'Straight net-onion bag' style collector, which is a net put over a long straight net, is also being used in Funka Bay. When the spat reach 3–4 mm in size, juvenile starfishes are removed, and the temporary seed collection or 'Karisaibyo' (spat being put in the pearl net of small mesh size) is carried out.

If the mesh size of the outer net is small, the collector becomes dirty immediately due to the mud and growth of fouling organisms. When the mesh size is too large, the spat fall off. Despite this, the larger size is favourable since the spat inside the bag grow fast. Megalopa larvae of crabs sometimes stick to collectors in the Nemuro Strait. Larvae enter the bag and metamorphose into juvenile crabs, using scallop spat inside the bag as feed.

1116

One juvenile crab is able to eat all scallop spat in a collector. Because of that, onion bags with mesh size smaller than the size of megalopa larva are used for the outer net in the Nemuro Strait. On the other hand, in Mutsu Bay, the crustacean, *Cymodoce japonica*, called 'Nihonkotubumusi', can enter the bag, when the mesh size of the outer net is large, and eat all the scallop spat. One type of 'Nihonkotubumusi' eats 20 to 30 scallop spat everyday. If over twenty crustaceans get into the collector, all scallop spat will be eaten in a month. The collector whose mesh size is small is favourable even if the growth of spat is reduced. On the Japan Sea side of Hokkaido, the larval density is very low and it seems as though the outer net does not become dirty, and so here too an onion bag is generally used for the outer net. Therefore, the mesh size of the outer net of a collector should be determined by the specific characteristics of its deployment.

22.2.6 Intermediate Culture

The rearing of spat from the collector up to a shell length of 3–5 cm in a pearl net is called intermediate culture. Intermediate culture is conducted by the process of sorting, caging and hanging: (1) "Bunsan" or sorting: sorting large spat with a sieve. (2) 'Shuyo' or caging: caging in nets at low density. (3) 'Suika' or hanging: hanging in water. Moreover, 'Bunsan' means the changing process from spat collection to intermediate culture, i.e., the whole process from (1) to (3). The concept of "bunsan" is to change a situation to increase productivity.

The hardware for intermediate culture is 'Zabutonkago', 'Paru-netto' or pearl net (Fig. 22.13). The sections of pearl net separate into single units, net cages, with a 30 cm square floor, and it is still used commonly in Japan due to low cost and the fact that it was originally introduced from the pearl culture. The mouth of a net is stitched closed using artificial gut and is therefore slightly troublesome to handle. Ten strings of net are linearly linked together with both ends of each singular stalk rope tied through the net centre. The stalk rope in a net becomes a vertical core prop to keep the net horizontal. Sometimes it causes crowding of scallops into a net corner because of irregular slanting due to shortening of the core rope. The seed produced by the above-mentioned process is used for both hanging and sowing culture. Generally the density of spat in the net for sowing culture is higher than for hanging culture. These cultures are described below.

22.2.7 Hanging Culture

Seeds are caged or suspended for hanging culture or 'Yoshoku'. Cultivation from seed to harvest is called 'Yosei' or 'Hon-yosei'. The fouled net is exchanged for a new net and the stocking density of scallops is reduced in the process of hanging culture. If stocking densities inside a net are too high, scallops sometimes bite each other, and damage mantles. This leads to slow growth and high mortality rates (Kosaka et al. 1996). To achieve normal growth, hanging culture is carried out according to the process shown in Table 22.6. It takes about 1.5–2.5 years to raise juvenile scallops to a marketable size. This process varies with the area or method of culture.

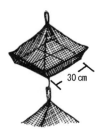

"Zabuton kago" or "Paru-net"
Pearl net

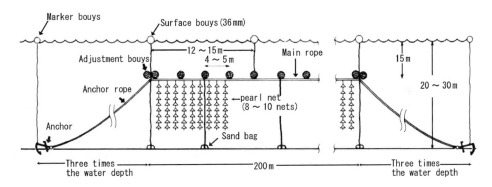

Figure 22.13. Facilities for intermediate culture of Japanese scallop, *Patinopecten* (*Mizuhopecten*) *yessoensis* Jay.

Table 22.6

Basic process of scallop culture in Japan.

July-Sept.	Sept.-Oct.	Feb.-May	Sept.-Nov.	Dec.-
Seed collection	First reducing density	Ear hanging or Second reducing density		Market
			Second reducing density	Market

Methods for culturing vary (as follows) (Fig. 22.14): (1) The 'Mimi-zuri', or ear suspended method; a hole of 1.3–1.5 mm drilled at the front eared beak of left valve through the right valve notch or through both front ears near a byssal notch. The 'Tegusu' method was the most commonly used method preceding the ear suspended culture method. In this method two scallops are suspended by #30 artificial strings, tied at every 12–15 cm of 6 mm polyethylene rope for hanging. Within the genre of ear suspended culture, there are methods called the 'age-pin' method and 'looped' method. The 'Age-pin' method is a method to put two scallops on a rope with a plastic clip called an 'Age-pin'. Another method is called the 'loop' method and was developed recently. The loop made of plastic string is made by the interval on the rope 12–15 cm, and an artificial string is passed through the rope between loops, and two scallops are suspended between those loops by the artificial string (Fig. 22.14). One hanging contains 130 individuals per twine. This method has become the most common recently since the materials are cheap and it is easier to harvest. This makes overcrowding imminent in Funka Bay. (2) The 'Maru-kago', 'Andon-kago', or lantern net: composed of short cylindrical units 50 cm in diameter and 15 cm high and framed by #8 vinyl covered wire. Cylinders are linearly connected together. The cage is made of a polyethylene mesh of 7 'Bu' or 21 mm knots and 5 mm hanging twine. Five to ten scallops per unit are caged. (3) The 'Poketto-kago' or pocket net: 60 cm wide and 20 cm high and is constructed from double nets and #8 vinyl covered wire, stitching two face-to-face sheets of net to each other into a wallet shape. This creates thin pocket net shapes, joining 10–25 units. The mesh is 7 "bu", the same as the lantern net. Six scallops per unit slit are inserted. (4) The 'Hausu-kago' or house cage: a cylindrical plastic sub-unit of 15 cm diameter divided into 12 rooms, five or six subunits making one unit. Sometimes these units are hung together vertically. One individual is placed in each room. This cage was only used at a part of Okhotsk. (5) The 'Paru-netto' or pearl net: coarse mesh is used for growing cultures. Though generally a long line system is used for the culture facilities, the facilities being called 'set' in Funka Bay were used too, which is that several long lines are connected together and concrete-block, sandbag and super-anchor are used for the fixation, and buoys are attached to beam rope and main rope and it has the structure to be able to stand a storm.

Technical research on cultured scallops concerning cultivation care by Aoyama et al. (1982) and Ito et al. (1986) has shown (1) lower densities in both intermediate and hanging (growing) cultures keeps a higher growth rate and a lower mortality by reducing crowding effects; (2) less time sorting scallops and changing cages and cleaning all of which lessens handling shock to the scallops; (3) stabilising the long line facilities and the hung scallops which also lessens wave-induced shock to a scallop. Wave-induced shock damages the scallop mantles and leads to high mortality and slow growth (Kosaka et al. 1996c). These measures produce a stable harvest without mass mortalities.

22.2.8 Fouling

Several organisms, e.g., barnacles, sea anemone, hydrozoas, sea squirts, bryozans, mussels, other bivalves and algae settle on the nets and grow (Kurata et al. 1996b). Heavy fouling prevents the growth of scallops inside the net. Most fouling attached to the culture net and the scallops is mussels. Because mussels often attach at a shallow depth, the settlement of mussels can be prevented when the culture facility is dropped deeper, at more than 20 m deep from April to July. The facility is then returned to its former depth after July (Hirano 1983).

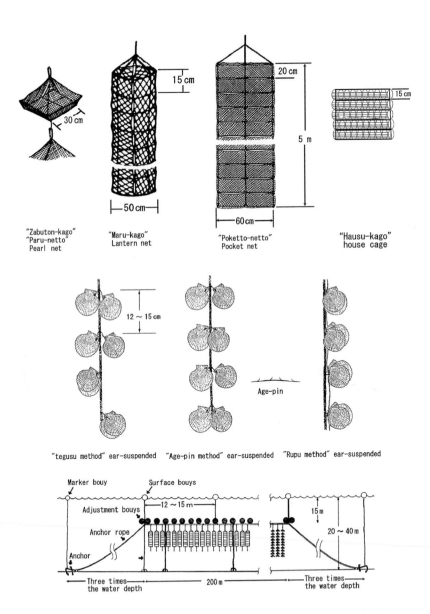

Figure 22.14. Facilities for hanging culture and a variety of hanging hardware for Japanese scallop, *Patinopecten* (*Mizuhopecten*) *yessoensis*, Jay.

22.2.9 Sowing Culture

Scallop seeds were sown for investigative purposes off Okhotsk Sea coast of Hokkaido in 1937 (Kinoshita 1949). Release of spat without intermediate culture was the practise until about 1940. Natural juvenile scallops and seeds imported from Hokkaido were sown in Mutsu Bay of Aomori Prefecture, north Honshu, before 1969 (Tsubata 1982). In Mutsu Bay, some stocks were recaptured; however, almost nothing resulted in Hokkaido after many efforts for first 32 years.

After the development of the collector and hanging culture, one-year-old scallop and mass seed could be produced and intermediate culture could be carried out under the frozen lake. One-year-old seeds could also be produced in Saroma Lake from approximately 1965. Hence, more than ten million seed scallops have been released on the cultural sea bottom since 1967. Survival of the sown scallop population was observed in the Okhotsk Sea coasts of Tokoro, the Abashiri region, Sarufutsu, and the Soya region during the period from the late 1960's to the early 1970's. On the Tokoro coast, 23 million seeds of one-year-old scallops were sown on the culture bottom in 1967. The sowed scallop population at the age of two years old was estimated at the high survival rate of 54.7–98.0% in the next year (Okesaku and Tanaka 1969). The seed input number in Sarufutsu coast was 14–60 million individuals in 1971–1973. Three populations of three-year-old scallops were observed to have survival rates of 5.5–28.9% in 1973–1975 (Hayashi et al. 1976).

22.2.10 Enemy Clearance

There are many animals which prey on or compete with scallops, such as starfishes, sea urchins, octopuses, crabs and fishes. Starfishes and sea urchins are removed from the culture ground by scallop dredges prior to seed sowing. A mop dredge made of hemp ropes and a baited trap of the flat basket type is used for starfish capture in supplemental cases. Sometimes octopuses are fished by "tako-bako" (wooden pot), long line or hook with barrel in culture grounds. Other predators and competitors are left intact due to lack of suitable methods to removal them economically.

22.2.11 Seed Sowing

Seeds are sown on a prepared bottom at a density of 5–6 individuals per square meter. The density value is based on natural stock density (Kinoshita et al. 1944). The optimum stock density is about 4 individuals per square meter in the Okhotsk Sea (Kurata 1999). The maximum production per square meter in Mutsu Bay is 1,200 g; therefore, less than 6 individuals per square meter should be released to achieve normal growth until the market size of 200 g is reached (Yamamoto 1971, 1973; Aoyama 1989). For these reasons the density is regulated at this same value in most locations; however, different densities have been applied locally as a new venture.

The seeds which grow to a shell height of more than 3 cm after intermediate culture are released in Okhotsk Sea coast from late May to early June in the next year (Nishihama

1994). In Mutsu Bay, seeds after intermediate culture are released in December of the same year or the following May or April when they reach a shell length of 3–5 cm (Aoyama 1989).

Annual seed input sown in Japan reached 2,876 million individuals from 1985 to 2000, a 1.5 fold increase in fifteen years (Fig. 22.15). Annual seed input sown off Hokkaido coasts amounted to 2,834 million individuals in 2000. Most of the seeds were sown off the coasts of north and east Hokkaido. In order of quantity: Abashiri (1,444 million individuals, 48%), Soya (729 million individuals, 24%), Nemuro (423 million individuals, 14%). Also, the subtotal seed input number of 1,342 million individuals was produced with the independent intermediate culture in respective areas; however, 1,254 million individuals were introduced from other areas.

Seed sowed in Abashiri is almost self-sufficient because the brackish lakes of Saroma and Notoro are good for natural seed production. In the other districts almost all seed is produced from other areas; nevertheless, seed supply has been chronically insufficient due to recent excess demand. To fill the seed deficiency a new technology for offshore spat collection and offshore intermediate culture in open sea coasts was developed rapidly in the 1980's and the self-sufficient rate of the seed seedling rose.

Mass seeds in marketing are introduced from other areas (Fig. 22.16). The seeds for sowing culture were produced in the Japan Sea region of Hokkaido, and have been sold to Okhotsk Sea region frequently since approximately 1975. But, because spat collection was not successful in the Japan Sea region in approximately 1985, a large quantity of spat were imported from Funka Bay or the Nemuro Strait (Nishihama 1994). Recently, seeds is in imported into Okhotsk mainly from the Japan Sea region because of bad harvest of spat in Funka Bay (Fig. 22.11). The half shared seeds for sowing cultures are produced in the Rumoi, Soya and Nemuro regions. These regions face the open seas of the Japan Sea and Okhotsk Sea, and so the seed production depends on offshore spat collection. However, because the spat collection is fundamentally unstable in offshore regions, this shows that offshore seed production is as indispensable as production in embayment for sowing culture.

22.2.12 Care

Density and growth of sowed scallops are checked regularly. If problems are noted, countermeasures are taken. In the case of overcrowding leading to a low growth rate, scallops are removed from the culture area to reduce the density.

22.2.13 Recapture

The cultured population is usually recaptured for 'harvesting' at three or four years old in Sarufutsu and Notori Lake at 5-years-old. The four circle harvest system called 'Yon rinsai' is being adopted in the Okhotsk Sea: the culture grounds are divided into four areas, and seeds are released on one area, and released on another area in the next year. After three years, four-year-old scallops are recaptured on the culture ground where seeds are first released. Seeds are released again in the next year when scallops are recaptured.

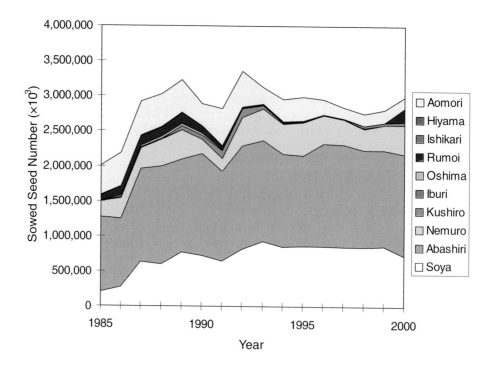

Figure 22.15. Changes of annual sown seed of Japanese scallop, *Patinopecten* (*Mizuhopecten*) *yessoensis* Jay, Source: Fisheries Agency, Government of Japan and Japan Sea-Farming Association, 'Actual results of aquaculture seed production, acquisition and release (the whole country), materials edition'.

This is repeated. This system can define the position of the grounds where seeds should be released and scallops should be recaptured, and makes a premeditated harvest possible (Nishihama 1994; Kurata 1996).

Relationships between seed input and production in sowing districts are significant (Fig. 22.17). The seed input has a highly significant effect on the harvest on a large scale, such as the Soya and Abashiri regions, though no significance is estimated on a small scale, such as Kushiro, Oshima and Iburi districts. In the Soya and Abashiri regions, the catch was extremely high. This scallop stock level was high due to the wild population in Sarufutsu coast of that region (Tomita et al. 1982). No more than about ten percent of the sowing culture contributed to Sarufutsu's scallop production. Moreover, the seed sowing had been stopped in 1980–1985 because of high growth of the wild scallop population on this coast.

The recent increase in catches in the Okhotsk Sea has occurred due to amount of seeds released and the increased reproduction from released seeds that have mature and added to the reproduction stock. The reproduction effect was markedly different between locations of sowing culture ground (Kitada et al.1996). Because an operating plan and a processing shipping plan are made based on the number of seeds released and the stock investigation made by many fishing cooperatives, the development of the wild population that compose year classes different from the sowed population will obstruct the production planning in many cases (Kurata 1996).

The statistical scallop harvest record in Japan inconveniently includes production from the wild population and masks the positive effect of the seed input estimates. Thus, accurate analyses have not reached completion for lack of separate harvest data on sowing culture. Statistical analyses are the primary and important step to investigating the effects of scallop culture technology.

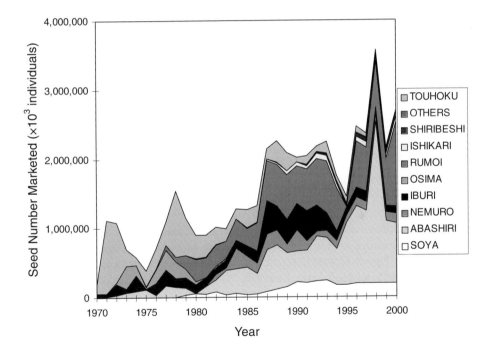

Figure 22.16. Changes in annual seed marketing of Japanese scallop, *Patinopecten* (*Mizuhopecten*) *yessoensis* Jay, in every region of Japan, 1970–2000. (Source: Ministry of Agriculture, Forestry and Fisheries, Government of Japan, 'Annual report of statistics of fishery and aquaculture production').

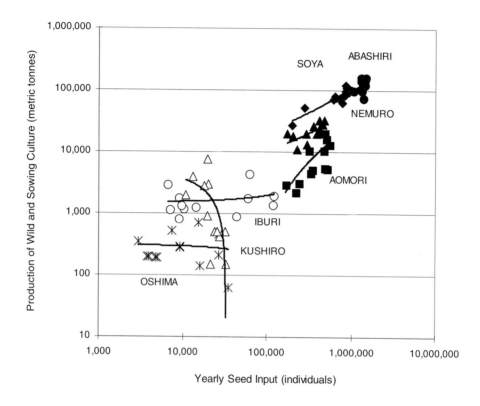

Figure 22.17. Relationship between seed input and production (wild and sowing culture) of Japanese scallop, *Patinopecten* (*Mizuhopecten*) *yessoensis* Jay, in the five main culture regions around Hokkaido. Figures of 1985–2000 are years of catch. Seed input numbers represent the corresponding values in assumed previous spat collection time (generating time of the scallop), four years traced back before recapture for catch.

22.2.14 Value

The scallop industry increased 32.1 times in value and 23.1 times in production over 30 years from 1970 to 2000 and was valued at more than 80 billion yen in 1998–2000 (Fig. 22.18). This value supports the economic activity of coastal fisheries in north Japan.

If one observes the relations between the amount of production of the scallop in Japan and the price of scallops it can be seen that the price decline began after 1987, when a total production exceeded 300,000 mt, and that tendency continues now. A tendency for

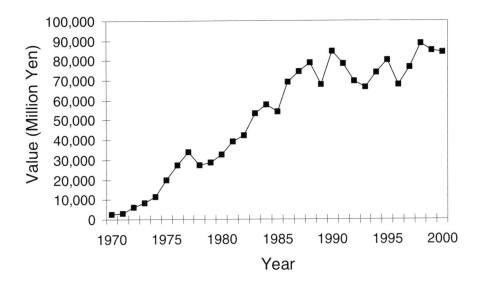

Figure 22.18. Changes in annual value of Japanese scallop, *Patinopecten* (*Mizuhopecten*) *yessoensis* Jay, in Japan, 1970–2000. Source: Ministry of Agriculture, Forestry and Fisheries, Government of Japan, 'Annual report of statistics of fishery and aquaculture production.

price decrease is seen after 1988 when total production exceeded 80,000 mt and continues to do so to this date in Aomori. It is considered that the decreasing price is caused by an increase in the amount of production in Japan on the whole and also imports from China.

The scallop price fluctuated in the ranges of 173–288 yen kg[-1] in 1971–1986 and has declined gradually since then. It had a peak of 288 yen kg[-1] in 1976. There was much anxiety about production due to the initial occurrence of a mass mortality in the hanging culture in northeast Honshu (east Tohoku) areas of Miyagi, Iwate and Aomori prefectures in the mid-1970's. In Aomori, the price soared suddenly in the next year after mass mortality occurred in 1975 (Fig. 22.19). Additionally, the culture cost rose suddenly at a remarkable rate due to increased oil prices in 1974 making the culture materials from petrochemical products expensive. A secondary peak of 264 yen kg[-1] appeared in 1980, because the hanging culture production touched bottom. In 1984 a third peak of 275 yen kg[-1] came of the brisk marketability for increasing the export of frozen muscle after the scallop fishery in USA closed. The unit price of yen kg[-1] at harvest from hanging culture has been higher than that of catch from sowing culture and wild scallops. In addition, the yearly price on harvest is inversely proportional to the quantity. The market for total harvest is presumed to depend directly on the current marketability and on catch hardly shows a tendency to be directly influenced by the current marketability. The scallop is an export item even now, though export of scallops to France was prohibited in 1990 and domestic consumption had tended to increase. In 1985–1986 the price declined due to

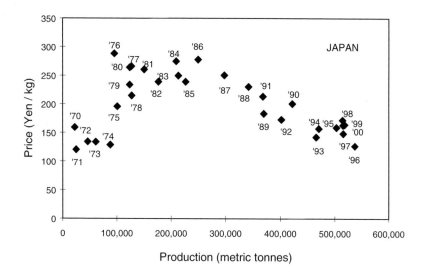

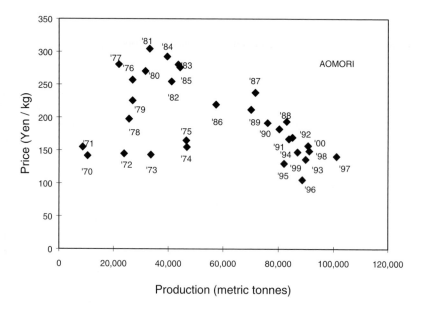

Figure 22.19. Relationship between annual production and price of Japanese scallop, *Patinopecten* (*Mizuhopecten*) *yessoensis* Jay, in Japan, 1970–2000. Source: Ministry of Agriculture, Forestry and Fisheries, Government of Japan, 'Annual report of statistics of fishery and aquaculture production.

both the increase in hanging culture production and the good Japanese yen exchange rate. Furthermore, that unit price fell gradually and reached the low level of 129 yen kg^{-1} in 1996 as the amounts of production of scallop in Japan increased greatly.

22.2.15 Processing

Though scallops produced in Iwate and Miyagi are almost all distributed fresh, most of the scallop is processed for marketing, with 16.3% of fresh scallops in Hokkaido and 2.3% in Aomori consumed in the fiscal year 2001–2002 (Fig. 22.20). Processed scallop products include the frozen, dried, canned, adductor muscle, and boiled meat (soft body). The distribution by scallop processing product in Hokkaido is calculated at four times [A-D] separated as follows: averages for 27 years throughout 1975/76–2001/2002 for the fiscal year (April-March) are presented (1) 22.4% for fresh scallop, (2) 35.5% for frozen muscle, (3) 20.8% for boiled meat, (4) 15.3% for dried muscle and (5) 6.2% for canned muscle and others. [A] Averages for first three years of 1975/76–1977/78 are (1)57.8%, (2) 4.1%, (3) 17.5%, (4) 16.5% and (5) 4.1%. [B] Averages for second 5 years of 1978/79–1982/83 are (1) 30.7%, (2) 20.9%, (3) 24.4%, (4) 19.9% and (5) 4.1%. [C] Averages for third ten years of 1983/84–1992/93 are (1) 16.8%, (2) 32.6%, (3) 26.6%, (4) 16.3% and (5) 7.7%. [D] Averages for fourth nine years of 1993/94–2001/2002 are (1) 20.1%, (2) 39.3%, (3) 18.8%, (4) 15.7% and (5) 6.0%. Consequently, each time span has a peculiar progress. [A] New processes: where 58% of the production was fresh scallop, and the processed products had a small share. Boiled meat and dried muscle were 17% at the same rate; however, the frozen muscle remained at an especially low rate. Production overwhelmed the processing capacity. In that period, the harvest from hanging culture increased rapidly toward a peak of 1977/78, though the catch was low. The annual harvest and catch were in the ratio of 66.5% to 33.5% and the total reached the first record, more than 100 thousand mt in 1977/78. [B] Developing stage: The fresh scallop decreased to the quantitative bottom in 1982/83; still the yearly average ratio was 30.7%. Frozen and dried muscles and boiled meat increased, especially the boiled meat, which reached a first peak over the harvest amount in 1982/83. The harvest touched bottom due to mass mortality in 1979/80; however, the catch increased rapidly. [C] Growing stage: absolute quantity for processing of every product increased with time. The boiled meat and fresh scallop suddenly increased from 1987 to 1989, the frozen muscle remained high. The frozen muscle and boiled meat increased in the average ratio, and the fresh scallop and the dried muscle decreased. The harvest recovered and exceeded 100,000 mt, sown culture exceeded 200,000 mt too; therefore, the total increased and surpassed 300,000 mt in 1990/91. [D] Mass processing stage: Processing of the frozen and the dried adductor muscle has grown rapidly by having increased the production of wild and sown culture scallop in the Okhotsk Sea. The frozen muscle

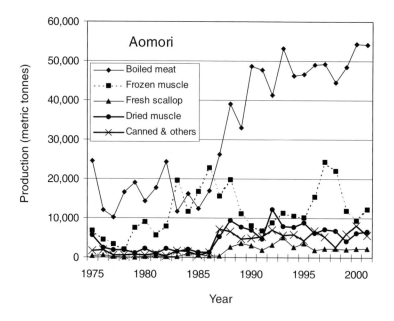

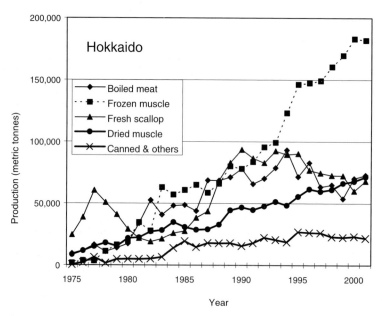

Figure 22.20. Changes in annual processing production of Japanese scallop, *Patinopecten* (*Mizuhopecten*) *yessoensis* Jay, in Japan, 1975–2000 of fiscal year (April-March). (Source: Hokkaido and Aomori Federation of Fisheries Co-optations survey).

especially came to occupy about 40% of the total production. On the other hand, the production of boiled and fresh scallop decreased gradually because the production of hanging culture scallop remained constant. The wild and sown culture comes to exceed 300,000 mt; therefore, the total increased and surpassed 400,000 mt in 1999/2000. Throughout this period of 27 years, the fresh scallop fluctuated randomly, and the canned muscle and other products decreased.

The distribution by the scallop processing product in Aomori has been calculated by the same method at four times [A-D] separated as follows: averages for 27 years throughout 1975/76–2001/2002 for the fiscal year are presented (1) 2.9% for fresh scallop, (2) 21.5% for frozen muscle, (3) 59.8% for boiled meat, (4) 9.1% for dried muscle and (5) 6.7% for canned muscle and others. [A] Averages for the same products over the first three years of 1975/76–1977/78 are (1) 1.3%, (2) 19.1%, (3) 61.2%, (4) 12.9% and (5) 5.4%. [B] Averages for the next 5 years, 1978/79–1982/83 are (1) 0.6%, (2) 23.6%, (3) 67.5%, (4) 6.2% and (5) 2.2%. [C] Averages for the next nine years, 1983/84–1990/91 are (1) 2.9%, (2) 30.9%, (3) 50.3%, (4) 8.8% and (5) 7.1%. [D] Averages for the next eleven years of 1991/92–2001/2002 are (1) 3.5%, (2) 16.9%, (3) 63.0%, (4) 9.3% and (5) 7.3%. Consequently, each epoch has a peculiar process. [A] New processes: most (60%) of the production was consumed as the boiled scallop, with processed products being a small share. A mass mortality occurred and the production of scallop in Mutsu Bay rapidly decreased. Frozen muscle was 21%; however, the fresh scallop, dried and canned muscle remained at an especially low rate. The annual harvest and catch were in the ratio of 57.2% to 42.7%. [B] Developing stage: Fresh scallop decreased to its ultimate quantitative, 44 mt in 1978/79. Production of frozen muscle and boiled meat increased, especially the boiled meat, which increased rapidly. At that time, the production of roe-on scallop to export to France increased, too. The harvest recovered and exceeded 20,000 mt in 1978/79, surpassing 30,000 mt in 1982/83. [C] Growing processing stage: The boiled meat suddenly increased from 1988 to 1990. Because export to America increased rapidly, the production of frozen muscle remained high and increased in accordance with average ratio, with the other products increasing in the late 1980's as well. The harvest recovered and exceeded 50,000 mt, with sowing culture coming to exceed 200,000 mt; hence the total increased and surpassed 70,000 mt in 1990/91. [D] Mass processing stage: The production of boiled scallop maintained a high level. The production rate of boiled scallop was more than 60% of total production in Mutsu Bay; especially the boiled small one-year-old scallop called 'Hansei-gai' which came to occupy more than 25%. Production of frozen muscle fluctuated at this stage because although export of the frozen muscle to America was favourable from 1997 to 1998, export to France was prohibited in 1990. The production of sowing culture decreased to less than 10,000 mt; therefore, the hanging culture came to occupy about 90% of the total production.

The boiled meat is a significant part of the harvest. The productions of frozen and dried muscles were significantly related to the catch, because scallop body size determines processing methods. Boiled meat is processed from a small scallop harvested within two years of age, while the frozen and dried muscles need to be good quality, large scallops caught at more than three-years-old.

22.2.16 Occurrence of Shellfish Poisoning

Paralytic shellfish poisoning (PSP) and diarrhetic shellfish poisoning (DSP) occurred in Hokkaido in 1978 (Nishihama 1994). Occurrences of PSP and DSP in the scallop industry change the harvest period and the array of processed products. Various countermeasures were taken (Okaichi et al. 1982; Nishihama 1980). A series of food poisonings resulting from mussel and scallop occurred in the Tohoku region in 1976 and the summer of 1977 and the presence of an unknown fat-soluble toxin was revealed (Yasumoto et al. 1978). In Mutsu Bay the regulation for DSP was applied to scallops and mussel in 1988 (Thubata 1982). The toxin levels of PSP in Funka Bay, Pacific Ocean and the Nemuro Strait are extremely high, but in the Japan Sea and northern part of the Okhotsk Sea in Hokkaido are not detected, or are very low (Tables 22.7 and 22.8; Nishihama 1994). Harvest was continued year round until 1977 when the shellfish poisoning limited the harvest. Both poisonings occur from spring to autumn, the catch is limited by PSP, though DSP changes the disposition to processing. The scallop can be distributed when the digestive gland is removed even if DSP has occurred, because DSP is lipid soluble toxin and does not spread to other tissues except for the digestive gland. The digestive glands of scallops are disposed of, and scallops in Mutsu Bay are distributed even if DSP occurs from spring to autumn, because PSP never occurred.

22.3 FUTURE

22.3.1 Technology

There is an unstable element in the natural spat collection in the bay as well as offshore. Spat collection was depressed several times during these ten years in Funka Bay. The cause of the natural spat collection depression should be detected, and a concrete countermeasure must be taken to maintain scallop industries' steady production. It is necessary to develop sustainable seed supplies to maintain production. Seed production in embayment areas such as Saroma Lake and Mutsu Bay may meet the demand and avoid the spat collection depression in the future; therefore, offshore spat collection has to reliably provide more seed for sowing culture.

Recently, there has been a decline in the growth of sowing cultured and hanging cultured scallop, probably caused by overpopulation and environmental conditions (Kurata 1999; Shimada et al. 2000; Kosaka et al. 2002).

New technology and continuous monitoring will be required, such as controlling cultured populations, culturing at low cost, and producing goods of quality, to develop and maintain scallop culture industries in Japan.

Table 22.7

Regulation period of PSP for Japanese scallop, *Patinopecten (Mizuhopecten) yessoensis* Jay, in Japan, 1991–2000. (Source: Hokkaido Prefecture Government survey).

Area	1991	1992	1993	1994	1995	1996	1997	1998	1999
East Funka Bay	4-Apr~12-Dec 4-Apr~12-Dec			1-Jul~8-Oct	25-Apr~10-Nov				
West Funka Bay	13-Mar~12-Dec 13-Mar~12-Dec		24-Mar~	10-Jun~13-Dec	21-Apr~14-Dec				
Entrance Funka Bay	~ ~	~27-Feb	19-Apr~23-Dec	28-Apr~6-Dec	17-Mar~2-Dec				25-Apr~1-Nov
Tsugaru Strait		6-Mar~2-Dec							
South Japan Sea									7-Apr~17-Dec
Mid Japan Sea									
Ishikari Bay									
North Japan Sea									
North Okhotsk Sea									
South Okhotsk Sea						9-Aug~26-Nov	11-Sep~21-Nov		
Saroma Lake									
NOTORI LAKE									
Nemuro Strait	20-Jul~8-Nov								
East Pacific Ocean	~7-Feb 20-Jun~							14-Jul~6-Nov	
Mid Pacific Ocean	17-May~	~ ~	~17-Dec						
West Pacific Ocean	11-Jul~	~10-Feb							

1131

Table 22.8

Regulation period of DSP for Japanese scallop, *Patinopecten (Mizuhopecten) yessoensis* Jay, in Japan, 1991–2000. (Source: Hokkaido and Aomori Prefecture Government survey).

Area	1991	1992	1993	1994	1995	1996	1997	1998	1999	2000
East Funka Bay	20-Apr~12-Dec	18-Apr~26-Sep	27-Apr~25-Sep	17-May~8-Oct	30-May~9-Dec	15-May~12-Oct	23-May~4-Oct	28-Oct~24-Dec	22-Apr~22-Oct 22-Nov~23-Dec	15-May~13-Oct 2-Dec~8-Jan
West Funka Bay	26-Apr~8-Feb		24-Apr~1-Oct	26-May~13-Dec	~	27-May~9-Nov	2-Jun~22-Oct	14-Nov~18-Dec	24-Apr~1-Nov 2-Dec~8-Jan	15-May~2-Nov
Entrance Funka Bay	11-Nov~	16-May~2-May	19-Apr~23-Dec 15-Jun 16-Jul	26-May~6-Dec	26-May~2-Dec				9-Apr~17-Dec 31-Jul~18-Aug	
Tsugaru Strait							11-Apr~28-Apr			
South Japan Sea					21-Apr~2-Jun					
Mid Japan Sea										
Ishikari Bay	16-May~18-Oct		29-Apr~9-Nov	9-Jun~26-Nov			23-May~19-Jul		8-May~3-Jun 24-Jun~20-Jul	4-Apr~2-May
North Japan Sea	11-Jul~16-Oct	21-May~5-Sep	18-May~1-Oct	29-Apr~15-Oct						
North Okhotsk Sea		20-Apr~24-Aug	28-Apr~4-Sep	22-Apr~12-Aug	25-Apr~8-Aug				22-May~10-Jul	10-May~13-Jul
South Okhotsk Sea		1-May~5-Sep	9-May~4-Sep 10-Dec~	~12-Sep						12-May~1-Jul
Saroma Lake	16-May~28-Sep	6-Jun~31-Aug	3-Jul~4-Sep	~13-Sep		23-Dec~26-Nov				
Nemuro Strait		6-Jun~31-Oct	17-Dec~	25-Feb~12-Apr 3-Jun~29-Oct	23-Dec	26-Jun~2-Nov			31-Jul~11-Sep	
East Pacific Ocean		6-Jun~31-Oct		~26-Nov						
Mid Pacific Ocean	16-May~	~ ~	~17-Dec	28-May~	~ ~18-Nov	24-May~14-Nov	16-Sep~1-Nov			
West Pacific Ocean	31-May~	~10-Jan 6-Jun~10-Oct								
Aomori	23-Mar~21-Sep	1-Apr~4-Sep	19-Mar~1-Oct	8-Apr~16-Sep	24-Mar~18-Sep	22-May~4-Oct	9-Apr~12-Sep	2-Apr~8-Oct	27-May~19-Aug	27-Apr~27-Jul

22.3.2 Control of Culture

Until now, the study of a carrying capacity level using primary production and feed amount of the scallop in Mutsu Bay, Funka Bay and Saroma Bay, a closed area, were made to prevent overcrowded culture. Similar studies were conducted in the same way even in Tokoro, Saroma and Yubetsu, which are open areas. Limiting mass of the hanging culture will be required with basic consensus of the producers discussed (Aomori Prefecture 1976, 2000; Hokkaido Fish. Res. Tec. Dev. Asso. 1977; 1978; Hokkaido Aquaculture Public Corp. 1983; Offshore Scallop Fishery Synthetic Investigation Special Commissioner 1997a, 1997b). Densities of seeds and surviving scallops have to be controlled in order to increase productivity (Ito et al. 1988b). The cultured and sown level is regulated in accordance with the results of studies in Hokkaido and Aomori. The density of seed sown on culture grounds has been kept at the same level everywhere because erroneous hypotheses that survival rates and cultured standing crops are the same everywhere are still believed. A profitable control for creating a systematic industry based on scientific fact will be developed. The density reduction by removal of scallops has been experimentally managed to avoid over crowded populations caused by high survival rates, concentration, and natural mass recruitment. Furthermore, for sustainable production of a high quality scallop population, introducing a concept of breeding management on scallop culture and genetic management should be carried out (Kosaka 1997b; 1999).

22.3.3 Cost-reduction

The net profit from scallop culture is decreasing because the price is declining rapidly, the materials and equipment have become more expensive and Japanese labour wages have risen. The indispensable sales strategy for the market of not only Japan but also the world is to maintain the price of the scallop. Moreover to keep its profitability, cost-reduction is necessary. In culture, the survival rate has to be improved and the quality must also be improved for a higher price. Furthermore, efficient and cheap materials or equipment must be developed. The labour involved in sowing culture is higher than that of hanging culture because the sowing culture system runs by low handling frequency and highly concentrated operations; therefore, automating sowing culture would be a great advantage in cost-reduction.

22.3.4 Concept

To develop the technology, the concept of scallop mariculture is indispensable. Culture systems need completion by elements of 'Tane-zukuri [making seed]', 'Gyoba-zukuri [making culture ground]' and 'Hito-zukuri [making person]': meaning seed production, culture control, and manpower. Every elemental technical method, know-how and manpower relates and responds to one another.

22.4 *PECTEN ALBICANS*

The bay scallop, *Pecten* (*Notovala*) *albicans* (Schröter, 1802), is called "Itaya-gai" in Japanese, the baking-scallop.

22.4.1 Biology

Bay scallops live on the sandy sea bottom in coastal areas of the temperate zones of China, Taiwan, Korea, and Japan. This scallop is distributed in shallow seas around Japan such as Kyushu, Shikoku, Honshu, and south Hokkaido except the Pacific coasts of east Honshu. In the Japan Sea coasts of Kyushu and Honshu, high density populations have sometimes appeared at depths of 20–120 m off Fukuoka, Yamaguchi, Shimane, and Tottori Prefecture (Hiramatsu 1960; Koga 1963).

The fan shaped valves have symmetrical ear beaks at the umbones. The shell shape is convex in the right valve with a white surface and the left one is flat with a slightly hollowed area below the umbo and brown colouration. The radiating rib number is between 8 and 10 on a right valve. The morphological development of the bay scallop is as follows (Hotta 1977): fertilised demersal eggs of 79 microns diameter reach D-shaped larvae of early veligers of 102 µm in length at two days. Larvae grow into the subsequent developmental stages of umbo 120–160 µm by 10 days, early full-grown 190–210 µm by 14 days and full-grown 154–270 µm with eye spots by 20 days. The pediveliger attaches to the substratum with byssus by 25 days. Attached spat produce the dissoconch of 230–250 µm length by 28 days. The spat grow to 315–356 µm in one month with a daily growth rate of 8.6–16 µm and 1.5–3.6 mm by two months at the rate of 20–140 µm. It becomes a benthic spat which parts from the substratum by three months at 10 mm length at the rate of 2.0–3.0 mm. Juveniles have an adult shell form at 5 cm length and 14 g weight at 6 months. Length and weight are 8 cm and 60 g at one year, 10 cm and 100 g at 1.5 years, 11 cm and 150 g at two years, and 11 cm and 150 g at 2.5 years.

The spawning season lasts from late-January to early-May with spawning occurring actively from mid-February to early-March (Satake et al. 1981). Bay scallops of just one-year-old matures and spawns from April to early June, but tends to spawn later than adult scallop (Moriwaki et al. 1982).

22.4.2 Fishery

The bay scallop fishery has been recorded in the Japan Sea coasts off Tottori Prefecture, San-in district of west Honshu, for more than 180 years (Tanaka 1984). The catch was below 200 mt per year before 1920 as the fishery was conducted by a hand trawl from a rowboat or sailboat and was regulated by scallop size and fishing period. A 1925 catch of 28,000 mt from a large recruitment was harvested after the introduction of powered boats and improvements in catch efficiency. The catch remained below 200 mt in the 40 years since that 1925 peak. A secondary high catch period occurred in 1965–67, and the catch ranged from 13,000–26,000 mt off Shimane and Ishikawa Prefecture. The catch decreased again in the 1980's.

22.4.3 Culture

Natural spat collection was tried in the Shimane Peninsula, Shimane Prefecture from 1978 to 1980 (Oshima 1978) and hanging culture of bay scallops was examined at Oki Island, Shimane Prefecture, in 1977 (Ayama 1986). After this promising result the culture extended to the fishery level. Two and a half million seed were harvested by about 150 fishermen in Shimane Prefecture by 1984. While that harvest slightly increased in the mid 1980's, the culture has become a fluctuating activity of late. After 1985 natural spat collection resulted in low level production (Yoshio et al. 1987; Seimura 1994). This decline caused the seed production levels to fluctuate. Recently the natural stocks have occurred at lower levels. The artificial spat production of bay scallop have been made by the public research organisation in Shimane, Miyazaki, Ishikawa, Hiroshima and private hatchery in Wakayama, but that method is not yet established. The hanging cultures were carried out with the increase in the natural spat collection in Shimane Prefecture at each place until 1985, but declined with the decrease in natural spat collection (Seimura 1991). Enhancement of the bay scallop was planned for 1980–1988 (Tanaka 1984, 1987); however, sowing culture has not developed.

22.5 CHLAMYS (MIMACHLAMYS) NOBILIS

The noble scallop, *Chlamys (Mimachlamys) nobilis* (Reeve, 1852), is called "hiougi-gai" in Japanese.

22.5.1 Biology

The noble scallops are distributed on the rocky or sandy sea bottom from the littoral zone to several fathoms depth in southwest Japan. Reproductive cycle of the noble scallop is biannual; the first spawning season from June to July is followed by the second spawning season from October to November. Most male juveniles matured at 8 and 9 months, whereas a few mature females were observed at 11 months of age (Komaru 1988). Fertilised eggs develop into D-shaped larvae by 20–24 hours. The pediveliger of 200 µm length attaches to the substratum with byssus by 2–3 weeks. The scallop grows to 7–8 cm by 1–1.5 years after fertilisation.

The shell colour of noble scallop is controlled genetically. It was argued that blue purple and red purple were dominant to brown, blue purple to red purple and lemon yellow, and blue purple were incomplete dominance to red purple (Nanba 1975).

22.5.2 Culture

The hanging culture is industrialised but the sowing culture is undeveloped (Shiihara 1986). Most fishermen culture scallops as a subsidiary job. So each harvest is kept small to preserve it as a unique local product. The culture is supported by seed from natural spat and artificially raised spat. To stabilise a seed supply, some experiments on artificial seed production have been carried out (Nanba and Nishiyama 1973; Hotta 1973; Shiihara and

Takeda 1978). Artificially produced larvae grow to attached spat at 1 mm shell length by about 40 days in tanks, and then they are caged for intermediate culture in the sea. Shell length becomes 3–5 mm at 2 months and 6–12 mm at three months. From September to November seeds of more than 10 mm are caged for hanging culture and they grow to 10–20 mm at 4 months, 13–31 mm by 5 months and 7–8 cm at 1.5 years. Scallops more than 7 cm are harvested from the next autumn to the following spring.

ACKNOWLEDGMENTS

The authors wish to thank Messrs. Mamoru Kurata and Akira Miyazono for their friendship and constructive discussion. The authors also wish to express sincere thanks to the Hokkaido and Aomori Federation of Fisheries Cooperatives for providing the scallop statistics.

REFERENCES

Aomori Prefecture, 1976. "Mutsu-wan gyogyo kaihatu kihon keikaku chosa saisyu hokokusyo" [Final survey report on environmental conditions of the fishing ground and basic plan for the fishery development in Mutsu Bay]. [in Jpn.] Aomori, Aomori Prefecture. 372 p.

Aomori Prefecture, 2000. "Mutsu-wan niokeru hotategai no souryo kisei" [Regulation on mass production of scallop in Mutsu Bay]. [in Jpn.] Aomori, Aomori Prefecture. 16 p.

Aoyama, S., et al., 1982. "Hotate-gai yoshoku gijutsu kenkyu review, Syowa 51–55" [Technical research review on hanging culture of Japanese scallop in 1976–1980]. [in Jpn.] Aquaculture Center of Aomori Prefecture. 196 p.

Aoyama, S., 1989. The Mutsu Bay scallop fisheries: Scallop culture, Stock enhancement, and resource management. In J. F. Caddy (Ed.). Marine Invertebrate Fisheries: Their Assessment and Management. John Wiley and Sons, New York. pp. 525–539.

Ayama, T., 1986. "Itaya-gai" [Bay scallop]. [in Jpn.] Senkai yoshoku (Tokyo). pp. 446–454.

Fuji, A. and Hashizume, M., 1974. Energy budget for a Japanese common scallop, *Patinopecten yessoensis* (Jay), in Mutsu Bay. Bull. Fac. Fish. Hokkaido Univ. 25(1):7–19.

Habe, T., 1977. Systematics of mollusca in Japan. Bivalvia and scaphopoda. (Tokyo). 372 p.

Hayashi, T., Tomita, K., Wakui, T., Ito, H. and Matsuya, M., 1976. Propagation of scallop by transplantation of seed in the Okhotsk Sea coast of northern Hokkaido. [in Jpn.] Jour. Hokkaido Fish. Exp. Stn. (Hokusuisi Geppo) 33(9):1–16.

Hiramatsu, T., 1960. "Itaya-gai no sigen-gaku-teki kosatsu" [Studies on bay scallop resources]. [in Jpn.] Ann. Rept. Fukuoka Fish. Exp. Stn. (F.Y. 1959). pp. 153–157.

Hirano, T., 1983. "Mutsu-Wan niokeru nimaigai to hitode no fuchaku no syunen henka" [Seasonal change of attachment of bivalves and sea stars in Mutsu Bay]. [in Jpn., Engl. abstr.] Sci. Rep. Aquaculture Center, Aomori Pref. 2:1–8.

Hokkaido Aquaculture Public Cooperation, 1983. "Saroma-ko niokeru hotategai no yousyoku kyoyoryo chosa hokokusyo" [Survey report of cultured capacity in Saroma Lake]. Hokkaido Aquaculture Public Cooperation. 55 p.

Hokkaido Fish. Res. Tec. Dev. Assoc., 1977. "Saroma-ko niokeru hotategai no yousyoku kyoyoryo chosa hokokusyo" [Survey Report of Cultured capacity in Saroma Lake]. Sapporo, Hokaido. Fish. Res. Tec. Dev. Asso. 48 p.

Hokkaido Fish. Res. Tec. Dev. Assoc., 1978. "Funka-wan hotategai no yousyoku kyoyoryo chosa hokokusyo" [Survey report of cultured capacity in Funka Bay]. Sapporo, Hokaido. Fish. Res. Tec. Dev. Assoc. 53 p.

Hotta, M., 1973. Artificial mass production of commercially important shellfish seedlings. [in Jpn.] Ann. Rept. Hiroshima Fish. Exp. Stn. (F.Y. 1972). pp. 7–13.

Hotta, M., 1977. On rearing the larvae and youngs of Japanese scallop, Pecten (Notovola) albicans (Shorter). (Preliminary report). [in Jpn.].

Ito, H., 1984. "Hokkaido no hotate-gai shubyo-horyu no gaiyo" [Outlines on seed sowing of Japanese scallop in Hokkaido] [in Jpn.] Hokusuiken news (31). Hokkaido Reg. Fish. Res. Lab. pp. 2–4.

Ito, H., 1985. "Hotate-gai fuyuyosei no kenkyo-sokutei notameno kanben-sochi nitsuite" [Simple computer devices for microscopic measuring of Japanese scallop larvae] [in Jpn.] Hokusuiken news 33:7–9.

Ito, H., 1986. "Hotate-gai ten-nen saibyo nikansuru flouchaat nitsuite" [A flowchart for natural spat collection of Japanese scallop]. [in Jpn.] Hokusuiken news 34:3–5.

Ito, H. and Hamaoka, S., 1988b. Abundance of net plankton in Nemuro Straights, east Hokkaido, in the season of spat collection of Japanese scallop, Patinopecten yessoensis, in 1984. Scallop (Kushiro). pp. 61–68.

Ito, H., Moriya, H. and Sasaki, T., 1988a. "Hokkaido enganiki niokeru hotate-gai fuyuyosei no bunpu to gaikai-saibyo gijutsu" [Larval distribution and offshore spat collection technology of Japanese scallop in Hokkaido's coasts]. [in Jpn.] Spring Meeting Japan. Sci. Soc. Fish. pp. 255.

Ito, H., Tanaka, S. and Wakui, T., 1986. "Hotate-gai" [Japanese scallop]. [in Jpn.] Senkai yoshoku (Tokyo). pp. 419–445.

Ito, H., Yano, Y. and Sakai, Y., 1988c. An analysis of seed and ground effects on sowing culture of Japanese scallop, Patinopecten yessoensis, in Nemuro coasts of east Hokkaido. [in Jpn.] Progress report of marine ranching project (9):121–160.

Ito, S., 1964. On the scalloping in Okhotsk Sea. Suisan zoyoshuku sosho (Tokyo). 7. 40 p.

Kanno, H. and Sato, S., 1980. "hotategai-zouyoushoku no jidai" [Present aspects of commercial scallop culture in Japan]. [in Jpn.] Suisan-gaku series (Tokyo) 31:11–25.

Kanno, H. and Wakui, T., 1978. "Hotate-gai bo-gai shudan no zosei" [Cultivation of mother shell population of the Japanese scallop]. [in Jpn.] Suisan-gaku series (Tokyo) 23:79–89.

Kawamata, K., 1983. Reproduction cycle of the scallop, Patinopecten yessoensis (JAY), planted in Funka Bay, Hokkaido. [in Jpn., Engl. abstr.] Sci. Rep. Hokkaido Fish. Exp. Stn. 25:15–20.

Kawamata, K., 1988. Gonadal development of the cultured scallops, Patinopecten yessoensis (JAY), off Rumoi, Hokkaido [in Jpn., Engl. abstr.] Sci. Rep. Hokkaido Fish. Exp. Stn. 31:9–13.

Kawamata, K., 2002. "housayroku de aterareruka? Hotaegai no sachi sono 2 zenkoku oyobi Hokkaido-hokubu-san ~ nemurokaikyo-san no maki" [Can you show the correct production area of the scallop by the number of radiating ribs? Volume of whole country and northern part of Hokkaido ~ the Nemuro Channel]. [in Jpn.] Hokusuiken news (55):1–7.

Kawamata, K., Tamaoki, Y. and Fuji A., 1981. Gonad development of the cultured scallops in Funka Bay. [in Jpn.] Jour. Hokkaido Fish. Exp. Stn. (Hokusuisi Geppo) 38:132–146.

Kinoshita, T., 1935a. "Hotate-gai saibyo shiken" [A test for natural spat collection of the Japanese scallop]. [in Jpn.] "Hokusui junpo" (Report of the Hokkaido Fish Res. Stn.] 273:1–8.

Kinoshita, T., 1935b. "Hokkaido san hotate-gai no kara no hosharoku-su no chihoteki heni" [Local variations in radiating rib number of Japanese scallop shells from Hokkaido]. [in Jpn.] Venus 5(4):223–229.

Kinoshita, T., 1949. "Hotate-gai no zoshoku ni kansuru kenkyu" [Studies on the enhancement of the Japanese scallop]. [in Jpn.] Suisan kagaku sosho 3 (Sapporo, Hokkaido). 106 p.

Kinoshita, T., Shibuya, S. and Shimizu, Z., 1943. Induction of spawning of the scallop, *Pecten* (*Patinopecten*) *yessoensis* Jay. [in Jpn., Engl. synop.] Bull. Japan. Soc. Sci, Fish. 11(5–6):168–170.

Kinoshita, T., Shimizu, Z. and Shibuya, S., 1944. "Hotate-gai gyojo sensui chosakekka (yoho)" [A survey on the Japanese scallop bed by diving (primary report)]. [in Jpn.] Jour. Hokkaido Fish. Exp. Stn. (Hokusuisi Geppo) 1(6):322–326.

Kitada S. and Fujishima, H., 1997. "Hokkaido oinker hetaerae no syubyohoryu- koka no kento" [The stocking effectiveness of scallop in Hokkaido]. Nippon Suisan Gakkaishi 63(5):686–693.

Koga, F., 1963. "Genkai-nada niokeru itaya-gai sigen chosa" [A survey on bay scallop resources in Genkai-nada waters]. [in Jpn.] Ann. Rep. Fukuoka Fish. Exp. Stn. (F.Y. 1962). pp. 22–40.

Komaki, S., 1975. "Senkai gyojo kaihatsu to gaikai joken" [Development of the coastal fishing ground and conditions of offshore water]. [in Jpn.] Suri-kogaku1 series. 75-B--5.

Komaru, A., 1988. Seasonal changes of gonad in the cultured scallops, *Chlamys nobilis*. Bull. Natl. Res. Inst. Aquaculture 14:125–132.

Kosaka, Y., Aisaka, K. and Kawamura, K., 1996a. "Hotategai no seicho to sono-tokucho". [The growth and those characteristics of Scallop] [in Jpn.] Annual report of the Aquaculture Center, Aomori Prefecture (26):130–136.

Kosaka, Y., Aisaka, K. and Kawamura, K., 1996b. "Mutsu-wan oinker kotuku -hetaerae no Shinjuku saran". [Maturation and spawning of the hanging cultured scallop in Mutsu Bay]. [in Jpn.] Annual report of the Aquaculture Center, Aomori Prefecture (26):121–129.

Kosaka, Y., Aisaka, K. and Takarada, M., 1996c. "Mimizuri-hotategai no syunki heisi-genin". [Cause of mortality of ear-hanging culture scallop in Spring]. [in Jpn.] Annual report of the Aquaculture Center, Aomori Prefecture (26):140–148.

Kosaka, Y., Kawamura, K., Kudo, T. and Tamura, W., 1997a. "Mutsu-wan niokeru yoshoku-hotategai no seijuku san-ran". [Maturation and spawning of the hanging cultured scallop in Mutsu Bay] [in Jpn.] Annual Report of the Aquaculture Center, Aomori Prefecture (27):173–182.

Kosaka, Y., 1997b. Genetic studies on scallop culture in Mutsu Bay. [in Jpn., Engl. abstr.] Sci. Rep. Aomori Pref. Aquacult. Res. Cen. (8):1–47.

Kosaka, Y., 1999. Breeding management of scallop culture in Mutsu Bay. [in Jpn., Engl. abstr.] Fish Genet. Breed. Sci. (27):57–66.

Kosaka, Y., Kimura, H., Yosida, M., Omizu, M. and Kawamura, K., 2002. "hotategai yuika-yosyoku jittai-tyosa". [Study of status of hanging cultured scallop]. Annual Report of the Aquaculture Center, Aomori Prefecture (31):155–170.

Kurata, M., 1996a. Oxygen consumption of Japanese common scallop, *Patinopecten yessoensis* (Jay) in the Okhotsk Sea. [in Jpn., Engl. abstr.]. Sci. Rep. Hokkaido Fish. Exp. Stn. 49:7–13.

Kurata, M., 1996. "Hokkaido tokoro, saroma-kaiiki deno hotategai syubyo horyu no seikou jirei". [The success case of the seed sowing of the scallop in Tokoro and Saroma distinct of sea Hokkaido. [in Jpn.] Saibai 80:17–23.

Kurata, M., 1999. On the decline in the growth of maricultured scallop, *Patinopecten yessoensis*, in the okhotsk coastal area of Hokkaido. [in Jpn., Engl. abstr.] Sci. Rept. Hokkaido Fish. Exp. Stn. 54:25–32.

Kurata, M., Hoshikawa, H. and Nishida, Y., 1991. Feeding rate of the Japanese scallop *Patinopecten yessoensis* in suspended cages in Lagoon Saroma-ko. [in Jpn., Engl. abstr.] Sci. Rep. Hokkaido Fish. Exp. Stn. 37:37–57.

Kurata, M., Nishida, Y. and Mizushima, T., 1996b. Fouling organisms attached on culture scallop, *Patinopecten yessoensis*, in Funka Bay. [in Jpn., Engl. abstr.] Sci. Rep. Hokkaido Fish. Exp. Stn. 49:15–22.

Kurata, M. and Yoshida, M., 2001. "hotategai fuyu-yousei bunpu chosa". [Study on the distribution of scallop larvae]. [in Jpn.] Annual Report of Abashiri Fish. Exp. Stn. Hokkaido Pref. (F.Y. 2000). pp. 101–113.

Maru, K., 1972. Morphological observations on the veliger larvae of a scallop, *Patinopecten yessoensis* (JAY). [in Jpn., Engl. abstr.] Sci. Rept. Hokkaido Fish. Exp. Stn. 14:55–62.

Maru, K., 1976. Studies on the reproduction of a scallop, *Patinopecten yessoensis* (JAY) - 1. Reproductive cycle of the cultured scallop. [in Jpn., Engl. abstr.] Sci. Rep. Hokkaido Fish. Exp. Stn. 18:9–26.

Maru, K., 1978a. Studies on the reproduction of a scallop, *Patinopecten yessoensis* (JAY) - 2. Gonad development in 1-year-old scallops. [in Jpn., Engl. aster.] Sic. Rep. Hokkaido Fish. Exp. Stn. 20:13–26.

Maru, K., 1978b. Studies on the reproduction of a scallop, *Patinopecten yessoensis* (JAY) - 3. Observations on hermaphroditic gonads. [in Jpn., Engl. abstr.] Sci. Rep. Hokkaido Fish. Exp. Stn. 20:27–33.

Maru, K., 1985a. Ecological studies on seed production of a scallop, *Patinopecten yessoensis* (JAY). [in Jpn., Engl. abstr.] Sci. Rep. Hokkaido Fish. Exp. Stn. 27:1–53.

Maru, K., 1985b. Tolerance of a scallop, *Patinopecten yessoensis* (JAY) to temperature and specific gravity during early development stage. [in Jpn., Engl. abstr.] Sci. Rep. Hokkaido Fish. Exp. Stn. 27:55–63.

Masuda, K., 1963. The so-called *Patinopecten* of Japan. Trans. Proc. Palaeont. Soc. Japan, N.S. (52):145–153.

Miyazono, A., 2000. Influence of the intensity of water motion on growth and physiological conditions of scallops cultured in flow tanks. [in Jpn., Engl. abstr.] Sci. Rep. Hokkaido Fish. Exp. Stn. 58:41–47.

Mori, K., 1975. Seasonal variation in physiological activity of scallop under culture in the coastal water of Sanriku district, Japan, and a physiological approach of possible cause of their mass mortality. Bull. Mar. Biol. St. Asamushi Tohoku Univ. 15(2):59–79.

Moriwaki, S., 1982. "Suika-siiku sita itayagai tonengai no seisyokusou no Hattatu" [Development of gonad on one-year-old bay scallop reared by suspend culture]. The Aquaculture 31:57–62.

Nanba, T. and Nishiyama, T., 1973. "Hiougi shubyo seisan kenkyu" [Experiments on artificial seed production of noble scallop] [in Jpn.] Ann. Rep. Wakayama Aquac. Exp. Stn. Hokkaido Pref (F.Y. 1972). pp. 88–95.

1140

Nanba, T. and Nishiyama, T., 1975. "Hiougi shubyo seisan kenkyu" [Experiments on artificial seed production of noble scallop] [in Jpn.] Ann. Rep. Wakayama Aquac. Exp. Stn. (F.Y. 1974) pp. 147–158.

Nishihama, Y., 1980. "kaidouka no kannkyou -youin" [Environmental estimate of carrying capacity for culturing scallop in coastal water]. [in Jpn.] Suisan-gaku series (Tokyo) 31:40–52.

Nishihama, Y., 1994. "Ohotuku no hotate gyogyou" [Scallop fisheries in Okhotsk.][in Jpn.] Hokkaido Daigaku Tosyo Kankokai (Sapporo). 218 p.

Offshore Scallop Fishery Synthetic Investigation Special Commissioner, 1997a. "Gaikai hotategai gyogyo sogo chosa (Tokoro, Saroma kyoyu kaiiki) hokokusyo (yoyakubann)". [Synthetic Investigation (Tokoro, Saroma common sea area) on scallop fishery (summary edition)]. Hokkaido,. Offshore Scallop Fishery Synthetic Investigation Special Commissioner. 41 p.

Offshore Scallop Fishery Synthetic Investigation Special Commissioner, 1997b. "Gaikai hotategai gyogyo sogo chosa (Yubetsu kaiiki) hokokusyo (yoyakubann)". [Synthetic Investigation on scallop fishery (summary edition)]. Hokkaido. Offshore Scallop Fishery Synthetic Investigation Special Commissioner. 53 p.

Okaichi, T., et al., 1982. Toxic phytoplankton-occurrence, mode of action, and toxins. [in Jpn.] Suisan-gaku series 42:135 p.

Okesaku, H. and Tanaka, S., 1969. "hotate-gai chi-gai no gaikai horyu koka ni tsuite" [On the effects of seed sowing of the Japanese scallop in the open sea coast]. [in Jpn.] Jour. Hokkaido Fish. Exp. Stn. (Hokusuisi Geppo) 26(3):32–49.

Osanai, K., et al., 1980. Sexual differentiation in the juveniles of the scallop, *Patinopecten yessoensis*. Bull. Mar. Biol. St. Asamushi Tohoku Univ. 16(4):21–320.

Oshima, T., 1978 "Itayagai no tennen saibyo nituite". Ann. Rept. Shimane Fish, Exp. Stn. (FY. 1977). pp. 288–292.

Satake, T. and Moriwaki, S., 1981. Maturing process in Japanese bay scallop-Seasonal change in gonads. Rept. Shimane Pref. Fish. Exp. Stn. 3:36–43.

Sato, K., Aisaka, K., Ebina, M., Nagamine, F. and Tanaka, S., 1993. Mutsu-wan niokeru yosyoku-hotategai no seicho to kankyo-youin. [Growth and environmental factor for hanging culture scallop in Mutsu Bay]. [in Jpn.] Annual report of the Aquaculture Center, Aomori Prefecture (22):185–203.

Seimura, H., 1991. "Shimaneken no itayagai yousyoku" [Bay scallop culture in Shimane prefecture]. Contribution Fish. Res. Jap. Sea Book. 23:65–72.

Shiihara, H., 1986. "Hiougi-gai" [Noble scallop]. [in Jpn.] Senkai yoshoku (Tokyo). pp. 455–472.

Shiihara, H. and Takeda, T., 1978. Studies on the artificial seed production of *Chlamys (Mimachlamys) nobilis* - 1. [in Jpn.] Bull. Oita. Pref. Fish. Exp. Stn. (10):59–66.

Shimada, H., Nishida, Y., Ito, Y. and Mizushima, T., 2000. Relationship among growth and survival of cultured scallops (*Patinopecten yessoensis* JAY), and environmental conditions in the coastal area of Yakumo, Funka Bay, Hokkaido, Japan. Sci. Rep. Hokkaido Fish. Exp. Stn. 18:9–26.

Tanaka, K., 1984. Increase of survival rate of bay scallop, *Pecten (Motorola) albcans* (Schrooner). [in Jpn.] A new approach to the enhancement of coastal fisheries resources. Results obtained from the study of Marine Ranching Program, Part-I. Ministry of Agriculture, Forestry and Fisheries. pp. 63–93.

Tanaka, K., 1987. High level stabilization of the population of bay scallop and ark-shell in Japan [Population increase with total management of whole life of bivalve]. [in Jpn.] A new approach to the enhancement of coastal fisheries - Results obtained from the study of Marine Ranching Program, Part-II (F.Y.1983–1985). Ministry of Agriculture, Forestry and Fisheries. pp. 101–135.

Tanaka, S., 1963. "Hokkaido niokeru hotate-gai gyogyo to kenkyu-jyo no sho-mondai" [Some problems on fishery and researches of the Japanese scallop around Hokkaido] [in Jpn.] Symposium at 1963 autumn meeting of Japan Soc. Sci. Fish. pp. 142–150.

Tanaka, S., et al., 1977. "Hotate-gai no subyo seisan" [Artificial seed production of Japanese scallop] [in Jpn.] Annual report of the Aquaculture Center, Aomori Prefecture. (6):65–78.

Tomita, K., Tajima, K., Uchida, M., Mon, M. and Wakui, T., 1982. On the population of scallop, *Patinopecten yessoensis* (JAY), in Sarufutsu, Hokkaido. [in Jpn.] Jour. Hokkaido Fish. Exp. Stn. (Hokususi Geppo) 39(6):111–125.

Tsubata, F., 1982. "Mutsu-wan hotate-gai gyogyo kenkyu-shi" [A history of the fisheries researches of the Japanese scallop in Mutsu Bay]. [in Jpn.] Aomori Prefecture. 120 p.

Wakui, T. and A. Obara. 1967. On the seasonal change of the gonads of scallop, *Patinopecten yessoensis* (JAY), in lake Saroma, Hokkaido. [in Jpn., Engl. abstr.] Bull. Hokkaido Reg. Fish. Res. Lab. 32:15–22.

Yamamoto, G., 1943. Gametogenesis and the breeding of the Japanese common scallop, *Pecten (Patinopecten) yessoensis* JAY. [in Jpn., Engl. synopsis.] Bull. Japan. Soc. Sci. Fish. 12:21–26.

Yamamoto, G., 1964. Studies on the propagation of the scallop, *Pecten (Patinopecten) yessoensis* (JAY), in Mutsu Bay. [in Jpn.] Suisan Zoyoshoku Sosho (Tokyo). 6:77 p.

Yamamoto, G., 1973. Biotic production of benthos. [in Jpn.] In: G. Yamamoto (Ed.). Marine Ecology. Tokyo. 213 p.

Yamamoto, G. and Nishioka, C., 1943. "Jinko-jusei niyoru hotate-gai no hassei nituite" [Development of Japanese scallop by artificial insemination]. [in Jpn.] Bull Japan. Soc. Sci. Fish. 11(5–6):219.

Yamamoto, G., et al., 1971. Through mariculture in shallow waters. [in Jpn.] (Tokyo). pp. 187–263.

Yasumoto, T., Oshima, Y. and Yamaguchi, M., 1978. Occurrence of a new type of shellfish poisoning in Tohoku distinct. Bull. Japan. Soc. Sci. Fish. 44:1249–1255.

Yoshio, J., Ishida, K., Hino, Y. and Hattori, M., 1987. Technology of multiple productions of shellfish in fishing ground of ivory shell. [in Jpn.] Progress report of marine ranching project (8). Japan Sea Reg. Fish. Res. Lab. pp. 97–105.

AUTHORS ADDRESSES

Yoshinobu Kosaka – Aomori Prefecture Fisheries Research Center, Aquaculture Institute, 10, Tsukidomari, Moura, Hiranai-cho, Aomori, 039–3381 Japan (E-mail: yoshinobu_kosaka@yahoo.co.jp or yoshinobu_kosaka@ags.pref.aomori.jp)

Hiroshi Ito – Hokkaido National Fisheries Research Institute, Fisheries Research Agency, Katsuarkoi, Kushiro, Hokkaido, 085–0802 Japan

Scallops: Biology, Ecology and Aquaculture
S.E. Shumway and G.J. Parsons (Editors)
© 2006 Elsevier B.V. All rights reserved.

Chapter 23

Scallop Culture in China

Ximing Guo and Yousheng Luo

23.1 INTRODUCTION

Scallops are among the most valuable marine resources in China. Scallops support a major aquaculture industry that is important to economies of China's coastal regions. In the past decade and particularly after the crash of the shrimp culture, scallop farming has expanded rapidly and become the largest mariculture industry in China. Aquaculture production increased by 41 folds between 1986 and 1997 (Fig. 23.1). In 1997, China produced one million metric tons of scallops from aquaculture (MAC 1998), with an estimated value of over US$1 billion. Compared with scallop aquaculture, scallop landing from wild fishery is negligible. Almost all scallops produced in China (>99%) are now from aquaculture production.

Scallops are found throughout China's 18,000-km coast, on rocky and sandy bottoms and at depths of 5–360 meters. There are over 44 species of scallops naturally occurring in coastal waters of China (Table 23.1). Most of the scallops are found along the southern coast (below 25°N) and in South China Sea (Wang 1983; Qi et al. 1989; Xu 1997). Only a few species are of commercial importance. The zhikong scallop (*Chlamys farreri* Jones et Preston) in the north is by far the most important commercial species. It is the dominant aquaculture species and accounts for over 60% of the total scallop production in China. The huagui scallop (*Chlamys nobilis* Reeve) in the south is also cultured commercially, but the production is limited to only about 2% of the total production.

Several non-native species have been introduced for aquaculture production, including the bay scallop (*Argopecten irradians* Lamarck) from North America, the Japanese scallop (*Patinopecten yessoensis* Jay) from Japan, and the great scallop (*Pecten maximus* Linnaeus) from Europe. The bay and Japanese scallops are cultured commercially in China. Aquaculture production of bay scallops increased considerably in recent years due to severe summer mortalities of zhikong scallops since 1997. The great scallop is being tested for aquaculture in the Dalian area.

This chapter focuses on the biology of the native zhikong scallop, and the current status and practices of the scallop aquaculture industry in China.

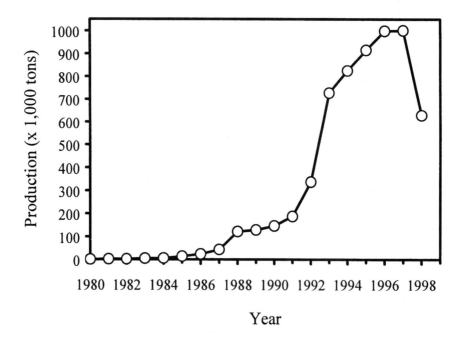

Figure 23.1. Aquaculture production of scallops (fresh whole weight) in China from 1980 to 1998. The sharp decrease in 1998 is caused by massive summer mortalities of the native zhikong scallop.

Table 23.1

A list of scallop species naturally occurring in coastal waters of China.

Species	Distribution
Amussium japonicum formosum Habe	18°N – 25°N
Amussium japonicum balloti Bernardi	8°N – 25°N
Amussium pleuronectes pleuronectes Linnaeus	8°N – 20°N
Amussium pleuronectes nanshaensis Wang	5°N – 12°N
Annachlamys reevei Adams et Reeve	22°N – 25°N
Annachlamys macassarensis Chenu	22°N – 25°N
Anquipecten superbus Sowerby	22°N – 25°N
Anquipecten lamberi Souverbie	22°N – 25°N
Bathyamussium jeffreysi Smith	10°N – 30°N
Bractaechlamys elegans sp. nov.	18°N – 19°N
Bractaechlamys quadrilirata Lischke	19°N
Bractaechlamys schmeltzii Kobelt	8°N – 18°N
Bractaechlamys teramachii Webb	20°N – 25°N

Chlamys albolineata Sowerby	18°N – 22°N
Chlamys asperulata Adams and Reeve	18°N – 22°N
Chlamys farreri Jones et Preston	25°N – 40°N
Chlamys gloriosa Reeve	18°N – 22°N
Chlamys irregularis Sowerby	20°N – 25°N
Chlamys jousseaumei Bavay	15°N – 22°N
Chlamys larvatus Reeve	18°N – 22°N
Chlamys lemniscata Reeve	18°N – 28°N
Chlamys lentiginosa Reeve	20°N – 25°N
Chlamys madreporarum Sowerby	18°N – 22°N
Chlamys miniacea Lamarck	20°N – 25°N
Chlamys nobilis Reeve	19°N – 22°N
Chlamys squamata Gmelin	18°N – 25°N
Chlamys valdecostatas Melvill	20°N – 22°N
Comptopallium radula Linnaeus	22°N – 25°N
Ctenamussium sinensis Wang	25°N – 32°C
Cryptopecten bullatus Cautzenberg et Bavay	10°N – 22°N
Cryptopecten complanus Wang	25°N – 30°N
Cryptopecten inaequivalvis Sowerby	20°N – 25°N
Cryptopecten nux Reeve	15°N – 22°N
Cryptopecten oweni De Gregorio	20°N – 25°N
Cryptopecten vesiculosus Dunker	22°N – 32°N
Decatopecten plicus Linnaeus	22°N – 25°N
Decatopecten striatus Schumacher	22°N – 25°N
Decatopecten amiculum Philippi	22°N – 25°N
Delectopecten macrocheiricola Habe	25°N – 32°N
Gloriopallium pallium Linnaeus	22°N – 25°N
Pecten albicans Schroter	22°N – 33°N
Pecten excavatus Anton	25°N – 32°N
Pecten pyxidatus Born	5°N – 28°N
Polynemamussium intuscostatum Yokoyama	5°N – 22°N
Propeamussium sibogai Dautzenberg et Bavay	5°N – 22°N
Propeamussium stella Wang	5°N – 22°N
Propeamussium gracilis Wang	5°N – 22°N
Propeamussium caducum Smith	5°N – 22°N
Propeamussium watsoni Smith	22°N – 32°N
Semipallium fulvicostum Adams et Reeve	19°N – 21°N
Semipallium spectabilis Reeve	19°N – 25°N
Semipallium tigris Lamarck	18°N – 19°N
Semipallium xishaensis sp. nov.	15°N – 20°N
Serratovola tricarinata Anton	8°N – 25°N
Volachlamys singaporinus Sowerby	18°N – 28°N
Volachlamys hirasei Bavay	30°N – 41°N

23.2 ZHIKONG SCALLOP

23.2.1 Shell Morphology

Shell morphology and anatomy of the zhikong scallop have been described in details elsewhere (DFC 1979; Wang 1983; Qi et al. 1989; Wang et al. 1993). The two valves of zhikong scallop are nearly equal. The external colour ranges from orange, light brown to purplish brown. The left valve is more convex than the right valve. The anterior auricle is larger than the posterior auricle on both valves. The anterior and posterior auricles of the left valve, and the posterior auricle of the right valve are triangular. The anterior auricle of the right valve is rectangular and forms the byssal opening. Six to ten denticles occur on the ventral side of the byssal opening. There are 10 major radian ribs (ranging from 8 to 13) on the left valve, with minor ribs between them. There are 20 to 30 less convex ribs on the right valve. Sharply pronounced scales occur on the ventral side of both valves. The interior surface of both valves is usually smooth and glossy white or pink in colour.

23.2.2 Distribution

Zhikong scallops are found from the coasts of Liaoning province, latitude 40°N, to the shores of Fujian province, latitude 25°N. Outside China, zhikong scallops are found in waters of Korea and Japan. Distribution is patchy throughout the range and no extensively abundant beds occur. Major populations occur on southern Liaoning peninsula, and the north and south coasts of Shandong peninsula, at depths of 5–60 m. At these waters the current is unblocked, the nutrients rich and the bottom solid. Zhikong scallop has well-developed byssus and uses it to attach onto rocks and other hard surfaces. Zhikong scallop can sever byssus and relocate under adverse conditions. Young scallops are particularly mobile.

During 1980 and 1984, the National Fishery Resource Survey showed that population density of zhikong scallop in Jin county on eastern Liaoning peninsula was 4.4 scallops (or 228 g) per square meter. In Changhai county, the average density was 2.3 scallops (or 126 g) per square meter. Total scallop resources in the above areas were estimated as between 200 and 750 t. In Lushun area at the western end of Liaoning peninsula, the average density was 1.8 scallops (or 98 g) per square meter. Total biomass of Lushun area was about 230 t. In Rongcheng county at the eastern end of Shandong peninsula, population density was about 0.8 scallops (or 43.7 g) per square meter. In Changdao county of the northern Shandong peninsula, population density was estimated as between 0.25 and 1.0 scallops (or 17–50 g) per square meter. Total resource in the above two areas was about 400 tons. In other areas such as Qingdao and Weihai waters of Shandong peninsula, zhikong scallops also occur, but no statistics is available.

Adult zhikong scallops are generally between 55 and 85 mm in shell height. The maximum shell height is about 100 mm. Figure 23.2 shows the size frequency distribution of scallops collected from Shandong waters by Zhang et al. (1956). In general, the distribution is skewed towards larger animals. The majority of scallops are

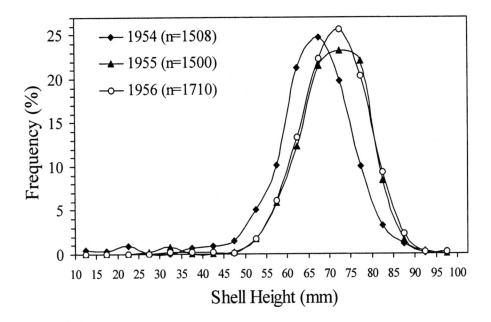

Figure 23.2. Size frequency distribution of zhikong scallops collected from Shandong waters in three years (Zhang et al. 1956).

between 60 and 80 mm. Scallops smaller than 50 mm or larger than 85 mm are rare and account only about 2% of the population. Similar distributions are found in scallops from other areas. Scallops from Jin county of Liaoning province have an average shell height of 72 mm and an average weight of 62 g. In Lushun area, scallops are between 33 and 87 mm in shell height, with an average of 74 mm, and 54.4 g in whole body weight.

23.2.3 Growth

Zhikong scallop grows fast during the first two years. Growth slows down considerably after two years of age. Zhang et al. (1956) studied the growth of scallops maintained in baskets in Qingdao area over a five-year period. Their results show that scallops reach a shell height of 23 mm in the first year (0.5 years in age), 50 mm in the second year, 64 mm in third year, 70 mm in the fourth year, and 76 mm in the fifth year (Fig. 23.3). The percent shell growth attained in Year 1 to 5 is 30% (first half year), 35%, 19%, 8% and 8%, respectively.

Growth rate of Zhikong scallops is strongly influenced by water temperature. Although zhikong scallop tolerates a wide range of water temperatures (-1.5–30°C), extreme temperatures greatly reduce growth. Growth is nearly absent during winter months when water temperature is below 5°C. Growth rate increases rapidly with

Figure 23.3. Growth in shell height of experimental zhikong scallops maintained in baskets in Qingdao waters (Zhang et al. 1956).

increasing temperature in the spring, and the fastest growth is found at 16–18°C (Yang et al. 1999). Growth slows down sharply with increasing temperatures beyond 23°C. Zhikong scallops in Dalian area have two growing periods. The first period is between May and July when the water temperature is at 14–21°C (Fig. 23.4). The second period of growth is between September and October when water temperature is between 22 and 14°C. Under optimal conditions, zhikong scallops can grow up to 9 mm per month. In northern Shandong peninsula, the main growing season is from early May to early July, and the second growing period is relatively insignificant (Fig. 23.5).

Growth of cultured scallops depends on culture density. In 1979, the Fisheries Institute of Jin County tested effects of density on growth at the same cultural depth. The results are summarised in Table 23.2. Growth of scallops decreased with increased density throughout the testing period. Depth may also affect the growth of cultured zhikong scallop. Generally, scallops suspended near the surface grow faster than those at deep waters, when there is little fouling. However, seaweeds and other fouling organisms, which grow well near the water surface, may seriously block water flow to lantern net close to the surface and limit food supply. Near the sea floor, the current is week and the presence of mud and sand is unfavourable for the growth of scallops. Lantern nets are often deployed close to the surface (1–3 m) and lowered to deeper waters (3–5 m) in seasons when fouling organisms reproduce. They are always above the sea floor.

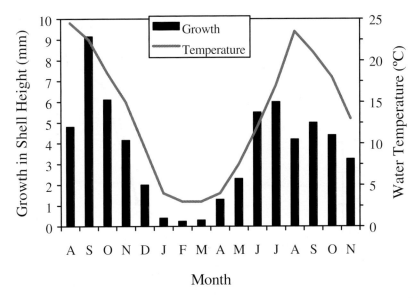

Figure 23.4. Relationship between water temperature and monthly increase in shell height of zhikong scallop during a cultural cycle in Dalian area.

23.2.4 Reproduction

Apart from occasional hermaphroditism, sexes in zhikong scallop are separate with an approximately 1:1 sex ratio. Zhang et al. (1956) examined the sex ratio of 6,025 scallops from a natural population, and found 52.4% females and 47.6% males. The colour of female gonad is reddish orange and the male is milky white. Zhikong scallops can reach sexual maturation in their first year, but the size at maturation depends on growth rate.

Gonad development and reproductive cycle of zhikong scallop depends on water temperature and food supply. From histological examinations, five main stages of gonadal development have been recognised in scallops from Qingdao area: 1) proliferation stage, from January to March; 2) growing stage, from April to May; 3) mature stage, from May to September; 4) spawning stage, from May to October; and 5) spent stage in November and December (Liao 1983).

Zhikong scallops usually have two spawning seasons in a year: one in mid-May to June and the other in late September to October. The gonad index (gonad weight/body weight x 100) of zhikong scallops in Dalian area increases sharply in late April (at 8–9°C) and reaches its first peak in mid June (around 16–19°C) (Fig. 23.6). After spawning, the gonad index drops to its lowest value in late July - August (22–24°C). The gonad index rises to its second peak in mid-September, and declines again in early October. Spawning time varies annually due to variations in water temperature. For example, the period of first spawning around Haiyang Island of Liaoning province occurred in late June/early

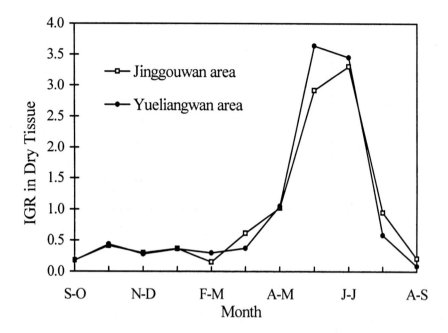

Figure 23.5. Annual variation of instantaneous growth rate (IGR) in dry-tissue weight of zhikong scallop in Jinggouwan and Yueliangwan areas of northern Shandong peninsula (Yang et al. 1999).

July in 1983, through out in July in 1984, and from early July to early August in 1985. Spawning time may differ by up to 30 days from year to year.

Mature eggs of zhikong scallop are about 65–72 μm in diameter. Sperm remain active for 12 hours after being released (at 19°C). At 18–20°C, fertilised eggs release polar body I at 15–20 min, polar body II at 25 min, and reach D-stage at 24–28 hours post-fertilisation. The average size of D-stage larvae is about 110 μm in length x 87 μm in height. Development to metamorphosis and settlement takes about 15 days at 20–23°C.

Table 23.2

Mean shell height (mm) of zhikong scallops cultured at different densities.

Duration (days)	Cultural density (scallops/basket)				
	50	100	200	400	600
25	6.6	6.5	5.5	5.0	4.7
43	7.4	7.0	6.7	6.9	4.7
76	8.5	6.6	6.7	6.4	6.1

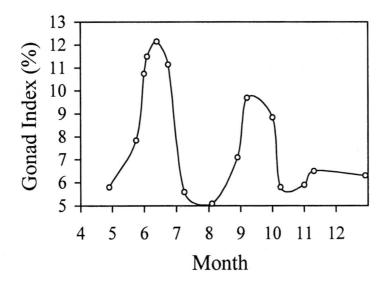

Figure 23.6. Seasonal changes in gonad index of zhikong scallop in Dalian waters, showing two spawning seasons (Month 4 = April; 5 = May; 6 = June, etc.).

Traditionally, recruitment rate of zhikong scallops is low in most areas of Bohai and Yellow Seas. Strong one-year classes (<30 mm) have been observed in Chrysanthemum Island waters, off Liaoning coast. Because of large-scale aquaculture since late 1980's, recruitment from cultured populations has become extremely abundant in several areas (Guo et al. 1999). The cultured populations are derived from hatchery-produced seed in early 1980's. There are concerns that the cultured populations are inbred because of the founding population effect. The loss of the genetic diversity is considered as one of the potential causes for the heavy mortality experienced by zhikong scallops since 1996.

Natural predators of zhikong scallop include fish, echinoderms, crabs and snails, which primarily feed on young scallops less than 10 mm. Pests and fouling organisms include the pea crab, *Pinnotheres sinensis* Shen, boring worms such as *Polydora*, sponges, barnacles, oysters and mussels.

23.3 OTHER SCALLOP SPECIES

23.3.1 Huagui Scallop

Huagui scallop is a warm water species. It has a larger adult-size and faster growth than zhikong scallop. It is the main commercial species in southern China.

Shells of the Huagui scallop are nearly circular, with the upper (left) valve more convex than the lower (right) one. The external colour of left valve appears pinkish brown, yellowish brown or light red with cloudy grains. Both valves have 23 ribs. The distance between rows of ribs is smaller than the width of the ribs. The anterior and posterior auricles of the left valve are triangular, with the anterior auricle being slightly larger and having 7–8 coarse ribs on the surface. The anterior auricle of the right valve is smaller than the posterior article and has four coarse ribs. The posterior auricle is triangular with the byssal opening on its ventral side and numerous fine ribs. The internal surface of both shells is generally yellowish brown and has as many ribs or grooves as the external surface.

Huagui scallops are primarily distributed in the South China Sea, and from eastern Japan as far south as Indonesia. Its distribution in China is mostly limited to coasts of Zhaoyan, Haimen, Jilang, Aotau, Guanghai, Jiapo of Guangdong province and Xincun of Hainan province. Abundant areas of huagui scallop are found along the coast of Daya Bay in Guangdong. Huagui scallops are usually distributed from the lower intertidal zone to subtidal depths of more than 360 m. They occur on subtidal rocky or sandy bottoms with good current and clear water. Usually they are attached by byssus to rocks or other substratum. Sometimes, they also lay freely on the right valve. Huagui scallops have a patchy distribution and are not particularly abundant in any one locality. Population density is generally not more than one scallop per square meter, and the biomass may occasionally reach 38 g per square meter.

Temperature range in natural habitats of huagui scallop is between 8 and 32°C. The optimum temperature of growth is between 20 and 25°C. In Guangdong province, huagui scallops exhibit rapid growth from May to September. Hatchery produced seed may grow to a shell height of 68 mm in one year in waters of Guangdong province. The largest individual grown in suspended culture may reach 120 mm in three years.

Sexes of huagui scallop are separate. Mature eggs are about 65 μm in diameter. Huagui scallops have high fecundities, although it depends on the size of the individual. Females usually produce about 3–7 million eggs, and the largest individuals may produce up to 15 million eggs. Huagui scallops have two spawning seasons on the northern coast of South China Sea. In Sunwei port of Guangdong province, the first peak of spawning occurs between late April and the first ten days of May when water temperature reaches 25 to 28°C. The second peak is between August and September when water temperature drops from 29 to 21°C. In Daya Bay, the first peak occurs between April and June, and the second peak between October and December (Cui 1980).

Larval development of huagui scallop has been studied in the laboratory (Zhang 1982). At 26–29.5°C and salinities of 27–29 ‰, development from fertilised egg to D-stage takes about 22 hours (shell length 101 μm x shell height 82 μm). Larvae usually settle at 230 μm in about 14 days at 26°C. At temperatures of above 28°C, larval development is shortened to 9 days (Zhang 1982).

23.3.2 Non-native Species

At least three species have been introduced to China for aquaculture production. The first species introduced is the Japanese scallop. It was introduced from Japan to Qingdao and Dalian areas in the late 1970's. The Japanese scallop is a cold-water species and cultured only in northern provinces, Liaoning and the north side of Shandong peninsula. Japanese scallops do poorly in regular lantern nets. They grow well when cultured using the ear-hanging method. They have difficulties with high summer temperatures in Bohai Sea, which often reach 25–28°C. The yield of Japanese scallop is probably less than 1% of the total scallop production in China. Although its yield is low, Japanese scallop is larger in size and enjoys a higher market price than zhikong and bay scallops.

The bay scallop was introduced from the United States to Qingdao in 1982 (Zhang et al. 1986). Of the original shipment, twenty-six bay scallops survived and spawned in January 1983, producing the first generation of bay scallops in China. Juvenile scallops reached an average 6.9 mm in May and were transferred to culture sites in Shandong and Fujian provinces. Scallops grew to 50 mm by September and 59 mm by December 1993. Market size, 50–60 mm, can therefore be reached within a year. This is a major advantage over the zhikong scallops, which usually take 1.5–2.0 years to reach market size. The shorter turn-around time of bay scallops is partly due to its faster growth, and partly to the fact that they are spawned in the early spring (or late winter) so that they catch a full growing season. Because zhikong scallop seeds are collected in the fall, they miss most of the first growing season. Because of the short grow-out time, bay scallops quickly gained acceptance by scallop farmers, and aquaculture expanded rapidly after 1984. By early 1990, the annual production of bay scallops had reached to about 200,000 tons. Bay scallop culture production has declined somewhat in early 1990's. It accounted for about 15–18% of the total scallop production. Because of the recent mortality problem in zhikong scallops, bay scallop culture is on the increase again.

The great scallop was introduced from France to Dalian in 1999. About three million F1 seed have been reared to 20–30 mm. They are being evaluated in Dalian area.

23.4 FISHERY

Scallops have been harvested in China for centuries, but the yield has always been limited. Scallop fishery reached its highest level in the 1950's, with an annual yield of about 3,000 tons of whole scallops. Scallop resource and landing declined in the 1950's because of over-fishing.

Zhikong scallop is the main species targeted by the fishery. Scallops were harvested by divers and trawl nets in depth less than 30 meters. Before the 1960's, scallops were mainly harvested by divers equipped with man-powered wooden boats, air compressors and diving suits. Air-compression and navigation were all done by hand. Dives were generally carried out near shores or around isles in depths less than 20 m. In the early 1970's, diesel-powered vessels were introduced, and fishing grounds were extended. Nevertheless, the fishing depth was never beyond 30 m. By the 1980's, small dredges were employed in some areas, and the fishing depth increased to more than 30 m.

Fishing grounds for zhikong scallops are mainly in Bohai Sea and the northern end of Yellow Sea. The number of divers back in the 1950's was about 150 and by the 1960's, the number of scallop fishing boats was declining. There were about 100 boats in operation in 1990. Scallops are not the only species harvested by divers. Other species harvested include abalone, sea cucumber and snails. In the 1950's, the annual yield of zhikong scallop was between 1,000–1,500 t (fresh) in Shandong province and about 2,000 t in Liaoning province. By the 1960's, the yield was absent in Shandong and in Liaoning had declined to a few hundred tons. Now, due to the development of raft culture, scallop resources in Liaoning and Shandong waters have recovered. Current landing from scallop fishery is not reported in official statistics and negligible compared to aquaculture production.

For zhikong scallops, a seasonal restriction, from May 15–July 31, has been introduced in the northern parts of Yellow Sea, and a size limit of 70 mm in shell height is also in place. No area closures have been enacted.

Huagui scallops are also harvested by divers and trawl nets. Fishing grounds are usually within 1–5 km from the shore. Fishery landing of huagui scallop has been limited since ancient times. Annual production has never reached over 100 t fresh weight. Most huagui scallops are marketed live.

23.5 AQUACULTURE

23.5.1 Species and History

China has a long history of culturing marine molluscs. Traditional molluscan culture dates back to over 2,000 years ago, but it is primarily limited to the "traditional four" species: oysters, blood cockle, razor clam and *Ruditapes* clams (Guo et al. 1999). Modern molluscan aquaculture developed after 1949 and went through a period of rapid expansion since the early 1980's. In 1998, China produced 7 million metric tons of marine molluscs from aquaculture, which accounts for 81% of mariculture production (MAC 1999).

Compared with the culture of other molluscs, scallop aquaculture is relatively new. Over-fishing and the decline in wild fishery in the 1950's led to studies on scallop biology and aquaculture. Studies in the 1950's and 1960's focused on taxonomy, distribution and reproductive characteristics of native scallops. Experimental culture of scallops began in 1968. Key technologies for hatchery production, seed collection and grow-out were developed between 1973 and 1977 (DFC 1979; Wang et al. 1993). Scallop aquaculture became a mature industry in the early 1980's and expanded rapidly between 1986 and 1997 (Fig. 23.1). In 1979, China produced only 35 tons of scallops. In 1985, scallop production exceeded 10,000 t for the first time. In 1997, over 1 million tonnes of scallops were produced from aquaculture, and fishery production became negligible in comparison (MAC 1998). In 1998, however, massive summer mortality of zhikong scallop caused a 37% decline in aquaculture production (MAC 1999).

Scallops are mainly cultured in Bohai and Yellow Seas in the north. Shandong province alone produced 84% of the national total in 1996 (Guo et al. 1999). Liaoning province, with 14% of the national production, was the second largest producer. Four southern provinces, Fujian, Zhejiang, Guangdong and Hainan, produced about 2% of the national total.

Currently there are four major species cultured in China: the zhikong scallop, bay scallop, huagui scallop and Japanese scallop (Fig. 23.7A and B). The native zhikong scallop is the most important species for aquaculture, accounting for over 80% of the total scallop production. The introduced bay scallops account for about 15–18% of the total production. Huagui and Japanese scallops together may account 1–3% of the total. Zhikong scallop is primarily cultured in north China, particularly in Shandong and Liaoning provinces. Rizhao, in southern Shandong, probably represents the southern-most site for zhikong scallop culture.

23.5.2 Collection of Natural Seed for Zhikong Scallop

Zhikong scallop culture was first developed using hatchery-produced seed, which were first developed in 1974. For many years after 1974, zhikong scallop culture used exclusively hatchery-produced seed. As scallop aquaculture grew, cultured populations became so large that they provided abundant recruitment for seed collection. In 1984, annual collection in Shandong waters reached 2.5 billion seed. In 1986, collection from Liaoning waters also reached 0.5 billion seed. In 1996, one of the most productive bays in north Shandong produced about 130 billion scallop seed, all from cultured populations. High levels of reproduction of cultured populations provided sufficient natural seed for aquaculture use in recent years. There has been virtually no hatchery production of seed for zhikong scallops.

Protocols for seed collection have been well established through years of research and experience (DFC 1979; Wang et al. 1993). Successful collection involves site selection, preparation of collection material, and forecasting collection dates by observing gonadal development, and larval stages and density at the collection site. Seed are usually collected using spat bags (30 x 40 cm, 1.2–1.5 mm mesh) stuffed with nylon screens (about 100 g). Collectors are hung on the floating raft below the surface. A string of collectors contains about 10–12 bags. A typical raft is about 60 m long and carries an average of 500–600 strings. Each bag may collect 100 to 1,000 spat depending on the location, season and year. The depth for deploying collectors varies among species and location. For zhikong scallops, seeds can be collected at depths of 2 and 9 m, usually best at 6–8 m (Wang et al. 1993).

Zhikong scallop culture primarily uses summer seed from the first spawning season, which set between late June and mid-July (Table 23.3). Spat in the collectors are left undisturbed until harvest in early October at a commercial size of about 5–10 mm. Commercial seed are over-wintered at nursery sites on the sea until the following March when they reach about 30 mm and are ready for grow-out. The period of culturing 10-mm seed to 30-mm juveniles is referred to as seed nursery. It is conducted at selected areas on the sea using lantern nets with a small mesh (4–8 mm).

23.5.3 Hatchery Production of Bay Scallop

Because the bay scallop is a non-native species, seed are exclusively hatchery-produced. Hatchery production of bay scallops is similar to that for other marine bivalves. Conditioning of brood stock begins in February (Table 23.3), by raising water temperature

1156

Table 23.3

Milestones of typical culture cycles for zhikong and bay scallops.

Month	Zhikong Scallop	Bay Scallop
Feb.		Broodstock conditioning
March		Spawning, hatchery rearing
April		Spawning, hatchery rearing
May	Forecast larval settlement	Hatchery rearing to 1 mm, spat nursery
June	Deploy spat collectors	Spat nursery to 5 mm, seed nursery
July	Deploy collectors	Seed nursery
Aug.	Monitoring spat growth in collectors	Seed nursery to 20 mm
Sep.	Monitoring spat growth in collectors	Grow-out
Oct.	Harvest 10 mm seed, seed nursery	Grow-out
Nov.	Over-winter seed nursery	Harvest at 60 mm
Dec.	Over-winter seed nursery	Harvest at 60 mm
Jan.	Over-winter seed nursery	
Feb.	Over-winter seed nursery	
March	Grow-out with 30 mm juveniles	
April	Grow-out	
May	Grow-out	
June	Grow-out	
July	Grow-out	
Aug.	Grow-out	
Sep.	Grow-out	
Oct.	Grow-out	
Nov.	Harvest at 60–70 mm	
Dec.	Harvest at 60–70 mm	

and intensive feeding. Hatchery production usually occurs between March and May. Bay scallops are hermaphroditic. They are air-dried and then immersed in warm seawater to induce spawning. Thousands of mature scallops, placed in dozens of lantern nets, are induced to spawn in large concrete tanks, ranging from 10 to 100 cubic meters. When the desired egg density is reached (about 50 mL^{-1}), brood scallops are moved to the next tank to continue spawning. The optimal temperature for larval rearing is about 22–23°C. The first water change is made as soon as the larvae reach D-stage, usually in 24 hours. The optimal culture density is between 4 and 10 larvae mL^{-1}.

Figure 23.7. Scallop culture in China. A, the introduced bay scallop (upper) and the native zhikong scallop (lower) cultured in Shandong. B, the Japanese scallop cultured in Dalian. C, ropes made from palm tree fibres used for the collection of scallop spat and seaweed seedlings in hatcheries. D, suspended longlines used for the culture of scallops, as well as abalone, Pacific oyster and seaweeds in Rongcheng, Shandong province. E, lantern nets used for scallop culture. F, zhikong scallops are harvested at 1.5 years of age. Photos are reprinted from Guo et al. (1999).

Scallop larvae are fed with unicellular algae. Commonly used algae species include *Platymonas subcordiformis, Isochrysis galbana, Dunalliella* sp., *Phaeodactylum tricornutum, Nitzschia closterium, Chaetoceros muelleri, Dicrateria zhangjianensis* Hu. var. sp., *Isochrysis galbana, Heterogloea* sp., and *Chlorella* sp. The species of algae are selected according to the age/size of the scallop larvae. Earlier stages are fed with small species, later stage are usually fed with large species. At the present, most hatcheries use a mixture of several species for feeding. Vitamins and antibiotics are commonly used during larval culture.

At 22–23°C, bay scallop larvae usually reach the eyed-stage (170–190 μm) in 10 to 11 days at which time spat collectors are placed in the tanks. Two common types of spat collectors are used. One is a rope-curtain made from natural palm tree fibre (Fig. 23.7C). The other is polyethylene or nylon nets/screens. Spat, which attach to the collectors by byssal threads, are cultured in the hatchery until they reach 500–600 μm, after which they are moved out of the hatchery for spat nursery. The best spat nurseries are large shrimp ponds. Commercial seed are sold at a size of 5–10 mm. They go through a seed nursery period before reaching 20 mm and being deployed for grow-out. Scallops are harvested between November and December (Table 23.3).

Most of the bay scallop industry was developed using the 26 scallops first introduced in 1982, and there had been signs of inbreeding such as larval and juvenile mortality. Several new introductions have been made to expand the gene pool. Both the northern (*Argopecten irradians irradians*) and southern (*Argopecten irradians concentricus*) subspecies of bay scallops have been introduced to China. The population introduced from eastern Canada, is valued in China for its higher meat ratio. The Canadian population is an F3 generation of scallops introduced from Massachusetts (Scarratt 2000).

23.5.4 Grow-out

Although several other culture methods (such as ear-hanging and tube-cages) have been used in the past, lantern nets on suspended longlines are now the dominant form of grow-out for all scallops cultured in China (Fig. 23.7D). Lantern nets are multi-layered cylinders, with 8–10 layers of discs (35 cm in diameter), connected and coated with a mesh screen (Fig. 23.7E). For zhikong scallops, commercial seed, about 10 mm in size, are deployed in lantern nets in October for seed nursery and over-wintering (Table 23.3). About 200–300 seed scallops are stocked in one layer, or 2000–3000 per cage. Next spring, scallops are thinned to grow-out densities of 50–80 per layer or 400–500 per cage for grow-out. The recommended grow-out density is 30–35 per layer for zhikong scallops, 25–30 per layer for bay scallops, and 15–20 per layer for Japanese scallops (Wang et al. 1993). In realty, higher densities are often used by farmers. The lantern nets are hung on longlines, which are usually 80–100 m long and supported by rubber floats. Fouling of scallop cages is a common problem. Fouling organisms and mud must be cleaned regularly. Zhikong scallops usually reach market size, 60–70 mm, by December or 1.5 years old (Fig. 23.7F). Bay scallops are harvested in December of the same year (or 8 months old) at 60 mm.

Lantern nets are not well suited for culturing Japanese scallops. Because adult Japanese scallops have no byssus and behave differently, they are easily injured or killed from cutting into each other. Single-compartment cages have been developed in Weihai area and greatly improved the survival and growth of Japanese scallops.

Polyculture systems of scallops and other species are developed to fully utilise the primary productivity and maximise income. Polyculture of scallops and kelp are commonly practised. In Liaoning waters, for example, 50 vertical strings of scallops and 34 horizontal strings of kelps are simultaneously hung between two longlines 5–6 m apart. Bay scallops are also cultured on the bottom of shrimp ponds. Scallops and sea cucumbers are sometimes cultured in the same lantern nets. Sea cucumbers feed on seaweed and other fouling organisms on the cage. They are also highly priced seafood in China.

Zhikong scallops have been suffering from massive summer mortalities since 1996. Mortality in 1998 caused a 37% decline in production, with an estimate loss of about US$360 million. Mortality generally begins in late July or early August as the water temperature reaches and exceeds 28°C. It lasts for about 20 days and ends when the temperature begins to decrease. The summer mortality was first observed in 1994. It has worsened in recent years. Both 1994 and 1995 were warmer (by 2°C and 1°C, respectively) than normal. Temperatures in 1996 were more typical and the death rate lessened. Mortality worsened in 1997–98, reaching 60–80% or more in many areas throughout Shandong and Liaoning provinces. On June 3, 2000, we deployed zhikong scallops at three sites along Shandong coast: Jiaonan on the south side, Penglai and Yantai on the north side of Shandong peninsula. All three sites experienced heavy mortalities in late July and August. Mortality at Jiaonan occurred earlier than the other two sites because of its southern locality and higher water temperatures. By August 29, 2000, cumulative mortality reached 92% at Jiaonan, 86% at Yantai and 88% at Penglai (Xiao 2002). Bay scallops are generally not affected.

There are several suspected causes for the summer mortality, although they have not been studied extensively. Most scientists and farmers in China believe that the mortalities are caused by a combination of over-crowding, high summer temperature and deteriorating water quality. Scallop farmers often culture scallop at 2–3 times the density (30–35/layer) recommended by local scientists. The number of longlines and culture plots (not just for scallop culture) has been increasing rapidly in recent years, and may have exceeded the carrying capacity of many coastal areas. Over-crowding at both cage and bay level may have added considerable stress to the culture environment. A haplosporidan parasite of the type responsible for extensive mortalities in oysters in the United States was identified in bay scallops in China, but there was no evidence that it was causing mortalities in that species or that it had been transferred to zhikong scallops (Chu et al. 1996). Finally, there is also the suspicion that the scallop stock is genetically inbred. Because all seed are collected from cultured populations, which were originated from hatchery production during the late 1970's and early 1980's, it is possible that some inbreeding occurred. Hybrids between the Chinese and Korean or Japanese varieties of zhikong scallop tend to survive better than local seed.

23.6 HARVEST, PROCESSING AND MARKETING

Harvest season generally occurs between autumn and winter. Commercial size varies between species: 60 mm in height for zhikong and bay scallops; 65 mm for huagui scallop, and 70 mm for Japanese scallop. Some scallops are sold live to local seafood markets. The price of live scallops varies greatly depending on the season and year. Most of the cultured scallops are processed.

Traditionally, most of scallops are processed into "dry shellfish", a traditional name for dried scallop adductor muscles. "Dry shellfish" is processed in tree ways:

1. Fresh process: After harvest and cleaning in seawater, adductor muscles are placed in seawater containing 2% fine salt for 10 minutes, then scooped up from the water and dried in the sun.
2. Cooked process: Fresh adductor muscles are placed in 80°C seawater with 2% fine salt for 5 minutes, immediately cooled in flowing water, and then air-dried.
3. Steamed process: Whole scallops are steamed lightly for the valves to open, the adductor muscles removed, cleaned with seawater and dried in the sun. It is important that scallops are not over-cooked.

"Dry shellfish" are marketed in China and other parts of Asia. In China, the price of "dry shellfish" is about 60–165 RMB (US$7–20) per kg depending on the size and quality.

Since the late 1980's, scallops are processed into frozen products for export to North America and other regions. During the 1990's, many modern seafood-processing plants have been built with mechanised lines and according to international standards. Currently most of the scallops harvested are processed into Individually-Quick-Frozen or IQF adductor muscles. During the early 1990's, a large fraction (>60%) of the scallop production is exported. The domestic market has been growing rapidly. Because of the strong domestic demand and the summer mortality problem, only a small fraction of scallops is currently exported. China also imports sea scallops from North America.

ACKNOWLEDGMENTS

The authors sincerely thank all colleagues and friends for their valuable input and comments, particularly Professors Fengshan Xu, Hongsheng Yang, Ms. Jie Xiao and Dr. Yongping Wang of the Institute of Oceanology, Chinese Academy of Sciences; Professors Ziniu Yu and Zhaoping Wang of Qingdao Ocean University; Professors Rofei Wei, Zhenlong Xu and others of Liaoning Marine Fisheries Research Institute; and Dr. Huiping Yang of HSRL, Rutgers University. This work is partly supported by a grant from National Natural Science Foundation of China (No. 39825121), NOAA's US-China Joint Program in Marine Living Resources and Rutgers University (contribution No. IMCS-2003-24).

REFERENCES

Cui, R., 1980. Preliminary observation on the growth and reproduction of scallop, *Chlamys nobilis*, in Daya Bay. Bull. Zhangjiang Fish. College, No. 1. (In Chinese with English abstract)

Chu, F.-L.E., Burreson, E.M., Zhang, F. and Chew, K.K., 1996. An unidentified haplosporidian parasite of bay scallop *Argopecten irradians* cultured in the Shandong and Liaoning provinces of China. Dis. Aquat. Org. 25:155–158.

DFC (Dalian Fishery College), 1979. Molluscan Aquaculture. Agriculture Press, Beijing, China. (In Chinese)

Guo, X., Ford, S.E. and Zhang, F., 1999. Molluscan aquaculture in China. J. Shellfish Res. 18:19–31.

Liao, C.Y., Xu, Y.F. and Wang, Y.L., 1983. Reproductive cycle of the scallop, *Chlamys farreri*, at Qingdao. J. Fish. China 7(1):1–13. (In Chinese with English abstract)

MAC (Ministry of Agriculture of China, Bureau of Aquatic Products), 1986–1999. China Fishery Annual Statistics. MAC Bureau of Aquatic Products, Beijing, China. (In Chinese)

Qi, Z., Ma, X., Wang, Z., Lin, G., Xu, F., Dong, Z., Li, F. and Lu, R., 1989. Mollusca of Huanghai and Bohai. Agricultural Publishing House, Beijing, China. (In Chinese)

Scarratt, D., 2000. Welcome to Shellfish World. Shellfish World 1: pp. 3.

Wang, R., Wang, Z. and Zhang, J., 1993. Marine Molluscan Culture. Qingdao Ocean University Press, Qingdao, China. (In Chinese)

Wang, Z., 1983. Studies on Chinese species of the Families Pectinidae. III. Chlamydinae (1. *Chlamys*). Trans. Chinese Soc. Malacol. 1:47–55. (In Chinese with English abstract)

Xiao, J., 2002. Studies on Reasons for Massive Mortality of Cultured *Chlamys farreri* Along the Coast of Shandong Province. MS Thesis. Institute of Oceanology, Chinese Academy of Sciences, Qingdao, China. (In Chinese)

Xu, F., 1997. Bivalve Molluscs of China. Science Press, Beijing, China. (In Chinese)

Yang, H., Zhang, T., Wang, J., Wang, P., He, Y. and Zhang, F., 1999. Growth characteristics of *Chlamys farreri* and its relation with environmental factors in intensive raft-culture area of Shiliwan Bay, Yatai. J. Shellfish. Res. 18:71–76.

Zhang, D., 1982. An experiment of artificial seeding on scallop *Chlamys nobilis*. Marine Fisheries 4:67–70. (In Chinese with English abstract)

Zhang, F., He, Y., Liu, X., Ma, J., Li, S. and Qi, L., 1986. The introduction, hatchery rearing and culture of bay scallops. Oceanol. Limnol. Sinica 17:367–374. (In Chinese with English abstract)

Zhang, X., Qi, Z. and Li, J., 1956. Reproduction and growth of zhikong scallop. J. Zool. China 8:235–257. (In Chinese with French abstract)

AUTHORS ADDRESSES

Ximing Guo - Haskin Shellfish Research Laboratory, Institute of Marine and Coastal Sciences, Rutgers, the State University of New Jersey, 6959 Miller Avenue, Port Norris, New Jersey, USA 08349 (E-mail: xguo@vertigo.hsrl.rutgers.edu)

Yousheng Luo - Marine Fisheries Research Institute, No. 5-19 Xinghai Park, Dalian, Liaoning 116023, PRC

Chapter 24

Scallops Fisheries and Aquaculture of Northwestern Pacific, Russian Federation

Victor V. Ivin, Vasily Z. Kalashnikov, Sergey I. Maslennikov
and Vitaly G. Tarasov

24.1 INTRODUCTION

Scallops are the most intensively consumed and fished bivalve molluscs in Russia. There are more than ten scallop species found within the seas of the Russian Federation. The best known is the Yesso scallop, *Mizuhopecten yessoensis*, also referred to as the Ezo scallop, giant scallop, Japanese scallop, Russian scallop, Primorsky scallop and common scallop. For a long time these molluscs have been the focus of traditional fishing in the coastal waters of the Sea of Japan, Southern Sakhalin, and the Southern Kurile shoal. Commercial scallop beds, averaging 4,000 tons per year, can be found in the Barents Sea (*Chlamys islandica*), the Bering Sea (*C. behringiana*), the Kurile Islands and northern Primorye (*C. albida* and *C. chosenica*), and in Peter the Great and Posjet Bays of the Sea of Japan (*C. farreri*). This chapter reviews the biology and ecology of eight scallop species from the Russian part of the northern Pacific Ocean. Fishery statistics and typical technologies used at these commercial aquaculture farms are also presented.

24.2 TAXONOMIC STATUS

According to (Kafanov 1991; Kafanov and Lutaenko 1998) there are eight species of Pectinidae in the Russian part of northern Pacific Ocean. Their taxonomic status follows:

Class Bivalvia Linné, 1758
    Order Pectinida H. Adams et A. Adams, 1857
        Family Pectinidae Rafinesque, 1815
          Genus *Chlamys* Röding, 1798
            Chlamys (*Chlamys*) *albida* Arnold, 1906 (ex Dall, MS)
               Until recently, this species was confused with *C. islandica* (Müller, 1776)
            C. (*C.*) *asiatica* Scarlato, 1981
            C. (*C.*) *behringiana* (Middendorff, 1849)
            C. (*C.*) *chosenica* Kuroda, 1932
               Until recently, this species was confused with *C. rosealba* Scarlato, 1981
            C. (*Azumapecten*) *farreri* (Jones et Preston, 1904)

Until recently, this species was confused with *C. farreri nipponensis* Kuroda, 1932 or *C. nipponensis* Kuroda, 1932
C. (*Swiftopecten*) *swifti* (Bernardi, 1858)
Genus *Delectopecten* Stewart, 1930
Delectopecten randolphi (Dall, 1897)
Genus *Mizuhopecten* Masuda, 1963
Mizuhopecten yessoensis (Jay, 1857)

24.3 BIOLOGY AND ECOLOGY

24.3.1 *Chlamys albida* (Common names: white scallop and commercial scallop)

Chlamys albida is a widespread, high-boreal, Pacific species (Fig. 24.1). It occurs from Middle Primorye (Lutaenko 1999) up to the northern part of the Sea of Japan (Tatar Strait), along the northern coastline of the Sea of Okhotsk, and near the Kurile (Paramushir and Iturup), Commodore and Aleut Islands. It lives in muddy-sands with pebbles at depths between 36–398 m (Scarlato 1981). Environmental ranges are typical for near-bottom waters: temperature -0.79 to 4.79°C, salinity 33.05–33.43‰ and oxygen concentrations from 5.67 to 6.40 mg L^{-1} (Myasnikov 1985). At Kurile, white scallops were more frequently fouled by sponges *Mycale adhaerens* and *Myxilla parasitica* and rarely inhabited by various hydrozoa, barnacles, bryozoa, algae, polychaetes, actinias and juveniles of bivalve molluscs (Myasnikov 1986).

The growth rates of scallops from different regions of the Kurile Ridge differ greatly from each other (Silina and Pozdnyakova 1986) but approximately equal and high rates of linear growth (up to 13.5 mm yr^{-1}) are observed in all regions during the first three years (Fig. 24.2; Table 24.1). After this age, the most intensive growth is observed in scallops living along the Sea of Okhotsk side of Onekotan Island, where shell growth is up to 18 mm·yr^{-1}. Minimum shell growth was observed at Simushir Island (middle Kuriles) where the annual rates did not exceed 13 mm yr^{-1} and in Pacific Ocean waters of Onekotan Island. According to Zolotarev (1979) these molluscs have three strongly pronounced stages of linear growth: juvenile, mature and senile. For *C. albida*, juvenile or fast growth stage finishes when the age of puberty is reached. The next stage (mature) is limited to 10 years. Senile stage starts at 10 years when annual shell growth does not exceed 0.5–1.5 mm. Scallops reach marketable size (60 mm) at about 5 years (Table 24.1).

Scallop weights also change irregularly with age. In scallops up to 50–60 mm, weight increases by equal rates. After this size, the rate of weight increase in scallops inhabiting the Sea of Okhotsk side of Onekotan Island differs from that of scallops from Pacific Ocean waters. Maximum weight increases (up to 23 g yr^{-1}) are observed at the age of 5 years.

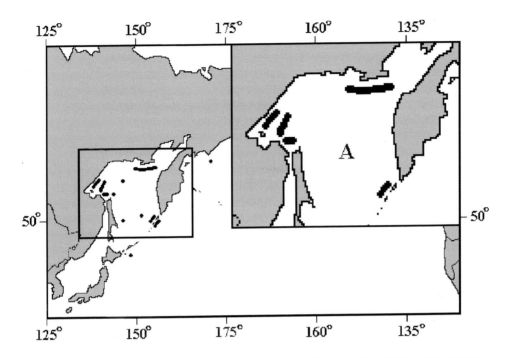

Figure 24.1. Natural habitat of the White scallop *Chlamys albida* Arnold, 1906 (ex Dall, MS) (by Scarlato 1981). A – map showing known commercial assemblages.

Table 24.1

Shell height (mm) and ages of commercial *Chlamys* scallops in the northwest Pacific, mm ± s.d.

Age (years)	*C. albida* (Silina and Pozdnyakova 1991)	*C. behringiana*	*C. chosenica* (Silina and Pozdnyakova 1990)
0.5	7.5 ± 0.6	8.6 ± 0.9	9.6 ± 0.5
1.5	19.7 ± 1.0	19.0 ± 1.9	23.5 ± 0.5
2.5	31.6 ± 1.4	31.3 ± 1.9	35.1 ± 0.6
3.5	43.5 ± 1.5	44.0 ± 2.0	45.0 ± 0.6
4.5	54.9 ± 1.5	55.9 ± 2.0	53.2 ± 0.6
5.5	64.5 ± 1.5	65.4 ± 2.2	59.2 ± 0.6
6.5	71.4 ± 1.5	72.5 ± 2.4	63.5 ± 0.6
7.5	77.0 ± 1.8	76.9 ± 2.4	66.7 ± 0.6
8.5			69.6 ± 0.6
9.5			72.0 ± 0.6
10.5			74.4 ± 0.7
11.5			76.2 ± 0.8
12.5			77.4 ± 1.2

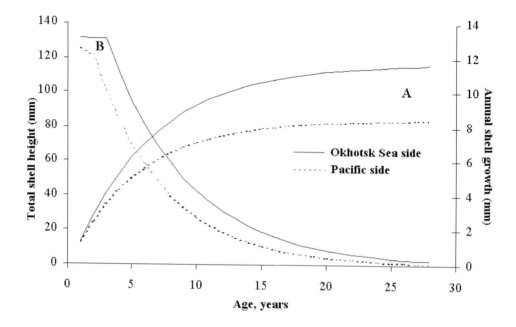

Figure 24.2. Growth rates of the White scallop *Chlamys albida* at Okhotsk Sea and Pacific sides of Onekotan Island, Kurile Islands (by data of Myasnikov, Kochnev 1988). A - Total shell height, mm. B - Annual shell growth, mm.

According to Myasnikov (1988), the size range in the population of *C. albida* from the northern part of the Sea of Okhotsk includes scallops with shell heights between 18–93 mm (average of 68 mm) but individuals within the size of 70–80 mm (about 40%) were most common. The average shell height correlates with depth. As depth increases from 50 to 125 meters the average shell height increases from 41 to 73 mm. At depths of over 125 m, average shell heights decreased to 62 mm.

The maximum age of these scallops does not exceed 28–30 yr. The age of sexual maturity is 3 to 5 years with shell heights between 40–70 mm. The sex ratio (male to female) changes from 0.6:1.0 to 2.0:1.0. At North Kuriles, mass spawning in population starts in June (Myasnikov and Kochnev 1988).

There are four commercial concentrations of *C. albida* along the southern and northern coastlines of Northern Kurile Islands (Sea of Okhotsk side and Pacific Ocean side), at Simushir Island and in the Northern part of the Sea of Okhotsk (Myasnikov and Hen 1990; Myasnikov et al. 199?).

24.3.2 *Chlamys asiatica* (Common name: Asiatic scallop)

Chlamys asiatica is a high-boreal, Asian-Pacific species (Fig. 24.3). The Asiatic scallop occurs on Kurile Island, on the eastern coast of Kamchatka Peninsula and in the

Bering Sea (Anadyr Bay). It lives in sandy substrates mixed with shingle and muddy sands at depths between 80–120 m (Scarlato 1981). It was rarely found in its natural habitat.

24.3.3 *Chlamys behringiana* (Common name: Bering Sea's scallop)

Chlamys behringiana is a widespread, high-boreal, Pacific species (Fig. 24.4). It occurs within the Sea of Okhotsk at Sakhalin and Aniva Bays, the Strait of Laperuz, the southern and eastern waters of Kamchatka Peninsula, and the Kurile Islands (Paramushir and Shikotan). It is also found in the Bering Sea and in the Arctic Ocean at the southern part of the Chuckchee Sea and Sea of Beaufort. It is found at depths between 24–200 m in muddy sands mixed with shingle and gravel and in gravel with pebbled substrates (Scarlato 1981). Environmental ranges have been reported by Myasnikov (1985) for temperature (0.79 to 4.79°C), salinity (33.05–33.43‰) and oxygen concentrations (5.67 to 6.40 mg L^{-1}). Bering Sea's scallops are often inhabited by hydroids, *Eunephtya* sp., and ascidians, *Pyaridae* sp. (Myasnikov 1986).

According to Buyanovsky (1999), larger scallops are observed at Karaginsky Island (size: 72.7 mm, age: 10–15 yr) and at Olyutorsky Bay (size: 83.3 mm, age: 20–25 yr). After these ages, shell height does not change. Higher growth rates in scallops from Olyutorsky Bay are probably linked to the more intensive water exchange via the main branch of Kamchatka Current.

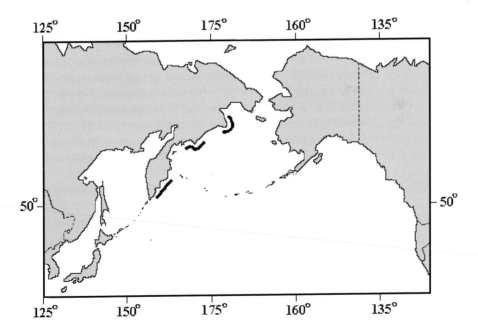

Figure 24.3. Natural habitat of the Asiatic scallop, *Chlamys asiatica* Scarlato, 1981 (Scarlato 1981).

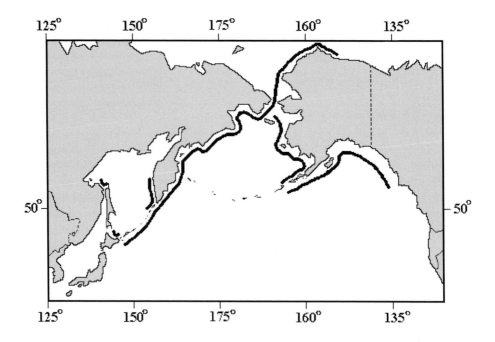

Figure 24.4. Natural habitat of the Bering Sea's scallop *Chlamys behringiana* (Middendorff, 1849) (Scarlato 1981).

The maximum age of the Bering Sea's scallop is 35 years old but most of the older population is 25–28 years old. The most intensive growth period is observed during the first two years and can be as high as 15 mm yr^{-1}. Between the ages of 3–6 yrs, the shell typically grows 4.2–11.5 mm yr^{-1}. Then, at the age of sexual maturity, growth rates decrease considerably until they reach 1.6–1.2 mm yr^{-1} in 10-year-old scallops (Silina and Pozdnyakova 1991). Scallops reach marketable size (60 mm) in about 6 years (Table 24.1). Scallop weights increase almost in proportion to their age (Buyanovsky 1999).

24.3.4 *Chlamys chosenica* (Common names: Pink scallop and White-pink scallop)

Chlamys chosenica is a low-boreal, Asian Pacific species (Fig. 24.5). It occurs in the Sea of Japan along the Primorye coastline, at the western Sakhalin Island and the north western part of Hokkaido Island, in waters of the small Kurile Ridge and at Iturup Island. Pink scallops are found at depths between 13–2,030 m in muddy sands mixed with shingle and gravel and sometimes in sand or shell-rock substrates (Scarlato 1981). The pink scallop is a eurybiontic species occurring at temperatures ranging from -0.09 to 13.03°C, salinities ranging of 33.34–34.15‰ and oxygen concentrations ranging from 5.49 to 7.49 mg L^{-1} (Myasnikov 1985). Along the coast of Northern Primorye, pink scallops are mainly colonised by the gastropod *Vetulina* sp. (Myasnikov 1986).

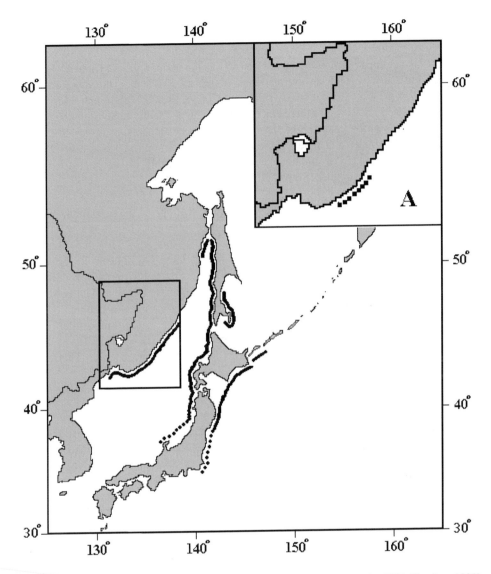

Figure 24.5. Natural habitat of the Pink scallop, *Chlamys chosenica* Kuroda, 1932 (Scarlato 1981).
A – map showing known commercial populations.

According to Silina and Pozdnyakova (1990), the population from the northern Primorye is characterised by scallops with shell heights up to 90 mm. The size distribution is a uni-modal curve with most individuals measuring 68–81 mm (peak at 77–76 mm). The maximum age of scallops is 22 years but most of the population is 11–12 years old.

The most intensive time of growth is observed within the first two years and can be up to 16 mm yr^{-1}. Between the ages of 3 to 6 years the shell grows by 5.2–10.8 mm yr^{-1}. At the age of sexual maturity, growth rates decrease to 2.5–1.0 mm yr^{-1} in 10-year-old scallops. Pink scallops reach marketable size (60 mm) in about 6 years (Table 24.1). In 9 to 14 year old scallops, with an average shell height of 70–79 mm, the muscle weight ranges from 7.4–10.4 g. In four years, from 10 to 14 yr, muscle weight increases by not more than 25% and shell weight increases by least by 50%. At the same time, the muscle comprises 20–25% of the total weight of the mollusc whenever shells comprise 44–52%.

The sex ratio (male to female) in these populations is approximately 1.0:1.2. The domination of females (55%) implies stable status. Spawning of the population takes place in the first half of summer (Silina and Pozdnyakova 1990).

24.3.5 *Chlamys farreri* (Common names: Japanese scallop, Chinese scallop, Farrer's scallop, Akazara scallop)

Chlamys farreri is a subtropical, Asian-Pacific species (Fig. 24.6). In the Sea of Japan it is widely distributed along the southern coasts. This is the northern border of its natural habitat. It also occurs at Middle Primorye and Japanese Island at depths from 0.5 to 24 m (Scarlato 1981). The scallop is mainly found in gravel and pebbled habitats. Frequently these scallops form many-tiered reefs in the rocks and oyster banks. Settlement densities can vary between 150–180 specimens m^{-2} (in reefs) at depths of 1.5–2.5 m. At other depths, their density decreases to 5–10 specimens m^{-2}. This species occurs at a temperature range of 19–22°C and a salinity range of 32–34‰.

This species is a candidate for mariculture and fishery in East Asian countries (Wang and Shieh 1991). In Russia, the Japanese scallop is one of the most prominent species for commercial fishing and mariculture (Bregman 1982; Afreichuk 1992a).

According to Afreichuk (1992a), the population of *C. farreri* from Posjet Bay includes scallops with shell heights of up to 112 mm but most individuals measure 70–80 mm. The maximum age of this species is 9 years but most of the population is three (35%) or four (27%) years old. After age 6, natural mortality increases and results in only 12% of population being older than 6 years.

The most intensive growth period is observed in the first three years and can be as high as 19 mm·yr^{-1}. Between the ages of 4 and 7, the shell grows by 5.8–11.5 mm yr^{-1}. After this, growth rates decrease (Afreichuk 1992b). These scallops reach marketable size (72 mm) in about 3–4 years. At that time the muscle weighs about 5.5 g. The muscle weight of a 5-year-old scallop with average shell height of 90 mm is 11 g. In total, the muscle comprises 12–14% of the total weight of the scallop and the shells amounts to 63% (Afreichuk 1990).

The age of sexual maturity is 2 years old with shell heights typically 40–45 mm (Table 24.2). The approximate sex ratio (male to female) in population is 1:1.36. At Posjet Bay spawning of the population starts at the end of June when the temperature of near-bottom waters approaches 16°C. The mass spawning (average 65 days) begins in the middle of July at temperatures of 18–20°C and ends at the end of August or early in September (Afreichuk 1992a).

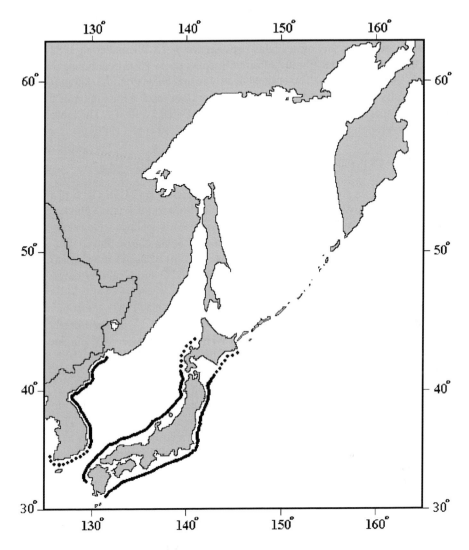

Figure 24.6. Natural habitat of the Japanese scallop, *Chlamys farreri* (Jones and Preston, 1904) (Scarlato 1981).

Table 24.2

Shell height (mm) of different ages of Japanese scallop *Chlamys farreri* at southern part of the Sea of Japan (Bregman 1982), mm ± s.d.

Age (years)	1	2	3	4	5
Shell height (mm)	2.70 ± 0.96	4.30 ± 1.50	5.85 ± 0.90	7.00 ± 0.96	7.80 ± 0.96

At Peter the Great Bay larvae occur in plankton during the warmest months (July and August) at water temperatures between 15–20°C (Kas'yanov et al. 1980). The maximum density of larvae occurs in shallower bights between 5–7 m (Afreichuk et al. 1988). The period of settling on hard and fibrous substrate (depths up to 23 m) continues from the middle of July to early August. The optimal depth for spat collection is 5–8.5 m (Gabaev 1988). The density of juvenile settlement on collectors can reach a maximum of 229 and an average of 48 specimens m^{-2} (Afreichuk et al. 1988; Gabaev 1990b). The morphology of larvae and larval shell structure were described by Kulikova et al. (1981).

24.3.6 *Chlamys swifti* (Common name: Swift's scallop)

Chlamys swifti is a low-boreal, Asian-Pacific species (Fig. 24.7). It is distributed along the southern coasts of the Sea of Japan, in Western Sakhalin, Hokkaido and northern waters of Honshu Island. In the Okhotsk Sea Swift's scallops have been found in southern Sakhalin (Aniva Bay) and in shallow waters of South Kurile Shoal. The scallop mainly inhabits the gravel, pebbled and shell areas at depths of 2–143 m (Scarlato 1981). This species occurs within a temperature range of 9 to 22°C and a salinity range of 32–34‰.

According to Ponurovsky (1982) and Ponurovsky and Silina (1983), the population of *C. swifti* from the Northern Primorye is characterised by scallops with shell heights up to 121 mm, but individuals measuring between 65–90 mm predominate. The maximum age of scallops is 13 years, but the majority of the population is five years old.

Swift's scallops grow throughout their lifetime. However, the greatest growth period is observed during the first three years after settling on the bottom and can be as high as 21–26 mm yr^{-1}. After 4 years old, linear growth rates decline and are only 1.0 mm yr^{-1} in 8-year-old individuals (Ponurovsky 1977; Ponurovsky and Silina 1983). The highest growth rates were found in regions of the Sea of Japan, Petrov and Putyatin Islands, and Vostok Bay (Table 24.3). These areas are more favourable for intensive growth of Swift's scallop (Ponurovsky 1982).

The sexual maturity is reached at 3 years when shell heights are 50–70 mm. The approximate sex ratios (male to female to hermaphrodite) in the population are 1.0:0.67:0.02 (Denisova 1981). By the age of 3, males predominate (86.3%) in the population. Later, the percentage of males declines. By the age of 10, females predominate (81.2%) the population.

At Peter the Great Bay and Bousse lagoon (southern Sakhalin Island) larvae are abundant in the plankton from August to September in water temperatures between 15–20°C. The maximum density of larvae occurs in near-bottom waters at depths of 10–20 m (Kas'yanov et al. 1983). The morphology of larvae and larval shell structure were described by V. A. Kulikova with co-authors (1981). The optimal depth for spat collection is greater than 15 m (Gabaev 1988).

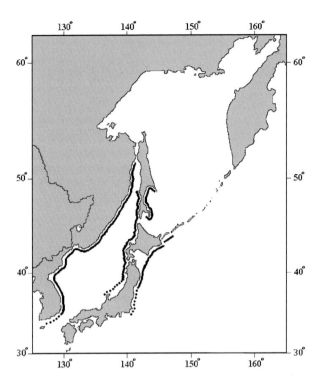

Figure 24.7. Natural habitat of the Swift's scallop, *Chlamys swifti* (Bernardi, 1858) (Scarlato 1981).

Table 24.3

Shell height (mm) of Swift's scallop, *Chlamys swifti* in the northwest part of the Sea of Japan (Ponurovsky 1982), mm ± s.d.

Age years	Region (from southern to northern) of scallops collection							
	Furugelm Island	Vityaz Inlet	Stenin Island	Klykov Island	Putyatin Island	Vostok Bay	Melkovod-naya Inlet	Petrov Island
1	29.7±0.5	26.7±0.4	21.2±0.5	24.3±0.4	29.3±0.2	21.3±0.2	24.2±0.4	29.8±0.3
2	44.1±1.0	37.4±0.7	32.3±1.0	34.9±0.8	45.8±0.4	45.8±0.6	40.0±0.7	47.3±0.5
3	62.3±1.7	52.8±1.5	48.9±1.6	49.4±1.4	67.2±0.6	70.3±0.7	60.58±0.7	69.8±0.6
4	87.4±1.8	70.2±20.1	67.6±2.0	69.1±1.7	90.6±0.6	88.9±0.7	81.3±0.7	90.5±0.5
5	95.1±1.4	87.5±2.0	86.1±1.9	87.9±1.7	106.4±0.4	100.3±0.7	94.7±0.6	101.6±0.5
6	101.5±1.5	97.8±1.9	96.9±1.4	99.8±1.7	112.9±0.4	105.7±0.8	98.8±0.6	105.1±0.5
7	102.3±1.8	104.4±1.7	102.1±1.2	105.3±1.3	116.8±0.5	106.7±0.9	102.0±0.6	107.5±0.6
8	104.5±2.5	107.1±1.4	105.3±1.2	109.1±1.3	118.6±0.6	108.7±1.1	104.0±0.8	109.6±0.8
9	105.6±2.7	107.4±1.6	106.5±1.3	111.3±1.5	120.6±0.7	109.0±1.5	105.6±0.9	110.8±0.9
10	No data	No data	107.2±1.4	112.1±1.4	121.1±1.0	No data	106.2±1.7	110.5±0.8
n	22	33	30	43	243	160	51	219

1174

24.3.7 *Delectopecten randolphi* (Common name: Randolph's scallop)

Delectopecten randolphi is a widespread North Pacific species (Fig. 24.8). It is found in the Sea of Japan at Posjet Bay, Peter the Great Bay, and the eastern coastline waters of Honshu Island (Sagami Bay). It is also found in the Okhotsk Sea along eastern and western coastlines and in the Bering Sea along the eastern coastline. The Randolph's scallop inhabits muddy areas at depths of 418–3,080 m (Scarlato 1981). Its temperature range is 0.2–2.5°C. The largest specimen known had dimensions of 23.5 x 27.0 mm and was found in Peter the Great Bay.

24.3.8 *Mizuhopecten yessoensis* (Common Names: Yesso scallop, Ezo scallop, Giant scallop, Japanese scallop, Russian scallop, Primorsky scallop and Common scallop)

The Yesso scallop, *Mizuhopecten yessoensis,* is a low-boreal, Asian-Pacific species (Fig. 24.9) with the highest commercial value of all the Pectinidae. It is found along the northern coastline of the Korean peninsula, the coastline of Primorye, near the shores of the Sakhalin islands, South Kuriles and Hokkaido and on the northern coastline of Honshu Island (Scarlato 1981).

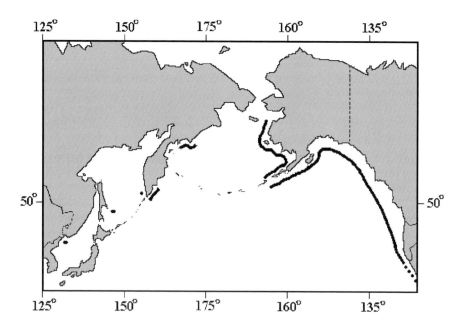

Figure 24.8. Natural habitat of Randolph's scallop, *Delectopecten randolphi* (Dall, 1897) (Scarlato 1981).

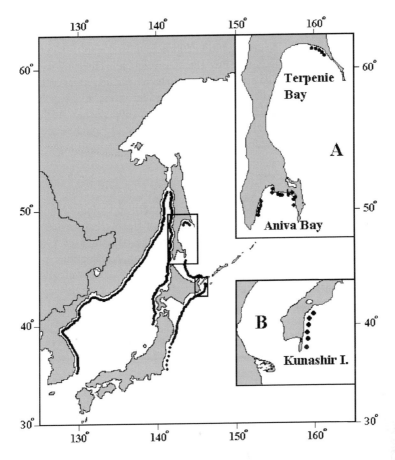

Figure 24.9. Natural habitat of the Yesso scallop, *Mizuhopecten yessoensis* (Jay, 1857) (Scarlato 1981 with additions). A and B – maps show known commercial assemblages at Sakhalin and Kunashir Islands.

24.3.8.1 Total populations and biomass

The most exhaustive studies of the Yesso scallop were completed along the coasts of Primorye by Razin (1934), Biryulina and Rodionov (1972), Markovskaya (1951), Bregman (1979), and Kalashnikov (1986, 1991).

In 1932, an estimated 40 million scallops inhabited some of the 16,000–17,000 hectares along the coast of Primorye. Between the 1940's-1960's, the scallop population in most areas decreased and in some places completely disappeared. From 1932 to 1959, the scallop stocks in Peter the Great Bay had reduced thrice. In the following decade, the abundance remained almost the same at 5,703 million scallops on 906 hectares with a biomass of 1,708 tons (Biryulina and Rodionov 1972).

At the same time, a significant drop in the abundance of scallops was observed in the northern areas of Peter the Great Bay (Olga Bay) where the scallop population in 1932 was four times as high as it was in 1975 (Silina and Bregman 1986). According to Skalkin (1971), the biomass in Aniva Bay (southern Sakhalin) in 1969 was twice as low as compared to 1961–1962. The scallop populations became ten times as low in Terpenie Bay (northern Sakhalin) and some areas of the Kuriles. Most investigators believe that intensive industrial fishing caused the overall drop in scallop stocks. At the present time, commercial reserves of wild Yesso scallop along the coast of Primorye are exhausted due to overfishing during the last decade.

24.3.8.2 Distribution in Primorye

The Yesso scallop was widely distributed in southern and middle Primorye where it is found in bays and coves and forms aggregations at depths between 6–30 m. These scallops were an object of traditional catching until the beginning of 1970's. The map (Fig. 24.10) shows the locations of existing and potential plantations sites for bottom cultivation of this scallop species. Commercial populations were well known in Bays of Posjet, Ussuri, Amur, Vladimir, Vostok and Strelok. The average density of settlements on bottom grounds increased from 0.05 to 1.0 specimen m^{-2} (average 0.1 specimen m^{-2}). Average individual biomass was 0.2–0.4 kg. At the present time, the average density does not exceed 0.001 specimen m^{-2} at sites described by other authors.

24.3.8.3 Distribution over depths

In the shallows, the Yesso scallop occurs between depths of 0.5–1.0 m in small inlets protected from wind and waves. Minimum depth corresponds to the winter time water level under the ice cover. Most of the scallops were found in the range of 4–10 m in closed inlets and at depths between 20–25 m in open and relatively deep-water sites of bays and inlets. According to Razin (1934) this species is typically found between 14–30 m on the open coastline of Primorye, however some specimens are found at 48 m.

Most of the scallops were found along the coastline at a depth of 20–25 m near rugged shores. This is apparently due to the corresponding range of spat settlement in the region, i.e., underwater rocks are substrate carriers for attaching scallop larvae. In Peter the Great Bay, scallops were recorded at a maximum depth of 82 m (Scarlato 1981).

24.3.8.4 Age structure of scallop settlements

Biologists believe that this species does not live more than an average of eleven years (Tibilova and Bregman 1975) with a typical life expectancy ranging from 7 to 9 years (Makarova 1985). The most extensively studied populations in Primorye have no 1–2 year old specimens. Eleven-to-twelve year-old scallops are frequent but older specimens are rare (see Table 24.4). In some settlements, 70 percent of the individuals range in age from 8 to 16 years. Generally, the age structure of the scallop population of various

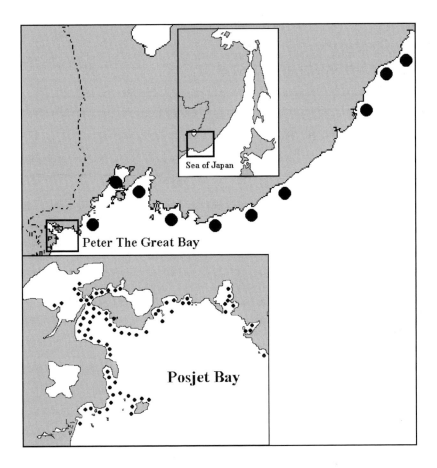

Figure 24.10. Locations of existing and potential plantations for bottom cultivation of the Yesso scallop *M. yessoensis* in Primorye.

regions reflects the randomness of replenishment and presence of abundant and non-abundant generations (Fig. 24.11).

24.3.8.5 Scallop growth

Fertilised eggs (average 60–70 micrometers in diameter) develop into larvae, which settle onto substrates (shell averages 260–285 μm) in 20–40 days, depending on temperature. Definitive development of scallops terminates on the substrate. The scallop detaches from the substrate in 3–4 months when the shell height is 10–30 mm. Subsequent scallop growth rate depends on temperature, feeding, water exchange, and many other conditions on the sea floor.

Table 24.4

Age composition of Yesso scallop settlements at Peter the Great Bay (Silina and Bregman 1986). Number of specimens from respective age groups.

Region	1–2	3–4	5–6	7–8	9–10	11–12	13–14
Vityaz Inlet	--	1	8	14	10	7	1
Stenin Island	--	1	1	5	11	7	5
Priboinaya Inlet	--	10	30	6	1	1	--
Shkota Inlet	--	2	16	6	6	6	2
Andreev Inlet	15	10	14	21	1	--	--
Putyatin Island	--	22	10	4	4	4	--
Lake Vtoroe	18	32	101	45	7	--	--
Melkovodnaya Inlet	2	13	20	11	6	5	1
Olga Bay	--	--	1	23	16	7	1

The scallop shell grows isometrically to retain its initial form. Makarova (1985) calculated the general equation for the linear growth rate of the Yesso scallop:

$$H_t = (160.92 \pm 18.7) \cdot (1 - e^{(-0.378 \pm 0.04)} \cdot t), \tag{1}$$

where H_t is the shell height in mm and t is scallops age in years.

The scallop grows at temperatures ranging from -2°C to 26°C. The optimal growth temperature is 4–6°C. In Primorye, the temperature optimum occurs in May-June and in September-October (Silina and Pozdnyakova 1986). Within the initial three years, the scallop height reaches 90–110 mm and then its growth slows exponentially (Table 24.5). The largest individuals occur at depths of 20 m in populations located on silty-sandy soils with good water exchange and relatively stable temperatures. These specimens reach 190–195 mm shell height or longer at the age of 16 and older. In the South Kurile shallows, specimens older than 20 years with shell heights of 220 mm are found (Skalkin 1966). In shallow silty inlets, scallops seldom exceed 150 mm and live for not more than 10–12 years. In Posjet Bay, we recorded the largest scallop specimen, whose dimensions were as follows: shell height 222 mm, length 202 mm, and width 37 mm. Scallop weights vary proportionally to its linear dimensions (Fig. 24.12; Silina and Pozdnyakova 1986). The proportion of muscle in scallop total mass amounts to 10–18% in various settlements of Peter the Great Bay (Belogrudov 1981).

24.3.8.6 Sex structure of settlements

The average (male to female) sex ratio within populations is 1.0:1.0. Hermaphrodites were rarely observed (i.e., not more than 0.3–0.4% of all cases; Bregman 1979). In these populations, males dominated in younger generations and females dominated in older ones. By age 1–3 the sex ratio was 2.0:1.0, by age 5–6 it was 1.9:1.0 and by the age of 7–8 the sex ratio was 1.0:1.9. This could be due to either sex specific different death rates or hermaphroditism.

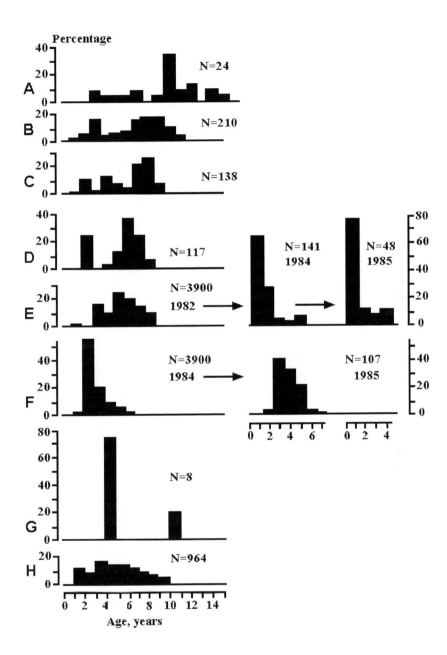

Figure 24.11. The age structure of the Yesso scallop population of various regions of Primorye. A – dispersal settlement from deep and open region of Posjet Bay; B – Olga Bay ribbon settlement; C – settlement of closed Vladimir Bay; D – settlement under kelp farm; E – shallow-water settlement in surf bay; F – shallow-water settlement in closed inlet; G – single scallops near surf shore; H – age composition (Razin 1934). (From Kalashnikov 1991)

Table 24.5

Linear growth of Yesso scallop *Mizuhopecten yessoensis* from various regions of natural habitat (by Silina and Pozdnyakova 1986).

Region	Shell height at corresponding age (in numerator) and annual shell growth (in denominator), mm (mm±s.d.)									
	1	2	3	4	5	6	7	8	9	10
Furugel'm Island	50.2±1.2	101.4±1.2	125.9±1.3	141.4±1.9	151.8±1.8	156.4±2.5	---	---	---	---
	50.2±1.2	51.2±0.9	24.5±1.0	15.5±1.0	10.4±0.8	4.6±0.8				
Bol'soi Pelis Island	---	89.8±4.2	120.4±3.2	139.1±2.1	147.8±2.4	151.5±2.6	151.8±2.8	---	---	---
		no data	30.5±1.0	18.7±2.0	8.7±1.8	no data	no data			
Andreev Inlet	46.9±0.8	85.8±1.0	107.0±1.0	118.8±1.1	124.7±1.3	128.3±1.1	130.4±1.5	133.5±1.7	---	---
	46.9±0.8	38.9±1.0	21.2±1.0	11.8±1.0	5.6±0.9	3.6±0.8	2.4±0.6	3.0±0.8		
Putyatin Island	50.7±1.6	90.8±1.2	115.3±1.0	128.5±1.1	135.7±1.8	141.1±1.0	---	---	---	---
	50.7±1.6	40.2±1.4	24.5±1.2	13.2±1.1	7.3±1.2	no data				
Olga Bay	36.6±1.1	83.2±1.4	114.6±1.4	133.8±1.0	145.4±0.9	152.0±1.1	154.5±1.1	155.4±1.5	158.8±1.7	---
	36.6±1.1	43.8±2.2	30.4±2.0	14.0±1.6	11.0±1.3	no data	no data	no data	no data	
Vladimir Bay	33.6±2.6	77.4±2.8	107.8±2.4	121.8±2.8	132.8±2.8	---	---	---	---	---
	33.6±2.6	43.8±2.2	30.4±2.0	14.0±1.6	11.0±1.3					
Aniva Bay	26.1±2.4	56.2±2.1	86.9±2.0	110.9±1.42	125.5±1.6	134.2±2.1	143.6±2.2	148.4±2.3	156.2±2.3	160.8±2.5
	26.1±2.0	30.1±2.0	30.7±1.8	24.0±1.6	14.6±1.4	8.7±1.0	9.4±0.8	4.8±0.6	no data	no data
Izmena Strait	24.9±5.8	75.2±8.4	97.9±10.4	119.9±8.2	130.7±8.4	138.4±5.2	141.2±3.9	143.0±4.8		
	24.9	50.3	22.7	22.0	10.8	7.7	2.8	1.8		
South Kuril Strait	20.7±5.5	56.9±6.8	89.0±11.5	119.1±12.2	141.3±11.1	150.3±8.7	155.3±7.7	158.4±7.5	161.3±8.0	---
	20.7	36.2	32.1	30.1	22.2	9.0	5.0	3.1	2.9	

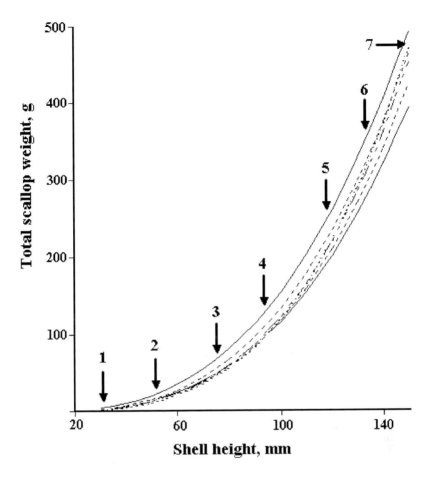

Figure 24.12. Relationship between Yesso scallop weight and shell height in various settlements in Peter the Great Bay. Numbers represent scallop age in years. (From Salina and Pozdnyakova 1986)

24.3.8.7 Replenishment

Scallop populations replenish annually owing to spawning, subsequent development of larvae in plankton, their settling on substrate and juveniles' transition to free life on the seabed.

24.3.8.8 Spawning

Spawning in scallop populations starts at temperatures ranging from 7–9°C. (Belogrudov 1981). In the waters around Primorye, spawning begins in the middle of May and spawning ends at the end of June. Spawning begins in shallow waters of southern regions. As the seawater warms up individuals from the deeper and more northerly settlements begin to spawn. Absolute fertility varies from 25–30 to 180 million eggs (Yamamoto 1964) depending on age and size.

24.3.8.9 Larvae morphology

The morphology of larvae and larval shell structure of three widespread scallops: Yesso scallop *M. yessoensis*, Japanese scallop *C. farreri* and Swift's scallop *C. swifti* have been described by Kulikova et al. (1981). All of the larvae are of triangular form and the anterior end is the apex of the triangle. The larvae are inequivalve. The umbos are low, rounded and poorly defined. The taxodonte hinge has several teeth at each side of the hinge line. The central hinge area is undifferentiated. Some specific characteristics were distinguished among the scallop larvae of Peter the Great Bay. The main differences are in the shell form, umbo form, shell size, and number of teeth at each side of the hinge line (Kulikova et al. 1981).

24.3.8.10 Development in plankton

The duration of larval growth in the plankton lasts from 20 to 40 days (Kas'yanov et al. 1980). Current-induced distribution disseminates larvae to create new local settlements. Plankton surveys (Belogrudov 1981) show that larvae can form dense concentrations in Posjet and Peter the Great Bays over several square kilometers. In that case, higher densities are noted at areas with abounding adult scallops. In bonanza years, larval density reached 200–300 specimens per cubic meter in the closed inlets and at the same time, it did not exceed 20–30 specimens per cubic meter in the adjacent open inlets and bays. Long-term studies in Posjet Bay show the absence of direct spatial relations between parents and new scallop generations.

On the eve of scallop cultivation, the number of sexually mature specimens in Minonosok Inlet in 1972 was 70,000. Ten years later, the number has increased up to 650,000 because of sowing culture (Fig. 24.13). The number of annually collected spat showed instability of larvae settling throughout all periods. Nor was there any sign of increases in the number of spat during the year. Apparently, the parent-larvae relationship was indirect because of great dilution of larvae pool by the water mass (Kalashnikov 1986). Water exchange with adjacent bodies of water (about 10% of waters every day is replaced by tidal) disturbed the parent-larvae relationships.

A comparison of the age composition (Fig. 24.13) and the results of spat collection in various bays and inlets showed that intensity of replenishment changes asynchronously. Bonanza generations in one bay do not necessarily correspond to the same in another bay. For instance in 1980, the lowest spat settling rates (2–20 specimens per collector) were noted in Posjet Bay, but in Vostok Bay the collectors recorded 400–500 specimens each. It is of interest to note that during that year, red tides were absent in Vostok Bay but observed in Posjet Bay. However, these distinctions were also noted in other bays without red tides, and this suggests that the intensity of replenishment is a local characteristic.

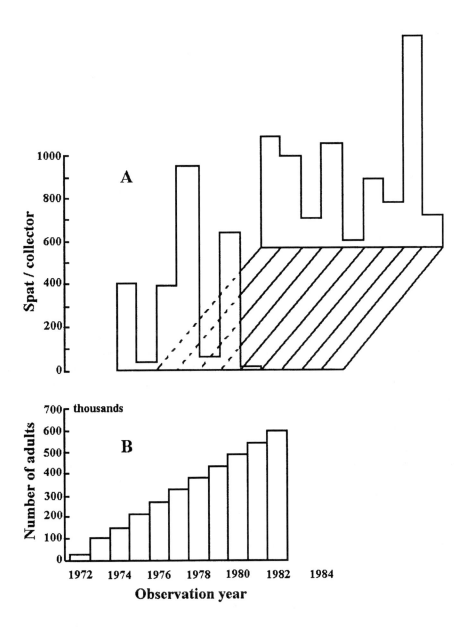

Figure 24.13. Dynamics of spat collection (A) and size of the Yesso scallop population of Minonosok Inlet (B). (From Kalashnikov 1991)

24.3.8.11 Migration behaviour

Yesso scallops can freely move along the sea floor. The mechanism of reactive movement of freely living scallop is widely known (for example: Dautov and Karpenko 1983). Researchers have noted random movement in natural conditions or individual scallop behaviour in aquarium. These observations support the idea that the scallop is a migrating species. However, the observation that populations are present for many years in specific sites shows that ability of scallops to migrate is limited.

24.3.8.12 Risk factors

24.3.8.12.1 Abiotic factors

Survival rates of pelagic larvae at metamorphosis depend on water temperature (ranges for survival is 5–20°C, with an optimum at 10–15°C), salinity (30–40‰, optimum at 33‰), density (optimum 8 specimens mL^{-1}), quantity and composition of food and on the abundance of predators (Belogrudov 1973; Bregman and Guida 1983; Chan 1989). At attachment, mortality is mainly caused by the absence of suitable substrate and intolerance to changing environments (Yamamoto 1964; Belogrudov 1973). Bregman and Guida (1983) reported that the number of attached juveniles resulted from only 5–8% of all fertilised eggs. Settling on the bottom is the next critical period in the life history of the scallop and here the mortality increases sharply. According to Yamamoto (1964), only 5–10% or sometimes none of the settled juveniles survive. Golikov and Scarlato (1970) reported that only 4% of the settled juveniles survived as long as 6 months. Only Golikov and Scarlato (1970) have reported on the mortality of older scallops. These researchers conjectured that after 6 months the quantity of survived scallops decreased gradually. Moreover, they noted that winter was the period during which the highest mortality of scallops occurred off the north western coastline of the Sea of Japan. In contrast, in Honshu Island, the most difficult season for scallop survival is summer and early autumn when water temperature increases to above 15°C (Yamamoto 1960).

Recent investigations (Silina 1996) also show the highest mortality when scallops were less than 2 years. Mortality of 2 to 5 year-old scallops is minimal. By 6–7 years of age (probably the beginning of the senile period of scallop development) and upwards to 9–10 (transition to the old-aged stage) scallops mortality increased sharply.

24.3.8.12.2 Storms

Storms are another factor, which cause mass mortalities of scallops in coastal shallow settlements. Depending on local topography, storms can increase settlement dispersion over vast flat areas or make them denser at the foothills of cliffs. In either case, a considerable number of scallops are buried under moving soil. Near shallow coasts, particularly beaches, the majority of scallops die in storm debris. Even small storms deform settlements but the loss is usually compensated by annual replenishment. While the typhoons (annual frequency about 50%) are natural calamities for the benthic coastline populations, including the scallop, they destroy most of the species around open waters at 20 m deep. By counting the number of shells in the breaker zone of only one beach following typhoon "Ellis" in 1983, researchers revealed the simultaneous death of 10,000 specimens of different ages. In a similar count in 1986 after typhoon "Vera", we discovered as many as 72,000 specimens in debris (Kalashnikov 1984). The joint effect of various factors on the sea floor populations shows perennial changes in density of artificial scallop populations, which regularly declined in all cases during the first cultivation season. The decline was greater in more open and unprotected waters (Fig. 24.14; Kalashnikov 1985).

24.3.8.12.3 Predators

The first hours and days on the sea floor after the scallops become one year old (shell height about 30 mm) appeared to be the most dangerous. During this period, natural death is maximal and young scallops are preyed upon by various starfish species such as *Asterias amurensis* Lütken, which can grow up to 165 mm (radius) and weigh up to 450 g. Another species, *Distolasterias nipon* (Döderlein) is even larger growing up to 250 mm with an average weight of 1,000 g. These starfish species attack scallops of the same or younger ages. The one-time ration of one starfish increases from 1 up to 8.5 g of fish and annually amounts to 400–450 g (Biryulina 1972). In view of the great abundance of these predators (density can be up to 15 specimens m^{-2}), the damage to scallop populations can be considerable. The death rate of young scallops located in super dense aggregations (over 100 specimens m^{-2}) in which predators temporarily eat only scallops is especially high.

When storms are so strong that they reach the sea floor scallops become weaker and readily accessible to predators. The joint effects of storms and starfishes have destroyed an artificial scallop settlement (about two hectares) with a population of over 200,000 specimens. Under stable conditions, starfish and scallops were noted to co-exist when they occupied a single habitat for several years in succession. Generally, however, the number of cultivated molluscs declines more in sites where starfish are more abundant.

Sowing spat are also preyed upon by various benthic fishes such as flounders and bullheads. Other predators of sown seed and adult scallops are of lesser importance but nevertheless pose a threat to small seed and juveniles. These include the octopuses, king crabs, *Paralithodes camtschatica* (Tilesius), and hermit crabs (Kalashnikov 1986). Crabs predate mainly on seed scallops present in large numbers. For example, during seasonal migrations, crabs can greatly denude newly seeded grounds. Some predatory gastropod molluscs pose a threat to both juveniles and adult scallops. Belogrudov (1973) reported that the drilling Muricidae gastropod *Boreotrophon candelabrum* (Adams et Reeve) and *Tritonia japonica* (Dunker) could attack and eat the scallops. At natural population levels, 14–27% of adult scallops had drilling marks on the shells. Gabaev and Kolotukhina (1999) reported that two-year scallops (shell height up to 73 mm) in the cages are preyed upon the gastropod *Nucella heyseana* (Dunker).

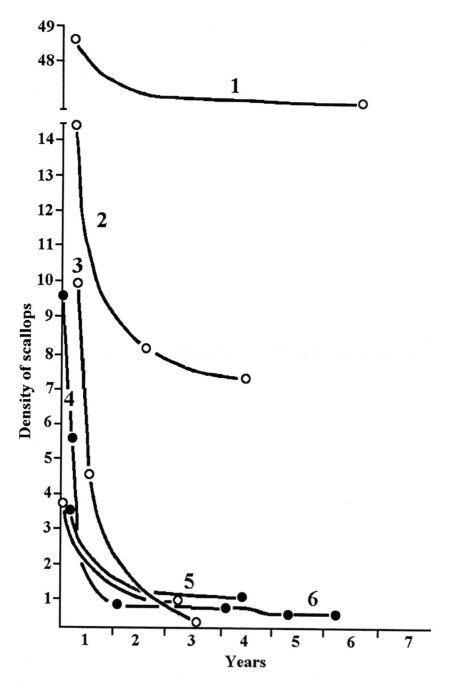

Figure 24.14. Density of Yesso scallops plantations in Posjet Bay (number m^{-2}). 1 = closed lagoon; 2 = protected island terrace; 3–6 = sites of open waters. (From Kalashnikov 1991)

24.3.8.12.4 Parasites

In comparison to other cultured bivalves, such as oysters and mussels, little is known about the parasites and diseases of scallops. Epizootic diseases, like those that have devastated the oyster culture industry in parts of the world, have not been encountered by the scallop culture industry. The relative lack of information on parasites and diseases in scallops may be attributed to less intensive cultures and comparatively fewer investigations. Infection of scallops by parasites is low, as the parasitic fauna is scarce and includes only a few potentially pathogenic species. Mass scallop deaths caused by parasites have never been recorded. There are only 17 parasites and commensals now associated with scallops (Kurochkin et al. 1986; Kovalenko 1990; Plyusnin 1990; Rakov 1990; Didenko 1996).

Sirolpidium zoophthorum Vishniac, 1955 (Lagenidiales)
This fungus was found in 1972 in juveniles at a scallop farm in Posjet Bay.

Myxosporidia gen. sp. (Myxosporea)
Local pestholes, which probably were caused by unknown Myxosporidia, were noted at a scallop farm in Minonosok Inlet (Posjet Bay) in April of 1996.

Perkinsus sp. (Sporozoa)
They were unknown until 1979 when a spherical cyst (0.2–03 mm in diameter) of *Perkinsus* sp. was found in 86% of discovered scallops (intensity of invasion by 1–2 cysts). Occasionally they can pose serious threats to scallop spat. Conceivably it was introduced with scallop seed from Aomori prefecture, Japan.

Pectenita golikowi Jankowski, 1973 (Ciliophora)
Almost all scallops are affected by this endoparasitic infusoria. Intensity of invasion is several tens of individuals within a scallop's intestine. Pathogenic effects on the host are unknown.

Trichodina pectenis Stein, 1974 (Ciliophora)
See below.

Trichodina sp. Stein, 1974 (Ciliophora)
These two species of endoparasitic infusoria were described in the mantle of Yesso scallops. Intensity of invasion is 20–100% and by several tens of specimens. Pathogenic effects on the host are unknown but probably caused by a secondary parasite.

Cliona sp. (Porifera)
This drilling parasitic sponge causes extensive invasions of up to 70% on lower (right) valve and up to 10% on upper (left) one.

Hirudinea gen. sp. (Hirudinea)
Leeches were found in the mantle on isolated instances.

Podocotype spp. (Trematoda)
Approximately 1% of all scallops are affected by trematode metacercaria, which were found in various tissues (including adductor muscle). Intensity of invasion is only one specimen per scallop.

Anisakidae gen. sp. (Nematoda)
Larvae of these nematodes were found in the digestive system on isolated instances.

Ohridiidae gen. sp. (Nematoda)
Ohridiidae affects approximately 73% of scallops. Larvae of these nematodes were found in the mantle with intensity of invasion up to 17 specimens per scallop.

Polydora ciliata (Johnston, 1838) (Polychaeta)
This widespread and well-known drilling polychaete is responsible for loss of market quality among cultivated and natural Yesso scallops. It has a larger effect on the upper (left) valve. The burrows excavated by *Polydora* in scallop shells cause unsightly blisters containing compacted mud. Approximately 75% of scallops have wormholes and blisters. Since 1974 infestation of scallop shells by the boring polychaetes has increased as siltation of the bottom increases (Silina et al. 2000).

Polydora websteri Hartmann, 1943 (Polychaeta)
Burrows and blisters of this species are indistinguishable from ones of *P. ciliata*.

Dodecaceria concharum Oersted (Polychaeta)
These drilling polychaete have wormholes similar with ones of *Polydora*. They can use old holes created by other polychaetes and sponges.

Herrmannella longicaudata G. Avdeev, 1975 (Copepoda)
These small (maximal length up to 2.2 mm) cyclopoid copepods were found in approximately 73% of Giant Yesso and Swift scallops. Pathogenic effects on the host are unknown even in the case of great abundances. Average intensity of invasion by commensals is nine specimens in the mantle.

Odostomia fujitanii Yokogawa, 1927 (Gastropoda)

O. (Evalea) culta Dall et Bartsch, 1906 (Gastropoda)
These small (maximal shell height up to 5 mm) littoral gastropods often prey on various bivalves as temporary parasites. Gastropods feed with using a long proboscis, which they introduce between shell valves.

24.3.8.12.5 Bacterial contamination

Numerous species of bacterial contaminants have been identified from cultivated scallops (Table 24.6; Avdeeva and Filipchuk 1988, Kovalenko 1989, Plyusnin 1990, Plyusnin and Cherkashin 1991, Kovalenko 1994). A total of 29 species were identified from scallop farms. Most of them are Gram-negative bacteria (22 species). Some of them (e.g., *Aeromonas*, *Vibrio* and *Pseudomonas*) are potentially pathogenic in situations where environmental culture conditions are poor.

24.3.8.12.6 Epibionts

Natural and farmed scallops are an excellent substrate for the settlement of many other organisms (collectively called fouling communities). Marine organisms that occur on scallop shells may be competitors for space and food. Epizoans may also reduce water flow and food accessibility.

24.4 FISHING AND AQUACULTURE

24.4.1 Fishing

24.4.1.1 History

Paleontological and archaeological studies reveal that the people inhabiting the coastal areas of the Far East had, from time immemorial, used marine organisms, including bivalve molluscs, to develop their national economy (Krasnov et al. 1977). A unique Yankovsky culture of shell mounds has been found in many regions (Okladnikov and Derevyanko 1973) dating from the Paleolithic age (25,000–30,000 years BC). Numerous shell concentrations suggest that the inhabitants of the coasts of Sakhalin, the Kuriles, Japan, Primorye and Korea preferred gastropods, mussels, and scallops.
We still do not know how ancient people obtained the scallops. Without any suitable gear and boats they may have only been able to collect them from storm debris. The subsequent history of scallop fishing in Primorye is known from descriptions by pioneer explorers of Ussuri Region, including N. M. Przevalsky and V. K. Arseniev. Russian merchants recall that during the last century Yesso scallops were exported as seasoned meats (muscles). Scallops were fished in Vladimir Bay and sold to China. The second half of the 19th century saw further development of scallop fishery along with the settlement of the Russian Far East. This was supported by its value and by the boundless demand on the local market. By the 1920's, the price for Yesso scallop in Vladivostok was as high as 10 rubles per 100 molluscs. During that time, they fished scallops in shallow bights using one-or-two-pronged lances and scoop nets (while watching it from the surface through a glass-bottom box), primitive dredges and cords (Razin 1934).

1190

Table 24.6

Known bacteria of Yesso scallop *Mizuhopecten yessoensis* cultivated in Peter the Great Bay at cages and bottom ground (compiled from Avdeeva and Filipchuk 1988; Kovalenko 1989; Plyusnin 1990; Avdeeva et al. 1991; Plyusnin and Cherkashin 1991; Kovalenko 1994) Cg = Cages; Gr = Ground

Species	1987	1988		1989		1990		1991	
	C	Gr	Cg	Gr	Cg	Gr	Cg	Gr	Cg
Gram-negative									
Pseudomonas sp.	+	+	--	+	+	+	+	--	--
P. putida	--	--	--	--	+	--	--	+	--
Yersinia ruckeri	--	--	+	+	+	--	--	+	--
Vibrio sp.	+	+	+	+	+	+	+	--	+
V. anguillarium	--	--	--	--	--	--	--	+	--
V. parahaemolyticus	--	--	--	+	--	+	+	+	+
Aeromonas hydrophila	+	--	--	--	--	+	--	+	+
A. punctata	--	--	--	--	--	--	--	+	--
A. salmonicida	--	--	--	+	--	--	--	--	--
A. s. achromogenes	--	+	+	+	+	+	--	--	--
A. s. masoucida	--	--	--	+	+	--	--	--	--
A. dourgesii	--	--	--	+	+	+	--	+	+
Aeromonas sp.	+	+	+	+	+	+	+	--	--
Plesiomonas sp.	+	+	--	--	+	--	--	+	--
Bacteroides sp.	+	+	--	--	--	+	--	--	--
Moraxella sp.	--	--	--	--	--	+	--	--	--
Acinetobacter sp.	--	+	--	--	--	--	--	--	--
Alcaligenes sp.	+	+	+	+	--	+	+	+	+
Chromobacterium sp.	--	+	--	+	--	--	--	--	--
Flavobacterium breve	+	+	+	+	--	+	--	+	+
F. halmephilum	+	--	--	--	--	--	--	--	--
Enterobacteriaceae g. sp.	+	--	--	+	--	--	--	--	--
Gram-positive									
Micrococcus sp.	--	+	+	+	--	+	+	--	+
Leuconosfoe sp.	--	+	+	--	--	+	--	+	--
Streptococcus sp.	--	+	--	+	+	+	--	--	--
Listeria sp.	--	+	+	--	+	+	+	--	--
Corynebacterium sp.	+	--	--	--	+	--	--	+	--
Artrobacter sp.	+	--	--	+	--	+	--	+	--
Lactobacter sp.	--	--	--	+	--	--	--	--	--

24.4.1.2 Fishing gear

Dip-nets were used to fish Yesso scallops in clear water and quiet weather. The pole was 8 m long and the average haul was 50–60 specimens per day, though it could be much higher in shallower waters with excellent weather. Following rain storms the turbid water made this virtually impossible.

Initially only simple dredges (about 75 cm wide) were used for fishing scallops. They were towed by small sampans, whose prototypes were brought by members of an expedition led by N. N. Muraviev-Amursky, the first governor-general of the Ussuri Territory, together with boats from neighbouring countries. Their displacement was 1–1.5 t, and they used sails or one oar, the so-called "yula". Industrial fishing by means of cords was unique at that time. It was performed in the following way: an anchored buoy was placed in the centre of the catching site and a cord was attached to it. The cord had length of 200 m and was 1–3 mm in diameter. The small lead loads (11–15 g each) were fixed to the cord at intervals of every 2 m. The catcher would sail in a boat to drag the cord and start making circles near the buoy. When the cord was trapped by the open scallop valves, the molluscs would abruptly close them and affix themselves. The more often the catcher would pull the cord, the tighter the bivalves would grip the gear. On even ground, a good catcher would haul several thousand scallops. In 1919–1920, Yesso scallop in Ussuri Territory was fished by professional women-divers with the ability to stay underwater for long periods of time even in the cold autumn months. At the end of 1920's when diving gear and motorboats equipped with dredges were introduced, greater scallop hauls were caught. When the Association for Exploiting Marine Resources was especially active the stocks rapidly diminished. After World War II industrial fishing was resumed for a short period of time but then totally banned in 1960.

Yesso scallops and commercial scallops (*C. albida, C. behringiana* and *C. chosenica*) were fished using small seiners (about 300 tons displacement) on the coasts of northern Primorye, southern Sakhalin, Kurile Islands and the Bering Sea. This involves a steel dredge, which is 1.5–3.0 m wide. The dredge is towed for 5–30 min. In Kuriles, one such haul would yield 0.39–1.28 tons of shells (Kochnev 1987). In Peter the Great Bay, divers use free diving down to 5 m and SCUBA diving down to 30 m to gather Yesso scallops.

24.4.1.3 Yesso scallop landings

Scallops are the most intensively consumed and fished bivalve molluscs. The best known is the Yesso scallop, *M. yessoensis*. For a long time these molluscs were an object of traditional catching (Tables 24.7 and 24.8).

24.4.1.3.1 Primorsky territory

Yesso scallop landings in southern Primorye were apparently not recorded for a long period of time. Starting in 1919, when diver boats were introduced, scallop hauls reached the impressive figure of 400 tons. By 1920, industrial fishing grew threefold, but in subsequent years, it sharply declined and then almost stopped. A Trust for Marine

Fisheries was organised in 1933 in Vladivostok with a network of enterprises all over the Soviet Far East. The new outfit resumed industrial fishing of the Yesso scallop to raise the average hauls for 1933–1937 up to 900 tons (Table 24.7). In subsequent years, the scallop was not fished and only 160 tons were landed in 1948–1949.

24.4.1.3.2 Sakhalin-Kurile region

In addition to the fishing areas in southern Primorye, the Yesso scallop was harvested in Sakhalin-Kurile areas. The commercial fishing of Yesso scallops at Sakhalin and Kurile Islands by Japanese fishers occurred from the 1930's to 1945. The main fishing area at that time was Aniva Bay. Between 1933 and 1943, annual yield was 1,000–2,300 tons. At Southern Kuriles values for landings were significantly greater (Skalkin 1966).

Commercial exploitation of scallop populations after World War II was founded by Russian fishers in 1961. The scallops were only fished with dredges from small seiners (Kochnev 1993). One year later in 1962 the catch of molluscs peaked at 5,230 tons. In the four following years, because of excessive catching, the annual yields in Aniva Bay decreased to only 30 tons (Table 24.8). Although it was a stable but poorly maintained population of scallops, the commercial stock of this population was evaluated at about 3,600 t. The reason for such a decrease was the dredging method of catching scallops and in 1967 catching of scallops in Aniva Bay was banned. In 2000, after a long prohibition, commercial catching using dredging was reinstated again in Aniva Bay (Shpakova 2001b).

Several years later commercial catching was banned at Southern Kuriles. At Terpenie Bay scallop landings endured for only one more year. As in Aniva Bay, annual yields were decreased because of excessive and irrational catching (Table 24.8). Between 1976 and 1984, commercial catching was reinstated and the annual yields were 9.5–282.9 tons. From 1985 to the present catching of scallops has been banned (Kochnev 1993) in Southern Kuriles.

24.4.1.4 Yesso scallop commercial stock

24.4.1.4.1 Primorsky territory

The commercial reserves of natural Yesso scallops along the coast of Primorye are presently exhausted after the last decade of irrational catching and poaching.

24.4.1.4.2 Sakhalin-Kurile region

Commercial populations of Yesso scallops along the coast of Sakhalin Island at Aniva and Terpenie Bays are shown on the map (Fig. 24.9). Commercial stocks at Tartar Strait have not been estimated.

Table 24.7

Annual catch (metric tons*) of Yesso scallop in Primorsky Territory in 1919–1937 (Belogrudov 1981).

Years	1919	1920	1923–1926	1933–1937	1948–1949
Catch	400	1,200	35	900	160

* To remove the confusion originated with using out-of-date units in first edition (Kalashnikov 1991) we shall use metrical ones.

Table 24.8

Annual catch (metric tons) of Yesso scallop at Sakhalin and Kurile Islands in 1961–1985 (by Kochnev 1993).

Years	Regions			
	Aniva Bay	South Kuriles	Terpenie Bay	Total catch
1961	200	---	---	200
1962	1,800	1,230	2,200	5,230
1963	260	2,010	---	2,270
1964	150	5,070	100	5,320
1965	110	2,460	banned	2,570
1966	30	1,400	---	1,430
1967	banned	1,220	---	1,220
1968	---	1,410	---	1,410
1969	---	1,200	---	1,200
1970	---	600	---	600
1971	---	banned	---	banned
1972	---	---	---	---
1973	---	---	---	---
1974	---	---	---	---
1975	---	---	---	---
1976	56	141	---	197
1977	70	102	---	172
1978	67	99	---	166
1979	169	97	16.9*	282.9
1980	114	12	---	126
1981	117	12	0.5	129.5
1982	64	36	---	100
1983	3	6.5	---	9.5
1984	28	11	---	39
1985	banned	banned	banned	---

*Experimental fishing

24.4.1.4.2.1 Aniva Bay

In Aniva Bay, Yesso scallops are found along western and north eastern coasts at depths of 8–30 meters. Their distribution has an irregular and mosaic pattern. The total stock in Aniva Bay is estimated at 16,030 t (79.37 million scallops) with 4,970 t (15.63 million scallops) of them being of commercial stock (Shpakova 2001a). Settlement densities on bottom grounds in the bathymetric range of 7–21 meters average 4.33 specimens m^{-2} with an average biomass up to 0.3 kg m^{-2}. Average shell height of the commercial molluscs (14–99%) is between 139–156 mm (Shpakova 2001a, b).

24.4.1.4.2.2 Terpenie Bay

In Terpenie Bay, Yesso scallops are distributed along the north eastern coast in waters 11–20 m deep. Total stock in Terpenie Bay is estimated at 1,300 tons (2.1 million scallops) but only 600 tons of them are of commercial stock. Settlement densities on bottom grounds average 0.03 specimens m^{-2} with an average biomass of 0.008 kg m^{-2}. Average shell height of the commercial molluscs is about 167.1 ± 6.9 mm at individual biomass of 518.7 ± 9.8 g. Mean age of population is about 6.9 years.

24.4.1.4.2.3 Kuriles

In Kuriles, commercial assemblages of Yesso scallop are known in shallow waters of Kunashir Island and on the South Kuriles Shoal (Fig. 24.9). Total stock is specified as 40,000 t (200–300 million scallops) with about 18,000 tons of commercial stock (Ponurovsky et al. 2000; Ponurovsky and Brykov 2001). Settlement densities on bottom grounds in the bathymetric range of 5–20 m average 0.5 specimens m^{-2}. Average shell height of the commercial molluscs (44.8% of total population) is 126.8 ± 1.1 mm. Mean age of population is 3.5 years with the maximum lifetime of 12 yr.

24.4.1.5 Commercial Chlamys scallops

In addition to *M. yessoensis*, some scallops from the genus *Chlamys* have trade significance. These are less known because of their smaller size and their occurrence at depths greater than 50 m. The most abundant of them in north western Pacific are white scallops, *C. albida*, pink scallops, *C. chosenica,* and Bering Sea's scallops, *C. behringiana*. The Asiatic scallops, *C. asiatica,* rarely occur here.

24.4.1.5.1 Primorye

In Primorye, *Chlamys* scallops, mainly *C. chosenica*, form concentrations at depths between 25–250 m. Myasnikov (1982) reported three large populations of pink scallops along the northern coast of Primorye (stretching from Cape Povorotnyj to Cape Zolotoj). The highest density of settlements reaches up to 25 specimens m^{-2} with an average of five specimens m^{-2}. Total stock of scallops is estimated at 420,000 tons (10.6 million scallops) located at depths of 90–100 m. Since 1990 these reserves have provided annual yields of 1,000 t (Myasnikov and Hen 1990).

24.4.1.5.2 Kurile Islands

On the Kurile Islands, commercial fishing of *Chlamys* scallops by Japanese fishers occurred from the 1930's to World War II (Skalkin 1975). After 1972, there was a renewal of commercial fishing. From 1972 to 1975 annual yields did not exceed 140–1,170 t. After 1976 annual yields increased up to 1,500–3,050 t with an average of 1,990 tons. Major scallop landings (about 75% of annual yield) are derived from the Sea of Okhotsk side of Onekotan Island at depths between 50–200 m (Kochnev 1987). It is now the most stable scallop fishery in the Russian Far East region (Myasnikov et al. 1992). Two species (*C. albida* and *C. chosenica*) occur in mixed settlements. Due to an intense fishing increase the annual yields was 3,462–7,198 t with an average of 4,693 t (Table 24.9).

Populations of *Chlamys* scallops on Onekotan Island are found on both the Sea of Okhotsk side and at pacific side of the island. On the Sea of Okhotsk side of the island scallops are distributed over depths of 40–140 m. Density of settlements on the bottom grounds are 90.0 specimens m^{-2} with a biomass of 6.0 kg m^{-2}. Average density and average biomass are 2.7 specimens m^{-2} and 0.25 kg m^{-2}, respectfully. On the pacific side of the island, the largest density of scallops was found between 40–100 m. Densities of settlements on bottom grounds on this side are as high as 175 specimens m^{-2}. Total stock of *Chlamys* scallops in the region is estimated at 64,400 t (730.5 million scallops) with about 42,000 t of commercial stock.

Table 24.9

Annual catch (metric tons) of *Chlamys* scallops near Onekotan Island (Kurile Islands) in 1976–1997 (by Kochnev 1993).

Year	Catch	Year	Catch
1976	1,601	1987	2,700
1977	3,050	1988	2,000
1978	2,392	1989	2,898
1979	2,317	1990	1,754
1980	1,501	1991	1,494
1981	1,625	1992	2,400
1982	1,543	1993	3,462
1983	2,101	1994	7,198
1984	1,413	1995	3,574
1985	2,370	1996	4,963
1986	2,945	1997	4,269

24.4.1.5.3 In Bering Sea

In the Bering Sea commercial concentrations of Bering Sea's scallop *C. behringiana* are found (Myasnikov 1992). Commercial stock of scallops is estimated at 3,000 t within bathymetric range of 110–120 m.

24.4.1.6 Other Chlamys species

Along with the commercial importance of *Chlamys* species described above there are other potential species such as the Japanese scallop, *C. farreri,* and the Swift's scallop, *C. swifti.*

24.4.1.6.1 Chlamys farreri

The Japanese scallop, *C. farreri,* is the most thermophilic scallop in Russian waters. It only occurs in southern Primorye (Fig. 24.6). Afreichuk (1992b) reported concentrations in Posjet Bay. The total stock is estimated at several thousand tons located at depths between 3–5 m. Commercial stock has not been estimated. This species is an object for mariculture and fishery in East Asian countries (Wang and Shieh 1991). In Russia, mainly in southern Primorye, the Japanese scallop is one of most promising species for fishery and mariculture (Bregman 1982; Afreichuk 1992a).

24.4.1.6.2 Chlamys swifti

Stocks of Swift's scallop, *C. swifti,* in Primorye have not been evaluated. In Aniva Bay, Swift scallops are found between 2–19 m. Average density of settlements on bottom grounds ranges from 0.04 to 5.50 specimen m^{-2} with an average of 0.17 specimen m^{-2}. Biomass ranges from 3.8 to 498.8 g m^{-2}, with an average of less than 16 g m^{-2}. Shell height ranges between 32–114 mm with an average of 86.8 mm. Individual biomass ranges from 24 to 208 g with an average of 94.1 g. Although this species is widespread in the bay it does not form commercial aggregations because of its low density.

24.4.2 Aquaculture

The Yesso scallop, *M. yessoensis,* is the only scallop species cultured in Russia. It is cultured in the coastal waters of Primorye in the north western part of the Sea of Japan.

24.4.2.1 History

After the ban on scallop fishing in Primorye in 1962 stocks were replenished very slowly. The first steps in the mariculture of Yesso scallops in the Russian Far East occurred after 1968. In 1971, after several years of research, the first scallop farm was organised in Posjet Bay (southern Primorye). This industry reached its greatest development in Primorye in the 1980's. In addition to the farm in southern Primorye, experimental work for spat collection was attempted in the 1970's in Sakhalin in Bousse lagoon but further cultivation was discontinued. Mariculture at this time had financial assistance from the largest fishing enterprises of region, such as "Dal'ryba" and "Primorrybprom" and from Ministry of Fishery of the USSR.

By the mid-1980's production reached over 10 million spat a year and at least 40 million one-year-olds were settled in various sites within Posjet Bay. Since 1977, in addition to its own plantations, the Posjet farm transferred 2–10 million young scallops from other bays to other marine farms. Starting in 1983, scallop cultivators in Posjet Bay collected 30 million spat, while industrial production was only 20–50 tons. It was not until 1989 that production output exceeded 100 t (Table 24.10).

In the 1990's, during the crash of the socialist system and the disintegration of the USSR, all the created farms were bankrupted or crisis-ridden. Since the end of the 1990's with the formation of the exchange relations in the Russian Federation, a new period in mariculture development has started. After the economic depression in 1997 the interest in mariculture has increased and a new period of economic expansion has started.

24.4.2.2 Present situation

The number of scallop farms in Primorye is quickly growing (Table 24.10). There are now 20 Yesso scallop farms with a total area of more than 700 hectares of single-crop area. There are 125 hectares of hanging culture and about 600 hectares under sowing culture. The development of mariculture in the Sakhalin region has been completely terminated.

Table 24.10

Twenty years (1981–2000) annual trend in number of organisations involved in Yesso scallop culture and total production (metric tons) in Primorsky Territory. Figures have been compiled from reports of fishery organisations.

Years	Number of farms	Yield	Years	Number of farms	Yield
1981	3	9.0	1991	5	153.0
1982	3	4.5	1992	5	150.0
1983	3	18.1	1993	5	155.0
1984	3	38.0	1994	5	110.0
1985	4	10.4	1995	6	113.0
1986	4	48.8	1996	8	22.0
1987	4	62.3	1997	9	60.0
1988	4	64.0	1998	10	131.0
1989	5	196.0	1999	18	99.6
1990	5	122.5	2000	20	91.2

24.4.2.3 Marketing

Almost all of scallop production from scallop farms is unprocessed. Unprocessed Yesso scallops are frozen as packed meats and a very small percentage of scallops (within 1%) are cooled in the shells. As for commercial scallops (*Chlamys* spp.), in most cases they are processed by canning. In Vladivostok fish stores and supermarkets, frozen meat may be sold for 280–340 rubles per kilogram (equivalent of \$9.3–11.3). Cooled whole scallops are sold for 20 rubles per shell (equivalent of \$0.7). For comparison, the cost of frozen beef is equivalent to \$2.5–3.0 (75–90 rubles) per kg. Canned scallops (in sauce, oil and smoked) are sold for the equivalent of \$2.0–4.0 (30–60 rubles) per 8 oz. In other Russian regions prices are a little bit higher. It is profitable to cultivate scallops at this price. In Vladivostok, several dozen tons are sold every year, which is highly insufficient even for the small local market. Frozen scallop meat is in high demand in spite of a high price. Yet, the scales of cultivation do not match the levels of need.

24.4.2.4 The culture methods

All the methods used in the cultivation of Yesso scallops can be assigned to one of two categories: off-bottom cultivation or on-bottom cultivation. Experiments were conducted to harvest spat in closed and running controlled systems involving artificial spawning and larvae feeding. This technique is not popular because it is too complicated and expensive. Attempts were also made to create commercial scallop concentrations by using artificial reefs. The results concluded that this was an inefficient process.

There are two main Japanese methods for commercial cultivation: hanging culture (cultivation of the scallop in cages) and sowing culture (cultivation on the bottom substrates of bays and inlets; Ventilla 1982).

24.4.2.4.1 Spat collection

Long-line structures with plates or bags as collectors are commonly used for spat collection in Primorye. Long-lines are a series of floats (Ø 240–300 mm) connected together by horizontal lines (Ø 19–22 mm) that supports a large number of vertical ropes with attached collectors. There are two types of collectors:

1 - Conical plates of perforated plastic (Ø 250 mm) covered by mesh stockings (7–12 mm). Twenty-five collector plates are strung onto the rope (Ø 6–8 mm) as garland up to 2.5 m long.

2 - Commercial onion bags with mesh or monofilament filling (capron or polyethylene). The bags are attached in sets of ten to the rope (Ø 6–8 mm) as garland up to 5 m long.

Prepared garlands of collectors are placed at depths between 5–10 m in semi-closed bays and inlets and at depths between 15–20 m in open waters. Mass scallop spawning in southern Primorye starts in mid-May. In order not to miss the settling peak, collectors are hung deep in the water for 15–20 days in early June, so the substrates would be covered with a bacterial-algal film. This is used to promote spat attachment prior to larvae settling (Belogrudov 1986). In southern Primorye settlement occurs in mid-June, 22–30 days after spawning begins. The size range of settling larvae is 250–275 μm. By the beginning of August, the spat average 3 mm in size and have growth rates of 3–4 mm per month.

There is a direct relationship between larval concentrations in the plankton and spat abundance found on the collectors. At larval concentrations in the range of 20–30 larvae m^{-3} spatfall is about 100–400 spat per collector and at 50–100 larvae m^{-3} it is up to 500–1,500 spat per collector. The long-term maximal larval concentration in plankton is 600 larvae m^{-3} (Belogrudov 1981).

One month after settling spat size averages 10 mm. At least 200 spat have to settle in the collector in order for the result to be considered commercially profitable. In bonanza years 400–600, and even as many as 1,000 larvae can settle in a collector. In lean years, settling numbers from several to several dozen spat will make processing commercially unprofitable.

24.4.2.4.2 Intermediate culture

The long-line for intermediate culture is set up similar to that of spat collector lines. Horizontal lines support a large number of vertical garlands of cages.

In autumn, the collected spat (200–250 samples) are manually removed to hanging multi-tier cages (0.12 m^{-2}). Spat are collected and directly placed in cages from a raft above the plantation. The collectors are lifted from the water to remove the substrate and spat are scraped off into vats filled with water. All foreign organisms (mostly mussels) and empty shells are removed during the transfer process. The necessary numbers of scallops are poured into cages and instantly placed back into the water. When performing this work, the scallop farmer will put a tent over the raft to protect the scallops from desiccation and the sun. In wintertime, the floating structures on which the cages are hung are placed under ice to protect them from destruction by moving ice floes. All structures are preferentially placed under the water to protect from heavy waves.

By April and May, scallops grown in these cages average 25–30 mm in height. When cultivation is performed correctly the survival rate is approximately 90%. Most mortality occurs at the beginning and is due to the stress of the transfer and to the fragility of the thin shell. The viability of these scallops is dozens of times higher than in the autumn, and several methods are used to subsequently cultivate them to marketable size. At 30 mm the juveniles can be sowed in open waters. About 70% are used for sowing culture and the rest are used for hanging culture in lower density cages.

24.4.2.4.3 Transport of scallop seed

When scallop seed are sold to other farms, or delivered for sowing on the bottom, they are packed up in boxes (volume 20–60 L) with perforated walls and bottoms. The boxes are put under the canvas on board a transport deck and should be instantly filled with molluscs in layers of 20–30 cm. Scallop seed should be covered with moist algae or seaweed leaves after being taken out of the water. During transportation, the boxes are replenished by outboard water every 30 minutes. Transportation is preferable when the water and air temperatures are approximately the same (about 5–10°C). In this manner, the scallops can survive for 24 hours (Table 24.11).

24.4.2.4.4 Sowing or on-bottom culture

Sowing cultivation is based on the principle of transferring scallop juveniles from areas where they settled in great abundance, to bottom grounds, where they can be spread at lower density in order to obtain better growth rates and weights. Sowing culture is practised more widely in Primorye.

Scallops are usually transferred after intermediate culture to bottom grounds in May and June when they are about one year old. Specific sites should be pre-selected before the sowing on sandy-silty sediments or on shingle-shell mixtures and should be without starfish (or at most 0.5 specimens m^{-2}). The content of finely dispersed silt (particles smaller than 130 µm) shall not exceed 30%. Bottom grounds shall be at least 3 m deep in inlets protected from wave action. They should be over 10 m deep in open waters, partially protected from prevailing winds and over 20 m deep at bays opened to all winds, so that storms will not ruin the benthic habitat. Areas with natural scallop accumulations (past or present) are preferable for sowing culture. The selected sites should be established by taking bearings on shore to reference marks and then mapped. Prior to sowing the water space should be marked with buoys. The vessel should sail between the buoys at low speeds, and the one-year-old scallops will be evenly scattered all along the bottom grounds with a density of 10–20 specimens m^{-2}. Sowed scallops usually weigh less and are 10–15 mm smaller in shell height than native scallops (Silina 1994). This difference usually continues throughout the lifetime of scallops. It is thought that the slower growth of sowed scallops from collectors or cages unfavourably affects the growth during winter and spring when the densities are high. Additionally, some of the growth inhibition results from shell breakage during transportation and sowing on the bottom. Growth of sowed scallops is further reduced while the animals adapt to their new habitat and generate new shell at the ventral margins. Depending on the temperature and other ground conditions, the scallops will grow to marketable sizes in two to four years. Survival on the sea floor will depend on predation and wave intensity and may vary from 5 to 20% within the same grounds during different cultivation cycles.

In two to four years after sowing, SCUBA divers collect marketable scallops. Their productivity on shallow grounds (3–5 m deep) with densities over 10 samples m^{-2} amounts to 1,500–3,000 scallops hr^{-1}. In cases of lower densities or larger depths and in turbid waters divers' productivity will progressively decline down to only 300 scallops hr^{-1}.

24.4.2.4.5 Hanging or off-bottom culture

Hanging cultivation adds a third dimension to the essentially two-dimensional sowing culture. The crop is spread over and can utilise a much greater proportion of the water column. The long-line for intermediate culture is set up in the same manner as that for the

Table 24.11

Survival rates (%) of Yesso scallops juveniles during transportation depending on temperature (Bregman 1987).

Duration of transportation, h	Temperature °C			
	5	10	15	20
3	98	98	95	90
5	97	96	93	88
10	95	94	91	85
15	93	92	85	80
20	92	91	83	75
25	90	88	80	70

intermediate culture. They are usually set deep (5–15 m from the surface) to escape wave action and the thermocline in the summer months. There are two schemes for hanging culture:

Young scallops with shell heights of 10–15 mm are placed into cages of 200–250 specimens. One-year old scallops (20–30 mm) are transferred into cages of 20–25 individuals and two-year-olds (50–70 mm) scallops of 5–7 individuals. In three years the scallops will reach the marketable size of 100 mm in height.

Young scallops with shell heights of 10–15 mm are placed into cages of 20–30 individuals and in 1.5 years are transferred to cages of 5–7 specimens. The number of transfers in the second case is smaller, but there is strong fouling on cages due to the longer period of time between operations. Fouling has a negative effect on scallop growth and survival.

All scallop transfer operations are performed in the spring and autumn at relatively low temperatures (about 10°C). When the work is performed correctly, survival rates average 90%. Most mortality occurs at the beginning since it is due to the stress of the transfer and to the fragility of the thin shell.

Harvesting methods require lifting of cages and collecting the scallops of marketable size. Manual harvesting is time-consuming and labour intensive. It is assisted by mechanical winches for lifting lines.

24.4.2.4.6 Obstacles to mariculture development

There are several obstacles to the development of mariculture in Russia:

- Financial problems of protracted payback periods in mariculture and the necessity of a huge investment in equipment. Lump-sum investments for the creation of mariculture farms exceed $200,000. It is often very difficult to find investors for such a long-term project.
- Ethnic and gastronomic problems because shellfish are not a traditional nourishment of the Russian people.

1202

- Legal problems arise from the absence of the laws regulating sea farming.
- Socio-economic problems concerned with undeveloped infrastructure of inshore population centres.

24.4.2.4.7 Ecological constraints associated with cultivation

24.4.2.4.7.1 Predation

As mentioned earlier, young scallops are often preyed upon by various starfish species. In mariculture, when their bipinnaria larvae reach concentrations of 20 specimens m^{-3} in plankton or when inside collectors they develop twice as fast as the scallop spat they may cause 100% mortality (Belogrudov 1981; Gabaev 1981). During settling starfish larvae attach to the same collectors as the scallop juveniles. The average abundance of starfishes and scallops on spat collectors is shown in table 24.12. The relationship between the abundance dynamics of scallops and starfishes can be described by following equation (Gabaev 1990b):

$$A_{scallops} = 128.7 + 39.4 \cdot A_{starfishes} \ (r = 0.82; \ p < 0.05), \tag{2}$$

Where $A_{scallops}$ is the abundance of scallops (specimens m^{-2}) and $A_{starfishes}$ is the abundance of starfishes (specimens m^{-2}).

The relationship between the number of attached scallop juveniles and number of scallops consumed by starfish can be described by the following equation (Gabaev 1990b):

$$N_{eating} = 175.0 + 2.3 \cdot A_{scallops} \ (r = 0.74; \ p < 0.05), \tag{3}$$

Where N_{eating} is the amount of eaten scallops (specimens) and $A_{scallops}$ is the abundance of scallops on spat collectors (specimens m^{-2}). In other words the larger the abundance of scallop juveniles, the faster it decreases.

The gastropod *Nucella heyseana* is also a predator of scallops, which can cause similar problems (Gabaev and Kolotukhina 1999).

24.4.2.4.7.2 Epibionts

Farmed scallops constitute an excellent substrate for the settlement of many epifaunal and epifloral organisms. Almost 60 species of algae and invertebrates were noted on the shells of cultured scallops. In hanging culture, scallops are most frequently inhabited by other bivalve molluscs and barnacles (Silina and Ovsyannikova 2000). The dominant epibiotic species are mussels *Mytilus trossulus* (abundances up to 110 individuals per shell and biomass estimates of up to 46 g shell^{-1}), *Modiolus kurilensis* (up to 32 specimens) and *Hiatella arctica* (up to 27 specimens). Subdominants of epibiotic communities include barnacles, *Balanus improvisus* (up to 268 specimens per shell), *B. crenatus* (up to 189

Table 24.12

Average abundance (mean ± s.d.; specimens m^{-2}) of starfish and scallop juveniles on spat collectors at a scallop farm in Posjet Bay, Sea of Japan (Gabaev 1990).

Year	Species			
	Yesso scallop *Mizuhopecten yessoensis*	Japanese scallop *Chlamys farreri*	Swift's scallop *C. swifti*	Starfish *Asterias amurensis*
1977	479 ± 49	13 ± 1	0	9.5 ± 3.1
1978	69 ± 3.0	16 ± 8	0	0.9 ± 0.7
1979	327 ± 89	2 ± 1	2 ± 1	2.0 ± 0.3
1980	22 ± 0.4	103 ± 25	3 ± 1	0.2 ± 0.1
1981	259 ± 56	84 ± 43	16 ± 9	1.1 ± 0.6
1982	160 ± 18	229 ± 65	1	0.1 ± 0.0
1983	1,060 ± 232	16 ± 11	29 ± 5	0.8 ± 0.1
1984	193 ± 16	66 ± 40	9 ± 7	0.0
1985	109 ± 1	74 ± 62	5 ± 2	0.0
1986	541 ± 337	0	5 ± 3	0.2 ± 0.1
1987	176 ± 36	1 ± 0.5	0.3	0.1 ± 0.05
1988	458 ± 165	0	0.4	0.02

specimens) and *Hesperibalanus hisperius* (up to 27 specimens). Scallops from wild populations have epibionts only on the upper valve. But in cultured scallops the fouling is presented on both valves and in cage culture epibionts are more abundant on the lower valve than on the upper valve. Because of cage overcrowding and immobility, epibionts in hanging culture are more extensive than in sowing culture.

In sowing culture, epibionts on scallop shells are mainly the barnacle *H. hisperius*. Their abundance can reach up to 77 specimens per a shell on two-year scallops and up to 337 specimens on three-year scallops (Silina et al. 2000; Silina and Ovsyannikova 2000).

Excessive fouling is a scallop-farming problem mainly by virtue of the extra effort required to clean the crop for marketing. It may also have some influence on growth rate and productivity through competition for space and reduction of water circulation. Again, farm management provides the best means of alleviating this problem, i.e., using optimal techniques for cultivation tending to sustain less fouling.

24.4.2.4.7.3 Biofouling of cultivation structures

According to the techniques of hanging culture existing in Russia, the cages are placed at depths of 5–15 m. These depths teem with the larvae of various organisms, many of which may form abundant fouling communities on all the parts of the structures. Maslennikov and Kashin (1993) reported strong fouling occurs on cages during cultivation. The fouling biomass reached 5.7 kg m^{-2} at a depth of 5–7 m. The dominant species, accounting for more than 84% of the total biomass, is *M. trossulus* (4.9 kg m^{-2} at a population density of over 10,000 specimens m^{-2}). The fouling biomass on collectors at this depth horizon was 6.5 kg m^{-2}, of which *M. trossulus* and *B. improvisus* contributed 61.2 and 36.4%, respectively.

Fouling causes the weight of the structures to sharply increase and their storm resistance to decrease. This reduces the life span of these structures (Bregman and Kalashnikov 1983). When strong, fouling may also make the collectors and cages an inhospitable environment by restricting water flow through the bags, depleting nutrients, and hindering the growth of scallops within the collectors and cages. The rate of growth of the scallops (height and mass) is also strongly affected by fouling organisms.

Many ways of removing fouling organisms on sea farms (mechanical, physical, chemical and biological) are now known. Chemical methods are not suitable for scallop cultivation because they use toxic substances. The application of physical and mechanical methods is considerably limited because they damage cultivated molluscs or are very labour consuming.

24.4.2.4.8 Effect of scallop mariculture on coastal ecosystems

This section considers the effects of plantations for the cultivation of bivalve molluscs on the hydrological and chemical conditions of the marine environment and on the communities of marine organisms in coastal, semi-closed bays of the Russia seas. Due to eutrophication (from rearing molluscs in suspended culture), the species composition, structural and functional parameters of pelagic and benthic communities have changed. The rate of eutrophication depends on the configuration of the bays, intensity of the water exchange and the turbulence of bottom layer of water. After liquidation of mariculture facilities the biological components of ecosystems of semi-closed bays were restored within 5–10 yr.

The main scallop species cultivated in the Russian seas is the Yesso scallop, *Mizuhopecten yessoensis*. Molluscs are farmed in monoculture mostly in semi-closed bays. The common trend of rearing technologies for molluscs is the use of mariculture installations constructed of numerous ropes, supporting floats and suspended elements. The maintenance of these structures demands a great deal of manual work. For this reason, farming molluscs is often carrying out in areas protected from wind and storms. These technologies have a series of essential drawbacks which affect productivity and result in the high costs of production. In addition, mariculture structures become part of the environment. They become a substrate for the formation of settlements of many other organisms (fouling communities), change water exchange and hydrochemical conditions in bays, and affect the composition and structure of pelagic and bottom communities in adjacent water areas.

The fouling of the structures of a mariculture installations is one of most vivid examples of the interactions between mariculture facilities and the environment. Heavy fouling increases their gross weight and reduces their sustainability to storm action (Bregman and Kalashnikov 1983; Kotlovskaya 1984; Stotsenko 1984, Kashin and Maslennikov 1993). A population of fouling organisms is a large source of a great number of larvae, settling on collectors, cages and other anthropogenic substrata. Fouling animals rival cultivated molluscs for the area, nutrition, oxygen and significantly affect production. Additionally, the fouling organisms aggravate the problems of fecal pollution due to the feces and pseudofeces of invertebrates (both fouling and cultivated). Organic matter accumulates on the bottom underneath plantations and stimulates aerobic and anaerobic bacterial decomposition.

Research assessing the effect of mariculture on coastal ecosystems is mixed. Influence of plantations was considered positive (Golikov and Scarlato 1979; Pereladov and Sergeeva 1989; Pogrebov et al. 1990) and negative (Levin 1981; Kucheryavenko 1986a; Zolotarev and Gubanov 1989; Gabaev 1990a; Kucheryavenko and Bregman 1995). An augmentation of the total biomass of macrobenthos (Golikov and Scarlato 1979; Golikov et al. 1986; Pogrebov et al. 1990) and of meiobenthos (Gal'tsova and Pavlyuk 1987), an increase in species diversity of the benthic population and the abundant growth of some hydrobionts in the areas of mariculture facilities (Pereladov and Sergeeva 1989; Pogrebov et al. 1990) are considered as arguments for a positive effect. Accumulation of biodeposits under plantations (Levin 1981; Zolotnitsky 1988; Sedova and Koshkareva 1989), changes in the biochemical composition of water (Kucheryavenko 1986a, b; Kucheryavenko and Bregman 1995), disruption of interspecific relations within communities (Gabaev 1990a), development of epizootic loci (Zolotarev and Gubanov 1989), and the decrease in the intensity of water exchange (Grigoryeva et al. 1996; Kucheryavenko and Bregman 1995) are referred to as negative effects. Bottom deposits directly under plantations may become completely lifeless within one and a half to two years or a burst of reproduction of an opportunist species followed by a vanishing of unsustainable species to organic pollution occurs during mariculture operations. Accumulation of rich organic bottom sediments and concomitant physical and chemical changes of substrate invoke deterioration of water quality and result in a sharp fall in the productivity of the mariculture facilities. The microbial decomposition of bottom organic matter occurs in the season of highest water temperatures which often causes oxygen deficiency. In places of artificial rearing of molluscs, the biochemical composition of the dissolved organic matter (DOM) differs from that in habitats of natural populations. This so-called "biological pollution" is observed in the areas of plantations (Levin 1981; Agatova et al. 1983; Development 1989; Sadykhova 1995).

The greatest number of publications concerning the Russian part of the Sea of Japan was devoted to functioning of mariculture plantations located in South Primorye in Posjet Bay and Alekseeva Bight (Popov Island) of the Peter the Great Bay (Fig. 24.15). Farming of the Yesso scallop, *Mizuhopecten yessoensis*, and of the Pacific mussel, *Mytilus trossulus*, in shallow and semi-closed bays of the Posjet Bay (Minonosok Inlet and Reid Pallada Bay) was implemented in the early 1970's. The total area of facilities for spat collecting and cultivating of molluscs was about 16 ha in Minonosok Bay and about 10 ha in Reid Pallada Bay.

Minonosok Bay is a semi-closed basin. It has an area of about 11 km^2 characterised by low water exchange. Its hydrological regime is regulated by tidal flows and no more than 10% of its water is exchanged per day (Kucheryavenko 1986b). The average speed of currents in the bay, due to installations for rearing of molluscs, has decreased from

1206

6.9 cm s^{-1} in the mouth to 3.8 cm s^{-1} in the central part. At the innermost part of the bay the current speed is only 2.6 cm s^{-1}. The direction of the currents at various depths differs by up to tens degrees and becomes inverse under the influence of some of the elements of structures (Kucheryavenko and Bregman 1995). Nutrition requirements of farmed molluscs require about 56% of the potential trophic base of the bay (Sedova and Kucheryavenko 1995). In the same study it was recorded that molluscs filtered 5.5% of the total volume of water of the bay per day. Significant differences in DOM composition excreted by organisms at plantations and beyond were recorded (Agatova et al. 1983). It was revealed that cultivated molluscs significantly affect the biochemical composition of the water especially in close vicinity of the installations. By autumn the content of organic carbon was almost 2 times higher in Minonosok Bay due to the spawning of molluscs (May - August) (Kucheryavenko and Bregman 1995).

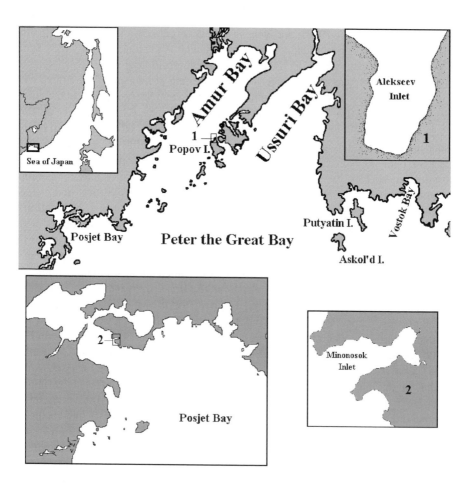

Figure 24.15. Map of sites mentioned in the text.

On surfaces of 1 m^2 located under the water column (volume of 2.5 m^3) there are about 450 scallops existing by the technology of rearing (Sedova and Kucheryavenko 1995). The molluscs filter that volume of water in 2 h. The content of suspended organic matter (SOM) is reduced with water motion from one collector string to another but the concentration of dissolved metabolites is increased. Installations reduce water exchange in the bay by decreasing the speed of the tidal currents. These combined effects slow the growth rates of cultivated scallops in the direction from the mouth to the innermost parts of the bay (Sedova and Kucheryavenko 1995). With a large number of farmed molluscs silting of the bays is promoted by the accumulation of biodeposits.

The effect of the biodeposits on the ecosystem of the bay has been discussed in series of works (Agatova et al. 1983; Kucheryavenko and Rakov 1983; Kucheryavenko 1986a, b, 1988, 1989; Rakov 1986). It has been shown that significant amount of feces and pseudo-feces are discharged from plantations. However, there are no an unambiguous answers concerning the effects of this process on the ecosystem (Kucheryavenko 1986b). Abundance of the fecal material dropping to the bottom may cause the development of anaerobic zones in the bottom sediments and the formation of hydrogen sulfide. Calculations show that cultivated molluscs excrete 10–15 tons of biodeposits per month. The actual amount of biodeposits may be even higher due to the additional feces of the fouling animals (Sedova and Kucheryavenko 1995). The maximum amount of biodeposits is discharged from plantations into adjacent environments during August and September and the minimum during the winter months (Kucheryavenko and Rakov 1983). Another study of Kucheryavenko (1988) describes the absence of a layer of biodeposits underneath plantations in Minonosok Inlet. This was attributed to high enzyme activity of the discharged suspension that resulted in its relatively quick decomposition. Gabaev (1990), summarising the operation of mariculture farm in Minonosok Inlet, drew conclusions on the degradation of benthic communities in the bay. Our examination of bottom (in 2000) confirmed the presence of a one-meter layer of silt deposits under mariculture plantations in the Minonosok Inlet.

In the semi-closed Alekseeva Bight (Popov Island) with the total area of 9 km^2, the rearing of Yesso scallops and mussels was conducted from 1978 to 1988. The mariculture farm in the Alekseeva Bight was liquidated in 1989. The area of suspended cultures varied from 2 to 6 ha during various years. The plantation reached its maximum area in 1986–87 of 14 ha. In addition to the scallop facilities there were also 500 collectors (about 3 ha) for mussel cultivating (Gal'tsova and Pavlyuk 1987).

Many authors have investigated the various aspects of the interactions of plantations and bay ecosystems. According to these assessments, molluscs from the Alekseeva Bight plantation discharged 17 t of biodeposits per year. They filtered 17% of the total volume of water per day and their nutrition requirement was 18% of all suspended organic matter of the bay (Sedova and Koshkareva 1989; Sedova and Kucheryavenko 1995). Growth rates of molluscs (*Mytilus trossulus*) were almost twice as high at the installations compared to their bottom-inhabiting vicariates (Soldatova et al. 1985).

Heavy silting of bottom and the depression and destruction of benthic communities was reported during the period of exploitation of the bay. Near the plantation there were changes in the microbial community. The number of benthic bacteria increased 5–10 times compared with that in a control area (Shcheglova and Bregman 1988). Meiobenthic researches have shown that there was a depletion of taxonomic structure and abnormal increases in populations of animals such as nematodes, harpacticids, and ostracods under the plantation (Gal'tsova and Pavlyuk 1987). The species composition of foraminiferans was also depleted under installations with the population density one order of magnitude lower than at areas away from installations (Tarasova and Preobrazhenskaya 2000).

Histopathological changes in gonads and disturbances of the development of progeny were recorded in sea urchins from the Alekseeva Bight (Vashchenko et al. 1992). An increase of heavy metal content in soft tissues of molluscs sampled from cages was observed (Chernova et al. 1988). Ichthyofauna of the bay has also changed. The fish species that are sensitive to the oxygen content of the water and to its overall quality have disappeared and the number of eurybiontic forms has increased (Gomelyuk et al. 1990). An increase in the abundance of the diatom, *Skeletonema costatum*, which is the indicator of eutrophication of a water basin, was observed during plankton research (Pautova 1987). In 1980 and 1982 there was a decrease of larval abundance of Yesso scallop and poor spat settling (Guida 1983). A trend of annual decreases in the number of trepang (*Apostichopus japonicus*) larvae was also recorded (Development 1984). Thus, in 1984 the number of trepang larvae was the lowest, compared to previous years. Only larvae of early developmental stages were observed in the plankton (Mokretsova and Gavrilova 1986). Significant decreases in the abundance and biodiversity of larval plankton were recorded in the meroplankton of the bay, compared to the mid 1970's (Korn and Kashin 1989). During its operation, the mollusc plantation rendered strong negative effects on the ecosystem of the bay (Pautova 1987).

Comparing data collected from 1986–1990 (Maslennikov et al. 1994) with data from 1973–74 (Mikulich and Biryulina 1977) reveals a sharp reduction in the abundance of larvae of common groups of invertebrates during the period of the maximum operations at the mariculture facilities in the Alekseeva Bight. The total number in the summer season decreased from several thousands of individuals down to only 500–800 individuals per m^3. In September of 1986, the maximal density of bivalve larvae was 785 individuals m^{-3} (Mikulich and Biryulina 1977). It had reached 5,260 individuals m^{-3} prior to installation of mariculture facilities (1973–74). In October of 1987, the number of polychaete larvae recorded was about 200 individuals m^{-3}. The average density in 1973–74 was 6,300 individuals m^{-3} with a maximum density of 22,760 individuals m^{-3}. In April of 1987, the maximum density of decapod larvae was only 20 individuals m^{-3}. Prior to the operations at the mariculture farm, the number of shrimp larvae was 170 individuals m^{-3} and that of crabs was 68 individuals m^{-3}. Only the number of gastropod larvae has comparable values with the previous data. The greatest changes took place for echinoderm larvae. In 1986–88 the brittle star, starfish and trepang larvae did not occur but individual sea urchin larvae were recorded. According to 1973–74 data the maximum total number of echinoderms reached 5,500 individuals m^{-3}. In 1974, of the 1,700 individuals m^{-3} only brittle star larvae were recorded in May and in September their density exceeded 400 individuals m^{-3}.

The effect of pollution on larvae may be mediated through the disturbances of the development of sex cells in mature animals giving rise to nonviable progeny of single individuals and, as a consequence, the disturbance of reproduction of the entire population (Vashchenko et al. 1992).

During the period of maximum activity at the mariculture facilities, the density of meroplankton could have been abnormally low but within one year after liquidation of the facilities the picture had changed significantly (Fig. 24.16). In May and June of 1990 the density of larvae was already 3,000–4,000 individuals m^{-3}. At the same time the ratio of larvae of fouling organisms sharply decreased and the abundance of naupli and ciprid larvae of benthic forms, e.g., *Balanus rostratus* increased (Korn 1991). According to our data, the same has occurred with polychaete larvae (Maslennikov et al. 1994). The most drastic result was the sharp increase in the number of echinoderm larvae in 1989–1990. These are the most sensitive of the benthic invertebrate larvae to pollution (Kobayashi 1979).

The similar picture of restoration was observed in meiobenthic communities in the Alekseeva Bight. In the seven years (1995) after liquidation of the scallop plantation the silting of bottom sediments decreased. That resulted in the appearance of a great number of foraminiferans that prefer sandy sediments and in the changes of the dominant and subdominant species of foraminiferans (Tarasova and Preobrazhenskaya 2000). In the meiobenthic community the number of taxonomic groups and the species diversity of foraminifers increased, the population density of meiobenthic animals decreased, and their dominating groups changed. Nematodes that dominated in 1985 were replaced by foraminiferans (Pavlyuk et al. 2001). The restoration of biological components of the ecosystem of the Alekseeva Bight to the status before the mariculture production took 5–10 years after the liquidation of the mariculture facilities.

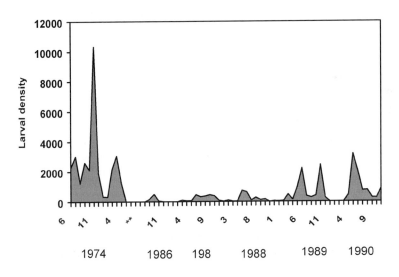

Figure 24.16. Multi-year variations in overall number of larval plankton in Alekseev Inlet in 1973–1974 (Mikulich, Biryulina 1977) and in 1986–1990 (our data). Date; month, year. Larval density; individuals per m^3.

The use of natural biotopes for artificial farming should not result in exhaustion of bioproduction potential. Cultivating of mass of molluscs is based on the use of this potential. Thus, artificially high densities of hydrobionts of one or several species, not intrinsic to the natural biotope, are created in a small area. Such modified communities should be balanced by the preservation of bioproduction potential, i.e., all the effects of industrial population of hydrobionts (seaweed, molluscs, fishes, etc.) on environment should not fall outside the limits of those "reparations" which are capable of providing the natural community. The community of organisms as a balanced system may only exist under conditions when products of metabolism of its incorporated populations do not accumulate in the biotope and do not vary in parameters corresponding to the structural and functional traits of community. Pollution results in a decrease in the total number of invertebrates, a decrease of indexes of species diversity and of total number of taxa in community. Thus, at places for the cultivation of molluscs there are all the prerequisites for the destruction of the bottom ecosystems eventually leading to the general decrease of output of production of mariculture (Gabaev 1990a).

Elements and structures of facilities that are in the water column significantly decrease water exchange at plantations (Babkov 1989; Kulakovsky and Sukhotin 1989; Silkin and Khailov 1988). The amount of nutrients coming into the mariculture farm, the growth rate of molluscs at the installations, the biomass of fouling communities, and the removal of metabolites all depend on the magnitude of this water exchange (Khalaman and Kulakovsky 1989; Khalaman and Sukhotin 1989). Reared molluscs and fouling species transform almost all seston into bottom sediments (except for organic material used for their feeding). Thus, bays with facilities for mollusc farming are traps for seston. Eutrophication results from the decrease in water exchange by exceeding inputs of biogenic elements and of organic matter from the plantations. It is further aggravated by high mortalities during seasons of the highest water temperatures and formation of bottom hypoxia. The sedimentation of seston and the low oxygen concentrations result in the degradation of benthic communities. On the other hand, excretions from molluscs may enhance the development of phytoplankton. At high concentrations, the relatively large phytoplankton cells sediment to the bottom in the form of pseudofeces. The rate of consumption at plantations (in water column) is higher than at natural substrates (Mikulich 1986). Nutritional requirements of molluscs farmed in bays vary from 20% to 59% of the trophic potential of the water basin (Alimov 1989; Sedova and Koshkareva 1989). As a result of using existing technologies for cultivating molluscs in semi-closed bays with low water exchange, silting of the bottom and biogenic pollution of the water area occur. This results in the depression of the development of one of the organisms and the enhanced reproduction of the other (opportunistic species) which disrupts the dynamic equilibrium of the ecosystem.

The generalised pattern of the interactions between the mariculture facilities and the natural ecosystem is shown in figure 24.17. Mollusc plantations in the waters of shallow bays may result in the following effects:

- Changes in the hydrological regime and decreases of water exchange;
- Changes in the specific communities of fouling organisms and farmed molluscs on mariculture facilities in water column;

- Cultivated molluscs and fouling organisms from mariculture installations filter a significant volume of water of a bay, thus their nutrition requirements consume a great part of the trophic potential of bays which mariculture facilities are located;
- Changes in the biochemical composition of the adjacent water areas due to the effects of exometabolites discharged from the plantation;
- The structure and dynamics of phyto- and zooplankton changes, species diversity and number of larvae of benthic invertebrate decreases;
- Settling of juveniles of commercial invertebrates (spat) on collectors decreases;
- Biodeposits from plantations cause silting of ground sediments;
- Number of heterotrophic microorganisms increases under plantations;
- At plantation areas a decrease in species diversity of macro- and meiofauna including changes in benthic communities;
- Changes of biota many times exceeding the size of plantations themselves.

To mitigate these potential effects technology of scallop rearing in open water areas has been developed at the Institute of Marine Biology of Far East Branch of Russian Academy of Sciences. These areas are less affected and at the same time have sufficient space to avoid competition with traditional species. That technology provides joint farming of Yesso scallop, *Mizuhopecten yessoensis*, and Red King crab, *Paralitodes camtschaticus*, in polyculture (Maslennikov 1998; Maslennikov and Kashin 1998). Thus, not only is the commercial cultivation of scallops and Japanese kelp achieved, the operations are also restoring the wild populations of Red King crab in the Far East seas (Maslennikov 1999).

24.4.2.5 Future prospects

The future of Yesso scallop farming is very promising. The market is constantly growing as demand continues to exceed supply. There is some potential for increased farming. Production of scallop farms might be increased through better scallop farm management and the application of technological innovations for mariculture.

There are prerequisites for future mariculture development in Primorye and the northern part of the Russian Far East including historical, organisational, market opportunities, resource environment and socio-economic (Arzamastsev 2000). According to experts, the inshore waters suitable for development of maricultural farms includes approximately 6,000–10,000 hectares in Primorye alone. This provides commercial output of 700,000 tons (Mokretsova 1996). In coastal waters of the Sakhalin and Kurile Islands, natural resources and environmental conditions are also favourable. The optimum place for mariculture development was found in shoaling waters of southern Sakhalin (Kochnev 2000) and South Kurile Shoal (Ponurovsky et al. 2000).

One of the promising trends of mariculture is polyculture. Scallop populations that spontaneously developed under kelp plantations at kelp farms, are well known in practice (Shaldybin 1983). Moreover, joint cultivation of kelps and scallops will help protect against excessive eutrophication, which often occurs at mollusc and kelp farms from overcrowding (Arakawa et al. 1971; Ventilla 1982; Rosenthal et al. 1988; Ivin 1999).

1212

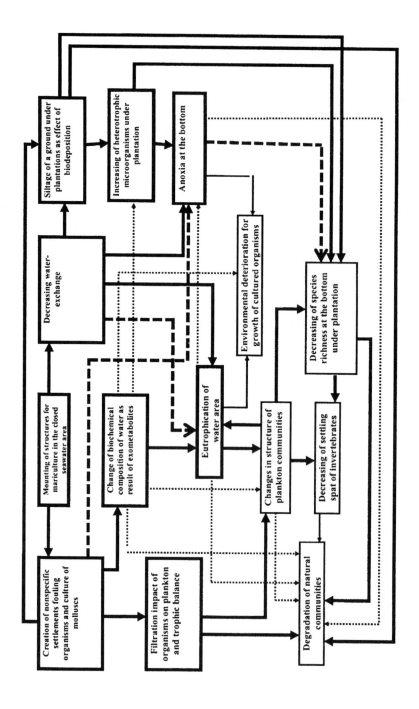

Figure 24.17. A schematic of scallop farming impact on coastal ecosystem.

Not long ago a project on the joint cultivation of Yesso scallops *M. yessoensis*, Japanese kelp *Laminaria japonica* and common sea cucumbers *Apostichopus japonicus* was conducted (Ivin and Maslennikov 2000; 2001). This project started in 2000 by one of the biggest joint stock companies of aquaculture and recently about 100 million spat were collected and more than 40 million scallop seed and 30,000 juvenile sea cucumbers were sowed. The methods for cultivation and construction of structures applied in the project were originally developed and protected by a patent of the Russian Federation No. 2149541 (Maslennikov and Kashin 2000). One hectare of sea farms can yield up to 400,000 market-sized scallops. This is equal to 8 tons of scallop meat. Kelp productivity is not less than 50 ton in wet weight or about 10 tons in dry weight. Technology of sea cucumber cultivation can yield up to 10,000 market-sized individuals. The projected profitability of the sea farms is about 15–18%.

World experiences with mariculture are driving sea farms transferred from closed bays to open waters. The technology uses open seawaters with depths of more than 30 m which are unsuitable for traditional mariculture. The shallow bays with depths of 10–15 m are used only during a short period (not longer than 6 months) for the collection of scallop spat and sea cucumber juveniles. Areas which are occupied by maricultural structures will cover no more than 33% of the general area for sea farms. The remaining water space is used as a reserve for the rotation of areas occupied by the cultivation of sea organisms. After termination of each culture cycle, the plantations will be relocated to a new area. This allows areas that are used in mariculture to be maintained in a natural state. Turnover of used areas allows long-term usage of sea farms without negative environmental impact. This rotational scheme has long been applied in agriculture.

ACKNOWLEDGMENTS

This work was partially supported by Grant 01–04–96922–p2001–Primorye from the Russian Foundation for Basic Research and Grant 00–15–97890 from the Foundation "Leading Scientific Schools of Russia".

REFERENCES

Afreichuk, L.S., 1990. Analysis of the size-age structure of Japanese scallop (*Chlamys farreri*) population for it cultivating. In: B.G. Ivanov (Ed.). Abstracts of the 5th All-Union Conference on Commercial Invertebrates, 9–13 October 1990, Minsk-Naroch, USSR. VNIRO, Moscow. pp. 102–103. (In Russian)

Afreichuk, L.S., 1992a. Growth pattern of the Japanese scallop (*Chlamys farreri*) in Posjet Bay. In: S.Yu. Shershenkov (Ed.). Ecology of the Sea Organisms: Sea Ecosystems. Abstracts of the Conference, 19–21 May 1992. TINRO-center, Vladivostok. pp. 61–62. (In Russian)

Afreichuk, L.S., 1992b. Reproduction of the Japanese scallop (*Chlamys farreri*) in Posjet Bay. In: S.Yu. Shershenkov (Ed.). Ecology of the Sea Organisms: Sea Ecosystems. Abstracts of the Conference, 19–21 May 1992. TINRO-center, Vladivostok. pp. 62–64. (In Russian)

Afreichuk, L.S., Gabaev, D.D. and Rakov, V.A., 1988. Characteristic of reproduction in Japanese scallop *Chlamys farreri nipponensis* in shallower bights of the Peter the Great Bay. In: V.E.

1214

Rodin (Ed.). Resources of Commercial Invertebrates and Biological Basis of their Rational Using. Abstracts of the All Union Meeting, 22–24 November 1988., TINRO, Vladivostok. pp. 111–112. (In Russian)

Agatova, A.I., Andreyeva, N.M., Kucheryavenko, A.V. and Torgunova, N.I., 1983. Biochemical composition of dissolved, suspended and sediment organic matter and its rate of destruction in Posjet Bay in places of artificial rearing and of natural populations of marine invertebrates. In: V.G. Markovtsev (Ed.). IV All-Union Conference on Scientific-technological Problems of Mariculture. Abstracts of papers, 23–28 October 1983. TINRO, Vladivostok. pp. 141–142. (In Russian)

Alimov, A.F., 1989. Introduction to production hydrobiology. Leningrad, Gidrometeoizdat. 159 p. (In Russian)

Arakawa, K.Y., Kusuki, Y. and Kamigaki, M., 1971. Studies on biodeposition in oyster beds. I. Economic density for ouster culture. Japan J. Malacology 30(3):113–128.

Arzamastsev, I.S., 2000. Mariculture in Primorye. Ecological Bulletin of Primorye 4:3–6. (In Russian)

Avdeeva, N.V. and Filipchuk, E.S., 1988. Conditionally pathogenous microflora in scallops from commercial farms in the Peter the Great Bay. In: Abstracts of the 3rd All-Union Conference on Marine Biology, 18–20 October 1988, Sevastopol, Kiev, INBYuM, Part 2. pp. 47. (In Russian)

Avdeeva, N.V., Kovalenko, L.M. and Matrosov, V.V., 1991. Epizootic situation at the commercial farms for the mollusk cultivation in the Peter the Great Bay. In: O.A. Bulatov (Ed.). Rational using of Pacific Biological Resources. Abstracts of the All-Union Conference, 8–10 October 1991. TINRO-center, Vladivostok. pp. 175–177. (In Russian)

Babkov, A.I., 1989. The registration of parameters of water exchange at organization and functioning mussel farm. In: V.G. Markovtsev (Ed.). Scientific and Technical Problems of Mariculture in the Country. Abstracts of the All-Union Conference, 23–28 October 1989, Vladivostok. TINRO. pp. 70–71. (In Russian)

Belogrudov, E.A., 1973. On the character of settlement and growth of the larvae of the sea scallop on different substrates. In: I.V. Kizevetter (Ed.). Studies on Fish Biology and Commercial Oceanography. TINRO, Vladivostok 4:87–90. (In Russian)

Belogrudov, E.A., 1981. Biological bases of cultivation of the Japanese scallop *Patinopecten yessoensis* (Jay) (Molluska, Bivalvia) in Posjet Bay, Sea of Japan. Dissertation, Candidate of Science in Biology (PhD Thesis). Vladivostok: Inst. Mar. Biol., Far East Center, Acad. Sciences USSR. 134 p. (In Russian)

Belogrudov, E.A., 1986. Scallops cultivation. 1. Settling of larvae on collectors. In: P.A. Motavkin (Ed.). Japanese Scallop *Mizuhopecten yessoensis* (Jay). Vladivostok, FESC Acad. Sciences USSR. pp. 201–203. (In Russian)

Biryulina, M.G., 1972. Starfishes of the Peter the Great Bay and their influence on the quantity of the food invertebrates. In: L.V. Mikulich (Ed.). Problems of the Hydrobiology of the Some Regions of the Pacific Ocean. Vladivostok, Pacific Branch of Oceanol. Inst. pp. 42–51. (In Russian)

Biryulina, M.G. and Rodionov, N.A., 1972. Distribution, stocks and age identification of the scallop in Peter the Great Bay, Sea of Japan. In: L.V. Mikulich (Ed.). Problems of the Hydrobiology of the Some Regions of the Pacific Ocean. Vladivostok, Pacific Branch of Oceanol. Inst. pp. 33–41. (In Russian)

Bregman, Yu.E., 1979. Population-genetic structure of the bivalve mollusk *Patinopecten yessoensis*. TINRO, Izvestiya 103:66–78. (In Russian)

Bregman, Yu.E., 1982. Biology and cultivation of scallop *Chlamys farreri nipponensis*. In: Biology of Continental Shelf Zones of the World Ocean. In: A.I. Kafanov (Ed.). Biology of Shelf Zones of the World Ocean. Abstracts of the 2nd All-Union Conference on Marine Biology, September 1982, Vladivostok. Inst. Mar. Biol. FESC USSR Acad. Sci., Part 3. pp. 58–60. (In Russian)

Bregman, Yu.E., 1987 (Ed.). Tentative manual on sowing cultivation of Yesso scallop after yearly intermediate culture. TINRO, Vladivostok. 25 p. (In Russian)

Bregman, Yu.E. and Guida, G.M., 1983. Ecology and development of the larvae of Yesso (Japanese) scallop in laboratory culture. In: Ya.I. Starobogatov (Ed.). Mollusks: Systematics, Ecology and Regularities of Distribution. Leningrad, Nauka 7:191–192. (In Russian)

Bregman, Yu.E. and Kalashnikov, V.Z., 1983. State, problems and prospects of culturing of bivalve mollusks in Primorye. In: V.G. Markovtsev (Ed.). IV All-Union Conference on scientific-technological problems of mariculture. Abstracts of papers, 23–28 October 1983. TINRO, Vladivostok. pp. 144–145. (In Russian)

Buyanovsky, A.I., 1999. Scallops ecology in the western part of Bering Sea. In: M.V. Pereladov (Ed.). Inshore Hydrobiological Investigations. VNIRO, Moscow. pp. 184–190. (In Russian)

Chan, G.M., 1989. Study on resistance of the Japanese scallop spat to reduced salinity. In: V.G. Markovtsev (Ed.). Scientific and Technical Problems of Mariculture in the Country. Abstracts of the All-Union Conference, 23–28 October 1989. TINRO, Vladivostok. pp. 126–127. (In Russian)

Chernova, E.I., Kavun, V.Ya. and Khristoforova, N.K., 1988. The assessment of chemico-ecological conditions in areas of mollusk rearing by microelement composition of Pacific mussel. Biologiya Morya (Marine Biology, Vladivostok) 4:71–73. (In Russian with English abstract)

Dautov, S.Sh. and Karpenko, A.A., 1983. Behavior and motion mechanisms in two scallop species in the Sea of Japan. Biologiya Morya (Marine Biology, Vladivostok) 5:40–44. (In Russian with English abstract)

Denisova, L.A., 1981. Sexual inversion in Swift's scallops *Swiftopecten swifti*. In: A.I. Kafanov (Ed.). Biological resources of continental shelf zone, theirs rational using and preservation. Abstracts of the Regional Conference, September 1981, Vladivostok. Inst. Mar. Biol. FESC USSR Acad. Sci. pp. 26–27. (In Russian)

Development of Mariculture in Primorsky Territory, 1989. In: A.M. Zarubova (Ed.). School of Progressive Experience in Mariculture. Abstracts of reports. 5–8 December 1989. Primorrybprom, Vladivostok. 30 p. (In Russian)

Development, Perfection and Introduction of Biotechniques of Rearing of Mollusks and Trepang, 1984. Results of investigations and application of the achievements of science to production in biotechnology for cultivation of mollusks and trepang. The report about RD of TINRO. Vladivostok. The copy dep. VNTITs. No. 01826005266. 92 p. (In Russian)

Didenko, E.M., 1996. Danger of epizooty occurrence at mariculture farms of Primorye. In: L.A. Dushkina (Ed.). The Status and Perspective of Scientific and Practical Works in Mariculture of Russia: Proceedings of conference, August 1996, Rostov-on-Don. VNIRO, Moscow. pp. 83–86. (In Russian)

1216

Gabaev, D.D., 1981. Settling of the larvae of bivalve mollusks and starfishes on suspended collectors in Posjet Bay (Sea of Japan). Biologiya Morya (Marine Biology, Vladivostok), 4:59–65. (In Russian with English abstract)

Gabaev, D.D., 1988. Dynamic of abundance of the commercial bivalve mollusks on collectors. In: Abstracts of the 3rd All-Union Conference on Marine Biology, 18–20 October 1988, Sevastopol. Kiev, INBYuM, Part 2. pp. 230–231. (In Russian)

Gabaev, D.D., 1990a. Changes of interspecific relations of invertebrates as a result of effect of mariculture. In: B.G. Ivanov (Ed.). Abstracts of the 5th All-Union Conference on Commercial Invertebrates, 9–13 October 1990, Minsk-Naroch. VNIRO, Moscow. pp. 7–8. (In Russian)

Gabaev, D.D., 1990b. The biological substantiation of new methods for cultivation of some commercial bivalves mollusks in Primorye. Abstract of Dissertation, Candidate of Science in Biology (PhD Thesis). Vladivostok: Inst. Mar. Biol., Far East Branch, Acad. Sciences USSR. 30 p. (In Russian)

Gabaev, D.D. and Kolotukhina, N.K., 1999. Influence of predatory gastropod *Nucella* (*Thais*) *heyseana* on population of scallop *Mizuhopecten yessoensis* (Jay). Russian J. Ecology 2:153–156. (In Russian)

Gal'tsova, V.V. and Pavljuk, O.N., 1987. Meiobenthos of Alekseeva Bight (Peter the Great Bay, Sea of Japan) in conditions of mariculture of Japanese scallop. Preprint No 20, Vladivostok, Institute Mar. Biol., FESC Ac. Sci. USSR. 49 p. (In Russian)

Golikov, A.N. and Scarlato O.A., 1970. Abundance, dynamics and production of populations of edible bivalves *Mizuhopecten yessoensis* and *Spisula sachalinensis* related to the problem of organization of controllable submarine farms at the western shores of the Japan Sea. Helgol. Wiss. Meeresunters. 20:493–513.

Golikov, A.N. and Scarlato, O.A., 1979. Effect of mussel rearing on adjacent water area in the White Sea. Biologiya Morya (Marine Biology, Vladivostok) 4:68–73. (In Russian with English abstract)

Golikov, A.N., Scarlato, O.A., Buzhinskaya, G.N., Vasilenko, S.V., Golikov, A.A., Perestenko, L.P. and Sirenko, B.N., 1986. Changes of benthos of the Posjet Bay (Sea of Japan) for last 20 years as result of accumulation of organic matter in bottom deposits. Okeanologia (Oceanology) 26(1):131–135. (In Russian with English abstract)

Gomelyuk, V.E., Kondrashev, S.L. and Levin, A.V., 1990. Ichthyofauna of Alekseeva Bight, Popov Island (Peter the Great bay, Sea of Japan) and effect of rearing of Japanese scallop on it. Biology of shelf zone and anadromous fishes. Vladivostok, FEB Ac. Sci. USSR. pp. 5–8. (In Russian)

Grigoryeva, N.I., Kucheryavenko, A.V., Novozhilov, A.V. and Vyshkvartsev, D.I., 1996. Speed of flows determining allocation of plantations in water area of shallow bays of the Posjet Bay (Sea of Japan). In: L.M. Sushenya (Ed.). Proceedings of the 7th Meeting. Hydrobiological Society of Russian Academy of Sciences, 8–11 October 1996, Kazan', Part 1. pp. 110–111. (In Russian)

Guida, G.M., 1983. Reproduction and dynamics of number of larvae of Japanese scallop *Patinopecten yessoensis* (Jay) in the Aleseyev Bay (Amursky bay, Sea of Japan). In: V.G. Markovtsev (Ed.). Mariculture at the Far East. TINRO, Vladivostok. pp. 25–28. (In Russian)

Ivin, V.V., 1999. Present situation, problems and perspective in Japanese kelp cultivation in Russia. In: K. Tazaki and H. Ikariyama (Eds.). Earth - Water – Humans. Proceedings of the

International Symposium, 30 May - 1 June 1999., Kanazawa University, Kanazawa, Japan. pp. 109–114.

Ivin, V.V. and Maslennikov, S.I., 2000. Seafarms for joint cultivation of invertebrates and seaweeds. Ocean & Business: Pacific Business Herald 4–5:44–45. (In Russian)

Ivin, V.V. and Maslennikov, S.I., 2001. Commercial technology of polyculture at the open waters of Russian Far East seas. In: E.N. Naumenko (Ed.). Proceedings of the 8th Meeting. Hydrobiological Society of Russian Academy of Sciences, 16–23 September 2001, Kaliningrad, Part 2. pp. 37–38. (In Russian)

Kafanov, A.I., 1991. Shelf and continental slope bivalve mollusks of the Northern Pacific Ocean: A checklist. Ja.I. Starobogatov (Ed.). Vladivostok, Far Eastern Branch, USSR Academy of Sciences. 200 p. (In Russian)

Kafanov, A.I. and Lutaenko K.A., 1998. New data on the bivalve mollusk fauna of the North Pacific Ocean. 5. The status of some scallops of the subfamily Chlamydinae von Teppner, 1922 and notes on the genus *Mizuhopecten* Masuda, 1963 (Pectinidae). Ruthenica (Russian Malacological Journal) 8(1):65–73. (In Russian with English abstract)

Kalashnikov, V.Z., 1984. Effect of typhoon Ellis on population of the Japanese scallop *Patinopecten yessoensis* in Posjet Bay, Sea of Japan. Biologiya Morya (Marine Biology, Vladivostok) 1:55–59. (In Russian with English abstract)

Kalashnikov, V.Z., 1985. Dynamics of population density of the Japanese scallop culture on the bottom of Posjet Bay, Sea of Japan. Biologiya Morya (Marine Biology, Vladivostok) 5:58–63. (In Russian with English abstract)

Kalashnikov, V.Z., 1986. Spatial structure and conditions of formation of settlements of the scallop *Mizuhopecten yessoensis* (Jay) in the north western part of the Sea of Japan. Abstract of Dissertation, Candidate of Science in Biology (PhD Thesis). Vladivostok: Inst. Mar. Biol., Far East Scientific Center, Acad. Sciences USSR. 23 p. (In Russian)

Kalashnikov, V.Z., 1991. Fisheries and aquaculture. Soviet Union. In: S.E. Shumway (Ed.). Scallops: Biology, Ecology and Aquaculture. Developments in Aquaculture and Fisheries Science, Vol. 1, Elsevier Science Publishers, Amsterdam, Netherlands. pp. 1057–1082.

Kashin, I.A. and Maslennikov, S.I., 1993. Fouling of mid-water hydrobiotechnical installations for rearing of Japanese scallop. Biologiya Morya (Marine Biology, Vladivostok) 4:90–97. (In Russian with English abstract)

Kas'yanov, V.L., Kryuchkova, G.A., Kulikova, V.A. and Medvedeva, L.A., 1983. Larvae of Marine Bivalves and Echinoderms. Nauka, Moscow. 215 p. (In Russian)

Kas'yanov, V.L., Medvedeva, L.A., Yakovlev, S.N. and Yakovlev, Yu.M., 1980. Reproduction of Echinoderms and Bivalves. Nauka, Moscow. 208 p. (In Russian)

Khalaman, V.V. and Kulakovsky, E.E., 1989. Formation of mussel biocenosis in conditions of mariculture in the White Sea. In: V.G. Markovtsev (Ed.). Scientific and Technical Problems of Mariculture in the Country. Abstracts of the All-Union Conference, 23–28 October 1989. TINRO, Vladivostok. pp. 122–123. (In Russian)

Khalaman, V.V. and Sukhotin, A.A., 1989. Variations of ecological parameters of fouling community at mussel farm in the White Sea. In: V.G. Markovtsev (Ed.). Scientific and Technical Problems of Mariculture in the Country. Abstracts of the All-Union Conference, 23–28 October 1989. TINRO, Vladivostok. pp. 123–124. (In Russian)

Kobayashi, N., 1979. Barnacle larvae and rotifers as indicatory materials for marine pollution bioassay, preliminary experiments. The Science and Engineering Review of Doshisha University 19(4):61–67.

Kochnev, Yu.R., 1987. Dynamics of resources and fishery of white scallop at the Onekotan Island. In: A.V. Ivanov (Ed.). Results of Investigations of Rational Using and Preservation of the Environment at Sakhalin and Kurile Isles. Abstracts of the 3rd Scientific Conference, 27–28 March 1987, Yuzhno-Sakhalinsk. Geographic Society of the USSR. pp. 130–131. (In Russian)

Kochnev, Yu.R., 1993. Sea scallops. In: V. Semenchik (Ed.). Commercial Fishes, Invertebrates and Algae of Seawaters of Sakhalin and Kurile Islands. SakhNIRO, Yuzhno-Sakhalinsk. pp. 49–60. (In Russian)

Kochnev, Yu.R., 2000. Distribution of *Mizuhopecten yessoensis* in the Izmenchivoe Lake (South Sakhalin). Bull. Russian Far East Malacology Society 4: 82–84. (In Russian)

Korn, O.M., 1991. Long term dynamics of barnacle larvae at the region hydrobiotechnical installations for scallop rearing. Vladivostok, Dep. VINITI 29.12.91. No 4865–B991. 21 p. (In Russian)

Korn, O.M. and Kashin, I.A., 1989. Effect of industrial rearing of bivalve mollusks on larval plankton of the water area. In: V.G. Markovtsev (Ed.). Scientific and Technical Problems of Mariculture in the Country. Abstracts of the All-Union Conference, 23–28 October 1989. TINRO, Vladivostok. pp. 97–98. (In Russian)

Kotlovskaya, T.P., 1984. Means of Protection from Fouling. The review. Preprint ONTI TsPKTB Dalryby. Vladivostok. 18 p. (In Russian)

Kovalenko, L.M., 1989. Investigations of microflora inhabited the scallops and environments at bottom plantation at Popov Island. In: V.G. Markovtsev (Ed.). Scientific and Technical Problems of Mariculture in the Country. Abstracts of the All-Union Conference, 23–28 October 1989, TINRO, Vladivostok. pp. 175–176. (In Russian)

Kovalenko, L.M., 1990. Some data on parasites of scallop *Mizuhopecten yessoensis* in the Peter the Great Bay. In: B.G. Ivanov (Ed.). Abstracts of the 5th All-Union Conference on Commercial Invertebrates, 9–13 October 1990, Minsk-Naroch. VNIRO, Moscow. pp. 177–178. (In Russian)

Kovalenko, L.M., 1994. The sanitary-microbiological state of mollusk farms in Peter the Great Bay. TINRO, Izvestiya. 113:33–37. (In Russian)

Krasnov, E.V., Evseev, G.A., Tatarnikov, V.A., Shavkunov, E.V., Besednov, K.N. and Diyakova, O.V., 1977. Sea organisms and their using by the human. Biologiya Morya (Marine Biology, Vladivostok) 1:81–90. (In Russian with English abstract)

Kucheryavenko, A.V., 1986a. Materials of research of trophic resources of bays of the Posjet Bay. In: Yu.E. Bregman and N.D. Mokretsova (Eds.). Results of Investigations and Application of the Achievements of Science to Production in Biotechnology for Cultivation of Mollusks and Trepang. The report about RD of TINRO. Vladivostok. The copy dep. VNTITs. pp. 62–67. (In Russian)

Kucheryavenko, A.V., 1986b. Variation of biochemical parameters of environment under effect of reared. Anthropogenic impact on coastal sea ecosystems. Papers collection. VNIRO, Moscow. pp. 148 – 154. (In Russian)

Kucheryavenko, A.V., 1988. Effect of reared mollusks on quantity and composition of organic matter in areas of mariculture. In: Abstracts of the 3rd All-Union Conference on Marine Biology, 18–20 October 1988, Sevastopol. Kiev, INBYuM, Part 2. pp. 54–56. (In Russian)

Kucheryavenko, A.V., 1989. Concerning biodeposits of reared mollusks. In: V.G. Markovtsev (Ed.). Scientific and Technical Problems of Mariculture in the Country. Abstracts of the All-Union Conference, 23–28 October 1989. TINRO, Vladivostok. pp. 100–101. (In Russian)

Kucheryavenko, A.V. and Bregman, Yu.E., 1995. Studying of biochemical composition of water and parameters of current in the Posjet Bay (Sea of Japan) by industrial rearing of mollusks. In: International Symposium on Mariculture. Abstracts. VNIRO, Moscow. pp. 30. (In Russian)

Kucheryavenko, A.V. and Rakov, V.A., 1983. The rate of accumulation of biodeposits by reared Pacific oyster in Novgorodskaya Bay (Posjet Bay, Sea of Japan). In: V.G. Markovtsev (Ed.). Mariculture at the Far East. TINRO, Vladivostok. pp. 14–19. (In Russian)

Kulakovsky, E.E. and Sukhotin, A.A., 1989. Results of 4-year rearing of mussels in conditions of pilot farm in the White Sea. In: V.G. Markovtsev (Ed.). Scientific and Technical Problems of Mariculture in the Country. Abstracts of the All-Union Conference, 23–28 October 1989. TINRO, Vladivostok. pp. 98–99. (In Russian)

Kulikova, V.A., Medvedeva, L.A. and Guida, G.M., 1981. Morphology of pelagic larvae of three bivalve species of the family Pectinidae from Peter the Great Bay (Sea of Japan). Biologiya Morya (Marine Biology, Vladivostok) 4:75–77. (In Russian with English abstract)

Kurochkin, Yu.V, Tsimbalyuk, E.M. and Rybakov, A.V., 1986. Parasites and diseases. In: P.A. Motavkin (Ed.). Japanese Scallop *Mizuhopecten yessoensis* (Jay). Vladivostok, FESC Acad. Sciences USSR. pp. 174–182. (In Russian)

Levin, V.S., 1981. Technology of development and reproduction of marine biological resources. In: I.B. Ikonnikov (Ed.). Underwater Technology. Leningrad, Korablestroyeniye. pp. 159–196. (In Russian)

Lutaenko, K.A., 1999. Additional data on the fauna of bivalve mollusks of the Russian continental coast of the Sea of Japan: Middle Primorye and Nakhodka Bay. Publications of Seto Marine Biological Laboratory 38(5/6):255–286.

Makarova, L.G., 1985. Production characteristics of the Japanese scallop *Mizuhopecten yessoensis* (Jay) as object of mariculture. Abstract of Dissertation, Candidate of Science in Biology (PhD Thesis). Leningrad: Zoological Inst. of Acad. Sciences USSR. 20 p. (In Russian)

Markovskaya, E.B., 1951. On distribution of the scallop in Peter the Great Bay, Sea of Japan. TINRO, Izvestiya. 35:199–200. (In Russian)

Maslennikov, S.I., 1998. The technology crab farming on a water area of the Far East seas. Far East of Russia: Economy, Investment, Conjuncture, 1:34–39. (In Russian)

Maslennikov, S.I., 1999. Artificial reproduction of kings crabs. Proceedings of the International Symposium "Earth, Water, Humans", 30 May - 1 June 1999, Kanazawa, Japan. pp. 142–145.

Maslennikov, S.I. and Kashin, I.A., 1993. Fouling of constructions for mariculture of *Mizuhopecten yessoensis*. Biologiya Morya (Marine Biology, Vladivostok) 4:90–97. (In Russian with English abstract)

Maslennikov, S.I. and Kashin, I.A., 1998. Culture of crabs on mariculture installations in an open water area. PISCES Seventh Annual Meeting October 14–22, 1998. Fairbanks Alaska USA. 1 p.

Maslennikov, S.I. and Kashin, I.A., 2000. Technique for cultivation of aquatic organisms in polyculture. The patent for an invention of Russian Federation, registered by 27 May 2000, No. 2149541. Invention Bulletin 15:1–18. (In Russian)

1220

Maslennikov, S.I., Korn, O.M., Kashin, I.A. and Martynchenko, J.N., 1994. Long-term variations of number of larvae of bottom invertebrates in the Alekseeva Bight, Popov Island, Sea of Japan. Biologiya Morya (Marine Biology, Vladivostok) 2:107–114. (In Russian with English abstract)

Mikulich, L.V., 1986. Composition of food. In: P.A. Motavkin (Ed.). Japanese Scallop *Mizuhopecten yessoensis* (Jay). Vladivostok, FESC Acad. Sciences USSR. pp. 95–99. (In Russian)

Mikulich, L.V. and Biryulina, N.G., 1977. Plankton of the Alekseeva Bight (Peter the Great Bay). In: Investigations of Oceanological Fields at Indian and Pacific Oceans. Vladivostok, FESC Acad. Sciences USSR. pp. 103–136. (In Russian)

Mokretsova, N.D., 1996. Present situation and perspectives of mariculture development in Far East Seas. In: L.A. Dushkina (Ed.). The Status and Perspective of Scientific and Practical Works in Mariculture of Russia. Proceedings of conference, August 1996, Rostov-on-Don. VNIRO, Moscow. pp. 203–209. (In Russian)

Mokretsova, N.D. and Gavrilova, G.S., 1986. Research on effect of ammonium nitrogen on Far East trepang during its rearing. In: V.G. Markovtsev (Ed.). Mariculture at the Far East. TINRO, Vladivostok. pp. 89–93. (In Russian)

Myasnikov, V.G., 1982. Distribution of the scallop *Chlamys rosealbus* (Scarlato, sp. nov.) along the north coastline of Primorye (Sea of Japan). TINRO, Izvestiya. 106:70–73. (In Russian)

Myasnikov, V.G., 1985. Distribution and ecology of scallops from genus *Chlamys* (Bivalvia, Pectinida) in Far East seas. In: V.N. Akulin (Ed.). Investigations and rational using of the bioresources of Far East and Northern seas of USSR. Abstracts of the All-Union Meeting, 15–17 October 1985. TINRO, Vladivostok. pp. 96–98. (In Russian)

Myasnikov, V.G., 1986. Some data on ecology of *Chlamys* scallops (Bivalvia, Pectinidae) from the Far East Seas. In: B.G. Ivanov and K.N. Nessis (Eds.). Abstracts of the 4th All-Union Conference on Commercial Invertebrates, April 1986, Sevastopol. Kiev, INBYuM, Part 2. pp. 265–266. (In Russian)

Myasnikov, V.G., 1988. Distribution and some biological particulars of white scallop *Chlamys albidus* (Bivalvia, Pectinidae) in the northern part of the Sea of Okhotsk. In: V.E. Rodin (Ed.). Resources of Commercial Invertebrates and Biological Basis of their Rational Using. Abstracts of the All-Union Meeting, 22–24 November. TINRO, Vladivostok. pp. 55–56. (In Russian)

Myasnikov, V.G., 1992. Commercial scallops of genera *Chlamys* (Bivalvia, Pectinidae) from temperate waters of north western Pacific: distribution, growth and resources. Abstract of Dissertation, Candidate of Science in Biology (PhD Thesis). St. Petersburg Zoological Inst. Russian Acad. Sciences. 22 p. (In Russian)

Myasnikov, V.G. and Hen, G.V., 1990. Conditions for the forming of commercial aggregations of *Chlamys* scallops in the Pacific north western. In: B.G. Ivanov (Ed.). Abstracts of the 5th All-Union Conference on Commercial Invertebrates, 9–13 October 1990, Minsk-Naroch. VNIRO, Moscow. pp. 122–123. (In Russian)

Myasnikov, V.G. and Kochnev, Yu.P., 1988. Life span, growth and sex structure of scallops (*Chlamys albidus*) off the Kurile Islands. In: B.G. Ivanov (Ed.). Marine Commercial Invertebrates. Collected papers. Moscow, VNIRO. pp. 153–166. (In Russian)

Myasnikov, V.G., Zgurovsky, K.A. and Temnych, O.S., 1992. Morphological differentiation of commercial scallops of the genus *Chlamys* (Bivalvia, Pectinidae) in the north western Pacific Ocean. Zoologichesky zhurnal 71(9):22–31. (In Russian with English abstract)

Okladnikov, A.R and Derevyanko, A.P., 1973. Distant Past of Primorye and Priamurye. Vladivostok. 439 p. (In Russian)

Pautova, L.A., 1987. Phytoplankton structure and the role dinoflagellates in coastal waters of the Peter the Great Bay of the Sea of Japan (at Popov Island). Abstract of Dissertation, Candidate of Science in Biology. PhD Thesis. IBSS, Sevastopol. 23 p. (In Russian)

Pavlyuk, O.N., Preobrazhenskaya, T.V. and Tarasova, T.S., 2001. Long-term variations in structure of meiobenthic communities of the Alekseeva Bight, Sea of Japan. Biologiya Morya (Marine Biology, Vladivostok) 27(2):127–132. (In Russian with English abstract)

Pereladov, M.V. and Sergeeva, Z.M., 1989. A tentative estimation of influence mussel farms on environment. In: International Symposium on Modern Problems of Mariculture in the Socialist Countries. Abstracts. VNIRO, Moscow. pp. 43–45. (In Russian)

Plyusnin, V.V., (Editor) 1990. Diseases of mollusks and algae, nidus and methods of preventive measures. Report on surveys in 1990 supported by a grant-in-aid for scientific research from the Ministry of Fishery, USSR, State reg. No. 01890027883. TINRO, Vladivostok. 39 p. (In Russian)

Plyusnin, V.V. and Cherkashin, S.A. (Eds.), 1991. Sanitarium and toxicological status of maricultural farms. Intervening Report on surveys in 1991 supported by a grant-in-aid for scientific research from the Ministry of Fishery, USSR. TINRO, Vladivostok. 36 p. (In Russian)

Pogrebov, V.B., Revkov, N.K. and Ryabushko, L.I., 1990. Effect of mussel farm on macrobenthos of the Laspi Bay, Black Sea. In: B.G. Ivanov (Ed.). Abstracts of the 5th All-Union Conference on Commercial Invertebrates, 9–13 October 1990, Minsk-Naroch, USSR. VNIRO, Moscow. pp. 21–22. (In Russian)

Ponurovsky, S.K., 1977. Growth of Swift's scallop at Vostok Bay, Sea of Japan. In: Matters of All-Union Scientific Conference on Using of Commercial Invertebrates for Food, Forage and Industry, 1977, Odessa. Moscow, TsNIIEIERH. pp. 74–75. (In Russian)

Ponurovsky, S.K., 1982. Swift's scallop as possible object for cultivation. In: A.I. Kafanov (Ed.). Biology of Shelf Zones of the World Ocean. Abstracts of the 2nd All-Union Conference on Marine Biology, September 1982. Vladivostok, Institute of Marine Biology, USSR Academy of Sciences, Part 3. pp. 82–84. (In Russian)

Ponurovsky, S.K. and Brykov, V.A., 2001. Present state in population of scallops *Mizuhopecten yessoensis* at South Kuriles Shoal. In: Near-shore fishery – XXI century. Abstracts of the International Scientific and Practical Conference, 19–21 September 2001, Yuzhno-Sakhalinsk, Russia. SakhNIRO. pp. 97–98. (In Russian)

Ponurovsky, S.K., Brykov, V.A. and Evseev, G.A., 2000. Population structure and growth of the scallop *Mizuhopecten yessoensis* in shallow waters of Kunashir Island South Kurile Shoal (Kurile Islands). Bull. Russian Ear East Malacology Soc. 4:93–94. (In Russian)

Ponurovsky, S.K. and Silina, A.V., 1983. The determination of age and linear growth rate in scallop *Swiftopecten swifti*. Biologiya Morya (Marine Biology, Vladivostok) 1:20–24. (In Russian with English abstract)

Rakov, V.A., 1990. Parasitic *Odostomia* (Gastropoda, Pyramidellidae) of commercial mollusks in Peter the Great Bay. In: B.G. Ivanov (Ed.). Abstracts of the 5th All-Union Conference on Commercial Invertebrates, 9–13 October 1990, Minsk-Naroch. VNIRO, Moscow. pp. 184–185. (In Russian)

Razin, A.I., 1934. Marine commercial mollusks of the Southern Primorye. Izvestiya TIRKh 8:48–63. (In Russian)

Rosenthal, H., Weston, D., Lowen, R. and Black, E., (Eds.), 1988. Environmental impact of mariculture. Report of the ad hoc study group. Cooperative Research Report. No. 154. 83 p.

Sadykhova, I.A., 1995. Mariculture of mollusks and its effect on coastal ecosystem. In: International Symposium on Mariculture. Abstracts. VNIRO, Moscow. p. 36–37. (In Russian)

Scarlato, O.A., 1981. Bivalve Mollusks of Temperate Latitudes of the Western Pacific. Nauka Press, Leningrad. 480 p. (In Russian)

Sedova, L.G. and Koshkareva, L.N., 1989. Nutrition requirements of reared mollusks in the Alekseeva Bight. In: V.E. Rodin (Ed.). Resources of Commercial Invertebrates and Biological Basis of their Rational Using. Abstracts of the All Union Meeting, 22–24 November 1988. TINRO, Vladivostok. pp. 106–107. (In Russian)

Sedova, L.G. and Kucheryavenko, A.V., 1995. Effect of mollusk rearing on ecology of two bays of the Peter the Great Bay (Sea of Japan). In: International Symposium on Mariculture. Abstracts. VNIRO, Moscow. pp. 38–39. (In Russian)

Shaldybin, S.L., 1983. Growth of the Japanese scallop in Kit Inlet. In: V.G. Markovtsev (Ed.). IV All-Union Conference on scientific-technological problems of mariculture. Abstracts of papers. 23–28 October 1983. TINRO, Vladivostok. pp. 201–202. (In Russian)

Shcheglova, I.K. and Bregman, Yu.E., 1988. Trophic activity of reared scallops, reforming the microbic coenosis of environment and biota. In: V.E. Rodin (Ed.). Resources of Commercial Invertebrates and Biological Basis of their Rational Using. Abstracts of the All Union Meeting, 22–24 November 1988. TINRO, Vladivostok. pp. 107–108. (In Russian)

Shpakova, T.A., 2001a. Distribution and resources of scallop *Mizuhopecten yessoensis* (Jay) in Aniva Bay (Eastern Sakhalin). In: Abstracts of the Conference, 19–21 September 2001, Yuzhno-Sakhalinsk. SakhNIRO. pp. 125–126. (In Russian)

Shpakova, T.A., 2001b. Dynamics of size composition in populations of scallop *Mizuhopecten yessoensis* (Jay) in Aniva Bay (Eastern Sakhalin). In: Abstracts of the Conference, 19–21 September 2001, Yuzhno-Sakhalinsk. SakhNIRO. pp. 126–127. (In Russian)

Silina, A.V., 1994. Growth of the scallop *Mizuhopecten yessoensis* cultured in the coastal waters of Primorye Province, Russia. In: N.F. Bourne (Ed.). Proceedings of the 9th International Pectinid Workshop, B.C., Canada, 22–27 April 1993. Volume 2. Can. Tech. Rep. Fish. Aquat. Sci. 1994:99–103.

Silina, A.V., 1996. Mortality of late juvenile and adult stages of the scallop *Mizuhopecten yessoensis* (Jay). Aquaculture 141:97–105.

Silina, A.V. and Bregman, Yu.E., 1986. Total populations and biomass. In: P.A. Motavkin (Ed.). Japanese Scallop *Mizuhopecten yessoensis* (Jay). Vladivostok, FESC Acad. Sciences USSR. pp. 190–200. (In Russian)

Silina, A.V. and Ovsyannikova, I.I., 2000. Yesso scallop and its epibioses in cage and bottom cultures in the Alekseeva Bight (Sea of Japan). Bull. Russian Far East Malacological Soc. 4:103–105. (in Russian)

Silina, A.V. and Pozdnyakova, L.A., 1986. Linear growth of the light scallop *Chlamys albidus* (Pectinida, Pectinidae). Zoologichesky zhurnal 65(5):741–746. (In Russian with English abstract)

Silina, A.V. and Pozdnyakova, L.A., 1990. Growth of the sea scallop *Chlamys rosealbus* in the Sea of Japan. Biologiya Morya (Marine Biology, Vladivostok) 1:37–42. (In Russian with English abstract)

Silina, A.V. and Pozdnyakova, L.A., 1991. Microsculpture of the shell and growth rates of three scallop species of the genus *Chlamys* at Onekotan Island, the Kurile Islands. Biologiya Morya (Marine Biology, Vladivostok) 5:23–30. (In Russian with English abstract)

Silina, A.V., Pozdnyakova, L.A. and Ovsyannikova, I.I., 2000. Population conditions of Japanese scallop in southwest part of the Peter the Great Bay. In: V.L. Kas'yanov (Ed.). Ecological condition and biota of a southwest part of the Peter the Great Bay and mouth of the Tumannaya River. Vladivostok: Dal'nauka. pp. 168–185. (In Russian)

Silkin, V.A. and Khailov, K.M., 1988. Bioecological Mechanisms of Control in Aquaculture. Nauka, Leningrad. 230 p. (In Russian)

Skalkin, V.A., 1966. Biology and Fishing of the Sea Scallop. Far East State Publisher, Vladivostok. 30 p. (In Russian)

Skalkin, V.A., 1971. Distribution, stocks and fishing of the sea scallop in the Sakhalin-Kurile region. In: I.M. Likharev (Ed.). Mollusks: Approaches, Methods and Results of their Study. Abstracts of the Meeting on the Investigations of Mollusks, March 1971. Nauka Press, Leningrad 4:56–57. (In Russian)

Skalkin, V.A., 1975. White scallops *Chlamys albidus* (Dall) (Molluska, Bivalvia, Pectinidae) at Onekotan Island (Kurile Ridge). Izvestiya TINRO 95:69–77. (In Russian)

Soldatova, I.N., Reznichenko, O.G. and Tsikhon-Lukanina, E.A., 1985.Specific of fouling on installation of mariculture of Japanese scallop. Okeanologia (Oceanology) 25(3):513–518.(In Russian)

Stotsenko, A.A., 1984. Hydrobiotechnical structures. DVGU, Vladivostok. 136 p. (In Russian)

Tarasova, T.S. and Preobrazhenskaya, T.V., 2000. Effect of mariculture on foraminifera complexes of the Alekseeva Bight of the Sea of Japan. Biologiya Morya (Marine Biology, Vladivostok) 3:166–174. (In Russian with English abstract)

Tibilova, T.Kh. and Bregman, Yu.E., 1975. Growth of the bivalve mollusk *Mizuhopecten yessoensis* in Troitsy Inlet (Posjet Bay, Sea of Japan). Russian J. Ecology 2:65–72. (In Russian)

Vashchenko, M.A., Zhadan, P.M., Kovaleva, A.L. and Chekmasova, N.M., 1992. Morphological and histological analysis of gonads of sea urchin *Strongylocentrotus intermedius*, inhabiting in conditions of anthropogenic pollution (Peter the Great Bay, Sea of Japan) Ekologia (Ecology) 1:45–54. (In Russian with English abstract)

Ventilla, R.F., 1982. The scallop industry in Japan. Adv. Mar. Biol. 20:309–382.

Wang, R. and Shieh, L.C., 1991. Culture of Farrer's scallops *Chlamys farreri* (Jones and Preston) in China. In: W. Menzel (Ed.). Estuarine and Marine Bivalve Mollusks Culture. CRC Press, Boston. pp. 277–281.

Yamamoto, G., 1960. Mortalities of the scallop during its life cycle. Bulletin of the Marine Biological Station Asamushi 10(2):149–152. (In Japanese)

Yamamoto, G., 1964. Studies on the propagation of the scallop *Patinopecten yessoensis* (Jay) in Mutsu Bay. Fish. Res. Board Can. Trans. Ser. No. 1054. 68 p.

Zolotarev, V.N., 1979. Investigations of the individual growth of the marine bivalves. In: I.M. Likharev (Ed.). Mollusks: Main Results of their Study. Abstracts of the 6th Meeting on the

Investigations of Mollusks, 7–9 February 1979, Leningrad. Zoological Institute, USSR Academy of Sciences. pp. 93–95. (In Russian).

Zolotarev, V.N. and Gubanov, V.V., 1989. Problems of development of mussel mariculture in the northwest part of Black sea. In: International Symposium on Modern Problems of Mariculture in the Socialist Countries. Abstracts. VNIRO, Moscow. pp.136–137. (In Russian)

Zolotnitsky, A.P., 1988. Effect of mariculture of mussel (*Mytilus galloprovincialis* Lam.) on secondary pollution of rearing areas. In: V.E. Rodin (Ed.). Resources of commercial invertebrates and biological basis of their rational using. Abstracts of the All Union Meeting, 22–24 November 1988, TINRO, Vladivostok. pp. 102–103. (In Russian)

AUTHORS ADDESSES

Victor V. Ivin - Institute of Marine Biology, Far East State Branch of the Russian Academy of Sciences, Vladivostok 690041, Russia (E-mail: ivin@hotbox.ru)

Vasily Z. Kalashnikov - Institute of Marine Biology, Far East State Branch of the Russian Academy of Sciences, Vladivostok 690041, Russia (E-mail: vassilikalashnikov@yahoo.com)

Sergey I. Maslennikov - Institute of Marine Biology, Far East State Branch of the Russian Academy of Sciences, Vladivostok 690041, Russia (E-mail: imb.dvo.ru)

Vitaly G. Tarasov - Institute of Marine Biology, Far East State Branch of the Russian Academy of Sciences, Vladivostok 690041, Russia (E-mail: vtarasov@imb.dvo.ru)

Scallops: Biology, Ecology and Aquaculture
S.E. Shumway and G.J. Parsons (Editors)
© 2006 Elsevier B.V. All rights reserved.

Chapter 25

Scallop Aquaculture and Fisheries in Brazil

Guilherme S. Rupp and G. Jay Parsons

25.1 INTRODUCTION

The Brazilian coast extends for more than 8,000 km along the Western Atlantic Ocean and ranges latitudinally from about 5° N to 33° S, being the largest tropical and subtropical coastline in the world (Fig. 25.1). Approximately 45% of the Brazilian population live within 200 km of the coast, which corresponded to approximately 70 million inhabitants in 1999 (Anonymous 1999). Although there are several well-preserved areas, anthropological pressure is causing serious environmental impacts on the coastal zone, such as marine pollution and damage to mangroves, estuaries, and other important ecosystems (Ab'Sáber 2001). The degradation of these nursery grounds of fishery resources along with overfishing and mismanagement by public agencies, are contributing to a decline of artisanal fishery production, which is an important socio-economic activity in several states. Recently, bivalve aquaculture has started to emerge as an important economic alternative to the declining artisanal fisheries in different areas of Brazil. As a result, the demand for unpolluted waters for bivalve growth is contributing to increased environmental awareness in coastal communities, as well as eliciting public action towards improvement of seawater quality and proper coastal zone management. Furthermore, along the Brazilian coast there are vast marine areas with pristine waters potentially available for future development of bivalve aquaculture, including scallops.

The Pectinidae is represented in Brazilian waters by 6 genera and 16 species (Rios 1994). Among them, two species are notable for their economic interest: the lion's paw scallop (Fig. 25.2), *Nodipecten* (= *Lyropecten*) *nodosus* (Linnaeus, 1758) and the zigzag scallop *Euvola* (= *Pecten*) *ziczac* (Linnaeus, 1758). *N. nodosus* is locally known as "vieira", "pata-de-leão" or in some areas "coquille", as a gallic derivation from a European scallop. *E. ziczac* is locally known as "vieira". The former is the largest scallop occurring in Brazilian tropical and subtropical waters, and has recently been the focus of increasing interest for aquaculture development (Rupp 1994, 1997, 2001, Uriarte et al. 2001). The latter has previously been the subject of a significant fishery, but stocks have recently declined and landings have dwindled. *Aequipecten* (= *Chlamys*) *tehuelchus* (Orbigny, 1846) is another species having potential economic interest in Brazil. Although it is an important fishery resource in Argentina and has aquaculture potential (Narvarte 1995), the absence of commercially significant stocks in Brazil precludes, at this time, any economic utilisation of *A. tehuelchus*. Other edible scallops from waters off south Brazil are *Chlamys felipponei* (Dall, 1922) and *Chlamys patagonicus* (King, 1832) (Rios 1994), which reaches about 7 cm in shell height, but no economically valuable stocks are known

1226

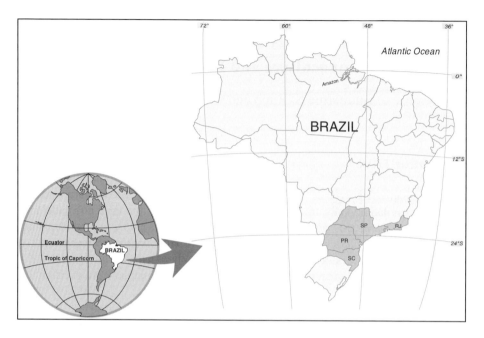

Figure 25.1. Map showing location of Brazil in the Americas, and in detail the States of Santa Catarina (SC), Paraná (PR), São Paulo, (SP) and Rio de Janeiro (RJ), where there are activities of scallop fishery and aquaculture.

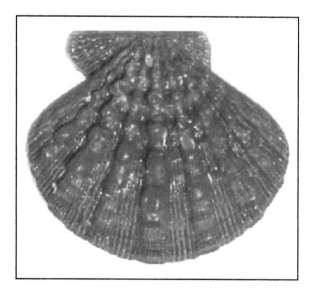

Figure 25.2. Left (superior) valve of *Nodipecten nodosus* (Linnaeus, 1758). Scale: shell length = 145 mm.

in the area. The remaining species, due to small sizes, have no potential for aquaculture or fisheries.

The next section of this review focuses on aspects of the biology and aquaculture of *Nodipecten nodosus*. As well, a small recreational fishery based upon this species will be described. In the last section, the information available on the biology, aquaculture and fishery of *Euvola ziczac* are discussed.

25.2 *Nodipecten nodosus* – BIOLOGY, AQUACULTURE AND FISHERIES

25.2.1 Taxonomy and Distribution

As an historic remark, the first record for *Nodipecten nodosus* in Brazilian waters was made by Haas (1953) who reported that *Pecten (Nodipecten) nodosus* Linnaeus, was collected by the Brazilian naturalist Helmut Sick off Ilha Grande, Rio de Janeiro (RJ), "almost under the Tropic of Capricorn" and sent to the Chicago Natural History Museum.

In recent literature, certain controversy persists with regard to the taxonomy of the genera *Nodipecten* and *Lyropecten*. Smith (1991) presents the distinction between both genera and characterises the different species. *Lyropecten* is an extinct genus in the Atlantic Ocean, and the only living species is *L. magnificus*, found in the Pacific Ocean off the Galapagos Islands. *Nodipecten* has extant representatives in the Atlantic and Pacific. In the Western Atlantic two holocenic species of *Nodipecten* are found, *N. fragosus*, which inhabits warm-temperate waters from the Gulf of Mexico to Cape Hatteras (North Carolina) in depths from 15 to 82 m, and *N. nodosus*, which is found from the Caribbean southwards. In the Pacific Ocean, *N. subnodosus* occurs off the west coast of Mexico and Baja California and *N. arthriticus* occurs from Mexico to northern Peru. According to Smith (1991), *N. nodosus* is the ancestor of *N. subnodosus*, but they no longer live adjacent to each other, speciation having occurred after the closure of the Panama Isthmus. The presence of bulbous hollow nodes is a distinctive characteristic of the genus *Nodipecten*. Taxonomic distinction between *N. nodosus* and *N. fragosus* presented by Smith (1991) is based on shell features (mainly number of ribs: 9 or 10 for *N. nodosus* and 7 or 8 for *N. fragosus;* finer macrosculpture and greater tendency to develop nodes on both valves of *N. fragosus*, and umbonal angle) (Table 25.1). In this manner, descriptions given by Abbott (1974), Eisenberg (1981), Abbott and Dance (1990), Rios (1994) and Lodeiros et al. (1999) included what is presently recognised as two distinct species (*N. nodosus* and *N. fragosus*) as being "*Lyropecten nodosus*". These studies also report the geographic distribution of "*L. nodosus*" combining the range of both Atlantic species of *Nodipecten*, in depths from 35 to 150 m. Specimens of *N. nodosus* collected in Brazil follow the descriptions given by Smith (1991), except that specimens displaying bulbous nodes also on the right valve were occasionally observed (Rupp, pers. obs.).

In the present review, we follow the taxonomy according to Smith (1991), also in agreement with Waller (pers. comm., Smithsonian Institution) and Peña (2001). Note, however, should be taken that taxonomic distinction of both species has been made solely on shell morphology. Further studies on genetic characterisation of different

Table 25.1

Morphological characteristics of *Nodipecten nodosus* and *N. fragosus* (from Smith 1991).

Feature	*N. nodosus*	*N. fragosus*
Number of ribs	9–10	7–8
Valve microsculpture	coarser	finer
Umbonal angle	95° in juveniles 105° in adults	86–94°
Nodes	Less tendency to develop large bulbous nodes; right valves lack nodes	Tendency to develop nodes on all ribs of both valves
Exterior colours	Red, orange, brown, purplish or mottled	Orange to reddish brown

species of *Nodipecten* and *Lyropecten* using molecular techniques may change the current view.

According to Smith (1991) *N. nodosus* is widespread in shallow tropical waters in the Caribbean, south of the Greater Antilles, the Virgin Islands, eastern Antilles, eastern Central America south of the Yucatan Peninsula; eastern Panama to Colombia and Venezuela, and discontinuously as far south as Rio de Janeiro state (Brazil). However, the southern distribution limit of *N. nodosus* is further south than Rio de Janeiro, reaching warm-temperate waters off Santa Catarina State (Rupp, pers. obs.). Except for a reference to Baía de Todos os Santos (Bahia, Brazil) (Smith 1991), and Recife (E. Rangel, pers. comm.) there are no other references for *N. nodosus* in the North and Northeast regions of Brazil. Rios (1994) also includes Ascension Is. in the geographic distribution of "*Lyropecten nodosus*". Bathymetric data are few for *N. nodosus*. According to Smith (1991), live organisms have been collected from 10–15 m. Rupp (1994, 1997) have collected *N. nodosus* in depths ranging from 6 to 30 m off Arvoredo and other islands in the vicinity of Santa Catarina island, (SC) Brazil (Fig. 25.3).

25.2.2 Ecology

Aspects of biology and ecology of *Nodipecten* spp. have been the subjects of few studies. According to Smith (1991), specimens of *Nodipecten* spp. were rarely observed alive, and most of her suppositions about this genus have been deduced from shell features or obtained through studies with other pectinids. Since Smith's (1991) publication, a few studies have been undertaken on *N. nodosus* (Brazil, Colombia and Venezuela) and *N. subnodosus* (Mexico), but mainly pertaining to issues related to aquaculture. For *N. nodosus*, no data are available about growth patterns in wild populations, life span, recruitment, mortality, or spatial distribution. The physiology and ecology have not yet been subjects of detailed studies.

Wild *Nodipecten nodosus*, as large as 17.8 cm, have been collected in Brazil (Rupp, pers. obs.) and it is considered within the group of living giant pectinids (Smith 1991). Considering the patterns for scallop life history described by Orensanz et al. (1991),

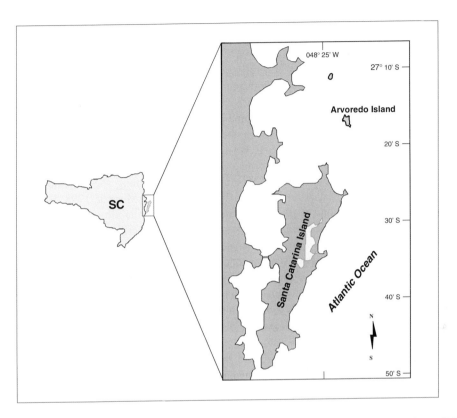

Figure 25.3. Map locating Arvoredo and Santa Catarina island, off Santa Catarina State (SC), Brazil.

N. nodosus can probably be included within the group of long-lived, iteroparous, large size species, which includes most of the commercially important scallops from temperate regions. Yet, it should be noted that, with regard to longevity, no published information is available.

In Brazil, *N. nodosus* inhabits mainly subtropical waters, and there is no information of populations occurring in the North and Northeast regions. The southern limit of distribution (Santa Catarina State - Lat. 27° S; Long. 048° W) is about 450 nautical miles southwest of Cabo Frio (RJ) (Lat. 23° S; Long. 042° W), which is considered the biogeographic boundary of the Western Atlantic Tropical Zone, and the distribution limit for many tropical marine species (Briggs 1995). The Brazilian southeast coast (Fig. 25.1), between Cape São Tomé (Lat. 22° S; Long 041° W) (RJ) and Cape Santa Marta (Lat. 28° 37' S Long. 048° 50' W) (SC) (Brandini 1990), is the region where populations of *N. nodosus* have been recognised, and where fisheries and aquaculture have been carried out, being included within the Western Atlantic warm-temperate biogeographic Region (cf. Briggs 1995).

The Brazilian southeast coast is seasonally influenced by different water masses. The oceanographic structure of this region is quite complex, but macro-scale oceanographic patterns have been identified (Emilsson 1961; Matsuura 1986; Castro and Miranda 1998, Sunyé and Servain 1998). As a generalisation, during summer to mid-autumn neritic waters are influenced by warm and oligitrophic waters of Brazil Current (tropical waters), which penetrates over the continental shelf. During winter, the area is mainly influenced by coastal advection of cold, low salinity and phytoplankton-rich waters of sub-Antarctic origin. Furthermore, from October–November to April, intrusions of cold and nutrient rich South Atlantic Central Water (SACW) into the bottom layers over the continental shelf have been reported to create a strong stratification in the water column influencing the inner shelf. This event leads to the formation of a strong thermocline creating subsurface phytoplankton maxima (Brandini 1990; Castro and Miranda 1998; Borzone et al. 1999). This temporal and spatial pattern in which food availability tend to vary in an opposite manner with temperature, is different from those described for most temperate and boreal regions, and can affect coastal aquaculture sites in Santa Catarina, having important influences on growth of scallops under culture conditions (Rupp 2003).

The main shell features of *Nodipecten* spp. are the prominent nodes and ledges. Under culture conditions in subtropical waters of southern Brazil, no evidence was found to support a relationship of node formation to variability in environmental factors or reproductive events, and the ecological significance of the hollow nodes is presently unclear. Furthermore, shell growth rings, commonly used as an age indicator in temperate scallops, are not clearly distinguishable in *N. nodosus*, except for disturbance rings caused by handling during thinning procedures, under culture conditions. Therefore, shell rings conveniently used as age estimators for temperate scallops (e.g., MacDonald and Thompson 1985a) are not useful for *N. nodosus*.

Unlike other scallops, which aggregate in economically valuable beds or banks (*Pecten maximus*, *Placopecten magellanicus*, *Argopecten irradians*, among others), *N. nodosus* is not gregarious. Rupp (pers. obs.) has found adult *N. nodosus* off Arvoredo Island (Brazil) at the limit between rocks and sandy or gravel bottoms, often byssally attached to the rocks. They are found particularly inside small rocky caves and crevices, or on sandy or calcareous patches adjacent to rocks. *N. nodosus* are usually found far from each other, but clumps of up to 3 or 4 animals have been observed, sometimes within the same cave. Shells are typically covered by epibionts and sediment, which provide camouflage, making them difficult to be found by inexperienced eyes. Juvenile *N. nodosus* have not been found in the same water depths as adults (Rupp, pers. obs.), as it is also reported for *N. subnodosus* (Smith 1991). Juvenile *N. subnodosus*, for example, have been found attached to spiny scallops (*Spondyllus princeps*) collected with deep trawl nets (40 m) (Félix-Pico et al. 1999). For *N. nodosus*, the question still remains whether the juveniles are highly cryptic or whether they occur in deeper waters.

A natural predator of *N. nodosus* in South Brazil is *Octopus vulgaris*. Octopus caves are often surrounded by several empty scallop shells, and octopuses preying upon scallops have been observed during underwater surveys (Rupp, pers. obs.). A characteristic of shells preyed by octopus is the presence of one or both auricles broken.

With regard to adductor muscle mass, wild *N. nodosus* of about 120–140 mm in shell height, from a population off Arvoredo Island, yield muscle wet mass of about 30–45 g (Rupp, unpubl.) and reach a maximum size of 17.8 cm. In the southern Caribbean region wild *N. nodosus* of about 120–140 mm yield muscle wet mass of about 18 g (Lodeiros et al. 1998) and the largest size reported is about 15 cm (Lodeiros et al. 1999). In southeastern Brazil, wild *N. nodosus* tend to have a larger shell size and adductor muscle weight and have heavier and thicker shells than *N. nodosus* from Margarita Islands, Venezuela (Rupp, pers. obs.). According to Sebens' model (Sebens 1982) relating the maximum size of suspension feeders and habitat suitability, animals living in environments resulting in lower physiological stress and/or supplying higher food availability, display a larger maximum size. This suggests that conditions in the subtropical environment of southern Brazil may be more suitable for scallop growth than the tropical areas of the Caribbean. However, genetic differences between these latitudinally separated populations should not be ruled out.

25.2.3 Reproduction

Information on the reproduction of *N. nodosus* is limited. *N. nodosus* is a simultaneous functional hermaphrodite (Rupp 1994). A study of the variation of the gonosomatic index (GSI) (wet weight) of a population off Arvoredo Island (Brazil) suggested a partial and asynchronous pattern of gamete release throughout the year, and higher GSI was recorded during late spring (November–December) (Manzoni et al. 1996) (Fig. 25.4). Lower reproductive activity was recorded during the winter (June–August), when a lower GSI was associated with periods of lower water temperatures. On the other hand, Manzoni and Banwart (2000) found higher gonosomatic indices of *N. nodosus* under culture conditions during winter months concomitant with lower water temperatures, but the asynchronous pattern of reproduction was again evidenced. It is possible that the previous results of Manzoni et al. (1996) were affected by the fact that temperature was recorded at 8 m but scallops were collected in a wide range of depths, from 8 to 30 m. From late spring to early autumn waters off Santa Catarina State can be influenced by subsurface intrusion of cold and phytoplankton-rich South Atlantic Central Water (SACW), which can reach depths up to 15 m (Borzone et al. 1999). Therefore, it is possible that scallops were collected below the thermocline, where they were exposed to temperatures lower than those recorded at 8 m, but higher phytoplankton abundance. As the reproductive cycle of scallops is a complex function of temperature and food availability (MacDonald and Thompson 1985b), the relationship among these factors and reproductive activity of this subtropical population of *N. nodosus* requires further investigation.

In the Caribbean region, Vélez et al. (1987) studied the seasonal variation in biochemical composition of Venezuelan scallops and suggested that *Nodipecten (= Lyropecten) nodosus* spawns throughout the year in an asynchronous fashion, with the period of greatest reproductive activity between August and February. In this geographic region, higher temperatures and lower phytoplankton abundance are generally found from August to December (Lodeiros and Himmelman 1994, 2000; Vélez et al. 1995).

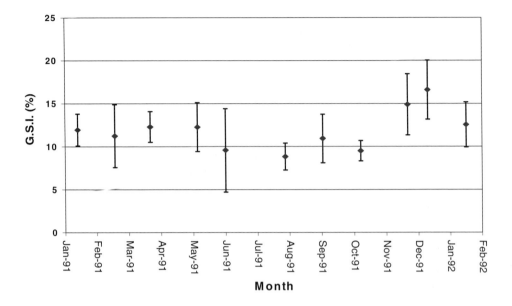

Figure 25.4. Gonosomatic index (G.S.I. ±SD) of *Nodipecten nodosus* sampled at Arvoredo Island, Santa Catarina State, Brazil, from January 1991 to January 1992, (From Manzoni et al. 1996).

Coronado et al. (1991) studied the genetic variability of some Venezuelan scallops, and suggested that *N. nodosus* has a much lower genetic variability compared to other scallops. Self-fertilisation is hypothesised to account for the low variability in natural populations (Coronado et al. 1991). For *N. nodosus*, the diploid number of chromosomes is reported to be 38, with 4 pairs metacentric, 5 pairs submetacentric, 7 pairs subtelocentric and 3 pairs telocentric chromosomes (Basoa et al. 2000). Thirty eight chromosomes, also reported for *Euvola ziczac,* agrees with the diploid numbers reported in other pectinid species such as *Pecten maximus, Placopecten magellanicus* and *Patinopecten yessoensis* among others (citations in Basoa et al. 2000). A comparison of self-fertilised orange *N. nodosus* and cross-fertilised orange and brown specimens undertaken by Alfonsi and Pérez (1998) indicated that self-fertilisation adversely affected survival, but, interestingly, favoured growth.

25.2.4 Aquaculture

25.2.4.1 Status

The recent success of bivalve mariculture in Santa Catarina State (for oysters, *Crassostrea gigas* and mussels, *Perna perna*), is increasing expectations for the scallop *Nodipecten nodosus* as the next candidate for the aquaculture industry. Although scallop culture is a promising activity, resources to study and promote its establishment have been limited, and to date there has been no official plan to foster its development.

Although initial studies on feasibility of culture of *N. nodosus* date from 1990 (Manzoni 1994; Rupp 1994), only recently, through internationally funded projects [European Union: Project INCO Scallop (1997–2000) and Canadian International Development Agency (Brazilian Mariculture Linkage Project - BMLP): Project Biology and Mariculture of *Nodipecten nodosus* 2000–2002)] has it been possible to advance the studies at the Laboratory for the Culture of Marine Molluscs (Federal University of Santa Catarina). The main objectives have been to overcome the primary constraint to the development of scallop culture: a sustained seed supply. Studies aiming to optimise larval and postlarval survivorship and growth in the hatchery, as well as the assessment of appropriate environmental conditions for sea-based nursery have been recently carried out (Pereira 2000, Bem et al. 2001a,b, Rupp et al. 2003a,b).

There are presently four institutions involved with aquaculture of *N. nodosus* in Brazil. At the Federal University of Santa Catarina (UFSC) research focusing on development of hatchery techniques, ecophysiology and experimental culture have been carried out. The UNIVALI (Universidade do Vale do Itajaí), a private university located in Santa Catarina State, is involved in experimental growout, and the establishment of an experimental hatchery is underway. Also in Santa Catarina, the State Government Agricultural Extension Agency (EPAGRI) is developing experimental cultivation of scallops in collaboration with mollusc growers. In Angra dos Reis, State of Rio de Janeiro, the IEDBIG (Instituto de Ecodesenvolvimento da Baía de Ilha Grande), a private foundation, funded by PETROBRAS (Brazilian Petroleum Company) and FURNAS (nuclear power plant), have been attempting commercial hatchery production of *N. nodosus* seed (Avelar 2000). Also, in Angra dos Reis, two private foundations have previously attempted to produce seed of *N. nodosus* (1995–1997), but activities were discontinued. Presently there are no commercial ventures growing *N. nodosus* in Brazil, and scallop aquaculture is still in its infancy at an experimental to pilot-scale.

25.2.4.2 Culture technology

25.2.4.2.1 Wild seed collection

Settlement of post-larval *N. nodosus* on artificial collectors deployed at Arvoredo Island (1991–1992) took place throughout the year, but in fairly low numbers (total 100 spat) compared with another pectinid (*Leptopecten bavahyi* – 18,315 spat) that also settled on the same collectors (Manzoni 1994). As well, *Euvola ziczac* and *Aequipecten tehuelchus* settled in low numbers (87 and 14, respectively). The low spat settlement associated with partial and asynchronous pattern of reproduction, the low-density populations widespread in a large geographic area, along with the complex hydrodynamics in the region, suggest that wild spat collection is unlikely to be a reliable source of *N. nodosus* seed for aquaculture. Therefore, hatchery production was proposed as the alternative source of spat to sustain potential aquaculture activity, and studies were oriented in that direction.

25.2.4.2.2 Hatchery production

Initial results with regard to larval production of *N. nodosus* in hatchery, were presented at the 9th International Pectinid Workshop, in Nanaimo, Canada, in 1993 (Rupp and Poli 1994). Induction of spawning using thermal stimulation and UV treated seawater, usually results in high quantities of oocytes (2.2–21.6 million per scallop, depending on size), within one to two hours, when gonads are sufficiently ripe. Sperm is usually released prior to oocytes. In four spawnings, when a total of 27 scallops were induced to spawn, 72.7% initially released sperm. Subsequently, 31.5% started to release oocytes, 22.7% released exclusively sperm, and 18.5% alternated release of oocytes and sperm, or released both gametes simultaneously. Of the ones that started releasing oocytes (27.3%), 18.2% subsequently released sperm and 9.1% released oocytes only (Rupp 1996). Considering that 68.2% of the induced scallops released gametes of both sexes and the switch from one sexual product to another is seldom preceded by an interval, a certain degree of self-fertilisation is difficult to avoid (Rupp 1996).

The ontogenic development of larvae follows a similar pattern as described for other pectinids (Table 25.2). Mean oocyte diameter ranges from 60–68 μm. "D" shape veligers are completely formed at 22 to 24 h after fertilisation at 25 to 27°C, and shell lengths are about 100 μm (Rupp 1994). The growth relationship between larval shell dimensions can be expressed by the allometric equation $H = 0.322 L^{1.198}$, where H = shell height (μm) and L = shell length (μm) ($r^2 = 0.99$) (Rupp 1994). Larval period ranges from 14 to 23 days and average larval growth rates vary from about 6–8 μm day^{-1} depending on culture conditions. Larval cultures are carried out using "standard" hatchery techniques, in tanks from 100 to 2,500 L depending on the experiment, and water changes are undertaken daily or every other day. Feeding typically consists of a mixed diet of flagellates and diatoms in concentrations ranging from 10–30 cells μL^{-1}. At the end of the larval culture, pediveligers settle and attach to artificial substrates (Netlon™, polycarbonate, and polyethlene) when dissoconch shell begins to form (Rupp 1994, 1997). Larval culture techniques for *N. nodosus* and other scallops from the Iberoamerican region have been recently reviewed by Uriarte et al. (2001).

Hatchery production of spat on an experimental-pilot scale is currently feasible on a routine basis. At the Laboratory for the Culture of Marine Molluscs, Federal University of Santa Catarina (Brazil) larval retrievals from "D" larvae to metamorphosis typically ranged from 0 to 14%, until 1997. The term retrieval here reflects not only the number of surviving larvae, but also excludes live larvae intentionally eliminated to reduce culture densities and select fast growing larvae. Table 25.3 shows the percent retrievals of "D" larvae, and pediveligers in selected larval cultures. The high larval mortalities, and sometimes total loss, had been associated with bacterial pathogenicity (Bem 1999). Isolation and identification of *Vibrio* strains from larval cultures have been carried out, as well as the development of alternatives to increase survivorship in larval cultures. Several bacterial strains were isolated from healthy and diseased larval cultures of *N. nodosus*. A highly pathogenic strain of *Vibrio* sp., (identification in progress), resulted in 100% mortality within 48 hours of inoculation during pathogenicity tests against *N. nodosus* larvae. Further, two novel *Vibrio* species (*V. neptunius* and *V. brasiliensis*) isolated

Table 25.2

Larval development of *Nodipecten nodosus* (from Rupp 1994).

	Temperature (°C)	Mean dimensions (μm)	Period after fertilisation
Oocytes	27	60–68	
Trochophore	27		10–12 hours
"D" larvae	27	length: 98.65 height: 76.52	22 hours
Umbo stage	26–27	length: 146.18 height: 128.01	7 days
Eyed pediveligers	23–26	length: 186.51 height: 167.77	14 days
Settlement	23–26	length: 207–214	14–19 days

from larval cultures of *N. nodosus*, at LCMM have been recently described (Thompson et al. 2003). The assessment of pathogenic activity of these new *Vibrio* spp. on scallop larvae should be the subject of further investigations. As a result of the identification of the major sources of larval mortality, strict hatchery sanitation procedures have recently resulted in an increase of larval survivorship leading to retrievals of up to 22% (Table 25.3), noting that a large percentage of live larvae were eliminated to reduce densities. The high susceptibility of scallop larvae to vibriosis in relation to other bivalves has been proposed to be caused by the relatively immature immune system in larval pectinids (Luna-Gonzales et al. 2002). Furthermore, granular hemocytes, the main cells actively involved in the immune defence in oysters and mussels (Cheng 1981; Hine 1999), have not been detected in adult lion's paw scallop *Nodipecten nodosus* (Vargas-Albores and Barraco 2001), which may also explain the high susceptibility of post-metamorphic life stages.

The artificial collectors with small spat attached can be transferred to the field 2 to 5 weeks after settlement, and 4–6 weeks later the spat attain a size of about 5–7 mm. The percent retrieval of 1–2 mm post-larvae and 5–10 mm juveniles are shown in Table 25.4. Land-based nursery techniques are costly, labour intensive and demand large volumes of cultured microalgae, due to the increasing feeding demand of growing scallops. Improvements are needed during nursery culture for a successful commercial-scale production, and the approach of early transfer of the spat to the field-based nursery is being investigated. Studies on the effect of the size at deployment, depth and seasonal variability of environmental factors during sea-based nursery on growth and survivorship of spat are underway (Rupp et al. 2003a, b).

With regard to broodstock conditioning, Rupp et al. (1997) concluded that *N. nodosus* maintained in the laboratory at temperatures of 17–18°C resulted in a higher percentage of spawning and higher quantities of oocytes released than the ones maintained at 25°C. Broodstock conditioning at low temperatures has been successfully accomplished in several trials for purposes of experimental seed production (Rupp, unpubl.).

Table 25.3

Number of oocytes released, "D" larvae and settling pediveligers and respective percent retrievals, obtained in some experimental larval cultures at the Laboratory for the Culture of Marine Molluscs, UFSC, Brazil.

Year	Number of oocytes	Number of "D" larvae recovered and used	% retrieval of "D" larvae in relation to oocytes	Number of settling pediveligers	% retrieval of pediveligers in relation to "D" larvae	Reference
1992	1.44×10^6	1.26×10^6	87.5%	76,000	6.03%	Rupp 1994
1996		11.06×10^6		211,000	1.9%	Rupp 1997
1996	12.6×10^6	9.52×10^6 used: 6.00×10^6	75.5%	750,000	12.5%	Rupp 1997
1996		5.58×10^6		350,000	6.27%	Rupp 1997
2000		49.2×10^6		8,775,000	17.84%	Unpubl.
2001		8.6×10^6		1,683,000	22.14%	Unpubl.
2001		17.1×10^6		2,366,000	13.18%	Unpubl.

Table 25.4

Number of 1–2 mm post-larvae and 5–10 mm juveniles, and respective percent retrievals, obtained in experimental cultures at the Laboratory for the Culture of Marine Molluscs, UFSC, Brazil.

Number of settling pediveligers	Number of 1–2 mm post-larvae	% retrieval of post-larvae in relation to pediveligers	Number of juveniles (5–10 mm)	% retrieval of juveniles in relation to post-larvae	Reference
76,000	2,286	3%	-	-	Rupp 1994
750,000	77,171	10.3%	21,274	27.5%	Rupp 1997

25.2.4.2.3 Growout

Growout experiments with *Nodipecten nodosus* in the subtropical marine environment of Santa Catarina have been limited. From initial suspended culture trials held in 1991 and 1992, in Arvoredo Island, growth rates varied from 5.5 to 16.7 mm month^{-1} depending on scallop size, with individuals reaching the size of 104 mm (from 10 mm spat) within 10 month of culture, resulting in an average growth rate of 9.4 mm month^{-1} (Manzoni 1994). These are amongst the fastest growth rates recorded for scallops (cf. Bricelj and Shumway 1991). Growth was determined by means of shell dimensions and 100% survival was attained. It should be noted that these experiments were carried out at depths of 8 m, with a limited number of scallops (2–10) and at a reduced density per layer in the lantern nets. In spite of low sample size, this study gives an indication of the high growth potential of *N. nodosus*, if suitable environmental conditions are provided. Manzoni and Marenzi (1997) reported an experimental culture of *N. nodosus* in Itapocoroi Cove (Santa Catarina) in which scallops reached 70 mm within 9 month (from 10 mm spat), indicating an average growth of about 6.6 mm month^{-1}, but mortality data were not presented. As well, in growout trials held at Pinheira Cove (Santa Catarina) using surface long lines in 1996 by Rupp (unpubl.) scallops reached about 66 mm within 10 month of culture (average growth rate 5.2 mm month^{-1}, and mortalities of approximately 60%). In this trial, heavy fouling colonisation on scallop shells and culture devices was an important factor affecting growth and survival of cultured scallops. A list of the most conspicuous epibionts observed on scallops cultured at Pinheira Cove is presented in Table 25.5. Although environmental parameters other than temperature had not been assessed in those studies, it is clear that the more oceanic conditions of waters off Arvoredo Island (located about 10 miles offshore) are more suitable for scallop growth than coastal embayments. Studies on the effects of environmental factors on growth and survivorship of *N. nodosus* on a coastal aquaculture site off Porto Belo island, are currently underway (Rupp et al. 2004a, b).

In Rio de Janeiro state (Brazil), scallop growth trials have been attempted but to date, limited information is available. Avelar (2000) reports that *N. nodosus* cultured at different densities off Ilha Grande, reached a mean shell height of 83 to 93 mm (from 10 mm spat) within one year. Survivorship ranged between 84 and 90%. In Ilha Grande Bay high mortalities have also been recorded in some periods of the year and it has been suggested to be caused by high temperatures concurrent with spawning events and possible low phytoplankton abundance (Rangel, pers. comm., Aquacultura do Norte Nordeste Ltda).

Lodeiros et al. (1998) reports experimental suspended culture of *Nodipecten (Lyropecten) nodosus* in the Caribbean Sea (Venezuela) in which growth rates progressively decreased with depth. At 8 m, scallops attained fast growth rates, but total mortality after 7 months of culture was attributed to heavy fouling colonisation on shells. In addition, other factors such as high temperature and harmful algal blooms may have accounted for the high mortality. At 31 m, slow growth and total mortality was associated with low phytoplankton abundance. The best results were recorded at 21 m, where 50% mortality was attained and the maximum size after 10 months of culture was about 5.5 cm. Vélez et al. (1997) reports a cumulative growth of 50 mm within 8 months for *N. nodosus* in suspended culture in Venezuela, but high mortality at the end of the experiment was associated with harmful algal blooms.

As previously mentioned, culture trials of *N. nodosus* on an experimental – pilot-scale have been undertaken in coastal embayments in Santa Catarina State, but growth rates were reduced when compared to growth trials undertaken in more oceanic conditions (Arvoredo Island). Even so, the growth rates obtained in inshore areas are still higher than those reported for other scallops currently cultured at a commercial level elsewhere. The catarina scallop, *Argopecten ventricosus*, reaches the commercial size of 5.7 cm in 11 months in Mexico (Maeda-Martinez et al. 1997) and bay scallop, *Argopecten irradians*, reaches a market size of 50 to 60 mm within one year in China (Guo et al. 1999).

Table 25.5

Conspicuous epibionts observed on shells of *Nodipecten nodosus* cultured in suspension at Pinheira Cove in 1996 (Rupp, unpubl.).

Phylum	Class	Species	Remarks
Porifera		Not identified	Yellow colour, shell-borer
		Not identified	Purple colour, tends to cover shells
Cnidaria	Anthozoa	*Bunodosoma* sp.	
	Hydrozoa	*Ectopleura* sp.	
Platyhelmintes	Turbellaria	*Stylochoplana* sp.	
Mollusca	Bivalvia	*Ostrea equestris*	
		Crassostrea rhizophorae	
		Perna perna	
		Leptopecten bavayi	
		Lithophaga patagonica	
	Gastropoda	*Thais haemastoma*	
		Cymatium parthenopeum	
		Egg masses of gastropods	
Annelida	Polychaeta	*Polidora websterii*	
		Not identified	Green colour
		Not identified	Red colour
Crustacea	Cirripedia	*Balanus* sp.	
	Malacostraca		Several species of the Orders Decapoda, Amphipoda and Isopoda
Bryozoa		*Bugula* sp. and *Schizoporella* sp.	
Chordata	Ascidiacea	*Clavelina* sp. and *Styela* sp.	

Further, the large size and growth potential of *N. nodosus* indicates that when cultured under more suitable environmental conditions, and as soon as experimental work proceeds leading to adequate culture strategies, this species may yield scallops of 10 cm, in a period much shorter than the large and high valued pectinids cultured in temperate waters (e.g., *Placopecten magellanicus, Pecten maximus* and *Patinopecten yessoensis*) which take 2 to 5 years to reach commercial size. Therefore, developing new strategies for culturing *N. nodosus* outside shallow protected areas should be pursued. In southeastern Brazilian coast, the continental shelf is very wide and subsurface long-lines could be deployed in depths from 20 to 50 m without interfering with navigation, providing vast areas for aquaculture developments. Furthermore, if adequate strategies are employed, especially to reduce fouling, the production of smaller size scallops (5–6 cm) in a short period of time (8–9 months) in coastal protected areas, may be commercially feasible. Bottom culture is another potential alternative for culture of *N. nodosus*, but to date, no information is available on this approach in Brazil. Further studies on the cost-effectiveness of the different growout methods are needed to determine the economic feasibility of culture of *N. nodosus* in Brazil.

25.2.4.2.4 Constraints

Seed supply has been the major constraint to the development of scallop culture in Brazil. Unlike most of the cultured scallops, on which industry can rely upon wild-collected and hatchery-produced seeds, the only source of juvenile *N. nodosus* to sustain aquaculture is hatchery production. The total dependence on hatcheries makes essential the establishment of a reliable and consistent technology in order to supply seed in high quantity and quality. Although *N. nodosus* larvae are amenable to culture, they are very sensitive to culture conditions and may be subject to mass mortalities due to pathogenic bacteria, as also reported in other pectinids. Developments in hatchery technology to improve water quality and produce high quality algal cultures, along with adequate hatchery sanitation, tend to lead to such reliable seed production. For sea-based cultivation the influence of environmental factors on growth and survivorship of *N. nodosus* during nursery culture and growout have to be considered, and site selection criteria have to be established based on ecophysiological parameters, in order to maximise production. Such constraints can be overcome in the near future if research efforts and institutional support are continued, and resources committed to the development of scallop culture in Brazil.

25.2.5 Fisheries and Marketing

The current market is mainly based on a small-scale fishery by recreational SCUBA divers or occasionally by fisherman. *N. nodosus* is the subject of anecdotal capture off coastal islands in the States of Rio de Janeiro, São Paulo, Paraná, Santa Catarina and Espirito Santo, supplying small local markets and fine restaurants. In some instances, the restaurant owners are the ones who dive to collect the scallops. There are no capture/production statistics available, but considering the low densities and patchy distribution of populations, wild stocks would not sustain any large-scale exploitation. *N. nodosus* is mentioned by Ostini and Poli (1990) as having threatened stocks due to intensive capture during summer tourist seasons. Although it may be true in some areas, to date no stock assessment of wild *N. nodosus* populations has been carried out and no conclusive statement can be made considering the spasmodic fluctuations characteristic of scallop populations. *N. nodosus* is the mollusc having the highest market value in Brazil and unit price can surpass US$ 5.00 for large whole scallops (>12 cm), including shells, depending on the region and season. The capture is more intense during the tourist season (summer), and its high value is stimulated not only by the large adductor muscle, which is served as a fine delicacy, but also by the ornamental appeal of the shells along with their scarcity of supply. There is also a parallel market for the large shells, which are refractory and used to prepare and serve seafood dishes in restaurants. Rare specimens with yellow or orange shells reach high prices in the collectors' market. Occasional small-scale aquaculture production yielded scallops from 6–7 cm, which were sold for about US$ 1.00 per whole scallop. No marketing surveys have been undertaken to date for the potential of *N. nodosus*, but considering the large market for seafood in general and the high demand for other bivalves (oysters, mussels and clams), in which the demand has been higher than the supply, there is a large potential market for high quality cultured *N. nodosus* in Brazil. This underdeveloped market has been constrained by the limited availability of the product and high prices of imported scallops.

An area with one of the most notable stocks of *N. nodosus* is Arvoredo Island, Santa Catarina State (Lat. 27° 17' S, Long. 048° 22' W) (Fig. 25.3), where capture by recreational divers was common until recently. Following an increase in environmental awareness, and the need to protect coastal environments, Arvoredo Island and its adjacent waters have been declared a Biological Preservation Area by the Instituto Brasileiro do Meio Ambiente (IBAMA), and the capture of organisms is now prohibited (BRASIL 1990), unless for research purposes.

25.2.6 Future Prospects

Hatchery technology for production of *Nodipecten nodosus* on an experimental – pilot scale is well advanced in Brazil. The transition to a sustainable commercial production will mainly depend upon continued research in order to overcome some remaining constraints and on logistic improvements in the Brazilian hatcheries. Hence, hatchery technology should not preclude further developments of scallop culture in Brazil if appropriate steps are taken to establish this new and promising activity, and private ventures are created to increase production. If future developments of scallop aquaculture are directed to the utilisation of open marine areas on the continental shelf at depths from 20 to 50 m, then legal aspects of coastal zone management will have to be addressed. Likewise, before large scale aquaculture is attempted in this area, research is needed to assess the effects of the oceanographic dynamics on scallop growth and survival. The recent success and development of oysters and mussels culture in Brazil created a positive political and social climate for further development of bivalve aquaculture, and *N. nodosus* is the next obvious candidate. The large size, commercial value and potential market are increasing interest of mollusc growers wanting to diversify their products and of new entrepreneurs wanting to carry out bivalve aquaculture. Thus, it appears that the commercial development of *N. nodosus* culture is imminent.

25.3 *Euvola (Pecten) ziczac* – BIOLOGY, AQUACULTURE AND FISHERIES

25.3.1 Taxonomy and Distribution

Among the marine molluscs collected by Helmut Sick off Ilha Grande, Brazil, in 1943–44, is *Pecten (Euvola) ziczac* (Haas 1953). The zigzag scallop, *Euvola ziczac* (Linnaeus, 1758) (cf. Waller 1991) is known in Portuguese as "vieira". The upper valve (left) is flat and displays fine zigzag markings of white and brown. The lower valve is very deep and convex, with very low ribs and radial and concentric grooves, displaying colours varying from white to brown or orange brown (Abbott 1974; Dance 1990). *Euvola* (= *Pecten*) *ziczac* occurs in the Western Atlantic, from North Carolina and Bermuda, to Santa Catarina State, Brazil (Rios 1994).

25.3.2 Ecology

Wild *Euvola* (= *Pecten*) *ziczac* are reported to reach up to 100 mm in shell height (Dance 1990). Its depth distribution varies according to the geographic area. It is reported to range from 15 to 75 meters in sandy and calcareous algae bottoms by Rios (1994). On the other hand, Abbott and Dance (1990) report its depth range from 1 to 40 m. Vélez and Lodeiros (1990) report natural beds of *E. ziczac* between 6 and 10 m in Venezuela, particularly in areas affected by coastal upwelling. In the southeastern Brazilian coast, *E. ziczac* aggregates in beds between 30 and 50 m, off the States of São Paulo, Paraná and northern Santa Catarina in sandy bottoms or sandy and broken shell bottoms, being absent on muddy substrates (BRASIL 1976). In this area, the most frequent size of the scallops was about 85 mm of shell height, but individuals up to 95 mm have been collected (BRASIL 1976). To date, no data are available on longevity of *E. ziczac* in wild populations, but Vélez et al. (1995) suggests, from a projection of growth curves for scallops cultured in partly buried cages, that *E. ziczac* could attain 70–80 mm in about 1.5 years in the wild. *E. ziczac* is a recessing scallop, and in its natural habitat is usually buried with a thin layer of sand covering the upper valve, which is slightly concave, preventing the colonisation of biofouling (Vélez et al. 1995).

25.3.3 Reproduction

Euvola ziczac is a hermaphroditic bivalve, which in southeastern Brazil, reproduces throughout the year in an asynchronous and continuous manner, but with variable intensities (Peres 1981). Higher spawning activity was recorded during summer, autumn and spring. On the other hand, Vélez et al. (1987) studying the biochemical composition of the gonads of *E. ziczac* of a population off Venezuela, report a synchronic bimodal pattern of reproduction, with two peak periods of spawning: from April to May, and from August to September, displaying a pattern typically different of most tropical bivalves, which tend to reproduce continuously. In this Venezuelan population, gametogenesis, accumulation of reserves and gonad growth are associated with periods of minimum temperature and increased primary productivity associated with upwelling events (Vélez et al. 1993). Peres (1981), in a histological study of the reproductive cycle of *E. ziczac* from a population off southeastern Brazil, indicated that *E. ziczac* show a high percentage of protandric individuals in spring, while in summer and autumn proterogynous individuals predominate. In this manner, although Vélez et al. (1990) indicate that *E. ziczac* from a Venezuelan population is a simultaneous hermaphrodite, Peres (1981), suggest that simultaneous hermaphroditism was not recorded in the population from southeastern Brazil. Such reproductive behaviour is not common in monoecious bivalves (Fretter and Graham 1964). Furthermore, Peres (1981), hypothesise that, as a consequence of this phenomenon, self-fertilisation is infrequent in this species.

Inductions of spawning of *E. ziczac* from the Brazilian population in the laboratory should clarify whether simultaneous hermaphroditism and self-fertilisation can occur.

25.3.4 Aquaculture

25.3.4.1 Status

To date no studies have focused on aquaculture of *E. ziczac* in Brazil. Such studies have been primarily oriented towards *Nodipecten nodosus*, due to the larger size and higher commercial interest, as well as due to the availability of broodstock to induce spawning in the hatchery. While *N. nodosus* can be readily collected by diving on coastal islands and transported quickly to the hatchery (Rupp 1994), *E. ziczac* is harvested as a by-catch product of commercial shrimp trawlers, being stored on ice where they dehydrate and die quickly (Morais and Kai 1980), thus, making it difficult to obtain live broodstock for husbandry practices.

25.3.4.2. Culture technology

25.3.4.2.1 Wild seed collection

Occasional settlement of spat of *E. ziczac* (up to 10–20 per collector) have been recorded on monofilament collectors holding hatchery produced spat of *N. nodosus*, after being deployed to the nursery phase in the sea (Rupp, pers. obs.). In addition, Manzoni (1994) reported settlement of 87 individuals on monofilaments off Arvoredo Island during one year. Due to the anecdotal nature of these results, no seasonal pattern can be discerned. Although low numbers for commercial purposes, wild collected spat could be grown until adult size and further used as broodstock for reproduction in the hatchery and seed production. In Venezuela, Vélez and Lodeiros (1990) indicated that wild spat collection of *E. ziczac* was also difficult to obtain. In Bermuda, Manuel (1999) reports that settlement of *E. ziczac* on artificial collectors takes place between October and June with evidence of a bimodal pattern, the larger peak occurring between February and April and the other peak in November–December.

25.3.4.2.2 Hatchery production

Studies pertaining to hatchery seed production of *E. ziczac* have been undertaken in Venezuela (Vélez and Freites 1993) and in Bermuda (Sarkis 1995). Larvae are amenable to culture using standard hatchery techniques and the larval period takes about 10 to 12 days at 26°C and densities of 5–10 larvae/mL (Vélez and Freites 1993). Settlement can be carried out on plastic filaments and plastic plates deployed in upwelling or downwelling tanks. Spat were successfully transferred to the field when they reached at sizes of 1 and 3 mm. Whereas there are presently no commercial enterprises in Venezuela producing *E. ziczac* (Lodeiros, pers. comm.), commercial hatchery production of zigzag scallop is under way in Bermuda (Sarkis, pers. comm.). In Brazil, *E. ziczac* is considered a potential candidate species for aquaculture, and as soon as culture of *Nodipecten nodosus* becomes established commercially, the hatchery technology available would be readily adapted to produce seeds of *E. ziczac* if a demand is raised from growers wanting to diversify their products.

25.3.4.2.3 Growout

Aquaculture studies focusing on growout stage of *E. ziczac* have not been undertaken in Brazil. In Venezuela, several studies have evaluated the biological feasibility of culturing *E. ziczac* in a tropical marine environment (See Lodeiros et al. Chapter 28).

25.3.4.2.4 Constraints

Several factors causing major constraints to culture of *E. ziczac* in suspension have been identified in a tropical environment, and bottom culture seems to be more suitable than suspension culture (See Lodeiros et al. Chapter 28). In the subtropical marine environment of Brazil, the effect of environmental factors on growth and survivorship of *E. ziczac* should be evaluated before commercial operations are attempted. As for *Nodipecten nodosus* aquaculture, development of *E. ziczac* will have to rely on hatchery production of seeds.

25.3.5 Fisheries

Fisheries targeting the zigzag scallop began in Brazil in 1972, when 12 t of scallops were landed in the State of São Paulo (BRASIL 1976). The identification of this unexploited resource gained the attention of the commercial shrimp fleet establishing a non-traditional fishery, which in subsequent years displayed variable yields, collapsing after 1986. Surveys undertaken in 1974 identified scallop beds off southern State of São Paulo, Paraná State and northern Santa Catarina, between latitudes 24° 26'S and 26° 20'S, in depths ranging from 20 to 50 m (BRASIL 1976).

From being a minor by-catch product of shrimp trawlers before 1972, *E. ziczac* started to be caught commercially and stimulated the possibilities of additional profits obtained by exporting this valuable product (Anonymous 1975). As soon as the scallop beds were identified, fishermen used the same type of shrimp otter-trawl nets with a few modifications to target scallops (Rebelo Neto 1980). Fishery statistics, fleet characteristics and destination of production of *Euvola ziczac* have been reviewed by Pezzuto and Borzone (1997). According to official fishery statistics, a peak production reached about 3,799.0 tonnes in 1975, subsequently declining to 8.7 tonnes in 1980. Maximum landings of zigzag scallop were reported to occur in 1981, when 8,845.3 t were harvested, followed again by a decrease in production subsequent years. In 1986, production reached about 1.9 t. Since then, landings have been insignificant, returning to the condition of minor by-catch product of shrimp fisheries, but no official statistics are available (Pezzuto and Borzone 1997). Considering the period over which the fishery operated, Pezzuto and Borzone (1997) suggest that there is a cycle of 3–5 years in scallop

1244

production. Recent surveys reported by Pezzuto et al. (1998) undertaken in the area where commercial scallop fisheries once took place, indicated that the yields were still low and the stocks are in a fragile situation. In nine cruises, yields ranged from 0 to a maximum of 35 scallops h^{-1}.

Scallop production in Brazil was completely directed to international markets, mainly as frozen muscle, generating an export value from 1973 to 1981 of US$ 6 million according to official statistics from the Ministry of Agriculture. The price of muscle varied from US$ 2.73 kg^{-1} in 1973 to US$ 6.55 kg^{-1} in 1981. The main importers were the United States (58%) and France (36%) (Pezzuto and Borzone 1997).

25.3.6 Future Prospects

Interest in culturing *Euvola ziczac* has been limited in Brazil compared to *Nodipecten nodosus*, a species with high interest in commercial culture. The low interest in *E. ziczac* is due to its smaller size and lower commercial value, along with a lack of seed supply and constraints to culture *E. ziczac* in suspension, which is the main system currently used by mollusc growers in Brazil. If aquaculture of *E. ziczac* is to be developed in Brazil, it will follow advances on culture of *N. nodosus* as a demand from growers is created in order to diversify their product, and if trends are oriented towards the use of bottom culture techniques.

Considering that most of scallop stocks display a pattern of fluctuation classified as spasmodic (Orensanz et al. 1991), characterised by irregular pulses of high abundance followed by periods of scarcity or collapse, it is possible that the *E. ziczac* beds off the southeastern coast of Brazil could re-establish in the future. Although stocks of *E. ziczac* have fallen below levels commercially exploitable, pulses of successful recruitment could occur, leading to population levels attained when scallop fisheries once operated in Brazil. It is necessary, however, to protect the natural grounds from the fishing fleet, along with regulations and adequate fishery management, otherwise a complete collapse of this resource will be unavoidable.

ACKNOWLEDGMENTS

Results on aquaculture of *Nodipecten nodosus* summarised in the present review were obtained under a M.Sc. Scholarship (CNPq – Conselho Nacional de Desenvolvimento Científico e Tecnológico, Brazil) (1992–1994) and a RHAE/PIBIO – CNPq Technology Development Fellowship (DTI) granted to GSR (1994–1996). Research has also been supported by European Union - Project INCO Scallop (1997–2000) and Canadian International Development Agency - Brazilian Mariculture Linkage Project (BMLP) (2000–2002). Compilation and writing of this article was undertaken under a Ph.D. scholarship (CNPq) granted to GSR. T. Waller and E. Rangel kindly provided information for this manuscript. We thank the personnel of the Laboratory for the Culture of Marine Molluscs (LCMM) - Department of Aquaculture - Universidade Federal de Santa Catarina, for the co-operation and assistance on experimental work.

REFERENCES

Abbott, R.T., 1974. American Seashells. Second edition. Van Nostrand Reinhold Company, NY. 663 p.

Abbott, R.T. and Dance, S.P., 1990. Compendium of Seashells. American Malacologists Inc., Melbourne, Florida. 411 p.

Ab'Sáber, A.N., 2001. Litoral do Brasil. Metalivros, São Paulo. 287 p.

Anonymous, 1975. Scallops e atum, novas opções de pesca. Revista Nacional da Pesca. São Paulo 16(143):5–6.

Anonymous, 1999. Almanaque Abril, 1999. CD Rom. Sixth Edition. Brasil. Editora Abril.

Alfonsi, C. and Pérez, J.E., 1998. Growth and survival in the scallop *Nodipecten nodosus* related to self-fertilization and shell colour. Bol. Inst. Oceanogr. Venezuela, Univ. Oriente 37(1–2):69–73.

Avelar, J.L., 2000. O cultivo de vieiras no Estado do Rio de Janeiro. Panorama da Aquicultura 62:41–47.

Basoa, E., Alfonsi, C., Perez, J.E. and Cequea, H., 2000. Karyotypes of the scallops *Euvola ziczac* and *Nodipecten nodosus* from the Golfo de Cariaco, Sucre State, Venezuela. Bol. Inst. Oceanogr. Venezuela, Univ. Oriente 39(1–2):49–54.

Bem, M.M., 1999. Efeito da adição de antibióticos no cultivo de larvas de *Nodipecten nodosus* (L. 1758) (Bivalvia: Pectinidae). Dissertação de Mestrado, Departamento de Aquicultura, Universidade Federal de Santa Catarina, Florianópolis (SC). 89 p.

Bem, M.M., Rupp, G.S., Pereira, A. and Poli, C.R., 2001a. Effect of antibiotics on larval survival and microbial levels during larval rearing of the tropical scallop *Nodipecten nodosus*. 13th International Pectinid Workshop. Universidad Católica del Norte, Coquimbo, Chile. Book of abstracts. pp. 114–116.

Bem, M.M., Rupp, G.S., Vandenberghe, J., Sorgeloos, P. and Ferreira, J., 2001b. Effects of marine bacteria on larval survival of the tropical scallop *Nodipecten nodosus* (Linnaeus, 1758) (Bivalvia : Pectinidea). 13th International Pectinid Workshop. Universidad Católica del Norte, Coquimbo, Chile. Book of abstracts. pp. 111–113.

Borzone, C.A., Pezzuto, P.R. and Marone, E., 1999. Oceanographic characteristics of a multi-specific fishing ground of the central south Brazil bight. Mar. Ecol. 20:132–146.

Brandini, F.P., 1990. Hydrography and characteristics of the phytoplankton in shelf and oceanic waters off southeastern Brazil during winter (July/August 1982) and summer (February/March 1984). Hydrobiologia 195:111–148.

BRASIL, 1976. Descrição sumária das principais realizações do Programa de Pesquisa e Desenvolvimento Pesqueiro do Brasil, PDP/1970/1976. Programa de Pesquisa e Desenvolvimento Pesqueiro do Brasil. Série Documentos Ocasionais 23. Brasília (DF).

BRASIL, 1990. Decreto no 99143, 12 março 1990. Diário Oficial da União. Brasília, 13 março 1990. Brasília, DF.

Bricelj, V.M. and Shumway, S., 1991. Physiology: Energy acquisition and utilization. In: S.E. Shumway (Ed.). Scallops: Biology, Ecology and Aquaculture. Developments in Aquaculture and Fisheries Science, Vol. 21. Elsevier, Amsterdam. pp. 305–376.

Briggs, J.C., 1995. Global Biogeography. Developments in Paleontology and Stratigraphy, Vol. 14. Elsevier. 452 p.

Castro, B.M. and Miranda, L.B., 1998. Physical oceanography of the western Atlantic Continental shelf located between 4° N and 34° S. In: A.R. Robinson and K.H. Brink (Eds.). The Global Coastal Ocean. The Sea, No. 11. pp. 209–251.

Coronado, M.C., Gonzales, P. and Perez, J.E., 1991. Genetic variation in Venezuelan molluscs: *Pecten ziczac* and *Lyropecten nodosus* (Pectinidae). Caribb. J. Sci. 27(1/2):71–74.

Cheng, T.C., 1981. Bivalves. In: N.A. Ratcliffe and A.F. Rowley (Eds.). Invertebrate Blood Cells. Vol. 1. Academic Press, London. pp. 223–300.

Dance, P.S., 1990. The Collectors Encyclopedia of Seashells. Zachary Kwintner Books, Singapore. 288 p.

Eisenberg, G.J.M., 1981. A Collector's Guide to Seashells of the World. McGraw-Hill Book Company, New York. 239 p.

Emilsson, I., 1961. The shelf and coastal waters of Southern Brazil. Bolm. Inst. Oceanogr., S. Paulo 11:101–112.

Félix-Pico, E.F., Villalejo-Fuerte, M., Tripp-Quezada, A. and Holguin-Quinones, O., 1999. Growth and survival of *Lyropecten subnodosus* (Sowerby, 1835) in suspended culture at the national marine park of Bahia de Loreto, B.C.S., México. 12th International Pectinid Workshop, Bergen, Norway. Book of abstracts. pp. 39–40.

Fretter, V. and Graham, A., 1964. Reproduction. In: K.M. Wilbur and C.M. Yonge (Eds.). Physiology of Mollusca. 1. Academic Press, New York. pp. 127–164.

Guo, X., Ford, S.E. and Zhang, F., 1999. Molluscan aquaculture in China. J. Shellfish Res. 18 (1):19–31.

Haas, F., 1953. Mollusks from Ilha Grande, Rio de Janeiro, Brazil. Fieldiana Zoology 34(20):203–209.

Hine, P.M., 1999. The inter-relationships of bivalve hemocytes. Fish Shellf. Immunol. 9:367–385.

Lodeiros, C.J. and Himmelman, J.H., 1994. Relations among environmental conditions and growth in the tropical scallop *Euvola* (*Pecten*) *ziczac* (L.) in suspended culture in the Golfo de Cariaco, Venezuela. Aquaculture 119(4):345–358.

Lodeiros, C.J. and Himmelman, J.H., 2000. Identification of factors affecting growth and survival of the tropical scallop *Euvola* (*Pecten*) *ziczac* in the Golfo de Cariaco, Venezuela. Aquaculture 182(1–2):91–114.

Lodeiros, C.J., Marín, B. and Prieto, A., 1999. Catálogo de moluscos marinos de las costas nororientales de Venezuela: Classe Bivalvia. Edición APUDONS. 109 p., 9 laminas.

Lodeiros, C.J., Rengel, J.J., Freites, L., Morales, F. and Himmelman, J.H., 1998. Growth and survival of the tropical scallop *Lyropecten* (*Nodipecten*) *nodosus* maintained in suspended culture at three depths. Aquaculture 165:41–50.

Luna-Gonzales, A., Maeda-Martinez, A.N., Sainz, J.C. and Ascencio-Valle, F., 2002. Comparative susceptibility of veliger larvae to four bivalve mollusks to a *Vibrio alginolyticus* strain. Dis. Aquat. Org. 49:221–226.

MacDonald, B.A. and Thompson, R.J., 1985a. Influence of temperature and food availability on the ecological energetics of the giant scallop *Placopecten magellanicus*. I. Growth rates of shell and somatic tissue. Mar. Ecol. Prog. Ser. 25:279–294.

MacDonald, B.A. and Thompson, R.J., 1985b. Influence of temperature and food availability on the ecological energetics of the giant scallop *Placopecten magellanicus*. II. Reproductive output and total production. Mar. Ecol. Prog. Ser. 25:295–303.

Maeda-Martinez, A., Reynoso-Granados, T., Monsalvo-Spencer, P., Sicard, M.T., Mazón-Suastegui, J.M., Hernandez, O., Segovia, E. and Morales, R., 1997. Suspension culture of catarina scallop *Argopecten ventricosus* (= *circularis*) (Sowerby II, 1842) in Bahia Magdalena, México, at different densities. Aquaculture 158:235–246.

Manuel, S., 1999. Spat settlement patterns of the scallops *Euvola ziczac* and *Argopecten gibbus* in Harrington Sound, Bermuda. Book of Abstracts: 12th International Pectinid Workshop, 12th Int. Pectinid Workshop, Bergen (Norway), 5–11 May 1999.

Manzoni, G.C., 1994. Aspectos da biologia de *Nodipecten nodosus* (Linnaeus, 1758) (Mollusca: Bivalvia), nos arredores da Ilha do Arvoredo (Santa Catarina – Brasil), com vista à utilização na aquicultura. Dissertação de Mestrado, Departamento de Aquicultura, Universidade Federal de Santa Catarina, Florianópolis (SC). 98 p.

Manzoni, G.C. and Banwart, J.P.F., 2000. Aspectos da biologia reprodutiva da vieira *Nodipecten nodosus*, cultivadas na enseada da Armação do Itapocoroy (26° 46' S - 48° 37' W) (Penha - SC). Anais XIII Semana Nacional da Oceanografia – 29 de Outubro a 03 de Novembro 1997. Itajai, Santa Catarina. pp. 537–539.

Manzoni, G.C. and Marenzi, A.W.C., 1997. Crescimento da viera *Nodipecten nodosus* (Linneaus, 1758) (Mollusca: Pectinidae) em cultivo experimental na enseada da armação do Itapocoroy (26° 46' S – 48° 37' W), Penha (SC). Anais X Semana Nacional da Oceanografia – 05 a 10 de Outubro 1997. Itajai, Santa Catarina. pp.178–180.

Manzoni, G.C., Poli, C.R. and Rupp, G.S., 1996. Período reproductivo del pectinido *Nodipecten nodosus* (Mollusca: Bivalvia) en los Alrededores de la Isla do Arvoredo (27° 17'S – 48° 22'W) – Santa Catarina – Brasil. In: A. Silva and G. Merino (Eds.). Acuicultura en Latinoamérica. Universidad Católica del Norte. Asociación Latinoamericana de Acuicultura, Coquimbo, Chile. pp. 197–201.

Matsuura, Y., 1986. Contribuição ao estudo da estrutura oceanográfica da região sudeste entre Cabo Frio (RJ) e Cabo de Santa Marta Grande (SC). Ciencia e Cultura 38(8):1439–1450.

Morais, C. and Kai, M., 1980.Considerações gerais sobre o manuseio e processamento de moluscos vieras. Bol. ITAL. Campinas. 17(3):253–273, jul./ set.

Narvarte, M.A., 1995. Spat collection and growth to commercial size of the tehuelche scallop *Aequipecten tehuelchus* (D'Orb.) in the San Matías Gulf, Patagonia, Argentina. J. World Aquacul. Soc. 26(1):59–64.

Orensanz, J.M., Parma, A.M. and Iribarne, O.O., 1991. Population dynamics and management of natural scallops. In: S.E. Shumway (Ed.). Scallops: Biology, Ecology and Aquaculture. Developments in Aquaculture and Fisheries Science, Vol. 21. Elsevier, Amsterdam. pp. 625–713.

Ostini, S. and Poli, C.R., 1990. A situação do cultivo de moluscos no Brasil. In: A. Hernandez (Ed.). Cultivo de Moluscos en America Latina. Memorias Segunda Reunión Grupo de Trabajo Tecnico. Chile. Centro International de Investigaciones para el Desarrollo CIID, Bogotá, Colombia. pp. 137–170.

Peña, J.B., 2001. Taxonomía, morfología, distribuición y hábitat de los pectínidos Iberoamericanos. In: A.N. Maeda-Martinez (Ed.). Los Moluscos Pectínidos de Iberoamérica: Ciencia y Acuicultura. Editorial LIMUSA, México. pp. 1–25.

1248

Pereira, A., 2000. Estudo da flora bacteriana associada a larvicultura de *Nodipecten nodosus* (Linnaeus, 1758) (Mollusca: Bivalvia). Dissertação de Mestrado, Departamento de Aquicultura, Unversidade Federal Santa Catarina, Florianópolis (SC).

Peres, S., 1981. Estudo do ciclo reprodutivo de *Pecten ziczac* (Linne, 1758) (Mollusca:Bivalvia). Dissertação de Mestrado, Instituto de Biociências, Unversidade de São Paulo, São Paulo (SP). 124 p.

Pezzuto, P.R. and Borzone, C.A., 1997. The scallop *Pecten ziczac* (Linnaeus, 1758) fishery in Brazil. J. Shellfish Res. 16(2):527–532.

Pezzuto, P.R., Borzone, C.A., Abrahão, R.L.B.E., Brandini, F. and Machado, E.C., 1998. Relatório técnico dos cruzeiros do projeto vieira. III. Cruzeiros iV (maio de 1996) a XiV (maio de 1997). Notas técnicas da FACIMAR 2:109–129.

Rebelo Neto, J.E., 1980. Considerações sobre as vieiras (*Pecten ziczac*) na região sudeste-sul do Brasil. Informe técnico Nº 4, Base de operações – PDP/SC. 16 p.

Rios, E.C., 1994. Seashells from Brazil. Fundação Cidade do Rio Grande. Museu Oceanográfico, Rio Grande, Brasil. 328 p.

Rupp, G.S., 1994. Obtenção de reprodutores, indução a desova, cultivo larval e pós larval de *Nodipecten nodosus* (Linnaeus, 1758) (Mollusca: Bivalvia). Dissertação de Mestrado, Departamento de Aquicultura, Unversidade Federal Santa Catarina, Florianópolis (SC). 132 p.

Rupp, G.S., 1996. Desenvolvimento de tecnologia de produção de sementes *Nodipecten nodosus* (Linnaeus, 1758) (BIVALVIA:PECTINIDAE). Relatório Parcial. Programa RHAE / PIBIO – UFSC. 71 p.

Rupp, G.S., 1997. Desenvolvimento de tecnologia de produção de sementes *Nodipecten nodosus* (Linnaeus, 1758) (BIVALVIA:PECTINIDAE). Relatório Final. Programa RHAE / PIBIO – UFSC. 57 p.

Rupp, G.S., 2001. O cultivo da vieira *Nodipecten nodosus* no Brasil: subsídios para o desenvolvimento sustentado. Panorama da Aquicultura 66:48–53.

Rupp, G.S., 2003. Influences of food availability and abiotic factors on growth and survival of the lion's paw scallop *Nodipecten nodosus* (Linnaeus, 1758) from a subtropical environment. PhD Thesis. Memorial University of Newfoundland, Canada. 203 p.

Rupp, G.S., Parsons, G.J., Thompson, R.J. and Bem M.M., 2003a. Growth and retrieval of postlarval lion's paw scallop *Nodipecten nodosus* in a subtropical environment: influence of environmental factors and size at deployment. 14th International Pectinid Workshop, 23–29 April 2003. St. Petersburg, Florida. Book of abstracts. pp. 20–21.

Rupp, G.S., Parsons, G.J., Thompson, R.J. and Bem M.M., 2003b. Growth and retrieval of postlarval lion's paw scallop *Nodipecten nodosus* in a subtropical environment: Influence of depth and density. 14th International Pectinid Workshop, 23–29 April 2003. St. Petersburg, Florida. Book of abstracts. pp. 65–66.

Rupp, G.S., Parsons, G.J., Thompson, R.T. and de Bem, M.M., 2004a. Effect of depth and stocking density on growth and retrieval of the postlarval lion's paw scallop, *Nodipecten nodosus* (Linnaeus, 1758). J. Shellfish Res. 23(2):473–482.

Rupp, G.S. and Poli, C.R., 1994. Spat production of the sea scallop *Nodipecten nodosus* (Linnaeus, 1758), in the hatchery: initial studies in Brazil. In: N.F. Bourne, B.L. Bunting, and L.D. Townsend (Eds.). Proceedings of the 9th International Pectinid Workshop, Nanaimo, B.C. Canada, April 22–27, 1993. Can. Tech. Rep. Fish. Aquat. Sci. 1994:91–96.

Rupp, G.S., Thompson, R.J. and Parsons G.J., 2004b. Influence of food supply on post-metamorphic growth and survival of hatchery-produced lion's paw scallop *Nodipecten nodosus* (Linnaeus, 1758). J. Shellfish Res. 23(1):5–13.

Rupp, G.S., Vélez, A., Bem, M.M. and Poli, C.R., 1997. Effect of temperature on conditioning and spawning of the tropical scallop *Nodipecten nodosus* (Linnaeus, 1758). Eleventh International Pectinid Workshop, 10–15 April 1997, La Paz, BCS, Mexico. Book of abstracts. E.F. Félix-Pico (Ed.). Centro Interdiscipinario de Ciencias Marinas, La Paz, México. Abril de 1997. pp. 132–133.

Uriarte, I., Rupp, G. and Abarca, A., 2001. Producción de juveniles de pectínidos Iberoamericanos bajo condiciones controladas. In: A.N. Maeda-Martinez (Ed.). Los Moluscos Pectínidos de Iberoamérica: Ciencia y Acuicultura. Editorial LIMUSA, México. pp. 147–171.

Sarkis, S., 1995. Scallop culture in Bermuda: A saga. In: J. Barret, J-C. Dao and P. Lubet (Eds.). Fisheries, Biology and Aquaculture of Pectinids. 8th Int. Pectinid Workshop, Cherbourg (France), 22–29 May 1991. Actes Colloq. IFREMER 17:115–122.

Sebens, K.P., 1982. The limits to indeterminate growth: an optimal size model applied to passive suspension feeders. Ecology 63:209–222.

Smith, J.T., 1991. Cenozoic giant pectinids from California and the Tertiary Caribbean Province: *Lyropecten*, "*Macrochlamis*," *Vertipecten*, and *Nodipecten* species. U.S. Geological Survey Professional Paper 1391:136 p.

Sunyé, P.S., and Servain, J., 1998. Effects of seasonal variations in meteorology and oceanography on the Brazilian sardine fishery. Fish. Oceanogr. 7:89–100.

Thompson, F.L., Li, Y., Gomez-Gil, B., Thompson, C.C., Hoste, B., Vandemeulebroecke, K., Rupp, G.S., Pereira, A., De Bem, M.M., Sorgeloos, P. and Swings, J., 2003. *Vibrio neptunius* sp. nov., *Vibrio brasiliensis* sp. nov. and *Vibrio xuii* sp. nov., isolated from the marine aquaculture environment (bivalves, fish, rotifers and shrimps). Int. J. Syst. Evol. Microbiol. 53:245–252.

Vargas-Albores, F. and Barraco, M.A., 2001. Mecanismos de defensa de los moluscos bivalvos, com énfasis in pectínidos. In: A.N. Maeda-Martinez (Ed.). Los Moluscos Pectínidos de Iberoamérica: Ciencia y Acuicultura. Editorial LIMUSA, México. pp. 127–146.

Vélez, A., Alifa, E. and Azuaje, O., 1990. Induction of spawning by temperature and serotonin in the hermaphroditic tropical scallop, *Pecten ziczac*. Aquaculture 84(3–4):307–313.

Vélez, A., Alifa, E. and Freites, L., 1993. Inducción de la reproducción en la vieira *Pecten ziczac*. I. Maduración y desove. Carib. J. Sci. 29:209–213.

Vélez, A. and Freites, L., 1993. Cultivo de semillas de *Pecten ziczac* (L), bajo condiciones ambientales controladas ("Hatchery"). Mem. IV Congr. Latinoam. Ciencias del Mar. Serie Ocasional No 2. Universidad Católica del Norte, Facultad de Ciencias del Mar, Coquimbo, Chile. pp. 311–317.

Vélez, A., Freites, L., Himmelman, J.H., Senior, W. and Marin, N., 1995. Growth of the tropical scallop, *Euvola* (*Pecten*) *ziczac*, in bottom and suspended culture in the Golfo de Cariaco, Venezuela. Aquaculture 136(3–4):257–276.

Vélez, A. and Lodeiros, C.J., 1990. Cultivo de moluscos en Venezuela. In: A. Hernandez, (Ed.). Cultivo de Moluscos en America Latina. Memorias Segunda Reunión Grupo de Trabajo Tecnico. Chile. Centro International de Investigaciones para el Desarrollo CIID, Bogotá, Colombia. pp. 345–368.

Vélez, A., Morales, F. and Jordan, N., 1997. Growth and survival of the tropical scallop *Nodipecten nodosus* (Linnaeus, 1758) in suspended culture at two areas of the eastern coast of Venezuela.

1250

Eleventh International Pectinid Workshop, 10–15 April 1997, La Paz, BCS, Mexico. Book of abstracts. E.F. Félix-Pico (Ed.). Centro Interdiscipinario de Ciencias Marinas, La Paz, México. Abril de 1997. pp. 33–34.

Vélez, A., Sotillo, F. and Pérez, J., 1987. Variación estacional de la composición química de los pectinídos *Pecten ziczac* y *Lyropecten nodosus*. Bol. Inst. Oceanog. Univ. Oriente 26(1:2):67–72.

Waller, T.R., 1991. Evolutionary relationships among commercial scallops (Mollusca: Bivalvia: Pectinidae). In: S.E. Shumway (Ed.). Scallops: Biology, Ecology and Aquaculture. Developments in Aquaculture and Fisheries Science, Vol. 21. Elsevier, Amsterdam. pp. 1–73.

AUTHORS ADDRESSES

Guilherme S. Rupp - Department of Biology, Memorial University, St. John's, Newfoundland, Canada A1B 3X9 [Current Address: EPAGRI, Centro de Desenvolvimento em Aquicultura e Pesca, Rod. Admar Gonzaga 1188, Itacorubi, P. O. Box 502, Florianópolis, SC, 88034–901, Brazil] (E-mail: rupp@epagri.rct-sc.br)

G. Jay Parsons - Marine Institute, Memorial University, P. O. Box 4920, St. John's, Newfoundland, Canada A1C 5R3 [Current Address: Fisheries and Oceans Canada, Aquaculture Science Branch, 200 Kent St., Ottawa, Ontario, Canada K1A 0E6] (E-mail: ParsonsJa@dfo-mpo.gc.ca)

Scallops: Biology, Ecology and Aquaculture
S.E. Shumway and G.J. Parsons (Editors)

1251

Chapter 26

Argentina

Néstor F. Ciocco, Mario L. Lasta, Maite Narvarte, Claudia Bremec,
Eugenia Bogazzi, Juan Valero and J.M. (Lobo) Orensanz

26.1 INTRODUCTION

When the first edition of this book was published over a decade ago, the only pectinid species commercially exploited in the southwestern Atlantic was the Tehuelche scallop, *Aequipecten tehuelchus* (d'Orbigny, 1846) (Orensanz et al. 1991). In the intervening time an industrial fishery developed for the Patagonian scallop, *Zygochlamys patagonica* (King & Broderip, 1832), whose potential value was highlighted in the first edition.

The fisheries supported by these two species are radically different from each other. The Tehuelche scallop is the target of small inshore fisheries in the gulfs of northern Patagonia, and involves dredging and commercial diving. In spite of the small volumes landed, these fisheries are of considerable significance for the local economies. The Patagonic scallop fishery, by contrast, is an industrial operation conducted by three or four factory trawlers that process the catch at sea. Catches in the order of 50,000 tons yr^{-1}, now rank this species among the most important scallop fisheries in the World.

Below we review knowledge about these two species, their fisheries, and experimental results that may have significance in the future development of aquaculture.

26.2 THE TEHUELCHE SCALLOP, *Aequipecten tehuelchus*

26.2.1 Biology

The Tehuelche scallop, *Aequipecten tehuelchus,* is distributed in the southwest Atlantic, from Río de Janeiro (23° SL, Brazil) to northern San Jorge Gulf (45° SL, Argentine Patagonia), inhabiting sandy bottoms at depths below 130 m. This is one of the most studied bivalves from that region. Its anatomy has been minutely described (Ciocco 1992a, b, 1995a, 1998). The shell has 14–19 conspicuous ribs (Fig. 26.1).

Main annual growth rings (corresponding to periods of slow growth) are marked in late spring/early summer, and are related to spawning; winter rings are always weaker (Orensanz et al. 1985). Rings marked during the first year of life are often inconspicuous. Growth rings apparently facilitate settlement of the spionid polychaete worm *Polydora websteri* (Orensanz 1986). Ciocco (1990) confirmed that *Polydora* infection depresses growth rate, and showed that the intensity of the infection increases with age. Slow growth rate favours settlement of *Polydora* near the border of the shell; as a result, growth

Figure 26.1. Left: *Zygochlamys patagonica*; right: *Aequipecten tehuelchus.*

rings tend to be 'highlighted' by shell-borers (Orensanz et al. 1991; Fig. 26.3). Maximum reported shell height is 102 mm, but specimens larger than 90 mm are uncommon. Growth rings are also readable in the ligament (Astort and Borzone 1980). Maximum age recorded from growth rings is 11 years (San José Gulf), but scallops older than 8 years are rare.

The Tehuelche scallop is a simultaneous hermaphrodite (Christiansen and Olivier 1971), and is iteroparous. First spawning occurs at the age of one year, although energy allocation to reproduction is proportionally smaller in first-time spawners. Gonadal development is controlled by temperature (Orensanz 1986), with spawning occurring when temperature reaches 14–15°C (de Vido de Mattio 1984; Ciocco 1985). In both the San Matías and José Gulfs (Fig. 26.2) there is a partial spawning pulse in spring; pre- and post-spawning stages coexist over the summer. Complete spawning occurs by late summer-early autumn, and is followed by rapid phagocytosis of the residual material; oocyte proliferation occurs over the entire winter (Christiansen et al. 1974; Lasta and Calvo 1978; Narvarte and Kroeck 2000). Fecundity (estimated by stereometric techniques) ranges from 2 to 17 million eggs over a size-range of 35–90 mm shell height (Christiansen et al. 1973, 1974). Energy for gonadal development appears to come directly from feeding (de Vido de Mattio 1984; Orensanz 1986), not from reserves stored in the adductor muscle or other tissues.

Size of the larvae at settlement is 162–201 µm (Orensanz 1976). Settlement occurs primarily on shell-hash in San Matías Gulf, and on the macroalga *Ulva* in San José Gulf. Two peaks of settlement may occur in San Matías, matching the two spawning pulses. Only one major settlement season has been observed, however, in San José Gulf, where the late summer spawning never resulted in important settlement (Ruzzante and Zaixso 1985; Ciocco 1985; Orensanz 1986). During the first 1.5 yr of life growth (shell height) is quasi-linear (constant growth rate), with a seasonal component over imposed. Growth in weight, also seasonal, is related to the reproductive cycle (and thus to temperature) and food availability. Growth rate decelerates (the growth curve becomes asymptotic) past the second winter. Figure 26.3 shows growth pattern in size (height) of three cohorts in one site (top) and of a single cohort in three sites (bottom). Parameters of the von Bertalanffy model (H_∞ and k) have been estimated for several data sets (Table 26.1).

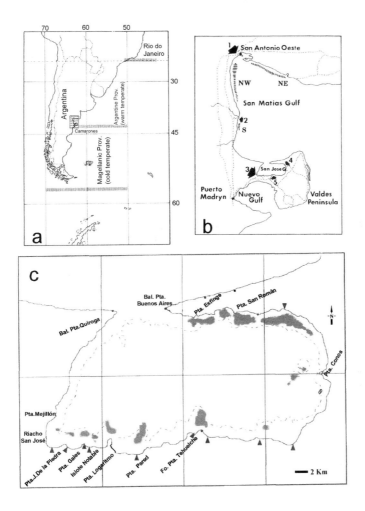

Figure 26.2. *Aequipecten tehuelchus.* a: Geographic distribution (dotted area expanded in b). b: North Patagonic Gulfs; arrows indicate main scallop landing sites (1: San Antonio, 2: Puerto Lobos, 3: El Riacho, 4: Los Abanicos, 5: Tehuelche); dashed lines indicate main roads; shaded areas (NW, NE and S) indicate fishing grounds in San Matías Gulf. c: San José Gulf; shaded areas indicate scallops grounds; black triangles: landing sites (depth in m). (Modified from Orensanz 1986 and Ciocco et al. 1998).

1254

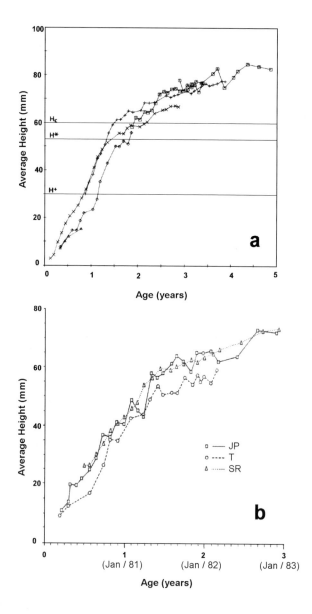

Figure 26.3. *Aquipecten tehuelchus.* Cohort-specific growth patterns in San José Gulf, assessed with periodic samples. a: Growth of six year-classes during a 3-year study, San Román Ground; horizontal lines: average sizes at which the cohorts reach full detection by the quantitative sampling technique (H+), to size-dependent mortality (H*), and to commercial size (Hc). Symbols for the year-classes are (O) 1972, (□) 1973, (+) 1974, (X) 1975, (◊) 1976, (Δ) 1977. b: Growth of the 1980 year-class in three grounds (JP: Juan de la Piedra; T: Tehuelche; SR: San Román) (From Orensanz 1986 and Ciocco 1991).

Table 26.1

Aequipecten tehuelchus. Parameters estimated for the von Bertalanffy growth model by different authors.

Region	Ground	H_∞ (mm)	$K \, yr^{-1}$	t_o (yr)	Source
San Matías Gulf	Bajo Oliveira	85.16	0.40	-0.88	Olivier and Capítoli 1980
San José Gulf	Gales	75.24	0.70	-0.03	Ciocco 1985
"	Logaritmo	82.12	0.57	-0.10	"
"	San Román	82.86	0.75	0.19	Orensanz 1986
"	San Román	77.17	0.98	-0.12	Ciocco 1991
"	J. de la Piedra	79.19	0.86	-0.07	"
"	Tehuelche	71.68	0.84	-0.09	"

Figure 26.4 shows growth in weight of different body parts, over two years (Nov. 1974-Nov. 1976) and for 3 year-classes (1973–1975), in San Román Ground (San José Gulf). Adductor muscle grew rapidly during the autumn, and the gonad during late-winter/early-spring of 1974, suggesting that energy from food was sequentially diverted to reserve storage (autumn) and gonadal maturation (spring). During 1976 (a year in which density was high) there was not a period of fast muscle growth. Other seasonal changes seemed to occur in phase (growth rates of the different body fractions were not negatively correlated). The weight of muscle and other meats drop synchronously with the gonad after spawning (de Vido de Mattio 1984; Orensanz 1986). This drop occurred for all year-classes and for all the spawning peaks detected, and is consistent with a rapid decline in content of glycogen (de Vido de Mattio 1984). Meats lose commercial value after spawning. Relative yield in muscle weight varies with season, area and density. In the San Roman ground it was 20% of total weight during the winter of 1975 (when density was low), but dropped to 15% one year later, after mean biomass increased to above 1 kg m^{-2} (Fig. 26.4).

Diet consists mainly of benthic microscopic algae, probably resuspended by tide or wind generated near bottom currents or turbulence (Vernet 1977). Only on one occasion (January), during one year of monthly sampling, did planktonic algae (largely the dinoflagellate *Prorocentrum micans)* constitute the main component of the diet. This was reflected in an increase of 22:6ω-3 fatty acid (a major component of dinoflagellate lipids) in the gonad and meats during that particular month (Pollero et al. 1979).

The Tehuelche scallop may do quick and brief swimming movements as escape responses to predators or other disturbances. Frequency of byssal attachment to the substrate vanishes gradually with age, but the capacity to form a byssus is not lost in the adults (Ciocco 1992a). Attachment rate of adult scallops (age 2 + and older) in the field increases with current velocity, from 65% at 4–5 m min^{-1} to 90% at 7–8 m min^{-1} (Ciocco et al. 1983).

1256

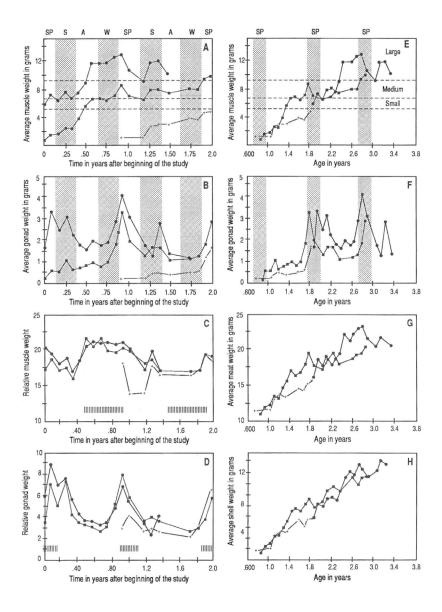

Figure 26.4. *Aquipecten tehuelchus*. San Román Ground, San José Gulf. Growth of adductor muscle (a, c, e), gonad (b, d, f), meats (g) and shell (h) between November, 1974, and November, 1976. Year-classes: (o) 1973, (□) 1974, (+) 1975. Relative weights (c, d) are expressed as percentages of total weight. Horizontal dashed lines in a and e separate commercial standards for muscle size (large, medium, small). Shaded bands in a indicate summers (S) and winters (W). Shaded areas in e and horizontal bars in d indicate springs (SP). Horizontal bars in c indicate fishing seasons (From Orensanz et al. 1991).

26.2.2 Population Dynamics

The stock from San José Gulf (the only one on which studies of population dynamics have been conducted) is structured as a metapopulation whose component sub-populations ('grounds') are inter-connected by larval dispersal (Fig. 26.2c). Location of grounds has been consistent through time (some have been known for more than 30 years). In some cases this may reflect the presence of eddies (produced by wind or tides, and coastal accidents) that function as areas of larval retention. Each ground is exposed to a distinct regime of food availability, predation and disturbance. Inter-populational gradients in growth rate correspond largely to gradients in nitrogen availability, primary production and carrying capacity (all decreasing from northwest to southeast; Orensanz 1986; Ciocco 1991, 1992c). Intra-populational differences in individual growth rate have been attributed to density-dependence. Fluctuations in abundance in the different grounds are largely out of phase, although some degree of synchrony points to large-scale modulation of recruitment. Settlement, for example, was strong everywhere in 1975 and insignificant in all the grounds in 1976. Spatial distribution within the grounds is contiguous; most individual scallops experience high concentration in their neighbourhoods.

Stranding, induced by storms and strong winds, seems to be the main source of mortality (Orensanz 1986; Orensanz et al. 1991). Sporadically it can reach catastrophic proportions, wiping out entire grounds. Stranding mortality seems to increase with size, perhaps because byssal attachment gets weaker while epibiotic load increases the flow-induced forces encountered by the scallop. Pulses of mortality were observed following minima in condition index: late-winter/early spring (minima of food availability and temperature) and after spawning. Main predators include a starfish (*Cosmasterias lurida*), a volutid snail (*Odontocymbiola magellanica*), an octopus *(Enteroctopus megalocyathus)*, and a ratfish (*Callorhynchus callorhynchus*) (Orensanz 1986, Di Giácomo et al. 1994; Ciocco and Orensanz 2001).

The San Roman Ground population (Fig. 26.2c) appears to be regulated by a combination of density-dependent growth and settlement, and size-dependent mortality (Orensanz 1986). Reproductive output was also depressed as a result of density-dependent effects. Natural mortality is very low within the size range [30–52 mm], and increases with size above 52–54 mm height. For that reason, mortality rate was higher in cohorts which grew faster: mean instantaneous per-capita mortality rate (above a mean size of 52 mm) was 1.98 yr^{-1}, 0.99 yr^{-1} and 0.62 yr^{-1} for the year classes 1973, 1974 and 1975 (respectively) which grew under increasingly high density conditions (Orensanz 1986). Ciocco (1996) reported significant geographic and temporal variation in "in situ" natural mortality, estimated from a combination of data on relative clucker abundance and mean clucker "life" (Tehuelche Ground: 0.21–1.56 yr^{-1}; San Román Ground: 0.27–1.22 yr^{-1}).

Average biomass and production are locally constrained by carrying capacity. In San Román, biomass reached an average of 1.3 kg m^{-2} during peaks of abundance, and a maximum (for individual patches) of 1.5 kg m^{-2}. Compensatory effects became detectable above 1 kg m^{-2}. Age at recruitment (60 mm, legal size) ranged from 1.5 to 2.2 yr, depending upon density conditions. Production in the core bed of the San Román ground was similar for two years (1975, 1976) that differed in average biomass, reflecting compensation. Annual production of meat was 382 g m^{-2} yr^{-1}, very close to the 379 g m^{-2} yr^{-1} that Charpy-Roubaud et al. (1982) calculated that could be sustained by primary production. In an independent study conducted on the same ground a few years later, Ciocco (1991) estimated values remarkably similar: 359 g m^{-2} yr^{-1} and 368 g m^{-2} yr^{-1}, respectively. Corresponding values estimated in the same study for the Tehuelche Ground were, as expected, lower (respectively 210 g m^{-2} yr^{-1} and 238 g m^{-2} yr^{-1}).

Year-class strength and average biomass showed a quasi 5-year cycle over a 20-year period in San Román Ground (1969–1990) for which information is available (Orensanz 1986; Ciocco and Aloia 1991). The cycle was produced by five evenly spaced, exceptionally strong year classes (1970, 1975, 1980, 1984, 1987). Abundance of year classes varies among populations in the San José Gulf. While the San Román sub-population (on the north coast) was dominated by one strong year class, the Tehuelche sub-population (on the south coast) has shown, over the years, a multi-cohort composition. Because of this the Tehuelche area has been suggested as a reasonable location for a reproductive reserve (Orensanz et al. 1997).

26.2.3 Fisheries

26.2.3.1 The inshore dredge fishery of San Matías Gulf (Rio Negro Province)

The main fishing areas of San Matías Gulf (Fig. 26.2b) are located on the northwest (Bajo Oliveira-El Sótano) and north-northeast (Orengo beach and surroundings). Most of the beds in these fishing grounds occur at depths greater than 20 m, and for that reason are largely inaccessible to commercial divers. Starting in 1995 sporadic harvests have been made by commercial divers near Puerto Lobos, on the S coast, and near El Fuerte (NW) of San Matías Gulf (Fig. 26.2b), at depths of 12–15 m. On that year 10 boats fished over 100 tons.

The dredge fishery for scallops, which is centred in San Antonio Oeste (Río Negro Province), boomed in 1969 due to favourable conditions in the international market. Boats that arrived in San Antonio from other areas of the country numbered 18 by 1972. Catch peaked in 1970 at 4,530 MT (Fig. 26.6a). Olivier and Capitoli (1980) reported a catch of 13,904 MT for the same year, but their data correspond to the unsorted catch; catch figures reported here correspond to scallops only. The CPUE dropped rapidly during the 1971 fishing season, following three years of intense harvesting.

The fleet was composed of stern-trawlers, 12–22 m long (Olivier et al. 1970). The Argentine scallop dredge (Fig. 26.5), developed in the thirties for mussel fishing, is built of iron pipe, 3 cm mesh chicken wire, and strips of spare tires; a smaller and lighter design was introduced during the 2000 fishing season (Morsan et al. 2000; Narvarte and Morsan 2000). The catch was not sorted on board during the 1968–1972 fishing period. The landed catch was composed of scallops (normally 30–40% of gross yield), other benthic invertebrates, and shell hash. Efficiency of the larger dredge has been estimated at 11–16% (Vacas et al. 1984) and 15–21% (Iribarne et al. 1991); it is completely unselective

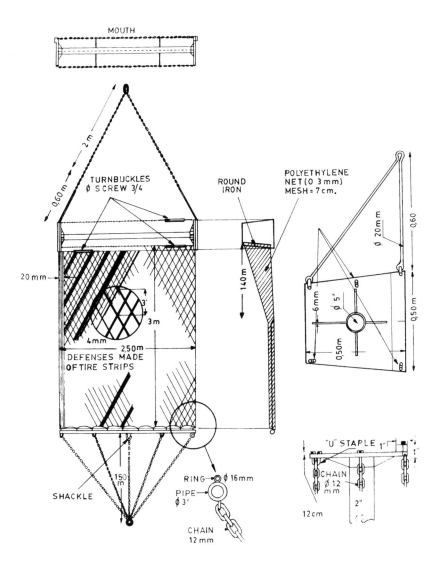

Figure 26.5. Dredge design traditionally used in the Argentine Tehuelche scallop fishery (From Orensanz et al. 1991).

due to clogging of the bag. No studies on the selectivity of the new design were carried out yet. Since 1990, including the last 2000 season, the catch was sorted on board.

Studies were conducted for the first time during the summer of 1970 (Olivier et al. 1970), and were followed by surveys of the fishing grounds in 1971–1973 (Olivier 1972; Olivier et al. 1971, 1973, 1974; Olivier and Capítoli 1980). Management recommendations during the first period included selection of the catch onboard, a

1260

minimum size (height) of 60 mm, and the closure of the fishery at the beginning of the spawning season (spring). The provincial fishery administration was, however, unable to enforce all of these measures. The fishery was closed at the end of the 1971 season, reopened briefly in 1972 (1,700 MT landed over three months), closed again, and reopened for one month in 1975. The remnants of the 1970 year-class (the last successful one) were wiped out by fishing and a storm (Orensanz et al. 1982). Afterwards, the fishery remained closed for six consecutive years. Its 'first pulse' was over.

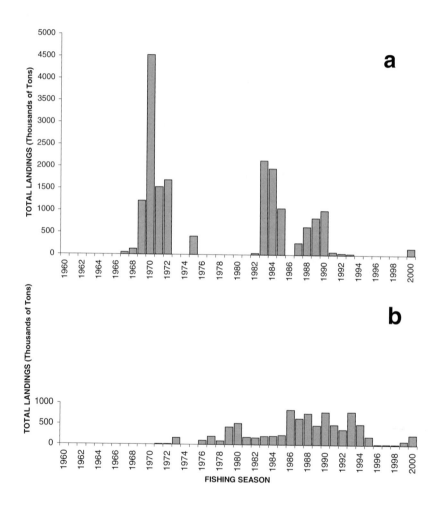

Figure 26.6. *Aequipecten tehuelchus*. Landings by year. a: San Matías Gulf dredge fishery; b: San José Gulf diving fishery (modified from Ciocco et al. 1998).

The 1980 spawning season was followed by strong settlement over the traditional fishing grounds (NW of the gulf; Fig. 26.2b), and the 'second pulse' of the fishery started in 1982, with a small-scale experimental fishing program. A management program was implemented in 1983 that included regulation of fishing effort (fleet size). Catch information was improved by cross checking data from the fleet and the buyers. Landings rose to 2,336 MT in 1983 (Fig. 26.6a), about half the peak of the 'first pulse' (1970). New fishing beds were discovered in the N-NE part of the gulf (Fig. 26.2b), allowing rotation of areas in 1984. The traditional NW ground was closed, and the N-NE ground (populated by a mixture of year classes) opened. Effort regulations were relaxed in 1984, leading to a rapid increase of fleet size. Although both grounds were opened in 1985, total landings declined (Fig. 26.6a). The NW ground was depleted; no significant settlement occurred until 1999, almost 20 years later.

Management again became conservative starting in 1986, in the face of a new collapse of the fishery. Significant settlement in the N-NE area in 1986 supported the fishery through 1990, this becoming the third pulse of the fishery. Only 4–6 boats operated, and catches were comparatively small (Figure 26.6a). An area-rotation program was effectively implemented during those four years by the provincial fisheries administration to protect areas where juvenile scallops predominated (Vacas et al. 1984). Yet, after the pulse was over following the 1990 season, no signs of recovery were detected in the N-NE grounds until 2000.

Landings were insignificant during the 1990s (Figure 26.6a) due to poor recruitment. In 1991–1992 scallops were landed only as by-catch (Morsan and Narvarte 2000). In 1992 two trawlers fished in the NE area, and during 1993 and 1994 two diving teams fished in shallow waters at the NW sector (Ciocco et al. 1998). The fishery was completely closed from late 1994 to 1999, except for illegal diving operations in 1995. In October, 2000, the fishery administration opened the Bajo Oliveira bed (NW fishing ground) during a 45-day season and with a 350-ton quota; participants, only artisanal boats, included both dredgers (6 boats) and commercial divers (5 boats) (Morsan et al. 2000; Narvarte and Morsan 2000).

In summary, just a few year-classes (1970, 1980, 1986 and 1999) have supported the fishery over the last 30 years. The fisheries authority considers that the only conceivable strategy to manage this spasmodic stock is an opportunistic allocation of effort (in space and time), in order to maximise the yield of unpredictable settlement pulses, with specific measures based on pre-season surveys and yield-per-recruit analysis. Surveys may allow the forecast of recruitment pulses up to two years in advance, facilitating a better spatial allocation of effort.

26.2.3.2 The commercial diving fishery of San José Gulf (Chubut Province)

While the San Matías fishery boomed in 1968–1970, some sporadic dredging was also done in the San José Gulf. Good fishing grounds were located by 1970, but the Chubut Province fishing authority, having witnessed the collapse of the fishery in neighbouring San Matías Gulf (Rio Negro Province), banned the use of dredges. Preliminary diving operations started in 1970–1971, and by 1972 the first efficient boat geared for commercial diving started to operate (Picallo 1979; Ciocco et al. 1998). Commercial diving is possible in this area because scallop beds are generally found at depths below 20 m.

The development of the fishery has been chronicled by Ciocco (1995b) and Ciocco et al. (1998); landings are summarised in Fig. 26.6b. Activity between 1971 and 1973 was incipient and irregular. The first commercial diving teams used inner engine boats, with 4–6 divers each. CPUE was high, in the order of 600 kg diver^{-1} d^{-1}, and there were no regulations. The fishery was closed shortly in 1972–1973, and then in 1974–1975, in both cases as preventive measures while a management plan was put together. Surveys conducted in 1973/1975 showed high scallop abundance (Olivier et al. 1974; Picallo 1980).

The fishery was opened every year during the 20-year period 1975–1995, to commercial divers only. Initial regulations (Olivier et al. 1974) included: i) a closed season, ii) ad-hoc area closures for the protection of sub-legals, iii) a size limit (60 mm), iv) catch only by diving (ban on dredging), and v) quotas. These regulations have continued with little change. The 60 mm legal size, originally established to allow one spawning before harvest, turned out to be a reasonable choice for other reasons: (1) maximum biomass of a cohort is reached around the age of 1.5 yr, shortly before 50% of the individuals reach legal size, and (2) average adductor muscle weight entered the smallest marketable size (in the export market) about two weeks after scallops (on average) reached legal size (Orensanz 1986). Diving fishermen soon developed a search image for size, and selected large scallops for economic reasons, whether a minimum size was enforced or not. Given these conditions, and the fact that harvestable beds were usually dominated by one strong year class, growth overfishing was not a main management concern in this fishery. Divers selected beds with the highest densities (40–60 scallops m^{-2}). Threshold density was around 20 scallops m^{-2}. Since growth and recruitment rates are thought to be density-dependent, it was claimed that the diving fishery had the potential for self regulation (Orensanz 1986).

The fishery consolidated between 1976 and 1985. Outboard motor boats (ca. 8-m long, with 35–110 HP motors) became the standard; the fleet consisted of 5–6 teams (one team per boat) fishing during each season. Effort concentrated on the S-SW of the gulf, and catch ranged from 100 to 500 MT yr^{-1} (Fig. 26.6b). CPUE declined from 600–800 kg diver^{-1} d^{-1} between 1976 and 1980, to 300–400 kg diver^{-1} d^{-1} between 1981 and 1985. After 1981 fishing was affected by red tide events during the warm season. In 1986 the fleet operated mostly on the north coast, and to a lesser extent to the east of the Gulf. The number of boats grew to 12–30, while CPUE declined from ca. 600 kg diver^{-1} d^{-1} in 1986 to less than 150 kg diver^{-1} d^{-1} in 1995 (a historical minimum for this fishery); catch, which fluctuated between 370 and 850 MT yr^{-1} during the period 1985–1994, dropped to 100 MT in 1995.

The 1995 fishing season was closed in advance due to an excess of sub-legal size scallops in the landings. Subsequently the fishery was closed for three years (1996–1998) due to low abundance (Ciocco et al. 1996, 1997). Several factors led to the collapse of the fishery in 1995: i) a domestic market developed for sub-legal size scallops, ii) the efficiency of the fishing units improved, iii) due to economic hardship in the country divers fished at progressively lower density thresholds, and risked fishing deeper beds, iv) fishing effort increased in the absence of a limited-entry program, v) lack of monitoring and enforcement. Removal of previous reproductive refugia (low density and deep beds, sub-legal size scallops) probably led to poor settlement. The fishery was reopened partially in1999 and normally in 2000 (both seasons with quota; around 100 MT fished in 1999 and 240 MT in 2000; Fig. 26.6b).

26.2.4 Aquaculture

26.2.4.1 Spat collection

Except for some trials in San José Gulf (Zaixso 1980; Zaixso and Spindola 1981; Zaixso and Toyos de Guerrero 1982), studies on spat collection in Argentina have been conducted mostly in San Matías Gulf (Río Negro Province; Fig. 26.2b), where experiments began in 1980. These were continued over twelve experimental seasons (1980–1985, 1989–1993 and 1996–1999), yielding results relevant for the development of culture techniques for *Aequipecten tehuelchus*. These are outlined below.

26.2.4.1.1 Vertical distribution of settlement

A significant correlation was observed between settlement on spat collectors and depth. Abundance of postlarvae per collector was highest (1973 'seed'/collector) in the collector closest to the bottom (1 m), and declined progressively towards the surface (Narvarte 1995) (Fig. 26.8). The settlement pattern observed in San José Gulf (Fig. 26.1) was less clear: in one experiment settlement was highest near the bottom, but in other occasions no correlation with depth was observed (Ciocco and Monsalve 1999a, b), or even the reverse pattern was observed (Ruzzante and Zaixso 1985).

26.2.4.1.2 Temporal variation in settlement

In San Matías Gulf the maximum intensity of larval settlement occurred in January-February in 9 out of 12 experimental seasons. In the other three seasons the number of settlers per collector reached its maximum after February (Orensanz et al. 1991, Narvarte 1995, 1998). Settlement on artificial substrates varied considerably, seasons with successful spat collection alternating with others in which it was virtually absent (Table 26.2).

26.2.4.1.3 Type of collector

Four types of collectors were tried in San Matías Gulf on longline system (Fig. 26.7): a) Japanese "ren", b) branches of "jarilla" (*Larrea divaricata*, a shrub common in the region) packed in onion bags (6 mm mesh), c) Japanese Netlon collectors with an external bag (2 mm mesh), and d) onion bags packed into external bags with a finer mesh (2 mm). With the first type the average number of postlarvae collected varied between 1 and 33

Table 26.2

Aequipecten tehuelchus. Interannual variability of settlement on spat collectors, San Matías Gulf.

Study periods	Successful seasons	Seasons with no settlement
1980–1985 [a]	1980–1981; 1982–1983	1984–1985
1989–1993	1989–1990; 1991–1992	1990–1991; 1992–1993
1996–1999	1996–1997; 1998–1999	1997–1998; 1999–2000

(a) from Orensanz et al. (1991)

per 100 mm^2 (Orensanz et al. 1991). These values are hardly comparable with those obtained in nets. The number of settlers per Netlon collector was far higher than in the other two types of net-mesh collectors, although collectors with an external fine mesh gave yields adequate for a commercial operation (Narvarte 1998, 2001). Highest average number of juveniles per "jarilla" collector was 146.4 (SE = 45.7), while the average for Netlon collectors was 589.5 (SE = 86.8).

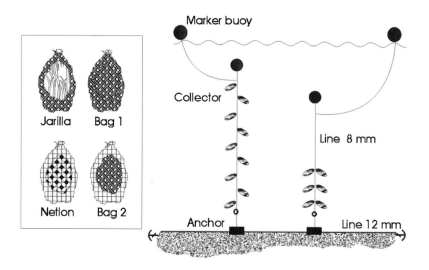

Figure 26.7. *Aequipecten tehuelchus.* Longline spat-collection system used at different sites in San Matías Gulf (schematic, collector types described in the text).

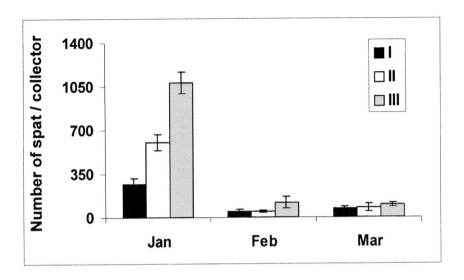

Figure 26.8. *Aequipecten tehuelchus.* Average number (± SE) of postlarvae/collector obtained at different depths (expressed as distance from the bottom: I = 3 m, II = 2 m and III = 1 m) during the 1991–1992 season.

26.2.4.1.4 Location

Experiences in San Matías Gulf were conducted at three different locations (Narvarte 2001): Playa Orengo (NE), El Sótano (NW) and El Fuerte (NW) (Fig. 26.2b). The number of settlers obtained in El Sótano was significantly higher than values for the other sites (Table 26.3). Three sets of lines were placed at each of the sites with the purpose of comparing spat collection at a smaller spatial scale (300 m on average). At this scale no significant differences were observed. Ciocco and Monsalve (1999a, b), in experiments conducted to assess larval availability in San José Gulf, also observed differences between sites. Spat density in Tehuelche (S) and San Román (N) was higher than in Mendioroz (W), although Fracasso (SE) was the location with the highest observed levels (Table 26.3).

26.2.4.1.5 Influence of depth, type of collector and location on the average size of the 'seed'

Significant differences in average seed size were observed between depths, collector types, and collection locations (Narvarte 2001). Differences associated with the first two factors may be related to the differences in the loss-rate of attached seed, while differences between sites may be attributable to the settlement of different larval cohorts.

Table 26.3

Aequipecten tehuelchus. Number of juveniles per spat collector in different locations of the Argentine coast, 1996 and 1999.

Region	Site	Mean	SD
San Matías Gulf	El Sótano	590	86.8
	El Fuerte	174.9	82.84
	Orengo	246	74.4
San José Gulf	Tehuelche	134	40.9
	Fracasso	786	260.92
	Mendioroz	14	2.83
	San Román	107.5	0.71

26.2.4.2 Growth in suspended structures of spat obtained from collectors

Three experiments were conducted in the region between 1982 and 1990. Zaixso (1982) utilised lanterns for intermediate culture (up to 30 mm shell height), and devices with rigid mesh and individual pockets for larger individuals, in two locations of San José Gulf. Experiments conducted in San Matías Gulf utilised plastic boxes suspended in mid-water. In the first (1985), a rigid plastic mesh enclosed the juveniles, which were spread over hard-plastic trays. An average size of 57 mm (shell height) was reached after 24 months (Orensanz et al. 1991). A second experiment employed the same plastic box, but the net enclosing the scallops was lighter and flexible. The same size was reached after only 13 months (Narvarte 1995).

Starting in 1996 the growth of juveniles was studied in three different suspended structures, all of them widely used in Japan: 1) cage (identical to the one used before), built with five rigid plastic trays, and covered inside with onion bag, 2) Japanese lantern, with 10 levels and a mesh size (6 mm) similar to that utilised in the cage, and 3) ear suspended (starting with scallops 37-mm high) (Narvarte 1999). The three structures were suspended vertically from a longline. The size of the juveniles was followed over 13 months (Fig. 26.9). The ear-suspension method was abandoned after three months due to the large volume of epibionts that attached to the valves (Fig. 26.10). In the case of the other two structures (cage and lantern), the season of fast growth went from April to December, when spawning takes place (Narvarte and Kroeck 2000). Growth rate slowed down afterwards (Fig. 26.9).

The von Bertalanffy growth model was fitted to data from the three experiments, and the different curves were compared using maximum likelihood methods. Commercial size was reached fastest in the lanterns, which were also easier to handle on board of small boats (Narvarte 1999). The relative mass of epibionts was comparable in the two structures and the difference between these and ear-hanging was significant (Fig. 26.10).

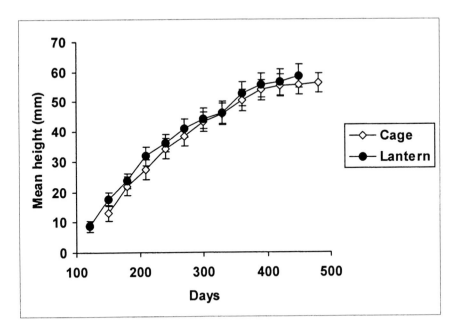

Figure 26.9. *Aequipecten tehuelchus.* Growth of juveniles kept in cages and lanterns, San Matías Gulf.

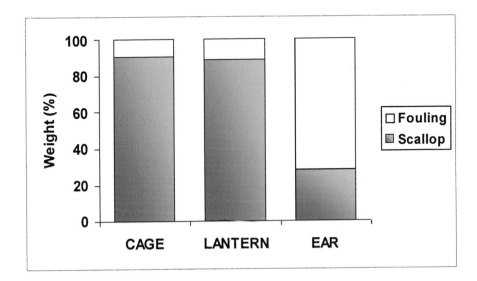

Figure 26.10. *Aequipecten tehuelchus.* Yield (weight of the final product, as a percentage of total weight with epibionts included) obtained from juveniles grown in cages, lanterns, and ear-hanging.

26.2.4.3 Larval culture in the laboratory

Experiments designed to optimise conditions for fertilisation and larval development included: a) fertilisation utilising different concentrations of sperm, b) incubation of embryos at different densities, c) feeding of the larvae with different algal diets, and d) settlement of the larvae on different substrates.

Spawning was induced by thermal stimulation. The number of D-veliger larvae obtained with 30–50 spermatozoids per egg was significantly higher to that obtained with 10 spermatozoids per egg. The 10:1 ratio, frequently used in aquaculture operations, yielded relatively few larvae.

Fertilised eggs were placed in glass containers at three different densities: 350, 700 and 1050 eggs mm^{-2}. No difference was found in the number of larvae obtained in these treatments, indicating that those are below the level at which density-dependence begins to operate, and that the highest density assayed could be used for incubation (Narvarte and Pascual, unpubl.).

Larval growth was assessed under three diets: a) *Isochrysis galbana* + *Chaetoceros calcitrans*, b) *Pavlova lutheri* + *C. calcitrans*, and c) *I. galbana* + *P. lutheri* + *C. calcitrans* (Narvarte and Pascual 2000). The diets containing *P. lutheri* yielded a significantly higher growth rate. Final average larval size varied among diets (Narvarte and Pascual 2000):

    a)   Iso + Ch: 161.77 µm (SE = 19.07)
    b)   Pav + Ch: 175.59 µm (SE = 21.66)
    c)   Iso + Pav + Ch: 200.48 µm (SE = 21.13)

The effect on settlement of adult shells (natural settlement substrate in San Matías Gulf) and young thalli of *Ulva lactuca* (natural settlement substrate in San José Gulf; Ciocco 1985) was investigated by observing the density (number of postlarvae per cm^2 of substrate) on experimental collectors. There was no difference between the two treatments; density of settlers was correlated with substrate surface (r = 0.74; Narvarte and Pascual, unpubl.).

26.2.4.4 Prospects and problems

Given that successful spat collection is obtained (on average) every two years, and considering that settlement in the natural environment is highly variable (episodes spaced 5–10 years in San Matías Gulf), the restocking of fishing grounds with seed from collectors is presently under consideration. However, a retrospective evaluation of research conducted over more than a decade highlights several problems that require further attention:

    i.   Year-to-year variation in spat collection.
    ii.   High cost of structures used for aquaculture, so that they cannot be left in exposed or unprotected locations.

The development of technology for the mass production of hatchery larvae will be conditioned by the knowledge obtained from the experiments under way. The following problems require further scrutiny:

  i.   In the case of hatchery work, the partial spawnings of the Tehuelche scallop provide small quantities of viable eggs at a time. This makes it necessary to stimulate a large number of spawners.

  ii. High mortality (up to 90%) during the first week of larval culture in tanks.

26.3 THE PATAGONIAN SCALLOP, *Zygochlamys patagonica*

26.3.1 Biology and Ecology

Zygochlamys patagonica (Fig. 26.1) is a species typical of the Magellanic Biogeographic Province (Waloszek 1984). It is distributed around southern South America, reaching 42° SL on the Pacific and 36° 15' SL on the Atlantic, between 40 to 200 m depth (Lasta and Zampatti 1981; Waloszek and Waloszek 1986; Defeo and Brazeiro 1994; Fig. 26.11) although some individuals have been reported from down to 960 m deep (Waloszek 1991). While grounds on the Pacific side occur in relatively shallow waters (Andrade et al. 1991), Atlantic grounds are oceanic, the most important beds being located along the 100-m isobath.

The spatial distribution and biological characteristics of scallops are probably related to hydrographic regimes. The continental shelf of the southwestern Atlantic is dominated by: (1) subantarctic waters, mainly the upper layer of the Malvinas Current, that flow through the west of Burdwood Bank (Guerrero et al. 1996), and (2) continental runoff that enters the shelf through the Magellan Strait (Krepper 1977; Bianchi et al. 1982) and the Tierra del Fuego Channels (Lusquiños and Valdes 1971). The Malvinas Current flows towards the north along the continental slope, transporting subantarctic waters characterised by high salinity and low temperature (Piola and Gordon 1989), converges between 35° and 40° SL with the Brazil Current, then flows poleward, generating a strong frontal zone between Subantarctic and Subtropical waters (Reid et al. 1977; Legeckis and Gordon 1982; Gordon and Green Grove 1986). The differential heating between shelf and slope regimes during the warm seasons develops a shelf-break front observed as a horizontal temperature gradient over the 90–100m isobath (Martos and Piccolo 1988; Baldoni 1993). This highly productive front (Brandhorst and Castello 1971; Carreto et al. 1981, 1995; Podestá and Esaías 1988, Negri 1993; Bertolotti et al. 1996) may be the factor that influences the area of distribution of beds located at these depths (Fig. 26.12). Tres Puntas and Sea Bay beds (Fig. 26.12) are influenced by a low salinity coastal system of continental origin, mainly via Magellan Strait and Tierra del Fuego Channels runoff (Brandhorst and Castello 1971; Lusquiños and Valdes 1971; Krepper and Rivas 1979; Guerrero and Piola 1997). The Tres Puntas Bed is associated with the coastal regime, characterised by water mixed due to tidal effect, with high nutrient concentration throughout the year and high chlorophyll concentration (Bertolotti et al. 1996). The Sea Bay Bed is associated with the intermediate shelf regime, characterised by lower productivity (Carreto and Benavides 1990; Carreto et al. 1995) due to strong seasonal

1270

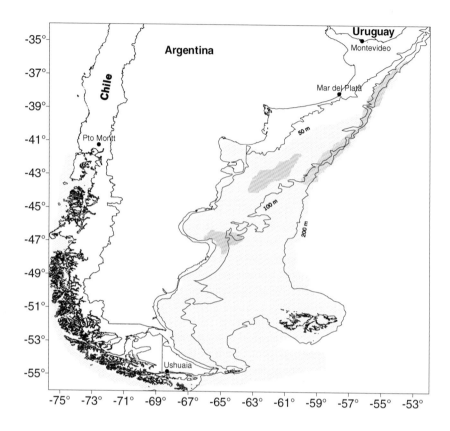

Figure 26.11. *Zygochlamys patagonica*. Geographic distribution. Dark areas represent main fishing scallop grounds in Uruguay, Argentina and Chile.

variation in the vertical structure of the water column, with an homogeneous layer in winter and a two-layer structure during warm months resulting in a strong thermocline (Krepper 1977; Bianchi et al. 1982). It is likely that currents associated with these frontal systems increase the food supply and/or helps retain drifting larvae (Natural Resources Consultants 1995).

Larval interconnection between the beds was inferred from current meter information, through simulation of the progressive trajectory of an ideal particle during a 30-day period, corresponding to the presumed duration of the pelagic larval stage (López 1998, Lasta et al. 1998). Mean vector speed, at 17 and 67 m depth was estimated at 5 cm s^{-1} with N-NE direction. Under these conditions, beds influenced by the shelf-break front and separated by distances fewer than 90 km could be interconnected by larval drift, assuming that larvae behave as passive particles. This pattern is more unrealistic in beds subject to coastal and intermediate shelf regimes, due to horizontal stratification.

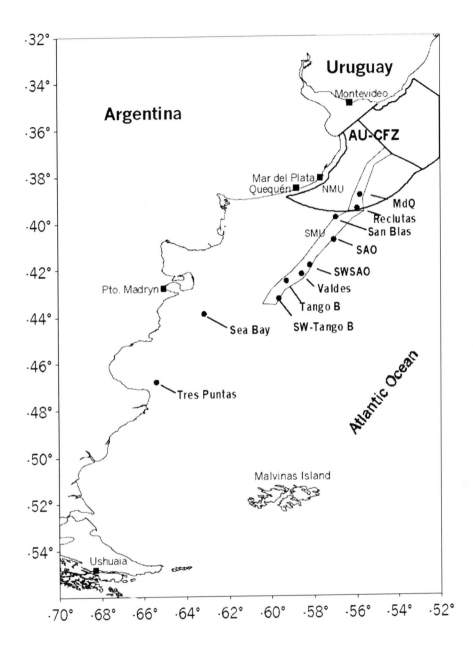

Figure 26.12. *Zygochlamys patagonica*. General location of scallop beds. CFZ: Common Fishing Zone (Uruguay-Argentina), NMU and SMU: Northern and Southern Management Units.

Studies on the benthic assemblages of Patagonian scallop beds were conducted in 1995, before they were disturbed by fishing operations. A total of 82 invertebrate taxa, representing 10 phylla, was recorded (Bremec and Lasta 1997, 1998). Although dominance (in terms of weight) in the epibenthos varied much among beds, its components characterise a distinctive assemblage quantitatively dominated by *Zygochlamys patagonica* over the investigated latitudinal range. The undisturbed beds were distributed along the shelf-break front (39°–43°S), and in regions under the influence of low salinity coastal regimes (44°–47°S). *Z. patagonica* was the dominant species (21%–58% of total biomass), followed by other suspension feeders (Porifera), predators (Cnidaria, Asteroidea, Gastropoda, Crustacea) and grazers-detritivores (Echinoidea, Ophiuroidea). The Patagonic scallop, the ophiuroids *Ophiuroglypha lymanii*, *Ophiactis asperula* and *Ophiacanta vivipara*, the echinoids *Austrocidaris canaliculata* and *Sterechinus agassizi*, and the anemone *Actinostola crassicornis* contribute most of the between-sites similarity (57.7) (Bremec and Lasta 2002). The most frequent epibionts are *Iophon* sp. (Porifera), *Alcyonium* sp. (Coelenterata), *Serpula narconensis* and *Idanthyrsus armatus* (Polychaeta), *Magellania venosa* and *Terebratella dorsata* (Brachiopoda), *Hiatella solida* and *Calyptraea pileolus* (Mollusca), and *Ornatoscalpellum* sp. (Crustacea).

Sexes are separate. Sequential protandric hermaphroditism was suggested based on observations of the size frequency distribution (SFD) by sexes (Orensanz et al. 1991), and on histological indication of a possible sexual inversion (Calvo et al. 1998). However, continuous studies conducted in the Reclutas Bed do not indicate significant differences in the SFDs of the sexes (sexing from frotis preparations; Bigatti and Bonard 2000). Only one hermaphrodite (61 mm shell height) was found among approximately 400 individuals histologically examined (E. Christiansen, pers. comm.).

Sexual maturity is reached at ca. 45 mm (shell height), which correspond to an approximate age of 2 years (Waloszek and Waloszek 1986), although a mature female 15 mm high was found (Calvo et al. 1998). There is no external dimorphism. Spawning takes place mainly during spring, although a second pulse has been reported in late summer to early autumn (Waloszek and Waloszek 1986).

Field observations showed that living scallops (mean shell height 57.2 mm) constitute the main settlement substrate (98.8% of total records) for juveniles between 3–30 mm shell height. Settlement experiments confirmed that individuals between 3–13 mm shell height are always attached (Bogazzi and Lasta 2000).

Size composition shows that recruitment varies between beds (Lasta and Bremec 1998). In 1995 unimodal size distributions characterised the Valdés and Sea Bay Beds, mainly composed by adults. However, size composition in the Reclutas Bed was consistent with intense recruitment during 1994, probably from early spring spawning but also showing a summer-autumn 1995 pulse. Successful recruitment has been registered in Reclutas Bed over the years (hence its name). Cohorts identified by tracking SFDs over time corresponded to the 1992, 1994, and 1995 cohorts; a weaker one settled in 1996–97 (Valero 1999). Recently (cruises from 2000), massive recruitment was observed in all beds (Lasta 2000). The assessment of biomass between 1995 and 2000 does not indicate a shortage of spawning stock. Large-scale hydrographic or ecological phenomena may govern the recruitment pattern observed (Lasta et al. 2001b).

Gut contents (35–50 mm-high scallops) are presently being studied on the basis of a study spanning one year cycle (Schejter, unpubl.). Preliminary results from spring 1996 in Reclutas Bed showed 60% of diatoms (mainly *Paralia sulcata, Nitzschia coarctata* and *Thalassiosira* spp.) and 20% of dinoflagellates (*Dinophysis mawsonii* and *D. rotundata*), plus prasinophytes, tintinids, foraminifers, silicoflagellates, and rests of sponges and arthropods (Schejter et al. 2000).

Soft tissues collected at Valdés Bed (n = 50; size range 30–78 mm) were not parasitised by metazoans (Cremonte 1999).

Analyses of heavy metals (Zn, Cu, total Hg and Cd) concentrations did not indicate accumulation in any tissue (muscle, mantle, viscera and gills). The lowest concentration of the four metals was found in muscle (Gerpe et al. 1995). Paralytic shellfish poisoning (PSP) was never detected through the control of meats (adductor muscle) landed by the commercial fishery (Sancho, pers. comm.). Separate analyses showed positive values of PSP in rests of viscera adhered to muscles from Tres Puntas Bed (Glaciar Pesquera, S.A., pers. comm.).

Enzymatic activity and reduced viscosity in adductor muscles are reduced by 35–40% after freezing at 2–4°C, mainly during the first two days (Paredi and Crupkin 2000).

26.3.2 Population Dynamics and Stock Assessment

Bimodal SFDs of the first annual ring (shell and ligament) has been interpreted as reflecting the two spawning peaks mentioned earlier. The first ring is assumed to be laid at a variable age, ranging from 0.5 to 1.5 years (Waloszek and Waloszek 1986). Size at settlement (inferred from the change in sculpture in the transition from prodissoconch to dissoconch in shells of adult individuals) is around 0.2 mm shell height (Waloszek 1984, Fig. 26.13). Individual growth in the Argentine Shelf was estimated by analysis of annual rings on the left valves and ligament by Waloszek (1979, 1991) and Waloszek and Waloszek (1986). These studies assumed that annual rings are formed during the winter, and that they are not related to spawning (Waloszek 1991). Recorded maximum size is around 90 mm shell height, and recorded maximum age is 10 years (few individuals are older than 8 years or higher than 80 mm). The von Bertalanffy growth model was fitted to size-at-age data by Waloszek and Waloszek (1986), Orensanz et al. (1991) and Valero (unpubl.). Estimated values of asymptotic height are in the range 53–79 mm, and estimated k in the range 0.35–0.67 yr^{-1} (Fig. 26.13a). There is a significant decrease in asymptotic height as latitude increases (Fig. 26.13b). Legal commercial size (55 mm, Figure 26.13a) is reached at ages varying from 3 to 5 years over much of the latitudinal range, but in some areas the scallops hardly reach legal size.

Lowest and highest values of muscle condition index were observed in September and May, respectively, whereas the index of gonad condition was lowest during March-April and highest during August-September in samples taken between 1995–1998 from the Reclutas Bed (Lasta and Bremec 1999; Lasta et al. 2001b).

1274

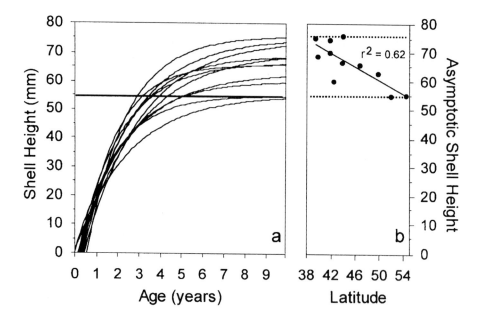

Figure 26.13. *Zygochlamys patagonica.* a: Von Bertalanffy growth curves fitted to size-at- age data from Waloszek and Waloszek (1986) and Valero (unpubl.). Solid horizontal line represents minimum commercial size (55 mm). b: estimated asymptotic shell height plotted against latitude. Dashed lines represent the range of estimates.

Total mortality rate estimated from SFDs (Reclutas Bed, 1995–1998) was 1.039 yr^{-1} (Lasta et al. 2001b).

Data obtained from 1995 to 1998 in Reclutas Bed show that the annual P/B ratio of the population was 0.679 yr^{-1}, corresponding to an annual total production of 45.5 g SFWM m^{-2} yr^{-1} (= 4.02 g C$_{org}$ m^{-2} yr^{-1}) (Lasta et al. 2001b). Muscle, soft tissue and gonad contributed 37.3%, 56.2% and 6.5% to total production, respectively. The major reason for this high P/B ratio is the low mean body mass in this population over the period of investigation. Estimates of productivity of the epifaunal community conducted in Reclutas Bed, based on fewer samples and on a P/B ratio derived from an empirical relation, resulted in lower production figures (Bremec et al. 2000). However, *Z. patagonica* dominated the epibenthic community in 1995, contributing 26% of the community biomass (3.15 g C m^{-2}) and 31% of community production. In 1998, community biomass (5.06 g C m^{-2}) as well as production were higher, and the share of *Z. patagonica* rose to 62% in both cases. A conspicuous decrease in biomass and annual production between 1995 and 1998, observed mostly among sessile taxa (Porifera and Cnidaria), contributed to reduce their share in biomass from 30% to 7%, and their share in production from 14% to 2%.

Table 26.4

Zygochlamys patagonica. Dimensions of beds (km^2), sample size (N), total and commercial scallop relative density (Tsd, Csd, expressed in tons km^{-2}) from evaluation cruises conducted during 2000. (*): Swept area method. CV = Coefficient of variation.

Bed	Area (km^2)	N	Tsd (t km^{-2})	CV (%)	Csd (t km^{-2})	CV (%)
MdQ	2,834	59	27.55	85.05	13.15	81.11
Reclutas	1,174	32	28.26	63.93	16.11	55.89
San Blas	355	18*	15.05	76.00	12.68	74.94
SWSAO	1,419	54*	24.13	65.58	16.67	54.42
Valdés	75	14	12.18	72.48	11.19	75.64
Tango B	456	9	15.14	126.95	13.72	129.18
SWTango B	274	13	24.06	90.26	18.62	102.46

Between 1995 and 1997 the stock was assessed using an index of abundance (CPUE, expressed as kg of commercial scallops/towing time) obtained from data recorded by the commercial fleet. Research cruises were started by Argentina's Institute for Fisheries Research and Development (INIDEP) in 1998 with the purpose of assessing absolute biomass in two management units (North and South). Sampling stations are spaced regularly at 5–6 nm intervals. The distribution of sampling stations is planned to cover the areas exploited since the beginning of the fishery. A non-selective 2.5-m wide dredge (efficiency 16–21%, Iribarne et al. 1991) is used; standard towing time and speed are 10 minutes and 3 knots, respectively. SCANMAR equipment is used to control the performance of the dredge over the bottom. Total catch is weighted and sub-samples are taken to estimate the percentage of commercial-size scallops, and to obtain SFDs. Geostatistical methods were recently introduced to estimate total and commercial abundance of scallops (Table 26.4) and to investigate changes in spatial distribution (Lasta and Hernández 1999; Lasta et al. 2001a).

26.3.3 The Fishery

26.3.3.1 Exploratory surveys and experimental fishing programs

The first commercial fishing operations were conducted in 1989 by the F/V "Sea Bay Alpha", a 64 m-long factory vessel. Although originally planned to extend for one year, the project was suspended after 24 days due to legal litigation. Most of the catch came from the Sea Bay Bed area. The muscle obtained was small in size (200 units per pound), but it was concluded that catch and muscle production were economically viable if the appropriate technologies were developed (Lasta 1992).

In 1991 an exploratory fishing trip off the coast of Uruguay was conducted by an Uruguayan company to locate scallop resources. The objectives were to test catch gear, to determine appropriate methods to harvest and handle the catch at sea, to develop a methodology for processing and packaging, and to collect data to assess the status of the stocks (DuPaul and Smolowitz 1994). An exploratory survey designed to assess the spatial distribution and composition of scallop stocks off Uruguay was conducted in the course of 17 trips during 1993 and 1994 (196 effective fishing day) by the F/V "Erin Bruce", a 56 m-long factory vessel (Defeo and Brazeiro 1994; Riestra and Barea 2000). The size of the muscles was small and the beds located were not extended; 772 MT of meats were landed (Riestra and Barea 2000). Vessel activity was suspended until January 1995.

Based on scattered data from exploratory and experimental fishing surveys (FRV "Prof. Siedlecki"/1973, Orensanz et al. 1991; FRV "Walther Herwig"/1978–79, Waloszek and Waloszek 1986; FRV "Walther Herwig"/1978–79 and FRV "Shinkai Maru"/1978, Lasta and Zampatti 1981, and FV "Sea Bay Alpha"/ 1989, Lasta 1992), a joint state-industry program was conducted in 1995 with the F/V "Erin Bruce" (which had been operating before off Uruguay) to evaluate the status of the resource across the Argentine Shelf. Data were collected by onboard observers in the course of 15 trips (Table 26.6). Results confirmed that the Patagonian scallop has a wide geographical distribution, with dense grounds between 50 and 130 m depth from 38° to 48° S (Lasta and Bremec 1998).

The F/V "Erin Bruce" was equipped with non-selective bottom otter trawls directly attached to the doors (otter board) (Lasta and Bremec 1997), 13 m in total length, a mesh size of 4 inches, and head and foot rope 15 m long. Estimated gear efficiency was 21–31% (Lasta and Iribarne 1997). The catch, composed of scallops, other benthic invertebrates and shell hash (Bremec and Lasta 1997), was mechanically processed onboard. By-catch and non-commercial size scallops were separated by drums and discarded, while commercial size scallops were processed as follows: (1) Steaming to open and separate soft tissue from the valves, (2) peeling to remove soft tissues and obtain muscles, (3) mechanical and/or manual control, and (4) freezing of the adductor muscle (IQF), grading in plates, and packing (Ciocco et al. 1998).

26.3.3.2 Management

A management plan was first outlined by INIDEP scientists following the experimental fishery of 1995, when basic information was collected (Lasta and Bremec 1997). The plan was conceived to incorporate adaptive criteria. The Argentine fisheries authority gave permits to two companies, each with 2 factory vessels, to start fishing commercially in 1996. Minimum legal size was set at 55 mm (total height), corresponding to an age of 3–4 years. No fishing season was imposed. Quotas were allocated to the fishing companies, even in the absence of strong information on stock size. A harvest rate of 40% of legal size biomass, and maximum meat counts (118 muscles pound^{-1}), were initially introduced, but subsequently suspended. Reproductive closures were demarcated within each bed. Fishing was also banned in areas selected to conduct research (Reclutas Bed, Fig. 26.12, Lasta and Bremec 1997). The companies were required to provide detailed fishing logs; these recorded fishing effort, positions, total catch, muscle yield, trawling time, etc.

Following consultation between the fishing authority, the industry and the scientists, the plan took final shape in 1999, structured for 4 year plus 1.

Two "Management Units" (MUs) were defined: "North", between 36° 45' and 39° 30' SL (within the Common Fishing Zone-CFZ- shared by Uruguay and Argentina), and "South", between 41° and 43° 30' SL. The North MU (including the MDQ and Reclutas beds) has an extension of 13,181 km², and the South MU (including the San Blas, SAO, SWSAO, Valdés, TangoB and SW-TangoB beds) is 25,858 km² (Fig. 26.12).

During 1999, a Uruguayan factory vessel (F/V "Holberg", 50 m long, see Riestra and Barea 2000; Riestra 2000) began operating at the CFZ (Fig. 26.12), in the Argentine sector. A quota by country criteria was recently agreed upon.

26.3.3.3 Development of the fishery

Information on fleet activity from the start of the fishery is summarised in Table 26.6 (see also Lasta and Bremec 1999). Fishing time was relatively constant over the years (Table 26.3). Although swept area and number of tows increased after 1995, fishing pulses occurred on different areas (Fig. 26.14, Table 26.6). One of the vessels (F/V "Atlantic Surf 2", Table 26.5) left the fishery in October 1997; this was reflected in a reduction of the landings in 1998 (Table 26.6). However, muscle production rose that year due to increased activity in the South MU, improving factory efficiency (Table 26.3), and enhanced ability of the skipper to detect the densest patches. Differences in landed muscle and daily muscle production between 1997 and 1999 (two years comparable in swept area and total number of tows) were due to improved processing efficiency (Table 26.6 and Fig. 26.15c).

Towing time increased from the beginning of the program (1995) through 1999 (Fig. 26.15a), and scallop CPUE show progressive depletion (Fig. 26.15b). After five years the fleet has trawled all over the shelf, following a rotating pattern and shifting the allocation of fishing effort (Fig. 26.16). This pattern follows from skipper decisions, based on prior knowledge as well as information gathered while fishing.

The ratio between landed muscle and commercial scallop processed (without epibionts), or "muscle yield" (MY) is used as an index of performance. MY was never higher than 20%. Performance has increased progressively (Table 26.3) as a result of improved technology on board. These results were corroborated by observers. An increasing trend in muscle production per fishing day is, thus, a result of increased efficiency at the factory as well as improved knowledge to locate and harvest dense patches. Epibionts make up ca. 7% of total scallop weight.

The development of this fishery was conditioned by two facts: the stocks are far from fishing harbours, and the size of the muscles is relatively small, rendering hand-processing impractical. The only option was mechanical processing on factory vessels. This led to the implementation of the current strategy. It is illustrative to notice that the Patagonian scallop fishery (including Uruguay) has involved five of the few scallop factory vessels built around the world (Table 26.5).

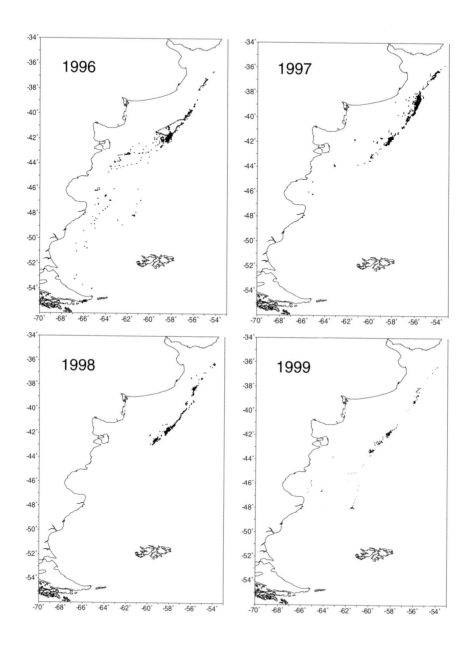

Figure 26.14. *Zygochlamys patagonica.* Activity of the fleet during 1996–1999. Each point represents approximately four tows.

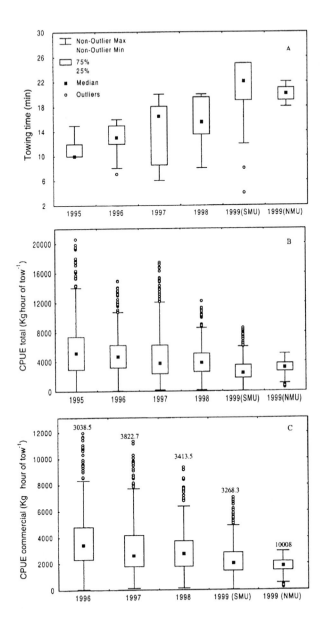

Figure 26.15. *Zygochlamys patagonica.* A: Median towing time per year (in minutes) for all vessels. B: Total CPUE (Catch-per-unit-effort, as kg per towing hour). C: CPUE, commercial size catch only (total height >55 mm), number above box indicates weight of landed muscles, in tons. SMU and MNU: Southern and Northern Management Units.

Table 26.5

Scallop factory vessels built around the world, and their current status.

Year	Factory vessel	Country of origin	Presently operating in
1986	Holberg	Norway	Uruguay
	Sea bay Alpha	Norway	Inactive
1987	Atlantic Surf 2 [1]	Norway	Russia
1988	Scalloper	Russia	Russia
	NN	USA	Never active
	NN	Norway	?
1990	Arctic Rose	USA	Never active
1993	Erin Bruce	USA	Argentina
1994	Alexygutnetsous	Russia	Russia
1996	Atlantic Surf 1	Canada	Argentina
	Mr. Big	USA	Argentina

[1] Active in Argentina between 1996–1997.

Vessels in the fleet that presently operates in Argentine waters have features similar to the F/V "Erin Bruce" (described above; see also Lasta and Bremec 1998). Average sailing speed is 9–10 knots and average towing speed is 4 knots. Bottom otter trawl was established as the regular fishing gear. The New Bedford dredge (4.5 m wide) was used for a short time (1996–1997) in south most beds. Trawl nets are operated with booms ("tangones"); the fishing activity goes on 24 hours a day. Depending on speed and other technical characteristics, the vessels can make 40–60 tows d^{-1} net^{-1}. The storage-room has a capacity up to 210 MT, which, together with storage capacity for fresh water (used in steam production) and fuel, limits the duration of the fishing trips to 20–40 days. The main landing harbours are Mar del Plata and Quequén, and occasionally Ushuaia and Puerto Madryn (Table 26.6).

Besides the industrial fishery, a small artisanal fleet lands small commercial catches in San Jorge Gulf. The whole meats are sold to local consumers.

26.3.3.4 Observers program

The management plan requires the companies to pay technical observers onboard scallopers on at least 50% of the trips to collect information. Vessel activity (position, towing speed and time), data on catches (total and commercial scallop catch, size frequency distribution total scallop), and factory process (estimated losses in the steam and peeling processes) are obtained by observers. Relative to community composition and gear disturbance, a daily catch monitoring sampling records abundance, biomass and minimum and maximum sizes of the main associated invertebrate species.

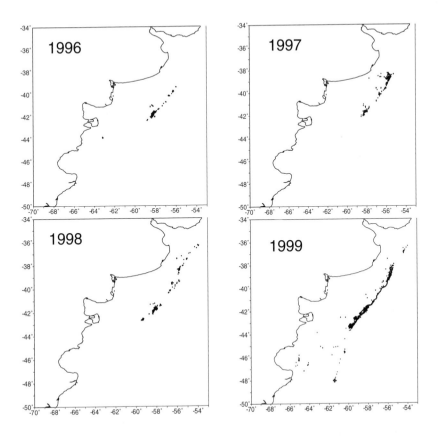

Figure 26.16. *Zygochlamys patagonica*. F/V "Atlantic Surf 1", activity between 1996 and 1999. Each point represents approximately four tows.

Processing onboard problems are frequently detected and tested by observers, mainly through evaluation of drum losses, steaming and peeling. Results are used for back-calculation to control catches from the amounts of muscle landed.

26.3.3.5 Ecological effects of fishing

The scallop community is affected by fishing operations (including both bottom trawling and onboard factory work) through: (1) sediment removal, (2) habitat modification, (3) damage to benthic organisms, (4) removal of target species and by-catch, (5) discards of scallops <55 mm (minimum legal size), (6) discards of shell and viscera, and (7) discards of non-target invertebrates. Invertebrates caught as by-catch (including mobile species and scallop epibionts) are discarded at sea after sorting in the factory, where they are injured or killed.

Table 26.6

Zygochlamys patagonica. Characteristics of the Argentine fishery, 1995–1999. Total time of each trip detailed as: fishing, sailing and searching time. SMU: Southern Management Unit, NMU: Northern Management Unit.

Year	Experimental		Commercial			
	1995	1996	1997	1998	1999	
Management Unit					SMU	NMU
Number of vessels	1	4	4–3	3	3	
Fishing days (%)[1]	200 (81.3)	658 (82)	845 (87.7)	652 (82.1)	783 (83.5)	
Total fleet tows	14565	51654	65809	46704	14980	48807
Mean Total Scallop Yield (SD)[2]	69.36 (12.31)	75.78 (8.57)	65.4 (12.74)	70.29 (10.82)	73.18 (10.77)	57.64 (16.25)
Mean Commercial Scallop Yield (SD)[2]	51.49 (10.12)	56.26 (7.9)	47.01 (12.74)	52.05 (10.02)	42.46 (9.72)	52.69 (15.89)
Swept area (km^2)	182	774	1157	835	1241	
Total scallop catch (t)	13581	49767	55386	47082	57500	
Total commercial scallop catch (t)	10592	36952	39817	28441	10008	32683
Mean Muscle by fishing day (t) (SD)[2]	6.8 (2.51)	4.78 (1.5)	4.68 (1.38)	5.41 (1.28)	6.69 (2.39)	
Commercial Muscle yield (%)	12.39 (1.97)	8.78 (2.54)	10.79 (2.26)	12.72 (3.96)	11.74 (3.27)	
Main mother-ports (%)	MDP (100)	MDP-USHUAIA	MDP-QN	MDP-QN	MDP (100)	
Observer coverage (%)	100	38	21	22	17	

[1] percentage of total time at the sea
[2] SD = Standard deviation

To test community responses to different levels of fishing effort, a depletion experiment was conducted in 1995 (undisturbed condition) in an extensive portion of Valdés Bed (Lasta and Iribarne 1997). The area swept amounted to 2.2 times the size of the experimental plot (which was not fished afterwards) and 1.15 times the size of the rest of the bed (which was subsequently opened to the fishery); both sectors were sampled again in 1997 (Bremec and Lasta 1998, 1999). Main effects on the fauna were damage of non-target invertebrates (short-term response), reduced of biomass of sessile filter-feeders like sponges and coelenterates (long-term response), and increment in the amount of empty shells on the bottom (a result of on-board processing). Scallops survive at least one removal from the bottom (Lasta and Bremec, unpubl.). Samples taken two years after depletion showed a higher relative biomass of scallops (60% of the total catch) than places with lower but persistent fishing effort (40% of total catch).

26.3.3.6 Marketing

Fishing production (adductor muscles only) is frozen both in block and IQF and exported to U.S.A, Canada and the European Union (EU). Exports through one company (Glaciar Pesquera S.A. pers. comm.) were destined as follows: 61% to the EU, 28% to Canada, and 10% to the USA, and the rest to the domestic market of Argentina. Although technology to produce scallop hamburgers (under various commercial names) was developed, this product found no significant acceptance in the market. Average calibre is 160 to 220 muscles per kg. Current (2000) value of the product is U$S 7.1 kg^{-1} (Fob) in Buenos Aires (Argentina), although prices fluctuate due to high supplies of scallops produced in China through aquaculture (Glaciar Pesquera, S.A., pers. comm.).

ACKNOWLEDGMENTS

Juan Valero was supported by scholarships from the Fulbright Program and the University of Washington. Maite Narvarte acknowledges Sandro Acosta and Natalia Saiz for assistance in sampling work, to Marcela Pascual, Eduardo Zampatti, Myriam Elvira and Ignacio Agulleiro for hatchery work and Enrique Morsan and Raúl González for their comments on individual growth in culture experiences. Norberto de Garín assisted in the preparation of some *A. tehuelchus* figures.

REFERENCES

Andrade, B.S., Bornicioli, D., Valladares, M.C., Scabini, V.V., Cormarck, A.B., Parada, G.J., Matsumoto, H., Perez, C., Pinto, M.E. and Tecay, V.A., 1991. Estudios de repoblamiento del ostión del sur, XII ª Region de Magallanes, Chile. Universidad de Magallanes (Punta Arenas, Chile), Informe de la Facultad de Ciencias No. 2/91. 184 p.

Astort, E.D. and Borzone, C.A., 1980. Edad y crecimiento de la vieira *Chlamys tehuelchus* (d'Orb.) en el golfo San Matías, y algunas observaciones sobre su ecología. Unpublished ms. 29 p.

Baldoni, A., 1993. Frente del talud. In: Seminario-taller sobre la dinámica marina y su impacto en la productividad de las regiones frontales del Mar Argentino. INIDEP, Technical Report No. 1 (Mar del Plata, Argentina). pp. 8–10.

Bertolotti, M.I., Brunetti, N.E., Carreto, J.I., Prenski, L.B. and Sánchez, R.P., 1996. Influence of shelf-break fronts on shellfish and fish stocks off Argentina. ICES C.M. 1996/S, 41. 15 p.

Bianchi, A., Massoneau, M. and Olivera, R.M., 1982. Análisis estadístico de las características T-S del sector austral de la plataforma continental argentina. Acta Oceanographica Argentina 3:93–118.

Bigatti, G. and Bonard, A., 2000. Diferenciación sexual en *Zygochlamys patagonica* (King y Broderip, 1832) (Mollusca, Bivalvia, Pectinidae) en el banco "Reclutas" del Mar Argentino. Seminario Curso de Oceanografía Biológica, Universidad de Buenos Aires. 20 p.

Bogazzi, E. and Lasta, M., 2000. Capacidad bisógena y natatoria de la vieira patagónica *Zygochlamys patagonica*. Resúmenes IV Jornadas Nacionales de Ciencias del Mar, Puerto Madryn, Argentina, 11–15 September 2000. pp. 42.

Brandhorst, W. and Castello, J.P., 1971. Evaluación de los recursos de anchoíta (*Engraulis anchoita*) frente a la Argentina y Uruguay. I. Las condiciones oceanográficas, sinopsis del conocimiento actual sobre la anchoíta y el plan para su evaluación. Proyecto Desarrollo Pesquero, FAO/UNDP (Mar del Plata, Argentina), Publicación 29. 63 p.

Bremec, C., Brey, T., Lasta, M., Valero, J. and Lucifora, L., 2000. *Zygochlamys patagonica* beds on the Argentinian shelf. Part I: Energy flow through the scallop bed community. Archive of Fishery and Marine Research 48:295–303.

Bremec, C.S. and Lasta, M.L., 1997. Macrobenthic bycatch associated with the scallop (*Zygochlamys patagonica* King & Broderip, 1832) assemblage in the argentine continental shelf: a baseline study. 11th International Pectinid Workshop, La Paz, México, 10–15 April 1997. pp. 145–147.

Bremec, C.S. and Lasta, M.L., 1998. Experimental study on macrobenthic community structure of Patagonian scallops (*Zygochlamys patagonica* King & Broderip, 1832) beds affected by fishing disturbances. ICES Symposium on Marine Benthic Dynamics: Environmental and Fisheries Impacts, Crete, Greece, 5–7 October 1998. pp. 42–43.

Bremec, C.S. and Lasta, M.L., 1999. Effects of fishing on faunistic composition of scallop beds in the Argentine Sea. 12th International Pectinid Workshop, Bergen, Norway, 5–11 May 1999. pp. 152–153.

Bremec, C.S. and Lasta, M.L., 2002. Epibenthic assemblage associated with scallop (*Zygochlamys patagonica*) beds in the Argentinian shelf. Bull. Mar. Sci. 70:89–105.

Calvo, J., Morriconi, E. and Orler, P.M., 1998. Estrategias reproductivas de moluscos bivalvos y equinoideos. In: E. Boschi (Ed.). El Mar Argentino y sus Recursos Pesqueros. INIDEP (Mar del Plata, Argentina). 2:195–231.

Carreto, J.I. and Benavides, H.R., 1990. Synopsis on the reproductive biology and early life history of *Engraulis anchoita*, and related environmental conditions in Argentine waters. Phytoplankton, IOC Workshop, Report No. 65, Annex V. pp. 2–5.

Carreto, J.I., Lutz, V., Carignan, M.O., Cucchi Colleoni, A.D. and De Marco, S.G., 1995. Hydrography and chlorophyll a in a transect from the coast to the shelf-break in the Argentinian Sea. Cont. Shelf Res. 15:315–336.

Carreto, J.I., Negri, R.M. and Benavides, H.R., 1981. Fitoplancton, pigmentos y nutrientes. Resultados de las campañas III y VI del B/I "Shinkai Maru", 1978. Campañas de Investigación Pesquera realizadas en el Mar Argentino por los B/I "Shinkai Maru" y "Walther Herwig" y el B/P "Marburg", años 1978 y 1979. Resultados de la parte argentina. INIDEP (Mar del Plata, Argentina), Contribución 383. pp. 181–201.

Charpy-Roubaud, C.J., Charpy, L.J. and Maestrini, S.Y., 1982. Fertilité des eaux cótieres nord-patagoniques: facteurs limitant la production du phytoplancton et potentialités d'exploitation mytilicole. Oceanol. Acta 5:179–188.

Christiansen, H.E., Brodsky, S.R. and Cabrera, M.E., 1973. Aplicación de una técnica histométrica en la determinación de la fecundidad en invertebrados marinos. Physis (Buenos Aires) 32(84):121–135.

Christiansen, H.E., Cabrera, M.E. and Brodsky, S.R., 1974. Ecología de las poblaciones de vieiras (Chlamys tehuelcha d'Orb., 1846) en el Golfo San Matías (Río Negro, Argentina). I. Estudio histológico del ciclo reproductivo. Instituto de Biología Marina (Mar del Plata, Argentina), Contribución No. 225. 17 p.

Christiansen, H.E. and Olivier, S.R., 1971. Sobre el hermafroditismo de Chlamys tehuelcha d'Orb. 1846 (Pelecypoda, Filibranchia, Pectinidae). Anales de la Sociedad Científica Argentina (Buenos Aires) 191:115–127.

Ciocco, N.F., 1985. Biología y ecología de Chlamys tehuelchus d'Orbigny en el golfo San José (Pcia. del Chubut, Republica Argentina) (Pelecypoda, Pectinidae). Doctoral Diss. Universidad Nacional de La Plata, Argentina. 406 p.

Ciocco, N.F., 1990. Infestación de la vieira tehuelche (Chlamys tehuelcha (d'Orbigny)) por Polydora websteri Hartman (Polychaeta: Spionidae) en el Golfo San José (Chubut, Argentina): un enfoque cuantitativo. Biología Pesquera (Chile) 19:9–18.

Ciocco, N.F., 1991. Differences in individual growth rate among scallop (Chlamys tehuelcha (d'Orb.)) populations from San José Gulf (Argentina). Fish. Res. 12:31–42.

Ciocco, N.F., 1992a. Anatomía de la vieira tehuelche, Chlamys tehuelcha (d'Orb.). I. Valvas, ligamento y manto (Pelecypoda, Pectinidae). Neotrópica (Argentina) 38(99):21–34.

Ciocco, N.F., 1992b. Anatomía de Chlamys tehuelcha (d'Orb.). II. Organización general, sistema branquial, pie y aparato bisógeno. (Bivalvia, Pectinidae). Revista de Biología Marina (Chile) 27:17–35.

Ciocco, N.F., 1992c. Differences in individual growth rate among scallop (Chlamys tehuelcha (d'Orb.)) populations from the San José Gulf (Argentina): experiments with transplanted individuals. J. Shellfish Res. 11:27–30.

Ciocco, N.F., 1995a. Anatomía de la vieira tehuelche Aequipecten tehuelchus (d'Orbigny, 1846). III. Sistemas digestivo, cardio-vascular y excretor (Bivalvia, Pectinidae). Revista de Biología Marina (Chile) 30:135–153.

Ciocco, N.F., 1995b. La marisquería mediante buceo en el Golfo San José (Chubut, Argentina). Informes Técnicos del Plan de Manejo Integrado de la Zona Costera Patagónica (PNUD-GEF, FPN) 2(1):1–39.

Ciocco, N.F., 1996. "In situ" natural mortality of the Tehuelche scallop, Aequipecten tehuelchus (d'Orb., 1846), from the San José Gulf (Argentina). Scientia Marina (Spain) 60:461–468.

Ciocco, N.F., 1998. Anatomía de la vieira tehuelche *Aequipecten tehuelchus* (d'Orbigny, 1846). IV. Sistema nervioso y estructuras sensoriales (Bivalvia, Pectinidae). Revista de Biología Marina, (Valparaíso, Chile) 33(1):25–42.

Ciocco, N.F. and Aloia, D.A., 1991. La pesquería de vieira tehuelche, *Chlamys tehuelcha* (d'Orb.), del golfo San José (Chubut, Argentina): vigor de clases anuales. Scientia Marina (Spain) 55:569–575.

Ciocco, N.F., Borzone, C.A. and Ruzzante, D.E., 1983. Observaciones sobre el comportamiento de fijación de *Chlamys tehuelchus* d'Orbigny en bancos naturales. Memorias de la Asociación Latinoamericana de Acuicultura 5:271–275.

Ciocco, N.F., Gosztonyi, A.E., Galvan, D., Monsalve, M.A., Diaz, M.A., Vera, R., Ibañez, J., Ascorti, J., Signorelli, J.C. and Berón, J.C., 1996. La vieira tehuelche del golfo San José: primeros resultados de la campaña de relevamiento SANJO/95. LAPEMAR, Centro Nacional Patagónico (Puerto Madryn, Argentina), Technical Report 1. 33 p.

Ciocco, N.F., Lasta, M.L. and Bremec, C.S., 1998. Pesquerías de bivalvos: mejillón, vieiras (tehuelche y patagónica) y otras especies. In: E. Boschi (Ed.). El Mar Argentino y sus recursos pesqueros. INIDEP, Mar del Plata, Argentina 2:143–166.

Ciocco, N.F. and Monsalve, M.A., 1999a. The Tehuelche scallop, *Aequipecten tehuelchus*, from San José Gulf (Argentina): spat settlement during the collapse of the fishery. 12th International Pectinid Workshop, 5–11 May, 1999, Bergen, Norway.

Ciocco, N.F. and Monsalve, M.A., 1999b. La vieira tehuelche, *Aequipecten tehuelchus* (d'Orb., 1846), del golfo San José (Argentina): captación de postlarvas durante el colapso de la pesquería. Biología Pesquera (Chile) 27:23–36.

Ciocco, N.F., Monsalve, M.A., Diaz, M.A., Vera, R., Signorelli, J.C. and Diaz, O., 1997. La vieira tehuelche del golfo San José: primeros resultados de la campaña de relevamiento SANJO/96. LAPEMAR, Centro Nacional Patagónico, Puerto Madryn, Argentina, Technical Report 1. 30 p.

Ciocco, N.F. and Orensanz, J.M., 2001. Depredación. In: A. Maeda-Martínez (Ed.). Los Moluscos Pectínidos de Iberoamérica: Ciencia y Acuicultura. Editorial LIMUSA. Mexico City, Mexico. pp. 267–284.

Cremonte, F., 1999. Estudio parasitológico de bivalvos que habitan ambientes marinos y mixohalinos en Argentina. Doctoral Diss. Universidad Nacional de La Plata, Argentina. 196 p.

Defeo, O. and Brazeiro, A., 1994. Distribución, estructura poblacional y relaciones biométricas de la vieira *Zygochlamys patagonica* en aguas uruguayas. Comunicaciones de la Sociedad Malacológica del Uruguay 7(66–67):362–367.

De Vido de Mattio, N., 1984. Variacion estacional de la composición bioquímica de la vieyra *Chlamys tehuelcha* (d'Orbigny) en el golfo San José. Centro Nacional Patagónico (Puerto Madryn, Argentina), Contribución 92. 22 p.

Di Giácomo, E., Parma, A.M. and Orensanz, J.M., 1994. Food consumption by the cock fish, *Callorhynchus callorhynchus* (Holocephali: Callorhynchidae), from Patagonia (Argentina). Env. Biol. Fishes 40:199–211.

DuPaul, W.D. and Smolowitz, R.J., 1994. A report on the field investigation into commercial potential of the Patagonian scallop, *Chlamys* sp. of Uruguay. BIVAR S.A., unpublished technical report. 22 p.

Gerpe, M., Aizpun de Moreno, J. and Moreno, V., 1995. Distribución de metales pesados en *Chlamys lischkei*. Congreso Latinoamericano Ciencias del Mar, Mar del Plata, Argentina, 23–27 October 1995, Resumenes. pp. 91.

Guerrero, R. and Piola, A.R., 1997. Masas de agua en la plataforma continental. In: E. Boschi (Ed.). El Mar Argentino y sus Recursos Pesqueros, INIDEP (Mar del Plata, Argentina). 1:107–118.

Guerrero, R., Baldoni, A. and Benavides, H., 1996. Oceanographic conditions at the southern end of the Argentine Continental Slope. In: Reproductive habitat, biology and acoustic biomass estimates of the Southern blue whiting (*Micromesistius australis*) in the Sea off Southern Patagonia. INIDEP (Mar del Plata, Argentina), unpublished ms. pp. 9.

Gordon, A.L. and Green Grove, C.L., 1986. Geostrophic circulation of the Brazil-Falkland Confluence. Deep-sea Res. 33:573–585.

Iribarne, O.O., Lasta, M., Vacas, H., Parma, A.M. and Pascual, M.S., 1991. Assessment of abundance, gear efficiency and disturbance in a scallop dredge fishery: results of a depletion experiment. In: S.E. Shumway and P.A. Sandifer (Eds.). Scallop Biology and Culture. World Aquaculture No. 1. The World Aquaculture Society, Baton Rouge, LA. pp. 244–248.

Krepper, C.M., 1977. Difusión del agua proveniente del Estrecho de Magallanes en las aguas de la plataforma continental. Acta Oceanographica Argentina 1:49–65.

Krepper, C.M. and Rivas, A.L., 1979. Análisis de las características oceanográficas de la zona austral de la plataforma continental argentina y aguas adyacentes. Acta Oceanographica Argentina 2:55–82.

Lasta, M.L., 1992. *Chlamys patagonica*: resultados del primer crucero de pesca experimental. 9° Simposio Cientifico de la Comision Tecnica Mixta del Frente Marítimo, Mar del Plata, Argentina, Nov. 30-Dec. 3, 1992. pp. 13 (abstract).

Lasta, M.L., 2000. Informe de campaña CC-06-2000. Evaluación de biomasa de vieira patagónica. Unidad Sur de Manejo: bancos San Blas, SWSAO, Valdés, Tango B y SW-Tango B. Complementariamente: Estación Fija banco Reclutas. Informe Interno INIDEP 181 (Mar del Plata). 49 p.

Lasta, M.L. and Bremec, C., 1997. *Zygochlamys patagonica* (King and Broderip, 1832): development of a new scallop fishery in the Southwestern Atlantic Ocean. 11th International Pectinid Workshop, La Paz, México, 10–15 Apr., 1997. pp. 138–139 (abstract).

Lasta, M.L. and Bremec, C., 1998. *Zygochlamys patagonica* in the Argentine Sea: a new scallop fishery. J. Shellfish Res. 17:103–111.

Lasta, M.L. and Bremec, C., 1999. Development of the scallop fishery (*Zygochlamys patagonica*) in the Argentine sea. 12th International Pectinid Workshop, Bergen, Norway, May 5–11, 1999. pp. 154–155. (abstract)

Lasta, M.L. and Calvo, J., 1978. Ciclo reproductivode la vieira *(Chlamys tehuelcha)* del golfo San José. Comunicaciones de la Sociedad Malacológica del Uruguay 5:1–42.

Lasta, M.L. and Hernández, D., 1999. Geostatistical techniques to estimate scallop (*Zygochlamys patagonica*) biomass and to evaluate associated uncertainty. 17th Lowell Wakefield Fisheries Symposium, Spatial Processes and Management of Fish Populations, Anchorage (Alaska), Oct. 27–30, 1999. pp. 13. (abstract)

Lasta, M.L, Hernández, D.R., Bogazzi, E., Burgos, G.E., Valero, J.E. and Lucifora, L., 2001a. Uso de técnicas geoestadísticas en la estimacion de la abundancia de vieira patagónica

1288

(*Zygochlamys patagonica*). Revista de Investigación y Desarrollo Pesquero (INIDEP, Argentina) 14:95–108.

Lasta, M.L. and Iribarne, O.O., 1997. Southern Atlantic scallop (*Zygochlamys patagonica*) fishery: assessment of gear efficiency through a depletion experiment. J. Shellfish Res. 16:59–62.

Lasta, M.L., López, F., Guerrero, R. and Bremec, C., 1998. Distribución espacial de vieira patagónica (*Zygochlamys patagonica*) en la plataforma continental argentina: régimen oceanográfico y deriva larval. Resúmenes XIII Simposio Científico-Técnológico, Comisión Técnica Mixta del Frente Marítimo, Mar del Plata, Argentina, 23–25 de November 1998:49–50.

Lasta, M., Valero, J., Brey, T. and Bremec, C., 2001b. *Zygochlamys patagonica* beds on the Argentinian shelf. Part 2. Population dynamics of *Z. patagonica*. Archive of Fishery and Marine Research 49:125–137.

Lasta, M.L. and Zampatti, E., 1981. Distribución de capturas de moluscos bivalvos de importancia comercial en el mar argentino. Resultados de las camapañas de los B/I " Walther Herwig" y "Shinkai Maru", años 1978 y 1979. INIDEP, Mar del Plata, Argentina Contribución 383:128–130.

Legeckis, R. and Gordon, A.L., 1982. Satellite observation of the Brazil and Falkland Currents 1975 to 1976 and 1978. Deep-sea Res. 29:375–402.

López, F., 1998. Caracterización oceanográfica de la Plataforma Continental Argentina entre los 40° y 48° de latitud sur: causalidad de la distribución espacial de bancos de vieira patagónica (*Zygochlamys patagonica* (King y Broderip, 1832)). Tesis Lic. Cs. Biológicas. Universidad Nacional de Mar del Plata, Argentina. 44 p.

Lusquiños, A. and Valdes, A.J., 1971. Aportes al conocimiento de las masas de agua del Atlántico Sudoccidental. Servicio de Hidrografía Naval, Buenos Aires, Argentina, Publicación H-659. 48 p.

Martos, P. and Piccolo, M.C., 1988. Hydrography of the Argentine continental shelf between 38° and 42° S. Cont. Shelf Res. 8:1043–1056.

Morsan, E. and Narvarte, M., 2000.Consideraciones sobre la situación actual del recurso vieira, *Aequipecten tehuelchus*, ante la factible explotación durante el año 2000. Technical Report IBMP "Alte. Storni", April 2000. 9 p.

Morsan, E.M., Narvarte, M.A. and González, R.A., 2000. Campaña de prospección y evaluación de vieira (*Aequipecten tehuelchus*) en el Golfo San Matías. Informe Técnico IBMP "Alte. Storni". unpublished technical report. 20 p.

Narvarte, M.A., 1995. Spat collection and growth to commercial size of the Tehuelche scallop *Aequipecten tehuelchus* (d'Orb.) in the San Matías Gulf, Argentina. J. World Aqua. Soc. 26:59–64.

Narvarte, M.A., 1998. Settlement of larvae of the Tehuelche scallop, *Aequipecten tehuelchus*, on artificial substrata in the San Matías Gulf (Argentina). Actas II Congreso Sudamericano de Acuicultura, Recife, Brasil. (abstract)

Narvarte, M.A., 1999. Cultivo de vieira tehuelche (*Aequipecten tehuelchus*) en el Golfo San Matías, Río Negro. Final Technical Report, Argentina-UE Fishery Agreement. 28 p.

Narvarte, M.A., 2001. Settlement of tehuelche scallop, *Aequipecten tehuelchus* D'Orb, larvae on artificial substrata in San Matías Gulf (Patagonia, Argentina). Aquaculture 96:55–65.

Narvarte, M.A. and Kroeck, M., 2000. Reproducción de la vieira tehuelche, *Aequipecten tehuelchus*, en el Golfo San Matías. Jornadas Nacionales de Ciencias del Mar, Puerto Madryn (Argentina), Sept. 11–15, 2000 (abstract).

Narvarte, M. and Morsan, E., 2000. Explotación pesquera de la vieira tehuelche, *Aequipecten tehuelchus*, en el Golfo San Matías. Evaluación de la temporada de pesca 2000. Technical Report IBMP "Alte. Storni", November 2000. 17 p.

Narvarte, M.A. and Pascual, M.S., 2000. Efecto de diferentes dietas algales sobre el crecimiento de larvas de vieira tehuelche (*Aequipecten tehuelchus* D´Orb.) en condiciones de cultivo. Jornadas Nacionales de Ciencias del Mar, Puerto Madryn (Argentina), Sept. 11–15, 2000 (abstract).

Natural Resources Consultants, 1995. A plan for development of scallops fishery off Argentina. NRC (Seattle, WA), report to the industry. 81 p.

Negri, R.M., 1993. Fitoplancton y producción primaria en el área de la plataforma bonaerense próxima al talud continental. Seminario-taller sobre la dinámica marina y su impacto en la productividad de las regiones frontales del Mar Argentino. INIDEP (Mar del Plata, Argentina), Technical Report No. 1. 7 p.

Olivier, S.R., 1972. La pesca de vieira en el golfo San Matías. Estado actual y perspectivas. Resultados de las investigaciones sobre la biología y ecología de la vieira tehuelche (*Chlamys tehuelcha*). Pautas para su explotación racional. Technical report prepared for the Río Negro Province fishery administration. 6 p.

Olivier, S.R. and Capítoli, R., 1980. Edad y crecimiento en *Chlamys tehuelcha* (d'Orbigny) (Mollusca, Pelecypoda, Pectinidae) del golfo San Matías (Pcia. de Río Negro, Argentina). Anales del Centro de Ciencias del Mar y Limnología, UNAM, Mexico 7:129–140.

Olivier, S.R., Christiansen, H.E. and Capítoli, R., 1970. Notas preliminares sobre la vieira tehuelche del golfo San Matías (Prov. de Río Negro). Proyecto de Desarrollo Pesquero, UNDP/FAO (Mar del Plata, Argentina), Doc. Inf. 30 p.

Olivier, S.R., Marziale, R.O. and Capítoli, R., 1971. Recursos malacológicos del golfo San Matías, con algunas observaciones realizadas en la campaña exploratoria 'SAO- 1/71'. CARPAS/5/Doc. Téc. 14. 20 p.

Olivier, S.R., Orensanz, J.M., Capítoli, R. and Quesada-Allue, L.A., 1973. Estado actual de las poblaciones de vieira tehuelche (*Chlamys tehuelcha*) en el sector comprendido entre el Bajo Oliveira y el Fuerte Argentino (golfo San Matías). Proyecto de Desarrollo Pesquero, FAO/PNUD (Mar del Plata, Argentina), unpublished technical report. 16 p.

Olivier, S.R., Orensanz, J.M., Capítoli, R. and Quesada-Allue, L.A., 1974. Estado actual de las poblaciones de vieiras, *Chlamys tehuelcha*, en las costas norte y sur del golfo San José, Pcia. del Chubut. Centro Nacional Patagónico, Puerto Madryn, Argentina. Inf. Cient. 2. 14 p.

Orensanz, J.M., 1976. Estado actual de las poblaciones de vieira tehuelche (*Chlamys tehuelcha* (d'Orbigny, 1846)) en el golfo San José. Pautas para su explotación racional. Centro Nacional Patagónico (Puerto Madryn, Argentina), unpublished technical report prepared for the Chubut Province fishery administration. 29 p.

Orensanz, J.M., 1986. Size, environment, and density: the regulation of a scallop stock and its management implications. Can. Spec. Publ. Fish. Aquat. Sci. 92:195–227.

Orensanz, J.M., Parma, A.M. and Ciocco, N.F., 1997. Reproductive reserves and zoning of uses as the only viable framework to prevent overfishing and protect wildlife in San José Gulf Marine

Park (Argentine Patagonia). Fisheries Centre Research Report Series (University of British Columbia, Canada) 5(1):21–22.

Orensanz, J.M., Parma, A.M. and Iribarne, O., 1991. Population dynamics and management of natural stocks. In: S.E. Shumway (Ed.). Scallops: Biology, Ecology and Aquaculture. Developments in Aquaculture and Fisheries, Vol. 21. Elsevier, Amsterdam. pp. 625–714.

Orensanz, J.M., Parma, A.M. and Vacas, H.C., 1982. The scallop (*Chlamys tehuelcha* (d'Orbigny, 1846)) stock of San Matías Gulf (Argentina). Observations on a non-resilient resource following intense harvesting. Unpublished ms. 24 p.

Orensanz, J.M., Pascual, M.S. and Fernandez, M., 1991. Fisheries and aquaculture: Argentina. In: S.E. Shumway (Ed.). Scallops: Biology, Ecology and Aquaculture. Developments in Aquaculture and Fisheries, Vol. 21. Elsevier, Amsterdam. pp. 981–1000.

Orensanz, J.M., San Román, N.A. and Ré, M.E., 1985. Results concerning the scallop *Chlamys tehuelcha* (d'Orb.) stocks of the North-Patagonic Gulfs (Argentina). School of Fisheries, University of Washington, Processed Report.

Paredi, M. and Crupkin, M., 2000. Propiedades bioquímicas y funcionales de actomiosina de músculo aductor de vieira (*Zygochlamys patagonica*). Almacenamiento a 2–4°C. Jornadas nacionales de Ciencias del Mar, Puerto Madryn (Argentina), Sept. 11–15, 2000, abstracts. pp. 98.

Picallo, S., 1979. Consideraciones sobre el empleo adecuado de equipos de suministro de aire para buceo con narguile. Centro Nacional Patagónico (Pto. Madryn, Argentina), unpublished technical report. 42 p.

Picallo, S., 1980. Sobre bancos de mariscos de explotación comercial potencialmente rentable en el golfo San José. Centro Nacional Patagónico (Puerto Madryn, Argentina), Contrib. No. 39. 20 p.

Piola, A.R. and Gordon, A.L., 1989. Intermediate waters in the southwest South Atlantic. Deep-sea Res. 36:1–16.

Podestá, G.P. and Esaías, W.E., 1988. Satellite-derived phytoplankton pigment concentration along the shelf-break off Argentina, 1979–1980. EOS 69:1144.

Pollero, R.I., Ré, M.E. and Brenner, R.R., 1979. Seasonal changes of the lipids of the mollusc *Chlamys tehuelcha*. Comp. Biochem. Physiol. 64A: 257–263.

Reid, J.L., Nowlin, W.D. and Patzert, W.C., 1977. On the characteristics and circulation of the southwestern Atlantic Ocean. J. Phys. Oceanogr. 7:62–91.

Riestra, G., 2000. Análisis de la fauna acompañante asociada a la pesquería de *Zygochlamys patagonica* en aguas uruguayas. In: M. Rey (Ed.). Recursos Pesqueros no Tradicionales: Moluscos Bentonicos Marinos. INAPE/ PNUD (Uruguay), Technical Report. pp. 153–157.

Riestra, G. and Barea, L., 2000. La pesca exploratoria de la vieira *Zygochlamys patagoinica* en aguas uruguayas. In: M. Rey (Ed.). Recursos Pesqueros no Tradicionales: Moluscos Bentonicos Marinos. INAPE/ PNUD (Uruguay), Technical Report. pp. 145–152.

Ruzzante, D.E. and Zaixso, H.E., 1985. Settlement of *Chlamys tehuelchus* (d'Orb.) on artificial collectors. Seasonal changes in spat settlement. Mar. Ecol. Progr. Ser. 26:195–197.

Schejter, L., Bremec, C., Akselman, R. and Hernández, D., 2000. Composición cualitativa de la dieta de *Zygochlamys patagonica* durante la primavera de 1996 en banco Reclutas (39°S – 55°W). IV Jornadas Nacionales de Ciencias del Mar, Puerto Madryn (Argentina), Sept. 11–15, 2000, abstracts. pp. 114.

Vacas, H.C., Pascual, M.S., Iribarne, O.O. and Lasta, M.L., 1984. Campaña de pesca experimental 'Vieira 83'. Unpublished technical report prepared for the Río Negro Province fisheries administration. 36 p.

Valero, J., 1999. Variación estacional, espacial e interanual en el crecimiento de vieira patagónica (*Zygochlamys patagonica*) en la plataforma argentina. INIDEP (Mar del Plata, Argentina), IICA, Final Report. 75 p.

Vernet, M., 1977. Alimentación de la vieira tehuelche *(Chlamys tehuelchus)*. Unpublished ms. 26 p. (20 figs. and tables).

Waloszek, D., 1979. Untersuchungen zum Wachstum, zur Variabilität an *Chlamys*-Fängen des Fischerei-forschungsschiffes "Walther Herwig" des Jahres 1978 vom Argentinischen Schelf mit Klärung des systematischen Status, nebst einigen Angaben zur Biologie und möglichen kommerziellen Nutzbarkeit. Diplomarbeit im Fachbereich Biologie, Hamburg. 160 p.

Waloszek, D., 1984. Variabilität, taxonomie und Verbreitung von *Chlamys patagonica* (King & Broderip, 1832) und Abnerkungen zu weiteren *Chlamys*-Arten von der Südspitze Süd-Amerikas (Mollusca, Bivalvia, Pectinidae). Verhandlungen naturwiss. Ver. Hamburg 27:207–276.

Waloszek, D., 1991. *Chlamys patagonica* (King and Broderip, 1832), a long "neglected" species from the shelf off the Patagonian Coast. In: S.E. Shumway and P.A. Sandifer (Eds.). An International Compendium of Scallop Biology and Culture. World Aquaculture No. 1. The World Aquaculture Society, Baton Rouge, LA. pp. 256–263.

Waloszek, D. and Waloszek, G., 1986. Ergebnisse der Forschungsreisen des FFS 'Walther Herwig' nach Südamerika, LXV. Vorkommen, Reproduktion, Wachstum und mogliche Nutzbarkeit von *Chlamys patagonica* (King and Broderip, 1832) (Bivalvia, Pectinidae) auf dem Schelf von Argentinien. Arch. Fish. Wiss. 37:69–99.

Zaixso, H.E., 1980. Captación de *Chlamys tehuelchus* (d'Orb.) sobre colectores. I. Observaciones preliminares. Centro Nac. Patagónico, Contrib. 37. 20 p.

Zaixso, H.E., 1982. Cultivo de *Chlamys tehuelchus*. I. Observaciones preliminares sobre el crecimiento en soportes de malla rígida. Centro Nacional Patagónico (Puerto Madryn, Argentina), Contrib. No. 74. 13 p.

Zaixso, H.E. and de Espíndola, J.A., 1981. Captación de *Chlamys tehuelchus* sobre colectores. II. Cantidad de material colector. Centro Nac. Patagónico, Contrib. 50, 11 p.

Zaixso, H.E. and Toyos de Guerrero, A., 1982. Captación de *Chlamys tehuelchus* (d'Orb.) sobre colectores. III. Observaciones sobre el nivel de colocación. Centro Nac. Patagónico, Contr. 58.

AUTHORS ADDRESSES

Néstor F. Ciocco - Centro Nacional Patagónico (CONICET), Bvd. Alte. Brown s/n, (9120) Puerto Madryn, Chubut, Argentina (E-mail: ciocco@cenpat.edu.ar)

Mario L. Lasta - INIDEP (Instituto Nacional de Investigacion y Desarrollo Pesquero), Paseo Victoria Ocampo 1, CC 175, Mar del Plata (7600), Argentina (E-mail: mlasta@inidep.edu.ar)

Maite Narvarte - Instituto de Biología Marina y Pesquera "Alte. Storni", Costanera s/n, CC 104, (8520) San Antonio Oeste, Río Negro, Argentina (E-mail: maitenarvarte@canaldig.com.ar)

Claudia Bremec - INIDEP (Instituto Nacional de Investigacion y Desarrollo Pesquero), Paseo Victoria Ocampo 1, CC 175, Mar del Plata (7600), Argentina (E-mail: cbremec@inidep.edu.ar)

Eugenia Bogazzi - INIDEP (Instituto Nacional de Investigacion y Desarrollo Pesquero), Paseo Victoria Ocampo 1, CC 175, Mar del Plata (7600), Argentina (E-mail: ebogazzi@inidep.edu.ar)

Juan Valero - School of Aquatic and Fishery Sciences, University of Washington, Box 355020, Seattle, WA 98195, USA (E-mail: juan@u.washington.edu)

J.M. (Lobo) Orensanz - Centro Nacional Patagónico (CONICET), Bvd. Alte. Brown s/n, (9120) Puerto Madryn, Chubut, Argentina (E-mail: lobo@cenpat.edu.ar)

Scallops: Biology, Ecology and Aquaculture
S.E. Shumway and G.J. Parsons (Editors)

Chapter 27

Scallop Fishery and Aquaculture in Chile

Elisabeth von Brand, German E. Merino, Alejandro Abarca and Wolfgang Stotz

27.1 INTRODUCTION

The total Chilean scallop production has reached the third place worldwide in 1997, and has maintained this position since (Table 27.1).

Three scallop species are of commercial interest in Chile: *Argopecten purpuratus* (Lamarck, 1819*), Zygochlamys patagonica* (King & Broderip, 1831) and *Chlamys vitrea* (King & Broderip, 1831). A fourth species, *Chlamys amandi* is also mentioned, but doubts exist if this is a true species or represents populations of *Z. patagonica*. *A. purpuratus* has a higher commercial value and it is the only scallop being intensively cultivated in bays located in Northern Chile. *Z. patagonica* and *C. vitrea*, are only fished from natural populations in Southern Chile.

This chapter includes a description of the scallop species in Chile with commercial value, development and collapse of the fisheries, and as a result the start of scallop aquaculture, with a brief history of its development, its production methods, and an example of recovery of an overfished bed.

27.2 SPECIES DESCRIPTION

For a better understanding of the scallop distribution in Chile, we have to explain briefly that Chile is divided into 12 regions from north to south, and a metropolitan area, located between the regions V and VI, which is the only one without a coast (Fig. 27.1).

While three scallop species are exploited in Chile: *Argopecten purpuratus, Zygochlamys patagonica* and *Chlamys vitrea,* official governmental sources only distinguish two resources, the "Northern scallop" (*Argopecten purpuratus)*, and the "Southern scallop" (*Z. patagonica* and *C. vitrea*) (Sernapesca 1999). In the latter case, this "Southern scallop", as described in the fishing statistics, is mainly *C. vitrea*, because for many years the minimum size of capture for the Southern scallop was set at 75 mm, which is a size *Z. patagonica* almost never reaches. Just recently the minimum size of capture for *Z. patagonica* was set at 55 mm, but still both species are not separated in the statistics, thus the values represent the sum of the landings of both species. While *A. purpuratus* is only cultivated in the northern part of the country in Regions I – IV (Fig. 27.1), the two latter species are fished exclusively in the far south (Region XII). A fourth species, *Chlamys amandi,* is mentioned for regions X and XI, but doubts exist if it is a true species. Nevertheless, no fishery exists in regions X and XI.

1294

Table 27.1

Main scallop producing countries (amounts are given in metric tons, t).

Year	Country		
	Chile	Japan	China
1989	1,123	369,373	129,605
1991	2,215	367,911	188,832
1993	6,015	465,270	728,411
1995	9,629	502,702	916,514
1997	14,084	515,250	1,001,539
1999	22,383	515,645	712,442
2000	19,370	514,989	919,681
2001	18,947	526,587	960,341
2002	15,552	578,662	935,610
2003[1]	15,109	N.D.	N.D.
2004[1]	24,577	N.D.	N.D.

N.D. = no data available. Source: FAO (1999), Sernapesca[1] (2003, 2004); FAOSTAT data, 2005 (http://faostat.fao.org/faostat/collections?version=ext&hasbulk=0&subset=fisheries)

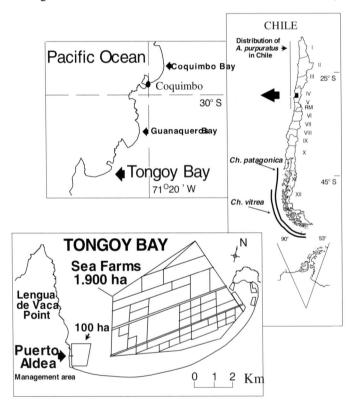

Figure 27.1. Map of Chile showing the distribution of scallop species and location of Tongoy Bay and sea farms within that bay.

27.2.1 *Argopecten purpuratus* (Lamarck 1819) the Northern Scallop

Grau (1959) changed *Argopecten purpuratus* into *Chlamys (Argopecten) purpurata*, and several authors have used this classification. But Waller (1969) states that this scallop species is the southernmost representative of the *Argopecten* group originated in the tropical / subtropical region of the Caribbean and Atlantic. Wolff (1987) studied the population dynamics of this scallop species during the El Niño phenomenon in Perú, and he states that these animals resemble very much the bay scallop *Argopecten irradians irradians* from the Northern Pacific in its population parameters, and he argues that his findings support the conclusion made by Waller (1969). On the other hand, von Brand et al. (1990) studying the chromosomes of the Northern Chilean scallop also agree with Waller (1969) and Wolff (1987) that this species belongs to the genus *Argopecten* and not to *Chlamys* using again the original species name proposed by Lamark 1819, *Argopecten purpuratus*.

Some authors have described the distribution of this species along the Pacific coast from Corinto, Nicaragua to Valparaíso Chile (Navarro et al. 1990), and from Panamá or Nicaragua to the X Region in Chile (40 °S). However, this information is quite inexact, because this species has never been found in Panamá nor any other country besides Perú and Chile (Villalaz, pers. comm.). The most important banks are located between Bahía Sechura - Perú (6° S) and Bahía Tongoy - Chile (31° S) (Vildoso and Chirichigno 1956; Marincovich 1973, Peña 2001) (Fig. 27.1). Individuals have been found as far south as the V Region, south of Valpararaíso. The main banks in Chile are actually found in Bahía San Jorge, La Rinconada (Antofagasta, region II) and Bahía Tongoy (region IV). It is important to point out that in the last decade various specimens were transferred to the South of Chile (Puerto Montt and Chiloé, region X) for aquaculture purposes, but no natural populations exist there.

Natural populations of *A. purpuratus* are found in sheltered zones from 5 to 40 m depth, especially in protected inshore bays with a low water exchange rate (Illanes 1990). Reproductively the northern scallop is a functional hermaphrodite with external fertilisation. The female and male gonads have simultaneous maturation, and it is very common that spermatozoa are released before oocytes during spawning periods (Uriarte et al. 2001). Partial or total spawnings occur all year round, with peaks in spring - summer (November-March). *A. purpuratus* has ideal characteristics for cultivation, which favoured its development as a new species to be reared by aquaculture methods in the early 1980's. This species has a fast growth, reaching the commercial size of about 75–80 mm shell height in about 18 months. Another advantage is that it is a native species, which shows a good survival in its original distribution area and has high commercial value.

27.2.2 *Zygochlamys patagonica* (King and Broderip 1831) the Southern Scallop

Zygochlamys patagonica was originally named *Chlamys patagonica* by King and Broderip (1831), and was transferred into the *Zygochlamys* genus by Waller (1991). It can be found in Chile (Fig. 27.1) from Puerto Montt (region X) to Punta Arenas (region

XII), and to the Magellan Strait (56° S) and Puerto Quequen in Argentina (Navarro et al. 1990). This scallop lives in protected bays, inlets, and channels between 2 to 40 meters deep, as well as in depths over 100 m on the Argentinean platform, where this species sustains a very important fishery in the last few years (Lasta and Bremec 1999). This species has separate sexes, without sexual dimorphism. The salinity in their natural areas fluctuates between 21 to 29‰, and the water temperature between 6 to 10°C. This species is not under commercial aquaculture production so far, therefore all of the captures are obtained from fisheries.

27.2.3 *Chlamys vitrea* (King & Broderip, 1831) the Southern Scallop

This bivalve has separate sexes, being a very mobile species, which in its early life swims around in the seaweeds (Walossek 1991). The shell is translucent whitish or orange in individuals up to 50 mm shell height, later on the shell increases in height and convexity, as well as thickness, attaining a dark brown or violet colour, even in the interior of the valves. Considering these dramatic changes in morphology, it is not surprising that this species has been described as various different species (Walossek 1991). *C. vitrea* is a species, which lives in the shallower parts of fjords, closely related to glaciers, between 48° and 55° S. Guzmán et al. (1999) observed that beds are between 2–17 m deep, with densities between 5–20 individuals m^{-2}. The growth parameters estimated by Guzmán et al. (1999) are L_α = 100–107 and K = 0.23–0.27, demonstrating to be a slow growing species.

27.3 FISHERIES

27.3.1 *Argopecten purpuratus*, the Northern Scallop

Historically, production of the Northern scallop was an extractive activity carried out by artisanal fishermen from the natural populations or "banks" located in North and South Chile. These have been exploited since the pre-Columbian times, and shells have been found at archaeological sites. In the period from 1945 to 1980, the fishery of *A. purpuratus,* the only valuable scallop species at that time, remained at a low level, with landings not exceeding 500 metric tons (t) per year (Fig. 27.2 and 27.3). This volume was used only for internal markets. The first fishery collapse occurred during early 1950, when annual landings did not surpass 400 t. Natural stocks of *A. purpuratus* were already overexploited in 1958, and therefore Chilean authorities applied successive administrative measures trying to avoid a total depletion of the resource, which ended with an indefinite ban of extraction from Arica to Valparaíso to protect the species (Lorenzen et al. 1979; Avendaño and Cantillanez 1996). These management measures were not, however, enough to recover the natural scallop stock. During 1962, only 11 mt of scallops were landed (Hancock 1969). These low captures forced to impose a new period of prohibition in 1971 that was extended until 1975. In the late 1970's, the government developed an active policy to supply international markets with fishery products. As a result, a sharp increase was observed after 1981, and in 1982, regular exports of

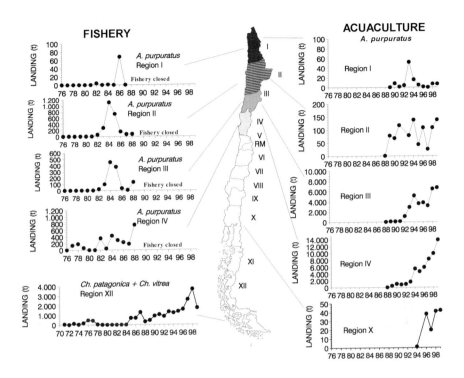

Figure 27.2. Scallop landings of fishery and aquaculture in different regions of Chile.

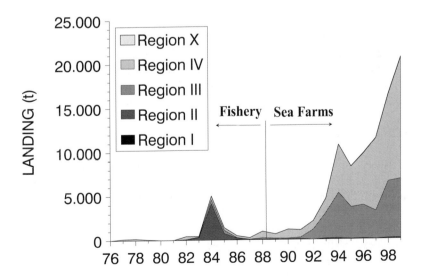

Figure 27.3. Comparison of total landings of fishery and aquaculture in Chile.

A. purpuratus began (Avendaño and Cantillanez 1996), with scallop landings reaching 5,276 t yr^{-1} in 1984. This increase was favoured by a good recruitment in northern Chile in 1983 due to the El Niño, a phenomenon well described for scallops on the Peruvian coast by Wolff (1987). But the fishery could not be sustained at that level, and landings decreased in the following years. In order to prevent another collapse, several regulations, including minimum size of capture and closed seasons were decreed by fishery authorities. Nevertheless, as no recovery was registered, the fishery was completely closed in 1988, and has remained closed since then, with a prohibition renewed on a two-year basis.

As shown in Fig. 27.2, the landings fluctuated heavily among years, which is probably caused by strong variations in the magnitude of settlement and recruitment. The strong recruitment during the 1983 "El Niño", as already mentioned, sustained high landings during the years 1984–1985 in most parts of the country (Fig. 27.2). However, this high fishing effort (and landings) could not be sustained during the following years. The heavily overexploited stocks were unable to produce a minimum amount, and the fishery had to be closed completely. The closure of the northern scallop fishery was the main incentive to develop and initiate the aquaculture of this species, an activity that began experimentally around 1985. The first harvests of cultured scallops were registered in 1989. However, as pointed out by Stotz and González (1997), at this time the aquaculture represented to some extent also the landing of illegal extraction of individuals from the natural scallop beds. These individuals collected were considered as "seed" for cultures, understanding all available sizes as "seed".

A. purpuratus is a fast growing and productive species. In Tongoy Bay the main settlement period occurs at the end of summer or beginning of autumn (February-April). The individuals grow to 90 mm shell height (minimum legal size for fisheries) in 18 months; the growth parameters being L_α = 124.58 and K = 0.84 (Stotz and González 1997). The production estimated for Tongoy Bay, with a low density of 1.64 individuals m^{-2} due to fishing, was of 0.528 ± 0.238 kg m^{-2} yr^{-1} (P/B = 2.08). With optimal management, maintaining a standing stock of 10 individuals m^{-2}, a production of 32 t ha^{-1} yr^{-1} is estimated, allowing an annual yield of 3.2 t ha^{-1} (Stotz and González 1997).

Scallops are collected by hand. The fishermen dive to a maximum depth of 20–25 m, with the aid of an air compressor on a boat ("Hooka" diving) using boats of 5–7 m length, with outboard motors of 15–45 HP).

27.3.2 *Chlamys vitrea* and *Zygochlamys patagonica*, the Southern Scallop

The fishery of the Southern scallop species was assumed to be capturing *Zygochlamys patagonica*. Only individuals bigger than the minimum size of capture of 75 mm were fished. This differed significantly from what was known for the species from the Argentinean Continental Shelf, where the maximum sizes (L_α) were below this level (54.4–74.6 mm). Doubts regarding the identity of this species fished in the Patagonian region appeared. Valladares and Stotz (1996) demonstrated through an experimental research that the growth of *Z. patagonica* in the Patagonian region was similar to that described for the Argentinean fishery, with an $L_\propto$ = 63 and K = 0.75. Thus, the fishery was really capturing another species, which was identified as *Chlamys vitrea*.

C. vitrea is a species, which lives in the shallower parts of fjords, closely related to glaciers, between 48° and 55° S. Guzmán et al. (1999) observed that the beds are located between 2–17 m depths, with densities between 5–20 individuals m^{-2}. The growth parameters estimated by Guzmán et al. (1999) are $L_\propto$ = 100–107 and K = 0.23–0.27, demonstrating it to be a slow growing species.

Z. patagonica lives in greater depth, but shallow beds can also be observed. Guzmán et al. (1999) sampled beds between 12–25 m, with densities between 3–50 individuals m^{-2}. Valladares and Stotz (1996) worked in a bed with a distribution between 5 to 25 m depth. On the Argentinean Continental Shelf, the species lives at a depth around the 100-m isobath (Lasta and Bremek 1999). Probably similar deep beds of the species exist in fjords in the Patagonian region, but no survey at such depth is known.

The minimum legal size of capture of 75 mm was only maintained for *C. vitrea*, being lowered for *C. patagonica* to 55 mm. Nevertheless, the fishery statistics still does not distinguish between both, giving only one figure for both species together. The landings were increasing until 1998, when they reached the level of 3,662 metric tons, but in the following year they decreased to 1,715 t suggesting that the maximum productive capacity of both species have already been reached. Extensive beds of *Z. patagonica* may exist at greater depths than those where diving is allowed (20 m). These could probably be fished using dredges. The Chilean law, however, completely prohibits fishing with dredges, as it is assumed to damage the benthic communities. Scallops are only allowed to be hand collected by divers, which work from 7–9 m boats equipped with air compressor (Hookah diving). Those beds, probably existing at greater depths are protected from fishery and will always serve as a reserve, which can provide larvae and settlers to the beds that are exploited. On the other hand, *C. vitrea*, which only lives in shallow water in fjords, is more exposed as a species to be harmed by fisheries. First attempts were made by the Universidad de Magallanes (region XII) to culture this species, and in this way assure its production. These attempts are still at a laboratory scale, with no commercial production until now.

27.4 HOW SCALLOP AQUACULTURE STARTED IN CHILE

The scallop aquaculture in Chile has a history of approximately 28 years starting with the first studies carried out by Hogg (1977), who made the first growth experiments in natural environment. One year later, Bustos (1978) obtained more information about seed collection. The first laboratory results were recorded by Sanzana (1978) studying the cycle of gonadal maturation and cultivating the first larval stages. Padilla (1979) followed with the culture of larvae in laboratory. While these were the first attempts at learning about the scallop *Argopecten purpuratus,* it was really in the early 1980's when research began with the main purpose to find a technical viability to produce the Northern scallop using aquaculture methods (Akaboshi and Illanes 1983). DiSalvo et al. (1984) were able for the first time to produce Chilean scallop larvae under controlled conditions. The next step was to find a way to produce them on a commercial scale, and the answer came from

Japan. The technology used to farm Japanese scallop (*Patinopecten yessoensis*) was introduced in Chile and applied successfully to the Northern scallop, *Argopecten purpuratus*. Three different organisations joined their efforts to develop commercial scallop farming, the Universidad [Católica] del Norte (Coquimbo, Chile), JICA (Japan International Cooperation Agency) and the Subsecretaría de Pesca (Governmental Agency for Fisheries and Aquaculture, Chile). The purpose of this joint effort was to start the necessary basic and applied research to develop the Northern scallop aquaculture under controlled conditions and in natural environment. Once the Japanese technology was adapted to the Northern scallop and tested in Tongoy Bay, the knowledge and technology developed was transferred to the productive sector (Illanes 1990; Vega 1996; Uriarte et al. 2003). In the following year, the OFCF (Overseas Fishery Corporation Foundation), the Subsecretaría de Pesca and SERNAP (Servicio Nacional de Pesca – National Fisheries Service, Chile) also started a project in Tongoy Bay to develop a new culture technology for *Argopecten purpuratus*. In 1984, the first controlled scallop production was 57 t from 6 culture centres (Vega 1996; Uriarte et al. 2003). But Stotz and González (1997) state that this production was possibly due to illegal landings because the production process still had several flaws. During the first attempts to start the commercial aquaculture production, the main restriction happened to be the seed source. Therefore a hatchery technology to produce good quality seed was developed as a response to the need of regular seed supplies for growers. Nowadays, the scallop producers in Chile, mainly private producers, have two seed sources: collecting them from the natural environment (80%) and hatchery production (20%) (Illanes 1988; Bourne et al. 1989; Abarca 2001; Uriarte et al. 2001).

The results from the first experimental-pilot hatcheries showed problems related to larval survival, settlement, metamorphosis and postmetamorphic culture or nursery. During the decade of 1980, the biggest constraint to the acceleration of the Northern scallop aquaculture development was the high degree of investment required and difficulties to determine or adapt different ranges of scallop production.

In 1985, the Universidad [Católica] del Norte and the Japanese Government through JICA built the Coastal Center for Aquaculture and Marine Research, which meant a big step forward in research and development of hatcheries for marine invertebrates, specially bivalves. The Coastal Center was built with state of the art of Japanese hatchery technology with a pumping capacity up to 50 m^3 h^{-1}, becoming the biggest and most technologically up-to-date facility of its kind in Latinamerica. The same year, CICUM (Centro de Investigacionesde Cultivos Marinos) belonging to a private fishing company, Pesquera Guanaye, built a hatchery and started its own scallop production in Antofagasta (Chile). Later Novamar (1986) appeared in Arica, and Biomar (1987) in Coquimbo. In 1989, six scallop hatcheries and 26 culture centres were operating, between region I and IV, with a production of 200 t (Illanes 1990). The commercial activities transformed themselves into one of the most important industries, mainly in the regions III and IV, and these regions were once considered the poor or marginal regions of Chile. Now they produce about 98% of the national scallop production (Sernapesca 2003).

Another big group of commercial and pilot hatcheries, Cultivos Marinos Internacionales was built in 1990 in region III, becoming the largest commercial hatchery in the country. Later, Pesquera Camanchaca (1992), Cultivos Marinos Guayacán (1993), Aquaingeniería (1994), Pesquera San Jose S.A., Cultivos Flamenco and Hidrocultivos (1995) were added, located in the regions III and IV. In addition, in the south of Chile (region X, 40° S) the hatchery of Cultivos Marinos Bahia Yal (1996) started culturing *A. purpuratus*; a region where the Northern scallop is not naturally distributed. Lately, new information has been obtained on growth and survival of *A. purpuratus* larvae from hatcheries in Southern Chile (Uriarte et al. 1996). Between 1985–1995, at least twelve to fourteen hatcheries were created, but several of these industries declined due to technical and probably economical problems (Vega 1996). In 2001, approximately twelve scallop hatcheries of different types and sizes produced *A. purpuratus* spat between regions II and X in Chile (Abarca 2001).

Most of the scallop production is for export, reaching markets such as France, USA, Belgium, Germany, Argentina and Asia. The production in Chile has gradually increased up to twenty times between 1989 and 1999, and now the country is considered the third scallop aquaculture producer worldwide, after China and Japan (Table 27.1). In 1999 a total of 22,383 t of scallops were harvested and 92% came from cultivation and extraction by artisanal fishermen of the Northern scallop. It seems that Northern scallop production reached a plateau after 1999 with a production range between 15,000 to 19,000 t (Table 27.1). The landings of the Southern scallop (*Zygochlamys patagonica* and probably *Chlamys vitrea*), come exclusively from artisanal fishery and not from aquaculture (Fig. 27.2). The highest capture peak has been of 3,662 tons in 1998, and the capture descended later to 1,715 t in 1999 (Sernapesca 1999). This predicts that Chilean aquaculture follows the activity's world trend of production.

27.5 AQUACULTURE PRODUCTION

The scallop hatchery culture involves broodstock conditioning, spawning, fertilisation, larval culture, settlement and metamorphosis and seed growth up to commercial size (Navarro et al. 1991; Merino et al. 2001). Each one of these stages of the production cycle has different time lengths (Table 27.2), depending on the biology of the species (Narvarte et al. 2001), physical and chemical conditions of the water (Uribe and Blanco 2001) and other factors such as biofouling (Uribe et al. 2001).

The culture technology used in Chile is based on the Japanese system developed in the 1960's and it is basically a suspended culture on floating structures called "long-lines" (Merino et al. 2001). To collect seed, "collector bags" are used to capture natural larvae in the wild prior to settlement (Narvarte et al. 2001) (Fig. 27.4). Trays known as "pearl nets" for the intermediate culture and structures for the final culture step called "lanterns" (Fig. 27.4). All the procedures described below are focused exclusively on the Northern scallop *Argopecten purpuratus*. This scallop is the only one being intensively produced in Chile by the aquaculture industry (Merino 1997).

Table 27.2

Culture systems and densities used in some scallop production centres in Tongoy Bay, Chile.

Culture system	Density (scallops / system)	Scallop size (mm)	Time (days)
Collector bag	variable	7–11 (out)	60–90
Pearl net 4.5 mm	150	7–11 (in)	90–120
Pearl net 9 mm	30–40	20–30 (in)	120
Lantern net 12 mm	30	30–40 (in)	60–90
Lantern net 21 mm	20–30	50–60 (in)	90–120

Figure 27.4. Different culture systems used for *Argopecten purpuratus* in Tongoy Bay, Chile. From left to right: collector bags, small lantern net, pearl nets and large lantern net.

The production cycle is usually between 14 to 18 months from fertilisation, reaching a commercial size of 75 to 85 mm in shell height. It has been established that scallops older than 18 months do not increase significantly in meat yield, because beyond this age they devote most of their energy to reproduction, showing more maturation peaks than younger individuals, not leaving much energy for meat growth.

27.5.1 Production Stages

27.5.1.1 Hatchery broodstock conditioning and spawning

Adults used as broodstock are brought into the hatchery, cleaned from biofouling and placed into clean running seawater in separate tanks for spawning. Depending on the developmental stage of the gonad, an adequate amount of microalgae must be given to reach full ripeness (Uriarte et al. 2001). The assessment of maturity, before the spawning stimulation starts, can be made in several ways. A simple, non-destructive method is a visual inspection of the state or condition of the gonads. The more conventional, destructive method is the gonad index (wet weight of gonad in relation to the total wet weight of the soft parts x 100). Generally, a full and bright orange gonad or a gonad index over 20% is desirable for successful spawning. A constant water flow is needed to maintain a suitable level of dissolved oxygen in the seawater (>7–8 mg L^{-1}). Bourne et al. (1989) mentioned water flow rates of 10 L min^{-1} for *Patinopecten yessoensis*. Seawater used for all phases of spawning must be filtered to at least 5 μm. The most used technique to induce spawning is thermal stimulation by increasing the temperature between 5 to 7°C (Bourne et al. 1989). If thermal stimulation does not work, as alternative solution chemical inducers like serotonin applied by intramuscular injection to promote sperm release can be used (Bariles and Gaete 1991). Chemical induction does not have widespread use in commercial hatcheries where thermal induction is preferred. To have successful fertilisation, DiSalvo et al. (1984) suggest the use of a proportion of 10 spermatozoa per oocyte for *A. purpuratus* to avoid polyspermy. Some small experiments using a much higher number of sperm cells did not show polyspermy in this species (Bellolio and Lohrmann, pers. comm.).

During the fertilisation period, it is advisable to use only 1 μm filtered and UV treated seawater. During this time no feeding and additional air supply are required. Approximately 48 h after fertilisation, the first larval stage, "D" larva, is reached.

27.5.1.2 Larval stage

The scallop begins its life as a free swimming or planktonic larva, but with a totally different behaviour and physical appearance compared to juvenile and adult scallops. The larvae can be cultivated in differently shaped tanks like conical cylinders, rectangular or cylindrical tanks. The most used tanks are the conical - cylinder with a drainage exit at the bottom. Water should be filtered, at least to 5 μm and UV treated. During this time the larvae require a fair amount of microalgae as live food. Many species of microalgae have been cultivated and used in hatcheries (Table 27.3). They can be divided in two groups: flagellated and diatoms. Most of the species commonly used in the hatcheries are known worldwide and recognised as species of easy culture and nutrition value (Uriarte et al. 2001). For the water changes, sieves of nylon mesh are used to retain the larvae. As far as possible, it is advisable to record some physical, chemical and biological factors like temperature, pH, salinity, dissolved oxygen, the amount of food, bacteria and others.

The larval period can last between 16 to 25 days depending upon water temperature, and the amount and quality of microalgae used. The final step is the pediveliger or eyed larva with a size of 230 μm. The morphological characteristics of this larva are the presence of a very well developed "foot" and an eyespot.

27.5.1.3 Settlement, metamorphosis and postlarval stage

The larval stage ends when the larva has settled on a prepared substrate, and it begins metamorphosis that will end as a small juvenile scallop. This new stage is not planktonic, and therefore the animal remains at the bottom of the sea or lightly attached on a prepared substrate within the culture tank, but eventually "swims" shaking its valves. During the metamorphic process, the animal stops food intake and starts using the energetic reserves accumulated during the larval stage. Under laboratory conditions, the process of settlement and metamorphosis takes between 2 and 5 days at 18–20°C. In some cases the substrate (collector bags) is kept for 3 to 5 days before larval settlement in seawater with microalgae to generate a microbial film on the surface to facilitate a natural induction to settlement (Narvarte et al. 2001). The use of artificial inducers at an industrial scale is still under evaluation from an economic and environmental point of view. After larval settlement is completed, the collector bags are transferred from the hatchery to a long-line (Illanes 1990; Merino 1997; Merino et al. 2001). Collector bags remain about 60 to 90 days in the sea, depending on the time of the year, until the scallops reach 6 to 8 mm in length (Abarca, pers. comm.).

Table 27.3

Common microalgae species used in scallop hatcheries.

Species	Class	Scallop size	Recommended use
Chaetoceros calcitrans	Baccilariophyceae	2.5–3.75	Larvae
C. simples	Baccilariophyceae	8.0–9.0	Juveniles, adults
C. gracilis	Baccilariophyceae	5.0–10.0	Juveniles, adults
Thalassiosira pseudonana	Baccilariophyceae	2.5–10.0	Juveniles, adults
Skeletonema costatum	Baccilariophyceae	6.3–10.0	Juveniles, adults
Isochrysis sp. *(var. Tahiti)*	Prymnesiophyceae	3.3–5.75	Larvae, juveniles and adults
Pavlova lutherii	Prymnesiophyceae	3.5–6.5	Larvae, juveniles
Tetraselmis sp.	Prasinophyceae	8.0–12.0	Juveniles, adults

27.5.1.4 Seed supply

Seed can be obtained from two sources: natural populations and hatchery. Natural stocks provide either eyed larvae (D larvae) or juveniles. The collector system consists of plastic mesh bags (generally Netlon™), which are suspended from a long-line to form a curtain or barrier to the passing of the swimming larvae in the water column. The time between settlement and metamorphosis in the collector bags will be mainly regulated by environmental temperature and occurs at least between 15 to 20 days after spawning in the wild (Illanes 1990). Each scallop farming company or culture centre usually has an exhaustive monitoring system to sample and record the quantity and size of Northern scallop larvae in the plankton to determine the moment when collector bags have to be deployed in the ocean. Collector bags are hung from a long-line for 2 to 3 months, which allow the settled scallops to reach around 7 to 9 mm before being transferred to the next culture stage in an intermediate culture systems (Table 27.2, Fig. 27.4). Keeping collector bags for periods over 3 months in the sea is not recommended since high mortality by predation, fouling and loss of the collectors due to excessive weight have been reported. Wild seed collection shows seasonal and interannual variability in the abundance of seed due to fluctuations of environmental conditions (Abarca et al. 1999). The presence of scallop larvae in collector bags are not time-constant, because they are influenced by environmental conditions like larval density, weather conditions, water temperature, transparency, salinity, currents and other biotic and abiotic factors of the natural spawning area. The quantity of seed captured in collector bags is highly variable from one year to another (Pereira and Molina 1997; Narvarte et al. 2001). The predictability of larvae in natural environment was the first and main bottleneck to enhance the industrial farming of the Northern scallop. Generally, during spring and summer all the scallop farming companies deploy collectors in the oceans on long-lines to catch natural larvae. If there are not sufficient larvae available in the wild, then hatchery practices are required to supply them. Hatcheries can get their broodstock from their own on-growing systems or from other producers. In Chile, a total of 13 commercial scallop hatcheries exist (Abarca, pers. comm.). Chilean Northern scallop hatcheries can produce almost all year around, which guarantees a constant supply of seed to the growers during the year and allows the possibility to have a planned production system. Natural seed sources are not constant in spawning time or amount of seed yielded.

27.5.1.5 Ongrowing stage

During the intermediate culture step, the seed between 7 and 9 mm are selected and transferred from the collector bags to pearl-net systems for at least 3 to 4 months, depending on the type of pearl nets used and the conditions of the culture area, near or far from the beach. The final stage of culture begins with the transfer of scallops from pearl nets to lantern nets for a period of 5 to 7 months (Table 27.2, Fig. 27.4) (Merino et al. 2001).

There is another culture system which is sometimes used for the final stage that is called "ear-hole", where a hole is drilled into the "ear" of the scallop and a plastic filament is used to hang the animal to a rope (Fig. 27.5). The "ear-hole" method is inexpensive, but has the risk of serious biofouling, which might lead to partial or complete loss of scallops due to a rupture of the "ear" when oceanographic conditions are not calm.

The ongrowing stage takes approximately 12–15 months after the seed stage is completed (out of collector bags). At the end of the first year starting from fertilisation, at about 40 mm shell height, they reach first reproductive maturity, and during their second year they reach the commercial size of 75 to 85 mm, depending upon the natural conditions of the growing area. Cultured animals can be harvested before, however, with smaller sizes if the market requires them.

27.5.2 Constraints and Opportunities for Northern Scallop Hatcheries

The development of scallop hatcheries has been marked by the following aspects:
a) hatcheries are not a high-priority as long as natural collected seed are available, cheaper and able to sustain the business;
b) the evolution of scallop prices on the international markets has not favoured new investments;

Figure 27.5. "Ear hole" culture system used in Tongoy Bay, Chile.

c) not many adequate and useful coastal areas exist with basic conditions such as electrical energy, fresh water or communications to install hatcheries;

d) the cyclic oceanographic fluctuation of El Niño and La Niña phenomena are responsible for the variation in collection of seed from nature. The El Niño phenomenon with its increase of water temperature produces an increase in spawning peaks, and finally a high seed production (Illanes et al. 1985; Merino et al. 2001). The La Niña with a decrease in water temperature produces exactly the opposite.

Nevertheless, some private companies have decided, on a long-term vision, to maintain or create new hatcheries to provide stability in seed availability, but with expensive seed. Scallop hatchery information is often limited for many reasons. The results have not always been better and finally the cost of hatchery seed is higher than those collected naturally (Illanes 1990). Considering the latter statement, hatchery-produced seed can have the following advantages:

a) to complement the natural collection;

b) to maintain a constant supply, even when environmental conditions are not favourable;

c) to allow early genetic selection and / or manipulation to produce better offspring (von Brand and Ibarra 2001).

Lastly, in a three-year study Abarca et al. (1999) found that the performance of hatchery seed compared with those from wild collection showed that 46.4% and 56.65% survive until harvest, respectively. Hatchery seed appears to be weaker than wild seed due to its "artificial nature". Some private companies that run hatcheries obtain up to 20 to 40% of the total amount of seed from hatcheries.

27.5.3 Interaction between Natural Beds and Aquaculture

Aquaculture of Northern scallop began through a special regulation, which guarantees farming companies some kind of ownership of a given seawater area to fulfil aquaculture activities. The seawater aquaculture concessions were granted in the same bays in which natural beds exist or existed. For example, in Tongoy Bay (which has a total surface of 3,600 ha) 1,900 ha were given to private enterprises to develop scallop culture. In this way, the space for recovery of wild scallop beds was greatly reduced. Moreover, on the open grounds, scallops were subjected to an intensive clandestine fishery and then sold as seed for ongrowing culture (Navarro et al. 1991; Wolff and Alarcón 1993; Stotz and González 1997). Thus, in spite of the closure of the fishery, beds did not recover. On the contrary, they were almost completely destroyed (Stotz 2000).

In the meantime, the cultures developed very rapidly, the harvests of cultured Northern scallops in the regions III and IV in 1999 already representing almost 10–20 times the volumes landed by fishery in the past (Fig. 27.3). A survey of stock sizes in natural beds and in cultures, shows that the wild stock already represents only 2–9% of the total stock in Tongoy Bay, or at best 10% in the entire country, being most of individuals in sea farms. This leads to the conclusion that A. purpuratus already represents a "domesticated" species (Stotz 2000). In this context, and considering the occurrence and development of selection practices in sea farms, genetic variability within the species appears to be at risk and already problems are showing up. As the faster growing individuals are harvested up to a year earlier than the slowest growing individuals, they contribute less to reproduction, with the result of an involuntary selection in favour of slower growing individuals. As a consequence, the increasing occurrence of dwarfs (very slow growers) in culture is mentioned as a problem by some sea farmers (Stotz 2000). For proper conservation of genetic variability, one possibility is to protect and restrain management of some wild populations. One reserve has already been established in region II (Avendaño et al. 1997). Other reserves in the more productive regions III and IV have not been considered, yet.

27.5.4 Recovery of a Natural Bed of *Argopecten purpuratus*

The Chilean fishery law of 1992 incorporates the possibility for fishermen to get exclusive fishing rights in limited coastal zones, called "management areas". In 1991, while the law was still in discussion, the Fishermen Union of Puerto Aldea, (region IV), a small fishing village on the southern corner of Tongoy Bay (Fig. 27.1), decided to protect the area in front of the town. Within few years the Puerto Aldea scallop bed increased its density and biomass, reaching by now a medium density of 4.74 individuals m^{-2} and a biomass of 0.572 kg m^{-2} (results of a direct stock assessment done in August 2000 by Stotz). Nevertheless, this recovery probably is closely related to the reproduction of the scallops maintained in sea farms, which, as mentioned before, represent between 91–97% of the scallops in Tongoy bay. Thus, we cannot really talk of the recovery of a true wild stock at Puerto Aldea. Therefore, and with the aim of conservation of genetic variability, natural beds in bays or sites not related to scallop sea farms need to be found. But the "management area" system represents a way to recover the productivity of scallop stocks in open areas within bays with sea farms. According to figures given by Stotz and González (1997) to the present, production of region IV, which amounts in 1999 to 13,793 t, the well managed "management areas" within Tongoy Bay could add another 11,200 t, thus completing a total production of around 25,000 t for the region. The management should include a restocking to an average density of 10 individuals m^{-2}, and limit the fishing rate to 10% of the total production of the stock. For restocking only protection of the bed is needed, but the introduction of seed collection techniques used in sea farms could improve the restocking. In the latter case, genetic selection for restocking areas should be considered, to avoid a loss of genetic variability due to the use of inadequate seed.

27.5.5 Final Overview and Projections of Chilean Scallop Farming

Actually, 27 companies are dedicated to the culture of this scallop species, 13 hatcheries producing seed and approximately 8 institutes and universities are involved in basic research like biology, reproduction physiology, nutrition, genetics, pathology (parasitology and microbiology), and technological aspects and bioengineering for farming facility systems design. The interaction among institutes, universities and the production sector is becoming more important, and many research projects are increasingly involving companies, trying to solve "real" farming and operational problems.

About 3,000 permanent and 1,000 temporary jobs are associated with the Chilean scallop industry (Fig. 27.6). On the other hand, the great development of this industry has allowed the direct or indirect development of new supply companies, which are producing pools, boats (Fig. 27.7), operational task rafts (Fig. 27.8) and operational docking areas. Also, an improvement of quality in products like plastic mesh, ropes, etc. has occurred.

Figure 27.6. An important amount of workers are required to support daily farming operations both in sea farming systems and land operation systems. Northern scallops are being introduced into pearl-nets in a land docking management area.

Figure 27.7. Aquaculture boats equipped with hydraulic cranes are being used to lift lantern nets from the long-line.

Figure 27.8. Task raft developed for scallop culture in Tongoy Bay, Chile. On this raft, culture steps such as selection, cleaning and change from one culture system to another are performed.

ACKNOWLEDGMENTS

We want to thank Leandro Sturla for his comments and Dr. Karin Lohrmann for her help with the manuscript.

REFERENCES

Abarca, A., 2001. Scallop hatcheries of *Argopecten purpuratus* (Lamarck, 1819) in Chile. A survey of the present situation. Book of abstracts 13th International Pectinid Workshop, Coquimbo, Chile. pp. 52–53.

Abarca, A., Fierro, J., Retamal, A. and Monsalve, H., 1999. Growth and survival of *Argopecten purpuratus* (Lamarck, 1819) juveniles originated from hatchery and natural collectors. A comparison of three years data. 12th International Pectinid Workshop, Bergen, Norway. Abstract. pp. 52–53.

Akaboshi, S. and Illanes, J.E., 1983. Estudio experimental sobre la captación, precultivo, y cultivo en ambiente natural de *Chlamys (Argopecten) purpurata,* Lamarck 1819, en Bahía Tongoy, IV Región, Coquimbo. Proceedings Symposium Internacional: Avances y Perspectivas de la Acuacultura en Chile. Coquimbo, Chile. pp. 233–254.

Avendaño, M. and Cantillanez, S.M., 1996. Efectos de la pesca clandestina, sobre *Argopecten purpuratus* (Lamarck, 1819) en el banco de la Rinconada, II Región. Ciencia y Tecnología del Mar (Chile) 19:57–65.

Avendaño, M. and Cantillánez, M., 1997. Necesidad de crear una Reserva Marina de ostiones en el banco de La Rinconada (Antofagasta, II Región, Chile). Estudios Oceanológicos 16:109–113.

Avendaño, M., Cantillánez, M., Rodríguez, L., Zuñiga, O., Escribano, R. and Oliva, M., 2000. Estudio de línea base proyecto "Conservación y protección Reserva Marina La Rinconada, Antofagasta". Informe Final, Proyecto Código BIP No. 20127869–0. Universidad de Antofagast. 185 p.

Bariles, S. and Gaete, U., 1991. Induction of sperm release in the scallop *Argopecten purpuratus* (Bivalvia: Pectinidae) using serotonin (5-hidroxytriptamine). Malacol. Rev. 24(1–2):19–24.

Bourne, N., Hodgson, C.A. and Whyte, J., 1989. A manual for scallop culture in British Columbia. Canadian Technical Report of Fisheries and Aquatic Sciences No. 1694.

Bustos, C., 1978. Investigaciones del ostión en la IV Región. Informe. Centro de Investigaciones Marinas. Universidad del Norte, Coquimbo, Chile. 44 p.

DiSalvo, L., Alarcón, E., Martínez, E. and Uribe, E., 1984. Progress in mass culture of *Chlamys (Argopecten) purpurata* Lamarck (1819) with notes on its Natural History. Rev. Chil. Hist. Nat. 57:34–45.

FAO, 1999. FAO Fisheries Circular / FAO Circulaire sur les Pêches / FAO Circular de Pesca. No. 815, rev. 11. Rome, Italy. 203 p.

Grau, G., 1959. Pectinidae of the eastern Pacific. Plates 1–57. Allen Hancock Pacific Expeditions. University of Southern California Press, Los Angeles, USA 23:103–105.

Guzmán, L., Brown, D., González, M., Cornejo, S. and Almonacid, E., 1999. Investigación biológica pesquera en ostiones en la XII Región. Informe Final FIP No. 97–27, Subsecretaría de Pesca, Ministerio de Economía, Fomento y Reconstrucción de Chile. 80 p.

1312

Hancock, D., 1969. La pesquería de mariscos en Chile. Published by Instituto de Fomento Pesquero de Chile (IFOP) 45: 94 p.

Hogg, D., 1977. Natural History of the Northern Chilean scallop. Informe. Centro de Investigaciones Submarinas. Universidad del Norte, Coquimbo, Chile. 19 p.

Illanes, J.E., 1988. Experiencias de captación de larvas de ostión (*Argopecten purpuratus*) en Chile, IV Región. En: E. Uribe (Ed.). Producción de larvas y juveniles de especies marinas. Universidad del Norte, Coquimbo, Chile. pp. 53–57.

Illanes, J.E., 1990. Cultivo del Ostión del Norte, *Argopecten purpuratus*. In: A. Hernández (Ed.). Cultivo de Moluscos en América Latina. 405 p.

Illanes, J.E., Akaboshi, S. and Uribe, E., 1985. Efectos de la temperatura en la reproducción del Ostión del Norte *Chlamys (Argopecten) purpuratus* en la Bahía de Tongoy durante el fenómeno El Niño 1982–83. Investigaciones Pesqueras (Chile) 32:167–173.

Illanes, J.E., Pereira, L. and López, A., 1993. Natural seed collection of the scallop *Argopecten purpuratus*, a case of technology transfer to artisanal fishermen. Proceedings of the 9th International Pectinid Workshop, Nanaimo, B.C., Canada. Can. Tech. Rep. Fish. Aquat. Sci. 1994(1):162–178.

Jaramillo, R., Winter, J., Valencia, J. and Rivera, A., 1993. Gametogenic cycle of the Chiloé scallop (*Chlamys amandi*). J. Shellfish Res. 12(1):59–64.

Lasta, M.L. and Bremec, C.S., 1999. Development of the scallop fishery (*Zygochlamys patagonica*) in the Argentine Sea. 12th International Pectinid Workshop 5–11 May 1999, Bergen, Norway. Abstract. pp. 154–155.

Lorenzen, S., Gallardo, C., Jara, C., Clasing, E., Pequeño, G. and Moreno, C., 1979. Mariscos y Peces de Importancia Comercial en el Sur de Chile. Universidad Austral de Chile, Valdivia, Chile. 131 p.

Marincovich, L.J., 1973. Intertidal mollusks of Iquique, Chile. Natural History Museum, Los Angeles County (USA), Science Bulletin No. 16: 49 p.

Merino, G., 1997. Considerations for longline culture system design: scallop production. In: C.E. Helsley (Ed.). Open Ocean Aquaculture 97. Charting the future of ocean farming. Proceedings of an International Conference. Maui, Hawaii. April 23-25, 1997. pp. 145–154.

Merino, G., Barraza, J. and Cortez-Monroy, J., 2001. Capitulo 19: Ingenieria y dimensionamiento de sistemas de cultivos en moluscos. In: A.N. Maeda-Martinez (Ed.). Los Moluscos Pectinidos de IberoAmerica: Ciencia y Acuicultura. Editorial Limusa, México. pp . 375–404

Narvarte, M., Félix-Pico, E. and Ysla-Chee, L., 2001. Capítulo 9: Asentamiento larvario de Pectínidos en colectores artificiales. In: A.N. Maeda-Martinez (Ed.). Los Moluscos Pectinidos de IberoAmerica: Ciencia y Acuicultura. Editorial Limusa, México. pp. 173–192.

Navarro, R., Sturla, L., Cordero, O. and Avendaño, M., 1991. Chile. In: S.E. Shumway (Ed.). Scallops: Biology, Ecology and Aquaculture. Developments in Aquaculture and Fisheries Science, Vol. 21. Elsevier, Amsterdam. pp. 1001–1015.

Peña, J.B., 2001. Capítulo 1: Taxonomía, Morfología, Distribución y Hábitat de los pectínidos Iberoamericanos. In: A.N. Maeda-Martinez (Ed.). Los Moluscos Pectínidos de Iberoamérica: Ciencia y Acuicultura. Editorial Limusa, Mexico. pp.1–25.

Pereira, L. and Molina, G., 1997. Development and present stage of the seed collection of the Northern Chilean scallop *Argopecten purpurata* (Lamarck) in Tongoy Bay, IV Región, Chile. 11th International Pectinid Workshop, La Paz, Mexico. Abstract. pp. 18–19.

Sanzana, J., 1978. Estudios preliminares del ostión *Chlamys purpurata* Lamarck (1819) en la zona de Valparaíso. Tesis, Escuela de Ciencias del Mar, Universidad Católica de Valparaíso, Chile. 57 p.

SERNAPESCA, 1999. Anuario Estadístico de Pesca. Servicio Nacional de Pesca, Ministerio de Economía, Fomento y Reconstrucción de Chile. 291 p.

SERNAPESCA, 2003. Anuario Estadístico de Pesca. Servicio Nacional de Pesca, Ministerio de Economía, Fomento y Reconstrucción de Chile. 78 p.

Stotz, W., 1999. Does the scallop fishery in northern Chile has a future? 12th International Pectinid Workshop 5–11 May 1999, Bergen, Norway. Abstract. pp. 143–144.

Stotz, W., 2000. When aquaculture restores and replaces a overfished stock: is the conservation of the species assured? The case of the scallop *Argopecten purpuratus* (Lamarck, 1819) in northern Chile. Aquaculture International 8(2–3):237–247.

Stotz, W. and González, S., 1997a. Abundance, growth, and production of the sea scallop *Argopecten purpuratus* (Lamarck 1819): bases for sustainable exploitation of natural scallop beds in north - central Chile. Fisheries Research 32:173–183.

Stotz, W. and González, S., 1997b. Recovery of an overfished scallop bed (*Argopecten purpuratus* (Lamarck 1819)) in Tongoy Bay (Chile) by the use of a new management tool. 11th International Pectinid Workshop, 10–15 April 1997, La Paz, Baja California Sur, México. Abstract. pp. 54–55.

Uriarte, I., Farias, A. and Muñoz, C., 1996. Hatchery culture and grow-out of the Chilean scallop, *Argopecten purpuratus* (Lamarck, 1819), in the South of Chile. Rev. Biol. Mar. (Valparaíso, Chile) 31(2):81–90.

Uriarte, I., Rupp, G. and Abarca, A., 2001. Capitulo 8: Produccion de juveniles de pectinidos iberoamericanos bajo condiciones controladas. In: A.N. Maeda-Martinez (Ed.). Los Moluscos Pectinidos de IberoAmerica: Ciencia y Acuicultura. Editorial Limusa, México. pp. 147–172.

Uribe, E. and Blanco, J., 2001. Capítulo 12: Capacidad de los sistemas acuáticos para el sostenimiento de los cultivos de pectínidos: el caso de Argopecten purpuratus en la Bahía Tongoy, Chile. In: A.N. Maeda-Martinez (Ed.). Los Moluscos Pectinidos de IberoAmerica: Ciencia y Acuicultura. Editorial Limusa, México. pp. 233–248.

Uribe, E., Lodeiros, C., Felix-Pico, E. and Etchepare, Y., 2001. Capítulo 13: Epibiontes de Pectínidos de Iberoamerica. In: A.N. Maeda-Martinez (Ed.). Los Moluscos Pectinidos de IberoAmerica: Ciencia y Acuicultura. Editorial Limusa, México. pp. 249–266.

Valladares, C. and Stotz, W., 1996. Crecimiento de *Chlamys patagonica* (Bivalvia: Pectinidae) en dos localidades de la región de Magallanes. Rev. Chil. Historia Natural 69:321–338.

Vega, G., 1996. Principales dificultades en la producción industrial de semilla de ostión del norte en ambiente controlado. En: Seminario: Desarrollo y perspectivas de la pectinicultura en Chile. 74 p.

Vega, R. and León, J., 1990. Evaluación del banco de ostiones de Bahía Coquimbo. Informe Técnico, Instituto de Fomento Pesquero (Chile). 24 p.

Vildoso, A. and Chirichigno, N., 1956. Contribución al estudio de la concha de abanico *Argopecten purpuratus* en el Perú. Pesca y Caza (Perú) 7: 26 p.

von Brand, E., Bellolio, G. and Lohrmann, K., 1990. Chromosome number of the Chilean scallop *Argopecten purpuratus*. Tohoku J. Agri. Res. 40(3–4):91–95.

1314

Von Brand-Skopnik, E. and Ibarra-Humphries, A.M., 2001. Capítulo 6: Genética de Pectínidos Iberoamericanos. In: A.N. Maeda-Martinez (Ed.). Los Moluscos Pectínidos de Iberoamérica: Ciencia y Acuicultura. Editorial Limusa, Mexico. pp. 105–126.

Waller, T.R., 1969. The evolution of the *Argopecten gibbus* stock (Mollusca : Bivalvia), with emphasis on the Tertiary and Quaternary species of eastern North America. J. Paleont., 43, part II of II, supplement No. 5.

Waller, T.R., 1991. Evolutionary relationships among commercial scallops (Mollusca: Bivalvia: Pectinidae). In: S.E. Shumway (Ed.). Scallops: Biology, Ecology and Aquaculture. Developments in Aquaculture and Fisheries Science, Vol. 21. Elsevier, Amsterdam. pp. 1–73.

Walossek, D., 1991. *Chlamys patagonica* (King & Broderip, 1832), a long "neglected" species from the shelf off the Patagonia Coast. In: S.E. Shumway and P.A. Sandifer (Eds.). Scallop Biology and Culture. Word Aquaculture Workshops, Vol. 1. World Aquaculture Society, Baton Rouge, LA. pp. 256–263.

Wolff, M., 1987. Population dynamics of the Peruvian scallop (*Argopecten purpuratus*) during the El Niño phenomenon of 1983. Can. J. Fish. Aqua. Sci. 44:1684–1691.

Wolff, M., 1994. A trophic model for Tongoy Bay - a system exposed to suspended scallop culture (Northern Chile). J. Exp. Mar. Biol. Ecol. 182:149–168.

Wolff, M. and Alarcón, E., 1993. Structure of a scallop *Argopecten purpuratus* (Lamarck, 1819) dominated subtidal macro-invertebrate assemblage in northern Chile. J. Shellfish Res. 12(2):295–304.

AUTHORS ADDRESSES

Elisabeth von Brand - Universidad Católica del Norte, Departamento de Biologia Marina, Facultad de Ciencias del Mar, Casilla 117, Coquimbo, Chile (E-mail: evonbran@ucn.cl)

German E. Merino - Universidad Católica del Norte, Departamento de Acuicultura, Facultad de Ciencias del Mar, Casilla 117, Coquimbo, Chile (E-mail: gmerino@ucn.cl)

Alejandro Abarca - Pesquera San José, S.A., Cultivo de Ostiones, Casilla 342, Tongoy, Chile (E-mail: aabarca@coloso.cl)

Wolfgang Stotz - Universidad Católica del Norte, Departamento de Biologia Marina, Facultad de Ciencias del Mar, Casilla 117, Coquimbo, Chile (E-mail: wstotz@ucn.cl)

Chapter 28

Venezuela

César J. Lodeiros, Luis Freites, Maximiano Nuñez, Anibal Vélez and John H. Himmelman

28.1 INTRODUCTION

The Venezuelan coast, in the southwestern Caribbean Sea (Fig. 28.1), is a tropical region that supports a high diversity of bivalve molluscs. To date, more than 200 species of bivalves have been catalogued (Lodeiros et al. 1999). These include ten species of pectinids: *Chlamys ornata* (L., 1758), *Chlamys muscosus* (Wood, 1828), *Aequipecten lineoralis* (Lamark, 1819), *Euvola* (*Pecten*) *ziczac* (L., 1758), *Leptopecten bavayi* (Dautzenber, 1900), *Argopecten nucleus* (Born, 1780), *Lyropecten* (*Nodipecten*) *antillarum* (Réluz, 1853), *Nodipecten* (*Lyropecten*) *nodosus* (L., 1758), *Amusium papiraceum* (Gabb, 1873) and *Amusium laurenti* (Gmelin, 1791). Most of these scallops are small (<40 mm shell height) and only *Nodipecten nodosus* and *Euvola ziczac* are exploited by coastal populations for local consumption and *Amusium papiraceum*, *Amusium laurenti* and *Argopecten nucleus* are exploited by a drag fishery.

Following the increase in demand for scallops in the 1980s, the Instituto Oceanográfico de Venezuela of the Universidad de Oriente began experimental studies on Caribbean scallops, principally in the Golfo de Cariaco (Fig. 28.1).

28.2 DISTRIBUTION, HABITAT AND REPRODUCTION

28.2.1 *Euvola* (*Pecten*) *ziczac*

This large semicircular species can attain up to 90 mm in shell height. The upper (right) valve is almost flat (slightly concave in the central region) and the lower (left) valve is strongly convex. The upper valve is ornamented with zigzag lines. Both shells are usually brown but occasionally purple or orange, and white for rare albino individuals (Lodeiros et al. 1999). *Euvola ziczac* is found in the western Atlantic, from North Carolina (eastern United States) and Bermuda to Santa Catarina in southern Brazil. In Venezuela, it is most common along the northeastern coast.

Euvola ziczac forms small beds on sandy bottoms in areas of upwelling. In contrast to the other scallops in the region, *E. ziczac* is an infaunal species. It buries itself flush with sand bottoms, with its upper valve covered with sand. As a result it cannot be easily seen, although the semicircle of fine tentacles, which extend above the substratum can be seen upon close inspection (Fig. 28.2). The burying behaviour of *E. ziczac* potentially camouflages it from predators and it readily swims when touched by certain predators

(Brokordt et al. 2000). Its main predators are gastropods of the genus *Cymatium* and the octopus *Octopus vulgaris* (shells are often abundant near octopus dens).

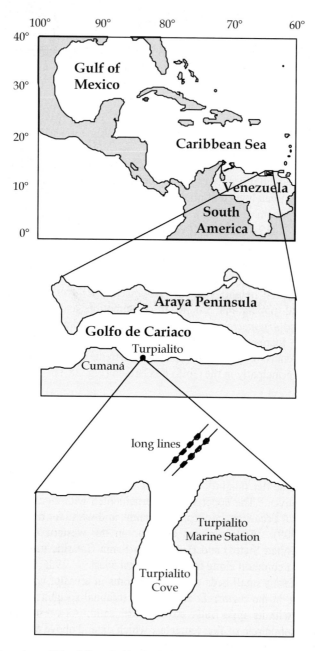

Figure 28.1. Location of Turpialito, Gulfo de Cariaco, where most of the studies on Venezuelan scallops were performed.

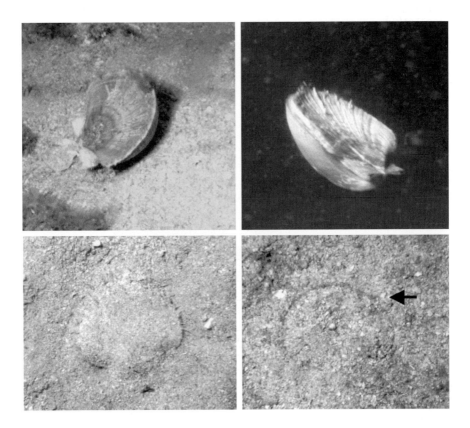

Figure 28.2. An unburied scallop, *Euvola ziczac*, a swimming scallop, and two scallops in their usual position, buried flush with the sediment surface (only a circle of tentacles is visible).

In natural beds of *E. ziczac*, most individuals measure between 60 and 88 mm in height (Fig. 28.3) and juveniles are found in extremely low densities. This suggests that scallops settle in areas outside of beds and later move to the beds.

Euvola ziczac is a functional hermaphrodite and is unusual among tropical bivalves because of its markedly seasonal reproduction. Active gonadal development and spawning occur between January and September and is followed by a period of reproductive inactivity between October and December (Brea 1986; Vélez et al. 1987). The reproductive period coincides with lower temperatures and high primary productivity, due to upwelling, and the period of reproductive inactivity with high temperatures and low productivity, associated with stratification of the water column (Lodeiros and Himmelman 1994). Natural populations of *E. ziczac* spawn twice, first in April to May and secondly in August to September, and during each spawning most individuals release gametes synchronously (Brea 1986; Vélez et al. 1987; Brokordt et al. 2000). Although Vélez et al. (1993) indicate that temperature is the major factor controlling gonadal maturation and

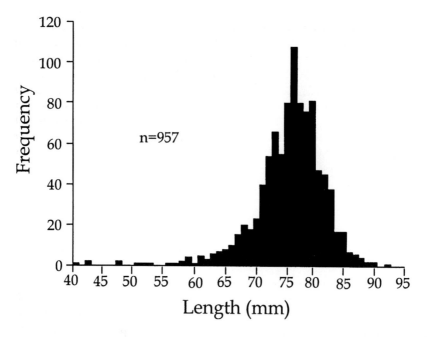

Figure 28.3. Size composition of a population of *Euvola ziczac* at Chacopatica in the Golfo de Cariaco, Venezuela, based on all individuals encountered during 8 hours of diving (two divers). No individuals were found measuring <40 cm.

spawning in *E. ziczac*, an inconsistent relationship between temperature and spawning has often been observed during culture studies (Vélez et al. 1995; Lodeiros and Himmelman 2000). For example, spawning may occur in one culture treatment but not in others, even though all are exposed to the same temperature. This suggests that the conditions in the immediate vicinity of the scallops, rather than large-scale factors, strongly influence reproduction. Spawnings outside of spawning periods on natural beds have also been documented for scallops in culture (Vélez et al. 1995; Lodeiros and Himmelman 2000). Finally, Lodeiros and Himmelman (2000) show the size (shell length or height) at which gonads first develop in juveniles in suspended culture varies with both culture depth and environmental conditions (including the abundance of fouling organisms).

Several studies indicate that reproductive investment and high temperatures may lead to stress in *E. ziczac*. For example, Boadas et al. (1997) documented a decreased oxidative capacity of mitochondria from the adductor muscle in the major reproductive period in May, which appeared to be related to a depletion of muscle reserves by gametogenesis during the period of low food availability and high temperatures. Further, Brokordt et al. (2000) showed that although *E. ziczac* maintains its clapping capacity in response to predators (which allows it to escape by swimming) during the various reproductive phases, its ability to recover from exhaustive clapping decreases during the course of gonadal maturation and spawning. This decrease is greater during the first reproduction (May) than the second reproduction (August), and this is likely explained by the greater gonadal production, coincident with lower food availability, during the first reproduction. Finally, a number of culture studies provide evidence of stress and mortality of adult *E. ziczac* during periods of high temperature and spawning activity (Lodeiros and Himmelman 1994, 2000; Boadas et al. 1997).

Natural populations of *E. ziczac* along the northeast coast of Venezuela show little genetic variability compared to most other pectinids, as Coronado et al. (1991) reported that the frequency of the polymorphic loci is 17.6% and mean heterozygosity is 0.074. The diploid number of chromosomes is 38 (Alfonsi et al. 1998) and studies by Pérez et al. (2000) of allozyme and biochemical variation at the octopine dehydrogenase (ODH) locus indicate that heterozygous superiority in fitness is the most likely explanation for an apparent over dominance at this locus.

28.2.2 *Argopecten nucleus*

This is a medium-sized scallop, up to 50–55 mm in shell height, with two similar convex valves, each with 10 to 22 robust radiating ridges. The colour is variable, but usually cream with brown patches (fewer brown patches on the lower valve). The juveniles attach bysally. *Argopecten nucleus* is similar to *Argopecten gibbus* (Linné, 1758), except that it has few additional ribs and a more globular shape (Lodeiros et al. 1999). *A. nucleus* occurs in southeastern Florida, the southern part of the Gulf of Mexico, and in the Caribbean and Surinam Seas and southward to the coasts of Guyana and Surinam. There are few data on its distribution in Venezuela, but it does settle on artificial collectors in the Golfo de Cariaco and the northern part of Sucre State and is occasionally captured in fishing drags.

Argopecten nucleus is a functional hermaphrodite. The one culture study on this species by Lodeiros et al. (1993) indicates that it is a short-lived scallop (8–10 months), which develops mature gonads at an age of about 6 months (at 38–48 mm in shell length). Thereafter, it reproduces continuously, even through periods of marked environmental changes, until about a month before it dies. In natural populations, individuals probably reproduce throughout the year, as larvae settle onto artificial collectors in all months (Lodeiros et al. 1997).

28.2.3 *Nodipecten* (*Lyropecten*) *nodosus*

This is the largest Caribbean scallop and attains sizes up to 150 mm in shell height. It has a strong rough shell with 7 to 9 wide nodular ribs and numerous longitudinal ribs. The auricles are asymmetrical and the colour is most often brown, but can also be purple to brownish red or orange. Juveniles can be confused with *Lyropecten antillarum*, which is a smaller scallop with a more fragile shell that does not have nodules (Lodeiros et al. 1999). *Nodipecten nodosus* is distributed from North Carolina to Brazil (Abbott 1974) but does not occur in Bermuda (Lodeiros et al. 1999). It occurs in low densities between 2 and 15 m in depth on coarse sand, or associated with corals, along the northeastern coast of Venezuela, but not in the Golfo de Paria.

Nodipecten nodosus is a functional hermaphrodite. Individuals in different reproductive phases are present throughout the year in populations along the Venezuelan Caribbean suggesting that it is constantly reproducing. Vélez et al. (1987) indicated a period of increased reproductive activity between August and December, but this was inferred only from changes in the biochemical composition of all tissues together (lipid content decreased during this period). Subsequent studies of individuals maintained in suspended and bottom culture showed gonadal decreases between May and July (Lodeiros et al. 1998; Acosta et al. 2000) and between October and November (Lodeiros et al. 1998; Freites et al. 2001).

Populations of *N. nodosus* along the northeast coast of Venezuela show even lower generic variability than *E. ziczac*, as the frequency of polymorphic loci is 1.92% and mean heterozygosity is 0.0016 (Coronado et al. 1991). The diploid number of chromosomes is 38 (Alfonsi et al. 1998). Alfonsi et al. (1998) report that consanguinity probably affects growth rate but not survival and suggest that individuals produced by self-fertilisation are more sensitive than cross-fertilised animals to environmental conditions.

28.2.4 *Amusium papyraceum* and *Amusium laurenti*

Amusium papyraceum has slightly convex symmetrical shells that are fragile and attains up to 90 mm in shell height. The internal surface of both shells has 30–40 fine radiating ribs, whereas the external surface has a shiny periostracum and no ribs (Lodeiros et al. 1999). The upper valve is brownish red in colour and the lower valve is white. *Amusium laurenti* is similar to *A. papyraceum* except that it is smaller, up to 60 mm in shell height, has less prominent lateral auricles and only 20–22 fine ribs are visible on the inner surface of the shell (Díaz and Puyana 1994). Both *Amusium* species occur in the tropical western Atlantic, *A. papyraceum* from the Golfo of Mexico to northern Brazil, and *A. laurenti* only in the Caribbean, from Honduras to the north coast of South America (Díaz and Puyana 1994). *Amusium papyraceum* inhabits the eastern and northeastern parts of Venezuela and the main beds are found north of Isla de Margarita and on the coast of Sucre State. *Amusium laurenti* is only found in eastern Venezuela (Penchazade et al. 2000, Fig. 28.1). Both *A. papyraceum* and *A. laurenti* occur between 10 and 50 m in depth (with peak abundance at 30–36 m) mainly on sandy bottoms but also on sand-mud bottoms (Salaya 1977; Díaz and Puyana 1994).

Both *A. papyraceum* and *A. laurenti* are hermaphrodites and a portion of the populations have mature gonads throughout the year, although a major spawning takes place in June to July and some smaller spawnings in December to January and in March to April (Penchazade and Salaya 1980; Penchazade et al. 2000). Gonadal development of *A. papyraceum* begins when it attains 35 mm in shell height (Salaya 1977).

28.3 FISHERIES

Amusium papyraceum forms the major scallop beds in Venezuela and is fished commercially. Ninety-two percent of the landings from the industrial trawl fishery are from the northeastern region of the country, including Isla de Margarita and northern

Sucre State, and 8% from the Golfo de Venezuela (Novoa et al. 1998). Fisheries landings have varied greatly over the past 30 years (Mendoza et al. 1994, Fig. 28.4), from 10 to 750 t yr^{-1} to peaks of 1,252 t yr^{-1} in 1972 and 1,873 t yr^{-1} in 1989 (data INIA 2001– Instituto Nacional de Investigaciones Agropecuarias de Venezuela).

Amusium papyraceum is part of a multi-species bottom trawl fishery, which targets shrimps, bottom fish and molluscs (octopus, squid and scallops). This mollusc first became important in the Venezuelan fisheries in 1972, when boats started dragging for scallops off northern Isla de Margarita and along the northern coast of Sucre State, in response to increased demand for scallops from the United States. The stocks do not appear to be overexploited in spite of the intensive fishing (Freddy Arocha, Fisheries Department of Instituto Oceanográfico de Venezuela, pers. comm.).

Amusium laurenti and *Argopecten nucleus* are captured by shrimp trawl in the northeastern and central parts of Venezuela, especially in the Golfo de Venezuela, but the landings are small and not recorded. Although *Euvola ziczac* and *Nodipecten nodosus* are the largest scallops in Venezuela, their densities are too small to support an organised fishery. They are only captured by individuals for local consumption, and occasionally appear in shrimp trawls.

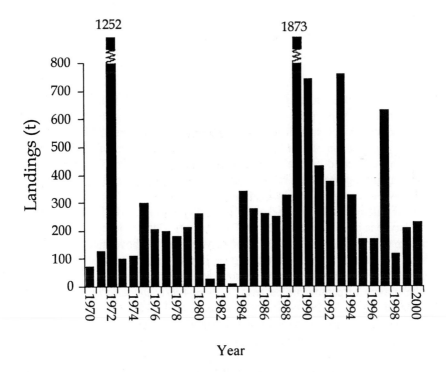

Figure 28.4. The number of metric tons of the scallop *Amusium papyraceum* landed in northeastern Venezuela between 1970 and 2000 (Data from J. Alió, Instituto Nacional de Investigaciones Agropecuarias, Sucre–Nueva Esparta, Venezuela).

28.4 AQUACULTURE

The large size of the adductor muscle of *Euvola ziczac* and *Nodipecten nodosus* suggests that they should be good species for aquaculture and a number of studies have evaluated their growth and survival in suspended and bottom culture. In addition, one study has examined growth of the smaller scallop *Argopecten nodosus* in suspended culture.

28.4.1 *Euvola ziczac*

28.4.1.1 Gonad maturation and spawning

Laboratory studies by Vélez et al. (1993) suggest that during periods of gonadal growth in the field, between February and July, maturation can be induced by exposing scallops in the laboratory to temperatures 3–4°C above those in the natural beds. Alternatively, maturation can be stimulated by placing scallops in small bays where temperatures are naturally warmer. Once the gonads are mature, release of male and female gametes can be induced using an abrupt temperature increase, from 20 to 29°C (Vélez et al. 1990). Mature scallops do not release gametes when injected with serotonin, and only male gametes are released when scallops are subjected to both serotonin injections and abrupt temperature increases (Vélez et al. 1990).

28.4.1.2 Spat production

As the numbers of *E. ziczac* spat settling on collectors is low in the Golfo de Cariaco (the monthly rate is <5 spat per 30 x 60 cm onion bag filled with mono-filament netting), natural spat collection is insufficient to supply commercial culture (Lodeiros et al. 1997). Thus, hatchery production of spat is required. The existing studies of larval development of *E. ziczac* have focused on evaluating the nutritional value of several microalgal diets and in determining optimal food rations, larval culture densities and water temperatures. The best growth and survival is obtained at 27°C with a diet of *Chaetoceros gracilis* and *Isochrysis aff galbana* (T-ISO) at a density of 30,000 cells mL^{-1} (Rojas et al. 1988). The density of larvae is initially maintained at 10 larvae mL^{-1}, later reduced to five larvae mL^{-1} and finally to three larvae mL^{-1}. In the laboratory, where the above studies were done, the water was filtered and treated with ultraviolet light to reduce the abundance of organic particles, competitors, predators and pathogenic microorganisms. It is important to control the growth of pathogenic bacteria, such as *Vibrio anguillarum*, *V. alginolyticus* and *Pseudomona* sp. (Lodeiros et al. 1992; Freites et al. 1993). When the above variables are adequately controlled, the larvae settle in 10 to 12 d with a survival rate of >50%. Settlement is greatest on the upper surfaces of roughened plexiglass collectors (angles from 0 to 45°) and when the water is maintained in motion (Vélez and Freites 1993).

A major factor in producing spat is the cost of growing phytoplanktonic food for the postlarvae. Vélez and Ortega (1988) studied the feasibility of producing microalgae in outside basins, using agricultural fertilisers as the source of nutrients and the sun as the source of light. Trials with *Tetraselmis chuii, T. suecica* and *Nannochloris oculata* produced favourable results. However, these species do not promote rapid growth of *E. ziczac* post-larvae. Slow larval growth increases costs because of the postlarvae must be manipulated more often and greater supplies of high quality water are needed. Vélez and Ortega (1988) could not reliably produce high quality diets, such as *Chaetoceros gracilis* and *Isochrysis aff galbana* (species characteristic of cooler regions). We have recently isolated two species of Venezuelan microalgae of the genus *Chaetoceros* and *Tetraselmis* that have been a good diet for culturing the larvae of several marine invertebrates, such as *Brachionus plicatilis* and *Litopenaeus vannamei,* but which we have not yet been tried as a diet for *E. ziczac* and other bivalves larvae and seed. Once *E. ziczac* spat attain 2 mm in shell length they are readily grown to 6 mm within 15 d (with >90% survival) in small 1-mm mesh lantern nets suspended in the sea (unpublished data). They can then be transferred to 3-mm mesh pearl nets.

28.4.1.3 Grow-out

The first study examining growth of *E. ziczac* in suspended culture showed a rapid increase in length during the first four month (from 3.3 to 44 mm) and then growth slowed coincident the an increase in gonadal size (Lodeiros and Himmelman 1994). A projection of the growth curve suggested it would take >3 years to attain the size generally found on natural beds (70–80 mm). We subsequently compared growth in three situations, (1) in cages suspended from a long line, (2) in cages at the same depth but placed on the bottom and (3) in partly buried cages at the same depth. The latter permitted the scallops to recess into the sand as they do in natural habitats. Shell and tissue growth was markedly greater in the partly buried cages than in the cages in suspension and on the bottom (A projection of the shell length curve indicated it would take about 1.5 years to attain the size on natural beds.). Further, survival was greatest in the partly buried cages (in the other treatments mortality may have been caused by organisms colonising the scallop shells). We hypothesised that *E. ziczac* grew better in the partly buried cages (1) because this scallop is better adapted to feeding on organic material at the sediment-water interface, or (2) because it is stressed when it cannot bury itself. A 5-month study evaluating the first hypothesis confirmed the greater growth in partly buried cages, and also showed an increase in the energetic content of sestonic food at the sediment/water interface compared to in suspension (Hunauld et al., in press). However, this difference could at most account for one third of the increase in growth in the partly buried cages.

In concurrent studies, Lodeiros and Himmelman (2000) ran four growth trials (starting with 10-mm scallops) to further evaluate factors determining growth and survival in suspended culture. Three trials followed growth of a cohort from spawning in August 1992, which was grown at three depths, 8, 21 and 34 m. The fourth trial used an April 1993 cohort, which was grown at 21 m. An exhaustive list of environmental variables was monitored during these growth trials (including physical and chemical factors, sestonic factors in two size fractions and fouling). Application of the stepwise multiple regression procedure to all trials together indicated that shell fouling was the main factor predicting somatic growth, and its affect was negative. Also, adding gonadal presence or absence as a "dummy" variable showed that gonadal production negatively affected somatic growth (Somatic tissue growth stopped or slowed at the initiation of gonadal development in the three trials with the 1992 cohort). Likely the strong effects of fouling and gonadal production masked the influence of the other environmental factors. This was indicated by analyses applied separately to the different growth trials that showed that phytoplankton abundance was a good predictor of growth in the two fastest growing trials and temperature variability negatively affected growth in the two slow-growing trials. The variations in food availability and physiochemical conditions in the Golfo de Cariaco did not appear to limit the growth of juveniles (as no environmental factors predicted the tissue increases in juveniles). The growth of the April 1993 cohort suspended at 21 m was higher than that observed by Vélez et al. (1995) for scallops in partly buried cages. In both cases there was low shell fouling. The rapid growth of the April 1993 cohort contradicted our previous hypothesis that *E. ziczac* is stressed when maintained in suspended culture.

A 32 day experiment (too short to be affected by fouling organisms) examining the relationship between RNA/DNA ratios and growth for three size groups of *E. ziczac* maintained in suspension at 8, 21 and 34 m in depth further indicated that food availability is a major factor determining growth (Lodeiros et al. 1996). Juvenile (without gonads) and maturing scallops (with developing gonads) showed a decrease in growth with depth and fully mature individuals (with full-sized gonads) showed a decrease in reserves with increasing depth (Fig. 28.5). The decrease in growth with depth was correlated with seston quality, as indicated by the decrease in chlorophyll-*a* values and the energy content of the seston with depth (temperature only decreased by ~1.3°C with depth).

The negative effect of fouling was examined further using an experimental design, which permitted separating the effects of fouling on the pearl nets and on the scallops shells (Lodeiros and Himmelman 1996). Growth was least when there was both net and shell fouling. Growth in shell dimensions and mass were strongly affected by organisms colonising the pearl nets and only weakly affected by organisms colonising the shells. Fouling organisms on the shells only slightly affected growth in tissue mass but probably accounted for increased mortality. The mass of organisms growing on the upper valve was much greater than in the longer-term growth trials of Lodeiros and Himmelman (1996) and could have impeded the ability of the ligament to open the valves.

The rapid growth of *Euvola ziczac* under good conditions (e.g., April 1993 cohort and partly buried cages in 1992) shows than commercial-sized scallops (>50 mm) can be produced in 10 months (Lodeiros and Himmelman 2000). Growth of juveniles in suspended culture is rapid and seems little affected by environmental conditions, including fouling. Grow trials in pearl nets by Freites et al. (1993), starting with 9.6-mm juveniles, indicate that a good culture density would be 800 individuals m^{-2} up to a length of 25 mm, 200 individuals m^{-2} until the scallops attain 40 mm, and 100 individuals m^{-2} for larger sizes. As the growth of somatic tissues markedly slows with the onset of gonadal production (Lodeiros and Himmelman 2000), techniques for preventing gonadal development (e.g., producing triploids) or delaying gonadal development (e.g., moving the scallops to greater depths) could be used to extend the rapid growth phase. Once the scallops reach 30–40 mm in shell length, steps need to be taken to limit fouling. Nets

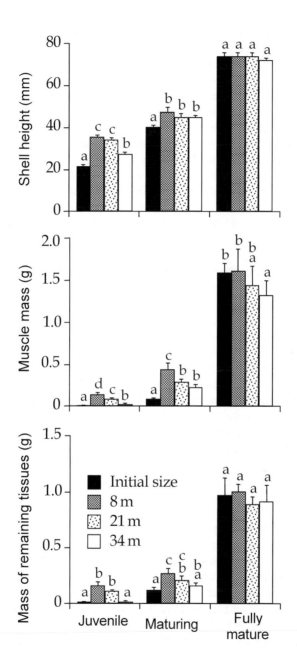

Figure 28.5. Mean of shell height and dry mass of muscle and remaining of tissues of *Euvola ziczac*, at the beginning of the study (initial size) compared to values after 32 days of culture at 8, 21 and 34 m in depth in the Golfo de Cariaco. Values sharing the same letter within each group are not significantly different, Scheffè test, P >0.05).

need to be changed periodically. Greater culture depths would both reduce fouling and retard gonadal development.

Bottom culture after 30–40 mm, in enclosures that permit the scallops to bury themselves, would eliminate the negative effects of fouling and not diminish growth. Freites et al. (2000b) compared three bottom culture techniques, (1) cages without bottoms, (2) corrals, and (3) being placed freely on the bottom. In all three treatments, the scallops were set out at a density of 15 individuals m^{-2} on a sandy bottom (1 m^2 areas). The high dispersion rate in the third treatment (12 scallops m^{-2} disappeared in the first month) prevented comparisons with the other treatments. In the corrals, the escape rate increased progressively with time, from 1 to 5 scallops m^{-2} month^{-1}, likely due to increased swimming ability with increasing size. In the treatment with cages, no scallops escaped, but survival was less than in the corrals (even when escaped scallops were counted as not surviving). This indicated that corrals were a better culture method than cages. Bottom culture density could likely be higher than 15 individuals m^{-2}, as we observed no differences in growth trials in corrals at 15, 35 and 70 individuals m^{-2} (unpublished data, this growth trial was during a period of high primary productivity).

28.4.2 *Nodipecten nodosus*

28.4.2.1 *Gonad maturation and spawning*

Gonadal growth and active gametogenesis occur at temperatures of 26 to 28°C during the principal period of reproduction from August to November (Vélez et al. 1987). Between February and July, gonadal growth is occurring in the field and maturation can be induced in the laboratory by maintaining the scallops at 7 to 8°C above the field temperature and providing them with phytoplanktonic food. Once the gonads are mature, spawning can be introduced by an abrupt temperature increase from 21 to 28°C (De La Roche et al. 2002).

28.4.2.2 *Spat production*

As for *E. ziczac*, hatchery production of spat will also be required for *N. nodosus* in the Golfo de Cariaco, because of the low numbers of spat settling on collectors (monthly rate of <5 spat per 30 x 60 cm onion bag filled with mono-filament netting, Lodeiros et al. 1997). We have cultured larvae and postlarvae using the same conditions (diet, food ration, density, temperature, water-changing schedule) as described above for *E. ziczac* and obtained larvae with an eyespot after 8 days and the post-larval dissoconch after 13 days (Vélez and Freites 1993). This compares with 11 and 14 days for the larval eyespot as observed for *N. nodosus* by De La Roche et al. (2002) and Marín (1994), respectively.

28.4.2.3 *Grow-out*

The first grow-out trials with *N. nodosus* by Lodeiros et al. (1998) were made in pearl nets suspended at 8, 21 and 34 m in depth at Turpialito, and started with scallops measuring

9.4 mm in shell height. The growth rate of the shell and somatic tissues decreased progressively with depth (Fig. 28.6) and appeared to be explained by food availability, as phytoplankton biomass decreased proportionally with depth (and temperature only decreased slightly with depth). Muscle mass attained 5–6 g after 5 months and at this size could be sold commercially. Mortality was low throughout the study at 21 m, whereas at 8 m major losses occurred after 6 months that appeared to be due to heavy shell fouling. The reduced growth rate, low survival and near lack of gonadal development at 34 m indicates that conditions at this depth are unfavourable for *N. nodosus*. We subsequently quantified the biochemical constituents of the major body components of *N. nodosus* for the scallops in the above experiment (Lodeiros et al. 2001). Protein in the digestive gland, and carbohydrates in the muscle and remaining somatic tissues, dropped during maximal gonadal growth suggesting transfer of nutrients from these tissues to gonadal production. Further, the greater use of lipids in the digestive gland during the reproductive period for individuals at 21 m than at 8 m was possibly because phytoplanktonic food was less abundant at 21 m. When conditions were likely stressful (low food availability, high temperatures, colonisation of fouling organisms and exhaustion from reproduction), protein levels in the remaining somatic tissues decreased and lipid levels stabilised.

Acosta et al. (2000) ran intermediate growth trials at 8 m in depth, starting with 9.4 mm juvenile *N. nodosus*, to examine the effect of culture density on growth and survival (6 densities: 7, 15, 30, 60, 125 and 250 scallops/pearl net). They show that increasing density decreases growth but not survival and recommend a schedule for reducing densities that should permit producing 50 mm scallops (ready for final growout) in 6 months.

Freites et al. (2001) compared three bottom culture methods for final growout (starting with 34 mm *N. nodosus*), (1) in corrals, (2) in pockets attached to an anchored line, and (3) in anchored sleeves. Survival was similar for the three methods, but all body components showed that growth was greatest in corrals and least in sleeves. The increased growth in the corrals was possibly because the scallops bysally attached to the corral walls so that they had greater access to suspended food particles (or to better quality particles), as for scallops in suspended culture. Finally, Mendoza et al. (2003) compared growth and survival of scallops, measuring 40 mm in shell length, maintained in four types of enclosures, in sacs, cones and lantern nets suspended at 6 m in depth, and in 1 x 1 m corrals set up at 6 m in depth on the bottom. The scallops covered about a third of the bottom for each treatment. The increase in tissue growth during the 4.5-month study was 36 to 48% greater in the suspended culture enclosures than in the bottom corrals and cumulative mortality was greater in the bottom corrals (45%) than in suspended culture treatments. There were no significant differences between lantern nets (33%), sacs (22%) and cones (20%). In contrast to the previous study by Lodeiros et al. (1998), this study did not show a negative effect of fouling, even though the study lasted for 4.5 months.

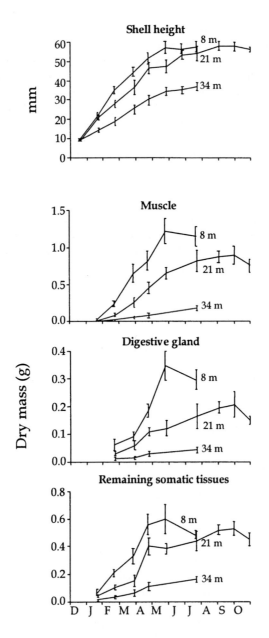

Figure 28.6. Changes in shell height and dry mass of the muscle, digestive gland, and remaining somatic tissues of the scallop *Nodipecten nodosus* maintained from December 1993 to October 1994 in suspended culture at 8, 21 and 34 m in depth at Turpialito, Golfo de Cariaco, Venezuela. Vertical bars represent 95% confidence intervals.

28.4.3 *Argopecten nucleus*

The numbers of natural spat settling onto onion bags filled with mono-filament netting is greater for *A. nucleus* than for *E. ziczac* and *N. nodosus* (monthly rate of ~20 spat per 30 x 60 cm onion bag; Lodeiros et al. 1997). Nevertheless, hatchery production will be needed to support commercial production.

The one growth study of *A. nucleus* (Lodeiros et al. 1993) started with individuals measuring 10 mm in shell length (~1.5 months in age), which were maintained in suspended pearl nets at 15–20 m in depth. The scallops attained 45–50 mm in length (approaching maximum size), with the adductor muscle weighing about 0.8 g in dry mass (3–4 g in wet mass), 6–7 months later (Fig. 28.7). At this size they could be sold commercially, either the muscle alone or all tissues together. This growth rate exceeds that reported for *Argopecten irradians* in North Carolina and *A. gibbus* in Florida. *Argopecten nucleus* appears to be suitable for suspended culture because it tends to resist colonisation by fouling organisms (at least when individuals are transferred to new pearl nets after the 1st and 6th month of culture, at the same time as reducing densities.

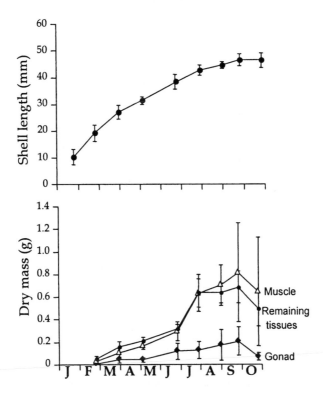

Figure 28.7. Mean of shell length and dry mass of tissues of *Argopecten nucleus* maintained in pearl nets at Turpialito, Golfo de Cariaco, Venezuela. Vertical bars indicate 95% confidence intervals.

28.5 PERSPECTIVES FOR CULTURE

For *Euvola ziczac* and *Nodipecten nodosus*, the low numbers of juveniles in natural beds and the low numbers of spat recruiting on suspended structures, as noted at Turpialito, indicates that commercial culture of these scallops will require the development of cost-effective hatchery techniques. As stated above, hatchery methods also need to be developed for *Argopecten nucleus*. It will likely be advantageous to grow the post larvae using natural sources of phytoplankton as early as possible. For *E. ziczac*, we have had >95% survival when growing 2 mm spat in suspended culture for 15 days (unpublished data).

The rapid growth and short life span of *A. nucleus* means that this scallop can be cultured in <9 months. Production of this scallop can likely be increased by varying culture depth and the type of enclosure employed. Growth trials need to be made starting in different seasons. Using a combination of suspended and bottom culture, it should be possible to produce commercial-sized *E. ziczac* (70–80 mm in length) and *N. nodosus* (80–90 mm) in 12 to 15 months. The culture will involve a first period of suspended culture (up to 30–40 mm in length for *E. ziczac* and to a larger size for *N. nodosus*), as the small individuals are easily handled, tend to be little colonised by fouling organisms, and have high rates of growth and survival. Both species tolerate relative high culture densities (Campos 1991; Freites et al. 1995; Acosta et al. 2000, unpublished data). The data available on the growth rates related to culture depth and density provide a good basis for developing a suspended culture strategy.

Although rapid growth rates provide a distinct advantage for suspended culture of scallops in the Golfo de Cariaco, and probably in many tropical regions, two factors, that may be specific to tropical regions, will limit production and increase costs. The first is fouling. A large variety of organisms, notably barnacles, anemones, other bivalves (e.g., oysters and pearl oysters), bryozoans, and tube forming amphipods and polychaetes, can colonise and grow profusely on scallops and suspended structures. Our growth trials document decreased rates of growth and survival of *E. ziczac* coincident with increases in fouling (Lodeiros and Himmelman 1994, 1996, 2000; Vélez et al. 1995; Freites et al. 1996). Inversely, prolonged rapid growth of *E. ziczac* (equivalent to that in bottom culture where the burying behaviour of the scallops prevents fouling) was observed in the one growth trial during a period when there was little colonisation by fouling organisms (August 1993 cohort, Lodeiros and Himmelman 2000). Fouling organisms may negatively affect production in suspended culture by reducing the flow of food particles to the scallops, by directly competing for food particles, and by encrusting the scallop shells so as to interfere with opening and closing of the valves. Mortality is most likely attributed or organisms colonising the shells (Lodeiros and Himmelman 1996). Fouling organisms will also decrease profitability by greatly adding to the costs of maintaining culture lines and enclosures in suspension. *N. nodosus* is much more resistant to fouling than *E. ziczac* (Lodeiros et al. 1998; Mendoza et al. 2003; Freites and Nuñez 2001). Nevertheless, Lodeiros et al. (1998) noted mortality and decreased growth coincident with heaving fouling in individuals grow in shallow water (8 m). In this case, the dry mass of fouling organisms on some dead scallops exceeded the mass of the shell.

The second factor that could limit suspension culture of scallops in Venezuela is the impact of predation by invertebrates settling on the suspended culture enclosures. Freites et al. (2000a) found that six carnivores (crabs and gastropods) that recruited onto pearl nets that had been immersed during 2 months (mid February to mid April 1994) caused significant mortality when subsequently added to pearl nets stocked with 25 *E. ziczac* (32 mm in shell length). One individual gastropod *Cymatium poulseni* or crab *Pilumnus caribbaeus* (both species settled onto about half of the pearl nets) killed 22–25 of the scallops in 30 days. Together, these predators could cause 50% mortality in pearl nets stocked with 25 scallops in 6 days.

Appropriate steps, such as periodically transferring the bivalves to new enclosures, removing predators from enclosures every 15 days (mainly during the first 2–3 months) or developing biological controls of predators, will be needed to limit predation losses but will add to production costs. Later, recruiting predators should be removed at about 2-month intervals, thus before they attain a size where then can attack the scallops. This may be achieved at the same time as changing nets to reduce fouling. The culture strategy should be flexible and involve monitoring fouling organisms and predators to determine the when control measures are needed.

Finally, a third factor that is critical in producing *E. ziczac* in suspended culture is the effect of wave action. Freites et al. (1999) demonstrated that the survival and growth rates of all body components are markedly reduced when scallops are maintained in enclosures that move about by wave action. The smooth convex lower shell and lack of byssal attachment, causes *E. ziczac* to readily slide around in suspended enclosures, and this causes damage to the shell and mantle edge. Thus, *E. ziczac* must be cultured in enclosures that are suspended in a way so that wave action is not transmitted from surface buoys, or be cultured in protected sites (if this does not reduce the carrying capacity of the enclosures). In contrast to *E. ziczac*, *N. nodosus* is not negatively affected by wave action because it bysally attaches to the sides of enclosures (Freites et al. 1999). This may also be the case for *A. nucleus* since it attaches bysally.

After a certain size, bottom culture will likely be a more appropriate culture strategy for both *E. ziczac* (at 30–40 mm) and *N. nodosus* (at 50–70 mm). Adult *E. ziczac* show high growth rates in bottom culture if the culture method permits the scallops to bury themselves as in natural beds (Vélez et al. 1995; Hunauld et al., in press). Whereas growth rates of adult *N. nodosus* are less in bottom than suspended culture, bottom culture may nevertheless be advantageous because of reduced equipment and maintenance costs. Seeding of the bottom needs to be explored as a culture technique for both scallops. Studies are particularly needed to evaluate losses due to predation and dispersion. Dispersion may be limited for *N. nodosus* because it attaches bysally and is not a good swimmer (especially individuals measuring >50 mm in shell length). Experimental studies should first be done at a small scale and then extended to trials at a large enough scale to mimic a commercial operation. There is likely a size at which losses from predation and dispersion decrease, and this size should be optimal for bottom seeding. Future studies should involve an economic analysis to evaluate the factors affecting profitability of different culture methods and to identify the research directions most likely

1332

to aid in establishing culture operations. An economic analysis could also be used to choose the optimal size for harvesting the scallops.

REFERENCES

Abbott, R.T., 1974. American Seashells. 2nd ed. Van Nostrand Reinhold Company. pp. 203–206.

Acosta, V., Freites, L. and Lodeiros, C.J., 2000. Densidad, crecimiento y supervivencia de juveniles de *Lyropecten* (*Nodipecten*) *nodosus* (Pteroida: Pectinidae), en cultivo suspendido en el Golfo de Cariaco, Venezuela. Revista de Biología Tropical 48:799–806.

Alfonsi, C., Basoa, E. and Pérez, J., 1998. Obtención de cromosomas de vieras a partir de ovocitos. Acta Científica Venezolana 49:198–200.

Alfonsi, C. and Pérez, J., 1998. Growth and survival in the scallop *Nodipecten nodosus* as related to self-fertilization and shell color. Boletín del Instituto Oceanográfico de Venezuela, Universidad de Oriente 37:69–73.

Boadas, M., Nusetti, O., Mundaraim, F., Lodeiros, C. and Guderley, H., 1997. Seasonal variation in the properties of mitochondria isolated from the phasic adductor muscle of the tropical scallop *Euvola* (*Pecten*) *ziczac* (L. 1758). Mar. Biol. 128:247–255.

Brea, J.M., 1986. Variaciones estacionales en la composición bioquímica de *Pecten ziczac* (Linnaeus 1758) en relación al metabolismo energético, reproducción y crecimiento. Licenciatura Thesis. Biology Department, Universidad de Oriente, Cumaná, Venezuela. 75 p.

Brokordt, K.B., Himmelman, J.H., Nusetti, O.A. and Guderley, H.E., 2000. Reproductive investment reduces recuperation from exhaustive escape activity in the tropical scallop *Euvola ziczac*. Mar. Biol. 137:857–865.

Campos, Y., 1991. Efecto de la densidad sobre la producción secundaria de la vieira *Pecten ziczac*. M.Sc. Thesis. Postgrado en Ciencias Marinas, Instituto Oceanográfico de Venezuela, Universidad de Oriente, Cumaná, Venezuela. 98 p.

Coronado, M., Gonzalez, P. and Pérez, J., 1991. Genetic variation in Venezuelan mollusks. *Pecten ziczac* and *Lyropecten nodosus* (Pectinidade). Caribb. J. Sci. 27:71–74.

De la Roche, J.P., Marín, B., Freites, L. and Vélez, A., 2002. Embryonic development and larval and post-larval growth of the tropical scallop *Nodipecten* (= *Lyropecten*) *nodosus* (L. 1758) (Mollusca:Pectinidae). Aquaculture Res. 33:819–827.

Díaz, J.M. and Puyana, M., 1994. Moluscos del Caribe Colombiano. Un catálogo Ilustrado. COLCIENCIAS, Fundación Natura e Instituto de Investigaciones Marina, Bogotá. 367 p.

Freites, L., Cote, J., Himmelman, J.H. and Lodeiros, C.J., 1999. Effect of wave action on the growth and survival of scallops *Euvola ziczac* and *Lyropecten nodosus* in hanging culture. J. Exp. Mar. Biol. Ecol. 239:47–59.

Freites, L., Himmelman, J.H., Babarro, J.M., Lodeiros, C.J. and Vélez, A., 2001. Bottom culture of the tropical scallop *Lyropecten* (*Nodipecten*) *nodosus* (L.) in the Golfo de Cariaco, Venezuela. Aquaculture International 9:45–60.

Freites, L., Lodeiros, C.J. and Himmelman, J.H., 2000a. Impact of predation by gastropods and decapods recruiting onto culture enclosures on the survival of the scallop *Euvola ziczac* (L.) in suspended culture. J. Exp. Mar. Biol. Ecol. 244:297–303.

Freites, L., Lodeiros, C.J. and Vélez, A., 2000b. Evaluation of three methods of bottom culture of the tropical scallop *Euvola* (*Pecten*) *ziczac*. (L. 1758). J. Shellfish Res. 19:77–83.

Freites, L., Lodeiros, C., Vélez, A. and Bastardo, J., 1993. Vibriosis tropical en ("Hatchery") del pectínido *Pecten ziczac* (L). Carib. J. Sci. 29:89–98.

Freites, L. and Nuñez, P., 2001. Cultivo suspendido de *Lyropecten* (*Nodipecten*) *nodosus* (L., 1758), en los métodos bolsas y aurículas ("ear hanging"). Boletín del Instituto Oceanográfico de Venezuela, Universidad de Oriente 40:21–29.

Freites, L., Vélez, A. and Hurtado, L., 1996. Crecimiento y producción secundaria en *Euvola* (*Pecten*) *ziczac* (L), en cultivo suspendido a tres profundidades. Boletín del Instituto Oceanográfico de Venezuela, Universidad de Oriente 35:17–26.

Freites, L, Vélez, A. and Lodeiros, C., 1993. Crecimiento y productividad de la vieira *Pecten ziczac* bajo varios sistemas de cultivos suspendidos. Mem. IV Congreso Ciencias del Mar, Serie Ocasional No. 2, Facultad de Ciencias del Mar, Universidad Católica del Norte. Coquimbo, Chile. pp. 259–269.

Freites, L., Vera, B., Lodeiros, C. and Vélez, A., 1995. Efecto de la densidad sobre el crecimiento y la producción secundaria de los juveniles de *Pecten ziczac* (L), bajo condiciones de cultivo suspendido. Ciencias Marinas 21:361–372.

Hunauld, P., Vélez, A., Jordan, N., Himmelman, J.H., Morales, F., Freites L. and Lodeiros, C., Contribution of food availability to the more rapid growth of the scallop, *Euvola ziczac*, in bottom than in suspended culture. Revista de Biología Tropical (in press).

Lodeiros, C.J., Fernández, R.I., Bonmati, A., Himmelman, J.H. and Chung, K.S., 1996. Relations of RNA/DNA ratios to growth for the scallop *Euvola* (*Pecten*) *ziczac* in suspended culture. Mar. Biol. 126:245–251.

Lodeiros, C., Freites, L., Nuñez, P. and Himmelman, J.H., 1993. Growth of the scallop *Argopecten nucleus* (Born, 1780) in suspended culture. J. Shellfish Res. 12:291–294.

Lodeiros, C., Freites, L. and Vélez, A., 1992. Necrosis bacilar en larvas del bivalvo *Euvola* (*Pecten*) *ziczac* (Linneo, 1758) causada por una *Pseudomona* sp. Acta Científica Venezolana 43:154–158.

Lodeiros, C.J. and Himmelman, J.H., 1994. Relations among environmental conditions and growth in the tropical scallop *Euvola* (*Pecten*) *ziczac* (L.) in suspended culture in the Golfo de Cariaco, Venezuela. Aquaculture 119:345–358.

Lodeiros, C. and Himmelman, J.H., 1996. Influence of fouling on the growth and survival of the tropical scallop *Euvola* (*Pecten*) *ziczac* in suspended culture. Aquaculture Research 27:749–756.

Lodeiros, C.J. and Himmelman, J.H., 2000. Identification of factors affecting growth and survival of the tropical scallop *Euvola* (*Pecten*) *ziczac* in the Golfo de Cariaco, Venezuela. Aquaculture 182:91–114.

Lodeiros, C., Marín, B. and Prieto, A., 1999. Catálogo de Moluscos del Nororiente de Venezuela: Clase Bivalvia. Edición APUDONS (Asociación Profesores, Universidad de Oriente, Núcleo Sucre), Venezuela. 110 p.

Lodeiros, C., Narváez, N., Rengel, J., Marquez, B., Jimenez, M., Marín, N. and Freites, L., 1997. Especies de bivalvos marinos con potencialidad para ser cultivados en el nororiente de Venezuela. Un estudio preliminar. XLVII Asociación Venezolana para el Avance de la Ciencia (AsoVac), Valencia, Venezuela. 5p.

Lodeiros, C., Rengel, J., Freites, L., Morales, F. and Himmelman, J.H., 1998. Growth and survival of the tropical scallop *Lyropecten* (*Nodipecten*) *nodosus* maintained in suspended culture at three depths. Aquaculture 165:41–50.

Lodeiros, C., Rengel, J.J., Guderley, H., Nusseti, O. and Himmelman, J.H., 2001. Biochemical composition and energy allocation in the tropical scallop *Lyropecten* (*Nodipecten*) *nodosus* during the months leading up to and following the development of gonads. Aquaculture 199:63–72.

Marín, B., 1984. Desove y desarrollo larval de la vieira *Lyropecten nodosus* L. en laboratorio. Licenciatura Thesis. Biology Department, Universidad de Oriente, Cumaná, Venezuela. 125 p.

Marín, N., Lodeiros, C. and Verginelli, R., 1994. Cultivo de microalgas y el rotífero *Brachionus pliacatilis* en gran escala. Acta Científica Venezolana 45:226–230.

Mendoza, Y., Freites, L., Lodeiros, C., López, J. and Himmelman, J.H., 2003 Evaluation of biological and economical aspects of the culture of the scallop *Lyropecten* (*Nodipecten) nodosus* in suspended and bottom culture. Aquaculture 221:207–219.

Mendoza, J., Sánchez, L. and Marcano, L., 1994. Variaciones en la distribución y abundancia de los principales recursos demersales del nororiente de Venezuela. II. Invertebrados. Memorias de la Sociedad de Ciencias Naturales La Salle, Vol. LIV (142):65–81.

Novoa, D., Mendoza, J., Marcano, L. and Cardenas, J., 1998. El atlas pesquero marítimo de Venezuela. Ministerio de Agricultura y Cría-servicio Autónomo de los Recursos Pesqueros y Acuícolas-Programa de Cooperación Técnica VECEP-Unión Europea. 197 p.

Penchazade, P., Paredes, C. and Salaya, J.J., 2000. Reproductive cycle of South American scallop *Amusium laurenti* (Gmelin, 1791) (Bivalvia, Pectinidae). Aquaculture International 8:227–235.

Penchazade, P. and Salaya, J.J., 1980. Contribución al conocimiento de la vieira *Pecten payraceum* en Golfo Triste, Venezuela. In: II Simpiosium, Asociación Latinoamericana de Acuicultura, México. pp. 846–870.

Pérez, J., Nusetti, O., Ramirez, N. and Alfonsi, C., 2000. Allozyme and biochemical variation at the octopine dehydrogenase locus in the scallop *Euvola ziczac*. J. Shellfish Res. 19:85–88.

Rojas, L., Azuaje, O. and Vélez, A., 1988. Efecto combinado de la ración y la densidad larvaria sobre la supervivencia y crecimiento de larvas de *Pecten ziczac* (Linné, 1758). Boletín del Instituto Oceanográfico de Venezuela, Universidad de Oriente 27:57–62.

Salaya, J.J., 1977. Contribución a la biología y pesquería de la vieira *Pecten payraceum* en Venezuela. M.Sc. Thesis. Postgrado en Ciencias Marinas, Instituto Oceanográfico de Venezuela, Universidad de Oriente. 140 p.

Vélez, A., Alifa, E. and Azuaje, O., 1990. Induction of spawning by temperature and serotonin in the hermaphroditic tropical scallop, *Pecten ziczac*. Aquaculture 84:307–313.

Vélez, A., Alifa, E. and Freites, L., 1993. Inducción de la reprodución el la vieira *Pecten ziczac*. I. Maduración y desove. Carib. J. Sci. 29:209–213.

Vélez, A. and Freites, L., 1993. Cultivo de "semillas" de la vieira *Pecten ziczac* bajo condiciones ambientales controladas (hatchery). Mem. IV Congreso Ciencias del Mar, Serie Ocasional No. 2, Faculdad de Ciencias del Mar, Universidad Católica del Norte. Coquimbo, Chile. pp. 311–317.

Vélez, A., Freites, L., Himmelman, J.H., Senior, W. and Marin, N., 1995. Growth of the tropical scallop, *Euvola* (*Pecten*) *ziczac* (L.), in bottom and suspended culture in the Golfo de Cariaco, Venezuela. Aquaculture 136:257–276.

1335

Vélez, A. and Ortega, L., 1988. Cultivo de microalgas en gran escala en el trópico para alimento de larvas de bivalvos. Revista Latinoamericana de Acuicultura 36:79–86.

Vélez, A., Sotillo, R. and Pérez, J., 1987. Variación estacional de la composición química de los pectínidos *Pecten ziczac* y *Lyropecten nodosus* y su relación con los períodos de desove. Boletín del Instituto Oceanográfico de Venezuela, Universidad de Oriente 26:56–61.

AUTHORS ADDRESSES

César J. Lodeiros - Departamento de Biología Pesquera, Instituto Oceanográfico de Venezuela, Universidad de Oriente, Apdo. Postal 245, Cumaná 6101, Estado Sucre, Venezuela (E-mail: clodeiro@sucre.udo.edu.ve)

Luis Freites - Departamento de Biología Pesquera, Instituto Oceanográfico de Venezuela, Universidad de Oriente, Apdo. Postal 245, Cumaná 6101, Estado Sucre, Venezuela (E-mail: lfreites@sucre.udo.edu.ve)

Maximiano Nuñez - Departamento de Biología Pesquera, Instituto Oceanográfico de Venezuela, Universidad de Oriente, Apdo. Postal 245, Cumaná 6101, Estado Sucre, Venezuela (E-mail: maxpaulinon@cantv.net)

Anibal Vélez - Departamento de Biología Pesquera, Instituto Oceanográfico de Venezuela, Universidad de Oriente, Apdo. Postal 245, Cumaná 6101, Estado Sucre, Venezuela (E-mail: avelez@sucre.edo.edu.ve)

John H. Himmelman - Département de Biologie, Université Laval, Quebec City, Canada G1K 7P4 (E-mail: John.Himmelman@bio.ulaval.ca)

Scallops: Biology, Ecology and Aquaculture
S.E. Shumway and G.J. Parsons (Editors)

1337

Chapter 29

Mexico

Esteban Fernando Félix-Pico

29.1 FISHERY

29.1.1 Introduction

Scallops are an important fishery and under intensive exploitation in México. Commercial fishing activity began in the Gulf of California in 1970, and at the end of the 1980's annual landings averaged about 32,000 t. The scallop fishery is notoriously variable, so by 1991 to 1995 landings had declined to about 900 t. This decline has served as an impetus for fishery enhancement through aquaculture. The State of Baja California Sur is the production leader with more than 95% of the Mexican production (Félix-Pico 1991a, b).

The Mexican species have been assigned new scientific names. Almeja Catarina or Pacific Calico Scallop (Poutiers 1995) was named *Argopecten circularis* (Sowerby I, 1835), but is now known as *Argopecten ventricosus* (Sowerby II, 1842) because the holotype was misidentified (Waller 1991, 1995). The Mexican lion's paw was known as *Lyropecten (Nodipecten) subnodosus*, but was replaced by *Nodipecten subnodosus* (Sowerby I, 1835) because the genus *Lyropecten* is extinct in the Pacific Ocean (Smith 1991). The Voladora scallop has been renamed from *Pecten (Oppenheimopecten) vogdesi* to *Euvola vogdesi* Arnold, 1906; reviewed by Waller (1991). These species are listed in order of their importance to the fishery. Pectinids are among the most attractive of all bivalves, both in shape and colours, and they have been used as the emblem of the shield of the State of Baja California Sur. Scallops inhabit only shallow tropical and subtropical Eastern Pacific waters, including the Gulf of California. The commercial fishing area for each species shows the geographic distribution of fishing and aquaculture (Fig. 29.1). As fisheries have developed, particularly over the last three decades, researchers and fishermen have discovered much about the species inhabiting Mexican waters; their basic biology, habits, physiology, ecology, genetics (for use in aquaculture) and sustainable production potential. More than 100 scientific papers have been published in the last 10 years on these subjects.

Currently, experimental and commercial efforts to develop further the aquaculture of scallops in México are being made by Secretaría del Medio Ambiente Recursos Naturales y Pesca (SEMARNAP), Centro Interdisciplinario de Ciencias Marinas (CICIMAR), Centro de Investigaciones Biológicas del Noroeste (CIBNOR), Universidad Autónoma de Baja California Sur (UABCS), Universidad Autónoma de Baja California (UABC),

1338

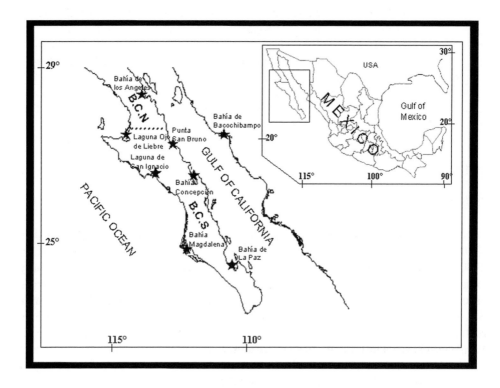

Figure 29.1. Principal scallop fishing grounds and aquaculture areas around the Península of Baja California and Sonora State, México.

Centro de Investigación Científica y Educación Superior de Ensenada (CICESE), Instituto Tenológico y de Estudios Superiores de Monterey (ITESM), Centro de Investigación en Alimentación y Desarrollo, A.C. (CIAD), fishing cooperatives, and aquaculture companies in México.

29.1.2 Species and Distribution of Scallops

Scientific name: *Argopecten ventricosus* (Sowerby II, 1842)
Common names: Almeja Catarina (México), Conchuela (Panamá), Pacific Calico scallop (FAO).
Mexican species code: 0281428H.
Diagnostic features: Solid medium to large shell, very inflated form, with both valves convex, sturdy shells, strongly ribbed with about 21 ribs (17 to 22 each valve), anterior and posterior ears about equal in size, raised concentric growth lines in spaces between ribs. A wide variety of colour and marking is seen, from almost pure white through blotched and streaked patterns to solid dark orange and brown or purple (Keen

1971; Keen and Coan 1974; Poutiers 1995). Usually has internally a wide chocolate-brown band or whitish background, varying in width and intensity and is similar in appearance to its twin species from the Gulf of México, *Argopecten gibbus*. Length is about 60 mm (maximum size 81 mm). Whole weight about 70 g and adductor muscle 8 g.

Distribution: From Isla Cedros, Baja California, through the Gulf of California and southern México to Paita, Perú. The northern subspecies, *A. ventricosus aequisulcatus* (Carpenter, 1864), ranges from Santa Barbara, California, USA, to Baja California Sur (Bahía de La Paz and Bahía Magdalena) (Poutiers 1995).

Scientific name: *Nodipecten subnodosus* (Sowerby I, 1835)
Common names: Almeja Garra or mano de león (México), Pacific lion's paw scallop (FAO).
Mexican species code: 0241026H
Diagnostic features: The largest and heaviest of the West American tropical species, grows to 170 mm. Thick, solid, convex equivalve and fan-shaped, 10 to 11 ribs are very conspicuous, some of them with large knobs or nodules, which makes the species easy to recognise (Rombouts 1991). Shell sturdy in texture, valves strongly convex, shell heavy, ribs subdivided by small, overlying riblets. Colours ranging from dull purple or white with purple lines to brilliant shades of orange and magenta. A similar species is *Nodipecten nodosus* (Linnaeus, 1758), from the Atlantic, more nodose and with one less rib (Keen 1971). Length about 175 mm (maximum size 218 mm), diameter about 75 mm. Whole weight is about 1,400 g and adductor muscle 160 g.

Distribution: From Laguna Ojo de Liebre (Guerrero Negro), Baja California Sur, through the Gulf of California and southern México to Perú.

Scientific name: *Euvola vogdesi* Arnold, 1906
Common names: Almeja Voladora (México), Vogde's scallop (FAO).
Mexican species code: 0401422H
Diagnostic features: Medium to large, strong shell. Right valve convex with about 21 (20–22) ribs, rounded ribs, left valve flat or slightly concave with about 20 ribs, darker colour, from buff to bright reddish brown, brighter on left valve, interior white to pinkish brown (Keen 1971; Rombouts 1991). Ears nearly equal in size, byssal notch small, ribs strong, smooth. Primarily a tropical genus, one species occurs in México. Length and height about 100 mm. Whole weight about 115 g and adductor muscle 12 g.

Distribution: From Punta Eugenia, Baja California Sur, through the Gulf of California and south at least to Panamá.

29.1.3 History of the Fishery

This fishery has been sustained by private investment authorised by the SEMARNAP, México. After the abalone fishery, the pectinid fisheries have been the most important economic factor in Baja California shellfish during the last decade. Within nine years (1989–1997), national scallop production rose from 1,000 t to over 32,000 t year^{-1}, caused entirely by the control and administration of the seabed stock. Scallop diving has become an efficient and well-organised form of fishing, and encouragement of scallop diving can help conserve an important natural resource for future harvests because the diver sees scallops in their natural surrounding and his observations can contribute to a better understanding of their habits and life cycle.

The Peninsula of Baja California is where the main exploitation grounds and culture areas are to be found (Fig. 29.1). They are Bahía de Los Angeles in Baja California, Lagunas Ojo de Liebre, Guerrero Negro, and San Ignacio, and Bahías Magdalena, Concepción, and de La Paz in Baja California Sur, and Bahía de Bacochibampo in Sonora, on the Gulf of California. It is difficult to obtain accurate data of landings in the official Mexican Sea Fishery statistical tables before the early 1970's, because before that landings of voladora and catarina scallops were so small they were included with other shellfish. Data for both species were combined until the 1970's. The information in the Mexican Annual Fishery Statistics (captured production) has tables named "Almeja". Although the term "Almeja" means "clam", it is normally used to refer to all bivalves, including all species of scallops, arks, clams and pens (SEMARNAP 1999).

At the beginning, in the 1970's, scallops were collected by divers operating from boats using hookah gear, the same equipment now still used. Dredges were never used. At each fishing ground, 50 to more than 800 boats worked.

The fishery of Bahía de los Angeles, Baja California collapsed suddenly after 4 years of intensive harvest. The fishery peaked in 1972, when about 4,400 t were landed in Bahía de Los Angeles and Punta San Bruno, Baja California Sur. The fishery of Bahía de La Paz peaked in 1973, with about 90 t landed. These beds were the first exploited on a commercial scale and the fishery expanded rapidly. In 1975, a fishery started at Bahía Magdalena and Laguna San Ignacio, but they have never reached the size of the Gulf fisheries. Between 1976 to 1980, there was a radical change in the fishery, with the catch falling in 1977 to only 20 t. As the fishery expanded the number of boats increased. The number of fishing boats had risen to more than 92 for each fishing season. In 1983, scallop landings reached 2,500 t (value $3.2 million US dollars). The discovery of a lucrative market in the United States of America caused the major part of the production to be exported. In 1990, scallop landings reached more than 30,000 t, with the number of fishing boats increased to 871.

29.1.4 Status of the Resource

29.1.4.1 Reproductive biology

29.1.4.1.1 Size at maturity

Argopecten ventricosus is a functional hermaphrodite and generally protandrous (Villalaz 1994) with two spawning periods during the year. The gonad is cream coloured in the male portion and orange reddish to vermilion in the female portion (Tripp-Quezada 1991; Félix-Pico et al. 1995). Catarina scallops reach sexual maturity at 8 months to one year. The minimum size at sexual maturity is approximately 30 mm (Baqueiro et al. 1981) or 35 mm (Villalejo-Fuerte and Ochoa Baez 1993).

Nodipecten subnodosus and *Euvola vogdesi* are also functional (Carvajal-Rascón 1987; Ruiz-Verdugo and Cáceres-Martínez 1991; Gutiérrez Villaseñor and Chee Barragán 1997. Under natural conditions, *N. subnodosus* is mature at one year or more, when the organism is approximately 80 mm for the male and 93 mm for the female (Reinecke 1981).

29.1.4.1.2 Fecundity

Argopecten ventricosus: Fecundity increases with shell height, but sizes between 58 and 77 mm have maximum reproductive potential (Fig. 29.2) (Villalejo-Fuerte and Ochoa-Baez 1993). The mean fecundity was calculated by regression analysis during spawning in February-March (12 million oöcytes). Using a sample of 101 individuals gave by the equation:

$$Y = 0.662 \ X^4 \text{ with } r = 0.70 \tag{1}$$

At the end of the reproductive season (Villalejo-Fuerte 1995), the relationship was described by the equation:

$$Y = 1.057 \ X + 146,000 \text{ with } r = 0.1 \tag{2}$$

Under laboratory conditions, when scallops have been stimulated, spawning fecundity varied between 2 and 3 million oöcytes per 45 mm shell height. For the male portion, there were an estimated 100 to 250 million spermatocysts (Monsalvo-Spencer 1998).

29.1.4.1.3 Maturity and spawning

Argopecten ventricosus: Spontaneous natural spawning occurs throughout the year, indicating there are at least some ripe individuals in the population at all times. Baqueiro and Arana (2000) studied the reproductive cycle in Ensenada de La Paz (Golfo de California), reporting three high intensity spawning periods with 30% in February, 90% in June, and 70% during August. Cáceres-Martínez et al. (1990) determined, by gonadic index from Bahía de La Paz, two intense gonadal activities between February-March and September-October. Luna-González et al. (2000) reported, also for Bahía de La Paz, ripe stages throughout the year were present in February to May, July-August (maximum 50%), and October to December. The resting stage was seen only in June and September (47 to 91%). The spent gonadal stage may be partial or complete. It has a flabby appearance of dull orange and the testis is yellowish white.

The reproductive season of *Argopecten ventricosus* in Bahía Magdalena is slightly different than that of the Gulf coastal lagoons. Tripp-Quezada (1991) reported that in northern Bahía Magdalena (Estero Sto. Domingo), spawning occurs sooner than in Bahía Concepción. There was a massive spawning from January to March and between October and November. Félix-Pico et al. (1995) determined two major spawnings a year, during April-May and in August in central Bahía Magdalena.

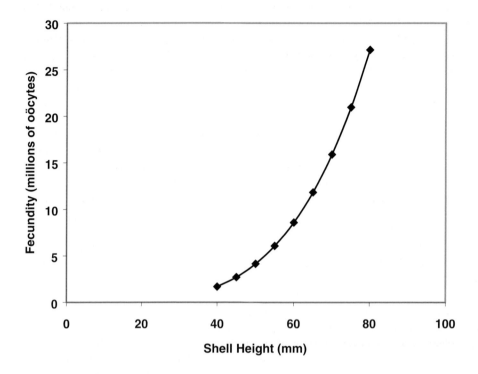

Figure 29.2. Relationship between shell height and potential fecundity in *Argopecten ventricosus*. Estimated by the equation No. 1. (From Villalejo-Fuerte and Ochoa Baez 1993).

Spawning intensity can be correlated with water temperatures. In central Bahía Magdalena, spawning occurs (March-April) when the bottom temperature reaches 15°C to 17.5°C, at about 20-m depth (Félix-Pico et al. 1995). In northern Bahía Magdalena, spawning occurs when temperature drops from 30°C to 22°C, with a minimum value of 19°C in January (Tripp-Quezada 1991). Villalejo-Fuerte and Ochoa-Baez (1993) reported that in Bahía Concepción the highest frequencies of mature scallop were caused by decreasing the temperature from 30°C to 21°C during autumn. The spawning season began in January, when the water temperature was decreasing, and that maximum spawning coincided with lowest temperature (16°C) recorded during the annual cycle, and continued until April, when water temperature began to increase. Reproduction occurred after a process of energy transfer from the adductor muscle to the gonad. This was attributed to the seasonal fall of temperature, which had a favourable influence for the energy transfer used in gametogenic development and spawning (Villalejo-Fuerte and Ceballos-Vázquez 1996).

For Ensenada de La Paz, scallops spawn from late February to March when the water temperature rises from 19°C to 21°C (Félix-Pico et al. 1989). Cáceres-Martínez et al.

(1990) determined ripe gonads are associated with a decrease in temperature from 31°C to 26°C in Bahía de La Paz.

Lora-Vilchis et al. (1997) studied the physiological mechanisms for spawning process control. When the sperm is ejected, the female part of the gonad begins to show the following changes. The nucleolus begins to disappear as do the endoplasmic reticulum and the Balbiani complex. Sperm emission usually begins before oocyte emission. Muscle fibres inside the gonad probably contribute to the spawning process. Avilés-Quevedo (1990) observed total evacuations of gametes under laboratory conditions and also indicated spermatozoids are released before the ovules.

Other exogenous factors have been related to the spawning of *A. ventricosus* in natural conditions. Baqueiro-Cárdenas and Aldana-Aranda (2000) associated a spawning period in June to a change in salinity, from 34.5‰ to 38‰, but Luna-González et al. (2000) found the salinity does not affect the reproductive cycle. Félix-Pico et al. (1989) mentioned that spawning is directly related to the lunar phases and it occurs during the full moon. In Bahía Concepción, Villalejo-Fuerte and Ochoa-Baez (1993) believed the proportion of mature scallops in relation to the photoperiod suggests the synchronisation of gametogenic development and maturity through an undescribed neuroendocrine component.

The gonadosomatic index is not a good indicator of the reproductive cycle because the gonads tend to accumulate water when they are in the resting stage Luna-González et al. (2000). But other authors have reported that the index is a good indicator because reproduction was believed to occur after a process of energy transfer from the adductor muscle to the reproductive system (Tripp-Quezada 1991; Villalejo-Fuerte and Ochoa-Baez 1993; Villalejo-Fuerte 1995; Villalejo-Fuerte and Ceballos-Vázquez 1996; Félix-Pico et al. 1989).

The influence of food, biochemical composition, and abundance of particles also must be considered. The phytoplankton community of Bahía Concepción was represented by diatoms (131 spp.), dominated by species of *Chaetoceros, Rhizosolenia* and *Nitzchia*. The carbohydrate/protein ratio and chlorophyll concentration were monitored during the cold season, and they have a better nutritional quality in terms of POM (Particulate Organic Matter) during the cold season (Verdugo-Díaz 1997). The inorganic seston/organic seston ratio (IS/OS ratio) influences the spawning stage, and with food and adductor muscle reserves, influences the developing and ripening of gametes. The IS/OS ratio had no clear relation to the reproductive cycle. Low quality and quantity of food in May led to the resting stage in June in Bahía de La Paz (Luna-González et al. 2000). In Bahía Concepción, Martínez-López and Garate-Lizárra (1994) reported an abundance of diatoms during the winter season (1994), related to maturity and spawning. This contribution of high-energy-content particles coincides with the maximum phytoplanktonic production cycle. This represents an important contribution of additional food, considerably reduced, that would cause a decline in production within the bay. These observations suggest that decreased food quantity may affect the catch of this scallop, which was low in 1994. On an interannual scale, in 1994 the particulate organic matter production decreased during a good part of the reproductive period of the catarina scallop (Palomares-García and Martínez-López 1997).

Nodipecten subnodosus: Under natural conditions *Nodipecten subnodosus*, spawning occurs when temperature decreased from 22°C to 17°C during October, with a maximum value of 84% for males and 56% for females (Reinecke 1981).

29.1.4.2 Population biology

The age composition of voladora and catarina scallop populations depend heavily on year-class strength. Because of the short life span of *Euvola vogdesi* and *Argopecten ventricosus*, age composition data generally show the presence of only two year-classes in the population. The nature of the fisheries is such that relative age composition data are not very important and hence largely unavailable. Age at first capture of catarina and voladora scallops is about 1 year, and for *Nodipecten subnodosus*, at least 2 years.

Size composition data for voladora and catarina scallop populations are scarce in the literature. Aguilar-Ruiz (1975) produced details of the size-frequency distribution of a population of *E. vogdesi* from Bahía de Los Angeles, Baja California. There was a compact size-group with a well defined modal peak at 72–81 mm length. The samples were collected in May 1971, and a unimodal year-class was present.

Yoshida-Yoshida and De Alba-Pérez (1977) collected samples of catarina scallop from Ensenada de La Paz, Baja California Sur, between May 1976 and May 1977. They demonstrated a bimodal distribution from May until July because of the presence of the two year-classes; the younger having been spawned in early spring. *A. ventricosus* has a distribution in the southern part of the Ensenada. There are distinct populations with similar size composition. The smallest individuals are 19–22 mm in length, the largest scallops are 51–69 mm, and the averages for the three separate populations are 44, 53 and 58 mm in length.

Baqueiro et al. (1981) continued with those studies in Ensenada de La Paz between December 1977 and August 1978. They demonstrated that the 0 and 1 (age-structure) groups are truly representative of their relative proportions in the population and found that groups are not unimodal; probably a result of the two spawnings in one year. They found evidence of a seasonal change in size composition. From March until June, there were no small (21–40 mm) catarina scallop in their samples, but in July small scallops constituted 5% to 8% of the total in the samples. In August, this increased to more than 60%, indicating the entry of another year-class into the fishery. The average length was between 51.5–57.7 mm in large scallops and 30 mm for small scallops. The population decreased because of overfishing and adverse environmental factors. This area is now closed to fishing.

Tripp-Quezada (1985) has monitored Bahía Magdalena at an eelgrass meadow in Estero Santo Domingo. The size at the first capture of the catarina scallop is about 42 mm (after the first spawning which takes place in the first half of the year), the average size is about 50 mm, and the maximum size is 63 mm. He also established condition factors and demonstrated the commercial size of 45 mm and the total weight of about 27 g.

León-Carballo et al. (1991) has evaluated Bahía Concepción and defined an adult population group between 51–71 mm with maximum concentrations between 55 and 63 mm. The average size was 59 mm for the catarina scallop.

Yoshida-Yoshida and De Alba-Pérez (1977) produced a length/weight relationship for *A. ventricosus* from Ensenada de La Paz. This relationship from a sample of 112 individuals gave:

$$Y = 0.78 X - 25.8 \text{ with } r = 0.56 \tag{3}$$

Tripp-Quezada (1985) produced total weight/visceral weight and visceral weight/adductor muscle weight relationships for catarina scallops from Santo Domingo. This first relationship for a sample with 75 individuals was:

$$Y = 0.964 X + 0.329 \text{ with } r = 0.98 \tag{4}$$

The second relationship can be described by the equation:

$$Y = 0.968 X + 0.096 \text{ with } r = 0.75 \tag{5}$$

These data represent proportional relationships of 1:1/4 and 1:1/3. Félix-Pico (1987) obtained a length/weight relationship for catarina scallop from the Bahía Concepción. This relationship for 151 individuals can be described by the equation:

$$W = 0.0024 L^{2.4} \text{ with } r = 0.67 \tag{6}$$

At the beginning of the season, the values are 58 mm and 40 g. In the last month of the season, we found low values between 46 mm and 50 mm with total weight 36 g. Félix-Pico (1993) has monitored central Bahía Magdalena at depths between 10 and 25 m. The size at the first capture of the catarina scallops was about 47 mm, the average size was about 56 mm, and the maximum size was 70 mm in height at the beginning of the fishing season. During April, the average size was 48 mm and in May the average size was 50 mm in height. In summer, the average size decreased from 54 to 51 mm and again increased in winter to 59 mm and 61 mm during December 1989 and January 1990.

The sampling site at Bahía Magdalena was assumed to represent monthly size increases during the 5 month period from April to August 1990 (Félix-Pico et al. 1994). The Ford-Walford method was used to estimate L_∞, the asymptotic shell height. A von Bertalanffy growth curve was fitted (Fig. 29.3) and the estimated parameters were $L_\infty = 71.6$, $K = 0.117$ (1.4 per year) and $t_0 = -6.23$, with age groups between 51 and 68 mm. The equation obtained was:

$$L_t = 71.63 \left[1 - e^{-0.1167 (t + 6.2311)}\right] \tag{7}$$

29.1.4.3 Abundance and density

Félix-Pico (1991a) reported the densities of *A. ventricosus* in Ensenada de La Paz in August 1975 averaged 4 organisms m^{-2} and in November to December 1975, density averaged 10 individuals m^{-2}. Yoshida-Yoshida and De Alba-Pérez (1977) assessed the

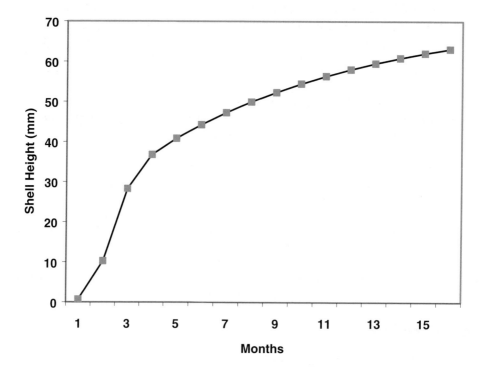

Figure 29.3. Growth curve calculated by the equation No. 7, from data of the natural population in Bahía Magdalena. (From Félix Pico et al. (1994)).

density of catarina scallops on the same beds. The sampling method was by diving using a one square meter area and random sampling. The density estimated during May to July 1976 averaged 6 individuals m^{-2}. Baqueiro et al. (1981) reported the densities of catarina scallops also on the Ensenada de La Paz beds from December 1977 averaged 13 individuals m^{-2} and in March 1978 averaged 3 individuals m^{-2}. They also estimated biomass ranged 545 g m^{-2} and 125 g m^{-2} for the same dates. Average biomass production was 55 g m^{-2} month during this period, and reported 48 g produced during this period.

The study of population size of *Argopecten ventricosus* was made by León-Carballo et al. (1991) who estimated abundance of Bahía Concepción stocks in 1988. An estimated abundance of 61.7 million scallops was calculated, with 85% above the minimum size limit for commercial catch (56 mm). A total biomass of 385 t of adductor muscles with an exploitable potential of 213 t was also calculated.

The population density during 1989 and 1990 had a maximum value of 26 scallops m² in Bahía Magdalena, decreasing to 4 scallops m^{-2} in April 1990 at the end of the fishing season (Félix-Pico 1993). A total biomass of 2,880 t of adductor muscle was taken during 1989 and 2,820 t in 1990, which represents about 79% of total production.

An intensive capture of the scallop *Euvola vogdesi* was made (Pineda-Barrera and López-Salas 1972; Aguilar-Ruiz 1975). They estimated the standing crop of voladora scallop at Isla Angel de La Guarda (Puerto Refugio, northern part) to be about 4 million, and at the eastern part of the island (Cantiles) to be about 3 million. Aguilar-Ruiz (1975) reported the standing crop of voladora scallops in Bahía de Los Angeles to be about 2 million. From June to August 1972, in the area among San Marcos de Tierra, San Rafael, and the southeast of Isla San Marcos, B.C.S., a dense bank of this scallop was exploited to its maximum level (Holguin-Quiñones 1997). In September 1972, Pineda-Barrera and López-Salas (1972) obtained densities of 16 and 26 organisms m^{-2} in an area of 2,240,000 m^2 at Isla Angel de la Guarda. At the end of summer, Holguin-Quiñones (1997) in San Marcos, when the fishing activity was declining, obtained an average density of 5 organisms m^{-2} covering an area of 12 km^2. Aguilar-Ruiz (1975) recorded densities of 15 and 16 organisms m^{-2} in two locations in Bahía de Los Angeles.

During 1989 and 1990, 750 million scallops were gathered at the mouth of Bahía Magdalena, at depths between 12 and 25 m, yielding 5,190 t of meat (Maeda-Martínez et al. 1993). Apparently exploited beds are formed only when temperatures are 16°C or lower deep inside the bay for at least 2 months. Leija-Tristán et al. (1996) estimated biomass of *A. ventricosus* varied from 11 kg ha^{-1} (depth to 15 m) and 0.9 kg ha^{-1} at greater depths (29 m deep). Capture of scallops over the continental shelf indicated that this species was associated with temperatures between 14°C and 15.9°C, with salinity approximately 35‰.

29.2 EXPLOITATION OF THE RESOURCE

29.2.1 Fishing Areas

The most important fished stocks of *Argopecten ventricosus* and *Euvola vogdesi* are along the coast of the Baja California Península - notably along the Pacific coast. The fishing grounds for *Nodipecten subnodosus* are in Lagunas de Ojo de Liebre and Guerrero Negro. Figure 29.1 shows the main stocks of scallops around the Baja California Peninsula.

There are five important catarina scallop grounds inside the bays of Baja California Sur. The smaller grounds on the East Coast are in Bahía de La Paz (closed) and Bahía Concepción. The largest areas containing commercial concentrations of catarina scallops are Laguna de San Ignacio, Lagunas Ojo de Liebre-Guerrero Negro and Bahía Magdalena on the Pacific coast. These are between 28°N 114°W and 24°N 111°W on the west coast of the peninsula. The grounds in the Gulf of California are at approximately 26°N 110°W (Bahía Concepción) and 24°N 110°W (Ensenada de La Paz).

Extinct beds have been found of *E. vogdesi* in Bahía de Los Angeles and at Isla Angel de La Guarda in Baja California (29°N 113°W), and Punta San Bruno, San Rafael and Isla San Marcos (27°N 112°W) in Baja California Sur.

The fishing grounds of *Nodipecten subnodosus* are in Lagunas de Ojo de Liebre and Guerrero Negro at approximately 28°N 114°W.

Bay scallops, by the nature of their habitat, are not found concentrated in large grounds. The catarina scallop beds have been estimated to cover some 200 hectares, which are distributed in patches inside the bays. Nine fishing areas were marked off over 374 ha of distribution sites in Bahía Concepción (León-Carballo 1991). Eighteen fishing areas were located ranging from Estero Santo Domingo (Sta. Helena), Bahía Magdalena (Pto. San Carlos), Bahía Almejas (Pto. Chale), to Rancho Bueno. These grounds were evaluated in the 114,000 ha of distribution area in Bahía Magdalena (Félix-Pico 1993).

Bay scallops are harvested from shallow water between 6 m and 35 m and often 3 m or less in the tidal channels. The maximum depth range for catarina scallops in the open sea is 150 m (not commercial grounds) (Maeda-Martínez 1993; Leija-Tristán et al. 1996).

In the northern part of the península, beds of voladora scallop in Bahía de Los Angeles were 15 m to 40 m. These grounds occur <1 km from the coast. The mano de león scallop beds had a depth range of 6 m to 14 m inside channels.

The bottom of the catarina scallop beds along the Pacific coast is medium hard, with shells and eelgrass under mud or sand substrate. In Bahía Concepción, the catarina beds are harder, with shells, pieces of coral and gravel mixed with sand or mud. The beds in the Gulf of California are covered with brown and red algae. Santamaría-Gallegos et al. (1999) determined the coincidence in the timing of appearance of *Zostera marina* meadow and the reproductive time of scallops, and juvenile scallop settlement over the eelgrass meadow at Punta Arena, Bahía Concepción. The eelgrass meadow existing from December to May can provide scallops with a habitat to avoid predators (i.e., a refuge). In Ensenada de La Paz, beds are medium hard, with shells, sponges, soft corals and algae under mud.

The voladora scallop (*Euvola vogdesi*) beds in the Gulf of California around Isla Angel de La Guarda are usually sandy or alluvium substrate.

29.2.2 Fishing Seasons

The development of catarina and voladora scallop fishing because of the advent of the processing plant has meant scallops can be fished all year. Most scallops are exported as frozen muscle, mainly to the United States. The establishment of processing factories in 1974 revolutionised the scallop fishery and enabled the catarina scallop fishery to operate throughout the year, even in the warmest months.

Aguilar-Ruiz (1975) reported that in the Bahía de Los Angeles for *E. vogdesi* the 1972–1973 season lasted from January to January, with more than half the catch taken in the summer months. Félix-Pico (1991b) reported in Ensenada de La Paz the bay scallop season is from May to September with a peak (in 1975) from July to August. According to Baqueiro et al. (1981), the 1977–1978 season was from January to April and September to November with two peaks in 1977 in March and in October.

According to the statistical fishery data from the Secretaria de Pesca (Polanco-Jaime et al. 1987), the catch can be all year; however, there are defined dates of beginning and end of season for each fishing ground. In Lagunas Ojo de Liebre and Guerrero Negro, the bay scallop season is from May to December. In Laguna de San Ignacio, this season was from June to December 1981. In Bahía Magdalena, the season was, during 1980 to 1984, from June to December. In 1983–1986, in Bahía Concepción, the maximum monthly catch was May to December.

The Ministry of Fisheries (SEMARNAP) is the legal authority. Survey and stock management is done by CRIP (Regional Fishery Centre). After 1990, the fishing season was set to be between May to June for *A. ventricosus* (Diario Oficial de La Federación 1990, 1993).

N. subnodosus had a fishing season during May to July and November and December of 1999.

29.2.3 Fishing Operations and Results

Unified effort data for scallops are scarce in México. The ideal measurement of effort would involve some measure of diving gear or number of divers per compressor and the time of diving, e.g., catch /No. divers h^{-1}. The catch is then reported as whole shellfish, meat, and muscle in kilograms. But usually effort data consist at best of catch per boat per day.

Aguilar-Ruiz (1975) reported that in Bahía de Los Angeles the recorded landings of *Pecten vogdesi* scallop from January 1972 to January 1973 varied from 5.3 kg muscle/boat day (1 kg muscle = 6.9 kg whole shellfish or 110 individuals) to 120 kg muscle/boat day^{-1}. The highest recorded catch in one month was 1,640 kg muscle in 18 fishing days - an average of 91 kg/boat day^{-1}. The average catch/boat day for the boats fishing during this period was 29–69 kg muscle/boat day^{-1}. The number of boats working Bahía de Los Angeles beds varied from 7 in January to 13 in June.

For the *Argopecten ventricosus* fishery, Félix-Pico (1991b) reported that at the Ensenada de La Paz grounds, catch averaged 29–35 kg muscle/boat day^{-1} during the 1975 season. Baqueiro et al. (1981) on the same grounds reported the average catch was 25 kg muscle/boat day^{-1} during the 1977–78 season. The number of boats working was 35 for 20 days. The highest recorded catch in one month was 5,000 kg muscle in 20 fishing days with all boats (1 kg muscle = 13 kg whole shellfish or 330 individuals).

The catch statistics for catarina and voladora scallop have been included amongst those for other shellfish, but in reality is almost entirely catarina scallop (Mexican Fisheries Statistics). The landings of scallop from 1981–1986 averaged 3,240 t year^{-1}. The highest recorded catch per year was 5,750 t. Scallops represented between 16% and 78% of total shellfish caught along the Pacific coast of México. In the Gulf of México, the recorded landings of shellfish from 1981–1984 was 1,500–2000 t year^{-1}, but there are no separate data on scallops. The data is in total fresh weight (Polanco-Jaime et al. 1987).

29.2.4 The Market Value-landings Data from 1970 to 2000

Landings of scallops in Baja California and Baja California Sur first became available in 1970; before this they had been included with other shellfish. As the voladora scallop landings increased in the late 1960's and early 1970's, the catarina scallop landings decreased from the 1975 peak to only 2.5 t in 1979, but they rose again as voladora scallop decreased and have since been about 224 t. The voladora scallop landings achieved their greatest weight (4,380 t) and value ($4,820,000 US dollars) in 1987.

The catch in México from 1970 to 2000 is given in Table 29.1. Catarina scallop catch is divided between those taken north of Baja California (Northwest Gulf of California) and those taken south of Baja California (Pacific coast and Gulf of California). Landings off northern Baja California were low in the late 1970's and early 1980's, and landings up to 1983 averaged 120 t year[-1]. The highest recorded catch was 1,140 t (value $6,400,000 US dollars) in 1987. The scallop catch off southern Baja California was low in 1979 with 2.5 t, but reached 900 t (value $1 million US dollars) in 1981. It has been estimated that annual landings had risen to some 885 t by 1981–1987 and in 1986 they were 1,785 t (value $4,460,000 US dollars). The number of boats in the scallop fishery had reached 95 in 1986. The data were taken from the statistical tables in the official bureau of SEMARNAP in La Paz and Ensenada. The values represent muscle weight in kilograms.

Major landings of catarina scallop occurred in 1989 and 1990, when 32,000 t and 28,000 t (whole weight) valued at $12 million US dollars were landed. At present they are harvesting only 130 t (Fig. 29.4).

The fishery for *Nodipecten subnodosus* has harvested more than 1,400 t valued at $2 million US dollars.

29.3 CONSTRAINTS

29.3.1 Pollution

In the Baja California Península, there is domestic and cannery sewage flowing into the bays, but there are no industrial wastes. There are small agriculture areas using pesticides.

Diseases of the scallop related to pollutants occurred in Ensenada de La Paz from 1977 until 1981. The rich organic soups created in inshore areas by domestic sewage outflows and sludge dumped into the bay contributed to increases in populations of facultative bacteria such as *Vibrio*, *Pseudomonas* and *Aeromonas*. Concentrations of such bacteria may provide sufficient infection pressure on scallops so that disease and mortality results. Evidence for this sequence of events can be found in studies done in polluted waters. Oseguera-Green (1977) reported scallops, water and sediments polluted by fecal coliform bacteria. At the present time, La Paz has a sewage processing plant and the pollution of Bahía de La Paz is controlled.

The implications of aquaculture development for human health assume importance in some coastal lagoons of Baja California Sur, but may gain further significance in the future. In some coastal regions, there is existing monitoring for regulatory control called the Mexican monitoring protocol for sanitation of mollusc bivalves (P.M.S.M.B.). This protocol began in 1973, and has been used in monitoring Laguna San Ignacio, Estero El Coyote (Pta. Abreojos), Bahía San Hipólito, Bahía Tortugas, Bahía Asunción and Estero Santo Domingo in Bahía Magdalena. Those sites have been shown to be of good quality for use of sea water in aquaculture (Treviño-Gracia 1995).

Table 29.1

Historical catch statistics for scallop species (*Argopecten ventricosus, Euvola vogdesi* and *Nodipecten subnodosus*), for the period 1970–2000. (Data from SEMARNAP, office in La Paz).

Year	Magdalena (t)	Guerrero N. (t)	La Paz (t)	Concepción (t)	Total Weight (t)	Price ($US/kg)	Value ($US/1000)
1970	235		224		459	0.07	$32
1971	1,907		110		2,017	0.07	$141
1972	4,415		186		4,601	0.10	$460
1973	434		449		883	0.12	$106
1974	69		351		420	0.11	$46
1975			390	20	410	0.08	$33
1976			304	15	319	0.11	$35
1977			240		240	0.12	$29
1978			82	216	298	0.13	$39
1979			1	4	5	0.12	$1
1980		699	308	139	1,146	0.11	$126
1981		1,288	1,532		2,820	0.13	$367
1982	37	1,125	1,351		2,476	0.16	$396
1983	77	973	998		2,085	0.14	$292
1984	92	1,901	313		2,306	0.17	$392
1985		2,203	90	135	2,428	0.36	$874
1986	84	2,841		2,705	5,693	0.27	$1,537
1987	11	2,599		3,020	5,630	0.62	$3,491
1988	437	5,040		5,472	10,949	0.82	$8,978
1989	25,290	1,205		5,531	32,026	2.41	$77,183
1990	24,673	2,040		2,507	29,220	2.12	$61,946
1991	2,539	1,289		3,089	6,917	0.51	$3,528
1992	848			100	948	0.07	$66
1993	270			1,030	1,300	0.11	$143
1994	472				472	0.03	$14
1995	922				922	0.06	$55
1996	14,474	2,390			16,864	1.27	$21,417
1997	1,263	296			1,559	0.9	$1,403
1998	1,707	200			1,907	0.26	$496
1999	109	3		27	130	1.07	$139
2000	120			60	180	1.3	$234

1352

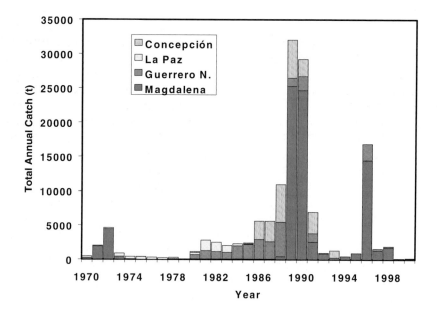

Figure 29.4. Landings of scallop species (*Argopecten ventricosus, Euvola vogdesi* and *Nodipecten subnodosus.* Data from SEMARNAP, office La Paz and Ensenada, B.C.

29.3.2 Biological Constraints

Natural biotoxins also pose public health problems, in particular, paralytic shellfish poisoning. The toxic dinoflagellates from the Pacific coast of México are *Gonyaulax polyedra, G. polygramma, Gymnodinium catenatum* and *Pyrodinium bahamense.* Some species in tropical and subtropical waters have long been known to be toxic. In the southern Gulf of California (Mazatlán, Sinaloa), dinoflagellate blooms or red tides have been reported during spring and summer causing toxic effects in fish and bivalves (Morquecho-Escamilla 1996). In the summer of 1986 in Bahía Concepción, blooms of dinoflagellates were noted but did not have any toxic effects on scallops and other bivalves. Morquecho-Escamilla (1996) established that during the highest peaks of dinoflagellate abundance, the potentially toxic species were *Dinophysis caudata, Ceratium furca, Prorocentrum micans* and *P. compressum.* By extracting the paralytic toxins (using mouse bioassays), Morquecho-Escamilla (1996) found the toxic activity in whole catarina scallop exceeded 400 MU, the maximum level allowed for human consumption of shellfish. Other toxic species; *Alexandrium catenella, Gonyaulax molinata, Gymnodinium catenatum* and *P. mexicanum* were identified in Bahía Concepción, Bahía de La Paz, and Laguna San Ignacio, B.C.S. (Garate-Lizarraga 1997). Possible pollution-mediated biotoxins should be an important consideration in planning aquaculture operations in these tropical waters.

29.3.3 Predation

The principal cause of mortality of *Argopecten ventricosus* is man, but natural mortality is caused by gastropods *Hexaplex nigritus*, *H. princeps* and *Phyllonotus erythrostomus*, fish (*Sphoeroides annulatus*, *Hoplopagrus guntheri*), stingray (*Heterodontus fransisci*) and the crustaceans *Callinectes bellicosus* and *Cronius ruber* (Tripp-Quezada 1985; Félix-Pico 1991a).

Maeda-Martínez et al. (1992) determined survival using the potential predator impact (PPI) in four zones of Rancho Bueno and Bahía Almejas, which had previously been sown with tagged spat of catarina scallops (sizes 3 to 4 cm). Results show an inverse correlation with scallop size up to 4 cm. Scallops larger than these were not affected. The average PPI in Rancho Bueno was 80% in 48 h and in Bahía Almejas was only 25%.

A similar experiment has been made in Ensenada de La Paz, Lango-Reynoso (1994) released three thousand tagged scallops (average size 43 mm). The results show the number of scallops decreased from 106 to 5 organisms m^{-2} in 56 days because of predator influence (96% mortality), with a predation rate of -0.19 x organisms eaten day^{-1}.

The catarina scallop community is composed of a large number of marine invertebrates that live on and within the shell. Although members of this community may indirectly affect the fishery, few of the species can be considered potential predators of the scallop. Pea crabs interfere with the normal feeding of scallops by ingesting strings of food filtered by the scallop, and cause chronic stress and reduce growth. In Bahía Magdalena, pea crab *Tumidotheres margarita*, (Crustacea: Pinnoteridae) infestation was found as high as 31% in adult *A. ventricosus*, between 47 and 56 mm shell height, and coincided with the spawning season of the scallop (Félix-Pico 1992).

In Bahía Magdalena, more than seventy-five species of invertebrates associated with scallops have been recorded, representing almost every major taxonomic group but crustaceans, polychaetes and molluscs are dominant. Heavy fouling covered 100% of upper valve surfaces and 30% of lower valve surfaces, adding considerably to the volume and weight of the catarina scallops. Three filter-feeders encrusting invertebrate species were dominant, *Megabalanus tintinnabulum*, *Tetraclina squamosa* and *Hydroides recurvispina*. The polychaetes *Nereis pelagica* and *Polydora rickettsi*, which inhabit space between the mantle and shell, cause the scallop to deposit extra shell layers. Consequent demands upon the scallop's energy may reduce growth or reproduction and may leave the weakened scallop prey to disease or to predators (Félix-Pico and García-Domínguez 1993).

In Laguna Ojo de Liebre and Estero El Coyote, Gómez del Prado (1984) reported more than 20 associated species of invertebrates, represented by the crustaceans *Fabia subcuadrata* and *Pinnotheres concharum* and molluscs. Other invertebrates of special interest because of their relation to the catarina scallop include *Echinocephalus seudouncinatus* (Nematoda: Gnathostomidae), which infects one host species, the horn shark, *Heterodontus francisci* (Pisces: Elasmobranchia). This final host eats scallops and becomes infected by ingesting the larval phase or cysts. The larvae occur in a high percentage of scallops seasonally, between 20% and 80%. The larvae also infect *Nodipecten subnodosus* as an intermediate host (Gómez del Prado 1983).

In Bahía Concepción, Gómez del Prado et al. (1992) reported *Phyllobotrium* sp. (Cestode) and a metacercariae (Family Fellodistomidae) infested 65% of catarina scallops. There is even less resemblance between species composition of the shallow areas of Bahía de La Paz and Bahía Magdalena, as described by Garduño-Méndez (1999). More than ten species were found in scallops in bottom culture, the copepod *Pseudomyticola spinosus*, a cestode larva (Family Phyllobothridae), and the turbellarian worms were prevalent. The coincident erosion of gill tissue and labial palps in some infested scallops suggests that *T. margarita* is a common parasite.

29.3.4 Resource Management Constraints

No legislative restrictions were placed on total catch of catarina and voladora scallops or on the efficiency or number of fishing units in scallop beds from 1970 to 1986. In the Baja California states, there are no limitations on daily total catch. Licenses are issued to those wishing to take catarina scallops commercially without any additional restriction of fishing grounds or number of fishing gears per ground. Management policies for the fisheries include licenses for both boats and personnel (3 boats and one diver per boat). The number of licenses for Bahía Magdalena and Bahía Concepción were 102 with 280 gears in recent years.

The capture of small scallops tends to be limited by the minimum size of shucked meat accepted by the processors. They accept 250–330 scallops kg^{-1}as a maximum. The maximum acceptable count per kg for *Argopecten ventricosus* is 330 scallops kg^{-1}, but the recommended is an average of 130 scallops kg^{-1} and the best size is 80 scallops kg^{-1} (Baqueiro et al. 1981; León-Carballo 1991; Félix-Pico et al. 1994), and for *E. vogdesi* is 173 scallops kg^{-1} with the minimum being 106 scallops kg^{-1} (Aguilar-Ruiz 1975). In addition, there are some minimum size limits imposed. In the Pacific coast beds, the recommended minimum size is 60 mm in the longest diameter, and along the Gulf of California coast is 56 mm (Diario Oficial de La Federación 1993).

The fishery for *Nodipecten subnodosus* is controlled, with a minimum size limit of 150 mm shell height, or 9 to 10 scallop muscles kg^{-1} and quotas limited to 1–2 t per license (25 kg muscle day^{-1}). The number of licenses per season was between 30 and 44 (Morales-Hernández and Cáceres-Martínez 1996).

29.4 HARVESTING AND TRANSPORTING

29.4.1 Methods

In the Baja California states, catarina and voladora scallops are usually consigned to a particular dealer, with scallops being sold by muscle weight or viscera weight. Many landings are made in remote places, and it is necessary to transport them by refrigerated trucks over considerable distances. Scallops landed in Lagunas Ojo de Liebre or Guerrero Negro might be taken to be sold in Ensenada or Tijuana, Baja California. Scallops landed in the Laguna San Ignacio, Bahía Concepción, and La Paz are sent to the continent, especially Mazatlán, Guadalajara, Monterrey and México City. Most scallops are exported as frozen muscle to the United Sates and some are marketed as canned.

Meat extraction, known as shucking, is a manual operation for catarina, mano de león and voladora scallops. The shell is opened by inserting a sharp knife along the inside of the flat valve. The viscera are then removed from the round valve, leaving the adductor muscle and roe attached and easily cut away. The whole operation, usually performed by a team of women, is done with amazing speed. The catarina scallops are opened, the flesh is then detached by shaking or by pushing with the finger, and the adductor muscle is pushed out from the rest of the viscera.

The smaller size meat is scalded in boiling water for few minutes; this usually to be sold to the home market.

The muscle meat is shucked, washed, and packed in plastic bags (5 pounds) for freezing and then mostly exported to the United States. New packaging is now used. Seal-top is a hermetically sealed, resistant, and reclosure overcap with printed specifications.

The food processing technology centre in Hermosillo, Sonora, studies quality conditions of frozen scallop muscles and have made flavoured, smoked, marinated, lyophilised, and cold binding or blending scallops (Ocaño-Higuera 1999).

29.4.2 Marketing

The home market for scallops is small because of the high price, though it is growing. Most, however, are exported to the United States of America. In addition to scallops sold as frozen meat, most scallops are exported as frozen muscle. There is no market in Baja California Sur for live scallops in the shell. The scallops are sold by the hundredweight and are usually consigned to a particular dealer or processor. The price in 1987 was $0.62 US dollars kg^{-1} and the price of scallop meat (muscle plus roe) varies according to the season from $2.25 to $3.41 US dollars kg^{-1} (Pérez-Concha 1990). The price of muscle only was $5.00 US dollars kg^{-1} and whole weight was $0.40 to $1.82 US dollars kg^{-1} in recent years (1997–1999). All the prices were paid by the dealer to fishermen at the beach.

The fishery statistics in Figure 29.4 and Table 29.2 illustrate how the states of Baja California contribute to the national production. Approximately 80% of scallop production from México is processed (muscle), and 82% of this processing is imported by The United States of America. In 1987, 380,000 kg of frozen muscles were exported. In the first months of 1988, 245,000 pounds of frozen muscle were exported (from Fishery Market News Report, January 1987 to March 1988; U.S. Department of Commerce).

The best prices are usually obtained in August to January in the U.S.A. market and Table 29.2 shows recent market prices for frozen muscle, an almost doubling in price. The Mexican scallops are small and cheap. There were plenty of Florida calicos and Chinese bays at prices averaging between $1.80 and $2.50 US dollars lb^{-1}, depending on size, through the winter of 1995.

Table 29.2

Historic number of licenses (2 or 3 boats), landed weight of sucked meats, price per kg, total values and percentage of exported to USA for 1978–2000. (From SEMARNAP (1999), Pérez-Concha (1990) and Massó-Rojas (1996)).

Year	No. of Licenses	Landings Muscles (t)	Price ($US/kg)	Total value ($US)	Exported to USA (%)
1978	7	33	2.85	$94,367	30
1979	12	1	3.25	$1,806	40
1980	12	127	4.32	$550,080	50
1981	11	313	4.12	$1,290,933	55
1982	14	275	3.77	$1,037,169	40
1983	23	232	3.89	$901,183	55
1984	20	256	4.90	$1,255,489	55
1985	13	270	4.52	$1,219,396	64
1986	17	633	4.02	$2,542,873	90
1987	10	626	3.61	$2,258,256	90
1988	63	1,217	3.41	$4,148,454	91
1989	135	3,558	3.25	$11,564,944	92
1990	300	3,247	4.10	$13,311,333	85
1991	230	769	3.85	$2,958,939	82
1992	59	105	4.30	$452,933	55
1993	32	144	4.88	$704,889	60
1994	45	52	5.86	$307,324	58
1995		102	6.30	$645,400	70
1996		1,874	4.50	$8,432,000	85
1997	102	173	5.40	$935,400	90
1998		212	4.20	$889,933	68
1999		14	5.25	$75,833	45
2000		20	6.40	$128,000	55

29.5 EXPECTED FUTURE

The Mexican scallop fishery has come a long way in three decades, from a dwindling overfished resource to a major fishery and culture in terms of tonnage and value. This success is caused by many factors integral to the Mexican society; the government fishery bodies, cooperatives, fishermen and scientists reacting quickly and sensibly to the problem of dwindling stocks and catch. There were also well established legal precedents for the exploitation of grounds, and for the development of culture in coastal areas where the long line installations were protected from industrial and other developments.

The growth of the scallop diving industry has been spectacular over the past few years, and its role in scallop conservation can be credited with creating employment through processing and handling of scallops. Practical culture depends on family labour of all ages. The balance of payments of México is helped through foreign earnings on exports. Methods of working scallops are now becoming standard, and the divers involved are putting a great deal of thought into running cost conscious and profitable operations. The new awareness of fishery bodies of the importance of environmental quality for the future of their scallop industry should result in a stable industry with healthy market outlets in the years to come.

29.6 AQUACULTURE

29.6.1 Introduction

At the present time, scallop aquaculture consists of culturing local species. Satisfactory results have been obtained by several culture facilities but there have not been any great advances in production. Aquaculture of pectinids, although rapidly developing, has production volumes <0.5% of the total landings (Table 29.3). Progress to massive production of pectinid species seed in hatcheries and their transference to the natural environment for growout has been successful. Future emphasis will be in selecting scallop species or producing hybrids that have improved qualities over local species. Strain evaluation and selection have been done, mass and family selection programs have been initiated, and intraspecies crossbreeding established (Cruz-Hernández and Ibarra 1997b; Ibarra 1999; Ruiz-Verdugo et al. 2000).

Table 29.3

Summary of the scallop aquaculture production, for the period 1988–1999. (Data from Maeda-Martínez et al. (1996), Mazón-Suástegui (1996) and SEMARNAP (1999)).

Year	Production (t) FAO (1999)	Production (t) SEMARNAP	Number of Spat Cultured	Origin
1988	20	15	340.000	Natural. hatchery
1989	20	30	60,000,000	Natural
1990	30	1,800	92,000,000	Natural
1991	10	576	478,000,000	Natural
1992	60	72	172,000,000	Natural
1993	60	47	2,000,000	Natural
1994	41	55	600,000	Natural, hatchery
1995	14	180	9,000,000	Natural, hatchery
1996	10	10	900,000	Natural
1997	10	10	200,000	Natural
1998	-	10	250,000	Natural
1999	-	200	9,000,000	Natural, hatchery
2000	-	200	10,000,000	Natural, hatchery

29.6.2 Species

The catarina scallop, *Argopecten ventricosus,* is the most studied and cultured scallop species in México. Other scallop species, *Nodipecten subnodosus* and *Euvola vogdesi,* are under study using some of the techniques developed and with the availability of hatchery-reared seed scallop.

29.6.3 History

Scallop culture research has just begun (Amador-Buenrostro 1978; Félix-Pico et al. 1980; Siewers 1982; Baqueiro 1984; Reyes-Sosa 1985; Tripp-Quezada 1985; Félix-Pico et al. 1989). The catarina scallop, *A. ventricosus* is cultivated because of its high value and fast growth. Many studies have been made on the techniques for the natural seed collection, broodstock conditioning, larvae rearing, and setting and growout of the spat at coastal sites. Investigations into hatchery cultivation of scallops have also been done and there has been successful work in management to obtain maximum production levels in suspended and bottom culture.

Under natural conditions, the catarina scallop spawns from late February to March when the water temperature rises from 19°C to 21°C (Félix-Pico et al. 1980). The pelagic larvae attach to spat collectors twenty days after spawning and are allowed to grow to about 8–12 mm. The juvenile scallops are put in plastic cages containing about 500 individuals at the start of culture, then reduced to a density of 50 individuals in the final culture. Market size (5–6 cm) is reached in about 6–8 months (Félix-Pico et al. 1989). The bottom culture practice in enclosed areas is adapted to shallow waters, where juvenile scallops are protected by plastic nets from benthic predators (Baqueiro 1984; Cáceres-Martínez et al. 1987; Singh-Cabanillas 1987).

Attempts have been made to rear only the bay scallop to market size, with spat produced in the laboratory reared to metamorphosis in laboratory pilot testing (Mazón-Suástegui 1986; Coronel-Solorzano et al. 1987; Avilés-Quevedo and Muciño-Díaz 1988a; Avilés-Quevedo and Muciño-Díaz 1988b). The first commercial hatchery production at the Centro de Investigaciones Biológicas del Noroeste (CIBNOR) was around 20,000 spat by Maeda-Martínez et al. (1989). The government hatchery at Bahía Magdalena (SEMARNAP) and this produced 325,000 spat (Mazón-Suástegui et al. 1991), and the state hatchery called Centro Reproductor de Especies Marinas del Estado de Sonora (CREMES) and this at Sonora began to sell seed at $1 US dollar per 1,000 (Robles-Mungaray and Serrano-Guzmán 1993, 1995). Cage culture can produce scallops in eight to 14 months, making the scallop a prime candidate for aquaculture.

No previous research on aquaculture for the commercial scallop *Nodipecten subnodosus* is known, just two works on larval rearing and wild spat. Carvajal-Rascón (1987) reported spat production to describe the larval cycle in laboratory and this species has been reared to metamorphosis in pilot testing. Another is about spat settlement and early growth in Laguna Ojo de Liebre (García-Domínguez et al. 1992).

There is a similar situation for *Euvola vogdesi,* with observations on spat settlement in collectors and growout in Bahía de La Paz (Cáceres-Martínez et al. 1989; Ruiz-

Verdugo and Cáceres-Martínez 1991) and in Bahía Bacochibampo (Carvajal-Rascón and Farell 1984). Attempts were made to produce juveniles from the laboratory pilot testing (Monsalvo-Spencer 1994), and a few thousand spat have been raised in the hatchery (personal communication, Miguel Robles).

29.7 HATCHERY TECHNIQUES

29.7.1 Conditioning

Laboratory experiments were done to determine the effect different temperatures and food levels might have on the timing of various stages of the reproductive cycle. Avilés-Quevedo and Muciño-Diaz (1988b) also reported diets to feed *Argopecten ventricosus* scallops to maturation. Gonads ripen when ambient water temperature is 17°C to 19°C, and with a food level increase from 4,000 to 5,000 million cells scallop^{-1} day^{-1} or at a concentration between 1 to 1.5 million cells mL^{-1} of *I. galbana*. Completely mature specimens were produced in 4 weeks. A broodstock-conditioning method was described by Mazón-Suastegui et al. (1986), when they were fed with a mixture of cornstarch and microalgae and conditioned in flow-through flumes provided with raw seawater at 20°C. The method was enhanced by Avilés-Quevedo et al. (1991) by optimising the clearance rate to maximise food assimilation and avoid pseudofeces production.

Catarina scallop have been cultured successfully at a commercial level from spawning to harvest size. Results indicate that recently spawned broodstock can be matured to spawning in 4 weeks at 23 ± 1°C and with a food ration of 3,900 million cells scallop^{-1} day^{-1} of a mixture (6:3:1) of *I. galbana*, *Chaetoceros* sp. and *Tetraselmis suecica* (Maeda-Martínez et al. 1989, 1995; Maeda-Martínez and Ormart-Castro 1995; Lora-Vilchis et al. 1997).

Scallops can be conditioned out of season in the laboratory by holding them in a cold room (20 ± 1°C) for 5 weeks with daily additions of *I. galbana* and *C. gracilis* (50:50), with the ration at 9% of the initial dry weight of the gonad or 50,000 cells mL^{-1} day^{-1} (Millán-Tovar 1997).

Racotta et al. (1998) investigated whether continuous rather than discontinuous feeding could improve reproductive performance of *A. ventricosus* during conditioning in the hatchery. The gonadal biochemical composition was analysed. Protein was 35% to 45% of dry weight or 75 mg g^{-1}, acylglycerols were 4 to 18 mg g^{-1}, and cholesterol (3.8 to 4.9 mg g^{-1}) and glucose (5 to 5.6 mg g^{-1}) were slightly higher. They concluded a continuous feeding system resulted in a better spawning capacity and increased rematuration ability than discontinuous feeding.

The continuous feeding system was optimised by using an agricultural drip system and a permanent supply of food in multiple rearing tanks. That result contrasted with the discontinuous system for conditioning scallop broodstock, for which the large amounts of food required, added in intervals, resulted in pseudofeces being produced and in a large variation in food availability and therefore of the microalgae removal rate (Ramírez and Ibarra 1999).

Carvajal-Rascón (1987) reported a broodstock of *Nodipecten subnodosus* conditioned for 8 weeks at 19°C, and fed a mixture of *Pavlova lutheri* and *Pseudoisochrysis* sp. algae, keeping concentrations at 1 to 2 million cells mL^{-1}. Gutierrez-Villaseñor and Chi-Barragán (1997) kept a population of the same species 6 weeks at 17°C to 18°C and then gradually raised temperature 2°C every four weeks, with a microalgal diet of *Monochrysis lutheri* and *Isochrysis galbana*. The scallops were also given the microalgae T-Iso daily so that it represented 3% or 1.7% of the dry weight of each organism. In a commercial hatchery (CIBNOR) at La Paz, a broodstock of *A. ventricosus* had been conditioned by maintaining the animals at 24 ± 1.5°C, and fed daily on a diet of *I. galbana, M. lutheri, Thalassiosira pseudonana,* and *C. gracilis* for 5 weeks at a concentration at 150,000 cells mL^{-1} in the proportion of 3:3:2:2 (Villavicencio-Peralta 1997).

29.7.2 Induction of Spawning and Fertilisation

Gametes are usually obtained by inducing ripe adult scallops (*A. ventricosus*) to spawn. Ripe scallops are placed in a tank or trough. The scallops are held at 30 °C for about 1 h and then the temperature decreased to 25°C until the scallops appear to be pumping vigorously, first male spawning and then after 20 min the females (Avilés-Quevedo and Muciño-Díaz 1990). Additionally, other authors induced spawning using the high algal density method (Mazón-Suástegui et al. 1991).

The spawning process has been induced by thermal shock and chemical (serotonin) stimulation for a broodstock of *A. ventricosus* (Ibarra et al. 1995; Lora-Vilchis et al. 1997; Monsalvo-Spencer et al. 1997). A broodstock of mature scallops collected from Bahía Magdalena was induced to spawn by each of the scallops being injected with 0.2 mL of serotonin (5-hydroxytryptamine, 0.5 nM) in the gonad (Ibarra et al. 1995). Adult specimens of *A. ventricosus* were collected from Bahía de La Paz, B.C.S. and were conditioned to gonadic maturity in the laboratory for 25 to 30 days. Two spawning procedures were used. Thermal shock consisted of a sudden increase from 24°C to 26°C, held for 45 min and then raised to 27°C and held 15 to 20 min until the sperm was observed. For the serotonin procedure, 0.4 mL of a 0.125 mM solution of serotonin diluted in filtered seawater was used (Lora-Vilchis et al. 1997).

For *A. ventricosus,* Monsalvo-Spencer et al. (1997) induced spawning with thermal shock (increasing 3°C/min from 18°C to 30°C) combined with addition of sexual products from another individual. This was effective as a spawning inducer in 50% of the breeders and was the only treatment that induced oöcyte release after 3 to 5 h. Spawning was also induced with injection of serotonin (5-hydroxytryptamine at 0.025, 0.25 and 2.5 nM) and was effective as a sperm releaser only. Fertilisation was made with a proportion of 7 sperm per oöcyte and 90% fertilisation obtained. Oocyte size averaged 53 μm.

Broodstock of *A. ventricosus* and *N. subnodosus* were conditioned at the CREMES at Bahía Kino, Sonora, and were successfully induced to spawn with thermal shock (ΔT from 18°C to 29°C). "D" larvae of *A. ventricosus* and *N. subnodosus* were found between 20 to 24 h after fertilisation with a length of 70 to 90 μm and height of 60 to 75 μm for the former, and a length from 80 to 100 μm and height from 65 to 70 μm for the latter (Serrano-Guzmán et al. 1997).

Spawning was obtained with thermal shock for *N. subnodosus* (Gutierrez-Villaseñor and Chee-Barragán 1997). More than half of the organisms spawned only as males, followed by those that spawned as both, and only a few spawned only as females. A similar behaviour has been found in other pectinids. Other authors used thermal shock (ΔT from 18 to 28°C) with successful induction (Velasco-Blanco 1997; Villavicencio-Peralta 1997).

29.7.3 Larval Culture and Metamorphosis

The catarina scallop has been reared to metamorphosis in pilot testing. The veliger larvae were cultured at 24°C and 35‰ salinity. After 10 days the velum (size 194 x 228 μm) was absorbed and a big foot developed. Between 14 to 16 days of fast growth, there was a shell without ribs (Avilés-Quevedo and Muciño-Diaz 1988a). When 60% of the larvae reached the eyed pediveliger stage, two basic arrangements for setting larvae were immersing a plastic mesh as substrate into the tank (bag collector), and depositing larvae in a continuous flow fibreglass-sieve containers, filled with or without plastic mesh collectors. Filtered seawater and microalgae were supplied twice a day for the culture tanks and continuously in the sieve containers (Mazón-Suástegui et al. 1991).

Monsalvo-Spencer (1998) placed free-swimming veliger "D" larvae after hatch-out (24 h) in a 100-L larval rearing tank with rearing densities of 10 larvae mL^{-1}. Daily feeding consisted of keeping a concentration of 60,000 cells mL^{-1} of a mixture of *I. galbana, Chaetoceros* sp. and *T. suecica* (6:3:1) over 6 days. After the shell height was 85 μm, density was reduced to 6 larvae mL^{-1} and they were fed with 100,000 cells mL^{-1}. Finally, density was reduced to 3 larvae mL^{-1} and fed with 150,000 cells mL^{-1}. The larval settlement (176–188 μm) occurred after 15 days and the temperature raised from 23 to 24°C. This period can be adjusted with the linear equation:

$$\text{Shell height} = 54.12 + 6.1 * \text{days} \tag{8}$$

Metamorphosis and setting were allowed to proceed on a nylon screen (30-cm diameter and 140-μm mesh), immersed in a 70-L container with running seawater at 0.5 L min^{-1}. The screens were changed to 160, 180, 212 and 250 μm as the larval size increased (Fig. 29.5). Juveniles were maintained until reaching 2 mm at 45 days. This period can be adjusted with the exponential equation:

$$\text{Shell height} = e^{(5.04 + 0.055 * \text{days})} \tag{9}$$

A successful culture and high yield is assured by changing water every three days until larval development is complete and the larvae are ready for settlement (Monsalvo-Spencer (1998). The larval rearing tanks are maintained with no water flow and soft aeration to avoid injuring the larvae. The larval rearing apparatus represents an advancement in scallop cultivation procedures, and the authors have obtained a patent (No. 180212) for this container (Maeda-Martínez et al. 1995; Maeda-Martínez 1996).

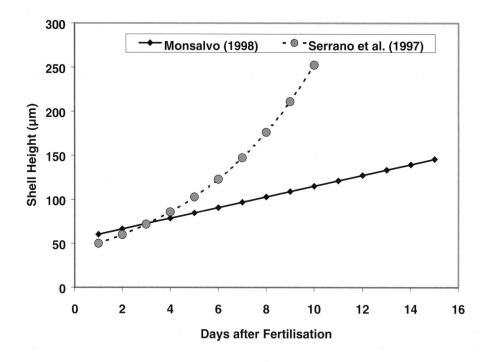

Figure 29.5. Laboratory growth of *Argopecten ventricosus* from fertilised egg to 15 day post fertilisation. Calculated by equations number 8 and 12. (From Serrano-Guzmán et al. (1997) and Monsalvo-Spencer (1998)).

Ibarra et al. (1995) studied the effects of inbreeding on growth and survival of self-fertilised catarina scallop larvae. This study was done at the Genetic Laboratory in La Paz (CIBNOR). The larvae were placed in 20-L experimental tanks. Daily feeding consisted of adding *I. galbana* at varying concentrations depending on age; 75,000 cells mL^{-1}, 100,000 cells mL^{-1}, and 150,000 cells mL^{-1} for larvae up to 3 days old, up to 6 days old, and up to 18 days old. The larvae produced from exclusive self-fertilisation and grown over a period of 18 days are smaller (112 µm) and have lower survival (31%) than larvae produced from either pair-mating (121 µm; 54%) or mass-spawning (118 µm; 56%), which latter two were not significantly different. At the end of rearing period, larvae with an inbreeding coefficient of F = 0.5 showed a decreased length of 12% and 11% and a reduced survival of 43% and 41% compared to mass-spawned and pair-mated groups.

Another factor that has an effect on growth and survival is the stocking density. Ibarra et al. (1997), with two populations of catarina scallop, tested larvae from Bahía Magdalena at densities of 10 to 20 larvae mL^{-1}, concluding the lower density had higher growth but equal survival. Larvae from a Bahía Concepción population grew better at densities of 15 and 20 larvae mL^{-1} than densities of 5 and 10 larvae mL^{-1}, although survival was lower at higher densities. The average seawater temperature was 21 to 22°C.

In both experiments, feeding was by adding, once a day, a mixture of *I. galbana* and *M. lutheri* (1:1) in concentrations of 30 cells μL^{-1} for 2–5 days, 50 cells μL^{-1} for 6–7 days, 60 cells μL^{-1} for 8–9 days, and 80 cells μL^{-1} for 10–14 days.

Several studies had been done on algal diets that yield fast growth and high survival in scallop larvae. Lora-Vilchis and Maeda-Martínez (1997) determined the nutritional values of ten microalgae species by studying the ingestion and digestion index of *A. ventricosus* veliger larvae during 7 days. Results showed a significant positive correlation between the smaller geometric dimension of the algae and the larval age and size of initial ingestion. Only *I. galbana*, *M. lutheri*, *C. calcitrans*, *C. muelleri* and *Thalassiosira pseudonana* clone 3H were ingested, and *Nannochloris oculata* and *Phaeodactylum tricornutum* were ingested but not digested. No ingestion was observed in *Tretaselmis suecica*, *Dunaliella tertiolecta* and *Thalassiosira pseudonana* clone s78 during the experiment.

Millán-Tovar (1997) found larvae of *A. ventricosus* could be cultured with a diet of *I. galbana* and *C. gracilis* (1:1) with the best growth rate of 11 μm day^{-1} and larval settlement after 16 days with the highest percentage.

The state hatchery CREMES at Sonora has cultured catarina and mano de león scallops in 3.5-m^3 tanks, in one trial in 1992 and two in 1993. Both culture systems followed the methods described by Robles-Mungaray and Serrano-Guzmán (1993, 1995) until larval settlement. "D" larvae of *A. ventricosus* and *N. subnodosus* between 20 to 24 h after fertilisation had a length from 70 to 90 μm and height from 60 to 75 μm for *A. ventricosus*. *N. subnodosus* larvae were 80 to 100 μm long and from 65 to 70 μm high. The prodissoconch II of catarina scallop (90 to 95 μm long) were observed after the second day and *N. subnodosus* prodissoconch II were present after the third day. The umbo stage for *A. ventricosus* (120 to 160 μm long and 105 to 150 μm height) appeared at the third day and that of *N. subnodosus* (150 to 185 μm long and 120 to 150 μm height) on the eighth day. Pediveliger with eyespot of catarina scallop (240 to 250 μm long and 215 to 220 μm height) were seen on day eight and for *N. subnodosus* (190 to 210 μm long and 165 to 180 μm height) on day eleven. *A. ventricosus* dissoconch starts between the eighth and the ninth days (>250 to 255 μm long and >220 to 230 μm high), and *N. subnodosus* dissoconch begin between the 12th and the 13th days (>210 to 215 μm long and 180 to 190 μm height).

Isometric larval growth occurred in both species following a linear model described by the equations:

$$L = -3.1 + 0.9 * \text{height with } R = 0.99 \text{ for } A.\ ventricosus, \tag{10}$$

$$L = -11.5 + 0.9 * \text{height with } R = 0.99 \text{ for } N.\ subnodosus. \tag{11}$$

A. ventricosus grew exponentially with time, leads however to giving the equation:

$$L = 60.5\ e^{\ 0.18\ \cdot\ days} \text{ with } R = 0.99, \tag{12}$$

Whereas *N. subnodosus* grew linearly with time described by the equation (Fig. 29.6):

$$L = 71.8 + 10.8\,1 * \text{days with } R = 0.99. \tag{13}$$

The highest growth rate in *A. ventricosus* was between the third and the seventh days (Serrano-Guzmán et al. 1997).

Carvajal-Rascón (1987) has reared *N. subnodosus* to metamorphosis in pilot testing, and described the morphological changes. Veliger larvae were observed between 21 to 24 h after fertilisation, with a density between 30 to 150 eggs mL^{-1}, and were fed algae at a concentration of 30,000 to 300,000 cells mL^{-1} under ambient temperature that averaged 26°C. After 10 days of culture, they reached a length 197 μm, with a growth rate of 6 μm day^{-1}. Larvae were dead before metamorphosis.

The CIBNOR, Centro de Investigación Científica y de Educación Superior de Ensenada, B.C. (CICESE), and Universidad Autónoma of Baja California (UABC) laboratories have cultured *N. subnodosus* (Ortiz-Cuel 1994; Velasco-Blanco 1997; Villavicencio-Peralta 1997). The larvae were reared at densities of 2 to 7 larvae mL^{-1} under ambient temperature that averaged 19.5°C for the first week and 24.5°C for the second week. In one experiment, four different rations of the microalgae *Isochrysis* var. *galbana* aff. *thaitiana* were used (1,250, 2,500, 3,750 and 5,000 cells larvae^{-1} day^{-1} in the first week, and the second week at 2,500, 5,000, 7,500 and 10,000 cells larvae^{-1} day^{-1}). During the second experiment, four larval densities were used (2, 5, 7 and 10 larvae mL^{-1}) with a constant ration of 3,000 cells larva^{-1} day^{-1}. The average larval growth rate at the higher rations averaged 6.1 μm day^{-1} during the two week period. Larval survival was reduced from 63% to 34% and from 84% to 45% by the increased culture density the first and second week of the experiment. Temperature had an appreciable effect on the larvae. Results showed that low to medium rations, low densities, and temperatures above 20°C are the most favourable for the larval culture of this species (Ortiz-Cuel 1994; Ortiz-Cuel et al. 1997).

Similar results were obtained by Velasco-Blanco (1997). Using a mixture of microalgae *I. galbana* and *C. calcitrans*, growth and survival rates were an optimum. Chemical analysis showed that microalgae *I. galbana* are higher in protein, lipid and carbohydrate and *C. calcitrans* was a complementary food. Villavicencio Peralta (1997) made a mixture with both algal species and *Nannochloropsis oculata*. These species produced an average larval growth rate between 0.7 and 1.6 μm day^{-1}. The temperature was about 27°C. The veliger "D" larvae occurred 18 h after fertilisation and larval settlement was at 16 days.

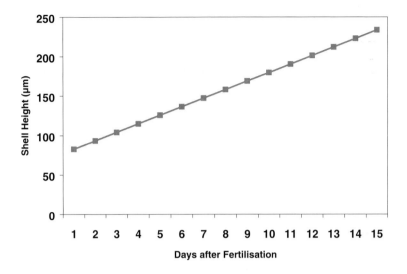

Figure 29.6. Larval growth of *Nodipecten subnodosus* from fertilised egg to 15-day post fertilisation. Calculated by equation number 13. (From Robles-Mungaray and Serrano-Guzmán (1993)).

Larvae of *N. subnodosus* were reared to evaluate the performance of downwelling screen as a settlement system in the laboratory (acrylic laminate of 112 x 27.5 x 29 cm). Each chamber received 30,000 mature larvae with a mean length of 190 μm. At the beginning of the experiment, the food provided was 30,000 cells mL[-1] of a 1:1 mixture of *Isochrysis var. galbana aff. thaitiana* and *T. pseudonana,* which was gradually increased to 50,000 cells mL[-1] after 4 weeks. The settlement of the larvae occurred during a protracted period of 15 days after which no swimming larvae were observed. Survival to the fourth week, as reflected in spat numbers, was significantly higher in the chambers without additional substrate (González-Ramos et al. 1997).

Larvae of voladora scallop (*Euvola vogdesi*) had been cultured in 4000–L rearing containers in the CIBNOR hatchery. The rearing and settlement methods used are similar to the process adopted for *A. ventricosus,* and preceded routinely along lines established by other workers (Maeda-Martínez et al. 1989; Avilés-Quevedo and Muciño-Díaz 1990; Robles-Mungaray and Serrano-Guzmán 1993, 1995; Monsalvo-Spencer 1994). Under optimum conditions, a high percentage of larvae metamorphosed with low mortality to 11 to 13 days after fertilisation. Successful sets were produced with 3 larvae mL[-1]. The best results in management of these spat were obtained by rearing them to 3 to 4 mm in the hatchery. The first attempt was with 60,000 juveniles, and the second near one million juveniles (Miguel Robles, personal communication).

29.7.4 Genetic Enhancement

29.7.4.1 Selection

The practice of selection relies on using the most desirable individuals within a population to produce the subsequent generation. The principle of selection is based on heritability of the trait and its relation to the intensity of selection. In bivalves, these are represented as shell traits; total length, height and width, and as whole weight. These growth variables can be measured without sacrificing the individuals. Ibarra et al. (1999) did an experiment by using one-generation selection for total weight and shell width at age 1 year in catarina scallop *A. ventricosus*. These selected lines were used to estimate direct and correlated responses to selection, produced heritabilities, and produced genetics. An experimental base generation was produced under laboratory conditions from a population at Laguna Rancho Bueno, in Bahía Magdalena. Growout was in the same lagoon. Larvae and spat were reared in the CIBNOR Genetics Laboratory at La Paz, B.C.S. Results were that two traits have high genetic correlation, evidenced by the high correlated response of shell width when selecting on total weight.

Similar results were obtained by Ibarra (1999) with the total weight. Correlated responses of all growth traits were large at the age of the selection (1 year), and at 5 months. For the most important trait in scallop, adductor muscle weight, and correlated gains of 13% to 19% are expected when selecting the heaviest 10% individuals. The same correlated response is expected for tissue weight. However, that was not true for shell width. No significant positive correlated responses at 5 months were seen for any trait, but negative correlated responses for shell height and shell length were significant at 1 year. Further studies need to be done to establish whether negative correlation of these traits, and for scallops, whether this results in a negative effect on muscle weight and on reproductive performance. The importance of stock selection is emphasised not only for juvenile or adult phenotypic traits of artificially-produced progeny, but also on the maternal quality of the spawners producing that progeny (Cruz-Hernández and Ibarra 1997a).

29.7.4.2 Crossbreeding

The evaluation of crosses among the populations is also important, because it might indicate whether the best breeding method would be crossbreeding. Crossbreeding would be the preferred method when heterosis for production traits occurs. There are two important catarina scallop fisheries along the Baja California Península, Bahía Magdalena and Bahía Concepción. Intra-and intercrosses between scallop populations from those two fisheries were produced and simultaneously grown under those two different environments. Cruz-Hernández and Ibarra (1997b) determined environmental effects on growth, and whether a genotype by environmental interaction is present that will cause the best genotype to be different for each environment.

During the larval stages, there was a clear maternal effect for survival and length, evidenced by a difference between the reciprocal crosses and a similarity of each reciprocal cross to its maternal parent. The presence of maternal effects was explained as egg energy reserves being different between the two populations, possibly caused by different environmental conditions during the initial gametogenesis process of spawners. Larvae from Magdalena population and the reciprocal cross produced with those eggs had a higher survival (46% and 47%) than larvae from the Concepción population and its maternal reciprocal cross (32% and 29%). The best larval growth rate was from Magdalena, reaching 167 µm, higher than Concepción at 140 µm. Similar results for the reciprocal crosses (160 and 153 µm) were obtained. The presence of maternal effects is explained because egg energy reserves are different between the two populations despite the simultaneous conditioning of 15 days for both groups of spawners prior to spawning induction (Cruz-Hernández and Ibarra 1997a).

During the simultaneous growout at Bahía Magdalena and Bahía Concepción to age 4 months, the environmental effect was the main determinant for growth and survival of the genetic groups. At Bahía Magdalena, there was a better performance of all genetic groups than at Bahía Concepción, which is explained as a consequence of a larger environmental stress in the latter site (low productivity and higher average annual temperature). These results suggest that Bahía Magdalena is a growout area with better environmental conditions for the growth of catarina scallops. The populations were differentiated during the last month in both environments, but only by shell width when grown in Bahía Magdalena, whereas when grown in Bahía Concepción genetic differences between the populations were seen for all growth traits and survival. For shell width, the best population in each environment was the one native to that bay, which resulted in significant of the interaction between genotypes and environments for this trait (Cruz-Hernández et al. 1998).

For survival, the genotype by environment interaction resulted not only from the poor performance of the Magdalena population when grown at Bahía Concepción, but also from a poor performance of the reciprocal crosses. At the end of the culture, the largest heterosis for growth was seen in Bahía Concepción (>15%) for the biomass trait. There was no heterosis for survival in either of the environments evaluated. For both hatchery and growout stages, and despite some growth heterosis, this cannot be considered a useful heterosis at the production level, because the reciprocal crosses were similar to the population with the best growth, never exceeding either of them significantly (Cruz-Hernández et al. 1998).

29.7.4.3 Triploidy

Production of triploid bivalves has been successfully achieved with many commercially important species of scallops. This practice had been used for the catarina scallop *Argopecten ventricosus* using two concentrations (0.1 and 0.5 mg L^{-1}) of cytochalasin-B (CB), applied for 15 min to inhibit extrusion of the second polar body (PBII). Results showed that the triploid percentage average ranged from 8.3% to 57.5%. Triploid individuals were obtained in both CB concentrations, but the triploid induction was low (8.3%) when using 0.1 mg L^{-1}. No triploid spat were detected in the controls. The highest level of triploidy (57.5%) was obtained with 0.5 mg L^{-1}. Cytochalasin B treatment of scallop eggs resulted in a lower survival than controls for both larval and spat stages (2% to 16%). One difficulty with triploid induction in this species is associated with the characteristic of hermaphroditism (Ruiz-Verdugo et al. 1997).

Growth of triploid scallops exceeds that of diploids in all evaluated traits. The largest percent difference between the diploid control group and the treated groups were seen after diploid scallops reached the peak of sexual maturation and began spawning. The gonad of the triploid had a brownish discolouration, few oöcytes and sperms developed, and male acini were replaced by female acini. Gametogenesis and the condition of hermaphroditism in this normally functional hermaphrodite were greatly affected by the triploid condition. At 280 days, the muscle index was larger than the control group, with an inverse relationship between the muscle index and gonad index in the control group, which when compared with diploid scallops from within the same treatment, had a significantly larger muscle weight than the differences seen between the treated and control group (Ruiz-Verdugo et al. 2000).

29.7.5 Antibiotics

Sick larvae in a culture exhibit slow growth, poor colour, slow ciliary movement, loss of velum, accumulation of lipid granules in the digestive system, empty stomachs and weak swimming activity (finally sinking to the bottom and dying). Sick larvae usually stop feeding, so a low rate of algal consumption will also indicate a problem. Sainz-Hernández et al. (1998) experimentally induced vibriosis with *Vibrio alginolyticus* in larvae of *Argopecten ventricosus* in a laboratory hatchery and determined that these bacterial infections came from seawater and freshwater effluents, and the broodstock. Under concentrations of 0.5×10^5 cells mL^{-1} or lower, *V. alginolyticus* was innocuous (80% survival), but at 5×10^5 cells mL^{-1}, there was no larval activity by day 10, nor had any larvae survived by day 12.

Adding chloramphenicol and erythromycin, effective antibiotics against both Gram-negative and Gram-positive bacteria, to the larval rearing water helps suppress bacterial growth. Both antibiotics helped increase survival in *Argopecten ventricosus*. The results obtained with the food algae (*Isochrysis galbana* and *Chaetoceros gracilis*) showed growth rates were not affected (Campa-Córdova 1997).

Chloramphenicol and erythromycin are added to the seawater in the rearing container to obtain the desired concentrations, 0.5 to 6 mg L^{-1}. The highest concentration gave best larval survival rates, 69% for chloramphenicol and 68% for erythromycin. There is no significant difference in survival with two antibiotics, but the author recommended erythromycin at 6 mg L^{-1} (Campa-Córdova 1997).

29.8 NATURAL SPAT

29.8.1 Methods Employed

In Bahía de La Paz, different kinds of structures were built for supporting gear and for collecting spat as described by Félix-Pico (1991a). The spat collectors were suspended from a long line held with buoys and anchors. For bottom culture, a pond was delimited with mangrove sticks and plastic nets in shallow areas. Three types of spat collectors were built; onion sacks (40 x 70 cm, 0.8-cm mesh) filled with shrubs and polypropylene nets, wood boxes filled with shells and chivato (*Atriplex* spp.) or chamizo (*Batis maritima*) shrubs, and Nestier plastic cages filled with polyfilm and shells. For rearing, the spat were placed in lantern type baskets and in Nestier plastic cages. Two hundred baskets were installed on each long line. Details of construction and specification of materials are described by Vicencio-Aguilar et al. (1988) and Vicencio-Aguilar et al. (1994).

29.8.2 Spat Collection

The timing of spawning and distribution of pectinid larvae are thought to depend upon environmental conditions. The natural spawning season was established by sampling the zooplankton each week and by installing spat collectors simultaneously. Also, the gonadal index was estimated in adult specimens, as were the number spat per collector, type of collector, spat sizes, depths, and the relationship between abiotic and biotic variables. Table 29.4 summarises the achievements in wild spat collection of *A. ventricosus* on both coasts of Baja California Sur between 1977 and 1999. Average figures of seed per collector are noticeably different. The incidence of pelagic larvae collected was associated, together with temperature at 3-m depth and number of seed per collector, with lunar phases, blooming of palo verde (*Cercidium peninsularae*), spawning season, and probable settlement period (Félix-Pico et al. 1980, 1989). A program was begun in the Gulf of California in 1977 and the Pacific lagoons in 1981. Previous results showed that 1977 settlement of pectinid spat on artificial collectors were good in Ensenada de La Paz. Monitoring of settlement continued until the population of catarina scallops declined in 1981. Results have shown great variation in number of spat per collector. From 90 to 635 scallop spat per collector were found (Félix-Pico 1991a).

Ruiz-Verdugo and Cáceres-Martínez (1990, 1991) deployed experimental collectors (25 x 25 cm, 2-mm mesh) at Bahía Falsa (Bahía de La Paz) during 1987 and 1988. Maximum spat collection occurred between 15 to 17 m depth during April to June. The range of spat collected was 85–217 per collector. For the first 5 weeks after settlement, the growth rate was 0.34 mm day $^{-1}$, and can be described by the linear equation:

$$\text{Shell height} = 0.2 + 0.34 * \text{days} \tag{14}$$

In Bahía Magdalena, B.C.S., Tripp-Quezada (1985) recorded from 100 to 300 spat per collector in April of 1981. Almost one year later, spat collection was more successful (up to 2,212 spat per collector) but the local scheme of abundance was somehow reversed. Results have shown that the best collecting season was between January to March when temperature had a minimum value of 19°C, and the best sites were Santa Elena and San Vicente (Estero Santo Domingo) in the northern part of Bahía Magdalena. In the south much lower numbers of spat were obtained per collector with a maximum of 100. The most efficient spat collectors were onion bags filled with plastic mesh (Tripp-Quezada

Table 29.4

Summary of wild spat collected from catarina scallops, for the period 1977 to 1999 at the aquaculture areas. (Data from Félix-Pico (1991a), Tripp-Quezada (1991), Félix-Pico et al. (1995), Ruiz-Verdugo and Cáceres-Martínez (1991), Félix-Pico et al. (1989), Tripp-Quezada (1985), Maeda-Martínez et al. (1993), Félix-Pico et al. (1997), Félix-Pico et al. (1980), Singh-Cabanillas (1987), García-Domínguez et al. (1992), Millán-Torvar (1997), Ruiz-Verdugo and Cáceres-Martínez (1990), Félix-Pico et al. (1992), Maeda-Martínez et al. (2000), Maeda-Martínez et al. (1996), Castro-Moroyoqui (1992), Félix-Pico and Villalejo-Fuerte (1999), Tobias-Sánchez and Cáceres-Martínez (1994)).

Geographical Areas, Lagoons, Sites	Years	Type of Collector	Spatfall per Collector	Season
Gulf of California				
Baja California Sur				
Bahía de La Paz				
Ensenada de La Paz	1977–1991	Onion bags, plastic cages	80–635	Feb-March
Bahía Falsa	1987–1988	Experimental collectors	10–217	Feb-April
Bahía Concepción				
El Remate	1988	Onion bags, plastic cages	25,000	Jan-April
	1989–1993	Onion bags, plastic cages	4,000–20,000	Jan-April
Pta. Coloradito	1994	Onion bags	315	Jan-April
Pta. Coloradito	1995	Onion bags	150	Jan-April
Sonora, Bahía Bacochibampo	1983–1984	Onion bags	1770	Feb-may
Pacific Ocean				
Bahía Magdalena (BCS)				
San Vicente	1981–1982	Onion bags, plastic cages	500–1,000	March-May
Los Prados	1981–1982	Onion bags, plastic cages	600–1,100	March-May
Santa Helena	1981–1982	Onion bags, plastic cages	300–2,000	Jan-May
San Buto	1981–1982	Onion bags, plastic cages	50–100	Jan-May
La Florida	1989–1991	Onion bags	500–9,000	
Puerto Magdalena	1989–1991	Onion bags	200–9,000	Feb-Oct.
Rancho Bueno	1989–1999	Onion bags	150–10,000	Jan-May
Laguna Guerrero Negro	1989–1990	Onion bags, plastic cages	4,000	Jun-Aug.
Laguna Ojo de Liebre		Onion bags, plastic cages	1,500–8,000	Jun-Aug.
Laguna San Ignacio	1990	Onion bags, plastic cages	500	Jun-Aug.

1991). In 1989 and 1990 when the megapopulation was 750 million scallops, the spat collection pattern changed, producing a longer spat collection season, and a fourfold increase in the number of spat collected. In central Bahía Magdalena, bottom water temperature was a minimum of 14°C during April and May. From July to August, the temperature increased from 21°C to 25°C, with the major summer settlement between

10,000 to 12,000 spat per collector (Félix-Pico et al. 1992; Santamaría-Gallegos et al. 1999; Maeda-Martínez et al. 2000).

Monitoring of settlement of catarina scallop at Bahía Magdalena continued between 1989 and 1999 extending the area studied south to include Laguna Rancho Bueno. Results have shown high numbers of spat in 1990 when more than 2,000 to 6,000 spat per bag were collected (Félix-Pico et al. 1992). Up to 300 to 5,000 scallop spat per collector were found. In 1995 when the population under culture was of 3.5 million scallops, the spat settlement increased to a maximum of 10,000 scallop spat per collector during January (Maeda-Martínez et al. 1996, 2000). In recent years, lower numbers of the spat in 1999 were found with only 700 scallop spat per collector (Tizoc Moctezuma, personal communication).

Spat collection was studied at Lagunas Guerrero Negro, San Ignacio, and Ojo de Liebre in Baja California Sur; from July 1985 to August 1986. The results of 950, 680 and 500 seed per collector were found in each lagoon. The highest incidence of spat was recorded at the end of summer. It was suggested that scallops spawn when water temperature is between 20°C and 22°C (Singh-Cabanillas 1987). Monitoring of settlement of catarina scallop continued between 1989 and 1992 in Laguna Ojo de Liebre, with higher numbers of spat obtained per collector, a maximum of 353, 6,300 and 8,100 between May and July during these years. A similar result was obtained at Laguna Guerrero Negro of 2,600 to 6,300 spat per collector. In the south (Laguna San Ignacio), much lower numbers of spat were obtained per collector, a maximum of 500 (Castro-Moroyoqui 1992).

In Bahía Bacochibampo, Sonora, Reyes-Sosa (1985, 1990) plotted abundance data as the mean number of spat per collector by month and found three peaks in November, February, and May. Those peaks were followed by an abrupt decrease of settlement in the following months. The largest peak was recorded in May when 177,000 spat were collected. In 1995, the area studied was extended north into the upper Gulf of California. Attempts were made to collect juveniles from a natural set of *Argopecten ventricosus* but results were disappointing. A storm caused the loss of the culture (Carvajal-Rascón and Farell 1984) (Douglas McLaurin, personal communication).

Singh-Cabanillas and Bojórquez-Verástica (1987) searched for commercial species for aquaculture purposes at Bahía Concepción, B.C.S. and found three species present on their collectors; *Argopecten ventricosus, Euvola vogdesi* and *Pinna rugosa*. Catarina scallop spat were monitored in Bahía Concepción using onion bags that were deployed, based on seasonal abundance of spat, from February 1987 to March 1994. During this period, scallop showed a high spawning intensity in March (peak), and lower spawning intensity continuing through April and May. In general, settlement was poor in the north but was good in the southern part of the bay. In 1987, the settlement resulting from a reduced winter-spring spawning (broodstock estimated at 70 million scallops) was 500 spat per bag. In 1988, the settlement increased to 40,000 spat per bag and the broodstock was estimated at 120 million. For 1991, the settlement decreased to 18,000 spat per bag and the broodstock was about 50 million scallops. For 1994, few juveniles were caught in spat-collecting operations (157 spat per bag). It was decided to stop the spat-collection program (Félix-Pico et al. 1995, 1997; Bojórquez-Verástica 1997).

Mano de león (*Nodipecten subnodosus*) in Laguna Ojo de Liebre can be a potential spat production bank. Observations on spat settlement in artificial collectors indicate that major settlement takes place during 2 months, September–October, and that the exact time of settlement varies from year to year (1990–1991) occurring from September to November. Numbers of spat per collecting onion bag varied between 4 to 74 spat per collector. Maximum spat collected occurred below 5-m depth (García-Domínguez et al. 1992). In Bahía de Loreto, Gulf of California, settlement was found on spiny shells (*Spondylus princeps*) in March and May 1994 and 1995. Adult of *N. subnodosus* spawn between December and February, indicating the major settlement takes place during a period of about 3 months. Numbers of spat per collecting shell (1–2) were relatively constant but varied with depth, ranging between 20 to 40 m (Félix-Pico and Villalejo-Fuerte 1999). Some authors reported spat settled on collectors in Bahía Concepción and Bahía de La Paz, but a low number of spat were collected there (Ruiz-Verugo and Cáceres-Martínez 1991; Bojórquez-Verástica 1997; Wright-López 1997) and at Bahía Bacochibampo, Sonora (Carvajal-Rascón 1987).

Settlement of voladora scallop (*Euvola vogdesi*) spat on experimental collectors (25 x 25 cm, 2-mm mesh) was studied from 1987 to 1988 at Bahía Falsa (Bahía de La Paz). Ruiz-Verdugo and Cáceres (1990, 1991) found maximum spat collection occurred between 7 to 17 m depth during June and July. The range of spat numbers collected was 7–21 per collector. For the first 5 weeks after settlement, growth rate was 0.29 mm day^{-1} and can be described by the linear equation:

$$\text{Shell height} = 0.21 + 0.29 * \text{days} \tag{15}$$

Tobías-Sánchez and Cáceres-Martínez (1994) tested the feasibility of collecting spat of *Euvola vogdesi* in Bahía de La Paz. Experimental collectors were made by Vexar filament net, monofilament, plastic mesh, coconut fibre, polypropylene bands and black polyethylene. After 5 weeks at 3-m depth, the spat were recorded from April-July. The largest number of spat were collected on Vexar filament and monofilament (44% of total collected), with 274 to 468 seed per collector in April 1988. The linear equation for the first five weeks was:

$$\text{Shell height} = 0.21 + 0.35 * \text{days} \tag{16}$$

29.9 GROWTH

29.9.1 Suspended Cultures

In 1977, the scallops (*A. ventricosus*) were used in a first attempt at growout trials in suspension culture grown from wild spat. Nearly one thousand specimens were obtained in 2.5 m^2. In March, they had a mean size of 0.95 cm. Two hundred scallops were put into each cage. Their mean size in May had reached 2.31 cm with 45% of scallops in the large-size category. Scallops were rearranged in several lots with different densities. In August, lot I had a mean size of 36 mm with a density of 250 individuals per cage and

35% of them in the large-size category. In October, the mean size was 40 mm with 35% of scallops in the large-size category. Results in other lots were not significantly different. Lot II reached a mean of 35 mm with 30% in the large-size category at a density of 100 scallops per cage. Lot III with same density reached a mean of 34 mm with 45% in the large-size category. In lot IV the mean size was 37 mm with 200 scallops per cage. Data of the monthly increase shell length were used to determine parameters of a von Bertalanffy growth curve. Growth response at density (50 scallops per cage) was estimated as $L_\infty = 4.9$ cm, and $K = 0.03$ (cm per month). This year larger sizes were recorded because of a more intensive handling (i.e., more frequent changes of cages) (Félix-Pico et al. 1989; Félix-Pico 1991a). The condition index for scallops can be described by the exponential equation

Meat weight = $0.43 \, L^{11.9}$ (17)

In 1979, the suspended culturing allowed a commercial-scale forecast by means of an analysis of the sensitivity in the production. Bioeconomic analysis of the catarina scallop has been made by Negrete-Martínez (1982) and Hernández-Llamas (1989). Analyses were made at low and high stocking rates. For plant sizes larger than 1.2 million juveniles, using high densities improved the internal rate of return but had a higher risk because of critical inconsistencies in survival. The internal rate of return was particularly sensitive to the maximum size attainable by the scallops, mortality, sale price and cost of equipment for growout. The importance of considering uncertainty in the variable estimates when projecting a pilot plant is addressed (Hernández-Llamas 1997).

México's first private scallop aquaculture company (Cultivos Marinos S.A.) was formed in La Paz in 1980. A local bay scallop, *Argopecten ventricosus*, was grown in lantern nets suspended from long lines. Scallops are stocked at a density of 25 scallops per 0.1 m^2 (50 per level) for the final growth stage. Market size (5 to 6 cm) was reached in 7 months. Four metric tons of scallops were marketed in México City in the first year of production (Siewers 1982). In 1982, when the scallop ground was overexploited, the company was closed because the collectors only caught 3 juveniles.

Tripp-Quezada (1985) reported growth for *Argopecten ventricosus* as an 11 month increase in the size expressed at San Vicente, 58.8 mm, Los Prados, 46 mm, and San Buto, 45 mm, at the northern part of Bahía Magdalena.

Reyes-Sosa (1988) reported growth for catarina scallop from Bahía de Bacochibampo experimental hanging culture for 27 weeks. The size increase was length 50 mm and total weight 35.9 g and recommended the period of culture from November to June. The growth observed on cultured scallops during 7 months can be described by the equation (Fig. 29.7):

Shell height = $62.36 \, (1 - e^{(-0.057 \, (t + 1.95))})$ (18)

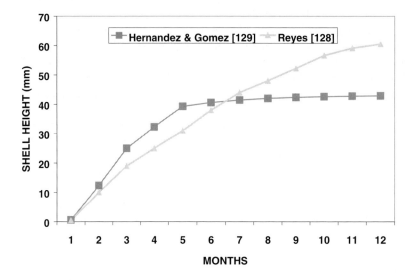

Figure 29.7. Growth curves calculated by the equations number 18 and 19, from data of suspended culture of *Argopecten ventricosus*. (From Reyes-Sosa (1988) and Hernández-Llamas and Gómez-Muñoz (1996)).

Hernández-Llamas and Gómez-Muñoz (1996) in experimental culture of catarina scallop at Ensenada de La Paz determined growth in relation to stocking density and length of culture period. In this study, a significant growth response to stocking rate was observed. In general, large sizes were produced when density decreased. The common allometric weight-length coefficients for the whole data set were estimated as a = 0.073 and b = 2.93 (P <0.05). Growth response at density (50% of available area occupied) was estimated as L_∞ = 4.31 cm, K = 0.42 (month) and t_o = -0.78 (P <0.005). The von Bertalanffy growth equation (Fig. 29.7) was:

$$L_t = 4.31(1 - e^{-0.42 (t - 0.78)})$$ (19)

The growth rates of juvenile catarina scallops (spat from hatchery) suspended in plastic trays was evaluated at different densities in Laguna Rancho Bueno (CULTEMAR, Co.), Bahía Magdalena (Maeda-Martínez et al. 1997). Three densities were tested at each of the culture stages. The first as nursery was 1,500, 2,500 and 4,000 scallops m^{-2} for 80 days, the second phase as intermediate was 400, 700 and 1,000 scallops m^{-2} for 80 to 136 days, and as growout was 150, 250 and 400 scallops m^{-2} for 136 to 320 days. In the first two stages, growth rates were higher at low densities, but growth rates were the same at low and medium densities in the growout stage. Mean final shell heights were 5.7, 5.6 and 5.4 cm for each density. Muscle weight was affected by density, estimating a 21% reduction caused by crowding. A cost analysis indicates that, regardless of the longer time, culturing scallops at high density would be less expensive than at low density, taking into account the muscle-weight loss caused by crowding.

An experiment to test feasibility of growing 400 juveniles originated in the laboratory of the Universidad Autón de Baja California in Ensenada. Scallops, *N. subnodosus*, were sown into triangular structures (100 x 40 x 50 cm) covered with a plastic mesh (8-mm opening), and suspended from a long line system off Isla Gaviota, Bahía de La Paz. The scallops reached 57-mm shell height in the first year of culture (Barrios-Ruiz et al. 1997). These results are similar to those found by García et al. (1992) for the same species in the Laguna Ojo de Liebre, where the scallops reached 58 mm in the first year of life. Some other experiments related to *N. subnodosus* report 35 mm in 107 days of culture in Bahía de La Paz (Morales-Hernández and Cáceres-Martínez 1996) and 43 mm in four months off the coast of Sonora (Carvajal-Rascón 1987). The instantaneous daily growth rate found in the experiment was 0.143 mm, which is similar to that found by García et al. (García-Domínguez et al. 1992)), but is less than half found by Morales-Hernández and Cáceres-Martínez (1996), 0.327 mm, and Carvajal-Rascón (1987), 0.352 mm.

Félix-Pico and Villalejo-Fuerte (1999) in experimental culture of *N. subnodosus* at Bahía Juncalito determined growth over one year. The instantaneous daily growth rate found was 0.23 mm day^{-1}.

The voladora scallop *Euvola vogdesi* was grown to determine its growth rate using organisms 20-mm high collected in Bahía Juncalito (Loreto), Gulf of California, observing them for twelve months in conditions of extensive culture. Through the summer and autumn the growth was fastest, obtaining a mean height of 46.9 mm. In winter, the mean height was 52.4 mm, with then a steady growth to the summer (Villalejo-Fuerte and Tripp-Quezada 1997).

29.9.2 Bottom Cultures

The growth rates of juvenile *A. ventricosus* in an enclosure called a "parque" were compared at different densities. The enclosure is a stationary structure of wooden bars buried in the substrate, to which plastic or anchovy fishing nets are fixed. Juvenile scallops with mean initial height of 28 mm and weight of 5.6 g were sown at densities of 100, 500, 1,000 and 2,000 scallops m^{-2} for 93 days. The average final heights were 47, 39, 32 and 30 mm, and weights were 33, 16, 9 and 7 g (Ramírez-Filippini and Cáceres-Martínez 1994). Another modular system was tested, which is based on individual modules that can be replaced for maintenance. The juveniles are sown inside the enclosures over a sand substrate and remain for 5 to months until they are harvested. The average growth is 5 to 6 mm per month using a density of 100 scallops m^{-2} (Ramírez-Filippini et al. 1990; Ramírez-Filippini and Cáceres-Martínez 1994).

Maeda-Martínez et al. (2000) used a bottom-culture system (tube or sack) on a commercial level to grow *A. ventricosus* in the Rancho Bueno (CULTEMAR, Co.) tidal channel in Bahía Magdalena. The system consisted of a 50 x 1 m sleeve of 19-mm mesh polyethylene netting placed on the sea floor of selected growout areas. The bottom culture gear represents an advancement in scallop cultivation procedures, and the authors obtained a patent (No. 180211) on this system (Maeda-Martínez and Ormart-Castro

1995). A total of 448 sleeves divided into 12 lots or groups of scallops were deployed from October 1994 to April 1995. Each sleeve contained 10,000 scallops with average size 32-mm shell height. A total of 19,300 kg of meat was harvested at the end of culture (between 109 to 319 days). Mean final size was 56.2 ± 3.2 mm with a minimum of 52.4 ± 2.9 mm in 118 days and a maximum of 58.2 ± 3.7 mm in 240 days. Final adductor muscle wet weight varied from 4.4 g at 118 days to 8.5 g at 319 days. Average survival in the sleeves was $71.2 \pm 3.8\%$ with a minimum of 60.4% and maximum of 86.5%.

In Bahía Concepción during February 1993, spat were counted and measured, and a bimodal recruitment pattern obtained (1.5 cm on average). The spat settled in the bottom and samples were taken and measured each month. The dates were fitted to the von Bertalanffy equation with $L_\infty = 78$ mm and $K = 1.4$. These results are in agreement with previous studies in other areas (Villalejo-Fuerte 1997). On 15 June 1994, about 933,000 juvenile scallops were placed in an enclosure with a density of 330 scallops m^{-2}. The scallops were held until 10 August, when they were released. Growth was estimated by measurements of the shell length and wet weights of different body portions. The culture period was 17 weeks. The final size of the scallops was 43.7 mm, total wet weight was 24.5 g, and meat weight was 9.0 g. The weekly size increase was 1.1 mm, the wet weight increase 1.24 g, and the muscle weight increase 0.1 g. In a protected bottom culture, the growth was at 6-m depth. The culture period was 25 weeks beginning in June. Shell length growth rate was 0.65 mm week^{-1}, total body weight growth was 0.62 g week^{-1}, and adductor muscle growth was 0.07 g week^{-1}. Adductor muscle production was 910 g m^{-2} (Félix-Pico et al. 1997).

Bottom culture of voladora scallop *E. vogdesi* was done in plastic cages at 10-m depth by Ruiz-Verdugo and Cáceres Martínez (1994). A total of 2,300 scallops 12 to 21 mm in height were grown at 450 scallops m^{-2} during 135 days. For this period, the size increase was estimated as 57 to 60 mm. The results of growth were adjusted to Von Bertalanffy equation as:

$$L_t = 74.19 \left(1 - e^{-0.003 (t-45)}\right) \tag{20}$$

29.9.3 Factors Influencing Growth

Temperature has been widely acknowledged as important in controlling growth rate. Sicard et al. (1999) found that growth of *A. ventricosus* (10–12 mm shell height) occurred between 12 and 28°C, with an optimum between 19°C and 22°C. Growth declined sharply at 28°C after 54 days of experiments. Studies of activity, ingestion and clearance rates, and the irrigation efficiency were made to obtain the optimum temperature. Results indicate that activity increased from 12°C to a maximum at 19°C, and then declined to 28°C. Mean ingestion and clearance rates were higher at 19 and 22°C and lower at 16, 25 and 28°C. Irrigation efficiency followed the same trend as activity. A similar experiment was done by Carvalho-Saucedo (1999), who determined the response of activity and standard metabolic rates to temperature, at 22°C and 28°C. The clearance rate was determined using two populations from Bahía Magdalena and Bahía Concepción, and reciprocal crosses and inbreeding lines. At the end of the experiment of clearance and metabolic activity, there were no significant differences for both populations and the reciprocal crosses, but the standard metabolic rate was higher at 28°C. A similar result was obtained with the inbreeding lines, when they were compared to a control.

Silva-Loera (1986) found the catarina scallop had a higher metabolic rate at a comparable temperature ranges (20°C to 30°C). The temperature coefficient ($Q_{10} = K_1/K_2$ when oxygen uptake at temperatures T_1 and T_2) for oxygen consumption was 1.94 and 2.77 mL h^{-1} g^{-1} for scallops 1 and 0.1 g dry tissue weight, respectively.

In addition to the factors discussed above, there are probably many others that influence growth. The coastal lagoons along the Pacific coast of México are exposed to heavy rainfall and freshwater runoff, however, the Baja California Peninsula and Sonora state have hypersaline lagoons where scallop grounds are found. Signoret-Brailovsky et al. (1996) found the species is a perfect osmoconformer by exposing specimens from normal salinity (37‰) up to 57‰ and down to 17‰ at a constant 28°C. Hemolymph osmolality followed the same trend as the changing external media throughout the salinity levels tested. Survival records indicate that salinity tolerance is restricted to 27–47‰.

Cruz-Hernández et al. (2000) evaluated two different populations and their cross (F1) for the age at first sexual maturity or the effects of different native environments. In Bahía Magdalena, a site characterised by high productivity, lower average temperatures (14–27°C), and a high chlorophyll-a concentration (1.5–5.1 mg m^{-3}). The environment conditions of Bahía Concepción are characterised by low productivity, higher average temperatures (17–32°C), and a low chlorophyll-a concentration (0.3–1.6 mg m^{-3}). Differences between the two populations were clearly evident when grown at Bahía Magdalena but not at Bahía Concepción. At Bahía Concepción, sexual maturity was not reached by any of the groups. However, at 7 months, some mature individuals were already present for Magdalena population, but not for the Concepción population. Results suggest that a triggering mechanism exists in catarina scallop for the initiation of sexual maturation, whereas the differences between environments suggest that regardless of the mechanism, environmental conditions have a significant role in further maturation processes.

29.10 MORTALITY

29.10.1 Suspended Culture

Natural mortality estimates have been calculated (Félix-Pico et al. 1980) by counting dead scallops found in each basket and those missing in destroyed baskets. In early stages (i.e., scallops up to three months old), mortality was between 6% and 12%. The highest figures (30%–40%) were recorded in summer (August and September) when the mean size of scallops was 3.8 cm. Mortality was lower in the baskets built with synthetic net, but some were destroyed by fish (*Spheroides annulatus*, *Diodon hystrix* and *Balistes polylepis*) and the crustaceans (*Callinectes bellicosus*). Siewers (1982) using the Japanese lantern net observed a pufferfish *Spheroides annulatus* preying on cultured scallops by chewing open the bottom compartments of some o the lantern nets. Shortening the lantern net by 3 levels alleviated this.

In Bahía Bacochibampo, Sonora, Reyes-Sosa (1988) recorded the highest rates of mortality in summer by comparing different baskets having a density of 150 scallops per cage.

Tripp-Quezada (1985) studied mortality in three different structures and found no significant difference; concluding mortality was related instead to the particular area of culturing.

Hernández-Llamas and Gómez-Muñoz (1996) in experimental culture of catarina scallop at Ensenada de La Paz identified two patterns of the mortality, both during the culture period and among stocking densities. One was associated with the post-spawning period and second occurred at high densities with increased temperatures and overstocking (at 66% to 100% of occupied area).

Maeda-Martínez et al. (1997) reported 9% mortality during the 320 day experiment of culture of *A. ventricosus* and did not correlate with density. Zero mortality was recorded during the first 8 months, and then began to occur during the spawning season at the onset of summer.

Félix-Pico and Villalejo-Fuerte (1999) made an experimental culture of *N. subnodosus* at Bahía Juncalito. Variations of the shell height achieved in spring had significant consequences on survival during the subsequent summer (survival rate 75%). Mortality was caused by high temperature over 29°C. Barrios-Ruiz et al. (1997) reported a high mortality during summer at Bahía de La Paz, when the temperature reached more than 30°C. Survival at the end of the first year was about 38%.

29.10.2 Bottom Culture

The natural mortality of catarina scallop, *A. ventricosus,* was estimated from organisms cultured in an enclosure called a "parque" at different densities. Results suggested that densities of 500 and 2000 scallops m^{-2} gave a declining growth rate and mortality increased. Mean final mortality was 26% and 37% at these densities (Ramírez-Filippini and Cáceres-Martínez 1994). Low densities of 40 scallops m^{-2} gave 14% mortality (Cáceres-Martínez et al. 1987). Another modular system was tested based on individual modules that can be replaced during maintenance. The juveniles are sown inside the enclosures at a density of 100 scallops per m^{-2} and remained 7 months. The mortality was estimated at 15% by the end of the culture period (Ramírez-Filippini et al. 1990; Ramírez-Filippini and Cáceres-Martínez 1994).

An experimental culture was developed, using a 25 m^2 surface mesh cage. A total of 2,000 catarina scallops of 30 to 40 mm in height were placed at 75 scallops/ m^2 for 60 days. For this period, a natural mortality about 10% was estimated (López-Contreras and Cáceres-Martínez 1994).

Maeda-Martínez et al. (2000) using a tube system had a mean survival of 71%. Survival figures are low when compared with results in suspension culture (>90%) and were probably caused by the escape of scallops when the sleeves were cast into the growing area.

Villalejo-Fuerte (1997) estimated a natural mortality M = 1.6 of catarina scallop spat when they were released on the bottom without protection at Bahía Concepción. Félix-

Pico et al. (1997) using an enclosure, had a mass mortality of scallops in September, with more than 80% mortality, increasing to 99% in October. The final survival was only 1%. The possible causes of mortality were heavy predation and high water temperatures of 29°C to 33°C. The final density was 3 scallops m^{-2}. When other methods was used, bottom cages with light plastic frame, the final survival was 40%, with a final density of 300 scallops m^{-2}.

Ruiz-Verdugo and Cáceres-Martínez (1994) used a bottom-cage system for *E. vogdesi* in Bahía de La Paz. Mean final survival was 6% to 10% after 135 days. A similar result was obtained, with mortality of 68% during the summer (Cáceres-Martínez et al. 1989).

29.11 CONSTRAINTS

Natural Seed Collection. Scallop supplies are erratic and usually insufficient from the natural stocks. Relatively little work has been done with these species, although it is known that spawning can be induced and larvae of some scallops can be raised to adult size.

Hatchery Production. Pilot testing of laboratory procedures is required to improved technology for rearing cultured *Euvola vogdesi* during the early life cycle stages.

Transporting Spat. Maeda-Martínez et al. (2000) tested the feasibility of transporting seed to growout areas. Spat were placed in sandwich layers of wet sponge and plywood lids inside vinyl bags, with these packed in styrofoam coolers with ice. Results indicated that seed can be successfully shipped out of water over long periods of time (>37 h). The highest survival was 98% at 17°C.

Disease. In aquaculture there so many disease problems that the use of bacteria as probiotics is gaining in importance as an alternative to antibiotics. There is some available information about probiotics in aquaculture.

Production and Cost. The economic analyses are also insufficient (Negrete-Martínez 1982; Hernández-Llamas 1997). Montaño and Pérez (1988) showed project finances and programming. Maeda-Martínez et al. (2000) determined the economic yield of the bottom culture system. The storage, processing, marketing, value, and future prospects are included in fishery charts.

29.12 FUTURE PROSPECTS

The decline of the commercial pectinid fishery in Baja California Sur, because of multiple factors, has given an impetus to the development of a scallop cultivation industry. The growth of this new industry has been slow. The reasons for this slow growth range from the difficulties of securing prime oceanfront sites for hatchery and growout operations at reasonable costs, to regulatory constraints, and to the lack of research and development funds. Considerable progress has been made in scallop cultivation practices, particularly in the areas of broodstock management, spawning induction, larval rearing, and growout. Further refinements are needed, principally in area of nutrition, genetics and diseases.

1380

REFERENCES

Aguilar-Ruiz, F., 1975. Disponibilidad de almeja voladora *Pecten vogdesi* en Bahía de Los Angeles, B.C. en primavera de 1971. Tesis de Licenciatura. Escuela Superior de Ciencias Marinas. Universidad Autónoma de Baja California. 51 p.

Amador-Buenrostro, J., 1978. Cultivo experimental de almeja catarina *Argopecten circularis* en la Ensenada de La Paz, B.C.S. Tesis de Licenciatura. Escuela Superior de Ciencias Marinas, Universidad Autónoma de Baja California. 44 p.

Avilés-Quevedo, M.A., 1990. Crecimiento de la almeja catarina (*Argopecten ventricosus*) en función del alimento, con anotaciones sobre su biología y desarrollo. Tesis de Maestría. CICIMAR-IPN, La Paz, B.C.S., México. 81 p.

Avilés-Quevedo, A. and Muciño-Díaz, M., 1988a. Desarrollo embrionario y morfología externa de la larva de *Argopecten circularis* (Sowerby, 1835). In: VII Simposio Internacional de Biologia Marina. U.A.B.C.S., La Paz, B.C.S. Junio 1–3. 3 p.

Avilés-Quevedo, A. and Muciño-Díaz, M., 1988b. Racion máxima de dos dietas microalgales para la almeja catarina *Argopecten circularis* (Sowerby, 1835). In: VII Simposio Internacional de Biología Marina, U.A.B.C.S., La Paz, B.C.S. Junio 1–3. 5 p.

Avilés-Quevedo, A. and Muciño-Díaz, M., 1990. Acondicionamiento gonádico y desove de la almeja catarina *Argopecten circularis* (Sowerby, 1835) en condiciones de laboratorio. Ciencia Pesquera, Inst. Nal.de la Pesca, Sria. De Pesca, México 7:9–15.

Avilés-Quevedo, A., Muciño-Díaz, M. and Mazón-Suastegui, M., 1991. Estimación de tasa de filtración de la almeja catarina (*Argopecten circularis*), a partir del consumo de *Isochrysis galbana* and *Tetraselmis chui*. Revista de Investigación Científica, U.A.B.C.S., La Paz, B.C.S. 25 p.

Baqueiro, E., 1984. Status of molluscan aquaculture on the Pacific coast of México. In: Recent Advances in Molluscan Diseases and Aquaculture. Elsevier Science Publishers B.V. pp. 83–93.

Baqueiro, E., Peña, I. and Massó J.A., 1981. Análisis de una población sobreexplotada de *Argopecten circularis* (Sowerby, 1835) en la Ensenada de La Paz, B.C.S., México. Ciencia Pesquera. Inst. Nal. Pesca, México I (2):57–65.

Baqueiro Cárdenas, E.C. and Aldana Aranda, D., 2000. A review of reproductive patterns of bivalve mollusks from México. Bull. Mar. Science 66(1):13–27.

Barrios-Ruiz, D., Chávez-Villalba, J. and Cáceres-Martínez, C., 1997. First trial of growth *Lyropecten subnodosus* under culture conditions in Bahía de La Paz, B.C.S., México. In: E. Félix-Pico (Ed.). Books of Abstracts, 11th International Pectinid Workshop, La Paz, México, 10–15 April. pp. 107–110.

Bojórquez-Verástica, G.J., 1997. Reclutamiento, crecimiento y supervivencia de *Argopecten circularis* (Sowerby, 1835), *Argopecten ventricosus* (Sowerby II, 1842), en bahía Concepción, B.C.S. Tesis Profesional. Universidad Autónoma de Baja California Sur, La Paz, México. 72 p.

Cáceres-Martínez, C., Ramírez-Filippini, D. and Chavez-Villalba, J., 1987. Cultivo en parques de la almeja catarina (*Argopecten circularis*). Revista Latinoamericana de Acuicultura 34:26–32.

Cáceres-Martínez, C., Ruiz-Verdugo, C. and De Silva-Davila, R., 1989. Preengorda de almeja voladora *Pecten vogdesi* (Arnold, 1906) en un sistema submarino. Revista Latinoamericana de Acuicultura 39:38–44.

1381

Cáceres-Martínez, C., Ruiz-Verdugo, C. and Rodríguez-Jaramillo, M.C., 1990. Variaciones estacionales del índice gonádico y muscular de *Argopecten circularis* (Sowerby 1835) en la Ensenada de La Paz, B.C.S., México. Investigaciones Marinas CICIMAR 5(1):1–6.

Campa-Córdova, A.I., 1997. Estudios encaminados al mejoramiento de la supervivencia larvaria de *Argopecten ventricosus* (Sowerby II, 1842), mediante el uso de antibióticos como tratamiento preventivo. Tesis de Maestría. UABCS, La Paz, BCS, México. 78 p.

Carvajal-Rascón, M.A., 1987. Cultivo larvario de la almeja mano de león (*Lyropecten subnodosus* Sowerby, 1835) a partir del crecimiento y maduración gonadal de los reproductores. Tesis de Maestria. Instituto Tecnológico y de Estudios Superiores, Campus Guaymas, Sonora.

Carvajal-Rascón, M.A. and Farell, S.C., 1984. Crecimiento intermedio de dos especies de escalopa (*Argopecten circularis* and *Pecten vogdesi*) dentro de módulos de crecimiento en la Bahía de Bacochibampo, Guaymas, Son., México. Reporte Interno de Investigación. Instituto Tecnologico y de Estudios Superiores, Campus Guaymas, Sonora.

Carvalho-Saucedo, L., 1999. Tasa metabólica y tasa de aclaramiento entre grupos genéticos con grado de heterocigosidad variable: poblaciones, cruza recíproca y líneas endogámicas de almeja catarina *Argopecten ventricosus* (Sowerby II, 1842). Tesis profesional de Biólogo. UABCS, La Paz, México. 60 p.

Castro-Moroyoqui, P., 1992. Colecta de semilla de almeja catarina *Argopecten circularis* (Sowerby) en las Lagunas de Ojo de Liebre, Guerrero Negro y San Ignacio, B.C.S. Informe Técnico Interno. Secretaría de Pesca, B.C.S, México.

Coronel-Solorzano, J.S., Porras, S. and Ormart, P., 1987. Metodología para la cría de larvas de almeja catarina *Argopecten circularis* en laboratorio. In: Resúmenes del Segundo Congreso de la Asociación Mexicana de Acuicultores AMAC'87. La Paz, B.C.S., México, del 24–28 de noviembre.

Cruz Hernández, P. and Ibarra, A.M., 1997a. Two catarina scallop *Argopecten ventricosus* (Sowerby II, 1842) populations as potential stock resources for a genetic improvement program. In: E. Félix-Pico (Ed.). Books of Abstracts, 11th Inter. Pectinid Workshop, La Paz, México, 10–15 April. pp. 67–69.

Cruz Hernández, P. and Ibarra, A.M., 1997b. Larval growth and survival of two catarina scallop (*Argopecten circularis,* Sowerby, 1835) populations and their reciprocal crosses. J. Exp. Mar. Biol. Ecol. 212:95–110.

Cruz Hernández, P., Ramírez, J.L., Guy García, A. and Ibarra, A.M., 1998. Genetic differences between two populations of catarina scallop (*Argopecten ventricosus*) for adaptations for growth and survival in a stressful environment. Aquaculture 166:321–335.

Cruz Hernández, P., Rodríguez-Jaramillo, C.L. and Ibarra, A.M., 2000. Environment and population origin effects on first sexual maturity of catarina scallop, *Argopecten ventricosus* (Sowerby II, 1842). J. Shellfish Res. 19(1):89–93.

Diario Oficial de La Federación, 1990. Acuerdo que establece veda para la almeja catarina (*Argopecten circularis*) en aguas litorales de los Estados de Baja California y Baja California Sur, durante el período comprendido entre el 15 de diciembre al 31 de marzo de cada año. México, D.F., 11 de enero de 1990. 2 p.

Diario Oficial de La Federación, 1993. Acuerdo que regula el aprovechamiento de la almeja catarina en aguas de juridisción federal de la Península de Baja California. (004-PESC-1993), México, D.F., 21 de diciembre de 1993. 3 p.

Félix-Pico, E.F., 1987. Scallop fishing in Concepcion Bay, Baja California Sur, México. In: Abstracts, 6th International Pectinid Workshop. Menai Bridge, Wales, UK, April 9–14 (1987).

Félix-Pico, E.F., 1991a. México: fishery and aquaculture. In: S.E. Shumway (Ed.). Scallops: Biology, Ecology and Aquaculture. Volume 21. Elsevier, Amsterdam. pp. 943–980.

Félix-Pico, E.F., 1991b. Scallop fisheries and mariculture in México. In: S.E. Shumway and P.A. Sandifer (Eds.). An International Compendium of Scallop Biology and Culture. World Aquaculture Workshops Number 1. World Aquaculture Society, Baton Rouge, LA. pp. 287–292.

Félix-Pico, E.F., 1992. La biología de el cangrejo chícharo, *Tumidotheres margarita* Smith, 1869 (Decapoda; Brachyura; Pinnotheridae) en las costas de Baja California Sur, México., In: R. Brusca and M. Hendrickx (Eds.). Memorias del I coloquio sobre macro-crustáceos bentónicos del Pacífico este tropical, UNAM, marzo 28–30 (1990), Mazatlán, Sin. Proceedings of the San Diego Natural History Museum No. 25:1–6.

Félix-Pico, E.F., 1993. Estudio Biológico de la Almeja Catarina, *Argopecten circularis* (Sowerby, 1835) en Bahía Magdalena, B.C.S., México. Tesis de Maestría en Ciencias. CICIMAR-IPN, La Paz, B.C.S., México. 89 p.

Félix-Pico, E.F., Castro-Ortíz, J.L. and Tripp-Quezada, A., 1994. Some aspects of the growth, recruitment and mortality of the catarina scallop, *Argopecten circularis* (Sowerby, 1835) in Bahía Magdalena, Baja California Sur, México. In: N.F. Bourne, B.L. Bunting and L.D. Townsend (Eds.). Proc. 9th International Pectinid Workshop. Nanaimo, B.C., Canada, April 22–27, 1993. Canadian Technical Report of Fisheries and Aquatic Sciences 1994(2):145–153.

Félix-Pico, E.F. and García-Domínguez, F.A., 1993. Macrobentos sublitoral de bahía Magdalena, B.C.S. En: Biodiversidad Marina y Costera de México. In: S.I. Salazar-Vallejo y N.E. González (Eds.). Com. Nal. Biodiversidad y CIQRO, México. pp. 389–410.

Félix-Pico, E.F., García-Domínguez, F.A. and Morales-Hernández, R., 1992. Fijación y reclutamiento de *Argopecten circularis* (Sowerby, 1835) en la pesquería de Bahía Magdalena, B.C.S., México. In: S.A. Guzmán del Proó (Ed.). Memorias del taller México-Australia sobre reclutamiento de recursos bentónicos de Baja California. Sría. Pesca e IPN. pp. 143–149.

Félix-Pico, E.F., Ibarra-Cruz, M.T., Merino-Márquez, R.E., Levy-Pérez, V.A., García-Domínguez, F.A. and Morales-Hernández, R., 1995. Reproductive cycle of *Argopecten circularis* in Magdalena Bay, B.C.S., México. In: IFREMER (Ed.). Actes de Colloques, No. 17, VIII International Pectinid Workshop. Cherbourg, France, 22–29 May 1991. pp. 151–155.

Félix-Pico, E.F., Morales, R., Cota, A., Singh, C.J. and Verdugo, J., 1980. Cultivo piloto de almeja catarina (*Argopecten circularis*) en la Ensenada de La Paz, B.C.S., México. In: Memorias del 2° Simposio - Latinoamericano de Acuacultura. México, D.F., Noviembre de 1978. Tomo I. pp. 823–844.

Félix Pico, E.F., Tripp-Quezada, A., Castro-Ortíz, J.L., Serrano-Casillas, G., González-Ramírez, P.G., Villalejo-Fuerte, M., Palomares-García, R., García Domínguez, F.A., Mazón-Suástegui, M., Bojórquez-Verástica, G. and López-García, G., 1997. Repopulation and culture of the scallop *Argopecten circularis* in Bahía Concepción, B.C.S., México. Aquaculture International 5(6):551–563.

Félix-Pico, E.F., Tripp-Quezada, A. and Singh-Cabanillas, J., 1989. Antecedentes en el cultivo de *Argopecten circularis* (Sowerby) en Baja California Sur, México. Revista Investigaciones Marinas, CICIMAR 4(1):75–92.

Félix-Pico, E.F. and Villalejo-Fuerte, M. 1999. Growth and survival of *Lyropecten subnodosus* (Sowerby, 1835) in suspended culture at the National Marine Park of Bahía de Loreto, B.C.S., México. In: 12th International Pectinid Workshop. Bergen, Norway, 5–11 May 1999. pp. 39–40.

Garate-Lizarraga, I., 1997. A review of the toxic phytoplankton species in distribution areas of pectinid bivalves along the Baja California peninsula. In: E. Félix-Pico (Ed.). Books of Abstracts, 11th Inter. Pectinid Workshop. La Paz, México, 10–15 April. pp. 184.

García-Domínguez, F., Castro-Moroyoqui, P. and Félix-Pico, E.F., 1992. Spat settlement and early growth of *Lyropecten subnodosus* (Sowerby, 1835) in Laguna Ojo de Liebre B.C.S., México, 1989–1990. In: Abstracts of the Aquaculture'92. Orlando, Fl., USA, May 21–25, 1992. pp. 101.

Garduño Méndez, M.L., 1999. Simbiontes de la almeja catarina *Argopecten ventricosus* (Sowerby, 1842), en Baja California Sur, México. Tesis de Maestría. Universidad Autónoma de Baja California Sur, La Paz, México.

Gómez del Prado, R.M. del C., 1983. Hallazgo de una forma larvaria de *Echinocephalus* sp. (Nemátodo: Gnathostomidae) en *Argopecten circularis* y *Lyropecten subnodosus* (Mollusca: Lamellibranchia) de la Laguna Ojo de Liebre, Guerrrero Negro, B.C.S. An. Inst. Biol. Univ. Nal. Autón. de Méx. Ser. Zool. 53(1):421–431.

Gómez del Prado, R.M. del C., 1984. *Echinocephalus seudouncinatus* nemátodo parásito de *Argopecten circularis* (Mollusca: Bivalvia) y *Heterodontus francisci* (Pisces: Elasmobranchia) en la costa occidental de Baja California Sur, México. Tesis de Maestria. Inst. de Cienc. del Mar y Limn. Univ. Nac. Autón. de Méx. 125 p.

Gómez del Prado, R.M. del C., Alvarez Torres, S. and Pérez Urbiola, J.C., 1992. Algunos parásitos de almeja catarina, *Argopecten circularis*, en bahía Concepción, B.C.S., México. An. Inst. Biol. Univ. Nal. Autón. de Méx. Ser. Zool. 63(2):265–271.

González Ramos, H., García Pámanes, F., García Pámanes, L., Chee Barragán, G., García Pámanes, J. and Medina Hurtado, O., 1997. Larval settlement of the scallop *Lyropecten subnodosus* in downwellers. In: E. Félix-Pico (Ed.). Books of Abstracts, 11th Inter. Pectinid Workshop. La Paz, México, 10–15 April. pp. 118–119.

Gutierrez Villaseñor, C.E. and Chee Barragán, G., 1997. Effect of temperature and feeding ratio on the conditioning of *Lyropecten subnodosus* (Sowerby, 1835). In: E. Félix-Pico (Ed.). Books of Abstracts, 11th Inter. Pectinid Workshop. La Paz, México, 10–15 April. pp. 73–75.

Hernández-Llamas, A., 1989. Modelo para la optimación del diseño y simulación de un cultivo de almeja catarina (*Argopecten circularis*). Tesis de Maestría. CICIMAR, La Paz, B.C.S., México. 85 p.

Hernández-Llamas, A., 1997. Management strategies of stocking density and length of culture period for Catarina scallop *Argopecten circularis* (Sowerby): a bioeconomic approach. Aquaculture Research 28:223–229.

Hernández-Llamas, A. and Gómez-Muñoz, V.M., 1996. Growth and survival response for the Catarina scallop *Argopecten circularis* (Sowerby) to stocking density and length of culture period. Aquaculture Research 27:711–719.

Holguin-Quiñones, O.E., 1997. The collapse of scallop fishery *Euvola vogdesi* Arnold, 1906 in the Gulf of California in 1972–1973. In: E. Félix-Pico (Ed.). Books of Abstracts, 11th Inter. Pectinid Workshop. La Paz, México, 10–15 April. pp. 137.

Ibarra, A.M., 1999. Correlated responses at age 5 months and 1 year for a number of growth traits to selection for total weight and shell width in catarina scallop (*Argopecten ventricosus*). Aquaculture 175:243–254.

Ibarra, A.M., Cruz, P. and Romero, B.A., 1995. Effects of inbreeding on growth and survival of self-fertilized catarina scallop larvae, *Argopecten circularis*. Aquaculture 134:37–47.

Ibarra, A.M., Ramírez, J.L. and Gracía, G.A., 1997. Stocking density effects on larval growth and survival of two catarina scallop, *Argopecten ventricosus* (= *circularis*) (Sowerby II, 1842). Aquaculture Research 28:443–451.

Ibarra, A.M., Ramírez, J.L., Ruiz, C.A., Cruz, P. and Avila, S., 1999. Realized heritabilities and genetic correlation for total weight and shell width in catarina scallop (*Argopecten ventricosus*). Aquaculture 175:227–241.

Keen, M.A., 1971. Sea Shells of Tropical West America. 2nd Ed. Stanford Univ. Press. Stanford. 1064 p.

Keen, A.M. and Coan, E., 1974. Marine Molluscan Genera of Western North America, an Illustrated Key. Stanford University Press. 208 p.

Lango-Reynoso, F., 1994. Estudios básicos sobre depredadores activos y potenciales, para el desarrollo del cultivo extensivo de *Argopecten circularis*. Tesis de Maestría. CICIMAR-IPN, La Paz, BCS, México. 99 p.

Leija-Tristán, A., Solís Marín, A., Aurioles-Gamboa, D. and Amador-Silva, E.S., 1996. Natural stocks of the scallop *Argopecten circularis* and relationships with the galatheid crab *Pleuroncodes planipes* in the Pacific coast of Baja California Sur, México. Cah. Biol. Mar. 37:153–157.

León-Carballo, G., Reinecke-Reyes, M.A. and Ceseña-Espinoza, N., 1991. Abundancia y estructura poblacional de los bancos de almeja catarina *Argopecten circularis* (Sowerby, 1835) durante abril de 1988, en Bahía Concepción, B.C.S. Ciencia Pesquera 8:35–40.

López-Contreras, L. and Cáceres-Martínez, C., 1994. Estudios de distribución espacial de *Argopecten circularis* (Sowerby, 1835). I. Experimento sin obstáculos. Rev. Inv. Cient., (No. Esp. AMAC), UABCS 1:18–26.

Lora-Vilchis, M.C. and Maeda-Martínez, A.N., 1997. Ingestion and digestion index of catarina scallop *Argopecten ventricosus* = *circularis*, Sowerby II, 1842, veliger larvae with ten microalgae species. Aquaculture Research 28:905–910.

Lora-Vilchis, M.C., Rodriguez Jaramillo, M.C. and Maeda-Martínez, A., 1997. Spawning process in *Argopecten ventricosus* = *circularis* (Sowerby II, 1842): histological description. In: E. Félix-Pico (Ed.). Books of Abstracts, 11th Inter. Pectinid Workshop. La Paz, México, 10–15 April. pp. 178–179.

Luna-González, A., Cáceres-Martínez, C., Zuñiga-Pacheco, C., López-López S. and Ceballos-Váquez, B.P., 2000. Reproductive cycle of *Argopecten ventricosus* (Sowerby 1842) (Bivalvia: Pectinidae) in the Rada del Puerto de Pichilingue, B.C.S., México and its relation to temperature, salinity, and food. J. Shellfish Res. 19(1):107–112.

Maeda-Martínez, A.N., 1996. Almejas cultivadas: Investigación e industria protegida por patentes. Urania (Méx.) 2(10):32–34.

Maeda-Martínez, A., Monsalvo, P. and Reynoso, T., 1989. Tecnología para la producción intensiva de semillas de almeja catarina *Argopecten circularis*. Publicación interna del Centro de Investigaciones Biológicas del Noroeste, La Paz, BCS, México 87 p.

Maeda-Martínez, A.N., Monsalvo-Spencer, P. and Reynoso-Granados, T., 1995. Sistema para crianza intensiva en su etapa juvenil de almeja catarina (*Argopecten circularis*). Título de Patente No. 180212. Titular Centro de Investigaciones Biológicas del Noroeste, S.C.

Maeda-Martínez, A.N. and Ormart-Castro, P., 1995. Sistema marino para crecimiento y engorda hasta la fase adulta de almeja catarina (*Argopecten circularis*). Título de Patente No. 180211. Titular Centro de Investigaciones Biológicas del Noroeste, S.C.

Maeda-Martínez, A.N., Ormart-Castro, P., Mendez, L., Acosta, B. and Sicard, M.T., 2000. Scallop growout using a new bottom-culture system. Aquaculture 189:73–84.

Maeda-Martínez, A.N., Ormart-Castro, P., Moctezuma Cano, T. and Osorio, V., 1996. Cultivo de scallop en Ecuador: una alternativa en desarrollo. Acuacultura del Ecuador. pp. 29–36.

Maeda-Martínez, A.N., Ormart-Castro, P., Polo, V., Reynoso-Granados, T., Monsalvo, P., Avila, S. and Espinosa, M., 1992. The potential predator impact on bottom cultures Mexican catarina scallops (*Argopecten circularis*). In: Abstracts of the Aquaculture '92. Orlando, Fl., USA, May 21–25, 1992. pp. 151.

Maeda-Martínez, A.N., Reynoso-Granados, T., Monsalvo-Spencer, P., Sicard, M.T., Mazón-Suástegui, J.M., Hernández, O., Segovia, E. and Morales, R., 1997. Suspension culture of catarina scallop *Argopecten ventricosus* (= *circularis*) (Sowerby II, 1842), in Bahía Magdalena, México, at different densities. Aquaculture 158:235–246.

Maeda-Martínez, A.N., Reynoso-Granados, T., Solís-Marin, F., Leija-Tristán, A., Aurioles-Gamboa, D., Salinas-Zavala, C., Lluch-Cota, D., Ormart-Castro, P. and Félix-Pico, E.F., 1993. Model to explain the formation of catarina scallop *Argopecten circularis* beds, in Magdalena bay, México. Aquaculture and Fisheries Management 24:399–415.

Maeda-Martínez, A.N., Sicard, M.T., Carvalho, L., Lluch-Cota, S.E. and Lluch-Cota, D.B., 2000. Las poblaciones de almeja catarina *Argopecten ventricosus* en el centro de actividad biológica de Bahía Magdalena, México. In: D. Lluch-Belda, J. Elorduy-Garay and G. Ponce-Díaz (Eds.). Centros de actividad Biológica del Pacífico Mexicano. CIBNOR/CONACyT., La Paz,, B.C.S., México.

Maeda-Martínez, A.N., Sicard, M.T. and Reynoso-Granados, T.A., 2000. Shipment method for scallop seed. J. Shellfish Res. 19:765–770.

Martínez-López, A. and Gárate-Lizárraga, I., 1994. Cantidad y calidad de la materia orgánica particulada en Bahía Concepción en la temporada de reproducción de almeja catarina *Argopecten circularis* (Sowerby, 1835). Ciencias Marinas 20(3):301–320.

Massó Rojas, J.A., 1996. Pesquería de almeja catarina. In: M. Casas and G. Ponce (Eds.). Estudio Potencial Pesquero y Acuícola de Baja California Sur, SEMARNAP, La Paz, Mex. Vol. I. pp. 71–85.

Mazón-Suástegui, J.M., 1986. Acondicionamiento gonádico y desove de cuatro especies de moluscos bivalvos, alimentados con dietas artificiales. In: (Resúmes) Primer Simposio Nacional de Acuacultura, Pachuca, Hgo., México.

Mazón-Suástegui, J.M., 1996. Cultivo de almeja catarina *Argopecten circularis*. In: M. Casas and G. Ponce (Eds.). Estudio Potencial Pesquero y Acuícola de Baja California Sur, Vol. II. SEMARNAP, La Paz, Mexico. pp. 513–544.

Mazón-Suástegui, J.M., Avilés-Quevedo, M.A., Rivera-Lucero, J.R. and Ríos-Arias, V.A., 1991. Advances on the pilot production of "catarina scallop" *Argopecten circularis* seed, in a Mexican shellfish hatchery. In: (Abstracts) 22th Ann. Conf. World Aquacult. Soc.

1386

Millán-Tovar, M.M., 1997. Experimentos de inducción a la maduración gonádica de *Argopecten ventricosus* (Sowerby, 1842) y estudio del valor nutricional de *Isochrysis galbana* y *Chaetoceros gracilis*, durante su crianza larvaria. Tesis de Maestría. UABCS, La Paz, BCS, México. 76 p.

Monsalvo-Spencer, P., 1994. Acondicionamiento gonádico, desarrollo embrionario y cultivo de larvas y juveniles de almeja voladora (*Pecten vogdesi*) (Mollusca: Pectinidae). Escuela Nacional de Estudios Profesionales-Iztacala. Universidad Nacional Autónoma de México. 78 p.

Monsalvo-Spencer, P., 1998. Biología reproductiva y cultivo de las primeras fases del desarrollo de la almeja catarina (*Argopecten circularis* = *ventricosus*). Tesis de Maestría. CICIMAR-IPN, La Paz, BCS, México. 70 p.

Monsalvo-Spencer, P., Maeda-Martínez, A.N. and Reynoso-Granados, T., 1997. Reproductive maturity and spawning induction in the Catarina scallop *Argopecten ventricosus* (= *circularis*) (Sowerby II, 1842). J. Shellfish Res. 16(1):67–70.

Montaño A., M.A. and Pérez C., J.C., 1988. Análisis económico de un proyecto de cultivo de almeja catarina en parques. In: (Resúmenes) VII Simposio Internacional de Biología Marina, U.A.B.C.S., La Paz, B.C.S., México, junio 1–3. 8 p.

Morales-Hernández, R. and Cáceres-Martínez, C., 1996. Pesquería de almeja mano de león, *Lyropecten subnodosus*. In: M. Casas and G. Ponce (Eds.). Estudio Potencial Pesquero y Acuícola de Baja California Sur, SEMARNAP, La Paz, México. Vol. I. pp. 87–100.

Morquecho-Escamilla, M.L., 1996. Fitoplancton tóxico y actividad de ficotoxinas en la almeja catarina *Argopecten circularis* (Sowerby, 1835) en Bahía Concepción, Golfo de California. Tesis de Maestría, CICIMAR-IPN, La Paz, BCS, México. 74 p.

Negrete-Martínez, J., 1982. El Análisis de Sensibilidad en la Producción Biológica. Instituto de Investigaciones Biomedicas. Biomatemáticas, Universidad Nacional Autónoma de México, D.F. 77 p.

Ocaño-Higuera, V.M., 1999. Caracterización parcial del comportamiento bioquímico posmortem y desarrollo de productos a partir del callo de almeja catarina (*Argopecten ventricosus*) y almeja mano de león (*Nodipecten subnodosus*) de Baja California, México. Tesis de Maestría. Centro de Investigación en Alimentación y Desarrollo, A.C., Hermosillo, México. 98 p.

Ortiz-Cuel, G., 1994. Efecto de la ración alimenticia y la densidad de cultivo sobre el desarrollo larval de *Lyropecten subnodosus*. Tesis Profesional de Oceanólogo. UABC, Ensenada, B.C., México. 58 p.

Ortiz-Cuel, G., García Pámanes, F., Chee Barragán, G., García Pámanes, L., García Pámanes, J. and Medina Hurtado, O., 1997. Food ration and culture density effect on the larval development of the scallop *Lyropecten subnodosus* (Sowerby, 1835). In: E. Félix-Pico (Ed.). Books of Abstracts, 11th Inter. Pectinid Workshop. La Paz, México, 10–15 April. pp. 164–165.

Oseguera-Green, V.M., 1977. Contribución al estudio de la contaminación por bacterias en almeja catarina (*Argopecten circularis*) en la Ensenada de la Bahía de La Paz, B.C.S. Tesis de Licenciatura. Facultad de Ciencias Biologicas, U.N.A.M., México, D.F. 57 p.

Palomares-García, R. and Martínez López, A., 1997. Food variability during the spawning peak of the scallop *Argopecten ventricosus* = *circularis* (soweby, 1835) in Bahía Concepción, Gulf of California. In: E. Félix-Pico (Ed.). Books of Abstracts, 11th Inter. Pectinid Workshop. La Paz, México, 10–15 April. pp. 79–81.

Pérez-Concha, J.C., 1990. Perspectivas de mercado en relación a la acuacultura de almeja catarina en Baja California Sur. Revista del Area Interdisciplinaria de Ciencias Sociales y Humanidades, U.A.B.C.S., Semestre 2:21–29.

Pineda-Barrera, J. and López-Salas, F., 1972. Prospección de la almeja voladora *Pecten vogdesi*. Instituto Nacional de Pesca, Ensenada, B.C., Boletín informativo, octubre de 1972. 5 p.

Polanco-Jaime, E., Mimbela, R., Belandez, L., González, P., Flores, M., Pérez, A., Aguilar, N., Pérez, R., Calderón, R., Guerra, J.L., Romo, J., Gómez, H., Mimbela, J., Cabrera, H., Peralta, M.D., García, J. and Ochoa, J.G., 1987. Pesquerías Mexicanas: Estrategias para su Administración. Secretaria de Pesca, México, D.F. 1061 p.

Poutiers, J.M., 1995. Bivalvos. In: W. Fisher, F. Krupp, W. Schneider, C. Sommer, K.E. Carpenter and V.H. Niem (Eds.). Guía FAO para la identificación de especies para fines de la pesca. Pacífico centro-oriental. Vol. I Plantas e invertebrados. Roma, FAO. pp. 99–222.

Racotta, I.S., Ramírez, J.L., Avila, S. and Ibarra, A.M., 1998. Biochemical composition of gonad and muscle in the catarina scallop, *Argopecten ventricosus*, after reproductive conditioning under two feeding systems. Aquaculture 163:111–122.

Ramírez, J.L. and Ibarra, A.M., 1999. Optimization of forage in two food-filtering organisms with the use of a continuous, low-food concentration, agricultural drip system. Aquacult. Eng. 20:175–189.

Ramírez-Filippini, D.H. and Cáceres Martínez, C., 1994. Colecta, preengorda y cultivo de *Argopecten circularis* en parques bajo diferentes densidades, en la Bahía de La Paz, B.C.S., México. Rev. Inv. Cient., (No. Especial AMAC) 1:35–44.

Ramírez-Filippini, D.H, Cáceres-Martínez, C. and Chávez-Villalba, J., 1990. Parque modular para el cultivo de la almeja catarina *Argopecten circularis*: Un diseño alternativo. Investigaciones Marinas CICIMAR 5(1):7–13.

Reinecke R., M.A., 1981. Madurez y desove de *Lyropecten subnodosus*, Sowerby, 1835 (Bivalvia; Pectinidae) en Bahía Ojo de Liebre, B.C.S., México. Informe Técnico Interno. CRIP La Paz, BCS, México. 10 p.

Reyes-Sosa, C., 1985. Experimental spat collection of the Catarina scallop, *Argopecten circularis* in Bacochibampo Bay, Guaymas, Sonora, México. MSc Thesis. The University of Miami, Coral Gables, Florida. 68 p.

Reyes-Sosa, C., 1988. Crecimiento en cultivo suspendido de escalopa catarina *Argopecten circularis* en la Bahía de Bacochibampo, Guaymas, Sonora, México. In: VII Simposio Internacional de Biología Marina, U.A.B.C.S., del 1–5 de junio, La Paz, B.C.S. 26 p.

Reyes-Sosa, C., 1990. El cultivo de pectínidos en México. Serie Científica, U.A.B.C.S. No. Esp. 1, AMAC. pp. 25–29.

Robles Mungaray, M. and Serrano Guzmán, S.J., 1993. Producción de semilla de almeja catarina *Argopecten circularis* (Mollusca: Bivalvia) en el Centro Ostrícola del Estado de Sonora. I Encuentro Regional para el cultivo de almeja catarina (*Argopecten circularis*), Guaymas, Sonora, México. In: Instituto de Acuacultura (Ed.). Gobierno de Sonora.

Robles Mungaray, M. and Serrano Guzmán, S.J., 1995. Laval growth, survival and spat production of *Argopecten circularis* (Sowerby, 1835) in a Mexican commercial hatchery. Rivista Italiana Acquacoltura 30:187–193.

Rombouts, A., 1991. Guidebook to Pecten Shells. Recent Pectinidae and Propeamussiidae of the World. Universal Book Services, Dr. W. Backhuys, Oegstgeest, The Netherlands. xiii + 157 p.

Ruiz-Verdugo, C.A. and Cáceres-Martínez, C., 1990. Estudio preliminar de captación de juveniles de moluscos bivalvos en la Bahía de La Paz, B.C.S. (septiembre 86 a abril 87). Investigaciones Marinas CICIMAR 5(1):29–38.

Ruiz-Verdugo, C.A. and Cáceres-Martínez, C., 1991. Experimental spat settlement of the Catarina scallop, *Argopecten circularis* (Sowerby, 1835) and *Pecten vogdesi* (Arnold, 1906) on a filament substrate in Falsa bay, B.C.S., México. In: S.E. Shumway and P.A. Sandifer (Eds.). An International Compendium of Scallop Biology and Culture. World Aquaculture Workshops No. 1. World Aquaculture Society, Baton Rouge, LA. pp. 21–27.

Ruiz-Verdugo, C.A. and Cáceres-Martínez, C., 1994. Cultivo sobre el fondo de la almeja voladora *Pecten vogdesi* en la Bahía de La Paz, Baja California Sur, México. Rev. Inv. Cient. (No. Esp. AMAC), UABCS 1:61–66.

Ruiz-Verdugo, C.A., Ramírez, J.L., Standish, S.K. and Ibarrra, A.M., 2000. Triploid Catarina scallop (*Argopecten ventricosus,* Sowerby II, 1842): growth, gametogenesis, and suppression of functional hermaphroditim. Aquaculture 186:13–32.

Ruiz-Verdugo, C.A., Standish, S.K. and Ibarrra, A.M., 1997. Experimental triploid induction in Catarina scallop (*Argopecten ventricosus = circularis,* Sowerby II, 1842). In: E. Félix-Pico (Ed.). Books of Abstracts, 11th Inter. Pectinid Workshop. La Paz, México, 10–15 April. pp. 159–161.

Sainz Hernández, J.C., Maeda-Martínez, A.N. and Ascencio, F., 1998. Experimental vibriosis induction with *Vibrio alginolyticus* on larvae of catarina scallop (*Argopecten ventricosus = circularis*) (Sowerby, 1842). Microbial Ecology 35:188–192.

Santamaría-Gallegos, N.A., Félix-Pico, E.F., Sánchez-Lizaso, J.L., Palomares-García, J.R. and Mazón Suástegui, M., 1999. Temporal coincidence of the annual eelgrass *Zostera marina* and juvenile scallops *Argopecten ventricosus* (Sowerby II, 1842) in bahía Concepción, México. J. Shellfish Res. 18(2):415–418.

SEMARNAP, 1999. Anuario Estadístico de Pesca 1998. Secretaría del Medio Ambiente Recursos Naturales y Pesca, México. 244 p.

Serrano Guzmán, S.J., Robles Mungaray, M., Velasco Blanco, G., Voltolina, L.D. and Hoyos Chairez, F., 1997. Larval culture of Mexican pectinid *Argopecten ventricosus = circularis* (Sowerby II, 1842) and *Lyropecten subnodosus* (Sowerby I, 1835) in a commercial hatchery. In: E. Félix-Pico (Ed.). Books of Abstracts, 11th Inter. Pectinid Workshop. La Paz, México, 10–15 April. pp. 25–27.

Sicard, M.T., Maeda-Martínez, A.N., Ormart, P., Reynoso-Granados, T. and Carvalho, L., 1999. Optimum temperature for growth in the catarina scallop (*Argopecten ventricosus-circularis* [Sowerby II, 1842]). J. Shellfish Res. 18(2):385–392.

Siewers, A.K., 1982. Commercial mariculture of a bay scallop *Argopecten circularis* (Sowerby) in the Ensenada of La Paz, Baja California Sur, México. J. Shellfish Res. 3(1):114.

Signoret-Brailovsky, G., Maeda-Martínez, A.N., Reynoso-Granados, T., Soto-Galera, E., Monsalvo-Spencer, P. and Valle-Meza, G., 1996. Salinity tolerance of the Catarina scallop *Argopecten ventricosus-circularis* (Sowerby II, 1842). J. Shellfish Res. 15(3):623–626.

Silva-Loera, A., 1986. Efecto del tamaño corporal, tensión de oxígeno y temperatura sobre la tasa de consumo de oxígeno en la escalopa *Argopecten circularis* (Sowerby) (Mollusca: Lamellibranchia). Tesis de Maestría. Instituto Tecnológico de Estudios Superiores de Monterrey, México. 93 p.

Singh-Cabanillas, J., 1987. Cultivo experimental de almeja catarina en corrales *Argopecten circularis*. Acuavisión, Revista Mexicana de Acuacultura 2(7):4–6.

Singh-Cabanillas, J. and Bojorquez-Verástica, G., 1987. Fijación de moluscos bivalvos en colectores artificiales en Bahía Concepción, Baja California Sur, un avance. In: Segundo Congreso Nacional de Acuacultura, AMAC, La Paz, B.C.S., noviembre 24–28 de 1987. 3 p.

Smith, J.T., 1991. Cenozoic giant pectinids from California and the Tertiary Caribbean Province: *Lyropecten*, "*Macrochlamys*", *Vertipecten*, and *Nodipecten* species. U.S. Geological Survey Prof. Paper 139.

Treviño-Gracia, E., 1995. Clasificación de la calidad sanitaria del agua en el área de cultivo de moluscos bivalvos en la zona de la Cooperativa Bahía Tortugas, Bahía Tortugas, B.C.S. Tesis Profesional. CICIMAR-IPN, La Paz, B.C.S. 49 pp.

Tripp-Quezada, A., 1985. Explotación y cultivo de la almeja catarina *Argopecten circularis* en Baja California Sur. Tesis de Maestria. CICIMAR-IPN, La Paz, B.C.S.

Tripp-Quezada, A., 1991. Spawning and spat settlement of the Catarina scallop *Argopecten circularis* (Sowerby, 1835) in bahía Magdalena, B.C.S., México. In: S.E. Shumway and P.A. Sandifer (Eds.). An International Compendium of Scallop Biology and Culture. World Aquaculture Workshops No. 1. World Aquaculture Society, Baton Rouge, LA. pp. 43–46.

Tobias-Sánchez, M. and Cáceres Martínez, C., 1994. Colecta experimental de juveniles de *Pecten vogdesi* sobre diferentes sustratos artificiales en la Bahía de La Paz, B.C.S., México. Rev. Inv. Cient., (No. Especial AMAC), UABCS 1:45–53.

Velasco-Blanco, G., 1997. Cultivo larvario a nivel piloto del callo de hacha *Atrina maura* Sowerby y de la almeja mano de león *Lyropecten subnodosus* Sowerby, con dos especies de microalgas. Tesis de Maestría. Centro de Invest. Cient. y de Educación Superior de Ensenada, B.C., México.

Verdugo-Díaz, G., 1997. Cambios estacionales del fitoplancton y de la composición bioquímica del material orgánico particulado en Bahía Concepción, B.C.S. Tesis de Maestría. CICIMAR-IPN, La Paz, B.C.S., México. 100 pp.

Vicencio Aguilar, A.M, Singh Cabanillas, J. and Amador Buenrostro, J., 1988. Guía Práctica para el Cultivo de almeja catarina. Secretaría de Pesca, La Paz, B.C.S., México. 20 p.

Vicencio Aguilar, A.M., Singh Cabanillas, J., Bojórquez Verástica, G.J., López García, G. and Llamas Hernández, A., 1994. Cultivo de la almeja catarina. Secretaría de Pesca, La Paz, B.C.S., México. 25 p.

Villalaz, G., J.R., 1994. Laboratory study of food concentration and temperature effect on the reproductive cycle of *Argopecten ventricosus*. J. Shellfish Res. 13(2):513–519.

Villalejo-Fuerte, M.T., 1995. Fecundidad en *Argopecten circularis* (Sowerby 1835) (Bivalvia: Pectinidae) de Bahía Concepción, B.C.S., México. An. Inst. Mar. Punta Betín 24:185–189.

Villalejo-Fuerte, M.T., 1997. Growth of *Argopecten ventricosus* (Sowerby II, 1842) in Bahía Concepción, Gulf of California, México. In: E. Félix-Pico (Ed.). Books of Abstracts, 11th International Pectinid Workshop. La Paz, México, 10–15 April. pp. 134.

Villalejo-Fuerte, M.T. and Ceballos-Vázquez, B.P., 1996. Variación de los índices de condición general, gonádico y de rendimiento muscular en *Argopecten circularis* (Bivalvia: Pectinidae). Rev. Biol. Trop. 44(2):571–575.

Villalejo-Fuerte, M.T. and Ochoa Baez, R.I., 1993. El ciclo reproductivo de la almeja catarina, *Argopecten circularis* (Sowerby 1835), en relación con temperatura y fotoperíodo, en Bahía Concepción, B.C.S., México. Ciencias Marinas 19(2):181–202.

Villalejo-Fuerte, M.T. and Tripp-Quezada, A., 1997. Growth of Vogde's *Euvola vogdesi* Arnold, 1906, in culture at Bahía Juncalito, B.C.S., México. In: E. Félix-Pico (Ed.). Books of Abstracts, 11th International Pectinid Workshop. La Paz, México, 10–15 April. pp. 29.

Villavicencio Peralta, G., 1997. Acondicionamiento gonadal, desarrollo embrionario y cultivo de larvas de almeja mano de león, *Lyropecten subnodosus* (Sowerby, 1835), alimentadas con cuatro especies de microalgas. Tesis Profesional. Inst. Tecnológico del Mar, Guaymas Unidad-La Paz, México. 93 p.

Waller, T.R., 1991. Evolutionary relationships among commercial scallops (Mollusca: Bivalvia: Pectinidae). In: S.E. Shumway (Ed.). Scallops: Biology, Ecology and Aquaculture. Vol. 21. Elsevier, Amsterdam. pp. 1–73.

Waller, T.R., 1995. The misidentified holotype of *Argopecten circularis* (Bivalvia: Pectinidae). The Veliger 38(4):298–303.

Wright López, B.M., 1997. Ecología de la captación de la semilla de madreperla *Pinctanda mazatlanica* y concha nácar *Pteria sterna* (Bivalvia: Pteriidae), en la isla Gaviota, Bahía de La Paz, B.C.S., México. Tesis de Maestría. CICIMAR-IPN, La Paz, B.C.S., México. 139 p.

Yoshida-Yoshida, M.K. and de Alba-Pérez, C.R., 1977. Densidad y distribución de la almeja catarina en la Ensenada de La Paz, B.C.S. Inf. de los Resultados de las Investigaciones 1977, CIB, La Paz, B.C.S., México. pp. 91–109.

AUTHOR'S ADDRESS

Esteban Fernando Félix-Pico - Centro Interdisciplinario de Ciencias Marinas, Departamento de Pesquerías y Biología Marina, Apartado Postal No. 592, La Paz, Baja California Sur, C.P. 23000, México (E-mail: efelix@ipn.mx)

Scallops: Biology, Ecology and Aquaculture
S.E. Shumway and G.J. Parsons (Editors)
© 2006 Elsevier B.V. All rights reserved.

Chapter 30

Scallop Fisheries, Mariculture and Enhancement in Australia

Mike Dredge

30.1 INTRODUCTION

At least 50 species of scallop have been described from Australian coastal waters (Lamprell and Whitehead 1992), although the taxonomy of many is confused and contradictory. There are, however, appreciable fisheries directed towards only two of these species. These are a trawl fishery for saucer scallops, *Amusium balloti*, in tropical and sub-tropical waters off Queensland and Western Australia, and a dredge fishery for commercial scallops, *Pecten fumatus*, in temperate waters off Victoria, Tasmania and, rarely, off New South Wales.

Pecten fumatus was originally described as three species - (*Notavola alba* (Tate) in Victorian waters, *N. meridionalis* (Tate) off Tasmania, and *N. fumatus* (Reeve) off New South Wales, but Woodburn (1990) identified these taxa as a single species on the basis of electrophoretic data. A separate species, *Pecten modestus*, occurs along the southern and south-western coasts of Western Australian. *Amusium balloti* was classified as *A. japonicum balloti* in its Queensland distribution, but the Western Australian distribution of the species was not recognised or referred to by Habe (1964). Until the taxonomy of the species is further investigated, the species will be referred to as *A. balloti* throughout its Australian distribution.

Queen scallops*, (Equichlamys bifrons)* and doughboys *(Mimachlamys asperrima)* are taken as a byproduct or minor component of southern Australian fisheries. *Pecten modestus*, the Western Australian equivalent of the commercial scallop, has supported intermittent commercial fishing operations off the south-west Western Australia coastline and the mud scallop *Amusium pleuronectes* is sometimes taken in small quantities by Queensland and Northern Australian trawlers. There are minor commercial and recreational dive fisheries for *P. fumatus* and *P. modestus* in South Australia, Victoria and Western Australia, Dive and dredge fishing are permitted in Tasmania's recreational scallop fishery (Anonymous 2000).

Australia's scallop fisheries have produced an average of 3,000 tonnes of meat, with a landed value of approximately Au $60m annually, which represents about 4% of Australia's total commercial fisheries production value (Caton and McLoughlin 2000).

Managers and resource users of all Australian fisheries, including those for scallops, have become increasingly aware of sustainability issues and the environmental impacts of fisheries. This awareness is partly attributable to legislative requirements. Most Australian fisheries statutes require fisheries to conform to the principals of ecologically sustainable development. Changes to Australian Commonwealth legislation, under the *Environmental Protection and Biodiversity Conservation Act 2000* that will effectively prohibit export of products from fisheries that cannot demonstrate target species sustainability, ecological sustainability and negligible impact on endangered and threatened species, are creating substantial changes in the way that Australian fisheries are currently being assessed and managed.

This review will describe the biology, fisheries, management and mariculture potential of *P. fumatus* and *A. balloti*, and the social environment in which the fisheries exist.

30.2 HISTORY OF THE FISHERIES

The Australian scallop fishery for *P. fumatus* commenced in Tasmania's Derwent River estuary and D'Entrecastaux Channel (Fig. 30.1) in the early to mid 1910's and peaked in the latter area in 1962, when production was in the order of 550 tonnes of meat (Table 30.1, Fig. 30.2). During the fishery's early development, an unrecorded but appreciable proportion of the catch consisted of doughboy scallop (*Mimachlamys asperrimus* (Fairbridge 1953). The D'Entrecastaux Channel population collapsed in about 1965, and appears not to have recovered since then (Harrison 1965, Young and Martin 1989, Zacharin 1990a). Vessels moved from the Channel to scallop beds that were discovered along the eastern Tasmanian coast over a period of 10–15 years. There appears to have been a history of the coastal Tasmanian grounds being serially depleted and not recovering for extended periods, if at all (Young et al. 1990). The Tasmanian fleet first fished grounds in Bass Straits in the early 1970's, following successful survey work undertaken by staff from Tasmania's Department of Fisheries (Grant and Alexander 1973). Catches from the Bass Strait grounds have varied considerably since the fishery commenced, although catches in recent years have been taken from relatively small areas in the south-eastern areas of the Strait. Proportions of the Bass Strait catch landed in Victorian and Tasmanian ports can no longer be separated.

The combination of reduced Tasmanian catches, growing Australian markets and increased publicity about the existence of a scallop resource in Port Phillip Bay (Victoria) led to an explosive growth of fishing operations in this area. Upward of 300 vessels, many from Tasmania, joined the fishery between 1963 and 1966 (Sanders 1970). Production exceeded 2,000 tonnes of meat in the peak production years of 1966 and 1967. The Port Phillip Bay scallop population declined in 1969 and much of the fleet moved to newly discovered beds off Lakes Entrance and other areas in Bass Strait. The Lakes Entrance and Bass Strait fisheries have persisted since that time, despite having to cope with irregular recruitment and intermittent closures caused by recruitment failure. The Port Phillip Bay fishery was closed in 1996, in response to community and political concerns about the impact of the fishery on benthic habitat and recreational fisheries resources (McCormack and McLoughlin 1999).

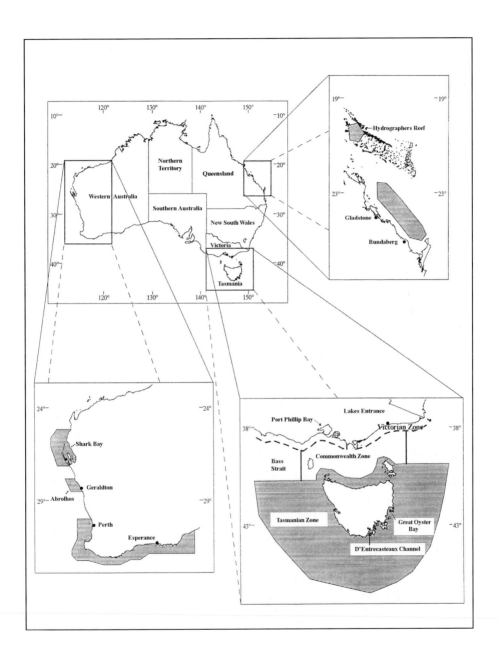

Figure 30.1. Fishery grounds and management areas of Australian scallop fisheries.

1394

Table 30.1

Production of *P. fumatus* and *P. modestus* in Australian waters (meat weight, tonnes).

Year	Pecten fumatus Tasmania (a)	Pecten fumatus Total
1922	60	60
1923	75	75
1924	N/A	N/A
1925	N/A	N/A
1926	N/A	N/A
1927	N/A	N/A
1928	N/A	N/A
1929	80	80
1930	100	100
1931	75	75
1932	115	115
1933	155	155
1934	180	180
1935	210	210
1936	235	235
1937	260	260
1938	330	330
1939	315	315
1940	255	255
1941	340	340
1942	205	205
1943	225	225
1944	260	260
1945	300	300
1946	390	390
1947	435	435
1948	375	375
1949	235	235
1950	205	205
1951	165	165
1952	145	145
1953	335	335
1954	405	405
1955	520	520
1956	535	535
1957	430	430
1958	415	415
1959	420	420
1960	415	415

Year	Pecten fumatus Tasmania (a)	Pecten fumatus Victoria (b)	Pecten fumatus New South Wales (c)	Pecten fumatus Total	Pecten modestus Western Australia (d)
1961	475			475	
1962	555			555	
1963	440	730		1170	
1964	385	1495		1880	
1965	410	1950		2360	
1966	50	2010		2060	
1967	35	2040		2075	
1968	20	710		730	
1969	10	280		290	5
1970	0	660	220	880	260
1971	10	1155	50	1215	5
1972	80	1885	20	1985	10
1973	180	1110	1	1290	5
1974	195	560		755	
1975	105	740		845	
1976	75	450		525	
1977	60	285		345	
1978	165	1670		1835	
1979	590	1735	15	2340	5
1980	520	2530	180	3230	
1981	1165	2540	435	4140	5
1982	1780	1815	105	3700	5
1983	1340	2290	20	3650	15
1984	355	1035		1390	10
1985	465	2100*		2565	80
1986	770	300*		1070	2
1987	75	1600*	2	1675	1
1988	Closed	220*	5	225	65
1989	Closed	100*	3	105	
1990	Closed	90		90	
1991	Closed	650		650	
1992	1	1980		1980	
1993	Closed	1335		1335	
1994	Closed	75		75	
1995	245	410		655	
1996	40	90		130	
1997	Closed	10		10	
1998	540				
1999	700				

Source: a) Fairbridge (1953), Australian Fisheries Newsletter (1962–1980), Harrison (1965), Anonymous (2000): b) Australian Fisheries Newsletter (1962–1980), Gwyther (1990), Anonymous (1999): c) Stewart et al. (1991): d) Stewart et al. (1991). * Data may be incomplete.

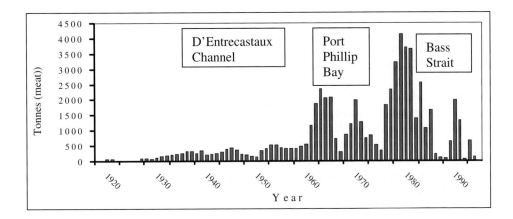

Figure 30.2. Annual production of *Pecten fumatus* in Australia.

The trawl fishery for *A. balloti* commenced in the mid 1950's off the central Queensland coast, between 23°S and 25°S (Ruello 1975), and in the early 1960's in Shark Bay (25°S–26°S), off the Western Australian coast (Harris et al. 1999) (Fig. 30.1). Both began as secondary operations to existing prawn fisheries but have since become substantial fisheries in their own right. The fisheries benefited and expanded when meat quality was recognised on international markets. The bulk of the Australian *A. balloti* catch is now exported to markets in southeast Asia. A number of the trawlers that fish for *A. balloti* in Shark Bay also work in the vicinity of the Abrolhos Islands (28°S), and, occasionally, in waters south and west of Perth (32°S–34°S), where recruitment is far less consistent than in Shark Bay. While more than 90% of average annual landings from the Queensland fishery have been taken between 23°S and 25°S, grounds in the vicinity of Hydrographer's Passage (22°S) receive intermittent recruitment and occasionally produce substantial quantities of saucer scallops (Fig. 30.1).

30.3 BIOLOGY OF TARGET SPECIES

30.3.1 *Pecten fumatus*

30.3.1.1 Distribution and life cycle

Pecten fumatus has been recorded along the southern and eastern Australian coastline between western South Australia (133°E) and central New South Wales (30°S) (Woodburn 1990). *Pecten modestus* occurs in waters off southern and southwestern Western Australian, and a form of *Pecten*, described as *Pecten jacobaeus byronensis* (perhaps in error) by Fleming (1955), recruits intermittently in waters off southern Queensland and northern New South Wales.

Pecten fumatus is a functional hermaphrodite that normally spawns in winter and spring (June to November) in Tasmanian and Victorian waters (Harrison 1961; Sause et al. 1987a; Young et al. 1999), and in late winter and early spring towards the northern limits of its distribution (Jacobs 1983). The species normally attains sexual maturity in its second year. Under hatchery conditions, the species has a 10–12 day larval phase (Dix and Sjardin 1975). There have been a number of studies directed towards understanding the timing of larval settlement in the species' natural environment, particularly in the context of predicting recruitment levels (Hortle and Cropp 1987; Sause et al. 1987b; Coleman 1990; Young et al. 1992). These studies have shown that maximum larval settlement onto spat collectors typically occurs in spring and early summer.

Hortle and Cropp (1987) and Sause et al. (1987b) recorded spat growth of 6–10 mm month^{-1}. Growth rates of juveniles and adults have been described by Fairbridge (1953), Sanders (1970), and Dix (1981), using tagging, shell check interpretation and size frequency analysis. These studies indicate that there is considerable variation in growth rates, with an average size of 30–40 mm shell length (maximum diameter) being attained in the first year, but in some circumstances attaining 60 mm in a year (Sause et al. 1987b). *Pecten fumatus* typically attains 70–80 mm in the second year of life and has an average maximum shell length (von Bertalanffy's L_∞) of 86–93 mm. There is some evidence of a change in the species' growth rates in its Port Phillip Bay distribution (Gwyther and McShane 1988), with growth rates increasing appreciably in a 20-year period. Whilst populations of *P. fumatus* including up to 16 year classes were recorded in the fishery's early history (Fairbridge 1953), the fishery is now almost entirely dependent upon 2+ and 3+ year old animals (McLoughlin 1994), which has caused concern about the potential for recruitment overfishing.

30.3.1.2 Mortality

While there have been a number of studies on predators and parasites that might cause mortalities to *P. fumatus*, there appears to be only one detailed study of population natural mortality rates for the species. Gwyther and McShane (1988) reported an annual instantaneous natural mortality rate (M) of 0.52, which is equivalent to an annual survival rate of approximately 60%. Starfish (*Costinasterias calamaria*) and whelks (*Fasciola australasia*) were reportedly responsible for the death of as much as 80% of a scallop population in the D'Entrecastaux Channel, in southern Tasmania, over a four year period (Olsen 1955). A number of parasites, including mudworm (*Polydora websteri*) (Dix 1981), bucephalid trematodes (Sanders and Lester 1981) and nematodes (McShane and Lester 1984) have been reported to cause mortality of *P. fumatus* but the impact of these species on population levels have not been determined.

Population levels of *P. fumatus* in Port Phillip Bay were at historically near-low levels in 1999, despite the termination of the commercial fishery in 1996 (Coleman, pers. comm.). There is evidence that interactions between a number of introduced species, particularly the clam *Corbula gibba*, which is now very abundant in Port Phillip Bay, may be responsible for a decline in scallop abundance (Currie and Parry 1999, Talman and Keough 2001).

30.3.1.3 Monitoring, abundance and population dynamics

Population levels and population status of *P. fumatus* are monitored largely through fisheries statistics. Data on landings, effort distribution and catch rates for catches made in state waters are collected and collated by fisheries management agencies in Tasmania and Victoria, and by the Australian Fisheries Management Authority (AFMA), for catches made in the central, Commonwealth managed, zone of Bass Strait. Data are sourced from daily or monthly fisher returns. AFMA produces an annual assessment report (Caton and McLoughlin 2000), which includes a summary on the state of the Bass Strait scallop population, but the respective state authorities do not currently produce population models or assessment reports other than the production of fishery data as annual catch and effort summaries.

The Tasmanian Government authorises annual exploratory fishing operations by licensed scallop boats. Results from operations carried out by these boats are used to determine whether or not the fishery is opened in Tasmanian waters.

Although the Victorian State Government does not conduct stock assessments on scallops in Bass Strait, annual recruitment surveys and population assessments using diver observations were conducted in Port Phillip Bay in the period 1982 to 1996 (Coleman 1998), prior to the fishery's closure in this area. Survey data were used to estimate total population numbers and have served as catch predictors in the Port Phillip Bay area. The surveys demonstrated that recruitment variation within the Bay was considerable. High recruitment levels did not invariably lead to successful fishing seasons, as juvenile populations occasionally suffered high mortality levels. While the annual survey was discontinued in 1996, it has been repeated irregularly since then. The exceptionally low recruitment levels recorded in 1999, in the absence of commercial fishing, (Coleman, pers. comm.), may be a consequence of competition from exotic benthic organisms.

Sporic et al. (1997) attempted to relate commercial scallop recruitment and meat condition to environmental parameters, including nutrient and trace element levels, chlorophyll levels and coarser measures such as river flows, salinity and water temperatures. Whilst adductor condition of scallops could be related to some of the variables used in analysis (often on a time lagged basis), the only environmental variable correlated to catch rate was water temperature. The authors concluded that there was little scope for predicting scallop catches on the basis of environmental parameters. Recruitment and population levels in waters administered by the Victorian Government are now monitored through exploratory fishing operations. The results of these surveys are used to determine when and where fisheries for scallops are allowed.

Total catch data (Table 30.1, Fig. 30.2), which are the only readily available, comprehensive and long-term information series on the population of *P. fumatus,* do not reflect the true state of the population. Rather, they reflect a complex series of fish-down events, some of which occurred over many years (as in the D'Entrecastaux Channel), discoveries of new beds, and irregular recruitment in areas such as Port Phillip Bay and the Bass Strait (Young et al. 1990). Caton and McLoughlin (2000) describe the species as being over-exploited. While detailed population models are not available, the general trends of effort distribution and total catch data support this conclusion.

30.3.2 *Amusium balloti*

30.3.2.1 Distribution and life cycle

Amusium balloti is a predominantly sub-tropical species that occurs in water depths of 15–50 m, typically between 18°S and 25°S along the Queensland continental shelf (Dredge 1988). Outliers have been recorded as far south as Jervis Bay (35°S), in New South Wales (Smith 1991). The species' Western Australian distribution extends between18°S and 35°S (Harris et al. 1999).

The species is gonochoristic and normally spawns in winter and spring (Dredge 1981, Joll and Caputi 1995a). It has a 12–25 day larval phase under hatchery conditions, with settlement of post larvae occurring at a size of approximately 200 µm. There is no evidence of more than a very transient byssal phase (Rose et al. 1988, Cropp 1994). This implies that conventional spat catching techniques are unlikely to collect *Amusium* spat. This observation is consistent with findings reported by Sumpton et al. (1990) and Robins-Troeger and Dredge (1993) in their studies on scallop spat collection in areas where *A. balloti* normally occurs. Successful settlement appears to create aggregations or beds of scallops. The only such beds that have been examined in detail were oblong in shape, with the longer axis lying parallel to the direction of tidal flow (Dredge 1985a).

Amusium balloti is characterised by exceptionally rapid growth. Joll (1988) has described the formation of daily growth rings in the Western Australian population of the species, and described growth rates of up to 2.2 mm week^{-1}. Williams and Dredge (1981) and Dredge (1985a) reported von Bertalanffy parameters for k (instantaneous growth rate) in the range 0.052–0.059 week^{-1} and L_∞ in the range 101–108 mm, using both tagging and sequential size frequency composition data. These growth rate parameters imply the species can attain a shell height of 90 mm in 6–12 months. While maximum lifespan may be 3 years, the bulk of the fished population consists of 0+ and 1+ animals (Dredge 1994, Harris et al. 1999). The species is processed in roe-off form, and the adductor meat displays seasonal variation in condition, which has significant implications in terms of yield optimisation.

30.3.2.2 Mortality

Amusium balloti has the relatively high natural mortality rate consistent with a short-lived species. The only published report on natural mortality rates for *A. balloti* in its normal distribution estimated the instantaneous rate for M of 0.02–0.025 week^{-1} for scallops in the size range 50–110 mm (Dredge 1985b). This is equivalent to an annual mortality rate of approximately 60% for scallops in this size range. There are no data available on natural mortality rates of juvenile saucer scallops. Studies of predators and parasites that may be responsible for natural mortality are limited. Jones (1988) described the scyllarid lobster *Thenus orientalis* as a predator of *A. balloti*. Other known predators include octopus, a number of turtle species and the sparid *Pagrus auratus* (Harris et al. 1999). There have been no comprehensive studies on the parasitology of *A. balloti*, but the species is known to act as the intermediate host for an ascaroid nematode, *Sulcascaris sulcuta*, which uses loggerhead turtles, *Caretta caretta* as a final host (Berry and Cannon 1981). There are no data on the impact of this species on survival rates of *A. balloti*. The top shell *Capulus dilatata* parasitises A. *balloti* after boring into the shell, but its impact upon population levels is again unknown.

30.3.2.3 Monitoring, abundance and population dynamics

The major *A. balloti* resources of Western Australia and Queensland are monitored and assessed by means of annual recruitment surveys and review of commercial catch and effort data collected by fishers. Each state fisheries authority conducts a comprehensive annual survey to monitor recruitment levels and provide predictions of subsequent stock levels available to the fishery. Reviews of catch, effort and trends in both fisheries are published in annual or irregular fisheries status reports (Penn 2000; Williams2002).

Annual landings from the Western Australian fishery are strongly correlated with recruitment levels as estimated through the annual survey (Joll and Caputi 1995b), and annual catch predictions are now offered on the basis of this survey (Penn 2000). The main determinant of recruitment levels in the Western Australian population appears to be prevailing environmental conditions driven by the el Nino – Southern Oscillation (ENSO) phenomenon. This, in turn, is thought to be related to the behaviour of the Leeuwin current system off the Western Australian coastline, although the exact mechanism driving recruitment levels remains unknown (Joll 1994; Caputi and Joll 1995b). Recruitment and catch levels were an order of magnitude greater than the long term average in 1991–1993, years in which the ENSO event had major impacts on water movements and temperatures off the Western Australian coastline (Table 30.2, Fig. 30.3). There appears to be no definable relationship between parent stock size and subsequent recruitment levels in the Western Australian population of *A. balloti*.

The Queensland fishery for *A. balloti* appears to exploit the resource far more heavily than is the case for the Western Australian population. There have been concerns about potential recruitment overfishing (Dredge 1988; Dichmont et al. 1999), which have lead to the introduction of precautionary management measures, including spatial closures to maintain spawner levels. Population assessments for the Queensland stock of *A. balloti* are now based upon reviews of commercial catch and effort data and information derived from an annual recruitment survey conducted in October each year (Dichmont et al. 2000). The recruitment survey has been conducted for three years and is not yet capable of giving meaningful predictions of annual landings. Dichmont et al. (1999) have developed models to examine the status of the Queensland saucer scallop resource, but their value as a management tool or predictor for landings has not yet been tested.

Table 30.2

Production of *A. balloti* in Australian waters (meat weight, tonnes).

Year	*A. balloti* Queensland	*A. balloti* Western Australia	Year	*A. balloti* Queensland	*A. balloti* Western Australia
1956	110		1978	645	109
1957	5		1979	280	57
1958	5		1980	520	113
1959	145		1981	660	169
1960	145		1982	1220	482
1961	35		1983	880	863
1962	60		1984	900	650
1963	75		1985	660	244
1964	40		1986	700	336
1965	65		1987	450	583
1966	40	1	1988	989	768
1967	30	N/A	1989	826	171
1968	55	69	1990	1689	557
1969	465	273	1991	1008	2567
1970	350	83	1992	983	4511
1971	430	NA	1993	2163	2233
1972	815	22	1994	1263	1488
1973	670	57	1995	1776	916
1974	300	36	1996	592	597
1975	180	33	1997	987	338
1976	100	111	1998	928	
1977	550	159	1999		

(Source – Queensland: Australian Fisheries 1956–1980, Stewart et al. 1991; Williams 2002; Western Australia: Harris et al. 1999)

30.4 FISHERIES AND THEIR MANAGEMENT

30.4.1 *Pecten fumatus*

30.4.1.1 Regulation

The major Australian fishery for *P. fumatus* is most readily described in terms of a single fishery managed by three jurisdictions. The Victorian Government is responsible for the species to a distance approximately 20 n. miles seaward of the Victorian shoreline. Tasmania has responsibility for management of scallop fisheries south of a line established under Australia's Offshore Constitutional Settlement that follows Latitude 40°45'S to about 144°E, and then approximately 20 n. miles seaward of the northern

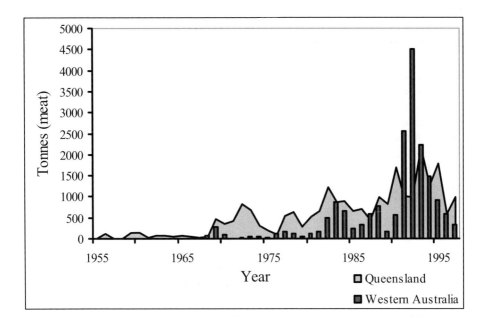

Figure 30.3. Annual production of *Amusium balloti* in Queensland and Western Australia.

Tasmanian coastline. The Commonwealth of Australian is responsible for fishing operations in the area between the two jurisdictions (Fig. 30.1). These arrangements can be linked to the fishery's history. The fisheries originated in coastal waters of Tasmania and Victoria, and moved to the oceanic waters of Bass Strait, the fisheries of which have historically been managed by a Commonwealth agency. Arrangements under which the three agencies co-manage the resource were formalised in 1986.

All three management agencies dealing with the major *P. fumatus* population in Victorian, Tasmanian and Bass Strait waters have had to deal with "boom and bust" fisheries that have attracted excess fishing capacity during boom years.

Managers administer the fishery in accordance with the objectives and requirements of three statutes, the Commonwealth Fisheries Management Act 1991, the Tasmanian Living Marine Resources Act 1995, and the Victorian Fisheries Act 1995. These statutes have some similarities in that they require sustainable use of fisheries resources amongst other objectives based on social and economic imperatives. All three jurisdictions have imposed limited entry into scallop fisheries, and many of the boats involved in the fishery have access rights to all three administrative zones.

While there has been no clear demonstration of a stock-recruitment relationship in the species, Young et al. (1990) presents limited data in support of the concept that spat settling in given areas of Bass Strait may originate from concentrations of spawners in the same areas. Depletion of spawners from such areas could therefore lead to localised recruitment failure. Managers in all three management areas support a policy of allowing scallops to spawn twice before being harvested, in order to minimise the risk of recruitment overfishing (McLoughlin 1994). In years of low recruitment, fisheries management agencies have closed the fisheries.

Tasmanian government policy on scallop resources now only allows for the harvesting scallops when there are adequate stock levels, at a time and size that will allow yield to be optimised (Anonymous 2000). The fishery has been closed for 6 of the past 10 years in Tasmanian waters, due to low recruitment levels. Consequently, the fishery has become a secondary operation for the 117 vessels that currently hold licenses to fish for scallops. All vessels that are so licensed have access to other fisheries and fish for scallops irregularly.

The fishery operates on the basis of marketing a roe-on product, both to local and European markets, and is administered to maximise yield and value by allowing only winter – spring fishing, when gonads are at peak condition. Environmental impacts of the fishery are managed by prohibiting scallop dredging in sensitive shallow areas, by limiting dredge numbers and by requiring the use of dredges that minimise impact on the seabed.

The Victorian scallop fishery's management arrangements have been profoundly affected by the legislated closure of Port Phillip Bay to scallop dredging in 1996. The fishery was permanently closed on the basis of social and ecological arguments, following jurisdictional procedures that had important implications for the nature of property rights in Australian fisheries (McCormack and McLoughlin 1999). The Victorian fishery is now restricted to coastal and oceanic waters to the south of the state, with an appreciable proportion of the state's landing being taken from Commonwealth-managed waters and, intermittently, from Tasmanian waters. Ninety-four operators are licensed to take scallops in Victorian and adjacent Commonwealth-managed waters. Vessels are allocated annual quotas that vary with available stock levels. The quotas, which are transferable, are further regulated on a monthly basis (currently ca. 3 tonnes, whole animal weight) to ensure an orderly supply of product to processing facilities and markets. The season is subject to seasonal closures that allow scallops to be taken in the period May-October, when the animals are in optimal condition.

Additional to this fishery, a minor fishery for *P. fumatus* in South Australian waters is managed as a limited entry dive fishery. The fishery currently includes only 4 licensed operators who have produced insignificant amounts of scallop in the past 10 years. The highly intermittent New South Wales and Western Australian *Pecten* fisheries have been managed on an "as-needs" basis in the past. Concerns about the impact of dredging on benthic communities may restrict use of these resources should future recruitment be sufficient to stimulate commercial interest.

30.4.1.2 Environmental management and sustainability issues

There have been detailed studies of the impact of scallop dredging, particularly in Port Phillip Bay. Currie and Parry (1996, 1999) concluded that dredging had a clearly detectable effect on benthic fauna and topography in the short term, but longer term impacts were far less noticeable. Following the closure of the Port Phillip Bay scallop fishery, environmental management issues were largely directed to the impact of the fishery on the target species. In addition to the 'two spawning' rule, scallop fishing is not allowed when a pre-determined proportion of the catch is smaller than the prevailing size limit in order to minimise wastage. Reducing the impact of scallop fishing on associated benthic communities has been considered in the context of using physically less impacting catch gear (Zacharin 1990b) but still remains a major issue for the industry to address.

30.4.2 *Amusium balloti*

30.4.2.1 *Regulation*

There are marked differences between the management of saucer scallop (*A. balloti*) fisheries in Western Australia and Queensland.

The Western Australian fishery occurs in spatially discrete areas near Shark Bay, off the Abrolhos Islands and, intermittently, to the west and south of Perth. Each area is managed as a separate entity. The collective fishery is managed under a basic tenet that the residual stock levels should not be fished below a level required to maintain long-term recruitment levels despite the absence of any apparent relationship between breeding stock levels and recruitment. Consequently, management of the fishery has been largely focused towards maximising yield, maintaining economic efficiency and minimising conflict with other fisheries that occur in the same locations. This is achieved through limited entry and seasonal closures (Harris et al. 1999). As all product is shucked at sea size limits are not a feasible option for yield optimisation. Entry to the Shark Bay fishery is restricted, with a total of 41 trawlers being authorised to take saucer scallops in this area. Of these, 27 are also entitled to take penaeid prawns and consequently fish for saucer scallops on a part time basis. The saucer scallop fishery is seasonal, with a regulated opening normally occurring in April - May and the season closing in November to coincide with the seasonal closure of a major prawn fishery. In reality, the saucer scallop fishery normally ceases well before November as numbers decline to economically unviable levels. The Abrolhos fishery is likewise subject to restricted entry and a seasonal fishery regime. Recruitment in this area is far less consistent than in the Shark Bay area, and most of the 16 trawlers that are licensed to fish in this area have access to alternative fisheries. A restricted entry fleet of four trawlers fishes saucer scallops that recruit intermittently on grounds off the state's southern coastline, typically off Esperence, and three trawlers are licensed to take scallops and other trawlable species in grounds west of Perth (34°). Effort in all these fishing grounds is constrained by vessel dimensions and trawl gear size, with trawl size being restricted to a maximum of nets with headrope length no greater than 18 m.

Saucer scallops taken in the Western Australian fishery are shucked as roe-off meat and frozen at sea. Almost all product produced by the fishery is exported, mostly to south-east Asia, where the market offers premium prices for saucer scallop meat (Hart 1994).

The Queensland fishery for saucer scallops is managed as an element of the Queensland East Coast trawl fishery. This complex fishery is largely directed towards the capture of 9 species of penaeid prawn, but also takes crabs, scyllarid lobsters and fish as byproduct, in addition to an average 1100 tonnes of scallop meat a year (Table 30.2, Fig. 30.3). While entry to the composite fishery is limited and boat numbers are being reduced, there are still more than 700 licensed trawlers that fish along the eastern Queensland seaboard between about 10°S and 28°S. All of these boats have access to the saucer scallop resource, but only 200–300 trawlers report landing scallops in any one year.

The resource is fished heavily. Management procedures are aimed both at yield optimisation and minimising the risk of recruitment overfishing. A seasonally varying size limit and a late winter-early spring closure are used to maximise yield (Dredge 1994). Trawl gear size is limited to a maximum 109 m of (combined) headrope and footrope in order to constrain effort, and three 10'*10' areas that have historically held high densities of saucer scallops have been closed to fishing as a means of maintaining spawner population levels. Saucer scallops are required to be processed ashore, or in a number of designated inshore shucking areas, in order that size limits can be policed.

The bulk of Queensland's saucer scallop production is exported, again largely to south-east Asian markets (Hart 1994).

The marked difference between the Queensland and Western Australian entry policies for saucer scallop fisheries reflect the fundamentally different approaches taken by fisheries managers during the development of Australian trawl fisheries in the 1960's and 1970's. Western Australian managers took a highly precautionary attitude, focused on species-specific management and were committed to a philosophy of developing profitable, flexible fisheries that could fully meet monitoring and management requirements (Hancock 1975). Queensland managers adopted a much more open market philosophy that eventually lead to Queensland becoming a 'dumping ground' for trawlers from a number of other Australian jurisdictions. At one point in time almost 2,000 trawlers were authorised to fish in Queensland waters (Hill and Pashen 1986). There has been a slow reduction of trawler numbers, but the fishery still has excess fishing effort.

30.4.2.2 Environmental management and sustainability issues

Both the Western Australian and Queensland saucer scallop fisheries have taken action in recognition of environmental management and conservation issues.

The Queensland scallop fishery is administered under a trawl fisheries management plan that is subsidiary legislation to the State's Fisheries Act. This plan includes trigger points for management intervention, based upon prescribed declines in catch rates of the target species. It includes a number of conservation measures, including limitations on ground chain size, compulsory use of turtle excluding devices, and the (temporary) closure of broodstock protection areas to ensure population levels are maintained. There is a requirement for reduction in bycatch levels, again with associated limit reference points, and a requirement for the fishery to have minimal impact on threatened and endangered species.

Western Australian fisheries also operate under management plans that are subsidiary to that State's Fisheries Act. These plans are essentially a set of rules for the fishery's operation. They allow a high degree of flexibility for management responses to changing conditions in the fishery. Bycatch action plans designed to minimise impacts on non-target species are in the process of being developed. More general regional environmental management plans for the fishery are also in the process of being finalised.

30.5 CULTURE OF SCALLOPS IN AUSTRALIA

While there has been intermittent interest in the culture of scallops in Australian waters, serious commitment to development of mariculture and marine ranching has been limited.

30.5.1 *Pecten fumatus*

The first recorded investigations into the culture of scallops began in 1917, in Tasmania, when a committee was appointed to inquire into the depletion of scallop beds in the Derwent River estuary (Flynn 1918). The potential for mid-water culture was recognised by this committee, as were the predation problems caused by starfish and crabs. There appears, however, to have been no serious attempt to carry out any form of enhancement until the mid 1970's, when the Tasmanian government established a program designed to assess the feasibility of reseeding depleted scallop beds.

The Victorian, South Australian and New South Wales Governments have invested in pilot studies on the feasibility of scallop mariculture or enhancement. Projects will be described on a state by state basis.

30.5.1.1 Tasmania

30.5.1.1.1 Spat production

The feasibility of developing cage culture and enhancement operations using both hatchery reared and wild-caught spat was investigated during the course of a program commenced in 1975. Dix and Sjardin (1975) reported successful rearing of larvae through to settlement, and their techniques were successfully applied to other scallop species (Dix 1976, Rose and Dix 1984). Whilst hatchery experience and competence have been developed to the point that commercial oyster and abalone operations rely on their output, there is currently no hatchery producing scallop spat at commercial levels (Crawford, Tasmanian Scallops Pty Ltd., pers. comm.). Spat collection rates in mesh bag collectors placed off the Tasmanian east coast were initially very low (Dix 1981; Hortle 1983), but improved with time and experience to reach up to 400 spat per collector (Hortle and Cropp 1987).

30.5.1.1.2 Culture operations

The Tasmanian Government entered into a joint-venture agreement with the Japanese Overseas Fisheries Foundation in 1986. Under this agreement, the Japanese organisation supplied technical expertise and capital for scallop mariculture, while the Tasmanian government provided research capacity, infrastructure, staff resources and an exclusive lease area in and adjacent to the Great Oyster Bay (Fig. 30.1). Thompson (1990) gives a brief account of the joint venture's progress.

The joint venture initially used open water spat catching techniques to obtain spat, and followed well described procedures of hanging cage culture for rearing spat, using cage culture and intermediate culture techniques as described by Ventilla (1982). The operations were not immediately successful. Short falls in spat supply and cage fouling was reported to be major constraints to growth and economic production of scallops.

The operation changed organisational structure and direction, being largely taken over by a Japanese consortium and moving towards an open water seeding operation. Open water seeding had limited success, as spat supply from both open water collectors and hatcheries was limited to between 5 and 20 million spat per year, and natural mortality rates appear to have been very high. Annual production has not exceeded 50 tonnes. Following further restructure, the operating company is now re-directing efforts towards hanging cage culture, and is investigating the feasibility of mariculture for other molluscan species, as spat supply continues to be a limiting factor for scallop culture.

30.5.1.2 Victoria

While there has been intermittent interest in the potential of cage culture within the protected waters of Port Phillip Bay, there has been little serious commitment of money or material to such operations. Spat catchers were deployed in the Bay during the 1980's as a means of monitoring spatfall for predictive fishing models (Coleman 1990). Mercer (Victorian Marine and Freshwater Institute, pers. comm.) has renewed spat collection in a pilot project to assess the potential for supplying a scallop aquaculture industry in the Bay. He obtained catches of up to 400 spat per bag in areas of eastern Port Phillip Bay in 1998 and 1999. An existing mussel farming operation has at least some of the infrastructure requirements for scallop culture work in Port Phillip Bay. Progress of this work to a pilot scale operation will depend upon State Government policy development in relation to the desirability of mariculture operations in what is regarded as an environmentally and socially sensitive area.

30.5.1.3 New South Wales

Pecten fumatus intermittently settles in large numbers along the southern coast of New South Wales. Short-lived fisheries in areas such as Jervis Bay (Fuentes 1994) have occurred as a consequence of these settlements.

There has been some interest in establishing whether the population could be stabilised through enhancement procedures, using hatchery-reared spat. There has been considerable progress in developing hatchery procedures for mass rearing of *P. fumatus* spat in hatcheries (Heasman et al. 1994; Heasman et al. 1998), but there has been little progress towards the development of a commercial cage culture or enhancement. An early (pilot) attempt to establish a bottom seeding program was unsuccessful, with predators, including rays, causing substantial losses to the stocked out population. Concerns about the environmental consequences of dredging seeded scallops may limit future restocking operations.

30.5.1.4 South Australia

There have been two commercial attempts to develop *Pecten* mariculture operations in Southern Australian waters. Both occurred in the period between 1996 and 2000, and were based on the concept of using spat collected from open waters in cage culture operations, in waters to the north of Kangaroo Island. Little has been documented about either operation, but the initial attempt apparently closed following two seasons of unacceptably low spat collection (Young, pers. comm.). An attempt to revive this operation is in an early stage of development.

30.5.2 *Amusium balloti*

The rapid growth rate and high market value of *A. balloti* suggests the species has inherent potential for mariculture and enhancement operations. There has been some research but little commercial commitment to translate this potential to reality. The apparent absence of anything more than a transient byssal attachment in this species (Robins-Troeger and Dredge 1993; Cropp 1994) limits the scope for wild spat collection. Consequentially any fisheries enhancement or mariculture operation would be reliant on hatchery reared spat.

Rose et al. (1988) reported on initial attempts to develop hatchery techniques for breeding *A. balloti* under controlled conditions, and Cropp (1992, 1994) attempted to expand on this work to develop an operational model for saucer scallop enhancement in Western Australian waters. While his work was not immediately productive in a commercial sense, there have been further attempts to enhance saucer scallop populations using hatchery-reared spat. The Western Australian government have granted an exclusive-access seabed lease in coastal waters at about 29°S, inshore of Abrolhos Islands, and a private venture company has established a hatchery as part of an program to generate a commercial bottom seeding operation (Joll, pers. comm.). There appears to be some scope for companies currently raising pearl oyster (*Pinctada maxima*) spat to diversify into *Amusium* culture should the present program achieves commercial success.

30.6 SUMMARY

Australian fisheries for scallops have been based on two species groupings and the fisheries that have evolved from these two resources have taken very different paths. The southern Australian dredge fishery for *Pecten* spp. has gone through the 'booms and busts' that appear to characterise many scallop fisheries. This fishery, however, appears to face difficulties and challenges of unprecedented magnitude. Recruitment in some previously productive areas has failed for an extended period of time, and the fishery has been based on scallop populations with a much reduced age structure than has been the case in historical times. Some parts of the fishery have been closed to meet social and environmental requirements, and the surviving elements still face major challenges in meeting the stringent environmental and sustainability standards associated with Australia's recently developed fisheries management standards.

The trawl fisheries for *A. balloti*, on the other hand, have been far more stable in terms of landings. Apart from the three years of massive landings from Western Australia when environmental conditions apparently improved recruitment, and a marked decline in Queensland in 1996, possibly associated with overfishing, annual landings have remained relatively stable by the standards of most scallop fisheries. These fisheries still need to address environmental sustainability issues, but have gone some way to meet these requirements.

By global standards, aquaculture and enhancement operations in Australia have received little commitment and met with little success to this time.

ACKNOWLEDGMENTS

While not wishing to ignore the help given to me by a great many friends and colleagues – too numerous to be named – I must specifically acknowledge the help of Lindsay Joll and Nic Caputi in Western Australia, Hilary Revill in Tasmania, Noel Coleman, Greg Parry and David Molloy in Victoria, Mike Heasman in New South Wales, and staff from Southern Fisheries Centre at Deception Bay, who have suffered endless talks on scallops and then had to review the written end product. Kate Yeomans put much of the art work together. My thanks to you all.

REFERENCES

Anonymous, 1999. Fisheries Victoria; catch and effort. Information Bulletin 1998, Marine and Freshwater Institute, Queenscliffe, Victoria.

Anonymous, 2000. Draft policy document and draft fisheries management plan for the Tasmanian scallop fishery. Department of Primary Industries, Water and Environment, Tasmania.

Berry, G.N. and Cannon, L.R.G., 1981. The life history of *Sulcraris sulcata* (Nematoda, Ascaridoidea), a parasite of marine molluscs and turtles. Int. J. Parasitol. 11:43–54.

Caton, A.E. and McLoughlin, K., 2000. Fishery Status Reports. Bureau of Rural Science, Canberra, Australia.

Coleman, N., 1990. Spat catches as an indication of recruitment to scallop populations in Victorian waters. In: M.L.C. Dredge, W.F. Zacharin and L.M. Joll (Eds.). Proceedings of the Australasian Scallop Workshop. Tasmanian Government Printer, Hobart, Australia. pp. 51–60.

Coleman, N., 1998. Counting scallops and managing the fishery in Port Phillip Bay, south-east Australia. Fish. Res. 38:145–157.

Cropp, D.A., 1992. Aquaculture of the saucer scallop *Amusium balloti*. Final project report to the Fisheries Research and Development Corporation 89/58, Canberra, Australia.

Cropp, D.A., 1994. Hatchery production of Western Australian scallops. Mem. Qld. Mus. 36(2):269–276.

Currie, D.R. and Parry, G.D., 1996. Effects of scallop dredging on a soft sediment community: a large scale experimental study. Mar. Ecol. Prog. Ser. 134:131–150.

Currie, D.R. and Parry, G.D., 1999. Impacts and efficiency of scallop dredging on different soft substrates. Can. J. Fish. Aquat. Sci. 56:539–550.

Dichmont, C.M, Dredge, M.C.L. and Yeomans, K., 2000. The first large-scale fishery-independent survey of the saucer scallop, *Amusium japonicum balloti* in Queensland, Australia. J. Shellfish Res. 19(2):731–739.

Dichmont, C.M., Haddon, M., Yeomans, K. and Kelly, K., 1999. Proceedings of the South-East Queensland stock assessment review workshop. Qld. D.P.I. Conference and Workshop Series QC99003.

Dix, T.G., 1976. Larval development of the queen scallop, *Equichlamys bifrons*. Aust. J. Mar. Freshw. Res. 35:399–403.

Dix, T.G., 1981. Preliminary experiments in commercial scallop (*Pecten meridionalis*) culture in Tasmania. Tas. Fish. Res. 23:18–24.

Dix, T.G. and Sjardin, M.J., 1975. Larvae of the commercial scallop (*Pecten meridionalis)* from Tasmania, Australia. Aust. J. Mar. Freshw. Res. 26:109–112.

Dredge, M.C.L., 1981. Reproductive biology of the saucer scallop *Amusium japonicum balloti* (Bernardi) in central Queensland waters. Aust. J. Mar. Freshw. Res. 32:775–787.

Dredge, M.C.L., 1985a. Growth and mortality in an isolated bed of saucer scallops, *Amusium japonicum balloti* (Bernardi). Queensl. J. Agric. Anim. Sci. 42:11–21.

Dredge, M.C.L., 1985b. Estimates of natural mortality for *Amusium japonicum balloti* (Bernardi) (Pectinidae) based on tag recoveries. J. Shellfish Res. 5:103–109.

Dredge, M.C.L., 1988. Recruitment overfishing in a tropical scallop fishery? J. Shellfish Res. 7:233–239.

Dredge, M.C.L., 1994. Modelling management measures in the Queensland scallop fishery. Mem. Qld. Mus. 36(2):277–282.

Fairbridge, W.S., 1953. A population study of the Tasmanian "commercial" scallop, *Notovola meridionalis* (Tate) (Lamellibranchiata, Pectinidae). Aust. J. Mar. Freshw. Res. 4:1–40.

Fleming, C.A., 1955. A new subspecies of scallop from Byron Bay, New South Wales. Aust. Zool. 12:108–109.

Flynn, T.T., 1918. Scallop investigation: preliminary report by Professor Flynn. Tasmania Parl. Pap. 79(59):6.

Fuentes, H.R., 1994. Population biology of the commercial scallop (*Pecten fumatus*) in Jervis Bay, NSW. Mem. Qld. Mus. 36(2):247–261.

Grant, J. and Alexander, K.R., 1973. The scallop resources of Bass Strait below Latitude 39°12'S, 1972/3. Tas. Fish. Res. 7(2):1–14.

Gwyther, D., 1990. History of management in the Victorian scallop fishery. In: M.L.C. Dredge, W.F. Zacharin and L.M. Joll (Eds.). Proceedings of the Australasian Scallop Workshop. Tasmanian Government Printer, Hobart, Australia. pp. 12–20.

Gwyther, D. and McShane, P.E., 1988. Growth rate and natural mortality of the scallop *Pecten alba* (Tate) in Port Phillip Bay, Australia, and evidence for changes in growth rate after a 20 year period. Fish. Res. 6:347–361.

Habe, T., 1964. Notes on the species of the genus *Amusium* (Mollusca). Bull. Nat. Sci. Mus. Tokyo 7:1–5.

Hancock, D.A., 1975. The basis for management of Western Australian prawn fisheries. In: P.C. Young (Ed.). Proceedings of the First Australian National Prawn Seminar. A.G.P.S., Canberra, Australia. pp. 252–269.

Harris, D.C., Joll, L.M. and Watson, R.A., 1999. The Western Australian scallop industry. Fisheries Research Report No. 114, Western Australian Marine Research Laboratories, Perth.

Harrison, A.J., 1961. Annual reproductive cycles in the Tasmanian commercial scallop *Notovola meridionalis*. BSc (Hons) Thesis. University of Tasmania, Hobart.

Harrison, A.J., 1965. Tasmanian scallop fishery and its future. Australian Fisheries Newsletter, June 1965:9–13.

Hart, B., 1994. Dilemma of the boutique Queensland scallop. Mem. Qld. Mus. 36(2):373–376.

Heasman, M.P., O'Connor, W.A. and Frazer, A.W., 1994. Improved hatchery and nursery rearing techniques for *Pecten fumatus* Reeve. Mem. Qld. Mus. 36(2):351–356.

Heasman, M.P., O'Connor, W.A., O'Connor, S.J. and Walker, W.W., 1998. Enhancement and farming of scallops in NSW using hatchery produced seedstock. Final project report to the Fisheries Research and Development Corporation 94/084, Canberra, Australia.

Hill, B.J. and Pashen, A.J., 1986. Management of the Queensland east coast otter trawl fishery: an historical review and future options. In: T.J.A. Hunloe (Ed.). Fisheries Management. Theory and Practice in Queensland. Griffith University Press, Brisbane, Australia. pp. 146–167.

Hortle, M.E., 1983. Scallop recruitment may be estimated. FINTAS 6(3):1–37.

Hortle, M.E. and Cropp, D.A., 1987. Settlement of the commercial scallop, *Pecten fumatus* (Reeve 1855), on artificial collectors in eastern Tasmania. Aquaculture 66:79–95.

Jacobs, N.E., 1983. The growth and reproductive biology of the scallop *Pecten fumatus (alba)* in Jervis Bay, N.S.W. and the hydrology of Jervis Bay. BSc (Hons) Thesis. University of New South Wales, Sydney.

Joll, L.M., 1988. Daily growth rings in saucer scallops, *Amusium balloti* (Bernardi). J. Shellfish Res. 7:73–76.

Joll, L.M., 1994. Unusually high recruitment in the Shark Bay saucer scallop (*Amusium balloti*) fishery. Mem. Qld. Mus. 36:261–268.

Joll, L.M. and Caputi, N., 1995a. Geographic variation in the reproductive cycle of the saucer scallop *Amusium balloti* (Bernardi 1861) (Mollusca: Pectinidae), along the Western Australian coast. Mar. Freshw. Res. 46:779–792.

Joll, L.M. and Caputi, N., 1995b. Environmental influences on recruitment in the saucer scallop (*Amusium balloti*) fishery of Shark Bay, Western Australia. ICES Mar. Sci. Symp. 199:47–53.

Jones, C.K., 1988. The biology and behaviour of bay lobsters, *Thenus* spp. (Decapoda: Scyllaridae), in northern Queensland, Australia. PhD Thesis. Department of Zoology, University of Queensland, Brisbane, Australia.

Lamprell, K. and Whitehead, T., 1992. Bivalves of Australia. Crawford House Press, Bathurst.

McCormack, S. and McLoughlin, R.J., 1999. The closure of the Port Phillip Bay scallop fishery. Mimeo, paper delivered to the Australian Society of Fisheries Managers, Perth, Western Australia.

McLoughlin, R.J., 1994. Sustainable management of Bass Strait scallops. Mem. Qld. Mus. 36:307–314.

McShane, P.E. and Lester, R.J.G., 1984. The occurrence of a larval nematode in Port Phillip Bay scallops. Mar. Sci. Lab. Queenscliff Tech. Rep. 37:1–6.

Olsen, A.M., 1955. Underwater studies on the Tasmanian commercial scallop, *Notovola meridionalis* (Tate) (Lamellibranchiata: Pectinidae). Aust. J. Mar. Freshw. Res. 6:392–409.

Penn, J.W. (Ed.), 2000. State of the Fisheries Report 1998/1999. Fisheries Research Division, Fisheries Laboratory, North Beach, Western Australia.

Robins-Troeger, J.B. and Dredge, M.C.L., 1993. Seasonal and depth characteristics of scallop spatfall in an Australian subtropical embayment. J. Shellfish Res. 12:285–290.

Rose, R.A., Campbell, G.R. and Saunders, S.G., 1988. Larval development of the saucer scallop *Amusium balloti* (Bernardi) (Mollusca: Pectinidae). Aust. J. Mar. Freshw. Res. 39:153–160.

Rose, R.A. and Dix, T.G., 1984. Larval and juvenile development of the doughboy scallop *Chlamys asperrimus* (Lamark) (Mollusca: Pectinidae). Aust. J. Mar. Freshw. Res. 35:315–323.

Ruello, N.V., 1975. An historical review and annotated bibliography of prawns and the prawning industry in Australia. In: P.C. Young (Ed.). Proceedings of the First Australian National Prawn Seminar. A.G.P.S., Canberra, Australia. pp. 305–341.

Sanders, M.J., 1970. The Australian scallop industry. Australian Fisheries, May 1970:3–11.

Sanders, M.J. and Lester, R.J.G., 1981. Further observations on a bucephalid trematode infection in scallops (*Pecten alba*) in Port Phillip Bay, Victoria. Aust. J. Mar. Freshw. Res. 32(3):475–478.

Sause, B.L., Gwyther, D. and Burgess, D., 1987b. Larval settlement, juvenile growth and the potential use of spatfall indices to predict recruitment of the scallop *Pecten alba* Tate in Port Phillip Bay, Victoria, Australia. Fish. Res. 6:81–92.

Sause, B.L., Gwyther, D., Hanna, P.J. and O'Conner, N.A., 1987a. Evidence for winter-spring spawning of the scallop *Pecten alba* (Tate) in Port Phillip Bay, Victoria. Aust. J. Mar. Freshw. Res. 38:329–337.

Smith, A.K., 1991. Tropical scallop found in Jervis Bay. Dive Log Australia, Jan. 1991, 3.

Sporic, M., Coleman, N. and Gason, A., 1997. Development of a predictive model for the scallop fishery in south-eastern Australia. Final Report to the Fisheries Research and Development Corporation, Queenscliff, Australia.

Stewart, P., Kailola, K. and Ramirez, C., 1991. Twenty-five Years of Australian Fisheries Statistics. Bureau of Rural Resources, Canberra, Australia.

Sumpton, W.D, Brown, I.W. and Dredge, M.C.L., 1990. Settlement of bivalve spat on artificial collectors in a subtropical embayment in Queensland, Australia. J. Shellfish Res. 9:227–231.

Talman, S.G. and Keough, M.J., 2001. The impact of an exotic clam, *Corbula gibba*, on the commercial scallop, *Pecten fumatus*, in Port Phillip Bay, south-east Australia: Evidence of resource-restricted growth in a subtidal environment. Mar. Ecol. Prog. Ser. 221:135–143.

Thompson, J., 1990. Administrative, legal and sociological difficulties of scallop culture and enhancement. In: M.L.C. Dredge, W.F. Zacharin and L.M. Joll (Eds.). Proceedings of the Australasian Scallop Workshop. Tasmanian Government Printer, Hobart, Australia. pp. 264–271.

Ventilla, R.F., 1982. The scallop industry in Japan. Advances in Marine Biology 20:309–82.

Williams, L.E., 2002. Queensland's Fisheries Resources. Queensland Department of Primary Industries Information Series QI02012, Brisbane, Australia.

Williams, M.J. and Dredge, M.C.L., 1981. Growth of the saucer scallop *Amusium japonicum balloti* Habe in central eastern Queensland. Aust. J. Mar. Freshw. Res. 32:657–666.

1412

Woodburn, L., 1990. Genetic variation in the southern Australasian *Pecten*. In: M.L.C. Dredge, W.F. Zacharin and L.M. Joll (Eds.). Proceedings of the Australasian Scallop Workshop. Tasmanian Government Printer, Hobart, Australia. pp. 226–240.

Young, P.C. and Martin, R.B., 1989. The scallop fisheries of Australia and their management. Rev. Aquatic Sci. 1:615–638.

Young, P.C., Martin, R.B., McLoughlin, R.J. and West, G.J., 1990. Variability in spatfall and recruitment of commercial scallops (*Pecten fumatus*) in Bass Strait. In: M.L.C. Dredge, W.F. Zacharin and L.M. Joll (Eds.). Proceedings of the Australasian Scallop Workshop. Tasmanian Government Printer, Hobart, Australia. pp. 80–91.

Young, P.C., West, G.J., McLoughlin, R.J. and Martin, R.B., 1999. Reproduction of the commercial scallop, *Pecten fumatus*, Reeve, 1852 in Bass Strait, Australia. Mar. Freshw. Res. 50:417–425.

Zacharin, W.F., 1990a. Scallop fisheries management: the Tasmanian experience. In: M.L.C. Dredge, W.F. Zacharin and L.M. Joll (Eds.). Proceedings of the Australasian Scallop Workshop. Tasmanian Government Printer, Hobart, Australia. pp. 1–11.

Zacharin, W.F., 1990b. Alternative dredge designs and their efficiency. In: M.L.C. Dredge, W.F. Zacharin and L.M. Joll (Eds.). Proceedings of the Australasian Scallop Workshop. Tasmanian Government Printer, Hobart, Australia. pp. 92–110.

AUTHOR'S ADDRESS

Mike Dredge - Queensland Fisheries Service, 40 Hall Road, Narangba, Queensland 4504, Australia (E-mail: flatcalm@ozemail.com.au)

Scallops: Biology, Ecology and Aquaculture
S.E. Shumway and G.J. Parsons (Editors)
© 2006 Elsevier B.V. All rights reserved.

1413

Chapter 31

New Zealand

Islay D. Marsden and Michael F. Bull[†]

31.1 INTRODUCTION

At least 18 different species of Pectinidae have been recorded from New Zealand waters (Powell 1979). Of these, only two (*Pecten novaezelandiae* and *Chlamys delicatula*) are large enough or occur at sufficient density to have encouraged any commercial exploitation. A fishery for *P. novaezelandiae* has been in existence for over 40 years and production from this fishery has averaged 660 tonnes meat weight (approx. 5,280 tonnes green or live weight) per annum between 1994 and 2002. Fishing for *Chlamys delicatula* on the other hand was carried out on an experimental basis in the late 1980's. However, because of its small size (60–70 mm) and deep water habitat (80–200 m) it did not develop into a significant fishery. This chapter has therefore been restricted to a discussion of the biology, fishery and aquaculture of *P. novaezelandiae*.

31.2 BIOLOGY

31.2.1 Morphology

Pecten novaezelandiae has the typical asymmetric shape of the genus with convex right valve and flat or slightly concave left valve and is also typical in being a functional hermaphrodite with separate male and female sections of the gonad. The shell is usually white, sometimes tinted with pink or brown and both valves are strongly ribbed with rib numbers ranging from 17–22 per valve. It is very similar in appearance to *P. fumatus* of Southern Australia but electrophoretic studies of protein similarities support its status as a separate species (Woodburn 1990).

31.2.2 Distribution

The species occurs sporadically around the whole New Zealand coastline including Stewart Island (Fig. 31.1) and the Chatham Islands (44°S, 176°W). It is found on a wide variety of substrates in semi-estuarine and coastal waters from low tide to depths of up to 60 m (up to 90 m at Chatham Islands). Individuals can be clumped in patches of conspecifics (McKinnon 1992; Morrisson 1999) but they are not associated exclusively with any single bottom community (Bull 1976). The main commercial beds at the northern tip of the South Island (Challenger fishery) occur in an open bay situation in 10–25 m on a soft muddy substrate while on the majority of North Island beds the substrate is

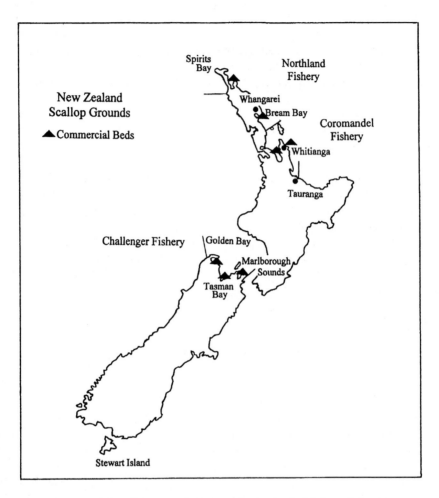

Figure 31.1. Commercial scallop grounds in New Zealand. Solid lines define the commercial fishing regions.

predominantly (hard) shell gravel and coarse sand (Morrisson 1999).

31.2.3 Life Cycle

The life cycle of *Pecten novaezelandiae* is similar to that of other members of the genus. Juveniles usually become mature at the end of their first year but probably contribute little to the spawning pool until the end of their second year. Fecundity is very high (15–50 million eggs/female) and variability in mortality of larvae and pre-recruits leads to great fluctuations in annual recruitment.

Observation of recently spent adults, presence of *P. novaezelandiae* larvae in plankton samples and records of spat settlement on artificial collectors indicate that in the area of the Challenger fishery some scallops spawn from August to April. In the Coromandel fishery Morrisson (1999) reported two highly synchronised spawning events in late October and mid January with some spawning in between. Fertilisation is external and larvae are thought to remain in the plankton for approximately 3–4 weeks (Bull 1976), based on a comparison with other temperate species.

Spat settlement in the Challenger area is usually highest from November to January and the data available for other areas of the country suggest a peak of settlement in early summer is probably widespread (Nicholson 1978; Bull 1980; McKinnon 1992; Morrisson 1999). At metamorphosis, late stage larvae (220 µm, Bull 1976) attach by byssus threads to filamentous material such as *Zostera* debris, algae and hydroids on the seabed and can also be caught on artificial collectors such as synthetic rope and mesh. This attachment is usually retained until the spat are at least 5 mm in shell length although smaller spat are sometimes found free living on the seabed. Juvenile scallops (2–9 mm) are quite mobile making post settlement movements by swimming or drifting at speeds equivalent to several hundreds of body lengths per second. In contrast, adult scallops are fairly sedentary and studies by Morrisson (1999) recorded movements of less than 10 m over a 14-month period from tagged individuals.

31.2.4 Growth Rates

Studies of size frequencies and measurement of tagged scallops indicate growth rates to be very variable among populations in different areas and among different depths in the same area and among different years. In the Challenger fishery area it is estimated that the majority of scallops reach 100 mm in length about 2.5 years after settlement. Large scallops in excess of 160 mm are occasionally taken but these are probably the result of especially rapid growth rather than extreme old age. Hauraki Gulf scallops reach 100 mm length in 18 months whereas scallops from the Coromandel Peninsula may take 3 years. Generally scallops grow faster in the Northland than the Coromandel fishery (Table 31.1) and the average size is larger further north. In contrast scallops from Paterson Inlet, Stewart Island have high longevity (6–7 years), and reach a larger maximal shell length, up to 170 mm (McKinnon 1992).

31.2.5 Mortality

The longevity of *P. novaezelandiae* in the wild is uncertain as is not possible to age scallops from rings on the shells (Bull 1976). In the Challenger fishery area, however, strong year classes are usually only traceable for three or four years suggesting relatively high rates of natural mortality. This ties in with estimates of the natural mortality rate obtained for the area from analysis of clucker ratios and tagging studies, which have ranged from 21% to 46% per year (Bull 1976; Bull and Drummond 1994). While rates of this order may be the norm, incidents of mass mortality have been reported on numerous occasions both in the Challenger and North Island fisheries (Table 31.2, Morrison 1999).

Table 31.1

Growth (estimated annual increments) from tagging and other studies in the Coromandel and Northland scallop fisheries (Cryer 1999) for scallops with initial shell lengths of 40 mm and 95 mm.

Growth increment	40 mm	95 mm
All data (1992–1997)	51.4	11.1
Coromandel (1992–1997)	49.9	10.3
Whitianga (1982, 1983, 1984)	52.4, 47.5, 30.3	13.8, 9.1, 6.0
Whitianga (depth 10.6 m)	60.1	15.1
Whitianga (depth 29.7 m)	31.2	6.8
Northland (1992–1977)	52.8	11.7
Bream Bay (depth 18.4 m)	53.0	8.8
Bream Bay (depth 21.4 m)	50.7	11.4

Potential causes of mortality include disease, predation, dredging, and environmental factors including storms, algal blooms, reduced salinity and excessive silt levels (Teaioro 1999). Histological examination (TEM) of tissue from wild and cultured, apparently diseased, specimens revealed virus-like particles associated with the digestive gland epithelium (Hine and Wesney 1997). Although the cytology changes were similar to those associated with enteroviruses and calciviruses, it could not be concluded that these caused the disease in the scallops. More recently, Hine and Diggles (2002) reported intracellular prokaryote infections in the digestive diverticular epithelial cells and branchial cells of both *P. novaezelandiae* and *C. delicatula*. Other potentially damaging parasites and commensals include the copepod *Pseudomyicola spinosus,* which is abundant in and around the stomach in some scallop populations, and the polychaete *Polydora hoplura*, which bores through the shell leading to weakening of the attachment of the adductor muscle. A tubellarian rhabdocoel *Paravortex* sp. has been reported from within the intestine of cultured scallops showing gaping and poor swimming ability (Woods and Hayden 1998). During the summer of 1999–2000, many poor condition shellfish, including scallops were collected from the Coromandel. This "black gill" condition was investigated histologically and bacteriologically but no bacteria were isolated and there was no indication that the tissues had been damaged by bacteria, parasites or viruses (Smith and Diggles 1999). It was suggested that this condition may have been the result of low upwelling nitrate levels over winter, warm high salinity water on the outer continental shelf, together with la Nina weather patterns..

The main predators of *P. novaezelandiae* appear to be starfish (especially *Coscinasterias calamaria),* fish including eagle rays *(Myliobatus tenuicaudatus)*, snapper (*Chrysophrys auratus*) and blue cod (*Parapercis colias*), and octopus (*Octopus maorum*). In many countries, crabs are a common predator of scallops but although the voracious paddle crab *Ovalipes catharus* is abundant along the sandy margins of such areas as Golden Bay, it is not commonly found on the scallop grounds further out. Hermit crabs are, however, abundant on some parts of the fishery and they are probably predators of small scallops as

Table 31.2

Mortality estimates for *P. novaezelandiae* from different parts of New Zealand, the method used to determine the estimate; Z, instantaneous mortality, which includes fishing mortality and M, natural mortality.

Locality	Method		Estimate y^{-1}	Reference
Coromandel	Density	M	0.45–0.50	Walshe 1984
Coromandel	Density	Z	1.06, 1.09	Osborne 1990
Challenger	Cluckers	M	0.29, 0.49	Bull 1976
Challenger	Tagging	M	0.21, 0.46	Bull and Drummond 1994
Hauraki Gulf				
Ti Point	Density	Z	3–11.15	Morrisson 1999
Omaha	Density	M	1.10–2.63	Morrisson 1999
Omaha (algal bloom)	Density	M	4.66	Morrisson 1999
Pukenihinihini Pt	Density	M	9.56–10.59	Morrisson 1999
Motuketekete	Density	Z	1.15	Morrisson 1999
Motuketekete	Tagging	Z	1.16	Morrisson 1999

seeding of young spat in areas where the crabs are abundant has usually been unsuccessful.

While fishing is likely to be the major cause of mortality in some areas, dredging activity can significantly affect scallop mortality and cause habitat disturbance (Thrush et al. 1995). Cryer and Morrison (1997) tested three designs of scallop dredge and found that despite the high incidental mortality, the current box dredge was best for the Coromandel scallop fishery because of its higher catching efficiency.

Scallops are thought to be sensitive to environmental factors such as turbidity and suspended silt that can result from dredging, bottom trawling, storms and run-off. Stevens (1987) reported that small *P. novaezelandiae* (20 mm shell height) rapidly expel water from the mantle cavity or swim when subjected to high silt levels then, after 30 min, settle with their valves open. Mortality (LD$_{50}$) occurred within 7–13 h and increased with increasing silt concentration (0.05–6% dry wt) and decreasing silt particle size.

In addition, anecdotal evidence from the North Island suggests that toxic algal blooms may have affected scallop recruitment and juvenile recruitment in the past. However, there have been no reported incidents of deaths of adult scallops due to harmful algal blooms. Commercial scallops are tested weekly for marine biotoxins and to date there has only been one precautionary closure of the fishery, which was due to low levels of ASP toxins in the roe. Suzuki et al. (2001) have investigated DSP toxins comparing the pectenotoxin profiles of scallops (*P. novaezelandiae)* with that of the toxic dinoflagellate *Dinophysis acuta* in the Marlborough Sounds. The results suggest that the New Zealand scallop has the ability to reduce the cytotoxicity of the pectenotoxin (PTX) found in *D. acuta* (PTX2) by conversion to a less toxic form (PTX2SA).

31.3 COMMERCIAL FISHING

Commercial fishing for scallops in New Zealand currently occurs in three main areas (Fig. 31.1): the Challenger fishery, previously known as the southern scallop fishery (Golden Bay, Tasman Bay and the Marlborough Sounds), the Coromandel fishery (= the Bay of Plenty and the Hauraki Gulf) and the Northland fishery (= Bream Bay to Spirits Bay). In 1997, landings totalled 519 tonnes meat weight with the three areas contributing 57.4%, 26.2% and 16.4%, respectively.

In 1988 commercial fishing also began at the Chatham Islands 850 km east of the New Zealand mainland and in that year 121 tonnes meat weight was landed. However, in the following year landings dropped severely and subsequent landings have been less than 50 tonnes per annum (Bull 1990).

31.4 DEVELOPMENT OF THE SCALLOP FISHERY

The scallop fishery in New Zealand began with an experimental dredging operation in Tasman Bay in 1959. Over the following ten years boat numbers gradually increased to 26 vessels and dredging operations spread to cover grounds in nearby Golden Bay and the Marlborough Sounds.

Production from the Challenger fishery reached a peak of about 1,200 t meat weight in 1975 from a fleet of 190 vessels. In 1976 there were 245 vessels but over the following five years the fishery suffered a dramatic decline (Table 31.3) and in 1981 and 1982 the whole fishery area was completely closed to fishing. Stocks have since recovered and the Challenger fishery in 1994/1995 season harvested 850 t meat weight.

Fishing in the Coromandel and Northland areas began in the early 1970's and the catches over the following ten years showed somewhat more stable production levels than the Challenger fishery (Table 31.3). From the mid 1980's the landings were recorded as meat weight (adductor muscle and roe), which reflects the commercial value. In recent years, these fisheries have yielded over 200 tons meat weight per annum but there have been significant declines over the past two years. Between 1995 and 1997, the Northland Fishery and Coramandel Fishery together contributed at times up to 47% of the total NZ catch.

31.4.1 Capture Methods

The commercial fishery for *P. novaezelandiae* is entirely a dredge fishery with the fleet currently consisting of about 100 vessels most of which are in the 12–20 m length range. Because of the relatively short season length (2–7 months) nearly all these vessels are involved with some other fishing activity for part of the year such as trawling, gill netting, long lining or rock lobster fishing.

Scallop fishing, however, generally provides 50% or more of these fishermen's income. Most vessels are privately owned and are run by a skipper and one or two crew. The type of dredge used depends largely on the substrate of the grounds concerned. In the Challenger fishery, where soft muddy substrate predominates and where scallop densities

Table 31.3

Landings (tonnes meat weight) from NZ scallop fisheries 1970–1999. Early records are based on estimates using the gazetted conversion factor from live weight to meat weight.

Season	Marlborough Sounds	Tasman and Golden Bays	Total Challenger	Northland	Coromandel controlled
1970	7	72	80		
1971	9	2	215		
1972	46	190	236		
1973	124	197	321		
1974	21	585	606	0.5	5
1975	72	1174	1246	15	9
1976	79	468	547	10	17
1977	63	512	574	93	17
1978	76	91	167	171	14
1979	39	65	105	107	19
1980	27	14	41	30	110
1981	0	0	0	70	146
1982	0	0	0	99	131
1983	61	164	225	146	194
1984	138	229	367	68	140
1985	100	145	245	43	110
1986	117	238	355	114	162
1987	105	114	160	183	384
1988	72	150	222	171	182
1989	95	110	205	164	104
1990	80	160	240	114	153
1991	30	642	672	158	203
1992	160	550	710	135	147
1993	135	670	805	114	62
1994	8	842	850	205	49
1995	76	445	521	208	88
1996	61	170	231	129	78
1997	58	240	298	136	85
1998	117	430	547	31	37
1999	7	669	676	18	8
2000	16	320	336	25	7
2001	30	686	716	55	22
2002	62	409	471	69	32

From: King (1985), King and McKoy (1984), Annala (1994), Cryer (2000), Drummond (pers. comm.), Sylvester (pers. comm.).

are often quite low (<1 per 5 m^2), most vessels use two ring bag dredges up to 2.4 m in width with heavy tickler chains. On the hard sand of North Island beds a self-tipping "box" dredge with rigid tooth bar is more common.

The duration of dredge tows is largely dependent on the amount of scallops or other material being caught but is commonly about 30 minutes. On recovering the dredge, legal size scallops are sorted from the catch and undersized scallops and shell returned to the sea. Marketable scallops are landed on a daily basis and trucked to nearby processing facilities where they are hand-shucked. Only the adductor muscle and gonad are retained and the shell and viscera are usually dumped. Meat weight recoveries are generally 12–20% of total weight with individual meats weighing 10–25 g of which the gonad may account for up to 30% or more. After shucking, the meats are usually chilled, frozen in layer packs or free flow form, although some are further processed to a crumbed or battered form. The local market currently consumes approximately half the catch with the remainder going to a variety of markets including France, Australia and USA. The quantity and destination of exports is very dependent on availability and demand and in some years, New Zealand has imported scallops from Japan and Australia.

31.5 MANAGEMENT

31.5.1 The Challenger Fishery

This fishery was one of the most highly Government regulated fisheries in New Zealand. Within the first ten years of its existence, controls had been introduced establishing a minimum size (4", later 100 mm), restrictions on number and size of dredges and an annual five months closed season (February to July).

In the late 1970's, strict controls were placed on boat numbers (245 reduced to 136) through the introduction of non transferable licenses under the Controlled Fisheries legislation and in the next 10 years controls including day light only and five day fishing weeks, daily quotas and shortened season lengths were introduced. In addition annual preseason stock assessment surveys were carried out and an upper limit on seasonal catch imposed by way of either competitive TAC's (1979 and 1980), short season length and daily catch limits (1983–1985) or individual seasonal allocations (1986–1991). These regulations were primarily aimed at prevention of biological overfishing, maximising the long-term economic return from the fishery and providing some stability for participants in the fishery.

In 1983, a three year trial on the feasibility of enhancing the fishery through the release of naturally-occurring spat caught on artificial collectors was set up. These initial seeding trials were successful and the first seeded scallops were harvested in 1986 (Clement and Bull 1988; Bull 1994). Funds generated from sale of crop from the trial seedings were used to keep the enhancement programme going for the next two years, but by 1989 an agreement had been made with the licence holders to fund seeding operations through a levy on annual catch and since that time the continuing enhancement operations in the fishery have been fully industry funded.

In 1989, in association with the development of the successful seeding operation, a switch was also made to the management system for the catching operation. A rotational system was set up in which the grounds in Golden Bay and Tasman Bay were divided into 9 sectors, with an agreed number of these sectors being opened for commercial fishing each year, followed by seeding and subsequent closure until seeded stock had reached takeable size. Under this system, in any given year, adequate stocks of potential spawners could be protected in the closed sectors and therefore there was little need to conserve spawning stock in the open sectors. Previous measures aimed at such protection such as a strict size limit and annual quotas could therefore be relaxed. As a result, the previous minimum size of 100 mm was dropped to 90 mm for the commercial fleet (= approximately the minimum size for marketing) and annual quotas were now used for allocation rather than stock conservation purposes.

In 1992, the fishery was introduced into the Quota Management System (QMS) system with 10% of the available take being allocated to Maori and the remainder being divided equally between the 48 licence holders. Between 1992 and 1995 there was a reorganisation of responsibilities resulting in a shift from Government agencies to the private sector (Arbuckle and Drummond 2000). Maori were granted an additional 10% of quota. Also, quota holdings were changed from being expressed as a set tonnage to proportional holdings so that the benefits of any increased production resulting from Industry funded enhancement and management initiatives would be returned directly to Industry.

In 1994 the Challenger Company consisting of all the quota holders in the fishery was incorporated and given the responsibility of providing a plan for enhancement, collecting research information and providing advice on management. Since 1996 there have been further management measures introduced: a compulsory levy on scallop quota holders, a daily catch balancing system and a dockside monitoring programme with penalties for exceeding the limit. In addition, Challenger Company could exclude vessels if yields were low or in the case of a biotoxin event. The Company commissioned a purpose designed enhancement vessel and has increased research effort aimed at improving the success of seeding and management of the fishery. Although there is a defined fishing season, this can be delayed to allow scallops to reach a better size before collection. However, the rules for an extended season still include an overall total allowable catch of 720 t, restricted fishing zones and biological and effort controls. In 2000 the delayed start to dredging together with variability in scallop size resulted in a lower than usual catch (Table31.3).

31.5.2 The Northland and Coromandel Scallop Fisheries

The Northland fishery consisted of 38 vessels up until 1997 when it was introduced into the Quota Management System (QMS). The number of vessels has since declined (Cryer 1999) and the Northland Scallop Enhancement Company includes 33 quota holders. The main commercial beds used to be Bream Bay and bays approximately 110 km to the north. However, in 1994, beds in the Spirits Bay area at the tip of the North Island were opened up and since then, this area has contributed more than 50% of the

Northland catch. Destruction of other benthic species by dredging in this area (especially sponges, hydroids and bryozoans) has however been of concern and a significant part of the grounds has subsequently been closed to scallop fishing.

The Coromandel fishery was a controlled fishery with 22 licence holders before it was included into the QMS in 2002. As with the Northland fishery there have been marked fluctuations in productivity among different beds over the years and yield from the fishery is currently at a very low level. Management controls on the fishery include controls on seasonal catch (based preseason surveys), on dredge size, fishing hours, non-fishing days and voluntary closed areas. In 1995 the previous minimum size of 100 mm was reduced to 90 mm in exchange for a package, which included further closed areas and reduced catch limits in an attempt to address concerns about the impacts of dredging on juvenile scallops.

31.6 RECREATIONAL FISHING

There is intense interest in non-commercial scallop fishing throughout New Zealand by recreational fishers. Scallops are taken by SCUBA diving and the use of small dredges and in a few areas by intertidal collections. Scallop divers are required by law to measure and count the shellfish on the seabed. They are not permitted to do this on the boat. Divers that fail to follow these rules could have their vessels confiscated. Daily bag limits vary among areas depending on scallop availability and recreational fishing pressure. At present, they are 10 per person per day in Southland, 50 per person in the Challenger area and 20 per day in the other 3 Fishery Management Areas. Because of depleted stocks, shellfish dredging was prohibited inside Paterson in 2003 and there is a temporary ban on collecting or possessing scallops until 30th September 2007. There are limited data on the extent of the recreation catch. Bradford (1997) estimated recreational take made up 20% of the total take from the Coromandel fishery and 5% for the Northland fishery in the 1994/95 season.

In the Challenger Fishery recreational fishers have access to both wild stocks and those seeded by the commercial industry but are not confined to the rotational zones that the commercial fishers are restricted to. Again, the extent of recreational take for this fishery is uncertain, although a national diary survey conducted by the Ministry of Agriculture and Fisheries gave an estimate of only 22 tonnes (= 3% of the take) for the area in the 1992/93 season.

31.7 AQUACULTURE AND ENHANCEMENT

In the late 1970's and early 1980's there was considerable local interest in farming scallops using Japanese hanging culture techniques and a number of small scale trials were carried out. While collection of spat in Japanese type spat collectors proved relatively easy, attempts at on-growing scallops using Japanese pearl net, lantern cage and ear hanging systems were plagued by fouling especially by mussels (*Perna canaliculus)* and also by unexpected mortality. Because of these problems, and relatively high costs of equipment and labour in New Zealand, it was concluded quite early on that scallops operations based on Japanese hanging culture techniques were unlikely to be economic in New Zealand (Cameron 1983). More recent attempts at scallop culture have suffered from the same problems and interest in such ventures appears to have now declined, especially as they would have to compete with the success of scallop enhancement operations in the Challenger fishery and the booming New Zealand mussel industry.

In addition to the early hanging culture trials, trials on the feasibility of enhancing scallop production by seeding the spat caught in collector bags onto the seabed were begun in 1982. In 1983, the New Zealand Ministry of Agriculture and Fisheries, in conjunction with the Overseas Fisheries Cooperation Foundation of Japan expanded this work and embarked on a three year pilot scale trial in Golden Bay (Challenger area). In its first year, the catch in the 44,000 collectors resulted in a release of approximately 35 million spat three to four months after settlement. A similar scale operation was carried out in the second and third years.

Follow-up work on the seeding trials indicated that rates of survival to harvest could be in excess of 20%. It also became apparent that survival of spat that settled on the outside of the collector bags and eventually released themselves naturally onto the seabed was quite high. Further these spat could also be used for seeding by dredging them from the seabed and redistributing them when they reached a suitable size (>25 mm).

On completion of the three year trial, the spat catching and seeding operations were expanded as quickly as possible leading to an estimated release in 1992 of more than 1 billion spat from collector bags and 204 million individuals transplanted from the seabed at the spat catching site (Bradford-Grieve 1994). Although survival rates for spat seeded in such large quantities has often been much lower than that found in the original seeding trials, research has shown that up to 80% of the commercial catch comes from seeded individuals (Annala 1994). The seeding operations are now an integral part of the management of the Challenger scallop fishery (Bull 1994).

Following the success of the enhancement trials in the Challenger area in the mid 1980's, similar trials were conducted in the Coromandel fishery and small scale spat catching trials have also been made in the Northland fishery (Morrisson 1999) and at the Chatham Islands. Catches in excess of 500 spat per collector bag were made in the Coromandel trials but subsequent survival of spat released onto the seabed was negligible. This was possibly due to extreme weather conditions at the time of release in the first year and heavy predation by fish on the clean sandy substratum at the release site chosen in the second year (Bartrom 1989).

31.8 FUTURE DEVELOPMENTS

In the Challenger scallop fishery it seems likely that there will be a gradual increase in yield and improved stability of the resource as seeding operations are refined and the seeding area is expanded to cover the Marlborough Sounds section of the fishery. In recent years, the Challenger Scallop Enhancement Company has placed more emphasis on spat survival and quality rather than the size of the spat catch. They have also moved to reduce the risk of spat fall failure by targeting more than a single period for spat collection.

In the Coromandel and Northland areas, increasing demand for amateur only fishing zones as the recreational fishing population expands is likely to lead to a long-term decline in the commercial fisheries in those areas. The shift to the QMS system in the Northland fishery and the success of self-management as shown by the Challenger Scallop Enhancement Company has improved the chances of an enhancement operation being established in that area. However, the feasibility of catching sufficient spat within the Northland fishery on a regular basis has yet to be demonstrated, as has the likely hood of an adequate survival rate of seeded spat. There are significant differences in substrate type and associated fauna between the Northland and Challenger scallop grounds and factors influencing survival are likely to be quite different between the two areas.

Potential threats to the industry are the apparently increasing incidence of toxic algal blooms in New Zealand waters and sporadic outbreaks of unexplainable shellfish disease. There are demands by conservationists for increased areas of coastal marine reserves and establishment of non-dredging zones. Finally, there is increasing pressure from other sectors of the marine industry for permits to farm within the Challenger fishery area.

ACKNOWLEDGMENTS

We would like to thank members the scallop industry who shared with us their knowledge and provided access to unpublished reports. Thanks especially to Kim Drummond, Martin Cryer, Graeme McGreggor, Martin Workman, Todd Sylvester and Russell Mincher. Mike Bull died in May 2002 and IDM would like to thank Sue Lindsay (Mike's widow) for her support in the publication of this chapter.

REFERENCES

Annala, J.H., (Comp.) 1994. Report from the Mid-year Fishery Assessment Plenary, November 1994: stock assessments and yield estimates. 37 p.

Arbuckle, M. and Drummond, K.L., 2000. Evolution of self-governance within a harvesting system governed by individual transferable quota. FishRights 99 Conference proceedings. Fisheries Western Australia.

Bartrom, A., 1989. The Coromandel scallop enhancement project 1987–1989. Unpublished Report, MAF Fisheries North. 22 p.

Bradford, E., 1997. Estimated recreational catches from Ministry of Fisheries North region marine recreational fishing surveys, 1993–4. NZ Fisheries Assessment Document 97/7:16 p.

Bradford-Grieve, J., (Comp.) 1994. Summary of knowledge of the Tasman and Golden Bay marine environment relevant to fisheries enhancement. Report produced by NIWA, MAF Fisheries and Cawthron Institute for Southern Scallop Advisory Committee, MAF Fisheries and Tasman District Council.

Bull, M.F., 1976. Aspects of the biology of the New Zealand scallop, *Pecten novaezelandiae* Reeve 1853, in the Marlborough Sounds. PhD Thesis. Victoria University of Wellington, New Zealand.

Bull, M.F., 1980. Scallop farming studies. In: Proceedings of the Aquaculture Conference. MAF Fisheries Research Division Occasional Publication No. 27:16–20.

Bull, M.F., 1988. New Zealand scallop. N.Z. Fisheries Assessment Document 88/25:16 p.

Bull, M.F., 1990. Report on a survey of the scallop resource at the Chatham Islands 22 April–5 May 1990. Central Fisheries Internal Report 16: 29 p.

Bull, M.F., 1994. Enhancement and management of New Zealand's southern scallop fishery. In: N.F. Bourne, B.L. Bunting and L.D. Townsend (Eds.). Proceedings of the 9th International Pectinid Workshop. Nanaimo, B.C., Canada. Canadian Technical Reports in Fisheries and Aquatic Science 1994(2):131–136.

Bull, M.F. and Drummond, K.L., 1994. Report on Tasman Bay and Golden Bay scallop mortality trials. MAF Central Region Internal Report 24:17 p.

Cameron, M.L., 1983. A preliminary investigation of the economic viability of scallop farming in New Zealand. MAF Economics Division, Unpublished Discussion Paper 15/83:33 p.

Clement, G. and Bull, M.F., 1988. New Zealand scallop fisheries proposals for future management. Internal report for MAFFish, March 1988:44 p.

Cryer, M., 1999. Coromandel and Northland scallop stock assessment for 1998. New Zealand Fisheries Assessment Research Document 99/37:14 p.

Cryer, M., 2000. Northern Coromandel and Northland scallops (Pecten novaezelandiae). Working Group Shellfish Report. 10 p.

Cryer, M. and Morrison, M., 1997. Incidental effects of commercial scallop dredges. Report for Ministry of Fisheries, Project AKSC03.

Hine, P.M. and Diggles, B.K., 2002. Prokaryote infections in the New Zealand scallop Pecten novaezelandiae and Chlamys delicatula. Diseases of Aquatic Organisms 50:137–144.

Hine, P.M. and Wesney, B., 1997. Virus-like particles associated with cyropathology in the digestive gland epithelium of scallops Pecten novaezelandiae and toheroa Paphies ventricosum. Diseases of Aquatic Organisms 29:197–204.

King, M.R., 1985. Fish and shellfish landings by domestic fishermen, 1974–1982. MAF Fisheries Research Division Occasional Publications: Data Series No. 20:88 p.

King, M.R. and McKoy, J.L., 1984. Scallop landings in the Nelson-Marlborough dredge fishery 1959–80. MAF Fisheries Research Division Occasional Publications, Data Series No. 14:12 p.

McKinnon, S.L.C., 1992. Review of the Paterson Inlet scallop fishery. New Zealand Fisheries Management Regional Series 2:30 p.

Morrison, M., 1999. Population dynamics of scallops, Pecten novaezelandiae, in the Hauraki Gulf. Unpublished PhD Thesis. University of Auckland.

Nicholson, J., 1978. Feeding and reproduction of the New Zealand scallop Pecten novaezelandiae. MSc Thesis. Auckland University of New Zealand.

Osborne, T., 1990. Assessment of northern scallop stocks 1990. Unpublished Draft Report. Ministry of Agriculture and Fisheries.

Powell, A.W.B., 1979. New Zealand Mollusca. William Collins Publishers Ltd, Auckland.

Smith, P. and Diggles, B., 1999. Coromandel shellfish update, November 1999. Unpublished NIWA Wellington Report. 2 p.

Stevens, P.M., 1987. Response of excised gill tissue from the New Zealand scallop Pecten novaezelandiae to suspended silt. NZ J. Mar. FW Res. 21:605–614.

Suzuki, T., Mackenzie, L. and Adamson, J., 2001. Conversion of pectenotoxin-2 to pectenotoxin-2 seco acid in the New Zealand scallop, Pecten novaezelandiae. Fisheries Science 67:506–510.

1426

Teaioro, I., 1999. The effects of turbidity on suspension feeding bivalves. Unpublished MSc Thesis. University of Waikato. 109 p.

Thrush, S.F., Hewitt, J.E., Cummings, V.J. and Dayton, P.K., 1995. The impact of habitat disturbance by scallop dredging on marine benthic communities: what can be predicted from the results of experiments? Mar. Ecol. Prog. Ser. 129:141–150.

Walshe, K.A.R., 1984. A study to determine the optimum number of licences for the Tauranga commercial scallop fishery based on an Optimum Yield estimate. Unpublished Dissertation for Dip. Bus. Admin. Massey University. 93 p.

Woodburn, L., 1990. *Pecten* in Southern Australasia. In: Proceedings of the Australasian Workshop on Scallop Fisheries. Tasmanian Government Printer, Hobart, Australia.

Woods, C.M.C. and Hayden, B.J., 1998. An observation of the turbellarian *Paravortex* sp. in the New Zealand scallop *Pecten novaezelandiae* (Bivalvia: Pectinidae). NZ J. Mar. FW Res. 32:551–553.

AUTHORS ADDRESSES

Islay D. Marsden - School of Biological Sciences, University of Canterbury, Christchurch, New Zealand (E-mail: i.marsden@zool.canterbury.ac.nz)

Michael F. Bull[†] - Nelson, New Zealand

[†] Deceased.

Species Index

A

Adamussium, 4-5, 14, 50, 487
Adamussium colbecki, 4-5, 34, 47, 83, 93, 106-107, 468, 475, 477, 481, 487-488, 580, 583-584, 596, 636, 639, 666, 674, 683, 693, 699, 708, 714, 719, 738, 747, 762-763
Aequipecten, 1-2, 4-5, 7, 11, 22-23, 27, 32, 54, 189, 547, 567, 582, 588, 596, 608, 623, 636, 650, 667, 728-729, 735, 737, 856, 991, 1006, 1037, 1047, 1058-1059, 1063, 1066, 1225, 1315
Aequipecten irradians, 107, 184, 196, 224, 301, 303, 307, 311, 354-355, 407, 411, 414, 477-478, 484, 584, 619, 720, 731-732, 737, 962, 964
Aequipecten irradians concentricus, 119, 176, 224, 414, 964
Aequipecten irradians irradians, 409
Aequipecten opercularis, 5, 11, 47, 105, 157, 184, 192, 219, 435, 485, 545-546, 558, 563, 587, 619, 632, 637, 645, 651-652, 655-656, 667-668, 670-679, 681-682, 685, 690, 693-696, 699, 701-711, 714, 716-717, 726-728, 731, 757, 774, 780-781, 796, 826, 844, 851, 931, 962, 991-992, 1000, 1004, 1006-1007, 1011, 1014-1017, 1019, 1023, 1031-1032, 1035-1040, 1044, 1047-1048, 1050-1051, 1054-1055, 1062-1063, 1067, 1085-1086
Aequipecten tehuelchus, 104, 115-116, 652, 665, 670, 687, 719, 770-771, 775, 782, 784, 793-794, 796, 806, 826, 834, 841, 848, 1233, 1247, 1251-1253, 1255, 1260, 1263-1267, 1285-1286, 1288-1289
Aeromonas, 1189-1190, 1350
Aeromonas hydrophila, 103, 118, 648, 1190

Aeromonas salmonicida, 639, 1084, 1086
Alexandrium fundyensein, 476
Alexandrium tamarense, 103, 121, 448-449, 451, 492
Amphidinium carterae, 435
Amusium, 2, 22, 31, 629, 664, 704, 711, 729, 770, 774, 777, 784, 805, 825, 829, 837, 840, 1315, 1320-1321, 1334, 1391, 1398, 1409
Amusium balloti, 47, 52, 61, 118, 370, 385, 409, 596, 629, 652, 664, 673-674, 676, 687, 690, 699, 702-703, 725-727, 742, 773, 780, 824, 853, 855, 1391, 1398, 1401, 1403, 1407-1408, 1410-1411
Amusium japonicum, 2, 375, 384
Amusium japonicum balloti, 362, 368, 371, 376, 408, 478, 491, 721, 742, 776, 850, 1409, 1411
Amusium papyraceum, 659, 735, 1320-1321
Amusium pleuronectes, 47, 52, 106-107, 355, 362, 384, 410, 596, 632, 652, 664, 678, 693, 699, 701-703, 707, 733, 806, 901
Amussiopecten, 22-23, 25
Amussiopecten burdigalensis, 23
Amussiopecten subpleuronectes, 23
Anachlamys leopardus, 629
Anatipopecten, 28, 39
Anguipecten, 28
Annachlamys, 22, 27-28, 31
Annachlamys iredalei, 27
Annachlamys okinawaensis, 27
Annachlamys reeve, 27
Antipecten pseudopandorae, 27
Apostichopus japonicus, 1208, 1213
Arca, 2, 82, 421
Argopecten, 1-2, 4, 6-7, 23, 32, 77, 88, 109, 111, 196, 214, 242, 253-262, 284, 329, 341-342, 345, 426, 478, 545, 585, 598, 606, 680, 723, 726, 927, 949-950, 1295, 1311-1312,

1430

General Index

A

D

L

M

N

W

X

Y

Z